FEATURES AND BENEFITS

Pre-Algebra: An Integrated Transition to Algebra and Geometry

1. **Integration** lessons apply mathematical concepts to all areas of mathematics (pages 139–144). Integration also appears *within* lessons (pages 134–135).

   Integration helps students to view mathematics as a whole—not as isolated areas of instruction.

2. **Applications** appear in every lesson to show where and when math is used in the real world. *Real-World Applications* are used to introduce lessons (page 134), as well as within lessons, (page 141), and in the exercises (pages 143–144).

   Realistic and relevant applications enhance a student's ability to grasp concepts and skills.

3. **Connections** to other disciplines are found in lesson openers (Lesson 4-5, page 190) and exercises (page 133, exercise 39). The features *The Shape of Things to Come* (page 40) and *Earth Watch* (page 214) are both excellent examples of interdisciplinary connections.

   Connection features help students to relate math skills to other subjects as well as enhancing learning and increasing student interest.

4. **Technology** is a powerful tool for exploring mathematics. *Mathematics Labs* (page 145), *Graphing Calculator Exercises* (page 132), and *Modeling with Technology* activities (page 124) encourage students to investigate using programming, a scientific or graphing calculator, or computer software.

   Technology enhances instruction and learning by engaging students in interactive activities and investigations.

5. **Modeling** activities open every lesson: *Modeling with Technology* (page 124), *Modeling a Real-World Application* (page 129), or *Modeling with Manipulatives* (page 146). Additional opportunities for modeling are found in *Modeling Mathematics* exercises (page 120).

   Modeling helps students move from concrete to abstract thinking through realistic and relevant applications, up-to-date technology, and hands-on manipulatives.

6. **Problem-Solving Strategy** lessons teach the strategies necessary to successfully solve problems (page 118). Students apply these strategies in every lesson in the *Applications and Problem Solving* section of the exercises (pages 136 and 137).

   Problem solving offers students an opportunity to learn the strategies and apply the skills learned in real-world situations, as well as answer the question, "Why should I study Pre-Algebra?"

7. **Investigations** are on-going, long-term projects that are revisited in the *Working on the Investigation* and *Closing the Investigation* features (pages 62–63, 128, 160).

   The Investigation is the ultimate in integration and application; it provides students the opportunity to apply important mathematics in interesting and relevant real-world settings. In addition, students develop important group skills as they work collaboratively.

Teacher's Wraparound Edition

GLENCOE

Pre-Algebra

An Integrated Transition to Algebra & Geometry

GLENCOE

McGraw-Hill

New York, New York Columbus, Ohio Mission Hills, California Peoria, Illinois

Visit the Glencoe Mathematics Internet Site for
Pre-Algebra at

www.glencoe.com/sec/math/prealg/mathnet

You'll find:

Group Activities

Games

interNET CONNECTION links to websites relevant to Chapter Projects, Investigations, and exercises

and much more!

Glencoe/McGraw-Hill
A Division of The McGraw-Hill Companies

Send all inquiries to:
Glencoe/McGraw Hill
8787 Orion Place
Columbus, OH 43240

ISBN: 0-02-833240-7 (Student Edition)
0-02-833241-5 (Teacher's Wraparound Edition)

4 5 6 7 8 9 10 027/043 06 05 04 03 02 01 00 99

Teacher's Wraparound Edition

Table of Contents

Pre-Algebra

An Integrated Transition to Algebra and Geometry

Glencoe Brings Pre-Algebra to Life! .T2-T3
Glencoe's Dynamic Lesson Structure T4-T5
Glencoe's ***Pre-Algebra*** Exemplifies the NCTM Standards . .T6-T7
Learning Objectives & NCTM Standards Correlation T8-T11
Planning Your Pre-Algebra Course T12-T13
Classroom Vignettes T14
Math in the News T15
Internet Connections T15
Glencoe's ***Pre-Algebra*** Research Activities T16
Table of Contents, Student Edition . . . . .iii-xx
Teacher's Wraparound Edition 1-739

Chapter Overviews
- Chapter 1 . . . . .4a-4d
- Chapter 2 . . . . .64a-64d
- Chapter 3 . . . . .116a-116d
- Chapter 4 . . . . .168a-168d
- Chapter 5 . . . . .222a-222d
- Chapter 6 . . . . .272a-272d
- Chapter 7 . . . . .328a-328d
- Chapter 8 . . . . .370a-370d
- Chapter 9 . . . . .430a-430d
- Chapter 10 . . . . .484a-484d
- Chapter 11 . . . . .546a-546d
- Chapter 12 . . . . .608a-608d
- Chapter 13 . . . . .662a-662d
- Chapter 14 . . . . .704a-704d

Extra Practice 740-773
Chapter Tests 774-787
Glossary 788-795
Spanish Glossary 796-803
Selected Answers 804-828
Photo Credits 830-832
Applications Index 834-835
Teacher's Wraparound Edition Index 836-845

Glencoe Brings *Pre-Algebra* to Life!

An Integrated Transition :

Students relate and apply algebraic concepts to geometry, statistics, data analysis, probability, and discrete mathematics.

Realistic and relevant applications from areas such as sports and travel are a hallmark of Glencoe's ***Pre-Algebra***.

Interdisciplinary connections to science, geography, music, and other curriculum areas help answer the question "When am I ever going to use this stuff?"

Every lesson begins with a modeling activity. There are three types:

- Modeling a Real-World Application
- Modeling with Manipulatives
- Modeling with Technology

lgebra and Geometry

Hands-On Activities

Students become more involved in their study of mathematics through Mathematics Labs and Modeling Math Activities.

Technology

The use of graphing technology and spreadsheets as problem-solving tools is integrated throughout Glencoe's ***Pre-Algebra***.

Assessment

The wide variety of assessment options includes multiple-choice and free-response tests as well as performance and portfolio assessment.

Comprehensive Program

Glencoe's ***Pre-Algebra*** has the built-in flexibility that gives teachers the tools to help make mathematics accessible to students with a wide range of mathematics achievement.

NCTM Standards

The sequence of topics, the content emphases, and dynamic feaures combine to form a program that supports and reflects the NCTM Curriculum, Teaching, and Assessment Standards.

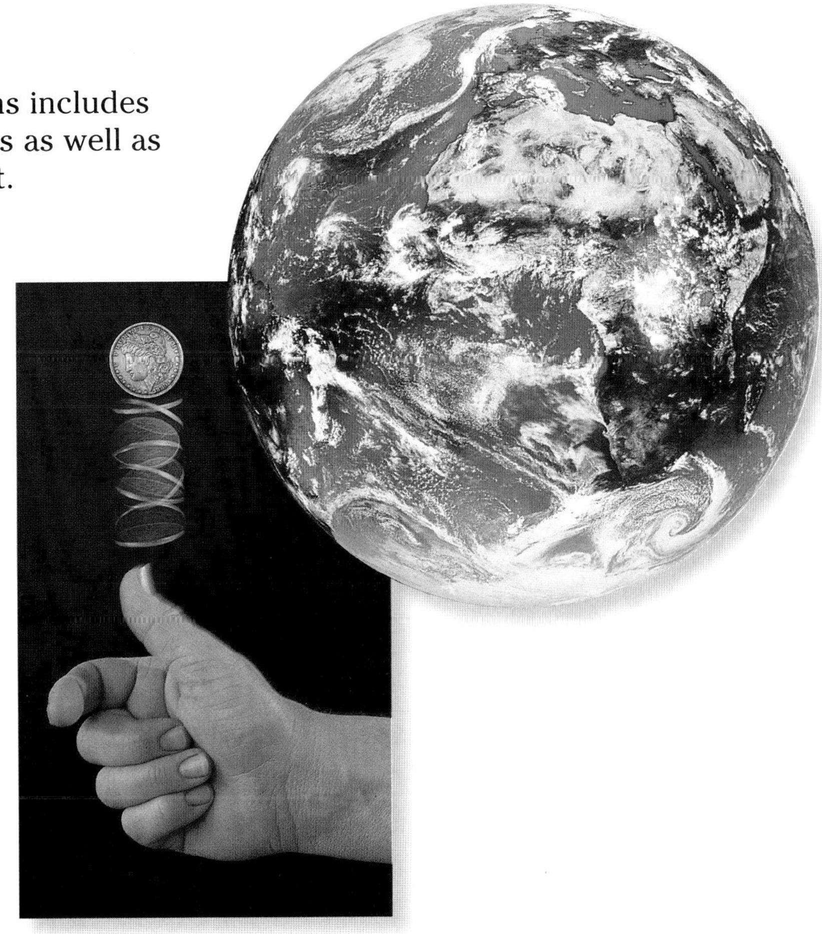

Glencoe's Dynamic Lesson Structure

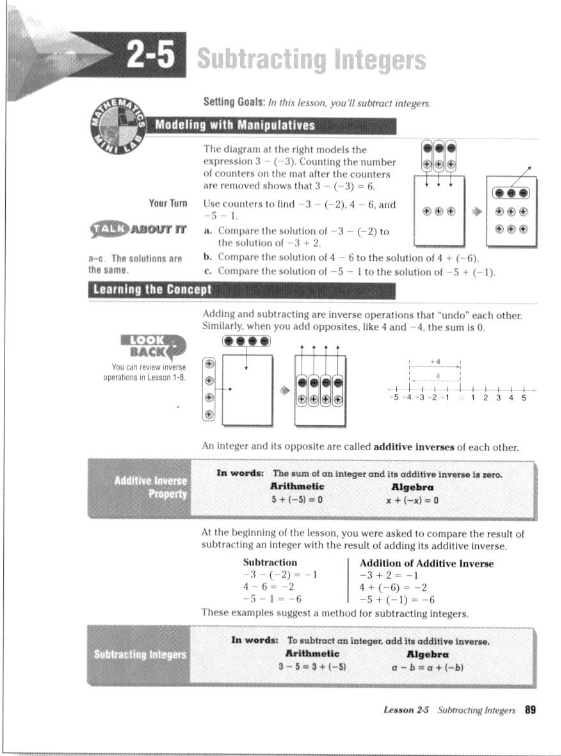

2-5 Subtracting Integers

Setting Goals: *In this lesson, you'll subtract integers.*

MATHEMATICS MINI-LAB — Modeling with Manipulatives

The diagram at the right models the expression $3 - (-3)$. Counting the number of counters on the mat after the counters are removed shows that $3 - (-3) = 6$.

Your Turn Use counters to find $-3 - (-2)$, $4 - 6$, and $-5 - 1$.

TALK ABOUT IT

a. Compare the solution of $-3 - (-2)$ to the solution of $-3 + 2$.
b. Compare the solution of $4 - 6$ to the solution of $4 + (-6)$.
c. Compare the solution of $-5 - 1$ to the solution of $-5 + (-1)$.

a–c. The solutions are the same.

Learning the Concept

Adding and subtracting are inverse operations that "undo" each other. Similarly, when you add opposites, like 4 and −4, the sum is 0.

LOOK BACK You can review inverse operations in Lesson 1-8.

An integer and its opposite are called **additive inverses** of each other.

Additive Inverse Property
In words: The sum of an integer and its additive inverse is zero.

Arithmetic	**Algebra**
$5 + (-5) = 0$	$x + (-x) = 0$

At the beginning of the lesson, you were asked to compare the result of subtracting an integer with the result of adding its additive inverse.

Subtraction	**Addition of Additive Inverse**
$-3 - (-2) = -1$	$-3 + 2 = -1$
$4 - 6 = -2$	$4 + (-6) = -2$
$-5 - 1 = -6$	$-5 + (-1) = -6$

These examples suggest a method for subtracting integers.

Subtracting Integers
In words: To subtract an integer, add its additive inverse.

Arithmetic	**Algebra**
$3 - 5 = 3 + (-5)$	$a - b = a + (-b)$

Lesson 2-5 Subtracting Integers **89**

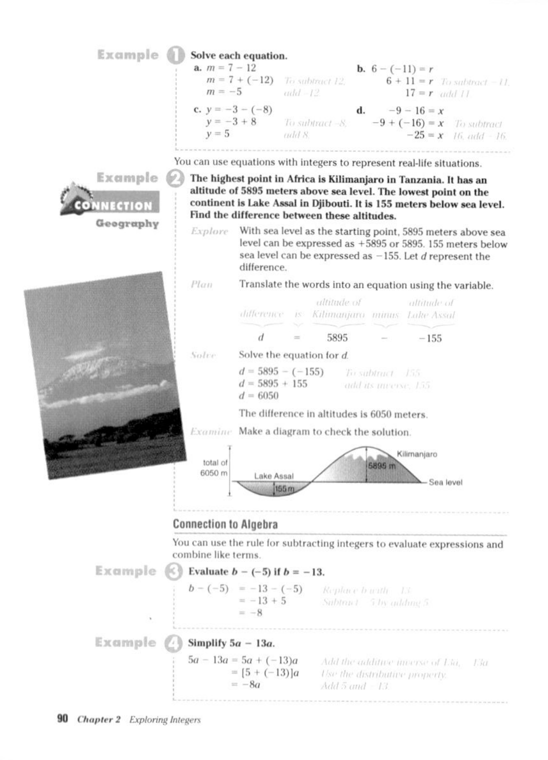

Example 1 **Solve each equation.**

a. $m = 7 - 12$
$m = 7 + (-12)$ *To subtract 12, add −12.*
$m = -5$

b. $6 - (-11) = r$
$6 + 11 = r$ *To subtract −11, add 11.*
$17 = r$

c. $y = -3 - (-8)$
$y = -3 + 8$ *To subtract −8, add 8.*
$y = 5$

d. $-9 - 16 = x$
$-9 + (-16) = x$ *To subtract 16, add −16.*
$-25 = x$

You can use equations with integers to represent real-life situations.

Example 2 **CONNECTION Geography** **The highest point in Africa is Kilimanjaro in Tanzania. It has an altitude of 5895 meters above sea level. The lowest point on the continent is Lake Assal in Djibouti. It is 155 meters below sea level. Find the difference between these altitudes.**

Explore With sea level as the starting point, 5895 meters above sea level can be expressed as +5895 or 5895. 155 meters below sea level can be expressed as −155. Let *d* represent the difference.

Plan Translate the words into an equation using the variable.

difference *is* *altitude of Kilimanjaro* *minus* *altitude of Lake Assal*
$d = 5895 - (-155)$

Solve Solve the equation for *d*.

$d = 5895 - (-155)$ *To subtract −155, add its inverse, 155.*
$d = 5895 + 155$
$d = 6050$

The difference in altitudes is 6050 meters.

Examine Make a diagram to check the solution.

Connection to Algebra

You can use the rule for subtracting integers to evaluate expressions and combine like terms.

Example 3 **Evaluate $b - (-5)$ if $b = -13$.**

$b - (-5) = -13 - (-5)$ *Replace b with −13.*
$= -13 + 5$ *Subtract −5 by adding 5.*
$= -8$

Example 4 **Simplify $5a - 13a$.**

$5a - 13a = 5a + (-13)a$ *Add the additive inverse of 13a, −13a.*
$= [5 + (-13)]a$ *Use the distributive property.*
$= -8a$ *Add 5 and −13.*

90 *Chapter 2 Exploring Integers*

Each lesson follows a straightforward, format —

> **Setting Goals**
> **Modeling**
> **Learning the Concept**
> **Checking for Understanding**
> **Exercises: Practicing and Applying the Concept**

— that focuses students on mathematical concepts and applications.

Setting Goals

Each lesson centers around the lesson goals or objectives. Students quickly see what performance is indicated and expected. "In this lesson, you'll ... ."

Modeling

Every lesson begins with a modeling activity designed to motivate students to learn the concept. The three types of modeling activities are Modeling a Real-World Application, Modeling with Manipulatives, and Modeling with Technology.

Learning the Concept

Completely worked-out examples with clear explanations support the objectives and match the exercises in the Guided Practice and Independent Practice. Examples include not only standard mathematical concepts, but also applications, connections, and integration.

Checking Your Understanding

This section ensures that all students are engaged in the lesson and understand the concepts.

Communicating Mathematics

Students work in small groups or as a whole class to define, explain, describe, decide, make lists, draw models, use symbols, or create graphs.

Modeling Mathematics and **Math Journal** activities, which further strengthen communication skills, appear in many lessons.

Guided Practice

Keyed to the lesson objectives, Guided Practice presents a sample of the exercises in the Practice section.

Exercises: Practicing and Applying the Concept

The Exercises always include **Independent Practice, Critical Thinking, Applications** and **Problem Solving**, and **Mixed Review**.

Independent Practice

The Independent Practice exercises are separated into A, B, and C sections, indicated *only* in the Teacher's Wraparound Edition. The A and B sections generally match the Guided Practice exercises in a 3:1 ratio. The C section offers challenging, thought-provoking questions that are not always matched by examples in the text.

Critical Thinking

Each lesson contains critical thinking exercises, in which students explain, evaluate, and justify mathematical concepts and relationships.

Applications and Problem Solving

Students find many opportunities to apply concepts to both real-life and mathematical problem situations.

Mixed Review

This spiraled, cumulative review comprises about 15% of the total number of exercises in each lesson and includes at least one application, connection, or integration problem. Glencoe's Mixed Reviews are designed to prepare students for success on proficiency tests and end-of-course exams.

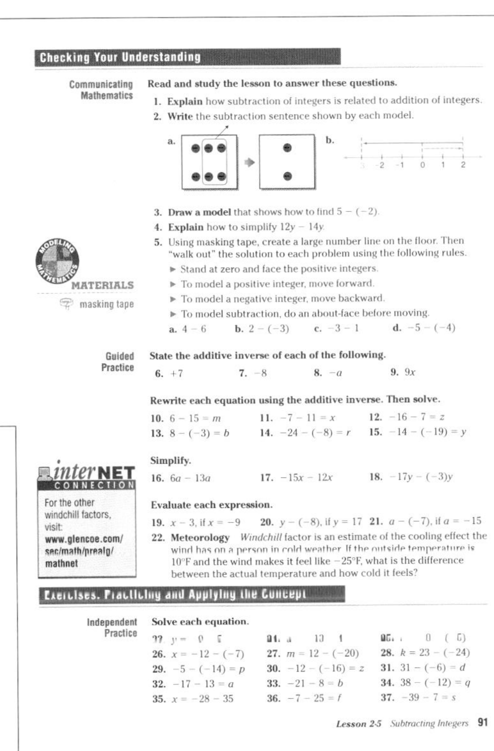

Checking Your Understanding

Communicating Mathematics

Read and study the lesson to answer these questions.

1. **Explain** how subtraction of integers is related to addition of integers.
2. **Write** the subtraction sentence shown by each model.

a. b.

3. **Draw a model** that shows how to find $5 - (-2)$.
4. **Explain** how to simplify $12y - 14y$.
5. Using masking tape, create a large number line on the floor. Then "walk out" the solution to each problem using the following rules.
   - Stand at zero and face the positive integers.
   - To model a positive integer, move forward.
   - To model a negative integer, move backward.
   - To model subtraction, do an about-face before moving.

   a. $4 - 6$ b. $2 - (-3)$ c. $-3 - 1$ d. $-5 - (-4)$

MODELING MATHEMATICS
MATERIALS
masking tape

Guided Practice

State the additive inverse of each of the following.

6. $+7$ 7. -8 8. $-a$ 9. $9x$

Rewrite each equation using the additive inverse. Then solve.

10. $6 - 15 = m$ 11. $-7 - 11 = x$ 12. $-16 - 7 = z$
13. $8 - (-3) = b$ 14. $-24 - (-8) = r$ 15. $-14 - (-19) = y$

Simplify.

16. $6a - 13a$ 17. $-15x - 12x$ 18. $-17y - (-3)y$

interNET CONNECTION
For the other windchill factors, visit: www.glencoe.com/sec/math/prealg/mathnet

Evaluate each expression.

19. $x - 3$, if $x = -9$ 20. $y - (-8)$, if $y = 17$ 21. $a - (-7)$, if $a = -15$
22. **Meteorology** *Windchill* factor is an estimate of the cooling effect the wind has on a person in cold weather. If the outside temperature is 10°F and the wind makes it feel like −25°F, what is the difference between the actual temperature and how cold it feels?

Exercises: Practicing and Applying the Concept

Independent Practice

Solve each equation.

23. [illegible] 24. [illegible] 25. [illegible]
26. $x = -12 - (-7)$ 27. $m = 12 - (-20)$ 28. $k = 23 - (-24)$
29. $-5 - (-14) = p$ 30. $-12 - (-16) = z$ 31. $31 - (-6) = d$
32. $-17 - 13 = a$ 33. $-21 - 8 = b$ 34. $38 - (-12) = q$
35. $x = -28 - 35$ 36. $-7 - 25 = f$ 37. $-39 - 7 = s$

Lesson 2-5 Subtracting Integers 91

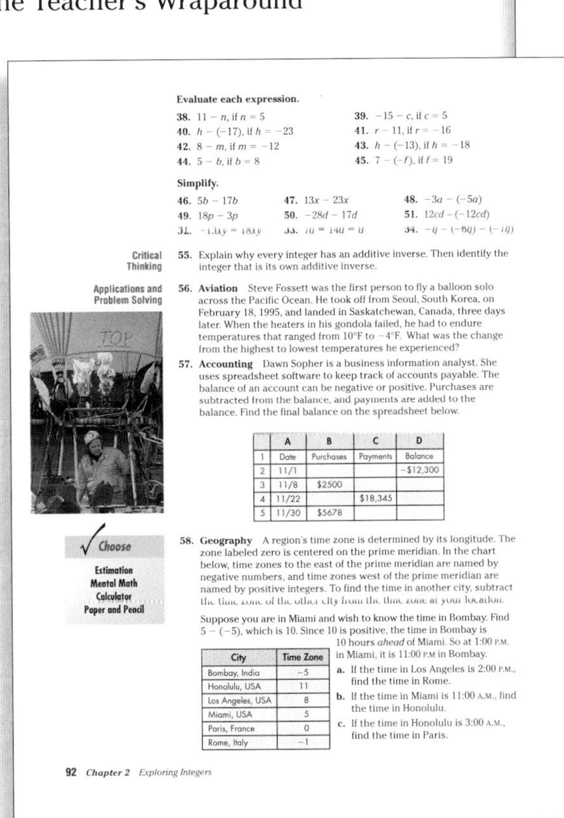

Evaluate each expression.

38. $11 - n$, if $n = 5$ 39. $-15 - c$, if $c = 5$
40. $h - (-17)$, if $h = -23$ 41. $r - 11$, if $r = -16$
42. $8 - m$, if $m = -12$ 43. $h - (-13)$, if $h = -18$
44. $5 - b$, if $b = 8$ 45. $7 - (-f)$, if $f = 19$

Simplify.

46. $5b - 17b$ 47. $13x - 23x$ 48. $-3a - (-5a)$
49. $18p - 3p$ 50. $-28d - 17d$ 51. $12cd - (-12cd)$
52. $-13xy - 18xy$ 53. $7a - 14a - a$ 54. $-q - (-6q) - (-7q)$

Critical Thinking

55. Explain why every integer has an additive inverse. Then identify the integer that is its own additive inverse.

Applications and Problem Solving

56. **Aviation** Steve Fossett was the first person to fly a balloon solo across the Pacific Ocean. He took off from Seoul, South Korea, on February 18, 1995, and landed in Saskatchewan, Canada, three days later. When the heaters in his gondola failed, he had to endure temperatures that ranged from 10°F to −4°F. What was the change from the highest to lowest temperatures he experienced?
57. **Accounting** Dawn Sopher is a business information analyst. She uses spreadsheet software to keep track of accounts payable. The balance of an account can be negative or positive. Purchases are subtracted from the balance, and payments are added to the balance. Find the final balance on the spreadsheet below.

	A	B	C	D
1	Date	Purchases	Payments	Balance
2	11/1			−$12,300
3	11/8	$2500		
4	11/22		$18,345	
5	11/30	$5678		

Choose
Estimation
Mental Math
Calculator
Paper and Pencil

58. **Geography** A region's time zone is determined by its longitude. The zone labeled zero is centered on the prime meridian. In the chart below, time zones to the east of the prime meridian are named by negative numbers, and time zones west of the prime meridian are named by positive integers. To find the time in another city, subtract the time zone of the other city from the time zone at your location.

Suppose you are in Miami and wish to know the time in Bombay. Find $5 - (-5)$, which is 10. Since 10 is positive, the time in Bombay is 10 hours *ahead* of Miami. So at 1:00 P.M. in Miami, it is 11:00 P.M in Bombay.

City	Time Zone
Bombay, India	−5
Honolulu, USA	11
Los Angeles, USA	8
Miami, USA	5
Paris, France	0
Rome, Italy	−1

a. If the time in Los Angeles is 2:00 P.M., find the time in Rome.
b. If the time in Miami is 11:00 A.M., find the time in Honolulu.
c. If the time in Honolulu is 3:00 A.M., find the time in Paris.

92 Chapter 2 Exploring Integers

Mixed Review

59. Simplify $9m + 43m + (-16m)$. (Lesson 2-4)
60. **Oceanography** A submarine at 1300 meters below sea level descends an additional 1150 meters. What integer represents the submarine's position with respect to sea level? (Lesson 2-4)

61. Write the inequality for *4 feet is shorter than 6 feet.* (Lesson 2-3)
62. Name the absolute value of −15. (Lesson 2-1)
63. **Geometry** Graph (2, 1), (2, 4), and (5, 1) on a coordinate plane. Connect the points with line segments. What figure is formed? (Lesson 1-7)
64. Solve $6b = 90$ mentally. (Lesson 1-6)
65. Simplify $y + 9 + 14 + 2y$. (Lesson 1-5)
66. Rewrite $(a + 3) + 4$ using the associative property. (Lesson 1-4)

Statistics Use the information in the graph at the right to answer each question. (Lesson 1-3)

67. If r represents the number of no-hitters that Nolan Ryan pitched, who can be represented by $r - 3$?
68. If y represents the number of no-hitters Cy Young pitched, which other pitcher can be represented by the expression y?
69. If k represents the number of no-hitters Sandy Koufax pitched, write an expression that represents the number of no-hitters Nolan Ryan pitched.

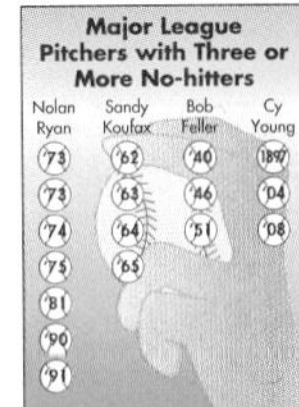

Self Test

Simplify. (Lesson 2-1)

1. $|-8|$ 2. $|-20| + |-19|$
3. Name the coordinate of each point graphed on the number line. (Lesson 2-1)

−7 −6 −5 −4 −3 −2 −1 0 1 2 3 4 5 6 7

On graph paper, draw coordinate axes. Then graph and label each point. (Lesson 2-2)

4. $Q(4, 3)$ 5. $R(-4, -3)$ 6. $C(-3, 0)$

Replace each ● with <, >, or =. (Lesson 2-3)

7. −9 ● −15 8. −11 ● 5 9. |−4| ● 4

Solve each equation. (Lesson 2-4, 2-5)

10. $-7 + 20 = a$ 11. $-12 + 14 + 8 = d$ 12. $-42 - 38 = q$ 13. $b = -34 - (-19)$
14. **Business** The formula $P = I - E$ is used to find the profit (P) when income (I) and expenses(E) are known. Find P if $I = \$15,525$ and $E = \$17,000$.

Lesson 2-5 Subtracting Integers 93

Glencoe's *Pre-Algebra* Exemplifies the NCTM Standards

CONTENT EMPHASIS

Pre-Algebra: An Integrated Transition to Algebra and Geometry implements the shift in perspective from just manipulative skills to a broader curriculum that is integrated and reflects the vital nature of mathematics.

"A broad range of topics should be taught, including number concepts, computation, estimation, functions algebra, statistics, probability, geometry, and measurement. Although each of these areas is valid mathematics in its own right, they should be taught as an integrated whole, not as isolated topics; the connections among them should be a prominent feature of the curriculum." NCTM Standards, page 67, 1989.

In each chapter, students explore the language of algebra, geometry, and other areas of mathematics in verbal, tabular, graphical, and symbolic form. Problem-solving activities and applications encourage students to model patterns and relationships with variables and functions and to construct, draw, measure, and classify geometric figures.

"Understanding the concept of variable is crucial to the study of algebra; a major problem in students' efforts to understand and so algebra results from their narrow interpretation of the term..." NCTM Standards, page 103, 1989.

"The study of geometry helps students represent and make sense of the world. Geometric models provide a perspective from which students can analyze and solve problems, and geometric interpretations can help make an abstract (symbolic) representation make more sense." NCTM Standards, page 112, 1989.

Each lesson is designed to motivate students to master the content they need to solve the problems involving applications, connections, and integration.

Glencoe's ***Pre-Algebra*** incorporates graphing-calculator and computer-software activities for discovery, problem solving, and modeling. The program balances modeling activities by providing students plenty of opportunity to develop critical paper-and-pencil skills such as computation involving integers and rationals and solving equations.

"Technology, including calculators, computers, and videos, should be used when appropriate. These devices and formats free students from tedious computations and allow them to concentrate on problem solving and other important content." NCTM Standards, page 67, 1989.

ASSESSMENT STANDARDS

The wide and varied collection of assessment tools and materials that accompany Glencoe's ***Pre-Algebra***, support the NCTM Assessment Standards. [Assessment Standards for School Mathematics, NCTM, 1995.]

Note: The bulleted headings and the quotations below are taken from the NCTM Assessment Standards, pages 11, 13, 15, 17, 19, and 21.

- **The Mathematics Standard** — "Assessment should reflect the mathematics that all students need to know and be able to do."
  Skills and problem solving are assessed through a broad range of short response exercises and longer-term, open-ended activities.

- **The Learning Standard** — "Assessment should enhance mathematics learning."
  Assessment is an integral part of the Communicating Mathematics, Guided Practice, and Exercises sections of each of Glencoe's ***Pre-Algebra*** lessons.

- **The Equity Standard** — "Assessment should promote equity."
  The wide variety of assessment tools enable all students to demonstrate what they know and can do.

- **The Openness Standard** — "Assessment should be an open process."
  Setting Goals tells students what they are expected to learn in each lesson. The examples, activities, and assessment match those objectives.

- **The Inferences Standard** — "Assessment should promote valid inferences about mathematics learning."
  The assortment of assessment tools provide many sources of evidence. Exercises, Mixed Review, and Chapter Tests are all correlated to the learning objectives and practice activities.

- **The Coherence Standard** — "Assessment should be a coherent process."
  The numerous, varied assessment tools are carefully designed to complement each other. They enable you to easily construct the assessment program that best fits your needs.

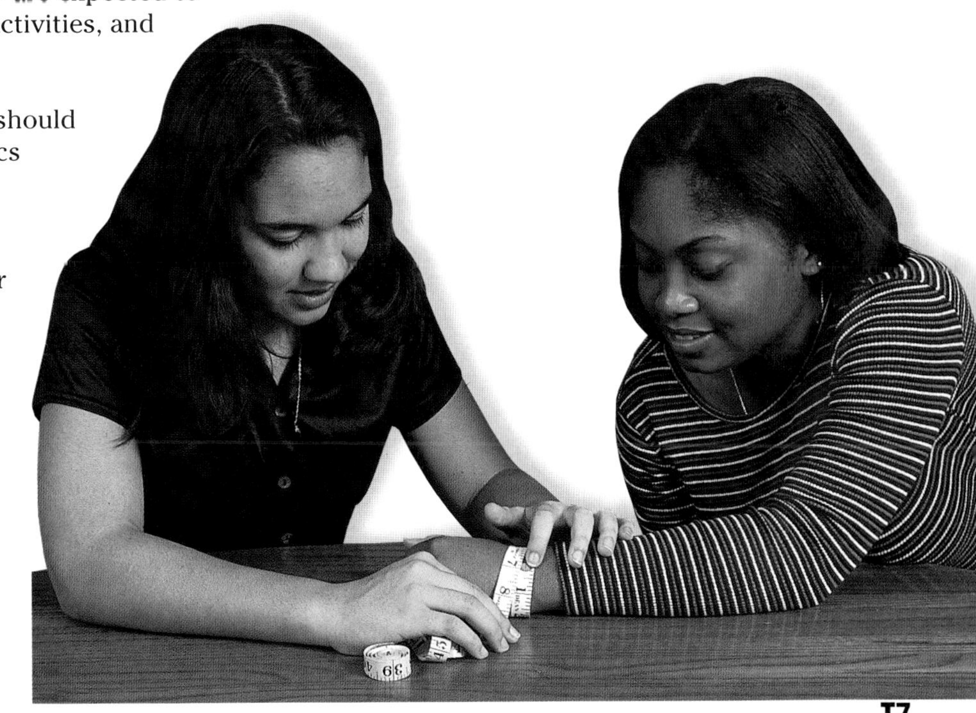

Learning Objectives & NCTM Standards Correlation

Lesson	Lesson Objectives	NCTM Standards
1-1	Use the four-step plan to solve real-life, math related problems. Choose an appropriate method of computation.	1-4, 7, 8
1-2	Use the order of operations to evaluate expressions.	1-5, 7, 8
1-3	Evaluate expressions containing variables. Translate verbal phrases into algebraic expressions.	1-4, 9
1-3B	Evaluate algebraic expressions using a graphing calculator.	1-4, 9
1-4	Identify properties of addition and multiplication. Use properties to solve problems.	1-4, 6, 7, 9
1-5	Simplify algebraic expressions using the distributive property.	1-4, 6, 7, 9
1-5B	Visualize the distributive property by modeling a geometric interpretation.	1-4, 6, 9
1-6	Identify and solve open sentences.	1-4, 7, 9
1-7	Use ordered pairs to locate points, to organize data, and to explore how certain quantities in expressions are related to each other.	1-4, 9, 12
1-8	Use inverse operations to solve equations. Solve problems using equations.	1-4, 6, 8, 9
1-9	Write and solve inequalities.	1-4, 9
1-10A	Gather and record data.	1-4, 10
1-10	Gather and record data in a frequency table or bar graph.	1-4, 10
2-1	Graph integers on a number line. Find the absolute value of an expression.	1-5, 12, 14
2-1B	Organize data into line plots.	1-4, 10, 12
2-2	Graph points in all quadrants of the coordinate plane.	1-5, 8, 12
2-2B	Plot points on a graphing calculator.	1-5, 12
2-3	Compare and order integers.	1-5, 12
2-4A	Use counters to model addition of integers.	1-5, 7, 12
2-4	Add integers.	1-5, 7, 12
2-5A	Use counters to model subtraction of integers.	1-5, 7, 12
2-5	Subtract integers.	1-5, 7, 12
2-6	Solve problems by looking for a pattern.	1-5, 8
2-7A	Use counters to model multiplication of integers.	1-5, 7, 12
2-7	Multiply integers.	1-5, 7, 12
2-8	Divide integers.	1-5, 7, 12
3-1	Solve problems by eliminating possibilities.	1-4
3-2A	Use cups and counters to solve equations.	1-4, 9
3-2	Solve equations by using the addition and subtraction properties of equality.	1-6, 9
3-3	Solve equations by using the multiplication and division properties of equality.	1-5, 9
3-4	Solve problems by using formulas.	1-4, 9
3-5A	Discover the greatest possible perimeter for a given area.	1-4, 8, 12, 13
3-5	Find the perimeter and area of rectangles and squares.	1-4, 9, 12, 13
3-5B	Use a graphing calculator to explore areas of irregular figures on a coordinate plane.	1-4, 12, 13

Lesson	Lesson Objectives	NCTM Standards
3-6	Solve inequalities by using the addition and subtraction properties of inequalities. Graph the solution sets of inequalities.	1-5, 9
3-7	Solve inequalities by using the multiplication and division properties of inequalities.	1-5, 9
3-8	Solve verbal problems by translating them into equations and inequalities. Write verbal problems for equations and inequalities.	1-4, 9
4-1	Determine whether one number is a factor of another.	1-6, 8, 9
4-2	Use powers and exponents in expressions and equations.	1-8
4-2B	Use a graphing calculator to evaluate expressions involving exponents.	2, 3, 5, 9
4-3	Solve problems by drawing diagrams.	1-6, 8
4-4	Identify prime and composite numbers. Write the prime factorizations of composite numbers.	1-6, 8
4-4B	Use a graphing calculator to investigate factor patterns.	1-3, 6, 8
4-5	Find the greatest common factor for two or more integers or monomials.	1-6, 8, 9
4-6A	Use models to investigate fractions.	1-6
4-6	Simplify fractions using the GCF.	1-6
4-7	Find the least common multiple of two or more numbers. Compare fractions.	1-6, 9
4-8	Multiply and divide monomials.	1-9
4-9	Use negative exponents in expressions and equations.	1-6, 9
5-1	Identify and compare rational numbers. Rename decimals as fractions.	1-6
5-2	Estimate sums and differences of decimals and fractions.	1-4, 7, 10
5-3	Add and subtract decimals.	1-4, 7
5-4	Add and subtract fractions with like denominators.	1-4, 7, 10
5-5	Add and subtract fractions with unlike denominators.	1-4, 7
5-6	Solve equations with rational numbers.	1-4, 7, 9
5-7	Solve inequalities with rational numbers.	1-4, 7, 9
5-8	Use deductive and inductive reasoning	1-4, 6-8
5-9	Find the terms of arithmetic sequences. Represent a sequence algebraically.	1-4, 7, 8
5-9B	Investigate patterns in the Fibonacci sequence.	1-4, 6, 8
6-1	Write fractions as terminating or repeating decimals. Compare rational numbers.	1-6, 8
6-2	Estimate products and quotients of rational numbers.	1-5, 7
6-3	Multiply fractions.	1-5, 7
6-4	Divide fractions using multiplicative inverses.	1-7, 9
6-5A	Multiply and divide decimals using decimal models.	1-5
6-5	Multiply and divide decimals.	1-8
6-6A	Use the mean, median, and mode of numbers to analyze data.	1-5, 10, 12
6-6	Use the mean, median, and mode as measures of central tendency.	1-5, 10
6-6B	Use a graphing calculator to find the mean and median of a set of data.	1-5, 10, 12
6-7	Solve equations and inequalities containing rational numbers.	1-5, 8, 9
6-8	Recognize and extend geometric sequences and represent them algebraically.	1-5, 8
6-9	Write numbers in scientific notation.	1-5, 7
7-1	Solve problems by working backward.	1-4
7-2A	Use cups and counters and to model and solve two-step equations.	1-4, 9
7-2	Solve equations that involve more than one operation.	1-4, 9, 12
7-3	Solve verbal problems by writing and solving equations.	1-4, 9
7-4	Find the circumference of a circle.	1-4, 9, 12
7-5A	Use cups and counters to model and solve equations with the variable on each side.	1-4, 9

Lesson	Lesson Objectives	NCTM Standards
7-5	Solve equations with the variable on each side.	1-4, 9
7-6	Solve inequalities that involve more than one operation.	1-5, 9
7-7	Solve verbal problems by writing and solving inequalities.	1-4, 9
7-8	Convert measurements in the metric system.	1-4, 13
8-1	Use tables and graphs to represent relations and functions.	1-4, 6, 8, 10, 12
8-2A	Graph number pairs on a coordinate system to determine if there is a relationship.	1-6, 10, 12
8-2	Construct and interpret scatter plots.	1-5, 8, 10
8-3	Find solutions for relations with two variables and graph the solutions.	1-5, 8, 9
8-3B	Investigate graphs that are nonlinear.	1-6, 8
8-4	Determine whether an equation represents a function. Find functional values for a given function.	1-5, 8, 9
8-5	Solve problems using graphs.	1-5, 8, 9, 12
8-6	Find the slope of a line.	1-5, 8-10
8-6B	Use cylindrical objects to discover a connection between algebra and geometry.	1-6, 8
8-7	Graph a linear equation using the *x*- and *y*-intercepts or using the slope and *y*-intercept.	1-5, 8, 9
8-7B	Use a graphing calculator to graph several functions to determine if any family traits exist.	1-6, 8
8-8	Solve systems of linear equations by graphing.	1-5, 8, 9
8-9A	Use a graphing calculator to investigate the graphs of inequalities.	1-6, 8
8-9	Graph linear inequalities.	1-5, 8, 9
9-1	Write ratios as fractions in simplest form and determine unit rates.	1-5, 7, 12
9-2	Solve problems by making a table, chart, or list.	1-5, 10, 12
9-3A	Determine whether a game is mathematically fair or unfair.	1-5, 10
9-3	Find the probability of simple events.	1-4, 10
9-4	Use proportions to solve problems.	1-5
9-4B	Use proportions to estimate a population.	1-5, 10
9-5	Use the percent proportion to write fractions as percents. Solve problems involving percents.	1-5, 9
9-6	Use a sample to predict the actions of a larger group.	1-5, 10
9-7	Express decimals and fractions as percents and vice versa.	1-5
9-8	Use percents to estimate.	1-5, 7, 10
9-9	Solve percent problems using percent equations.	1-5, 7
9-10	Find percent of increase or decrease.	1-5, 7
10-1	Display and interpret data in stem-and-leaf plots.	1-4, 10
10-2	Use measures of variation to compare data.	1-4, 10
10-3	Use box-and-whisker plots, pictographs, and line graphs to display data.	1-4, 10
10-3B	Use a graphing calculator to construct statistical graphs.	1-4, 10
10-4	Recognize when statistics are misleading.	1-4, 10
10-5	Use tree diagrams or the Fundamental Counting Principal to count outcomes.	1-4, 11
10-6A	Explore and use permutations and combinations.	1-4, 11
10-6	Use permutations and combinations.	1-4, 11
10-7	Find the odds of a simple event.	1-4, 11
10-8	Investigate problems using simulations.	1-4, 11
10-8B	Conduct experiments to find experimental probability.	1-4, 11
10-9	Find the probability of independent and dependent events.	1-4, 11
10-10	Find the probability of mutually exclusive or inclusive events.	1-4, 11
11-1	Identify points, lines, planes, rays, segments, angles, and parallel, perpendicular, and skew lines. Classify angles as acute, right, or obtuse.	1-4, 12
11-1B	Construct congruent line segments and angles with a compass and straightedge.	1-4, 12

Lesson	Lesson Objectives	NCTM Standards
11-2	Make a circle graph to illustrate data.	1-4, 10, 12
11-3	Identify the relationships of vertical, adjacent, complementary, and supplementary angles, and angles formed by two parallel lines and a transversal.	1-4, 12
11-3B	Use a graphing calculator to determine whether two lines are parallel.	1-4, 12
11-4	Find the missing angle measure of a triangle. Classify triangles by angles and sides.	1-4, 12
11-5	Identify congruent triangles and corresponding parts of congruent triangles.	1-4, 12
11-6	Identify corresponding parts and find missing measures of similar triangles. Solve problems involving indirect measurement using similar triangles.	1-4, 12
11-7	Find the missing angle measure of a quadrilateral. Classify quadrilaterals. Solve problems involving similar quadrilaterals.	1-4, 12
11-8	Classify polygons and determine the sum of the measures of the interior and exterior angles of a polygon.	1-4, 12
11-8B	Investigate tessellations using regular polygons.	1-4, 12
11-9	Identify and draw reflections, translations, and rotations and symmetric figures.	1-4, 12
12-1A	Use a geoboard to find areas of shapes that are not rectangular.	1-4, 12, 13
12-1	Find areas of parallelograms, triangles, and trapezoids.	1-4, 12, 13
12-1B	Explore and construct fractals.	1-4, 12, 13
12-2	Find areas of circles.	1-4, 12, 13
12-3	Find probabilities using area models.	1-4, 11, 12, 13
12-3B	Use a graphing calculator to experiment with area and probability.	1-4, 11, 12, 13
12-4	Solve problems using a model or a drawing.	1-4, 12, 13
12-5	Find surface areas of triangular and rectangular prisms and circular cylinders.	1-4, 12, 13
12-6	Find surface areas of pyramids and cones.	1-4, 12, 13
12-7A	Investigate volume.	1-4, 12, 13
12-7	Find volumes of prisms and circular cylinders.	1-4, 12, 13
12-8	Find volumes of pyramids and cones.	1-4, 12, 13
12-8B	Investigate how changing the dimensions of a solid affects its surface area and volume.	1-4, 12, 13
13-1	Find and use squares and square roots.	1-4, 6, 12, 13
13-2	Use Venn diagrams to solve problems.	1-4, 12
13-3	Identify numbers in the real number system. Solve equations by finding square roots.	1-4, 6, 12, 13
13-4	Use the Pythagorean Theorem to find the length of the side of a right triangle and to solve problems.	1-4, 12, 13
13-4B	Use the Pythagorean Theorem to graph irrational numbers.	1-4, 12, 13
13-5	Find missing measures in 30°-60° and 45°-45° right triangles.	1-4, 12, 13
13-6A	Explore the relationships of the sides of right triangles.	1-4, 12, 13
13-6	Find missing sides and angles of triangles using the sine, cosine, and tangent ratios.	1-4, 12, 13
13-6B	Use a graphing calculator to explore the slope of a line.	1-4, 12, 13
13-7	Solve problems by using the trigonometric ratios.	1-4, 12, 13
14-1	Identify and classify polynomials. Find the degree of a polynomial.	1-4, 9
14-1B	Use algebra tiles to model polynomials.	1-4, 9
14-2	Add polynomials.	1-4, 9
14-3	Subtract polynomials.	1-4, 9
14-4	Find powers of monomials.	1-4, 9
14-5A	Use algebra tiles to multiply a monomial and a polynomial.	1-4, 9
14-5	Multiply a polynomial by a monomial.	1-4, 9
14-6	Multiply binomials.	1-4, 9
14-6B	Use algebra tiles to factor simple trinomials.	1-4, 9

Planning Your *Pre-Algebra* Course

40- to 50-Minute Class Periods

The chart below gives suggested pacing guides for two options, Standard and Honors, for four 9-week grading periods. The Standard option covers Chapters 1-13 while the Honors option covers Chapters 1-14. Generally, one day is allotted for each lesson including the Mathematics Labs, one day for the Chapter Study Guide and Assessment, and one day for a chapter test. *A pacing guide that addresses both of these options is provided in each lesson of the Teacher's Wraparound Edition.*

The total days suggested for each option is about 165 days. This allows for teacher flexibility in planning due to school cancellation or shortened class periods.

Grading Period	Standard		Honors	
	Chapter	Days	Chapter	Days
1	1	14	1	13
	2	12	2	11
	3	12	3	11
	4-1 to 4-4	4	4-1 to 4-6	6
2	4-5 to end	9	4-7 to end	5
	5	13	5	12
	6	13	6	12
	7-1 to 7-5	7	7	11
3	7-6 to end	5	8	12
	8	13	9	13
	9	14	10	13
	10-1 to 10-8B	10	11-1 to 11-2	3
4	10-9 to end	4	11-3 to end	10
	11	13	12	11
	12	13	13	10
	13	11	14	9

For day-by-day lesson plans, please refer to Glencoe's *Pre-Algebra* Lesson Planning Guide.

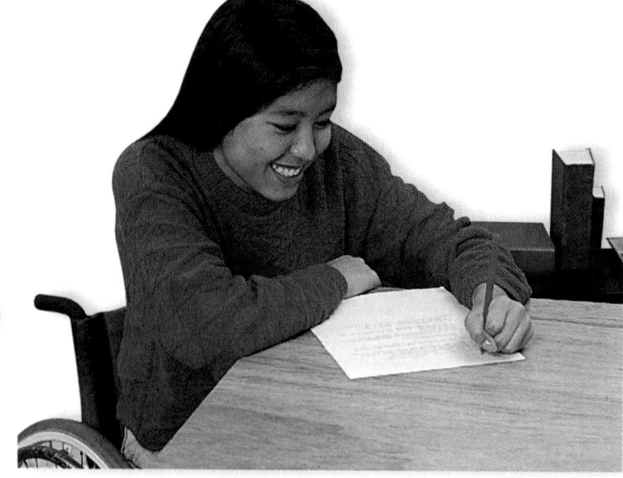

Block Schedule, 90-Minute Class Periods

The chart below gives a suggested pacing guide for teaching Glencoe's ***Pre-Algebra*** using block scheduling in 1 semester (classes meet every day) or 1 year (classes meet about 85 days). The guide is based on teaching Chapters 1-13. It is suggested that Mathematics Labs be paired with their corresponding lessons for one class period while most other lessons be taught individually. *The pacing guide in each lesson of the Teacher's Wraparound Edition also addresses block scheduling.*

Chapter	Class Periods
1	7
2	7
3	7
4	6
5	6
6	7
7-1 to 7-5	4
7-6 to end	3
8	7
9	7
10	7
11	7
12	7
13	6

For day-by-day lesson plans, please refer to Glencoe's *Pre-Algebra* Block Scheduling Booklet.

Classroom Vignettes

Glencoe's unique **Classroom Vignettes** are classroom-tested teaching suggestions made by experienced mathematics educators. Each vignette was written by a teacher who uses Glencoe's ***Pre-Algebra*** or by a Glencoe reviewer, consultant, or author. Glencoe thanks these outstanding mathematics educators for their unique contributions.

Yuria Alban, p. 171
Dade County Public Schools
Miami, Florida

Cynthia Anderson, p. 358
Trion High School
Trion, Georgia

Stacey Bondon, p. 67
Amos Alonzo Stagg High School
Palos Hills, Illinois

Tricia Burgess, p. 549
St. Paul's Episcopal School
Mobile, Alabama

K. Caliendo, p. 289
Hamden Middle School
Hamden, Connecticut

Niva Costa, p. 613
Ambridge Area High School
Ambridge, Pennsylvania

Naomi Harwood, p.462
Rutherfordton Spindle High School
Rutherfordton, North Carolina

Nancy E. Herald, p. 6
Wolfe County High School
Campton, Kentucky

Loren Holthaus, p. 556
Big Lake Middle School
Big Lake, Minnesota

Bonnie Langan, pp. 234, 433, 445
Northern Middle School
Accident, Maryland

William D. Leschensky, p. 691
Author

Robert Mandel, p. 83
Hamden Middle School
Hamden, Connecticut

Cleo M. Meek, p. 375, 715
North Carolina Department of
Instruction, Retired
Raleigh, North Carolina

Phyllis Plasky, p. 140
Camden High School
Camden, New Jersey

Dawn M. Rueter, p. 541
Amos Alonzo Stagg High School
Palos Hills, Illinois

Darlene Thurner, p. 104
Pine-Richland Middle School
Gibsonia, Pennsylvania

Paul Todd, p. 574
Onondaga Hill Middle School
Syracuse, New York

Denny Wagoner, pp. 27, 517
Cowan High School
Muncie, Indiana

Richard Whited, p 229
Kecoughtan High School
Hampton, Virginia

Math in the News

Glencoe's unique **Math in the News** features an issue, such as fashion, recycling, and computers, that young people can to. Each of these are based on actual articles that have appeared in a newspaper or other publication.

The **Employment Opportunity** and **Statistical Snapshot** are related to the content of **Math in the News**.

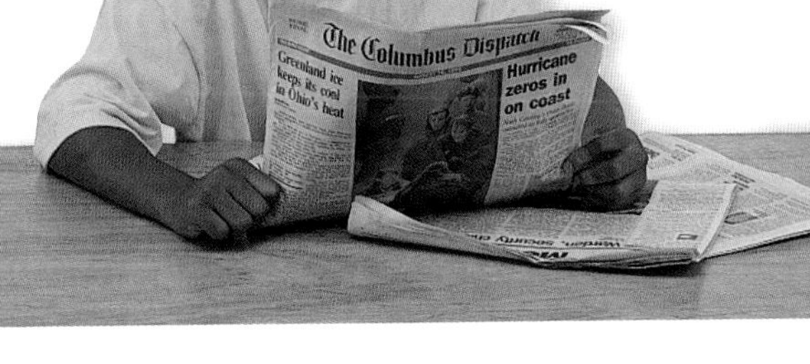

Fashion
Chapter 1, p. 4

Recycling
Chapter 2, p. 64

Braille
Chapter 3, p. 116

Computers
Chapter 4, p. 168

Fitness
Chapter 5, p. 222

Computers
Chapter 6, p. 272

Entertainment
Chapter 7, p. 328

Transportation
Chapter 8, p. 370

Genetics
Chapter 9, p. 430

Marketing
Chapter 10, p. 484

Sailing
Chapter 11, p. 546

Archaeology
Chapter 12, p. 608

Technology
Chapter 13, p. 662

Games
Chapter 14, p. 704

Internet Connections

The sites referenced in Glencoe's Internet Connections are not under the control of Glencoe. Therefore, Glencoe can make no representation concerning the content of these sites. Extreme care has been taken to list only reputable links by using educational and government sites whenever possible. Internet searches have been used that return only sites that contained no content apparently intended for mature audiences. The following list of mathematics internet sites may prove helpful. Useful search tools include http://webcrawler.com/, http://www.yahoo.com/search.html, and http://www.microsoft.com/search.

Internet Site	Comments
http://www.yahoo.com/science/mathematics	An ideal starting point that has links to other math sites.
http://www.tc.cornell.edu/Edu/MathSciGateway	Provides links to math and science resources. For grades 9-12.
http://forum.swarthmore.edu/dr.math/	Students in grades K-12 can ask Dr. Math their own questions.
http://www.enc.org/	This is the Eisenhower clearinghouse for K-12 math and science instructional materials.
http://forum.swarthmore.edu/mathmagic	K-12 telecommunications project in Texas that uses computers to increase problem-solving and communication skills.

Glencoe's *Pre-Algebra* Research Activities

Glencoe's ***Pre-Algebra: An Integrated Transition to Algebra and Geometry***, as well as the entire Glencoe Mathematics Series, is the product of ongoing, classroom-oriented research that involves students, teachers, curriculum supervisors, administrators, parents, and college-level mathematics educators.

The programs that make up the Glencoe Mathematics Series are currently being used by millions of students and tens of thousands of teachers. The key reason that Glencoe Mathematics programs are so successful in the classroom is the fact that each Glencoe author team is a mix of practicing classroom teachers, curriculum supervisors, and college-level mathematics educators. Glencoe's balanced author teams help ensure that Glencoe Mathematics programs are both practical and progressive.

Prior to publication of a Glencoe program, typical research activities include:

- a review of educational research and recommendations made by groups such as the NCTM
- mail surveys of mathematics educators
- discussion groups involving mathematics teachers, department heads, and supervisors
- focus groups involving mathematics educators
- face-to-face interviews with mathematics educators
- telephone surveys of mathematics educators
- in-depth analyses of manuscript by a wide range of reviewers and consultants
- field tests in which students and teachers use pre-publication manuscript in the classroom

Feedback from teachers, curriculum supervisors, and even students who currently use Glencoe mathematics programs is also incorporated as Glencoe plans and publishes new and revised programs. Classroom Vignettes, which are printed in the Teacher's Wraparound Editions, are one result of this feedback.

All of this research and information is used by Glencoe's authors and editors to publish the best instructional resources possible.

WHY IS MATHEMATICS IMPORTANT?

Why do I need to study mathematics? When am I ever going to have to use math in the real world?

Many people, not just mathematics students, wonder why mathematics is important. ***Pre-Algebra*** is designed to answer those questions through ***integration***, ***applications***, and ***connections***.

Did you know that algebra and geometry are closely related? Topics from all branches of mathematics, like geometry, probability, and statistics, are integrated throughout the text.

You'll learn how to use proportions to relate the measures of similar triangles. (Lesson 11-6, pages 578–583)

How did they make those dinosaurs in *Jurassic Park?* They used rational numbers! In ***Pre-Algebra***, you'll study many ways that math is used in the real world.

The size of a model is related to the size of a real object by using a rational number. (Lesson 5-1, page 224)

What does mathematics have to do with biology? Mathematical topics are connected to the other subjects that you study.

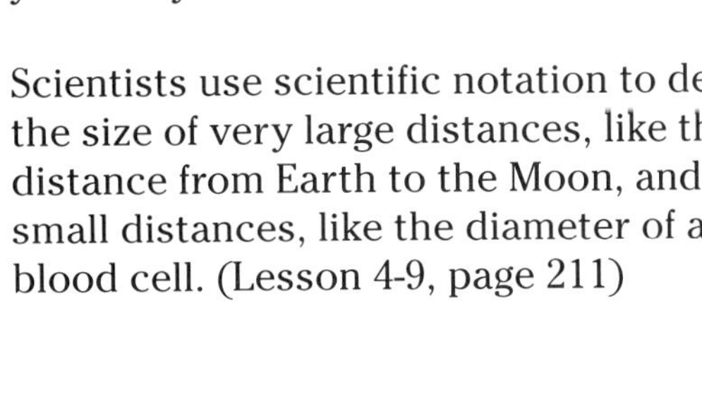

Scientists use scientific notation to describe the size of very large distances, like the distance from Earth to the Moon, and very small distances, like the diameter of a red blood cell. (Lesson 4-9, page 211)

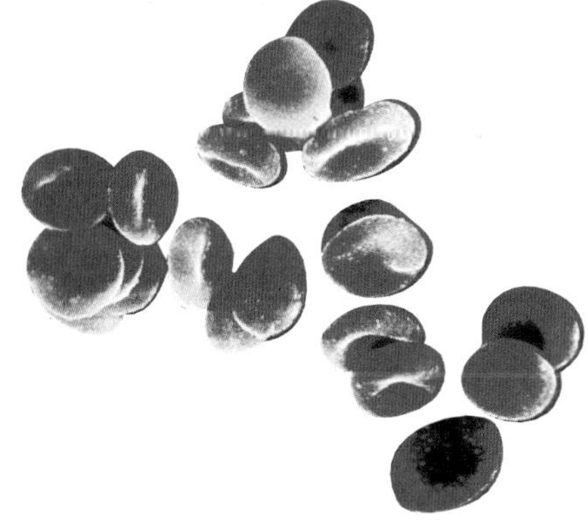

Authors

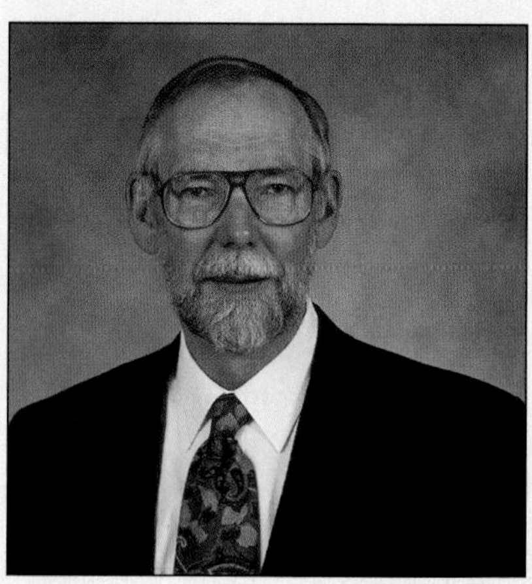

This program builds on the mathematical background students obtain in earlier grades and prepares them to succeed in algebra. Technology is included wherever appropriate, and links between mathematics and life situations abound. Math is fun!

William Leschensky is a former teacher of mathematics at Glenbard South High School and the College of DuPage in Glen Ellyn, Illinois. He has taught mathematics at every level of the high school curriculum, served as mathematics department chairperson, and is a former textbook editor and editorial director. Mr. Leschensky received his B.A. in mathematics from Cornell College and his M.A. in mathematics from the University of Northern Iowa. He has participated in special institutes in mathematics, physics, and computer science and is a frequent speaker at local, state, and regional conferences. Mr. Leschensky is a question writer for MATHCOUNTS, a nationwide series of competitions designed to stimulate student interest and achievement in mathematics.

Wm D Leschensky

Carol Malloy is an assistant professor of mathematics education at the University of North Carolina at Chapel Hill. Dr. Malloy previously taught middle and high school mathematics. She received her B.S. in mathematics and education from West Chester State College, her M.S.T. in mathematics from Illinois Institute of Technology, and her Ph.D. in curriculum and instruction from the University of North Carolina at Chapel Hill. Dr. Malloy is an active member of numerous local, state, and national professional organizations and boards, including National Council of Teachers of Mathematics (NCTM) and Association for Supervision and Curriculum Development (ASCD). Dr. Malloy also serves as an editor of the Benjamin Banneker Association newsletter. She frequently speaks at conferences and schools on the topics of equity and mathematics for all students.

Carol E. Malloy

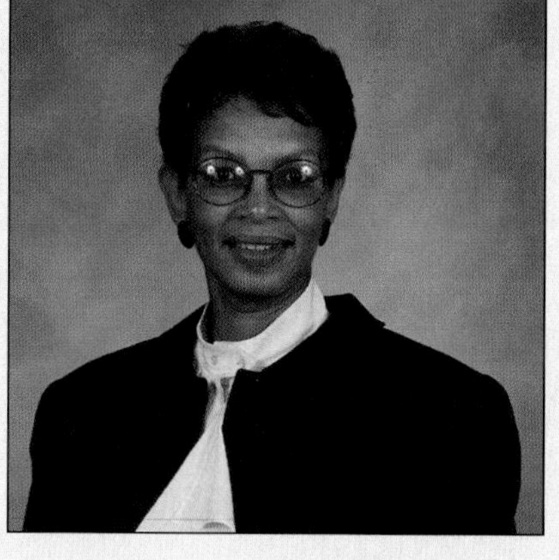

Mathematics teachers are excited about implementing the NCTM Standards in their mathematics classrooms. Glencoe's Pre-Algebra affords students with a strong foundation to high school mathematics through comprehensive real-world explorations, technology, and hands-on activities.

With the study of pre-algebra, you are taking your first steps toward a mathematics education that will be valuable to you throughout your lifetime. I know you will find it exciting and challenging.

Jack Price received his B.A. from Eastern Michigan University and his Doctorate in Mathematics Education from Wayne State University. Dr. Price has been active in mathematics education for over 40 years, 38 of those years at grades K through 12. In his current position as co-director of the Center for Education and Equity in Mathematics, Science, and Technology at California State Polytechnic University, he teaches mathematics and methods courses for preservice teachers and consults with school districts on curriculum change. Dr. Price is a past president of the National Council of Teachers of Mathematics, is a frequent speaker at professional conferences, conducts many in-service workshops, and is author of numerous mathematics and science instructional materials. Dr. Price is co-author on Glencoe's *Mathematics: Applications and Connections* and *Interactive Mathematics: Activities and Investigations*.

Jim Rath has 30 years of classroom experience in teaching mathematics at every level of the high school curriculum. He is a former mathematics teacher and department chairperson at Darien High School in Darien, Connecticut. Mr. Rath earned his B.A. in philosophy from the Catholic University of America and his M.Ed. and M.A. in mathematics from Boston College. He held an appointment as a Visiting Fellow in the Mathematics Department at Yale University in New Haven, Connecticut for the academic year 1994-1995. Mr. Rath is a co-author on Glencoe's *Algebra 1* and *Algebra 2*.

What you learn today is a building block for what comes tomorrow. That is the reason why a student needs to do his or her best each day . . . what you learn in pre-algebra is building a strong foundation for you.

Mathematics today is a thrilling mix of present reality and investigation. Students will travel on paths of exploration to reach into the future. This textbook will make the journey a unique and exciting experience.

Yuria Alban received her B.S. in Mathematics from Armstrong State College, her M.A. in Mathematics from the University of Georgia, and her Educational Specialist degree in Computer Education from Nova University. Ms. Alban has more than twenty years of teaching experience in middle school, high school, and college mathematics. She is currently a District Mathematics Supervisor for the Dade County Public Schools in Florida. Ms. Alban is a past President of the Dade County Council of Teachers of Mathematics and has received numerous teaching awards at the local and state levels. Recently, Ms. Alban participated in U.S. Justice Department programs to assess students' needs and advise on mathematics curriculum in Guantanamo Bay, Cuba and Venezuela.

Consultants and Reviewers

Consultants

Gilbert J. Cuevas
Professor of Mathematics Education
University of Miami
Coral Gables, Florida

Dr. Luis Ortiz-Franco
Consultant, Diversity
Associate Professor of Mathematics
Chapman University
Orange, California

David Foster
Glencoe Author and Mathematics Consultant
Morgan Hill, California

Daniel Marks
Consultant, Real-World Applications
Associate Professor of Mathematics
Auburn University at Montgomery
Montgomery, Alabama

Melissa McClure
Consultant, Tech Prep
Mathematics Consultant
Teaching for Tomorrow
Fort Worth, Texas

Cleo M. Meek, Ed. D.
Mathematics Consultant
NC Dept. of Public Instruction, Retired
Raleigh, North Carolina

Jim Zwick
Mathematics Department Chairman
Guion Creek Middle School—MSD of Pike Township
Indianapolis, Indiana

Reviewers

Robert M. Allen
Mathematics Teacher
Vista Verde Middle School
Phoenix, Arizona

Bruce I. Althouse, Sr.
Mathematics Department Chairperson
Northern Lebanon Junior/Senior High School
Fredericksburg, Pennsylvania

Glenn Aston-Reese
Mathematics Department Chairperson
Trinity Middle School
Washington, Pennsylvania

Mary R. Beard
Mathematics Teacher
Southeast Guilford Middle School
Greensboro, North Carolina

Ronald L. Bow
Mathematics Teacher
Henry Clay High School
Lexington, Kentucky

Alan Brown
Registrar, Adult Education
Springfield Adult Learning Center
Springfield, Massachusetts

Sharon A. Cichocki
Secondary Math Coordinator
Hamburg High School
Hamburg, New York

Debra M. Cline
Mathematics Teacher
Atkins Middle School
Winston-Salem, North Carolina

Linda Coutts
Mathematics Coordinator
Columbia Public Schools
Columbia, Missouri

Jean E. Ellis
Mathematics Teacher
Ruddiman Middle School
Detroit, Michigan

Gary A. Graves
Coordinator of Mathematics K-12
St. Joseph Public Schools
St. Joseph, Missouri

Earlene J. Hall
Administrator
Detroit Public Schools
Detroit, Michigan

Eileen D. Harris, Ph.D.
Mathematics/Science K-12 Specialist
Charlotte County School District
Port Charlotte, Florida

Richard Hedlund
Mathematics Teacher
South Broward High School
Hollywood, Florida

James P. Herrington
Mathematics/Science Department Chairperson
O'Fallon Township High School
O'Fallon, Illinois

Deborah D. Johnson
Supervisor of Mathematics
Camden City Public Schools
Camden, New Jersey

Nancy L. Jones
Mathematics Teacher
Aycock Middle School
Greensboro, North Carolina

Stephen R. Lightner
Mathematics Department Chairperson
D.S. Keith Junior High School
Altoona, Pennsylvania

Gaye S. McKinnon
Mathematics Teacher
Columbiana Middle School
Columbiana, Alabama

Bernadette M. Meglino
Mathematics Teacher
Lake Weir High School
Ocala, Florida

Nance Marie Minnick
Mathematics Teacher
Denver Public Schools
Denver, Colorado

Rhonda C. Niemi
Mathematics Teacher
Meyzeek Middle School
Louisville, Kentucky

Georgia Thielen O'Connor
Mathematics Instructor
Stevens High School
Rapid City, South Dakota

Pamela H. Ostrosky
Mathematics Teacher
Dake Junior High School
Rochester, New York

Marcia Perry
Mathematics Teacher
Palm Beach Gardens High School
Palm Beach Gardens, Florida

Odelia Schrunk
Mathematics Department Chairperson
Clinton High School
Clinton, Iowa

Sharon Ann Schueler
Mathematics Department Chairperson
Roosevelt High School
Sioux Falls, South Dakota

Sara P. Shaw
Mathematics Teacher
Adams Middle School
Saraland, Alabama

Mary Eleanor Wood Smith
Mathematics Teacher
North Augusta High School
North Augusta, South Carolina

Diane Stilwell
Mathematics Department Chairperson
South Junior High School
Morgantown, West Virginia

Janice M. Stricko
Mathematics Teacher
Gateway Middle School
Monroeville, Pennsylvania

Jeanette Tomasullo
Assistant Principal Supervision—Mathematics
Eastern District High School
Brooklyn, New York

Ivan T. Van Dyke
Mathematics Teacher
Norfolk Junior High School
Norfolk, Nebraska

Kathy Vielhaber
Mathematics Teacher
Parkway East Middle School
Creve Coeur, Missouri

Marlene T. Vitko
Mathematics Department Chairperson
Godwin Middle School
Dale City, Virginia

Carol Welsch
Learning Coordinator
Toki Middle School
Madison, Wisconsin

Table of Contents

Introduction to the Graphing Calculator xx

Chapter 1 Tools for Algebra and Geometry 4

1-1 Problem-Solving Strategy: Make a Plan 6
1-2 Order of Operations 11
1-3 Variables and Expressions 16
1-3B **Math Lab: Graphing Calculator Activity** Evaluating Expressions 21
1-4 Properties 22
1-5 The Distributive Property 26
1-5B **Math Lab: Hands-On Activity** Distributive Property 31
1-6 Variables and Equations 32
1-7 **Integration: Geometry** Ordered Pairs ... 36
1-8 Solving Equations Using Inverse Operations 41
1-9 Inequalities 46
1-10A **Math Lab: Hands-On Activity** Gathering Data 50
1-10 **Integration: Statistics** Gathering and Recording Data 51
Chapter 1 Highlights 57
Chapter 1 Study Guide and Assessment 58

In addition to the Graphing Calculator Activities, Modeling with Technology is included on page 11.

In addition to the Hands-On Activities, Modeling Math exercises are included on pages 14, 18, 24, 34, 47, and 54.

Up, Up, and Away! 2
- *Working On the Investigation* 10, 35
- *Closing the Investigation* 56

Chapter 2 Exploring Integers 64

2-1 Integers and Absolute Value 66
2-1B **Math Lab: Hands-On Activity** 71
2-2 **Integration: Geometry**
The Coordinate System 72
2-2B **Math Lab: Graphing Calculator Activity** . . 77
2-3 Comparing and Ordering 78
2-4A **Math Lab: Hands-On Activity** 82
2-4 Adding Integers 83
2-5A **Math Lab: Hands-On Activity** 88
2-5 Subtracting Integers 89
2-6 Problem-Solving Strategy: Look for a Pattern 94
2-7A **Math Lab: Hands-On Activity** 98
2-7 Multiplying Integers 99
2-8 Dividing Integers 104
Chapter 2 Highlights 109
Chapter 2 Study Guide and Assessment 110
Ongoing Assessment, Chapters 1–2 114

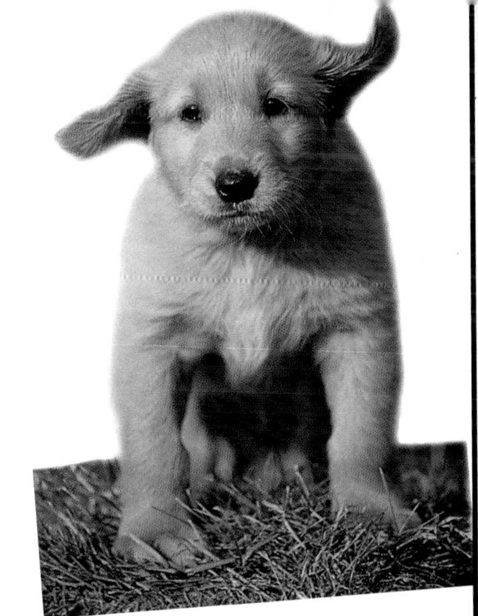

In addition to the Graphing Calculator Activities, Modeling with Technology is included on pages 83, 94, and 124.

In addition to the Hands-On Activities, Modeling Math exercises are included on pages 68, 91, 102, 120, and 135.

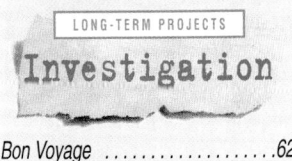

LONG-TERM PROJECTS

Investigation

Bon Voyage 62
• Working On the Investigation 70, 97, 128
• Closing the Investigation 160

Chapter 3 Solving One-Step Equations and Inequalities 116

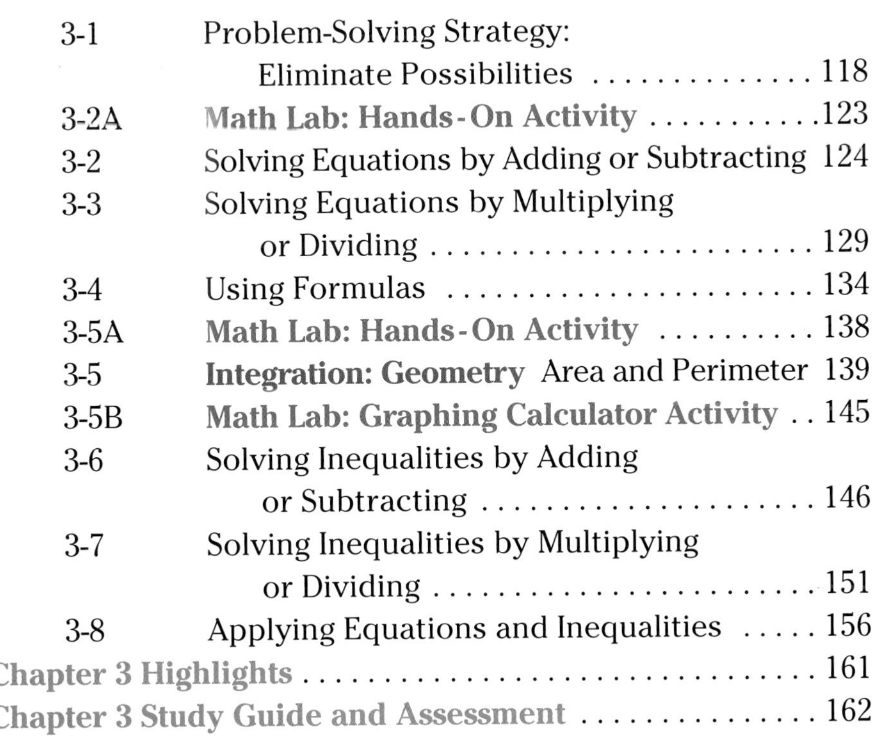

3-1 Problem-Solving Strategy:
Eliminate Possibilities 118
3-2A **Math Lab: Hands-On Activity** 123
3-2 Solving Equations by Adding or Subtracting 124
3-3 Solving Equations by Multiplying
or Dividing 129
3-4 Using Formulas 134
3-5A **Math Lab: Hands-On Activity** 138
3-5 **Integration: Geometry** Area and Perimeter 139
3-5B **Math Lab: Graphing Calculator Activity** . . 145
3-6 Solving Inequalities by Adding
or Subtracting 146
3-7 Solving Inequalities by Multiplying
or Dividing 151
3-8 Applying Equations and Inequalities 156
Chapter 3 Highlights 161
Chapter 3 Study Guide and Assessment 162

Content Integration

Why are topics like geometry and statistics in a pre-algebra book? Believe it or not, you can usually study an algebra topic from a geometric point of view and vice versa. Some examples are described below.

INTEGRATION

Probability You'll learn more about your chances of making that winning move in Backgammon. (Lesson 9-3, pages 440-443)

Geometry You'll use ordered pairs to develop algebraic expressions. (Lesson 1-7, pages 36-40)

Statistics
You'll learn how to use mean, median, and mode. (Lesson 6-6, pages 301-306)

Discrete Mathematics
You'll learn to represent an arithmetic sequence algebraically. (Lesson 5-9, pages 258-262)

Problem Solving
You'll learn how to solve problems by looking for a pattern. (Lesson 2-6, pages 94-97)

You can review dividing fractions in Lesson 6-4.

Look Back features refer you to skills and concepts that have been taught earlier in the book. Lesson 9-7, page 459

Chapter 4 Exploring Factors and Fractions 168

4-1 Factors and Monomials . . . 170
4-2 Powers and Exponents . . . 175
4-2B **Math Lab: Graphing Calculator Activity** Evaluating Expressions with Exponents 180
4-3 Problem-Solving Strategy: Draw a Diagram 181
4-4 Prime Factorization . . . 184
4-4B **Math Lab: Graphing Calculator Activity** Factor Patterns . . . 189
4-5 Greatest Common Factor (GCF) . . . 190
4-6A **Math Lab: Hands-On Activity** Equivalent Fractions . . . 195
4-6 Simplifying Fractions . . . 196
4-7 Using the Least Common Multiple (LCM) 200
4-8 Multiplying and Dividing Monomials . . . 205
4-9 Negative Exponents . . . 210
Chapter 4 Highlights . . . 215
Chapter 4 Study Guide and Assessment . . . 216
Ongoing Assessment, Chapters 1–4 . . . 220

Chapter 5 Rationals: Patterns in Addition and Subtraction 222

5-1 Rational Numbers . . . 224
5-2 Estimating Sums and Differences . . . 229
5-3 Adding and Subtracting Decimals . . . 234
5-4 Adding and Subtracting Like Fractions . . . 239
5-5 Adding and Subtracting Unlike Fractions . . 244
5-6 Solving Equations . . . 248
5-7 Solving Inequalities . . . 251
5-8 Problem-Solving Strategy: Using Logical Reasoning . . . 255
5-9 **Integration: Discrete Mathematics** Arithmetic Sequences . . . 258
5-9B **Math Lab: Hands-On Activity** Fibonacci Sequence . . . 263
Chapter 5 Highlights . . . 265
Chapter 5 Study Guide and Assessment . . . 266

In addition to the Graphing Calculator Activities, Modeling with Technology is included on pages 205, 210, 234, and 258.

In addition to the Hands-On Activities, Modeling Math exercises are included on pages 173, 182, 192, 198, 202, 227, 246.

LONG-TERM PROJECTS

Investigation

Taking Stock . . . 166
- *Working On the Investigation . . . 174, 209, 238*
- *Closing the Investigation . . . 264*

TABLE OF CONTENTS

Real-Life Applications

Every lesson in this book is designed to help show you where and when math is used in the real world. Since you'll explore many interesting topics, we believe you'll discover that math is relevant and exciting. Here are some examples.

APPLICATIONS

Recycling You'll discuss recycling as you learn about solving inequalities. (Lesson 3-7, page 151)

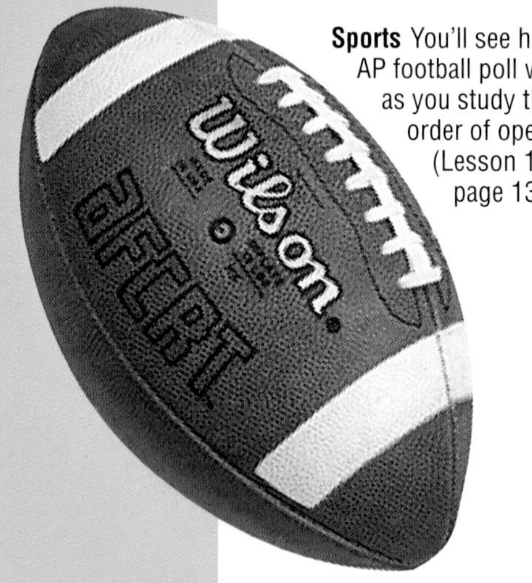

Sports You'll see how the AP football poll works as you study the order of operations. (Lesson 1-2, page 13)

Mae Jemison was the first African-American woman to go into space. Her first mission was a cooperative effort between U.S. and Japan which focused on experiments involving space travel and biology.

FYI features contain interesting facts that enhance the applications. (Lesson 2-3, page 81)

Games
Rolling dice in the game of parch which originated in India, is relat exponents. (Lesson 4-2, page 17

When is mathematics funny?
An actual reprinted *Fox Trot* comic illustrates how mathematics is a part of our society. (Lesson 1-3, page 20)

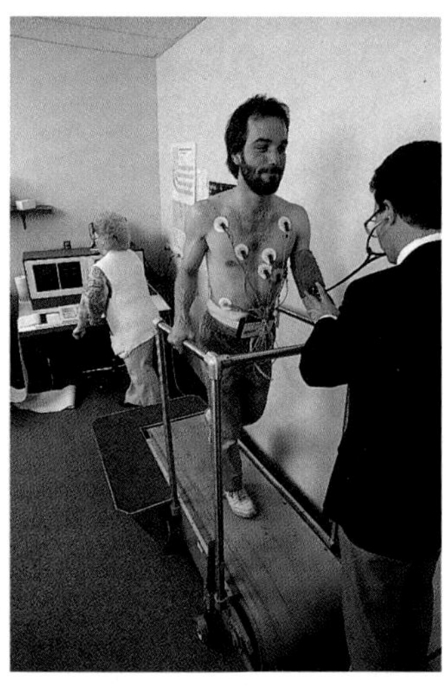

Health and Medicine You'll study a health-related application that involves heart rates and algebraic expressions. (Lesson 1-3, pages 16-17)

COOPERATIVE LEARNING PROJECT

THE SHAPE OF THINGS TO COME

Living Batteries

You'll do research about the possible future use of bacteria to generate electricity and power things like a watch. (Lesson 6-9, page 320)

Chapter 6 Rationals: Patterns in Multiplication and Division 272

6-1 Writing Fractions as Decimals 274
6-2 Estimating Products and Quotients 280
6-3 Multiplying Fractions 284
6-4 Dividing Fractions 289
6-5A **Math Lab: Hands-On Activity**
Multiplying and Dividing Decimals ... 294
6-5 Multiplying and Dividing Decimals 295
6-6A **Math Lab: Hands-On Activity**
Mean, Median & Mode 300
6-6 **Integration: Statistics**
Measures of Central Tendency 301
6-6B **Math Lab: Graphing Calculator Activity**
Finding Mean and Median 307
6-7 Solving Equations and Inequalities 308
6-8 **Integration: Discrete Mathematics**
Geometric Sequences 312
6-9 Scientific Notation 317
Chapter 6 Highlights 321
Chapter 6 Study Guide and Assessment 322
Ongoing Assessment, Chapters 1–6 326

Chapter 7 Solving Equations and Inequalities 328

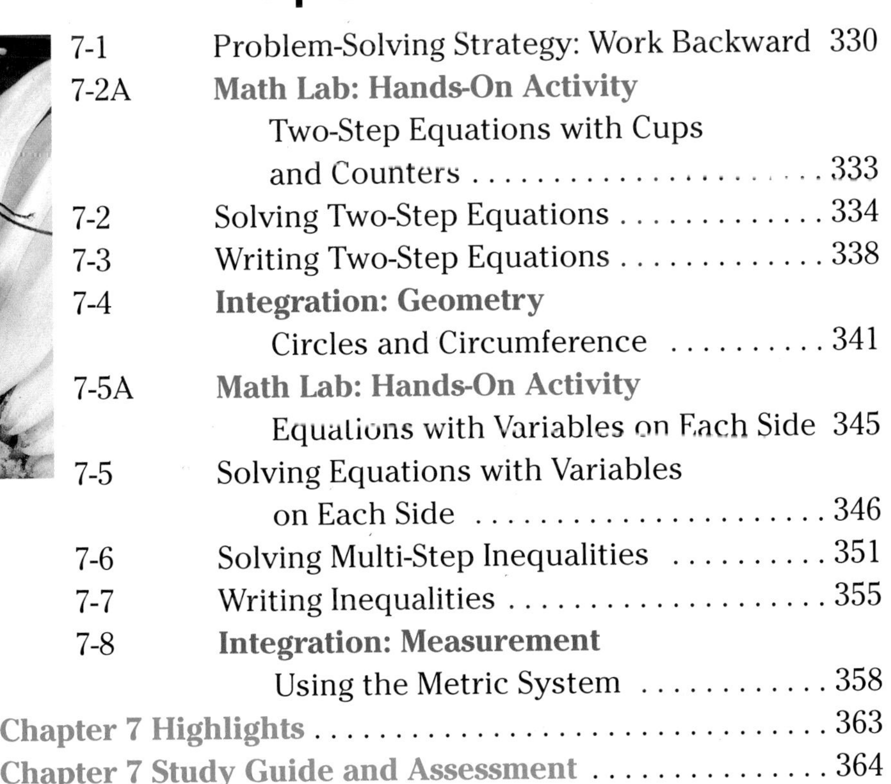

7-1 Problem-Solving Strategy: Work Backward 330
7-2A **Math Lab: Hands-On Activity**
Two-Step Equations with Cups and Counters 333
7-2 Solving Two-Step Equations 334
7-3 Writing Two-Step Equations 338
7-4 **Integration: Geometry**
Circles and Circumference 341
7-5A **Math Lab: Hands-On Activity**
Equations with Variables on Each Side 345
7-5 Solving Equations with Variables on Each Side 346
7-6 Solving Multi-Step Inequalities 351
7-7 Writing Inequalities 355
7-8 **Integration: Measurement**
Using the Metric System 358
Chapter 7 Highlights 363
Chapter 7 Study Guide and Assessment 364

In addition to the Graphing Calculator Activities, Modeling with Technology is included on pages 280, 295, 330, 346.

In addition to the Hands-On Activities, Modeling Math exercises are included on pages 277, 286, 297, 304, 315, 331, 335, 348, 353.

Roller Coaster Math 270
• Working On the Investigation 279, 293, 337
• Closing the Investigation 362

TABLE OF CONTENTS

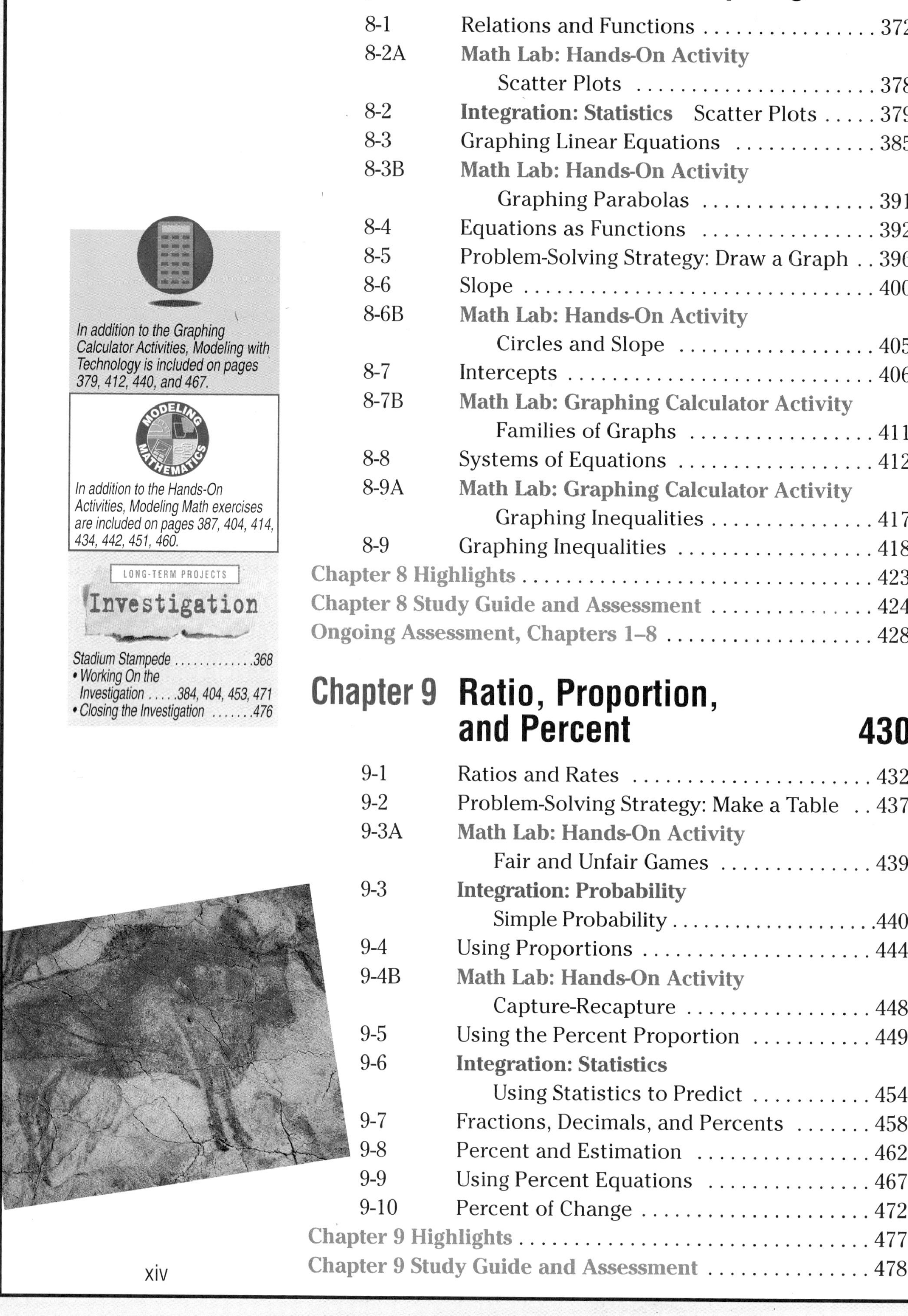

Chapter 8 Functions and Graphing 370

8-1 Relations and Functions 372
8-2A **Math Lab: Hands-On Activity**
Scatter Plots 378
8-2 **Integration: Statistics** Scatter Plots 379
8-3 Graphing Linear Equations 385
8-3B **Math Lab: Hands-On Activity**
Graphing Parabolas 391
8-4 Equations as Functions 392
8-5 Problem-Solving Strategy: Draw a Graph .. 396
8-6 Slope 400
8-6B **Math Lab: Hands-On Activity**
Circles and Slope 405
8-7 Intercepts 406
8-7B **Math Lab: Graphing Calculator Activity**
Families of Graphs 411
8-8 Systems of Equations 412
8-9A **Math Lab: Graphing Calculator Activity**
Graphing Inequalities 417
8-9 Graphing Inequalities 418
Chapter 8 Highlights 423
Chapter 8 Study Guide and Assessment 424
Ongoing Assessment, Chapters 1–8 428

Chapter 9 Ratio, Proportion, and Percent 430

9-1 Ratios and Rates 432
9-2 Problem-Solving Strategy: Make a Table .. 437
9-3A **Math Lab: Hands-On Activity**
Fair and Unfair Games 439
9-3 **Integration: Probability**
Simple Probability 440
9-4 Using Proportions 444
9-4B **Math Lab: Hands-On Activity**
Capture-Recapture 448
9-5 Using the Percent Proportion 449
9-6 **Integration: Statistics**
Using Statistics to Predict 454
9-7 Fractions, Decimals, and Percents 458
9-8 Percent and Estimation 462
9-9 Using Percent Equations 467
9-10 Percent of Change 472
Chapter 9 Highlights 477
Chapter 9 Study Guide and Assessment 478

In addition to the Graphing Calculator Activities, Modeling with Technology is included on pages 379, 412, 440, and 467.

MODELING MATHEMATICS

In addition to the Hands-On Activities, Modeling Math exercises are included on pages 387, 404, 414, 434, 442, 451, 460.

LONG-TERM PROJECTS

Investigation

Stadium Stampede 368
• Working On the Investigation 384, 404, 453, 471
• Closing the Investigation 476

Connections

Did you realize that math can be used to model many real-world applications? Do you know that mathematics is used in biology? in history? in geography? Yes, it may be hard to believe, but mathematics is frequently used in subjects besides math.

Connections to Algebra

You'll use algebra as a tool to solve problems from math topics such as geometry and arithmetic. (Lesson 7-4, page 342)

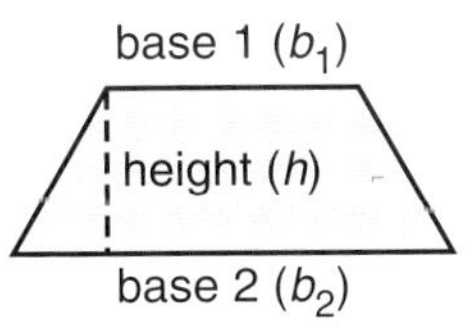

CULTURAL CONNECTIONS

People of different cultures have been playing games for centuries. One of the oldest games, Senet, made its first known appearance in Egypt in about 2686 B.C. Senet was played by people of all social classes and had both religious and secular forms.

Cultural Connections introduce you to a variety of world cultures. (Lesson 1-6, page 32)

Connections to Geometry

You'll learn how ratios can be used to describe the dimensions of the pyramids of Egypt. (Lesson 9-1, page 434)

CONNECTIONS Interdisciplinary Connections

Geography You'll study how to write and solve an equation involving the highest and lowest points in Africa. (Lesson 2-5, page 90)

Ecology You'll learn to model data about household waste as ordered pairs. (Lesson 8-1, page 372)

Music You'll solve equations that involve the number of vibrations on a guitar for a particular musical note. (Lesson 6-7, page 308)

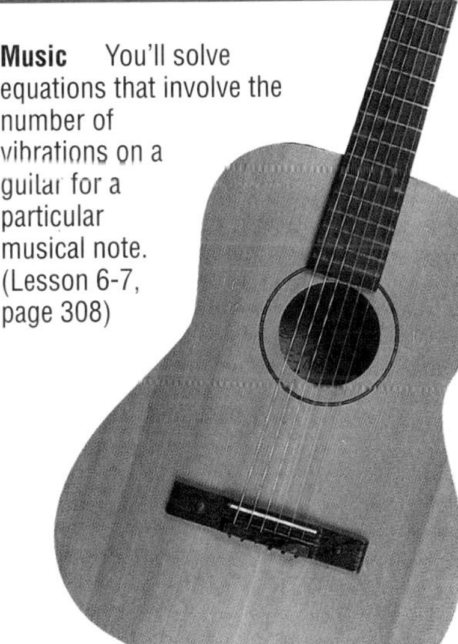

HELP WANTED

Are you interested in the patterns and changes of the weather? For more information on this field, contact:

American Meteorological Society
45 Beacon St.
Boston, MA 02108

Help Wanted features include information on interesting careers. (Lesson 5-6, page 249)

Health You'll relate pollen count to comparing integers on a number line. (Lesson 2-3, page 78)

MATH JOURNAL

Math Journal exercises give you the opportunity to assess yourself and write about your understanding of key math concepts. (Lesson 5-2, page 231)

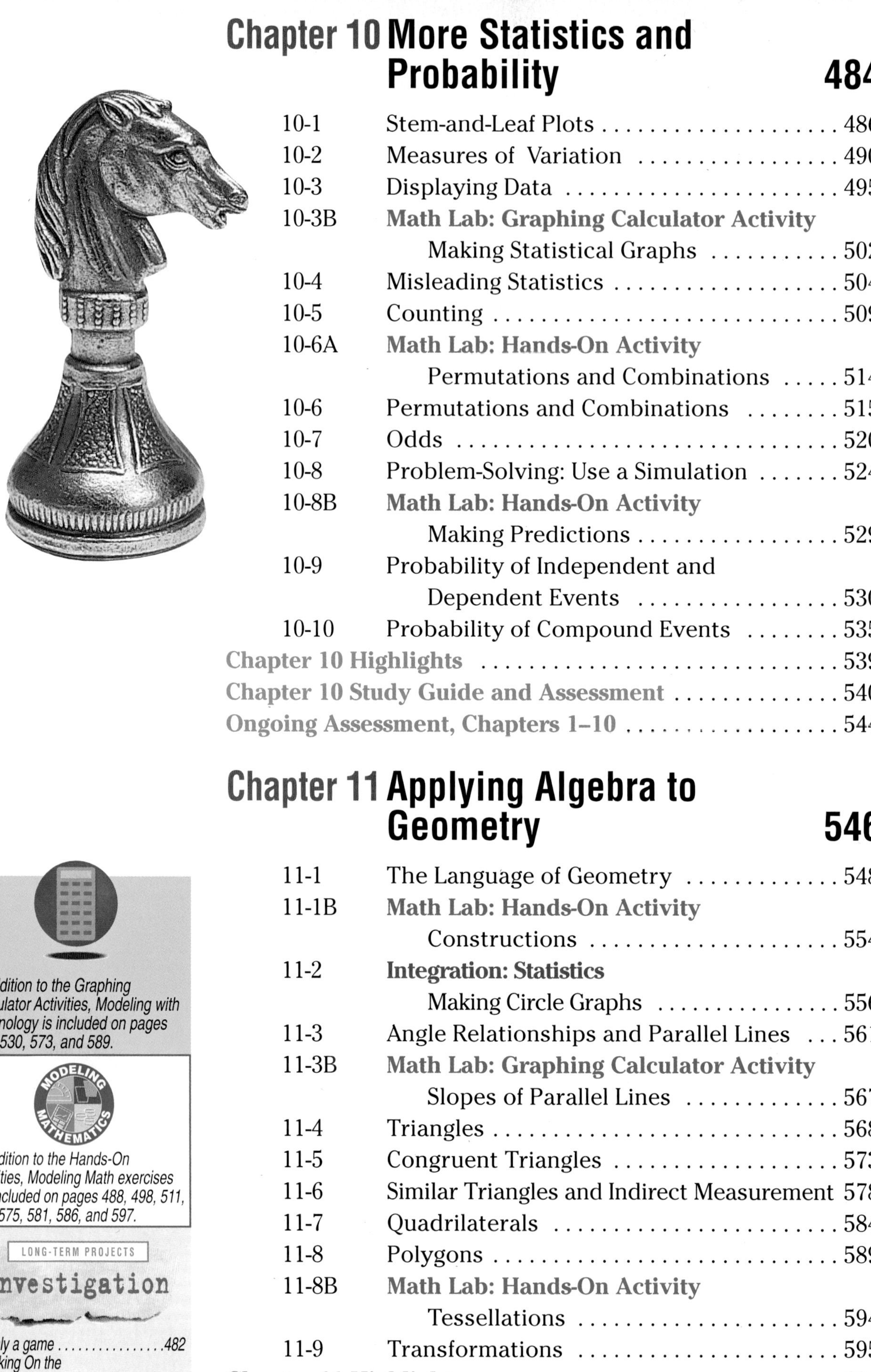

In addition to the Graphing Calculator Activities, Modeling with Technology is included on pages 490, 530, 573, and 589.

In addition to the Hands-On Activities, Modeling Math exercises are included on pages 488, 498, 511, 552, 575, 581, 586, and 597.

It's only a game 482
• Working On the Investigation 508, 528, 588
• Closing the Investigation 600

Chapter 10 More Statistics and Probability 484

10-1 Stem-and-Leaf Plots 486
10-2 Measures of Variation 490
10-3 Displaying Data 495
10-3B **Math Lab: Graphing Calculator Activity** Making Statistical Graphs 502
10-4 Misleading Statistics 504
10-5 Counting 509
10-6A **Math Lab: Hands-On Activity** Permutations and Combinations 514
10-6 Permutations and Combinations 515
10-7 Odds 520
10-8 Problem-Solving: Use a Simulation 524
10-8B **Math Lab: Hands-On Activity** Making Predictions 529
10-9 Probability of Independent and Dependent Events 530
10-10 Probability of Compound Events 535
Chapter 10 Highlights 539
Chapter 10 Study Guide and Assessment 540
Ongoing Assessment, Chapters 1–10 544

Chapter 11 Applying Algebra to Geometry 546

11-1 The Language of Geometry 548
11-1B **Math Lab: Hands-On Activity** Constructions 554
11-2 **Integration: Statistics** Making Circle Graphs 556
11-3 Angle Relationships and Parallel Lines ... 561
11-3B **Math Lab: Graphing Calculator Activity** Slopes of Parallel Lines 567
11-4 Triangles 568
11-5 Congruent Triangles 573
11-6 Similar Triangles and Indirect Measurement 578
11-7 Quadrilaterals 584
11-8 Polygons 589
11-8B **Math Lab: Hands-On Activity** Tessellations 594
11-9 Transformations 595
Chapter 11 Highlights 601
Chapter 11 Study Guide and Assessment 602

Modeling Mathematics

Using Manipulatives and Technology

One of the best ways to learn is to learn by doing. Here are two examples.

Hands-On Activities

In Lesson 9-4B, you'll learn how to use the capture-recapture method of counting deer.

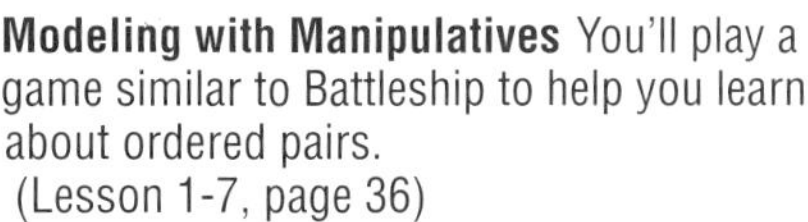

Modeling with Manipulatives You'll play a game similar to Battleship to help you learn about ordered pairs. (Lesson 1-7, page 36)

Do you know how to use high-tech machines like graphing calculators? We know that such knowledge is vital to your future success. That is why graphing calculator technology is integrated throughout this book.

There are several ways in which graphing calculators are integrated.

- **Introduction to Graphing Calculators** On page xx and page 1, you'll get acquainted with the basic features and functions of a graphing calculator.
- **Graphing Calculator Activities** In Lesson 3-5B, you'll use a program to find the area of quadrilaterals given the coordinates of the vertices.
- **Graphing Calculator Exercises** Many exercises are designed to be solved using a graphing calculator. For example, Exercise 23 in Lesson 8-2 involves making a scatter plot of keyboarding speeds.
- **Modeling with Technology** On page 467, Lesson 9-9 begins with an application involving a spreadsheet.

Option	Minimum Investment	Monthly Interest Rate (Percent)	Term (Months)	Interest for Term
1	$2500	0.375	6	$56.25
2	$2500	0.45	12	$135.00
3	$5000	0.54	36	$972.00
4	$7500	0.625	60	$2812.50

Techno Tips are designed to help you make more efficient use of technology through practical hints and suggestions.

Some calculators use a +/− key, called the "change sign" key instead of the (−) key.

(Lesson 2-4, page 83)

TABLE OF CONTENTS

Chapter 12 Measuring Area and Volume 608

12-1A **Math Lab: Hands-On Activity**
Areas and Geoboards 610
12-1 Area: Parallelograms, Triangles, and Trapezoids 612
12-1B **Math Lab: Hands-On Activity**
Fractals 618
12-2 Area: Circles 619
12-3 **Integration: Probability**
Geometric Probability 623
12-3B **Math Lab: Graphing Calculator Activity**
Geometric Probability 628
12-4 Problem-Solving Strategy: Make a Model or a Drawing 629
12-5 Surface Area: Prisms and Cylinders 632
12-6 Surface Area: Pyramids and Cones 638
12-7A **Math Lab: Hands-On Activity**
Volume 643
12-7 Volume: Prisms and Cylinders 644
12-8 Volume: Pyramids and Cones 649
12-8B **Math Lab: Hands-On Activity**
Similar Solid Figures 654

Chapter 12 Highlights 655
Chapter 12 Study Guide and Assessment 656
Ongoing Assessment, Chapters 1–12 660

In addition to the Graphing Calculator Activities, Modeling with Technology is included on pages 644 and 688.

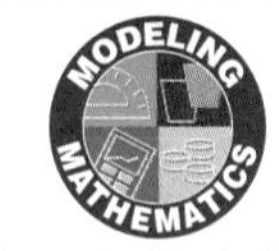

In addition to the Hands-On Activities, Modeling Math exercises are included on pages 625, 635, 651, 667, and 679.

LONG-TERM PROJECTS

Investigation

Oh, Give Me a Home! 606
• *Working On the Investigation* 637, 653, 671
• *Closing the Investigation* 698

Chapter 13 Applying Algebra to Right Triangles 662

13-1 Squares and Square Roots 664
13-2 Problem-Solving Strategy: Use Venn Diagrams 669
13-3 The Real Number System 672
13-4 The Pythagorean Theorem 676
13-4B **Math Lab: Hands-On Activity**
Graphing Irrational Numbers 682
13-5 Special Right Triangles 683
13-6A **Math Lab: Hands-On Activity**
Ratios in Right Triangles 687
13-6 The Sine, Cosine, and Tangent Ratios 688

13-6B Math Lab: Graphing Calculator Activity
Slope and Tangent 693
13-7 Using Trigonometric Ratios 694
Chapter 13 Highlights 699
Chapter 13 Study Guide and Assessment 700

Chapter 14 Polynomials 704

14-1 Polynomials 706
14-1B Math Lab: Hands-On Activity
Representing Polynomials with
Algebra Tiles 710
14-2 Adding Polynomials 711
14-3 Subtracting Polynomials 715
14-4 Powers of Monomials 719
14-5A Math Lab: Hands-On Activity
Multiplying Polynomials 724
14-5 Multiplying a Polynomial by a Monomial . . 725
14-6 Multiplying Binomials 728
14-6B Math Lab: Hands-On Activity
Factoring 732
Chapter 14 Highlights 733
Chapter 14 Study Guide and Assessment 734
Ongoing Assessment, Chapters 1–14 738

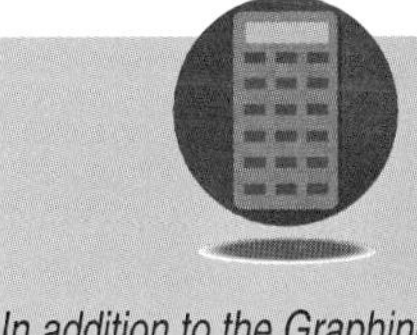

In addition to the Graphing Calculator Activities, Modeling with Technology is included on page 719.

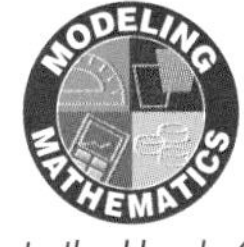

In addition to the Hands-On Activities, Modeling Math exercises are included on pages 713, 717, 726, and 730.

Student Handbook

Extra Practice 740
Chapter Tests 774
Glossary 788
Spanish Glossary 796
Selected Answers 804
Photo Credits 830
Applications Index 834
Index 836

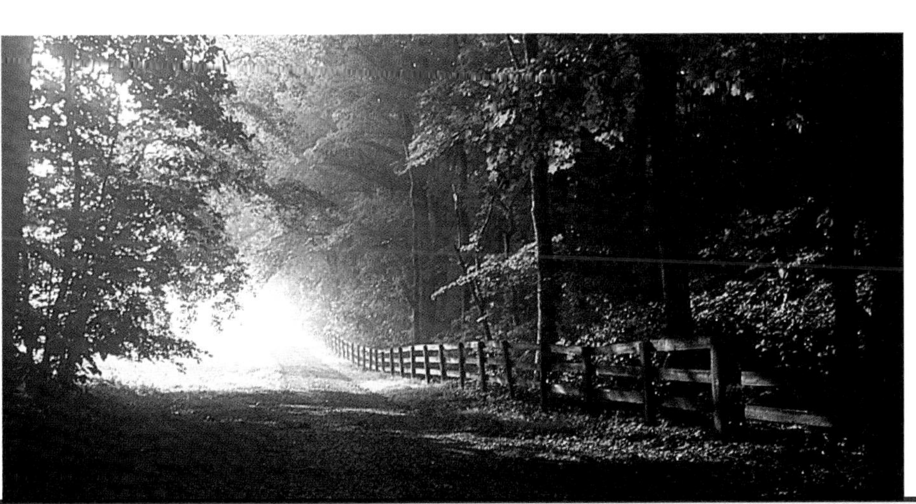

Introduction to the Graphing Calculator

What is it?
What does it do?
How is it going to help me learn math?

These are just a few of the questions many students ask themselves when they first see a graphing calculator. Some students may think, "Oh, no! Do we *have* to use one?", while others may think, "All right! We get to use these neat calculators!" There are as many thoughts and feelings about graphing calculators as there are students, but one thing is for sure: a graphing calculator *can* help you learn mathematics.

So what is a graphing calculator? Very simply, it is a calculator that draws graphs. This means that it will do all of the things that a "regular" calculator will do, *plus* it will draw graphs of simple or very complex equations. In pre-algebra, this capability is nice to have as you learn to graph equations and solve equations by graphing.

But a graphing calculator can do more than just calculate and draw graphs. For example, you can program it and work with data to make statistical graphs and computations. If you need to generate random numbers, you can do that on the graphing calculator. If you need to find the absolute value of numbers, you can do that too. It's really a very powerful tool—so powerful that it is often called a pocket computer. A graphing calculator can save you time and make doing mathematics easier.

As you may have noticed, graphing calculators have some keys that other calculators do not. The Texas Instruments TI-82 will be used in this text for graphing and programming. The keys located on the bottom half of the calculator are probably familiar to you as they are the keys found on basic scientific calculators. The keys located just below the screen are the graphing keys. You will also notice the up, down, left, and right arrow keys. These allow you to move the cursor around on the screen, to "trace" graphs that have been plotted, and to choose items from the menus. The other keys located on the top half of the calculator access the special features such as statistical computations and programming features.

There are some keystrokes that can save you time when using the graphing calculator. A few of them are listed below.

- Any light blue commands written above the calculator keys are accessed with the [2nd] key, which is also blue. Similarly, any gray characters above the keys are accessed with the [ALPHA] key, which is also gray.
- [2nd] [ENTRY] copies the previous calculation so you can edit and use it again.
- Pressing [ON] while the calculator is graphing stops the calculator from completing the graph.
- [2nd] [QUIT] will return you to the home (or text) screen.
- [2nd] [A-LOCK] locks the [ALPHA] key, which is like pressing "shift lock" or "caps locks" on a typewriter or computer. The result is that all letters will be typed and you do not have to repeatedly press the [ALPHA] key. (This is handy for programming.) Stop typing letters by pressing [ALPHA] again.
- [2nd] [OFF] turns the calculator off.

Some commonly-used mathematical functions are shown in the table below. As with any scientific calculator, the graphing calculator observes the order of operations.

Mathematical Operation	Example	Keys	Display
evaluate expressions	Find 2 + 5.	2 [+] 5 [ENTER]	2+5 7
exponents	Find 3^5.	3 [^] 5 [ENTER]	3^5 243
multiplication	Evaluate 3(9.1 + 0.8).	3 [×] [(] 9.1 [+] .8 [)] [ENTER]	3*(9.1+.8) 29.7
roots	Find $\sqrt{14}$.	[2nd] [√] 14 [ENTER]	√14 3.741657387
opposites	Enter −3.	[(−)] 3	−3

Graphing on the TI-82

Before graphing, we must instruct the calculator how to set up the axes in the coordinate plane. To do this, we define a **viewing window**. The viewing window for a graph is the portion of the coordinate grid that is displayed on the **graphics screen** of the calculator. The viewing window is written as [left, right] by [bottom, top] or [Xmin, Xmax] by [Ymin, Ymax]. A viewing window of $[-10, 10]$ by $[-10, 10]$ is called the **standard viewing window** and is a good viewing window to start with to graph an equation. The standard viewing window can be easily obtained by pressing ZOOM 6. Try this. Move the arrow keys around and observe what happens. You are seeing a portion of the coordinate plane that includes the region from -10 to 10 on the x-axis and from -10 to 10 on the y-axis. Move the cursor, and you can see the coordinates of the point for the position of the cursor.

Any viewing window can be set manually by pressing the WINDOW key. The window screen will appear and display the current settings for your viewing window. First press ENTER. Then, using the arrow and ENTER keys, move the cursor to edit the window settings. Xscl and Yscl refer to the x-scale and y-scale. This is the number of units between tick marks on the x- and y-axes. Xscl = 1 means that there will be a tick mark for every unit of one along the x-axis. The standard viewing window would appear as follows.

Xmin = -10
Xmax = 10
Xscl = 1
Ymin = -10
Ymax = 10
Yscl = 1

Programming on the TI-82

The TI-82 has programming features that allow us to write and execute a series of commands to perform tasks that may be too complex or cumbersome to perform otherwise. Each program is given a name. Commands begin with a colon (:), followed by an expression or an instruction. Most of the features of the calculator are accessible from program mode.

When you press PRGM, you see three menus: EXEC, EDIT, and NEW. EXEC allows you to execute a stored program, EDIT allows you to edit or change a program, and NEW allows you to create a program. The following tips will help you as you enter and run programs on the TI-82.

- To begin entering a new program, press PRGM ▶ ▶ ENTER.
- After a program is entered, press 2nd QUIT to exit the program mode and return to the home screen.
- To execute a program, press PRGM. Then use the down arrow key to locate the program name and press ENTER, or press the number or letter next to the program name.
- If you wish to edit a program, press PRGM ▶ and choose the program from the menu.
- To immediately re-execute a program after it is run, press ENTER when Done appears on the screen.
- To stop a program during execution, press ON.

While a graphing calculator cannot do everything, it can make some things easier. To prepare for whatever lies ahead, you should try to learn as much as you can. The future will definitely involve technology, and using a graphing calculator is a good start toward becoming familiar with technology. Who knows? Maybe one day you will be designing the next satellite, building the next skyscraper, or helping students learn mathematics with the aid of a graphing calculator!

Investigation

TEACHER NOTES

This Investigation is designed to be completed over several days or weeks. It may be considered optional. You may want to assign the Investigation and the follow-up activities to be completed at the same time.

NCTM Standards: 1-4, 6-10

Objective

Use mathematics to analyze a trip on an ultralight airplane.

Mathematical Overview

This Investigation will use the following skills and concepts from Chapter 1.

- Use the four-step, problem-solving plan to solve real-life, math-related problems.
- Solve problems using equations.
- Gather and record data.

Recommended Time		
Part	Pages	Time
Investigation	2–3	1 class period
Working on the Investigation	10, 35	20 minutes each
Closing the Investigation	56	1 class period

Instructional Resources

Investigations and Projects Masters, pp. 1–4

A recording sheet, teacher notes, and scoring guide are provided for each Investigation in the *Investigation and Projects Masters*.

1 MOTIVATION

This Investigation uses research to investigate ultralight airplanes. Ask students if any of them have ever heard of or ridden in an *ultralight airplane*. Discuss reasons why careful planning must be done before going on a trip in an ultralight.

LONG-TERM PROJECT

Investigation

Up, Up and Away!

Have you ever dreamed of breaking free of the ties of gravity and soaring among the clouds? Maybe you should try an ultralight airplane! An ultralight is a small one-person airplane that is very light, about the size and weight of a hang glider. Unlike a hang glider, an ultralight is propelled by a small engine and can be enclosed or open to the air.

The Federal Aviation Administration, or FAA, governs all types of aircraft in the United States. They set forth the following restrictions on ultralights.

Notice! All ultralight airplanes must:

- *be single seat aircrafts*
- *weigh 254 pounds or less*
- *have a top speed of 55 knots (63 miles per hour)*
- *stall at 24 knots (28 miles per hour)*
- *carry no more than 5 gallons of fuel*
- *never fly over towns or settlements*
- *never fly at night or around airports without special permission*
- *always yield the right of way to all other aircraft*

Special 2-seat exemptions, for training only, may weigh up to 496 pounds and carry 10 gallons of fuel.

Veteran fliers say that building and flying an ultralight is inexpensive when compared to other types of airplanes. A number of kits are available for you to put together yourself for between \$3000 and \$6000. Plan on spending 6 months to two years on construction. Flight training, while not required, is highly recommended. Instructors charge about \$600 to \$1200. Once the initial expenses of plane and training are completed, flying is very inexpensive. The engine of an ultralight usually burns only 2 to 3 gallons of fuel an hour. Many ultralights can be stored at home. If you need to store your plane in a hangar, budget about \$30 to \$90 a month.

Starting Your Investigation

Each year, the Experimental Aircraft Association, or EAA, gathers in Oshkosh, Wisconsin, for their annual convention. EAA members are concerned with all types of home-built, antique, and ultralight aircraft. Suppose you and the other members of your group are on a committee chosen by the local EAA chapter to make plans to fly the group to the convention. There are many things you must consider as you make your plans. Look over the following list of considerations.

Think About . . .

- *Will you be able to fly there in one day?*
- *How much fuel will you need? Can you make it there on one tank of fuel?*
- *How should you plan your route?*
- *Are there places to store the planes while you are at the convention?*
- *Must each person fly his or her own plane, or can two-seat ultralights be used?*

Work with your group to add to the list of questions to be considered. Then reread the information about ultralights on the previous page to find significant information. You may wish to look in some other sources.

You will continue working on this Investigation throughout Chapter 1. Be sure to keep your materials in your Investigation Folder.

interNET CONNECTION

For current information on ultralights, visit:
www.glencoe.com/sec/math/prealg/mathnet

Investigation
Up, Up, and Away!

Working on the Investigation
Lesson 1-1, p. 10

Working on the Investigation
Lesson 1-6, p. 35

Closing the Investigation
End of Chapter 1, p. 56

2 SETUP

You may wish to have a student read the first page of the Investigation to provide information about ultralight airplanes. You may then wish to read the first paragraph on the second page, which introduces the activity. Discuss the activity with your students. Then separate the class into groups of four.

3 MANAGEMENT

Each member of the groups should be responsible for a specific task.

Recorder Collects data and records the group discussion.

Gatekeeper Makes sure that there is participation by each group member.

Advocate Makes sure the contributions of each group member are appreciated.

Administrator Guides the discussion to keep the group on task.

At the end of the activity, each member should write a paragraph to evaluate the group's performance as a team.

Investigations and Projects Masters, p. 4

NAME ______________________ DATE ________

CHAPTER 1 Investigation
Up, Up, and Away!

Student Edition
Pages 2–3

Work with your group to add to the list of questions to be considered.

Will you be able to fly there in one day?

How much fuel will you need?

Can you make it there on one tank of fuel?

How should you plan your route?

Are there places to store the planes while you are there?

Must each person fly his or her own plane, or can two-seat ultralights be used?

Reread the information about ultralights on page 2 to find significant information. You may wish to look in some other sources. List your findings below.

Please keep this page and any other research in your Investigation Folder.

Cooperative Learning

This Investigation offers an excellent opportunity for using cooperative learning groups. For more information on cooperative learning strategies and group management, see *Cooperative Learning in the Mathematics Classroom*, one of the titles in the Glencoe Mathematics Professional Series.

Sample Answers

Answers will vary as they are based on the research of each group.

CHAPTER 1

Tools for Algebra and Geometry

PREVIEWING THE CHAPTER

This chapter lays the foundation for the studies of algebra, geometry, and statistics in the rest of the book. Students will learn the four-step, problem-solving plan and how to use it to solve all types of problems. They will evaluate expressions using the order of operations and use variables and expressions to solve real-world problems. The most important algebraic property introduced is the distributive property. Students will also see the connection between algebraic expressions and the coordinate plane. In the last lesson, students will begin to gather and record data.

Lesson (Pages)	Lesson Objectives	NCTM Standards	State/Local Objectives
1-1 (6–10)	Use the four-step plan to solve real-life, math-related problems. Choose an appropriate method of computation.	1-4, 7, 8	
1-2 (11–15)	Use the order of operations to evaluate expressions.	1-5, 7, 8	
1-3 (16–20)	Evaluate expressions containing variables. Translate verbal phrases into algebraic expressions.	1-4, 9	
1-3B (21)	Evaluate algebraic expressions using a graphing calculator.	1-4, 9	
1-4 (22–25)	Identify properties of addition and multiplication. Use properties to solve problems.	1-4, 6, 7, 9	
1-5 (26–30)	Simplify algebraic expressions using the distributive property.	1-4, 6, 7, 9	
1-5B (31)	Visualize the distributive property by modeling a geometric interpretation.	1-4, 6, 9	
1-6 (32–35)	Identify and solve open sentences.	1-4, 7, 9	
1-7 (36–40)	Use ordered pairs to locate points, to organize data, and to explore how certain quantities in expressions are related to each other.	1-4, 9, 12	
1-8 (41–45)	Use inverse operations to solve equations. Solve problems using equations.	1-4, 6, 8, 9	
1-9 (46–49)	Write and solve inequalities.	1-4, 9	
1-10A (50)	Gather and record data.	1-4, 10	
1-10 (51–55)	Gather and record data in a frequency table or bar graph.	1-4, 10	

A complete, 1-page lesson plan is provided for each lesson in the ***Lesson Planning Guide***. Answer keys for each lesson are available in the ***Answer Key Masters***.

ORGANIZING THE CHAPTER

You may want to refer to the **Course Planning Calendar** on page T12 for detailed information on pacing.
PACING: Standard—15 days; **Honors**—13 days; **Block**—6 days

LESSON PLANNING CHART

			BLACKLINE MASTERS												
Lesson (Pages)	**Materials/ Manipulatives**	**Extra Practice (Student Edition)**	**Study Guide**	**Practice**	**Enrichment**	**Assessment and Evaluation**	**Math Lab and Modeling Math**	**Multicultural Activity**	**Tech Prep Applications**	**Graphing Calculator**	**Activity**	**Real-World Applications**	**Interactive Mathematics Tools Software**	**Teaching Transparencies**	**Group Activity Cards**
1-1 (6–10)		p. 740	p. 1	p. 1	p. 1			p. 1						1-1A 1-1B	1-1
1-2 (11–15)	calculator index cards	p. 740	p. 2	p. 2	p. 2									1-2A 1-2B	1-2
1-3 (16–20)	cups and counters*	p. 740	p. 3	p. 3	p. 3	p. 15			p. 1		p. 15		1-3	1-3A 1-3B	1-3
1-3B (21)	graphing calculator									p. 15					
1-4 (22–25)	counters*	p. 741	p. 4	p. 4	p. 4									1-4A 1-4B	1-4
1-5 (26–30)		p. 741	p. 5	p. 5	p. 5	pp. 14, 15	p. 76							1-5A 1-5B	1-5
1-5B (31)	algebra tiles*						p. 27								
1-6 (32–35)	TI-82 graphing calculator index cards	p. 741	p. 6	p. 6	p. 6						p. 1	1		1-6A 1-6B	1-6
1-7 (36–40)	grid paper	p. 742	p. 7	p. 7	p. 7									1-7A 1-7B	1-7
1-8 (41–45)	cups and counters*	p. 742	p. 8	p. 8	p. 8	p. 16			p. 2		p. 29	2	1-8	1-8A 1-8B	1-8
1-9 (46–49)	index cards reference books	p. 742	p. 9	p. 9	p. 9					p. 1				1-9A 1-9B	1-9
1-10A (50)							p. 28								
1-10 (51–55)	M & Ms®	p. 743	p. 10	p. 10	p. 10	p. 16		p. 2						1-10A 1-10B	1-10
Study Guide/ Assessment (57–61)						pp. 1–13, 17–19									

*Included in Glencoe's *Student Manipulative Kit* and *Overhead Manipulative Resources*.

ORGANIZING THE CHAPTER

All of the blackline masters in the Teacher's Classroom Resources are available on the ***Electronic Teacher's Classroom Resources*** CD-ROM.

Other Chapter Resources

Student Edition
Chapter Opener, pp. 4–5
Investigation: Up, Up, and Away!, pp. 2–3
From the Funny Papers, p. 20
Working on the Investigation, pp. 10, 35
Closing the Investigation, p. 56
Cooperative Learning Project: The Shape of Things to Come, p. 40

Teacher's Classroom Resources
Investigations and Projects Masters, pp. 29–32

Pre-Algebra Overhead Manipulative Resources, pp. 1–3

Technology
Test and Review Software (IBM & Macintosh)
CD-ROM Activities

Professional Publications
Block Scheduling Booklet
Glencoe Mathematics Professional Series

Outside Resources

Books/Periodicals
Codes, Ciphers, and Secret Writing, Dover.
Mathematics and Humor, NCTM.

Videos/CD-ROMs
Math Who Needs It with Jaime Escalante, FASE Productions.
Becoming Successful Problem Solvers, HRM Video.

Software
Teasers by Tobbs with Whole Numbers, Sunburst
Geometric Connectors: Coordinates, Sunburst
Data Insights, Sunburst

Assessment Resources

Student Edition
Math Journal, pp. 9, 53
Mixed Review, pp. 15, 20, 25, 30, 35, 40, 45, 49, 55
Self Test, p. 30
Chapter Highlights, p. 57
Chapter Study Guide and Assessment, pp. 58–60
Alternative Assessment, p. 61
Portfolio, p. 61

MindJogger Videoquiz, 1

Teacher's Wraparound Edition
5-Minute Check, pp. 6, 11, 16, 22, 26, 32, 36, 41, 46, 51
Checking Your Understanding, pp. 9, 14, 19, 24, 28, 34, 39, 43, 48, 53
Closing Activity, pp. 10, 15, 20, 25, 30, 35, 40, 45, 49, 55
Cooperative Learning, pp. 12, 34

Assessment and Evaluation Masters
Multiple-Choice Tests, Forms 1A (Honors), 1B (Average), 1C (Basic), pp. 1–6
Free-Response Tests, Forms 2A (Honors), 2B (Average), 2C (Basic), pp. 7–12
Performance Assessment, p. 13
Mid-Chapter Test, p. 14
Quizzes A–D, pp. 15–16
Standardized Test Practice, p. 17
Cumulative Review, pp. 18–19

ENHANCING THE CHAPTER

Examples of some of the materials for enhancing Chapter 1 are shown below.

Multicultural Activity Masters, pp. 1, 2

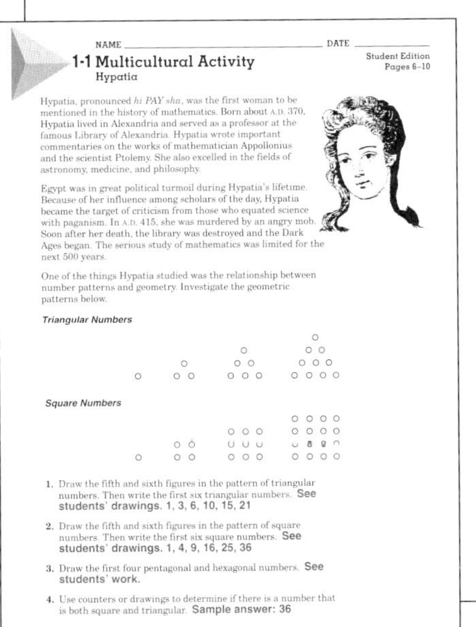

NAME ______ DATE ______

1-1 Multicultural Activity
Hypatia

Student Edition Pages 6–10

Hypatia, pronounced *hi PAY sha*, was the first woman to be mentioned in the history of mathematics. Born about A.D. 370, Hypatia lived in Alexandria and served as a professor at the famous Library of Alexandria. Hypatia wrote important commentaries on the works of mathematician Appollonius and the scientist Ptolemy. She also excelled in the fields of astronomy, medicine, and philosophy.

Egypt was in great political turmoil during Hypatia's lifetime. Because of her influence among scholars of the day, Hypatia became the target of criticism from those who equated science with paganism. In A.D. 415, she was murdered by an angry mob. Soon after her death, the library was destroyed and the Dark Ages began. The serious study of mathematics was limited for the next 500 years.

One of the things Hypatia studied was the relationship between number patterns and geometry. Investigate the geometric patterns below.

Triangular Numbers

Square Numbers

1. Draw the fifth and sixth figures in the pattern of triangular numbers. Then write the first six triangular numbers. **See students' drawings. 1, 3, 6, 10, 15, 21**
2. Draw the fifth and sixth figures in the pattern of square numbers. Then write the first six square numbers. **See students' drawings. 1, 4, 9, 16, 25, 36**
3. Draw the first four pentagonal and hexagonal numbers. **See students' work.**
4. Use counters or drawings to determine if there is a number that is both square and triangular. **Sample answer: 36**

Real-World Applications, 1, 2

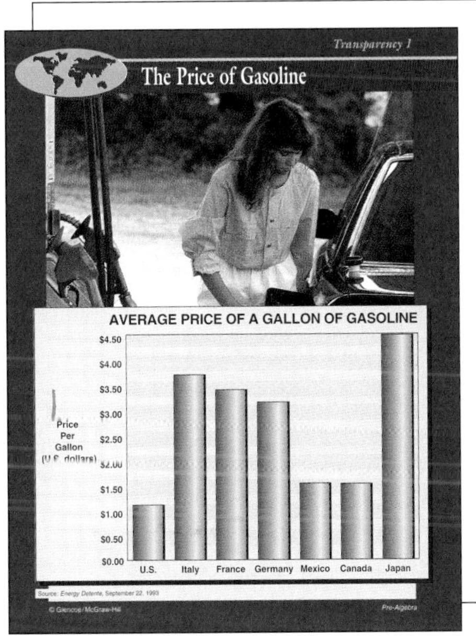

Graphing Calculator Masters, p. 1

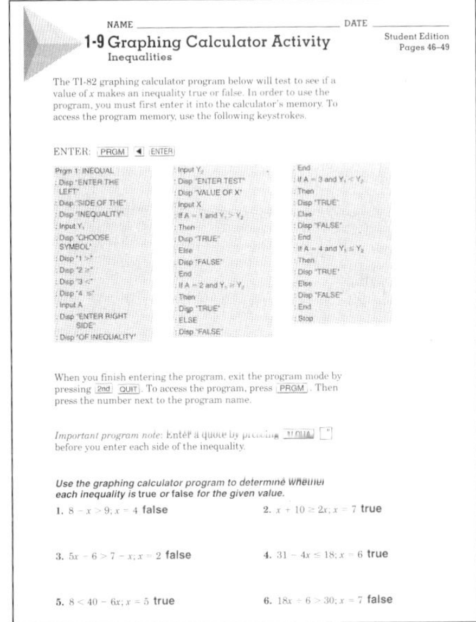

NAME ______ DATE ______

1-9 Graphing Calculator Activity
Inequalities

Student Edition Pages 46–49

The TI-82 graphing calculator program below will test to see if a value of x makes an inequality true or false. In order to use the program, you must first enter it into the calculator's memory. To access the program memory, use the following keystrokes.

ENTER: PRGM ◄ ENTER

```
Prgm 1: INEQUAL
:Disp "ENTER THE
  LEFT"
:Disp "SIDE OF THE"
:Disp "INEQUALITY"
:Input Y1
:Disp "CHOOSE
  SYMBOL"
:Disp "1 >"
:Disp "2 ≥"
:Disp "3 <"
:Disp "4 ≤"
:Input A
:Disp "ENTER RIGHT
    SIDE"
:Disp "OF INEQUALITY"
:Input Y2
:Disp "ENTER TEST"
:Disp "VALUE OF X"
:Input X
:If A = 1 and Y1 > Y2
:Then
:Disp "TRUE"
:Else
:Disp "FALSE"
:End
:If A = 2 and Y1 ≥ Y2
:Then
:Disp "TRUE"
:ELSE
:Disp "FALSE"
:End
:If A = 3 and Y1 < Y2
:Then
:Disp "TRUE"
:Else
:Disp "FALSE"
:End
:If A = 4 and Y1 ≤ Y2
:Then
:Disp "TRUE"
:Else
:Disp "FALSE"
:End
:Stop
```

When you finish entering the program, exit the program mode by pressing 2nd QUIT. To access the program, press PRGM. Then press the number next to the program name.

Important program note: Enter a quote by pressing ALPHA " before you enter each side of the inequality.

Use the graphing calculator program to determine whether each inequality is* true *or* false *for the given value.

1. $8 - x > 9; x = 4$ **false**
2. $x + 10 \geq 2x; x = 7$ **true**
3. $5x - 6 > 7 - x; x = 2$ **false**
4. $31 - 4x \leq 18; x = 6$ **true**
5. $8 < 40 - 6x; x = 5$ **true**
6. $18x \div 6 > 30; x = 7$ **false**

Tech Prep Applications Masters, pp. 1, 2

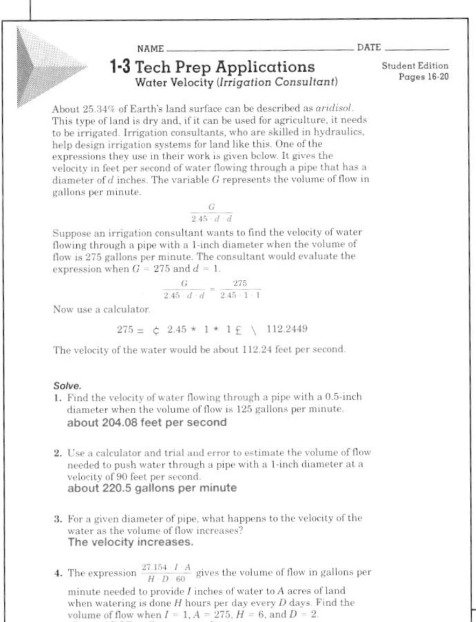

NAME ______ DATE ______

1-3 Tech Prep Applications
Water Velocity (*Irrigation Consultant*)

Student Edition Pages 16-20

About 25.34% of Earth's land surface can be described as *aridisol*. This type of land is dry and, if it can be used for agriculture, it needs to be irrigated. Irrigation consultants, who are skilled in hydraulics, help design irrigation systems for land like this. One of the expressions they use in their work is given below. It gives the velocity in feet per second of water flowing through a pipe that has a diameter of d inches. The variable G represents the volume of flow in gallons per minute.

$$\frac{G}{2.45 \cdot d \cdot d}$$

Suppose an irrigation consultant wants to find the velocity of water flowing through a pipe with a 1-inch diameter when the volume of flow is 275 gallons per minute. The consultant would evaluate the expression when $G = 275$ and $d = 1$.

$$\frac{G}{2.45 \cdot d \cdot d} = \frac{275}{2.45 \cdot 1 \cdot 1}$$

Now use a calculator.

275 = ¢ 2.45 * 1 * 1 £ \ 112.2449

The velocity of the water would be about 112.24 feet per second.

Solve.

1. Find the velocity of water flowing through a pipe with a 0.5-inch diameter when the volume of flow is 125 gallons per minute. **about 204.08 feet per second**
2. Use a calculator and trial and error to estimate the volume of flow needed to push water through a pipe with a 1-inch diameter at a velocity of 90 feet per second. **about 220.5 gallons per minute**
3. For a given diameter of pipe, what happens to the velocity of the water as the volume of flow increases? **The velocity increases.**
4. The expression $\frac{27.154 \cdot I \cdot A}{H \cdot D \cdot 60}$ gives the volume of flow in gallons per minute needed to provide I inches of water to A acres of land when watering is done H hours per day every D days. Find the volume of flow when $I = 1$, $A = 275$, $H = 6$, and $D = 2$. **about 10.37 gallons per minute**

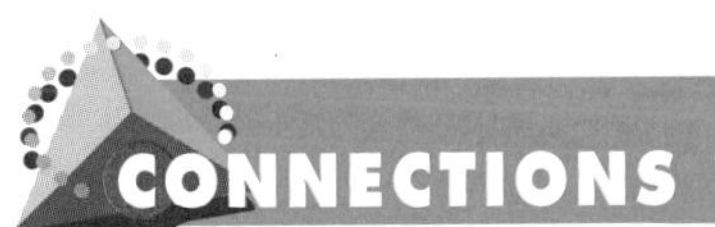

Activity Masters, pp. 1, 15, 29

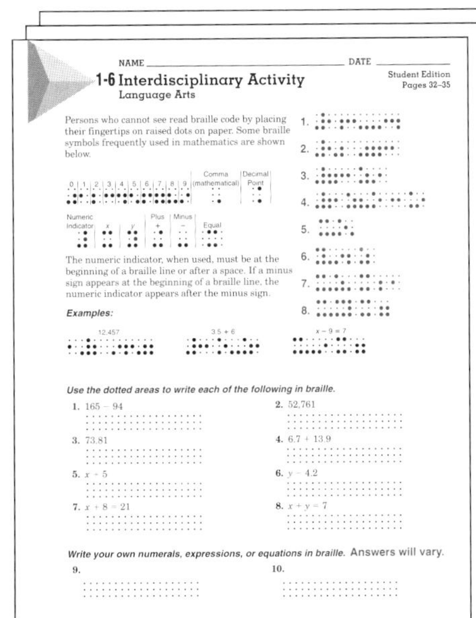

NAME ______ DATE ______

1-6 Interdisciplinary Activity
Language Arts

Student Edition Pages 32–35

Persons who cannot see read braille code by placing their fingertips on raised dots on paper. Some braille symbols frequently used in mathematics are shown below.

0 | 1 | 2 | 3 | 4 | 5 | 6 | 7 | 8 | 9 | Comma (mathematical) | Decimal Point

Numeric Indicator | x | y | Plus + | Minus − | Equal

The numeric indicator, when used, must be at the beginning of a braille line or after a space. If a minus sign appears at the beginning of a braille line, the numeric indicator appears after the minus sign.

Examples:

12.457   3.5 + 6 $x - 9 = 7$

Use the dotted areas to write each of the following in braille.

1. 165 − 94
2. 52,761
3. 73.81
4. 6.7 + 13.9
5. $x + 5$
6. $y - 4.2$
7. $x + 8 = 21$
8. $x + y = 7$

Write your own numerals, expressions, or equations in braille. Answers will vary.

9.
10.

Math Lab and Modeling Math Masters, p. 76

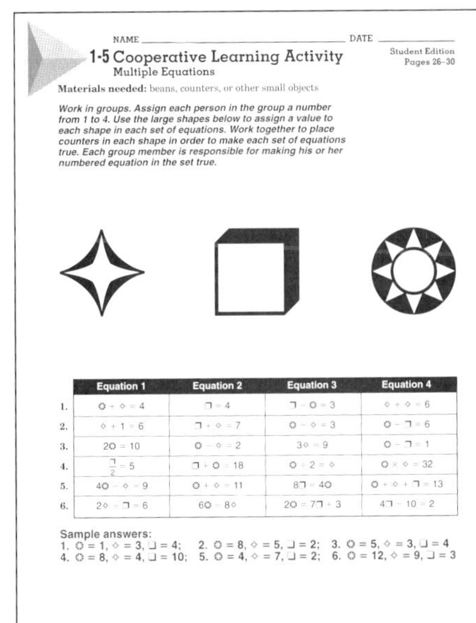

NAME ______ DATE ______

1-5 Cooperative Learning Activity
Multiple Equations

Student Edition Pages 26–30

Materials needed: beans, counters, or other small objects

Work in groups. Assign each person in the group a number from 1 to 4. Use the large shapes below to assign a value to each shape in each set of equations. Work together to place counters in each shape in order to make each set of equations true. Each group member is responsible for making his or her numbered equation in the set true.

	Equation 1	Equation 2	Equation 3	Equation 4
1.	○ + ◇ = 4	□ = 4	□ − ○ = 3	◇ + ◇ = 6
2.	◇ + 1 = 6	□ + ◇ = 7	○ − ◇ = 3	○ − □ = 6
3.	2○ = 10	○ − ◇ = 2	3◇ = 9	○ − □ = 1
4.	□/2 = 5	□ + ○ = 18	○ ÷ 2 = ◇	○ × ◇ = 32
5.	4○ − ◇ = 9	○ + ◇ = 11	8□ = 4○	○ + ◇ + □ = 13
6.	2◇ ÷ □ = 6	6○ = 8◇	2○ = 7□ + 3	4□ − 10 = 2

Sample answers:
1. ○ = 1, ◇ = 3, □ = 4; 2. ○ = 8, ◇ = 5, □ = 2; 3. ○ = 5, ◇ = 3, □ = 4
4. ○ = 8, ◇ = 4, □ = 10; 5. ○ = 4, ◇ = 7, □ = 2; 6. ○ = 12, ◇ = 9, □ = 3

MAKING MATHEMATICS RELEVANT

This two-page introduction to the chapter provides students with an opportunity to explore other disciplinary topics and their applications to mathematics.

Background Information

Math and Fashion in the News
Levi Strauss left the East Coast for San Francisco in 1853. His product was strong canvas cloth he thought miners would use for tents and wagons. When he saw that the miners needed strong pants that could stand up to the elements, Strauss made pants from the brown fabric. In the early 1860s, he switched to indigo-dyed denim from the Amoskeag Mill in New Hampshire.

CHAPTER 1 Tools for Algebra and Geometry

TOP STORIES in Chapter 1

In this chapter, you will:

- solve problems using the four-step problem-solving plan,
- choose the appropriate method of computation,
- use variables and algebraic properties,
- graph points in a coordinate system, and
- gather and record data.

MATH AND FASHION IN THE NEWS

The search is over . . . Jeans that fit!

Source: The Columbus Dispatch, December 27, 1994

Too long, too short, too big, too small . . . While jeans may be the most popular fashion of all time, sometimes finding a pair that fits is like searching for buried treasure. But Levi Strauss, the original maker of jeans, is coming to the rescue. You can now go to a store and get a pair of jeans made to measure. A salesperson takes four measurements on you and enters them into a computer. After some fancy figuring, the computer chooses a trial pair of jeans for you to try on. Any necessary adjustments are made, and your order is placed through the computer directly to the factory. In three weeks, your dream jeans arrive. Being in fashion has never been so easy!

Putting It into Perspective

1848 Discovery of gold in California hills draws Americans westward.

1853 Levi Strauss begins producing pants from canvas-like cloth for gold miners.

1873 U.S. patent is granted for use of rivets to reinforce stress points in jeans.

1926 First jeans with a zipper instead of buttons

1850 1880 1910

Putting It into Perspective

In 1897, German chemist Adolf von Baeyer discovered a formula for synthetic indigo. Since the 1920s, synthetic indigo has accounted for 95% of the nine thousand tons of indigo produced each year.

EMPLOYMENT OPPORTUNITIES

Shape the Fads and Fashions of Your Generation

DO YOU LIKE dealing with people? Do you have a flair for fashion? We are looking for sharp, energetic people to work as **retail salespeople** and **clothing buyers**. Salespeople assist customers in choosing clothing and offer advice on fit, styles, colors, and fabrics. Buyers work with manufacturers to choose merchandise for our retail outlets. A high-school education is required for salespeople, college degree preferred for buyer's position. Additional training or experience is a plus.

For more information, contact:
National Retail Federation
701 Pennsylvania Avenue NW, Suite 701
Washington, D.C. 20004

interNET CONNECTION For up-to-date information on selling fashions, visit:
www.glencoe.com/sec/math/prealg/mathnet

Statistical Snapshot

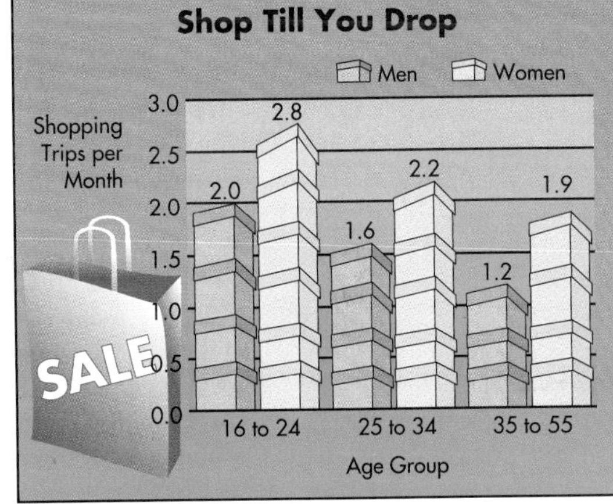

Source: *Cotton Incorporated*, 1994 survey

1939 Lady Levis, the first name-brand jeans made specifically for women introduced

1940

1965 Invention of the mini-skirt

1970

1974 Beverly Johnson, the first black model on the cover of a major fashion magazine

1986 Acid-washed jeans are introduced by an Italian jeansmaker.

1994 Made-to-measure jeans introduced

2000

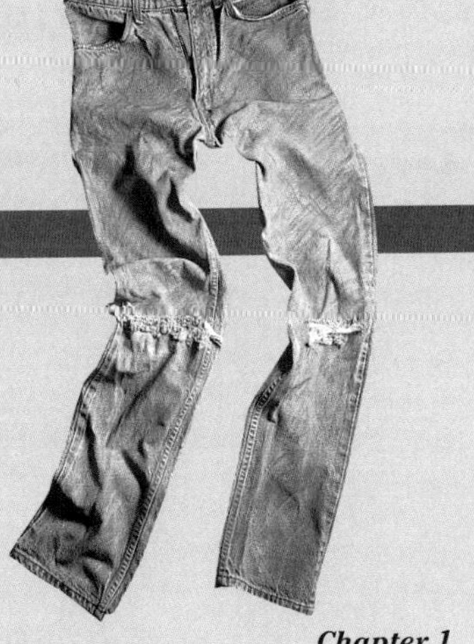

EMPLOYMENT OPPORTUNITIES

Some colleges offer bachelor's degree programs in fashion merchandising. Students who are interested in the field of fashion may wish to contact a school for information.

Statistical Snapshot

Point out to students that the graph shows that older shoppers tend to make fewer shopping trips per month than younger shoppers. Female shoppers make more shopping trips than males in all age groups.

Investigations and Projects Masters, p. 29

NAME ______________________ DATE ____________

CHAPTER 1 Project A
The Right Fit

Jeans have come a long way since their beginnings as sturdy pants for miners. Now, several different styles and shapes make jeans the right fit for a variety of occasions. They are a staple of many wardrobes, and can be seen as often in restaurants as in rodeos. Where do you buy jeans in your area?

1. Visit several stores that sell jeans, or contact the stores by phone. Find out about jeans sales by asking questions like: What sizes, styles, and colors of jeans do you carry? What prices do you charge? Which brands, styles, and sizes are your best sellers? Which don't sell well? Do sales of jeans vary during the year? If so, when is the busiest time? Who are your best customers?

2. Interview neighbors, relatives, and families of friends. Find out what brands and styles of jeans they own and which they plan to buy. Ask them how often they shop for jeans, when they usually shop, and where they shop. Use this information to figure out what a family typically spends on jeans in a year. Then find out what they think of the idea of made-to-measure jeans, and ask them to describe any experiences they've had ordering a pair.

3. Think about the process of making made-to-measure jeans. First, predict what the four measurements might be taken. Then speak with salespeople who have had experience with made-to-measure jeans. Ask them to describe the process. Try to determine if all stores take the same measurements the same way. Find out if prices for "dream jeans" are the same at each store.

4. Organize your findings to prepare an article for your school newspaper about shopping for jeans in your neighborhood. Provide a table to display the differences in brands and styles carried, and prices for jeans and made-to-measure jeans in the stores you contacted. Include a description of the process of getting made-to-measure jeans and a sketch to show what parts of the jeans are measured. Also include a summary of what you learned about family jeans shopping habits.

Cooperative Learning

Chapter Projects Two chapter projects are included in the *Investigations and Projects Masters*. In Chapter Project A, students extend the topic in the Chapter Opener. In Chapter Project B, students explore passing a basketball across the United States. A student page and a parent letter are provided for each Chapter Project.

interNET CONNECTION

Glencoe has made every effort to ensure that the website links for *Pre-Algebra* at www.glencoe.com/sec/math/prealg/mathnet are current and contain appropriate content. However, these website links are not under Glencoe's control.

1-1 LESSON NOTES

NCTM Standards: 1-4, 7, 8

Instructional Resources

- Study Guide Master 1-1
- Practice Master 1-1
- Enrichment Master 1-1
- Group Activity Card 1-1
- Multicultural Activity Masters, p. 1

Transparency 1-1A contains the 5-Minute Check for this lesson; **Transparency 1-1B** contains a teaching aid for this lesson.

Recommended Pacing	
Standard Pacing	Day 1 of 14
Honors Pacing	Day 1 of 13
Block Scheduling*	Day 1 of 7 (along with Lesson 1-2)

*For more information on pacing and possible lesson plans, refer to the ***Block Scheduling Booklet***.

1 FOCUS

5-Minute Check

Find the value of each expression.

1. 18 + 6 **24**
2. 85 + 12 **97**
3. 114 − 29 **85**
4. 8 × 47 **376**
5. 72 ÷ 4 **18**
6. The sales figures for a business for the first quarter months were as follows.
   January $12,000
   February $15,000
   March $10,000
   What were total sales for the first quarter? **$37,000**

1-1 Problem-Solving Strategy: Make a Plan

Setting Goals: *In this lesson, you'll use the four-step plan to solve real-life, math-related problems and choose an appropriate method of computation.*

Modeling a Real-World Application: Health and Exercise

Venus Williams studies French and Spanish so that she can talk to the tennis players who she meets from other countries. After her professional tennis career is over, Venus hopes to become a paleontologist.

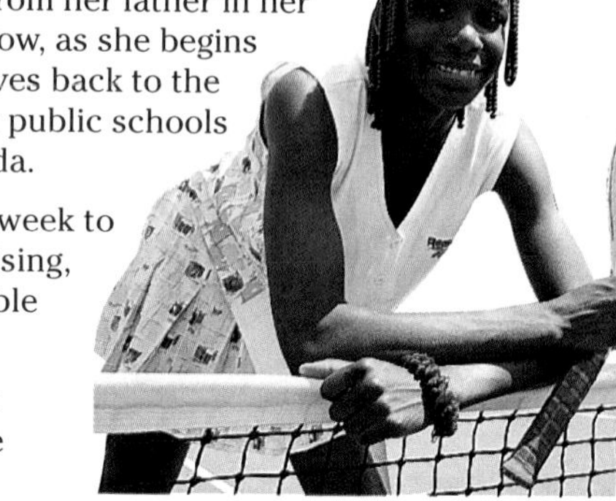

At 15, Venus Williams is the newest superstar in the world of tennis. Venus learned to play tennis from her father in her home town of Compton, California. Now, as she begins her professional tennis career, she gives back to the community by giving tennis clinics at public schools near her home in Delray Beach, Florida.

Venus spends hours exercising each week to remain in top condition. When exercising, it is important to maintain a reasonable heart rate.

Based on the table below, what heart rate should Venus maintain when she exercises?

Guide to Reasonable Heart Rate (85% capacity)											
Age	20	25	30	35	40	45	50	55	60	65	70
Heart Rate (beats per minute)	174	170	166	162	157	153	149	145	140	136	132

This problem will be solved in Example 1.

Learning the Concept

You can use four steps to solve real-life, math-related problems.

1. **Explore**
   - Read the problem carefully.
   - Ask yourself questions like "What facts do I know" and "What do I need to find out?"
2. **Plan**
   - See how the facts relate to each other.
   - Make a plan for solving the problem.
   - Estimate the answer.
3. **Solve**
   - Use your plan to solve the problem.
   - If your plan does not work, revise it or make a new plan.
4. **Examine**
   - Reread the problem.
   - Ask, "Is my answer close to my estimate?"
   - Ask, "Does my answer make sense for the problem?
   - If not, solve the problem another way.

Classroom Vignette

To start the school year, I ask teams of students to determine approximately how many days it will take a line of people (standing 10 feet apart) to pass a basketball from San Francisco, California, to Washington, D.C. To solve this problem, students will need a basketball, stopwatch, U.S. map, ruler, calculator, and masking tape to mark off 10-foot intervals on the gym floor.

Nancy E. Herald

Nancy E. Herald
Wolf County High School
Campton, Kentucky

Example 1

CONNECTION

Health and Exercise

Refer to the application at the beginning of the lesson. Find a reasonable heart rate for exercise for 15-year-old Venus.

Use the problem-solving plan to find the heart rate.

Explore — *What facts do we know and what are we trying to find out?*

The table shows the suggested heart rates for several other ages. We are trying to find the rate for a 15 year old.

Plan — *Make a plan for attacking the problem and estimate the answer.*

Since we know the heart rates for several ages, it makes sense to look for a pattern to extend to find the rate for a 15 year old. The heart rate decreases as the age increases. So the heart rate for a 15 year old should be higher than that for a 20 year old.

Solve — *Use the plan to solve the problem.*

Find the differences between consecutive ages and heart rates.

+5 +5 +5 +5 +5 +5 +5 +5 +5 +5

Age (years)	20	25	30	35	40	45	50	55	60	65	70
Heart Rate (beats per minute)	174	170	166	162	157	153	149	145	140	136	132

−4 −4 −4 −5 −4 −4 −4 −5 −4 −4

Each consecutive age increases 5 years. The pattern for the heart rates is −4, −4, −4, −5. Extend this pattern as follows.

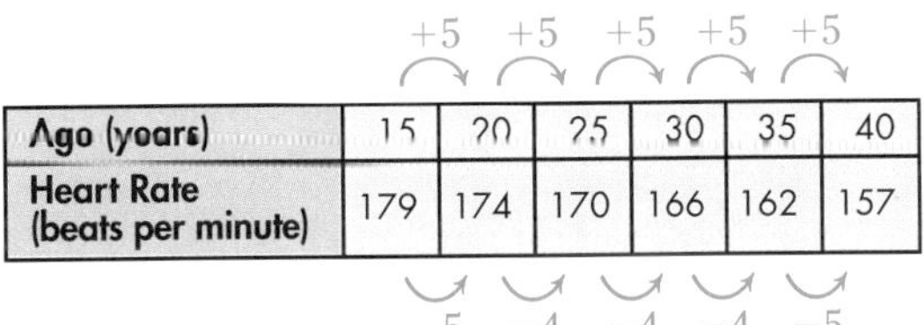

+5 +5 +5 +5 +5

Age (years)	15	20	25	30	35	40
Heart Rate (beats per minute)	179	174	170	166	162	157

−5 −4 −4 −4 −5

A reasonable heart rate for exercise for a 15 year old is 174 + 5 or 179.

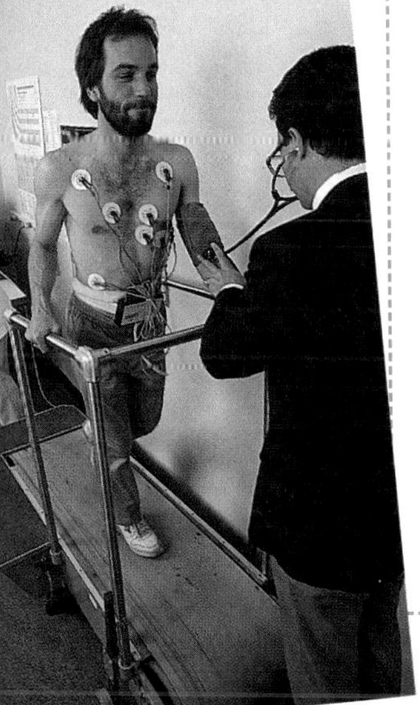

Examine — *Compare the answer to the problem and the estimate.*

The heart rate of 179 that we found for a 15 year old makes sense with the estimate that we made.

Motivating the Lesson

Questioning Ask students if they have ever read a mystery book or seen a mystery movie. Have them explain how the characters solved the mystery. What steps did they follow? How did they verify their theory? Explain that solving a problem is like solving a mystery and then discuss the problem-solving plan.

2 TEACH

In-Class Example

For Example 1

Jerry Gardner owns an auto service shop. The chart below hangs on the wall of the shop to indicate the prices for different amounts of time spent on a repair. Find the labor cost of a repair that takes $4\frac{1}{2}$ hours. **$152.00**

All charges are determined by the labor time. The cost of parts and sales tax are added to the total.

Number of hours	Labor cost
0–1 hours	$36.00
1–2 hours	$65.00
2–3 hours	$94.00
3–4 hours	$123.00

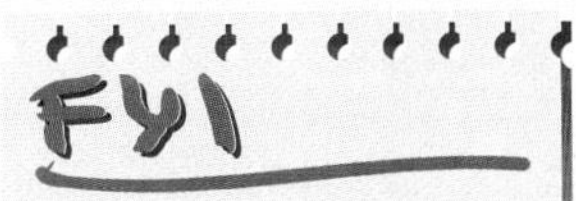

Venus William's younger sister Serena is also a promising tennis player.

Teaching Tip You may wish to discuss the factors other than age that could affect the optimal heart rate for exercising. Physical fitness, injuries, weight, or fitness goals may all make a person adjust his or her heart rate during exercise.

Alternative Learning Styles

Kinesthetic Label each corner of the room with the four steps of the problem-solving plan. Allow students to move from one corner to the next as they complete each step of solving the problem below.

In 1994, the United States had 2496 radio stations that specialized in country music. There were 2307 adult contemporary stations, 1220 religious stations, 972 top-40 stations, 723 oldies stations, 491 album-rock stations, and 460 news stations. About how many radio stations were in the United States in 1994? **about 8700 stations**

One of the important steps in solving problems is choosing the method of computation. The diagram below can help you decide which method is most appropriate.

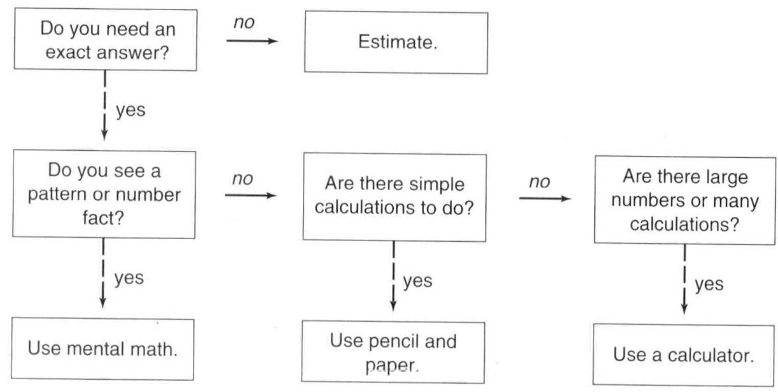

Modern medicine has made amazing advances in human transplants. The numbers of different types of transplants that were performed in the United States in 1992 are shown at the right. Find the total number of transplants performed in 1992.

Transplant	Number
heart	2172
liver	3059
kidney	10,210
heart-lung	48
lung	535
pancreas	557

Source: U.S. Department of Health and Human Services

Estimation
Mental Math
Calculator
Paper and Pencil

Explore We are given the number of each type of transplant and wish to find the total.

Plan Add the numbers of transplants to find the total. Use the diagram above to find the appropriate method of computation.

Since we need an exact answer, we will not solve using estimation. There is no pattern in the numbers and there are many calculations to be performed, so using a calculator may be the best choice.

Even though we will solve using a calculator, it is a good idea to estimate first so that we know whether the answer we find is reasonable. Round each number to the nearest hundred and then add mentally.

Estimate:

2172	→	2200
3059	→	3100
10,210	→	10,200
48	→	0
535	→	500
557	→	600
		16,600

Do you think the estimate is higher or lower than the actual total?

Teaching Tip Discuss the reasoning behind the choices indicated by the chart. Stress that the chart is only a suggestion and alternate methods of computation may be used to solve a problem.

Teaching Tip Point out the prompt for students to choose the appropriate method of computation. This prompt will be used throughout the book to indicate times when students need to choose their computation method.

Teaching Tip You may wish to review the rules of rounding with students as you discuss Example 2.

In-Class Example

For Example 2

Big Ben is the largest bell located in London's House of Parliament. The bell weighs about 27,000 pounds. This is about 68 times as much as the hammer that strikes the bell weighs. *About* how much does the hammer weigh? **about 400 pounds**

Study Guide Masters, p. 1

NAME ________________ DATE ________

1-1 Study Guide
Problem-Solving Strategy: Make a Plan

Student Edition Pages 6–10

The Four-Step Plan

1. **Explore** the problem. Read and ask questions about what you know and what you need to know.
2. **Plan** your method of solution. Estimate the answer and decide how you will solve the problem.
3. **Solve** the problem using your plan.
4. **Examine** your solution. Ask yourself whether it answers the question or makes sense for the problem.

Solve each problem using the four-step plan.

1. **Engineering** Geothermal energy is heat from inside the earth. Underground temperatures usually rise 9°C for each 300 feet of depth. For the ground temperature to rise 90°C, how deep would you have to dig? **3,000 feet**
2. **Transportation** A DC-11 jumbo jet carries 342 passengers with 36 in first class seating and the rest in coach. On a certain day, a first class ticket from Los Angeles to Chicago costs $750, and a coach ticket costs $450. What will be the ticket sales for the airline if the flight is full? **$164,700**
3. **Retail** At the Woodward Park School bookstore, a ball point pen costs 28¢ and a notepad costs 23¢. What could you buy and spend exactly 74¢? **2 tablets and 1 pen**
4. **Geography** On Kenny's map, each inch represents 30 miles. Raleigh, NC, is about 12 inches from Atlanta, GA. About how many miles is this? **about 360 miles**
5. **Fundraising** Emily sold 332 boxes of candy in two weeks for her band trip. If she sold the same number of boxes each day, how many boxes of candy did Emily sell each day? **23 boxes**
6. **Catering** Swan Catering Services offers banquet facilities and food service. They charge $6 per person for a cold buffet. If the Everetts invite 75 people to a retirement party, how much should they budget for the cold buffet? **$450**

Reteaching

Using Cooperative Groups Display and discuss the four-step, problem-solving plan. Then have students work together to solve the following problem. Each group will need several pennies and a ruler or measuring tape. *Mrs. Nall, a teacher in North Carolina, challenged her students to determine how many pennies are in a mile. Use the four-step plan to solve.* **about 84,480 pennies**

Solve	*Use the plan to solve the problem.* Use a calculator to find the total. 2172 [+] 3059 [+] 10210 [+] 48 [+] 535 [+] 557 [=] *16581*
Examine	The total we found with the calculator is very close to our estimate, so the answer is reasonable.

Checking Your Understanding

Communicating Mathematics

Read and study the lesson to answer these questions. 1. **See margin.**

1. **Describe** each step of the problem-solving plan in your own words.
2. **Explain** why it is important to examine your solution after you solve a problem. **See margin.**
3. **Demonstrate** how to use the diagram on page 8 to determine the best method of computation. **See students' work.**

4. Begin your Math Journal by writing two or three sentences that describe what you like and dislike about mathematics. **See students' work.**

Guided Practice

5. **Sales** John Diaz sells large computer software packages to accounting corporations for Object Systems, Inc. Each month, his supervisor writes a sales plan for the number of software packages that Mr. Diaz is supposed to sell. Mr. Diaz receives a salary, plus a bonus for any software packages he sells in excess of the plan. The table at the right shows the bonus amount for different levels of sales. Determine Mr. Diaz's bonus if he sells 16 software packages more than the plan. **See margin.**

Sales above plan	Bonus
2	$100
4	$125
6	$150
8	$175

   a. Write the *Explore* step. What do you know and what do you need to find?
   b. Write the *Plan* step. What strategy will you use? What do you estimate the answer to be?
   c. *Solve* the problem using your plan. What is your answer?
   d. *Examine* your solution. Is it reasonable? Does it answer the question?

6. **Hygiene** A survey performed by the Lever Brothers Corporation found that, on average, people washed their hands or face 6.8 times per day. Based on this finding, about how many times does a person wash their hands or face in one year? **See margin.**
   a. Which method of computation do you think is most appropriate? Justify your choice.
   b. Solve the problem using the four-step plan. Be sure to examine your solution.
   c. Do the results of the survey imply that each person actually washes his or her hands 6.8 times each day? Explain.

3 PRACTICE/APPLY

Checking Your Understanding

Exercises 1–6 are designed to help you assess your students' understanding through reading, writing, speaking, and modeling. You should work through Exercises 1–4 with your students and then monitor their work on Exercises 5–6.

Additional Answers

1. Explore—read and decide what is given and what is to be found
   Plan—plan a method for solution
   Solve—execute the plan
   Examine—reflect on whether the solution answers the question and makes sense
2. Sample answer: To make sure that it answers the question and makes sense.
5. Sample answers:
   a. We know the bonus for 2, 4, 6, and 8 over plan. We need to know the bonus for 16 over plan.
   b. Extend the pattern. The bonus should be about $250.
   c.

10	$200
12	$225
14	$250
16	$275

Practice Masters, p. 1

NAME ______________________ DATE ________

1-1 Practice
Problem-Solving Strategy: Make a Plan

Student Edition Pages 6–10

Solve each problem using the four-step plan.

1. **Art** Eric Walton uses pewter to make keychains to sell at arts and crafts fairs. His best-selling keychain has a dragon-shaped ornament that takes 2 ounces of pewter to make. If Mr. Walton has 2 pounds of pewter on hand, how many dragons can he make? (16 ounces = 1 pound) **16 dragons**

2. **Transportation** Helen Westman takes public transportation to and from work each day. The blue line bus stops at State Street and Main every twenty minutes. The red line bus stops there every half hour. If both buses were at the stop at 1:10 P.M., what is the next time that Ms. Westman will be able to change buses at the State Street stop without waiting? **2:10 P.M.**

3. **Music** The Grandview High School Music department is organizing an autumn concert involving the choruses, the orchestra, and the symphonic band. The 10th grade girls' chorus has 15 minutes to fill in the concert. If each of the songs they are considering performing is about 4 minutes long, about how many songs can they plan to sing? **3 or 4**

4. **Geometry** What is the total number of rectangles in the figure below? (*Hint:* There are more than 6.) **18**

5. **Music** In the 1940s, record players were made to spin at 78 revolution per minute (rpm). A more modern record player spins singles at a rate of 45 rpm. If both turntables spin for 9 minutes, find the difference in the number of turns they make. **297 revolutions**

Group Activity Card 1-1

Guess My Number **Group Activity 1-1**

In a group of two or more, one member of the group secretly chooses a number between 1 and 250 and writes it on a piece of paper without showing it to the other players. The other players then try to guess the number by asking questions that can be answered with a "yes" or "no." A player's turn consists of asking a question and making a guess as to the number. In groups of three or more, the players try to be the first player to guess the number. If only two people play the game, each player tries to guess the number in the least number of attempts.

Sample questions: Is the number greater than 100? Is the number a multiple of 5? Does the number end in 6?

 Pre-Algebra

Additional Answers

5d. The answer is reasonable and answers the question.
6a. Sample answer: estimation. The question asks for "about how many times", so an exact answer is not necessary.
b. $6.8 \times 365 \rightarrow 7 \times 350 = 2450$
c. Sample answer: No, it is an average. A person can only wash his or her hands a whole number of times.

Error Analysis

Students may be tempted to skip steps in the four-step plan. It may be helpful to ask students to write out each step they use to solve the problem as they complete the exercises.

Assignment Guide

Core: 7–12
Enriched: 7–12

For **Extra Practice**, see p. 740.

The red A, B, and C flags, printed only in the Teacher's Wraparound Edition, indicate the level of difficulty of the exercises.

4 ASSESS

Closing Activity

Writing Have students write a sentence or two describing the purpose of each step in the four-step, problem-solving plan.

Additional Answers

11a.

b. 2, 4, 6, 8, 10, 12
c. 1, 2, 3, 4, 5, 6
d. The number of cuts is half the number of pieces.

Enrichment Masters, p. 1

NAME ______________________ DATE __________

1-1 Enrichment
Cyclic Numbers

Student Edition
Pages 6–10

Look closely at the products below. Do you see a pattern in the products?

$1 \times 142{,}857 = 142{,}857$
$2 \times 142{,}857 = 285{,}714$
$3 \times 142{,}857 = 428{,}571$
$4 \times 142{,}857 = 571{,}428$
$5 \times 142{,}857 = 714{,}285$
$6 \times 142{,}857 = 857{,}142$

The same six digits repeat in the same order in the products. Numbers like 142,857 are called *cyclic numbers*.

1. Cyclic numbers are related to prime numbers. A prime number is a number that has exactly two factors, 1 and itself. You can use a calculator and the decimal equivalents of fractions of the form $\frac{1}{p}$, where p is a prime number to find cyclic numbers. Use a calculator to find the decimal equivalent of each fraction below.
   a. $\frac{1}{2}$ **0.5** b. $\frac{1}{3}$ **0.3333 · · ·** c. $\frac{1}{5}$ **0.2** d. $\frac{1}{7}$ **0.142857· · ·**

2. Study the decimal equivalents you found. Do you observe a pattern in any of the digits? **The equivalent of $\frac{1}{7}$ contains the same digits as the cyclic number.**

3. The cyclic number 142,857 has six digits. The next largest cyclic number has sixteen digits. What fraction do you think might help you find the next cyclic number? Explain. **$\frac{1}{17}$; The digits of the equivalent of $\frac{1}{7}$ were related to a cyclic number with 6 digits, so it follows that a cyclic number with 16 digits is related to $\frac{1}{17}$.**

4. Find the next largest cyclic number. **0,588,235,294,117,647**

Exercises: Practicing and Applying the Concept

Independent Practice

Solve each problem using the four-step plan.

7. **Money Management** The Westerville North High School Band plans to travel to Miami, Florida, to perform in the Orange Bowl Parade. Ana Vera needs to save $300 for the trip. She has $265 in her bank account and earns $20 shoveling snow and $25 babysitting for the neighbors. Does Ana have enough for the trip? **yes**

8. Find the total cost of building supplies that cost $2.34, $9.28, $92.34, $501.38 and $0.34. **$605.68**

Choose

Estimation
Mental Math
Calculator
Paper and Pencil

9. **Postal Service** The chart at the right shows the cost of mailing a first-class letter of different weights. How much would it cost to send a letter that weighs 8 ounces? **$2.07**

Weight (ounces)	Cost (1995)
1	$0.32
2	$0.57
3	$0.82
4	$1.07
5	$1.32

10. **Space Exploration** The space shuttle can carry a payload of about 65,000 pounds. A compact car weighs about 2450 pounds. About how many compact cars could be carried on the space shuttle? **about 26**

11. **Geometry** Study the pattern of circles. **See margin.**

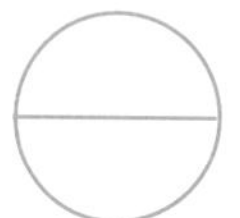

   a. Draw the next two figures in the sequence.
   b. Write a sequence of numbers for the number of pieces in each circle.
   c. Write a sequence of numbers for the number of cuts in each circle.
   d. What is the relationship between the number of cuts and the number of pieces in each circle?

12. When Gabriella divided 45,109.5 by 1479 the calculator showed 305. Is this a reasonable answer? Explain. **No, it should be about 30.**

WORKING ON THE
Investigation

Up, Up, and Away!

Refer to the Investigation on pages 2–3.

Look over your research and the list of considerations that you placed in your Investigation Folder.

- Work with the members of your group to choose one of the questions to consider first.
- Explore the question. Decide what you know and what you need to know to find the answer. You may need to do some more research.
- Make a plan for solving. If possible, make an estimate.
- Execute your plan to solve. If your plan doesn't work, make another plan and solve again.
- Examine your answer. Does it come close to the estimate? Can you justify it?

Add the results of your work to your Investigation Folder.

WORKING ON THE
Investigation

The Investigation on pages 2 and 3 is designed to be a long-term project that is completed over several days or weeks. Encourage students to keep their materials in their Investigation Folder as they work on the Investigation.

Extension

Business Connection Have students ask adults how they solve problems or make decisions at work. As a class activity, list the different strategies. Discuss how the strategies could be used to solve mathematical problems.

1-2 Order of Operations

Setting Goals: *In this lesson, you'll use the order of operations to evaluate expressions.*

Modeling with Technology

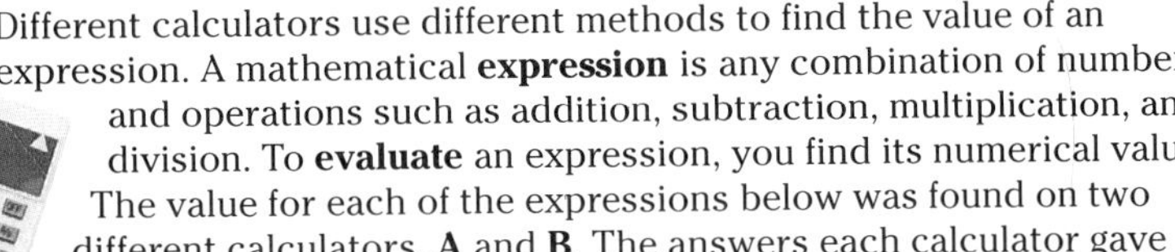

Different calculators use different methods to find the value of an expression. A mathematical **expression** is any combination of numbers and operations such as addition, subtraction, multiplication, and division. To **evaluate** an expression, you find its numerical value. The value for each of the expressions below was found on two different calculators, **A** and **B**. The answers each calculator gave are shown.

Expression	$7 + 2 \times 3$	$8 + 4 \div 2$	$12 \div 6 + 16 \div 4$	$19 - 7 + 12 \times 2 \div 8$
Calculator A	13	10	6	15
Calculator B	27	6	4.5	6

Your Turn Use a calculator to evaluate each expression above, entering the numbers and symbols in the order shown.

a–d. See students' work.

a. Did the calculator you used agree with the answers given by calculator A or calculator B?

b. Compare the answers you found to the ones that the rest of the class found. Did everyone agree with either calculator A or B?

c. Discuss reasons that the calculators may have given different answers. How did each calculator find the answer?

d. Which value do you think is correct for each expression? Defend your answer.

Learning the Concept

There are two methods that someone might use to find the value of an expression that contains more than one operation. Consider the expression $8 + 5 \times 2$.

Method 1		**Method 2**	
$8 + 5 \times 2 = 8 + 10$	*Multiply,*	$8 + 5 \times 2 = 13 \times 2$	*Add, then*
$8 + 5 \times 2 = 18$	*then add.*	$8 + 5 \times 2 = 26$	*multiply.*

In order to avoid confusion, mathematicians established the **order of operations** to tell us how to find the value of an expression.

First: Do all multiplications and divisions from left to right.
Second: Do all additions and subtractions from left to right.

Tech Prep

General Contractor For students who are interested in construction, you may wish to point out that contractors use steps to plan building projects. They must complete tasks in a certain order as they plan and construct a building. First, the design is chosen. Then, the site is prepared, the foundation laid, the walls and the roof are added, plumbing and electrical work is done, and finally finish work. For more information on tech prep, see the *Teacher's Handbook*.

1-2 LESSON NOTES

NCTM Standards: 1-5, 7, 8

Lesson Resources

- Study Guide Master 1-2
- Practice Master 1-2
- Enrichment Master 1-2
- Group Activity Card 1-2

Transparency 1-2A contains the 5-Minute Check for this lesson; **Transparency 1-2B** contains a teaching aid for this lesson.

Recommended Pacing	
Standard Pacing	Day 2 of 14
Honors Pacing	Day 2 of 13
Block Scheduling*	Day 1 of 7 (along with Lesson 1-1)

*For more information on pacing and possible lesson plans, refer to the ***Block Scheduling Booklet***.

1 FOCUS

5-Minute Check
(over Lesson 1-1)

Solve using the four-step plan.

According to the Florida Department of Citrus, 734 million gallons of orange juice and 213 million gallons of apple juice are sold each year. How much more orange juice than apple juice is sold each year?

1. What method of computation is most appropriate? **Sample answer: pencil & paper**
2. Solve for the difference. **521 million gallons**

Motivating the Lesson

Questioning Write the problem below on the chalkboard or overhead projector with the answers 5 and 8.

$2 \times 3 + 4 \div 2 = 5 \text{ or } 8$

Ask students which answer is correct. Discuss with students why they chose their answer.

2 TEACH

In-Class Examples

For Example 1
Find the value of each expression.

a. $8 \times 3 + 9$ **33**

b. $20 - 6 + 4 \times 3$ **26**

c. $12 \div 4 + 5 \times 2 - 24 \div 6$ **9**

For Example 2
Find the value of $(18 + 6) \div (9 - 6)$. **8**

Teaching Tip Ask students how they could correctly evaluate an expression like $4.5 + 3.1 \times 2.0$ on a non-scientific calculator.

THINK ABOUT IT

Calculators that follow the order of operations are called scientific calculators. Is your calculator a scientific or non-scientific calculator?

The order of operations guarantees that each numerical expression has a *unique* value. Using the order of operations, the correct value of the expression $8 + 5 \times 2$ is 18.

Example 1 **Find the value of each expression.**

a. $7 \times 9 + 3 = 63 + 3$ *Multiply 7 and 9.*
$= 66$ *Add 63 and 3.*

b. $18 + 6 - 8 \div 2 = 18 + 6 - 4$ *Divide 8 by 2.*
$= 24 - 4$ *Add 18 and 6.*
$= 20$ *Subtract 4 from 24.*

c. $144 \div 16 + 9 \div 3 + 12 = 9 + 9 \div 3 + 12$ *Divide 144 by 16.*
$= 9 + 3 + 12$ *Divide 9 by 3.*
$= 12 + 12$ *Add 9 and 3.*
$= 24$ *Add 12 and 12.*

TECHNO TIP

Many calculators allow you to evaluate expressions with parentheses. Consult your User's Guide to see if the calculator you use has this capability.

The order of operations can be changed by using grouping symbols like **parentheses**, ( ), and **brackets**, [ ]. The value of the expression is found by performing the operation in the grouping symbols first.

Example 2 **Find the value of each expression.**

a. $45 - 2 \times (15 - 3)$
$45 - 2 \times (15 - 3) = 45 - 2 \times (12)$ *Do the operations in parentheses first.*
$= 45 - 24$
$= 21$ *Multiply 2 by 12.*

b. $(22 + 3) \div (9 - 4)$
$(22 + 3) \div (9 - 4) = 25 \div 5$ *Do the operations in parentheses first.*
$= 5$ *Divide 25 by 5.*

The order for performing operations is as follows.

Order of Operations

1. **Do all operations within grouping symbols first; start with the innermost grouping symbol.**
2. **Next, do all multiplications and divisions from left to right.**
3. **Then, do all additions and subtractions from left to right.**

Cooperative Learning

Round Table Have groups of students make a list of expressions that have a value of 18. One student states and writes an expression on an answer sheet and passes the answer sheet to the next person for an answer. Allow students to work for a few minutes. Then have each group write their solutions on the chalkboard or overhead projector for the rest of the class to check. For more information on the round table strategy, see *Cooperative Learning in the Mathematics Classroom*, one of the titles in the Glencoe Mathematics Professional Series, p. 21.

Example 3 **APPLICATION** Sports

In each week of the football season, the Associated Press poll ranks football teams using votes from a group of sports reporters. A team gets 25 points for each first-place vote, 24 points for each second-place vote, 23 points for each third-place vote, and so on, to one point for each twenty-fifth place vote. The team with the highest total is ranked number one. Suppose that in one week the Miami Hurricanes received 10 first-place votes, 4 second-place votes, 2 fourth-place votes, and 1 tenth-place vote.

a. Write an expression for the number of points that the Hurricanes earned.

$$\begin{pmatrix}\text{points for}\\ 10 \times \text{first-place}\\ \text{vote}\end{pmatrix} + \begin{pmatrix}\text{points for}\\ 4 \times \text{second-place}\\ \text{vote}\end{pmatrix} + \begin{pmatrix}\text{points for}\\ 2 \times \text{fourth-place}\\ \text{vote}\end{pmatrix} + \begin{pmatrix}\text{points for}\\ 1 \times \text{tenth-place}\\ \text{vote}\end{pmatrix}$$

$$(10 \times 25) \quad + \quad (4 \times 24) \quad + \quad (2 \times 22) \quad + \quad (1 \times 16)$$

b. Find the total number of points for the Hurricanes.

$(10 \times 25) + (4 \times 24) + (2 \times 22) + (1 \times 16)$

$= 250 + 96 + 44 + 16$ *Do operations in parentheses.*

$= 406$ *Add.*

The Hurricanes earned 406 points.

There are many ways used to indicate multiplication and division in algebra.

A raised dot or parentheses can be used to indicate multiplication.

$$5 \cdot 6 \xrightarrow{\text{means}} 5 \times 6$$

$$7(3), (7)3, \text{ or } (7)(3) \xrightarrow{\text{means}} 7 \times 3$$

A fraction bar can be used to indicate division.

$$\frac{37 + 38}{30 - 5} \xrightarrow{\text{means}} (37 + 38) \div (30 - 5)$$

Example 4 **Find the value of each expression.**

a. $5(7 + 4) - 8 \cdot 4 = 5(11) - 8 \cdot 4$

$= 55 - 32$ *5(11) means 5 × 11 and*

$= 23$ *8 · 4 means 8 × 4*

b. $5[(5 + 14) - 2(7)] = 5[19 - 2(7)]$ *Do operations in innermost grouping symbols first.*

$= 5[19 - 14]$

$= 5[5]$

$= 25$

c. $\frac{72 + 12}{35 + 7} = (72 + 12) \div (35 + 7)$ *Rewrite division using parentheses.*

$= 84 \div 42$

$= 2$

In-Class Examples

For Example 3

VideoCity rents videotapes and VCRs. A newly released tape rents for $2.75 for 2 days, other tapes rent for $1.25 for 2 days, and a VCR rental is $8.00 per day.

a. In one month, the Harnapp family rented 2 newly released tapes for 2 days, 4 regular tapes for 4 days, and a VCR for a week. Write an expression for the price of the rentals. **2(2.75) + 4(2 × 1.25) + 7(8.00)**

b. Find the cost of the rentals before sales tax is added. **$71.50**

For Example 4
Find the value of each expression.

a. $3(6 + 8) - 9 \cdot 2$ **24**

b. $2[(10 - 2) + 2(9)]$ **52**

c. $\frac{84 - 12}{6 + 2}$ **9**

Study Guide Masters, p. 2

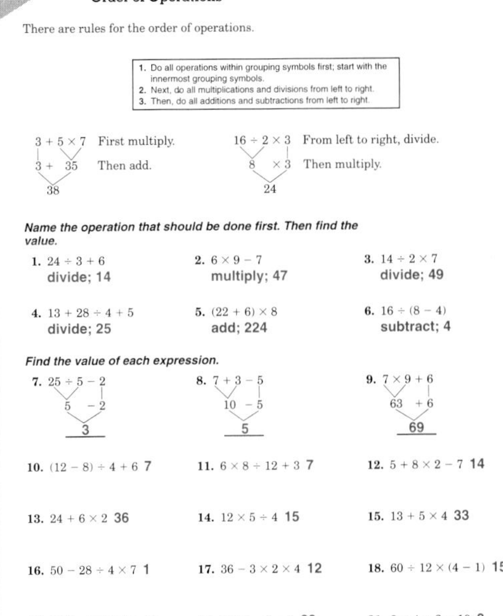

NAME ______ DATE ______

1-2 Study Guide Student Edition Pages 11–15

Order of Operations

There are rules for the order of operations.

1. Do all operations within grouping symbols first; start with the innermost grouping symbols.
2. Next, do all multiplications and divisions from left to right.
3. Then, do all additions and subtractions from left to right.

3 + 5 × 7 → 3 + 35 → 38 First multiply. Then add.

16 ÷ 2 × 3 → 8 × 3 → 24 From left to right, divide. Then multiply.

Name the operation that should be done first. Then find the value.

1. 24 ÷ 3 + 6 divide; 14
2. 6 × 9 − 7 multiply; 47
3. 14 ÷ 2 × 7 divide; 49
4. 13 + 28 ÷ 4 + 5 divide; 25
5. (22 + 6) × 8 add; 224
6. 16 ÷ (8 − 4) subtract; 4

Find the value of each expression.

7. 25 ÷ 5 − 2 → 5 − 2 → 3
8. 7 + 3 − 5 → 10 − 5 → 5
9. 7 × 9 + 6 → 63 + 6 → 69
10. (12 − 8) ÷ 4 + 6 7
11. 6 × 8 ÷ 12 + 3 7
12. 5 + 8 × 2 − 7 14
13. 24 + 6 × 2 36
14. 12 × 5 ÷ 4 15
15. 13 + 5 × 4 33
16. 50 − 28 ÷ 4 × 7 1
17. 36 − 3 × 2 × 4 12
18. 60 ÷ 12 × (4 − 1) 15
19. (100 − 25) × 2 + 25 175
20. 3 × 7 − 5 + 4 20
21. 9 × 4 ÷ 2 − 10 8

Reteaching

Using Lists Have students use three numbers in the same order to list as many expressions as possible; for example, 4 × 2 + 1 and 4 − 2 × 1. Evaluate each expression. Repeat using expressions involving parentheses, for example, 4 × (2 + 1).

3 PRACTICE/APPLY

Checking Your Understanding

Exercises 1–12 are designed to help you assess your students' understanding through reading, writing, speaking, and modeling. You should work through Exercises 1–4 with your students and then monitor their work on Exercises 5–12.

Additional Answers

2. No; to evaluate 9(4 + 3), you would add 4 and 3 and then multiply the sum by 9; to evaluate 9 · 4 + 3, you would multiply 9 and 4 and then add 3 to the product.
31. 71 − (17 + 4) = 50
32. (8 − 5) × (4 + 2) = 18
33. 18 ÷ (3 + 6) + 12 = 14

Practice Masters, p. 2

NAME ______________________ DATE ____________

1-2 Practice
Order of Operations

Student Edition
Pages 11–15

Find the value of each expression.

1. 8 + 9 − 3 + 5 **19**	2. 7 · 5 + 2 · 3 **41**
3. 18 − 5 · 2 **8**	4. (9 + 4)(8 − 7) **13**
5. (16 + 5) − (13 + 2) **6**	6. 24 ÷ 6 + 2 **6**
7. 32 · 4 ÷ 2 **64**	8. 18 − (9 + 3) + 2 **8**
9. 6 + 5 · 2 + 3 **19**	10. 18 + 24 ÷ 12 + 3 **23**
11. 67 + 84 − 12 · 4 ÷ 16 **148**	12. 75 ÷ 15 · 6 **30**
13. 34 + 8 ÷ 2 + 4 · 9 **74**	14. 6 · 3 ÷ 9 · 2 + 1 **5**
15. (15 + 21) ÷ 3 **12**	16. (45 + 21) ÷ 11 **6**
17. 5 · 6 − 25 ÷ 5 − 2 **23**	18. (84 ÷ 4) ÷ 3 **7**
19. $\frac{15 + 35}{21 + 4}$ **2**	20. 6(38 − 12) + 4 **160**
21. (13 + 4) + (17 · 4) **120**	22. $\frac{18 + 66}{35 - 14}$ **4**
23. 10[8(15 − 7) − (4 · 3)] **520**	24. 8[(26 + 10) − 4(3 + 2)] **128**

State whether each equation is true or false.

25. 16 + 24 ÷ 8 − 4 = 1 **false**	26. 39 − 9 · 3 + 6 = 18 **true**
27. 5(35 − 18) + 1 = 102 **false**	28. 60 ÷ 6 + 4 · 3 = 2 **false**
29. 25 ÷ 5 · 4 = 20 **true**	30. 17 − 4 + 8 · 4 = 45 **true**
31. 28 ÷ 7 · 5 ÷ 5 = 4 **true**	32. 2(3 + 4) − 2 · 3 = 8 **true**

Checking Your Understanding

Communicating Mathematics

Read and study the lesson to answer these questions.

1. **Identify** the first operation you perform when evaluating the expression (13 + 11) ÷ 4. **Add 13 and 11.**
2. Are the expressions 9(4 + 3) and 9 · 4 + 3 equivalent? Explain. **2. See margin.**
3. **Make a diagram** to show the order of steps you should follow when evaluating an expression. **See students' work.**
4. **Write** an expression involving multiplication and subtraction in which the first step when you evaluate would be to subtract. **4. Sample answer: 7 × (9 − 5).**
5. Play this game with a partner.
   - Write each of the digits 1–9 on a card and place cards face down.
   - Draw six cards from the pile. Out of the six numbers chosen, choose one to act as the answer.
   - Then work with your partner to write an expression with the other five numbers so that the result is the chosen answer.
   - You may use any of the operations—addition, subtraction, multiplication, and division—and add grouping symbols as needed. You may also rearrange the numbers. **See students' work.**

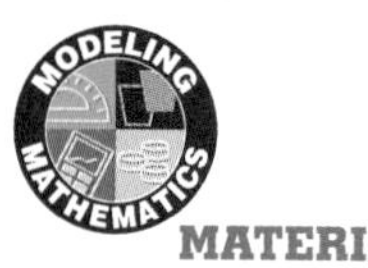

MATERIALS

index cards
pencil

Guided Practice

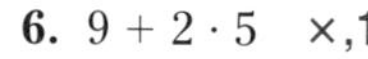

Name the operation that should be performed first. Then find the value of each expression.

6. 9 + 2 · 5 **×,19**
7. 65 − 32 ÷ 8 **÷,61**
8. 33 ÷ 3 + 6 **÷,17**
9. 17 − 6(2) **×,5**
10. 8(2 + 4) **+,48**
11. $\frac{20 - 8}{9 - 5}$ **−,3**
12. **Retail Sales** Mediaphile CDs and Books has a new pricing policy on CDs. Every single album CD is $11.99 or less. Anthony rang up four $11.99 CDs and a double-album CD that sells for $18.99.
   a. Write an expression for the total cost of the merchandise before sales tax. **4($11.99) + $18.99**
   b. What was the total before sales tax was added? **$66.95**

Exercises: Practicing and Applying the Concept

Independent Practice

Find the value of each expression.

13. 15 ÷ 3 + 12 ÷ 4 **8**
14. 3 · (4 + 5) − 7 **20**
15. 15 ÷ 5 × 3 **9**

16. 40 · (6 − 2) **160**
17. 12(7 − 2 × 3) **12**
18. 36 − 6 · 5 **6**
19. $\frac{18 + 66}{35 - 14}$ **4**
20. $\frac{16 + 8}{15 - 7}$ **3**
21. $\frac{2(14 - 6)}{4}$ **4**
22. 96 ÷ (12 · 4) ÷ 2 **1**
23. 72 ÷ (9 × 4) ÷ 2 **1**
24. (30 × 2) − (6 · 9) **6**
25. 3[6(12 − 3)] − 17 **145**
26. 7[5 + (13 − 4) ÷ 3] **56**
27. 4[3(21 − 17) + 3] **60**
28. 5[(12 + 5) − 3(19 − 14)] **10**
29. 8[(26 + 10) − 4(3 + 2)] **128**
30. 10[8(15 − 7) − 4 · 3] **520**

Calculators

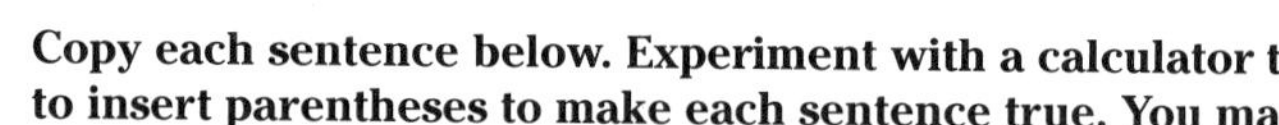

Copy each sentence below. Experiment with a calculator to find where to insert parentheses to make each sentence true. You may need to use the parentheses keys. **See margin.**

31. 71 − 17 + 4 = 50
32. 8 − 5 × 4 + 2 = 18
33. 18 ÷ 3 + 6 + 12 = 14

Group Activity Card 1-2

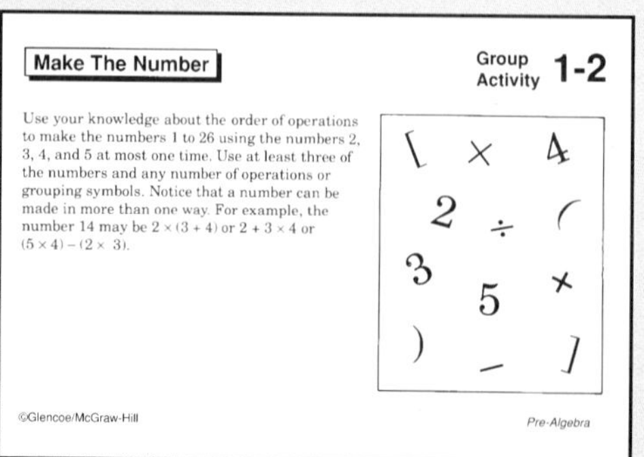

Make The Number

Group Activity 1-2

Use your knowledge about the order of operations to make the numbers 1 to 26 using the numbers 2, 3, 4, and 5 at most one time. Use at least three of the numbers and any number of operations or grouping symbols. Notice that a number can be made in more than one way. For example, the number 14 may be 2 × (3 + 4) or 2 + 3 × 4 or (5 × 4) − (2 × 3).

Pre-Algebra

Critical Thinking

34. Only the 1, [+], [−], [×], [÷], [(], [)] and [=] keys on a scientific calculator are working. How can a result of 75 be reached by pushing these keys less than 20 times? **Answers may vary. Sample answer 111 − (1 + 1 + 1) × (11 + 1)**

Applications and Problem Solving

35. **Business** Phyllis Sokol leads seminars on computer technology. On an upcoming trip, she will conduct a three-day seminar in New York City followed by a two-day seminar in Washington, D.C., and then a four-day seminar in Chicago. *Nation's Business* magazine estimates that a business traveler will spend $330 on food and lodging per day in New York City, $247 per day in Washington, D.C., and $229 per day in Chicago. **35a. 3($330) + 2($247) + 4($229)**

   a. Write an expression for the amount of money that Ms. Sokol can expect to spend on food and lodging on this trip.

   b. Find the amount of money that Ms. Sokol should expect to spend on food and lodging. **$2400** **35c. See students' work.**

   c. List some other items that Ms. Sokol will have to pay for on her trip. Estimate the cost of each item and the total cost of the trip.

36. **Nutrition** The table below shows the number of Calories in one serving of different snack foods movie-goers love.

Snack	Calories per serving
Junior Mints (3 ounces)	360
Raisinets (2.3 ounces)	270
Buttered popcorn, medium bucket	1221
Coca-Cola (16 ounces)	205
Diet 7•Up (16 ounces)	0

   a. David and Marta ordered 2 boxes of Junior Mints, a bucket of popcorn, and four Cokes to share with friends. Find the total number of Calories in all the food. **2761**

   b. Yolanda bought two Diet 7•Ups and a box of Raisinets to share with Curtiss. How many Calories did each person consume? **135**

37. **History** You can plant a bit of history in your yard. A non-profit conservation group called American Forests collects seeds from trees at historic homes, grows them into saplings, and sells them to the public. Each sapling costs $35 and $7 is added to each order for shipping and handling. How much would an order for seven saplings cost? **$252**

Mixed Review

38. **Sports** According to *Women's Sports and Fitness* magazine, a good swimmer can swim the length of a standard-sized pool in 24 or fewer strokes. If 1470 laps in a pool is equivalent to swimming the distance across the English Channel, about how many strokes would a good swimmer use to swim across the English Channel? (Lesson 1-1)

   a. Which method of computation do you think is most appropriate to solve this problem? **Sample answer: estimate**

   b. Solve the problem using the four-step plan. **37,500**

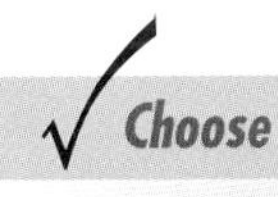

Choose

Estimation
Mental Math
Calculator
Paper and Pencil

39. **Number Theory** Write a sentence or two about the pattern at the right. Find the products 44 × 46, 54 × 56, and 84 × 86 by extending the pattern. (Lesson 1-1) **2024, 3024, 7224**

$4 \times 6 = 24$
$14 \times 16 = 224$
$24 \times 26 = 624$
$34 \times 36 = 1224$

Extension

Using Technology Have students work with a calculator to find expressions that are equal to the whole numbers from 1 to 5. Each expression should use each of the numbers 1, 2, 3, and 4 once. Students may use grouping symbols and the operations of addition, subtraction, multiplication, and division.

Sample answer: $(4 + 2) \div 3 - 1 = 1$; $4 + 2 - 3 - 1 = 2$; $2 \times 3 - 4 + 1 = 3$; $4 \div 2 + 3 - 1 = 4$; $4 + 3 - 2 \div 1 = 5$

Error Analysis

Students may try to do too many steps at one time as they evaluate expressions. Have students underline the first operation to be performed, then write the second line with only that operation completed. Repeat until the problem is completed.

Assignment Guide

Core: 13–33 odd, 34, 35, 37–39
Enriched: 14–32 even, 34–39

For **Extra Practice**, see p. 740.

The red A, B, and C flags, printed in the Teacher's Wraparound Edition, indicate the level of difficulty of the exercises.

4 ASSESS

Closing Activity

Speaking Have students describe an activity in real life that has an order of operations. For example, socks are always put on before shoes.

Enrichment Masters, p. 2

NAME ________ DATE ________

Student Edition Pages 11–15

1-2 Enrichment
Operations Search

This is a fun activity that you can try, and then make up similar problems for your family or classmates.

In each exercise below, you are given some numbers. Insert operations symbols (+, −, ×, ÷) and parentheses so that a true mathematical sentence is formed.

1. You cannot change the order of the numbers and you must observe the order of operations.

   a. 5 4 3 2 1 = 3 $(5 \times 4) \div (3 + 2) - 1 = 3$
   b. 5 4 3 2 1 = 0 $(5 + 4) \div 3 - 2 - 1 = 0$
   c. 5 4 3 2 1 = 1 $(5 + 4) \div 3 - 2 \div 1 = 1$
   d. 5 4 3 2 1 = 50 $5 \times (4 + 3 + 2 + 1) = 50$

2. You cannot change the order of the numbers, but you may put two numbers together to form a two-digit number. You must observe the order of operations.

   a. 1 2 3 4 5 6 7 8 = 90 $1 + 23 + 45 + 6 + 7 + 8 = 90$
   b. 8 7 6 5 4 3 2 1 = 25 $(8 + 7 + 65) \div 4 + 3 + 2 \times 1 = 25$

3. You can change the order of the numbers and you may put numbers together to form a two- or three-digit number. You must observe the order of operations.

   Use these four digits: 4 3 2 1

   Make these totals:
   a. 1312 $41 \times 32 = 1312$
   b. 2 $(4 \times 2) \div (3 + 1) = 2$
   c. 16 $4 \times (3 + 2 - 1) = 16$
   d. 1 $(4 + 2) \div 3 - 1 = 1$

1-3 LESSON NOTES

NCTM Standards: 1-4, 9

Instructional Resources

- Study Guide Master 1-3
- Practice Master 1-3
- Enrichment Master 1-3
- Group Activity Card 1-3
- Activity Masters, p. 15
- Assessment and Evaluation Masters, p. 15
- Tech Prep Applications Masters, p. 1

Transparency 1-3A contains the 5-Minute Check for this lesson; **Transparency 1-3B** contains a teaching aid for this lesson.

Recommended Pacing	
Standard Pacing	Day 3 of 14
Honors Pacing	Day 3 of 13
Block Scheduling*	Day 2 of 7 (along with Lesson 1-3B)

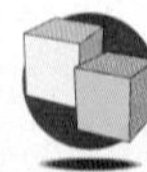

*For more information on pacing and possible lesson plans, refer to the ***Block Scheduling Booklet***.

1 FOCUS

5-Minute Check
(over Lesson 1-2)

Find the value of each expression.

1. $54 - 42 \div 7$ **48**
2. $36 \div 9 \cdot 3$ **12**
3. $(40 \cdot 2) - (6 \cdot 11)$ **14**
4. $\frac{37 + 38}{30 - 5}$ **3**
5. $3[(18 - 3) + 4(5 + 7)]$ **189**
6. Write an expression that would help you find the total cost of twenty adult zoo tickets at $6 each and ten children's tickets at $2 each. Then find the total cost. **20(6) + 10(2); $140**

1-3 Variables and Expressions

Setting Goals: *In this lesson, you'll evaluate expressions containing variables and translate verbal phrases into algebraic expressions.*

Modeling a Real-World Application: Medicine

How much do you know about your hardworking heart?

- An adult heart is about the size of a clenched fist.
- In one year, a human heart beats over 30 million times. The heart of a 70 year old has beaten over 2.5 billion times!
- In an average lifetime, the heart pumps 1 million barrels of blood.

One way that people maintain good health is by monitoring their blood pressure. Blood pressure is the pressure that your blood exerts on your arteries. The *systolic pressure* is the pressure when the heart contracts, and the *diastolic pressure* is the pressure when the heart is between contractions.

As people age, their blood pressure rises. You can approximate a person's normal systolic pressure by dividing his or her age by 2 and then adding 110.

Learning the Concept

Algebra, like any language, is a language of symbols. You know the symbols for division and addition, so you can write the blood-pressure relationship as $age \div 2 + 110$.

In arithmetic, you could write ● ÷ 2 + 110, where ● is the person's age in years. The ● serves as a placeholder.

In algebra, we use placeholders called **variables**. Variables are usually letters. The letter x is used very often as a variable, but it is also common to use the first letter of the value you are representing. For example, we can use a to represent age in the blood-pressure relationship. The relationship can then be written as $a \div 2 + 110$.

$age \div 2 + 110$	*Words and symbols*
● ÷ 2 + 110	*Arithmetic*
$a \div 2 + 110$	*Algebra*

$a \div 2 + 110$ is called an **algebraic expression** because it is a combination of variables, numbers, and at least one operation.

Expressions like $a \div 2 + 110$ can be evaluated by replacing the variables with numbers and then finding the numerical value of the expression.

Tech Prep

Medical Technician For students who are interested in medicine, you may wish to point out that medical technicians often perform medical tests involving blood pressure and medical conditions. For more information on tech prep, see the *Teacher's Handbook*.

If Malcolm is 18 years old, he could estimate his normal blood pressure by evaluating the expression $18 \div 2 + 110$.

$$a \div 2 + 110 = 18 \div 2 + 110 \quad \textit{Replace a with 18.}$$
$$= 9 + 110 \quad \textit{Divide first, then add.}$$
$$= 119$$

Malcolm's normal systolic blood pressure is approximately 119.

Replacing a with 18 demonstrates an important property of numbers.

Substitution Property of Equality	**For all numbers a and b, if $a = b$, then a may be replaced with b.**

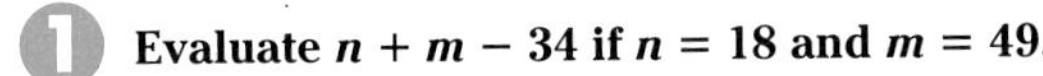

Example 1 **Evaluate $n + m - 34$ if $n = 18$ and $m = 49$.**

$$n + m - 34 = 18 + 49 - 34 \quad \textit{Replace n with 18 and m with 49.}$$
$$= (18 + 49) - 34 \quad \textit{Follow the order of operations.}$$
$$= 67 - 34 \quad 18 + 49 = 67$$
$$= 33 \quad 67 - 34 = 33$$

THINK ABOUT IT

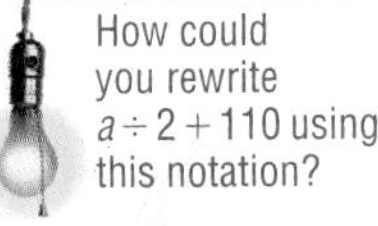

How could you rewrite $a \div 2 + 110$ using this notation?

$\frac{a}{2} + 110$

As with numerical expression, there is special notation for multiplication and division with variables.

$3d$ *means* $3 \times d$

xy *means* $x \times y$

$7st$ *means* $7 \times s \times t$

$\frac{q}{4}$ *means* $q \div 4$

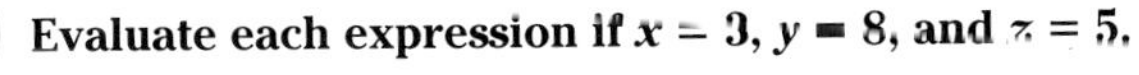

Example 2 **Evaluate each expression if $x = 3$, $y = 8$, and $z = 5$.**

a. $7x - 4z = 7(3) - 4(5)$ *Replace x with 3 and z with 5.*
$= 21 - 20$ *Use the order of operations.*
$= 1$

b. $\frac{yz}{4} = (8)(5) \div 4$ $\quad \frac{yz}{4} = yz \div 4$
$= 40 \div 4$ *Use the order of operations.*
$= 10$

c. $3x + (z + 2y) - 12 = 3(3) + (5 + 2 \cdot 8) - 12$
$= 3(3) + (5 + 16) - 12$
$= 9 + 21 - 12$
$= 18$

You can use variables to show patterns in expressions. For example, a babysitter may earn $4 per hour. The chart at the right shows several possibilities for number of hours and earnings.

Hours of babysitting	Money earned
2	$4 \cdot 2$ or 8
5	$4 \cdot 5$ or 20
8	$4 \cdot 8$ or 32
h	$4 \cdot h$ or $4h$

The variable h and the algebraic expression $4h$ summarize the relationship between number of hours of babysitting and earnings.

Motivating the Lesson

Situational Problem Ask students to describe times where substitutes are used. Two examples are margarine for butter, or a relief pitcher. Ask when substitutions are made. Emphasize the fact that when a substitution is made the substitute is equal to the original.

2 TEACH

In-Class Examples

For Example 1
Evaluate $q + r - 15$ if $q = 21$ and $r = 18$. **24**

For Example 2
Evaluate each expression if $a = 4$, $b = 2$, and $c = 3$.

a. $6a - bc$ **18**

b. $\frac{6a}{c}$ **8**

c. $2c + 3a + 6b$ **30**

Teaching Tip Point out that an algebraic expression is different from a numerical expression because it contains a variable.

Teaching Tip Remind students that the order of operations still applies when evaluating algebraic expressions.

Alternative Teaching Strategies

Reading Mathematics When the order of operations is used to evaluate an algebraic expression for given values of the variables, we always get the same value. This is called the *unique answer*. Discuss what *unique* means.

GLENCOE Technology

Interactive Mathematics Tools Software

In this interactive computer lesson, students use the computer to investigate evaluating expressions. A **Computer Journal** gives students the opportunity to write about what they have learned.

For Windows & Macintosh

Teaching Tip Discuss the different words that indicate mathematical operations. Have students make up sentences that use the words.

In-Class Examples

For Example 3
Translate each phrase into an algebraic expression.

a. 3 years younger than Quint **$q - 3$**

b. twenty minus r **$20 - r$**

c. Marty's total divided by four **$m \div 4$**

For Example 4
A taxicab charges $2.00 plus $0.80 for each mile traveled for a ride. Write an expression for the price of a cab ride. **2.00 + 0.80*m***

3 PRACTICE/APPLY

Checking Your Understanding

Exercises 1–15 are designed to help you assess your students' understanding through reading, writing, speaking, and modeling. You should work through Exercises 1–3 with your students and then monitor their work on Exercises 4–15.

Study Guide Masters, p. 3

NAME ______________________ DATE __________

1-3 Study Guide
Variables and Expressions

Student Edition
Pages 16–20

A variable is used to represent some number. The value of an expression may be changed by replacing a variable with different numbers. An expression may contain more than one variable. Remember to use the order of operations when evaluating expressions.

Example: Evaluate $5a - 2b$ if $a = 3$ and $b = 2$.
$5a - 2b = 5(3) - 2(2)$
$= 15 - 4$
$= 11$

Example: Evaluate $5a - 2b$ if $a = 10$ and $b = 0$.
$5a - 2b = 5(10) - 2(0)$
$= 50 - 0$
$= 50$

Evaluate each expression if a = 1, b = 2, x = 5, and y = 10.

1. $y - a$ **9**
2. $y - b$ **8**
3. $x + y$ **15**
4. $a + b$ **3**
5. ab **2**
6. ax **5**
7. bx **10**
8. xy **50**
9. $ab + y$ **12**
10. $ax - b$ **3**
11. $bx - y$ **0**
12. $xy + ab$ **52**
13. $\frac{y}{b}$ **5**
14. $\frac{y}{x}$ **2**
15. $\frac{x}{a} + x$ **10**
16. $\frac{y}{a} - bx$ **0**
17. $5a$ **5**
18. $5b$ **10**
19. $10y + y$ **110**
20. $10b - a$ **19**

Evaluate each expression if a = 3, b = 5, x = 2, and y = 4.

21. $a \times 7$ **21**
22. $b \times 8$ **40**
23. $2 \times x + y$ **8**
24. $a + 6 \times b$ **33**
25. $5x - y$ **6**
26. $2b - 2a$ **4**
27. $6x \div y$ **3**
28. $x + y \times 7 \div 2$ **16**
29. $a \times x + b \times y$ **26**
30. $2ab$ **30**
31. $b + x \times a$ **11**
32. $4x \div y$ **2**

When reading a verbal sentence and writing an algebraic expression to represent it, there are many words and phrases that suggest the operation to use. The following chart shows some common phrases and operations.

Addition	Subtraction	Multiplication	Division
plus sum more than increased by total in all	minus difference less than subtract decreased by	times product multiplied each of	divided quotient

Example  **Translate each phrase into an algebraic expression.**

a. ten more points than Marisa scored
Let p represent the points Marisa scored.
The words *more than* suggest addition.
The algebraic expression is $p + 10$ or $10 + p$.

b. three times last year's total
Let t represent last year's total.
The words *three times* suggest multiplication.
The algebraic expression is $3t$ or $3 \cdot t$.

Example 4

Business

Warnet Cable charges $19.95 a month for basic cable television service. Each premium channel selected costs an additional $4.95 per month. Write an expression for the cost of a month of cable service.

Let n represent the number of premium channels selected.

Since you are paying $4.95 *for each* premium channel, the total for premium channels will be $\$4.95 \times n$ or $\$4.95n$.

The word *additional* suggests addition. You will pay for basic service and the premium channels.

The algebraic expression is $19.95 + 4.95n$.

Checking Your Understanding

Communicating Mathematics

Read and study the lesson to answer these questions.

1. **Define** a variable in your own words. **See Solutions Manual.**
2. **Write** two different expressions that are the same as $7a$. **$7 \times a$, $7 \cdot a$**
3. If m is the number of games the Jets won, what does $m - 5$ represent? **5 less than the number of games the Jets won**
4. Cups and counters can be used to model algebraic expressions. The cup represents the variable, and the counters are one unit each. The diagram models $2x + 3$.
   a. Use cups and counters to model the expression $4x + 2$.
   b. Model the phrase *four plus three times a number.* Then write an expression to represent the phrase. **4. See Solutions Manual.**

MATERIALS

 cups

 counters and mat

Reteaching

Using Models Model the examples using different-sized or different-colored cups to represent the different variables in each expression. Use the substitution property to replace each cup with the number of counters each variable represents.

Group Activity Card 1-3

Guess My Expression — Group Activity **1-3**

MATERIALS: 0-9 spinner • clock

SETUP: Each member of the group writes four algebraic expressions using the numbers 1-9 and a variable. For example, one such expression is $3n - 5$. The expression $\frac{3(n+9)}{n+4}$ is not acceptable since the variable "n" is used more than one time.

Players take turns spinning the spinner. They substitute the spinner result into their expression and announce the value to the other players. The other players must then guess the expression. If no one can guess the expression in one minute, the player spins again and announces another value. The first player to get the correct expression scores one point. If no player gets the expression after 4 spins, the writer of the expression scores one point and the spinner is passed to the next player. When all players have used all four of their expressions, the person with the greatest number of points wins.

 Pre-Algebra

Guided Practice

Evaluate each expression if $m = 4$, $n = 11$, $p = 2$, and $q = 5$.

5. $8 + n$ **19**
6. $10 - q$ **5**
7. $3m$ **12**
8. $24 - 4q$ **4**
9. $\frac{5m}{2}$ **10**
10. $n + 110 - 2p$ **117**

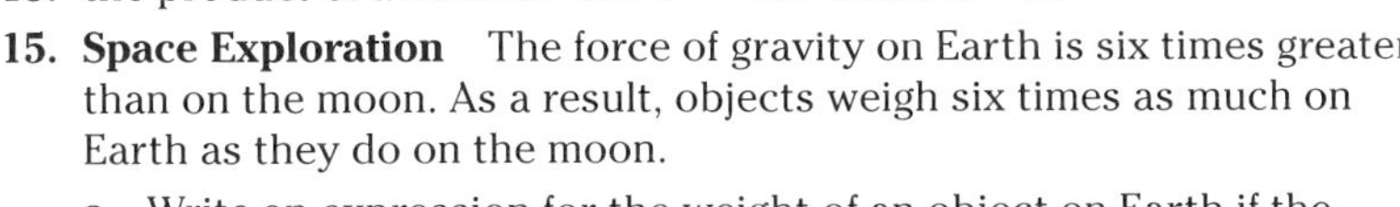

Translate each phrase into an algebraic expression. **13. $6n$**

11. three more than v **$3 + v$**
12. seven less than h **$h - 7$**
13. the product of a number and 6
14. twice w **$2w$**
15. **Space Exploration** The force of gravity on Earth is six times greater than on the moon. As a result, objects weigh six times as much on Earth as they do on the moon.
    a. Write an expression for the weight of an object on Earth if the weight on the moon is x. **$6x$**
    b. A scientific instrument weighs 34 pounds on the moon. How much does the instrument weigh on Earth? **204 pounds**

Exercises: Practicing and Applying the Concept

Independent Practice

Evaluate each expression if $a = 6$, $b = 3$, and $c = 7$.

16. $a + 22$ **28**
17. $2b - 4$ **2**
18. $ca - ab$ **24**

A ▶

19. $4a - (b + c)$ **14**
20. $5c + 5b$ **50**
21. $\frac{8b}{a}$ **4**

B ▶

22. $2c + 3a + 6b$ **50**
23. $\frac{6a}{b}$ **12**
24. $9a - (4b + 2c)$ **28**
25. $\frac{3(4a - 3c)}{c - 4}$ **3**
26. $12b - \frac{2c - 4}{a + 4}$ **35**
27. $\frac{15ab}{3c + 6}$ **10**

Translate each phrase into an algebraic expression. **29. $92 \div c$**

28. a number divided by 8 **$x \div 8$**
29. the quotient of ninety-two and c
30. the sum of g and 8 **$g + 8$**
31. p decreased by 5 **$p - 5$**
32. eleven increased by t **$11 + t$**
33. the difference of u and 10 **$u - 10$**
34. twice a number **$2n$**
35. the quotient of eighty-eight and b
36. Gil's salary plus a $500 bonus
37. three times as many fouls as Sumi

C ▶

38. three times a number decreased by 18 **$3n - 18$**
39. two more than twice the previous month's sales **$2s + 2$**

35. $88 \div b$ 36. $g + \$500$ 37. $3k$

Write a verbal phrase for each algebraic expression. **See margin.**

40. $x + 4$
41. $16 - b$
42. $7n$
43. $v \div 5$
44. $2(a + 1)$
45. $2a + 1$

Critical Thinking

46. Write an expression for the value of a three-digit number whose hundreds, tens, and units digits are x, y, and z, respectively. **$100x + 10y + z$**

Applications and Problem Solving

47. **Meteorology** You can estimate the temperature in degrees Fahrenheit by counting the number of times a cricket chirps in one minute. Count the number of chirps, divide by 4, and then add 37.

    a. Write an expression for the temperature relationship. **$c \div 4 + 37$**
    b. Find the approximate temperature when a cricket chirps 124 times in a minute. **68°**

Extension

Using Problem Solving Each symbol represents a single digit (0–9). Determine the value of each symbol.

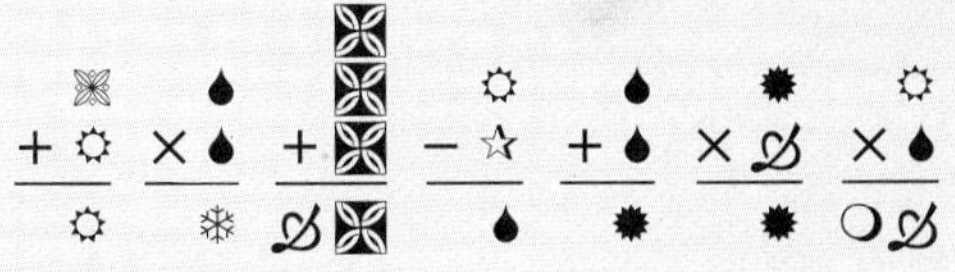

✱ = 0; ☼ = 7; ♦ = 3; ❄ = 9; ⊠ = 5; ∅ = 1; ☆ = 4; ✹ = 6; ❍ = 2; ⅄ = 8

Assignment Guide

Core: 17–45 odd, 46, 47, 49–52
Enriched: 18–44 even, 46–52

For **Extra Practice**, see p. 740.

The red A, B, C flags, printed only in the Teacher's Wraparound Edition, indicate the level of difficulty of the exercises.

Additional Answers

40. some number plus 4
41. 16 less some number
42. seven times some number
43. the quotient of some number and 5
44. twice the sum of some number and one
45. twice some number plus one

Practice Masters, p. 3

NAME ______ DATE ______

1-3 Practice Student Edition Pages 16–20
Variables and Expressions

Evaluate each expression if $x = 7$, $y = 10$, $r = 15$, $t = 3$, and $c = 8$.

1. $x + y - r$ **2**
2. $c + c + y + y$ **36**
3. $(r + t) + c$ **26**
4. $x + t + c - r + y$ **13**
5. $y + 15 - c + 12 - x$ **22**
6. $85 - 17 + t - x + c$ **72**
7. $72 - r + y - c$ **59**
8. $t + t + t + t + t$ **15**
9. $y + y + c - 10 + x$ **25**
10. $125 + x - y + r - t$ **134**

Evaluate each expression if $x = 3$, $y = 4$, and $z = 5$.

11. $6x - 3y$ **6**
12. $6(x + y)$ **42**
13. $\frac{y - x}{z - y}$ **1**
14. $2x + 3z + y$ **25**
15. $14x - (2y + z)$ **29**
16. $2(x + z) - y$ **12**
17. $4z - (2y + x)$ **9**
18. $x(y + z + 4)$ **39**
19. $\frac{10(z - x)}{z}$ **4**
20. $\frac{21xy}{x + y}$ **36**
21. $\frac{4z + 2y}{7}$ **4**
22. $\frac{y(z + x + y)}{y}$ **12**

Translate each phrase into an algebraic expression.

23. six minutes less than Bob's time **$t - 6$**
24. four points more than the Bearcubs scored **$s + 4$ or $4 + s$**
25. Joan's temperature increased by two degrees **$t + 2$ or $2 + t$**
26. the cost decreased by ten dollars **$c - 10$**
27. seven times a certain number **$7n$ or $n \cdot 7$**
28. twice a number decreased by four **$2n - 4$**
29. twice the sum of two and y **$2(2 + y)$ or $2(y + 2)$**
30. the quotient of x and 2 **$x \div 2$ or $\frac{x}{2}$**

4 ASSESS

Closing Activity

Writing Have students write a mathematical expression using the letters *k*, *m*, or *n* and any numbers. Then have them exchange their expressions with another student and evaluate if $k = 1$, $m = 2$, and $n = 0$.

Chapter 1, Quiz A (Lessons 1-1 through 1-3) is available in the *Assessment and Evaluation Masters*, p. 15.

Additional Answer

52a. An estimate is appropriate because an exact answer is not needed.

Enrichment Masters, p. 3

NAME ______________________ DATE ____________

1-3 Enrichment
Sets

Student Edition
Pages 16–20

A **set** is a collection or group of objects. The members of a set are called **elements**. A set can be named by listing the elements within a pair of braces, { }. Here are two sets:

A = {0.5, 1.5, 2.5, 3.5}   B = {0.01, 0.02, 0.03, 0.04}

The following symbols are frequently used with sets. Note that every element of a subset of a set must be an element of the set.

Symbol	Meaning	Example
$\in$	is an element of	$2.5 \in \{0.5, 1.5, 2.5, 3.5\}$
$\notin$	is not an element of	$0.2 \notin \{0.01, 0.02, 0.03, 0.04\}$
$\subseteq$	is a subset of	$\{2.5, 3.5\} \subseteq \{0.5, 1.5, 2.5, 3.5\}$
$\not\subseteq$	is not a subset of	$\{3, 4\} \not\subseteq \{0.5, 1.5, 2.5, 3.5\}$
$\varnothing$ or { }	empty set (null set)	A set having no elements.

List the following sets. Use braces.

1. set C, decimals in tenths between 2.7 and 3.4
   C = {2.8, 2.9, 3.0, 3.1, 3.2, 3.3}
2. set D, whole numbers between 8.3 and 13.7 D = {9, 10, 11, 12, 13}
3. set E, whole numbers between 2.1 and 2.5 E = { } or $\varnothing$
4. set F, decimals in hundredths between 4.56 and 4.62
   F = {4.57, 4.58, 4.59, 4.60, 4.61}
5. set G, names of the first 3 place values to the right of the decimal point G = {tenths, hundredths, thousandths}

Write true or false for each of the following.

6. $3.5 \in \{1.5, 2.5, 3.5, 4.5\}$ true
7. $0.1 \notin \{0.1, 0.01, 0.001\}$ false
8. $\{0.4, 0.6\} \subseteq \{0.2, 0.4, 0.6, 0.8\}$ true
9. $0.8 \in \{0.8, 0.9\}$ true
10. $5.1 \subseteq \{4.1, 5.1, 6.1, 7.1\}$ false
11. $0.5 \in \{0.2, 0.4, 0.6, 0.8\}$ false
12. $\{1.3\} \not\subseteq \{0.3, 1, 1.3, 2, 2.3\}$ false
13. $\{+, -, \times\} \subseteq \{+, -, \times\ \div\}$ true
14. tenths $\notin$ {tenths, hundredths} false
15. $\{2, 3\} \not\subseteq \{1.5, 2.5, 3.5\}$ true

48. **Statistics** The *mean* is a number that is often used to represent a set of numbers. To find the mean of three numbers, find the sum of the numbers and then divide by 3.
    a. Write an expression for the mean of *a*, *b*, and *c*. $\frac{a+b+c}{3}$
    b. Find the mean of 51, 66, and 63. **60**

Mixed Review

49. Find the value of the expression $6 \times 8 - (9 + 12)$. (Lesson 1-2) **27**
50. **Retail Sales** Jamie bought four adult tickets at \$6.25 each and two children's tickets at \$3.75 each. (Lesson 1-2)
    a. Write an expression for the total cost of the tickets.
    b. Find the cost. **\$32.50**
    c. Write your own problem involving movie tickets and solve.

50a. 4(\$6.25) + 2(\$3.75)
50c. See students' work.

51. **Patterns** Use the four-step problem-solving plan to find the next number in the sequence. (Lesson 1-1)
    1, 2, 4, 7, 11, 16, 22, __?__ **29**

52. **Art** A reproduction of Vincent van Gogh's *Sunflowers* was constructed in a wheat field near Duns, Scotland. The "painting" covered a 46,000-square foot area and contained 250,000 plants and flowers. About how many flowers and plants were planted per square foot? (Lesson 1-1)
    a. Which method of computation do you think is most appropriate for solving this problem? Explain. **See margin.**
    b. Solve the problem. **about 5**

TALK ABOUT IT

1. Explain why the comic is funny. **See students' work.**
2. What difference would it make if Paige mistook the *x* in $3x(2 + 8)$ for a multiplication symbol?
3. Write an explanation that Paige could use to distinguish between an *x* and a multiplication symbol. **See students' work.**

2. She would multiply 3 and (2 + 8) instead of multiplying 3 times *x* by (2 + 8).

From the FUNNY PAPERS

Ask students to write an expression that would have the same value if the variable *x* was mistaken for a multiplication symbol.

Sample answer:
$2x(4 - 2)$ if $x = 1$.

GRAPHING CALCULATOR ACTIVITY

1-3B Evaluating Expressions

An Extension of Lesson **1-3**

Graphing calculators observe the order of operations. Since you can specify the value of a variable, evaluating expressions is as simple as entering the expression with the calculator keys. The calculators also have parentheses.

Activity

Keystrokes are shown for the TI-82 graphing calculator. If you are using a different graphing calculator, consult your User's Guide.

Evaluate $3 + (2x - y)$ if $x = 7$ and $y = 5$ on a graphing calculator.

ENTER: 7 [STO▸] [X,T,θ] [2nd] [:] — *Enters the value of x.*

5 [STO▸] [ALPHA] [Y] [2nd] [:] — *Enters the value of y.*

3 [+] [(] 2 [X,T,θ] [−] [ALPHA] [Y] [)] [ENTER] — *Enters and evaluates the expression.*

DISPLAY: 12

If you get an error message or discover that you entered the expression incorrectly, you can use the REPLAY feature to correct your error and reevaluate without reentering the expression. Follow the steps below to use the REPLAY feature.

- Press [2nd] [ENTRY] to display your expression.
- Use the arrow keys to move to the location of the correction. Then type over, use [INS], or use [DEL] to make the correction.
- Press [ENTER] to reevaluate.

Your Turn

Evaluate each expression with a graphing calculator if $a = 9$, $b = 0$, $c = 7$, and $d = 3$.

1. $5d - 1$ 14
2. $b(a - c)$ 0
3. $3(b + c) \div d$ 7
4. $4a \div d$ 12
5. $9a - (4d + 2c)$ 55
6. $a(b + c) - 7$ 56

7. How can you use the REPLAY feature to evaluate several expressions for variables of the same value? **Use REPLAY then change the expression.**
8. What do you observe about the display when you press the [÷] and the [×] keys? **Display shows / for [÷] and * for [×].**

Technology

This lesson offers an excellent opportunity for using technology in your pre-algebra classroom. For more information on using technology, see *Graphing Calculators in the Mathematics Classroom*, one of the titles in the Glencoe Mathematics Professional Series.

1-3B LESSON NOTES

NCTM Standards: 1-4, 9

Objective

Use a graphing calculator to evaluate algebraic expressions.

Recommended Time

15 minutes

Instructional Resources

Graphing Calculator Masters, p. 15

This master provides keystroking instruction for this lesson for the TI-81 and Casio graphing calculators.

1 FOCUS

Motivating the Lesson

If this is the first time your students have used a graphing calculator, allow them some time to explore the calculators before beginning.

2 TEACH

Teaching Tip Question 2 will apply only if students are using Texas Instruments calculators.

3 PRACTICE/APPLY

Assignment Guide

Core: 1–8
Enriched: 1–8

4 ASSESS

Observing students working with technology is an excellent method of assessment.

1-4 LESSON NOTES

NCTM Standards: 1-4, 6, 7, 9

Instructional Resources

- Study Guide Master 1-4
- Practice Master 1-4
- Enrichment Master 1-4
- Group Activity Card 1-4

Transparency 1-4A contains the 5-Minute Check for this lesson; **Transparency 1-4B** contains a teaching aid for this lesson.

Recommended Pacing	
Standard Pacing	Day 5 of 14
Honors Pacing	Day 5 of 13
Block Scheduling*	Day 3 of 7 (along with Lesson 1-5)

*For more information on pacing and possible lesson plans, refer to the ***Block Scheduling Booklet***.

1 FOCUS

5-Minute Check
(over Lesson 1-3)

Evaluate each expression if $a = 4$, $b = 2$, and $c = 3$.

1. $a + bc$ **10**
2. $4a + b \cdot b$ **20**
3. $\frac{ac}{b}$ **6**
4. $7a - (2c + b)$ **20**
5. $\frac{15ac}{3c + 6}$ **12**
6. The number of fat Calories in a food item can be calculated by multiplying the number of fat grams by 9.
   a. Write an expression for the relationship between fat grams and Calories. **$c = 9g$**
   b. Find the number of fat Calories in a bag of pretzels with 6 grams of fat. **54**

1-4 Properties

Setting Goals: *In this lesson, you'll identify properties of addition and multiplication and use these properties to solve problems.*

Modeling with Manipulatives

MATERIALS

counters, beans, or other small objects

Small objects like counters or beans can be used to model operations with whole numbers. The diagram at the right models the expression $5 + 3$. Counting the number of counters in the group after the additions shows that $5 + 3 = 8$.

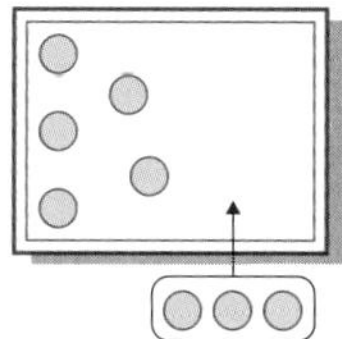

Your Turn Use counters to model $3 + 5$.

TALK ABOUT IT

b. They are the same.

a. What is the value of $3 + 5$ that you found using the counters? **8**
b. What do you observe about the values of $5 + 3$ and $3 + 5$?
c. Work with a partner to write and solve other problems like $a + b$ and $b + a$. Summarize your results. **See students' work.**
d. Investigate $3 \cdot 5$ and $5 \cdot 3$ with counters by making 3 groups of 5 counters and 5 groups of 3 counters. Then explore ab and ba. What do you observe? **$ab = ba$**

Learning the Concept

The charts below and on the next page summarize some of the properties of addition and multiplication.

Commutative Properties of Addition and Multiplication	
The order in which numbers are added does not change the sum. $5 + 3 = 3 + 5$ For any numbers a and b, $a + b = b + a$	The order in which numbers are multiplied does not change the product. $2 \cdot 4 = 4 \cdot 2$ For any numbers a and b, $a \cdot b = b \cdot a$
Associative Properties of Addition and Multiplication	
The way in which addends are grouped does not change the sum. $(2 + 4) + 6 = 2 + (4 + 6)$ For any numbers a, b, and c, $(a + b) + c = a + (b + c)$	The way in which factors are grouped does not change the product. $(6 \cdot 3) \cdot 7 = 6 \cdot (3 \cdot 7)$ For any numbers a, b, and c, $(a \cdot b) \cdot c = a \cdot (b \cdot c)$

Alternative Learning Styles

Visual Have students write three examples of each of the properties shown in the chart on pages 22 and 23. Then use colored pencils or markers to color each number or variable in a problem the same color each time it appears. Then look for the patterns in the examples of each property.

Identity Properties of Addition and Multiplication	
The sum of an addend and zero is the addend. $6 + 0 = 6$ For any number a, $a + 0 = a$.	The product of a factor and one is the factor. $6 \cdot 1 = 6$ For any number a, $a \cdot 1 = a$.

Multiplicative Property of Zero
The product of a factor and zero is zero. $5 \cdot 0 = 0$ For any number a, $a \cdot 0 = 0$

These properties can be helpful when you find sums and products mentally.

Example  **Find the value of each expression mentally.**

a. $17 + (13 + 15) = (17 + 13) + 15$ *Regroup addends.*
$= 30 + 15$
$= 45$

b. $2 \cdot 16 \cdot 5 = (2 \cdot 5) \cdot 16$ *Rearrange and group factors.*
$= 10 \cdot 16$ or 160

Example 2

Entertainment

On a recent tour, Elton John and Billy Joel visited Knoxville, Raleigh, Richmond, and Washington, D.C. The mileages between the cities are shown in the table below. If the equipment truck traveled from each city to the next in consecutive days, how many miles did it travel?

City	Destination	Mileage
Knoxville, TN	Raleigh, NC	362
Raleigh, NC	Richmond, VA	156
Richmond, VA	Washington, D.C.	108

They would travel the total of the mileages between the cities.

Rearrange and group addends so addition is easier.

$362 + 156 + 108 = (362 + 108) + 156$
$= 470 + 156$
$= 626$

Connection to Algebra

Variables represent numbers, so the properties can be used to rewrite and simplify algebraic expressions.

Motivating the Lesson

Hands-On Activity Have students look up the words *commute, associate*, and *identity* in a dictionary and relate their definitions to the mathematical meanings.

2 TEACH

In-Class Examples

For Example 1
Find the value of each expression mentally.

a. $37 + 45 + 53 + 65$ **200**

b. $2 \cdot 8 \cdot 50$ **800**

For Example 2
According to *Health* magazine, the average American spends 15.1 hours each week watching television, 2.2 hours playing sports, and 4.9 hours visiting with friends. How many hours each week are spent on these three activities? **22.2 hours**

Teaching Tip Point out that the last digit of the numbers that you wish to add mentally can help determine the order to add. Have students make a list of the pairs of last digits they would combine first.

Study Guide Masters, p. 4

NAME ____________ DATE ________

1-4 Study Guide
Properties

Student Edition
Pages 22–25

Three Properties of Addition

Commutative	The order of adding does not change the sum. $15 + 23 = 23 + 15$
Associative	Grouping numbers differently does not change the sum. $(18 + 2) + 8 = 18 + (2 + 8)$
Identity	Adding zero to a number does not change its value. $68 + 0 = 68$

Four Properties of Multiplication

Commutative	The order of multiplying does not change the product. $3 \times 5 = 5 \times 3$
Associative	Grouping numbers differently does not change the product. $(3 \times 2) \times 4 = 3 \times (2 \times 4)$
Identity	Multiplying a number by 1 does not change its value. $5 \times 1 = 5$
Zero	Zero times any number equals zero. $16 \times 0 = 0$

Name the addition property shown by each statement.

1. $300 + 0 = 300$ **Identity**
2. $96 + 4 = 4 + 96$ **Commutative**
3. $0 + 36 = 36$ **Identity**
4. $(6 + 2) + 4 = 6 + (2 + 4)$ **Associative**
5. $20 + (12 + 2) = (20 + 12) + 2$ **Associative**
6. $27 + 45 = 45 + 27$ **Commutative**
7. $75 + 25 = 25 + 75$ **Commutative**
8. $2115 = 2115 + 0$ **Identity**
9. $(3 + 7) + 6 = 3 + (7 + 6)$ **Associative**
10. $x + y = y + x$ **Commutative**
11. $y + 0 = y$ **Identity**
12. $x + (y + z) = (x + y) + z$ **Associative**

Name the multiplication property shown by each statement.

13. $3 \times 2 = 2 \times 3$ **Commutative**
14. $(6 \times 3) \times 4 = 6 \times (3 \times 4)$ **Associative**
15. $7 \times 1 = 7$ **Identity**
16. $0 \times 8 = 0$ **Zero**
17. $76 \times 13 = 13 \times 76$ **Commutative**
18. $42 \times 0 = 0$ **Zero**
19. $xy = yx$ **Commutative**
20. $x(yz) = (xy)z$ **Associative**
21. $1 \times y = y$ **Identity**

Reteaching

Using Manipulatives Use base-ten blocks or counters to illustrate the commutative, associative, and identity properties for whole-number expressions. For example, $5 \times 3 = 3 \times 5$; $3 \times (4 \times 5) = (3 \times 4) \times 5$; and $3 \times 1 = 3$.

In-Class Examples

For Example 3
Rewrite each expression using a commutative property.

a. $8 + r$ **$r + 8$**

b. $f \cdot g$ **$g \cdot f$**

For Example 4
Rewrite each expression using an associative property. Then simplify.

a. $(y + 3) + 12$
$y + (3 + 12)$; $y + 15$

b. $4(13b)$ **$(4 \cdot 13)b$; $52b$**

For Example 5
Evaluate fgh mentally if $f = 3$, $g = 8$, and $h = 2$. **48**

3 PRACTICE/APPLY

Checking Your Understanding

Exercises 1–15 are designed to help you assess your students' understanding through reading, writing, speaking, and modeling. You should work through Exercises 1–4 with your students and then monitor their work on Exercises 5–15.

Practice Masters, p. 4

NAME ______________________ DATE ____________

1-4 Practice
Properties

Student Edition Pages 22–25

Name the property shown by each statement.

1. $4 + (9 + 6) = (4 + 9) + 6$ Associative Property of Addition
2. $x + 12 = 12 + x$ Commutative Property of Addition
3. $(3 + y) + 0 = 3 + y$ Identity Property of Addition
4. $(x + y) + z = x + (y + z)$ Associative Property of Addition
5. $(15 + x) + 2 = 2 + (15 + x)$ Commutative Property of Addition
6. $x \cdot 1 = x$ Identity Property of Multiplication
7. $14xy = 14yx$ Commutative Property of Multiplication
8. $(3 + 5) + c = 3 + (5 + c)$ Associative Property of Addition
9. $(2 \cdot 5) \cdot 0 = 0$ Multiplicative Property of Zero
10. $6 \cdot (8 + c) = (8 + c) \cdot 6$ Commutative Property of Multiplication
11. $6 \cdot (4 \cdot 3) = (6 \cdot 4) \cdot 3$ Associative Property of Multiplication
12. $(3 \cdot 9) \cdot 1 = 3 \cdot 9$ Identity Property of Multiplication
13. $(a + b) + c = c + (a + b)$ Commutative Property of Addition
14. $(x + y) \cdot 5 = (y + x) \cdot 5$ Commutative Property of Addition
15. $ab + 0 = ab$ Identity Property of Addition
16. $a \cdot b = b \cdot a$ Commutative Property of Multiplication
17. $(x \cdot y) \cdot z = x \cdot (y \cdot z)$ Associative Property of Multiplication
18. $(7 \cdot 3) \cdot 5 = 7 \cdot (3 \cdot 5)$ Associative Property of Multiplication
19. $(2 + x) \cdot 0 = 0$ Multiplicative Property of Zero
20. $(8 + 5) + 3 = 3 + (8 + 5)$ Commutative Property of Addition
21. $(a + b) \cdot 1 = a + b$ Identity Property of Multiplication

Example 3 Rewrite each expression using a commutative property.

a. $n + 6$
$n + 6 = 6 + n$ *Change the order.*

b. $4 \cdot s$
$4 \cdot s = s \cdot 4$

Example 4 Rewrite each expression using an associative property. Then simplify.

THINK ABOUT IT
Is subtraction commutative or associative? **no, no**

a. $(m + 5) + 4$
$(m + 5) + 4 = m + (5 + 4)$
$= m + 9$

b. $3(7t)$
$3(7t) = (3 \cdot 7)t$ *Regroup.*
$= 21t$

Example  Evaluate xyz if $x = 2$, $y = 8$, and $z = 15$ mentally.

$xyz = 2 \cdot 8 \cdot 15$ *Replace x with 2, y with 8, and z with 15.*
$= 2 \cdot 15 \cdot 8$ *Use the Commutative property to rearrange factors.*
$= (2 \cdot 15) \cdot 8$ *Group 2 and 15 to make multiplication easier.*
$= 30 \cdot 8$
$= 240$

Checking Your Understanding

Communicating Mathematics

Read and study the lesson to answer these questions. **1. See margin.**

1. **Compare and contrast** the associative and commutative properties.
2. **Write** mathematical sentences to illustrate each of the seven properties described in the tables on pages 22 and 23. **See margin.**
3. **Demonstrate** how the properties can be used to perform calculations mentally. **See students' work.** **4a–b. See Solutions Manual.**
4. Use counters to model the expressions $(2 + 5) + 6$ and $2 + (5 + 6)$.
   a. What do you observe about the values of each expression?
   b. Model several more expressions of the form $(a + b) + c$ and $a + (b + c)$. Explain your results in terms of one of the properties of addition.

MODELING MATHEMATICS

MATERIALS
counters

Guided Practice

Name the property shown by each statement. **5–8. See Solutions Manual.**

5. $(3 + 8) + 7 = 3 + (8 + 7)$ **6.** $8 + 9 = 9 + 8$
7. $(5 + 7) + 2 = (7 + 5) + 2$ **8.** $0 + 9 = 9$

Find each sum or product mentally.

9. $12 + 9 + 8$ **29** **10.** $5 \cdot 3 \cdot 8$ **120** **11.** $4 + 13 + 26 + 5$ **48**

Rewrite each expression using an associative property. Then simplify.

12. $(y + 9) + 8$ **13.** $7(6z)$ **14.** $(k \cdot 8)3$ **$k(8 \cdot 3)$; $24k$**

12. $y + (9 + 8)$; $y + 17$
13. $(7 \cdot 6)z$, $42z$

15. Food Service For lunch, Juana ordered an iced tea, a bowl of clam chowder, and a small Caesar salad. If the prices of the items are \$1.25, \$3.35, and \$3.75 respectively, how much should Juana's bill be before tax and tip? **\$8.35**

Group Activity Card 1-4

Property Rummy Group Activity **1-4**

12+34 15x87 72

MATERIALS: Set of "Property Rummy" cards (see back)

The object of this card game is to form sets of three cards each. Each set must contain one property name card and two expression cards that illustrate the property. To begin, each player is dealt eight cards. The remaining cards are placed face down, and the top card is turned face up. You then take turns either drawing the top face up or face down card and then discarding one of your cards. When you can form two sets from the cards in your hand, place them in front of you. The first person to do this wins the game.

 Pre-Algebra

Additional Answer

1. Sample answer: Both properties involve multiplication and addition. The associative property involves grouping numbers that are being added or multiplied. The commutative property involves the order in which numbers are added or multiplied.

Exercises: Practicing and Applying the Concept

Independent Practice

Name the property shown by each statement. 16–27. See margin.

16. $22 \cdot 1 = 22$
17. $19 \cdot 12 = 12 \cdot 19$
18. $6 \cdot 33 = 33 \cdot 6$
19. $3 \cdot 7 \cdot 0 = 0$
20. $8cd = 8dc$
21. $x + (2y + z) = (2y + z) + x$
22. $0 + 9 = 9 + 0$
23. $5s + 9 = 9 + 5s$
24. $(7a)(b) = 7(ab)$
25. $8t + 6 = 6 + 8t$
26. $1m = m$
27. $7 + (2 + u) = (7 + 2) + u$

Find each sum or product mentally. 30. 200

28. $16 + 7 + 14$ 37
29. $2 \cdot 9 \cdot 20$ 360
30. $82 + 58 + 23 + 37$
31. $18 \cdot 6 \cdot 0$ 0
32. $7 + 99 + 123$ 229
33. $2 \cdot 13 \cdot 5$ 130
34. $7 + 2 + 13 + 18$ 40
35. $6 \cdot 11 \cdot 10$ 660
36. $129 \cdot 8 \cdot 0$ 0

Rewrite each expression using a commutative property.

37. $5 + 9$ $9 + 5$
38. $8a + 6$ $6 + 8a$
39. $9 + 18w$ $18w + 9$

Rewrite each expression using an associative property. Then simplify.

40. $(y + 7) + 6$
41. $(b \cdot 6) \cdot 5$
42. $2(8f)$

43. $13 + (11 + m)$
44. $3(2z)$
45. $(p \cdot 7) \cdot 4$
46. $(u + 8) + 16$
47. $(15 + 4w) + w$
48. $0(3x)$

40–48. See Solutions Manual.

Critical Thinking

49. **Calculator** Use a calculator to find $28 \div 7$ and $7 \div 28$.
    a. Experiment with other expressions of the form $a \div b$ and $b \div a$.
    b. Write a statement about division and the commutative property.
    49a–b. See Solutions Manual.

Applications and Problem Solving

50. **Sports** In the game of volleyball, the net is 3 feet 3 inches tall. The bottom of the net is to be set 4 feet 8 inches from the floor.
    a. Write an expression for the distance from the floor to the top of the net. (3 feet + 3 inches) + (4 feet + 8 inches)
    b. Find the distance from the floor to the top of the net. Explain how you used the associative and commutative properties.

51. **Science** Mr. Rex instructed students in his class to pour acid in their flasks into water in their beakers. He warned that pouring water into acid could cause spattering and burns. These actions are not commutative. Give another example of actions that are not commutative. See students' work.

Mixed Review

50b. 7 feet 11 inches; See students' explanations.

52. Write an expression for *the product of a number and 18.* (Lesson 1-3) $18n$
53. Evaluate the expression $8xy + 6$ if $x = 0$ and $y = 2$. (Lesson 1-3) 6
54. Find the value of the expression $7 \cdot (9 - 3) + 4$. (Lesson 1-2) 46
55. **Medicine** You can find the approximate number of pints of blood in your body by dividing your weight by 16. (Lesson 1-2)
    a. Write an expression for the blood relationship. $w \div 16$
    b. JT weighs 144 pounds. Find the approximate number of pints of blood in JT's body. 9 pints
56. **Patterns** Find the next number in the sequence 15, 14, 12, 9, 5, __?__. (Lesson 1-1) 0

Assignment Guide

Core: 17–51 odd, 52–56
Enriched: 16–48 even, 49–56

For **Extra Practice**, see p. 741.

The red A, B, and C flags, printed in the Teacher's Wraparound Edition, indicate the level of difficulty of the exercises.

4 ASSESS

Closing Activity

Act It Out Have students work in groups to create a skit to show one of the properties they have studied in this lesson.

Additional Answers

16. identity multiplication
17. commutative multiplication
18. commutative multiplication
19. multiplicative zero
20. commutative multiplication
21. commutative addition
22. commutative addition
23. commutative addition
24. associative multiplication
25. commutative addition
26. identity multiplication
27. associative addition

Enrichment Masters, p. 4

NAME ________________ DATE ________

1-4 Enrichment Student Edition Pages 22–25

Algebraic Proof

Axioms are statements assumed to be true without being proven. They are used in the proofs of theorems. The following properties are examples of algebraic axioms. Abbreviations for these properties are used in the examples below.

Commutative Property of Addition (CPA)
Associative Property of Addition (APA)
Subtraction Property of Equality (SubPE)
Addition Property of Equality (APE)
Substitution Property of Equality (SPE)
Additive Identity Property (AIP)

Example: Prove: $a + (b + c) = c + (a + b)$

Statement	Reason
$a + (b + c) = (a + b) + c$	APA
$= c + (a + b)$	CPA

Example: Prove: $5 + (x + 2) = 7 + x$

Statement	Reason
$5 + (x + 2) = 5 + (2 + x)$	CPA
$= (5 + 2) + x$	APA
$= 7 + x$	SPE

Write the reason for each statement.

1. Prove: $(9 + 6) + (x + 3) = x + 18$

Statement	Reason
$(9 + 6) + (x + 3) = 15 + (x + 3)$	a. SPE
$= (15 + x) + 3$	b. APA
$= (x + 15) + 3$	c. CPA
$= x + (15 + 3)$	d. APA
$= x + 18$	e. SPE

Prove each of the following. Identify a reason for each statement.

2. $9 + (x + 4) = 13 + x$
   $9 + (x + 4) = 9 + (4 + x)$ CPA
   $= (9 + 4) + x$ APA
   $= 13 + x$ SPE

3. $5 + (x + 11) = x + (10 + 6)$
   $5 + (x + 11) = (x + 11) + 5$ CPA
   $= x + (11 + 5)$ APA
   $= x + 16$ SPE
   $= x + (10 + 6)$ SPE

Extension

Using Number Sense Have students experiment with a calculator and the problems below to determine whether subtraction and division are associative.

$14 - (6 - 4) \stackrel{?}{=} (14 - 6) - 4$
$18 - (9 - 5) \stackrel{?}{=} (18 - 9) - 5$
$24 \div (6 \div 2) \stackrel{?}{=} (24 \div 6) \div 2$
$81 \div (9 \div 3) \stackrel{?}{=} (81 \div 9) \div 3$

They are not.

Additional Answer

2. Sample answer:
   Commutative addition: $8 + 9 = 9 + 8$
   Commutative multiplication: $6 \times 3 = 3 \times 6$
   Associative addition: $3 + (5 + 4) = (3 + 5) + 4$
   Associative multiplication: $7 \times (3 \times 1) = (7 \times 3) \times 1$
   Identity addition: $8 + 0 = 8$
   Identity multiplication: $17 \times 1 = 17$
   Multiplicative of zero: $12 \times 0 = 0$

1-5 LESSON NOTES

NCTM Standards: 1-4, 6, 7, 9

Instructional Resources

- Study Guide Master 1-5
- Practice Master 1-5
- Enrichment Master 1-5
- Group Activity Card 1-5
- Assessment and Evaluation Masters, pp. 14, 15
- Math Lab and Modeling Math Masters, p. 76

Transparency 1-5A contains the 5-Minute Check for this lesson; **Transparency 1-5B** contains a teaching aid for this lesson.

Recommended Pacing	
Standard Pacing	Day 6 of 14
Honors Pacing	Day 6 of 13
Block Scheduling*	Day 3 of 7 (along with Lesson 1-4)

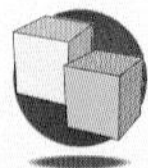

*For more information on pacing and possible lesson plans, refer to the ***Block Scheduling Booklet***.

1 FOCUS

5-Minute Check
(over Lesson 1-4)

Name the property shown by each statement.

1. $(4b)c = 4(bc)$ **associative, ×**
2. $(7 + 4)0 = 0$ **mult. prop. of 0**

Find each sum or product mentally.

3. $18 + 13 + 2 + 7$ **40**
4. $6 \cdot 7 \cdot 5$ **210**
5. A recent study showed that 63 million Americans choose swimming as their favorite sport, 55 million chose hiking, and 47 million chose camping. How many people chose one of these three sports as their favorite? **165 million**

1-5 The Distributive Property

Setting Goals: *In this lesson, you'll simplify algebraic expressions using the distributive property.*

Modeling a Real-World Application: Travel

HELP WANTED

Energetic people who enjoy working with people will find rewards in the travel industry. If you are interested in finding out more about this exciting field, contact:

The Travel Industry Association of America
1133 21st Street NW
Suite 800
Washington, D.C. 20036

Sunny beaches, exciting ski lodges, and exotic experiences in far-off lands await you! How do you get there? Call a travel agent! Elena Chávez is a travel agent for Carlson Travel. She helps people plan vacations and business trips by making arrangements for hotels, airline tickets, rental cars, and any other services her clients need.

Ms. Chávez received the brochure at the right for a travel package to Barbados. Her clients, Mr. and Mrs. Yoder, are considering Barbados for a summer vacation. If the round-trip airfare to Barbados from Columbus is \$679, how much should Ms. Chávez tell the Yoders that the airfare and package trip would cost the couple? *This problem will be solved in Example 2.*

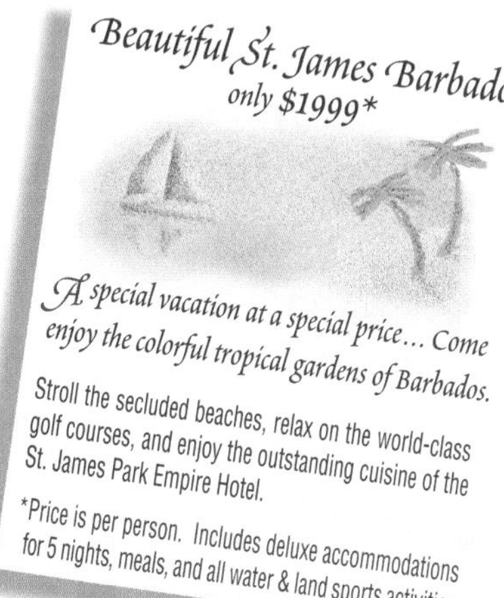

Beautiful St. James Barbados
only $1999*
A special vacation at a special price... Come enjoy the colorful tropical gardens of Barbados.
Stroll the secluded beaches, relax on the world-class golf courses, and enjoy the outstanding cuisine of the St. James Park Empire Hotel.
*Price is per person. Includes deluxe accommodations for 5 nights, meals, and all water & land sports activities.

Learning the Concept

In Lesson 1-4, you learned some of the properties of addition and multiplication with whole numbers. The **distributive property** ties addition and multiplication together.

Distributive Property	
In words:	The sum of two addends multiplied by a number is the sum of the product of each addend and the number.
In symbols:	For any numbers a, b, and c, $a(b + c) = ab + ac$ and $(b + c)a = ba + ca$.

The expression $a(b + c)$ is read "a times the quantity b plus c" or "a times the sum of b and c."

The distributive property allows you to write expressions in different forms.

HELP WANTED

Many technical schools offer associate's degree programs in travel. Interested students may wish to contact a local community college for more information.

Example  **Find 14 · 12 mentally using the distributive property.**

$14 \cdot 12 = 14(10 + 2)$ *Replace 12 with 10 + 2*
$= 14 \cdot 10 + 14 \cdot 2$ *Think of 140 + 28.*
$= 140 + 28$ or 168

Example 

Travel

Refer to the application at the beginning of the lesson.

a. Write two different expressions for the price of the trip for two people.

b. Find the cost of the trip.

a. There are two methods of writing the total cost: find the total cost for one person and multiply by 2, or find the cost of each part of the trip for two people and then add.

Method 1			**Method 2**		
number of people	*times*	*total trip cost*	*airfare for two*	*plus*	*package for two*
2	×	($679 + $1999)	2($679)	+	2($1999)

Notice that if we apply the distributive property to the expression on the left, the result is the expression on the right.

b. Evaluate either expression to find the cost of the trip.
2($679 + 1999) = 2($2678)
= $5356
Check the result by evaluating the expression 2($679) + 2($1999).

Connection to Algebra

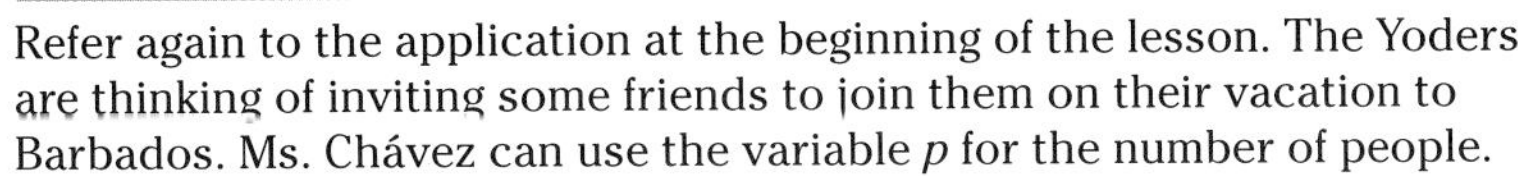

Refer again to the application at the beginning of the lesson. The Yoders are thinking of inviting some friends to join them on their vacation to Barbados. Ms. Chávez can use the variable p for the number of people.

If there are p people traveling, the airfare will be $679p$ dollars.
For the same number of people, the package cost will be $1999p$ dollars.

The total cost for p people could be represented as $679p + 1999p$ dollars.

THINK ABOUT IT

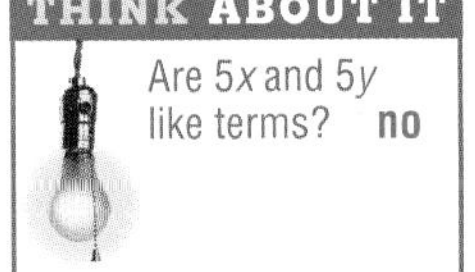

Are $5x$ and $5y$ like terms? **no**

The expression $679p + 1999p$ has two **terms** with the same variable. These terms are called **like terms**. Some other pairs of like terms are $6d$ and $17d$, $3st$ and $5st$, and x and $4x$. The distributive property can be helpful in simplifying expressions that have like terms. Example 3a shows how to simplify $679p + 1999p$.

Example 3 **Simplify each expression.**

a. $679p + 1999p$
$679p + 1999p = (679 + 1999)p$ *p is a factor of both 679p and 1999p.*
$= 2678p$ *Use the distributive property.*

b. $x + 12x$
$x + 12x = 1x + 12x$ *$x = 1 \cdot x$ by multiplicative identity*
$x + 12x = (1 + 12)x$ *Distributive property*
$= 13x$

Motivating the Lesson

Situational Problem Ask students if they know anyone who works in distribution. Have students discuss what distribution means in business.

2 TEACH

In-Class Examples

For Example 1
Find 12 · 15 mentally.
12(10 + 5) = 180

For Example 2
North Country Rivers of York, Maine, offers one-day white-water rafting trips on the Kennebec River. The trip costs $69 per person and optional wet suits are $15 each.

a. Write two different expressions for the cost of taking a family of four on a trip if each person used a wet suit. **4(69 + 15) or 4(69) + 4(15)**

b. Find the cost of the trip. **$336**

For Example 3
Simplify each expression.

a. $185a + 332a$ **$517a$**

b. $24v + v$ **$25v$**

Teaching Tip Remind students that a variable represents a number. So any operation or property that applies to a number also applies to a variable.

Classroom Vignette

To teach combining like terms, I simulate an inventory problem in a retail outlet. Students must physically collect packages of golf balls, tennis balls, or eggs from various parts of the classroom and put them in the same place to determine how many they have, and ultimately, the cost of the items.

Dennis L. Wagoner
Cowan High School
Muncie, IN

An expression is in **simplest form** when it has no like terms and no parentheses.

Example  **Write each expression in simplest form.**

a. $\mathbf{16k + 9 + 5k}$

$16k + 9 + 5k = 16k + 5k + 9$ — *Commutative property of addition*
$= (16 + 5)k + 9$ — *Distributive property*
$= 21k + 9$

b. $\mathbf{m + 5(n + 7m)}$

$m + 5(n + 7m) = m + 5n + 5 \cdot 7m$ — *Distributive property*
$= m + 5n + 35m$ — *Substitution property*
$= m + 35m + 5n$ — *Commutative property of addition*
$= 1m + 35m + 5n$ — *Multiplicative identity*
$= (1 + 35)m + 5n$ — *Distributive property*
$= 36m + 5n$

Checking Your Understanding

Communicating Mathematics

Read and study the lesson to answer these questions. **1–3. See margin.**

1. **Rewrite** the expression $7(5 + 3)$ in three different ways using the distributive property.
2. **Give an example** of two like terms. Explain why they are like terms.
3. **Make a table** of expressions that are and are not in simplest form. Explain how you know whether an expression is in simplest form.

MATH JOURNAL

4. Make up a method for remembering the distributive property. Write your method and a description of how it works in your journal. **See student's work.**

Guided Practice

Restate each expression using the distributive property. Do not simplify.

5. $5(7 + 8)$ **5(7) + 5(8)**
6. $6(2 + 4)$ **6(2) + 6(4)**
7. $4(x + 3)$ **4(x) + 4(3)**

Simplify each expression. **9. 13a + 7 12. 38c + 8 13. 20ab**

8. $6 \cdot 5 + 9 \cdot 6$ **84**
9. $12a + a + 7$
10. $x + 9x$ **10x**
11. $6c + c + 7$ **7c + 7**
12. $22c + 4(2 + 4c)$
13. $4ab + 6ab + 10ab$
14. **Money** Carlos works part time at The Arizona Pizzeria to save for college. He earns \$5.25 an hour plus tips.
    a. Carlos worked 4 hours on Friday and 8 hours on Saturday. Write two different expressions for his wages that weekend.
    b. Find the amount Carlos earned if his tips totaled \$35.50.

Exercises: Practicing and Applying the Concept

14a. \$5.25(4 + 8) or 4 · \$5.25 + 8 · \$5.25 14b. \$98.50

Independent Practice

Restate each expression using the distributive property. Do not simplify. **See margin.**

 A

15. $3(11 + 12)$
16. $(5 + 7)x$
17. $6t + 11t$
18. $2a + 4b$
19. $2r + 12s$
20. $6x + 6y + 1$

B

21. $(9 + 8)v$
22. $12r + 12s$
23. $(4x + 9y)2$

In-Class Example

For Example 4
Write each expression in simplest form.

a. $11b + 3 + 26b$ **37b + 3**

b. $c + 4(d + 8c)$ **33c + 4d**

3 PRACTICE/APPLY

Checking Your Understanding

Exercises 1–14 are designed to help you assess your students' understanding through reading, writing, speaking, and modeling. You should work through Exercises 1–4 with your students and then monitor their work on Exercises 5–14.

Additional Answers

1. $7(5) + 7(3)$; $7 \cdot 5 + 7 \cdot 3$; $7 \times 5 + 7 \times 3$
2. Sample answer: $8x$ and $2x$ both contain only the variable x
3. Sample answer: Simplest form: $7x$, $9a + 2$; $u + uv$; Not in simplest form: $12x + x$; $7 + 6$; $2(7y)$
   Expressions are in simplest form when no terms can be combined.

Study Guide Master, p. 5

NAME ________________ DATE ________

1-5 Study Guide
The Distributive Property

Student Edition Pages 26–30

The distributive property involves multiplication and addition. It states that the sum of two products with a common factor is equal to the sum of the other factors times the common factor.

$$(12 \times 5) + (4 \times 5) = (12 + 4) \times 5$$

Notice that the two products on the left have a common factor, 5.

The distributive property is used to combine like terms in algebraic expressions. Two or more terms with the same variables are combined by adding the coefficients.

$$12a + 4a = (12 + 4)a = 16a$$
$$b + 6b = (1 + 6)b = 6b$$

Use the distributive property to find the value of each of the following.

1. $(3 \times 8) + (3 \times 2)$ **30**
2. $(2 \times 9) + (2 \times 11)$ **40**
3. $(13 \times 6) + (7 \times 6)$ **120**
4. $(59 \times 5) + (41 \times 5)$ **500**
5. $(25 \times 731) + (25 \times 129)$ **21,500**
6. $(486 \times 40) + (144 \times 40)$ **25,200**
7. $(6 \times 31) + (6 \times 147)$ **1068**
8. $(7 \times 4) + (7 \times 9) + (7 \times 8)$ **147**

Simplify each expression.

9. $8y + 2y$ **10y**
10. $9m + 2m$ **11m**
11. $13a + 7a$ **20a**
12. $5x + x$ **6x**
13. $d + 10d$ **11d**
14. $z + z$ **2z**
15. $8y + 2y + 3y$ **13y**
16. $9m + 2m + 3m$ **14m**
17. $13a + 7a + 2a$ **22a**
18. $8y + 2y + y$ **11y**
19. $m + 2m + m$ **4m**
20. $a + a + 2a$ **4a**

Reteaching

Using Manipulatives Have students represent distributive problems with manipulatives. For example, if a red cube represents a 69¢ can of cola and a green cube represents an 89¢ bottle of cola, represent an order of six 69¢ colas and six 89¢ colas. Have students solve several problems with the manipulatives.

Additional Answers

15. $3(11) + 3(12)$
16. $5(x) + 7(x)$
17. $t(6 + 11)$
18. $2(a + 2b)$
19. $2(r + 6s)$
20. $6(x + y) + 1$
21. $9(v) + 8(v)$
22. $12(r + s)$
23. $2(4x) + 2(9y)$

Simplify each expression. 24–35. See margin.

24. $6d + d + 15$ **25.** $17a + 21a + 45$ **26.** $m + m$
27. $c + 24c + 16$ **28.** $q + 4 + 11 + 4q$ **29.** $18y + 5(7 + 3y)$
30. $14(b + 3) + 8b$ **31.** $30(b + 2) + 2b$ **32.** $3(8 + a) + 7(6 + 4a)$
C **33.** $8(9 + 3f) + f$ **34.** $x + 5x + 8(x + 2)$ **35.** $3(x + y) + 4(2x + 3y)$

Critical Thinking

36. The distributive property given on page 26 is often called the *distributive property of multiplication over addition.* Is there a distributive property of multiplication over subtraction? Evaluate several different expressions of the form $a(b - c)$ and $ab - ac$ and explain your findings. See students' work.

Applications and Problem Solving

37. **Engineering** The foundation of a building is critical to ensuring that the building is safe. Before the foundation can be poured, an engineer calculates the volume of material to be removed. The expression used to find the volume of earth to remove for an opening of the shape shown is $\frac{d(a + b)}{2} \times L$. Find the volume of earth to remove if $a = 52$ m, $b = 48$ m, $d = 6$ m, and $L = 40$ m.

37. 12,000 cubic meters

38. **Business Operations** Most products sold in the United States carry a universal product code (UPC) like the one at the right. The last digit of the code, the check digit, is used to ensure that the numbers of the code are entered correctly. The computer uses the steps below to verify the check digit. a. $3(0 + 6 + 0 + 3 + 6 + 5) + (3 + 8 + 0 + 0 + 8)$

1. Add the digits in the odd-numbered positions. Multiply the total by 3.
2. Add the digits in the even-numbered positions. Then add to the previous total.
3. Subtract the result from the next highest multiple of 10. For example, a result of 57 would be subtracted from 60. The answer is the check digit.

a. Write an expression for the first two steps of checking UPC 0 36800 30685 1. Be sure that you don't include the check digit in the calculation.

b. Evaluate the expression you wrote for part a. Then subtract from the next highest multiple of 10. Does the check digit verify the code? 79, 80 − 79 = 1; yes

Mixed Review

39. Rewrite $(8 + 19) + 17$ using an associative property. (Lesson 1-4)

40. Name the property shown by the statement $12 \times 1 = 12$. (Lesson 1-4)

41. Translate the phrase *the quotient of a number and 9* into an algebraic expression. (Lesson 1-3) $n \div 9$

39. $8 + (19 + 17)$
40. Multiplicative Identity

Assignment Guide

Core: 15–35 odd, 36, 37, 39–45
Enriched: 16–34 even, 36–45
All: Self Test, 1–10

For **Extra Practice**, see p. 741.

The red A, B, and C flags, printed in the Teacher's Wraparound Edition, indicate the level of difficulty of the exercises.

Additional Answers

24. $7d + 15$ **25.** $38a + 45$
26. $2m$ **27.** $25c + 16$
28. $5q + 15$ **29.** $33y + 35$
30. $22b + 42$ **31.** $32b + 60$
32. $66 + 31a$ **33.** $72 + 25f$
34. $14x + 16$ **35.** $11x + 15y$

Practice Masters, p. 5

NAME ______ DATE ______

1-5 Practice
The Distributive Property

Student Edition Pages 26–30

Use the distributive property to compute each of the following.

1. $8(50 + 4)$ 432 **2.** $(20 + 9)5$ 145 **3.** $2 \cdot 60 + 2 \cdot 4$ 128
4. $7(40 - 2)$ 266 **5.** $4 \cdot 400 - 4 \cdot 2$ 1592 **6.** $67 \cdot 40$ 2680
7. $501 \cdot 11$ 5511 **8.** $210 \cdot 800$ 168,000 **9.** $89 \cdot 12$ 1068

Simplify each expression.

10. $5a + a$ $6a$ **11.** $k - k$ 0
12. $m + 3m + 8$ $4m + 8$ **13.** $10b - b + 1$ $9b + 1$
14. $9ab + 8ab - 7ab$ $10ab$ **15.** $6x + 3y + 6y - 2x$ $4x + 9y$
16. $3xy + 2xy - xy$ $4xy$ **17.** $18 + 7x - 12 + 5x$ $12x + 6$
18. $12a + 3 + 18 - 9a$ $3a + 21$ **19.** $5(x + 2y) + 6x$ $11x + 10y$
20. $5(r + 2)7r$ $12r + 10$ **21.** $x + 5x + 8(x + 2)$ $14x + 16$
22. $4(x + 2) + 3(x + 5)$ $7x + 23$ **23.** $8(r + 15) + 7(2r + 10)$ $22r + 190$
24. $2(r + 3) + 3(r + 7) - 10$ $5r + 17$ **25.** $12(c + 3d + 4f) + 2(2c + d + 6f)$ $16c + 38d + 60f$
26. $5 \cdot 4a + 6(5a + 2)$ $50a + 12$ **27.** $4 \cdot 8 + 9(3a + 5) + 8(2a + 1)$ $43a + 85$
28. $4 \cdot 3a + 2(a + 6b)$ $14a + 12b$ **29.** $10r + 100s + 50r$ $60r + 100s$
30. $9[5 + 3(x + 2)]$ $27x + 99$ **31.** $3[9(x + 4) + 2(x + 1)]$ $33x + 114$

Group Activity Card 1-5

Mental Manipulations Group Activity **1-5**

MATERIALS: 0-9 spinner

PROCEDURE: You and a partner take turns thinking of a two-digit number between 10 and 40. After you think of a number, you say it out loud. Your partner then decides how to break up the number to form a sum that can easily be used to do mental computations. He or she then spins the spinner to get another number and then mentally calculates the product of the sum and the number using the distributive property.

EXAMPLE: Suppose your number is 27. Your partner could break up the number into (20 + 7) or (25 + 2). If the spinner stops on 6, then your partner would mentally calculate either 120 + 42 = 162 or 150 + 12 = 162.

 Pre-Algebra

Extension

Using Technology Have students experiment with a calculator to determine whether division and addition are distributive.

$(16 + 4) \div 2 \stackrel{?}{=} 16 \div 2 + 4 \div 2$

$(18 + 15) \div 3 \stackrel{?}{=} 18 \div 3 + 15 \div 3$

Yes they are.

4 ASSESS

Closing Activity

Writing Have students write a few sentences to explain the distributive property to an absent classmate.

Chapter 1, Quiz B (Lessons 1-3 through 1-5) is available in the *Assessment and Evaluation Masters*, p. 15.

Mid-Chapter Test (Lessons 1-1 through 1-5) is available in the *Assessment and Evaluation Masters*, p. 14.

Additional Answer

44.

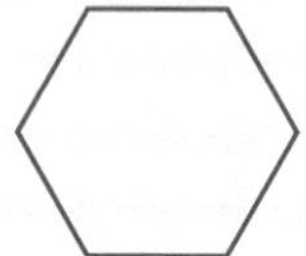

Enrichment Masters, p. 5

NAME ______________________ DATE __________

1-5 Enrichment
Finite and Infinite Sets

Student Edition
Pages 26–30

A **set** is a collection of objects. The members of a set are called **elements**. A set can be named by listing the elements within a pair of braces, { }. Here are two sets:

A = {1, 2, 3, 4, 5}

B = {+, −, ÷, ×}

In many sets, all the elements cannot be listed (or cannot conveniently be listed). When this is the case, three dots inside the braces indicate that not all the elements are listed, but the pattern shown by those elements listed continues.

set of even numbers less than 100: E = {2, 4, 6, 8, · · ·, 98}

set of whole numbers: W = {0, 1, 2, 3, · · ·}

The set E is an example of a **finite** set. That is, all the elements could be listed. The set W is an example of an **infinite** set. That is, not all the elements can be listed without using the three dots.

The **cardinal number** of a set is the number obtained by counting the elements in the set. The cardinal number for set E is 49. Note that each element of a set is listed only once.

State whether the following sets are finite or infinite. For each finite set, indicate the cardinal number.

1. {3, 6, 9, 12} **finite, 4**
2. {0} **finite, 1**
3. {2, 4, 6, 8, · · ·} **infinite**
4. {2, 4, 6, · · ·, 30} **finite, 15**
5. the set of all counting numbers **infinite**

Equal sets have the same elements. Equivalent sets have the same number of elements (same cardinal number). Each element is listed only once and order is not important.

Complete. **Answers will vary. Sample answers given.**

6. List two sets equivalent to {+, −, ×}.
{A, B, C}, {×, −, +}
7. List two infinite sets that are equivalent (the elements can be matched one-to-one) but not equal. **{even numbers}, {odd numbers}**

42. Evaluate $75 - 10n$ if $n = 7$. (Lesson 1-3) **5**
43. **Fundraising** The local chapter of the Cancer Society sells Valentine candy to raise funds for research. Cindy sold 15 white chocolate hearts at \$4.25 each, 36 milk chocolate hearts at \$3.75 each, and 22 milk chocolate assortments at \$7.45 each. How much money did Cindy raise? (Lesson 1-2) **\$362.65**
44. **Geometry** Draw the next figure in the pattern. (Lesson 1-1)
See margin.

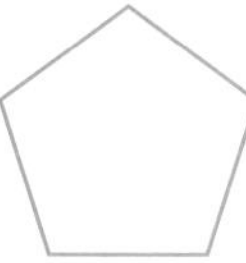

45. Bandi used a calculator to find $528 \times 7 = 3696$. Without actually calculating, determine if this answer is reasonable. (Lesson 1-1) **yes**

Self Test

1. **Number Theory** The pattern at the right is known as Pascal's Triangle. Find the pattern and complete the 6th and 7th rows. (Lesson 1-1) **See margin.**

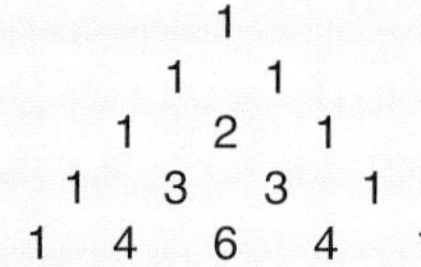

2. **Construction** Lillehammer, Norway hosted the 1994 Winter Olympic Games. Gjovick Olympic Cavern Hall, the site of the hockey competition, was carved out of Hovdeteoppen Mountain. 29,000 truckloads of rock were removed. If a truck holds about 5 cubic yards of rock, about how much rock was removed to build the hall? (Lesson 1-1)
   a. Which method of computation do you think is most appropriate for this problem? Justify your choice. **Estimate. It says *about*, so an exact answer is not needed.**
   b. Solve the problem using the four-step plan. Be sure to examine your solution. **about 150,000 cubic yards**

Find the value of each expression. (Lesson 1-2)

3. $9(2 + 5) - 4$ **59**
4. $\frac{19 - 11}{2(4)}$ **1**

Evaluate each expression if $a = 3$, $b = 8$, and $c = 12$. (Lesson 1-3)

5. $\frac{ab}{c}$ **2**
6. $8c - (b + 3a)$ **79**

Name the property shown by each statement. (Lesson 1-4) **See margin.**

7. $0 + c = c + 0$
8. $19 + 7 + 1 = 19 + 1 + 7$
9. **Music** Enrique's piano teacher charges \$12.50 for a lesson. If Enrique has 5 lessons in one month, how much does he owe his piano teacher? Solve mentally. (Lesson 1-5) **\$62.50**
10. Simplify $2n + 4(n + 8n)$. (Lesson 1-5) **38*n***

Self Test

The Self Test provides students with a brief review of the concepts and skills in Lessons 1-1 through 1-5. Lesson numbers are given at the end of problems or instruction lines so students may review concepts not yet mastered.

Additional Answers Self Test

1.
1 5 10 10 5 1
1 6 15 20 15 6 1

7. commutative addition
8. commutative addition

HANDS-ON ACTIVITY

1-5B Distributive Property

An Extension of Lesson **1-5**

MATERIALS

algebra tiles

Many algebraic principles can be modeled using geometry. You can look at the distributive property by modeling with algebra tiles of different sizes.

- Let [1 × 1 tile] be a 1×1 square. That is, the length and width are each 1 unit. If the area of a square or a rectangle is found by multiplying the length and the width, what is the area of the square? **1 square unit**

- Let [x × 1 tile] be a $1 \times x$ rectangle. The width is 1 unit, and the length is x units. What is the area of the rectangle? **x square units**

Then [x and 1 tiles] is a rectangle that has a width of 1 unit and a length of $x + 1$ units. What is the area of the rectangle? **$x + 1$ square units**

Your Turn Using your tiles, make rectangles with areas of $x + 2$, $2x$, $2x + 1$, and $2x + 2$ square units. **See students' work.**

Explore You can use geometric models to check the distributive property. Is it true that $2(x + 1) = 2x + 1$?

$2(x + 1)$ means [tiles] *Does it matter how the tiles are arranged?*

$2x + 1$ means [tiles]

Therefore, $2(x + 1)$ is not the same as $2x + 1$. From the drawing above, $2(x + 1) = 2x + 2$.

Tell whether each of the following statements is *true* or *false*. Justify your answers with algebra tiles or a drawing.

1. $2x + 3 = 6x$ F **2.** $2x = x + x$ T **3.** $3x + 3 = 3(x + 1)$ T

4. $3x + 2x = 6x$ F **5.** $3x + 3 = 3(x + 3)$ F **6.** $3x + 2x = x(3 + 2)$ T

7. You Decide Steve says that $3(x + 2) = 3x + 2$. Linh disagrees. She thinks that $3(x + 2) = 3x + 6$. Write a paragraph with drawings to explain who you think is correct. **See students' work.**

1-5B LESSON NOTES

NCTM Standards: 1-4, 6, 9

Objective

Visualize the distributive property by modeling a geometric interpretation.

Recommended Time

Demonstration and discussion: 15 minutes; Exercises: 30 minutes

Instructional Resources

For each student or group of students

Student Manipulative Kit
- algebra tiles

Math Lab and Modeling Math Masters
- p. 1 (algebra tiles)
- p. 27 (worksheet)

For teacher demonstration

Overhead Manipulative Resources

1 FOCUS

Motivating the Lesson

A 4-by-5 vegetable garden and a 3-by-5 herb garden are side by side. If the length of each garden is extended 2 feet, how could that be represented geometrically and algebraically?

2 TEACH

Teaching Tip Model each example for students. Remind students that the area of a rectangle is found by multiplying the length by the width.

3 PRACTICE/APPLY

Assignment Guide
Core: 1–7
Enriched: 1–7

4 ASSESS

Observing students working in cooperative groups is an excellent method of assessment.

1-6 LESSON NOTES

NCTM Standards: 1-4, 7, 9

Instructional Resources

- Study Guide Master 1-6
- Practice Master 1-6
- Enrichment Master 1-6
- Group Activity Card 1-6
- Activity Masters, p. 1
- Real-World Applications, 1

Transparency 1-6A contains the 5-Minute Check for this lesson; **Transparency 1-6B** contains a teaching aid for this lesson.

Recommended Pacing	
Standard Pacing	Day 7 of 14
Honors Pacing	Day 7 of 13
Block Scheduling*	Day 4 of 7 (along with Lesson 1-7)

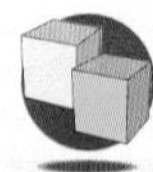

*For more information on pacing and possible lesson plans, refer to the ***Block Scheduling Booklet***.

1 FOCUS

5-Minute Check
(over Lesson 1-5)

Simplify each expression.

1. $4q + q$ **$5q$**
2. $4m + 6m + 3$ **$10m + 3$**
3. $24a + a + 16$ **$25a + 16$**
4. $9(r + 7) + 12r$ **$21r + 63$**
5. $6(x + y) + 4(2x + 3y)$ **$14x + 18y$**
6. A two-day pass for Colonial Williamsburg costs \$24 and admission to the Governor's Palace is an additional \$5.
   a. Write two expressions for the cost of taking a family of three to Williamsburg if each person goes to the palace also. **3(24 + 5); 3(24) + 3(5)**
   b. Find the total cost. **\$87**

1-6 Variables and Equations

Setting Goals: *In this lesson, you'll identify and solve open sentences.*

Modeling a Real-World Application: Entertainment

CULTURAL CONNECTIONS
People of different cultures have been playing games for centuries. One of the oldest games, Senet, made its first known appearance in Egypt in about 2686 B.C. Senet was played by people of all social classes and had both religious and secular forms.

How do you like to spend a rainy Saturday afternoon? With a good book, an old movie, or how about a nice long board game? People have been playing board games for centuries and even with the advent of computer games, they still enjoy the real thing.

Parcheesi was one of the first games to receive a copyright in the United States. It was copyrighted in 1874. Parcheesi originated in India more than 1200 years ago.

The most famous board game, Monopoly, was invented by Charles Darrow during the Great Depression. Darrow was an unemployed engineer who longed to be rich and spend his days on the boardwalks of Atlantic City, New Jersey. Darrow's game may be based on an earlier game called The Landlord's Game invented by Elizabeth J. Magie in 1904. The copyright for Monopoly was issued in 1933.

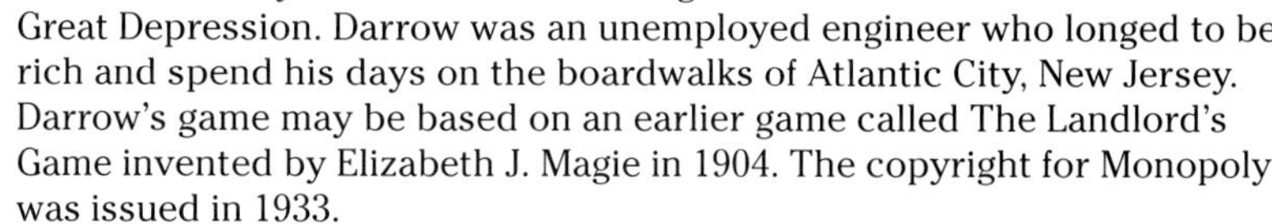

The number of years between the copyrights of Parcheesi and Monopoly can be represented by $1933 - 1874 = y$.

Learning the Concept

A mathematical sentence like $1933 - 1874 = y$ is called an **equation**. Some examples of equations are $6 + 8 = 14$, $3(6) - 9 = 9$, and $6x - 2 = 11$. Equations that contain variables are **open sentences**. When the variables in an open sentence are replaced with numbers, the sentence may be true or false.

$1933 - 1874 = y$

$1933 - 1874 \stackrel{?}{=} 67$ *Replace y with 67.*
This sentence is false because $1933 - 1874 \neq 67$.

$1933 - 1874 \stackrel{?}{=} 59$ *Replace y with 59.*
This sentence is true.

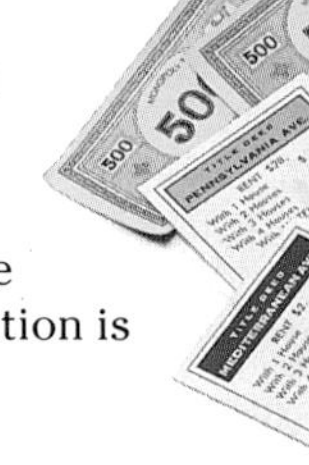

A value for the variable that makes an equation true is called a **solution** of the equation. The process of finding a solution is called **solving the equation**.

The solution of $1933 - 1874 = y$ is 59.

Reteaching

Using Manipulatives Separate students into groups, giving each group a balance scale. Have students use different weights to illustrate how to solve equations. The students should write down the combinations of weights used.

CULTURAL CONNECTIONS
Two other games you may wish to discuss with your students are the African game of strategy *wari* and *ko-no*, a checker-type game from China and Korea.

Example 1 **Which of the numbers 16, 24, or 32 is the solution of $66 + x = 98$?**

Replace x with each of the possible solutions to solve the equation.

$66 + 16 = 98$ *Replace x with 16.*
$82 = 98$
This sentence is false.

$66 + 24 = 98$ *Replace x with 24.*
$90 \neq 98$
This sentence is false.

$66 + 32 = 98$ *Replace x with 32.*
$98 = 98$

This sentence is true. The solution is 32.

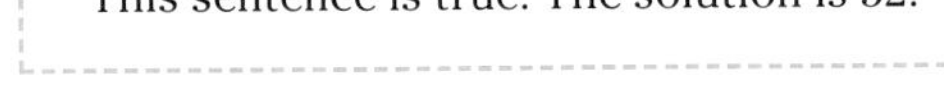

THINK ABOUT IT

How could you find the solution to the equation in Example 1 without substituting each of the possible solutions?

$x = 98 - 66$

You will learn many ways to solve equations in this course. Some equations can be solved mentally by using basic facts or arithmetic skills.

Example 2

APPLICATION

Sports

Basketball great Shaquille O'Neal is an imposing 7' 1" tall and 300 pounds. O'Neal has appeared in Reebok commercials since 1992. When O'Neal began working with Reebok, his shoe size was 19. By 1995, O'Neal's size had risen to a whopping size 22. How many shoe sizes did Shaquille O'Neal increase from 1992 to 1995?

Since the numbers are simple, this equation can be solved mentally.

$19 + s = 22$
$19 + 3 = 22$ *Replace s with 3 since this will result in a true sentence. The solution is 3.*
$s = 3$

Shaquille O'Neal increased 3 shoe sizes from 1992 to 1995.

Equations involving products and quotients can also be found mentally.

Example 3 **Solve each equation mentally.**

a. $6x = 36$
$6 \cdot 6 = 36$ *Think "What number times 6 is 36?"*
$x = 6$

b. $\frac{45}{n} = 9$
$\frac{45}{5} = 9$ *Think "What number times 9 is 45?"*
$n = 5$

Checking Your Understanding

Communicating Mathematics

Read and study the lesson to answer these questions.

1. **In your own words,** define an equation. **See students' work.**
2. **Write** an equation that is always true and another equation that is always false. **Sample answers: $5 + 2 = 7$; $5 + 2 = 9$**

Motivating the Lesson

Situational Problem Ask students whether each is true or false.

- $35 in wages and $20 in tips total $55 earned.
- $200 in salary plus a bonus total $350.

The first is true. The second may be true or false depending upon the amount of the bonus.

2 TEACH

In-Class Examples

For Example 1
Which of the numbers 29, 30, or 31 is the solution of $y + 27 = 58$? **31**

For Example 2
Earnest Lorenz learned to be a cobbler in his native Austria. He now runs the shoe-repair concession at a Lazarus Department store in Columbus, Ohio. If he had been a cobbler for 48 years in 1995, in what year did he begin his career?
$x + 48 = 1995$; 1947

For Example 3
Solve each equation mentally.
a. $10y = 60$ **6**
b. $\frac{72}{m} - 9$ **8**

Study Guide Masters, p. 6

NAME ____________ DATE ________

1-6 Study Guide
Variables and Equations

Student Edition Pages 32–35

The solution of an equation is the number or numbers that make it true. When you have found the solution of an equation, you have solved it.

One way to solve an equation is to guess a number and then check to see if your guess is correct.

Example: Solve $m + 13 = 20$

First guess: 33 $33 + 13 = 46$
So 33 is not the solution.

Next guess: 10 $10 + 13 = 23$
So 10 is not the solution.
But, 10 is closer than 33.

Next guess: 7 $7 + 13 = 20$
So 7 is the solution to the equation.

Name the number that is a solution of the given equation.

1. $19 \times p = 95$; 0, 1, 2, 3, 4, 5, 6 **5**
2. $11 - q = 8$; 1, 3, 5, 7 **3**
3. $51 \div r = 17$; 0, 3, 6, 9, 12 **3**
4. $21 + j = 31$; 0, 5, 10, 15, 20 **10**
5. $28 = 4 \times s$; 1, 3, 5, 7, 9 **7**
6. $13 - t = 9$; 0, 2, 4, 6, 8, 10 **4**
7. $100 = h + 25$; 0, 25, 50, 75 **75**
8. $g \times 9 = 0$; 0, 3, 6, 9, 12 **0**

Solve each equation mentally.

9. $m + 21 = 35$ **$m = 14$**
10. $17 - c = 2$ **$c = 15$**
11. $150 = 15 \times s$ **$s = 10$**
12. $9 = a \div 2$ **$a = 18$**
13. $h + 23 = 32$ **$h = 9$**
14. $c - 12 = 50$ **$c = 62$**
15. $81 = 9 \times t$ **$t = 9$**
16. $3 = 12 \div t$ **$t = 4$**
17. $82 + a = 102$ **$a = 20$**
18. $11 = h - 7$ **$h = 18$**
19. $6 \times a = 48$ **$a = 8$**
20. $u \div 4 = 20$ **$u = 80$**

Group Activity Card 1-6

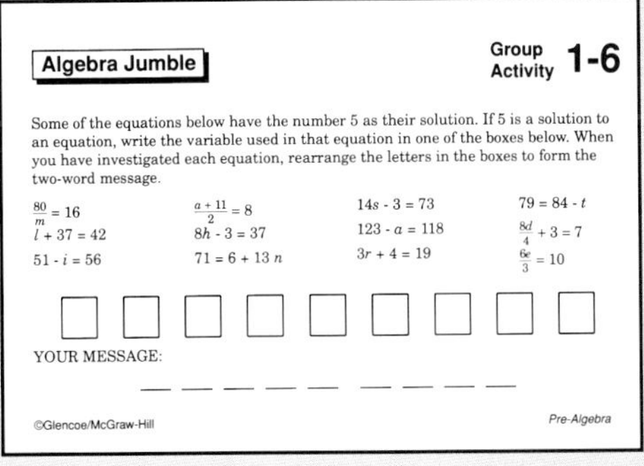

Algebra Jumble

Group Activity **1-6**

Some of the equations below have the number 5 as their solution. If 5 is a solution to an equation, write the variable used in that equation in one of the boxes below. When you have investigated each equation, rearrange the letters in the boxes to form the two-word message.

$\frac{80}{m} = 16$	$\frac{a + 11}{2} = 8$	$14s - 3 = 73$	$79 = 84 - t$
$l + 37 = 42$	$8h - 3 = 37$	$123 - a = 118$	$\frac{8d}{4} + 3 = 7$
$51 - i = 56$	$71 = 6 + 13n$	$3r + 4 = 19$	$\frac{6e}{3} = 10$

□ □ □ □ □ □ □ □ □

YOUR MESSAGE:

__ __ __ __ __   __ __ __ __

 Pre-Algebra

3 PRACTICE/APPLY

Checking Your Understanding

Exercises 1–13 are designed to help you assess your students' understanding through reading, writing, speaking, and modeling. You should work through Exercises 1–4 with your students and then monitor their work on Exercises 5–13.

3. **Find** the solution to the equation $8 + x = 12$ using the four-step problem-solving plan. **4**
4. **Write** an open sentence and then rewrite it so that it is false. **Sample answer: $3 + x = 6$; $3 + 8 = 6$**
5. (MODELING MATHEMATICS) You can use a TI-82 graphing calculator to solve equations if the right side of the equation is zero. You need to enter the left side of the equation, the variable you wish to solve for, and a guess at the solution. Use the keystrokes below to solve $x - 5 = 0$. We guessed that the solution was 5.

   [MATH] 0 [X,T,θ] [−] 5 [,] [X,T,θ] [,] 5 [)] [ENTER]

   The solution of the equation, 5, is displayed.

   Use a TI-82 to solve each equation below.

   a. $x - 9 = 0$ **9** **b.** $2x = 0$ **0** **c.** $3x - 6 = 0$ **2**

Guided Practice

Identify the solution to each equation from the list given.

6. $7 - x = 4$; $\underline{3}, 4, 6$
7. $y + 19 = 32$; $9, \underline{13}, 19$
8. $2x + 1 = 7$; $\underline{3}, 4, 5$

Solve each equation mentally.

9. $t + 5 = 10$ **5**
10. $19 - q = 11$ **8**
11. $78 - r = 28$ **50**
12. $\frac{12}{x} = 3$ **4**
13. **Food** If you could choose only one snack food to eat for the rest of your life, what would it be? In a survey of 1000 people, 220 said hamburgers and 360 said pizza.

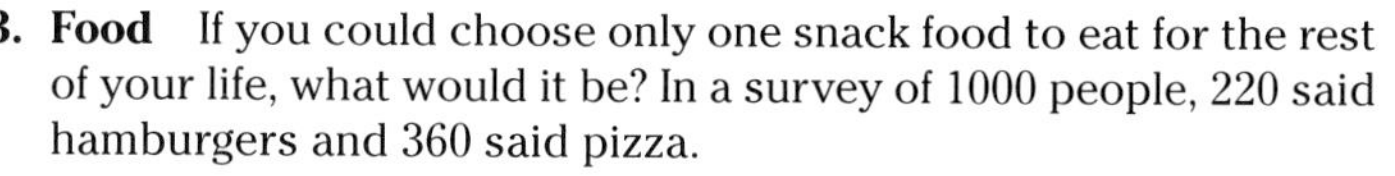

   a. Write an equation to find how many more people said pizza than hamburgers. **$220 + d = 360$**

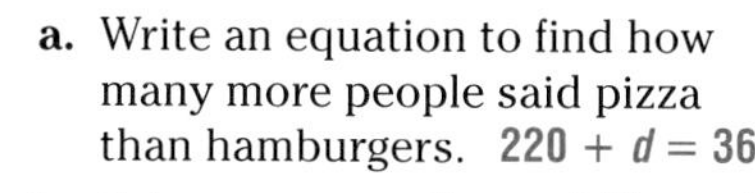

   b. Solve your equation. **140 people**

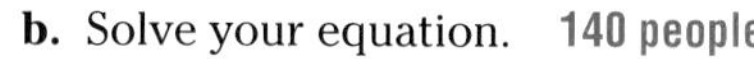

Exercises: Practicing and Applying the Concept

Independent Practice

Identify the solution to each equation from the list given.

14. $12 - c = 8$; $2, \underline{4}, 7$
15. $t + 33 = 72$; $29, 34, \underline{39}$

A
16. $8 = \frac{16}{a}$; $1, \underline{2}, 4$
17. $8 = \frac{g}{4}$; $\underline{32}, 24, 16$

B
18. $3x + 1 = 10$; $2, \underline{3}, 4$
19. $7 = 5b + 2$; $0, \underline{1}, 2$
20. $110 = 145 - t$; $\underline{35}, 40, 45$
21. $8 = \frac{48}{m}$; $3, 4, \underline{6}$

Solve each equation mentally.

22. $m + 8 = 10$ **2**
23. $129 - q = 9$ **120**
24. $17 = k - 3$ **20**
25. $8x = 64$ **8**
26. $\frac{50}{h} = 5$ **10**
27. $63 = 7j$ **9**

C
28. $56 - s = 0$ **56**
29. $99 = 9x$ **11**
30. $\frac{72}{c} = 8$ **9**
31. $6x = 42$ **7**
32. $18 = 2w$ **9**
33. $\frac{21}{y} = 3$ **7**

Practice Masters, p. 6

NAME ______________________ DATE ____________

1-6 Practice
Variables and Equations

Student Edition Pages 32–35

Solve each equation mentally.

1. $8c = 24$ **$c = 3$**	2. $14 - 10 = y$ **$y = 4$**	3. $24 = 16 + b$ **$b = 8$**
4. $8 = \frac{x}{5}$ **$x = 40$**	5. $\frac{z}{15} = 2$ **$z = 30$**	6. $30 = 3w$ **$w = 10$**
7. $32 + p = 50$ **$p = 18$**	8. $\frac{r}{7} = 10$ **$r = 70$**	9. $21 - d = 5$ **$d = 16$**
10. $x + 13 = 22$ **$x = 9$**	11. $\frac{m}{5} = 20$ **$m = 100$**	12. $72 = 9k$ **$k = 8$**
13. $t - 25 = 25$ **$t = 50$**	14. $5m = 0$ **$m = 0$**	15. $12 + a = 29$ **$a = 17$**
16. $33 - h = 13$ **$h = 20$**	17. $44 = p - 1$ **$p = 45$**	18. $\frac{n}{8} = 0$ **$n = 0$**
19. $10 + q = 10$ **$q = 0$**	20. $66 - 33 = f$ **$f = 33$**	21. $\frac{t}{7} = 7$ **$t = 49$**
22. $\frac{u}{15} = 1$ **$u = 15$**	23. $36 - k = 0$ **$k = 36$**	24. $\frac{28}{x} = 4$ **$x = 7$**
25. $48 = t - 2$ **$t = 50$**	26. $17 = r + 7$ **$r = 10$**	27. $8 = \frac{32}{s}$ **$s = 4$**

Extension

Using Puzzles This number puzzle is similar to a word search puzzle. Have students find and circle as many number combinations as they can that, when linked together with symbols like +, −, ×, ÷, and =, will form a correct mathematical equation. One combination is circled. Have students find at least two more.

66	14	42	10	5	15
38	56	3	9	6	7
(28 − 18 =	10)		42	7	3
20	31	2	49	5	63
4	43	32	12	13	7
1	9	6	3	9	1

Critical Thinking

34. Write two different open sentences that fit each description.

a. The solution is 2. **Sample answer: $5 - x = 3$; $2x = 4$**

b. There is no solution that is a whole number. *Hint: The whole numbers are 0, 1, 2, 3, . . .* **$5 + x = 4$; $8 - x = 9$**

Applications and Problem Solving

CONNECTION

For the latest Dow Jones Industrial Average, visit: **www.glencoe.com/sec/math/prealg/mathnet**

35. Finance Stock market performance helps economists determine the health of the economy. One of the indicators of stock market performance is the Dow Jones Industrial Average. On February 21, 1995, the Dow Jones Industrial Average climbed above 4000 for the first time ever. That day, it closed at 4003.33. On February 20, 1995, the closing mark was 3973.05. **35a. $3973.05 + c = 4003.33$**

a. Write an open sentence that describes the climb in the average.

b. Is the solution to the open sentence 29.92, 30.28, or 33.32? **30.28**

36. Geometry The sum of the measures of the angles in any triangle is 180°.

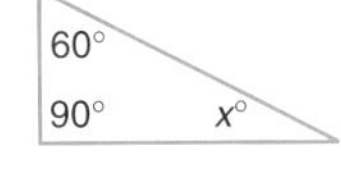

a. Write an equation for the sum of the angle measures in the triangle at the right. **$60° + x° + 90° = 180°$**

b. Solve your equation for x. **30°**

Mixed Review

37. Simplify the expression $5x + 2x + 1$. (Lesson 1-5) **$7x + 1$**

38. Rewrite the expression $8(3 \cdot 6)$ using an associative property. (Lesson 1-4) **$(8 \cdot 3)6$**

39. $1943 + n$

39. Entertainment The game Chutes and Ladders was copyrighted in the United States in 1943. Clue was copyrighted n years later. Write an expression for the year that Clue was copyrighted. (Lesson 1-3)

40. Write a verbal phrase for $y - 9$. (Lesson 1-3) **nine less than a number**

41. Evaluate $8 + 9 - 6 \cdot 2$. (Lesson 1-2) **5**

42. Patterns What is the total number of rectangles, of all sizes, in the figure below? (Lesson 1-1) **21**

WORKING ON THE

Investigation

Up, Up, and Away!

Refer to the Investigation on pages 2–3.

Look over your research and the list of considerations that you placed in your Investigation Folder at the beginning of the Investigation.

You probably listed time and fuel consumption as two of the things you need to consider as you plan your trip. You can use the formulas *speed* × *time* = *distance* and *miles traveled* ÷ *miles per gallon* = *gallons of fuel needed* to determine your time and fuel needed.

- Use a map to determine the approximate distance to Oshkosh.
- Use the information you found to write an equation involving the speed of an ultralight and the distance you will travel. Use a calculator to solve for the time needed.
- Then write an equation with the information about fuel. How many gallons of fuel will you need to make the trip?

Add the results of your work to your Investigation Folder.

WORKING ON THE

Investigation

The Investigation on pages 2 and 3 is designed to be a long-term project that is completed over several days or weeks. Encourage students to keep their materials in their Investigation Folder as they work on the Investigation.

Assignment Guide

Core: 15–33 odd, 34, 35, 37–42
Enriched: 14–34 even, 35–42

For **Extra Practice**, see p. 741.

The red A, B, and C flags, printed in the Teacher's Wraparound Edition, indicate the level of difficulty of the exercises.

4 ASSESS

Closing Activity

Modeling Give each student a card with a different number on it. Have each student create an equation with their number as the solution.

Enrichment Masters, p. 6

NAME ______________________ DATE ____________

1-6 Enrichment Student Edition Pages 32–35

Solution Sets

Consider the following open sentence.

It is a robot that starred in STAR WARS.

You know that a replacement for the word *It* must be found in order to determine if the sentence is true or false. If *It* is replaced by either *R2D2* or *C3PO*, the sentence is true.

The set {*R2D2, C3PO*} can be thought of as the *solution set* of the open sentence given above. This set includes all replacements for the word that make the sentence true.

Write the solution set of each open sentence.

1. It is the name of a state beginning with the letter *A*. **{Alabama, Alaska, Arkansas, Arizona}**
2. It is a primary color. **{red, yellow, blue}**
3. Its capital is Harrisburg. **{Pennsylvania}**
4. It is a New England state. **{Connecticut, Maine, Massachusetts, New Hampshire, Rhode Island, Vermont}**
5. In 1964, he was one of the Beatles. **{John Lennon, George Harrison, Paul McCartney, Ringo Starr}**
6. It is the name of a month that contains the letter *q*. **∅**
7. During the 1970s, she was the wife of a U.S. President. **{Patricia Nixon, Betty Ford, Rosalynn Carter}**
8. It is an even number between 1 and 13. **{2, 4, 6, 8, 10, 12}**
9. $x + 4 = 10$ **{6}**
10. $31 = 72 - k$ **{41}**
11. It is the square of 2, 3, or 4. **{4, 9, 16}**

Write a description of each set.

12. {A, E, I, O, U} **the vowels**
13. {1, 3, 5, 7, 9} **the odd numbers between 0 and 10**
14. {June, July, August} **the months between May and September**
15. {Atlantic, Pacific, Indian, Arctic} **the oceans**

1-7 LESSON NOTES

NCTM Standards: 1-4, 9, 12

Instructional Resources

- Study Guide Master 1-7
- Practice Master 1-7
- Enrichment Master 1-7
- Group Activity Card 1-7

Transparency 1-7A contains the 5-Minute Check for this lesson; **Transparency 1-7B** contains a teaching aid for this lesson.

Recommended Pacing	
Standard Pacing	Day 8 of 14
Honors Pacing	Day 8 of 13
Block Scheduling*	Day 4 of 7 (along with Lesson 1-6)

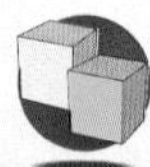

*For more information on pacing and possible lesson plans, refer to the ***Block Scheduling Booklet***.

1 FOCUS

5-Minute Check *(over Lesson 1-6)*

Identify the solution to each equation from the list given.

1. $5 - x = 2$; 3, 5, 7, **3**
2. $11 = \frac{b}{3} + 2$; 27, 33, 36 **27**

Solve each equation mentally.

3. $5x = 25$ **5**
4. $19 - w = 5$ **14**
5. The average baseball player weighed 171 pounds in the 1880s. In the 1980s, the average baseball player weighed 16 pounds more. How much did the average baseball player weigh in the 1980s? **187 pounds**

1-7 Integration: Geometry
Ordered Pairs

Setting Goals: *In this lesson you'll use ordered pairs to locate points, to organize data, and to explore how certain quantities in expressions are related to each other.*

Modeling with Manipulatives

MATERIALS

grid paper

You may have played a board game called Battleship. In this game, players locate their battleships on grids like the one shown here. One player tries to locate and "sink" the other player's battleships by calling out a letter-number combination like F2. If a ship is located at this point, the player says *hit*. If not, the player says *miss*. The object of the game is to sink all of your opponent's ships before he or she sinks yours.

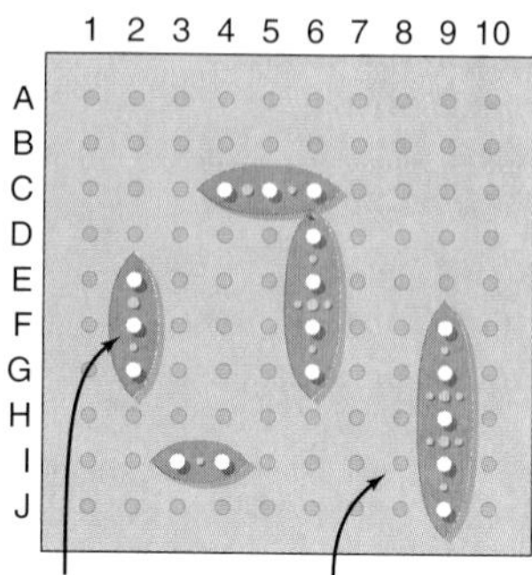

Your Turn TALK ABOUT IT

Play a game of Battleship on grid paper.

a. Name an advantage of using a letter-number instead of a number-number combination to locate points on a grid.

b. Is the point named by C3 the same as 3C? If the locations were both named by numbers, would (3, 4) be different than (4, 3)?

c. Name a disadvantage of using a letter-number combination to locate points on a grid.

Learning the Concept

In mathematics, we can locate a point by using a **coordinate system**. The coordinate system is formed by the intersection of two number lines that meet at their zero points. This point is called the **origin**. The horizontal number line is called the ***x*-axis**, and the vertical number line is called the ***y*-axis**.

Alternative Learning Styles

Kinesthetic Use the grid created by a tile floor or use masking tape to lay out a grid pattern and axes on the floor. Give each student a card with a different ordered pair. Have students start at the origin and step to the right and up to model graphing their ordered pair on the grid.

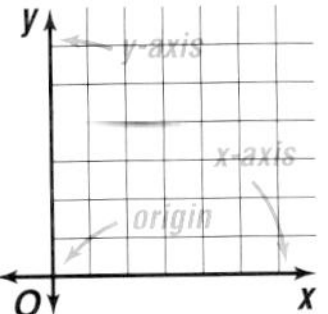

You can graph any point on a coordinate system by using an **ordered pair** of numbers. The first number in the pair is called the ***x*-coordinate**. The second number is called the ***y*-coordinate**. The coordinates are your directions to find the point.

Example 1 **Graph each ordered pair.**

a. (5, 3)
Begin at the origin. The x-coordinate is 5. This tells you to go 5 units right of the origin.

The y-coordinate is 3. This tells you to go up 3 units.

Draw a dot. You have now graphed the point whose coordinates are (5, 3).

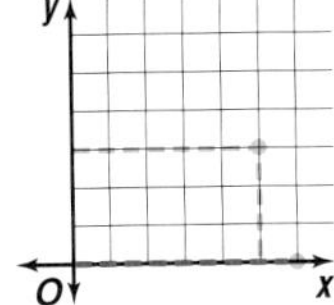

THINK ABOUT IT

Are the numbers in an ordered pair commutative?
no

b. (6, 0)
Begin at the origin again. The x-coordinate is 6, so go 6 units right of the origin. The y-coordinate is 0, so you will not go up to place the dot.

Sometimes points are named by using letters. The symbol $B(3, 4)$ means point B has an x-coordinate of 3 and a y-coordinate of 4.

The coordinate plane allows you to locate points.

Example 2 **Name the ordered pair for each point.**

a. ***C***
Go right on the x-axis to find the x-coordinate of point C, which is 2. Go up along the y-axis to find the y-coordinate, which is 2.

The ordered pair for point C is (2, 2).

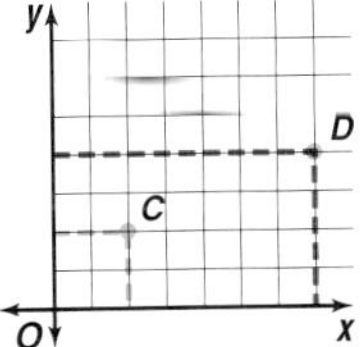

b. ***D***
The x-coordinate of D is 7, and the y-coordinate is 4.

The ordered pair for point D is (7, 4).

Ordered pairs can also be used to show how the value substituted for a variable is related to the value of an expression containing the variable.

Motivating the Lesson

Situational Problem Ask students if they have ever been to a concert or sporting event where they had reserved seat tickets. Have them explain how they found their seats. Relate the seating system to a coordinate system.

2 TEACH

In-Class Examples

For Example 1
Graph each point.

a. $S(5, 7)$

b. $N(0, 5)$

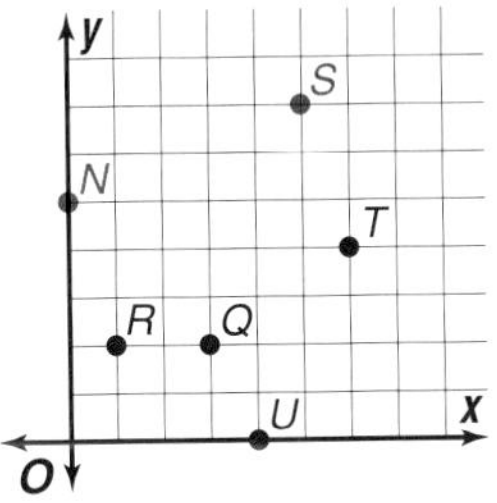

For Example 2
Name the ordered pair for each point on the grid above.

a. Q (3, 2)

b. T (6, 4)

Teaching Tip Point out to students that the axes are not always labeled x and y. For example, the horizontal axis could be labeled t for time and the vertical axis could be labeled d for distance.

Reteaching

Using Models Have students use dominoes to generate ordered pairs. Have one student pick a domino and state an ordered pair. Have the other students locate and graph the point. Be sure that students alternate activities.

In-Class Example

For Example 3

A survey by *SmartMoney* magazine found that the average cost of a baby-sitter in Atlanta in 1994, was \$3 per hour. If a \$2 tip is given and x represents the number of hours, the expression $3x + 2$ represents the cost of a babysitter.

a. Make a list of ordered pairs for 3, 6, 9, and 12 hours of babysitting.

x	$3x + 2$	y	(x, y)
3	$3(3) + 2$	11	(3, 11)
6	$3(6) + 2$	20	(6, 20)
9	$3(9) + 2$	29	(9, 29)
12	$3(12) + 2$	38	(12, 38)

b. Graph the ordered pairs.

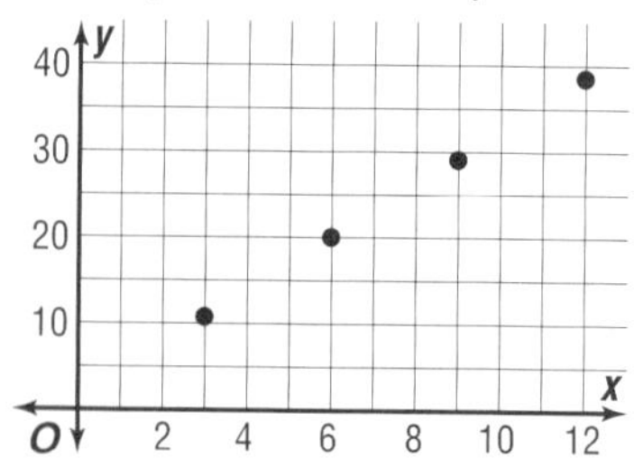

Study Guide Masters, p. 7

NAME ______________ DATE ______

1-7 Study Guide
Integration: Geometry
Ordered Pairs

Student Edition Pages 36–40

An ordered pair of numbers, such as the point (5, 2), can be graphed as follows.

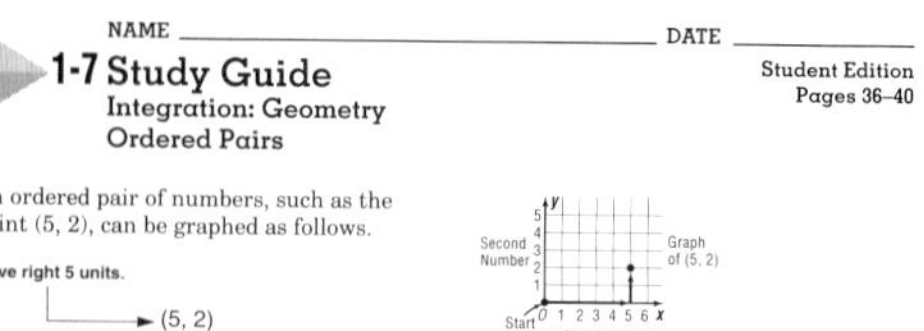

Describe the location of each point with respect to the point (0, 0).

1. (4, 4) **4 units right, 4 units up**
2. (1, 4) **1 unit right, 4 units up**
3. (6, 5) **6 units right, 5 units up**
4. (5, 3) **5 units right, 3 units up**
5. (0, 2) **2 units up**
6. (2, 1) **2 units right, 1 unit up**
7. (3, 0) **3 units right**
8. (2, 6) **2 units right, 6 units up**
9. (3, 3) **3 units right, 3 units up**

Use the grid at the right to name the point for each ordered pair.

10. (4, 4) **A**
11. (5, 3) **D**
12. (3, 0) **C**
13. (1, 6) **B**
14. (2, 2) **F**
15. (3, 4) **E**

Use the grid at the right to find the ordered pair for each labeled point.

16. *A* **(1, 5)**
17. *G* **(7, 5)**
18. *B* **(5, 1)**
19. *H* **(2, 7)**
20. *C* **(4, 4)**
21. *I* **(4, 7)**
22. *D* **(0, 3)**
23. *J* **(0, 6)**

Example 3 APPLICATION Retailing

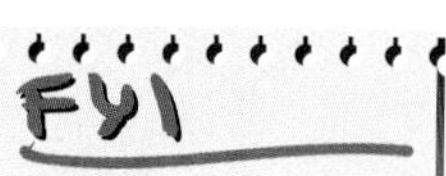

Pocahontas was a Powhatan Indian. President Woodrow Wilson was one of her descendants.

At Movie Madness Video Store, newly-released videos rent for \$2 a day. The store paid \$18 for a copy of *Pocahontas*. If x represents the number of times the video is rented in the first month, the expression $2x - 18$ gives the amount of money the store will make on this video.

a. Make a list of ordered pairs in which the x-coordinate represents the number of rentals in the first month and the y-coordinate represents the money made for 10, 15, 20, 25, or 30 rentals.

b. Then graph the ordered pairs.

a. Evaluate the expression $2x - 18$ for 10, 15, 20, 25, and 30. A table of ordered pairs that result is shown below.

x (rentals)	$2x - 18$	y (money made)	(x, y)
10	$2(10) - 18$	2	(10, 2)
15	$2(15) - 18$	12	(15, 12)
20	$2(20) - 18$	22	(20, 22)
25	$2(25) - 18$	32	(25, 32)
30	$2(30) - 18$	42	(30, 42)

b.

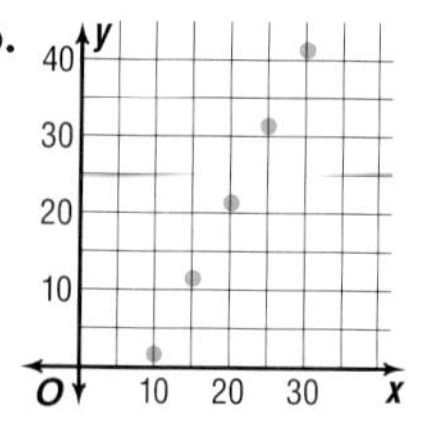

Checking Your Understanding

Communicating Mathematics

Read and study the lesson to answer these questions.

1. **Draw** a coordinate system and label the origin, x-axis, and y-axis.
2. **Tell** three different uses of ordered pairs. **See students' work.**
3. **Describe** a coordinate system that you have seen used in everyday life.
4. **You Decide** Takala says that the points (4, 7) and (7, 4) are the same. Jamal disagrees. Who is correct and why? **See margin.**

1. See Solutions Manual. 3. seats at a concert hall

Guided Practice

Use the map at the right to answer each question.

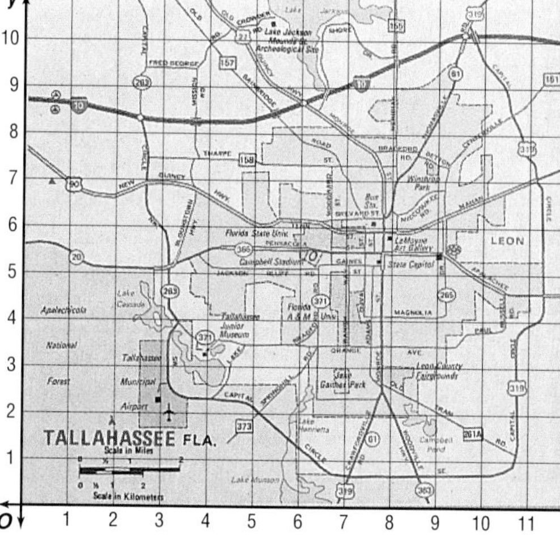

5. Which ordered pair indicates the location of the Tallahassee Municipal Airport? **5. (3, 2) and (3, 3)**
6. Is (5, 6) the correct location for Campbell Stadium? **no**
7. What location does (8, 5) indicate? **7. State Capitol**
8. Which ordered pair indicates the location of the intersection of routes 319 and 10? **8. (10,10)**
9. Is (3, 7) the correct ordered pair for the Lake Jackson Mounds State Archeological site? **no**

Group Activity Card 1-7

Name That Point — Group Activity 1-7

MATERIALS: Enlarged picture of cat on grid

Each group needs a grid with a picture of a cat on it as shown. To begin, the leader names parts of the picture (such as top of left foot) and determines if answers are correct. Everyone tries to be the first to identify the ordered pair associated with the feature. The leader calls on the first person to raise their hand. The person who names the correct ordered pair scores one point. If an ordered pair is incorrectly identified, the person loses one point and someone else responds.

Play continues until everyone has had an opportunity to be the leader.

 Pre-Algebra

Additional Answer

4. Sample answer: Jamal; to graph (4, 7) you move 4 units right and 7 up and to graph (7, 4) you move 7 units right and 4 up.

Exercises: Practicing and Applying the Concept

Independent Practice

Use the grid at the right to name the point for each ordered pair.

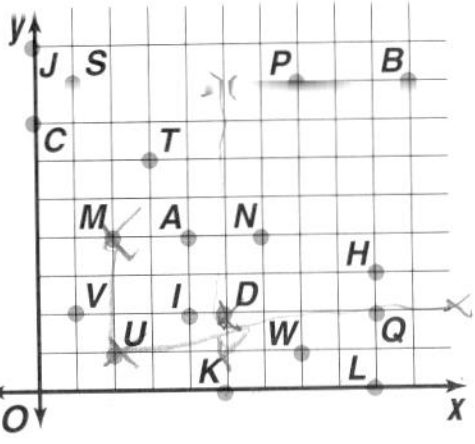

10. (5, 2) D
11. (2, 4) M
12. (1, 8) S
13. (9, 0) L
14. (4, 4) A
15. (3, 6) T
16. (9, 3) H
17. (0, 9) J

Use the grid to name the ordered pair for each point.

18. V (1, 2)
19. N (6, 4)
20. P (7, 8)
21. K (5, 0)
22. W (7, 1)
23. Q (9, 2)
24. C (0, 7)
25. B (10, 8)
26. U (2, 1)

27. a. Write an ordered pair for the coordinates of a point on the x-axis.
    b. Write an ordered pair for the coordinates of a point on the y-axis.
    c. What point lies on both the x- and y-axes? origin
    27a. (a, 0) 27b. (0, b)

Critical Thinking

28. **Geometry** Graph (2, 1), (2, 4), and (5, 1) on a coordinate system.
    a. Connect the points with line segments. What figure is formed?
    b. Multiply each number in the set of ordered pairs by 2. Graph and connect the new ordered pairs. What figure is formed?
    c. Compare the two figures you drew. Write a sentence that tells how the figures are the same and how they are different.
    28a–c. See Solutions Manual.

Communicating Mathematics

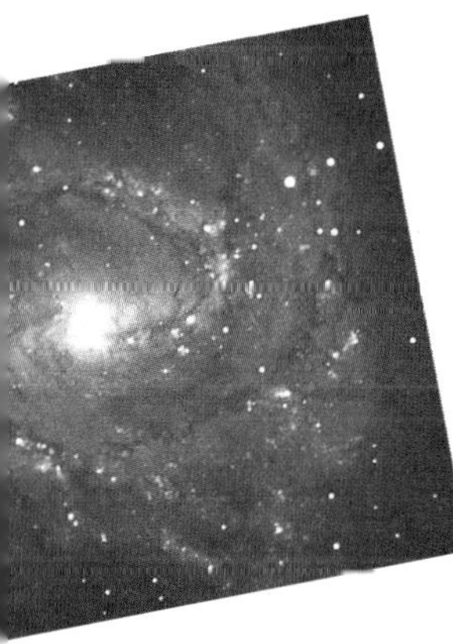

29. **Astronomy** The table at the right contains data about the planets of our solar system. The x-value is the planet's mean distance from the sun in millions of miles. The y-value is how many years it takes each planet to complete its orbit around the Sun. a. See Solutions Manual.
    a. Graph each ordered pair.
    b. Suppose a planet was discovered 1.5 billion miles from the sun. Use the graph to estimate how long it would take to orbit the sun. about 78 years

Planet	x	y
Mercury	36.0	0.241
Venus	67.0	0.615
Earth	93.0	1.000
Mars	141.5	1.880
Jupiter	483.0	11.900
Saturn	886.0	29.500
Uranus	1782.0	84.000
Neptune	2793.0	165.000
Pluto	3670.0	248.000

30. **Geology** The underground temperature of rocks varies with their depth below the surface. The temperature in degrees Celsius is estimated by the expression $35x + 20$, where x is the depth in kilometers. a–b. See Solutions Manual.
    a. Make a list of ordered pairs in which the x-coordinate represents the depth and the y-coordinate represents the temperature for depths of 0, 2, 4, 6, and 8 kilometers.
    b. Graph the ordered pairs.

3 PRACTICE/APPLY

Checking Your Understanding

Exercises 1–9 are designed to help you assess your students' understanding through reading, writing, speaking, and modeling. You should work through Exercises 1–4 with your students and then monitor their work on Exercises 5–9.

Assignment Guide

Core: 11–27 odd, 28, 29, 31–37
Enriched: 10–26 even, 28–37

For **Extra Practice**, see p. 742.

The red A, B, C flags, printed only in the Teacher's Wraparound Edition, indicate the level of difficulty of the exercises.

Exercise Note You may wish to have students use a graphing calculator to graph the points in Exercise 29 since the numbers are so large.

Practice Masters, p. 7

NAME ______ DATE ______

1-7 Practice
Integration : Geometry
Ordered Pairs

Student Edition Pages [illegible]

Use the grid below to name the point for each ordered pair. Write the letter directly below the ordered pair. After completing all the exercises, read the message formed by the letters.

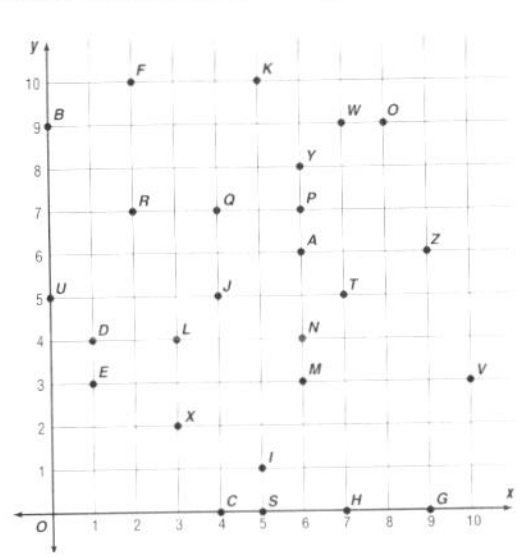

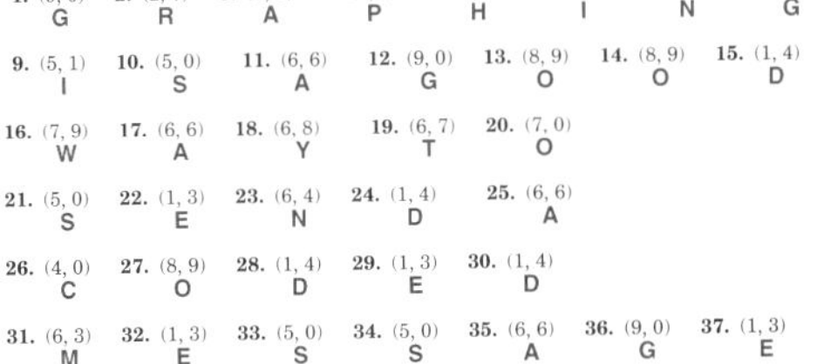

1. (9, 0) G
2. (2, 7) R
3. (6, 6) A
4. (6, 7) P
5. (7, 0) H
6. (5, 1) I
7. (6, 4) N
8. (9, 0) G
9. (5, 1) I
10. (5, 0) S
11. (6, 6) A
12. (9, 0) G
13. (8, 9) O
14. (8, 9) O
15. (1, 4) D
16. (7, 9) W
17. (6, 6) A
18. (6, 8) Y
19. (6, 7) T
20. (7, 0) O
21. (5, 0) S
22. (1, 3) E
23. (6, 4) N
24. (1, 4) D
25. (6, 6) A
26. (4, 0) C
27. (8, 9) O
28. (1, 4) D
29. (1, 3) E
30. (1, 4) D
31. (6, 3) M
32. (1, 3) E
33. (5, 0) S
34. (5, 0) S
35. (6, 6) A
36. (9, 0) G
37. (1, 3) E

Extension

Using Graphing Draw two large grids on the chalkboard or overhead transparency. Have students create an ordered pair to represent their height in inches as the x-coordinate and their shoe size as the y-coordinate. Then have the male students plot their points on one grid while female students plot their points on the other. Discuss the graphs. What observations about the graphs can students make?

4 ASSESS

Closing Activity

Speaking Have students explain the difference between the ordered pairs (3, 2) and (2, 3).

Additional Answer

31a. 50 cars are waiting at the light at 7:00 A.M. and 10 cars are waiting at 9:00 A.M.

b. The workday started at most businesses in the area.

c. Sample answer: Add point (5:15, 50). About the same number of cars will take people away from work that took them there in the morning.

31. See margin.

31. Traffic Control The graph at the right shows some information about the number of cars at the traffic light at State Street and Schrock Road at two different times of a certain weekday.

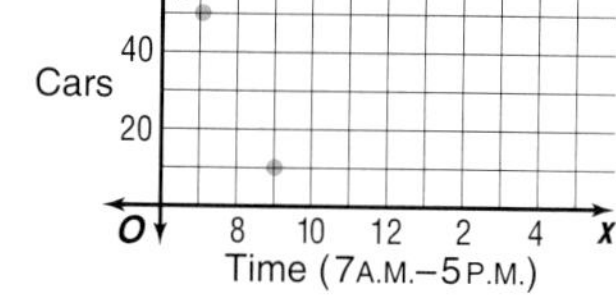

a. What information is given by the two points?

b. What do you think happened between the two times?

c. If most of the local businesses close at 5:00 P.M., copy the graph and place a point for the number of cars you think would be at the intersection at 5:15 P.M. Explain how you chose your point.

Mixed Review

32. Solve the equation $7a = 28$ mentally. (Lesson 1-6) **4**

33. Simplify the expression $3m + 7m + 1$. (Lesson 1-5) **$10m + 1$**

34. Medicine There are three-times more women enrolled in U.S. medical schools today than in the mid-70s. Write an expression for the number of women in medical schools today if x represents the number in the 1970s. (Lesson 1-3) **$3x$**

35. Evaluate the expression $4r$ if $r = 3$. (Lesson 1-3) **12**

36. Evaluate $4c + 2a$ if $a = 2$ and $c = 7$. (Lesson 1-3) **32**

37. Find the value of the expression $6 \cdot 3 \div 9 - 1$. (Lesson 1-2) **1**

COOPERATIVE LEARNING PROJECT

THE SHAPE OF THINGS TO COME

The Global Positioning System

Soon you will be able to hit the open road without stopping to consult a road map. The Global Positioning System, or GPS, is becoming more available to the consumer market, with some guidance systems mounted in cars and other hand-held models as small as a candy bar. The GPS uses a system of 24 satellites to determine your exact location. The monitor receives signals from the satellites and then finds the distances to three or four different satellites to tell your location anywhere on Earth. The display tells you your latitude, longitude, altitude, and direction and speed of travel.

Latitude and longitude lines form a coordinate system for locating any point on Earth. Early sailors developed the latitude system from observations of the North Star. Latitude lines tell the location north or south of the equator. Longitude is used to determine east-west distance. Longitude lines tell the location east or west of the prime meridian in Greenwich, England. Both latitude and longitude are measured in degrees.

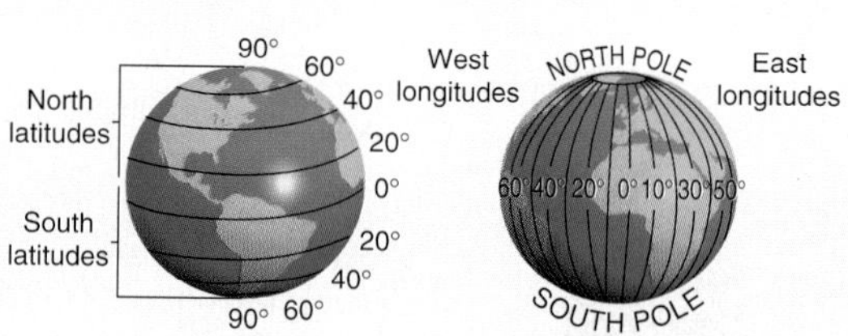

See For Yourself **See Solutions Manual.**

Research the Global Positioning System and the latitude and longitude system.

- How does the GPS use the satellite distances to determine position?
- What are some of the ways consumers are using the GPS? What are the plans for the GPS?
- Why was the GPS first developed?
- What is the approximate latitude and longitude of your hometown?

COOPERATIVE LEARNING PROJECT

THE SHAPE OF THINGS TO COME

The system of longitude and latitude arose from early sailors who used the sun and stars to guide them.

Enrichment Masters, p. 7

NAME ______ DATE ______

1-7 Enrichment
The Hidden Animal

Student Edition Pages 36–40

Graph the following sets of points. Join successive points by a line segment. Begin a new line segment with each numbered set of ordered pairs. When you finish, you will have a picture of an animal.

1. (10, 12), (8, 12), $\left(2, 11\frac{1}{2}\right)$, $\left(\frac{1}{2}, 10\right)$, $\left(\frac{1}{2}, 5\right)$
2. $\left(\frac{1}{2}, 8\right)$, $\left(1, 6\frac{1}{2}\right)$, $\left(1, 5\frac{1}{2}\right)$, $\left(1\frac{1}{2}, 3\right)$, $\left(1\frac{1}{2}, 1\right)$, (4, 1), $\left(3\frac{1}{2}, 2\right)$, $\left(3\frac{1}{2}, 5\frac{1}{2}\right)$
3. $\left(8\frac{1}{2}, 6\frac{1}{2}\right)$, (8, 6), $\left(8\frac{1}{2}, 4\right)$, $\left(8\frac{1}{2}, 1\right)$, $\left(10\frac{1}{2}, 1\right)$, (10, 2), (10, 6)
4. (10, 5), (12, 3), $\left(12, 1\frac{1}{2}\right)$, $\left(10\frac{1}{2}, 1\frac{1}{2}\right)$, $\left(10\frac{1}{2}, 2\frac{1}{2}\right)$, (10, 3)
5. (8, 6), $\left(6, 5\frac{1}{2}\right)$, $\left(3, 5\frac{1}{2}\right)$, (3, 2), $\left(3\frac{1}{2}, 1\right)$
6. (12, 12), $\left(11, 12\frac{1}{2}\right)$, $\left(9, 11\frac{1}{2}\right)$, $\left(8\frac{1}{2}, 10\frac{1}{2}\right)$, $\left(8\frac{1}{2}, 9\frac{1}{2}\right)$, $\left(9, 8\frac{1}{2}\right)$, (10, 8), (11, 9)
7. (12, 12), (13, 12), (14, 11), (14, 9), $\left(13\frac{1}{2}, 8\right)$, $\left(13\frac{1}{2}, 7\right)$, (13, 3), $\left(12\frac{1}{2}, 2\right)$, $\left(12, 2\frac{1}{2}\right)$, $\left(12\frac{1}{2}, 3\right)$, $\left(12\frac{1}{2}, 6\right)$, (12, 7), $\left(11, 7\frac{1}{2}\right)$, $\left(10\frac{1}{2}, 7\right)$, (10, 6)
8. Suppose you multiply both coordinates of each ordered pair by 2 and graph the resulting pairs on graph paper using the same scale on the axes as for the drawing above. How would the drawings compare? **The new figure would be twice as large.**

1-8 Solving Equations Using Inverse Operations

Setting Goals: *In this lesson, you'll use inverse operations to solve equations and solve problems using equations.*

Modeling a Real-World Application: Transportation

The New York City subway system is among the oldest and longest in the world. It has 461 stations, 230 miles of track, and most trains run 24 hours a day, 365 days a year.

When the New York City subway service began in 1904, passengers bought tickets from an attendant for five cents. Another attendant tore the ticket as you passed through the turnstile. Soon the turnstiles were made to operate on coins. When the subway fare was changed to a dime in 1948, the turnstiles could still operate on a coin. However, when the fare went to fifteen cents in July of 1953, there was a problem. That's when the New York City subway token was introduced.

The "bulls-eye" tokens used in New York subways today have brass surrounding a steel center. Since the turnstiles read tokens with a magnet, using bulls-eye tokens helps reduce the use of slugs, which are worthless pieces of metal used instead of tokens. Before the bulls-eyes, New York took in about 165,000 slugs each month. Now slugs total about 20,000 a month.

In 1995, New York subways lost about \$25,000 a month to people using slugs. What was the subway fare in 1995? *This problem will be solved in Example 3.*

Learning the Concept

You have solved some equations using mental math or patterns, but sometimes the solution is not easy to see. Another way to solve equations is to use **inverse operations**. Inverse operations "undo" each other. For example, to "undo" addition you would subtract. These can be shown by related sentences.

The sentences below show how addition and subtraction are related.

$$9 + 12 = 21 \quad \begin{array}{l} 9 = 21 - 12 \\ 12 = 21 - 9 \end{array}$$

$$9 + n = 18 \quad \begin{array}{l} 9 = 18 - n \\ n = 18 - 9 \end{array}$$

1-8 LESSON NOTES

NCTM Standards: 1-4, 6, 8, 9

Instructional Resources

- Study Guide Master 1-8
- Practice Master 1-8
- Enrichment Master 1-8
- Group Activity Card 1-8
- Activity Masters, p. 29
- Assessment and Evaluation Masters, p. 16
- Real-World Applications, 2
- Tech Prep Applications Masters, p. 2

Transparency 1-8A contains the 5-Minute Check for this lesson; **Transparency 1-8B** contains a teaching aid for this lesson.

Recommended Pacing	
Standard Pacing	Day 9 of 14
Honors Pacing	Day 9 of 13
Block Scheduling*	Day 5 of 7 (along with Lesson 1-9)

*For more information on pacing and possible lesson plans, refer to the ***Block Scheduling Booklet***.

1 FOCUS

5-Minute Check
(over Lesson 1-7)

Use the grid below to name the point for each ordered pair.

1. (2, 5) *P*
2. (4, 3) *R*
3. (1, 3) *V*

Use the grid at the right to name the ordered pair for each point.

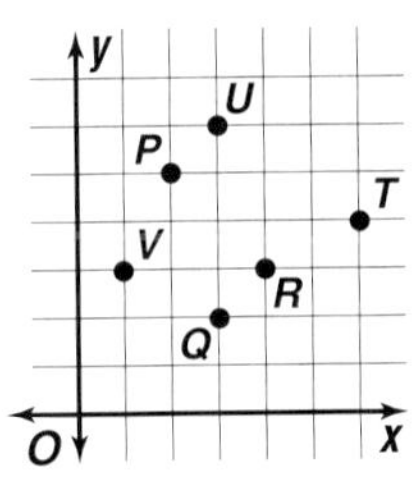

4. *Q* (3, 2)
5. *T* (6, 4)
6. *U* (3, 6)

GLENCOE Technology

Interactive Mathematics Tools Software

In this interactive computer lesson, students use computer manipulatives to solve equations. A **Computer Journal** gives students the opportunity to write about what they have learned.

For Windows and Macintosh

The first subway in the United States opened on Tremont Street in Boston on September 1, 1897. The first subway in the world opened in London in 1863.

Motivating the Lesson

Hands-On Activity Draw the maze shown below on the chalkboard or overhead transparency and work backward with your students to solve it. Discuss why it was easier to solve it backward. *You can only go one way if you work backward, but you can go three ways if you go forward.* You can relate this to solving equations since it is easier to solve an equation by "undoing" the operations than by guessing the solution.

2 TEACH

In-Class Examples

For Example 1
Solve each equation by using inverse operations.

a. $x - 8 = 29$ **37**

b. $m + 16 = 51$ **35**

c. $\frac{a}{7} = 34$ **238**

d. $15b = 210$ **14**

For Example 2
The product of a number and 16 is 80. Find the number. **5**

Multiplication and division are also inverse operations.

$$3 \cdot 5 = 15 \quad \begin{cases} 3 = 15 \div 5 \text{ or } 3 = \frac{15}{5} \\ 5 = 15 \div 3 \text{ or } 5 = \frac{15}{3} \end{cases}$$

$$12t = 48 \quad \begin{cases} 12 = 48 \div t \text{ or } 12 = \frac{48}{t} \\ t = 48 \div 12 \text{ or } t = \frac{48}{12} \end{cases}$$

You can use inverse operations to solve equations.

Example 1 **Solve each equation by using the inverse operations.**

a. $\boldsymbol{b - 6 = 42}$
$b = 42 + 6$ *Write a related addition sentence.*
$b = 48$

b. $\boldsymbol{c + 16 = 54}$
$c = 54 - 16$ *Write a related subtraction sentence.*
$c = 38$ *You can check your answer by replacing c with 38 in the original equation.*

c. $\boldsymbol{\frac{v}{9} = 16}$
$v = 16 \cdot 9$ *Write a related multiplication sentence.*
$v = 144$

d. $\boldsymbol{2.5s = 37.5}$
$s = 37.5 \div 2.5$ *Write a related division sentence.*
$s = 15$ *Use a calculator.*

THINK ABOUT IT
Study each of the rewritten sentences. Why do you think each one is written with the variable alone on one side?
to find the value of the variable

Using an equation to solve a problem is an important problem-solving strategy. The first step in writing an equation is to choose a variable and a quantity for the variable to represent. This is called **defining a variable**.

Example 2 **The sum of 215 and a certain number is equal to 266. Find the number.**

Let n represent the number. Now that you have a variable, translate the words into an equation using the variable.

215	*plus*	*number*	*equals*	*266*
215	+	n	=	266

$215 + n = 266$
$n = 266 - 215$ *Write the related subtraction sentence.*
$n = 51$ The number is 51.

Check: Since 215 + 51 = 266, the solution is correct.

Alternative Learning Styles

Auditory Think aloud as you demonstrate a problem for the class. For example to solve $16r = 48$ you might say "I could try to guess the solution by substituting different values for r, but that might take a while. I know that multiplying and dividing are opposites, so if I divide 48 by 16, that should give me the value of r. 48 divided by 8 is 6 and 16 divided by 8 is 2, so 48 divided by 16 must be 6 divided by 2 or 3. I'll check by multiplying. 3 times 16 is 48. It checks." Allow students to think aloud as they solve problems.

Example 3 APPLICATION Transportation

Refer to the application at the beginning of the lesson. Use the four-step problem-solving plan to solve.

Explore Read the problem for the general idea. You know that about 20,000 slugs are used each month for a loss of \$25,000. You need to find the subway fare. Each slug is equal to one fare. Let f represent the fare.

Plan Translate the words into an equation using the variable. The total amount of money lost to people using slugs is the number of slugs multiplied by the fare.

fare	*times*	*number of slugs*	*equals*	*total lost*
f	$\times$	20,000	=	\$25,000

Solve Solve the equation for the fare.

$f \times 20{,}000 = \$25{,}000$
$f = \$25{,}000 \div 20{,}000$
Write a related division sentence.

25000 [÷] 20000 [=] 1.25
Use a calculator to solve.

The fare is \$1.25.

Examine The answer is close to the estimate. Since $20{,}000 \times \$1.25$ equals \$25,000, the answer checks.

THINK ABOUT IT

Always estimate before you solve the equation. THINK: Since 25,000 is a little more than 20,000 the quotient will be a little over 1.

Checking Your Understanding

Communicating Mathematics

Read and study the lesson to answer these questions.

1. **Write** an addition sentence and then write a related subtraction sentence. Sample answer: $8 + 7 = 15$, $15 - 8 = 7$
2. What is the inverse operation of gaining five yards in football? 2. losing 5 yards
3. **Explain** what it means to *define a variable*. See margin.
4. **Transportation** Refer to the application at the beginning of the lesson.
   a. Write an equation and find the amount of money lost each month if the New York Subway System was still taking in 165,000 slugs per month and the fare is \$1.25. \$206,250 each month
   b. Check your answer using the words of the problem. Write a convincing argument for checking the answer of a problem against the words of the problem instead of the equation that you wrote. See students' work.

Guided Practice

For each sentence, write a related sentence using the inverse operation. 5–8. See margin.

5. $3 + 9 = 12$
6. $13 \cdot t = 39$
7. $16 - 12 = 4$
8. $44x \div 4x = 11$

In-Class Example

For Example 3
The Humane Society of the United States estimates that it costs \$7800 to care for a dog for 12 years. Use the four-step plan to find how much it costs to care for a dog for one year. **\$650**

3 PRACTICE/APPLY

Checking Your Understanding

Exercises 1–16 are designed to help you assess your students' understanding through reading, writing, speaking, and modeling. You should work through Exercises 1–4 with your students and then monitor their work on Exercises 5–16.

Additional Answers

3. Choosing a variable and a quantity for it to represent.

5–8. Sample answers given.

5. $12 - 9 = 3$
6. $39 \div 3 = t$
7. $4 + 12 = 16$
8. $44x = 4x \cdot 11$

Study Guide Masters, p. 8

NAME ______ DATE ______

1-8 Study Guide Student Edition Pages 41–45
Solving Equations Using Inverse Operations

Addition and subtraction are **inverse** operations. Because of this, these four equations give the same information.

$10 + 5 = 15$ $5 + 10 = 15$ $15 - 5 = 10$ $15 - 10 = 5$

Multiplication and division are also inverse operations.

$4 \times 5 = 20$ $5 \times 4 = 20$ $20 \div 5 = 4$ $20 \div 4 = 5$

To solve a single-operation equation, write and solve the related equation that has the variable all by itself on one side of the equal sign.

$x + 5 = 15 \rightarrow 15 - 5 = x$ $4 \times n = 20 \rightarrow 20 \div 4 = n$

Write the three related equations for each given equation.

1. $7 \times 6 = 42$ — $6 \times 7 = 42$, $42 \div 7 = 6$, $42 \div 6 = 7$
2. $50 - 30 = 20$ — $50 - 20 = 30$, $20 + 30 = 50$, $30 + 20 = 50$
3. $56 \div 8 = 7$ — $56 \div 7 = 8$, $7 \times 8 = 56$, $8 \times 7 = 56$
4. $13 + 3 = 16$ — $3 + 13 = 16$, $16 - 3 = 13$, $16 - 13 = 3$
5. $6 + m = 10$ — $m + 6 = 10$, $10 - m = 6$, $10 - 6 = m$
6. $12 = 3y$ — $12 = y \times 3$, $y = 12 \div 3$, $3 = 12 \div y$
7. $d = 32 \div 8$ — $8 = 32 \div d$, $8 \times d = 32$, $d \times 8 = 32$
8. $12 = t - 3$ — $t - 12 = 3$, $3 + 12 = t$, $12 + 3 = t$

Solve each equation by using the inverse operation. Use a calculator where necessary.

9. $6x = 42$ — $x = 7$
10. $42 = 7y$ — $y = 6$
11. $42 \div z = 6$ — $z = 7$
12. $13 + a = 23$ — $a = 10$
13. $b - 13 = 23$ — $b = 36$
14. $23 - c = 13$ — $c = 10$
15. $56 - p = 0$ — $p = 56$
16. $56 + q = 56$ — $q = 0$
17. $r - 0 = 56$ — $r = 56$
18. $30 = 10k$ — $k = 3$
19. $30 \div h = 10$ — $h = 3$
20. $j \times 3 = 30$ — $j = 10$

Reteaching

Using Cooperative Groups Have students work in groups of three. Give students several problems to solve. To reinforce the importance of writing each step in the solution, have one student write the original problem, another write the middle step, and the third student write the solution.

Error Analysis

Students may use the operation in the problem instead of the inverse operation. For example, they may solve $x + 3 = 5$ by adding 3 to each side. It may be helpful to have students estimate the solution before solving or by checking the solution with substitution.

Assignment Guide

Core: 17–45 odd, 46–53
Enriched: 18–42 even, 43–53

For **Extra Practice**, see p. 742.

The red A, B, C flags, printed only in the Teacher's Wraparound Edition, indicate the level of difficulty of the exercises.

Additional Answers

14–15. Sample answers given.
14. Dana earns \$6 an hour. How many hours must she work to earn \$135?
15. Evan missed 7 questions on the quiz. If his score was 28, how many points were possible?

17–22. Sample answers given.
17. $14 = 11 + 3$
18. $23 - 17 = n$
19. $14 = 28 \div 2$
20. $x = 12 \cdot 11$
21. $h = 48 \div 6$
22. $k = 13 \cdot 7$

Solve each equation using the inverse operation. Use a calculator when needed.

9. $8 + y = 22$ 14
10. $77 = 7h$ 11
11. $5.4 = \frac{w}{2.2}$ 11.88

Translate each sentence into an equation. Then solve.

12. Twice a certain number is 6. $2x = 6$; $x = 3$
13. Twenty-seven more than a number is thirty-one. $n + 27 = 31$; $n = 4$

14–15. See margin.

Write a problem that could be solved using each equation.

14. $\$6x = \$135$
15. $x - 7 = 28$
16. **Sports** In April 1994, Jose Maria Olazabal won the \$360,000 prize of the Masters Tournament in Augusta, Georgia. The following week, Hale Irwin took the prize of the Heritage Classic. Irwin won \$135,000 less than Olazabal. How much was the Heritage Classic prize?
    a. Define a variable for this problem. h = Heritage Classic prize
    b. Write an equation. $h + 135{,}000 = 360{,}000$
    c. Solve for the amount of the Heritage Classic prize. $h = \$225{,}000$
    d. Check your solution against the words of the problem. \$225,000 is \$135,000 less than \$360,000.

Exercises: Practicing and Applying the Concept

Independent Practice

For each sentence, write a related sentence using the inverse operation. 17–22. See margin.

17. $14 - 3 = 11$
18. $23 = 17 + n$
19. $14 \cdot 2 = 28$
20. $\frac{x}{11} = 12$
21. $6h = 48$
22. $k \div 13 = 7$

Solve each equation using the inverse operation. Use a calculator when needed.

23. $6 + n = 28$ 22
24. $g + 11 = 15$ 4
25. $86 = c - 2$ 88
26. $144 = 12h$ 12
27. $8 = \frac{v}{15}$ 120
28. $210 = 15w$ 14
29. $5.56 - q = 4.73$ 0.83
30. $3.7 = \frac{r}{5.4}$ 19.98
31. $f + 15.98 = 19.75$ 3.77
32. $6.8b = 34.68$ 5.1
33. $1.5 + b = 2.1$ 0.6
34. $\frac{p}{2} = 9$ 18

Translate each sentence into an equation. Then solve.

35. Three times a certain number is 18. $3x = 18$; 6
36. Ten less than a number is 19. $n - 10 = 19$; 29
37. When a number is decreased by seven, the result is 22. $n - 7 = 22$; 29
38. The product of a number and eight is 88. $8t = 88$; 11
39. A number divided by four is 14. $n \div 4 = 14$; 56
40. A number divided by three is thirteen. $x \div 3 = 13$; 39

41–42. See margin.

Write a problem that could be solved using each equation.

41. $n + 2 = 16$
42. $\$30n = \$90$

Practice Masters, p. 8

NAME ______________________ DATE __________

1-8 Practice
Solving Equations Using Inverse Operations

Student Edition
Pages 41–45

Solve each equation by using the inverse operation. Use a calculator where necessary.

1. $9 + x = 16$ $x = 7$
2. $\frac{v}{16} = 1$ $v = 16$
3. $k - 13 = 18$ $k = 31$
4. $378 = 18z$ $z = 21$
5. $55 = 5c$ $c = 11$
6. $32 = \frac{f}{8}$ $f = 256$
7. $z - 5 = 19$ $z = 24$
8. $m + 15 = 20$ $m = 5$
9. $6c = 54$ $c = 9$
10. $\frac{x}{6} = 19$ $x = 114$
11. $73 = b - 42$ $b = 115$
12. $155 = n + 137$ $n = 18$
13. $\frac{h}{10} = 100$ $h = 1000$
14. $27d = 945$ $d = 35$
15. $94 - p = 12$ $p = 82$
16. $98 = 38 + c$ $c = 60$
17. $201 = \frac{t}{10}$ $t = 2010$
18. $1479 = 17c$ $c = 87$
19. $145 = s - 121$ $s = 266$
20. $12 + r = 54$ $r = 42$
21. $\frac{b}{4.9} = 4.9$ $b = 24.01$

Define a variable, write an equation, then solve.

22. Earth is about 93,000,000 miles from the sun. When Venus is on the opposite side of the sun from Earth, it is about 69,000,000 miles from the sun. What is the distance from Earth to Venus? Let d = distance from Earth to Venus; $d = 93{,}000{,}000 + 69{,}000{,}000$; 162,000,000 miles
23. Mrs. Walsh plans to drive from New York to Chicago, a distance of 850 miles. How long will it take her to make the trip if she averages 50 miles per hour? Let t = time; $850 = 50t$; 17 hours

Group Activity Card 1-8

Does This Event Have An Inverse?

Group Activity **1-8**

With a partner, decide which of the events below have an inverse and which do not. For those that do have an inverse, state what its inverse would be. Then, together, write a paragraph explaining why some events have inverses and others do not. In your paragraph, include another event that has an inverse and one that does not.

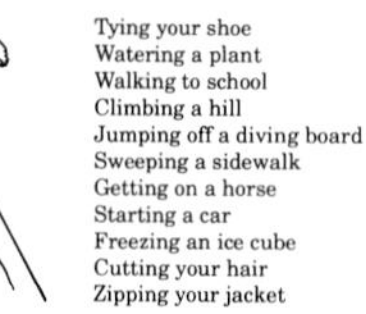

Tying your shoe
Watering a plant
Walking to school
Climbing a hill
Jumping off a diving board
Sweeping a sidewalk
Getting on a horse
Starting a car
Freezing an ice cube
Cutting your hair
Zipping your jacket

 Pre-Algebra

Additional Answers

41. Sample answer: Jack scored 16 points in this week's game. That is two more than what he scored last week. How many points did Jack score last week?
42. Sample answer: Service Plus charges \$30 per hour for a service call. The total for the service call on the furnace was \$90, how many hours of labor were there?

Critical Thinking

43. Consider the division sentence $a = \frac{5}{0}$ and the related multiplication sentence $5 = a \cdot 0$. Use these statements to explain why division by zero is not allowed. **See margin.**

Applications and Problem Solving

44. **Entertainment** In the movie *Maverick*, Bret Maverick worked to raise enough money to enter a card championship. Twenty players entered the tournament, each one paying the same entrance fee. The entrance fees totaled $500,000. How much did each player pay to play?

a. Define a variable. **f = entrance fee**

b. Write an equation for the problem.

44b. $20f = 500,000$

c. Solve for the entrance fee. **$25,000**

d. Check your solution against the words of the problem. **See students' work.**

45. **Sports** *Monday Night Football* has been thrilling fans and topping the television ratings since 1969. As of the 1993 season, the San Francisco 49ers had appeared on Monday Night Football 35 times. The Miami Dolphins had appeared the most times, with 11 more appearances than the 49ers. How many times had the Dolphins appeared? **46 times**

Mixed Review

46. **Geometry**

a. Which of the points on the graph at the right has the coordinates (5, 8)? (Lesson 1-7) **C**

b. Write the coordinates of the point E. (Lesson 1-7) **(5, 0)**

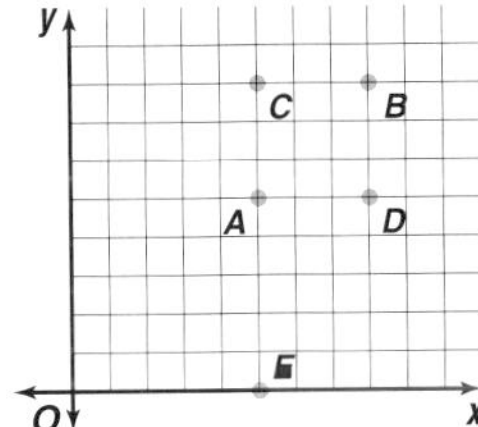

47. Which of the numbers 13, 16, or 19 is the solution of $77 + b = 96$? (Lesson 1-6) **19**

48. Find $13 \cdot 13$ mentally using the distributive property. (Lesson 1-5) **169**

49. Name the property shown by the statement $(9 + 4) + 5 = 9 + (4 + 5)$. (Lesson 1-4) **Associative property of addition**

50. Evaluate $b + cd$ if $b = 2$, $c = 5$, and $d = 7$. (Lesson 1-3) **37**

51. Translate the phrase *five times last year's total* into an algebraic expression. (Lesson 1-3) **$5 \cdot t$**

52. Find the value of $(18 + 6) \div (8 - 2)$. (Lesson 1-2) **4**

√ Choose

Estimation
Mental Math
Calculator
Paper and Pencil

53. **Demographics** *American Demographics* magazine reported that American men drive an average of 16,500 miles each year. How many miles does an average American man drive in a week? (Lesson 1-1)

a. Which method of computation do you think is most appropriate for this problem? Justify your choice. **Sample answer: calculator**

b. Solve the problem using the four-step plan. Be sure to examine your solution. **about 317 miles**

4 ASSESS

Closing Activity

Modeling Have students use cups and counters to model simple equations. First model the equation. Then model the related sentence and solve.

Chapter 1, Quiz C (Lessons 1-6 through 1-8) is available in the *Assessment and Evaluation Masters*, p. 16.

Additional Answer

43. Sample answer: The product of any number and 0 is 0 according to the Multiplicative property of zero. Thus, we could rewrite $5 = a \cdot 0$ as $5 = 0$. $5 \neq 0$, so division by zero does not make sense.

Enrichment Masters, p. 8

NAME ____________ DATE ______

1-8 Enrichment Student Edition Pages 41–45

Mental Math: Compensation

To add or subtract in your head, work with multiples of 10 (20, 30, 40, ...) or 100 (200, 300, 400, ...) and then adjust your answer.

To add 52, first add 50, then add 2 more.

To subtract 74, first subtract 70, then subtract 4 more.

To subtract 38, first subtract 40, then add 2.

To add 296, first add 300, then subtract 4.

Write the second step you would use to do each of the following.

1. Add 83. 1) Add 80. 2) **Add 3 more.**
2. Add 304. 1) Add 300. 2) **Add 4 more.**
3. Subtract 62. 1) Subtract 60. 2) **Subtract 2 more.**
4. Add 27. 1) Add 30. 2) **Subtract 3.**
5. Subtract 79. 1) Subtract 80. 2) **Add 1.**
6. Subtract 103. 1) Subtract 100. 2) **Subtract 3 more.**
7. Add 499. 1) Add 500. 2) **Subtract 1.**
8. Add 294. 1) Add 300. 2) **Subtract 6.**
9. Subtract 590. 1) Subtract 600. 2) **Add 10.**

Use the method above to add 59 to each of the following.

10. 40 **40 + 60 − 1 = 99**
11. 72 **72 + 60 − 1 = 131**
12. 53 **53 + 60 − 1 = 112**
13. 15 **15 + 60 − 1 = 74**

Use the method above to subtract 18 from each of the following.

14. 96 **96 − 20 + 2 = 78**
15. 45 **45 − 20 + 2 = 27**
16. 71 **71 − 20 + 2 = 53**
17. 67 **67 − 20 + 2 = 49**

Extension

Using Cooperative Groups Have small groups of students discuss the equation $3x - 2 = 7$ and devise a method to solve the equation. Encourage them to check their solution to verify their method. *Add 2 to each side, then divide each side by 3.* Then have the groups generalize their methods and use them to solve other equations.

1-9 LESSON NOTES

NCTM Standards: 1-4, 9

Instructional Resources

- Study Guide Master 1-9
- Practice Master 1-9
- Enrichment Master 1-9
- Group Activity Card 1-9
- Graphing Calculator Masters, p. 1

Transparency 1-9A contains the 5-Minute Check for this lesson; **Transparency 1-9B** contains a teaching aid for this lesson.

Recommended Pacing	
Standard Pacing	Day 10 of 14
Honors Pacing	Day 10 of 13
Block Scheduling*	Day 5 of 7 (along with Lesson 1-8)

*For more information on pacing and possible lesson plans, refer to the ***Block Scheduling Booklet***.

1 FOCUS

5-Minute Check
(over Lesson 1-8)

Solve each equation using the inverse operation. Use a calculator when needed.

1. $5 + r = 12$ **7**
2. $34 = 2s$ **17**
3. $7 = \frac{g}{12}$ **84**
4. $4.37 = y - 9.32$ **13.69**
5. Astronomers keep track of about 7000 objects. One out of every ten of these is a satellite. The rest are "space trash." About how many satellites are astronomers tracking?
   a. s = satellites
   a. Define a variable.
   b. Write an equation. **$10s = 7000$**
   c. Solve for the number of satellites. **700**
   d. Check your solution.

1-9 Inequalities

Setting Goals: *In this lesson, you'll write and solve inequalities.*

Modeling a Real-World Application: Health

FYI

23% of Americans say that they dream in black and white.

Insomnia, the inability to get enough sleep, affects millions of Americans each year. Researchers have come up with some suggestions on how to avoid insomnia.

- Maintain a temperature of between 68 and 75°F in your bedroom.
- Stop eating at least 2 hours before bedtime.
- Avoid reading or watching television in bed.
- Go to bed and get up at the same times each day, even when you can sleep in.

Learning the Concept

Let the variable t represent the bedroom temperature. We can write mathematical sentences to describe the temperature that physicians recommend, which is between 68°F and 75°F.

Say: t is less than 75
Write: $t < 75$

Say: t is greater than 68
Write: $t > 68$

A mathematical sentence that contains < or > is called an **inequality**. Inequalities, like equations, can be true, false, or open.

$18 > 13$ *This sentence is true.*
$9 < 5$ *This sentence is false.*
$t < 75$ *This sentence is open. It is neither true nor false until t is replaced with a number.*

Some inequalities contain ≤ or ≥ symbols. They are the combinations of the equals sign and the inequality symbols.

Say: n is less than or equal to 4
Write: $n \leq 4$

Say: m is greater than or equal to 8
Write: $m \geq 8$

Alternative Teaching Strategies

Reading Mathematics Point out to students that there are several ways to read an inequality. Use the phrases in the chart on page 47 to read $4c > 16$. Then have groups write and read their own inequalities, using each of the symbols <, >, ≤, and ≥ at least once.

FYI

Well-known psychologist Sigmund Freud was the first to interpret dreams as a part of a study of the unconscious mind.

Example 1 For the given value, state whether each inequality is *true* or *false*.

a. $r - 8 > 7, r = 12$

$12 - 8 > 7$

$4 > 7$

This sentence is false.

b. $14 \leq \frac{3s}{2} + 5, s = 10$

$14 \leq \frac{3 \times 10}{2} + 5$

$14 \leq 15 + 5$

$14 \leq 20$

This sentence is true.

Many situations in real life can be described using inequalities. The table below shows some common phrases and corresponding inequalities.

<	>	≤	≥
• less than • fewer than	• greater than • more than • exceeds	• less than or equal to • no more than • at most	• greater than or equal to • no less than • at least

Example 2

Agriculture

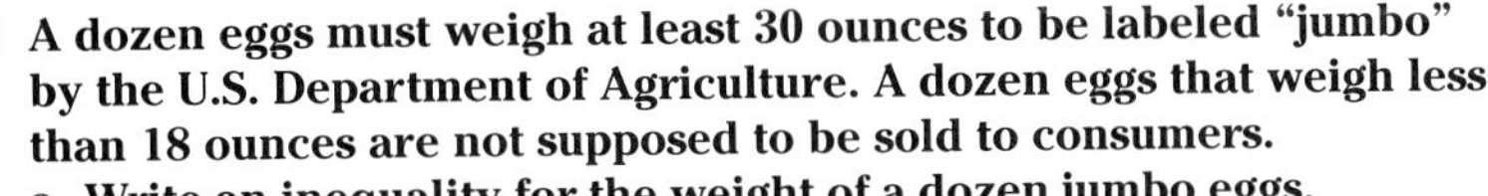

A dozen eggs must weigh at least 30 ounces to be labeled "jumbo" by the U.S. Department of Agriculture. A dozen eggs that weigh less than 18 ounces are not supposed to be sold to consumers.

a. Write an inequality for the weight of a dozen jumbo eggs.

b. Write an inequality for the weight of eggs not sold.

a. The phrase *at least* implies a greater than or equal relationship. Let j represent the weight of a dozen jumbo eggs.
$j \geq 30$

b. Eggs that weigh less than 18 ounces per dozen are not sold. Let u represent the weight of unsold eggs.
$u < 18$

Checking Your Understanding

Communicating Mathematics

Read and study the lesson to answer these questions.

1. **Explain** the difference between $x > 6$ and $x \geq 6$. See Solutions Manual.
2. **Find** a value for which $6x > 12$ is true. Sample answer: 3
3. **Compare and contrast** equations and inequalities. See margin.
4. Work with a partner. See students' work.
   - ▶ Write each of the digits 0–9 on a card and place cards face down.
   - ▶ Then write each of the inequality symbols on a card and place these cards face down in a separate pile. (Mark the bottom of the symbol cards so you can tell which direction the symbol should face.)
   - ▶ Choose one card from the digits pile.
   - ▶ Then choose a card from the symbol pile.
   - ▶ Choose another card from the digits pile.
   - ▶ Determine if the inequality you created is true or false.
   - ▶ Then try to rearrange the cards to make a true inequality.
   - ▶ Shuffle the cards and draw again to create several inequalities.

MODELING MATHEMATICS

MATERIALS

index cards

Motivating the Lesson

Situational Problem Service technicians often set appointments for in-home repairs by saying they will arrive "between 1:00 and 4:00." Ask students to name some possible arrival times. Then represent the situation with an inequality.

2 TEACH

In-Class Examples

For Example 1

For the given value, state whether each inequality is *true* or *false*.

a. $y - 7 \leq 9, y = 15$ **true**

b. $23 \geq \frac{4x}{2} + 6, x = 8$ **true**

For Example 2

Under normal conditions, water is in the form of steam at temperatures greater than 212°F. When the temperature is less than 32°F, water is in the form of ice.

a. Write an inequality for the temperature if water is steam. **$t > 212$°F**

b. Write an inequality for the temperature if water is ice. **$t < 32$°F**

Study Guide Masters, p. 9

NAME ______________ DATE ______________

1-9 Study Guide
Inequalities

Student Edition
Pages 46–49

An **inequality** is a number sentence which states that two expressions are *not* equal. Four symbols for inequality are >, <, ≥, and ≤. Notice that the point of the V-shaped symbol points toward the lesser expression.

$2 + 2 > 3$ Two plus two is greater than 3.
$2 + 2 \geq 3$ Two plus two is greater than or equal to 3.
$2 + 2 \geq 4$ Two plus two is greater than or equal to 4.
$2 + 2 < 5$ Two plus two is less than 5.
$2 + 2 \leq 5$ Two plus two is less than or equal to 5.
$2 + 2 \leq 4$ Two plus two is less than or equal to 4.

Translate each statement into an algebraic inequality.

1. x is less than 10. $x < 10$
2. 20 is greater than or equal to y. $20 \geq y$
3. 14 is greater than a. $14 > a$
4. b is less than or equal to 8. $b \leq 8$
5. 6 is less than the product of f and 20. $6 < 20f$
6. The sum of t and 9 is greater than or equal to 36. $t + 9 \geq 36$
7. 7 more than w is less than or equal to 10. $w + 7 \leq 10$
8. 19 decreased by p is greater than or equal to 2. $19 - p \geq 2$

Name the numbers that are a solutions of the given inequality.

9. $r > 10$; 5, 10, 15, 20 — 15, 20
10. $t \geq 10$; 5, 10, 15, 20 — 10, 15, 20
11. $2 + n < 5$; 0, 1, 2, 3, 4 — 0, 1, 2
12. $2 + m \leq 5$; 0, 1, 2, 3, 4 — 0, 1, 2, 3
13. $6 + m \geq 10$; 3, 4, 5, 6, 7 — 4, 5, 6, 7
14. $6 + m \leq 10$; 3, 4, 5, 6, 7 — 3, 4
15. $30 \leq 5d$; 4, 5, 6, 7, 8 — 6, 7, 8
16. $30 \geq 5d$; 4, 5, 6, 7, 8 — 4, 5, 6

Reteaching

Using Manipulatives Have students use counters to solve inequalities like $3n \leq 15$. Have one student solve the inequality and others use counters to illustrate the solution. For example, the solution of 5 can be shown by using 5 groups of 3 counters.

Additional Answer

3. Sample answer: Equations and inequalities are both number sentences. Equations contain "=", and inequalities contain a symbol indicating something other than equality, like > or ≤.

3 PRACTICE/APPLY

Checking Your Understanding

Exercises 1–12 are designed to help you assess your students' understanding through reading, writing, speaking, and modeling. You should work through Exercises 1–3 with your students and then monitor their work on Exercises 4–12.

Assignment Guide

Core: 13–33 odd, 34, 35, 37–44
Enriched: 14–32 even, 34–44

For **Extra Practice**, see p. 742.

The red A, B, and C flags, printed in the Teacher's Wraparound Edition, indicate the level of difficulty of the exercises.

Practice Masters, p. 9

NAME ______________ DATE ________

1-9 Practice
Inequalities

Student Edition
Pages 46–49

State whether each inequality is true or false for the given value.

1. $b + 10 < 12, b = 4$ false
2. $3 < x - 8, x = 12$ true
3. $6m + 3 \leq 8, m = 1$ false
4. $12 \leq 2p - 6, p = 9$ true
5. $k - 12 < 18, k = 31$ false
6. $13 > 4 + c, c = 9$ false
7. $15 + n \geq 15, n = 6$ true
8. $2 \geq t - 3, t = 3$ true
9. $4t - 4 < 20, t = 7$ false
10. $29 < 24 + a, a = 6$ true
11. $10 \geq 2a + 4, a = 4$ false
12. $5v > 25, v = 4$ false
13. $21 < \frac{r}{3}, r = 66$ true
14. $\frac{s}{8} \geq 4, s = 32$ true
15. $5w + 8 \leq 12, w = 0$ true
16. $2y - 7 < 41, y = 16$ true
17. $3z + z - 6 < 11, z = 4$ true
18. $5f - 2f + 3 \geq 9, f = 2$ true
19. $6h - 3 > 15, h = 2$ false
20. $81 + 3d \geq 90, d = 2$ false
21. $7g - 14 > 0, g = 3$ true
22. $9 \leq 5j - 6, j = 3$ true

Evaluate each expression if a = 2, b = 4, and c = 6. Then write >, <, or = in the box to make a true sentence.

23. bc [>] ac
24. $c + 6$ [<] $3a + 2c$
25. $5b - 2a$ [=] $4b$
26. $3c$ [=] $2b + 4a + 2$
27. $4c - 5b$ [>] $b - a$
28. $5c - 3b - a + 16$ [>] 0

Guided Practice

State whether each inequality is *true*, *false*, or *open*.

5. $9 < 3$ false
6. $8 < 8$ false
7. $12 > 4$ true

State whether each inequality is *true* or *false* for the given value.

8. $n + 2 > 4, n = 8$ true
9. $15 \leq 2q, q = 9$ true

Translate each sentence into an inequality.

10. A number is less than 6. $n < 6$
11. A dozen eggs must weigh 27 ounces or more to be labeled "extra large." $d \geq 27$
12. **Civics** On July 1, 1971, the 26th amendment to the U.S. Constitution was ratified. This amendment states "The right of citizens of the United States, who are 18 years of age or older, to vote shall not be denied or abridged by the United States or any state on account of age." Write an inequality showing the age of a voter. $v \geq 18$

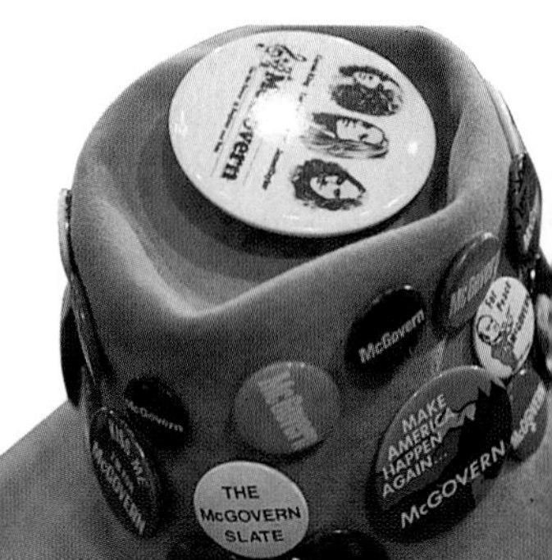

Exercises: Practicing and Applying the Concept

Independent Practice

State whether each inequality is *true*, *false*, or *open*.

13. $7 > 6$ true
14. $40 < 28$ false
15. $2n > 4$ open
16. $18 > 6m$ open
17. $0 \geq 0$ true
18. $j + 7 < 34$ open
19. $16 \leq 16$ true
20. $17 < p$ open
21. $y - 1 < 4$ open

State whether each inequality is *true* or *false* for the given value.

22. $22 - w > 4, w = 11$ true
23. $24 \geq 3c, c = 6$ true
24. $h - 9 > 6, h = 10$ false
25. $2x - 7 \leq 7, x = 7$ true
26. $4x + 8x < 7, x = 1$ false
27. $5z - 16 \leq 11, z = 4$ true

Translate each sentence into an inequality.

28. A number exceeds 24. $n > 24$
29. Tertius earned \$100 or more. $t \geq \$100$
30. At most, 75 people applied. $a \leq 75$

31. It takes no less than three years to complete law school. $l \geq 3$
32. Julita scored at least 16 points. $k \geq 16$
33. Fewer than 15 people attended. $p < 15$

Critical Thinking

34. Suppose $x > 6$ and $x < 18$. a. x is between 6 and 18.
    a. Explain what the open sentence $6 < x < 18$ means.
    b. Find a value of x that makes the sentence $6 < x < 18$ true. Sample answers: 7, 8, 9, 10, 11, 12, . . . , or 17

Group Activity Card 1-9

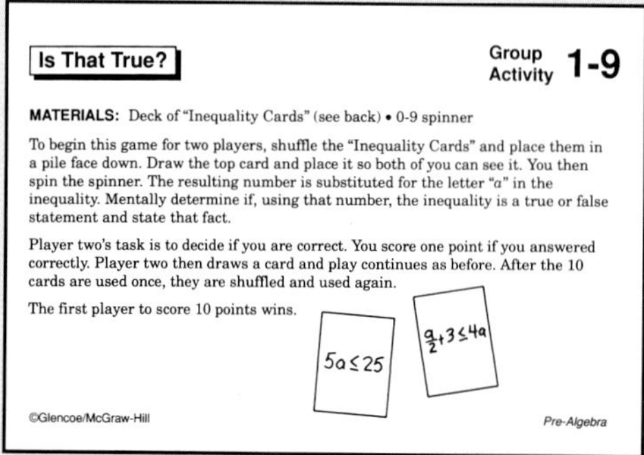

Is That True?

Group Activity **1-9**

MATERIALS: Deck of "Inequality Cards" (see back) • 0-9 spinner

To begin this game for two players, shuffle the "Inequality Cards" and place them in a pile face down. Draw the top card and place it so both of you can see it. You then spin the spinner. The resulting number is substituted for the letter "a" in the inequality. Mentally determine if, using that number, the inequality is a true or false statement and state that fact.

Player two's task is to decide if you are correct. You score one point if you answered correctly. Player two then draws a card and play continues as before. After the 10 cards are used once, they are shuffled and used again.

The first player to score 10 points wins.

Pre-Algebra

Applications and Problem Solving

35. **Sports** On April 9, 1993, the Colorado Rockies played their first home opener in Mile High Stadium in Denver. More than 80,000 fans were in attendance. Write an inequality for the number of people who attended. **$f > 80{,}000$**

36. **Finance** Some companies provide 401(k) plans that help their employees save for retirement. The Internal Revenue Service sets a limit on the maximum amount that an employee may invest in a 401(k) plan in any given year. In 1995, the limitation was changed to $9240. Write an inequality for the amount that an employee may invest in a 401(k) plan. **$a \leq \$9240$**

37. **Health** Some of the benefits of quitting smoking are listed in the chart at the right.
   a. Write an inequality for the amount of time in which a person who quits smoking can expect lung cancer death risk reduced to that of a nonsmoker. **$t \leq 10$ years**
   b. Write an inequality for the amount of time it takes for a person who quits smoking to cut their risk of heart disease to half that of a smoker. **$t \leq 1$ year**

Benefits of Quitting Smoking	
Within	**Benefit**
2 weeks to 3 months	• circulation improves • walking becomes easier • lung function increases 38%
1 year	• excess risk of heart disease is half that of a smoker
10 years	• lung cancer death risk is similar to that of a nonsmoker • precancerous cells are replaced • risk of various cancers decreases

Mixed Review

38. **Charity** Many airlines donate change found in seats and foreign coins that travelers no longer need to charity. TWA gives the foreign coins donated on international flights to UNICEF. The donations total $14,400 in an average year. How much does TWA collect for UNICEF in a month? (Lesson 1-8)
   a. Define a variable. **m = monthly donations**
   b. Write an equation for the problem. **$12m = 14{,}400$**
   c. Solve for the monthly donations. **$1200**
   d. Check your solution against the words of the problem.

38d. $12(1200) = 14{,}400$

39. Solve $7b = 77$ mentally. (Lesson 1-6) **11**

40. Use the distributive property to simplify $4 \cdot 8 + 4 \cdot 2$. (Lesson 1-5) **40**

41. Rewrite the expression $8(7c)$ using an associative property. Then simplify. (Lesson 1-4) **$(8 \cdot 7)c$; $56c$**

42. Evaluate the expression $9t - 2s$ if $s = 8$ and $t = 3$. (Lesson 1-3) **11**

43. Find the value of $\frac{84 - 12}{14 - 6}$. (Lesson 1-2) **9**

44. **Patterns** Find the next number in the pattern 8, 10, 14, 22, 38, __?__ (Lesson 1-1) **70**

4 ASSESS

Closing Activity

Writing Give pairs of students reference books like *The Guinness Book of World Records* or an almanac. Have students use facts from the books to write questions that can be answered using inequalities. For example, how long of a jump would set a record for women's long jump?

Extension

Using Logical Reasoning Three numbers x, y, and z satisfy the conditions below. Which is the greatest, x, y, or z? **x**

$y - z < 0$
$x - y > 0$
$z - x < 0$

Enrichment Masters, p. 9

NAME ______ DATE ______

1-9 Enrichment Student Edition Pages 46–49
Make Up a Problem

You have seen that some problem situations can be solved using variables and open sentences. Usually in math, you are asked to solve problems. In this activity, you will be writing the problems.

Use the open sentence, story idea, and your imagination to write an interesting word problem. Answers will vary. Sample answers given.

1. $x + 4 = 83$
scores on a chapter test Betty outscored Sam by four points on the chapter test. She scored 83. What was Sam's score?

2. $15 - x \leq 10$
money spent on clothes Jill has $15 to spend on a belt and earrings. She does not want to spend more than $10 on the belt. What is the most she can spend on the earrings?

3. $12 + v = 20$
a team's won/loss record The Bulldogs have played 20 games this season. They have lost 12 games. How many have they won?

4. $9 > p - 10$
people at a party A certain number of people were at a party. Then 10 people left. If there were less than 9 people left at the party, how many were there at the start?

5. $13 + y \geq 21$
ages Bill will graduate from college when he is at least 21. If Bill is 13 now, in how many years will he finish college?

6. $36 + h = 47$
temperatures When I got up this morning, the temperature was 36°F. It warmed up to 47°F by noon. How many degrees had the temperature risen by noon?

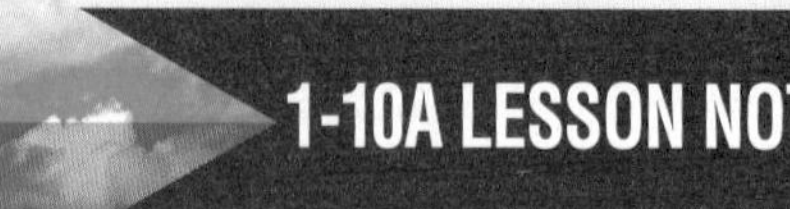

1-10A LESSON NOTES

NCTM Standards: 1-4, 10

Objective
Gather and record data.

Recommended Time
Demonstration and discussion: 15 minutes; Exercises: 30 minutes

Instructional Resources
For each student or group of students
Math Lab and Modeling Math Masters
- p. 28

For teacher demonstration
Overhead Manipulative Resources

1 FOCUS

Motivating the Lesson
Tell students that the number one reason that people leave their jobs is the lack of praise or recognition. Ask students how they think statistics like these are discovered.

2 TEACH

Teaching Tip Ask students to brainstorm reasons why they might be asked for their opinion. Discuss how advertisers try to convince people to use their product.

3 PRACTICE/APPLY

Assignment Guide
Core: 1–5
Enriched: 1–5

4 ASSESS

Observing students working in cooperative groups is an excellent method of assessment.

MATHEMATICS LAB

HANDS-ON ACTIVITY

1-10A Gathering Data

A Preview of Lesson **1-10**

Advertisers, political parties, and media such as television stations and newspapers all want to find out about public opinion. **Demographics** is the study of the characteristics of a population. **Surveys** are often used to study a population. The results of surveys are often modeled by an equation.

Your Turn Work in small groups. Find out about the opinions of students in your school by conducting a survey. The pieces of information that you collect are called **data**.

a. Your group has been hired by a manufacturer to develop an ad campaign to convince students in your school to buy their product. Decide on a product to advertise.

b. How many people will you survey? You probably can't ask every person in your school, so you will need to ask a small group of students. A group that is used to represent a whole population is called a **sample**.

c. Write your questions.

d. Make a plan for asking students. Make sure that the group of students you will ask represents the whole school population.

e. Display your survey results in a way that is easily understood. The results of a sample survey question about television-watching habits is shown below.

How many hours of television do you watch in a week?		
Number of hours watched	**Tally**	**Frequency**
0–2	𝍸	5
3–5	𝍸 𝍸 ||	12
6–8	𝍸 𝍸 𝍸 𝍸 ||||	24
9–11	𝍸 𝍸 𝍸 𝍸 𝍸 𝍸 |||	33
12–14	𝍸 𝍸 𝍸 𝍸 𝍸	25
15–17	𝍸 |	6

TALK ABOUT IT

1. Use the table above to find how many hours of television most of the students surveyed watch in a week. **9–11 hours**
2. Write a few sentences analyzing what you found out about the students in your school through your survey. **See students' work.**
3. Explain how your advertising campaign can convince students to buy the product. **See students' work.**
4. Prepare a report, with charts and graphs, that summarizes your findings. **See students' work.**

Extension

5. Survey the cars in a parking lot. Determine what characteristics you will study. Then collect the data and analyze it. **See students' work.**

Cooperative Learning
This lesson offers an excellent opportunity for using cooperative learning groups. For more information on cooperative learning strategies, see *Cooperative Learning in the Mathematics Classroom*, one of the titles in the Glencoe Mathematics Professional Series.

1-10 Integration: Statistics
Gathering and Recording Data

Setting Goals: *In this lesson, you'll gather and record data in a frequency table or bar graph.*

Modeling a Real-World Application: Entertainment

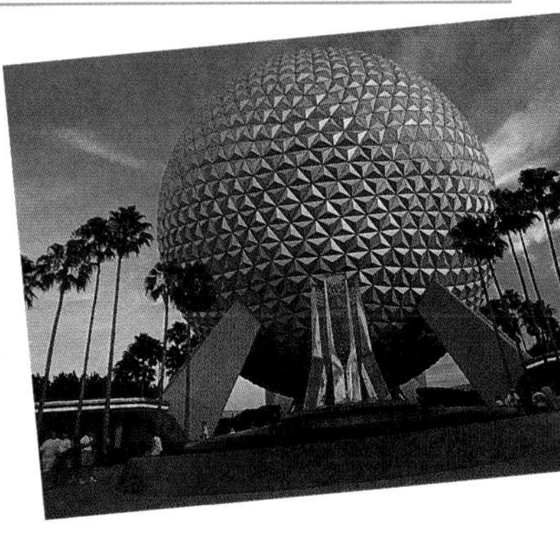

"What do you think?" Opinion polls often ask that question. The results of an opinion poll are reported using **statistics**. Statistics involves collecting, analyzing, and presenting data.

Disney's Epcot Center in Orlando, Florida, was the dream of founder Walt Disney. Disney wanted Epcot to be an open forum for people of the world to meet and talk. As part of that dream, Epcot Center conducts one of the world's largest ongoing public opinion polls. People visiting the park can gather in a 172-seat theater to participate in polls on everything from vegetables to rock videos.

> **FYI**
> On January 1, 1990, Epcot Center began a survey on "The Person of the Century" that will last until the year 2000. Who do you think is the most influential person of the 1900s? So far in the survey, the leaders are Albert Einstein, Martin Luther King, Jr., and Thomas Edison.

The **frequency table** below shows the results of one Epcot poll on movie tastes.

What type of movie do you like most?		
	Men	Women
Comedy	39	47
Action/Adventure	40	18
Drama	6	20
Horror	2	2
Other	5	5
No response	8	8

The poll was conducted with 5174 adults at the Epcot Center. Results are shown as if groups of 100 men and 100 women were surveyed.

Learning the Concept

Obviously, it would be very difficult to ask every man and woman in the United States which type of movie they prefer. Epcot Center surveys smaller groups, called **samples**. The sample group is assumed to be representative of the entire population. However, since the group of people could not be chosen randomly but only from the group at Epcot Center, the results may not be representative.

Cooperative Learning

Send-A-Problem Have students work in groups with a reference book or a newspaper to create data problems. Have groups trade problems and solve. Each group may present the problem and their solution to the class. For more information on the send-a-problem strategy, see *Cooperative Learning in the Mathematics Classroom*, p. 23.

> **FYI**
> Martin Luther King, Jr. led a march on Washington, D.C., of about 200,000 people in August of 1963. They gathered to protest violence and discrimination.

1-10 LESSON NOTES

NCTM Standards: 1-4, 10

Instructional Resources

- Study Guide Master 1-10
- Practice Master 1-10
- Enrichment Master 1-10
- Group Activity Card 1-10
- Assessment and Evaluation Masters, p. 16
- Multicultural Activity Masters, p. 2

Transparency 1-10A contains the 5-Minute Check for this lesson; **Transparency 1-10B** contains a teaching aid for this lesson.

Recommended Pacing	
Standard Pacing	Day 12 of 14
Honors Pacing	Day 11 of 13
Block Scheduling*	Day 6 of 7 (along with Lesson 1-10A)

*For more information on pacing and possible lesson plans, refer to the ***Block Scheduling Booklet***.

1 FOCUS

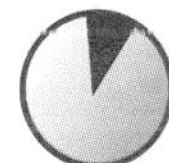

5-Minute Check
(over Lesson 1-9)

State whether each inequality is *true* or *false* for the given value.

1. $5v + 10 < 20, v = 2$ **false**
2. $3x - 12 \geq 0, x = 4$ **true**
3. $7m > 4m, m = 8$ **true**

Translate each sentence into an inequality. **5.** $y > 350$

4. More than 500 people attended the rally. $r > 500$
5. Mathematicians tried to prove Fermat's Last Theorem for more than 350 years when Andrew Wiles presented his effort in 1994.

Motivating the Lesson

Questioning How did students spend last Saturday afternoon? After a few students share, write general activities like "Sports" on the chalkboard and take a survey. Display in a frequency table.

2 TEACH

In-Class Examples

For Example 1
Use the frequency table below.

Number of hours of TV watched in one week	Tally	Frequency
0–2	||||	4
3–5	卌 |||	8
6–8	卌 卌 卌 卌 ||	22
9–11	卌 卌 卌 卌 卌 卌 ||	32
12–14	卌 卌 卌 卌 卌 卌	30
15–17	||||	4

a. How many hours of television did the most students watch? **9-11**

b. How many people were surveyed? **100**

For Example 2
The table below shows the heights of the tallest free-standing statues in the world. Make a bar graph of the data.

Statue	Height (m)
Chief Crazy Horse, USA	172
Buddha, Japan	120
The Indian Rope Trick, Sweden	103
Motherland, Russia	82
Buddha, Afghanistan	53

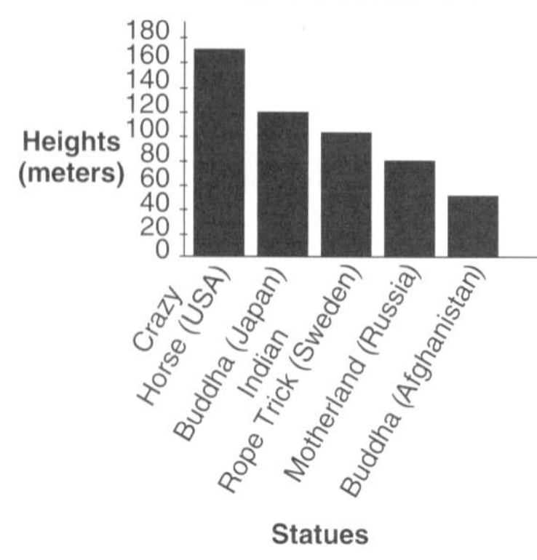

Example 1 APPLICATION Entertainment

Refer to the frequency table at the beginning of the lesson.

a. Which type of movie did most men in the group prefer?
The greatest number of men, 40, chose action/adventure. Comedy was a close second with 39 men choosing it.

b. Overall, which type of movie is most popular?
In order to determine the type of movie most preferred, find the sum of the numbers of men and women who prefer each type.

	men	+	*women*	=	*total*
Comedy	39	+	47	=	86
Action/Adventure	40	+	18	=	58
Drama	6	+	20	=	26
Horror	2	+	2	=	4
Other	5	+	5	=	10
No response	8	+	8	=	16

Comedy is the most popular type of movie overall.

THINK ABOUT IT

The Epcot poll involved adults. How different would the results be if teens were asked? Conduct your own survey on movie preferences.

See students' work

Statistical information can be displayed in a **bar graph**.

Example 2 CONNECTION Social Studies

In 1995, the Gallup Organization conducted a poll on the women that Americans admire most. The results are shown in the frequency table at the right. Make a bar graph to display the data.

Each of the categories in a bar graph has a bar to represent it. The vertical scale shows the number of votes.

Women Americans Admire	
Woman	**Votes**
Queen Elizabeth II	28
Jacqueline Kennedy Onassis	27
Mamie Eisenhower	22
Helen Keller	18
Margaret Chase Smith	18
Claire Booth Luce	17
Mother Teresa	16
Margaret Thatcher	16

Source: The Gallup Organization

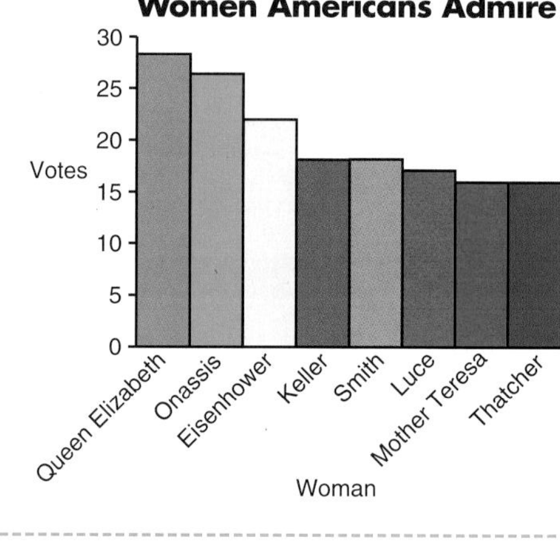

FYI

Claire Boothe Luce was an American playwright and politician noted for her sense of humor. Ms. Luce served as a U.S. Congressional Representative from Connecticut in the 1940s, as an ambassador to Italy in the 1950s, and on the Foreign Intelligence Advisory Board in the 1980s.

Reteaching

Using Modeling Have students roll pairs of number cubes and create a frequency distribution of the data they create. Make sure students record each sum with a tally mark until they have at least twenty pieces of data. Then find the total for each sum and create a bar graph of the data.

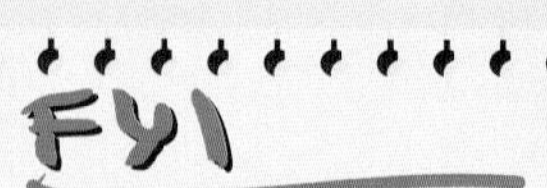

FYI

Margaret Chase Smith was the first woman to be elected to both houses of the United States Congress.

Some graphing calculators allow you to create bar graphs.

Sports

Since the first modern Olympics in 1896, the number of events at each Summer Olympic Games has grown. The table at the right shows the number of events at recent Olympic Games. Use a TI-82 graphing calculator to make a bar graph of the data.

Olympic Year	Number of Events
1968	172
1972	196
1976	199
1980	200
1984	223
1988	237
1992	257
1996	271

Begin by accessing and clearing lists L_1 and L_2 in the calculator's memory.

Enter: STAT ENTER ▲ CLEAR ENTER ▶ ▲ CLEAR ENTER

Then enter the data. Use list L_1 for the years and L_2 for the events.

Enter: ◀ 1968 ENTER 1972 ENTER *and so on* 1996 ENTER ▶ 172 ENTER 196 ENTER *and so on* 271 ENTER

Next set the range of the data for the viewing window.

Enter: WINDOW ENTER 1968 ENTER 2000 ENTER 4 ENTER 0 ENTER 300 ENTER 25 ENTER

Now choose the type of graph and construct the graph.

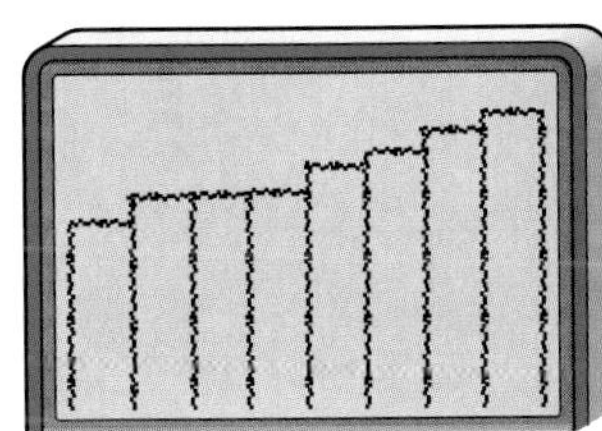

Enter: 2nd STAT PLOT ENTER

Use the arrow and ENTER keys to highlight "On", the bar graph, "L_1", and "L_2". Then press GRAPH to display the graph.

Checking Your Understanding

Communicating Mathematics

Read and study the lesson to answer these questions. 1–3. See margin.

1. **Describe** the kind of information a frequency table shows in your own words.
2. **Explain** why a poll taken at Epcot Center may not be representative of the entire United States population.
3. **Statistics** Find a newspaper or magazine article that contains a graph or table. Explain how you think the results were found. Do you think the display of the information is effective? Explain.

4. **Assess Yourself** When do you use statistics in your life? Do you ever make decisions based on statistics? See students' work.

In-Class Example

For Example 3

Use a TI-82 graphing calculator to make a bar graph of the data on the number of cable television systems.

Year	Number of Cable Systems
1988	8500
1989	9050
1990	9575
1991	10,704
1992	11,075
1993	11,385

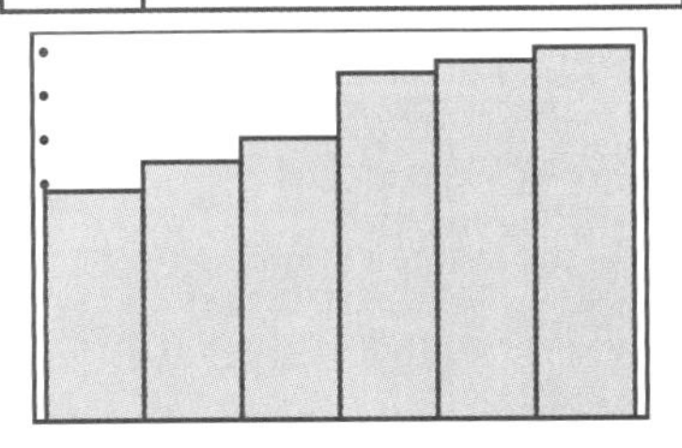

3 PRACTICE/APPLY

Checking Your Understanding

Exercises 1–10 are designed to help you assess your students' understanding through reading, writing, speaking, and modeling. You should work through Exercises 1–4 with your students and then monitor their work on Exercises 5–10.

Study Guide Masters, p. 10

NAME ______________________ DATE ____________

1-10 Study Guide
Integration: Statistics
Gathering and Recording Data

Student Edition
Pages 51–55

To make a frequency table:
a. collect the data.
b. tally the results.
c. count the tallies.

Birthdays

Month	Tally	Frequency
Jan	卌	5
Feb	卌 |||	8
March	||||	4
April	卌 卌		11
May	卌 |||	8
June	卌 卌 ||	12
July	卌 卌		11
Aug	卌 卌 |||	13
Sept	卌 ||||	9
Oct	卌 ||	7
Nov	卌 卌	10
Dec	卌 |||	8

1. Complete the frequency table for the number of birthdays per month.
2. Which four months had the greatest number of birthdays? April, June, July, August
3. Compare the number of birthdays in the first six months to the number in the second six months. 48 < 58

Favorite books	Tally	Frequency
Science Fiction	卌 卌 卌		16
Mystery	卌 卌 ||	12
Biography	卌 卌 |||	13
Romance	卌 卌 卌	15
Sports	卌 ||||	9
Historical Novel	卌 卌 卌 ||||	19

4. Complete the frequency table for the number of books read.
5. How many people responded to this survey? 84
6. What percent chose mysteries? (Round to the nearest whole percent.) 14%
7. What is the ratio of those who chose romances to those who chose sports? 15:9 or 5:3
8. Complete the frequency table for these scores.

Math Quiz Scores

82	78	91	67	72
75	81	68	90	75
83	72	77	71	65
55	93	73	76	81
90	88	78	75	60

Range	Tally	Frequency
95–100		0
90–94	||||	4
85–89			1
80–84	||||	4
75–79	卌 ||	7
70–74	||||	4
65–69	|||	3
below 65	||	2

Additional Answers

1. Sample answer: A frequency table contains categories and the number of times that the answers occur.
2. Only people that visit Epcot Center can be included in the survey. These people have chosen to come there and may not represent people who did not.
3. A frequency table shows each response and a bar graph shows the responses in a visual way. A frequency table is good when you want to know specific numbers. A bar graph is good for giving a general idea quickly.

Assignment Guide

Core: 11–16, 20, 21, 23–27
Enriched: 17–27

For **Extra Practice**, see p. 743.

The red A, B, and C flags, printed in the Teacher's Wraparound Edition, indicate the level of difficulty of the exercises.

Additional Answers

10.

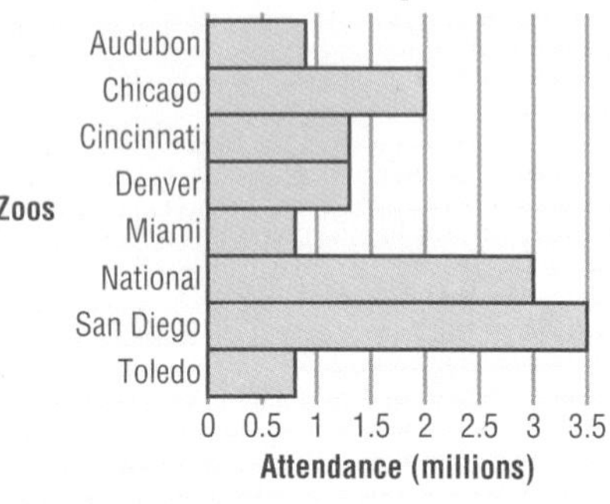

15. Most states will issue a driver's license to a 16-year-old, all to an 18-year-old.

MATERIALS

M & M's or other colored candies

5. Bar graphs are excellent tools for displaying data. Examine the colors of the candies in a bag of M & Ms or a similar product. As you remove each candy from the bag, place it in a column to form a bar graph like the one at the right.

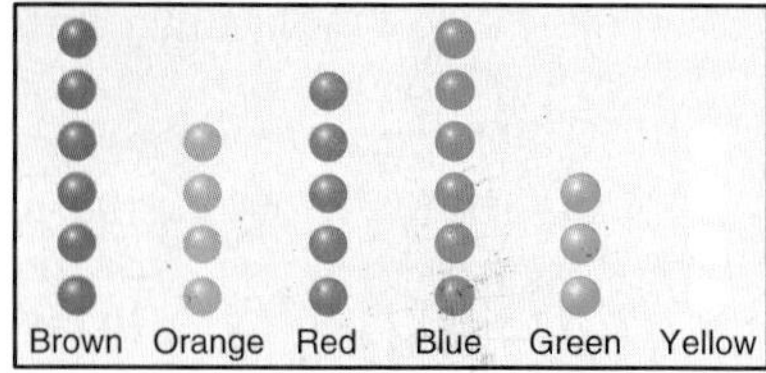

a. Use your data to make a bar graph. **See students' work.**

b. How does your graph compare to the ones made by others in your class? Did each package of candy have the same color distribution and the same number of candies? **See students' work.**

Guided Practice

The frequency table at the right contains data about the 1993 attendance at some of the major zoos in the United States.

Attendance of Selected Zoos in 1993	
Zoo	Attendance (millions)
Audubon (New Orleans)	0.9
Chicago	2.0
Cincinnati	1.3
Denver	1.3
Miami Metrozoo	0.8
National (Wash. D.C.)	3.0
San Diego	3.5
Toledo	0.8

6. Which zoo was visited by the most people? **San Diego**
7. Which zoo was visited least frequently?
8. How many people visited the Denver Zoo in 1993?
9. How many times were all of these zoos visited in 1993?
10. Make a bar graph of the zoo attendance figures. **See margin.**

7. Miami Metrozoo and Toledo
8. 1.3 million
9. 13.6 million

Exercises: Practicing and Applying the Concept

Independent Practice

The table below shows the number of states with different laws about driving ages. The table shows the lowest age at which a person can receive a restricted or unrestricted driver's license.

Driver's License Laws	
Law	States
Restricted at 15	3
Unrestricted at 15	2
Restricted at 16	20
Unrestricted at 16	18
Restricted at 17	3
Unrestricted at 17	2
Unrestricted at 18	3

11. What is the lowest age at which someone can obtain a driver's license in any state? **15**
12. When can a person get a driver's license in most states? Is it a restricted or unrestricted license? **16, restricted**

13. Are all of the states and the District of Columbia represented in the table? How can you tell? **yes, the total of the states is 51**
14. Make a bar graph of the data.
15. Write a sentence to describe the driver's age laws in the United States. **See margin.**
16. Where in the table does your state fall? **Answers will vary.**

14. See Solutions Manual.

Practice Masters, p. 10

NAME ________________ DATE ________

1-10 Practice
Integration : Statistics
Gathering and Recording Data

Student Edition
Pages 51–55

The scores on an English test were 80, 95, 60, 75, 80, 70, 65, 70, 95, 45, 55, 60, 65, 90, 75, 65, and 80.

Score	Tally	Frequency
95	||	2
90	|	1
85		0
80	|||	3
75	||	2
70	||	2
65	|||	3
60	||	2
Below 60	||	2

1. Complete the frequency table for this set of data.
2. What is the highest score? **95**
3. What is the lowest score? **45**
4. What is the frequency of the score that occurred least often? **3**
5. What is the frequency of the score that occurred least often? **0**
6. How many scores are 75 or higher? **8**
7. Write a sentence that describes the test-score data. **Answers may vary. Sample answer: The test scores range from 45 to 95.**

The scores on a mathematics test were 75, 80, 85, 70, 95, 80, 100, 95, 80, 60, 85, 85, 70, 90, 85, 80, 80, 75, 75, 50, 100, 85, 50, 95, and 80.

Score	Tally	Frequency
100	||	2
95	|||	3
90	|	1
85	~~||||~~	5
80	~~||||~~ |	6
75	|||	3
70	||	2
65		0
60	|	1
Below 60	||	2

8. Complete the frequency table for this set of data.
9. What is the highest score? **100**
10. What is the frequency of the score that occurred most often? **6**
11. How many scores are 90 or better? **6**
12. If 70 is the lowest passing score, how many scores are not passing scores? **3**
13. Write a sentence that describes the test-score data. **Answers may vary. Sample answer: 88% of the students passed the test.**

Group Activity Card 1-10

To New Heights — **Group Activity 1-10**

How tall are the students in your pre-algebra class?

Record the heights of all your classmates. Then group the data in reasonable intervals. For example, you may want to use 2" intervals such as 4′10" to 5′, 5′ to 5′2", 5′2" to 5′4", and so on. Next, make a frequency table from which you can then make a histogram. Finally, answer the following questions about the data for your class:

1. What height interval is the most common?
2. What height interval is the least common?
3. What is the range of heights?
4. Is the frequency of the heights evenly spread across the range of heights?
5. Can you tell where the data for girls or boys lie in the histogram? How could you change the graph to show this?

 Pre-Algebra

A list of the heights of the tall buildings in Seattle, Washington, is given below. Heights are given in feet.

17. Make a frequency table for the building heights. Use intervals of 100 feet. **See margin.**

954	740	730	722	609	605
580	574	543	514	500	498
493	487	466	456	454	409
397	389	379	371		

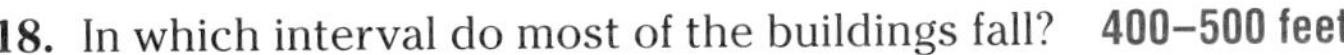

18. In which interval do most of the buildings fall? **400–500 feet**

19. A histogram is a special bar graph that displays the frequency of data in equal intervals. Use your frequency table to make a histogram of the data. **See margin.**

Critical Thinking

20. a. Find a bar graph in a newspaper or magazine.
    b. Does the graph effectively communicate the data? When would a frequency distribution be more useful? **See students' work.**

Applications and Problem Solving

21. **Social Studies** The heights of the U.S. presidents are given in the table at the right.

U.S. Presidents' Heights	
Height (in.)	Frequency
63–65	1
66–68	9
69–71	12
72–74	18
75–77	1

a. How many different presidents have there been? **41**

b. What range of heights describes the most presidents? **72–74 in.**

c. How many presidents were exactly 6 feet tall? Explain your answer. **See margin.**

d. Why do you think so many presidents were or are fairly tall? **Leadership is associated with tallness.**

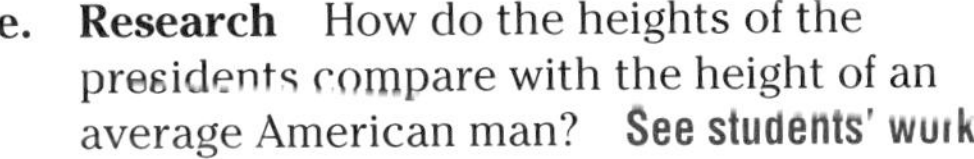

e. **Research** How do the heights of the presidents compare with the height of an average American man? **See students' work.**

22. **Family Activity** Write your own survey question. You might ask family and friends about their favorite color, the number of people living in their home, or how many languages they speak. Make a frequency distribution and a bar graph of your findings. **See students' work.**

Mixed Review

23. **Family Management** The Family Economics Research Group, which is a division of the U.S. Department of Agriculture, estimates that it costs a middle-income family $210,070 or more to raise a child to age 17. Write an inequality for the amount a middle income family will spend. (Lesson 1-9) $s \geq \$210,070$

24. Solve the equation $t - 8 = 42$ using an inverse operation. (Lesson 1-8) **50**

25. **Geometry** Graph the points $D(6, 3)$, $F(2, 2)$, $G(4, 5)$, and $H(5, 10)$ on a coordinate system. (Lesson 1-7) **See Solutions Manual.**

26. Rewrite the expression $(t + 17) + 6$ using an associative property. Then simplify. (Lesson 1-4) $t + (17 + 6); t + 23$

27. Evaluate each expression if $g = 9$ and $h = 4$. (Lesson 1-3)
    a. $6g - 12h$ **6**
    b. $\frac{gh}{3}$ **12**

4 ASSESS

Closing Activity

Speaking Have students state the month of their birthday. Each student should create a frequency table and bar graph of the data.

Chapter 1, Quiz D (Lessons 1-9 and 1-10) is available in the *Assessment and Evaluation Masters*, p. 16.

Additional Answers

17.

Height	0–100	101–200	201–300	301–400	401–500
Buildings	0	0	0	4	8

Height	501–600	601–700	701–800	801–900	901–1000
Buildings	4	1	3	0	1

19.

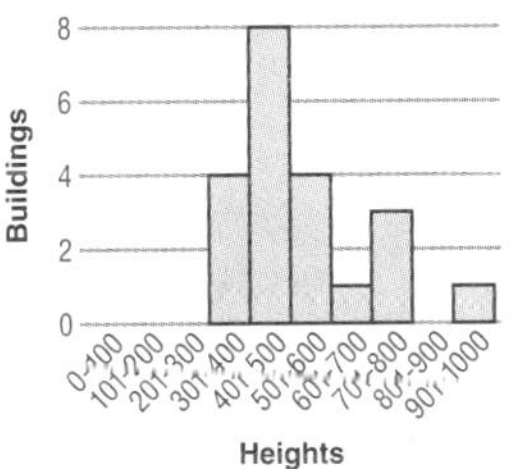

21c. You cannot tell from the frequency table how many presidents were exactly six feet tall. The table shows how many presidents fell in different height ranges, not exact heights.

Enrichment Masters, p. 10

NAME ______ DATE ______

1-10 Enrichment
Histograms

Student Edition Pages 51–55

A histogram is a type of bar graph that displays the frequencies in a distribution of data that has been divided into equal intervals. Notice that there is no space between the bars. (A boundary grade, like 200 or 240, is included in the lower interval. 200 would be included in the 180–200 interval.)

Example: Weight of the players on the Mather Jr. High football team

Weight	Frequency
180–200	2
200–220	10
220–240	9
240–260	7
260–280	4
280–300	2

Use the graph to answer each of the following.

1. Give the histogram a name. **Weight of Players Mather Jr. High Football Team**
2. What does the vertical scale represent? **number of players**
3. What does the horizontal scale represent? **weight of the players**
4. How many weigh between 220 and 240? **9**
5. How many players are there on the team? **34**
6. How many weigh over 240 lb? **13**
7. How many weigh 227 lb? **Cannot be determined**

Make a histogram using the given data. **Check students' graphs.**

8. Grades in Mr. Miner's Math Class

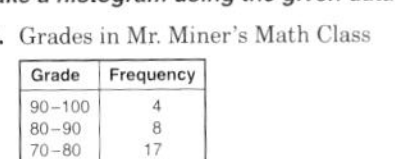

Grade	Frequency
90–100	4
80–90	8
70–80	17
60–70	5
0–60	2

9. Grades in Mrs. Colburn's Math Class

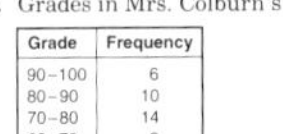

Grade	Frequency
90–100	6
80–90	10
70–80	14
60–70	8
0–60	3

10. How many students in each class received a grade above 70? **Miner: 29; Colburn: 30**
11. How many students in each class had a grade of 60 or below? **Miner: 2; Colburn: 3**

Extension

Using Cooperative Groups Give students a newspaper page or use a page of their textbook to make a frequency table for the letters in the words on the page. Have students make a bar graph showing the frequencies of the ten most-used letters. Discuss how the frequency of letters in words is used in solving word puzzles like those on the game show *Wheel of Fortune*.

CLOSING THE
Investigation

This activity provides students an opportunity to bring their work on the Investigation to a close. For each Investigation, students should present their findings to the class. Here are some ways students can display their work.

- Conduct and report on an interview or survey.
- Write a letter, proposal, or report.
- Write an article for the school or local newspaper.
- Make a display, including graphs and/or charts.
- Plan an activity.

Assessment

To assess students' understanding of the concepts and topics explored in this Investigation and its follow-up activities, you may wish to examine students' Investigation Folders.

The scoring guide provided in the *Investigations and Projects Masters*, p. 3, provides a means for you to score students' work on this Investigation.

Investigations and Projects Masters, p. 3

CHAPTER 1 Investigation
Up, Up, and Away!

Scoring Guide

Level	Specific Criteria
3 Superior	• Shows thorough understanding of the concepts of *the four-step problem solving plan, variables, equations*, and *gathering and recording data.* • Uses appropriate strategies to solve problems. • Computations are correct. • Written explanations are exemplary. • Charts, graphs, and brochure are appropriate and sensible. • Goes beyond the requirements of the Investigation.
2 Satisfactory, with minor flaws	• Shows understanding of the concepts of *the four-step problem solving plan, variables, equations*, and *gathering and recording data.* • Uses appropriate strategies to solve problems. • Computations are mostly correct. • Written explanations are effective. • Charts, graphs, and brochure are appropriate and sensible. • Satisfies all requirements of the Investigation.
1 Nearly Satisfactory, with serious flaws	• Shows understanding of most of the concepts of *the four-step problem solving plan, variables, equations*, and *gathering and recording data.* • May not use appropriate strategies to solve problems. • Computations are mostly correct. • Written explanations are satisfactory. • Charts, graphs, and brochure are mostly appropriate and sensible. • Satisfies most requirements of the Investigation.
0 Unsatisfactory	• Shows little or no understanding of the concepts of *the four-step problem solving plan, variables, equations*, and *gathering and recording data.* • May not use appropriate strategies to solve problems. • Computations are incorrect. • Written explanations are not satisfactory. • Charts, graphs, and brochure are not appropriate or sensible. • Does not satisfy requirements of the Investigation.

CLOSING THE
Investigation

Up, Up, and Away!

Refer to the Investigation on pages 2–3.

Your chapter of the EAA is about to meet to finalize arrangements for the trip to Oshkosh. Your committee will be called upon to present its plan for the trip at the meeting next week. Complete your plan and make a presentation to the club.

- Suppose that your school represents the members of the club. Take a survey about the type of place they would like to stay on a trip. Would they like to camp? How about a hotel? Sometimes people rent rooms in their homes to travelers. Would that be an option? Use your results to choose a place to stay on your trip.
- Use the results of your study of the time and fuel consumption to choose the time of departure and to estimate the cost of the trip.
- Map out the route of the trip so that no Federal regulations are violated and the trip is as enjoyable as possible.
- Compile all of your results into a brochure to give to members interested in making the trip. Include your research and reasons for your choices.

PORTFOLIO ASSESSMENT

You may want to keep your work on this Investigation in your portfolio.

CHAPTER 1 Highlights

Highlights

The Chapter Highlights begins with a listing of the new terms, properties, and phrases that were introduced in this chapter. Have students define each term and provide an example or two of it, if appropriate.

Vocabulary

After completing this chapter, you should be able to define each term, property, or phrase and give an example or two of each.

Algebra

algebraic expression (p. 16)
associative property of addition (p. 22)
associative property of multiplication (p. 22)
brackets (p. 12)
commutative property of addition (p. 22)
commutative property of multiplication (p. 22)
defining a variable (p. 42)
distributive property (p. 26)
equation (p. 32)
evaluate (p. 11)
evaluating an expression (p. 16)
expression (p. 11)
identity property of addition (p. 23)
identity property of multiplication (p. 23)
inequality (p. 16)
inverse operations (p. 41)
like terms (p. 27)
multiplicative property of zero (p. 23)
open sentence (p. 32)
order of operations (p. 11)
parentheses (p. 12)
simplest form (p. 28)
solution (p. 32)
solving an equation (p. 32)
substitution property of equality (p. 17)
terms (p. 27)
variable (p. 16)

Geometry

coordinate system (p. 36)
ordered pair (p. 37)
origin (p. 36)
x-axis (p. 36)
x-coordinate (p. 37)
y-axis (p. 36)
y-coordinate (p. 37)

Statistics

bar graph (p. 52)
data (p. 50)
demographics (p. 50)
frequency table (p. 51)
sample (pp. 50, 51)
statistics (p. 51)
surveys (p. 50)

Problem-Solving

method of computation (p. 8)
problem-solving plan (p. 6)

Understanding and Using Vocabulary

Choose the term that best completes each statement.

1. $8x + 6$ is an example of an __?__. (algebraic expression, inequality)
2. "For any numbers a and b, $a + b = b + a$" is a statement of the __?__ property of addition. (commutative, associative)
3. An expression is __?__ when it has no like terms and no parentheses. (an algebraic expression, in simplest form)
4. In (5, 6), 5 is the __?__. (x-coordinate, y-coordinate)
5. The inverse operation for subtraction is __?__. (addition, division)
6. A group used to represent a larger population is a __?__. (survey, sample)
7. $7y$ and $12y$ are __?__. (like terms, ordered pairs)

Instructional Resources

Three multiple-choice tests and three free-response tests are all provided in the *Assessment and Evaluation Masters*. Forms 1A and 2A are for honors pacing, Forms 1B and 2B are for average pacing, and Forms 1C and 2C are for basic pacing. Chapter 1 Test, Form 1B is shown at the right. Chapter 1 Test, Form 2B is shown on the next page.

Assessment and Evaluation Masters, pp. 3–4

NAME ____________ DATE ______

CHAPTER 1 Test, Form 1B

1. Find the next number in the pattern 3, 7, 15, 31, 63, __?__.
   A. 126 B. 127 C. 94 D. 52 — 1. B
2. What is the value of $12 \cdot 4 + 3$?
   A. 51 B. 144 C. 84 D. 12 — 2. A
3. What is the value of $6(8 - 2)$?
   A. 36 B. 46 C. 4 D. 28 — 3. A
4. Find the value of $7(2 + 6) - 18 \div 3$.
   A. 1 B. 16 C. 50 D. 56 — 4. C
5. What is the value of $ab - 5$ if $a = 7$ and $b = 4$?
   A. 37 B. 33 C. 28 D. 23 — 5. D
6. Evaluate $r + 2s - 9$ if $r = 8$ and $s = 5$.
   A. 1 B. 9 C. 17 D. 24 — 6. B
7. Evaluate $u + 4v - w$ if $u = 9$, $v = 4$, and $w = 6$.
   A. 19 B. 25 C. 40 D. 18 — 7. A
8. Translate the phrase *n increased by 13* into an algebraic expression.
   A. $n - 13$ B. $10n$ C. $n + 13$ D. $13 - n$ — 8. C
9. Choose a phrase that represents the expression $6x + 1$.
   A. x increased by 6
   B. the sum of six times x and 1
   C. 6 more than x
   D. six times the sum of x and 1 — 9. B
10. What property is shown by the statement $4 \cdot 1 = 1 \cdot 4$?
    A. identity property of addition
    B. commutative property of multiplication
    C. identity property of multiplication
    D. associative property of multiplication — 10. B
11. Rewrite $5(2 + 6)$ using the distributive property.
    A. $5 \cdot 2 + 5 \cdot 6$ B. $5(6 + 2)$
    C. $(2 + 6)5$ D. $5 \cdot 2 + 6$ — 11. A
12. Simplify the expression $12n + 2n - 6$.
    A. $12n - 6$ B. 20 C. $14n - 6$ D. $6 - 2n$ — 12. C

NAME ____________ DATE ______

Chapter 1 Test, Form 1B (continued)

13. Simplify the expression $5(m + n) + 2(4n + 2m)$.
    A. $10n + 4m$ B. $10(5n + 3m)$
    C. $13n + 9m$ D. $10m + 12n$ — 13. C
14. Choose the solution for $x - 44 = 57$.
    A. 85 B. 101 C. 13 D. 39 — 14. B
15. What is the solution for y if the difference of y and 19 is 16?
    A. 35 B. 3 C. 19 D. 16 — 15. A
16. Choose the point for the ordered pair (7, 3).
    A. M
    B. N
    C. S
    D. V — 16. B
17. Choose the ordered pair for the point labeled T.
    A. (2, 5) B. (0, 7) C. (6, 9) D. (7, 7) — 17. B
18. Which word describes the inequality $5p + 7 \geq 25$ if $p = 5$?
    A. open B. true C. false D. addition — 18. B

The table at the right shows the average snowfall amounts in the winter in Brighton, Utah.

Month	Snowfall (in inches)
October	21
November	50
December	66
January	69
February	63
March	70
April	51
May	16

19. Choose the month with the greatest snowfall.
    A. January
    B. February
    C. March
    D. April — 19. C
20. How much snow falls in an average winter in Brighton?
    A. 406 inches B. 70 inches
    C. 21 inches D. 451 inches — 20. A

CHAPTER 1 Study Guide and Assessment

Using the Study Guide and Assessment

Skills and Concepts Encourage students to refer to the objectives and examples on the left as they complete the review exercises on the right.

Assessment and Evaluation Masters, pp. 9–10

NAME ______________________ DATE __________

CHAPTER 1 Test, Form 2B

1. **Physical Science** If gravity is the only force acting on a falling object, its speed will increase 32 feet per second each second. Use the four-step problem solving plan to find the speed in feet per second per second of a dropped ball after 4 seconds. — 1. 128 ft/sec

Find the value of each expression.

2. $16 \div 4 + 20$ — 2. 24
3. $(18 - 9) \div 3$ — 3. 3
4. $9(22 - 18)$ — 4. 38
5. $2(5 \cdot 4) - 3 \cdot 6$ — 5. 22

Evaluate each expression if n = 7, m = 9, and p = 2.

6. $12 - m$ — 6. 3
7. $m + 3n$ — 7. 30
8. $\frac{mn}{3}$ — 8. 21
9. $p + 7 - n$ — 9. 2

Translate each phrase into an algebraic expression.

10. two more than k — 10. $k + 2$
11. twice f — 11. $2f$

Name the property shown by each statement.

12. $(2 \cdot 7) \cdot 3 = (7 \cdot 2) \cdot 3$ — 12. commutative
13. $4(5 + 6) = 4 \cdot 5 + 4 \cdot 6$ — 13. distributive

Simplify each expression.

14. $8c + c + 5$ — 14. $9c + 5$
15. $4(5 + 3x)$ — 15. $20 + 12x$
16. $2z + 6(z + 4)$ — 16. $8z + 24$
17. $7(v + w) + 2(2v + w)$ — 17. $11v + 9w$

Identify the solution to each equation from the list given.

18. $15 - h = 11$; 3, 4, 5 — 18. 4
19. $9 = \frac{18}{y}$; 2, 4, 6 — 19. 2
20. $8 = 4x - 28$; 7, 9, 11 — 20. 9
21. $114 = 150 - 3w$; 10, 12, 14 — 21. 12

Solve each equation mentally.

22. $w - 8 = 11$ — 22. 19
23. $32 = 8k$ — 23. 4

NAME ______________________ DATE __________

Chapter 1 Test, Form 2B (continued)

24. Use the grid at the right to name the point for the ordered pair (5, 8). — 24. R
25. Use the grid at the right to name the ordered pair for the point labeled N. — 25. (4, 7)

Solve each equation using the inverse operation.

26. $12 + x = 29$ — 26. 17
27. $\frac{36}{y} = 4$ — 27. 9
28. $12w = 48$ — 28. 4
29. $180 - n = 92$ — 29. 88

State whether each inequality is true *or* false *for the given value.*

30. $22 + x \geq 49$, $x = 32$ — 30. true
31. $2c + 4 \leq 26$; $c = 11$ — 31. true

The table at the right shows the amount of gold produced by selected countries in 1992. Amounts are given in hundreds of thousands of troy ounces.

Country	Production
Australia	77
Canada	51
China	45
Colombia	12
Ghana	10
Mexico	3
Philippines	77
South Africa	197
United States	106
USSR	81
Zaire	3

32. Which country produced the most gold in 1992? — 32. South Africa
33. Make a bar graph of the data on another sheet of paper. — 33. See students' work.

CHAPTER 1 Study Guide and Assessment

Skills and Concepts

Objectives and Examples | **Review Exercises**

Upon completing this chapter, you should be able to:

Use these exercises to review and prepare for the chapter test.

▸ **use the order of operations to evaluate expressions** (Lesson 1-2)

$$\begin{aligned} 3[(7-3)+8\cdot 2] &= 3[(4)+8\cdot 2] \\ &= 3[4+16] \\ &= 3[20] \\ &= 60 \end{aligned}$$

Find the value of each expression.

8. $8(16 + 14)$ **240**
9. $3(9 - 3 \cdot 2)$ **9**
10. $14 \div 2 + 4 \cdot 3$ **19**
11. $12 \div 3 \cdot 4$ **16**
12. $\frac{14 + 4}{11 - 2}$ **2**
13. $\frac{3(2 + 6)}{4}$ **6**
14. $9(12 - 5) - 11(21 - 8 \cdot 2)$ **8**
15. $12[3(17 - 6) - 7 \cdot 4]$ **60**

▸ **evaluate expressions containing variables** (Lesson 1-3)

Evaluate $8x - 2y$ if $x = 3$ and $y = 7$.

$$\begin{aligned} 8x - 2y &= 8(3) - 2(7) \\ &= 24 - 14 \\ &= 10 \end{aligned}$$

Evaluate each expression if $a = 10$, $b = 8$, $c = 6$, and $d = 12$.

16. $14 - a + b$ **12**
17. $2b - d$ **4**
18. $d - c$ **6**
19. $c + (d - b)$ **10**
20. $5a + 4b$ **82**
21. $\frac{a}{b + 2}$ **1**
22. $ab - cd$ **8**
23. $6(d - b)$ **24**

▸ **translate verbal phrases into algebraic expressions** (Lesson 1-3)

Forty-five fewer fans attended this football game than attended last week's game.

Let a represent last week's attendance. Then this week's attendance is $a - 45$.

Translate each phrase into an algebraic expression. 27. $b + 10$ 28. $4 + 2w$

24. seven more than some number $n + 7$
25. the product of a number and ten $10n$
26. twice as many as last week $2n$
27. Joi's score plus 10 points extra credit
28. four more than twice last week's earnings
29. the difference of r and 12 $r - 12$

▸ **use properties of addition and multiplication** (Lesson 1-4)

Name the property shown by the statement $a + 12 = 12 + a$.

commutative property of addition

Name the property shown by each statement.

30. $22 + 0 = 22$ **additive identity**
31. $9 + 3t = 3t + 9$ **commutative, +**
32. $8 + (2 + y) = (8 + 2) + y$ **associative, +**
33. $4(6 \cdot 3) = (4 \cdot 6) \cdot 3$ **associative, ×**
34. $0(b) = 0$ **mult. property of zero**
35. $1(m + 5) = m + 5$ **mult. identity**

GLENCOE Technology

Test and Review Software

You may use this software, a combination of an item generator and an item bank, to create your own tests or worksheets. Types of items include free response, multiple choice, short answer, and open ended.

For IBM & Macintosh

Chapter 1 •
Study Guide and Assessment

Objectives and Examples

▶ **simplify expressions using the distributive property** (Lesson 1-5)

Simplify the expression $6(8 + n) - 6$.

$$\begin{aligned} 6(8 + n) - 6 &= 6(8) + 6(n) - 6 \\ &= 48 + 6n - 6 \\ &= 48 - 6 + 6n \\ &= 42 + 6n \end{aligned}$$

Review Exercises

Simplify each expression. **38. $19b - 10$**

36. $7 \cdot 8 + 7 \cdot 3$ **77** **37.** $n + 7n$ **8*n***

38. $7b + 12b - 10$ **39.** $16x + 2x + 8$ **18*x* + 8**

40. $8(5ab + 4) + 3(2ab + 2)$ **46*ab* + 38**

41. $2(x + y) + 8(x + 3y)$ **10*x* + 26*y***

42. $10(u + 3) + 4(5 + u)$ **14*u* + 50**

▶ **solve open sentences** (Lesson 1-6)

Which of the numbers 8, 12, or 16 is the solution of $84 - n = 72$?

$84 - 8 = 72$
$76 = 72$ *false*

$84 - 12 = 72$
$72 = 72$ *true*

$84 - 16 = 72$
$68 = 72$ *false*

The solution is 12.

Identify the solution to each equation from the list given.

43. $15 - h = 9$; 4, <u>6</u>, 7 **44.** $7 = \frac{m}{4}$; 21, <u>28</u>, 35

45. $5b = 45$; 7, <u>9</u>, 11 **46.** $18 - n = 18$; 18, 1, <u>0</u>

47. $12 = 9x - 15$; <u>3</u>, 6, 9

48. $16 = 144 \div k$; <u>9</u>, 12, 16

49. $\frac{35}{r + 2} = 5$; 3, <u>5</u>, 7

▶ **use ordered pairs** (Lesson 1-7)

Graph the point (7, 3).

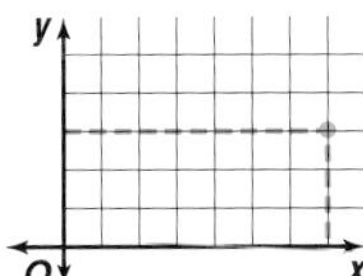

Use the grid to name the point for each ordered pair.

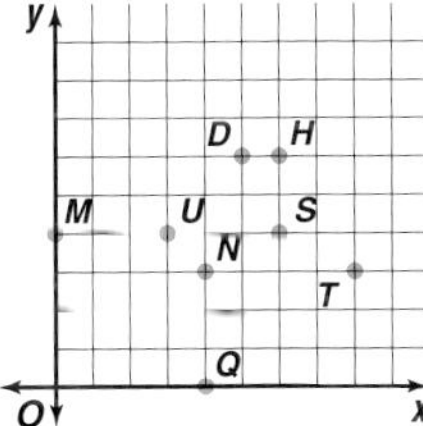

50. (8, 3) ***T***

51. (4, 3) ***N***

52. (3, 4) ***U***

Use the grid to name the ordered pair for each point.

53. *D* **(5, 6)**

54. *H* **(6, 6)**

55. *M* **(0, 4)**

▶ **solve equations using inverse operations** (Lesson 1-8)

$\frac{h}{7} = 13$

$h = 7 \cdot 13$ *Write a related multiplication sentence.*

$h = 91$

Solve each equation using the inverse operation. Use a calculator when needed.

56. $72 = 8n$ **9** **57.** $19 - g = 7$ **12**

58. $6n = 84$ **14** **59.** $\frac{y}{4.5} = 6.3$ **28.35**

60. $19.9 - x = 17.6$ **61.** $9 = \frac{w}{12}$ **108**

62. The product of a number and seven is 84. Find the number. **12** **60. 2.3**

Applications and Problem Solving

Encourage students to work through the exercises in the Applications and Problem Solving section to strengthen their problem-solving skills.

Additional Answers

69.

Representatives	States
1–10	38
11–20	7
21–30	3
31–40	1
41–50	0
51–60	1

70.

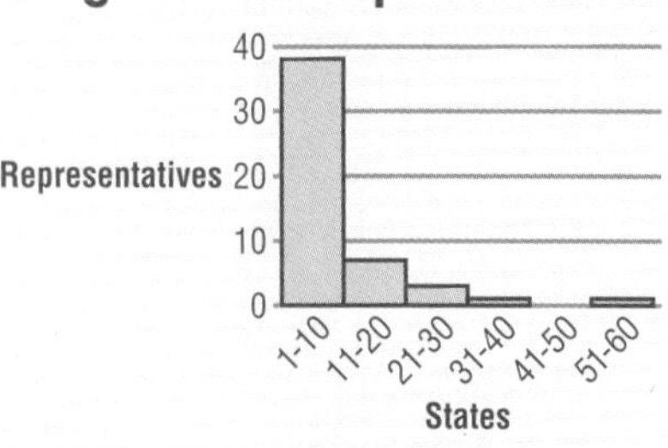

71a. Sample answer: Estimation because an exact answer is not needed.

72a. 125 pounds

b. 190 pounds

c. males: $w = 6h - 260$; females: $w = 5h - 200$

Chapter 1 • Study Guide and Assessment

Objectives and Examples

▶ **write and solve inequalities** (Lesson 1-9)

Translate "*A number is greater than or equal to fifteen.*" into an inequality. Is the inequality *true* or *false* for the value 15?

Let n represent the number.
$n \geq 15$

$15 \geq 15$ is true.

Review Exercises

State whether each inequality is *true* or *false* for the given value.

63. $37 - z \leq 9, z = 22$ **false**
64. $5x + 8 > 0; x = 0$ **true**
65. $7c + 9c \geq 17c, c = 8$ **false**

Translate each sentence into an inequality.

66. A number exceeds 25. $n > 25$
67. Fewer than 60 points were scored. $p < 60$
68. It takes no less than seven hours to drive to Chicago from here. $h \geq 7$

▶ **gather and record data in a frequency table or bar graph** (Lesson 1-10)

Age	Freq.
0–10	2
11–20	8
21–30	11
31–40	6

Frequency
15
10
5
0
0–10 11–20 21–30 31–40
Age

The list below shows the numbers of U.S. congressional representatives from each state for the years 1990–2000. **See margin.**

7 1 6 4 52 6 6 1 23 11
2 2 20 10 5 4 6 7 2 8
10 16 8 5 9 1 3 2 2 13
3 31 12 1 19 6 5 21 2 6
1 9 30 3 1 11 9 3 9 1

69. Make a frequency table of the numbers of representatives. Use intervals of 10.
70. Make a bar graph of the data.

Applications and Problem Solving

71. **Space Exploration** The longest flight ever made by a space shuttle ended on March 18, 1995. The astronauts aboard *Endeavour* spent 16 days, 15 hours in flight, traveling 6.9 million miles and circling Earth 262 times. About how many miles did the shuttle travel on each trip around Earth? (Lesson 1-1) **71a. See margin.**
   a. Which method of computation do you think is most appropriate for this problem? Justify your choice.
   b. Solve the problem using the four-step plan. Be sure to examine your solution. **about 28,000 miles**

72. **Health and Fitness** One health magazine gives the equations below for finding your ideal weight, where w represents weight in pounds and h is height in inches. **See margin.**

males: $w = 100 + 6(h - 60)$
females: $w = 100 + 5(h - 60)$

   a. Find the ideal weight of Cassie, who is 5 feet 5 inches tall. (Lesson 1-2)
   b. What is Rituso's ideal weight if he is 6 feet 3 inches tall? (Lesson 1-2)
   c. Rewrite each formula using the distributive property. (Lesson 1-5)

A practice test for Chapter 1 is available on page 774.

Alternative Assessment

Performance Task

Demonstrate your knowledge by giving a clear, concise solution to each problem. Be sure to include all relevant drawings and justify your answers. You may show your solutions in more than one way or investigate beyond the requirements of the problem.

Vitality magazine reports on issues of health and wellness. The table of Calories and grams of fat in different snack foods shown at the right accompanied an article on nutrition.

Snack	Calories	Fat grams
KitKat Bar (4 oz)	588	33
Strawberry Twizzlers (5 oz)	500	5
Butterfinger (4 oz)	492	21
Reeses' Peanut Butter Cups (3.2 oz)	380	22
M & M's peanut (2.6 oz)	363	20
Goobers (2.2 oz)	320	21
Skittles, fruit flavor (2.6 oz)	286	2
Plain air-popped popcorn (med. bucket)	180	trace
Coca-Cola (16 oz)	205	0
Diet Pepsi (16 oz)	<1	0

1. About how many Calories would you consume if you ate one candy item and drank a soft drink?
2. Write an expression for the number of grams of fat in two packages of peanut M & Ms and a Coca-Cola. Then find the number of Calories.
3. There are 192 Calories in a 16-ounce 7•Up.
   a. Write two expressions for the number of calories in a 7•Up and a package of candy. **$192 + C$ or $C + 192$**
   b. Find the number of calories in a 7•Up and a KitKat. **780**
4. Explain what the entry in the Calories column for Diet Pepsi means. Could there be 6 Calories in a 16-ounce Diet Pepsi? **See margin.**
5. You have been asked to make a presentation on nutrition to your health class. Represent the data in the table in a different way to show the nutritious value of snack foods. **See students' work.**

1. 300 to 800
2. 2(20) + 0; 40

Thinking Critically

► Graph several points of the form (a, b) and (b, a). What do you observe? When are the graphs of (a, b) and (b, a) the same point? Write an equation that has no solution. Explain why it has no solution. **See Solutions Manual.**

Portfolio

► A portfolio is representative samples of your work, collected over a period of time. Begin your portfolio by selecting an item that shows something new that you learned in this chapter. **See students' work.**

Self Evaluation

► A good problem solver makes a plan before attacking a problem. The first step toward success is planning the route to get there.

See students' work.

Assess yourself. How do you start the process of solving a problem? Do you make a plan before you begin? Choose a problem from the chapter or a task in your life and make a plan for completing it. Then execute your plan and describe how the plan helped you as you worked.

Alternative Assessment

The Alternative Assessment section provides students with the opportunity to assess their own work by thinking critically, working with others, keeping a portfolio, and honestly evaluating their own progress. For more information on alternative forms of assessment, see *Alternative Assessment in the Mathematics Classroom*.

Performance Assessment

Performance Assessment tasks for this chapter are included in the *Assessment and Evaluations Masters*. A scoring guide is also provided.

Alternative Assessment

Additional Answer

4. There is less than 1 Calorie in 16 ounces of Diet Pepsi. No, since $6 > 1$ there could not be 6 Calories.

Assessment and Evaluation Masters, pp. 13, 25

NAME ____________ DATE ________

CHAPTER 1 Performance Assessment

Instructions: Demonstrate your knowledge by giving a clear, concise solution to each problem. Be sure to include all relevant drawings and justify your answers. You may show your solution in more than one way or investigate beyond the requirements of the problem.

1. State University requires that football fans join the Touchdown Club in order to be eligible to buy season football tickets. The annual fee for a family membership in the Touchdown Club is $150. It cost the Brewtons $480 for their club membership and tickets. How much did the tickets cost?

   Solve the problem in at least two different ways. For each solution, explain the strategy you used. Also, show how you can check each solution to make sure it is reasonable.

2. Ben is 3 years older than his sister Rachel.
   a. Make a table that shows five different ages for Rachel and the corresponding ages for Ben. Write an algebraic expression that represents Ben's age if the variable r is used to represent Rachel's age. Explain why r is called a variable.
   b. Suppose the variable b is used to represent Ben's age. Write a verbal expression for what the algebraic expression $b + 20$ might represent.
   c. Make up a problem based on Ben and Rachel's ages that can be solved by using an equation. Then write two or three sentences describing how to solve the problem.

3. Explain why the expressions $2 \cdot 4 + 6$ and $2(4 + 6)$ have different values even though they have the same numbers and operations.

4. Which statement is correct?
   a. $2(x + 1) = 2x + 1$
   b. $2(x + 1) = 2x + 2$

   Explain your reasoning by writing a paragraph complete with drawings, if appropriate.

CHAPTER 1 Scoring Guide
Performance Assessment

Level	Specific Criteria
3 Superior	• Shows thorough understanding of the concepts *variable, algebraic expression*, and *equation*. • Uses appropriate strategies to find the cost of the tickets. • Computations to solve problems and make table are correct. • Written explanations are exemplary. • Word problem concerning Ben and Rachel's ages is appropriate and makes sense. • Table is accurate. • Goes beyond requirements of some or all problems.
2 Satisfactory, with minor flaws	• Shows understanding of the concepts *variable, algebraic expression*, and *equation*. • Uses appropriate strategies to find the cost of the tickets. • Computations to solve problems and make table are mostly correct. • Written explanations are effective. • Word problem concerning Ben and Rachel's ages is appropriate and makes sense. • Table is mostly accurate. • Satisfies all requirements of problems.
1 Nearly Satisfactory, with serious flaws	• Shows understanding of most of the concepts *variable, algebraic expression*, and *equation*. • May not use appropriate strategies to find the cost of the tickets. • Computations to solve problems and make table are mostly correct. • Written explanations are satisfactory. • Word problem concerning Ben and Rachel's ages is appropriate and makes sense. • Table is mostly accurate. • Satisfies most requirements of problems.
0 Unsatisfactory	• Shows little or no understanding of the concepts *variable, algebraic expression*, and *equation*. • May not use appropriate strategies to find the cost of the tickets. • Computations to solve problems and make table are incorrect. • Written explanations are not satisfactory. • Word problem concerning Ben and Rachel's ages is not appropriate or sensible. • Table is not accurate. • Does not satisfy requirements of problems.

Investigation

TEACHER NOTES

NCTM Standards: 1-4

This Investigation is designed to be completed over several days or weeks. It may be considered optional. You may want to assign the Investigation and the follow-up activities to be completed at the same time.

Objective

- Prepare to be a foreign exchange student by completing appropriate research and making cost estimates.

Mathematical Overview

This Investigation will use the following skills and concepts from Chapters 2 and 3.

- estimate with whole numbers and decimals
- use coordinate systems to read maps
- read tables and graphs about environmental data
- estimate equivalent Fahrenheit and Celsius temperatures
- multiply decimals to compare foreign currency

Recommended Time		
Part	Pages	Time
Investigation	62–64	1 class period
Working on the Investigation	70, 97, 128	20 minutes each
Closing the Investigation	160	1 class period

Instructional Resources

Investigations and Projects Masters, pp. 5-8

A recording sheet, teacher notes, and scoring guide are provided for each Investigation in the *Investigation and Projects Masters.*

1 MOTIVATION

In this Investigation, students use research to find out about being an exchange student in a foreign country. Ask if any students have lived in or visited another country. What did they find the same or different there?

LONG-TERM PROJECT

Investigation

BON VOYAGE

You and another member of your class have been selected to participate in a student foreign exchange program. You won't be leaving the country until June, so there is plenty of time to prepare. The first thing you should do is obtain a passport.

MATERIALS

travel guides

reference books or magazines

To apply for a passport, you must:

- *complete an application in person before an official designated to accept passport applications. (A parent or legal guardian must complete the application for children less than 13 years old.)*
- *provide proof of U.S. citizenship (usually a previously-issued passport or a birth certificate).*
- *submit two identical photographs that are 2 inches by 2 inches in size. (The photographs should not be more than 6 months old and must be clear, front view, full face, with a white background.)*
- *pay $30. (The fee is $55 for passports issued to persons age 18 and older.)*

Work with your partner to prepare an estimate of your pre-flight expenses. Be sure to include costs associated with obtaining your passport. Be prepared to explain your list of pre-trip expenses to the class.

Starting the Investigation

Once passports have been obtained, another concern is what items to pack. There are many things to consider as you make your plans. Look over the following list of questions to consider.

Think About . . .

- *From what airport will you depart?*
- *How much luggage can you take?*
- *Are there restrictions on items you can take into the foreign country?*
- *Are there any restrictions on items you can bring or ship back?*
- *Will you have enough luggage space to pack purchases, gifts, and souvenirs?*
- *What are the costs and quality of the communication services (phone, mail) between your final destination and your home?*
- *What's the time difference between your final destination and your home?*

Brainstorm with your partner to choose a country of interest. Use this country as your destination as an exchange student. Be prepared to share the reasoning behind your choice with the class. Then work together to add to the list of questions to be considered. You may want to write to the U.S. Customs Service for more information.

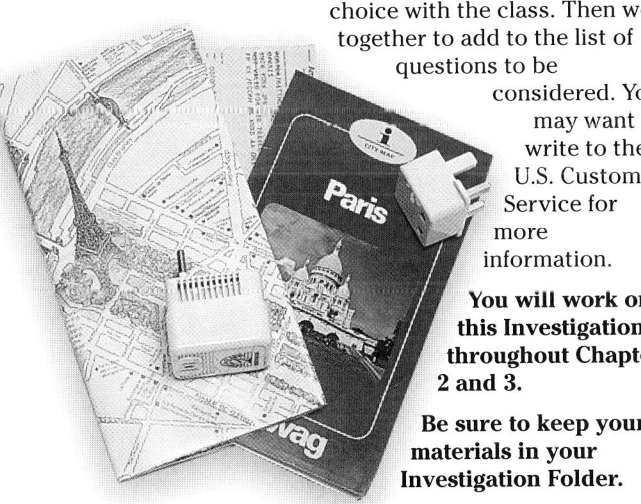

You will work on this Investigation throughout Chapters 2 and 3.

Be sure to keep your materials in your Investigation Folder.

For current information on being an exchange student, visit: **www.glencoe.com/sec/math/prealg/mathnet**

Investigation: Bon Voyage!

Working on the Investigation
Lesson 2-1, p. 70

Working on the Investigation
Lesson 2-6, p. 97

Working on the Investigation
Lesson 3-2, p. 128

Closing the Investigation
End of Chapter 3, p. 160

Cooperative Learning

This Investigation offers an excellent opportunity for using cooperative learning groups of two. For more information on cooperative learning strategies and group management, see *Cooperative Learning in the Mathematics Classroom,* one of the titles in the Glencoe Mathematics Professional Series.

2 SETUP

Have students work in pairs. Provide travel guides, reference books, and magazines about various countries around the world. You may also want to review sources of finding further materials, such as the local library or a travel agent. For further information, students may use the following resource.

Association for Childhood Education International through Council on Standards for International Educational Travel
3 Loudoun St., SE
Leesburg, VA 22075

3 MANAGEMENT

Each partner should be equally involved in this project. Partners should decide how tasks will be divided. A record needs to be made of all research completed and data collected. At the end of the activity, each partner should write a paragraph to evaluate the pair's performance as a team.

Sample Answers

Answers will vary as they are based on the research of each pair of students.

Investigations and Projects Masters, p. 8

NAME ______________ DATE ______

CHAPTERS 2 and 3 Investigation
Bon Voyage

Student Edition
Pages 62–63

Work with your partner to add to the lists of questions to be considered when planning your trip.

What country did you and your partner choose as your destination, and why?

Discuss travel plans in detail with your partner. List the questions you will want answered as you plan for your student foreign exchange program. Be sure to consider all travel plans, not just the flight. Think about passports, climate, travel time, food, and possible health requirements such as inoculations.

Where will you stay, and what will it cost?

Will you do any local traveling once you are there? List below the points of interest you will want to visit.

Please keep this page and any other research in your Investigation Folder.

CHAPTER 2

Exploring Integers

PREVIEWING THE CHAPTER

In this chapter, students explore integers through graphing, patterns, and manipulatives. In the beginning of the chapter, students will locate integers on a number line and will explore the absolute values of integers. Students will extend the number-line concepts to completing line plots for collected data. In the third lesson, students will compare and order integers. Students will first explore and model addition of integers and then will formally add integers. Then students explore, model, and formally subtract, multiply, and divide integers. Students also will explore number and other logical patterns.

Lesson (pages)	Lesson Objectives	NCTM Standards	State/Local Objectives
2-1 (66–70)	Graph integers on a number line. Find the absolute value of an expression.	1-5, 12, 14	
2-1B (71)	Organize data into line plots.	1-4,10, 12	
2-2 (72–76)	Graph points in all quadrants of the coordinate plane.	1-5, 8, 12	
2-2B (77)	Plot points on a graphing calculator.	1-5, 12	
2-3 (78–81)	Compare and order integers.	1-5, 12	
2-4A (82)	Use counters to model addition of integers.	1-5, 7, 12	
2-4 (83–87)	Add integers.	1-5, 7, 12	
2-5A (88)	Use counters to model subtraction of integers.	1-5, 7, 12	
2-5 (89–93)	Subtract integers.	1-5, 7, 12	
2-6 (94–97)	Solve problems by looking for a pattern.	1-5, 8	
2-7A (98)	Use counters to model multiplication of integers.	1-5, 7, 12	
2-7 (99–103)	Multiply integers.	1-5, 7, 12	
2-8 (104–108)	Divide integers.	1-5, 7, 12	

A complete, 1-page lesson plan is provided for each lesson in the ***Lesson Planning Guide***. Answer keys for each lesson are available in the ***Answer Key Masters***.

ORGANIZING THE CHAPTER

You may want to refer to the **Course Planning Calendar** on page T12 for information on pacing.
PACING: Standard—12 days; **Honors**—11 days; **Block**—7 days

LESSON PLANNING CHART

			Blackline Masters												
Lesson (Pages)	**Materials/ Manipulatives**	**Extra Practice (Student Edition)**	**Study Guide**	**Practice**	**Enrichment**	**Assessment and Evaluation**	**Math Lab and Modeling Math**	**Multicultural Activity**	**Tech Prep Applications**	**Graphing Calculator**	**Activity**	**Real-World Applications**	**Interactive Mathematics Tools Software**	**Teaching Transparencies**	**Group Activity Cards**
2-1 (66–70)	counters* tape, string	p. 743	p. 11	p. 11	p. 11						p. 16			2-1A 2-1B	2-1
2-1B (71)	rulor						p. 29								
2-2 (72–76)	Battleship™	p. 743	p. 12	p. 12	p. 12	p. 43					p. 30		2-2	2-2A 2-2B	2-2
2-2B (77)	graphing calculator									p.16					
2-3 (78–81)	graph paper	p. 744	p. 13	p. 13	p. 13				p. 3	p. 2				2-3A 2-3B	2-3
2-4A (82)	counters* integer mat*						p. 30						2-4A		
2-4 (83–87)	calculator	p. 744	p. 14	p. 14	p. 14	pp. 42, 43					p. 31			2-4A 2-4B	2-4
2-5A (88)	counters* integer mat*						p. 31						2-5A		
2-5 (89–93)	counters*, mat* masking tape	p. 744	p. 15	p. 15	p. 15			p. 3	p. 4			3		2-5A 2-5B	2-5
2-6 (94–97)		p. 744	p. 16	p. 16	p. 16	p. 44					p. 32			2-6A 2-6B	2-6
2-7A (98)	counters* integer mat*						p. 32						2-7A		
2-7 (99–103)		p. 745	p. 17	p. 17	p. 17			p. 4			p. 2			2-7A 2-7B	2-7
2-8 (104–108)	counter* integer mat*	p. 745	p. 18	p. 18	p. 18	p. 44	p. 77				p. 33	4		2-8A 2-8B	2-8
Study Guide/ Assessment (119–115)						pp. 29–41, 45–47									

*Included in Glencoe's *Student Manipulative Kit* and *Overhead Manipulative Resources*.

ORGANIZING THE CHAPTER

All of the blackline masters in the Teacher's Classroom Resources are available on the ***Electronic Teacher's Classroom Resources*** CD-ROM.

Other Chapter Resources

Student Edition
Chapter Opener, pp. 64–65
Investigation, Bon Voyage, pp. 62–63
Working on the Investigation, pp. 70, 97
Earth Watch, pp. 80–81

Teacher's Classroom Resources
Investigations and Projects Masters, pp. 33–36
Pre-Algebra Overhead Manipulative Resources, pp. 3–7

Technology
Test and Review Software (IBM & Macintosh)
CD-ROM Activities

Professional Publications
Block Scheduling Booklet
Glencoe Mathematics Professional Series

Outside Resources

Books/Periodicals
Activities for Junior High School and Middle School Mathematics, K. Easterday, L. Henry, and F. M. Simpson
Exploring Pre-Algebra and Algebra with Manipulatives, Don Balka
Math Motivators!! Investigations in Pre-Algebra, Alfred S. Posamentier and Gordon Sheridan
Using the Math Challenger Calculator, A Sourcebook for Teachers, Chapter 4

Videos/CD-ROMs
Algebra Math Video Series: Algebraic Terms and Operations, ETA
Basic Word Problems, Video ETA
Teaching Mathematics Effectively, ASCD

Software
Teasers by Tobbs with Integers, Sunburst

Assessment Resources

Student Edition
Math Journal, pp. 79, 85, 106
Mixed Review, pp. 70, 76, 81, 87, 93, 97, 103, 108
Self Test, p. 93
Chapter Highlights, p. 109
Chapter Study Guide and Assessment, pp. 110–112
Alternative Assessment, p. 113
Portfolio, p. 113
Ongoing Assessment, pp. 114–115
MindJogger Videoquiz, 2

Teacher's Wraparound Edition
5-Minute Check, pp. 66, 72, 78, 83, 89, 94, 99, 104
Checking Your Understanding, pp. 68, 74, 80, 86, 91, 95, 101, 106
Closing Activity, pp. 70, 76, 81, 87, 93, 97, 103, 108
Cooperative Learning, pp. 75, 100

Assessment and Evaluation Masters
Multiple-Choice Tests, Forms 1A (Honors), 1B (Average), 1C (Basic), pp. 29–34
Free-Response Tests, Forms 2A (Honors), 2B (Average), 2C (Basic), pp. 35–40
Performance Assessment, p. 41
Mid-Chapter Test, p. 42
Quizzes A–D, pp. 43–44
Standardized Test Practice, p. 45
Cumulative Review, pp. 46–47

ENHANCING THE CHAPTER

Examples of some of the materials for enhancing Chapter 2 are shown below.

DIVERSITY

Multicultural Activity Masters, pp. 3, 4

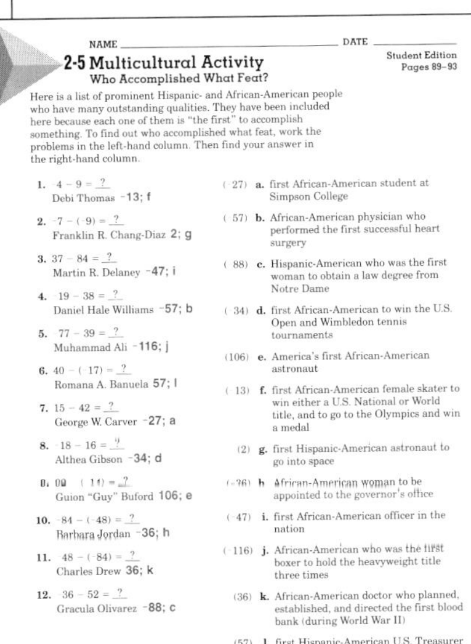

NAME ______________ DATE ________

2-5 Multicultural Activity
Who Accomplished What Feat?

Student Edition Pages 89–93

Here is a list of prominent Hispanic- and African-American people who have many outstanding qualities. They have been included here because each one of them is "the first" to accomplish something. To find out who accomplished what feat, work the problems in the left-hand column. Then find your answer in the right-hand column.

1. -4 − 9 = ? Debi Thomas **-13; f**
2. -7 − (-9) = ? Franklin R. Chang-Diaz **2; g**
3. 37 − 84 = ? Martin R. Delaney **-47; i**
4. -19 − 38 = ? Daniel Hale Williams **-57; b**
5. -77 − 39 = ? Muhammad Ali **-116; j**
6. 40 − (-17) = ? Romana A. Banuela **57; l**
7. 15 − 42 = ? George W. Carver **-27; a**
8. -18 − 16 = ? Althea Gibson **-34; d**
9. [illegible] = ? Guion "Guy" Buford **106; e**
10. -84 − (-48) = ? Barbara Jordan **-36; h**
11. -48 − (-84) = ? Charles Drew **36; k**
12. -36 − 52 = ? Gracula Olivarez **-88; c**

(-27) a. first African-American student at Simpson College
(-57) b. African-American physician who performed the first successful heart surgery
(-88) c. Hispanic-American who was the first woman to obtain a law degree from Notre Dame
(-34) d. first African-American to win the U.S. Open and Wimbledon tennis tournaments
(106) e. America's first African-American astronaut
(-13) f. first African-American female skater to win either a U.S. National or World title, and to go to the Olympics and win a medal
(2) g. first Hispanic-American astronaut to go into space
(-36) h. African-American woman to be appointed to the governor's office
(-47) i. first African-American officer in the nation
(-116) j. African-American who was the first boxer to hold the heavyweight title three times
(36) k. African-American doctor who planned, established, and directed the first blood bank (during World War II)
(57) l. first Hispanic-American U.S. Treasurer

APPLICATIONS

Real-World Applications, 3, 4

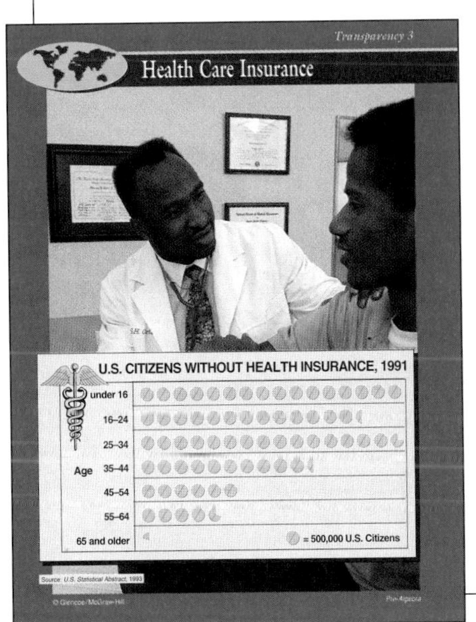

TECHNOLOGY

Graphing Calculator Masters, p. 2

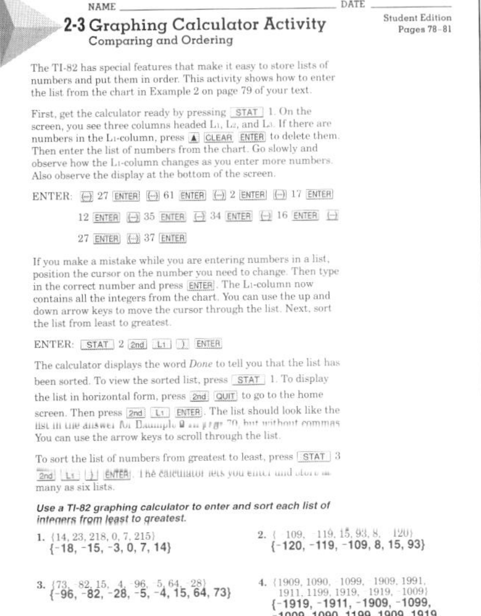

NAME ______________ DATE ________

2-3 Graphing Calculator Activity
Comparing and Ordering

Student Edition Pages 78–81

The TI-82 has special features that make it easy to store lists of numbers and put them in order. This activity shows how to enter the list from the chart in Example 2 on page 79 of your text.

First, get the calculator ready by pressing [STAT] 1. On the screen, you see three columns headed L_1, L_2, and L_3. If there are numbers in the L_1-column, press [▲] [CLEAR] [ENTER] to delete them. Then enter the list of numbers from the chart. Go slowly and observe how the L_1-column changes as you enter more numbers. Also observe the display at the bottom of the screen.

ENTER: [(-)] 27 [ENTER] [(-)] 61 [ENTER] [(-)] 2 [ENTER] [(-)] 17 [ENTER] 12 [ENTER] [(-)] 35 [ENTER] [(-)] 34 [ENTER] [(-)] 16 [ENTER] [(-)] 27 [ENTER] [(-)] 37 [ENTER]

If you make a mistake while you are entering numbers in a list, position the cursor on the number you need to change. Then type in the correct number and press [ENTER]. The L_1-column now contains all the integers from the chart. You can use the up and down arrow keys to move the cursor through the list. Next, sort the list from least to greatest.

ENTER: [STAT] 2 [2nd] [L1] [)] [ENTER]

The calculator displays the word *Done* to tell you that the list has been sorted. To view the sorted list, press [STAT] 1. To display the list in horizontal form, press [2nd] [QUIT] to go to the home screen. Then press [2nd] [L1] [ENTER]. The list should look like the list in the answer for Example [illegible] on page 79, but without commas. You can use the arrow keys to scroll through the list.

To sort the list of numbers from greatest to least, press [STAT] 3 [2nd] [L1] [)] [ENTER]. The calculator lets you enter and store as many as six lists.

Use a TI-82 graphing calculator to enter and sort each list of integers from least to greatest.

1. {14, 23, 218, 0, 7, 215} **{-18, -15, -3, 0, 7, 14}**
2. {-109, -119, 15, 93, 8, -120} **{-120, -119, -109, 8, 15, 93}**
3. {73, -82, 15, -4, -96, -5, 64, -28} **{-96, -82, -28, -5, -4, 15, 64, 73}**
4. {1909, 1090, -1099, -1909, 1991, -1911, 1199, 1919, -1919, -1009} **{-1919, -1911, -1909, -1099, -1009, 1090, 1199, 1909, 1919, 1991}**

TECH PREP

Tech Prep Applications Masters, pp. 3, 4

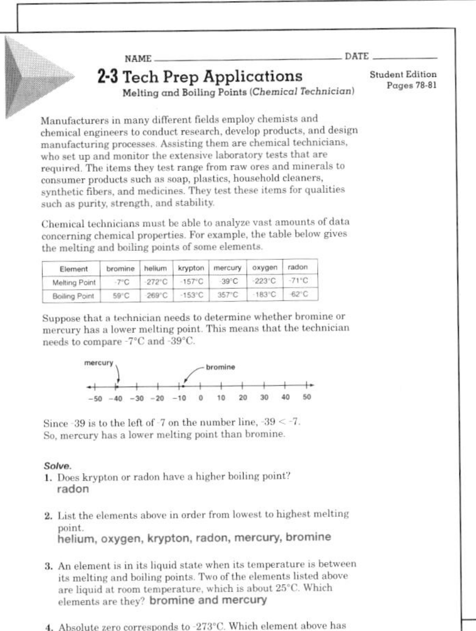

NAME ______________ DATE ________

2-3 Tech Prep Applications
Melting and Boiling Points (*Chemical Technician*)

Student Edition Pages 78-81

Manufacturers in many different fields employ chemists and chemical engineers to conduct research, develop products, and design manufacturing processes. Assisting them are chemical technicians, who set up and monitor the extensive laboratory tests that are required. The items they test range from raw ores and minerals to consumer products such as soap, plastics, household cleaners, synthetic fibers, and medicines. They test these items for qualities such as purity, strength, and stability.

Chemical technicians must be able to analyze vast amounts of data concerning chemical properties. For example, the table below gives the melting and boiling points of some elements.

Element	bromine	helium	krypton	mercury	oxygen	radon
Melting Point	-7°C	-272°C	-157°C	-39°C	-223°C	-71°C
Boiling Point	59°C	-269°C	-153°C	357°C	-183°C	-62°C

Suppose that a technician needs to determine whether bromine or mercury has a lower melting point. This means that the technician needs to compare -7°C and -39°C.

mercury, bromine
−50 −40 −30 −20 −10 0 10 20 30 40 50

Since -39 is to the left of -7 on the number line, -39 < -7. So, mercury has a lower melting point than bromine.

Solve.

1. Does krypton or radon have a higher boiling point? **radon**
2. List the elements above in order from lowest to highest melting point. **helium, oxygen, krypton, radon, mercury, bromine**
3. An element is in its liquid state when its temperature is between its melting and boiling points. Two of the elements listed above are liquid at room temperature, which is about 25°C. Which elements are they? **bromine and mercury**
4. Absolute zero corresponds to -273°C. Which element above has the boiling point closest to absolute zero? **helium**

CONNECTIONS

Activity Masters, pp. 2, 16, 30, 31, 32, 33

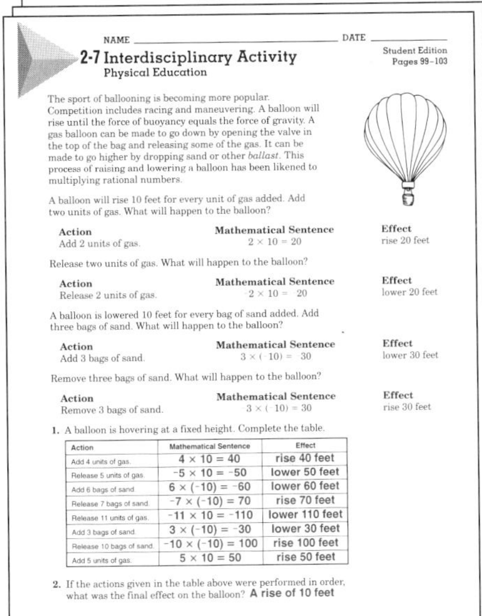

NAME ______________ DATE ________

2-7 Interdisciplinary Activity
Physical Education

Student Edition Pages 99–103

The sport of ballooning is becoming more popular. Competition includes racing and maneuvering. A balloon will rise until the force of buoyancy equals the force of gravity. A gas balloon can be made to go down by opening the valve in the top of the bag and releasing some of the gas. It can be made to go higher by dropping sand or other *ballast*. This process of raising and lowering a balloon has been likened to multiplying rational numbers.

A balloon will rise 10 feet for every unit of gas added. Add two units of gas. What will happen to the balloon?

Action	Mathematical Sentence	Effect
Add 2 units of gas.	2 × 10 = 20	rise 20 feet

Release two units of gas. What will happen to the balloon?

Action	Mathematical Sentence	Effect
Release 2 units of gas.	-2 × 10 = -20	lower 20 feet

A balloon is lowered 10 feet for every bag of sand added. Add three bags of sand. What will happen to the balloon?

Action	Mathematical Sentence	Effect
Add 3 bags of sand.	3 × (-10) = -30	lower 30 feet

Remove three bags of sand. What will happen to the balloon?

Action	Mathematical Sentence	Effect
Remove 3 bags of sand.	-3 × (-10) = 30	rise 30 feet

1. A balloon is hovering at a fixed height. Complete the table.

Action	Mathematical Sentence	Effect
Add 4 units of gas.	**4 × 10 = 40**	**rise 40 feet**
Release 5 units of gas.	**-5 × 10 = -50**	**lower 50 feet**
Add 6 bags of sand.	**6 × (-10) = -60**	**lower 60 feet**
Release 7 bags of sand.	**-7 × (-10) = 70**	**rise 70 feet**
Release 11 units of gas.	**-11 × 10 = -110**	**lower 110 feet**
Add 3 bags of sand.	**3 × (-10) = -30**	**lower 30 feet**
Release 10 bags of sand.	**-10 × (-10) = 100**	**rise 100 feet**
Add 5 units of gas.	**5 × 10 = 50**	**rise 50 feet**

2. If the actions given in the table above were performed in order, what was the final effect on the balloon? **A rise of 10 feet**

COOPERATIVE LEARNING

Math Lab and Modeling Math Masters, p. 77

NAME ______________ DATE ________

2-8 Cooperative Learning Activity
Negative Numbers in History

Student Edition Pages 104–108

Work in groups. Research the history of negative numbers. Find answers for the following exercises. Share the results of your group's study with the entire class. **Answers will vary. Sample answers are given.**

1. Before the 16th century in Europe, what was the "image" of negative numbers? Find some examples in history to support your answer. **A bad image; Europeans refused to accept negative numbers. Descartes said a negative solution of a problem was false.**
2. What happened around the 16th century that caused people to need to use negative numbers? **In the Renaissance, explorers in all fields became more creative, and new discoveries expanded the practical requirements of mathematics.**

After your research is completed, answer the following questions.

3. What kinds of practical problems have you met that required the use of negative numbers? **Answers will vary. Sample answers: Money problems involving debts and credits, temperature problems, measurements above and below sea level, weight gains and losses, football gains and losses of yards, card games with negative scores.**
4. A company is said to be operating "in the red" if it is spending more money than it is making. Find out where the expression "in the red" came from. **It comes from the practice of writing losses in company accounting records in red ink. It started in the Renaissance period, before the use of the minus sign, to indicate a negative number.**
5. A man died in A.D. 5 at the age of 51 years. In what year was he born? Can you use the rules for adding and subtracting integers to solve this problem? Explain. **47 B.C.; No, there is no zero year, so A.D. 1 comes immediately after 1 B.C.**
6. In telling time, how is the idea of a negative number often used? **To express time before the hour, such as "15 minutes before 3 o'clock."**

MAKING MATHEMATICS RELEVANT

This two-page introduction to the chapter provides students with an opportunity to explore other disciplinary topics and their applications to mathematics.

Background Information

Math and Recycling in the News
More than half of the waste produced in the U.S. is dumped into landfills, while only about one-fourth is recycled or composted. The percentage recycled or composted has been growing for several years. Although there appears to be more than enough open space for more landfills, in fact, their number has been decreasing. The government has imposed strict rules for landfill operation and it is becoming much more difficult, socially and politically, to find acceptable sites for new landfills.

CHAPTER 2 Exploring Integers

TOP STORIES in Chapter 2

In this chapter, you will:

- graph on number lines and coordinate planes,
- compare and order integers,
- add, subtract, multiply, and divide integers,
- solve problems looking for a pattern, and
- solve problems with integers.

MATH AND RECYCLING IN THE NEWS

Where does all my money go?!

Source: New York Times, May 22, 1994

It goes to the landfill. At least for now. For decades, worn out paper money has been shredded and sent to a landfill. But since landfill space is becoming scarce, the Federal Reserve is looking for ways to recycle its tired money. About 715 million bills weighing 7000 tons and having a face value of nearly $10 billion are shredded each year. The shredded bills are then pressed into bricks that weigh 2.2 pounds each. A number of companies have expressed interest in using the recycled money, including companies making roofing shingles, particle board, fuel pellets, stationery, packing material, and artwork. So where *does* all the money go? All over!

Putting It into Perspective

500 B.C.
Athens, Greece organizes the first municipal dump in the western world, 1 mile outside the city walls.

A.D. 1388
English Parliament bans waste disposal in waterways and ditches.

500 B.C. — A.D. 1400 — 1600

1690
The first paper mill in the U.S. opens near Philadelphia.

Japan creates 160,000 tons of household waste per day. Tokyo, one of the world's largest cities, has built artificial islands made of waste material in Tokyo Bay. However, this space is rapidly filling up. Some of the legislative debate centers around the issue of whether to do more recycling or to implement other programs aimed at reducing the amount of waste produced.

interNET CONNECTION For up-to-date information on printing money, visit:
www.glencoe.com/sec/math/prealg/mathnet

Statistical Snapshot

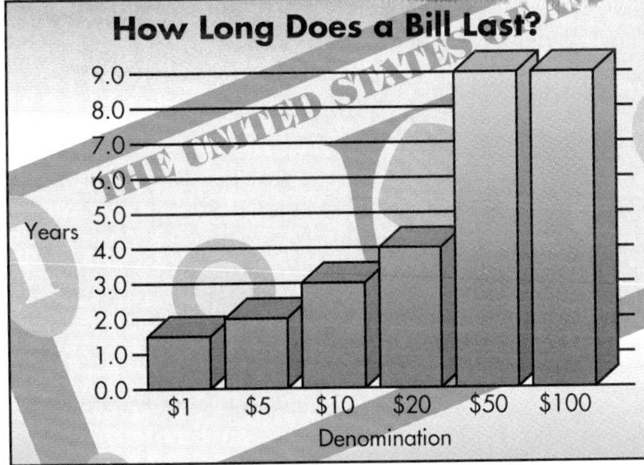

Source: Federal Reserve Bank of Los Angeles

On the Lighter Side

Frank & Ernest

I'M ENVIRONMENTALLY FRIENDLY – I DON'T MOP IT, SWEEP IT, DUST IT OR WAX IT.

1942–45
Americans collect paper, rubber, and tin to recycle for the war effort.

1994
U.S. Federal Reserve begins recycling money.

1900 | 1950 | 2000

1904
The first two major aluminum recycling plants open in Cleveland and Chicago.

1970
The U.S. Environmental Protection Agency is formed.

On the Lighter Side

Ask students if they have heard of things being called environmentally friendly. What does it mean to be environmentally friendly?

Statistical Snapshot

Have students take a survey to find the years on 100 random dollar bills still in circulation. Students can decide how to collect and display the data.

Cooperative Learning

Chapter Projects Two chapter projects are included in the *Investigations and Project Masters*. In Chapter Project A, students extend the topic in the Chapter Opener. In Chapter Project B, students explore tornadoes and tornado safety. A student page and a parent letter are provided for each Chapter Project.

interNET CONNECTION

Glencoe has made every effort to ensure that the website links for *Pre-Algebra* at www.glencoe.com/sec/math/prealg/mathnet are current and contain appropriate content. However, these website links are not under Glencoe's control.

Investigations and Projects Masters, p. 33

NAME ________________ DATE ________

CHAPTER 2 Project A
Recycling In Your Community

Have you ever noticed the sparkle in some roads and in some sidewalks? These surfaces sparkle because of the recycled porcelain from sinks, tubs, and toilets that is used in their construction. Porcelain is one of the many materials communities recycle. Other recyclable materials include glass, paper, aluminum, tin, plastic, batteries, tires, and motor oil. What else gets recycled in your community?

1. Find out what you can about recycling where you live. Which common materials get recycled? Which don't?
2. Next, choose a category of recycled goods, such as paper, glass, or aluminum. Form a group with others interested in that same category. Work cooperatively to further investigate the recycling process for that category of material.
3. For your material, find answers to questions like these.
   - Where is the recycling done?
   - What is done to the material, and how is it done?
   - How much is done in a day, a week, a year?
   - What uses does the processed material have?
   - Who pays for recycled materials?
   - How much do they pay for a ton of it?

   Call or visit a recycling site to get answers. For example, to research aluminum recycling, contact a scrap dealer to see how the metal arrives, how it is separated, repackaged, stored, priced, and then shipped.
4. How big is a ton? Figure out how much space is needed to hold a ton of recycled material. For instance, how many classrooms or closets would be needed to store a ton of paper?
5. Make graphs or tables to display information you've gathered. Draw pictures or take photos of what you've seen and prepare descriptive labels for each.
6. Contribute your written descriptions, graphs, tables, and photos to a class recycling bulletin board. Make a brief presentation to accompany your contribution to the bulletin board.

2-1 LESSON NOTES

NCTM Standards: 1-5, 12, 14

Instructional Resources

- Study Guide Master 2-1
- Practice Master 2-1
- Enrichment Master 2-1
- Group Activity Card 2-1
- Activity Masters, p. 16

Transparency 2-1A contains the 5-Minute Check for this lesson; **Transparency 2-1B** contains a teaching aid for this lesson.

Recommended Pacing	
Standard Pacing	Day 1 of 12
Honors Pacing	Day 1 of 11
Block Scheduling*	Day 1 of 7 (along with Lesson 2-2)

*For more information on pacing and possible lesson plans, refer to the ***Block Scheduling Booklet***.

1 FOCUS

5-Minute Check
(over Chapter 1)

Find the value of each expression.

1. $3 \times 5 - 4 \times 2$ **7**
2. $19 + 3(1 + 5)$ **37**
3. $48 \div 6 \times 5$ **40**

Simplify each expression.

4. $(x - 7) + 3x$ **$4x - 7$**
5. $2(y - 3) + 3(2y + 2)$ **$8y$**

CULTURAL CONNECTIONS
The Dead Sea is actually a salt lake. More than half of the sea's water evaporates each year, which causes a thick mist to hover over the lake. The Dead Sea is 25% salt, or 7 times as salty as the ocean, making it easy for bathers to float. The salinity prevents most animal and vegetable life except for bacteria and brine shrimp.

2-1 Integers and Absolute Value

Setting Goals: *In this lesson, you'll graph integers on a number line and find absolute value.*

Modeling a Real-World Application: Geography

CULTURAL CONNECTIONS
For centuries, people have traveled miles to bathe in the Dead Sea. They believed that the minerals in the water were healthful. Several resorts are still in operation on the shores today.

The shores of the Dead Sea, between Israel and Jordan, are 1312 feet *below* sea level. If you use the number 1312 to represent 1312 feet *above* sea level, what number could be used to represent 1312 feet below sea level?

Learning the Concept

Study the pattern of the following subtraction sentences.

$$5 - 1 = 4$$
$$5 - 2 = 3$$
$$5 - 3 = 2$$
$$5 - 4 = 1$$
$$5 - 5 = 0$$
$$5 - 6 = \underline{\ ?\ }$$

From the pattern, it appears that $5 - 6$ has an answer less than 0. Study the diagram below.

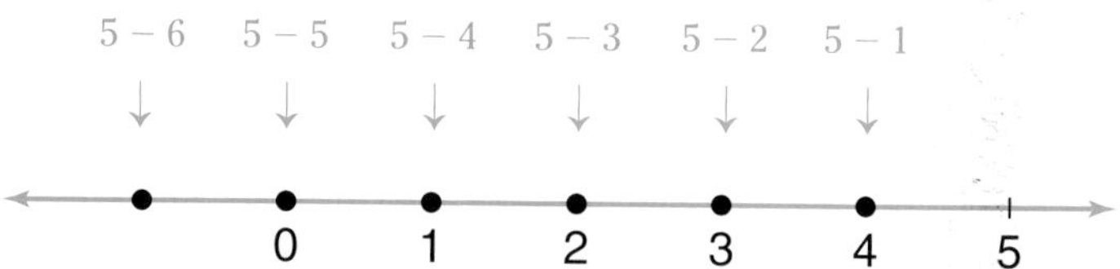

The answer for $5 - 6$ should be *one less than zero*. The number *one less than zero* is written as -1. So $5 - 6 = -1$. This is an example of a **negative** number. A negative number is a number less than zero.

You can use a negative number to write 1312 feet below sea level, the elevation of the shores of the Dead Sea. With sea level as the starting point or 0, you can express the elevation as $0 - 1312$, or -1312. So, the elevation of the Dead Sea is -1312 feet.

Negative numbers, like -1312, are members of the set of **integers**. Integers can also be represented as points on a number line.

Alternative Teaching Strategies

Student Diversity Use a local or state map to locate the town where students live. Locate a town about 25 miles to the east of where students live. Locate another town about 25 miles to the west of where students live.

a. If someone asked you how far away the first town is, would you reply 25 miles? **yes**

b. If someone asked you how far away the second town is, would you reply 25 miles? **yes**

c. How do you distinguish between the location of the two towns? **You need a second locator, such as east or west.**

Numbers to the left of zero are less than zero.

negative positive

-6 -5 -4 -3 -2 -1 0 1 2 3 4 5 6

Numbers to the right of zero are greater than zero.

The numbers −1, −2, −3, . . . are called negative integers. The number negative 3 is written −3.

Zero is neither negative nor positive.

The numbers 1, 2, 3, . . . are called positive integers. The number positive 4 is written +4 or 4.

THINK ABOUT IT

Which integer is neither positive nor negative? **zero**

This set can be written {. . ., −3, −2, −1, 0, 1, 2, 3, . . .} where . . . means continues indefinitely.

To graph a particular set of integers, locate the integer points on a number line. The number that corresponds to a point on the number line is called the **coordinate** of the point.

Example 

a. Name the coordinates of *D*, *E*, and *B*.

The coordinate of *D* is 4, *E* is −3, and B is 2.

b. Graph points *F*, *U*, and *N* on a number line if *F* has coordinate 1, *U* has coordinate −3, and *N* has coordinate 4.

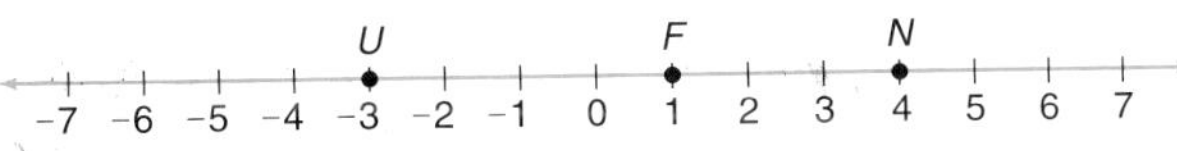

To graph a number, find its location on the number line and draw a dot. Write the letter above the dot.

Looking at the number line shown below, you can see that 4 and −4 are on opposite sides of the starting point, zero. However, they are the same distance from zero. In mathematics, we say that they have the same **absolute value**, 4.

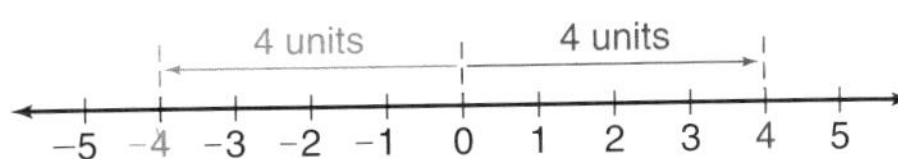

The symbol for the absolute value of a number is two vertical bars on either side of the number. $|-4| = 4$ is read *The absolute value of −4 is 4.*

Absolute Value		
	In words:	**The absolute value of a number is the distance the number is from the zero point on the number line.**
	In symbols:	$|4| = 4$ **and** $|-4| = 4$

Motivating the Lesson

Situational Problem The temperature is 2°F and then goes down 5°. What is the new temperature? Ask students what is different about this problem. **The answer is a negative number.** Have students brainstorm other applications of negative numbers.

2 TEACH

In-Class Example

For Example 1

a. Name the coordinates of *A*, *D*, and *E*. **A, −2; D, 3; E, 5**

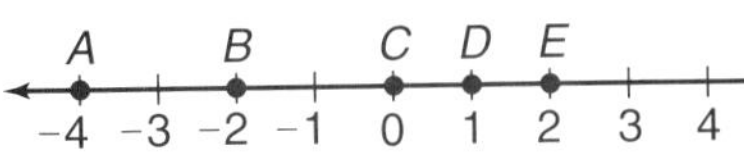

b. Graph points *H*, *Q*, and *T* on a number line, if *H* has coordinate 2, *Q* has coordinate −4, and *T* has coordinate −1.

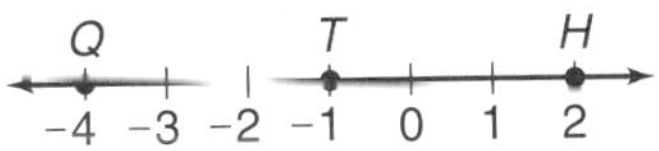

Teaching Tip Ask students what they think the opposite of the absolute value of 7 is. How would they write it? $-|7| = -7$

Classroom Vignettes

"I use Wilbur the Worm to help with introducing negative numbers, as well as adding and subtracting integers. Ground level is 0. When Wilbur digs down in the ground, the number is negative. When he climbs up a pile of dirt, the number is positive."

Stacey Bondon
Amos Alonzo Stagg High School
Palos Hills, IL

In-Class Examples

For Example 2
Simplify.

a. $|-7| + |-5|$ 12

b. $|-6| - |-4|$ 2

For Example 3
Evaluate the expression $|a| \times |b| - 12$, if $a = -5$ and $b = 8$. 28

3 PRACTICE/APPLY

Checking Your Understanding

Exercises 1–16 are designed to help you assess your students' understanding through reading, writing, speaking, and modeling. You should work through Exercises 1–4 with your students and then monitor their work on Exercises 5–17.

Additional Answers

1. Sample answer: Draw a number line. Locate -7. Draw a dot there.
2. Sample answer: temperatures below zero, loss of money, below par scores in golf
4.

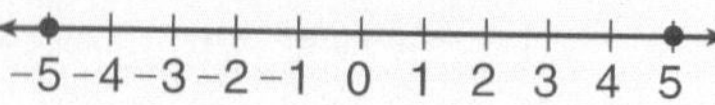

Study Guide Masters, p. 11

NAME ______________________ DATE __________

2-1 Study Guide
Integers and Absolute Value

Student Edition Pages 66–70

Situations that involve growth or increase are usually represented by positive integers.

Situations Represented by Positive Integers
Profit of \$50 > +50
Deposit of \$400 > +400
Increase of 20 > +20

Situations that involve decline or decrease are usually represented by negative integers.

Situations Represented by Negative Integers
Loss of \$30 > −30
Withdrawal of \$250 > −250
Decrease of 40 > −40

Write the integer that describes the situation.

1. loss of 8 yards −8
2. 4° rise in temperature +4
3. 50-foot drop in altitude −50
4. debt of \$500 −500
5. deposit of \$70 +70
6. gain of 10 pounds +10

Graph each set of numbers on the number line provided.

7. {−4, −1, 3} -5 -4 -3 -2 -1 0 1 2 3 4 5
8. {−2, 0, 5} -5 -4 -3 -2 -1 0 1 2 3 4 5

Write the absolute value of each integer.

9. −3 3
10. 14 14
11. 20 20
12. −5 5

Write the two integers that have the given absolute value.

13. 6 −6, 6
14. 1 −1, 1
15. 15 −15, 15
16. 8 −8, 8
17. 3 −3, 3
18. 12 −12, 12
19. 30 −30, 30
20. 21 −21, 21

Simplify.

21. $|4| - |-2|$ 2
22. $|-8| + |-3|$ 11
23. $|-15| - |6|$ 9
24. $|-7| \cdot |-11|$ 77
25. $|12| \cdot |-4|$ 48
26. $|-36| \div |-9|$ 4

Example 2 **Simplify.**

a. $|9| + |-9|$

$|9| + |-9| = 9 + 9$ *The absolute value of −9 is 9.*
$= 18$ *The absolute value of 9 is 9.*

b. $|13| - |-2|$

$|13| - |-2| = 13 - 2$ *The absolute value of 13 is 13.*
$= 11$ *The absolute value of −2 is 2.*

Connection to Algebra

Since variables represent numbers, you can use absolute value notation with algebraic expressions involving variables.

Example  **Evaluate the expression $|x| - 7$ if $x = -13$.**

$|x| - 7 = |-13| - 7$ *Replace x with −13.*
$= 13 - 7$ *The absolute value of −13 is 13.*
$= 6$

Checking Your Understanding

Communicating Mathematics

Read and study the lesson to answer these questions.

1. **Explain** how you would graph −7. **See margin.**
2. **Describe** some situations in the real world where negative numbers are used. **See margin.**
3. **Identify** the integers graphed on the number line at the right. **−3, −1, 1, 3**
   -4 -3 -2 -1 0 1 2 3 4
4. **Graph** two points on a number line so that the coordinates of both points have an absolute value of 5. **See margin.**
5. You can use counters to model integers. Each yellow counter represents a positive unit, and each red counter represents a negative unit. **a–b. See margin.**

MODELING MATHEMATICS

MATERIALS
counters
integer mat

Place three yellow counters on the mat to represent +3.

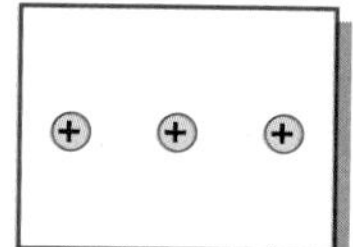

Place eight red counters on the mat to represent −8.

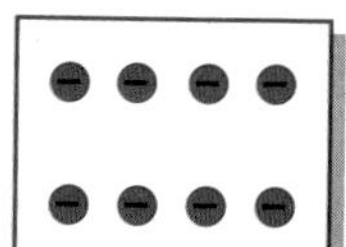

a. Model −4 using counters. **b.** Model +5 using counters.

Guided Practice

6. Name the coordinate of each point graphed on the number line.

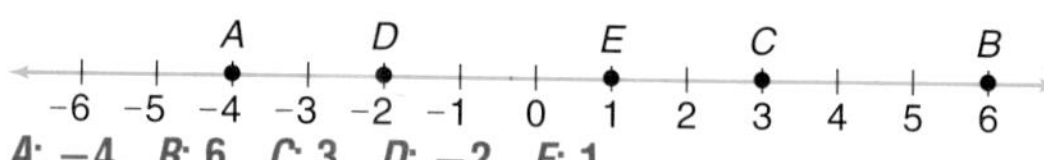

A: −4, B: 6, C: 3, D: −2, E: 1

Reteaching

Using Practice Partners Have students work in pairs. One student names an integer between −10 and 10. The other student locates that number on the number line and states its absolute value. Then students switch roles and repeat the procedure.

Additional Answers

5a.

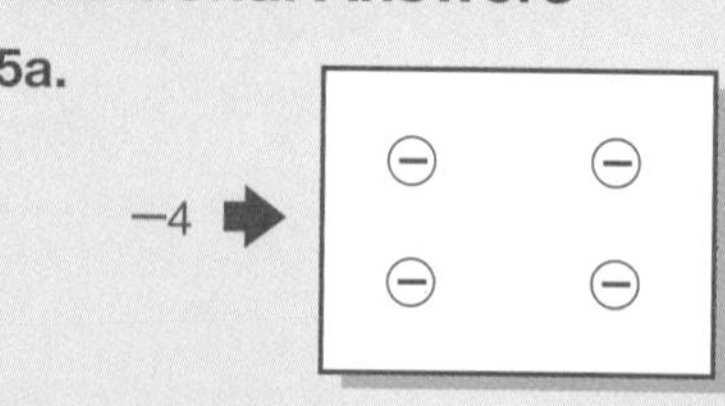

5b.

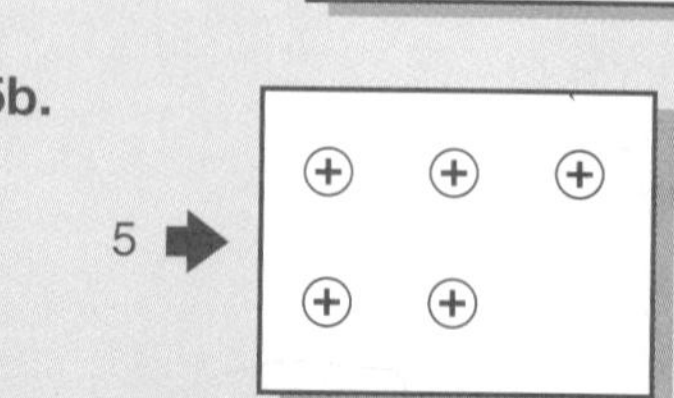

Write an integer for each situation. Then graph on a number line.

7. 5°F below zero -5
8. a 12-pound loss -12
9. 13 yards gained $+13$
10. a bank deposit of \$1200 $+1200$

7–10. See Solutions Manual for graphs.

Simplify.

11. $|-5|$ 5
12. $|8|$ 8
13. $|13| - |-3|$ 10
14. $|15| + |-8|$ 23

Evaluate each expression if $d = -5$ and $r = 3$.

15. $7 + |d|$ 12
16. $|d| - r$ 2
17. **Scuba Diving** Two divers are descending to a coral reef off the coast of Madagascar. One diver is 15 meters below the water's surface. The other diver is 9 meters lower.
    a. Represent the depths of the divers as integers. $-15, -24$
    b. Graph each point on a number line. See margin.

Exercises: Practicing and Applying the Concept

Independent Practice

Graph each set of numbers on a number line. See margin.

A
18. $\{0, -1, 3\}$
19. $\{-2, -5, 4\}$
20. $\{-4, 1, 3, 8\}$
21. $\{-3, -5, -7, -9\}$
22. Name the coordinates of each point graphed on the number line.

B D C A
-6 -4 -2 0 2 4 6 8

A: 5, B: −4, C: 2, D: −1

Write an integer for each situation.

23. a salary increase of \$400 $+400$
24. a bank withdrawal of \$50 -50
25. 3 seconds before liftoff -3
26. a gain of 9 yards $+9$

B
27. 5° above zero $+5$
28. a loss of 6 pounds -6

Simplify. 36. 27 40. −25

29. $|-11|$ 11
30. $-|24|$ -24
31. $|0|$ 0
32. $|7|$ 7
33. $|-5| + |3|$ 8
34. $|-8| + |-10|$ 18
35. $|14| - |-5|$ 9
36. $|-15| + |-12|$
37. $|0 + 12|$ 12
38. $|16 - 2|$ 14
39. $-|-36|$ -36
40. $-||-4| + |-21||$

Evaluate each expression if $a = 0$, $b = 2$, and $c = -4$.

41. $|c| - 2$ 2
42. $12 - |b|$ 10
43. $b + a + |c|$ 6

44. $ac + |-30|$ 30
45. $|c| - b$ 2
46. $|b| \cdot |c| + |a|$ 8

Critical Thinking

Find the next three numbers in each pattern. 48. −8, −3, 2

47. 32, 24, 16, _?_, _?_, _?_ 8, 0, −8
48. −23, −18, −13, _?_, _?_, _?_
49. **Logical Reasoning** Name a negative number that is not an integer. Sample answer: −1.5

Applications and Problem Solving

50. **Game Shows** On *Jeopardy!*, contestants earn points for each correct response and lose points for each incorrect response. Questions are valued at 100, 200, 300, 400, 500, 600, 800, or 1000 points. Describe a situation where a contestant would have a score of −300. See margin.

Error Analysis

Some students may write the absolute value as the opposite of the number. Stress the definition of absolute value and opposites.

Assignment Guide

Core: 19–51 odd, 52–60
Enriched: 18–46 even, 47–60

For **Extra Practice**, see p. 743.

The red A, B, and C flags, printed only in the Teacher's Wraparound Edition, indicate the level of difficulty of the exercises.

Additional Answers

17b.

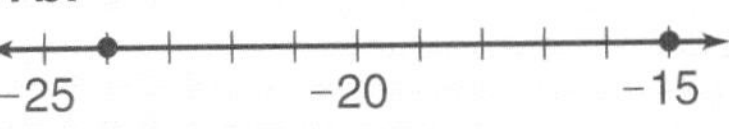

18.

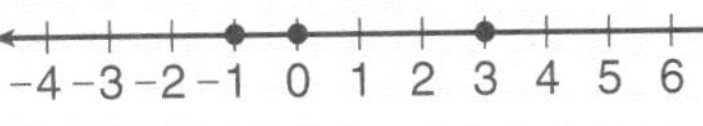

19.

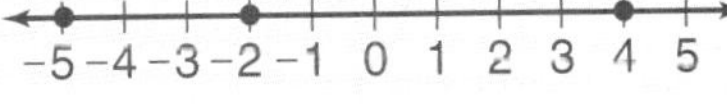

20.

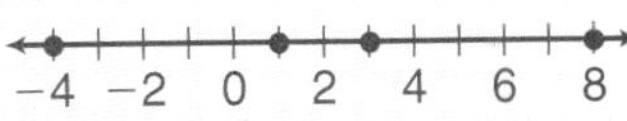

21.

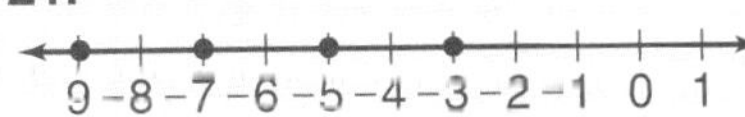

Practice Masters, p. 11

NAME ______ DATE ______

2-1 Practice
Integers and Absolute Value

Student Edition Pages 69–70

Graph each set of numbers on the number line provided.

1. $\{0, 1, 5\}$
2. $\{-1, 0, 3\}$
3. $\{-4, -2, 2\}$
4. $\{-3, 0, 4\}$

Write an integer for each situation.

5. a gain of 8 pounds $+8$
6. 21° below zero -21
7. a loss of three yards -3
8. a bank deposit of \$120 $+120$
9. 10 meters below sea level -10
10. a loss of \$10 -10

Simplify.

11. $|1|$ 1
12. $|-10|$ 10
13. $|-8|$ 8
14. $|10|$ 10
15. $|4| + |-4|$ 8
16. $|9| - |-5|$ 4
17. $0 + |-1|$ 1
18. $|-6| + |-5|$ 11
19. $|-8| - |-8|$ 0
20. $|12| + |-3|$ 15
21. $|-15| - |6|$ 9
22. $|-13| + |-7|$ 20

Evaluate each expression if $a = -3$, $b = 0$, and $c = 1$.

23. $|a| - |c|$ 2
24. $2|c| + |a|$ 5
25. $|a|c + |-12|$ 15
26. $3a - |b|$ −9
27. $|a| \cdot |c| + |b|$ 3
28. $|14| - |a|$ 11

Group Activity Card 2-1

Birthday Number Line Group Activity **2-1**

Draw a calendar line like the one shown below. Make the space between the marks for the months about two centimeters. Notice that January is the zero point on this line and July through December are represented by negative integers. February through June are represented by positive integers.

Survey your classmates to find out the month in which each of them was born. When you have done this, record your data on the calendar line by putting an x above the month for each student. If two or more students have a birthday falling in the same month, line up the additional x's above the first one in a column. Then answer the questions on the back based on the information from your calendar line.

July August September October November December January February March April May June
-6 -5 -4 -3 -2 -1 0 1 2 3 4 5

 Pre-Algebra

Additional Answer

50. Sample answer: The contestant had 500 points and answered two 400-point questions incorrectly.

4 ASSESS

Closing Activity

Act It Out Make a large walk-on number line with tape. Anchor a piece of string or rope to zero. Have a student walk to -5 and put a knot in the string to show the location of -5. Now have a student walk to 5 and check whether the distance from 0 to -5 is the same as from 0 to 5.

Additional Answers

51.

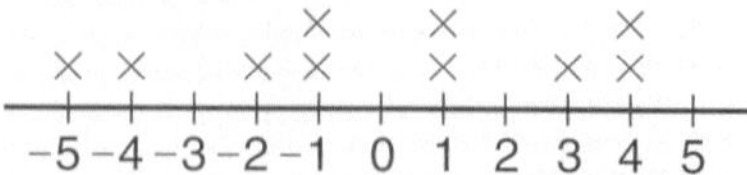

55.

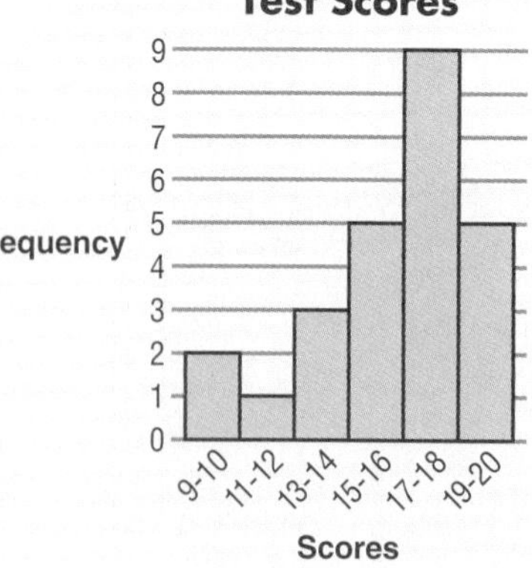

Enrichment Masters, p. 11

NAME ______________ DATE ______

2-1 Enrichment Student Edition Pages 66–70
Integers on the Number Line

We can graph integers on the number line, describe them in words, or describe them in mathematical sentences. Sometimes it is important to be able to think of them in several ways at the same time. The graphs can help us "check" if our problem solutions make sense. The mathematical sentences can let us express answers in a form of shorthand. The word description can help us use integers in a variety of problem situations.

In this activity, you will be completing the table below. Each column shows different ways of thinking about the same integers. You need to find ways to fill in the blanks. You can use the completed ones to help you figure out the ones that are missing.

	Integers	Open Sentence	Word Description	Number Line Graph
	-2, 2	$\lvert x \rvert = 2$	Integers 2 units from zero	-4 -2 0 2 4
1.	-3, -2	$-4 < x < -1$	Integers between -4 and -1	-4 -2 0 2 4
2.	**-2, -1, 0, 1, 2**	$\lvert x \rvert < 3$	Integers less than 3 units from zero	-4 -2 0 2 4
3.	-1, 1	**$\lvert x \rvert = 1$**	Integers 1 unit from zero	-4 -2 0 2 4
4.	1, 2, 3	$0 < x < 4$	**Integers between 0 and 4**	-4 -2 0 2 4
5.	**-3, 3**	**$\lvert x \rvert = 3$**	Integers 3 units from zero	-4 -2 0 2 4
6.	**0, 1**	**$-1 < x < 2$**	**Integers between -1 and 2**	-4 -2 0 2 4
7.	**-1, 3**	$\lvert x - 1 \rvert = 2$	Integers 2 units from 1	-4 -2 0 2 4
8.	-2, 4	$\lvert x - 1 \rvert = 3$	**Integers 3 units from 1**	-4 -2 0 2 4
9.	**0, 4**	$\lvert x - 2 \rvert = 2$	**Integers 2 units from 2**	-4 -2 0 2 4

51. See margin.
52. 25
53. 17–18
54. 11–12
55. See margin.

51. **Sports** Golf scores are reported in reference to *par*. A score of +2 means two strokes over par. The leaders in the 1995 U.S. Open in Southampton, New York, had the scores −2, 3, 4, 1, 4, −1, −1, −5, −4, and 1 in the final round. Graph the scores on a number line.

Mixed Review

The table shows test scores for a physical science class. Use the table to answer each question. (Lesson 1-10)

Score	Test Score Tally	Frequency
19–20	𝍸	5
17–18	𝍸 IIII	9
15–16	𝍸	5
13–14	III	3
11–12	I	1
9–10	II	2

52. How many students are in the class?
53. What scores occurred most often?
54. What scores occurred least often?
55. Make a bar graph of the data.
56. Solve $16 = z - 25$. (Lesson 1-8) **41**
57. **Geometry** The area of a rectangle is given by $A = \ell w$, where ℓ is the length and w is the width. Find the area of a rectangular garden if its length is 15 feet and its width is 12 feet. (Lesson 1-6) **180 square feet**

58. Simplify $5a + 3(7 + 2a)$. (Lesson 1-5) **$11a + 21$**
59. Write an algebraic expression for the phrase *three times a number decreased by 8.* (Lesson 1-3) **$3x - 8$**
60. **Consumer Awareness** The Columbus Association for the Performing Arts runs a summer movie series in the restored Ohio Theater. Tickets in the 1995 season were $3.00 for adults and $2.50 for senior citizens. (Lesson 1-2)
    a. Write an expression that would help you find the total cost of fifteen adult tickets and eight senior citizen tickets. **15 × $3.00 + 8 × $2.50**
    b. Then find the total cost. **$65.00**

WORKING ON THE
Investigation
BON VOYAGE

Refer to the Investigation on pages 62–63.

Materials: map or globe

Lines of longitude indicate degrees east or west of the prime meridian. The prime meridian is at 0°. The prime meridian is a semicircle passing through Greenwich, England.

A city's longitude determines its time zone. The zone labeled zero is centered at the prime meridian.

- How does the way the zones are numbered differ from the integer number line?

$37\frac{1}{2}$W $22\frac{1}{2}$W $7\frac{1}{2}$W $7\frac{1}{2}$E $22\frac{1}{2}$E $37\frac{1}{2}$E
Zone 3 | Zone 2 | Zone 1 | Zone 0 | Zone −1 | Zone −2 | Zone −3
30W 15W 0 15E 30E

- Compare your time zone to that of the country you chose at the beginning of this Investigation. How would you use this information when traveling abroad?
- Work with the members of your group to develop reasons why time zones exist.

Add the results of your work to your Investigation Folder.

Extension

Using Lists Have students divide a sheet of paper into two columns. In one column, have students list sentences indicating integers, such as "Mary put $4 in the bank." In the second column, have students write a sentence that is the opposite of the first. For example, "Mary took $4 out of the bank."

The Investigation on pages 62 and 63 is designed to be a long-term project that is completed over several days or weeks. Encourage students to keep their materials in their Investigation Folder as they work on the Investigation.

HANDS-ON ACTIVITY

2-1B Statistical Line Plots

An Extension of Lesson **2-1**

MATERIALS

ruler

In the previous lesson, you saw how number lines helped to visualize integers. In statistics, data is also organized and presented on a number line. A picture of information on a number line is called a **line plot**.

Population figures from the census are used to determine the number of House of Representative members from each state of the United States.

Work with a partner.

- This chart shows the states that had a change in the number of House members after the 1990 census. Determine the change for each state. Express a decrease as a negative integer and an increase as a positive integer.

State	1980	1990	Change
Arizona	5	6	+1
California	45	52	+7
Florida	19	23	+4
Georgia	10	11	+1
Illinois	22	20	−2
Iowa	6	5	−1
Kansas	5	4	−1
Kentucky	7	6	−1
Louisiana	8	7	−1
Massachusetts	11	10	−1
Michigan	18	16	−2
Montana	2	1	−1
New Jersey	14	13	−1
New York	34	31	−3
North Carolina	11	12	+1
Ohio	21	19	−2
Pennsylvania	23	21	−2
Texas	27	30	+3
Virginia	10	11	+1
Washington	8	9	+1
West Virginia	4	3	−1

FYI

Patsy Mink, a Democrat from Hawaii, became the first Japanese American woman to serve in the U.S. Congress on January 4, 1965.

New York

ornia

- Draw a number line. Since the largest and smallest changes were 7 and −3 for California and New York, respectively, use a scale of −4 to +8 and intervals of 1.

-4 -3 -2 -1 0 1 2 3 4 5 6 7 8

- To make the line plot, put an "x" above the number that represents the change in House members for each state. The graph below shows the first five values from the table. **See Solutions Manual for completed graph.**

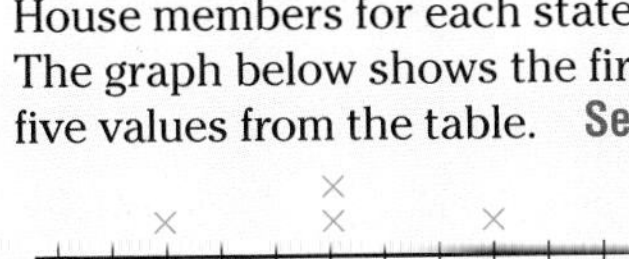

TALK ABOUT IT

1. What number occurs most frequently? **−1**
2. A number that is far apart from the rest of the data is called an **outlier**. Identify any numbers that appear to be outliers.
3. Data that are grouped closely together are called a **cluster**. Identify any clusters.
4. The *net change* is the change in the total number of House members. What was the net change in the number of House members? **no net change**
5. What general geographic patterns do you notice from the information in the table? In which areas of the country are states gaining or losing population? **See Solutions Manual.**

Extension

2. Sample answer: 7
3. Sample answer: There seems to be a cluster between −2 and 1.

2-1B LESSON NOTES

NCTM Standards: 1-4, 10, 12

Objective

Organize data into line plots.

Recommended Time

Demonstration and discussion: 15 minutes; Exercises: 30 minutes

Instructional Resources

For each student or group of students
Student Manipulative Kit
- ruler

Math Lab and Modeling Math Masters
- p. 12 (number lines)
- p. 29 (worksheet)

For teacher demonstration
Overhead Manipulative Resources

1 FOCUS

Motivating the Lesson

Ask students whether they think they learn better by seeing or hearing information. Explain that this lesson shows one way to present a table of information so that a visual display helps some people interpret the data quicker and easier.

2 TEACH

Teaching Tip Remind students to stack and space their "xs" evenly so that the display will be an accurate visual indicator of how much greater one number is than another.

3 PRACTICE/APPLY

Assignment Guide
Core: 1–5
Enriched: 1–5

4 ASSESS

Observing students working in cooperative groups is an excellent method of assessment.

Cooperative Learning

This lesson offers an excellent opportunity for using cooperative learning groups. For more information on cooperative learning strategies and group management, see *Cooperative Learning in the Mathematics Classroom*, one of the titles in the Glencoe Mathematics Professional Series.

2-2 LESSON NOTES

NCTM Standards: 1-5, 8, 12

Instructional Resources

- Study Guide Master 2-2
- Practice Master 2-2
- Enrichment Master 2-2
- Group Activity Card 2-2
- Assessment and Evaluation Masters, p. 43
- Activity Masters, p. 30

Transparency 2-2A contains the 5-Minute Check for this lesson; **Transparency 2-2B** contains a teaching aid for this lesson.

Recommended Pacing	
Standard Pacing	Day 2 of 12
Honors Pacing	Day 2 of 11
Block Scheduling*	Day 1 of 7 (along with Lesson 2-1)

*For more information on pacing and possible lesson plans, refer to the ***Block Scheduling Booklet***.

1 FOCUS

5-Minute Check
(over Lesson 2-1)

Graph each set on a number line.

1. {-2, 0, 2}

-3 -2 -1 0 1 2 3

2. {-5, -3, 1}

-5 -4 -3 -2 -1 0 1

Simplify.

3. $|4| - |-2|$ **2**

4. $|-3| + |-1|$ **4**

5. Evaluate $|x| - 3$ if $x = -3$. **0**

Motivating the Lesson

Hands-On Activity Bring the game Battleship™ to class or make a similar game and play it with your students. Explain that the numbers and letters form an ordered pair and represent a location on the grid.

2-2 Integration: Geometry
The Coordinate System

Setting Goals: *In this lesson, you'll graph points in all quadrants of the coordinate plane.*

Modeling a Real-World Application: Cartography

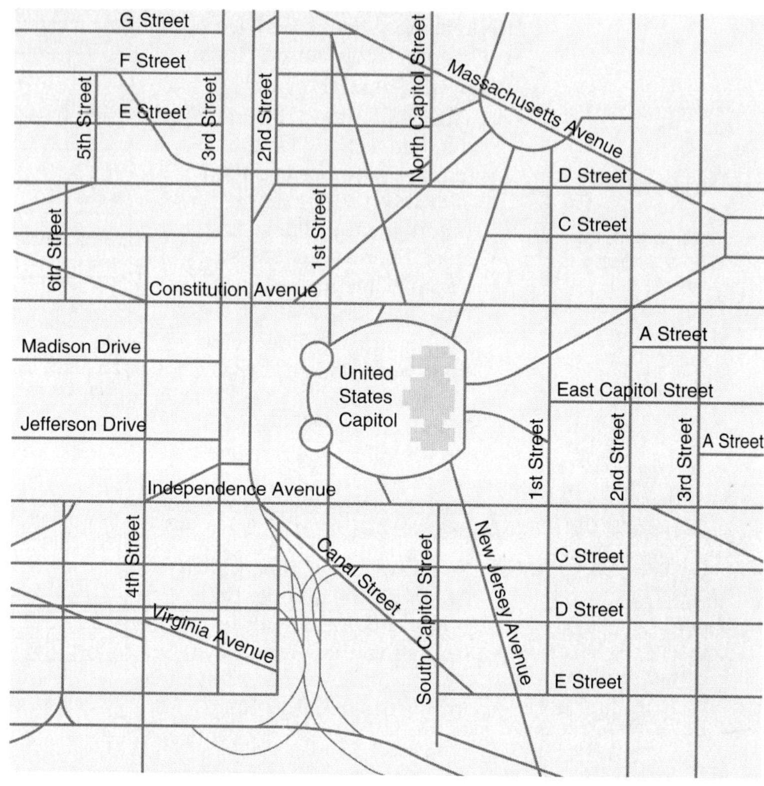

CULTURAL CONNECTIONS
Benjamin Banneker, the grandson of a slave, was part of a team charged with designing Washington, D.C. When Pierre L'Enfant resigned and took all the plans with him, Banneker reproduced the entire set of plans from memory.

Our nation's capital has a unique system for naming its streets. The Capitol is at the center. Streets that lie east and west of the Capitol are named numerically: 1st, 2nd, 3rd, 4th, and so on. Streets that lie north and south of the Capitol are named alphabetically: A, B, C, D, . . . through W, skipping the letter J. This means that there are *two* sets each of the numerical and alphabetical street names.

Each location on the map could be indicated by a pair of numbers and letters, such as (3, D). This would represent the intersection of 3rd and D streets. However, there are *four* places where 3rd and D streets meet. The problem is resolved by separating Washington into four different sections. The sections are northeast (NE), southeast (SE), southwest (SW), and northwest (NW).

CULTURAL CONNECTIONS
Banneker was born in 1731. In his early schooling, he demonstrated his mathematical abilities. He also taught himself about astronomy and started writing local almanacs in 1771. In 1789, he accurately predicted a solar eclipse.

Learning the Concept

You can review the coordinate system in Lesson 1-7.

A mathematical model of the city map of Washington, D.C., could be represented by a **coordinate system**. Remember that a coordinate system is formed by the intersection of two number lines, called **axes**. The axes separate the coordinate plane into four regions called **quadrants**, like the four sections of Washington, D.C.

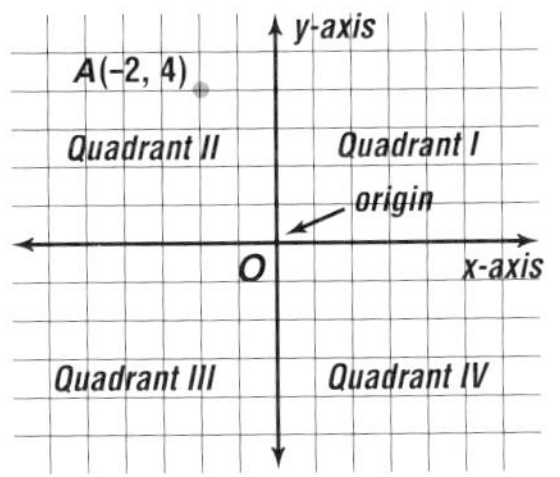

Axes is the plural of axis.

Notice that the numbers to the left of zero on the x-axis are negative as are the numbers below zero on the y-axis.

The *x-coordinate* of the ordered pair $(-2, 4)$ is -2, and the *y-coordinate* is 4.

ordered pair

$(-2, 4)$

x-coordinate ↑ ↑ *y-coordinate*

The dot at $(-2, 4)$ is the **graph** of point *A*.

The origin and the two axes do not lie in any quadrant.

Example 1

Cartography

In 1992, Ben Nighthorse Campbell became the first Native American to serve as a U.S. Senator.

a. Write the ordered pair for NASA.
The *x*-coordinate is -6, and the *y*-coordinate is -3. Thus, the ordered pair is $(-6, -3)$.

b. What is located at (2, 3)?
Count 2 units to the right and 3 units up. The point for the Senate Offices lies at (2, 3).

c. What is located at the origin?
The origin is at (0, 0). The Capitol is at the origin.

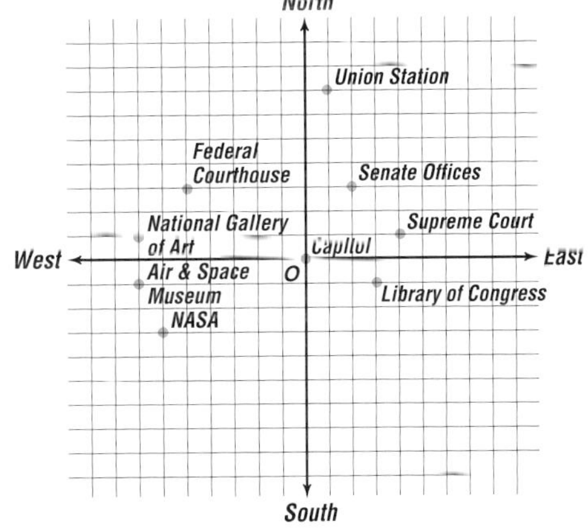

Remember that to graph a point means to place a dot at the point named by the ordered pair. This is sometimes called **plotting the point**.

2 TEACH

In-Class Example

For Example 1
Using the map on page 73, answer these questions.

a. What ordered pair shows the location of the Federal Courthouse? **(−5, 3)**

b. What is located at (3, −1)? **Library of Congress**

c. What is located at (−7, 1)? **National Gallery of Art**

Teaching Tip Ask students these questions.

- How do you know which quadrant contains a point, if you know the coordinates of the point? **If both *x* and *y* are positive, the point is in Quadrant I; if *x* is negative and *y* is positive, the point is in Quadrant II; if both *x* and *y* are negative, the point is in Quadrant III; if *x* is positive and *y* is negative, the point is in Quadrant IV.**
- How do you know when a point is on the *x*-axis? on the *y*-axis? **If the *y*-coordinate is 0, the point is on the *x*-axis; if the *x*-coordinate is 0, the point is on the *y*-axis.**

Mr. Campbell is a Northern Cheyenne Indian chief. He was a member of the U.S. judo team in the 1964 Olympic Games. In the 1960s and 1970s, Mr. Campbell owned a jewelry design business. He was elected to the Colorado House of Representatives in 1983, and to the U.S. House of Representatives in 1986.

GLENCOE Technology

Interactive Mathematics Tools Software

In this interactive computer lesson, students will explore graphing ordered pairs. A **Computer Journal** gives students an opportunity to write about what they have learned.

For Windows & Macintosh

Alternative Learning Styles

Kinesthetic Make a large grid on the floor. Name an ordered pair and have a student move to that location on the grid. If necessary, have the student start at the origin and show how to move to the left or right and then up or down to locate the point. Repeat as needed.

In-Class Example

For Example 2
Graph each point on a coordinate plane. Then name the quadrant in which each point lies.

a. $A(-2, -3)$ **quadrant III**

b. $B(3, 0)$ **not in any quadrant, on y-axis**

c. $C(-1, 2)$ **quadrant II**
See students' graphs.

3 PRACTICE/APPLY

Checking Your Understanding

Exercises 1–13 are designed to help you assess your students' understanding through reading, writing, speaking, and modeling. You should work through Exercises 1–4 with your students and then monitor their work on Exercises 5–13.

Additional Answer

2. Sample answer: (4, 8) represents a point 4 units to the right and 8 units up from the origin. (8, 4) represents a point 8 units to the right and 4 units up from the origin.

Study Guide Masters, p. 12

NAME ______________________ DATE ____________

2-2 Study Guide
Integration: Geometry
The Coordinate System

Student Edition
Pages 72–76

A horizontal number line and a vertical number line that meet at zero are shown in the figure at the right.

Such a **coordinate system** is used for maps and graphs, and has many uses in mathematics.

Point *A* in the figure is the point for the **ordered pair** (2, 4).

To find point *A*, start at 0 and move to the point for 2 on the horizontal number line. Then move *up* 4 units. To find point *B*, start at 0 and move to the point for −3 on the horizontal number line. Then move *down* 5 units.

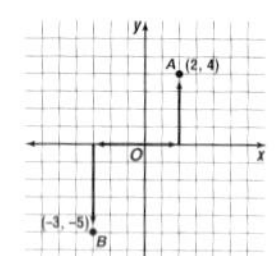

Use the coordinate system at the right to name the ordered pair for each point.

1. *A* **(1, 4)**   2. *B* **(3, 3)**   3. *C* **(−5, −3)**
4. *D* **(−3, −2)**   5. *E* **(−3, 2)**   6. *F* **(3, −3)**
7. *G* **(2, −5)**   8. *H* **(−4, 4)**   9. *J* **(1, −2)**

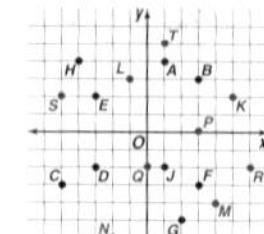

Use the coordinate system at the right. Graph and label each point. Name the quadrant in which each point is located.

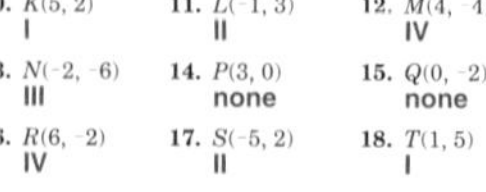

10. *K*(5, 2) **I**   11. *L*(−1, 3) **II**   12. *M*(4, −4) **IV**
13. *N*(−2, −6) **III**   14. *P*(3, 0) **none**   15. *Q*(0, −2) **none**
16. *R*(6, −2) **IV**   17. *S*(−5, 2) **II**   18. *T*(1, 5) **I**

Example 2 **Graph each point on a coordinate plane. Then name the quadrant in which each point lies.**

a. $D(3, -4)$

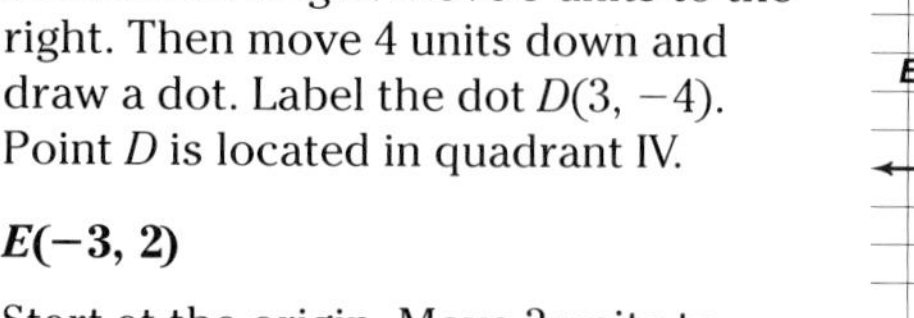

Start at the origin. Move 3 units to the right. Then move 4 units down and draw a dot. Label the dot $D(3, -4)$. Point *D* is located in quadrant IV.

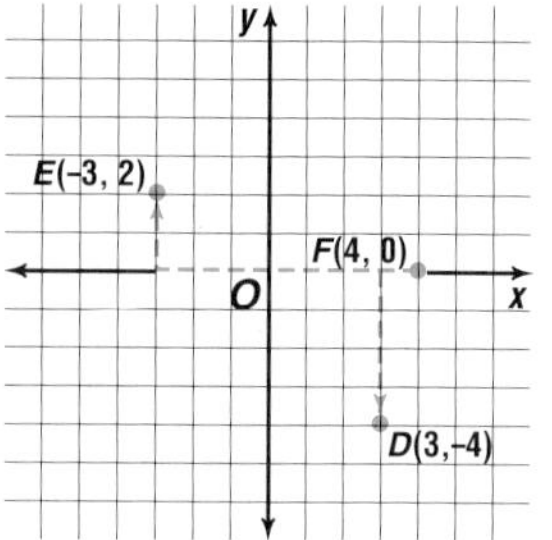

b. $E(-3, 2)$

Start at the origin. Move 3 units to the left. Then move 2 units up and draw a dot. Label this dot $E(-3, 2)$. Point *E* is located in quadrant II.

c. $F(4, 0)$

Start at the origin. Move 4 units to the right.The graph is on the *x*-axis. Label this dot $F(4, 0)$. Since point *F* lies on the *x*-axis, it is not in any quadrant.

Checking Your Understanding

Communicating Mathematics

Read and study the lesson to answer these questions. **2. See margin.**

1. **Draw** a coordinate system and label the origin, the *x*- and *y*-axes, and the four quadrants. **See students' work.**
2. **Explain** why the point (4, 8) is different from the point (8, 4).
3. **Name** two ordered pairs whose graphs are *not* located in one of the four quadrants. **Sample answer: (0, 0), (9, 0).**
4. **You Decide** Danny said that if you interchange the coordinates of any point in Quadrant I, the new point still would be in Quadrant I. Cherita disagrees. She says the new point would be in Quadrant 3. Who is right? Explain. **Danny; coordinates of points in Quadrant I are both positive.**

Guided Practice

Use the coordinate grid at the right to name the point for each ordered pair.

5. $(-3, 4)$ ***B***
6. $(1, 4)$ ***H***
7. $(-2, -3)$ ***D***
8. $(-4, 3)$ ***F***

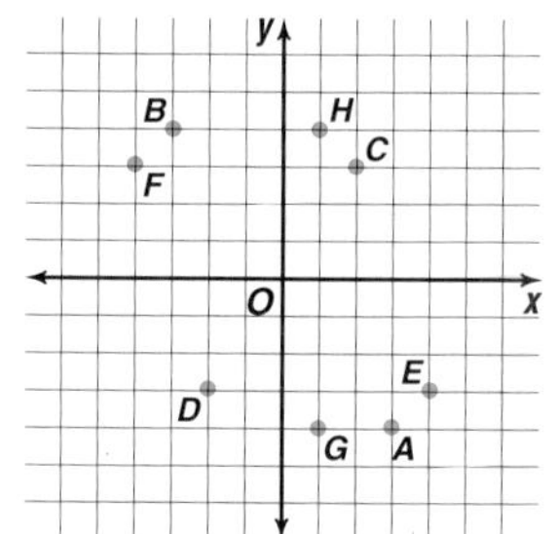

On graph paper, draw coordinate axes. Then graph and label each point. Name the quadrant in which each point is located.

9–12. See margin for graphs.

9. $I(4, -7)$ **IV**
10. $J(-4, 8)$ **II**
11. $K(0, 7)$ **none**
12. $L(-6, -3)$ **III**
13. **Logical Reasoning** If the graph of $A(x, y)$ satisfies the given condition, name the quadrant in which point *A* is located.
a. $x > 0, y > 0$ **I** **b.** $x < 0, y < 0$ **III** **c.** $x < 0, y > 0$ **II**

Reteaching

Using Writing Give students a four-quadrant graph with each letter of the alphabet located at a point somewhere on the graph. Give them a short, coded message by using the coordinates of the letters needed. After decoding the message, have each student write a coded message of his or her own. Trade and decode.

Additional Answers

9–12.

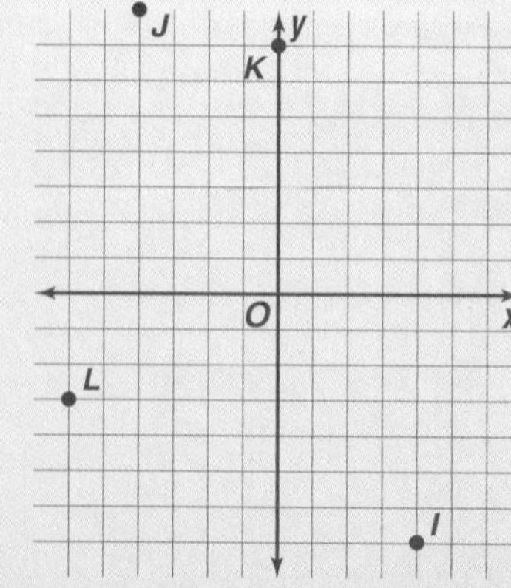

Exercises: Practicing and Applying the Concept

Independent Practice

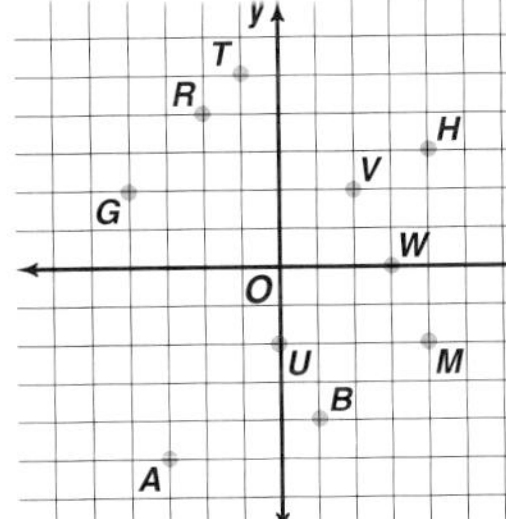

Name the ordered pair for each point graphed on the coordinate plane at the right.

14. R (−2, 4)	**15.** M (4, −2)
16. A (−3, −5)	**17.** B (1, −4)
18. G (−4, 2)	**19.** H (4, 3)
20. U (0, −2)	**21.** V (2, 2)
22. W (3, 0)	**23.** T (−1, 5)

On graph paper, draw coordinate axes. Then graph and label each point. Name the quadrant in which each point is located.

24. $A(3, 5)$ I	**25.** $K(-6, 1)$ II	**26.** $M(2, -4)$ IV
27. $B(-7, -4)$ III	**28.** $R(-1, 3)$ II	**29.** $S(3, -5)$ IV
30. $Q(-3, -5)$ III	**31.** $D(4, 2)$ I	**32.** $E(7, -8)$ IV
33. $F(-4, -7)$ III	**34.** $G(1, 0)$ none	**35.** $H(-7, 0)$ none

For graphs to Exercises 24–35, see margin.

Graph each point. Then connect the points in alphabetical order and identify the figure. 36–37. See Solutions Manual.

36. $A(0, 6), B(4, -6), C(-6, 2), D(6, 2), E(-4, -6), F(0, 6)$

37. $A(5, 8), B(1, 13), C(5, 18), D(9, 13), E(5, 8), F(5, 6), G(3, 7), H(3, 5), I(7, 7), J(7, 5), K(5, 6), L(5, 3), M(3, 4), N(3, 2), P(7, 4), Q(7, 2), R(5, 3), S(5, 1)$

If x and y are integers, graph all ordered pairs that satisfy the given conditions. 38–39. See margin.

38. $|x| < 4$ and $|y| < 3$ **39.** $|x| \leq 2$ and $|y| \leq 1$

Critical Thinking

40. Describe the possible location in terms of quadrants or axes, for the graph of (x, y) if x and y satisfy the condition $xy = 0$. origin or x- or y-axes

Applications and Problem Solving

41a. $P(2, 2)$, $Q(2, 5)$, $R(7, 5)$, $S(7, 2)$
41b. Rectangle $PQRS$ shifted 4 units down.
41c. A rectangle shaped like $PQRS$ in quadrant III.

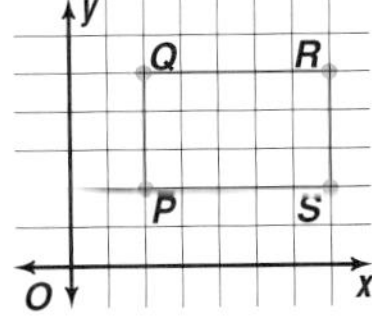

41. Geometry A **vertex** of a polygon is a point where two sides of the polygon intersect.

a. Identify the coordinates of the vertices in the rectangle at the right.

b. Subtract 4 from each y-coordinate. Graph the new ordered pairs. What figure is formed if the points are connected?

c. Multiply each x- and y-coordinate of the vertices of $PQRS$ by −1. Graph the new ordered pairs. Describe the figure that is formed if the points are connected.

Error Analysis

Students may incorrectly plot (x, y) as (y, x), especially when x or $y = 0$. Emphasize the order in such a pair by plotting any ordered pair incorrectly and then correctly.

Assignment Guide

Core: 15–39 odd, 40–41, 43–49
Enriched: 14–38 even, 40–49

For **Extra Practice**, see p. 743.

The red A, B, and C flags, printed only in the Teacher's Wraparound Edition, indicate the level of difficulty of the exercises.

Additional Answers

24–35.

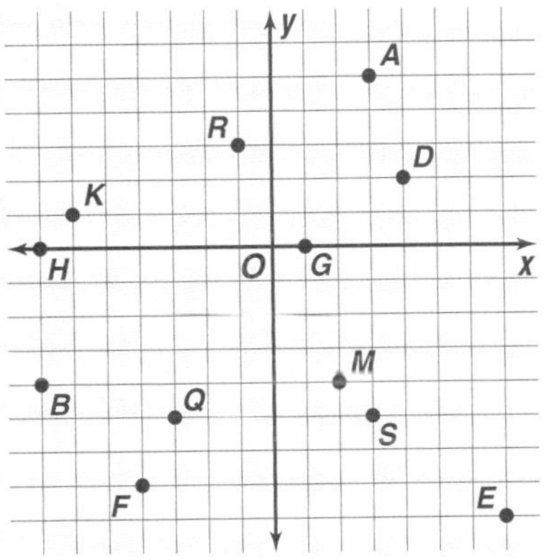

Practice Masters, p. 12

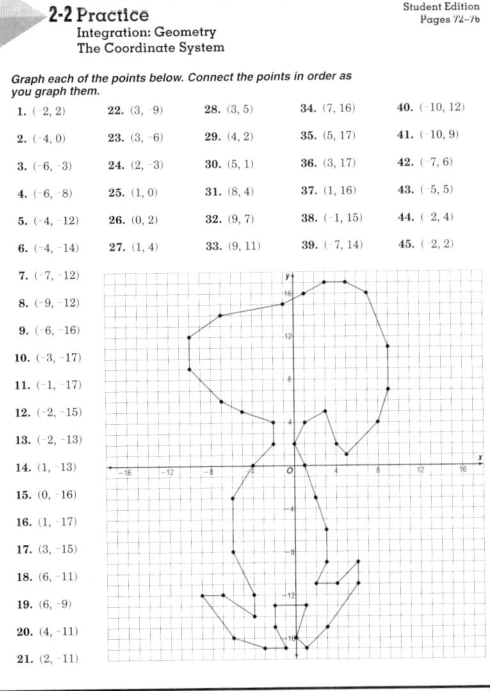

NAME ______________ DATE ______

2-2 Practice
Integration: Geometry
The Coordinate System

Student Edition
Pages 72–76

Graph each of the points below. Connect the points in order as you graph them.

1. (−2, 2)	**22.** (3, −9)	**28.** (3, 5)	**34.** (7, 16)	**40.** (−10, 12)
2. (−4, 0)	**23.** (3, −6)	**29.** (4, 2)	**35.** (5, 17)	**41.** (−10, 9)
3. (−6, −3)	**24.** (2, −3)	**30.** (5, 1)	**36.** (3, 17)	**42.** (−7, 6)
4. (−6, −8)	**25.** (1, 0)	**31.** (8, 4)	**37.** (1, 16)	**43.** (−5, 5)
5. (−4, −12)	**26.** (0, 2)	**32.** (9, 7)	**38.** (−1, 15)	**44.** (−2, 4)
6. (−4, −14)	**27.** (1, 4)	**33.** (9, 11)	**39.** (−7, 14)	**45.** (−2, 2)

7. (−7, −12)
8. (−9, −12)
9. (−6, −16)
10. (−3, −17)
11. (−1, −17)
12. (−2, −15)
13. (−2, −13)
14. (1, −13)
15. (0, −16)
16. (1, −17)
17. (3, −15)
18. (6, −11)
19. (6, −9)
20. (4, −11)
21. (2, −11)

Cooperative Learning

Pairs Check Use dominoes and two coins to generate number pairs. One student picks a domino and then flips the coins to determine whether the numbers are positive, negative, or one of each—heads, positive and tails, negative. That student then states an ordered pair. Other students locate and graph the point. An alternate activity is for one student to pick a domino and graph the point. Other students state the ordered pair.

For more information on this strategy, see *Cooperative Learning in the Mathematics Classroom*, one of the titles in the Glencoe Mathematics Professional Series, p. 12

4 ASSESS

Closing Activity

Modeling Make a treasure map, but do not tell the students where the treasure is. Have them find the treasure by laying a grid over your map and asking which quadrant the treasure is in. Then have them name coordinates until they land on the treasure. (Hints such as hot—within 2 units for both *x* and *y* coordinates—or cold—in wrong quadrant, etc. may speed up the game.)

Chapter 2, Quiz A (Lessons 2-1 and 2-2) is available in the *Assessment and Evaluation Masters*, p. 43.

Additional Answers

38.

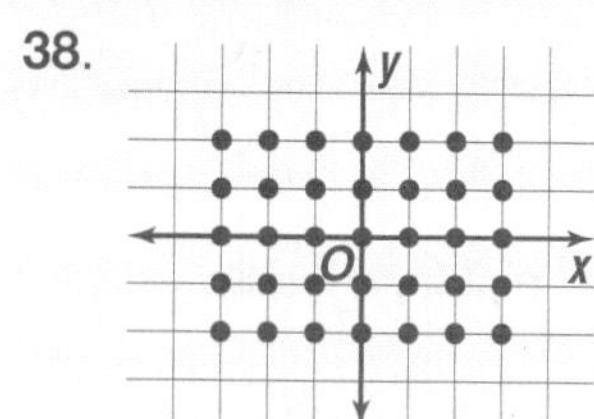

39.

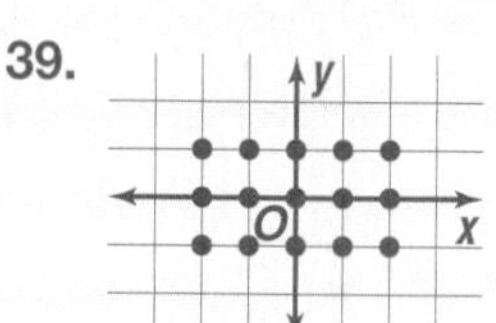

44.

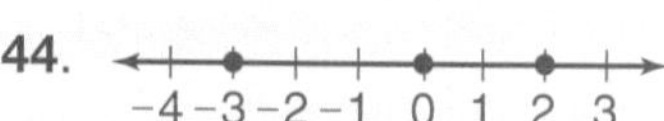

Enrichment Masters, p. 12

NAME ____________ DATE ____________

2-2 Enrichment
Polar Coordinates

Student Edition Pages 72–76

In a rectangular coordinate system, the ordered pair (x, y) describes the location of a point x units from the origin along the x-axis and y units from the origin along the y-axis.

In a polar coordinate system, the ordered pair (r, θ) describes the location of a point r units from the pole on the ray (vector) whose endpoint is the pole and which forms an angle of θ with the polar axis.

The graph of (2, 30°) is shown on the polar coordinate system at the right below. Note that the concentric circles indicate the number of units from the pole.

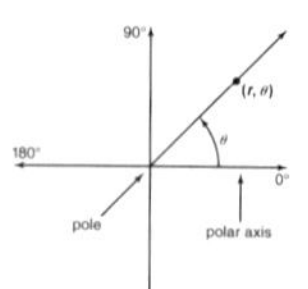

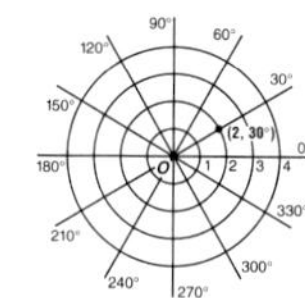

Locate each point on the polar coordinate system below.

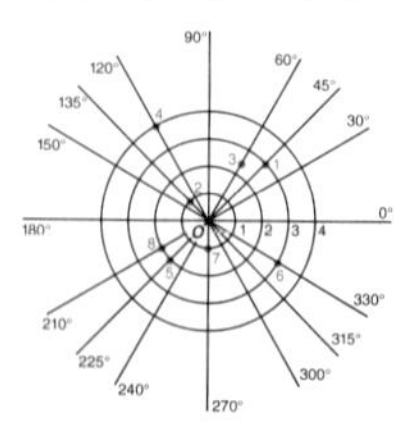

1. (3, 45°)
2. (1, 135°)
3. $(2\frac{1}{2}, 60°)$
4. (4, 120°)
5. (2, 225°)
6. (3, −30°)
7. (1, −90°)
8. (−2, 30°)

42. **Geography** Latitude and longitude lines are used to locate places on a map or globe. Latitude is measured in degrees north or south of the equator while longitude is measured in degrees east or west of the prime meridian. When describing a location, the latitude is usually given first. For example, the location of 30°N, 90°W is New Orleans, Louisiana. Find the approximate latitude and longitude of the following locations.
   a. Sydney, Australia **30°S, 150°E**
   b. Porto Alegre, Brazil **30°S, 50°W**
   c. Johannesburg, Republic of South Africa **30°S, 30°E**

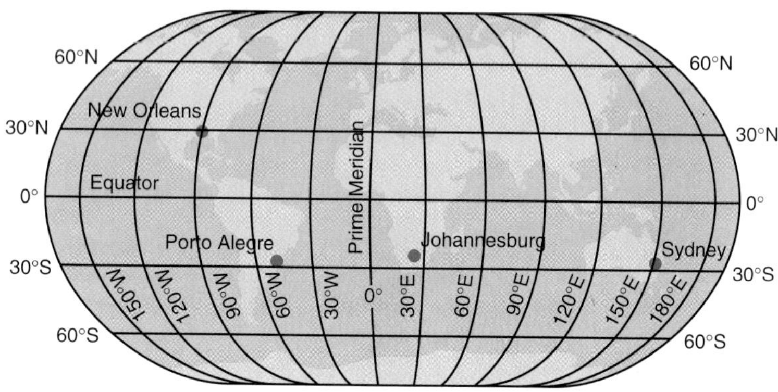

Mixed Review

43. Find the value of $|15| - |-3|$. (Lesson 2-1) **12**

44. Graph {−3, 0, 2} on a number line. (Lesson 2-1) **See margin.**

45. **Recreation** During the last few years, in-line skating has become a popular pastime. Today there are over 500,000 people who participate in this sport. This is more than twice as many people as there were several years ago. Which inequality, $s > 250{,}000$ or $s < 250{,}000$, describes how many skaters there were several years ago? (Lesson 1-9) $s < 250{,}000$

Choose
Estimation
Mental Math
Calculator
Paper and Pencil

46. What property is shown by $(6 \cdot 15) \cdot 3 = 6 \cdot (15 \cdot 3)$? (Lesson 1-4)

47. Evaluate each expression if $a = 4$, $b = 2$, and $c = 3$. (Lesson 1-3)
   a. $9a - (4b + 2c)$ **22**
   b. $\frac{6(a + b)}{3c}$ **4**

46. associative property of multiplication

48. Find the value of $7[(12 + 5) - 3(19 - 14)]$. (Lesson 1-2) **14**

49. **Sports** The Middletown Junior High School basketball team made 21 baskets to score 36 points. If two of the baskets were 3-point shots, how many baskets were worth 2 points and how many were worth 1 point? (Lesson 1-1) **eleven 2-point; eight 1-point**

Group Activity Card 2-2

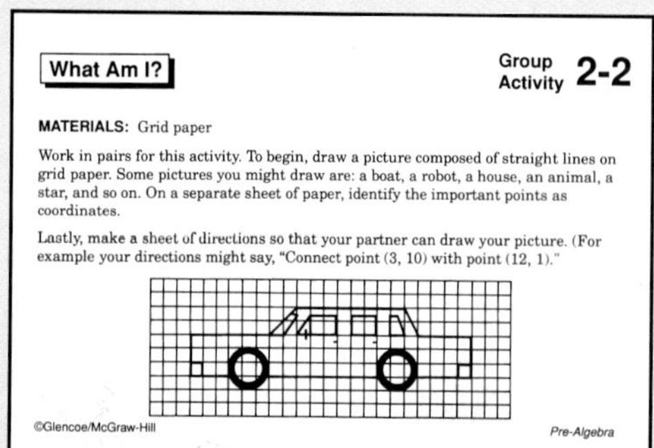

What Am I?

Group Activity 2-2

MATERIALS: Grid paper

Work in pairs for this activity. To begin, draw a picture composed of straight lines on grid paper. Some pictures you might draw are: a boat, a robot, a house, an animal, a star, and so on. On a separate sheet of paper, identify the important points as coordinates.

Lastly, make a sheet of directions so that your partner can draw your picture. (For example your directions might say, "Connect point (3, 10) with point (12, 1).")

Pre-Algebra

Extension

Using Connections Have each student draw a picture on a four-quadrant graph. Only straight lines are used. Turns are made only at integral points. Write, in order, coordinates for each point in the drawing. Then exchange coordinates and re-create the partner's drawing in a dot-to-dot fashion.

GRAPHING CALCULATOR ACTIVITY

2-2B Plotting Points

An Extension of Lesson **2-2**

You can plot points on a graphing calculator just as you do on grid paper.

Activity **Plot the ordered pairs (6, 9), (27, 13), (15, –13), (–33, 24), and (–12, –8) on a graphing calculator.**

First, set the viewing window. The viewing window is the range of x- and y-values that are visible on the graphics screen. The viewing window [−47, 47] by [−31, 31], with scale factors of 10 on both axes works well for these points. This means that the scale will be from −47 to 47 on the x-axis and from −31 to 31 on the y-axis, with tick marks every 10 units.

ENTER: WINDOW ENTER (–) 47 ENTER 47 ENTER 10 ENTER (–) 31 ENTER 31 ENTER 10 GRAPH

Now draw the points.

ENTER: 2nd DRAW ▶ ENTER

Use the arrow keys to move the cursor to the first point at (6, 9) and press ENTER. As you move the cursor to the next point, at (27, 13), you will see that the first point was plotted. After you position the cursor at the next point, press ENTER again. Continue until all points are plotted. To see the last point, press CLEAR.

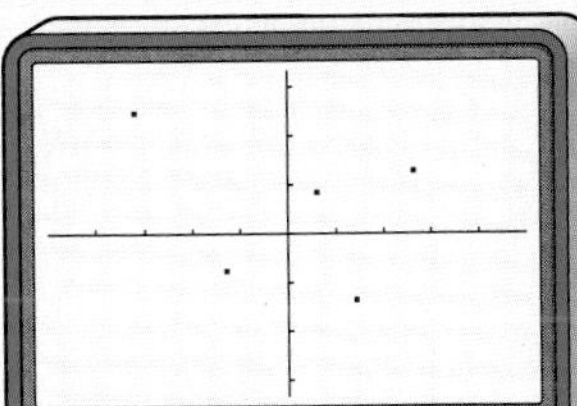

Your Turn **Clear the graphics screen by pressing 2nd DRAW ENTER. Then graph each set of ordered pairs on a graphing calculator.**

1–4. See Solutions Manual.

1. (8, –3), (9, 7), (–3, 4), (–7, 4), (–6, –5)
2. (–17, 22), (8, 19), (–12, 1), (–21, –7), (16, 23)
3. (–32, 4), (26, 16), (–3, –8), (–7, –11)
4. (18, –30), (39, 17), (–33, 24), (–27, 14), (32, –21)

TALK ABOUT IT

5. The *standard viewing window* can be set automatically by pressing ZOOM 6. **b. [−47, 47] by [−31, 31]**
   a. What are the minimum and maximum values of the standard viewing window? **[−10, 10] by [−10, 10]**
   b. What are the minimum and maximum values of the viewing window after pressing ZOOM 6 ZOOM 8 ENTER ?
6. You should choose the viewing window to best display each graph.
   a. How would you set the viewing window to display a graph of Quadrant II? Quadrant IV? **See Solutions Manual.**
   b. If the last ordered pair in Exercise 4 was (–49, 39), how could you make sure it was visible in the viewing window? **See Solutions Manual.**

Technology

This lesson offers an excellent opportunity for using technology in your pre-algebra classroom. For more information on using technology, see *Graphing Calculators in the Mathematics Classroom*, one of the titles in the Glencoe Mathematics Professional Series.

2-2B LESSON NOTES

NCTM Standards: 1-5, 12

Objective

Use a graphing calculator to plot points on a coordinate grid.

Recommended Time

15 minutes

Instructional Resources

Graphing Calculator Masters, p. 16
This master provides keystroking instruction for this lesson for the TI-81 and Casio graphing calculators.

1 FOCUS

Motivating the Lesson

Tell students that the calculator screen can model a coordinate grid.

2 TEACH

If students have used their graphing calculator for graphing equations, drawing, or graphing statistical graphs, they will need to clear the graphics screen before they begin. Use the following keystrokes for a TI-82.

2nd Y-VARS 5 2 ENTER
Turns off functions in the Y – list.

2nd DRAW ENTER ENTER
Clears all drawn elements.

2nd STAT PLOT 4 ENTER
Turns the STAT PLOTS off.

3 PRACTICE/APPLY

Assignment Guide
Core: 1–6
Enriched: 1–6

4 ASSESS

Observing students working with technology is an excellent method of assessment.

2-3 Comparing and Ordering

Setting Goals: *In this lesson, you'll compare and order integers.*

Modeling a Real-World Application: Health

Nature's way of using pollen to produce seeds in plants is remarkable. However, pollen is the enemy of millions of people who suffer from allergies. Pollen from trees or weeds can cause sneezing and eye irritation.

During allergy season, local television stations or newspapers may give daily pollen counts to warn people. As the pollen count increases, so does the discomfort level as seen moving left to right on the scale below.

Pollen Count	0–100	101–600	601–1200	over 1200
Discomfort Level	low	uncomfortable	high	very high

Learning the Concept

Just like the pollen scale, the numbers on a number line increase as you move to the right. On a number line, it is easy to determine which of two numbers is greater. –3 and 4 are graphed on the number line below.

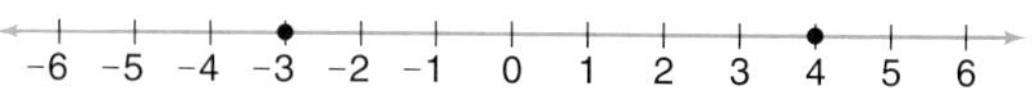

Say: 4 is greater than −3 *The number to the right is always greater.*
Write: $4 > -3$ *The number to the left is always less.*

You might also conclude that −3 is to the left of 4.

Say: −3 is less than 4 *Remember, the symbol points*
Write: $-3 < 4$ *to the lesser number.*

Remember that any mathematical sentence containing < or > is called an inequality. In inequalities, numbers are compared.

Example 1 **Use the integers graphed on the number line below for each question.**

-6 -5 -4 -3 -2 -1 0 1 2 3 4 5 6

a. Write two inequalities involving −5 and −1.
Since –5 is to the left of –1, write $-5 < -1$.
Since –1 is to the right of –5, write $-1 > -5$.

b. State which number is greater.
−1 is greater since it lies to the right of –5.

c. State which number has the greater absolute value.
The absolute value of –5 is greater than the absolute value of –1 because it lies farther from the origin.

2-3 LESSON NOTES

NCTM Standards: 1-5, 12

Instructional Resources

- Study Guide Master 2-3
- Practice Master 2-3
- Enrichment master 2-3
- Group Activity Card 2-3
- Graphing Calculator Masters, p. 2
- Tech Prep Applications Masters, p. 3

Transparency 2-3A contains the 5-Minute Check for this lesson; **Transparency 2-3B** contains a teaching aid for this lesson.

Recommended Pacing	
Standard Pacing	Day 3 of 12
Honors Pacing	Day 3 of 11
Block Scheduling*	Day 2 of 7

*For more information on pacing and possible lesson plans, refer to the ***Block Scheduling Booklet***.

1 FOCUS

5-Minute Check
(over Lesson 2-2)

On graph paper, draw coordinate axes. Then graph and label each point. Name the quadrant in which each point is located.

1. $B(1, -3)$ 2. $N(-4, 2)$
3. $E(0, -2)$ 4. $W(2, 0)$
5. $R(-3, -4)$

See Transparency 2-3A for answers.

Motivating the Lesson

Using Questioning Ask the following questions.

- "If Juan has $2 and Karen has $3, who has more?" **As students answer, write 3 > 2.**
- "If it is 8°C outside today, and was 10°C outside yesterday, which day was cooler or had the lower temperature?" **8 < 10.**

Alternative Teaching Strategies

Reading Mathematics When students work on comparing and ordering integers, encourage the use of correct vocabulary when reading inequalities. Provide an opportunity for students to practice reading inequalities. One such way would be to have students read their answers to Exercises 15–26 as inequalities or equations, not just naming the symbol that makes a true statement.

Integers are used to compare numbers in many everyday applications.

The chart at the right shows the record low temperatures for selected states through 1992. Order the temperatures from least to greatest.

Record Low Temperatures for Selected States (°F)	
State	Temperature
Alabama	−27
California	−61
Florida	−2
Georgia	−17
Hawaii	12
Indiana	−35
Kentucky	−34
Louisiana	−16
Oklahoma	−27
West Virginia	−37

Graph each integer on a number line.

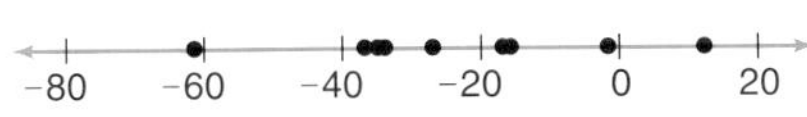

Write the integers as they appear on the number line from left to right.

−61, −37, −35, −34, −27, −27, −17, −16, −2, 12 are in order from least to greatest.

- "If it is −3°C outside today, and was −5°C outside yesterday, which day was cooler or had the lower temperature?" **yesterday**

Tell students they will learn to solve the last problem in this lesson.

2 TEACH

In-Class Examples

For Example 1

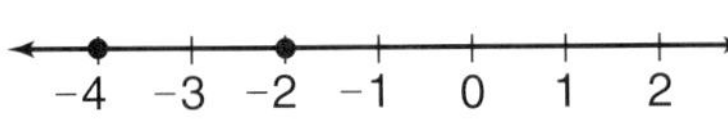

a. Write two inequalities using the numbers shown on the graph. **$-4 < -2$; $-2 > -4$**

b. Which number is greater? **−2**

c. Which number has the greater absolute value? **−4**

For Example 2

One town had these low temperatures for a week: −5°F, 0°F, 4°F, 2°F, −3°F, −4°F, −1°F. Graph each integer on a number line. Then write the integers in order from least to greatest. **−5, −4, −3, −1, 0, 2, 4**

Checking Your Understanding

Communicating Mathematics

Read and study the lesson to answer these questions.

1. **Write** two inequalities that show how the highest and lowest temperatures in Example 2 are related.
2. **List** the integers graphed on the number line from least to greatest.

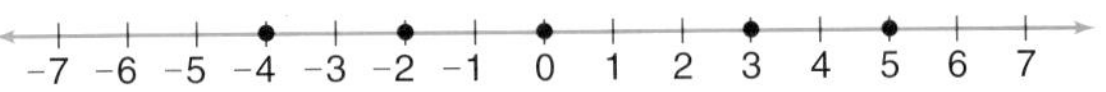

3. Graph two integers on a number line. Then write two inequalities that show how the integers are related. **See students' work.**

4. Write a sentence that explains how to determine when one integer is less than another integer. **See margin.**

Guided Practice

Write an inequality using the numbers in each sentence. Use the symbols $<$ or $>$.

5. 2° is warmer than −5°. **$2 > -5$**
6. 5 is greater than −3. **$5 > -3$**

Replace each ● with $<$, $>$, or $=$.

7. −19 ● −9 **$<$**
8. 0 ● −8 **$>$**
9. 4 ● |−4| **$=$**
10. |3| ● |8| **$<$**
11. Order the integers {43, 0, −9, −4, 3, −88, 234} from least to greatest.

Write two inequalities for each graph.

12. −7 −6 −5 −4 −3 −2 −1 0
13. −3 −2 −1 0 1 2 3 4

14. **Sports** Patty Sheehan won the \$82,500 first prize in the Rochester International LPGA tournament in June, 1995. The final round scores of the top ten finishers were −2, +1, −5, +3, −2, −1, +2, +4, 0, and +1.
    a. Order the scores from lowest to highest.
    b. If the lowest score wins, what was the winning score for this round? **−5**

Answers (margin):
1. Sample answer: $-61 < 12$, $12 > -61$
2. −4, −2, 0, 3, 5
11. {−88, −9, −4, 0, 3, 43, 234}
12. $-6 < -2$, $-2 > -6$
13. $-2 < 3$, $3 > -2$
14a. −5, −2, −2, −1, 0, +1, +1, +2, +3, +4

Reteaching

Using Models Make a large thermometer on the board or overhead. Have students locate various pairs of temperatures and write an inequality to compare them.

Additional Answer

4. Sample answer: The number farthest to the left on a number line is the lesser number. Inequality symbols always point to the lesser number.

Study Guide Masters, p. 13

NAME ____________ DATE ________

2-3 Study Guide
Comparing and Ordering Integers

Student Edition Pages 78–81

The graphs of −4, −1, 3, and 5 are shown on the number line.

−6 −5 −4 −3 −2 −1 0 1 2 3 4 5 6

The following statements can be made about the numbers and their graphs.

−4 is graphed to the left of −1, so $-4 < -1$.
−1 is graphed to the left of 5, so $-1 < 5$.
3 is graphed to the right of −1, so $3 > -1$.
5 is graphed to the right of −4, so $5 > -4$.

Write $>$, $<$, or $=$ in each □.

1. 4 [$>$] −4
2. 8 [$<$] 12
3. −7 [$<$] −5
4. 2 [$>$] −1
5. −8 [$=$] −8
6. −4 [$<$] 3
7. −3 [$>$] −8
8. −11 [$<$] −10
9. 6 [$>$] −9
10. −12 [$<$] 7
11. 9 [$>$] −9
12. 5 [$>$] −5

Order the numbers in each set from least to greatest.

13. {4, −4, 2, −1} **{−4, −1, 2, 4}**
14. {−5, 0, −3, 1} **{−5, −3, 0, 1}**
15. {−9, −6, −12, −7} **{−12, −9, −7, −6}**
16. {−1, 3, −5, −7} **{−7, −5, −1, 3}**
17. {−5, 5, −4, 4} **{−5, −4, 4, 5}**
18. {2, −3, 0, 4} **{−3, 0, 2, 4}**

Cross out one number in each group so that the rest of the numbers are in order. More than one answer is possible.

19. −3, −2, 0, −1, 3, 4 **0 or −1**
20. 5, 4, 3, 1, 0, −3, −2, −4 **−3 or −2**
21. −6, −4, −2, 0, 4, 2, 6 **4 or 2**
22. 15, 10, 0, 5, −5, −10, −15 **0 or 5**
23. −1, −2, 0, 1, 2, 3, 4 **−1 or −2**
24. 9, 7, 5, −5, −9, −7, −11 **−7 or −9**

3 PRACTICE/APPLY

Checking Your Understanding

Exercises 1–14 are designed to help you assess your students' understanding through reading, writing, speaking, and modeling. You should work through Exercises 1–4 with your students and then monitor their work on Exercises 5–14.

Error Analysis

Students may say that, for example, $^-7$ is greater than $^-5$, or $^-7$ is greater than 5 because 7 is greater than 5. Have students locate both numbers on a number line. Point out that the number to the right is always the greater number.

Assignment Guide

Core: 15–43 odd, 44–45, 47–53
Enriched: 16–42 even, 44–53

For **Extra Practice,** see p. 744.

The red A, B, and C flags, printed only in the Teacher's Wraparound Edition, indicate the level of difficulty of the exercises.

Practice Masters, p. 13

NAME ______________________ DATE __________

2-3 Practice
Comparing and Ordering Integers

Student Edition Pages 78–81

Write <, >, or = in each ☐.

1. 5 [>] $^-8$ 2. $^-3$ [<] 2 3. $^-4$ [>] $^-7$ 4. 4 [>] $^-6$
5. 0 [>] $^-12$ 6. 35 [>] $^-21$ 7. $^-55$ [<] 65 8. $^-16$ [>] $^-40$
9. 3 [>] 0 10. $^-5$ [<] 5 11. $^-7$ [<] 6 12. $^-10$ [<] 12
13. $^-10$ [<] 9 14. 6 [>] 4 15. $^-3$ [<] 0 16. $^-1$ [<] 1
17. $|^-4|$ [<] 5 18. $|^-1|$ [=] 1 19. $^-2$ [<] $|2|$ 20. 5 [>] $|^-3|$

Order the numbers in each set from least to greatest.

21. {7, 0, 4} {0, 4, 7} 22. {9, $^-7$, $^-3$} {$^-7$, $^-3$, 9} 23. {11, 0, $^-2$} {$^-2$, 0, 11}
24. {$^-3$, 1, $^-5$, 2} {$^-5$, $^-3$, 1, 2} 25. {$^-4$, $^-6$, 0, $^-2$} {$^-6$, $^-4$, $^-2$, 0} 26. {$^-7$, 5, $^-9$, 4} {$^-9$, $^-7$, 4, 5}
27. {3, $^-6$, 6, $^-3$} {$^-6$, $^-3$, 3, 6} 28. {10, $^-7$, 8, $^-9$} {$^-9$, $^-7$, 8, 10} 29. {$^-4$, 3, 0, $^-2$} {$^-4$, $^-2$, 0, 3}

Write an inequality using the numbers in each situation. Use the symbols < or >.

30. Yesterday's high wind speed was 25 mph. The low wind speed was 12 mph. 25 > 12, or 12 < 25
31. Rebecca made 7 foul shots in a game. In the same game, she missed 5 foul shots. 7 > $^-5$, or $^-5$ < 7
32. The team gained 15 yards. Then the team lost 6 yards. 15 > $^-6$, or $^-6$ < 15
33. Lucille spent \$10. She had earned \$9. $^-10$ < 9, or 9 > $^-10$

Exercises: Practicing and Applying the Concept

Independent Practice

Replace each ● with <, >, or =.

A
15. 9 ● −11 > 16. −3 ● −5 > 17. |−3| ● 3 =
18. −10 ● −3 < 19. −7 ● −13 > 20. −|15| ● −15 =
21. −4 ● 0 < 22. 0 ● −2 > 23. 0 ● |−8| <
24. −33 ● 5 < 25. −439 ● −23 < 26. −10 ● −3 <

Write an inequality using the numbers in each sentence. Use the symbols < or >. 29. 60 < 75

27. 3 m is taller than 2 m. 3 > 2 **28.** \$20 is more than \$15. 20 > 15
29. 60 kilometers per hour is slower than 75 kilometers per hour.
B
30. Yesterday's high temperature was 41°F. The low temperature was −3°F. 41 > −3
31. Today's pollen count is 565. Yesterday's count was 344. 565 > 344
32. Water boils at 212°F, and it freezes at 32°F. 212 > 32
33. One cup of green bean casserole has 265 Calories. The same amount of fresh green beans has 66 Calories. Sample answer: 66 < 265

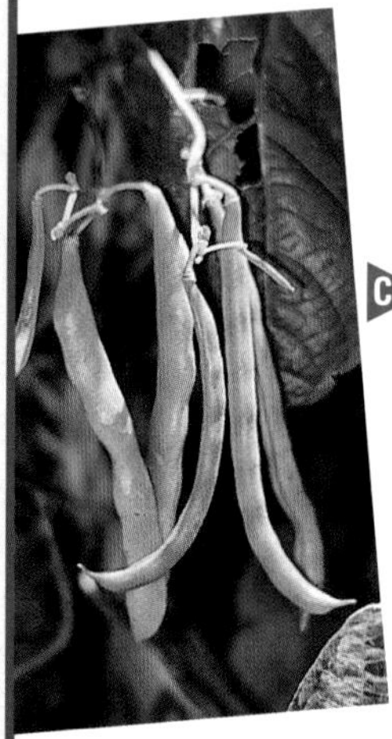

Write two inequalities for each graph.

C
34. 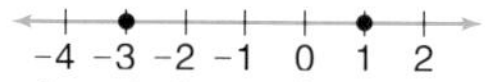

−3 < 1; 1 > −3

35. 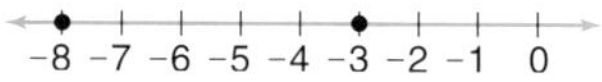

−8 < −3; −3 > −8

36. 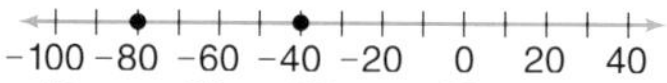

−40 > −80; −80 < −40

37. 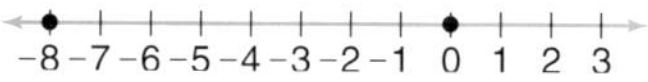

0 > −8; −8 < 0

Order the integers in each set from least to greatest. 40–43. See margin.

38. {7, 0, −5} {−5, 0, 7} 39. {−11, 8, −3} {−11, −3, 8}
40. {0, −5, −8, −1} 41. {−6, 56, 29, 1, −65}
42. {33, 9, −99, 7, 0, −4} 43. {87, −65, −53, 48, 199}

Critical Thinking

44. Graph all integer solutions for $|x| < 4$ on a number line. See margin.

EARTH WATCH

With the human population exploding, competition for Earth's dwindling resources is accelerating. This means that many plants and animals are in an almost impossible struggle for survival. An *endangered species* is one whose numbers have been reduced so that there is a real danger of it becoming extinct. While extinction is a natural result of changes in the environment, humans have had a hand in destroying habitat and overusing resources to make some species endangered. The table at the right shows the number of wild buffalo living in the United States in different years.

Year	Population
1851	50,000,000
1865	15,000,000
1872	7,000,000
1900	<1000
1972	30,000
1993	120,000

Source: *Audubon*

Additional Answers

40. {−8, −5, −1, 0}
41. {−65, −6, 1, 29, 56}
42. {−99, −4, 0, 7, 9, 33}
43. {−65, −53, 48, 87, 199}
44. 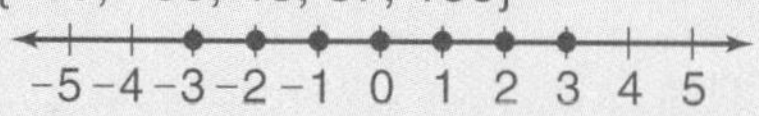

EARTH WATCH

The northern spotted owl was mistakenly listed as endangered. In the early 1990s, the U.S. government spent over \$20 million to save it from extinction. Within 5 years, over 5000 such owls were counted in Washington and Oregon alone.

Applications and Problem Solving

45a. See students' work.

45. **Chemistry** The chart at the right shows the freezing/melting points of water and selected elements.

Element	Freezing/melting point (°C)
Water	0
Helium	−272
Lead	328
Mercury	−39
Oxygen	−219
Sodium	98
Tin	232

a. Graph this data on a number line.

b. Which of these elements has the highest freezing/melting point?

c. Which has the lowest freezing/melting point? Helium 45b. Lead

46. **Weather** The table below shows the average monthly temperatures (°F) in Fairbanks, Alaska, for the 30-year period 1961–1990. Graph these temperatures on a number line. See students' work.

Jan.	Feb.	Mar.	Apr.	May	June	July	Aug.	Sept.	Oct.	Nov.	Dec.
−10	−4	11	31	49	60	63	57	46	25	3	−7

46a. Sample answer: $-58 < -4$

a. In 1993, the lowest temperature in Fairbanks was −58°F on February 1. Write an inequality to compare this with the February average.

b. The highest temperature in 1993 was 93°F on July 15. Write an inequality to compare this with the July average. Sample answer: $93 > 63$

Mixed Review

On graph paper, draw and label coordinate axes. Then graph and label each point. (Lesson 2-2)

47. $L(2, -8)$ See margin.

48. $M(-5, -5)$ See margin.

49. **Space Exploration** About six seconds before liftoff, the three main shuttle engines start. About 120 seconds after liftoff, the solid rocket boosters burn out. Use integers to describe these events. (Lesson 2-1)

50. State whether $15 < 3m + 7$ is *true* or *false* when $m = 4$. (Lesson 1-9)

51. Restate 6×26 using the distributive property. (Lesson 1-5)

52. Evaluate $4a + b \cdot b$, if $a = 4$ and $b = 2$. (Lesson 1-3) 20

53. **Consumer Awareness** The Jacksonville Middle School chorale bought 30 tickets to a musical play. The total cost of the tickets is $98. If student tickets cost $3 each and adult tickets cost $5 each, how many adults are going with the chorale? (Lesson 1-1) 4 adults

49. −6; 120 50. true 51. Sample answer: $(6 \times 20) + (6 \times 6)$

Mae Jemison was the first African-American woman to go into space. Her first mission was a cooperative effort between U.S. and Japan that focused on experiments involving space travel and biology.

See for Yourself

There was a time when the American buffalo was in danger of extinction. Thanks to conservation, they have made a comeback.

1. The change in population from 1851 to 1865 was 15,000,000 − 50,000,000 or −35,000,000. Use a calculator to find the change in the population from each year to the next. See margin.

2. Which years showed the greatest loss in the population of buffalo? Which years showed the most growth in the population of buffalo? See margin.

3. **Research** Investigate the efforts to save the northern spotted owl. Does the program appear to be successful? See students' work.

4 ASSESS

Closing Activity

Speaking Tell about examples where negative numbers are used to compare scientific data. For example, telling temperature in degrees below zero.

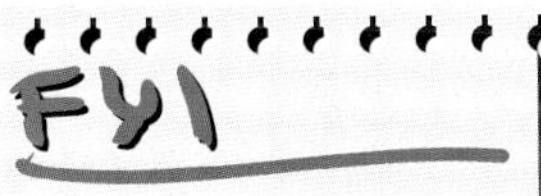

Ms. Jenson worked as a staff physician for the Peace Corps in Sierra Leone before joining the space program.

Additional Answers

47-48.

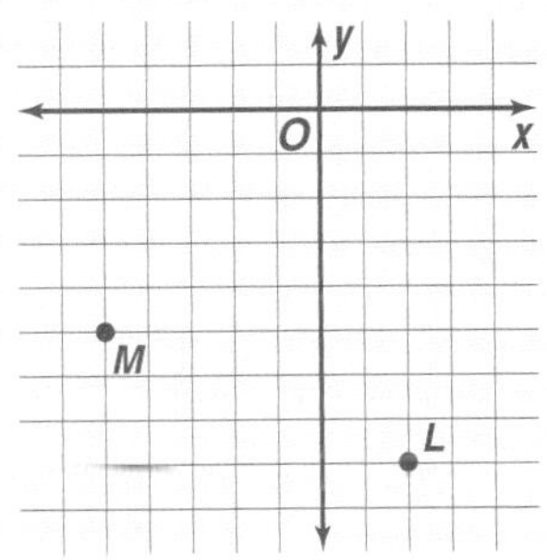

Earth Watch

1. −8,000,000; > −6,999,000; > 29,000; 90,000
2. 1851 to 1865; 1972 to 1993

Enrichment Masters, p. 13

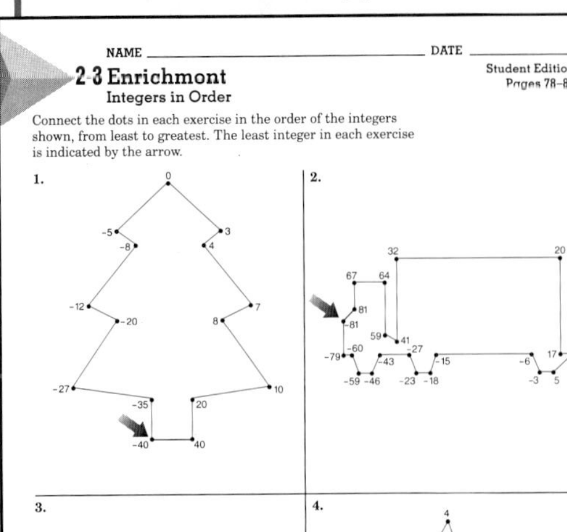

NAME ______ DATE ______ Student Edition Pages 78–81

2-3 Enrichment

Integers in Order

Connect the dots in each exercise in the order of the integers shown, from least to greatest. The least integer in each exercise is indicated by the arrow.

1.

2.

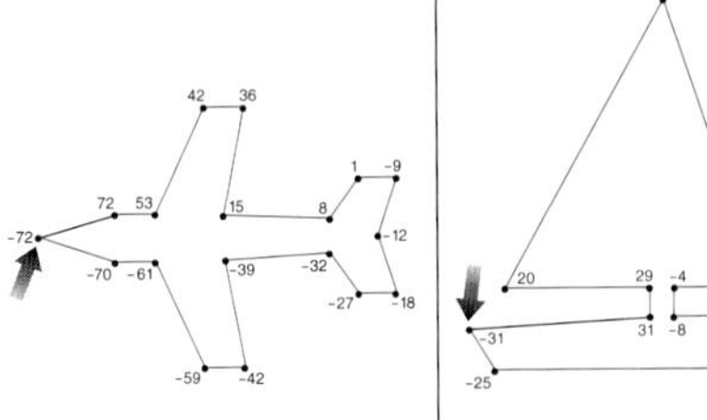

3.

4.

Extension

Using Connections Two kilometers above ground, the temperature is 0 degrees Celsius. For each kilometer of altitude, the temperature drops 7 degrees. What is the temperature 9 kilometers above ground? −63°C

Group Activity Card 2-3

Don't Be The Joker Group Activity 2-3

MATERIALS: 2–10 cards in diamonds and clubs from a deck of cards • one joker card • three 3x5 cards, with either ≤, ≥, or "has the same absolute value as" written on them

Shuffle the 19 cards and deal them. One player gets 9 cards and the other 10. Red cards represent negative integers. Black cards represent positive integers. The joker represents zero.

The player with 10 cards lays a card on the table, draws a card from the other player's hand, and lays that card beside the first card. Lastly, the player makes a true statement about those numbers using one of the 3x5 cards. If correct, the player discards the pair of cards. If incorrect, the player keeps the cards.

It is now the other player's turn. If a joker is drawn, the player must put the played card and the joker back into his or her hand, and it is the other player's turn. The person who runs out of cards wins, causing the other player to be the joker.

 Pre-Algebra

2-4A LESSON NOTES

NCTM Standards: 1-5, 7, 12

Objective

Use counters to model addition of integers.

Recommended Time

Demonstration and discussion: 15 minutes; Exercises: 30 minutes

Instructional Resources

For each student or group of students
Student Manipulative Kit
- counters
- integer mat

Math Lab and Modeling Math Masters
- p. 2 (integer mat)
- p. 5 (integer counters)
- p. 30 (worksheet)

For teacher demonstration
Overhead Manipulative Resources

1 FOCUS

Motivating the Lesson

Ask students if they have ever taken clothes out of a dryer and found a sock sticking to a shirt. The clothes stick together because one has a positive charge and the other has a negative charge. In this activity, the idea of opposite charges is used as a model for adding integers.

2 TEACH

Teaching Tip You might want to assemble and store sets of 10 positive counters and 10 negative counters in small plastic bags. Students could oversee and be responsible for the assembling, distribution, return, and storage of these materials.

3 PRACTICE/APPLY

Assignment Guide
Core: 1–9
Enriched: 1–9

MATHEMATICS LAB

HANDS-ON ACTIVITY

2-4A Adding Integers

A Preview of Lesson **2-4**

MATERIALS

- counters
- integer mat

You can use counters to model operations with integers. In this book, yellow counters will represent positive integers, and red counters will represent negative integers.

There are two important properties to keep in mind when modeling integers.

- When one positive counter is paired with one negative counter, the result is called a **zero pair**.
- You can add or remove zero pairs from a mat because removing or adding zero does not change the value of the counters on the mat.

Activity 1 **Find $-2 + (-4)$.**

Remember that $2 + 4$ means *combine a set of two items with a set of four items.* The expression $-2 + (-4)$ means something similar. It tells you to combine a set of two negative items with a set of four negative items.

- Place 2 negative counters and 4 negative counters on the mat.

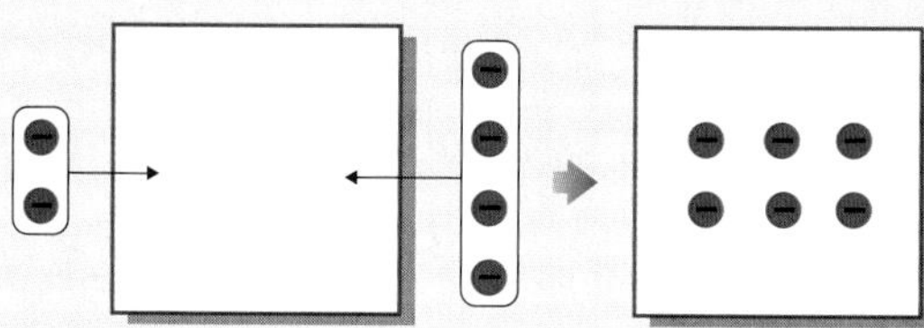

- Since there are 6 negative counters on the mat, the sum is -6. Therefore, $-2 + (-4) = -6$.

Activity 2 **Find $-4 + 2$.**

Remember that removing zero pairs does not change the value on the mat.

Place 4 negative counters and 2 positive counters on the mat. It is possible to remove 2 zero pairs.

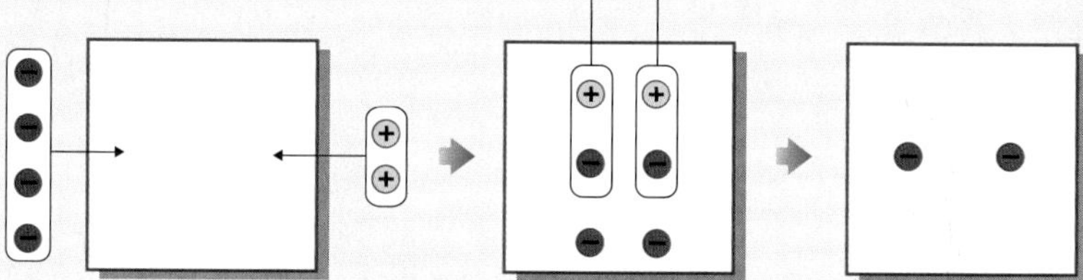

- Since there are 2 negative counters remaining, the sum is -2. Therefore, $-4 + 2 = -2$.

Your Turn **Find each sum using counters.**

1. $3 + 2$ **5** **2.** $3 + (-2)$ **1** **3.** $-3 + 2$ **−1** **4.** $-3 + (-2)$ **−5**
5. $1 + (-4)$ **−3** **6.** $-3 + 7$ **4** **7.** $-4 + (-4)$ **−8** **8.** $-4 + 4$ **0**

Write About It **9.** Suppose you add a positive integer and a negative integer. Explain how you use a zero pair to find the sum. **See Solutions Manual.**

4 ASSESS

Observing students working in cooperative groups is an excellent method of assessment. You could also ask students to write a rule for adding integers.

GLENCOE Technology

Interactive Mathematics Tools Software

In this interactive lesson, students use an equation mat, cups, and counters to solve equations. A **Computer Journal** gives students the opportunity to write about what they have learned.

For Windows & Macintosh

2-4 Adding Integers

Setting Goals: *In this lesson, you'll add integers.*

Modeling with Technology

ne calculators use a +/− key, called the ge sign" key instead of the (−) key.

You already know that the sum of two positive integers is a positive integer. What is the sign of the sum of two negative integers? The keystrokes for finding the sum of −5 and −8 on a scientific calculator are as follows. *Notice that the negative key is different than the subtraction key and that the negative is entered after the number.*

Enter: 5 (−) + 8 (−) = **Display:** -13

Your Turn Use a calculator to study the patterns of the sums in each column.

A	B	C	D
4 + 8 = **12**	−3 + (−4) = **−7**	17 + (−17) = **0**	−18 + 2 = **−16**
12 + 3 = **15**	−4 + (−6) = **−10**	13 + (−2) = **11**	−5 + 12 = **7**
29 + 17 = **46**	−13 + (−2) = **−15**	12 + (−15) = **−3**	−8 + 21 = **13**
25 + 93 = **118**	−7 + (−3) = **−10**	11 + (−18) = **−7**	−7 + 7 = **0**

a. Suggest a rule for adding integers with the same signs, as in Columns A and B. **add the absolute values and keep the sign**

b. Suggest a rule for adding integers with different signs, as in Columns C and D. **See margin.**

c. Compare your rules to the rules of two of your classmates. **c. See students' work.**

Learning the Concept

Another way to add integers is by using a number line.

Example 1 **Find −5 + (−2).**

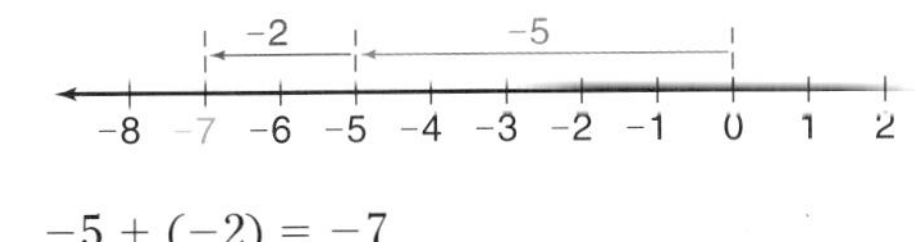

Start at zero. Move 5 units to the left. From there, move 2 more units to the left.

−5 + (−2) = −7

The results of this example and those in Columns A and B above suggest the following rule.

Adding Integers with the Same Signs

To add integers with the same sign, add their absolute values. Give the result the same sign as the integers.

Study Example 2 to see if you can discover the rule for adding integers that have different signs.

2-4 LESSON NOTES

NCTM Standards: 1-5, 7, 12

Instructional Resources

- Study Guide Master 2-4
- Practice Master 2-4
- Enrichment Master 2-4
- Group Activity Card 2-4
- Assessment and Evaluation Masters, pp. 42-43
- Activity Masters, p. 31

Transparency 2-4A contains the 5-Minute Check for this lesson; **Transparency 2-4B** contains a teaching aid for this lesson.

Recommended Pacing	
Standard Pacing	Day 5 of 12
Honors Pacing	Day 4 of 11
Block Scheduling*	Day 3 of 7

*For more information on pacing and possible lesson plans, refer to the ***Block Scheduling Booklet***.

1 FOCUS

5-Minute Check
(over Lesson 2-3)

Replace each ■ with <, > or =.

1. −9 ■ 8 <
2. 0 ■ −4 >

Write an inequality using the numbers in each sentence. Use the symbols < or >.

3. A steak sandwich costs $6. A steak dinner costs $11. **6 < 11**
4. Yesterday's low temperature was −4°F. The temperature now is −5°F. **−4 > −5**

Additional Answer Talk About It

b. Subtract the lessor absolute value from the greater absolute value. Keep the sign of the number with the greater absolute value.

Classroom Vignette

"You can use a staircase to represent a number line. Number each of the steps. Then have a student stand on the step representing the first addend and move up or down the stairs as determined by the second addend. The sum of the two integers is determined by the last step on which they land."

Robert Mandel
Hamden Middle School,
Hamden, CT

Motivating the Lesson

Situational Problem Use your students to model or act out the following situation: One student has $5 but owes another student $3. How much is left once the debt is paid? **$2** Continue with similar situations.

2 TEACH

In-Class Examples

For Example 1
Find −4 + (−3). **−7**

For Example 2
Find each sum.

a. 3 + (−7) **−4**

b. −5 + 5 **0**

For Example 3
Solve each equation.

a. $-2 + 9 = y$ **7**

b. $b = 3 + (-5) + (-6)$ **−8**

For Example 4
The temperature outside the space shuttle was −125°F. The temperature inside the shuttle was 193°F warmer. Find the temperature inside the space shuttle. **68°F**

Example  **Find each sum.**

a. 8 + (−4)

Recall that a zero pair is one positive and one negative counter.

Use counters. Put 8 positive counters on the mat. Add 4 negative counters. Remove all zero pairs.

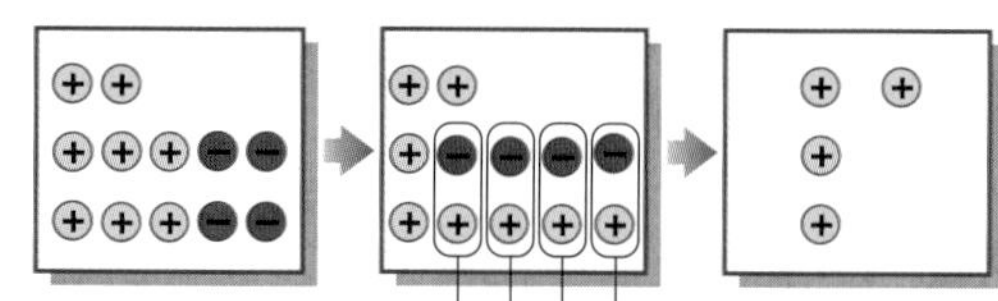

$8 + (-4) = 4$

b. −7 + 4

Use a number line. Start at zero. Move 7 units to the left. From there, move 4 units to the right.

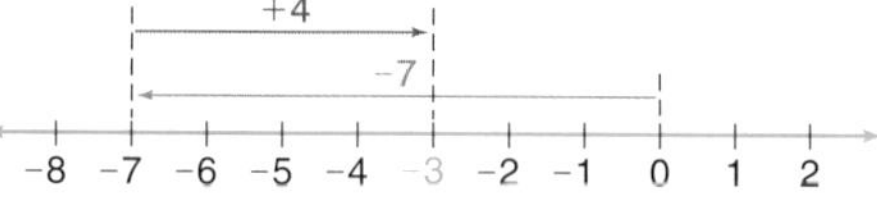

$-7 + 4 = -3$

The results in these examples and those in Columns C and D on page 83 suggest the following rule.

Adding Integers with Different Signs — **To add integers with different signs, subtract their absolute values. Give the result the same sign as the integer with the greater absolute value.**

Example  **Solve each equation.**

a. $x = -9 + 6$

$x = -(|-9| - |6|)$ *Subtract absolute values. The result is negative because the integer with the greater absolute value, −9, is negative.*
$x = -(9 - 6)$
$x = -3$

b. $8 + (-3) + 4 = a$

$8 + 4 + (-3) = a$ *Commutative property*
$12 + (-3) = a$ *Add 8 and 4.*
$+(12 - 3) = a$ *The result is positive because $|12| > |3|$.*
$9 = a$ *Check your result with a calculator.*

Example 4

Space Exploration

During a spacewalk on February 9, 1995, astronauts Bernard Harris Jr. and Michael Foale worked in the shadow of the shuttle *Discovery*. The temperature outside the cargo bay was −125°F. If the temperature inside the cargo bay was 111 degrees warmer, what was the temperature inside?

Explore The temperature outside the cargo bay was −125°F. Inside, it was 111 degrees warmer. You need to find the temperature inside. Let t represent the temperature inside.

Plan Translate the words into an equation using the variable. The temperature inside is the sum of −125 and 111.

Bernard Harris Jr. became the first African-American to walk in space on this *Discovery* flight.

Solve inside temperature is outside temperature plus 111°

$$t = -125 + 111$$

Solve the equation for t.

$t = -125 + 111$

125 [(-)] [+] 111 [=] -14 *Use a calculator to solve.*

The temperature inside the cargo bay was -14°F.

Examine The amount of increase was less than the absolute value of -125, so the temperature inside the cargo bay should still be negative. The answer is reasonable.

Connection to Algebra

You can use the rules for adding integers and the distributive property to combine like terms.

Example  **Simplify each expression.**

a. $-15n + 22n$

$-15n + 22n = (-15 + 22)n$ *Distributive property*

$= (7)n$ *Find the sum of −15 and 22.*

$= 7n$

b. $9x + (-18x) + 8x$

$9x + (-18x) + 8x = [9 + (-18) + 8]x$ *Distributive property*

$= -1x$ *Add 9, −18, and 8.*

$= -x$ *−1x can be written as −x.*

Checking Your Understanding

Communicating Mathematics

Read and study the lesson to answer these questions.

1. **Explain** how to determine whether to add or subtract the absolute values to find the sum of two integers. **See margin.**
2. **Write** the addition sentence shown by each model.

a. 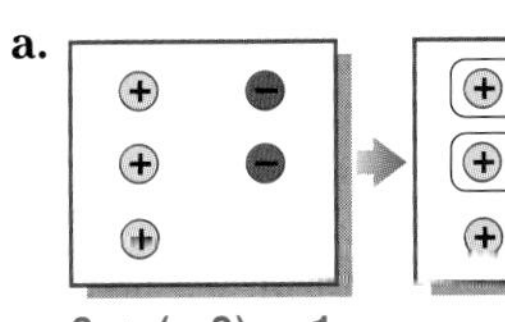

$3 + (-2) = 1$

b. 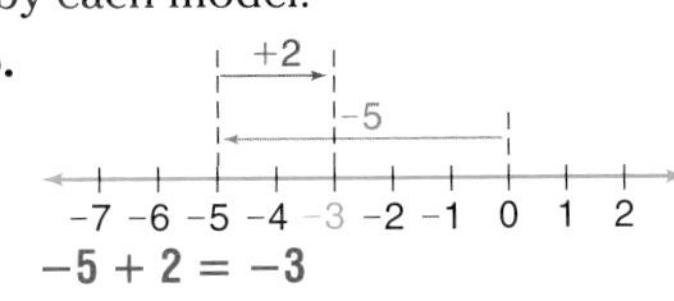

$-5 + 2 = -3$

3. **Draw a model** that shows how to find the sum of 5 and -2. **See margin.**
4. **Explain** how to simplify $-3a + 12a + (-14a)$. **4–5. See margin.**

5. Write, in your own words, how to add pairs of integers with the same sign and with different signs. Include an example for each situation.

Guided Practice

Write an addition sentence for each model.

6. 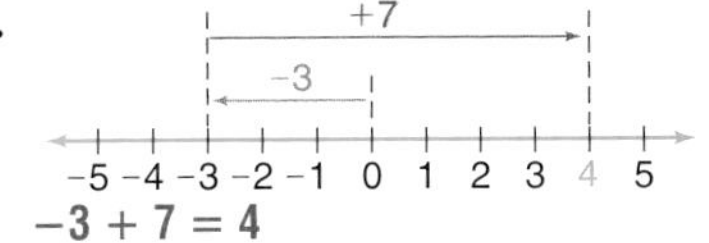

$-3 + 7 = 4$

7. 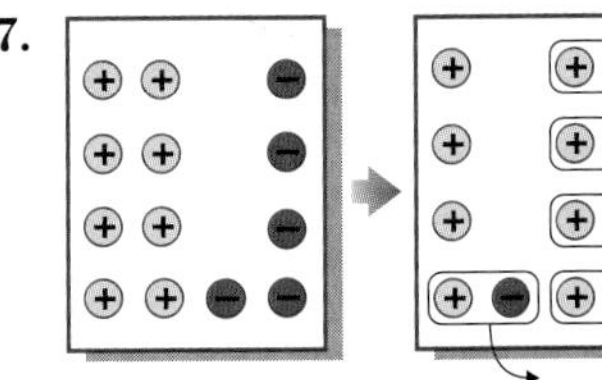

7. $8 + (-5) = 3$

In-Class Example

For Example 5
Simplify each expression.

a. $9d + (-5d)$ **4d**

b. $-3y + (-y) + 9y$ **5y**

Additional Answers

1. Sample answer: If the signs are the same, add. If the signs differ, subtract.
3. 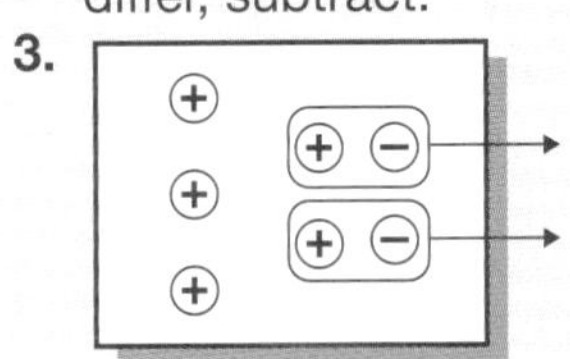
4. Sample answer: Use the distributive property to write the expression as $[-3 + 12 + (-14)]a$. Then use the commutative property to write it as $[-3 + (-14) + 12]a$. Add −3 and −14 to get −17. Then subtract the absolute value of 12 from the absolute value of −17 and give the result a negative sign. The simplified expression is $-5a$.
5. If the signs are alike just add and keep the sign. For example, $-4 + (-3) = -7$. If the signs differ, subtract the integer with the smaller absolute value from the integer with the larger absolute value and give the result with the sign of the integer with the larger absolute value. For example, $-6 + 4 = -2$.

Study Guide Masters, p. 14

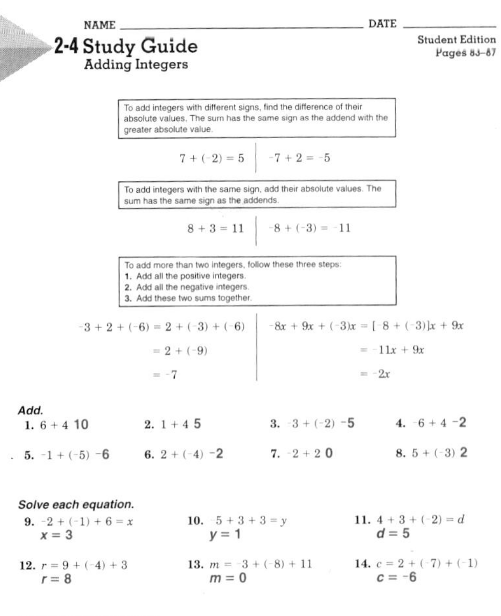

NAME ______ DATE ______

2-4 Study Guide
Adding Integers

Student Edition Pages 83–87

To add integers with different signs, find the difference of their absolute values. The sum has the same sign as the addend with the greater absolute value.

$7 + (-2) = 5$ | $-7 + 2 = -5$

To add integers with the same sign, add their absolute values. The sum has the same sign as the addends.

$8 + 3 = 11$ | $-8 + (-3) = -11$

To add more than two integers, follow these three steps:
1. Add all the positive integers.
2. Add all the negative integers.
3. Add these two sums together.

$-3 + 2 + (-6) = 2 + (-3) + (-6)$
$= 2 + (-9)$
$= -7$

$-8x + 9x + (-3)x = [-8 + (-3)]x + 9x$
$= -11x + 9x$
$= -2x$

Add.

1. $6 + 4$ **10**
2. $1 + 4$ **5**
3. $-3 + (-2)$ **−5**
4. $-6 + 4$ **−2**
5. $-1 + (-5)$ **−6**
6. $2 + (-4)$ **−2**
7. $-2 + 2$ **0**
8. $5 + (-3)$ **2**

Solve each equation.

9. $-2 + (-1) + 6 = x$ **x = 3**
10. $-5 + 3 + 3 = y$ **y = 1**
11. $4 + 3 + (-2) = d$ **d = 5**
12. $r = 9 + (-4) + 3$ **r = 8**
13. $m = -3 + (-8) + 11$ **m = 0**
14. $c = 2 + (-7) + (-1)$ **c = −6**

Simplify each expression.

15. $-4x + 7x + (-6)x$ **−3x**
16. $8f + (-2) + (-5)f$ **1f or f**
17. $-7e + (-10)e + 7e$ **−10e**
18. $3y + 6y + (-10)y$ **−1y or −y**
19. $-10t + 9t + 3t$ **2t**
20. $5d + (-1)d + (-8)d$ **−4d**

Reteaching

Using Models Have students sketch a model of a twenty-yard football field. Modify the field by making the 50-yard line the 0-yard line. Using a coin or counter to represent a football, have students place the ball on the 0-yard line. Describe two plays at a time and ask where the ball ends up in relation to the starting point.

FYI: During the five-hour space walk, Harris and Foale tested thermal improvements in their space suits and practiced lifting a 2,800-pound telescope to help NASA plan for future expeditions.

3 PRACTICE/APPLY

Checking Your Understanding

Exercises 1–18 are designed to help you assess your students' understanding through reading, writing, speaking, and modeling. You should work through Exercises 1–5 with your students and then monitor their work on Exercises 6–18.

Error Analysis

Some students may use the sign of the first addend for the sign of the sum. Emphasize the addition process by using a number line.

Assignment Guide

Core: 19–55 odd, 57–65
Enriched: 20–52 even, 53–65

For **Extra Practice**, see p. 744.

The red A, B, and C flags, printed only in the Teacher's Wraparound Edition, indicate the level of difficulty of the exercises.

Teaching Tip For Exercises 19–33, allow students to use counters or draw a numbers line to check their answers.

Practice Masters, p. 14

NAME ______________________ DATE ____________

2-4 Practice
Adding Integers

Student Edition
Pages 83–87

Solve each equation.

1. $x = -7 + (-5)$ **−12**
2. $10 + 9 = n$ **19**
3. $w = -12 + (-5)$ **−17**
4. $t = -13 + (-3)$ **−16**
5. $-10 + 12 = z$ **2**
6. $-7 + 8 = k$ **1**
7. $m = -11 + (-6)$ **−17**
8. $0 + (-21) = b$ **−21**
9. $72 + (-10) = c$ **62**
10. $d = 72 + 10$ **82**
11. $-13 + (-11) = h$ **−24**
12. $f = -52 + 52$ **0**
13. $6 + 5 + (-4) = t$ **$t = 7$**
14. $-4 + (-5) + 6 = m$ **$m = -3$**
15. $k = -3 + 8 + (-9)$ **$k = -4$**
16. $a = -6 + (-2) + (-1)$ **$a = -9$**
17. $10 + (-5) + 6 = n$ **$n = 11$**
18. $c = -8 + 8 + (-10)$ **$c = -10$**
19. $36 + (-28) + (-16) + 24 = y$ **$y = 16$**
20. $x = -31 + 19 + (-15) + (-6)$ **$x = -33$**

Simplify each expression.

21. $6y + (-13y)$ **$-7y$**
22. $-12z + (-9z)$ **$-21z$**
23. $-8x + 9x + (-3x)$ **$-2x$**
24. $18e + (-7e) + (-14e)$ **$-3e$**
25. $5m + 29m + (-15m)$ **$19m$**
26. $-3d + (-8d) + (-17d)$ **$-28d$**
27. $12n + (-25n) + 20n$ **$7n$**
28. $-9t + (-9t) + 17t$ **$-t$**

8. 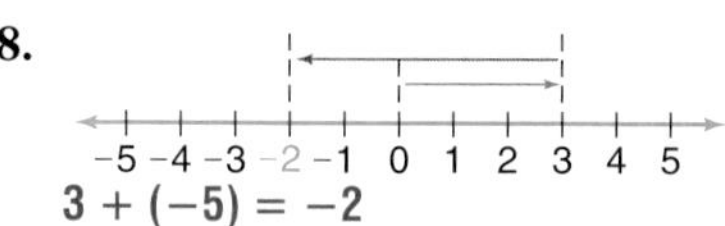

$3 + (-5) = -2$

9. 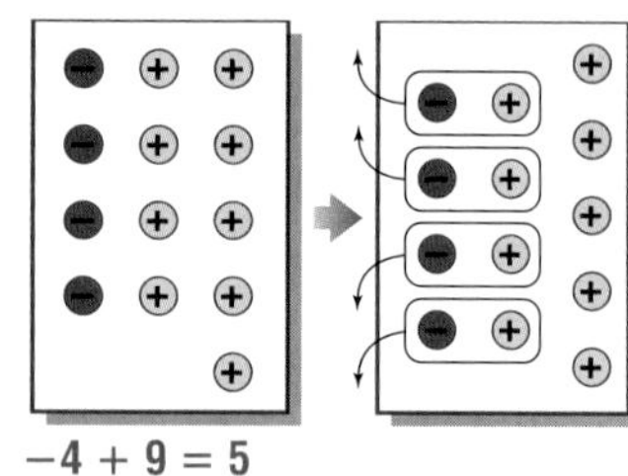
$-4 + 9 = 5$

State whether each sum is positive or negative. Then find the sum.

10. $-6 + (-9)$ **negative, −15**
11. $13 + (-4)$ **positive, 9**
12. $-8 + 12$ **positive, 4**
13. $-3 + 7 + (-8)$ **negative, −4**

Solve each equation.

14. $r = -13 + (-5) + 7$ **−11**
15. $(-83) + (-21) + (-7) = z$ **−111**

Simplify each expression.

16. $-3z + (-17z)$ **$-20z$**
17. $-8a + 14a + (-12a)$ **$-6a$**
18. **Golf** In golf, a score of 0 is called *even par.* A score of 3 under par is written as −3. A score of 2 over par is written +2 or 2. In 1995, Jose Maria Olazabal shot 6 under par, 2 over par, even par, and even par for the four rounds of the Masters Tournament. What was his final score? **−4 or 4 under par**

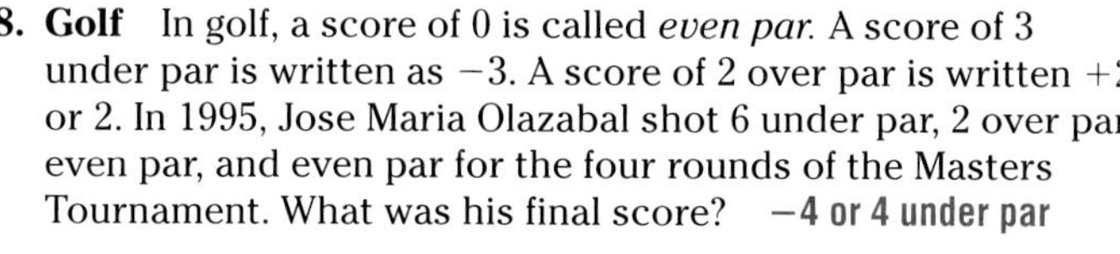

Exercises: Practicing and Applying the Concept

Independent Practice

Solve each equation. **20. −11 22. −16 26. −26 32. −17**

19. $18 + (-5) = x$ **13**
20. $b = -8 + (-3)$
21. $16 + (-9) = m$ **7**
22. $v = -12 + (-4)$
23. $k = 9 + (-13)$ **−4**
24. $-15 + 6 = q$ **−9**
25. $g = 19 + (-7)$ **12**
26. $-11 + (-15) = t$
27. $-23 + (-43) = h$ **−66**
28. $-3 + 18 = n$ **15**
29. $-5 + 31 = r$ **26**
30. $y = 6 + (-16)$ **−10**
31. $m = 8 + (-17)$ **−9**
32. $z = -12 + (-5)$
33. $-12 + 5 = s$ **−7**

34. 500 + (−700); −200

Write an addition sentence for each situation. Then find the sum.

34. Maria has $500 in the bank. She owes $700 on her car.
35. In Monday night's football game, the Jacksonville Jaguars lost 7 yards on one play. They gained 12 yards on the next. **−7 + 12; 5**
36. The research submarine *Alvin* is at 1500 meters below the sea level. It descends another 1250 meters to the ocean floor.

B

37. **Make Up a Problem** Write a problem that can be solved using the addition sentence $-15 + 25 = c$. **See students' work.**

36. −1500 + (−1250); −2750

Solve each equation.

38. $m = 3 + (-11) + (-5)$ **−13**
39. $k = 14 + 9 + (-2)$ **21**
40. $-18 + (-23) + 10 = c$ **−31**
41. $-16 + (-6) + (-5) = a$ **−27**
42. $-12 + (17) + (-7) = x$ **−2**
43. $y = 47 + 32 + (-16)$ **63**

Group Activity Card 2-4

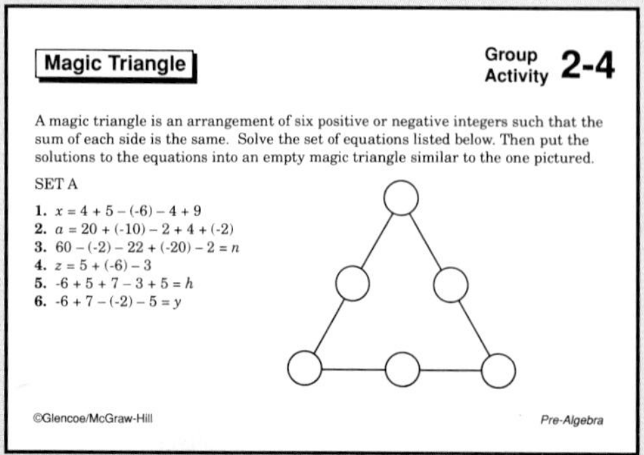

Magic Triangle

Group Activity **2-4**

A magic triangle is an arrangement of six positive or negative integers such that the sum of each side is the same. Solve the set of equations listed below. Then put the solutions to the equations into an empty magic triangle similar to the one pictured.

SET A

1. $x = 4 + 5 - (-6) - 4 + 9$
2. $a = 20 + (-10) - 2 + 4 + (-2)$
3. $60 - (-2) - 22 + (-20) - 2 = n$
4. $z = 5 + (-6) - 3$
5. $-6 + 5 + 7 - 3 + 5 = h$
6. $-6 + 7 - (-2) - 5 = y$

Pre-Algebra

Simplify each expression.

44. $6x + (-15)x$ **$-9x$**
45. $-11y + 14y$ **$3y$**
46. $-14k + (-7)k + 15k$ **$-6k$**
47. $8m + (-23)m$ **$-15m$**
48. $14b + (-21b) + 37b$ **$30b$**
49. $16d + (-9d) + (-27d)$ **$-20d$**

Calculator

Use a calculator to find each sum.

50. $139 + (-316)$ **-177**
51. $-249 + 7915$ **7666**
52. $-3916 + (-8128)$ **$-12{,}044$**

Critical Thinking

53. *True* or *false*: $-n$ names a negative number. **false**

Applications and Problem Solving

54. **Astronomy** At noon, the average temperature on the moon is 112°C. During the night, the average temperature drops 252°C. What is the average temperature on the moon's surface during the night? **−140°C**

55. **Pets** The chart at the right shows the numbers of dogs of different breeds registered in the American Kennel Club in 1992 and 1993.

Breed	1992 Registration	1993 Registration
Akita	11,574	11,383
Beagle	61,051	60,661
Chow Chow	33,824	42,670
Dachshund	48,573	50,046
Labrador Retriever	124,899	120,879
Pug	15,722	16,008

a. Describe the change in the number of dogs of each breed registered from 1992 to 1993. **See margin.**

b. What was the total change in the number of dogs of these breeds registered from 1992 to 1993? **+6004**

Mixed Review

56. Order the numbers {0, −4, −8, −3} from least to greatest. (Lesson 2-3)

57. *True* or *false*: The graph of a point whose coordinates are both negative integers would be located in Quadrant IV. (Lesson 2-2) **false**

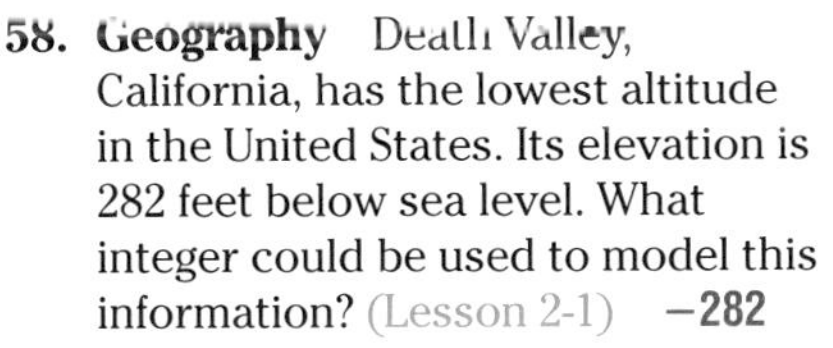

58. **Geography** Death Valley, California, has the lowest altitude in the United States. Its elevation is 282 feet below sea level. What integer could be used to model this information? (Lesson 2-1) **−282**

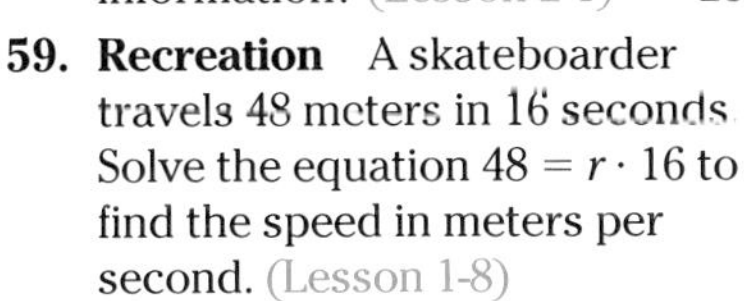

59. **Recreation** A skateboarder travels 48 meters in 16 seconds. Solve the equation $48 = r \cdot 16$ to find the speed in meters per second. (Lesson 1-8)

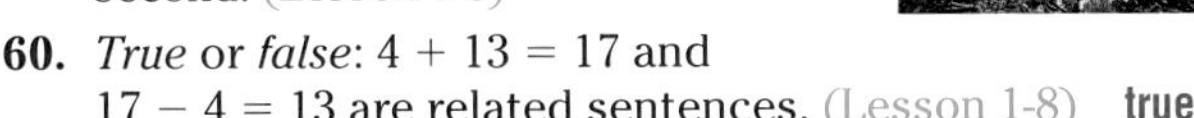

60. *True* or *false*: $4 + 13 = 17$ and $17 - 4 = 13$ are related sentences. (Lesson 1-8) **true**

61. Write an equation and solve: Antonia paid $15 for a shirt on sale. It was reduced by $9. What was the regular price? (Lesson 1-8)

62. Of 19, 29, or 39, which is the solution of $3x = 87$? (Lesson 1-6) **29**

63. Rewrite $(n + 8) + 9$ using an associative property. Then simplify. (Lesson 1-4) **$n + (8 + 9)$; $n + 17$**

64. Write a verbal phrase for the expression $2x + 3$. (Lesson 1-3)

65. Determine whether $6 \div 2 + 5 \times 4 = 32$ is *true* or *false*. (Lesson 1-2) **false**

56. $\{-8, -4, -3, 0\}$
59. 3 m/s
61. $r - 9 = 15$; $24

64. See margin.

4 ASSESS

Closing Activity

Writing Have students write a descriptive phrase for what $-6 + 8$ could represent. Then explain how to find the sum.

Chapter 2, Quiz B (Lessons 2-3 and 2-4) is available in the *Assessment and Evaluation Masters*, p. 43.

Mid-Chapter Test (Lessons 2-1 through 2-4) is available in the *Assessment and Evaluation Masters*, p. 42.

Additional Answers

55a.

Breed	Change
Akita	−191
Beagle	−390
Chow Chow	+8846
Dachshund	+1473
Labrador Retriever	−4020
Pug	+286

64. Sample answer: two times a number plus three

Enrichment Masters, p. 14

NAME ______________________ DATE __________

Student Edition Pages 83–87

2-4 Enrichment
Modular Arithmetic

In modular arithmetic, there is a finite set of numbers. An example of a finite set would be the numbers on the face of a clock. No matter how many numbers are added together, the answer will still be one of the twelve numbers found on the face of the clock.

Example: In mod 12, $5 + 8 + 2 = 3$.

To find the sum, start at 0. Count 5 units in a clockwise direction, then 8 units, and then 2 units. You should end at the number 3.

Find each sum in mod 12.

1. $6 + 8 + 4$ **6**
2. $7 + 9 + 11 + 5$ **8**
3. $6 + 4 + 11 + 12 + 5$ **2**
4. $1 + 4 + 9 + 11 + 6$ **7**
5. $7 + 10 + 9 + 12 + 10$ **0**
6. $6 + 12 + 5 + 12 + 8 + 12$ **7**
7. $12 + 12 + 12 + 12 + 6$ **6**
8. $4 + 9 + 3 + 7 + 8 + 11 + 5$ **11**
9. $6 + 10 + 7 + 8 + 4 + 3 + 9$ **11**
10. $9 + 2 + 4 + 3 + 10 + 12 + 4 + 11$ **7**

Example: In mod 7, $3 + 5 + 2 = 3$.

To find the sum, start at 0. Count 3 units in a clockwise direction, then 5 units, and then 2 units. You should end at the number 3.

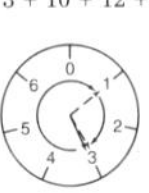

Find each sum in mod 7.

11. $2 + 4 + 3$ **2**
12. $5 + 1 + 6 + 3$ **1**
13. $2 + 2 + 4 + 6 + 7$ **0**
14. $5 + 3 + 4 + 6 + 1$ **5**
15. $7 + 7 + 7 + 7 + 5$ **5**
16. $6 + 5 + 1 + 2 + 4$ **4**
17. $4 + 3 + 1 + 5 + 6 + 2$ **0**
18. $4 + 2 + 1 + 6 + 7 + 5 + 4 + 3$ **4**
19. $6 + 5 + 4 + 3 + 2 + 1 + 7 + 6$ **6**
20. $5 + 7 + 4 + 4 + 2 + 6 + 1 + 2$ **3**
21. Can you find a pattern to these exercises? Explain. **Yes; the sum in mod x is the remainder when the sum is divided by x.**

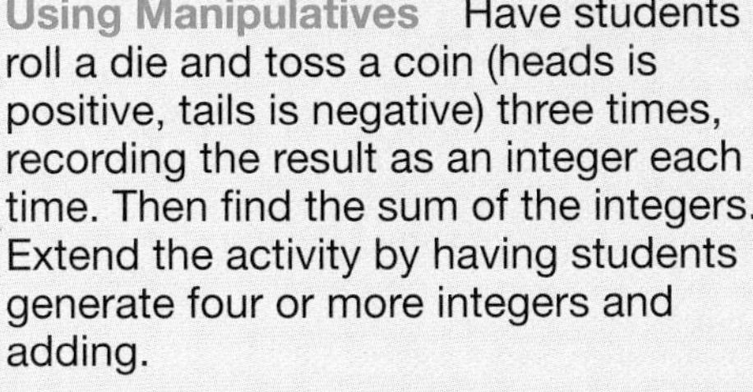

Extension

Using Manipulatives Have students roll a die and toss a coin (heads is positive, tails is negative) three times, recording the result as an integer each time. Then find the sum of the integers. Extend the activity by having students generate four or more integers and adding.

Tech Prep

Veterinary Technician In exercise 55, some students may be interested in a career as a veterinary technician. They keep records, take specimens, assist in surgery, and dress wounds. For more information on tech prep, see the *Teacher's Handbook*.

2-5A LESSON NOTES

NCTM Standards: 1-5, 7, 12

Objective

Use counters to model subtraction of integers.

Recommended Time

Demonstration and discussion: 15 minutes; Exercises: 30 minutes

Instructional Resources

For each student or group of students

Student Manipulative Kit
- counters
- integer mat

Math Lab and Modeling Math Masters
- p. 2 (integer mat)
- p. 5 (integer counters)
- p. 31 (worksheet)

For teacher demonstration

Overhead Manipulative Resources

1 FOCUS

Motivating the Lesson

Ask students to consider the following situation. Your friend owes you $12, but he only has $10 in his wallet. He gives you the $10. How would you describe the situation and end result mathematically? **−12 + 10 = −2**

2 TEACH

Teaching Tip Students who have trouble understanding the results of Activity 2 may benefit from a different approach). Have students find the differences and identify a pattern for 4 − 4, 4 − 3, 4 − 2, 4 − 1, 4 − 0. Then ask them to continue the pattern and predict the next two differences.

3 PRACTICE/APPLY

Assignment Guide
Core: 1–10
Enriched: 1–10

MATHEMATICS LAB

HANDS-ON ACTIVITY

2-5A Subtracting Integers

A Preview of Lesson **2-5**

MATERIALS

 counters

integer mat

You can use counters to model the subtraction of integers.

Activity 1 **Find −6 − (−3).**

- Start with 6 negative counters. Remove 3 negative counters.

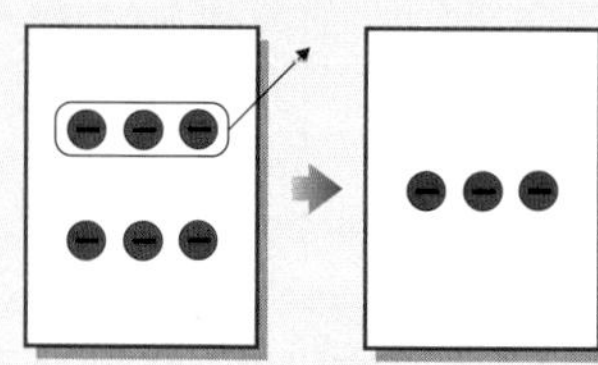

- Since there are 3 negative counters left, the difference is −3. Therefore, −6 − (−3) = −3.

Activity 2 **Model 4 − (−2).**

- Start with 4 positive counters. There are no negative counters, so you can't remove 2 negative counters. Add 2 zero pairs to the mat. Remember adding zero pairs does not change the value of the set. Now remove 2 negative counters.

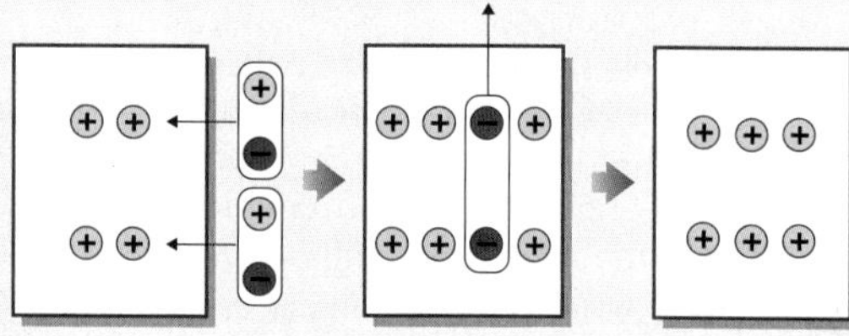

- Since there are 6 positive counters remaining, the difference is +6 or 6. Therefore, 4 − (−2) = 6.

Your Turn **Find each difference using counters.**

1. 2 − (−3) **5**
2. 2 − 3 **−1**
3. −2 − (−3) **1**
4. −2 − 3 **−5**
5. −6 − 9 **−15**
6. 13 − 4 **9**
7. −8 − (−12) **4**
8. −4 − (−4) **0**

TALK ABOUT IT

9. Compare Exercises 5–7 above with Exercises 10–12 on page 86. How are they alike? How are they different? **See Solutions Manual.**
10. If you subtract a negative integer from a lesser negative integer, is the difference positive or negative? Justify your answer with a drawing. **See Solutions Manual.**

4 ASSESS

Observing students working in cooperative groups is an excellent method of assessment. As you circulate, you may want to ask individuals to summarize their modeling procedures or results.

GLENCOE Technology

 **Interactive Mathematics Tools Software**

In this interactive lesson, students use an equation mat, cups, and counters to solve equations. A **Computer Journal** gives students the opportunity to write about what they have learned.

For Windows & Macintosh

2-5 Subtracting Integers

Setting Goals: *In this lesson, you'll subtract integers.*

Modeling with Manipulatives

The diagram at the right models the expression $3-(-3)$. Counting the number of counters on the mat after the counters are removed shows that $3-(-3)=6$.

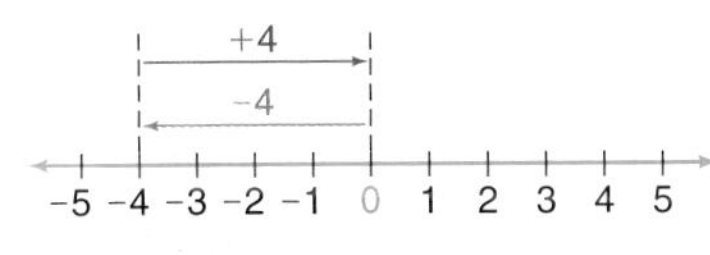

Your Turn Use counters to find $-3-(-2)$, $4-6$, and $-5-1$.

TALK ABOUT IT

a. Compare the solution of $-3-(-2)$ to the solution of $-3+2$.

b. Compare the solution of $4-6$ to the solution of $4+(-6)$.

c. Compare the solution of $-5-1$ to the solution of $-5+(-1)$.

a–c. The solutions are the same.

Learning the Concept

Adding and subtracting are inverse operations that "undo" each other. Similarly, when you add opposites, like 4 and -4, the sum is 0.

LOOK BACK You can review inverse operations in Lesson 1-8.

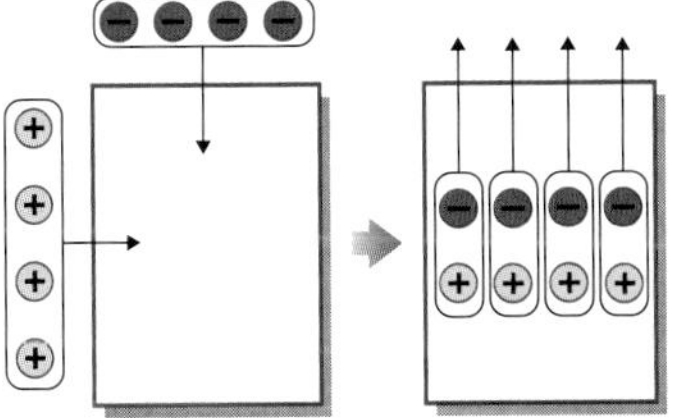

An integer and its opposite are called **additive inverses** of each other.

Additive Inverse Property

In words: The sum of an integer and its additive inverse is zero.

Arithmetic	**Algebra**
$5+(-5)=0$	$x+(-x)=0$

At the beginning of the lesson, you were asked to compare the result of subtracting an integer with the result of adding its additive inverse.

Subtraction	**Addition of Additive Inverse**
$-3-(-2)=-1$	$-3+2=-1$
$4-6=-2$	$4+(-6)=-2$
$-5-1=-6$	$-5+(-1)=-6$

These examples suggest a method for subtracting integers.

Subtracting Integers

In words: To subtract an integer, add its additive inverse.

Arithmetic	**Algebra**
$3-5=3+(-5)$	$a-b=a+(-b)$

Alternative Learning Styles

Visual As each example is discussed, have students use a number line to show what is being done and why. Be sure students understand that subtracting a number is the same as adding the opposite of a number.

2-5 LESSON NOTES

NCTM Standards: 1-5, 7, 12

Instructional Resources

- Study Guide Master 2-5
- Practice Master 2-5
- Enrichment Master 2-5
- Group Activity Card 2-5
- Multicultural Activity Masters, p. 3
- Real-World Applications, 3
- Tech Prep Applications Masters, p. 4

Transparency 2-5A contains the 5-Minute Check for this lesson; **Transparency 2-5B** contains a teaching aid for this lesson.

Recommended Pacing	
Standard Pacing	Day 7 of 12
Honors Pacing	Day 6 of 11
Block Scheduling*	Day 4 of 7

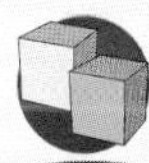

*For more information on pacing and possible lesson plans, refer to the ***Block Scheduling Booklet***.

1 FOCUS

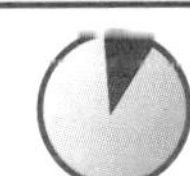

5-Minute Check
(over Lesson 2-4)

Solve each equation.

1. $-4+(-3)=a$ **−7**
2. $b=2+(-6)$ **−4**
3. $-1+4+(-6)=c$ **−3**

Simplify each expression.

4. $15x+(-22x)+2x$ **$-5x$**
5. $-2x+(-6x)+(-5x)$ **$-13x$**

Motivating the Lesson

Questioning Tell students that you borrowed $5 from a friend and later borrowed another $3. Ask students:

- How much do I owe my friend? **$8**
- How can this example be written as an addition problem? **-5 + (-3) = -8**
- How can this example be written as a subtraction problem? **-5 − 3 = -8**

2 TEACH

In-Class Examples

For Example 1
Solve each equation.

a. $m = -3 - 4$ **-7**

b. $n = 3 - 4$ **-1**

c. $-3 - (-4) = p$ **1**

d. $m = 3 - (-4)$ **7**

For Example 2
The outside temperature was 8°F. The windchill factor made it feel like it was -12°F. Find the difference between the real temperature and the apparent temperature. **-20°F**

For Example 3
Evaluate $d - (-12)$ if $d = -9$. **3**

For Example 4
Simplify $7b - 11b$. **-4b**

3 PRACTICE/APPLY

Checking Your Understanding

Exercises 1–22 are designed to help you assess your students' understanding through reading, writing, speaking, and modeling. You should work through Exercises 1–5 with your students and then monitor their work on Exercises 6–22.

Example  **Solve each equation.**

a. $m = 7 - 12$
$m = 7 + (-12)$ *To subtract 12, add −12.*
$m = -5$

b. $6 - (-11) = r$
$6 + 11 = r$ *To subtract −11, add 11.*
$17 = r$

c. $y = -3 - (-8)$
$y = -3 + 8$ *To subtract −8, add 8.*
$y = 5$

d. $-9 - 16 = x$
$-9 + (-16) = x$ *To subtract 16, add −16.*
$-25 = x$

You can use equations with integers to represent real-life situations.

Example 2

Geography

The highest point in Africa is Kilimanjaro in Tanzania. It has an altitude of 5895 meters above sea level. The lowest point on the continent is Lake Assal in Djibouti. It is 155 meters below sea level. Find the difference between these altitudes.

Explore With sea level as the starting point, 5895 meters above sea level can be expressed as +5895 or 5895. 155 meters below sea level can be expressed as −155. Let d represent the difference.

Plan Translate the words into an equation using the variable.

difference	*is*	*altitude of Kilimanjaro*	*minus*	*altitude of Lake Assal*
d	$=$	5895	$-$	-155

Solve Solve the equation for d.

$d = 5895 - (-155)$ *To subtract −155*
$d = 5895 + 155$ *add its inverse, 155.*
$d = 6050$

The difference in altitudes is 6050 meters.

Examine Make a diagram to check the solution.

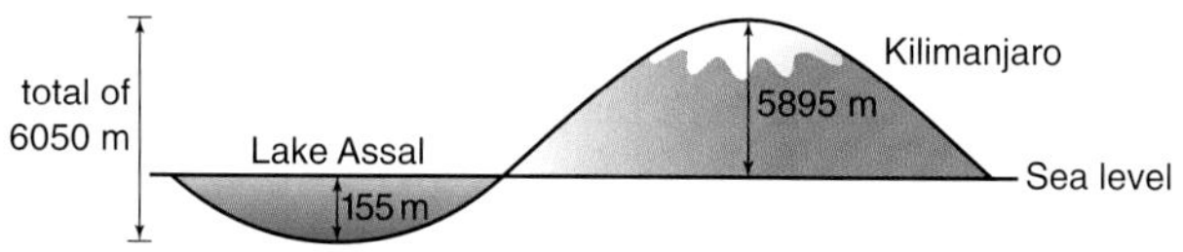

Connection to Algebra

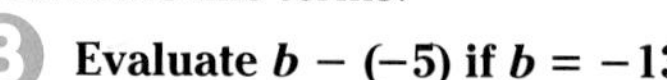

You can use the rule for subtracting integers to evaluate expressions and combine like terms.

Example 3 **Evaluate $b - (-5)$ if $b = -13$.**

$b - (-5) = -13 - (-5)$ *Replace b with −13.*
$= -13 + 5$ *Subtract −5 by adding 5.*
$= -8$

Example  **Simplify $5a - 13a$.**

$5a - 13a = 5a + (-13)a$ *Add the additive inverse of 13a, −13a.*
$= [5 + (-13)]a$ *Use the distributive property.*
$= -8a$ *Add 5 and −13.*

Checking Your Understanding

Communicating Mathematics

Read and study the lesson to answer these questions. 1. See margin.

1. **Explain** how subtraction of integers is related to addition of integers.
2. **Write** the subtraction sentence shown by each model.

a. 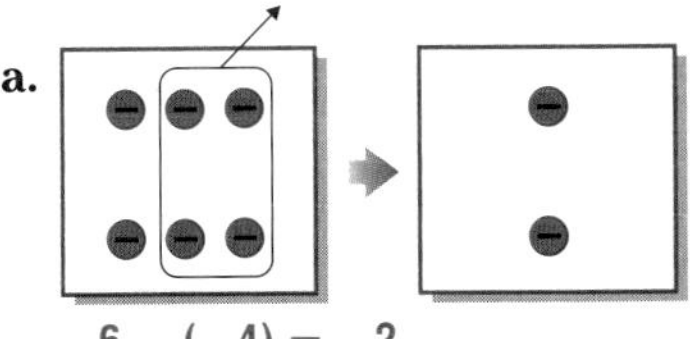

$-6 - (-4) = -2$

b. $2 - 5 = -3$

3. **Draw a model** that shows how to find $5 - (-2)$. See margin.
4. **Explain** how to simplify $12y - 14y$. See margin.
5. Using masking tape, create a large number line on the floor. Then "walk out" the solution to each problem using the following rules.
   - Stand at zero and face the positive integers.
   - To model a positive integer, move forward.
   - To model a negative integer, move backward.
   - To model subtraction, do an about-face before moving.

   a. $4 - 6$ **−2** **b.** $2 - (-3)$ **5** **c.** $-3 - 1$ **−4** **d.** $-5 - (-4)$ **−1**

MODELING MATHEMATICS

MATERIALS

masking tape

Guided Practice

State the additive inverse of each of the following.

6. $+7$ **−7**
7. -8 **8**
8. $-a$ ***a***
9. $9x$ **−9*x***

Rewrite each equation using the additive inverse. Then solve.

10. $6 + (-15) = m$; −9
11. $-7 + (-11) = x$; −18
12. $16 + (-7) = z$; −23

10. $6 - 15 = m$
11. $-7 - 11 = x$
12. $-16 - 7 = z$
13. $8 - (-3) = b$ **$8 + 3 = b$; 11**
14. $-24 - (-8) = r$ **$-24 + 8 = r$; −16**
15. $-14 - (-19) = y$ **$-14 + 19 = y$; 5**

Simplify.

16. $6a - 13a$ **−7*a***
17. $-15x - 12x$ **−27*x***
18. $-17y - (-3)y$ **−14*y***

interNET CONNECTION

For the other windchill factors, visit: **www.glencoe.com/sec/math/prealg/mathnet**

Evaluate each expression. 19. −12 20. 25 21. −8

19. $x - 3$, if $x = -9$
20. $y - (-8)$, if $y = 17$
21. $a - (-7)$, if $a = -15$
22. **Meteorology** *Windchill* factor is an estimate of the cooling effect the wind has on a person in cold weather. If the outside temperature is 10°F and the wind makes it feel like −25°F, what is the difference between the actual temperature and how cold it feels? **−35°F**

Exercises: Practicing and Applying the Concept

Independent Practice

Solve each equation.

A

23. $y = -9 - 5$ **−14**
24. $a = -13 - 4$ **−17**
25. $r = -8 - (-5)$ **−3**
26. $x = -12 - (-7)$ **−5**
27. $m = 12 - (-20)$ **32**
28. $k = 23 - (-24)$ **47**
29. $-5 - (-14) = p$ **9**
30. $-12 - (-16) = z$ **4**
31. $31 - (-6) = d$ **37**
32. $-17 - 13 = a$ **−30**
33. $-21 - 8 = b$ **−29**
34. $38 - (-12) = q$ **50**

B

35. $x = -28 - 35$ **−63**
36. $-7 - 25 = f$ **−32**
37. $-39 - 7 = s$ **−46**

Teaching Tip Exercises 11–16 require students to rewrite the subtraction equation as an addition equation. Monitor work carefully. If necessary, have students rewrite Exercises 24–38 as addition problems before they begin working independently.

Assignment Guide

Core: 23–57 odd, 59–69
Enriched: 24–54 even, 55–69
All: Self Test, 1–14

For **Extra Practice**, see p. 744.

The red A, B, and C flags, printed only in the Teacher's Wraparound Edition, indicate the level of difficulty of the exercises.

Additional Answers

1. Sample answer: subtraction is adding the opposite
3. 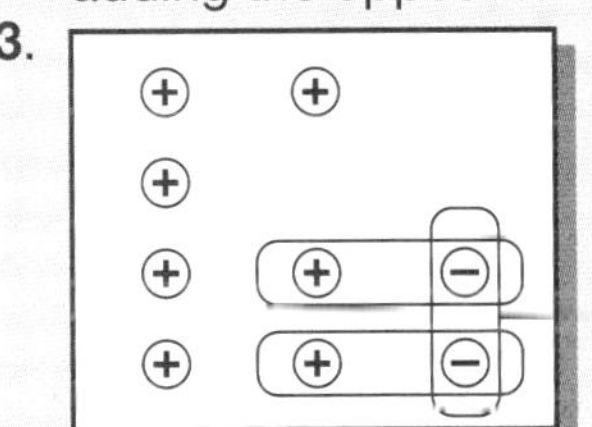
4. Sample answer: First, subtract $14y$ by adding $-14y$. Then use the distributive property to rewrite the expression. Find the sum of 12 and −14.

Reteaching

Using Models Have each student write two positive integers on cards. Have each student write all the subtraction problems that would use the integers or their opposites. For example, if the cards show 2 and 3, write the problems $2 - 3$, $2 - (-3)$, $-2 - 3$, and $-2 - (-3)$. Solve. To check the solutions, model the four problems on a number line or with counters.

Study Guide Masters, p. 15

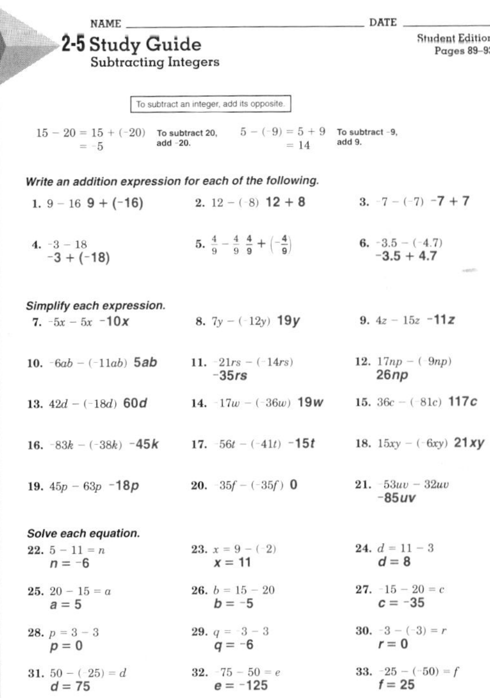

NAME ______________________ DATE __________

2-5 Study Guide
Subtracting Integers

Student Edition Pages 89–93

To subtract an integer, add its opposite.

$15 - 20 = 15 + (-20) = -5$ To subtract 20, add −20.

$5 - (-9) = 5 + 9 = 14$ To subtract −9, add 9.

Write an addition expression for each of the following.

1. $9 - 16$ **$9 + (-16)$**
2. $12 - (-8)$ **$12 + 8$**
3. $-7 - (-7)$ **$-7 + 7$**
4. $-3 - 18$ **$-3 + (-18)$**
5. $\frac{4}{9} - \frac{4}{9}$ **$\frac{4}{9} + \left(-\frac{4}{9}\right)$**
6. $-3.5 - (-4.7)$ **$-3.5 + 4.7$**

Simplify each expression.

7. $-5x - 5x$ **−10*x***
8. $7y - (-12y)$ **19*y***
9. $4z - 15z$ **−11*z***
10. $-6ab - (-11ab)$ **5*ab***
11. $-21rs - (-14rs)$ **−35*rs***
12. $17np - (-9np)$ **26*np***
13. $42d - (-18d)$ **60*d***
14. $-17w - (-36w)$ **19*w***
15. $36c - (-81c)$ **117*c***
16. $-83k - (-38k)$ **−45*k***
17. $-56t - (-41t)$ **−15*t***
18. $15xy - (-6xy)$ **21*xy***
19. $45p - 63p$ **−18*p***
20. $-35f - (-35f)$ **0**
21. $-53uv - 32uv$ **−85*uv***

Solve each equation.

22. $5 - 11 = n$ **$n = -6$**
23. $x = 9 - (-2)$ **$x = 11$**
24. $d = 11 - 3$ **$d = 8$**
25. $20 - 15 = a$ **$a = 5$**
26. $b = 15 - 20$ **$b = -5$**
27. $-15 - 20 = c$ **$c = -35$**
28. $p = 3 - 3$ **$p = 0$**
29. $q = -3 - 3$ **$q = -6$**
30. $-3 - (-3) = r$ **$r = 0$**
31. $50 - (-25) = d$ **$d = 75$**
32. $-75 - 50 = e$ **$e = -125$**
33. $-25 - (-50) = f$ **$f = 25$**

Teaching Tip You may wish to supply students with a check register from a checking account and a starting balance. Have students list checks and deposits, including amounts that a business of their choice would make. Have each student determine a final balance. (Emphasize that negative balances are not acceptable.)

Additional Answer

55. Sample answer: Every integer has an additive inverse because every integer has an opposite. The sum of an integer and its opposite is zero. Zero is its own additive inverse because $0 + 0 = 0$.

Evaluate each expression.

38. $11 - n$, if $n = 5$ **6**
39. $-15 - c$, if $c = 5$ **−20**
40. $h - (-17)$, if $h = -23$ **−6**
41. $r - 11$, if $r = -16$ **−27**
42. $8 - m$, if $m = -12$ **20**
43. $h - (-13)$, if $h = -18$ **−5**
44. $5 - b$, if $b = 8$ **−3**
45. $7 - (-f)$, if $f = 19$ **26**

Simplify. **52. −33*xy* 54. 12*q***

46. $5b - 17b$ **−12*b***
47. $13x - 23x$ **−10*x***
48. $-3a - (-5a)$ **2*a***
49. $18p - 3p$ **15*p***
50. $-28d - 17d$ **−45*d***
51. $12cd - (-12cd)$ **24*cd***
52. $-15xy - 18xy$
53. $7a - 14a - a$ **−8*a***
54. $-q - (-6q) - (-7q)$

Critical Thinking

55. Explain why every integer has an additive inverse. Then identify the integer that is its own additive inverse. **See margin.**

Applications and Problem Solving

56. **Aviation** Steve Fossett was the first person to fly a balloon solo across the Pacific Ocean. He took off from Seoul, South Korea, on February 18, 1995, and landed in Saskatchewan, Canada, three days later. When the heaters in his gondola failed, he had to endure temperatures that ranged from 10°F to −4°F. What was the change from the highest to lowest temperatures he experienced? **−14°F**

57. **Accounting** Dawn Sopher is a business information analyst. She uses spreadsheet software to keep track of accounts payable. The balance of an account can be negative or positive. Purchases are subtracted from the balance, and payments are added to the balance. Find the final balance on the spreadsheet below. **−$2133**

	A	B	C	D
1	Date	Purchases	Payments	Balance
2	11/1			−$12,300
3	11/8	$2500		
4	11/22		$18,345	
5	11/30	$5678		

Estimation
Mental Math
Calculator
Paper and Pencil

58. **Geography** A region's time zone is determined by its longitude. The zone labeled zero is centered on the prime meridian. In the chart below, time zones to the east of the prime meridian are named by negative numbers, and time zones west of the prime meridian are named by positive integers. To find the time in another city, subtract the time zone of the other city from the time zone at your location.

Suppose you are in Miami and wish to know the time in Bombay. Find $5 - (-5)$, which is 10. Since 10 is positive, the time in Bombay is 10 hours *ahead* of Miami. So at 1:00 P.M. in Miami, it is 11:00 P.M in Bombay.

City	Time Zone
Bombay, India	−5
Honolulu, USA	11
Los Angeles, USA	8
Miami, USA	5
Paris, France	0
Rome, Italy	−1

a. If the time in Los Angeles is 2:00 P.M., find the time in Rome. **11:00 P.M.**
b. If the time in Miami is 11:00 A.M., find the time in Honolulu. **5:00 A.M.**
c. If the time in Honolulu is 3:00 A.M., find the time in Paris. **2:00 P.M.**

Practice Guide Masters, p. 15

NAME ______________________ DATE __________

2-5 Practice
Subtracting Integers

Student Edition
Pages 89–93

Rewrite each equation using the additive inverse. Then solve.

1. $39 - 18 = x$
   39 + (−18) = x
   21 = x
2. $65 - 72 = y$
   65 + (−72) = y
   −7 = y
3. $-85 - (-42) = z$
   −85 + 42 = z
   −43 = z
4. $-15 - (-86) = a$
   −15 + 86 = a
   71 = a
5. $-21 - 24 = b$
   −21 + (−24) = b
   −45 = b
6. $-16 - (-57) = c$
   −16 + 57 = c
   41 = c
7. $84 - 92 = t$
   84 + (−92) = t
   −8 = t
8. $-32 - 74 = w$
   −32 + (−74) = w
   −106 = w
9. $-74 - (-21) = d$
   −74 + 21 = d
   −53 = d

Simplify each expression.

10. $-124k - (-65k)$ **−59k**
11. $15x - 21x$ **−6x**
12. $-32y - (-15y)$ **−17y**
13. $65x - (-12x)$ **77x**
14. $-74a - 56a$ **−130a**
15. $-21xy - 32xy$ **−53xy**
16. $-95ab - (-16ab)$ **−79ab**
17. $84ac - 15ac$ **69ac**
18. $124ad - (-203ad)$ **327ad**
19. $56xy - 83xy$ **−27xy**
20. $-453ab - (-675ab)$ **222ab**
21. $2045m - (-3056m)$ **5101m**

Solve each equation.

22. $-4 - 1 = f$ **f = −5**
23. $h = -5 - (-7)$ **h = 2**
24. $z = 9 - 12$ **z = −3**
25. $a = -765 - (-34)$ **a = −731**
26. $652 - (-57) = b$ **b = 709**
27. $c = 346 - 865$ **c = −519**
28. $d = -136 - (-158)$ **d = 22**
29. $x = 342 - (-456)$ **x = 798**
30. $y = -684 - (-379)$ **y = −305**
31. $b = -658 - 867$ **b = −1525**
32. $657 - 899 = t$ **t = −242**
33. $3004 - (-1007) = r$ **r = 4011**

Tech Prep

Accounting Clerk In exercise 59, point out that accounts payable and accounts receivable clerks are responsible for a specific portion of the accounting operations of a business. For more information on tech prep, see the *Teacher's Handbook*.

Group Activity Card 2-5

Amazing Difference — Group Activity **2-5**

MATERIALS: Copies of the number array

The object of this activity is to be the first player or group of players to find a track from a circled number on the left to a boxed number on the right or from a circled number on the top to a boxed number on the bottom.

Beginning with the circled number, mentally subtract each successive number on your track. Your final difference must be the ending boxed number. Your path must consist of only six moves.

START

		(20)	(8)	(24)	(18)	
START	(10)	−2	10	30	26	[−8]
	(15)	20	8	−7	−10	[−21]
	(4)	−13	−20	−14	1	[7]
	(−6)	4	−9	5	−12	[6] FINISH
		[12]	[5]	[14]	[−9]	

FINISH

 Pre-Algebra

Mixed Review

61. $4 < 6$

59. Simplify $9m + 43m + (-16m)$. (Lesson 2-4) **36*m***

60. **Oceanography** A submarine at 1300 meters below sea level descends an additional 1150 meters. What integer represents the submarine's position with respect to sea level? (Lesson 2-4) **−2450**

61. Write the inequality for *4 feet is shorter than 6 feet.* (Lesson 2-3)

62. Name the absolute value of −15. (Lesson 2-1) **15**

63. **Geometry** Graph (2, 1), (2, 4), and (5, 1) on a coordinate plane. Connect the points with line segments. What figure is formed? (Lesson 1-7) **triangle**

64. Solve $6b = 90$ mentally. (Lesson 1-6) **15**

65. Simplify $y + 9 + 14 + 2y$. (Lesson 1-5) **3*y* + 23**

66. Rewrite $(a + 3) + 4$ using the associative property. (Lesson 1-4) **$a + (3 + 4)$**

Statistics Use the information in the graph at the right to answer each question. (Lesson 1-3)

67. Sandy Koufax

67. If *r* represents the number of no-hitters that Nolan Ryan pitched, who can be represented by $r - 3$?

68. If *y* represents the number of no-hitters Cy Young pitched, which other pitcher can be represented by the expression *y*? **Bob Feller**

69. If *k* represents the number of no-hitters Sandy Koufax pitched, write an expression that represents the number of no-hitters Nolan Ryan pitched. **$k + 3$**

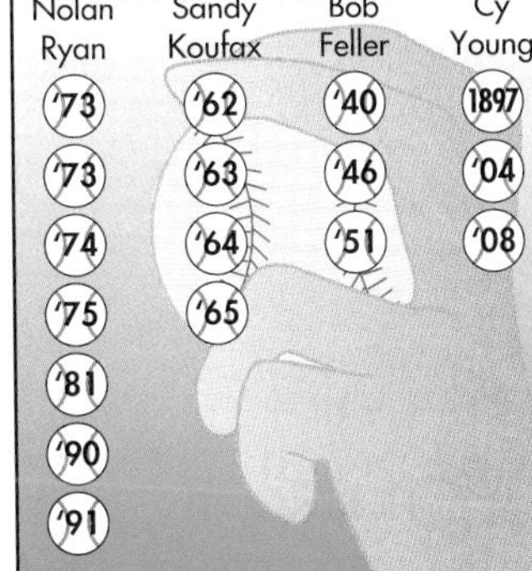

Self Test

Simplify. (Lesson 2-1)

1. $|-8|$ **8**

2. $|-20| + |-19|$ **39**

3. Name the coordinate of each point graphed on the number line. (Lesson 2-1) **−3, 0, 2, 5**

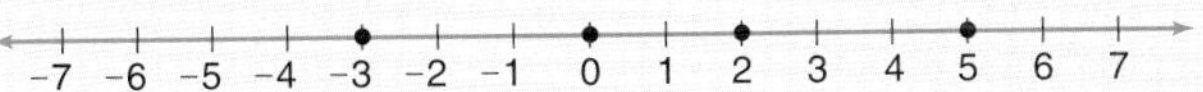

On graph paper, draw coordinate axes. Then graph and label each point. (Lesson 2-2)

4. $Q(4, 3)$

5. $R(-4, -3)$

6. $C(-3, 0)$ **See Solutions Manual.**

Replace each ● with <, >, or =. (Lesson 2-3)

7. −9 ● −15 **>**

8. −11 ● 5 **<**

9. $|-4|$ ● 4 **=**

Solve each equation. (Lesson 2-4, 2-5)

10. $-7 + 20 = a$ **13**

11. $-12 + 14 + 8 = d$ **10**

12. $-42 - 38 = q$ **−80**

13. $b = -34 - (-19)$ **−15**

14. **Business** The formula $P = I - E$ is used to find the profit (P) when income (I) and expenses(E) are known. Find P if $I = \$15,525$ and $E = \$17,000$. **−$1475**

Extension

Using Models Have students write a positive integer on each of six cards. Students then draw three cards without looking and write six addition, subtraction, and/or combination problems using those three integers or their opposites.

Self Test

The Self Test provides students with a brief review of the concepts and skills in Lessons 2-1 through 2-5. Lesson numbers are given to the right of exercises or instruction lines so students can review concepts not yet mastered.

4 ASSESS

Closing Activity

Modeling Have students model the following problem situation using play money. A business did $5000 worth of business in one week. If their expenses were $2750 for supplies, how much profit did they make? **$2250**

Enrichment Masters, p. 15

NAME ________________ DATE ________

Student Edition Pages 89–93

2-5 Enrichment

Subtracting in a Finite Number System

To subtract in a finite number system, add a number's additive inverse. The additive inverse of a number in a finite number system is that number which, when added to the first number, results in a sum that is the mod or identity.

Example: What is the additive inverse of 4 in mod 7?

$-(4) = 3$ What is added to 4 to equal 7? The answer is 3. The additive inverse of 4 in mod 7 is 3.

The additive inverses for 1, 2, 3, 5, and 6 are 6, 5, 4, 2, and 1, respectively.

In mod 7, subtraction is the same as adding the additive inverse.

Example: $3 - 4 = 3 + (-4) = 3 + 3 = 6$ Subtract 4 by adding its additive inverse, 3.

$1 - 6 = 1 + (-6) = 1 + 1 = 2$ Subtract 6 by adding its additive inverse, 1.

Solve each equation in mod 7.

1. $x = 1 - 3$ — $x = 1 + 4 = 5$
2. $5 - 6 = y$ — $y = 5 + 1 = 6$
3. $2 - 5 = z$ — $z = 2 + 2 = 4$
4. $q = 6 - 4$ — $q = 6 + 3 = 2$
5. $b = 2 - 6$ — $b = 2 + 1 = 3$
6. $3 - 4 = z$ — $z = 3 + 3 = 6$
7. $y = 0 - 4$ — $y = 0 + 3 = 3$
8. $3 - 6 = x$ — $x = 3 + 1 = 4$
9. $4 - 5 = p$ — $p = 4 + 2 = 6$

Solve each equation in mod 12.

10. $y = 2 - 9$ — $y = 2 + 3 = 5$
11. $x = 1 - 8$ — $x = 1 + 4 = 5$
12. $3 - 5 = r$ — $r = 3 + 7 = 10$
13. $z = 0 - 4$ — $z = 0 + 8 = 8$
14. $q = 4 - 10$ — $q = 4 + 2 = 6$
15. $6 - 11 = p$ — $p = 6 + 1 = 7$
16. $5 - 3 = k$ — $k = 5 + 9 = 2$
17. $7 - 11 = z$ — $z = 7 + 1 = 8$
18. $2 - 3 = b$ — $b = 2 + 9 = 11$
19. $8 - 10 = r$ — $r = 8 + 2 = 10$
20. $9 - 7 = f$ — $f = 9 + 5 = 2$
21. $4 - 7 = w$ — $w = 4 + 5 = 9$
22. $m = 1 - 11$ — $m = 1 + 1 = 2$
23. $z = 3 - 8$ — $z = 3 + 4 = 7$
24. $g = 5 - 9$ — $g = 5 + 3 = 8$

2-6 LESSON NOTES

NCTM Standards: 1-5, 8

Instructional Resources

- Study Guide Master 2-6
- Practice Master 2-6
- Enrichment Master 2-6
- Group Activity Card 2-6
- Assessment and Evaluation Masters, p. 44
- Activity Masters, p. 32

Transparency 2-6A contains the 5-Minute Check for this lesson; **Transparency 2-6B** contains a teaching aid for this lesson.

Recommended Pacing	
Standard Pacing	Day 8 of 12
Honors Pacing	Day 7 of 11
Block Scheduling*	Day 5 of 7

*For more information on pacing and possible lesson plans, refer to the ***Block Scheduling Booklet***.

1 FOCUS

5-Minute Check
(over Lesson 2-5)

Solve each equation.

1. $n = 9 - (-1)$ **10**
2. $x = -3 - (21)$ **-24**
3. $t = -8 - (-3)$ **-5**

Simplify.

4. $8m - (-6m)$ **14*m***
5. $-15c - 17c$ **-32*c***

Motivating the Lesson

Situational Problem Ask students which they would rather have—$1000 now, or 1 cent on the first day, 2 cents on the second day, 4 cents on the third day, 8 cents on the fourth day, and so on for 20 days. **Second option yields $10,485.76 after 20 days.**

2-6 Problem-Solving Strategy: Look for a Pattern

Setting Goals: *In this lesson, you'll solve problems by looking for a pattern.*

Modeling with Technology

Carl F. Gauss (1777–1855), a famous German mathematician, taught himself to read and to make mathematical calculations. When he was 10 years old, his teacher asked the students in his class to find the sum of the numbers from 1 to 100. The teacher thought that this problem would keep the students busy for a long time, but young Gauss wrote the answer almost immediately!

Your Turn

The notation "· · ·" means continue the pattern.

You can write this sum as $1 + 2 + 3 + \cdots + 98 + 99 + 100$. The sum of the least and greatest numbers is $1 + 100$ or 101.

- Find the sum of the second least and second greatest number in the set.
- Find the sum of the third least and third greatest number in the set.

a. What do you notice about the sums? **all 101**

b. How many of these sums are in the set of numbers from 1 to 100? **50**

Use a calculator to find 101×11, 101×12, 101×13, and 101×14.

c. What do you notice about the products? **1111, 1212, 1313, . . .**

d. Why do you think this happens? **See students' work.**

e. How did Gauss use this pattern to amaze his teacher? **See students' work.**

Learning the Concept

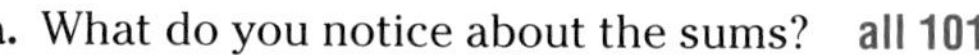

Looking for, and then extending, patterns is a good problem-solving strategy. Sometimes, solving one or more simpler problems allows you to see a pattern.

Example  **Use the pattern at the right to find 74 × 76.**

$4 \times 6 = 24$
$14 \times 16 = 224$
$24 \times 26 = 624$
$34 \times 36 = 1224$

As you look at the products of these simpler problems, observe the pattern in the factors. In each successive product, each factor is increased by 10. To find the product of 74 and 76, we will need to extend the pattern to include four more products.

Now look at the products. Each product has "24" as the last two digits. The digits before "24" follow the pattern 0, 2, 6, 12. Take a close look at this pattern and extend it four times.

0 2 6 12 20 30 42 56

+2 +4 +6 +8 +10 +12 +14

Use the pattern to write the next four products.

$$4 \times 6 = 24$$
$$14 \times 16 = 224$$
$$24 \times 26 = 624$$
$$34 \times 36 = 1224$$
$$44 \times 46 = 2024$$
$$54 \times 56 = 3024$$
$$64 \times 66 = 4224$$
$$74 \times 76 = 5624$$

The pattern shows that $74 \times 76 = 5624$. Use a calculator to verify the result.

Mathematicians have been studying patterns for centuries. Example 2 involves a pattern first made famous in a book published in 1202!

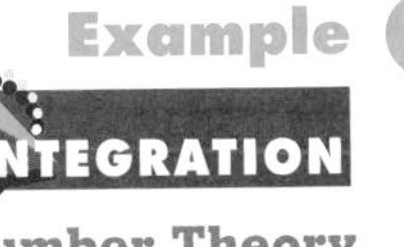

Number Theory

Example 2 **The pattern of numbers 1, 1, 2, 3, 5, 8, . . . is called the *Fibonacci sequence*. The terms of this sequence are called *Fibonacci numbers*. List the first ten Fibonacci numbers.**

What do you add to each term to get the succeeding term?

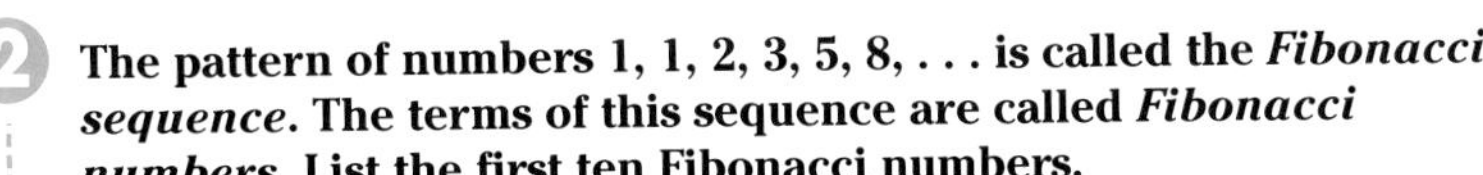

1 + 1 1 + 2 2 + 3 3 + 5

1 1 2 3 5 8

To continue the pattern, add the previous two terms. So, $5 + 8 = 13$, $8 + 13 = 21$, $13 + 21 = 34$, and $21 + 34 = 55$.

The first ten Fibonacci numbers are 1, 1, 2, 3, 5, 8, 13, 21, 34, and 55.

Checking Your Understanding

Communicating Mathematics

Read and study the lesson to answer these questions.

1. **Write** the sum of the numbers from 1 to 160. **12,880**
2. **Explain** how $\frac{n(n+1)}{2}$ relates to the pattern at the beginning of the lesson. **See margin.**
3. **Write** a rule to find the successive terms in the pattern 63, 48, 35, 24, ... **See margin.**
4. **You Decide** Julita says that the next number in the pattern below is 28. Jack thinks it is 27. Who do you think is correct? Explain.

   1 3 6 10 15 21 __?__

   Julita; the numbers are increasing by 2, 3, 4, . . .

Guided Practice

Solve. Look for a pattern. **5. 120, 720**

5. Find the next two integers in the pattern 1, 1, 2, 6, 24, __?__, __?__.
6. Use the simpler problems at the right to find 8888×5. Explain your thinking. **44,440**
7. How many squares of any size are there in an 8-by-8 checkerboard? **204**
8. **Carpentry** If shorter boards are cut from one long board, how many cuts will it take to have 12 shorter boards if there is no waste? **11 cuts**

$8888 \times 1 = 8888$
$8888 \times 2 = 17,776$
$8888 \times 3 = 26,664$
$8888 \times 4 = 35,552$

2 TEACH

In-Class Examples

For Example 1
Extend the pattern in Example 1 to find 94×96. **9024**

For Example 2
Use the diagram to find a pattern for triangular numbers. Then state the next three triangular numbers. **15, 21, 28**

3 PRACTICE/APPLY

Checking Your Understanding

Exercises 1–8 are designed to help you assess your students' understanding through reading, writing, speaking, and modeling. You should work through Exercises 1–4 with your students and then monitor their work on Exercises 5–8.

Study Guide Masters, p. 16

NAME ______________________ DATE ____________

Student Edition Pages 94–97

2-6 Study Guide
Problem Solving Strategy: Look for a Pattern

Example: At Fairview High School the bell rings at 7:35, 8:22, 8:25, 9:12, and 9:15 each weekday morning. When do the next three bells ring?

Bell	1	2	3	4	5
Time	7:35	8:22	8:25	9:12	9:15

Explore The chart shows the times that the bell rings.

Plan Since bell schedules often follow patterns, look for a pattern. Once you find a pattern, you can determine the next three bells.

Solve Notice that each class period is forty-seven minutes long and there is three minutes between each class. According to this pattern, the next three bells will be at 10:02, 10:05, and 10:52.

Examine The first class begins at 7:35. The class periods are forty-seven minutes in length, so dismissal bell will ring at 8:22. Having three minutes between classes, you must be in the next class by 8:25. Continuing this pattern, you will see that the sixth, seventh, and eighth bells ring at 10:02, 10:05, and 10:52.

Solve. Look for a pattern.

1. Find the next two integers in the set. {3, 5, 9, 17, 33, **65**, **129**}
2. Jon ran 1 lap on the first day. He ran 2 laps on the second day, 4 laps on the third day, and 8 laps on the fourth day. How many laps should he run on the seventh day? **64**
3. The following chart shows the scores at the baseball game.

Cubs	0	1	2	4	0	1	3
Reds	4	3	2	0	4	3	?

   If the pattern continues, how many runs will the Reds score in the bottom of the 7th inning? **1**

4. Ellie is using the following chart to help her calculate the discount on T-shirts.

T-shirts	1	2	3
Discount	$1.75	$3.25	$4.75

   Mr. Smith would like to order 8 T-shirts. How much of a discount should Ellie give him? **$12.25**

Reteaching

Using Comparisons Give students who are having trouble several simple patterns to identify. For example, use multiples of 3: 3, 6, 9, 12, 15,... . Have students find the difference between terms in the series. Continue with other simple patterns.

Additional Answers

2. Sample answer: The formula $\frac{n(n+1)}{2}$ is another rule for finding the sum of the first n integers.
3. Sample answer: Subtract 2 less each time; first, subtract 9, then subtract 7, and then 5.

Error Analysis

Students may become discouraged easily when looking for a pattern in a sequence. Suggest that students make a diagram or a list. Also remind students that more than one arithmetic operation may be involved and they should look for all four operations.

Assignment Guide

Core: 9–21 odd, 22–26
Enriched: 10–20 even, 21–26

For **Extra Practice**, see p. 744.

The red A, B, and C flags, printed only in the Teacher's Wraparound Edition, indicate the level of difficulty of the exercises.

Additional Answers

15a.

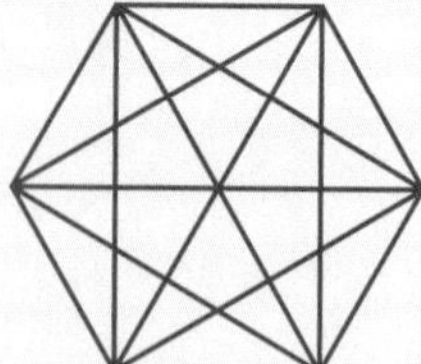
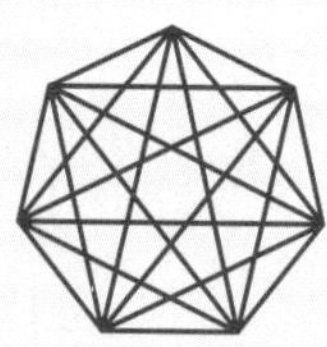

Practice Masters, p. 16

NAME ______________________ DATE ____________

2-6 Practice
Problem-Solving Strategy: Look for a Pattern

Student Edition Pages 94–97

Solve. Look for a pattern.

1. Ralph and Ella are playing a game called "Guess My Rule." Ralph has kept track of his guesses and Ella's responses in this table.

Ralph	0	1	2	3	4	5	6
Ella	10	9	8	7	6	5	**4**

Look for a pattern and predict Ella's response for the number 6. Describe this pattern. **Ella's response is Ralph's number subtracted from 10.**

2. Mollie is using the following chart to help her calculate prices for tickets.

Tickets	1	2	3	4
Price	$7.50	$12.50	$17.50	$22.50

A customer came in and ordered 10 tickets. How much should Mollie charge for this ticket order? **$52.50**

3. Brad needs to set up a coding system for files in the library using two-letter combinations. He has begun this table.

Letters	1	2	3	4	5
Combinations	1	4	9	16	**25**

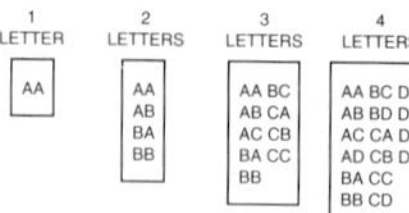

How many files can Brad code using the letters A, B, C, D, and E?

4. If the library has 400 items to code, how many letters will the librarian need if she uses Brad's system? **20 letters**

5. Billie needs to make a tower of soup cans as a display in a grocery store. Each layer of the tower will be in the shape of a rectangle. The length and the width of each layer will be one less than the layer below it.

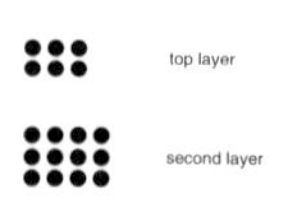

   a. How many cans will be needed for the fifth layer of the tower? **42 cans**

   b. How many total cans will be needed for a 10-layer tower? **570 cans**

Exercises: Practicing and Applying the Concept

Solve. Use any strategy.

9. Use the pattern at the right to find 11,111 × 11,111. **123,454,321**

$1 \times 1 = 1$
$11 \times 11 = 121$
$111 \times 111 = 12{,}321$
$1111 \times 1111 = 1{,}234{,}321$

10. Find the next two integers in the pattern 2, 5, 9, 14, 20, __?__, __?__. **27, 35**

✓ Choose

Estimation
Mental Math
Calculator
Paper and Pencil

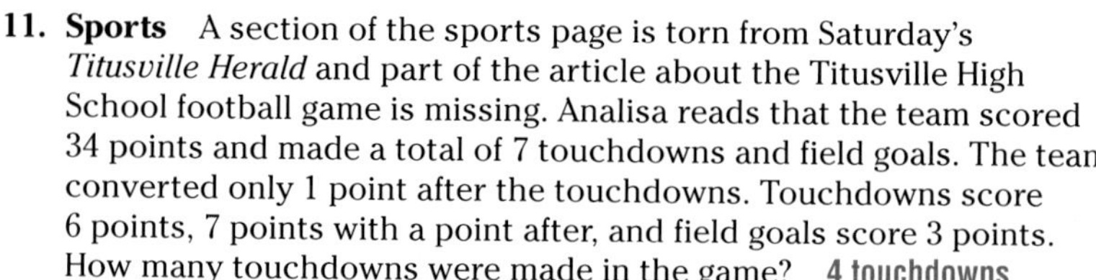

11. **Sports** A section of the sports page is torn from Saturday's *Titusville Herald* and part of the article about the Titusville High School football game is missing. Analisa reads that the team scored 34 points and made a total of 7 touchdowns and field goals. The team converted only 1 point after the touchdowns. Touchdowns score 6 points, 7 points with a point after, and field goals score 3 points. How many touchdowns were made in the game? **4 touchdowns**

12. Brian travels south on his bicycle riding 8 miles per hour. One hour later, his friend Maya starts riding her bicycle from the same location. If she travels south at 10 miles per hour, how long will it take her to catch Brian? **4 hours**

13. A ferry service is available to take passengers and cars to South Bass Island in Lake Erie. Michelle has part of a ferry schedule. She wishes to take the ferry to the island, but she cannot leave until 1:30 P.M. Assuming the ferry departs at regular intervals, what is the earliest time Michelle can catch it? **1:33 P.M.**

South Bass Island Ferry Schedule

Departures	Arrivals
8:45 A.M.	9:
9:33 A.M.	1
10:21 A.M.	
11:09 A.M.	

14. Two hot dogs and one hamburger cost $3.39. Three hot dogs and three hamburgers cost $6.57. What is the cost of one hot dog? **$1.20**

15. **Geometry** The green line segments in the figures at the right are called the *diagonals* of the polygon. The square has two diagonals, and the pentagon has five diagonals.

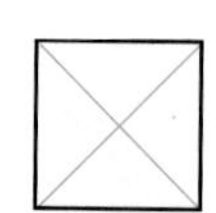
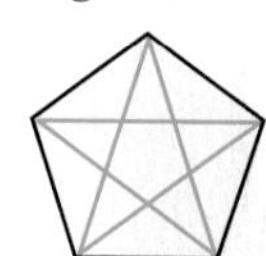

   a. Draw the next two figures in the pattern. **See margin.**

   b. Write a sequence to represent the number of diagonals in successive figures. How many diagonals does a 9-sided polygon have?

15b. 2, 5, 9, 14; 27

16. What is the sum of the first 50 odd numbers? **2500**

17. **Geometry** Find the number of line segments determined by nine points on a line. **36 segments**

18. A college student sent home the message at the right. If each letter stands for one digit 0–9, how much money did she ask for? **$10,652**

SEND
+MORE
MONEY

19. **Patterns** What is the next number in the pattern 1, 5, 14, 30, . . . ? **55**

20. **Family Activity** Survey your family members about their hair and eye color. Do you see a pattern? What color hair and eyes do you think your children will have? Explain. **See students' work.**

Alternative Learning Styles

Auditory For students who have trouble finding patterns, start by having them identify and extend a rhythmic clapping pattern. Patterns might include soft clap, loud clap, soft clap, repeated several time or clap hands, slap knee, clap hands.

Group Activity Card 2-6

Climb Those Stairs Group Activity **2-6**

You and a partner will investigate how many blocks are needed to make staircases with increasing numbers of steps.

Consider a staircase of just one step. It is made from one square. Now draw a staircase with two steps, as shown below. It is made from three squares. If you draw a staircase of three steps, how many squares are needed? (6) How many squares would be needed to draw a staircase of 12 steps?

Look for a pattern in the number of squares needed by recording some information you know. Make a table with two columns, Number of Steps and Number of Squares, and your data for staircases of 1, 2, 3, 4, 5, and 6 steps. Do you see a pattern in the Number of Squares column? How could you determine the number of squares needed for 7 steps? Then for 8 steps? Write a sentence describing the pattern(s) you and your partner found.

Refer to the back of the card for more variations.

?

 Pre-Algebra

Critical Thinking

21. The product of a 2-digit number and 101 contains the digits of the 2-digit number repeated twice. For example, $42 \times 101 = 4242$. What should you multiply a three-digit number by in order to have the product repeat the digits? **1001**

Mixed Review

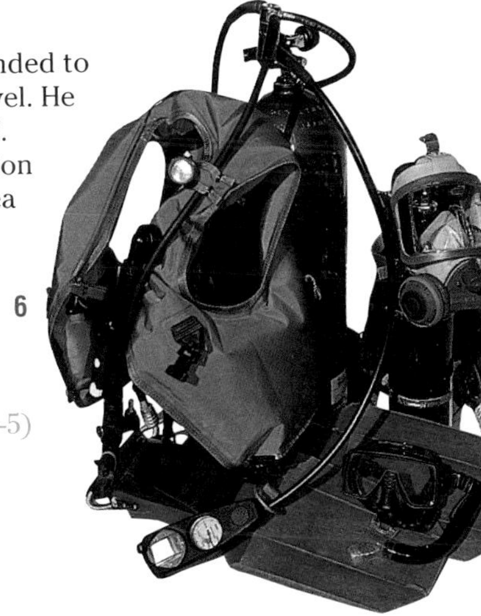

22. Recreation A scuba diver descended to a depth of 75 meters below sea level. He then rose 30 meters to a coral reef. What integer represents the position of the coral reef with respect to sea level? (Lesson 2-5) **−45 meters**

23. Solve the equation $h = 8 + (-15) + 13$. (Lesson 2-4) **6**

24. true

24. State whether $3x + 2x - 9 \geq 17$ is *true* or *false* if $x = 6$. (Lesson 1-9)

25. $11y + 21$

25. Simplify $5y + 3(7 + 2y)$. (Lesson 1-5)

26. Find the value of $2[5(4 + 6) - 3]$. (Lesson 1-2) **94**

WORKING ON THE Investigation

Refer to the Investigation on pages 62 63.

Depending on which season it is in the country you chose at the beginning of the Investigation, you may want to consider the *windchill factor* or the *heat index* when you are deciding what clothes to pack. If the country chosen is south of the equator, it will be winter when you visit that country. You may need to prepare for windchill factors. It may be summer when you visit and you need to be aware of the heat index.

The *windchill factor* is an estimate of the cooling effect the wind has on a person in cold weather. Increased wind velocity causes the body to lose heat more rapidly. Therefore, the chilling effect of cold increases as the speed of the wind increases.

The *heat index* is a measure of how high temperatures accompanied by high humidity affect the body's ability to cool itself. Increased humidity makes abnormally high temperatures "feel" even hotter to the average person.

- Determine which season it will be in June in the country you chose at the beginning of the Investigation. Then find a windchill table (winter) or a heat index table (summer).
- Explain why you would use the heat index table if the country is between the Tropic of Cancer and the Tropic of Capricorn.
- Find the average daily temperature for the country for a month in the appropriate season (winter or summer). Then find the average wind velocity (winter) or average relative humidity (summer) for the country.
- Look for a pattern in the table you are using. Use the pattern to estimate an extreme measure that is not listed in the table. For example, if your table for windchill goes to 0 degrees and to a wind of 40 mph, estimate the windchill for 0 degrees and a 50-mph wind.
- Write some general safety guidelines with respect to windchill factors or heat index.

Add the results of your work to your Investigation Folder.

4 ASSESS

Closing Activity

Speaking Have students brainstorm ways they look for a pattern. Make a list of these ways on the board. Encourage students to give as many different responses as possible.

Chapter 2, Quiz C (Lessons 2-5 and 2-6) is available in the *Assessment and Evaluation Masters*, p. 44.

Extension

Using Diagrams Have students draw a diagram and use patterns to solve the following problem: A painter can climb a ladder by going up one or two rungs. How many different ways can he reach the first rung? **1** the second rung? **2** the fifth rung? **8** the twelfth rung? **233**

WORKING ON THE Investigation

The Investigation on pages 62 and 63 is designed to be a long-term project that is completed over several days or weeks. Encourage students to keep their materials in their Investigation Folder as they work on the Investigation.

Enrichment Masters, p. 16

NAME ______________ DATE ______

2-6 Enrichment Student Edition Pages 94–97

Mind Reading

Make four cards having the following numbers and letters.

A	B	C	D
1 3 5 7 9 11 13 15	2 3 6 7 10 11 14 15	4 5 6 7 12 13 14 15	8 9 10 11 12 13 14 15

Ask someone to think of a number from 1 to 15. Then ask the person which cards the number is on.

For example, suppose the number appears on cards A, B, and C. Add the least number on each card. This sum is 1 + 2 + 4, or 7. The person's number is 7.

2. C: Beginning with 4, there are four consecutive numbers, four missing numbers, four consecutive numbers; D: Beginning with 8, there are eight consecutive numbers.

Answer these questions.

1. What is the pattern of the numbers on card A? **They are all odd numbers.**
2. The least number on card B is 2. Note that the next counting number, 3, is also on the card. The next two counting numbers, 4 and 5, are skipped. Then the next two counting numbers, 6 and 7, do appear on the card. This pattern is repeated up to 15. Can you explain the patterns on cards C and D? **See above.**
3. The least numbers on card A through card D are 1, 2, 4, 8. How are these numbers related to the base-two numeration system? **Since $1 = 2^0$, $2 = 2^1$, $4 = 2^2$, and $8 = 2^3$, then $1 = 1_2$, $2 = 10_2$, $4 = 100_2$, and $8 = 1000_2$.**

Identify the number chosen when it appears on the following cards.

4. C and D **12**
5. A, B, and D **11**
6. A, B, C, and D **15**

Each of the following numbers is chosen. Write the corresponding 4-digit base-two numeral. Arrange the cards in base-two place-value position from left to right, D (eights), C (fours), B (twos), A (ones). Interpret 1 as on the card and 0 as not on the card. Use the base-two numeral to determine on which cards each given number appears.

7. 6 **0110_2; DCBA; B and C**
8. 14 **1110_2; DCBA; B, C, and D**
9. 10 **1010_2; DCBA; A and C**

Suppose you want to do this trick with five cards, A through E. Which numbers should appear on each of the following cards?

10. card A
11. card B
12. card C
13. card D
14. card E

10. 1, 3, 5, 7, 9, 11, 13, 15, 17, 19, 21, 23, 25, 27, 29, 31
11. 2, 3, 6, 7, 10, 11, 14, 15, 18, 19, 22, 23, 26, 27, 30, 31
12. 4, 5, 6, 7, 12, 13, 14, 15, 20, 21, 22, 23, 28, 29, 30, 31
13. 8, 9, 10, 11, 12, 13, 14, 15, 24, 25, 26, 27, 28, 29, 30, 31
14. 16, 17, 18, 19, 20, 21, 22, 23, 24, 25, 26, 27, 28, 29, 30, 31

2-7A LESSON NOTES

NCTM Standards: 1-5, 7, 12

Objective

Use counters to model multiplication of integers.

Recommended Time

Demonstration and discussion: 15 minutes; Exercises: 30 minutes

Instructional Resources

For each student or group of students

Student Manipulative Kit
- counters
- integer mat

Math Lab and Modeling Math Masters
- p. 2 (integer mat)
- p. 5 (integer counters)
- p. 32 (worksheet)

For teacher demonstration

Overhead Manipulative Resources

1 FOCUS

Motivating the Lesson

Discuss the meaning of 5×4. Then ask students to find a pattern for the products 5×3, 5×2, 5×1, and 5×0 and predict the next two products.

2 TEACH

Teaching Tip As you observe students modeling Activity 2, you could informally assess their understanding by asking them to model the following situations.
- removing four sets of two negative counters;
- subtracting two positive counters from four negative counters.

3 PRACTICE/APPLY

Assignment Guide
Core: 1–10
Enriched: 1–10

MATHEMATICS LAB

HANDS-ON ACTIVITY

2-7A Multiplying Integers

A Preview of Lesson **2-7**

MATERIALS
- counters
- integer mat

You can use counters to model multiplying integers. Remember that 2×4 means *two sets of four items*. Using models, 2×4 means ***put in*** *two sets of four* ***positive*** *counters.*

Activity  **Find 2 × (−4).**

▸ Put in two sets of four negative counters.

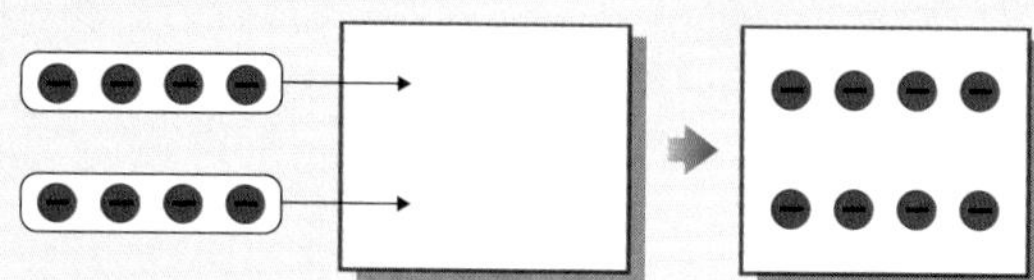

Therefore, $2 \times (-4) = -8$.

Activity 2 **Find −2 × (−4).**

▸ Since one meaning of the negative sign is *the opposite of*, $-2 \times (-4)$ means to ***remove*** *two sets of four* ***negative*** *counters.* How can you take out two sets? First, put in as many zero pairs as you need. Then, remove 2 sets of 4 negative counters.

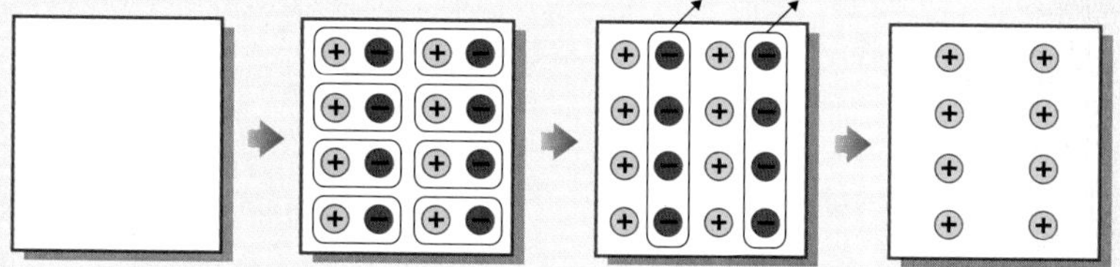

Therefore, $-2 \times (-4) = 8$.

Your Turn **Find each product using counters.**

1. $2 \times (-3)$ **−6**
2. 2×3 **6**
3. $-2 \times (-3)$ **6**
4. -2×3 **−6**
5. -3×2 **−6**
6. $-3 \times (-6)$ **18**
7. -4×4 **−16**
8. $-4 \times (-4)$ **16**

TALK ABOUT IT

9. What does -2×4 mean? Write your explanation in paragraph form. Model this operation. **See Solutions Manual.**
10. How are the expressions $2 \times (-3)$ and -3×2 alike? How do they differ? **See Solutions Manual.**

4 ASSESS

Observing students working in cooperative groups is an excellent method of assessment.

GLENCOE Technology

Interactive Mathematics Tools Software

In this interactive computer lesson, students use an equation mat, cups, and counters to multiply integers. A **Computer Journal** gives students the opportunity to write about what they have learned.

For Windows & Macintosh

2-7 Multiplying Integers

Setting Goals: *In this lesson, you'll multiply integers.*

Modeling a Real-World Application: Oceanography

Robert Ballard and his team from Woods Hole Oceanographic Institution are scientists who have discovered such famous shipwrecks as the *Titanic* and the *Lusitania*. In 1989, they discovered the sunken remains of the German battleship *Bismarck*, which sank in 1942 during World War II.

The *Bismarck* sank in 16,000 feet of water in the North Atlantic. The team used the research submarine, *Alvin*, to locate the hulk. It took *Alvin* over 2 hours to reach the *Bismarck* on the ocean floor. Its rate of descent was about 100 feet per minute. At this rate, how far would *Alvin* descend in 3 minutes?

Learning the Concept

The submarine descends 100 feet per minute, or it moves -100 feet. In 3 minutes, it will move $3 \times (-100)$ feet. You can find this product by using a number line or by observing a pattern.

Method 1: Number Line

-100 -100 -100

-300 -200 -100 0 100

Method 2: Patterns

$3 \cdot 200 = 600$
) -300
$3 \cdot 100 = 300$
) -300
$3 \cdot 0 = 0$
) -300
$3 \cdot (-100) = -300$

So, $3 \times (-100) = -300$. In three minutes, Alvin would descend 300 feet.

If $3 \times (-100) = -300$, then $-100 \times 3 = -300$ by the commutative property of multiplication. Notice that in each sentence one integer is positive, one integer is negative, and the product is negative. The pattern above and other similar examples suggest the following rule.

Multiplying Integers with Different Signs	The product of two integers with different signs is negative.

2-7 LESSON NOTES

NCTM Standards: 1-5, 7, 12

Instructional Resources

- Study Guide Master 2-7
- Practice Master 2-7
- Enrichment Master 2-7
- Group Activity Card 2-7
- Activity Masters, p. 2
- Multicultural Activity Masters, p. 4

Transparency 2-7A contains the 5-Minute Check for this lesson; **Transparency 2-7B** contains a teaching aid for this lesson.

Recommended Pacing	
Standard Pacing	Day 9 of 12
Honors Pacing	Day 8 of 11
Block Scheduling*	Day 6 of 7 (along with Lesson 2-8)

*For more information on pacing and possible lesson plans, refer to the ***Block Scheduling Booklet***.

1 FOCUS

5-Minute Check
(over Lesson 2-6)

1. Use the pattern below to find the product of 48×52. **2496**
   $8 \times 12 = 96$
   $18 \times 22 = 396$
   $28 \times 32 = 896$
   $38 \times 42 = 1596$

Find the next two integers in each pattern.

2. 5, 10, 20, 40, ... **80, 160**
3. −2, 6, −18, 54, ... **−162, 486**

Give the next two letters in each pattern.

4. N, O, R, S, V, ... **W, Z**
5. J, F, M, A, M, J, J, A, ... **S, O (September, October)**

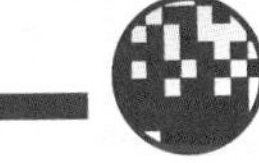

Alternative Teaching Strategies

Student Diversity Have students complete the first quadrant in the multiplication grid. Then have them find a pattern for the 3rd row (from right to left) and 3rd column (top to bottom). Have them continue the pattern and complete the table for quadrants 2 and 4. Then use patterns in the rows and columns of those quadrants to complete quadrant 3.

	−12			4				12	
−12	−9	−6	−3	3	0	3	6	9	12
	−6			2				6	
	−3			1				3	
	0			0				0	
−4	−3	−2	−1	×	0	1	2	3	4
	3			−1				−3	
	6			−2				−6	
12	9	6	3	−3	0	−3	−6	−9	−12
	12			−4				−12	

Motivating the Lesson

Questioning In a golf match, what integer is used to represent par? **0** How is three above par represented? **+3** three below par? **−3** If a player had a score of −2 on each of the first three holes, what would the score be at the end of the third hole? **−6**

2 TEACH

In-Class Examples

For Example 1
Solve each equation.

a. $r = -9(12)$ **−108**

b. $s - 8(-18)$ **−144**

For Example 2
Solve each equation.

a. $t = -6(-13)$ **108**

b. $-3 \cdot (-8) \cdot (-10) = v$ **−240**

Teaching Tip Write examples of two positive integers, one positive and one negative, one negative and one positive, and two negative integers on the board. Have students state the rule for multiplying each. Then have students alternate, repeating the appropriate rule to a partner while working Exercises 5–10.

Example  **Solve each equation.**

a. $m = -5(18)$

The two factors have different signs. The product is negative.

$m = -5(18)$
$m = -90$

b. $6(-12) = n$

The two factors have different signs. The product is negative.

$6(-12) = n$
$-72 = n$

You already know that the product of two positive integers is positive.

Both factors are positive. $3 \cdot 7 = 21$ The product is positive.

How would you find $-3(-2)$? You can solve this problem by again observing a pattern or by using counters.

Method 1: Patterns

$-3(2) = -6$
$-3(1) = -3$ (+3)
$-3(0) = 0$ (+3)
$-3(-1) = 3$ (+3)
$-3(-2) = 6$ (+3)

Method 2: Counters

$-3(-2)$ means that you will remove 3 sets of 2 negative counters.

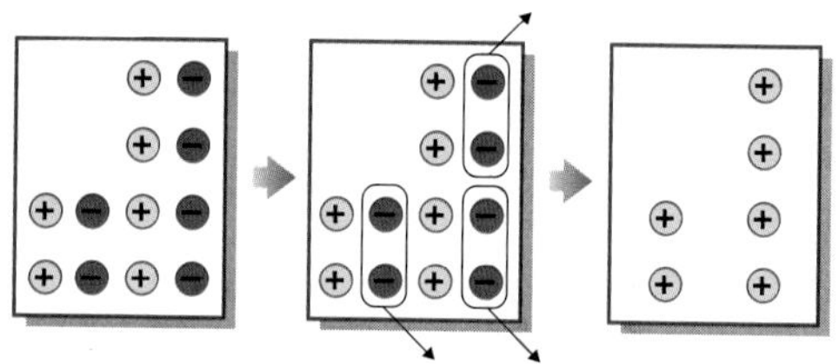

So, $-3(-2) = 6$. This suggests the following rule.

Multiplying Integers with the Same Sign — **The product of two integers with the same sign is positive.**

Example  **Solve each equation.**

a. $m = -8(-15)$

The two factors have the same sign. The product is positive.

$m = -8(-15)$
$m = 120$

b. $-7(-5)(-4) = q$

$[-7(-5)](-4) = q$ *Associative property*
$35(-4) = q$ *$-7(-5) = 35$*
$-140 = q$

Many real-life situations can be modeled using products of integers.

Cooperative Learning

Trade-A-Problem Have students work in small groups to create word problems that can be solved by adding, subtracting, or multiplying integers. After a problem has been created and solved, it is traded with another group. Each group solves the new problem. Finally, solutions are compared and errors are worked out by both groups until they are satisfied that both word problems are good ones.
For more information on this strategy, see *Cooperative Learning in the Mathematics Classroom*, one of the titles in the Glencoe Mathematics Professional Series, p. 25.

Meteorology

Example 3 **The temperature between the ground and 11 kilometers above the ground drops about 7°C for each kilometer of increase in altitude. Suppose the temperature 2 kilometers above the ground is 0°C. What would the ground temperature be?**

Explore The temperature 2 km above ground is 0°C, and the temperature drops 7°C for each km of increase in altitude. You need to find the ground temperature. Let t represent the ground temperature, -2 represent 2 km lower, and -7 represent a drop of 7°C.

Plan Write an equation.

Solve *ground temperature = temperature at 2 km + (drop of 7° per km)(2 km lower)*

$$t = 0 + (-7)(-2)$$
$$t = 0 + 14$$
$$t = 14$$

The ground temperature is 14°C.

Examine Since the altitude is decreasing, the temperature rises. The answer is reasonable.

Connection to Algebra

You can use the rules for multiplying integers to evaluate algebraic expressions and to simplify expressions.

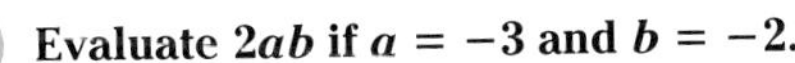

Example 4 **Evaluate $2ab$ if $a = -3$ and $b = -2$.**

$2ab = 2(-3)(-2)$	*Replace a with −3 and b with −2.*
$= 2[(-3)(-2)]$	*Associative property of multiplication*
$= 2(6)$	*The product of −3 and −2 is positive.*
$= 12$	

Example 5 **Find the product of -3 and $8m$.**

$-3(8m) = (-3 \cdot 8)m$ *Associative property of multiplication*
$= -24m$

The product is $-24m$.

Checking Your Understanding

Communicating Mathematics **Read and study the lesson to answer these questions.**

1. **Explain** why the solution of $-3(-4) = p$ must be 12. Use a pattern in your explanation. **See margin.**
2. **Explain** how to multiply 12 and $-4y$. **See margin.**

In-Class Examples

For Example 3
The price of a stock dropped $3 per day for 5 days. What was the total change in the value of that stock after those 5 days? **-$15**

For Example 4
Evaluate $3abc$ if $a = -4$, $b = -2$, and $c = -3$. **-120**

For Example 5
Find the product of -9 and -3n. **27n**

3 PRACTICE/APPLY

Checking Your Understanding

Exercises 1–19 are designed to help you assess your students' understanding through reading, writing, speaking, and modeling. You should work through Exercises 1–4 with your students and then monitor their work on Exercises 5–19.

Additional Answer

1. Sample answer:
   -4 · 2 = -8
   -4 · 1 = -4
   -4 · 0 = 0
   -4 · (-1) = 4
   -4 · (-2) = 8
   -4 · (-3) = 12

Study Guide Masters, p. 17

NAME ______ DATE ______

2-7 Study Guide
Multiplying Integers

Student Edition Pages 99–103

The product of two integers with different signs is negative.

8 × -2 = -16 | -8 × 2 = -16

The product of two integers with same signs is positive.

5 × 6 = 30 | -5 × (-6) = 30

State whether each statement is true or false.

1. The product of two positive integers is positive. **true**
2. The product of two negative integers is negative. **false**
3. The product of a negative and a positive integer is positive. **false**
4. The product of one negative and two positive integers is negative. **true**

State whether each product is positive or negative.

5. 6 × 7 **positive**
6. -3 × 4 **negative**
7. -5 × (-2) **positive**
8. 8 × (-8) **negative**
9. -7 × (-9) **positive**
10. 11 × 4 **positive**
11. -3 × (-12) **positive**
12. 2 × 7 **positive**
13. 3 × (-8) **negative**

Solve each equation

14. x = -4 × (-15) **x = 60**
15. y = -8 × 7 **y = -56**
16. x = 3 × (-6) **x = -18**
17. -4 × 5 × 2 = c **c = -40**
18. 3 × (-9) × (-2) = d **d = 54**
19. 2 × (-5) × (-5) = n **n = 50**
20. -22 × (-12) = t **t = 264**
21. s = -18 × 32 **s = -576**
22. w = 15 × (-25) **w = -375**

Reteaching

Using Models Students can practice multiplying integers by telling a story and using play money. For example, you had some money in your pocket. You paid for lunch with three five-dollar bills and got back no change . How does the amount in your pocket now compare with the amount you had before lunch? **-$15**

Additional Answer

2. Use the associative property of multiplication to rewrite the expression as [12 · (-4)]y. Then multiply. Since the integers have different signs, the product is -48y.

Error Analysis

Students may interchange the rules for adding integers with the rules for multiplying integers. Have students write the rules at the top of the paper with examples. Then refer to the rules as problems are being worked.

Assignment Guide
Core: 21–49 odd, 50–58
Enriched: 20–46 even, 47–58

For **Extra Practice**, see p. 745.

The red A, B, and C flags, printed only in the Teacher's Wraparound Edition, indicate the level of difficulty of the exercises.

Additional Answer

4.

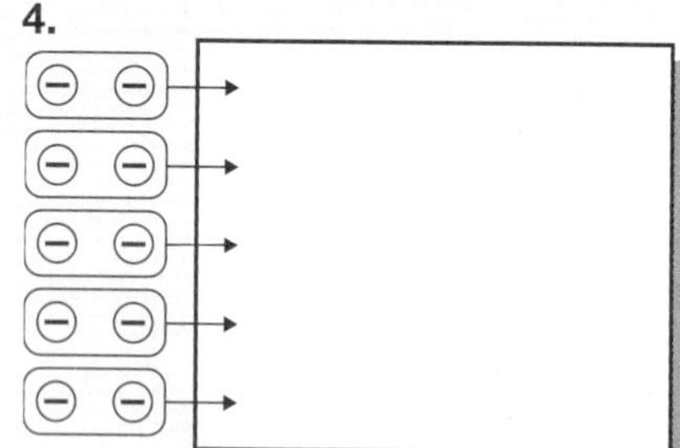

3. **Write** the multiplication sentence shown by each model.

a. 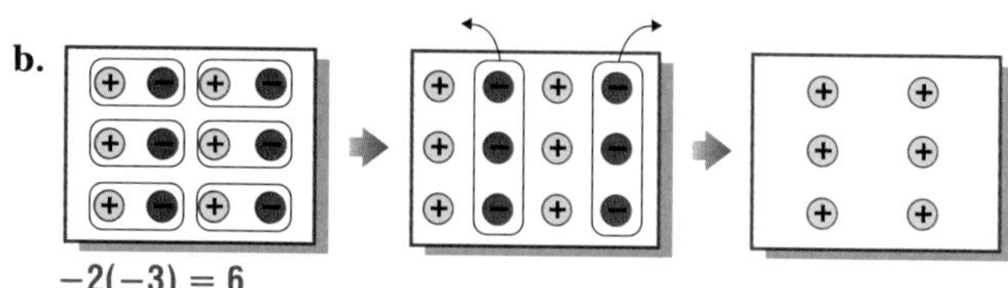

$3(-4) = -12$

b. $-2(-3) = 6$

4. **Draw a model** that shows the solution for $q = 5(-2)$. See margin.

Guided Practice

State whether each product is positive or negative. Then find the product.

5. $-8 \cdot 7$ −, −56
6. $5(-9)$ −, −45
7. $14 \cdot 37$ +, 518
8. $-35(-57)$ +, 1995
9. $(-9)(2)(-3)$ +, 54
10. $(-3)(-5)(-7)$ −, −105

Solve each equation.

11. $t = -9 \cdot 3$ −27
12. $-7(-3) = u$ 21
13. $8(-10)(4) = v$ −320

Evaluate each expression.

14. $4m$, if $m = -12$ −48
15. $-3gh$, if $g = -4$, $h = -11$ −132

Multiply.

16. $-5 \cdot 3w$ −15w
17. $8(-5x)$ −40x
18. $-4(-3y)$ 12y

19. **Geology** In December of 1994, geologists found that the Bering Glacier had come to a stop. Until that time, the glacier had been rushing forward at a rate of up to 300 feet per day. Before the surge began, the glacier was retreating at a rate of about 2 feet per day. If the retreat resumes at the old rate, what integer represents how far the glacier will have advanced after 28 days? −56 feet

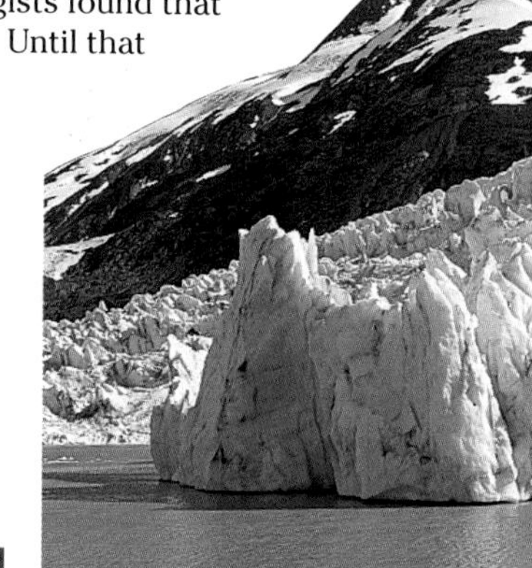

Exercises: Practicing and Applying the Concept

Independent Practice

Solve each equation. 26. 120 27. −585 30. 147

20. $m = -6(-8)$ 48
21. $x = -7(-4)$ 28
22. $a = -3(12)$ −36
23. $r = 13(-2)$ −26
24. $-5 \cdot 7 = b$ −35
25. $-15(-5) = y$ 75
26. $-4(-5)(6) = z$
27. $9(-13)(5) = q$
28. $6(-5)(-8) = g$ 240
29. $h = 2 \cdot 7 \cdot 12$ 168
30. $-7(-7)(3) = p$
31. $11(-4)(7) = s$ −308

Practice Masters, p. 17

NAME ________________ DATE ________

2-7 Practice
Multiplying Integers

Student Edition
Pages 99–103

Multiply.

1. $-2 \cdot 3x$ **−6x**
2. $-4 \cdot 5y$ **−20y**
3. $9 \cdot (-2z)$ **−18z**
4. $-5 \cdot (-6a)$ **30a**
5. $8t \cdot (-3)$ **−24t**
6. $2n \cdot (-1)$ **−2n**
7. $-5 \cdot 2w$ **−10w**
8. $8c \cdot (-2)$ **−16c**
9. $-3c \cdot (-5d)$ **15cd**
10. $4r \cdot 7s$ **28rs**
11. $-3x \cdot (-z)$ **3xz**
12. $-4ab \cdot (-6)$ **24ab**
13. $(-3)(4)(-x)$ **12x**
14. $-3(5)(-y)$ **15y**
15. $(-6)(-2)(8r)$ **96r**
16. $-5(0)(-xy)$ **0**
17. $5(-7)(4w)$ **−140w**
18. $(-8)(-4)(m)$ **32m**
19. $(-3)(6n)(-2p)$ **36np**
20. $(3)(9)(-d)$ **−27d**
21. $(0)(6m)(-10f)$ **0**
22. $7k(-3)(-5t)$ **105kt**
23. $(7)(-x)(-y)$ **7xy**
24. $(-5)(-8g)(-h)$ **−40gh**

Solve each equation.

25. $x = -6 \cdot -8$ **x = 48**
26. $y = -12 \cdot 4$ **y = −48**
27. $x = -9 \cdot (-11)$ **x = 99**
28. $y = (-7)(17)$ **y = −119**
29. $-14(-4) = h$ **h = 56**
30. $-15(10) = k$ **k = −150**
31. $-22(-3) = c$ **c = 66**
32. $7(-24) = d$ **d = −168**
33. $p = -21(13)$ **p = −273**
34. $(-5)(-6)(-4) = m$ **m = −120**
35. $(10)(-8)(-2) = r$ **r = 160**
36. $(-3)(3)(-10) = t$ **t = 90**
37. $w = (-12)(-1)(6)$ **w = 72**
38. $y = (20)(-5)(-5)$ **y = 500**
39. $x = (4)(-16)(-6)$ **x = 384**
40. $n = (16)(9)(-2)$ **n = −288**
41. $z = (-11)(-4)(-7)$ **z = −308**
42. $f = (21)(-7)(-2)$ **f = 294**

Evaluate each expression if x = −5 and y = −6.

43. $3y$ **−18**
44. $-8x$ **40**
45. $-4y$ **24**
46. $12x$ **−60**
47. $-15x$ **75**
48. $-19y$ **114**
49. $-6xy$ **−180**
50. $4xy$ **120**

Group Activity Card 2-7

Powerful Integer Products Group Activity **2-7**

MATERIALS: One calculator • a die • a coin • paper • pencils

Three students should do this activity with one of them being the leader. Another player uses the calculator to compute products, and the other player is to use mental computation or paper and pencil. The object is to be first in computing the product of given integers.

To begin, the leader writes the first expression given on the back of this card without showing it to the two players. The leader also states which side of the coin indicates positive or negative, rolls the die, and flips the coin. This integer will replace the variable in the expression the leader has written on the paper. The leader next shows the players the expression. They are to compute the product as quickly as possible and show it to the leader. The leader decides who gave the first correct answer, and the winner earns one point. The first player to earn five points wins the game. The leader should watch the player with the calculator to make sure that all the necessary strokes and not just the final answer are input.

 Pre-Algebra

Evaluate each expression.

32. $5n$, if $n = -7$ **35**
33. st, if $s = 7, t = 8$ **56**
34. $12ab$, if $a = -3$, $b = -7$ **252**
35. $-6b$, if $b = 8$ **−48**
36. $-15c$, if $c = -1$ **15**
37. $12x$, if $x = -3$ **−36**

Find each product. **44. −24*ab*** **45. −3*ab***

38. $8(-5x)$ **−40*x***
39. $-4 \cdot 3b$ **−12*b***
40. $12(-5n)$ **−60*n***
41. $(-12)(-4b)$ **48*b***
42. $(7y)(5z)$ **35*yz***
43. $(-10r)(5s)$ **−50*rs***
44. $2ab \cdot 6 \cdot (-2)$
45. $-3(-a)(-b)$
46. $7(-m)(k)$ **−7*mk***

Critical Thinking

47. Write a rule that will help you determine the sign of the product if you are multiplying two or more integers. **See margin.**

Applications and Problem Solving

48. **Geometry** On a sheet of graph paper, copy the figure at the right. **See margin.**
 a. Name the ordered pairs for the coordinates of the vertices of ΔPQR.
 b. Multiply each coordinate in the ordered pairs by −1. Label the new ordered pairs as P', Q', and R'.
 c. Now graph $\Delta P'Q'R'$.
 d. How do the two triangles compare?

49. **Weather** In the great flood of 1993, the Mississippi River was so high that it caused the Illinois River to flow backward. If the Illinois River flowed at the rate of −1500 feet per hour, how far would the water travel in 24 hours? **−36,000 feet or about −6.8 miles**

Mixed Review

50. **Patterns** Look at the pattern of numbers below. **no; 7 ≠ 4 × 2**

 1 2 4 7

 Is the next number found by multiplying by 2? Explain. (Lesson 2-6)

51. **Weather** In Spearfish, South Dakota, on January 22, 1943, the temperature was −4°F at 7:30 A.M. At 7:32 A.M., it was 45°F. How much did the temperature rise in two minutes? (Lesson 2-5) **49°F**

52. State whether the sum $-10 + 9 + 8$ is positive or negative. Then find the sum. (Lesson 2-4) **positive, 7**

53. Write an inequality that compares yesterday's high temperature of 65°F with yesterday's low temperature of 34°F. (Lesson 2-2) **Sample answer: 65 > 34**

54. Graph the numbers {−1, 1, 4} on a number line. (Lesson 2-1) **See margin.**

55. 39,177

55. **Sports** The attendance at Shea Stadium was 35,892 on Friday. This was 3285 less than the attendance Saturday. Solve the equation $35{,}892 = a - 3285$ to find Saturday's attendance. (Lesson 1-8)

56. Solve $\frac{18}{a} = 9$ mentally. (Lesson 1-6) **2**

57. What property is shown by the statement $9a + b = b + 9a$? (Lesson 1-4) **commutative, +**

58. Find the value of $21 \div 7 + 4 \cdot 11$. (Lesson 1-2) **47**

Extension

Using Cooperative Groups Give each group a bag with five red and five blue number cubes in it. Red cubes are positive, and blue cubes are negative. Draw four cubes. Find the product and record the results. Play five rounds. Have each group find the sum of the absolute values of all results. The group with the highest total wins.

4 ASSESS

Closing Activity

Speaking Have students brainstorm and provide as many models as possible for the expression −6 · 4.

Additional Answers

47. If there is an even number of negative numbers being multiplied, the product will be positive. If there is an odd number of negative numbers being multiplied, the product will be negative.

48a. *P*(−1, 0), *Q*(−3, −2), *R*(−2, −4)

48b. *P′*(1, 0), *Q′*(3, 2), *R′*(2, 4)

48c.

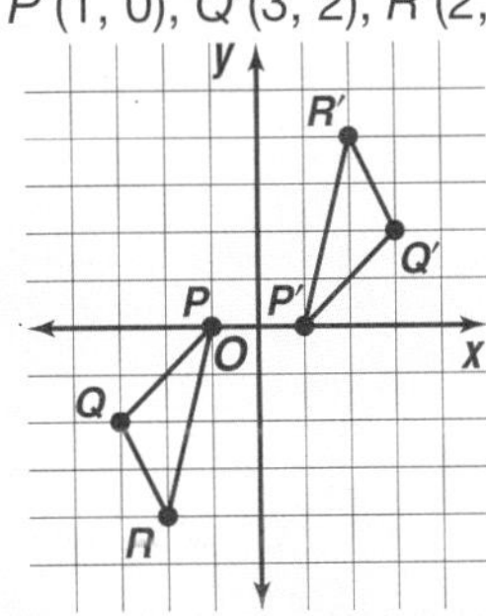

48d. The triangles are the same size and shape. $\Delta P'Q'R'$ is rotated from ΔPQR.

54.

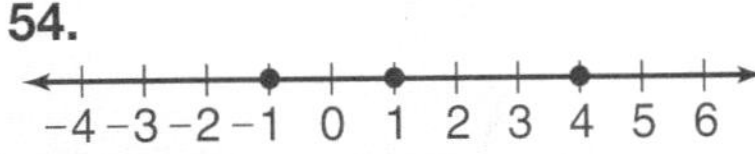

Enrichment Masters, p. 17

NAME ____________ DATE ____________

2-7 Enrichment Student Edition Pages 99–103

Carpeting

Suppose you want to figure the cost of carpeting a room. Room sizes are usually given in square feet. Carpet prices are usually given in square yards. You need to apply three formulas.

The formula $a = \ell w$ is used to compute area, a, of a rectangle in square feet.
a (area in square feet) = ℓ (length in feet) · w (width in feet)

The formula for the area in square yards in given below.
A (area in square yards) = $\frac{a \text{ (area in square feet)}}{9}$

The following formula can be used to compute the cost of carpeting.
c (cost) = A (Number of square yards) · p (price per square yard)

Example: Find the cost of carpeting a room 21 feet long and 15 feet wide if the carpeting costs \$7.70 per square yard.

$a = \ell \cdot w$
$= 21 \cdot 15$
$= 315$ square feet

$A = \frac{a}{9}$
$= \frac{315}{9}$
$= 35$ square yards

$c = A \cdot p$
$= 35 \cdot \$7.70$
$= \$269.50$

The cost of carpeting the room is \$269.50.

Find the cost of carpeting each room listed in the chart below.

	Room	Length	Width	Price Per Square Yard	Cost Per Room
1.	Living Room	18	24	\$17.00	**\$816**
2.	Dining Room	15	15	\$22.50	**\$562.50**
3.	Den	12	9	\$15.00	**\$180**
4.	Bedroom A	12	18	\$18.50	**\$444**
5.	Bedroom B	9	15	\$21.00	**\$315**
6.	Bathroom	6	6	\$ 8.95	**\$35.80**

7. Find the total carpeting bill. **\$2353.30**

2-8 LESSON NOTES

NCTM Standards: 1-5, 7, 12

Instructional Resources

- Study Guide Master 2-8
- Practice Master 2-8
- Enrichment Master 2-8
- Group Activity Card 2-8
- Assessment and Evaluation Masters, p. 44
- Activity Masters, p. 33
- Math Lab and Modeling Math Masters, p. 77
- Real-World Applications, 4

Transparency 2-8A contains the 5-Minute Check for this lesson; **Transparency 2-8B** contains a teaching aid for this lesson.

Recommended Pacing	
Standard Pacing	Day 10 of 12
Honors Pacing	Day 9 of 11
Block Scheduling*	Day 6 of 7 (along with Lesson 2-7)

*For more information on pacing and possible lesson plans, refer to the ***Block Scheduling Booklet***.

1 FOCUS

5-Minute Check
(over Lesson 2-7)

Solve each equation.

1. $(-5)(-3)(4) = a$ **60**
2. $(20)(-6)(2) = b$ **−240**

Find each product.

3. $(-8x)(-9)$ **72x**
4. $(3xy)(-3)(7)$ **−63xy**
5. $-9(-m)(-n)$ **−9mn**

Motivating the Lesson

Hands-On Activity Have students use a drawing of a football field to represent the following problem. A football team was penalized a total of 35 yards in one game. If each penalty was 5 yards, how many penalties did they receive? **−35 ÷ (−5) = 7**

2-8 Dividing Integers

Setting Goals: *In this lesson, you'll divide integers.*

Modeling with Manipulatives

Remember that you used counters in Lesson 2-7 to understand the rules for multiplying integers. You can also use counters to model division. The diagram at the right shows 12 negative counters separated into 4 equal groups. Counting the number of counters in each group shows that $-12 \div 4 = -3$.

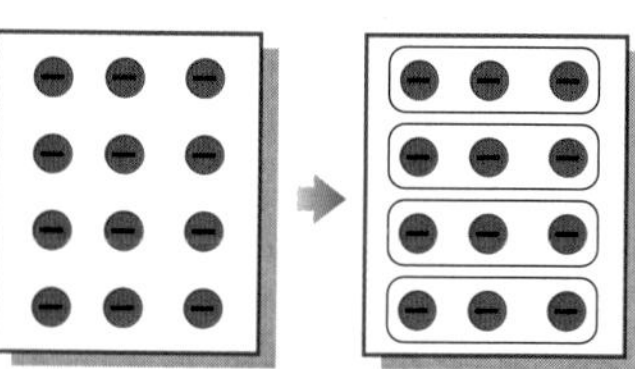

Your Turn Use counters to solve $-10 \div 5 = t$.

a. How is $-12 \div 4 = -3$ related to $-3 \times 4 = -12$?

b. What do you think is the quotient of 12 and -4? **−3**

a. ÷ and × are inverse operations.

Learning the Concept

You could also find $-12 \div 4$ by looking for a pattern.

$$4 \div 4 = 1$$
$$0 \div 4 = 0$$
$$-4 \div 4 = -1$$
$$-8 \div 4 = -2$$
$$-12 \div 4 = -3$$

(Each step: −1)

THINK ABOUT IT
What is the next division sentence in the pattern? **−16 ÷ 4 = −4**

In the example above, the dividend is negative and the divisor is positive. Let's try a case where the signs are reversed. How would you solve $f = 12 \div (-4)$?

Use a calculator. 12 [÷] 4 [(−)] [=] -3

So, the solution of $f = 12 \div (-4)$ is -3.

These and other similar examples suggest the following rule.

Dividing Integers with Different Signs	The quotient of two integers with different signs is negative.

Classroom Vignette

"After practicing solving problems with integers, I separate students into groups of three. Each group is given a legal-size sheet of paper. Students fold the paper into three columns and select one of the problems from the chapter to solve. Column 1 is used to write the problem. Column 2 is used to show a picture explaining the problem, and Column 3 is used to show the numerical solution. Each member of the group should be able to explain how the problem is solved."

Darlene Thurner
Darlene Thurner
Pine-Richland Middle School
Gibsonia, PA

Example 1 **Solve each equation.**

a. $-34 \div 2 = x$

The dividend and the divisor have different signs. The quotient is negative.

$-34 \div 2 = x$

$-17 = x$

b. $y = \frac{42}{-3}$

The dividend and the divisor have different signs. The quotient is negative.

$y = \frac{42}{-3}$

$y = -14$

You already know that the quotient of two positive integers is positive.

$$21 \div 3 = 7$$

Both the dividend and the divisor are positive. The quotient is positive.

How would you solve $z = -27 \div (-9)$? You can solve this problem by again looking for a pattern.

$$\begin{aligned} 18 \div (-9) &= -2 \\ 9 \div (-9) &= -1 \\ 0 \div (-9) &= 0 \\ -9 \div (-9) &= 1 \\ -18 \div (-9) &= 2 \\ -27 \div (-9) &= 3 \end{aligned}$$

(each step +1)

So, the solution to $z = -27 \div (-9)$ is 3. This pattern suggests the following rule.

Dividing Integers with the Same Sign The quotient of two integers with the same sign is positive.

Example  **Solve each equation.**

a. $m = -40 \div (-8)$

The dividend and the divisor have the same sign. The quotient is positive.

$m = -40 \div (-8)$

$m = 5$

b. $\frac{90}{15} = r$

The dividend and the divisor have the same sign. The quotient is positive.

$\frac{90}{15} = r$

$6 = r$

2 TEACH

In-Class Examples

For Example 1
Solve each equation.

a. $-54 \div 3 = b$ **−18**

b. $\frac{86}{(-2)} = d$ **−43**

For Example 2
Solve each equation.

a. $x = 72 \div 9$ **8**

b. $\frac{-144}{-12} = b$ **12**

In-Class Examples

For Example 3

Evaluate $rs \div t$, if $r = -3$, $s = -4$ and $t = -6$. **-2**

For Example 4

In one partnership, the monthly net profit or loss is divided equally among the three partners and recorded in their accounts. If their partnership lost $621 one month, how much should be entered in each partner's account? **-$207**

3 PRACTICE/APPLY

Checking Your Understanding

Exercises 1–19 are designed to help you assess your students' understanding through reading, writing, speaking, and modeling. You should work through Exercises 1–4 with your students and then monitor their work on Exercises 5–19.

Additional Answer

1. Sample answer:

4 ÷ 2 = 2	2 ÷ 2 = 1
0 ÷ 2 = 0	-2 ÷ 2 = -1
-4 ÷ 2 = -2	-6 ÷ 2 = -3
-8 ÷ 2 = -4	-10 ÷ 2 = -5

Study Guide Masters, p. 18

NAME ______________ DATE ________

2-8 Study Guide
Dividing Integers

Student Edition
Pages 104–108

When dividing two integers with different signs, the result is negative.	When dividing two integers with the same sign, the result is positive.
Examples: 35 ÷ (-7) = -5	***Examples:*** -35 ÷ (-7) = 5
$-\frac{30}{5} = -6$	$\frac{30}{5} = 6$

State whether each quotient is positive or negative. Then find the quotient

1. 132 ÷ 11 **positive; 12**	2. -108 ÷ 12 **negative; -9**	3. -80 ÷ (-16) **positive; 5**
4. 98 ÷ (-14) **negative; -7**	5. 30 ÷ (-3) **negative; -10**	6. -88 ÷ 8 **negative; -11**
7. -120 ÷ (-15) **positive; 8**	8. 196 ÷ 14 **positive; 14**	9. 81 ÷ (-3) **negative; -27**

State whether each statement is true or false.

10. The quotient of two negative numbers is positive. **true**

11. The quotient of one positive and one negative number is negative. **true**

Divide.

12. 12 ÷ (-6) **-2**	13. -15 ÷ 3 **-5**	14. 14 ÷ 2 **7**
15. (-21) ÷ (-7) **3**	16. 30 ÷ (-5) **-6**	17. 0 ÷ 6 **0**
18. 64 ÷ 8 **8**	19. -81 ÷ 9 **-9**	20. -49 ÷ (-7) **7**

Connection to Algebra

You can find the value of expressions involving division by using the same rules as division with integers.

Example  **Evaluate $cd \div (-3)$, if $c = -5$ and $d = -9$.**

$cd \div (-3) = -5(-9) \div (-3)$ *Replace c with −5 and d with −9.*
$= 45 \div (-3)$ *The product of −5 and −9 is positive.*
$= -15$ *The quotient of 45 and −3 is negative.*

Meteorology

Example 4 **People in Duluth, Minnesota, find ways to enjoy their lengthy cold spells. Energy providers use *degree days* to estimate the energy that will be needed for heating in these cold days. The degree-day number indicates how far the day's average temperature was above or below 65°F. The formula $d = 65 - \frac{h + \ell}{2}$ can be used to find degree days. In the formula, *d* represents degree days, *h* represents the high temperature of the day, and ℓ represents the low temperature of the day. If *d* is positive, it represents heating degree days. If *d* is negative, it represents cooling degree days. If the high temperature in Duluth was −2°F and the low temperature was −16°F, find the degree days.**

$$d = 65 - \frac{h + \ell}{2}$$
$$= 65 - \frac{-2 + (-16)}{2}$$
$$= 65 - \frac{-18}{2}$$
$$= 65 - (-9)$$
$$= 74$$

The day had 74 heating degree days.

Checking Your Understanding

Communicating Mathematics

1. See margin.

Read and study the lesson to answer these questions.

1. **Explain** why the solution of $q = -10 \div 2$ must be −5. Use a pattern.
2. **Draw a model** that shows the solution for $15 \div 3 = r$. **See margin.**
3. **Write** the division sentence shown at the right. **−8 ÷ 4 = −2**

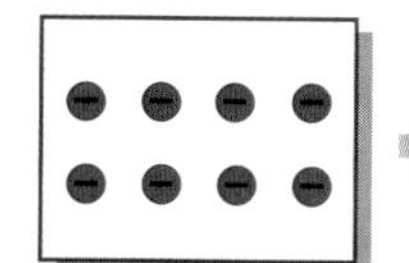
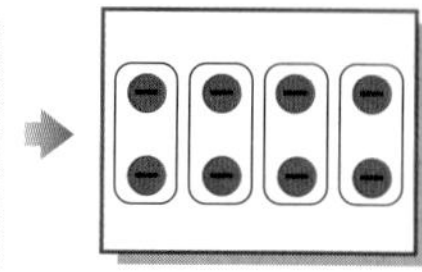

4. **Assess Yourself** In your own words, describe a situation in your life where you use negative numbers. **See students' work.**

Reteaching

Using Writing Have students rewrite a division problem as a multiplication problem. Then solve.
■ ÷ (-8) = 7 → 7 × (-8) = ■ **-56**

Additional Answer

2.

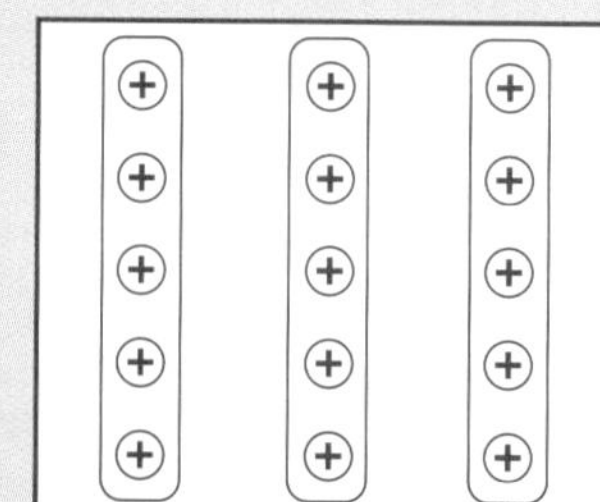

Guided Practice State whether each quotient is positive or negative. Then find the quotient.

5. $-56 \div 8$ **−; −7**
6. $63 \div 9$ **+; 7**
7. $42 \div (-7)$ **−; −6**
8. $\frac{91}{7}$ **+; 13**
9. $\frac{105}{-7}$ **−; −15**
10. $\frac{222}{-37}$ **−; −6**

Solve each equation.

11. $d = \frac{-240}{-6}$ **40**
12. $-96 \div 24 = k$ **−4**
13. $\frac{-105}{-15} = m$ **7**
14. $n = \frac{-804}{67}$ **−12**

Evaluate each expression.

15. $\frac{x}{-5}$, if $x = -65$ **13**
16. $\frac{y}{5}$, if $y = -50$ **−10**
17. $\frac{42}{z}$, if $z = -14$ **−3**
18. $\frac{b}{-7}$, if $b = -98$ **14**

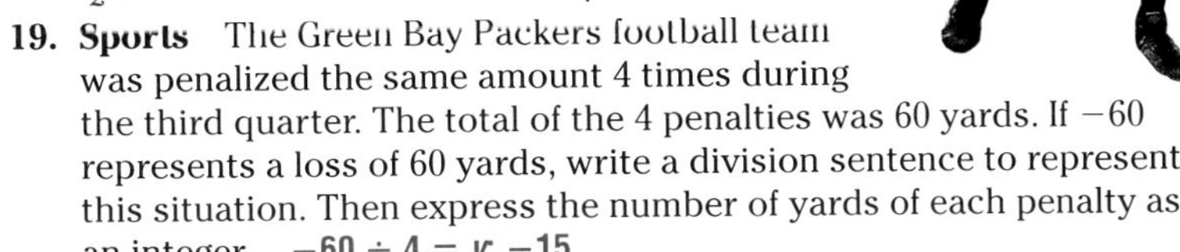

19. **Sports** The Green Bay Packers football team was penalized the same amount 4 times during the third quarter. The total of the 4 penalties was 60 yards. If −60 represents a loss of 60 yards, write a division sentence to represent this situation. Then express the number of yards of each penalty as an integer. **$-60 \div 4 = y$; −15**

Exercises: Practicing and Applying the Concept

Independent Practice **Divide.**

20. $56 \div (-7)$ **−8**
21. $48 \div (-12)$ **−4**
22. $-72 \div 8$ **−9**
23. $-36 \div (-6)$ **6**
24. $84 \div (-12)$ **−7**
25. $-52 \div (-4)$ **13**
26. $64 \div 4$ **16**
27. $-555 \div 5$ **−111**
28. $\frac{24}{6}$ **4**
29. $\frac{27}{-3}$ **−9**
30. $\frac{-570}{19}$ **−30**
31. $\frac{-804}{67}$ **−12**
32. Divide −54 by 9. **−6**
33. Divide 63 by −7. **−9**
34. Find the quotient of 121 and 11. **11**
35. Find the quotient of −28 and −7. **4**

Solve each equation.

36. $m = \frac{91}{-7}$ **−13**
37. $r = \frac{76}{-4}$ **−19**
38. $\frac{-63}{-21} = a$ **3**
39. $\frac{-90}{18} = z$ **−5**
40. $\frac{45}{-15} = k$ **−3**
41. $\frac{-119}{-17} = x$ **7**
42. $p = \frac{-143}{13}$ **−11**
43. $b = \frac{-144}{16}$ **−9**
44. $t = \frac{-175}{-25}$ **7**

Evaluate each expression. **45. −17 46. −5 48. −16 49. −7**

45. $\frac{x}{6}$, if $x = -102$
46. $\frac{105}{m}$, if $m = -21$
47. $\frac{y}{-9}$, if $y = -108$ **12**
48. $\frac{112}{z}$, if $z = -7$
49. $\frac{126}{a}$, if $a = -18$
50. $\frac{-144}{p}$, if $p = 9$ **−16**
51. $\frac{v}{q}$, if $v = -24$, and $q = 8$ **−3**
52. $\frac{n \times (-n)}{m}$, if $m = 27$, and $n = -9$ **−3**

Error Analysis

Students may find the wrong sign for the quotient because

- they use rules for addition and subtraction.
- they always use the sign of the first number (or the last number).
- they use no sign, thus making all answers positive.

Have students make a reference card containing the rules for division of integers and use that when completing exercises.

Assignment Guide

Core: 21–51 odd, 53, 55–67
Enriched: 20–52 even, 53–67

For **Extra Practice**, see p. 745.

The red A, B, and C flags, printed only in the Teacher's Wraparound Edition, indicate the level of difficulty of the exercises.

Teaching Tip In Exercises 5–10, students are to state the sign of the quotient. If students are still having difficulty with the signs, continue to practice identifying the sign of the quotient using the Independent Practice exercises.

Practice Masters, p. 18

NAME ______________ DATE ________

2-8 Practice
Dividing Integers

Student Edition
Pages 104–108

Divide.

1. $16 \div 4$ **4**
2. $-27 \div 3$ **−9**
3. $25 \div (-5)$ **−5**
4. $63 \div (-9)$ **−7**
5. $-15 \div (-3)$ **5**
6. $14 \div (-7)$ **−2**
7. $-124 \div 4$ **−31**
8. $60 \div 15$ **4**
9. $28 \div (-4)$ **−7**
10. $-56 \div (-8)$ **7**
11. $72 \div 8$ **9**
12. $-21 \div (-7)$ **3**
13. $\frac{-32}{4}$ **−8**
14. $\frac{45}{9}$ **5**
15. $\frac{-45}{3}$ **−15**
16. $\frac{-25}{-5}$ **5**
17. $\frac{35}{-7}$ **−5**
18. $\frac{-63}{-7}$ **9**
19. $\frac{-144}{12}$ **−12**
20. $\frac{48}{-6}$ **−8**

Evaluate each expression if $x = -8$ and $y = -12$.

21. $x \div 2$ **−4**
22. $x \div (-4)$ **2**
23. $36 \div y$ **−3**
24. $0 \div y$ **0**
25. $\frac{y}{-6}$ **2**
26. $\frac{x}{4}$ **−2**
27. $\frac{-144}{y}$ **12**
28. $\frac{-136}{x}$ **17**

Solve each equation.

29. $x = \frac{-150}{-25}$ **$x = 6$**
30. $k = \frac{-98}{14}$ **$k = -7$**
31. $m = \frac{-144}{-16}$ **$m = 9$**
32. $y = \frac{243}{-81}$ **$y = -3$**
33. $\frac{-208}{-26} = t$ **$t = 8$**
34. $\frac{-180}{15} = n$ **$n = -12$**
35. $\frac{-189}{-21} = p$ **$p = 9$**
36. $\frac{288}{-18} = d$ **$d = -16$**
37. $z = \frac{930}{-30}$ **$z = -31$**
38. $w = \frac{-312}{24}$ **$w = -13$**
39. $b = \frac{-396}{-36}$ **$b = 11$**
40. $c = \frac{-336}{12}$ **$c = -28$**

Group Activity Card 2-8

Division Rummy Group Activity **2-8**

MATERIALS: Deck of "Integer Division" cards, with expressions given on back

This is a game for two or three players. To begin, the deck of "Integer Division" cards is shuffled and five cards are dealt to each player, with the remaining cards placed face down in a pile in the middle of the table. The top card is removed, turned over, and becomes the discard pile. The rules and procedures for rummy are used for this game.

Players try to make a "book" of three cards, all of which have the same quotient represented by the division problem on them.

Players take turns drawing a card from the face-down pile or the discard pile. After drawing a card, a player may lay down a book or play a card on a book already on the table before discarding a card from his or her hand. The first player to run out of cards wins.

 Pre-Algebra

4 ASSESS

Closing Activity

Writing Have students write a problem that can be solved by using the equation $-36 \div (-3) = 12$.

Chapter 2, Quiz D (Lessons 2-7 and 2-8) is available in the *Assessment and Evaluation Masters*, p. 44.

Additional Answers

60.

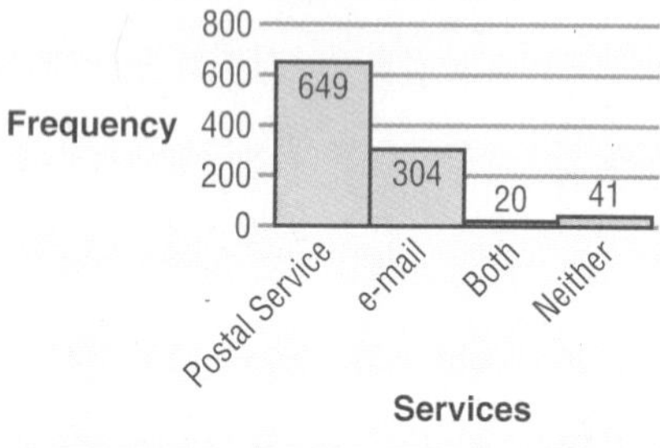

Enrichment Masters, p. 18

NAME ______________________ DATE ____________

2-8 Enrichment
Cross-Number Puzzle

Student Edition Pages 104–108

Use the clues below to complete the following cross-number puzzle. Any negative signs should be placed in the same box as the first digit of the integer.

<table>
<tr><td></td><td></td><td></td><td>1 −1</td><td>2 2</td><td>3 4</td><td>4 2</td><td></td><td></td><td></td></tr>
<tr><td>5 −7</td><td>6 1</td><td></td><td>7 3</td><td>0</td><td>3</td><td>3</td><td></td><td>8 −3</td><td>9 6</td></tr>
<tr><td>10 4</td><td>0</td><td></td><td>11 4</td><td>4</td><td>5</td><td>6</td><td></td><td>12 8</td><td>0</td></tr>
<tr><td>8</td><td></td><td></td><td></td><td></td><td></td><td></td><td></td><td></td><td>8</td></tr>
<tr><td></td><td></td><td>13 −7</td><td>14 2</td><td>1</td><td></td><td>15 −2</td><td>16 4</td><td></td><td></td></tr>
<tr><td>17 −7</td><td></td><td>18 5</td><td>3</td><td></td><td>19 −9</td><td>0</td><td>5</td><td></td><td>20 8</td></tr>
<tr><td>21 1</td><td>22 9</td><td></td><td>23 2</td><td>24 4</td><td>7</td><td>0</td><td></td><td>25 −5</td><td>8</td></tr>
<tr><td>26 1</td><td>8</td><td></td><td>27 9</td><td>6</td><td>4</td><td>0</td><td></td><td>28 1</td><td>6</td></tr>
<tr><td></td><td></td><td></td><td>29 6</td><td>0</td><td>5</td><td>0</td><td></td><td></td><td></td></tr>
</table>

Across

1. $23 \times (-54)$
5. $-56 - 15$
7. $-4283 + 7316$
8. $46.8 \div (-1.3)$
10. $-3840 \div (-96)$
11. $-1.6 \times (-2785)$
12. $91.5 + (-11.5)$
13. $-42 - 679$
15. $50 \times (-0.48)$
18. $-53{,}000 \times (-0.001)$
19. -181×5
21. $-42 + 61$
23. $10 \times (-13) \times (-19)$
25. $26.1 \div (-0.45)$
26. $-750 \times (-0.024)$
27. $-38.56 \div (-0.004)$
28. $-38 - (-54)$
29. $11{,}260 + (-5210)$

Down

1. $-718 + 584$
2. $-0.8 \times (-255)$
3. $198 - (-237)$
4. $-118 \div (-0.5)$
5. $-2 \times (-22) \times (-17)$
6. $-1000 \div (-100)$
8. $-6.46 \div 0.17$
9. $1547 + (-939)$
13. -5000×0.015
14. $-832 \times (-28)$
15. $-40 \times (-20) \times (-25)$
16. $-9 \times (-5)$
17. $-79 \times 3 \times 3$
19. $-13{,}406 + 3661$
20. $-649 + 1535$
22. $46 - (-52)$
24. $-10 \times (-46)$
25. $-0.663 \div 0.013$

Critical Thinking

53. Find values for x, y, and z, so that all of the following statements are true.
- ▶ $y > x$, $z < y$, and $x < 0$.
- ▶ $x \div z = -z$
- ▶ z is divisible by 2 and 3.
- ▶ $x \div y = z$ **Sample answer: $x = -144$; $y = 12$; $z = -12$**

Applications and Problem Solving

54. Aerospace To simulate space travel, NASA's Lewis Research Center in Cleveland, Ohio, uses a 430-foot shaft. The air is pumped out of the shaft to create a near-vacuum for free-fall experiments. If the free fall of an object in the shaft takes 5 seconds to travel the −430 feet, on average how far does the object travel in each second? **−86 feet per second**

55. Economics At the end of 1992, the United States Federal Government had a debt of 4065 billion dollars. By the end of 1993, the debt had grown to more than 4351 billion. The term *per capita* means per person. If there were about 261,000,000 U.S. citizens in 1993, how much did the per capita debt grow from 1992 to 1993? (*Hint*: Find the amount of change in the debt, then divide by the number of citizens.) **about $1096**

Mixed Review

56. Solve $y = 7(-11)$. (Lesson 2-7) **−77**

57. Education At Clay Battelle Junior High School, the bell rings at 8:05, 8:51, 8:55, 9:41, and 9:45 each weekday morning. When do the next three bells ring? (Lesson 2-6) **10:31, 10:35, 11:21**

The frequency table at the right contains data from a survey on message service. (Lesson 1-10)

Who do you trust more with a message?	Frequency
Postal Service	649
e-mail	304
Both Postal Service and e-mail equally	20
Neither Postal Service nor e-mail/Don't know	41

Source: ICR Survey Research Group

58. What message service was chosen most often? **Postal Service**

59. How many people responded that they didn't trust the Postal Service or e-mail, or didn't know which one they trusted more? **41**

60. Make a bar graph of the data in the table. **See margin.**

61. How many people gave information for this survey? **1014**

62. Personal Finance Silvia Vidal just landed her first job as a veterinary assistant. The paycheck for her first week was $280 before taxes were deducted. If Ms. Vidal worked 35 hours that week, how much did she earn per hour? (Lesson 1-8) **$8**

63. Solve $r + 11 = 17$ mentally. (Lesson 1-6) **6**

64. Name the property shown by $(9 + 7) + 4 = 4 + (9 + 7)$. (Lesson 1-4) **commutative, +**

Find the value of each expression. (Lesson 1-2)

65. $12 \div 3 + 12 \div 4$ **7** **66.** $(25 \cdot 3) + (15 \cdot 3)$ **120** **67.** $144 \div 16 \cdot 9 \div 3$ **27**

Extension

Using Manipulatives Have students make a number cube with the sides marked with 12, 24, 36, 48, 60, and 72. Students then roll that number cube and toss a coin to establish the dividend. If the coin is heads, the dividend is positive; if it is tails, the dividend is negative. Then toss the coin again and roll a regular die to find the divisor. If a 5 is rolled, roll again. Then the student is to find the quotient of the two numbers.

CHAPTER 2 Highlights

Vocabulary

After completing this chapter, you should be able to define each term, property, or phrase and give an example or two of each.

Algebra

absolute value (p. 67)
adding integers with different signs (p. 84)
adding integers with the same sign (p. 83)
additive inverses (p. 89)
dividing integers with different signs (p. 104)
dividing integers with the same signs (p. 105)
integers (p. 66)
multiplying integers with different signs (p. 99)
multiplying integers with the same sign (p. 100)
negative (p. 66)
subtracting integers (p. 89)
zero pair (p. 82)

Geometry

axes (p. 73)
coordinate (p. 67)
coordinate system (p. 73)
graph (p. 73)
plotting a point (p. 73)
quadrants (p. 73)
vertex (p. 75)

Problem Solving

look for a pattern (p. 94)

Statistics

cluster (p. 71)
line plot (p. 71)
outlier (p. 71)

Understanding and Using Vocabulary

Choose the letter of the term that best matches each statement or phrase.

1. an integer less than zero F
2. the four regions separated by the axes on a coordinate plane G
3. the number that corresponds to a point on the number line C
4. an integer and its opposite B
5. positive and negative whole numbers E
6. a dot placed at the point named by the ordered pair D
7. the distance a number is from the zero point on the number line A

a. absolute value
b. additive inverses
c. coordinate
d. graph
e. integers
f. negative integer
g. quadrants

Highlights

Using the Chapter Highlights

The Chapter Highlights begins with a listing of the new terms, properties, and phrases that were introduced in this chapter.

Instructional Resources

Three multiple-choice tests and three free-response tests are provided in the *Assessment and Evaluation Masters*. Forms 1A and 2A are for honors pacing, Forms 1B and 2B are for average pacing, Forms 1C and 2C are for basic pacing. Chapter 2 Test, Form 1B is shown at the right. Chapter 2 Test, Form 2B is shown on the next page.

Assessment and Evaluation Masters, p. 31–32

NAME ______________________ DATE __________

CHAPTER 2 Test, Form 1B

Simplify.

1. |-18|
   A. -18 B. 18 C. 180 D. 0 — 1. B
2. |-12| + |18|
   A. 30 B. -30 C. 0 D. 6 — 2. A
3. |15| − |-7|
   A. 22 B. 8 C. -22 D. -8 — 3. B
4. Use the coordinate grid at the right to choose the point for the ordered pair (2, 3).
   A. A B. B C. E D. F — 4. A
5. Use the coordinate grid at the right to choose the ordered pair for the point labeled C.
   A. (0, 1) B. (1, 1) C. (1, 0) D. (-1, 1) — 5. B

Choose the symbol to replace ● to make a true sentence.

6. -7 ● -10
   A. < B. = C. > D. + — 6. C
7. 4 ● -15
   A. < B. = C. > D. + — 7. C
8. Choose the set of numbers that is correctly ordered from greatest to least.
   A. {-3, 2, 1, 0} B. {5, -4, 0, -1} C. {-4, -3, -2, 0} D. {7, 4, -1, -3} — 8. D

Simplify.

9. 19 + 21
   A. -40 B. -2 C. 2 D. 40 — 9. D
10. -5 + 12
    A. 17 B. -7 C. 7 D. -17 — 10. C
11. -17 + (-12)
    A. 29 B. -5 C. 5 D. -29 — 11. D
12. -8 − (-23)
    A. -31 B. 31 C. 15 D. -15 — 12. C

NAME ______________________ DATE __________

Chapter 2 Test, Form 1B (continued)

Find the difference.

13. 14 − 24
    A. 38 B. -10 C. 10 D. -38 — 13. B
14. -16 − 15
    A. -31 B. 31 C. -1 D. 1 — 14. A

Find the product.

15. -7 · 5
    A. 12 B. -35 C. 35 D. -12 — 15. B
16. -8 · (-7)
    A. -15 B. -56 C. 56 D. -1 — 16. C
17. 11 · (-6)
    A. -66 B. 5 C. -17 D. 66 — 17. A

Find the quotient.

18. -15 ÷ (-5)
    A. -3 B. 3 C. -10 D. -20 — 18. B
19. 12 ÷ (-2)
    A. -6 B. 6 C. -10 D. 14 — 19. A
20. -36 ÷ 4
    A. -40 B. -32 C. 9 D. -9 — 20. D

Evaluate each expression if $a = -3$, $b = 4$, and $c = -6$.

21. $b \cdot c - a$
    A. -21 B. -27 C. 6 D. 72 — 21. A
22. $(b \cdot c) \div a$
    A. -21 B. -8 C. 27 D. 8 — 22. D

Find the next number in each pattern.

23. 15, 17, 18, 20, 21, · · ·
    A. 23 B. 22 C. 19 D. 24 — 23. A
24. 5, 25, 20, 100, 95, · · ·
    A. 90 B. 475 C. 500 D. 70 — 24. B
25. An elevator was at floor 24. It went down 6 floors, up 10 floors, and then down 8 floors. At what floor did it stop?
    A. 28 B. 26 C. 20 D. 40 — 25. C

CHAPTER 2 Study Guide and Assessment

Using the Study Guide and Assessment

Skills and Concepts Encourage students to refer to the objectives and examples on the left as they complete the review exercises on the right.

Assessment and Evaluation Masters, p. 37–38

NAME ______ DATE ______

CHAPTER 2 Test, Form 2B

Simplify.

1. $|-15|$ — 1. 15
2. $23 + 13$ — 2. 36
3. $43 - 19$ — 3. 24
4. $14 \cdot 6$ — 4. 84
5. $564 \div 4$ — 5. 141
6. $|-18| + |9|$ — 6. 27
7. $-16 - (-12)$ — 7. −4
8. $-24 + (-25)$ — 8. −49

Solve each equation.

9. $a = -38 \cdot (-3)$ — 9. $a = 114$
10. $-72 \div (-12) = b$ — 10. $6 = b$
11. $|20| - |-5| = c$ — 11. $15 = c$
12. $m = -67 + 18$ — 12. $m = -49$
13. $p = -14 - 12$ — 13. $p = -26$
14. $-9 \cdot 23 = x$ — 14. $-207 = x$
15. $y = -150 \div 5$ — 15. $y = -30$

Replace each ● with <, >, or =.

16. -4 ● 1 — 16. <
17. 7 ● -6 — 17. >
18. $6 \cdot (-3)$ ● $-2 \cdot (-9)$ — 18. <
19. $3 \cdot 8$ ● $-4 \cdot (-6)$ — 19. =
20. $x - (-5)$ ● 12, if $x = 9$ — 20. >
21. Write two inequalities for the graph below. — 21. $-4 < -1$, $-1 > -4$

−7 −6 −5 −4 −3 −2 −1 0

NAME ______ DATE ______

Chapter 2 Test, Form 2B (continued)

Order the integers in each set from least to greatest.

22. {1, −2, 7, 4, −5} — 22. {−5, −2, 1, 4, 7}
23. {5, −3, 0, 3, −1} — 23. {−3, −1, 0, 3, 5}

Evaluate each expression if r = 2, s = −3, and t = 5.

24. $2r + 3t$ — 24. 19
25. $4s - 3r$ — 25. −18
26. $12r \div 2s$ — 26. −4
27. $rst + st$ — 27. −45
28. Write the next two integers in the pattern 7, 9, 13, 19, 27, ⋯. — 28. 37, 49

Use the coordinate grid to name the ordered pair for the following points.

29. E — 29. (2, 3)
30. F — 30. (4, −2)

On graph paper, draw coordinate axes. Then graph and label each of the following points.

31. T(−3, −4) — 31. See students' graphs.
32. U(−4, −3) — 32. See students' graphs.
33. At the New England Professional Golfer's Association Pro-Am Tourney in Hollis, New Hampshire, three golfers scored 72, 74, and 75 on October 4, 1995 on a par 71 course. What was the total points scored? How many points over par did each golfer score? — 33. 221; 1; 3; 4

CHAPTER 2 Study Guide and Assessment

Skills and Concepts

Objectives and Examples

Upon completing this chapter, you should be able to:

▶ **graph integers on a number line** (Lesson 2-1)

Graph points *A*, *B*, and *C* on a number line if *A* has coordinate −2, *B* has coordinate 3, and *C* has coordinate −1.

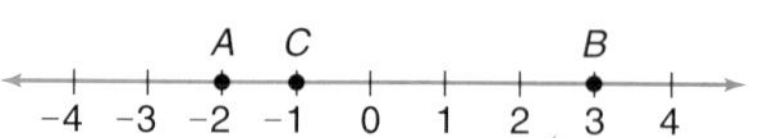

▶ **find absolute value** (Lesson 2-1)

$|-3| = 3$

$|4| - |-9| = 4 - 9 = -5$

▶ **graph points in all quadrants of the coordinate plane** (Lesson 2-2)

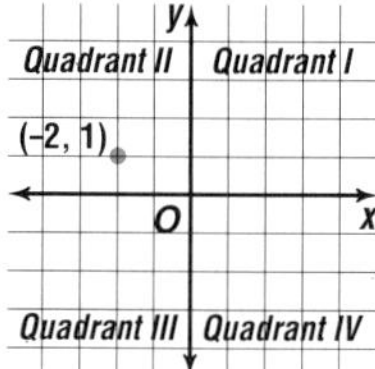

▶ **compare integers** (Lesson 2-3)

Compare −4 and −6.

Since −6 is to the left of −4 on the number line, $-4 > -6$.

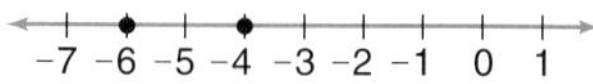

Review Exercises

Use these exercises to review and prepare for the chapter test. **8–11. See margin.**

Graph each set of numbers on a number line.

8. {−3, 6, −4, −6, 11, −1}
9. {22, 74, −30, 82, −10, −12}
10. {112, 98, −21, −41, 67}
11. {−1, −16, −4, −6, 4}

Simplify.

12. $|-31|$	**31**	13. $|24|$	**24**
14. $-|18|$	**−18**	15. $-|40|$	**−40**
16. $-|6 - 2|$	**−4**	17. $|-8| + |-21|$	**29**
18. $|17 - 29|$	**12**	19. $||7| - |-19||$	**12**

On graph paper, draw coordinate axes. Then graph and label each point. Name the quadrant in which each point is located.

20. $A(4, 3)$	**I**	21. $R(-1, -1)$	**III**
22. $W(-5, 3)$	**II**	23. $G(-6, 5)$	**II**
24. $M(-4, -1)$	**III**	25. $L(3, -2)$	**IV**

20–25. See margin for graph.

Replace each ● with <, >, or =.

26. 9 ● −9	**>**	27. −3 ● −3	**=**
28. 6 ● 13	**<**	29. −12 ● −21	**>**
30. −2 ● 0	**<**	31. −43 ● 43	**<**

Write an inequality using the numbers in each sentence. Use the symbols < or >.

32. The population of Detroit, Michigan, is 1,027,974. The population of Ann Arbor, Michigan, is 109,592.
33. The winning team in Super Bowl XXVIII won by 17 points. In Super Bowl XXVII, the winning team won by 35 points.

32–33. See margin.

GLENCOE Technology

Test and Review Software

You use this software, a combination of an item generator and an item bank, to create your own tests and worksheets. Types of items include free response, multiple choice, short answer, and open ended.

For IBM & Macintosh

Additional Answers

8.
9.

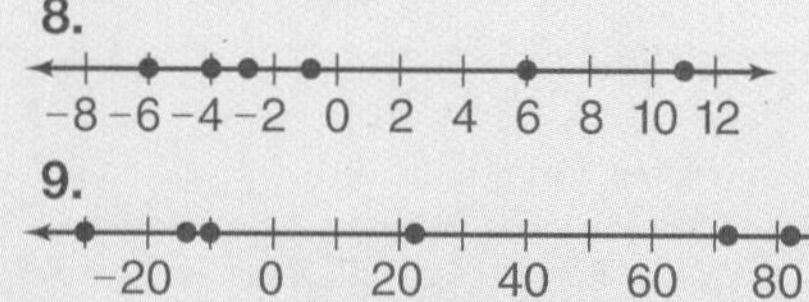

Objectives and Examples

▶ **order integers** (Lesson 2-3)

Order the integers $\{-3, 5, -1, 8, -6\}$ from least to greatest.

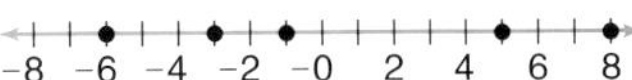

$-6, -3, -1, 5, 8$ are in order from least to greatest.

▶ **add integers** (Lesson 2-4)

$d = 4 + (-5)$
$d = -1$

$r = -4 + (-5)$
$r = -9$

$m = -4 + 5 + (-7)$
$m = 1 + (-7)$
$m = -6$

▶ **subtract integers** (Lesson 2-5)

$r = 5 - 12$
$r = 5 + (-12)$
$r = -7$

$-8p - (-11p) = -8p + 11p$
$= 3p$

▶ **multiply integers** (Lesson 2-7)

$m = -7 \cdot (-4)$
$m = +(7 \cdot 4)$
$m = 28$

$-4a(3b) = (-4 \cdot 3)(a \cdot b)$
$= -12ab$

Review Exercises

Order the integers in each set from least to greatest.

34. $\{8, 0, -3, 6, 5\}$ **−3, 0, 5, 6, 8**
35. $\{-16, 3, -4, 2\}$ **−16, −4, 2, 3**
36. $\{-75, 21, 4, -34, -88\}$ **−88, −75, −34, 4, 21**
37. $\{210, -300, 124, 9, -33\}$ **−300, −33, 9, 124, 210**

Solve each equation.

38. $a = -4 + (-5)$
39. $8 + (-11) = s$ **−3**
40. $-12 + 4 = t$ **−8**
41. $k = -15 + 9$ **−6**
42. $g = 18 + (-25)$
43. $h = -13 + (-2)$
44. $-6 + (-18) + 40 = q$ **16**
45. $a = 7 + (-3) + (-14)$ **−10**

38. −9 42. −7 43. −15

Solve each equation. **47. −30 48. −25**

46. $g = -8 - 7$ **−15**
47. $-33 - (-3) = s$
48. $-17 - 8 = w$
49. $t = 15 - (-6)$ **21**

Simplify. **52. 26m**

50. $2a - 5a$ **−3a**
51. $-12b - (-5b)$ **−7b**
52. $14m - (-12m)$
53. $-15r - (-21r)$ **6r**

Solve each equation. **56. 165**

54. $z = -9 \cdot 4$ **−36**
55. $(-4)(2) = k$ **−8**
56. $(-11)(-15) = y$
57. $p = -5 \cdot (-13)$ **65**

Find each product. **58. −48z 61. −300yz**

58. $(-2)(8z)(3)$
59. $6 \cdot (-6c) \cdot (-d)$ **36cd**
60. $-15x(5y)$ **−75xy**
61. $(-25y) \cdot (-6) \cdot (-2z)$

Additional Answers

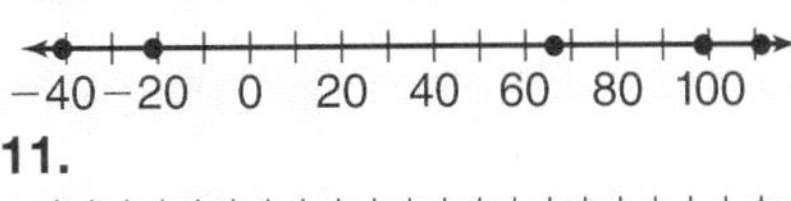

20-25.

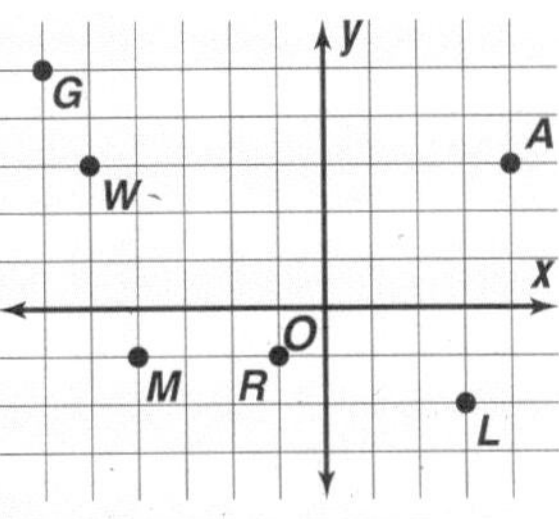

32. 1,027,974 > 109,592 or 109,592 < 1,027,974
33. 17 < 35 or 35 > 17

Applications and Problem Solving

Encourage students to work through the exercises in the Applications and Problem Solving section to strengthen their problem-solving skills.

Additional Answer

70.

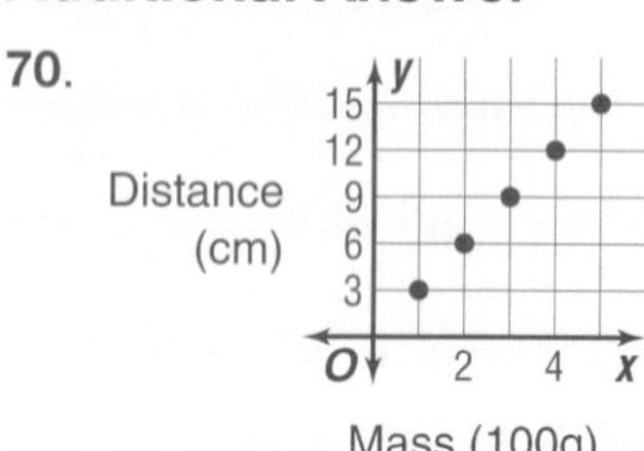

Objectives and Examples

▶ **divide integers** (Lesson 2-8)

$u = -105 \div 15$
$u = -(105 \div 15)$
$u = -7$

Evaluate $t \div 18$, if $t = -108$.

$t \div 18 = -108 \div 18$
$= -6$

Review Exercises

Solve each equation. 62. −22 64. −9

62. $k = 88 \div -4$ **63.** $-14 \div (-2) = h$ 7
64. $99 \div -11 = n$ **65.** $q = -45 \div (-3)$ 15

Evaluate each expression.

66. $d \div 8$, if $d = 120$ 15
67. $64 \div m$, if $m = \;\;4$ −16
68. $-700 \div k$, if $k = -10$ 70
69. $z \div -12$, if $z = -228$ 19

Applications and Problem Solving

70. Physics Jalisa did an experiment to see how the mass of an object attached to a spring affected the distance a spring stretched. The data are shown at the right. Graph the ordered pairs (mass, distance) in the coordinate plane. (Lesson 2-2) **See margin.** **72. 5, −6**

Stretching of a Spring	
Mass	**Distance**
100 g	3 cm
200 g	6 cm
300 g	9 cm
400 g	12 cm
500g	15 cm

71. Weather A combination of wind and cold makes you feel colder than the actual temperature. According to the National Weather Service, a temperature of 20 degrees Fahrenheit with a wind of 30 miles per hour makes you feel like it is 38 degrees colder than it is. How cold does it feel when it is 20 degrees with a wind of 30 mph? (Lesson 2-4) **−18°F**

72. Patterns Find the next two integers in the pattern 1, −2, 3, −4, __?__, __?__. (Lesson 2-6)

73. Patterns Use the pattern at the right to find $1089 \cdot 7$. (Lesson 2-6) **7623**

$1089 \cdot 1 = 1089$
$1089 \cdot 2 = 2178$
$1089 \cdot 3 = 3267$
$1089 \cdot 4 = 4356$

74. Physics When an object that is traveling increases its speed it is *accelerating*, when it slows, it is *decelerating*. Acceleration and deceleration are described by positive and negative numbers respectively. Suppose a car is decelerating at a rate of 1.2 meters per second per second. This is represented by $-1.2\ \text{m/s}^2$.

a. How much will a car's speed change if it decelerates at a rate of $-1.2\ \text{m/s}^2$ for 11 seconds? (Lesson 2-7) **−13.2 m/s**

b. If a motorcycle slows down its rate of speed by 3.6 meters per second in 12 seconds, what is its rate of deceleration? (Lesson 2-8) **$-0.3\ \text{m/s}^2$**

A practice test for Chapter 2 is available on page 775.

Alternative Assessment

Cooperative Learning Project

Temperature changes occur hourly, daily, weekly, and so on. For this project, you will need to be aware of this change and record it as follows.

1–6. See students' work.

1. Pick three cities in the world and record their temperature at the same time each day for a period of five days.
2. Make a table showing the time of day, name of the city, and the temperature for each day.
3. Describe the change in temperature for each city from day 1 to day 2, day 2 to day 3, day 3 to day 4, and day 4 to day 5.
4. On a coordinate graph, plot the daily temperature change for each city. The x-coordinate should be the first day of the series. The y-coordinate should be the change in temperature.
5. What was the total change in temperature for each city from day 1 to day 5?
6. Find the average change in temperature for each city. (Step 1: Add the change in temperature from day to day. Step 2: Divide that number by the number of temperature changes.)

Thinking Critically

▶ The point (a, b) is located in quadrant IV. What are all the possible values for a and b? **$a > 0$; $b < 0$**

▶ Describe all the points located in each quadrant using + (for positive numbers) and − (for numbers) for the values of the x-coordinate and the y-coordinate. (*Hint*: You should have four ordered pairs as your answer.) **I: (+, +); II: (−, +); III (−, −); IV (+, −)**

Portfolio

▶ Select an item from this chapter that you feel shows your best work and place it in your portfolio. **See students' work.**

Self Assessment

▶ Use the talents that you possess, for the trees would be quiet if no birds sang but the best.

Assess yourself. Are you putting your best foot forward? Does your work reflect the best that you can be? Is your work correct and presentable? Make a decision to complete your school work or a task in your life to the best of your ability. Never settle for less than your best. **See students' work.**

Assessment and Evaluation Masters, pp. 41, 53

NAME ________________ DATE ________

CHAPTER 2 Performance Assessment

Instructions: Demonstrate your knowledge by giving a clear, concise solution to each problem. Be sure to include all relevant drawings and justify your answers. You may show your solution in more than one way or investigate beyond the requirement of the problem.

1. Lincoln High School is playing Irmo High School in football. Lincoln High gained 6 yards and 8 yards on their first two plays. They lost yards on their next play. What was the total yards gained or lost in the first three plays?
   a. Solve the problem in at least two ways. Explain each strategy. Which strategy do you like best? Why?
   b. Irmo High had the ball for eight plays before fumbling. Their yardage on each play was +7, +7, ⁻4, +7, +7, +7, ⁻4, and ⁻4. Choose what you consider to be the easiest strategy to find Irmo's total yardage. Tell why your strategy is easier than another possible strategy.
2. A professional golfer had scores of 4 under par (⁻4) and 3 over par (+3), on the first two days of the tournament.
   a. Find the difference between her second day and first day scores. Use a number line in explaining the meaning of your answer.
   b. Her average score for each day of the four-day tournament was ⁻2. Choose third-day and fourth-day scores that would result in such a four-day average. Tell how you determined your scores.
3. Write a paragraph explaining why the product of a positive integer and a negative integer is a negative integer. Use a number line or models in your exploration.

CHAPTER 2 Scoring Guide

Level	Specific Criteria
3 Superior	• Shows thorough understanding of the concepts *positive and negative integers*, *average*, and *sum*, *difference*, *product of integers*, and *number patterns*. • Clearly demonstrates ability to create and interpret coordinate grids. • Uses appropriate strategies to solve problems. • Computations to find total yardage, difference in scores, and average score are correct. • Written explanations are exemplary. • Number lines or models are accurate. • Goes beyond requirements of some or all problems.
2 Satisfactory, with Minor Flaws	• Shows understanding of the concepts *positive and negative integers*, *average*, and *sum*, *difference*, *product of integers*, and *number patterns*. • Demonstrates ability to create and interpret coordinate grids. • Uses appropriate strategies to solve problems. • Computations to find total yardage, difference in scores, and average score are mostly correct. • Written explanations are effective. • Number lines or models are mostly accurate. • Satisfies all requirements of problems.
1 Nearly Satisfactory, with Serious Flaws	• Shows understanding of most of the concepts *positive and negative integers*, *average*, and *sum*, *difference*, *product of integers*, and *number patterns*. • May not be able to create and interpret coordinate grids. • May not uses appropriate strategies to solve problems. • Computations to find total yardage, difference in scores, and average score are mostly correct. • Written explanations are satisfactory. • Number lines or models are mostly accurate. • Satisfies most requirements of problems.
0 Unsatisfactory	• Shows little or no understanding of most of the concepts *positive and negative integers*, *average*, and *sum*, *difference*, *product of integers*, and *number patterns*. • May not be able to create and interpret coordinate grids. • May not use appropriate strategies to solve problems. • Computations to find total yardage, difference in scores, and average score are incorrect. • Written explanations are not satisfactory. • Number lines or models are not accurate. • Does not satisfy requirements of problems.

Alternative Assessment

The Alternative Assessment section provides students with the opportunity to assess their own work by thinking critically, working with others, keeping a portfolio, and honestly evaluating their own progress. For more information on alternative forms of assessment, see Alternative *Assessment in the Mathematics Classroom*.

Performance Assessment

Performance Assessment tasks for this chapter are included in the *Assessment and Evaluation Masters*. A scoring guide is also provided.

Using the Ongoing Assessment

These two pages review the skills and concepts presented in Chapters 1–2. This review is formatted to reflect new trends in standardized testing. A more traditional cumulative review, shown below, is provided in the *Assessment and Evaluation Masters*, pp. 46–47.

Assessment and Evaluation Masters, pp. 46–47

NAME ______________________ DATE __________

CHAPTERS 1–2 Cumulative Review

1. A school is having a candy sale. The table shows the values of the prize categories that students can choose from at various levels of candy sales. Use the table to find the prize category if a student sells \$175 worth of candy. (Lesson 1-1) — 1. \$85

Value of Candy Sold	Prize Category
\$25	\$5
\$50	\$20
\$75	\$35
\$100	\$50

Find the value of each expression. (Lessons 1-2, 1-3)

2. $12 + 12 \div 3$ — 2. 16
3. $x - y + 7$ if $x = 20$, and $y = 9$ — 3. 18
4. $m \cdot p + 5 \cdot m$ if $m = 6$, and $p = 7$ — 4. 72

Name the property shown by each statement. (Lessons 1-4, 1-5)

5. $7(3 + 8) = 7 \cdot 3 + 7 \cdot 8$ — 5. Distributive
6. $(9 + 5) + 11 = 9 + (5 + 11)$ — 6. Associative

Identify the solution to each equation from the list given. (Lessons 1-6)

7. $y + 9 = 36$; 25, 27, 29 — 7. 27
8. $24 = n - 6$; 18, 24, 30 — 8. 30
9. $\frac{36}{t} = 9$; 3, 4, 5 — 9. 4

Translate each sentence into an equation. Then solve. (Lesson 1-8)

10. Four times a number is thirty-two. — 10. $4x = 32$; $x = 8$
11. Twenty more than a number is ninety. — 11. $x + 20 = 90$; $x = 70$

State whether each inequality is true, false, or open. (Lesson 1-9)

12. $5x + 2 < 12$ when $x = 5$ — 12. false
13. $4c - 8 > 10$ when $c = 5$ — 13. true

Use the frequency table to answer each question. (Lesson 1-10)

Precipitation in Rockland, U.S.A., 1995	
Month	Precipitation (in.)
January	3.4
February	2.9
March	4.2
April	3.0
May	1.8
June	2.1

14. In which month was there the greatest precipitation? — 14. March
15. On another sheet of paper, make a bar graph of the data. — 15. See students' work.

NAME ______________________ DATE __________

Chapters 1–2, Cumulative Review (continued)

Simplify. (Lesson 2-1)

16. $|^-27| - |^-12|$ — 16. 15
17. $^-45 + (^-9)$ — 17. $^-54$

On graph paper, draw coordinate axes. Then graph and label each point. (Lesson 2-2)

18. $F(^-4, 0)$ — 18. See students' work.
19. $G(3, ^-4)$ — 19. See students' work.

Replace each ● with > or <. (Lesson 2-3).

20. $^-6 + (^-7)$ ● $^-12$ — 20. <
21. $17 - j$ ● $^-20$ if $j = {^-5}$ — 21. >

Simplify. (Lessons 2-4, 2-5, 2-7, 2-8)

22. $18 - 35$ — 22. $^-17$
23. $^-8 + (^-19)$ — 23. $^-27$
24. $44 + (^-67)$ — 24. $^-23$
25. $^-4 - (^-16)$ — 25. 12
26. $^-18 \cdot 3$ — 26. $^-54$
27. $(^-15)(^-23)$ — 27. 345
28. $24 \div (^-8)$ — 28. $^-3$
29. $^-364 \div 14$ — 29. $^-26$

Find the next two integers in each pattern. (Lesson 2-6)

30. 7, 10, 15, 22, 31, ... — 30. 42, 55
31. 1, 3, 9, 27, 81, ... — 31. 243, 729
32. At midnight, the temperature was $^-15$°F. By the next morning, the temperature had gone up 20 degrees. What was the temperature then? (Lesson 2-4) — 32. 5°F
33. From 8 A.M. to noon, the temperature change was $^-16$°F. Find the average change per hour. (Lesson 2-8) — 33. $^-4$°F

CHAPTERS 1–2 Ongoing Assessment

Section One: Multiple Choice

There are ten multiple-choice questions in this section. After working each problem, write the letter of the correct answer on your paper.

1. Which is equivalent to $3 \times 8 - 6 \div 2$? D
   A. 3
   B. 9
   C. 13
   D. 21

2. You have two more brothers than sisters. If you have b brothers, which sentence could be used to find s, the number of sisters you have? B
   A. $b = s - 2$
   B. $b - 2 = s$
   C. $s = b + 2$
   D. $s = 2b$

3. The perimeter of a rectangle is $2(\ell + w)$, where ℓ is the length and w is the width. What is the perimeter of a rectangle with length 10 cm and width 8 cm? A
   A. 36 cm
   B. 28 cm
   C. 20 cm
   D. 18 cm

4. If a diver descends at a rate of 5 meters per minute, at what depth will she be after 12 minutes? D
   A. +60 m
   B. −17 m
   C. −50 m
   D. −60 m

5. If $36 - b = 20$, what is the value of b? C
   A. 56
   B. 26
   C. 16
   D. 6

6. $7 \times (10 + 5) =$ C
   A. $(7 + 10) + 5$
   B. $(7 + 10) \times 5$
   C. $(7 \times 10) + (7 \times 5)$
   D. $(7 + 10) \times (7 + 5)$

Use the following graph for exercises 7 and 8.

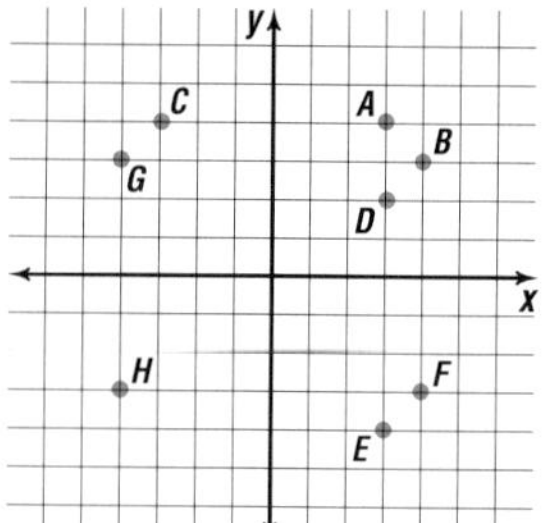

7. Which letter represents the ordered pair (4, 3)? B
   A. A
   B. B
   C. C
   D. D

8. Which letter represents the ordered pair (−4, 3)? C
   A. E
   B. F
   C. G
   D. H

9. The frequency table below contains students' test scores on a 25-point test.

Score	18	19	20	21	22	23	24	25
Number of Students	2	1	5	4	5	6	4	1

   How many students had a score greater than 20? C
   A. 4
   B. 5
   C. 20
   D. 25

10. If $a = 5$ and $b = 20$, what is the value of $2a + b$? D
   A. 50
   B. 45
   C. 40
   D. 30

Standardized test practice questions are also provided in the *Assessment and Evaluation Masters*, p. 45.

Section Two: Free Response

This section contains nine questions for which you will provide short answers. Write your answer on your paper.

11. Rewrite $(x + 2) + 12$ using an associative property. Then simplify. $x + (2 + 12); x + 14$

12. Many of the roller coasters at Cedar Point in Sandusky, Ohio, require the rider to be taller than 48 inches. Write an inequality for the height requirement. $h > 48$

13. A bolt of fabric contains about 18 yards of fabric. You want to make T-shirts for your friends. You need 1.875 yards for each T-shirt. About how many T-shirts can you make from one bolt of fabric? about 10

14. During a cold spell in January, 1994, the low temperatures were −22°F, −20°F, and −18°F. The average is the sum of these temperatures divided by 3. What was the average low temperature for those three days? −20°F

15. Translate *sixteen less than a number is twelve* into an equation and solve. $x - 16 = 12; 28$

16. During one of the Chicago Bears' offensive drives, they gained 12 yards on one play. On the next play, the quarterback was sacked for a loss of 6 yards. Write and solve an addition sentence representing the net yards for these two plays. $12 + -6$; 6 yards

17. Write a verbal phrase for $6u + 7$. Answers will vary.

18. The Branson Middle School football team is 2 yards from their opponent's goal line. On the next two plays, their offense loses 17 yards and then gains 11 yards. How many yards are they from the goal line now? 8

19. The temperature at 6:00 A.M. was −5°F. What was the temperature at 8:00 A.M. if it had risen 7 degrees? 2°F

Test-Taking Tip

You can prepare for taking standardized tests by working through practice tests like this one. The more you work with questions in a format similar to the actual test, the better you become in the art of test taking.

Do not wait until the night before taking a test to review. Allow yourself plenty of time to review the basic skills and formulas that are tested. Find your weaknesses so that you can ask for help.

Section Three: Open-Ended

This section contains four open-ended problems. Demonstrate your knowledge by giving a clear, concise solution to each problem. Your score on these problems will depend on how well you do the following.

- Explain your reasoning.
- Show your understanding of the mathematics in an organized manner.
- Use charts, graphs, and diagrams in your explanation.
- Show the solution in more than one way or relate it to other situations.
- Investigate beyond the requirements of the problem.

20. Show that $7ac + 14ab + 28a = 7a(c + 2b + 4)$. Use values for a, b, c, and d to justify your answer. See students' work.

21. You are at a party with 9 other guests. If everyone is to shake hands with every other guest, how many handshakes will occur? 45

22. If y is an integer, what is the least value that $y \cdot y$ could have? Explain your answer. 0

23. Rivera High School will graduate 268 seniors on June 1. The ceremony will be held in the school gymnasium. The gymnasium holds 1400 people in addition to the graduates. How many tickets should be offered to each graduate for family and friends? 5

CHAPTER 3

Solving One-Step Equations and Inequalities

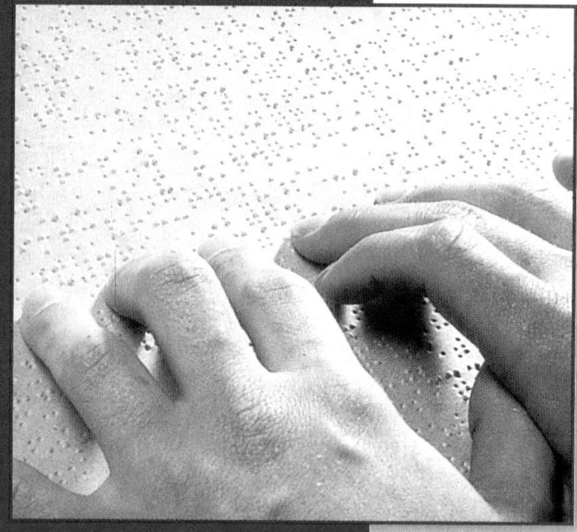

PREVIEWING THE CHAPTER

Students begin the chapter by solving mathematical and logic problems using the strategy of eliminating possibilities. Then students will learn to solve one-step equations by using addition, subtraction, multiplication, and division. Using formulas to solve real-world and geometrical problems, allows students to utilize their equation-solving skills. The study will then be extended to solving one-step inequalities. Finally, students will use the skills they have developed to solve real-world application problems.

Lesson (Pages)	Lesson Objectives	NCTM Standards	State/Local Objectives
3-1 (118–122)	Solve problems by eliminating possibilities.	1-4	
3-2A (123)	Use cups and counters to solve equations.	1-4, 9	
3-2 (124–128)	Solve equations by using the addition and subtraction properties of equality.	1-6, 9	
3-3 (129–133)	Solve equations by using the multiplication and division properties of equality.	1-5, 9	
3-4 (134–137)	Solve problems by using formulas.	1-4, 9	
3-5A (138)	Discover the greatest possible perimeter for a given area.	1-4, 8, 12, 13	
3-5 (139–144)	Find the perimeter and area of rectangles and squares.	1-4, 9, 12, 13	
3-5B (145)	Use a graphing calculator to explore the area of irregular figures on a coordinate plane.	1-4, 12, 13	
3-6 (146–150)	Solve inequalities by using the addition and subtraction properties of inequalities. Graph the solution set of inequalities.	1-5, 9	
3-7 (151–155)	Solve inequalities by using the multiplication and division properties of inequalities.	1-5, 9	
3-8 (156–159)	Solve verbal problems by translating them into equations and inequalities. Write verbal problems for equations and inequalities.	1-4, 9	

A complete, 1-page lesson plan is provided for each lesson in the ***Lesson Planning Guide***. Answer keys for each lesson are available in the **Answer Key Masters**.

ORGANIZING THE CHAPTER

You may want to refer to the **Course Planning Calendar** on page T12 for detailed information on pacing.
PACING: Standard—12 days; **Honors**—11 days; **Block**—7 days

LESSON PLANNING CHART

Lesson (Pages)	Materials/ Manipulatives	Extra Practice (Student Edition)	Blackline Masters: Study Guide	Blackline Masters: Practice	Blackline Masters: Enrichment	Blackline Masters: Assessment and Evaluation	Blackline Masters: Math Lab and Modeling Math	Blackline Masters: Multicultural Activity	Blackline Masters: Tech Prep Application	Blackline Masters: Graphing Calculator	Blackline Masters: Activity	Real-World Applications	Interactive Mathematics Tools Software	Teaching Transparencies	Group Activity Cards
3-1 (118–122)	balance scale	p. 745	p. 19	p. 19	p. 19									3-1A 3-1B	3-1
3-2A (123)	balance scale cups and counters*						p. 33						3-2A		
3-2 (122–128)	balance scale pennies thermometer (°F and °C)	p. 745	p. 20	p. 20	p. 20	p. 71			p. 5				3-2	3-2A 3-2B	3-2
3-3 (129–133)	calculator	p. 746	p. 21	p. 21	p. 21			p. 5					3-3.1 3-3.2 3-3.3	3-3A 3-3B	3-3
3-4 (134–137)		p. 746	p. 22	p. 22	p. 22	pp. 70–71		p. 6			pp. 3, 17	5	3-4	3-4A 3-4B	3-4
3-5A (138)	grid paper or tiles*						p. 34								
3-5 (139–144)	paper rectangles of different sizes	p. 746	p. 23	p. 23	p. 23		pp. 59, 78							3-5A 3-5B	3-5
3-5B (145)	calculator grid paper									p. 17					
3-6 (146–150)	balance scale small objects	p. 747	p. 24	p. 24	p. 24	p. 72			p. 6					3-6A 3-6B	3-6
3-7 (151–155)	balance scale counters*	p. 747	p. 25	p. 25	p. 25					p. 3		6		3-7A 3-7B	3-7
3-8 (156–159)		p. 747	p. 26	p. 26	p. 26	p. 72						7		3-8A 3-8B	3-8
Study Guide/ Assessment (156–159)						pp. 57–69, 73–75									

*Included in Glencoe's *Student Manipulative Kit* and *Overhead Manipulative Resources*.

ORGANIZING THE CHAPTER

All of the blackline masters in the Teacher's Classroom Resources are available on the ***Electronic Teacher's Classroom Resources*** CD-ROM.

OTHER CHAPTER RESOURCES

Student Edition
Chapter Opener, pp. 116–117
Working on the Investigation, p. 128
From the Funny Papers, p. 144
Closing the Investigation, p. 160

Teacher's Classroom Resources
Investigations and Projects Masters, pp. 37–40

Pre-Algebra Overhead Manipulative Resources, pp. 7–10

Technology
Test and Review Software (IBM and Macintosh)
CD-ROM Activities

Professional Publications
Block Scheduling Booklet
Glencoe Mathematics Professional Series

OUTSIDE RESOURCES

Books/Periodicals
Quincy Market, Regional Math Network
Problem-Solving Experiences in Pre-Algebra, Randall I. Charles and Carne S. Barnett
Math Connections, David J. and Joyce Glatzer

Videos/CD-ROMs
Pre-Algebra Video, ETA
Algebra Math Video Series: Solving Algebraic Equations & Inequalities, ETA
Mathematical Eye: Equations and Formula, Journal Films, Multimedia and Videodisc Compendium

Software
Quincy Market, Regional Math Network
Mastering Equations, Roots and Exponents, Phoenix/BFA Films and Videos

 ASSESSMENT RESOURCES

Student Edition
Math Journal, pp. 148, 153
Mixed Review, pp. 122, 128, 133, 137, 144, 150, 155, 159
Self Test, p. 137
Chapter Highlights, p. 161
Chapter Study Guide and Assessment, pp. 162–164
Alternative Assessment, p. 165
Portfolio, p. 165

MindJogger Videoquiz, 3

Teacher's Wraparound Edition
5-Minute Check, pp. 118, 124, 129, 134, 139, 146, 151, 156
Checking Your Understanding, pp. 120, 126, 131, 135, 141, 148, 153, 157
Closing Activity, pp. 122, 128, 133, 137, 144, 150, 155, 159
Cooperative Learning, pp. 136, 152

Assessment and Evaluation Masters
Multiple-Choice Tests, Forms 1A (Honors), 1B (Average), 1C (Basic), pp. 57–62
Free-Response Tests, Forms 2A (Honors), 2B (Average), 2C (Basic), pp. 63–68
Performance Assessment, p. 69
Mid-Chapter Test, p. 70
Quizzes A–D, pp. 71–72
Standardized Test Practice, p. 73
Cumulative Review, pp. 74–75

ENHANCING THE CHAPTER

Examples of some of the materials for enhancing Chapter 3 are shown below.

Multicultural Activity Masters, pp. 5, 6

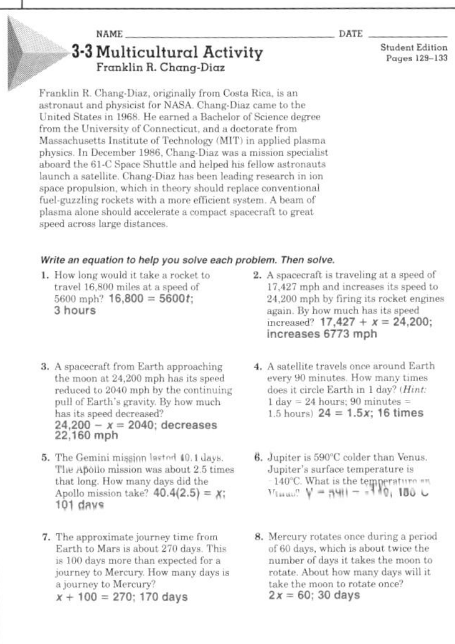

NAME ____________ DATE ________

3-3 Multicultural Activity
Franklin R. Chang-Diaz

Student Edition Pages 129–133

Franklin R. Chang-Diaz, originally from Costa Rica, is an astronaut and physicist for NASA. Chang-Diaz came to the United States in 1968. He earned a Bachelor of Science degree from the University of Connecticut, and a doctorate from Massachusetts Institute of Technology (MIT) in applied plasma physics. In December 1986, Chang-Diaz was a mission specialist aboard the 61-C Space Shuttle and helped his fellow astronauts launch a satellite. Chang-Diaz has been leading research in ion space propulsion, which in theory should replace conventional fuel-guzzling rockets with a more efficient system. A beam of plasma alone should accelerate a compact spacecraft to great speed across large distances.

Write an equation to help you solve each problem. Then solve.

1. How long would it take a rocket to travel 16,800 miles at a speed of 5600 mph? **$16,800 = 5600t$; 3 hours**
2. A spacecraft is traveling at a speed of 17,427 mph and increases its speed to 24,200 mph by firing its rocket engines again. By how much has its speed increased? **$17,427 + x = 24,200$; increases 6773 mph**
3. A spacecraft from Earth approaching the moon at 24,200 mph has its speed reduced to 2040 mph by the continuing pull of Earth's gravity. By how much has its speed decreased? **$24,200 - x = 2040$; decreases 22,160 mph**
4. A satellite travels once around Earth every 90 minutes. How many times does it circle Earth in 1 day? (*Hint:* 1 day = 24 hours; 90 minutes = 1.5 hours) **$24 = 1.5x$; 16 times**
5. The Gemini mission lasted 40.1 days. The Apollo mission was about 2.5 times that long. How many days did the Apollo mission take? **$40.4(2.5) = x$; 101 days**
6. Jupiter is 590°C colder than Venus. Jupiter's surface temperature is -140°C. What is the temperature on Venus? [illegible]
7. The approximate journey time from Earth to Mars is about 270 days. This is 100 days more than expected for a journey to Mercury. How many days is a journey to Mercury? **$x + 100 = 270$; 170 days**
8. Mercury rotates once during a period of 60 days, which is about twice the number of days it takes the moon to rotate. About how many days will it take the moon to rotate once? **$2x = 60$; 30 days**

Real-World Applications, 5, 6, 7

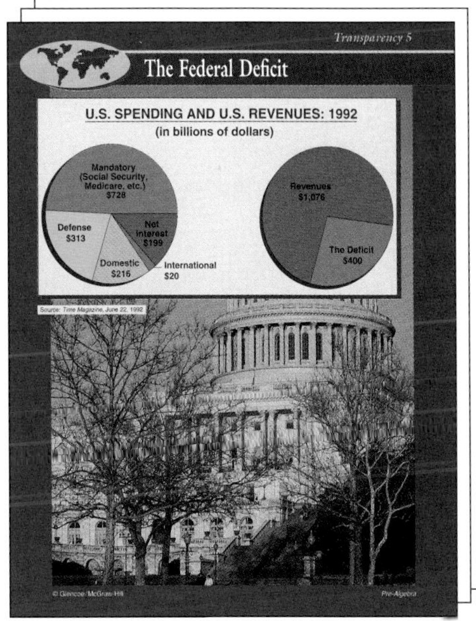

Graphing Calculator Masters, p. 3

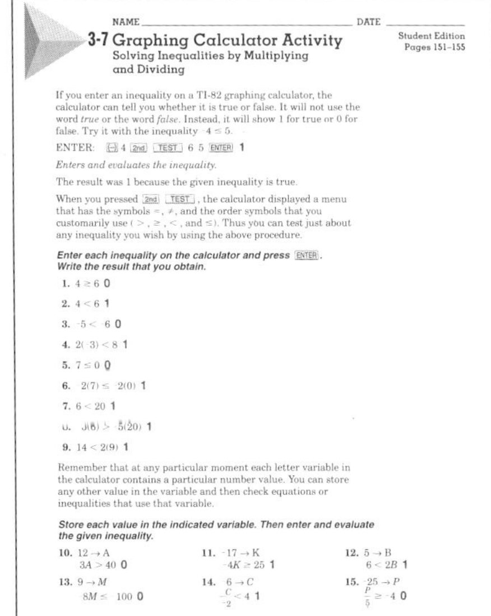

NAME ____________ DATE ________

3-7 Graphing Calculator Activity
Solving Inequalities by Multiplying and Dividing

Student Edition Pages 151–155

If you enter an inequality on a TI-82 graphing calculator, the calculator can tell you whether it is true or false. It will not use the word *true* or the word *false*. Instead, it will show 1 for true or 0 for false. Try it with the inequality $^-4 \leq 5$.

ENTER: (-) 4 [2nd] [TEST] 6 5 [ENTER] 1

Enters and evaluates the inequality.

The result was 1 because the given inequality is true.

When you pressed [2nd] [TEST], the calculator displayed a menu that has the symbols =, ≠, and the order symbols that you customarily use (>, ≥, <, and ≤). Thus you can test just about any inequality you wish by using the above procedure.

Enter each inequality on the calculator and press [ENTER]. Write the result that you obtain.

1. $4 \geq 6$ **0**
2. $4 < 6$ **1**
3. $^-5 < {}^-6$ **0**
4. $2(^-3) < 8$ **1**
5. $7 \leq 0$ **0**
6. $^-2(7) \leq {}^-2(0)$ **1**
7. $6 < 20$ **1**
8. $3(6) > {}^-5(20)$ **1**
9. $14 < 2(9)$ **1**

Remember that at any particular moment each letter variable in the calculator contains a particular number value. You can store any other value in the variable and then check equations or inequalities that use that variable.

Store each value in the indicated variable. Then enter and evaluate the given inequality.

10. $12 \to A$; $3A > 40$ **0**
11. $^-17 \to K$; $^-4K \geq 25$ **1**
12. $5 \to B$; $6 < 2B$ **1**
13. $9 \to M$; $8M \leq {}^-100$ **0**
14. $6 \to C$; $\frac{C}{^-2} < 4$ **1**
15. $^-25 \to P$; $\frac{P}{5} \geq {}^-4$ **0**

Tech Prep Applications Masters, pp. 5, 6

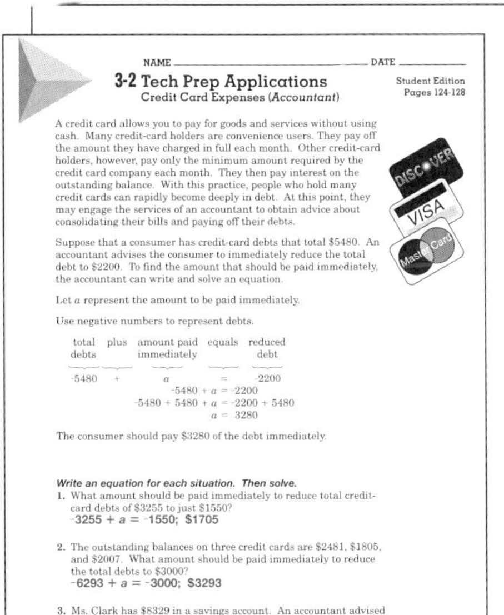

NAME ____________ DATE ________

3-2 Tech Prep Applications
Credit Card Expenses (*Accountant*)

Student Edition Pages 124-128

A credit card allows you to pay for goods and services without using cash. Many credit-card holders are convenience users. They pay off the amount they have charged in full each month. Other credit-card holders, however, pay only the minimum amount required by the credit card company each month. They then pay interest on the outstanding balance. With this practice, people who hold many credit cards can rapidly become deeply in debt. At this point, they may engage the services of an accountant to obtain advice about consolidating their bills and paying off their debts.

Suppose that a consumer has credit-card debts that total \$5480. An accountant advises the consumer to immediately reduce the total debt to \$2200. To find the amount that should be paid immediately, the accountant can write and solve an equation.

Let a represent the amount to be paid immediately.

Use negative numbers to represent debts.

total debts	plus	amount paid immediately	equals	reduced debt
$^-5480$	+	a	=	$^-2200$

$$^-5480 + a = {}^-2200$$
$$^-5480 + 5480 + a = {}^-2200 + 5480$$
$$a = 3280$$

The consumer should pay \$3280 of the debt immediately.

Write an equation for each situation. Then solve.

1. What amount should be paid immediately to reduce total credit-card debts of \$3255 to just \$1550? **$^-3255 + a = {}^-1550$; \$1705**
2. The outstanding balances on three credit cards are \$2481, \$1805, and \$2007. What amount should be paid immediately to reduce the total debts to \$3000? **$^-6293 + a = {}^-3000$; \$3293**
3. Ms. Clark has \$8329 in a savings account. An accountant advised her to leave \$2500 in the savings account and withdraw the rest to pay off her credit card debt. What amount should she withdraw from the savings account? **$8329 + a = 2500$; $^-$\$5829**

Activity Masters, pp. 3, 17

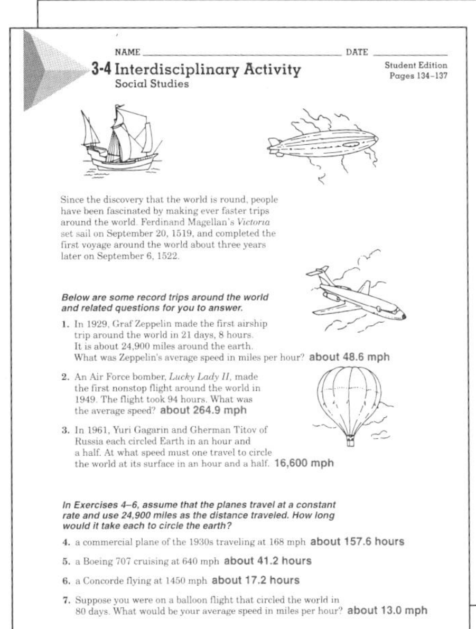

NAME ____________ DATE ________

3-4 Interdisciplinary Activity
Social Studies

Student Edition Pages 134–137

Since the discovery that the world is round, people have been fascinated by making ever faster trips around the world. Ferdinand Magellan's *Victoria* set sail on September 20, 1519, and completed the first voyage around the world about three years later on September 6, 1522.

Below are some record trips around the world and related questions for you to answer.

1. In 1929, Graf Zeppelin made the first airship trip around the world in 21 days, 8 hours. It is about 24,900 miles around the earth. What was Zeppelin's average speed in miles per hour? **about 48.6 mph**
2. An Air Force bomber, *Lucky Lady II*, made the first nonstop flight around the world in 1949. The flight took 94 hours. What was the average speed? **about 264.9 mph**
3. In 1961, Yuri Gagarin and Gherman Titov of Russia each circled Earth in an hour and a half. At what speed must one travel to circle the world at its surface in an hour and a half. **16,600 mph**

In Exercises 4–6, assume that the planes travel at a constant rate and use 24,900 miles as the distance traveled. How long would it take each to circle the earth?

4. a commercial plane of the 1930s traveling at 168 mph **about 157.6 hours**
5. a Boeing 707 cruising at 640 mph **about 41.2 hours**
6. a Concorde flying at 1450 mph **about 17.2 hours**
7. Suppose you were on a balloon flight that circled the world in 80 days. What would be your average speed in miles per hour? **about 13.0 mph**

Math Lab and Modeling Math Masters, pp. 59, 78

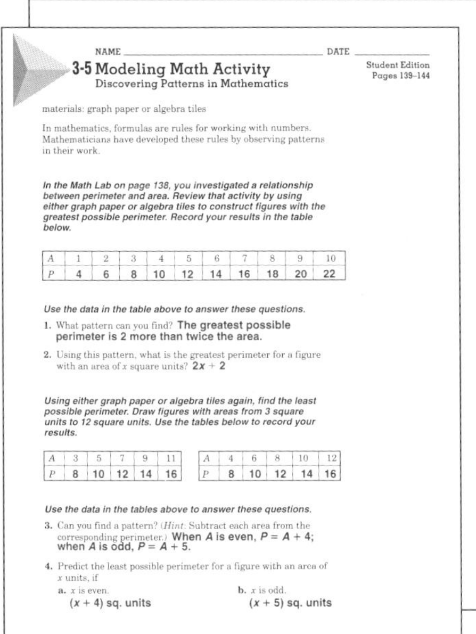

NAME ____________ DATE ________

3-5 Modeling Math Activity
Discovering Patterns in Mathematics

Student Edition Pages 139–144

materials: graph paper or algebra tiles

In mathematics, formulas are rules for working with numbers. Mathematicians have developed these rules by observing patterns in their work.

In the Math Lab on page 138, you investigated a relationship between perimeter and area. Review that activity by using either graph paper or algebra tiles to construct figures with the greatest possible perimeter. Record your results in the table below.

A	1	2	3	4	5	6	7	8	9	10
P	4	6	8	10	12	14	16	18	20	22

Use the data in the table above to answer these questions.

1. What pattern can you find? **The greatest possible perimeter is 2 more than twice the area.**
2. Using this pattern, what is the greatest perimeter for a figure with an area of x square units? **$2x + 2$**

Using either graph paper or algebra tiles again, find the least possible perimeter. Draw figures with areas from 3 square units to 12 square units. Use the tables below to record your results.

A	3	5	7	9	11
P	8	10	12	14	16

A	4	6	8	10	12
P	8	10	12	14	16

Use the data in the tables above to answer these questions.

3. Can you find a pattern? (*Hint:* Subtract each area from the corresponding perimeter.) **When A is even, $P = A + 4$; when A is odd, $P = A + 5$.**
4. Predict the least possible perimeter for a figure with an area of x units, if
   a. x is even. **$(x + 4)$ sq. units**
   b. x is odd. **$(x + 5)$ sq. units**

MAKING MATHEMATICS RELEVANT

This two-page introduction to the chapter provides students with an opportunity to explore other disciplinary topics and their applications to mathematics.

Background Information

Math and Braille in the News
Prior to the Dots Plus system, students had to use one braille system for math, another for literary publications, and another for computers. The existing math braille code has been around for about 40 years and is called the Nemeth code, after its inventor, Abraham Nemeth. Nemeth found it necessary to invent a "math" system because Louis Braille provided for the use of numbers by using the same symbols as letters A through I and preceding them with a symbol for "number".

CHAPTER 3 Solving One-Step Equations and Inequalities

TOP STORIES in Chapter 3

In this chapter, you will:

- solve problems by eliminating possibilities,
- solve equations and inequalities using addition, subtraction, multiplication and division, and
- use formulas to solve real-world and geometry problems.

MATH AND BRAILLE IN THE NEWS

Building Better Braille

Source: Physics Today, March, 1995

Do you ever think of math as a language? Imagine how difficult it might be for someone who cannot see letters, numbers and symbols. Physicist John Gardner and mathematician Norberto Salinas, both of whom are blind, have teamed up to develop a better method for writing mathematics for the blind using a new eight-dot code called Dots Plus. The Dots Plus system allows for four times as many symbols as the traditional six-dot system. Other new developments like speech synthesizers that "read" are opening new doors to those with visual impairments.

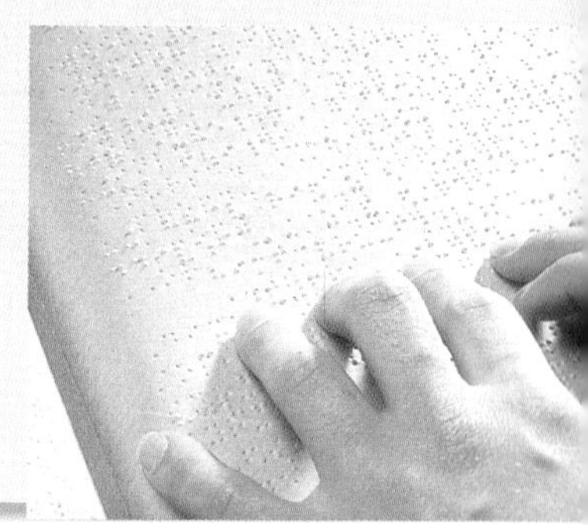

Putting It into Perspective

1800 — 1840 — 1880

1829 Louis Braille develops the Braille system, and Laura Dewey Bridgman, the first blind and deaf child to be successfully educated in the U.S., is born.

1832 The first school for the blind in the U.S. opens in Boston, MA.

1858 The American Printing House for the Blind, a major publisher, opens in Louisville, KY.

Putting It into Perspective

There are other ways for blind people to study, such as using recordings or computers. However, scientific and higher-level mathematics situations provide difficult, if not impossible, problems to be translated. Gardner's Dots Plus system includes the use of many symbols, such as integrals and summations. The symbols are greatly enlarged.

EMPLOYMENT OPPORTUNITIES

We See the Possibilities!

WE KNOW THAT new technologies are making it easier than ever for the visually impaired to choose any field of work. We are looking for talented engineers, entertainers, factory workers, teachers, social workers, computer programmers, lawyers, and anything else your imagination can conceive!

For more information, contact:
National Blindness Information Center
1800 Johnson Street
Baltimore, MD 21230

Statistical Snapshot

The Braille System

The braille cell is three dots high and two dots wide. This means that 63 different characters can be formed.

A B C D E F G H I J

The braille alphabet starts by using 10 combinations of the top 4 dots. The same 10 characters, when preceded by a special number sign, are used to express the numbers 1 to 0.

K L M N O P Q R S T

Adding the lower left-hand dot makes the next 10 letters. Adding the lower right-hand dot makes the last 5 letters of the alphabet (except *w*) and five word symbols, *below*.

U V X Y Z and for of the with

Omitting the lower left-hand dot forms nine *digraphs*, or speech sounds, and the letter *w*. This construction continues until all possible combinations have been used.

ch gh sh th wh ed er ou ow W

interNET CONNECTION For up-to-date information on DotsPlus, visit: **www.glencoe.com/sec/math/prealg/mathnet**

1920 **1960** **2000**

1930 Famed blind blues singer Ray Charles is born in Albany, GA.

1944 The first eye bank is formed in New York City.

1968 Helen Keller, a famous blind and deaf writer and speaker, dies. Keller was the subject of the movie *The Miracle Worker*.

1995 Dots Plus system of writing mathematics is developed.

EMPLOYMENT OPPORTUNITIES

The American Association for the Advancement of Science, throughout its Project on Science, Technology and Disability, maintains a resource directory of scientists and engineers with disabilities. Contact AAAS, 1333 H Street NW, Washington, DC 20005.

Statistical Snapshot

What patterns do students see in the Braille letters and numbers?

Cooperative Learning

Chapter Projects Two chapter projects are included in the *Investigations and Projects Masters.* In Chapter Project A, students extend the topic in the Chapter Opener. In Chapter Project B, students explore nutrition. A student page and a parent letter are provided for each Chapter Project.

interNET CONNECTION

Glencoe has made every effort to ensure that the website links for *Pre-Algebra* at www.glencoe.com/sec/math/prealg/mathnet are current and contain appropriate content. However, these website links are not under Glencoe's control.

Investigations and Projects Masters, p. 37

NAME ______________ DATE ________

CHAPTER 3 **Project A**
Designing a Work Station

In 1812, Louis Braille lost his sight at the age of three. His father, determined that his son would not suffer a life of begging or coal shoveling—typical fates for the physically handicapped in France at that time—made sure his son got an education. Louis made good use of this opportunity. He developed a tactile alphabet system. Examine the Braille system shown in your book on page 117. This system helped to open up a new world of opportunities for the blind. Over time, the world has come to be much more sensitive to the needs of the physically challenged.

1. Many devices help physically-challenged people negotiate life, such as wheelchair ramps and hearing aid devices for theater-goers. With your group, brainstorm a list of other devices that are available. You can contact a hospital or rehabilitation clinic for ideas. Share your findings with the class.

2. Today, people with physical challenges work productively in many challenging fields. Often, accommodations in the design of the workplaces are needed. For instance, people in wheelchairs need a wider space to maneuver by their desks and shelving low enough to be useful to them. Learn about and think about the special work needs of a person with a particular physical handicap. Then design a work station that helps that person do his or her job.

3. To help get ideas, arrange an interview with a person who has experience with specially-designed workspaces. Or speak with a human resources administrator at a company that employs physically-challenged people. Tape these interviews or take notes. Additionally, you can talk with family members or knowledgeable neighbors, or contact a local physical rehabilitation facility for suggestions. Also, you can visit the library to find magazines or journals that may be sources of ideas.

4. Draw a picture of your new work station design. Include labels and all measurements. Accompany your design with a report that describes all the features and the reasons for each. Then make a presentation to the class. Afterwards, you and your classmates may wish to present your completed works to local companies, clinics, or even to the newspaper.

3-1 LESSON NOTES

NCTM Standards: 1-4

Instructional Resources

- Study Guide Master 3-1
- Practice Master 3-1
- Enrichment Master 3-1
- Group Activity Card 3-1

Transparency 3-1A contains the 5-Minute Check for this lesson; **Transparency 3-1B** contains a teaching aid for this lesson.

Recommended Pacing	
Standard Pacing	Day 1 of 12
Honors Pacing	Day 1 of 11
Block Scheduling*	Day 1 of 7

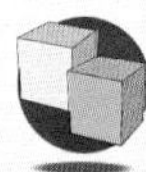

*For more information on pacing and possible lesson plans, refer to the ***Block Scheduling Booklet***.

1 FOCUS

5-Minute Check
(over Chapter 2)

Solve each equation.

1. $x = 4 - (-5)$ **9**
2. $n = (4)(-5)$ **−20**
3. $-8 \div -2 = m$ **4**

Simplify each expression.

4. $8t - 3t + 2t$ **$7t$**
5. $-9k - (-9k)$ **0**

Motivating the Lesson

Situational Problem Pose this problem, filling in the blanks with information about a student. I am taller than _?_. I am shorter than _?_. I (*do/do not*) wear glasses. My shoes are _?_ in color. I am wearing a _?_-colored shirt. I am sitting in the _?_ row. Ask students who the mystery student is. Ask how they solved the problem. Tell students they will learn more about eliminating possibilities in this lesson.

3-1 Problem-Solving Strategy: Eliminate Possibilities

Setting Goals: *In this lesson, you'll solve problems by eliminating possibilities.*

Modeling A Real-World Application: Education

Was there ever a question on a multiple-choice test for which you weren't quite sure of the answer? How did you go about choosing an answer? Eliminating some of the possible answers may help you choose the correct answer without having to guess.

Learning the Concept

This problem-solving strategy, **eliminating possibilities**, is helpful in solving many types of problems.

Example 1 APPLICATION Education

The question below appeared on a multiple choice test. Use the strategy of eliminating possibilities to choose the correct answer.

For what values of y will $2y - 4$ be equal to $2y + 6$?
a. all negative numbers
b. 0
c. all positive numbers
d. no value

Explore You know that y can represent any number. You want to find the values of y, if any, for which $2y - 4 = 2y + 6$.

Plan Look at the choices of values of y to see which ones are impossible and eliminate those.

Solve Choose a test value of y for each of the possible answers. Then test to see if $2y - 4 = 2y + 6$ for that value. If $2y - 4 \neq 2y + 6$ for that value, then we can eliminate that answer choice.

Choice	Test value for y	$2y - 4$	$2y + 6$	Is $2y - 4 = 2y + 6$?
a	−3	$2(-3) - 4 = -6 - 4$ or -10	$2(-3) + 6 = -6 + 6$ or 0	no
b	0	$2(0) - 4 = 0 - 4$ or -4	$2(0) + 6 = 0 + 6$ or 6	no
c	2	$2(2) - 4 = 4 - 4$ or 0	$2(2) + 6 = 4 + 6$ or 10	no

We can eliminate a, b, and c as possible answers. Therefore, d must be the correct choice.

Examine Study the equation $2y - 4 = 2y + 6$. Notice that $2y$ appears on both sides. Since $2y$ will have the same value for any given value of y, the equation will only be true when $-4 = 6$. But $-4 \neq 6$, so there is no value of y that makes the two expressions equal.

The process of elimination is also useful in solving some logic problems. Making a chart will help you organize the facts. The process in Example 2 is called **matrix logic**.

Example  **Shalana, Ken, Rolon, and Erica each take their pets to Dr. McDermott for veterinary care. They each have a different pet: a dog, a cat, a gerbil, or a parrot. Use the following information to match each person with his or her pet.**

- **Rolon lives next door to the person with the gerbil.**
- **Erica and Ken frequently care for the dog when its owner is out of town.**
- **Shalana cannot have a dog or cat because of allergies.**
- **Erica is teaching her pet how to talk.**

Make a chart to organize the information. Mark an X whenever you eliminate a possibility and draw a circle when you find a match.

	Shalana	Ken	Rolon	Erica
Dog				
Cat				
Gerbil				
Parrot				

Since Rolon cannot live next door to himself, the first clue tells you that he does not own the gerbil. The second clue tells us that neither Ken nor Erica owns the dog. And the third clue tells us that Shalana does not own the cat or the dog.

	Shalana	Ken	Rolon	Erica
Dog	X	X		X
Cat	X			
Gerbil			X	
Parrot				

So Rolon must own the dog. Draw a circle to indicate the match. Then mark X's for the other parts of Rolon's column.

(continued on the next page)

2 TEACH

In-Class Examples

For Example 1
For what values of x will $3x - 3$ be equal to $3(x - 1)$? **C**

A. all negative numbers

B. 0

C. all numbers

D. no numbers

For Example 2
Kalere, Hernando, Jessica, and Bryan played in a video-game tournament. Use the information below to determine who placed first and second.

- Kalere finished lower than Hernando.
- Jessica beat Bryan's score.
- Bryan finished last.
- A girl won.

First—Jessica; Second—Hernando

Teaching Tip Many crossword puzzle magazines contain logic problems similar to those found in this lesson.

Alternative Learning Styles

Kinesthetic Have students act out this logic problem. *Five students are standing in line. Kendra is not last. Melissa is directly ahead of Arnie. Juan has a girl directly in front of him and directly behind him. Samuel is not last. Juan is somewhere ahead of Arnie.* Name the students in order from first to last. **Samuel, Kendra, Juan, Melissa, Arnie**

3 PRACTICE/APPLY

Checking Your Understanding

Exercises 1–6 are designed to help you assess your students' understanding through reading, writing, speaking, and modeling. You should work through Exercises 1–3 with your students and then monitor their work on Exercises 4–6.

Additional Answers

1. Sample answer: It can help narrow down the acceptable solutions.
2. Sample answer: "Guess-and-check" method takes a solution and tests it. "Elimination" method narrows down the possible answers to a select few before the answer is tested.

Study Guide Masters, p. 19

NAME ______________ DATE ________

3-1 Study Guide
Problem-Solving Strategy: Eliminate Possibilities

Student Edition Pages 118–122

Dan, Ellie, and Ian each went on a vacation. One went to Mexico, one went to Florida, and one went to Canada. Ian did not go to Florida or Canada. Ellie did not go to Florida. Find where each went for vacation.

Explore We need to find out where each person went on vacation.

Plan Make a chart to organize the information. Mark an X to eliminate a possibility.

Solve Since Ian did not go to Florida or Canada, he must have gone to Mexico. By using the second clue and eliminating possibilities, we find that Ellie went to Canada and that Dan went to Florida.

	Mexico	Canada	Florida
Dan	X	X	O
Ellie	X	O	X
Ian	O	X	X

Examine Check the results with the clues. There is no conflict with any of the given clues.

Pick an answer by eliminating as many possibilities as possible.

1. 462 765 612 87 223 492
534 591 688 638 333 692
   a. It is an even number divisible by 6.
   b. It is larger than 400 and smaller than 600.
   c. The sum of the digits is 15. **492**

2. 12 248 76 3 240
504 7 26 48 25
86 592 50 9 72
   a. It is not divisible by 6.
   b. It is even.
   c. It is divisible by 4.
   d. The sum of its digits is 14. **248**

3. Theo, Sue, and Lenita each had breakfast. One had toast, one had pancakes, and one had eggs. Use the clues to find out who had what for breakfast.
   a. Lenita did not have pancakes.
   b. Theo did not have eggs or pancakes.
   Theo-toast; Lenita-eggs; Sue-pancakes

4. Ross is twice as tall as Enrico. Enrico is 6 inches taller than Tony. Which answer could be the heights of the three boys?
   a. Ross, 64 in.; Enrico, 29 in.; Tony, 32 in.
   b. Ross, 32 in.; Enrico, 64 in.; Tony, 67 in.
   c. Ross, 64 in.; Enrico, 32 in.; Tony, 26 in. **c**

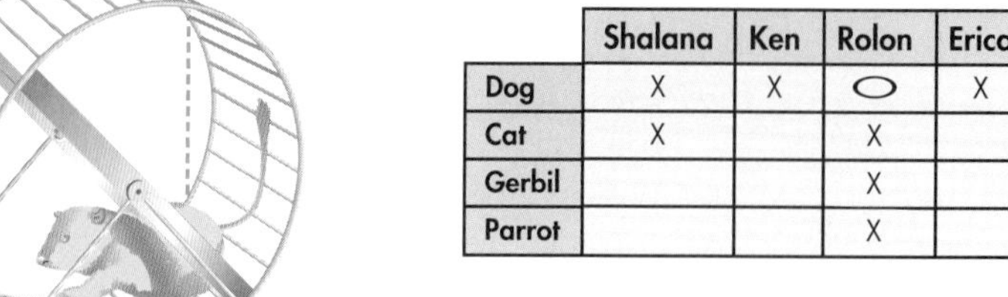

	Shalana	Ken	Rolon	Erica
Dog	X	X	O	X
Cat	X		X	
Gerbil			X	
Parrot			X	

Since the only pet that could possibly talk is a parrot, the fourth clue tells us that Erica must own the parrot. Draw a circle to indicate the match. Then mark Xs for each of the other choices of pets for Erica and owners for the parrot.

Complete the table to determine the answers.

	Shalana	Ken	Rolon	Erica
Dog	X	X	O	X
Cat	X	O	X	X
Gerbil	O	X	X	X
Parrot	X	X	X	O

Shalana owns the gerbil, Ken owns the cat, Rolon owns the dog, and Erica owns the parrot.

Checking Your Understanding

Communicating Mathematics

Read and study the lesson to answer these questions.

1. **Explain** why eliminating possibilities can be helpful in solving problems. See margin.
2. **Explain** how eliminating possibilities is different from guessing and checking to find the answer. See margin.

MATERIALS

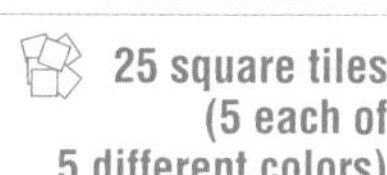

25 square tiles (5 each of 5 different colors)

3. Arrange the tiles in a 5-by-5 square so that no row, column, or diagonal has more than one of the same color. See students' work.
   a. Compare your arrangement with others in your class.
   b. Does the order of the first row make a difference to your final arrangement?
   c. Is there more than one way to arrange the tiles?

Guided Practice

Solve by eliminating possibilities.

4. Matt, Stacey, and Keisha each ate either cheese pizza, a hot dog, or a hamburger for dinner. Use the clues to find each person's dinner.
   - Keisha did not have any meat for dinner. Keisha – pizza;
   - Stacey did not have a hot dog. Stacey – hamburger; Matt – hot dog
5. Use the following clues to eliminate all but the correct number.

4235	8765	4223	6087	6444	5163
9635	7123	6668	5228	8224	9296

   - It is an odd number that is divisible by 3.
   - It is greater than 6000 and less than 7000.
   - The sum of the digits is 21. 6087

Reteaching

Using Discussion Choose four students from the class. Place their names across the top of a chart. List four activities (band, baseball, art, and so on) on the left of the chart, being certain each activity applies to only one student. Brainstorm with students to create a list of clues that would eliminate possibilities and help match each students with his or her activity.

6. If a number is increased by 5 and the result is multiplied by 8, the product is 168. What is the original number? **c**

a. 42 **b.** 26 **c.** 16 **d.** 128

Exercises: Practicing and Applying the Concept

Independent Practice

Solve using any strategy.

A ▶ 7. Study the list of numbers below. Using the clues, find the mystery number. **24**

- It is not divisible by 5.
- It is even.
- It is divisible by 3.
- It has 2 digits.
- The sum of its digits is 6.

6	124	76	25	7
13	252	321	60	9
24	120	15	27	8
43	561	18	78	5

8. How many packages of hot dogs balance with the ketchup? **3**

ketchup + mustard = 4 packages ketchup = mustard + 2 packages

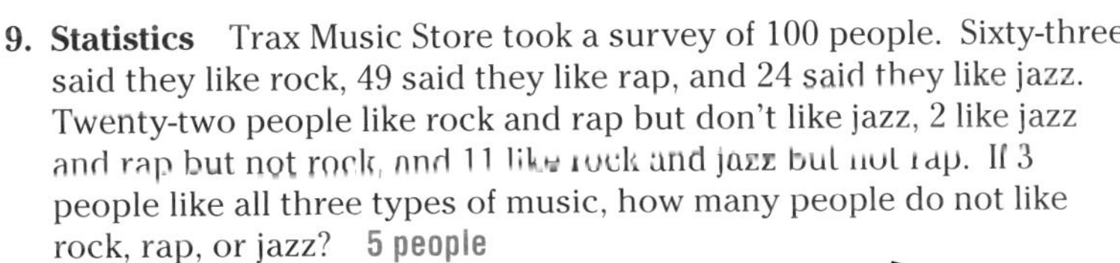

B ▶ 9. **Statistics** Trax Music Store took a survey of 100 people. Sixty-three said they like rock, 49 said they like rap, and 24 said they like jazz. Twenty-two people like rock and rap but don't like jazz, 2 like jazz and rap but not rock, and 11 like rock and jazz but not rap. If 3 people like all three types of music, how many people do not like rock, rap, or jazz? **5 people**

10. If 9 less than the product of a number and −4 is greater than 7, which of the following could be that number? **d**

a. 5 **b.** 4 **c.** −3 **d.** −5

11. What is the least positive number that you can divide by 7 and get a remainder of 4, divide by 8 and get a remainder of 5, and divide by 9 and get a remainder of 6? **501**

12. **Business** A concession stand at the football game sold 500 candy bars for a total of $255.00. A small candy bar sells for 40¢ and a large candy bar sells for 65¢. How many of the small candy bars were sold? **b**

a. 320 **b.** 280 **c.** 220 **d.** 77

C ▶ 13. **Consumerism** According to *Money* magazine, most Americans choose to buy shoes that cost $70 and last two years instead of shoes that cost $200 and last seven years. Is this good money management? Explain. **See margin.**

Assignment Guide

Core: 7–20
Enriched: 7–20

For **Extra Practice**, see p. 745.

The red A, B, and C flags, printed only in the Teacher's Wraparound Edition, indicate the level of difficulty of the exercises.

Additional Answer

13. You would buy 4 pairs of $70 shoes at a cost of $280 before the $200 pair wore out. It is less expensive to buy the $200 pair of shoes that lasts longer. But, if the shoes are uncomfortable or go out of style, it may make more sense to buy the less expensive shoes.

Group Activity Card 3-1

Master Eliminator Group Activity 3-1

Solve the following perplexing problem:

On an airplane, Mr. Smith, Mr. Robinson, and Ms. Jones are the pilot, copilot, and navigator — but *not* necessarily in that order. Three passengers who have names of Mr. Smith, Mr. Robinson, and Ms. Jones are also on the airplane.

Use the following clues to determine the name of the navigator:

CLUES:

1. Mr. Robinson lives in New York.
2. The copilot lives halfway between New York and Los Angeles.
3. Ms. Jones earns exactly $45,000 per year.
4. The copilot's nearest neighbor, one of the passengers, earns exactly three times as much as the copilot.
5. Mr. Smith beats the pilot when playing cards.
6. The passenger whose name is the same as the copilot lives in Los Angeles.

 Pre-Algebra

Practice Masters, p. 19

NAME ______________________ DATE ____________

3-1 Practice
Problem-Solving Strategy: Eliminate Possibilities

Student Edition Pages 118–122

Solve by eliminating possibilities.

1. The number is odd.
The number has two digits.
The sum of the digits is nine.
The product of the digits is twenty.
The ten's digit is one less than the unit's digit.
What is the number? **45**

2. Pencils cost $0.05.
Notebooks cost $0.30.
Henry spent $1.40.
How many of each did he buy if he bought the same number of pencils and notebooks? **B**
A. 3
B. 4
C. 6
D. 8

3. A number is between 300 and 400. If it is divided by 2, the remainder is 1. If it is divided by 4, 6, or 8, the remainder is 3. If it is divided by 10, the remainder is 5. If it is divided by 3, 5, 7, or 9, the remainder is zero. What is the number? **315**

4. Harry, Merrie, Sherrie, Larry, and Carrie live on the same street. Their houses are white, yellow, tan, green, and blue. One of them has a dog, one has a cat, one has two goldfish, one has a hamster, and one doesn't have a pet. Follow the clues to determine who lives in which house and what pet that person has.
a. The white house is farthest to the right on the street.
b. Larry lives between Merrie and Harry.
c. Harry lives in the middle house which is blue.
d. The house farthest on the left has a dog.
e. A hamster lives in the white house.
f. The yellow house is next to the white house and no pet lives there.
g. Larry has a cat.
h. Sherrie doesn't live next to Harry.
i. The green house is to the right of the tan house.
Merrie, tan, dog; Larry, green, cat; Harry, blue, two goldfish; Carrie, yellow, no pet; Sherrie, white, hamster

4 ASSESS

Closing Activity

Writing Have students write the steps they would take to create a table for use in solving a problem that involves eliminating the possibilities.

Enrichment Masters, p. 19

NAME ______________________ DATE ____________

3-1 Enrichment
Four Fours

Student Edition Pages 118–122

In mathematics, there are many ways to express numbers. For instance, the number five can be expressed as 5, V, 4 + 1, 10 ÷ 2, 2 + 3, 8 − 3, 20 ÷ 4, and so on.

Operations can be combined to produce a desired result. For example, the number one can be expressed using four fours as shown below.

$\frac{4+4}{4+4} = 1 \quad \frac{4 \times 4}{4 \times 4} = 1 \quad \frac{4}{4} \times \frac{4}{4} = 1 \quad \frac{44}{44} = 1$

Use four fours to express each result. Other answers are possible.

1. 2 $\frac{4}{4} + \frac{4}{4}$
2. 5 $\frac{4 \times 4 + 4}{4}$
3. 7 $4 + 4 - \frac{4}{4}$
4. 6 $\frac{4+4}{4} + 4$
5. 9 $4 + 4 + \frac{4}{4}$
6. 3 $\frac{4+4+4}{4}$
7. 8 $4 + 4 + 4 - 4$
8. 17 $4 \times 4 + \frac{4}{4}$
9. 16 $4 + 4 + 4 + 4$
10. 0 $4 - 4 + 4 - 4$
11. 4 $\frac{4-4}{4} + 4$
12. 15 $4 \times 4 - \frac{4}{4}$
13. 10 $\frac{44-4}{4}$
14. 12 $\frac{44+4}{4}$

14. **Horse Riding** Five members of the Riding Club at Kingston High School went for a two-day riding trip to Manchester Lodge. They rode Sommie, Samantha, Friar, Claude, and Blaze on the first day. The stable staff wanted the club members to ride the same horses on the second day, but they could not remember who was on which horse. Use the clues below to match the rider with the horse.
    - Blaze is the fastest horse and always the first horse on the trail.
    - Friar is the most patient and gentle horse.
    - Sommie is old so she always has the lightest rider, and on the trail she liked to be near the rear.
    - Samantha and Claude are spirited and always start off together with the most experienced riders.
    - Thomas was nervous because this was his first riding trip.
    - Amber and Cynthia were the first two riders back to the stable.
    - Amber taught Cynthia how to ride, but her horse had trouble keeping up with Cynthia's horse.
    - Vernita was afraid that she was not tall enough to handle the horses she saw in the stable, so she rode behind the experienced riders.
    - Carlos rode a male horse. He started out with Amber, but stopped on the trail to check his horse's hooves. He was one of the most experienced riders, so the rest of the group knew he would catch up. **Thomas – Friar, Cynthia – Blaze, Vernita – Sommie, Carlos – Claude, Amber – Samantha**

Critical Thinking

15. Three students are blindfolded and stand in a single-file line. Kieko takes three hats from a box containing three blue hats and two green hats and places one on each of the students. She tells the students how many hats of each color were in the box and removes the blindfolds. The student in the back looks at the hats on the two students in front of him and says "I don't know what color hat I am wearing." The student in the middle hears him, looks at the hat on the student in front of her and says the same thing. The student in front says "I *do* know what color hat I am wearing." What color hat is he wearing? Explain how he knows. **Blue; see Solutions Manual for explanation.**

Mixed Review

16. Solve $x = \frac{-120}{-15}$. (Lesson 2-8) **8**
17. **Patterns** Find the next number in the pattern 0, 1, 3, 6, 10, 15, . . . (Lesson 2-6) **21**
18. **Sports** In Saturday's game, the Warriors lost 3 yards on a first down play. On second down, they gained 7 yards. What was the gain or loss for the two plays? (Lesson 2-4) **gain of 4 yards**

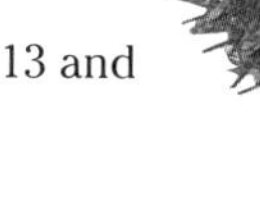

19. Simplify $21k + 3(k + 1)$. (Lesson 1-5) **$24k + 3$**
20. Evaluate $35 - c - b$ if $b = 13$ and $c = 7$. (Lesson 1-3) **15**

Extension

Using Manipulatives Have students work in small groups and use classroom materials (erasers, pencils, books, and so on) and a balance scale to make up problems similar to Exercise 8. Then have students exchange problems and solve them.

HANDS-ON ACTIVITY

3-2A Solving Equations with Cups and Counters

A Preview of Lesson **3-2**

MATERIALS

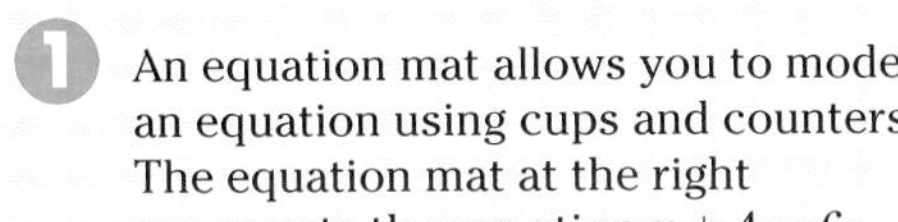

cups and counters

equation mat

You can use cups and counters as a model for solving equations.

Activity 1

An equation mat allows you to model an equation using cups and counters. The equation mat at the right represents the equation $x + 4 = 6$.

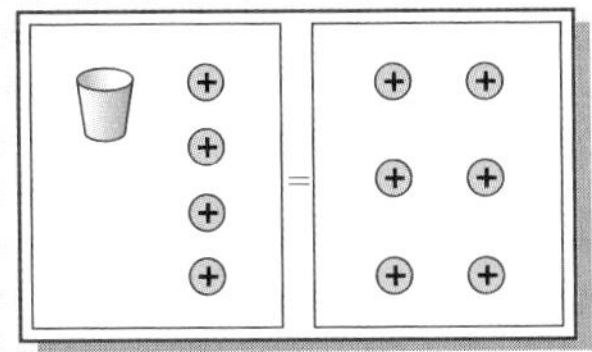

$x + 4 = 6$

To solve $x + 4 = 6$, find the value of x that makes the equation true. Think of this as finding the number of counters in the cup. To do that, get the cup by itself on one side of the mat.

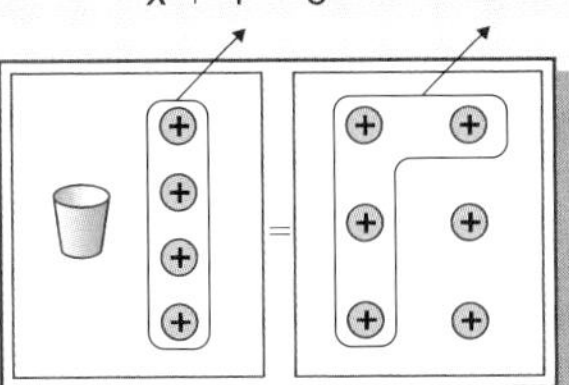

$x + 4 - 4 = 6 - 4$

If you remove four counters from each side of the mat, then you can see that the cup is equal to two counters. So, $x = 2$. Since $2 + 4 = 6$, our solution is correct.

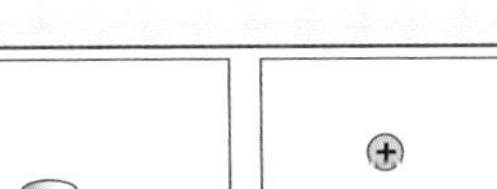

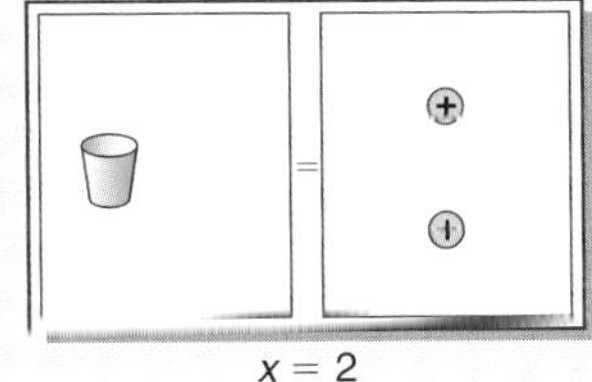

$x = 2$

Activity 2

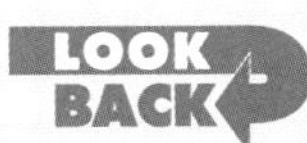

You can review adding integers with counters in Lesson 2-4A.

Some equations involve subtraction. The mat at the right represents $7 = y - 5$. To get the cup by itself, remove 5 negative counters from each side. Since there are no negative counters on the left side, add 5 positive counters to each side to make 5 zero pairs on the right side.

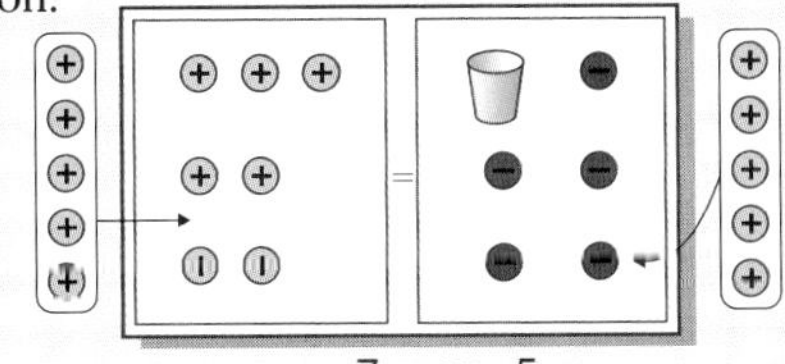

$7 = y - 5$

$7 + 5 = y - 5 + 5$

Now remove the zero pairs. The remaining cup and counters show that $y = 12$. Check by substituting 12 for y in the original equation.

$12 = y$

Your Turn **Use cups and counters to model and solve each equation.**

1. $3 + x = 7$ **4**
2. $n + 6 = 9$ **3**
3. $y + 15 = 18$ **3**
4. $x + 6 = 10$ **4**
5. $t + 2 = 9$ **7**
6. $4 + w = 5$ **1**
7. $x - 2 = 10$ **12**
8. $n - 4 = 9$ **13**
9. $6 = q - 5$ **11**

3-2A LESSON NOTES

NCTM Standards: 1-4, 9

Objective

Use cups and counters to solve equations.

Recommended Time

Demonstration and discussion: 15 minutes; Exercises: 30 minutes

Instructional Resources

For each student or group of students

Student Manipulative Kit
- cups and counters

Math Lab and Modeling Math Masters
- p. 3 (equation mat)
- p. 5 (integer counters)
- p. 6 (cup pattern)
- p. 33 (worksheet)

For teacher demonstration

Overhead Manipulative Resources

1 FOCUS

Motivating the Lesson

Have students discuss solutions to this problem. *Andre and Maria stopped at the playground on their way home from the grocery store. Maria held the sack of groceries and sat on one end of the teeter-totter. Andre sat on the other end and the teeter-totter balanced. Maria put the groceries down, but now the teeter-totter did not balance. Why?* **A weight was removed from only one side of the teeter-totter.**

2 TEACH

Teaching Tip Relate the removal or addition of counters to each side of the equation to keeping a balance scale in balance.

3 PRACTICE/APPLY

Assignment Guide
Core: 1–9
Enriched: 1–9

4 ASSESS

Observing students working in cooperative groups is an excellent method of assessment.

GLENCOE Technology

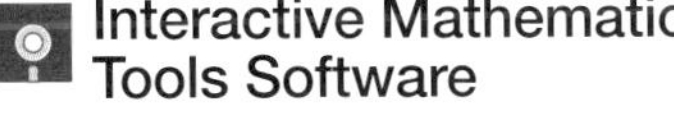

Interactive Mathematics Tools Software

In this interactive computer lesson, students add and subtract to solve equations. A **Computer Journal** gives students the opportunity to write about what they have learned.

For Windows & Macintosh

3-2 LESSON NOTES

NCTM Standards: 1-6, 9

Instructional Resources

- Study Guide Master 3-2
- Practice Master 3-2
- Enrichment Master 3-2
- Group Activity Card 3-2
- Assessment and Evaluation Masters, p. 71
- Tech Prep Applications Masters, p. 5

Transparency 3-2A contains the 5-Minute Check for this lesson; **Transparency 3-2B** contains a teaching aid for this lesson.

Recommended Pacing	
Standard Pacing	Day 3 of 12
Honors Pacing	Day 2 of 11
Block Scheduling*	Day 2 of 7

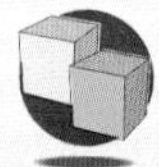

*For more information on pacing and possible lesson plans, refer to the ***Block Scheduling Booklet***.

1 FOCUS

5-Minute Check
(over Lesson 3-1)

Mr. Brown, Ms. Red, and Ms. Black each have a dog. The dogs' colors are brown, red, and black. A woman owns the black dog. No owner's name is the same as the color of his or her dog. What is the color of each person's dog? **Mr. Brown—red; Ms. Red—black; Ms. Black—brown**

Motivating the Lesson

Hands-On Activity Place an equal amount of pennies on each pan of a two-pan balance. Remove some pennies from one pan. Ask students how to get the scale to balance. **Remove the same number of pennies from the other pan.** Repeat by adding some pennies to one pan. Ask students how to get the scale to balance. **Add the same number of pennies to the other pan.**

3-2 Solving Equations by Adding or Subtracting

Setting Goals: *In this lesson, you'll solve equations by using the addition and subtraction properties of equality.*

Modeling with Technology

See page 1 for information on programming the T1-82.

Graphing calculators are computers, so you can run programs. The T1-82 program at the right allows you to enter an equation of the form $x + a = b$ or $x - a - b$ and guess the solution. Using this program and eliminating possibilities, you can solve equations.

```
PROGRAM: ONESTEP
: Disp "PRESS 1: X + A = B"
: Input "PRESS 2: X – A =
  B", N
: Input "ENTER A", A
: Input "ENTER B", B
: Lbl 1
: Input "GUESS X", X
: If N = 1 and X + A = B or
  N = 2 and X – A = B
: Then
: Disp "CORRECT"
: Stop
: Else
: Disp "TRY AGAIN"
: Goto 1
```

Your Turn Use the program to choose the solution of each equation from the possible values. Record the solutions.

$x + 5 = 9$	$x = 1, 3, 14, \underline{4}$
$x - 9 = 6$	$x = 3, 6, \underline{15}, 18$
$x + 7 = 8$	$x = \underline{1}, 7, 8, 15$
$x - 2 = 14$	$x = 2, 12, 14, \underline{16}$
$x + 10 = 15$	$x = 2, \underline{5}, 10, 20$
$x - 8 = 8$	$x = 0, \underline{16}, 1, 64$

a. They are $b - a$.
b. They are $b + a$.
c. 9; 20

a. What do you observe about the solutions of equations like $x + a = b$?
b. What do you observe about the solutions of equations like $x - a = b$?
c. Solve $x + 3 = 12$ and $x - 5 = 15$. Use the program to verify the solutions.

Learning the Concept

Consider $x + 10 = 14$. One way to solve is to mentally find the number that makes the mathematical sentence true. Since $4 + 10 = 14$, the solution is 4.

Not all equations are easy to solve mentally. You can subtract to isolate the variable. If you subtract 10 from the left side of the equation, you must also subtract 10 from the right side to keep the equation balanced.

$$x + 10 = 14$$
$$x + 10 - 10 = 14 - 10 \quad \text{Subtract 10 from each side.}$$
$$x + 0 = 4 \quad 10 - 10 = 0 \text{ and } 14 - 10 = 4.$$
$$x = 4$$

These are **equivalent equations** because they all have the same solution, 4. We used the **subtraction property of equality** to solve this equation.

Subtraction Property of Equality

In words: If you subtract the same number from each side of an equation, the two sides remain equal.

In symbols: For any numbers a, b, and c, if $a = b$, then $a - c = b - c$.

Alternative Teaching Strategies

Student Diversity Show students how to solve the equation at the bottom of the page using a vertical format.

$$\begin{array}{rcr} x + 10 &=& 14 \\ -10 &=& -10 \\ \hline x + 0 &=& 4 \text{ or } x = 4 \end{array}$$

Example 1 Solve $r + 56 = -5$.

$$r + 56 = -5$$
$$r + 56 - 56 = -5 - 56 \quad \textit{Subtract 56 from each side.}$$
$$r = -61 \quad \textit{Check your solution by replacing r with −61.}$$

Some equations can be solved by using the **addition property of equality**.

Addition Property of Equality		
	In words:	If you add the same number to each side of an equation, the two sides remain equal.
	In symbols:	For any numbers a, b, and c, if $a = b$, then $a + c = b + c$.

Example  Solve $y - 9 = -17$.

$$y - 9 = -17$$
$$y - 9 + 9 = -17 + 9 \quad \textit{Add 9 to each side.}$$
$$y = -8 \quad \textit{Check your solution.}$$

You can often use an equation to represent a situation in real life.

Example 3

Entertainment

***Forrest Gump* is the only one of the top ten money-making films to win an Oscar for Best Picture. It earned \$317 million at the box office. This is \$83 million less than *E.T. the Extra-Terrestrial*, which was the all-time top money maker. How much did *E.T.* earn?**

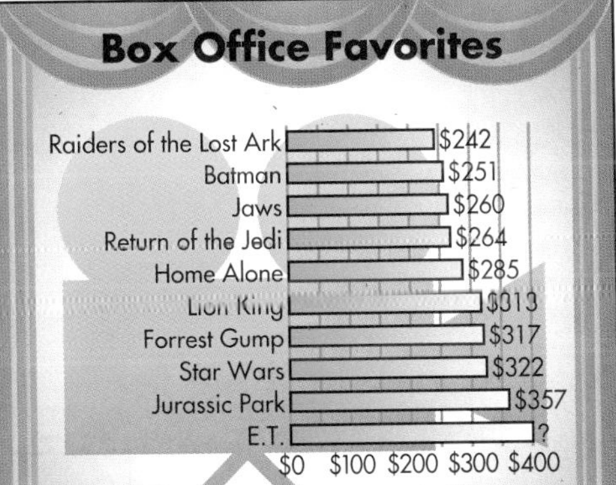

Source: The Associated Press

Write an equation.

Forrest Gump revenue	*equals*	*E.T. revenue*	*minus*	*\$83 million*
317	=	r	−	83

$$317 + 83 = r - 83 + 83 \quad \textit{Add 83 to each side.}$$
$$400 = r \quad \textit{Check your solution.}$$

E.T. the Extra-Terrestrial earned \$400 million at the box office.

LOOK BACK

You can review additive inverses in Lesson 2-5.

Recall that to subtract an integer, you can add its inverse.

subtraction		addition
	additive inverse	
$8 - (-2) = 10$		$8 + 2 = 10$
	same result	

2 TEACH

In-Class Examples

For Example 1
Solve $x + 82 = 23$. **−59**

For Example 2
Solve $m - 18 = 3$. **21**

For Example 3
The Lion King earned \$313 million at the box office. That is \$44 million less than *Jurassic Park* earned. Write and solve an equation to find out how much *Jurassic Park* earned. **$J - 44 = 313$; 357; *Jurassic Park* earned \$357 million at the box office.**

Teaching Tip You may relate solving these types of equations to this situation: There is to be a tug-of-war contest at an outdoor fun day. Each team has the same number of people. Just before the tug-of-war starts, two brothers from one team must leave. What can be done to make the contest fair? How does this relate to the lesson?

GLENCOE Technology

Interactive Mathematics Tools Software

In this interactive computer lesson, students explore equations and the number line. A **Computer Journal** gives students an opportunity to write about what they have learned.

For Windows & Macintosh

In-Class Examples

For Example 4

Solve $x + (-2) = -3$. **−1**

For Example 5

Solve $x - (-5) = 3$. Graph the solution on a number line. **−2**

3 PRACTICE/APPLY

Checking Your Understanding

Exercises 1–13 are designed to help you assess your students' understanding through reading, writing, speaking, and modeling. You should work through Exercises 1–6 with your students and then monitor their work on Exercises 7–13.

Additional Answers

1. Sample answer: The addition property of equality can be used to solve an equation like $x + a = b$ or $x - a = b$. Add the inverse of a to each side of the equation to solve $x + a = b$ or add a to each side to solve $x - a = b$.
2. the addition property of equality or the subtraction property of equality

Study Guide Masters, p. 20

NAME ______________________ DATE ____________

3-2 Study Guide
Solving Equations by Adding or Subtracting

Student Edition
Pages 124–128

Method: 1. Identify the variable.
2. To get the variable by itself, add the same number to or subtract the same number from each side of the equation.
3. Check the solution.

Example: Solve $x + (-2) = 6$.

$x + (-2) = 6$
$x + (-2) - (-2) = 6 - (-2)$ Subtract −2 from each side. The solution is 8.
$x = 8$

Check: $x + (-2) = 6$ In the original equation, replace x with 8.
$8 + (-2) \stackrel{?}{=} 6$
$6 = 6$ ✓

Example: Solve $x - 9 = -13$.

$x - 9 = -13$
$x - 9 + 9 = -13 + 9$ Add 9 to each side. The solution is −4.
$x = -4$

Check: $x - 9 = -13$ In the original equation, replace x with −4.
$-4 - 9 \stackrel{?}{=} -13$
$-13 = -13$ ✓

Solve each equation and check your solution. Then graph the solution on the number line.

1. $x + 5 = 2$ **$x = -3$**
2. $11 + w = 10$ **$w = -1$**
3. $a - 7 = -5$ **$a = 2$**
4. $b + (-13) = -13$ **$b = 0$**
5. $-3 + h = -7$ **$h = -4$**
6. $y - (-9) = 12$ **$y = 3$**

You can also use the additive inverse to solve equations involving addition.

Example 4 **Solve $c + 8 = 12$.**

Method 1: Subtracting

$$c + 8 = 12$$
$$c + 8 - 8 = 12 - 8$$
$$c = 4$$

Method 2: Adding Inverse

$$c + 8 = 12$$
$$c + 8 + (-8) = 12 + (-8)$$
$$c + 0 = 4$$
$$c = 4$$

Check: $c + 8 = 12$
$4 + 8 \stackrel{?}{=} 12$ *Replace c with 4.*
$12 = 12$ ✓ The solution is 4.

Solutions to equations can be represented on a *number line*. Each solution is a **coordinate** of a point. The coordinate tells its distance and direction from the 0-point on the line. The dot marking the point is called the **graph** of the number.

The solution in the example above, 4, is represented by placing a dot above 4 on a number line.

Example 5 **Solve $y - 24 = 16$. Graph the solution on a number line.**

$y - 24 = 16$
$y - 24 + 24 = 16 + 24$ *Add 24 to each side.*
$y = 40$ The solution is 40.

Graph the solution, 40, on a number line.

Checking Your Understanding

Communicating Mathematics

Read and study the lesson to answer these questions. **1–5. See margin.**

1. **Explain** when you use the addition property of equality to solve an equation.
2. **Tell** what property you would use to solve $x + 34 = -6$.
3. **Write** two equations that are equivalent. Then write two equations that are not equivalent.
4. **Write** an equation in the form $x - a = b$ where the solution is -12.
5. **You Decide** Ellen and Tertius solved $x - (-25) = 56$ in two different ways. Who is correct? Explain.

Ellen

$x - (-25) = 56$
$x + 25 = 56$
$x + 25 - 25 = 56 - 25$
$x = 31$

Tertius

$x - (-25) = 56$
$x - (-25) + (-25) = 56 + (-25)$
$x = 56 - 25$
$x = 31$

MATERIALS

cups and counters

6. Use cups and counters to model the solution to $9 = w + (-5)$. **See margin.**

Reteaching

Using Modeling Separate students into small groups to play "I Am Thinking of a Number". Have one student state an addition equality. For example, "When I add −5 to my number, I get 13. What is my number?" Other students can model the equation and use the model to solve the equation.

Additional Answers

3. Sample answers: $2 + x = 6$ and $x = 4$; $x - 7 = 14$ and $x = 14$
4. Sample answer: $x - 5 = -17$
5. Both Ellen and Tertius are correct. Ellen rewrote the equation and used the subtraction property of equality. Tertius used the addition property of equality.

Guided Practice **Solve each equation and check your solution. Then graph the solution on a number line.** 7–12. See margin for graphs.

7. $w + 8 = -17$ **−25**
8. $25 = -4 + y$ **29**
9. $m - (-3) = 40$ **37**
10. $y + (-7) = 9$ **16**
11. $k - 36 = -37$ **−1**
12. $h - 8 = -22$ **−14**
13. **Entertainment** *Star Wars* and its sequel, *Return of the Jedi*, were both blockbusters. Refer to the graph in Example 3 for the revenue earned by each of these movies. How much more did *Star Wars* earn? **$322 = 264 + e$, $58 million**

Exercises: Practicing and Applying the Concept

Independent Practice **Solve each equation and check your solution. Then graph the solution on a number line.** See Solutions Manual for graphs.

A

14. $y + 7 = 21$ **14**
15. $k - 6 = 32$ **38**
16. $19 = g - 5$ **24**
17. $-24 = h - 22$ **−2**
18. $x + 49 = 13$ **−36**
19. $f + 34 = 2$ **−32**

B

20. $d - (-3) = 2$ **−1**
21. $a - (-26) = 5$ **−21**
22. $12 + p = -14$ **−26**
23. $x + 18 = 14$ **−4**
24. $59 = s + 95$ **−36**
25. $x - 27 = 63$ **90**
26. $-7 = z - (-12)$ **−19**
27. $34 + r = 84$ **50**
28. $-234 = x + 183$ **−417**
29. $y - 94 = 562$ **656**
30. $-591 = m - (-112)$ **−703**
31. $846 + t = -538$ **−1384**

Choose the equation whose solution is graphed.

32. 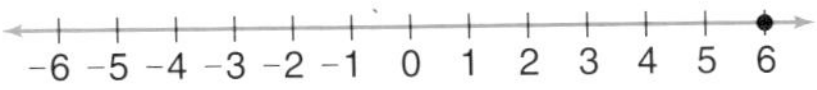

c
a. $x + 5 = -11$ b. $x - 5 = 11$ c. $x + 5 = 11$ d. $x - 5 = -11$

33. 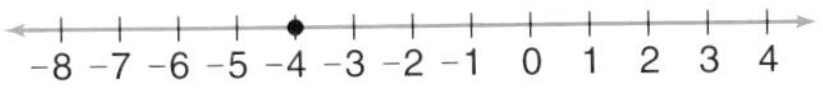

a
a. $-8 = t - 4$ b. $-8 = t + 4$ c. $8 = t - 4$ d. $8 = t + 4$

Solve each equation. Check each solution.

C

34. $(f + 5) + (-2) = 6$ **3**
35. $[y + (-3)] + 2 = 4$ **5**
36. $16 = [n - (-2)] + (-3)$ **17**
37. $-10 = [b + (-4)] + 2$ **−8**

Critical Thinking

38. Write an equation whose solution is represented by the number line. **Sample answer: $x + 8 = 3$**
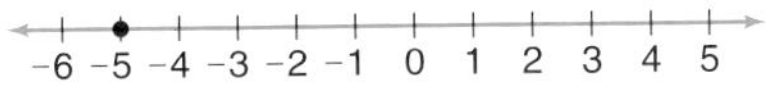

Applications and Problem Solving

39. 12

39. **Advertising** Morris has been the official mascot for 9-Lives® Cat Food for 26 years. Actually, there have been two cats named Morris. Morris I died unexpectedly at the age of 19. If he was the official mascot for 14 years, how long has Morris II been the official mascot?

40. **Meteorology** Weather fronts usually move across the United States from west to east and can bring with them astounding changes in temperatures.
    a. On January 23–24, 1916, the temperature in Browning, Montana, dropped 100° in 24 hours. If the temperature started at 44°F, how cold did it get? **−56°F**
    b. In 1892, at Fort Assinaborn, Montana, the temperature rose 43° in 15 minutes. At 2:00 A.M., the temperature was −5°F. If that is when the temperature began to rise, what was the temperature at 2:15 A.M.? **38°F**

Error Analysis

Some students will always add if the equation contains an addition symbol and subtract if the equation contains a subtraction symbol. Review the rules for solving equations and remind students to check their solutions for all equations.

Assignment Guide

Core: 15–37 odd, 38, 39, 41–48
Enriched: 14–38 even, 39–48

For **Extra Practice**, see p. 745.

The red A, B, and C flags, printed only in the Teacher's Wraparound Edition, indicate the level of difficulty of the exercises.

Additional Answers

7.
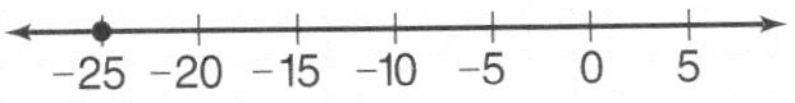

8.
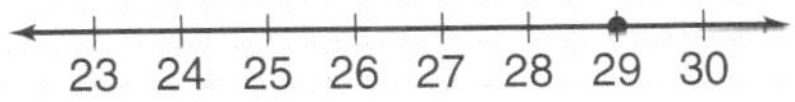

9.
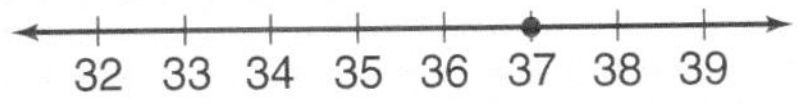

10.

11.

12.
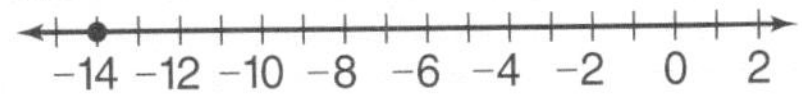

Practice Masters, p. 20

NAME ______________________ DATE ____________

3-2 Practice
Solving Equations by Adding or Subtracting

Student Edition
Pages 124–128

Solve each equation and check your solution. Then graph the solution on the number line.

1. $m + 7 = 12$ **$m = 5$**
2. $x - 12 = -10$ **$x = 2$**
3. $y + 19 = 15$ **$y = -4$**
4. $14 = y - (-13)$ **$y = 1$**
5. $11 = t + 16$ **$t = -5$**
6. $n - 13 = -11$ **$n = 2$**
7. $13 = z + 18$ **$z = -5$**
8. $z + (-6) = -7$ **$z = -1$**
9. $m - (-14) = 17$ **$m = 3$**
10. $-31 = c - 33$ **$c = -2$**
11. $35 = w + 35$ **$w = 0$**
12. $0 = j - 4$ **$j = 4$**
13. $-15 = -18 + f$ **$f = -3$**
14. $-7 + r = -11$ **$r = -4$**
15. $z + (-7) = -8$ **$z = -1$**
16. $n + 25 = 26$ **$n = 1$**

Additional Answer

6.
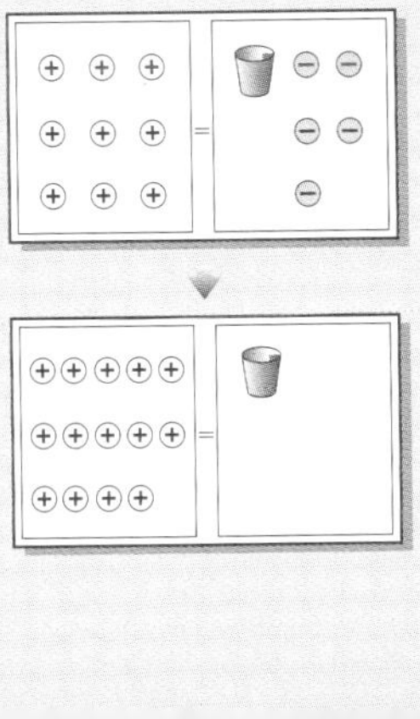

Group Activity Card 3-2

Equation Creation Group Activity **3-2**

MATERIALS: 0-9 spinner • coin • paper • pencils

Two to four players create equations for others in the group to solve. To begin, someone in the group spins the spinner four times and flips the coin two times. The four digits and the two operations (+ or −) are recorded on a piece of paper for each player to see. Everyone in the group now writes an equation of the form $x + a = b$ or $b = a + x$ using all four digits and the two operations. The digits may be used to form 1-, 2-, or 3-digit numbers in the equation.

Everyone then exchanges their equation with someone in the group and solves the given equation. The player who wrote the equation checks to be sure the equation is solved correctly.

Each player who solved an equation correctly earns a point. The winners are the players who score the most points after five rounds.

Pre-Algebra

4 ASSESS

Closing Activity

Speaking Have the student give an explanation for solving equations involving subtracting negative numbers, such as $r - (-3) = 3$.

Chapter 3, Quiz A (Lessons 3-1 and 3-2) is available in the *Assessment and Evaluation Masters*, p. 71.

Additional Answer

46.

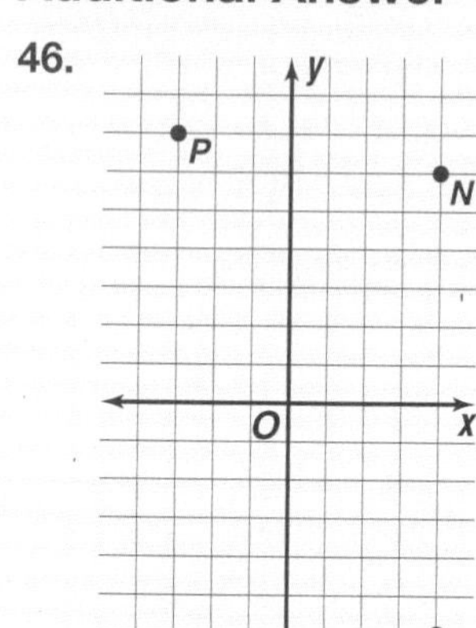

Enrichment Masters, p. 20

NAME ______________________ DATE ____________

3-2 Enrichment — Solving Equations Using Addition and Subtraction — Student Edition Pages 124–128

Write an equation and then solve.

1. The sum of a number and 8 is 15. Find the number. **$x + 8 = 15$; 7**
2. The difference of a number and 12 is 15. Find the number. **$x - 12 = 15$; 27**
3. If 17 less than a number is 25, find the number. **$x - 17 = 25$; 42**
4. 29 is 12 more than a number. What is the number? **$29 = 12 + x$; 17**
5. The sum of a number and $6\frac{1}{2}$ is $12\frac{3}{4}$. Find the number. **$x + 6\frac{1}{2} = 12\frac{3}{4}$; $6\frac{1}{4}$**
6. A number decreased by 6.2 is equal to 10.9. Find the number. **$x - 6.2 = 10.9$; 17.1**
7. The difference of some number and 36 is $12\frac{1}{2}$. What is the number? **$x - 36 = 12\frac{1}{2}$; $48\frac{1}{2}$**
8. If $2\frac{2}{3}$ more than a number is $6\frac{1}{2}$ find the number. **$2\frac{2}{3} + x = 6\frac{1}{2}$; $3\frac{5}{6}$**
9. 14.32 is 15 less than x. Find x. **$14.32 = x - 15$; 29.32**
10. $3\frac{1}{2}$ increased by m is $12\frac{3}{4}$. Find m. **$3\frac{1}{2} + m = 12\frac{3}{4}$; $9\frac{1}{4}$**
11. The sum of y and 25.8 is 36.5. Find y. **$y + 25.8 = 36.5$; 10.7**
12. The difference of 35.9 and p is 12.7. Find p. **$35.9 - p = 12.7$; 23.2**

Mixed Review

46. N – I, P – II, Q – IV; see margin for graph.
47a. v = Calories burned in hour of volleyball
47c. $v = 264$

41. **Illustration** In 1995, Charles Schulz, the creator of Peanuts comic strip, was 70 years old. He created Peanuts in 1950. How old was he when he started drawing Peanuts? **25 years old**

42. **Travel** On a trip to Florida, a bus traveled 600 miles in 10 hours. At that rate, how far could the bus travel in 4 hours? (Lesson 3-1) **c**
    a. 2400 miles **b.** 400 miles **c.** 240 miles **d.** 150 miles

43. Solve $y = (-9)(18)$. (Lesson 2-7) **-162**

44. Solve $7 - 25 = h$. (Lesson 2-5) **-18**

45. Order the integers in the set {3, −1, 15, −3} from least to greatest. (Lesson 2-3) **−3, −1, 3, 15**

46. **Geometry** On graph paper, draw coordinate axes. Then graph and label $N(4, 6)$, $P(-3, 7)$, and $Q(4, -6)$. Name the quadrant in which each point is located. (Lesson 2-2)

47. **Exercise** A 150-pound person burns 384 Calories an hour jogging at a rate of 6 miles per hour. That is 120 Calories more than the same person burns playing volleyball for one hour. How many Calories does a 150-pound person burn playing volleyball for one hour? (Lesson 1-8)
    a. Define a variable for this problem.
    b. Write an equation. **$v + 120 = 384$**
    c. Solve for the number of Calories burned in an hour of volleyball.

48. Find the value of $7 + 8 \div 2 \cdot 6 - 1$. (Lesson 1-2) **30**

WORKING ON THE

Investigation

BON VOYAGE

Refer to the Investigation on pages 62–63.

Look over the list of considerations that you placed in your Investigation Folder at the beginning of Chapter 2.

You probably listed weather as one thing to consider before packing clothes. You don't want to pack shorts and sandals if temperatures are going to be in the 20s—or do you?

Unless you selected Great Britain, Australia, New Zealand, or Ireland, as your destination, your hosts will refer to temperatures measured in *degrees Celsius.*

Like the Fahrenheit scale, the Celsius scale has two fixed points: the boiling point of water and the freezing point of water (or the melting point of ice).

On the Celsius scale, the boiling point of water was set at 100 and the freezing point of water was set at 0 so that the scale would have 100 equal divisions (called degrees Celsius) between them.

If you've grown up using the Fahrenheit scale, you can probably translate temperatures into "feel". You know, for example, that 85°F feels very warm.

- Work with your partner to choose at least four reference Fahrenheit temperatures that you "feel."
- Find a thermometer that has both scales on it. Use it to estimate the equivalent Celsius temperature for each Fahrenheit temperature you selected.
- If you haven't already done so, obtain environmental data for the country you and your partner selected at the beginning of the investigation. Then provide general guidelines for your wardrobe.

Add your results to your Investigation Folder.

Extension

Consumer Awareness Provide students with this information. Kenneth wrote these checks this month: #1233 for $150 and #1234 for $49.50. His check register shows a balance of $515.34. The bank statement shows a balance of $665.34. Explain how the bank and Kenneth can both be correct. **The check for $150 has not cleared the bank yet.**

The Investigation on pages 62 and 63 is designed to be a long-term project that is completed over several days or weeks. Encourage students to keep their materials in their Investigation Folder as they work on the Investigation.

3-3 Solving Equations by Multiplying or Dividing

Setting Goals: *In this lesson, you'll solve equations by using the multiplication and division properties of equality.*

Modeling a Real-World Application: Toys

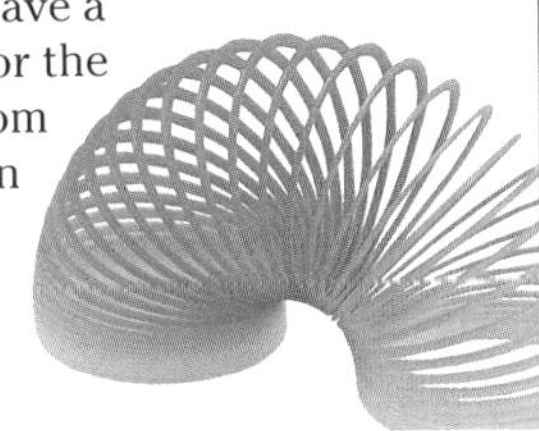

Popular toys are often created by individuals who have a simple idea. Richard James came up with the idea for the Slinky® toy in 1943 when he watched a spring fall from his desk to the floor. He introduced the first Slinky in 1945 at the annual Toy fair. The Slinky is a wiry coil, that with a nudge, appears to walk down the stairs. After more than fifty years, the Slinky is still a "walking" success.

Millions of Slinkys have been sold over the past 50 years. They are still made using the same machines designed years ago. One Slinky is produced every 8 seconds, 24 hours a day. How many Slinkys are made each day?

Learning the Concept

To find out how many Slinkys are made in a day, translate what we know into an equation.

First find the number of seconds in a day. There are 60 seconds in a minute, 60 minutes in an hour, and 24 hours in a day.

$$\left(\frac{60 \text{ s}}{\text{min}}\right)\left(\frac{60 \text{ min}}{\text{hr}}\right)\left(\frac{24 \text{ hr}}{\text{day}}\right) = 86{,}400 \frac{\text{s}}{\text{day}}$$

Let d represent the number of Slinkys made in a day. If each Slinky takes 8 seconds to make, we can write an equation to represent the daily production of Slinkys.

seconds per Slinky	*times*	*Slinkys made in a day*	*equals*	*seconds in a day*
8	×	d	=	86,400

To solve this equation, undo the multiplication by dividing each side by the same number.

$8d = 86{,}400$

$\frac{8d}{8} = \frac{86{,}400}{8}$ *Divide each side by 8 to undo the multiplication $8 \cdot d$.*

$d = 10{,}800$

Check: $8d = 86{,}400$

$8(10{,}800) \stackrel{?}{=} 86{,}400$ *Replace d with 10,800.*

$86{,}400 = 86{,}400$ ✓ The solution is 10,800.

Therefore, 10,800 Slinkys are made in a day.

3-3 LESSON NOTES

NCTM Standards: 1-5, 9

Instructional Resources

- Study Guide Master 3-3
- Practice Master 3-3
- Enrichment Master 3-3
- Group Activity Card 3-3
- Multicultural Activity Masters, p. 5

Transparency 3-3A contains the 5-Minute Check for this lesson; **Transparency 3-3B** contains a teaching aid for this lesson.

Recommended Pacing	
Standard Pacing	Day 4 of 12
Honors Pacing	Day 3 of 11
Block Scheduling*	Day 3 of 7 (along with Lesson 3-4)

*For more information on pacing and possible lesson plans, refer to the ***Block Scheduling Booklet***.

1 FOCUS

5-Minute Check *(over Lesson 3-2)*

Solve each equation.

1. $x - 3 = 5$ **8**
2. $y + 9 = 7$ **−2**
3. $13 = m - 6$ **19**
4. $n - (-1) = 1$ **0**
5. $f + 4 = -5$ **−9**

Motivating the Lesson

Hands-On Activity Measure the hallway outside your classroom and have students find how many "hallways" are in a mile. Then time a student while he or she walks the hallway and have students find how long it would take him or her to walk a mile.

Alternative Learning Styles

Visual Provide a drawing to help students visualize how many Slinkys are made in a day.

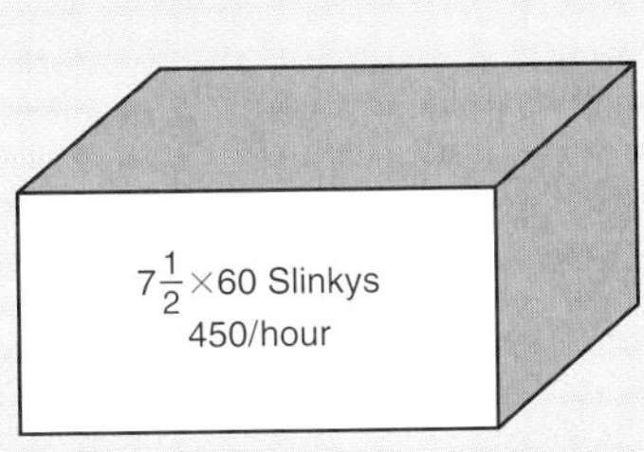

2 TEACH

In-Class Examples

For Example 1
Solve 28 = −4*t*. Check the solution and graph it on a number line. **−7**

−9 −8 −7 −6 −5 −4

For Example 2
Solve $\frac{w}{-12} = 24$. Check the solution and graph it on a number line. **288**

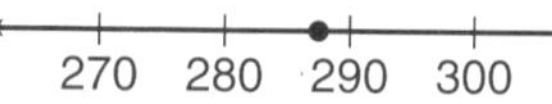

Teaching Tip Watch for students who say the answer to Example 2 above is −2 simply because 12 is a factor of 24.

Division Property of Equality

In words: If you divide each side of an equation by the same nonzero number, the two sides remain equal.

In symbols: For any numbers *a*, *b*, and *c*, where $c \neq 0$, if $a = b$, then $\frac{a}{c} = \frac{b}{c}$.

Example 1 **Solve $-68 = -4m$. Check the solution and graph it on a number line.**

$-68 = -4m$

$\frac{-68}{-4} = \frac{-4m}{-4}$ *Divide each side by −4.*

$17 = m$ *−68 ÷ (−4) = 17*

Check: $-68 = -4m$

$-68 \stackrel{?}{=} -4(17)$ *Replace m with 17.*

$-68 = -68$ ✓

The solution is 17.

13 14 15 16 17 18 19 20

Just as division can be used to solve equations in some situations, multiplication can also be used to solve equations. This makes use of the multiplication property of equality.

Multiplication Property of Equality

In words: If you multiply each side of an equation by the same number, the two sides remain equal.

In symbols: For any numbers *a*, *b*, and *c*, if $a = b$, then $a \cdot c = b \cdot c$.

Example 2 **Solve $\frac{d}{3} = -3$. Check the solution and graph it on a number line.**

$\frac{d}{3} = -3$

$\frac{d}{3}(3) = -3(3)$ *Multiply each side by 3 to undo the division in $\frac{d}{3}$.*

$d = -9$

Check: $\frac{d}{3} = -3$

$\frac{-9}{3} \stackrel{?}{=} -3$ *Replace d with −9.*

$-3 = -3$ ✓

The solution is −9.

−12 −11 −10 −9 −8 −7 −6 −5

GLENCOE Technology

Interactive Mathematics Tools Software

In this interactive computer lesson, students explore solving equations. A **Computer Journal** gives students an opportunity to write about what they have learned.

For Windows & Macintosh

Example 3 APPLICATION Sports

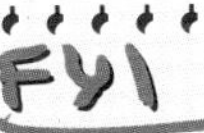

FYI

The tradition of the seventh-inning stretch started in 1910, when President William Taft stood to stretch his legs during a game in Washington, D.C. The fans stood up out of respect because they thought he was leaving.

In the 1993 season, the fans at the Baltimore Orioles' Camden Yards bought 232,618 bags of peanuts at ball games. There were 80 games in the park that season.

a. On average, how many bags of peanuts were sold at each game?

b. How many bags of peanuts should the concession manager order for the 1997 season if there will be 86 games in the park that year?

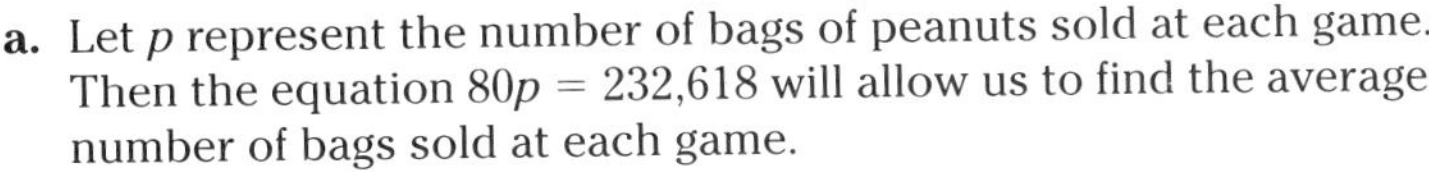

a. Let p represent the number of bags of peanuts sold at each game. Then the equation $80p = 232{,}618$ will allow us to find the average number of bags sold at each game.

$$80p = 232{,}618$$

$$\frac{80p}{80} = \frac{232{,}618}{80} \quad \textit{Divide each side by 80.}$$

$$p = \frac{232{,}618}{80} \quad \textit{Estimate: } 240{,}000 \div 80 = 3000$$

Use a calculator to find p.

232618 [÷] 80 [=] 2907.725

On average, about 2908 bags of peanuts were sold at each game.

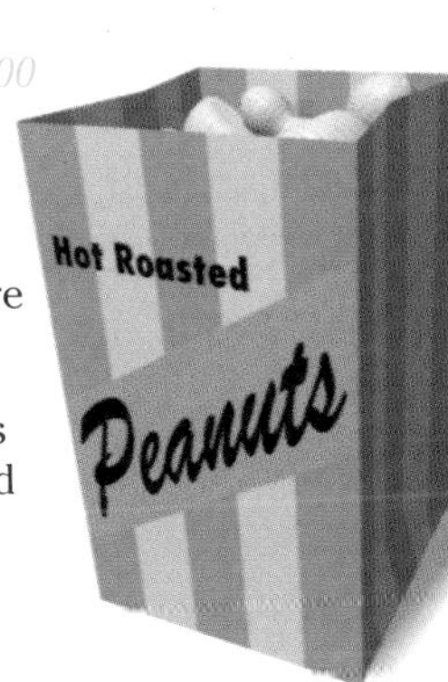

b. You can estimate to find out how many bags of peanuts the manager should order. Round to 3000 bags per game and 90 games in the season. So, the manager should order 3000×90 or 270,000 bags of peanuts for a season with 86 games.

Checking Your Understanding

Communicating Mathematics

Read and study the lesson to answer these questions. 1–4. See margin.

1. **In your own words, explain** how you would solve $\frac{x}{4} = -3$.
2. **Compare and contrast** solving equations using the multiplication property of equality and solving equations using the division property of equality.
3. **Write** an equation in the form $ax = c$ where the solution is -5.
4. In an equation of the form $\frac{x}{a} = b$, explain why a cannot be 0.

Guided Practice

Solve each equation and check your solution. Then graph the solution on a number line. See Solutions Manual for graphs.

5. $5x = 45$ **9**
6. $8y = -64$ **−8**
7. $-\frac{p}{3} = 4$ **−12**
8. $\frac{a}{-5} = 8$ **−40**
9. $-3g = -51$ **17**
10. $-2 = \frac{p}{6}$ **−12**

In-Class Example

For Example 3

Josh spent \$112 on boxes of baseball cards. He paid \$14 per box. Each box contained 540 cards.

a. How many boxes did Josh buy? **8 boxes**

b. How many cards did he get? **4320 cards**

3 PRACTICE/APPLY

Checking Your Understanding

Exercises 1–11 are designed to help you assess your students' understanding through reading, writing, speaking, and modeling. You should work through Exercises 1–4 with your students and then monitor their work on Exercises 5–11.

Additional Answers

1. Multiply each side by 4.
2. Sample answer: The division property is similar to the multiplication property since multiplying by the reciprocal is the same as dividing. Use the division property when a number is multiplied by x and the multiplication property when x is divided by a number.
3. Sample answer: $8x = -40$
4. Division by zero is undefined.

Study Guide Masters, p. 21

NAME ________ DATE ________

3-3 Study Guide
Solving Equations by Multiplying or Dividing

Student Edition Pages 129–133

Method: 1. Identify the variable.
2. Multiply or divide each side of the equation by the same nonzero number to get the variable by itself.
3. Check the solution.

Example: Solve $-7x = 42$.

$-7x = 42$

$\frac{-7x}{-7} = \frac{42}{-7}$ Divide each side by −7.

$x = -6$ The solution is −6.

−8 −7 −6 −5 −4 −3 −2 −1 0 1 2

Check: $-7x = 42$

$-7(-6) \stackrel{?}{=} 42$

$42 = 42$ ✓

Example: Solve $\frac{y}{2} = -2$.

$\frac{y}{2} = -2$

$\frac{y}{2} \cdot (2) = -2 \cdot (2)$ Multiply each side by 2.

$y = -4$ The solution is −4.

−6 −5 −4 −3 −2 −1 0 1 2 3 4

Check: $\frac{y}{2} = -2$

$\frac{-4}{2} \stackrel{?}{=} -2$

$-2 = -2$ ✓

Solve each equation and check your solution. Then graph the solution on the number line.

1. $-3a = 15$ $a = -5$
2. $-t = 5$ $t = -5$
3. $-1 = \frac{n}{4}$ $n = -4$
4. $7r = 28$ $r = 4$
5. $0 = \frac{h}{7}$ $h = 0$
6. $24 = -8m$ $m = -3$
7. $-11b = 44$ $b = -4$
8. $\frac{a}{-2} = -1$ $a = 2$

 T 21 *Pre-Algebra*

FYI

At Camden Yards on September 6, 1995, Cal Ripken, Jr. played in his 2131st consecutive baseball game. In doing so, he broke Lou Gehrig's record for playing in the most consecutive games, 2130.

Reteaching

Working Backward Have students count the number of letters in their last name. Then students are to create five multiplication and five division equations that have that number as the solution.

Error Analysis

Watch for students who have difficulty simplifying fractions. For instance, students may have difficulty simplifying $x = \frac{24}{-6}$. Review simplifying fractions, reminding students that the fraction bar indicates division.

Assignment Guide

Core: 13–41 odd, 42–49
Enriched: 12–36 even, 37–49

For **Extra Practice**, see p. 746.

The red A, B, and C flags, printed only in the Teacher's Wraparound Edition, indicate the level of difficulty of the exercises.

interNET CONNECTION

For the latest exchange rates, visit: **www.glencoe.com/sec/math/prealg/mathnet**

11. **Traveling** Paloma's social studies class was planning an imaginary trip to Kenya. Paloma had to find out about the rate of exchange for money in Kenya. She learned that money in Kenya is based on shillings and that the current exchange rate was one U.S. dollar for 40 shillings. If one night at a hotel cost 960 shillings, how much would that be in U.S. dollars? **24 dollars**

Exercises: Practicing and Applying the Concept

Independent Practice

Solve each equation and check your solution. Then graph the solution on a number line. **See Solutions Manual for graphs.**

A 12. $8x = 72$ **9** 13. $-5y = 95$ **−19** 14. $\frac{k}{-2} = 7$ **−14**

B 15. $-\frac{h}{7} = 20$ **−140** 16. $86 = 2v$ **43** 17. $-21 = -\frac{g}{8}$ **168**

18. $\frac{a}{45} = -3$ **−135** 19. $\frac{b}{-9} = -4$ **36** 20. $672 = -21t$ **−32**

21. $-56 = -7p$ **8** 22. $-116 = -4u$ **29** 23. $-\frac{y}{11} = 132$ **−1452**

24. $\frac{f}{-34} = -14$ **476** 25. $\frac{y}{8} = 117$ **936** 26. $17r = -357$ **−21**

27. $-18p = 306$ **−17** 28. $-144 = 8x$ **−18** 29. $-384 = -3m$ **128**

C 30. $\frac{b}{46} = 216$ **9936** 31. $-71 = \frac{x}{31}$ **−2201** 32. $-171 = \frac{x}{-12}$ **2052**

33. $584 = -\frac{s}{23}$ **−13,432** 34. $\frac{p}{47} = 123$ **5781** 35. $-67w = -5561$ **83**

Graphing Calculator

36. You can check your solutions to equations using a T1-82 graphing calculator. Enter one side of the equation as an expression using your solution and see if the value is the same as the other side of the equation.

For example, if $\frac{x}{5} = -2$, then $x = -10$.

Enter: [(−)] 10 [STO▸] [X,T,θ] [ENTER] *Enters the value of x.*

[X,T,θ] [÷] 5 [ENTER] -2 *Enters equation and evaluates it.*

The solution checks.

Solve each equation. Then check your solution using a graphing calculator.

a. $3x = 126$ **42** b. $\frac{x}{6} = -24$ **−144** c. $-4x = 88$ **−22**

Critical Thinking

37. Write a word problem that can be solved using the equation $4x = 13$. **Answers will vary.**

Applications and Problem Solving

38. **Manufacturing** One Slinky is made from 80 feet of wire. If a standard spool of wire contains 4000 feet, how many Slinkys can be made from one spool? **50**

Practice Masters, p. 21

NAME ______________ DATE ________

3-3 Practice
Solving Equations by Multiplying or Dividing

Student Edition Pages 129–133

Solve each equation and check your solution. Then graph the solution on the number line.

1. $-4 = 4t$ **t = −1**
2. $\frac{u}{-4} = 0$ **u = 0**
3. $5x = -15$ **x = −3**
4. $28 = -7f$ **f = −4**
5. $-1 = \frac{n}{-5}$ **n = 5**
6. $2 = \frac{k}{-2}$ **k = −4**
7. $0 = \frac{y}{-36}$ **y = 0**
8. $0 = -9r$ **r = 0**
9. $-1 = \frac{m}{-3}$ **m = 3**
10. $-4x = -12$ **x = 3**
11. $\frac{c}{1} = -2$ **c = −2**
12. $-12p = -48$ **p = 4**
13. $3 = \frac{t}{-1}$ **t = −3**
14. $-9r = -27$ **r = 3**
15. $35 = 7y$ **y = 5**
16. $1 = \frac{n}{1}$ **n = 1**

Group Activity Card 3-3

Algebra Bingo Group Activity **3-3**

MATERIALS: Paper • pencils

A group of two to four students can play Algebra Bingo.

To begin, each player must copy the Bingo Card shown below. Make the playing area large enough to write a number in the small box in each square. Each player uses a copy of the same card.

To win at Algebra Bingo, a player must cross out all the numbers in any row, column or diagonal. In order to cross out a number, the player must get that number as the solution to one of the equations given on the back of this card. The equations do not need to be solved in a certain order.

For each solution found on the Bingo Card, each player must write the number of the equation in the corner box for that square. The winner is the first player who gets a Bingo and has the equation numbers in the corners verified by the other players.

Algebra Bingo Card

-84	-91	-33	-90
16	-12	-4	33
90	15	39	135
-22	4	11	12

 Pre-Algebra

39. **Biology** The mola is an interesting fish because it appears to have no body. Weighing more than two tons, it looks like a large head with two perpendicular fins. Female molas can have 300 million eggs. This is 6 to 15 times the number in most species of fish.

a. Write an equation to represent 6 times the number of eggs is 300 million eggs. **$6x = 300{,}000{,}000$**

b. Write an equation to represent 15 times the number of eggs is 300 million eggs. **$15x = 300{,}000{,}000$**

c. Write a sentence that states the range of eggs produced by most species of fish. **See margin.**

40. **Finance** Refer to Exercise 11. On the trip to Kenya, the students in Paloma's class wanted to take a safari to the Masai Mara. They estimated that the cost of the safari would be $300 a person.

a. Research to find out how many shillings are exchanged for a U.S. dollar today. **Answers will vary.**

b. Write an equation to find out how much the safari would cost in shillings. **Answers will vary.**

c. Solve the equation. **Answers will vary.**

41. **Geometry** The perimeter of any square is 4 times the length of one of its sides. If the perimeter of a square is 72 centimeters, what is the length of each side of the square? **18 cm**

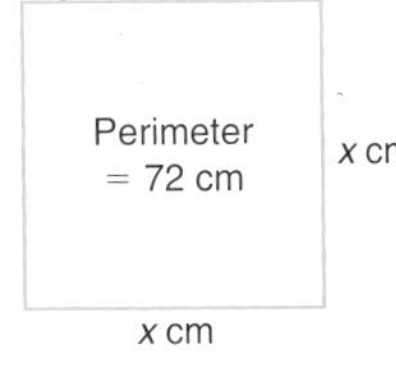

Mixed Review

42. Solve $121 = k - 34$ and check your solution. Then graph the solution on a number line. (Lesson 3-2) **155; see margin for graph.**

43. If a number is multiplied by 4 and the result is subtracted from 52, the difference is -12. What is the original number? (Lesson 3-1) **d**

a. -3 b. 0 c. 12 d. 16

44. Find the product of -7 and -15. (Lesson 2-7) **105**

45. Simplify $|-5| + |12|$. (Lesson 2-1) **17**

46. State whether the inequality $7 > 2$ is *true*, *false*, or *open*. (Lesson 1-9)

46. true

47. Solve $7n = 70$ mentally. (Lesson 1-6) **10**

48. **Careers** Curt Harris received a raise. He now makes $150 less than twice his old salary. Write an expression for Mr. Harris's current salary if his old salary was *s*. (Lesson 1-3) **$2s - 150$**

49. **Shopping** Is $7 enough money to buy a loaf of bread for $0.98, a pound of hamburger for $2.29, and a pound of roast turkey for $3.29? (Lesson 1-1)

a. Which method of computation do you think is most appropriate? Explain.

b. Solve the problem using the four-step plan. **Yes, $7 is enough.**

49a. Estimation; an exact answer is not needed.

Choose
- Estimation
- Mental Math
- Calculator
- Paper and Pencil

4 ASSESS

Closing Activity

Modeling Have students make up two word problems; one that requires a multiplication equation and one that requires a division equation. Then have students model how to write and solve the equations.

Additional Answers

39c. Most species lay between 20 and 50 million eggs.

42. Number line from 135 to 165 with a point at 155: 135 140 145 150 155 160 165

Extension

Using Logic Have students create a crossword puzzle where a number rather than a letter fits in each square and where the clues are addition or subtraction equations rather than definitions.

Enrichment Masters, p. 21

NAME ______________ DATE ______

3-3 Enrichment
Puzzle

Student Edition Pages 129–133

Solve each equation. The first one is completed.

1. $\frac{m}{12} = 13$	156	U
2. $17v = -578$	-34	E
3. $\frac{c}{75} = 18$	1350	R
4. $-252d = -5796$	23	H
5. $64 \cdot w = 5568$	87	O
6. $g \div 29 = 61$	1769	B
7. $p(85) = -7225$	-85	T
8. $39x = 663$	17	I
9. $\frac{k}{18} = 30$	540	A
10. $\frac{z}{-94} = -32$	3008	N
11. $-112q = 1456$	-13	F
12. $201y = -1608$	-8	S

Put the letter that is next to each solution in the box below the corresponding solution shown above.

What is a RINGLEADER?

-85	23	-34		-13	17	1350	-8	-85		87	3008	-34
T	H	E		F	I	R	S	T		O	N	E

17	3008		-85	23	-34		1769	540	-85	23	-85	156	1769
I	N		T	H	E		B	A	T	H	T	U	B

3-4 LESSON NOTES

NCTM Standards: 1-4, 9

Instructional Resources

- Study Guide Master 3-4
- Practice Master 3-4
- Enrichment Master 3-4
- Group Activity Card 3-4
- Assessment and Evaluation Masters, pp. 70, 71
- Multicultural Activity Masters, p. 6
- Activity Masters, pp 3, 17
- Real-World Applications, 5

Transparency 3-4A contains the 5-Minute Check for this lesson; **Transparency 3-4B** contains a teaching aid for this lesson.

Recommended Pacing	
Standard Pacing	Day 5 of 12
Honors Pacing	Day 4 of 11
Block Scheduling*	Day 3 of 7 (along with Lesson 3-3)

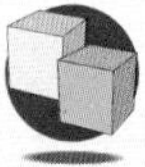

*For more information on pacing and possible lesson plans, refer to the ***Block Scheduling Booklet***.

1 FOCUS

5-Minute Check
(over Lesson 1-2)

Solve each equation. Check your solution.

1. $\frac{x}{3} = -6$ **−18**
2. $-3x = -6$ **2**
3. $\frac{x}{-2} = -16$ **32**
4. $-2x = 16$ **−8**
5. $12x = -156$ **−13**

Motivating the Lesson

Questioning Ask students each question. "If you drive 50 miles each hour, how far can you drive in 2 hours?" **100 miles** "If you drove 225 miles in 5 hours, what was your average speed?" **45 mph** Ask how students found each answer. Show how the equation $d = rt$ can be used to solve each problem.

3-4 Using Formulas

Setting Goals: *In this lesson, you'll solve problems by using formulas.*

Modeling a Real-World Application: Travel

You slide behind the wheel of a red convertible. The engine roars as you squeal away down the road. With the top down, the sound of the whipping wind drowns out the world until suddenly the engine stops.

"Just how far back was that gas station? Wish I would have figured the gas mileage on this thing before I left home."

You can use a formula to compute gas mileage.

Learning the Concept

A **formula** shows the relationship among certain quantities. The formula below can be used to find the miles per gallon achieved by a car.

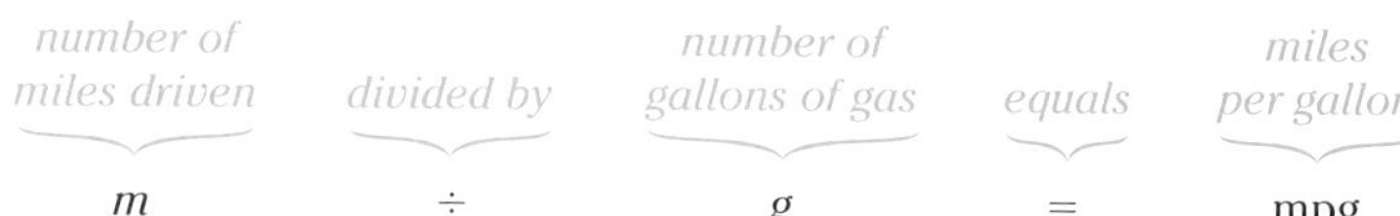

number of miles driven
divided by
number of gallons of gas
equals
miles per gallon
m
÷
g
=
mpg

Example 1 **You drove 192.8 miles before your convertible ran out of gas. The gas tank holds 11 gallons and was full when you left. What gas mileage does the car get?**

$m \div g = \text{mpg}$

$192.8 \div 11 = \text{mpg}$ *Replace m with 192.8 and g with 11.*

192.8 [÷] 11 [=] 17.527272 *Estimate: 200 ÷ 10 = 20*

The car gets about 17.5 miles per gallon.

Connection to Geometry

There are formulas associated with almost every field of study. Example 2 uses a formula from geometry.

Example 2 *Geometry*

A pentagon is a polygon with five sides. In a regular pentagon, all of the sides are the same length, and all of the angles have the same measure. The area, *A*, of a regular pentagon can be found using the formula $A = \frac{5}{2}sa$, where *s* is the length of a side of the pentagon and *a* is the length of the apothem. Find the area of the pentagon shown at the right.

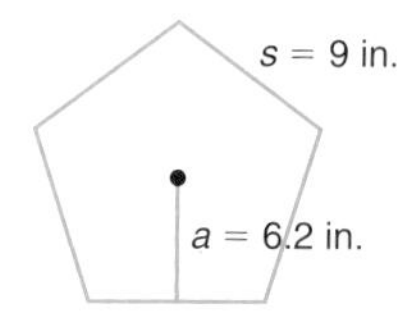

HELP WANTED

There is a wide variety of opportunities and places to learn about machines. You may wish to contact a local machine shop, automotive service center, or junior college for information on training that is available locally.

Interactive Mathematics Tools Software

In this interactive computer lesson, students explore acceleration. A **Computer Journal** gives students an opportunity to write about what they have learned.

For Windows & Macintosh

$A = \frac{5}{2}sa$

$A = \frac{5}{2}(9)(6.2)$ Replace s with 9 and a with 6.2.

$A = 139.5$

The area of the pentagon is 139.5 square inches.

Checking Your Understanding

Communicating Mathematics Read and study the lesson to answer these questions.

1–2. See Solutions Manual.

1. **Explain** how to use a formula.
2. **Write** a formula that you have used in your life.
3. **Draw** polygons with the numbers of sides in the table below. Pick a vertex and draw all the diagonals from this vertex. Count the number of triangles formed. Copy and complete the table. See Solutions Manual.

MODELING MATHEMATICS

MATERIALS

ruler

Number of sides	3	4	6	8	10
Number of triangles	1	2	4	6	8

Can you find a formula that will predict how many triangles will be formed in a polygon with 100 sides? 1250 sides? x sides? 98; 1248; $x - 2$

Guided Practice Solve by replacing the variables with the given values.

4. $d = rt$, if $d = 315$, $r = 45$ 7
5. $s = l - d$, if $s = \$35$, $d = \$10$ 45

Solve. Use the correct formula.

6. **Travel** Find the miles per gallon for a car that travels 271.7 miles on 0.6 gallons of fuel. 20.3 mpg
7. **Transportation** The formula $d = rt$ relates distance, d, rate, r, and time, t, traveled. Find the distance you travel if you drive at 55 miles per hour for 3 hours. 165 miles
8. **Sports** In league bowling, each team has an equal chance of winning because team members are given a handicap based on their averages. They use a formula to find the handicap score. The formula is as follows.

   handicap score(s) = game score(g) + handicap(h) or $s = g + h$

   Monique has a game score of 145 and a handicap of 35. Dennis has a game score of 153 and a handicap of 30. Who has the higher handicap score? Dennis

Exercises: Practicing and Applying the Concept

Independent Practice Solve by replacing the variables with the given values.

9. $d = rt$, if $r = 120$ and $t = 3$ 360
10. $F = \frac{9}{5}C + 32$, if $C = 30$ 86

11. $P = 5s$, if $P = 65$ 13
12. $h = 69 + 2.2F$, if $F = 14$ 99.8

13. $d = 2r$, if $r = 1.7$ 3.4
14. $A = \frac{b + d}{2}$, if $b = 11$ and $d = 13$ 12

2 TEACH

In-Class Examples

For Example 1

You car averages 24.5 miles per gallon. If you fill up your 12-gallon gas tank, how far can you drive on that fuel? **294 miles**

For Example 2

The area of a triangle can be found using the formula $A = \frac{1}{2}(bh)$, where b is the length of the base and h is the height of the triangle. Find the area of a triangle with a base of 10 cm and a height of 12 cm. **60 square centimeters**

3 PRACTICE/APPLY

Checking Your Understanding

Exercises 1–8 are designed to help you assess your students' understanding through reading, writing, speaking, and modeling. You should work through Exercises 1–3 with your students and then monitor their work on Exercises 4–8.

Study Guide Masters, p. 22

NAME ______________ DATE ______

3-4 Study Guide
Using Formulas

Student Edition Pages 134–137

A **formula** shows the relationship between certain quantities. The formula for the distance traveled by a moving object is $d = rt$. In the formula, d represents distance in kilometers (km), r represents the rate in kilometers per hour (km/h), and t represents the time in hours (h).

Example: Suppose r is 40 kilometers per hour and t is 3 hours. Find the distance traveled (d).

$d = rt$

$d = 40 \times 3$ Replace r with 40 and t with 3.

$d = 120$ The distance traveled is 120 kilometers.

Use the formula d = rt to find the indicated variables.

1. $r = 60$ km/h; $t = 4$ h; d **d = 240 km**
2. $d = 100$ km; $t = 2$ h; r **r = 50 km/h**
3. $r = 55$ km/h; $d = 110$ km; t **t = 2 h**
4. $r = 35$ km/h; $t = 3$ h; d **d = 105 km**
5. $d = 210$ km; $t = 7$ h; r **r = 30 km/h**
6. $r = 80$ km/h; $d = 320$ km; t **t = 4 h**

The formula $I = \frac{V}{R}$ shows the relationship between the current in amperes (I), the voltage in volts (V), and the resistance in ohms (R) in an electrical circuit.

Use the formula $I = \frac{V}{R}$ to find the current for each of the following. (Current is measured in amperes.)

7. V: 60 volts; R: 3 ohms **20 amperes**
8. V: 90 volts; R: 3 ohms **30 amperes**
9. V: 100 volts; R: 2 ohms **50 amperes**
10. V: 120 volts; R: 3 ohms **40 amperes**

Reteaching

Using Tables Allow students to discuss the formula $d = rt$. Have them copy this table.

d	r	t
120	40	?
320	?	8
1400	350	?
?	48	4

Have students use the formula to complete the table. Discuss their work.

Group Activity Card 3-4

Formulating A Formula Group Activity **3-4**

How many lines can you draw to connect two dots any distance apart? (1, of course.) How many lines can you draw between three dots? (3) How many lines can be drawn between 15 dots?

Write a formula that will tell you the number of lines that can be drawn between a given number of dots. Make a table to record the data. Complete the table for 8 or 9 or more dots.

Number of dots	Number of lines
2	1
3	3
4	?
.	.
.	.
.	.

1. Can you add the same value to the number of dots to get the number of lines?
2. Can you always multiply the number of dots by the same value to get the number of lines?
3. What is the relationship between a pair of numbers in the first column and the "Number of Lines" in the second column?
4. Write a formula to show the relationship. Let N = the number of dots, and L = the number of lines.

 Pre-Algebra

Error Analysis

Some students obtain incorrect answers when working with formulas because they try to take shortcuts or work too many steps at one time. Have students work in pairs and take turns writing all the steps needed to solve an equation.

Assignment Guide

Core: 9–21 odd, 22–23, 25–30
Enriched: 10–20 even, 22–30
All: Self Test, 1–10

For **Extra Practice**, see p. 746.

The red A, B, and C flags, printed only in the Teacher's Wraparound Edition, indicate the level of difficulty of the exercises.

Practice Masters, p. 22

NAME ______ DATE ______

3-4 Practice
Using Formulas

Student Edition
Pages 134–137

Solve. Use the correct formula.

1. A salesclerk must put a \$4 markup on a shirt that costs \$12.00 wholesale. What should the retail price be? **\$16**

The formula for the retail price is given below.

Retail Price	=	Wholesale Price	+	Markup
p	=	*w*	+	*m*

2. A pair of boots has a retail price of \$75. The store's markup is \$12. What is the wholesale price? **\$63**

3. A cassette that regularly sells for \$8.99 has a discount of \$2.50. What is the sale price? **\$6.49**

The following is the formula for the sale price.

Sale Price	=	Regular Price	–	Discount (markdown)
s	=	*p*	–	*d*

4. A book that regularly sells for \$14.50 was marked \$11.95. How much of a discount was there? **\$2.55**

5. An account opened three years ago with a principal of \$250 now has \$300.50. Find the amount of interest. **\$50.50**

The formula for adding principal and interest is given below.

Amount	=	Principal	+	Interest
a	=	*p*	+	*i*

6. After 4 years interest, an account has \$884. The interest is \$234. Find the principal. **\$650**

Solve. Use the correct formula.

15. **Ballooning** What is the speed in miles per hour of a balloon that travels 56 miles in 4 hours? **14 mph**

16. **Transportation** A 1992 Ford Taurus can travel an average of 464 miles on one tank of gas. If the tank holds 16 gallons, how many miles per gallon does it get? **29 mpg**

17. **Travel** The Flynn family plans to drive 600 miles from Harrisburg, Pennsylvania to Chicago, Illinois for their summer vacation. The speed limit on the highways they plan to use is 55 miles per hour. How long will the trip take if they don't exceed the speed limit? **10.9 hrs**

18. **Air Travel** How long does it take an Air Force jet fighter to fly 5200 miles at 650 miles per hour? **8 hours**

Translate each sentence into a formula.

19. The sale price of an item, s, is equal to the list price, ℓ, minus the discount, d. $s = \ell - d$

20. **Geometry** In a circle, the diameter, d, is twice the length of the radius, r. $d = 2r$

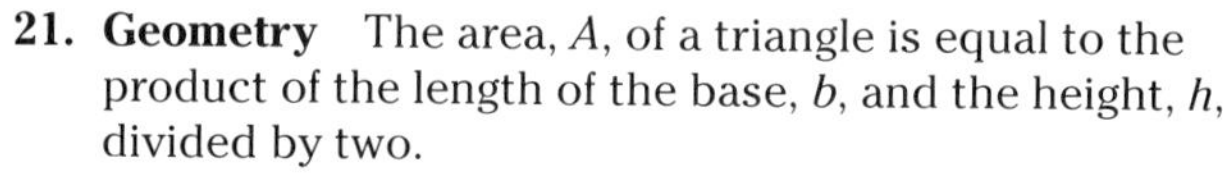

21. (C) **Geometry** The area, A, of a triangle is equal to the product of the length of the base, b, and the height, h, divided by two.
    a. What is the area of a triangle with a base of 12 inches and a height of 5 inches? **30 sq. in.**

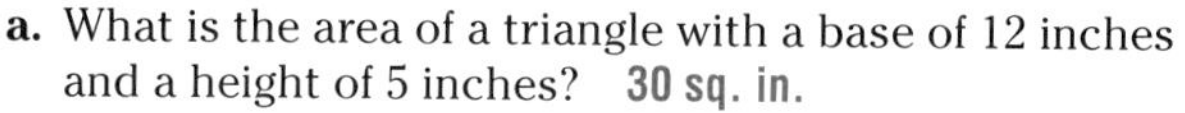

    b. What is the length of the base of a triangle with an area of 24 square centimeters and a height of 8 centimeters? **6 cm**

Critical Thinking

22. Measure the height and head circumference of five people. Use these measures to write a formula to predict a person's height from the size of their head. **Answers will vary.**

Applications and Problem Solving

23. **Jobs** Dorinda Pinel found a job about 8 miles from home. For a few weeks, she rode the bus to and from work. But the connections were poor, and it took an hour to get home in the afternoon. She thought that she could get home earlier if she rode her bicycle to and from work. Use a calculator to answer the following questions. **23b. 8.9 mph**
    a. If she rode at an average speed of 10 miles an hour, how long would it take for her to get home? **0.8 hours or 48 minutes**
    b. The trip to work was uphill most of the way. It took Dorinda 54 minutes, or 0.9 hours, to get to work. How fast was she riding?

24. **Physical Therapy** Physical therapists work to help people regain strength after surgery or an accident. The formula they use to determine a person's arm strength, S, is as follows.

$$S = \frac{D + P}{\frac{W}{10} + H - 60}$$

Find the arm strength of a person with the following information.

D(dips on a parallel bar) = 4 dips
P(pull-ups) = 8 pull-ups
W(weight in pounds) = 120 pounds
H(height in inches) = 60 inches **1**

Cooperative Learning

Corners Provide each group of four students with four equations to solve. Assign each student a number from 1 to 4. Student 1 solves equation 1, student 2 solves 2, and so on. After solving the equation, students go to the appropriate corner of the room to compare their answer with other students having the same number. Then they return to their group and share the results.

For more information on this strategy, see *Cooperative Learning in the Mathematics Classroom*, one of the titles in the Glencoe Mathematics Professional Series, pp. 17–20.

25. **Physical Science** Acceleration is the rate at which velocity is changing with respect to time. To find the acceleration, first find the change in velocity by subtracting the starting velocity, s, from the final velocity, f. Then divide by the time, t, it took to make the change.

   a. Write the formula for acceleration, a. $a = \frac{f - s}{t}$

   b. A motorcycle goes from 2 m/s to 14 m/s in 6 seconds. Find its acceleration. **2 m/s²**

   c. Find the acceleration of a skateboard that goes from 6 m/s to 0 m/s in 3 seconds. **−2 m/s²**

   d. When something slows down, it *decelerates*. Describe the value of the acceleration when an object decelerates. **negative**

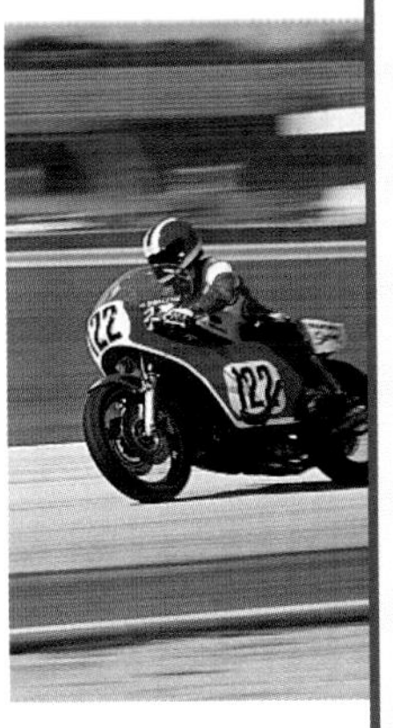

Mixed Review

26. Solve $17p = 119$ and check your solution. Then graph the solution on a number line. (Lesson 3-3) **7; see margin for graph.**
27. Find the quotient of −81 and 3. (Lesson 2-8) **−27**
28. Solve $n = 8 + (-21)$. (Lesson 2-4) **−13**
29. **Sports** There are 25 players on the Lady Scot soccer team. There are 11 players on the field at one time. Write an equation for the number of players on the bench. Then solve. (Lesson 1-8) **$x + 11 = 25$; 14**
30. Name the property shown by $(2 + 6) + 9 = 2 + (6 + 9)$. (Lesson 1-4) **Associative, +**

Self Test

1. **Entertainment** David Ogden Stiers, Mel Gibson, Russell Means, Linda Hunt, and Irene Bedard all lent their voices to characters in Disney's *Pocahontas*. The characters they played were Governor John Ratcliffe, Captain John Smith, Pocahontas, Grandmother Willow, and Powhatan. Use the following information to match each person with his or her character. (Lesson 3-1) **See margin.**
   - Irene Bedard is part Inupiat and was proud to play the part of a Native American heroine.
   - The character of Grandmother Willow is a tree.
   - Governor John Ratcliffe and the character played by Mel Gibson came to America from England. The character played by Russell Means is Native American.
   - Powhatan is Pocahontas' father and chief of the Powhatan tribe.

Solve each equation and check your solution. Then graph the solution on a number line. (Lessons 3-2 and 3-3) **See Solutions Manual for graphs.**

2. $18 + m = -57$ **−75**
3. $v - 11 = -5$ **6**
4. $67 = h + 38$ **29**
5. $d - (-16) = 9$ **−7**
6. $4h = -52$ **−13**
7. $\frac{x}{5} = 20$ **100**
8. $\frac{m}{11} = -10$ **−110**
9. $-2b = -13$ **$\frac{13}{2}$**

10. **Electronics** Ohm's Law states that the relationship between the current in amperes, I, the voltage in volts, V, and the resistance in ohms, R, in an electrical circuit is $I = \frac{V}{R}$. Current is usually stated in amperes. An automobile has a 12-volt system. What is the current in an automobile system that has a resistance of 3 ohms? (Lesson 3-4) **4 amps**

4 ASSESS

Closing Activity

Speaking Have students choose a formula, write it on the board, and explain how it is used to solve a problem. You may have the student actually make up a problem and demonstrate the use of the formula.

Chapter 3, Quiz B (Lessons 3-3 and 3-4) is available in the *Assessment and Evaluation Masters,* p. 71.

Mid-Chapter Test (Lessons 3-1 through 3-4) is available in the *Assessment and Evaluation Masters,* p. 70.

Additional Answers

26.

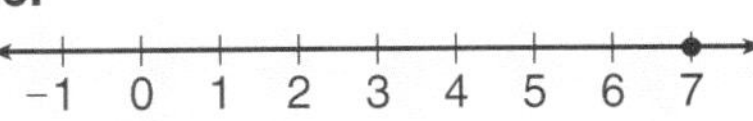

Self Test

1. Stiers-Ratcliffe, Gibson-Smith, Means-Powhatan, Hunt-Willow, Bedard-Pocahontas

Enrichment Masters, p. 22

NAME ____________ DATE ________

3-4 Enrichment
Formulas

Student Edition Pages 134–137

The following formula can be used to determine the specific day of the week on which a date occurred.

$$s = d + 2m + [(3m + 3) \div 5] + y + \left[\frac{y}{4}\right] - \left[\frac{y}{100}\right] + \left[\frac{y}{400}\right] + 2$$

s = sum
d = day of the month, using numbers from 1–31
m = month, beginning with March is 3, April is 4, and so on, up to December is 12, January is 13, and February is 14
y = year except for dates in January or February when the previous year is used

For example, for February 13, 1985, $d = 13$, $m = 14$, and $y = 1984$; and for July 4, 1776, $d = 4$, $m = 7$, and $y = 1776$

The brackets, [ ], mean you are to do the division inside them, discard the remainder, and use only the whole number part of the quotient. The next step is to divide s by 7 and note the remainder. The remainder 0 is Saturday, 1 is Sunday, 2 is Monday, and so on, up to 6 is Friday.

Example: What day of the week was October 3, 1854?

For October 3, 1854, $d = 3$, $m = 10$, and $y = 1854$.

$$s = 3 + [\,2(10)\,] + [\,(3 \times 10 + 3) \div 5\,] + 1854 + \left[\frac{1854}{4}\right] - \left[\frac{1854}{100}\right] + \left[\frac{1854}{400}\right] + 2$$
$$= 3 + 20 + 6 + 1854 + 463 - 18 + 4 + 2$$
$$= 2334$$

$$s \div 7 = 2334 \div 7$$
$$= 333 \text{ R3}$$

Since the remainder is 3, the day of the week was Tuesday.

Solve.

1. See if the formula works for today's date. **Answers will vary.**
2. On what day of the week were you born? **Answers will vary.**
3. What will be the day of the week on April 13, 2006?
   $s = 13 + 2(4) + [(3 \times 4 + 3) \div 5] + 2006 + \left[\frac{2006}{4}\right] - \left[\frac{2006}{100}\right] + \left[\frac{2006}{400}\right] + 2$
   $= 13 + 8 + 3 + 2006 + 501 - 20 + 4 + 2 = 2518$; $2518 \div 7 = 359$ R5 → Thursday
4. On what day of the week was July 4, 1776?
   $s = 4 + 2(7) + [(3 \times 7 + 3) \div 5] + 1776 + \left[\frac{1776}{4}\right] - \left[\frac{1776}{100}\right] + \left[\frac{1776}{400}\right] + 2$
   $= 4 + 14 + 4 + 1776 + 444 - 17 + 4 + 2 = 2231$; $2231 \div 7 = 318$ R5 → Thursday

Extension

Using Connections Have students research other textbooks to find formulas that are used in physics, interior decorating, auto mechanics, etc. Have students report on the formulas they found and use them to make up problems to exchange with other students.

Self Test

The Self Test provides students with a brief review of the concepts and skills in Lessons 3-1 through 3-4. Lesson numbers are given to the right of exercises or instruction lines so students can review concepts not yet mastered.

3-5A LESSON NOTES

NCTM Standards: 1-4, 8, 12, 13

Objective

Discover the greatest possible perimeter for a given area.

Recommended Time

Demonstration and discussion: 15 minutes; Exercises: 30 minutes

Instructional Resources

For each student or group of students

Student Manipulative Kit
- square tiles (centimeter cubes)

Math Lab and Modeling Math Masters
- p. 7, 8 (grid paper)
- p. 34 (worksheet)

For teacher demonstration

Overhead Manipulative Resources

1 FOCUS

Motivating the Lesson

Pose this situation: *For a walk-a-thon, students were to walk around the outside of an area that contained 16 square city blocks. How many blocks did the students walk?* **16 blocks to 34 blocks**

2 TEACH

Teaching Tip Encourage students to communicate their reasoning process by drawing pictures and describing patterns.

3 PRACTICE/APPLY

Assignment Guide
Core: 1–6
Enriched: 1–6

4 ASSESS

Observing students working in cooperative groups is an excellent method of assessment.

MATHEMATICS LAB **3-5A**

HANDS-ON ACTIVITY

Area and Perimeter

A Preview of Lesson **3-5**

MATERIALS
- grid paper or
- 15 square tiles

The distance around a geometric figure is called its **perimeter**. The measure of the surface enclosed by a geometric figure is its **area**. You can investigate the relationship between perimeter and area by drawing figures on grid paper or making models with square tiles.

Activity

The figures below show several figures with an area of 4 square units.

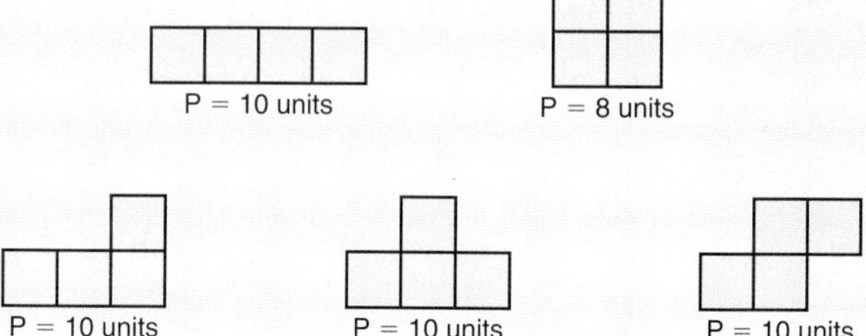

It appears that for an area of 4 square units, the greatest possible perimeter is 10 units.

Your Turn Repeat the activity for areas of 1, 2, 3, 5, and 6 square units. Find the greatest possible perimeter for each area. Record your results in a table.

For each given area, find the greatest possible perimeter. Draw a figure to show the greatest possible perimeter.

1–3. See Solutions Manual for drawings.

1. 8 square units **18 units**
2. 10 square units **22 units**
3. 15 square units **32 units**

TALK ABOUT IT

4. A figure has an area of 50 square units. Predict the greatest possible perimeter. **102 units**

5. See margin.

5. Explain why the two figures at the right have the same perimeter. Draw another figure with the same perimeter.
6. Suppose a figure has an area of x square units. Write an expression for the greatest possible perimeter. **$2x + 2$ units**

Additional Answer

5. The two figures have the same perimeter because the squares still share the same number of sides.

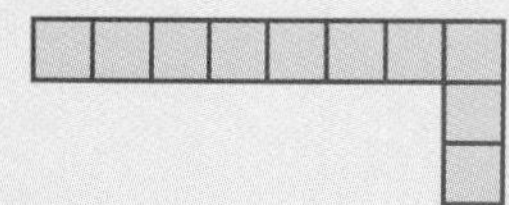

3-5 Integration: Geometry
Area and Perimeter

Setting Goals: *In this lesson, you'll find the perimeter and area of rectangles and squares.*

Modeling with Manipulatives

MATERIALS

16 square tiles

or grid paper

Below are all the different rectangles that have sides with measures that are whole numbers and a perimeter of 12 units. Model each with tiles or by drawing on grid paper.

A 

B 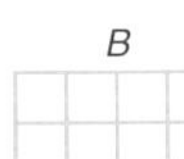

C 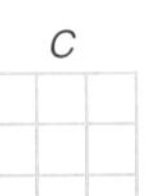

On a paper, copy and complete the table at the right using the information from figures A, B, and C.

Rectangle	Length of Rectangle	Width of Rectangle	Perimeter of Rectangle
A	5	1	12
B	4	**2**	12
C	**3**	**3**	12

Your Turn Use grid paper or tiles to make as many different rectangles as you can with a perimeter of 16 units. Make and complete a table similar to the one above.

a–c. See margin.

a. What observations can you make about the tables?

b. Work with a partner to write a rule on how to find the perimeter of a rectangle.

c. Will your rule also work for squares?

Learning the Concept

One method of finding the perimeter of a rectangle is to add the measures of the four sides. Let P represent the measure of the perimeter of a rectangle. Let ℓ represent the measure of the length and w the measure of the width.

$P = \ell + w + \ell + w$ *Definition of perimeter*
$P = \ell + \ell + w + w$ *Commutative property of addition*
$P = 2\ell + 2w$
$P = 2(\ell + w)$ *Distributive property*

Perimeter of a Rectangle

In words: If a rectangle has a length of ℓ units and a width of w units, then the perimeter is twice the sum of the length and width.

In symbols: $P = 2(\ell + w)$

$2(\ell + w)$ is read two times the quantity of ℓ plus w.

3-5 LESSON NOTES

NCTM Standards: 1-4, 9, 12, 13

Instructional Resources

- Study Guide Master 3-5
- Practice Master 3-5
- Enrichment Master 3-5
- Group Activity Card 3-5
- Math Lab and Modeling Math Masters, pp. 59, 78

Transparency 3-5A contains the 5-Minute Check for this lesson; **Transparency 3-5B** contains a teaching aid for this lesson.

Recommended Pacing	
Standard Pacing	Day 7 of 12
Honors Pacing	Day 6 of 11
Block Scheduling*	Day 4 of 7

*For more information on pacing and possible lesson plans, refer to the ***Block Scheduling Booklet***.

1 FOCUS

5-Minute Check
(over Lesson 3-4)

Solve by replacing the variables with the given values.

1. $d = rt$, if $r = 50$ and $t = 3$ **150**
2. $A = \ell w$, if $\ell = 23$ and $w = 11$ **253**
3. $A = \ell w$, if $A = 500$ and $w = 10$ **50**
4. $F = \frac{9}{5}C + 32$, if $C = 20$ **68**
5. $a = \frac{(f - s)}{t}$, if $f = 18$, $s = 2$, and $t = 8$ **2**

Motivating the Lesson

Situational Problem Tell students that you want to put new tiles on the floor and new edging around the base of the room. Allow students to suggest ways of finding how much flooring and edging will be needed.

Alternative Learning Styles

Auditory Have students discuss perimeter and area. The discussion should include the fact that perimeter is measured in linear units and area is measured in square units.

Have students discuss how to find the perimeter and area of various squares and rectangles.

Additional Answers Talk About It

a. Answers may vary.
b. Sample answer: Add the lengths of all the sides.
c. Yes

2 TEACH

In-Class Examples

For Example 1
Find the perimeter of each rectangle.

a. a 4-foot-by-8-foot rectangle **24 feet**

b. a 6-centimeter square **24 centimeters**

For Example 2
Find the area of a rectangle with a length of 90 feet and a width of 120 feet. **10,800 square feet**

Teaching Tip You may want to challenge students to find all possible *pentominoes*. A pentomino is made up of 5 squares so that every square has at least one edge in common with another of the 5 squares. Two sample pentominoes are shown below.

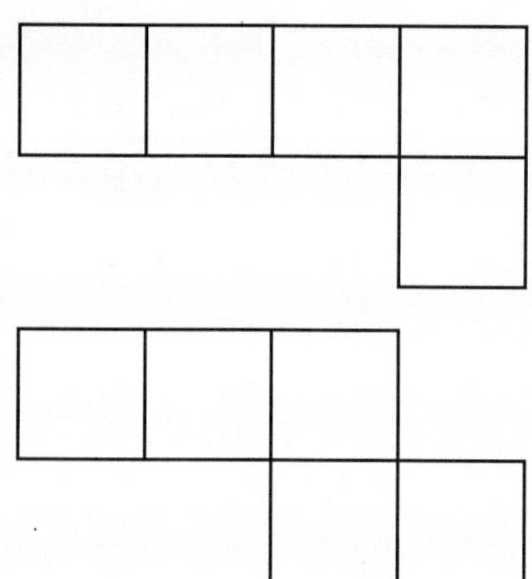

Example 1 **Find the perimeter of each rectangle.**

a. 9 in. by 4 in.

$P = 2(\ell + w)$
$P = 2(9 + 4)$ $\ell = 9, w = 4$
$P = 2(13)$ Add 9 and 4.
$P = 26$ inches

The perimeter is 26 inches.

b. 8 m by 8 m

$P = 2(\ell + w)$
$P = 2(8 + 8)$ $\ell = 8, w = 8$
$P = 2(16)$
$P = 32$ m

The perimeter is 32 meters.

Another measurement of a rectangle is its **area**. By counting the number of 1 unit-by-1 unit squares, you can find the area of the rectangle shown at the right. The area of the rectangle at the right is 12 square units.

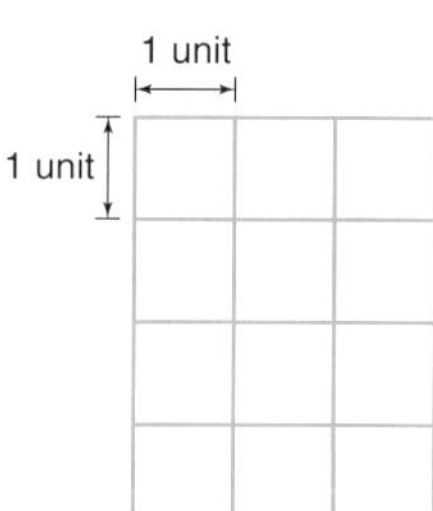

Look back at the models of rectangles that you investigated at the beginning of this lesson. Add another column called Area of Rectangle to your table. Count the squares in each rectangle to fill in the column. Do you notice a pattern? The area can also be found by multiplying the measures of the length and the width.

Area of a Rectangle

In words: If a rectangle has a length of ℓ units and a width of w units, then the area is $\ell \cdot w$ square units.

In symbols: $A = \ell w$

Example 2 **Find the area of a rectangle with length 15 meters and width of 8 meters.**

First draw a picture of the rectangle and label the length and width.

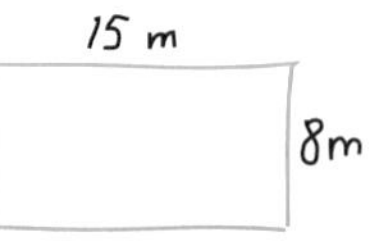

$A = \ell \cdot w$ formula for area of a rectangle
$A = 15 \cdot 8$ Replace ℓ with 15 and w with 8.
$A = 120$

The area is 120 square meters.

Many real-life problems involve finding the area or perimeter of a rectangle.

Classroom Vignette

"I have students use square chips to form rectangles on their desks. The students can visually find the perimeters and areas. I then ask students to determine the formulas on their own. If students do not discover them after several attempts, I give the formulas and have students practice finding area and perimeter."

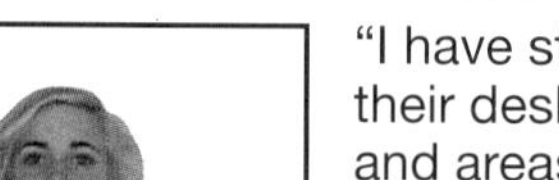

Phyllis Plasky
Camden High School
Camden, NJ

Example 3

Sumi has just come home from the plant nursery with 20 outdoor plants. The man at the plant nursery told her that each of these particular types of plants needs 24 square feet of space to grow, so that it won't be choked off by its neighbors.

a. How many square feet does Sumi need in a garden plot so all the plants have enough room?

square feet per plant	*times*	*number of plants*	*equals*	*square feet*
24	$\cdot$	20	=	480

b. Sumi has decided to use old railroad ties to enclose a rectangular space along the side of her house for the plants. The space is 16 feet wide, so how long should the garden be?

$A = \ell \cdot w$ *formula for area of a rectangle*

$480 = \ell \cdot 16$ *Replace A with 480 and w with 16.*

$\frac{480}{16} = \frac{\ell \cdot 16}{16}$ *Divide each side by 16.*

$30 = \ell$ The length should be 30 feet.

Example 4

APPLICATION

Real Estate

Real estate agents help their clients choose a home that meets their needs. One of the ways that real estate agents compare homes is by their total square footage of living space. Find the square footage of each room of the home whose floor plan is shown at the right.

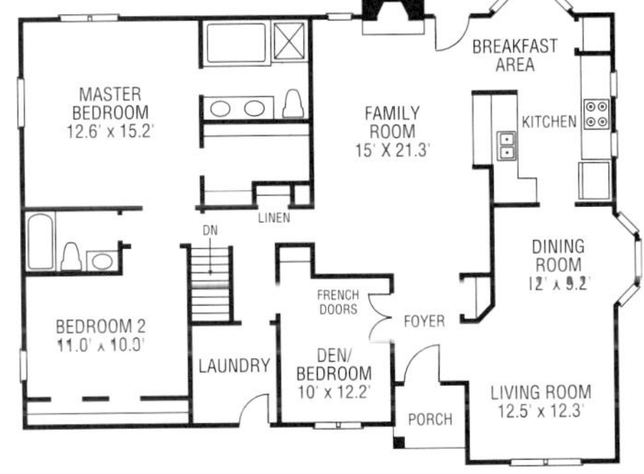

Room	$\ell \times w$	Area
master bedroom	12.6×15.2	191.52
family room	15×21.3	319.5
dining room	12×9.2	110.4
living room	12.5×12.3	153.75
den/bedroom	10×12.2	122
bedroom 2	11.6×10.3	119.48

Checking Your Understanding

Communicating Mathematics

2–4. See margin.

Read and study the lesson to answer these questions. 1. Answers will vary.

1. **Restate** the definition of perimeter in your own words.
2. **Compare and contrast** the perimeter and area of a rectangle
3. **Draw and label** a rectangle that has a perimeter of 18 inches.
4. **Explain** how to find the perimeter and area of a rectangle that is 9 cm long and 5 cm wide.

In-Class Examples

For Example 3

For a 4-H project, Javier is to have a garden with an area equivalent to a 15-foot-by-16-foot lot.

a. What must the area of Javier's garden be? **240 feet**

b. The space Javier has available for his garden is only 12 feet wide. How long will his garden have to be? **20 feet**

For Example 4

Four families are involved in a garden co-op in their city. Using this sketch of their garden space, find the garden area available to each family.

12 ft, 8 ft, 6 ft, Smith, Mays, 8 ft, Kolwalski, 8 ft, Shin, 6 ft

Smith, 72 square feet; Mays, 64 square feet; Kolwalski, 96 square feet; Shin, 48 square feet

Additional Answers

2. Sample answer: Perimeter and area are the same in the respect that they both deal with the dimensions and the shape of an object. They are different in the respect that perimeter deals with the outer boundary of an object whereas area deals with the space an object occupies.
3. Sample answer:

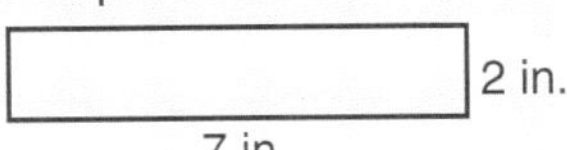

4. To find the perimeter, multiply each dimension by 2 and then find the sum. To find the area, multiply the two dimensions.

3 PRACTICE/APPLY

Checking Your Understanding

Exercises 1–10 are designed to help you assess your students' understanding through reading, writing, speaking, and modeling. You should work through Exercises 1–4 with your students and then monitor their work on Exercises 5–10.

Error Analysis

Students may confuse the terms *perimeter* and *area* or use the wrong units to label answers involving perimeter or area. Review the meaning of each term and the linear units used for perimeter and the square units used for area.

Study Guide Masters, p. 23

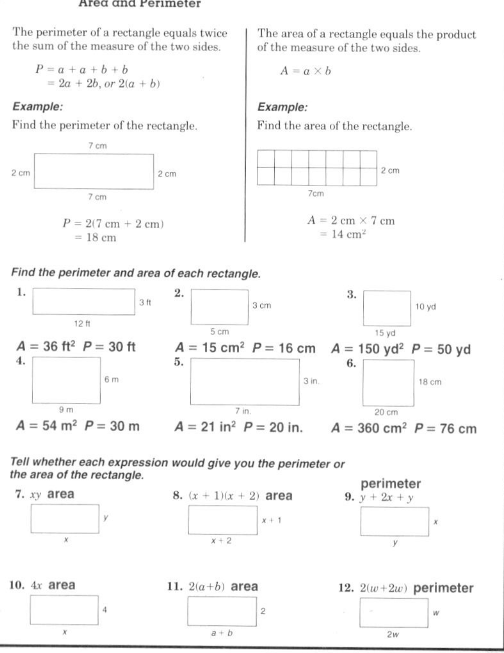

NAME ______________ DATE ______

3-5 Study Guide
Integration: Geometry
Area and Perimeter

Student Edition Pages 139–144

The perimeter of a rectangle equals twice the sum of the measure of the two sides.

$P = a + a + b + b$
$= 2a + 2b$, or $2(a + b)$

Example:
Find the perimeter of the rectangle.

$P = 2(7 \text{ cm} + 2 \text{ cm})$
$= 18 \text{ cm}$

The area of a rectangle equals the product of the measure of the two sides.

$A = a \times b$

Example:
Find the area of the rectangle.

$A = 2 \text{ cm} \times 7 \text{ cm}$
$= 14 \text{ cm}^2$

Find the perimeter and area of each rectangle.

1. $A = 36 \text{ ft}^2$ $P = 30$ ft
2. $A = 15 \text{ cm}^2$ $P = 16$ cm
3. $A = 150 \text{ yd}^2$ $P = 50$ yd
4. $A = 54 \text{ m}^2$ $P = 30$ m
5. $A = 21 \text{ in}^2$ $P = 20$ in.
6. $A = 360 \text{ cm}^2$ $P = 76$ cm

Tell whether each expression would give you the perimeter or the area of the rectangle.

7. xy area
8. $(x + 1)(x + 2)$ area
9. $y + 2x + y$ perimeter
10. $4x$ area
11. $2(a+b)$ area
12. $2(w+2w)$ perimeter

Guided Practice

Find the perimeter and area of each rectangle.

5.

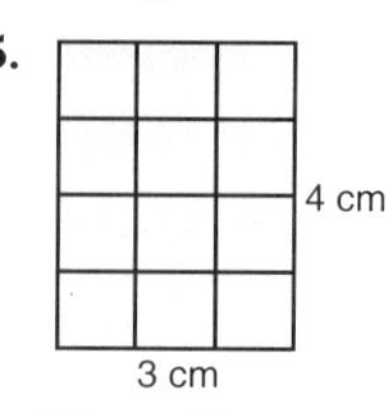

14 cm, 12 sq cm

6.

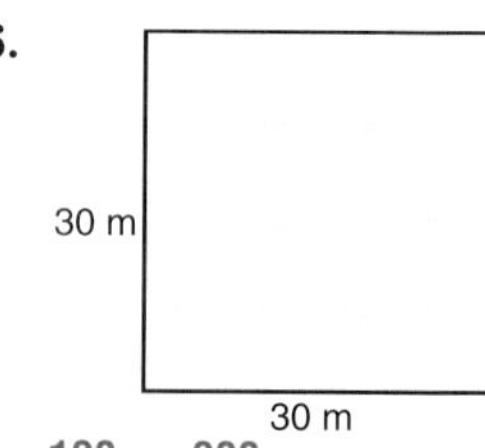

120 m, 900 sq m

7. rectangle with length of 15 feet and a width of 6 feet **42 ft; 90 sq ft**

Find the missing dimension of each rectangle.

8. 12 in.; Perimeter = 32 in.; *w* **4 in.**

9.

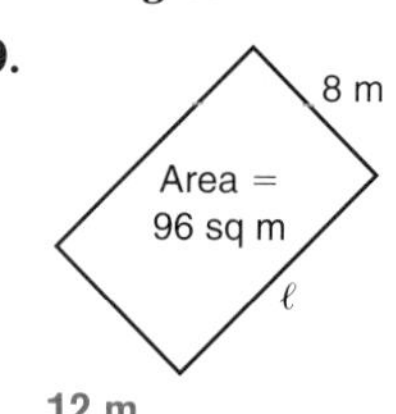

12 m

10. Community Project A neighborhood uses an empty city lot for a community vegetable garden. Each participant is allotted a space of 18 feet by 90 feet. What is the perimeter and area of each plot? **216 ft; 1620 sq ft**

Exercises: Practicing and Applying the Concept

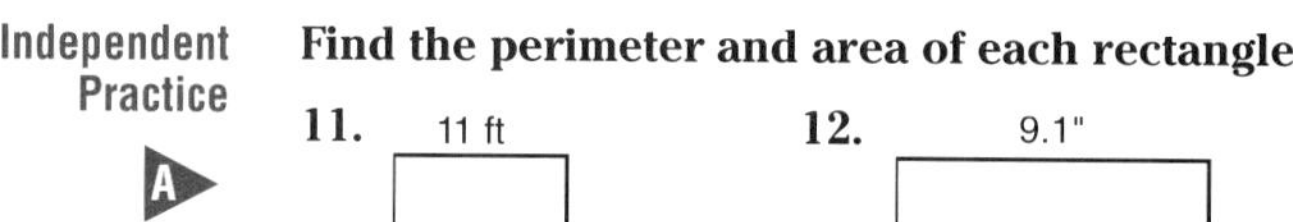

Independent Practice

Find the perimeter and area of each rectangle.

A

11.

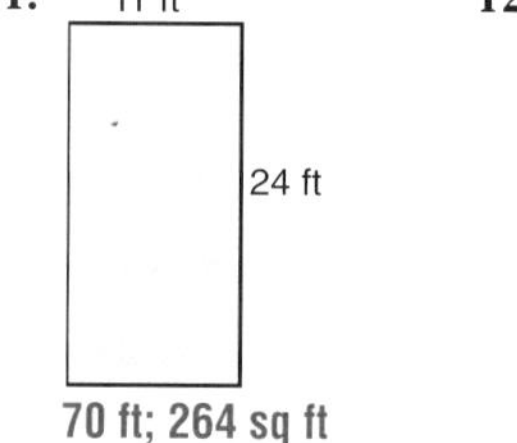

70 ft; 264 sq ft

12.

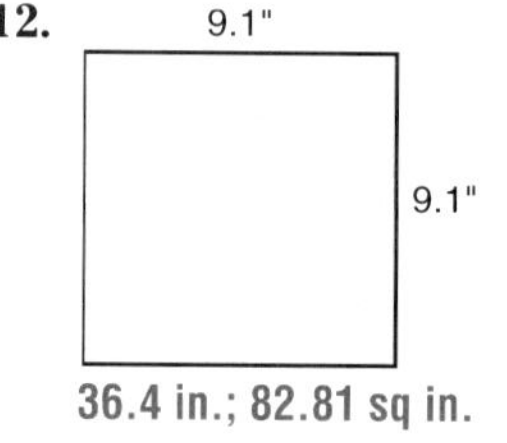

36.4 in.; 82.81 sq in.

13.

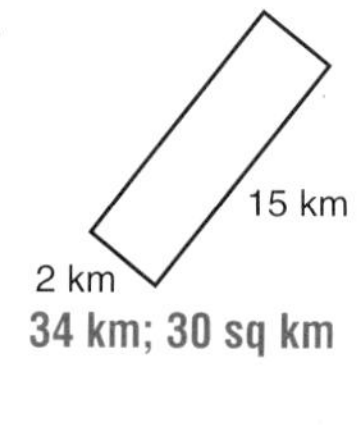

34 km; 30 sq km

B

14.

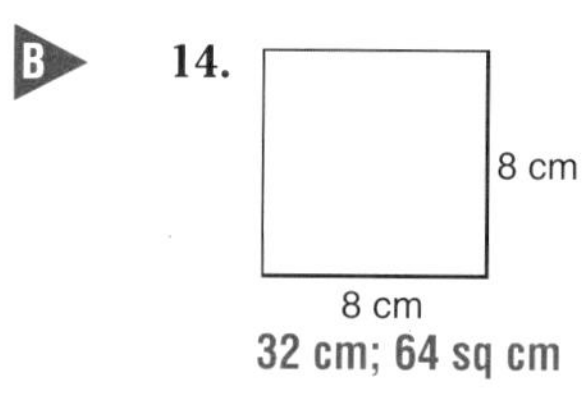

32 cm; 64 sq cm

15.

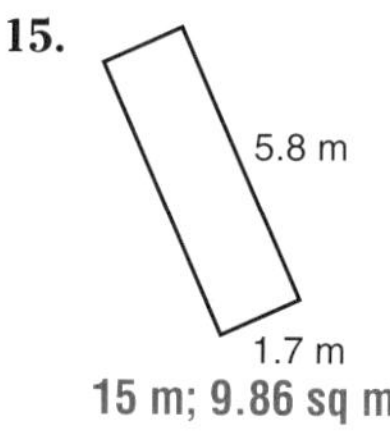

15 m; 9.86 sq m

16.

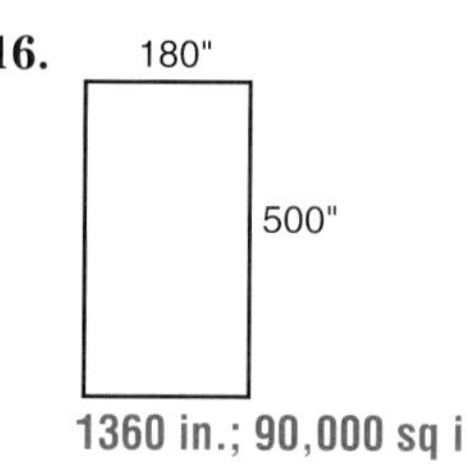

1360 in.; 90,000 sq in.

17.

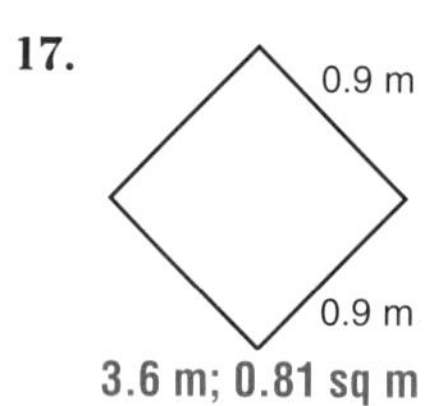

3.6 m; 0.81 sq m

18.

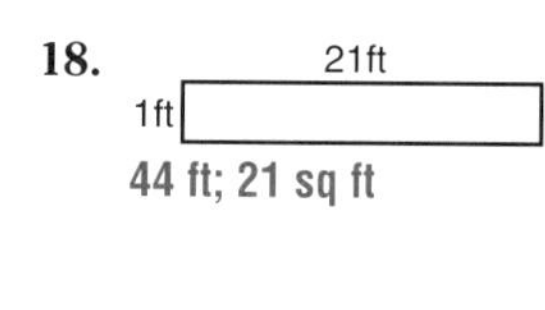

44 ft; 21 sq ft

19.

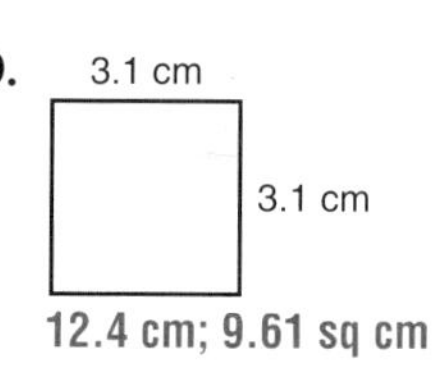

12.4 cm; 9.61 sq cm

Reteaching

Using Models On graph paper, have students design a small, one-level house with rectangular rooms. Use a scale where one grid unit represents one foot. Ignoring doorways, determine the amount of molding needed for baseboards and the area of carpet needed for flooring.

20. a rectangle that is 3.8 meters long and 1.1 meters wide **9.8 m, 4.18 sq m**

21. a square that is 4.8 millimeters on each side **19.2 mm, 23.04 sq mm**

22. an $8\frac{1}{2}$-inch by 11-inch rectangle **39 in.; 93.5 sq in.**

Find the missing dimension of each rectangle.

	Length	Width	Perimeter	Area
23.	7 m	**5 m**	24 m	35 sq m
24.	**18 cm**	15 cm	66 cm	270 sq cm
25.	16 yd	**11 yd**	54 yd	176 sq yd
26.	11 km	**24 km**	70 km	264 sq km
27.	**39 ft**	12 ft	102 ft	468 sq ft
28.	**17 yd**	91 yd	216 yd	1547 sq yd

Find the area of the blue part of each rectangle.

C 29.

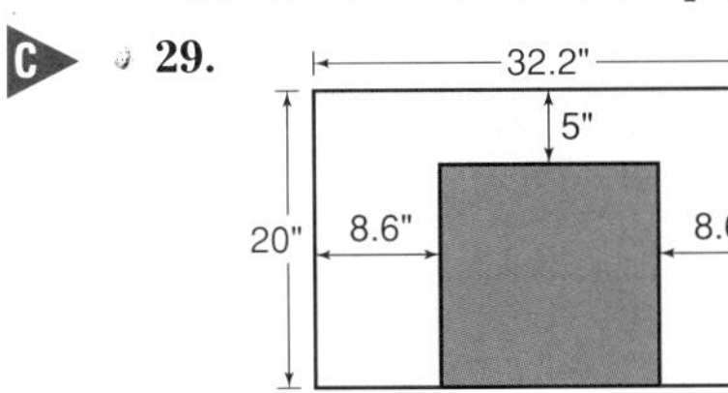

225 sq in.

30.

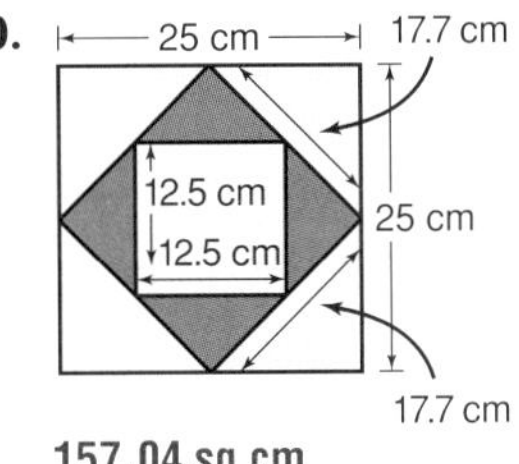

157.04 sq cm

Critical Thinking 31. Copy and complete the following table that compares changes in dimension, perimeter, and area of a square.

Dimension of square	Change in each from original	Perimeter	Change in perimeter from original	Area	Change in area from original
2 by 2	—	8	—	4	—
4 by 4	multiplied by 2	**16**	multiplied by 2	**16**	multiplied by 4
6 by 6	**multiplied by 3**	**24**	**multiplied by 3**	**36**	**multiplied by 9**
8 by 8	**multiplied by 4**	**32**	**multiplied by 4**	**64**	**multiplied by 16**

a. Show a relationship between the length of a side of a square and its perimeter using a graph. **See margin.**

b. Show a relationship between the length of a side of a square and its area using a graph. **See margin.**

c. The change in each square from the original is called the *scale factor*. Suppose the scale factor is x. What are the new perimeter and area? **perimeter: $8x$ units; area: $4x^2$ square units**

Applications and Problem Solving 32. **Landscaping** Tachiyuki Mizumoto wishes to fertilize his lawn. Fertilizer comes in bags that cover 5000 square feet. His yard is 110 by 210 feet. His house occupies a rectangle of 30 by 45 feet and the driveway is 18 by 50 feet.

a. What is the area he must fertilize? **20,850 sq ft**

b. How many bags should Mr. Mizumoto buy? **5 bags**

Assignment Guide

Core: 11–35 odd, 36–42
Enriched: 12–30 even, 31–42

For **Extra Practice**, see p. 746.

The red A, B, and C flags, printed only in the Teacher's Wraparound Edition, indicate the level of difficulty of the exercises.

Additional Answers

31a.

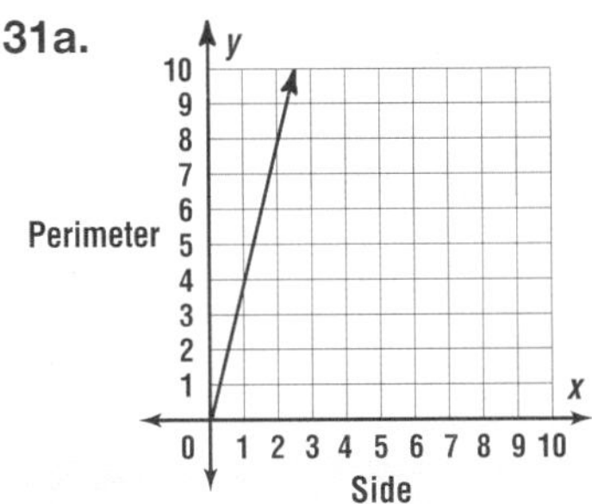

The perimeter is four times the length of a side.

31b.

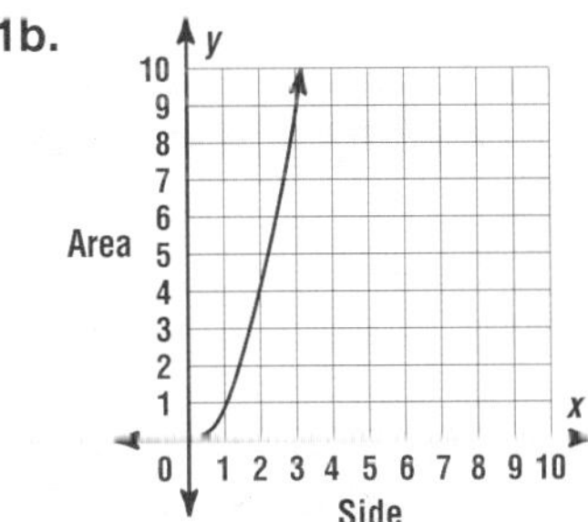

The area is the square of the length of a side.

Practice Masters, p. 23

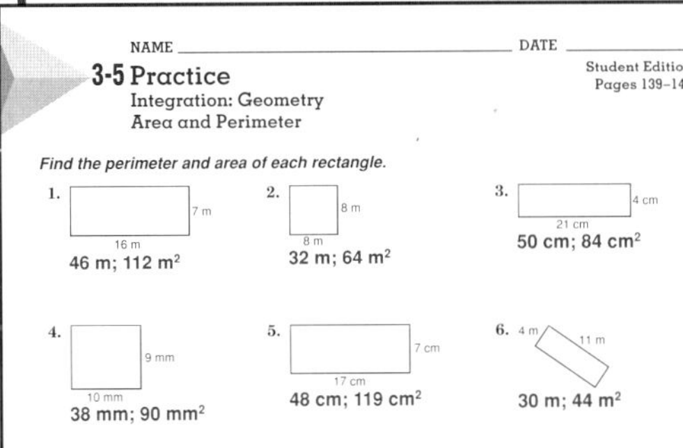

NAME ______ DATE ______

3-5 Practice
Integration: Geometry
Area and Perimeter

Student Edition Pages 139–144

Find the perimeter and area of each rectangle.

1. 16 m, 7 m **46 m; 112 m²**
2. 8 m, 8 m **32 m; 64 m²**
3. 21 cm, 4 cm **50 cm; 84 cm²**
4. 10 mm, 9 mm **38 mm; 90 mm²**
5. 17 cm, 7 cm **48 cm; 119 cm²**
6. 4 m, 11 m **30 m; 44 m²**

7. a square with each side 15 meters long **60 m; 225 m²**

8. a rectangle with a length of 27 meters and a width of 8 meters **70 m; 216 m²**

9. a square with each side 21 centimeters long **84 cm; 441 cm²**

10. a rectangle, 13 m by 11 m **48 m; 143 m²**

11. a square with each side 2 miles long **8 mi; 4 mi²**

Given each area, find the missing dimensions of each rectangle.

12. $A = 225$ m², $\ell = 17$ m, $w =$? **15 m**

13. $A = 216$ cm², $\ell =$?, $w = 12$ cm **18 cm**

14. $A = 250$ km², $\ell = 25$ km, $w =$? **10 km**

15. $A = 45$ yd², $\ell =$?, $w = 3$ yd **15 yd**

16. $A = 105$ mm², $\ell = 15$ mm, $w =$? **7 mm**

17. $A = 3055$ m², $\ell = 65$ m, $w =$? **47 m**

Extension

Using Diagrams Have students draw as many rectangles as possible that have an area of 24 square units. Have students order these rectangles according to perimeter, least to greatest.

Group Activity Card 3-5

Fence Me In — Group Activity 3-5

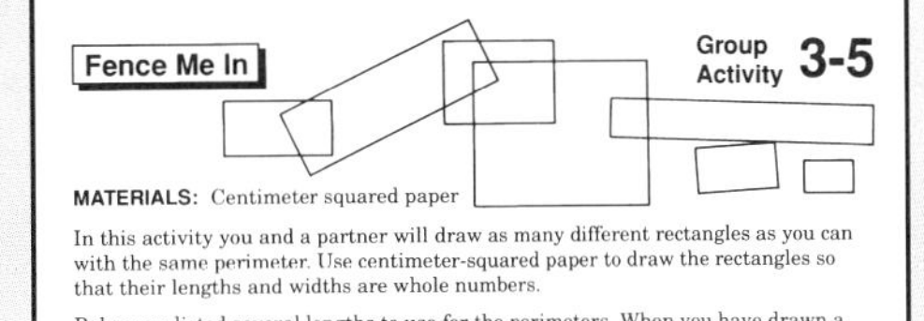

MATERIALS: Centimeter squared paper

In this activity you and a partner will draw as many different rectangles as you can with the same perimeter. Use centimeter-squared paper to draw the rectangles so that their lengths and widths are whole numbers.

Below are listed several lengths to use for the perimeters. When you have drawn a rectangle with a given perimeter, you are then to determine its area. It would be useful for this activity to record the areas and dimensions of the rectangles with a given perimeter.

When you have worked with all of the perimeters given, answer the questions given on the back.

PERIMETERS TO USE: 22 cm, 24 cm, 28 cm and 34 cm

 Pre-Algebra

4 ASSESS

Closing Activity

Modeling Cut out rectangles of various sizes and shapes and give one to each student. Then have each student find the perimeter and area of his or her rectangle by measuring the sides with a ruler and applying the perimeter and area formulas.

Additional Answer

37.

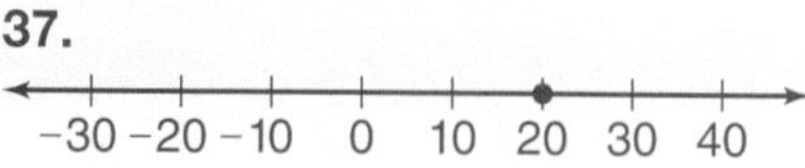

Enrichment Masters, p. 23

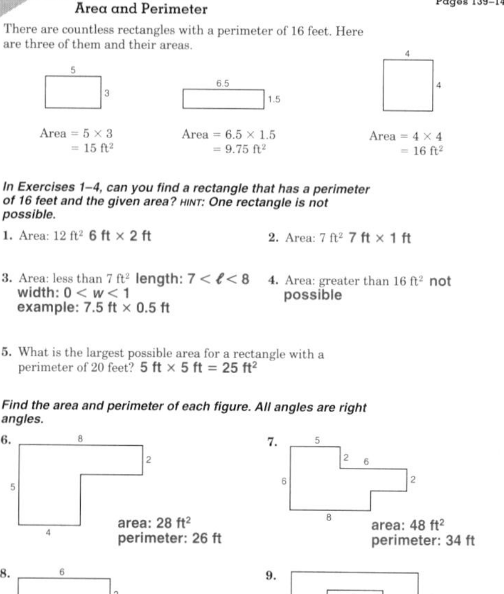

NAME ______________ DATE ______

3-5 Enrichment
Area and Perimeter

Student Edition
Pages 139–144

There are countless rectangles with a perimeter of 16 feet. Here are three of them and their areas.

Area = 5×3 = 15 ft²   Area = 6.5×1.5 = 9.75 ft²   Area = 4×4 = 16 ft²

In Exercises 1–4, can you find a rectangle that has a perimeter of 16 feet and the given area? HINT: One rectangle is not possible.

1. Area: 12 ft² **6 ft × 2 ft**
2. Area: 7 ft² **7 ft × 1 ft**
3. Area: less than 7 ft² **length: $7 < \ell < 8$ width: $0 < w < 1$ example: 7.5 ft × 0.5 ft**
4. Area: greater than 16 ft² **not possible**
5. What is the largest possible area for a rectangle with a perimeter of 20 feet? **5 ft × 5 ft = 25 ft²**

Find the area and perimeter of each figure. All angles are right angles.

6. **area: 28 ft² perimeter: 26 ft**
7. **area: 48 ft² perimeter: 34 ft**
8. **area: 42 ft² perimeter: 30 ft**
9. **area: 60 ft² perimeter: 48 ft**

33. **Kennels** Liam raises Irish setters at his home in the country. He wants to enclose an area for a run in which his dogs can exercise.
    a. If the dimensions of the run are 60 by 15 feet, how much fencing would he need? **150 ft**
    b. What is the area enclosed? **900 sq ft**
34. **Construction** Jackie Rockford wishes to add a 15-by-40-foot rectangular deck to her house. However, she needs to allow for two square wells for the large trees in the deck area. The wells are 5-by-5 feet each. How much area must Jackie allow for decking materials? **550 sq ft**
35. **Family Activity** Suppose you are going to add a deck to the back of your home or have a garden in a community plot. Design two different plans and give the area and perimeter of each option. **See student's work.**

Mixed Review

36. **Travel** Ana Perez travels 371 miles in 7 hours on a bus trip to a concert. What is the average rate of speed for the bus? (Lesson 3-4) **53 mph**
37. Solve $b + (-14) = 6$ and check your solution. Then, graph the solution on a number line. (Lesson 3-2) **20; see margin for graph.**
38. Evaluate $-6ab$, if $a = -3$ and $b = -5$. (Lesson 2-7) **−90**
39. **Cooking** A casserole made with ground beef has 375 Calories per serving. The same recipe made with ground turkey has 210 Calories per serving. Write an inequality using the numbers and $<$ or $>$. (Lesson 2-3) **$375 > 210$**
40. Solve $121 = 11x$ mentally. (Lesson 1-6) **11**
41. Simplify $6d + 14(d + 2)$. (Lesson 1-5) **$20d + 28$**
42. Evaluate $8 + 9 \cdot (7 - 4)$. (Lesson 1-2) **35**

From the FUNNY PAPERS

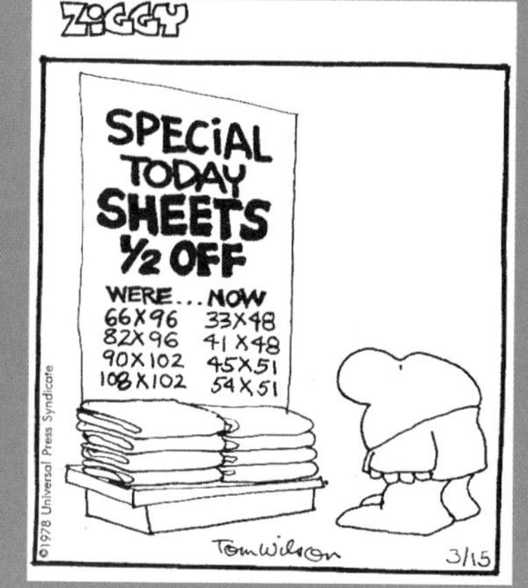

1. What does "half off" usually mean at a sale?
2. What is actually half off in this sale?
3. Draw a diagram of one of the sheets. Is the sheet half of the size that it was when it was not on sale? Explain. **1–3. See margin.**

From the FUNNY PAPERS

Have students create their own cartoon about a funny use of "half." Ideas might include children talking about getting the bigger half or about a toll bridge that is half done so it costs only half the usual toll.

Additional Answers From the Funny Papers

1. half off the price
2. the sizes of the sheets
3. See students' diagrams. The areas are one fourth of the areas of the sheets that are not on sale.

3-5B Area and Coordinates

An Extension of Lesson **3-5**

You have learned about rectangles and squares, but many figures have irregular shapes. Coordinates can be used to find the area of these shapes. Aerial surveyors use this technique to find areas of plots of land. Use the TI-82 graphing calculator program below to find the area of the figure at the right.

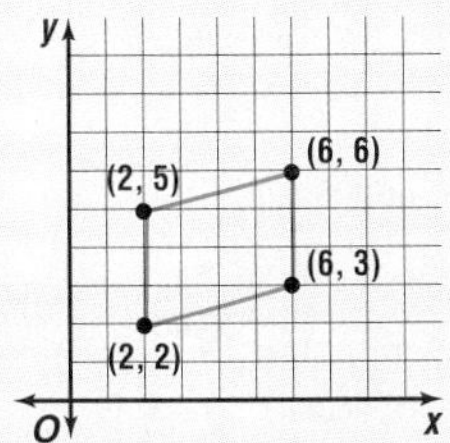

Activity

Before you run the program, choose one vertex as the first and list the ordered pairs for the vertices in counterclockwise order. Repeat the first ordered pair at the end of the list.

See page 1 for information on programming the TI-82.

(2, 2)
(6, 3)
(6, 6)
(2, 5)
(2, 2)

```
PROGRAM:AREA
: ClrList L1, L2
: 0 → D
: 0 → U
: 0 → X
: Input "VERTICES", N
: Disp "X COORDINATES"
: Input L1
: Disp "Y COORDINATES"
: Input L2
: For (X, 1, N, 1)
: L1(X)*L2(X+1)+D → D
: L2(X)*L1(X+1)+U → U
: End
: (D−U)/2 → A
: Disp "AREA ="
: Disp A
: Stop
```

The program will ask for the number of vertices in the figure. For this figure, press 4 and ENTER.

When asked for the *x*- and *y*-coordinates, enter the lists in counterclockwise order. Use braces, { }, before and after the lists.

The program will display area in square units.

Your Turn **Draw a coordinate plane and graph each figure whose vertices are listed. Then use the graphing calculator program to find the area of each figure.** **See Solutions Manual for graphs.**

1. (2, 3), (4, 1), (6, 8) **9 sq units**
2. (−4, 2), (5, 2), (5, 5), (−4, 5) **27 sq units**
3. (0, 4), (4, 0), (−4, 0), (0, −4) **32 sq units**
4. (0, 0), (8, 0), (9, 5), (6, 8), (−1, 4) **57 sq units**

Extension

5. Graph a rectangle or square on a coordinate grid and determine the coordinates of its vertices. Use the formula and the graphing calculator program to find its area. How do the methods compare? **Both give the same result.**

Technology

This lesson offers an excellent opportunity for using technology in your pre-algebra classroom. For more information on using technology, see *Graphing Calculators in the Mathematics Classroom,* one of the titles in the Glencoe Mathematics Professional Series.

3-5B LESSON NOTES

NCTM Standards: 1-4, 12, 13

Objective

Use a graphing calculator to explore the area of irregular figures on a coordinate plane.

Recommended Time

15 minutes

Instructional Resources

Graphing Calculator Masters, p. 17

This master provides keystroking instruction for this lesson for the TI-81 and Casio graphing calculators.

1 FOCUS

Motivating the Lesson

Have students draw a rectangle of any size on grid paper. Ask them how they could find the area of their rectangle. **Sample answers: use a formula, count the squares**

2 TEACH

Teaching Tip Some students may find it helpful to draw a flowchart of the steps the calculator uses to find the area of the parallelogram.

3 PRACTICE/APPLY

Assignment Guide

Core: 1–5
Enriched: 1–5

4 ASSESS

Observing students working with technology is an excellent method of assessment.

3-6 LESSON NOTES

NCTM Standards: 1-5, 9

Instructional Resources

- Study Guide Master 3-6
- Practice Master 3-6
- Enrichment Master 3-6
- Group Activity Card 3-6
- Assessment and Evaluation Master, p. 72
- Tech Prep Applications Masters, p. 6

Transparency 3-6A contains the 5-Minute Check for this lesson; **Transparency 3-6B** contains a teaching aid for this lesson.

Recommended Pacing	
Standard Pacing	Day 8 of 12
Honors Pacing	Day 7 of 11
Block Scheduling*	Day 5 of 7 (along with Lesson 3-7)

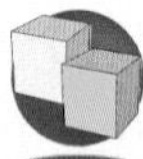

*For more information on pacing and possible lesson plans, refer to the **Block Scheduling Booklet**.

1 FOCUS

5-Minute Check
(over Lesson 3-5)

Find the perimeter and area of each rectangle.

1. rectangle with length of 10 feet and width of 5 feet **30 ft; 50 sq ft**
2. rectangle with length of 8 cm and width of 7 cm **30 cm; 56 sq cm**
3. square with each side 11 inches long **44 in.; 121 sq in.**
4. square with sides 1 kilometer long **4 km; 1 sq km**
5. rectangle 200 meters wide and 300 meters long **1000 m; 60,000 sq m**

3-6 Solving Inequalities by Adding or Subtracting

Setting Goals: *In this lesson, you'll solve inequalities by using the addition and subtraction properties of inequalities and graph the solution set.*

Modeling with Manipulatives

MATERIALS
- balance
- blocks
- paper cups

Small objects and a balance can be used to represent mathematical inequalities.

The diagram at the right shows $9 > 5$ since 9 objects are heavier than 5. The balance scale models an *inequality* because the two sides are not equal.

Let's see if the addition and subtraction properties of equality apply to inequalities.

CULTURAL CONNECTIONS
Balance scales were first used in Mesopotamia and Egypt around 2000 B.C.

First, set up your balance like the diagram. Remove 4 blocks from the left side. Remove 4 blocks from the right side. The scale returns to the position it was originally. This means that subtracting the same amount from each side does not change the inequality.

Your Turn Work with a partner. Model the inequality $2 < 5$ and draw a picture of your model. Add 6 blocks to the side with 2. Draw a picture of what happens to the balance. Next add 6 blocks to the side with 5. Draw a picture of the result. Write down your observations.

Use a paper cup and blocks with a balance to model the inequality $x + 3 > 6$.

a. How many blocks have to be in the paper cup to model this inequality? **4 or more**

b. Is this the only answer? **4, 5, and so on**

c. Remove the 3 blocks on the side with the variable and draw a picture of what happened. **See students' drawings.**

d. Next, remove 3 blocks from the other side.

e. What does this tell you about the variable? Is there more than one answer to this inequality? **$x > 3$; yes**

Learning the Concept

To solve an inequality that involves addition, you can use subtraction just as you did to solve equations. Likewise, you can use addition to solve an inequality that involves subtraction.

Alternative Teaching Strategies

Reading Mathematics Students may still confuse the < and > symbols. Remind them that the symbol always points to the lesser value. You may also use the elementary reminder that the symbol is like an open mouth and always opens toward the greater value.

CULTURAL CONNECTIONS
The first true precision balance was built in the early 1770s by British clockmaker John Harrison.

Addition and Subtraction Properties of Inequalities

In words: Adding or subtracting the same number from each side of an inequality does not change the truth of the inequality.

In symbols: For all numbers a, b, and c:
1. If $a > b$, then $a + c > b + c$ and $a - c > b - c$.
2. If $a < b$, then $a + c < b + c$ and $a - c < b - c$.

The rules are similar for $a \geq b$ and $a \leq b$.

Example  **Solve $m + 4 < 9$. Check the solution.**

$m + 4 < 9$
$m + 4 - 4 < 9 - 4$ *Subtract 4 from each side.*
$m < 5$

Check: Try −3, a number less than 5.

$-3 + 4 \overset{?}{<} 9$ *Replace m with −3.*
$1 < 9$ ✓ *This statement is true, so it checks.*

Any number less than 5 will make the statement true. Therefore, the solution is $m < 5$, all numbers less than 5.

Like solutions to equations, solutions to inequalities can be graphed. The solutions to $m < 5$ are graphed on the number line below.

The arrow represents all numbers to the left of (less than) 5. *The point, 5, is not included. Thus, the circle is open.*

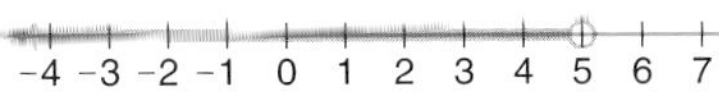

Example  **Solve $-13 \geq x - 8$ and check the solution. Graph the solution.**

$-13 \geq x - 8$
$-13 + 8 \geq x - 8 + 8$ *Add 8 to each side.*
$-5 \geq x$

Check: Try −5 and −7, numbers less than or equal to −5.

$-13 \overset{?}{\geq} -5 - 8$ $\quad$ $-13 \overset{?}{\geq} -7 - 8$
$-13 \geq -13$ ✓ *Both are true, so they check.* $-13 \geq -15$ ✓

The solution is $-5 \geq x$. This means that the inequality is true for all numbers less than or equal to −5. The solution may also be written as $x \leq -5$. The graph of the solution set is shown below.

The point, −5, is included. Thus, the circle is solid.

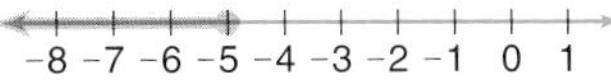

Motivating the Lesson

Questioning Draw the graphs of $y = -4$, $y < -4$, $y > -4$, and $y > -4$ on the board or overhead. Ask questions to elicit from students how each graph should be labeled with an inequality. Ask students which graph or graphs have −4 for a solution.

2 TEACH

In-Class Examples

For Example 1
Solve $s - 6 > -3$. $\mathbf{s > 3}$

For Example 2
Solve $12 \geq t + (-8)$. Graph the solution. $\mathbf{t \leq 20}$

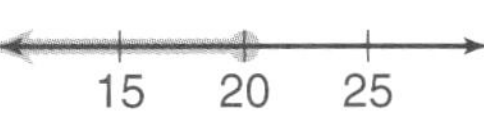

Teaching Tip If students have trouble working with the variable on the right side of the inequality, you may wish to suggest that they rewrite the problem with the variable on the left side before they begin solving. However, be sure to caution students to change the order of the inequality symbol as they change the variables and numbers. Point out that "2 is greater than x" and "x is less than 2" are equivalent statements.

In-Class Example

For Example 3
Alexandra knows that each box holds 800 baseball cards. She has 479 cards in one box. How many more cards will the box hold? **321 more cards**

1 FOCUS

Checking Your Understanding

Exercises 1–16 are designed to help you assess your students' understanding through reading, writing, speaking, and modeling. You should work through Exercises 1–5 with your students and then monitor their work on Exercises 6–16.

Additional Answers

1. Answers will vary. Sample answer: Add 6 to each side of the inequality to get m alone. The answers is $m < 9$.
2. Sample answer: You can add the same number to each side of an inequality and the inequality is still true.
3. The solution of $n + 5 = -3$ is $n = -8$. The solution of $n + 5 > -3$ is $n > -8$. This is all numbers greater than -8.

Study Guide Masters, p. 24

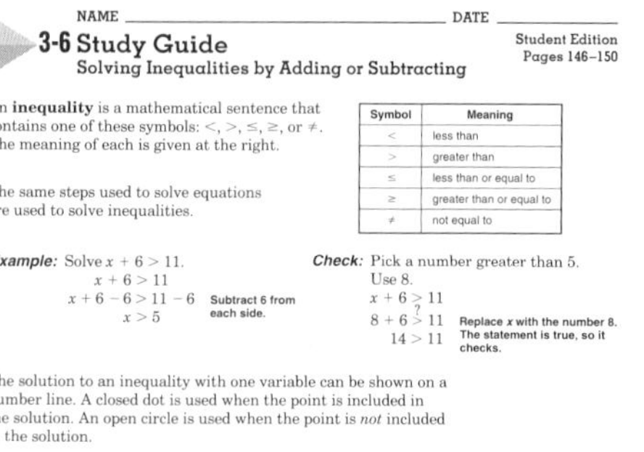

NAME ______________ DATE ________

3-6 Study Guide
Solving Inequalities by Adding or Subtracting

Student Edition
Pages 146–150

An **inequality** is a mathematical sentence that contains one of these symbols: $<$, $>$, $\leq$, $\geq$, or $\neq$. The meaning of each is given at the right.

Symbol	Meaning
$<$	less than
$>$	greater than
$\leq$	less than or equal to
$\geq$	greater than or equal to
$\neq$	not equal to

The same steps used to solve equations are used to solve inequalities.

Example: Solve $x + 6 > 11$.

$x + 6 > 11$
$x + 6 - 6 > 11 - 6$ **Subtract 6 from each side.**
$x > 5$

Check: Pick a number greater than 5. Use 8.

$x + 6 > 11$
$8 + 6 \overset{?}{>} 11$ **Replace x with the number 8.**
$14 > 11$ **The statement is true, so it checks.**

The solution to an inequality with one variable can be shown on a number line. A closed dot is used when the point is included in the solution. An open circle is used when the point is *not* included in the solution.

Examples: $x - 3 \leq -6$, $x \leq -3$ $x - 3 > -6$, $x > -3$

Solve each inequality and check your solution. Then graph the solution on the number line.

1. $-1 > x + 3$ **$x < -4$**
2. $t + 9 \geq 6$ **$t \geq -3$**
3. $-5 \leq r - 3$ **$r \geq -2$**
4. $-7 + m \geq -9$ **$m \geq -2$**
5. $b - 14 < -10$ **$b < 4$**
6. $8 + y \geq 6$ **$y \geq -2$**

 T 24 *Pre-Algebra*

Example 3 APPLICATION **Hobbies**

Christopher has \$30 to buy new baseball cards for his collection. He has chosen a Greg Maddux rookie card that costs \$8. What is the most he can spend on other baseball cards?

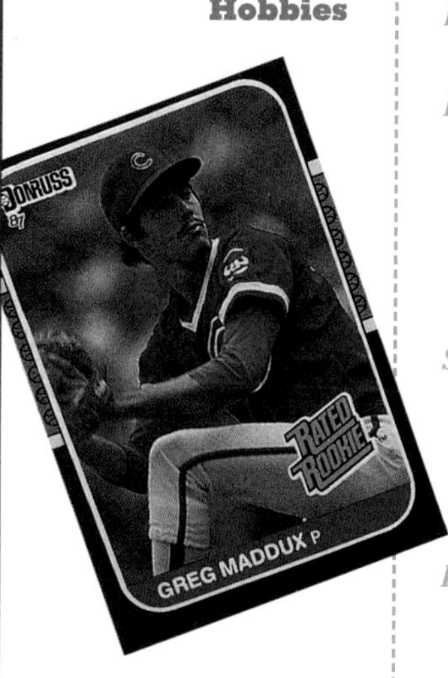

Explore We need to find out how much money he can spend on other new cards.

Plan Translate the words into an equation. Let s represent the amount he can spend on the other cards.

Cost of Maddux card	*plus*	*cost of other cards*	*must be less than or equal to*	*\$30*
8	+	s	$\leq$	30

Solve

$8 + s \leq 30$
$8 - 8 + s \leq 30 - 8$ *Subtract 8 from each side.*
$s \leq 22$

Christopher can spend no more than \$22 on other baseball cards.

Examine Suppose Christopher spent \$20, a number less than \$22, on other baseball cards. Then he would have spent \$20 + \$8 or \$28 in all. \$28 $\leq$ \$30, so the answer checks.

Checking Your Understanding

Communicating Mathematics

Read and study the lesson to answer these questions. **1–3. See margin.**

1. **Describe** how you would explain to a friend how to solve the inequality $m - 6 < 3$.
2. **Restate** in your own words the addition property of inequality.
3. **Compare and contrast** the solutions of $n + 5 = -3$ and $n + 5 > -3$.
4. **You Decide** Kelli translated the sentence *"Rachel scored no less than 8 points."* as $p < 8$. Derrick translated it as $p \geq 8$. Who is correct and why? **Derrick; *no less* means that many or more.**

5. **Assess Yourself** Measure your height and a friend's height. Write an inequality that compares your heights. Suppose that each of you were to grow 3 more inches. Now write an inequality that compares your new heights. **See students' work.**

6. $x > -2$
7. $y \leq -1$

Guided Practice

Write an inequality for each solution set graphed below.

8–9. See margin for graphs.

Write an inequality for each sentence. Then draw its graph.

8. Roberto spent more than \$7 for lunch. **$d > 7$**
9. Liz drove less than 75 miles today. **$d < 75$**

Reteaching

Using Models On the chalkboard or overhead, list several equations that involve only addition or subtraction. Beneath each equation, rewrite it as an inequality. For each pair, solve both the equation and the inequality, comparing the methods and solutions. Then graph the solutions for each pair.

Additional Answers

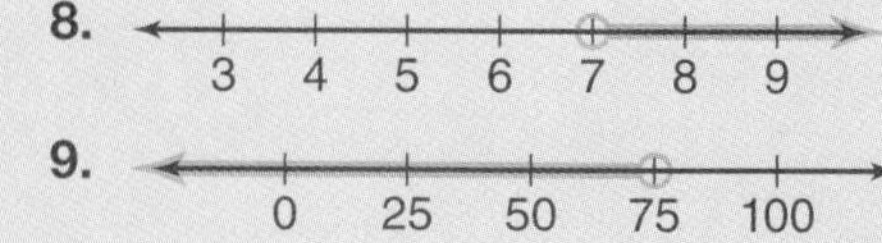

Solve each inequality and check your solution. Then graph the solution on a number line. 10–15. See margin for graphs.

10. $z + (-5) > -3$ $z > 2$
11. $m + 4 < -9$ $m < -13$
12. $9 < x - 7$ $x > 16$
13. $-6 > y - 2$ $y < -4$
14. $k + (-1) \geq -6$ $k \geq -5$
15. $18 + a \leq -13$ $a \leq -31$
16. **Money** Miguel spent more than \$6 at a fast food restaurant. If he had \$25 to start, what was his remaining pocket money? $25 - 6 > x$; $x < 19$

Exercises: Practicing and Applying the Concept

Independent Practice

Write an inequality for each solution set graphed below.

17. 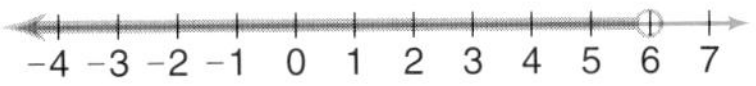 (number line −4 to 7)
18. 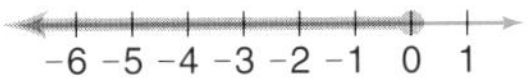 (number line −6 to 1)
19. 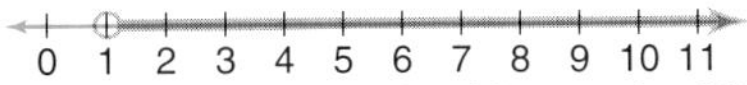 (number line 0 to 11)
20. 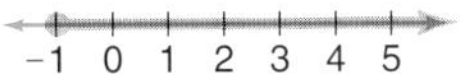 (number line −1 to 5)

17. $x < 6$ 18. $y \leq 0$ 19. $y > 1$ 20. $x \geq -1$

See margin for graphs.

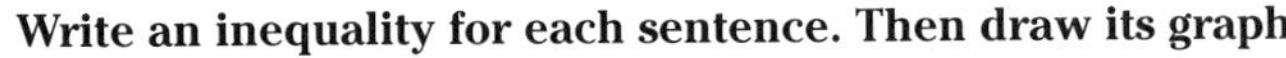

Write an inequality for each sentence. Then draw its graph.

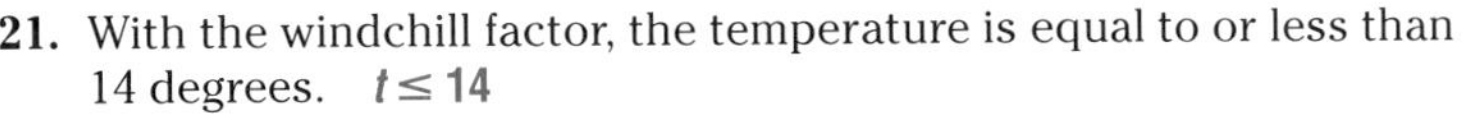

21. With the windchill factor, the temperature is equal to or less than 14 degrees. $t \leq 14$
22. Each of the NFL team rosters must have no more than 45 members at the beginning of the season. $r \leq 45$
23. Maria Vasquez is less than 5 ft. 4 in. (64 in.) tall. $m < 64$
24. The U.S. Constitution is more than 200 years old. $c > 200$

Solve each inequality and check your solution. Then graph the solution on a number line. See margin for solutions. See Solutions Manual for graphs.

25. $m - 6 < 13$
26. $r \leq -12 - 8$
27. $y + 7 > 13$
28. $k + 9 \geq 21$
29. $-4 + x > 23$
30. $14 \geq a + -2$
31. $-7 + x < -3$
32. $-13 + y \geq 8$
33. $m + (-2) > -11$
34. $f + (-5) \geq 14$
35. $19 \geq z + (-9)$
36. $-41 < m - 12$
37. $73 + k < 47$
38. $33 < m - (-6)$
39. $-31 \geq x + (-5)$
40. $22 < n - (-16)$
41. $-30 < c + (-5)$
42. $56 > w + 72$

Critical Thinking

43. Graph the solutions for the compound inequality $y < -1$ or $y > 3$. (*Hint:* An **or** in a sentence means that either part is true.) See Solutions Manual.
44. Graph the solutions for the compound inequality $x > -5$ and $x \leq 4$. (*Hint:* An **and** in a sentence means that both parts are true.) See Solutions Manual.

Applications and Problem Solving

45. **Shopping** Martin is saving money to buy a new mountain bike. Bikes that he likes start at \$375. If he already has saved \$285, what is the least amount he must still save? $x + 285 \geq 375$; $x \geq \$90$

Assignment Guide

Core: 17–47 odd, 48–55
Enriched: 18–42 even, 43–55

For **Extra Practice**, see p. 747.

The red A, B, and C flags, printed only in the Teacher's Wraparound Edition, indicate the level of difficulty of the exercises.

Additional Answers

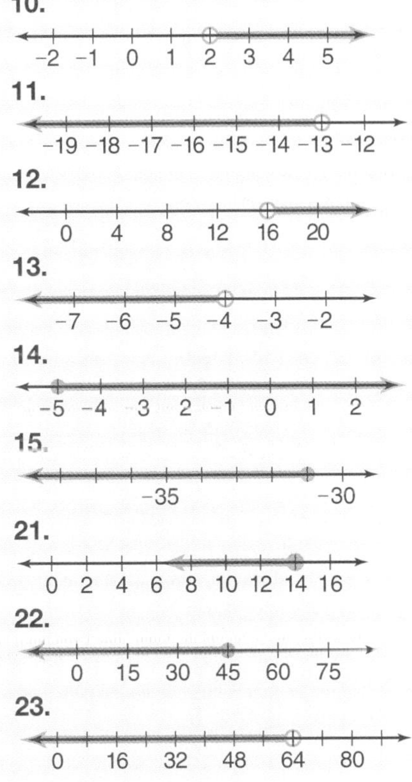

Practice Masters, p. 24

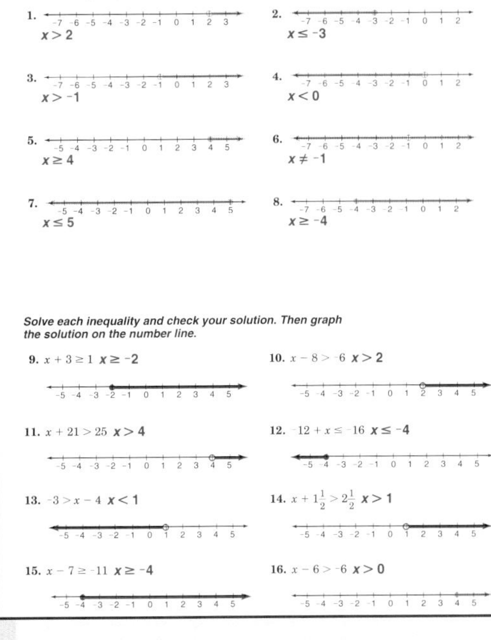

NAME ______ DATE ______

3-6 Practice
Solving Inequalities by Adding or Subtracting

Student Edition Pages 146–150

Write an inequality for each solution set graphed below.

1. $x > 2$
2. $x \leq -3$
3. $x > -1$
4. $x < 0$
5. $x \geq 4$
6. $x \neq -1$
7. $x \leq 5$
8. $x \geq -4$

Solve each inequality and check your solution. Then graph the solution on the number line.

9. $x + 3 \geq 1$ $x \geq -2$
10. $x - 8 > -6$ $x > 2$
11. $x + 21 > 25$ $x > 4$
12. $-12 + x \leq -16$ $x \leq -4$
13. $-3 > x - 4$ $x < 1$
14. $x + 1\frac{1}{2} > 2\frac{1}{2}$ $x > 1$
15. $x - 7 \geq -11$ $x \geq -4$
16. $x - 6 > -6$ $x > 0$

Group Activity Card 3-6

Inequality Detective Group Activity **3-6**

MATERIALS: 0-9 spinner • three dice • coin

Write inequalities for others in your group to solve. The inequalities you write should use only the operations of addition and subtraction and the symbols >, <, ≥, ≤.

To begin, group members roll three dice and flip the coin. From these, each person writes a 1-digit and a 2-digit number. If the coin is heads, both numbers are positive. If the coin is tails, one number is positive and the other number is negative. Next, spin the spinner and flip the coin to obtain the integer value that will be substituted for the variable in your inequalities. Using these values, write one correct inequality and one incorrect inequality. As a group, determine which are correct.

For example, if a 3, 5, and 6 are obtained on the dice and the coin gives a negative, you make an inequality from these numbers. If a 9 is the number from the spinner and the second flip of the coin is heads (positive), then +9 is the value to be substituted for the variable in the inequality. Hence, a correct inequality you could write is $x + (-35) < 6$, and an incorrect inequality could be $x + 6 < -35$.

 Pre-Algebra

Additional Answers

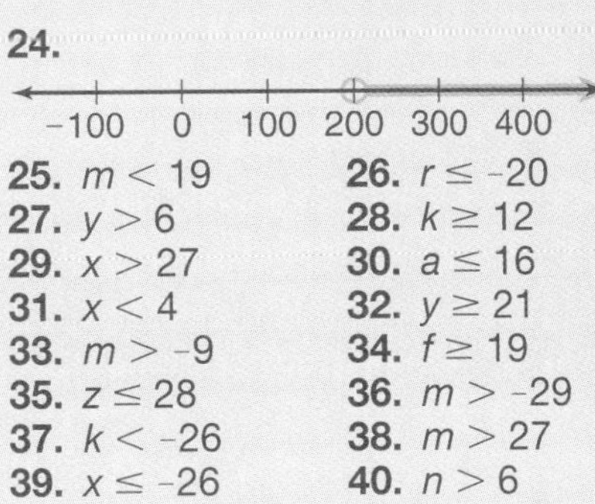

25. $m < 19$
26. $r \leq -20$
27. $y > 6$
28. $k \geq 12$
29. $x > 27$
30. $a \leq 16$
31. $x < 4$
32. $y \geq 21$
33. $m > -9$
34. $f \geq 19$
35. $z \leq 28$
36. $m > -29$
37. $k < -26$
38. $m > 27$
39. $x \leq -26$
40. $n > 6$
41. $c \geq -25$
42. $w < -16$

4 ASSESS

Closing Activity

Speaking Have students tell what number should be added to or subtracted from each side of the following inequalities to solve.

1. $y - 8 < 12$ **add 8**
2. $x + 3 > 1$ **subtract 3**
3. $n - (-2) < -4$ **add -2**
4. $7 > v + (-3)$ **subtract -3 or add 3**

Chapter 3, Quiz C (Lessons 3-5 and 3-6) is available in the *Assessment and Evaluation Masters*, p. 72.

Additional Answer

53b.

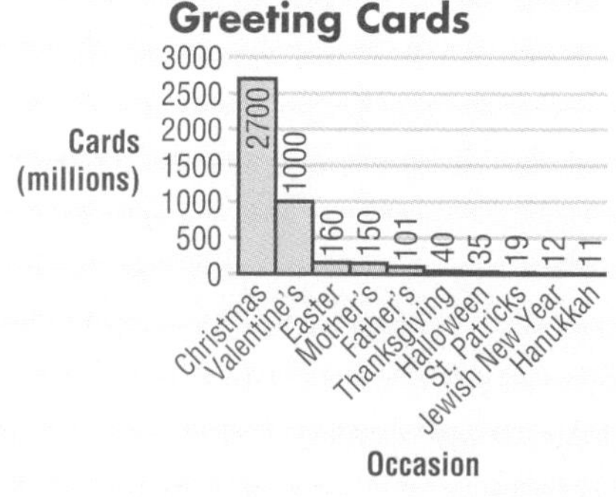

Enrichment Masters, p. 24

NAME ______________________ DATE __________

3-6 Enrichment — Student Edition Pages 146–150

Solving Inequalities Using Addition and Subtraction

Inequalities can be written many different ways in English.

English Phrase	Mathematical Phrase
at most x	$\leq x$
at least x	$\geq x$
5 less than x	$x - 5$
5 more than x	$x + 5$
5 is less than x	$5 < x$
x is between 4 and 6	$4 < x < 6$ or $x > 4$ and $x < 6$

1. Write a mathematical phrase for each of the following English phrases.
   a. 8 is no more than x. **$8 \leq x$**
   b. x is no less than 3. **$x \geq 3$**
   c. x is not between 6 and 12. **$x \leq 6$ or $x \geq 12$**
   d. x cannot exceed -5. **$x \leq -5$**
2. For what values of x, if any, is the mathematical phrase $5 > x > 7$ true? **none**

To solve compound inequalities, you must perform the same operation on each of the three parts of the inequality.

Example: Solve the compound inequality $3 < x - 2 < 5$.

$3 < x - 2 < 5$
$3 + 2 < x - 2 + 2 < 5 + 2$ — To get x by itself, add 2 to each part.
$5 < x < 7$

Thus, x is all numbers between 5 and 7.

Solve each compound inequality.

3. $-2 < x + 1 < 4$ **$-3 < x < 3$**
4. $-4 \leq x - 1 \leq 0$ **$-3 \leq x \leq 1$**
5. $3 < 15 + x < 10$ **$-12 < x < -5$**
6. $-12 < x - 3 \leq 1$ **$-9 < x \leq 4$**

46. **Grades** On three extra credit problems on a math test, students are given up to 4 points each, provided that their scores do not exceed 100 (points). What is the minimum raw score students could have and still score 100 with the extra credit? **$x + 3(4) \geq 100$; $x \geq 88$**

47. **Taxes** Residents of New York and Connecticut pay the highest taxes in the country. They must work 3 hours 9 minutes (189 minutes) of each 8-hour day to earn enough to pay their federal, state, and local taxes. The inequality $x + 52 \leq 189$ compares the time in these states with x, where x is the minimum time required in Alaska. What is the time needed in Alaska to pay off taxes? **$x \leq 137$ minutes or 2 hours 17 minutes**

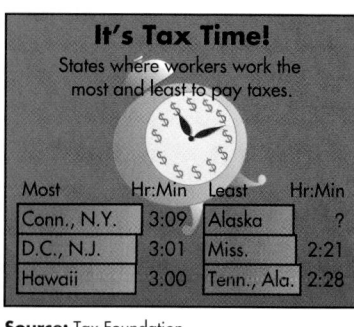

It's Tax Time!
States where workers work the most and least to pay taxes.

Most	Hr:Min	Least	Hr:Min
Conn., N.Y.	3:09	Alaska	?
D.C., N.J.	3:01	Miss.	2:21
Hawaii	3:00	Tenn., Ala.	2:28

Source: Tax Foundation

Mixed Review

48. **Landscaping** Cherita has a rectangular garden that is 96 square feet. If the garden is 8 feet long, what is its width? (Lesson 3-5) **12 feet**

49. Solve $f = t - h$, if $t = 125$ and $h = 25$ by replacing the variables with the given values. (Lesson 3-4) **100**

50. **Business** Jennifer made a \$150 down payment for soccer camp. Her unpaid balance was \$300. What is the total fee for the soccer camp? (Lesson 3-2) **\$450**

51. Simplify $10a - 12a$. (Lesson 2-5) **$-2a$**

52. Find the sum of -12 and 7. (Lesson 2-4) **-5**

53. **Communication** The table at the right shows the numbers of millions of different types of greeting cards sent in the United States each year. (Lesson 1-10) **53a. 49 million** **53b. See margin.**
    a. How many more cards are sent for Mother's Day than are sent for Father's Day?
    b. Make a bar graph of the data.

Occasion	Estimated number sent (millions)
Christmas	2700
Valentine's Day	1000
Easter	160
Mother's Day	150
Father's Day	101
Thanksgiving	40
Halloween	35
St. Patrick's Day	19
Jewish New Year	12
Hanukkah	11

54. Solve $\frac{36}{b} = -4$. (Lesson 1-6) **-9**

55. Name the property shown by $8 \cdot 1 = 8$. (Lesson 1-4) **Identity, ×**

Extension

Using Guess and Check Have students try to graph the solutions to each of these inequalities. Have them test various points on their graphs to check their answers.

a. $|x| > 2$ b. $|x| < 2$
c. $|x| < -2$

3-7 Solving Inequalities by Multiplying or Dividing

Setting Goals: *In this lesson, you'll solve inequalities by using the multiplication and division properties of inequalities.*

Modeling a Real-World Application: Recycling

Environmental protection is an ever-growing concern for the world. When we pollute the environment, we are polluting *our* living space. Recycling materials is a good way to keep our habitat cleaner.

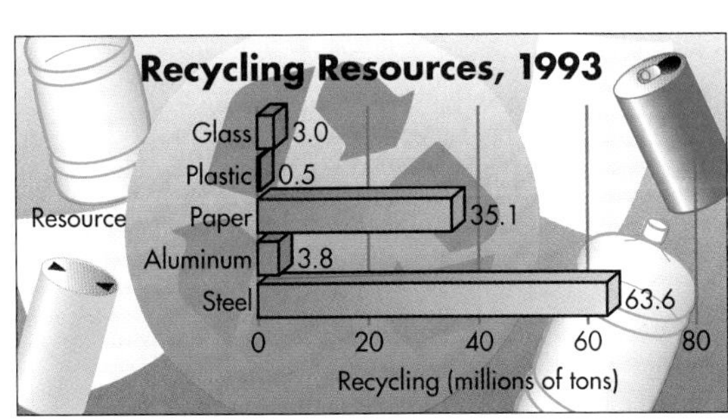

Source: Bureau of Mines

By weight, the reuse of steel and paper lead the way. The recycling of steel, by weight, is greater than that of paper.

Learning the Concept

The comparison of paper to steel recycling provides a way to look at inequalities. Rounding the weights, 35 million tons of paper being recycled is less than 64 million tons of steel being recycled. In symbols, $35 < 64$.

It is estimated that people use four times as much paper and steel as they recycle. Does the inequality between paper and steel still keep its order for the amount used?

$$35 \times 4 \stackrel{?}{<} 64 \times 4$$
$$140 < 256 \checkmark$$ *The inequality is still true.*

Another way people can help the environment is to use as little of these products as possible. What if people cut the amount of paper and steel that they used in half? Would this affect the inequality?

$140 < 256$ — *The original amount of tons of paper and steel used.*

$\frac{140}{2} \stackrel{?}{<} \frac{256}{2}$ — *Each amount is divided by 2.*

$70 < 128 \checkmark$ — *The inequality is still true.*

These examples suggest the properties on the next page.

3-7 LESSON NOTES

NCTM Standards: 1-5, 9

Instructional Resources

- Study Guide Master 3-7
- Practice Master 3-7
- Enrichment Master 3-7
- Group Activity Card 3-7
- Graphing Calculator Masters, p. 3
- Real-World Applications, 6

Transparency 3-7A contains the 5-Minute Check for this lesson; **Transparency 3-7B** contains a teaching aid for this lesson.

Recommended Pacing	
Standard Pacing	Day 9 of 12
Honors Pacing	Day 8 of 11
Block Scheduling*	Day 5 of 7 (along with Lesson 3-6)

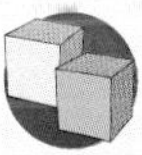

*For more information on pacing and possible lesson plans, refer to the ***Block Scheduling Booklet***.

1 FOCUS

5-Minute Check *(over Lesson 3-6)*

Solve each inequality.

1. $r - 9 < 4$ **$r < 13$**
2. $s + 24 > 23$ **$s > -1$**
3. $t - (-30) \leq 40$ **$t \leq 10$**
4. $u + (-14) \geq -29$ **$u \geq -15$**
5. $19 < v - 9$ **$v > 28$**

Motivating the Lesson

Questioning Ask which is heavier, a pound of feathers or a pound of iron. After allowing for any discussion that occurs, point out that of course a pound is a pound, so both weigh the same. Ask which would take more space, a pound of feathers or a pound of iron. **feathers**

2 TEACH

In-Class Example

For Example 1
Solve each inequality and check the solution. Then graph the solution on a number line.

a. $-56 \geq 7x$ $x \leq -8$

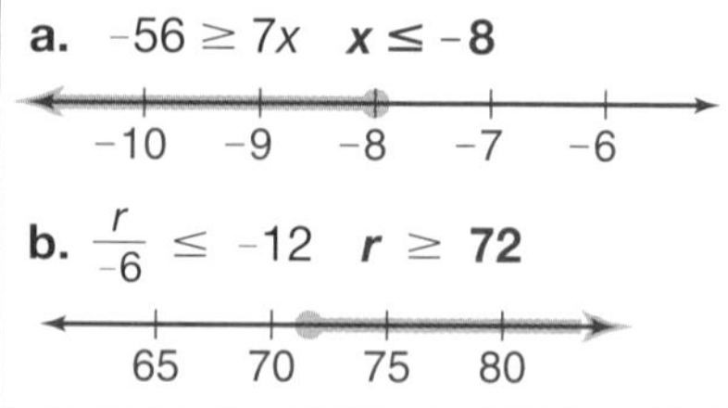

b. $\frac{r}{-6} \leq -12$ $r \geq 72$

Teaching Tip Suggest that students choose a negative value, a positive value, and 0 when checking solutions. Values that do not make the original inequality true should not be included in the solution set.

Multiplication and Division Properties of Inequalities

In words: When you multiply or divide each side of a true inequality by a *positive* integer, the result remains true.

In symbols: For all integers a, b, and c, where $c > 0$, if $a > b$, then $a \cdot c > b \cdot c$ and $\frac{a}{c} > \frac{b}{c}$.

The rule is similar for $a < b$, $a \geq b$, and $a \leq b$.

Next, we need to consider if an inequality remains true if we multiply or divide each side by a negative number. Consider the inequality $6 < 10$.

Multiply each side by −1.

$6 < 10$
$6(-1) \overset{?}{<} 10(-1)$
$-6 < -10$ **False**

Divide each side by −2.

$6 < 10$
$\frac{6}{-2} < \frac{10}{-2}$
$-3 < -5$ **False**

The inequalities $-6 < -10$ and $-3 < -5$ are both false. However, notice that both inequalities would be true if you reverse the order symbol. That is, change $<$ to $>$.

$-6 > -10$ and $-3 > -5$ are both true.

These examples suggest the following properties.

Multiplication and Division Properties of Inequalities

In words: When you multiply or divide each side of an inequality by a *negative* integer, you must *reverse the order symbol*.

In symbols: For all integers a, b, and c, where $c < 0$, if $a > b$, then $a \cdot c < b \cdot c$ and $\frac{a}{c} < \frac{b}{c}$.

The rule is similar for $a < b$, $a \geq b$, and $a \leq b$.

Example 1 **Solve each inequality and check the solution. Then graph the solution on a number line.**

a. $4x < -28$

$4x < -28$
$\frac{4x}{4} < \frac{-28}{4}$ *Divide each side by 4.*
$x < -7$

Check: Try −10, a number less than −7.
$4(-10) \overset{?}{<} -28$
$-40 < -28$  *This is true.*

The solution is $x < -7$, all numbers less than −7. The graph is shown below.

Cooperative Learning

Pairs Check Have students work in pairs to complete Exercises 7–12. One student works one problem while the other coaches and observes. When the problem is solved to the satisfaction of both students, the roles are reversed. After all problems are completed, answers can be checked with another pair of students.

For more information on this strategy, see *Cooperative Learning in the Mathematics Classroom*, one of the titles in the Glencoe Mathematics Professional Series, p. 12.

b. $\frac{m}{-3} \leq 4$

$$\frac{m}{-3} \leq 4$$

$$\frac{m}{-3} \times (-3) \geq 4 \times (-3) \quad \textit{Multiply each side by } -3 \textit{ and reverse the order symbol.}$$

$$m \geq -12$$

Check: Try 0, a number greater than -12.

$$\frac{m}{-3} \leq 4$$

$$\frac{0}{-3} \stackrel{?}{\leq} 4 \quad \textit{Replace m with 0.}$$

$$0 \leq 4 \checkmark$$

The solution is all numbers greater than or equal to -12. The graph is shown below.

−15 −14 −13 −12 −11 −10 −9 −8

Inequalities are often used to solve problems involving money.

Example 2 APPLICATION

Personal Finance

Carlissa earns $5.00 per hour busing tables at The Italian Eatery Restaurant. How many hours must she work to earn at least $110 each week?

Let x represent the hours required to earn at least $110.

$$5x \geq 110$$

$$\frac{5x}{5} \geq \frac{110}{5} \quad \textit{Divide each side by 5.}$$

$$x \geq 22$$

Carlissa must work at least 22 hours to earn at least $110 a week.

In-Class Example

For Example 2

Maria saves $6 each week. How many weeks must she save money to be able to buy a coat that costs at least $48.95?

at least 9 weeks

Checking Your Understanding

Communicating Mathematics

Read and study the lesson to answer these questions.

1. **Write** an inequality that is always true. **Sample answer: $5 > 2$**
2. **Write** an inequality that would be solved using division where the solution is $x < 4$. **Sample answer: $5x < 20$**
3. **Compare and contrast** solving an inequality involving multiplication and solving an inequality involving division. **See margin.**

MATH JOURNAL

4. **Write** a note to a classmate about how to solve the inequalities $5x < 35$ and $\frac{y}{3} \geq 12$. **See students' work.**

3 PRACTICE/APPLY

Checking Your Understanding

Exercises 1–14 are designed to help you assess your students' understanding through reading, writing, speaking, and modeling. You should work through Exercises 1–4 with your students and then monitor their work on Exercises 5–14.

Additional Answer

3. When you solve inequalities involving multiplication or division, you must do the same operation to each side of the equation. When you solve an inequality involving multiplication, you divide each side by the coefficient. When you solve an inequality involving division, you multiply each side by the divisor.

Reteaching

Using Manipulatives To help students remember to change the order of the inequality symbol when multiplying or dividing by a negative number, provide them with cards as shown. Then, when students multiply or divide by a negative number, they physically turn the card upside down.

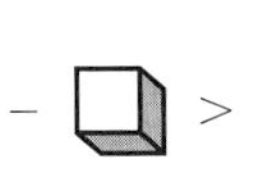

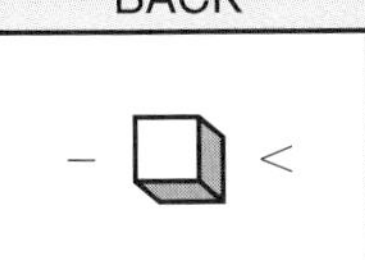

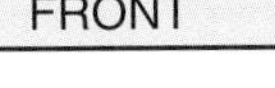

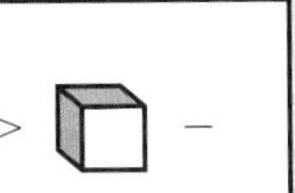

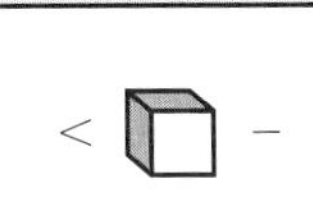

FRONT BACK

FRONT BACK

Study Guide Masters, p. 25

NAME ______________ DATE ______

3-7 Study Guide
Solving Inequalities by Multiplying or Dividing

Student Edition Pages 151–155

When you multiply or divide each side of an inequality by a *positive* number, you get a new inequality with the same solutions.

$$3h < -12 \qquad \frac{h}{5} > 10$$
$$3h \div 3 < -12 \div 3 \qquad \frac{h}{5} \cdot 5 > 10 \cdot 5$$
$$h < -4 \qquad h > 50$$

When you multiply or divide each side by a *negative* number, you must reverse the inequality symbol. Otherwise, the new inequality will not have the same solutions.

$$-3h < -12 \qquad \frac{h}{-5} > 10$$
$$-3h \div (-3) > -12 \div (-3) \qquad \frac{h}{-5} \cdot (-5) < 10 \cdot (-5)$$
$$h > 4 \qquad h < -50$$

Do the two inequalities have the same solutions? Write yes or no.

1. $2x < 14$, $x > 7$ **no**
2. $-x < 0$, $x > 0$ **yes**
3. $3x < 9$, $x < 3$ **yes**
4. $-5x > 0$, $x > 0$ **no**
5. $-4x < 4$, $x > -1$ **yes**
6. $-3x > -3$, $x > 1$ **no**

Solve each inequality and check your solution.

7. $7x < 84$ **$x < 12$**
8. $9x > 81$ **$x > 9$**
9. $\frac{h}{3} < -10$ **$h < -30$**
10. $6p < 12$ **$p < 2$**
11. $\frac{h}{4} > -7$ **$h > -28$**
12. $0 > -5c$ **$c > 0$**
13. $-2d > 4$ **$d < -2$**
14. $-2d > -4$ **$d < 2$**
15. $-2d < -4$ **$d > 2$**
16. $\frac{a}{-3} < 9$ **$a > -27$**
17. $\frac{a}{-3} > -9$ **$a < 27$**
18. $\frac{a}{3} < -9$ **$a < -27$**

Assignment Guide

Core: 15–37 odd, 39–45
Enriched: 16–32 even, 33–45

For **Extra Practice**, see p. 747.

The red A, B, and C flags, printed only in the Teacher's Wraparound Edition, indicate the level of difficulty of the exercises.

Additional Answers

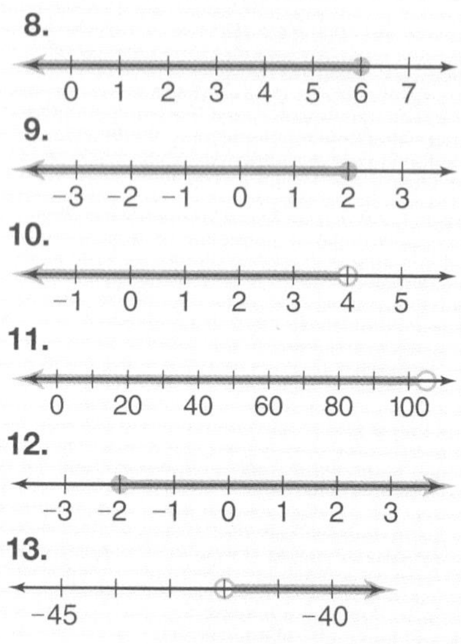

Guided Practice

State the number needed to multiply or divide each side by to solve each inequality. Then tell whether the order symbol should be reversed.

5. $-15 \leq 5a$ 5, no **6.** $\frac{y}{4} > -20$ 4, no **7.** $-3m < -33$ −3, yes

Solve each inequality and check your solution. Then graph the solution on a number line. See margin for graphs.

8. $-3y \geq -18$ $y \leq 6$ **9.** $-8 \leq -4x$ $x \leq 2$ **10.** $\frac{z}{-2} > -2$ $z < 4$

11. $\frac{p}{-3} > -35$ $p < 105$ **12.** $13y \geq -26$ $y \geq -2$ **13.** $6 > \frac{a}{-7}$ $a > -42$

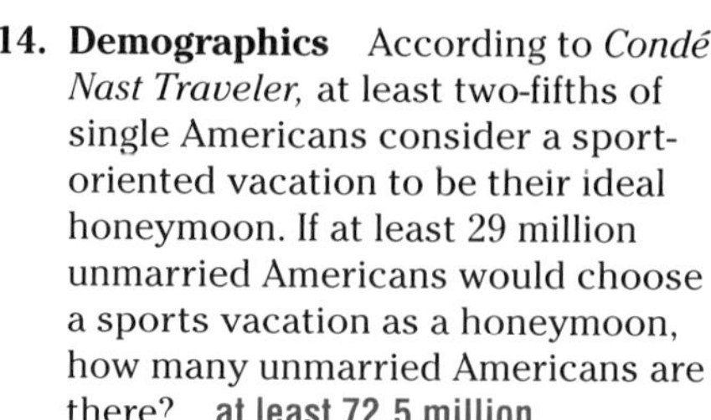

14. Demographics According to *Condé Nast Traveler,* at least two-fifths of single Americans consider a sport-oriented vacation to be their ideal honeymoon. If at least 29 million unmarried Americans would choose a sports vacation as a honeymoon, how many unmarried Americans are there? at least 72.5 million

Exercises: Practicing and Applying the Concept

Independent Practice

Solve each inequality and check your solution. Then graph the solution on a number line.

15–32. See Solutions Manual for graphs.

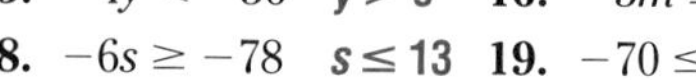

A **15.** $-4y < -36$ $y > 9$ **16.** $-3m \geq -24$ $m \leq 8$ **17.** $-72 > -4x$ $x > 18$

18. $-6s \geq -78$ $s \leq 13$ **19.** $-70 \leq 5p$ $p \geq -14$ **20.** $9x \geq -72$ $x \geq -8$

B **21.** $\frac{x}{3} < -7$ $x < -21$ **22.** $-8 > \frac{y}{4}$ $y < -32$ **23.** $\frac{m}{-2} \leq -7$ $m \geq 14$

24. $\frac{p}{-4} < 20$ $p > -80$ **25.** $-7 \leq \frac{z}{14}$ $z \geq -98$ **26.** $-13a \geq -273$ $a \leq 21$

C **27.** $-378 < 18r$ $r > -21$ **28.** $\frac{a}{14} \leq -17$ $a \leq -238$ **29.** $-14 \leq \frac{s}{-23}$ $s \leq 322$

30. $\frac{g}{-21} > 33$ $g < -693$ **31.** $-4 > \frac{x}{3}$ $x < -12$ **32.** $-25a \leq 400$ $a \geq -16$

Critical Thinking

33. The product of an integer and −4 is greater than −24. Find the greatest integer that meets this condition. 5

Applications and Problem Solving

34. Statistics The Boston Marathon had more than 2,600,000 spectators along its 26-mile route. What was the average number of spectators per mile? $26x > 2{,}600{,}000$; $x > 100{,}000$

35. Paper Recycling A giant roll of paper towels made from recycled paper can be cut into 80,000 small rolls for retail sale. If the smallest carton for shipment to a retail store holds 100 small rolls, what is the maximum number of cartons required to package and ship the rolls of towels made from one giant roll of towels? 800

Practice Masters, p. 25

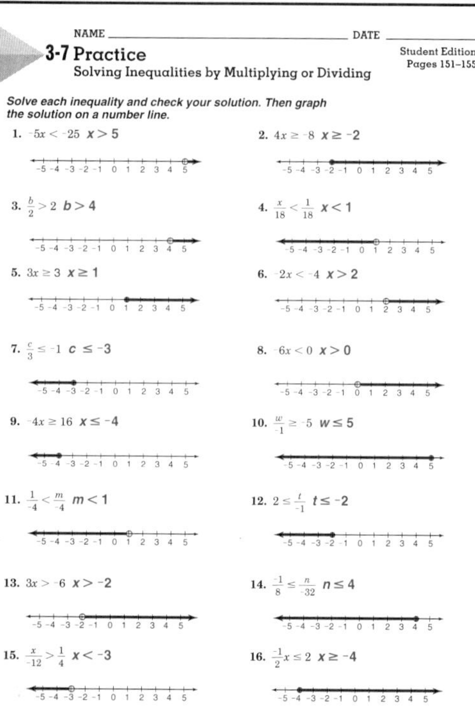

NAME ______ DATE ______

3-7 Practice
Solving Inequalities by Multiplying or Dividing

Student Edition Pages 151–155

Solve each inequality and check your solution. Then graph the solution on a number line.

1. $^-5x < {}^-25$ $x > 5$
2. $4x \geq {}^-8$ $x \geq -2$
3. $\frac{b}{2} > 2$ $b > 4$
4. $\frac{x}{18} < \frac{1}{18}$ $x < 1$
5. $3x \geq 3$ $x \geq 1$
6. $^-2x < {}^-4$ $x > 2$
7. $\frac{c}{3} \leq {}^-1$ $c \leq -3$
8. $^-6x < 0$ $x > 0$
9. $^-4x \geq 16$ $x \leq -4$
10. $\frac{w}{^-1} \geq {}^-5$ $w \leq 5$
11. $\frac{1}{^-4} < \frac{m}{^-4}$ $m < 1$
12. $2 \leq \frac{t}{^-1}$ $t \leq -2$
13. $3x > {}^-6$ $x > -2$
14. $\frac{^-1}{8} \leq \frac{n}{^-32}$ $n \leq 4$
15. $\frac{x}{^-12} > \frac{1}{4}$ $x < -3$
16. $\frac{^-1}{2}x \leq 2$ $x \geq -4$

Group Activity Card 3-7

Hit That Range Group Activity **3-7**

MATERIALS: Three dice • a coin

To begin the activity roll the three dice and flip the coin (to get a positive or negative sign). Using the numbers and sign obtained, each group member writes an inequality involving multiplication or division and one of the inequality symbols (>, <, ≥, ≤.). The three digits are to be used to make a 1-digit and a 2-digit number in the expression. If the flip of the coin indicates a negative, then one of the numbers must be negative; otherwise they are both positive.

Everyone must write an inequality that satisfies either of the following conditions: the solution set for the inequality begins at a number between -5 and 5, or the solution set for the inequality begins at a number whose absolute value is greater than 100.

After writing your inequality, exchange it for someone else to solve. That person must check to see if the solution to the inequality satisfies the conditions given above.

 Pre-Algebra

36. **Pets** For those who like a well-dressed pet, NuBreed Pet Hats in Batavia, Illinois, makes baseball, cowboy, beach, and other hats for dogs and cats. The hats cost \$10–20 in retail stores.
   a. Write an inequality for the lowest possible retail cost of 15 pet hats. **$x \geq \$150$**
   b. Write an inequality for the greatest possible retail cost of 15 pet hats. **$x \leq \$300$**

37. **Lifestyles** According to a study by *Men's Fitness* magazine, the average American spends six months of his or her life waiting at stop lights. More than twelve times that amount of time is spent eating. How long does the average American spend eating in a lifetime? **$t > 6$ years**

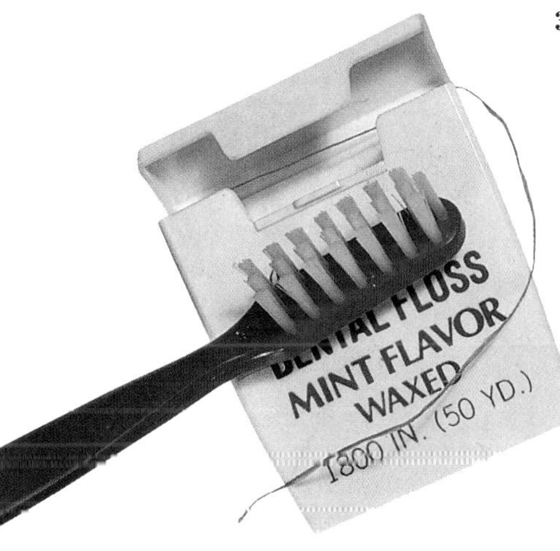

38. **Health** Americans bought almost 59,000 *miles* of dental floss in 1992.
   a. There are 5280 feet in a mile. How many feet of dental floss were purchased in 1992? **$f < 311{,}520{,}000$**
   b. If there were about 248,710,000 Americans in 1992, how much dental floss did the average American buy? **$p < 1.25$ feet**
   c. Dentists recommend that people floss their teeth once each day. Based on your answer to part b, do you think most Americans follow this recommendation? Explain. **See margin.**

Mixed Review

39. Solve $m + (-5) < 23$. (Lesson 3-6) **$m < 28$**
40. **Pets** Find the perimeter and area of a square dog kennel that is 6.5 feet on each side. (Lesson 3-5) **26 feet, 42.25 sq ft**
41. If $x + 5 = \frac{1}{3}(3x - 5)$, what is the value of x? (Lesson 3-1) **d**
   a. 5　　b. any real number
   c. 3　　d. no real number
42. Find the quotient of -126 and 3. (Lesson 2-8) **-42**
43. Find the product $8 \cdot (-7y)$. (Lesson 2-7) **$-56y$**
44. Find the value of $-|-4|$. (Lesson 2-1) **-4**
45. **Grades** When finding a grade point average, an A is worth 4 points, a B is worth 3 points, a C is worth 2 points, a D is 1 point, and an F is 0. After totaling the points, the sum is divided by the number of grades. (Lesson 1-2)
   a. Write an expression for the grade point average of a student who received 3 As, 2 Bs, 1 C, and 1 D.
   b. Find the grade point average. **3.0**

45a. $[(3 \times 4) + (2 \times 3) + (1 \times 2) + (1 \times 1)] \div 7$

Extension

Using Mental Math Have nine students stand side-by-side in a line at the front of the room to form a human number line from -4 to 4. Call out an inequality. Students in the line raise their hand if they think they are a solution to the inequality. Other students solve the inequality and check that students have correctly raised their hands. After solving a few inequalities, have new students form the number line.

4 ASSESS

Closing Activity

Modeling Have students model inequalities on a balance scale by placing unequal, but even, numbers of counters in each pan. Demonstrate how the multiplication and division properties of equality are true for inequalities by doubling and halving the counters in each pan.

Additional Answer

38c. No. The average American buys less than 1.25 feet of floss each year. That amount would allow you to floss once or twice, but not each day.

Enrichment Masters, p. 25

NAME ________________ DATE ________

3-7 Enrichment
Hidden Word

Student Edition
Pages 151–155

In each group of five inequalities, only two have the same solution set. For each group, write the solution of each inequality and then circle the letters of the two inequalities having the same solution set. After completing all four groups, use the circled letters to form a one-word answer to the question at the bottom of the page.

GROUP 1

(P.) $3x < 30$	B. $-3x < -30$	F. $-30 > 3x$	(O.) $-30 < -3x$	D. $-3x < 30$
$x < 10$	$x > 10$	$-10 > x$	$10 > x$	$x > -10$

GROUP 2

M. $\frac{x}{5} \leq -2$	N. $\frac{-x}{5} \leq 2$	(I.) $\frac{x}{-5} \leq -2$	(H.) $\frac{-x}{2} \leq -5$	R. $\frac{-x}{5} < -2$
$x \leq -10$	$x \geq -10$	$x \geq 10$	$x \geq 10$	$x > 10$

GROUP 3

Q. $\frac{3x}{2} < 0$	K. $0 \leq -12x$	L. $\frac{-1}{x} \leq 0$	(C.) $0.73x \geq 0$	(N.) $0 \geq -912x$
$x < 0$	$0 \geq x$	$x > 0$	$x \geq 0$	$0 \leq x$

GROUP 4

L. $200 > 0.05x$	(A.) $0.05x < 20$	D. $\frac{1}{5}x > 20$	E. $200 > 5x$	(M.) $\frac{x}{0.1} < 4000$
$4000 > x$	$x < 4000$	$x > 100$	$40 > x$	$x < 400$

QUESTION: What was the name of Gene Autry's horse on *The Gene Autry Show*? **Champion**

3-8 LESSON NOTES

NCTM Standards: 1-4, 9

Instructional Resources

- Study Guide Master 3-8
- Practice Master 3-8
- Enrichment Master 3-8
- Group Activity Card 3-8
- Assessment and Evaluation Masters, p. 72
- Real-World Applications, 7

Transparency 3-8A contains the 5-Minute Check for this lesson; **Transparency 3-8B** contains a teaching aid for this lesson.

Recommended Pacing	
Standard Pacing	Day 10 of 12
Honors Pacing	Day 9 of 11
Block Scheduling*	Day 6 of 7

*For more information on pacing and possible lesson plans, refer to the ***Block Scheduling Booklet***.

1 FOCUS

5-Minute Check
(over Lesson 3-7)

Solve each inequality and check the solution. Then graph the solution on a number line.

1. $-2x > 4$ $\mathbf{x < -2}$

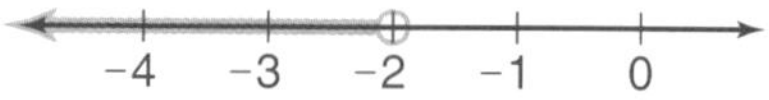

2. $3x < -15$ $\mathbf{x < -5}$

3. $\frac{x}{-4} \leq 0$ $\mathbf{x \geq 0}$

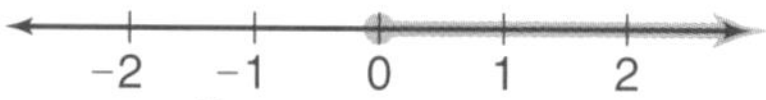

4. $6 \leq \frac{x}{3}$ $\mathbf{x \geq 18}$

Motivating the Lesson

Situational Problem Have students write and solve an inequality for the problem.

- My age plus 2 is greater than Bill's. Bill is 10 years old. What can you say about my age? **age > 8 years**

3-8 Applying Equations and Inequalities

Setting Goals: *In this lesson, you'll solve verbal problems by translating them into equations and inequalities, and write verbal problems for equations and inequalities.*

Modeling a Real-World Application: World Records

What takes 400 pounds of butter, 500 volunteers, and 3 blocks of tables? The World's Biggest Breakfast!

Springfield, Massachusetts, and Battle Creek, Michigan, compete every year to serve the world's largest pancake breakfast. In 1994, 66,344 people were served 132,688 pancakes in two hours in Springfield. Each person was served the same number of pancakes. How many pancakes were served to each person?

Learning the Concept

We can use the four-step plan to solve this problem.

Explore We know how many pancakes were served to how many people. We want to find out how many pancakes each person had.

Plan Translate the words into an equation. Let p represent the number of pancakes served per person.

number of people served	*times*	*number of pancakes per person*	*equal*	*total number of pancakes*
66,344	×	p	=	132,688

Solve

$$66{,}344p = 132{,}688$$

$$\frac{66{,}344p}{66{,}344} = \frac{132{,}688}{66{,}344} \quad \textit{Estimate: } \frac{130{,}000}{65{,}000} = 2.$$

$$p = 2 \quad \textit{Use a calculator.}$$

Examine If each of the 66,344 people were served 2 pancakes, $66{,}344 \times 2$ or 132,688 pancakes were served. The answer checks.

Each person was served 2 pancakes.

Example 1 APPLICATION Sports

Special Olympics athlete, Andy Leonard, is one of the few American athletes who can lift up to four times his own body weight. The most weight he has lifted is 385 pounds. What is his body weight, rounded to the nearest whole number?

Explore Mr. Leonard's maximum lift is 385 pounds, which is up to four times his weight.

Plan First, estimate the solution by rounding.

$\frac{400}{4} = 100$ pounds

Translate the words into an inequality. Let w represent Mr. Leonard's body weight.

lifting weight	*is less than or equal to*	*4 times body weight*
385	$\leq$	$4w$

Solve

$385 \leq 4w$

$\frac{385}{4} \leq \frac{4w}{4}$ *Divide each side by 4.*

$96.25 \leq w$ *Use a calculator to divide 385 by 4.*

Examine Compare to the estimate. The answer appears to be reasonable. Therefore, Mr. Leonard's weight is about 96 pounds or less.

Many types of situations can be represented using an equation.

Example 2 APPLICATION Entertainment

Write a realistic problem that could be solved using the equation $25x = 325$.

There is an endless number of possible situations that this equation could describe. One situation is as follows.

The pep band went on a trip to Six Flags Amusement Park. Admission was \$25 per person. If the total price of admission was \$325, how many people went to Six Flags?

Checking Your Understanding

Communicating Mathematics

Read and study the lesson to answer these questions. 1–3. See margin.

1. **Explain** the advantages in using the four step problem-solving plan on word problems using equations or inequalities.
2. **List** as many word phrases as you can think of for which you would use the symbol $\leq$.
3. **You Decide** Over 2100 pounds of pancake mix were used to make the largest breakfast in 1994. Rosa solved the inequality $5b > 2100$ to find the number of 5-pound bags of mix that were used. Kiele used the inequality $5b < 2100$. Who is correct and why?

Tech Prep

Sports Medicine Point out that equations and inequalities are in sports medicine used in many ways. A goal of running a certain distance within a certain time is common. Have students suggest an inequality that could be used in this situation. For more information on tech prep, see the *Teacher's Handbook*.

Additional Answers

1. Sample answer: You can solve the problem effectively and you check the answer to make sure it is accurate.
2. less than or equal to, no more than, at most
3. Rosa, more than 2100 pounds of mix was used.

2 TEACH

In-Class Examples

For Example 1

Marta is graduating from college. Her parents reserved a dining room for 25 people for a reception. The cost for the room is \$20. Each meal will cost at least \$8. What is the total cost of the dinner? **cost $\geq$ \$220**

For Example 2

Write a realistic problem that could be solved using the equation $12x > 100$. **Answers will vary. Sample answer: A dozen CDs cost more than \$100. How much does one CD cost?**

3 PRACTICE/APPLY

Checking Your Understanding

Exercises 1–7 are designed to help you assess your students' understanding through reading, writing, speaking, and modeling. You should work through Exercises 1–3 with your students and then monitor their work on Exercises 4–7.

Study Guide Masters, p. 26

NAME ______ DATE ______

3-8 Study Guide Student Edition Pages 156–159

Applying Equations and Inequalities

Compare these two examples. On the left, an equation is used to find the solution. On the right, an inequality is used.

Example: The record low temperature for Bakersville is $^-35$°F. This is 5 times the record low for Springtown. What is the record low for Springtown?

Explore Bakersville's low is 5 times Springtown's.

Plan $B = 5 \times S$

Solve $^-35 = 5 \times S$

$\frac{^-35}{5} = S$

$^-7 = S$

Springtown's low is $^-7$°F.

Examine Check the answer in the original problem.

Example: The record high temperature for Bakersville is 120°F. This is more than 2 times the record high for Springtown. What is the record high for Springtown?

Explore Bakersville's high is *more than* 2 times Springtown's.

Plan $B > 2 \times S$

Solve $120 > 2 \times S$

$\frac{120}{2} > S$

$60 > S$

Springtown's high is *less than* 60°F.

Examine Check whether the answer should include *less than* or *more than*.

Choose the equation or inequality that could be used to solve each problem.

1. On the way to school it was 10°F. It dropped 13 degrees by the end of the day. What is the temperature at the end of the day?
   A. $10 + t = ^-13$ (B.) $10 - 13 = t$
   C. $10 + t < ^-13$ D. $10 - 13 > t$
2. Football practice begins in 6 weeks. Ted wants to gain at least 12 pounds. How much must he gain per week?
   A. $g \leq 6 \times 12$ B. $g \leq 12 \div 6$
   C. $g \geq 6 \times 12$ (D.) $g \geq 12 \div 6$
3. The radio Andrea wants costs \$16 less than a cassette player. How much does the radio cost?
   A. $r = c + 16$ (B.) $r = c - 16$
   C. $r > c + 16$ D. $r < c + 16$
4. Randy invested \$2000 in 150 shares of stock last year. This is twice what the stock is worth today. What is the present value of the stock?
   A. $p < 2000 \times 2$ B. $p > 2000 \div 2$
   C. $p = 2000 \times 2$ (D.) $p = 2000 \div 2$

Assignment Guide

Core: 9–25 odd, 27–31
Enriched: 8–16 even, 17, 18, 20, 22–31

For **Extra Practice**, see p. 747.

The red A, B, and C flags, printed only in the Teacher's Wraparound Edition, indicate the level of difficulty of the exercises.

Additional Answers

4. x = integer; $x + 12 = 28$; 16
5. n = number; $4n < 96$; $n < 24$
6a. What is greatest possible cost of the labor?
6b. What is the greatest possible cost?
6c. $39 + x \leq 63$
6d. $x \leq 24$ dollars
7. Sample answer: The total cost of a car repair was $187. If the labor cost $169, how much did the parts cost?
8. h = hours worked; $13h \geq 52$; $h \geq 4$
9. n = number; $\frac{n}{-3} > 5$; $n < -15$
10. t = winner's time; $88 = 2t$; 44
11. a = account balance; $a + 50 > 400$; $a > 350$
12. n = number; $7n < 84$; $n < 12$
13. p = original purchase price; $4p \leq 85{,}000$; $p \leq 21{,}250$

Practice Masters, p. 26

NAME ______________________ DATE ____________

3-8 Practice
Applying Equations and Inequalities

Student Edition
Pages 156–159

Define a variable and translate each sentence into an equation or inequality. Then solve.

1. The sum of 39 and some number is 103. What is the number? **Let n be the number. $39 + n = 103$; $n = 64$ The number is 64.**
2. An unknown number less 7 is 19. What is the number? **Let n be the number. $n - 7 = 19$; $n = 26$ The number is 26.**
3. Six times a number is $^-84$. What is the number? **Let n be the number. $6n = {^-84}$; $n = {^-14}$ The number is $^-14$.**
4. Some number divided by $^-8$ is equal to $^-15$. What is the number? **Let n be the number. $\frac{n}{^-8} = {^-15}$; $n = 120$ The number is 120.**
5. The product of a number and 6 is less than 36. Find the number. **Let n be the number. $6n < 36$; $n < 6$ The number is any number less than 6.**
6. A store makes a profit of $25 on each moon watch it sells. How many of these must it sell to make a profit of at least $275? **Let w be the number of watches. $25w \geq 275$; $w \geq 11$ They must sell at least 11 moon watches.**
7. Kim bought a new fishing pole. It was on sale for $35. She saved $8. What was the original price? **Let p be the original price. $p - 8 = 35$; $p = 43$ The original price was $43.**
8. Leif's score on his second test was 87. This was 14 points more than his score on the first test. What was his score on the first test? **Let s be Leif's score on the first test. $s + 14 = 87$; $s = 73$ Leif's score on the first test was 73.**
9. Carol sold 20 shares of stock for a total of $2980. What was the value of one share? **Let v be the value of one share. $20v = 2980$; $v = 149$ The value of one share was $149.**
10. Jake worked a total of 38 hours last week. His earnings for the week were more than $228. What is his hourly rate of pay? **Let r be his hourly rate. $38r > 228$; $r > 6$ His hourly rate of pay is more than $6.**
11. The sum of two integers is at most $^-57$. One integer is 33. What is the other integer? **Let z be the other integer. $z + 33 \leq {^-57}$; $z \leq {^-90}$ The number is any integer less than or equal to $^-90$.**
12. The difference between two integers is at least 12. The smaller integer is 2. What is the larger integer? **Let z be the larger integer. $z - 2 \geq 12$; $z \geq 14$ The number is any integer greater than or equal to 14.**

Guided Practice

Define a variable and translate each sentence into an equation or inequality. Then solve. 4–7. See margin.

4. An integer increased by 12 is 28.
5. Four times some number is less than 96.
6. **Cycling** Dwain took his bike to the Leshnock Bike Shop to have the gear-shift mechanism repaired. At the shop, Dwain was told the total bill for parts and labor would be at most $63. The cost of the parts was $39. How much could Dwain expect to pay for labor?
   a. What is being asked?
   b. What does the phrase *at most* mean?
   c. Write an equation or inequality that describes the problem.
   d. Solve the problem.
7. Write a realistic problem that could be solved using the equation $169 + x = 187$.

Exercises: Practicing and Applying the Concept

Independent Practice

Define a variable and translate each sentence into an equation or inequality. Then solve. 8–13. See margin.

8. The bill for labor at $13 per hour was at least $52.
9. The quotient when dividing by -3 is greater than 5.
10. The 10-km race time of 88 minutes was twice as long as the winner's time.

11. A savings account increased by $50 is now more than $400.
12. Seven times a number is less than 84.
13. The house sold for $85,000. That is at least four times the original purchase price.

Write a realistic problem that could be solved using each equation or inequality. 14–16. See Solutions Manual.

14. $9 + 3x = 24$
15. $b + 9 \leq 16$
16. $30m > 300$

Critical Thinking

17. Write an inequality to represent the numbers whose absolute values are less than 4. Find all the integer solutions to the inequality. $|x| < 4$; $-3, -2, -1, 0, 1, 2, 3$

Applications and Problem Solving

18–19. See Solutions Manual.

Define a variable and write an equation or inequality. Then solve.

18. Brisa delivers pizza for Mama's Pizzeria. Her average tip is $1.50 for each pizza that she delivers. How many pizzas must she deliver to earn at least $20 in tips?
   a. What is being asked?
   b. What does the phrase *at least $20* mean?
   c. Write an equation or inequality that describes the problem. Solve.
   d. Can your answer be anything other than a whole number? Explain.

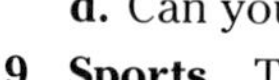

19. **Sports** The University of Connecticut Huskies women's basketball team was the undefeated NCAA Champion for the 1994–95 season. The team played 35 games, counting both regular and post-season play. Rebecca Lobo, one of Connecticut's star players, scored almost 600 points for the season. What was her average per game?

Reteaching

Using Lists List symbols $+, -, \times, \div, <, >, \leq, \geq$, and $=$ on the chalkboard. Read word problems from the lesson and pick out words or terms that have the same meaning as each listed symbol. List each under the symbol. Add other words or terms. Write problems using these terms.

Group Activity Card 3-8

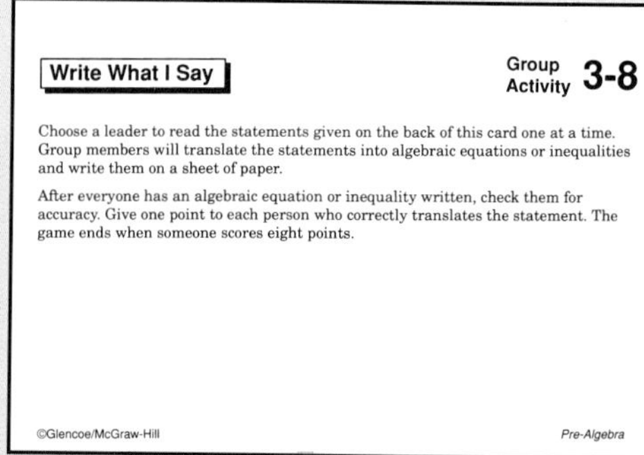

Write What I Say

Group Activity **3-8**

Choose a leader to read the statements given on the back of this card one at a time. Group members will translate the statements into algebraic equations or inequalities and write them on a sheet of paper.

After everyone has an algebraic equation or inequality written, check them for accuracy. Give one point to each person who correctly translates the statement. The game ends when someone scores eight points.

©Glencoe/McGraw-Hill *Pre-Algebra*

20–26. See margin.

20. Transportation A minivan is rated for maximum carrying capacity of 900 pounds.

a. If the luggage weighs 100 pounds, what is the maximum weight allowable for passengers?

b. What is the maximum average weight allowable for 5 passengers?

21. Postage In 1971, first-class postage stamps cost 8 cents for the first ounce. In 1995, stamps for the same first-class mail cost 32 cents for the first ounce. How many times more did stamps cost in 1995 than in 1971?

22. Air Travel In 1903, the Wright brothers' first flight was at 30 mph. A Boeing 747 cruises at 600 mph. How many times faster is the flight speed of a 747?

23. Geometry The volume of a rectangular solid is found by multiplying its length, width, and depth. If the volume of the solid shown at the right is 440 cubic centimeters, what is its height?

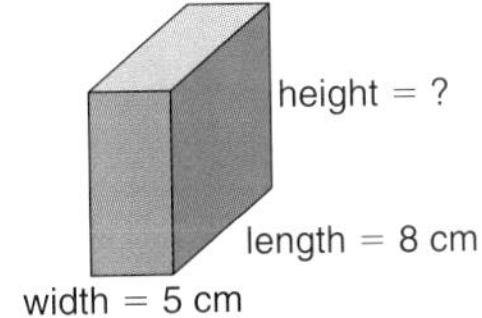

24. Agriculture Between 1987 and 1992, the number of female farmers increased 14,386 to 145,156. What was the number of female farmers in 1987?

25. Sports Danika is on her high school golf team. Her personal goal is to shoot 80. Her practice scores during the week were within 6 strokes above her desired score. What could be some of her scores?

26. Tournaments In the NBA playoffs, a team advances from the first round if they win the "best of five." Best of five means that they win more than half of the five games played between the teams. The semifinal, conference final, and final rounds are all best of seven.

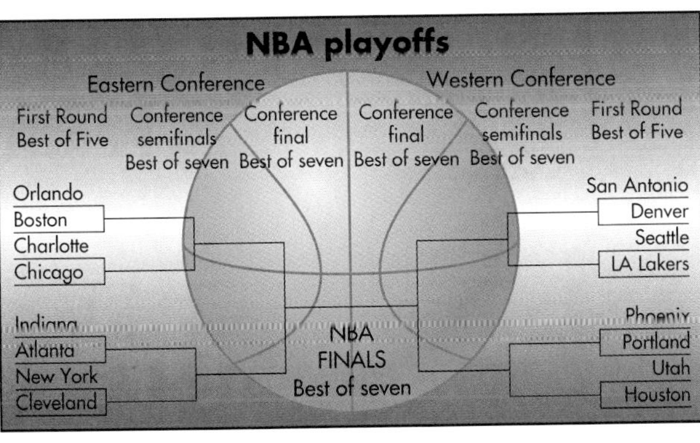

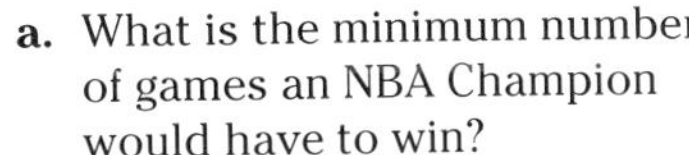

a. What is the minimum number of games an NBA Champion would have to win?

b. How many games could an NBA Champion lose? Express your answer as an inequality.

Mixed Review

27. Solve $-3h < -39$ and check your solution. Then graph the solution on a number line. (Lesson 3-7) **$h > 13$, See margin for graph.**

28. Science The magnification of lenses in a compound microscope, M, can be found by using the formula $E \times O = M$, where E is the power of the eyepiece lens and O is the power of the objective lens. If an eyepiece lens of a microscope has a power of 10 and its objective lens has a power of 43, what is the magnification of the microscope? (Lesson 3-4) **430**

29. Simplify $-12y + 5y$. (Lesson 2-4) **$-7y$**

30. Name the absolute value of -9. (Lesson 2-1) **9**

31. Solve $\frac{t}{3} = 8$ mentally. (Lesson 1-6) **24**

4 ASSESS

Closing Activity

Speaking Have students work in groups of four to solve a problem similar to those in this lesson. Each student in the group then explains to you or to the class one step in the solution of the problem. The steps should include *Explore, Plan, Solve,* and *Examine* as shown on pages 156 and 157.

Chapter 3, Quiz D (Lessons 3-7 and 3-8) is available in the *Assessment and Evaluation Masters,* p. 72.

Additional Answers

20a. w = weight; $w + 100 \leq 900$; $w \leq 800$ lb

20b. $w \leq 160$ lb average for 5 passengers

21. c = changes in price; $8c = 32$; $c = 4$ times the 1971 price/oz

22. t = changes in speed; $30t = 600$; $t = 20$ times faster

23. h = height; $5(8)h = 440$; $h = 11$ centimeters

24. f = female farmers in 1987; $145{,}156 = f + 14{,}386$; $f = 130{,}770$ female farmers

25. x = sample scores; $80 - 6 \leq x \leq 80 + 6$; $74 \leq x \leq 86$ strokes

Enrichment Masters, p. 26

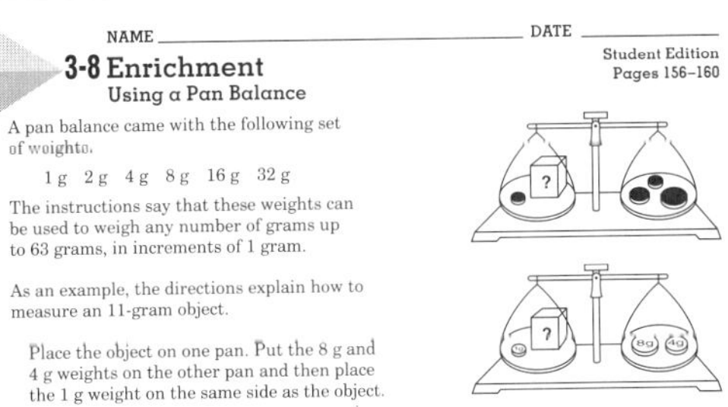
NAME ____ DATE ____

3-8 Enrichment Student Edition Pages 156–160

Using a Pan Balance

A pan balance came with the following set of weights.

1 g 2 g 4 g 8 g 16 g 32 g

The instructions say that these weights can be used to weigh any number of grams up to 63 grams, in increments of 1 gram.

As an example, the directions explain how to measure an 11-gram object.

Place the object on one pan. Put the 8 g and 4 g weights on the other pan and then place the 1 g weight on the same side as the object.

1. The equation $x + 1 = 8 + 4$ can be used to represent this situation. Can you find another way to weigh an 11 g object? **Place weights of 8 g, 2 g, and 1 g opposite the 11 g object.**

In Exercises 2–7, the weight of an object is given. Imagine the circles are the two pans of the balance discussed above. Show how the balance can be used to weigh the object. Use a square to represent the object and write the numbers for the gram weights given above. The first one has been started. Answers may vary.

2. 15 g — [15] 1 / 16

3. 29 g — [29] 4 / 32 1

4. 43 g — [43] / 32 2 8 1

5. 60 g — [60] / 32 16 8 4

Another balance comes with this set of weights: 1 g, 3 g, 9 g, and 27 g. See if you can determine how to weigh an object having each of the given weights.

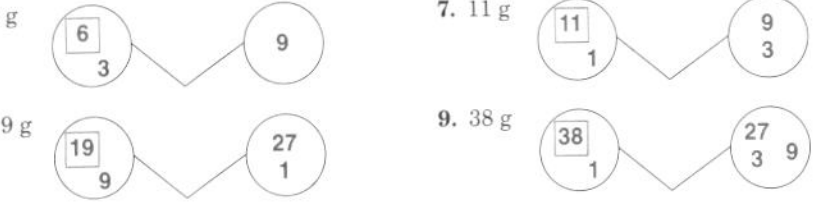
6. 6 g — [6] 3 / 9

7. 11 g — [11] 1 / 9 3

8. 19 g — [19] 9 / 27 1

9. 38 g — [38] 1 / 27 3 9

Extension

Using Research Have students research a topic, such as prices of cassettes and CDs, and write equations and inequalities to be solved using that information.

Additional Answers

26a. $g = 15$ minimum number of games (15 wins and 0 losses)

26b. $g \leq 11$ losses

27. (number line: open circle at 13, shaded to the right; labels 10 11 12 13 14 15 16)

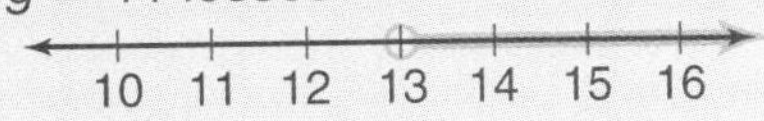

CLOSING THE Investigation

This activity provides students with an opportunity to bring their work on the Investigation to a close. For each Investigation, students should present their findings to the class. Here are some ways students can display their work.

- Conduct and report on an interview or survey.
- Write a letter, proposal, or report.
- Write an article for the school or local paper.
- Make a display, including graphs and/or charts.
- Plan an activity.

Assessment

To assess students' understanding of the concepts and topics explored in this Investigation and its follow-up activities, you may wish to examine students' Investigation Folders.

The scoring guide provided in the *Investigations and Projects Masters*, p. 7, provides a means for you to score students' work on this Investigation.

Investigations and Projects Masters, p. 7

CHAPTERS 2 and 3 Investigation
Bon Voyage

Scoring Guide

Level	Specific Criteria
3 Superior	• Shows thorough understanding of *graphing on numberlines and coordinate planes, comparing and ordering integers, operations with integers, solving problems looking for a pattern, solving problems with integers, solving problems by eliminating possibilities, solving equations and inequalities,* and *using formulas to solve real-world and geometry problems.* • Uses appropriate strategies to solve problems. • Computations are correct. • Written explanations are exemplary. • Goes beyond the requirements of the Investigation.
2 Satisfactory, with minor flaws	• Shows understanding of *graphing on numberlines and coordinate planes, comparing and ordering integers, operations with integers, solving problems looking for a pattern, solving equations and inequalities,* and *using formulas to solve real-world and geometry problems.* • Uses appropriate strategies to solve problems. • Computation are mostly correct. • Written explanations are effective. • Satisfies all requirements of the Investigation.
1 Nearly Satisfactory, with serious flaws	• Shows understanding of most of the concepts of *graphing on numberlines and coordinate planes, comparing and ordering integers, operations with integers, solving problems looking for a pattern, solving equations and inequalities,* and *using formulas to solve real-world and geometry problems.* • May not use appropriate strategies to solve problems. • Computations are mostly correct. • Written explanations are satisfactory. • Satisfies most requirements of the Investigation.
0 Unsatisfactory	• Shows little or no understanding of *graphing on numberlines and coordinate planes, comparing and ordering integers, operations with integers, solving problems looking for a pattern, solving equations and inequalities,* and *using formulas to solve real-world and geometry problems.* • May not use appropriate strategies to solve problems. • Computations are incorrect. • Written explanations are not satisfactory. • Does not satisfy requirements of the Investigation.

CLOSING THE Investigation

Refer to the Investigation on pages 62–63.

You are ready to finalize arrangements for your trip. You and your partner will be called upon to present your plan for the trip at the parents' meeting next week. Complete your research and make your presentation to the parents.

- Find the most current foreign exchange rate for the currency used in the country you chose at the beginning of the investigation. In other words, how much is *one* franc (or one krone, one peso, one rupee or one yen) worth in American dollars? These can be found in *The Wall Street Journal* and in the business section of other major newspapers.
- Explain how to find the dollar value of the foreign currency. Provide both an exact formula (that could be done on a calculator) and a simple way to estimate the value. You should be able to determine, for example, what a pair of jeans would cost in the United States if they cost 45,280 lire in Rome, Italy.
- Furnish the time zone data you found when working on the Investigation in Lesson 2-1. Work with your family to determine a convenient time to make phone calls.
- Supply environmental data for the country. Then provide the data you found when working on the Investigation in Lesson 3-2.
- Provide a short refresher course on metric units of length, weight, and liquid volume.
- Review any important business and social customs. If you can also explain the rationale behind them, it may be easier for you to remember. For example, in some Middle Eastern countries, it is considered unclean to eat or drink with your left hand!

You may wish to organize all of the information on the country you studied in a travel scrapbook to be displayed in your classroom.

PORTFOLIO ASSESSMENT

You may want to keep your work on this Investigation in your portfolio.

CHAPTER 3 Highlights

Vocabulary

After completing this chapter, you should be able to define each term, property, or phrase and give an example or two of each.

Algebra
addition property of equality (p. 125)
addition property of inequality (p. 147)
division properties of inequality (p. 152)
division property of equality (p. 130)
equivalent equations (p. 124)
formula (p. 134)
multiplication properties of inequality (p. 152)
multiplication property of equality (p. 130)
subtraction property of equality (p. 124)
subtraction property of inequality (p. 147)

Geometry
area (p. 138)
area of a rectangle (p. 140)
coordinate (p. 126)
graph (p. 126)
perimeter (p. 138)
perimeter of a rectangle (p. 139)

Problem Solving
eliminating possibilities (p. 118)
matrix logic (p. 119)

Understanding and Using Vocabulary

Determine whether each statement is *true* or *false*.

1. When two numbers that are additive inverses of each other are added, the sum is zero. **true**
2. The formula *rate* · *time* = *distance* can be used to find the rate when only the distance is known. **false**
3. The coordinates of the points on a number line are greater as you move from right to left. **false**
4. The area of a geometric figure is the measure of the surface it encloses. **true**
5. A solid dot on a number line indicates that the point is included in the solution set. **true**
6. The perimeter of a geometric figure is the distance around it. **true**
7. The division property of inequality says that if you divide both sides of an inequality by the same number, the order of the inequality symbol must be reversed. **false**

Instructional Resources

Three multiple-choice tests and three free-response tests are provided in the *Assessment and Evaluation Masters*. Forms 1A and 2A are for honors pacing, Forms 1B and 2B are for average pacing, Forms 1C and 2C are for basic pacing. Chapter 3 Test, Form 1B is shown at the right. Chapter 3 Test, Form 2B is shown on the next page.

Highlights

Using the Chapter Highlights

The Chapter Highlights begins with an alphabetical listing of the new terms, properties, and phrases that were introduced in this chapter.

Assessment and Evaluation Masters, pp. 59–60

NAME ______________________ DATE __________

CHAPTER 3 Test, Form 1B

Solve by eliminating possibilities.

1. Roger, Stephen, Kathie, and Diane are planning a dinner. Each person will bring a dish he or she made at home. Use the clues to find out who brings the salad.

   Roger uses a grill to prepare his dish.
   Stephen preheats the oven at 350°F for his dish.
   Diane's oven is broken.
   Kathie had to wait one hour while the dish she prepared rises.

   A. Diane **B.** Roger **C.** Stephen **D.** Kathie — 1. A

2. For what value of x is $15x < x^2 + 3$?
   A. 3 **B.** $\frac{1}{5}$ **C.** 5 **D.** 1 — 2. B

3. Which property of equality is used to solve the equation $x + 7 = -15$?
   A. addition **B.** subtraction
   C. multiplication **D.** division — 3. B

4. Solve $p + (-12) = -14$.
   A. -2 **B.** 2 **C.** 26 **D.** -26 — 4. A

5. Solve $w - (-15) = -12$.
   A. 3 **B.** -3 **C.** -27 **D.** 27 — 5. C

6. Solve $x - 26 = -32$.
   A. 6 **B.** -6 **C.** 58 **D.** -58 — 6. B

7. Choose the equation whose solution is graphed below.

   −7 −6 −5 −4 −3 −2 −1 0 1 2 3 4 5 6 7

   A. $x + 90 = 100$ **B.** $2x + 90 = 100$
   C. $x - 90 = 100$ **D.** $9x - 10 = 100$ — 7. A

Solve each equation.

8. $-8m = -24$
   A. -16 **B.** -3 **C.** 3 **D.** -32 — 8. C

9. $\frac{m}{-3} = 24$
   A. 8 **B.** -8 **C.** 72 **D.** -72 — 9. D

10. $\frac{a}{-6} = -12$
    A. 18 **B.** -18 **C.** 72 **D.** -72 — 10. C

NAME ______________________ DATE __________

Chapter 3 Test, Form 1B (continued)

Solve by replacing the variables with the given values.

11. Find the distance traveled if you drive at 55 miles per hour for $3\frac{1}{2}$ hours.
    A. 165 **B.** 192.5 **C.** 137.5 **D.** 15.7 — 11. B

12. $d = 2r$, if $r = 8.5$
    A. 10.5 **B.** 4.25 **C.** 16.5 **D.** 17 — 12. D

13. $a = \frac{f - s}{t}$, if $f = 28$ m/s, $s = 10$ m/s, and $t = 10$ s
    A. 18 m/s **B.** 1.8 m/s **C.** -18 m/s **D.** -1.8 m/s — 13. B

14. Find the perimeter of a square measuring 12 meters on each side.
    A. 16 m **B.** 32 m **C.** 48 m **D.** 144 m — 14. C

15. Find the area of the rectangle at the right.
    A. 168 cm^2
    B. 158 cm^2
    C. 52 cm^2
    D. 26 cm^2 — 15. A

    14 cm
    12 cm

Solve each inequality.

16. $x + 4 < 12$
    A. $x < 8$ **B.** $x < 16$
    C. $x > 8$ **D.** $x > 16$ — 16. A

17. $y - 3 > -16$
    A. $y < -13$ **B.** $y > -19$ **C.** $y < -19$ **D.** $y > -13$ — 17. D

18. $6x > -12$
    A. $x > -2$ **B.** $x > 2$ **C.** $x > -6$ **D.** $x > -18$ — 18. A

19. Choose the inequality whose solution is graphed below.

    −8 −7 −6 −5 −4 −3 −2 −1 0 1 2 3 4 5 6 7 8

    A. $-3x - 2 \leq -17$ **B.** $3x - 2 \leq -17$
    C. $3x - 2 \geq -17$ **D.** $3x - 2 \geq 17$ — 19. B

20. Charles and Betty spent more than $230 while shopping. Betty spent $138. Choose the inequality that shows what Charles spent.
    A. C > $108 **B.** C < $92 **C.** C = $92 **D.** C > $92 — 20. D

Using the Study Guide and Assessment

Skills and Concepts Encourage students to refer to the objectives and examples on the left as they complete the review exercises on the right.

Assessment and Evaluation Masters, pp. 65–66

NAME ______________________ DATE ________

CHAPTER 3 Test, Form 2B

Solve by eliminating possibilities.

1. The town library surveyed visitors to determine their reading preferences. Fifty people said they like mysteries, 30 said they like history, and 6 said they like science fiction. However, of these, 5 like mysteries and history, 3 like history and science fiction, 2 like mysteries and science fiction, and 1 likes all three. Find the number of visitors surveyed. — 1. **24**

2. Examine the list of numbers at the right. Using the clues, find the correct number.

The sum of its digits is 12.
It is an odd number divisible by 3 and 5.
It is not divisible by 9.
It is $\frac{1}{7}$ another number.

18	54	42	39
525	65	78	325
57	75	225	30
45	81	15	95

2. **75**

Solve each equation.

3. $a + (-18) = -42$ — 3. **−24**
4. $c - 12 = 31$ — 4. **43**
5. $d - (-18) = -27$ — 5. **−45**
6. $e - 18 = -16$ — 6. **2**
7. Solve the equation $3x - 4 = 5$ and graph its solution on the number line. — 7. **$x = 3$; See students' graphs.**

Solve each equation.

8. $5f = -35$ — 8. **−7**
9. $-7g = -56$ — 9. **8**
10. $-11h = 66$ — 10. **−6**

Tell how you would solve each equation.

11. $\frac{k}{4} = -16$ — 11. **multiply both sides by 4**
12. $m - 82 = -24$ — 12. **add 82 to both sides**

NAME ______________________ DATE ________

Chapter 3 Test, Form 2B (continued)

Solve by replacing the variables with the values given.

13. $P = 2\ell w$, if $\ell = 1$ and $w = 15$ — 13. **P = 30**
14. $H = m(T_2 - T_1)$, if $m = 100$, $T_2 = 25$, and $T_1 = 10$ — 14. **H = 1500**
15. Anita wants a table cloth for a square table that is 4 feet on each side. She wants the cloth to hang 1 foot off each side. Find the area of the cloth she needs. — 15. **36 sq. ft**

Solve each problem.

16. Find the perimeter of a square with sides 12 centimeters long. — 16. **48 cm**
17. Find the area of a rectangle with a length of 14 centimeters and a width of 20 centimeters. — 17. **280 cm²**

Solve each inequality.

18. $x + 8 < 16$ — 18. **$x < 8$**
19. $y - 12 \geq 15$ — 19. **$y \geq 27$**
20. $-12y < -108$ — 20. **$y > 9$**
21. $-5z \geq 50$ — 21. **$z \leq -10$**
22. Write an inequality that describes the graph. — 22. **$x \geq -3$**

−5 −4 −3 −2 −1 0 1 2 3 4 5

23. Leon has reduced his intake of saturated fat to 20 grams each day. If he is planning to have a 6-ounce serving of chicken wings for dinner, and a 4-ounce serving has 6 grams of saturated fat, write an inequality that describes how much saturated fat he can have at other times of the day. — 23. **$sf \leq 11$ g**

Solve.

24. Fred needs at least 86 points to pass Ellen's score. Ellen has 631 points. How many points does Fred have? — 24. **$f \leq 545$**
25. Donna received a shipment of 38 grapefruit, and wants to use them in fruit baskets for family gifts. How many baskets can Donna make if she puts at least 6 grapefruit in each basket? — 25. **$b \leq 6$**

CHAPTER 3 Study Guide and Assessment

Skills and Concepts

Objectives and Examples

Upon completing this chapter, you should be able to:

Review Exercises

Use these exercises to review and prepare for the chapter test.

▶ **solve equations using the addition and subtraction properties of equality.** (Lesson 3-2)

Solve $g - (-19) = 23$.

$g - (-19) = 23$ *Rewrite as addition sentence.*
$g + 19 = 23$
$g + 19 - 19 = 23 - 19$ *Subtract 19 from each side.*
$g = 4$

Solve each equation and check your solution. Then graph the solution on a number line. **See Solutions Manual for graphs.**

8. $f + 31 = -5$ **−36**
9. $r + (-11) = 40$ **51**
10. $12 = k - 7$ **19**
11. $-9 = r + (-15)$ **6**
12. $16 = h - (-4)$ **12**
13. $7 = q - 12$ **19**
14. $m - 3 = -10$ **−7**
15. $z - (-51) = 36$ **−15**

▶ **solve equations using the multiplication and division properties of equality.** (Lesson 3-3)

Solve $\frac{r}{-7} = 15$.

$\frac{r}{-7} = 15$
$(-7) \cdot \frac{r}{-7} = 15 \cdot (-7)$ *Multiply each side by (−7).*
$r = -105$

Solve each equation and check your solution. Then graph the solution on a number line. **See Solutions Manual for graphs.**

16. $\frac{c}{6} = -29$ **−174**
17. $\frac{p}{-3} = -42$ **126**
18. $840 = -28r$ **−30**
19. $13b = -299$ **−23**
20. $14x = 224$ **16**
21. $-\frac{y}{10} = -33$ **330**
22. $\frac{g}{-9} = 21$ **−189**
23. $25d = -1775$ **−71**

▶ **solve problems using formulas.** (Lesson 3-4)

The Herrs flew 360 miles to Tallahassee in 2 hours. What was the speed of the plane? Use $d = rt$.

$d = rt$
$360 = r(2)$ *Replace d with 360 and t with 2.*
$\frac{360}{2} = \frac{r(2)}{2}$ *Divide each side by 2.*
$180 = r$

Solve by replacing the variables with the given values.

24. $A = \ell \cdot w, A = 105, \ell = 7$ **15**
25. $I = p \cdot r \cdot t, p = 200, r = 0.06, t = 2$ **24**
26. $A = \frac{b \cdot h}{2}, A = 72, b = 6$ **24**
27. $C = 2\pi r, \pi = 3.14, r = 9$ **56.52**

GLENCOE Technology

Test and Review Software

You may use this software, a combination of an item generator and item bank, to create your own tests or worksheets. Types of items include free response, multiple choice, short answer, and open ended.

For IBM & MacIntosh

Objectives and Examples

▶ **find the perimeter and area of rectangles and squares.** (Lesson 3-5)

Find the area and perimeter of a rectangle with length 11 cm and width 7 cm.

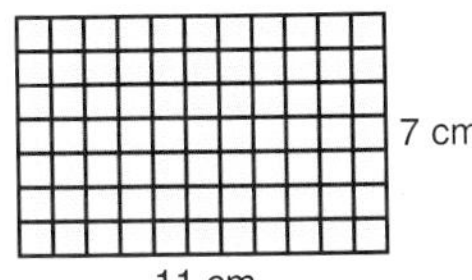

$P = 2(\ell + w)$
$= 2(11 + 7)$
$= 2(18)$ or 36 cm

$A = \ell \cdot w$
$= 11 \cdot 7$
$= 77$ sq cm

Review Exercises

Find the perimeter and area of each rectangle.

28. **29.**

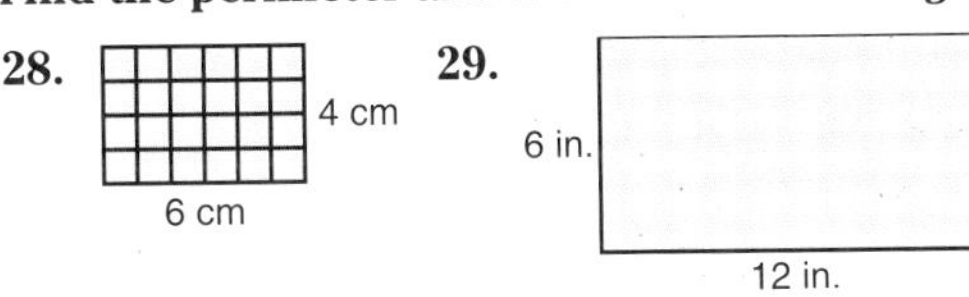

20 cm, 24 sq cm **36 in., 72 sq in.**

30. a rectangle that is 4 meters wide and 12 meters long **32 m, 48 sq m**

31. a 15 foot-by-5 foot rectangle **40 ft; 75 sq ft**

32. a square with side 22 yards **88 yd; 484 sq yd**

▶ **solve inequalities by using the addition and subtraction properties and graph the solution set of the inequalities.** (Lesson 3-6)

Solve $-6 < n - 6$. Then graph the solution.

$-6 < n - 6$
$-6 + 6 < n - 6 + 6$ *Add 6 to each side.*
$0 < n$

−2 −1 0 1 2 3 4 5 6

Solve each inequality and check your solution. Then graph the solution on a number line. **See margin.**

33. $x - 2 < -14$ **34.** $12 + a > -9$

35. $-7 + b < -5$ **36.** $4 < y - 23$

37. $f + (-8) > -12$ **38.** $w + 9 \leq -5$

▶ **solve inequalities by using the multiplication and division properties.** (Lesson 3-7)

Solve $-7s > 70$. Then graph the solution.

$-7s > 70$
$\frac{-7s}{-7} < \frac{70}{-7}$ *Divide each side by −7 and*
$s < -10$ *reverse the order symbol.*

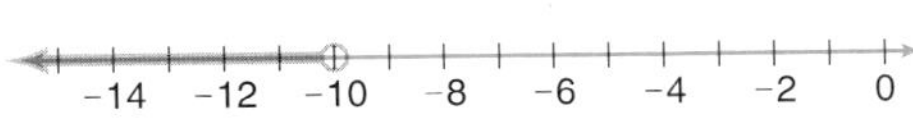

Solve each inequality and check your solution. Then graph the solution on a number line. **See margin for graphs.**

39. $\frac{m}{-15} < 9$ **40.** $-72 \geq -4k$ **$k \geq 18$**

41. $\frac{t}{12} < 11$ **$t < 132$** **42.** $144 < 9k$ **$k > 16$**

43. $88 > -8p$ **44.** $\frac{s}{-22} < -7$ **$s > 154$**

45. $285 \leq 19f$ **$f \geq 15$** **46.** $18 \leq \frac{r}{6}$ **$r \geq 108$**

39. $m > -135$ **43.** $p > -11$

Additional Answers

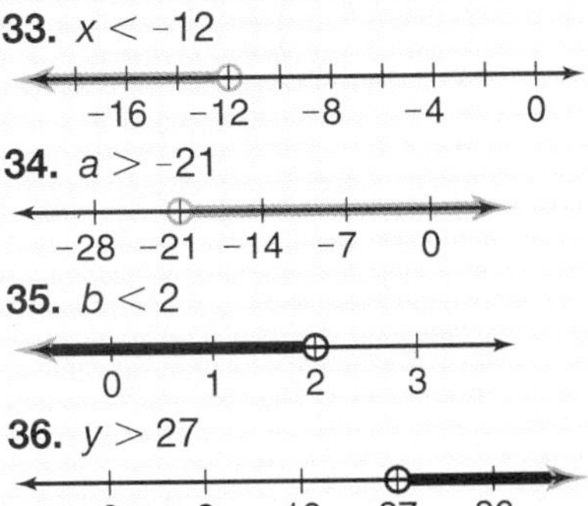

37. $f > -4$

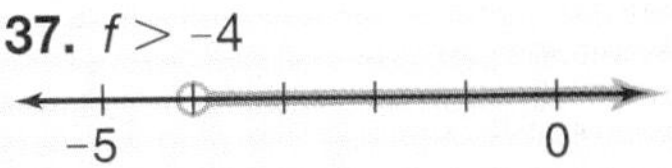

38. $w \leq -14$

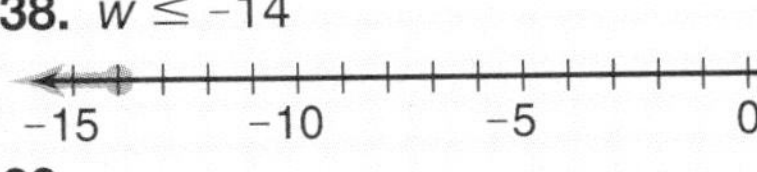

39.

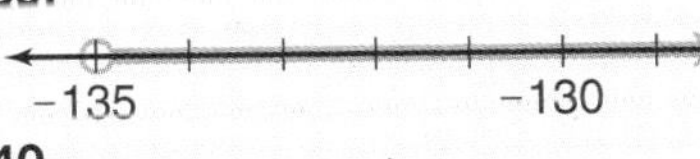

40.

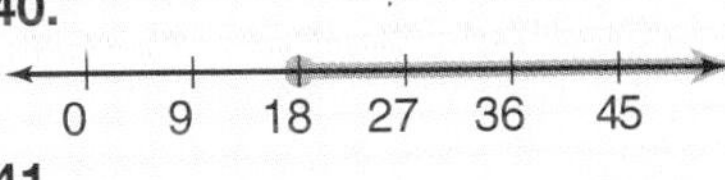

41.

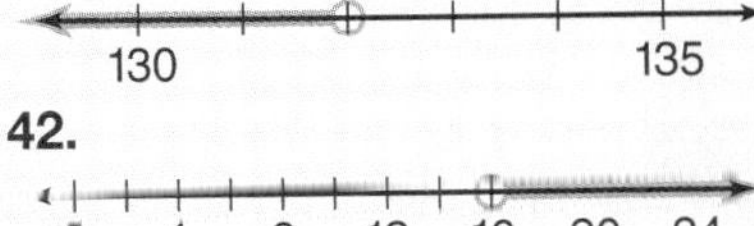

42.

0 4 8 12 16 20 24

43.

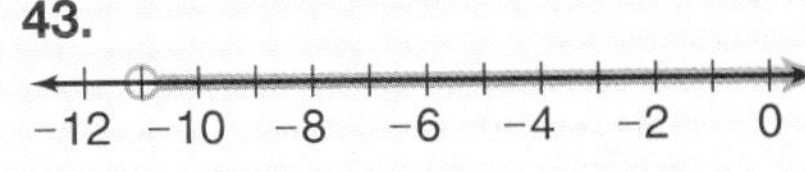

44.

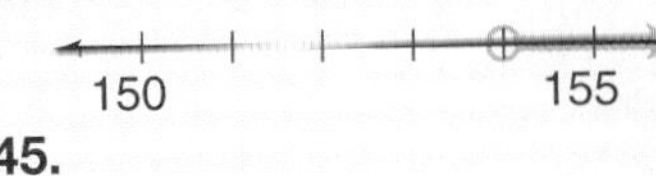

45.

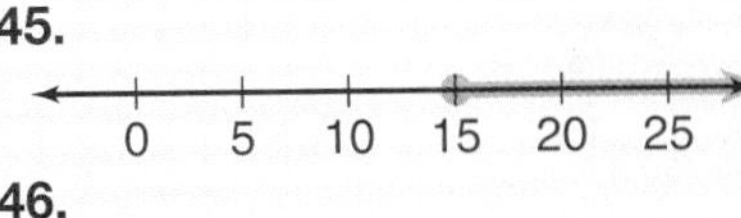

46.

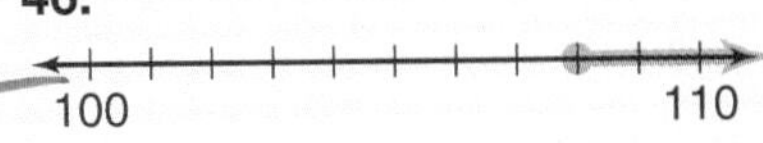

Applications and Problem Solving

Encourage students to work through the exercises in the Applications and Problem Solving section to strengthen their problem-solving skills.

Objectives and Examples

- **solve verbal problems by translating them into equations and inequalities.** (Lesson 3-8)

After spending \$17 to get into the amusement park, Marcus had \$21 left. How much money did he have before he bought his ticket?

Let x be the amount of money Marcus had at the beginning.

money at start	*minus*	*ticket price*	*equals*	*money left*
x	$-$	17	$=$	21

$$x - 17 = 21$$
$$x - 17 + 17 = 21 + 17$$
$$x = 38$$

Marcus had \$38 before buying his ticket.

Review Exercises

Define a variable and write an equation or inequality. Then solve.

47. Three friends went to dinner to celebrate one of them getting a job promotion. The total check was less than \$36. If the cost was shared evenly, how much did each person pay for dinner? **$3x < 36$; less than \$12**

48. Four times a number is greater than 76. What is the number? **$4n > 76$; $n > 19$**

49. A Girl Scout troop must sell more than 800 boxes of cookies to earn a trip to camp. If troop #332 has sold 576 boxes, how many boxes must they still sell to win a trip? **more than 224 boxes**

Applications and Problem Solving

50. Four runners represented South, Kennedy, Lincoln, and Anthony High Schools in an 880-meter race. The runner from Lincoln was faster than the runner from Kennedy. The runner from South beat the runner from Anthony, but lost to the runner from Kennedy. Which school did the runner who came in last represent? (Lesson 3-1) **d**
    - **a.** Lincoln
    - **b.** Kennedy
    - **c.** South
    - **d.** Anthony

51. **Statistics** For many years, immigrants to the United States arrived at Ellis Island in New York. The flow of immigrants peaked in 1907 with more than 1 million. During that year, 37,807 people arrived from Germany. Find the average number of Germans who arrived daily at Ellis Island in 1907. (Lesson 3-3) **103**

52. **Physics** A 110-car train goes from a velocity of 1.2 miles per hour to 6.8 miles per hour in 4 minutes. Use the formula $a = \frac{f - s}{t}$, where a is acceleration, f is final velocity, s is starting velocity, and t is time it took to make the change in velocity to find the acceleration of the train. (Lesson 3-4) **1.4 miles per hour per minute**

A practice test for Chapter 3 is available on page 776.

Additional Answer Alternative Assessment

5. The label states that there is less than 1 gram of sugar in one serving of the product. If x represents the number of grams of sugar in the product, then $x < 1$. Thus, six servings of the product will contain less than 6 grams of sugar. Since $5 < 6$, it is possible that there will be less than 5 grams of sugar in 6 servings of the product.

Alternative Assessment

Performance Task **Demonstrate your knowledge by giving a clear, concise solution to each problem. Be sure to include all relevant drawings and justify your answers. You may show your solutions in more than one way or investigate beyond the requirements of the problems.**

	Oatmeal Creme Pies	Peanut Butter Cookies	Chocolate-Chip Cookies
Calories per serving	170	**190**	90
Grams of Fat	8 g	**16 g**	**4 g**
Sodium	190 mg	130 mg	80 mg

Define a variable, write an equation or inequality, and then solve. Then copy and complete the table.

1. Chocolate-chip cookies have 100 Calories less than peanut butter cookies. **$c - 100 = 90$; 190**
2. Oatmeal Creme Pies have half as much fat as peanut butter cookies.
3. Chocolate-chip cookies have 20 times more sodium than grams of fat.
4. Write a realistic problem involving the numbers given in the chart.

2. $\frac{f}{2} = 8$; 16

Nutrition Facts	Amount/serving	%DV	Amount/serving	%DV
Serv. Size 43 g	**Total fat** 12 g	19%	**Total Carb.** 23 g	8%
Serv. per package 1	Sat. fat 2.5 g	13%	Fiber 1 g	5%
Calories 220	**Cholest.** 5 mg	1%	Sugars < 1 g	
Fat Cal. 110	**Sodium** 340 mg	14%	**Protein** 6 g	
* Percent Daily Values (DV) are based on a 2000 calorie diet.	Vitamin A 0% • Vitamin C 0% • Calcium 6% • Iron 10%			

5. Examine the nutrition label at the left. Could you eat six servings of this product and still consume less than 5 grams of sugar? Explain.

3. $20s = 80$; 4
4. See students' work.
5. See margin.

Thinking Critically

▶ A rectangle is 7 inches by w inches. Suppose you increase the width by 1 inch. **a. increases 2 inches; increases 7 sq in.**

a. How do the perimeter and area change?
b. If the length was 10 inches instead of 7 inches, would the change in the perimeter be different? Explain.
c. If the length was 10 inches instead of 7 inches, would the change in the area be different? Explain.

b. No; see Solutions Manual for explanation.
c. Yes; see Solutions Manual for explanation.

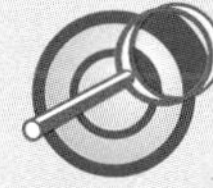

Portfolio

▶ Place your favorite word problem from this chapter in your portfolio and attach a note explaining why it is your favorite.

Self Evaluation

▶ Don't follow the paved road. Instead leave the beaten path and start a new trail. In other words, be yourself!

Assess yourself. When you try to solve a challenging problem, do you quit if the method you know doesn't work, or do you branch off and try a new way? Sometimes all you need to do is to look at a problem in a different way in order to be successful.

Alternative Assessment

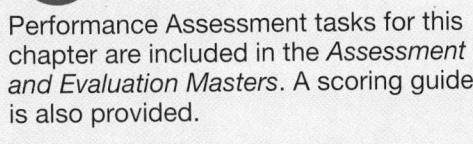

The Alternative Assessment section provides students with the opportunity to assess their own work by thinking critically, working with others, keeping a portfolio, and honestly evaluating their own progress. For more information on alternative forms of assessment, see *Alternative Assessment in the Mathematics Classroom*.

Performance Assessment

Performance Assessment tasks for this chapter are included in the *Assessment and Evaluation Masters*. A scoring guide is also provided.

Alternative Assessment

Assessment and Evaluation Masters, pp. 69, 81

NAME ________________ DATE ________

CHAPTER 3 Performance Assessment

Instructions: Demonstrate your knowledge by giving a clear, concise solution to each problem. Be sure to include all relevant drawings and justify your answers. You may show your solution in more than one way or investigate beyond the requirements of the problem.

1. Make up a problem that can be solved by using each equation or inequality. Then solve each problem and explain the meaning of the answer.
   a. $x - 14 = 9$
   b. $2x \geq 250$
2. Draw pictures of cups and counters for each equation. Then solve. Explain each step.
   a. $k + 3 = 7$
   b. $a - 4 = 2$
3. Carlos is planning a new vegetable garden. He has 70 feet of fencing with which to enclose the garden. If he wants to obtain the greatest possible area, what are the dimensions of the garden? Explain your reasoning.
4. Alecia, Terri, Maria, and Carla baby-sit for families in the neighborhood. They take care of different numbers of children: twin boys, three girls, two girls and a boy, and one girl and two boys. Use the following information to match each person with her baby-sitting job.

   Maria baby-sits next to the home with twin boys.
   Terri baby-sits for two children.
   Alecia baby-sits for children of the same gender.
   Carla shows the girls how to help take care of their baby brother.

CHAPTER 3 SCORING GUIDE

Level	Specific Criteria
3 Superior	• Shows thorough understanding of the concepts *variable, equation, inequality, area, perimeter,* and *matrix logic.* • Uses appropriate strategies of reason. • Computations to solve equations and inequalities are correct. • Written explanations are exemplary. • Word problems for equations and inequalities are appropriate and make sense. • Modeling is accurate and clear. • Goes beyond requirements of some or all problems.
2 Satisfactory, with minor flaws	• Shows understanding of the concepts *variable, equation, inequality, area, perimeter,* and *matrix logic.* • Uses appropriate strategies or reason. • Computations to solve equations and inequalities are mostly correct. • Written explanations are effective. • Word problems for equations and inequalities are appropriate and make sense. • Modeling is mostly accurate and clear. • Satisfies all requirements of problems.
1 Nearly Satisfactory, with serious flaws	• Shows understanding of most of the concepts *variable, equation, inequality, area, perimeter,* and *matrix logic.* • May not use appropriate strategies of reason. • Computations to solve equations and inequalities are mostly correct. • Written explanations are satisfactory. • Word problems for equations and inequalities are appropriate and make sense. • Modeling is mostly accurate and clear. • Satisfies most requirements of problems.
0 Unsatisfactory	• Shows little or no understanding of the concepts *variable, equation, inequality, area, perimeter,* and *matrix logic.* • May not use appropriate strategies of reasoning. • Computations to solve equations and inequalities are incorrect. • Written explanations are not satisfactory. • Word problems for equations and inequalities are not appropriate or sensible. • Modeling is not accurate or clear. • Does not satisfy requirements of problems.

Investigation

TEACHER NOTES

NCTM Standards: 1-8, 10

This Investigation is designed to be completed over several days or weeks. It may be considered optional. You may want to assign the Investigation and the follow-up activities to be completed at the same time.

Objective

Use mathematics to "purchase" and manage a stock portfolio.

Mathematical Overview

This Investigation will use the following skills and concepts from Chapters 4 and 5.

- simplifying fractions
- comparing fractions
- adding and subtracting fractions

Recommended Time		
Part	Pages	Time
Investigation	166–167	1 class period
Working on the Investigation	174, 209, 240	20 minutes each
Closing the Investigation	264	1 class period

Instructional Resources

Investigations and Projects Masters, pp. 9–12

A recording sheet, teacher notes, and scoring guide are provided for each Investigation in the *Investigations and Projects Masters.*

1 MOTIVATION

This Investigation uses research to study the stock market. Ask students to share what they know about stocks and the stock market.

LONG-TERM PROJECT

Investigation

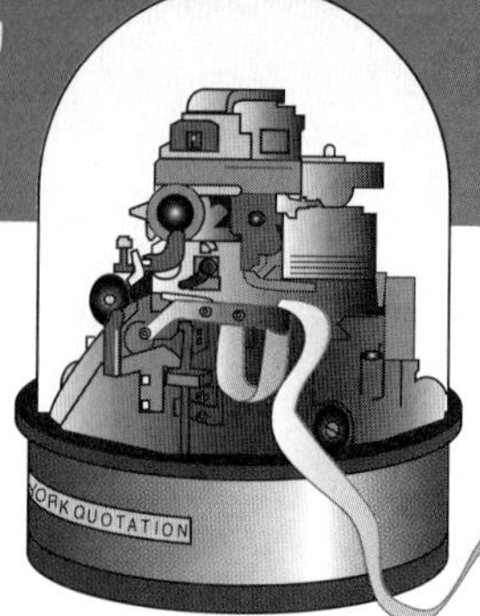

Good news: You have just inherited $10,000!

Bad news: You can't claim it until your 21st birthday.

More good news: By investing wisely, you'll have even more money in a few years!

There are several kinds of investments. You are probably most familiar with savings accounts, stocks, and bonds. Before making any kind of investment, you should learn as much as possible about investing in general.

MATERIALS

newspaper

Some important investment terms are:

- ***Capital Gain*** – *profit made by buying something at one price and selling it at a higher price*
- ***Dividend*** – *a payment made by a corporation to its stockholders*
- ***Market Price*** – *the price per share that a stock may be purchased for or sold for at a particular time*
- ***Security*** – *a document specifying the right to property in the form of stocks or bonds*
- ***Share*** – *unit used for owning stock*
- ***Stock Exchange*** – *place where the buying and selling of stocks takes place*

Upon the advice of a relative who is a financial counselor, you decide to invest your inheritance in the stock market.

When you purchase stock, you become part owner of a corporation. The stockholders of a company share in the profits and losses of the firm. If a company's profits are high, stockholders may receive a dividend.

How to Read a Newspaper Stock Report

Numbers in this column show the current annual dividend paid per share.

These columns show, in dollars, the highest and lowest prices paid in the last 52 weeks, excluding yesterday's trading.

Bold face type denotes something special about the stock. In this case, it shows stock that rose 3% or more.

This column lists the name of the stock or an abbreviation for it. You can find the full name for an abbreviation at the library or from a broker.

52-Week High	52-Week Low	Stock	Div	Sales 100s	Cls	Chg
19 1/4	5 1/8	Corprp		177	5 5/8	
26 3/4	**12 3/8**	**CntwCrd**	**.32**	**7171**	**21**	+7/8
19 5/8	16	CoutsPr	1.08f	102	19 1/8	+1/8
21 1/2	14	CosCm		3266	18 3/8	+1/8
39 1/2	25 3/4	Crane	.75	480	37 1/8	+1/8
17 3/4	14 3/8	CrwfdA	.58	27	16 1/8	+1/8
17 3/4	14 3/8	CrwfdB	.54	47	16 1/4	-3/8
29 1/4	14 5/8	CrayRs		947	23 7/8	+3/8
18 1/2	14 3/4	Credicp n		764	17 1/4	+1/4
35 7/8	25	CrescRE	2.20f	552	33 7/8	
61	36 3/4	Crestar	1.80	661	59 1/8	+1/8
9 1/4	6 1/2	CrimMa	.94f	215	8 7/8	+1/8
20	12	CrmpKno	.54	2238	12 3/4	-1/4
11 7/8	**9 1/2**	**CrosTim**	**1.00e**	**316**	**10 1/4**	**+3/8**
14 1/8	6 1/2	CrwnAm	.80	1357	7 5/8	
50 5/8	33 1/2	CwnCork		5613	41 1/8	+3/8
18	11 1/2	CwnCr	.12	119	12	
21 5/8	17	CwnPac	2.04	657	18 1/4	-1/8
24	23 1/4	CullgW		2216	24	
48 5/8	34	CumEng	1.00	1400	37 5/8	+1/8
27 3/4	10 3/4	CySem		38153	12 1/2	-1/2
32 1/8	24 1/4	Cyprus	.80a	2596	26 5/8	-1/4
64 1/4	31 3/4	Cytec		379	61 3/4	+1/2
14	10 1/8	CzechFd	.10e	172	13 1/2	

This column lists the number of *round lots* (group of 100s) traded during the day. *Odd lots* (groups with less than 100) are not included.

This is the *closing price*, the price per share in the last sale of the day.

This indicates that the closing price was one-half dollar or 50 cents less than on the preceding business day.

Before you decide in which stocks to invest your money, you should familiarize yourself with how a stock exchange reports its daily activity. These reports use several columns of numbers, symbols, and abbreviations.

Starting the Investigation

A list of investments, such as stocks, that a person or a company owns is called a **portfolio**. Stockbrokers recommend selecting stock from several companies so that if the price of one company's stock drops, it will not cause a large decrease in the total value of your portfolio.

Create your portfolio by "spending" no more than $10,000 on at least six different stocks. You will need to track your holdings and periodically analyze their net worth. You may change stock holdings by selling and buying other stocks at any time during this Investigation. Your goal is to have the largest capital gain possible by the end of this Investigation.

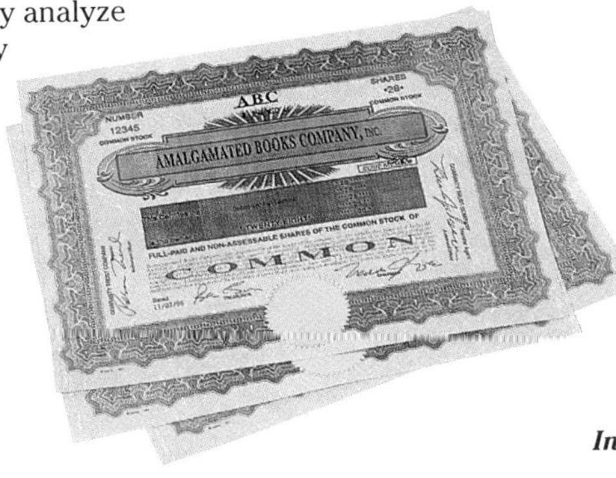

For current information on the stock market, visit: **www.glencoe.com/sec/math/prealg/mathnet**

Investigation
Taking Stock

Working on the Investigation
Lesson 4-1, p. 174

Working on the Investigation
Lesson 4-8, p. 209

Working on the Investigation
Lesson 5-3, p. 238

Closing the Investigation
End of Chapter 5, p. 264

Cooperative Learning

This Investigation offers an excellent opportunity for using cooperative learning groups. For more information on cooperative learning strategies and group management, see *Cooperative Learning in the Mathematics Classroom*, one of the titles in the Glencoe Mathematics Professional Series.

2 SETUP

You may wish to have a student read the Investigation to give information about stocks and the stock market and describe the activity. You may wish to bring some newspaper stock market reports to have students "purchase" their stocks. Students may complete the Investigation individually, in pairs, or in groups.

3 MANAGEMENT

Provide students with background information and materials that discuss different strategies for investing in the stock market.

Encourage students to follow the performance of the stocks they choose by reading newspaper reports or watching television newscasts.

Sample Answers

Answers will vary as they are based on the stock purchases of each student or group.

Investigations and Projects Masters, p. 12

NAME ________________ DATE ________

CHAPTERS 4 and 5 Investigation
Taking Stock

Student Edition
Pages 166–167

Complete the following table for your portfolio.

Name of Stock	Number of Shares	Price Per Share	Total Value
1.			
2.			
3.			
4.			
5.			
6.			

Is your stock portfolio diversified? Explain.

Explain why you chose each stock in your portfolio.

Did you spend the entire $10,000 on stocks? If not, you have some cash that is not invested in stocks. How will you invest the cash?

Please keep this page and any other research in your Investigation Folder.

CHAPTER

4 Exploring Factors and Fractions

PREVIEWING THE CHAPTER

This chapter relates the concepts of number theory to algebra. Students learn to find factors of numbers, to factor a number into primes, and evaluate expressions with powers and exponents. They find the greatest common factor and least common multiple and use these concepts to compare and simplify fractions. Finally, students multiply and divide monomials and learn how to apply negative exponents.

Lesson (Pages)	Lesson Objectives	NCTM Standards	State/Local Objectives
4-1 (170–174)	Determine whether one number is a factor of another.	1-6, 8, 9	
4-2 (175–179)	Use powers and exponents in expressions and equations.	1-8	
4-2B (180)	Use a graphing calculator to evaluate expressions involving exponents.	2, 3, 5, 9	
4-3 (181–183)	Solve problems by drawing diagrams.	1-6, 8	
4-4 (184–188)	Identify prime and composite numbers and write the prime factorizations of composite numbers.	1-6, 8	
4-4B (189)	Use a graphing calculator to investigate factor patterns.	1-3, 6, 8	
4-5 (190–194)	Find the greatest common factor for two or more integers or monomials.	1-6, 8, 9	
4-6A (195)	Use models to investigate fractions.	1-6	
4-6 (196–199)	Simplify fractions using the GCF.	1-6	
4-7 (200–204)	Find the least common multiple of two or more numbers and compare fractions.	1-6, 9	
4-8 (205–209)	Multiply and divide monomials.	1-9	
4-9 (210–214)	Use negative exponents in expressions and equations.	1-6, 9	

A complete, 1-page lesson plan is provided for each lesson in the ***Lesson Planning Guide***. Answer keys for each lesson are available in the ***Answer Key Masters***.

ORGANIZING THE CHAPTER

You may want to refer to the **Course Planning Calendar** on page T12 for detailed information on pacing.
PACING: **Standard—**13 days; **Honors—**12 days; **Block—**6 days

LESSON PLANNING CHART

Lesson (Pages)	Materials/ Manipulatives	Extra Practice (Student Edition)	Blackline Masters									Real-World Applications	Interactive Mathematics Tools Software	Teaching Transparencies	Group Activity Cards
			Study Guide	Practice	Enrichment	Assessment and Evaluation	Math Lab and Modeling Math	Multicultural Activity	Tech Prep Applications	Graphing Calculator	Activity				
4-1 (170–174)	grid paper 36 cubes* counters*	p. 748	p. 27	p. 27	p. 27		p. 79	p. 7						4-1A 4-1B	4-1
4-2 (175–179)		p. 748	p. 28	p. 28	p. 28			p. 8						4-2A 4-2B	4-2
4-2B (180)	graphing calculator									p. 18					
4-3 (181–183)	grid paper	p. 748	p. 29	p. 29	p. 29	p. 99								4-3A 4-3B	4-3
4-4 (184–188)	grid paper	p. 749	p. 30	p. 30	p. 30		p. 60				p. 34	8		4-4A 4-4B	4-4
4-4B (189)	graphing calculator									p. 19					
4-5 (190–194)	colored pencils tennis shoes	p. 749	p. 31	p. 31	p. 31	pp. 98, 99							4-5	4-5A 4-5B	4-5
4-6A (195)	ruler*						p. 35								
4-6 (196–199)		p. 749	p. 32	p. 32	p. 32					p. 4	p. 4			4-6A 4-6B	4-6
4-7 (200–204)		p. 749	p. 33	p. 33	p. 33	p. 100			p. 7					4-7A 4-7B	4-7
4-8 (205–209)		p. 750	p. 34	p. 34	p. 34				p. 8		p. 18	9	4-8	4-8A 4-8B	4-8
4-9 (210–214)		p. 750	p. 35	p. 35	p. 35	p. 100								4-9A 4-9B	4-9
Study Guide/ Assessment (216–219)						pp. 85–97, 101–103									

*Included in Glencoe's *Students Manipulative Kit* and *Overhead Manipulative Resources*.

ORGANIZING THE CHAPTER

All of the blackline masters in the Teacher's Classroom Resources are available on the ***Electronic Teacher's Classroom Resources*** CD-ROM.

Other Chapter Resources

Student Edition
Investigation, *Taking Stock* pp. 166–167
Chapter Opener, pp. 168–169
Working on the Investigation, p. 174, 209
From the Funny Papers, p. 179
Earth Watch p. 214

Teacher's Classroom Resources
Investigations and Projects Masters, pp. 41–44
Pre-Algebra Overhead Manipulative Resources, pp. 10–12

Technology
Test and Review Software (IBM & Macintosh)
CD-ROM Activities

Professional Publications

Block Scheduling Booklet
Glencoe Mathematics Professional Series

Outside Resources

Books/Periodicals
Middle Grade Mathematics Project Series: Factors and Multiples, Fitzgerald, et al
Middle Grade Mathematics Project Series: Similarity and Equivalent Fractions, Fitzgerald, et al
Key to Fractions, Key Curriculum Press

Videos/CD-ROMs
Fractions Video, ETA
Basic Number Concepts Video, ETA
Real Life Math Fractions Series, ETA

Software
Mastering Fractions, Phoenix/BFA Films and Videos

Assessment Resources

Student Edition
Math Journal, pp. 177, 207, 212
Mixed Review, pp. 174, 179, 183, 188, 194, 199, 204, 208, 214
Self Test, p. 194
Chapter Highlights, p. 215
Chapter Study Guide and Assessment, pp. 216–218
Alternative Assessment, p. 219
Portfolio, p. 219
Ongoing Assessment, pp. 220–221
MindJogger Videoquiz, 4

Teacher's Wraparound Edition
5-Minute Check, pp. 170, 175, 181, 184, 190, 196, 200, 205, 210
Checking Your Understanding, pp. 172, 177, 182, 187, 192, 198, 202, 207, 212
Closing Activity, pp. 174, 179, 183, 188, 194, 199, 204, 209, 214
Cooperative Learning, pp. 176, 191

Assessment and Evaluation Masters
Multiple-Choice Tests, Forms 1A (Honors), 1B (Average), 1C (Basic), pp. 85–90
Free-Response Tests, Forms 2A (Honors), 2B (Average), 2C (Basic), pp. 91–96
Performance Assessment, p. 97
Mid-Chapter Test, p. 98
Quizzes A-D, pp. 99–100
Standardized Test Practice, p. 101
Cumulative Review, pp. 102–103

ENHANCING THE CHAPTER

Examples of some of the materials for enhancing Chapter 4 are shown below.

Multicultural Activity Masters, pp. 7, 8

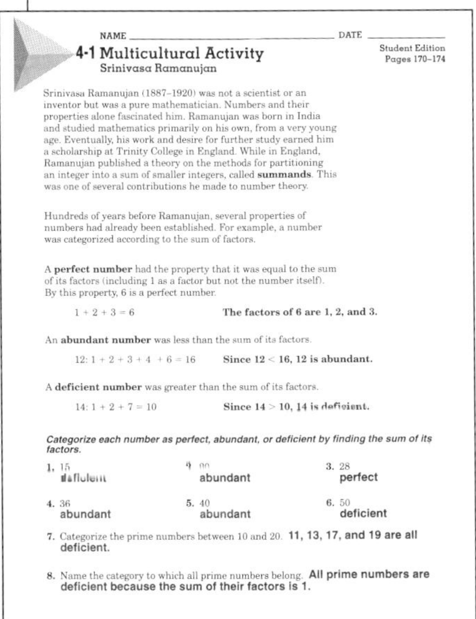

NAME ______________________ DATE __________

4-1 Multicultural Activity
Srinivasa Ramanujan

Student Edition
Pages 170–174

Srinivasa Ramanujan (1887–1920) was not a scientist or an inventor but was a pure mathematician. Numbers and their properties alone fascinated him. Ramanujan was born in India and studied mathematics primarily on his own, from a very young age. Eventually, his work and desire for further study earned him a scholarship at Trinity College in England. While in England, Ramanujan published a theory on the methods for partitioning an integer into a sum of smaller integers, called **summands**. This was one of several contributions he made to number theory.

Hundreds of years before Ramanujan, several properties of numbers had already been established. For example, a number was categorized according to the sum of factors.

A **perfect number** had the property that it was equal to the sum of its factors (including 1 as a factor but not the number itself). By this property, 6 is a perfect number.

$1 + 2 + 3 = 6$ **The factors of 6 are 1, 2, and 3.**

An **abundant number** was less than the sum of its factors.

$12: 1 + 2 + 3 + 4 + 6 = 16$ **Since 12 < 16, 12 is abundant.**

A **deficient number** was greater than the sum of its factors.

$14: 1 + 2 + 7 = 10$ **Since 14 > 10, 14 is deficient.**

Categorize each number as perfect, abundant, or deficient by finding the sum of its factors.

1. 15 **deficient**
2. [illegible] **abundant**
3. 28 **perfect**
4. 36 **abundant**
5. 40 **abundant**
6. 50 **deficient**
7. Categorize the prime numbers between 10 and 20. **11, 13, 17, and 19 are all deficient.**
8. Name the category to which all prime numbers belong. **All prime numbers are deficient because the sum of their factors is 1.**

Real-World Applications, 8, 9

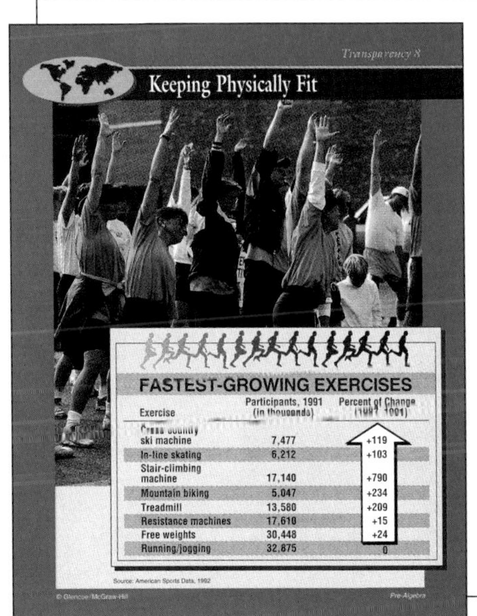

Exercise	Participants, 1991 (in thousands)	Percent of Change (1987–1991)
Cross-country ski machine	7,477	+119
In-line skating	6,212	+103
Stair-climbing machine	17,140	+790
Mountain biking	5,047	+234
Treadmill	13,580	+209
Resistance machines	17,610	+15
Free weights	30,448	+24
Running/jogging	32,875	0

Graphing Calculator Masters, p. 4

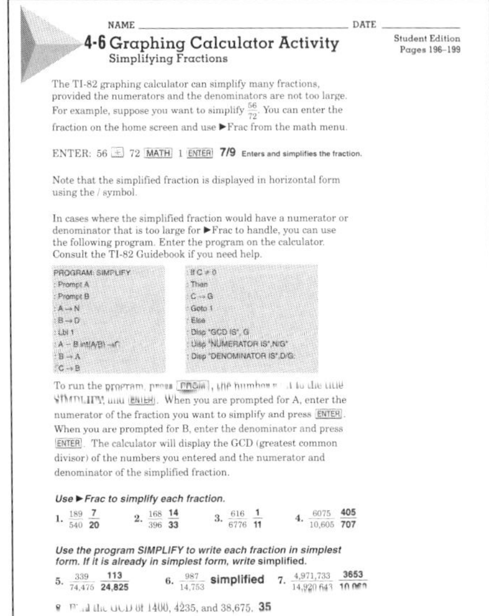

NAME ______________________ DATE __________

4-6 Graphing Calculator Activity
Simplifying Fractions

Student Edition
Pages 196–199

The TI-82 graphing calculator can simplify many fractions, provided the numerators and the denominators are not too large. For example, suppose you want to simplify $\frac{56}{72}$. You can enter the fraction on the home screen and use ►Frac from the math menu.

ENTER: 56 [÷] 72 [MATH] 1 [ENTER] **7/9** Enters and simplifies the fraction.

Note that the simplified fraction is displayed in horizontal form using the / symbol.

In cases where the simplified fraction would have a numerator or denominator that is too large for ►Frac to handle, you can use the following program. Enter the program on the calculator. Consult the TI-82 Guidebook if you need help.

```
PROGRAM: SIMPLIFY          :If C ≠ 0
:Prompt A                  :Then
:Prompt B                  :C → G
:A → N                     :Goto 1
:B → D                     :Else
:Lbl 1                     :Disp "GCD IS", G
:A – B int(A/B) → C        :Disp "NUMERATOR IS",N/G
:B → A                     :Disp "DENOMINATOR IS",D/G
:C → B
```

To run the program, press [PRGM], the number next to the title SIMPLIFY, and [ENTER]. When you are prompted for A, enter the numerator of the fraction you want to simplify and press [ENTER]. When you are prompted for B, enter the denominator and press [ENTER]. The calculator will display the GCD (greatest common divisor) of the numbers you entered and the numerator and denominator of the simplified fraction.

Use ►Frac to simplify each fraction.

1. $\frac{189}{540}$ $\mathbf{\frac{7}{20}}$
2. $\frac{168}{396}$ $\mathbf{\frac{14}{33}}$
3. $\frac{616}{6776}$ $\mathbf{\frac{1}{11}}$
4. $\frac{6075}{10{,}605}$ $\mathbf{\frac{405}{707}}$

***Use the program SIMPLIFY to write each fraction in simplest form. If it is already in simplest form, write* simplified.**

5. $\frac{339}{74{,}475}$ $\mathbf{\frac{113}{24{,}825}}$
6. $\frac{987}{14{,}753}$ **simplified**
7. $\frac{4{,}971{,}733}{14{,}920{,}643}$ **3653/[illegible]**
8. Find the GCD of 1400, 4235, and 38,675. **35**

Tech Prep Applications Masters, pp. 7, 8

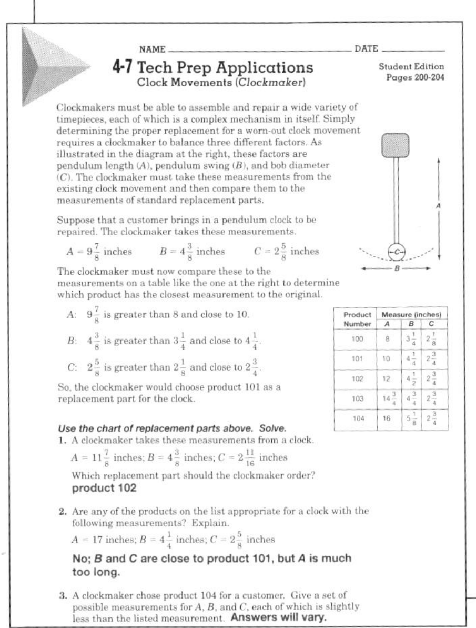

NAME ______________________ DATE __________

4-7 Tech Prep Applications
Clock Movements (*Clockmaker*)

Student Edition
Pages 200-204

Clockmakers must be able to assemble and repair a wide variety of timepieces, each of which is a complex mechanism in itself. Simply determining the proper replacement for a worn-out clock movement requires a clockmaker to balance three different factors. As illustrated in the diagram at the right, these factors are pendulum length (A), pendulum swing (B), and bob diameter (C). The clockmaker must take these measurements from the existing clock movement and then compare them to the measurements of standard replacement parts.

Suppose that a customer brings in a pendulum clock to be repaired. The clockmaker takes these measurements.

$A = 9\frac{7}{8}$ inches $\quad B = 4\frac{3}{8}$ inches $\quad C = 2\frac{5}{8}$ inches

The clockmaker must now compare these to the measurements on a table like the one at the right to determine which product has the closest measurement to the original.

A: $9\frac{7}{8}$ is greater than 8 and close to 10.

B: $4\frac{3}{8}$ is greater than $3\frac{1}{4}$ and close to $4\frac{1}{4}$.

C: $2\frac{5}{8}$ is greater than $2\frac{1}{8}$ and close to $2\frac{3}{4}$.

So, the clockmaker would choose product 101 as a replacement part for the clock.

Product Number	Measure (inches) A	B	C
100	8	$3\frac{1}{4}$	$2\frac{1}{8}$
101	10	$4\frac{1}{4}$	$2\frac{3}{4}$
102	12	$4\frac{1}{2}$	$2\frac{3}{4}$
103	$14\frac{3}{4}$	$4\frac{3}{4}$	$2\frac{3}{4}$
104	16	$5\frac{1}{8}$	$2\frac{3}{4}$

Use the chart of replacement parts above. Solve.

1. A clockmaker takes these measurements from a clock.
   $A = 11\frac{7}{8}$ inches; $B = 4\frac{3}{8}$ inches; $C = 2\frac{11}{16}$ inches
   Which replacement part should the clockmaker order? **product 102**
2. Are any of the products on the list appropriate for a clock with the following measurements? Explain.
   $A = 17$ inches; $B = 4\frac{1}{4}$ inches; $C = 2\frac{5}{8}$ inches
   No; *B* and *C* are close to product 101, but *A* is much too long.
3. A clockmaker chose product 104 for a customer. Give a set of possible measurements for A, B, and C, each of which is slightly less than the listed measurement. **Answers will vary.**

Activity Masters, pp. 4, 18, 34

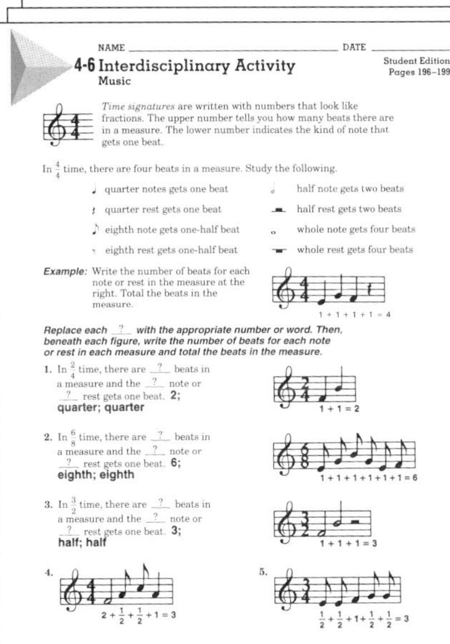

NAME ______________________ DATE __________

4-6 Interdisciplinary Activity
Music

Student Edition
Pages 196–199

Time signatures are written with numbers that look like fractions. The upper number tells you how many beats there are in a measure. The lower number indicates the kind of note that gets one beat.

In $\frac{4}{4}$ time, there are four beats in a measure. Study the following.

- quarter notes gets one beat
- quarter rest gets one beat
- eighth note gets one-half beat
- eighth rest gets one-half beat
- half note gets two beats
- half rest gets two beats
- whole note gets four beats
- whole rest gets four beats

Example: Write the number of beats for each note or rest in the measure at the right. Total the beats in the measure.

$1 + 1 + 1 + 1 = 4$

Replace each ? with the appropriate number or word. Then, beneath each figure, write the number of beats for each note or rest in each measure and total the beats in the measure.

1. In $\frac{2}{4}$ time, there are __?__ beats in a measure and the __?__ note or __?__ rest gets one beat. **2; quarter; quarter**
   $1 + 1 = 2$
2. In $\frac{6}{8}$ time, there are __?__ beats in a measure and the __?__ note or __?__ rest gets one beat. **6; eighth; eighth**
   $1 + 1 + 1 + 1 + 1 + 1 = 6$
3. In $\frac{3}{2}$ time, there are __?__ beats in a measure and the __?__ note or __?__ rest gets one beat. **3; half; half**
   $1 + 1 + 1 = 3$
4. $2 + \frac{1}{2} + \frac{1}{2} + 1 = 3$
5. $\frac{1}{2} + \frac{1}{2} + 1 + \frac{1}{2} + \frac{1}{2} = 3$

Math Lab and Modeling Math Masters, pp. 60, 79

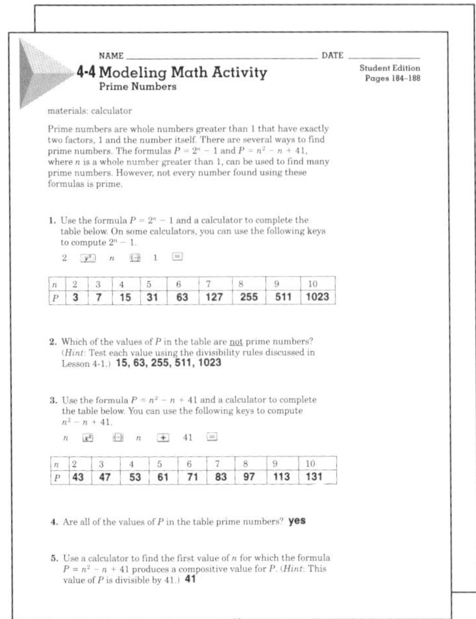

NAME ______________________ DATE __________

4-4 Modeling Math Activity
Prime Numbers

Student Edition
Pages 184–188

materials: calculator

Prime numbers are whole numbers greater than 1 that have exactly two factors, 1 and the number itself. There are several ways to find prime numbers. The formulas $P = 2^n - 1$ and $P = n^2 - n + 41$, where n is a whole number greater than 1, can be used to find many prime numbers. However, not every number found using these formulas is prime.

1. Use the formula $P = 2^n - 1$ and a calculator to complete the table below. On some calculators, you can use the following keys to compute $2^n - 1$.
   2 [y^x] n [–] 1 [=]

n	2	3	4	5	6	7	8	9	10
P	**3**	**7**	**15**	**31**	**63**	**127**	**255**	**511**	**1023**

2. Which of the values of P in the table are not prime numbers? (*Hint*: Test each value using the divisibility rules discussed in Lesson 4-1.) **15, 63, 255, 511, 1023**
3. Use the formula $P = n^2 - n + 41$ and a calculator to complete the table below. You can use the following keys to compute $n^2 - n + 41$.
   n [x^2] [–] n [+] 41 [=]

n	2	3	4	5	6	7	8	9	10
P	**43**	**47**	**53**	**61**	**71**	**83**	**97**	**113**	**131**

4. Are all of the values of P in the table prime numbers? **yes**
5. Use a calculator to find the first value of n for which the formula $P = n^2 - n + 41$ produces a compositive value for P. (*Hint*: This value of P is divisible by 41.) **41**

MAKING MATHEMATICS RELEVANT

This two-page introduction to the chapter provides students with an opportunity to explore other disciplinary topics and their applications to mathematics.

Background Information

Math and Computers in the News The digital technology used in CD-ROMs means that information is stored in the form of discrete values or numbers. To record, sound is converted into billions of binary numbers (1 or 0) and stored. A laser beam is used to "read" the stored data and convert it back into sound. A typical CD-ROM can hold about 800 megabytes (800 million bytes) of data.

CHAPTER 4 Exploring Factors and Fractions

TOP STORIES in Chapter 4

In this chapter, you will:

- use powers and exponents,
- solve problems by drawing a diagram,
- find the greatest common factor of two or more numbers or monomials,
- simplify and compare fractions, and
- multiply and divide monomials.

MATH AND COMPUTERS IN THE NEWS

The Incredible Shrinking CD-ROM

Source: CD-ROM News Extra, August, 1994

How many CDs can you fit on a CD? Sound like "Who's buried in Grant's tomb?" Not really. Scientists are working on a method for expanding the amount of information that can be stored on a CD by stacking superthin recording surfaces on top of each other on one disk. This new storage system will make the CD a three-dimensional system. Each layer contains data which is read by a movable lens. The lens will move up and down to choose the layer it will access. The number of possible data layers is limited only by the power of the CD-ROM drive's scanning laser, the transparency of the storage layers, and the production cost. So how many CDs *can* you fit on a CD? . . . who knows?!

Putting It into Perspective

1921 Pace Phonograph Company, the first African-American-owned record company, sells over 500,000 records under the Black Swan label.

1850 1880 1910

1877 Thomas Edison invents the phonograph using a wax cylinder covered with tin foil.

1887 Emile Berliner invents the audio disk.

Putting It into Perspective

The *Xplora 1* CD-ROM produced by singer Peter Gabriel provides a comprehensive interactive experience. It is part-video game, part musical encyclopedia, and part multimedia rock concert. Users can sample live performances by numerous artists, sounds of different musical instruments, and CD recordings of songs from around the world. Both this disc and the disc produced by Todd Rundgren, *No World Order*, allow the user to actually remix and change the sound of some of the music.

EMPLOYMENT OPPORTUNITIES

Many Opportunities in a Fast-Growing Field

DO YOU LIKE to work in a fast-paced, ever-changing environment? You may be suited for a career in **multimedia development**. You will work on developing laserdisc, video, and CD-ROM materials for entertainment, education, and business applications. College degree in computer science desirable, but not necessary.

For more information, contact:
Interactive Multimedia Association
3 Church Circle, Suite 800
Annapolis, MD 21401-1933

Statistical Snapshot

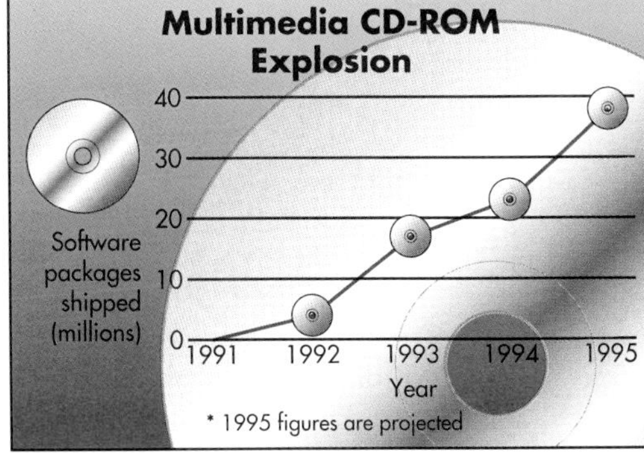

Source: Information and Interactive Services Report, Dataquest

interNET CONNECTION For up-to-date information on multimedia development, visit: **www.glencoe.com/sec/math/prealg/mathnet**

1979 A prototype of the compact disc player is demonstrated by the N.V. Phillips company.

1994 Multilevel CDs developed

1940 — 1970 — 2000

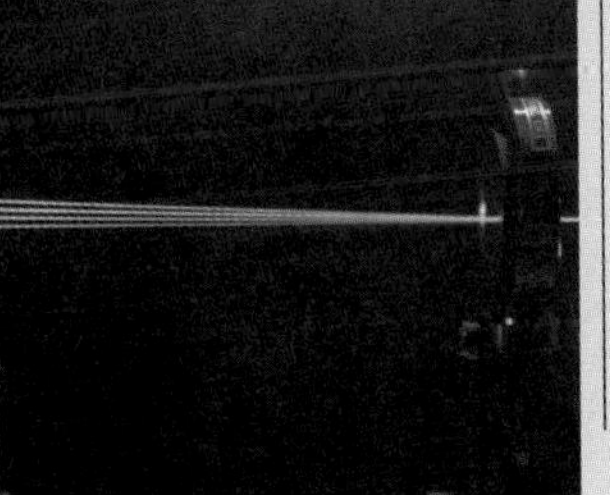

1958 Invention of the laser beam

1993 Todd Rundgren and Peter Gabriel produce the first interactive CD-ROM for rock music fans.

EMPLOYMENT OPPORTUNITIES

Multimedia employers are often looking for people with broad interests and backgrounds. If they could, they would like to hire people who are creative,. technically oriented, capable of programming and communicating well, and who have a good understanding of both the book publishing and electronic publishing industries. Because the multimedia field is so new, people like this are hard to find.

Statistical Snapshot

The graph shows the amazing growth in the number of CD-ROMs produced in recent years. Ask students why they think the number of CD-ROMs produced was zero in 1991.

Investigations and Projects Masters, p. 41

NAME ______________________ DATE ____________

CHAPTER 4 **Project A**
Comparing Information Sources

Each day brings new advances in CD-ROM technology. Will this electronic interactive tool make other sources of information obsolete? Maybe. But, maybe not. Think about calculators. Has the use of calculators caused paper and pencil calculations and mental math to fall by the wayside? Suppose you want to find $\frac{1}{4}$ of 2000. Which is quicker, mental math, pencil and paper, or a calculator?

Consider that despite the wonders of interactive multimedia technology, there may still be a place for "old fashioned" print materials like books, magazines, and newspapers. Explore this possibility by looking into how using CD-ROM reference sources compares with using traditional sources.

1. Choose a CD-ROM reference source, such as an encyclopedia, a thesaurus, a dictionary, a sports encyclopedia, or even a movie guide. Examine the interactive tool selected. Browse through it. Look something up. Think about how easy it is to get started, to move around within the reference, and to locate information.

2. Select a traditional source to compare with the CD-ROM product. Visit the library and look for similar information in an encyclopedia or other source. Compare the traditional research process with the experience using the CD-ROM resource. Think about which source of data is easier to use and which provides more information.

3. Focus on other aspects of using a reference source. Compare the CD-ROM product with the traditional source by investigating and thinking about questions like these: How do the two products compare in price? How do they compare in size? Which is easier to store? Which is easier to update? Which do you prefer, and why?

4. Use your experiences with both electronic and print reference materials to prepare a chart or poster that compares the pros and cons of using each kind of tool. Use fractions to compare sizes and costs. Prepare a short speech to accompany the visual presentation in which you summarize what you discovered and draw conclusions from your findings.

Cooperative Learning

Chapter Projects Two chapter projects are included in the *Investigations and Projects Masters.* In Chapter Project A, students extend the topic in the Chapter Opener. In Chapter Project B, students explore music. A student page and a parent letter are provided for each Chapter Project.

interNET CONNECTION

Glencoe has made every effort to ensure that the website links for *Pre-Algebra* at www.glencoe.com/sec/math/prealg/mathnet are current and contain appropriate content. However, these website links are not under Glencoe's control.

4-1 LESSON NOTES

NCTM Standards: 1-6, 8, 9

Instructional Resources

- Study Guide Master 4-1
- Practice Master 4-1
- Enrichment Master 4-1
- Group Activity Card 4-1
- Math Lab and Modeling Math Masters, p. 79
- Multicultural Activity Masters, p. 7

Transparency 4-1A contains the 5-Minute Check for this lesson; **Transparency 4-1B** contains a teaching aid for this lesson.

Recommended Pacing	
Standard Pacing	Day 1 of 13
Honors Pacing	Day 1 of 12
Block Scheduling*	Day 1 of 6 (along with Lesson 4-2)

*For more information on pacing and possible lesson plans, refer to the ***Block Scheduling Booklet***.

1 FOCUS

5-Minute Check
(over Chapter 3)

Solve. Check each solution.

1. $x - 3 = 9$ **12**
2. $2x = -16$ **-8**
3. $-12 = -\frac{x}{6}$ **72**
4. $5 + x > -2$ **$x > -7$**
5. $24 \leq 6x$ **$x \geq 4$**

4-1 Factors and Monomials

Setting Goals: *In this lesson, you'll determine whether one number is a factor of another.*

Modeling with Manipulatives

MATERIALS

grid paper

Although it may seem like they are not at all related, algebra and geometry are so interconnected that you can learn about algebra by exploring geometric figures.

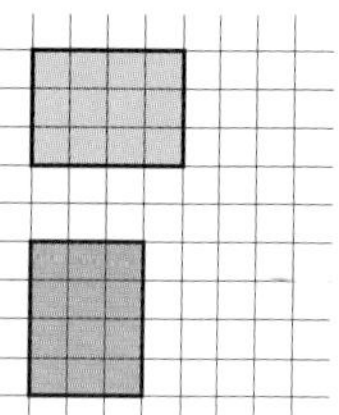

Your Turn

Two different rectangular arrangements of 12 squares are shown at the right. There are six possible rectangular arrangements of 12 squares. Use grid paper to draw the remaining four arrangements.

TALK ABOUT IT

a. 1 × 12, 12 × 1, 2 × 6, 6 × 2, 3 × 4, 4 × 3

c. You can find the pairs of factors of a given number by using rectangular arrangements of that number of squares.

a. You know that $3 \times 4 = 12$. List all of the pairs of whole numbers that have a product of 12. **b. They are the same.**

b. Compare your list of whole number pairs to the lengths of the sides of the rectangles that you drew. What do you observe?

c. Write a sentence about the relationship between rectangular arrangements of squares and pairs of whole number factors of a number.

d. Can you arrange 14 squares in a rectangle so that one side of the rectangle is 2 units long? 3 units long? Did you know whether or not it could be done before you tried? **yes; no; See students' work.**

Learning the Concept

The **factors** of a whole number divide that number with a remainder of 0. For example, 2 is a factor of 12 because $12 \div 2 = 6$, and 5 is not a factor of 12 because $12 \div 5 = 2$ with a remainder of 2.

Another way of saying that 6 is a factor of 12 is to say that 12 is **divisible** by 6.

Sometimes you can test for divisibility mentally. These rules will help you determine whether a number is divisible by 2, 3, 5, 6, or 10.

A number is divisible by:

- 2 if the ones digit is divisible by 2.
- 3 if the sum of its digits is divisible by 3.
- 5 if the ones digit is 0 or 5.
- 6 if the number is divisible by 2 and 3.
- 10 if the ones digit is 0.

A number that is divisible by 2 is called ***even***. *An* ***odd*** *number is not divisible by 2.*

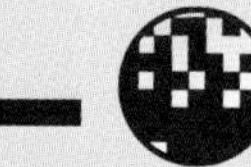

Alternative Teaching Strategies

Student Diversity Provide students with 40 coins, centimeter cubes, or other objects. Have them randomly choose a number of objects and write the number in the first column of a chart like the one used in the Motivating the Lesson activity. Students separate the objects to see if they can be shared by 2, 3, 4, 5, 6, and 10 people. Then they write Y for yes or N for no to show if the number can be equally shared by that number of people.

Example 1 **Determine whether 146 is divisible by 2, 3, 5, 6, or 10.**

Number	Divisible?	Reason
2	yes	The ones digit is 6 and 6 ÷ 2 = 3.
3	no	The sum of the digits is 1 + 4 + 6 or 11. 11 is not divisible by 3.
5	no	The ones digit of 146 is 6, not 0 or 5.
6	no	146 is divisible by 2 but not by 3.
10	no	The ones digit of 146 is not 0.

Use a calculator to verify the divisibility of each number.
So, 146 is divisible by 2, but not 3, 5, 6, or 10.

THINK ABOUT IT

Explain why you can say that a number is divisible by 10 if it is divisible by 2 and 5.

Because 2 and 5 are factors of 10.

As the rules for divisibility suggest, a pattern can be found in many numbers that share factors. Looking for a pattern can help you determine other rules for divisibility.

Example 2 **Look for a pattern in the digits of numbers to write a rule for determining if a number is divisible by 4.**

PROBLEM SOLVING
Look for a Pattern

Explore We know a rule to determine if a number is divisible by 2. We need to find a rule for numbers divisible by 4.

Plan Use a calculator to determine many numbers that are divisible by 4. Then look for a pattern among those numbers.

Solve Since 2 is a factor of 4, any number divisible by 4 must also be divisible by 2. So, let's look at some even numbers.

Number	Last Two Digits	Divisible by 4?
16	16	yes
28	28	yes
112	12	yes
148	48	yes
254	54	no
284	84	yes
1016	16	yes
2026	26	no

In all of the numbers that are divisible by 4, the last two digits form a number that is divisible by 4. For example in 112, 12 is divisible by 4. In 254, 54 is not divisible by 4, so neither is 254 divisible by 4.

A possible rule for determining whether a number is divisible by 4 is that if the last two digits of a number are divisible by 4, then the number is divisible by 4.

Examine According to the rule, 524, 816, and 1248 are divisible by 4. Use a calculator to check.

Being able to determine whether one number is divisible by another is a useful skill that can be applied in everyday tasks.

Motivating the Lesson

Hands-On Activity Place 30 counters on the overhead. Ask students if it is possible to distribute the counters evenly to three students without having any left over. **yes** Then ask for other numbers of students that you can distribute an equal amount of counters to and not have any left over. **1, 2, 5, 6, 10, 15, 30**

Number of Objects	Can the objects be shared equally by this many people?					
	2	3	4	5	6	10
30	Y	Y	N	Y	Y	Y

2 TEACH

In-Class Examples

For Example 1
Determine whether 255 is divisible by 2, 3, 5, 6, or 10. **2, no; 3, yes; 5, yes; 6, no; 10, no**

For Example 2
Look for a pattern in the sum of digits to write a rule for determining whether a number is divisible by 9. **If the sum of the digits of a number is divisible by 9, then the number is divisible by 9.**

Classroom Vignette

"I like to allow students to discover the rules for divisibility. They can use calculators to explore numbers that are divisible by a given factor and then use a pattern to write the divisibility rule."

Yuria R. Alban

Yuria Alban
Author

In-Class Examples

For Example 3
Determine whether the years 1990, 1900, 1880, and 1800 were leap years. **1990, no; 1900, no; 1880, yes; 1800, no.**

For Example 4
Determine whether each expression is a monomial.

a. $6(y - 3)$ **no**

b. $-903mnp$ **yes**

Teaching Tip Clarify that a monomial is the product of variables and rational numbers and may have a number as a denominator, but not a variable. For example, $\frac{x}{2}$ is a monomial, but $\frac{2}{x}$ is not a monomial.

3 PRACTICE/APPLY

Checking Your Understanding

Exercises 1–14 are designed to help you assess your students' understanding through reading, writing, speaking, and modeling. You should work through Exercises 1–5 with your students and then monitor their work on Exercises 6–14.

Study Guide Masters, p. 27

NAME ______________________ DATE ____________

4-1 Study Guide
Factors and Monomials

Student Edition Pages 170–174

The **factors** of a whole number divide that number with no remainder.

3 is a factor of 12 because $12 \div 3 = 4$ with no remainder.
3 is not a factor of 16 because $16 \div 3 = 5$ with a remainder of 1.

Because $12 \div 3 = 4$, with no remainder, we say that 12 is divisible by 3.

A number is divisible by:
- 2 if the ones digit is divisible by 2.
- 3 if the sum of the digits is divisible by 3.
- 5 if the ones digit is 0 or 5.
- 6 if the number is divisible by 2 *and* 3.
- 10 if the ones digit is 0.

A **monomial** is an integer, a variable, or a product of integers or variables.

Expressions like -537 and $8ac$ are monomials.
Expressions like $4t + 5$ and $-2(3x - 3)$ are not monomials.

Using divisibility rules, state whether each number is divisible by 2, 3, 5, 6, or 10.

1. 1060 **2, 5, 10**
2. 996 **2, 3, 6**
3. 285 **3, 5**
4. 705 **3, 5**
5. 32 **2**
6. 64,230 **2, 3, 5, 6, 10**
7. 1645 **5**
8. 3241 **none**
9. 42,246 **2, 3, 6**

Determine whether each expression is a monomial. Explain why or why not.

10. $-3mn$ **Yes; it is the product of an integer and a variable.**
11. $-4x + y$ **No; it involves addition.**
12. 672 **Yes; it is an integer.**
13. $65cde$ **Yes; it is the product of an integer and variables.**
14. k **Yes; it is a variable.**
15. $-9(3x - 2)$ **No; it involves subtraction.**

Example 3

CONNECTION
Social Studies

In the Gregorian Calendar, each "common year" is 365 days long. Years that are divisible by 4, called "leap years" are 366 days long. Also, years ending in "00" that are divisible by 400 are leap years and those that are not are common. Determine whether 1998, 2000, 2066, and 2116 are leap years.

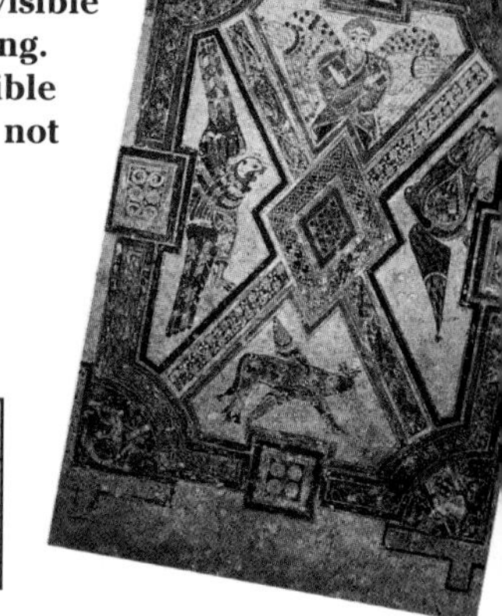

Use the rule for divisibility by 4 that we discovered in Example 2.

CULTURAL CONNECTIONS
One of the earliest calendars was created by the Babylonians. Their year started at the vernal equinox, which occurs near mid-March in our calendar.

Year	Divisible by 4?	Leap Year?
1998	no, 98 is not divisible by 4	no
2066	no, 66 is not divisible by 4	no
2116	yes, 16 is divisible by 4	yes

Since 2000 ends in "00", it must be divisible by 400 to be a leap year. The quotient $2000 \div 400$ is 5, so 2000 is a leap year.

The years 2000 and 2116 are leap years, but 1998 and 2066 are not.

Connection to Algebra

We know that 5 and 12 are factors of 60 because $5 \cdot 12 = 60$. Consider the algebraic expression $5x$. Since this notation means $5 \cdot x$, it follows that 5 and x are factors of $5x$.

An expression like $5x$ is called a **monomial**. A monomial is an integer, a variable, or a product of integers or variables. Other monomials are z, 16, mn, and $6ab$. Expressions like $4x - 2$ and $x + 3$ are not monomials since they involve addition or subtraction.

Example 

Determine whether each expression is a monomial.

a. $5(x + y)$

This expression is not a monomial because it involves addition.

b. $-25xyz$

This expression is a monomial because it is the product of integers and variables.

Checking Your Understanding

Communicating Mathematics

Read and study the lesson to answer these questions. **1–3. See margin.**

1. **Explain** why the diagram at the right demonstrates that 8 is a factor of 40.

8
5
Area = 40 sq. units

2. **Determine** whether 6 is a factor of 78. Explain.
3. Is $7n$ a monomial? Justify your answer.
4. What number is a factor of every nonzero number? **1**
5. You know that the factors of a number can be modeled using rectangles. Arrange 36 small cubes in a shape like a rectangular box in as many ways as you can. Write a multiplication sentence for each model you created. **See margin.**

MATERIALS

36 cubes

CULTURAL CONNECTIONS
The Hindu calendar is circular and based on lunar months.

Additional Answers

1. Because a solid rectangular shape can be made.
2. Yes; with 78 squares, a 6×13 rectangle can be made.
3. Yes; it is a product of an integer and a variable.

Guided Practice

Using divisibility rules, state whether each number is divisible by 2, 3, 5, 6, or 10. 9. 2, 3, 5, 6, 10

6. 39 3 **7.** 118 2 **8.** 876 2, 3, 6 **9.** 5010

Determine whether each expression is a monomial. Explain why or why not. See Solutions Manual for explanations.

10. 24 yes **11.** $-3m$ yes **12.** $4y + 2$ no **13.** $5xyz$ yes

14. Music The 78 members of the Brookhaven High School Marching Band will be performing in the Independence Day parade on Saturday. Due to the width of the street they will be marching down, the parade director has recommended that all bands march in rows of 6. Can the whole Brookhaven band be arranged in rows of 6? yes

Exercises: Practicing and Applying the Concept

Independent Practice

Using divisibility rules, state whether each number is divisible by 2, 3, 5, 6, or 10. 22. 2, 3, 5, 6, 10

A

15. 111 3 **16.** 11,222 2 **17.** 5050 2, 5, 10 **18.** 4444 2

19. 10,505 5 **20.** 24,640 2, 5, 10 **21.** 117 3 **22.** 330

23. 49 none **24.** 10,523 none **25.** 434 2 **26.** 378 2, 3, 6

Determine whether each expression is a monomial. Explain why or why not. See Solutions Manual for explanations.

27. -234 yes **28.** $3c - 3$ no **29.** $-4b$ yes **30.** $6(xy)$ yes

31. $4(3x - 1)$ no **32.** v yes **33.** $q + r$ no **34.** $166h$ yes

35. Replace the _?_ in 82_?_ so that the number is divisible by 4. 0, 4, or 8

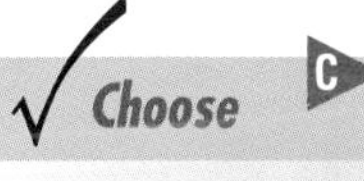

Estimation
Mental Math
Calculator
Paper and Pencil

C

Use mental math, paper and pencil, or a calculator to find at least one number that satisfies each condition. Sample answers given. 37. 1035

36. is divisible by 2, 3, and 5 30

37. has four digits and is divisible by 3 and 5, but is not divisible by 10

38. has 3 digits and is not divisible by 2, 3, 5, or 10 343

39. has four equal digits and is divisible by 5 5555

Critical Thinking

40. a. What is the greatest three-digit number that is not divisible by 2, 3, 5, or 10? 997

b. What is the least three-digit number that is not divisible by 2, 3, 5, or 10? 101

41a. Answers will vary.

Applications and Problem Solving

41. Social Studies Chinese New Year, the most important holiday of the Chinese year, comes in late January or early February. The Chinese calendar follows an annual cycle, which repeats every twelve years. For example, 1996 was the Year of the Rat, so 1996 + 12 or 2008 and 1996 − 12 or 1984 are also the Year of the Rat.

Boar–2007, Ox–1997, Tiger–1998, Hare–1999, Dragon–2000, Snake–2001, Horse–2002, Sheep–2003, Monkey–2004, Rooster–2005, Dog–2006; 1996 Rat

a. 1998 is the Year of the Tiger. How old is the youngest person born in a Year of the Tiger?

b. List five years that are Years of the Rat. 1996, 2008, 2020, 2032, 2044

Error Analysis

Students may think that a number is divisible by 3 if the last digit is divisible by 3. Provide students with examples that show this is not true, such as 13, 26, and 49.

Assignment Guide

Core: 15–39 odd, 40–41, 43–50
Enriched: 16–38 even, 40–50

For **Extra Practice**, see p. 748.

The red A, B, and C flags, printed only in the Teacher's Wraparound Edition, indicate the level of difficulty of the exercises.

Additional Answers

5. Possible answers: $2 \cdot 3 \cdot 6 = 36$; $1 \cdot 1 \cdot 36$; $3 \cdot 3 \cdot 4$; $2 \cdot 2 \cdot 9$; $1 \cdot 3 \cdot 12$; $1 \cdot 2 \cdot 18$

Practice Masters, p. 27

NAME ______________ DATE ________

4-1 Practice
Factors and Monomials

Student Edition Pages 170–174

Using divisibility rules, state whether each number is divisible by 2, 3, 5, 6, or 10.

1. 39 3	2. 82 2	3. 157 none	4. 56 2
5. 315 3, 5	6. 30 2, 3, 5, 6, 10	7. 81 3	8. 105 3, 5
9. 136 2	10. 195 3, 5	11. 75 3, 5	12. 29 none
13. 350 2, 5, 10	14. 42 2, 3, 6	15. 50 2, 5, 10	16. 86 2
17. 72 2, 3, 6	18. 88 2	19. 90 2, 3, 5, 6, 10	20. 27 3
21. 70 2, 5, 10	22. 45 3, 5	23. 96 2, 3, 6	24. 100 2, 5, 10
25. 69 3	26. 74 2	27. 85 5	28. 78 2, 3, 6
29. 1025 5	30. 969 3	31. 805 5	32. 888 2, 3, 6
33. 177 3	34. 1046 2	35. 282 2, 3, 6	36. 1010 2, 5, 10
37. 6237 3	38. 3762 2, 3, 6	39. 2367 3	40. 7623 3

Determine whether each expression is a monomial. Explain why or why not.

41. $21abc$ Yes; it is the product of an integer and variables.	42. $-3(x + y)$ No; it involves addition.	43. $4n - 7$ No; it involves subtraction.	44. -512 Yes; it is just an integer.
45. z Yes; it is just a variable.	46. $-16p$ Yes; it is the product of an integer and a variable.	47. $r - st$ No; it involves subtraction.	48. $35df$ Yes; it is the product of an integer and variables.

Group Activity Card 4-1

Factor To Win Group Activity **4-1**

MATERIALS: Pencil • paper

In this game you try to get more points than your opponent by choosing the best numbers from those that are less than the starting number. You get as many points as the number you choose, but your opponent gets the total from the factors of the number you chose. Make your best choice, but remember each number may be used only once.

The winner is the person who has the highest score after all numbers are used.

Example: If the given number is 25 and you start out by choosing number 24, you will score 24 points; however, your opponent will score the sum of 1, 2, 4, 6, 8, and 12 since these are all factors of 24. He or she will be ahead of you, and all the numbers listed may not be counted again this game. On your opponent's turn if the number 16 is picked, you will score 0 points since all the factors of 16 have been used.

Starting Numbers: 39, 42, 56, 122, 339, 57, 79, 81, 184, 49, 67, 228

 Pre-Algebra

Reteaching

Using Tables Display a chart of the first 20 multiples of 2, 3, 4, 5, 6, and 10 on the chalkboard or overhead. Have students describe the similarities and patterns in each sequence.

Teaching Tip After students complete Exercise 43, they might ask about numbers that are not perfect. If the sum of the factors is more than the number, the number is *abundant*. If the sum of the factors is less than the number, the number is *deficient.*

Additional Answer

42b. See students' work. It could be a 1-by-51 or a 3-by-17 rectangle, but neither of these is well suited to the space available.

4 ASSESS

Closing Activity

Writing Have students make study cards by writing divisibility rules in their own words.

Enrichment Masters, p. 27

NAME ______________________ DATE __________

4-1 Enrichment Student Edition Pages 170–174
Divisibility

Divisibility rule for 7

Determine the divisibility by 7 for 4032.

- 4032 Cross out the ones digit.
  − 4 Subtract twice the value of the ones digit from the rest of the number.
  399 If the difference is a number that you know is divisible by 7, stop. If not, repeat.
  − 18
  21 Since 21 is divisible by 7, 4032 is divisible by 7.

Divisibility rule for 11

Determine the divisibility by 11 for 5159.

- 5159 Cross out the ones digit.
  − 9 Subtract the value of the ones digit from the rest of the number.
  506 If the difference is a number that you know is divisible by 11, stop. If not, repeat.
  − 6
  44 Since 44 is divisible by 11, 5159 is divisible by 11.

- 5159 Note: The 1st digit is 5, 2nd digit is 1, 3rd digit is 5, 4th digit is 9.
  5 + 5 = 10 Find the sum of the values of the odd-numbered digits.
  1 + 9 = 10 Find the sum of the values of the even-numbered digits.
  0 Subtract the sums. If the difference is divisible by 11, the number is divisible by 11.
  Since 0 is divisible by 11, 5159 is divisible by 11.

Determine the divisibility by 11 for 6238.

- 6238
  2 + 8 = 10
  6 + 3 = 9
  1 Since 1 is not divisible by 11, 6238 is not divisible by 11.

Determine the divisibility by 7 and 11 for each of the following.

1. 266 divisible by 7, not divisible by 11
2. 4312 divisible by 7 and 11
3. 8976 not divisible by 7, divisible by 11
4. 936 not divisible by 7 or 11
5. 13,293 divisible by 7, not divisible by 11
6. 7085 not divisible by 7 or 11
7. 2957 not divisible by 7 or 11
8. 3124 not divisible by 7, divisible by 11
9. 6545 divisible by 7 and 11

42. **Civics** Each state has a star on the U.S. flag and the stripes represent the thirteen British colonies that became the United States.
    a. How do you think the 48 stars were arranged before Alaska and Hawaii joined the Union?
    b. Suggest an arrangement for the stars if another state were to be added. Could it be rectangular? **See margin.**

42a. in a 6-by-8 arrangement

43. **Number Theory** A perfect number is equal to the sum of its factors, except itself. There are two perfect numbers less than 30. The third perfect number is 496, since 496 = 1 + 2 + 4 + 8 + 16 + 31 + 62 + 124 + 248.
    a. Find a perfect number between 1 and 10. **6**
    b. Find a perfect number between 20 and 30. **28**

Mixed Review

44. **Geography** Colorado is less than twice as large as Arkansas. Colorado is 103,730 square miles. How large is Arkansas? (Lesson 3-8) **more than 51,865 square miles**
45. Solve $89 = \frac{x}{-3}$. (Lesson 3-3) **−267**
46. Solve $f = -64 \div -4$. (Lesson 2-8) **16**
47. Solve $t = -6 + (-14) + 12$. (Lesson 2-4) **−8**
48. Solve $5b = 95$ mentally. (Lesson 1-6) **19**
49. Evaluate $ac - 3b$ if $a = 7$, $b = 3$, and $c = 6$. (Lesson 1-3) **33**
50. Find the value of $3 \cdot (3 + 5) \div 2$. (Lesson 1-2) **12**

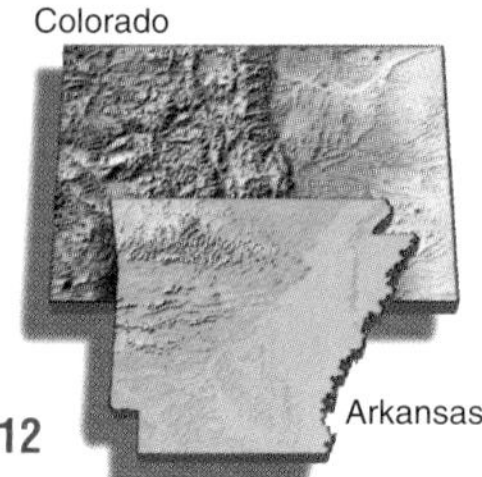

WORKING ON THE

Investigation

TAKING STOCK

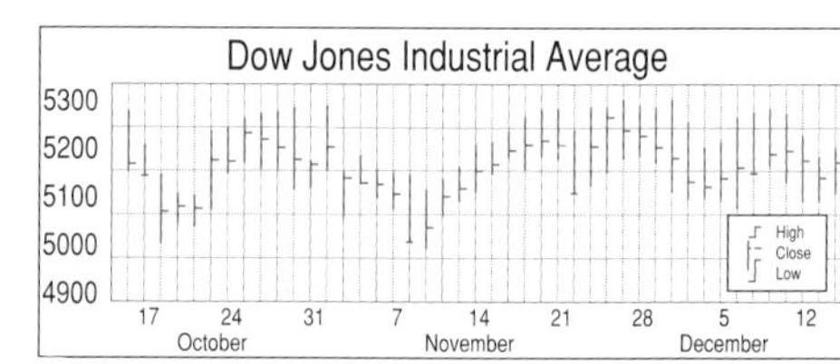

Refer to the Investigation on pages 166–167.

One of the factors to consider when investing in stock is the trends in the prices. You may see something that looks like the diagram at the right in the financial pages of a newspaper. This diagram 'charts' 30 large industrial companies whose stocks are traded on the New York Stock Exchange. This is called the *Dow Jones Industrial Average*.

You can also chart an individual stock. When you chart a stock's activity for a day, the top of the vertical bar represents the highest price at which the stock traded, while the bottom of the bar represents the lowest price at which the stock traded. The horizontal bar indicates the closing price.

- Go to the library and use the last three months' editions of a local or national newspaper to chart the price of one of the stocks you have chosen. If the paper you are using doesn't provide the daily highs and lows, prepare a line graph of the closing prices.
- Write a brief analysis of your stocks' recent performance.

Add the results of your work to your Investigation Folder.

Extension

Using Cooperative Groups Have students work in groups to develop divisibility rules for 15 and 25.

The Investigation on pages 166 and 167 is designed to be a long-term project that is completed over several days or weeks. Encourage students to keep their materials in their Investigation Folder as they work on the Investigation.

4-2 Powers and Exponents

Setting Goals: *In this lesson, you'll use powers in expressions and equations.*

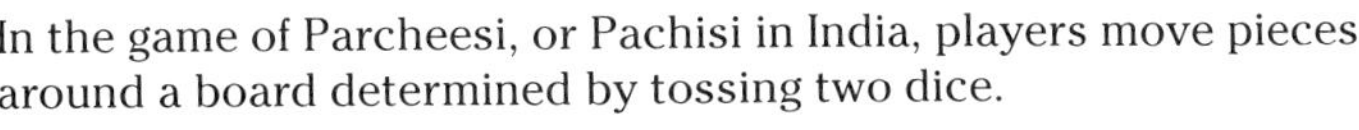

Modeling a Real-World Application: Games

Think about India 1200 years ago. What did the people there have in common with people in the United States today? Parcheesi!

In the game of Parcheesi, or Pachisi in India, players move pieces around a board determined by tossing two dice.

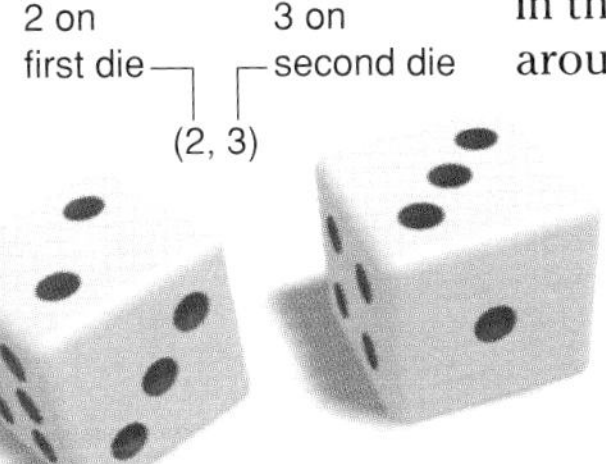

Since each die has six faces, we know that in rolling one die we can come up with six different results. How many different results are there with two dice? Make a table of the possible results.

First Die

Second Die	1	2	3	4	5	6
1	(1, 1)	(2, 1)	(3, 1)	(4, 1)	(5, 1)	(6, 1)
2	(1, 2)	(2, 2)	(3, 2)	(4, 2)	(5, 2)	(6, 2)
3	(1, 3)	(2, 3)	(3, 3)	(4, 3)	(5, 3)	(6, 3)
4	(1, 4)	(2, 4)	(3, 4)	(4, 4)	(5, 4)	(6, 4)
5	(1, 5)	(2, 5)	(3, 5)	(4, 5)	(5, 5)	(6, 5)
6	(1, 6)	(2, 6)	(3, 6)	(4, 6)	(5, 6)	(6, 6)

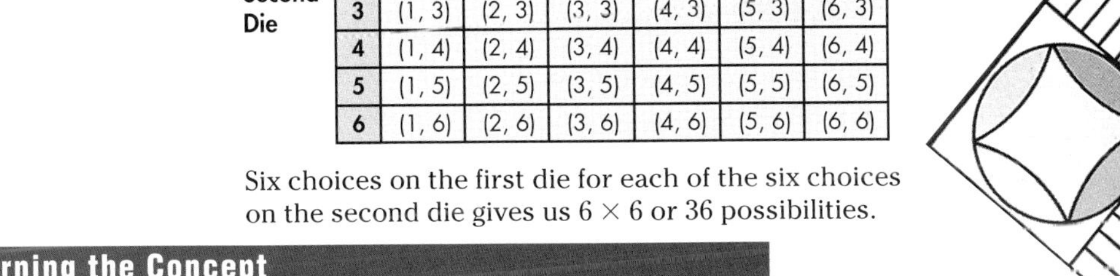

Six choices on the first die for each of the six choices on the second die gives us 6×6 or 36 possibilities.

Learning the Concept

The expression 6×6 can be written in a shorter way using exponents. An **exponent** tells how many times a number, called the **base**, is used as a factor. Numbers that are expressed using exponents are called **powers**.

The expression 6×6 can be written as 6^2.

base → 6^2 ← *exponent*
power

The value of any number to the first power is the number. For example, $6^1 = 6$.

The powers 7^2, 6^3, 10^4 and $(-2)^5$ are read as follows.

7^2 is seven to the second power or seven squared.
6^3 is six to the third power or six cubed.
10^4 is ten to the fourth power.
$(-2)^5$ is negative two to the fifth power.

Any number, except 0, raised to the zero power, like 5^0, is defined to be 1.

Example  **Write each multiplication expression using exponents.**

a. $3 \cdot 3 \cdot 3 \cdot 3$

The base is 3. It appears as a factor 4 times. The exponent is 4.

$3 \cdot 3 \cdot 3 \cdot 3 = 3^4$

b. $y \cdot y \cdot y \cdot y \cdot y$

The base is y. It appears as a factor 5 times. The exponent is 5.

$y \cdot y \cdot y \cdot y \cdot y = y^5$

Alternative Learning Styles

Auditory Have students recite this phrase as a reminder of the proper order of operations: **P**lease **E**xcuse **M**y **D**ear **A**unt **S**ally. The first letters of the words indicate the correct order of **P**arentheses, **E**xponents, **M**ultiplication, **D**ivision, **A**ddition, **S**ubtraction.

4-2 LESSON NOTES

NCTM Standards: 1-8

Instructional Resources

- Study Guide Master 4-2
- Practice Master 4-2
- Enrichment Master 4-2
- Group Activity Card 4-2
- Multicultural Activity Masters, p. 8

Transparency 4-2A contains the 5-Minute Check for this lesson; **Transparency 4-2B** contains a teaching aid for this lesson.

Recommended Pacing	
Standard Pacing	Day 2 of 13
Honors Pacing	Day 2 of 12
Block Scheduling*	Day 1 of 6 (along with Lesson 4-1)

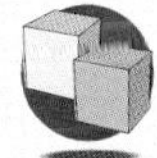

*For more information on pacing and possible lesson plans, refer to the ***Block Scheduling Booklet***.

1 FOCUS

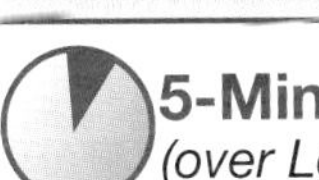

5-Minute Check
(over Lesson 4-1)

Using the divisibility rules, state whether each number is divisible by 2, 3, 5, 6, or 10.

1. 48 **2, yes; 3, yes; 5, no; 6, yes; 10, no**
2. 155 **2, no; 3, no; 5, yes; 6, no; 10, no**
3. 543 **2, no; 3, yes; 5, no; 6, no; 10, no**
4. 1230 **2, yes; 3, yes; 5, yes; 6, yes; 10, yes**
5. 345,678 **2, yes; 3, yes; 5, no; 6, yes; 10, no**

Motivating the Lesson

Situational Problem For winning a contest, Jacob had two choices for this prize. He could either accept \$250 or receive \$2 the first day, and on each successive day, receive double the amount received on the previous day for 7 days. Which prize should Jacob choose so that he gets as much money as is possible? **the second option**

2 TEACH

In-Class Examples

For Example 1
Write each product using exponents.

a. $5 \cdot 5 \cdot 5 \cdot 5 \cdot 5 \cdot 5$ **5^6**

b. $x \cdot x \cdot x$ **x^3**

c. 3 **3^{-1}**

d. $(-3)(-3)(-3)(-3)$ **$(-3)^4$**

For Example 2
Write each power as the product of the same factor.

a. 7^3 **$7 \cdot 7 \cdot 7$**

b. m^5 **$m \cdot m \cdot m \cdot m \cdot m$**

For Example 3
Write 8725 in expanded form. **$(8 \times 10^3) + (7 \times 10^2) + (2 \times 10^1) + (5 \times 10^0)$**

For Example 4
Evaluate $5a - b^2$, if $a = 6$ and $b = -5$. **5**

c. 5
The base is 5. It appears once. The exponent is 1.
$5 = 5^1$

d. $(-2)(-2)(-2)(-2)$
The base is -2. It appears as a factor 4 times. The exponent is 4.
$(-2)(-2)(-2)(-2) = (-2)^4$

Example  **Write each power as a multiplication expression.**

a. 5^4
The base is 5. The exponent 4 means 5 is a factor 4 times.
$5^4 = 5 \cdot 5 \cdot 5 \cdot 5$

b. b^2
The base is b. The exponent 2 means b is a factor 2 times.
$b^2 = b \cdot b$

The number 12,496 is in **standard form**. You can use exponents to express a number in **expanded form**.

Example  **Express 12,496 in expanded form.**

To express a number in expanded form, use place value to write the value of each digit in the number.

$$12{,}496 = 10{,}000 + 2000 + 400 + 90 + 6$$
$$= (1 \times 10{,}000) + (2 \times 1000) + (4 \times 100) + (9 \times 10) + (6 \times 1)$$

Then write the multiples of 10 using exponents.

$$12{,}496 = (1 \times 10^4) + (2 \times 10^3) + (4 \times 10^2) + (9 \times 10^1) + (6 \times 10^0)$$

Recall that $10^0 = 1$.

Since powers are forms of multiplication, they need to be included in the rules for order of operations.

Order of Operations

1. **Do all operations within grouping symbols first; start with the innermost grouping symbols.**
2. ***Evaluate all powers in order from left to right.***
3. **Next do all multiplications and divisions in order from left to right.**
4. **Then do all additions and subtractions in order from left to right.**

Connection to Algebra

Follow the order of operations as you evaluate algebraic expressions.

Example  **Evaluate $3a + b^3$ if $a = 2$ and $b = 5$.**

$3a + b^3 = 3(2) + 5^3$	*Replace a with 2 and b with 5.*
$= 3(2) + 125$	*Evaluate the power.*
$= 6 + 125$	*Find the product of 3 and 2.*
$= 131$	*Find the sum.*

Many scientific formulas involve evaluating expressions containing exponents.

Cooperative Learning

Trade-A-Problem Have each student research formulas that contain exponents that can be found in other texts, such as science books. Then use the formula to write and solve a problem. Have students exchange and solve. Finally, compare solutions and edit problems as needed.

For more information on the trade-a-problem strategy, see *Cooperative Learning in the Mathematics Classroom*, one of the titles in the Glencoe Mathematics Professional Series, p. 25.

Example 5

APPLICATION

Space Flight

At liftoff, the space shuttle *Discovery* has a constant acceleration, a, of 16.4 ft/s². The initial velocity, v, is 1341 ft/s. Use the formula $d = vt + \frac{1}{2}at^2$, where d represents distance and t represents time in seconds, to find the distance that the shuttle has traveled after 30 seconds.

$d = vt + \frac{1}{2}at^2$

$= (1341)(30) + \frac{1}{2}(16.4)(30)^2$ *Replace v with 1341, t with 30, and a with 16.4.*

Estimate: *40,000 + 8 · 900 is about 47,000.*

Evaluate this expression by using a calculator. Many calculators have a special key labeled x^2 for finding squares.

Enter: 1341 [×] 30 [+] .5 [×] 16.4 [×] 30 [x^2] [=] *47610*

The shuttle has traveled 47,610 feet after 30 seconds.

TECHNO TIP

You can evaluate expressions involving exponents by using a calculator. Enter the base first, press the [y^x] key, and then enter the exponent.

In-Class Example

For Example 5

Use the formula $d = vt + \frac{1}{2}at^2$ to find the distance the shuttle has traveled after 15 seconds. Use v = 1341 ft/s and a = 16.4 ft/s². **21,960 ft**

Teaching Tip After using Example 5 given above, ask why the distance is not half the distance found in Example 5 in the text, since the time is half as long. Emphasize that exponential values do not increase linearly.

Checking Your Understanding

Communicating Mathematics

Read and study the lesson to answer these questions. **1–4. See margin.**

1. **Explain** how to find the value of the expression *3 squared*.
2. **Write** 10^5 as a product of the same factor. What is the value of 10^5?
3. Regardless of the value of n, what can you say about 1^n? Explain.
4. **You Decide** Nichelle says that for every value of n, $n^2 > n$. Cardida disagrees. Who is correct? Explain.

MATH JOURNAL

5. **Assess Yourself** Why do you think that exponents were invented? **See students' work.**

Guided Practice

Write each multiplication expression using exponents.

6. $10 \cdot 10 \cdot 10$ **10^3**
7. $m \cdot m \cdot m$ **m^3**
8. $(3)(3)(3)$ **3^3**

Write each power as a multiplication expression.

9. 2^3 **$2 \cdot 2 \cdot 2$**
10. $(-8)^4$ **$(-8)(-8)(-8)(-8)$**
11. 10^6 **$10 \cdot 10 \cdot 10 \cdot 10 \cdot 10 \cdot 10$**

Evaluate each expression if $x = 5$, $y = 9$, and $z = -3$.

12. x^4 **625**
13. $y^2 + z^3$ **54**
14. Write 598 in expanded form. **14. $(5 \times 10^2) + (9 \times 10^1) + (8 \times 10^0)$**
15. **Geometry** The edges of the cube shown at the right are 5 inches long.

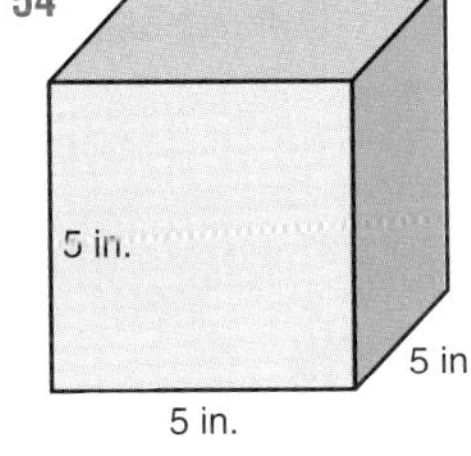

a. The surface area of a cube is the sum of the areas of the faces. Use exponents to write an expression for the surface area of the cube. **$6(5^2)$ or 150 in.²**

b. The *volume* of a cube is the amount of space that it occupies. The volume can be found by multiplying the length, width, and height of the cube. Write an expression for the volume of the cube. **15b. 5^3 or 125 in.³**

c. If you double the length of each edge of the cube, what is the effect on the surface area and volume? **See margin.**

3 PRACTICE/APPLY

Checking Your Understanding

Exercises 1–15 are designed to help you assess your students' understanding through reading, writing, speaking, and modeling. You should work through Exercises 1–5 with your students and then monitor their work on Exercises 6–15.

Additional Answers

1. Multiply 3 and 3, or use 3 as a factor twice to obtain 9.
2. 10 × 10 × 10 × 10 × 10; 100,000

Study Guide Masters, p. 28

NAME ______ DATE ______

4-2 Study Guide
Powers and Exponents

Student Edition Pages 175–179

Expressions such as 4^2, a^3, 2^n, and $(x + 3)^5$ are written using exponents. In 4^2, the base is 4, and the exponent is 2.

The exponent tells you how many times to use the base as a factor.

$3^4 = 3 \cdot 3 \cdot 3 \cdot 3$, or 81 The number named by 3^4 is 81.

Write each product using exponents.

1. $5 \cdot 5 \cdot 5$ **5^3**
2. $6 \cdot 6 \cdot 6 \cdot 6 \cdot 6$ **6^5**
3. $7 \cdot 7 \cdot 7 \cdot 7 \cdot 7 \cdot 7 \cdot 7 \cdot 7$ **7^8**
4. $2 \cdot 2 \cdot 2 \cdot 3 \cdot 3$ **$2^3 \cdot 3^2$**
5. $2 \cdot 2 \cdot 2 \cdot 2 \cdot 2 \cdot 2 \cdot 2$ **2^7**
6. $2 \cdot 2 \cdot 2 \cdot 2 \cdot 2 \cdot 5 \cdot 5 \cdot 5$ **$2^5 \cdot 5^3$**
7. $m \cdot m \cdot m \cdot m \cdot m \cdot m$ **m^6**
8. $n \cdot n \cdot n \cdot n \cdot n \cdot y \cdot y \cdot y \cdot y$ **$n^5 \cdot y^4$**
9. $q \cdot q \cdot q \cdot p \cdot p \cdot p \cdot p$ **$q^3 \cdot p^4$**
10. $2 \cdot t \cdot t \cdot t \cdot t \cdot t \cdot t$ **$2 \cdot t^6$**
11. $9 \cdot 9 \cdot 9 \cdot 9 \cdot 9 \cdot 9$ **9^6**
12. $4 \cdot 4 \cdot 6 \cdot 6 \cdot x \cdot x \cdot x \cdot v \cdot v$ **$4^2 6^2 x^3 v^2$**

Write each power as the product of the same factor.

13. 5^3 **$5 \cdot 5 \cdot 5$**
14. a^5 **$a \cdot a \cdot a \cdot a \cdot a$**
15. 1^4 **$1 \cdot 1 \cdot 1 \cdot 1$**
16. $(-j)^2$ **$(-j)(-j)$**
17. $(y-2)^2$ **$(y-2)(y-2)$**
18. 7^2 **$7 \cdot 7$**
19. $(-4)^3$ **$(-4)(-4)(-4)$**
20. 5^{22} **$5 \cdot 5 \cdot \ldots \cdot 5$ £ 22 factors**

Reteaching

Using Models Have students construct different-sized squares on graph paper. Then have them express the number of enclosed smaller squares using exponents. For example, a 3-by-3 square encloses 3^2 or 9 smaller squares.

Additional Answers

3. Since 1 multiplied by itself is always 1, no matter how many times 1 appears as a factor, the value of this expression is always 1.
4. Cardida is correct. If n is between 0 and 1, then $n^2 < n$.

15c. The surface area is multiplied by 4 and the volume is multiplied by 8.

Teaching Tip For exercise 44, you may include a discussion of the word *googol*, which is a 1 followed by 100 zeros, or 10^{100}. The word *googol* was coined by 9-year-old Milton Sirotta in 1938.

Assignment Guide

Core: 17–49 odd, 50–57
Enriched: 16–44 even, 45–57

For **Extra Practice**, see p. 748.

The red A, B, and C flags, printed only in the Teacher's Wraparound Edition, indicate the level of difficulty of the exercises.

Additional Answers

40. $(5 \times 10^1) + (6 \times 10^0)$
41. $(1 \times 10^2) + (4 \times 10^1) + (9 \times 10^0)$
42. $(2 \times 10^3) + (5 \times 10^1) + (3 \times 10^0)$
43. 23,405

Exercises: Practicing and Applying the Concept

Independent Practice

Write each multiplication expression using exponents.

16. $(5)(5)(5)$ **5^3**
17. $n \cdot n \cdot n \cdot n$ **n^4**
18. $1 \cdot 1 \cdot 1 \cdot 1 \cdot 1$ **1^5**
19. 14 **14^1**
20. $(p \cdot p)(p \cdot p)$ **p^4**
21. $\underbrace{3 \cdot 3 \cdot \cdots \cdot 3 \cdot 3}_{15 \text{ factors}}$ **3^{15}**
22. $6 \cdot 6 \cdot 6 \cdot 6 \cdot 6$ **6^5**
23. $b \times b \times b \times b$ **b^4**
24. $(-5)(-5)$ **$(-5)^2$**

Write each power as a multiplication expression of the same factor.

25. 12^2 **$12 \cdot 12$**
26. y^6 **$y \cdot y \cdot y \cdot y \cdot y \cdot y$**
27. $(-7)^3$ **$(-7)(-7)(-7)$**
28. $(-f)^2$ **$(-f)(-f)$**
29. 6^{20}
30. $(x+1)^2$ **$(x+1)(x+1)$**
31. 1^{55}
32. q^3 **$q \cdot q \cdot q$**
33. $(-1)^4$

29. $\underbrace{6 \cdot 6 \cdot \ldots \cdot 6 \cdot 6}_{20 \text{ factors}}$
31. $\underbrace{1 \cdot 1 \cdot \ldots \cdot 1 \cdot 1}_{55 \text{ factors}}$
33. $(-1)(-1)(-1)(-1)$

Evaluate each expression if $a = 3$, $b = 9$, and $c = -2$.

34. $a^4 + b$ **90**
35. $c^2 + ab$ **31**
36. $b^0 - 10$ **−9**
37. $2b^2$ **162**
38. $10a^5 + c^2$ **2434**
39. $a - b^2$ **−78**

Write each number in expanded form. **40–43. See margin.**

40. 56
41. 149
42. 2053
43. Write $(2 \times 10^4) + (3 \times 10^3) + (4 \times 10^2) + (5 \times 10^0)$ in standard form.

44. **Patterns** Study the pattern at the right. Extend the pattern to find the value of each expression.

$10^0 \cdot 10^2 = 1 \cdot 100$	$= 100$	or 10^2
$10^1 \cdot 10^2 = 10 \cdot 100$	$= 1000$	or 10^3
$10^2 \cdot 10^2 = 100 \cdot 100$	$= 10{,}000$	or 10^4
$10^3 \cdot 10^2 = 1000 \cdot 100$	$= 100{,}000$	or 10^5

a. $10^5 \cdot 10^2$ **10^7** b. $10^{10} \cdot 10^2$ **10^{12}** c. $10^x \cdot 10^2$ **10^{x+2}**

Critical Thinking

45. **Patterns** Study the pattern of exponents at the right.

$2^4 = 16$
$2^3 = 8$
$2^2 = 4$
$2^1 = 2$
$2^0 = 1$
(÷ 2 between each value)

a. If you continue the pattern, what is the next exponent for 2? **−1**
b. What is the next value in the pattern? **$\frac{1}{2}$**
c. Use a calculator to verify your results. **See students' work.**
d. Make a conjecture about the value of 3^{-1}. **$\frac{1}{3}$**

Applications and Problem Solving

46. **Electricity** Niagara Falls and Hoover Dam can produce electric energy with little pollution. As the power goes to its point of use, energy is lost to the resistance in the wires. The amount of power lost can be found by using the formula $P = I^2R$, where P is power in watts, I is current in amps, and R is resistance in ohms. **a. $P = (41)^2(2)$**
    a. The resistance of the wire leading from the source of power to a home is 2 ohms. If an electric stove causes a current of 41 amps to flow through the wire, write an equation for the power lost.
    b. Find the power lost from the wire powering the stove. **3362 watts**
47. **Geometry** The second and third powers have special names related to geometry. Since the area of a square whose side is s units long is $s \cdot s$ or s^2, s^2 is often read as "s squared." A third power like s^3 is often read as "s cubed." Can you explain why? **The volume of a cube with sides s units long is s^3.**

Practice Masters, p. 28

NAME ______ DATE ______

4-2 Practice
Powers and Exponents

Student Edition Pages 175–179

Write each product using exponents.

1. $2 \cdot 2 \cdot 2 \cdot 3 \cdot 3 \cdot 7$ **$2^3 \cdot 3^2 \cdot 7$**
2. $2 \cdot 3 \cdot 3 \cdot 7 \cdot 7 \cdot 7 \cdot 11$ **$2 \cdot 3^2 \cdot 7^3 \cdot 11$**
3. $3 \cdot 3 \cdot 5 \cdot 7 \cdot x \cdot x \cdot x$ **$3^2 \cdot 5 \cdot 7 \cdot x^3$**
4. $5 \cdot 7 \cdot 7 \cdot r \cdot r \cdot t \cdot t \cdot t$ **$5 \cdot 7^2 \cdot r^2 \cdot t^3$**
5. $a \cdot a \cdot b \cdot b \cdot c \cdot c \cdot c$ **$a^2 \cdot b^2 \cdot c^3$**
6. $2 \cdot n \cdot p \cdot p \cdot s \cdot s \cdot s \cdot s$ **$2 \cdot n \cdot p^2 \cdot s^4$**
7. 8 to the fourth power **8^4**
8. m to the third power **m^3**
9. n to the seventh power **n^7**
10. h cubed **h^3**

Write each power as the product of the same factor.

11. x^7 **$x \cdot x \cdot x \cdot x \cdot x \cdot x \cdot x$**
12. $(-2)^4$ **$(-2)(-2)(-2)(-2)$**
13. 6^1 **6**
14. $(y+3)^2$ **$(y+3)(y+3)$**
15. 13^3 **$13 \cdot 13 \cdot 13$**
16. 5^{25} **$\underbrace{5 \cdot 5 \cdot \cdots \cdot 5}_{25 \text{ factors}}$**
17. 1^6 **$1 \cdot 1 \cdot 1 \cdot 1 \cdot 1 \cdot 1$**
18. $(cd)^4$ **$(cd)(cd)(cd)(cd)$**
19. $(-g)^5$ **$(-g)(-g)(-g)(-g)(-g)$**

Evaluate each expression if $p = 1$, $m = 6$, $r = 2$, $y = 3$, and $z = 5$.

20. $3ry$ **18**
21. p^2m^2 **36**
22. $(rm)^2$ **144**
23. $p^2(ry)$ **6**
24. $2zy^2$ **90**
25. $y^3r^3p^3$ **216**
26. $5p^8$ **5**
27. $p^{10}y^4$ **81**
28. $r^2y^2z^2$ **900**

Group Activity Card 4-2

Make An Exponential Score — Group Activity 4-2

MATERIALS: Deck of "Exponential Expressions" cards using expressions given on back • 0-9 spinner • paper • pencil

Before play begins, shuffle the "Exponential Expression" cards and deal four cards to 2-4 players.

Spin the spinner and use that number to evaluate the exponential expression on one of your cards, which you then lay down in front of you. If evaluated correctly, you then record that value on your piece of paper. If evaluated incorrectly, you get no points.

Play continues around the table, with players adding the score they get on each round to their previous total. Each player spins the spinner once on each of four turns. After players have played all of their cards, the person with the highest final total wins the game.

Play the game several times to see if you can develop a strategy to help you get a high score.

 Pre-Algebra

Extension

Using Manipulatives Each student in a group, in turn, tosses two standard dice. Each student chooses how to use the two numbers rolled to write an exponential expression, such as 2^3 or 3^2. Then the student finds the value of the expression and scores that many points in the round. The student with the highest total at the end of five rounds wins.

48. **Entertainment** In the movie *I.Q.*, Albert Einstein tries to make his niece Catherine Boyd fall in love with auto mechanic Ed Walters. In one scene, Ed asks Catherine how many stars she thinks are in the sky. Her answer is $10^{12} + 1$. Write this number in standard form.

49. **Literature** There is a story of a knight who slew a dragon that was destroying a small kingdom. The knight asked the king to give him reward money for a month following this pattern. The first day of the month, he would get 1 cent, the second day, 2 cents, the third day, 4 cents, and so on, continuing to double the amount for 30 days.
   a. Express his reward on each of the first three days as a power of 2.
   b. What is his reward on the fourth day? Express it as a power of 2. **2^3**
   c. How much would the knight be getting from the king on the 30th day? Express it as a power of 2. **2^{29} cents**
   d. Write your answer to part c in standard form. **536,870,912**
   e. Write an expression using a sum and exponents for the amount of money the knight would receive for the whole month.
   f. The king agreed to the knight's request because he thought it would be a cheap reward. Was it? Explain. **See margin.**

Mixed Review

48. 1,000,000,000,001
49a. $2^0, 2^1, 2^2$
49e. $2^0 + 2^1 + 2^2 + \ldots + 2^{29}$

50. State whether 945 is divisible by 2, 3, 5, 6, or 10. (Lesson 4-1) **3, 5**
51. Solve and graph $-7x < 84$. (Lesson 3-7) **$x > -12$; see margin for graph.**
52. **Geometry** Find the perimeter and area of a rectangle that is 4 meters wide and 12 meters long. (Lesson 3-5) **32 m; 48 m^2**
53. **Meteorology** A maximum and minimum thermometer records both the high and low temperatures of the day. If the difference between one day's high and low was 49 degrees and the high was 67°F, what was the low temperature? (Lesson 3-2) **18°F**
54. Solve $m = \frac{126}{-21}$. (Lesson 2-8) **−6**
55. Simplify $18p - 26p$. (Lesson 2-5) **$-8p$**
56. **Education** The class of 1988 had 536 graduates, and 497 students graduated in the class of 1995. Write and solve an equation to find how many more students graduated in 1988. (Lesson 1-6) **39 students**
57. Write an expression to represent the phrase *four more than the Bears scored in the last game.* (Lesson 1-3) **$b + 4$**

From the FUNNY PAPERS

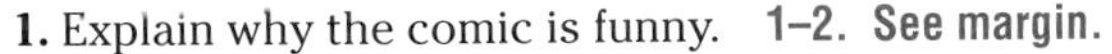

1. Explain why the comic is funny. **1–2. See margin.**
2. Explain the differences in the formulas on the chalkboard.
3. Einstein's famous formula $E = mc^2$ relates energy, E, mass, m, and the speed of light, c. The speed of light is about 300,000 km/s. Use a calculator to find the value of c^2. **90,000,000,000 km^2/s^2**

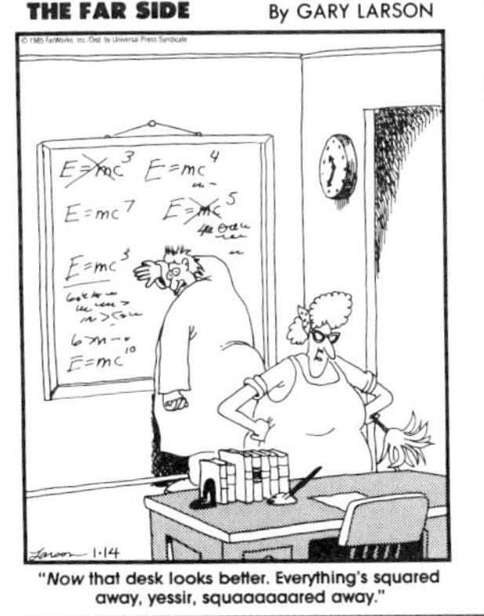

"Now that desk looks better. Everything's squared away, yessir, squaaaaaared away."

From the FUNNY PAPERS

Two aliens are discussing American schools.
"No wonder Earthlings have trouble learning. In one class they learn pie are square, and in another they learn pies are round and baked."
Allow students to discuss this joke.

Additional Answers From the Funny Papers

1. Sample answer: The man has been working on the problem for a long time and the cleaning person gives him the answer by saying "squared".
2. The exponent on c is different in each one.

4 ASSESS

Closing Activity

Modeling Have students model each expression symbolically, using exponents.
1. r squared **r^2**
2. six times n to the fourth power **$6n^4$**
3. five times the sum of c and d cubed **$5(c + d)^3$**

Additional Answers

49f. No; the reward will cost $10,737,418.23 over the 30-day period.

51.

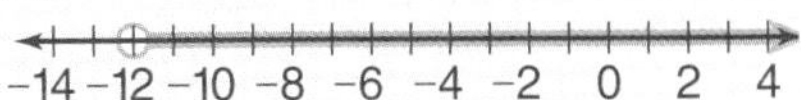

Enrichment Masters, p. 28

NAME ______________ DATE ________

4-2 Enrichment
Exponents

Student Edition Pages 175–179

Numbers can be expressed in several ways. Some numbers are expressed as sums. Some numbers are expressed as products of factors, while other numbers are expressed as powers.

Two ways to express 27 are $3 \cdot 3 \cdot 3$ and 3^3.

The number 1 million can be expressed in the following ways.

1,000,000	$1000 \cdot 1000$	$100 \cdot 100 \cdot 100$	$10^2 \cdot 10^2 \cdot 10^2$
$1{,}000{,}000^1$	1000^2	100^3	10^6

Write names for each number below using the given exponents.

1. 16, exponents: 2 and 4 **$4^2, 2^4$**
2. 81, exponents: 2 and 4 **$9^2, 3^4$**
3. 64, exponents: 2 and 6 **$8^2, 2^6$**
4. 256, exponents: 2 and 8 **$16^2, 2^8$**
5. 625, exponents: 2 and 4 **$25^2, 5^4$**
6. 729, exponents: 2 and 6 **$27^2, 3^6$**
7. 2401, exponents: 2 and 4 **$49^2, 7^4$**
8. 4096, exponents: 2 and 12 **$64^2, 2^{12}$**
9. 6561, exponents: 2 and 8 **$81^2, 3^8$**
10. 390,625, exponents: 2 and 8 **$625^2, 5^8$**

Numbers that can be named as powers with like bases can be multiplied by adding the exponents.

$$8 \cdot 8 = 2^3 \cdot 2^3 = 2^{3+3} = 2^6$$

Write the product of each pair of factors in exponential form.

11. $9 \cdot 9$ **$3^2 \cdot 3^2 = 3^4$**
12. $4 \cdot 4$ **$2^2 \cdot 2^2 = 2^4$**
13. $16 \cdot 8$ **$2^4 \cdot 2^3 = 2^7$**
14. $125 \cdot 25$ **$5^3 \cdot 5^2 = 5^5$**
15. $27 \cdot 9$ **$3^3 \cdot 3^2 = 3^5$**
16. $81 \cdot 27$ **$3^4 \cdot 3^3 = 3^7$**
17. $49 \cdot 49$ **$7^2 \cdot 7^2 = 7^4$**
18. $121 \cdot 121$ **$11^2 \cdot 11^2 = 11^4$**

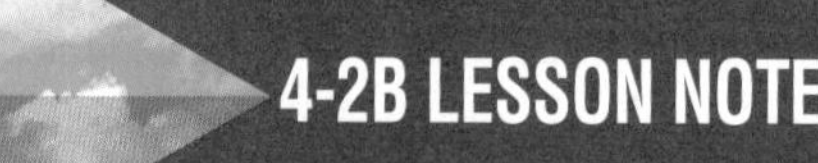

4-2B LESSON NOTES

NCTM Standards: 2, 3, 5, 9

Objective
Use a graphing calculator to evaluate expressions with exponents.

Recommended Time
15 minutes

Instructional Resources
Graphing Calculator Masters, p. 18

This master provides keystroking instruction for this lesson for the TI-81 and Casio graphing calculators.

1 FOCUS

Motivating the Lesson
Have students predict whether the expressions $(-2)^6$ and -2^6 are equivalent. Then have them evaluate both expressions.

2 TEACH

Teaching Tip Point out that the calculator will evaluate an expression according to order of operations. Expressions such as $(-2)^6$ and -2^6 are not the same.
$(-2)^6 = (-2)(-2)(-2)(-2)(-2)(-2)$
$-2^6 = (-1)(2)(2)(2)(2)(2)(2)$

3 PRACTICE/APPLY

Assignment Guide
Core: 1–7
Enriched: 1–7

4 ASSESS

Observing students working with technology is an excellent method of assessment.

GRAPHING CALCULATOR ACTIVITY

4-2B Evaluating Expressions with Exponents

An Extension of Lesson **4-2**

You learned how to evaluate expressions with a graphing calculator in Lesson 1-3B. You can also use a graphing calculator to evaluate expressions involving exponents.

Storing the values of the variables allows you to evaluate complicated expressions accurately. Also, it will save time when you want to evaluate several expressions for the same values.

Activity  **Evaluate $\frac{x}{(y-2)^2}$ on a graphing calculator if $x = 24$ and $y = 4$.**

Keystrokes are shown for the TI-82 graphing calculator. If you are using a different graphing calculator, consult your User's Guide.

ENTER:
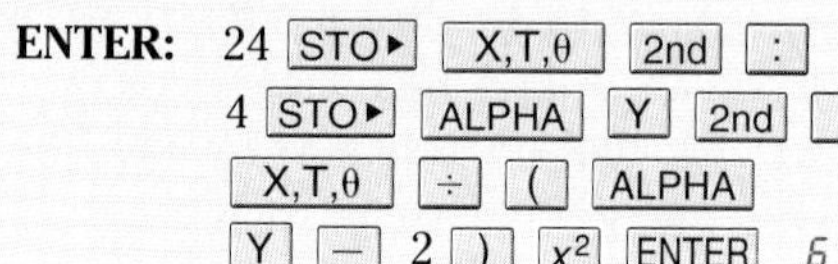

Enters the value of x.
Enters the value of y.
Enters and evaluates the expression.

Expressions involving exponents other than 2 can be evaluated using the ^ key.

Activity  **Use a graphing calculator to evaluate $3ab^5$ if $a = -4$ and $b = 2$.**

ENTER: (−) 4 STO▸ ALPHA A 2nd : *Enters the value of a.*
2 STO▸ ALPHA B 2nd : *Enters the value of b.*
3 ALPHA A ALPHA B *Enters and evaluates*
^ 5 ENTER -384 *the expression.*

Notice that the calculator was able to evaluate the expression without entering multiplication symbols between 3, a, and b^5.

Your Turn **Use a graphing calculator to evaluate each expression if $a = 9$, $b = -1$, $c = 7$, and $d = 3$.**

1. $7d^2 - 1$ **62**
2. $\frac{32}{(a-c)^3}$ **4**
3. $3(b + c) \div d^2$ **2**
4. $12b^4 \div d$ **4**
5. $(2c - 4d)^{10}$ **1024**
6. $11 - a^2(b + c)$ **−475**

7. Evaluate $12 - 4d^2$ and $(12 - 4d)^2$ if $d = 3$. Are the values the same? Explain. **See margin.**

Technology
This lesson offers an excellent opportunity for using technology in your pre-algebra classroom. For more information on using technology, see *Graphing Calculators in the Mathematics Classroom,* one of the titles in the Glencoe Mathematics Professional Series.

Additional Answer
7. No. Sample answer: The order of operations will be different. The value of the first expression is -24 and the value of the second expression is 0.

4-3 Problem-Solving Strategy: Draw a Diagram

Setting Goals: *In this lesson, you'll solve problems by drawing diagrams.*

Modeling a Real-World Application: Entertainment

The game of Totolospi was invented by the Hopi Indians. You play the game with three cane dice, a counting board inscribed in stone, and a counter for each player. Each cane die can land round side up, (R), or flat side up, (F). When two players play Totolospi, each player places a counter on the nearest circle. The moves in the game are determined by tossing three cane dice.

- Toss three round sides up, (RRR), and advance 2 lines.
- Toss three flat sides up, (FFF), and advance 1 line.
- No advance with any other combination.

The first player to reach the opposite side wins.

If you are playing Totolospi, how many combinations of tosses will allow an advance?

Learning the Concept

When we are able to draw a diagram for a given situation, we often are able to understand it better. Drawing a diagram is a powerful problem-solving strategy.

Explore There are 3 dice being tossed. We need to know how many combinations of round and flat side up can be tossed.

Plan We can draw a diagram showing how each of the dice lands to determine the number of possible combinations.

Solve

First Die	Second Die	Third Die	
R	R	R (RRR)	← *advance 2*
		F (RRF)	*no advance*
	F	R (RFR)	*no advance*
		F (RFF)	*no advance*
F	R	R (FRR)	*no advance*
		F (FRF)	*no advance*
	F	R (FFR)	*no advance*
		F (FFF)	← *advance 1*

There are 8 possible outcomes. Two of them, RRR and FFF, allow you to advance. So 8 − 2 or 6 tosses allow no advance.

Examine Each die can land two different ways. So there should be $2 \cdot 2 \cdot 2$ or 8 possible tosses.

4-3 LESSON NOTES

NCTM Standards: 1-6, 8

Instructional Resources

- Study Guide Master 4-3
- Practice Master 4-3
- Enrichment Master 4-3
- Group Activity Card 4-3
- Assessment and Evaluation Masters, p. 99

Transparency 4-3A contains the 5-Minute Check for this lesson; **Transparency 4-3B** contains a teaching aid for this lesson.

Recommended Pacing	
Standard Pacing	Day 3 of 13
Honors Pacing	Day 3 of 12
Block Scheduling*	Day 2 of 6 (along with Lesson 4-4)

*For more information on pacing and possible lesson plans, refer to the ***Block Scheduling Booklet***.

1 FOCUS

5-Minute Check
(over Lesson 4-2)

Write using exponents.

1. (−3)(−3) $(-3)^2$
2. 19 19^1
3. $m \cdot m \cdot m \cdot m \cdot m \cdot m$ m^6

Evaluate, if $a = -3$ and $b = -1$.

4. a^3b 27
5. $a^2 - 3b$ 12

Motivating the Lesson

Questioning Ask students why diagrams are included in programs for athletic events or plays. **It helps people find seats and other areas, such as restrooms and exits.**

Study Guide Masters, p. 29
The Study Guide Master provides a concise presentation of the lesson along with practice problems.

Alternative Learning Styles

Kinesthetic Have students volunteer to show the class the steps to a favorite line dance. Challenge the class to draw a diagram to show the progression of the steps through a part of the dance.

2 TEACH

In-Class Example

A tennis class of 16 students will participate in a single elimination tournament; that is, only the winners continue to play. How many games will be played? **15 games**

3 PRACTICE/APPLY

Checking Your Understanding

Exercises 1–5 are designed to help you assess your students' understanding through reading, writing, speaking, and modeling. You should work through Exercises 1–3 with your students and then monitor their work on Exercises 4 and 5.

Assignment Guide
Core: 6–20
Enriched: 6–20

For **Extra Practice**, see p. 748.

The red A, B, and C flags, printed only in the Teacher's Wraparound Edition, indicate the level of difficulty of the exercises.

Practice Masters, p. 29

NAME ______________________ DATE ____________

4-3 Practice
Problem-Solving Strategy: Draw a Diagram

Student Edition
Pages 181–183

Solve. Use any strategy.

1. A sandwich shop has 7 kinds of sandwiches and 4 kinds of drinks. How many different orders of one sandwich and one drink could you order? **28**

2. There are 16 golfers in a single-elimination tournament. How many golf matches will be played during the tournament? **15**

3. Ethel, Mike, Pete, and Gail wanted to go to the movies. In how many different ways could they stand in line to buy their tickets? **24**

4. If you have 4 pairs of jeans, 3 shirts, and 2 pairs of running shoes, how many different outfits can you make? Each outfit contains one pair of running shoes. **24**

5. There are 5 members in the Washington family. Suppose each member hugs every other member. How many hugs take place? **10**

6. Show how you can cut this cake into sixteenths with exactly 5 cuts.

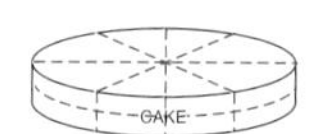

Checking Your Understanding

Communicating Mathematics

1–2. See Solutions Manual.

MATERIALS

grid paper

Read and study the lesson to answer these questions.

1. **Explain** why a diagram is a good strategy for problem solving.
2. Are diagrams helpful in solving *every* type of problem?
3. The diagram at the right shows a model of a five-step staircase built with concrete blocks. Use grid paper to draw a staircase with seven steps. How many blocks would be required to build a staircase with seven steps? **28 blocks**

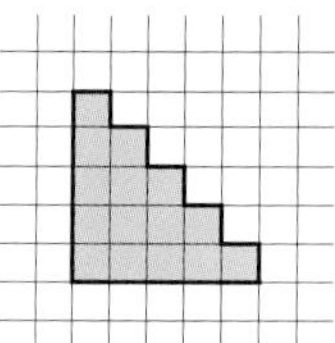

Guided Practice

4. **Draw** a diagram to determine how many flights an airline would have to schedule if it wants to provide nonstop flights between Miami, Ft. Lauderdale, Tallahassee, Tampa, Orlando, and Daytona. **See Solutions Manual.**

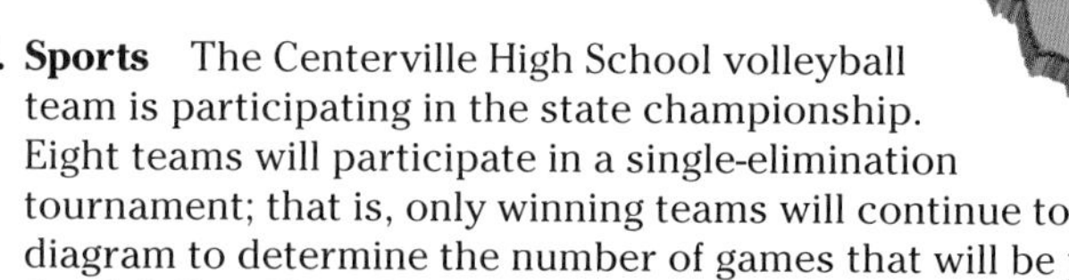

5. **Sports** The Centerville High School volleyball team is participating in the state championship. Eight teams will participate in a single-elimination tournament; that is, only winning teams will continue to play. Draw a diagram to determine the number of games that will be played. **See Solutions Manual.**

Practicing and Applying the Concept

Independent Practice

Solve. Use any strategy.

6. The West Peoria Baseball League has 7 teams and the East Peoria League has 8 teams. Every team from the West League must play every team from the East League at least once. What is the least number of games that must be played? **56 games**
7. Replace each ● with an operation symbol to make the equation true. Add grouping symbols if necessary. **(4 + 3) × 6 + 3 = 45**

4 ● 3 ● 6 ● 3 = 45

8. During the first minute of a game of tug-of-war, the Red Team pulled the Blue Team forward 2 feet. During the second minute, the Blue Team pulled the Red Team forward 1 foot. During the third minute, the Red Team pulled the Blue Team forward 2 feet and during the fourth minute, the Blue Team pulled the Red Team forward 1 foot. If this pattern continues, how long will it take the Red Team to pull the Blue Team forward a total of 10 feet to win the game? **17 minutes**
9. Meg is conditioning for the start of soccer season. She does seven minutes of stretching, followed by an 18-minute run. Then she cools down with 8 minutes of walking and 7 minutes of leg lifts. Finally, she finishes with 4 minutes of stretching. If Meg finishes her workout at 2:07 P.M., when did she begin? **1:23 P.M.**

Group Activity Card 4-3

How Can I Get To My Friend's House? Group Activity **4-3**

In this activity, work with a partner to solve the following dilemma.

My brother and I went to Phoenix to visit our friend. However, when we got two blocks east of his house there was an accident, and we had to go south for four blocks and then turn west. This street had an overpass, so we had to go three blocks before we could turn north. We went two blocks on that street and came to a dead end sign. The sign on the street to the right says "No Outlet."

What directions would you give my brother and me that would get us to our friend's house by the shortest way?

 Pre-Algebra

Reteaching

Using Questioning Help students draw a diagram to solve this problem. *A ferry carrying 37 cars, each containing two people, is crossing Lake Erie. Each person has three suitcases. The captain and his assistant each have two suitcases. Each of the rest of the crew of 15 has one suitcase. How many suitcases are there on the ferry?* **241 suitcases**

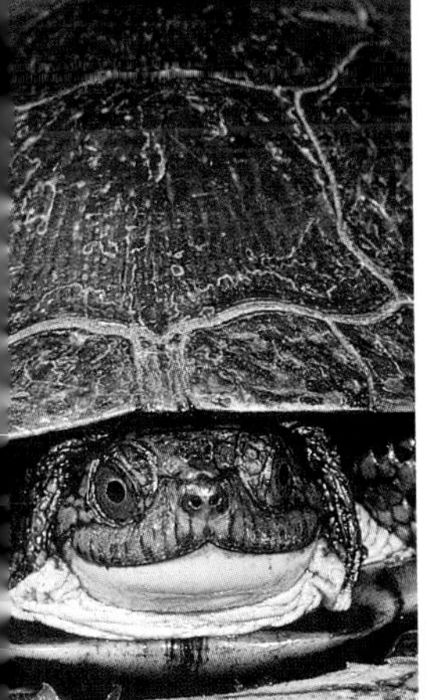

10. History Jorge has to write a paper and complete a family tree for history class. If part of the assignment was to find the names and birthdates of the five generations that preceded him, direct relatives only, no step-parents, how many people will Jorge have to investigate? **62 people**

11. Magic squares are arrays of numbers in which the sum of the numbers on each row, column, and diagonal are the same. According to legend, the most famous magic square, the *lo-shu*, was discovered on the back of a divine tortoise by China's Emperor Yu in about 2200 B.C. Copy and complete the magic square pictured at the right.

4	15	1	14
9	6	12	7
16	3	13	2
5	10	8	11

12. Geometry A diagonal is a line segment that can be drawn from one vertex of a polygon to another vertex that does not share a side. The diagram at the right shows the two diagonals of a rectangle. **12a. 5 diagonals**

a. A pentagon has five sides. How many diagonals does it have?

b. How many diagonals can be drawn in a hexagon? (*Hint:* A hexagon has six sides.) **9 diagonals**

c. Use a pattern to predict how many diagonals a heptagon, 7 sides, has. Check your answer by drawing a diagram. **14 diagonals**

d. In your own words, write a description of the pattern. **See margin.**

13. Charlie traded in his old car which averaged 22 miles per gallon. The EPA sticker on the new car stated that it should average 37 miles per gallon. If Charlie drives about 12,000 miles per year and gasoline is about $1.10 per gallon, how much should he expect to save on gasoline in the first year of owning the new car? **about $243**

14. At Grandma's Bakery, muffins are sold in boxes of 4, 6, or 13. You can buy 8 muffins by choosing two boxes of 4 muffins, but you can't buy 9 muffins with any combination of boxes. Find all of the numbers of muffins less than 25 that you can't buy. **1, 2, 3, 5, 7, 9, 11, 15**

Critical Thinking

15. City code requires that a party house provide 9 square feet for each person on a dance floor. If the owners of the Westman Hotel want to provide a square dance floor that is large enough for 100 people, how long should each side be? **30 feet**

Mixed Review

16. Write $x \cdot x \cdot x \cdot x \cdot x$ using exponents. (Lesson 4-2) x^5

17. Solve $x - 8 \leq -17$. (Lesson 3-6) $x \leq -9$

18. Patterns What is the next number in the pattern 3, 3, 6, 18, . . . ? (Lesson 2-6) **72**

19. Find the sum $5 + (-8) + 7$. (Lesson 2-4) **4**

20. Name the property shown by $(3 \cdot 2) \cdot 8 = 3 \cdot (2 \cdot 8)$. (Lesson 1-4) **Assoc, ×**

Teaching Tip Students may wonder why Exercise 4 has a greater answer than Exercise 5, even though Exercise 4 involves 6 cities and Exercise 5 involves 8 teams. Point out that the questions are not asking for the same information—these problems cannot be generalized into one rule. A diagram can help them to think about a question properly.

4 ASSESS

Closing Activity

Speaking Have students orally describe personal situations when they used a diagram to solve a problem. Sample answers could include travel, geometry, or finding patterns.

Chapter 8, Quiz A (Lessons 4-1 through 4-3) is available in the *Assessment and Evaluation Masters*, p. 99.

Additional Answer

12d. A polygon with n sides has $(2 + 3 + \ldots + (n - 2))$ diagonals.

Extension

Using Models Have students model the solution to this problem by drawing a series of diagrams.

A man has a wolf, a chicken, and some plant seeds on one side of the river. He needs to get everything to the other side of the river. His small boat will only hold himself and one other item. But, if left together, the wolf will eat the chicken and the chicken will eat the seed. How can the man safely get everything to the other side of the river? **Diagrams should show: Take the chicken across the river. Go back empty-handed. Take the plant seed across, and return with the chicken. Take the wolf across, and return empty-handed. Take the chicken across the river.**

Enrichment Masters, p. 29

NAME ____________ DATE ________

4-3 Enrichment
Intersection and Union of Sets

Student Edition Pages 181–183

The set of all factors of 8 and 12 can be listed in set notation as follows.

factors of 8 A = {1, 2, 4, 8}

factors of 12 B = {1, 2, 3, 4, 6, 12}

The **intersection** of sets A and B is the set of elements that are in *both* A and B. The diagram shows the intersection of A and B.

A intersection B = the set containing 1, 2, and 4

$A \cap B = \{1, 2, 4\}$

The **union** of sets A and B is the set of elements that are in A *or* in B *or* in both.

A union B = the set containing 1, 2, 3, 4, 6, 8, and 12

$A \cup B = \{1, 2, 3, 4, 6, 8, 12\}$

Indicate the intersection and union for each of the following.

1. $X \cap Y$ **{6, 8}**

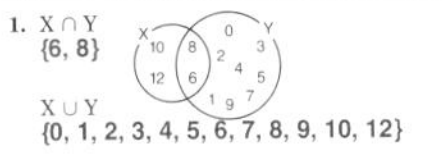

$X \cup Y$ **{0, 1, 2, 3, 4, 5, 6, 7, 8, 9, 10, 12}**

2. $M \cap N$ **{1, 2, 6}**

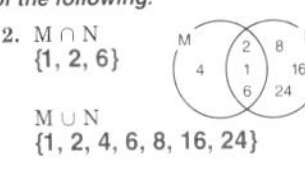

$M \cup N$ **{1, 2, 4, 6, 8, 16, 24}**

3. C = {2, 4, 6, 8, 10, 12}
D = {0, 3, 6, 9, 12, 15, 18}
C ∩ D = {6, 12}
C ∪ D = {0, 2, 3, 4, 6, 8, 9, 10, 12, 15, 18}

4. E = {a, e, i, o, u}
F = {c, a, k, e}
E ∩ F = {a, e}
E ∪ F = {a, c, e, i, k, o, u}

5. G = {1, 2, 3, 4, 5}
H = {2, 3, 5, 7, 11}
G ∩ H = {2, 3, 5}
G ∪ H = {1, 2, 3, 4, 5, 7, 11}

6. Let J = {3, 12, 21, 30} and R = {2, 11, 20}. What is J ∩ R?
Hint: The empty set has no elements. The empty set is written ∅. **∅**

Let K be the set of all counting numbers less than 60. Let L be the set of all counting numbers greater than 50.

7. Find K ∩ L. **{51, 52, 53, ⋯, 59}**

8. Find K ∪ L. **{1, 2, 3, ⋯}**

9. Describe K ∩ L using the words *greater than* and *less than*.
The set of counting numbers greater than 50 and less than 60

 T 29 *Pre-Algebra*

4-4 LESSON NOTES

NCTM Standards: 1-6, 8

Instructional Resources

- Study Guide Master 4-4
- Practice Master 4-4
- Enrichment Master 4-4
- Group Activity Card 4-4
- Activity Masters, p. 34
- Math Lab and Modeling Math Masters, p. 60
- Real-World Applications, 8

Transparency 4-4A contains the 5-Minute Check for this lesson; **Transparency 4-4B** contains a teaching aid for this lesson.

Recommended Pacing	
Standard Pacing	Day 4 of 13
Honors Pacing	Day 4 of 12
Block Scheduling*	Day 2 of 6 (along with Lesson 4-3)

*For more information on pacing and possible lesson plans, refer to the ***Block Scheduling Booklet***.

1 FOCUS

5-Minute Check
(over Lesson 4-3)

In the N.I.T. basketball tournament, 32 teams are invited to participate in a single-elimination format. How many games will be played before a winner is determined? **31 games**

Motivating the Lesson

Questioning Is it possible to write the numbers 30, 75, and 180 as products, using only the factors 2, 3, and 5? If so, how? **Yes; $30 = 2 \cdot 3 \cdot 5$; $75 = 3 \cdot 5 \cdot 5$; $180 = 2 \cdot 2 \cdot 3 \cdot 3 \cdot 5$**

4-4 Prime Factorization

Setting Goals: *In this lesson, you'll identify prime and composite numbers and write the prime factorizations of composite numbers.*

Modeling with Manipulatives

MATERIALS

grid paper

As you learned in Lesson 4-3, drawing a diagram can be very helpful as you solve problems.

Your Turn

The grid at the right shows one way of drawing a rectangle consisting of 10 squares. Use grid paper to draw as many different rectangular arrangements of 1, 2, 3, 4, 5, 6, 7, 8, 9, and 10 squares as possible.

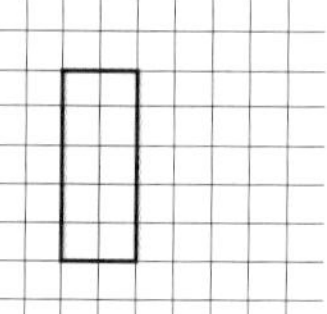

a. Which numbers of squares could only be arranged in either one row or one column? **1, 2, 3, 5, 7**

b. Which numbers of squares could be arranged in a way other than one row or column? **4, 6, 8, 9, 10**

c. Choose one of the numbers that could only be arranged in one row or one column. Write as many multiplication sentences with the number as the product and two whole numbers as factors as you can. What do you observe? **There are only two: $n \times 1 = n$ and $1 \times n = n$.**

Learning the Concept

In the activity above, the numbers of squares that can be arranged in only one row or one column are prime numbers. A **prime number** is a whole number greater than one that has *exactly* two factors, 1 and itself.

Numbers of squares like 4, 6, 8, 9, and 10 that can be arranged in rectangles other than a row or a column are called composite numbers. A **composite number** is a whole number greater than one that has more than two factors. A composite number can always be expressed as a product of two or more primes.

- ▶ The numbers 0 and 1 are considered *neither* prime *nor* composite.
- ▶ Every number is a factor of 0, since any number multiplied by zero is zero.
- ▶ The number 1 has only one factor, itself.
- ▶ Every whole number greater than 1 is either prime or composite.

Sometimes a number can be factored in several ways. Three different ways to find all of the prime factors of 24 using diagrams called **factor trees** are shown below.

$24 = 2 \cdot 12 = 2 \cdot 3 \cdot 4 = 2 \cdot 3 \cdot 2 \cdot 2$

$24 = 3 \cdot 8 = 3 \cdot 2 \cdot 4 = 3 \cdot 2 \cdot 2 \cdot 2$

$24 = 4 \cdot 6 = 2 \cdot 2 \cdot 2 \cdot 3$

The factors are in a different order, but the result is the same. The order in which you factor is not important. The factoring process ends when all of the factors are prime. When a positive integer (other than 1) is expressed as a product of factors that are all prime, the expression is called the **prime factorization**.

Example  **Use a factor tree to factor 64 completely.**

Begin with any two whole number factors of 64.

$64 = 8 \cdot 8$

$= 4 \cdot 2 \cdot 4 \cdot 2$ *Factor each 8.*

$= 2 \cdot 2 \cdot 2 \cdot 2 \cdot 2 \cdot 2$ *Factor each 4.*

The factorization is complete when all the factors are prime. The prime factorization of 64 is $2 \cdot 2 \cdot 2 \cdot 2 \cdot 2 \cdot 2$ or 2^6.

The prime factorization of a negative integer (other than -1) contains a factor of -1, and the rest of the factors are prime.

Example  **Use a factor tree to factor −72 completely.**

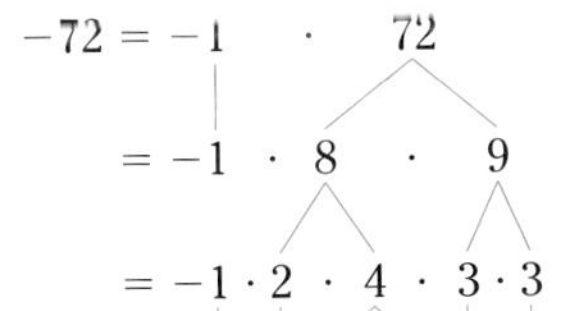

$-72 = -1 \cdot 72$ *A negative integer may be expressed as the product of −1 and a whole number.*

$= -1 \cdot 8 \cdot 9$ *Begin with any two whole number factors of 72.*

$= -1 \cdot 2 \cdot 4 \cdot 3 \cdot 3$ *Factor 8 and 9.*

$= -1 \cdot 2 \cdot 2 \cdot 2 \cdot 3 \cdot 3$ *Factor 4.*

You are finished factoring when all of the factors, other than -1, are prime. The factorization of -72 is $-1 \cdot 2 \cdot 2 \cdot 2 \cdot 3 \cdot 3$. You can write this product as $-1 \cdot 2^3 \cdot 3^2$.

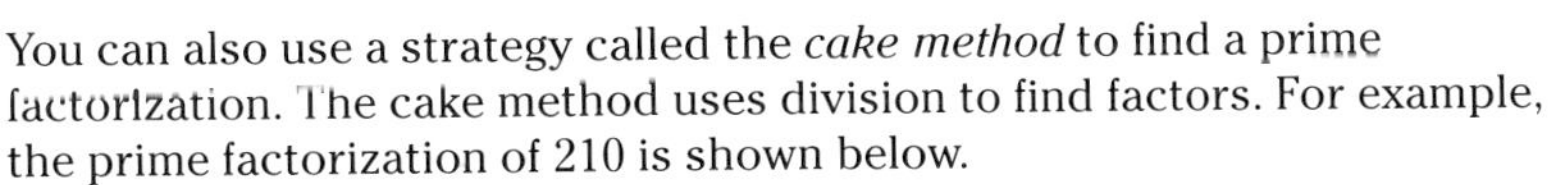

You can also use a strategy called the *cake method* to find a prime factorization. The cake method uses division to find factors. For example, the prime factorization of 210 is shown below.

- Begin with the smallest prime that is a factor.
- Then divide the quotient by the smallest possible prime factor.
- Repeat until the quotient is prime.

```
   7
5)35
3)105
2)210
```

The prime factorization of 210 is $2 \cdot 3 \cdot 5 \cdot 7$. *Use a tree diagram to check the result.*

2 TEACH

In-Class Example

For Example 1

Use a factor tree to factor −60 completely. **$-1 \cdot 2^2 \cdot 3 \cdot 5$**

Teaching Tip Some students may have seen the cake method used, but written going downward, as shown below.

```
2)308
2)154
 7)77
   11
```

Alternative Teaching Strategies

Student Diversity Have students construct the Sieve of Eratosthenes using numbers 1–100 as shown at the right.

Tell students to cross out 1, then circle 2 and cross out all its multiples. Proceed to the next number that is not circled or crossed out (3) and repeat the process. Continue until all numbers are circled (prime) or crossed out (composite).

1 2 3 4 5 6 7 8 9 10
11 12 13 14 15 16 17 18 19 20
21 22 23 24 25 26 27 28 29 30
31 32 33 34 35 36 37 38 39 40
41 42 43 44 45 46 47 48 49 50
51 52 53 54 55 56 57 . . .

In-Class Examples

For Example 2

The key to decoding an RSA cryptoalgorithm is $n = pq$. Find p and q, if $n = 1457$. **31 and 47**

For Example 3

Factor each monomial completely.

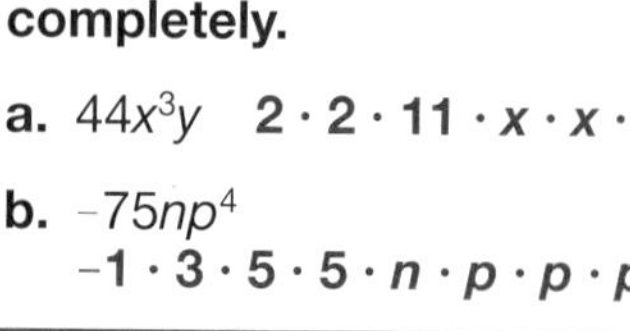

a. $44x^3y$ **$2 \cdot 2 \cdot 11 \cdot x \cdot x \cdot x \cdot y$**

b. $-75np^4$ **$-1 \cdot 3 \cdot 5 \cdot 5 \cdot n \cdot p \cdot p \cdot p \cdot p$**

Example 3 **Cryptography**

Prime numbers are the key to coding and decoding information using the Rivest-Shamirs-Adleman (RSA) cryptoalgorithm. Two prime numbers p and q are chosen and the key to the code is $n = pq$. Find p and q if $n = 1073$.

The greater the prime numbers, the more difficult it is to break the code.

Some possibilities can be eliminated by using the divisibility rules you learned in Lesson 4-1.

1073 is not divisible by 2 since the ones digit, 3, is not divisible by 2.
1073 is not divisible by 3 since $1 + 0 + 7 + 3$ or 11 is not divisible by 3.
1073 is not divisible by 5 since the ones digit, 3, is not 0 or 5.

Now use a calculator and the guess-and-check problem-solving strategy to continue checking primes.

x	1073 ÷ x	factor of 1073?
7	153.286	no
11	97.5455	no
13	82.5385	no
17	63.1176	no
19	56.4737	no
23	46.6522	no
29	37	yes

If $n = 1073$, then p and q are 29 and 37.

FYI

Dr. Arjen Lenstra and a team of volunteers spent eight months in 1993 and 1994 using calculations to factor RSA 129, a 129-digit number for encoding. The message they found was "The magic words are squeamish ossifrage."

Connection to Algebra

A monomial can also be written in factored form as a product of prime numbers, -1, and variables with no exponent greater than 1.

Example 4 **Factor each monomial completely.**

a. $28x^2y$

$$28x^2y = 2 \cdot 14 \cdot x^2 \cdot y$$
$$= 2 \cdot 2 \cdot 7 \cdot x \cdot x \cdot y$$

b. $64ab^3$

$$64ab^3 = 8 \cdot 8 \cdot a \cdot b^3$$
$$= 2 \cdot 4 \cdot 2 \cdot 4 \cdot a \cdot b \cdot b \cdot b$$
$$= 2 \cdot 2 \cdot 2 \cdot 2 \cdot 2 \cdot 2 \cdot a \cdot b \cdot b \cdot b$$

Study Guide Masters, p. 30

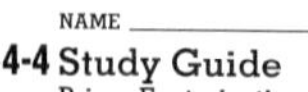

NAME ______________________ DATE __________

4-4 Study Guide
Prime Factorization

Student Edition Pages 184–188

A **factor tree** can be used to find the **prime factorization** of a composite number. Test prime numbers as factors in order from least to greatest. Test 2, 3, 5, 7, and so on.

Are all the factors prime?

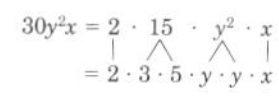

Complete each factor tree.

1. 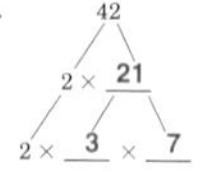

2. 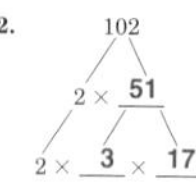

3. 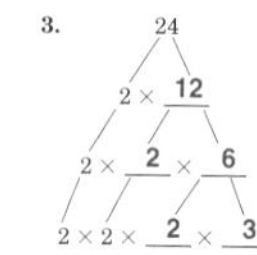

4. 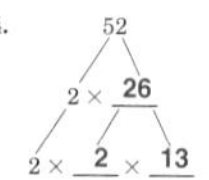

5. 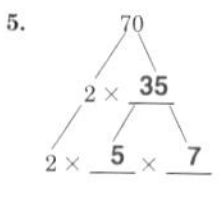

6. 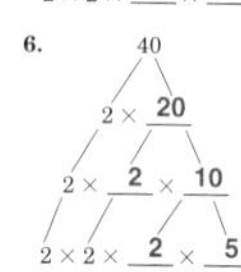

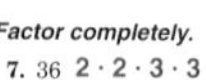

Factor completely.

7. 36 $2 \cdot 2 \cdot 3 \cdot 3$
8. -28 $-1 \cdot 2 \cdot 2 \cdot 7$
9. 50 $2 \cdot 5 \cdot 5$
10. $54xyz$ $2 \cdot 3 \cdot 3 \cdot 3 \cdot x \cdot y \cdot z$
11. $81m^2n$ $3 \cdot 3 \cdot 3 \cdot 3 \cdot m \cdot m \cdot n$
12. $100a^2b^2$ $2 \cdot 2 \cdot 5 \cdot 5 \cdot a \cdot a \cdot b \cdot b$
13. $-164st^3$ $-1 \cdot 2 \cdot 2 \cdot 41 \cdot s \cdot t \cdot t \cdot t$
14. $102f^3g^2$ $2 \cdot 3 \cdot 17 \cdot f \cdot f \cdot f \cdot g \cdot g$
15. $-72rt$ $-2 \cdot 2 \cdot 2 \cdot 3 \cdot 3 \cdot r \cdot t$
16. $200xy^3$ $2 \cdot 2 \cdot 2 \cdot 5 \cdot 5 \cdot x \cdot y \cdot y \cdot y$
17. $-225w^2vu$ $-3 \cdot 3 \cdot 5 \cdot 5 \cdot w \cdot w \cdot v \cdot u$
18. $140cd^2$ $2 \cdot 2 \cdot 5 \cdot 7 \cdot c \cdot d \cdot d$

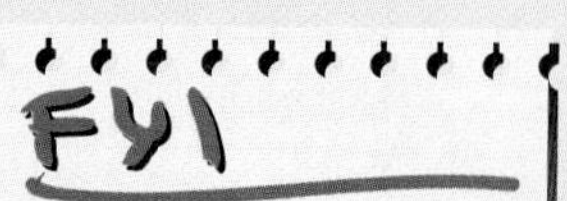

FYI

Volunteers worked on computers to complete 100 quadrillion calculations. The authors of RSA 129 thought it would take 40 quadrillion years to factor the number with methods available at that time.

Checking Your Understanding

Communicating Mathematics

1–2. See Solutions Manual.

Read and study the lesson to answer these questions.

1. **Explain** the differences between prime and composite numbers.
2. **Find** the prime factorization of 84 by using both a factor tree and the cake method. Are the results the same? Explain.
3. **You Decide** Mansi and Wes each factored -124. Whose factorization is complete and why? **Mansi; 4 is not prime**

Mansi	Wes
$-124 = -1 \cdot 124$ $= -1 \cdot 4 \cdot 31$ $= -1 \cdot 2 \cdot 2 \cdot 31$	31 $4\overline{)124}$ $-1\overline{)-124}$ $-124 = -1 \cdot 4 \cdot 31$

Guided Practice

Determine whether each number is *prime* or *composite*.

4. 17 **prime**
5. 9 **composite**
6. 27 **composite**

Factor each number or monomial completely.

7. 38
8. 66
9. 56
10. $8x^2y^3$
11. $-42abc$
12. $25(xy)^2$
13. **Entertainment** In the movie *Little Man Tate*, Damon is called "the Mathemagician" because he can perform complicated calculations mentally. In a contest, he is asked how many factors the number 3067 has. Damon answers "Come on, guys. There are no factors of 3067. The number is prime." Is Damon correct? Explain.

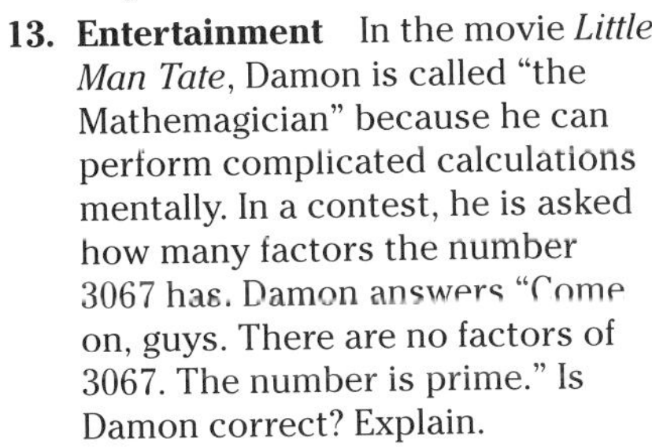

7. $2 \cdot 19$
8. $2 \cdot 3 \cdot 11$
9. $2 \cdot 2 \cdot 2 \cdot 7$
10. $2 \cdot 2 \cdot 2 \cdot x \cdot x \cdot y \cdot y \cdot y$
11. $-1 \cdot 2 \cdot 3 \cdot 7 \cdot a \cdot b \cdot c$
12. $5 \cdot 5 \cdot x \cdot x \cdot y \cdot y$
13. Damon is correct that 3067 is prime, but it has two factors—1 and 3067.

Exercises: Practicing and Applying the Concept

Independent Practice

14. composite
16. composite
17. prime

A **Determine whether each number is prime or composite.**

14. 35
15. 19 **prime**
16. 55
17. 101
18. 2 **prime**

Factor each number or monomial completely. **19–38. See margin.**

19. 51
20. -63
21. 41
22. -95
23. 110
24. -333
25. 81
26. 59
27. 13
28. 1024

C

29. $42xy^2$
30. $38mnp$
31. $21xy^3$
32. $560x^4y^2$
33. $28f^2g$
34. $275st^3$
35. $210mn^3$
36. $-8a^3b^2$
37. $75m^2k$
38. $-400a^2b^3$

Critical Thinking

39. Find the least number that gives you a remainder of 1 when you divide it by 2 or by 3 or by 5 or by 7. **211**
40. February the third is a *prime day* because the month and day (2/3) are represented by prime numbers. How many prime days are there in a leap year? **53 days**

3 PRACTICE/APPLY

Checking Your Understanding

Exercises 1–13 are designed to help you assess your students' understanding through reading, writing, speaking, and modeling. You should work through Exercises 1–3 with your students and then monitor their work on Exercises 4–13.

Assignment Guide

Core: 15–43 odd, 45–52
Enriched: 14–38 even, 39–52

For **Extra Practice**, see p. 749.

The red A, B, and C flags, printed only in the Teacher's Wraparound Edition, indicate the level of difficulty of the exercises.

Additional Answers

19. $3 \cdot 17$
20. $-1 \cdot 3 \cdot 3 \cdot 7$
21. $1 \cdot 41$
22. $-1 \cdot 5 \cdot 19$
23. $2 \cdot 5 \cdot 11$
24. $-1 \cdot 3 \cdot 3 \cdot 37$
25. $3 \cdot 3 \cdot 3 \cdot 3$
26. $1 \cdot 59$
27. $1 \cdot 13$
28. $2 \cdot 2 \cdot 2 \cdot 2 \cdot 2 \cdot 2 \cdot 2 \cdot 2 \cdot 2 \cdot 2$

Practice Masters, p. 30

NAME ______________________ DATE __________

4-4 Practice
Prime Factorization

Student Edition Pages 184–188

Factor each number or monomial completely.

1. 16 $2 \cdot 2 \cdot 2 \cdot 2$	2. 72 $2 \cdot 2 \cdot 2 \cdot 3 \cdot 3$	3. 75 $3 \cdot 5 \cdot 5$
4. -80 $-1 \cdot 2 \cdot 2 \cdot 2 \cdot 5$	5. -55 $-1 \cdot 5 \cdot 11$	6. 44 $2 \cdot 2 \cdot 11$
7. -60 $-1 \cdot 2 \cdot 2 \cdot 3 \cdot 5$	8. 54 $2 \cdot 3 \cdot 3 \cdot 3$	9. 96 $2 \cdot 2 \cdot 2 \cdot 2 \cdot 2 \cdot 3$
10. 98 $-1 \cdot 2 \cdot 7 \cdot 7$	11. 105 $3 \cdot 5 \cdot 7$	12. 125 $5 \cdot 5 \cdot 5$
13. 144 $2 \cdot 2 \cdot 2 \cdot 2 \cdot 3 \cdot 3$	14. -110 $-1 \cdot 2 \cdot 5 \cdot 11$	15. -123 $-1 \cdot 3 \cdot 41$
16. -200 $-1 \cdot 2 \cdot 2 \cdot 2 \cdot 5 \cdot 5$	17. 275 $5 \cdot 5 \cdot 11$	18. -280 $-1 \cdot 2 \cdot 2 \cdot 2 \cdot 5 \cdot 7$
19. 297 $3 \cdot 3 \cdot 3 \cdot 11$	20. -900 $-1 \cdot 2 \cdot 2 \cdot 3 \cdot 3 \cdot 5 \cdot 5$	21. 108 $2 \cdot 2 \cdot 3 \cdot 3 \cdot 3$
22. -1500 $-1 \cdot 2 \cdot 2 \cdot 3 \cdot 5 \cdot 5 \cdot 5$	23. 1521 $3 \cdot 3 \cdot 13 \cdot 13$	24. -1600 $-1 \cdot 2 \cdot 2 \cdot 2 \cdot 2 \cdot 2 \cdot 2 \cdot 5 \cdot 5$
25. $35xy^2$ $5 \cdot 7 \cdot x \cdot y \cdot y$	26. $12x^2z^2$ $-1 \cdot 2 \cdot 2 \cdot 3 \cdot x \cdot x \cdot z \cdot z$	27. $32pq$ $2 \cdot 2 \cdot 2 \cdot 2 \cdot 2 \cdot p \cdot q$
28. $-42mn^3$ $-1 \cdot 2 \cdot 3 \cdot 7 \cdot m \cdot n \cdot n \cdot n$	29. $51e^2f$ $3 \cdot 17 \cdot e \cdot e \cdot f$	30. $-64jk$ $-1 \cdot 2 \cdot 2 \cdot 2 \cdot 2 \cdot 2 \cdot 2 \cdot j \cdot k$
31. $98r^2t^3$ $2 \cdot 7 \cdot 7 \cdot r \cdot r \cdot t \cdot t \cdot t$	32. $-27v^3w$ $-1 \cdot 3 \cdot 3 \cdot 3 \cdot v \cdot v \cdot v \cdot w$	33. $90t^3m^2$ $2 \cdot 3 \cdot 3 \cdot 5 \cdot t \cdot t \cdot t \cdot m \cdot m$
34. $105ab^2$ $3 \cdot 5 \cdot 7 \cdot a \cdot b \cdot b$	35. $143m^2p$ $11 \cdot 13 \cdot m \cdot m \cdot p$	36. $525ac^2$ $3 \cdot 5 \cdot 5 \cdot 7 \cdot a \cdot c \cdot c$
37. $-150c^2d^3$ $-1 \cdot 2 \cdot 3 \cdot 5 \cdot 5 \cdot c \cdot c \cdot d \cdot d \cdot d$	38. $600xy$ $2 \cdot 2 \cdot 2 \cdot 3 \cdot 5 \cdot 5 \cdot x \cdot y$	39. $-450s^2t^3$ $-1 \cdot 2 \cdot 3 \cdot 3 \cdot 5 \cdot 5 \cdot s \cdot s \cdot t \cdot t \cdot t$
40. $100kt^3$ $2 \cdot 2 \cdot 5 \cdot 5 \cdot k \cdot t \cdot t \cdot t$	41. $500hj^2$ $2 \cdot 2 \cdot 5 \cdot 5 \cdot 5 \cdot h \cdot j \cdot j$	42. $-625b^3c$ $-1 \cdot 5 \cdot 5 \cdot 5 \cdot 5 \cdot b \cdot b \cdot b \cdot c$

Group Activity Card 4-4

Mind Those Prime Factors — Group Activity **4-4**

MATERIALS: 0-9 spinner

In this activity each group member tries to create the three-digit number with the greatest number of *prime* factors and the greatest prime factor.

To begin each round of the activity, the spinner is spun four times to determine the four digits each person can use to make a number. Each member is to use any three of the four digits to make any three-digit number, but it should have 1) as many prime factors as possible and 2) the largest prime factor possible.

For example, suppose the digits 0, 2, 7, 8 are spun. You could make the number 208 which factors into (24) (13) or 782 which factors into (2) (391). If these two were the only two numbers made, 208 would have the most prime factors (5), while 782 would have the greater prime factor (391). Thus, each player would "win" a part of that round.

A calculator could be used to check the factorization given by each member.

 Pre-Algebra

Additional Answers

29. $2 \cdot 3 \cdot 7 \cdot x \cdot y \cdot y$
30. $2 \cdot 19 \cdot m \cdot n \cdot p$
31. $3 \cdot 7 \cdot x \cdot y \cdot y \cdot y$
32. $2 \cdot 2 \cdot 2 \cdot 2 \cdot 5 \cdot 7 \cdot x \cdot x \cdot x \cdot x \cdot y \cdot y$
33. $2 \cdot 2 \cdot 7 \cdot f \cdot f \cdot g$
34. $5 \cdot 5 \cdot 11 \cdot s \cdot t \cdot t \cdot t$
35. $2 \cdot 3 \cdot 5 \cdot 7 \cdot m \cdot n \cdot n \cdot n$
36. $-1 \cdot 2 \cdot 2 \cdot 2 \cdot a \cdot a \cdot a \cdot b \cdot b$
37. $3 \cdot 5 \cdot 5 \cdot m \cdot m \cdot k$
38. $-1 \cdot 2 \cdot 2 \cdot 2 \cdot 2 \cdot 5 \cdot 5 \cdot a \cdot a \cdot b \cdot b \cdot b$

4 ASSESS

Closing Activity

Modeling Have students use a calculator to make factor trees for the following numbers.

1. 500 $2 \cdot 2 \cdot 5 \cdot 5 \cdot 5$
2. 288 $2 \cdot 2 \cdot 2 \cdot 2 \cdot 3 \cdot 3$
3. 2261 $7 \cdot 17 \cdot 19$
4. 3080 $2 \cdot 2 \cdot 2 \cdot 5 \cdot 7 \cdot 11$

Additional Answer

42. 3 and 5, 5 and 7, 11 and 13, 17 and 19, 29 and 31, 41 and 43, 59 and 61, 71 and 73

Enrichment Masters, p. 30

NAME ______________________ DATE __________

4-4 Enrichment Student Edition Pages 184–188

Prime Pyramid

A **prime number** is a whole number that has exactly two factors—itself and 1. A pyramid composed of prime numbers is called a *prime pyramid*. Each row of the prime pyramid below begins with 1 and ends with the number of that row. In each row, the numbers from 1 to the row number are arranged such that the sum of any two adjacent numbers is a prime.

For example, look at row 4:

It must contain the numbers 1, 2, 3, and 4.
It must begin with 1 and end with 4.
The sum of adjacent pairs must be a prime number:
$1 + 2 = 3,\ 2 + 3 = 5,\ 3 + 4 = 7$

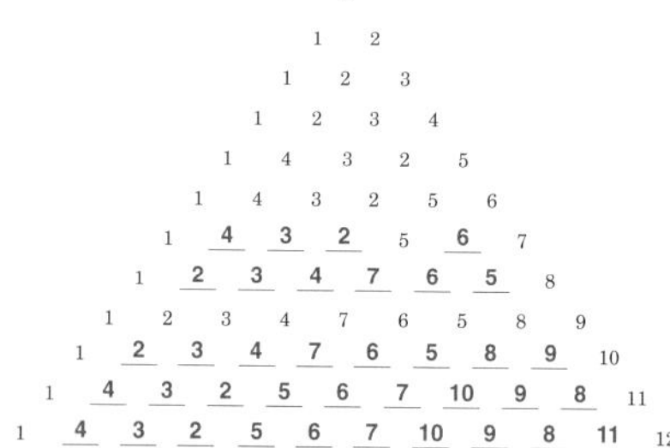

1. Complete the pyramid by filling in the missing numbers.
2. Extend the pyramid to row 13.
   1, 4, 3, 2, 5, 6, 7, 12, 11, 8, 9, 10, 13
3. Explain the patterns you see in the completed pyramid.
   Answer will vary. Each row alternates odd and even numbers.

Applications and Problem Solving

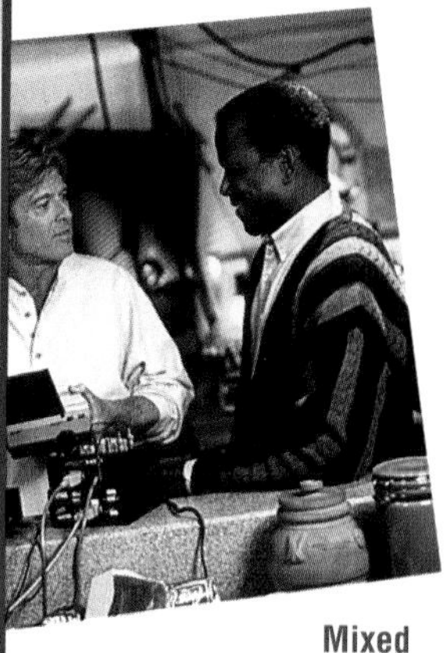

41. Cryptography Use the formula in Example 3 to find p and q if $n = 1643$ for an RSA code. **31 and 53**

42. Number Theory Mathematicians have many theories that are unproved. One theory is that there is an infinite number of *twin primes*. Twin primes are prime numbers that differ by 2, like 3 and 5. List all the twin primes that are less than 100. **See margin.**

43. History Chinese mathematician Sun-Tsŭ lived in the 1st century A.D. He studied *Chinese remainders*. This involves the remainders left when a number is divided by different primes.

a. Find the least positive integer having the remainders 2, 3, 2 when divided by 3, 5, and 7 respectively. **23**

b. Is the least integer described in **part a** prime? **Yes**

44. Family Activity The movie *Sneakers* involves a supposed invention that would break codes. Find the movie in a library or video store and watch it to see what the invention did. **It factors large numbers.**

Mixed Review

46. $(-8)(-8)(-8)(-8)(-8)(-8)(-8)$

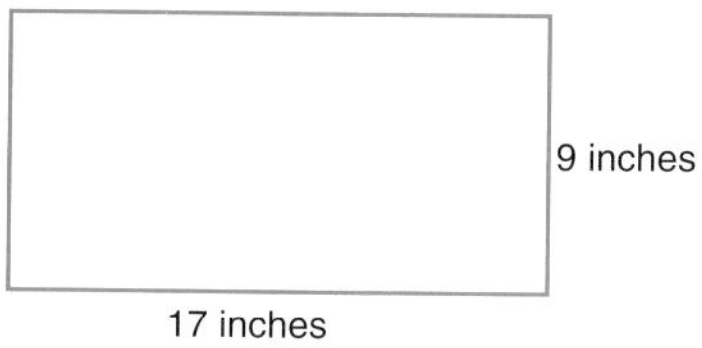

47. 52 inches; 153 square inches

50. $3 \cdot 12 - 3 \cdot 4$

52a. Sample answer: calculator because of large numbers.

Estimation
Mental Math
Calculator
Paper and Pencil

45. Probability If you toss a coin 4 times, how many different combinations of outcomes are possible? (Lesson 4-3) **16**

46. Write $(-8)^7$ as the product of the same factor. (Lesson 4-2)

47. Geometry Find the perimeter and area of the rectangle at the right. (Lesson 3-5)

[Rectangle: 17 inches by 9 inches]

48. Solve $\frac{x}{-7} = -28$. (Lesson 3-3) **196**

49. Replace ● in 6 ● −6 with <, >, or = to make a true sentence. (Lesson 2-3) **>**

50. Rewrite $3 \cdot (12 - 4)$ using the distributive property. (Lesson 1-5)

51. Exercise *Healthy Woman* magazine reports that you burn 15 additional calories each hour that you spend standing instead of sitting. Write an expression for the number of calories you burn in an hour of standing if you burn x calories per hour while sitting. (Lesson 1-3) $x + 15$

52. Demographics The average American earns $1,235,720 and pays $178,364 in taxes in his or her lifetime. The amount of money that is left in a paycheck after taxes is called take-home pay. What is the average American's lifetime take-home pay? (Lesson 1-1)

a. Which method of computation do you think is most appropriate? Justify your choice.

b. Solve the problem using the four-step plan. Be sure to examine your solution. **$1,057,356**

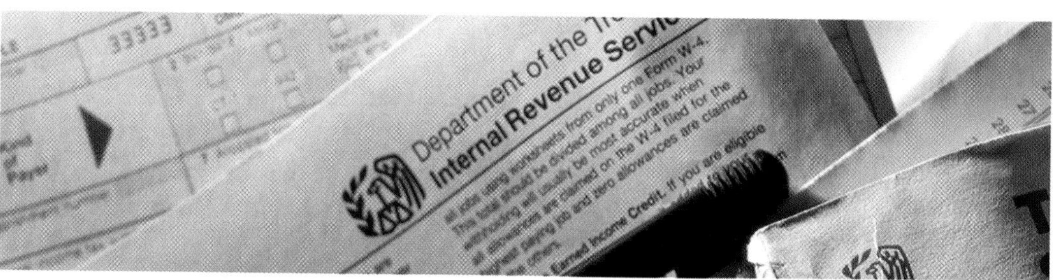

Reteaching

Using Discussion Discuss the use of factor trees and division and see which method is more easily understood by the students. Emphasize that method, working toward the more complex problems.

Extension

Using Logical Reasoning Have students solve this problem.
I am a three-digit number. I am divisible by nine. Two and seven are two of my six prime factors. My digits are consecutive numbers, but not in order. What number am I? **756**

GRAPHING CALCULATOR ACTIVITY

4-4B Factor Patterns

An Extension of Lesson **4-4**

You already know that prime numbers have exactly two factors. Which numbers have exactly three factors? or four factors? In this Math Lab, you will investigate factor patterns.

The chart below shows the numbers 2 through 18 and their factors arranged according to the number of factors.

Exactly 2 Factors	Exactly 3 Factors	Exactly 4 Factors	Exactly 5 Factors	Exactly 6 Factors
2: 1, 2 **3:** 1, 3 **5:** 1, 5 **7:** 1, 7 **11:** 1, 11 **13:** 1, 13 **17:** 1, 17 **19, 23, 29, 31, 37, 41, 43, 47**	**4:** 1, 2, 4 **9:** 1, 3, 9 **25, 49**	**6:** 1, 2, 3, 6 **8:** 1, 2, 4, 8 **10:** 1, 2, 5, 10 **14:** 1, 2, 7, 14 **15:** 1, 3, 5, 15 **21, 22, 26, 27, 33, 34, 35, 38, 39, 46**	**16:** 1, 2, 4, 8, 16	**12:** 1, 2, 3, 4, 6, 12 **18:** 1, 2, 3, 6, 9, 18 **20, 28, 32, 44, 45, 50** **More than 6 factors: 24, 30, 36, 40, 42, 48**

Your Turn Copy the chart above. Then use the TI-82 graphing calculator program below to find the factors of the numbers 19 through 50. Place each number in the correct column of the chart.

Begin entering the graphing calculator program by pressing PRGM ◄ ENTER. Then type in the program title, FACTOR. Consult your User's Guide if you need help finding the keys for the functions. When you finish entering the program, press 2nd QUIT to exit the programming mode. To run the program, press PRGM, the number next to the title FACTOR, and ENTER.

```
PROGRAM: FACTOR
: Input "ENTER NUMBER ", N
: For (D, 1, N)
: If iPart (N/D) – (N/D)
: Disp D
: End
```

TALK ABOUT IT

1. Predict a number from 50 to 150 that can be placed in each column. Check your prediction by using the graphing calculator program.
2. Write a paragraph that describes the pattern in each column.

1. Sample answer: 67, 121, 55, 81, 63, 66 2. See Solutions Manual.

Technology

This lesson offers an excellent opportunity for using technology in your pre-algebra classroom. For more information on using technology, see *Graphing Calculators in the Mathematics Classroom,* one of the titles in the Glencoe Mathematics Professional Series.

4-4B LESSON NOTES

NCTM Standards: 1-3, 6, 8

Objective

Use the graphing calculator to investigate factor patterns.

Recommended Time

15 minutes

Instructional Resources

Graphing Calculator Masters, p. 19

This master provides keystroking instruction for this lesson for the TI-81 and Casio graphing calculators.

1 FOCUS

Motivating the Lesson

Ask students to consider whether the number of factors a number has is arbitrary, or are there certain kinds of numbers that have the same number of factors.

2 TEACH

Teaching Tip The programming mode is different than the normal computational mode. When you press ENTER, the program moves to the next line instead of giving an answer. When the calculator is turned off, the program is **not** forgotten!

3 PRACTICE/APPLY

Assignment Guide

Core: 1–2
Enriched: 1–2

4 ASSESS

Observing students working with technology is an excellent method of assessment.

4-5 LESSON NOTES

NCTM Standards: 1-6, 8, 9

Instructional Resources

- Study Guide Master 4-5
- Practice Master 4-5
- Enrichment Master 4-5
- Group Activity Card 4-5
- Assessment and Evaluation Masters, pp. 98, 99

Transparency 4-5A contains the 5-Minute Check for this lesson; **Transparency 4-5B** contains a teaching aid for this lesson.

Recommended Pacing	
Standard Pacing	Day 5 of 13
Honors Pacing	Day 5 of 12
Block Scheduling*	Day 3 of 6 (along with Lesson 4-6)

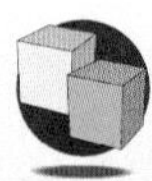

*For more information on pacing and possible lesson plans, refer to the ***Block Scheduling Booklet***.

1 FOCUS

5-Minute Check
(over Lesson 4-4)

Factor each number or monomial completely.

1. 108 **$2 \cdot 2 \cdot 3 \cdot 3 \cdot 3$**
2. 117 **$3 \cdot 3 \cdot 13$**
3. -88 **$-1 \cdot 2 \cdot 2 \cdot 2 \cdot 11$**
4. $72x^2y$ **$2 \cdot 2 \cdot 2 \cdot 3 \cdot 3 \cdot x \cdot x \cdot y$**
5. $-25xy^4$ **$-1 \cdot 5 \cdot 5 \cdot x \cdot y \cdot y \cdot y \cdot y$**

Motivating the Lesson

Hands-On Activity Bring in two different tennis shoes. Have students examine the shoes and list things that are common to both shoes and things that are different.

4-5 Greatest Common Factor (GCF)

Setting Goals: *In this lesson, you'll find the greatest common factor of two or more integers or monomials.*

Modeling a Real-World Application: History

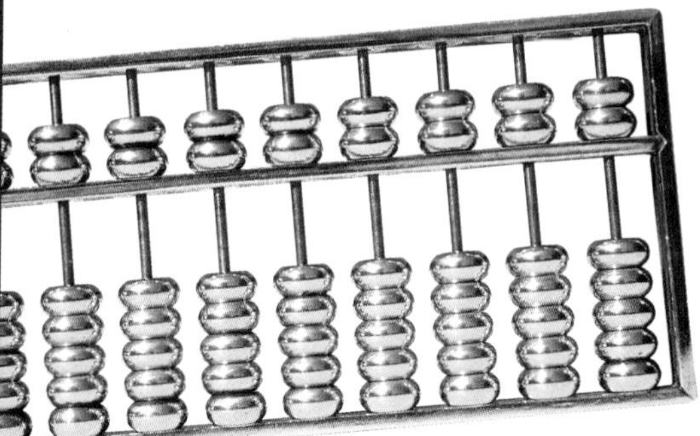

Very little of the mathematics of ancient China remains known to us today. The Chinese recorded information on bamboo, which did not last very long. Also, many of the books were destroyed during the reign of Emperor Shǐ Huang-ti. But the *Jiuzhang Suanshu*, or *The Nine Chapters on the Mathematical Art* survived to tell us a great deal about the mathematical accomplishments of the culture.

One of the chapters in the *Jiuzhang Suanshu* contains information on field measurement. In it, a procedure for finding the **greatest common factor (GCF)** of two numbers is described.

In Exercise 40, you will use the Chinese method to find the GCF.

Learning the Concept

The greatest of the factors of two or more numbers is called the greatest common factor. There are many methods for finding the greatest common factor. One method is to simply list the factors of each number and identify the greatest of the factors common to the numbers.

For example, consider finding the GCF of 78 and 91.

Factors of 78: 1, 2, 3, 6, 13, 26, 39, 78
Factors of 91: 1, 7, 13, 91

The common factors of 78 and 91, shown in blue, are 1 and 13. The GCF of 78 and 91 is 13.

Another method for finding the greatest common factor of two or more numbers is to find the prime factorization of the numbers and then find the product of their common factors.

Example 1 **Use prime factorization to find the GCF of each set of numbers.**

a. 315 and 135

First find the prime factorization of each number.

$$315 = 3 \cdot 105 = 3 \cdot 5 \cdot 21 = 3 \cdot 5 \cdot 3 \cdot 7$$

$$135 = 5 \cdot 27 = 5 \cdot 3 \cdot 9 = 5 \cdot 3 \cdot 3 \cdot 3$$

Then find the common factors.

$315 = 3 \cdot 5 \cdot 3 \cdot 7$ or $3 \cdot 3 \cdot 5 \cdot 7$ *The loops indicate each common factor.*

$135 = 5 \cdot 3 \cdot 3 \cdot 3$ or $3 \cdot 3 \cdot 5 \cdot 3$ *They are 3, 3, and 5.*

The greatest common factor of 315 and 135 is $3 \cdot 3 \cdot 5$ or 45.

b. 66, 90, and 150

Express each number as a product of prime factors.

$66 = 2 \cdot 3 \cdot 11$

$90 = 2 \cdot 3 \cdot 3 \cdot 5$ *The common factors are 2 and 3.*

$150 = 2 \cdot 3 \cdot 5 \cdot 5$

The GCF is $2 \cdot 3$ or 6.

Some real-life problems can be solved using the GCF.

Example 2

Hobbies

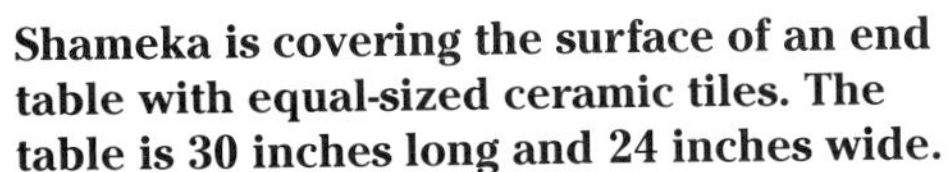

Shameka is covering the surface of an end table with equal-sized ceramic tiles. The table is 30 inches long and 24 inches wide.

a. What is the largest square tile that Shameka can use and not have to cut any tiles?

b. How many tiles will Shameka need?

a. The size of the tile is the greatest common factor of 30 and 24.

$30 = 2 \cdot 3 \cdot 5$

$24 = 2 \cdot 2 \cdot 2 \cdot 3$

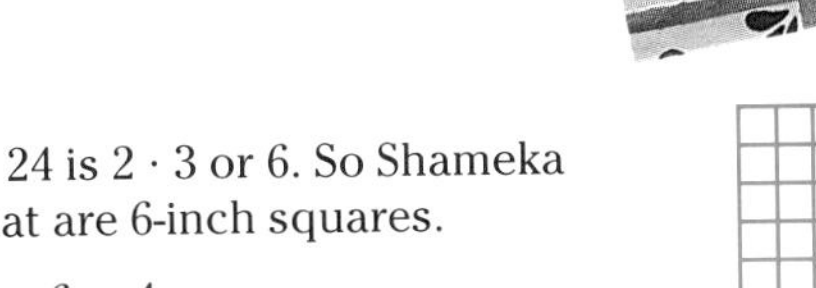

The GCF of 30 and 24 is $2 \cdot 3$ or 6. So Shameka should use tiles that are 6-inch squares.

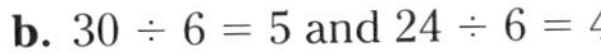

b. $30 \div 6 = 5$ and $24 \div 6 = 4$

So Shameka will need $5 \cdot 4$ or 20 tiles to cover the table.

Connection to Algebra

The product of the common prime factors of two or more monomials is their GCF.

Example  **Find the greatest common factor of $36x^3y$ and $56xy^2$.**

$36x^3y = 2 \cdot 2 \cdot 3 \cdot 3 \cdot x \cdot x \cdot x \cdot y$

$56xy^2 = 2 \cdot 2 \cdot 2 \cdot 7 \cdot x \cdot y \cdot y$

The GCF of $36x^3y$ and $56xy^2$ is $2 \cdot 2 \cdot x \cdot y$ or $4xy$.

2 TEACH

In-Class Examples

For Example 1

Use prime factorization to find the GCF of each set of numbers.

a. 108 and 126 **18**

b. 30, 150, 315 **15**

For Example 2

A bicycle path is being planned from Alendale to Barker to Compton. It is 42 miles from Alendale to Barker and 70 miles from Barker to Compton. Rest stops are to be located at each city and at equal distances along the entire path. What is the greatest possible distance between rest stops? **14 miles**

For Example 3

Find the greatest common factor of $28mn^3$ and $36m^2n^2$. **$4mn^2$**

Cooperative Learning

Inside-Outside Circle Have students form two concentric circles with students in each circle facing those in the other circle. There should be a one-to-one correspondence of students. Students rotate after each question is answered. Start by giving the inside students a set of cards with two numbers less than 100 on one side of the card. Have students find the GCF. Then show the card to a student in the outside ring and that student names the GCF. Give new cards to students in the outside circle and repeat. For more information on the inside-outside circle strategy, see *Cooperative Learning in the Mathematics Classroom*, p. 33.

3 PRACTICE/APPLY

Checking Your Understanding

Exercises 1–11 are designed to help you assess your students' understanding through reading, writing, speaking, and modeling. You should work through Exercises 1–4 with your students and then monitor their work on Exercises 5–11.

Teaching Tip Remind students that for two numbers that do not appear to have any common factors, the greatest common factor is 1, not 0.

Additional Answer

1. Find the prime factorizations for each number. Multiply all the common prime factors.

Assignment Guide

Core: 13–35 odd, 36, 37, 39, 41–48
Enriched: 12–34 even, 36–48
All: Self Test 1–13

For **Extra Practice**, see p. 749.

The red A, B, and C flags, printed only in the Teacher's Wraparound Edition, indicate the level of difficulty of the exercises.

Study Guide Masters, p. 31

NAME ______ DATE ______

4-5 Study Guide
Greatest Common Factor (GCF)

Student Edition Pages 190–194

Step 1	Step 2	Step 3
Factor each number completely.	Circle all pairs of factors that the numbers have in common.	Find the product of the common factors circled in Step 2.

Find the GCF of 24 and 56.

$24 = 2 \cdot 12 = 2 \cdot 2 \cdot 6 = 2 \cdot 2 \cdot 2 \cdot 3$	$56 = 2 \cdot 28 = 2 \cdot 2 \cdot 14 = 2 \cdot 2 \cdot 2 \cdot 7$	$24 = 2 \cdot 2 \cdot 2 \cdot 3$ $56 = 2 \cdot 2 \cdot 2 \cdot 7$	$2 \cdot 2 \cdot 2 = 8$ 8 is the GCF.

Find the GCF of $15xy^2$ and $18x^2y$.

$15xy^2 = 3 \cdot 5 \cdot x \cdot y \cdot y$ $18x^2y = 2 \cdot 9 \cdot x \cdot x \cdot y = 2 \cdot 3 \cdot 3 \cdot x \cdot x \cdot y$	$15xy^2 = 3 \cdot 5 \cdot x \cdot y \cdot y$ $18x^2y = 2 \cdot 3 \cdot 3 \cdot x \cdot x \cdot y$	$3 \cdot x \cdot y = 3xy$ $3xy$ is the GCF.

Find the GCF of each set of numbers or monomials.

1. 8, 10 2	2. 15, 24 3	3. 42, 54 6
4. 22, 55 11	5. 21, 49 7	6. 75, 100 25
7. $8n$, $18n$ $2n$	8. $15vw^2$, $27v^3w$ $3vw$	9. $125ab$, $200a^2b^2$ $25ab$
10. $26r^2$, $91s^2$, $13rs$ 13	11. $48xy$, $72x^2y^3$ $24xy$	12. $15m^3n^2$, $30mn$, $135m^2n^2$ $15mn$

Checking Your Understanding

Communicating Mathematics

Read and study the lesson to answer these questions. 1–2. See margin.

1. **Explain** how to find the greatest common factor of two or more numbers.
2. **Find** the common factors of 21 and 45 by using factor trees.
3. Name two numbers whose GCF is 15. **Sample answer: 15 and 30.**

4. In 1880, English mathematician John Venn developed the use of diagrams to show the relationships between collections of objects. The diagram at the right shows the prime factors of 16 and 28. The common factors are in both circles, so the GCF is $2 \cdot 2$ or 4. Find the greatest common factor of 36 and 48 by making a Venn diagram. **See margin.**

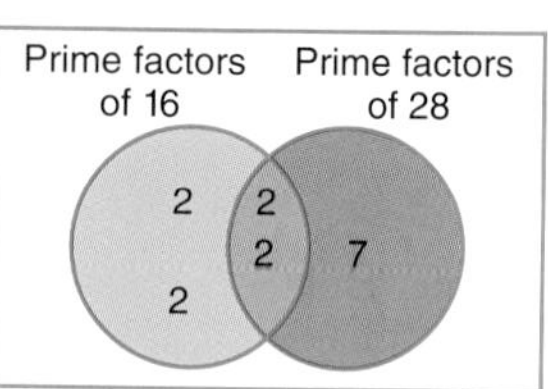

Guided Practice

Find the GCF of each set of numbers or monomials.

5. 6, 8 **2**
6. 12, 8 **4**
7. 1, 20 **1**
8. $12x$, $40x^2$ **$4x$**
9. $-5ab$, $6b^2$ **b**
10. $15a^2b^2$, $27a^3b^3$

10. $3a^2b^2$

11. **Interior Design** Mrs. Garcia has picked out two different fabrics to use to make square pillows for her living room. One fabric comes in a width of 48 inches, and the other comes in a width of 60 inches. How long should each side of the squares for the pillows be if all the pillows are the same size and there is no fabric wasted? **12 inches**

Exercises: Practicing and Applying the Concept

Independent Practice

Find the GCF of each set of numbers or monomials.

12. 16, 56 **8**
13. 24, 40 **8**
A
14. 6, 8, 12 **2**
15. 12, 24, 36 **12**
16. 20, 21, 25 **1**
17. 108, 144 **36**
B
18. $14n$, $42n^2$ **$14n$**
19. $40x^2$, $16x$ **$8x$**
20. $24a^2$, $-60a$ **$12a$**
21. $14b$, $-56b^2$ **$14b$**
22. $33y^2$, $44y$ **$11y$**
23. -18, $45mn$ **9**
24. $32mn^2$, $16n$, $12n^3$ **$4n$**
25. $18a$, $30ab$, $42b$ **6**
26. $15v^2$, $36w^2$, $70vw$ **1**
27. $24x^3$, $36y^3$, $18xy$ **6**
28. $15r^2$, $35s^2$, $70rs$ **5**
29. $18ab$, $6a^2$, $42a^2b$ **$6a$**

Reteaching

Using Calculators Use calculators and short division to write the prime factorization of numbers.

3)819 819 [÷] 3 [=] 273
3)273 273 [÷] 3 [=] 91
7)91 91 [÷] 7 [=] 13
13
$819 = 3 \cdot 3 \cdot 7 \cdot 13$

GLENCOE Technology

Interactive Mathematics Tools Software

In this interactive computer lesson, students explore factoring integers. A **Computer Journal** gives students an opportunity to write about what they have learned.

For Windows & Macintosh

Two numbers are *relatively prime* if their only common factor is 1. Determine whether the numbers in each pair are relatively prime. Write *yes* or *no*.

30. 8 and 9 **yes** **31.** 11 and 13 **yes** **32.** 21 and 14 **no**
33. 25 and 30 **no** **34.** 21 and 23 **yes** **35.** 9 and 12 **no**

Critical Thinking

36. Can the GCF of any set of numbers be greater than any one of the numbers? Explain. **See margin.**

Applications and Problem Solving

37. Patterns What is the GCF of all the numbers in the sequence 15, 30, 45, 60, 75, . . . ? **15**

38. 6; See Solutions Manual for explanation and sieve.

38. History Eratosthenes was a Greek mathematician who lived during the third century B.C. One of his contributions to the field was the *sieve of Eratosthenes*. Follow the steps below to complete a sieve.

- ▶ Write the numbers 2 to 50 in a list.
- ▶ Circle 2, the first prime number. Then cross out every second number after 2.
- ▶ Using a different color, circle 3, the second prime number. Then cross out every third number after 3.
- ▶ Continue using different colors to circle prime numbers and cross out their multiples until all the numbers in the list have been circled or crossed out.

Determine the greatest common factor of 30 and 42 using the sieve. Explain how you found the GCF.

MATERIALS

colored pencils or crayons

39. Carpentry Elena and her father are making shelves to store sports equipment and garden supplies in the garage. They would like to make the best use of a 48 in. by 72 in. piece of plywood. How many shelves measuring 12 in. by 16 in. could be cut from the plywood if there were no waste? **18 shelves**

40. History Follow the steps outlined below to find the GCF of 42 and 86 by using the Chinese method mentioned at the beginning of the lesson.

- ▶ Subtract the lesser number, *a*, from the greater number, *b*.
- ▶ If the result is a factor of both numbers, it is the greatest common factor. If the result is not a factor of both numbers, subtract the result from *a* or subtract *a* from the result so that the difference is a positive number.
- ▶ Continue subtracting and checking the results until you find a number that is a factor of both numbers. **2**

Mixed Review

41. Find the prime factorization of 3080. (Lesson 4-4) $2 \cdot 2 \cdot 2 \cdot 5 \cdot 7 \cdot 11$

42. Solve and graph $\frac{r}{4} \geq -14$. (Lesson 3-7) **See margin.**

43. Evaluate $|a| - |b| \cdot |c|$ if $a = -16$, $b = 2$, and $c = 3$. (Lesson 2-1) **10**

Error Analysis

Some students cannot differentiate between a list of factors and prime factorization. Emphasize that the product of all factors in a prime factorization should be the original number.

Additional Answers

2.

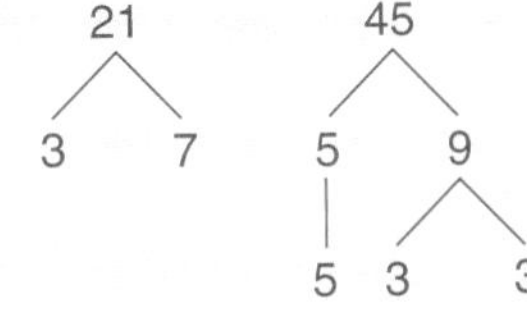

The common factor is 3.

4.

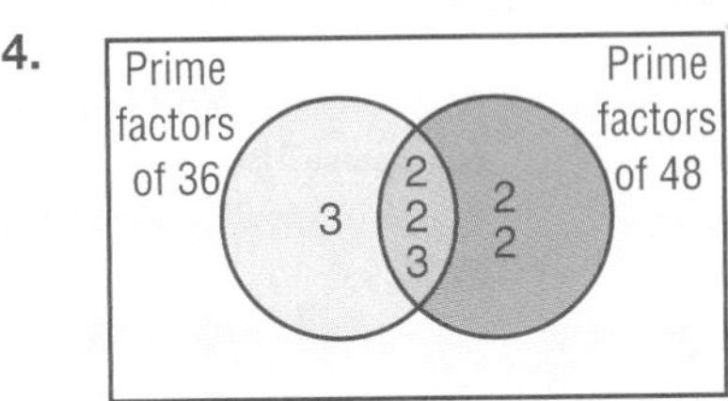

36. No; because a number cannot be a factor of a smaller number.

42. $r \geq -56$

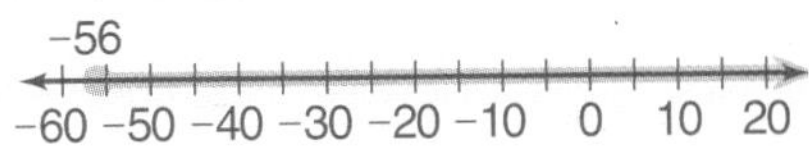

Extension

Using Cooperative Groups

Demonstrate Euclid's method for finding the GCF of 20 and 56.

Divide the lesser number into the greater number (20 into 56 goes 2 times, with 16 as a remainder).

Divide the remainder into the previous divisor (16 into 20 goes 1 time, with 4 as a remainder).

Continue until there is a remainder of zero (4 into 16 goes 4 times, with 0 as a remainder).

The last divisor is the GCF.
The GCF of 20 and 56 is 4.

Have students use this method to find the GCF of 8 and 34, 125 and 500, and 16 and 57. **2, 125, 1**

Practice Masters, p. 31

NAME ____________ DATE ________

4-5 Practice
Greatest Common Factor (GCF)

Student Edition Pages 190–194

Find the GCF of each set of numbers or monomials.

1. 14, 21 **7**	**2.** 15, 18 **3**	**3.** -14, 28 **14**
4. 36, 45 **9**	**5.** -28, 32 **4**	**6.** 48, 56 **8**
7. 25, -30 **5**	**8.** 25, 27 **1**	**9.** -60, 24 **12**
10. 32, 48, 96 **16**	**11.** 20, 28, 36 **4**	**12.** -72, 84, 132 **12**
13. 10, 25, 30 **5**	**14.** -14, 28, 42 **14**	**15.** 40, 60, 180 **20**
16. -42, 105, 126 **21**	**17.** 33, 198, 330 **33**	**18.** -126, 168, 210 **42**
19. $15ab$, $10ac$ $5a$	**20.** $14xy$, 28 **14**	**21.** $17xy$, $15x^2z$ x
22. $12am^2$, $18a^3m$ $6am$	**23.** $-120x^2$, $150xy$ $30x$	**24.** $105x^3y^2$, $165x^2y^4$ $15x^2y^2$
25. $9r^2t^2$, $12r^2$ $3r^2$	**26.** $-160zw$, $240w^2$ $80w$	**27.** $280ac^3$, $320a^3c$ $40ac$
28. $14m$, $21ny$, 28 **7**	**29.** $21pt$, $49p^2t$, $42pt^2$ $7pt$	**30.** $-5m^2$, $10m$, $15m^3$ $5m$
31. $5a^2$, $25b^2$, $50ab$ **5**	**32.** $9x$, $30xy$, $42y$ **3**	**33.** $15np$, $6n^2$, $39n^2p$ $3n$

44. **Sports** The table at the right shows the average cost of equipping an athlete for different sports. Make a bar graph of the costs. (Lesson 1-10) **See margin.**

Sport	Cost
Bowling	\$100
Golf	\$400
Racquetball	\$60
Skiing	\$500
Tennis	\$75

45. Solve $c - 15 = 120$ using the inverse operation. (Lesson 1-8) **45. 135**

46. **Business** Book Distributors adds a \$1.50 shipping and handling charge to the total price of every order. If the cost of a book is c, write an expression for the total cost. (Lesson 1-3) **c + \$1.50**

47. **Merchandising** Manufacturers often deliver small items in a quantity called a great gross. There are 12 items in a dozen, 12 dozens in a gross, and 12 gross in a great gross. How many items are in a great gross? (Lesson 1-2) **1728**

48. **Automobiles** Mechanics recommend changing the motor oil in your car every 3000 miles in order to keep your car running its best. If you just changed the oil and the odometer reads 56,893 miles, at what odometer reading should you change the oil next? (Lesson 1-1)
   a. Which method of computation do you think is most appropriate? Justify your choice. **Sample answer: mental math, simple numbers**
   b. Solve the problem using the four-step plan. Be sure to examine your solution. **59,893 miles**

Self Test

Using divisibility rules, state whether each number is divisible by 2, 3, 5, 6, or 10. (Lesson 4-1)

1. 117 **3**
2. 1002 **2, 3, 6**
3. 57 **3**

Write each product using exponents. (Lesson 4-2)

4. $9 \cdot 9 \cdot 9 \cdot 9 \cdot 9$ **9^5**
5. $2(2 \cdot 2)$ **2^3**
6. $\underbrace{k \cdot k \cdot k \cdot \ldots \cdot k \cdot k \cdot k}_{28 \text{ factors}}$ **k^{28}**

7. How many cubes are in the twenty-fifth building in the sequence at the right? (Lesson 4-3) **121**

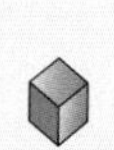
Building 1

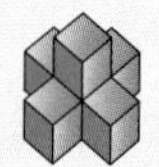
Building 2

Building 3

Factor each number or monomial completely. (Lesson 4-4)

8. 80 **$2 \cdot 2 \cdot 2 \cdot 2 \cdot 5$**
9. −26 **$-1 \cdot 2 \cdot 13$**
10. $42xy^2$ **$2 \cdot 3 \cdot 7 \cdot x \cdot y \cdot y$**

Find the GCF of each set of numbers or monomials. (Lesson 4-5)

11. 42, 56 **14**
12. 9, 15, 24 **3**
13. −18, $45xy$ **9**

4 ASSESS

Closing Activity

Modeling Have students demonstrate how to find the greatest common factor for each of the following.

1. 24 and 45 **3**
2. 190 and 380 **190**
3. 17 and 23 **1**
4. $6x^3$ and $15xy^2$ **$3x$**

Chapter 4, Quiz B (Lessons 4-4 and 4-5) is available in the *Assessment and Evaluation Masters*, p. 99.

Mid-Chapter Test (Lessons 4-1 through 4-5) is available in the *Assessment and Evaluation Masters*, p. 98.

Additional Answer

44.

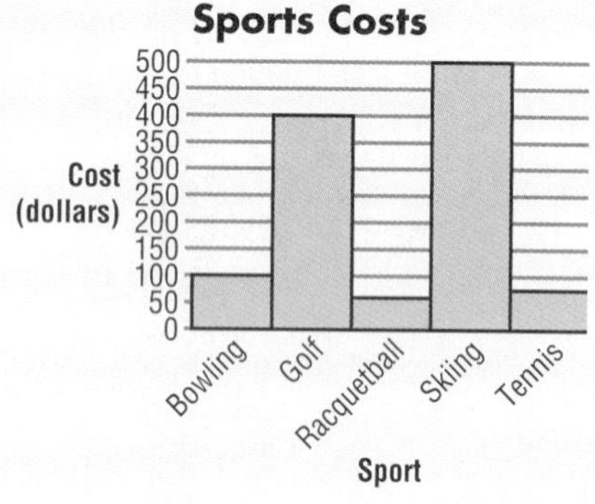

Enrichment Masters, p. 31

NAME ________________ DATE ________

4-5 Enrichment
Perfect and Imperfect Numbers

Student Edition Pages 190–194

In addition to being classified as composite or prime, numbers can be classified as perfect or imperfect.

A **perfect number** is one that is equal to the sum of all of its factors, except itself. There are two perfect numbers less than 30. The third perfect number is 496.

$$496 = 1 + 2 + 4 + 8 + 16 + 31 + 62 + 124 + 248$$

An **imperfect number** is one that is not equal to the sum of its factors, except itself. When the sum of its factors is greater than the number, it is called an **abundant number**. When the sum of its factors is less than the number, it is called a **deficient number**.

Solve.

1. Find a perfect number between 1 and 10. **6 = 1 + 2 + 3**
2. Find a perfect number between 20 and 30. **28 = 1 + 2 + 4 + 7 + 14**
3. The formula $N = (2^{n+1} - 1) \cdot 2n$ can be used to find perfect numbers. N is a perfect number if $(2^{n+1} - 1)$ is a prime number. Find N when $n = 1$. **6**
4. Find N when $n = 2$. Is N a perfect number? **28; yes**
5. Find N when $n = 3$. Is N a perfect number? **120; no**
6. Find the value of n that produces a value of 496 for N. **4**
7. Find the fourth perfect number. **when $n = 6$, $N = 8128$**
8. Find an abundant number between 20 and 30. **24**
9. Find a deficient number between 15 and 20. **16, 17, or 19**

Group Activity Card 4-5

Quick Draw Group Activity **4-5**

MATERIALS: Deck of regular playing cards

In this activity, you and your partner will use a regular deck of playing cards. The cards are shuffled and placed face down between you. Each person draws six or less cards from the deck. Next, each of you determines the total number represented by the cards (all face cards count ten).

Write these two numbers down and then work together to find their greatest common factor. You may want to start out drawing fewer cards and work your way up to greater numbers.

 Pre-Algebra

Self Test

The Self Test provides students with a brief review of the concepts and skills in Lessons 4-1 through 4-5. Lesson numbers are given to the right of exercises or instruction lines so students can review concepts not yet mastered.

HANDS-ON ACTIVITY

4-6A Equivalent Fractions

A Preview of Lesson **4-6**

MATERIALS

ruler

It is often helpful to use a model to help you understand a concept. In this lab, you will use models to help you understand fractions.

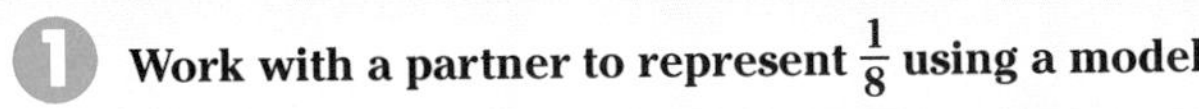

Activity 1 Work with a partner to represent $\frac{1}{8}$ using a model.

Use a ruler to draw a rectangle and separate the rectangle into eight equal parts. Shade one part. Since one part in eight is shaded, $\frac{1}{8}$ of the rectangle is shaded.

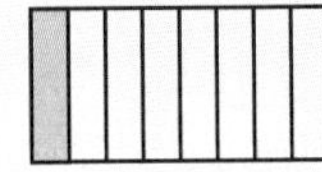

Your Turn Draw a model to represent $\frac{1}{6}$.

TALK ABOUT IT

1. How would you represent $\frac{2}{6}$ using a model? **Shade 2 of 6 sections.**
2. If you shaded five parts of a rectangle with eight equal parts, what fraction is represented? $\frac{5}{8}$

Activity  Compare $\frac{2}{8}$ and $\frac{3}{12}$ using models.

Draw two identical rectangles and separate one rectangle into eight equal parts. Separate the other into twelve equal parts. Shade two parts of the rectangle that has eight sections and shade three parts of the rectangle that has twelve sections.

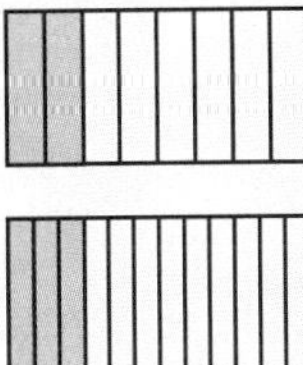

TALK ABOUT IT

3. What do you notice about the shaded regions in the two rectangles in Activity 2? **They are equal in size.**
4. Do the fractions $\frac{2}{8}$ and $\frac{3}{12}$ name the same number? Explain. **yes**
5. Are there other fractions that name the same number? If so, draw a diagram to show an example. **See Solutions Manual.**

Extension **Use models to determine whether each pair of fractions represents the same number.**

6. $\frac{1}{2}, \frac{5}{10}$ **yes**
7. $\frac{1}{3}, \frac{3}{9}$ **yes**
8. $\frac{1}{7}, \frac{2}{12}$ **no**

4-6A LESSON NOTES

NCTM Standards: 1-6

Objective

Use models to recognize equivalent fractions.

Recommended Time

Demonstration and discussion: 15 minutes; Exercises: 30 minutes

Instructional Resources

For each student or group of students
Math Lab and Modeling Math Masters
• p. 35 (worksheet)
For teacher demonstration
Overhead Manipulative Resources

1 FOCUS

Motivating the Lesson

Have students write a solution to this problem: *Fred and Ethel decided to run home from school. Fred tired out $\frac{7}{10}$ of the way home, and Ethel tired out $\frac{2}{3}$ of the way home. Who ran farther?* **Fred**

2 TEACH

Teaching Tip Emphasize the importance of carefully drawn models—the rectangles have to be the same size in order to make accurate comparisons.

3 PRACTICE/APPLY

Assignment Guide
Core: 1–8
Enriched: 1–8

4 ASSESS

Observing students working in cooperative groups is an excellent method of assessment.

4-6 LESSON NOTES

NCTM Standards: 1-6

Instructional Resources

- Study Guide Master 4-6
- Practice Master 4-6
- Enrichment Master 4-6
- Group Activity Card 4-6
- Activity Masters, p. 4
- Graphing Calculator Masters, p. 4

Transparency 4-6A contains the 5-Minute Check for his lesson; **Transparency 4-6B** contains a teaching aid for this lesson.

Recommended Pacing	
Standard Pacing	Day 7 of 13
Honors Pacing	Day 7 of 12
Block Scheduling*	Day 3 of 6 (along with Lesson 4-5)

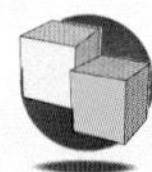

*For more information on pacing and possible lesson plans, refer to the ***Block Scheduling Booklet***.

1 FOCUS

5-Minute Check
(over Lesson 1-2)

Find the GCF of each set of numbers or monomials.

1. 22, 55 **11**
2. 106, 318 **106**
3. 15, 18, 30 **3**
4. $12ab^3$, $16a^4b^2$ **$4ab^2$**
5. $15mnp$, $45mp^3$, $90m^3n^4$ **$15m$**

Motivating the Lesson

Hands-On Activity Have students write the number 2 on several slips of paper. Repeat with other prime numbers through 11. Draw two groups of five slips of paper each. One group represents factors of the numerator of a fraction and the other group, the denominator. Remove the common factors and name the fraction in simplest form.

4-6 Simplifying Fractions

Setting Goals: *In this lesson, you'll simplify fractions using the GCF.*

Modeling a Real-World Application: Forestry

When the Rolling Stones, Eric Clapton, and the Allman Brothers need a keyboardist, they turn to Chuck Leavell. When the National Arbor Day Foundation, the world's largest tree-planting advocacy group, needs a farmer to speak on their behalf, they turn to Chuck Leavell, too.

Fifteen years ago, Mr. Leavell was a rock 'n' roller who had no experience with tree farming. His life changed when his wife's grandparents left her a 1200-acre farm in Twiggs County, Georgia. "The more I learned, the more I fell in love with it," said Leavell. Now the Leavells divide their time between their two loves, music and forestry. Their farm has now grown to 1500 acres. You can compare the original size of their farm to its current size by using a ratio.

Learning the Concept

A **ratio** is a comparison of two numbers by division. A ratio can be expressed in several ways. The expressions below all represent the same ratio.

$$2 \text{ to } 3 \qquad 2:3 \qquad \frac{2}{3} \qquad 2 \div 3$$

A ratio is most often written as a fraction in **simplest form**. A fraction is in simplest form when the GCF of the numerator and the denominator is 1.

Example 1 APPLICATION Forestry

Refer to the application at the beginning of the lesson. Write the ratio of the amount of land that Mrs. Leavell inherited to the size of the farm today in simplest form.

Mrs. Leavell inherited 1200 acres, and the farm is 1500 acres today. So the ratio is $\frac{1200}{1500}$.

First find the GCF of the numerator and denominator.

$1200 = 2 \cdot 2 \cdot 2 \cdot 2 \cdot 3 \cdot 5 \cdot 5$ *The GCF of 1200 and*
$1500 = 2 \cdot 2 \cdot 3 \cdot 5 \cdot 5 \cdot 5$ *1500 is $2 \cdot 2 \cdot 3 \cdot 5 \cdot 5$ or 300.*

Alternative Learning Styles

Kinesthetic Have one student pour two half cups of water into a clear jar. Have another student pour four quarter cups of water into an identical container. Point out that the same amount of water was poured into each container. Relate this to simplifying fractions.

Now divide the numerator and the denominator by 300 to write the fraction in simplest form.

$$\frac{1200}{1500} \overset{\div 300}{=} \frac{4}{5} \quad (\div 300)$$

Since the GCF of 4 and 5 is 1, the fraction $\frac{4}{5}$ is in simplest form.

The division in Example 1 can be represented in a different way.

$$\frac{1200}{1500} = \frac{\overset{1}{\cancel{2}} \cdot \overset{1}{\cancel{2}} \cdot 2 \cdot 2 \cdot \overset{1}{\cancel{3}} \cdot \overset{1}{\cancel{5}} \cdot \overset{1}{\cancel{5}}}{\underset{1}{\cancel{2}} \cdot \underset{1}{\cancel{2}} \cdot \underset{1}{\cancel{3}} \cdot 5 \cdot \underset{1}{\cancel{5}} \cdot \underset{1}{\cancel{5}}} = \frac{2 \cdot 2}{5} \text{ or } \frac{4}{5}$$

The slashes indicate that the numerator and denominator are both divided by 2 · 2 · 3 · 5 · 5, the GCF.

Example  **Write $\frac{12}{40}$ in simplest form.**

$$\frac{12}{40} = \frac{\overset{1}{\cancel{2}} \cdot \overset{1}{\cancel{2}} \cdot 3}{\underset{1}{\cancel{2}} \cdot \underset{1}{\cancel{2}} \cdot 2 \cdot 5}$$

Divide both the numerator and denominator by 2 · 2.

$$= \frac{3}{10}$$

Connection to Algebra

A fraction with variables in the numerator or denominator is called an **algebraic fraction**. Algebraic fractions can also be written in simplest form.

Example  **Simplify $\frac{16x^2y^3}{12xy^4}$. Assume that x and y are not equal to zero.**

$$\frac{16x^2y^3}{12xy^4} = \frac{\overset{1}{\cancel{2}} \cdot \overset{1}{\cancel{2}} \cdot 2 \cdot 2 \cdot \overset{1}{\cancel{x}} \cdot x \cdot \overset{1}{\cancel{y}} \cdot \overset{1}{\cancel{y}} \cdot \overset{1}{\cancel{y}}}{\underset{1}{\cancel{2}} \cdot \underset{1}{\cancel{2}} \cdot 3 \cdot \underset{1}{\cancel{x}} \cdot \underset{1}{\cancel{y}} \cdot \underset{1}{\cancel{y}} \cdot \underset{1}{\cancel{y}} \cdot y}$$

Divide both numerator and denominator by 2 · 2 · x · y · y · y.

$$= \frac{4x}{3y}$$

THINK ABOUT IT

What is the GCF of $16x^2y^3$ and $12xy^4$? $4xy^3$

Checking Your Understanding

Communicating Mathematics

Read and study the lesson to answer these questions.

1. **Explain** what is meant by expressing a fraction in simplest form.
2. **Express** the ratio 12:14 as a fraction in simplest form. $\frac{6}{7}$
3. **Give examples** of fractions in simplest form and fractions that are not in simplest form.
4. Why is it important to be able to simplify fractions?

1. The GCF of the numerator and denominator is 1.
3. Sample answers: $\frac{1}{3}, \frac{3}{4}, \frac{2}{4}, \frac{8}{12}$
4. See margin.

2 TEACH

In-Class Examples

For Example 1
Ms. Thompkins earned $40,000 this year. Last year she earned $35,000. Write the ratio of this year's salary to last year's salary in simplest form. $\frac{8}{7}$

For Example 2
Write $\frac{32}{72}$ in simplest form. $\frac{4}{9}$

For Example 3
Simplify $\frac{18x^3y^5}{15xy^8}$. Assume that x and y are not equal to zero. $\frac{6x^2}{5y^3}$

Teaching Tip For students who do not completely simplify fractions, tell them that for a fraction to be in simplest form the GCF of the numerator and denominator must be 1.

Study Guide Masters, p. 32

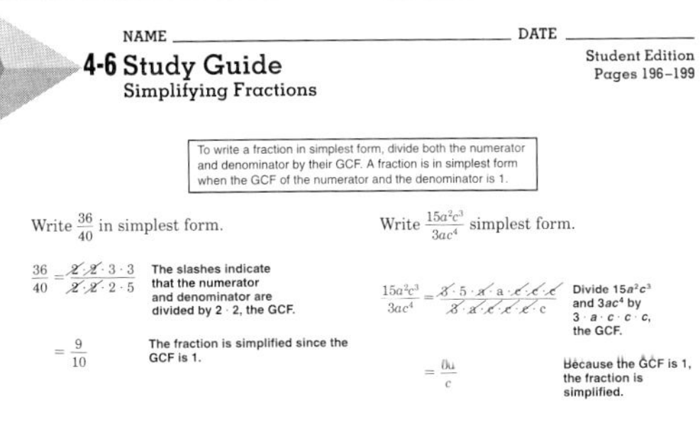
NAME ______________________ DATE ________

4-6 Study Guide
Simplifying Fractions

Student Edition Pages 196–199

To write a fraction in simplest form, divide both the numerator and denominator by their GCF. A fraction is in simplest form when the GCF of the numerator and the denominator is 1.

Write $\frac{36}{40}$ in simplest form.

$\frac{36}{40} = \frac{\cancel{2} \cdot \cancel{2} \cdot 3 \cdot 3}{\cancel{2} \cdot \cancel{2} \cdot 2 \cdot 5}$ The slashes indicate that the numerator and denominator are divided by 2 · 2, the GCF.

$= \frac{9}{10}$ The fraction is simplified since the GCF is 1.

Write $\frac{15a^2c^3}{3ac^4}$ simplest form.

$\frac{15a^2c^3}{3ac^4} = \frac{\cancel{3} \cdot 5 \cdot \cancel{a} \cdot a \cdot \cancel{c} \cdot \cancel{c} \cdot \cancel{c}}{\cancel{3} \cdot \cancel{a} \cdot \cancel{c} \cdot \cancel{c} \cdot \cancel{c} \cdot c}$ Divide $15a^2c^3$ and $3ac^4$ by $3 \cdot a \cdot c \cdot c \cdot c$, the GCF.

$= \frac{5a}{c}$ Because the GCF is 1, the fraction is simplified.

Write each fraction in simplest form. If the fraction is already in simplest form, write simplified.

1. $\frac{14}{20}$ $\frac{7}{10}$
2. $\frac{16}{35}$ simplified
3. $\frac{16}{20}$ $\frac{4}{5}$
4. $\frac{10}{40}$ $\frac{1}{4}$
5. $\frac{16}{36}$ $\frac{4}{9}$
6. $\frac{45}{48}$ $\frac{15}{16}$
7. $\frac{23}{55}$ simplified
8. $\frac{49}{56}$ $\frac{7}{8}$
9. $\frac{13}{26}$ $\frac{1}{2}$
10. $\frac{3x^3}{12x^4}$ $\frac{1}{4x}$
11. $\frac{9x^2}{16y^3}$ simplified
12. $\frac{4p^2q^3}{12pq}$ $\frac{pq^2}{3}$
13. $\frac{3ab}{15a^2b^3}$ $\frac{1}{5ab^2}$
14. $\frac{-3r^2s^5}{-12r^3s^6}$ $\frac{1}{4rs}$
15. $\frac{7m^3n^{12}}{15mn^8}$ $\frac{7m^2n^4}{15}$
16. $\frac{-13p^4q^3}{26p^5q^7}$ $-\frac{1}{2pq^4}$
17. $\frac{12a^4m^3}{16a^3m^8}$ $\frac{3a}{4m^5}$
18. $\frac{6q^3v^3}{18q^2v^4}$ $\frac{q}{3v}$

Reteaching

Using Manipulatives Provide students with circular pieces of paper with edges marked eighths and sixths. Have students fold the circles and shade to show that $\frac{4}{8} = \frac{2}{4} = \frac{1}{2}$ and that $\frac{3}{6} = \frac{1}{2}$.

Additional Answer

4. Sample answer: Simplified fractions are the most easily understood in everyday usage.

3 PRACTICE/APPLY

Checking Your Understanding

Exercises 1–18 are designed to help you assess your students' understanding through reading, writing, speaking, and modeling. You should work through Exercises 1–5 with your students and then monitor their work on Exercises 6–18.

Error Analysis

When simplifying fractions, students may use a factor other than the GCF, resulting in an answer that can still be simplified. Other students may always write the answer as a proper fraction even though they started with an improper fraction. Emphasize that students need to check all answers to see that they are indeed in simplest form.

Assignment Guide

Core: 19–43 odd, 44, 45, 47–54
Enriched: 20–42 even, 44–54

For **Extra Practice**, see p. 749.

The red A, B, and C flags, printed only in the Teacher's Wraparound Edition, indicate the level of difficulty of the exercises.

Practice Masters, p. 32

NAME ____________ DATE ________

4-6 Practice
Simplifying Fractions

Student Edition Pages 196–199

rite each fraction in simplest form. If the fraction is ready in simplified form, write simplified.

$\frac{3}{9}$ $\frac{1}{3}$	2. $\frac{6}{10}$ $\frac{3}{5}$	3. $\frac{12}{18}$ $\frac{2}{3}$	4. $\frac{5}{20}$ $\frac{1}{4}$
$\frac{9}{12}$ $\frac{3}{4}$	6. $\frac{15}{20}$ $\frac{3}{4}$	7. $\frac{3}{7}$ simplified	8. $\frac{28}{32}$ $\frac{7}{8}$
$\frac{10}{35}$ $\frac{2}{7}$	10. $\frac{24}{30}$ $\frac{4}{5}$	11. $\frac{49}{98}$ $\frac{1}{2}$	12. $\frac{28}{48}$ $\frac{7}{12}$
$\frac{14}{26}$ $\frac{7}{13}$	14. $\frac{11}{88}$ $\frac{1}{8}$	15. $\frac{45}{81}$ $\frac{5}{9}$	16. $\frac{27}{45}$ $\frac{3}{5}$
$\frac{16}{48}$ $\frac{1}{3}$	18. $\frac{47}{99}$ simplified	19. $\frac{13}{91}$ $\frac{1}{7}$	20. $\frac{30}{42}$ $\frac{5}{7}$
$\frac{84}{140}$ $\frac{3}{5}$	22. $\frac{96}{112}$ $\frac{6}{7}$	23. $\frac{53}{78}$ simplified	24. $\frac{62}{66}$ $\frac{31}{33}$
$\frac{15}{90}$ $\frac{1}{6}$	26. $\frac{56}{84}$ $\frac{2}{3}$	27. $\frac{105}{175}$ $\frac{3}{5}$	28. $\frac{258}{387}$ $\frac{2}{3}$
$\frac{296}{375}$ simplified	30. $\frac{240}{255}$ $\frac{16}{17}$	31. $\frac{64}{776}$ $\frac{8}{97}$	32. $\frac{1320}{1650}$ $\frac{4}{5}$
$\frac{5x^2y}{30xy}$ $\frac{x}{6}$	34. $\frac{25ab^2}{36c^2d}$ simplified	35. $\frac{15ef^2}{35ef}$ $\frac{3f}{7}$	36. $\frac{3r^2s^2}{27rs}$ $\frac{4s}{9}$
$\frac{15wx}{45w}$ $\frac{x}{3}$	38. $\frac{36l^2m}{81lm^2}$ $\frac{4l}{9m}$	39. $\frac{27m^2n^2}{45mn}$ $\frac{3mn}{5}$	40. $\frac{24xz}{64x^2z}$ $\frac{3}{8x}$
$\frac{18h^2d}{54d}$ $\frac{h^2}{3}$	42. $\frac{36st^2}{72t^2}$ $\frac{s}{2}$	43. $\frac{19x^2y^2}{38xy}$ $\frac{xy}{2}$	44. $\frac{48a^2b^2}{64a^2b}$ $\frac{3b}{4}$

Glencoe/McGraw-Hill T 32 *Pre-Algebra*

5. The fractions $\frac{3}{4}$ and $\frac{6}{8}$ are graphed on the number lines at the right. Notice that the fractions have the same graph. Thus, $\frac{3}{4}$ and $\frac{6}{8}$ are equivalent. Use number lines to determine whether $\frac{3}{9}$ and $\frac{1}{3}$ are equivalent. yes

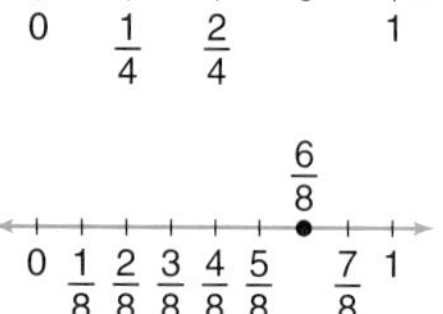

Guided Practice

Express each ratio as a fraction. Then if the fraction is not in simplest form, write it in simplest form.

6. $\frac{6}{10}, \frac{3}{5}$

6. 6 to 10 **7.** $1 \div 7$ $\frac{1}{7}$ **8.** $20 : 100$ $\frac{1}{5}$ **9.** eleven to fifteen $\frac{11}{15}$

Write each fraction in simplest form. If the fraction is already in *simplest* form, write simplified. 17. simplified

10. $\frac{10}{37}$ simplified **11.** $\frac{15}{21}$ $\frac{5}{7}$ **12.** $\frac{51}{60}$ $\frac{17}{20}$ **13.** $\frac{18}{44}$ $\frac{9}{22}$
14. $\frac{x}{x^3}$ $\frac{1}{x^2}$ **15.** $\frac{11t}{121t^2}$ $\frac{1}{11t}$ **16.** $\frac{8z^2}{16z}$ $\frac{z}{2}$ **17.** $\frac{8ab}{15cd}$

interNET CONNECTION
For the latest on government spending, visit: **www.glencoe.com/sec/math/prealg/mathnet**

18. Economics How does Uncle Sam spend a dollar? The graph at the right shows how each dollar spent by the Federal Government is used.

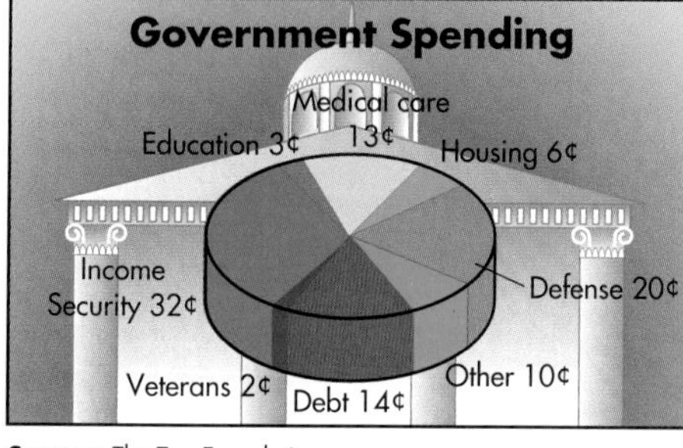

Source: The Tax Foundation

a. Write the ratio of the amount spent on housing assistance to total spending as a fraction in simplest form. $\frac{3}{50}$

b. Write the ratio of the amount spent on the national debt to total spending as a fraction in simplest form. $\frac{7}{50}$

Exercises: Practicing and Applying the Concept

Independent Practice

20. simplified
22. simplified
27. simplified
29–30. simplified
36. simplified
41. simplified
43. simplified

Write each fraction in simplest form. If the fraction is already in simplest form, write *simplified*.

A
19. $\frac{2}{14}$ $\frac{1}{7}$ **20.** $\frac{17}{20}$ **21.** $\frac{34}{38}$ $\frac{17}{19}$ **22.** $\frac{20}{53}$ **23.** $\frac{17}{51}$ $\frac{1}{3}$

B
24. $\frac{9}{15}$ $\frac{3}{5}$ **25.** $\frac{25}{40}$ $\frac{5}{8}$ **26.** $\frac{30}{51}$ $\frac{10}{17}$ **27.** $\frac{30}{37}$ **28.** $\frac{16}{64}$ $\frac{1}{4}$

29. $\frac{11}{13}$ **30.** $\frac{8}{27}$ **31.** $\frac{124}{222}$ $\frac{62}{111}$ **32.** $\frac{12m}{15m}$ $\frac{4}{5}$ **33.** $\frac{12x}{15y}$ $\frac{4x}{5y}$

34. $\frac{40a}{42a}$ $\frac{20}{21}$ **35.** $\frac{82t}{14t}$ $\frac{41}{7}$ **36.** $\frac{12cd}{19ef}$ **37.** $\frac{xyz^3}{x^2y}$ $\frac{z^3}{x}$ **38.** $\frac{30x^2}{51xy}$ $\frac{10x}{17y}$

C
39. $\frac{40p^2q}{52pq^2}$ $\frac{10p}{13q}$ **40.** $\frac{17k^2z}{51z}$ $\frac{k^2}{3}$ **41.** $\frac{31gh}{14}$ **42.** $\frac{52rst}{26rst}$ 2 **43.** $\frac{12x^2z}{23y^3}$

Group Activity Card 4-6

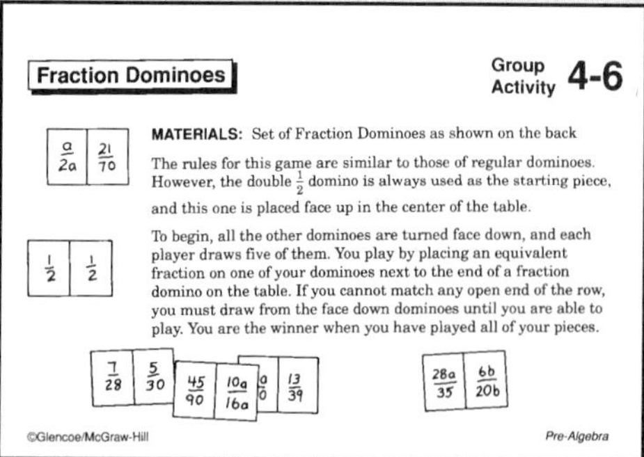
Fraction Dominoes Group Activity 4-6

MATERIALS: Set of Fraction Dominoes as shown on the back

The rules for this game are similar to those of regular dominoes. However, the double $\frac{1}{2}$ domino is always used as the starting piece, and this one is placed face up in the center of the table.

To begin, all the other dominoes are turned face down, and each player draws five of them. You play by placing an equivalent fraction on one of your dominoes next to the end of a fraction domino on the table. If you cannot match any open end of the row, you must draw from the face down dominoes until you are able to play. You are the winner when you have played all of your pieces.

 Pre-Algebra

Critical Thinking

44. Suppose $(1 \times 10^a) + (2 \times 10^b) + (3 \times 10^c) + (4 \times 10^d) = 24{,}130$, and a, b, c and d are all positive integers. Find the value of $\frac{a+b+c+d}{16}$. **$\frac{5}{8}$**

Applications and Problem Solving

45. Transportation According to *American Demographics* magazine, the average American worker spends 22 minutes traveling to work. What fraction of the day is this? **$\frac{22}{60 \cdot 24}$ or $\frac{11}{720}$**

46. Demographics The United States is aging. Not only is the country getting older, but the number of older Americans increases each year. There are 2,500,000 Americans over the age of 85 and 30,000 of them are 100 or older.

a. What is the ratio of the number of Americans age 100 or over to the number of Americans over age 85? **30,000 : 2,500,000**

b. Write the ratio in simplest terms. **$\frac{3}{250}$**

c. Explain what the ratio in part b means.

47. Medicine The University of California recommends using a soft gel pack to relieve muscle pain. To make a soft gel pack, fill a plastic bag with 2 ounces of rubbing alcohol and 6 ounces of water. Seal the bag and put it in a freezer. Apply to the injured area.

a. What is the ratio of rubbing alcohol to water for the gel pack? **2 : 6**

b. Find a fraction in simplest form to represent the ratio. **$\frac{1}{3}$**

c. Find other amounts of rubbing alcohol and water that could be used to make a gel pack. **Sample answer: 1 oz. alcohol, 3 oz. water**

46c. For every 250 Americans over age 85, 3 are over the age of 100.

Mixed Review

48. Find the GCF of $42x^2y$ and $38xy^2$. (Lesson 4-5) **$2xy$**

49. Solve $v = \ell \cdot w \cdot h$, by replacing the variables with the values $\ell = 6$, $w = 3$, and $h = 6$. (Lesson 3-4) **108**

50. Solve $180 = 15r$. (Lesson 3-3) **12**

51. Find the product of -4 and $-17d$. (Lesson 2-7) **$68d$**

52. History Ohio was declared a state in 1803. Hawaii became a state in 1959. Write a mathematical statement comparing the ages of these states in 1996. (Lesson 2-3) **193 > 37**

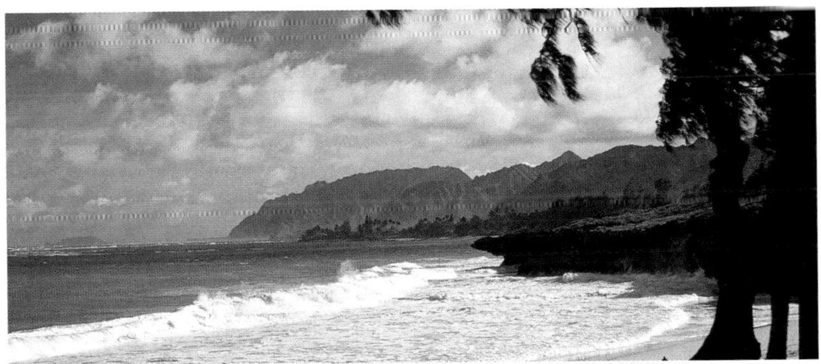

53. Solve $\frac{144}{x} = 12$ mentally. (Lesson 1-6) **12**

54. Evaluate $324 \div (2 \times 3 + 6)$. (Lesson 1-2) **27**

Extension

Using Surveys Have students take a survey that requires yes or no answers to three questions. Then have students report their findings, using as many fractions as possible. For example, $\frac{2}{5}$ of the people I surveyed were girls. $\frac{7}{10}$ of all people said they like to swim.

4 ASSESS

Closing Activity

Speaking Have students explain how to simplify each fraction.

1. $\frac{15}{20}$ **divide numerator and denominator by 5**
2. $\frac{3}{6}$ **divide numerator and denominator by 3**
3. $\frac{33}{22}$ **divide numerator and denominator by 11**

Enrichment Masters, p. 32

NAME ______________________ DATE ____________

4-6 Enrichment
Matching Equivalent Fractions

Student Edition Pages 196–199

Cut out the pieces below and match the edges so that equivalent fractions meet. The pieces form a rectangle. The outer edges of the rectangle formed will have no fractions on them.

Piece	Top	Left	Right	Bottom
15	9/10	15/35	7/9	3/24
18	1/7	18/42	6/8	2/10
18	11/12	10/28	3/8	45/50
10	12/13	8/20	4/7	40/45
14	8/9	5/15	3/7	3/21
8	9/11	15/21		40/45
5	3/10		1/2	52/56
23	1/9	35/77	7/13	
19	1/8	18/48	4/9	2/18
4		8/10		27/33
20	1/12	24/54		2/20
24	1/10	49/91		
3		6/8	4/5	21/33
13	7/8		1/3	3/18
7	7/11	9/15	5/7	44/48
2		4/6	3/4	9/33
22	1/5	14/56	5/11	
21	1/4		2/8	
17	1/6		2/7	2/8
9	13/14		2/5	35/40
1			2/3	6/20
16	6/7	25/45		3/36
6	3/11	3/6	3/5	48/52
12	10/11	20/32		30/35

1	2	3	4
5	6	7	8
9	10	11	12
13	14	15	16
17	18	19	20
21	22	23	24

4-7 LESSON NOTES

NCTM Standards: 1-6, 9

Instructional Resources

- Study Guide Master 4-7
- Practice Master 4-7
- Enrichment Master 4-7
- Group Activity Card 4-7
- Assessment and Evaluation Masters, p. 100
- Tech Prep Applications Masters, p. 7

Transparency 4-7A contains the 5-Minute Check for this lesson; **Transparency 4-7B** contains a teaching aid for this lesson.

Recommended Pacing	
Standard Pacing	Day 9 of 13
Honors Pacing	Day 8 of 12
Block Scheduling*	Day 4 of 6

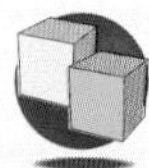

*For more information on pacing and possible lesson plans, refer to the ***Block Scheduling Booklet***.

1 FOCUS

5-Minute Check
(over Lesson 4-6)

Write each fraction in simplest form.

1. $\frac{8}{24}$ $\frac{1}{3}$
2. $\frac{15}{25}$ $\frac{3}{5}$
3. $\frac{100}{40}$ $\frac{5}{2}$
4. $\frac{15x^2}{18xy}$ $\frac{5x}{6y}$
5. $\frac{17abc^4}{a^3b}$ $\frac{17c^4}{a^2}$

4-7 Using the Least Common Multiple (LCM)

Setting Goals: *In this lesson, you'll find the least common multiple of two or more numbers and compare fractions.*

Modeling a Real-World Application: Astronomy

Some clear night, go to a park or out in the country and spend some time looking at the night sky. In ideal conditions, you will be able to see about 2500 stars, although there are many more than this. Some nights you will even be able to see some planets.

The chart at the right shows the number of Earth years it takes for some of the planets to revolve around the Sun. In 1982, the planets Jupiter and Saturn were in conjunction; that is, they appeared very close to each other in the sky. How often do these planets line up in these positions? *This problem will be solved in Example 1.*

Planet	Revolution Time (in Earth Years)
Earth	1
Jupiter	12
Saturn	30
Uranus	84

Learning the Concept

A **multiple** of a number is a product of that number and any whole number. *Recall that the whole numbers are 0, 1, 2, 3, . . .*

multiples of 2: **0**, 2, 4, 6, 8, **10**, 12, 14, 16, 18, **20**, . . .
multiples of 5: **0**, 5, **10**, 15, **20**, 25, 30, 35, 40, . . .

Multiples that are shared by two or more numbers are called **common multiples**. Some of the common multiples of 2 and 5, shown in blue, are 0, 10, and 20.

The least of the nonzero common multiples of two or more numbers is called the **least common multiple (LCM)** of the numbers. The least common multiple of 2 and 5 is 10.

Example 1

APPLICATION

Astronomy

Refer to the application at the beginning of the lesson.

a. How often will Jupiter and Saturn appear close in the sky?
b. When will this happen next?
c. How often do Saturn, Jupiter, and Uranus coincide?

a. According to the table, Jupiter revolves around the Sun every 12 Earth years and Saturn every 30 years. The time required for the planets to coincide in this way again will be a multiple of 12 and 30.

multiples of 12: 0, 12, 24, 36, 48, **60**, . . .
multiples of 30: 0, 30, **60**, 90, 120, . . .

The least common multiple of 12 and 30 is 60. Jupiter and Saturn coincide every 60 years.

Tech Prep

Mason A mason is constructing a brick wall. Two sizes of bricks are being used. The regular bricks are 8 inches long. Every fifth row of bricks is made of decorative bricks that are 10 inches long. Give three possible lengths for the wall if all the edges are to be aligned. **Sample answers: 40 inches, 80 inches, 120 inches**

For more information on tech prep, see the *Teacher's Handbook*.

b. The last time Jupiter and Saturn coincided was 1982. The next time will be 1982 + 60 or 2042.

c. Find the LCM of 12, 30, and 84 to find the number of years for Saturn, Jupiter, and Uranus to coincide in the same way.

A second method for finding an LCM is to find the prime factorization of the numbers. A common multiple of the numbers contains *all* of the prime factors of each number. The LCM contains *each* factor the greatest number of times it appears for any of the numbers.

$$12 = 2 \cdot 2 \cdot 3$$
$$30 = 2 \cdot 3 \cdot 5$$
$$84 = 2 \cdot 2 \cdot 3 \cdot 7$$

Find the common factors. Then multiply all of the factors, using the common factors only once.

$2 \cdot 2 \cdot 3 \cdot 5 \cdot 7 = 420$

Jupiter, Saturn, and Uranus will coincide in this way every 420 years.

Connection to Algebra

The LCM of two or more monomials is found in the same way as the LCM of two or more numbers.

Example 2 **Find the LCM of $12a^2b$ and $8ac$.**

THINK ABOUT IT

Why are all of the prime factors of 12 also factors of 24?

24 is a multiple of 12.

$$12a^2b = 2 \cdot 2 \cdot 3 \cdot a \cdot a \cdot b$$
$$8ac = 2 \cdot 2 \cdot 2 \cdot a \cdot c$$

The LCM is $2 \cdot 2 \cdot 2 \cdot 3 \cdot a \cdot a \cdot b \cdot c$ or $24a^2bc$.

One way to compare fractions is to write them with the *same* denominator. While any common denominator could be used, the least common multiple of the denominators, or the **least common denominator** (LCD) is usually the most convenient.

Motivating the Lesson

Situational Problem Tell students that you asked two people to give you numbers related to the number 18. One person named 1, 2, 3, 6, 9, and 18. The other person named 18, 36, 54, 72, and 180. Ask students to classify the numbers named in each answer. **The first person listed factors of 18. The second person listed multiples of 18.**

2 TEACH

Teaching Tip Many students find that listing the multiples is the most efficient way for them to find the least common multiple of numbers. This method should be encouraged.

In-Class Examples

For Example 1

Janet works every other day, Lynnette every third day, and Mark every fifth day. All three worked together on Tuesday, August 1. They all work at the same company.

a. How often will Janet and Mark work together? **every 10 days**

b. On what date will Janet and Mark work together? **Friday, August 11**

c. On what date will all three work together again? **Thursday, August 31**

For Example 2

Find the LCM of $10x^2y^2$ and $6x^3y$. $\mathbf{30x^3y^2}$

In-Class Examples

For Example 3

Which is less, $\frac{7}{8}$ or $\frac{9}{10}$? $\frac{7}{8}$

For Example 4

Find the LCD for $\frac{6}{5m^2n}$ and $\frac{3}{8n^4}$. $\mathbf{40m^2n^4}$

3 PRACTICE/APPLY

Checking Your Understanding

Exercises 1–17 are designed to help you assess your students' understanding through reading, writing, speaking, and modeling. You should work through Exercises 1–4 with your students and then monitor their work on Exercises 5–17.

Additional Answers

1. If the definition did not say nonzero, zero would always be the LCM of any set of numbers.
3. The LCM involves the common multiples of a set of numbers, the LCD is the LCM of the denominators of two fractions.

Study Guide Masters, p. 33

NAME ______________________ DATE __________

4-7 Study Guide
Using the Least Common Multiple (LCM)

Student Edition Pages 200–204

Follow these steps to find the **least common multiple (LCM)** of two or more numbers or algebraic expressions.

Step 1	Step 2	Step 3
Factor each number or monomial completely.	Write the prime factorization as powers.	Multiply the greatest power of each number or variable to find the LCM.

Find the LCM of $30a$ and $45b^2$.

$30a = 2 \cdot 3 \cdot 5 \cdot a$	$30a = 2^1 \cdot 3^1 \cdot 5^1 \cdot a^1$	LCM $= 2^1 \cdot 3^2 \cdot 5^1 \cdot a^1 \cdot b^2$
$45b^2 = 3 \cdot 3 \cdot 5 \cdot b \cdot b$	$45b^2 = 3^2 \cdot 5^1 \cdot b^2$	$= 2 \cdot 9 \cdot 5 \cdot a \cdot b^2$
		$= 90ab^2$

You can use the LCM to compare two fractions. To compare two fractions, write them as equal fractions with the same denominator by multiplying or dividing the numerator and denominator by the same *nonzero* number.

Which is greater, $\frac{3}{5}$ or $\frac{2}{3}$?

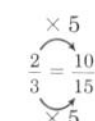

$\frac{3}{5} = \frac{9}{15}$ (×3) $\frac{2}{3} = \frac{10}{15}$ (×5) Since $\frac{9}{15} < \frac{10}{15}$, $\frac{3}{5} < \frac{2}{3}$.

Find the LCM of each set of numbers or algebraic expressions.

1. 4, 6 **12**	2. 6, 9 **18**	3. 21, 9 **63**	4. 6, 10 **30**
5. 20, 25 **100**	6. 15, 20 **60**	7. 12, 36 **36**	8. 20, 30 **60**
9. $4x$, $3x$ **$12x$**	10. $3c^2$, $7c$ **$21c^2$**	11. $16w^2$, 72 **$144w^2$**	12. $4f$, $10f^2$, $12f^2$ **$60f^2$**

Write < or > in each box to make a true statement.

13. $\frac{2}{3} > \frac{6}{24}$	14. $\frac{2}{5} < \frac{3}{4}$	15. $\frac{5}{8} < \frac{5}{6}$
16. $\frac{7}{10} < \frac{3}{4}$	17. $\frac{4}{6} > \frac{3}{12}$	18. $\frac{6}{9} > \frac{2}{7}$

Example 3 Which is greater, $\frac{5}{6}$ or $\frac{7}{9}$?

THINK ABOUT IT

Multiplying the numerator and denominator by the same number is like multiplying the fraction by what number? 1

First, find the LCM of 6 and 9.

$6 = 2 \cdot 3$
$9 = 3 \cdot 3$

The LCM is $2 \cdot 3 \cdot 3$ or 18. Find equivalent fractions with 18 as the denominator.

$\frac{5}{6} = \frac{\bullet}{18}$ (×3) → $\frac{5}{6} = \frac{15}{18}$ $\frac{7}{9} = \frac{\bullet}{18}$ (×2) → $\frac{7}{9} = \frac{14}{18}$

Since $\frac{15}{18} > \frac{14}{18}$, $\frac{5}{6} > \frac{7}{9}$.

Connection to Algebra

You can also find the LCD for algebraic fractions with different denominators.

Example 4 **Find the LCD for $\frac{5}{6k^2}$ and $\frac{3}{8km}$.**

$6k^2 = 2 \cdot 3 \cdot k \cdot k$
$8km = 2 \cdot 2 \cdot 2 \cdot k \cdot m$

The LCM of $6k^2$ and $8km$ is $2 \cdot 2 \cdot 2 \cdot 3 \cdot k \cdot k \cdot m$ or $24k^2m$.

Thus, the LCD of $\frac{5}{6k^2}$ and $\frac{3}{8km}$ is $24k^2m$.

Checking Your Understanding

Communicating Mathematics

Read and study the lesson to answer these questions. 1, 3. See margin.

1. **Explain** why the definition of the least common multiple says that it is the least *nonzero* multiple common to the numbers involved.
2. **Demonstrate** how to find the LCM of two numbers by using the prime factorizations. See students' work.
3. **Compare and contrast** the LCM and the LCD.

4. The Venn diagram at the right shows the prime factorizations of 12 and 16. Using each number shown as a factor will give you the LCM of the numbers. The LCM of 12 and 16 is $2 \cdot 2 \cdot 2 \cdot 2 \cdot 3$ or 48. Use a Venn diagram to find the LCM of 12, 28, and 48. See Solutions Manual for diagram. LCM = 336

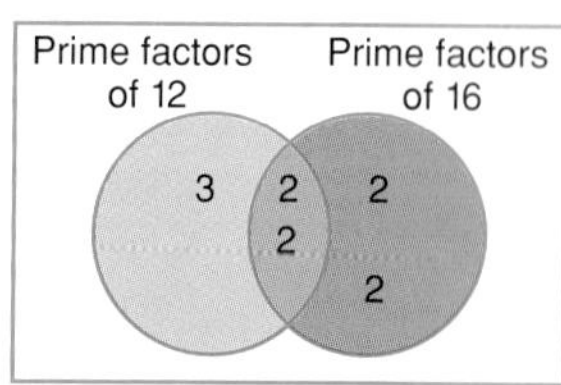

Guided Practice

Find the least common multiple (LCM) of each set of numbers or algebraic expressions.

5. 10, 14 **70**	6. 7, 9 **63**	7. $20a$, $12a$ **$60a$**
8. 12, 16, 24 **48**	9. 45, 30, 35 **630**	10. a, b **ab**

Reteaching

Using Manipulatives Provide students with large circular sheets of paper marked in 24ths. Have students use these to compare fractions with denominators that are factors of 24.

Find the least common denominator (LCD) for each pair of fractions.

11. $\frac{4}{5}, \frac{1}{2}$ **10**
12. $\frac{3}{4}, \frac{1}{8}$ **8**
13. $\frac{3a}{25}, \frac{17a}{20}$ **100**

Replace each ● with <, >, or = to make a true statement.

14. $\frac{5}{12}$ ● $\frac{1}{2}$ **<**
15. $\frac{5}{7}$ ● $\frac{7}{9}$ **<**
16. $\frac{14}{15}$ ● $\frac{9}{10}$ **>**

17. **Food** Americans ate about 606 million pounds of potato chips in the summer of 1994. That is about the weight of six ships! The graph at the right shows the regions where most of the chips were eaten. **a. East Central**

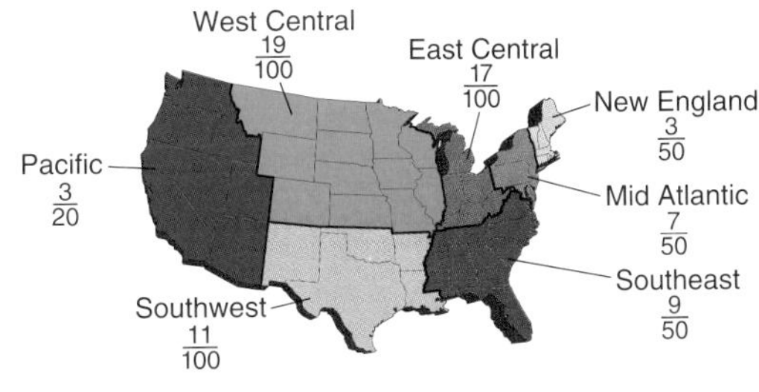

a. Were more chips eaten in the Pacific or in the East Central region?
b. In which region were the most chips eaten? **West Central**
c. In which region were the fewest chips eaten? **New England**

Exercises: Practicing and Applying the Concept

Independent Practice

Find the least common multiple (LCM) of each set of numbers or algebraic expressions.

18. 20, 12 **60**
19. 2, 9 **18**
20. 15, 75 **75**
21. 18, 32 **288**
22. 24, 32 **96**
23. 21, 28 **84**

B

24. $6c, 8cd$ **$24cd$**
25. $7y, 12y$ **$84y$**
26. $20cd, 50d$ **$100cd$**
27. 10, 20, 40 **40**
28. 7, 21, 84 **84**
29. 9, 12, 15 **180**
30. $3b, 5c, 2$ **$30bc$**
31. $3t, 5t^2, 7$ **$105t^2$**
32. $3a^2, 6a^3, 9a^4$ **$18a^4$**

Find the least common denominator (LCD) for each set of fractions.

33. $\frac{1}{4}, \frac{3}{16}$ **16**
34. $\frac{5}{6}, \frac{3}{8}$ **24**
35. $\frac{3}{5}, \frac{4}{7}$ **35**
36. $\frac{2}{9}, \frac{5}{12}$ **36**
37. $\frac{5}{11}, \frac{11}{20}$ **220**
38. $\frac{1}{2}, \frac{1}{3}, \frac{2}{5}$ **30**
39. $\frac{6c}{5b}, \frac{7c}{25b}$ **$25b$**
40. $\frac{7}{8a}, \frac{2}{15a}$

40. $120a$

Replace each ● with <, >, or = to make a true statement.

41. $\frac{5}{6}$ ● $\frac{9}{11}$ **>**
42. $\frac{3}{5}$ ● $\frac{4}{7}$ **>**
43. $\frac{35}{51}$ ● $\frac{12}{17}$ **<**
44. $\frac{10}{12}$ ● $\frac{9}{11}$ **>**
45. $\frac{15}{32}$ ● $\frac{1}{2}$ **<**
46. $\frac{21}{100}$ ● $\frac{1}{5}$ **>**
47. $\frac{8}{9}$ ● $\frac{19}{21}$ **<**
48. $\frac{7}{9}$ ● $\frac{8}{10}$ **<**
49. Use $\frac{3}{8}, \frac{5}{12}$, and < to write a true statement. **$\frac{3}{8} < \frac{5}{12}$**

Critical Thinking

50. Two numbers have a LCM of $2^2 \cdot 3 \cdot 5^2$. Their GCF is $2 \cdot 5$. If one of the numbers is $2 \cdot 3 \cdot 5$, what is the other number? **$2^2 \cdot 5^2$ or 100**

Applications and Problem Solving

51. **Veterinary Medicine** According to the Humane Society, caring for a dog costs an average of \$7815 over 12 years or $\frac{\$7815}{12}$ a year. Caring for a cat costs \$4723 over 15 years or $\frac{\$4723}{15}$ a year. Which is more costly, caring for a cat or a dog? Explain.

51. dog, $\frac{7815}{12} > \frac{4723}{15}$

Error Analysis

Some students may multiply common factors more than once when using prime factorizations to determine the LCM. Have students circle factors as they multiply them.

Assignment Guide

Core: 19–49 odd, 50, 51, 53–62
Enriched: 18–48 even, 50–62

For **Extra Practice**, see p. 749.

The red A, B, and C flags, printed only in the Teacher's Wraparound Edition, indicate the level of difficulty of the exercises.

Practice Masters, p. 33

NAME ______________________ DATE ____________

4-7 Practice
Using the Least Common Multiple (LCM)

Student Edition
Pages 200–204

Find the least common multiple (LCM) of each set of numbers or algebraic expressions.

1. 4, 5 **20**
2. 10, 15 **30**
3. 5, 8 **40**
4. 8, 20 **40**
5. $5x, 12x$ **$60x$**
6. $15x, 45y$ **$45xy$**
7. $15k, 35k^2$ **$105k^2$**
8. $12h^2, 28$ **$84h^2$**
9. $6p, 8p, 12p$ **$24p$**
10. $3x, 15x^2, 30$ **$30x^2$**
11. $8k, 20k, 24k^2$ **$120k^2$**
12. $3c, 5c^2, 7c$ **$105c^2$**

Find the least common denominator (LCD) for each set of fractions.

13. $\frac{1}{2}, \frac{1}{3}$ **6**
14. $\frac{1}{4}, \frac{3}{8}$ **8**
15. $\frac{3}{4}, \frac{7}{9}$ **36**
16. $\frac{5}{8}, \frac{1}{6}$ **24**
17. $\frac{1}{4}, \frac{5}{7}$ **28**
18. $\frac{1}{8}, \frac{1}{7}$ **56**
19. $\frac{3}{5}, \frac{3}{10}$ **10**
20. $\frac{5}{12}, \frac{1}{5}$ **60**
21. $\frac{5}{4a}, \frac{7}{5a^2}$ **$20a^2$**
22. $\frac{1}{7x}, \frac{1}{9x}$ **$63x$**
23. $\frac{3}{8m}, \frac{7}{9k}$ **$72mk$**
24. $\frac{3}{10x}, \frac{7}{20x^2}$ **$20x^2$**

Write < or > in each box to make a true statement.

25. $\frac{2}{3}$ [>] $\frac{5}{9}$
26. $\frac{1}{3}$ [<] $\frac{5}{6}$
27. $\frac{2}{3}$ [>] $\frac{3}{5}$
28. $\frac{2}{7}$ [<] $\frac{1}{3}$
29. $\frac{3}{5}$ [<] $\frac{2}{3}$
30. $\frac{1}{5}$ [<] $\frac{3}{7}$
31. $\frac{7}{9}$ [<] $\frac{5}{6}$
32. $\frac{3}{4}$ [<] $\frac{9}{10}$
33. $\frac{5}{9}$ [<] $\frac{7}{12}$

Group Activity Card 4-7

Making Multiples — Group Activity **4-7**

MATERIALS: Deck of regular playing cards • pencil • paper

The object of this game is to be the first player to determine the least common multiple of a set of numbers.

Each player draws three cards from a deck of regular playing cards. The player announces the numbers to the other players. (Face cards are counted as 10.) Everyone records the numbers and figures out the least common multiple of these numbers.

The first player to arrive at the correct answer scores one point. Players who give incorrect least common multiples must subtract one point from their score.

The game finishes when one player scores five points.

 Pre-Algebra

4 ASSESS

Closing Activity

Act It Out Show students a drawing of a grocery display like the one below.

←7 in.→	←7 in.→			
←10 in.→	←10 in.→			

Have students use strips of paper that are 7 inches and 10 inches long to show how long the display will be when the ends of the two sizes of boxes coincide. **70 in.**

Chapter 4, Quiz C (Lessons 4-6 and 4-7) is available in the *Assessment and Evaluation Masters*, p. 100.

Additional Answer

61.

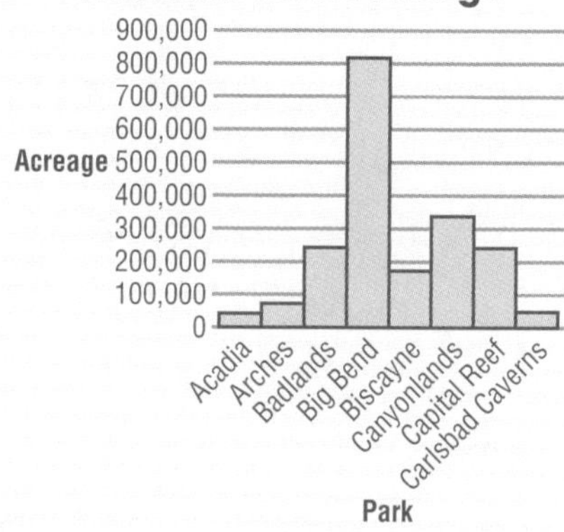

***Enrichment Masters*, p. 33**

NAME ______________________ DATE __________

4-7 Enrichment
GCF, LCM, and Ladders

Student Edition Pages 200–204

When two or more numbers have common factors, the greatest of these is termed the **greatest common factor (GCF)**. When two or more numbers have nonzero common multiples, the least of these is called the **least common multiple (LCM)** of the numbers. A method of determining the GCF and LCM for two or more numbers is the *ladder.*

Example: Find the GCF and LCM of 24 and 36.

```
6 | 24  36   Find any number that will divide into both numbers.
2 |  4   6   Both 24 and 36 can be divided evenly by 6.
1 |  2   3   Continue to choose divisors until the only
     2   3   remaining divisor is 1.
```

To find the GCF, multiply all the divisors.

$6 \cdot 2 \cdot 1 = 12$ The GCF is 12.

To find the LCM, multiply the GCF and the numbers remaining on the bottom of the ladder.

$12 \cdot 2 \cdot 3 = 72$ The LCM is 72.

Find the GCF and LCM for each pair of numbers by using a ladder. The numbers used as divisors in the ladder may vary.

1. 18, 27 GCF: 9 LCM: 54
2. 24, 42 GCF: 6 LCM: 168
3. 44, 66 GCF: 22 LCM: 132
4. 36, 81 GCF: 9 LCM: 324
5. 42, 48 GCF: 6 LCM: 336
6. 100, 125 GCF: 25 LCM: 500
7. 64, 72 GCF: 8 LCM: 576
8. 90, 150 GCF: 30 LCM: 450
9. 250, 300 GCF: 50 LCM: 1500

53a. video game player

52. **Politics** In the United States, presidents are elected every four years. Senators are elected every six years. In 1992, a presidential election year, Carol Moseley Braun of Illinois became the first African-American woman elected to the U.S. Senate. If she continues to run and win each time her term expires, when is the next presidential election year in which Senator Braun will run? **2004**

53. **Technology** Chilton Research studied the amount of technology that American teens use. The table at the right shows the fractions of the teens surveyed who had different technology at home in 1995.
    a. Did more teens say that they had cable TV or a video game player?
    b. Which technology did most teens use? **VCR**
    c. Which technology was used the least? **on-line services**

Technology	Users
Cable TV	$\frac{2}{3}$
Cellular phone	$\frac{7}{20}$
Computer	$\frac{23}{50}$
On-line services	$\frac{17}{100}$
Video game player	$\frac{77}{100}$
VCR	$\frac{24}{25}$

Source: Chilton Research

Mixed Review

54. Write $\frac{30}{36}$ in simplest form. (Lesson 4-6) $\frac{5}{6}$
55. If the quotient of x and 8 is greater than 72, find x. (Lesson 3-8) $x > 576$
56. Solve $-13 < x - (-5)$. (Lesson 3-6) $x > -18$
57. Solve $14 = \frac{a}{-7}$. (Lesson 3-3) -98
58. Evaluate $a \cdot b$ if $a = -3$ and $b = 14$. (Lesson 2-7) -42

The table at the right shows the sizes of several national parks in acres. (Lesson 1-10) **61. See margin.**

59. Which state has the most parks listed? **Utah**
60. What is the difference between the size of the largest and smallest parks? **759,212 acres**
61. Make a bar graph of the data.
62. Solve $5g = 65$ mentally. (Lesson 1-6) **13**

Park	Size (acres)
Acadia, ME	41,951
Arches, UT	73,374
Badlands, SD	242,756
Big Bend, TX	801,163
Biscayne, FL	172,925
Canyonlands, UT	337,570
Capital Reef, UT	241,904
Carlsbad Caverns, NM	46,766

Source: *World Almanac*, 1995

Extension

Using Models Use a set of dominoes to generate problems. Have a student draw two dominoes and write proper fractions to represent the dominoes. Then have students write an inequality to compare the fractions.

4-8 Multiplying and Dividing Monomials

Setting Goals: *In this lesson, you'll multiply and divide monomials.*

Modeling with Technology

In dealing with large numbers and small numbers, calculators can simplify and speed our computation.

Your Turn Copy the tables below. Then use a calculator to find each product and quotient and complete the table.

Factors	Product	Product written as a power
$10^1 \cdot 10^1$	100	10^2
$10^1 \cdot 10^2$	1000	10^3
$10^1 \cdot 10^3$	10,000	10^4
$10^1 \cdot 10^4$	100,000	10^5
$10^1 \cdot 10^5$	1,000,000	10^6

Division	Quotient	Quotient written as a power
$10^5 \div 10^1$	10,000	10^4
$10^4 \div 10^1$	1000	10^3
$10^3 \div 10^1$	100	10^2
$10^2 \div 10^1$	10	10^1
$10^1 \div 10^1$	1	10^0

TALK ABOUT IT

a. The exponents are added.
b. The exponents are subtracted.

a. Compare the exponents of the factors to the exponent in the products. What do you observe?

b. Compare the exponents of the division expressions to the exponents in the quotients. What do you observe?

c. Write a rule for determining the exponent of the product when you multiply powers. Test your rule by multiplying $2^2 \cdot 2^4$ using a calculator. See students' work.

d. Write a rule for determining the exponent in the quotient when you divide powers. Test your rule by dividing 8^6 by 8^4 on a calculator. See students' work.

Learning the Concept

The pattern you observed in the products table above leads to the following rule for multiplying powers that have the same base.

Product of Powers

In words: You can multiply powers *that have the same base* by adding their exponents.

In symbols: For any number a and positive integers m and n, $a^m \cdot a^n = a^{m+n}$.

Alternative Teaching Strategies

Reading Mathematics When studying monomials, encourage students to use the correct vocabulary, such as *m squared* or *m to the second power*, rather than *m two*.

4-8 LESSON NOTES

NCTM Standards: 1-9

Instructional Resources

- Study Guide Master 4-8
- Practice Master 4-8
- Enrichment Master 4-8
- Group Activity Card 4-8
- Activity Masters, p. 18
- Real-World Applications, 9
- Tech Prep Applications Masters, p. 8

Transparency 4-8A contains the 5-Minute Check for this lesson; **Transparency 4-8B** contains a teaching aid for this lesson.

Recommended Pacing	
Standard Pacing	Day 10 of 13
Honors Pacing	Day 9 of 12
Block Scheduling*	Day 5 of 6 (along with Lesson 4-9)

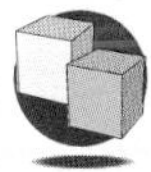

*For more information on pacing and possible lesson plans, refer to the ***Block Scheduling Booklet***.

1 FOCUS

5-Minute Check
(over Lesson 4-7)

Find the LCM of each set of numbers or algebraic expressions.

1. 8, 20 **40** **2.** 11, 12 **132**
3. $9d$, $12df$ **$36df$**
4. 6, 15, 17 **510**
5. $6a^3c$, $8abc^2$ **$24a^3bc^2$**

Motivating the Lesson

Situational Problem Under ideal conditions, one pair of fleas can produce 10^5 fleas in just 30 days. Ask students to use a power to express the number of fleas there would be in 30 days if you started with 10^3 parent fleas. 10^8 fleas

2 TEACH

In-Class Examples

For Example 1
Find $6^4 \cdot 6^3$. **6^7**

For Example 2
Find each product.

a. $r^3 \cdot r^6$ **r^9**

b. $(-3m^7)(-8m^3)$ **$24m^{10}$**

For Example 3
Find each quotient.

a. $\frac{4^5}{4^3}$ **4^2**

b. $\frac{n^9}{n^8}$ **n^1 or n**

Teaching Tip Students may multiply or divide the bases when simplifying. Emphasize writing out the factors to show that this is not the case.

Example 1 **Find $8^4 \cdot 8^5$.**

$8^4 \cdot 8^5 = 8^{4+5}$ or 8^9

Check: $8^4 \cdot 8^5 = (8 \cdot 8 \cdot 8 \cdot 8)(8 \cdot 8 \cdot 8 \cdot 8 \cdot 8)$
$= 8 \cdot 8 \cdot 8 \cdot 8 \cdot 8 \cdot 8 \cdot 8 \cdot 8 \cdot 8$ or 8^9

Connection to Algebra

Some monomials can also be multiplied using the rule for the product of powers.

Example  **Find each product.**

a. $n^2 \cdot n^4$

$n^2 \cdot n^4 = n^{2+4}$ or n^6

b. $(-6x^2)(5x^3)$

$(-6x^2)(5x^3) = (-6 \cdot 5)(x^2 \cdot x^3)$ *Commutative and associative properties*
$= (-30)(x^{2+3})$ *Product of powers*
$= -30x^5$

You can also write a rule for finding quotients of powers. Study the pattern of division sentences below. Each dividend, divisor, and quotient has been replaced with a power of 2.

$16 \div 2 = 8$	$64 \div 8 = 8$	$256 \div 8 = 32$
↓ ↓ ↓	↓ ↓ ↓	↓ ↓ ↓
$2^4 \div 2^1 = 2^3$	$2^6 \div 2^3 = 2^3$	$2^8 \div 2^3 = 2^5$

$2^1 = 2$
$2^2 = 4$
$2^3 = 8$
$2^4 = 16$
$2^5 = 32$
$2^6 = 64$
$2^7 = 128$
$2^8 = 256$

Look at the exponents only. Do you see a pattern?

Quotient of Powers

In words: You can divide powers *that have the same base* by subtracting their exponents.

In symbols: For any nonzero number a and whole numbers m and n, $\frac{a^m}{a^n} = a^{m-n}$.

Example  **Find each quotient.**

a. $\frac{9^7}{9^5}$

$\frac{9^7}{9^5} = 9^{7-5}$ or 9^2

Check: $\frac{9^7}{9^5} = \frac{\overset{1}{\cancel{9}} \cdot \overset{1}{\cancel{9}} \cdot \overset{1}{\cancel{9}} \cdot \overset{1}{\cancel{9}} \cdot \overset{1}{\cancel{9}} \cdot 9 \cdot 9}{\underset{1}{\cancel{9}} \cdot \underset{1}{\cancel{9}} \cdot \underset{1}{\cancel{9}} \cdot \underset{1}{\cancel{9}} \cdot \underset{1}{\cancel{9}}}$
$= 9 \cdot 9$
$= 9^2$

b. $\frac{c^8}{c^8}$

$\frac{c^8}{c^8} = c^{8-8}$
$= c^0$ or 1

Check: $\frac{c^8}{c^8} = \frac{\overset{1}{\cancel{c}} \cdot \overset{1}{\cancel{c}} \cdot \overset{1}{\cancel{c}} \cdot \overset{1}{\cancel{c}} \cdot \overset{1}{\cancel{c}} \cdot \overset{1}{\cancel{c}} \cdot \overset{1}{\cancel{c}} \cdot \overset{1}{\cancel{c}}}{\underset{1}{\cancel{c}} \cdot \underset{1}{\cancel{c}} \cdot \underset{1}{\cancel{c}} \cdot \underset{1}{\cancel{c}} \cdot \underset{1}{\cancel{c}} \cdot \underset{1}{\cancel{c}} \cdot \underset{1}{\cancel{c}} \cdot \underset{1}{\cancel{c}}}$
$= 1$

GLENCOE Technology

Interactive Mathematics Tools Software

In this interactive computer lesson, students explore monomials by measuring sound. A **Computer Journal** gives students an opportunity to write about what they have learned.

For Windows & Macintosh

Many applications involve calculations with very large or very small numbers. The quotient of powers rule often makes these calculations less complicated.

The intensity of an earthquake is usually measured on the Richter scale. On this scale the number designation given to an earthquake represents the intensity as a power of 10. For example, an earthquake measuring 8 on the Richter scale has an intensity of 10^7. In April, 1992, an earthquake measuring 5 on the Richter scale shook the Netherlands. Twelve days later, northern California experienced an earthquake measuring 7. How many times more intense was the California earthquake?

The Netherlands earthquake had an intensity of 10^4, and the California earthquake had an intensity of 10^6. Divide 10^6 by 10^4 to find out how many times more intense the California earthquake was.

$$\frac{10^6}{10^4} = 10^{6-4} \quad \textit{Quotient of powers}$$
$$= 10^2 \text{ or } 100$$

The California earthquake was 100 times more intense.

In-Class Example

For Example 4

Sound is measured in decibels. Normal conversation has an intensity of 50 decibels, which is expressed as 10^5. A motorcycle has an intensity of 110 decibels or 10^{11} and a jet airplane 150 decibels or 10^{15}. How many times more intense is a jet airplane than a motorcycle? **10^4 or 10,000**

Communicating Mathematics

Read and study the lesson to answer these questions. **1–3. See margin.**

1. **Explain** how to find the product $6^6 \cdot 6^3$.
2. **Justify** the product of powers rule by multiplying 2^2 and 2^5 using the definition of exponents.
3. Make up a division problem whose solution is 4^4.

4. Describe, in your own words, the relationship between the rules for multiplying powers and dividing powers. **See students' work.**

Guided Practice

Choose
Estimation
Mental Math
Calculator
Paper and Pencil

Find each product or quotient. Express your answer in exponential form.

5. $10^4 \cdot 10^3$ **10^7**
6. $\frac{2^5}{2^2}$ **2^3**
7. $3^6 \cdot 3^4$ **3^{10}**
8. $\frac{8^4}{8^3}$ **8^1 or 8**
9. $8^2 \cdot 8^3 \cdot 8$ **8^6**
10. $m \cdot m^6$ **m^7**
11. $\frac{y^{11}}{y^9}$ **y^2**
12. $\frac{ab^4}{b^2}$ **ab^2**
13. **Seismology** On June 6, 1994, an earthquake measuring 6.8 on the Richter scale struck southwest Colombia. Three days later an earthquake measuring 8.2 hit La Paz, Bolivia. *Approximately* how many times more intense was the earthquake in Bolivia? **about 10 times more intense**

3 PRACTICE/APPLY

Checking Your Understanding

Exercises 1–13 are designed to help you assess your students' understanding through reading, writing, speaking, and modeling. You should work through Exercises 1–4 with your students and then monitor their work on Exercises 5–13.

Additional Answers

1. Since the numbers have the same base, add the exponents: $6^{6+3} = 6^9$.
2. $2^2 = 4$ and $2^5 = 32$. $4 \cdot 32 = 128$, which is 2^7.
3. Sample answer: $4^7 \div 4^3$

Reteaching

Using Calculators Have students multiply 2^3 and 2^4. The calculator will display 128. Suggest that students evaluate 2^7 and 4^7 and compare to the product $2^3 \cdot 2^4$. For exercises that are numerical, have students predict what the answer will be. Then let them check their answers with the calculator.

Study Guide Masters, p. 34

NAME ______________________ DATE ____________

4-8 Study Guide
Multiplying and Dividing Monomials

Student Edition
Pages 205–209

To multiply powers with the same base, add the exponents.

Examples: $x^4 \cdot x^8 = x^{4+8} = x^{12}$ $(4x^2)(2x^5) = (4 \cdot 2)(x^2 \cdot x^5) = 8(x^{2+5}) = 8x^7$

To divide powers with the same base, subtract the exponents.

Examples: $\frac{4^8}{4^5} = 4^{8-5} = 4^3$ $\frac{x^5}{x^3} = x^{5-3} = x^2$

Find each product or quotient. Express your answer in exponential form.

1. $8^2 \cdot 8^3$ **8^5**
2. $x^5 \cdot x^1$ **x^6**
3. $10^2 \cdot 10^7$ **10^9**
4. $(3x^2)(2x^4)$ **$6x^6$**
5. $y^3(y^2x)$ **y^5x**
6. $a^3 \cdot a^3$ **a^6**
7. $n^1 \cdot n^3$ **n^4**
8. $20^3 \cdot 20^5$ **20^8**
9. $(4x^3)(-2x^5)$ **$-8x^8$**
10. $\frac{a^5}{a^2}$ **a^3**
11. $\frac{10^4}{10^2}$ **10^2**
12. $\frac{h^8}{h^4}$ **h^4**
13. $\frac{t^9}{t^3}$ **t^6**
14. $\frac{(-d)^3}{(-d)^2}$ **$-d$**
15. $\frac{c^3}{c^1}$ **c^2**
16. $\frac{x^4x^3}{x^5}$ **x^2**
17. $\frac{5^6}{5^3}$ **5^3**
18. $\frac{6^7}{6^2}$ **6^5**

Error Analysis

In multiplication or division exercises, students may actually multiply or divide exponents rather than add or subtract them. Review the rules with students and encourage them to check that their answers are reasonable.

Assignment Guide

Core: 15–43 odd, 44–45, 47–55
Enriched: 14–42 even, 44–55

For **Extra Practice**, see p. 750.

The red A, B, and C flags, printed only in the Teacher's Wraparound Edition, indicate the level of difficulty of the exercises.

Exercises: Practicing and Applying the Concept

Independent Practice

Find each product or quotient. Express your answer in exponential form.

14. $10^5 \cdot 10^5$ **10^{10}** 15. $w \cdot w^5$ **w^6** 16. $3^3 \cdot 3^2$ **3^5** 17. $b^5 \cdot b^2$ **b^7**

18. $\frac{5^5}{5^2}$ **5^3** 19. $\frac{10^{10}}{10^3}$ **10^7** 20. $\frac{z^3}{z}$ **z^2** 21. $\frac{a^{10}}{a^6}$ **a^4**

B

22. $\frac{(-2)^6}{(-2)^5}$ **-2** 23. $(5x^3)(4x^4)$ **$20x^7$** 24. $\frac{(-x)^4}{(-x)^3}$ **$-x$** 25. $\frac{f^{20}}{f^8}$ **f^{12}**

26. $(3x^4)(-5x^2)$ **$-15x^6$** 27. $3a^2 \cdot 4a^3$ **$12a^5$** 28. $(-10x^3)(2x^2)$ **$-20x^5$**

29. $a^2b \cdot ab^3$ **a^3b^4** 30. $x^4(x^3y^2)$ **x^7y^2** 31. $t^3t^2 \div t^4$ **t**

32. $\frac{y^{100}}{y^{100}}$ **1** 33. $k^3m^2 \div km$ **k^2m** 34. $(-3y^3z)(7y^4)$ **$-21y^7z$**

35. $w^5xy^2 \div wx$ **w^4y^2** 36. $7ab(a^{16}c)$ **$7a^{17}bc$** 37. $15n^9q^2 \div 5nq^2$ **$3n^8$**

Find each missing exponent.

38. $(5^{\bullet})(5^3) = 5^{11}$ **8** 39. $x(x^2)(x^6) = x^{\bullet}$ **9** 40. $\frac{t^{\bullet}}{t} = t^{14}$ **15**

41. $3^5 \cdot 3^{\bullet} = 3^8$ **3** 42. $\frac{12^5}{12^{\bullet}} = 1$ **5** 43. $\frac{2^5}{4^2} = 2^{\bullet}$ **1**

Critical Thinking

44. If $10^n \div 10^m = 1$, what can you conclude about n and m? **$n = m$**

Applications and Problem Solving

45. **Chemistry** The pH of a solution is a measure of its acidity. A low pH indicates an acid and a high pH indicates a base. Neutral water has a pH of 7; this is the dividing line between acids and bases. Each one-unit increase or decrease in the pH means that the intensity of the acid or base is changed by a factor of ten. For example, a pH of 4 is ten times more acidic than a pH of 5. **b. base**
    a. Acid rain is an environmental problem related to the burning of fossil fuels. Suppose the pH of a lake is 5 due to acid rain. How much more acidic is the lake than neutral water? **100 times**
    b. Human blood usually has a pH of about 7.4. Is it an acid or a base?

46. **Biology** When bacteria reproduce, they split so that one cell becomes two. The number of cells after t time periods is 2^t.
    a. *E. coli* reproduce very quickly, about every 15 minutes. If there are 100 *e. coli* in a dish now, how many will there be in 30 minutes? **400**
    b. How many more *e. coli* are there in a population after 3 hours than there were after 1 hour? **256 times as many**

Mixed Review

47. Replace ● with $<$ or $>$ to make $\frac{5}{8} \bullet \frac{8}{11}$ a true inequality. (Lesson 4-7) **$<$**

48. Find the prime factorization of $16ab^3$. (Lesson 4-4) **$2 \cdot 2 \cdot 2 \cdot 2 \cdot a \cdot b \cdot b \cdot b$**

49. Use divisibility rules to determine whether 298 is divisible by 2, 3, 5, 6, or 10. (Lesson 4-1) **2**

Practice Masters, p. 34

NAME ______________________ DATE __________

4-8 Practice Student Edition Pages 205–209
Multiplying and Dividing Monomials

Find each product or quotient. Express your answer in exponential form.

1. $2^2 \cdot 2^4 \cdot 2^1$ **2^7**
2. $x^4 \cdot x^2 \cdot x^5$ **x^{11}**
3. $(3x^2)(-2xy)$ **$-6x^3y$**
4. $x \cdot y \cdot z \cdot x \cdot y \cdot x \cdot z$ **$x^3y^2z^2$**
5. $(x^2y)(-4x^6y^3)$ **$-4x^8y^4$**
6. $(-5a^2m^7)(-3a^5m)$ **$15a^7m^8$**
7. $(-x^2z)(-xyz)$ **x^3yz^2**
8. $(-2n^2)(y^4)(-3n)$ **$6n^3y^4$**
9. $x^3(x^4y^2)$ **x^7y^2**
10. $(-5r^2s)(-3rs^4)$ **$15r^3s^5$**
11. $(a^2b^2)(a^3b)$ **a^5b^3**
12. $(2n^3)(-6n^4)$ **$-12n^7$**
13. $(5wz^2)(8w^4z^3)$ **$40w^5z^5$**
14. $(c^2d)(-10c^3d)$ **$-10c^5d^2$**
15. $5^9 \div 5^2$ **5^7**
16. $\frac{x^5}{x^1}$ **x^4**
17. $10^{10} \div 10^3$ **10^7**
18. $\frac{m^7}{m^4}$ **m^3**
19. $w^6 \div w^1$ **w^5**
20. $\frac{y^4}{y^2}$ **y^2**
21. $\frac{a^7}{a^6}$ **a**
22. $\frac{6^8}{6^3}$ **6^5**
23. $8^4 \div 8^3$ **8**
24. $\frac{(-3)^9}{(-3)^8}$ **-3**
25. $\frac{r^6r^4}{r^8}$ **r^2**
26. $\frac{a^{40}}{a^{16}}$ **a^{24}**
27. $\frac{b^7}{b^7}$ **1**
28. $\frac{(-z)^{12}}{(-z)^{10}}$ **$-z^2$**
29. $\frac{f^2f^2}{f^3}$ **f**

Group Activity Card 4-8

More Power To You Group Activity **4-8**

For this activity students work in groups.

To begin, each student writes down two simplified monomials that include constants, variables, and exponents. Then they exchange papers with any other student, and that student must rewrite each expression as the product of two or three other monomials in two different ways, one set correct and the other incorrect.

Lastly, they give the papers to someone else, and that person checks to see which expression is correct.

The first one to determine the correct expressions scores one point. The winner is the first person to score five points.

 Pre-Algebra

Extension

Using Patterns Have students use patterns to find the difference between the product of powers and the power of a power. For example, $3^3 \times 3^2 = 243$, but $(3^3)^2 = 729$.

50. **Art** A photograph measuring 9" by 11" is matted and placed in a 10" by 12" frame. Find the area of the matting. (Lesson 3-5) **21 sq in.**

51. Solve $5 + n = -2$ and check your solution. Graph the solution on a number line. (Lesson 3-2) **−7; See margin for graph.**

52. Solve $s = 40 - (-17)$. (Lesson 2-5) **57**

53. In which quadrant is the graph of $(-3, 4)$ located? (Lesson 2-2) **II**

54. Is $4x < 16$ *true*, *false*, or *open*? (Lesson 1-9) **open**

55. **Music** A survey of 6500 Top-10 songs released between 1900 and 1975 showed that "I" is the most frequent first word of a song. "My" was the second most popular first word. "I" was the first word of four less than twice as many songs as the word "my." Write an expression for the number of songs that begin with "my" if x songs started with "I." (Lesson 1-3) **$2x - 4$**

WORKING ON THE

Investigation

TAKING STOCK

Refer to the Investigation on pages 166–167.

You can use a **spreadsheet** to find the value of your portfolio on any given day. Using a spreadsheet, it is possible to project results, make calculations, and print almost anything that can be arranged in a table.

The basic unit of a spreadsheet is called a **cell**. A cell may contain numbers, words (called labels), or a formula. Each cell is named by the column and row that describes its location. The cell C2 is the box at the intersection of column C and row 2. *Cell C2 in the spreadsheet at the right contains the number 20.*

As you know, stock prices are listed as mixed numbers. You can enter a mixed number in a spreadsheet as an expression. For example, the price of Ameritech's stock was listed as $48\frac{3}{8}$. It can be entered in cell B2 of the spreadsheet as $48 + \left(\frac{3}{8}\right)$ or 48.375.

When cells are related, a formula can be used to generate this relationship. If you type B2*C2 into cell D2, the cell will not show the formula, but the result of the multiplication of the contents of B2 and C2.

	A	B	C	D
1	Company	Price	Shares	Value
2	Ameritech	48.375	20	
3	Exxon	70.375	10	
4	GE	50	100	
5	GM	48	40	
6	IBM	109.625	10	
7	Sears	32.75	10	
8	Total Value			

- Write spreadsheet formulas for cells D3, D4, D5, D6, D7, and D8. **See margin.**
- Create your own spreadsheet to find the value of your portfolio. Update the value using share prices in a local or national newspaper each week. Also document any changes in your portfolio such as selling old stock holdings and buying new ones.

Add the results of your work to your Investigation Folder.

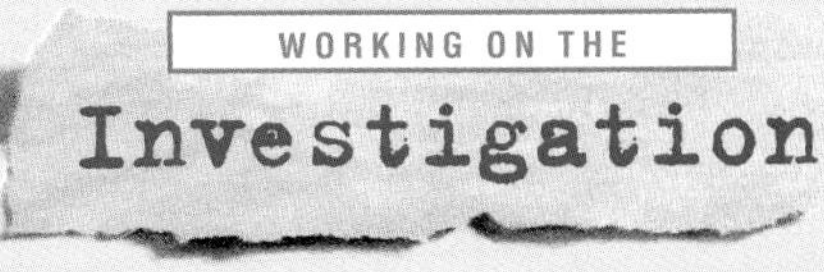

The Investigation on pages 166 and 167 is designed to be a long-term project that is completed over several days or weeks. Encourage students to keep their materials in their Investigation Folder as they work on the Investigation.

Additional Answer Working on the Investigation

D3 = B3 * C3; D4 = B4 * C4; D5 = B5 * C5; D6 = B6 * C6; D7 = B7 * C7; D8 = D2 + D3 + D4 + D5 + D6 + D7

4 ASSESS

Closing Activity

Writing Have students use exponents to write five multiplication equations and five division equations, such as $(-4)^2 \cdot (-4)^5 = (-4)^7$ and $\frac{w^8}{w^3} = w^5$.

Have students use calculators to verify equations involving integers.

Additional Answer

51.

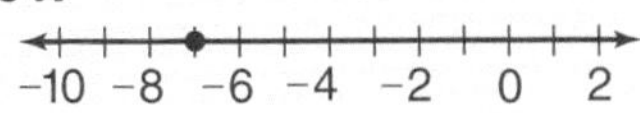

Enrichment Masters, p. 34

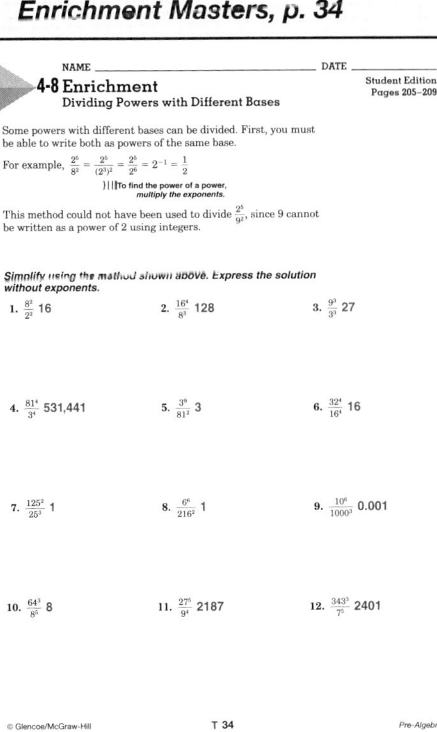

NAME ____________ DATE ________

4-8 Enrichment
Dividing Powers with Different Bases

Student Edition
Pages 205–209

Some powers with different bases can be divided. First, you must be able to write both as powers of the same base.

For example, $\frac{2^5}{8^2} = \frac{2^5}{(2^3)^2} = \frac{2^5}{2^6} = 2^{-1} = \frac{1}{2}$

To find the power of a power, *multiply the exponents.*

This method could not have been used to divide $\frac{2^5}{9^2}$, since 9 cannot be written as a power of 2 using integers.

Simplify using the method shown above. Express the solution without exponents.

1. $\frac{8^2}{2^2}$ 16
2. $\frac{16^4}{8^3}$ 128
3. $\frac{9^3}{3^3}$ 27
4. $\frac{81^4}{3^4}$ 531,441
5. $\frac{3^9}{81^2}$ 3
6. $\frac{32^4}{16^4}$ 16
7. $\frac{125^2}{25^3}$ 1
8. $\frac{6^6}{216^2}$ 1
9. $\frac{10^6}{1000^3}$ 0.001
10. $\frac{64^3}{8^5}$ 8
11. $\frac{27^5}{9^4}$ 2187
12. $\frac{343^3}{7^5}$ 2401

 T 34 *Pre-Algebra*

4-9 LESSON NOTES

NCTM Standards: 1-6, 9

Instructional Resources

- Study Guide Master 4-9
- Practice Master 4-9
- Enrichment Master 4-9
- Group Activity Card 4-9
- Assessment and Evaluation Masters, p. 100

Transparency 4-9A contains the 5-Minute Check for this lesson; **Transparency 4-9B** contains a teaching aid for this lesson.

Recommended Pacing	
Standard Pacing	Day 11 of 13
Honors Pacing	Day 19 of 12
Block Scheduling*	Day 5 of (along with Lesson 4-8)

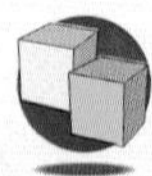

*For more information on pacing and possible lesson plans, refer to the ***Block Scheduling Booklet***.

1 FOCUS

5-Minute Check
(over Lesson 4-8)

Find each product or quotient. Express your answer in exponential form.

1. $10^9 \cdot 10^4$ **10^{13}**
2. $a^6 \cdot a^6$ **a^{12}**
3. $\frac{10^9}{10^4}$ **10^5**
4. $\frac{a^6}{a^6}$ **a^0 or 1**
5. $\frac{-12a^7h^2}{4a^3h}$ **$-3a^4h$**

4-9 Negative Exponents

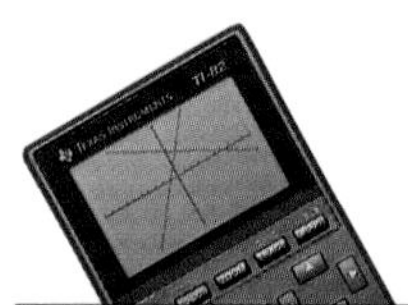

Setting Goals: *In this lesson, you'll use negative exponents in expressions and equations.*

Modeling with Technology

You can often discover important concepts by looking for a pattern.

Your Turn Copy the table at the right. Then use a calculator to complete the second column.

Expression	Value
2^6	64
2^5	32
2^4	16
2^3	8
2^2	4
2^1	2
2^0	1

TALK ABOUT IT

a. Describe how successive expressions differ. **The exponents decrease by 1.**

b. Describe how successive values differ. **b. Each value is half of the previous.**

c. If the pattern continues, what is the next expression? **2^{-1}**

d. If the pattern continues, what is the next value? **$\frac{1}{2}$**

e. Use a calculator to verify that the expression you wrote for part c is equal to the value you wrote for part d. **See students' work.**

Learning the Concept

Extending the pattern above suggests that $2^{-1} = \frac{1}{2}$. You can use the quotient of powers rule and the definition of a power to simplify the expression $\frac{x^3}{x^6}$ and write a general rule about negative powers.

Method 1: Quotient of Powers Rule

$$\frac{x^3}{x^6} = x^{3-6}$$
$$= x^{-3}$$

Method 2: Definition of Powers

$$\frac{x^3}{x^6} = \frac{\overset{1}{\cancel{x}} \cdot \overset{1}{\cancel{x}} \cdot \overset{1}{\cancel{x}}}{\underset{1}{\cancel{x}} \cdot \underset{1}{\cancel{x}} \cdot \underset{1}{\cancel{x}} \cdot x \cdot x \cdot x}$$
$$= \frac{1}{x \cdot x \cdot x}$$
$$= \frac{1}{x^3}$$

Since $\frac{x^3}{x^6}$ cannot have two different values, you can conclude that x^{-3} is equal to $\frac{1}{x^3}$. This and other examples suggest the following definition.

Negative Exponents — For any nonzero number a and any integer n, $a^{-n} = \frac{1}{a^n}$.

Example  **Write each expression using positive exponents.**

a. $\mathbf{10^{-5}}$

$10^{-5} = \frac{1}{10^5}$ *Definition of negative exponents*

b. $\boldsymbol{ab^{-3}}$

$ab^{-3} = a \cdot b^{-3}$

$= a \cdot \frac{1}{b^3}$ *Definition of negative exponents*

$= \frac{a}{b^3}$

Negative exponents are often used in science when dealing with very small numbers.

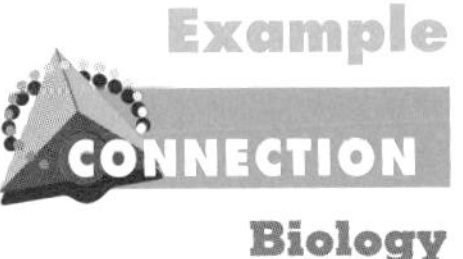

CONNECTION Biology

Example  **Red blood cells are circular-shaped cells that carry oxygen through your bloodstream. A red blood cell is about 7.75×10^{-7} meters across.**

a. Write the width of a red blood cell using positive exponents.

b. Write the width of a red blood cell as a decimal.

a. $7.75 \times 10^{-7} = 7.75 \times \frac{1}{10^7}$ or $\frac{7.75}{10^7}$ meters

b. Use a calculator to write the number as a decimal.

7.75 [÷] 10 [y^x] 7 [=] 0.000000775

You can use the prime factorization of a number to write a fraction as an expression with negative exponents.

Example  **Write each fraction as an expression using negative exponents.**

a. $\frac{1}{8}$

$\frac{1}{8} = \frac{1}{2^3}$

$= 2^{-3}$ *Definition of negative exponents*

b. $\frac{x}{y^2}$

$\frac{x}{y^2} = x \cdot \frac{1}{y^2}$

$= x \cdot y^{-2}$ or xy^{-2}

Connection to Algebra

Expressions involving variables can also contain negative exponents.

Motivating the Lesson

Questioning Ask students how to write the number of dollars in each of the following in exponential form and in standard form.

dollars in one thousand dollars **10^3 or 1000**

dollars in one hundred dollars **10^2 or 100**

dollars in ten dollars **10^1 or 10**

dollars in a dollar **10^0 or 1**

dollars in a dime **$10^?$**

Tell students they will learn to answer the last problem in this lesson.

2 TEACH

In-Class Examples

For Example 1

Write each expression using positive exponents.

a. 10^{-8} $\frac{1}{10^8}$ **b.** $m^{-2}n^3$ $\frac{n^3}{m^2}$

For Example 2

The length of a microchip is 3×10^{-7} mm.

a. Write that length using positive exponents. $\frac{3}{10^7}$

b. Write that length as a decimal. **0.0000003**

For Example 3

Write each fraction as an expression using negative exponents.

a. $\frac{1}{81}$ 3^{-4}

b. $\frac{m^2}{n^5}$ m^2n^{-5}

Teaching Tip In Example 2 in the text, help students understand how the decimal is written by showing the movement of the decimal point to the left as shown below.

0.000000775

Alternative Learning Styles

Visual Provide students with a large square separated into four smaller squares. Tell students to write an exponential number to show how many squares there are. **2^2** Have students shade $\frac{1}{2}$ of the large square. Tell students to write an exponential number to show how many squares are not shaded. **2^1** Have students shade $\frac{1}{2}$ of the unshaded squares and write an exponential number to show how many squares are not shaded. **2^0 or 1** Continue the process for 2^{-1} and 2^{-2}.

Example  **Evaluate the expression $4x^{-3}$ if $x = 3$.**

$4x^{-3} = 4(3)^{-3}$ *Replace x with 3.*

$= 4\left(\frac{1}{3^3}\right)$ *Definition of negative exponents*

$= 4\left(\frac{1}{27}\right)$ *Find 3^3.*

$= \frac{4}{27}$ *Multiply.*

Checking Your Understanding

Communicating Mathematics

Read and study the lesson to answer these questions.

1. **Express** x^{-3} with a positive exponent. $\frac{1}{x^3}$
2. **Write** a convincing argument that $2^0 = 1$ using the fact that $2^4 = 16$, $2^3 = 8$, $2^2 = 4$, and $2^1 = 2$. **See margin.**

3. Write in your own words the relationship between the rules for multiplying powers and dividing powers. **See students' work.**

Guided Practice

Write each expression using positive exponents.

4. 3^{-3} $\frac{1}{3^3}$
5. 15^{-5} $\frac{1}{15^5}$
6. a^{-12} $\frac{1}{a^{12}}$

Write each fraction as an expression using negative exponents.

7. $\frac{1}{10^7}$ 10^{-7}
8. $\frac{1}{4^3}$ 4^{-3}
9. $\frac{1}{9}$ 3^{-2}

Evaluate each expression.

10. 2^x if $x = -2$ $\frac{1}{4}$
11. $3n^{-2}$ if $n = 4$ $\frac{3}{16}$
12. **Biology** Deoxyribonucleic acid, or DNA, contains the genetic code of an organism. The length of a DNA strand is about 10^{-7} meters.
    a. Write the length of a DNA strand using positive exponents. $\frac{1}{10^7}$ **meters**
    b. Write the length of a DNA strand as a decimal. **0.0000001 meters**

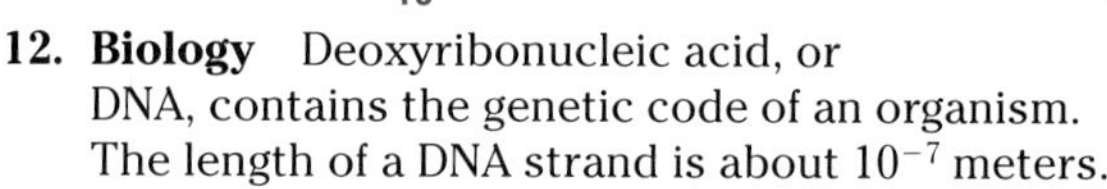

Practicing and Applying the Concept

Independent Practice

Write each expression using positive exponents.

13. 4^{-1} $\frac{1}{4}$
14. 5^{-2} $\frac{1}{5^2}$
15. $(-2)^{-5}$ $\frac{1}{(-2)^5}$
16. x^{-1} $\frac{1}{x}$
17. $s^{-2}t^{-1}$ $\frac{1}{s^2t}$
18. mn^{-2} $\frac{m}{n^2}$
19. $2(xy)^{-2}$
20. $\frac{1}{2^{-2}}$ 2^2

19. $\frac{2}{(xy)^2}$

In-Class Example

For Example 4

Evaluate the expression $-3n^{-4}$ if $n = 2$. $-\frac{3}{16}$

3 PRACTICE/APPLY

Checking Your Understanding

Exercises 1–12 are designed to help you assess your students' understanding through reading, writing, speaking, and modeling. You should work through Exercises 1–3 with your students and then monitor their work on Exercises 4–12.

Additional Answer

2. Sample answer: Look for a pattern.

$2^4 = 16$
$\quad \div 2$
$2^3 = 8$
$\quad \div 2$
$2^2 = 4$
$\quad \div 2$
$2^1 = 2$
$\quad \div 2$
$2^0 = 1$

Each successive power is half of the previous power. Therefore, $2^0 = 2^1 \div 2$ or 1.

Study Guide Masters, p. 35

NAME __________ DATE ________

4-9 Study Guide
Negative Exponents

Student Edition Pages 210–214

Definition of Negative Exponents
For any nonzero number a and any integer n, $a^{-n} = \frac{1}{a^n}$

Using the definition, $10^{-6} = \frac{1}{10}$ and $x^{-2} = \frac{1}{x}$.

The definition also shows that $\frac{1}{10} = 10^{-6}$ and $\frac{1}{x} = x^{-2}$.

Represent each expression using positive exponents.

1. 3^{-1} $\frac{1}{3}$
2. $e^{-4}f$ $\frac{f}{e^4}$
3. w^{-2} $\frac{1}{w^2}$
4. $\frac{1}{5^{-2}}$ 5^2
5. $(-4)^{-3}$ $\frac{1}{(-4)^3}$
6. $7(xy)^{-1}$ $\frac{7}{(xy)}$

Write each fraction as an exponent with negative exponents.

7. $\frac{b}{a^7}$ ba^{-7}
8. $\frac{6}{2^3}$ $6 \cdot 2^{-3}$
9. $\frac{1}{6^3}$ 6^{-3}
10. $\frac{1}{100}$ 10^{-2}
11. $\frac{1}{u}$ u^{-1}
12. $\frac{s}{r^3t^2}$ $sr^{-3}t^{-2}$

Evaluate each expression.

13. 2^x if $x = -4$ $\frac{1}{16}$
14. $(3b)^{-3}$ if $b = -2$ $\frac{-1}{216}$
15. $5w^{-2}$ if $w = 3$ $\frac{5}{9}$

Reteaching

Using Manipulatives Have students roll two dice. Then have two students write an exponential expression with those two numbers and find the value of that expression. Then have students change the exponent to a negative exponent and write the value of the new expression. For instance, a roll of 2 and 5 would result in either $2^5 = 32$ and $2^{-5} = \frac{1}{32}$ or $5^2 = 25$ and $5^{-2} = \frac{1}{25}$.

Tech Prep

Astronomy Technician For Exercise 39, students who are interested in astronomy or physics may want to do research on this career. Astronomy technicians often assist astronomers doing research. For more information on tech prep, see the *Teacher's Handbook*.

Write each fraction as an expression using negative exponents.

21. $\frac{1}{5^3}$ 5^{-3}
22. $\frac{1}{a}$ a^{-1}
23. $\frac{1}{16}$ 2^{-4} or 16^{-1}
24. $\frac{1}{x^6}$ x^{-6}
25. $\frac{2}{3^2}$ $2 \cdot 3^{-2}$
26. $\frac{a}{b^6}$ ab^{-6}
27. $\frac{f}{g^3}$ fg^{-3}
28. $\frac{3m}{n^2}$ $3mn^{-2}$

Evaluate each expression.

29. 3^n if $n = -3$ $\frac{1}{27}$
30. $4x^{-3}y^2$ if $x = 2$ and $y = 6$ 18
31. $6t^{-2}$ if $t = 3$ $\frac{2}{3}$
32. $(2b)^{-3}$ if $b = -2$ $-\frac{1}{64}$

Find each product or quotient. Express using positive exponents.

33. $(a^6)(a^{-3})$ a^3
34. $(b^{-10})(b^5)$ $\frac{1}{b^5}$
35. $\frac{c^5}{c^{-2}}$ c^7
36. $\frac{d^2}{d^3}$ $\frac{1}{d}$

Critical Thinking

37. Simplify $\frac{a^b}{a^{a-b}}$. a^{2b-a}

Applications and Problem Solving

38. **Physical Science** Electromagnetic waves are used to transmit television and radio signals. The diagram at the right shows the electromagnetic wave spectrum.

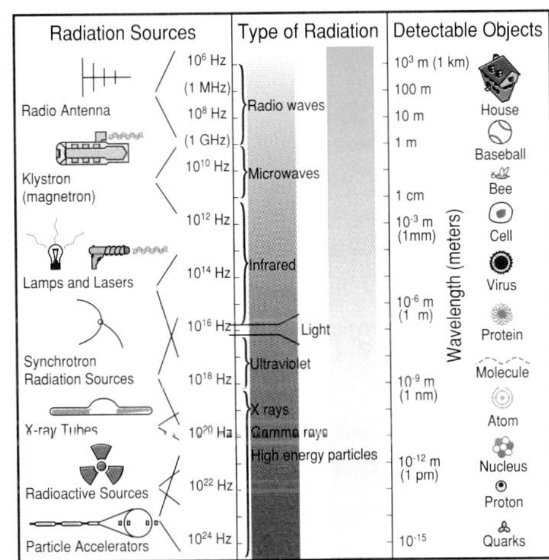

38a. $\frac{1}{10^6}$ m

a. Write the length of a light wave using a positive exponent.

b. What is the range of lengths of infrared waves written in decimal form? 0.000001 to 0.001 m

39. **Astronomy** Any two objects in space have an attraction that can be calculated by using a formula written by English scientist Henry Cavendish. The formula uses the universal gravitational constant, which is 6.67×10^{-11} Nm^2/kg^2. (N is newtons.)

39a. $\frac{6.67}{10^{11}}$ Nm^2/kg^2

a. Write the universal gravitational constant by using positive exponents.

b. Write the constant as a decimal. 0.0000000000667 Nm^2/kg^2

40. **Physics** An object has energy if it is able to produce a change in itself or its surroundings. For example, people use energy to run or to digest food. Scientists use the joule as the unit in which they measure energy. A joule is $1 \text{ kg} \cdot \text{m}^2 \cdot \text{s}^{-2}$. Write this unit as a fraction using positive exponents. $\frac{1 \text{ kg} \cdot \text{m}^2}{\text{s}^2}$

Mixed Review

41. Find the product of $(10x)$ and $(-5x^3)$. (Lesson 4-8) $-50x^4$

42. **Sports** Calida made 8 of 14 free throws in her last basketball game. How would you describe her success as a fraction in simplest form? (Lesson 4-6) $\frac{4}{7}$

Error Analysis

When evaluating x^{-2}, students make the value of the expression negative by writing $\frac{-1}{x^2}$. Remind students that the negative exponent does not change the sign of the expression.

Assignment Guide

Core: 13–39 odd, 41–48
Enriched: 14–36 even, 37–48

For **Extra Practice**, see p. 750.

The red A, B, and C flags, printed only in the Teacher's Wraparound Edition, indicate the level of difficulty of the exercises.

Practice Masters, p. 35

NAME ______ DATE ______

4-9 Practice
Negative Exponents

Student Edition Pages 210–214

Write each expression using positive exponents.

1. 6^{-3} $\frac{1}{6^3}$
2. 8^{-5} $\frac{1}{8^5}$
3. $(-3)^{-2}$ $\frac{1}{(-3)^2}$
4. $c^{-6}d^{-1}$ $\frac{1}{c^6d}$
5. $a^{-4}b$ $\frac{b}{a^4}$
6. $2(mn)^{-4}$ $\frac{2}{(mn)^4}$
7. 3^{-1} $\frac{1}{3}$
8. $\frac{1}{3^{-3}}$ 3^3
9. y^{-1} $\frac{1}{y}$
10. $\frac{s^{-3}}{r^{-2}}$ $\frac{r^2}{s^3}$
11. $4xy^{-3}$ $\frac{4x}{y^3}$
12. $\frac{1}{2^{-4}}$ 2^4

Write each fraction as an expression with negative exponents.

13. $\frac{v}{w^2}$ vw^{-2}
14. $\frac{1}{6^4}$ 6^{-4}
15. $\frac{a}{b^5}$ ab^{-5}
16. $\frac{1}{25}$ 5^{-2}
17. $\frac{3}{2^3}$ $3 \cdot 2^{-3}$
18. $\frac{1}{t}$ t^{-1}
19. $\frac{-5}{4^2}$ $-5 \cdot 4^{-2}$
20. $\frac{1}{x^3y^9}$ $x^{-3}y^{-9}$
21. $\frac{7}{cd}$ $7c^{-1}d^{-1}$
22. $\frac{jk}{t^7}$ jkt^{-7}
23. $\frac{-1}{13^5}$ -13^{-5}
24. $\frac{-4}{(xy)^4}$ $-4(xy)^{-4}$

Evaluate each expression.

25. $4t$ if $t = -2$ $\frac{1}{16}$
26. $3y^{-1}$ if $y = 3$ 1
27. $(5w)^{-3}$ if $w = -1$ $\frac{1}{125}$
28. $6z^x$ if $x = -3$ and $z = 4$ $\frac{3}{32}$
29. $2a^{-3}b^1$ if $a = 2$ and $b = 12$ 3
30. $5g^{-2}h^1$ if $g = 6$ and $h = -3$ $\frac{-5}{12}$

Extension

Using Manipulatives Show students a standard sheet of typing paper. Ask how thick they think the paper is. Allow students to brainstorm ideas for finding the thickness. **Sample answer: Find the thickness of 100 sheets of paper and divide by 100.** Challenge students to write the thickness of a sheet of paper as a decimal and as a number between 1 and 10 rounded to the nearest hundredth times a power of 10. **Sample answer: 100 sheets are 1.25 cm thick, so each sheet is 1.25×10^{-2} cm.**

4 ASSESS

Closing Activity

Speaking Have students orally explain how they can show that $\frac{x^2}{x^6} = x^{-4}$ or $\frac{1}{x^4}$.

Chapter 4, Quiz D (Lessons 4-8 and 4-9) is available in the *Assessment and Evaluation Masters*, p. 100.

43. Write the product $7 \cdot 7 \cdot 7 \cdot 7 \cdot 7$ using exponents. (Lesson 4-2) **7^5**

44. Solve $V = \frac{4}{3}\pi r^3$ for V if $\pi = 3.14$ and $r = 5$. (Lesson 3-4) **523.3**

45. **Personal Finance** When Chandra and Scott got married, they put their savings into one account. Chandra had $2230 and together they had $4750. Write and solve an equation to find the amount of money Scott had. (Lesson 3-2)

45. $2230 + x = 4750$; $2520

46. Evaluate $3c - ac$ if $a = -4$ and $c = 2$. (Lesson 2-7) **14**

47. Write and solve an equation for *eight less than a number is three.* (Lesson 1-8) **$n - 8 = 3$; 11**

48. **Exercise** According to one medical study, you can add 21 minutes to your life for every mile that you walk. If you walked 2 miles each day for a year, how much time would you add to your life? (Lesson 1-1)
   a. Which method of computation do you think is most appropriate? Justify your choice. **Sample answer: calculator**
   b. Solve the problem using the four-step plan. Be sure to examine your solution. **15,330 minutes or 10 days, 15.5 hours**

EARTH WATCH

GLOBAL WARMING

You have heard about the dreadful winter suffered by General George Washington and his troops at Valley Forge. At that time, Earth was nearing the end of what has been called the "Little Ice Age." In fact, the temperatures on Earth have fluctuated from very warm to very cold many times.

Beginning in the 1980s, environmental advocates have been warning us about a global warming crisis. Carbon dioxide and other gases, called greenhouse gases, blanket Earth and keep it warm. It is said that the burning of fossil fuels adds too much carbon dioxide to the atmosphere, making Earth dangerously warm.

See for Yourself **See students' work.**

- The carbon dioxide that is released from human activity accounts for $\frac{174}{10^7}$ of all of the greenhouse gases released each year. Write this number as a decimal. **0.0000174**
- Some of the other greenhouse gases are chlorofluorocarbons, methane, nitrous oxide, and carbon monoxide. Research the activities that contribute to the production of these gases.
- Investigate the benefits of the greenhouse effect. Why is it necessary for life on Earth?
- Write a paragraph or two stating your opinion on the greenhouse effect. Is there any danger? If so, what should be done?

Enrichment Masters, p. 35

NAME ________________ DATE ________

4-9 Enrichment Student Edition Pages 210–214

Truth Tables

A **conditional statement** is a statement formed by two statements (*p and q*) connected by *if . . . , then*.

Statement: ABC is a triangle.

Statement: ABC has 3 sides.

Conditional: If ABC is a triangle, then ABC has 3 sides.

The logic symbol for a conditional is $p \to q$, read "if p, then q."

Three other statements can be formed by changing the order of p and q, or by *negating* p and q.

$q \to p$ **Converse:** If ABC has 3 sides, then ABC is a triangle.

$\sim p \to \sim q$ **Inverse:** If ABC is not a triangle, then ABC does not have 3 sides.

$\sim q \to \sim p$ **Contrapositive:** If ABC does not have 3 sides, then ABC is not a triangle.

The symbol ~ is used for *not.*

Write the converse, inverse, and contrapositive of each conditional. Then complete the truth table.

1. $p \to q$ If $x + 3 = 8$, then $x = 5$.
   $q \to p$ **If $x = 5$, then $x + 3 = 8$.**
   $\sim p \to \sim q$ **If $x + 3 \neq 8$, then $x \neq 5$.**
   $\sim q \to \sim p$ **If $x \neq 5$, then $x + 3 \neq 8$.**

$p \to q$	$q \to p$	$\sim p \to \sim q$	$\sim q \to \sim p$
True	T	T	T

2. $p \to q$ If $ABCD$ is a square, then $ABCD$ has 4 sides.
   $q \to p$ **If $ABCD$ has 4 sides, then $ABCD$ is a square.**
   $\sim p \to \sim q$ **If $ABCD$ is not a square, then $ABCD$ does not have 4 sides.**
   $\sim q \to \sim p$ **If $ABCD$ does not have 4 sides, then $ABCD$ is not a square.**

$p \to q$	$q \to p$	$\sim p \to \sim q$	$\sim q \to \sim p$
True	F	F	T

3. $p \to q$ If it is 9 A.M., it is morning.
   $q \to p$ **If it is morning, then it is 9 A.M.**
   $\sim p \to \sim q$ **If it is not 9 A.M., then it is not morning.**
   $\sim q \to \sim p$ **If it is not morning, then it is not 9 A.M.**

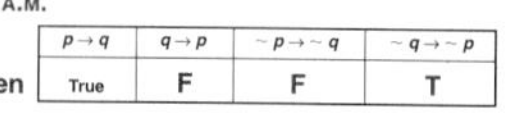

$p \to q$	$q \to p$	$\sim p \to \sim q$	$\sim q \to \sim p$
True	F	F	T

Group Activity Card 4-9

Plenty of Powers Group Activity **4-9**

In this activity you will work in pairs.

To begin, each student writes down two or three expressions that include constants, variables, and exponents. The same variables are to be used in each expression. The exponents may be negative, and the expressions may be in fraction form. You then exchange papers, and your partner is to simplify the expression into a monomial. Lastly, check the answer for its correctness.

An example is $(3x^3 y^{-2})(4x^{-1} y^5)(2x^{-4} y^2)$ which is simplified into $24 x^{-2} y^5$.

Another example is $\frac{(4m^{-6}n^5p^2)(5m^8n^{-3}p^4)}{(7m^{-4}n^3p^1)(8m^7n^2p^3)}$, which is simplified into $\frac{5}{14} m^{-1} n^{-3} p^2$.

 Pre-Algebra

EARTH WATCH

Since the late 19th century, the average global temperature has increased between 0.54°F and 1.08°F (0.3°C and 1.06°C). Internationally, 1990 was the hottest year on record since official weather records first started being kept by the British in 1860.

CHAPTER 4 Highlights

Vocabulary

After completing this chapter, you should be able to define each term, property, or phrase and give an example or two of each.

Algebra

algebraic fraction (p. 197)
base (p. 175)
common multiples (p. 200)
composite number (p. 184)
divisible (p. 170)
expanded form (p. 176)
exponent (p. 175)
factor trees (p. 185)
factors (p. 170)
greatest common factor (GCF) (p. 190)
least common denominator (LCD) (p. 201)
least common multiple (LCM) (p. 200)
monomial (p. 172)
multiple (p. 200)
negative exponents (p. 210)
order of operations (p. 176)
powers (p. 175)
prime factorization (p. 185)
prime number (p. 184)
product of powers (p. 205)
quotient of powers (p. 206)
ratio (p. 196)
simplest form (p. 196)
standard form (p. 176)

Problem Solving

draw a diagram (p. 181)
guess and check (p. 186)
look for a pattern (p. 171)

Understanding and Using Vocabulary

Complete each statement.

1. The greatest integer that is a factor of each of two or more integers is called their greatest common factor.
2. Numbers that are expressed using exponents are called powers.
3. The factors of a whole number divide that number with a remainder of zero.
4. A prime number is a whole number greater than one that has exactly two factors, 1 and itself.
5. The product of a number and any whole number is called a multiple of the original number.
6. Expanded form is writing numbers using place values and exponents.
7. The least common multiple of the denominators of two or more fractions is called the least common denominator.
8. A monomial is an integer, a variable, or a product of integers or variables.
9. A comparison of two numbers by division is called a ratio.

Instructional Resources

Three multiple-choice tests and three free-response tests are provided in the *Assessment and Evaluation Masters*. Forms 1A and 2A are for honors pacing, Forms 1B and 2B are for average pacing, Forms 1C and 2C are for basic pacing. Chapter 4 Test, Form 1B is shown at the right. Chapter 4 Test, Form 2B is shown on the next page.

Highlights

Using the Chapter Highlights

The Chapter Highlights begins with a listing of the new terms, properties, and phrases that were introduced in this chapter.

Assessment and Evaluation Masters, pp. 87–88

NAME ______________________ DATE ____________

CHAPTER 4 Test, Form 1B

Use the divisibility rules to answer each question.

1. Which is a factor of 8346?
   A. 4 B. 5 C. 6 D. 10 — 1. C
2. Which is a factor of 9039?
   A. 2 B. 3 C. 6 D. 10 — 2. B
3. Determine which expression is not a monomial.
   A. $3x - y$ B. $-13xyz$ C. 16 D. w — 3. A
4. Choose the product of $(w)(w)(w)(w)(w)$.
   A. $5 + w$ B. $5w$ C. 5^w D. w^5 — 4. D
5. Write 15^3 as a product of the same factor.
   A. $15 \cdot 15 \cdot 15$
   B. $1 \cdot 15 \cdot 15$
   C. $3 \cdot 3 \cdot 3 \cdot 3 \cdot 3 \cdot 3 \cdot 3 \cdot 3 \cdot 3 \cdot 3 \cdot 3 \cdot 3 \cdot 3 \cdot 3 \cdot 3$
   D. $3 \cdot 15$ — 5. A
6. Evaluate $3m^2$ if $m = 5$.
   A. 21 B. 30 C. 75 D. 225 — 6. C
7. In an invitational basketball championship tournament, 16 teams participate in a single-elimination format. That is, if a team loses a game, it is out of the tournament. How many games are played in this tournament?
   A. 8 B. 12 C. 15 D. 16 — 7. C
8. Determine the number of rounds required for 32 teams to play in a single-elimination tournament by drawing a diagram.
   A. 32 B. 31 C. 16 D. 5 — 8. D
9. Which is a prime number?
   A. 21 B. 23 C. 24 D. 27 — 9. B
10. Find the prime factorization of 36.
   A. $2 \cdot 2 \cdot 9$ B. $2 \cdot 2 \cdot 2 \cdot 2$ C. $6 \cdot 6$ D. $2 \cdot 2 \cdot 3 \cdot 3$ — 10. D
11. Factor $28xy^2$ completely.
   A. $4 \cdot 7 \cdot x \cdot y \cdot y$ B. $2 \cdot 2 \cdot 7 \cdot x \cdot y \cdot y$
   C. $1 \cdot 28 \cdot x \cdot y \cdot y$ D. $2 \cdot 2 \cdot 7 \cdot x \cdot y^2$ — 11. B

Find the GCF for each set of numbers or monomials.

12. 12, 20, and 36
   A. 720 B. 12 C. 3 D. 4 — 12. D

NAME ______________________ DATE ____________

Chapter 4 Test, Form 1B (continued)

13. $15a^2b^3$ and $20ab^4$
   A. $5ab^3$ B. $60a^3b^7$ C. $5a^2b^4$ D. $300a^2b$ — 13. A

Write each fraction in simplest form.

14. $\frac{36}{66}$
   A. $\frac{1}{2}$ B. $\frac{18}{33}$ C. $\frac{6}{11}$ D. $\frac{12}{22}$ — 14. C
15. $\frac{[illegible]}{50ab}$
   A. $\frac{2ab}{5}$ B. $\frac{2b}{5a}$ C. $\frac{2ab^2}{5ab}$ D. $\frac{2b}{5}$ — 15. D

Find the LCM for each set of numbers or monomials.

16. 14 and 21
   A. 7 B. 42 C. 147 D. 294 — 16. B
17. $18x$ and $24y$
   A. 6 B. $72x^2y^2$ C. $72xy$ D. $6xy$ — 17. C
18. Replace ● with $<, >, \geq,$ or $=$ to make $\frac{2}{3}$ ● $\frac{4}{5}$ a true statement.
   A. $=$ B. $>$ C. $<$ D. $\geq$ — 18. C

Find each product.

19. $10^3 \cdot 10^6$
   A. 100^9 B. 10^{36} C. 10^{18} D. 10^9 — 19. D
20. $(3a^3)(-5a^4)$
   A. $-15a^{12}$ B. $-15a^7$ C. $15a^4$ D. $15a^7$ — 20. B

Find each quotient.

21. $\frac{x^9}{x}$
   A. 1^9 B. 9 C. x^8 D. x^9 — 21. C
22. $\frac{a^6}{a^8}$
   A. a^{-2} B. a^2 C. $2a$ D. a^{14} — 22. A
23. Write 8^{-3} using a positive exponent.
   A. 2^3 B. $\frac{1}{8^3}$ C. $\frac{1}{3^8}$ D. 3^8 — 23. B
24. Write $\frac{1}{5^2}$ using a negative exponent.
   A. 2^{-5} B. 5^{-1} C. 5^{-2} D. -5 — 24. C
25. Evaluate $3x^{-4}$ if $x = 2$.
   A. $\frac{3}{16}$ B. $\frac{3}{32}$ C. -48 D. 48 — 25. A

CHAPTER 4 Study Guide and Assessment

Using the Study Guide and Assessment

Skills and Concepts Encourage students to refer to the objectives and examples on the left as they complete the review exercises on the right.

Assessment and Evaluation Masters, pp. 93–94

NAME ________________ DATE ________

CHAPTER 4 Test, Form 2B

State whether each number is divisible by 2, 3, 5, 6, or 10.

1. 1000 — 1. 2, 5, 10
2. 324 — 2. 2, 3, 6
3. 300 — 3. 2, 3, 5, 6, 10
4. Write the product (3)(3)(3)(3)(3)(3) using exponents. — 4. 3^6

Evaluate.

5. four cubed — 5. 64
6. $a^2 - b^2$ if $a = 6$ and $b = 3$ — 6. 27

Factor each number or monomial completely.

7. 36 — 7. $2 \cdot 2 \cdot 3 \cdot 3$
8. 84 — 8. $2 \cdot 2 \cdot 3 \cdot 7$
9. $24y$ — 9. $2 \cdot 2 \cdot 2 \cdot 3 \cdot y$

Find the GCF for each set of numbers or monomials.

10. 12 and 18 — 10. 6
11. 20, 30, and 45 — 11. 5
12. $16x^2y$ and $18xy^2$ — 12. $2xy$

Write each fraction in simplest form.

13. $\frac{18}{64}$ — 13. $\frac{9}{32}$
14. $\frac{16a^2b}{20ab^2}$ — 14. $\frac{4a}{5b}$

Find the LCM for each set of numbers or monomials.

15. 8 and 12 — 15. 24
16. 12, 18, 36 — 16. 36
17. $6x^2y$ and $8x^3y$ — 17. $24x^3y$

Replace each ● with >, <, or = to make a true sentence.

18. $\frac{2}{3}$ ● $\frac{3}{4}$ — 18. <
19. $\frac{6}{9}$ ● $\frac{8}{12}$ — 19. =

NAME ________________ DATE ________

Chapter 4 Test, Form 2B (continued)

Find each product.

20. $(^-12x^2y)(3x)$ — 20. $^-36x^3y$
21. $d^3 \cdot d^2 \cdot d$ — 21. d^6

Find each quotient.

22. $\frac{3^5}{3^2}$ — 22. 3^3
23. $\frac{a^{16}}{a}$ — 23. a^{15}
24. David is packing bundles of 3-inch-by-5-inch cards into a square box. He places the cards side-by-side so that there is no wasted space. Find the smallest possible measure for the edge of the box. — 24. 15 in.
25. Shareen is covering a tabletop with tiles measuring 3 inches by 6 inches. The table is 30 inches wide and 40 inches long. Can she cover the top without cutting some of the tiles? — 25. no

CHAPTER 4 Study Guide and Assessment

Skills and Concepts

Objectives and Examples	Review Exercises

Objectives and Examples

After completing this chapter, you should be able to:

▶ **determine whether one number is a factor of another.** (Lesson 4-1)

Is 342 divisible by 2, 3, 5, 6, or 10?

It is divisible by 2 since its last digit is 2.

It is divisible by 3 because 3 + 4 + 2 or 9 is divisible by 3.

The last digit is not 5 or 0, so it is not divisible by 5.

It is divisible by 6 since it is divisible by 2 and 3.

It is not divisible by 10 since its last digit is not 0.

Review Exercises

Use these exercises to review and prepare for the chapter test.

Using divisibility rules, state whether each number is divisible by 2, 3, 5, 6, or 10.

10. 863 none
11. 635 5
12. 4200 2, 3, 5, 6, 10
13. 14,577 3
14. 40, 180 2, 5, 10
15. 582 2, 3, 6
16. 5103 3
17. 1962 2, 3, 6

▶ **use powers and exponents in expressions and equations.** (Lesson 4-2)

Write a^5 as a product of the same factor.

$a^5 = a \cdot a \cdot a \cdot a \cdot a$

Write $3 \cdot 3 \cdot 3 \cdot 3$ using exponents.

$3 \cdot 3 \cdot 3 \cdot 3 = 3^4$

Write each multiplication expression using exponents. 18–21. See margin.

18. $6 \cdot 6 \cdot 6 \cdot 6 \cdot 6$
19. $(c + 2)(c + 2)(c + 2)$
20. $(-4)(-4)(-4)(-4)$
21. $(1 \cdot 1) \cdot 1$

Write each power as a multiplication expression. 22–25. See margin.

22. $(-16)^3$
23. k^9
24. $(-w)^4$
25. 32^{11}

▶ **write the prime factorization of composite numbers.** (Lesson 4-4)

Write the prime factorization of 648.

$$648 = 9 \cdot 72$$
$$= 3 \cdot 3 \cdot 8 \cdot 9$$
$$= 3 \cdot 3 \cdot 2 \cdot 4 \cdot 3 \cdot 3$$
$$= 3 \cdot 3 \cdot 2 \cdot 2 \cdot 2 \cdot 3 \cdot 3$$

Factor each number or monomial completely. 26–33. See margin.

26. 120
27. −114
28. $630x^3$
29. $-825x^2y$
30. 2805
31. $-1827jk^3$
32. $66g^5h^2$
33. $550mnq^2$

GLENCOE Technology

Test and Review Software

You may use this software, a combination of an item generator and item bank, to create your own tests or worksheets. Types of items include free response, multiple choice, short answer, and open ended.

For IBM & Macintosh

Additional Answers

18. 6^5
19. $(c + 2)^3$
20. $(-4)^4$
21. 1^3
22. $-16 \cdot -16 \cdot -16$
23. $k \cdot k \cdot k \cdot k \cdot k \cdot k \cdot k \cdot k \cdot k$
24. $(-w)(-w)(-w)(-w)$
25. $32 \cdot 32 \cdot 32 \cdot 32 \cdot 32 \cdot 32 \cdot 32 \cdot 32 \cdot 32 \cdot 32 \cdot 32$

Objectives and Examples

▶ **find the greatest common factor of two or more integers or monomials.** (Lesson 4-5)

Find the GCF of $42x^2$ and $110xy$.

$42x^2 = 2 \cdot 3 \cdot 7 \cdot x \cdot x$
$110xy = 2 \cdot 5 \cdot 11 \cdot x \cdot y$
The GCF is $2 \cdot x$ or $2x$.

Review Exercises

Find the GCF of each set of numbers or monomials. 36. $4x$ 37. $35a^2b^2$ 38. $16j^2$

34. 70, 66 2
35. 1092, 325 13
36. $100x, 84xy, -76x^2$
37. $210a^3b^2, 875a^2b^3$
38. $112j^5k, -144j^2$
39. $40x, 25y$ 5
40. $115wx, 224wz$ w
41. $441ac, 223bd$ 1

▶ **simplify fractions using the GCF.** (Lesson 4-6)

Write $\frac{6}{45}$ in simplest form.

$\frac{6}{45} = \frac{2}{15}$ (÷3) *The GCF of 6 and 45 is 3.*

Simplify $\frac{15ac^2}{24ab}$.

$$\frac{15ac^2}{24ab} = \frac{\overset{1}{\cancel{3}} \cdot 5 \cdot \overset{1}{\cancel{a}} \cdot c \cdot c}{2 \cdot 2 \cdot 2 \cdot \underset{1}{\cancel{3}} \cdot \underset{1}{\cancel{a}} \cdot b} = \frac{5c^2}{8b}$$

Write each fraction in simplest form. If the fraction is already in simplest form, write *simplified*.

42. $\frac{72}{88}$ $\frac{9}{11}$
43. $\frac{133}{140}$ $\frac{19}{20}$
44. $\frac{225}{315}$ $\frac{5}{7}$
45. $\frac{48}{66}$ $\frac{8}{11}$
46. $\frac{23a}{32b}$ simplified
47. $\frac{500x}{1000x}$ $\frac{1}{2}$
48. $-\frac{125mn}{625m}$ $-\frac{n}{5}$
49. $\frac{68b^2}{105c}$ simplified

▶ **find the least common multiple of two or more numbers.** (Lesson 4-7)

Find the LCM of 18 and 15.

$18 = 2 \cdot 3 \cdot 3$
$15 = 3 \cdot 5$
The LCM is $2 \cdot 3 \cdot 3 \cdot 5 = 90$

Find the least common multiple of each set of numbers or algebraic expressions.

50. $11ab, 6b$ $66ab$
51. 12, 4, 5 60
52. $7x^3, 49xy$ $49x^3y$
53. $10j, 6jk, 3$ $30jk$
54. $5f^4g, 13fg^2$ $65f^4g^2$
55. 12, 15, 18 180
56. $72x^3, 64x^5$ $576x^5$
57. $6ab, 18c, 36d$ $36abcd$

▶ **compare fractions.** (Lesson 4-7)

Which is greater, $\frac{3}{4}$ or $\frac{7}{12}$?

$\frac{3}{4} = \frac{9}{12}$ (×3) *The LCM of 4 and 12 is 12.*

$\frac{9}{12} > \frac{7}{12}$, so $\frac{3}{4} > \frac{7}{12}$.

Replace each ● with <, >, or = to make each statement true.

58. $\frac{6}{7}$ ● $\frac{5}{14}$ >
59. $\frac{9}{16}$ ● $\frac{8}{15}$ >
60. $\frac{9}{100}$ ● $\frac{1}{9}$ <
61. $\frac{1}{11}$ ● $\frac{2}{23}$ >

Chapter 4 • Study Guide and Assessment

Additional Answers

26. $2 \cdot 2 \cdot 2 \cdot 3 \cdot 5$
27. $-1 \cdot 2 \cdot 3 \cdot 19$
28. $2 \cdot 3 \cdot 3 \cdot 5 \cdot 7 \cdot x \cdot x \cdot x$
29. $-1 \cdot 3 \cdot 5 \cdot 5 \cdot 11 \cdot x \cdot x \cdot y$
30. $3 \cdot 5 \cdot 11 \cdot 17$
31. $-1 \cdot 3 \cdot 3 \cdot 7 \cdot 29 \cdot j \cdot k \cdot k \cdot k$
32. $2 \cdot 3 \cdot 11 \cdot g \cdot g \cdot g \cdot g \cdot g \cdot h \cdot h$
33. $2 \cdot 5 \cdot 5 \cdot 11 \cdot m \cdot n \cdot q \cdot q$

Applications and Problem Solving

Encourage students to work through the exercises in the Applications and Problem Solving section to strengthen their problem-solving skills.

Objectives and Examples	Review Exercises
▸ **multiply and divide monomials.** (Lesson 4-8) **Find each product or quotient.** $g^2 \cdot g^7 = g^{2+7}$ $= g^9$ $\frac{(-3)^5}{(-3)^2} = (-3)^{5-2}$ $= (-3)^3$	**Find each product or quotient. Express your answer in exponential form.** **62.** $12^4 \cdot 12^7$ **12^{11}** **63.** $\frac{d^9}{d^4}$ **d^5** **64.** $n^5 \cdot n^{14}$ **n^{19}** **65.** $\frac{(-4)^5}{(-4)}$ **$(-4)^4$** **66.** $(6w^3)(10w^4)$ **$60w^7$** **67.** $\frac{6x^{40}}{3x^{18}}$ **$2x^{22}$** **68.** $\frac{30c^2d}{15c}$ **$2cd$** **69.** $22ab^3 \cdot 5a^9$ **$110a^{10}b^3$**
▸ **use negative exponents in expressions.** (Lesson 4-9) **Write $-2x^3y^{-2}$ using positive exponents.** $-2x^3y^{-2} = -2x^3 \cdot y^{-2}$ $= -2x^3 \cdot \frac{1}{y^2}$ $= \frac{-2x^3}{y^2}$	**Write each expression using positive exponents.** **70.** $(-5)^{-3}$ **$\frac{1}{(-5)^3}$** **71.** $4c^{-2}d$ **$\frac{4d}{c^2}$** **72.** $\frac{1}{6^{-18}}$ **6^{18}** **73.** $4x(yz)^{-2}$ **$\frac{4x}{(yz)^2}$** **74.** $\left(\frac{2}{x}\right)^{-4}$ **$\left(\frac{x}{2}\right)^4$** **75.** $\frac{3}{(f)^{-9}}$ **$3f^9$**

Applications and Problem Solving

76. Business The Carpet Experts cleaning team can clean carpet in a room that is 10 feet by 10 feet in 20 minutes. Draw a diagram to find how long it would take them to do a walk-in closet that is 5 feet by 5 feet. (Lesson 4-3) **5 minutes**

77. Sports The City Monday-Night Softball League has 7 teams. The director is making a schedule where every team plays every other team once. How many games must be scheduled? (Lesson 4-3) **21 games**

78. Number Theory Numbers that have a greatest common factor of 1 are said to be **relatively prime**. Find the least two composite numbers that are relatively prime. (Lesson 4-5) **4, 9**

79. Literature *USA TODAY* polled Americans on the types of books that they buy. The graph below shows the results of the poll. Which type of book is more popular, travel or home improvement? (Lesson 4-7) **home improvement**

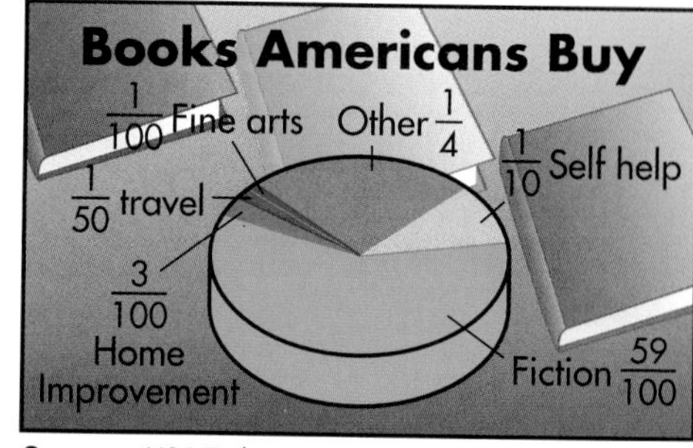

Source: *USA-Today*

A practice test for Chapter 4 is available on page 777 .

Alternative Assessment

Performance Task

Demonstrate your knowledge by giving a clear, concise solution to each problem. Be sure to include all relevant drawings and justify your answers. You may show your solutions in more than one way or investigate beyond the requirements of the problem.

1. See students' work.
6. 15: 0, 15, 30, 45, 60, 75, . . . ; 25: 0, 25, 50, 75, . . .
8. 75 feet; See students' explanations.

A planned community housing project is providing a gardening area for its residents. The area will be laid out in individual garden plots as shown.

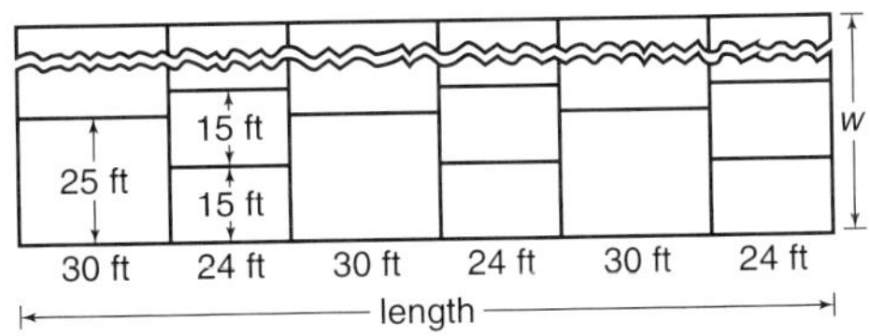

1. Explain how to find the prime factorization of a number.
2. Find the prime factorizations of 30 and 24. **$30 = 2 \cdot 3 \cdot 5$; $24 = 2^3 \cdot 3$**
3. Find the common factors of 30 and 24. **2, 3, 6**
4. The garden area will be fenced. The fence posts in front are to be equally spaced along the entire length with posts at the corners of each garden plot. Would a spacing of 3 feet between posts work? How do you know? **Yes; 3 is a factor of 30 and 24.**
5. What is the longest spacing between posts that can be used? What is this distance called? **6 feet; the greatest common factor of 24 and 30**
6. List the multiples of the widths of each size lot.
7. If the back fence of the overall gardening area is to be straight, give at least two possible widths, w, of the area. **75, 150**
8. What is the shortest possible width of the gardening area? Explain.
9. If the fence posts are to be equally spaced along the width of the garden area with posts at the corners of each garden plot, what spacing would you recommend? Why? **5 feet; 5 is the GCF of 15 and 25**

Thinking Critically

- ▶ Find the GCF and LCM of 45 and 72. **GCF: 9; LCM: 360**
- ▶ Explain why you think the greatest common factor is called that when it is less than the numbers given. **See students' work.**
- ▶ Explain why you think the least common multiple is called that when it is greater than the numbers given. **See students' work.**

Portfolio

Select an item from your work in this chapter that shows your creativity and place it in your portfolio.

Self Evaluation

It isn't climbing the hill that makes you tired. It's the rock in your shoe. You are able to reach the top of the hill, if you don't let anything stop you.

Assess yourself. When you feel that rock in your shoe, what do you do? Do you stop and take it out or do you hope it moves to a place where it won't bother you? In mathematics class or in life, stop and ask a question or ask for help so you can remove the rock and continue to the top.

Alternative Assessment

The Alternative Assessment section provides students with the opportunity to assess their own work by thinking critically, working with others, keeping a portfolio, and honestly evaluating their own progress. For more information on alternative forms of assessment, see *Alternative Assessment in the Mathematics Classroom*.

Performance Assessment

Performance Assessment tasks for this chapter are included in the *Assessment and Evaluation Masters*. A scoring guide is also provided.

Alternative Assessment

Assessment and Evaluation Masters, pp. 97, 109

NAME ______________________ DATE ____________

CHAPTER 4 Performance Assessment

Instructions: Demonstrate your knowledge by giving a clear, concise solution to each problem. Be sure to include all relevant drawings and justify your answers. You may show your solution in more than one way or investigate beyond the requirements of the problem.

1. The Burnsville High School band and chorus have been invited to participate in the Thanksgiving parade. There are 210 members in the band and 40 members in the chorus.
   a. Can the band march in a rectangular formation having 8 band members in each row? Why or why not? If not, draw a diagram of a rectangular formation that would be permissible.
   b. Twenty-six chorus members and 140 band members have not brought in their permission slips for the Thanksgiving trip. Explain how to tell whether a greater fractional part of the chorus or band has not brought in their slips. Find which part is greater.
2. To understand a communication, you must understand the language used in the communication. To understand mathematics, you must understand the language or symbols used.
   a. Explain the difference in the meanings of $4a$ and a^4.
   b. Explain the difference in the meanings of ^-4b and b^{-4}.
3. Write an argument or counter-example to support your answer to the following questions.
   a. Are all numbers that are divisible by 4 also divisible by 2?
   b. Are all numbers that are divisible by 2 also divisible by 4?
4. Methods of finding prime numbers have intrigued mathematicians for years. For example, Goldbach's conjecture states that every even number greater than 2 can be written as the sum of two prime numbers. Choose three even numbers greater than 20 and less than 100. Write each as the sum of two prime numbers.

CHAPTER 4 Scoring Guide

Level	Specific Criteria
3 Superior	• Shows thorough understanding of the concepts *factors, fraction, exponent, coefficient, prime,* and *composite.* • Uses appropriate strategies to compare fractions. • Computations to compare fractions and identify prime numbers are correct. • Written explanations are exemplary. • Diagram is accurate. • Goes beyond requirements of some or all problems.
2 Satisfactory, with minor flaws	• Shows understanding of the concepts *factors, fraction, exponent, coefficient, prime,* and *composite.* • Uses appropriate strategies to compare fractions. • Computations to compare fractions and identify prime numbers are mostly correct. • Written explanations are effective. • Diagram is mostly accurate. • Satisfies requirements of problems.
1 Nearly Satisfactory, with serious flaws	• Shows understanding of most of the concepts *factors, fraction, exponent, coefficient, prime,* and *composite.* • May not use appropriate strategies to compare fractions. • Computations to compare fractions and identify prime numbers are mostly correct. • Written explanations are satisfactory. • Diagram is mostly accurate. • Satisfies most requirements of problems.
0 Unsatisfactory	• Shows little or no understanding of the concepts *factors, fraction, exponent, coefficient, prime,* and *composite.* • May not use appropriate strategies to compare fractions. • Computations to compare fractions and identify prime numbers are incorrect. • Written explanations are not satisfactory. • Diagram is not accurate. • Does not satisfy requirements of problems.

Using the Ongoing Assessment

These two pages review the skills and concepts presented in Chapters 1–4. This review is formatted to reflect new trends in standardized testing.

Assessment and Evaluation Masters, pp. 102–103

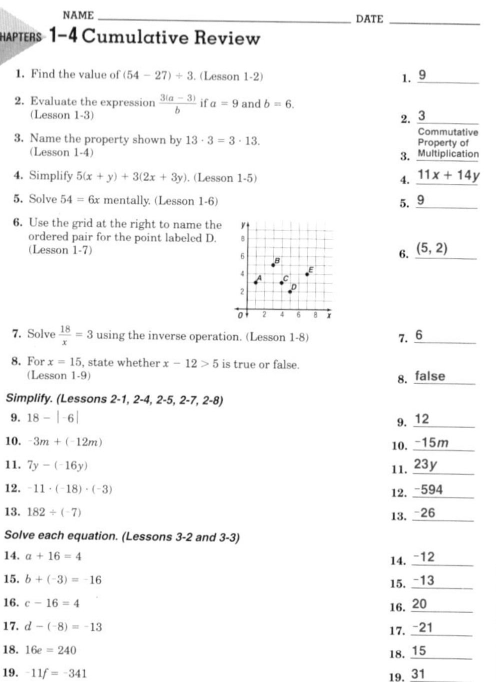

NAME ________________ DATE ________

CHAPTERS **1–4 Cumulative Review**

1. Find the value of $(54 - 27) \div 3$. (Lesson 1-2) — 1. 9
2. Evaluate the expression $\frac{3(a-3)}{b}$ if $a = 9$ and $b = 6$. (Lesson 1-3) — 2. 3
3. Name the property shown by $13 \cdot 3 = 3 \cdot 13$. (Lesson 1-4) — 3. Commutative Property of Multiplication
4. Simplify $5(x + y) + 3(2x + 3y)$. (Lesson 1-5) — 4. $11x + 14y$
5. Solve $54 = 6x$ mentally. (Lesson 1-6) — 5. 9
6. Use the grid at the right to name the ordered pair for the point labeled D. (Lesson 1-7) — 6. (5, 2)
7. Solve $\frac{18}{x} = 3$ using the inverse operation. (Lesson 1-8) — 7. 6
8. For $x = 15$, state whether $x - 12 > 5$ is true or false. (Lesson 1-9) — 8. false

Simplify. (Lessons 2-1, 2-4, 2-5, 2-7, 2-8)

9. $18 - |-6|$ — 9. 12
10. $-3m + (-12m)$ — 10. $-15m$
11. $7y - (-16y)$ — 11. $23y$
12. $-11 \cdot (-18) \cdot (-3)$ — 12. -594
13. $182 \div (-7)$ — 13. -26

Solve each equation. (Lessons 3-2 and 3-3)

14. $a + 16 = 4$ — 14. -12
15. $b + (-3) = -16$ — 15. -13
16. $c - 16 = 4$ — 16. 20
17. $d - (-8) = -13$ — 17. -21
18. $16e = 240$ — 18. 15
19. $-11f = -341$ — 19. 31
20. $\frac{m}{-12} = 16$ — 20. -192

NAME ________________ DATE ________

Cumulative Review (continued)

21. Paul and Tia have a square-shaped garden 50 feet on a side. Find the perimeter of the garden. (Lesson 3-5) — 21. 200 ft

Solve each inequality. (Lessons 3-6, 3-7, and 3-8)

22. $x + 9 > 4$ — 22. $x > -5$
23. $\frac{x}{-3} \geq 15$ — 23. $x \leq -45$
24. The Eagles need at least 7 more points to win the football game. If the opponent's score is 27, what is the Eagles' score? — 24. 21

Use divisibility rules to state whether each number is divisible by 2, 3, 5, 6, or 10. (Lesson 4-1)

25. 1340 — 25. 1340 is divisible by 2, 5, and 10
26. 150 — 26. 150 is divisible by 2, 3, 5, 6, and 10

Evaluate each expression. (Lesson 4-2)

27. n^4 if $n = 6$ — 27. 1296
28. $15r^3$ if $r = 4$ — 28. 960

Write the prime factorization of each number. (Lesson 4-4)

29. 891 — 29. $3^4 \cdot 11$
30. 864 — 30. $2^5 \cdot 3^3$
31. Find the GCF for 16 and 19. (Lesson 4-5) — 31. 1
32. Find the LCM for 12 and 16. (Lesson 4-7) — 32. 48
33. Simplify $\frac{x^7}{x^8}$. (Lesson 4-9) — 33. x^{-1}

CHAPTERS 1–4 *Ongoing Assessment*

Section One: Multiple Choice

There are eight multiple-choice questions in this section. After working each problem, write the letter of the correct answer on your paper.

1. The family sizes of students in a mathematics class are recorded in the following table.

Family Size	2	3	4	5	6	7	8
Number of Students	1	4	3	6	5	2	3

How many students have fewer than 5 members in their family? **B**

A. 6
B. 8
C. 9
D. 14

2. A group of divers needs to descend to a depth 3 times their present depth of -25 meters. At what depth do they need to be? **D**

A. 75 meters
B. 28 meters
C. -25 meters
D. -75 meters

3. Simplify $|-12| - |-3|$. **C**

A. -9
B. -15
C. 9
D. 15

4. An airplane descended 250 feet in 5 minutes. What was its average rate of descent? **A**

A. 50 ft/min
B. 250 ft/min
C. 1250 ft/min
D. Not here

5. Which of the following is equivalent to 4^{-3}? **D**

A. $3 \cdot 3 \cdot 3 \cdot 3$
B. $4 \cdot 4 \cdot 4$
C. $\frac{1}{3 \cdot 3 \cdot 3 \cdot 3}$
D. $\frac{1}{4 \cdot 4 \cdot 4}$

6. Which fraction is in simplest form? **C**

A. $\frac{12a^2}{15ab}$
B. $\frac{15a}{50}$
C. $\frac{12d^2}{25ab}$
D. $\frac{5b^2c}{7bh^2}$

7. Which set of numbers is in order from least to greatest? **A**

A. $\{-5, -3, 1, 14\}$
B. $\{-1, -6, 8, 11\}$
C. $\{15, -8, 1\}$
D. $\{10, 0, -5\}$

8. The distance traveled is given by the formula $d = rt$, where r is the rate of speed and t is the time spent traveling. How long would it take to ride a bike 9 miles at the rate of 6 miles per hour? **B**

A. 15 min
B. 1.5 hours
C. 54 hours
D. 54 miles

Standardized test practice questions are also provided in the *Assessment and Evaluation Masters*, p. 101.

Chapter 4 • Ongoing Assessment

Section Two: Free Response

This section contains eight questions for which you will provide short answers. Write your answer on your paper.

9. The temperature when you left for school this morning was 10°F. The wind speed was 20 mph, which produced a windchill of −24°F. What is the difference between the actual temperature and the windchill temperature? **34**

10. Before she paid the electric bill, Rocio's checking account had a balance of \$131. Her new account balance is \$109. How much was her electric bill? **\$22**

11. Juwan's yard is 10 yards wide and 20 yards long. His dad wants to build a rectangular swimming pool that has an area of 100 square feet. What are the possible dimensions of the pool? **Sample Answers: 20' × 5'; 25' × 4'; 10' × 10'; 40' × $2\frac{1}{2}$'**

12. Simplify $12a + 8b - 6a$. **$6a + 8b$**

13. A college basketball court is 94 feet long by 50 feet wide. What is the area of the court? **4700 sq ft**

14. Nituna must earn 250 points to win a prize. She has already earned 176 points. Write a sentence to find p, the number of points she still needs to earn. **$176 + p = 250$**

15. Carrie is cutting a round pizza. What is the greatest number of pieces she can get from 1 pizza with 5 straight cuts? **15**

16. In high school football, a team can score 6 points for a touchdown, 7 points for a touchdown and an extra point, 8 points for a touchdown and a two-point conversion, 3 points for a field goal, or 2 points for a safety. Can a team have a score of 23 points? Explain. **Yes; for example, three seven-point touchdowns and one safety are 23 points.**

Test-Taking Tip

You can solve many problems without much calculating if you understand the basic mathematical concepts. Always look carefully at what is asked, and think of possible shortcuts for solving the problem.

Section Three: Open-Ended

This section contains two open-ended problems. Demonstrate your knowledge by giving a clear, concise solution to each problem. Your score on these problems will depend on how well you do the following.

- Explain your reasoning.
- Show your understanding of the mathematics in an organized manner.
- Use charts, graphs, and diagrams in your explanation.
- Show the solution in more than one way or relate it to other situations.
- Investigate beyond the requirements of the problem.

17. Drina, Susan, Latisha, and Joshua are in the orchestra at school. Each student plays one of the following instruments: drums, violin, trumpet, or trombone. If each student plays a different instrument, use the following information to match each student with an instrument.
    — Drina and Joshua both play brass instruments.
    — The trumpet player and Joshua also play in the jazz band.
    — Susan lost one of her sticks last week.
    Drina–trumpet; Latisha–violin; Susan–drums; Joshua–trombone

18. Determine whether 435 is divisible by 2, 3, 5, 6, or 10. Justify your answers. **3, 5; see students' work.**

CHAPTER 5

Rationals: Patterns in Addition and Subtraction

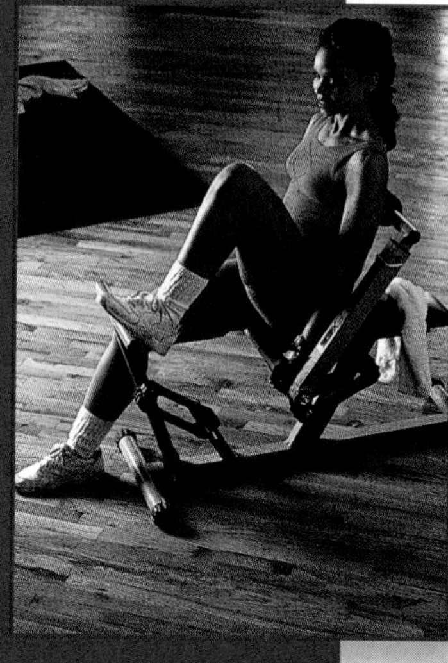

PREVIEWING THE CHAPTER

In this chapter, students explore and use patterns in addition and subtraction of rational numbers. Students explore our number system of rational numbers, including writing repeating and terminating decimals as fractions. Students then estimate sums and differences of rational numbers. Addition and subtraction of decimals is followed by addition and subtraction of fractions. The concepts are expanded to include simplifying algebraic expressions. Students then solve one-step equations and inequalities with addition and subtraction. Finally, students explore arithmetic sequences and the Fibonacci sequence.

Lesson (Pages)	Lesson Objectives	NCTM Standards	State/Local Objectives
5-1 (224–228)	Identify and compare rational numbers. Rename decimals as fractions.	1-6	
5-2 (229–233)	Estimate sums and differences of decimals and fractions.	1-4, 7, 10	
5-3 (234–238)	Add and subtract decimals.	1-4, 7	
5-4 (239–243)	Add and subtract fractions with like denominators.	1-4, 7, 10	
5-5 (244–247)	Add and subtract fractions with unlike denominators.	1-4, 7	
5-6 (248–250)	Solve equations with rational numbers.	1-4, 7, 9	
5-7 (251–254)	Solve inequalities with rational numbers.	1-4, 7, 9	
5-8 (255–257)	Use deductive and inductive reasoning.	1-4, 6-8	
5-9 (258-262)	Find the terms of arithmetic sequences. Represent a sequence algebraically.	1-4, 7, 8	
5-9B (263)	Investigate patterns in the Fibonacci sequence.	1-4, 6, 8	

A complete, 1-page lesson plan is provided for each lesson in the ***Lesson Planning Guide***. Answer keys for each lesson are available in the ***Answer Key Masters***.

ORGANIZING THE CHAPTER

You may want to refer to the **Course Planning Calendar** on page T12 for detailed information on pacing.
PACING: Standard—13 days; **Honors**—12 days; **Block**—6 days

LESSON PLANNING CHART

Lesson (Pages)	Materials/ Manipulatives	Extra Practice (Student Edition)	Blackline Masters: Study Guide	Blackline Masters: Practice	Blackline Masters: Enrichment	Blackline Masters: Assessment and Evaluation	Blackline Masters: Math Lab and Modeling Math	Blackline Masters: Multicultural Activity	Blackline Masters: Tech Prep Applications	Blackline Masters: Graphing Calculator	Blackline Masters: Activity	Real-World Applications	Interactive Mathematics Tools Software	Teaching Transparencies	Group Activity Cards
5-1 (224–228)	coins dominoes masking tape	p. 750	p. 36	p. 36	p. 36									5-1A 5-1B	5-1
5-2 (229–233)	empty boxes newspaper bar graphs	p. 751	p. 37	p. 37	p. 37				p. 9			10		5-2A 5-2B	5-2
5-3 (234–238)	bowls index cards play money (dollars, dimes, pennies)	p. 751	p. 38	p. 38	p. 38	p. 127		p. 9			p. 19		5-3	5-3A 5-3B	5-3
5-4 (239–243)	rulers*	p. 751	p. 39	p. 39	p. 39				p. 10		p. 5		5-4	5-4A 5-4B	5-4
5-5 (244–247)	index cards measuring cups (customary) rulers*	p. 751	p. 40	p. 40	p. 40	pp. 126–127				p. 5		11	5-5.1 5-5.2	5-5A 5-5B	5-5
5-6 (248–250)	balance scale* containers measuring cups	p. 752	p. 41	p. 41	p. 41			p. 10						5-6A 5-6B	5-6
5-7 (251-254)	masking tape	p. 752	p. 42	p. 42	p. 42	p. 128								5-7A 5-7B	5-7
5-8 (255–257)		p. 752	p. 43	p. 43	p. 43						p. 35			5-8A 5-8B	5-8
5-9 (258–262)	cubes* index cards periodic chart	p. 752	p. 44	p. 44	p. 44	p. 128	pp. 61–62, 80				p. 36		5-9	5-9A 5-9B	5-9
5-9B (263)	daisies						p. 36						5-9B		
Study Guide/ Assessment (266–269)						pp. 113–125, 129–131									

*Included in Glencoe's *Student Manipulative Kit* and *Overhead Manipulative Resources*.

ORGANIZING THE CHAPTER

All of the blackline masters in the Teacher's Classroom Resources are available on the ***Electronic Teacher's Classroom Resources*** CD-ROM.

Other Chapter Resources

Student Edition
Chapter Opener, pp. 222–223
Earth Watch, pp. 232–233
Working on the Investigation, p. 238
From the Funny Pages, p. 257
Closing the Investigation, p. 264

Teacher's Classroom Resources
Investigations and Projects Masters, pp. 45–48

Pre-Algebra Overhead Manipulative Resources, pp. 12–13

Technology
Test and Review Software (IBM & Macintosh)
CD-ROM Activities

Professional Publications
Block Scheduling Booklet
Glencoe Mathematics Professional Series

Outside Resources

Books/Periodicals
Trivia Math: Pre-Algebra, Katherine Pedersen, et al
Fascinating Fibonaccis, Mind Ware
Stories with Holes, Mind Ware

Videos/CD-ROMs
Algebra Math Video Series: Solving Algebraic Equations & Inequalities, ETA
Mathematical Eye: Decimal Investigations, Journal Films, Multimedia and Videodisc Compendium

Software
Conquering Fractions (+, −), MECC
Mastering Decimals, Phoenix/BFA Films and Video

Assessment Resources

Student Edition
Math Journal, pp. 231, 241, 256, 260
Mixed Review, pp. 228, 233, 238, 243, 247, 250, 254, 257, 262
Self Test, p. 247
Chapter Highlights, p. 265
Chapter Study Guide and Assessment, pp. 266–268
Alternative Assessment, p. 269

Portfolio, p. 61
MindJogger Videoquiz, 5

Teacher's Wraparound Edition
5-Minute Check, pp. 224, 229, 234, 239, 244, 248, 251, 255, 258
Checking Your Understanding, pp. 227, 230, 236, 242, 245, 250, 253, 257, 260
Closing Activity, pp. 228, 233, 238, 243, 247, 250, 254, 257, 262
Cooperative Learning, pp. 225, 259

Assessment and Evaluation Masters
Multiple-Choice Tests, Forms 1A (Honors), 1B (Average), 1C (Basic), pp. 113–118
Free-Response Tests, Forms 2A (Honors), 2B (Average), 2C (Basic), pp. 119–124
Performance Assessment, p. 125
Mid-Chapter Test, p. 126
Quizzes A-D, pp. 127–128
Standardized Test Practice, p. 129
Cumulative Review, pp. 130–131

ENHANCING THE CHAPTER

Examples of some of the materials for enhancing Chapter 5 are shown below.

Multicultural Activity Masters, pp. 9, 10

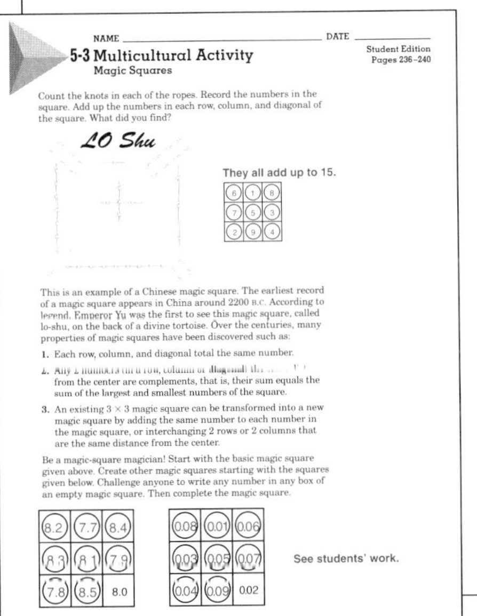

NAME ____________ DATE ________

5-3 Multicultural Activity
Magic Squares

Student Edition Pages 236–240

Count the knots in each of the ropes. Record the numbers in the square. Add up the numbers in each row, column, and diagonal of the square. What did you find?

LO Shu

They all add up to 15.

6	1	8
7	5	3
2	9	4

This is an example of a Chinese magic square. The earliest record of a magic square appears in China around 2200 B.C. According to legend, Emperor Yu was the first to see this magic square, called lo-shu, on the back of a divine tortoise. Over the centuries, many properties of magic squares have been discovered such as:

1. Each row, column, and diagonal total the same number.
2. Any 2 numbers in a row, column or diagonal that are equidistant from the center are complements, that is, their sum equals the sum of the largest and smallest numbers of the square.
3. An existing 3 × 3 magic square can be transformed into a new magic square by adding the same number to each number in the magic square, or interchanging 2 rows or 2 columns that are the same distance from the center.

Be a magic-square magician! Start with the basic magic square given above. Create other magic squares starting with the squares given below. Challenge anyone to write any number in any box of an empty magic square. Then complete the magic square.

8.2	7.7	8.4
8.3	8.1	7.9
7.8	8.5	8.0

0.08	0.01	0.06
0.03	0.05	0.07
0.04	0.09	0.02

See students' work.

Real-World Applications, 10, 11

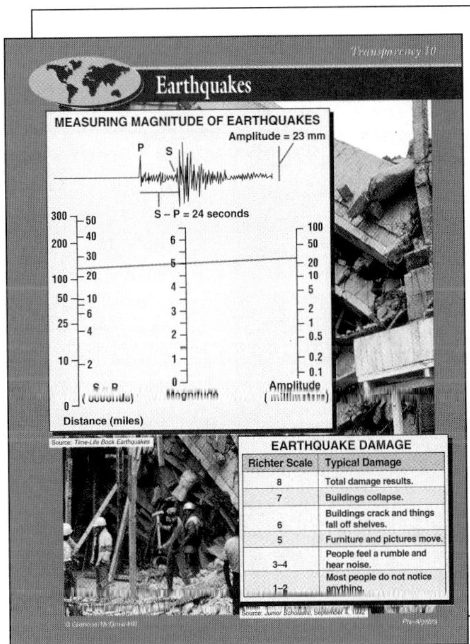

Transparency 10
Earthquakes
MEASURING MAGNITUDE OF EARTHQUAKES
Amplitude = 23 mm
P
S
S – P = 24 seconds
Distance (miles)
S – P (seconds)
Magnitude
Amplitude (millimeters)
EARTHQUAKE DAMAGE
Richter Scale
Typical Damage
8 Total damage results.
7 Buildings collapse.
6 Buildings crack and things fall off shelves.
5 Furniture and pictures move.
3–4 People feel a rumble and hear noise.
1–2 Most people do not notice anything.
Source: Time-Life Book Earthquakes
Pre-Algebra

Graphing Calculator Masters, p. 5

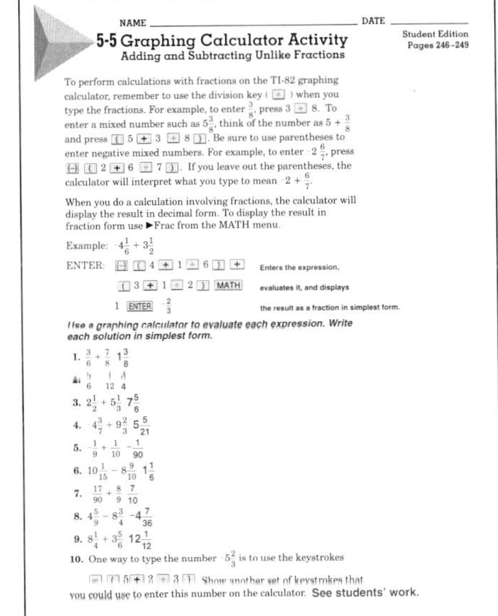

NAME ____________ DATE ________

5-5 Graphing Calculator Activity
Adding and Subtracting Unlike Fractions

Student Edition Pages 246–249

To perform calculations with fractions on the TI-82 graphing calculator, remember to use the division key (÷) when you type the fractions. For example, to enter $\frac{3}{8}$, press 3 ÷ 8. To enter a mixed number such as $5\frac{3}{8}$, think of the number as $5 + \frac{3}{8}$ and press (5 + 3 ÷ 8). Be sure to use parentheses to enter negative mixed numbers. For example, to enter $-2\frac{6}{7}$, press (–) (2 + 6 ÷ 7). If you leave out the parentheses, the calculator will interpret what you type to mean $-2 + \frac{6}{7}$.

When you do a calculation involving fractions, the calculator will display the result in decimal form. To display the result in fraction form use ▶Frac from the MATH menu.

Example: $4\frac{1}{6} + 3\frac{1}{2}$

ENTER: ((4 + 1 ÷ 6) + ... Enters the expression, ... MATH evaluates it, and displays ... ENTER $\frac{2}{3}$ the result as a fraction in simplest form.

Use a graphing calculator to evaluate each expression. Write each solution in simplest form.

1. $\frac{3}{6} + \frac{7}{8}$ $1\frac{3}{8}$
2. ...
3. $2\frac{1}{2} + 5\frac{1}{3}$ $7\frac{5}{6}$
4. $4\frac{3}{7} + 9\frac{2}{3}$ $5\frac{5}{21}$
5. $\frac{1}{9} + \frac{1}{10}$ $-\frac{1}{90}$
6. $10\frac{1}{15} - 8\frac{9}{10}$ $1\frac{1}{6}$
7. $\frac{17}{90} + \frac{8}{9}$ $\frac{7}{10}$
8. $4\frac{5}{9} - 8\frac{3}{4}$ $-4\frac{7}{36}$
9. $8\frac{1}{4} + 3\frac{5}{6}$ $12\frac{1}{12}$
10. One way to type the number $-5\frac{2}{3}$ is to use the keystrokes ... Show another set of keystrokes that you could use to enter this number on the calculator. See students' work.

Tech Prep Application Masters, pp. 9, 10

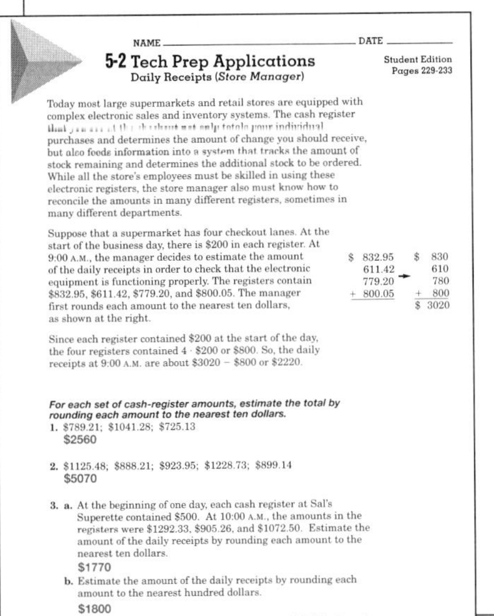

NAME ____________ DATE ________

5-2 Tech Prep Applications
Daily Receipts (Store Manager)

Student Edition Pages 229–233

Today most large supermarkets and retail stores are equipped with complex electronic sales and inventory systems. The cash register that you see at the checkout not only totals your individual purchases and determines the amount of change you should receive, but also feeds information into a system that tracks the amount of stock remaining and determines the additional stock to be ordered. While all the store's employees must be skilled in using these electronic registers, the store manager also must know how to reconcile the amounts in many different registers, sometimes in many different departments.

Suppose that a supermarket has four checkout lanes. At the start of the business day, there is $200 in each register. At 9:00 A.M., the manager decides to estimate the amount of the daily receipts in order to check that the electronic equipment is functioning properly. The registers contain $832.95, $611.42, $779.20, and $800.05. The manager first rounds each amount to the nearest ten dollars, as shown at the right.

$ 832.95	$ 830
611.42	610
779.20	780
+ 800.05	+ 800
	$ 3020

Since each register contained $200 at the start of the day, the four registers contained 4 · $200 or $800. So, the daily receipts at 9:00 A.M. are about $3020 – $800 or $2220.

For each set of cash-register amounts, estimate the total by rounding each amount to the nearest ten dollars.

1. $789.21; $1041.28; $725.13
   $2560
2. $1125.48; $888.21; $923.95; $1228.73; $899.14
   $5070
3. a. At the beginning of one day, each cash register at Sal's Superette contained $500. At 10:00 A.M., the amounts in the registers were $1292.33, $905.26, and $1072.50. Estimate the amount of the daily receipts by rounding each amount to the nearest ten dollars.
   $1770
   b. Estimate the amount of the daily receipts by rounding each amount to the nearest hundred dollars.
   $1800
   c. By how much do your estimates in parts a and b differ from the actual daily receipts? $0.09; $29.91

Activity Masters, pp. 5, 19, 35, 36

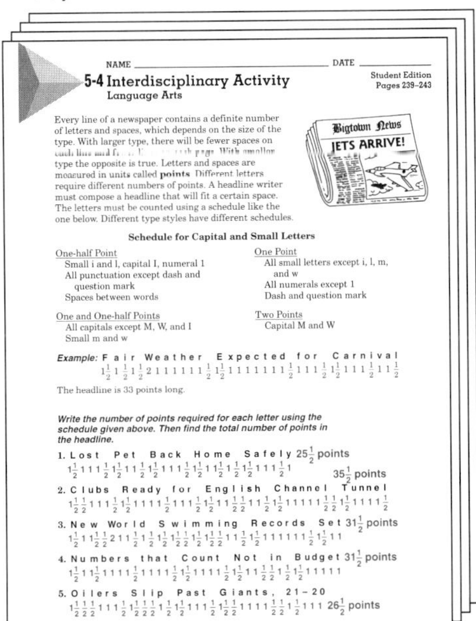

NAME ____________ DATE ________

5-4 Interdisciplinary Activity
Language Arts

Student Edition Pages 239–243

Every line of a newspaper contains a definite number of letters and spaces, which depends on the size of the type. With larger type, there will be fewer spaces on each line and fewer lines on each page. With smaller type the opposite is true. Letters and spaces are measured in units called **points**. Different letters require different numbers of points. A headline writer must compose a headline that will fit a certain space. The letters must be counted using a schedule like the one below. Different type styles have different schedules.

Bigtown News
JETS ARRIVE!

Schedule for Capital and Small Letters

One-half Point
Small i and l, capital I, numeral 1
All punctuation except dash and question mark
Spaces between words

One Point
All small letters except i, l, m, and w
All numerals except 1
Dash and question mark

One and One-half Points
All capitals except M, W, and I
Small m and w

Two Points
Capital M and W

Example: F a i r W e a t h e r E x p e c t e d f o r C a r n i v a l

The headline is 33 points long.

Write the number of points required for each letter using the schedule given above. Then find the total number of points in the headline.

1. L o s t P e t B a c k H o m e S a f e l y $25\frac{1}{2}$ points
2. C l u b s R e a d y f o r E n g l i s h C h a n n e l T u n n e l $35\frac{1}{2}$ points
3. N e w W o r l d S w i m m i n g R e c o r d s S e t $31\frac{1}{2}$ points
4. N u m b e r s t h a t C o u n t N o t i n B u d g e t $31\frac{1}{2}$ points
5. O i l e r s S l i p P a s t G i a n t s , 2 1 – 2 0 $26\frac{1}{2}$ points

COOPERATIVE LEARNING

Math Lab and Modeling Math Masters, pp. 61–62, 80

NAME ____________ DATE ________

5-9 Modeling Math Activity
Triangular and Square Numbers

Student Edition Pages 260–264

materials: geoboard pattern

Triangular and square numbers had great meaning for the ancient Greeks. Use the tables on this and the following page and the following notation to help discover the patterns for these numbers. Draw the figures on the geoboard pattern.

T refers to a triangular number.

S refers to a square number.

n refers to any term in the sequence. *n* – 1 is the term before the *n*th term. *n* + 1 is the term after the *n*th term.

1. Fill in the table to determine the first five triangular numbers.

n	Number of Dots Added	Total Number of Dots
1	1	T_1 = ? 1
2	2	T_2 = ? 3
3	3	T_3 = ? 6
4	4	T_4 = ? 10
5	5	T_5 = ? 15

2. What is the value of T_n? $T_{n-1} + n$

MAKING MATHEMATICS RELEVANT

This two-page introduction to the chapter provides students with an opportunity to explore other disciplinary topics and their applications to mathematics.

Background Information

Math and Fitness in the News
U. S. consumption of take-out food increased sharply (13%) in the 1980s as hectic lifestyles gave people less and less time to prepare meals. But problems arose. Take-out foods are usually high in fat, Calories and sodium; they are usually eaten quickly, so more is consumed; all the food tends to be eaten, instead of a portion being discarded or saved for later consumption. In 1994, Americans ate nearly 64 pounds of beef per person—the highest level in five years.

CHAPTER 5

Rationals: Patterns in Addition and Subtraction

TOP STORIES in Chapter 5

In this chapter, you will:

- estimate and find sums and differences of rational numbers,
- solve equations and inequalities involving rational numbers,
- solve problems using logical reasoning, and
- find terms of an arithmetic sequence.

MATH AND FITNESS IN THE NEWS

Is thin *really* in?

Source: TIME, January 16, 1995

The 1980s were not the healthy years that diet and fitness experts had hoped. According to a long-term federal study, Americans gained weight and increased their related health risks in that decade. The percentage of Americans who are seriously overweight jumped to about one-third during the 1980s. About *58 million* people in the U.S. are now classified as obese, that is they weigh at least 20% more than their ideal body weight. The percentage of overweight teenagers also rose, from 15% in the early 1970s to 21% in 1991, a 40% increase. As a nation, Americans are eating too much food, eating unhealthy food, and not exercising enough. Possible causes include heavy advertising by the food industry, eating out more often, eating more fast food, and spending more time in front of televisions and computers instead of exercising.

Putting It into Perspective
1750
1800
1850
1762
The fourth Earl of Sandwich orders slices of meat and cheese served between slices of bread; the sandwich is invented.
1853
The potato chip is invented by American Indian chef George Crum in Saratoga Springs, NY.
222 Chapter 5 Rationals: Patterns in Addition and Subtraction

Putting It into Perspective

Standardized nutrition labels appeared on packaged foods in 1994 as a result of federal law. The goal was to provide food consumers with more information so they could make informed purchase decisions. The labels include data about Calories, fiber content and several nutrients including fat, sodium and cholesterol.

On the Lighter Side

"Boy, Dad, you're sure being all you can be!"

interNET CONNECTION For up-to-date information on nutrition and weight, visit: **www.glencoe.com/sec/math/prealg/mathnet**

Statistical Snapshot

How hard do you exercise?
If you monitor how hard you exert yourself, you can estimate the number of calories you burn. The table below shows the number of calories a 130-pound person burns in an hour of different activities.

Exertion Level	Calories per hour	Type of activity
No exertion	60	sleeping, watching television reading
Almost no exertion	120	eating, doing desk work, driving washing dishes, strolling
Very light exertion	180	bowling, slowly walking walking down stairs
Light exertion	240	mopping, golfing with a cart leisurely biking, raking leaves
Moderate exertion	300	walking, riding a stationary bike playing softball, calisthenics
Vigorous exertion	360	Leisurely swimming, mowing grass shoveling light snow
Heavy exertion	420	jogging, shoveling deep snow playing tennis, casual racquetball
Extreme exertion	560 or more	sprinting, jogging uphill jumping rope

Source: *The New England Journal of Medicine*

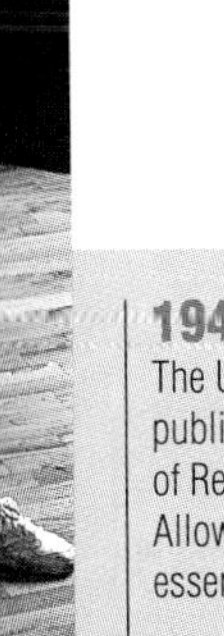

1900 — **1950** — **2000**

1904
The ice-cream cone is born at the St. Louis World's Fair when an ice cream vendor runs out of dishes and uses rolled waffles as a substitute.

1941
The United States publishes their first set of Recommended Daily Allowances (RDAs) of essential nutrients.

1994
Nutrition labels appear on all food packages as a result of U.S. federal law.

1995
Study shows alarming increase in number of overweight Americans.

On the Lighter Side

As people grow older, they tend to be less physically active and their bodies' metabolisms slow down. Unless they are careful about their food intake, they will find it easier to gain weight.

Statistical Snapshot

The chart shows how hard a person has to work or exercise to burn off various amounts of Calories. Students may be happy to know that they burn some Calories even when they are resting or sleeping. However, if they engage in some of the activities shown, they can burn off several times as many Calories.

Cooperative Learning

Chapter Projects Two chapter projects are included in the *Investigation and Project Masters*. In Chapter Project A, students extend the topic in the Chapter Opener. In Chapter Project B, students explore scoring rules in different sports. A student page and a parent letter are provided for each Chapter Project.

interNET CONNECTION

Glencoe has made every effort to ensure that the website links for *Pre-Algebra* at www.glencoe.com/sec/math/prealg/mathnet are current and contain appropriate content. However, these website links are not under Glencoe's control.

Investigations and Projects Masters, p. 45

NAME ____________ DATE ________

CHAPTER **5 Project A**
Feel the Burn

Everybody, from a fit muscular Olympic decathlon champion to a small child learning to walk, takes in calories by eating and burns them off through physical activity. Gaining and losing weight is a function of the kinds and quantities of foods you eat and the amount of exercise you get. On average, do you think you burn off more calories than you ingest in a day? Find out.

1. Begin by doing research to gather more information about the calories per hour burned doing various activities. Supplement the information in the Statistical Snapshot with data from other sources. You may find it useful to talk with a nutritionist or doctor to learn what factors other than weight can affect how fast a person burns calories.
2. Next, find out about the calorie content of the foods you commonly eat. You can find this information in several sources, including nutrition labels on foods. *The Wellness Encyclopedia of Food and Nutrition* (Random House, 1992) is one good source of nutrition information.
3. Now that you have your calorie data, you can begin. Choose a day and keep track of everything you eat that day. Be sure to include snacks and all drinks, too. Also, record all the activities you did that day, and the amount of time you spent doing them. Don't forget all the casual walking you do and any stairs you climb.
4. Make one table listing the foods you ate and the calories you took in. To estimate with any accuracy the number of calories you ingest, you'll need to gather information about serving sizes and figure out how these amounts compare with the portions of food you ate.
5. Make another table in which you list your activities and the calories you burned. Group each activity as light, moderate, vigorous, and so on. Use proportional reasoning to determine how many calories you, at your present weight, burned doing each activity. Estimate the number of calories you burned at activities you did that are not listed in the table you found.
6. Total the calories for both lists. Summarize the data your tables provide. Then organize the data the class has gathered to find facts on daily calorie intake *vs.* calories burned for (1) a typical class member, (2) a typical boy in the class, and (3) a typical girl in the class. Display the data in a double bar graph and prepare a summary of what the graph shows. What conclusions did you draw about calorie intake *vs.* calories burned?

Chapter 5 **223**

5-1 LESSON NOTES

NCTM Standards: 1-6

Instructional Resources

- Study Guide Master 5-1
- Practice Master 5-1
- Enrichment Master 5-1
- Group Activity Card 5-1

Transparency 5-1A contains the 5-Minute Check for this lesson; **Transparency 5-1B** contains a teaching aid for this lesson.

Recommended Pacing	
Standard Pacing	Day 1 of 13
Honors Pacing	Day 1 of 12
Block Scheduling*	Day 1 of 6 (along with Lesson 5-2)

*For more information on pacing and possible lesson plans, refer to the ***Block Scheduling Booklet***.

1 FOCUS

5-Minute Check
(over Chapter 4)

Find the greatest common factor and the least common multiple of each pair of numbers or algebraic expressions.

1. 12, 30 **GCF, 6; LCM, 60**
2. $9a$, $17ab$ **GCF, a; LCM, $153ab$**

Find each product or quotient. Express your answer in exponential form, using positive exponents.

3. $3^3 \cdot 3^4$ 3^7
4. $\dfrac{(-5a)^4}{(-15a^2b)^3}$ $\dfrac{(5)}{(-3b)^3a^2}$
5. $\dfrac{1}{(2b^{-2})^{-3}}$ $\dfrac{2^3}{b^6}$

5-1 Rational Numbers

Setting Goals: *In this lesson you'll identify and compare rational numbers and rename decimals as fractions.*

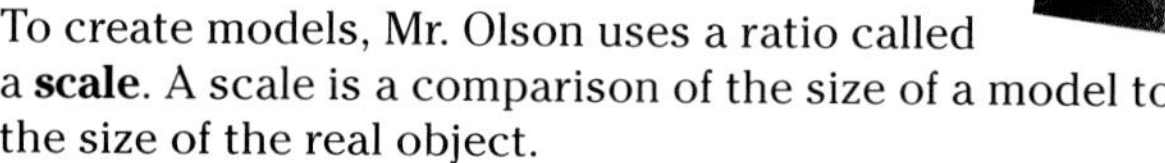

Modeling a Real-World Application: Movies

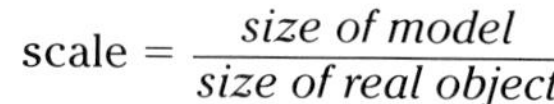

Were you frightened by the dinosaurs of *Jurassic Park*? Or amazed by the flight of the Enterprise in *Star Trek*? Then you have Jeff Olson to thank. Mr. Olson created miniature models for these and other blockbuster movies. Movie makers often use models to simulate situations that would be difficult or impossible to do in real life.

To create models, Mr. Olson uses a ratio called a **scale**. A scale is a comparison of the size of a model to the size of the real object.

$$\text{scale} = \frac{\textit{size of model}}{\textit{size of real object}}$$

The scale tells how much smaller or larger the model has to be than the real object. Suppose Mr. Olson was building a 15-inch model of a Bigfoot that would be 120 inches tall in real life. The scale would be $\frac{15}{120}$ or $\frac{1}{8}$. That means absolutely everything on the model is $\frac{1}{8}$ the size of the same item on the original Bigfoot. Numbers such as $\frac{1}{8}$ that can be written as fractions are called **rational numbers**.

Learning the Concept

Numbers can be organized into sets. One set is the **whole numbers**.

Whole Numbers
This set includes 0, 1, 2, 3, . . . It also includes any number that can be written as a whole number, such as $\frac{5}{5}$ or $\frac{9}{1}$.

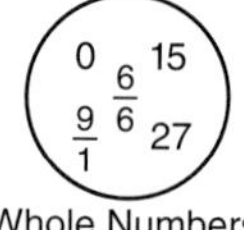

Whole Numbers

Another set of numbers is the **integers**.

Integers
This set includes . . . , −2, −1, 0, 1, 2, . . . Notice that all whole numbers are included in the set of integers.

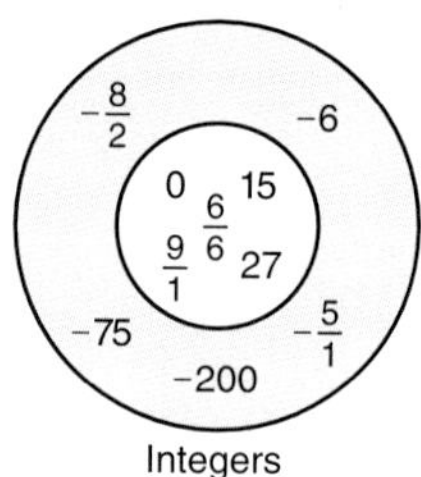

Integers

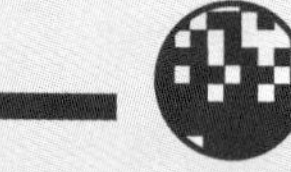

Alternative Teaching Strategies

Student Diversity Have students work in pairs. One student secretly creates a decimal on a calculator by dividing two numbers between 0 and 50. The other student looks at the display and gets three chances to divide two numbers to recreate the decimal. Reverse roles and repeat.

Refer to the application at the beginning of the lesson. The model of Bigfoot had a scale of $\frac{1}{8}$. Both 1 and 8 are integers, but $\frac{1}{8}$ is not. The number system is extended to include numbers like $\frac{1}{8}$. Numbers like $\frac{1}{8}$ are rational numbers.

Definition of a Rational Number

Any number that can be expressed in the form $\frac{a}{b}$, where a and b are integers and $b \neq 0$, is called a rational number.

Rational Numbers
This set includes common fractions, such as $\frac{1}{4}$. It also includes mixed numbers, decimals, integers, and whole numbers because all these numbers can be written in the form $\frac{a}{b}$.

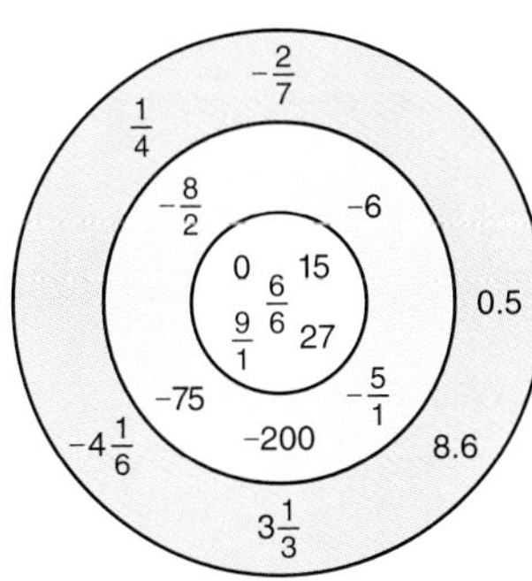

Rational Numbers

You can review simplifying fractions in Lesson 4-6.

Some decimals are rational numbers. Decimals either terminate or they go on forever. The decimal 3.74 is an example of a **terminating decimal**. Every terminating decimal can be written as a fraction with a denominator of 10, 100, 1000, and so on. For example, $0.75 = \frac{75}{100}$ or $\frac{3}{4}$, and $1.3 = 1\frac{3}{10}$ or $\frac{13}{10}$. So terminating decimals are always rational numbers.

Example 1 Connection — Biology

The average adult's fingernails grow 0.125 inch each month. Write 0.125 as a fraction.

$0.125 = \frac{125}{1000}$

$= \frac{1}{8}$ *Simplify. The GCF of 125 and 1000 is 125.*

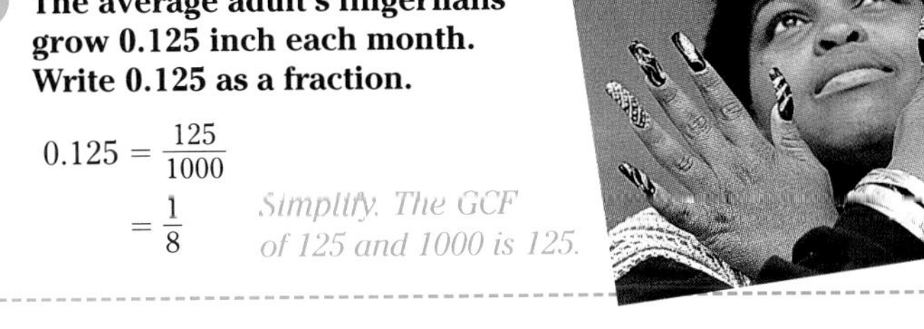

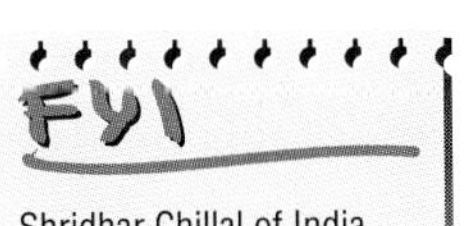

Shridhar Chillal of India has the longest fingernails in the world. On March 3, 1993, the nails on his left hand were a total of 205 inches long.

Decimals like 0.4444444444 . . . are called **repeating decimals**. Because it is inconvenient to write all of these digits, you can use the **bar notation** $0.\overline{4}$ to indicate that the 4 repeats forever. Here are some other examples.

$6.23232323232323\ldots = 6.\overline{23}$ The digits 23 repeat.

$8.4613613613613\ldots = 8.4\overline{613}$ The digits 613 repeat.

Repeating decimals can always be written as fractions. For example, $0.\overline{4}$ is equivalent to $\frac{4}{9}$. So, repeating decimals are always rational numbers.

Connection to Algebra

Example 2 shows how to rename repeating decimals as fractions.

Motivating the Lesson

Questioning Write the following rational numbers on the chalkboard: $-8, -3\frac{1}{3}, -3, -\frac{3}{5}, 0, 0.3, \frac{1}{2}, 2\frac{2}{3}, 8$.

- Ask students how these numbers can be classified into two different types of numbers. **Sample answer: integers and nonintegers**
- Ask if each of the two classifications of numbers can be organized in any way. **Sample answer: ordered from least to greatest**

2 TEACH

In-Class Example

For Example 1

Hair grows at a rate of about 0.4375 inch per month. Write 0.4375 as a fraction. $\frac{7}{16}$

The oldest fingernails were found on an intact body from the late Stone Age called the Iceman from the Similaun. The body was discovered in the Austrian Alps in 1991 and is estimated to be at least 5000 years old.

Cooperative Learning

Numbered Heads Together Assign each group of four students a number. Then have students in each group count off from one to four. Write a number on the board or the overhead, such as $-1\frac{1}{2}$ or 1.53. Each group classifies that number. Randomly choose a number from 1 to 4 and students with that number raise their hands. Then choose a number from 1 to x, where x is the number of groups. The person chosen from x group is expected to respond. For more information on this strategy, see *Cooperative Learning in the Mathematics Classroom*, one of the titles in the Glencoe Mathematics Professional Series, p. 9.

Example 2 **Express each repeating decimal as a fraction or mixed number in simplest form.**

a. $0.\overline{5}$

Let $N = 0.555\ldots$ Then $10N = 5.555\ldots$ *Multiply N by 10 because one digit repeats.*

Subtract N from $10N$ to eliminate the repeating part.

$$\begin{aligned} 10N &= 5.555\ldots \quad \text{Recall } 10N - N = 9N. \\ -\ 1N &= 0.555\ldots \\ \hline 9N &= 5 \\ N &= \frac{5}{9} \quad \text{Divide each side by 9.} \end{aligned}$$

Therefore $0.\overline{5} = \frac{5}{9}$.

b. 3.363636 . . .

Let $N = 3.363636\ldots$ Then $100N = 336.3636\ldots$. *Multiply N by 100 because two digits repeat.*

$$\begin{aligned} 100N &= 336.3636\ldots \\ -\ \ \ N &= \ \ \ 3.3636\ldots \\ \hline 99N &= 333 \\ N &= \frac{333}{99} \text{ or } 3\frac{4}{11} \end{aligned}$$

So $3.363636\ldots = 3\frac{4}{11}$.

THINK ABOUT IT

What multiplier should you use to express $0.\overline{718}$ as a fraction?

1000

Decimals that do not terminate and do not repeat such as 4.252627 . . . , cannot be written as fractions. So they are not rational numbers.

Thus, the set of rational numbers includes whole numbers, integers, fractions, terminating decimals, and repeating decimals.

Example 3 **Name the set(s) of numbers to which each number belongs.**

a. −7 −7 is an integer and a rational number.

b. $-3\frac{2}{5}$ Since $-3\frac{2}{5}$ can be written as $-\frac{17}{5}$, it is a rational number. It is not a whole number nor is it an integer.

c. 0.545556 . . . This is a non-terminating, non-repeating decimal. So it is not rational.

d. 1.469 1.469 is a terminating decimal. It is a rational number.

Like integers, rational numbers can be graphed on a number line. Examples of rational numbers graphed on a number line are shown below.

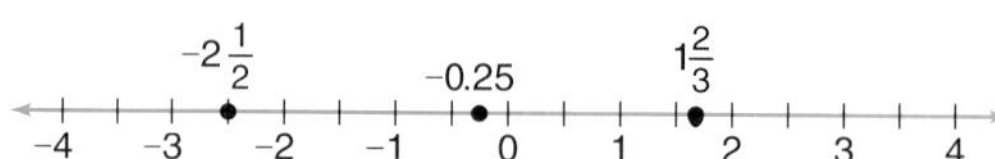

You can use a number line to compare rational numbers.

In-Class Examples

For Example 2

Express each repeating decimal as a fraction or mixed number in simplest form.

a. 0. **$\frac{7}{9}$**

b. 2.212121... **$2\frac{7}{33}$**

For Example 3

Name the set(s) of numbers to which each number belongs.

a. 9 **whole number, integer, rational number**

b. $-1\frac{1}{2}$ **rational number**

c. 0.121131114... **none of these**

d. 6.78 **rational number**

For Example 4

Use a number line to choose the greater number from each pair.

a. $-\frac{1}{3}, -\frac{1}{4}$ **$-\frac{1}{4}$**

−1 $-\frac{3}{4}$ $-\frac{1}{2}$ $-\frac{1}{3}$ $-\frac{1}{4}$ 0

b. 0.9, $\frac{4}{5}$ **0.9**

0 $\frac{4}{5}$ 0.9 1

Study Guide Masters, p. 36

NAME ______________________ DATE __________

5-1 Study Guide
Rational Numbers

Student Edition Pages 224–228

The set of **whole numbers** includes 0, 1, 2, 3, ...

The set of **integers** includes . . . , −2, −1, 0, 1, 2, ...

Any number that can be written as a fraction is called a **rational number**.

To express a decimal as a fraction or mixed number, write the digits of the decimal as the numerator and use the appropriate power of ten (10, 100, 1000, · · ·) as the denominator. Simplify if necessary.

Examples: Express each decimal as a fraction in simplest form.

a. $0.125 = \frac{125}{1000} = \frac{1}{8}$

So, $0.125 = \frac{1}{8}$.

b. $0.\overline{7}$

Let $N = 0.777\ldots.$ **Multiply *N* by 10 since one digit repeats.**

Then $10N = 7.777\ldots.$

$$\begin{aligned} 10N &= 7.777\ldots \\ -\ 1N &= 0.777\ldots \\ \hline 9N &= 7 \\ N &= \frac{7}{9} \end{aligned}$$

Subtract *N* from 10*N* to eliminate the repeating part.

Express each decimal as a fraction or mixed number in simplest form.

1. 0.8 **$\frac{4}{5}$**
2. −0.4 **$-\frac{2}{5}$**
3. 0.09 **$\frac{9}{100}$**
4. $0.\overline{48}$ **$\frac{16}{33}$**
5. 0.15 **$\frac{3}{20}$**
6. 0.25 **$\frac{1}{4}$**
7. $-0.\overline{81}$ **$-\frac{81}{99}$**
8. 0.88 **$\frac{22}{25}$**
9. $0.\overline{6}$ **$\frac{2}{3}$**
10. 0.845 **$\frac{169}{200}$**
11. $5.\overline{36}$ **$5\frac{12}{33}$**
12. $7.\overline{16}$ **$7\frac{16}{99}$**

Name the set(s) of numbers to which each number belongs. (Use the symbols W = whole numbers, I = integers, and R = rationals.)

13. 2 **W, I, R**
14. −4 **I, R**
15. 4.169 **R**
16. $-\frac{1}{4}$ **R**
17. $-2\frac{3}{5}$ **R**
18. 0.32 **R**
19. $\frac{18}{3}$ **W, I, R**
20. −8.0 **I, R**

Group Activity Card 5-1

Line Them Up Group Activity **5-1**

MATERIALS: Two dice • coin • colored, four-foot long piece of heavy paper.

SETUP: Draw a number line from -6 to +6 on the long piece of paper. (If your teacher wants to use the number line over and over, cover it with contact paper and use a special pen to write on it.) There should be a space of about three inches between each number.

To begin, roll the two dice and flip the coin. From the two dice, form a rational number. If the coin is heads, the number is positive. If it is tails, the number is negative. Next graph this number on the number line you have made. Below this point, give two other names for this rational number. Your partner checks to make sure these are correct.

 Pre-Algebra

Example 4 Use a number line to choose the greater number from each pair.

a. $-\frac{2}{3}, -\frac{1}{2}$

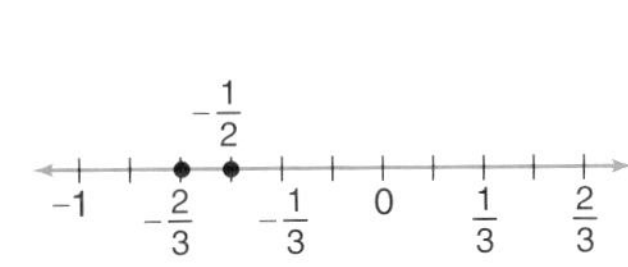

$-\frac{1}{2}$ is greater or $-\frac{1}{2} > -\frac{2}{3}$.

b. $0.3, \frac{1}{3}$

$0.3 = \frac{3}{10}$

0.3 $\frac{1}{3}$

0 $\frac{1}{10}$ $\frac{1}{5}$ $\frac{3}{10}$ $\frac{2}{5}$

$\frac{1}{3}$ is greater or $\frac{1}{3} > 0.3$.

Checking Your Understanding

Communicating Mathematics

Read and study the lesson to answer these questions. 1–2, 4. See margin.

1. **Explain** what a rational number is.
2. Are all integers also rational numbers? Explain.
3. Use $\frac{2}{3}, \frac{4}{5}$, and < to write a true statement. $\frac{2}{3} < \frac{4}{5}$
4. **Explain** why 0.13 and $0.\overline{13}$ are rational numbers.
5. Use a ruler to draw a number line. Mark tick marks that are 1 centimeter apart from −8 to 8. **5a. See margin.**
   a. Mark $3\frac{1}{5}$, $-1\frac{1}{3}$, 0.7, −5.6, $4\frac{1}{3}$, 4.3, and −2.9 on the number line.
   b. Which of the points indicates the greatest number? $4\frac{1}{3}$
   c. Which number in the list is the least? **−5.6**

MODELING MATHEMATICS

MATERIALS

ruler

Guided Practice

Express each decimal as a fraction or mixed number in simplest form.

6. 0.8 $\frac{4}{5}$
7. 0.05 $\frac{1}{20}$
8. −9.64 $-9\frac{16}{25}$
9. $0.\overline{2}$ $\frac{2}{9}$

For Exercises 10–13, W = whole numbers, I = integers, and R = rationals.

Name the set(s) of numbers to which each number belongs.

10. −3 **I, R**
11. $-8\frac{3}{4}$ **R**
12. 15 **W, I, R**
13. 0.414243 . . . **none**

interNET CONNECTION

For the latest on models, visit: **www.glencoe.com/sec/math/prealg/mathnet**

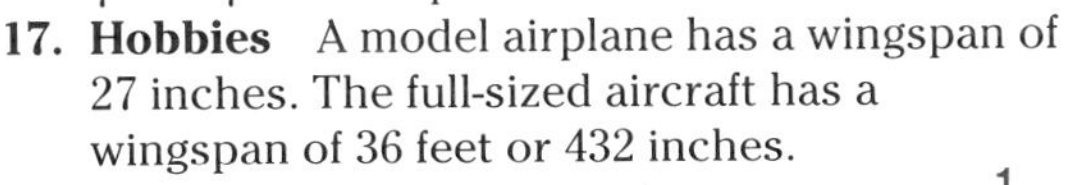

Replace each ● with <, >, or = to make each sentence true. Use a number line if necessary.

14. $\frac{3}{7}$ ● $-\frac{3}{7}$ **>**
15. $\frac{3}{4}$ ● 0.75 **=**
16. −3.78 ● −3.88 **>**
17. **Hobbies** A model airplane has a wingspan of 27 inches. The full-sized aircraft has a wingspan of 36 feet or 432 inches.
    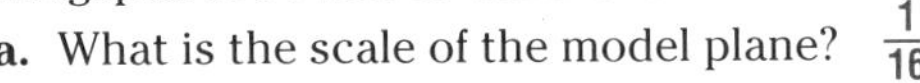
    a. What is the scale of the model plane? $\frac{1}{16}$
    b. Is the scale a whole number, an integer, or a rational number? Explain.

17b. Rational; it is a fraction.

Exercises: Practicing and Applying the Concept

Independent Practice

Express each decimal as a fraction or mixed number in simplest form.

A
18. 0.4 $\frac{2}{5}$
19. −0.7 $-\frac{7}{10}$
20. 0.23 $\frac{23}{100}$
21. 0.57 $\frac{57}{100}$
22. 0.48 $\frac{12}{25}$
23. 0.03 $\frac{3}{100}$
24. $0.\overline{8}$ $\frac{8}{9}$
25. −0.333 . . . $-\frac{1}{3}$

B
26. −0.51 $-\frac{51}{100}$
27. $0.\overline{25}$ $\frac{25}{99}$
28. 0.375 $\frac{3}{8}$
29. $2.\overline{34}$ $2\frac{34}{99}$

3 PRACTICE/APPLY

Checking Your Understanding

Exercises 1–17 are designed to help you assess your students' understanding through reading, writing, speaking, and modeling. You should work through Exercises 1–5 with your students and then monitor their work on Exercises 6–17.

Additional Answers

1. Sample answer: a rational number is a number that can be written in the form of a common fraction. Some sample rationals: 4.8, 7.1, $1\frac{3}{4}$.
2. Sample answer: Yes; integers are always written with a denominator of 1.
4. 0.13 is a rational number because it can be written in the form $\frac{13}{100}$. $0.\overline{13}$ is rational because it can be written as $\frac{13}{99}$.

5a. Sample answer:

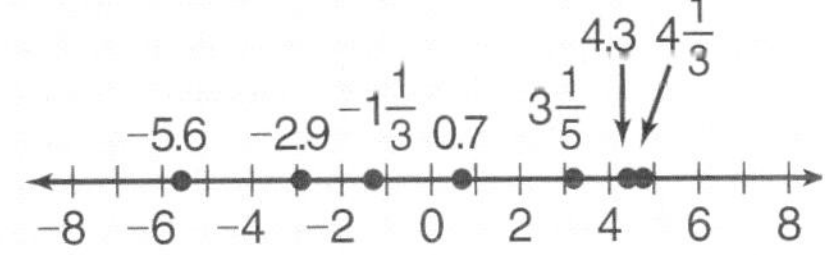

Practice Masters, p. 36

NAME ______ DATE ______

5-1 Practice
Rational Numbers

Student Edition Pages 224–228

Express each decimal as a fraction or mixed number in simplest form.

1. 0.4 $\frac{2}{5}$
2. −0.9 $-\frac{9}{10}$
3. 0.06 $\frac{3}{50}$
4. $-0.\overline{5}$ $-\frac{5}{9}$
5. 0.15 $\frac{3}{20}$
6. $0.\overline{48}$ $\frac{48}{99}$
7. 0.79 $\frac{79}{100}$
8. −0.755 $-\frac{151}{200}$
9. −0.125 $-\frac{1}{8}$
10. 0.64 $\frac{16}{25}$
11. −0.95 $-\frac{19}{20}$
12. 0.99 $\frac{99}{100}$
13. −1.5 $-1\frac{1}{2}$
14. $9.\overline{08}$ $9\frac{8}{99}$
15. 5.25 $5\frac{1}{4}$

Name the set(s) of numbers to which each number belongs. (Use the symbols W = whole numbers, I = integers, and R = rationals.)

16. 0 W, I, R
17. −0.15 R
18. −5 I, R
19. −10 I, R
20. $\frac{28}{4}$ W, I, R
21. $21\frac{1}{2}$ R
22. 0.13 R
23. $-1\frac{1}{3}$ R
24. −625.0 I, R
25. 0.14159 . . . none
26. $-\frac{2}{3}$ R
27. 2.11 R

Write >, <, or = in each box to make a true sentence.

28. $\frac{1}{3}$ > $-\frac{1}{3}$
29. $-\frac{5}{4}$ = −1.25
30. 0.6666 . . . > $\frac{3}{5}$
31. 0.26 = 0.26
32. $\frac{9}{11}$ < 0.9
33. 0.3 > $\frac{1}{8}$
34. −1.6 = $-1\frac{3}{5}$
35. 5.8 > 5.7
36. $\frac{10}{16}$ > $-\frac{5}{8}$

Reteaching

Using Models Create a large number line on the floor. Label the number line from −1 to 1, showing marks for thirds, fourths, and tenths. Write a rational number between −1 and 1 on the board. Choose a student to first tell whether the number is a whole number, integer, or rational number. Then have that student locate the number on the number line. Repeat for another rational number. Finally, have students compare the two rational numbers.

Assignment Guide

Core: 19–57 odd, 58–66
Enriched: 18–52 even, 53–66

For **Extra Practice**, see p. 750.

The red A, B, and C flags, printed only in the Teacher's Wraparound Edition, indicate the level of difficulty of the exercises.

4 ASSESS

Closing Activity

Writing Have students write each number as you name a negative integer, a positive fraction, a mixed number, a decimal, and two more numbers of your choice. Have students classify each number and then write them in order from least to greatest.

Enrichment Masters, p. 36

NAME ______________ DATE ________

5-1 Enrichment
Irrational Numbers

Student Edition
Pages 224–228

Decimals in which a digit or group of digits repeats are called **repeating decimals**.

Example: $0.333\ldots = 0.\overline{3}$
$1.454545\ldots = 1.\overline{45}$ The bar indicates the repeating digit or digits.

Decimals in which only a 0 repeats are called **terminating decimals**.

Example: $0.5000\ldots = 0.5$
$42.19500\ldots = 42.195$

Consider the decimal 0.757757775 . . . . It does not terminate in zero nor does it have a group of digits that repeat. Numbers that are represented by nonterminating, nonrepeating decimals are **irrational numbers**.

Example: $\pi = 3.14159\ldots$
$\sqrt{5} = 2.236068\ldots$

Determine whether the given decimal is repeating or nonrepeating.

1. 0.373373337 . . . nonrepeating
2. 24.15971597 . . . repeating
3. 5.71571571 . . . repeating
4. 0.5795579555 . . . nonrepeating
5. 8.3121121112 . . . nonrepeating

Name the next three digits in the following irrational numbers.

6. 0.13141516 . . . 171
7. 0.96796679666 . . . 796

Name an irrational number between the given numbers. Answers will vary. Sample answers given.

8. 6.7 and 6.8 6.7121221222 . . .
9. 17.3 and 17.4 $10\sqrt{3}$
10. 0.1231233 . . . and 0.1231134111 . . . 0.1231152552555 . . .
11. 2.333 . . . and 2.444 . . . 2.43344333444 . . .

Find each sum.

12. 0.232232223 . . . + 0.323323332 . . . 0.555555555 . . . or $0.\overline{5}$
13. 0.131131113 . . . + 0.868868886 . . . 0.999999999 . . . or $0.\overline{9}$

For Exercises 30–41, W = whole numbers, I = integers, and R = rationals.

Name the set(s) of numbers to which each number belongs. 35. none

30. -14 I, R **31.** $-\frac{6}{7}$ R **32.** $-3\frac{7}{8}$ R **33.** $\frac{35}{7}$ W, I, R
34. 0.12 R **35.** $-3.3738\ldots$ **36.** -0.17 R **37.** 3.11 R
38. -10 I, R **39.** $29\frac{1}{2}$ R **40.** -32.0 I, R **41.** $0.1234\ldots$ none

Replace each ● with >, <, or = to make a true sentence. Use a number line if necessary.

C

42. $\frac{3}{4}$ ● $-\frac{3}{8}$ > **43.** $-6\frac{1}{2}$ ● $-6.\overline{5}$ > **44.** 9.9 ● $9.\overline{8}$ >
45. $\frac{3}{5}$ ● 0.6 = **46.** $\frac{1}{7}$ ● 0.222 . . . < **47.** $\frac{13}{26}$ ● $-\frac{1}{2}$ >
48. $0.3\overline{4}$ ● $0.\overline{34}$ > **49.** $\frac{1}{11}$ ● 0.1 < **50.** $-1\frac{1}{3}$ ● $-1.\overline{3}$ =

51. Which number is the greatest, $\frac{2}{7}$, $\frac{2}{9}$, or $\frac{2}{11}$? $\frac{2}{7}$

52. Which number is the greatest, $-\frac{3}{5}$, $-\frac{5}{6}$, or $-\frac{6}{7}$? $-\frac{3}{5}$

54. yes, $2\frac{3}{8} > 2.37$

Critical Thinking

53. Does $\frac{6}{2.4}$ name a rational number? Explain. Yes; $\frac{6}{2.4} = \frac{5}{2}$ or $\frac{60}{24}$.

54. A machinist made a steel peg 2.37 inches in diameter for a $2\frac{3}{8}$-inch diameter hole. Will the peg fit? How do you know?

Applications and Problem Solving

55. Measurement A gauge measured the thickness of a piece of metal as 0.025 inch. What fraction of an inch is this? $\frac{1}{40}$

56. Manufacturing A GLAD Garbage Bag has a thickness of 0.8 mils. This is 0.0008 inch. What fraction of an inch is this?

57. Engineering Ingrid Proctor-Fridia works for the United States Department of Defense checking to see that airplane parts conform to the blueprints. Blueprints are drawings of objects that show their scale.

a. The wingspan of an airplane on a blueprint is 26.9 centimeters. If the wingspan of the actual airplane is 8.07 meters, or 807 centimeters, what is the scale of the blueprints?

b. Is the scale a rational number? Explain.

56. $\frac{1}{1250}$
57a. $\frac{1}{30}$
57b. Yes; it can be written as a fraction.

Mixed Review

58. Physical Science The wavelength of visible red light is 7.50×10^{-7} meters. (Lesson 4-9)

a. Write the size of the wavelength using positive exponents. $\frac{7.50}{10^7}$ m

b. Write the size of the wavelength using decimals. 0.00000075 m

59. Find the least common multiple of $4x^2$, $3x$, and 5. (Lesson 4-7) $60x^2$

60. Is 945 divisible by 2, 3, 5, 6, or 10? (Lesson 4-1) 3, 5

61. Solve $-4b > 72$. (Lesson 3-7) $b < -18$

62. Solve $\frac{y}{12} = -7$. (Lesson 3-3) -84

63. Solve $q = \frac{323}{-17}$. (Lesson 2-8) -19

64. Is $17 > 40$ a true, false, or open sentence? (Lesson 1-9) false

65. Food A carton of orange juice contains 64 ounces. If the nutritional information lists the serving size as 8 ounces, how many servings are in the carton? (Lesson 1-3) 8 servings

66. Find the value of $24 \div (9 - 3)$. (Lesson 1-2) 4

Extension

Using Manipulatives Have students play a game with dominoes and a coin. Each student draws a domino and forms a proper fraction or "1" from the numbers on the domino, then tosses a coin. If it lands heads, the number is positive; if it lands tails, the number is negative. Students then compare fractions. The one with the greater number keeps both dominoes. In case of a tie, students draw again and the one with the greater number gets all four dominos. For variation, you may have students write numbers as decimals.

5-2 Estimating Sums and Differences

Setting Goals: *In this lesson, you'll estimate sums and differences of decimals and fractions.*

Modeling a Real-World Application: Business

The exercise craze of the 1980s changed the world of fitness *and* the world of fashion. Once seen only in the locker room, athletic shoes are now a part of the everyday wardrobe. The graph at the right shows the athletic shoe sales made to women, men, and children in 1993. What were the total sales of athletic shoes in 1993?

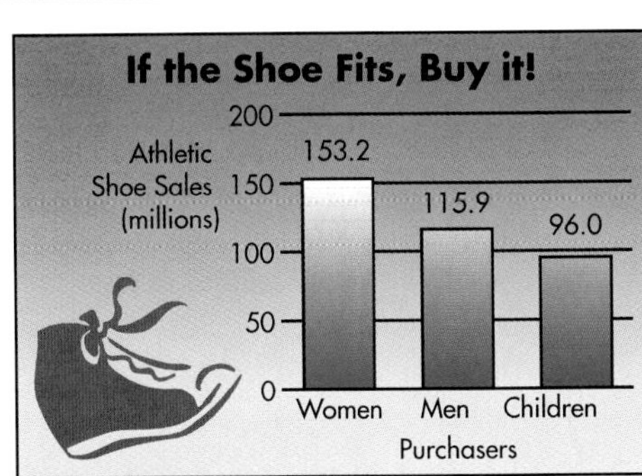

Source: Sporting Goods Manufacturers Association

Learning the Concept

Estimation is often used to provide a quick and easy answer when a precise answer is not necessary. It is also an excellent way to quickly see if your answer is reasonable or not.

To solve the problem about shoe sales, you can use *rounding* to estimate the answer. Round each number to a convenient place-value position. Often the greatest place-value position is used. Then complete the operation.

To find the total sales, round each of the sales figures to the greatest place value that they all share and then add.

153.2	→	150	*Round to the nearest ten.*
115.9		120	
+ 96.0		+100	
		370	

There were approximately 370 million pairs of athletic shoes sold in 1993.

Example  **Estimate using rounding.**

a. 8.890 + 15.98

8.890 + 15.98 → 9 + 16 = 25
8.890 + 15.98 is about 25.

b. 132.62 − 45.81

132.62 − 45.81 → 130 − 50 = 80
132.62 − 45.81 is about 80.

Before you use a calculator to compute an answer, it is a good idea to estimate the answer. This helps make sure that you entered the numbers correctly.

Classroom Vignette

"We play *Supermarket Challenge*. Put prices on about 12 items from the grocery store (empty boxes work forever). Then have several students each estimate which items could be purchased for a certain amount; for example, $8. Allow 30 seconds for the activity. Students enjoy competing to see who can get the closest."

Richard Whited
Kecoughtan High School
Hampton, VA

5-2 LESSON NOTES

NCTM Standards: 1-4, 7, 10

Instructional Resources

- Study Guide Master 5-2
- Practice Master 5-2
- Enrichment Master 5-2
- Group Activity Card 5-2
- Real-World Applications, 10
- Tech Prep Applications Masters, p. 9

Transparency 5-2A contains the 5-Minute Check for this lesson; **Transparency 5-2B** contains a teaching aid for this lesson.

Recommended Pacing	
Standard Pacing	Day 2 of 13
Honors Pacing	Day 2 of 12
Block Scheduling*	Day 1 of 6 (along with Lesson 5-1)

*For more information on pacing and possible lesson plans, refer to the ***Block Scheduling Booklet***.

1 FOCUS

5-Minute Check
(over Lesson 5-1)

Express each decimal as a fraction or mixed number in simplest form.

1. 0.3 $\frac{3}{10}$
2. −1.5 $-1\frac{1}{2}$
3. $5.\overline{4}$ $5\frac{4}{9}$

Replace each ● with <, >, or = to make a true statement.

4. -0.5 ● $-\frac{1}{2}$ =
5. $2\frac{2}{3}$ ● $2\frac{3}{4}$ <

Motivating the Lesson

Situational Problem Make a transparency of the bar graph below or one from a newspaper. Point out that exact numbers cannot be determined from the graph and ask questions such as "How many more students were there per computer in 1984–1985 than in 1994–1995?"

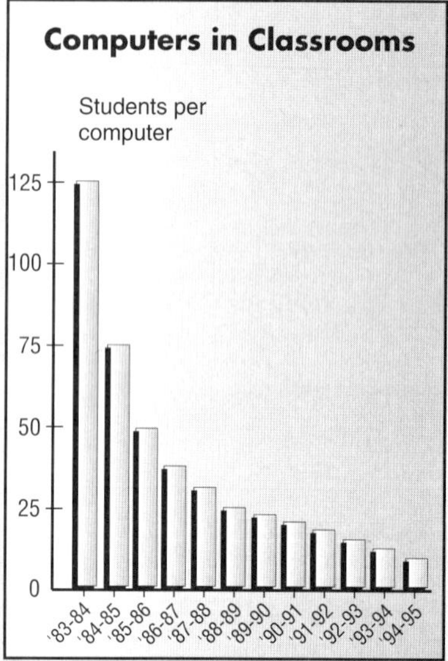

2 TEACH

In-Class Examples

For Example 1
Estimate using rounding.

a. 9.243 + 13.64 **23**
b. 96.9015 − 15.09 **82**

For Example 2
Suppose you choose a set of baseball cards for \$19.90, a notebook for \$5.95, and plastic sheets for \$9.25. The cashier tells you the total is \$35. Is that a reasonable cost? **yes**

3 PRACTICE/APPLY

Checking Your Understanding

Exercises 1–22 are designed to help you assess your students' understanding through reading, writing, speaking, and modeling. You should work through Exercises 1–4 with your students and then monitor their work on Exercises 5–22.

Example 2

Consumerism

Suppose you go to Miller Cycles to buy equipment for in-line skating. You purchase a helmet for \$29.89, a set of elbow pads for \$19.26, knee pads for \$21.67, and some energy bars for \$6.03. The cashier tells you that the total is \$82.88. Is this a reasonable cost?

Round each item to the nearest dollar amount.

29.89	→	30
19.26	→	19
21.67	→	22
+ 6.03	→	+ 6
		77

Obviously the amount charged is not correct. Probably the cashier accidentally charged you twice for the energy bars. The correct total is \$76.85, which is very close to the estimate.

You can estimate sums and differences of fractions and mixed numbers by rounding also. To estimate the sum or difference of mixed numbers, round each mixed number to the nearest whole number. To estimate the sum or difference of proper fractions, round each fraction to 0, $\frac{1}{2}$, or 1.

Example 3 **Estimate each sum or difference.**

a. $4\frac{1}{12} + 16\frac{19}{32}$

$4\frac{1}{12} + 16\frac{19}{32} \rightarrow 4 + 17 = 21$

The sum of $4\frac{1}{12}$ and $16\frac{19}{32}$ is about 21.

b. $\frac{7}{8} + \frac{7}{16}$

$\frac{7}{8} + \frac{7}{16} \rightarrow 1 + \frac{1}{2} = 1\frac{1}{2}$

The sum of $\frac{7}{8}$ and $\frac{7}{16}$ is about $1\frac{1}{2}$.

c. $\frac{4}{5} - \frac{3}{8}$

$\frac{4}{5} - \frac{3}{8} \rightarrow 1 - \frac{1}{2} = \frac{1}{2}$

$\frac{4}{5} - \frac{3}{8}$ is about $\frac{1}{2}$.

Checking Your Understanding

Communicating Mathematics

Read and study the lesson to answer these questions. **1–3. See margin.**

1. **Give** two examples of numbers that round to a nearest whole number. One should round up; the other should not.
2. **State** two reasons for using estimation.
3. **You Decide** Pascual estimates that 569 − 345 = 300. Nida estimates the difference as 220. Which is the better estimate? Explain.

Reteaching

Using Mental Math Have students estimate to solve each problem.

- I went to a store and bought a coat for \$79.95 and a scarf for \$8.95. About how much did I spend? **\$89**
- The bill for the meal was \$15.39. I gave the cashier a \$20 bill. About how much change should I get? **\$5**
- I had \$32.40 and spent \$16.94. About how much do I have left? **\$15**

Additional Answers

1. Sample answers: 4.8, 7.1
2. Sample answers: to give approximate answers, to verify calculations, to detect errors in calculations
3. Sample answer: 220 is the better estimate, since 569 and 345 were rounded to the nearest ten instead of to the nearest hundred.

4. **Assess Yourself** Write about a situation or an example in your life in which only an estimate is necessary, not an exact answer. **See students' work.**

Guided Practice

Round to the nearest whole number.

5. 6.19 **6** 6. 12.641 **13** 7. 803.487 **803**

8. $6\frac{3}{4}$ **7** 9. $11\frac{1}{6}$ **11** 10. $33\frac{7}{24}$ **33**

Round each fraction to 0, $\frac{1}{2}$, or 1.

11. $\frac{10}{11}$ **1** 12. $\frac{1}{9}$ **0** 13. $\frac{15}{32}$ **$\frac{1}{2}$**

Choose the best estimate.

14. $4\frac{1}{8} - 2\frac{9}{10}$ a. almost 1 b. more than 1 c. more than $1\frac{1}{2}$ **b**

15. $5\frac{2}{3} + 2\frac{1}{11}$ a. almost 8 b. more than 8 c. less than $7\frac{1}{2}$ **a**

Estimate each sum or difference. **Sample answers given.**

16. 34.32 + 19.51 **50** 17. 159.7 − 124.8 **40** 18. $6\frac{7}{10} - 3\frac{1}{6}$ **4**

19. $12\frac{1}{9} - 2\frac{9}{14}$ **9** 20. $1\frac{7}{12} + \frac{8}{19}$ **2** 21. $23\frac{8}{9} + 5\frac{4}{25}$ **29**

22. **Conservation** The chart at the right shows the amount of water used for average and quick showers with standard and low-flow showerheads. About how much water could be saved by changing from an average shower with a standard showerhead to a quick shower with a low-flow showerhead? **22,000 gallons per year**

Rub-A-Dub-Dub

Gallons used by one shower daily per year:

Standard head

Average shower[1] 26,718

Quick shower[2] 13,140

Low-flow head

Average shower[1] 11,132

Quick shower[2] 5,475

(one drop) = 2,000

1-National average 12.2 minutes, 2-Recommended 6.0 minutes

Source: Opinion Research Corp. for Teledyne Water Pik

Exercises: Practicing and Applying the Concept

Independent Practice

Estimate each sum or difference. **Sample answers given.**

23. 12.5 + 44.8 **60** 24. \$20.00 − \$15.34 **\$5** 25. 8.6 + 11.9 **21**

26. 32 − 29.75 **2** 27. $\frac{8}{9} + \frac{1}{5}$ **1** 28. $\frac{11}{15} + \frac{7}{8}$ **2**

B

29. $6\frac{9}{10} + \frac{2}{5}$ **$7\frac{1}{2}$** 30. $\frac{17}{20} + 8\frac{3}{4}$ **10** 31. $\frac{4}{5} - \frac{1}{10}$ **1**

32. $\frac{7}{9} - \frac{13}{18}$ **0** 33. $18\frac{1}{8} - 12\frac{1}{2}$ **5** 34. $11\frac{89}{100} - 4\frac{1}{7}$ **8**

35. 125.8 − 22.4 **110** 36. 16.432 + 11.910 **28** 37. $32\frac{8}{10} - 4\frac{3}{8}$ **$28\frac{1}{2}$**

38. $\frac{65}{131} + \frac{3}{35}$ **$\frac{1}{2}$** 39. $22\frac{2}{95} + 28\frac{3}{75}$ **50** 40. $145\frac{4}{5} - 121\frac{2}{15}$ **25**

Fast Food Supervisor When working as a supervisor of workers in a fast food restaurant, a person must be able to estimate whether the price of a meal is approximately correct and whether the change is correct. The supervisor may also have to estimate the amount of food to be prepared based on experience. For more information on tech prep, see the *Teacher's Handbook*.

In-Class Example

For Example 3
Estimate each sum or difference.

a. $5\frac{1}{5} + 8\frac{9}{10}$ **14**

b. $\frac{7}{8} + 1\frac{2}{5}$ **$2\frac{1}{2}$**

c. $\frac{5}{11} - \frac{1}{4}$ **$\frac{1}{4}$**

Teaching Tip Remind students that different techniques can be used when making an estimate so there is not one correct answer.

Error Analysis

Some students may round numbers incorrectly when the digit in the place to be rounded is a 9. For instance, students might round 39.53 to 30. Remind students that if they are rounding up, they must increase the number by one, not just the digit.

Assignment Guide

Core: 23–51 odd, 52–60
Enriched: 24–46 even, 47–60

For **Extra Practice**, see p. 751.

The red A, B, and C flags, printed only in the Teacher's Wraparound Edition, indicate the level of difficulty of the exercises.

Study Guide Masters, p. 37

NAME ______________________ DATE __________

5-2 Study Guide
Estimating Sums and Differences

Student Edition
Pages 229–233

You can use rounding to estimate sums and differences. Round each number to a convenient place-value position.

4.25 + 3.56
4.25 + 3.56 → 4 + 4 = 8
4.25 + 3.56 is about 8.

\$187.45 − \$53.81
\$187.45 − \$53.81 → 190 − 50 = 140
\$187.45 − \$53.81 is about \$140.

To estimate sums and differences of mixed numbers, round each mixed number to the nearest whole number. To estimate sums and differences of proper fractions, round each fraction to 0, $\frac{1}{2}$, or 1.

$15\frac{17}{19} - 7\frac{3}{43} \rightarrow 16 - 7 = 9$

$\frac{12}{25} + \frac{7}{65} + \frac{34}{35} \rightarrow \frac{1}{2} + 0 + 1 = 1\frac{1}{2}$

Round to the nearest whole number.

1. 9.23 **9** 2. 3.045 **3** 3. 17.792 **18** 4. 634.572 **635**

5. $37\frac{5}{29}$ **37** 6. $6\frac{5}{6}$ **7** 7. $13\frac{1}{7}$ **13** 8. $4\frac{8}{9}$ **5**

Round each fraction to 0, $\frac{1}{2}$, or 1.

9. $\frac{17}{35}$ **$\frac{1}{2}$** 10. $\frac{5}{9}$ **$\frac{1}{2}$** 11. $\frac{2}{11}$ **0** 12. $\frac{15}{16}$ **1**

Estimate each sum or difference. Sample answers are given.

13. 24.02 + 17.46 **40** 14. 74.63 − 65.89 **9** 15. $10\frac{17}{20} - 8\frac{2}{9}$ **3**

16. $28\frac{1}{6} - 4\frac{9}{17}$ **23** 17. $2\frac{4}{11} + \frac{6}{31}$ **2** 18. $21\frac{6}{7} + 5\frac{8}{27}$ **27**

 T 37 *Pre-Algebra*

Pirate Corner Store Price List	
felt-tip pen	$1.39
one inch binder	$2.89
dividers	$0.89
loose-leaf paper	$1.29
9 x 12 envelopes	$0.93
whiteout	$1.59
package of highliters	$2.79
TI-34 calculator	$13.89
composition books	
1 theme	$2.69
2 theme	$3.99
5 theme	$4.19
index tabs	$1.53
3 x 5 index cards	$0.79
ruler	$0.89
scotch tape	$1.69
can of soda or juice	$0.79
ice cream bar	$1.29
candy bars	$0.69
bag of popcorn	$1.74

The DECA club at Union High School runs a store in the school lobby as a fundraiser. Use their price list at the left to estimate each purchase price or change amount to the nearest dollar. Sample answers given. 44. $3

41. price of binder, loose-leaf paper, index cards **$5**
42. price of loose-leaf paper, dividers, white-out **$4**
43. change from $10 for purchase of 5-theme composition book, and package of hi-liters **$3**
44. change from $20 for purchase of TI-34 and one-inch binder

45. Is $8.15 the correct change from $10.00 for a ruler, ice cream bar, and a candy bar? **No, the change should be about $7.**
46. What supplies would you purchase for $10.00, allowing yourself a candy bar? Let your goal be to get back less than $0.50 change. **See students' work.**

Critical Thinking

47. Make up a problem using estimation in which the answer is $12\frac{1}{2}$. **Sample answer: $11\frac{1}{10} + 1\frac{7}{15}$**

Applications and Problem Solving

48. **Recycling** *National Geographic* magazine estimated the amounts of steel, paper, and aluminum available for recycling in 1994. Which material accounted for the greatest amount of recycling? **paper**

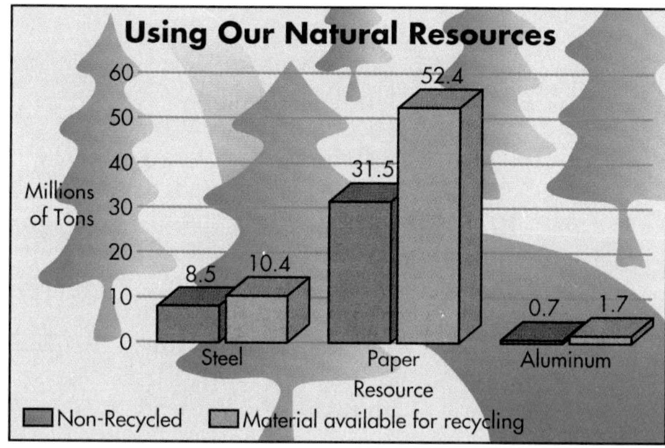

Source: *National Geographic*

EARTH WATCH

THE OZONE LAYER AND CFCs

You have probably heard people speak of the "hole in the ozone layer." Did you wonder why they were concerned? Ozone is a form of oxygen that is able to block out the harmful ultraviolet radiation that radiates from the Sun. A shield of ozone, called the ozone layer, surrounds Earth about 10 to 40 kilometers above the surface. This layer protects us from the cancer-causing UVA and UVB radiation.

Chlorofluorocarbons, or CFCs, are used as coolants in air conditioners, refrigerators, and other products. Scientists have found that CFCs are able to break down ozone and destroy the protective layer above Earth.

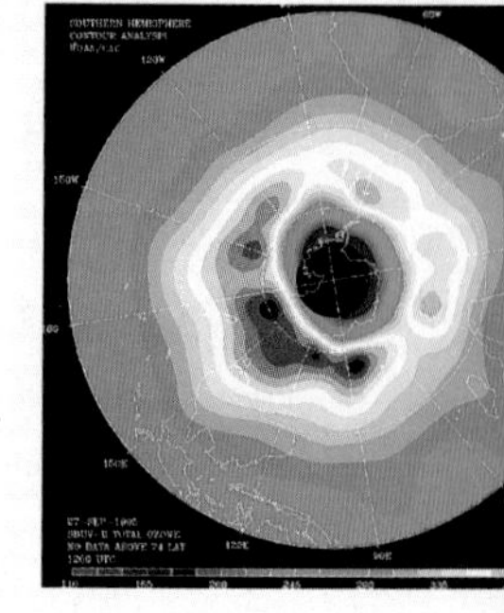

Practice Masters, p. 37

NAME ______________________ DATE ________

5-2 Practice
Estimating Sums and Differences

Student Edition
Pages 229–233

Estimate each sum or difference. Sample answers are given.

1. 13.4 + 27.9 **40**
2. $20.00 − $8.47 **$12.00**
3. 7.3 + 12.7 **20**
4. 24 − 17.25 **7**
5. $\frac{7}{9} + \frac{3}{4}$ **2**
6. $\frac{5}{6} + \frac{1}{9}$ **1**
7. $4\frac{7}{12} + \frac{1}{7}$ **5**
8. $2\frac{3}{4} + 4\frac{2}{9}$ **7**
9. $1\frac{7}{100} + 2\frac{3}{80}$ **3**
10. $\frac{6}{7} - \frac{13}{16}$ **0**
11. $\frac{5}{8} - \frac{1}{5}$ **1**
12. $13\frac{77}{100} - 2\frac{3}{80}$ **12**
13. 23.864 + 4.493 **29**
14. $4\frac{23}{48} - 2\frac{79}{100}$ **$1\frac{1}{2}$**
15. $212\frac{2}{3} - 122\frac{7}{13}$ **90**

Use the price list at the right to estimate each purchase price or change amount to the nearest dollar. Sample answers are given.

Movie Theater Price List	
hot dog	$2.25
slice of pizza	$1.75
nachos	$1.59
bag of popcorn	$1.85
small soda	$1.09
small candy bar	$0.89
large candy bar	$1.39
potato chips	$0.99

16. price of two slices of pizza, a bag of popcorn, and nachos **$7.50**
17. cost of a hot dog and three bags of popcorn **$8.00**
18. change from $15 for a large candy bar, a hot dog, a small candy bar, and a bag of potato chips **$9.50**
19. change from $20 for two hot dogs, a small soda, and a small candy bar **$14.00**
20. cost of three large candy bars, nachos, and a small candy bar **$7.00**
21. change from $5 for a slice of pizza and a bag of popcorn **$1.00**

Group Activity Card 5-2

Round 'em Up — Group Activity **5-2**

MATERIALS: 0-9 spinner

To begin, each person draws a place value board similar to the one pictured below. This has five connected squares with a decimal placed between the third and fourth squares.

Play two rounds, and the player with the most points is the overall winner. One round consists of five games. (See games on the back.)

To play, take turns spinning the spinner to get a digit. After each spin, place that number in an empty square on your board. After 5 spins each player rounds his or her number as indicated by the game number.

The numbers are compared, and one point is awarded to the persons whose numbers are correctly rounded and fit the requirements below. In case of ties, all who tie score one point.

 Pre-Algebra

EARTH WATCH

Ozone, or O_3, is made up of three oxygen atoms. Normally, oxygen atoms exist in pairs as O_2 molecules. Ultraviolet radiation, from the sun splits O_2 molecules into two separate oxygen atoms. Each oxygen atom immediately combines with another O_2 molecule and forms ozone, O_3.

49. **Retail Sales** Jaime bought a sweatshirt with his high school name and mascot for \$41.29 and a T-shirt with the track team logo for \$18.25. Estimate his cost to the nearest dollar. **\$59**

50. **Fundraising** The Freshman class needs to earn \$500 for their annual trip to Six Flags. They made \$95.60 at a car wash, \$41.75 at a cookie and bake sale, \$150.95 at a band concert, and \$125.16 selling candy bars.
    a. Estimate their total earnings. **\$414**
    b. Estimate, if necessary, the amount needed to reach their goal. **\$86**

51. **Business** Pepsi-Cola reported that $\frac{3}{10}$ of the sales of their soft drinks were made through fountains and restaurants in 1994. Another $\frac{2}{5}$ of the sales were through supermarkets and other retail stores. Did the sales through these channels account for more or less than half of the total soft drink sales for Pepsi in 1994? **more than half**

Mixed Review

52. Express -0.88 as a fraction in simplest form. (Lesson 5-1) $-\frac{22}{25}$
53. **Food** Tortilla chips account for $\frac{26}{100}$ of the snack chip retail sales. Write $\frac{26}{100}$ as a fraction in simplest form. (Lesson 4-6) $\frac{13}{50}$
54. Solve $364 = w - 84$. (Lesson 3-2) **448**
55. Divide $48 \div (-3)$. (Lesson 2-8) **−16**
56. In which quadrant is the graph of $(-6, 0)$ located? (Lesson 2-2) **none**
57. Is the inequality $4 < 3$ *true, false,* or *open*? (Lesson 1-9) **false**
58. Solve $12n = 48$ mentally. (Lesson 1-6) **4**
59. **Entertainment** When Bob and Neva went to the movies, the tickets were \$6.25 apiece. They also bought two small boxes of popcorn for \$2.25 each. (Lesson 1-5)
    a. Write two different expressions for the amount of money Bob and Neva spent. **2(6.25 + 2.25) or 2(6.25) + 2(2.25)**
    b. Find the amount of money they spent. **\$17**
60. Find the value of $(11 + 9) \div (3 - 1)$. (Lesson 1-2) **10**

See for Yourself **2800 million kilograms**

- The table at the right shows the amount of freon, the most common CFC, released into the atmosphere in different years. Estimate the amount released in the 1980s.
- A "hole" in the ozone layer is an area where the ozone is half as thick as it is in the rest of the layer. Every year, a hole appears over Antarctica. Investigate the reasons behind the appearance of the hole. Is there a similar hole over the North Pole?
- As the table of data suggests, several governments have recently passed new guidelines on the use of CFCs. What are the new guidelines? When do manufacturers have to comply? What are the expected results? **See Solutions Manual.**

Year	Released Freon (millions of kilograms)
1980	250.8
1981	248.2
1982	239.5
1983	252.8
1984	271.1
1985	280.8
1986	295.1
1987	310.6
1988	314.5
1989	265.2
1990	216.1
1991	188.3
1992	171.1

4 ASSESS

Closing Activity

Speaking Have students explain how they use estimation in various real-life situations.

Extension

Using Graphs Have students find bar graphs in newspapers and magazines and write problems that can be solved by estimating sums and differences. You may wish to have students prepare the graphs and related problems for display in the classroom.

Enrichment Masters, p. 37

NAME ______________________ DATE ____________

5-2 Enrichment
Using Rounding to Estimate with Fractions

Student Edition Pages 229–233

In Exercises 1 and 2, write an X above the problem that has the greatest result and circle the problem that has the least result. Round each mixed number to determine your answer. Do not find the exact results.

1. $3\frac{2}{3} + 5\frac{1}{8}$; $4\frac{1}{12} + 6\frac{2}{7}$; X $1\frac{6}{7} + 8\frac{3}{4}$; $2\frac{4}{5} + 7\frac{1}{9}$; $6\frac{3}{11} + 3\frac{7}{16}$; (circled) $1\frac{4}{10} + 7\frac{9}{36}$

2. (circled) $9\frac{1}{5} - 5\frac{3}{4}$; $8\frac{4}{7} - 3\frac{3}{5}$; X $7\frac{4}{5} - 2\frac{1}{3}$; $6\frac{3}{10} - 1\frac{3}{8}$; $5\frac{7}{8} - \frac{15}{16}$; $7\frac{3}{7} - 3\frac{5}{12}$

For each problem, use rounding to estimate the result. Then decide if the exact result will be greater than or less than your estimate. If the actual result will be greater than your estimate, write "up" below your estimate. If the actual result will be less than your estimate, write "down."

3. $4\frac{1}{3} + 7\frac{2}{8}$ — 11 up
4. $2\frac{2}{3} - 1\frac{1}{5}$ — 2 down
5. $6\frac{4}{7} + 3\frac{5}{6}$ — 11 down
6. $8\frac{1}{10} - 5\frac{3}{4}$ — 2 up
7. $9\frac{3}{7} - \frac{5}{8}$ — 8 up
8. $2\frac{6}{7} + 5\frac{9}{10}$ — 9 down
9. $3\frac{5}{7} + 7\frac{3}{5}$ — 12 down
10. $8\frac{7}{10} - 4\frac{1}{3}$ — 5 down
11. $6\frac{2}{9} - 2\frac{7}{8}$ — 3 up

5-3 LESSON NOTES

NCTM Standards: 1-4, 7

Instructional Resources

- Study Guide Master 5-3
- Practice Master 5-3
- Enrichment Master 5-3
- Group Activity Card 5-3
- Assessment and Evaluation Masters, p. 127
- Activity Masters, p. 19
- Multicultural Activity Masters, p. 9

Transparency 5-3A contains the 5-Minute Check for this lesson; **Transparency 5-3B** contains a teaching aid for this lesson.

Recommended Pacing	
Standard Pacing	Day 3 of 13
Honors Pacing	Day 3 of 12
Block Scheduling*	Day 2 of 6 (along with Lesson 5-4)

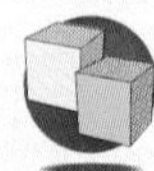

*For more information on pacing and possible lesson plans, refer to the ***Block Scheduling Booklet***.

1 FOCUS

5-Minute Check
(over Lesson 5-2)

Round to the nearest whole number.

1. 9.099 **9**

2. $15\frac{13}{16}$ **16**

Estimate each sum or difference.

3. 5.36 + 6.023 + 9.902 **21**

4. $\frac{5}{6} - \frac{3}{7}$ $\mathbf{\frac{1}{2}}$

5. 902.958 − 190.39 **700**

Motivating the Lesson

Hands-On Activity Have each of three students reach into a bowl of pennies and take as many pennies as possible. Write the amount each student got in dollars-and-cents form. Ask how you could find the total number of pennies the students pulled out.

5-3 Adding and Subtracting Decimals

Setting Goals: *In this lesson, you'll add and subtract decimals.*

Modeling with Technology

You have experience with adding and subtracting integers. Use a calculator to develop rules for adding and subtracting with decimals.

Your Turn Copy the table below. Use a calculator to find each sum or difference and complete the table.

Column A		Column B	
Sum or Difference	Answer	Sum or Difference	Answer
18 + 66	**84**	1.8 + 6.6	**8.4**
56 + (−78)	**−22**	5.6 + (−7.8)	**−2.2**
44 − 15	**29**	4.4 − 1.5	**2.9**
220 − (−104)	**324**	2.20 − (−1.04)	**3.24**

TALK ABOUT IT

a. Compare the sums and differences in Column A to the sums and differences in Column B. **They have the same digits, in different place values.**

b. Compare the answers in Column A to the answers in Column B.

b. They have the same digits, in different place values.

c. Describe how you could find the sum of 6.8 and 9.3 if you know that 68 + 93 = 161. **Move the decimal point in 161 one place to the left to get the sum 16.1.**

Learning the Concept

Adding and subtracting rational numbers follow the same principles as addition and subtraction of integers.

Properties of Addition	Examples
Commutative Property For any rational numbers a and b, $a + b = b + a$.	$-7 + 9 = 9 + (-7)$ $-8.1 + 9.6 = 9.6 + (-8.1)$
Associative Property For any rational numbers a, b, and c, $(a + b) + c = a + (b + c)$.	$(-3.5 + 5.2) + 3.1 = -3.5 + (5.2 + 3.1)$
Identity Property For every rational number a, $a + 0 = a$ and $0 + a = a$.	$8 + 0 = 8$ $0 + -5.2 = -5.2$
Inverse Property For every rational number a, $a + (-a) = 0$.	$6 + (-6) = 0$ $15.8 + (-15.8) = 0$

Classroom Vignette

"I use the names and examples of properties to create a game of Concentration. Write the property name on a card and an example on another. Mix and place cards face down on a desk. One student selects two cards. If the cards match, the student keeps them and continues. If they do not, the cards are returned face down, and another student selects. Play until all cards are matched. The one with the most cards wins."

Bonnie Langan
Northern Middle School, Accident, MD

You will use these properties and the rules you learned for adding and subtracting integers to add and subtract rational numbers. Estimating will help you determine the reasonableness of your answer.

Example 1 **Solve each equation**

a. $m = 13.2 + 11.7$ *Estimate: 13 + 12 = 25*

$$\begin{array}{r} 13.2 \\ +11.7 \\ \hline 24.9 \end{array}$$

$m = 24.9$ *Compare with the estimate.*

b. $g = 231 - 126.7$ *Estimate: 230 − 130 = 100*

Pencil and Paper

$$\begin{array}{r} 231.0 \\ -126.7 \\ \hline 104.3 \end{array}$$

Calculator

231  126.7 [=] 104.3

$g = 104.3$ *Compare with the estimate.*

c. $d = 8.3 + (-12.5)$ *Estimate: 8 − 13 = −5*

$d = 8.3 - 12.5$ *The addition can be written as subtraction.*

$d = -|12.5 - 8.3|$ *The difference will be negative since $|-12.5| > |8.3|$.*

$$\begin{array}{r} 12.5 \\ -\ 8.3 \\ \hline 4.2 \end{array}$$

$d = -4.2$ *Compare with the estimate.*

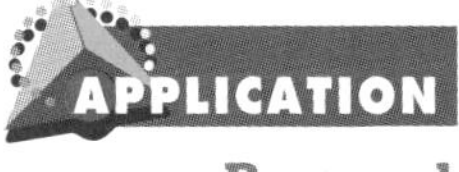

TECHNO TIP

You can save time by entering a sum involving a negative as a subtraction problem. For example, find 8.3 + (−12.5) by entering 8.3 − 12.5.

Connection to Algebra

Simplifying expressions may involve adding and subtracting decimals.

Example 2 **Simplify $4.7x + 6.3x - 13.7x$.**

$4.7x + 6.3x - 13.7x = (4.7 + 6.3 - 13.7)x$ *Distributive property*

$= (11.0 - 13.7)x$ *Associative property*

$= -2.7x$

You use sums and differences of rational numbers everyday when you deal with money.

Example 3 APPLICATION **Personal Finance**

Kiana uses a spreadsheet program to keep track of her money. The printout at the right shows her expenses and budgeted amounts for the month of January. Find the amount of money that Kiana is under or over budget for the month.

Item	Budget	Actual
Car payment	159.14	159.14
Car insurance	195.63	195.63
Fuel	35.00	37.32
Car repairs	150.00	135.67
Entertainment	60.00	52.44
Savings	50.00	50.00

(continued on the next page)

Teaching Tip Relate the names of the properties to common uses of similar words. "You can *commute* between two cities. You can *associate* with any person you wish." Allow students to create their own illustrative sentences.

2 TEACH

In-Class Examples

For Example 1
Solve each equation.

a. $x = 15.6 + 11.2$ **26.8**

b. $238 - 112.9 = y$ **125.1**

c. $z = 6.5 + (-8.4)$ **−1.9**

For Example 2
Simplify $6.3x - 3.7x - (-7.4x)$. **10x**

GLENCOE Technology

Interactive Mathematics Tools Software

In this interactive computer lesson, students use technology to explore the addition of decimals. A **Computer Journal** gives students an opportunity to write about what they have learned.

For Windows & Macintosh

In-Class Example

For Example 3
Cari's Carpenter Corner printed out the spreadsheet of its budgeted and actual expenses.

Item	Budget	Actual
Store Rent	\$1200.00	\$1200.00
Inventory Expense	\$3900.00	\$4506.94
Salaries & Commissions	\$2450.75	\$2690.50
Advertising Expense	\$800.00	\$636.47

Find the amount that the store is under or over budget. **\$683.16 over budget**

Additional Answers

1. commutative: $12.3 + (8.3 + 2.5) = 12.3 + (2.5 + 8.3)$
   associative: $12.3 + (8.3 + 2.5) = (12.3 + 8.3) + 2.5$
3. Both are correct because adding 2.9 is the same as subtracting -2.9.

Study Guide Masters, p. 38

NAME ______________ DATE ________

5-3 Study Guide
Adding and Subtracting Decimals

Student Edition Pages 234–238

To add or subtract decimals, first align the decimal points and place zeros where necessary. Then add or subtract as with whole numbers.

Solve $m = 38.5 + 52.12 \rightarrow 38.50 + 52.12$, $m = 90.62$ | Solve $d = 235 - 13.8 \rightarrow 235.0 - 13.8$, $d = 221.1$

Solve each equation.

1. $x = 3.2 + 4.7$ **7.9**
2. $-2.06 + 3.15 = m$ **1.09**
3. $y = -36.09 - (-7.01)$ **−29.08**
4. $47.9 + 3.24 = w$ **51.14**
5. $g = 0.5623 - 0.3541$ **0.2082**
6. $52.5 + 8.62 = k$ **61.12**
7. $h = -27.8 + (-14.32)$ **−42.12**
8. $3.09 - 0.05 = b$ **3.04**
9. $m = 16.2 - 5.59$ **10.61**
10. $58 - 0.232 = z$ **57.768**
11. $j = 23 + (-1.59)$ **21.41**
12. $15.6 - 0.423 = g$ **15.177**
13. $r = -8.52 + 2.43$ **−6.09**
14. $150 - 25.6 = f$ **124.4**
15. $v = -3.56 - 0.49$ **−4.05**

Simplify each expression.

16. $0.5y + 0.7y$ **1.2*y***
17. $9x - 2.5x$ **6.5*x***
18. $3.81n + 0.092n + 4.6n$ **8.502*n***
19. $3.56t - 0.59t + 46.08t$ **49.05*t***
20. $3.0w - 24.15w + 56.052w$ **34.902*w***
21. $215.5v - 36v - 4.63v$ **174.87*v***

Find the sums of the budget and the actual amounts. Use a calculator.

budget $= 159.14 + 195.63 + 35.00 + 150.00 + 60.00 + 50.00$
$= 649.77$
actual $= 159.14 + 195.63 + 37.32 + 135.67 + 52.44 + 50.00$
$= 630.20$

The actual total is less than the budgeted total, so Kiana is under budget by $649.77 - 630.20$ or \$19.57.

Checking Your Understanding

Communicating Mathematics

Read and study the lesson to answer these questions. **1, 3. See margin.**

1. Using the expression $12.3 + (8.3 + 2.5)$, illustrate the commutative property of addition and the associative property of addition.
2. **State** the additive inverse of 7.398. **−7.398**
3. **You Decide** Mariana and Leslie used different methods to solve $-3.8 - (-2.9)$. Who is correct? Explain.

Mariana	Leslie
3.8 [(−)] [−] 2.9 [(−)] [=] −.9	3.8 [(−)] [+] 2.9 [=] −.9

Guided Practice

State where the decimal point should be placed in each sum or difference.

4. $4.6 + 5.9 = 105$ **10.5**
5. $7.97 - 4.29 = 368$ **3.68**
6. $3.16 + 4.2 = 736$ **7.36**
7. $-8.47 + 3.56 = -491$ **−4.91**
8. $8.17 + 9.123 = 17293$ **17.293**
9. $70.3 + 7.03 = 7733$ **77.33**

Solve each equation.

10. $a = 41.3 + 0.28$ **41.58**
11. $b = -9.6 + 3.2$ **−6.4**
12. $34.2 - 43.0 = c$ **−8.8**
13. $t = 81.9 - 38$ **43.9**

Simplify each expression.

14. $4.7x + 2x$ **6.7*x***
15. $5.3m - 1.4m - 8m$ **−4.1*m***

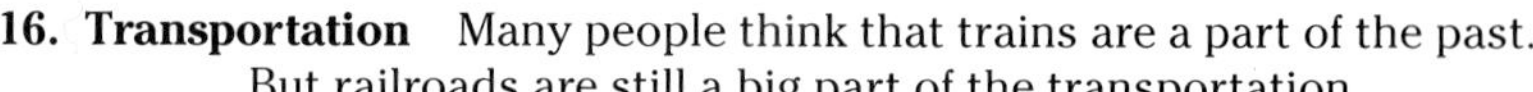

16. **Transportation** Many people think that trains are a part of the past. But railroads are still a big part of the transportation industry in the United States. Use the table below to find the change in the number of passengers on each type of public transportation listed from 1993 to 1994.

Type of Transportation	1993 (billions)	1994 (billions)	
Commuter rail	330.8	349.5	**+18.7**
Heavy rail	2189.0	2279.0	**+90.0**
Light rail	224.9	232.9	**+8.09**
Bus/trolley	5412.1	5403.4	**−8.7**

Source: *American Public Transit Association*

Reteaching

Using Manipulatives Give each group of four students play money including pennies, dimes, and dollar bills. Each student takes some dollars, dimes, and pennies. Then the group writes the decimal for the total amount of money taken by members of the group. For students needing help finding the sum, have them first count the total number of each type of coin, exchanging 10 pennies for a dime and 10 dimes for a dollar as needed. Relate that to adding the amounts.

Exercises: Practicing and Applying the Concept

Independent Practice

Solve each equation.

A

17. $a = 4.3 + 9.8$ **14.1**
18. $x = 31.92 + 14.2$ **46.12**
19. $15.3 - 13.8 = m$ **1.5**
20. $85.3 - 37.07 = r$ **48.23**

B

21. $72.47 - 9.039 = b$ **63.431**
22. $c = 0.735 - 0.3879$ **0.3471**
23. $-7.5 + 9.8 = x$ **2.3**
24. $7.4 + (-3.9) = y$ **3.5**
25. $z = -13.9 + (-12.5)$ **−26.4**
26. $k = -3.91 - (-0.6)$ **−3.31**
27. $a = -34.1 + (-17.63)$ **−51.73**
28. $y = -18.12 - (-7.3)$ **−10.82**

Simplify each expression.

29. $5.3m + 7m$ **12.3*m***
30. $47.9w - 31.8w$ **16.1*w***
31. $0.3y + 4.1y + 2.5y$ **6.9*y***
32. $27y - 4.7y - 13.8y$ **8.5*y***
33. $3.5x - 5 + 8x$ **11.5*x* − 5**
34. $6.93 + (3.1 + 4.07)m$ **6.93 + 7.17*m***

Evaluate each expression if $a = 5.3$, $b = 8.07$, $x = 21.33$, and $y = 0.7$.

35. $b + x$ **29.4**
36. $a - y$ **4.6**
37. $x - (b + y)$ **12.56**
38. $b + x + y$ **30.1**
39. $a + b - y$ **12.67**
40. $(x - b) + a$ **18.56**

Critical Thinking

41. Cassie has 45 feet of chicken wire for a fence. She wants to build a triangular pen that is 11.2 feet on one side and 9.35 feet on a second side. How much wire will she have left for the third side? Can she use all of it? Why or why not? **24.45 left; no; third side must be less than 20.55 feet**

Applications and Problem Solving

42. **Sports** The physical size of the top 10 men and women tennis players has changed in 20 years. How much more do the men and women, respectively, weigh now than 20 years ago?

42. men: 8.2 lb; women: 8.5 lb

Average weight of top ten players

	1974	1994
Men (lbs)	161.7	169.9
Women (lbs)	127.2	135.7

Source: *USA-Today* Research

43. **Cycling** Miguel Indurain of Spain won the Tour de France, the world's most prestigious bicycle race, for the fifth time in 1995. At the beginning of one of his workouts, the odometer on his bicycle reads 201.9 kilometers. If Miguel rides 176.6 kilometers that day, what will the odometer read at the end of the workout? **378.5**

44. **Recycling** It is estimated that for every 13.2 million tons of glass used per year in the U.S., 2.6 million tons are recycled. How many millions of tons are simply discarded? **10.6 million tons discarded**

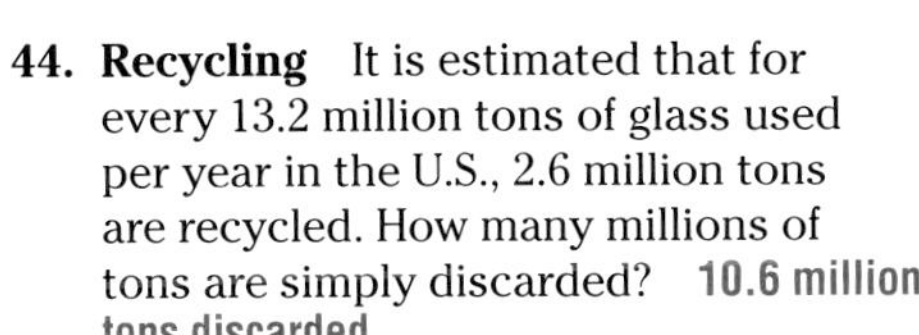

3 PRACTICE/APPLY

Checking Your Understanding

Exercises 1–16 are designed to help you assess your students' understanding through reading, writing, speaking, and modeling. You should work through Exercises 1–3 with your students and then monitor their work on Exercises 4–16.

Assignment Guide

Core: 17–45 odd, 47–54
Enriched: 18–40 even, 41–54

For **Extra Practice**, see p. 751.

The red A, B, and C flags, printed only in the Teacher's Wraparound Edition, indicate the level of difficulty of the exercises.

Practice Masters, p. 38

NAME ____________ DATE ________

5-3 Practice
Adding and Subtracting Decimals

Student Edition Pages 234–238

Solve each equation.

1. $x = 4.7 + 8.3$ **13**
2. $a = 14.1 - 7.2$ **6.9**
3. $-9.2 - (-6.03) = y$ **−3.17**
4. $q = -18.4 + (-28.7)$ **−47.1**
5. $23.1 + (-10.9) = m$ **12.2**
6. $n = -19.21 + 12.8$ **−6.41**
7. $-6.35 - (-0.9) = b$ **−5.45**
8. $m = -25.4 + (-18.93)$ **−44.33**
9. $8.56 - 3.492 = t$ **5.068**
10. $y = 0.834 - 0.54$ **0.294**
11. $x = 49.95 + 3.75$ **53.70**
12. $43.27 - 4.59 = r$ **38.68**
13. $425.9 - 173.2 = d$ **252.7**
14. $0.4999 - 0.375 = x$ **0.1249**

Simplify each expression.

15. $12w + 3.4w$ **15.4*w***
16. $87.5d - 3 + 15d$ **102.5*d* − 3**
17. $(0.04 + 9.2)p + 0.07$ **9.24*p* + 0.07**
18. $45.9m - 23.6m$ **22.3*m***
19. $0.2a + 1.4a + 4.3a$ **5.9*a***
20. $49x - 15.6x - 3.7x$ **29.7*x***

Evaluate each expression if a = 0.4, b = 3.5 c = 15.61, and d = 0.03.

21. $c + b$ **19.11**
22. $b + d$ **3.53**
23. $a - d$ **0.37**
24. $(b + c) + a$ **19.51**
25. $c - b$ **12.11**
26. $(a + c) - b$ **12.51**
27. $c - d - a$ **15.18**
28. $(b - d) + a$ **3.87**
29. $(c + b) - a$ **18.71**

Group Activity Card 5-3

Silly Skeleton Group Activity **5-3**

In pairs, solve the following equations. Next arrange the answers in order from least to greatest. Write the variables used in the equations below the answers. They should tell you the answer to this riddle:

Why didn't the skeleton cross the road?

1. -3.421 − (-15.13) = O
2. 53.2 + 8.14 − 15.261 = T
3. -24.56 + (-5.07) = H
4. -17.3 − 26.85 = I
5. -18.94 + 42.31 = G
6. 73.42 − 18.6 + (-4.382) = S
7. 13.7 − 22.06 − 4.36 = A
8. 14 − 9.67 + 3.2 = N
9. -51.6 + 4.31 + 6.93 = T
10. 82.73 − 39.97 = U
11. 61.08 − 93.75 + 32.67 = D

ARRANGEMENT OF ANSWERS FROM LEAST TO GREATEST:

ANSWER:

__ __ __ __ __ __ __ __ __ __ __

 Pre-Algebra

Extension

Using Logical Reasoning Have students play "Concentration" as described in the Classroom Vignette on page 234.

4 ASSESS

Closing Activity

Act It Out Provide students with play money (about $20 in dollars, dimes, and pennies). Name an amount between $1 and $19. Hand the student a $20 bill and have the student give you the correct amount of change.

Chapter 5, Quiz A (Lessons 5-1 through 5-3) is available in the *Assessment and Evaluation Masters*, p. 127.

Enrichment Masters, p. 38

NAME ______________________ DATE ____________

5-3 Enrichment
Adding and Subtracting Decimals

Student Edition Pages 234–238

1. The chart below shows Mrs. White's credit card statement for September. Find the total amount due.

PB Previous Balance $1252.86		
Merchandise Description		Amounts (Credit Items Followed by –)
CC	Stereo	786.95
PY	Payment — Thank You	420.00–
FC	Finance Charge	29.16
TT	Total Now Due	**$1648.97**

2. Mr. Fair's savings account statement shows the following transactions. What is his balance at the end of the month?

Previous Balance	8/1	$8792.50
Deposit	8/6	820.40
Deposit	8/9	120.50
Withdrawal	8/15	1260.00
Deposit	8/20	472.26
Withdrawal	8/26	2768.40
Interest	8/27	23.16
Balance	8/31	**$6200.42**

3. A weather forecaster kept track of the temperature change over a six-hour period. The temperature at 5:00 A.M. was −10.5°F. What was the temperature at the end of the six-hour period?

1st hour	rise	5.3°
2nd hour	drop	2.1°
3rd hour	drop	4.8°
4th hour	rise	3.5°
5th hour	rise	7.1°
6th hour	drop	2.1°

−3.6°F

4. For every 160 hours Marie works, she receives 11.5 hours of vacation. She would like to accumulate at least 50 hours of vacation so she could travel to Mexico. Using the chart, determine how many more hours she needs to work. She receives each additional 11.5 hours only after working a *full* 160 hours.

Month	Hours Worked
January	161.5
February	120
March	138.5
April	110.5
May	170.5

Since $50 \div 11.5 > 4$, Marie needs to work $5 \times 160 = 800$ hours. She has already worked 701 hours. Thus, she needs to work $800 - 701 = 99$ additional hours.

5. A submarine at 1300 meters below sea level descends an additional 1150 meters. How far below sea level is the submarine now? **2450 m**

6. Mike has a box 5.8 inches on each side. If he has a piece of rope 27.4 inches long, how much extra rope will he have after wrapping it around the box once? **4.2 in.**

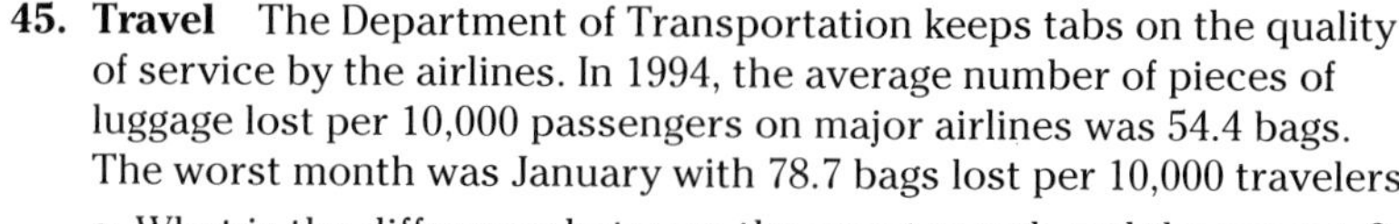

45a. 24.3 bags per 10,000 passengers
45b. Sample answer: Bad weather causes changes in flight schedules.

45. **Travel** The Department of Transportation keeps tabs on the quality of service by the airlines. In 1994, the average number of pieces of luggage lost per 10,000 passengers on major airlines was 54.4 bags. The worst month was January with 78.7 bags lost per 10,000 travelers.
    a. What is the difference between the worst month and the average?
    b. Why do you think January is the worst month for lost baggage?
46. **Geometry** Find the perimeter of the triangle shown at the right. **9.6 cm**

2.4 cm 4.0 cm 3.2 cm

Mixed Review

47. **Personal Finance** When packing for a trip, Deandra bought toothpaste for $1.89, suntan lotion for $4.39, hair spray for $1.27, and soap for $2.04. She estimated that the $10 she had would be enough to pay the bill. Do you agree? Explain. (Lesson 5-2) **yes; $2 + 4 + 1 + 2 = \$9$**
48. Write $p \cdot p \cdot p \cdot p$ using exponents. (Lesson 4-2) p^4
49. Solve $19 < 17 + a$. (Lesson 3-6) $2 < a$
50. **Fundraising** The freshman class earned $120 selling magazines. The class account now has a balance of $435. How much money was in the account before the magazine sale? (Lesson 3-2) **$315**
51. Solve $b = -121 \div 11$. (Lesson 2-8) **−11**
52. Replace ● with $<$, $>$, or $=$ in $|-10|$ ● 9 to make a true sentence. (Lesson 2-3) $>$
53. Find the value of $|8| + |-4|$. (Lesson 2-1) **12**
54. Evaluate $2m - n$, if $m = 8$ and $n = 5$. (Lesson 1-3) **11**

WORKING ON THE

Investigation

Refer to the Investigation on pages 166–167.

To get an overall feel of how the stock market is performing, you can consult one of the market indexes. An **index** is an analysis of a selected group of individual stocks such as communications, pharmaceuticals, technology, or utilities.

There are two types of indexes. An index that is calculated by adding the price per share of all the stocks in the index and dividing by the number of stocks in the index is called a **price-weighted index**. The *Dow Jones Industrial Average* is an example of a price-weighted index.

If an index tracks the total market value, it is called a **value-weighted index**. The total market value is the product of the price per share and the total number of shares outstanding. To evaluate this type of index, the total market value is divided by the total market value on the date when the index was first started. The quotient is then multiplied by 100. The *Standard and Poor's 500* is an example of a value-weighted index.

Refer to your work with a spreadsheet on page 209 of Lesson 4-8.

- Using the stocks in your portfolio, create a spreadsheet to calculate your own price-weighted stock market index.
- Update your index each day for a five-day period and note any changes.
- Show the changes in your index by drawing a line graph of the five-day period.

Add the results of your work to your Investigation Folder.

WORKING ON THE

Investigation

This Investigation on pages 166–167 is designed to be a long-term project that is completed over several days or weeks. Encourage students to keep their materials in their Investigation Folder as they work on the Investigation.

5-4 Adding and Subtracting Like Fractions

Setting Goals: *In this lesson you'll add and subtract fractions with like denominators.*

Modeling with Manipulatives

MATERIALS

colored pencils

ruler

As you learned in Chapter 4, a fraction can be represented using a model like the one at the right. Different colors of shading can be used to show addition of fractions. This model represents $\frac{1}{6} + \frac{2}{6}$. Since 3 of the 6 sections are shaded, the sum is $\frac{3}{6}$ or $\frac{1}{2}$.

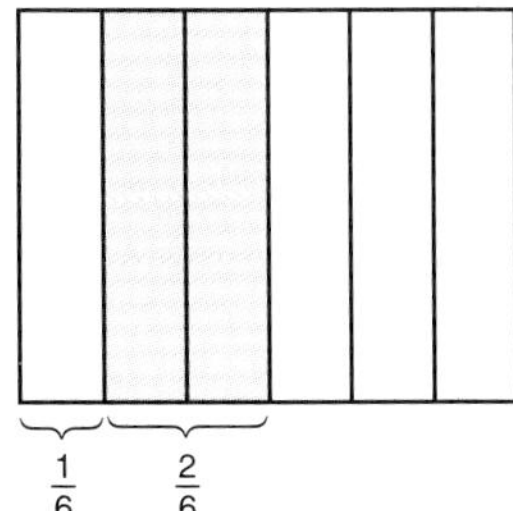
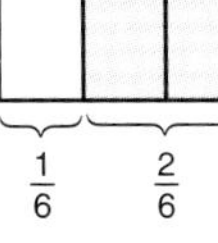

Your Turn Draw and shade a model that represents the sum of $\frac{1}{8}$ and $\frac{5}{8}$.

TALK ABOUT IT

a. How many sections of your model are shaded? How does this compare to the sum of the numerators? **6; They are the same.**

b. How many sections are there in your rectangle? How does this compare to the denominator of the addends? **8; They are the same.**

c. What fraction represents the shaded sections of your rectangle? $\frac{6}{8}$

d. Write a rule for finding the sum of fractions with the same denominator. **The numerator is the sum of the numerators and the denominator is the same as the denominators of the addends.**

Learning the Concept

Adding fractions with like denominators can be described by the following rule.

Adding Like Fractions

In words: To add fractions with like denominators, add the numerators and write the sum over the same denominator.

In symbols: For fractions $\frac{a}{c}$ and $\frac{b}{c}$, where $c \neq 0$, $\frac{a}{c} + \frac{b}{c} = \frac{a+b}{c}$.

When the sum of two fractions is greater than one, the sum is usually written as a mixed number in simplest form. A **mixed number** indicates the sum of a whole number and a fraction.

Alternative Learning Styles

Kinesthetic Provide students with rulers marked in eighths of an inch. Have students find the perimeter of various polygons whose sides are between $\frac{1}{2}$ inch and 2 inches long.

5-4 LESSON NOTES

NCTM Standards: 1-4, 7, 10

Instructional Resources

- Study Guide Master 5-4
- Practice Master 5-4
- Enrichment Master 5-4
- Group Activity Card 5-4
- Activity Masters, p. 5
- Tech Prep Applications, p. 10

Transparency 5-4A contains the 5-Minute Check for this lesson; **Transparency 5-4B** contains a teaching aid for this lesson.

Recommended Pacing	
Standard Pacing	Day 4 of 13
Honors Pacing	Day 4 of 12
Block Scheduling*	Day 2 of 6 (along with Lesson 5-3)

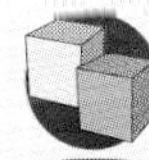

*For more information on pacing and possible lesson plans, refer to the ***Block Scheduling Booklet***.

1 FOCUS

5-Minute Check
(over Lesson 5-3)

Solve each equation.

1. $a = 6.4 + 16.3$ **22.7**
2. $96.34 - 63.74 = b$ **32.6**
3. $c = 5.3 + (-5.6) - (3.4)$ **−3.7**

Simplify each expression.

4. $7.4n - 9.3n$ **$-1.9n$**
5. $6.3m + 4 - 1.1m$ **$5.2m + 4$**

Motivating the Lesson

Situational Problem Relate the situation that occurs in *Ramona the Pest* when Ramona is late to school. Her mother tells her to leave for school at quarter after eight. Ramona reasons that a quarter is 25¢, so she leaves for school at 8:25 and is late. Have students discuss this situation and the various meanings of the word "quarter".

2 TEACH

In-Class Examples

For Example 1
Solve each equation.

a. $s = \frac{3}{5} + \frac{4}{5}$ $\mathbf{1\frac{2}{5}}$

b. $y = 16\frac{2}{3} + 28\frac{2}{3}$ $\mathbf{45\frac{1}{3}}$

For Example 2
Solve $t = \frac{6}{7} - \frac{16}{7}$. $\mathbf{-1\frac{3}{7}}$

For Example 3
Evaluate $b - c$, if $b = 2\frac{1}{8}$ and $c = -3\frac{3}{8}$. $\mathbf{5\frac{1}{2}}$

Teaching Tip Students may need extra help and practice in completing subtraction exercises that require renaming. You may wish to have students who understand the process act as peer tutors.

In-Class Example

For Example 4
A survey asked 100 people about their favorite ice cream flavors.

Flavor	Fraction
Butter pecan	$\frac{11}{100}$
Chocolate	$\frac{23}{100}$
Chocolate chip	$\frac{9}{100}$
Rocky road	$\frac{7}{100}$
Strawberry	$\frac{21}{100}$
Vanilla	$\frac{29}{100}$

a. What fraction of the people chose chocolate or chocolate chip ice cream? $\mathbf{\frac{8}{25}}$

b. What fraction of the people chose strawberry or vanilla ice cream? $\mathbf{\frac{1}{2}}$

Example 1 **Solve each equation.**

a. $s = \frac{3}{4} + \frac{3}{4}$ *Estimate: 1 + 1 = 2*

$s = \frac{6}{4}$ *Since the denominators are the same, add the numerators.*

$s = 1\frac{2}{4}$ *Rename $\frac{6}{4}$ as a mixed number.*

$s = 1\frac{1}{2}$ *Write the mixed number in simplest form.*

b. $y = 3\frac{2}{5} + 4\frac{3}{5}$ ***Estimate:** 3 + 5 = 8*

$y = (3 + 4) + \left(\frac{2}{5} + \frac{3}{5}\right)$ *Use the Associative and Commutative properties to add the whole numbers and fractions separately.*

$y = 7 + \frac{5}{5}$ *Add the numerators of the fractions*

$y = 7 + 1$ or 8 *Simplify.*

The rule for subtracting fractions with like denominators is similar to the rule for addition.

Subtracting Like Fractions

In words: To subtract fractions with like denominators, subtract the numerators and write the difference over the same denominator.

In symbols: For fractions $\frac{a}{c}$ and $\frac{b}{c}$, where $c \neq 0$, $\frac{a}{c} - \frac{b}{c} = \frac{a-b}{c}$.

Example 2 **Solve $y = \frac{7}{15} - \frac{28}{15}$. Write the solution in simplest form.**

$y = \frac{7}{15} - \frac{28}{15}$ *Since the denominators are the same, subtract the numerators.*

$y = \frac{-21}{15}$

$y = -1\frac{6}{15}$ or $-1\frac{2}{5}$ *Rename as a mixed number and simplify.*

Connection to Algebra

You can use the same rules for adding and subtracting like fractions when you evaluate algebraic expressions.

Example 3 **Evaluate $m + n$ if $m = 3\frac{1}{6}$ and $n = -1\frac{5}{6}$.**

$m + n = 3\frac{1}{6} + \left(-1\frac{5}{6}\right)$ *Replace m with $3\frac{1}{6}$ and n with $-1\frac{5}{6}$.*

$= 3\frac{1}{6} - 1\frac{5}{6}$

$= 2\frac{7}{6} - 1\frac{5}{6}$ *Rename $3\frac{1}{6}$ as $2\frac{7}{6}$.*

$= (2 - 1) + \left(\frac{7}{6} - \frac{5}{6}\right)$

$= 1\frac{2}{6}$ or $1\frac{1}{3}$

GLENCOE Technology

Interactive Mathematics Tools Software

In this interactive computer lesson, students explore adding like fractions. A **Computer Journal** gives students an opportunity to write about what they have learned.

For Windows & Macintosh

How do you like to spend time in the great outdoors? The results of a survey by Roper Starch Worldwide on that subject are shown in the graph at the right.

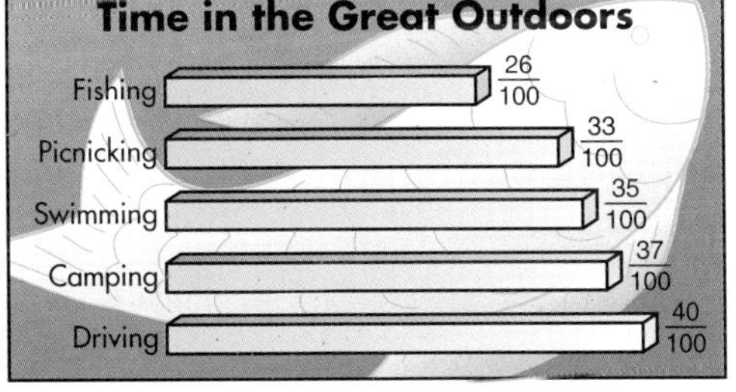

Source: Roper Starch Worldwide

a. What fraction of the people said that they enjoyed driving or swimming?

$$\frac{40}{100} + \frac{35}{100} = \frac{40 + 35}{100}$$
$$= \frac{75}{100} \text{ or } \frac{3}{4}$$

b. What fraction of the people chose camping or picnicking as favorite activities?

$$\frac{37}{100} + \frac{33}{100} = \frac{37 + 33}{100}$$
$$= \frac{70}{100} \text{ or } \frac{7}{10}$$

Checking Your Understanding

Communicating Mathematics

Read and study the lesson to answer these questions.

1. What number properties allow you to write $4\frac{2}{7} + 8\frac{3}{7}$ as $(4 + 8) + \left(\frac{2}{7} + \frac{3}{7}\right)$? **associative and commutative, +**
2. **Draw** and shade a model or use objects like paper plates to represent the sum of $\frac{3}{5}$ and $\frac{1}{5}$. **See margin.**
3. **Express** $\frac{33}{7}$ as a mixed number. **$4\frac{5}{7}$**

4. List several examples of like fractions and unlike fractions. Then in your own words state a simple rule for adding and subtracting like fractions. **See students' work.**

Guided Practice

Solve each equation. Write the solution in simplest form.

5. $\frac{4}{7} + \frac{5}{7} = m$ **$1\frac{2}{7}$**
6. $1\frac{2}{19} + \frac{17}{19} = a$ **2**
7. $\frac{19}{24} - \frac{11}{24} = r$ **$\frac{1}{3}$**
8. $-\frac{19}{30} + \frac{7}{30} = z$ **$-\frac{2}{5}$**

Evaluate each expression if $x = \frac{3}{8}$, $y = \frac{7}{8}$, and $z = \frac{5}{8}$. Write in simplest form.

9. $x + z$ **1**
10. $z - y$ **$-\frac{1}{4}$**

Reteaching

Using Models Use measuring cups and water to model addition and subtraction of fractions with like denominators, such as $\frac{1}{3} + \frac{1}{3} = \frac{2}{3}$ and $\frac{3}{4} - \frac{1}{4} = \frac{1}{2}$. Ask students how an answer is changed to simplest form.

Additional Answer

2.

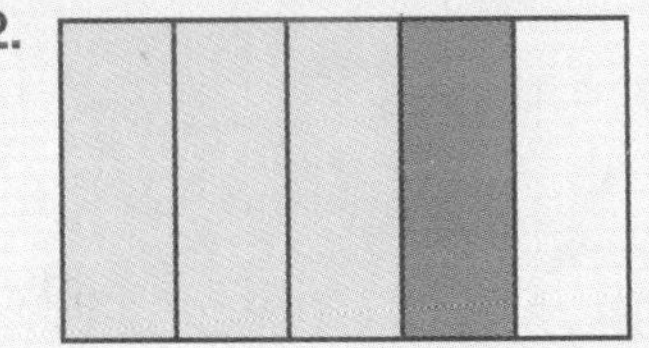

3 PRACTICE/APPLY

Checking Your Understanding

Exercises 1–11 are designed to help you assess your students' understanding through reading, writing, speaking, and modeling. You should work through Exercises 1–4 with your students and then monitor their work on Exercises 5–11.

Teaching Tip Remind students that answers are to be given in simplest form.

Study Guide Masters, p. 39

NAME ______________________ DATE ____________

5-4 Study Guide
Adding and Subtracting Like Fractions

Student Edition Pages 239–243

To add fractions with like denominators, add the numerators. Write the sum over the common denominator. When the sum of the two fractions is greater than 1, the sum is written as a mixed number.

$$w = \frac{4}{12} + \frac{10}{12} \qquad n = 2\frac{2}{5} + 1\frac{1}{5}$$
$$= \frac{14}{12} \qquad = 3\frac{3}{5}$$
$$= 1\frac{2}{12} \text{ or } 1\frac{1}{6}$$

To subtract fractions with like denominators, subtract the numerators. Write the difference over the common denominator.

$$x = \frac{7}{15} - \frac{1}{15} \qquad y = 4\frac{3}{4} - 2\frac{1}{4}$$
$$x = \frac{6}{15} \text{ or } \frac{2}{5} \qquad y = 2\frac{2}{4} - 2\frac{1}{2}$$

Solve each equation. Write the solution in simplest form.

1. $\frac{4}{7} + \frac{2}{7} = a$ **$\frac{6}{7}$**
2. $m = 1\frac{6}{9} + 2\frac{3}{9}$ **4**
3. $s = 1\frac{12}{20} + \frac{7}{20}$ **$1\frac{19}{20}$**
4. $\frac{15}{16} + \frac{7}{16} = n$ **$1\frac{3}{8}$**
5. $\frac{13}{20} - \frac{3}{20} = v$ **$\frac{1}{2}$**
6. $d = \frac{23}{18} - \frac{15}{18}$ **$\frac{4}{9}$**
7. $-\frac{12}{50} + \left(-\frac{2}{50}\right) = h$ **$-\frac{7}{25}$**
8. $j = 1\frac{13}{16} - \frac{7}{16}$ **$1\frac{3}{8}$**
9. $\frac{62}{52} - \frac{12}{52} = f$ **$\frac{25}{26}$**
10. $\frac{11}{8} - \frac{17}{8} = g$ **$-\frac{3}{4}$**
11. $\frac{17}{32} - \frac{5}{32} = d$ **$\frac{3}{8}$**
12. $b = 1\frac{15}{42} + \left(-\frac{11}{42}\right)$ **$1\frac{2}{21}$**
13. $2\frac{2}{30} - \frac{12}{30} = y$ **$1\frac{2}{3}$**
14. $c = 6\frac{6}{7} + \frac{6}{7}$ **$7\frac{5}{7}$**
15. $p = \frac{7}{8} - \frac{3}{8}$ **$\frac{1}{2}$**

Evaluate each expression if $x = \frac{1}{15}$, $y = \frac{8}{15}$, and $z = \frac{4}{15}$. Write in simplest form.

16. $x - y$ **$-\frac{7}{15}$**
17. $z - x$ **$\frac{1}{5}$**
18. $x + z$ **$\frac{1}{3}$**
19. $x + y$ **$\frac{3}{5}$**
20. $y - z$ **$\frac{4}{15}$**
21. $y + z$ **$\frac{12}{15}$**

Assignment Guide

Core: 13–41 odd, 42–48
Enriched: 12–34 even, 36–48

For **Extra Practice**, see p. 751.

The red A, B, and C flags, printed only in the Teacher's Wraparound Edition, indicate the level of difficulty of the exercises.

11. **Home Maintenance** The graph at the right shows the fraction of each dollar an owner spends on different expenses for a house.
    a. What fraction of each dollar is spent on utilities and property taxes? $\frac{9}{25}$
    b. How much more is spent on the mortgage than is spent on paying for housekeeping supplies? $\frac{1}{5}$ of each dollar

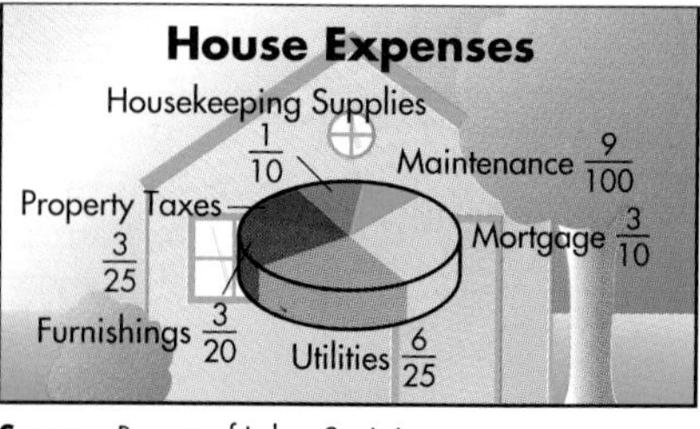

Source: Bureau of Labor Statistics

Exercises: Practicing and Applying the Concept

Independent Practice

Solve each equation. Write the solution in simplest form.

A

12. $\frac{5}{7} + \frac{1}{7} = b$ $\frac{6}{7}$
13. $m = \frac{11}{15} - \frac{8}{15}$ $\frac{1}{5}$
14. $a = \frac{19}{31} - \frac{8}{31}$ $\frac{11}{31}$
15. $\frac{19}{27} - \frac{7}{27} = r$ $\frac{4}{9}$
16. $1\frac{13}{16} - 1\frac{5}{16} = x$ $\frac{1}{2}$
17. $\frac{13}{18} + \frac{11}{18} = a$ $1\frac{1}{3}$

B

18. $m = 1\frac{9}{16} + \frac{15}{16}$ $2\frac{1}{2}$
19. $1\frac{16}{21} - \frac{9}{21} = y$ $1\frac{1}{3}$
20. $p = -\frac{19}{27} + \left(-\frac{7}{27}\right)$ $-\frac{26}{27}$
21. $1\frac{5}{12} + \frac{11}{12} = z$ $2\frac{1}{3}$
22. $\frac{9}{20} - \frac{17}{20} = c$ $-\frac{2}{5}$
23. $1\frac{19}{41} + \left(-\frac{19}{41}\right) = w$ 1

Evaluate each expression if $a = \frac{5}{12}$, $b = \frac{7}{12}$, and $c = \frac{1}{12}$. Write in simplest form.

24. $a + b$ 1
25. $b - c$ $\frac{1}{2}$
26. $a + c$ $\frac{1}{2}$
27. $b + c$ $\frac{2}{3}$
28. $c - a$ $-\frac{1}{3}$
29. $a - b$ $-\frac{1}{6}$

Simplify each expression.

30. $4\frac{1}{3}x + \frac{2}{3}x - 3\frac{1}{3}x$ $1\frac{2}{3}x$
31. $-4\frac{1}{4}r + \left(-2\frac{3}{4}\right)r + 5r$ $-2r$
32. $5\frac{1}{7}m + \left(-3\frac{3}{7}\right)m + 1\frac{2}{7}m$ $3m$
33. $6\frac{3}{5}a - 3\frac{4}{5}a - 2\frac{2}{5}a$ $\frac{2}{5}a$
34. $5\frac{5}{6}y + 3\frac{5}{6}y - 2\frac{1}{6}y$ $7\frac{1}{2}y$
35. $3\frac{5}{9}b - 2\frac{4}{9}b - \frac{7}{9}b$ $\frac{1}{3}b$

Critical Thinking

36. Write $\frac{x + y}{8}$ as the sum of two fractions. Sample answer: $\frac{x}{8} + \frac{y}{8}$
37. Write $\frac{2x - 1}{5}$ as the difference of two fractions. Sample answer: $\frac{2x}{5} - \frac{1}{5}$
38. Find the sum of $\frac{1}{20} + \frac{2}{20} + \frac{3}{20} + \cdots + \frac{17}{20} + \frac{18}{20} + \frac{19}{20}$. Describe any pattern or shortcut you used to find the sum. $9\frac{1}{2}$; See students' work.

Applications and Problem Solving

39. **Carpentry** A carpenter is installing a countertop in a new kitchen. The countertop will be between two walls that are $54\frac{5}{8}$ inches apart. If the piece of countertop material is five feet long, how much will have to be cut off before the countertop is installed? $5\frac{3}{8}$ inches

Practice Masters, p. 39

NAME ______________________ DATE ________

5-4 Practice
Adding and Subtracting Like Fractions

Student Edition Pages 239–243

Solve each equation. Write the solution in simplest form.

1. $\frac{2}{9} + \frac{2}{9} = x$ $\frac{4}{9}$
2. $\frac{2}{3} - \frac{1}{3} = y$ $\frac{1}{3}$
3. $2\frac{7}{10} - 2\frac{1}{10} = t$ $\frac{3}{5}$
4. $s = 1\frac{9}{16} + \frac{11}{16}$ $2\frac{1}{4}$
5. $r = \frac{13}{16} - \frac{5}{16}$ $\frac{1}{2}$
6. $v = 3\frac{7}{20} + \left(-\frac{9}{20}\right)$ $2\frac{9}{10}$
7. $1\frac{11}{24} - \frac{5}{24} = w$ $1\frac{1}{4}$
8. $b = \frac{11}{12} + 1\frac{7}{12}$ $2\frac{1}{2}$
9. $d = \frac{3}{12} - \frac{11}{12}$ $-\frac{2}{3}$
10. $p = \frac{7}{20} + \frac{17}{20}$ $1\frac{1}{5}$
11. $\frac{13}{18} - \frac{1}{18} = h$ $\frac{2}{3}$
12. $2\frac{11}{12} + \left(-\frac{11}{12}\right) = n$ 2
13. $j = \frac{8}{15} + \frac{11}{15}$ $1\frac{4}{15}$
14. $k = \frac{17}{24} - \frac{23}{24}$ $-\frac{1}{4}$
15. $\frac{11}{21} + \frac{17}{21} = n$ $1\frac{1}{3}$
16. $\frac{29}{6} - \frac{19}{6} = f$ $1\frac{2}{3}$
17. $a = \frac{1}{8} + \frac{57}{8}$ $7\frac{1}{4}$
18. $\frac{17}{18} + \frac{5}{18} = m$ $1\frac{2}{9}$

Evaluate each expression if $x = \frac{3}{8}$, $y = \frac{7}{8}$, and $z = \frac{1}{8}$. Write the solution in simplest form.

19. $x + y$ $1\frac{1}{4}$
20. $y - z$ $\frac{3}{4}$
21. $x + z$ $\frac{1}{2}$
22. $y + z$ 1
23. $z - x$ $-\frac{1}{4}$
24. $x - y$ $-\frac{1}{2}$

Simplify each expression.

25. $3\frac{1}{4}a + \frac{3}{4}a - 2\frac{1}{4}a$ $1\frac{3}{4}a$
26. $4\frac{3}{8}b - 1\frac{5}{8}b - \frac{7}{8}b$ $1\frac{7}{8}b$
27. $5\frac{2}{5}x - 2\frac{3}{5}x - 1\frac{1}{5}x$ $1\frac{3}{5}x$
28. $6\frac{1}{6}y + 2\frac{1}{6}y - 3\frac{1}{6}y$ $5\frac{1}{6}y$

Group Activity Card 5-4

Concentrate On Fractions Group Activity **5-4**

MATERIALS: Twenty 3x5 note cards • watch for keeping time or one minute timer • paper and pencil

SETUP: Copy the expressions given on the back onto note cards.

This is a concentration game using the deck of cards. Begin by placing all the cards face down in four rows of five cards each. Take turns being the timer.

The object of the game is to turn over two cards that represent the same expression. You have one minute to do this. In this concentration deck, one card states the exercise, and its pair gives the answer in simplest form.

If you turn over two cards that match you keep them, and you can turn over two more cards.

When all cards are taken, the person with the most cards wins the game.

 Pre-Algebra

41. $3\frac{1}{2}$ yards

40. **Home Projects** The Stadlers added an exercise room to the rear of their house. They installed $\frac{3}{8}$-inch thick paneling over a layer of dry wall $\frac{5}{8}$ inch thick. How thick are the wall coverings? **1 inch**

41. **Sewing** Nyoko is making a linen suit. The portion of the pattern envelope that shows the amounts of fabric needed for different sizes of skirts and jackets is shown at the right. If Nyoko is making a size 6 from 45-inch fabric, how much fabric should she buy for the jacket and skirt?

SIZE	(6	8	10)
JACKET			
45"	$2\frac{5}{8}$	$2\frac{3}{4}$	$2\frac{3}{4}$
60"	2	2	2
SKIRT			
45"	$\frac{7}{8}$	$1\frac{1}{8}$	$1\frac{1}{8}$
60"	$\frac{7}{8}$	$\frac{7}{8}$	$\frac{7}{8}$

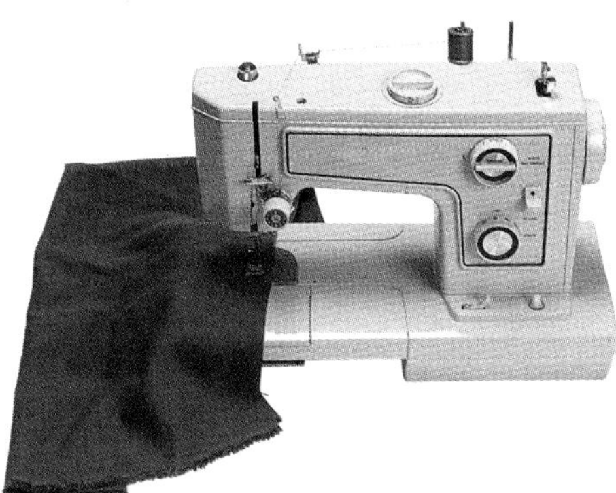

Mixed Review

42. **Shopping** Cheryl has \$38.78 in her wallet. If she buys a purse for \$25.59, how much will she have left? (Lesson 5-3) **\$13.19**

43. **Economics** The Valdosta Recreation Center needs \$17 million for renovations. The table at the right shows the contributions that have been made so far. (Lesson 4-6)

Contributor	Amount
City of Valdosta	\$2.5 million
First One Federal Bank	\$0.5 million
Montgomery County	\$0.5 million
State of Georgia	\$1.3 million
Recreation Center Fundraisers	\$3 million
Sale of tax credits	\$6.2 million

a. Write the ratio of the amount of money the city contributed to the total amount of money needed. $\frac{25}{170}$ or $\frac{5}{34}$

b. Write the ratio of the amount of money Montgomery County is contributing to the total amount of money received so far. $\frac{1}{28}$

44. Solve $\frac{d}{6} < -14$. (Lesson 3-7) $d < 84$

45. Solve $c - 12 \geq -25$. (Lesson 3-6) $c \geq -13$

46. **Geometry** Find the missing dimension in the rectangle at the right. (Lesson 3-5) **15 meters**

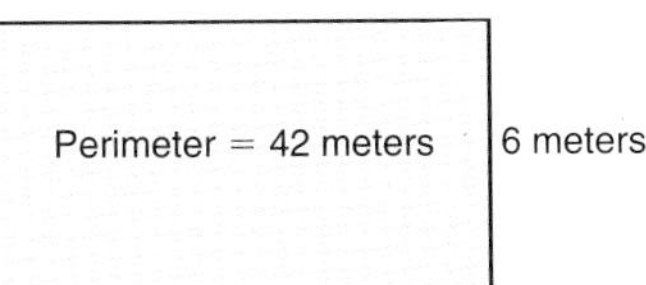

47. Simplify $-8a + (-17a) - (-3a)$. (Lesson 2-4) $-22a$

48. Name the property shown by $1z = z$. (Lesson 1-4) **Identity property, ×**

4 ASSESS

Closing Activity

Writing Have students evaluate each of the following, if $a = \frac{3}{8}$, $b = \frac{5}{8}$, and $c = 2\frac{1}{8}$. Tell students to write each answer in simplest form.

1. $a + b$ **1**
2. $b + a$ **1**
3. $a - b$ $-\frac{1}{4}$
4. $b - a$ $\frac{1}{4}$
5. $a + c$ $2\frac{1}{2}$
6. $a - b - c$ $-2\frac{3}{8}$

Enrichment Masters, p. 39

NAME ______________ DATE ________

5-4 Enrichment Student Edition Pages 239–243

Magic Squares

A magic square is a square array of numbers in which all rows, columns, and diagonals have the same sum, call the *magic sum*.

To make a magic square with fractions, follow these steps.

1. Draw a 3-by-3 square with 9 spaces. Then add a square on each side as shown by the dashed lines in Figure 1 below. Number the squares as shown.
2. Choose any fraction and put it in box 5. See Figure 2. The fraction $\frac{1}{24}$ was put in box 5.
3. Choose any fraction to add to the fraction in box 5 and put the answer in box 2. In the example, $\frac{3}{24}$ was chosen. Then add that same fraction to the result in box 2 and put this answer in box 1. See Figure 2.
4. Choose any fraction to add to the fraction in box 5 and put the answer in box 10. In the example, $\frac{5}{24}$ was chosen. Then add that same fraction to the result in box 10 and put this answer in box 13. See Figure 2.
5. Complete the upward diagonals by adding $\frac{3}{24}$ each time.
6. Check the downward diagonals by adding $\frac{5}{24}$ each time.
7. Now complete the magic square by putting the fraction in box 5 into box 8, the fraction in box 9 into box 6, the fraction in box 1 into box 11, and the fraction in box 13 into box 3. See Figure 3.

Figure 1

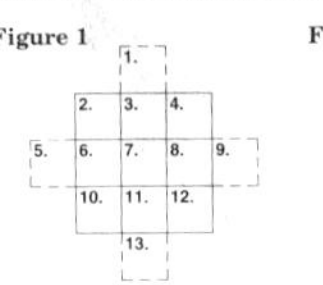

Figure 2

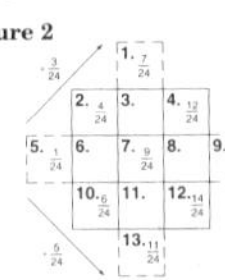

Figure 3

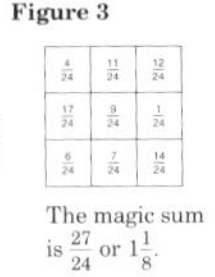

The magic sum is $\frac{27}{24}$ or $1\frac{1}{8}$.

Use the steps above to complete each exercise.

1. Complete the magic square below.

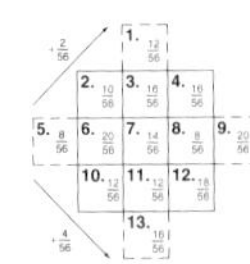

2. Make your own magic square. Check the rows and diagonals to be sure they have the same sum. **Answers will vary.**

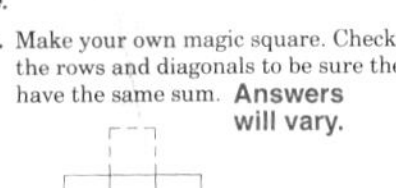

Extension

Using Patterns Write the equations at the right on the chalkboard or the overhead. Then have students explain why the sum of two identical fractions with an even denominator will always have the same numerator and the denominator is one half of the original denominator.

$$\frac{1}{4} + \frac{1}{4} = \frac{1}{2}$$
$$\frac{3}{4} + \frac{3}{4} = \frac{3}{2}$$
$$\frac{3}{8} + \frac{3}{8} = \frac{3}{4}$$
$$\frac{5}{6} + \frac{5}{6} = \frac{5}{3}$$

5-5 LESSON NOTES

NCTM Standards: 1-4, 7

Instructional Resources

- Study Guide Master 5-5
- Practice Master 5-5
- Enrichment Master 5-5
- Group Activity Card 5-5
- Assessment and Evaluation Masters, pp. 126, 127
- Graphing Calculator Masters, p. 5
- Real-World Applications, 11

Transparency 5-5A contains the 5-Minute Check for this lesson; **Transparency 5-5B** contains a teaching aid for this lesson.

Recommended Pacing	
Standard Pacing	Day 5–6 of 13
Honors Pacing	Day 5 of 12
Block Scheduling*	Day 3 of 6 (along with Lesson 5-6)

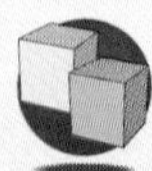

*For more information on pacing and possible lesson plans, refer to the ***Block Scheduling Booklet***.

1 FOCUS

5-Minute Check
(over Lesson 5-4)

Solve each equation. Write the solution in simplest form.

1. $\frac{1}{5} - \frac{4}{5} = x$ $\quad -\frac{3}{5}$
2. $2\frac{3}{8} + 1\frac{3}{8} = m$ $\quad 3\frac{3}{4}$
3. $y = 4\frac{1}{3} - 2\frac{2}{3}$ $\quad 1\frac{2}{3}$

Simplify each expression.

4. $2\frac{1}{3}t + 5\frac{2}{3}t - 3\frac{1}{3}t$ $\quad 4\frac{2}{3}t$
5. $\frac{11}{7}g + \left(-2\frac{3}{7}g\right) - 5\frac{1}{7}g$ $\quad -6g$

Motivating the Lesson

Questioning Ask students why a quarter and a half-dollar have the same total value as three quarters.

5-5 Adding and Subtracting Unlike Fractions

Setting Goals: *In this lesson, you'll add and subtract fractions with unlike denominators.*

Modeling with Manipulatives

MATERIALS

ruler
colored pencils

You can add or subtract fractions with unlike denominators by changing them into fractions with like denominators.

Your Turn

Make a model that represents the sum of $\frac{1}{4}$ and $\frac{1}{2}$. First shade $\frac{1}{4}$ of a rectangle like the one at the right. Then shade $\frac{1}{2}$ using a second color to represent $\frac{1}{4} + \frac{1}{2}$.

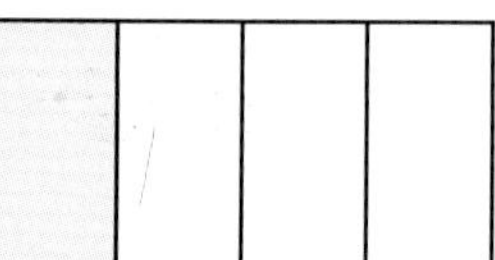

TALK ABOUT IT

a. What fraction of the model is shaded? $\frac{3}{4}$

b. What is the least common denominator of $\frac{1}{4}$ and $\frac{1}{2}$? 4

c. Write an addition sentence equivalent to $\frac{1}{4} + \frac{1}{2}$ by using the least common denominator. Then find the sum. $\frac{1}{4} + \frac{2}{4} = \frac{3}{4}$

Learning the Concept

The activity leads us to the following rule for adding and subtracting unlike fractions.

Adding and Subtracting Unlike Fractions

To find the sum or difference of two fractions with unlike denominators, rename the fractions with a common denominator. Then add or subtract and simplify.

One way to rename unlike fractions before you add or subtract is to use the LCD.

Example 1 **Solve each equation. Write the solution in simplest form.**

a. $m = \frac{7}{9} + \frac{11}{12}$ *Estimate: 1 + 1 = 2*

$m = \frac{7}{9} \cdot \frac{4}{4} + \frac{11}{12} \cdot \frac{3}{3}$ *The LCD is 3 · 3 · 2 · 2 or 36.*

$m = \frac{28}{36} + \frac{33}{36}$ *Rename each fraction with the LCD.*

$m = \frac{61}{36}$ or $1\frac{25}{36}$

Alternative Learning Styles

Visual Have students use a ruler to explain why $\frac{1}{2} + \frac{1}{4} = \frac{3}{4}$. To show the sum of other fractions, have students make drawings of rulers for other fractions with denominators of powers of two.

Additional Answer

2. Adding or subtracting the numerators without having a common denominator would be meaningless.

b. $\frac{1}{4} - \frac{2}{3} = a$ *Estimate: $\frac{1}{2} - 1 = -\frac{1}{2}$*

$\frac{3}{12} - \frac{8}{12} = a$ *The LCD is 12. $\frac{1}{4} = \frac{3}{12}, \frac{2}{3} = \frac{8}{12}$*

$-\frac{5}{12} = a$

To add and subtract mixed numbers with unlike denominators, rename using the LCD.

Example 2 **Solve each equation. Write the solution in simplest form.**

a. $n = 4\frac{1}{6} + 7\frac{11}{18}$ *Estimate: 4 + 8 = 12*

$n = 4\frac{3}{18} + 7\frac{11}{18}$ *Use the LCD to rename $\frac{1}{6}$ as $\frac{3}{18}$.*

$n = 11\frac{14}{18}$ or $11\frac{7}{9}$

b. $7\frac{2}{3} - 9\frac{1}{12} = y$ *Estimate: 8 − 9 = −1*

$7\frac{8}{12} - 9\frac{1}{12} = y$ *The LCD is 12. Rename the fractions.*

$7\frac{8}{12} - 8\frac{13}{12} = y$ *Rename $9\frac{1}{12}$ as $8\frac{13}{12}$.*

$-1\frac{5}{12} = y$

Stock market prices are expressed as mixed numbers.

Example 3

Business

News that Microsoft's Windows 95 software sales were brisk after its August, 1995, debut sent stocks of technological companies up. On Monday, September 11, 1995, Oracle Systems stock rose $\$1\frac{3}{8}$ to close at $\$45\frac{1}{4}$. What was the opening price for Oracle that day?

Let p represent the opening price. Then write an equation.

$p = 45\frac{1}{4} - 1\frac{3}{8}$

$p = 45\frac{2}{8} - 1\frac{3}{8}$ *The LCD is 8. Rename $\frac{1}{4}$ as $\frac{2}{8}$.*

$p = 44\frac{10}{8} - 1\frac{3}{8}$ *Rename $45\frac{2}{8}$ as $44\frac{10}{8}$.*

$p = 43\frac{7}{8}$

The opening price was $\$43\frac{7}{8}$.

Checking Your Understanding

Communicating Mathematics **Read and study the lesson to answer these questions.**

1. What is the first step in adding or subtracting fractions with unlike denominators? **Rename them using the LCD.**
2. **Explain** why you cannot add or subtract fractions without first finding a common denominator. **See margin.**

2 TEACH

In-Class Examples

For Example 1
Solve each equation.

a. $n = \frac{3}{8} + \frac{5}{6}$ $1\frac{5}{24}$

b. $\frac{2}{5} - \frac{1}{2} = p$ $-\frac{1}{10}$

For Example 2
Solve each equation.

a. $r = 6\frac{1}{3} + 5\frac{4}{9}$ $11\frac{7}{9}$

b. $8\frac{3}{4} - 9\frac{7}{12} = y$ $-\frac{5}{6}$

For Example 3
A stock opened the day at $32\frac{1}{8}$ and closed at $30\frac{1}{2}$. What was the change in the stock price during that day? $-1\frac{5}{8}$

3 PRACTICE/APPLY

Checking Your Understanding

Exercises 1–12 are designed to help you assess your students' understanding through reading, writing, speaking, and modeling. You should work through Exercises 1–5 with your students and then monitor their work on Exercises 6–12.

Study Guide Masters, p. 40

NAME ____________ DATE ____________

5-5 Study Guide
Adding and Subtracting Unlike Fractions

Student Edition Pages 244–247

To find the sum or difference of two fractions with unlike denominators, rename each fraction with a common denominator. The common denominator will be the least common multiple of the given denominators. This is called the **least common denominator (LCD)**.

	List the multiples to find the LCD.		Rename each fraction with the LCD.		Add or subtract. Simplify if necessary.
$\frac{1}{9}$	9: 9, 18, 27, . . .	→	$\frac{1}{9} = \frac{2}{18}$		$\frac{2}{18}$
$+\frac{2}{6}$	6: 6, 12, 18, . . .	→	$+\frac{2}{6} = \frac{6}{18}$	→	$+\frac{6}{18}$
	LCM: 18				$\frac{8}{18}$ or $\frac{4}{9}$

Solve each equation. Write the solution in simplest form.

1. $\frac{1}{3} - \frac{1}{6} = c$ $\frac{1}{6}$
2. $\frac{1}{5} + \frac{1}{7} = k$ $\frac{12}{35}$
3. $b = \frac{1}{8} + \frac{1}{9}$ $\frac{17}{72}$
4. $\frac{7}{16} - \frac{3}{8} = a$ $\frac{1}{16}$
5. $g = \frac{7}{10} + \frac{2}{5}$ $1\frac{1}{10}$
6. $\frac{3}{14} - \frac{1}{7} = h$ $\frac{1}{14}$
7. $2\frac{5}{12} + 1\frac{1}{3} = d$ $3\frac{3}{4}$
8. $6\frac{5}{4} + 3\frac{1}{2} = b$ $10\frac{3}{4}$
9. $4\frac{1}{6} - 3\frac{1}{8} = s$ $1\frac{1}{24}$
10. $a = 9\frac{1}{6} + 7\frac{4}{9}$ $16\frac{11}{18}$
11. $11\frac{3}{16} - 5\frac{1}{12} = m$ $6\frac{5}{48}$
12. $18\frac{7}{30} - 3\frac{1}{6} = y$ $15\frac{1}{15}$

Evaluate each expression if $c = -\frac{2}{3}$, $d = \frac{3}{4}$, and $f = 2\frac{5}{6}$. Write the solution in simplest form.

13. $f + c$ $2\frac{1}{6}$
14. $d + f$ $3\frac{7}{12}$
15. $c - d$ $-1\frac{5}{12}$
16. $f - c$ $3\frac{1}{2}$
17. $c + d$ $\frac{1}{12}$
18. $d + f + c$ $2\frac{11}{12}$
19. $f - d + c$ $1\frac{5}{12}$
20. $f + d - c$ $4\frac{1}{4}$
21. $c + f - d$ $1\frac{5}{12}$

Reteaching

Using Manipulatives Use measuring cups and water to model addition and subtraction of unlike fractions, such as $\frac{1}{2} + \frac{1}{4}$ and $\frac{3}{4} - \frac{1}{2}$. Have students explain the importance of renaming unlike fractions to add and subtract.

GLENCOE Technology

Interactive Mathematics Tools Software

In these interactive computer lessons, students explore adding unlike fractions and patterns in stock prices. A **Computer Journal** gives students an opportunity to write about what they have learned.

For Windows & Macintosh

Error Analysis

Some students may rename the denominator correctly, but not rename the numerator correctly. Some students may not rename the numerator at all, others may find the new denominator by adding rather than multiplying. Remind students to estimate answers before doing the computation and then to check their answer against the estimate.

Assignment Guide

Core: 13–39 odd, 40–47
Enriched: 14–36 even, 37–47
Self Test: 1–13

For **Extra Practice**, see p. 751.

The red A, B, and C flags, printed only in the Teacher's Wraparound Edition, indicate the level of difficulty of the exercises.

Additional Answers

3. Sample answer: A carpet layer needs $12\frac{1}{2}$ square yards of carpet for one room and $18\frac{3}{4}$ square yards for another room. How much carpet is needed in all?
4. Keri; Hank didn't find a common denominator.

Practice Masters, p. 40

NAME ______________________ DATE __________

5-5 Practice
Adding and Subtracting Unlike Fractions

Student Edition
Pages 244–247

Solve each equation. Write the solution in simplest form.

1. $\frac{1}{2} - \frac{1}{3} = x$ $\frac{1}{6}$
2. $y = \frac{3}{8} + \frac{1}{4}$ $\frac{5}{8}$
3. $3\frac{1}{3} - 2\frac{1}{2} = z$ $\frac{5}{6}$
4. $\frac{5}{12} + \frac{1}{3} = r$ $\frac{3}{4}$
5. $6\frac{3}{4} - 3\frac{5}{8} = d$ $3\frac{1}{8}$
6. $\frac{2}{3} + \frac{1}{6} = t$ $\frac{5}{6}$
7. $\frac{7}{8} - \frac{7}{12} = a$ $\frac{7}{24}$
8. $\frac{1}{6} + \frac{1}{2} = b$ $\frac{2}{3}$
9. $c = \frac{7}{9} - \frac{3}{5}$ $\frac{8}{45}$
10. $\frac{7}{8} + \frac{3}{4} = d$ $1\frac{5}{8}$
11. $10\frac{1}{6} + 2\frac{1}{18} = m$ $12\frac{2}{9}$
12. $n = \frac{1}{9} + \frac{2}{3}$ $\frac{7}{9}$
13. $x = 2\frac{1}{2} - 1\frac{3}{4}$ $\frac{3}{4}$
14. $7\frac{5}{6} - 2\frac{3}{4} = k$ $5\frac{1}{12}$
15. $9\frac{7}{8} + 2\frac{1}{6} = h$ $12\frac{1}{24}$
16. $\frac{7}{8} + \frac{1}{2} = m$ $1\frac{3}{8}$
17. $17\frac{4}{5} + 4\frac{5}{6} = j$ $22\frac{19}{30}$
18. $\frac{11}{12} - \frac{1}{16} = t$ $\frac{41}{48}$
19. $b = 3\frac{1}{8} - \frac{7}{8}$ $2\frac{1}{4}$
20. $\frac{1}{3} + \frac{5}{7} = r$ $1\frac{1}{21}$
21. $\frac{5}{8} - \frac{9}{16} = s$ $\frac{1}{16}$
22. $u = 3 - \frac{3}{8}$ $2\frac{5}{8}$
23. $1\frac{1}{3} + 2\frac{1}{6} = a$ $3\frac{1}{2}$
24. $6 - 2\frac{7}{8} = g$ $3\frac{1}{8}$

Evaluate each expression if $a = \frac{4}{9}$, $b = -\frac{2}{3}$, and $c = 3\frac{7}{18}$. Write the solution in simplest form.

25. $a + c$ $3\frac{5}{6}$
26. $b - a$ $-1\frac{1}{9}$
27. $c + b$ $2\frac{13}{18}$
28. $c - a + b$ $2\frac{5}{18}$
29. $a + b + c$ $3\frac{1}{6}$
30. $c + a - b$ $4\frac{1}{2}$
31. $a + b$ $-\frac{2}{9}$
32. $c - b$ $4\frac{1}{18}$
33. $c - a - b$ $3\frac{11}{18}$

3–5. See margin.

3. **Make up a problem** in which you would add $12\frac{1}{2}$ and $18\frac{3}{4}$.
4. **You Decide** Hank found the sum of $\frac{7}{9}$ and $\frac{11}{12}$ to be $\frac{18}{12}$ or $1\frac{1}{2}$. Keri found the sum to be $1\frac{25}{36}$. Who is correct? Explain.
5. **Draw** a model or use objects such as paper plates to represent the difference $\frac{3}{4} - \frac{1}{2}$.

Guided Practice

Solve each equation. Write the solution in simplest form.

6. $\frac{1}{3} + \frac{5}{6} = h$ $1\frac{1}{6}$
7. $m = \frac{3}{4} - \frac{5}{8}$ $\frac{1}{8}$
8. $t = \frac{3}{5} + \frac{3}{10}$ $\frac{9}{10}$
9. $n = \frac{3}{8} - \frac{1}{2}$ $-\frac{1}{8}$
10. $\frac{7}{12} - \frac{2}{3} = g$ $-\frac{1}{12}$
11. $\frac{3}{7} + \frac{5}{14} = w$ $\frac{11}{14}$
12. **Travel** Chilton Research studied the factors that travelers consider when choosing an airline.
    a. What fraction of those asked said they look at cost or safety record? $\frac{39}{50}$
    b. What is the difference between the fraction of people who said they considered time of arrival or departure and the number who chose based on the size or type of aircraft? $\frac{11}{100}$

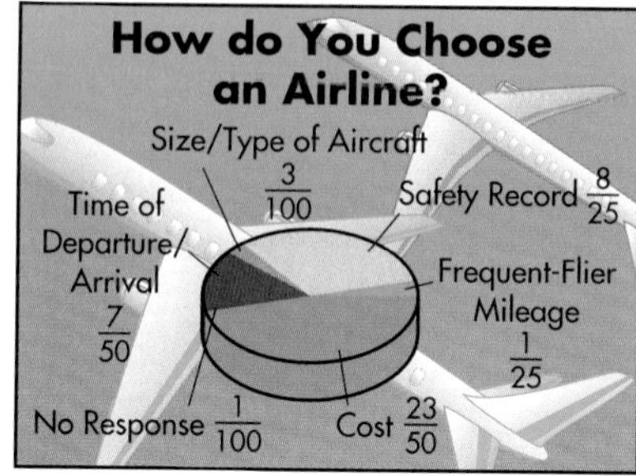

Source: Chilton Research

Exercises: Practicing and Applying the Concept

Independent Practice

Solve each equation. Write the solution in simplest form.

A

13. $x = \frac{1}{6} + \frac{7}{18}$ $\frac{5}{9}$
14. $m = \frac{4}{7} + \frac{9}{14}$ $1\frac{3}{14}$
15. $\frac{5}{12} - \frac{1}{2} = d$ $-\frac{1}{12}$
16. $y = \frac{6}{7} - \frac{5}{21}$ $\frac{13}{21}$
17. $r = \frac{9}{26} + \frac{3}{13}$ $\frac{15}{26}$
18. $b = \frac{11}{12} + \frac{3}{4}$ $1\frac{2}{3}$

B

19. $p = 4\frac{1}{3} - 2\frac{1}{2}$ $1\frac{5}{6}$
20. $k = \frac{7}{8} + \frac{3}{16}$ $1\frac{1}{16}$
21. $5\frac{1}{3} - 2\frac{1}{6} = m$ $3\frac{1}{6}$
22. $\frac{5}{7} - \frac{10}{21} = a$ $\frac{5}{21}$
23. $9\frac{3}{4} - 5\frac{1}{2} = d$ $4\frac{1}{4}$
24. $x = 7\frac{1}{3} - 3\frac{1}{2}$ $3\frac{5}{6}$
25. $y = 12\frac{3}{7} + 4\frac{1}{21}$ $16\frac{10}{21}$
26. $19\frac{3}{8} - 4\frac{3}{4} = s$ $14\frac{5}{8}$
27. $8\frac{9}{10} + 1\frac{1}{6} = r$ $10\frac{1}{15}$
28. $b = \frac{1}{3} + \frac{2}{35}$ $\frac{41}{105}$
29. $\frac{7}{10} - \frac{8}{9} = c$ $-\frac{17}{90}$
30. $18\frac{6}{7} + 2\frac{3}{5} = a$ $21\frac{16}{35}$

Evaluate each expression if $x = \frac{5}{8}$, $y = -\frac{3}{4}$, and $z = 2\frac{7}{12}$. Write in simplest form.

C

31. $x + z$ $3\frac{5}{24}$
32. $z + y$ $1\frac{5}{6}$
33. $y - x$ $-1\frac{3}{8}$
34. $x + y + z$ $2\frac{11}{24}$
35. $z - x + y$ $1\frac{5}{24}$
36. $z + x - y$ $3\frac{23}{24}$

Critical Thinking

37. Studies of unit fractions have been found in Greek papyrus dating back to 500 and 800 A.D. A *unit fraction* is a fraction that has a numerator of 1, such as $\frac{1}{5}$, $\frac{1}{7}$, or $\frac{1}{4}$. Greeks wrote all fractions as the sum of unit fractions. Express $\frac{2}{9}$ as the sum of two different unit fractions. Sample answer: $\frac{1}{18} + \frac{1}{6}$

Group Activity Card 5-5

Keep on Track

Group Activity **5-5**

MATERIALS: Copies of the number array below

The object of this activity is to find tracks from a circled number on the left to a boxed number on the right or from a circled number on the top to a boxed number on the bottom.

The sum of the starting circled number plus all of the numbers you cross on the track must add up to the ending boxed number.

You score one point for every correct track you draw and you lose one point for every incorrect track you draw. After ten minutes, total your points. The winner is the player or group of players who have the highest score.

NUMBER TRACK

	$-\frac{5}{6}$	$\frac{1}{2}$	$\frac{1}{6}$	$\frac{2}{3}$	
$-\frac{1}{4}$	$\frac{3}{4}$	$\frac{1}{3}$	$-\frac{7}{8}$	$\frac{5}{8}$	$\frac{23}{24}$
$-\frac{1}{3}$	$-\frac{1}{2}$	$\frac{5}{6}$	$-\frac{1}{4}$	$\frac{1}{8}$	$-\frac{3}{4}$
$\frac{3}{4}$	$-\frac{1}{6}$	$\frac{1}{2}$	$-\frac{1}{8}$	$-\frac{3}{4}$	$-\frac{1}{24}$
$\frac{7}{8}$	$-\frac{3}{8}$	$\frac{1}{4}$	$\frac{1}{6}$	$-\frac{2}{3}$	$\frac{1}{4}$
	$\frac{1}{2}$	1	$-\frac{11}{12}$	$-\frac{1}{8}$	

 Pre-Algebra

Extension

Make a Drawing Have students research the meaning of the word *tolerance* in measurement. Then have students draw line segments to show the shortest and longest acceptable segments for each of the following.

1. $\frac{1}{2}$ in. $\pm \frac{1}{16}$ in.
2. 3 cm $\pm$ 0.05 cm

Applications and Problem Solving

38. **Construction** The outside walls of a new home have $\frac{5}{8}$-inch drywall, $5\frac{1}{2}$ inches of insulation, $\frac{3}{4}$-inch outside wall plywood sheathing, and $\frac{7}{8}$-inch siding. How thick is the wall? **$7\frac{3}{4}$ in.**

39. **Nutrition** The Beverage Marketing Corporation found that $\frac{1}{6}$ of American households bought bottled water in 1992. Only $\frac{1}{17}$ of the households bought bottled water in 1985. What fraction of the population bought bottled water in 1992 that did not in 1985? **$\frac{11}{102}$**

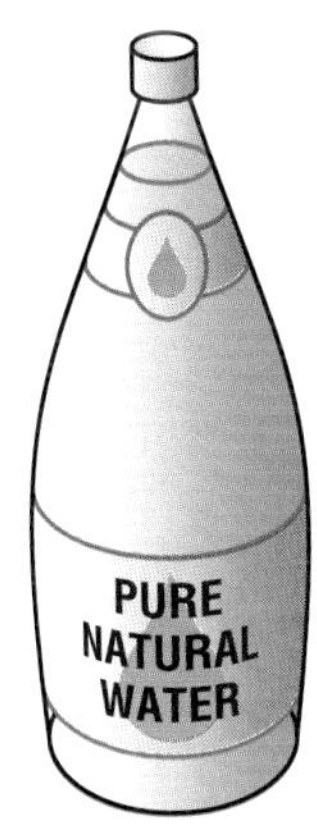

Mixed Review

40. Simplify $5\frac{1}{3}b + 3\frac{2}{3}b - 2\frac{1}{3}b$. (Lesson 5-4) **$6\frac{2}{3}b$**
41. Express $s^{-2}t^3$ with positive exponents. (Lesson 4-9) **$\frac{t^3}{s^2}$**
42. **Boating** The U.S. Coast Guard recommends using the formula $p = \frac{\ell w}{15}$, where ℓ is the length of the boat and w is the width, to determine the number of people, p, who can safely occupy a pleasure boat. If a Daysailer is 18 feet long and 5 feet wide, how many people can occupy it? (Lesson 3-4) **6 people**
43. Solve $b = -40 \div 8$. (Lesson 2-8) **−5**
44. Find $-8(3)(-2)$. (Lesson 2-7) **48**
45. Solve $y = 9 - (-4)$. (Lesson 2-5) **13**
46. Order the integers in the set {9, −3, −1, 12, −11} from least to greatest. (Lesson 2-3) **−11, −3, −1, 9, 12**
47. Simplify $3x + 15x$. (Lesson 1-5) **$18x$**

Self Test

Express each decimal as a fraction or mixed number in simplest form. (Lesson 5-1)

1. $-0.1666\ldots$ **$-\frac{1}{6}$**
2. 2.98 **$2\frac{49}{50}$**
3. $1.\overline{4}$ **$1\frac{4}{9}$**

Estimate each sum or difference. (Lesson 5-2) **Sample answers are given.**

4. \$50.00 − \$37.52 **\$12**
5. $\frac{8}{9} + \frac{1}{15}$ **1**
6. $\frac{9}{11} - \frac{5}{12}$ **$\frac{1}{2}$**

Solve each equation. (Lesson 5-3)

7. $r = 5.4 + 9.12$ **14.52**
8. $90.6 - 33.1 = k$ **57.5**
9. $m = 17.4 - (-13.2)$ **30.6**

Solve each equation. Write the solution in simplest form. (Lesson 5-4)

10. $\frac{8}{10} + \frac{3}{10} = a$ **$1\frac{1}{10}$**
11. $b = 4\frac{6}{7} + \frac{2}{7}$ **$5\frac{1}{7}$**
12. $8\frac{5}{35} - 2\frac{8}{35} = d$ **$5\frac{32}{35}$**

13. **Publishing** The length of a page in a yearbook is 10 inches. The top margin is $\frac{1}{2}$ inch, and the bottom margin is $\frac{3}{4}$ inch. What is the length of the page inside the margins? (Lesson 5-5) **$8\frac{3}{4}$ inches**

4 ASSESS

Closing Activity

Modeling Write the numbers −12 through 12 on blank index cards. Have students draw four cards and form two fractions. Then have students add or subtract the fractions and write the solution in simplest form.

Chapter 5, Quiz B (Lessons 5-4 and 5-5) is available in the *Assessment and Evaluation Masters*, p. 127.

Mid-Chapter Test (Lessons 5-1 through 5-5) is available in the *Assessment and Evaluation Masters*, p. 126.

Self Test

The Self Test provides students with a brief review of the concepts and skills in Lessons 5-1 through 5-5. Lesson numbers are given to the right of exercises or instruction lines so students can review concepts not yet mastered.

Additional Answer

5. $\frac{3}{4} - \frac{1}{2} = \frac{1}{4}$

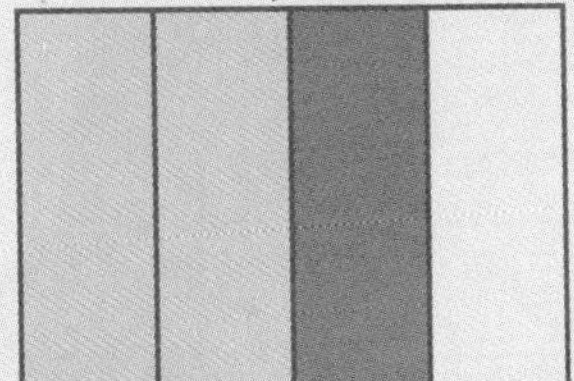

Enrichment Masters, p. 40

NAME ______ DATE ______

5-5 Enrichment
Adding and Subtracting Fractions

Student Edition Pages 244–247

Solve each problem. The first one is done for you.

1. $\frac{7}{14} + \frac{3}{14}$ $\frac{5}{7}$ I
2. $\frac{5}{6} + \frac{7}{8}$ $1\frac{17}{24}$ K
3. $\frac{3}{5} + \frac{7}{10} + \frac{1}{15}$ $1\frac{11}{30}$ U
4. $\frac{3}{8} - \frac{1}{8}$ $\frac{1}{4}$ A
5. $\frac{6}{10} - \frac{2}{5}$ $\frac{1}{5}$ G
6. $2\frac{2}{3} + \frac{3}{5}$ $3\frac{4}{15}$ D
7. $8\frac{1}{2} - 6\frac{1}{3}$ $2\frac{1}{6}$ H
8. $25\frac{1}{9} - 14\frac{5}{6}$ $10\frac{5}{18}$ B
9. $15\frac{3}{8} + 13\frac{1}{4}$ $28\frac{5}{8}$ L
10. $5\frac{3}{4} + 6\frac{2}{3}$ $12\frac{5}{12}$ F
11. $8 - 2\frac{2}{3}$ $5\frac{1}{3}$ E
12. $5\frac{5}{7} - 3\frac{1}{6}$ $2\frac{23}{42}$ N
13. $1\frac{5}{24} + \frac{11}{12}$ $2\frac{1}{8}$ O
14. $8\frac{1}{6} - 3\frac{5}{8}$ $4\frac{13}{24}$ T
15. $15\frac{11}{30} + 5\frac{4}{15}$ $20\frac{19}{30}$ C
16. $15 - 8\frac{2}{3}$ $6\frac{1}{3}$ S

Write the letter that is next to each solution above on the corresponding line shown below to complete the following question. The first letter is completed.

Did you hear about the football player who . . .

A S K E D   H I S   C O A C H   T O
$\frac{1}{4}$ $6\frac{1}{3}$ $1\frac{17}{24}$ $5\frac{1}{3}$ $3\frac{4}{15}$ $2\frac{1}{6}$ $\frac{5}{7}$ $6\frac{1}{3}$ $20\frac{19}{30}$ $2\frac{1}{8}$ $\frac{1}{4}$ $20\frac{19}{30}$ $2\frac{1}{6}$ $4\frac{13}{24}$ $2\frac{1}{8}$

F L O O D   T H E   F I E L D
$12\frac{5}{12}$ $28\frac{5}{8}$ $2\frac{1}{8}$ $2\frac{1}{8}$ $3\frac{4}{15}$ $4\frac{13}{15}$ $2\frac{1}{6}$ $5\frac{1}{3}$ $12\frac{5}{12}$ $\frac{5}{7}$ $5\frac{1}{3}$ $28\frac{5}{8}$ $3\frac{4}{15}$

S O   H E   C O U L D   G O   I N
$6\frac{1}{3}$ $2\frac{1}{8}$ $2\frac{1}{6}$ $5\frac{1}{3}$ $20\frac{19}{30}$ $2\frac{1}{8}$ $1\frac{11}{30}$ $28\frac{5}{8}$ $3\frac{4}{15}$ $\frac{1}{5}$ $2\frac{1}{8}$ $\frac{5}{7}$ $2\frac{23}{42}$

A S   A   S U B?
$\frac{1}{4}$ $6\frac{1}{3}$ $\frac{1}{4}$ $6\frac{1}{3}$ $1\frac{11}{30}$ $10\frac{5}{8}$

5-6 LESSON NOTES

NCTM Standards: 1-4, 7, 9

Instructional Resources

- Study Guide Master 5-6
- Practice Master 5-6
- Enrichment Master 5-6
- Group Activity Card 5-6
- Multicultural Activity Masters, p. 10

Transparency 5-6A contains the 5-Minute Check for this lesson; **Transparency 5-6B** contains a teaching aid for this lesson.

Recommended Pacing	
Standard Pacing	Day 7 of 13
Honors Pacing	Day 6 of 12
Block Scheduling*	Day 3 of 6 (along with Lesson 5-5)

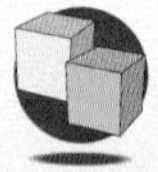

*For more information on pacing and possible lesson plans, refer to the ***Block Scheduling Booklet***.

1 FOCUS

5-Minute Check
(over Lesson 5-5)

Solve each equation. Write the solution in simplest form.

1. $\frac{7}{15} + \frac{3}{9} = y$ $\quad \frac{4}{5}$
2. $g = \frac{-2}{3} - \frac{5}{9}$ $\quad -1\frac{2}{9}$
3. $d = 6\frac{1}{4} + 11\frac{5}{6}$ $\quad 18\frac{1}{12}$
4. $-3\frac{3}{10} - \left(-6\frac{2}{5}\right) = v$ $\quad 3\frac{1}{10}$
5. $t = -9 - \left(-5\frac{1}{4}\right)$ $\quad -3\frac{3}{4}$

Motivating the Lesson

Hands-On Activity Place two identical containers with equal amounts of water on each pan of a two-pan scale. Remove $\frac{1}{3}$ cup of water from one container. Discuss what must be done to get the scale in balance again.

Study Guide Masters, p. 54 The Study Guide Master provides a concise presentation of the lesson along with practice exercises.

5-6 Solving Equations

Setting Goals: *In this lesson, you'll solve equations with rational numbers.*

Modeling a Real-World Application: Meteorology

Hurricanes can be devastating to life and property. The graph at the right compares the statistics for the five most costly hurricanes to strike the U.S. mainland. The two most costly hurricanes were Andrew and Hugo. Together, the damage they caused totaled $32.16 billion. How much did Hurricane Andrew cost?

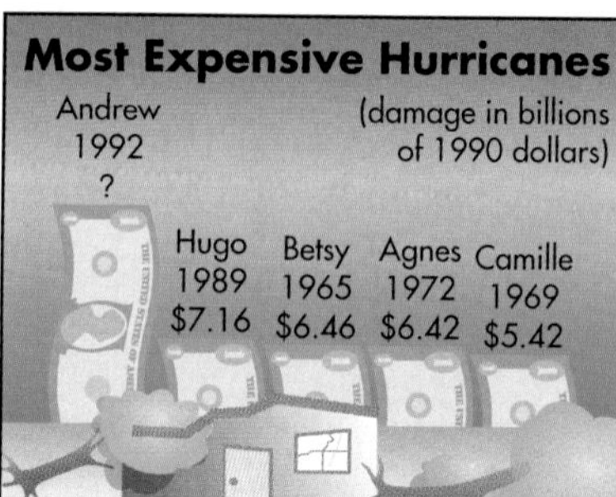

Source: National Hurricane Center

Learning the Concept

FYI

The earliest date for a hurricane to strike the U.S. was Hurricane Allison on June 5, 1995. The previous early hurricane record holder was Hurricane Alma on June 9, 1966, which like Allison, struck the Florida panhandle.

To find the cost of Hurricane Andrew, let c represent its cost and write an equation.

cost of Andrew	*plus*	*cost of Hugo*	*equals*	*total cost*
c	$+$	7.16	$=$	32.16

$$c + 7.16 = 32.16$$

$$c + 7.16 - 7.16 = 32.16 - 7.16 \quad \textit{Subtract 7.16 from each side.}$$

$$c = 25$$

Hurricane Andrew cost $25 billion.

You can solve rational number equations using the same skills you used to solve equations involving integers.

Example  **Solve each equation. Check the solution.**

a. $b + \frac{3}{5} = \frac{3}{2}$

$$b + \frac{3}{5} = \frac{3}{2}$$

$$b + \frac{3}{5} - \frac{3}{5} = \frac{3}{2} - \frac{3}{5} \quad \textit{Subtract } \tfrac{3}{5} \textit{ from each side.}$$

$$b = \frac{15}{10} - \frac{6}{10} \quad \textit{The LCD is 10.}$$

$$b = \frac{9}{10}$$

Group Activity Card 5-6

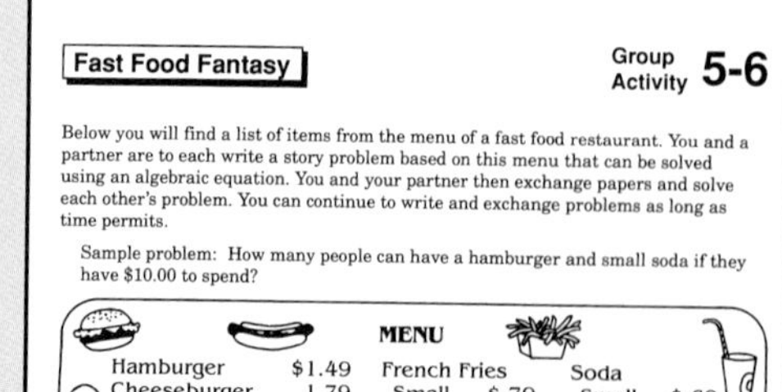

Fast Food Fantasy — Group Activity 5-6

Below you will find a list of items from the menu of a fast food restaurant. You and a partner are to each write a story problem based on this menu that can be solved using an algebraic equation. You and your partner then exchange papers and solve each other's problem. You can continue to write and exchange problems as long as time permits.

Sample problem: How many people can have a hamburger and small soda if they have $10.00 to spend?

MENU

Hamburger	$1.49	French Fries		Soda	
Cheeseburger	1.79	Small	$.79	Small	$.69
Hot dog	1.19	Large	.99	Medium	.89
Ice Cream Cone	.59				

 Pre-Algebra

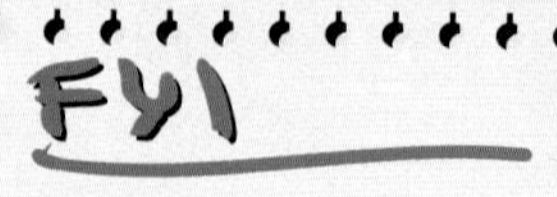

FYI

1995 was one of the busiest years for tropical storms, with 18 in the Atlantic Ocean. Hurricane Opal struck Florida on October 5, 1995, killing more than 20 people and causing more than $1.8 billion in damage.

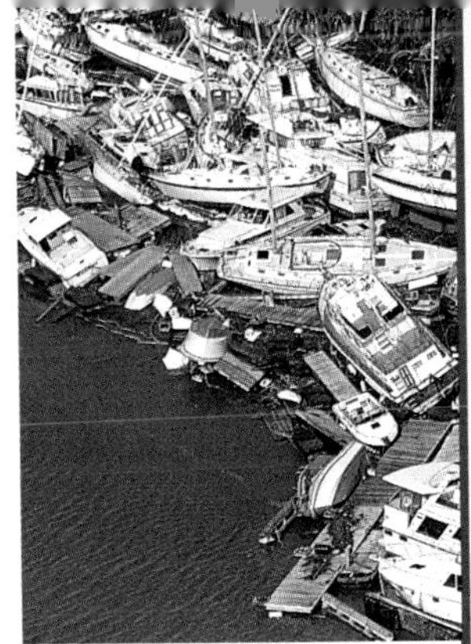

Check: $b + \frac{3}{5} = \frac{3}{2}$

$\frac{9}{10} + \frac{3}{5} \stackrel{?}{=} \frac{3}{2}$

$\frac{9}{10} + \frac{6}{10} \stackrel{?}{=} \frac{3}{2}$

$\frac{15}{10} \stackrel{?}{=} \frac{3}{2}$ ✓ The solution is $\frac{9}{10}$.

b. $a - 3.2 = 2.7$

$a - 3.2 = 2.7$

$a - 3.2 + 3.2 = 2.7 + 3.2$ *Add 3.2 to each side.*

$a = 5.9$

Check: $a - 3.2 = 2.7$

$5.9 - 3.2 \stackrel{?}{=} 2.7$

$2.7 = 2.7$ ✓ The solution is 5.9.

You can solve equations that represent real-world situations involving rational numbers.

Example 2 APPLICATION Meteorology

Meteorologists use barometers to measure the air pressure in the atmosphere. Changes in the pressure indicate that there will be a change in the weather. As a hurricane approaches, the barometric pressure at 7:00 P.M. was 28.65 inches. It had dropped 0.37 inches from the previous reading at noon. What was the noon reading?

Let p represent the barometric pressure at noon.

pressure at noon	*dropped*	*0.37 inches*	*to*	*28.65*
p		0.37	=	28.65

$p - 0.37 = 28.65$

$p - 0.37 + 0.37 = 28.65 + 0.37$ *Add 0.37 to each side.*

$p = 29.02$

The reading was 29.02 inches at noon.

HELP WANTED

Are you interested in the patterns and changes of the weather? For more information on this field, contact:

American Meteorological Society
45 Beacon St.
Boston, MA 02108

Checking Your Understanding

Communicating Mathematics

Read and study the lesson to answer these questions. 1–3. See margin.

1. What property of equality would you use to solve the equation $x - 5.3 = 6.8$?
2. **Explain** how to solve $n - 8.9 = 6.8$.
3. **Tell** how you know that 6.9 is not a solution of $n + 8.5 = 14.7$.

Guided Practice

Solve each equation. Check your solution. 7. −5.28 12. −6.34

4. $\frac{7}{6} = m + \frac{5}{12}$ $\frac{3}{4}$
5. $\frac{2}{3} + r = \frac{3}{5}$ $-\frac{1}{15}$
6. $7\frac{1}{2} = x - 5\frac{2}{3}$ $13\frac{1}{6}$
7. $m - 4.1 = -9.38$
8. $y + 7.2 = 21.9$ 14.7
9. $x - 1.5 = 1.75$ 3.25
10. $a - 1\frac{1}{3} = 4\frac{1}{6}$ $5\frac{1}{2}$
11. $-13.7 = b - 5$ −8.7
12. $y + 3.17 = -3.17$
13. **Driver's Education** The odometer on the driver's education car read 26,375.4 miles at the start of Rachel's first in-car class. After the class, the odometer read 26,397.3 miles. How far did Rachel drive?

13. 21.9 miles

2 TEACH

In-Class Examples

For Example 1

Solve each equation. Check the solution.

a. $\frac{5}{6} = n + \frac{4}{3}$ $-\frac{1}{2}$

b. $r - 8.2 = 2.5$ **10.7**

For Example 2

In the two hours before a storm struck Oceanside, the barometric pressure dropped 0.42 inches. When the storm struck, the pressure was 28.94 inches. What was the pressure before the storm?

29.36 inches

Additional Answers

1. addition property of equality
2. Sample answer: add 8.9 to each side to isolate the x.
3. Sample answer: $6.9 + 8.5 \neq 14.7$

Practice Masters, p. 41

NAME ______________________ DATE __________

5-6 Practice
Solving Equations

Student Edition Pages 248–250

Solve each equation. Check your solution.

1. $x + 7\frac{1}{2} = -8$ $-15\frac{1}{2}$
2. $y - 12 = -7\frac{1}{3}$ $4\frac{2}{3}$
3. $z - \left(-\frac{3}{4}\right) = 6\frac{1}{2}$ $5\frac{3}{4}$
4. $a - \frac{1}{6} = 2\frac{1}{3}$ $2\frac{1}{2}$
5. $b - 4.3 = 21.5$ 25.8
6. $c - \frac{4}{5} = 2\frac{1}{2}$ $3\frac{3}{10}$
7. $d + 2.4 = -15$ −17.4
8. $f + \frac{1}{3} = 7$ $6\frac{2}{3}$
9. $u - \frac{1}{4} = \frac{5}{4}$ $1\frac{1}{2}$
10. $v - 0.4 = 1.5$ 1.9
11. $54.7 + w = -6.72$ −61.42
12. $m + 4 = \frac{7}{8}$ $-3\frac{1}{8}$
13. $n - 1\frac{3}{4} = 3\frac{1}{6}$ $4\frac{11}{12}$
14. $p + \frac{7}{2} = -\frac{5}{4}$ $-4\frac{3}{4}$
15. $16\frac{4}{5} = 21\frac{3}{5} + q$ $-4\frac{4}{5}$
16. $x + 4.5 = 21.8$ 17.3
17. $t - 7.4 = -1.0$ 6.4
18. $b + 9.2 = 6.8$ −2.4
19. $m + 3.7 = 0.82$ −2.88
20. $g - (-23.6) = 18.3$ −5.3
21. $p + (-4.32) = 0.79$ 5.11
22. $w + 2\frac{1}{3} = 4\frac{1}{9}$ $1\frac{7}{9}$
23. $c - \frac{1}{7} = 2\frac{4}{5}$ $2\frac{33}{35}$
24. $u + \frac{2}{5} = 4\frac{1}{2}$ $4\frac{1}{10}$
25. $b - 3\frac{4}{7} = 7\frac{1}{5}$ $10\frac{27}{35}$
26. $n + 21.6 = 16.8$ −4.8
27. $z - \frac{5}{8} = 3\frac{1}{4}$ $3\frac{7}{8}$
28. $m + \frac{3}{2} = \frac{7}{2}$ 2
29. $\frac{2}{225} + z = \frac{5}{75}$ $\frac{13}{225}$
30. $b + 3\frac{1}{4} = 5$ $1\frac{3}{4}$

Reteaching

Using Properties Allow students to work in small groups and review the addition and subtraction properties for equations presented in Lessons 3–2 and 3–3. Then have students work together to solve the equations in Exercises 4–12.

HELP WANTED

Meteorologists make observations and analyses of past and current weather, extrapolate to find the future state of the atmosphere, and then use all that information to make a prediction.

3 PRACTICE/APPLY

Checking Your Understanding

Exercises 1–13 are designed to help you assess your students' understanding through reading, writing, speaking, and modeling. You should work through Exercises 1–3 with your students and then monitor their work on Exercises 4–13.

Assignment Guide

Core: 15–31 odd, 32, 33, 35, 37–43
Enriched: 14–30 even, 32–43

For **Extra Practice**, see p. 752.

The red A, B, and C flags, printed only in the Teacher's Wraparound Edition, indicate the level of difficulty of the exercises.

4 ASSESS

Closing Activity

Speaking Tell what needs to be added to or subtracted from each side of the equation to solve it.

1. $x + \frac{3}{4} = 2$ **subtract $\frac{3}{4}$**
2. $y - 2.4 = 2.4$ **add 2.4**

Enrichment Masters, p. 41

NAME ______________ DATE ______

5-6 Enrichment
Solving Equations Involving Addition and Subtraction

Student Edition Pages 248–250

Solve each equation.

1. $p + 5 = 8$ **3**
2. $16 + a = 20$ **4**
3. $d - 3\frac{2}{3} = 1\frac{1}{3}$ **5**
4. $j + 2\frac{1}{2} = 9\frac{1}{2}$ **7**
5. $8.2 + n = 14.2$ **6**
6. $g - 5 = 3$ **8**
7. $k + 1\frac{2}{3} = 5\frac{2}{3}$ **4**
8. $s - 3 = 4\frac{2}{5}$ **$7\frac{2}{5}$**
9. $x + 2\frac{1}{2} = 5$ **$2\frac{1}{2}$**
10. $15 = w + 7$ **8**
11. $r + 3.23 = 8.23$ **5**
12. $2\frac{4}{5} = t - 4\frac{3}{5}$ **$7\frac{2}{5}$**
13. $7 + y = 11$ **4**
14. $1.4 = x - 3.4$ **4.8**
15. $n + 5.3 = 8.7$ **3.4**
16. $g - 1\frac{2}{3} = 1\frac{1}{3}$ **3**
17. $n + 7 = 12.23$ **5.23**
18. $24 = x + 18$ **6**
19. $\frac{5}{6} = y - \frac{5}{6}$ **$1\frac{2}{3}$**
20. $7 + y = 12$ **5**
21. $t - 0.8 = 3.2$ **4**
22. $1\frac{2}{3} = n + 1\frac{2}{3}$ **0**

Match each solution above to the corresponding letter in the table below. Then place each letter on the appropriate blank at the bottom of the page to form the answer to this riddle.

Why didn't the man receive the basketballs he ordered?

A	B	C	D	E	H	I	K	N	O	S	U
7	3	5	0	4	$7\frac{2}{5}$	$2\frac{1}{2}$	3.4	$1\frac{2}{3}$	5.23	8	6

B	E	C	A	U	S	E		H	I	S
1	2	3	4	5	6	7		8	9	10

C	H	E	C	K		B	O	U	N	C	E	D
11	12	13	14	15		16	17	18	19	20	21	22

Exercises: Practicing and Applying the Concept

Independent Practice

Solve each equation. Check your solution.

A

14. $y + 3.5 = 14.9$ **11.4**
15. $r - 8.5 = -2.1$ **6.4**
16. $a - \frac{3}{5} = \frac{5}{6}$ **$1\frac{13}{30}$**
17. $b - 1\frac{1}{2} = 14\frac{1}{4}$ **$15\frac{3}{4}$**
18. $a + 7.1 = 4.7$ **−2.4**
19. $b - 5.3 = 8.1$ **13.4**

B

20. $y + 1\frac{1}{3} = 3\frac{1}{18}$ **$1\frac{13}{18}$**
21. $m + \frac{7}{12} = -\frac{5}{18}$ **$-\frac{31}{36}$**
22. $x + \frac{5}{8} = 7\frac{1}{2}$ **$6\frac{7}{8}$**
23. $k - \frac{3}{8} = 1\frac{3}{5}$ **$1\frac{39}{40}$**
24. $n + 1.4 = 0.72$ **−0.68**
25. $d - (-31.4) = 28.6$

C

26. $w - 0.04 = 1.2$ **1.24**
27. $d + (-7.03) = 0.98$
28. $5\frac{3}{10} + z = 2\frac{2}{3}$ **$-2\frac{19}{30}$**
29. $v - 4\frac{7}{9} = 8\frac{1}{6}$ **$12\frac{17}{18}$**
30. $501.1 = y - 9.32$
31. $\frac{1}{312} + t = \frac{5}{78}$ **$\frac{19}{312}$**

25. −2.8 27. 8.01 30. 510.42

Critical Thinking

32. Find two rational numbers, a and b, such that $a + b = ab$. **Sample answer: 6, 1.2**

Applications and Problem Solving

33. **Oil Production** In 1989, the top two oil-producing states were Texas and Alaska. Together they produced 1372.2 million barrels of oil. Alaska produced 684.0 million barrels. How many barrels of oil were produced in Texas? **688.2 million barrels**

34. **Geometry** The sum of the measures of the angles in a triangle is 180°. If two of the angles measure 75.5° and 60.3°, what is the measure of the third angle? **44.2°**

35. **Airports** The graph at the right gives information on the world's busiest cargo airports for 1994. Memphis tops the list at 1.65 million metric tons. O'Hare is the world's busiest passenger airport. What is the difference in cargo handling between Memphis and O'Hare? **0.39 million or 390,000 metric tons**

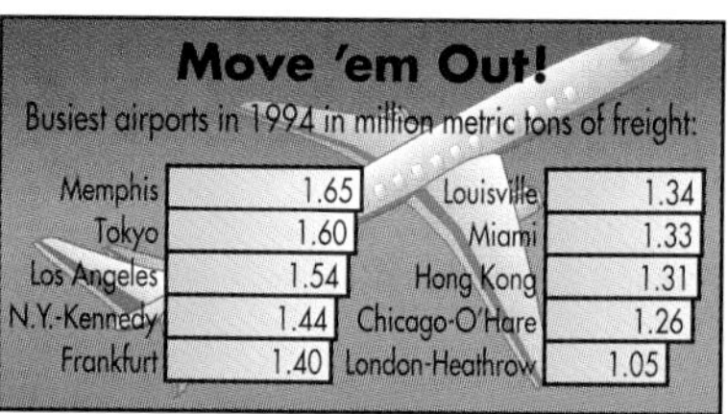

Move 'em Out!
Busiest airports in 1994 in million metric tons of freight:
Memphis 1.65
Tokyo 1.60
Los Angeles 1.54
N.Y.-Kennedy 1.44
Frankfurt 1.40
Louisville 1.34
Miami 1.33
Hong Kong 1.31
Chicago-O'Hare 1.26
London-Heathrow 1.05

Source: Airports Council International

36. **Trains** As a train begins to roll, the cars are "jerked" into motion, and the movement spreads like a ripple towards the back of the train. Slack is built into the couplings so that the engine does not have to move every single car all at once. If the slack built into each coupling is 3 inches or $\frac{1}{4}$ foot, how many feet of slack is there between two freight cars and the engine? **$\frac{1}{2}$ foot**

Mixed Review

37. Solve $y = \frac{7}{8} + 4\frac{1}{24}$. Write the solution in simplest form. (Lesson 5-5)
38. Factor $420ab^3$ completely. (Lesson 4-4)
39. Solve $\frac{d}{6} \geq -14$. (Lesson 3-7) **$d \geq -84$**
40. **Geometry** Find the area of a 4" by 6" rectangle. (Lesson 3-5) **24 in²**
41. Solve $x + 5 = 23$. (Lesson 3-2) **18**
42. **Medicine** When a baby is born, a doctor can take measurements and estimate the child's adult height. The doctor estimated that Tobi would be 5'11" tall. If Tobi is actually 5'9" tall, what is the difference between his actual height and his estimated height? (Lesson 2-5) **2 inches**
43. Solve $\frac{90}{x} = 6$ mentally. (Lesson 1-6) **15**

37. $4\frac{11}{12}$
38. $2 \cdot 2 \cdot 3 \cdot 5 \cdot 7 \cdot a \cdot b \cdot b \cdot b$

Extension

Using Lists Provide students with simple recipes such as the one at the right. Have students make up word problems involving the recipe. For each problem, students are to write an equation and then solve it.

Then have students exchange problems, solve, and check.

Honey-Fruit Salad
1 1/2 cups each, pears & peaches
1 1/4 cups orange sections
2 cups bananas
Syrup:
1 1/4 cup honey
1 1/2 T grated lemon peel
2 1/2 T lemon juice & orange juice
Heat syrup slowly for 5 minutes. Cool. Pour over fruit. Refrigerate.

5-7 Solving Inequalities

Setting Goals: *In this lesson, you'll solve inequalities with rational numbers.*

Modeling a Real-World Application: Sports

"So let's root, root, root, for the home team. . . ." Commentators and athletes often talk about the advantage of being the home team. Being the home team with a *new* home field may be even better.

CULTURAL CONNECTIONS
Baseball is very similar to the European game of cricket. Cricket was probably invented in England in the 1300s.

When a baseball team has a winning average of 0.600, this means they have won 0.600 or $\frac{3}{5}$ of their games. The table at the right lists the winning rates for teams in the first season in new homes. In 1995, the Colorado Rockies opened in a new park. As of May 22nd, their winning average at home was 0.600. How much would the Rockies have to raise their winning average to have the best new field record?

Home Sweet (New) Home

Field	Winning average
Jacobs Field (1994 Indians)	0.686
SkyDome (1989 Blue Jays)	0.618
Comiskey Park (1991 White Sox)	0.568
Camden Yards (1992 Orioles)	0.531
The Ballpark (1994 Rangers)	0.492
Metrodome (1982 Twins)	0.457

Source: Major League Baseball teams

Learning the Concept

The amount that the Rockies need to raise their average should be an inequality since anything greater than the present record will give them the best new field record.

$$\underbrace{0.600}_{\text{current average}} \underbrace{+}_{\text{plus}} \underbrace{x}_{\text{increase}} \underbrace{>}_{\text{is greater than}} \underbrace{0.686}_{\text{current record}}$$

$$0.600 - 0.600 + x > 0.686 - 0.600 \quad \textit{Subtract 0.600 from each side.}$$
$$x > 0.086$$

The Rockies would have to increase their winning record by more than 0.086 to establish a new home-field advantage record.

You can solve rational number inequalities using the same skills you used to solve inequalities involving integers.

Example 1 **Solve each inequality. Graph each solution on a number line.**

a. $a + \frac{3}{8} > 2$

$a + \frac{3}{8} - \frac{3}{8} > 2 - \frac{3}{8}$ *Subtract $\frac{3}{8}$ from each side.*

$a > \frac{16}{8} - \frac{3}{8}$ *Rename 2 as a fraction with a denominator of 8.*

$a > \frac{13}{8}$ or $1\frac{5}{8}$

(continued on the next page)

CULTURAL CONNECTIONS
In cricket, international matches lasting 5 days are called "test matches." The test match between England and Australia called "The Ashes" dates from 1882. In 1975, a World Cup of 1-day matches, played every 4 years, was introduced.

Alternative Teaching Strategies

Reading Mathematics Have students make lists of various words and phrases that can be used to write different inequalities, such as *at least, at most, not greater than, not less than,* and so on.

5-7 LESSON NOTES

NCTM Standards: 1-4, 7, 9

Instructional Resources

- Study Guide Master 5-7
- Practice Master 5-7
- Enrichment Master 5-7
- Group Activity Card 5-7
- Assessment and Evaluation Masters, p. 128

Transparency 5-7A contains the 5-Minute Check for this lesson; **Transparency 5-7B** contains a teaching aid for this lesson.

Recommended Pacing	
Standard Pacing	Day 8 of 13
Honors Pacing	Day 7 of 12
Block Scheduling*	Day 4 of 6 (along with Lesson 5-8)

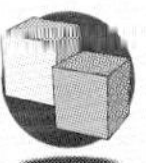

*For more information on pacing and possible lesson plans, refer to the ***Block Scheduling Booklet***.

1 FOCUS

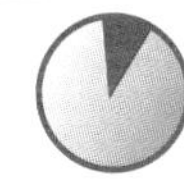

5-Minute Check *(over Lesson 5-6)*

Solve each equation. Check the solution.

1. $n + 6.3 = -2.5$ **−8.8**
2. $9.4 = w - 9.4$ **18.8**
3. $k - \frac{3}{4} = \frac{-2}{3}$ $\mathbf{\frac{1}{12}}$
4. $\frac{4}{5} = b - \left(\frac{-1}{2}\right)$ $\mathbf{\frac{3}{10}}$
5. $2\frac{1}{5} + p = 1\frac{7}{8}$ $\mathbf{-\frac{13}{40}}$

Motivating the Lesson

Situational Problem Use a catalog to select prices for merchandise. Then use the prices to make problems using decimals. For example, you have $15 and want to buy a necklace for $7.89. Can you buy earrings that cost $7.49? $\mathbf{x + 7.89 \le \$15;}$ $\mathbf{x \le \$7.11;}$ **no**

Teaching Tip You may wish to review the addition and subtraction properties for inequalities presented in Lessons 3-6 and 3-7.

2 TEACH

In-Class Examples

For Example 1

Solve each inequality. Graph each solution on a number line.

a. $n - \frac{3}{4} < 1\frac{1}{2}$ $\boldsymbol{n < 2\frac{1}{4}}$

0 1 2 3 4

b. $9.43 \geq 3.24 + w$ $\boldsymbol{w \leq 6.19}$

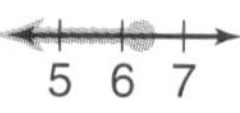

For Example 2

A large piece for a machine is to be 54.5 cm wide. The tolerance for the piece is ± 0.11 cm. What is the range of acceptable measurements for the piece? **53.4 cm ≤ *w* and *w* ≤ 55.6 cm**

Study Guide Masters, p. 42

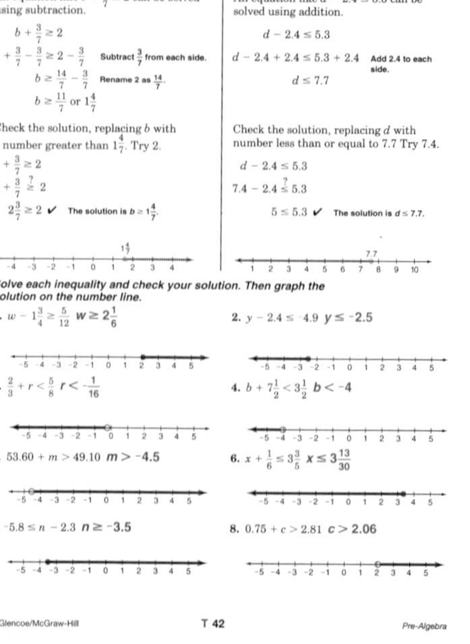

NAME ______ DATE ______

5-7 Study Guide
Solving Inequalities

Student Edition Pages 251–254

An equation like $b + \frac{3}{7} \geq 2$ can be solved using subtraction.

$b + \frac{3}{7} \geq 2$

$b + \frac{3}{7} - \frac{3}{7} \geq 2 - \frac{3}{7}$ Subtract $\frac{3}{7}$ from each side.

$b \geq \frac{14}{7} - \frac{3}{7}$ Rename 2 as $\frac{14}{7}$.

$b \geq \frac{11}{7}$ or $1\frac{4}{7}$

Check the solution, replacing b with a number greater than $1\frac{4}{7}$. Try 2.

$b + \frac{3}{7} \geq 2$

$2 + \frac{3}{7} \overset{?}{\geq} 2$

$2\frac{3}{7} \geq 2$ ✔ The solution is $b \geq 1\frac{4}{7}$.

An equation like $d - 2.4 \leq 5.3$ can be solved using addition.

$d - 2.4 \leq 5.3$

$d - 2.4 + 2.4 \leq 5.3 + 2.4$ Add 2.4 to each side.

$d \leq 7.7$

Check the solution, replacing d with number less than or equal to 7.7 Try 7.4.

$d - 2.4 \leq 5.3$

$7.4 - 2.4 \overset{?}{\leq} 5.3$

$5 \leq 5.3$ ✔ The solution is $d \leq 7.7$.

Solve each inequality and check your solution. Then graph the solution on the number line.

1. $w - 1\frac{3}{4} \geq \frac{5}{12}$ $\boldsymbol{w \geq 2\frac{1}{6}}$
2. $y - 2.4 \leq -4.9$ $\boldsymbol{y \leq -2.5}$
3. $\frac{2}{3} + r < \frac{5}{8}$ $\boldsymbol{r < -\frac{1}{16}}$
4. $b + 7\frac{1}{2} < 3\frac{1}{2}$ $\boldsymbol{b < -4}$
5. $53.60 + m > 49.10$ $\boldsymbol{m > -4.5}$
6. $x + \frac{1}{6} \leq 3\frac{3}{5}$ $\boldsymbol{x \leq 3\frac{13}{30}}$
7. $-5.8 \leq n - 2.3$ $\boldsymbol{n \geq -3.5}$
8. $0.75 + c > 2.81$ $\boldsymbol{c > 2.06}$

 T 42 *Pre-Algebra*

Check: Try 2, a number greater than $1\frac{5}{8}$.

$a + \frac{3}{8} > 2$

$2 + \frac{3}{8} \overset{?}{>} 2$

$2\frac{3}{8} > 2$ ✔ It checks.

The solution is $a > 1\frac{5}{8}$, all numbers greater than $1\frac{5}{8}$.

$1\frac{5}{8}$

0 1 2

b. $\boldsymbol{14.92 + r \geq 7.65}$

$14.92 + r \geq 7.65$

$14.92 - 14.92 + r \geq 7.65 - 14.92$ *Subtract 14.92 from each side.*

$r \geq -7.27$ 7.65 [−] 14.92 [=] -7.27

Check: Try −7, a number greater than −7.27.

$14.92 + r \geq 7.65$

$14.92 + (-7) \overset{?}{\geq} 7.65$

$7.92 \geq 7.65$ ✔ *It checks.*

The solution is $r \geq -7.27$, all numbers greater than or equal to −7.27.

−7.30 −7.20

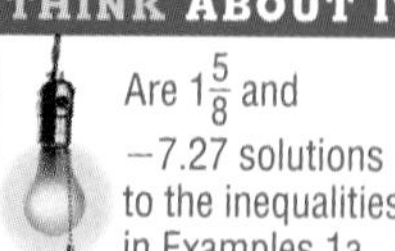

THINK ABOUT IT

Are $1\frac{5}{8}$ and −7.27 solutions to the inequalities in Examples 1a and 1b, respectively? Explain.

No; $1\frac{5}{8}$ and −7.27 are the boundaries of the solutions.

You can translate a phrase into an inequality involving rational numbers.

Example 2

Medicine

They say "If the shoe fits, wear it." But people often wear shoes that don't fit properly. Podiatrists recommend that a shoe be at least $\frac{3}{8}$ inch but no more than $\frac{1}{2}$ inch longer than your foot. If a shoe is $9\frac{1}{2}$ inches long, how long should a wearer's foot be?

Explore We know the length of the shoe. We need to find the length of the wearer's foot.

Plan This situation can be represented using two inequalities. Let f represent the length of the foot.

Solve Write the inequalities.

foot length	*plus*	*$\frac{3}{8}$ inch*	*no more than*	*$9\frac{1}{2}$ inches*
f	$+$	$\frac{3}{8}$	$\leq$	$9\frac{1}{2}$

$f + \frac{3}{8} \leq 9\frac{1}{2}$

$f + \frac{3}{8} - \frac{3}{8} \leq 9\frac{1}{2} - \frac{3}{8}$

$f \leq 9\frac{4}{8} - \frac{3}{8}$

$f \leq 9\frac{1}{8}$

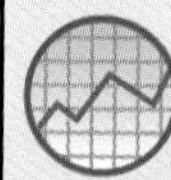

Tech Prep

Gymnastics Trainer A gymnastics trainer must know how to estimate a potential scoring range for his or her gymnasts as they complete their rotations. The trainer must also be able to tell athletes approximately what score they need in order to take the lead in a competition.

For more information on tech prep, see the *Teacher's Handbook.*

foot length	plus	$\frac{1}{2}$ inch	is no more than	$9\frac{1}{2}$ inches
f	$+$	$\frac{1}{2}$	$\geq$	$9\frac{1}{2}$

$$f + \frac{1}{2} \geq 9\frac{1}{2}$$

$$f + \frac{1}{2} - \frac{1}{2} \geq 9\frac{1}{2} - \frac{1}{2}$$

$$f \geq 9$$

The foot should be between 9 and $9\frac{1}{8}$ inches long.

Examine Since $9\frac{1}{8} + \frac{3}{8} = 9\frac{1}{2}$ and $9 + \frac{1}{2} = 9\frac{1}{2}$, the answer is reasonable.

Checking Your Understanding

Communicating Mathematics

Read and study the lesson to answer each question. 1–2. See margin.

1. **Compare and contrast** solving an inequality involving rational numbers to solving an inequality involving integers.
2. **Write a problem** that could be solved using the inequality $65.50 + x > 89.75$.
3. **Write** the inequality whose solution is graphed below. $n > 3.4$

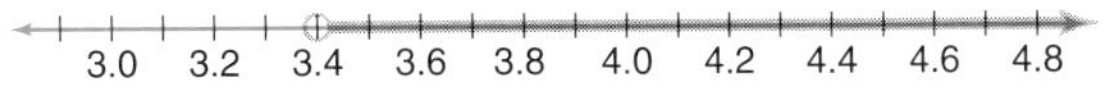

(number line: 3.0 3.2 3.4 3.6 3.8 4.0 4.2 4.4 4.6 4.8)

Guided Practice

Solve each inequality and check your solution. Graph the solution on a number line. See Solutions Manual.

4. $m - 2\frac{3}{4} \geq \frac{7}{12}$
5. $x + \frac{5}{6} \leq 3\frac{3}{5}$
6. $\frac{5}{9} < 1\frac{1}{2} + y$
7. $x - 1.4 \leq 7.9$
8. $n - 3.7 \geq -7.2$
9. $y + 5.2 \leq 7.12$
10. **Computers** The number of electronic mailboxes on private networks was 2.5 million in 1990. It is predicted that more than 36 million will exist in 1996. How many mailboxes will have been created between 1990 and 1996? more than 33.5 million

Exercises: Practicing and Applying the Concept

Independent Practice

Solve each inequality and check your solution. Graph the solution on a number line. See Solutions Manual.

11. $r + 7.5 \leq 13.2$
12. $x + 4.2 \geq -7.3$
13. $y + \frac{3}{4} < \frac{7}{12}$
14. $a - 2\frac{2}{3} \geq 8\frac{5}{6}$
15. $m - 3.1 < 7.4$
16. $b - 8.9 > -7.2$

17. $d - 1\frac{2}{3} < 1\frac{1}{6}$
18. $q - \frac{5}{12} \leq \frac{7}{18}$
19. $r + \frac{7}{8} \leq 2\frac{3}{4}$
20. $z - 1\frac{1}{2} \geq 4\frac{5}{9}$
21. $\frac{3}{4} \leq -1\frac{1}{2} + w$
22. $13.5 \leq 18.3 + k$
23. $h - 9.76 < 6.2$
24. $7\frac{6}{11} + f > 2\frac{14}{15}$
25. $u - 9.03 \leq 0.8$

C

26. $n - 289.90 < 479.21$
27. $\frac{17}{18} + t > \frac{107}{108}$
28. $0.027 + k \leq 0.00013$

Critical Thinking

29. If $\frac{1}{x} > x$, what can you say about the value of x? x is between 0 and 1 or x is less than -1.

3 PRACTICE/APPLY

Checking Your Understanding

Exercises 1–10 are designed to help you assess your students' understanding through reading, writing, speaking, and modeling. You should work through Exercises 1–3 with your students and then monitor their work on Exercises 4–10.

Assignment Guide

Core: 11–33 odd, 35–41
Enriched: 12–28 even, 29–41

For **Extra Practice**, see p. 752.

The red A, B, and C flags, printed only in the Teacher's Wraparound Edition, indicate the level of difficulty of the exercises.

Additional Answers

1. The same procedure is used to solve both types of inequalities. You may have to rename fractions in order to solve inequalities involving rational numbers.
2. Sample answer: Sharon earns \$65.50 plus tips each week delivering pizzas. How much will she have to make in tips to make more than \$89.75?

Practice Masters, p. 42

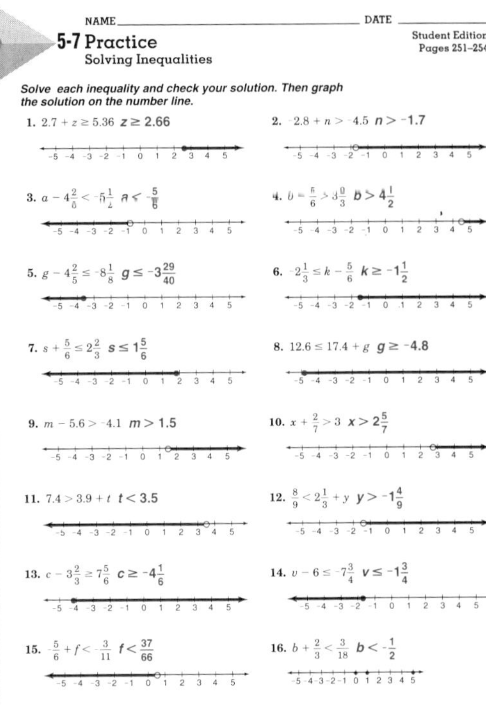

NAME ________ DATE ________

5-7 Practice
Solving Inequalities

Student Edition
Pages 251–254

Solve each inequality and check your solution. Then graph the solution on the number line.

1. $2.7 + z \geq 5.36$ $z \geq 2.66$
2. $-2.8 + n > -4.5$ $n > -1.7$
3. $a - 4\frac{2}{3} < -5\frac{1}{2}$ $a < -\frac{5}{6}$
4. $b - \frac{5}{6} > 3\frac{2}{3}$ $b > 4\frac{1}{2}$
5. $g - 4\frac{2}{5} \leq -8\frac{1}{8}$ $g \leq -3\frac{29}{40}$
6. $-2\frac{1}{3} \leq k - \frac{5}{6}$ $k \geq -1\frac{1}{2}$
7. $s + \frac{5}{6} \leq 2\frac{2}{3}$ $s \leq 1\frac{5}{6}$
8. $12.6 \leq 17.4 + g$ $g \geq -4.8$
9. $m - 5.6 > -4.1$ $m > 1.5$
10. $x + \frac{2}{7} > 3$ $x > 2\frac{5}{7}$
11. $7.4 > 3.9 + t$ $t < 3.5$
12. $\frac{8}{9} < 2\frac{1}{3} + y$ $y > -1\frac{4}{9}$
13. $c - 3\frac{2}{3} \geq 7\frac{5}{6}$ $c \geq -4\frac{1}{6}$
14. $v - 6 \leq -7\frac{3}{4}$ $v \leq -1\frac{3}{4}$
15. $-\frac{5}{6} + f < -\frac{3}{11}$ $f < \frac{37}{66}$
16. $b + \frac{2}{3} < \frac{3}{18}$ $b < -\frac{1}{2}$

Group Activity Card 5-7

Inequality Maker — Group Activity 5-7

MATERIALS: Three dice

Each player must make an inequality from the numbers rolled on three dice. You score a point if your inequality solution includes an integer rolled on one die.

To begin, someone rolls three dice, and all players make two different mixed numbers from combinations of them. One of the digits will be the whole number, one will be the numerator of the fraction, and the last will be the denominator of the fraction.

Next, everyone writes an inequality using these two mixed numbers, a variable, addition and subtraction, and one of the inequality signs. All players then solve their inequality.

Lastly, one die is rolled to get an integer. When players' solution sets include this number, they score one point. The first to score five points wins.

 Pre-Algebra

Reteaching

Using Connections Remind students that they can solve an inequality by solving the associated equation and then testing points on either side of the solution of the equation to find the solution to the inequality.

4 ASSESS

Closing Activity

Modeling Use masking tape to make a large number line on the floor. Have students solve an inequality and stand on the number line with arms extended in one direction to show the solution. Challenge students to create a way to show that the point they are standing on is not part of the solution, such as in $x < 2$.

Chapter 5, Quiz C (Lessons 5-6 and 5-7) is available in the *Assessment and Evaluation Masters*, p. 128.

Additional Answers

33a. Sample answer: The average salary of a high school graduate is slightly less than half of the average salary of a college graduate.

33b. Sample answer: No; if it was that low the inequality would be written differently. The inequality means that it is less than but close to one-half.

Enrichment Masters, p. 42

NAME ________________ DATE ________

5-7 Enrichment
Problem Solving: Gathering Data

Student Edition Pages 251–254

When investors decide to buy or sell stock, they usually base their decisions on data. They read stock market reports, investigate companies they are interested in, and look at economic trends. This activity focuses on reading stock market reports.

To read a stock market report, you need to become familiar with the terms that are used, the facts and figures that are published in the newspaper, and the methods used to keep track of the stocks.

Many newspapers print stock reports such as the New York Stock Exchange (NYSE) report. The final entry for each stock is a fraction or mixed number that is positive (indicating a gain in value) or negative (indicating a loss).

Get the business section of your local Sunday newspaper and find the New York Stock Exchange (NYSE) report. Using just those stocks that begin with A, B, or C, tally the number of stocks that had each of the gains and losses listed below.

Gain/Loss	Tally	Gain/Loss	Tally	Gain/Loss	Tally	Gain/Loss	Tally
+2 points or more		+1		$-\frac{1}{8}$		$-1\frac{1}{8}$	
$+1\frac{7}{8}$		$+\frac{7}{8}$		$-\frac{1}{4}$		$-1\frac{1}{4}$	
$+1\frac{3}{4}$		$+\frac{3}{4}$		$-\frac{3}{8}$		$-1\frac{3}{8}$	
$+1\frac{5}{8}$		$+\frac{5}{8}$		$-\frac{1}{2}$		$-1\frac{1}{2}$	
$+1\frac{1}{2}$		$+\frac{1}{2}$		$-\frac{5}{8}$		$-1\frac{5}{8}$	
$+1\frac{3}{8}$		$+\frac{3}{8}$		$-\frac{3}{4}$		$-1\frac{3}{4}$	
$+1\frac{1}{4}$		$+\frac{1}{4}$		$-\frac{7}{8}$		$-1\frac{7}{8}$	
$+1\frac{1}{8}$		$+\frac{1}{8}$		-1		-2 points or more	
		No change					

1. Did more stocks gain or lose? **Answers will vary.**
2. How did the market do as a whole? **Answers will vary.**
3. How did the newspapers describe the trading this week? **Answers will vary.**

Project

4. Watch two specific stocks each weekday for a month, and graph their daily closing price.

Applications and Problem Solving

30. Hobbies Susan plans to spend no more than \$30 on model airplanes and supplies. If she buys a model for \$19.95, how much more can she spend on supplies? Express your solution as an inequality. **$s \leq \$10.05$**

31. Figure Skating Michelle Kwan has scores of 9.4, 9.3, 9.8, and 8.9 in a figure skating competition. She has one more event to skate. Michelle will win the competition if her total score is greater than 47.1. What must Michelle's final event be rated in order for her to win? **$x > 9.7$**

32. Family Activity Imagine that you have won \$1000.00 in a radio contest. Choose one item that you would purchase for each person in your family including yourself. Research the prices in an advertisement or catalog. Would you have enough money? Write an inequality to show the amount of money you would have left over to put in savings. **See students' work.**

33. Employment According to the Census Bureau, a high school graduate earns less than half what a college graduate earns in a lifetime.

a. What do you think this statement means? **33a–33b. See margin.**

b. According to the statement, the average high school graduate could earn one-tenth of what an average college graduate does. Do you think that is true? Explain.

34. Geometry The length of each side of a triangle is always less than the sum of the lengths of the other two sides. Find s for the triangle shown at the right. **$s < 23\frac{7}{8}$**

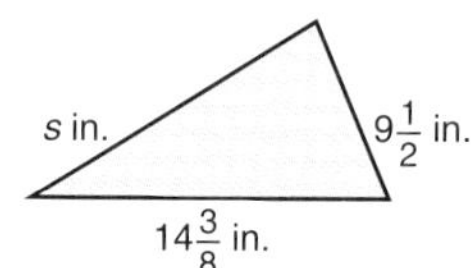

Mixed Review

35. Solve $x + 3\frac{1}{2} = 7\frac{1}{4}$. (Lesson 5-6) **$3\frac{3}{4}$**

36. Write $\frac{3}{a^2}$ as an expression using negative exponents. (Lesson 4-9) **$3a^{-2}$**

37. Find the GCF of 46 and 72. (Lesson 4-5) **2**

38. **Geometry** Find the perimeter and area of the rectangle shown at the right. (Lesson 3-5) **46 ft; 120 sq ft**

39. Simplify $2a(-3b)(5c)$. (Lesson 2-7) **$-30abc$**

40. **Sports** Plastic lenses and rubber eye pieces are used to make lightweight, shatterproof sunglasses for serious athletes. Ray-Ban's new sport-specific glasses, called Xrays, weigh less than 25 grams. Write an inequality for the weight of the glasses. (Lesson 1-9) **$w < 25$**

41. **Driving** The table at the right shows the cost of driving a car in different cities in cents per mile. The costs include insurance, depreciation of the car, license fees, taxes, fuel, oil, tires, and maintenance. (Lesson 1-10)

a. Which of the listed cities is the most expensive for driving?

b. What is the difference between the cost of driving in the most expensive city and in the least expensive city listed? **20 cents per mile** **41a. Los Angeles**

City	Cost (cents per mile)
Bismarck, ND	36.3
Boston, MA	49.8
Burlington, VT	36.4
Hartford, CT	48.0
Los Angeles, CA	55.8
Nashville, TN	37.1
Philadelphia, PA	49.0
Providence, RI	48.5
Raleigh, NC	37.1
Sioux Falls, SD	35.8

Source: Runzheimer International

Extension

Using Logical Reasoning Provide students with a short story. For example, Jason has scored 6.7, 7.0, and 6.6 in the first three rotations at a gymnastics meet. The leader has finished the fourth rotation and has an average score of 6.85. What score must Jason get on the fourth rotation to take the lead? **$s > 7.1$**

5-8 Problem-Solving Strategy: Using Logical Reasoning

Setting Goals: *In this lesson, you'll use deductive and inductive reasoning.*

Modeling a Real-World Application: Comics

Peppermint Patty has used **inductive reasoning**. Inductive reasoning *makes* a rule after seeing several examples. Patty made a conclusion based on what happened in the past.

Example  **What is the next term in the sequence 1, 2, 4, 8, 16, . . . ?**

Notice that every number is twice the previous number. For example, 4 is twice 2 and 8 is twice 4. The next term is probably twice 16, or 32.

Suppose Patty gets back her algebra paper and finds that this time $x = 9$ and $y = 15$. Then her method of inductive reasoning did not work. Since inductive reasoning is based on past evidence only, it may sometimes fail.

The Greek letter π, called pi, represents the number 3.14159. . . . If the question on Patty's paper had been "What is the value of π?", she would know that the answer was 3.14159. . . . This is an example of **deductive reasoning**. Deductive reasoning *uses* a rule to make a conclusion.

Example 2

APPLICATION

Traffic

All octagons have eight sides. A stop sign is shaped like an octagon. How many sides does a stop sign have?

You do not need to count the sides of a stop sign. Using deductive reasoning, you can conclude that all stop signs have eight sides.

Checking Your Understanding

Communicating Mathematics

Read and study the lesson to answer these questions. **1–2. See margin.**

1. **Compare and contrast** deductive reasoning and inductive reasoning.
2. Which type of reasoning are you using when you look for a pattern?

5-8 LESSON NOTES

NCTM Standards: 1-4, 6-8

Instructional Resources

- Study Guide Master 5-8
- Practice Master 5-8
- Enrichment Master 5-8
- Group Activity Card 5-8
- Activity Masters, p. 35

Transparency 5-8A contains the 5-Minute Check for this lesson; **Transparency 5-8B** contains a teaching aid for this lesson.

Recommended Pacing	
Standard Pacing	Day 9 of 13
Honors Pacing	Day 8 of 12
Block Scheduling	Day 4 of 6 (along with Lesson 5-7)

*For more information on pacing and possible lesson plans, refer to the ***Block Scheduling Booklet***.

1 FOCUS

5-Minute Check
(over Lesson 5-7)

Solve each inequality and check your solution. Graph the solution on a number line.

1. $n + 1 \geq -1$ $\quad n \geq -2$ (number line: −3 −2 −1 0)
2. $r - \left(-\frac{1}{2}\right) < 2$ $\quad r < 1\frac{1}{2}$ (number line: −1 0 1 2)
3. $2.5 + x > -0.5$ $\quad x > -3$ (number line: −4 −3 −2 −1)
4. $-\frac{1}{7} \geq y - \frac{5}{7}$ $\quad y \leq \frac{4}{7}$ (number line: 0 1)
5. $z - 0.253 \leq -0.253$ $\quad z \leq 0$ (number line: −1 0 1 2)

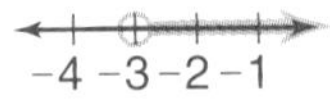

Study Guide Masters, p. 43
The Study Guide Master provides a concise presentation of the lesson along with practice problems.

Additional Answers

1. Inductive reasoning is based upon past experience; deductive reasoning is based upon a given rule.
2. inductive reasoning
3. deductive; This uses a rule to draw a conclusion about a specific case.

Motivating the Lesson

Questioning Ask students what the difference is between these two situations.

- The water froze when the temperature was −2°C, −8°C, −12°C, and −1°C. It did not freeze at 2°C, 10°C, or 4°C. So, water freezes when the temperature is below 0°C.
- I know water freezes at 0°C or below. The temperature is −2°C, so the water is frozen.

In the first situation, reasoning was from specific examples to a general conclusion. In the second situation, a conclusion was drawn based on a rule or law.

2 TEACH

In-Class Examples

For Example 1

What is the next term in the sequence 27, 9, 3, 1, ...? $\frac{1}{3}$

For Example 2

All squares have four congruent sides. The east side of my square garden is 10 feet long. How long is the north side of my garden? **10 ft**

Practice Masters, p. 43

NAME ______________ DATE ________

5-8 Practice
Problem-Solving Strategy: Using Logical Reasoning

Student Edition Pages 255–257

Use inductive reasoning to determine the next two numbers in each list.

1. 109, 110, 111, 112, ··· 113, 114
2. 80, 75, 70, 65, ··· 60, 55
3. 22, 32, 42, 52, 62, ··· 72, 82
4. 1, 5, 11, 15, 21, 25, ··· 31, 35
5. 2, 5, 4, 5, 6, 5, 8, 5, ··· 10, 5
6. 2, 3, 5, 8, 12, 17, ··· 23, 30
7. 8, 8, 10, 10, 12, 12, 14, ··· 14, 16
8. 1, 4, 8, 13, 19, ··· 26, 34
9. 1024, 512, 256, 128, 64, ··· 32, 16
10. 16, 16, 16, 16, 16, ··· 16, 16
11. 1, 2, 4, 5, 7, 8, ··· 10, 11
12. 1, 0.5, 0, −0.5, −1, ··· −1.5, −2
13. 1, 2, 3, 3, 4, 5, 6, 6, ··· 7, 8
14. 1, 4, 9, 16, 25, 36, ··· 49, 64
15. $\frac{1}{2}$, 1, 2, 4, 8, ··· 16, 32
16. 1, 3, 6, 10, 15, ··· 21, 28

State whether each is an example of inductive or deductive reasoning. Explain your answer. See students' explanations.

17. Numbers ending in zero are divisible by five. 25,893,690 is divisible by five. deductive
18. Everyone who came into the store today was wearing sunglasses. It is sunny today. inductive
19. Every student in class has a math book. This must be math class. inductive
20. Every triangle has 180° as the sum of its angle measures. Polygon *ABC* is a triangle. The sum of its angle measures must be 180°. deductive
21. If you are in first place, you will be able to go to the state tournament. You are in first place. You will be able to go to the tournament. deductive
22. It has rained every Monday for four weeks. Marsha says, "Tomorrow is Monday. I think it will rain." inductive

3. The law requires that a person be born in the United States in order to serve as president. What type of reasoning are you using when you say that President Clinton was born in the United States? Explain. See margin.

MATH JOURNAL

4. Make up your own examples of inductive and deductive reasoning. See students' work.

Guided Practice

State whether each is an example of *inductive* or *deductive* reasoning. Explain your answer. Sample answers given. See students' explanations.

5. Every year for the past five years it has rained during spring break. Jamal says that it will rain during spring break this year. inductive
6. If a student earns an "A" for each nine weeks in a class, then he or she will not have to take the final exam in that class. Stacey earned an A each nine weeks in Pre-Algebra, so she will not have to take the final. deductive
7. The citizens of San Juan Capistrano celebrate when the swallows leave and return. For many years, the swallows have left on October 13 and returned on March 19. The citizens are predicting the swallows will return this year on March 19. inductive

Exercises: Practicing and Applying the Concept

Independent Practice

8\. See Solutions Manual.

A

B

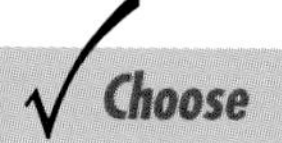

Estimation
Mental Math
Calculator
Paper and Pencil

15c. See Solutions Manual.

8. Is the statement *integers that are even are always divisible by 2* an example of inductive or deductive reasoning? Explain.
9. Find the next two numbers in the pattern 3, 6, 9, 12, 15, _?_, _?_. 18, 21
10. The product of a number and itself is 196. Find the number. 14
11. **Patterns** Write the equation that you think should come next in the pattern at the right. Check your answer with a calculator. $1111^2 = 1{,}234{,}321$

$$1^2 = 1$$
$$11^2 = 121$$
$$111^2 = 12{,}321$$

12. Insert one set of parentheses in $4 \times 5 - 2 + 7 = 19$ to make the equation true. $4 \times (5 - 2) + 7 = 19$
13. **School** If you do not pass your math test, then your parents will ground you for the weekend. 13b. No, there could be another reason.
    a. Suppose you do not pass your math test. What will your parents do? What type of reasoning did you use? ground you; deductive
    b. Suppose you do pass your math test. Does that mean you will definitely not be grounded? Why or why not?
14. At Chicken Little's, you can order Chicken Fingers in boxes of 6, 9, or 20. If you order a box of 6 and a box of 9, you can get 15 fingers. Since no combination of 6, 9, and 20 adds up to 13, you cannot order 13 fingers. What is the greatest number of fingers you *cannot* order? 43
15. **Patterns** Study the pattern at the right.

$$11 \times 1 = 11$$
$$11 \times 2 = 22$$
$$11 \times 3 = 33$$
$$11 \times 4 = 44$$

    a. If you were to continue the pattern, how would you write 11×10? 1010
    b. What is the actual value of 11×10? 110
    c. What does this tell you about inductive reasoning? Could you use inductive reasoning to prove something?
16. At Champion High School, if you take an art class you do not have to take a music class. Anastasia is taking a technical drawing course, so she does not have to take a music course. Is this an example of inductive or deductive reasoning? deductive

Reteaching

Using Lists Choose a topic. Divide a piece of paper into two columns. Write inductive statements about the topic in column 1 and deductive statements in column 2. For example, the topic is a basketball game. Column 1: Al will make this shot. Column 2: If made, this foul shot will raise the score by 1 point.

Group Activity Card 5-8

Think About It

Group Activity **5-8**

MATERIALS: Sets of football, baseball, or basketball cards

To begin, use one sport (baseball for example) and select 20 cards from this set. A player for each position should be selected. Use a rule such as "The best team will have the highest batting average," to select players for your baseball team. Now as a group, consider all the cards and select another team based upon the group's knowledge of the players. Tell why you think this team will be better (or worse) than the one originally selected. Tell which method used inductive reasoning and which method used deductive reasoning.

Work with other sets of cards to form teams of players. Determine a deductive and an inductive way to form a team.

 Pre-Algebra

17. What color piece is needed to complete the quilt at the right? **pink**

Critical Thinking

18. A set of dominoes has 28 rectangular pieces with two numbers of dots on each one. Each number, 0 through 6, is paired with every other number, including itself, on exactly one domino. The grid at the right was made by arranging dominoes and then recording the numbers. Four dominoes are shown. Copy the grid and use deductive reasoning to draw the positions of the remaining dominoes.

1	0	2	0	0	5	4	1
1	1	5	3	6	2	4	2
3	3	1	0	3	5	3	4
0	6	6	4	6	5	1	1
0	4	0	2	5	4	2	6
1	2	3	2	6	4	5	2
3	5	5	0	3	4	6	6

Mixed Review

19. Solve $z - 4.71 \leq -3.8$. (Lesson 5-7) $z \leq 0.91$

20. Solve $a - 9\frac{5}{6} = 2\frac{3}{24}$. (Lesson 5-6) $11\frac{23}{24}$

21. Find the GCF of 78 and 35. (Lesson 4-5) **1**

22. **Fashion** The formula that relates men's shoe size, *s*, and foot length, *f*, in inches is $s = 3f - 21$. What is the shoe size for a man whose foot is 11 inches long? (Lesson 3-4) **12**

23. Find $|-6| + |4|$. (Lesson 2-1) **10**

From the FUNNY PAPERS

"Water boils down to nothing . . . snow boils down to nothing . . . ice boils down to nothing . . . everything boils down to nothing."

1. What type of reasoning did the caveman use? Explain. **Inductive; See students' explanations.**
2. Is the conclusion the caveman reached correct? Give an example to justify your answer. **No; Milk doesn't boil down to nothing.**

3 PRACTICE/APPLY

Checking Your Understanding

Exercises 1–7 are designed to help you assess your students' understanding through reading, writing, speaking, and modeling. You should work through Exercises 1–4 with your students and then monitor their work on Exercises 5–7.

Assignment Guide

Core: 9–17 odd, 18–23
Enriched: 8–18 even, 19–23

For **Extra Practice**, see p. 752.

The red A, B, and C flags, printed only in the Teacher's Wraparound Edition, indicate the level of difficulty of the exercises.

4 ASSESS

Closing Activity

Writing Have students work in pairs and write one deductive and one inductive reasoning problem. Then have them exchange their problems and solve using inductive and deductive reasoning. Have the students tell which type of problem they are solving.

Enrichment Masters, p. 43

NAME ______________________ DATE __________

5-8 Enrichment
Sums of Even and Odd Numbers

Student Edition Pages 255–257

How do you find the sums of even and odd numbers?
The formula $k(k + 1)$ can be used to find the sum of the first k even numbers. | The sum of the first k odd numbers can be bound by using the formula k^2.

Example: Find the sum of the first 9 even numbers.

List the first nine even numbers. Then add them together.
$2 + 4 + 6 + 8 + 10 + 12 + 14 + 16 + 18 = 90$

Now use the formula $k(k + 1)$ if $k = 9$.

$$k(k+1) = 9(9+1) = 9(10) \text{ or } 90$$
The sum is the same.

Example: Find the sum of the first 8 odd numbers.

Use the formula $k \cdot k$ if $k = 8$ to find the sum.

$$k \cdot k = 8 \cdot 8 = 64$$

List the first eight odd numbers. Then add them together.
$1 + 3 + 5 + 7 + 9 + 11 + 13 + 15 = 64$ The sum checks.

Example: Find the sum of the first 10 even numbers plus the first 10 odd numbers.

$$k(k+1) + k^2 = 10(11) + 10^2 = 110 + 100 = 210$$

Solve. Use the correct formula to find the sum of each.

1. the first 10 even numbers **100**
2. the first 16 even numbers **272**
3. the first 100 even numbers **10,100**
4. the first 50 even numbers **2550**
5. the first 5 odd numbers **25**
6. the first 10 odd numbers **100**
7. the first 100 odd numbers **10,000**
8. the first 300 odd numbers **90,000**
9. the first 60 numbers $30(31) + 30^2 = 1830$
10. the first 75 numbers $75(76) + 75^2 = 11,325$
11. the first 50 odd numbers plus the first 50 even numbers $50^2 + 50(51) = 5050$
12. the first 20 even numbers minus the first 20 odd numbers $20(21) - 20^2 = 20$
13. the first 20 even numbers plus the first 10 odd numbers $20(21) + 10^2 = 520$

Extension

Using Connections Have students find an example of inductive reasoning and an example of deductive reasoning in a textbook for another subject. Have students present their findings to the class.

From the FUNNY PAPERS

Junior, "Boy, those big fish must really be smart."
Friend, "How do you know?"
Junior, "Pop always says 'You should have seen the big one that got away', so the big fish must be smart."

5-9 LESSON NOTES

NCTM Standards: 1-4, 7, 8

Instructional Resources

- Study Guide Master 5-9
- Practice Master 5-9
- Enrichment Master 5-9
- Group Activity Card 5-9
- Assessment and Evaluation Masters, p. 128
- Activity Masters, p. 36
- Math Lab and Modeling Math Masters, pp. 61, 62, 80

Transparency 5-9A contains the 5-Minute Check for this lesson; **Transparency 5-9B** contains a teaching aid for this lesson.

Recommended Pacing	
Standard Pacing	Day 10 of 13
Honors Pacing	Day 9 of 12
Block Scheduling*	Day 5 of 6

*For more information on pacing and possible lesson plans, refer to the ***Block Scheduling Booklet***.

1 FOCUS

5-Minute Check
(over Lesson 5-8)

State whether each is an example of *inductive* or *deductive* reasoning. Explain your answer.

1. Children ages 2 or younger get into the zoo free. Mr. Mitchel had to pay for his son, Mike. Mike must be older than 2 years. **deductive**
2. Every person walking to school had on a winter coat. Krista decided it must be cold outside. **inductive**

5-9 Integration: Discrete Mathematics
Arithmetic Sequences

Setting Goals: *In this lesson, you'll find the terms of arithmetic sequences and represent a sequence algebraically.*

Modeling with Technology

Have you ever seen a flash of lightning and heard the roll of thunder and wondered how far away it was? We can generate a table to estimate the distance to a lightning bolt for which you heard the thunder. Sound travels approximately 1100 feet or $\frac{1}{5}$ of a mile per second. For every second that passes, the soundwaves of the thunder travel an additional 1100 feet.

Time (seconds)	0	1	2	3	4	5	6
Distance (feet)	0	1100	2200	**3300**	**4400**	**5500**	**6600**
Distance (miles)	0	$\frac{1}{5}$	$\frac{2}{5}$	$\mathbf{\frac{3}{5}}$	$\mathbf{\frac{4}{5}}$	**1**	$\mathbf{1\frac{1}{5}}$

Your Turn Copy the table above. Use a calculator and a pattern to complete the table.

a. How far away is the flash of lightning if the thunderclap takes 4 seconds to reach the observer? **4400 feet or about $\frac{4}{5}$ miles**

b. What pattern do you see in the table? **See margin.**

c. Write a rule for finding the next numbers in the patterns. **See margin.**

Learning the Concept

A branch of mathematics called **discrete mathematics** deals with topics like logic and statistics. One topic of discrete mathematics is **sequences**. A sequence is a list of numbers in a certain order, such as 0, 1, 2, 3, or 2, 4, 6, 8. Each number is called a **term** of the sequence. A sequence like the distances for thunder, where the difference between any two consecutive terms is the same, is called an **arithmetic sequence**. The difference is called the **common difference**.

Arithmetic Sequence — **An arithmetic sequence is a sequence in which the difference between any two consecutive terms is the same.**

Additional Answers Talk About It

b. The distance increases by 1100 feet or $\frac{1}{5}$ mile with each second.

c. Add 1100 feet (or $\frac{1}{5}$ mile) to the previous distance.

GLENCOE Technology

Interactive Mathematics Tools Software

In this interactive computer lesson, students explore arithmetic sequences. A **Computer Journal** gives students an opportunity to write about what they have learned.

For Windows & Macintosh

Example  **State whether each sequence is arithmetic. Then write the next three terms of each sequence.**

a. 1.25, 1.45, 1.65, 1.85, 2.05

Find the difference between consecutive terms in the sequence.

1.25 1.45 1.65 1.85 2.05

+0.20 +0.20 +0.20 +0.20

Since the difference between any two consecutive terms is the same, the sequence is arithmetic.

Continue the sequence to find the next three terms.

. . . 1.85 2.05 2.25 2.45 2.65

+0.20 +0.20 +0.20 +0.20

The next three terms of the sequence are 2.25, 2.45, and 2.65.

b. 1, 3, 7, 13, 21, . . .

Since there is no common difference, the sequence is not arithmetic.

1 3 7 13 21

+2 +4 +6 +8

Although the sequence is not arithmetic, it is still a sequence. Notice that the differences are in a pattern of 2, 4, 6, 8. So, the next three differences will be 10, 12, 14. The next three terms are 21 + 10 or 31, 31 + 12 or 43, and 43 + 14 or 57.

c. 13, 8, 3, −2, −7, . . .

13 8 3 −2 −7 **−12** **−17** **−22**

−5 −5 −5 −5 **−5** **−5** **−5**

Since there is a common difference of −5, the sequence is arithmetic. The next three terms are −12, −17, and −22.

Connection to Algebra

If we know any term of an arithmetic sequence and the common difference, we can list the terms of the sequence. Consider the following sequence.

3, 7, 11, 15, 19, 23, . . .

We can write an expression that represents any term in the sequence. The first term is 3; call the first term a. The common difference between terms is 4; call the common difference d. Study this pattern.

1st term	a	3
2nd term	$a + d$	$3 + 4 = 7$
3rd term	$a + d + d$ or $a + 2d$	$3 + 2 \cdot (4) = 11$
4th term	$a + d + d + d$ or $a + 3d$	$3 + 3 \cdot (4) = 15$
⋮	⋮	
nth term	$a + (n - 1)d$	

Motivating the Lesson

Hands-On Activity Make stacks of small cubes to illustrate this situation.

- Start with one cube. Put 3 cubes in the next stack. Put 5 cubes in the next stack. Ask how many cubes you should put in the next stack. **7**

2 TEACH

In-Class Examples

For Example 1

State whether each sequence is arithmetic. Then write the next three terms of each sequence.

a. $\frac{1}{3}, \frac{5}{6}, 1\frac{1}{3}, 1\frac{5}{6}$ **arithmetic; $2\frac{1}{3}, 2\frac{5}{6}, 3\frac{1}{3}$**

b. 5, 10, 20, 40 **not arithmetic; 80, 160, 320**

c. 5.6, 4.2, 2.8, 1.4 **arithmetic; 0, −1.4, −2.8**

Teaching Tip You may wish to discuss the pronunciation of *arithmetic* (ar-ith-met′-ik) when it is used as an adjective rather than a noun.

Alternative Learning Styles

Auditory Have one student clap in a rhythmic pattern. Have other students try to continue the pattern. Many familiar songs have distinctive rhythmic patterns. You may wish to bring audio tapes to class for students to use in finding audio patterns.

Cooperative Learning

Send-A-Problem Have students work in small groups to create review problems involving arithmetic sequences. Then have groups exchange problems. For more information on this strategy, see *Cooperative Learning in the Mathematics Classroom*, p. 23.

In-Class Example

For Example 2

To send a fax, an office supply store charges $2.50 for the first page and $1.25 for each additional page. How much will it cost to send a 6-page fax? **$8.75**

3 PRACTICE/APPLY

Checking Your Understanding

Exercises 1–11 are designed to help you assess your students' understanding through reading, writing, speaking, and modeling. You should work through Exercises 1–4 with your students and then monitor their work on Exercises 5–11.

Additional Answers

1. Sample answer: 1, 3, 5, 7, 9, ...
2. Each term is the sum of the previous term and the common difference.
3. Find the difference between each pair of consecutive terms. If they are all the same, the sequence is arithmetic.

Study Guide Masters, p. 44

NAME ______________________ DATE __________

5-9 Study Guide

Integration: Discrete Mathematics

Arithmetic Sequences

Student Edition Pages 258–262

The list of numbers 4, 11, 18, 25, 32, 39, ··· is called a **sequence**.

Each term is 7 more than the previous term. That is, there is a **common difference** of 7.

$11 - 4 = 7,\ 18 - 11 = 7,\ 25 - 18 = 7, \cdots$

Since the difference between any two consecutive terms is the same, the sequence is called an **arithmetic sequence**.

State whether each sequence is an arithmetic sequence. Then write the next three terms of each sequence.

1. 5, 14, 23, 32, 41, ··· **yes; 50, 59, 68**
2. 24, 39, 54, 69, 84, ··· **yes; 99, 114, 129**
3. 16.5, 18.3, 21.1, 24.9, 29.7, ··· **no; 35.5, 42.3, 50.1**
4. 0.50, 0.54, 0.58, 0.62, 0.66, ··· **yes; 0.7, 0.74, 0.78**
5. $4\frac{1}{3}, 5\frac{2}{3}, 7, 8\frac{1}{3}, 9\frac{2}{3}, \cdots$ **yes; $11, 12\frac{1}{3}, 13\frac{2}{3}$**
6. $\frac{2}{3}, \frac{3}{4}, \frac{4}{5}, \frac{5}{6}, \frac{6}{7}, \cdots$ **no; $\frac{7}{8}, \frac{8}{9}, \frac{9}{10}$**
7. 112, 125, 139, 154, ··· **no; 170, 187, 205**
8. 0.3, 0.6, 0.9, ··· **yes; 1.2, 1.5, 1.8**
9. 4, 4, 4, 4, ··· **yes; 4, 4, 4**
10. 2.06, 2.12, 2.18, 2.24, ··· **yes; 2.3, 2.36, 2.42**
11. $53\frac{3}{10}, 55\frac{4}{5}, 58\frac{3}{10}, \cdots$ **no; $60\frac{4}{5}, 63\frac{3}{10}, 65\frac{4}{5}$**
12. $6\frac{1}{2}, 7\frac{1}{12}, 7\frac{2}{3}, \cdots$ **yes; $8\frac{1}{4}, 8\frac{5}{6}, 9\frac{5}{12}$**

Because n represents any term, you can use the expression $a + (n - 1)d$ to find any term in the sequence.

Example 2 APPLICATION **Postage**

In 1995, the first class postage rates were raised to 32 cents for the first ounce and 23 cents for each additional ounce. A chart showing the postage for weights up to 6 ounces is shown below. What was the cost for a 10-ounce letter?

Weight (ounces)	1	2	3	4	5	6
Postage (cents)	32	55	78	101	124	147

Use the expression $a + (n - 1)d$ to find the 10th term of the postage sequence.

The first term, a, is 32. The common difference, d, is 23. Since we are looking for the tenth term, n is 10.

$a + (n - 1)d = 32 + (10 - 1)(23)$ *Substitute 32 for a, 10 for n, and 23 for d.*

$= 32 + 9(23)$ *Use the order of operations.*

$= 32 + 207$

$= 239$ The 10th term is 239.

A 10-ounce letter would cost $2.39.

THINK ABOUT IT

Which would be easier to find the cost of a 16-ounce letter: extending the table to 16 ounces or using the formula?

Sample answer: formula

Checking Your Understanding

Communicating Mathematics

Read and study the lesson to answer these questions. **1–3. See margin.**

1. **Give** an example of an arithmetic sequence.
2. **Tell** the relationship between consecutive terms of an arithmetic sequence.
3. **Explain** how to determine if a sequence is arithmetic.

4. Given the expression $a + (n - 1)d$, record in your Math Journal the meaning of the variables a, n, and d. Illustrate the meaning with a numerical example. **See students' work.**

Guided Practice

State whether each sequence is arithmetic. Then write the next three terms of each sequence.

5. 2, 5, 8, 11, 14, . . . **Yes; 17, 20, 23**
6. 5, 9, 13, 17, 21, . . . **Yes; 25, 29, 33**
7. 17, 16, 14, 11, 7, . . . **7. No; 2, −4, −11**
8. $\frac{3}{2}, 2, \frac{5}{2}, 3, \ldots$ **Yes; $\frac{7}{2}, 4, \frac{9}{2}$**
9. 3, 5, 8, 12, 17, . . . **No; 23, 30, 38**
10. 15, 25, 40, 60, 85, . . . **No; 115, 150, 190**
11. **On-Line Service** In 1995, America Online charged $9.95 and $2.95 per hour after the first five hours to use their service. Let n represent the number of hours after five.
    a. Write the sequence of prices for $n = 0$ to 5. **11a. 9.95, 12.90, 15.85, 18.80, 21.75, 24.70**
    b. How much would it cost if you used America Online 9 hours more than the allotted five hours in one month? **$36.50**

Reteaching

Using Manipulatives In groups of three or four, have each student make up three arithmetic sequences with six terms each. Write the first three numbers in the sequence on one index card and the last three numbers on another card. Lay the cards face down in a 4-by-6 array. Have students take turns turning over two cards. If the cards form a complete sequence, then that student takes the pair of cards. The student with the most pairs after all cards are removed is the winner.

Exercises: Practicing and Applying the Concept

Independent Practice

State whether each sequence is arithmetic. Then write the next three terms of each sequence. 12. Yes; 3.75, 4.5, 5.25 13. Yes; 2.5, 2.0, 1.5

12. 0.75, 1.5, 2.25, 3, . . .
13. 4.5, 4.0, 3.5, 3.0, . . .
14. 0,1100, 2200, 3300
15. 1, 4, 9, 16, . . . No; 25, 36, 49

16. 1, 1, 1, 1, 1, . . . Yes; 1, 1, 1
17. 1, 2, 4, 8, 16, . . . No; 32, 64, 128
18. 7, 10, 13, 16, . . . Yes; 19, 22, 25
19. 91, 82, 73, 64, . . . Yes; 55, 46, 37
20. 83, 77, 71, 65, . . . Yes; 59,53,47
21. 19, 15, 11, 7, . . . Yes; 3, −1, −5
22. 0.1, 0.3, 0.9, 2.7, . . .
23. 5.47, 6.49, 7.51, 8.53, . . .
24. 10, 12, 15, 19, 24, . . . No; 30,37,45
25. 11, 14, 19, 26, 35, . . . No; 46,59,74
26. $1, \frac{1}{2}, \frac{1}{3}, \frac{1}{4}, \ldots$ No; $\frac{1}{5}, \frac{1}{6}, \frac{1}{7}$
27. $\frac{3}{4}, \frac{4}{5}, \frac{5}{6}, \frac{6}{7}, \ldots$ No; $\frac{7}{8}, \frac{8}{9}, \frac{9}{10}$

14. Yes; 4400, 5500, 6600
22. No; 8.1, 24.3, 72.9
23. Yes; 9.55, 10.57, 11.59

28. Write the first six terms in an arithmetic sequence with a common difference of −3. The first term is 35. 35, 32, 29, 26, 23, 20
29. Given the sequence 7, 18, 29, 40, . . . , name the first term and the common difference. 7; 11
30. Find the 25th even integer. 50
31. The first term of an arithmetic sequence is 100, and the common difference is 25. Write the first six terms of the sequence. 100, 125, 150, 175, 200, 225

Critical Thinking

32. Draw the next figure in the sequence below. See margin.

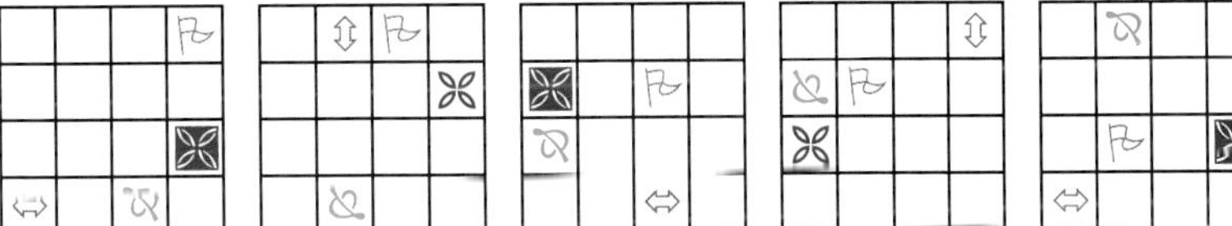

Applications and Problem Solving

33. **Meteorology** Refer to the application at the beginning of the lesson. Thunder can be heard from up to 10 miles away.
    a. How many seconds would it take for the thunder to travel 10 miles? 50 s
    b. Write a formula for finding the distance, d, in miles from lightning for which the thunder is heard t seconds after the lightning is seen. $d = \frac{1}{5}t$

34. **Geometry** Russian mathematician Sonya Kovalevsky (1850–1891) studied number sequences. One of the sequences she studied is represented by the shaded squares below. 34a–b. See margin.

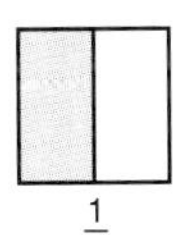
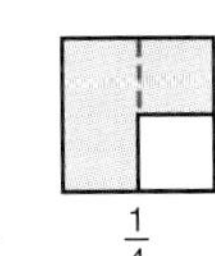
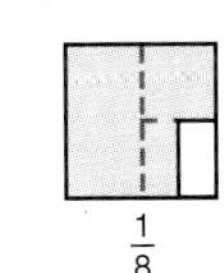
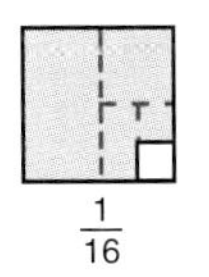
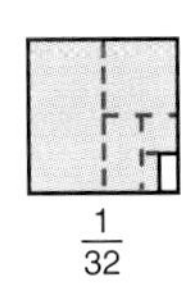
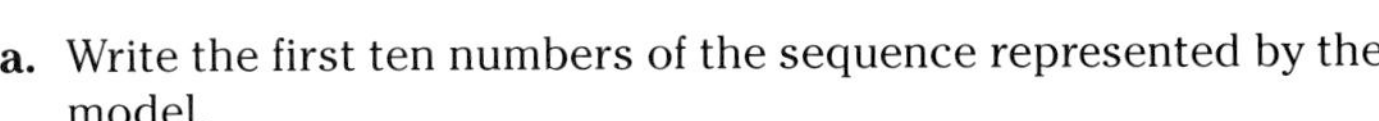

$\frac{1}{2} + \frac{1}{4} + \frac{1}{8} + \frac{1}{16} + \frac{1}{32}$

    a. Write the first ten numbers of the sequence represented by the model.
    b. Is the sequence arithmetic? Explain.
    c. What do you think the sum of the first ten terms is close to? Do not actually add. 1

ПОЧТА СССР 1951г. 40 коп. С.В. КОВАЛЕВСКАЯ-ВЫДАЮЩИЙСЯ РУССКИЙ МАТЕМАТИК

Error Analysis

Students may think that a sequence such as 100, 75, 50, 25, 0, −25 is not arithmetic. Point out to students that the common difference in an arithmetic sequence can be negative.

Assignment Guide

Core: 13–31 odd, 32, 33, 35, 37–42
Enriched: 12–30 even, 32–42

For **Extra Practice**, see p. 752.

The red A, B, and C flags, printed only in the Teacher's Wraparound Edition, indicate the level of difficulty of the exercises.

Additional Answers

32.

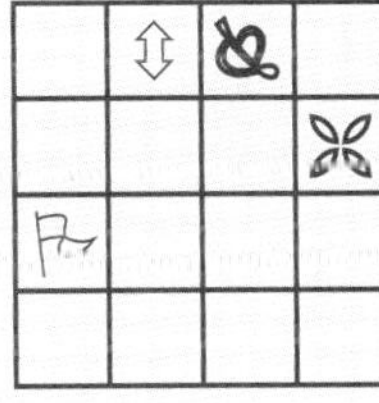

34a. $\frac{1}{2}, \frac{1}{4}, \frac{1}{8}, \frac{1}{16}, \frac{1}{32}, \frac{1}{64}, \frac{1}{128}, \frac{1}{256}, \frac{1}{512}, \frac{1}{1024}$

34b. No; each term is half of the previous term so there is no common difference.

Practice Masters, p. 44

NAME ______ DATE ______

5-9 Practice
Integration: Discrete Mathematics
Arithmetic Sequences
Student Edition Pages 258-262

State whether each sequence is an arithmetic sequence. Then write the next three terms of each sequence.

1. 6.2, 6.4, 6.6, 6.8, ··· yes; 7, 7.2, 7.4
2. −4, −1, 2, 5, 8, ··· yes; 11, 14, 17
3. 0, 3, 9, 12, 18, ··· no; 21, 27, 30
4. −5, −3, 0, 2, 5, ··· no; 7, 10, 12
5. 1, 2, 4, 7, 11, 16, ··· no; 22, 29, 37
6. 95, 85, 75, 65, ··· yes; 55, 45, 35
7. 5, 11, 17, 23, ··· yes; 29, 35, 41
8. −11, −15, −19, −23, ··· yes; −27, −31, −35
9. 6, 9, 12, 15, ··· yes; 18, 21, 24
10. −17, −16, −13, −8, −1, ··· no; 8, 19, 32
11. 3.6, 2.6, 3.6, 2.6, ··· no; 3.6, 2.6, 3.6
12. 0.8, 2.7, 4.6, 6.5, ··· yes; 8.4, 10.3, 12.2
13. 34, 26, 18, 10, ··· yes; 2, −6, −14
14. 206, 217, 228, 239, ··· yes; 250, 261, 272
15. 15, 8, 1, −6 ··· yes; −13, −20, −27
16. 20, 25, 35, 50, 70, ··· no; 95, 125, 160
17. 28, 29, 29, 30, 30, ··· no; 31, 31, 32
18. −8, −13, −18, −23, ··· yes; −28, −33, −38
19. Find the eighth number in the sequence 20, 10, 0, −10, ··· −50
20. Find the tenth number in the sequence 1, 1.25, 1.5, 1.75, 2, ··· 3.25
21. The fifth term of a sequence is 42. The common difference is −3. Find the first four terms. 54, 51, 48, 45
22. The seventh term of a sequence is 12. The common difference is 1.5. Find the first six terms. 3, 4.5, 6, 7.5, 9, 10.5

Group Activity Card 5-9

Powerful Patterns Group Activity 5-9

MATERIALS: Deck of "Pattern Maker" cards with the 40 numbers given on the back of this card

SETUP: The cards are shuffled and each player is dealt 7 cards. The rest are placed face down in the center of the table, and one card is turned face up.

The object of the game is to form as many four number arithmetic sequences as possible. In each turn draw one card from either the face-up or face-down pile. Then look in your hand to see if you can put down an arithmetic sequence of 4 cards. If you do so, then you may draw 4 more cards.

Lastly, you must discard one of the cards from your hand on the face-up pile. You may only play one sequence during your turn. Play continues until all cards are drawn from the face-down pile and everyone has one last turn to make any possible arithmetic sequences from their hand.

The winner is the player who has formed the most sequences.

 Pre-Algebra

4 ASSESS

Closing Activity

Speaking Have students state whether each sequence is arithmetic.

1. 7.2, 8.1, 9.0, ... **yes**
2. $\frac{1}{4}, \frac{1}{2}, \frac{3}{4}, 1, \ldots$ **yes**
3. 4, 6, 10, 16, ... **no**

Chapter 5, Quiz D (Lessons 5-8 and 5-9) is available in the *Assessment and Evaluation Masters*, p. 128.

Additional Answer

41.

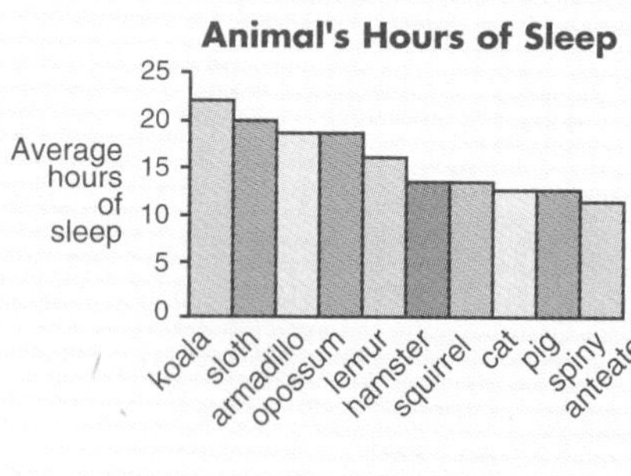

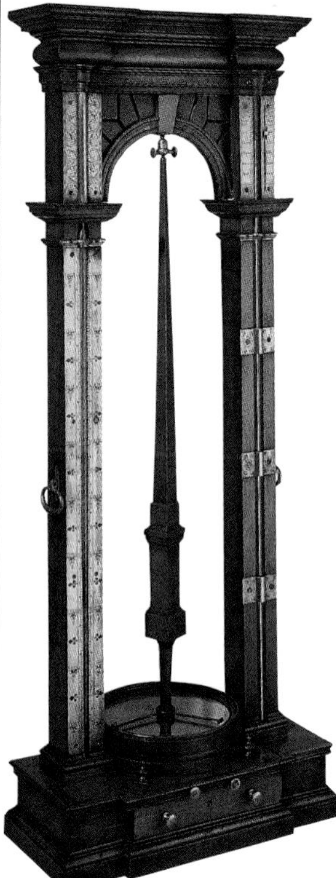

35. **Physics** While flying his hot-air balloon at 4000 feet, Rey dropped a sandbag in order to climb higher. A free-falling object increases its velocity by 32 feet per second every second. Use a sequence to determine how fast the sandbag will be falling at the end of 10 seconds. **320 feet/second**

36. **Meteorology** Television weather reporters frequently give the daily barometric pressure, such as 29.56 inches. This means that atmospheric pressure alone will hold up a column of mercury to a height of 29.56 inches. Centuries ago, scientists experimented with other materials such as water and discovered that atmospheric pressure would hold up a column of water more than thirty feet high.

Mercury (inches)	25	26	27	28	29	30	31	32
Water (feet)	27.2	28.33	29.46	30.59	**31.72**	**32.85**	**33.98**	**35.11**

a. Copy and complete the table of heights above.

b. What is a in the sequence of water heights shown? What is the common difference? **27.2; 1.13**

c. How many feet of water are held by the pressure that holds up thirty inches of mercury? **32.85 feet**

Mixed Review

37. **Business** The long-distance telephone company that Lathan uses charges $0.10 per minute for a long distance call. Lathan talked for one hour, so he knows his bill will be $6.00. Is this an example of inductive or deductive reasoning? (Lesson 5-8) **deductive**

38. Write $\frac{x}{125}$ as an expression with negative exponents. (Lesson 4-9)

38. $x \cdot 125^{-1}$ **or** $x \cdot 5^{-3}$

39. Find the missing exponent in $(x^{\bullet})(x^9) = x^{15}$. (Lesson 4-8) **6**

40. In which quadrant does the graph of (−17, −4) lie? (Lesson 2-2) **III**

41. **Statistics** Have you ever heard of "sleeping like a baby?" How about "sleeping like a koala!" A koala sleeps more than any other animal. The table at the right shows the average number of hours that different animals sleep each day. Make a bar graph of the data. (Lesson 1-10)

41. See margin.

Animal	Average hours of sleep
Koala	22
Sloth	20
Armadillo	19
Opossum	19
Lemur	16
Hamster	14
Squirrel	14
Cat	13
Pig	13
Spiny anteater	12

42. Write a related sentence for 35 = 16 + m using an inverse operation. (Lesson 1-8) **35 − m = 16 or 35 − 16 = m**

Extension

Using Science Connections Provide each student with a periodic table. Point out where atomic masses are located. Round each mass used to the nearest whole number. Point out that Li, Na, and K are similar and their masses form an arithmetic sequence. Have students find other arithmetic sequences on the chart.

Enrichment Masters, p. 44

NAME ______________________ DATE __________

5-9 Enrichment
Finding Sums of Arithmetic Series

Student Edition Pages 258–262

The numbers 7, 11, 15, 19, 23, 27, 31, 35, ⋯ form an arithmetic sequence. It might take a long time to find the sum of the **series** 7 + 11 + 15 + ⋯ + 35 even if you had a computer or a calculator. Formulas have been developed to make solving these problems much easier.

An **arithmetic series** is the sum of the first n terms of an arithmetic sequence, where n is the number of terms in the arithmetic series. To add an arithmetic series using a formula, you need to know the first term, the common difference, and the number of terms.

Example: Find the sum of the arithmetic series 7 + 11 + 15 + 19 + 23 + 27 + 31 + 35.

The first term, a, is 7.
The common difference, d, is 4.
The number of terms, n, is 8.

Use the formula $S_n = \frac{n}{2}[2a + (n - 1)d]$ and substitute to find the sum.

$S_8 = \frac{8}{2}[2(7) + (8 - 1)4]$ — S_8 means the sum of the first eight terms of the series.
$S_8 = 4[14 + (7)4]$
$S_8 = 4(42)$
$S_8 = 168$

The sum of the series is 168.

Use the formula to find the sum of each arithmetic series.

1. 7 + 11 + 15 + 19 + 23 **75**
2. 12 + 17 + 22 + 27 + 32 + 37 **147**
3. 0 + 6 + 12 + 18 + ⋯ + 36 **126**
4. 50 + 57 + 64 + ⋯ + 92 **497**
5. 5 + 10 + 15 + 20 + ⋯ + 65 **455**
6. 9 + 18 + 27 + 36 + ⋯ + 72 **324**
7. 9 + 20 + 31 + 42 + ⋯ + 86 **380**
8. 21 + 29 + 37 + 45 + 53 + ⋯ + 93 **570**

HANDS-ON ACTIVITY

5-9B Fibonacci Sequence

An Extension of Lesson **5-9**

As you have learned, there are sequences that are not arithmetic. One special sequence is called the **Fibonacci sequence**.

The Fibonacci sequence is not arithmetic, but its terms are found by adding. The sequence begins with 1, 1, and each term after the second is the sum of the previous two terms of the sequence.

$$\begin{array}{cccccc} & & 1+1 & 1+2 & 2+3 & 3+5 \\ 1, & 1, & 2, & 3, & 5, & 8, \ldots \end{array}$$

Your Turn List the first ten terms of the Fibonacci sequence. **1, 1, 2, 3, 5, 8, 13, 21, 34, 55**

There are many patterns in the Fibonacci sequence. Here are some of the patterns.

- Every third term is divisible by 2.
- Every fourth term is divisible by 3.
- Every fifth term is divisible by 5.
- Every sixth term is divisible by 8.

1. What number do you think may be a factor of every seventh term? Use a calculator to check your conjecture. **13; See students' work.**
2. There is only one perfect square and one perfect cube, except for 1, in the Fibonacci sequence. Name them. **144, 8**
3. If you square the Fibonacci numbers and then add adjacent squares, a new sequence is formed. Describe the sequence. **See Solutions Manual.**

Extension

4. Investigate how the Fibonacci numbers are related to a pineapple, a pine cone, and a sunflower. **See Solutions Manual.**

5-9B Lesson Notes

NCTM Standards: 1-4, 6, 8

Objective

Investigate patterns in the Fibonacci sequence.

Recommended Time

Demonstration and discussion: 15 minutes; Exercises: 30 minutes

Instructional Resources

For each student or group of students
Math Lab and Modeling Math Masters
- p. 36 (worksheet)

For teacher demonstration
Overhead Manipulative Resources

1 FOCUS

Motivating the Lesson

Bring in a bouquet of daisies and have students count the number of petals on each one. Most daisies have 13, 21, or 34 petals.

2 TEACH

Teaching Tip The pattern for divisibility can be continued. Every seventh number is divisible by 13 and every eighth number is divisible by 21.

3 PRACTICE/APPLY

Assignment Guide
Core: 1–4
Enriched: 1–4

4 ASSESS

Observing students working in cooperative groups is an excellent method of assessment.

GLENCOE Technology

Interactive Mathematics Tools Software

In this interactive computer lesson, students explore the Fibonacci sequence. A **Computer Journal** gives students an opportunity to write about what they have learned.

For Windows & Macintosh

CLOSING THE

Investigation

This activity provides students an opportunity to bring their work on the Investigation to a close. For each Investigation, students should present their findings to the class. Here are some ways students can display their work.

- Conduct and report on an interview or survey.
- Write a letter, proposal, or report.
- Write an article for the school or local paper.
- Make a display, including graphs and/or charts.
- Plan an activity.

Assessment

To assess students' understanding of the concepts and topics explored in this Investigation and its follow-up activities, you may wish to examine students' Investigation Folders.

The scoring guide provided in the *Investigation and Project Masters*, p. 11, provides a means for you to score students' work on this Investigation.

Investigations and Projects Masters, p. 11

CHAPTERS **4 and 5** Investigation
Taking Stock

Scoring Guide

Level	Specific Criteria
3 Superior	• Shows thorough understanding of the concepts of *solving problems by drawing a diagram, simplifying and comparing fractions, multiplying and dividing monomials, estimating and finding sums and differences of rational numbers*, and *solving problems using logical reasoning*. • Uses appropriate strategies to solve problems. • Written explanations are exemplary. • Tables, charts, and graphs are appropriate and sensible. • Goes beyond the requirements of the Investigation.
2 Satisfactory, with minor flaws	• Shows understanding of the concepts of *solving problems by drawing a diagram, simplifying and comparing fractions, multiplying and dividing monomials, estimating and finding sums and differences of rational numbers*, and *solving problems using logical reasoning*. • Uses appropriate strategies to solve problems. • Written explanations are effective. • Tables, charts, and graphs are appropriate and sensible. • Satisfies all requirements of the Investigation.
1 Nearly Satisfactory, with serious flaws	• Shows understanding of most of the concepts of *solving problems by drawing a diagram, simplifying and comparing fractions, multiplying and dividing monomials, estimating and finding sums and differences of rational numbers*, and *solving problems using logical reasoning*. • May not use appropriate strategies to solve problems. • Written explanations are satisfactory. • Tables, charts, and graphs are mostly appropriate and sensible. • Satisfies most requirements of the Investigation.
0 Unsatisfactory	• Shows little or no understanding of the concepts of *solving problems by drawing a diagram, simplifying and comparing fractions, multiplying and dividing monomials, estimating and finding sums and differences of rational numbers*, and *solving problems using logical reasoning*. • May not use appropriate strategies to solve problems. • Written explanations are not satisfactory. • Tables, charts, and graphs are not appropriate or sensible. • Does not satisfy requirements of the Investigation.

CLOSING THE

Investigation

TAKING STOCK

Refer to the Investigation on pages 166–167.

Throughout this investigation, the terms *capital gain*, *closing price*, and *share* were often used.

What do these terms mean?

What do these words mean in terms of the work you did in this investigation?

Plan and give an oral presentation to describe what you know about these and any other investment terms. Your presentation should include the following.

- Examples from your portfolio to help you describe and define the investment terms you used most often in this investigation.
- A table showing how you initially invested your $10,000 and the source of the prices.
- A stock chart or line graph showing a price-weighted index of your portfolio's performance over the course of this investigation.
- An explanation of the analysis of your portfolio's performance.
- A comparison of your portfolio's performance to how the *Standard and Poor's (S&P) 500* performed over the same time period.
- A statement of what your shares would be worth if you sold all of them today.
- An explanation of what you would do differently if you had to do this investigation over again.

Extension

Schedule an interview with a financial analyst or stock broker. Make a list of questions to ask him or her.

- Be sure to ask about the requirements and duties of their job.
- Include questions about investment strategies for teenagers and how investment strategies differ depending on the age of the investor.

PORTFOLIO ASSESSMENT

You may want to keep your work on this Investigation in your portfolio.

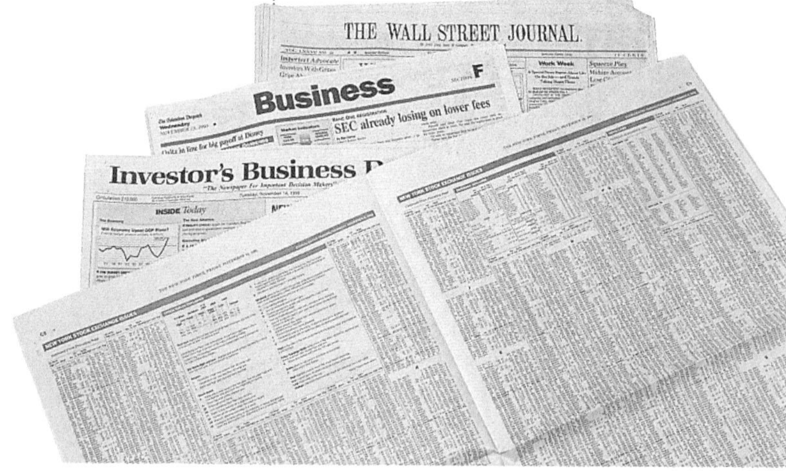

CHAPTER 5 Highlights

Vocabulary

After completing this chapter, you should be able to define each term, property, or phrase and give an example or two of each.

Algebra
adding like fractions (p. 239)
adding unlike fractions (p. 244)
associative property of addition (p. 234)
bar notation (p. 225)
commutative property of addition (p. 234)
identity property of addition (p. 234)
integers (p. 224)
inverse property of addition (p. 234)
mixed number (p. 239)
rational numbers (p. 224)
repeating decimals (p. 225)
subtracting like fractions (p. 240)
subtracting unlike fractions (p. 244)
terminating decimal (p. 225)
whole numbers (p. 224)

Geometry
scale (p. 224)

Problem Solving
deductive reasoning (p. 255)
inductive reasoning (p. 255)

Discrete Mathematics
arithmetic sequence (p. 258)
common difference (p. 258)
discrete mathematics (p. 258)
Fibonacci sequence (p. 263)
sequence (p. 258)
term (p. 258)

Understanding and Using Vocabulary

Choose the letter of the term that best matches each statement.

A. Identity property of addition
B. rational number
C. integers
D. Commutative property of addition
E. whole numbers
F. deductive reasoning
G. Inverse property of addition
H. inductive reasoning
I. scale
J. Associative property of addition

1. This is any number that can be expressed in the form $\frac{a}{b}$, where a and b are integers, and $b \neq 0$ *B*
2. This type of reasoning creates a rule based on consistent examples. *H*
3. A comparison of the size of a model to the size of the real object is called a __?__. *I*
4. This property justifies the statement $3 + 5 = 5 + 3$. *D*
5. This is any number which is positive, negative, or zero and whose absolute value is a whole number. *C*
6. This type of reasoning uses a given rule to make a conclusion or a decision. *F*
7. This property justifies the statement $4 + 0 = 4$ *A*

Instructional Resources

Three multiple-choice tests and three free-response tests are provided in the *Assessment and Evaluation Masters*. Forms 1A and 2A are for honors pacing, Forms 1B and 2B are for average pacing, Forms 1C and 2C are for basic pacing, Chapter 5 Test, Form 1B is shown at the right. Chapter 5 Test, Form 2B is shown on the next page.

Highlights

Using the Chapter Highlights

The Chapter Highlights begins with a listing of the new terms, properties, and phrases that were introduced in this chapter.

Assessment and Evaluation Masters, pp. 115–116

NAME ____________ DATE ________

CHAPTER 5 Test, Form 1B

1. Name the set(s) of numbers to which $-\frac{3}{5}$ belongs.
   A. whole numbers, integers, rational numbers
   B. rational numbers
   C. integers, rational numbers
   D. odd numbers, whole numbers, integers, rational numbers — 1. B
2. Choose the symbol to replace ● in $\frac{12}{-4}$ ● -3 to make a true sentence.
   A. $<$ B. $>$ C. $=$ D. cannot tell — 2. C
3. Choose the symbol to replace ● in $-\frac{1}{2}$ ● $\frac{3}{100}$ to make a true sentence.
   A. $>$ B. $<$ C. $=$ D. cannot tell — 3. B
4. Express $0.\overline{12}$ as a fraction in simplest form.
   A. $\frac{12}{20}$ B. $\frac{4}{33}$ C. $\frac{3}{25}$ D. $\frac{12}{100}$ — 4. B
5. Express 0.55 as a fraction in simplest form.
   A. $\frac{11}{2}$ B. $\frac{55}{10}$ C. $\frac{50}{9}$ D. $\frac{11}{20}$ — 5. D
6. Round $36.86 to the nearest dollar.
   A. $36.90 B. $40.00 C. $37.00 D. $36.00 — 6. C
7. Estimate the sum $503 + $98 + $106 + $2.
   A. $600 B. $700 C. $800 D. $900 — 7. B
8. Solve $r = -13.68 - (-14.7)$.
   A. -1.02 B. 1.02 C. -28.38 D. 28.38 — 8. B
9. Simplify $3.6x + 9.5x - 15.7x$.
   A. [illegible] B. $2.6x$ C. $-6.2x$ D. $-12.1x$ — 9. A

Solve each equation. Write the solution in simplest form.

10. $\frac{3}{11} + \frac{2}{11} = d$
    A. $\frac{5}{11}$ B. $\frac{1}{11}$ C. $\frac{3}{2}$ D. $\frac{2}{3}$ — 10. A
11. $s = 1\frac{2}{5} + 2\frac{1}{5}$
    A. 4 B. 3 C. $3\frac{3}{5}$ D. $\frac{17}{5}$ — 11. C
12. $\frac{1}{3} + \frac{3}{5} = h$
    A. $\frac{4}{8}$ B. $\frac{14}{15}$ C. $\frac{1}{5}$ D. $\frac{1}{2}$ — 12. B

NAME ____________ DATE ________

Chapter 5 Test, Form 1B (continued)

Evaluate each expression if $x = \frac{1}{3}$, $y = -\frac{3}{4}$, and $z = 1\frac{1}{2}$.

13. $y + z$
    A. $\frac{3}{4}$ B. $\frac{7}{4}$ C. $-\frac{3}{4}$ D. $\frac{1}{4}$ — 13. A
14. $x + y + z$
    A. $\frac{31}{12}$ B. $\frac{22}{12}$ C. $\frac{13}{12}$ D. $\frac{7}{12}$ — 14. C

Solve each equation.

15. $c - 3.8 = 7.7$
    A. 11.5 B. 11.5 C. -3.9 D. 3.9 — 15. B
16. $p - 0.36 = 27.93$
    A. 27.75 B. 28.36 C. 28.93 D. 28.29 — 16. D

Solve each inequality.

17. $d + \frac{2}{3} < \frac{7}{3}$
    A. $d < 3$ B. $d < -\frac{5}{3}$ C. $d < \frac{5}{3}$ D. $d < -3$ — 17. C
18. $y - 5 \geq -8.3$
    A. $y \geq 13.3$ B. $y \geq -3.3$ C. $y \geq 3.3$ D. $y \geq -13.3$ — 18. B
19. $-4 + a \geq -\frac{13}{2}$
    A. $a \geq \frac{9}{2}$ B. $a \geq -\frac{17}{2}$ C. $a \geq \frac{5}{2}$ D. $a \geq -\frac{5}{2}$ — 19. D
20. Use deductive reasoning to determine the next number in the pattern 2, 22, 222, . . .
    A. 242 B. 224 C. 2222 D. 2200 — 20. C
21. As cook at summer camp, Hector discovered he needed more than one egg per camper for his breakfast omelets. How many eggs does he need to make omelets for 30 campers?

Number of campers	Number of eggs
5	7
10	14
15	21
20	28

    A. 38 B. 40 C. 42 D. 46 — 21. C
22. Write the next three terms in the sequence 8, 2, -4, -10, ...
    A. $-14, -18, -22$ B. $-16, -20, -24$
    C. $-16, -26, -36$ D. $-16, -22, -28$ — 22. D
23. Find the 10th term of the sequence 54, 60, 66, 72, ...
    A. 108 B. 114 C. 120 D. 126 — 23. A
24. Find the common difference in the arithmetic sequence 11, 15, 19, 23, ...
    A. +2 B. -2 C. +4 D. -4 — 24. C
25. Find the 100th term in the arithmetic sequence 3, 6, 9, 12, ...
    A. 294 B. 300 C. 303 D. 297 — 25. B

CHAPTER 5 Study Guide and Assessment

Using the Study Guide and Assessment

Skills and Concepts Encourage students to refer to the objectives and examples on the left as they complete the review exercises on the right.

Assessment and Evaluation Masters, pp. 121–122

NAME ____________ DATE ______

CHAPTER 5 Test, Form 2B

Replace each ● with >, <, or = to make each sentence true.

1. $^{-}\frac{9}{3}$ ● $^{-}3$	1. =
2. $\frac{4}{5}$ ● $^{-}\frac{5}{4}$	2. >
3. Is the statement $^{-}6$ *is an integer* true or false?	3. true

Express each decimal as a fraction in simplest form.

4. 0.103	4. $\frac{103}{1000}$
5. $0.\overline{12}$	5. $\frac{4}{33}$

Round to the nearest whole number or dollar.

6. 38.19	6. 38
7. $6.49	7. $6
8. $42.51	8. $43

Estimate each sum or difference.

9. 815 + 415	9. 1200
10. $\frac{19}{24} - \left(^{-}\frac{5}{24}\right)$	10. 1

Find each sum or difference.

11. $^{-}0.386 + 4.23$	11. 3.844
12. $6.38 + (^{-}5.79)$	12. 0.59
13. 8.2 + 3.9	13. 12.1
14. $^{-}4.35 - (^{-}4.36)$	14. 0.01

Solve each equation. Write each solution in simplest form.

15. $\frac{6}{7} - \frac{4}{7} = p$	15. $\frac{2}{7}$
16. $\frac{6}{13} - \left(^{-}\frac{7}{13}\right) = r$	16. 1
17. $\frac{3}{4} + \frac{4}{5} = t$	17. $1\frac{11}{20}$
18. $\frac{1}{6} + \frac{3}{4} = u$	18. $\frac{11}{12}$
19. $4\frac{3}{4} + 2\frac{1}{8} = z$	19. $6\frac{7}{8}$

NAME ____________ DATE ______

Chapter 5 Test, Form 2B (continued)

Solve each equation.

20. $w + 2\frac{1}{3} = 3\frac{1}{2}$	20. $1\frac{1}{6}$
21. $c + \frac{5}{3} = \frac{6}{5}$	21. $^{-}\frac{7}{15}$
22. $b - 6.3 = 4.7$	22. 11
23. $p + (^{-}0.63) = 28.46$	23. 29.09

Solve each inequality.

24. $a + 2.4 > 3.7$	24. $a > 1.3$
25. $b - 4.3 \leq 6.2$	25. $b \leq 10.5$
26. $d - \frac{1}{7} \geq \frac{1}{4}$	26. $d \geq \frac{11}{28}$
27. $f + \left(^{-}\frac{3}{5}\right) \leq \frac{3}{4}$	27. $f \leq 1\frac{7}{20}$

Use deductive reasoning to determine the next three numbers in each pattern.

28. 13, 11.7, 10.4, 9.1, . . .	28. 7.8, 6.5, 5.2
29. $^{-}13, ^{-}8, ^{-}3, 2, \ldots$	29. 7, 12, 17

30. In the laboratory, you apply heat to different volumes of water. The table below shows the temperature increase in each volume of water after a given time. Use inductive reasoning to determine the temperature increase in 8 liters of water. — 30. 1.25

Volume (L)	Temperature Increase (°C)
1	10
2	5
4	2.5
8	

31. Find the next three terms of the arithmetic sequence 1.75, 3, 4.25, 5.5, ...	31. 6.75, 8, 9.25
32. Find the 13th term of the sequence 100, 97, 94, 91, ...	32. 64
33. State whether the sequence 24, 35, 46, 57, . . . is arithmetic. Then write the next three terms in the sequence.	33. yes; 68, 79, 90

CHAPTER 5 Study Guide and Assessment

Skills and Concepts

Objectives and Examples	Review Exercises
Upon completing this chapter, you should be able to:	Use these exercises to review and prepare for the chapter test.

▶ **identify and compare rational numbers** (Lesson 5-1)

0 is a rational number, an integer, and a whole number.

0.1234567 . . . is a non-terminating, non-repeating decimal. So it is not rational.

Name the set(s) of numbers to which each number belongs.

8. $\frac{9}{10}$ R
9. -4 I, R
10. 18 W, I, R
11. 0.122333 . . . none

For Exercises 8–11, W = whole numbers, I = integers, and R = rationals.

▶ **rename decimals as equivalent fractions.** (Lesson 5-1)

Let $N = 0.888\ldots$. Then $10N = 8.888\ldots$

$$\begin{aligned} 10N &= 8.8888\ldots \\ -\quad N &= 0.8888\ldots \\ \hline 9N &= 8 \\ N &= \frac{8}{9} \\ 0.\overline{8} &= \frac{8}{9} \end{aligned}$$

Express each decimal as a fraction or mixed number in simplest form.

12. 3.25 $3\frac{1}{4}$
13. -9.45 $-9\frac{9}{20}$
14. 0.40 $\frac{2}{5}$
15. -0.13 $-\frac{13}{100}$
16. $0.\overline{72}$ $\frac{8}{11}$
17. $4.\overline{24}$ $4\frac{8}{33}$
18. $0.0\overline{6}$ $\frac{1}{15}$
19. 8.89 $8\frac{89}{100}$

▶ **estimate sums and differences of decimals and fractions.** (Lesson 5-2)

$48.89 + 91.21 \rightarrow 50 + 90 = 140$

$6\frac{9}{10} - 2\frac{3}{8} \rightarrow 7 - 2\frac{1}{2} = 4\frac{1}{2}$

Estimate each sum or difference.

20. 53.6 + 41.2 90
21. 4.99 + 3.29 8
22. 325.44 − 249.25 75
23. 50.00 − 39.89 10
24. $\frac{11}{12} + \frac{6}{10}$ $1\frac{1}{2}$
25. $\frac{24}{25} + \frac{1}{11}$ 1
26. $18\frac{1}{10} - 3\frac{1}{9}$ 15
27. $24\frac{4}{7} - 22\frac{1}{6}$ $2\frac{1}{2}$

20–27. Sample answers are given.

▶ **add and subtract decimals.** (Lesson 5-3)

$f = 4.56 - 2.358$

$$\begin{array}{r} 4.560 \\ -2.358 \\ \hline 2.202 \end{array}$$

$f = 2.202$

Solve each equation. 28–33. See margin.

28. $t = 8.5 + 42.25$
29. $9.43 - (-1.8) = p$
30. $j = -7.43 + 5.34$
31. $m = 17.19 - 24.87$
32. $y = 13.983 + 4.52$
33. $-8.52 - 9.43 = d$

GLENCOE Technology

Test and Review Software

You may use this software, a combination of an item generator and item bank, to create your own tests or worksheets. Types of items include free response, multiple choice, short answer, and open ended.

For IBM & Macintosh

Additional Answers

28. 50.75
29. 11.23
30. −2.09
31. −7.68
32. 18.503
33. −17.95

Objectives and Examples	Review Exercises
▸ **add and subtract fractions with like denominators.** (Lesson 5-4) $n = 3\frac{3}{5} + 4\frac{4}{5}$ $n = (3 + 4) + \left(\frac{3}{5} + \frac{4}{5}\right)$ $n = 7 + \frac{7}{5}$ $n = 7 + 1\frac{2}{5}$ or $8\frac{2}{5}$	**Solve each equation. Write the solution in simplest form.** **34.** $w = \frac{4}{7} + \frac{6}{7}$ $1\frac{3}{7}$ **35.** $t - 5\frac{2}{9} = 8\frac{5}{9}$ $13\frac{7}{9}$ **36.** $7\frac{4}{7} + \left(-2\frac{5}{7}\right) = s$ $4\frac{6}{7}$ **37.** $\frac{7}{8} - \frac{5}{8} = b$ $\frac{1}{4}$ **38.** $k = 5\frac{1}{5} - 4\frac{4}{5}$ $\frac{2}{5}$ **39.** $\frac{4}{15} - \frac{13}{15} = y$ $-\frac{3}{5}$
▸ **add and subtract fractions with unlike denominators.** (Lesson 5-5) $r = \frac{2}{5} + \frac{7}{8}$ $r = \frac{2 \cdot 8}{5 \cdot 8} + \frac{7 \cdot 5}{8 \cdot 5}$ *The LCD is 40.* $r = \frac{16}{40} + \frac{35}{40}$ $r = \frac{51}{40}$ or $1\frac{11}{40}$	**Solve each equation. Write the solution in simplest form.** **40.** $\frac{3}{7} + \frac{11}{14} = q$ $1\frac{3}{14}$ **41.** $s = \frac{7}{15} - \frac{16}{30}$ $-\frac{1}{15}$ **42.** $6\frac{4}{5} + 1\frac{3}{4} = h$ $8\frac{11}{20}$ **43.** $4\frac{1}{6} - \left(-2\frac{3}{4}\right) = m$ $6\frac{11}{12}$ **44.** $1\frac{2}{5} + \frac{1}{3} = a$ $1\frac{11}{15}$ **45.** $9\frac{5}{12} - 4\frac{7}{18} = h$ $5\frac{1}{36}$
▸ **solve equations with rational numbers.** (Lesson 5-6) $y + 5.7 = 3.1$ $y + 5.7 - 5.7 = 3.1 - 5.7$ $y = -2.6$	**Solve each equation. Check your solution.** **46.** $v - 4.72 = 7.52$ **47.** $s + (-13.5) = -22.3$ **48.** $x + \frac{3}{4} = -1\frac{2}{5}$ **49.** $b - \frac{5}{8} = 1\frac{3}{16}$ **50.** $z - (-5.8) = 1.36$ **51.** $k - \frac{4}{9} = 1\frac{7}{18}$ **46–51. See margin.**
▸ **solve inequalities with rational numbers.** (Lesson 5-7) $p - 4\frac{9}{10} > -8\frac{6}{10}$ $p - 4\frac{9}{10} + 4\frac{9}{10} > -8\frac{6}{10} + 4\frac{9}{10}$ $p > -3\frac{7}{10}$	**Solve each inequality and check your solution. Graph the solution on a number line.** **52–59. See margin.** **52.** $w + \frac{7}{12} \geq \frac{5}{18}$ **53.** $f - 3\frac{1}{4} \leq \frac{7}{8}$ **54.** $31.6 < -5.26 + g$ **55.** $q - (-6.7) > 12$ **56.** $w - 4.32 \leq 1.234$ **57.** $m + 7.17 > 1.019$ **58.** $\frac{7}{6} + c \geq 3\frac{5}{24}$ **59.** $a - \left(\frac{-3}{10}\right) \leq 1\frac{1}{5}$

Chapter 5 • Study Guide and Assessment

Additional Answers

46. 12.24

47. −8.8

48. $-2\frac{3}{20}$

49. $1\frac{13}{16}$

50. −4.44

51. $1\frac{5}{6}$

52. $w \geq -\frac{11}{36}$

−1 0

53. $f \leq 4\frac{1}{8}$

3 4 5

54. $g > 36.86$

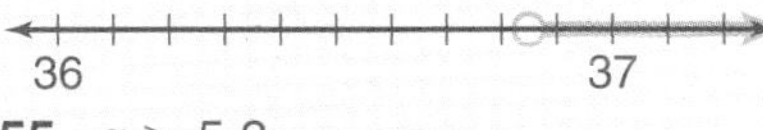

55. $q > 5.3$

5 6

56. $w \leq 5.554$

5 6

57. $m > -6.151$

−6 −5

58. $c \geq 2\frac{1}{24}$

2 3

59. $a \leq \frac{9}{10}$

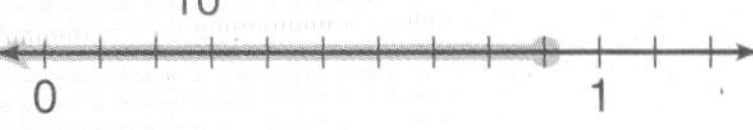

Applications and Problem Solving

Encourage students to work through the exercises in the Applications and Problem Solving section to strengthen their problem-solving skills.

Additional Answers

68. deductive reasoning, a rule used to make a conclusion about a specific case.
69b. The lengths are the squares of 1, 2, 3, ... It is not an arithmetic sequence because there is no common difference.

Objectives and Examples	Review Exercises

Objectives and Examples

▶ **find the terms of arithmetic sequences and represent a sequence algebraically.** (Lesson 5-9)

2 4 8 16
+2 +4 +8

This is not an arithmetic sequence because the difference between any two consecutive terms is not the same.

The next three terms are 16 + 16 or 32, 32 + 32 or 64, and 64 + 64 or 128.

Review Exercises

State whether each sequence is arithmetic. Then write the next three terms of each sequence.

60. 75, 90, 105, . . . **Yes; 120, 135, 150**
61. 45, 37, 29, 21, . . . **Yes; 13, 5, −3**
62. 0.0625, 0.125, 0.25, . . . **No; 0.5, 1, 2**
63. $\frac{10}{11}, \frac{8}{9}, \frac{6}{7}$ **No; $\frac{4}{5}, \frac{2}{3}, \frac{0}{1}$**
64. 67, 58, 49, 40, . . . **Yes; 31, 22, 13**

65a. $\frac{35}{1200}$ or $\frac{7}{240}$
65b. Rational, it is a fraction.

Applications and Problem Solving

65. Hobbies Tedrick is building a model yacht that is 35 centimeters long. The actual yacht is 12 meters long. (Lesson 5-1)
a. What is the scale of the model?
b. Is the scale a whole number, an integer, or a rational number? Explain.

66. Politics In the 1992 presidential election, 186.7 million people were eligible to vote. Only 113.9 million people actually voted. How many people were eligible to vote in 1992 but did not? (Lesson 5-3) **72.8 million**

67. Carpentry Currito is making a porch swing for his house. He bought a 4-foot piece of poplar for the seat. If the finished seat on the pattern is to be $41\frac{3}{8}$ inches long, how much should Currito cut off? (Lesson 5-5) **$6\frac{5}{8}$ inches**

68. Numbers that can be divided by both 2 and 3 are always divisible by 6. 18 is divisible by 2 and 3, so it is divisible by 6. Is this inductive or deductive reasoning? Explain. (Lesson 5-8) **See margin.**

69. Physical Science The Italian scientist Galileo discovered that there was a relationship between the time of the back-and-forth swing of a pendulum and its length. The table at the right shows the time for a swing and the length of several pendulums. (Lesson 5-9)
a. How long do you think a pendulum with a swing of 5 seconds is? **25 units**
b. Describe the sequence of lengths. Is it an arithmetic sequence? Explain. **See margin.**

Time of swing (seconds)	Pendulum length (units)
1	1
2	4
3	9
4	16

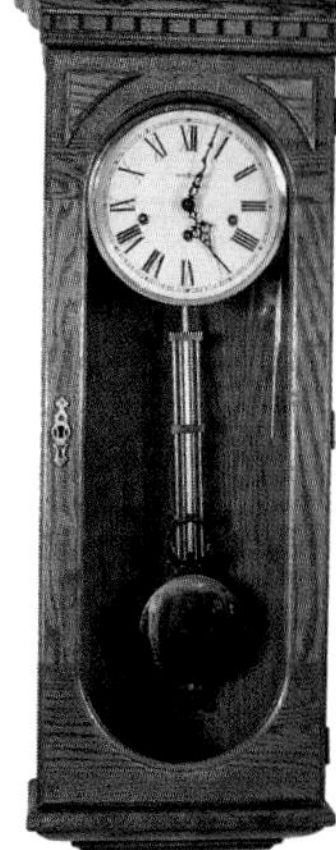

A practice test for Chapter 5 is available on page 778.

Additional Answers
Thinking Critically

Sample answer: If you forgot the amount of your paycheck that was deposited in your checking account, you would want to underestimate the amount to find the estimated balance.
Sample answer: You come to the actual answer when you round to the nearest tenth because the rounded number is closer to the actual number than if it were rounded to the nearest whole number.

Alternative Assessment

Performance Task **Demonstrate your knowledge by giving a clear, concise solution to each problem. Be sure to include all relevant drawings and justify your answers. You may show your solutions in more than one way or investigate beyond the requirements of the problem.** 1b. Rational, it is a fraction.

1. A house is 36 feet from front to back. On the blueprints, the drawing of the house is 18 inches from front to back.
   a. What is the scale factor of the blueprints to the house? $\frac{1}{24}$
   b. What type of number is the scale factor? Explain.
2. The windows on the side of the house are $2\frac{7}{8}$ feet from the ground and $7\frac{3}{4}$ feet from the roof. The house is 15 feet from the ground to the roof. 2b. $4\frac{3}{8}$ feet
   a. Approximately how tall are the windows? About 4 feet
   b. Exactly how tall are the windows on the side of the house?
3. The window on the front of the house is 6.72 feet wide, and there are 4.14 feet on either side of the window. What is the width of the front section of the house? 15 feet
4. The house covers 1023.5 square feet. The house and the yard together cover 1 acre, or 43,560 square feet. How large is the yard? (4. 42,536.5 square feet)
5. The law in Genoa Township states that an individual may erect a fence as long as it is no closer than 3 inches from the property line. Mr. Owens says that the fence he is erecting 5 inches from the property line is legal. Did Mr. Owens use inductive or deductive reasoning? Explain. Deductive; he used a rule to draw a conclusion about a specific case.
6. A series of bushes is to be planted along the foundation of the house. Devise a plan for the planting that uses an arithmetic sequence. Sample answer: 1, 3, 5, 7 feet from the front

Thinking Critically

- When you estimate an area in order to purchase wallpaper, it is wise to overestimate the area you need to cover. Give an example of a situation when it is better to estimate low. See margin.
- Do you come closer to the actual answer when you round to the nearest tenth or to the nearest whole number? Explain. See margin.

Portfolio Review the items in your portfolio. Make a list or table of contents of the items, noting why each item was chosen and place it in your portfolio.

Self Evaluation People who are willing to get their hands dirty can always find a sink to clean up. This is a way of saying that if you are willing to go out of your way, you can find a place to be helpful.

Assess yourself. Do you help others when they are having difficulties? Do you notice that people you have helped in the past are more likely to help you when you need some assistance? See students' work.

Alternative Assessment

The Alternative Assessment section provides students with the opportunity to assess their own work by thinking critically, working with others, keeping a portfolio, and honestly evaluating their own progress. For more information on alternative forms of assessment, see *Alternative Assessment in the Mathematics Classroom*.

Performance Assessment

Performance Assessment tasks for this chapter are included in the *Assessment and Evaluation Masters*. A scoring guide is also provided.

Alternative Assessment

Assessment and Evaluation Masters, pp. 125, 137

NAME ______________ DATE ______

CHAPTER 5 Performance Assessment

Instructions: Demonstrate your knowledge by giving a clear, concise solution to each problem. Be sure to include all relevant drawings and justify your answers. You may show your solution in more than one way or investigate beyond the requirements of the problem.

1. Mr. Chen is building a house in a new subdivision. The house plan for the first floor is shown below.

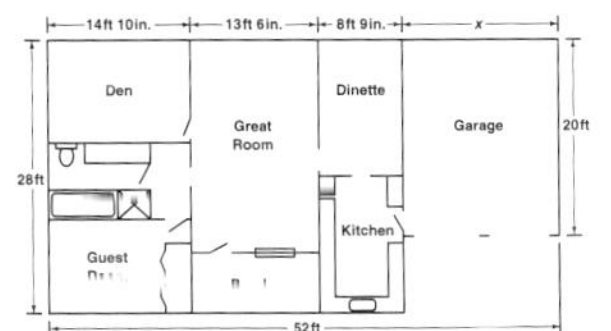

   a. Mr. Chen has a choice of two lots. One lot has an area of 1.7 acres and the other has an area of $1\frac{2}{3}$ acres. Which lot is larger? Explain your reasoning. Give at least three factors other than the area he might consider before buying a lot.
   b. Estimate the width of the garage. Is your estimate too large or too small? Why?
   c. Explain how to add dimensions given in feet and inches. Find the actual width of the garage.
2. In 1995, the cost of mailing a letter first class was 32¢ for the first ounce and 23¢ for each additional ounce.
   a. Make a table that shows the cost of mailing letters for 1, 2, 3, ⋯, 6 ounces. What is the pattern? What is this pattern called?
   b. Write a word problem that uses an arithmetic sequence. Solve the problem and explain the meaning of the answer.
3. Explain how you can use an estimate to place the decimal point in the sum 53.72 + 7.093.

CHAPTER 5 Scoring Guide

Level	Specific Criteria
3 Superior	• Shows thorough understanding of the concepts *estimation* and *arithmetic sequence*. • Uses appropriate strategies to estimate the width of the garage. • Computations are correct in adding dimensions and solving problem. • Written explanations are exemplary. • Word problem concerning arithmetic sequence is appropriate and makes sense. • Table is accurate. • Goes beyond requirements of some or all problems.
2 Satisfactory, with minor flaws	• Shows understanding of the concepts *estimation* and *arithmetic sequence*. • Uses appropriate strategies to estimate the width of the garage. • Computations are mostly correct in adding dimensions and solving problem. • Written explanations are effective. • Word problem concerning arithmetic sequence is appropriate and makes sense. • Table is mostly accurate. • Satisfies all requirements of problems.
1 Nearly Satisfactory, with serious flaws	• Shows understanding of most of the concepts *estimation* and *arithmetic sequence*. • May not use appropriate strategies to estimate the width of the garage. • Computations are mostly correct in adding dimensions and solving problem. • Written explanations are satisfactory. • Word problem concerning arithmetic sequence is appropriate and makes sense. • Table is mostly accurate. • Satisfies most requirements of problems.
0 Unsatisfactory	• Shows little or no understanding of the concepts *estimation* and *arithmetic sequence*. • May not use appropriate strategies to estimate the width of the garage. • Computations are incorrect. • Written explanations are not satisfactory. • Word problem concerning arithmetic sequence is not appropriate or sensible. • Table is not accurate. • Does not satisfy requirements of problems.

Investigation

TEACHER NOTES

NCTM Standards: 1-4

This Investigation is designed to be completed over several days or weeks. It may be considered optional. You may want to assign the Investigation and the follow-up activities to be completed at the same time.

Objectives

- Research the history of roller coasters.
- Create an original design for a roller coaster, including ride data and special interest items such as providing entertainment for people waiting in line.

Mathematical Overview

This Investigation will use the following skills and concepts from Chapters 6 and 7.

- estimate products and quotients of rational numbers
- divide fractions
- work backward
- write and solve equations
- create scale drawings

Recommended Time		
Part	**Pages**	**Time**
Investigation	270–271	1 class period
Working on the Investigation	279, 293, 337	20 minutes each
Closing the Investigation	362	1 class period

Instructional Resources

Investigations and Projects Masters, pp. 13–16

A recording sheet, teacher notes, and scoring guide are provided for each Investigation in the *Investigation and Projects Masters.*

1 MOTIVATION

In this Investigation, students use research and simulation to design a roller coaster. Allow time at the beginning of the Investigation for students to discuss roller coasters that they have ridden and what they feel makes for a great ride.

LONG-TERM PROJECT

Investigation

MATERIALS

- ruler
- protractor
- compass
- graph paper

"Please remain seated and keep your hands and feet inside the ride at all times. Enjoy your ride."

You are a member of a group of three engineers that has been hired to design and build a new roller coaster ride. You will oversee every detail of the design of the ride from the length of its track to how riders will wait in line and board the cars. You'll even name the ride. The amusement park where the roller coaster will be built wants you to design it so that it establishes at least one new record.

Some of the existing world records are:

Longest

- *1.42 miles*—The Ultimate *at Lightwater Valley Theme Park in Ripon, Great Britain*

Tallest

- *222 feet above its footings*—Pepsi Max-The Big One *at Blackpool Pleasure Beach, Great Britain*

Fastest

- *85 mph*—Desperado *at Buffalo Bill's, Jean, NV*

Greatest Drop

- *225 feet*—Steel Phantom *at Kennywood Amusement Park, West Mifflin, PA*

Coasters are classified as either steel or wooden thus allowing for separate records. Some records depend on whether passengers remain seated or are standing during the ride.

Starting the Investigation

To start the design, you will want to draw a detailed model of the path your roller coaster will take. Before you begin drawing, familiarize yourself with some roller coaster terms.

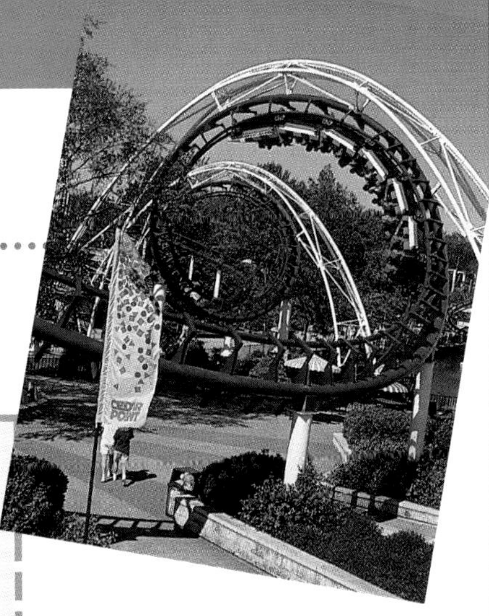

A few important roller coaster definitions are:

- ***Banked Turn***—*Turn in which the tracks are tilted to allow trains to turn at high speed.*
- ***Car***—*The part of a coaster train that carries between two and eight passengers.*
- ***Chain Lift***—*The rolling chain that carries the train to the crest of a hill.*
- ***Circuit***—*A completed journey on a coaster track.*
- ***First Drop***—*Usually the highest and fastest drop on a coaster. It most often follows immediately after the chain lift.*
- ***Gully Coaster***—*One that uses the natural terrain to give the added feeling of speed by keeping the track close to the ground.*
- ***Lift Hill***—*The hill that the train is carried up by the chain lift.*
- ***Loading Platform***—*The part of the station where passengers board the coaster trains.*
- ***Speed Dip***—*A small hill taken at high speeds usually lifting riders off their seats.*
- ***Train***—*A series of cars hooked together to travel the circuit of the coaster track.*

Since the design of a roller coaster can be very complex, you will need to include a scale drawing from several different angles including an overhead view. Be sure to include a sketch of the loading platform. Also include the number of cars in each train and how many passengers each car will carry.

Brainstorm with your group to decide what record you want to break with your design. You may wish to look in some other sources for other records and for guidance on how to design your coaster.

You will work on this investigation throughout Chapters 6 and 7. Be sure to keep your materials in your Investigation Folder.

For current information on roller coasters, visit:
www.glencoe.com/sec/math/prealg/mathnet

Investigation
Roller Coaster Math

Working on the Investigation
Lesson 6-1, p. 279

Working on the Investigation
Lesson 6-4, p. 293

Working on the Investigation
Lesson 7-2, p. 337

Closing the Investigation
End of Chapter 7, p. 362

2 SETUP

Have students work in small groups. For this Investigation, students will need rulers, protractors, compasses and graph paper.

For further information, students can use any of these resources.

World Wide Web at http: //sunsite.unc.edu/darlene/coaster/FAQ/FAQ.cover

American Coaster Enthusiasts; P.O. Box 8226; Chicago, IL 60680

Cartmell, Robert. *The Incredible Scream Machine: A History of the Roller Coaster.* Bowling Green State University Popular Press

Rollercoaster (quarterly magazine), American Coaster Enthusiasts

3 MANAGEMENT

Students should understand that the designs need to be neat and accurate. Each member of the group should check the final product before it is turned in.

Sample Answers

Answers will vary as they are based on the research of each group.

Investigations and Projects Masters, p. 16

NAME ______________________ DATE __________

CHAPTERS **6 and 7** **Investigation**
Roller Coaster Math

Student Edition Pages 270–271

Work with the other engineers in your group to answer the following questions about the design of your roller coaster.

What is the name of your ride?

____

Four records are discussed on page 270: longest, tallest, fastest, and greatest drop. For each category, explain the problems and limitations related to breaking that record.

____

What record does your group want to break with your roller coaster?

____

What details of your design would help you achieve your goal?

____

List below any additional questions that will need to be addressed concerning the new roller coaster, including details of the track, cars, train, loading platform, and materials.

____

Please keep this page and any other research in your Investigation Folder.

Cooperative Learning

This Investigation offers an excellent opportunity for using cooperative learning groups. For more information on cooperative learning strategies and group management, see *Cooperative Learning in the Mathematics Classroom*, one of the titles in the Glencoe Mathematics Professional Series.

CHAPTER 6

Rationals: Patterns in Multiplication and Division

PREVIEWING THE CHAPTER

This chapter begins with a study of fractions expressed as terminating or repeating decimals. The emphasis of the remainder of the chapter is on estimating and finding products and quotients of decimals and fractions. Students will determine the mean, median, and mode of a set of data. They will also learn to solve equations and inequalities that contain rational numbers. In addition, students will use algebra to find the terms of a geometric sequence and learn how to write large and small numbers using scientific notation.

Lesson (pages)	Lesson Objectives	NCTM Standards	State/Local Objectives
6-1 (274–279)	Write fractions as terminating or repeating decimals. Compare rational numbers.	1-6, 8	
6-2 (280–283)	Estimate products and quotients of rational numbers.	1-5, 7	
6-3 (284–288)	Multiply fractions.	1-5, 7	
6-4 (289–293)	Divide fractions using multiplicative inverses.	1-7, 9	
6-5A (294)	Multiply and divide decimals using decimal models.	1-5	
6-5 (295–299)	Multiply and divide decimals.	1-8	
6-6A (300)	Use the mean, median, and mode of numbers to analyze data.	1-5, 10, 12	
6-6 (301–306)	Use the mean, median, and mode as measures of central tendency.	1-5, 10	
6-6B (307)	Use a graphing calculator to find the mean, median, and mode of a set of data.	1-5, 10, 12	
6-7 (308–311)	Solve equations and inequalities containing rational numbers.	1-5, 8, 9	
6-8 (312–316)	Recognize and extend geometric sequences and represent them algebraically.	1-5, 8	
6-9 (317–320)	Write numbers in scientific notation.	1-5, 7	

A complete, 1-page lesson plan is provided for each lesson in the ***Lesson Planning Guide***. Answer keys for each lesson are available in the ***Answer Key Masters***.

ORGANIZING THE CHAPTER

You may want to refer to the **Course Planning Calendar** on page T12 for detailed information on pacing.
PACING: Standard—13 days; **Honors**—12 days; **Block**—7 days

LESSON PLANNING CHART

Lesson (Pages)	Materials/ Manipulatives	Extra Practice (Student Edition)	Blackline Masters									Real-World Applications	Interactive Mathematics Tools Software	Teaching Transparencies	Group Activity Cards
			Study Guide	Practice	Enrichment	Assessment and Evaluation	Math Lab and Modeling Math	Multicultural Activity	Tech Prep Applications	Graphing Calculator	Activity				
6-1 (274–279)	grid paper	p. 753	p. 45	p. 45	p. 45						p. 37		6-1	6-1A 6-1B	6-1
6-2 (280–283)	calculator	p. 753	p. 46	p. 46	p. 46									6-2A 6-2B	6-2
6-3 (284–288)	grid paper colored pencils ruler*	p. 753	p. 47	p. 47	p. 47	p. 155	p. 63		p. 11		p. 20			6-3A 6-3B	6-3
6-4 (289–293)	colored pencils	p. 754	p. 48	p. 48	p. 48			p. 11	p. 12					6-4A 6-4B	6-4
6-5A (294)	grid paper colored pencils						p. 37								
6-5 (295–299)	calculator grid paper colored pencils	p. 754	p. 49	p. 49	p. 49	pp. 154–155					p. 6		6-5	6-5A 6-5B	6-5
6-6A (300)	small boxes of raisins						p. 38								
6-6 (301–306)	roll of 50 pennies	p. 754	p. 50	p. 50	p. 50			p. 12				12	6-6	6-6A 6-6B	6-6
6-6B (307)	graphing calculator									p. 20					
6-7 (308–311)		p. 755	p. 51	p. 51	p. 51	p. 156	p. 64					13		6-7A 6-7B	6-7
6-8 (312–316)	$8\frac{1}{2}$" by 11" paper	p. 755	p. 52	p. 52	p. 52					p. 6			6-8	6-8A 6-8B	6-8
6-9 (317–320)		p. 755	p. 53	p. 53	p. 53	p. 156	p. 81							6-9A 6-9B	6-9
Study Guide/ Assessment (322–324)						pp. 141–153, 157–159									

*Included in Glencoe's *Student Manipulative Kit* and *Overhead Manipulative Resources*.

ORGANIZING THE CHAPTER

All of the blackline masters in the Teacher's Classroom Resources are available on the ***Electronic Teacher's Classroom Resources*** CD-ROM.

OTHER CHAPTER RESOURCES

Student Edition
Chapter Opener, pp. 272–273
Investigation, pp. 270–271
Working on the Investigation, pp. 279, 293
Cooperative Learning Project, The Shape of Things to Come, p. 320

Teacher's Classroom Resources
Investigations and Projects Masters, pp. 49–52
Pre-Algebra Overhead Manipulative Resources, pp. 13–16

Technology
Test and Review Software (IBM & Macintosh)
CD-ROM Activities

Professional Publications
Block Scheduling Booklet
Glencoe Mathematics Professional Series

OUTSIDE RESOURCES

Books/Periodicals
The Minneapolis General Math Project: A Reasonably Close Encounter (estimation), The Minneapolis Public Schools
Pre-Algebra with Pizzazz!, Steve and Janis Marcy

Videos/CD-ROMs
Real Life Decimals Series, ETA
Decimals Video, ETA

Software
Mastering Decimals, Phoenix/BFA Films and Videos

ASSESSMENT RESOURCES

Student Edition
Math Journal, pp. 282, 291, 319
Mixed Review, pp. 279, 283, 288, 293, 299, 306, 311, 316, 320
Self Test, p. 299
Chapter Highlights, p. 321
Chapter Study Guide and Assessment, pp. 322–324
Alternative Assessment, p. 325
Portfolio, p. 325
Ongoing Assessment, pp. 326–327
MindJogger Videoquiz, 6

Teacher's Wraparound Edition
5-Minute Check, pp. 274, 280, 284, 289, 295, 301, 308, 312, 317
Checking Your Understanding, pp. 277, 282, 286, 291, 297, 304, 310, 314, 319
Closing Activity, pp. 279, 283, 288, 293, 299, 306, 311, 316, 320
Cooperative Learning, pp. 276, 319

Assessment and Evaluation Masters
Multiple-Choice Tests, Forms 1A (Honors), 1B (Average), 1C (Basic), pp. 141–146
Free-Response Tests, Forms 2A (Honors), 2B (Average), 2C (Basic), pp. 147–152
Performance Assessment, p. 153
Mid-Chapter Test, p. 154
Quizzes A–D, pp. 155–156
Standardized Test Practice, p. 157
Cumulative Review, pp. 158–159

ENHANCING THE CHAPTER

Examples of some of the materials for enhancing Chapter 6 are shown below.

Multicultural Activity Masters, pp. 11, 12

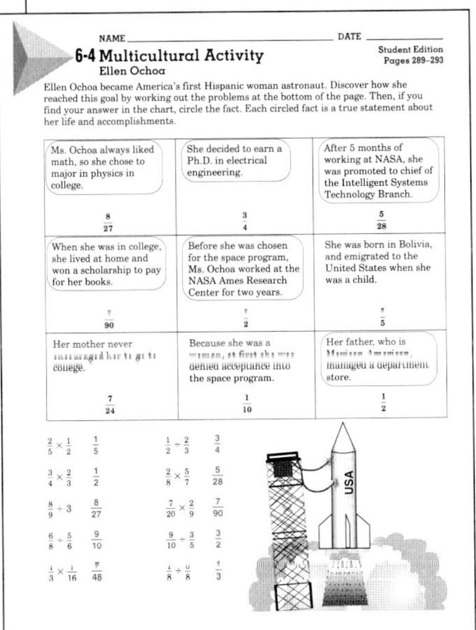

NAME ____________ DATE ______

Student Edition Pages 289–293

6-4 Multicultural Activity
Ellen Ochoa

Ellen Ochoa became America's first Hispanic woman astronaut. Discover how she reached this goal by working out the problems at the bottom of the page. Then, if you find your answer in the chart, circle the fact. Each circled fact is a true statement about her life and accomplishments.

Ms. Ochoa always liked math, so she chose to major in physics in college. $\frac{8}{27}$	She decided to earn a Ph.D. in electrical engineering. $\frac{3}{4}$	After 5 months of working at NASA, she was promoted to chief of the Intelligent Systems Technology Branch. $\frac{5}{28}$
When she was in college, she lived at home and won a scholarship to pay for her books. $\frac{7}{90}$	Before she was chosen for the space program, Ms. Ochoa worked at the NASA Ames Research Center for two years. $\frac{3}{2}$	She was born in Bolivia, and emigrated to the United States when she was a child. $\frac{7}{5}$
Her mother never encouraged her to go to college. $\frac{7}{24}$	Because she was a woman, at first she was denied acceptance into the space program. $\frac{1}{10}$	Her father, who is Mexican American, managed a department store. $\frac{1}{2}$

$\frac{2}{5} \times \frac{1}{2}$ $\frac{1}{5}$ $\qquad$ $\frac{1}{2} \div \frac{2}{3}$ $\frac{3}{4}$

$\frac{3}{4} \times \frac{2}{3}$ $\frac{1}{2}$ $\qquad$ $\frac{2}{8} \times \frac{5}{7}$ $\frac{5}{28}$

$\frac{8}{9} \div 3$ $\frac{8}{27}$ $\qquad$ $\frac{7}{20} \times \frac{2}{9}$ $\frac{7}{90}$

$\frac{6}{8} \div \frac{5}{6}$ $\frac{9}{10}$ $\qquad$ $\frac{9}{10} \div \frac{3}{5}$ $\frac{3}{2}$

$\frac{1}{3} \times \frac{1}{16}$ $\frac{1}{48}$ $\qquad$ $\frac{1}{8} \div \frac{9}{8}$ $\frac{1}{3}$

Real-World Applications, 12, 13

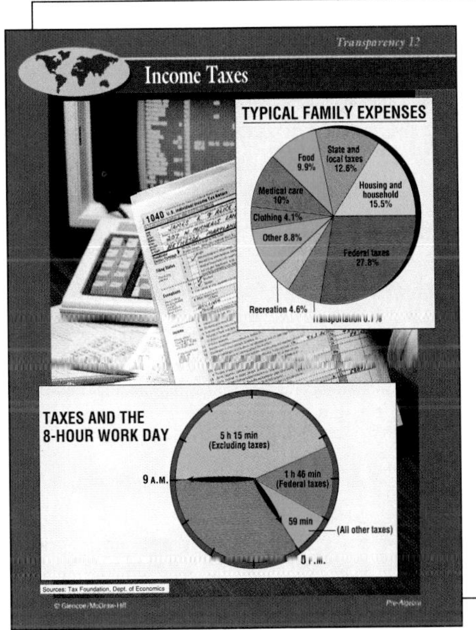

Graphing Calculator Masters, p. 6

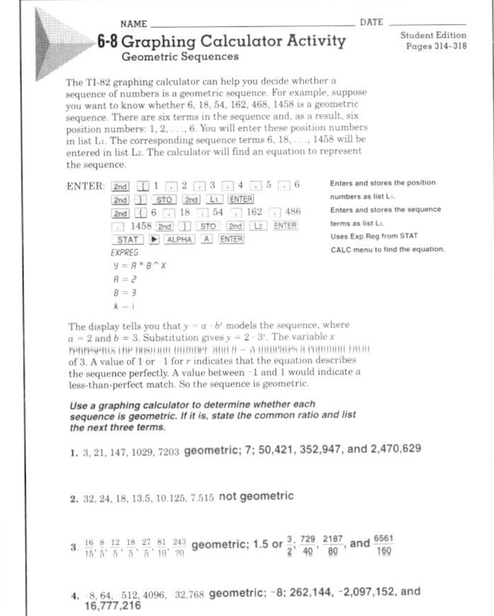

NAME ____________ DATE ______

Student Edition Pages 314–318

6-8 Graphing Calculator Activity
Geometric Sequences

The TI-82 graphing calculator can help you decide whether a sequence of numbers is a geometric sequence. For example, suppose you want to know whether 6, 18, 54, 162, 468, 1458 is a geometric sequence. There are six terms in the sequence and, as a result, six position numbers: 1, 2, …, 6. You will enter these position numbers in list L1. The corresponding sequence terms 6, 18, …, 1458 will be entered in list L2. The calculator will find an equation to represent the sequence.

ENTER: 2nd { 1 , 2 , 3 , 4 , 5 , 6 2nd } STO 2nd L1 ENTER — Enters and stores the position numbers as list L1.
2nd { 6 , 18 , 54 , 162 , 486 , 1458 2nd } STO 2nd L2 ENTER — Enters and stores the sequence terms as list L2.
STAT ▶ ALPHA A ENTER — Uses Exp Reg from STAT CALC menu to find the equation.

EXPREG
y = a * b ^ x
a = 2
b = 3
r = 1

The display tells you that $y = a \cdot b^x$ models the sequence, where $a = 2$ and $b = 3$. Substitution gives $y = 2 \cdot 3^x$. The variable x represents the position number and b = 3 indicates a common ratio of 3. A value of 1 or −1 for r indicates that the equation describes the sequence perfectly. A value between −1 and 1 would indicate a less-than-perfect match. So the sequence is geometric.

Use a graphing calculator to determine whether each sequence is geometric. If it is, state the common ratio and list the next three terms.

1. 3, 21, 147, 1029, 7203 **geometric; 7; 50,421, 352,947, and 2,470,629**
2. 32, 24, 18, 13.5, 10.125, 7.515 **not geometric**
3. $\frac{16}{15}, \frac{8}{5}, \frac{12}{5}, \frac{18}{5}, \frac{27}{5}, \frac{81}{10}, \frac{243}{20}$ **geometric; 1.5 or $\frac{3}{2}$; $\frac{729}{40}$, $\frac{2187}{80}$, and $\frac{6561}{160}$**
4. −8, 64, −512, 4096, −32,768 **geometric; −8; 262,144, −2,097,152, and 16,777,216**

Tech Prep Applications Masters, pp. 11, 12

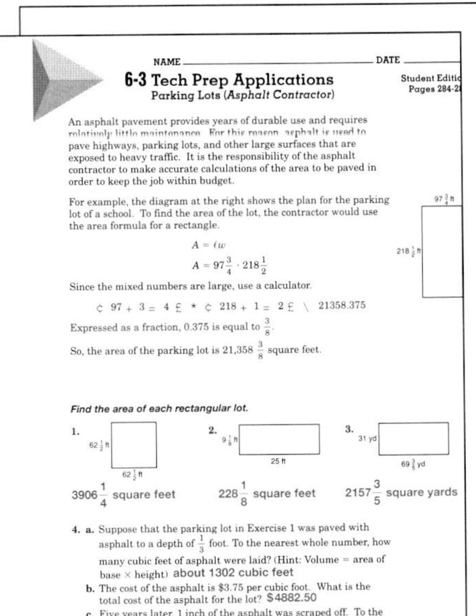

NAME ____________ DATE ______

Student Edition Pages 284–2…

6-3 Tech Prep Applications
Parking Lots (Asphalt Contractor)

An asphalt pavement provides years of durable use and requires relatively little maintenance. For this reason, asphalt is used to pave highways, parking lots, and other large surfaces that are exposed to heavy traffic. It is the responsibility of the asphalt contractor to make accurate calculations of the area to be paved in order to keep the job within budget.

For example, the diagram at the right shows the plan for the parking lot of a school. To find the area of the lot, the contractor would use the area formula for a rectangle.

$A = \ell w$

$A = 97\frac{3}{4} \cdot 218\frac{1}{2}$

Since the mixed numbers are large, use a calculator.

97 + 3 ÷ 4 = × 218 + 1 ÷ 2 = 21358.375

Expressed as a fraction, 0.375 is equal to $\frac{3}{8}$.

So, the area of the parking lot is $21{,}358\frac{3}{8}$ square feet.

Find the area of each rectangular lot.

1. $62\frac{1}{2}$ ft by $62\frac{1}{2}$ ft **$3906\frac{1}{4}$ square feet**
2. $9\frac{1}{8}$ ft by 25 ft **$228\frac{1}{8}$ square feet**
3. 31 yd by $69\frac{3}{5}$ yd **$2157\frac{3}{5}$ square yards**

4. a. Suppose that the parking lot in Exercise 1 was paved with asphalt to a depth of $\frac{1}{3}$ foot. To the nearest whole number, how many cubic feet of asphalt were laid? (Hint: Volume = area of base × height) **about 1302 cubic feet**
   b. The cost of the asphalt is $3.75 per cubic foot. What is the total cost of the asphalt for the lot? **$4882.50**
   c. Five years later, 1 inch of the asphalt was scraped off. To the nearest whole number, how many cubic feet was this? (Hint: What fraction of 1 foot is 1 inch?) **about 326 cubic feet**

Activity Masters, pp. 6, 20, 37

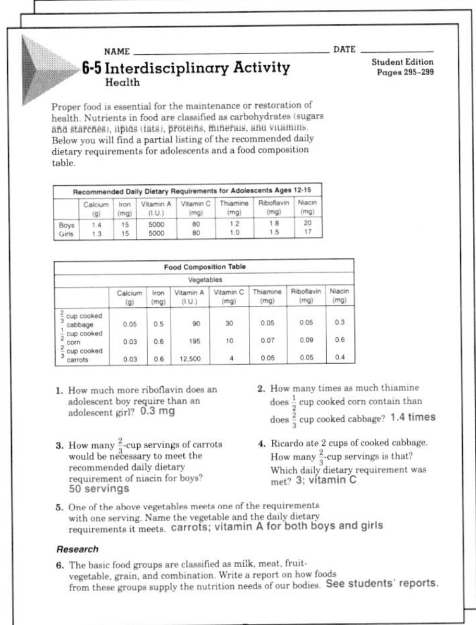

NAME ____________ DATE ______

Student Edition Pages 295–299

6-5 Interdisciplinary Activity
Health

Proper food is essential for the maintenance or restoration of health. Nutrients in food are classified as carbohydrates (sugars and starches), lipids (fats), proteins, minerals, and vitamins. Below you will find a partial listing of the recommended daily dietary requirements for adolescents and a food composition table.

Recommended Daily Dietary Requirements for Adolescents Ages 12-15

	Calcium (g)	Iron (mg)	Vitamin A (I.U.)	Vitamin C (mg)	Thiamine (mg)	Riboflavin (mg)	Niacin (mg)
Boys	1.4	15	5000	80	1.2	1.8	20
Girls	1.3	15	5000	80	1.0	1.5	17

Food Composition Table — Vegetables

	Calcium (g)	Iron (mg)	Vitamin A (I.U.)	Vitamin C (mg)	Thiamine (mg)	Riboflavin (mg)	Niacin (mg)
$\frac{2}{3}$ cup cooked cabbage	0.05	0.5	90	30	0.05	0.05	0.3
$\frac{2}{3}$ cup cooked corn	0.03	0.6	195	10	0.07	0.09	0.6
$\frac{2}{3}$ cup cooked carrots	0.03	0.6	12,500	4	0.05	0.05	0.4

1. How much more riboflavin does an adolescent boy require than an adolescent girl? **0.3 mg**
2. How many times as much thiamine does $\frac{2}{3}$ cup cooked corn contain than does $\frac{2}{3}$ cup cooked cabbage? **1.4 times**
3. How many $\frac{2}{3}$-cup servings of carrots would be necessary to meet the recommended daily dietary requirement of niacin for boys? **50 servings**
4. Ricardo ate 2 cups of cooked cabbage. How many $\frac{2}{3}$-cup servings is that? Which daily dietary requirement was met? **3; vitamin C**
5. One of the above vegetables meets one of the requirements with one serving. Name the vegetable and the daily dietary requirements it meets. **carrots; vitamin A for both boys and girls**

Research

6. The basic food groups are classified as milk, meat, fruit-vegetable, grain, and combination. Write a report on how foods from these groups supply the nutrition needs of our bodies. **See students' reports.**

COOPERATIVE LEARNING

Math Lab and Modeling Math Masters, pp. 63, 64, 81

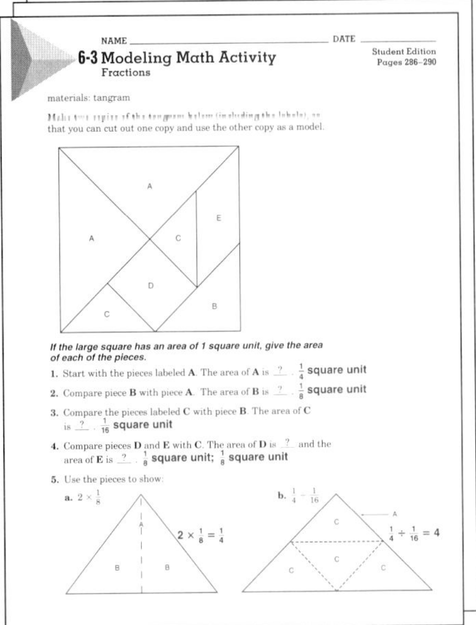

NAME ____________ DATE ______

Student Edition Pages 286–290

6-3 Modeling Math Activity
Fractions

materials: tangram

Make two copies of the tangram below (including the labels), so that you can cut out one copy and use the other copy as a model.

If the large square has an area of 1 square unit, give the area of each of the pieces.

1. Start with the pieces labeled A. The area of A is __?__. **$\frac{1}{4}$ square unit**
2. Compare piece B with piece A. The area of B is __?__. **$\frac{1}{8}$ square unit**
3. Compare the pieces labeled C with piece B. The area of C is __?__. **$\frac{1}{16}$ square unit**
4. Compare pieces D and E with C. The area of D is __?__ and the area of E is __?__. **$\frac{1}{8}$ square unit; $\frac{1}{8}$ square unit**
5. Use the pieces to show:
   a. $2 \times \frac{1}{8}$ **$2 \times \frac{1}{8} = \frac{1}{4}$**
   b. $\frac{1}{4} \div \frac{1}{16}$ **$\frac{1}{4} \div \frac{1}{16} = 4$**

MAKING MATHEMATICS RELEVANT

This two-page introduction to the chapter provides students with an opportunity to explore other disciplinary topics and their applications to mathematics.

Background Information

Math and Computers in the News One of the most famous virus creators is known as Dark Avenger. These viruses have been found in at least 10 countries. It has created viruses that change their appearance as they spread so they are harder to detect. It even has produced a time-release virus that is difficult to locate because the data on a disk is destroyed very slowly.

CHAPTER 6 Rationals: Patterns in Multiplication and Division

TOP STORIES in Chapter 6

In this chapter, you will:

- write fractions as decimals,
- estimate and find products and quotients of rational numbers,
- find the mean, median, and mode of a set of data,
- identify and extend geometric sequences, and
- write numbers in scientific notation.

MATH AND COMPUTERS IN THE NEWS

How do you vaccinate a computer?

Source: PC Computing, September, 1994

Feed a cold, starve a fever, but what do you do for a virus? Especially if the patient has circuits and electric power! Computer viruses are destructive programs that are placed on a person's computer without their knowledge. Once inside the system, a virus manipulates or destroys the programs and data. Viruses have been a problem to computer owners for some time now, but with the dawn of the Information Superhighway, safeguards are now more important than ever.

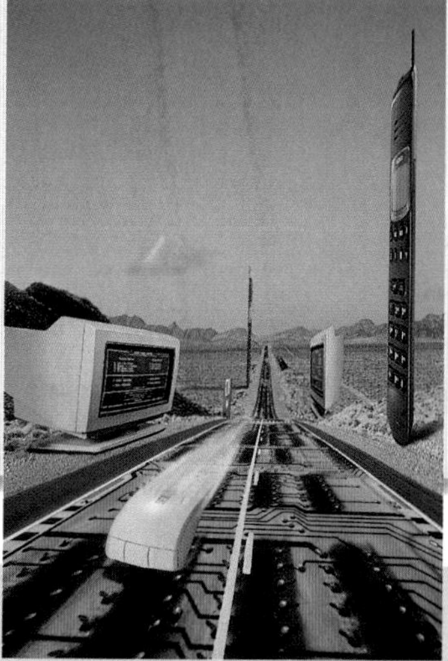

Putting It into Perspective

1850

1833
Charles Babbage designs the "analytical engine" with the programming help of Ada Byron Lovelace.

1880

1888
The first successful punch-card tabulating machine is invented by an American.

1910

Putting It into Perspective

The Internet is an informal system that connects 15–20 million users in more than 125 countries using more than a million computers. Users of Internet have access to millions of files, thousands of different discussion groups, and numerous databases and news sources.

On the Lighter Side

"Excuse me, I'm lost. Can you direct me to the information superhighway?"

Statistical Snapshot

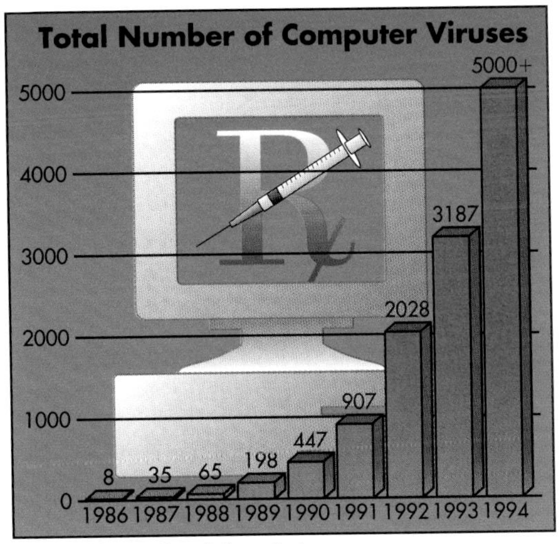

Source: Norman Data Defense Systems

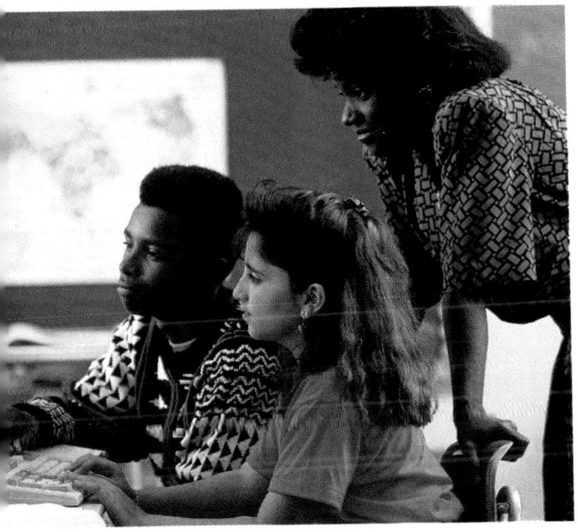

interNET CONNECTION For up-to-date information on computer viruses, visit: **www.glencoe.com/sec/math/prealg/mathnet**

1946 The first general purpose digital computer, the ENIAC, is constructed at the University of Pennsylvania.

1977 The Apple II is introduced as the first personal computer marketed on a large scale.

1995 The Information Superhighway, a worldwide computer network, grows.

1940 1970 2000

1968 Blockbuster movie *2001: A Space Odyssey* features the character HAL, a talking, thinking computer.

On the Lighter Side

The Information Superhighway, or Internet, began as a network of government and research computers. It continues to grow into a huge network with many products and services available.

Statistical Snapshot

The graph shows the rapid growth in the number of computer viruses since they were first discovered in 1986. Ask students to use the graph to estimate the number of new viruses produced in the two-year period between January 1, 1992 and December 31, 1993.

Cooperative Learning

Chapter Projects Two chapter projects are included in the *Investigations and Projects Masters*. In Chapter Project A, students extend the topic in the Chapter Opener. In Chapter Project B, students explore archaeology. A student page and a parent letter are provided for each Chapter Project.

interNET CONNECTION

Glencoe has made every effort to ensure that the website links for *Pre-Algebra* at www.glencoe.com/sec/math/prealg/mathnet are current and contain appropriate content. However, these website links are not under Glencoe's control.

Investigations and Projects Masters, p. 49

NAME ____________ DATE ________

CHAPTER 6 Project A

Keeping Information Safe

There are many things we all do in our daily lives to protect ourselves and our property. From locking doors and installing house and car alarms to padlocking bicycles, most people recognize the need to take security precautions. But now we face a new security challenge—how to protect private information that is transmitted electronically.

1. Interview a few adults you know to find out about ways they protect themselves from theft. Ask questions like
   - What security measures protect you during visits to ATM machines?
   - How do you know that you are the only one using your credit cards?
   - How is information about such things as your insurance policies and driving records kept away from anyone but yourself and the administrators?
2. Interview a person who works for a bank to find out what measures are taken to maintain the privacy of financial records. Or speak with a person who works for a hospital to learn how personal medical files are kept private.
3. Now focus on the privacy problems computer users face and the security measures that are currently available. Talk to on-line service users to find out about the role of passwords and codes. Read a computer or news magazine, or interview a knowledgeable computer programmer to learn more about the menace of computer viruses. Try to find out the names and the damage done by some of the more well-known viruses. Find out about some safeguards that can protect files from viruses.
4. Use your research to prepare a pamphlet about the rise of electronic theft and the ways in which we are combating the problem. Organize your findings to include an overview of the issue, how the possibility of theft affects people in their everyday lives, and how institutions deal with problems of monitoring electronic information. Be sure to include a section that addresses the rise of computer viruses and the ways that are being developed to "vaccinate" computers from them.

6-1 LESSON NOTES

NCTM Standards: 1-6, 8

Instructional Resources

- Study Guide Master 6-1
- Practice Master 6-1
- Enrichment Master 6-1
- Group Activity Card 6-1
- Activity Masters, p. 37

Transparency 6-1A contains the 5-Minute Check for this lesson; **Transparency 6-1B** contains a teaching aid for this lesson.

Recommended Pacing	
Standard Pacing	Day 1 of 13
Honors Pacing	Day 1 of 12
Block Scheduling*	Day 1 of 7 (along with Lesson 6-2)

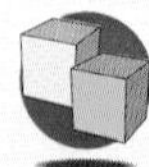

*For more information on pacing and possible lesson plans, refer to the ***Block Scheduling Booklet***.

1 FOCUS

5-Minute Check
(over Chapter 5)

Rename each decimal as a fraction or mixed number in simplest form.

1. 0.4 $\frac{2}{5}$

2. 1.666... $1\frac{2}{3}$

3. -2.45 $-2\frac{9}{20}$

Solve each equation.

4. $m = 2.9 - (-2.6)$ **5.5**

5. $-7.1 + 5.7 = x$ **−1.4**

FYI

In 1851, the first ice-cream factory opened in Baltimore, Maryland. The ice-cream soda was introduced in 1879 and the ice-cream cone in 1904.

6-1 Writing Fractions as Decimals

Setting Goals: *In this lesson, you'll write fractions as terminating or repeating decimals and compare rational numbers.*

Modeling a Real-World Application: Food

FYI

Finland produces more ice cream per citizen each year, 42 pints, than any other country in the world.

"You scream, I scream, we all scream for ice cream!" What is your favorite flavor? The International Ice Cream Association polled Americans about their favorite flavors. Some of the top finishers and the approximate fractions of the votes they received are shown in the graph below. If $\frac{3}{76}$ of those surveyed chose chocolate chip as their favorite, did more people choose cookies-and-cream or chocolate chip? *This problem will be solved in Example 3.*

Learning the Concept

You can review the definition of rational numbers in Lesson 5-1.

In the case above, it is convenient to write rational numbers as decimals instead of fractions because decimals are easier to compare. Any fraction can be written as a decimal.

Consider the fraction $\frac{3}{5}$. The fraction $\frac{3}{5}$ indicates $3 \div 5$.

Alternative Teaching Strategies

Student Diversity Have students compare $\frac{16}{19}$ and $\frac{6}{7}$ using cross products. Write the ratios as 16:19 and 6:7. Then multiply the first and last numbers and the two "middle" numbers so that the products are in the same order as the fractions.

For $\frac{16}{19}$ and $\frac{6}{7}$, students find $16 \cdot 7$ and $19 \cdot 6$. Compare 112 to 114. Since $112 < 114$, $\frac{16}{19} < \frac{6}{7}$.

Method 1
Divide using paper and pencil.

$$\frac{3}{5} \rightarrow \begin{array}{r} 0.6 \\ 5\overline{)3.0} \\ -3\,0 \\ \hline 0 \end{array}$$

Method 2
Divide using a calculator.

3  5 = .6

The fraction $\frac{3}{5}$ can be written as the decimal 0.6. This means that $\frac{3}{5}$ and 0.6 are **equivalent** rational numbers. Remember that a decimal like 0.6 is called a *terminating decimal* because the division ends, or terminates, when the remainder is zero.

Example 1 **Write $2\frac{3}{4}$ as a decimal.**

$2\frac{3}{4} = 2 + \frac{3}{4}$

Method 1
Use paper and pencil.

$$\frac{3}{4} \rightarrow \begin{array}{r} 0.75 \\ 4\overline{)3.00} \\ -2\,8 \\ \hline 20 \\ -20 \\ \hline 0 \end{array}$$

Consider only $\frac{3}{4}$.

Method 2
Use a calculator.

2 + 3 ÷ 4 = 2.75

Be sure your calculator follows the order of operations.

$2 + 0.75 = 2.75$

The mixed number $2\frac{3}{4}$ equals the decimal 2.75.

Not all fractions can be written as terminating decimals. Consider the fraction $\frac{2}{3}$. How can $\frac{2}{3}$ be expressed as a decimal?

Use paper and pencil.

$$\frac{2}{3} \rightarrow \begin{array}{r} 0.666 \\ 3\overline{)2.000} \\ -1\,8 \\ \hline 20 \\ -18 \\ \hline 20 \end{array}$$

Use a calculator.

2 ÷ 3 = .6666666

When you divide using paper and pencil, the remainder after each step is 2. If you continue dividing, the pattern repeats. The calculator display fills with 6s. Remember, a decimal like 0.6666666 . . . , or $0.\overline{6}$, is called a *repeating decimal.*

Some calculators round answers and others truncate answers. **Truncate** means to cut off at a certain place-value position, ignoring the digits that follow.

Motivating the Lesson

Questioning Have students brainstorm to answer these questions:

- What situation can you think of in which fractions are easy to use? **Sample answer: measuring length**
- What situation can you think of in which decimals are more practical than fractions? **Sample answer: working with money**

2 TEACH

In-Class Example

For Example 1

Write $1\frac{7}{8}$ as a decimal. **1.875**

Teaching Tip You may wish to review dividing a decimal by a whole number. Question students about how they determine the location of the decimal point in the answer.

Teaching Tip Relate the word *terminating* to other words such as *terminal* and titles such as *The Terminator*.

GLENCOE Technology

Interactive Mathematics Tools Software

In this interactive computer lesson, students use grids to explore and compare rational numbers. A **Computer Journal** gives students the opportunity to write about what they have learned.

For Windows & Macintosh

Teaching Tip Students may assume that if a decimal equivalent does not repeat after one or two numbers, it does not repeat. Using an example of sevenths is a good way to show that many digits can repeat.

In-Class Examples

For Example 2

Write $\frac{7}{12}$ as a decimal.
0.5833...

For Example 3

Seven sixteenths of the students chose blue and gold when surveyed about new school colors. Black and gold was chosen by $\frac{5}{12}$. Which colors were chosen by more students? **blue and gold**

For Example 4

Replace each ● with <, >, or = to make a true sentence.

a. $\frac{3}{4}$ ● $\frac{2}{3}$ **>**

b. -1.65 ● $-1\frac{2}{3}$ **>**

Example  **Write $\frac{5}{11}$ as a decimal.**

Some calculators round and others truncate. Find the decimal value of $\frac{5}{11}$ on your calculator to determine whether it rounds or truncates.

Use paper and pencil.

$$\frac{5}{11} \rightarrow 11\overline{)5.0000} = 0.4545$$

$$\begin{array}{r} 0.4545 \\ 11\overline{)5.0000} \\ -4\,4 \\ 60 \\ -55 \\ 50 \\ -44 \\ 60 \\ -55 \end{array}$$

$\frac{5}{11} = 0.\overline{45}$

Use a calculator.

5 [÷] 11 [=] .4545455
This calculator rounds.

5 [÷] 11 [=] .4545454
This calculator truncates.

Sometimes you need to compare numbers, as in the application at the beginning of the lesson. You can use what you have learned about writing fractions as decimals to compare fractions.

Example 3

APPLICATION
Food

Refer to the application at the beginning of the lesson. Which is more popular, chocolate chip or cookies-and-cream ice cream?

Write the fractions as decimals and then compare the decimals.

chocolate chip: $\frac{3}{76} \rightarrow 0.0394736842$

cookies-and-cream: $\frac{2}{55} \rightarrow 0.036363636$

In thousandths place, $9 > 6$.

Note that on a number line, 0.039 . . . is to the right of 0.036 . . . .

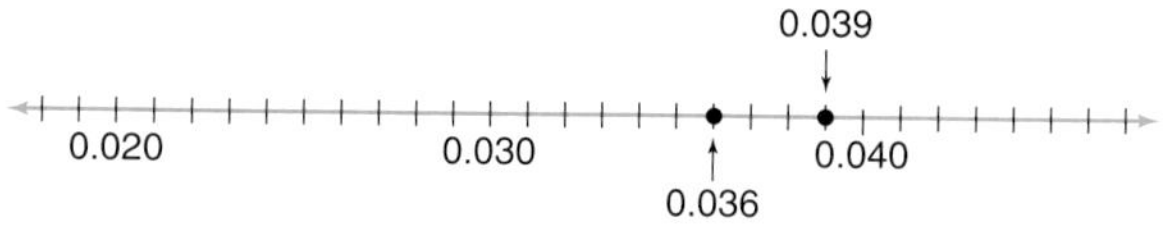

[Number line from 0.020 to beyond 0.040, with labels 0.020, 0.030, 0.040; points marked at 0.036 and 0.039]

Since $0.039 \ldots > 0.036 \ldots$, you know that $\frac{3}{76} > \frac{2}{55}$.

More people chose chocolate chip than chose cookies-and-cream.

Another way to compare two rational numbers is to write them as equivalent fractions with like denominators.

Example  **Replace each ● with <, >, or = to make a true sentence.**

a. $\frac{7}{8}$ ● $\frac{5}{6}$

$\frac{21}{24}$ ● $\frac{20}{24}$ *The LCM is 24. $\frac{7}{8} = \frac{21}{24}$ and $\frac{5}{6} = \frac{20}{24}$*

$\frac{21}{24} > \frac{20}{24}$ *Since $21 > 20$, $\frac{21}{24} > \frac{20}{24}$.*

Therefore, $\frac{7}{8} > \frac{5}{6}$.

Cooperative Learning

Group Discussion Have students work in groups of three. One student names two numbers, another locates the numbers on a number line, and the third names a number in between them. Start with decimals, then try fractions, and end using a combination of fractions and decimals.

For more information on this strategy, see *Cooperative Learning in the Mathematics Classroom*, one of the titles in the Glencoe Mathematics Professional Series, p. 31.

b. $-\frac{3}{5}$ ● -0.75

-0.6 ● -0.75 *Write $-\frac{3}{5}$ as a decimal.*

$-0.6 > -0.75$ *In tenths place, $-6 > -7$.*

Therefore, $-\frac{3}{5} > -0.75$.

Verify this answer using a number line.

Checking Your Understanding

Communicating Mathematics

Read and study the lesson to answer these questions. 1–3. See margin.

1. **Explain** why we use division to change a fraction to a decimal.
2. **Describe** the difference between terminating and repeating decimals. Give an example of each.
3. **Explain** the difference between 0.6 and $0.\overline{6}$. Which is greater?
4. **You Decide** Juanita says that $\frac{2}{3} > \frac{3}{5}$. Drew disagrees. Who is correct and why? Juanita, $\frac{2}{3} = \frac{10}{15}, \frac{3}{5} = \frac{9}{15}$ and $\frac{10}{15} > \frac{9}{15}$.
5. The grid at the right represents the number 0.45. Represent 0.5, $\frac{1}{5}$, $\frac{1}{3}$, and 0.25 by shading 10-by-10 grids.

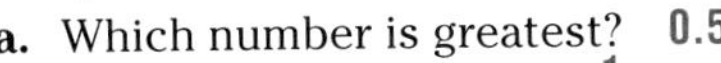

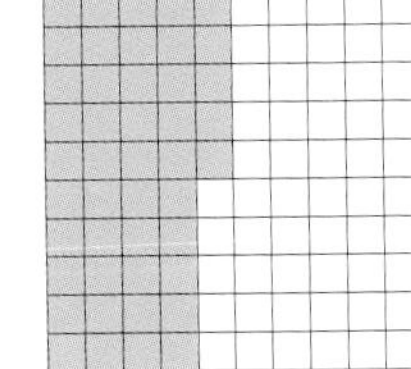

   a. Which number is greatest? 0.5
   b. Which number is least? $\frac{1}{5}$
   c. Order the numbers from least to greatest.

MATERIALS

grid paper

5c. $\frac{1}{5}$, 0.25, $\frac{1}{3}$, 0.45, 0.5

Guided Practice

Write each fraction as a decimal. Use a bar to show a repeating decimal.

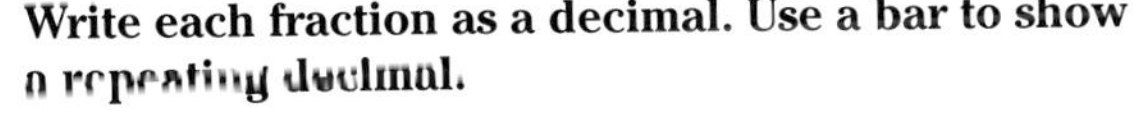

6. $\frac{5}{9}$ $0.\overline{5}$
7. $2\frac{2}{3}$ $2.\overline{6}$
8. $-3\frac{1}{8}$ -3.125
9. $\frac{1}{8}$ 0.125

State the greater number for each pair. 13. $-\frac{3}{4}$

10. $\frac{1}{2}, \frac{1}{3}$ $\frac{1}{2}$
11. $-\frac{3}{4}, -\frac{7}{8}$ $-\frac{3}{4}$
12. $\frac{3}{8}, \frac{10}{33}$ $\frac{3}{8}$
13. $-\frac{3}{4}, -0.\overline{75}$

14. **Business** On Monday, June 12, 1995, the price of Wendy's stock closed at $16\frac{5}{8}$. The previous Friday, the stock had closed at $16\frac{3}{4}$. On which day did the stock close at a higher price? Friday

Exercises: Practicing and Applying the Concept

Independent Practice

Write each fraction as a decimal. Use a bar to show a repeating decimal. 25. -4.3125

15. $\frac{7}{10}$ 0.7
16. $-\frac{1}{3}$ $-0.\overline{3}$
17. $\frac{1}{2}$ 0.5
18. $-1\frac{1}{4}$ -1.25
19. $\frac{14}{20}$ 0.7
20. $-\frac{2}{9}$ $-0.\overline{2}$
21. $\frac{5}{16}$ 0.3125
22. $\frac{21}{25}$ 0.84
23. $-3\frac{4}{11}$ $-3.\overline{36}$
24. $\frac{23}{45}$ $0.5\overline{1}$
25. $-4\frac{5}{16}$
26. $\frac{31}{40}$ 0.775

3 PRACTICE/APPLY

Checking Your Understanding

Exercises 1–14 are designed to help you assess your students' understanding through reading, writing, speaking, and modeling. You should work through Exercises 1–5 with your students and then monitor their work on Exercises 6–14.

Error Analysis

Some students may divide the denominator by the numerator when changing a fraction to a decimal. Have students practice reading fractions as division problems, emphasizing that the fraction is read from top to bottom.

Additional Answers

1. Fractions represent division problems.
2. Terminating decimals have a zero remainder; repeating decimals have a pattern that repeats without end. Check students' examples.
3. $0.6 = 0.60$ and $0.\overline{6} = 0.66...$; $0.\overline{6}$ is greater.

Reteaching

Hands-On Activity Have students roll a pair of dice and then write the ratio of the lower number to the higher number as a fraction in simplest form and as a decimal. Ask students what happens when the same number appears on both dice. Then have students write the ratio of the higher number to the lower number as a mixed number or whole number and as a decimal.

Study Guide Masters, p. 45

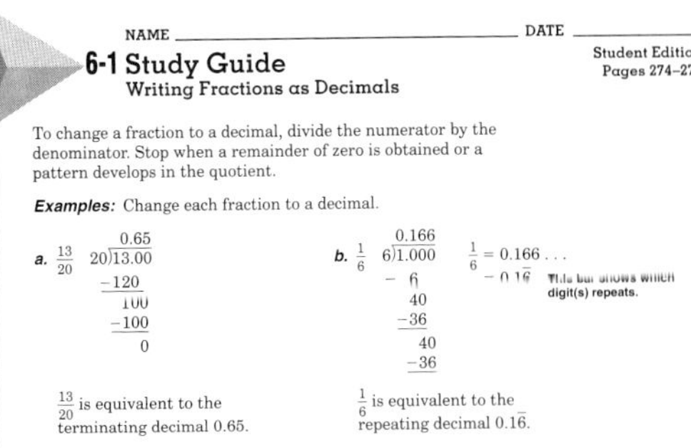

NAME ______ DATE ______

6-1 Study Guide
Writing Fractions as Decimals

Student Edition Pages 274–279

To change a fraction to a decimal, divide the numerator by the denominator. Stop when a remainder of zero is obtained or a pattern develops in the quotient.

Examples: Change each fraction to a decimal.

a. $\frac{13}{20}$ $\quad 20\overline{)13.00} = 0.65$; -120, 100, -100, 0

b. $\frac{1}{6}$ $\quad 6\overline{)1.000} = 0.166$; -6, 40, -36, 40, -36 $\quad \frac{1}{6} = 0.166\ldots = 0.1\overline{6}$ This bar shows which digit(s) repeats.

$\frac{13}{20}$ is equivalent to the terminating decimal 0.65.

$\frac{1}{6}$ is equivalent to the repeating decimal $0.1\overline{6}$.

On a number line, numbers are in order of size (magnitude).

$-\frac{4}{5} < \frac{5}{6}$

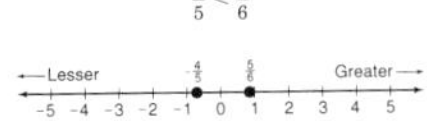

Write each fraction as a decimal. Use a bar to show a repeating decimal.

1. $\frac{1}{4}$ 0.25
2. $\frac{17}{20}$ 0.85
3. $-\frac{3}{8}$ -0.375
4. $\frac{2}{9}$ $0.\overline{2}$
5. $\frac{19}{25}$ 0.76
6. $-\frac{5}{6}$ $-0.8\overline{3}$
7. $-\frac{4}{5}$ -0.8
8. $\frac{5}{8}$ 0.625
9. $\frac{33}{40}$ 0.825
10. $\frac{7}{16}$ 0.4375
11. $-\frac{11}{15}$ $-0.7\overline{3}$
12. $-\frac{5}{12}$ $-0.41\overline{6}$

Write > or < in each blank to make a true sentence.

13. $4\frac{1}{4}$ $\underline{>}$ 4.13
14. $\frac{2}{3}$ $\underline{>}$ $\frac{3}{5}$
15. $3\frac{2}{7}$ $\underline{>}$ $-2\frac{1}{4}$
16. $\frac{4}{11}$ $\underline{<}$ $\frac{7}{12}$
17. $-7\frac{1}{8}$ $\underline{>}$ -7.34
18. $1\frac{6}{7}$ $\underline{<}$ 1.893

Assignment Guide

Core: 15–47 odd, 49–57
Enriched: 16–42 even, 43–57

For **Extra Practice**, see p. 753.

The red A, B, and C flags, printed only in the Teacher's Wraparound Edition, indicate the level of difficulty of the exercises.

Additional Answers

43. $0.\overline{3}, 0.25, 0.7, 0.8, 0.41\overline{6}, 0.\overline{571428}, 0.12$

43a. terminating: $\frac{1}{4}, \frac{7}{10}, \frac{4}{5}, \frac{3}{25}$; repeating: $\frac{1}{3}, \frac{5}{12}, \frac{4}{7}$

43b. $3 = 1 \times 3$; $4 = 2^2$; $10 = 2 \times 5$; $5 = 1 \times 5$; $12 = 2^2 \times 3$; $7 = 1 \times 7$; $25 = 5^2$

43c. 2 and 5; see students' work.

44. Sample answer: $\frac{3}{5}$ and $\frac{2}{3}$; see students' explanations.

Replace each ● with >, <, or = to make a true sentence.

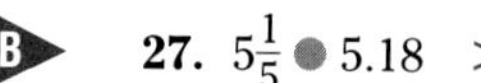

B
27. $5\frac{1}{5}$ ● 5.18 **>**
28. 7.56 ● $7\frac{12}{25}$ **>**
29. −3.45 ● $3\frac{2}{3}$ **<**
30. $-1\frac{1}{20}$ ● 1.01 **<**
31. $\frac{7}{8}$ ● $\frac{8}{9}$ **<**
32. $\frac{2}{5}$ ● $\frac{1}{3}$ **>**
33. $3\frac{4}{7}$ ● $3\frac{5}{8}$ **<**
34. $2\frac{5}{8}$ ● $2\frac{2}{3}$ **<**
35. $-\frac{11}{14}$ ● $-\frac{13}{16}$ **>**

C
36. $\frac{9}{11}$ ● $\frac{19}{23}$ **<**
37. $1\frac{4}{5}$ ● 1.857 **<**
38. $-5\frac{1}{12}$ ● −5.09 **>**

Calculators

Several different calculators' displays for 2 ÷ 3 are shown below. What would the display for 7 ÷ 9 be for each calculator?

39. .6666666 **.7777777**
40. 0.6666666 **0.7777777**
41. 0.6666667 **0.7777778**
42. .67 **.78**

Critical Thinking

43. Find the decimal equivalents for $\frac{1}{3}, \frac{1}{4}, \frac{7}{10}, \frac{4}{5}, \frac{5}{12}, \frac{4}{7}$, and $\frac{3}{25}$. **See margin.**
   a. Which are terminating decimals? Which are repeating decimals?
   b. Find the prime factorization of each denominator.
   c. What prime factors are present in the denominator of a fraction whose equivalent decimal is terminating? Write about what you observe.

44. Find a terminating and a repeating decimal between $\frac{1}{6}$ and $\frac{8}{9}$. Explain how you found them. **See margin.**

45. **Zoology** Monarch butterflies migrate up to 2000 miles from Canada and the northern United States to Florida, California, and Mexico. The fastest monarch butterfly can fly $\frac{1}{3}$ mile in one minute. Write $\frac{1}{3}$ as a decimal rounded to the nearest hundredth. **0.33**

Applications and Problem Solving

46. **Physics** For best performance, 0.91 to 0.93 of a rocket's total mass should be propellant. Aerospace engineers call the following ratio the *mass fraction (MF)* and usually express it as a decimal.

$$MF = \frac{\text{mass of propellant}}{\text{total mass}}$$

If the total mass of a rocket is 6500 pounds and the mass of the propellant is 6000 pounds, would this ratio fall into the range for best performance? Explain.
yes; $\frac{6000}{6500} = 0.9231$

Practice Masters, p. 45

NAME ______ DATE ______

6-1 Practice
Writing Fractions as Decimals
Student Edition Pages 274–279

Write each fraction as a decimal. Use a bar to show a repeating decimal.

1. $\frac{4}{5}$ 0.8
2. $-\frac{1}{9}$ $0.\overline{1}$
3. $\frac{1}{5}$ 0.2
4. $\frac{8}{9}$ $0.\overline{8}$
5. $-6\frac{5}{20}$ −6.25
6. $\frac{10}{11}$ $0.\overline{90}$
7. $\frac{5}{12}$ $0.41\overline{6}$
8. $\frac{2}{15}$ $0.1\overline{3}$
9. $-\frac{9}{16}$ −0.5625
10. $\frac{18}{25}$ 0.72
11. $-\frac{3}{11}$ $-0.\overline{27}$
12. $-5\frac{1}{9}$ $-5.\overline{1}$
13. $\frac{7}{33}$ $0.\overline{21}$
14. $-\frac{16}{45}$ $-0.3\overline{5}$
15. $8\frac{10}{32}$ 8.3125
16. $-\frac{21}{30}$ 0.7
17. $-2\frac{5}{22}$ $-2.2\overline{27}$
18. $-3\frac{3}{4}$ −3.75

Write > or < in each blank to make a true sentence.

19. 4.79 __>__ $4\frac{1}{8}$
20. 3.12 __<__ $3\frac{3}{17}$
21. −4.39 __<__ $4\frac{3}{4}$
22. $-2\frac{8}{50}$ __<__ 2.08
23. $\frac{6}{7}$ __<__ $\frac{7}{8}$
24. $\frac{5}{8}$ __>__ $\frac{1}{2}$
25. $\frac{2}{3}$ __>__ $\frac{4}{9}$
26. $2\frac{7}{8}$ __>__ $2\frac{9}{11}$
27. $\frac{5}{6}$ __>__ $\frac{4}{5}$
28. $6\frac{3}{5}$ __>__ $6\frac{4}{7}$
29. $-\frac{1}{3}$ __<__ −0.16
30. $2\frac{1}{9}$ __>__ $2\frac{1}{10}$
31. $-1\frac{1}{7}$ __<__ −1.143
32. $\frac{23}{27}$ __>__ $\frac{15}{19}$
33. $-7\frac{1}{14}$ __<__ −7.06

Group Activity Card 6-1

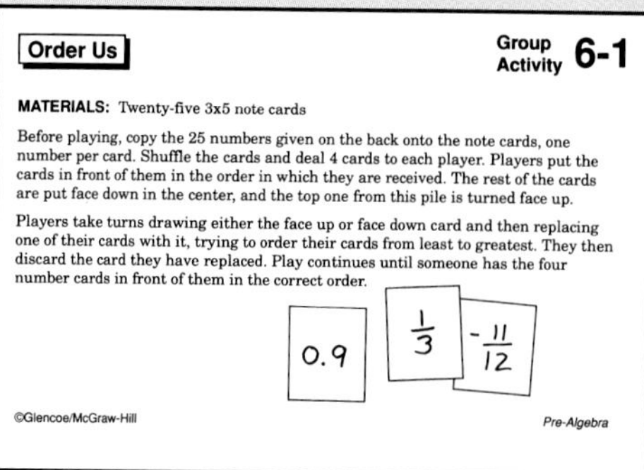

Order Us Group Activity **6-1**

MATERIALS: Twenty-five 3x5 note cards

Before playing, copy the 25 numbers given on the back onto the note cards, one number per card. Shuffle the cards and deal 4 cards to each player. Players put the cards in front of them in the order in which they are received. The rest of the cards are put face down in the center, and the top one from this pile is turned face up.

Players take turns drawing either the face up or face down card and then replacing one of their cards with it, trying to order their cards from least to greatest. They then discard the card they have replaced. Play continues until someone has the four number cards in front of them in the correct order.

 Pre-Algebra

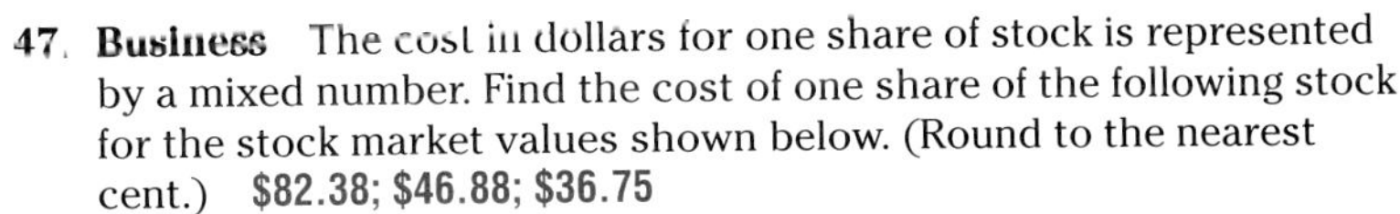

47. **Business** The cost in dollars for one share of stock is represented by a mixed number. Find the cost of one share of the following stock for the stock market values shown below. (Round to the nearest cent.) **$82.38; $46.88; $36.75**

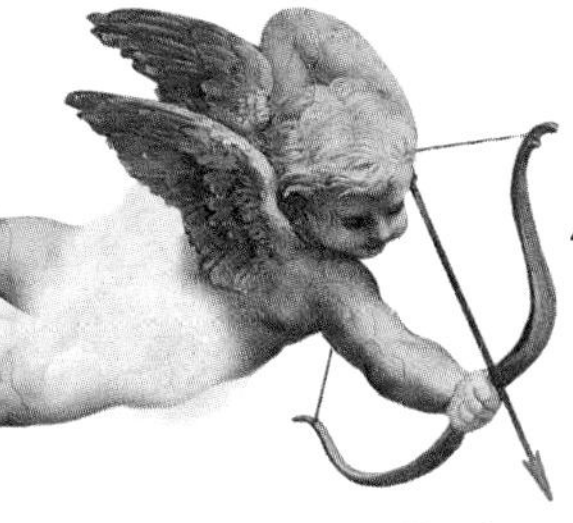

Nike 82 3/8	Pepsi 46 7/8	The Gap 36 3/4

48. **Demographics** Do you believe in love at first sight? In a survey by Roper Starch Worldwide, $\frac{2}{5}$ of those asked said "It happens," $\frac{7}{25}$ said "It could happen," and $\frac{8}{25}$ said "No way." What did most people respond? **It happens.**

Mixed Review

49. Write the next three terms of the sequence 88, 82, 76, 70, . . . (Lesson 5-9) **64, 58, 52**
50. Solve $t - 3\frac{1}{4} = \frac{7}{2}$. (Lesson 5-6) $6\frac{3}{4}$
51. Find the product of $b^5 \cdot b^6$. (Lesson 4-8) b^{11}
52. Find a three-digit number that is divisible by 2, 3 and 6. (Lesson 4-1)

52. Sample answer: 108

53. **Sewing** The Home Economics Class made a quilt measuring 4 feet by 6 feet to donate for a raffle. How much binding should they buy to trim the outer edges? (Lesson 3-5) **20 feet**
54. Order the integers {0, −7, 15, −6, 20} from least to greatest. (Lesson 2-3) **−7, −6, 0, 15, 20**
55. **Sports** The Cleveland Indians beat the California Angels by a score of 10 to 3. Write an inequality that would tell you how many more runs the Angels would have needed to win the game. (Lesson 1-9) $r > 7$ **or** $r \geq 8$
56. Solve $13 + y = 25$ by using the inverse operation. (Lesson 1-8) **12**
57. Solve $3t = 81$ mentally. (Lesson 1-6) **27**

WORKING ON THE Investigation

Refer to the Investigation on pages 270–271.

Begin working on the investigation by doing a group activity to determine the steepness of your lift hill.

- Build a scale model of your lift hill using a yardstick to represent the length of the track on the downhill side.
- While one member of your group holds the yardstick at the angle matching your design, another member measures how high the end of the yardstick is from the floor. Record the ratio of the height to the length of the yardstick. For example, if the height is 30 inches, you would write $\frac{30}{36}$ or $\frac{5}{6}$. Write the ratio as a decimal.
- Next, use the same method to find the ratio of an angle that is not as steep as yours and another angle that is steeper than yours.
- Write a short summary explaining the relationship between steepness and the decimal values of the ratios.

Add the results of your work to your Investigation Folder.

4 ASSESS

Closing Activity

Writing Express the following fractions as decimals. Use a calculator or paper and pencil.

1. $\frac{3}{10}$ 0.3
2. $-\frac{5}{8}$ −0.625
3. $\frac{2}{9}$ $0.\overline{2}$
4. $1\frac{2}{5}$ 1.4
5. $-3\frac{1}{3}$ $-3.\overline{3}$

Enrichment Masters, p. 45

NAME ______________________ DATE __________

6-1 Enrichment
Decimals in Sports

Student Edition Pages 274–279

Many sports use decimals to show how well a team or player is doing statistically. You will need to change fractions to decimals to answer the questions below.

Solve each problem. Use a bar to show a repeating digit.

1. Cindy made 12 of 15 free throw attempts in the last basketball game. In decimal form, what was her success rate? **0.8**
2. Andy carried the football eight times, but gained only five yards. How many yards per carry did he average? **0.625 yard**
3. Julio plays shortstop and had only seven errors in the 105 times the baseball was hit to him. Express his error rate as a decimal. $0.0\overline{6}$
4. Fay's tennis team won six of nine games in May, then only won four and lost twenty the rest of the season. Use a decimal to show the fraction of games won during the entire season. $\frac{10}{33} = 0.\overline{30}$
5. Arnie putted only 25 times in his last round of golf (18 holes). In decimal form, what was his average number of putts per hole? $1.3\overline{8}$ **putts**
6. Diana bowled 35 games last year and got a total of 203 strikes. How many strikes per game did she average? **5.8 strikes**

A baseball pitcher's earned run average (ERA) is the number of runs allowed per nine-inning game. To compute this average, the number of earned runs is divided by the quotient of the number of innings pitched and 9.

Example: Find Alex's ERA if he allowed 5 earned runs in 18 total innings.
$5 \div \frac{18}{9} = 5 \div 2 = 2.5$ runs per game

7. Find Gregg's earned run average if he allowed 44 earned runs in 200 innings. **1.98 runs per game**
8. What is Carla's earned run average if she allowed 35 earned runs in 150 innings? **2.1 runs per game**

Extension

Using Models Have students create a fraction for each domino in a set except ones containing blanks. Then have students rename each fraction as a decimal. Ask students to look for a pattern to determine whether a fraction is a terminating or repeating decimal. Extend the activity by having students draw two dominos, write the fraction for each and then determine which is greater.

WORKING ON THE Investigation

The Investigation on pages 270–271 is designed to be a long-term project that is completed over several days or weeks. Encourage students to keep their materials in their Investigation Folder as they work on the Investigation.

6-2 LESSON NOTES

NCTM Standards: 1-5, 7

Instructional Resources

- Study Guide Master 6-2
- Practice Master 6-2
- Enrichment Master 6-2
- Group Activity Card 6-2

Transparency 6-2A contains the 5-Minute Check for this lesson; **Transparency 6-2B** contains a teaching aid for this lesson.

Recommended Pacing	
Standard Pacing	Day 2 of 13
Honors Pacing	Day 2 of 12
Block Scheduling*	Day 1 of 7 (along with Lesson 6-1)

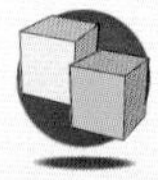

*For more information on pacing and possible lesson plans, refer to the **Block Scheduling Booklet**.

1 FOCUS

5-Minute Check
(over Lesson 6-1)

Write each fraction as a decimal. Use a bar to show a repeating decimal.

1. $\frac{9}{16}$ 0.5625
2. $-1\frac{3}{8}$ −1.375
3. $\frac{3}{7}$ $0.\overline{428571}$

Replace each ● with <, >, or = to make a true sentence.

4. $2\frac{1}{2}$ ● $2\frac{6}{13}$ >
5. −0.35 ● $-\frac{1}{3}$ <

Motivating the Lesson

Hands-On Activity Have students roll two dice and write a decimal using the numbers. For example, for a roll of 3 and 5, 5.3, 3.5, 0.35, or 0.53 could be written. Then the student rolls again and writes another decimal. Then the decimals are added. The student with a *sum* closest to 8 is the winner. Ask students how their strategy would change if the goal was a *product* of 15.

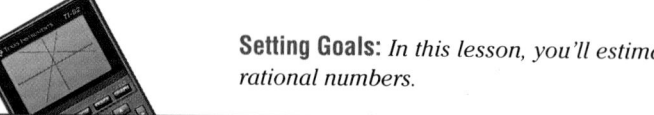

6-2 Estimating Products and Quotients

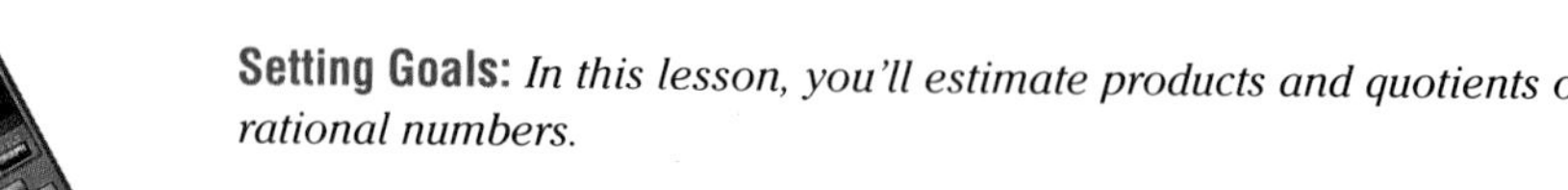

Setting Goals: *In this lesson, you'll estimate products and quotients of rational numbers.*

Modeling with Technology

In this activity, you and a partner will try to score as many points as possible by estimating decimal products. Products receive points as shown in the table below.

Your Turn Pick two numbers from the chart below and use your calculator to find the product. You may use each number only once.

Numbers:							
4.6	0.02	12.1	30.5	11.5	0.6	31.8	0.05
19.8	0.115	10.6	0.43	0.08	2.9	0.96	18.8
24.2	9.7	1.7	5.64				

Add the points for your products to find your score. You can do this activity more than once to see if you can increase your team score.

Product is between:	Points Scored
0 and 0.1	1
0.1 and 1	2
1 and 10	3
10 and 100	2
100 and 1000	1

TALK ABOUT IT

a. What strategy did you use to choose the numbers for each product?
b. How could you increase your score?
a–b. See students' work.

Learning the Concept

In the activity above, it is useful to know how to estimate the product of the decimals before actually multiplying them. It is also important to estimate products and quotients when using a calculator. We can make mistakes when entering numbers, and estimating will tell us if our result is reasonable.

Using rounding and **compatible numbers**, it is possible to estimate the products and quotients of rational numbers. Compatible numbers are rounded so it is easy to compute with them mentally.

47.5 ÷ 5.23 → 48 ÷ 6 or 8 *Even though 5.23 rounds to 5, 6 is a compatible number because 48 is divisible by 6.*

8 × 98.24 → 8 × 100 or 800.

Teaching Tip Remind students that estimates can vary, even when using compatible numbers. For example, for 47.5 ÷ 4.23, you can also use 45 ÷ 5 = 9 as an estimate.

Example 1 **Use rounding or compatible numbers to estimate 67.45 ÷ 7.6.**

$67.45 \div 7.6 \rightarrow 70 \div 7$ *Round 67.45 to 70 and round 7.6 to 7 since 70 and 7 are compatible numbers.*

Think: $70 \div 7 = 10$

$67.45 \div 7.6$ is about 10.

The strategy of using compatible numbers works for fractions as well as decimals.

Example 2 **Use rounding or compatible numbers to estimate $\frac{5}{14} \times 30$.**

$\frac{5}{14} \times 30 \rightarrow \frac{1}{3} \times 30$ *Replace $\frac{5}{14}$ with $\frac{1}{3}$ because $\frac{5}{14}$ is close to $\frac{5}{15}$ and $\frac{5}{15} = \frac{1}{3}$.*

Think: $\frac{1}{3} \times 30 = 10$

$\frac{5}{14} \times 30$ is about 10.

Being able to estimate products and quotients of rational numbers is very useful in everyday life.

Employment

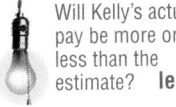

Example 3 **Kelly found a part-time job working at the Discus Record Store for \$5.85 an hour. The manager told Kelly that she will probably work about $7\frac{1}{2}$ hours each week. She will be paid every two weeks. Estimate Kelly's pay before taxes are deducted.**

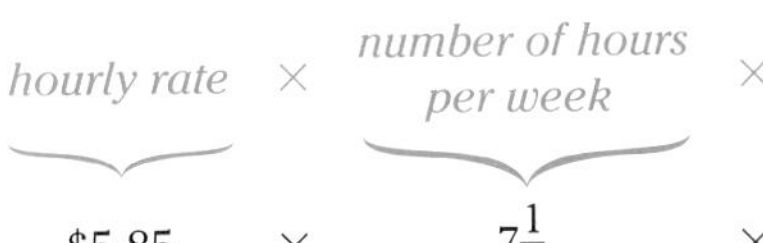

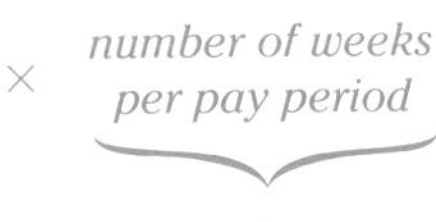

hourly rate	×	*number of hours per week*	×	*number of weeks per pay period*
\$5.85	×	$7\frac{1}{2}$	×	2

Think: \$5.85 rounds to \$6 and $7\frac{1}{2}$ hours rounds to 8 hours.

$\$5.85 \times 7\frac{1}{2} \rightarrow \$6 \times 8$ or \$48 a week

For two weeks, multiply by 2.
$\$48 \times 2 \rightarrow \$50 \times 2 = \$100$

Kelly will earn about \$100 every two weeks.

THINK ABOUT IT

Will Kelly's actual pay be more or less than the estimate? **less**

2 TEACH

Teaching Tip For Example 1, $64 \div 8 = 8$ is also a good estimate.

Teaching Tip For Example 2, $\frac{5}{14} \times 28 = 10$ is also a good estimate.

In-Class Examples

Estimation results may vary.

For Example 1
Use rounding or compatible numbers to estimate 375 ÷ 19.
400 ÷ 20 = 20

For Example 2
Use rounding or compatible numbers to estimate $\frac{11}{15} \times 44$.
$\frac{11}{15} \times 45 = 30$ or $\frac{2}{3} \times 45 = 30$

For Example 3
Jacob's recipe calls for $1\frac{3}{4}$ cups of flour. For the party, he is making 4 batches. Estimate how much flour he will need for all 4 batches. **8 cups**

Teaching Tip Encourage students to estimate *whenever they solve a problem*. Explain that by estimating each time, it will become a habit and the accuracy of their work will improve.

Study Guide Masters, p. 46

NAME ________________ DATE ________

6-2 Study Guide
Estimating Products and Quotients

Student Edition
Pages 280–283

Estimate the products and quotients of rational numbers by using rounding and compatible numbers.

$26.5 \div 4.16 \rightarrow 28 \div 4$ or 7 — Even though 26.5 rounds to 27, 28 is a compatible number because 28 is divisible by 4.

$3 \times 47.98 \rightarrow 3 \times 50$ or 150

$\frac{4}{9} \times 20 \rightarrow \frac{1}{2} \times 20$ or 10 — $\frac{4}{9}$ is close to $\frac{1}{2}$.

$25 \div 4\frac{5}{6} \rightarrow 25 \div 5$ or 5 — $4\frac{5}{6}$ is close to 5.

Estimate each product or quotient. Sample answers are given.

1. 27.2×8.1 240	2. 9.32×6.5 63	3. $19.1 \div 3.6$ 5
4. $(8.53)(4.86)$ 45	5. $75.61 \div 1.9$ 38	6. $24.6 \div 4.8$ 5
7. $\frac{1}{3} \times 23$ 8	8. $\left(\frac{1}{9}\right)(35)$ 4	9. $\frac{3}{7} \times 12$ 6
10. $\frac{7}{16} \times 240$ 120	11. $\frac{6}{10} \times 28$ 14	12. $16 \times \frac{21}{48}$ 8
13. $45 \div 8\frac{6}{7}$ 5	14. $315 \div 4\frac{11}{12}$ 63	15. $\frac{29}{54} \times 304$ 152
16. $400 \times \frac{11}{21}$ 52	17. $156 \div 12\frac{14}{15}$ 12	18. $46 \div 1\frac{32}{37}$ 23
19. $14.7 \div 5.03$ 3	20. $48.9 \div 24.8$ 2	21. $68.04 \div 0.96$ 68

3 PRACTICE/APPLY

Checking Your Understanding

Exercises 1–14 are designed to help you assess your students' understanding through reading, writing, speaking, and modeling. You should work through Exercises 1–4 with your students and then monitor their work on Exercises 5–14.

Assignment Guide

Core: 15–37 odd, 38–44
Enriched: 16–32 even, 33–44

For **Extra Practice**, see p. 753.

The red A, B, and C flags, printed only in the Teacher's Wraparound Edition, indicate the level of difficulty of the exercises.

Additional Answer

1. Sample answer: An estimate lets you check for possible errors in entering numbers into the calculator.

Practice Masters, p. 46

NAME ____________ DATE ________

6-2 Practice
Estimating Products and Quotients

Student Edition
Pages 280–283

Estimate each product or quotient. Sample answers are given.

1. 18.87×7.6 160	2. 3.19×2.6 9	3. $6.3 \div 3.05$ 2
4. 28.9×6.6 210	5. 8.29×7.1 56	6. 9.7×89.7 900
7. $47.56 \div 2.9$ 16	8. $10.4 \div 9.67$ 1	9. $6.82 \div 7.09$ 1
10. $29.61 \div 5.4$ 6	11. $56 \div 8.4$ 7	12. $80.3 \div 20.2$ 4
13. $(10.16)(8.8)$ 90	14. $(39.6)(9.6)$ 400	15. $(4.37)(64.5)$ 325
16. $\frac{1}{3} \times 8$ 3	17. $\frac{1}{4} \times 15$ 4	18. $\frac{1}{5} \times 29$ 6
19. $\frac{1}{6} \times 13$ 2	20. $\frac{1}{10} \times 19$ 2	21. $\frac{1}{6} \times 32$ 5
22. $\frac{4}{9} \times 19$ 8	23. $\frac{5}{6} \times 35$ 30	24. $\frac{4}{7} \times 61$ 36
25. $\frac{7}{8} \times 73$ 73	26. $12 \times \frac{2}{76}$ 0	27. $\frac{10}{19} \times 100$ 50
28. $16 \div 3\frac{4}{5}$ 4	29. $45 \div 8\frac{3}{4}$ 5	30. $26\frac{1}{2} \div 6$ 4
31. $179 \div 20\frac{3}{11}$ 9	32. $130 \div 12\frac{11}{14}$ 10	33. $66 \div 3\frac{10}{31}$ 22

Checking Your Understanding

Communicating Mathematics

Read and study the lesson to answer these questions.

1. **Explain** why estimation is more important in mathematics since calculator use has become more widespread. See margin.
2. **Give an example** of two compatible numbers. Explain why they are compatible. Sample answer: 36 and 6; $36 \div 6 = 6$
3. Choose the correct phrase: The product of a whole number and a fraction less than one is always (less than, greater than, equal to) the whole number. less than

4. **Assess Yourself** Describe a time in your daily life when estimation is useful. See students' work.

Guided Practice

Choose the best estimate for each product or quotient.

5. $\left(1\frac{1}{8}\right)\left(3\frac{1}{4}\right)$ — a. almost 3 b. less than 1 c. more than 3 d. none of these — c
6. $49 \times \frac{2}{5}$ — a. about 20 b. more than 25 c. less than 10 d. none of these — a
7. $\$24.99 \div 4$ — a. more than \$8 b. almost \$6 c. about \$5 d. none of these — d

Rewrite the problem using rounding and compatible numbers. Then estimate each product or quotient. See margin.

8. $40 \times \frac{2}{9}$
9. $17.8 \div 3.2$
10. $\frac{5}{12} \times 44$
11. $8.4 \div 0.95$
12. $75.4 \div 9.8$
13. $20\frac{1}{9} \times \frac{1}{4}$
14. **Geometry** The formula for the perimeter of a rectangle is $P = 2(\ell + w)$, where ℓ is the length and w is the width. Estimate the perimeter of the rectangle at the right. Is it greater than or less than 20 meters? less than

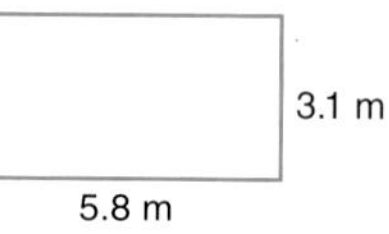

Exercises: Practicing and Applying the Concept

Independent Practice

Estimate each product or quotient. Sample answers given.

15. 15.93×9.8 160
16. $2.94 \cdot 1.8$ 6
17. $8.1 \div 2.2$ 4

A

18. $47.6 \div 7.8$ 6
19. $15.2 \div 2.7$ 5
20. $(84.2)(3.9)$ 320
21. $\frac{1}{4} \cdot 9$ 2
22. $\left(\frac{1}{3}\right)(14)$ 5
23. $\frac{3}{8} \times 17$ 6

B

24. $\frac{9}{19} \times 120$ 60
25. $\frac{11}{20} \times 41$ 20
26. $75 \div 6\frac{7}{8}$ 10
27. $\frac{31}{40} \times 200$ 150
28. $146 \div 13\frac{1}{15}$ 12
29. $43.8 \div 9.2$ 5
30. Estimate $\frac{5}{6}$ times 10. 8
31. Estimate the quotient of 27.26 and 2.6. 10

Group Activity Card 6-2

Up or Down — Group Activity 6-2

MATERIALS: Calculator

You and your partner can win points by giving the closer estimate of a product.

To begin, think of a number between 1 and 10 with one decimal place. Your partner thinks of a number between 20 and 80 with one decimal place. Then each person should estimate the product of the numbers. Compute the actual product using a calculator. The player whose estimate is closer to the exact product scores one point.

Play continues until one person is five points ahead of the other.

EXAMPLE: The problem is 9.8×54.3. You round both numbers up and mentally compute $10 \times 55 = 550$. You know this is high so you guess 545. Your partner estimates the product as 10×50 or 500 and makes an actual guess of 525. Using a calculator you get the exact product of 532.14. Your partner is closer; therefore, your partner scores a point.

 Pre-Algebra

Reteaching

Using Logical Reasoning Provide exercises and possible estimates such as 3.62×63.4 and 18, 180, 1800. Have students choose a reasonable answer and explain how they made their choice.

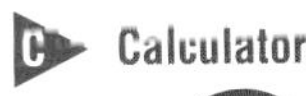

Calculators

32. Use a calculator to find the following quotients.

a.		**b.**		**c.**	
$80 \div 0.5$	**160**	$10 \div 0.1$	**100**	$4 \div 0.25$	**16**
$6 \div 0.5$	**12**	$2 \div 0.1$	**20**	$20 \div 0.25$	**80**
$1.5 \div 0.5$	**3**	$6.5 \div 0.1$	**65**	$1.1 \div 0.25$	**4.4**

d. What patterns do you observe? **See margin.**

Use the patterns to estimate each quotient.

e. $14 \div 0.48$ **28** **f.** $5 \div 0.12$ **50** **g.** $12 \div 0.23$ **48**

h. $2\frac{1}{5} \div 0.25$ **8** **i.** $21.2 \div 0.085$ **212** **j.** $3\frac{1}{4} \div 0.51$ **6**

Critical Thinking

33. Explain how you would know if an estimated answer is greater than or less than the actual answer. **See margin.**

Applications and Problem Solving

34. **Decorating** Jennifer Green ran out of wallpaper before she finished wallpapering her room. The remaining area is 12 feet by $10\frac{1}{2}$ feet and has one window, which is $3\frac{1}{2}$ feet by $4\frac{3}{4}$ feet.
   a. Estimate the amount (square feet) of wall paper Jennifer needs to cover the wall. **about 110 square feet**
   b. If the wallpaper comes on rolls that cover 51 square feet, how many rolls should she purchase? **3 rolls**

35. **Population** In 1994, Heather Whitestone became the first hearing-impaired Miss America. In the United States, $\frac{3}{50}$ of the people are hearing impaired or deaf. If 1990 census showed that the population of the United States was 248,709,873 at that time, estimate how many people were hearing impaired. **about 15,000,000**

36. **Stock Market** Stock prices are stated as mixed numbers. On July 18, 1995, a share of IBM stock was listed as $107\frac{1}{4}$. The cost of one share of stock is \$107.25. (The fraction indicates a part of a dollar.)
   a. Estimate the cost of 37 shares of stock listed at $107\frac{1}{4}$ a share. **\$4400**
   b. Suppose you had \$1000 to invest in IBM stock. About how many shares could you buy? **about 9**

37. **Biology** A dolphin can swim at a speed of 37 miles per hour. A human can swim about one-eighth that speed. About how fast can a human swim? **5 mph**

Mixed Review

38. Write $\frac{3}{8}$ as a decimal. (Lesson 6-1) **0.375**

39. Solve $\frac{6}{7} + \frac{5}{14} = s$. (Lesson 5-5) $\frac{17}{14}$ **or** $1\frac{3}{14}$

40. **Food** A foot-long submarine sandwich was cut into 6 equal pieces. There are 2 pieces left. Write the ratio of pieces eaten to total pieces as a fraction in simplest form. (Lesson 4-6) $\frac{2}{3}$

41. Solve $-14 + c = 13$. (Lesson 3-2) **27**

42. Find $-13 + 6$. (Lesson 2-4) **−7**

43. Simplify $|-6| - |-4|$. (Lesson 2-1) **2**

44. Solve $\frac{r}{9} = 15$ by using the inverse operation. (Lesson 1-8) **135**

Extension

Using Cooperative Groups Have students discuss Example 3 on page 281. Students could use a calculator to find Kelly's actual wages of \$87.55. Discuss whether Kelly might have been disappointed upon seeing her first paycheck. Have each group devise a method to give a more accurate estimate.

4 ASSESS

Closing Activity

Writing Have students write about a situation where they had to estimate the product or quotient of rational numbers.

Additional Answers

8–13. Sample answers given.

8. $40 \times \frac{1}{4} = 10$

9. $18 \div 3 = 6$

10. $\frac{1}{2} \times 44 = 22$

11. $8 \div 1 = 8$

12. $70 \div 10 = 7$

13. $20 \times \frac{1}{4} = 5$

32d. The quotients of the numbers divided by 0.5 are the dividend multiplied by 2. The quotients of the numbers divided by 0.1 are the dividend multiplied by 10. The quotients of the numbers divided by 0.25 are the dividend multiplied by 4.

33. If the numbers were rounded up, the actual answer is less than the estimate. If the numbers were rounded down, the product is greater than the estimate.

Enrichment Masters, p. 46

NAME ______________ DATE ______

6-2 Enrichment
Estimating Fractional Products

Student Edition Pages 280–283

Find the best estimate of the product from the list at the right. Write the corresponding letter to the right of each exercise.

1. a little more than $\frac{1}{3}$ of 1488 **E**
2. a little less than $\frac{3}{4}$ of 1635 **I**
3. a little less than $\frac{1}{5}$ of 462 **C**
4. a little more than $\frac{1}{7}$ of 1326 **O**
5. very close to $\frac{2}{3}$ of 922 **M**
6. a little less than $\frac{5}{6}$ of 1871 **H**
7. a little less than $\frac{5}{8}$ of 1729 **T**
8. a little more than $\frac{1}{9}$ of 861 **A**
9. a little more than $\frac{5}{3}$ of 1298 **F**
10. very close to $\frac{3}{7}$ of 2079 **K**
11. a little less than $\frac{1}{4}$ of 2866 **U**
12. a little more than $\frac{1}{10}$ of 1292 **N**
13. a little less than $\frac{1}{8}$ of 428 **S**
14. a little more than $\frac{2}{9}$ of 3549 **R**

A	100
C	90
E	500
F	2000
H	1500
I	1200
K	900
M	615
N	130
O	200
R	800
S	50
T	1000
U	700

Place the letters from the answers above on the blanks with the corresponding exercise number below them.

The best way to make a fire with two sticks is to . . .

M	A	K	E		S	U	R	E		O	N	E		O	F
5	8	10	1		13	11	14	1		4	12	1		4	9

T	H	E	M		I	S		A		M	A	T	C	H
7	6	1	5		2	13		8		5	8	7	3	6

Will Rogers

6-3 LESSON NOTES

NCTM Standards: 1-5, 7

Instructional Resources

- Study Guide Master 6-3
- Practice Master 6-3
- Enrichment Master 6-3
- Group Activity Card 6-3
- Assessment and Evaluation Masters, p. 155
- Activity Masters, p. 20
- Math Lab and Modeling Math Masters, p. 63
- Tech Prep Applications Masters, p. 11

Transparency 6-3A contains the 5-Minute Check for this lesson; **Transparency 6-3B** contains a teaching aid for this lesson.

Recommended Pacing	
Standard Pacing	Day 3 of 13
Honors Pacing	Day 3 of 12
Block Scheduling*	Day 2 of 7 (along with Lesson 6-4)

*For more information on pacing and possible lesson plans, refer to the ***Block Scheduling Booklet***.

1 FOCUS

5-Minute Check
(over Lesson 6-2)

Estimate each product or quotient.

Estimates may vary.

1. 53.64×8.95 450
2. $\frac{7}{15} \times 24$ 12
3. $104 \times 1\frac{1}{20}$ 104
4. $164.36 \div 81$ 2
5. $\frac{2}{9} \div \frac{4}{15}$ $\frac{5}{6}$

6-3 Multiplying Fractions

Setting Goals: *In this lesson, you'll multiply fractions.*

Modeling with Manipulatives

MATERIALS

You can use grid paper to make an **area model** for multiplication of fractions. The diagram at the right models the product $\frac{2}{3} \cdot \frac{5}{6}$. Notice that the rectangle has three rows to represent thirds. Two rows are shaded yellow to represent $\frac{2}{3}$. The rectangle has six columns to represent sixths. Five columns are shaded blue to represent $\frac{5}{6}$. Notice that the overlapping area, $\frac{10}{18}$ of the rectangle, is shaded green. The product of $\frac{2}{3}$ and $\frac{5}{6}$ is $\frac{10}{18}$, or $\frac{5}{9}$.

Your Turn Use grid paper to make an area model of $\frac{3}{5} \cdot \frac{1}{4}$.

a. How did you represent $\frac{3}{5}$ in your model?

b. How did you represent $\frac{1}{4}$?

c. What is the value of $\frac{3}{5} \cdot \frac{1}{4}$ that you found using the grid paper? $\frac{3}{20}$

d. Write a description of how an area model shows multiplication of fractions. **a, b, d. See students' work.**

Learning the Concept

An area model shows what happens when you multiply fractions. You will get the same results if you multiply fractions as shown below.

$\frac{2}{3} \cdot \frac{5}{6} = \frac{2 \cdot 5}{3 \cdot 6}$ *Multiply the numerators and multiply the denominators.*

$= \frac{10}{18}$ or $\frac{5}{9}$ *Simplify by dividing numerator and denominator by 2.*

Multiplying Fractions

In words: To multiply fractions, multiply the numerators and multiply the denominators.

In symbols: For fractions $\frac{a}{b}$ and $\frac{c}{d}$, where $b \neq 0$ and $d \neq 0$, $\frac{a}{b} \cdot \frac{c}{d} = \frac{ac}{bd}$.

If the fractions have common factors in the numerators and denominators, you can simplify before you multiply. Compare the two methods used in Example 1.

Example 1 Solve $x = \frac{5}{12} \cdot \frac{2}{9}$.

Method 1: Use the rule.

$x = \frac{5}{12} \cdot \frac{2}{9}$

$= \frac{5 \cdot 2}{12 \cdot 9}$ *Multiply.*

$= \frac{10}{108}$ or $\frac{5}{54}$

Method 2: Use the GCF.

$x = \frac{5}{12} \cdot \frac{2}{9}$

$= \frac{5 \cdot \overset{1}{\cancel{2}}}{\underset{6}{\cancel{12}} \cdot 9}$ *The GCF of 2 and 12 is 2. Divide 2 and 12 by 2.*

$= \frac{5 \cdot 1}{6 \cdot 9}$ or $\frac{5}{54}$

You can use the multiplication skills that you developed with integers and fractions to multiply negative fractions.

Example 2 **Solve each equation.**

a. $3\frac{3}{8}\left(-5\frac{1}{3}\right) = x$ *Estimate: 3(−5) = −15.*

$\frac{27}{8}\left(-\frac{16}{3}\right) = x$ *Rename $3\frac{3}{8}$ as $\frac{27}{8}$ and $-5\frac{1}{3}$ as $-\frac{16}{3}$.*

$\frac{\overset{9}{\cancel{27}}}{\underset{1}{\cancel{8}}}\left(\frac{\overset{-2}{\cancel{-16}}}{\underset{1}{\cancel{3}}}\right) = x$ *Simplify.*

$\frac{-18}{1} = x$

$-18 = x$ *Compare with the estimate.*

b. $a = \left(-1\frac{4}{9}\right)\left(-2\frac{3}{5}\right)$ *Estimate: (−1)(−3) = 3.*

$a = \left(\frac{-13}{9}\right)\left(\frac{-13}{5}\right)$ *Rename $-1\frac{4}{9}$ as $-\frac{13}{9}$ and $-2\frac{3}{5}$ as $-\frac{13}{5}$.*

$a = \frac{169}{45}$ or $3\frac{34}{45}$ *The product of two rational numbers with the same sign is positive.*

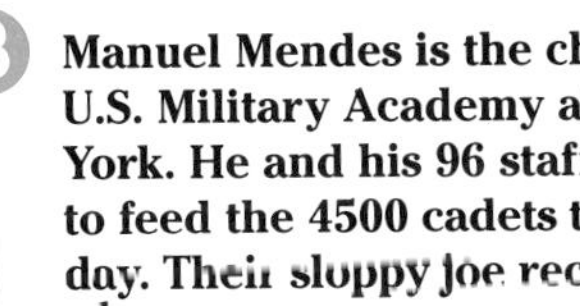

Example 3 APPLICATION **Cooking**

Manuel Mendes is the chief cook at the U.S. Military Academy at West Point, New York. He and his 96 staff members work to feed the 4500 cadets three times each day. Their sloppy joe recipe calls for $7\frac{1}{2}$ pounds of ketchup and serves 100 people. How many pounds of ketchup are needed for one meal of sloppy joes?

The recipe serves 100 people, so the staff will need to purchase enough ingredients for 4500 ÷ 100 or 45 batches of sloppy joes.

$45 \cdot 7\frac{1}{2} = \frac{45}{1} \cdot \frac{15}{2}$ *Rename 45 as $\frac{45}{1}$ and $7\frac{1}{2}$ as $\frac{15}{2}$.*

$= \frac{45 \cdot 15}{1 \cdot 2}$

$= \frac{675}{2}$ or $337\frac{1}{2}$ *Write as a mixed number.*

$337\frac{1}{2}$ pounds of ketchup are needed.

FYI

Long a staple in Latin countries, salsa has now passed ketchup in the race for most popular condiment in America.

Motivating the Lesson

Hands-On Activity Bring in a long stick or piece of rope or string. Have students estimate its length and then measure to find its actual length. Students could estimate how long half that length is. Have students devise a method of measuring to check their estimates.

2 TEACH

In-Class Examples

For Example 1

Solve $r = \frac{9}{10} \cdot \frac{2}{3}$. $\frac{3}{5}$

For Example 2

Solve each equation.

a. $m = -1\frac{1}{2} \cdot 2\frac{5}{6}$ $-4\frac{1}{4}$

b. $p = -3\frac{1}{3} \cdot \left(-2\frac{1}{10}\right)$ 7

For Example 3

One kilogram equals about $2\frac{1}{5}$ pounds. Misha weighs 75 kilograms. What is Misha's weight in pounds? **165 lb**

Alternative Learning Styles

Auditory When evaluating phrases, have students explain orally each required step. Prior to doing this, you may wish to review the order of operations with students.

FYI

In 1991, salsa outsold ketchup by $79 million. Sales of tortilla chips, often served with salsa, have grown three times as fast as sales of potato chips in the last few years.

In-Class Example

For Example 4
Evaluate m^2n if $m = \frac{3}{8}$ and $n = 1\frac{3}{5}$. $\frac{9}{40}$

3 PRACTICE/APPLY

Checking Your Understanding

Exercises 1–17 are designed to help you assess your students' understanding through reading, writing, speaking, and modeling. You should work through Exercises 1–4 with your students and then monitor their work on Exercises 5–17.

Connection to Algebra

You can evaluate algebraic expressions involving multiplication of fractions in the same way that you evaluated expressions using integers.

Example 4 **Evaluate the expression ab^2 if $a = 1\frac{1}{2}$ and $b = \frac{2}{3}$.**

$ab^2 = 1\frac{1}{2} \cdot \left(\frac{2}{3}\right)^2$ *Substitute $1\frac{1}{2}$ for a and $\frac{2}{3}$ for b.*

$= 1\frac{1}{2} \cdot \frac{2}{3} \cdot \frac{2}{3}$ *Rewrite the problem without exponents.*

$= \frac{3}{2} \cdot \frac{4}{9}$ *Write $1\frac{1}{2}$ as $\frac{3}{2}$.*

$= \frac{12}{18}$ or $\frac{2}{3}$ *Multiply and write in simplest form.*

Checking Your Understanding

Communicating Mathematics

Read and study the lesson to answer these questions.

1. **Explain** how the model at the right shows the product of $\frac{3}{4}$ and $\frac{1}{2}$. What is the product?

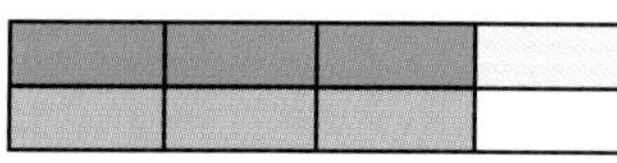

2. **You Decide** Jamal and Penny multiplied the fractions $\frac{18}{21}$ and $\frac{3}{14}$ in the following ways. Who is correct and why?

Jamal

$\frac{\overset{9}{18}}{\underset{7}{21}} \cdot \frac{\overset{1}{3}}{\underset{7}{14}} = \frac{9}{49}$

Penny

$\frac{\overset{6}{18}}{\underset{3}{21}} \cdot \frac{\overset{1}{3}}{\underset{2}{14}} = \frac{6}{6}$ or 1

3. **Identify** which point shown on the number line below could be the product of the numbers graphed at A and B. Explain your reasoning.

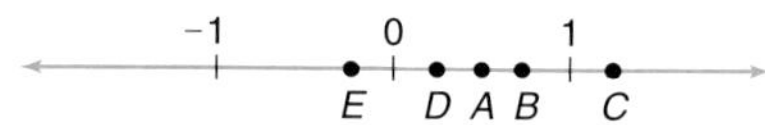

1. Yellow shading is $\frac{3}{4}$, blue shading is $\frac{1}{2}$, green shading is the product, $\frac{3}{8}$.
2. Jamal; Penny canceled incorrectly.
3. D; see students' work.

MATERIALS

construction paper
ruler

4. Fold a piece of $8\frac{1}{2}$" by 11" construction paper in half and then in half again as shown. Cut out a square from the open end of the folded paper. Unfold the paper. Then fold up the edges to form a box with no lid. Use your ruler to measure the length, width, and height of your box to the nearest sixteenth of an inch. **See students' work.**
   a. Find the total area of the bottom and four sides of your box.
   b. How does the area of your box compare with your neighbor's? Why do you think they are the same or different?

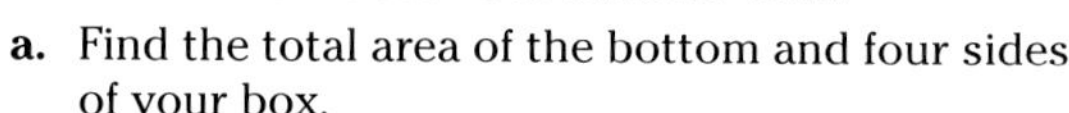

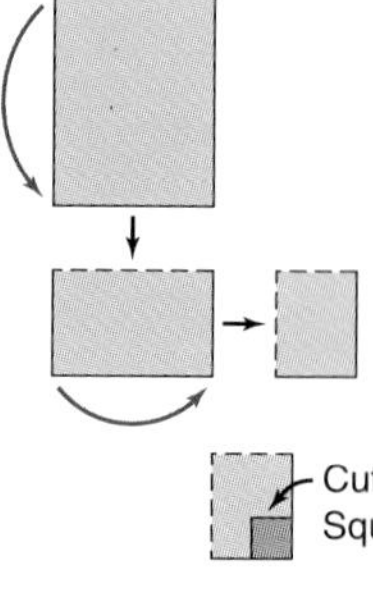

Guided Practice

Choose the correct product.

5. $\frac{1}{6} \cdot \left(-\frac{3}{5}\right)$ **a.** $\frac{2}{5}$ **b.** $-\frac{2}{5}$ **c.** $\frac{1}{10}$ **d.** $-\frac{1}{10}$ d

6. $\left(\frac{1}{2}\right)^2$ **a.** $\frac{1}{2}$ **b.** $\frac{1}{4}$ **c.** 1 **d.** 2 b

Study Guide Masters, p. 47

NAME ______________ DATE ________

6-3 Study Guide
Multiplying Fractions

Student Edition
Pages 284–288

To multiply two fractions, first multiply the numerators. Then, multiply the denominators. Write the product in simplest form.

$\frac{1}{2} \times \frac{3}{5} = \frac{1 \times 3}{2 \times 5} = \frac{3}{10}$

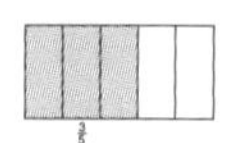

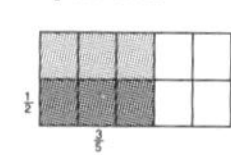

To multiply with mixed numbers, first change the mixed numbers to improper fractions. Then multiply the fraction.

Solve each equation. Write the solution in simplest form.

1. $t = \frac{1}{2} \times \left(-\frac{1}{3}\right)$ $-\frac{1}{6}$
2. $\frac{3}{4} \times \frac{1}{2} = f$ $\frac{3}{8}$
3. $c = -\frac{2}{3} \times \left(-\frac{1}{5}\right)$ $\frac{2}{15}$
4. $\frac{1}{3} \times \frac{2}{7} = d$ $\frac{2}{21}$
5. $n = \frac{1}{10} \times \left(-\frac{3}{6}\right)$ $-\frac{1}{20}$
6. $-\frac{4}{9} \times \frac{6}{7} = b$ $-\frac{8}{21}$
7. $\frac{6}{7} \times \frac{1}{12} = w$ $\frac{1}{14}$
8. $\frac{2}{9} \times \frac{4}{5} = q$ $\frac{8}{45}$
9. $m = 3\frac{1}{3} \times 2\frac{1}{4}$ $7\frac{1}{2}$
10. $y = -\frac{3}{4} \times \left(-1\frac{1}{2}\right)$ $1\frac{1}{8}$
11. $t = 5\frac{5}{6} \times \left(-\frac{1}{14}\right)$ $-\frac{5}{12}$
12. $4\frac{3}{4} \times 2\frac{2}{3} = r$ $12\frac{2}{3}$

Evaluate each expression if $a = \frac{3}{4}$, $x = \frac{1}{2}$, and $y = \frac{3}{7}$.

13. a^2 $\frac{9}{16}$
14. $2ax$ $\frac{3}{4}$
15. ay $\frac{9}{28}$
16. xy $\frac{3}{14}$
17. $\frac{1}{2}a$ $\frac{3}{8}$
18. $3y$ $\frac{9}{7}$

Reteaching

Using Problem Solving Use transparency overlays to illustrate multiplication. For example, illustrate $\frac{1}{4}$ of $\frac{1}{2}$ by drawing a rectangle, marking $\frac{1}{2}$ of it with diagonal stripes. Then have students shade $\frac{1}{4}$ of the striped half. Lead students to notice that $\frac{1}{8}$ of the entire rectangle has diagonal stripes and shading. Have them illustrate $\frac{1}{4}$ of $\frac{2}{3}$, $\frac{1}{2}$ of $\frac{3}{4}$, and $\frac{1}{4}$ of $\frac{1}{3}$.

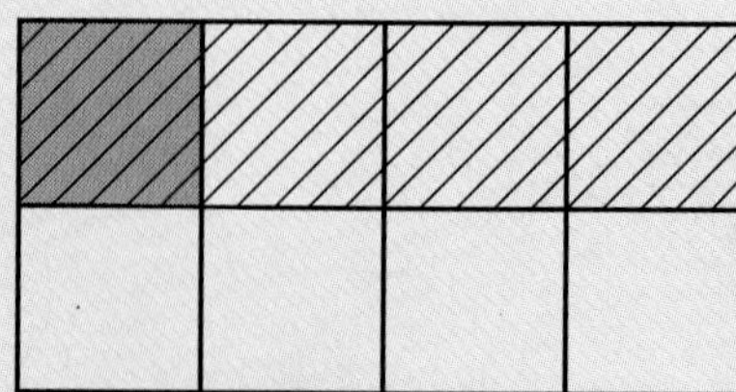

Solve each equation. Write each solution in simplest form.

7. $\frac{3}{4} \cdot \frac{2}{3} = a$ **$\frac{1}{2}$**
8. $b = \frac{1}{2}\left(-\frac{5}{6}\right)$ **$-\frac{5}{12}$**
9. $c = \frac{8}{12} \cdot \frac{4}{6}$ **$\frac{4}{9}$**
10. $x = \left(3\frac{1}{2}\right)4$ **14**
11. $m = \frac{8}{15}(-45)$ **-24**
12. $\left(-\frac{4}{5}\right)^2 = p$ **$\frac{16}{25}$**

Evaluate each expression if $a = \frac{2}{3}$, $x = \frac{3}{5}$, and $y = \frac{5}{8}$.

13. xy **$\frac{3}{8}$**
14. $2a$ **$1\frac{1}{3}$**
15. x^2 **$\frac{9}{25}$**
16. ay **$\frac{5}{12}$**

17. **Economics** The U.S. Congress is debating whether to replace paper one-dollar bills with coins in order to save money.
    a. It costs 8¢ to produce a one-dollar coin. The cost of producing a paper dollar is $\frac{19}{40}$ of that. How much does it cost to produce a paper dollar? **3.8¢**
    b. An average paper dollar lasts $1\frac{1}{4}$ years. A coin lasts about $22\frac{1}{2}$ times longer. How long does a coin last? **28 years, 1.5 months**
    c. Do you think it would save money to use coins instead of paper dollars? Explain. **Answers will vary.**

Exercises: Practicing and Applying the Concept

Independent Practice

Solve each equation. Write each solution in simplest form.

18. $\frac{3}{4} \cdot \left(-\frac{1}{3}\right) = b$ **$-\frac{1}{4}$**
19. $\frac{1}{2} \cdot \frac{2}{7} = t$ **$\frac{1}{7}$**
20. $k = -\frac{5}{6}\left(-\frac{2}{5}\right)$ **$\frac{1}{3}$**

21. $d = -4\left(\frac{3}{8}\right)$ **$-1\frac{1}{2}$**
22. $(-7)\left(-2\frac{1}{3}\right) = h$
23. $c = 1\frac{4}{5} \cdot \left(-2\frac{1}{2}\right)$ **$4\frac{1}{2}$**
24. $v = 2\frac{5}{6} \cdot 3\frac{1}{3}$ **$9\frac{4}{9}$**
25. $s = \left(2\frac{1}{4}\right)\left(-\frac{4}{3}\right)$ **-3**
26. $\left(-9\frac{3}{5}\right)\left(\frac{5}{12}\right) = y$ **-4**
27. $\left(-3\frac{1}{5}\right)\left(7\frac{1}{2}\right) = w$
28. $m = (9)\left(-1\frac{5}{6}\right)$
29. $p = \left(-\frac{3}{5}\right)^2$ **$\frac{9}{25}$**
30. $r = \left(\frac{7}{11}\right)^2$ **$\frac{49}{121}$**
31. $\left(-\frac{9}{10}\right)^2 \cdot 8 = f$
32. $\left(\frac{5}{7}\right)^2 = n$ **$\frac{25}{49}$**
33. What is the product of $\frac{4}{9}$ and $\frac{3}{5}$? **$\frac{4}{15}$**
34. What is $\frac{5}{8}$ of 36? **$22\frac{1}{2}$**

22. $16\frac{1}{3}$ **27.** -24
28. $-16\frac{1}{2}$ **31.** $6\frac{12}{25}$

Evaluate each expression if $a = -\frac{1}{3}$, $b = \frac{3}{4}$, $x = 1\frac{2}{5}$, and $y = -3\frac{1}{6}$.

35. ay **$1\frac{1}{18}$**
36. $2a^2$ **$\frac{2}{9}$**
37. $b(x + a)$ **$\frac{4}{5}$**
38. $3a - 5x$ **-8**
39. $a^2(b + 2)$ **$\frac{11}{36}$**
40. $-a(a - b)$ **$-\frac{13}{36}$**

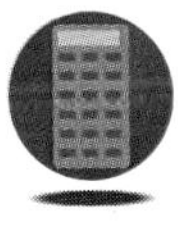

41. Use a fraction calculator and the numbers 2, 3, 4, 5, 6, and 7 to make two fractions with the greatest product. Use each digit only once. Explain your answer. **See margin.**

Critical Thinking

42. **Number Theory** Every whole number is a product of primes.
    a. Can you write $\frac{42}{45}$ as the product of fractions with prime numerators and denominators? You may use as many fractions as you need in the product. **42a. Sample answer: $\frac{2}{3} \times \frac{3}{5} \times \frac{7}{3}$**
    b. Is there more than one way to write the product? Explain. **Yes; rearrange the numerators and denominators.**

Error Analysis

Watch for students who use modified addition rules, such as finding a common denominator and multiplying only numerators in a multiplication problem. To avoid this, encourage students to estimate their products first.

Assignment Guide

Core: 19–41 odd, 42–43, 45, 47–54
Enriched: 18–40 even, 41–54

For **Extra Practice**, see p. 753.

The red A, B, and C flags, printed only in the Teacher's Wraparound Edition, indicate the level of difficulty of the exercises.

Additional Answer

41. $\frac{7}{2} \times \frac{6}{3}$ or $\frac{6}{2} \times \frac{7}{3}$; Use the greatest numbers as numerators and the least as denominators.

Group Activity Card 6-3

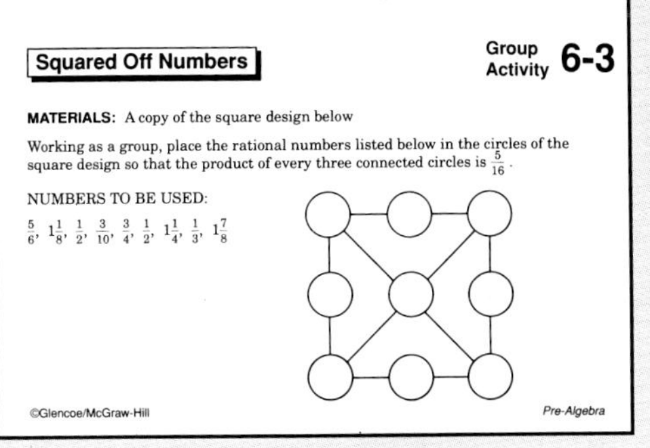

Squared Off Numbers Group Activity **6-3**

MATERIALS: A copy of the square design below

Working as a group, place the rational numbers listed below in the circles of the square design so that the product of every three connected circles is $\frac{5}{16}$.

NUMBERS TO BE USED:

$\frac{5}{6}$, $1\frac{1}{8}$, $\frac{1}{2}$, $\frac{3}{10}$, $\frac{3}{4}$, $\frac{1}{2}$, $1\frac{1}{4}$, $\frac{1}{3}$, $1\frac{7}{8}$

 Pre-Algebra

Practice Masters, p. 47

NAME ______________________ DATE ____________

6-3 Practice Student Edition Pages 284–288
Multiplying Fractions

Solve each equation. Write each solution in simplest form.

1. $a = -\frac{5}{7} \cdot \frac{14}{15}$ $-\frac{2}{3}$
2. $\frac{6}{11}\left(-\frac{33}{34}\right) = b$ $-\frac{9}{17}$
3. $c = \left(-6\frac{2}{3}\right)\left(-\frac{15}{16}\right)$ $6\frac{1}{4}$
4. $x = 5\left(-\frac{1}{10}\right)$ $-\frac{1}{2}$
5. $\frac{1}{7} \cdot \frac{1}{8} = y$ $\frac{1}{56}$
6. $z = -5\left(-\frac{21}{25}\right)$ $4\frac{1}{5}$
7. $m = \left(-9\frac{1}{5}\right)\left(\frac{10}{23}\right)$ -4
8. $\left(-\frac{3}{4}\right)\left(-\frac{8}{9}\right) = n$ $\frac{2}{3}$
9. $p = 4\frac{1}{2} \cdot 8$ 36
10. $\left(-\frac{8}{9}\right)\left(\frac{9}{8}\right) = q$ -1
11. $-5\frac{1}{3} \cdot 1\frac{4}{5} = r$ $-9\frac{3}{5}$
12. $9\left(-3\frac{1}{3}\right) = s$ -30
13. $t = \left(7\frac{7}{8}\right)\left(-\frac{5}{9}\right)$ $-4\frac{3}{8}$
14. $h = \left(1\frac{1}{9}\right)\left(\frac{27}{40}\right)$ $\frac{3}{4}$
15. $\left(-\frac{36}{50}\right)\left(\frac{75}{48}\right) = h$ $-1\frac{1}{8}$
16. $a = \left(-\frac{1}{2}\right)^2$ $\frac{1}{4}$
17. $b = 3\left(\frac{4}{5}\right)^2$ $1\frac{23}{25}$
18. $c = -1\left(-\frac{3}{5}\right)^2$ $-\frac{9}{25}$

Evaluate each expression if $a = -\frac{1}{4}$, $b = \frac{5}{6}$, $c = -1\frac{1}{2}$, and $d = 2\frac{1}{3}$.

19. $4c$ -6
20. bd $1\frac{17}{18}$
21. $3b - 4a$ $3\frac{1}{2}$
22. $18b - 6c$ 24
23. $a + cd$ $-3\frac{3}{4}$
24. $9d + \frac{7}{8}$ $21\frac{7}{8}$
25. $a(c + 4)$ $-\frac{5}{8}$
26. $b(a + 8)$ $6\frac{11}{24}$
27. $d(b + 6)$ $15\frac{17}{18}$

4 ASSESS

Closing Activity

Modeling Have students model the following products.

1. $\frac{1}{2} \times \frac{5}{6}$ $\frac{5}{12}$

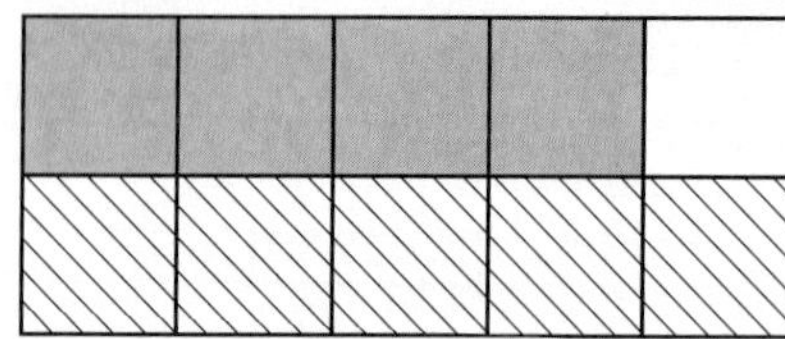

2. $\frac{3}{8} \times \frac{3}{4}$ $\frac{9}{32}$

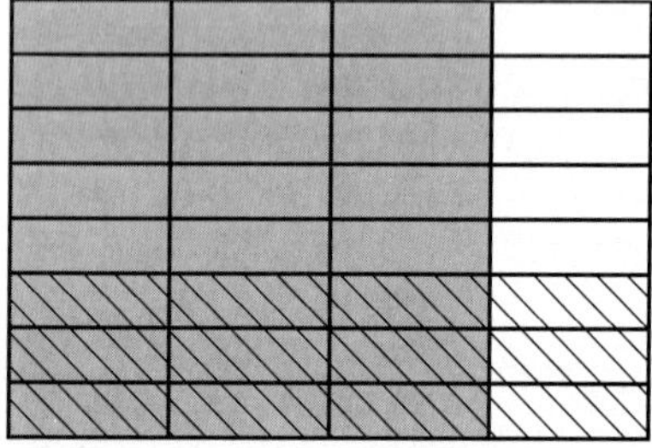

Chapter 6, Quiz A (Lessons 6-1 through 6-3) is available in the *Assessment and Evaluation Masters*, p. 155.

Additional Answer

46. African $= \frac{3}{20}$; modern $= \frac{2}{5}$; ballet $= \frac{1}{20}$

Enrichment Masters, p. 47

NAME ______________________ DATE ____________

6-3 Enrichment
Visualizing the Distributive Property

Student Edition Pages 284–288

Using the distributive property to estimate the product of fractions can be shown by using the area of a rectangle. For example, suppose you are working with this product:

$$3\frac{1}{2} \times 2\frac{1}{3}$$

This product is modeled by the rectangle at the right. The length is split into two parts, 3 units and $\frac{1}{2}$ unit. The width is split into 2 units and $\frac{1}{3}$ unit. The area of the rectangle is then divided into four parts.

3 | $\frac{1}{2}$ | 2 | $2 \times 3 = 6$ | $2 \times \frac{1}{2} = 1$ | $\frac{1}{3}$ | $3 \times \frac{1}{3} = 1$ | $\frac{1}{2} \times \frac{1}{3} = \frac{1}{6}$

Now, find the area of each part. The area of the largest part is just 2×3, or 6.

The areas of the two next-largest rectangles are $2 \times \frac{1}{2}$, or 1, and $3 \times \frac{1}{3}$, or 1.

The area of the smallest rectangle is $\frac{1}{2} \times \frac{1}{3}$, or $\frac{1}{6}$.
Thus, the product is $6 + 1 + 1 + \frac{1}{6}$, or $8\frac{1}{6}$.

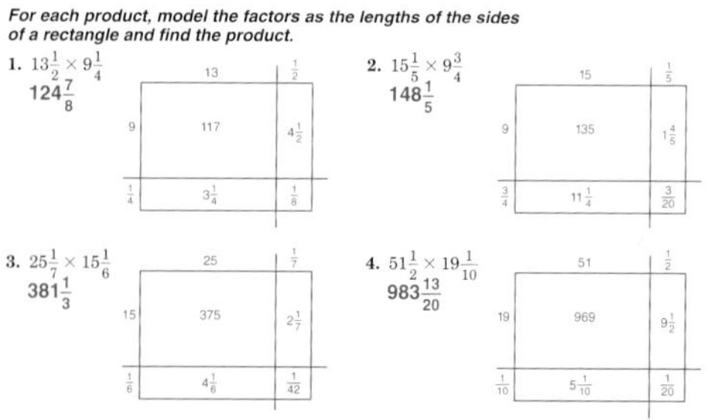

For each product, model the factors as the lengths of the sides of a rectangle and find the product.

1. $13\frac{1}{2} \times 9\frac{1}{4}$ $124\frac{7}{8}$

2. $15\frac{1}{5} \times 9\frac{3}{4}$ $148\frac{1}{5}$

3. $25\frac{1}{7} \times 15\frac{1}{6}$ $381\frac{1}{3}$

4. $51\frac{1}{2} \times 19\frac{1}{10}$ $983\frac{13}{20}$

Applications and Problem Solving

43. **Health** If you usually burn after $\frac{1}{2}$ hour exposure to sun, you can lengthen the time you are in the sun to three hours by applying SPF number 6. How long could a person who burns after $\frac{1}{4}$ hour stay in the sun using SPF 15 lotion? **$3\frac{3}{4}$ hours**

44. **Cooking** The West Point sloppy joe recipe mentioned in Example 3 uses $3\frac{1}{4}$ ounces of cider vinegar and $1\frac{2}{3}$ ounces of garlic powder. How much of these ingredients would be needed to make sloppy joes for 4500 people?

44. $146\frac{1}{4}$ oz cider, 75 oz garlic powder

45. **Measurement** People who design printed products use a system of measurement that is different from the customary and metric systems. The measurements used in this system are the *point* and the *pica*. One point equals $\frac{1}{72}$ inch, and 12 points make one pica.

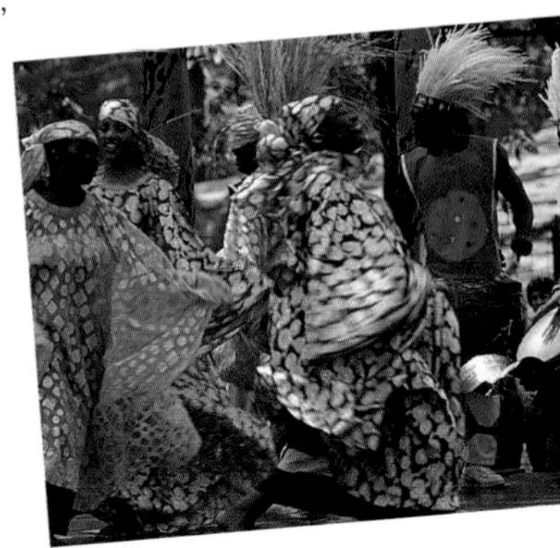

a. How many picas are in an inch? **6 picas**

b. The margin of a book is $\frac{3}{4}$ inches. How many picas is this?

45b. $4\frac{1}{2}$ picas

c. If a particular kind of type has an average measure of $2\frac{3}{4}$ characters per pica, about how many characters would fit across a $5\frac{1}{2}$-inch line?

45c. about 91 characters

46. **Statistics** At Westside High School, $\frac{3}{5}$ of the students are female. The dance preferences of those students are shown in the graph at the right. What part of the students are female and interested in each kind of dance? **See margin.**

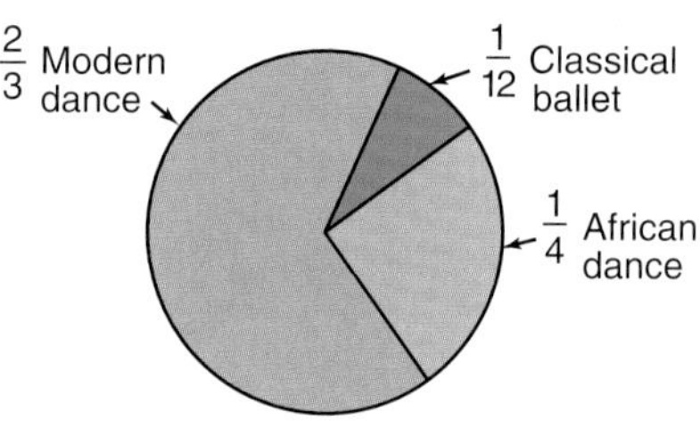

Mixed Review

47. Estimate $\frac{13}{28} \times 30$. (Lesson 6-2) **15**

48. **Consumerism** At the movies, it costs \$6.50 for a ticket, \$1.25 for a drink, and \$2.50 for popcorn. Will \$10 be enough to purchase one of each? Explain. (Lesson 5-2) **no; 7 + 1 + 3 = 11**

49. Find the prime factorization of 72. (Lesson 4-4) **$2^3 \cdot 3^2$**

50. Solve $f + (-3) > 8$. (Lesson 3-6) **$f > 11$**

51. Find the product $-4(-x)(-5y)$. (Lesson 2-7) **$-20xy$**

52. **Geometry** Supplementary angles are angles whose measures have a sum of 180°. One angle measures 57°. Solve the equation $y + 57 = 180$ to find the measure of a supplementary angle. (Lesson 1-8) **123°**

53. Find the value of $18 \div 3 + 18 \div 2$. (Lesson 1-2) **15**

54. When Kim divided 1033.5 by 26 the calculator showed 397.5. Is this a reasonable answer? Explain. (Lesson 1-1) **No, it should be about 40.**

Estimation
Mental Math
Calculator
Paper and Pencil

Extension

Using Logical Reasoning Write $-4\frac{7}{8}$, $1\frac{1}{6}$, $-6\frac{3}{10}$, $\frac{2}{3}$ on the board.

1. Ask students to mentally determine which two will give the greatest product. **$-4\frac{7}{8}$ and $-6\frac{3}{10}$**
2. Ask students to mentally determine which two will give the least product. **$1\frac{1}{6}$ and $-6\frac{3}{10}$**

Tech Prep

Electronic Pagination System Operator Persons in this occupation use the mathematics described in Exercise 45. They use a keyboard to enter and select the size and style of type, the column width, and the appropriate spacing, and then store it in a computer. For more information on tech prep, see the *Teacher's Handbook*.

6-4 Dividing Fractions

Setting Goals: *In this lesson, you'll divide fractions using multiplicative inverses.*

Modeling with Manipulatives

MATERIALS

colored pencils

One way to think of 6 ÷ 2 is to say "How many 2s are there in 6?" This way of thinking and an area model can help you divide fractions.

The drawing at the right shows a model for $\frac{3}{4} \div \frac{1}{2}$. To find the quotient, first draw a rectangle to represent one unit. Divide it into four equal parts and shade three to represent $\frac{3}{4}$. Then determine how many halves are contained in the shaded area. There is one half, and half of a second. So, $\frac{3}{4} \div \frac{1}{2} = 1\frac{1}{2}$.

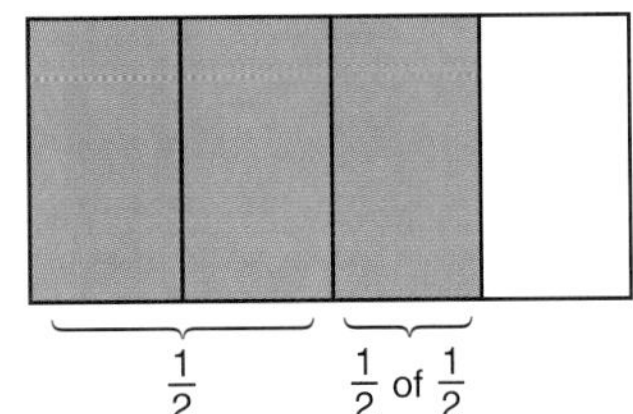

Your Turn

- Model $2\frac{2}{3} \div \frac{4}{9}$. Draw three rectangles and divide into thirds. Then shade $2\frac{2}{3}$.
- Add additional marks to show ninths in each rectangle.
- Mark as many groups of $\frac{4}{9}$ as you can.

TALK ABOUT IT

a. How many groups of $\frac{4}{9}$ did you find? 6

b. What is the quotient of $2\frac{2}{3} \div \frac{4}{9}$? 6

Learning the Concept

THINK ABOUT IT

Are $\frac{3}{4}$ and $-\frac{4}{3}$ multiplicative inverses?

no; $\frac{3}{4}\left(-\frac{4}{3}\right) = -1$

You can use the same skills to divide fractions that you used to divide integers. For example, dividing 40 by 2 is the same as multiplying 40 by $\frac{1}{2}$. In each case, the solution is 20. In other words, dividing by 2 is the same as multiplying by $\frac{1}{2}$.

Notice that $2 \times \frac{1}{2} = 1$. Two numbers whose product is 1 are **multiplicative inverses**, or **reciprocals**, of each other. So, 2 and $\frac{1}{2}$ are multiplicative inverses. In the same way, $-\frac{3}{2}$ and $-\frac{2}{3}$ are multiplicative inverses because $-\frac{3}{2} \cdot \left(-\frac{2}{3}\right) = 1$.

Inverse Property of Multiplication

In words: The product of a number and its multiplicative inverse is 1.

In symbols: For every nonzero number $\frac{a}{b}$, where $a, b \neq 0$, there is exactly one number $\frac{b}{a}$ such that $\frac{a}{b} \cdot \frac{b}{a} = 1$.

Classroom Vignette

"During this lesson, I introduce the concept of complex fractions. I use this concept to explain why we multiply by the reciprocal to divide fractions."

Example: $\frac{2}{3} \div \frac{5}{7}$

$$\frac{\frac{2}{3}}{\frac{5}{7}} \cdot \frac{\left(\frac{7}{5}\right)}{\left(\frac{7}{5}\right)} = \frac{\frac{2}{3} \cdot \frac{7}{5}}{\frac{5}{7} \cdot \frac{7}{5}} = \frac{\frac{2}{3} \cdot \frac{7}{5}}{1} \text{ or } \frac{2}{3} \cdot \frac{7}{5}$$

Kathryn A. Caliendo

Kathryn Caliendo
Hamden Middle School
Hamden, CT

6-4 LESSON NOTES

NCTM Standards: 1-7, 9

Instructional Resources

- Study Guide Master 6-4
- Practice Master 6-4
- Enrichment Master 6-4
- Group Activity Card 6-4
- Multicultural Activity Masters, p. 11
- Tech Prep Applications Masters, p. 12

Transparency 6-4A contains the 5-Minute Check for this lesson; **Transparency 6-4B** contains a teaching aid for this lesson.

Recommended Pacing	
Standard Pacing	Day 4 of 13
Honors Pacing	Day 4 of 12
Block Scheduling*	Day 2 of 7 (along with Lesson 6-3)

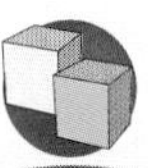

*For more information on pacing and possible lesson plans, refer to the ***Block Scheduling Booklet***.

1 FOCUS

5-Minute Check
(over Lesson 6-3)

Solve each equation. Write each solution in simplest form.

1. $n = \frac{3}{8} \cdot \frac{16}{21}$ $\frac{2}{7}$
2. $x = 5\frac{1}{2} \cdot 8$ 44
3. $\left(-\frac{3}{4}\right)^2 = q$ $\frac{9}{16}$

Evaluate each expression if $r = -\frac{2}{3}$ and $s = \frac{3}{8}$.

4. rs $-\frac{1}{4}$
5. $r^2(s + 3)$ $1\frac{1}{2}$

Motivating the Lesson

Situational Problem Inform students that a honeybee can produce $\frac{1}{10}$ of a pound of honey in its life time. Ask students how many honeybees it would take to produce a half pound of honey. 5

2 TEACH

Teaching Tip Write numbers on one set of cards and their multiplicative inverses on another set of cards. Give each student a card. Have students quietly find the person who has the number that, when multiplied by their number, equals 1.

In-Class Example

For Example 1
Jessica is designing a mural for a wall that is $10\frac{1}{2}$ feet high and $27\frac{3}{4}$ feet wide. It will have 4 panels of equal width. How wide will each panel be if there is a one-foot border around the mural and one foot between each panel? **$5\frac{11}{16}$ feet**

Remember that dividing by 2 was the same as multiplying by $\frac{1}{2}$, its multiplicative inverse. This is true for any rational number.

Division with Fractions

In words: To divide by a fraction, multiply by its multiplicative inverse.

In symbols: For fractions $\frac{a}{b}$ and $\frac{c}{d}$ where $b, c, d \neq 0$, $\frac{a}{b} \div \frac{c}{d} = \frac{a}{b} \cdot \frac{d}{c}$.

Example 1 APPLICATION Desktop Publishing

Robert G. Fernandez, a vice-president at Smith Barney financial corporation in Austin, Texas, wrote an article on money management in *Hispanic* magazine. *Hispanic* is printed on $8\frac{1}{2}$-by-11 inch pages, and the articles are printed in three columns. If the margins are $\frac{3}{8}$ inch on the left and right side of the page and there is $\frac{1}{4}$ inch of space between the columns, how wide should the columns on his computer be set?

Explore You know the total width of the page, the width of the margins, and the amount of space between each column. There are three equal columns. You need to know how wide to make each column.

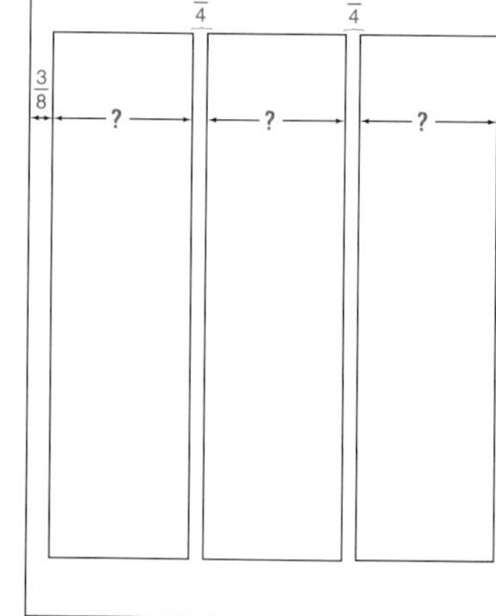

Plan Make a diagram and mark the widths you know. Subtract the known widths from the total width. Then divide the remaining width by 3.

Solve

$$8\frac{1}{2} - \frac{3}{8} - \frac{3}{8} - \frac{1}{4} - \frac{1}{4} = 7\frac{1}{4}$$

$$7\frac{1}{4} \div 3 = \frac{29}{4} \div \frac{3}{1} \quad \text{Rename } 7\frac{1}{4} \text{ as } \frac{29}{4} \text{ and 3 as } \frac{3}{1}.$$

$$= \frac{29}{4} \times \frac{1}{3} \quad \text{Multiply by the multiplicative inverse of 3, } \frac{1}{3}.$$

$$= \frac{29}{12} \text{ or } 2\frac{5}{12}$$

Each column should be $2\frac{5}{12}$ inches wide.

Examine Add the widths of all the margins and the three columns. The sum should be $8\frac{1}{2}$.

$$\frac{3}{8} + 2\frac{5}{12} + \frac{1}{4} + 2\frac{5}{12} + \frac{1}{4} + 2\frac{5}{12} + \frac{3}{8} \stackrel{?}{=} 8\frac{1}{2}$$

$$8\frac{1}{2} = 8\frac{1}{2} \checkmark \quad \text{It checks.}$$

Alternative Learning Styles

Kinesthetic Have students use quarters to show $4.50 ÷ 3 $\left(4\frac{1}{2} \div 3\right)$ by separating the quarters into 3 equal groups. Have students make up other problems that can be modeled using various coins.

You can use the division skills that you developed with integers and fractions to divide negative fractions.

Example 2 **Solve $p = -3\frac{3}{5} \div \frac{6}{7}$.**

$p = -3\frac{3}{5} \div \frac{6}{7}$ — *Estimate: $-4 \div 1 = -4$*

$= -\frac{18}{5} \div \frac{6}{7}$ — *Rename $-3\frac{3}{5}$ as $-\frac{18}{5}$.*

$= -\frac{18}{5} \times \frac{7}{6}$ — *Multiply by the multiplicative inverse of $\frac{6}{7}$, $\frac{7}{6}$.*

$= -\frac{21}{5}$ or $-4\frac{1}{5}$ — *Rename as a mixed number in simplest form. Compare to the estimate.*

Connection to Algebra

Evaluating algebraic expressions involving division of fractions uses the same procedures as evaluating algebraic expressions with integers.

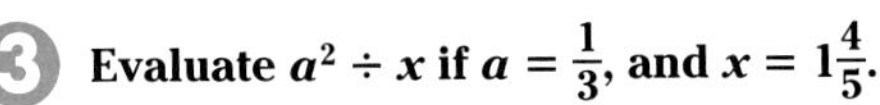

Example 3 **Evaluate $a^2 \div x$ if $a = \frac{1}{3}$, and $x = 1\frac{4}{5}$.**

$a^2 \div x = \left(\frac{1}{3}\right)^2 \div 1\frac{4}{5}$ — *Replace a with $\frac{1}{3}$ and x with $1\frac{4}{5}$.*

$= \left(\frac{1}{3}\right)\left(\frac{1}{3}\right) \div 1\frac{4}{5}$ — *Evaluate powers first.*

$= \frac{1}{9} \div 1\frac{4}{5}$

$= \frac{1}{9} \div \frac{9}{5}$ — *Rename $1\frac{4}{5}$ as $\frac{9}{5}$.*

$= \frac{1}{9} \times \frac{5}{9}$ — *Multiply by the multiplicative inverse of $\frac{9}{5}$, $\frac{5}{9}$.*

$= \frac{5}{81}$

Checking Your Understanding

Communicating Mathematics

Read and study the lesson to answer these questions. **1–3. See margin.**

1. In your own words, what are reciprocals? Give an example.
2. How and when do you use a reciprocal in division of fractions?
3. **Write** an equation that you could use the multiplicative inverse of $\frac{3}{8}$ to solve.
4. **Give a counter example** to the statement that every rational number has a reciprocal. **0**

5. A number is both multiplied and divided by the same rational number n, where $0 < n < 1$. Which is greater, the product or the quotient? Explain your reasoning. **See margin.**

Guided Practice

State whether each pair of numbers are multiplicative inverses. Write *yes* or *no*.

6. $7, \frac{1}{7}$ **yes**
7. $-\frac{2}{5}, -1\frac{1}{2}$ **no**
8. $5\frac{2}{3}, \frac{17}{3}$ **no**
9. $1.4, \frac{5}{7}$ **yes**

In-Class Examples

For Example 2

Solve $d = \frac{5}{8} \div 1\frac{3}{4}$. $-\frac{5}{14}$

For Example 3

Evaluate each expression if $p = -\frac{1}{4}$, $r = 1\frac{2}{3}$, and $s = -2\frac{1}{2}$.

a. $\frac{r}{s}$ $-\frac{2}{3}$

b. $p^2 \div s$ $-\frac{1}{40}$

3 PRACTICE/APPLY

Checking Your Understanding

Exercises 1–18 are designed to help you assess your students' understanding through reading, writing, speaking, and modeling. You should work through Exercises 1–5 with your students and then monitor their work on Exercises 6–18.

Additional Answers

1. Sample answer: Reciprocals are numbers whose product is 1. $\frac{4}{7}$ and $\frac{7}{4}$ are reciprocals.
2. Dividing by a number is the same as multiplying by the reciprocal.
3. Sample answer: $5 \div \frac{3}{8} = y$

Study Guide Masters, p. 48

NAME ______________ DATE ________

6-4 Study Guide
Dividing Fractions

Student Edition
Pages 289–293

When dividing with fractions, you will use the reciprocal of a number. Two numbers whose product is 1 are called **reciprocals**. The reciprocal is also called the **multiplicative inverse**.

The reciprocal of 4 is $\frac{1}{4}$ because $4 \times \frac{1}{4} = 1$.

The reciprocal of $-\frac{4}{5}$ is $-\frac{5}{4}$ because $-\frac{4}{5} \times -\frac{5}{4} = 1$.

To divide by a rational number, multiply by its multiplicative inverse (its reciprocal).

$4 \div \frac{1}{2} = 4 \times \frac{2}{1} = 8$

If dividing mixed numbers, first rename the mixed numbers as fractions. Then, divide.

$-1\frac{1}{5} \div 2\frac{2}{5} = -\frac{6}{5} \div \frac{12}{5} = -\frac{6}{5} \times \frac{5}{12} = -\frac{30}{60} = -\frac{1}{2}$

Name the multiplicative inverse for each rational number.

1. $\frac{4}{5}$ $\frac{5}{4}$
2. $\frac{8}{3}$ $\frac{3}{8}$
3. $-\frac{2}{7}$ $-\frac{7}{2}$
4. $\frac{5}{6}$ $\frac{6}{5}$
5. 3 $\frac{1}{3}$
6. -7 $-\frac{1}{7}$
7. $-\frac{1}{8}$ -8
8. $\frac{1}{10}$ 10

Solve each equation. Write the solution in simplest form.

9. $\frac{1}{4} \div \left(-\frac{2}{5}\right) = r$ $-\frac{5}{8}$
10. $\frac{5}{8} \div \frac{4}{5} = j$ $\frac{25}{32}$
11. $y = -\frac{2}{9} \div \frac{3}{5}$ $-\frac{10}{27}$
12. $m = 10 \div \frac{2}{3}$ 15
13. $\ell = -\frac{1}{3} \div (-4)$ $\frac{1}{12}$
14. $n = \frac{2}{3} \div (-\frac{1}{2})$ $-1\frac{1}{3}$
15. $-1\frac{3}{5} \div \frac{1}{2} = q$ $-3\frac{1}{5}$
16. $4\frac{1}{9} \div \frac{5}{6} = x$ $4\frac{14}{15}$
17. $\frac{3}{5} \div 1\frac{1}{2} = d$ $\frac{2}{5}$
18. $p = 1\frac{3}{4} \div 2\frac{2}{3}$ $\frac{21}{32}$
19. $f = -2\frac{1}{2} \div (-1\frac{4}{5})$ $1\frac{7}{18}$
20. $1\frac{1}{4} \div 4\frac{1}{8} = w$ $\frac{10}{33}$

Reteaching

Using Models Draw six identical circles on the board or overhead. Divide each circle into fourths. Ask students how many groups of $\frac{3}{4}$ they see. **8** Have students write the division problem that represents this model. Repeat this activity finding groups of $\frac{3}{8}$ after dividing the circles into eighths. **16**

Additional Answer

5. The quotient is greater because dividing by the number is the same as multiplying by the reciprocal. If a number is between 0 and 1, its reciprocal will be greater than it is.

Assignment Guide

Core: 19–45 odd, 46, 47, 49–57
Enriched: 20–44 even, 46–57

For **Extra Practice**, see p. 754.

The red A, B, and C flags, printed only in the Teacher's wraparound Edition, indicate the level of difficulty of the exercises.

Write each division expression as a multiplication expression. Then find its value.

10. $\frac{1}{2} \div \frac{6}{7}$ $\frac{7}{12}$ **11.** $-\frac{3}{4} \div \frac{3}{4}$ -1 **12.** $\frac{7}{9} \div \frac{2}{3}$ $1\frac{1}{6}$

13. $5 \div \left(-1\frac{1}{3}\right)$ $-3\frac{3}{4}$ **14.** $2\frac{3}{5} \div 3\frac{6}{7}$ $\frac{91}{135}$ **15.** $-8 \div \left(-22\frac{4}{5}\right)$ $\frac{20}{57}$

Evaluate each expression.

16. $x \div y$, if $x = 1\frac{1}{2}$ and $y = \frac{1}{2}$ 3

17. $c \div d + e$, if $c = \frac{2}{3}$, $d = 1\frac{1}{2}$, and $e = \frac{1}{6}$ $\frac{11}{18}$

18. Carpentry How many boards, each 2 feet 8 inches long, can be cut from a board 16 feet long? **6 boards**

Exercises: Practicing and Applying the Concept

Independent Practice

A

Name the multiplicative inverse for each rational number.

19. $-\frac{7}{3}$ $-\frac{3}{7}$ **20.** 8 $\frac{1}{8}$ **21.** $1\frac{3}{5}$ $\frac{5}{8}$ **22.** $-3\frac{2}{7}$ $-\frac{7}{23}$

23. 1.5 $\frac{2}{3}$ **24.** -0.4 $-\frac{5}{2}$ **25.** $\frac{x}{y}$ $\frac{y}{x}$ **26.** $\frac{m}{n}$ $\frac{n}{m}$

Estimate the solution to each equation. Then solve. Write the solution in simplest form. 31. $-1\frac{1}{2}$ 39. $6\frac{2}{3}$

27. $r = -\frac{3}{5} \div \frac{5}{9}$ $-1\frac{2}{25}$ **28.** $u = -1\frac{1}{9} \div \frac{2}{3}$ $-1\frac{2}{3}$ **29.** $-8 \div \frac{4}{5} = t$ -10

30. $6 \div \frac{1}{3} = m$ 18 **31.** $2\frac{1}{4} \div \left(-1\frac{1}{2}\right) = h$ **32.** $-2\frac{3}{7} \div \left(-4\frac{4}{7}\right) = d$ $\frac{17}{32}$

33. $12 \div \frac{4}{9} = c$ 27 **34.** $-10 \div \frac{3}{8} = f$ $-26\frac{2}{3}$ **35.** $z = 24 \div \frac{7}{10}$ $34\frac{2}{7}$

36. $q = -2 \div \left(-\frac{1}{3}\right)$ 6 **37.** $s = -3\frac{1}{4} \div 2\frac{1}{6}$ $-1\frac{1}{2}$ **38.** $7\frac{1}{2} \div 1\frac{1}{5} = n$ $6\frac{1}{4}$

39. $m = -\frac{16}{7} \div \left(-\frac{12}{35}\right)$ **40.** $a = \frac{21}{30} \div \frac{7}{15}$ $1\frac{1}{2}$ **41.** $12\frac{1}{4} \div \left(-\frac{14}{3}\right) = j$ $-2\frac{5}{8}$

Evaluate each expression.

42. $a \div b$, if $a = \frac{2}{3}$ and $b = 1\frac{1}{3}$ $\frac{1}{2}$

43. $r \div s$, if $r = -\frac{8}{9}$ and $s = \frac{7}{18}$ $-2\frac{2}{7}$

44. $a^2 \div b^2$, if $a = -\frac{3}{4}$ and $b = 1\frac{1}{3}$ $\frac{81}{256}$

45. $m + n \div p$, if $m = \frac{2}{3}$, $n = 1\frac{1}{3}$, and $p = \frac{1}{9}$ $12\frac{2}{3}$

Critical Thinking

46. a. Divide the fraction $\frac{3}{4}$ by $\frac{1}{2}$, $\frac{1}{4}$, $\frac{1}{8}$, and $\frac{1}{12}$. $\frac{3}{2}$, 3, 6, 9

b. What happens to the quotient as the divisor changes?

c. Make a conjecture about what happens when you divide $\frac{3}{4}$ by fractions that increase in size. Then give examples that support your conjecture. **The quotient would get smaller.**

46b. The quotient gets larger.

Practice Masters, p. 48

NAME ______________ DATE ______________

6-4 Practice
Dividing Fractions

Student Edition
Pages 289–293

Name the multiplicative inverse for each rational number.

1. 7 $\frac{1}{7}$ **2.** -10 $-\frac{1}{10}$ **3.** 1 1 **4.** 0.6 $\frac{5}{3}$

5. $\frac{1}{3}$ 3 **6.** $\frac{1}{5}$ 5 **7.** $-\frac{1}{12}$ -12 **8.** $\frac{1}{10}$ 10

9. $-\frac{4}{3}$ $-\frac{3}{4}$ **10.** $\frac{2}{3}$ $\frac{3}{2}$ **11.** $-\frac{8}{7}$ $-\frac{7}{8}$ **12.** 1.5 $\frac{2}{3}$

13. $-\frac{3}{23}$ $-\frac{23}{3}$ **14.** $\frac{6}{41}$ $\frac{41}{6}$ **15.** $3\frac{1}{5}$ $\frac{5}{16}$ **16.** $-3\frac{3}{4}$ $-\frac{4}{15}$

Solve each equation. Write the solution in simplest form.

17. $a = -8 \div (-12)$ $\frac{2}{3}$ **18.** $x = -15 \div \frac{3}{4}$ -20 **19.** $h = -\frac{3}{4} \div 3$ $-\frac{1}{4}$

20. $b = -5\frac{1}{2} \div \left(-2\frac{3}{4}\right)$ 2 **21.** $-8 \div \left(-\frac{4}{5}\right) = p$ 10 **22.** $c = 2\frac{1}{5} \div \left(-1\frac{7}{10}\right)$ $-1\frac{5}{17}$

23. $-1\frac{1}{9} \div 1\frac{17}{63} = d$ $-\frac{7}{8}$ **24.** $-10 \div \left(-5\frac{3}{4}\right) = k$ $1\frac{17}{23}$ **25.** $g = -\frac{1}{5} \div \frac{7}{8}$ $-\frac{8}{35}$

26. $-12\frac{1}{4} \div 4\frac{2}{3} = x$ $-2\frac{5}{8}$ **27.** $-5\frac{2}{7} \div \left(-\frac{3}{8}\right) = r$ $14\frac{2}{21}$ **28.** $-7\frac{3}{5} \div \left(-1\frac{9}{10}\right) = y$ 4

29. $6\frac{2}{3} \div \left(-\frac{10}{3}\right) = p$ -2 **30.** $-12\frac{1}{4} \div \left(-\frac{7}{8}\right) = k$ 14 **31.** $h = -5\frac{2}{3} \div \left(-2\frac{4}{15}\right)$ $2\frac{1}{2}$

Group Activity Card 6-4

Less Is Better Group Activity **6-4**

Work in groups of three. Complete the following multiplication table together, using as few new multiplications as possible. One person must write out the multiplications your group used, another person must list the properties you used, and the other person must keep track of the order of the multiplications you used and the number of multiplications. The group that finishes with the least number of new multiplications performed is the winner.

×	12	10	2	$\frac{5}{4}$	$\frac{3}{4}$
12					9
10				$\frac{25}{2}$	
2					$\frac{3}{2}$
$\frac{5}{4}$	15		$\frac{5}{2}$	$\frac{25}{16}$	
$\frac{3}{4}$		$\frac{15}{2}$		$\frac{15}{16}$	$\frac{9}{16}$

Use the commutative and distributive properties of multiplication. For example, your group might decide to start by doing the problem 12 × 12. You might rewrite it using the distributive property as 12 (10 + 2) and perform 2 new multiplications. Then the answer for the problems 12 × 10 and 12 × 2 would be known, so they would not be new multiplications.

 Pre-Algebra

Applications and Problem Solving

47. 21 uniforms plus extra fabric remaining

47. Business Bertha Thangelane of Soweto, South Africa used the political changes in South Africa to start her business. She is the co-owner and managing director of RNB Creations (Pty) Ltd., which makes school uniforms. If she can make an elementary school child's uniform with $2\frac{1}{3}$ yards of fabric, how many uniforms can she make with a 50-yard bolt of fabric?

48. Sequences What is the quotient if the eighth term of the sequence 1, $\frac{1}{2}$, $\frac{1}{4}$, $\frac{1}{8}$, ... is divided by the ninth term? Explain your method and strategy for solving this problem. **2; see student's work.**

49. Food According to the U.S. Department of Agriculture, the average young American woman drinks $1\frac{1}{2}$ cans of cola each day. At this rate, how long would a 12-pack of cola last? **8 days**

Mixed Review

50. Solve $\frac{3}{5} \cdot \frac{2}{7} = a$. Write the solution in simplest form. (Lesson 6-3) $\frac{6}{35}$

51. Which number is greater, $\frac{2}{5}$ or 0.25? (Lesson 6-1) $\frac{2}{5}$

52. 5 pounds

52. Health Teri lost $1\frac{3}{4}$ pounds the first week of her diet. The following week she lost $3\frac{1}{4}$ pounds. How much did Teri lose? (Lesson 5-4)

53. Find the greatest common factor (GCF) of 30 and 12. (Lesson 4-5) **6**

54. Personal Finance You received your paycheck on Friday. After paying your rent, which is \$395, you had \$510 left. How much was your paycheck? (Lesson 3-2) **\$905**

55. Geometry Name the quadrant in which $(-6, 18)$ is located. (Lesson 2-2) **II**

56. State whether $9 > 6$ is *true, false,* or *open.* (Lesson 1-9) **true**

57. Simplify $h + 12h + 23$. (Lesson 1-5) **$13h + 23$**

WORKING ON THE

Investigation

Refer to the Investigation on pages 270–271.

Look over the scale drawings that you placed in your Investigation Folder at the beginning of Chapter 6. Make any necessary revisions to them based on what you've learned so far.

- Using the scale drawings, determine the speed of your train for each section of the circuit you designed. A typical speed down the first drop is about 60 miles an hour (88 feet per second). The length of each section should be determined by the group and will, for the most part, be dictated by the design, that is, uphill, downhill, straight, or curved.
- Determine the time it will take for your coaster train to travel each section of the circuit you designed. This can be found by dividing the length of the section of track by the speed you think the train will be traveling over that section.
- Find the total time it will take for the coaster train to complete the circuit you designed.

Add the results of your work to your Investigation Folder.

4 ASSESS

Closing Activity

Writing Write a fraction or number on the board. Then have students write a division problem with that quotient. For example, write $\frac{4}{5}$ on the board and have students write a problem that will result in that answer, such as $\frac{1}{5} \div \frac{1}{4}$ or $2\frac{2}{3} \div 3\frac{1}{3}$.

Enrichment Masters, p. 48

NAME ____________ DATE ________

6-4 Enrichment
Dividing Fractions

Student Edition Pages 289–293

When multiplying fractions, you multiply the numerators and multiply the denominators. What would happen if, when dividing fractions, you would divide the numerators and divide the denominators?

$$\frac{5}{8} \div \frac{1}{4} = \frac{5 \div 1}{8 \div 4} = \frac{5}{2} = 2\frac{1}{2}$$

When adding and subtracting fractions, you first change the fractions to like fractions. What would happen if, when dividing fractions, you change to like fractions before you divide?

$$\frac{2}{3} \div \frac{1}{4} = \frac{8}{12} \div \frac{3}{12} = \frac{8}{12} \times \frac{12}{3} = \frac{8}{3} = 2\frac{2}{3}$$

Find the quotient by dividing numerators and dividing denominators. Check each answer by using the traditional method.

1. $\frac{1}{6} \div \frac{1}{3}$ $\frac{1}{2}$
2. $\frac{5}{12} \div \frac{1}{4}$ $\frac{5}{3} = 1\frac{2}{3}$
3. $\frac{3}{8} \div \frac{3}{4}$ $\frac{1}{2}$

4. Will this method always produce the correct quotient? Explain. **Yes; $\frac{A}{B} \div \frac{C}{D} = \frac{AD}{BC}$ and $\frac{A \div C}{B \div D} = \frac{A}{C} \div \frac{B}{D} = \frac{AD}{BC}$; therefore, $\frac{A}{B} \div \frac{C}{D} = \frac{A \div C}{B \div D}$**

5. Give an example when this method may be more difficult than the traditional method. **Examples will vary. One possible answer is $\frac{3}{4} \div \frac{5}{3} = \frac{3 \div 5}{4 \div 3}$.**

Find the quotient by first changing to like fractions. Check by using the traditional method.

6. $\frac{5}{6} \div \frac{3}{4}$ $\frac{10}{9} = 1\frac{1}{9}$
7. $\frac{4}{5} \div \frac{7}{10}$ $\frac{8}{7} = 1\frac{1}{7}$
8. $\frac{7}{8} \div \frac{11}{16}$ $\frac{14}{11} = 1\frac{3}{11}$

9. Will this method always produce the correct quotient? Explain. **Yes; when you change to like fractions, you are not changing the value of either fraction or their quotient.**

10. Name the method you like best and explain why. **Answers will vary.**

Extension

Using Models Using a set of dominoes, have students express a problem with a quotient of 1, such as $\frac{1}{2} \div \frac{2}{4}$. Have students find two dominoes that can express a problem with a quotient of 2, such as $\frac{4}{3} \div \frac{2}{3}$. Ask students if division is commutative. Have students explain using the models.

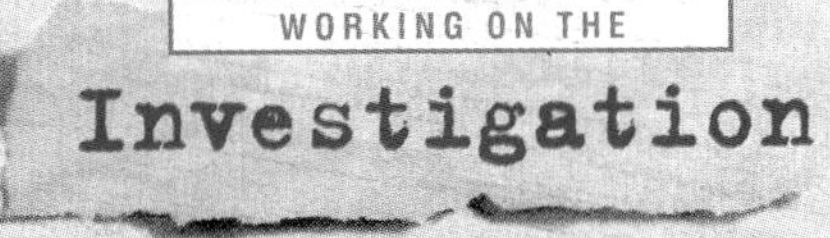

The Investigation on pages 270–271 is designed to be a long-term project that is completed over several days or weeks. Encourage students to keep their materials in the Investigation Folder as they work on the Investigation.

6-5A LESSON NOTES

NCTM Standards: 1-5

Objective

Use models to multiply and divide decimals.

Recommended Time

Demonstration and discussion: 15 minutes; Exercises: 30 minutes

Instructional Resources

For each student or group of students

- colored pencils

Math Lab and Modeling Math Masters

- p. 11 (Base-Ten Models)
- p. 37 (worksheet)

For teacher demonstration

Overhead Manipulative Resources

1 FOCUS

Motivating the Lesson

Have students explore the following pattern.

8×3 ones = 24 ones or 24
8×3 tenths = 24 tenths or 2.4
8×3 hundredths = 24 hundredths or 0.24

Ask students to compare the number of decimal places in the factors and the final product.

2 TEACH

Teaching Tip Students who have difficulty working with decimal models may prefer using fraction models to represent the multiplication of decimals.

3 PRACTICE/APPLY

Assignment Guide
Core: 1–6
Enriched: 1–6

MATHEMATICS LAB

HANDS-ON ACTIVITY

6-5A Multiplying and Dividing Decimals

A Preview of Lesson **6-5**

MATERIALS

grid paper

colored pencils

In the activities below, you will explore the multiplication and division of decimals using grid paper decimal models. The whole decimal model represents 1, and each row or column represents 0.1.

Activity 1 **Work with a partner.**

- Use a 10-by-10 decimal model to find 0.3×0.8.
- Color 3 rows of the model yellow to represent 0.3.
- Color 8 columns of the model blue to represent 0.8.

0.3
0.8

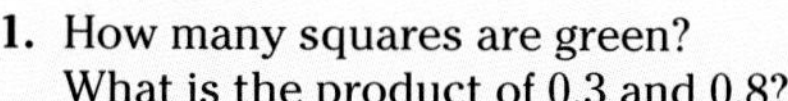

TALK ABOUT IT

1. How many squares are green? What is the product of 0.3 and 0.8? **24 squares; 24 hundredths or 0.24**

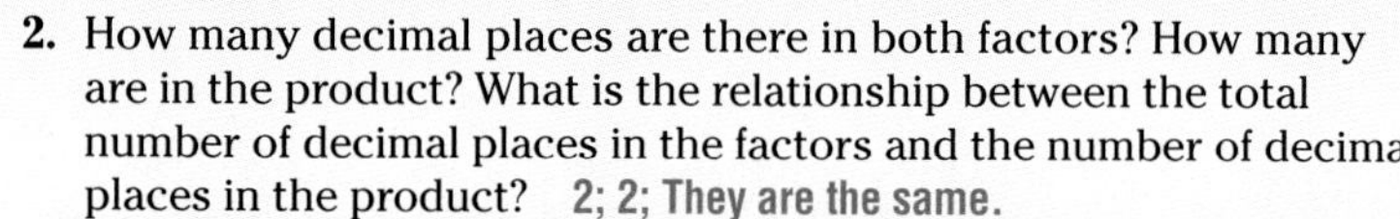

2. How many decimal places are there in both factors? How many are in the product? What is the relationship between the total number of decimal places in the factors and the number of decimal places in the product? **2; 2; They are the same.**
3. Use a model and a calculator to find the product 0.4×0.2. Explain your results. **$0.4 \times 0.2 = 0.08$, each square represents one one-hundredth**

Activity 2 **Use decimal models to find $1.5 \div 0.3$.**

- Outline one complete model and 5 columns in the second model to represent 1.5.

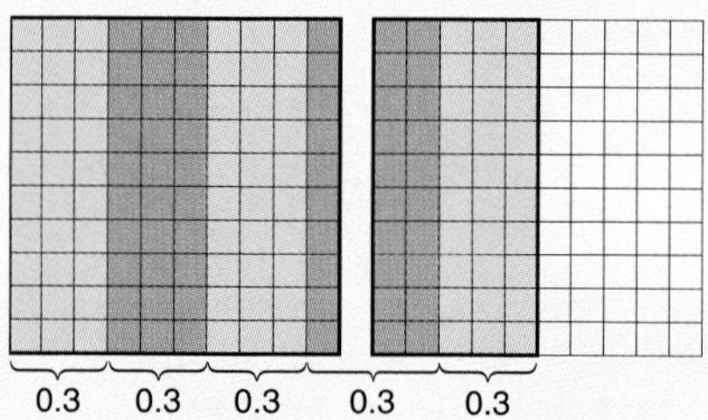

0.3 0.3 0.3 0.3 0.3

- Shade the first three columns yellow to represent 0.3.
- Shade the next three columns blue to represent another 0.3.
- Continue shading groups of three columns in alternating colors until you have filled the outline representing 1.5.

TALK ABOUT IT

4. How many groups of three tenths do you have? **5 groups**
5. What is the quotient of $1.5 \div 0.3$? **5**
6. Use decimal models to divide 2.4 by 0.4. **6**

4 ASSESS

Observing students working in cooperative groups is an excellent method of assessment. You may wish to ask a student at random from each group to explain how to model and solve a problem. Also watch for and acknowledge students who are helping others to understand the concept being taught.

6-5 Multiplying and Dividing Decimals

Setting Goals: *In this lesson, you'll multiply and divide decimals.*

Modeling with Technology

You can multiply decimals in the same way that you multiply whole numbers. Look for a pattern in the products below to help you write a rule for placing the decimal point in the product.

Your Turn Use a calculator to find each product.

$15.6 \times 38 =$ **592.8**

$15.6 \times 3.8 =$ **59.28**

$15.6 \times 0.38 =$ **5.928**

$15.6 \times 0.038 =$ **0.5928**

TALK ABOUT IT

a. Describe the pattern in the factors of the problems above.

b. Describe the pattern in the products. **a–d. See margin**

c. What do you think 15.6×0.0038 equals? Explain your answer.

d. Write a rule for placing the decimal point in a product of two decimals.

Learning the Concept

When you multiply decimals, it is helpful to use estimation to verify the placement of the decimal point.

Example 1 **Solve each equation.**

a. $c = (3.9)(8.2)$ *Estimate: $4 \times 8 = 32$*

```
   3.9    ←  one decimal place
 × 8.2    ←  one decimal place
   7 8
 312 0
 31.98    ←  two decimal places
```

$c = 31.98$ *Compare to the estimate.*

b. $x = (-6.302)(0.81)$ *Estimate: $-6 \times 1 = -6$*

```
  −6.302   ←  three decimal places
  × 0.81   ←  two decimal places
    6302
  504160
−5.10462   ←  five decimal places
```

$x = -5.10462$ *Compare to the estimate.*

Additional Answers Talk About It

a. The numbers are the same, but the decimal point moves one place to the left in the second factor in each successive problem.

b. The numbers are the same, but the decimal point moves one place to the left in each successive problem.

c. 0.05928; The decimal point moved one place to the left from the previous problem.

d. The number of digits after the decimal point in the product is the sum of the number of digits after the decimal point in the factors.

6-5 LESSON NOTES

NCTM Standards: 1-8

Instructional Resources

- Study Guide Master 6-5
- Practice Master 6-5
- Enrichment Master 6-5
- Group Activity Card 6-5
- Assessment and Evaluation Masters, pp. 154, 155
- Activity Masters, p. 6

Transparency 6-5A contains the 5-Minute Check for this lesson; **Transparency 6-5B** contains a teaching aid for this lesson.

Recommended Pacing	
Standard Pacing	Day 6 of 13
Honors Pacing	Day 5 of 12
Block Scheduling*	Day 3 of 7

*For more information on pacing and possible lesson plans, refer to the ***Block Scheduling Booklet***.

1 FOCUS

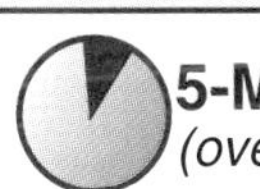

5-Minute Check *(over Lesson 6-4)*

1. Name the multiplicative inverse of $1\frac{2}{5}$. $\frac{5}{7}$

Estimate the solution to each equation. Then solve. Write the solution in simplest form.

2. $x = -6 \div 1\frac{1}{2}$ -4
3. $n = \frac{5}{12} \div \frac{7}{15}$ $\frac{25}{28}$

Evaluate each expression if $m = \frac{1}{3}$ and $n = -\frac{2}{9}$.

4. $m \div n$ $-1\frac{1}{2}$
5. $n \div m^2$ -2

Motivating the Lesson

Situational Problem Present students with these problems: Jessie earns $193.68 in a week for 30.5 hours of work. Find Jessie's hourly pay rate. **$6.35** At that rate, how much would Jessie earn for 37.5 hours of work? **$238.13**

2 TEACH

In-Class Examples

For Example 1
Solve each equation.

a. $d = (-2.3)(-1.9)$ **4.37**

b. $y = (-0.264)(0.025)$ **0.0066**

For Example 2
Solve each equation.

a. $f = 2.73 \div 1.3$ **2.1**

b. $g = 95.04 \div 0.6$ **158.4**

Teaching Tip Have students make index cards naming the property on one side and giving examples, one algebraic and one numeric, on the other side. Students can then use these as study cards and while completing exercises.

Teaching Tip List several decimal numbers on the board or overhead. For each, ask students what they would multiply each number by to make it a whole number.

All the properties that were true for multiplication of integers are also true for multiplication of rationals. The properties of rationals are summarized in the following chart.

Properties of Multiplication	Examples
Commutative Property For all rational numbers x and y, $x \cdot y = y \cdot x$.	$-5(-7) = -7(-5)$ $0.6 \cdot 1.5 = 1.5 \cdot 0.6$ $\frac{5}{6} \cdot \frac{8}{9} = \frac{8}{9} \cdot \frac{5}{6}$
Associative Property For all rational numbers x, y, and z, $(x \cdot y) \cdot z = x \cdot (y \cdot z)$.	$(-6 \cdot 5) \cdot 8 = -6 \cdot (5 \cdot 8)$ $(0.3 \cdot 4) \cdot 5.7 = 0.3 \cdot (4 \cdot 5.7)$ $\left(-\frac{1}{3} \cdot \frac{9}{10}\right) \cdot \frac{2}{5} = -\frac{1}{3} \cdot \left(\frac{9}{10} \cdot \frac{2}{5}\right)$
Identity Property For every rational number x, $x \cdot 1 = x$ and $1 \cdot x = x$.	$-8 \cdot 1 = -8$ $1 \cdot \frac{1}{4} = \frac{1}{4}$ $10.3 \cdot 1 = 10.3$
Inverse Property For every rational number $\frac{x}{y}$, where $x, y \neq 0$, there is a unique number $\frac{y}{x}$ such that $\frac{x}{y} \cdot \frac{y}{x} = 1$.	$6 \cdot \frac{1}{6} = 1$ $-\frac{5}{12}\left(-\frac{12}{5}\right) = 1$ $0.7 \cdot \frac{1}{0.7} = 1$

To divide by a decimal, think of the division as a fraction and write the divisor, or denominator, as a whole number. For example, consider $50 \div 0.26$.

Think of it as $\frac{50}{0.26}$.

$$50 \div 0.26 \rightarrow \frac{50}{0.26} \times \frac{100}{100} = \frac{5000}{26}$$

Since $\frac{100}{100} = 1$, the value of the fraction is unchanged. This becomes a division problem involving whole numbers.

$$\frac{50}{0.26} \rightarrow 0.26\overline{)50.00} \rightarrow 26\overline{)5000.00}$$

```
      192.30
26)5000.00
   26
    240
    234
      60
      52
       80
       78
       20
```

To the nearest tenth, the quotient is 192.3.

Example 2 **Solve $y = 42 \div (-0.8)$.** *Estimate: $42 \div -1 = -42$.*

$0.8\overline{)42.0}$ *Multiply 0.8 and 42 by 10 to get a whole number in the divisor.*

$8\overline{)420}$ with quotient 52.5, or 42 [÷] .8 [+/−] [=] -52.5

$y = -52.5$ *Compare to the estimate.*

GLENCOE Technology

Interactive Mathematics Tools Software

In this interactive computer lesson, students use decimal models to explore multiplication and division of decimals. A **Computer Journal** gives students the opportunity to write about what they have learned.

For Windows & Macintosh

Alternative Learning Styles

Visual Have students shade a 100-unit square to show 0.2 × 0.7. Ask how many squares are shaded and how many there are in all. **14; 100** Have students write a decimal to show the shaded part. **0.14** Then have students write a number sentence. **0.2 × 0.7 = 0.14** Repeat for other pairs of factors.

Decimals are used every day when people work with money.

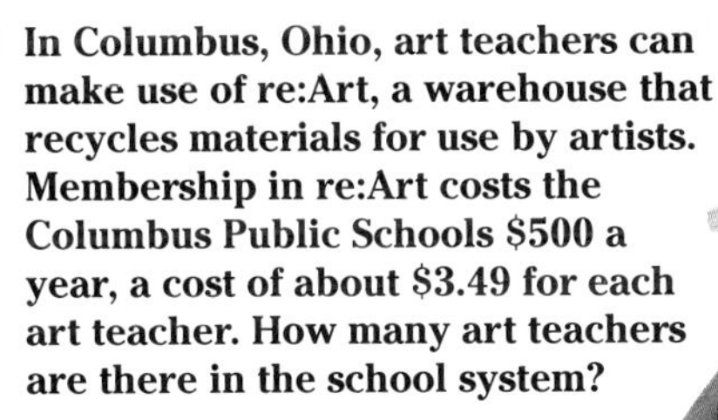

Example 3 APPLICATION Recycling

In Columbus, Ohio, art teachers can make use of re:Art, a warehouse that recycles materials for use by artists. Membership in re:Art costs the Columbus Public Schools \$500 a year, a cost of about \$3.49 for each art teacher. How many art teachers are there in the school system?

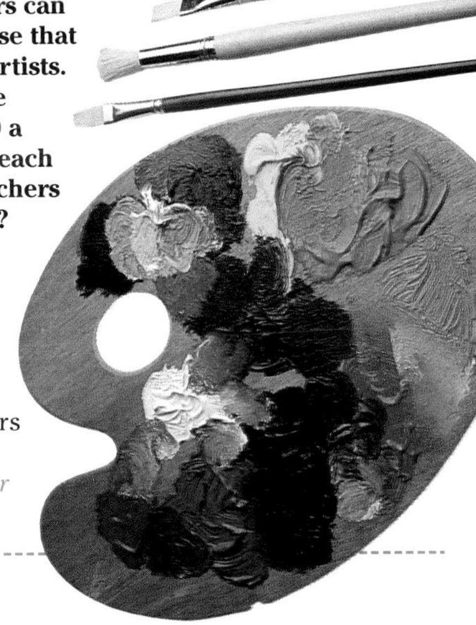

$500 ÷ $3.49 = ?

Estimate: 500 ÷ 4 = 125

500 [÷] 3.49 [=] 143.2664756

There are 143 or 144 art teachers in the system.

Check by comparing your answer to the estimate.

HELP WANTED

Many creative people find fulfillment as professional artists. If you would like to learn more, contact:

National Assn. of Schools of Art and Design
11250 Roger Bacon Dr.
Suite 21
Reston, VA 22090

Connection to Algebra

Evaluate expressions involving division of decimals in the same way you evaluate other algebraic expressions.

Example 4 **Evaluate $\frac{6m}{n}$ if $m = 0.5$ and $n = -3.2$.**

$\frac{6m}{n} = \frac{6(0.5)}{-3.2}$

6 [×] .5 [÷] 3.2 [(-)] [=] -0.9375

The value of the expression is −0.9375.

Checking Your Understanding

Communicating Mathematics

1. **Illustrate** or explain the rule for placement of the decimal point in (2.54)(0.067). **See margin.**
2. What is the first step you should take before you actually begin to multiply or divide decimals? **Find the estimate.**
3. **Decide** where to place a decimal point in each factor so that the product is correct. $493 \times 17 = 8.381$
4. In Lesson 6-5A, you used decimal models to represent multiplication and division problems. Work with a partner to represent these multiplication and division problems. **See Solutions Manual.**
   a. (1.2)(3) b. (1.2)(0.3) c. 4.8 ÷ 4 d. 4.8 ÷ 0.4

3. **Sample answer: 4.93 and 1.7**

MATERIALS

grid paper
colored pencils

Guided Practice

State where the decimal point should be placed in each product or quotient.

5. (7.2)(0.2) = 144 **1.44**
6. (0.06)(3) = 18 **0.18**
7. 0.63 ÷ 0.9 = 7 **0.7**
8. 8.4 ÷ 0.4 = 21 **21**

In-Class Examples

For Example 3

Several classes from Sterling Middle School are going to visit the art museum. There are 148 students and 12 adults going. Each bus can hold 48 passengers. How many buses will be needed to transport the students and adults to the museum? **4**

For Example 4

Evaluate $\frac{9.2a}{c}$, if $a = -5$ and $c = 0.4$. **115**

3 PRACTICE/APPLY

Checking Your Understanding

Exercises 1–17 are designed to help you assess your students' understanding through reading, writing, speaking, and modeling. You should work through Exercises 1–4 with your students and then monitor their work on Exercises 5–17.

Study Guide Masters, p. 49

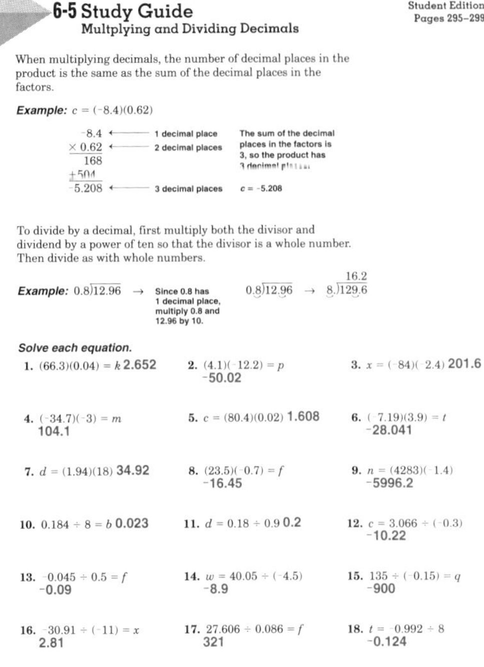

NAME ______________ DATE ________

6-5 Study Guide
Multplying and Dividing Decimals

Student Edition
Pages 295–299

When multiplying decimals, the number of decimal places in the product is the same as the sum of the decimal places in the factors.

Example: $c = (-8.4)(0.62)$

```
  -8.4  ←—— 1 decimal place
× 0.62  ←—— 2 decimal places
   168
 +504
-5.208  ←—— 3 decimal places
```

The sum of the decimal places in the factors is 3, so the product has 3 decimal places.

$c = -5.208$

To divide by a decimal, first multiply both the divisor and dividend by a power of ten so that the divisor is a whole number. Then divide as with whole numbers.

Example: 0.8)12.96 → Since 0.8 has 1 decimal place, multiply 0.8 and 12.96 by 10. 0.8)12.96 → 8.)129.6 = 16.2

Solve each equation.

1. $(66.3)(0.04) = k$ **2.652**
2. $(4.1)(-12.2) = p$ **−50.02**
3. $x = (-84)(-2.4)$ **201.6**
4. $(-34.7)(-3) = m$ **104.1**
5. $c = (80.4)(0.02)$ **1.608**
6. $(-7.19)(3.9) = t$ **−28.041**
7. $d = (1.94)(18)$ **34.92**
8. $(23.5)(-0.7) = f$ **−16.45**
9. $n = (4283)(-1.4)$ **−5996.2**
10. $0.184 \div 8 = b$ **0.023**
11. $d = 0.18 \div 0.9$ **0.2**
12. $c = 3.066 \div (-0.3)$ **−10.22**
13. $-0.045 \div 0.5 = f$ **−0.09**
14. $w = 40.05 \div (-4.5)$ **−8.9**
15. $135 \div (-0.15) = q$ **−900**
16. $-30.91 \div (-11) = x$ **2.81**
17. $27.606 \div 0.086 = f$ **321**
18. $t = -0.992 \div 8$ **−0.124**

HELP WANTED

Students might find it interesting to research Andy Warhol and discover the different media and subjects he used for his artwork.

Additional Answer

1. There will be five places. The first factor has hundredths (2 places), the second factor has thousandths (3 places). 2 + 3 = 5 places.

Assignment Guide

Core: 19–49 odd, 50, 51, 53–62
Enriched: 18–48 even, 50–62
All: Self-Test 1–10

For **Extra Practice**, see p. 754.

The red A, B, and C flags, printed only in the Teacher's Wraparound Edition, indicate the level of difficulty of the exercises.

Additional Answers

18. -0.845
19. 1.4147
20. 8.344
21. -5.5875
22. 0.00709
23. -6.592
24. 90
25. 0.32
26. 50
27. -17
28. 1.7
29. 1460
44. 24; 26.7
45. -2.5; -2.6
46. 0.24; 0.24
47. -10; -10.2
48. 47, 39.0
49. 0.1; 0.2

Solve each equation.

9. $x = (-0.2)(-3.1)$ 0.62
10. $y = (1.2)(-0.05)$ −0.06
11. $s = 27 \div (-0.3)$ −90
12. $t = 0.4 \div 2$ 0.2

Evaluate each expression if $a = 15.7$, $b = 0.4$, and $c = 1.6$.

13. ac 25.12
14. $5c \div b$ 20

Name the property shown by each statement. 16. Commutative, ×

15. $\left(-\frac{5}{6}\right)\left(-\frac{6}{5}\right) = 1$ Inverse, ×
16. $(-4 \cdot 5) \cdot \frac{1}{4} = [5 \cdot (-4)] \cdot \frac{1}{4}$

17. **Recycling** Sun Shares, a recycling company, is paid \$0.65 a pound for aluminum. How many pounds of aluminum would they have to collect a week to meet their weekly phone costs of \$45.50? 70 pounds

Exercises: Practicing and Applying the Concept

Independent Practice

Solve each equation. See margin.

18. $g = (-6.5)(0.13)$
19. $k = (0.47)(3.01)$
20. $-14.9(-0.56) = n$
21. $7.45(-0.75) = t$
22. $x = (0.001)(7.09)$
23. $b = (1.03)(-6.4)$
24. $r = 14.4 \div (0.16)$
25. $q = 0.384 \div 1.2$
26. $-85 \div (-1.7) = k$
27. $-0.51 \div 0.03 = g$
28. $s = -15.3 \div (-9)$
29. $h = 2.92 \div 0.002$
30. Find the product of 7.5 and 0.03. 0.225
31. What is the quotient of 52.64 and 9.4? 5.6

Evaluate each expression.

32. $3y^2$ if $y = 0.4$ 0.48
33. ab if $a = -1.5$, $b = 10$ −15
34. $\frac{5t}{s}$ if $t = 6.2$, $s = 2.5$ 12.4
35. $n^2 \div q$ if $n = 2.2$, $q = 4$ 1.21

36. $12y \div z$ if $y = 9.8$, $z = 6.4$ 18.375
37. $(x + 4)(x - 2)$ if $x = 1.8$ −1.16

Name the property shown by each statement.

38. $\left(2 \cdot \frac{5}{2}\right) \cdot \frac{3}{4} = \frac{3}{4}\left(2 \cdot \frac{5}{2}\right)$ Commutative, ×
39. $-\frac{8}{9} \cdot 1 = -\frac{8}{9}$ Identity, ×
40. $0.6\left(\frac{1}{0.6}\right) = 1$ Inverse, ×
41. $(-6.5 \cdot 9.3) \cdot 2.1 = -6.5 \cdot (9.3 \cdot 2.1)$ Associative, ×
42. $-9.3 = 1 \cdot -9.3$ Identity, ×
43. $(0.6 \cdot 3.2) \cdot 0.3 = (3.2 \cdot 0.6) \cdot 0.3$ Commutative, ×

Calculators

Estimate. Then compute with a calculator. Round quotients to the nearest tenth if necessary. See margin.

44. $(8.01)(3.33) = w$
45. $(-0.56)(4.59) = r$
46. $-8.022(-0.03) = y$
47. $c = 90.5 \div (-8.9)$
48. $f = 93.702 \div 2.4$
49. $z = -0.36 \div (-2.1)$

Practice Masters, p. 49

NAME ____________ DATE ________

6-5 Practice
Multiplying and Dividing Decimals

Student Edition Pages 295–299

Solve each equation.

1. $(0.32)(-1.4) = a$ -0.448
2. $b = (0.52)(4.07)$ 2.1164
3. $c = (0.01)(-15.8)$ -0.158
4. $(-21.04)(-4.2) = d$ 88.368
5. $t = (-8.61)(0.48)$ -4.1328
6. $(-3.2)(2.06) = f$ -6.592
7. $k = 400(-8.15)$ -3260
8. $(2.18)(3.4) = z$ 7.412
9. $(-0.111)(0.12) = p$ -0.01332
10. $2.413 \div (-0.019) = a$ -127
11. $b = -40.3 \div (-0.62)$ 65
12. $c = 0.3936 \div 4.8$ 0.082
13. $d = -0.672 \div (-67.2)$ 0.01
14. $0.2208 \div (-3) = k$ -0.0736
15. $-35 \div (-2.5) = q$ 14

Evaluate each expression.

16. $2m^2$ if $m = 0.6$ 0.72
17. xy if $x = 5.3$, $y = -4$ -21.2
18. $4c \div d$ if $c = 0.9$, $d = 1.2$ 3
19. $\frac{b}{k}$ if $b = 16.4$, $k = 1.6$ 10.25
20. $\frac{8h}{g}$ if $h = 3.8$, $g = 0.76$ 40
21. n^2w if $n = 1.1$, $w = 12.3$ 14.883

Group Activity Card 6-5

Disturbing Decimals Group Activity **6-5**

MATERIALS: Five dice • one coin

The object of this activity is to create two decimal numbers that when multiplied together will yield the greatest or the least product.

To begin, one person flips the coin. If the coin lands heads, the object is to form the greatest product. If the coin lands tails, the object is to form the least product.

Now another person rolls the dice. Each person in the group uses the numbers on the dice to form a three-digit number and a two-digit number. Each number must have a tenths digit.

Lastly, calculate the product. The person with the greatest or least product depending on the flip of the coin scores one point. The first person to score eight points wins the game.

 Pre-Algebra

Reteaching

Using Groups of Two Before reteaching, determine exact causes of errors. Students may have trouble with the numerical division process, placing the point, or rounding. Ask students to work a problem while you watch. Reteach, then pair up students and have them verbalize each step.

Critical Thinking

50. a. Name two decimals whose sum is 2.4 and whose product is 1.44.

b. Name two decimals whose sum is 2.7 and whose quotient is 2. **See margin.**

Applications and Problem Solving

51. Consumer Economics A 13.5 ounce box of muesli cereal costs \$2.45, and a 19.5 ounce box of frosted corn flakes costs \$3.59. Which costs less per ounce? **See margin.**

52. 67,567.5 pesos

52. Economics When Lilia's softball team visited Mexico, \$1 in American money could be exchanged for 2702.7 pesos. If Lilia exchanged \$25 in American money, how many pesos did she receive?

53. Sports American consumers spent \$0.8 billion dollars for golf equipment in a recent year. That same year they spent four times that much on athletic shoes. How much did consumers spend on athletic shoes? **\$3.2 billion**

Mixed Review

54. $\frac{5}{11} \cdot \frac{11}{10} = \frac{1}{2}$

54. Write $\frac{5}{11} \div \frac{10}{11}$ as a multiplication expression. Then find its value. (Lesson 6-4)

55. Solve $3\frac{4}{9} \cdot \frac{3}{4} = h$. (Lesson 6-3) **$2\frac{7}{12}$**

56. Solve $b - 1.6 \leq 4.3$. (Lesson 5-7) **$b \leq 5.9$**

57. integers, rationals

57. Name the set(s) of numbers to which -10 belongs. (Lesson 5-1)

58. Solve $8 = -\frac{b}{11}$. (Lesson 3-3) **-88**

59. Banking Account #8234620 had an opening balance of \$3245. Three checks cleared on Tuesday, leaving a balance of \$2931. What was the total of the checks? (Lesson 3-2) **\$314**

60. Meteorology The *heat index* is an estimate of the warming effect that humidity has on a person in hot weather. If the outside temperature is 86°F and the humidity makes it feel like 93°F, how much warmer does the humidity make it feel? (Lesson 2-5) **7°F**

61. commutative, +

61. Name the property shown by $3 + a = a + 3$. (Lesson 1-4)

62. Find the value of $18 \div (18 - 6 \cdot 2)$. (Lesson 1-2) **3**

Self Test

Express each fraction as a decimal. Use a bar to show a repeating decimal. (Lesson 6-1)

1. $\frac{7}{8}$ **0.875** **2.** $-1\frac{2}{3}$ **$-1.\overline{6}$** **3.** $-\frac{8}{11}$ **$-0.\overline{72}$**

4. Business On Tuesday, Motorola stock fell $2\frac{1}{8}$ points from the previous day's closing price. If the closing price on Monday was \$75.50, what was the closing price, in dollars, on Tuesday? (Lesson 6-1) **\$73.38**

Estimate each product or quotient. (Lesson 6-2)

5. $\frac{2}{3} \times 25$ **16** **6.** 3.85×6 **24** **7.** $6.2 \div 2.1$ **3**

Solve each equation. Write each solution in simplest form. (Lessons 6-3, 6-4, and 6-5)

8. $x = \frac{1}{4}\left(-\frac{3}{8}\right)$ **$-\frac{3}{32}$** **9.** $y = -9 \div \frac{3}{8}$ **-24** **10.** $w = (0.5)(1.46)$ **0.73**

4 ASSESS

Closing Activity

Modeling Write decimal division problems with blanks or "x"s for numbers, such as xx.x)x.xxx. Then use number cards 0–9 to fill in the blanks and solve.

Chapter 6, Quiz B (Lessons 6-4 and 6-5) is available in the *Assessment and Evaluation Masters*, p. 155.

Mid-Chapter Test (Lessons 6-1 through 6-5) is available in the *Assessment and Evaluation Masters*, p. 154.

Additional Answers

50a. 1.2, 1.2

50b. 1.8, 0.9

51. Muesli; muesli 18.1¢ per oz; frosted corn flakes 18.4¢ per oz

Enrichment Masters, p. 49

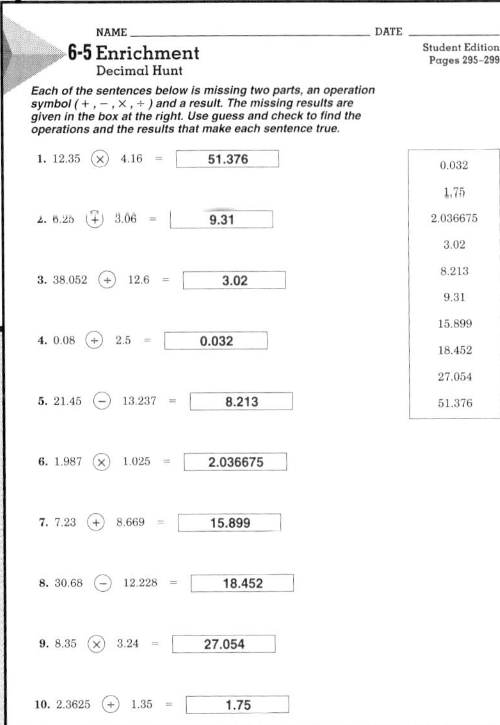

NAME ______ DATE ______

6-5 Enrichment Student Edition Pages 295–299

Decimal Hunt

Each of the sentences below is missing two parts, an operation symbol (+, −, ×, ÷) and a result. The missing results are given in the box at the right. Use guess and check to find the operations and the results that make each sentence true.

1. 12.35 (×) 4.16 = 51.376
2. 6.25 (+) 3.06 = 9.31
3. 38.052 (÷) 12.6 = 3.02
4. 0.08 (÷) 2.5 = 0.032
5. 21.45 (−) 13.237 = 8.213
6. 1.987 (×) 1.025 = 2.036675
7. 7.23 (+) 8.669 = 15.899
8. 30.68 (−) 12.228 = 18.452
9. 8.35 (×) 3.24 = 27.054
10. 2.3625 (÷) 1.35 = 1.75

0.032
1.75
2.036675
3.02
8.213
9.31
15.899
18.452
27.054
51.376

Self Test

The Self Test provides students with a brief review of the concepts and skills in Lessons 6-1 through 6-5. Lesson numbers are given to the right of exercises or instruction lines so students can review concepts not yet mastered.

Extension

Using Consumer Connections Have students write problems giving total cost and number of units purchased and ask for the unit cost. For example, Mark buys 2.83 pounds of steak for \$16.95. What is the cost per pound? **\$5.99** Have students use familiar purchases and make the problems realistic. Trade and solve.

6-6A LESSON NOTES

NCTM Standards: 1-5, 10, 12

Objective

Explore the mean, median, and mode using real data.

Recommended Time

Demonstration and discussion: 15 minutes; Exercises: 30 minutes

Instructional Resources

For each student or group of students
- small boxes of raisins
- calculator

Math Lab and Modeling Math Masters
- p. 38 (worksheet)

For teacher demonstration
Overhead Manipulative Resources

1 FOCUS

Motivating the Lesson

Inform the class that this activity involves having each group collect its own data by counting the number of raisins in a box. Then each group shares information with all the other groups to determine the measures of central tendency.

2 TEACH

Teaching Tip Caution groups to record all the data including their own. Make sure that the number of groups and the number of data match so they do not count one group twice or forget to include a group. Groups could be identified by letters to assure the correct one-to-one correspondence. Encourage groups to record the data in tables.

MATHEMATICS LAB **6-6A**

HANDS-ON ACTIVITY

Mean, Median, and Mode

A Preview of Lesson **6-6**

MATERIALS
- small boxes of raisins
- calculator

Measures of central tendency are numbers that represent a set of data. Researchers use these numbers when analyzing data.

Your Turn

Work in small groups.

- Get a small box of raisins for your group from your teacher.
- Estimate the number of raisins you think will be in the box.
- Count the number of raisins and record the number.
- Record the data for each group.
- Use a calculator to add the data. Then divide the sum by the number of groups. This number is called the **mean** of the data.
- List the data in order and mark the middle number. If there is no middle number in the set of data, circle the middle two numbers and find their mean. This number is called the **median**.
- If a number appears more often than the others in the data, this number is called the **mode**. **1–8. See students' work.**

TALK ABOUT IT

1. If you add the number 5 to your group's set of data, predict which measure of central tendency will be affected the most.
2. Find the mean, median, and mode of the new set of data.
3. Which measure of central tendency will be most representative of the data before adding 5? after adding 5?
4. If you add the number 400 to your group's set of data, predict which measure of central tendency will be affected the most.
5. Find the mean, median, and mode of the new set of data.
6. Which measure of central tendency will be most representative of the data after adding 400?

Extension

7. Suppose you count the number of raisins in a large box. If you add this number to your set, will it affect the mean? the median? the mode?
8. Choose your own numbers to add to your set of data. Study the effects on the mean, median, and mode.

3 PRACTICE/APPLY

Assignment Guide
Core: 1–8
Enriched: 1–8

4 ASSESS

Observing students working in cooperative groups is an excellent method of assessment.

6-6 Integration: Statistics
Measures of Central Tendency

Setting Goals: *In this lesson, you'll use the mean, median, and mode as measures of central tendency.*

Modeling a Real-World Application: Meteorology

"It's hot fun in the summertime today, folks! The high will reach a scorching 92 degrees, way above the normal of 84 for this date."

You have heard meteorologists report the normal high temperature for a given date. The normal high temperature is determined by averaging the high temperatures for a given date over several years. The high temperatures on March 1 for twelve consecutive years are recorded below. What is the normal high temperature for the date?

1985	1986	1987	1988	1989	1990	1991	1992	1993	1994	1995	1996
59°F	50°F	48°F	43°F	40°F	46°F	50°F	53°F	58°F	61°F	64°F	73°F

Learning the Concept

To analyze sets of data, researchers often try to find a number that can represent the whole set. These numbers or pieces of data are **measures of central tendency**. Three that we will study are the **mean**, the **median**, and the **mode**.

The **mean** is what people usually are talking about when they say average. It is the arithmetic average of the data. For the temperatures above, the mean is

$$\frac{59 + 50 + 48 + 43 + 40 + 46 + 50 + 53 + 58 + 61 + 64 + 73}{12} = \frac{645}{12} \text{ or } 53.75.$$

The mean is about 53.8°.

Notice that the mean may not be a member of the set of data.

Definition of Mean — **The mean of a set of data is the sum of the data divided by the number of pieces of data.**

The **median** is the middle number when the data are in order. Consider two cases.

6-6 LESSON NOTES

NCTM Standards: 1–5, 10

Instructional Resources

- Study Guide Master 6-6
- Practice Master 6-6
- Enrichment Master 6-6
- Group Activity Card 6-6
- Multicultural Activity Masters, p. 12
- Real-World Applications, 12

Transparency 6-6A contains the 5-Minute Check for this lesson; **Transparency 6-6B** contains a teaching aid for this lesson.

Recommended Pacing	
Standard Pacing	Day 8 of 13
Honors Pacing	Day 7 of 12
Block Scheduling*	Day 4 of 7

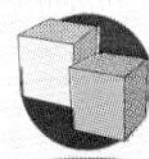

*For more information on pacing and possible lesson plans, refer to the ***Block Scheduling Booklet***.

1 FOCUS

5-Minute Check
(over Lesson 6-5)

Solve each equation.

a. $a = (1.2)(6)$ **7.2**

b. $(-1.64)(0.05) = b$ **−0.082**

c. $c = 105.4 \div 0.62$ **170**

d. $-7.56 \div 2.1 = d$ **−3.6**

e. $e = 13.10355 \div 5.55$ **2.361**

Motivating the Lesson

Questioning Have students describe the types of data that would be used for each type of survey. **Sample answers given.**

- How many people are in your family? **whole numbers**
- How tall are you? **measurements in feet and inches or in centimeters**
- How old are you? **rational numbers**
- What is your favorite color? **words that name colors**

Teaching Tip Ask students whether it is safe to dive into a lake that has an average depth of 9 feet. **No** Have them explain their reasoning.

Teaching Tip Pose this situation to students: Suppose that for this grading period I were to allow you to choose whether I should use the mean, the median, or the mode of your test and quiz scores to determine your grade. Which would you choose and why?

Case 1: An Odd Number of Data

Write the set of data in order from least to greatest. Consider the set of data below.

8 12 22 22 25 34 35 36 55

The middle number is 25.

Case 2: An Even Number of Data

If the number of data is even, there are two middle numbers. In this case, the median is the mean of the two numbers. Look back at the set of temperatures.

40 43 46 48 50 50 53 58 59 61 64 73

$$\frac{50 + 53}{2} = \frac{103}{2} \text{ or } 51.5$$

The median is 51.5°.

If there is an even number of data in a set, the median may not be a member of the data set.

Definition of Median — **The median is the number in the middle when the data are arranged in order. When there are two middle numbers, the median is their mean.**

The number or item in a data set that appears most often is called the **mode**. There can be one mode, more than one mode, or no mode in a data set.

In the temperatures above, the temperature 50° appears more times than any other (two times), so 50° is the mode.

The mode, if there is one, is always a member of the set of data. If there is no one piece of data that appears more often than the rest, the set of data has no mode.

Definition of Mode — **The mode of a set of data is the number or item that appears most often.**

Interactive Mathematics Tools Software

In this interactive computer lesson, students explore measures of central tendency. A **Computer Journal** gives students the opportunity to write about what they have learned.

For Windows & Macintosh

Tech Prep

Statistician Statisticians analyze and track statistics in many varied businesses, including professional sports. Oakland A's manager Tony LaRussa bases many game decisions on statistics provided by the team's statistician. For more information on tech prep, see the *Teacher's Handbook*.

Example 1

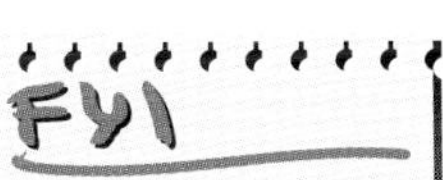

Social Studies

The table below shows the populations of several American Indian tribes as of the 1990 U.S. census. Organize the data and then find the mean, median, and mode.

Tribe	Population (thousands)
Apache	50
Cherokee	308
Chippewa	104
Choctaw	82
Creek	44
Iroquois	49
Lumbee	48
Navajo	219
Pueblo	53
Sioux	103

Median List the populations in order from greatest to least. The median is the middle number. There are 10 populations listed, so the average of the 5th and 6th populations is the median.

44 48 49 50 53 82 103 104 219 308

The median population is $\frac{53 + 82}{2}$ or 67.5 thousand.

Mode Since no population occurs more than the rest, there is no mode.

Mean To find the mean, add the populations and divide by 10.

$$\frac{50 + 308 + 104 + 82 + 44 + 49 + 48 + 219 + 53 + 103}{10}$$

$$= \frac{1060}{10} \text{ or } 106$$

The mean population is 106 thousand.

FYI

In 1985, Wilma Mankiller was elected chief of the Cherokee Nation. As chief, she placed a high priority on the economic growth of the nation.

You can use the mean, median, and mode to compare two different sets of data.

Example 2

Entertainment

Nielsen Media Research monitors the popularity of television programs and assigns a rating to each one. The Nielsen ratings for programs airing at 8:00 P.M. on CBS and NBC for each day of the week of July 3, 1995 are shown at the right.

CBS	NBC
7.2	6.1
5.5	5.7
7.1	4.1
8.4	9.6
5.6	5.6
5.1	5.3
9.9	12.5

a. Find the mean, median, and mode of the ratings for each network.

Begin by putting the data in order.

CBS	5.1	5.5	5.6	7.1	7.2	8.4	9.9
NBC	4.1	5.3	5.6	5.7	6.1	9.6	12.5

Median Because there are 7 pieces of data in each set, the medians are the 4th piece of data.

(continued on the next page)

2 TEACH

In-Class Examples

For Example 1

Suppose your quiz grades for one marking period were 80, 83, 99, 92, 95, 90, 80, 74, 80, 100. Organize the data and then find the mean, median, and mode. **mean, 87.3; median, 86.5; mode, 80**

For Example 2

Arejay Sports Store and Super Sports Store each have 5 full-time employees. Below is a list of the monthly salaries of the employees. Arejay Sports: $1290, $1400, $1400, $1600, $3650.
Super Sports: $1400, $1450, $1550, $1600, $2000.

a. Find the mean, median, and mode for each set of data. **Arejay Sports: mean, $1868; median, $1400; mode, $1400 Super Sports: mean, $1600; median, $1550; no mode**

b. Compare the averages to determine which store pays its employees better. **The extreme salary of 1 employee for Arejay Sports makes its mean higher. However, with the exception of that one person, the employees at Super Sports are better paid.**

FYI:

In the early 1800's, Chief Sequoyah developed an alphabet for the Cherokee and taught his people to read and write. The Cherokee nation had its own constitution and government.

3 PRACTICE/APPLY

Checking Your Understanding

Exercises 1–8 are designed to help you assess your students' understanding through reading, writing, speaking, and modeling. You should work through Exercises 1–3 with your students and then monitor their work on Exercises 4–8.

Additional Answer

1. Sample Answer: The mean of 7.0 for both networks best represents the data. It shows that, overall, the ratings for the two networks were about the same.

CBS The median is 7.1.

NBC The median is 5.7.

Mode None of the pieces of data occur more than once in either set of data. Thus, neither set has a mode.

Mean The means are the sums of the data divided by 7.

CBS $\frac{5.1 + 5.5 + 5.6 + 7.1 + 7.2 + 8.4 + 9.9}{7} = \frac{48.8}{7}$ or about 7.0

NBC $\frac{4.1 + 5.3 + 5.6 + 5.7 + 6.1 + 9.6 + 12.5}{7} = \frac{48.9}{7}$ or about 7.0

b. Compare the averages to determine whether CBS or NBC programs rated better for the week.

The median rating for CBS programs was higher than for NBC programs. The means for the two networks were about the same. It appears that the popularity of the CBS programs was more consistent. NBC had more programs that were rated very high or very low.

Checking Your Understanding

Communicating Mathematics

Read and study the lesson to answer these questions.

1. Which "average" do you think best represents the data in Example 2? Explain your answer. **See margin.**
2. **Write** a set of data that has a mean of 7. Is the median 7 also? **Sample answers: 6 7 7 8; yes; 5 6 8 9, no**
3. Complete this activity with a group of three or four students. Arrange the 50 pennies in a roll in order from oldest to newest.
   a. Find the mean, median, and mode for the year the coins were minted. **Answers will vary.**
   b. Based on your sample of pennies, what can you say about all pennies in circulation?
   c. Compare your results with the rest of the class and check your conclusion in **part b.** **Same as b.**
   3b. Recently minted coins should have greater number.

MODELING MATHEMATICS

MATERIALS

roll of 50 pennies

Guided Practice

List the data in each set from least to greatest. Then find the mean, median, and mode. When necessary, round to the nearest tenth.

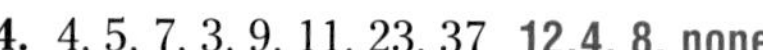

4. 4, 5, 7, 3, 9, 11, 23, 37 **12.4, 8, none**
5. 11, 45, 62, 12, 47, 8, 12, 35, 33 **29.4, 33, 12**
6. 25.98, 30.00, 45.36, 25.00, 45.36 **34.3, 30.0, 45.4**
7. 105, 116, 125, 78, 78 **100.4, 105, 78**
8. **Health** If 50 males have a resting heart rate of 59 beats a minute, 30 males have 72, 15 males have 84, and 5 males have 93, what is the mean resting heart rate for all 100 males represented? **68.35**

Study Guide Masters, p. 50

NAME ______________________ DATE __________

6-6 Study Guide
Integration: Statistics
Measures of Central Tendency

Student Edition Pages 301–306

Member	Height in Inches
JC	57
KL	60
MW	61
NL	62
PK	62
LV	70

The heights of the school ensemble members are listed at the left.

The **mean** height is the sum of all the heights divided by the number of addends.

$\frac{57 + 60 + 61 + 62 + 62 + 70}{6} = \frac{372}{6}$ or 62

The mean is 62.

The **median** height is the middle number when the data are listed in order. Since there are two middle numbers, 61 and 62, the median is the mean of these two. The median is 61.5.

The **mode** is the height that appears most often. The mode is 62.

List each set of data from least to greatest. Then find the mean, median, and mode. When necessary, round to the nearest tenth.

1. 98, 63, 51, 52, 99, 57, 54, 99 **51, 52, 54, 57, 63, 98, 99, 99; 60, 71.6, 99**
2. 69, 68, 65, 64, 68, 69, 68, 67 **64, 65, 67, 68, 68, 68, 69, 69; 68, 67.3, 68**
3. 73, 75, 71, 69, 72, 71, 73, 71 **69, 71, 71, 71, 72, 73, 73, 75; 71.5, 71.9, 71**
4. 14, 11, 12, 13, 14, 15, 16, 13, 12, 13 **11, 12, 12, 13, 13, 13, 14, 14, 15, 16; 13, 13.3, 13**

Solve.

5. Jim's math quiz scores were 87, 79, 100, 83, and 88. Find the mean of his scores. **87.4**
6. The high temperatures for a week in May were 68, 70, 68, 66, 70, 74, and 72. Find the median, mean, and mode. **70, ≈ 69.7, 68 and 70**

Reteaching

Using Models Give students an odd number of different-colored counters. Have students count the number of each color and list the number of each color. The data should be listed vertically from least to greatest. Ask students what number in their list occurred most often. Ask what this number represents. **the mode** Have students cross out the top and bottom numbers on their lists. Continue this process until one number is left. Ask students what this number represents. **the median**

Exercises: Practicing and Applying the Concept

Independent Practice

Find the mean, median, and mode for each set of data. When necessary, round to the nearest tenth.

9. 78, 45, 32, 64, 22, 63, 45 **49.9, 45, 45**

10. 22, 63, 53, 26, 41, 44, 12, 26 **35.9, 33.5, 26**

11. 3.6, 7.2, 9.0, 5.2, 7.2, 6.5, 3.6 **6.0, 6.5, 3.6 and 7.2**

12. 0.6, 0.7, 1.4, 0.9, 0.2, 0.7, 0.7, 1.4 **0.8, 0.7, 0.7**

13. 100, 113, 94, 86, 113, 120, 105 **104.4, 105, 113**

14. 2.3, 3.6, 4.1, 3.6, 2.9, 3.0 **3.3, 3.3, 3.6**

15. 0.4, 1.6, 0.8, 0.9, 0.7, 1.1 **0.92, 0.85, none**

16. 3.4, 5.6, 1.2, 4.8, 8.8, 4.5, 3.4, 7.6 **4.9, 4.7, 3.4**

Use the following spreadsheet to answer exercises 17–25.

Test Scores						
	A	B	C	D	E	F
1	Student	Test 1	Test 2	Test 3	Test 4	Test 5
2	Colleen	93	86	84	92	85
3	Lakesha	83	85	93	78	86
4	Bryan	88	92	73	78	86
5	Phongsava	90	88	83	76	90
6	Raul	74	90	86	89	88

17. What is the mean of all the test scores? **85.4**

18. What is the median of all the test scores? **86**

19. What is the mode of all the test scores? **86**

20. a. What is Colleen's mean on the five tests? **88**

b. How does Colleen's mean affect the class mean? **It raises it.**

21. a. What is Colleen's median on the first five tests? **86**

b. How does Colleen's median affect the class median? **no effect**

22. a. What is Phongsava's mean on the five tests? **85.4**

b. How does Phongsava's mean affect the class mean? **no effect**

23. a. Find Phongsava's mode on the first five tests. **90**

b. How does Phongsava's mode affect the class mode? **no effect**

24. If test 4 is removed, how will this change the mean? **It will be raised.**

25. The lowest average grade that receives a B on a report card is 86. If Lakesha wanted a B after the sixth test, what grade does she need? **91**

26. See margin.

Critical Thinking

26. Several years ago, the mayor of Glasshouse City quit and moved to the city of Stonethrowers. An elated citizen of Glasshouse City was heard to say that the move raised the average IQ in both cities. Suppose that the average referred to is the mean. Explain how the statement could be true.

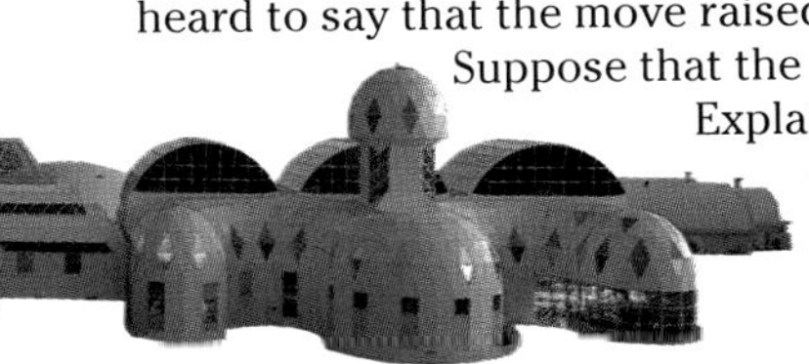

27. Construct a set of data with at least seven different items in which the mean, median, and mode are the same number. **See margin.**

Error Analysis

Students may use the terms mean, median, and mode interchangeably and incorrectly. Have students make up study index cards with the word and its definition on one side and an example on the other side.

Assignment Guide

Core: 9–25 odd, 26, 27, 29, 31–36

Enriched: 10–24 even, 26–36

For **Extra Practice**, see p. 754.

The red A, B, and C flags, printed only in the Teacher's Wraparound Edition, indicate the level of difficulty of the exercises.

Additional Answers

26. The mayor's IQ was less than the mean IQ in Glasshouse City and greater than the mean IQ in Stonethrowers.

27. Sample answer: 2, 4, 6, 8, 8, 10, 12, 14

Practice Masters, p. 50

NAME ______________________ DATE ____________

6-6 Practice
Integration: Statistics
Measures of Central Tendency

Student Edition
Pages 301–306

Find the mean, median, and mode for each set of data. When necessary, round to the nearest tenth.

1. 2.5, 2.4, 2.9, 2.7, 2.4, 2.3, 2.4, 2.9, 2.3, 2.4 **2.5; 2.4; 2.4**

2. 1, 5, 8, 3, 10, 7, 8, 10, 3, 8, 6, 3, 4, 9 **6.1; 6.5; 3 and 8**

3. 70, 85, 90, 65, 70, 85, 100, 60, 55, 95, 85, 70, 75 **77.3; 75; 70 and 85**

4. 80, 70, 85, 90, 75, 75, 90 **80.7; 80; 75 and 90**

5. 7.0, 6.3, 7.5, 6.4, 8.9, 5.4, 7.9, 6.8 **7.0; 6.9; none**

6. 5, 7, 7, 9, 10, 10, 12 **8.6; 9; 7 and 10**

Use the data at the right to answer Exercises 7–12.

Weights of Students in Class	
Name	Weight (kg)
Malissa	49
Marco	60
Tyrill	58
Gerd	73
Cierra	67
Maria	60
Dillon	63
Serena	60
Kelly	64
Amanda	68
Jason	60

7. What is the mode? **60**

8. What is the mean? **62**

9. What is the median? **60**

Suppose Sonya enrolls in the class and her weight is 51 kg. Without computing, answer these questions.

10. How will Sonya affect the new mean? **The new mean will be lower.**

11. How will Sonya affect the new median? **The median will not be affected, it will remain the same.**

Suppose Hector now joins the class. His weight is 70 kg.

12. After both Sonya and Hector join the class, what are the new mode, mean, and median? **60; 61.8; 60**

Group Activity Card 6-6

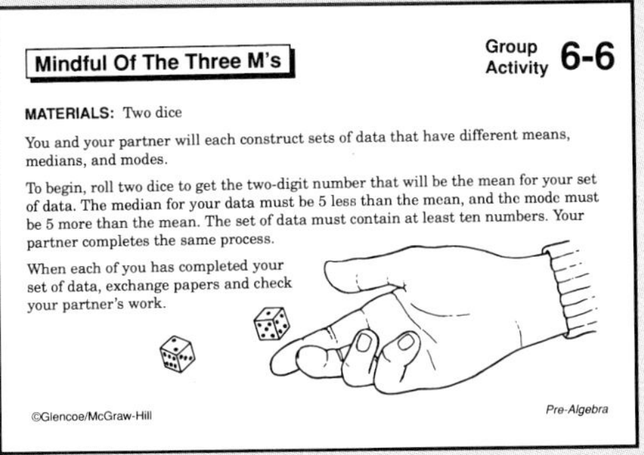

Mindful Of The Three M's Group Activity **6-6**

MATERIALS: Two dice

You and your partner will each construct sets of data that have different means, medians, and modes.

To begin, roll two dice to get the two-digit number that will be the mean for your set of data. The median for your data must be 5 less than the mean, and the mode must be 5 more than the mean. The set of data must contain at least ten numbers. Your partner completes the same process.

When each of you has completed your set of data, exchange papers and check your partner's work.

 Pre-Algebra

4 ASSESS

Closing Activity

Modeling Survey the number of hours of television each student watched the night before and list the data on the board or overhead. Then have students find the mean, median, and mode of that set of data.

Additional Answers

28a. 1983: mean, 90.6; median, 91.4; no mode
1992: mean, 96.5; median, 94.6; mode, 94.6

28b. Sample answers: The situation has improved from 1983 to 1992, but women still earn less than men for the same work.

Enrichment Masters, p. 50

NAME ______________________ DATE ____________

6-6 Enrichment
Mean Variation

Student Edition
Pages 301–306

Mean variation is the average amount by which the data differ from the mean.

Example: The mean for the set of data at the right is 17. 12, 16, 27, 16, 14

Find the mean variation as follows.

Step 1 Find the difference between the mean and each item in the set.

$17 - 12 = 5$
$17 - 16 = 1$
$27 - 17 = 10$
$17 - 16 = 1$
$17 - 14 = +3$
$\quad 20$

Step 2 Add the differences.

Step 3 Find the mean of the differences. This is the mean variation. $5\overline{)20}$ = 4

The mean variation is 4.

Find the mean variation.

1. 100, 250, 200, 175, 300
The mean is 205.
$205 - 100 =$ **105**
$250 - 205 =$ **45**
$205 - 200 =$ **5**
$205 - 175 =$ **30**
$300 - 205 =$ **95**
Add the differences above.
Then, divide by 5.
The mean variation is **56**.

2. 124, 128, 121, 123
The mean is 124.
$124 - 124 =$ **0**
$128 - 124 =$ **4**
$124 - 121 =$ **3**
$124 - 123 =$ **1**
Add the differences above.
Then, divide by 4.
The mean variation is **2**.

For each set of data, find the mean and the mean variation.

3. 68, 43, 28, 25 **41; 14.5**

4. 13, 18, 22, 28, 35, 46 **27; $9\frac{1}{3}$**

5. 68, 25, 36, 42, 603, 16, 8, 18 **102; 125.25**

6. 79, 81, 85, 80, 78, 86, 84, 83, 75, 88 **81.9; 3.3**

Applications and Problem Solving

28. Employment The table at the right shows the earnings for women for every $100 earned by a man in the same occupation for 1983 and 1992.

Occupation	1983	1992
Nurses	99.5	104.7
Teachers	88.6	90.3
Police	91.2	94.2
Food service	102.5	105.6
Packagers	91.6	94.6
Secretaries	76.7	91.6
Orderlies	81.0	96.0
Postal clerks	93.4	94.6

a. Find the mean, median, and mode of the earnings for each year. **See margin.**

b. Compare the averages for the two years. What might you conclude about the changes in women's employment situations from 1983 to 1992? Explain. **See margin.**

29. Family Activity For one hour while you are watching television or listening to the radio with your family, record the length of each commercial break. Find the mean, median, and mode. Compare your results with three other people in your class. Find the mean, median, and mode for the four sets of data. Write about your findings. **See students' work.**

30. Business The 1994 revenues of the top ten money-making businesses that are owned by women are listed in the chart below. **30a. $971.90, $807.50, none**

a. What are the mean, median, and mode revenues for the ten companies?

b. Which measure of central tendency would complete this statement correctly? "There are more companies below the ______ than above it." **mean**

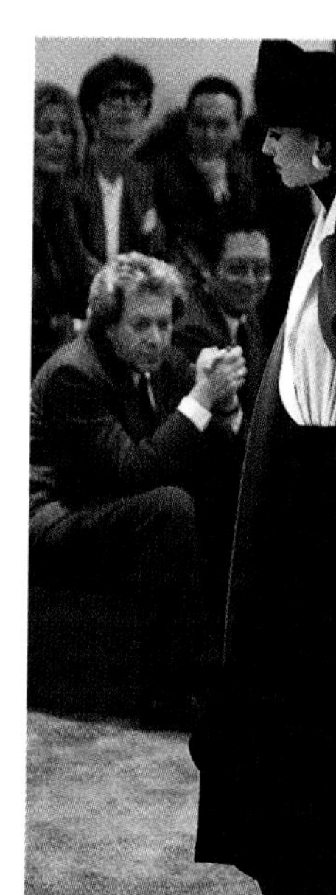

Rank	Company	Revenues*
1	TLC Beatrice	$1820
2	Raley's	$1800
3	Roll International	$1400
4	Little Caesar Enterprises	$1000
5	Axel Johnson	$815
6	Minyard Food Store	$800
7	Warnaco	$789
8	Donna Karan	$465
9	Jocket International	$450
10	Copley Press	$380

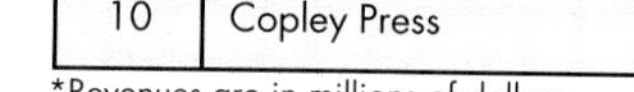

*Revenues are in millions of dollars.

Mixed Review

31. Solve $w = (-3.2)(4.057)$. (Lesson 6-5) **−12.9824**

32. Write $\frac{36a^3b}{52ab^2}$ in simplest form. (Lesson 4-6) $\frac{9a^2}{13b}$

33. Solve $-3y \geq 72$. (Lesson 3-7) $y \leq -24$

34. Oceanography A submarine at 2200 feet below sea level descends another 1325 feet. How far below sea level is the submarine? (Lesson 2-4) **3525**

35. Name the property shown by $3 \cdot 0 = 0$. (Lesson 1-4) **Mult. Prop. of 0**

36. Nutrition You can find the number of calories from fat in a food item by multiplying the number of grams of fat by nine. (Lesson 1-3)

a. Write an expression for the relationship. **36a. 9g 36b. 45**

b. Find the number of calories from fat in a snack with 5 grams of fat.

Extension

Work Backward Give an example of a set of data that has a mean of 5. Give several other examples, such as data with a mean of 100 or $5.50. Working in small groups, have one student give a mean and other students try to guess what the survey question might have been.

GRAPHING CALCULATOR ACTIVITY

6-6B Finding Mean and Median

An Extension of Lesson **6-6**

A graphing calculator is a powerful computer. You can use a graphing calculator to find the mean and median of a set of data.

Activity

Find the mean and median of the bicycle helmet prices (in dollars) shown below.

51 40 58 60 30 45 66 40 87 65 41 60 40 35 47 49 54 50 52 47

Begin by clearing list L1.

Enter: STAT 4 2nd L1 ENTER

Then enter the prices into list L1.

Enter: STAT ENTER 51 ENTER 40 ENTER . . . 52 ENTER 47 ENTER

Now compute the values.

Enter: STAT ▶ ENTER ENTER

The TI-82 displays many statistical values at one time. The first value, $\bar{x}$, is the mean. The calculator found the mean to be 50.85 or $50.85.

Use the ▼ key to scroll through the list to locate "Med". The value displayed is the median. The median is 49.5 or $49.50.

Your Turn

3. −16.36, −15

Clear list L1 and find the mean and median of each data set with a graphing calculator. Round to the nearest hundredth. 4. 0.29, 1.35

1. 6.4, 5.0, 7.3, 1.2, 5.7, 8.9, 3.9, 5.8, 6.0, 4.7 5.55, 5.75
2. 123, 423, 190, 289, 99, 178, 156, 217, 217 210.22, 190
3. −23, −13, −16, −21, −15, −34, −4, −6, −11, −15, −22
4. −8.4, 2.2, 7.3, −5.3, 6.7, −4.3, 5.1, 1.3, −1.1, −3.2, 2.2, 2.9, −2.8, 1.4

5. **Geography** The table at the right shows the populations per square mile in thousands for different regions as of 1994. Find the mean and median of the data. 73.6, 49.7

Region	Population
N. America	30.7
Latin America	68.7
Europe	133.9
Asia	192.2
Oceania	16.2
Antarctica	0

TALK ABOUT IT

6. when there is an odd number of data pieces

6. Look back at the medians that you found. When is the median a member of the data set?
7. When does the median best represent the data? When does the mean best represent the data? Explain. See Solutions Manual.

Technology

This lesson offers an excellent opportunity for using technology in your pre-algebra classroom. For more information on using technology, see *Graphing Calculators in the Mathematics Classroom*, one of the titles in the Glencoe Mathematics Professional Series.

6-6B LESSON NOTES

NCTM Standards: 1-5, 10, 12

Objective

Use a graphing calculator to explore the mean and median of given data.

Recommended Time

15 minutes

Instructional Resources

Graphing Calculator Masters, p. 20

This master provides keystroking instruction for this lesson for the TI-81 and Casio graphing calculators.

1 FOCUS

Motivating the Lesson

Write a long list of numbers and ask students to explain why they would or would not use a calculator to find the median, the mean, and the mode. Lead students to realize at times a calculator is useful, but at other times its use is too time consuming.

2 TEACH

Teaching Tip Students may try to enter a new set of data in the second column and ignore the instruction to clear list L1 first. Tell students that the setup must be changed in the STAT CALC menu for this method to work.

3 PRACTICE/APPLY

Assignment Guide

Core: 1–7
Enriched: 1–7

4 ASSESS

Observing students working with technology is an excellent method of assessment.

6-7 LESSON NOTES

NCTM Standards: 1-5, 8, 9

Instructional Resources

- Study Guide Master 6-7
- Practice Master 6-7
- Enrichment Master 6-7
- Group Activity Card 6-7
- Assessment and Evaluation Masters, p. 156
- Math Lab and Modeling Math Masters, p. 64
- Real-World Applications, 13

Transparency 6-7A contains the 5-Minute Check for this lesson; **Transparency 6-7B** contains a teaching aid for this lesson.

Recommended Pacing	
Standard Pacing	Day 9 of 13
Honors Pacing	Day 8 of 12
Block Scheduling*	Day 5 of 7

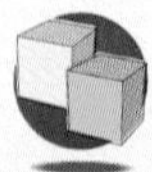

*For more information on pacing and possible lesson plans, refer to the ***Block Scheduling Booklet***.

1 FOCUS

5-Minute Check
(over Lesson 6-6)

Find the mean, median, and mode for each set of data. When necessary, round to the nearest tenth.

1. 52, 35, 63, 44, 23, 44, 52, 90, 44, 51 **mean, 49.8; median, 47.5; mode 44**
2. \$17,500; \$25,900; \$87,500; \$34,200; \$19,500 **mean, \$36,920; median, \$25,900; no mode**
3. 6, 7, 3, 3, 3, 6, 7, 4, 9, 1, 6, 6, 4 **mean, 5; median, 6; mode, 6**
4. 1.5, 2.5, 1.9, 2.7, 1.7, 3.1, 6.0, 5.4 **mean, 3.1; median, 2.6; no mode**

6-7 Solving Equations and Inequalities

Setting Goals: *In this lesson, you'll solve equations and inequalities containing rational numbers.*

Modeling a Real-World Application: Music

Whether it's rock'n'roll, jazz, rap, or classical, the art of making music is the same. About 2600 years ago, Pythagoras discovered the mathematical principles behind the musical scale. Musical sounds are made by vibrations. He discovered the relationships between the vibrations and the notes of the scale. If n represents the vibrations for *do*, or middle C, the vibrations for the other notes going up the scale are given below.

middle C — 1 octave up

do	re	mi	fa	so	la	ti	do
n	$\frac{9}{8}n$	$\frac{5}{4}n$	$\frac{4}{3}n$	$\frac{3}{2}n$	$\frac{5}{3}n$	$\frac{15}{8}n$	$\frac{2}{1}n$

A guitar string vibrates 352 times per second to produce fa. Find the number of vibrations per second needed to produce middle C.

Learning the Concept

CULTURAL CONNECTIONS
The guitar originated in Spain with a medieval *guitarra latina* of four strings. There are also Mexican 5-string *jaranas*, South American 5-string *charangos*, and Hawaiian steel guitars.

What you learned about solving equations with integers can be used to solve equations with rational numbers.

Let n represent vibrations for middle C.

vibrations for fa → $\frac{4}{3}n = 352$ ← *guitar string vibrations for fa*

$\frac{3}{4} \cdot \frac{4}{3}n = \frac{3}{4}(352)$ *Multiply each side by the multiplicative inverse of* $\frac{4}{3}$, $\frac{3}{4}$.

$n = 264$

The guitar vibrates 264 times a second to produce middle C.

CULTURAL CONNECTIONS
The guitar originated in Spain during the early 16th century. Modern classical guitar technique owes much to the Spaniard Francisco Tarrega (1852–1909), who transcribed the works of Bach, Mozart, and other composers for the guitar.

Example **1** **Solve each equation. Check your solution.**

a. $6.5y = -63.7$

$\frac{6.5}{6.5}y = \frac{-63.7}{6.5}$ *Divide each side by 6.5.*

$y = -9.8$

Check: $6.5y = -63.7$

$(6.5)(-9.8) \stackrel{?}{=} -63.7$ 6.5 [×] 9.8 [+/−] [=] −63.7

$-63.7 = -63.7$ ✓

b. $-\frac{y}{5.3} = 8.9$

$(-5.3)\left(-\frac{y}{5.3}\right) = (-5.3)(8.9)$ *Multiply each side by −5.3.*

$y = -47.17$ *Check by estimation: −5 × 9 = −45*

c. $\frac{7}{9}k = -\frac{5}{12}$

$\frac{9}{7} \cdot \frac{7}{9}k = \frac{9}{7} \cdot \left(-\frac{5}{12}\right)$ *Multiply each side by $\frac{9}{7}$. Why?*

$k = -\frac{15}{28}$

Check: $\frac{7}{9}k = -\frac{5}{12}$

$\frac{7}{9}\left(-\frac{15}{28}\right) \stackrel{?}{=} -\frac{5}{12}$

$-\frac{5}{12} = -\frac{5}{12}$ ✓

You can use the same skills you developed to solve inequalities with integers to solve inequalities with rational numbers.

Example **2** **Solve each inequality.**

LOOK BACK You can review graphing inequalities in Lesson 3-6.

a. $\frac{t}{4.3} \geq 5$

$(4.3)\frac{t}{4.3} \geq (4.3)5$ *Multiply each side by 4.3.*

$t \geq 21.5$ *Any number greater than or equal to 21.5 is a solution.*

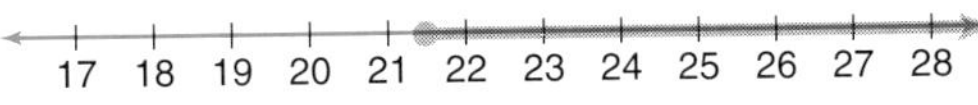

Check: Try 25, a number greater than 21.5.

$\frac{25}{4.3} \stackrel{?}{\geq} 5$

$5.8 \geq 5$ ✓

b. $-\frac{2}{3}b > -9\frac{1}{2}$

$\left(-\frac{3}{2}\right)\left(-\frac{2}{3}\right)b < \left(-\frac{3}{2}\right)\left(-\frac{19}{2}\right)$ *Multiply each side by $-\frac{3}{2}$ and reverse the order symbol.*

$b < \frac{57}{4}$ or $14\frac{1}{4}$ *Any number less than $14\frac{1}{4}$ is a solution.*

Motivating the Lesson

Questioning Give several situations and ask students to tell whether they represent inequalities or equations. Examples:

- The labor charge will be at least \$72. **inequality**
- The total number of players is 11. **equation**

2 TEACH

Teaching Tip You may want to introduce the use of rational numbers in equations and inequalities by first solving problems using whole numbers. Then replace whole numbers with decimals and fractions and solve.

In-Class Examples

For Example 1

Solve each equation. Check your solution.

a. $5.3x = 102.85$ **18.7**

b. $-36.8 = \frac{x}{9.2}$ **−338.56**

c. $\frac{3}{2}m = 1\frac{3}{8}$ $\mathbf{\frac{11}{12}}$

For Example 2

Solve each inequality.

a. $-\frac{3}{4} \leq \frac{7}{12}n$ $\mathbf{n \geq -1\frac{2}{7}}$

b. $-\frac{m}{9} > -5.76$ $\mathbf{m < 51.84}$

Study Guide Masters, p. 51

NAME ______________________ DATE ____________

6-7 Study Guide

Solving Equations and Inequalities

Student Edition Pages 308–311

To solve equations containing rational numbers, multiply each side by the same number to get the variable by itself.

The number you multiply each side by is the reciprocal of the number that is multiplied times the variable.

$-\frac{y}{5} = 8$

$-\frac{1}{5} \cdot y = 8$

$-\frac{5}{1} \cdot \left(-\frac{1}{5}\right) \cdot y = -\frac{5}{1} \cdot 8$

$y = -40$

Use the same method to solve inequalities containing rational numbers.

When you multiply or divide each side of an inequality by a negative number, you must reverse the inequality sign.

$-\frac{2}{3}x \leq \frac{1}{2}$

$-\frac{3}{2} \cdot -\frac{2}{3} \cdot x \leq \frac{1}{2} \cdot -\frac{3}{2}$

$x \geq -\frac{3}{4}$

Write the number you would multiply each side by in order to get the variable by itself.

1. $4n = 12$ $\frac{1}{4}$ **2.** $\frac{a}{2} = 50$ $\frac{2}{1}$ **3.** $-7t = 21$ $-\frac{1}{7}$ **4.** $-\frac{m}{6} = 2$ −6

5. $-35 < \frac{h}{7}$ $\frac{7}{1}$ **6.** $0.25b < 5$ $\frac{1}{0.25}$ or 4 **7.** $2.4 \leq -6x$ $-\frac{1}{6}$ **8.** $-\frac{c}{3} \geq -12$ −3

Solve each equation or inequality. Check your solution.

9. $4n = 12$ 3 **10.** $\frac{a}{2} = 50$ 100 **11.** $-7t = 21$ −3 **12.** $-\frac{m}{6} = 2$ −12

13. $4n \leq 12$ $n \leq 3$ **14.** $\frac{a}{2} > 50$ $a > 100$ **15.** $-7t > 21$ $t < -3$ **16.** $-\frac{m}{6} \leq 2$ $m \geq -12$

17. $-35 = \frac{h}{7}$ −245 **18.** $0.25b = 5$ 20 **19.** $2.4 = -6x$ −0.4 **20.** $-\frac{c}{3} = -12$ 36

21. $-35 \geq \frac{h}{7}$ $h \leq -245$ **22.** $0.25b < 5$ $b < 20$ **23.** $2.4 < -6x$ $x < -0.4$ **24.** $-\frac{c}{3} > -12$ $c < 36$

Teaching Tip Review reversing the order symbol in an inequality when multiplying or dividing by a negative number.

In-Class Example

For Example 3
The Square-Root Club is sponsoring a field trip. They have raised \$95. They estimate that this amount is a little less than $\frac{1}{5}$ of the money they need for the field trip. How much is their estimated cost? **$n >$ \$475**

3 PRACTICE/APPLY

Checking Your Understanding

Exercises 1–11 are designed to help you assess your students' understanding through reading, writing, speaking, and modeling. You should work through Exercises 1–4 with your students and then monitor their work on Exercises 5–11.

Additional Answers

1. Divide each side by 0.3.
2a. Answers will vary.
2b. Multiplication property of equality
3. Divide each side by $\frac{1}{4}$ or multiply each side by 4.

Practice Masters, p. 51

NAME ____________ DATE ________

6-7 Practice
Solving Equations and Inequalities

Student Edition Pages 308–311

Solve each equation or inequality. Check your solution.

1. $-8a = -5.68$ **0.71**	2. $\frac{b}{4.2} = -5$ **−21**	3. $27.44 = -4.9c$ **−5.6**
4. $\frac{d}{4} < -4.7$ **$d > 18.8$**	5. $-12.6 \leq 3n$ **$n \geq -4.2$**	6. $-5f > 35.5$ **$f < -7.1$**
7. $-125 = -\frac{5}{8}m$ **200**	8. $-2.3n = 0.805$ **−0.35**	9. $-\frac{t}{8.7} = -3.01$ **26.187**
10. $-8.4r \geq 5.88$ **$r \leq -0.7$**	11. $\frac{k}{1.5} < -4.5$ **$k < -6.75$**	12. $20.4 \leq -3.4d$ **$d \leq -6$**
13. $-18.24 = -6x$ **3.04**	14. $-\frac{z}{2.1} = -100$ **210**	15. $-1.27y = 0.0381$ **−0.03**
16. $\frac{r}{0.5} < -3.1$ **$r < -1.55$**	17. $-0.16s > -9.6$ **$s < 60$**	18. $-\frac{2}{3}t \leq -\frac{8}{9}$ **$t \geq 1\frac{1}{3}$**
19. $3.4j = 0.816$ **0.24**	20. $\frac{r}{7.4} = -0.5$ **−3.7**	21. $48.374 = -1.34w$ **−36.1**
22. $0.3u < -2.73$ **$u < -9.1$**	23. $-\frac{3}{4}v \geq -1\frac{11}{16}$ **$v \leq 2\frac{1}{4}$**	24. $0.025x \leq 5.25$ **$x \leq 210$**
25. $-\frac{z}{50.3} = 7.6$ **−382.28**	26. $-\frac{8}{9}k = -0.16$ **0.18**	27. $0.42y = -2.6166$ **−6.23**

Theater

A replica of the Globe Theater, where Shakespeare's plays were originally performed, is currently being constructed in London, England. Supporters of the project have raised almost \$13 million, which is over four-fifths of the budget. How much is the project budget?

Write an equation. Let b represent the budget.

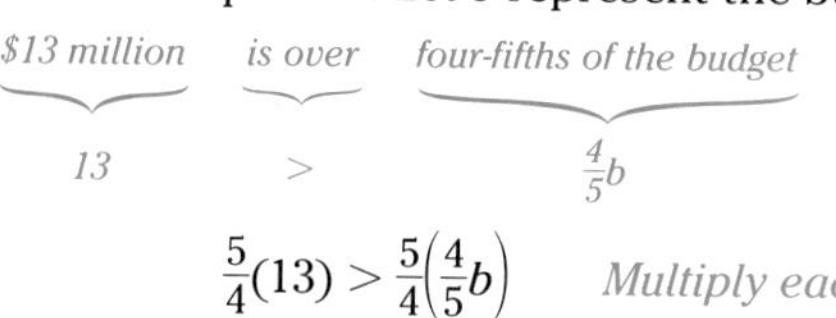

$$\frac{5}{4}(13) > \frac{5}{4}\left(\frac{4}{5}b\right) \quad \textit{Multiply each side by } \frac{5}{4}.$$

$$\frac{65}{4} > b$$

The budget is less than $\frac{65}{4}$ or \$16.25 million.

Checking Your Understanding

Communicating Mathematics

Read and study the lesson to answer these questions. **1–3. See margin.**

1. **Explain** how you would solve the equation $0.3x = 4.5$.
2. In the application at the beginning of the lesson, the equation was solved by multiplying each side by the multiplicative inverse, or reciprocal, of $\frac{4}{3}$.
   a. How does this help you solve the equation?
   b. Describe the property of equality that allows you to do this.
3. **State** two ways to solve the equation $\frac{1}{4}x = 28$.
4. **Write** an inequality that has the solution graphed below. **$x \leq -2$**

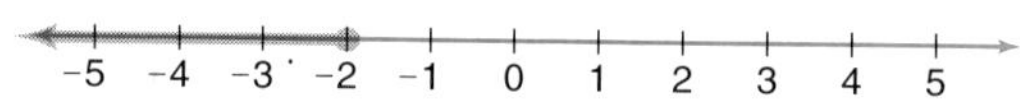

Guided Practice

Explain how to solve each equation or inequality. Then solve. Graph the solutions on a number line. **See Solutions Manual for graphs.**

5. $0.6n = -9$ **−15**
6. $4.1p = 16.4$ **4**
7. $-2.7 < 0.3f$ **$-9 < f$**
8. $-\frac{2}{3}x > 6$ **$x < -9$**
9. $\frac{5}{6}y = \frac{3}{5}$ **$\frac{18}{25}$**
10. $\frac{r}{-4.6} \geq 0.01$ **10. $r \leq -0.046$**

11. **Geometry** You are asked to draw a rectangle with one side measuring 2.5 cm and with an area of 15 cm^2. How long should you make the other side? **6 cm**

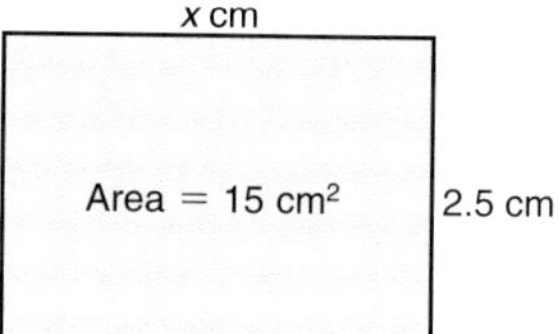

Group Activity Card 6-7

Comparing To One — Group Activity 6-7

MATERIALS: Four 0-9 spinners

You and a partner will investigate the solutions to equations involving rational numbers. To begin, spin the four 0-9 spinners. If a zero turns up, you will needed to spin again until you get all nonzero numbers.

From these four digits, form two fractions, both of which are less than one (that is, in both fractions the numerator is less than the denominator). Using the two fractions, write an equation of the form $ax = b$, replacing the a and b with the fractions.

Now decide if the solution to the equation is less than 1 or greater than 1. Solving the equation using algebraic techniques will allow you to tell if your decision was correct.

Do five more of these equations. Write a short paragraph explaining when the value of x will be less than one and when it will be greater than one.

 Pre-Algebra

Reteaching

Using Examples Have students rewrite Examples 1b and 2a without using the fraction bar. Ask students how they would solve these if the number on the same side as the variable were a whole number. Have students use calculators to apply their responses to the rewritten examples.

Exercises: Practicing and Applying the Concept

Independent Practice

Solve each equation or inequality. Check your solution.

12. $7y < 3.5$ **$y < 0.5$** 13. $-3y = 1.5$ **-0.5** 14. $5a > -7\frac{1}{2}$ **$a > -1\frac{1}{2}$**

15. $\frac{1}{4}x = 4\frac{1}{8}$ **$16\frac{1}{2}$** 16. $-\frac{1}{8}r = \frac{1}{4}$ **-2** 17. $\frac{x}{3.2} < -4.5$ **$x < -14.4$**

18. $\frac{7}{8}r \leq 1.4$ **$r \leq 1.6$** 19. $-15 = 2\frac{2}{5}x$ **$-6\frac{1}{4}$** 20. $-6.7b = 14.07$ **-2.1**

21. $\frac{p}{7.1} = -0.5$ **-3.55** 22. $\frac{1}{3}d = -0.36$ **-1.08** 23. $-1.3z \geq 1.69$ **$z \leq -1.3$**

24. $\frac{3}{4}y > 1\frac{4}{5}$ **$y > 2\frac{2}{5}$** 25. $-7\frac{1}{2}x = 5\frac{1}{4}$ **$\frac{-7}{10}$** 26. $-2h > 4.6$ **$h < -2.3$**

27. $-0.05x = -0.95$ **19** 28. $\frac{x}{-2.5} \leq 3.2$ **$x \geq -8$** 29. $7x < 1\frac{4}{10}$ **$x < \frac{1}{5}$**

Solve each equation or inequality and graph the solution on a number line. **See Solutions Manual for graphs.** **32. $p \leq -2.3$**

30. $-\frac{1}{4}c = 3.4$ **-13.6** 31. $\frac{4}{5}y < \frac{3}{8}$ **$y < \frac{15}{32}$** 32. $-12.6p \geq 28.98$

Critical Thinking

33. Write an inequality involving multiplication or division, using fractions, whose solution is $p < \frac{1}{3}$. **Answers will vary.**

Applications and Problem Solving

34. $h > 14.4$ inches

34. **Geometry** The formula for the area of a triangle is $A = \frac{1}{2}bh$. A triangle has an area greater than 18 square inches. If the base of the triangle is 2.5 inches, what is the height of the triangle?

35. **Business** The Rose Department Store chain was closing one of their stores so they had to sell most of the inventory in the store. In the final markdown, they marked all items $\frac{1}{3}$ off the ticketed price. How much would a shirt that was ticketed at \$24.99 cost? **\$16.66**

36. **Food** The National Restaurant Association polled a group of Americans about their food preferences. Less than $\frac{1}{3}$ of those polled had tried Indian or Thai food.
    a. If 332 people had tried Indian or Thai food, how large a group was polled? **more than 996 people**
    b. How large do you think the group polled was? Explain your answer. **See margin.**

Mixed Review

37. **Education** If you scored 93, 86, 84, 92, 85, and 93 on your first six pre-algebra tests, what is your mean test score? (Lesson 6-6) **88.8**

38. Write $\frac{28}{45}$ as a decimal. (Lesson 6-1) **$0.6\overline{2}$**

39. Solve $\frac{6}{7} + \frac{5}{14} = s$. (Lesson 5-5) **$1\frac{3}{14}$**

40. Find the LCM of $15c$ and $12cd$. (Lesson 4-7) **$60cd$**

41. Solve $a = (-12)(3)$. (Lesson 2-7) **-36**

42. Name the property shown by $5(6 \cdot 3) = (5 \cdot 6)3$. (Lesson 1-4) **Associative, ×**

Extension

Using Partners Have students work in pairs of two to solve inequalities with the variable on both sides of the inequality, such as $3x < x - 3.6$.

Assignment Guide

Core: 13–35 odd, 37–42
Enriched: 12–32 even, 33–42

For **Extra Practice**, see p. 755.

The red A, B, and C flags, printed only in the Teacher's Wraparound Edition, indicate the level of difficulty of the exercises.

4 ASSESS

Closing Activity

Modeling Have students make up problems involving inequalities. Another student models how to solve the problem. A third student models that solution on the number line.

Chapter 6, Quiz C (Lessons 6-5 and 6-6) is available in the *Assessment and Evaluation Masters*, p. 156.

Additional Answer

36b. Sample answer: 1000; because it is more than 996 and is a round number. Also, if the size of the group was much larger than 1000, say 1500, it would be reported that less than $\frac{1}{4}$ of those polled had tried Indian or Thai food.

Enrichment Masters, p. 51

NAME ______________________ DATE ____________

6-7 Enrichment
Multi-Step Problems

Student Edition
Pages 308–311

Solving an equation such as $5 + 3y = 17$ requires two steps. First, subtract 5 from each side of the equation. Then divide each side of the equation by 3.

$$5 + 3y = 17$$
$$-5 \quad\quad -5$$
$$3y = 12$$
$$\frac{3y}{3} = \frac{12}{3}$$
$$y = 4$$

Write the operation and number that would be used in each step of the following two-step equations. Use A for addition, S for subtraction, M for multiplication, and D for division.

	Equation	Step 1	Step 2	Solution
Example	$2y - 6 = 10$	A 6	D 2	$y = 8$
1.	$2a + 3.4 = 75.6$	S 3.4	D 2	$a = 36.1$
2.	$3.456 = 4b - 1$	A 1	D 4	$b = 1.114$
3.	$15 = 3.6 + \frac{c}{3}$	S 3.6	M 3	$c = 34.2$
4.	$0.4 = \frac{d}{12.6} + 2$	S 2	M 12.6	$d = {}^-20.16$
5.	$7 - 0.87e = 8.4355$	S 7	D −0.87	$e = {}^-1.65$
6.	$87 = 0.25f - 38.6$	A 38.6	D 0.25	$f = 502.4$

Write and solve a multi-step equation for each problem.

7. What are the length and width of a rectangular dog pen with a perimeter of 38 feet if the length is two feet less than twice the width? **$2(2w - 2) + 2w = 38$; 7 ft by 12 ft**

8. Apex Taxi charges \$2.80 per mile plus \$1.50 per passenger. How many miles did Mr. and Mrs. Kelly travel if the charge was \$31. **$3 + 2.8m = 31$; 10 mi**

9. Alice earned a 5% commission for selling encyclopedias. How many dollars in sales did she have if she still needed \$60 to buy a \$100 radio? **$0.05x + 60 = 100$; \$800**

6-8 LESSON NOTES

NCTM Standards: 1-5, 8

Instructional Resources

- Study Guide Master 6-8
- Practice Master 6-8
- Enrichment Master 6-8
- Group Activity Card 6-8
- Graphing Calculator Masters, p. 6

Transparency 6-8A contains the 5-Minute Check for this lesson; **Transparency 6-8B** contains a teaching aid for this lesson.

Recommended Pacing	
Standard Pacing	Day 10 of 13
Honors Pacing	Day 9 of 12
Block Scheduling*	Day 6 of 7 (along with Lesson 6-9)

*For more information on pacing and possible lesson plans, refer to the ***Block Scheduling Booklet***.

1 FOCUS

5-Minute Check
(over Lesson 6-7)

Solve each equation or inequality. Check your solution.

1. $3x = -36$ -12
2. $\frac{y}{3} > -7.4$ $y > -22.2$
3. $\frac{6}{7}c = \frac{3}{5}$ $\frac{7}{10}$
4. $\frac{r}{5} < -\frac{2}{3}$ $r < -3\frac{1}{3}$
5. $-\frac{2}{3}m \geq \frac{1}{9}$ $m \leq -\frac{1}{6}$

Motivating the Lesson

Hands-On Activity Hold a ball up in the air. Ask students how many times they think the ball will bounce if it is dropped. Drop the ball and have students count the number of bounces. Repeat the experiment from the same height and then from different heights. Examine the heights to look for a pattern.

6-8 Integration: Discrete Mathematics
Geometric Sequences

Setting Goals: *In this lesson, you'll recognize and extend geometric sequences and represent them algebraically.*

Modeling a Real-World Application: Genealogy

In 1974, Alex Haley finished his book, *Roots*, which traced his family's history. His story began in 1767 with the African Kunta Kinte. Alex Haley's great-great-great-great-great-grandfather, Kunta Kinte, was brought to America to be a slave. With each preceding generation that Alex Haley researched, he found more and more ancestors.

Learning the Concept

A TV miniseries of *Roots* made in 1977 won a record-setting 9 Emmys. A follow up series about Haley's grandmother, Queen, was made in 1994.

You can use a pattern to find the number of ancestors Alex Haley had back to Kunta Kinte's generation. The number of ancestors doubles as shown below.

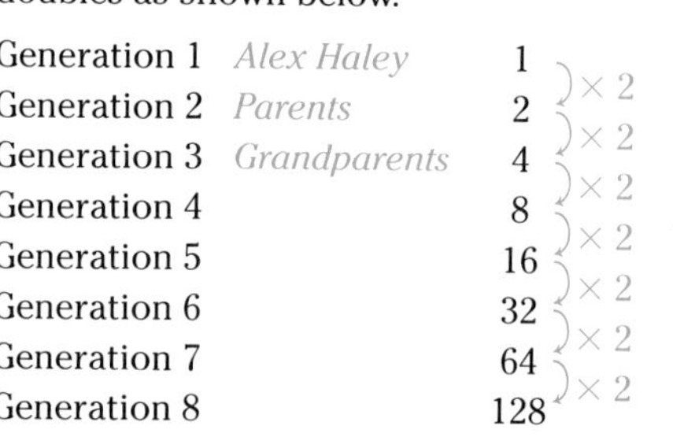

Generation 1	*Alex Haley*	1
Generation 2	*Parents*	2
Generation 3	*Grandparents*	4
Generation 4		8
Generation 5		16
Generation 6		32
Generation 7		64
Generation 8		128

(each step $\times 2$)

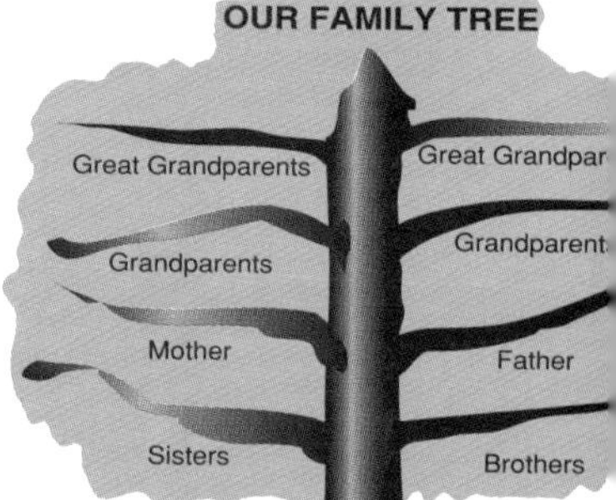

The list of numbers 1, 2, 4, 8, 16, 32, 64, 128 forms a **geometric sequence.** Each term in a geometric sequence increases or decreases by a common *factor*, called the **common ratio**. The number of ancestors doubles in each generation so the common ratio for this sequence is 2.

Geometric Sequence — **A geometric sequence is a sequence in which the ratio between any two successive terms is the same.**

Example  **State whether each sequence is geometric. If so, state the common ratio and list the next three terms.**

a. 3, 15, 75, 375, . . .

3, 15, 75, 375 ($\times 5$, $\times 5$, $\times 5$) *Notice* $\frac{15}{3} = 5$, $\frac{75}{15} = 5$, *and* $\frac{375}{75} = 5$.

The sequence has a common ratio so it is a geometric sequence. The common ratio is 5. The next three terms are 375×5 or 1875, 1875×5 or 9375, and 9375×5 or 46,875.

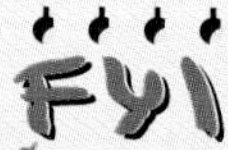

Alex Haley (1921-1992) wrote other historical fiction and nonfiction prior to *Roots*. His first major work was assisting on *The Autobiography of Malcom X* in 1965. For Roots, Haley received the Pulitzer Prize.

b. $-4, -2, -\frac{2}{3}, -\frac{1}{6}, \ldots$

$-4 \quad -2 \quad -\frac{2}{3} \quad -\frac{1}{6}$

$\times\frac{1}{2} \quad \times\frac{1}{3} \quad \times\frac{1}{4}$

Since there is no common ratio, the sequence is *not* geometric.

c. $-4, 1, -\frac{1}{4}, \frac{1}{16}, \ldots$

$-4 \quad 1 \quad -\frac{1}{4} \quad \frac{1}{16}$

$\times\left(-\frac{1}{4}\right) \quad \times\left(-\frac{1}{4}\right) \quad \times\left(-\frac{1}{4}\right)$

The sequence is geometric. The common ratio is $-\frac{1}{4}$.

$\frac{1}{16} \times \left(-\frac{1}{4}\right) = -\frac{1}{64} \qquad -\frac{1}{64} \times \left(-\frac{1}{4}\right) = \frac{1}{256} \qquad \frac{1}{256} \times \left(-\frac{1}{4}\right) = -\frac{1}{1024}$

The next three terms are $-\frac{1}{64}, \frac{1}{256}$, and $-\frac{1}{1024}$.

Connection to Algebra

If we know the first term of a geometric sequence and the common ratio, we can find any other term of the sequence. Consider the sequence 3, 12, 48, 192, ...

Here's how to write an expression that represents a term in the sequence. The first term is 3; call the first term a. The common ratio is 4; call the common ratio r.

Term	1st	2nd	3rd	4th	nth
Sequence	3	12	48	192	
Factors of term	3	$3 \cdot 4$	$3 \cdot 4 \cdot 4$	$3 \cdot 4 \cdot 4 \cdot 4$	$3 \cdot \underbrace{4 \cdot 4 \cdot 4 \cdot \ldots \cdot 4}_{n-1}$
Term using exponents	3	$3 \cdot 4^1$	$3 \cdot 4^2$	$3 \cdot 4^3$	$3 \cdot 4^{n-1}$
Variables	a	$a \cdot r^1$	$a \cdot r^2$	$a \cdot r^3$	$a \cdot r^{n-1}$

The exponent is one less than the number of the term.

Because n represents any term, you can use this expression to find any term in the sequence.

Example 2 **Use the expression $a \cdot r^{n-1}$ to find the eighth term in the sequence 4, 12, 36, 108, . . .**

The first term is 4, so $a = 4$. The common ratio is 3, so $r = 3$, and $n = 8$.

$a \cdot r^{n-1} = 4 \cdot 3^{8-1}$ or $4 \cdot 3^7$

4 [×] 3 [y^x] 7 [=] 8748

The eighth term is 8748.

2 TEACH

In-Class Examples

For Example 1

State whether each sequence is geometric. If so, state the common ratio and list the next three terms.

a. 4, 2, 1, $\frac{1}{2}$, ... **geometric, $\frac{1}{4}, \frac{1}{8}, \frac{1}{16}$**

b. −16, −8, 4, 2, ... **not geometric**

c. $\frac{1}{9}, -\frac{1}{3}$, 1, -3, ... **geometric, 9, −27, 81**

For Example 2

Use the expression $a \cdot r^{n-1}$ to find the ninth and twelfth terms in the sequence 1024, 512, 256, 128, ... **4, $\frac{1}{2}$**

Teaching Tip You may want to review powers of negative numbers and of fractions by having students use the formula developed here to check the answers found in Example 1.

GLENCOE Technology

Interactive Mathematics Tools Software

In this interactive computer lesson, students explore geometric sequences. A **Computer Journal** gives students the opportunity to write about what they have learned.

For Windows & Macintosh

In-Class Example

For Example 3

A ball bounces $\frac{2}{3}$ of its height after every fall. If it is dropped from a height of 48 feet, how high will it bounce at the end of the fourth bounce? $9\frac{13}{27}$ ft

3 PRACTICE/APPLY

Checking Your Understanding

Exercises 1–13 are designed to help you assess your students' understanding through reading, writing, speaking, and modeling. You should work through Exercises 1–5 with your students and then monitor their work on Exercises 6–13.

Additional Answers

1. The number by which a term is multiplied to get the next term.
2. Geometric sequences have common factors and arithmetic sequences have common addends.

Study Guide Masters, p.52

NAME ______________________ DATE ____________

6-8 Study Guide
Integration: Discrete Mathematics
Geometric Sequences

Student Edition
Pages 312–316

Two lists of numbers, or sequences, are shown below.

3, 6, 12, 24, 48, ···	Each number in the pattern is 2 times the number before it.
162, 54, 18, 6, 2, ···	Each number in the pattern is $\frac{1}{3}$ times the number before it.

In each sequence above, each number after the first can be obtained from the previous one by multiplying it by a fixed number called the **common ratio**. Such a sequence is called a **geometric sequence**. The common ratios above are 2 and $\frac{1}{3}$, respectively.

$\frac{6}{3} = 2$ $\quad \frac{12}{6} = 2$ $\quad \frac{24}{12} = 2$ $\quad \frac{48}{24} = 2$

$\frac{54}{162} = \frac{1}{3}$ $\quad \frac{18}{54} = \frac{1}{3}$ $\quad \frac{6}{18} = \frac{1}{3}$ $\quad \frac{2}{6} = \frac{1}{3}$

State whether each sequence is a geometric sequence. If so, state the common ratio and list the next three terms.

1. 1, 2, 4, 8, 16, ··· yes; 2; 32, 64, 128
2. $\frac{1}{3}, \frac{1}{9}, \frac{1}{27}, \frac{1}{81}, \frac{1}{243}, \cdots$ yes; $\frac{1}{3}$; $\frac{1}{729}, \frac{1}{2187}, \frac{1}{6561}$
3. $\frac{1}{2}, \frac{1}{3}, \frac{1}{4}, \frac{1}{5}, \frac{1}{6}, \cdots$ no

Complete each geometric sequence.

4. 8, 4, 2, __?__, __?__, __?__ 1; $\frac{1}{2}$; $\frac{1}{4}$
5. __?__, 7, 14, 28, __?__, __?__ $3\frac{1}{2}$; 56; 112
6. __?__, $\frac{1}{24}, \frac{1}{12}, \frac{1}{6}$, __?__, __?__ $\frac{1}{48}$; $\frac{1}{3}$; $\frac{2}{3}$
7. __?__, 10^{98}, 10^{96}, __?__, __?__, 10^{90} 10^{100}; 10^{94}; 10^{92}

Write the first five terms of each geometric sequence described below.

8. first term, 5; ratio, 2 5, 10, 20, 40, 80
9. first term, 8; ratio, $\frac{1}{2}$ 8, 4, 2, 1, $\frac{1}{2}$
10. first term, 81; ratio, $\frac{1}{3}$ 81, 27, 9, 3, 1
11. first term, $\frac{8}{9}$; ratio, $\frac{3}{2}$ $\frac{8}{9}, \frac{24}{18}, \frac{72}{36}, \frac{216}{72}, \frac{648}{144}$

Some real-life problems can be solved using geometric sequences.

Example 3 Sports

The NCAA Women's Basketball tournament is a single elimination tournament. After each game, the winning team progresses to the next level of competition and the loser is out of the tournament. How many teams are in the first round of play if there are 6 rounds in the tournament?

Explore With each round, half of the teams are out of the competition. So, this is a geometric sequence. If you start with the final game and work backward, each round had twice as many teams as the one after it. The first term is 2, the 2 teams who played in the final game of the tournament. The common ratio is 2 because there are twice as many games in each previous round. The number of terms is 6 because there were 6 rounds.

Plan Find out how many teams played in each round by starting with 2 and multiplying by 2 until you find the 6*th* term.

Solve 2 [×] 2 [×] 2 [×] 2 [×] 2 [×] 2 [=] 64

There were 64 teams.

Examine Another strategy is to use the formula $a \cdot r^{n-1}$

$a = 2$, $r = 2$, and $n = 6$.

$2 \cdot 2^{(6-1)} = 2 \cdot 2^5$ or 64

The solution checks.

Checking Your Understanding

Communicating Mathematics

Read and study the lesson to answer these questions.

1. **Describe** the common ratio in a geometric sequence. **See margin.**
2. You studied arithmetic sequences in Lesson 5-9. Compare and contrast geometric and arithmetic sequences. **See margin.**
3. **Give an example** of a geometric sequence with a first term of $\frac{1}{4}$.
4. **Find** the number of ancestors Alex Haley has in the generation before Kunta Kinte. **256**

3. Sample answer: $\frac{1}{4}, \frac{1}{40}, \frac{1}{400}$

Reteaching

Using Cooperative Groups Have a student name a number. A second student multiplies that number by a whole number and states the second term in the sequence. Each following student must decide what factor was used and give the next term. Rotate and repeat. Extend from whole numbers to other rational numbers.

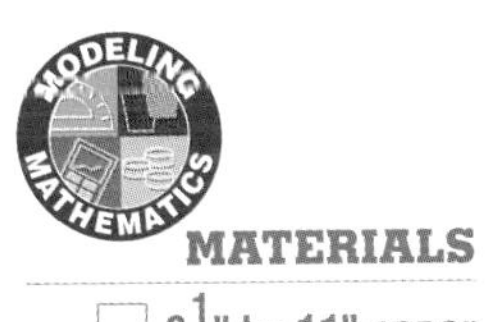

MATERIALS

☐ $8\frac{1}{2}$" by 11" paper

5. You can model a geometric sequence by folding a piece of paper and counting the number of sections.
   a. Start with an $8\frac{1}{2}$"-by-11" piece of paper. There are 0 folds and 1 section. The first term of the sequence is 1.
   b. Fold the paper in half. There is 1 fold and 2 sections. The second term of the sequence is 2. Continue to fold the paper in half. Keep a record of the number of folds and the number of sections.
   c. How many sections result from the 5th fold? **16 more, 32 total**
   d. Write the sequence. Does the expression $a \cdot r^{(n-1)}$ represent the terms in the sequence? Explain. **See margin.**

Guided Practice

State whether each sequence is a geometric sequence. If so, state the common ratio and list the next three terms. **See margin.**

6. 1, 3, 9, 27, ...
7. 2, 4, 6, 8, ...
8. 1, −3, 5, −7, ...
9. 81, 27, 9, 3, ...
10. 10, 14, 18, 22, ...
11. $\frac{1}{2}$, 1, 2, 4, ...

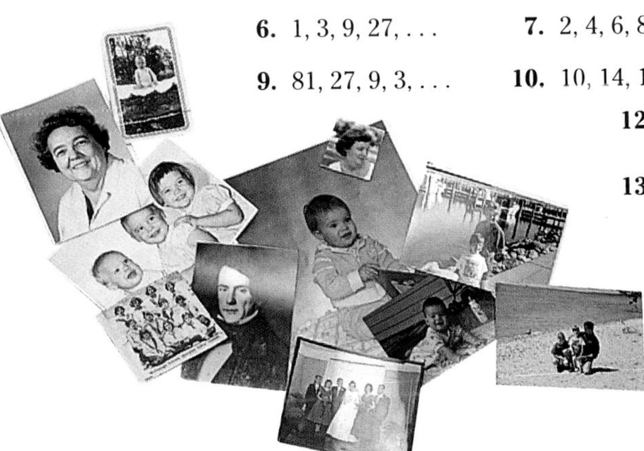

12. Find the fourth term of a geometric sequence if $a = -5$ and $r = -2$. **40**
13. **Genealogy** Find the number of ancestors in your family 11 generations ago. **2048**

Exercises: Practicing and Applying the Concept

Independent Practice

State whether each sequence is a geometric sequence. If so, state the common ratio and list the next three terms. **See margin.**

14. $-5, 1, -\frac{1}{5}, \frac{1}{25}, \ldots$
15. 24, 12, 6, 3, ...
16. 9, 3, −3, −9, ...
17. $144, 12, 1, \frac{1}{12}, \ldots$
18. $-14, 7, -1, -\frac{1}{7}, \ldots$
19. $\frac{1}{2}, \frac{1}{4}, \frac{1}{8}, \frac{1}{16}, \ldots$

20. −8, 16, −32, 64, ...
21. 7, −14, 28, −56, ...
22. $25, 5, 1, \frac{1}{5}, \ldots$
23. $2\frac{1}{2}, \frac{1}{2}, -1\frac{1}{2}, \ldots$
24. $-\frac{1}{4}, -\frac{1}{2}, -1, -2, \ldots$
25. 1, 1, 2, 3, 5, ...
26. $18, -6, 2, -\frac{2}{3}$
27. $24, 6, \frac{3}{2}, \frac{3}{8}, \ldots$
28. 5, 7, 9, 11, 13, ...

29. Write the first five terms in a geometric sequence with a common ratio of −2. The first term is 2. **2, −4, 8, −16, 32**

30–31. **See margin.**

30. Write the first four terms of a geometric sequence if $a = -6$ and $r = \frac{2}{3}$.
31. Write the first three terms of a geometric sequence if $a = 80$ and $r = \frac{5}{4}$.
32. Use the expression $ar^{(n-1)}$ to find the ninth term in the geometric sequence $-14, -\frac{7}{2}, -\frac{7}{8}, -\frac{7}{32}, \ldots$. **$-\frac{7}{128}$**

Assignment Guide

Core: 15–37 odd, 39–46
Enriched: 14–32 even, 33–46

For **Extra Practice**, see p. 755.

The red A, B, and C flags, printed only in the Teacher's Wraparound Edition, indicate the level of difficulty of the exercises.

Additional Answers

5d. 1, 2, 4, 8, 16, 32, ... Yes, the common ratio is 2.
6. yes, 3; 81, 243, 729
7. no
8. no
9. yes, $\frac{1}{3}$; 1, $\frac{1}{3}$, $\frac{1}{9}$
10. no
11. yes, 2; 8, 16, 32
14. yes, $-\frac{1}{5}$; $-\frac{1}{125}, \frac{1}{625}, -\frac{1}{3125}$
15. yes, $\frac{1}{2}$; $\frac{3}{2}, \frac{3}{4}, \frac{3}{8}$
16. no
17. yes, $\frac{1}{12}$; $\frac{1}{144}, \frac{1}{1728}, \frac{1}{20736}$
18. no
19. yes, $\frac{1}{2}$; $\frac{1}{32}, \frac{1}{64}, \frac{1}{128}$
20. yes, −2; −128, 256, −512
21. yes, −2; 112, −224, 448
22. yes, $\frac{1}{5}$; $\frac{1}{25}, \frac{1}{125}, \frac{1}{625}$
23. no
24. yes, 2; −4, −8, −16
25. no
26. yes, $-\frac{1}{3}$; $\frac{2}{9}, -\frac{2}{27}, \frac{2}{81}$
27. yes, $\frac{1}{4}$; $\frac{3}{32}, \frac{3}{128}, \frac{3}{512}$
28. no

Practice Masters, p. 52

NAME ____________ DATE ________

6-8 Practice — Student Edition Pages 312–316
Integration: Discrete Mathematics
Geometric Sequences

State whether each sequence is a geometric sequence. If so, state the common ratio and list the next three terms.

1. 2, 6, 18, 54, ⋯ yes; 3; 162, 486, 1458
2. 50, 46, 41, 37, ⋯ no
3. 8, 4, 2, 1, ⋯ yes; $\frac{1}{2}$; $\frac{1}{2}, \frac{1}{4}, \frac{1}{8}$
4. $0, 1, -\frac{1}{8}, \frac{1}{64}, \cdots$ yes; $-\frac{1}{8}$; $\frac{1}{512}, \frac{1}{4096}, \frac{1}{32,768}$
5. 64, 16, 4, 1, ⋯ yes; $\frac{1}{4}$; $\frac{1}{4}, \frac{1}{16}, \frac{1}{64}$
6. $\frac{2}{3}, \frac{2}{9}, \frac{2}{27}, \frac{2}{81}, \cdots$ yes; $\frac{1}{3}$; $\frac{2}{243}, \frac{2}{729}, \frac{2}{2187}$
7. 51, −5.1, 0.51, −0.051, ⋯ yes; $-\frac{1}{10}$; 0.0051, −0.00051, 0.000051
8. 3, 3, 3, 3, ⋯ yes; 1; 3, 3, 3
9. $\frac{1}{3}, \frac{1}{6}, \frac{1}{12}, \frac{1}{24}, \cdots$ yes; $\frac{1}{2}$; $\frac{1}{48}, \frac{1}{96}, \frac{1}{192}$
10. 2125, −425, 85, −17, ⋯ yes; $-\frac{1}{5}$; $\frac{17}{5}, \frac{17}{25}, \frac{17}{125}$
11. 10, 20, 60, 240, ⋯ no
12. $3\frac{1}{2}, 4\frac{1}{2}, 6\frac{1}{3}, 9\frac{1}{2}, \cdots$ no
13. $-289, 17, -1, \frac{1}{17}, \cdots$ yes; $-\frac{1}{17}$; $-\frac{1}{289}, \frac{1}{4913}, \frac{1}{83,521}$
14. $8, 6, 4\frac{1}{2}, 3\frac{3}{8}, \cdots$ yes; $\frac{3}{4}$; $\frac{81}{32}, \frac{243}{128}, \frac{729}{512}$
15. −1,000,000, −10,000, −100, −1, ⋯ yes; $\frac{1}{100}$; −0.01, −0.0001, −0.000001
16. $-\frac{2}{12}, -\frac{2}{24}, -\frac{2}{72}, -\frac{2}{288}, \cdots$ no
17. 3, 4, 7, 11, ⋯ no
18. $18, -3, \frac{1}{2}, -\frac{1}{12}, \cdots$ yes; $-\frac{1}{6}$; $\frac{1}{72}, -\frac{1}{432}, \frac{1}{2592}$
19. $36, 4, \frac{4}{9}, \frac{4}{81}, \cdots$ yes; $\frac{1}{9}$; $\frac{4}{729}, \frac{4}{6561}, \frac{4}{59,049}$
20. $-\frac{1}{7}, \frac{2}{21}, -\frac{4}{63}, \frac{8}{189}, \cdots$ yes; $-\frac{2}{3}$; $-\frac{16}{567}, \frac{32}{1701}, -\frac{64}{5103}$

Group Activity Card 6-8

Geometrically Speaking — Group Activity **6-8**

MATERIALS: 0-9 spinner

This activity works with geometric sequences. To begin, spin the spinner four times. Use these numbers to create a geometric sequence. One number needs to be the first term (a) of the sequence. Two of the numbers will be the common ratio (r), one of which is the numerator of the ratio, and the other is the denominator of the ratio. The last number will indicate how many terms of the geometric sequence you are to determine.

After everyone has created one sequence, exchange papers and see if each of you can compute the indicated number of terms of the sequence given to you. As a group, check the accuracy of these computations.

 Pre-Algebra

Additional Answers

30. $-6, -4, -\frac{8}{3}, -\frac{16}{9}, \ldots$
31. 80, 100, 125

4 ASSESS

Closing Activity

Speaking Name a ratio and the starting number for a geometric sequence. Then have the students give the next four numbers in the sequence.

Additional Answers

34a. The term increases.
34b. When you multiply by a number less than 1, the term decreases.

Critical Thinking

33\. 7, 9, 3, 1, 7, 9, 3, 1, . . .; 1

33. Use your calculator to find the first eight terms in the geometric sequence defined by $1 \cdot 7^n$, where n represents the number of the term. Determine the pattern that develops in the ones digits of each term. Use the pattern to find the ones digit of the 100th term.

34. Write the first five terms in a geometric sequence with a common ratio that is greater than 1. Then write the first five terms in a geometric sequence with a common ratio that is between 0 and 1.

a. What happens to the size of the terms in a geometric sequence if the common ratio is greater than 1? **See margin.**

b. What happens to the size of the terms in a geometric sequence if the common ratio is less than 1? **See margin.**

Applications and Problem Solving

35. Geography Between 1980 and 1990, the population of Florida increased from approximately 10 million to 13 million.

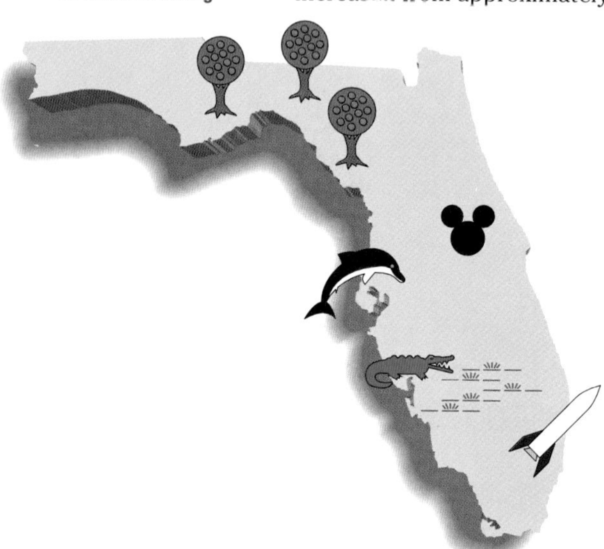

a. Find the common ratio between these numbers. **1.3**

b. If the population grows in a geometric pattern over each of the next four decades, what would the population of Florida be in 2030? **37,129,300**

36. Investment The face value of a high return bond increases to $1\frac{1}{10}$ of its previous value each year. If the starting value is $1250, what is the face value after 3 years? **$1663.75**

37. Patterns Jacob told three friends a secret. Each of these friends told three more friends, who each told three more friends. How many people in all knew about the secret? **40**

38. Games *Jeopardy* holds a tournament each year. Three people participate in each game of the competition and only the winners proceed to the next round. How many players participate if there are five rounds? **243**

Mixed Review

39. Solve $\frac{5}{9}k = 15$. (Lesson 6-7) **27**

40. Solve $d = 119 - 9.8$. (Lesson 5-3) **109.2**

41. Find the quotient $\frac{3^3}{3^7}$. (Lesson 4-9) $\frac{1}{3^4}$ *or* $\frac{1}{81}$

42. Geometry The area of a square is the product of its length and width. Use exponents to write an expression for the area of a square that has sides b units long. (Lesson 4-2) b^2

43. Solve $t = \frac{120}{24}$. (Lesson 2-8) **5**

44. Simplify $5x + 3x + 1$. (Lesson 1-5) $8x + 1$

45. Find the value of $6 \cdot 5 + 12 \div 3$. (Lesson 1-2) **34**

46. Patterns Find the next number in the pattern 8, 9, 11, 14, 18, 23, . . . (Lesson 1-1) **29**

Extension

Using Tables Draw three circles. Mark 2, 3, and 4 points, respectively on the circles. Using straight lines, connect points in all possible ways. Make a table that shows the number of points and the number of regions formed. Predict the number of regions formed by 5 and 6 points.

Enrichment Masters, p. 52

NAME ________________ DATE ________

6-8 Enrichment
Applications of Geometric Sequences

Student Edition
Pages 312–316

Populations often grow according to a geometric sequence. If the population of a country grows at the rate of 2% per year, then the common ratio, r, is 1.02. To find the population of a city of 100,000 after 5 years of 2% growth, use the formula ar^{n-1}, where r is the common ratio and n is the number of years.

$$ar^{n-1} = 100,000 \times 1.02^{5-1} \quad a = 100,000;\ r = 1.02;\ n = 5$$
$$= 100,000 \times 1.02^4$$
$$\approx 108,243$$

After a few years, a small change in the annual growth rate can cause enormous differences in the population.

Assume that the nation of Grogro had a population of one million in 1990. Using a calculator, find the population of the country in the years 2000, 2050, and 2100 at growth rates of 1%, 3%, and 5% per year. Record your results in the table below.

	Population of Grogro			
	Growth Rate	Year		
		2000	2050	2100
1.	1%	1,093,685	1,798,709	2,958,215
2.	3%	1,304,773	5,720,003	25,075,956
3.	5%	1,551,328	17,789,701	204,001,612

Suppose we want to find the *total* distance a bouncing ball has moved. We need a formula for the *sum* of the terms of a geometric sequence. This is called a **geometric series.**

$$S_n = \frac{a - ar^n}{1 - r} \quad \begin{array}{l} a = \text{first term} \\ r = \text{common ratio } (r \neq 1) \\ n = \text{number of terms} \end{array}$$

Use the formula above to find each sum. Then check your answer by adding.

4. 5 + 10 + 20 + 40 + 80 155

5. 80 + 240 + 720 + 2160 + 6480 9680

6-9 Scientific Notation

Setting Goals: *In this lesson, you'll write numbers in scientific notation.*

Modeling a Real-World Application: Geology

Dr. William Rose, a geologist, specializes in the study of volcanoes and their effect on the atmosphere. He and his team studied the Augustine volcano about 280 kilometers from Anchorage, Alaska. The Augustine volcano erupted twice in 1986, once on March 27 and again on April 4.

Even though these eruptions were relatively small, on March 27th and 28th, the volcano produced 75,000,000 metric tons of fallout per day. The individual particles in the fallout were very small. The diameter of a particle was 0.000022 meter.

Learning the Concept

It's easy to make mistakes when computing with small numbers such as 0.000022 and large numbers such as 75,000,000. People who deal regularly with such numbers use **scientific notation**.

Scientific Notation: **Numbers expressed in scientific notation are written as the product of a factor and a power of 10. The factor must be greater than or equal to 1 and less than 10.**

To write a large positive or negative number in scientific notation, move the decimal point to the right of the left-most digit, and multiply this number by a power of ten.

To find the power of ten, count the number of places you moved the decimal point.

Example 1 **Write each number in scientific notation.**

a. 687,000
$687{,}000 = 6.87 \times 10^5$ *Move the decimal point 5 places to the left. Multiply by 10^5.*

b. 50,000,000
$50{,}000{,}000 = 5 \times 10^7$ *Move the decimal point 7 places to the left. Multiply by 10^7.*

LOOK BACK You can review negative exponents in Lesson 4-9.

To write a small positive number in scientific notation, move the decimal point to the right of the first nonzero digit, and multiply this number by a power of ten. Remember that $10^{-n} = \frac{1}{10^n}$. Count the number of places you moved the decimal point and use the negative of that number as the exponent of ten.

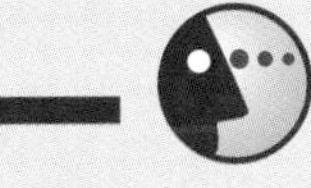

Alternative Learning Styles

Visual For students who have visual difficulty working with a series of zeros, suggest that they turn their lined paper sideways and write one digit in each column formed by the lines on the paper.

6-9 LESSON NOTES

NCTM Standards: 1-5, 7

Instructional Resources

- Study Guide Master 6-9
- Practice Master 6-9
- Enrichment Master 6-9
- Group Activity Card 6-9
- Assessment and Evaluation Masters, p. 156
- Math Lab and Modeling Math Masters, p. 81

Transparency 6-9A contains the 5-Minute Check for this lesson; **Transparency 6-9B** contains a teaching aid for this lesson.

Recommended Pacing	
Standard Pacing	Day 11 of 13
Honors Pacing	Day 10 of 12
Block Scheduling*	Day 6 of 7 (along with Lesson 6-8)

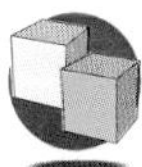

*For more information on pacing and possible lesson plans, refer to the ***Block Scheduling Booklet***.

1 FOCUS

5-Minute Check
(over Lesson 6-8)

State whether each sequence is a geometric sequence. If so, state the common ratio and list the next three terms.

1. 6, 18, 54, 162, ... **yes; 3; 486, 1458, 4374**
2. 8, 4, 0, −4, ... **no**
3. 1024, 256, 64, 16, ... **yes; $\frac{1}{4}$; 4, 1, $\frac{1}{4}$**
4. $-\frac{1}{2}, \frac{3}{2}, -\frac{9}{2}, \frac{27}{2}, \ldots$ **yes; −3; $-\frac{81}{2}, \frac{243}{2}, -\frac{729}{2}$**
5. $\frac{2}{3}, -\frac{1}{3}, -\frac{4}{3}, -\frac{7}{3}, \ldots$ **no**

Motivating the Lesson

Questioning Ask students to write the diameter of Saturn, 121 thousand kilometers. **121,000** Then ask students to express the distance from Saturn to the sun, 1 billion, 430 million kilometers. **1,430,000,000**

2 TEACH

In-Class Examples

For Example 1
Write each number in scientific notation.

a. 92,000 **9.2×10^4**

b. 634,800,000 **6.348×10^8**

For Example 2
Write 0.00002605 in scientific notation. **2.605×10^{-5}**

For Example 3
Write each number in standard form.

a. 3.253×10^6 **3,253,000**

b. 8.3004×10^{-5} **0.000083004**

For Example 4
The Atlantic Ocean has an area of 3.18×10^7 square miles, the Arctic Ocean, 5.4×10^6 square miles, the Indian Ocean, 2.89×10^7 square miles, and the Pacific Ocean, 6.4×10^7 square miles. Rank these oceans by area from least to greatest. **Arctic, Indian, Atlantic, Pacific**

Study Guide Masters, p. 53

NAME ________________ DATE ________

6-9 Study Guide
Scientific Notation

Student Edition
Pages 317–320

A convenient form for writing very large and very small numbers is called **scientific notation**. All numbers expressed in scientific notation are given as the product of a number between 1 and 10 and a power of 10.

Write 2.31×10^{-4} in standard form as follows: — To multiply by 10^{-4}, move the decimal point 4 places to the left.

$2.31 \times 10^{-4} \longrightarrow 0.000231$

Write 2.31×10^4 in standard form as follows: — To multiply by 10^4, move the decimal point 4 places to the right.

$2.31 \times 10^4 \longrightarrow 23,100$

Write 4000 in scientific notation as follows: — Move the decimal point 3 places to the left. Multiply by 10^3.

$4000 = 4 \times 10^3$

Write 0.00712 in scientific notation as follows: — Move the decimal point 3 places to the right. Multiply by 10^{-3}.

$0.00712 = 7.12 \times 10^{-3}$

Write each number in standard form.

1. 7×10^5 **700,000**
2. -3.2×10^3 **−3200**
3. 1.7×10^{-6} **0.0000017**
4. 6.366×10^{-4} **0.0006366**
5. 2.979×10^4 **29,790**
6. -5.09×10^{-7} **−0.000000509**

Write each number in scientific notation.

7. 50,000 **5×10^4**
8. −3700 **-3.7×10^3**
9. 0.000249 **2.49×10^{-4}**
10. 2030 **2.03×10^3**
11. 0.0000755 **7.55×10^{-5}**
12. 51,000 **5.1×10^4**
13. 0.0046 **4.6×10^{-3}**
14. −12,800 **-1.28×10^4**
15. −724 **-7.24×10^2**
16. 0.176 **1.76×10^{-1}**
17. −2156 **-2.156×10^3**
18. 0.03278 **3.278×10^{-2}**

Example  **Write 0.000003576 in scientific notation.**

$0.000003576 = 3.576 \times 10^{-6}$ *Move the decimal point 6 places to the right. Multiply by 10^{-6}.*

Remember that $10^{-6} = \frac{1}{1,000,000}$.

Example  **Write each number in standard form.**

TECHNO TIP

Use the EE key to enter numbers in scientific notation in a calculator. Enter the decimal portion, press EE, then enter the exponent.

a. 8.495×10^5

$8.495 \times 10^5 = 8.495 \times 100,000$ *Move the decimal point 5 places to the right.*

$= 849,500$ 8.49500

b. 3.08×10^{-4}

$3.08 \times 10^{-4} = 3.08 \times \left(\frac{1}{10}\right)^4$ *Remember that $10^{-4} = \left(\frac{1}{10}\right)^4$.*

$= 3.08 \times \frac{1}{10,000}$

$= 3.08 \times 0.0001$ *Move the decimal point 4 places to the left.*

$= 0.000308$ 00003.08

You can compare and order numbers in scientific notation. First compare the exponents. With positive numbers, any number with a greater exponent is greater. If the exponents are the same, compare the numbers.

Example 4 **CONNECTION** **Science**

The diameter of Venus is 1.208×10^4 km, the diameter of Earth is 1.276×10^4 km, and the diameter of Mars is 6.79×10^3 km. Rank these planets by diameter, from least to greatest.

$1.208 \times 10^4 < 1.276 \times 10^4$
The exponent is the same; $1.208 < 1.276$.
So Venus is smaller than Earth.

$6.79 \times 10^3 < 1.208 \times 10^4$ *$10^3 < 10^4$*
So Mars is smaller than Venus.

The rank from least to greatest is Mars, Venus, Earth.

Checking Your Understanding

Communicating Mathematics

1–2. See Solutions Manual.

Read and study the lesson to answer these questions.

1. In your own words, explain how to write these numbers in scientific notation.
   a. 634,930
   b. 0.0000407
2. **Explain** the relationship between a number in standard form and the sign of the exponent when the number is written in scientific notation.

Reteaching

Using Models Write numbers in standard form on a magnetic board or chalkboard. Use a magnet or a piece of masking tape as a decimal point. Have a student physically move the decimal to its new location. Have other students count places as the decimal is moved. Write the correct power of 10 beside the new number.

Cooperative Learning

Jigsaw Place books on different subjects in each corner of the room. Each student in a group of four is assigned a corner to find data that can be written in scientific notation. The group then uses the data to write word problems. For more information on this strategy, see *Cooperative Learning in the Mathematics Classroom*, p. 26.

3. **Assess Yourself** What career are you interested in pursuing after high school? How might you use scientific notation in that career? See students' work.

Guided Practice

State whether each number is in scientific notation. Write *yes* or *no*. If *no*, explain why not. 4. No; 23.45 is not between 0 and 10.

4. 23.45×10^5 5. 5.689×10^{-3} yes 6. 0.23×10^8
6. No; 0.23 is not between 0 and 10.

State where the decimal point should be placed in order to express each number in scientific notation. State the power of ten by which you should multiply. Then write the number in scientific notation.

7. 8490 8.490×10^3 8. 0.000045 4.5×10^{-5} 9. 847.9 8.479×10^2

Write each number in standard form. 11. 61,000

10. 5.2×10^5 520,000 11. 6.1×10^4 12. 9.34×10^{-2} 0.0934

13. **Physics** The speed of light is 3.00×10^5 kilometers per second. If a light-year is the distance light travels in one Earth year, how many kilometers are in one light-year? (One Earth year = 365.25 days. Remember to convert days to seconds) 31,557,600 seconds in a year; $(31,557,600)(3.00 \times 10^5)$ or about 9.47×10^{12} km

Exercises: Practicing and Applying the Concept

Independent Practice

Write each number in scientific notation. See margin.

14. 5,894,000 15. 0.0059 16. 269,000

17. 0.00015 18. 80,000,000 19. 0.0000000498
20. the possible hands in a 5-card game, 2,598,960
21. the population of Africa in 1994, 701,000,000
22. the diameter of a spider's thread, 0.001 inch

20. 2.59896×10^6
21. 7.01×10^8
22. 1.0×10^{-3}
24. 139.9
26. 780,000,000,000
28. 40.54

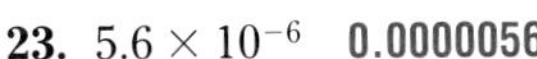

Write each number in standard form.

23. 5.6×10^{-6} 0.0000056 24. 1.399×10^2
25. 9.001×10^{-3} 0.009001 26. 7.8×10^{11}
27. 5.985×10^{-5} 0.00005985 28. 4.054×10^1
29. the height of the Sears Tower, 1.454×10^3 feet 1454 feet
30. the diameter of a uranium atom, 3.5×10^{-8} 0.000000035
31. the volume of a drop of liquid, 5×10^{-5} liter 0.00005

32. Write $(4 \times 10^4) + (8 \times 10^3) + (3 \times 10^2) + (9 \times 10^1) + (6 \times 10^0)$ in standard notation. 48,396
33. Write $(7 \times 10^8) + (8 \times 10^4) + (3 \times 10^0)$ in standard notation.
34. Write $(6 \times 10^0) + (4 \times 10^{-3}) + (3 \times 10^{-5})$ in standard notation. 6.00403

33. 700,080,003

Critical Thinking

35. In standard form, $3.14 \times 10^{-4} = 0.000314$, and $3.14 \times 10^4 = 31,400$. What is 3.14×10^0 in standard form? 3.14

Applications and Problem Solving

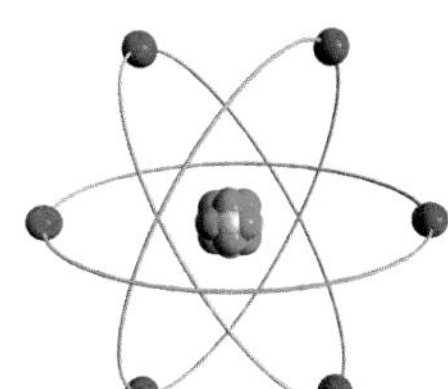

36. **Science** Everything around you is made up of atoms. Atoms are extremely small. An atom is about two millionths of an inch in diameter. Write this number in standard form and then in scientific notation. 0.000002 or 2.0×10^{-6}

3 PRACTICE/APPLY

Checking Your Understanding

Exercises 1–13 are designed to help you assess your students' understanding through reading, writing, speaking, and modeling. You should work through Exercises 1–3 with your students and then monitor their work on Exercises 4–13.

Assignment Guide

Core: 15–37 odd, 38–43
Enriched: 14–34 even, 35–43

For **Extra Practice**, see p. 755.

The red A, B, and C flags, printed only in the Teacher's Wraparound Edition, indicate the level of difficulty of the exercises.

Additional Answers

14. 5.894×10^6
15. 5.9×10^{-3}
16. 2.69×10^5
17. 1.5×10^{-4}
18. 8.0×10^7
19. 4.98×10^{-8}

Practice Masters, p. 53

NAME ______________________ DATE ____________

6-9 Practice
Scientific Notation

Student Edition
Pages 317–320

Write each number in standard form.

1. 8.2×10^3 8200	2. 6.4×10^2 640	3. 3.1×10^4 31,000
4. 9.03×10^{11} 903,000,000,000	5. -6.8×10^8 -680,000,000	6. 9.347×10^4 93,470
7. 1.5×10^{-1} 0.15	8. 7.3×10^{-3} 0.0073	9. 8.7×10^0 8.7
10. 2.9×10^{-2} 0.029	11. -3.07×10^{-4} -0.000307	12. -7.16×10^{-5} -0.0000716
13. 1.234×10^{-3} 0.001234	14. 5.008×10^4 50,080	15. -4.11×10^5 -411,000
16. -2.307×10^0 -2.307	17. 3.09×10^{-4} 0.000309	18. -1.4685×10^1 -14.685

Write each number in scientific notation.

19. 65,000,000 6.5×10^7	20. -9200 -9.2×10^3	21. 840,000 8.4×10^5
22. 0.0056 5.6×10^{-3}	23. 28,400,000 2.84×10^7	24. -5.65 -5.65×10^0
25. 5,620,800,000 5.6208×10^9	26. -0.00087 -8.7×10^{-4}	27. 769.5 7.695×10^2
28. 59,300 5.93×10^4	29. 9,000,000 9.0×10^6	30. -0.3054 -3.054×10^{-1}
31. 0.00001 1.0×10^{-5}	32. -8 -8.0×10^0	33. 89,000,000,000 8.9×10^{10}
34. -175 -1.75×10^2	35. 0.08792 8.792×10^{-2}	36. -31 -3.1×10^1
37. 0.0003141 3.141×10^{-4}	38. -1 -1.0×10^0	39. 6,801,700 6.8017×10^6
40. -5.001 -5.001×10^0	41. 1,000,000,000 1.0×10^9	42. 0.00000938 9.38×10^{-6}

Group Activity Card 6-9

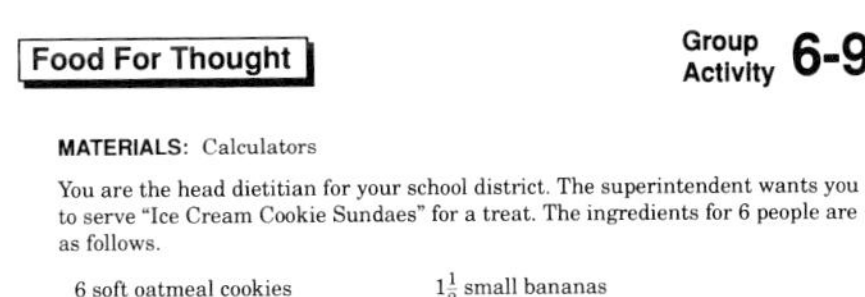

Food For Thought Group Activity **6-9**

MATERIALS: Calculators

You are the head dietitian for your school district. The superintendent wants you to serve "Ice Cream Cookie Sundaes" for a treat. The ingredients for 6 people are as follows.

6 soft oatmeal cookies
$\frac{1}{4}$ cup butterscotch topping
$\frac{1}{5}$ cup chocolate topping
$\frac{1}{3}$ cup caramel topping
$1\frac{1}{2}$ small bananas
$2\frac{7}{8}$ cups vanilla or chocolate ice cream
$\frac{1}{6}$ cup peanuts

Determine how much of each ingredient is needed to serve the 12,000 students in your district. Compute the amounts and express the answers in scientific notation.

 Pre-Algebra

Extension

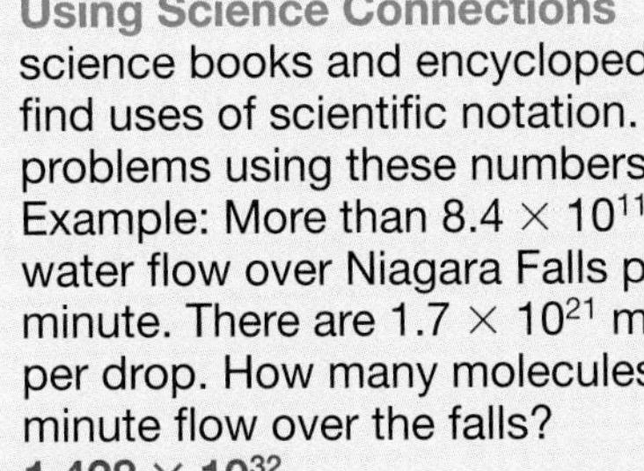

Using Science Connections Use science books and encyclopedias to find uses of scientific notation. Write problems using these numbers. Example: More than 8.4×10^{11} drops of water flow over Niagara Falls per minute. There are 1.7×10^{21} molecules per drop. How many molecules per minute flow over the falls? 1.428×10^{32}

4 ASSESS

Closing Activity

Speaking Have students explain how to write each of the following in standard form. Then have students name the numbers in order from least to greatest.

1.23×10^{-3}
1.23×10^{3}
1.23×10^{-6}
1.23×10^{0}
1.23×10^{5}
1.23×10^{2}

0.00000123, 0.00123, 1.23, 123, 1230, 123,000

Chapter 6, Quiz D (Lessons 6-8 and 6-9) is available in the *Assessment and Evaluation Masters*, p. 156.

Additional Answer

37c. Bezymianny; Santa Maria; Agung; Mount St. Helens tied with Hekla 1947; Hekla, 1970; Ngauruhoe

interNET CONNECTION

For the latest volcano eruption rates, visit: **www.glencoe.com/sec/math/prealg/mathnet**

37. Geology This graph below gives the maximum eruption rate in cubic meters per second of seven volcanoes in the last century.

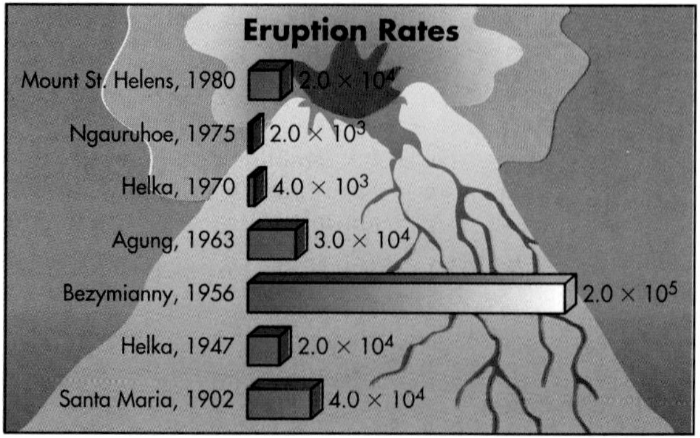

Source: University of Alaska

a. Which volcano had the greatest eruption rate? **Bezymianny**

b. Which had the least eruption rate? **Ngauruhoe**

c. Use standard form to rank the volcanoes in order from greatest to least eruption rate. **See margin.**

Mixed Review

38. State whether $9, -3, 1, -\frac{1}{3}, \ldots$ is a geometric sequence. If so, state the common ratio and list the next three terms. (Lesson 6-8)

38. yes; $-\frac{1}{3}$; $\frac{1}{9}, -\frac{1}{27}, \frac{1}{81}$

39. Solve $x + \frac{2}{3} \geq 3$. (Lesson 5-7) $x \geq 2\frac{1}{3}$

40. Solve $-3\frac{1}{5} - \left(-\frac{2}{5}\right) = a$. (Lesson 5-4) $-2\frac{4}{5}$

41. Evaluate $\frac{b}{7}$, if $b = -91$. (Lesson 2-8) **−13**

42. Solve $x + 7 = 19$ mentally. (Lesson 1-6) **12**

43. Sports Translate *three more than the Bears' score* into an algebraic expression. (Lesson 1-3) $b + 3$

COOPERATIVE LEARNING PROJECT

THE SHAPE OF THINGS TO COME

Living Batteries

Can you imagine taking your car to the service station for a fresh tank of sugar water? Or plugging your stereo into a lemon?! Scientists at King's College in London, England are learning to harness the energy produced by digestion to power living batteries that use sugar and bacteria.

Escherichia coli, or *E. coli*, bacteria break down carbohydrates as a part of the digestive process. This process frees up electrons, which generates electricity. As many as 100 billion, or 1×10^{11}, *E. coli* can exist in a cubic centimeter. One of the prototype living batteries is as small as a button and can run a watch for over a year.

See For Yourself

Research living batteries and the digestive system.

- How does the process of digestion work?
- A simple battery can be made using a zinc needle, a copper needle, wire, and a lemon. Place both needles into the lemon and connect them with the wire. Then place a small light bulb, like one from a flashlight, on the wire to see the result of the power flow. How does the battery work?
- How are the living batteries different than the alkaline batteries we usually use?

COOPERATIVE LEARNING PROJECT

THE SHAPE OF THINGS TO COME

Electrochemist Peter Bennetto and colleagues at King's College created a more complicated living battery than the lemon battery. Free electrons produced by *E. coli* digestion are collected and forced to travel along a copper wire to get to the oxygen they need. While rushing along the copper wire, they can be used to power electrical devices.

Enrichment Masters, p. 53

NAME ______________________ DATE __________

6-9 Enrichment
Scientific Notation

Student Edition Pages 317–320

It is sometimes necessary to multiply and divide numbers in scientific notation.

The following rule is used to multiply numbers in scientific notation.

For any numbers a and b, and any numbers c and d,
$(c \times 10^a)(d \times 10^b) = (c \times d) \times 10^{a+b}$

Example: $(3 \times 10^4)(^-5 \times 10^{-2}) = (3 \times {^-5})10^{4+(-2)}$
$= {^-15} \times 10^2$
$= {^-1500}$

The following rule is used to divide numbers in scientific notation.

For any numbers a and b, and any numbers c and d, $(d \neq 0)$,
$(c \times 10^a) \div (d \times 10^b) = (c \div d) \times 10^{a-b}$

Example: $(30 \times 10^{-4}) \div (6 \times 10^2) = (30 \div 6) \times 10^{-4-2}$
$= 5 \times 10^{-6}$
$= 0.000005$

Multiply. Express the product or quotient in scientific notation.

1. $(2.7 \times 10^9) \times (3.1 \times 10^2)$ 8.37×10^{11}
2. $(6.1 \times 10^{-2}) \times (1.3 \times 10^5)$ 7.93×10^3
3. $(5.4 \times 10^{-3}) \div (1.8 \times 10^2)$ 3×10^{-5}
4. $(6.9 \times 10^{-3}) \div (0.3 \times 10^{-7})$ 2.3×10^5
5. $(1.1 \times 10^{-5}) \times (9.9 \times 10^{-1})$ 1.089×10^{-5}
6. $(4 \times 10^0) \div (10^{-2})$ 4×10^2

Solve. Write your answers in standard form.

7. The distance from Earth to the moon is about 2×10^5 miles. The distance from Earth to the sun is about 9.3×10^7 miles. How many times farther is it to the sun than to the moon? 465
8. If each of the 3×10^4 people employed by Suny Motors earned 4×10^4 dollars last year, how much money did the company pay out to its employees? $1,200,000,000

CHAPTER 6 Highlights

Vocabulary After completing this chapter, you should be able to define each term, property, or phrase, and give an example or two of each.

Algebra
common ratio (p. 312)
compatible numbers (p. 280)
equivalent (p. 275)
geometric sequence (p. 312)
inverse property of multiplication (p. 289)
multiplicative inverses (p. 289)
reciprocal (p. 289)
repeating decimal (p. 275)
scientific notation (p. 317)
terminating decimal (p. 275)
truncate (p. 275)

Statistics
mean (p. 301)
measures of central tendency (p. 301)
median (p. 301)
mode (p. 301)

Geometry
area model (p. 284)

Problem Solving
work backward (p. 314)

Understanding and Using Vocabulary

Determine whether each statement is *true* or *false*. If the statement is false, rewrite it to make it true. See Solutions Manual for rewritten statements.

1. Two numbers whose product is 1 are multiplicative inverses, or reciprocals, of each other. true
2. The mode is the number in the middle when the data are arranged in order. false
3. Each term in a common ratio increases or decreases by a geometric sequence. false
4. Mean is what people are usually talking about when they say average. 4. true
5. A decimal is called a terminating decimal because the division ends when the divisor is zero. false
6. Equivalent numbers are rounded so that they fit together. false
7. Numbers written in scientific notation are expressed as the product of a factor and a power of 10. true
8. A calculator that truncates will display .66666666 when you divide 2 by 3. A calculator that rounds will display .66666667 when you divide 2 by 3. true
9. It is possible to have a set of numerical data with no mean. false
10. The inverse property of multiplication says that every number, including zero, has a multiplicative inverse. Zero is its own multiplicative inverse. false

Highlights

Using the Chapter Highlights

The Chapter Highlights begins with a listing of the new terms, properties, and phrases that were introduced in this chapter.

Assessment and Evaluation Masters, pp. 143-144

NAME ______________________ DATE __________

CHAPTER 6 Test, Form 1B

1. Write $\frac{2}{3}$ as a decimal. If necessary, use bar notation to show a repeating decimal.
   A. -0.6 B. 0.6 C. $-0.\overline{6}$ D. $0.\overline{6}$ — 1. D
2. Express $\frac{5}{16}$ as a decimal.
   A. 0.3125 B. 0.312 C. 5.16 D. 3 — 2. A
3. Replace ● with <, >, or = to make $\frac{2}{3}$ ● $\frac{3}{5}$ a true sentence.
   A. = B. < C. > D. none of these — 3. C
4. Estimate the product $\frac{1}{5}(16)$.
   A. 2 B. 3 C. 5 D. 80 — 4. B
5. Estimate the quotient $32 \div 3.1$.
   A. 96 B. 8 C. 12 D. 10 — 5. D
6. Evaluate ab, if $a = 2\frac{1}{2}$ and $b = \frac{2}{3}$.
   A. [illegible] B. $1\frac{2}{3}$ C. [illegible] D. $1\frac{5}{6}$ — 6. B
7. Evaluate $7\frac{x}{y}$, if $x = 7.12$, and $y = 3.2$.
   A. 4.413 B. 3.14 C. 15.575 D. 16.33 — 7. C

Choose the correct product or quotient.

8. $3\frac{1}{5} \cdot 3\frac{3}{4} = y$.
   A. 12 B. 10 C. $9\frac{4}{20}$ D. $9\frac{1}{5}$ — 8. A
9. $-\frac{2}{3} \div \left(-\frac{3}{2}\right)$
   A. 1 B. -1 C. $-\frac{4}{9}$ D. $\frac{4}{9}$ — 9. D
10. $(3.6)(-1.1)$
    A. -39.6 B. 4.7 C. -3.66 D. -3.96 — 10. D
11. $36.48 \div 8.0$
    A. 4.26 B. 4.06 C. 5.16 D. 4.56 — 11. D
12. Solve $2\frac{1}{4} \div 1\frac{1}{4} = y$.
    A. $1\frac{4}{5}$ B. 2 C. $2\frac{13}{16}$ D. $\frac{5}{9}$ — 12. A
13. Solve $(9.6)(0.4) = w$.
    A. 10 B. 3.44 C. 3.84 D. 38.4 — 13. C

NAME ______________________ DATE __________

Chapter 6 Test, Form 1B (continued)

14. Choose the quotient for $t = 17.92 \div 5.6$ in which the decimal is correctly placed.
    A. 0.032 B. 0.32 C. 3.2 D. 32.0 — 14. C
15. Name the property of multiplication shown by $(3.6)(4.8) = (4.8)(3.6)$.
    A. Identity B. Associative C. Commutative D. Inverse — 15. C
16. Name the property of multiplication shown by $(3.6)\frac{1}{3.6} = 1$.
    A. Identity B. Associative C. Commutative D. Inverse — 16. D
17. Find the median of 62, 63, 63, 64, 65, 70, and 71.
    A. 62 B. 63 C. 64 D. 65 — 17. C
18. Find the mean of 18, 20, 22, 25, and 30.
    A. 12 B. 22 C. 23 D. 115 — 18. C
19. Find the mode of 18, 20, 19, 18, 20, 18, 19, 19, 19, 20, and 21.
    A. 18 B. 19 C. 20 D. 21 — 19. B
20. Solve $\frac{2}{3}y = \frac{3}{7}$.
    A. $\frac{9}{14}$ B. $\frac{2}{7}$ C. $\frac{6}{7}$ D. $\frac{6}{14}$ — 20. A
21. Solve $\frac{x}{-3.2} < 12.8$.
    A. $x < -40.96$ B. $x > -40.96$ C. $x < -4$ D. $x > -4$ — 21. B
22. Find the next term of the geometric sequence 24, 12, 6, 3, $\cdots$.
    A. 1.5 B. 1 C. 0.5 D. 0.25 — 22. A
23. Use the expression $a \cdot r^{n-1}$ to find the sixth term in the sequence 3, 6, 12, 24, $\cdots$.
    A. 32 B. 48 C. 96 D. 192 — 23. C
24. Mary read that a red blood cell is about 7.5×10^{-4} centimeter long. Write this number in standard form.
    A. 0.0075 B. 0.07500 C. 7,500 D. 0.00075 — 24. D
25. The speed of sound waves in the tissues of the human body averages about 154,000 centimeters per second. Write this number in scientific notation.
    A. 15.4×10^5 B. 1.54×10^5 C. 0.154×10^5 D. 154×10^5 — 25. B

Instructional Resources

Three multiple-choice tests and three free-response tests are provided in the *Assessment and Evaluation Masters*. Forms 1A and 2A are for honors pacing, Forms 1B and 2B are for average pacing, Forms 1C and 2C are for basic pacing. Chapter 6 Test, Form 1B is shown at the right. Chapter 6 Test, Form 2B is shown on the next page.

CHAPTER 6 Study Guide and Assessment

Using the Chapter Study Guide and Assessment

A more traditional cumulative review, shown on the next page, is provided in the *Assessment and Evaluation Masters*, pp. 158–159.

Assessment and Evaluation Masters, pp. 149-150

NAME ______________ DATE ________

CHAPTER 6 Test, Form 2B

1. Write $\frac{4}{9}$ as a decimal. — 1. $0.\overline{4}$
2. Replace ● with >, <, or = to make $^-\frac{4}{5}$ ● $\frac{1}{7}$ a true sentence. — 2. <
3. Estimate the product $\frac{2}{3}(31)$. — 3. 20
4. Estimate the product (3.97)(1.9). — 4. 8
5. Solve $\left(^-2\frac{1}{5}\right)(5) = t$. — 5. $^-11$
6. Solve $\frac{2}{5}x = 3\frac{2}{5}$. — 6. $8\frac{1}{2}$
7. Evaluate $a^2(x - y) + 25$, if $a = 2$, $x = 4$, and $y = 7$. — 7. 13
8. Evaluate $\frac{a}{b}$, if $a = {}^-21.06$, and $b = {}^-8.1$. — 8. 2.6

Find the product or quotient.

9. (3.17)(1.02) — 9. 3.2334
10. ($^-6$)(4.37) — 10. $^-26.22$
11. $^-\frac{2}{3} \div \left(^-\frac{4}{5}\right)$ — 11. $\frac{5}{6}$
12. Name the property of multiplication shown by $[^-2(^-4)]3 = {}^-2(^-4 \cdot 3)$. — 12. Associative
13. Name the property of multiplication shown by $\left(\frac{4}{5}\right)\left(\frac{5}{4}\right) = 1$. — 13. Inverse
14. Solve $^-\frac{1}{3}x = \frac{3}{5}$. — 14. $^-1\frac{4}{5}$
15. Solve $3.5 > 7a$. — 15. $a < 0.5$
16. What do we call the number in the middle of a set of numbers arranged in order of size? — 16. median

NAME ______________ DATE ________

Chapter 6 Test, Form 2B (continued)

A new set of encyclopedias consists of seven volumes. In a grocery store promotion, the first volume costs 16¢, the second $1.99, and the remaining volumes cost $9.79 each.

17. What is the mean price of each book in the set? — 17. $7.30
18. What is the median price? — 18. $9.79
19. What is the mode of the book prices? — 19. $9.79
20. Solve $\frac{y}{2.5} = {}^-2.5$. — 20. $^-6.25$
21. Solve $\frac{x}{^-6.4} < 3.2$. — 21. $x > {}^-20.48$
22. Mal states that 1, 10, 2, 20, 3, 30, · · · is not a geometric sequence. Do you agree? Explain. — 22. Yes; there is no common ratio.
23. The game chips at the right illustrate the first three terms of a geometric sequence. What is the seventh term of the sequence? — 23. 192

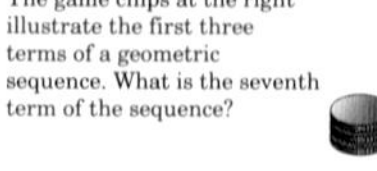
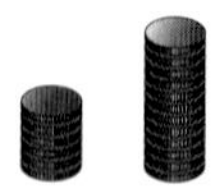

24. Robbi read that the total population of the six New England states in 1995 was 13.3 million people. Write this number in scientific notation. — 24. 1.33×10^7
25. How many microbes are on your skin if you have about 2×10^4 square centimeters of skin and there are 4×10^6 microbes on each square centimeter? Write your answer in scientific notation. — 25. 8×10^{10}

CHAPTER 6 Study Guide and Assessment

Skills and Concepts

Objectives and Examples	Review Exercises

Upon completing this chapter, you should be able to:

Use these exercises to review and prepare for the chapter test.

▶ **write fractions as terminating or repeating decimals.** (Lesson 6-1)

$\frac{2}{5} \rightarrow 5\overline{)2.0}$ with quotient 0.4 $\qquad \frac{4}{9} \rightarrow 4 \boxed{\div} 9 \boxed{=} .44444$

$\frac{2}{5} = 0.4 \qquad \frac{4}{9} = 0.\overline{4}$

Write each fraction as a decimal. Use a bar to show a repeating decimal.

11. $\frac{5}{8}$ **0.625** **12.** $\frac{6}{9}$ **$0.\overline{6}$** **13.** $\frac{17}{40}$ **0.425**

14. $\frac{2}{25}$ **0.08** **15.** $\frac{4}{15}$ **$0.2\overline{6}$** **16.** $\frac{7}{12}$ **$0.58\overline{3}$**

▶ **estimate products and quotients of rational numbers.** (Lesson 6-2)

$14\frac{1}{5} \times 4\frac{8}{9} \rightarrow 14 \times 5 = 70$

$84.73 \div 4.930 \rightarrow 85 \div 5 = 17$

Estimate each product or quotient.

17. $\frac{9}{10} \cdot 21$ **21** **18.** $8.36 \div 16.53$ **0.5**

19. $9 \div 1\frac{1}{5}$ **9** **20.** 18.17×6.19 **120**

▶ **multiply fractions.** (Lesson 6-3)

$a = \frac{4}{3} \cdot \frac{5}{12}$

$a = \frac{\overset{1}{\cancel{4}}}{3} \cdot \frac{5}{\underset{3}{\cancel{12}}}$

$a = \frac{5}{9}$

Solve each equation. Write each solution in simplest form.

21. $-\frac{5}{9} \cdot \frac{8}{25} = n$ $\mathbf{-\frac{8}{45}}$

22. $6\frac{2}{3} \cdot \frac{1}{2} = t$ $\mathbf{3\frac{1}{3}}$

23. $w = -3\left(\frac{6}{15}\right)$ $\mathbf{-1\frac{1}{5}}$

24. $-7\left(\frac{8}{21}\right) = d$ $\mathbf{-2\frac{2}{3}}$

25. $g = \left(\frac{4}{5}\right)^3$ $\mathbf{\frac{64}{125}}$

26. $q = \frac{2}{5} \cdot \frac{7}{8}$ $\mathbf{\frac{7}{20}}$

▶ **divide fractions using multiplicative inverses.** (Lesson 6-4)

Evaluate $c \div d$ if $c = \frac{1}{3}$ and $d = \frac{7}{10}$.

$c \div d = \frac{1}{3} \div \frac{7}{10}$

$= \frac{1}{3} \cdot \frac{10}{7}$ *Multiply by the inverse.*

$= \frac{10}{21}$

Evaluate each expression.

27. $a \div b$, if $a = \frac{1}{4}$ and $b = \frac{2}{6}$ $\mathbf{\frac{3}{4}}$

28. $bc \div f$, if $b = -\frac{8}{15}$, $c = 3$, and $f = 1\frac{1}{2}$ $\mathbf{-1\frac{1}{15}}$

29. $\frac{x}{y}$, if $x = \frac{8}{12}$ and $y = -6$ $\mathbf{-\frac{1}{9}}$

30. $\frac{gh}{8}$, if $g = 3\frac{2}{7}$ and $h = -5\frac{1}{3}$ $\mathbf{-2\frac{4}{21}}$

GLENCOE Technology

Test and Review Software

You may use this software, a combination of an item generator and an item bank, to create your own tests or worksheets. Types of items include free response, multiple choice, short answer, and open ended.

For IBM & Macintosh

Objectives and Examples	Review Exercises

▶ **multiply and divide decimals.** (Lesson 6-5)

Solve $w = -5.7(3.8)$.

$$\begin{array}{rl} -5.7 & \leftarrow \text{ one decimal place} \\ \times\, 3.8 & \leftarrow \text{ one decimal place} \\ 456 & \\ 1710 & \\ -21.66 & \leftarrow \text{ two decimal places} \end{array}$$

Solve each equation. See margin.

31. $4 \times 7.07 = y$ **32.** $k = -1.25 \times 12$
33. $b = 62.9 \div 1000$ **34.** $2.65 \cdot 3.46 = c$
35. $0.7 \div 2.4 = d$ **36.** $-25.9 \div 2.8 = s$
37. $n = 5.3 \cdot (-9)$ **38.** $f = 1.3 \div 100$

▶ **find the mean, median, and mode of a set of data.** (Lesson 6-6)

Find the mean, median, and mode of the test scores 90, 88, 95, 73, and 92.

mean $\frac{90 + 88 + 95 + 73 + 92}{5} = \frac{438}{5}$ or 87.6

median 73 88 90 92 95
↑ *The median is 90.*

mode Since no scores occur more than once, there is no mode.

Find the mean, median, and mode for each set of data. When necessary, round to the nearest tenth. See margin.

39. 21, 25, 16, 18, 18, 26, 21, 17, 19, 25
40. 2.6, 3.1, 6.8, 4.9, 5.7, 3.4, 4.3
41. 0, 9, 14, 22, 26, 14
42. 5.4, 4.6, 6.2. 2.7, 8.0

▶ **solve equations and inequalities containing rational numbers.** (Lesson 6-7)

Solve $-4.7x < 14.1$.

$-4.7x < 14.1$

$\frac{-4.7x}{-4.7} > \frac{14.1}{-4.7}$ *Reverse the order symbol.*

$x > -3$

Solve each equation or inequality. Check your solution. See margin.

43. $8.67k = 78.03$ **44.** $6.3a \leq 52.92$
45. $-\frac{1}{4}r = 3$ **46.** $4.5 < -\frac{5}{8}f$
47. $-25 \geq 6.25n$ **48.** $3x = 55.8$
49. $8x = 42.4$ **50.** $2.6v > 19.76$

▶ **recognize and extend geometric sequences.** (Lesson 6-8)

State whether 768, 192, 48, . . . is a geometric sequence. If so, state the common ratio and list the next three terms.

768, 192, 48, . . .
$\times \frac{1}{4}$ $\times \frac{1}{4}$

The sequence is geometric. The common ratio is $\frac{1}{4}$. The next three terms are $48 \times \frac{1}{4}$ or 12, $12 \times \frac{1}{4}$ or 3, and $3 \times \frac{1}{4}$ or $\frac{3}{4}$.

State whether each sequence is a geometric sequence. If so, state the common ratio and list the next three terms.

51. 6.6, 5.7, 4.9, . . . no
52. 3, 9, 27, . . . yes; 3; 81, 243, 729
53. 0.5, 2, 3.5, 5, . . . no
54. $\frac{1}{4}, \frac{1}{2}$, 1, 2, . . . yes; 2; 4, 8, 16
55. 17, 13, 18, 14, 19, . . . no

Additional Answers

31. 28.28
32. −15
33. 0.0629
34. 9.169
35. $0.291\overline{6}$
36. −9.25
37. −47.7
38. 0.013
39. 20.6; 20; 18, 21, and 25
40. 4.4; 4.3; none
41. 14.2; 14; 14
42. 5.4; 5.4; none
43. 9
44. $a \leq 8.4$
45. −12
46. $f < -7.2$
47. $-4 \geq n$
48. 18.6
49. 5.3
50. $v > 7.6$

Applications and Problem Solving

Encourage students to work through the exercises in the Applications and Problem Solving section to strengthen their problem-solving skills.

Additional Answers

56. 1.349×10^7
57. 6.74×10^{-6}
58. 3.2×10^{-4}
59. 5.81×10^3

Objectives and Examples

▶ **write numbers in standard form and scientific notation.** (Lesson 6-9)

Write 58,970,000,000 in scientific notation.

$58{,}970{,}000{,}000 = 5.897 \times 10^{10}$

Write 6.892×10^{-2} in standard form.

$$6.892 \times 10^{-2} = 6.892 \times \left(\frac{1}{10}\right)^2 = 6.892 \times 0.01 = 0.06892$$

Review Exercises

Write each number in scientific notation.

56. 13,490,000 **57.** 0.00000674
58. 0.00032 **59.** 5810
56–59. See margin.

Write each number in standard form.

60. 4.24×10^2 **424**
61. 5.72×10^4 **57,200**
62. 3.347×10^{-1} **0.3347**
63. 2.02×10^8 **202,000,000**

Applications and Problem Solving

64. Sports Of the medals that the United States' 1992 Summer Olympic Team earned, 0.34 were gold. $\frac{44}{112}$ of the medals earned by the Unified Team were gold. Which team had a greater portion of their medals as gold? (Lesson 6-1) **the Unified team**

65. Economics The sales tax for many cities is $5\frac{1}{2}$%. This means that for every dollar you spend, you pay $5\frac{1}{2}$¢ in tax. If you purchase an item for $24, how much will you owe in tax? (Lesson 6-3) **132¢ or $1.32**

66. Population Canada is divided into ten provinces. The table at the right shows the population in thousands of each province in 1991. (Lesson 6-6)

a. Find the mean population for the provinces. **2,721.2 thousand**

b. Find the median population for the provinces. **1,040.5 thousand**

c. Find the mode population for the provinces. **none**

Province	Population (thousands)
Alberta	2546
British Columbia	3282
Manitoba	1092
New Brunswick	724
Newfoundland	568
Nova Scotia	900
Ontario	10,085
Prince Edward Island	130
Quebec	6896
Saskatchewan	989

67. Astronomy The first new moon discovered by *Voyager* revolves around Uranus. Called Puck, this moon makes one orbit around Uranus every 18 hours. How many times has Puck circled Uranus in 273.6 hours? (Lesson 6-7) **15.2 times**

68. Forestry The U.S. National Park System covers a total of 80,663,217.42 acres. Write this number in scientific notation. (Lesson 6-9) **8.066321742×10^7**

A practice test for Chapter 6 is available on page 779.

Alternative Assessment

Performance Task **Demonstrate your knowledge by giving a clear, concise solution to each problem. Be sure to include all relevant drawings and justify your answers. You may show your solutions in more than one way or investigate beyond the requirements of the problem.**

Stock	Price per share
McDonalds	$39\frac{5}{8}$
Wendys	19
Sony	$55\frac{3}{8}$
Nike	$92\frac{3}{4}$

1. How much does one share of Nike stock cost in dollars? **$92.75**
2. How much would it cost if you bought 6 shares of Sony stock? **$332.25**
3. If you spent $951 buying shares of stock in McDonalds, how many shares would you have purchased? **24**
4. Find the mean of the stock prices listed. **$51.69**
5. Find the median stock price. **$47.50**
6. If you bought 32 shares of Wendy's at $17\frac{3}{4}$ per share and sold it all at the price listed, how much money would you make? **$40.00**
7. Find a list of stock prices in a local newspaper. Choose ten shares of at least three stocks to "sell". Record the prices and the number of shares of each stock that you chose in a table. Then look at a paper from one week ago and record the price of each stock then.
   a. If you bought the shares last week and sold this week, how much did you make or lose?
   b. Find the gain or loss on each stock. Then find the mean, median, and mode of the numbers. If you were bragging about your financial accomplishments, which number would you tell? If you were complaining about the stock market which number would you tell? Explain. **See students' work.**

7a. See students' work.

Thinking Critically

- Why is scientific notation important? **faster, smaller margin of error**
- One meter is 1×10^2 centimeters. The approximate size of an atom is 1×10^{-10} centimeters. How many meters is that? Write the number of meters in scientific notation and standard form. **1×10^{-12} or 0.000000000001 meters**

Portfolio Select some of your work from this chapter that shows how you used a calculator or a computer and place it in your portfolio. **See students' work.**

Self Evaluation Assess yourself. When you first started this chapter, what did you think was going to be the most difficult lesson? Did you let the lesson win, or did you work hard and learn how to do it? How will you react when faced with a similar situation next time? **See students' work.**

Alternative Assessment

The Alternative Assessment section provides students with the opportunity to assess their own work by thinking critically, working with others, keeping a portfolio, and honestly evaluating their own progress. For more information on alternative forms of assessment, see *Alternative Assessment in the Mathematics Classroom*.

Performance Assessment

Performance Assessment tasks for this chapter are included in the *Assessment and Evaluation Masters*. A scoring guide is also provided.

Alternative Assessment

Assessment and Evaluation Masters, pp. 153, 165

NAME ______________________ DATE __________

CHAPTER 6 Performance Assessment

Instructions: Demonstrate your knowledge by giving a clear, concise solution to each problem. Be sure to include all relevant drawings and justify your answers. You may show your solution in more than one way or investigate beyond the requirements of the problem.

1. a. Write the multiplication sentence modeled by the area diagram shown at the right. Find the product.

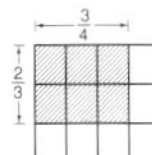

   b. Draw an area model to illustrate the product of $\frac{1}{2} \times \frac{3}{5}$. Explain your model.
   c. Make up a problem that involves the product of two fractions. Find the product and answer the problem.
   d. Write the division sentence modeled by the area diagram below. Find the quotient.

   $1\frac{1}{2}$ Hint: 8

   e. Draw an area model to illustrate the quotient of $2\frac{1}{2} \div \frac{1}{2}$. Explain your model.
2. Seams and Hems is a custom sewing center. The company does sewing for individuals and organizations. Suppose you are in charge of this company.
   a. The sewing center is making 7 cheerleading uniforms for Northmont High School. Each uniform will require about $4\frac{5}{8}$ yards of material. Estimate how much material will be required. Then compute the actual amount of material that will be required. Compare your answer with your estimate. Explain how you know your answer is reasonable.
   b. A customer has brought in 40 yards of material that he wants made into aprons. Each apron uses $1\frac{3}{4}$ yards of material. Write an equation and solve to find how many aprons can be made. Explain in your own words whether the answer needs to be rounded up or down and why. Include any diagrams that may help your explanation.
3. a. Explain why $1, \frac{1}{2}, \frac{1}{4}, \frac{1}{8}, \cdots$ is a geometric sequence.
   b. What is the quotient if the 99th term of the sequence is divided by the 100th term?

CHAPTER 6 Scoring Guide

Level	Specific Criteria
3 Superior	• Shows thorough understanding of the concepts *rounding, geometric sequence*, and *multiplication and division of fractions and mixed numbers*. • Uses appropriate strategies to solve problems. • Computations to multiply and divide fractions and mixed numbers are correct. • Written explanations are exemplary. • Word problem concerning the product of fractions is appropriate and makes sense. • Modeling is accurate and clear. • Goes beyond requirements of some or all problems.
2 Satisfactory, with minor flaws	• Shows understanding of the concepts *rounding, geometric sequence*, and *multiplication and division of fractions and mixed numbers*. • Uses appropriate strategies to solve problems. • Computations to multiply and divide fractions and mixed numbers are mostly correct. • Written explanations are effective. • Word problem concerning the product of fractions is appropriate and makes sense. • Modeling is mostly accurate and clear. • Satisfies requirements of problems.
1 Nearly Satisfactory, with serious flaws	• Shows understanding of most of the concepts *rounding, geometric sequence*, and *multiplication and division of fractions and mixed numbers*. • May not use appropriate strategies to solve problems. • Computations to multiply and divide fractions and mixed numbers are mostly correct. • Written explanations are satisfactory. • Word problem concerning the product of fractions is appropriate and makes sense. • Modeling is mostly accurate and clear. • Satisfies most requirements of problems.
0 Unsatisfactory	• Shows little or no understanding of the concepts *rounding, geometric sequence*, and *multiplication and division of fractions and mixed numbers*. • May not use appropriate strategies to solve problems. • Computations to multiply and divide fractions and mixed numbers are incorrect. • Written explanations are not satisfactory. • Word problem concerning the product of fractions is not appropriate or sensible. • Modeling is not accurate or clear. • Does not satisfy requirements of problems.

Chapter 6 • Ongoing Assessment

Using the Ongoing Assessment

These two pages review the skills and concepts presented in Chapters 1–6. This review is formatted to reflect new trends in standardized testing.

Assessment and Evaluation Masters, pp. 158–159

NAME ________________ DATE ________

CHAPTERS 1–6 Cumulative Review

Simplify each expression. (Lessons 1-2, 2-5, 2-8)

1. $16 \div 4 + 4$ — 1. 8
2. $8 + 8 \div 2$ — 2. 12
3. $^-8 - 15$ — 3. $^-23$
4. $^-12 - (^-12)$ — 4. 0
5. $^-64 \div (^-4)$ — 5. 16

Solve each equation or inequality. (Lessons 3-3, 3-6, 3-7)

6. $^-48 = 3y$ — 6. $^-16$
7. $\frac{x}{16} = 12$ — 7. 192
8. $x + 18 < 12$ — 8. $x < {^-6}$
9. $^-3y > {^-42}$ — 9. $y < 14$
10. Maria is waxing a dance floor shaped as a square 60 feet long on each side. Find the area of the floor. (Lesson 3-5) — 10. 3600 ft^2

Find the prime factorization of each number. (Lesson 4-5)

11. 128 — 11. 2^7
12. 1225 — 12. $5^2 \cdot 7^2$
13. Find the GCF of 24 and 40. (Lesson 4-6) — 13. 8
14. Find the LCM of 12 and 16. (Lesson 4-8) — 14. 48
15. Simplify $n^6 \cdot n \cdot n^{14}$. (Lesson 4-9) — 15. n^{21}

Write each decimal as a fraction. (Lesson 5-1)

16. 0.08 — 16. $\frac{2}{25}$
17. 0.45 — 17. $\frac{9}{20}$

Solve each equation or inequality. (Lessons 5-3, 5-7)

18. $18 - 3.64 = x$ — 18. 14.36
19. $f - 3.7 < 5.2$ — 19. $f < 8.9$

NAME ________________ DATE ________

Chapters 1–6, Cumulative Review (continued)

Solve each equation. (Lessons 6-3, 6-4, 6-5)

20. $\frac{6}{8} \cdot \frac{20}{32} = y$ — 20. $\frac{15}{32}$
21. $^-\frac{24}{17} \div \frac{36}{34} = x$ — 21. $^-1\frac{1}{3}$
22. $(8.36)(^-0.16) = w$ — 22. $^-1.3376$
23. $^-29.58 \div 3.4 = z$ — 23. $^-8.7$
24. Colleen is the freshman class treasurer. The class has made $380 from a bake sale, $425 from a car-wash, and $902 from a paper drive in the fall. She divides the earnings equally among three accounts. How much money is in each account? (Lesson 6-5) — 24. $569

On July 4th, several southern cities reported the following temperatures in degrees Fahrenheit: 102, 103, 99, 101, 102, 100, 102. (Lesson 6-6)

25. What was the median temperature? — 25. 102
26. What was the mode of the temperatures? — 26. 102
27. What was the mean temperature? — 27. 101.3

State whether each sequence is a geometric sequence. If it is, state the common ratio and list the next three terms. (Lesson 6-8)

28. 1, 4, 9, 16, 25, · · · — 28. no
29. 5, 25, 125, 625, · · · — 29. yes; 5; 3125, 15,625, 78,125
30. 1, 8, 15, 22, 29, · · · — 30. no

Write each number in scientific notation. (Lesson 6-9)

31. 0.0000302 — 31. 3.02×10^{-5}
32. 2850000000 — 32. 2.85×10^9
33. Write 5.032×10^{10} in standard form. (Lesson 6-9) — 33. 50,320,000,000

CHAPTERS 1–6 Ongoing Assessment

Section One: Multiple Choice

There are ten multiple-choice questions in this section. After working each problem, write the letter of the correct answer on your paper.

1. Carl bought a bike that cost $532.16. He paid for the bike in equal payments for 12 months. Estimate: the amount of each payment was between **C**

A. $10 and $20.
B. $25 and $35.
C. $40 and $50.
D. $55 and $65.

2. The Sanchez's added a 12-foot-by-15-foot rectangular deck to the back of their house. How much wood do they need for the railing? (The deck is attached to the house on one of the long sides.) **A**

A. 39 feet
B. 27 feet
C. 42 feet
D. 90 feet

3. Gladys bought grocery items for the following prices: $1.39, $2.89, $0.58, and $1.19. The best estimate of the total cost is **B**

A. $9.
B. $6.
C. $5.
D. $4.

4. If $m = 12$ and $p = 10$, what is the value of $m + mp$? **B**

A. 120
B. 132
C. 240
D. 1440

5. If $3.2y = 80$, what is the value of y? **A**

A. 25
B. 2.5
C. 256
D. 76.8

6. Which statement is *not* true? **C**

A. $\frac{1}{2} > \frac{3}{8}$
B. $\frac{4}{5} < \frac{5}{6}$
C. $\frac{8}{11} < \frac{7}{13}$
D. $\frac{1}{3} < \frac{10}{13}$

7. Brent knows that to have an average of more than 85 in biology, his two exam scores must total more than 170. If his first grade was 82, which inequality could he use to find g, his second grade needed for an average greater than 85? **A**

A. $82 + g > 170$
B. $82 > 170 + g$
C. $g + 82 < 170$
D. $\frac{82 + g}{2} > 170$

8. The mileage reading on the Garcia's car was 256.8 before they left on vacation. When the family returned, the reading was 739.4. How many miles did the Garcia family travel on their vacation? **D**

A. 996.2 mi
B. 583.5 mi
C. 483.6 mi
D. 482.6 mi

9. Which is equivalent to $\frac{a^2}{a^4}$? **A**

A. a^{-2}
B. a^2
C. a^6
D. a^8

10. The labor charge of repairing a car is $42.50 per hour. If it takes 2.5 hours to repair the car, what will be the charge for labor? **B**

A. $812.50
B. $106.25
C. $13.00
D. none of these

Standardized test practice questions are provided in the *Assessment and Evaluation Masters*, p. 157.

Section Two: Free Response

This section contains eight questions for which you will provide short answers. Write your answer on your paper.

11. Solve $\frac{a}{-5} < 1$ and graph on a number line. $a > -5$;

12. Express 0.36 as a fraction. $\frac{9}{25}$

13. Find the GCF of $60x^2y$, $24x^3y^2$, and $84xy^4$. $12xy$

14. Write the first five terms of a geometric sequence that has a common ratio of -2. See students' work.

15. Carlos used $2\frac{5}{8}$ yards of material to make a shirt. He began with $3\frac{1}{4}$ yards of material. How much material did he have left? $\frac{5}{8}$ yards

16. The difference between the average high temperature in January and July in New York City is 37°F. The average high in July is 85°F. What is the average high in January? 48°F

17. The dimensions of Adita's yard are 25 feet long by 40 feet wide. Express the ratio of the length to the width in simplest form. $\frac{5}{8}$

18. Ms. Clinton calculated that it would take $2\frac{5}{8}$ gallons of paint to paint one classroom at her school. She is going to paint 5 classrooms. How many gallons of paint will she need? $13\frac{1}{8}$ gallons

Test-Taking Tip

You may be able to eliminate all or most of the answer choices by estimating. Also, look to see which answer choices are not reasonable for the information given in the problem.

When questions involve variables, check your solutions by substituting values for the variables.

Section Three: Open-Ended

This section contains three open-ended problems. Demonstrate your knowledge by giving a clear, concise solution to each problem. Your score on these problems will depend on how well you do the following.

- Explain your reasoning
- Show your understanding of the mathematics in an organized manner.
- Use charts, graphs, and diagrams in your explanation.
- Show the solution in more than one way or relate it to other situations.
- Investigate beyond the requirements of the problem.

19. Give an example of either inductive or deductive reasoning. Explain why your example is either inductive or deductive reasoning. See students' work.

20. The high temperatures for a week in April were 56°F, 58°F, 60°F, 63°F, 58°F, 62°F, and 70°F. What were the mean and the median temperatures? Explain the difference between the two temperatures. Mean: 61°F, Median: 60°F; The mean is the arithmetic average and the median is the middle number.

21. Give an example of a decimal that neither terminates nor repeats. Sample answer: 0.01001000100001 . . .

CHAPTER 7

Solving Equations and Inequalities

PREVIEWING THE CHAPTER

In this chapter, students explore the solutions of equations and inequalities. Students start by solving problems using the strategy of working backward. This strategy is extended and applied throughout the chapter as students solve equations and inequalities. Next, students use concrete manipulatives to model and solve two-step equations. Students then learn to solve equations algebraically and to write and solve such equations in order to solve word problems. The circumference of a circle is presented as a real-world application of solving equations. Students extend their skills to solving inequalities. Finally, students use their knowledge of equations to convert measures in the metric system.

Lesson (Pages)	Lesson Objectives	NCTM Standards	State/Local Objectives
7-1 (330–332)	Solve problems by working backward.	1-4	
7-2A (333)	Use cups and counters to model and solve two-step equations.	1-4, 9	
7-2 (334–337)	Solve equations that involve more than one operation.	1-4, 9, 12	
7-3 (338–340)	Solve verbal problems by writing and solving equations.	1-4, 9	
7-4 (341–344)	Find the circumference of a circle.	1-4, 9, 12	
7-5A (345)	Use cups and counters to model and solve equations with the variable on each side.	1-4, 9	
7-5 (346–350)	Solve equations with the variable on each side.	1-4, 9	
7-6 (351–354)	Solve inequalities that involve more than one operation.	1-5, 9	
7-7 (355–357)	Solve verbal problems by writing and solving inequalities.	1-4, 9	
7-8 (358–361)	Convert measurements in the metric system.	1-4, 13	

A complete, 1-page lesson plan is provided for each lesson in the ***Lesson Planning Guide***. Answer keys for each lesson are available in the ***Answer Key Masters***.

ORGANIZING THE CHAPTER

You may want to refer to the **Course Planning Calendar** on page T12 for detailed information on pacing.
PACING: Standard—12 days; **Honors**—11 days; **Block**—7 days

LESSON PLANNING CHART

Lesson (Pages)	Materials/ Manipulatives	Extra Practice (Student Edition)	Blackline Masters									Real-World Applications	Interactive Mathematics Tools Software	Teaching Transparencies	Group Activity Cards
			Study Guide	Practice	Enrichment	Assessment and Evaluation	Math Lab and Modeling Math	Multicultural Activity	Tech Prep Application	Graphing Calculator	Activity				
7-1 (330–332)		p. 756	p. 54	p. 54	p. 54		p. 82				pp. 7, 38			7-1A 7-1B	7-1
7-2A (333)	cups* counters* equation mat*						p. 39						7-2A		
7-2 (334–337)	cups* counters* equation mat*	p. 756	p. 55	p. 55	p. 55	p. 183								7-2A 7-2B	7-2
7-3 (338–340)		p. 756	p. 56	p. 56	p. 56				p. 13					7-3A 7-3B	7-3
7-4 (341–344)	circular objects string metric ruler*	p. 757	p. 57	p. 57	p. 57	pp. 182, 183		p. 13						7-4A 7-4B	7-4
7-5A (345)	cups* counters* equation mat*						p. 40								
7-5 (346–350)	graphing calculator cups* counters* equation mat*	p. 757	p. 58	p. 58	p. 58		p. 65			p. 7		14	7-5	7-5A 7-5B	7-5
7-6 (351-354)	cups* counters*	p. 757	p. 59	p. 59	p. 59	p. 184		p. 14	p. 14					7-6A 7-6B	7-6
7-7 (355–357)		p. 757	p. 60	p. 60	p. 60									7-7A 7-7B	7-7
7-8 (358–361)		p. 758	p. 61	p. 61	p. 61	p. 184					p. 21	15		7-8A 7-8B	7-8
Study Guide/ Assessment (364–367)						pp. 169–181, 185–187									

*Included in Glencoe's *Student Manipulative Kit* and *Overhead Manipulative Resources*.

ORGANIZING THE CHAPTER

All of the blackline masters in the Teacher's Classroom Resources are available on the ***Electronic Teacher's Classroom Resources*** CD-ROM.

OTHER CHAPTER RESOURCES

Student Edition
Chapter Opener, pp. 328–329
Working on the Investigation, p. 337
Closing the Investigation, p. 362
Cooperative Learning Project: The Shape of Things to Come, p. 354
From The Funny Papers, p. 340

Teacher's Classroom Resources
Investigations and Projects Masters, pp. 53-56
Pre-Algebra Overhead Manipulative Resources, pp. 16–18

Technology
Test and Review Software (IBM & Macintosh)
CD-ROM Activities

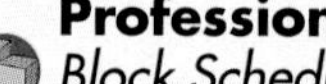

Professional Publications
Block Scheduling Booklet
Glencoe Mathematics Professional Series

OUTSIDE RESOURCES

Books/Periodicals
Algebra With Pizzazz! (Grades 8–12) Creative Publications

Videos/CD-ROMs
Mathematical Eye: Equations and Formula, Journal Films, Multimedia and Videodisc Compendium

Software
Quincy Market, Regional Math Network
Mastering Equations, Roots and Exponents, Phoenix/BFA Films and Videos

ASSESSMENT RESOURCES

Student Edition
Math Journal, pp. 343, 356, 360
Mixed Review, pp. 332, 337, 340, 344, 350, 354, 357, 361
Self Test, p. 350
Chapter Highlights, p. 363
Chapter Study Guide and Assessment, pp. 364–366
Alternative Assessment, p. 367
Portfolio, p. 367

MindJogger Videoquiz, 7

Teacher's Wraparound Edition
5-Minute Check, pp. 330, 334, 338, 341, 346, 351, 355, 358
Checking Your Understanding, pp. 331, 335, 340, 342, 348, 353, 357, 360
Closing Activity, pp. 332, 337, 340, 344, 350, 354, 357, 361
Cooperative Learning, pp. 355, 359

Assessment and Evaluation Masters
Multiple-Choice Tests, Forms 1A (Honors), 1B (Average), 1C (Basic), pp. 169–174
Free-Response Tests, Forms 2A (Honors), 2B (Average), 2C (Basic), pp. 175–180
Performance Assessment, p. 181
Mid-Chapter Test, p. 182
Quizzes A–D, pp. 183–184
Standardized Test Practice, p. 185
Cumulative Review, pp. 186-187

ENHANCING THE CHAPTER

Examples of some of the materials for enhancing Chapter 7 are shown below.

Multicultural Activity Masters, pp. 13, 14

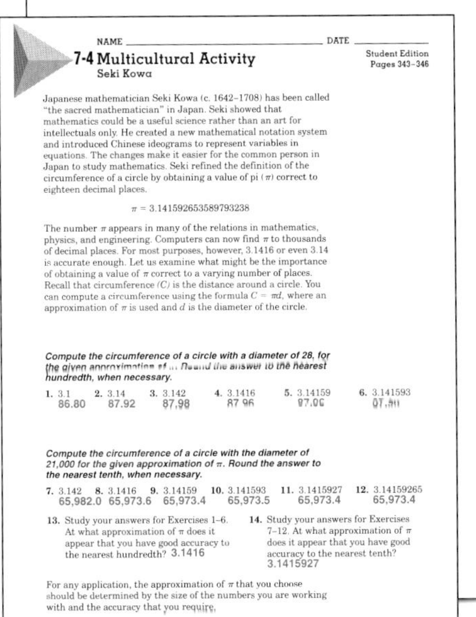

NAME ______________ DATE ________

7-4 Multicultural Activity
Seki Kowa

Student Edition Pages 343–346

Japanese mathematician Seki Kowa (c. 1642–1708) has been called "the sacred mathematician" in Japan. Seki showed that mathematics could be a useful science rather than an art for intellectuals only. He created a new mathematical notation system and introduced Chinese ideograms to represent variables in equations. The changes make it easier for the common person in Japan to study mathematics. Seki refined the definition of the circumference of a circle by obtaining a value of pi (π) correct to eighteen decimal places.

$$\pi = 3.141592653589793238$$

The number π appears in many of the relations in mathematics, physics, and engineering. Computers can now find π to thousands of decimal places. For most purposes, however, 3.1416 or even 3.14 is accurate enough. Let us examine what might be the importance of obtaining a value of π correct to a varying number of places. Recall that circumference (C) is the distance around a circle. You can compute a circumference using the formula $C = \pi d$, where an approximation of π is used and d is the diameter of the circle.

Compute the circumference of a circle with a diameter of 28, for the given approximation of π. Round the answer to the nearest hundredth, when necessary.

1. 3.1 86.80
2. 3.14 87.92
3. 3.142 87.98
4. 3.1416 87.96
5. 3.14159 [illegible]
6. 3.141593 [illegible]

Compute the circumference of a circle with the diameter of 21,000 for the given approximation of π. Round the answer to the nearest tenth, when necessary.

7. 3.142 65,982.0
8. 3.1416 65,973.6
9. 3.14159 65,973.4
10. 3.141593 65,973.5
11. 3.1415927 65,973.4
12. 3.14159265 65,973.4

13. Study your answers for Exercises 1–6. At what approximation of π does it appear that you have good accuracy to the nearest hundredth? 3.1416
14. Study your answers for Exercises 7–12. At what approximation of π does it appear that you have good accuracy to the nearest tenth? 3.1415927

For any application, the approximation of π that you choose should be determined by the size of the numbers you are working with and the accuracy that you require.

Real-World Applications, 14, 15

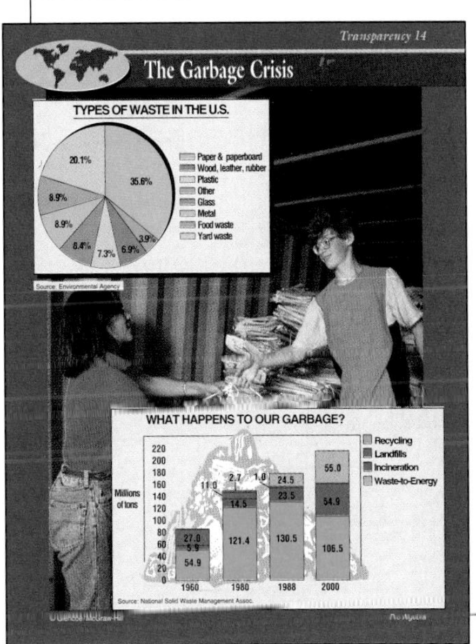

Graphing Calculator Masters, p. 7

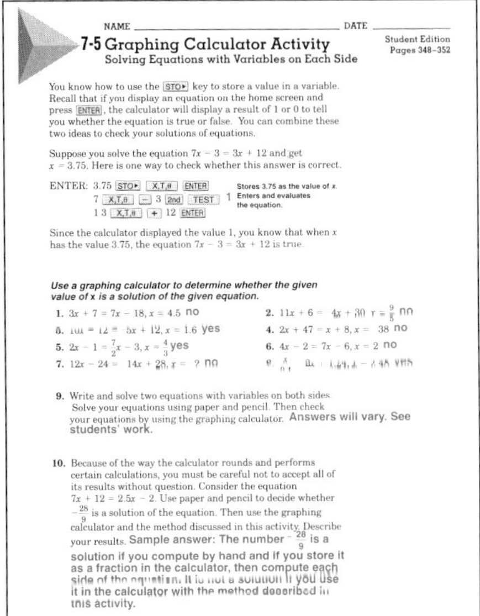

NAME ______________ DATE ________

7-5 Graphing Calculator Activity
Solving Equations with Variables on Each Side

Student Edition Pages 348–352

You know how to use the STO▸ key to store a value in a variable. Recall that if you display an equation on the home screen and press ENTER, the calculator will display a result of 1 or 0 to tell you whether the equation is true or false. You can combine these two ideas to check your solutions of equations.

Suppose you solve the equation $7x - 3 = 3x + 12$ and get $x = 3.75$. Here is one way to check whether this answer is correct.

ENTER: 3.75 STO▸ X,T,θ ENTER — Stores 3.75 as the value of x.
7 X,T,θ − 3 2nd TEST 1 3 X,T,θ + 12 ENTER — 1 — Enters and evaluates the equation.

Since the calculator displayed the value 1, you know that when x has the value 3.75, the equation $7x - 3 = 3x + 12$ is true.

Use a graphing calculator to determine whether the given value of x is a solution of the given equation.

1. $3x + 7 = 7x - 18, x = 4.5$ no
2. $11x + 6 = 4x + 30, x = \frac{9}{5}$ no
3. [illegible] $= 5x + 12, x = 1.6$ yes
4. $2x + 47 = x + 8, x = 38$ no
5. $2x - 1 = \frac{7}{2}x - 3, x = \frac{4}{3}$ yes
6. $4x - 2 = 7x - 6, x = 2$ no
7. $12x - 24 = 14x + 28, x = ?$ no
8. [illegible] yes

9. Write and solve two equations with variables on both sides. Solve your equations using paper and pencil. Then check your equations by using the graphing calculator. Answers will vary. See students' work.

10. Because of the way the calculator rounds and performs certain calculations, you must be careful not to accept all of its results without question. Consider the equation $7x + 12 = 2.5x - 2$. Use paper and pencil to decide whether $-\frac{28}{9}$ is a solution of the equation. Then use the graphing calculator and the method discussed in this activity. Describe your results. Sample answer: The number $-\frac{28}{9}$ is a solution if you compute by hand and if you store it as a fraction in the calculator, then compute each side of the equation. It is not a solution if you use it in the calculator with the method described in this activity.

Tech Prep Applications Masters, pp. 13, 14

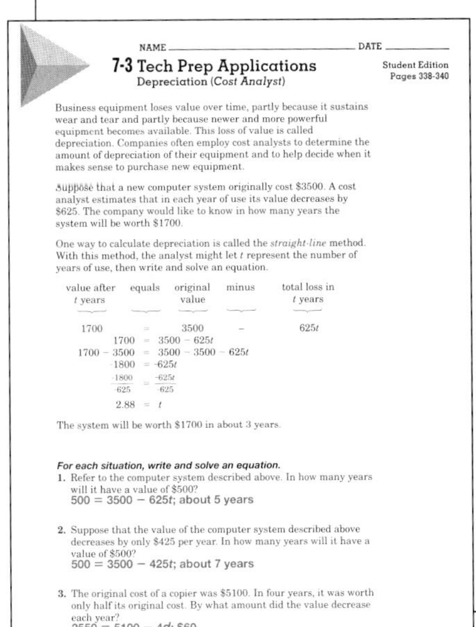

NAME ______________ DATE ________

7-3 Tech Prep Applications
Depreciation (*Cost Analyst*)

Student Edition Pages 338-340

Business equipment loses value over time, partly because it sustains wear and tear and partly because newer and more powerful equipment becomes available. This loss of value is called depreciation. Companies often employ cost analysts to determine the amount of depreciation of their equipment and to help decide when it makes sense to purchase new equipment.

Suppose that a new computer system originally cost $3500. A cost analyst estimates that in each year of use its value decreases by $625. The company would like to know in how many years the system will be worth $1700.

One way to calculate depreciation is called the *straight-line* method. With this method, the analyst might let t represent the number of years of use, then write and solve an equation.

value after t years	equals	original value	minus	total loss in t years
1700	=	3500	−	$625t$

$$1700 = 3500 - 625t$$
$$1700 - 3500 = 3500 - 3500 - 625t$$
$$-1800 = -625t$$
$$\frac{-1800}{-625} = \frac{-625t}{-625}$$
$$2.88 = t$$

The system will be worth $1700 in about 3 years.

For each situation, write and solve an equation.

1. Refer to the computer system described above. In how many years will it have a value of $500?
   $500 = 3500 - 625t$; about 5 years
2. Suppose that the value of the computer system described above decreases by only $425 per year. In how many years will it have a value of $500?
   $500 = 3500 - 425t$; about 7 years
3. The original cost of a copier was $5100. In four years, it was worth only half its original cost. By what amount did the value decrease each year?
   $2550 = 5100 - 4d$; $60

Activity Masters, pp. 7, 21, 38

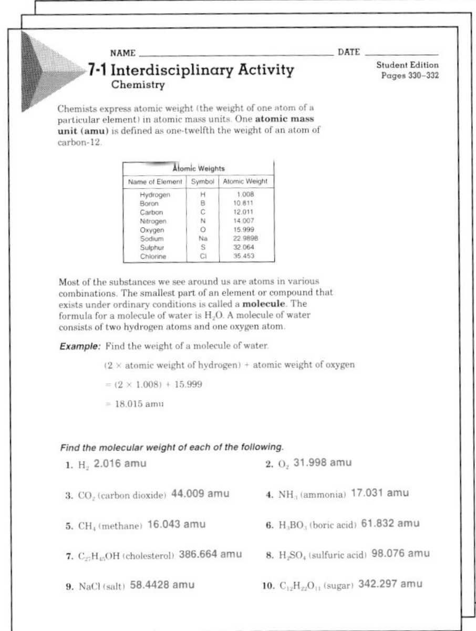

NAME ______________ DATE ________

7-1 Interdisciplinary Activity
Chemistry

Student Edition Pages 330–332

Chemists express atomic weight (the weight of one atom of a particular element) in atomic mass units. One **atomic mass unit (amu)** is defined as one-twelfth the weight of an atom of carbon-12.

Atomic Weights

Name of Element	Symbol	Atomic Weight
Hydrogen	H	1.008
Boron	B	10.811
Carbon	C	12.011
Nitrogen	N	14.007
Oxygen	O	15.999
Sodium	Na	22.9898
Sulphur	S	32.064
Chlorine	Cl	35.453

Most of the substances we see around us are atoms in various combinations. The smallest part of an element or compound that exists under ordinary conditions is called a **molecule**. The formula for a molecule of water is H_2O. A molecule of water consists of two hydrogen atoms and one oxygen atom.

Example: Find the weight of a molecule of water.

(2 × atomic weight of hydrogen) + atomic weight of oxygen
= (2 × 1.008) + 15.999
= 18.015 amu

Find the molecular weight of each of the following.

1. H_2 2.016 amu
2. O_2 31.998 amu
3. CO_2 (carbon dioxide) 44.009 amu
4. NH_3 (ammonia) 17.031 amu
5. CH_4 (methane) 16.043 amu
6. H_3BO_3 (boric acid) 61.832 amu
7. $C_{27}H_{45}OH$ (cholesterol) 386.664 amu
8. H_2SO_4 (sulfuric acid) 98.076 amu
9. NaCl (salt) 58.4428 amu
10. $C_{12}H_{22}O_{11}$ (sugar) 342.297 amu

Math Lab and Modeling Math Masters, pp. 65, 82

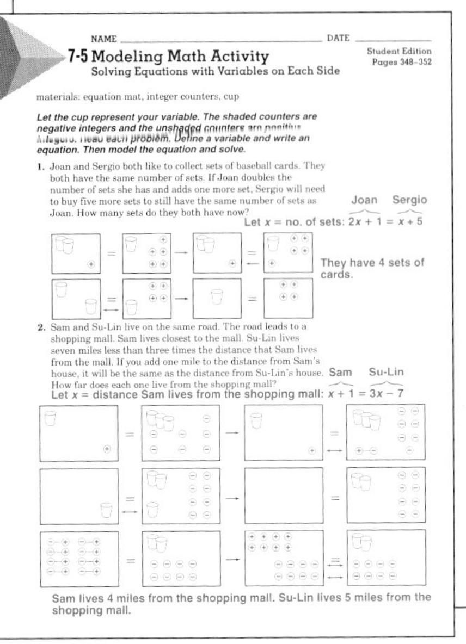

NAME ______________ DATE ________

7-5 Modeling Math Activity
Solving Equations with Variables on Each Side

Student Edition Pages 348–352

materials: equation mat, integer counters, cup

Let the cup represent your variable. The shaded counters are negative integers and the unshaded counters are positive integers. Read each problem. Define a variable and write an equation. Then model the equation and solve.

1. Joan and Sergio both like to collect sets of baseball cards. They both have the same number of sets. If Joan doubles the number of sets she has and adds one more set, Sergio will need to buy five more sets to still have the same number of sets as Joan. How many sets do they both have now?
   Let x = no. of sets: $2x + 1 = x + 5$ (Joan, Sergio)
   They have 4 sets of cards.
2. Sam and Su-Lin live on the same road. The road leads to a shopping mall. Sam lives closest to the mall. Su-Lin lives seven miles less than three times the distance that Sam lives from the mall. If you add one mile to the distance from Sam's house, it will be the same as the distance from Su-Lin's house. How far does each one live from the shopping mall?
   Let x = distance Sam lives from the shopping mall: $x + 1 = 3x - 7$ (Sam, Su-Lin)
   Sam lives 4 miles from the shopping mall. Su-Lin lives 5 miles from the shopping mall.

MAKING MATHEMATICS RELEVANT

This two-page introduction to the chapter provides students with an opportunity to explore other disciplinary topics and their applications to mathematics.

Background Information

Math and Entertainment in the News The unique *Jeopardy!* format, in which players are given the answers and asked to provide the correct questions, arose from the quiz-show scandals of the late Fifties. It was thought that this format might find more viewer acceptance. The most famous of the players who cheated was Charles Van Doren, a college professor who won $129,000 on Twenty-One. He was the subject of the 1994 movie *Quiz Show*.

CHAPTER 7 Solving Equations and Inequalities

TOP STORIES in Chapter 7

In this chapter, you will:

- solve problems by working backward,
- write and solve multi-step equations and inequalities,
- find the circumferences of circles, and
- convert measurements within the metric system.

MATH AND ENTERTAINMENT IN THE NEWS

Winning on a game show isn't all in what you know

Source: The Mathematics Teacher, May, 1994

Have you ever dreamed of hitting it big on a game show? Before you sign up, study up, on your math! Strategy is a big part of winning, and math is a big part of strategy. For example, on the show *Jeopardy!*, the winner is often determined by the last question, called *Final Jeopardy*. Before the question is asked, contestants are told the category and asked to wager any amount less than or equal to their current score. For a correct answer, the amount is added to their score. For an incorrect answer, it is subtracted. A mathematical analysis using inequalities shows that each player can carefully choose his or her wager to get the best chance of winning. So next time you play, take mathematics for $1000!

Putting It into Perspective

1928 Television recording system using aluminum gramophone records invented by John Logie Baird in London.

195 *The $64,00 Question* ushers era of big-mone quiz shows c televisio

1920 — 1935 — 1950

1941 The first television commercial airs on WBNT of New York.

Putting It into Perspective

The Game Show Network is produced by Sony Entertainment and reflects the belief that there is a significant untapped market for game shows in the U.S. The network broadcasts reruns of classic game shows from the past and continues to develop new shows, some of which allow viewers to play. The trend is toward more viewer interaction.

On the Lighter Side

THE FAR SIDE By GARY LARSON

JEOPARDY
EINSTEIN
57800
EDISON
54900
JONES
-100

"Excuse me . . . I know the game's almost over, but just for the record, I don't think my buzzer was working properly."

Statistical Snapshot

1994's Top Ten Syndicated Programs

Rank	Show	Rating	Stations
1	Wheel of Fortune	13.6	225
2	Jeopardy!	11.5	216
3	Star Trek	9.6	244
4	Oprah Winfrey Show	8.9	234
5	Buena Vista I	8.2	162
6	Entertainment Tonight	8.1	180
7	Star Trek: Deep Space Nine	7.5	234
8	National Geographic on Assignment	7.4	188
8	Roseanne	7.4	188
10	Inside Edition	6.7	166

Source: *Brandweek*, April 3, 1995

interNET CONNECTION For up-to-date information on game shows, visit: **www.glencoe.com/sec/math/prealg/mathnet**

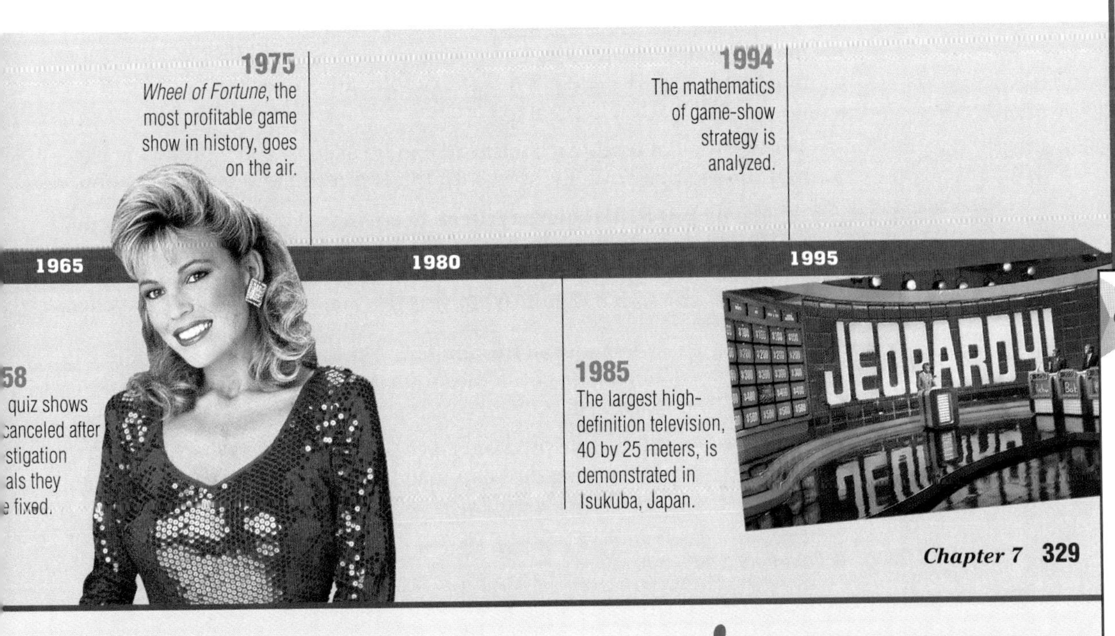
1975
Wheel of Fortune, the most profitable game show in history, goes on the air.
1994
The mathematics of game-show strategy is analyzed.
1965
1980
1995
58
quiz shows canceled after stigation als they e fixed.
1985
The largest high-definition television, 40 by 25 meters, is demonstrated in Isukuba, Japan.
JEOPARDY!

On the Lighter Side

Students may enjoy creating their own comics about game shows or other situations involving mathematics.

Statistical Snapshot

Syndicated programs are produced and sold to individual TV stations around the country. They can be a profitable source of programming for local stations. Students may note that the two top-rated programs are both game shows. (*Jeopardy!* is more technically a quiz show, a type of game show.) It might be interesting to ask students which of the shows listed they consider to be "educational" and why.

Cooperative Learning

Chapter Projects Two chapter projects are included in the *Investigations and Projects Masters*. In Chapter Project A, students extend the topic in the Chapter Opener. In Chapter Project B, students explore consumer issues. A student page and a parent letter are provided for each Chapter Project.

interNET CONNECTION

Glencoe has made every effort to ensure that the website links for *Pre-Algebra* at www.glencoe.com/sec/math/prealg/mathnet are current and contain appropriate content. However, these website links are not under Glencoe's control.

Investigations and Projects Masters, p. 53

NAME ______________________ DATE __________

CHAPTER 7 Project A
Class *Jeopardy!*

Most of you are familiar with how the TV game show *Jeopardy!* works: contestants answer questions from a category they select. All answers must be in the form of questions. Over the next few weeks, you and your classmates are going to work collaboratively to play *Class Championship Jeopardy!*

1. With the whole class, watch a videotape of a *Jeopardy!* show. Talk about how the game is played and scored. Review and discuss the strategies players use in *Final Jeopardy!*

2. Next, form small groups. Each group is responsible for writing questions and answers for one or more subjects such as math, social studies, science, language arts, health, music, art, or foreign language. Agree on the number of questions each group should produce. Write each question with its answer on an index card. Write the subject and point value for all questions (10, 20, 30, 40, 50). Be sure all subject areas and point values are represented.

3. Once questions have been written, have a group of volunteers collect them, separate them by subject, and then divide them into sets for use in different games.

4. Now get organized to play. Choose groups of three players by drawing names at random. As needed, some games can be played by either two or four players. Choose volunteers to act as "hosts," scorekeepers, and timers. Decide on a way for players to show that they want to answer. Determine a reasonable amount of time to answer a question and to play a game. Then make a schedule of games and choose locations for them.

5. Play enough games so that everyone gets to play. Rotate hosts, scorekeepers, and timers. Record all scores. You may choose to play additional games to determine a class champion.

6. Once all games have been completed, discuss and analyze the results. Talk about game-winning strategies, including those for *Final Class Jeopardy!* Then write a report in which you summarize and outline the strategies you think work best. Provide clear examples to support your conclusions.

7-1 Problem-Solving Strategy: Work Backward

Setting Goals: *In this lesson, you'll solve problems by working backward.*

Modeling with Technology

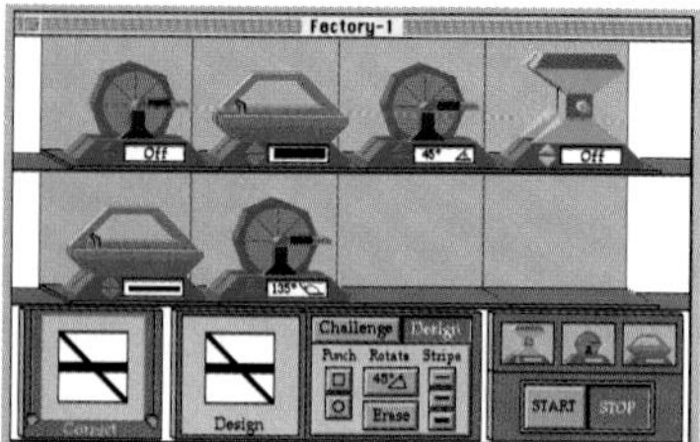

The computer software *The Factory* by Sunburst Communications simulates a factory assembly line. Machines rotate, punch holes, and paint a vertical stripe on geometric-shaped objects so that they match a given design.

Your Turn Copy and complete the chart.

Operation	Begin with a designed square.	Rotate 45° clockwise.	Remove thin line.	Rotate 45° clockwise.	Remove thick line.
Result					

a. What process is involved in completing the chart above? **See margin.**

b. Look at the two squares at the right. What would you do to the first square to get the square on the far right as the end result from *The Factory*? **See margin.**

Learning the Concept

In most problems, a set of conditions or facts is given and an end result must be found. However, *The Factory* software shows the result and asks for something that happened earlier.

The strategy of **working backward** can be used to solve problems like this. To use this strategy, start with the end result and *undo* each step.

Example APPLICATION **Personal Finance**

Marita put $10 of her paycheck in savings. Then she spent one-half of what was left on clothes. She paid $26 for a haircut and then spent one-half of what was left on a concert ticket. When she got home, she had $17 left. What was the amount of Marita's paycheck?

Explore You know all the amounts Marita spent and what she had when she got home. You want to find the amount of her paycheck.

Plan Since this problem gives the end result and asks for something that happened earlier, start with the result and work backward. *Undo* each step.

7-1 LESSON NOTES

NCTM Standards: 1-4

Instructional Resources

- Study Guide Master 7-1
- Practice Master 7-1
- Enrichment Master 7-1
- Group Activity Card 7-1
- Activity Masters, pp 7, 38
- Math Lab and Modeling Math Masters, p. 82

Transparency 7-1A contains the 5-Minute Check for this lesson; **Transparency 7-1B** contains a teaching aid for this lesson.

Recommended Pacing	
Standard Pacing	Day 1 of 12
Honors Pacing	Day 1 of 11
Block Scheduling*	Day 1 of 7

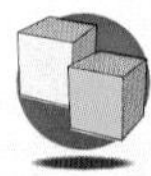

*For more information on pacing and possible lesson plans, refer to the ***Block Scheduling Booklet***.

1 FOCUS

5-Minute Check
(over Chapter 6)

Solve each equation or problem.

1. $n = 5.4 \cdot 9.2$ **49.68**
2. $\frac{3}{5} \cdot \left(-1\frac{1}{4}\right) = r$ $-\frac{3}{4}$
3. $-\frac{5}{6} \div -\frac{2}{3} = x$ $1\frac{1}{4}$
4. $w = 0.4 \div 2.5$ **0.16**
5. Jess had test scores of 96, 100, 80, and 88. What was his average score? **91**

Motivating the Lesson

Hands-On Activity Choose a student who is wearing tennis shoes and socks. Ask that student to take off one shoe and one sock. Then have the student put the shoe and sock back on. Make a list of the order in which each activity is completed. Point out that one undoes the other and that one way to undo something is to work backward.

Additional Answers Talk About it

a. Undoing the steps used to assemble the designed square.

b. Sample answer: Paint thick line. Rotate counterclockwise 45°. Paint thin line. Rotate counterclockwise 135°.

Study Guide Masters, p. 54 The Study Guide Master provides a concise presentation of the lesson along with practice problems.

Solve When Marita got home, she had $17. → 17

Undo the half that she spent on a concert ticket. → $\times 2$ = 34

Undo the $26 she spent on a haircut. → $+ 26$ = 60

Undo the half she spent for clothes. → $\times 2$ = 120

Undo the $10 she put in savings. → $+ 10$ = 130

The amount of Marita's paycheck was $130.

Examine Assume that Marita started with $130. After putting $10 in savings, she had $120. She spent half of $120, so she had $120 \div 2$ or $60. Then she spent $26, so she had $60 - 26$ or $34. Finally, she spent half of $34, so she had $34 \div 2$ or $17. In the end, Marita would have $17, so the answer is correct.

Checking Your Understanding

Communicating Mathematics

Read and study the lesson to answer these questions.

1. **Describe** the circumstances under which you could use the strategy of working backward. **See margin.**
2. **Explain** how to solve a problem by working backward. **See margin.**
3. **Write** a problem that can be solved by working backward. Trade problems with a classmate and solve his or her problem.
4. Create a design that could be made using *The Factory*. Record the procedure as you make it. Trade designs with a friend and reproduce your friend's design. **See students' work.**

3\. See students' work.

Guided Practice

Solve by working backward.

5. Mr. Fuentes uses half of a can of evaporated milk to make pumpkin pie. The can and the milk that is left weigh 9 ounces. If the can and the milk weighed 15 ounces before it was opened, how much does the can weigh? **3 ounces**
6. A certain number is divided by 5, and then 1 is subtracted from the result. The final answer is 32. Find the number. **165**

Exercises: Practicing and Applying the Concept

Independent Practice

Solve. Use any strategy.

7. Mr. and Mrs. Gentry each own an equal number of shares of IBM stock. Mr. Gentry sells one-third of his shares for $2700. What was the total value of Mr. and Mrs. Gentry's stock before the sale? **$16,200**
8. A certain number is multiplied by 3 and then 5 is added to the result. The final answer is 41. Find the number. **12**

2 TEACH

In-Class Example

For the Example

Lisa and Andre decided to trade CDs. Lisa gave Andre half of her CDs in exchange for 5 newer CDs. Then Lisa gave Andre's sister 2 CDs as a gift. If Lisa now has 18 CDs, how many did she have to start with? **30**

3 PRACTICE/APPLY

Checking Your Understanding

Exercises 1–6 are designed to help you assess your students' understanding through reading, writing, speaking, and modeling. You should work through Exercises 1–4 with your students and then monitor their work on Exercises 5–6.

Additional Answers

1. Sample answer: When you have the final result after a sequence of known operations and you want to know the starting point.
2. Sample answer: Reverse the order.

Practice Masters, p. 54

NAME ______________________ DATE __________

7-1 Practice — Student Edition Pages 330–332

Problem-Solving Strategy: Work Backward

Solve by working backward.

1. Bus #17 runs from Apple Street to Ellis Avenue, making 3 stops in between. At Bonz Avenue, 2 people got off and 5 people got on the bus. At Crump Road, half the people on the bus got off and 4 people got on. At Dane Square, 3 people got off and 1 person got on. At the final stop, the remaining 12 people got off. How many people were on the bus when it left Apple Street? **17 people**
2. Janette bought a share of stock in PRT Corporation. The first week, it increased in value 25%. During the second week, it decreased $1.40 in value. The next week it doubled in value, so she sold it. She got $27.20 for it. How much had she paid for it? **$12**
3. Hector's mother sent him on two errands. She gave him $5.00. He picked up clothes at the dry cleaners and later spent half the change on a loaf of bread. He returned 97¢ change to his mother. How much did the dry cleaning cost? **$3.06**
4. Dawn baked cookies and gave $\frac{3}{4}$ of them to Penny. Penny gave back a dozen. Dawn ended up with 18 cookies. How many had she baked originally? **24 cookies**

Solve. Use any strategy.

5. Guppies cost 20¢ less than swordtails. Three guppies and 4 swordtails cost $2.83. How much do guppies cost? **29¢**
6. A bus holds 40 people. At its first stop it picks up 8 people. At each stop after the first, 3 people get off and 7 get on. After which stop will the bus be full? **9th stop**
7. Bjorn has 5 coins with a total value of 50¢. Not all of the coins are dimes. What are the coins? **1 quarter, 1 dime, 3 nickels**
8. A certain number is added to 6 and the result is multiplied by 25. The final answer is 50. Find the number. **$^-4$**

Reteaching

Using Lists Provide students with a grocery cash register tape that has one price blacked out. Have students work backward to find the missing price.

Group Activity Card 7-1

Back Out — Group Activity **7-1**

Work with a partner to figure out each other's starting number by working backwards from the final number.

To begin, think of a number, but do not tell your partner what it is. Next, do an operation on that number, such as multiply it by 7. Take that product and do an operation on it, such as subtract 15. After doing at least three operations, tell your partner the final number and all of the operations you did to the original number in the order in which you did them.

It is your partner's task to determine your starting number.

You and your partner should take turns creating the problem and solving the problem.

6×7−15 =

 Pre-Algebra

Assignment Guide

Core: 7–19
Enriched: 7–19

For **Extra Practice**, see p. 756.

The red A, B, and C flags, printed only in the Teacher's Wraparound Edition, indicate the level of difficulty of the exercises.

4 ASSESS

Closing Activity

Speaking Have students explain how they know when to use the working backward strategy.

Enrichment Masters, p. 54

NAME ______________________ DATE ________

7-1 Enrichment
Fraction Detective

Student Edition
Pages 330–332

The fifteen mixed numbers in the table at the right below belong in five groups with three numbers in each group. They have been scrambled. Your job is to use the clues below to place these numbers in their proper groups.

Clue 1: Each group has only one number less than 1.

Clue 2: In each group, the greatest number is the quotient of its middle number divided by its least number.

Clue 3: Group A contains both the greatest number and the least number of all fifteen.

Clue 4: $3\frac{3}{8}$ and $2\frac{1}{4}$ are in the same group—group B.

Clue 5: The greatest number in group C is the number closest to 3.

Clue 6: $1\frac{7}{8}$ and $\frac{4}{5}$ are in the same group—group D.

Clue 7: Group E has no number greater than 2.

$\frac{1}{2}$	$\frac{5}{6}$	$\frac{4}{5}$
$1\frac{1}{3}$	$\frac{3}{4}$	$\frac{2}{3}$
$3\frac{1}{5}$	$3\frac{3}{8}$	$1\frac{7}{8}$
$2\frac{1}{4}$	$2\frac{11}{32}$	$3\frac{1}{3}$
$1\frac{3}{5}$	$2\frac{2}{5}$	$6\frac{2}{3}$

Use the chart below to record your findings.

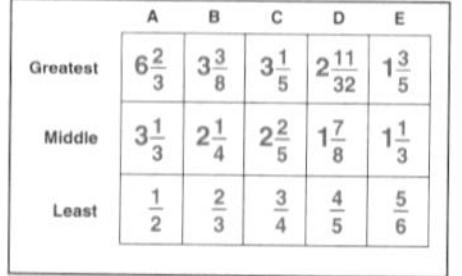

	A	B	C	D	E
Greatest	$6\frac{2}{3}$	$3\frac{3}{8}$	$3\frac{1}{5}$	$2\frac{11}{32}$	$1\frac{3}{5}$
Middle	$3\frac{1}{3}$	$2\frac{1}{4}$	$2\frac{2}{5}$	$1\frac{7}{8}$	$1\frac{1}{3}$
Least	$\frac{1}{2}$	$\frac{2}{3}$	$\frac{3}{4}$	$\frac{4}{5}$	$\frac{5}{6}$

Choose
Estimation
Mental Math
Calculator
Paper and Pencil

11. 729 people
13. 29.8.92 and 2.9.92

B

9. Chris has $1.50. Half of the money he had when he left home this morning was spent on lunch. He lent Bill a dollar after lunch. How much money did Chris start with? **$5**

10. Forty-four students took the bus to the St. Louis Science Center. Each student paid $1.25. How much did the driver collect? **$55**

11. On Monday, Fumiko told a joke to 3 of her friends. On Tuesday, each of those friends told the joke to 3 other friends. On Wednesday, each person who heard the joke on Tuesday told 3 other people. If this pattern continues, how many people will hear the joke on Saturday?

12. An ice sculpture is melting at the rate of half its weight every hour. After 8 hours, it weighs $\frac{5}{16}$ of a pound. How much did it weigh in the beginning? **80 pounds**

13. In Canada and other countries, dates are abbreviated as "date.month.year." For example, September 19, 1991 would be abbreviated as 19.9.91 and September 5, 1995 would be 5.9.95. Dates like these, where the pattern of numbers is the same when read forward and backward, are called *palindromic dates*. What are the two palindromic dates in the 1900s that are closest together?

Critical Thinking

14. At the end of the second round of a game, Bart's and Carlos' scores were double the score each had at the end of the first round. The total number of points they gained were subtracted from Al's first round score. After the third round, Al's and Carlos' scores were double their second round scores. The total number of points they gained were subtracted from Bart's second round score. After the fourth round, Al's and Bart's third round scores doubled. The total number of points they gained were subtracted from Carlos' third round score. Now each of them has 8 points. How many points did each of them have after the first round? **Al: 13 points; Bart: 7 points; Carlos: 4 points**

Mixed Review

16. 0.28 million or 280,000
17. whole, integer, rational

15. Write 6.789×10^{-7} in standard form. (Lesson 6-9) **0.0000006789**

16. **Entertainment** The table at the right shows the 1993 attendance in millions of people of the largest fairs in the United States. How many more people attended the New Mexico State Fair than attended the Los Angeles County Fair? (Lesson 5-3)

Fair	Attendance (millions)
State Fair of Texas, Dallas	3.15
State Fair of Oklahoma, Oklahoma City	1.79
New Mexico State Fair, Albuquerque	1.68
Minnesota State Fair, St. Paul	1.60
Houston (Texas) Livestock Show	1.57
Western Washington Fair, Puyallup	1.42
Los Angeles County Fair, Pomona	1.40
De Mar (California) Fair	1.11
Colorado State Fair, Pueblo	1.08
Tulsa (Oklahoma) State Fair	1.03

Source: *Top Ten of Everything*

17. Name the sets of numbers to which $\frac{42}{7}$ belongs—whole numbers, integers, or rationals. (Lesson 5-1)

18. Evaluate $ab \div (-7)$ if $a = 49$ and $b = -2$. (Lesson 2-8) **14**

19. Find the value of $[4(3 + 8) \div (4 - 2)] + 9$. (Lesson 1-2) **31**

Extension

Analyzing Mazes Have students bring in mazes from magazines or books. Have students analyze the mazes by making two lists. The first list is made by starting at the beginning of the maze. Each time a choice is to be made a number is written down that indicates how many different paths there are to choose from. The second list is made by starting at the end of the maze and working backward. Again, each time a choice is to be made the number of possible paths is written down. Students will probably find far fewer choices need to be made when working backward.

HANDS-ON ACTIVITY

7-2A Two-Step Equations

A Preview of Lesson **7-2**

MATERIALS

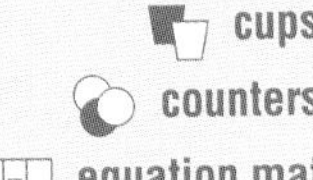

cups

counters

equation mat

In this activity, you will extend what you know about solving one-step equations to model and solve two-step equations.

Recall from Chapter 3 that you can use cups and counters to model one-step equations. You can model two-step equations also. A two-step equation contains two operations. For example, the equation $2x + 3 = 7$ involves multiplication and addition.

Activity

Model and solve the equation $2x + 3 = 7$.

- Place 2 cups and 3 positive counters on the left side of the mat. Then place 7 positive counters on the right side of the mat.

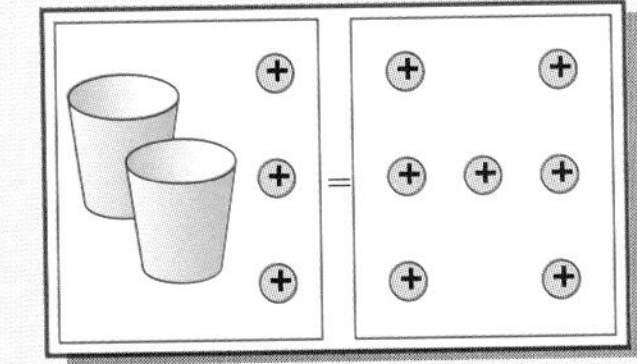

$2x + 3 = 7$

- To determine how many counters are in each cup, the goal is to isolate the cups on one side of the mat. Since the two sides of the mat represent equal quantities, you can remove an equal number of counters from each side without changing the value of the equation.

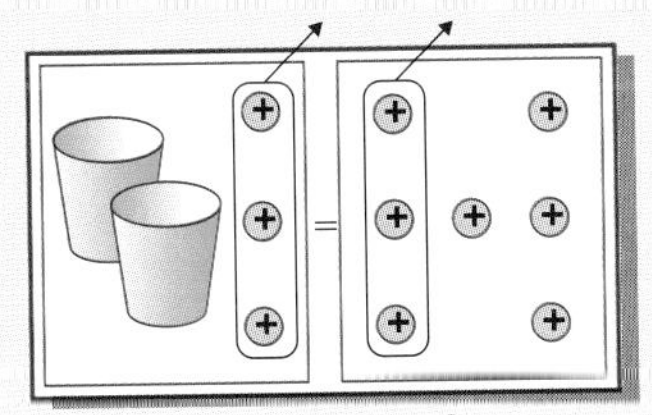

$2x + 3 - 3 = 7 - 3$

$2x = 4$

- Now the equation is $2x = 4$. Match an equal number of counters with each cup. Therefore, each cup must contain 2 counters, so $x = 2$.

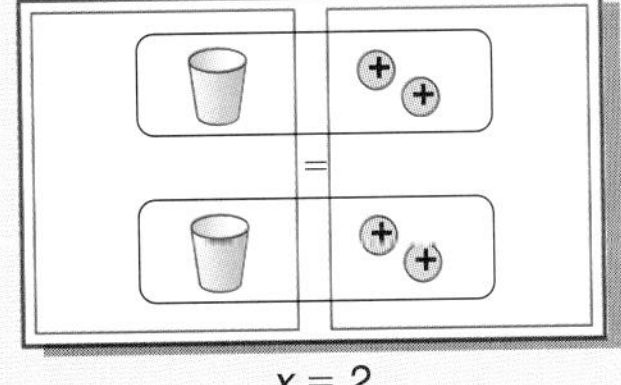

$x = 2$

Your Turn **Model each equation and solve.** See students' work. Solutions are given.

1. $2x + 2 = -4$ **−3**
2. $3x + 3 = 12$ **3**
3. $2x - 2 = 6$ **4**
4. $2x - 2 = -4$ **−1**
5. $4 + 2x = 8$ **2**
6. $3x - (-2) = -4$ **−2**

7. Explain how this activity models the work-backward strategy presented in Lesson 7-1. **See Solutions Manual.**
8. Explain how you can use models to solve $-2x + 1 = -5$. **See Solutions Manual.**

7-2A LESSON NOTES

NCTM Standards: 1-4, 9

Objective

Use cups and counters to model and solve two-step equations.

Recommended Time

Demonstration and discussion: 15 minutes; Exercises: 30 minutes

Instructional Resources

For each student or group of students

Student Manipulative Kit
- cups, counters, equation mats

Math Lab and Modeling Math Masters
- p. 3 (integer mat)
- p. 5 (integer counters)
- p. 6 (pattern for cup)
- p. 39 (worksheet)

For teacher demonstration

Overhead Manipulative Resources

1 FOCUS

Motivating the Lesson

Tell students that you started the day with some pennies. After going to the bank, you had twice as many pennies. Then you got two more pennies in change and had a total of 18 pennies. Ask students how working backward can be used to find how many pennies you had in the beginning.

2 TEACH

Teaching Tip Remind students that for equations such as $2x + 2 = -4$, positive counters cannot be removed from both sides. Suggest adding 2 negative counters to each side and using zero pairs to remove the 2 positive counters from the side with the $2x$.

3 PRACTICE/APPLY

Assignment Guide

Core: 1–8
Enriched: 1–8

4 ASSESS

Observing students working in cooperative groups is an excellent method of assessment.

GLENCOE Technology

Interactive Mathematics Tools Software

In this interactive computer lesson, students explore equations. A **Computer Journal** gives students the opportunity to write about what they have learned.

For Windows & Macintosh

7-2 LESSON NOTES

NCTM Standards: 1-4, 9, 12

Instructional Resources

- Study Guide Master 7-2
- Practice Master 7-2
- Enrichment Master 7-2
- Group Activity Card 7-2
- Assessment and Evaluation Masters, p. 183

Transparency 7-2A contains the 5-Minute Check for this lesson; **Transparency 7-2B** contains a teaching aid for this lesson.

Recommended Pacing	
Standard Pacing	Day 3 of 12
Honors Pacing	Day 3 of 11
Block Scheduling*	Day 2 of 7

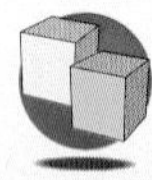

*For more information on pacing and possible lesson plans, refer to the ***Block Scheduling Booklet***.

1 FOCUS

5-Minute Check
(over Lesson 7-1)

1. Rupesh earned some money mowing lawns one month. He put half of his money into savings. With the rest, he spent \$15 on a new CD, \$6 to see a movie, \$3 on food, and he still had \$24 left. How much money did Rupesh earn mowing lawns that month? **\$96**
2. In an intramural basketball game, Annie made 6 points more than Roberto. Roberto made half as many points as Charlie. Charlie made 3 more points than Darla. If Darla made 9 points, how many points did Annie make? **12 points**

Motivating the Lesson

Questioning Ask students what you must do to check the oil in a car. You have to lift the hood and find the dipstick. Then ask what you do before you drive the car again. **You must work backward by replacing the dipstick before shutting the hood.**

7-2 Solving Two-Step Equations

Setting Goals: *In this lesson, you'll solve equations that involve more than one operation.*

Modeling a Real-World Application: Consumerism

"At Carpet World, we will not be undersold! Get 18 square yards of the best stain-resistant carpet installed for just \$305.89! That *includes* the \$54 installation charge. Hurry in! At this price, the carpet will not last!"

The Jackson family is comparing prices at different carpet stores. Most carpet stores price carpet by the square yard. What price should the Jacksons use to compare the price of one square yard of carpet at Carpet World to the prices at other stores? *You will solve this problem in Exercise 36.*

Learning the Concept

To solve this problem, you will use the equation $54 + 18p = 305.89$. Some equations, like this one, contain more than one operation. To solve an equation with more than one operation, use the work-backward strategy and undo each operation.

Example 1 **Solve each equation. Check your solution.**

a. $\frac{c}{4} - 19 = 17$

$\frac{c}{4} - 19 + 19 = 17 + 19$ *Add to undo the subtraction.*

$\frac{c}{4} = 36$

$4 \cdot \frac{c}{4} = 4 \cdot 36$ *Multiply to undo the division.*

$c = 144$

Check: $\frac{c}{4} - 19 = 17$

$\frac{144}{4} - 19 \stackrel{?}{=} 17$ *Replace c with 144.*

$36 - 19 \stackrel{?}{=} 17$ *Do the division first.*

$17 = 17$ ✓ The solution is 144.

Alternative Learning Styles

Auditory For students who need a reminder to reverse the order of operations when solving a multi-step equation, encourage them to use the acronym SADMP, which stands for *Simple As Doing Math Problems* and indicates that subtraction and addition are undone first, then division and multiplication, and finally operations within parentheses.

b. $-3n + 8 = -7$

$$-3n + 8 - 8 = -7 - 8 \quad \textit{Subtract 8 from each side.}$$
$$-3n = -15$$
$$\frac{-3n}{-3} = \frac{-15}{-3} \quad \textit{Divide each side by } -3.$$
$$n = 5$$

Check:
$$-3n + 8 = -7$$
$$-3(5) + 8 \stackrel{?}{=} -7$$
$$-15 + 8 \stackrel{?}{=} -7$$
$$-7 = -7 \checkmark$$
The solution is 5.

In Chapter 3, you learned that a formula is an algebraic expression that can be used to show the relationship among certain quantities. Some formulas involve more than one operation. You can use the work-backward strategy to solve problems involving formulas.

Example 2

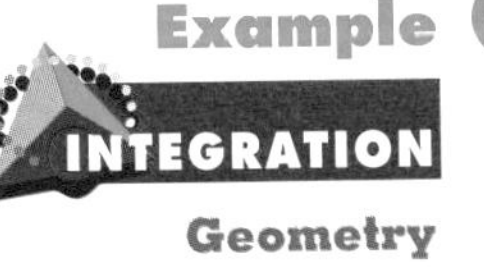

Geometry

The area A of a trapezoid can be found by multiplying the height h and one-half the sum of the lengths of the bases b_1 and b_2. The formula is $A = \frac{1}{2} \cdot h(b_1 + b_2)$. The area of the trapezoid at the right is 32 square inches. The trapezoid is 4 inches high, and the length of one of the bases is 9 inches. What is the length of the other base?

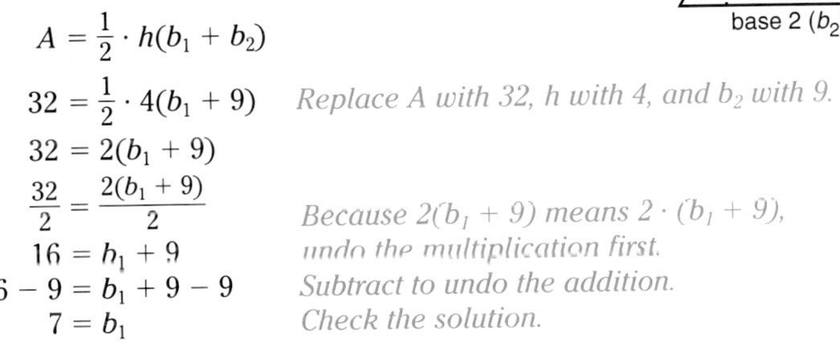

$$A = \frac{1}{2} \cdot h(b_1 + b_2)$$
$$32 = \frac{1}{2} \cdot 4(b_1 + 9) \quad \textit{Replace A with 32, h with 4, and } b_2 \textit{ with 9.}$$
$$32 = 2(b_1 + 9)$$
$$\frac{32}{2} = \frac{2(b_1 + 9)}{2} \quad \textit{Because } 2(b_1 + 9) \textit{ means } 2 \cdot (b_1 + 9)\textit{, undo the multiplication first.}$$
$$16 = b_1 + 9$$
$$16 - 9 = b_1 + 9 - 9 \quad \textit{Subtract to undo the addition.}$$
$$7 = b_1 \quad \textit{Check the solution.}$$

The second base is 7 inches long.

Checking Your Understanding

Communicating Mathematics

1–3. See Solutions Manual.

Read and study the lesson to answer these questions.

1. How is the order of operations used in solving two-step equations?
2. **You Decide** Kelsey says that to solve the equation $3x + 6 = 24$, the first step should be to subtract 6 from each side. Dana says the first step should be to divide each term by 3.
   a. Which method would you use and why?
   b. Whose method would be easier to use to solve the equation $2x + 3 = 5$? Explain.
3. **Write** a problem that could be solved using the equation $3x - 5 = 15$.

MATERIALS

cups
counters
equation mat

4. This diagram models the equation $4 + 2x = 8$.
   a. What is the solution of $4 + 2x = 8$? 2
   b. Model the equation $2x - 3 = 7$. Then solve the equation. See margin.

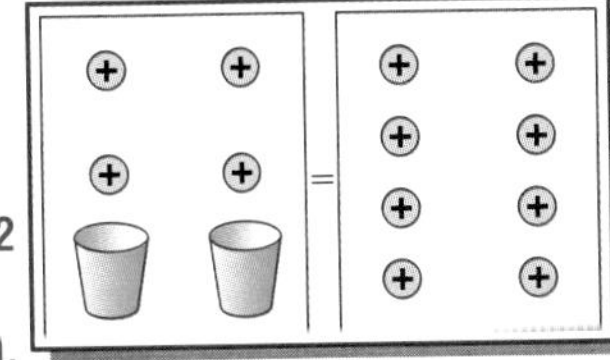

2 TEACH

In-Class Examples

For Example 1
Solve each equation. Check your solution.

a. $\frac{n}{5} + 6 = 2$ −20

b. $-23 = -8m + 17$ 5

For Example 2
Use the formula for the area of a trapezoid to find the length of the other base of a trapezoid with one base 7.2 centimeters long, a height of 2.4 centimeters, and an area of 25.2 square centimeters.
13.8 cm

3 PRACTICE/APPLY

Checking Your Understanding

Exercises 1–11 are designed to help you assess your students' understanding through reading, writing, speaking, and modeling. You should work through Exercises 1–4 with your students and then monitor their work on Exercises 5–11.

Study Guide Masters, p. 55

NAME ________ DATE ________

7-2 Study Guide
Solving Two-Step Equations

Student Edition
Pages 334–337

Some equations contain more than one operation. To solve an equation with more than one operation, use the work-backward strategy and undo each operation.

Example: Solve $\frac{c}{2} - 13 = 7$.

$$\frac{c}{2} - 13 + 13 = 7 + 13 \quad \text{Add to undo subtraction.}$$
$$\frac{c}{2} = 20$$
$$2 \cdot \frac{c}{2} = 20 \cdot 2 \quad \text{Multiply to undo division.}$$
$$c = 40$$

Check the solution.
$$\frac{c}{2} - 13 = 7 \quad \text{Replace } c \text{ with 40.}$$
$$\frac{40}{2} - 13 \stackrel{?}{=} 7$$
$$20 - 13 \stackrel{?}{=} 7$$
$$7 = 7 \checkmark \quad \text{The solution is 40.}$$

Solve each equation. Check your solution.

1. $4a - 10 = 42$ 13
2. $12 - 3m = 18$ −2
3. $-10 = -5w - 25$ −3
4. $\frac{m}{4} + 6 = 70$ 256
5. $-3 + \frac{c}{2} = 12$ 30
6. $\frac{-v}{3} + 8 = 22$ −42
7. $5.8t + 15 = -14$ −5
8. $8 - 6.2u = -23$ 5
9. $-4 - 2.4w = -16$ 5
10. $4(x + 6) = 12$ −3
11. $-13 = \frac{4 - b}{3}$ 43
12. $-16 = 4(2 - 2x)$ 3
13. $-1.4(a + 2) = 4.2$ −5
14. $7.7 = 2.1 - 7m$ −0.8
15. $\frac{a + 4}{2} = 10.8$ 17.6

Reteaching

Using Flowcharts Have students create a flowchart to solve the general equation in the form $ax + b = c$, where a, b, and c are any rational numbers.

Additional Answer

4b.

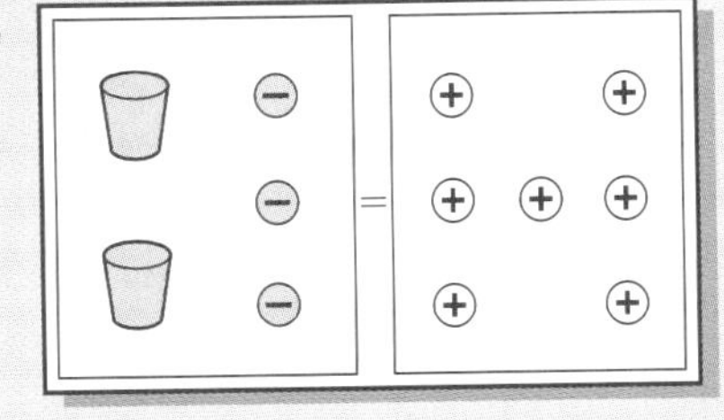

$x = 5$

Error Analysis

When solving equations similar to those in the exercises, students may try to multiply or divide first. Remind students of the need to use the order of operations *in reverse.* Thus, they should add or subtract first.

Assignment Guide

Core: 13–37 odd, 38–44
Enriched: 12–32 even, 33–44

For **Extra Practice**, see p. 756.

The red A, B, and C flags, printed only in the Teacher's Wraparound Edition, indicate the level of difficulty of the exercises.

Ralph J. Bunche was a United States delegate at the formation of the United Nations. In 1949, he persuaded Israel and the Arab League to agree to an armistice. That mediation role earned him the Nobel Peace Prize.

Guided Practice

Solve each equation. Check your solutions. 9. −3

5. $2n - 5 = 21$ **13**
6. $3 + \frac{t}{2} = 35$ **64**
7. $7.5r + 2 = -28$ **−4**
8. $3(x + 5) = 9$ **−2**
9. $-2.5(a + 2a) = 22.5$
10. $\frac{a+5}{2} = 10.5$ **16**
11. **Postage** Judi mailed some photographs to her friend Misae. She paid \$1.77 to send them first-class. The Post Office charges 32 cents for the first ounce and 29 cents for each additional ounce. Use the equation $32 + 29w = 177$ to find how much Judi's package weighs. **6 ounces**

Exercises: Practicing and Applying the Concept

Independent Practice

Solve each equation. Check your solution.

12. $85 = 4d + 5$ **20**
13. $2r - 7 = 1$ **4**
14. $4 - 2b = -8$ **6**

A
15. $-8 - t = -25$ **17**
16. $-4y + 3 = 19$ **−4**
17. $1.8 = 0.6 - y$ **−1.2**

B
18. $19 = 7 + \frac{b}{7}$ **84**
19. $-12 + \frac{j}{4} = 9$ **84**
20. $-3 = -31 + \frac{c}{6}$ **168**
21. $8 = \frac{h}{-3} + 19$ **33**
22. $\frac{-4x}{3} = 24$ **−18**
23. $13 + \frac{-p}{3} = -4$ **51**
24. $\frac{3}{4}n - 3 = 9$ **16**
25. $\frac{y}{3} + 6 = -45$ **−153**
26. $\frac{c}{-4} - 8 = -42$ **136**
27. $\frac{n-10}{5} = 2.5$ **22.5**
28. $16 = \frac{-8+s}{-7}$ **−104**
29. $\frac{-d-(-5)}{7} = 14$ **−93**

30. $\frac{6+c}{-13} = -3$ **33**
31. $\frac{n}{2} + (-3) = 5$ **16**
32. $4.7 = 1.2 - 7m$ **−0.5**

Critical Thinking

33. **Make Up a Problem** Write a two-step equation using the numbers 2, 3, and 6, in which the solution is $-\frac{1}{6}$. **Sample answer: $6x + 3 = 2$**

Applications and Problem Solving

34. **Medicine** Dr. Bell recommended that Marcie take eight tablets on the first day and then 4 tablets each day until the prescription was used. The prescription contained 28 tablets. Use the equation $8 + 4d = 28$ to find how many days Marcie will be taking pills after the first day. **5 more days**
35. **Chemistry** Chemical equations are like algebraic equations; they must be balanced on either side of the arrow. The symbol $3H_2O$ represents 3 molecules made up of 2 hydrogen and 1 oxygen atoms. Find the numbers that belong in the boxes in order for the equation $\square P + \square O_2 \rightarrow P_4O_{10}$ to be balanced. **4, 5**
36. **Consumerism** Refer to the application at the beginning of the lesson. How much is the carpet per square yard? **about \$13.99**
37. **Academics** Swedish scientist Alfred Nobel left a major part of his fortune to establish annual prizes to be awarded to those who have achieved the greatest common good in the fields of physics, chemistry, physiology or medicine, literature, peace, and economic science. People from the United States have won Nobel Prizes 213 times. That is 33 more than twice as many as the people from the United Kingdom, which is second in producing prize winners. How many people from the United Kingdom have won Nobel Prizes? **90**

FYI

In 1950, Ralph Johnson Bunche became the first African-American to win the Nobel Peace Prize. Mr. Bunche worked at the U.S. State Department and the United Nations.

Practice Masters, p. 55

NAME ______________________ DATE __________

7-2 Practice
Solving Two-Step Equations

Student Edition
Pages 334–337

Solve each equation. Check your solution.

1. $20 = 6x + 8$ **2**
2. $-10 - k = -36$ **26**
3. $2y - 7 = 15$ **11**
4. $15 - 4g = -33$ **12**
5. $2.1 = 0.8 - z$ **−1.3**
6. $-9x + 36 = 72$ **−4**
7. $-5c + 4 = 64$ **−12**
8. $8h + 7 = -113$ **−15**
9. $15d - 21 = 564$ **39**
10. $2x + 5 = 5$ **0**
11. $14 = 27 - x$ **13**
12. $44 = -4 + 8p$ **6**
13. $3 + 6u = -63$ **−11**
14. $33 = 5w - 12$ **9**
15. $19 = -3a - 5$ **−8**
16. $-21 - 15m = 219$ **−16**
17. $\frac{x}{12} - 15 = 31$ **552**
18. $\frac{5}{3}j - 6 = 94$ **60**
19. $-17 + \frac{t}{5} = 3$ **100**
20. $29 = \frac{b}{-4} + 15$ **−56**
21. $\frac{-2k}{7} = 36$ **−126**
22. $-30 = -37 + \frac{b}{15}$ **105**
23. $\frac{-c}{4} - 8 = -48$ **160**
24. $2.7 = 1.3 - 2d$ **−0.7**
25. $12 + \frac{-v}{5} = -3$ **75**
26. $9 = 14 + \frac{m}{2}$ **−10**
27. $\frac{z}{6} + 11 = -49$ **−360**
28. $\frac{t}{3} + (-2) = -5$ **−9**
29. $\frac{5+r}{-2} = -6$ **7**
30. $\frac{f-6}{5} = 3.2$ **22**
31. $\frac{s-8}{-8} = -1$ **16**
32. $\frac{-a-(-3)}{3} = 10$ **−27**
33. $16 = \frac{-6+c}{-3}$ **−42**

Group Activity Card 7-2

Seek And Find

Group Activity **7-2**

Work in small groups to find equations in the puzzle that have solutions from 1 through 20. In the puzzle these equations are written either horizontally, vertically, or along diagonals. Additionally, they may be written either forward or backward.

One member of your group is to keep the official group copy, listing the equations as you find them. Another member should circle the equation. Groups have ten minutes to find as many equations as possible.

2y	15	=	14	–	2x	+	5	=	19	14
+	9	–	3	3y	–	58	=	-4	=	x
8	-5	12	x	26	7	1	5x	6x	1	12
=	–	2x	–	8	=	18	=	6	+	m
16	-2x	+	10	-9	17	6	–	3x	2x	+
+	=	24	+	8	=	15	+	7	=	4a
7	-45	2a	–	30	=	4	+	4x	=	64
50	–	–	16	19	=	3x	+	-6	+	=
–	3y	11	21	10	+	1	7y	8	18	y
2x	=	=	+	-6	=	-5	–	x	9	+
=	4	9	=	13	–	2a	+	20	=	48
12	5	y	–	7	=	-1	+	5a	=	14

 Pre-Algebra

Extension

Using Connections At Pay-Alot Rentals, the cost of renting a car is \$52 per day. The first 75 miles driven are free, but any miles after that cost 33¢ per mile. Susan needs the car for three days and only has \$250 to spend. How many miles can she drive? **about 360 miles**

Mixed Review

38. **Personal Finance** Before Shantal left on her trip, she decided to plan carefully how much money she should set aside to cover the cost of the vacation. First, the plane ticket would cost $156. Shantal decided she would use three-fourths of the remaining money in her budget for a hotel, rental car, and meals. She added $30 for souvenirs and she wanted to have $100 left over for emergencies. How much money should Shantal start with in her vacation budget? (Lesson 7-1) **$676**

39. Replace ● with <, >, or = to make 16.43 ● $16\frac{3}{7}$ a true sentence. (Lesson 6-1) **>**

40. Solve $c - \frac{3}{5} > 14\frac{2}{9}$. (Lesson 5-7) $c > 14\frac{37}{45}$

41. Write $y \cdot y \cdot y$ as a product using exponents. (Lesson 4-2) y^3

42. Simplify $|18| - |-4|$. (Lesson 2-1) **14**

43. Is $x - (-17) > 29$ *true*, *false*, or *open* if $x = 11$? (Lesson 1-9) **false**

44. **Food** A survey performed by the *American Demographics* magazine found that the average American family spent $1664 eating out in 1994. Based on this finding, about how much did the average family spend on eating out in a month? (Lesson 1-1)
    a. Which method of computation do you think is most appropriate? Justify your choice.
    b. Solve the problem using the four step plan. Be sure to examine your solution.

✓ Choose

Estimation
Mental Math
Calculator
Paper and Pencil

44a. estimate; the problem says about
44b. about $140

WORKING ON THE

Investigation

Refer to the Investigation on pages 270–271.

Look at your design of the loading platform that you placed in your Investigation Folder at the beginning of Chapter 6.

Determine how long it would take to load a coaster train by doing a whole-class experiment.

- Simulate the unloading and loading of different numbers of passengers in different numbers and sizes of cars. For example, determine how long it would take to unload and load 24 passengers in six cars or 20 passengers in ten cars. Record the time it takes from when the train stops to the time it starts moving again.
- Using the total ride time you calculated at the end of Lesson 6-4, determine how many trains you can safely operate on the track at the same time using your loading time and total ride time.

Add the results of your work to your Investigation Folder.

WORKING ON THE

Investigation

The Investigation on pages 270 and 271 is designed to be a long-term project that is completed over several days or weeks. Encourage students to keep their materials in their Investigation Folder as they work on the Investigation.

Tech Prep

Advertising Display Designer Among other displays, an advertising designer must create window displays. The professional needs to know how to find area and perimeter to be able to calculate materials needed. For more information on tech prep, see the *Teacher's Handbook*.

4 ASSESS

Closing Activity

Modeling Have students use cups and counters to solve the following equations.

1. $2x + 4 = 8$ **2**
2. $3x - 12 = -9$ **1**
3. $8 = 2x + -6$ **7**

Chapter 7, Quiz A (Lessons 7-1 and 7-2) is available in the *Assessment and Evaluation Masters*, p. 183.

Enrichment Masters, p. 55

NAME ______________________ DATE __________

Student Edition Pages 334–337

7-2 Enrichment
Writing Word Problems

You have seen how using equations can help you solve word problems. Now it is your turn to write the word problems! When writers compose story problems, they usually have a particular type of equation in mind. Then they write about a situation that fits that equation. They try to make the story more interesting by inserting names and additional information that will help you understand the situation.

Here is a sample equation: $5 \times 3 + 2 = 17$

An author might think of this equation as showing 17 items separated into 5 groups of 3 items, with two items left over.

Here is one story that fits the equation.

Seventeen girls went on a canoeing trip. There were three girls in each canoe except the last one, which contained only two girls. How many canoes did they use?

Here is the second possible story.

Brad had some new baseball cards. He arranged them in 3 piles, with 5 cards in each pile. There were two cards left over. How many cards did Brad have in all?

Write a word problem for each of the following equations. Answers will vary. Sample answers are given.

1. $\frac{15 + 6}{7} = x$ Jenny had 15 baby gerbils and 6 adult gerbils. She put them in different cages with 7 in each cage. How many cages did she use?
2. $3 \times (6 + 4) = x$ Three tables were set for an awards dinner. Six athletes and 4 parents were seated at each table. How many people attended the dinner?
3. $19 - 5x = 9$ Marty started with 19 flowers. He gave 5 of his friends each an equal number of flowers. He has 9 left. How many flowers did he give each friend?
4. $6x + 3 = 45$ The math club wants to buy 45 posters to decorate the math classrooms. The posters come in sets of 6, or separately. If they buy 3 posters separately, how many sets do they need?

7-3 LESSON NOTES

NCTM Standards: 1-4, 9

Instructional Resources

- Study Guide Master 7-3
- Practice Master 7-3
- Enrichment Master 7-3
- Group Activity Card 7-3
- Tech Prep Applications Masters, p. 13

Transparency 7-3A contains the 5-Minute Check for this lesson; **Transparency 7-3B** contains a teaching aid for this lesson.

Recommended Pacing	
Standard Pacing	Day 4 of 12
Honors Pacing	Day 4 of 11
Block Scheduling*	Day 3 of 7 (along with Lesson 7-4)

*For more information on pacing and possible lesson plans, refer to the ***Block Scheduling Booklet***.

1 FOCUS

5-Minute Check
(over Lesson 7-2)

Solve each equation. Check your solution.

1. $-b + 5 = 12$ **−7**
2. $3m + 5 = 17$ **4**
3. $\frac{(w+2)}{7} = 3$ **19**
4. $\frac{-8+a}{3} = 6$ **26**
5. $\frac{(r-3)}{4} = 7$ **31**

Motivating the Lesson

Questioning Ask a student to give directions from his or her house to the house of another student in the class. Have the other student do the same. Ask the class if the second student worked backward from the first students' directions.

7-3 Writing Two-Step Equations

Setting Goals: *In this lesson, you'll solve verbal problems by writing and solving equations.*

Modeling a Real-World Application: Baseball

"It has been an interesting year for the Red Sox. Here at the 1995 All-Star Game, they stand at 10 games above 0.500. So 68 games into the season, they are atop the eastern division of the American League."

How many games had the Red Sox won before the All-Star Game?

Learning the Concept

If a baseball team is at 0.500, that means that they have won $\frac{500}{1000}$ or $\frac{1}{2}$ of their games. If a team wins more than half of their games, then their status is given as a number of games "over 0.500." At 10 over 0.500, the Red Sox have won 10 more games than they have lost. Let w represent the number of games the Red Sox have won. Then they have lost $w - 10$ games. Write an equation.

$$\underbrace{68}_{68}\ \underbrace{=}_{\text{equals}}\ \underbrace{w}_{\text{the number of wins}}\ \underbrace{+}_{\text{plus}}\ \underbrace{(w-10)}_{\text{the number of losses}}$$

Work backward to solve the equation.

$$68 = w + (w - 10)$$
$$68 = 2w - 10 \quad \textit{Simplify.}$$
$$68 + 10 = 2w - 10 + 10 \quad \textit{Add 10 to each side.}$$
$$78 = 2w$$
$$\frac{78}{2} = \frac{2w}{2} \quad \textit{Divide each side by 2.}$$
$$39 = w$$

The Boston Red Sox had won 39 games before the All-Star Game.

Example  **Five more than twice some number is 19. Find the number.**

First define a variable. Let n represent the number.

$$\underbrace{5}_{\text{five}}\ \underbrace{+}_{\text{more than}}\ \underbrace{2n}_{\text{twice some number}}\ \underbrace{=}_{\text{is}}\ \underbrace{19}_{19}$$
$$5 - 5 + 2n = 19 - 5 \quad \textit{Subtract 5 from each side.}$$
$$2n = 14$$
$$\frac{2n}{2} = \frac{14}{2} \quad \textit{Divide each side by 2.}$$
$$n = 7$$

The number is 7.

Alternative Teaching Strategies

Reading Mathematics Have students make a list of different phrases and how they are translated into mathematical expressions. For instance, *is increased by* usually indicates addition and *separated into equal parts* indicates division.

Study Guide Masters, p. 56
The Study Guide Master provides a concise presentation of the lesson along with practice problems.

"Play it Sam." is a well known line from the classic film *Casablanca*. Rick Blaine, played by Humphrey Bogart, spoke this line to Sam, played by Dooley Wilson. Humphrey Bogart earned \$36,667 for his role in the movie. This is \$1667 more than ten times the salary Dooley Wilson earned. How much did Mr. Wilson earn for *Casablanca*?

Explore You know Mr. Bogart's salary. You are looking for Mr. Wilson's salary.

Plan Let w represent Mr. Wilson's salary. Write an equation.

Solve *36,667 is 1667 more than ten times Mr. Wilson's salary*

$$36{,}667 = 1667 + 10w$$

$$36{,}667 - 1667 = 1667 - 1667 + 10w \quad \textit{Subtract 1667 from each side.}$$

$$35{,}000 = 10w$$

$$\frac{35{,}000}{10} = \frac{10w}{10} \quad \textit{Divide each side by 10.}$$

$$3500 = w$$

Examine Since $10(3500) + 1667 = 36{,}667$, the answer is reasonable. Mr. Wilson's salary for *Casablanca* was \$3500.

FYI

Many people misquote the line from *Casablanca* as "Play it *again*, Sam."

Checking Your Understanding

Communicating Mathematics

Read and study the lesson to answer these questions.

1. **List** the steps you would use to solve a verbal problem.
2. a. **Pick** any number. Double it and then add 30. Divide the sum by 2. Subtract the original number. What do you get?
   b. Repeat the exercise with two other numbers.
   c. Write an algebraic equation to explain your answers.

1–2. See Solutions Manual.

Guided Practice

Define a variable and write an equation for each situation. Then solve.

3. If 17 is decreased by twice a number, the result is 5. Find the number.
4. Three times a number plus twice the number plus one is 6. What is the number? $3n + 2n + 1 = 6$; 1
5. **Pets** According to a survey by the Pet Food Institute, about 2682 thousand dogs in the United States can "speak." This is 354 thousand fewer than four times the number of dogs that can "sing."
   a. Let d represent the number of dogs that can "sing." Write an equation for this situation. $2682 = 4d - 354$
   b. How many dogs can "sing"? 759 thousand

3. $17 - 2x = 5$; 6

Exercises: Practicing and Applying the Concept

Independent Practice

Match each sentence with the equation it represents.

6. Four plus three times a number is 18. A
7. Eighteen equals seven plus twice some number. C
8. Four times a number minus 15 is 92. B
9. Ten minus three times a number is forty-five. E

A. $4 + 3x = 18$
B. $4n - 15 = 92$
C. $18 = 7 + 2n$
D. $18 = 7n + 2$
E. $10 - 3x = 45$
F. $15 - 4n = 92$
G. $10x - 3 = 45$

2 TEACH

In-Class Examples

For Example 1
Twelve is three less than five times a number. Find the number. **3**

For Example 2
Ms. Allison earns \$48,400 per year. That is \$4150 more than three times as much as her daughter earns. How much does her daughter earn? **\$14,750**

FYI

Lincoln is often given credit for the phrase "Government of the people, by the people, for the people." However, Daniel Webster had previously talked about "The people's government, made for the people, made by the people, and answerable to the people."

Practice Masters, p. 56

NAME ____________ DATE ________

7-3 Practice
Writing Two-Step Equations

Student Edition Pages 338–340

Define a variable and write an equation for each situation. Then solve.

1. Find a number such that three times the number increased by 7 is 52. **Let n = the number; $3n + 7 = 52$; 15**
2. Five times a number decreases by 11 is 19. Find the number. **Let n = the number; $5n - 11 = 19$; 6**
3. Thirteen more than four times a number is $^-91$. Find the number. **Let n = the number; $4n + 13 = {^-91}$; $^-26$**
4. Find a number such that seven less than twice the number is 43. **Let n = the number; $2n - 7 = 43$; 25**
5. The length of a rectangle is four times its width. Its perimeter is 90 m. Find its dimensions. Use $P = 2\ell + 2w$. **Let w = width, then $4w$ = length; $90 = 2(4w) + 2w$; 9 m × 36 m**
6. The total cost of a suit and a coat is \$291. The coat cost twice as much as the suit. How much did the coat cost? **Let s = the price of suit, then $2s$ = the price of coat; $2s + s = 291$; \$194**
7. In one season, Kim ran 18 races. This was four fewer races than twice the number of races Kelly ran. How many races did Kelly run? **Let r = races run by Kelly; $18 = 2r - 4$; 11 races**
8. The perimeter of a triangle is 51 cm. The lengths of its sides are consecutive odd integers. Find the lengths of all three sides. **Let x = shortest side, then $x + 2$ and $x + 4$ are the remaining sides; $x + x + 2 + x + 4 = 51$; 15 cm, 17 cm, 19 cm**
9. The length of a rectangle is 5 more than twice its width. Its perimeter is 88 feet. Find its dimensions. Use $P = 2\ell + 2w$. **Let w = width, then $2w + 5$ = length; $88 = 2(2w + 5) + 2w$; 13 ft × 31 ft**
10. Steve hit four more home runs than twice the number of home runs Larry hit. Together they hit 10 home runs. How many home runs did Steve hit? **Let h = home runs hit by Larry, then $2h + 4$ = home runs hit by Steve; $h + 2h + 4 = 10$;**

Group Activity Card 7-3

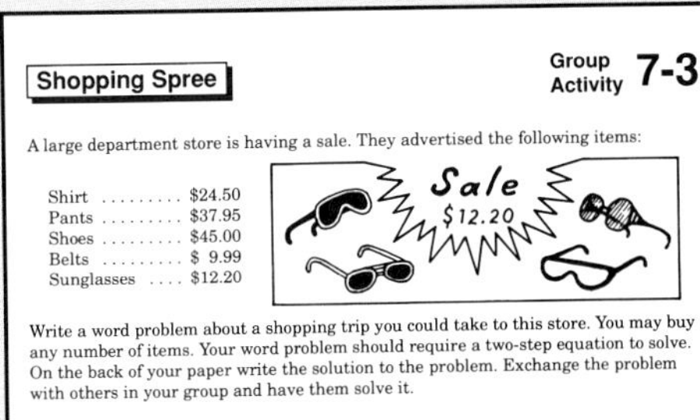

Shopping Spree Group Activity **7-3**

A large department store is having a sale. They advertised the following items:

Shirt	\$24.50
Pants	\$37.95
Shoes	\$45.00
Belts	\$ 9.99
Sunglasses	\$12.20

Write a word problem about a shopping trip you could take to this store. You may buy any number of items. Your word problem should require a two-step equation to solve. On the back of your paper write the solution to the problem. Exchange the problem with others in your group and have them solve it.

 Pre-Algebra

Reteaching

Act It Out Bring to class a vest labeled "multiply/divide" and a coat labeled "add/subtract". Show students that the normal order of operation is to put on the vest first (multiply/divide) and then the jacket (add/subtract). Removing the coat and vest is similar to solving an equation in that the order must be reversed.

3 PRACTICE/APPLY

Checking Your Understanding

Exercises 1–5 are designed to help you assess your students' understanding through reading, writing, speaking, and modeling. You should work through Exercises 1–2 with your students and then monitor their work on Exercises 3–5.

Assignment Guide

Core: 7–17 odd, 18–19, 21–25
Enriched: 6–16 even, 18–25

For **Extra Practice**, see p. 756.

The red A, B, and C flags, printed only in the Teacher's Wraparound Edition, indicate the level of difficulty of the exercises.

4 ASSESS

Closing Activity

Writing Have students solve the following problem. *Mary charges $3.50 per hour to baby-sit plus $3 for gas. If Mary made $11.75 for one night of baby-sitting, how many hours did she baby-sit?*
3 + 3.50x = 11.75; 2.5 hours

Enrichment Masters, p. 56

NAME ____________ DATE ________

7-3 Enrichment
Multi-Step Problems

Student Edition
Pages 338–340

Solve.

1. Danny bought a baseball glove for $22.50 and a pair of sneakers for $27.95. How much change did he receive if he paid with a $100 bill? **$49.55**
2. Artis bought four gifts. They cost $4.39, $3.25, $8.95, and $2.50. How much did he receive if he gave the clerk a $20 bill? **$0.91**
3. During the summer, Bob worked 35 hours per week for 8 weeks as a lifeguard. If he was paid $3.75 an hour, how much money did he earn for the summer? **$1050**
4. Mindy purchased 13 gallons of gasoline at $1.35 a gallon, 5 quarts of oil at $1.05 a quart, and a new tire for $84.75. What was her total bill? **$107.55**
5. Mrs. Clarke has a weekly budget of $95 for groceries. She spent $24.65 on Monday, $32.75 on Tuesday, $12.25 on Wednesday, $8.50 on Thursday, and $17.50 on Friday. Did she stay within her budget? Explain. **No; she spent a total of $95.65.**
6. Gloria agreed to buy a boat costing $1183.84 by paying $400 down and the balance in equal monthly payments for one year. How much did she pay each month? **$65.32**
7. A grocery store purchased 40 gallons of milk at $1.10 per gallon and sold it at 52 cents per quart. What profit was made on the 40 gallons of milk? **$39.20**
8. The telephone rate between two calling zones is $0.45 for the first three minutes and $0.10 for each additional minute. How long was a call for which the charge was $0.95? **8 minutes**

10. $3x - 4 = 17$; 7 **B**
11. $\frac{n+6}{7} = 5$; 29
12. $20 + 2c = -30$; -25
13. $\frac{c}{-4} - 8 = -42$; 136
14. $2x - 9 = 16$; 12.5
15. $5x - 10 = 145$; 31
C
16. Answers will vary.
18. $2x + (2x + 2)$

Define a variable and write an equation for each situation. Then solve.

10. Three times a number less four is 17. What is the number?

11. The sum of a number and 6, divided by 7 is 5. Find the number.

12. Twenty more than twice a number is -30. What is the number?

13. The quotient of a number and -4, less 8 is -42. What is the number?

14. The difference between twice a number and 9 is 16. Find the number.

15. The product of 5 and a number is 10 more than 145. Find the number.

16. Write a real-life problem that could be solved using $3x + 2 = 17$.

17. Geometry The perimeter of the triangle at the right is 27 yards. What are the lengths of the sides of the triangle? **7 yd, 9 yd, 11 yd**

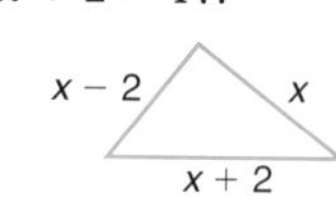

Critical Thinking

18. Numbers that are in order and differ by 1, such as 2, 3, 4, are *consecutive numbers*. If you begin with an even integer and count by two, you are counting *consecutive even integers*. Write an expression for the sum of two consecutive even integers where $2x$ is the lesser integer.

Applications and Problem Solving

19. Fencing Wanda used 130 feet of fence to enclose a rectangular flower garden. She also used the 50-foot wall of her house as one side of the garden. What was the width of the garden?

a. Write an equation that represents this situation. **$2w + 50 = 130$**

b. Solve the equation to find the width of the garden. **40 feet**

20. Running In cross-country, a team's score is the sum of the place numbers of the first five finishers on the team. The co-captains finished first and fourth in a meet. The next three finishers on the team placed in consecutive order. The team score was 35.

20a. $5 + n + (n + 1) + (n + 2) = 35$

a. Write an equation that represents this situation.

b. In what places did the other three members finish? **9th, 10th, 11th**

Mixed Review

21. Solve $8 + \frac{2}{5}n = 28$. (Lesson 7-2) **50**

22. Estimate $188.76 + 14.31 - 106.5$. (Lesson 5-2) **Sample answer: 110**

23. Solve $y - 19 = 53$. (Lesson 3-2) **72**

24. Money What is the least positive number of coins that is *impossible* to give as change for a dollar? (Lesson 3-1) **77**

25. assoc., +

25. Name the property shown by $(2 + 5) + c = 2 + (5 + c)$. (Lesson 1-4)

From the FUNNY PAPERS

AN INTEGRAL NUMBER OF YEARS AGO, THE INTEGER BEING SEVEN GREATER THAN FOUR TIMES THE LEAST COMMON MULTIPLE OF 4 AND 5, OUR FATHERS...

1. Explain why the comic is funny. **See students' work.**
2. Use the information in the comic to write an equation. Let y represent the number of years ago. **$y = 4(20) + 7$**
3. What is the value of y? **87**

Extension

Make Up a Problem Have students write a verbal problem that can be solved by the equation $2x - 5 = 19$. **Sample answer: Olga scored five less than twice as many points as Xena scored in the basketball game.**

From the FUNNY PAPERS

Legend has it that the Gettysburg Address was scribbled on the back of an envelope. However, several drafts have been found, including one on executive stationary.

7-4 Integration: Geometry
Circles and Circumference

Setting Goals: *In this lesson, you'll find the circumference of a circle.*

Modeling with Manipulatives

MATERIALS

4 circular objects
string

metric ruler

A **circle** is the set of all points in a plane that are the same distance from a given point in the plane. The given point is called the **center**. The distance from the center to any point on the circle is called the **radius (*r*)**. The distance across the circle through its center is its **diameter (*d*)**. The **circumference (*C*)** of a circle is the the distance around the circle.

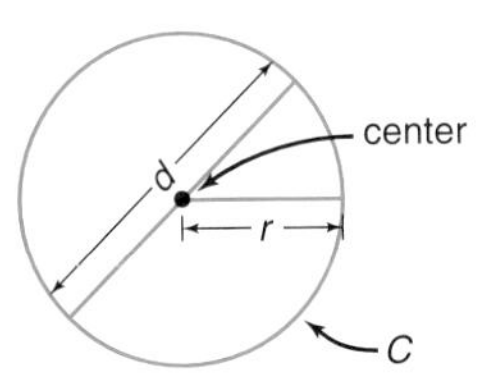

You can measure the circumference of a circle with a metric ruler.

Your Turn

- Use a metric ruler to measure the diameter of a circular object. Record your findings in a table like the one below.
- Wrap a string around the circular object once. Mark the string where it meets itself.
- Lay the string out straight and measure its length with your ruler. Record your finding. This is the circumference of the circular object.
- Use a calculator to divide the circumference by the diameter. Round to the nearest hundredth and record your finding.
- Repeat this activity with three other circular objects of various sizes.

THINK ABOUT IT

What is the relationship between the diameter and the radius of a circle?

The diameter is twice the radius.

	Object	Diameter	Circumference	$\frac{\text{Circumference}}{\text{Diameter}}$
#1				
#2				
#3				
#4				

TALK ABOUT IT

a. What is the mean of the quotients in the last column?

b. Press the  key on your calculator. How does your mean compare to this value? **Answers will vary. π and your mean should be close.**

a. about 3.14 Answers will vary

Learning the Concept

The relationship you discovered in the activity above is true for all circles. The circumference of a circle is always 3.1415926... times the diameter. The Greek letter π **(pi)** stands for this number. Although π is not a rational number, the rational numbers 3.14 and $\frac{22}{7}$ are two generally accepted approximations for π.

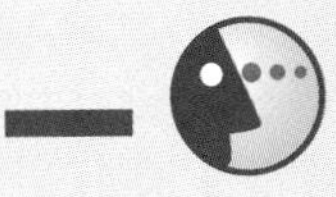

Alternative Learning Styles

Visual Arrange four circular objects with different diameters on a table. Provide at least eight pieces of string of different lengths. Have students take turns trying to match the proper string with each object by using visual estimation.

Alternative Teaching Strategies

Student Diversity To help students remember the terminology of the lesson, relate *circumference* to *circle* and *radius* to *radiate*, as in heat radiates from a central source.

7-4 LESSON NOTES

NCTM Standards: 1-4, 9, 12

Instructional Resources

- Study Guide Master 7-4
- Practice Master 7-4
- Enrichment Master 7-4
- Group Activity Card 7-4
- Assessment and Evaluation Masters pp. 182, 183
- Multicultural Activity Masters, p. 13

Transparency 7-4A contains the 5-Minute Check for this lesson; **Transparency 7-4B** contains a teaching aid for this lesson.

Recommended Pacing	
Standard Pacing	Day 5 of 12
Honors Pacing	Day 5 of 11
Block Scheduling*	Day 3 of 7 (along with Lesson 7-3)

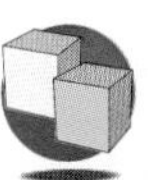

*For more information on pacing and possible lesson plans, refer to the ***Block Scheduling Booklet***.

1 FOCUS

5-Minute Check
(over Lesson 7-3)

Define a variable. Then write and solve an equation.

1. Shyam sold the family car for \$200 more than half of the original price. If he sold the car for \$6540, what was the original price? **$6540 = 200 + \frac{x}{2}$; \$12,680**
2. Erik scored 13 points more than twice the lowest score on a science test. If he scored 87 points, what was the lowest score on the test? **$87 = 13 + 2x$; 37**

Motivating the Lesson

Situational Problem Tell students that the exercise wheel in a hamster's cage has a diameter of 6 inches. Ask students how they would find how far the hamster runs on the wheel in a day.

2 TEACH

In-Class Examples

For Example 1
Find the circumference of each circle.

a. The diameter is 3.2 centimeters. Use 3.14 for π. Round to the nearest tenth. **10.0 cm**

b. The radius is $4\frac{2}{3}$ inches. Use $\frac{22}{7}$ for π. **$29\frac{1}{3}$ in**

For Example 2
Jenna has 30 feet of decorative fence that she wants to put around a circular flower garden. To the nearest tenth of a foot, what should the radius of her garden be? **9.5 feet**

3 PRACTICE/APPLY

Checking Your Understanding

Exercises 1–10 are designed to help you assess your students' understanding through reading, writing, speaking, and modeling. You should work through Exercises 1–5 with your students and then monitor their work on Exercises 6–10.

Study Guide Masters, p. 57

NAME ______________ DATE ______
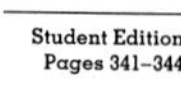

7-4 Study Guide
Integration: Geometry
Circles and Circumference

Student Edition
Pages 341–344

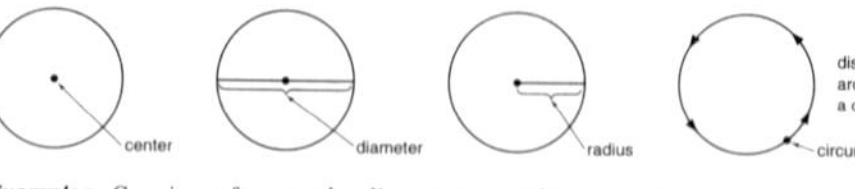

Examples: C = circumference; d = diameter; r = radius; use 3.14 for π.

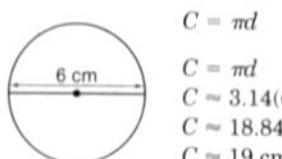

$C = \pi d$

$C = \pi d$
$C \approx 3.14(6)$
$C \approx 18.84$
$C \approx 19$ cm

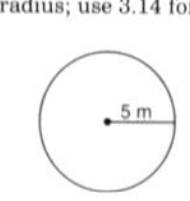
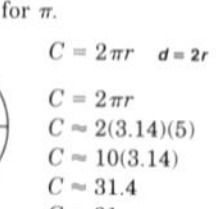

$C = 2\pi r$ $\quad d = 2r$

$C = 2\pi r$
$C \approx 2(3.14)(5)$
$C \approx 10(3.14)$
$C \approx 31.4$
$C \approx 31$ m

Find the circumference of each circle.

1. 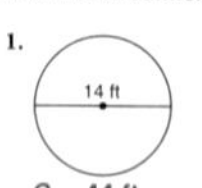

$C \approx$ 44 ft

2. 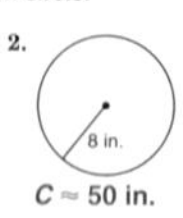

$C \approx$ 50 in.

3. 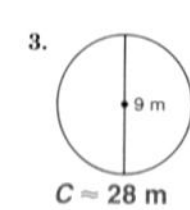

$C \approx$ 28 m

4. The radius is $6\frac{1}{5}$ feet. $C \approx$ 39 m

5. The diameter is 4.7 yards. $C \approx$ 15 yd

Solve. Round to the nearest inch.

6. What is the circumference of the top of an ice cream cone if its diameter is about $1\frac{7}{8}$ inches? $\left(\frac{7}{8} = 0.875\right)$ $C \approx$ 6 in.

7. The radius of the basketball rim is 9 inches. What is the circumference? $C \approx$ 57 in.

Circumference	
In words:	The circumference of a circle is equal to its diameter times π, or 2 times its radius times π.
In symbols:	$C = \pi d$ or $C = 2\pi r$

Example  **Find the circumference of each circle described below.**

a. The diameter is 8.25 cm. Use a calculator.

$C = \pi d$

$C = \pi \cdot 8.25$

[π] [×] 8.25 [=] *25.91813939*

The circumference is about 25.9 cm.

b. Use $\frac{22}{7}$ for π.

$C = 2\pi r$

$C \approx \frac{2}{1} \cdot \frac{22}{7} \cdot \frac{28}{3}$ (28 and 7 simplified to 4 and 1)

$C \approx \frac{176}{3}$ or $58\frac{2}{3}$

The circumference is about $58\frac{2}{3}$ in.

$9\frac{1}{3}$ in.

Connection to Algebra

Sometimes you'll know the circumference of a circle and need to find its radius or diameter. In this case, you can write an equation and solve it.

Example 2 APPLICATION Geography

Earth's circumference is approximately 25,000 miles. If you could dig a tunnel to the center of the Earth, how long would the tunnel be?

Explore You know the circumference of Earth. You want to know the distance to the center, or the radius of Earth.

Plan Replace C with 25,000 and solve $C = 2\pi r$ for r.

Solve $25{,}000 = 2\pi r$

$\frac{25{,}000}{2\pi} = \frac{2\pi r}{2\pi}$ *Divide each side by 2π.*

$\frac{25{,}000}{2\pi} = r$ ***Estimate:*** *24,000 ÷ 6 = 4000*

25,000  [÷] [(] 2 [×] [π] [)] [=]  *3978.8736*

The distance to the center of Earth is about 3979 miles.

Examine The radius is about 4000 and π is about 3. Estimate the circumference by finding their product.

$2 \cdot 3 \cdot 4000 = 24{,}000$

The answer is reasonable.

CULTURAL CONNECTIONS
The Native American culture has always valued the symbolism of a circle. Oglala Sioux writer Black Elk relates "The Power of the World always works in circles, and everything tries to be round."

Checking Your Understanding

Communicating Mathematics

Read and study the lesson to answer these questions. **1. See margin.**

1. **Explain** how to find the radius of a circle if you know its diameter.
2. **Name** two rational numbers that are approximations for π. **3.14, $\frac{22}{7}$**

Reteaching

Using Connections Write the formula for circumference on the board or overhead. Ask students how to find the value of π if the circumference and diameter are known. **divide each side by the diameter** Have students use calculators or computers to rename $\frac{22}{7}$ as a decimal.

CULTURAL CONNECTIONS
Many Native Americans historically constructed their homes and communities in circular shapes. Many of their religious objects were circular.

3. **Calculate** the circumference of a circle with a radius of 6 inches using the pi key on a calculator and each of the rational approximations. **See margin.**
4. A *chord* of a circle is a segment whose endpoints both lie on the circle. What is the longest chord you can draw? **a diameter**
5. Write a note to someone at home that explains how to estimate the circumference of a circle. **See margin.**

MATH JOURNAL

Guided Practice

6. 9.42 ft
7. 11.31 cm
9. 10.99 m

Find the circumference of each circle described below.

6.

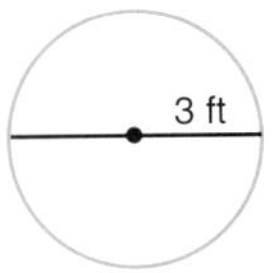

7.

8. The diameter is $5\frac{1}{4}$ inches. **$16\frac{1}{2}$ in.**
9. The radius is 1.75 meters.
10. **Recreation** A 10-speed bicycle tire has a diameter of 27 inches. Find the distance the bicycle will travel in 10 rotations of the tire. **about 848.2 in.**

Exercises: Practicing and Applying the Concept

Independent Practice

11. 12.57 cm
12. 44 m
13. 34.56 ft
14. 28.27 in.
15. 44 mm
16. 213.63 ft.
18. 3.14 m
20. 42.39 in.

A

Find the circumference of each circle.

11.

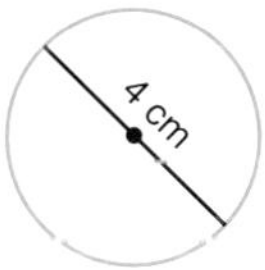

12.

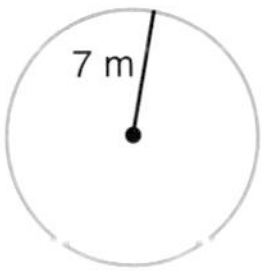

13.

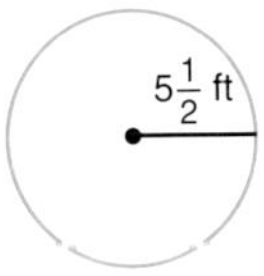

14.

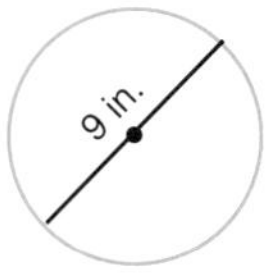

15.

16.

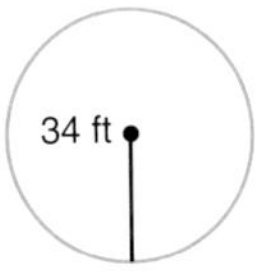

Choose
Estimation
Mental Math
Calculator
Paper and Pencil

17. The diameter is 18.8 m. **59.032 m**
18. The radius is 0.5 meters.
19. The radius is 1.3 yd. **8.164 yd**
20. The diameter is $13\frac{1}{2}$ in.
21. The diameter is $2\frac{1}{3}$ ft. **$7\frac{1}{3}$ ft**
22. The radius is $4\frac{1}{2}$ km. **28.27 km**

Match each circle described in the column on the left with its corresponding measurement in the column on the right.

B

23. $C = 628$ cm **C**	A. $d = 15.2$ cm
24. $r = 30$ cm **D**	B. $C = 87.92$ cm
25. $2r = 28$ cm **B**	C. $r = 100$ cm
26. $C = 47.728$ cm **A**	D. $C = 188.4$ cm

C

28. $\frac{1}{2}$ unit

27. The circumference of a circle is 4.082 meters. Find its radius. **0.65 m**
28. The circumference of a circle is π units. What is the radius of this circle?
29. The radius of a circle is $\frac{7}{22}$ units. Find its circumference. **2 units**

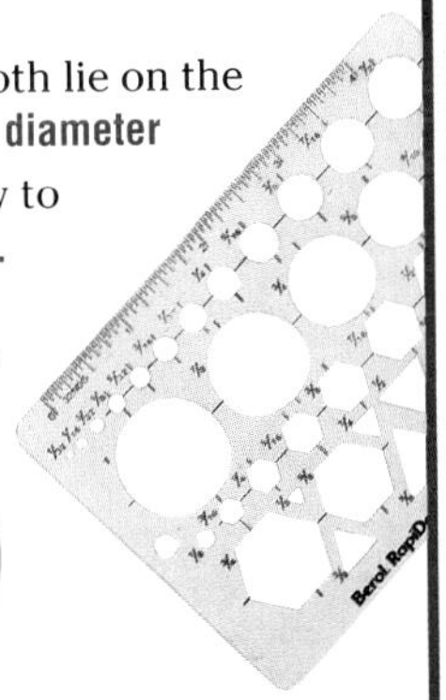

Error Analysis

When finding the circumference, students may forget to multiply by 2 when given the radius rather than the diameter. Encourage students to check their work by estimation or by working backward.

Assignment Guide

Core: 11–29 odd, 30–31, 33, 35–40
Enriched: 12–28 even, 30–40

For **Extra Practice**, see p. 757.

The red A, B, and C flags, printed only in the Teacher's Wraparound Edition, indicate the level of difficulty of the exercises.

Additional Answers

1. Sample answer: The radius of a circle is one-half its diameter.
3. Sample answer: Multiply the diameter times π, or 2 times the radius times π.
5. Sample answer: Multiply the diameter by 3 or the radius by 6.

Practice Masters, p. 57

NAME ____________ DATE ________

7-4 Practice
Integration: Geometry
Circles and Circumference

Student Edition
Pages 341–344

Find the circumference of each circle.

1.

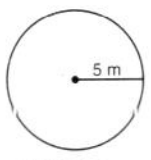

31.4 m

2.

25.12 mm

3.

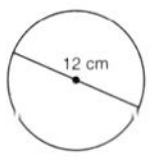

37.68 cm

4.

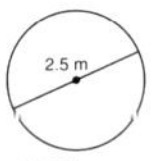

7.85 m

5.

12.56 ft

6.

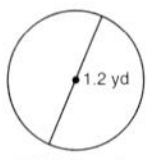

3.768 yd

7.

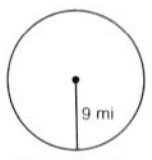

56.52 mi

8.

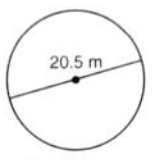

64.37 m

9. The diameter is 15.2 km. **47.728 km**
10. The radius is $\frac{13}{20}$ yd. **4.082 yd**
11. The diameter is $\frac{1}{2}$ ft. **1.57 ft**
12. The radius is $3\frac{3}{4}$ in. **23.55 in.**
13. The diameter is 25.6 cm. **80.384 cm**
14. The radius is 12 mm. **75.36 mm**

Match each circle described in the column on the left with its corresponding measurement in the column on the right.

15. $C = 43.96$ cm **C**	A. $d = 16.2$ cm
16. $r = 11.2$ cm **B**	B. $C = 70.336$ cm
17. $2r = 10.4$ cm **D**	C. $r = 7$ cm
18. $C = 50.868$ cm **A**	D. $C = 32.656$ cm

Group Activity Card 7-4

Curious Circumferences — Group Activity 7-4

In a small group, complete the following activity to discover an astounding fact about the distance around circles of different diameters.

Suppose you had a small plate and a bicycle tire. The diameter of the plate was 5 inches and the diameter of the tire was 30 inches. Compute the circumference of both the plate and the bicycle tire.

Pretend you had strings of these two lengths. Now suppose you added 9.42 inches to each string. If you put these new strings around the plate and the tire, how far outside each would the string be? You will be amazed when you compute the new diameters.

More amazing is the following: The diameter of Earth at the equator is about 8,000 miles. If you could put a string around Earth and then add 9.42 inches to this string, how far above the surface do you think the string would be? Change the diameter measurement to inches and do the computations to find the answer.

 Pre-Algebra

4 ASSESS

Closing Activity

Modeling Provide a variety of circular objects. Have students demonstrate how to find the circumference of two different objects.

Chapter 7, Quiz B (Lessons 7-3 and 7-4) is available in the *Assessment and Evaluation Masters*, p. 183.

Mid-Chapter Test (Lessons 7-1 through 7-4) is available in the *Assessment and Evaluation Masters*, p. 182.

Additional Answers

33b. Sample answer: The perimeter of a square is 4 times the length of one side. The circumference of a circle is approximately 3.14 times the diameter. $4 > 3.14$

38. $m \geq 1$

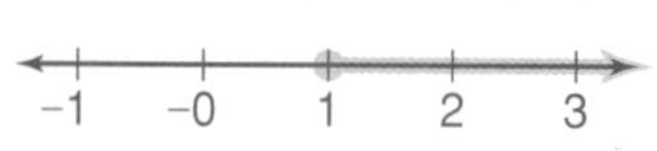

Enrichment Masters, p. 57

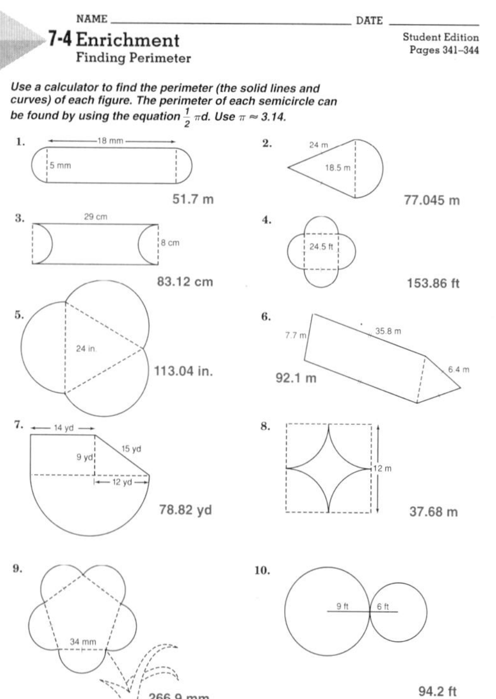

NAME ______ DATE ______

7-4 Enrichment
Finding Perimeter

Student Edition Pages 341–344

Use a calculator to find the perimeter (the solid lines and curves) of each figure. The perimeter of each semicircle can be found by using the equation $\frac{1}{2}\pi d$. Use $\pi \approx 3.14$.

1. 18 mm; 5 mm — **51.7 m**
2. 24 m; 18.5 m — **77.045 m**
3. 29 cm; 8 cm — **83.12 cm**
4. 24.5 ft — **153.86 ft**
5. 24 in. — **113.04 in.**
6. 7.7 m; 35.8 m; 6.4 m — **92.1 m**
7. 14 yd; 9 yd; 15 yd; 12 yd — **78.82 yd**
8. 12 m — **37.68 m**
9. 34 mm — **266.9 mm**
10. 9 ft; 6 ft — **94.2 ft**

Critical Thinking

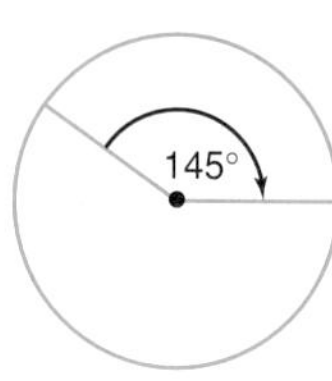

30. A *central angle* of a circle is an angle whose vertex is the center of a circle. A 90° angle intersects an arc of the circle that has a length that is $\frac{1}{4}$ the circumference. Find the length of the arc intersected by a 145° central angle of a circle that has a diameter of 8 feet. **about 10.12 ft**

Applications and Problem Solving

31. Entertainment The first Ferris wheel, designed by George Ferris, was built in 1893 in Paris, France. The diameter of that Ferris wheel was 76 meters. What distance would you travel in one revolution if you rode on the first Ferris wheel? **about 238.76 m**

32. Automotive The diameter of the wheels on many passenger cars is 14 inches. All four tires are inflated so that the weight of the car does not distort the shape of the tires. The cross section height of each 14-inch tire is 5.4 inches

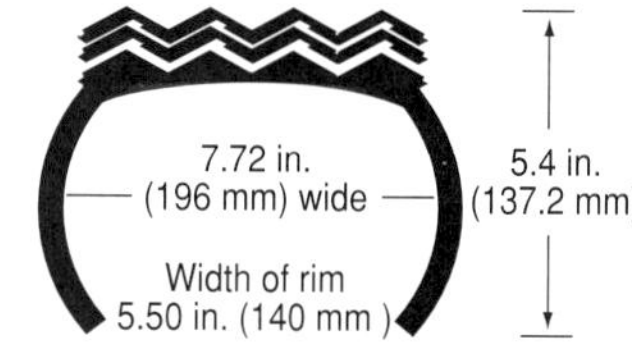

a. How far does a car with 14-inch tires travel in one complete turn of the tires? **32a. about 77.91 in. or about 6.49 ft**

b. How many turns of the tires would it take to travel 1 mile? **32b. about 813.56**

c. A well-cared-for tire will usually be able to travel about 50,000 miles before it wears out. How many turns of the tire would it take to travel 50,000 miles? **about 40,678,000**

33. Geometry Each side of a square is 7 cm long. A circle has a diameter that is 7 cm long. **33b. See margin.**

a. Which is longer, the perimeter of the square or the circumference of the circle? **The perimeter of the square is longer.**

b. Justify your answer to part a without doing any computation.

interNET CONNECTION

For the latest calculation of π, visit: **www.glencoe.com/sec/math/prealg/mathnet**

34. Research Mathematicians have worked for centuries to find an accurate value for π. Investigate the methods that have been used. Who was the first to calculate a value of π? How many decimal places are known now? **See Solutions Manual**

Mixed Review

35. Tourism The Louvre Museum, home to the *Mona Lisa*, is one of the most-visited museums in the world. The National Gallery of Art in Washington, D.C. is the most-visited art museum in the United States with 7.5 million visitors each year. This is 0.1 million more than twice what the second most visited museum, the Metropolitan Museum of Art, in New York, receives. How many visitors go to the Metropolitan Museum each year? (Lesson 7-3) **3.7 million**

36. Solve $\frac{x}{-1.7} \geq 6.8$. (Lesson 6-7) $x \leq -11.56$

37. Factor $220pq^2$ completely. (Lesson 4-4) $2 \cdot 2 \cdot 5 \cdot 11 \cdot p \cdot q \cdot q$

38. Solve $-4 + m \geq -3$ and graph the solution. (Lesson 3-6) **See margin.**

39. Replace ● with <, >, or = to make $|-13|$ ● $|7|$ a true sentence. (Lesson 2-3) **>**

40. Simplify $18z + 7(2 + 3z)$. (Lesson 1-5) $39z + 14$

Extension

Using Cooperative Groups Have small groups determine the answer to the following problem.

Two planes are to fly around Earth at the equator. One plane will fly at an altitude of 5 miles. The other will fly at an altitude of 6.5 miles. How much farther will the second plane travel? **about 9.4 miles farther**

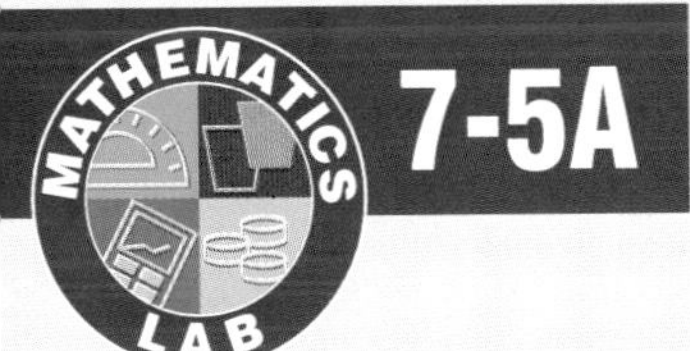

HANDS-ON ACTIVITY

7-5A Equations with Variables on Each Side

A Preview of Lesson **7-5**

In this activity, you will use cups and counters to model and solve equations with variables on each side.

Activity

MATERIALS

cups

counters

equation mat

Model and solve the equation $2x - 4 = x - 2$.

▶ Place 2 cups and 4 negative counters on the left side of the mat. Then place 1 cup and 2 negative counters on the right side of the mat.

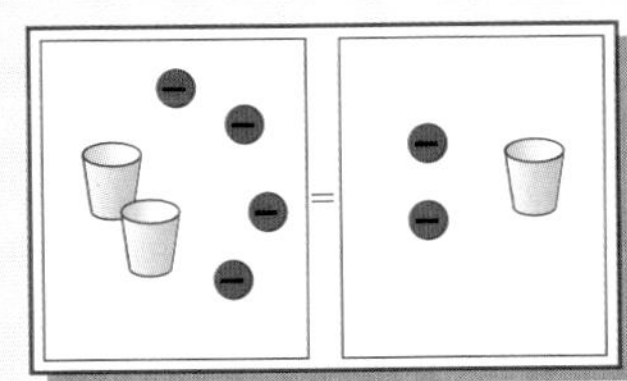

▶ Since the two sides of the mat represent equal quantities, you can remove an equal number of cups or counters from each side without changing the value of the equation.

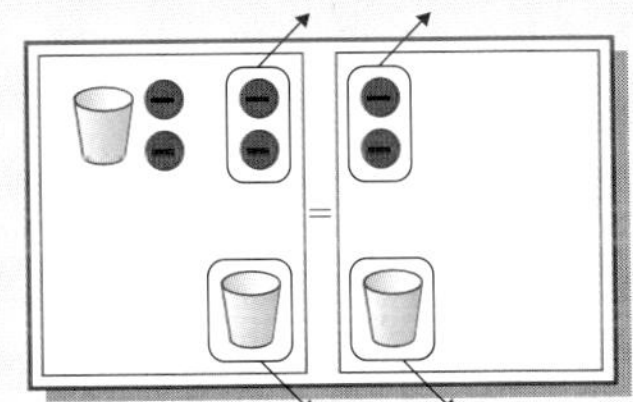

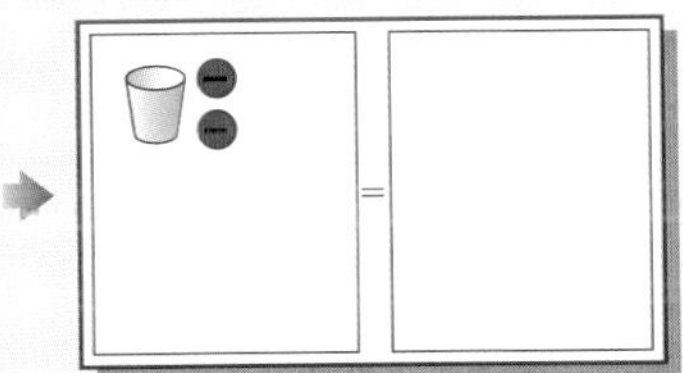

▶ Now the equation is $x - 2 = 0$. Since there are no counters on the right, add 2 zero pairs. Then you can remove 2 negative counters from each side.

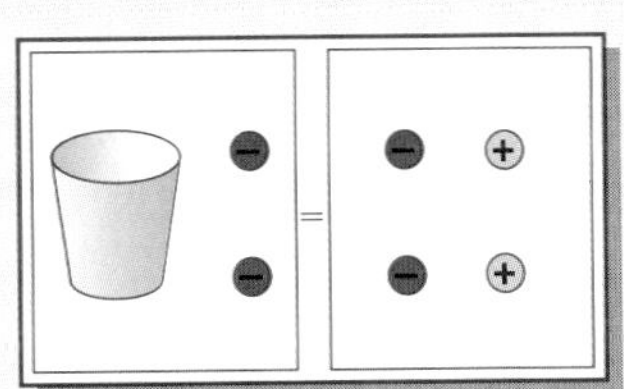

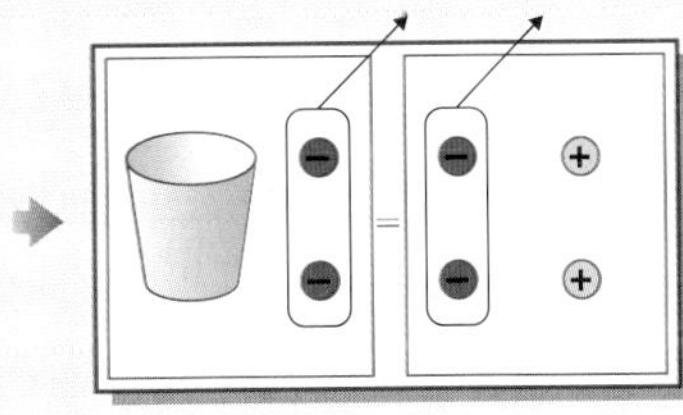

Therefore, each cup must contain 2 counters and $x = 2$.

Your Turn **Model and solve each equation.** **1–6. See students' work.**

1. $2x + 3 = x - 5$ **−8**
2. $2x + 3 = x + 1$ **−2**
3. $3x - 2 = x + 6$ **4**
4. $3x - 7 = x + 1$ **4**
5. $6 + x = 5x + 2$ **1**
6. $x - 1 = 3x + 7$ **−4**

TALK ABOUT IT

7. Does it matter whether you remove cups or counters first? Explain.
8. Explain how you could use models to solve $-2x + 3 = -x - 5$.

7–8. See Solutions Manual.

7-5A LESSON NOTES

NCTM Standards: 1-4, 9

Objective

Use cups and counters to model and solve equations with the variable on each side.

Recommended Time

Demonstration and discussion: 15 minutes; Exercises: 30 minutes

Instructional Resources

For each student or group of students

Student Manipulative Kit

- cups, counters, equation mats

Math Lab and Modeling Math Masters

- p. 3 (integer mat)
- p. 5 (integer counters)
- p. 6 (pattern for cup)
- p. 40 (worksheet)

For teacher demonstration

Overhead Manipulative Resources

1 FOCUS

Motivating the Lesson

Model several examples of equations with the variable on both sides. Help students write the corresponding equations. Model the example for the entire class. Ask students to identify the steps as you undo the operations.

2 TEACH

Teaching Tip Allow students to work in pairs or small groups to complete this activity. Circulate around the room as students model the problems. Have students check their answers by substituting them into the original equations.

3 PRACTICE/APPLY

Assignment Guide

Core: 1–8

Enriched: 1–8

4 ASSESS

Observing students working in cooperative groups is an excellent method of assessment.

7-5 LESSON NOTES

NCTM Standards: 1-4, 9

Instructional Resources

- Study Guide Master 7-5
- Practice Master 7-5
- Enrichment Master 7-5
- Group Activity Card 7-5
- Graphing Calculator Masters, p. 7
- Math Lab and Modeling Math Masters, p. 65
- Real-World Applications, 14

Transparency 7-5A contains the 5-Minute Check for this lesson; **Transparency 7-5B** contains a teaching aid for this lesson.

Recommended Pacing	
Standard Pacing	Day 7 of 12
Honors Pacing	Day 6 of 11
Block Scheduling*	Day 4 of 7

*For more information on pacing and possible lesson plans, refer to the ***Block Scheduling Booklet***.

1 FOCUS

5-Minute Check
(over Lesson 7-4)

Find the circumference of each circle.

1. The radius is 5 cm. **31.4 cm**
2. The diameter is 10.5 m. **32.97 m**
3. The diameter is $1\frac{3}{4}$ ft. **$5\frac{1}{2}$ ft**
4. The radius is $18\frac{2}{3}$ in. **$117\frac{1}{3}$ in.**

Solve.

5. A wheel has a circumference of 40.82 cm. Find the radius of the wheel. **6.5 cm**

7-5 Solving Equations with Variables on Each Side

Setting Goals: *In this lesson, you'll solve equations with the variable on each side.*

Modeling with Technology

Most countries use the Celsius scale to measure temperature. In the United States, we use the Fahrenheit scale most of the time. There is one temperature which is the same on both scales. You can use the T1-82 graphing calculator program at the right to find that temperature. In this program, C represents Celsius temperature, and F represents Fahrenheit temperature.

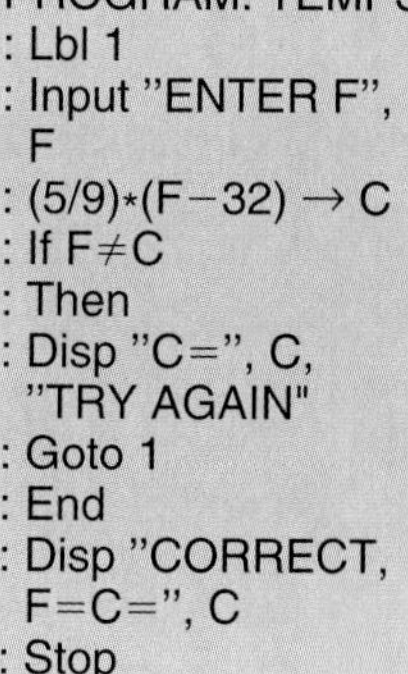

```
PROGRAM: TEMPS
: Lbl 1
: Input "ENTER F",
  F
: (5/9)*(F−32) → C
: If F≠C
: Then
: Disp "C=", C,
  "TRY AGAIN"
: Goto 1
: End
: Disp "CORRECT,
  F=C=", C
: Stop
```

Your Turn

▶ Enter the program into the calculator's memory.

▶ Run the program.

Enter a guess at the Fahrenheit temperature that has the same value on the Celsius scale. Continue guessing and checking until you determine the temperature.

TALK ABOUT IT

a. What temperature is the same on both scales? **−40**

b. Explain what the formula in the third line of the program does.

b. It converts the Fahrenheit temperature to Celsius.

Learning the Concept

The formula, $F = \frac{9}{5}C + 32$, is used for finding the Fahrenheit temperature when a Celsius temperature is known. When the temperature is the same on both scales, $F = C$. Replacing F with C results in the equation $C = \frac{9}{5}C + 32$.

In $C = \frac{9}{5}C + 32$, the variable is on each side of the equal sign. Use the properties of equality to eliminate the variable from one side. Then solve the equation.

$$C = \frac{9}{5}C + 32$$

$$C - \frac{9}{5}C = \frac{9}{5}C - \frac{9}{5}C + 32 \quad \text{Subtract } \tfrac{9}{5}C \text{ from each side.}$$

$$\frac{5}{5}C - \frac{9}{5}C = 32 \quad \text{Rename } C \text{ as } \tfrac{5}{5}C. \text{ Why?}$$

$$-\frac{4}{5}C = 32$$

$$\left(-\frac{\overset{1}{\cancel{5}}}{\underset{1}{\cancel{4}}}\right)\left(-\frac{\overset{1}{\cancel{4}}}{\underset{1}{\cancel{5}}}\right)C = \left(-\frac{5}{\underset{1}{\cancel{4}}}\right) \cdot \frac{\overset{8}{\cancel{32}}}{1} \quad \text{Multiply each side by } -\tfrac{5}{4}. \text{ Why?}$$

$$C = -40$$

Check: $C = \frac{9}{5}C + 32$

$-40 \stackrel{?}{=} \frac{9}{5}(-40) + 32$ *Replace C with −40.*

$-40 \stackrel{?}{=} -72 + 32$

$-40 = -40$ ✓

The temperature is the same on the Celsius and Fahrenheit scales at $-40°$.

Example 1 **Solve $7n + 12 = 2.5n - 2$.**

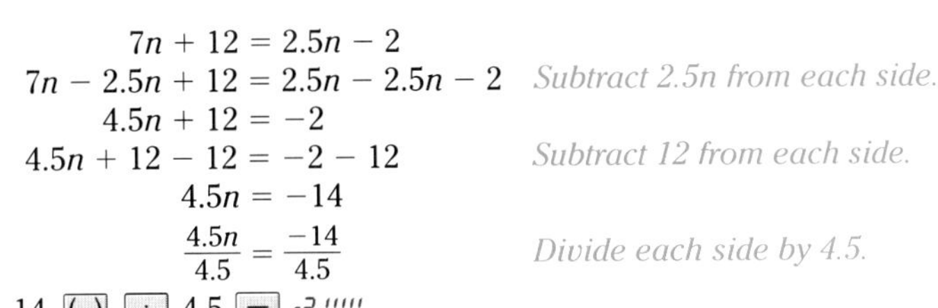

$7n + 12 = 2.5n - 2$

$7n - 2.5n + 12 = 2.5n - 2.5n - 2$ *Subtract 2.5n from each side.*

$4.5n + 12 = -2$

$4.5n + 12 - 12 = -2 - 12$ *Subtract 12 from each side.*

$4.5n = -14$

$\frac{4.5n}{4.5} = \frac{-14}{4.5}$ *Divide each side by 4.5.*

14 [(−)] [÷] 4.5 [=] -3.11111

$n = -3.\overline{1}$

Sometimes a rectangle is described in terms of only one of its dimension. To find the dimensions, you will have to solve an equation that contains grouping symbols. When solving equations of this type, first use the distributive property to remove the grouping symbols.

INTEGRATION Geometry

The perimeter of a rectangle is 74 inches. Find the dimensions if the length is 7 inches greater than twice the width.

Let w represent the width in inches. Then $2w + 7$ represents the length in inches. Recall that the formula for perimeter can be represented as $2w + 2\ell = P$.

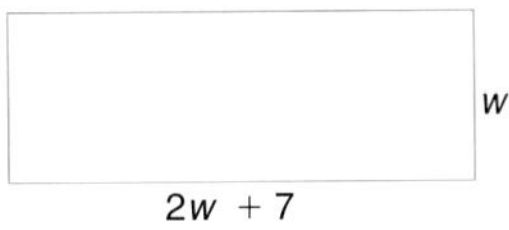

$2w + 2\ell = P$

$2w + 2(2w + 7) = 74$ *Replace ℓ with 2w + 7 and P with 74.*

$2w + 4w + 14 = 74$ *Use the distributive property.*

$6w + 14 = 74$ *Simplify.*

$6w + 14 - 14 = 74 - 14$ *Subtract 14 from each side.*

$6w = 60$

$\frac{6w}{6} = \frac{60}{6}$ *Divide each side by 6.*

$w = 10$

The width is 10 inches. Now evaluate $2w + 7$ to find the length.

$2w + 7 = 2(10) + 7$ or 27

The length is 27 inches. *Check to see if this length and width results in the correct perimeter.*

Some equations may have *no* solution. The solution set is the **null** or **empty set**. It is shown by the symbol { } or ∅. Other equations may always be true and have every number in their solution set.

Motivating the Lesson

Hands-On Activity Provide students with both Fahrenheit and Celsius thermometers. Have them read the classroom temperature on each and write the readings on the board. Have students read the temperature of cold water and of warm water and write those temperatures on the board. Ask whether anyone knows the relationship between the two temperature scales.

2 TEACH

In-Class Examples

For Example 1

Solve $5x - 6 = 3x + 9$. $7\frac{1}{2}$

For Example 2

The perimeter of a rectangle is 102 centimeters. Find the dimensions if the length is 5 centimeters less than 4 times the width. **11.2 cm**

Tech Prep

Landscaper A landscaper often may have to find the perimeter or circumference of an area to estimate costs of materials such as fencing, decorative rocks, shrubs, and so on. For more information on tech prep, see the *Teacher's Handbook*.

GLENCOE Technology

Interactive Mathematics Tools Software

In this interactive computer lesson, students explore altitudes and boiling points. A **Computer Journal** gives students an opportunity to write about what they have learned.

For Windows & Macintosh

In-Class Example

For Example 3
Solve each equation.

a. $6x + 4 = 2(3x - 9)$ **no solution**

b. $9 + 2n = 15 - (6 - 2n)$ **all numbers**

Teaching Tip Remind students that the symbol for no solutions is { } or ∅, but not {∅}.

3 PRACTICE/APPLY

Checking Your Understanding

Exercises 1–10 are designed to help you assess your students' understanding through reading, writing, speaking, and modeling. You should work through Exercises 1–3 with your students and then monitor their work on Exercises 4–10.

Additional Answer

1. Sample answer: Subtract $7p$ from each side. Divide each side by -5.

Study Guide Masters, p. 58

NAME ______________ DATE ______

7-5 Study Guide
Solving Equations with Variables on Each Side

Student Edition Pages 346–350

When an equation has the variable on each side, the first step is to write an equivalent equation with the variable on just one side.

$$4a - 25 = 6a + 50$$
$$4a - 25 + (-6a) = 6a + 50 + (-6a) \quad \text{Add } -6a \text{ to each side.}$$
$$-2a - 25 = 50$$

From this point, the equation is solved by first adding 25 to each side, and then dividing each side by -2.

1. Complete the solution of the equation in the example above.
   $-2a = 75$
   $a = -37.5$
2. Solve the equation in the example again. This time, first isolate the variable on the right side.
   $4a - 25 = 6a + 50$
   $-25 - 50 = 6a - 4a$
   $-75 = 2a$
   $-37.5 = a$

Solve each equation. Check your solution.

3. $4a + 26 = 50 + 6a$ **-12**
4. $2r + 36 = 6r - 12$ **12**
5. $4(b + 24) = 16b + 60$ **3**
6. $x + 21 = -x + 87$ **33**
7. $25c + 17 = -5c + 143$ **4.2**
8. $8v + 5 = 7v - 21$ **-26**
9. $6n - 42 = 4n$ **21**
10. $2x = 3x + 2$ **-2**
11. $5y = 2y - 12$ **-4**
12. $a = 5a - 28$ **7**
13. $2w + 3 = 5w$ **1**
14. $-45m + 68 = 84m - 61$ **1**

Example 3 Solve each equation.

a. $2x + 5 = 2x - 3$

$$2x + 5 = 2x - 3$$
$$2x - 2x + 5 = 2x - 2x - 3 \quad \textit{Subtract 2x from each side.}$$
$$5 = 3$$

This sentence is *never* true. The solution is ∅.

b. $3(x + 1) - 5 = 3x - 2$

$$3(x + 1) - 5 = 3x - 2$$
$$3(x + 1) - 3 = 3x \quad \textit{Add 2 to each side.}$$
$$3x + 3 - 3 = 3x \quad \textit{Distributive property}$$
$$3x = 3x \quad \textit{3 − 3 = 0}$$
$$\frac{3x}{3} = \frac{3x}{3} \quad \textit{Divide each side by 3.}$$
$$x = x$$

This sentence is *always* true. The solution set is all numbers.

Checking Your Understanding

Communicating Mathematics

MODELING MATHEMATICS

MATERIALS
- cups and counters
- equation mat

Read and study the lesson to answer these questions.

1. **Describe** the steps you would take to solve $2p = 15 + 7p$. **See margin.**
2. **Name** the property of equality that allows you to subtract the same term from each side of an equation.
3. The diagram at the right models the equation $2x - 1 = x - 4$.
   a. Solve $2x - 1 = x - 4$. **-3**
   b. Solve $3x - 1 = x + 3$ using cups and counters. **See students' work; $x = 2$**
   c. Solve $x + 5 = x + 7$ using cups and counters. **See students' work; ∅**

2. Sample answer: subtraction property of equality

Guided Practice

Name the first two steps you should take to solve each equation. Then solve. Check your solution. **4–9. See Solutions Manual for steps.**

4. $12k + 15 = 35 + 2k$ **2**
5. $3x + 2 = 4x - 1$ **3**
6. $3b + 8 = -10b + b$ **$-\frac{2}{3}$**
7. $3(a + 22) = 12a + 30$ **4**
8. $3k + 10 = 2k - 21$ **-31**
9. $n + 4 = -n + 10$ **3**
10. **Geometry** The perimeter of a rectangle is 32 feet. Find the dimensions if the length is 4 feet longer than three times the width. **3 ft, 13 ft**

Exercises: Practicing and Applying the Concept

Independent Practice

Solve each equation. Check your solution. **19. all numbers**

11. $6n - 42 = 4n$ **21**
12. $8 - 3g = -2 + 2g$ **2**
13. $3(k + 2) = 12$ **2**
14. $\frac{4}{7}y - 8 = \frac{2}{7}y + 10$ **63**

B

15. $6 - 8x = 20x + 20$ **-0.5**
16. $13 - t = -t + 7$ **∅**
17. $3n + 7 = 7n - 13$ **5**
18. $7b - 3 = -b + 4$ **0.875**
19. $2 + 7(d + 1) = 9 + 7d$
20. $2a - 1 = 3.5a - 3$ **$\frac{4}{3}$**
21. $5m + 4 = 7(m + 1) - 2m$ **∅**
22. $12x - 24 = -14x + 28$ **2**

Reteaching

Using Discussion Have students discuss the steps, including the proper order, they would use to solve each equation.

1. $2x - 4 = 10$ **Sample answer: Add -4 to each side, then divide each side by 2.**
2. $8 + 4x = x - 4$ **Sample answer: Subtract x from each side, add -8 to each side, divide by 3.**
3. $2(x - 3) = 2x - 6$ **Answers may vary. Some students may recognize by inspection that this is an identity and true for all values of x.**

25. all numbers

23. $2(f-3)+5=3(f-1)$ **2**

24. $4[z+3(z-1)]=36$ **3**

25. $-3(4b-10)=\frac{1}{2}(-24b+60)$

26. $\frac{3}{4}a+16=2-\frac{1}{8}a$ **−16**

27. $\frac{d}{0.4}=2d+1.24$ **2.48**

28. $\frac{a-6}{12}=\frac{a-2}{4}$ **0**

Find the dimensions of each rectangle. The perimeter is given.

29. $P = 460$ m

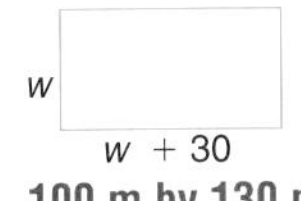

w

w + 30

100 m by 130 m

30. $P = 440$ yd

w

3w − 60

70 yd by 150 yd

31. $P = 110$ ft

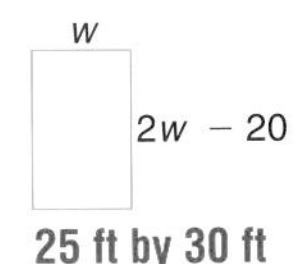

w

2w − 20

25 ft by 30 ft

Define a variable and write an equation for each situation. Then solve.

32. Twice a number is 220 less than six times the number. What is the number? **$2x = 6x - 220$; 55**

33. Fourteen less than three times a number equals the number. What is the number? **$3y - 14 = y$; 7**

34. apple: 50¢; banana: 35¢; orange: 25¢

Critical Thinking

34. An apple costs the same as 2 oranges. Together, an orange and a banana cost 10¢ more than an apple. Two oranges cost 15¢ more than a banana. What is the cost for one of each fruit?

Applications and Problem Solving

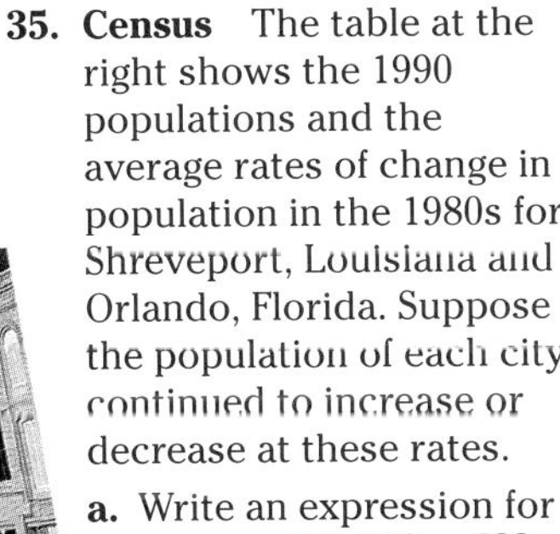

35. **Census** The table at the right shows the 1990 populations and the average rates of change in population in the 1980s for Shreveport, Louisiana and Orlando, Florida. Suppose the population of each city continued to increase or decrease at these rates.

City	Shreveport, LA	Orlando, FL
1990 Population	199,000	165,000
Annual Population Change in 1980s	−700	+3700

a. Write an expression for the population of Shreveport after x years. **199,000 − 700x**

b. Write an expression for the population of Orlando after x years.

c. Use the expressions in parts a and b to write an equation to find the number of years until the populations are the same.

d. In how many years would the population of the two cities be the same? **about 8 years**

35b. 165,000 + 3700x
35c. 199,000 − 700x = 165,000 + 3700x

36. **Cooking** Water boils at 212°F at sea level. However, as the altitude increases, the air pressure drops. The lower air pressure causes water to boil at a lower temperature. For every 550 feet above sea level, the boiling point is lowered 1°F. The boiling point rises 1°F for every 550 feet below sea level.

a. Write a formula for finding the boiling point b of water at any altitude A above sea level. **$b = 212 - \frac{A}{550}$**

b. Denver is 5280 feet above sea level. If you were boiling an egg in Denver, how hot would the water be? **202.4°F**

c. Find the temperature at which water would boil in Death Valley, which is 282 feet *below* sea level. **212.513°F**

Error Analysis

Watch for students who fail to multiply or divide *all* terms in order to clear parentheses or fraction bars. For example, in Exercise 13, students may write $3k + 2 = 12$ rather than $3k + 6 = 12$. Emphasize the need to check the solution to each equation by substituting back into the *original* equation.

Assignment Guide

Core: 11–33 odd, 34–35, 37–44
Enriched: 12–32 even, 34–44
All: Self Test 1–9

For **Extra Practice**, see p. 757.

The red A, B, and C flags, printed only in the Teacher's Wraparound Edition, indicate the level of difficulty of the exercises.

Practice Masters, p. 58

NAME ______ DATE ______

7-5 Practice Student Edition Pages 346–350
Solving Equations with Variables on Each Side

Solve each equation. Check your solution.

1. $3n - 21 = 2n$ 21
2. $-3b = 96 + b$ −24
3. $2(x + 4) = 6x$ 2
4. $12 - 6r = 2r + 36$ −3
5. $21 - y = 87 + 2y$ 36
6. $2v - 54 = -v + 21$ 25
7. $6 - y = -y + 2$ ∅
8. $25c + 17 = 5c - 143$ −8
9. $\frac{4}{3}u - 6 = \frac{7}{3}u + 8$ −14
10. $3k - 5 = 7k + 7$ −3
11. $7 + 6z = 8z - 13$ 10
12. $18d = 21 - 13d + 3$ 0
13. $12p = 6 - 3p$ $\frac{2}{5}$
14. $9 + 3k = 2k - 12$ −21
15. $\frac{5}{6}t + 4 = 2 - \frac{1}{6}t$ −2
16. $3 + 8(2m + 1) = 11 + 16m$ all numbers
17. $3 + 2(k + 1) = 6 + 3k$ −1
18. $3(z - 2) + 6 = 5(z + 4)$ −2
19. $6r + 5 = 8(r + 2) - 2r$ ∅
20. $28 - 14z = -24 + 12z$ 2
21. $-2(2c - 4) = \frac{1}{3}(-12c + 24)$ all numbers
22. $9[n + 2(n - 2)] = 45$ 3
23. $\frac{u}{0.3} = 4u + 6.28$ −9.42
24. $\frac{g-2}{5} = \frac{g+4}{7}$ 17
25. $0.3x - 15 = 0.2x - 5$ 100

Group Activity Card 7-5

Spaced Out — Group Activity 7-5

With a partner solve the following equations. To find the answer to the riddle below, put the variable above the indicated solution. Not all of the solutions are used in the answer.

RIDDLE: What do you call a satellite of Earth that has just eaten a big meal?

EQUATIONS:

1. $5F + 6 = 2F - 3$
2. $7(N - 3) = 2(2N + 1) + 4$
3. $48 - 5(B - 4) = 2(6B + 2) + 20$
4. $\frac{2}{9}L - 3 = \frac{1}{9}(L + 54)$
5. $2A + 12 = 5(4A - 12)$
6. $5H + 2(H - 1) = 7(H + 2) - 3H - 1$
7. $-6 + 4(2M + 4) = 3(2M + 1) + 5$
8. $8.9U + 64.8 = 3.5U$
9. $\frac{2}{5}E - 16\frac{1}{4} = \frac{1}{6}E + 12\frac{3}{4}$
10. $6(0 + 13) + 6 = 3(0 - 1)$

__ __ __ __ __ __ __ __ __
4 -3 -12 81 81 -1 -29 -29 9

 Pre-Algebra

Extension

Using Connections The chorale needs to have concert programs printed. Company A charges $30 plus 2¢ per program. Company B charges $10 plus 4¢ per program. How many programs would they need to buy for the prices to be equal? **1000** If they need 600 programs, from which company should they buy programs? **B**

4 ASSESS

Closing Activity

Speaking Have students explain, in their own words, the meaning of the null or empty set. **Sample answer: The solution set of an equation with no solution.**

37. **Sales** Shoe World offers Veronica a temporary job during her spring break. The manager gives her a choice as to how she wants to be paid, but she must decide before she starts working.
Plan 1 \$2 an hour plus 10¢ for every dollar's worth of shoes she sells.
Plan 2 \$3 an hour plus 5¢ for every dollar's worth of shoes she sells.
    a. How much would Veronica's sales need to be in one hour to earn the same amount in either plan? **\$20**
    b. If you were in Veronica's place, which method of payment would you choose? Explain your reasoning. **See students' work.**

Mixed Review

38. **Life Science** Sequoias are the largest trees on Earth. Botanists can estimate the age of a tree based on its diameter. If the circumference of a tree is 18 inches, what is its diameter? (Lesson 7-4) **5.7 in.**
39. Solve $3x + 3 = 99$. (Lesson 7-2) **32**
40. Solve $15 \div \frac{3}{7} = n$. (Lesson 6-4) **35**
41. Write $\frac{36}{126}$ in simplest form. (Lesson 4-5)
42. Solve $A = \ell w$ for w if $A = 65$ and $\ell = 13$. (Lesson 3-4) **5**
43. Solve $n - 10 = 42$ by using the inverse operation. (Lesson 1-8) **52**
44. Solve $7m = 56$ mentally. (Lesson 1-6) **8**

41. $\frac{2}{7}$

Self Test

1. **Banking** Dorinda is balancing her checkbook. Since the bank statement was mailed to her, she has written checks for \$13.85, \$19.72, and \$59.66 and made a deposit in the amount of \$497.92. If the balance she has written in her check register is \$973.88, how much should the bank statement say is in her account? (Lesson 7-1) **\$569.19**

Solve each equation. (Lesson 7-2)

2. $12 - z = 28$ **−16**
3. $9 - 4z = 57$ **−12**
4. $\frac{d-5}{7} = 14$ **103**

Define a variable and write an equation for each situation. Then solve. (Lesson 7-3)

5. Four times a number less twelve is 18. What is the number? **$4x - 12 = 18$; 7.5**
6. The product of a number and 6 is eight greater than 10. **$6x - 8 = 10$; 3**
7. The circumference of a circle is 6.398 yards. Find its radius. (Lesson 7-4) **1.02 yd**
8. The diameter of a circle is 16.2 inches. What is the circumference of the circle? (Lesson 7-4) **50.89 in.**
9. **Sports** A soccer field is 75 yards shorter than 3 times its width. If the perimeter is 370 yards, find the dimensions of the soccer field. (Lesson 7-5) **65 by 120 yards**

Enrichment Masters, p. 58

NAME ______________________ DATE ____________

7-5 Enrichment Student Edition Pages 346–350

Fractional Equations

To solve equations containing fractions, multiply both sides by the least common denominator. Then solve as usual.

Example: Solve $\frac{2x}{5} - \frac{x}{10} = 6$.

$$\frac{2x}{5} - \frac{x}{10} = 6 \quad \text{The least common denominator is 10.}$$
$$10\left(\frac{2x}{5}\right) - 10\left(\frac{x}{10}\right) = 10(6)$$
$$4x - x = 60$$
$$3x = 60$$
$$\frac{3x}{3} = \frac{60}{3}$$
$$x = 20$$

Solve each equation.

1. $\frac{3x}{2} - x = 1$ **2**
2. $\frac{3x}{8} = \frac{x}{3} + \frac{4}{3}$ **32**
3. $\frac{y}{6} - \frac{y}{4} = 5$ **−60**
4. $2a + \frac{a}{3} = \frac{a}{4} + 5$ **$\frac{12}{5}$**
5. $\frac{x-2}{3} = \frac{x+1}{4}$ **11**
6. $\frac{x-1}{2} + \frac{x-2}{3} = 1$ **$\frac{13}{5}$**
7. $\frac{x-3}{5} - \frac{x+2}{15} + \frac{2}{3} = 0$ **$\frac{1}{2}$**
8. $\frac{x+4}{3} - 4 = \frac{x-11}{4}$ **−1**
9. $-\frac{d}{4} + d = \frac{1}{8}$ **$\frac{1}{6}$**
10. $\frac{x-7}{5} + 2 = \frac{x+8}{10}$ **2**
11. $z + \frac{z}{4} = 14 - \frac{z}{2}$ **8**
12. $\frac{y+3}{16} - \frac{y-4}{6} = \frac{1}{3}$ **5**

Self Test

The Self Test provides students with a brief review of the concepts and skills in Lessons 7-1 through 7-5. Lesson numbers are given to the right of exercises or instruction lines so students can review concepts not yet mastered.

7-6 Solving Multi-Step Inequalities

Setting Goals: *In this lesson, you'll solve inequalities that involve more than one operation.*

Modeling with Manipulatives

MATERIALS

cups

counters

The diagram at the right models the inequality $4 + 2x > x + (-2)$. Notice that the side that is lower has the heavier weight. To solve an inequality, use the same process used for equations.

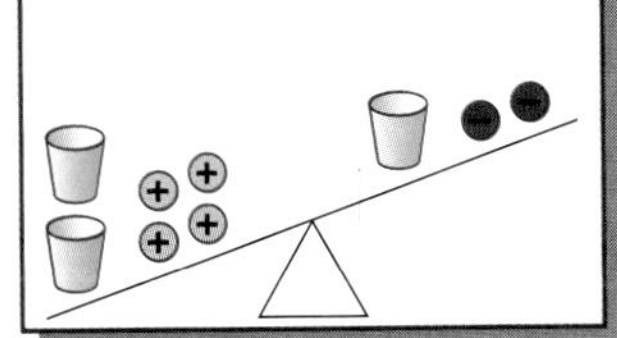

Your Turn

▶ Draw a diagram of the unbalanced scale on a sheet of paper and model the diagram shown above.

▶ Remove one cup from each side. *Perform operations on each side.*

▶ Place four zero pairs on the right side.

▶ Remove four positive counters from each side.

a. State the inequality after removing one cup from each side. **$4 + x > -2$**

b. Why did you have to place four zero pairs on the right side?

c. What is the solution to the inequality? **$x > -6$** **b. See margin.**

d. How could you check the solution? **Substitute numbers greater and less than −6 for each cup into the original inequality.**

Learning the Concept

Inequalities can be solved by applying the methods for solving equations. Remember that if each side of an inequality is multiplied or divided by a negative number, the sign of the inequality should be reversed.

Example 1 **Solve each inequality and check the solution. Graph the solution on a number line.**

a. $7x + 29 > 15$

$$7x + 29 > 15$$

$$7x + 29 - 29 > 15 - 29 \quad \textit{Subtract 29 from each side.}$$

$$7x > -14$$

$$\frac{7x}{7} > \frac{-14}{7} \quad \textit{Divide each side by 7.}$$

$$x > -2$$

Any number greater than −2 is a solution.

Check: Try 0, a number greater than −2.

$$7(0) + 29 \stackrel{?}{>} 15 \quad \textit{Replace x with 0.}$$

$$0 + 29 \stackrel{?}{>} 15$$

$$29 > 15 \checkmark$$

−5 −4 −3 −2 −1 0 1 2 3 4

Alternative Learning Styles

Kinesthetic Bring a balance scale to class and allow students to model the solution to $3 + 2x > x + 5$ using the scale, cups, and counters.

Additional Answer Talk About It

b. Sample answer: So that 4 positive counters can be removed from each side.

7-6 LESSON NOTES

NCTM Standards: 1-5, 9

Instructional Resources

- Study Guide Master 7-6
- Practice Master 7-6
- Enrichment Master 7-6
- Group Activity Card 7-6
- Assessment and Evaluation Masters, p. 184
- Multicultural Activity Masters, p. 14
- Tech Prep Applications Masters, p. 14

Transparency 7-6A contains the 5-Minute Check for this lesson; **Transparency 7-6B** contains a teaching aid for this lesson.

Recommended Pacing	
Standard Pacing	Day 8 of 12
Honors Pacing	Day 7 of 11
Block Scheduling*	Day 5 of 7 (along with Lesson 7-7)

*For more information on pacing and possible lesson plans, refer to the ***Block Scheduling Booklet***.

1 FOCUS

5-Minute Check *(over Lesson 7-5)*

Solve each equation.

1. $5a - 18 = 3a$ **9**
2. $-3b + 13 = -b - 3$ **8**
3. $4(2x - 1) - 3 = 5(x - 5)$ **−6**
4. $7(y - 3) - 2 = 3(2y - 5) + 4$ **12**
5. $\frac{4}{7}m + 5 = \frac{2}{7}m - 3$ **−28**

Motivating the Lesson

Situational Problem Dixon and Sterling had scored the same number of touchdowns. On each touchdown, Sterling scored 7 points, and Dixon scored 6 points. Sterling also scored a 2-point safety. Dixon also scored two 3-point field goals. What is the maximum number of touchdowns scored by each team if Dixon won? **3**

2 TEACH

In-Class Examples

For Example 1
Solve each inequality and check the solution. Graph the solution on a number line.

a. $8x - 52 < -100$ **$x < -6$**

−7 −6 −5 −4

b. $-19 + 4y \geq 6y - 17$ **$y \leq -1$**

−2 −1 0 1

For Example 2
Old Route 66 went from Chicago to Los Angeles. You rent a car in Chicago for $99.95 plus $0.28 per mile for each mile over 50 miles. Your budget for car rental while on vacation is $250. How many miles can you drive and stay within your budget? **585 miles**

Additional Answers

1. Sample answer: Substitute numbers greater and less than 340.2 for m.
2. Sample answer: Tinesha is correct because simplifying the grouping symbols will make it easier to multiply.

Study Guide Masters, p. 59

NAME ______________________ DATE __________

7-6 Study Guide
Solving Multi-Step Inequalities

Student Edition
Pages 351–354

The inequality $13 - 6y > 49 - 2y$ can be solved by applying the same steps as solving equations.

$$13 - 6y > 49 - 2y$$
$$13 - 6y - 13 > 49 - 2y - 13 \quad \text{Subtract 13 from each side}$$
$$-6y > 36 - 2y$$
$$-6y + 2y > 36 - 2y + 2y \quad \text{Add } 2y \text{ to each side.}$$
$$-4y > 36$$
$$\frac{-4y}{-4} < \frac{36}{-4} \quad \text{Divide each side by } -4. \text{ Reverse the order symbol.}$$
$$y < -9 \quad \text{Any number less than } -9 \text{ is a solution.}$$

Check: Replace y with a number less than −9. Try −10.

$$13 - 6y > 49 - 2y$$
$$13 - 6(-10) \overset{?}{>} 49 - 2(-10)$$
$$13 + 60 \overset{?}{>} 49 + 20$$
$$73 > 69 \checkmark$$

$y < -9$

−12 −11 −10 −9 −8 −7

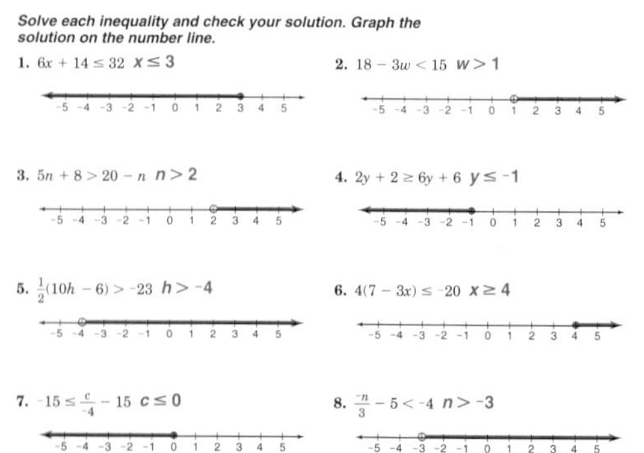
Solve each inequality and check your solution. Graph the solution on the number line.

1. $6x + 14 \leq 32$ **$x \leq 3$**
2. $18 - 3w < 15$ **$w > 1$**
3. $5n + 8 > 20 - n$ **$n > 2$**
4. $2y + 2 \geq 6y + 6$ **$y \leq -1$**
5. $\frac{1}{2}(10h - 6) > -23$ **$h > -4$**
6. $4(7 - 3x) \leq -20$ **$x \geq 4$**
7. $-15 \leq \frac{c}{-4} - 15$ **$c \leq 0$**
8. $\frac{-n}{3} - 5 < -4$ **$n > -3$**

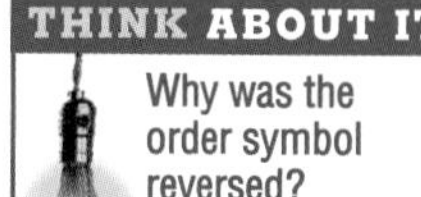

THINK ABOUT IT
Why was the order symbol reversed?

Each side was divided by a negative number.

b. $13 - 6n \leq 49 - 2n$

$$13 - 6n \leq 49 - 2n$$
$$13 - 6n + 2n \leq 49 - 2n + 2n \quad \textit{Add } 2n \textit{ to each side.}$$
$$13 - 4n \leq 49$$
$$13 - 13 - 4n \leq 49 - 13 \quad \textit{Subtract 13 from each side.}$$
$$-4n \leq 36$$
$$\frac{-4n}{-4} \geq \frac{36}{-4} \quad \textit{Divide each side by } -4 \textit{ and change } \leq \textit{ to } \geq.$$
$$n \geq -9$$

Any number greater than or equal to −9 is a solution.

Check: Try −5, a number greater than −9.

$$13 - 6n \leq 49 - 2n$$
$$13 - 6(-5) \overset{?}{\leq} 49 - 2(-5) \quad \textit{Replace } n \textit{ with } -5.$$
$$13 + 30 \overset{?}{\leq} 49 + 10$$
$$43 \leq 59 \checkmark$$

−10 −8 −6 −4 −2 0 2 4

When solving inequalities that contain grouping symbols, remember to first use the distributive property to remove the grouping symbols.

Example 2 APPLICATION
Tourism

Route 1A along the Maine coast is called one of the most scenic highways in America. Every year, thousands of people go there to see the ocean and the charming New England villages. Suppose you rented a car in Portland for $139.95 plus $0.25 per mile for each mile over 100 miles. If you had budgeted up to $200 for car rental in your vacation budget, how many miles could you drive and stay within the budget?

Let m represent the number of miles driven.

$139.95	*plus*	*$0.25 per mile over 100*	*is no more than*	*$200*
139.95	+	$0.25(m - 100)$	$\leq$	200

$$139.95 + 0.25(m - 100) \leq 200$$
$$139.95 + 0.25m - 25 \leq 200 \quad \textit{Simplify.}$$
$$114.95 + 0.25m \leq 200$$
$$114.95 - 114.95 + 0.25m \leq 200 - 114.95 \quad \textit{Subtract 114.95 from each side.}$$
$$0.25m \leq 85.05$$
$$\frac{0.25m}{0.25} \leq \frac{85.05}{0.25} \quad \textit{Divide each side by 0.25.}$$
$$m \leq 340.2 \quad \textit{Use a calculator.}$$

You could drive up to 340.2 miles and stay within your budget.

Additional Answers

3.

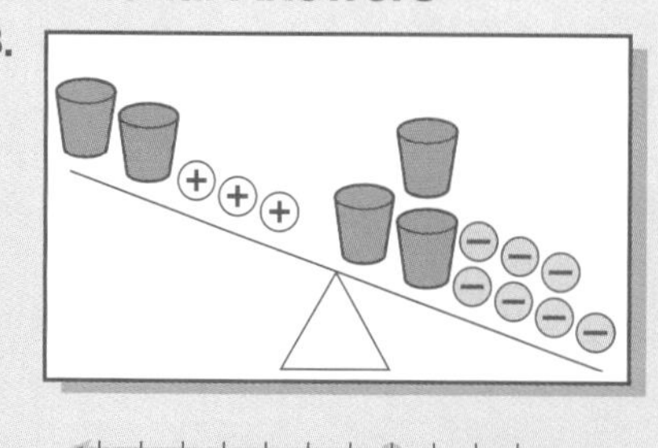

5. 0 5 10

6. −4 −3 −2 −1 0 1 2 3 4

Group Activity Card 7-6

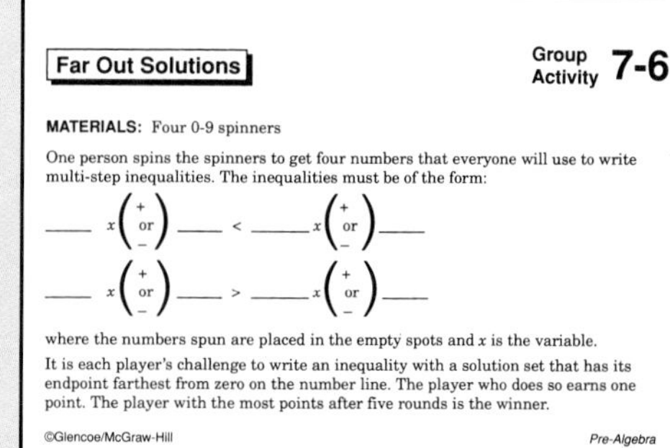
Far Out Solutions Group Activity **7-6**

MATERIALS: Four 0-9 spinners

One person spins the spinners to get four numbers that everyone will use to write multi-step inequalities. The inequalities must be of the form:

___ x (+ or −) ___ < ___ x (+ or −) ___

___ x (+ or −) ___ > ___ x (+ or −) ___

where the numbers spun are placed in the empty spots and x is the variable.

It is each player's challenge to write an inequality with a solution set that has its endpoint farthest from zero on the number line. The player who does so earns one point. The player with the most points after five rounds is the winner.

 Pre-Algebra

Checking Your Understanding

Communicating Mathematics

Read and study the lesson to answer these questions. 1–3. See margin.

1. **Explain** how to check the solution for Example 2.
2. **You Decide** Tinesha says that to solve the inequality $2(2y + 3) > y + 1$, the first step should be to divide each side by 2. Kiki says the first step should be to subtract 3 from each side. Who is correct, and why?

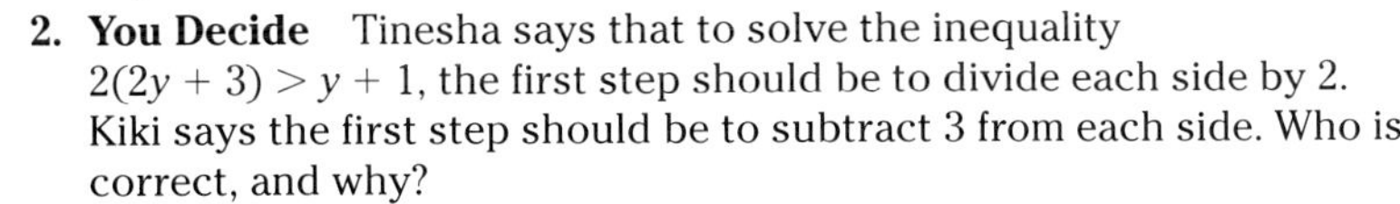

MATERIALS

cups and counters

3. **Draw a model** that you could use to solve the inequality $2x + 3 < 3x - 7$.
4. Write an inequality for the model at the right. Then use cups and counters to solve the inequality. $x + 1 < 2x + (-4)$; $5 < x$

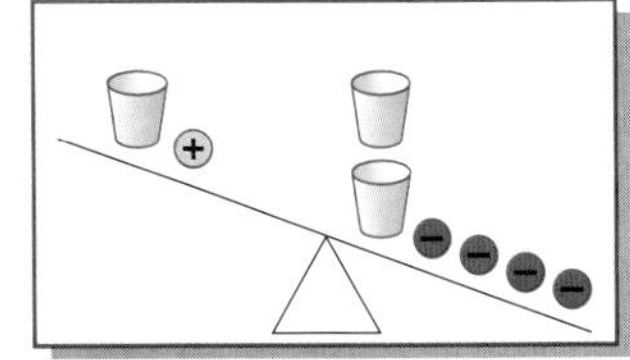

Guided Practice

Solve each inequality and check your solution. Graph the solution on a number line. See margin for graphs.

5. $3x + 2 \leq 23$ $x \leq 7$
6. $18 - 2v < 16$ $v > 1$
7. $2k + 7 > 13 - k$ $k > 2$
8. $y + 1 \geq 5y + 5$ $y \leq -1$
9. $-3(m - 2) > 12$
10. $-5 \leq \frac{x}{4} - 7$ $x \geq 8$
11. **Consumer Awareness** Heather has \$90 that relatives gave her for her birthday. She needs to buy a new pair of shoes and would also like to buy two CDs. The shoes she likes cost \$65. Use the inequality $2d + 65 \leq 90$ to determine how much she can spend on each CD.

9. $m < -2$

11. See Solutions Manual.

Exercises: Practicing and Applying the Concept

Independent Practice

Solve each inequality and check your solution. Graph the solution on a number line. See Solutions Manual for graphs. 13. $u \geq -4$ 16. $z > 30$

12. $2x + 9 > 25$ $x > 8$
13. $-2u + 3 \leq 11$
14. $16 < 18 - 2n$ $n < 1$
15. $2x - 5 < 2x - 9$ $\varnothing$
16. $1.2z - 2 > 7 + 0.9z$
17. $3(j - 1) \geq -12$ $j \geq -3$
18. $\frac{m}{3} - 7 > 11$ $m > 54$
19. $36 + \frac{k}{5} \geq 51$ $k \geq 75$
20. $\frac{2x}{9} - 2 < -4$ $x < -9$
21. $0.47 < \frac{t}{-9} + 0.6$ $1.17 > t$
22. $-4.4 > \frac{b}{-5} - 4.8$ $-2 < b$

B

23. $12 - \frac{5z}{4} < 37$ $z > -20$
24. $\frac{n - 11}{2} \leq -6$ $n \leq -1$
25. $1.3x + 6.7 \geq 3.1x - 1.4$ $x \leq 4.5$
26. $-5x + 3 < 3x + 23$ $x > -2.5$
27. $-5(k + 4) \geq 3(k - 4)$ $k \leq -1$
28. $8c - (c - 5) > c + 17$ $c > 2$

C

29. $\frac{c + 8}{4} < \frac{5 - c}{9}$ $c < -4$
30. $\frac{2(n + 1)}{7} \geq \frac{n + 4}{5}$ $n \geq 6$

Critical Thinking

31. The sum of three times a number and 5 lies between -10 and 8. Solve the *compound inequality* $-10 < 3x + 5 < 8$ to find the solution(s). (*Hint*: Any operation must be done to every part of the inequality.) $-5 < x < 1$

3 PRACTICE/APPLY

Checking Your Understanding

Exercises 1–11 are designed to help you assess your students' understanding through reading, writing, speaking, and modeling. You should work through Exercises 1–4 with your students and then monitor their work on Exercises 5–11.

Assignment Guide

Core: 13–33 odd, 34–39
Enriched: 12–30 even, 31–39

For **Extra Practice**, see p. 757.

The red A, B, and C flags, printed only in the Teacher's Wraparound Edition, indicate the level of difficulty of the exercises.

Additional Answers

7.

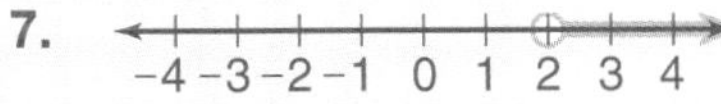

8.

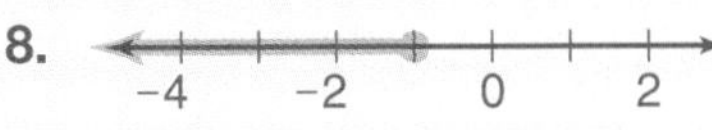

9.

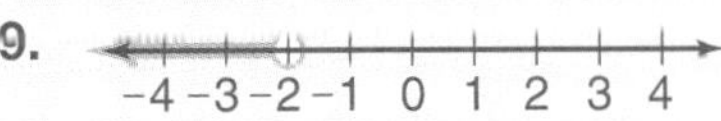

10.

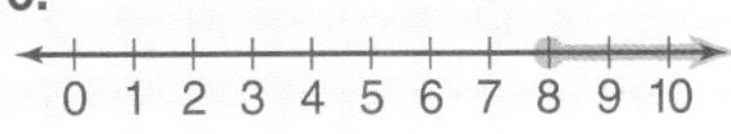

Practice Masters, p. 59

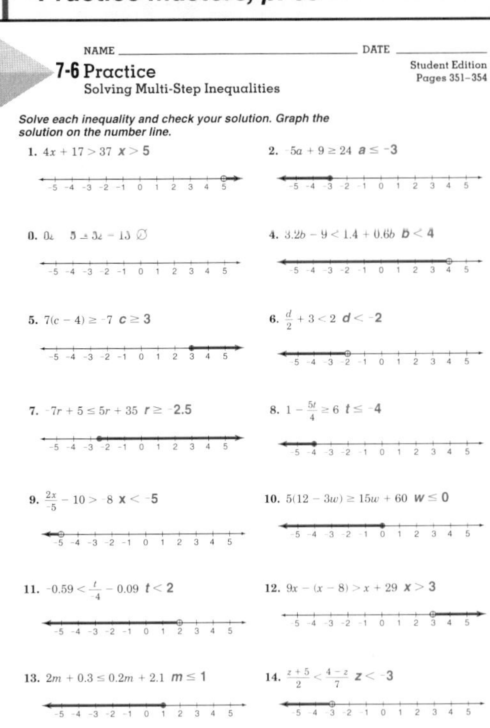

NAME ________________ DATE ________

7-6 Practice
Solving Multi-Step Inequalities

Student Edition Pages 351–354

Solve each inequality and check your solution. Graph the solution on the number line.

1. $4x + 17 > 37$ $x > 5$
2. $-5a + 9 \geq 24$ $a \leq -3$
3. $8z + 5 \leq 8z - 13$ $\varnothing$
4. $3.2b - 9 < 1.4 + 0.6b$ $b < 4$
5. $7(c - 4) \geq -7$ $c \geq 3$
6. $\frac{d}{2} + 3 < 2$ $d < -2$
7. $-7r + 5 \leq 5r + 35$ $r \geq -2.5$
8. $1 - \frac{5t}{4} \geq 6$ $t \leq -4$
9. $\frac{2x}{-5} - 10 > -8$ $x < -5$
10. $5(12 - 3w) \geq 15w + 60$ $w \leq 0$
11. $-0.59 < \frac{t}{-4} - 0.09$ $t < 2$
12. $9x - (x - 8) > x + 29$ $x > 3$
13. $2m + 0.3 \leq 0.2m + 2.1$ $m \leq 1$
14. $\frac{z + 5}{2} < \frac{4 - z}{7}$ $z < -3$

Reteaching

Using Flowcharts Have students draw a flowchart of the steps needed to solve a multi-step inequality. Be sure students include a step that checks whether they are multiplying or dividing by a negative number.

Extension

Using Manipulatives Have students toss four dice to determine the values of a, b, c, and d in $a(x - b) < cx - d$ and then solve it. Ask what values of a, b, c, and d are necessary for there to be no solutions or an infinite number of solutions. **no solutions: $a = c$ and $ab \geq d$; infinite solutions: $a = c$ and $ab < d$.**

4 ASSESS

Closing Activity

Modeling Have students use cups and counters to model and solve $3(x + 2) < x + 8$. $x < 1$

Chapter 7, Quiz C (Lessons 7-5 and 7-6) is available in the *Assessment and Evaluation Masters*, p. 184.

Additional Answers Cooperative Learning Project

- Aerodynamics is the study of the way gases interact with something in motion. If something is "more aerodynamic", it interacts better with the air than another object.
- Javelins were flying too far—into stands of spectators and out of stadiums. The new javelins were banned from competition.

Enrichment Masters, p. 59

NAME ______________________ DATE ____________

7-6 Enrichment

Conditional and Unconditional Inequalities

Student Edition Pages 351–354

When the replacement set is the set of real numbers, the inequality $2x < 16$ is called a **conditional inequality** because it is true for at least one but not all values of the replacement set. Other examples of conditional inequalities are $x + 5 > 8$ and $2y - 6 < 10$.

If the replacement set is the set of real numbers, $x + 5 > x$ is true for every element of the replacement set. Such an inequality is called an **unconditional inequality**. Other examples of unconditional inequalities are $2x + 9 > 2x$ and $x - 7 < x$.

Solve each inequality. Then determine whether each inequality is conditional or unconditional.

1. $x - 2 > 4$
   $x > 6$; conditional
2. $3x - 2 < 2x + 4$
   $x < 6$; conditional
3. $4x + 5 \geq 4x$
   $5 \geq 0$; unconditional
4. $2(3x + 5) > 6x + 5$
   $10 > 5$; unconditional
5. $7y - 4 > 6 + 2y$
   $y > 2$; conditional
6. $8y - 3y > 5y - 10$
   $0 > -10$; unconditional
7. $5x \leq 10 + 2(3x - 4)$
   $x \geq -2$; conditional
8. $x < 7 + x$
   $0 < 7$; unconditional
9. $2c + 5 < 8 + 2c$
   $5 < 8$; unconditional
10. $6x + 3x - 1 < 8x + x$
    $-1 < 0$; unconditional
11. $2x + 3x \geq 4x + 1$
    $x \geq 1$; conditional
12. $7x + x + 10 \geq 8x$
    $10 \geq 0$; unconditional
13. $x + 8 < 8$
    $x < 0$; conditional
14. $8x < 5(2x + 4)$
    $x > -10$; conditional

Applications and Problem Solving

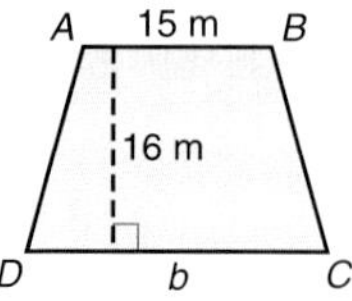

32. **Geometry** The area of trapezoid *ABCD* is at least 320 square meters. The height of the trapezoid is 16 meters and the length of one of the bases is 15 meters. Use the inequality $16\left(\frac{15 + b}{2}\right) \geq 320$ to find the length of the other base, b. $b \geq 25$ **meters**
33. **Statistics** Odina's scores on the first three of four 100-point tests were 89, 92, and 82. Use the inequality $\frac{89 + 92 + 82 + s}{4} \geq 90$ to determine what scores she could receive on the fourth test and have an average of at least 90 for all the tests. $s \geq 97$

Mixed Review

34. Solve $-75 \quad m = 25 + 4m$. (Lesson 7-5) -20
35. **Patterns** Duncan received a chain letter that said he should copy the letter and send it to six friends in order to have good luck. Each of Duncan's friends sent copies to six other friends, who each sent copies to six other friends. How many people have received the chain letter after Duncan? (Lesson 6-8) **258**
36. Simplify $7.5x - 31 + 13.78x$. (Lesson 5-3) $21.28x - 31$
37. Find the missing exponent $\frac{t^{17}}{t^{\bullet}} = t^9$. (Lesson 4-8) **8**
38. Find the value of $17 + -13 + 8$. (Lesson 2-4) **12**
39. Translate *3 more days of summer vacation than Northridge High School* into an algebraic expression. (Lesson 1-3) $n + 3$

COOPERATIVE LEARNING PROJECT

THE SHAPE OF THINGS TO COME

Technology in Sports

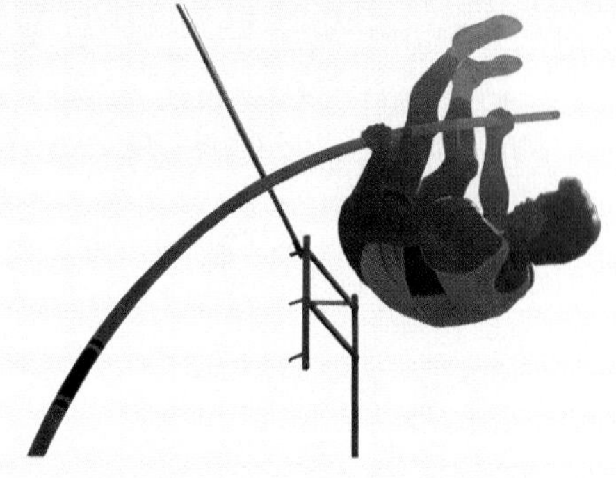

The Olympic motto is "Faster – Higher – Stronger." Thanks to technology, athletes are accomplishing things they never dreamed were possible before.

Divers will soon be reaching new heights with diving boards that provide more spring than those of the past. Baseball players and golfers can practice their swings inside with electronic sensors that analyze their strength and accuracy. Pole vaulters will use poles made specifically for their height, weight, speed, and hold technique. And cyclists will speed by on bicycles with less drag and more stability.

See for Yourself

- Many of the advances in sports technology have to do with making something more *aerodynamic*. Investigate what it means for something to be aerodynamic. **See margin.**
- In the 1980s, advances in aerodynamic technology made javelins *too* good. Research the situation. What was the final outcome? **See margin.**
- Choose a sport and investigate the changes that technological developments have made and will soon make in the sport. **See students' work.**

COOPERATIVE LEARNING PROJECT

THE SHAPE OF THINGS TO COME

Even the simple discus has undergone technological improvements. The traditional wooden discus had its weight concentrated in a metal central core. Now a discus has a plastic shell with lead weights on the perimeter of the discus. By moving the weights outward, the discus is more aerodynamically stable and could be thrown farther.

7-7 Writing Inequalities

Setting Goals: *In this lesson, you'll solve verbal problems by writing and solving inequalities.*

Modeling a Real-World Application: Backpacking

Ahh, the great outdoors. Nearly 10 million Americans hit the trail to go backpacking each year. State and national parks are among the most popular getaways. *Women's Sports and Fitness* magazine says that three times the weight of your backpack and its contents should be less than your body weight in order for you to avoid an injury. If you weigh 120 pounds and your empty backpack weighs 15 pounds, how much should the contents of your backpack weigh? *This problem will be solved in Example 2.*

Learning the Concept

LOOK BACK You can review writing simple inequalities in Lesson 1-9.

Verbal problems with phrases like *greater than* or *less than* describe inequalities. Other phrases that suggest inequalities are *at least, at most,* and *between.* Inequalities that correspond to each statement are below.

x is *at least* 3.	$x \geq 3$	*At least 3 means 3 or greater.*
y is *at most* 9.	$y \leq 9$	*At most 9 means 9 or less.*
z is *between* 2 and 4.	$2 < z < 4$	*Between 2 and 4 means greater than 2 and less than 4.*

Example 1 **Three times a number increased by eight is at least twenty-five. What is the number?**

Let x represent the number. Write an inequality.

three times a number $\;3x$ *increased by* $+$ *eight* 8 *is at least* $\geq$ *twenty-five* 25

$$3x + 8 \geq 25$$
$$3x + 8 - 8 \geq 25 - 8 \quad \text{Subtract 8 from each side.}$$
$$3x \geq 17$$
$$\frac{3x}{3} \geq \frac{17}{3} \quad \text{Divide each side by 3.}$$
$$x \geq \frac{17}{3} \text{ or } 5\frac{2}{3}$$

In many real-life problems, you will use a multi-step inequality to solve.

Example 2 APPLICATION Backpacking

Refer to the application at the beginning of the lesson. How much should the contents of your backpack weigh?

Explore You want to find what the backpack contents should weigh.

Plan Let c represent the weight of the backpack contents.

(continued on the next page)

Cooperative Learning

Send-A-Problem Have groups of students make up two verbal problems that can be solved by writing an inequality. Each is written on an index card, with the inequality on the back. Then send the cards to another group to solve. For more information on this strategy, see *Cooperative Learning in the Mathematics Classroom*, p. 23.

Study Guide Masters, p. 60
The Study Guide Master provides a concise presentation of the lesson along with practice problems.

7-7 LESSON NOTES

NCTM Standards: 1-4, 9

Instructional Resources

- Study Guide Master 7-7
- Practice Master 7-7
- Enrichment Master 7-7
- Group Activity Card 7-7

Transparency 7-7A contains the 5-Minute Check for this lesson; **Transparency 7-7B** contains a teaching aid for this lesson.

Recommended Pacing	
Standard Pacing	Day 9 of 12
Honors Pacing	Day 8 of 11
Block Scheduling*	Day 5 of 7 (along with Lesson 7-6)

*For more information on pacing and possible lesson plans, refer to the ***Block Scheduling Booklet***

1 FOCUS

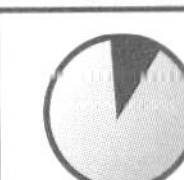

5-Minute Check *(over Lesson 7-6)*

Solve each inequality and check your solution. Graph each solution.

1. $21 - 4t \geq 25$ $\quad t \leq -1$ 
   (number line: -2 -1 0 1)
2. $18 + 6m \geq -24$ $\quad m \geq -7$ 
   (number line: -9 -8 -7 -6)
3. $-5.7 < \frac{a}{-20}$ $\quad a < 114$
   (number line: 113 114 115 116)
4. $13x - 8 \leq 15x + 4$ $\quad x \geq -6$
   (number line: -8 -7 -6 -5)

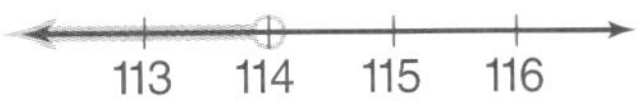

Motivating the Lesson

Questioning Write these scores on the board: 68, 66, 72. Ask students to find the average of these scores. Tell students that to qualify for the semifinals, you need an average score of 70. Ask what score this student needs in the next round to qualify for the semifinals.

2 TEACH

In-Class Examples

For Example 1

Four times a number decreased by six is at most thirty. What is the number?

$x \leq 9$

For Example 2

Write and solve an inequality. You must have an average score of at least 80 to get a B on your report card. You have scores of 61, 70, 99, and 70. What is the minimum score you must get on the last test to get a B on your report card?

$s > 100$; You must score 100 to get a B.

Solve

three times	weight of backpack	plus	weight of contents	is less than	your weight
3	(15	+	c)	<	120

$3(15 + c) < 120$

$\frac{3(15 + c)}{3} < \frac{120}{3}$ *Divide each side by 3.*

$15 + c < 40$

$15 - 15 + c < 40 - 15$ *Subtract 15 from each side.*

$c < 25$

The backpack contents should weigh less than 25 pounds.

Examine Suppose the contents of your backpack weigh 20 pounds, a number less than 25. The backpack and its contents would weigh 20 + 15 or 35 pounds. 3(35) = 105 and 105 < 120, so this would be a safe weight to carry.

Checking Your Understanding

Communicating Mathematics

Read and study the lesson to answer these questions.

1. a. **Write** an inequality for *six times a number decreased by four is at most 41.* **$6n - 4 \leq 41$**
   b. **Explain** how you would know if $4\frac{1}{2}$ is a member of the solution set to the inequality you wrote for part a. **Replace *n* with $4\frac{1}{2}$.**
2. **Write** an inequality for *a number is between −7 and −9.* **$-7 > x > -9$**
3. **Assess Yourself** Think of a time when you needed to use an inequality. Write an inequality about the situation. **See students' work.**

MATH JOURNAL

Guided Practice

Define a variable and write an inequality. Then solve.

4. Four less than a number is at most 10. What is the number?
5. Twice a number decreased by 9 is greater than 11. Find the number.
6. You earn $2.00 for every magazine subscription you sell plus a salary of $10 a week. How many subscriptions do you need to sell each week to earn at least $40 a week? **$2s + 10 \geq 40$; $s \geq 15$**
7. You buy some candy bars at 55¢ each and one newspaper for 35¢. How many candy bars can you buy if you only have $2?
8. **Sports** Gao Min represents China in springboard diving. Suppose Gao scored 68.2, 68.9, 67.5, and 71.7 for her first four dives and has one more dive remaining. If the diver in first place has a score of 345.4, what must Gao receive on her fifth dive to overtake her? **Gao will overtake the first-place diver if her fifth-dive score is greater than 69.1.**

4. $n - 4 \leq 10$; $n \leq 14$
5. $2n - 9 > 11$; $n > 10$
7. $55c + 35 \leq 200$; $c \leq 3$

Exercises: Practicing and Applying the Concept

Independent Practice

Define a variable and write an inequality. Then solve.

9. Four times a number increased by four is at least 16. Find the number.

10. Nine less than six times a number is at most 33. What is the number?

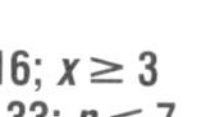

Write a problem that could be solved using each inequality.

11. $f + 115 \geq 260$
12. $\frac{87 + 92 + 89 + 97 + s}{5} \geq 90$

9. $4x + 4 \geq 16$; $x \geq 3$
10. $6n - 9 \leq 33$; $n \leq 7$
11–12. See margin.

Additional Answers

11. Sample answer: A stove and a freezer weigh at least 260 kg. The stove weighs 115 kg. What is the freezer's weight?
12. Sample answer: Jenny has scores of 87, 92, 89, and 97 on quizzes. There will be one more quiz. What must Jenny's fifth score be to have an average score of at least 90?

Reteaching

Paired Partners Have students work in pairs to write an "at least" problem, an "at most" problem, and a "between" problem. Trade with other paired partners and solve.

Practice Masters, p. 60

NAME ______________ DATE ________

7-7 Practice
Writing Inequalities

Student Edition Pages 355–357

Define a variable and write an inequality for each situation. Then solve.

1. Three times a number increased by 4 is at least 16. What is the number? Let n = the number; $3n + 4 \geq 16$; any number greater than or equal to 4
2. Five less than a number is at most 11. What is the number? Let n = the number; $n - 5 \leq 11$; any number less than or equal to 16
3. The sum of a number and 7 is less than 19. What is the number? Let n = the number; $n + 7 < 19$; any number less than 12
4. Twice a number decreased by 9 is greater than 11. What is the number? Let n = the number; $2n - 9 > 11$; any number greater than 10
5. The sum of two consecutive positive integers is less than 19. What are the integers? Let n = the lesser number, then $n + 1$ = the greater number; $n + n + 1 < 19$; any number less than 9
6. The sum of two consecutive positive odd integers is at most 16. What are the integers? Let n = the lesser number, then $n + 2$ = the greater number; $n + n + 2 \leq 16$; any number less than or equal to 7
7. Your test scores are 75, 93, 90, 82 and 85. What is the lowest score you can obtain on the next test to achieve an average of at least 86? Let t = next test score; $\frac{75 + 93 + 90 + 82 + 85 + t}{6} \geq 86$; 91
8. Juan spent at most $2.50 on apples and oranges. He bought 5 apples at $0.36 each. What is the most he spent on the oranges? Let p = the price of the oranges; $p + 5(0.36) \leq 2.50$; $0.70
9. Three times a number increased by twice the number is greater than 125. What is the number? Let n = the number; $3n + 2n > 125$; any number greater than 25
10. Five times a number decreased by 7 times the same number is at most 20. What is the number? Let n = the number; $5n - 7n \leq 20$; any number greater than or equal to −10

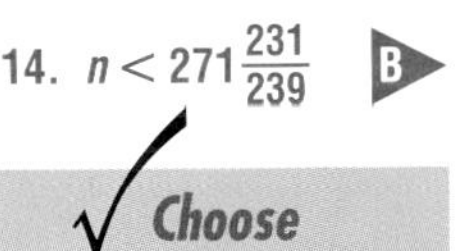

13. $3(h-1) \geq 18$; $h \geq 7$

14. $n < 271\frac{231}{239}$

Choose

Estimation
Mental Math
Calculator
Paper and Pencil

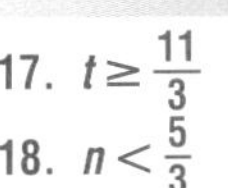

17. $t \geq \frac{11}{3}$

18. $n < \frac{5}{3}$

20b. $90 < m\angle A < 180$

25a. $\frac{1.7}{10^{24}}$

13. You can hike along the Appalachian Trail at 3 miles per hour. You will stop for one hour for lunch. You want to walk at least 18 miles. How many hours should you expect to spend on the trail?

B ▶ 14. Twelve times a number decreased by $\frac{1}{20}$ of the number is less than 3250. What is the number?

15. The sum of an integer and the next greater integer is at least 35. Find the least pair of such integers. **17, 18**

16. The sum of an odd integer and the next greater odd integer is more than 136. Find the least pair of such integers. **69, 71**

17. Three times a number t exceeds six by at least five. What are the possible values of t?

18. A number is less than the sum of twice its opposite and five. Find the number.

C ▶ 19. If 12.15 times an integer is increased by 9.348, the result is between 39 and 75. Find the integer. **3, 4, or 5**

Critical Thinking

20. A *compound inequality* is formed when two inequalities are connected by *and* or *or*. A compound inequality containing *and* is true only if *both* inequalities are true.
   a. Write an inequality equivalent to $2x - 5 < 3$ and $x + 5 > 4$ without using *and*. $-1 < x < 4$
   b. **Geometry** An obtuse angle is one that measures more than 90° but less than 180°. Write an inequality to represent the possible measures of angle A if you know that angle A is obtuse.

Applications and Problem Solving

21. **Geometry** The sum of the lengths of any two sides of a triangle is greater than the length of the third side. Triangle ABC has side lengths of 22 inches and 28 inches. Write a compound inequality to describe the length of the third side of triangle ABC. $50 > c > 6$

22. **Agriculture** Rockwell International and John Deere Precision Farming Group are working on a system that uses satellites to help dispense the correct amount of fertilizer on crops. The basic equipment costs about \$7000. Annual fees for signal reception will be about \$1200.
   a. How much additional revenue would the technology have to generate for a farmer in ten years for the investment to have been a good one? $r > \$19{,}000$

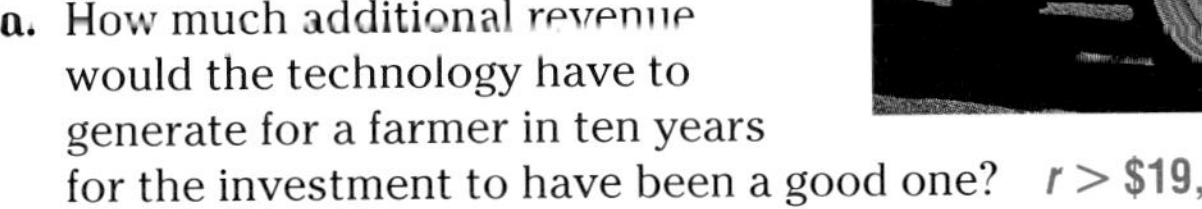

   b. One farm expert estimated that the system would produce an additional \$100 per acre each year on a certain crop. How large a farm should a farmer have in order to expect to make money over a five-year period using the system? **over 26 acres**

Mixed Review

23. Solve $\frac{2}{5}x - 1 > -3$. (Lesson 7-6) $x > -5$

24. Solve $\frac{20}{27} - \frac{11}{27} = b$. (Lesson 5-4) $\frac{1}{3}$

25. **Science** Neutrons and protons weigh about 1.7×10^{-24} grams each. (Lesson 4-9) **25b. 0.0000000000000000000000017**
   a. Write the weight of a proton or neutron using positive exponents.
   b. Write the weight of a proton or neutron as a decimal.

26. Solve $6 \leq \frac{n}{-7}$. (Lesson 3-7) $42 \geq n$

27. Find the product of -14 and $9b$. (Lesson 2-7) $-126b$

3 PRACTICE/APPLY

Checking Your Understanding

Exercises 1–8 are designed to help you assess your students' understanding through reading, writing, speaking, and modeling. You should work through Exercises 1–3 with your students and then monitor their work on Exercises 4–8.

Assignment Guide

Core: 9–19 odd, 20–21, 23–27
Enriched: 10–18 even, 20–27

For **Extra Practice**, see p. 757.

The red A, B, and C flags, printed only in the Teacher's Wraparound Edition, indicate the level of difficulty of the exercises.

4 ASSESS

Closing Activity

Writing Have students refer to the application at the beginning of the lesson and find how much they can safely carry in a 15-lb backpack. **Answers will vary, depending upon student weight.**

Enrichment Masters, p. 60

NAME ______ DATE ______

7-7 Enrichment Student Edition Pages 355–357

Consecutive Integers and Inequalities

Consecutive integers follow one after another. For example, 4, 5, 6, and 7 are consecutive integers, as are −8, −7, −6. Each number to the right in the series is one greater than the one that comes before it. If x = the first consecutive integer, then $x + 1$ = the second consecutive integer, $x + 2$ = the third consecutive integer, $x + 3$ = the fourth consecutive integer, and so on.

Example: Find three consecutive positive integers whose sum is less than 12.

first integer		second integer		third integer	
x	+	$x + 1$	+	$x + 2 < 12$	Simplify the expression by combining like terms.

$3x + 3 < 12$

$3x + 3 - 3 < 12 - 3$ — Subtract 3 from each side.

$3x < 9$

$\frac{3x}{3} < \frac{9}{3}$ — Divide each side by 3.

$x < 3$ — So x could equal 1 or 2.

If $x = 1$, then $x + 1 = 2$, $x + 2 = 3$, and {1, 2, 3} is one solution.
If $x = 2$, then $x + 1 = 3$, $x + 2 = 4$, and {2, 3, 4} is another solution.
Each of the two solutions must be considered in the answer. The solution set is {1, 2, 3; 2, 3, 4}.

Solve. Show all possible solutions.

1. Find three consecutive positive integers whose sum is less than 15. {1, 2, 3; 2, 3, 4; 3, 4, 5}

2. Find two consecutive positive even integers whose sum is less than 10. {2, 4}

3. Find three consecutive positive integers such that the second plus four times the first is less than 21. {1, 2, 3; 2, 3, 4; 3, 4, 5}

4. Find three consecutive positive even integers such that the third plus twice the second is less than 26. {2, 4, 6; 4, 6, 8}

Group Activity Card 7-7

Clue Me In — Group Activity 7-7

MATERIALS: One die

Work with a partner to give clues and make guesses about a secret number. Mentally choose a number less than 50 and keep it a secret. Your partner will try to guess your number from clues you give. Clues must be inequalities that use two operations (×, ÷, +, and −), a variable, a number determined by rolling the die, and other numbers you choose to use. Your secret number must be a solution to the inequality you make, so your partner can solve the inequality and get a clue about your number. (For example, if your secret number is 30 and the number rolled is 6, you could write $2x - 6 < 55$ or $\frac{x}{10} + 4 > 6$ as a clue.) Roll the die to get a new number for each clue. Your partner should guess your number using the least number of clues.

 Pre-Algebra

Extension

Using Logic After five dives in a diving competition, the first place diver has 350.5 points, and the fifth place diver has 338.25 points. If your first four dives earned scores of 66.2, 72, 70.1, and 60.8, what score must you get on your last dive to finish among the top five divers? $s \geq 69.2$ What score must you get to win? $s > 81.4$

7-8 LESSON NOTES

NCTM Standards: 1-4, 13

Instructional Resources

- Study Guide Master 7-8
- Practice Master 7-8
- Enrichment Master 7-8
- Group Activity Card 7-8
- Assessment and Evaluation Masters. p. 184
- Activity Masters, p. 21
- Real-World Applications, 15

Transparency 7-8A contains the 5-Minute Check for this lesson; **Transparency 7-8B** contains a teaching aid for this lesson.

Recommended Pacing	
Standard Pacing	Day 10 of 12
Honors Pacing	Day 9 of 11
Block Scheduling*	Day 6 of 7

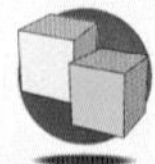

*For more information on pacing and possible lesson plans, refer to the ***Block Scheduling Booklet.***

1 FOCUS

5-Minute Check
(over Lesson 7-7)

Define a variable and write an inequality for each situation. Then solve.

1. The sum of two numbers is at most 48. If one number is 17, what is the other number? $n \leq 31$
2. Sixteen times a number increased by $\frac{3}{4}$ of the number is less than 536. What is the number? $n < 32$

HELP WANTED

Students may wish to discuss careers in science with a science teacher or guidance counselor in your school.

7-8 Integration: Measurement
Using the Metric System

Setting Goals: *In this lesson, you'll convert measurements within the metric system.*

Modeling a Real-World Application: Travel

You're finally here! Your dream vacation abroad. Noticing that the gas gauge is low, you pull into a gas station to buy some fuel. As you finish filling the tank, you notice that the pump stopped at 41.6!

How can the gas tank in your little rental car hold 41.6 gallons? It doesn't! The tank holds 41.6 *liters*. Most countries other than the United States use the **metric system of measurement**. They use liters to measure capacity instead of gallons.

Learning the Concept

HELP WANTED

The metric system was developed so that scientists in different countries could communicate. For more information on science-related careers, contact:

National Science Teachers Association
1840 Wilson Blvd.
Arlington, VA 22201

The metric system was developed by a group of French scientists in 1795. The **meter (m)** was defined as $\frac{1}{10{,}000{,}000}$ of the distance between the North Pole and the Equator. All units of length in the metric system are defined in terms of the meter. A prefix is added to indicate the decimal place-value position of the measurement. The chart below shows the relationships between the prefixes and the decimal place-value positions.

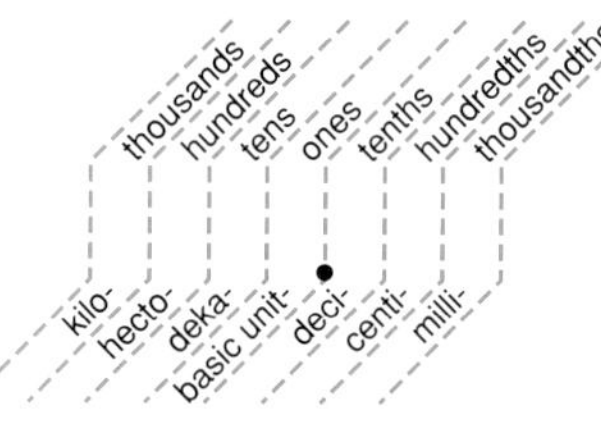

Note that each place value is 10 times the place value to its right.

Note that the value of each metric prefix is 10 times the value of the prefix to its right.

You can review multiplying and dividing by powers of ten in Lesson 6-9.

Converting units within the metric system follows the same procedure as multiplying or dividing by powers of ten.

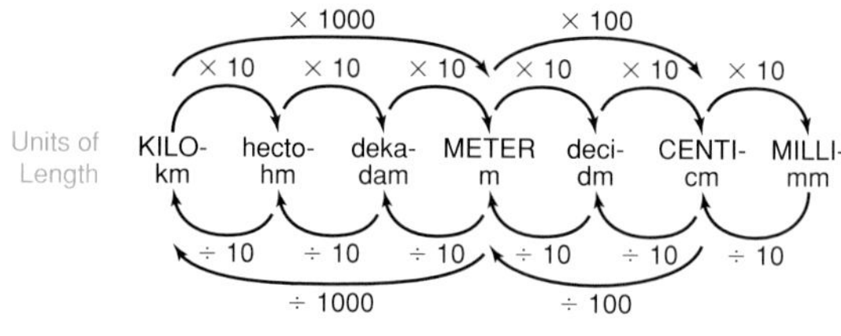

Classroom Vignette

"I give each student a metric cord, a Popsicle® stick, and different objects measuring under a decimeter. I ask as many students as possible to demonstrate measuring the lengths of various objects. Also, many students remember the order of the prefixes by using the phrase 'King Hector Died Monday Don't Call Me.'"

Cynthia Anderson
Cynthia Anderson
Trion High School
Trion, GA

Example  **Complete each sentence.**

a. 4.8 km = _?_ m

Kilometers are larger than meters. Converting to smaller units means there will be more units. Multiply by 1000.

$4.8 \times 1000 = 4800$
$4.8 \text{ km} = 4800 \text{ m}$

b. 9 mm = _?_ cm

Millimeters are smaller than centimeters. Converting to larger units means there will be fewer units. Divide by 10.

$9 \div 10 = 0.9$
$9 \text{ mm} = 0.9 \text{ cm}$

The application at the beginning of the lesson involved units of *capacity*. Capacity is the amount of liquid or dry substance a container can hold. The basic unit of capacity in the metric system is the **liter (L)**. A liter and milliliter are related in a manner similar to meter and millimeter.

$$1 \text{ L} = 1000 \text{ mL}$$

Example 2 APPLICATION Consumer Awareness

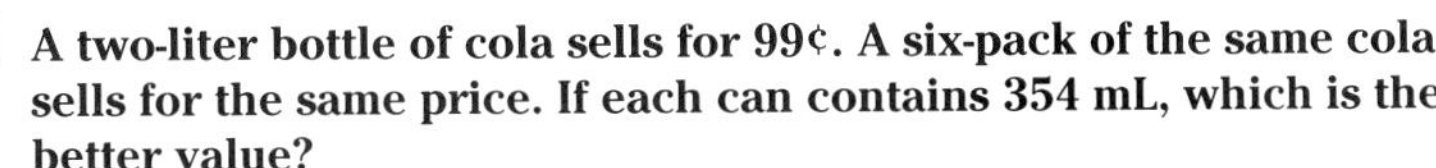

A two-liter bottle of cola sells for 99¢. A six-pack of the same cola sells for the same price. If each can contains 354 mL, which is the better value?

Explore Both packages sell for the same price. You want to determine which package gives you more cola for your money.

Plan To find the total capacity of the six-pack, multiply 354 times 6. To compare capacities, change from milliliters to liters. A milliliter is less than a liter, so divide by 1000.

$$\text{capacity of cans in liters} = \frac{(354 \text{ mL/can})(6 \text{ cans})}{1000 \text{mL/L}}$$

Estimate: $350 \times 6 \div 1000 = 2.1$

Solve 354 [×] 6 [÷] 1000 [=] 2.124 The six-pack holds 2.124 L.

The six-pack is the better value since you get 2.124 liters of cola for the same price as the 2-liter bottle.

Examine The capacity we found with the calculator is very close to our estimate, so the answer is reasonable.

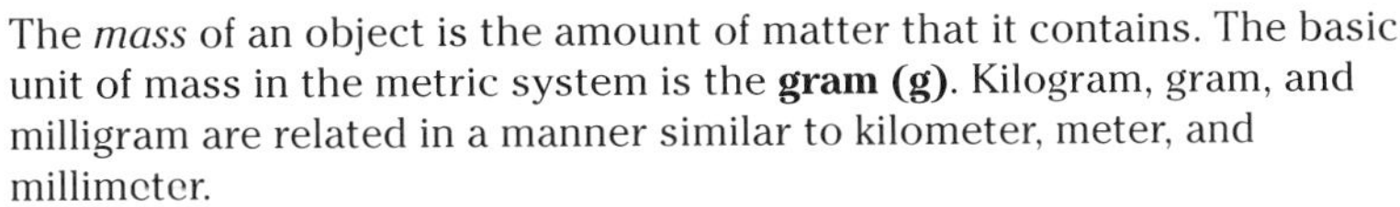

The *mass* of an object is the amount of matter that it contains. The basic unit of mass in the metric system is the **gram (g)**. Kilogram, gram, and milligram are related in a manner similar to kilometer, meter, and millimeter.

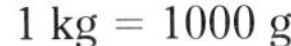

$$1 \text{ kg} = 1000 \text{ g} \qquad 1 \text{g} = 1000 \text{ mg}$$

  **Complete each sentence.**

a. 3.2 kg = _?_ g

Kilograms are larger than grams. Larger to smaller means more units. Multiply by 1000.

$3.2 \times 1000 = 3200$
$3.2 \text{ kg} = 3200 \text{ g}$

b. 200 mg = _?_ g

Milligrams are smaller than grams. Smaller to larger means fewer units. Divide by 1000.

$200 \div 1000 = 0.2$
$200 \text{ mg} = 0.2 \text{ g}$

Cooperative Learning

Brainstorming Have each small group of students choose a different topic from this chapter. Students are to create questions of various difficulties for a *Jeopardy!* game. The questions are written on blank index cards with the point values on the back. Set 1 has point values of 5, 10, 15, 20, 25. Set 2 has point values of 10, 20, 30, 40, 50. Answers are written on a sheet of paper. After all groups have turned in their cards, play a mini game of *Jeopardy!* with the class.

For more information on this strategy, see *Cooperative Learning in the Mathematics Classroom*, one of the titles in the Glencoe Mathematics Professional Series, p. 30

Motivating the Lesson

Hands-On Activity Have a "Guess the Size" contest. Place signs on several objects throughout your classroom asking students to guess the metric length, volume, or mass of the object.

2 TEACH

In-Class Examples

For Example 1
Complete each sentence.

a. 8.9 km = ● m **8900**

b. 15 mm = ● cm **1.5**

For Example 2
In preparing for a banquet, Misha found she could purchase vegetable oil in one 3.78-liter bottle or in four 950-ml bottles for the same price. Which is the better buy? **four 950 ml bottles**

Teaching Tip You may want to show the conversion in Example 2 using fractions to show how units cancel.

$$\frac{354 \text{ mL}}{1 \text{ can}} \cdot 6 \text{ cans} \div \frac{1000 \text{ mL}}{1 \text{ L}} =$$

$$\frac{354 \cancel{\text{mL}}}{1 \cancel{\text{can}}} \cdot 6 \cancel{\text{cans}} \cdot \frac{1 \text{ L}}{1000 \cancel{\text{mL}}} = \frac{2124}{1000}$$

$$= 2.124 \text{ L}$$

Study Guide Masters, p. 61

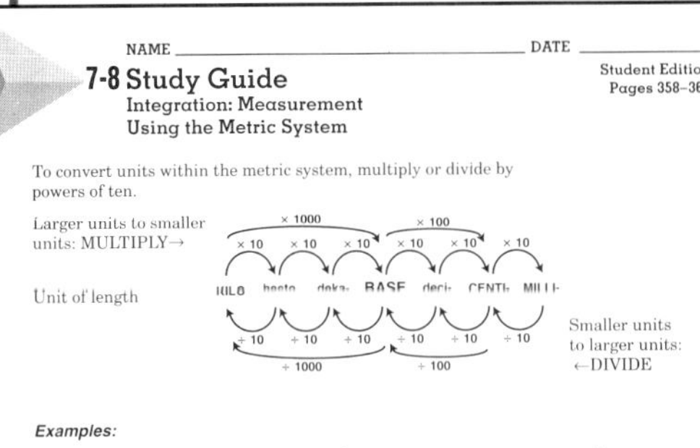

NAME ______ DATE ______

7-8 Study Guide Integration: Measurement Using the Metric System

Student Edition Pages 358–362

To convert units within the metric system, multiply or divide by powers of ten.

Larger units to smaller units: MULTIPLY→

Unit of length: KILO hecto deka- BASE deci- CENTI- MILLI-

(× 10 between each unit; × 1000 from KILO to BASE; × 100 from BASE to CENTI-)

(÷ 10 between each unit; ÷ 1000 from BASE to KILO; ÷ 100 from CENTI- to BASE)

Smaller units to larger units: ←DIVIDE

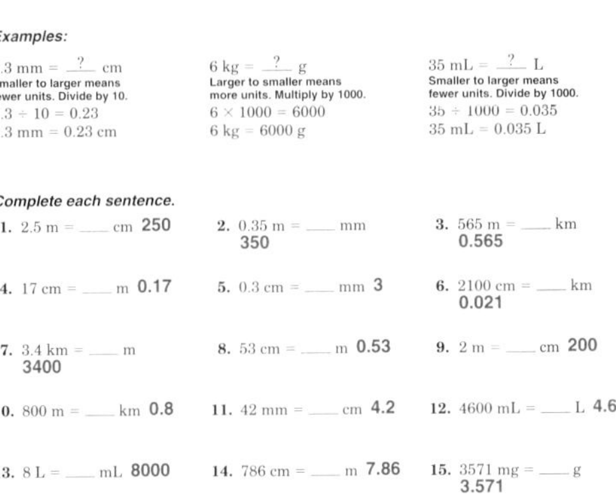

Examples:

2.3 mm = _?_ cm
Smaller to larger means fewer units. Divide by 10.
$2.3 \div 10 = 0.23$
2.3 mm = 0.23 cm

6 kg = _?_ g
Larger to smaller means more units. Multiply by 1000.
$6 \times 1000 = 6000$
6 kg = 6000 g

35 mL = _?_ L
Smaller to larger means fewer units. Divide by 1000.
$35 \div 1000 = 0.035$
35 mL = 0.035 L

Complete each sentence.

1. 2.5 m = ____ cm **250**
2. 0.35 m = ____ mm **350**
3. 565 m = ____ km **0.565**
4. 17 cm = ____ m **0.17**
5. 0.3 cm = ____ mm **3**
6. 2100 cm = ____ km **0.021**
7. 3.4 km = ____ m **3400**
8. 53 cm = ____ m **0.53**
9. 2 m = ____ cm **200**
10. 800 m = ____ km **0.8**
11. 42 mm = ____ cm **4.2**
12. 4600 mL = ____ L **4.6**
13. 8 L = ____ mL **8000**
14. 786 cm = ____ m **7.86**
15. 3571 mg = ____ g **3.571**
16. 3 kg = ____ g **3000**
17. 58 g = ____ mg **58,000**
18. 0.045 m = ____ mm **45**

In-Class Example

For Example 3
Complete each sentence.

a. 0.53 kg = ● g **530**

b. 140 mg = ● g **0.14**

3 PRACTICE/APPLY

Checking Your Understanding

Exercises 1–11 are designed to help you assess your students' understanding through reading, writing, speaking, and modeling. You should work through Exercises 1–3 with your students and then monitor their work on Exercises 4–11.

Additional Answers

1. Let y be in centimeters and x be in meters. $y = 100x$ for converting meters to centimeters; $x = \frac{y}{100}$ for converting centimeters to meters.
2. Sample answer: A kilometer is 1,000,000 times longer than a millimeter.

Checking Your Understanding

Communicating Mathematics

Read and study the lesson to answer these questions. **1–2. See margin.**

1. Write a formula for converting a measure in meters to an equivalent measure in centimeters. Then write a second formula for converting a measure in centimeters to an equivalent measure in meters.
2. **Show** how many times longer a kilometer is than a millimeter by using the diagram on page 358.

MATH JOURNAL

3. Which concept in this chapter was the most difficult for you? Why was it challenging? **See students' work.**

Guided Practice

State which metric unit you would probably use to measure each item.

4. width of a computer screen **cm**
5. juice in a small glass **mL**
6. thickness of a coin **mm**
7. a load of lumber **kg**

Complete each sentence.

8. 0.035 m = _?_ mm **35**
9. 40 mL = _?_ L **0.040**
10. 3 kg = _?_ g **3000 g**
11. **Running** Last week, Ramon ran 8 laps around a 400-meter track on Monday, 4 laps on Tuesday, 8 laps on Wednesday, 4 laps on Thursday, and 8 laps on Friday. How many kilometers did he run last week? **12.8 km**

Exercises: Practicing and Applying the Concept

Independent Practice

Write which metric unit you would probably use to measure each item.

12. distance between two cities **km**
13. height of a 15-year-old female
14. weight of an aspirin **mg**
15. water in a washing machine **L**

13. cm or m

Complete each sentence.

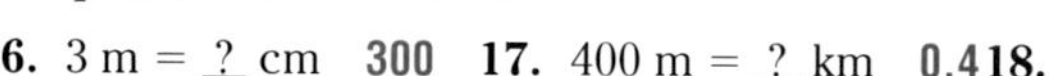

16. 3 m = _?_ cm **300**
17. 400 m = _?_ km **0.4**
18. 30 mm = _?_ cm **3**
19. 9400 mL = _?_ L **9.4**
20. 5 L = _?_ mL **5000**
21. 600 cm = _?_ m **6**
22. 5000 mg = _?_ g **5**
23. 8 kg = _?_ g **8000**
24. 67 g = _?_ mg **67,000**

B

25. How many millimeters are in a meter? **1000**
26. How many centimeters are in a kilometer? **100,000**
27. How many grams are in 0.316 kilograms? **316**
28. **Biology** The thickness of a leg of an ant is 0.035 centimeters. How many millimeters is this? **0.35 mm**
29. A bottle of ketchup holds 0.946 liters. How many milliliters does it hold? **946 mL**

C

30. How many milligrams are in 0.019 kilograms? **19,000 mg**
31. How many centimeters are in 6.032 kilometers? **603,200 cm**
32. The hair on your scalp grows about 13 millimeters per month. How long would it take for hair to grow 0.5 meter? **more than 38 months**

Practice Masters, p. 61

NAME ______ DATE ______

7-8 Practice
Integration: Measurement
Using the Metric System

Student Edition
Pages 358–362

Complete each sentence.

1. 3 m = ___ cm 300	2. 1.6 m = ___ cm 160	3. 0.9 m = ___ cm 90
4. 250 cm = ___ m 2.5	5. 60 cm = ___ m 0.6	6. 8 cm = ___ m 0.08
7. 2000 mm = ___ m 2	8. 15 mm = ___ m 0.015	9. 500 mm = ___ m 0.5
10. 5 m = ___ mm 5000	11. 12 cm = ___ mm 120	12. 3000 mm = ___ cm 300
13. 2 km = ___ m 2000	14. 10 km = ___ m 10,000	15. 0.8 kg = ___ g 800
16. 2000 mg = ___ g 2	17. 50 mg = ___ g 0.05	18. 6000 g = ___ kg 6
19. 2.9 kg = ___ g 2900	20. 0.004 kg = ___ g 4	21. 75 g = ___ kg 0.075
22. 1.5 kg = ___ g 1500	23. 0.008 kg = ___ mg 8000	24. 15,000 g = ___ kg 15
25. 3 L = ___ m 3000	26. 5000 mL = ___ L 5	27. 4.5 L = ___ mL 4500
28. 75 mL = ___ L 0.075	29. 7.5 mL = ___ L 0.0075	30. 390 mL = ___ L 0.39
31. 9.9 g = ___ kg 0.0099	32. 0.03 m = ___ mm 30	33. 0.2 L = ___ mL 200
34. 6 m = ___ mm 6000	35. 8 mg = ___ g 0.008	36. 2.48 L = ___ mL 2480
37. 7.8 kg = ___ g 7800	38. 43.2 L = ___ mL 43,200	39. 4569 g = ___ kg 4.569
40. 807 mL = ___ L 0.807	41. 5.8 km = ___ m 5800	42. 3751 m = ___ km 3.751

Reteaching

Using Models Use the model below to show how to make unit conversions.

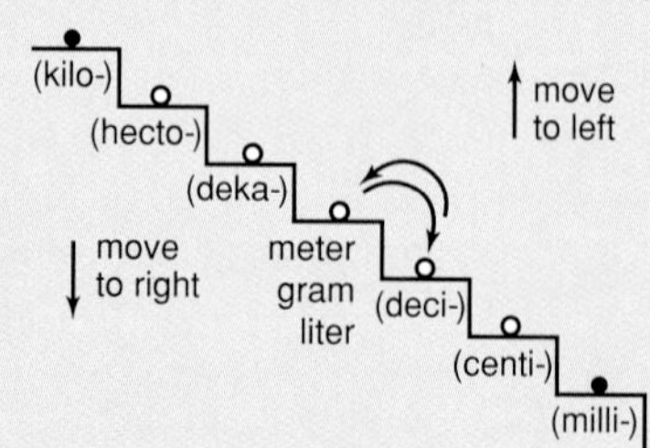

Group Activity Card 7-8

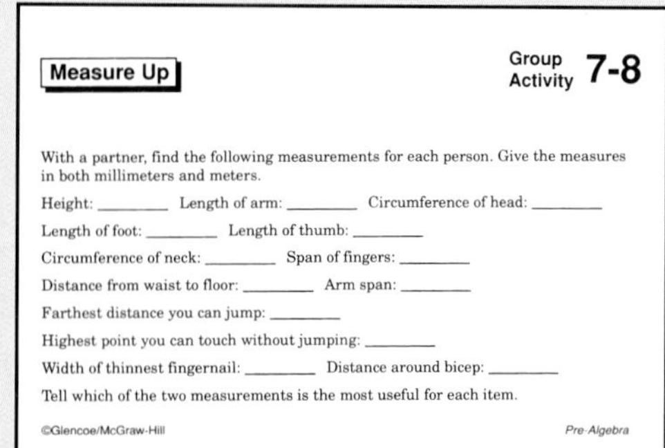
Measure Up

Group Activity 7-8

With a partner, find the following measurements for each person. Give the measures in both millimeters and meters.

Height: ______ Length of arm: ______ Circumference of head: ______

Length of foot: ______ Length of thumb: ______

Circumference of neck: ______ Span of fingers: ______

Distance from waist to floor: ______ Arm span: ______

Farthest distance you can jump: ______

Highest point you can touch without jumping: ______

Width of thinnest fingernail: ______ Distance around bicep: ______

Tell which of the two measurements is the most useful for each item.

 Pre-Algebra

Critical Thinking

34. about 312 trees

33. Four volumes of International Recipes are in order on a shelf. The total pages of each volume are 5 cm thick. Each cover is 5 mm thick. A bookworm started eating at page 1 of Volume I and stopped eating at the last page of Volume IV. Describe the amount the bookworm ate in terms of distance. **13 cm**

Applications and Problem Solving

Choose

Estimation
Mental Math
Calculator
Paper and Pencil

34. Ecology Every metric ton of recycled office paper saves about 19 trees. Glencoe Publishing's office in Westerville, Ohio, recycled 16,427 kilograms of paper in 1994. How many trees did this save?

35. Geometry The perimeter of a triangle is 1.6 meters. The length of the shortest side is 40 centimeters. Another side is 150 millimeters longer than the shortest side. Find the lengths of the other sides.

35. 65 cm, 55 cm

36. Refreshments To mix a punch, Lavonna starts with 3 liters of lemon-lime soda and adds 1893 milliliters of cranberry juice.

a. Estimate how many liters are in the punch bowl when the punch is completed.

b. Use your estimate to find how many 250-mL glasses of punch can be served.

37. Photography Most cameras use 35-millimeter film. This means that the film is 35 millimeters wide. Write the width of the film in meters.

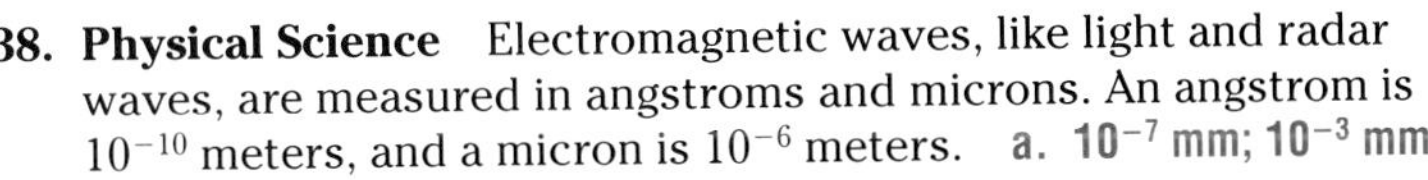

38. Physical Science Electromagnetic waves, like light and radar waves, are measured in angstroms and microns. An angstrom is 10^{-10} meters, and a micron is 10^{-6} meters. **a. 10^{-7} mm; 10^{-3} mm**

a. Write the length of an angstrom and a micron in millimeters.

b. The visible portion of the electromagnetic spectrum, including white light, has waves that are approximately 3900 to 7700 angstroms long. Write the lengths of these waves in millimeters.

c. The waves used for radio and television transmission are 1 to 10 meters long. Write these lengths in angstroms.

36a. 5 L
36b. about 20 glasses
37. 0.035 m
38b. 0.00039 to 0.00077 mm
38c. 10^{10} to 10^{11} angstroms

39. Family Activity Find at least five metric measures in your home. Take measurements or find some on packages of food or other products. Write each measure as you found it and in one other metric unit of measure. **See students' work.**

Mixed Review

40. The sum of three integers, x, $x + 1$, and $x + 2$, is at least 216. Find the three smallest such integers. (Lesson 7-7) **71, 72, 73**

41. Solve $7b - 39 > 2(b - 2)$. (Lesson 7-6) **$b > 7$**

42. Solve $2z - 500 = 100 - z$. (Lesson 7-5) **200**

43. Evaluate $\frac{5x}{y}$ if $x = 31.74$ and $y = 9.2$. (Lesson 6-5) **17.25**

44. 26.125 million; 16 million; 17 million

44. Agriculture The chart at the right shows the numbers of millions of 42-pound units of different types of apples produced in 1992. What is the mean, median, and mode of the numbers of apples? (Lesson 6-6)

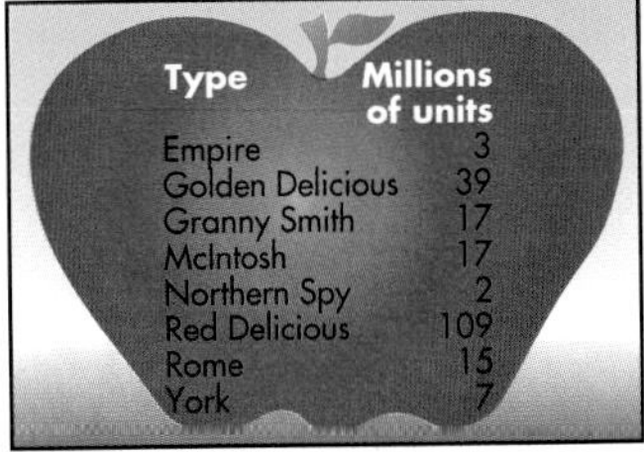

Type	Millions of units
Empire	3
Golden Delicious	39
Granny Smith	17
McIntosh	17
Northern Spy	2
Red Delicious	109
Rome	15
York	7

Source: International Apple Institute

45. $-20\frac{4}{5}$

45. Solve $\left(3\frac{1}{4}\right)\left(-6\frac{2}{5}\right) = r$. (Lesson 6-3)

46. 3

46. Solve $15x = 45$. (Lesson 3-3)

Assignment Guide

Core: 13–39 odd, 40–46
Enriched: 12–32 even, 33–46

For **Extra Practice**, see p. 758.

The red A, B, and C flags, printed only in the Teacher's Wraparound Edition, indicate the level of difficulty of the exercises.

4 ASSESS

Closing Activity

Act It Out Bring different sizes of grocery products to class. Mark them with prices so one large size product is equal in cost to a number of smaller sizes of the same product. Have students show how they would determine which is the better buy by showing their work on the board or overhead and then placing a sign indicating the better buy.

Chapter 7, Quiz D (Lessons 7-7 and 7-8) is available in the *Assessment and Evaluation Masters*, p. 184.

Extension

Using Manipulatives Using metric rulers, have students measure various objects in the classroom to the nearest centimeter. Have them convert each of these measurements to meters and millimeters.

Enrichment Masters, p. 61

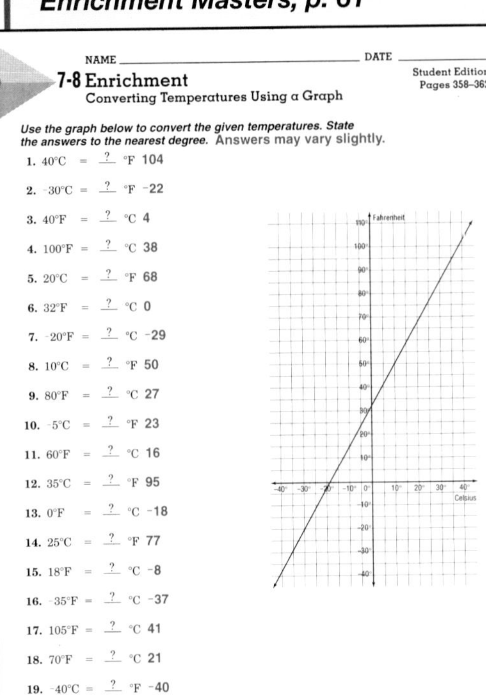

NAME ______________________ DATE ____________

7-8 Enrichment — Student Edition Pages 358–362

Converting Temperatures Using a Graph

Use the graph below to convert the given temperatures. State the answers to the nearest degree. Answers may vary slightly.

1. 40°C = ? °F 104
2. -30°C = ? °F -22
3. 40°F = ? °C 4
4. 100°F = ? °C 38
5. 20°C = ? °F 68
6. 32°F = ? °C 0
7. -20°F = ? °C -29
8. 10°C = ? °F 50
9. 80°F = ? °C 27
10. -5°C = ? °F 23
11. 60°F = ? °C 16
12. 35°C = ? °F 95
13. 0°F = ? °C -18
14. 25°C = ? °F 77
15. 18°F = ? °C -8
16. -35°F = ? °C -37
17. 105°F = ? °C 41
18. 70°F = ? °C 21
19. -40°C = ? °F -40
20. -40°F = ? °C -40

CLOSING THE

Investigation

This activity provides students an opportunity to bring their work on the Investigation to a close. For each Investigation, students should present their findings to the class. Here are some ways students can display their work.

- Conduct and report on an interview or survey.
- Write a letter, proposal, or report.
- Write an article for the school or local paper.
- Make a display, including graphs and/or charts.
- Plan an activity.

Assessment

To assess students' understanding of the concepts and topics explored in this Investigation and its follow-up activities, you may wish to examine students' Investigation Folders.

The scoring guide provided in the *Investigations and Projects Masters*, p. 15, provides a means for you to score students' work on this Investigation.

Investigations and Projects Masters, p. 15

CHAPTERS 6 and 7 Investigation
Roller Coaster Math

Scoring Guide

Level	Specific Criteria
3 Superior	• Shows thorough understanding of *writing fractions as decimals, estimating and finding products and quotients of rationals, solving problems by working backward, writing and solving multi-step equations and inequalities, finding circumferences*, and *converting metric measurements*. • Uses appropriate strategies to solve problems. • Computations are correct. • Written explanations are exemplary. • Tables and scale drawings are appropriate and sensible. • Goes beyond the requirements of the Investigation.
2 Satisfactory, with minor flaws	• Shows understanding of *writing fractions as decimals, estimating and finding products and quotients of rationals, solving problems by working backward, writing and solving multi-step equations and inequalities, finding circumferences*, and *converting metric measurements*. • Uses appropriate strategies to solve problems. • Computations are mostly correct. • Written explanations are effective. • Tables and scale drawings are appropriate and sensible. • Satisfies all requirements of the Investigation.
1 Nearly Satisfactory, with serious flaws	• Shows understanding of most of the concepts of *writing fractions as decimals, estimating and finding products and quotients of rationals, solving problems by working backward, writing and solving multi-step equations and inequalities, finding circumferences*, and *converting metric measurements*. • May not use appropriate strategies to solve problems. • Computations are mostly correct. • Written explanations are satisfactory. • Tables and scale drawings are mostly appropriate and sensible. • Satisfies most requirements of the Investigation.
0 Unsatisfactory	• Shows little or no understanding of *writing fractions as decimals, estimating and finding products and quotients of rationals, solving problems by working backward, writing and solving multi-step equations and inequalities, finding circumferences*, and *converting metric measurements*. • May not use appropriate strategies to solve problems. • Computations are incorrect. • Written explanations are not satisfactory. • Tables and scale drawings are not appropriate or sensible. • Does not satisfy requirements of the Investigation.

CLOSING THE

Investigation

Refer to the Investigation on pages 270–271.

You are ready to begin construction of your new roller coaster ride. The amusement park's information director is preparing a press release to announce when the new ride will open.

Using information from your design and the data from your class experiments, write a report for the information director. Your report should include the following.

- The name of your roller coaster and an explanation of why you chose it.
- Any records and track data such as height, speed, and so on.
- Ride data such as how long a ride lasts and how many passengers will be carried per hour.
- Any special-interest items like providing entertainment for people waiting in line.

Plan and give an oral presentation to describe what you know about roller coasters and designing roller coasters. Your presentation could include the following.

- A brief history of roller coasters.
- Examples from your portfolio to help you describe and define the roller coaster terms you used most often in this investigation.
- A table showing how you determined the most efficient number of passengers and trains for your design.
- An explanation of what you would do differently if you had to do this Investigation over again.

PORTFOLIO ASSESSMENT

You may want to keep your work on this Investigation in your portfolio.

Extension

Build a scale model of your roller coaster using an Erector Set®, or wooden craft sticks.

CHAPTER 7 Highlights

Vocabulary

After completing this chapter, you should be able to define each term, property, or phrase and give an example or two of each.

Algebra

empty set (p. 347)
null set (p. 347)

Geometry

center (p. 341)
circle (p. 341)
circumference (p. 341)
diameter (p. 341)
pi, π (p. 341)
radius (p. 341)

Measurement

gram (p. 359)
liter (p. 359)
meter (p. 358)
metric system (p. 358)

Problem Solving

working backward (p. 330)

Understanding and Using Vocabulary

Determine whether each statement is *true* or *false*.

1. The basic unit of capacity in the metric system is the gram. false
2. The distance around a circle is its circumference. true
3. In the metric system, the meter is the basic unit used for measuring distance. true
4. A radius of a circle is twice as long as its diameter. false
5. The null set, or the empty set, contains no elements. true
6. A segment from any point on a circle to the center of the circle is a radius. true
7. A formula for the circumference of a circle is $C = \pi d$. true

Instructional Resources

Three multiple-choice tests and three free-response tests are provided in the *Assessment and Evaluation Masters*. Forms 1A and 2A are for honors pacing, Forms 1B and 2B are for average pacing, Forms 1C and 2C are for basic pacing, Chapter 7 Test, Form 1B is shown at the right. Chapter 7 Test, Form 2B is shown on the next page.

Highlights

Using the Chapter Highlights

The Chapter Highlights begins with a listing of the new terms, properties, and phrases that were introduced in this chapter.

Assessment and Evaluation Masters, pp. 171–172

NAME ______________ DATE ______

CHAPTER 7 Test, Form 1B

1. A certain number is divided by 3 and then 2 is subtracted from the result. The final answer is 14. Find the number.
   A. 40 B. 36 C. 16 D. 48 — 1. D

2. Shareen earns one dollar per hour more baby-sitting for the Rothstein children than for the Miller children. If she earned $12 baby-sitting for the Miller children for three hours, how much would she earn baby-sitting for the Rothstein children for 4 hours and the Miller children for 2 hours?
   A. $28 B. $20 C. $21 D. $48 — 2. A

Solve each equation. Check your solution.

3. $3n + 2 = 26$
   A. 7 B. 8 C. 9 D. 6.5 — 3. B

4. $^-4m - (^-2) = 6$
   A. $^-2$ B. 2 C. $^-1$ D. 1 — 4. C

5. $\frac{x}{3} + 2 = 12$
   A. 10 B. 42 C. 36 D. 30 — 5. D

6. Faraz has $30 and saves an additional $6 per week. Choose the equation that would be used to find the number of weeks it will take him to save $96. Then solve.
   A. $30 + w = 96, w = 66$ B. $96 - w = 30, w = 66$
   C. $30 + 6w = 96, w = 11$ D. $6w - 30 = 96, w = 21$ — 6. C

7. Ami buys $80 worth of non-taxable items and other items that are taxable at 5%. Her total bill is $118.85. Choose the equation to find the cost of the taxable items. Then solve.
   A. $t + 0.05t + 80 = 118.85, t = \$37$
   B. $t + 0.5t + 80 = 118.85, t = \$25.90$
   C. $t + 5t - 80 = 118.85, t = \$6.48$
   D. $80 - 0.05t - t = 118.85, t = \$37$ — 7. A

8. Kareem wants to buy a portable CD player that costs $127. He has saved $57 from money he made by recycling. If he can save $14 per month, how long it will take him to save enough money to buy the CD player?
   A. 5 months B. 2 months C. 13 months D. 7 months — 8. A

9. Members of the pep club are making circular posters to advertise an upcoming basketball tournament. They want to put crepe paper streamers around the edge of each poster. What length of crepe paper is needed to go around a poster 4 feet in diameter?
   A. 6.28 ft B. 13.64 ft C. 12.56 ft D. 25.12 ft — 9. C

NAME ______________ DATE ______

Chapter 7 Test, Form 1B (continued)

10. Find the circumference of a circle with a radius of 3 yards.
    A. 9.42 yd B. 18.84 yd C. 37.68 yd D. 4.71 yd — 10. B

Solve each equation. Check your solution.

11. $4x + 8 = 3x - 1$
    A. $^-7$ B. 7 C. $^-9$ D. $^-1$ — 11. C

12. $3(a + 4) - 2 = 8a + 5$
    A. 1 B. 2 C. 0.4 D. 3 — 12. A

13. $7a - 3 = 5(a - 1)$
    A. 0.25 B. 1 C. $^-1$ D. 2 — 13. C

Solve each inequality and check your solution.

14. $4x - 2 > 10$
    A. $x < 3$ B. $x > 3$ C. $x < 2$ D. $x > 2$ — 14. B

15. $^-4p - 5 < {}^-25$
    A. $p > 5$ B. $p < 5$ C. $p > 4$ D. $p < 7.5$ — 15. A

16. $\frac{x}{2} - 2 < 2$
    A. $x < 4$ B. $x < 6$ C. $x > 8$ D. $x < 8$ — 16. D

17. Nichelle earns $7 per hour and gets a 10% commission on the sale price of each item she sells. She wants to work only 10 hours each week, and has a weekly earnings goal of $200. Which inequality would be used to find the total sales she must make to reach her goal?
    A. $(7)(10) - 1.10s \geq 200, s = 118.18$
    B. $(7)(10)(0.10)s \geq 200, s = 28.25$
    C. $(7)(10) + 1.10s \leq 200, s = 118.18$
    D. $7(10) + 0.10s \geq 200, s = 1300$ — 17. D

18. Five times a number decreased by 8 is less than 92. Choose the inequality to find the unknown number. Then solve.
    A. $5n \div 8 < 92, n < 147.2$
    B. $5n + 8 < 92, n < 16.8$
    C. $5n - 8 < 92, n < 20$
    D. $5n - 8 > 92, n > 20$ — 18. C

Complete each sentence.

19. 300 mg = _?_ g
    A. 0.3 B. 3 C. 30 D. 0.03 — 19. A

20. 60,000 m = _?_ km
    A. 6 B. 60 C. 600 D. 0.6 — 20. B

CHAPTER 7 Study Guide and Assessment

Using the Study Guide and Assessment

Skills and Concepts Encourage students to refer to the objectives and examples on the left as they complete the review exercises on the right.

Assessment and Evaluation Masters, pp. 177–178

NAME ______________ DATE ________

CHAPTER 7 Test, Form 2B

1. A box of 90 red, black, and blue map pins has twice as many red pins as blue, and half as many blue as black. How many pins are blue? — 1. 18
2. The diameter of a balloon doubles every time 10 pounds of pressure is added. The diameter was 2 inches before 30 pounds of pressure was added. Find the balloon's present diameter. — 2. 16 in.
3. Marcia read for 45 minutes, exercised for 60 minutes, and rested for 20 minutes. At the end of her rest period, it was noon. At what time did Marcia begin reading? — 3. 9:55 A.M.

Solve each equation. Check your solution.

4. $3x + 8 = 23$ — 4. 5
5. $6y - 12 = 30$ — 5. 7
6. $\frac{a + 4}{6} = 6$ — 6. 32
7. $\frac{b}{4} + 5 = 2$ — 7. -12
8. $-3m - 7 = 34$ — 8. $-13\frac{2}{3}$
9. $-6p - 15 = -63$ — 9. 8
10. $\frac{y}{2} + 9 = 7$ — 10. -4
11. $\frac{z}{3} - 10 = -8$ — 11. 6

Find the circumference of each circle described below.

12. The diameter is 12.5 centimeters. — 12. 39.25 cm
13. The radius is 4.9 inches. — 13. 30.772 in.
14. 

— 14. 81.64 m
15. 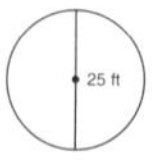

— 15. 78.5 ft

NAME ______________ DATE ________

Chapter 7 Test, Form 2B (continued)

Solve each equation. Check your solution.

16. $x + 9 = 2x + 3$ — 16. 6
17. $5a - 10 = 4a + 8$ — 17. 18
18. $9b - 4 = 12b - 8$ — 18. $1\frac{1}{3}$
19. $2.5d + 1 = 3d - 4$ — 19. 10
20. $8f + 14 = 12f + 2$ — 20. 3
21. $2(x - 5) = 20x$ — 21. $-\frac{5}{9}$
22. $4(y - 9) = 2(y + 21)$ — 22. 39

Solve each inequality and check your solution.

23. $7y - 5 < 30$ — 23. $y < 5$
24. $-\frac{1}{2}m + 3 < 23$ — 24. $m > -40$
25. $2p + 6 < 5p - 6$ — 25. $p > 4$
26. $-3(x - 2) > 12$ — 26. $x < -2$
27. $4y + 2 > 5y - 5$ — 27. $y < 7$
28. The sum of an integer and the next greater integer is at least 13. Find the lesser integer. — 28. $n \geq 6$
29. If 5 times an integer is decreased by 17, the result is less than 63. Find the integer. — 29. $n < 16$

Complete each sentence.

30. 6 m = _?_ cm — 30. 600
31. 8300 g = _?_ kg — 31. 8.3
32. 5.2 L = _?_ mL — 32. 5200
33. 400 cm = _?_ mm — 33. 4000

CHAPTER 7 Study Guide and Assessment

Skills and Concepts

Objectives and Examples

Upon completing this chapter, you should be able to:

▸ **solve equations that involve more than one operation.** (Lesson 7-2)

$$3x + 6 = 27$$
$$3x + 6 - 6 = 27 - 6 \quad \textit{Subtract 6 from each side.}$$
$$3x = 21$$
$$\frac{3x}{3} = \frac{21}{3} \quad \textit{Divide each side by 3.}$$
$$x = 7$$

Review Exercises

Use these exercises to review and prepare for the chapter test.

Solve each equation. Check your solution.

8. $54 = 7x - 9$ **9**
9. $\frac{g}{3} + 10 = 14$ **12**
10. $12 = 7 - m$ **−5**
11. $\frac{x}{4} - 17 = -2$ **60**
12. $24 - y = -9$ **33**
13. $4d - 8 = -88$ **−20**
14. $-2f - 37 = -11$ **−13**
15. $2.9 = 3.1 - t$ **0.2**

▸ **solve verbal problems by writing and solving equations.** (Lesson 7-3)

One more than twice a number is 17. What is the number?

one *more than* *twice a number* *is* *17*

$$1 \quad + \quad 2n \quad = 17$$
$$1 + 2n = 17$$
$$1 - 1 + 2n = 17 - 1$$
$$2n = 16$$
$$\frac{2n}{2} = \frac{16}{2}$$
$$n = 8$$

Define a variable and write an equation for each situation. Then solve.

16. The sum of six times a number and 4 is 52. What is the number? **8**
17. Twice a number less 7 is 19. Find the number. **13**
18. Six hundred is 15 more than 15 times a number. What is the number? **39**
19. Vashawn earns \$50 a week less than twice his old salary. If he earns \$450 a week now, what was his old salary? **\$250**

16–19. See Solutions Manual for equations.

▸ **find the circumference of a circle.** (Lesson 7-4)

Find the circumference of a circle with a diameter 12.3 cm long.

$$C = \pi d$$
$$C = \pi(12.3)$$
$$C \approx 38.6 \text{ cm}$$

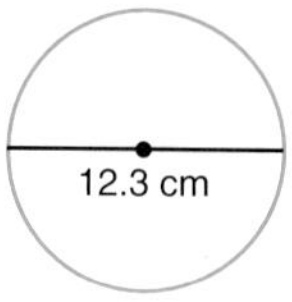

Find the circumference of each circle shown or described below.

20. 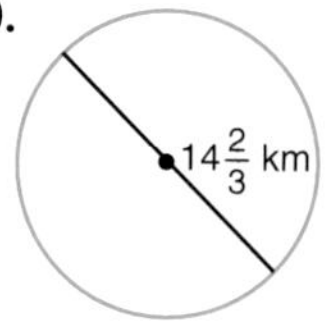

46.1 km

21. The radius is $3\frac{1}{2}$ inches. **22.0 in.**
22. The diameter is 18.45 centimeters. **58.0 cm**
23. The radius is 4.9 meters. **30.8 m**

GLENCOE Technology

Test and Review Software

You may use this software, a combination of an item generator and an item bank, to create your own tests or worksheets. Types of items include free response, multiple choice, short answer, and open ended.

For IBM & Macintosh

Objectives and Examples

▶ **solve equations with the variable on each side.** (Lesson 7-5)

$3t - 1 = t + 8$

$3t - t - 1 = t - t + 8$ *Subtract t from each side.*

$2t - 1 = 8$

$2t - 1 + 1 = 8 + 1$ *Add 1 to each side.*

$2t = 9$

$\frac{2t}{2} = \frac{9}{2}$ *Divide each side by 2.*

$t = \frac{9}{2}$ or $4\frac{1}{2}$

Review Exercises

Solve each equation. Check your solution.

24. $6n - 8 = 2n$ **2**
25. $4 + 2w = 8w + 16$ **−2**
26. $4m + 11 = 6 - m$ **−1**
27. $5(3 - k) = 20$ **−1**
28. $5(q + 2) = 2q + 1$ **−3**
29. $\frac{2}{7}h - 10 = h - 20$ **14**
30. $\frac{x - 5}{6} = \frac{x - 11}{2}$ **14**

▶ **solve inequalities that involve more than one operation.** (Lesson 7-6)

$5b - 10 < 7b + 4$

$5b - 5b - 10 < 7b - 5b + 4$ *Subtract 5b from each side.*

$-10 < 2b + 4$

$-10 - 4 < 2b + 4 - 4$ *Subtract 4 from each side.*

$-14 < 2b$

$\frac{-14}{2} < \frac{2b}{2}$ *Divide each side by 2.*

$-7 < b$

Solve each inequality and check your solution. Graph the solution on a number line.

31. $5n + 4 > 34$ **$n > 6$**
32. $-7m - 12 < 9$ **$m > -3$**
33. $\frac{r}{7} - 11 \leq 3$ **$r \leq 98$**
34. $16f > 13f + 45$ **$f > 15$**
35. $0.52c + 14.7 > 2.48c$ **$c < 7.5$**
36. $\frac{a - 5}{6} \leq \frac{a + 9}{8}$ **$a \leq 47$**

31–36. See margin for graphs.

▶ **solve verbal problems by writing and solving inequalities.** (Lesson 7-7)

Six times a number less 12 is at least twenty-two. What is the number?

six times a number	*less*	*12*	*is at least*	*twenty-two*
$6x$	$-$	12	$\geq$	22

$6x - 12 \geq 22$

$6x - 12 + 12 \geq 22 + 12$ *Add 12 to each side.*

$6x \geq 34$

$\frac{6x}{6} \geq \frac{34}{6}$ *Divide each side by 6.*

$x \geq \frac{34}{6}$ or $5\frac{2}{3}$

Define a variable and write an inequality for each situation. Then solve.

37. Eight times a number increased by two is at least 18. What is the number? **$n \geq 2$**
38. Twelve less than three times a number is at most 27. Find the number. **$n \leq 13$**
39. The sum of an integer and the next greater integer is more than 47. What is the integer? **$x > 23$**
40. Four times a number s exceeds 6 by at least 8. What are the possible values of s? **$s \geq \frac{7}{2}$**

37–40. See Solutions Manual for inequalities.

Additional Answers

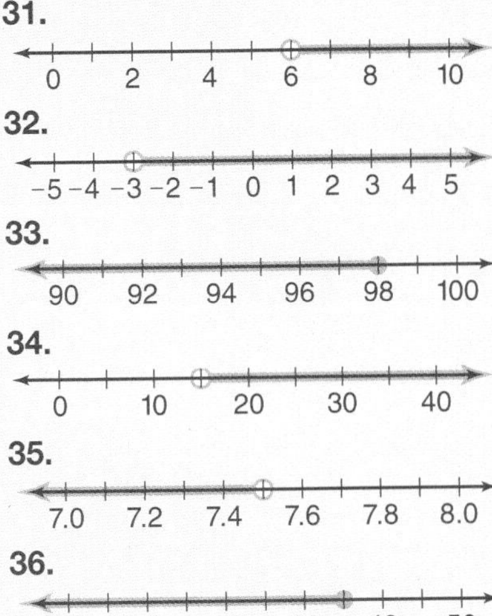

Applications and Problem Solving Encourage students to work through the exercises in the Applications and Problem Solving section to strengthen their problem-solving skills.

Objectives and Examples

- **convert measurements within the metric system.** (Lesson 7-8)

4.5 km = ? m

Kilometers are larger than meters, so there will be more units. There are 1000 meters in a kilometer.

$4.5 \times 1000 = 4500$
4.5 km = 4500 m

Review Exercises

Complete each sentence.

41. 6 mm = ? m **0.006**
42. 8.3 g = ? kg **0.0083**
43. 43 L = ? mL **43,000**
44. 560 mg = ? g **0.56**
45. 3 km = ? cm **300,000**
46. 40 mL = ? L **0.04**
47. 88 m = ? cm **8800**

Applications and Problem Solving

48. Patterns "One potato, two potato, . . ." says Alejándro as the children decide who is "it." At the beginning, six children were standing in a circle and each round of the rhyme eliminated the eighth child. Alejándro counts clockwise and each round begins with the person next to the one who was eliminated. Let A represent Alejándro and B through F the rest of the children clockwise around the circle. If E is declared "it," where did Alejándro begin the first rhyme? (Lesson 7-1) **C**

49. Health Doctors recommend that you not exceed a safe heart rate when you exercise. The formula $r = \frac{4(220 - a)}{5}$, gives the desirable heart rate, r, in beats per minute for a person a years old. Find the maximum safe heart rate for a 44-year old. (Lesson 7-2) **140.8 beats per minute**

50. Music Before valves were invented in the late 19th century, musicians used to adjust the scale of their brass instruments by adding a crook. A crook is a metal piece of tubing inserted between the mouthpiece and the main tubing of the instrument. If a circular crook has a diameter of 3.4 inches, how long was the piece of metal used to make it? (Lesson 7-4) **about 10.7 inches**

51. Aviation The pilot of a small plane wants to avoid going above the cloud ceiling that is currently at 12,000 feet. If the plane is ascending at a rate of 630 feet per minute, how long can the pilot ascend and still stay at least 1500 feet below the ceiling? (Lesson 7-7) **less than 16.67 minutes**

A practice test for Chapter 7 is available on page 780.

Performance Task **Demonstrate your knowledge by giving a clear, concise, solution to each problem. Be sure to include all relevant drawings and justify your answers. You may show your solutions in more than one way or investigate beyond the requirements of the problem.**

1. The table at the right shows the top five biggest money-losing films.

Film	Loss (millions)
The Adventures of Baron Münchhausen	48.1
Ishtar	47.3
Hudson Hawk	47.0
Inchon	44.1
The Cotton Club	38.1

Source: *Top Ten of Everything*

   a. *The Adventures of Baron Münchhausen* lost $10.3 less than twice what the British film *Raise the Titanic* lost. How much did *Raise the Titanic* lose? 29.2 million

   b. The sum of what *Billy Bathgate* and *The Cotton Club* lost is $5.1 million more than twice what *Billy Bathgate* lost. How much did *Billy Bathgate* lose? 33 million

   c. *Rambo III* lost just over $6.3 million more than half what *Ishtar* lost. How much did *Rambo III* lose? Write your answer as an inequality. $r >$ $29.95 million

   d. Use the numbers in the table to write a problem that could be solved using a two-step equation or inequality. See students' work.

2. Terrica is cutting fabric to make a tablecloth for a round table in her living room. The top of the table is 56.25 centimeters across. The table is 1.2 meters tall. 2a. 176.7 cm or 1.76 m

   a. What is the circumference of the top of the table?

   b. What should Terrica make the diameter of the tablecloth be if it goes to the floor on all sides? 296.25 cm or 2.9625 m

   c. If Terrica is going to put lace around the edge of the tablecloth, how much should she buy? about 9.3 meters

Thinking Critically Two formulas for the circumference of a circle are $C = 2\pi r$ and $C = \pi d$. If you could only remember one of them, could you always find the circumference of a circle? Explain. See margin.

 **Portfolio** Select one of the assignments from this chapter that you found especially challenging and place it in your portfolio.

Self Evaluation The only place where success comes before work is in the dictionary. In other words, if you want to succeed, you must first work!

Assess yourself. Are you willing to put forth the effort to succeed? Describe a situation that was difficult for you when you worked hard and succeeded or didn't work hard and failed. What did you learn from the situation? What would you do the same or differently if you could do it again? See students' work.

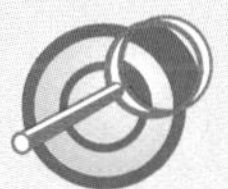

Alternative Assessment

The Alternative Assessment section provides students with the opportunity to assess their own work by thinking critically, working with others, keeping a portfolio, and honestly evaluating their own progress. For more information on alternative forms of assessment, see *Alternative Assessment in the Mathematics Classroom*.

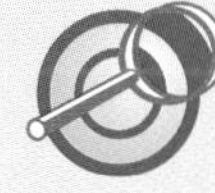

Performance Assessment

Performance Assessment tasks for this chapter are included in the *Assessment and Evaluation Masters*. A scoring guide is also provided.

Assessment and Evaluation Masters, pp. 181,193

NAME ______________________ DATE ____________

CHAPTER 7 Performance Assessment

Instructions: Demonstrate your knowledge by giving a clear, concise solution to each problem. Be sure to include all relevant drawings and justify your answers. You may show your solution in more than one way or investigate beyond the requirements of the problem.

1. a. Explain how you can use the strategy of working backward to solve a problem.

   b. Mrs. Grant withdrew money from her checking account in the bank. She gave half of it to her husband, put $7.50 worth of gasoline in the car, and found 80¢ in a chair when she returned home. If she had no money when she went to the bank and has $18.30 now, how much did she withdraw at the bank?

   c. Write a word problem that can be solved by working backward. Solve and explain each step. What does the answer represent?

2. A postal clerk uses mathematics when selling stamps, weighing packages, and determining the amount of postage needed.

   a. A customer bought an equal number of each of the following stamps: 5¢, 10¢, and 32¢. She also mailed a package that required $2.75 postage. The total bill was $12.15. Write an equation that describes this situation and solve. What does the answer represent?

   b. A customer wishes to keep the weight of a package he is sending to his sister and three children under 4 pounds. He wishes to send three identical bags of candy and a $1\frac{3}{4}$-pound present. Write an inequality that describes this situation and solve. Tell what the answer represents.

   c. Make up a problem that can be solved by using the equation below. Then solve and tell what the answer represents.

   $$\$8.46 = \$15 - 3x$$

   d. Make up a problem that can be solved by using the inequality below. Then solve and tell what the answer represents.

   $$\$10 - 2v \geq \$2.50$$

CHAPTER 7 Scoring Guide

Level	Specific Criteria
3 Superior	• Shows thorough understanding of the concepts *variable, equation, inequality*, and *working backward*. • Uses appropriate strategies to solve problems. • Computations to solve equations, inequalities, and problems involving working backward are correct. • Written explanations are exemplary. • Word problems concerning equations, inequalities, and working backward are appropriate and make sense. • Diagram/table/chart is accurate (as applicable). • Goes beyond requirements of some or all problems.
2 Satisfactory, with minor flaws	• Shows understanding of the concepts *variable, equation, inequality*, and *working backward*. • Uses appropriate strategies to solve problems. • Computations to solve equations, inequalities, and problems involving working backward are mostly correct. • Written explanations are effective. • Word problems concerning equations, inequalities, and working backward are appropriate and make sense. • Diagram/table/chart is mostly accurate (as applicable). • Satisfies all requirements of problems.
1 Nearly Satisfactory, with serious flaws	• Shows understanding of most of the concepts *variable, equation, inequality*, and *working backward*. • May not use appropriate strategies to solve problems. • Computations to solve equations, inequalities, and problems involving working backward are mostly correct. • Written explanations are satisfactory. • Word problems concerning equations, inequalities, and working backward are appropriate and make sense. • Diagram/table/chart is mostly accurate (as applicable). • Satisfies most requirements of problems.
0 Unsatisfactory	• Shows little or no understanding of the *variable, equation, inequality*, and *working backward*. • May not use appropriate strategies to solve problems. • Computations to solve equations, inequalities, and problems involving working backward are incorrect. • Written explanations are not satisfactory. • Word problems concerning equations, inequalities, and working backward are not appropriate or sensible. • Diagram/table/chart is not accurate (as applicable). • Does not satisfy requirements of problems.

Investigation

TEACHER NOTES

NCTM Standards: 1–4

This Investigation is designed to be completed over several days or weeks. It may be considered optional. You may want to assign the Investigation and the follow-up activities to be completed at the same time.

Objective

- Explore, conduct research, and make decisions about the operations of the stadium at a major university.

Mathematical Overview

This Investigation will use the following skills and concepts from Chapters 8 and 9.

- use scatter plots
- graph linear functions
- use slope
- make a table
- use statistics to predict
- find percents of change

Recommended Time		
Part	Pages	Time
Investigation	368-369	1 class period
Working on the Investigation	384, 404, 453, 471	20 minutes each
Closing the Investigation	476	1 class period

Instructional Resources

Investigations and Projects Masters, pp. 17–20

A recording sheet, teacher notes, and scoring guide are provided for each Investigation in the *Investigation and Projects Masters.*

1 MOTIVATION

In this Investigation, students explore what is involved in managing a large stadium at a university. Ask if students have ever attended a high school or university football game. What do they think the stadium manager has to decide or plan for a game?

LONG-TERM PROJECT

Investigation

You are the stadium manager for the West Prairie State University, home of the Mighty Mustangs. You oversee the operations of Mustang Stadium and help plan team-spirit activities for the home football games. The university has a large stadium with a seating capacity of 80,000. The stadium has two levels, an upper level and a lower level.

- *In each level, there are 26 sections. Each section has about the same number of seats.*
- *The sections are labeled A through Z, with section A situated at the center of the north end zone of the field.*
- *The rest of the sections are labeled in alphabetical order going around the stadium in a clockwise direction.*
- *In each section of the lower level, there are 90 rows numbered from 1 to 90. They are in sequential order starting with row 1 next to the field.*
- *Each section of the upper level has only 75 rows numbered from 1 to 75. Each seat in the row is also numbered.*

The following ticket stub shows a seat for the big game.

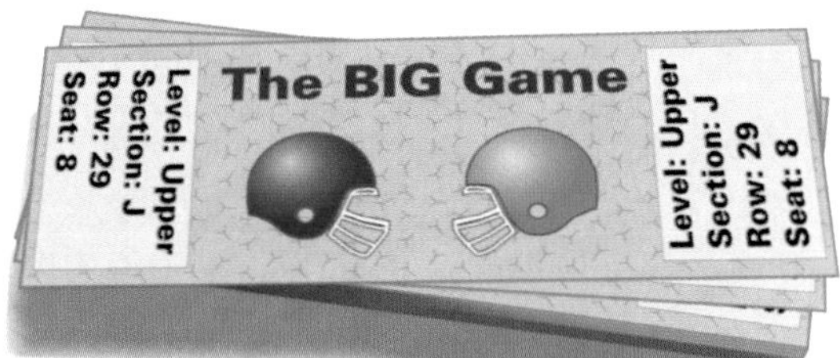

Various groups come to your office for research, advice, and permission for different projects. The spirit squad has been working on keeping the team spirit high this year for the football games. Also, the athletic director has been working on ticket pricing for stadium seating.

Starting Your Investigation

To start the research for either the spirit squad or athletic director, you will want to draw a model of the stadium. Since every row in every section of the stadium seats about the same number of people (plus or minus one person), how many seats are in a row? Explain your answer. What is the seating capacity of an upper section in the stadium? What is the seating capacity of a lower section?

On your model, label all the sections and levels in the stadium. Also, draw an enlargement of section A (both upper and lower levels), and label the number of each row in the section. Also, choose one row and label all the seat numbers.

Describe where the best seats are in the stadium. If you could sit wherever you want during the big game, what seat would you pick?

Think About . . .
The spirit squad wants to do the wave during a time-out. The wave is a type of stadium cheer where the people in each section of the stadium stand, raise their hands, and cheer in a sequence around the stadium. The spirit squad asks you a few questions:

- *How long does it take the crowd to do one rotation around the stadium?*
- *How many times can the wave go around the stadium during a two minute time-out?*
- *If the wave is done for two minutes during each of six time-outs in the game and for three minutes during each quarter and half-time, how many times would the average fan have to stand up and sit down during a game?*

Reread the above questions throughout this Investigation. As you come up with other questions, write these new questions in the list also.

interNET CONNECTION

For current information on football stadiums, visit: **www.glencoe.com/sec/math/prealg/mathnet**

You will continue working on this Investigation throughout Chapters 8 and 9.

Be sure to keep your materials in your Investigation Folder.

Investigation
Stadium Stampede

Working on the Investigation
Lesson 8-2, p. 384

Working on the Investigation
Lesson 8-6, p. 404

Working on the Investigation
Lesson 9-5, p. 453

Working on the Investigation
Lesson 9-9, p. 471

Closing the Investigation
End of Chapter 9, p. 476

Cooperative Learning

This Investigation offers an excellent opportunity for using cooperative learning groups. For more information on cooperative learning strategies and group management, see *Cooperative Learning in the Mathematics Classroom*, one of the titles in the Glencoe Mathematics Professional Series.

2 SETUP

Have students work in small groups. No special materials are needed. You may wish to set up groups so that each group has at least one sports enthusiast.

3 MANAGEMENT

After each group has completed its work, have groups exchange reports. Provide a copy of the scoring guide from the *Investigation and Projects Masters* and have each group grade each other group's project. Then have groups revise their final report before turning it in to the teacher.

Sample Answers

Answers will vary as they are based on the research of each pair of students.

Investigations and Projects Masters, p. 20

NAME ______________________ DATE __________

CHAPTERS 8 and 9 Investigation
Stadium Stampede

Student Edition
Pages 368–369

Answer the following questions as Stadium Manager.

How would you determine the pricing of seats?

What restrictions would influence seat prices?

How would you label seats?

Which sections would you reserve for students? Why?

Where would the press box be located?

Answer the following questions as Director of the Spirit Squad.

How would you promote school spirit?

What makes stadium cheers effective?

Where should the cheerleaders stand during the game to be seen by most of the crowd?

Would a marching band or team mascot provide team spirit?

Please keep this page and any other research in your Investigation Folder.

CHAPTER 8

Functions and Graphing

PREVIEWING THE CHAPTER

In this chapter, students explore functions and their graphs. In the first lesson, students use tables and graphs to represent relations and functions. Next, students explore and construct scatter plots. Students then determine whether an equation represents a function and use graphs to solve problems. Students continue their study of functions by finding the slope of a line and graphing equations using two intercepts or the slope and one intercept. Students then learn to solve systems of linear equations by graphing. The chapter concludes with students graphing linear inequalities.

Lesson (Pages)	Lesson Objectives	NCTM Standards	State/Local Objectives
8-1 (372–377)	Use tables and graphs to represent relations and functions.	1–4, 6, 8, 10, 12	
8-2A (378)	Graph number pairs on a coordinate system to determine if there is a relationship.	1–6, 10, 12	
8-2 (379–384)	Construct and interpret scatter plots.	1–5, 8, 10	
8-3 (385–390)	Find solutions for relations with two variables and graph the solutions.	1–5, 8, 9	
8-3B (391)	Investigate graphs that are nonlinear.	1–6, 8	
8-4 (392–395)	Determine whether an equation represents a function and find functional values for a given function.	1–5, 8, 9	
8-5 (396–399)	Solve problems using graphs.	1–5, 8, 9, 12	
8-6 (400–404)	Find the slope of a line.	1–5, 8–10	
8-6B (405)	Use cylindrical objects to discover a connection between algebra and geometry.	1–6, 8	
8-7 (406–410)	Graph a linear equation using the *x*- and *y*-intercepts or using the slope and *y*-intercept.	1–5, 8, 9	
8-7B (411)	Use a graphing calculator to graph several functions to determine if any family traits exist.	1–6, 8	
8-8 (412–416)	Solve systems of linear equations by graphing.	1–5, 8, 9	
8-9A (417)	Use a graphing calculator to investigate the graphs of inequalities.	1–6, 8	
8-9 (418–422)	Graph linear inequalities.	1–5, 8, 9	

A complete, 1-page lesson plan is provided for each lesson in the ***Lesson Planning Guide***. Answer keys for each lesson are available in the ***Answer Key Masters***.

ORGANIZING THE CHAPTER

You may want to refer to the **Course Planning Calendar** on page T12 for detailed information on pacing.
PACING: Standard—13 days; **Honors**—12 days; **Block**—7 days

LESSON PLANNING CHART

Lesson (Pages)	Materials/ Manipulatives	Extra Practice (Student Edition)	Blackline Masters: Study Guide	Practice	Enrichment	Assessment and Evaluation	Math Lab and Modeling Math	Multicultural Activity	Tech Prep Applications	Graphing Calculator	Activity	Real-World Applications	Interactive Mathematics Tools Software	Teaching Transparencies	Group Activity Cards
8-1 (372–377)	geoboard* rubber bands*	p. 758	p. 62	p. 62	p. 62									8-1A 8-1B	8-1
8-2A (378)	string or yarn pins or tacks						p. 41								
8-2 (379–384)	checkerboard and checkers	p. 758	p. 63	p. 63	p. 63			p. 15				16	8-2	8-2A 8-2B	8-2
8-3 (385–390)		p. 759	p. 64	p. 64	p. 64	p. 211							8-3	8-3A 8-3B	8-3
8-3B (391)							p. 42								
8-4 (392–395)	counters*	p. 759	p. 65	p. 65	p. 65					p. 8			8-4	8-4A 8-4B	8-4
8-5 (396–399)		p. 759	p. 66	p. 66	p. 66	pp. 210, 211			p. 15		p. 22			8-5A 8-5B	8-5
8-6 (400–404)	two different-colored dice*	p. 760	p. 67	p. 67	p. 67						p. 8		8-6	8-6A 8-6B	8-6
8-6B (405)							p. 43								
8-7 (406–410)		p. 760	p. 68	p. 68	p. 68	p. 212	p. 66						8-7	8-7A 8-7B	8-7
8-7B (411)										p. 21					
8-8 (412–416)	grid paper drinking straw	p. 760	p. 69	p. 69	p. 69							17	8-8	8-8A 8-8B	8-8
8-9A (417)										p. 22					
8-9 (418–422)		p. 761	p. 70	p. 70	p. 70	p. 212	p. 83	p. 16	p. 16					8-9A 8-9B	8-9
Study Guide/ Assessment (424–427)						pp. 197–209, 213–215									

*Included in Glencoe's *Student Manipulative Kit* and *Overhead Manipulative Resources*.

ORGANIZING THE CHAPTER

All of the blackline masters in the Teacher's Classroom Resources are available on the ***Electronic Teacher's Classroom Resources*** CD-ROM.

OTHER CHAPTER RESOURCES

Student Edition
Investigation, Stadium Stampede, pp. 370–371
Chapter Opener, pp. 372–373
Earth Watch, pp. 376–377
Working on the Investigation, p. 384, 404

Teacher's Classroom Resources
Investigations and Projects Masters, pp. 57–60
Pre-Algebra Overhead Manipulative Resources, pp. 19–23

Technology
Test and Review Software (IBM & Macintosh)
CD ROM Activities

Professional Publications
Block Scheduling Booklet
Glencoe Mathematics Professional Series

OUTSIDE RESOURCES

Books/Periodicals
Math-O-Graphs: Critical Thinking Through Graphing, Donna K. Buck and Francis Hildebrand
Graphing Power: Activities for the TI-81 and TI-82 (middle school), *The Graphing Technology in Mathematics Project*
EQUALS Investigations Project: Does Alcohol Cause Cancer? (data analysis), Creative Publications

Videos/CD-ROMs
Algebra Math Video Series: Solving Simultaneous Equations and Inequalities Algebraically & Geometrically, ETA

Software
The Algebra Sketchbook, Sunburst
The Function Supposer, Sunburst
Green Globs and Graphing Equations, Sunburst

ASSESSMENT RESOURCES

Student Edition
Math Journal, pp. 402, 408, 420
Mixed Review, pp. 377, 384, 390, 395, 399, 404, 410, 416, 422
Self Test, p. 399
Chapter Highlights, p. 423
Chapter Study Guide and Assessment, pp. 424–426
Alternative Assessment, p. 427
Portfolio, p. 479
MindJogger Videoquiz, 8

Teacher's Wraparound Edition
5-Minute Check, pp. 372, 379, 385, 392, 396, 400, 406, 412, 418
Checking Your Understanding, pp. 373, 381, 387, 394, 398, 402, 408, 414, 420
Closing Activity, pp. 377, 384, 390, 395, 399, 404, 410, 416, 422
Cooperative Learning, pp. 373, 388, 401

Assessment and Evaluation Masters
Multiple-Choice Tests, Forms 1A (Honors), 1B (Average), 1C (Basic), pp. 197–202
Free-Response Tests, Forms 2A (Honors), 2B (Average), 2C (Basic), pp. 203–208
Performance Assessment, p. 209
Mid-Chapter Test, p. 210
Quizzes A–D, pp. 211–212
Standardized Test Practice, p. 213
Cumulative Review, pp. 214–215

ENHANCING THE CHAPTER

Examples of some of the materials for enhancing Chapter 8 are shown below.

Multicultural Activity Masters, pp. 15, 16

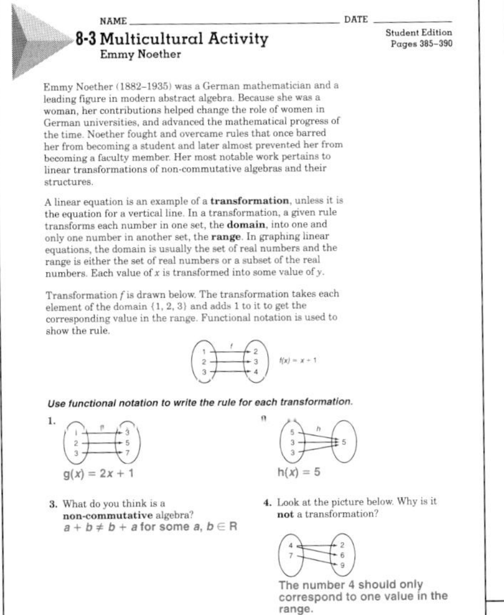

NAME ______ DATE ______

8-3 Multicultural Activity
Emmy Noether

Student Edition Pages 385–390

Emmy Noether (1882–1935) was a German mathematician and a leading figure in modern abstract algebra. Because she was a woman, her contributions helped change the role of women in German universities, and advanced the mathematical progress of the time. Noether fought and overcame rules that once barred her from becoming a student and later almost prevented her from becoming a faculty member. Her most notable work pertains to linear transformations of non-commutative algebras and their structures.

A linear equation is an example of a **transformation**, unless it is the equation for a vertical line. In a transformation, a given rule transforms each number in one set, the **domain**, into one and only one number in another set, the **range**. In graphing linear equations, the domain is usually the set of real numbers and the range is either the set of real numbers or a subset of the real numbers. Each value of x is transformed into some value of y.

Transformation f is drawn below. The transformation takes each element of the domain {1, 2, 3} and adds 1 to it to get the corresponding value in the range. Functional notation is used to show the rule.

$f(x) = x + 1$

Use functional notation to write the rule for each transformation.

1. $g(x) = 2x + 1$
2. $h(x) = 5$
3. What do you think is a **non-commutative** algebra? $a + b \neq b + a$ for some $a, b \in R$
4. Look at the picture below. Why is it **not** a transformation? The number 4 should only correspond to one value in the range.

Real-World Applications, 16, 17

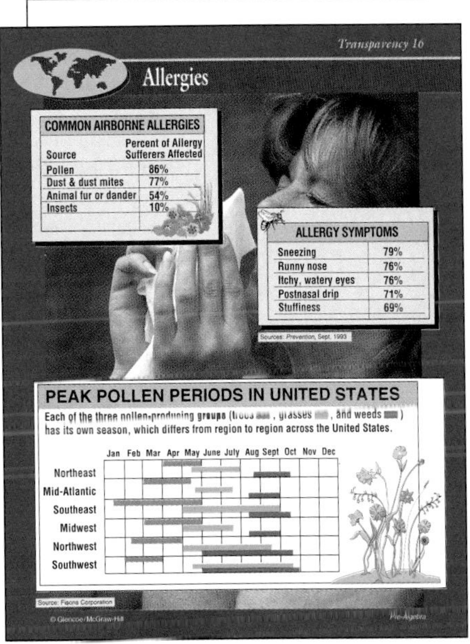

Graphing Calculator Masters, p. 8

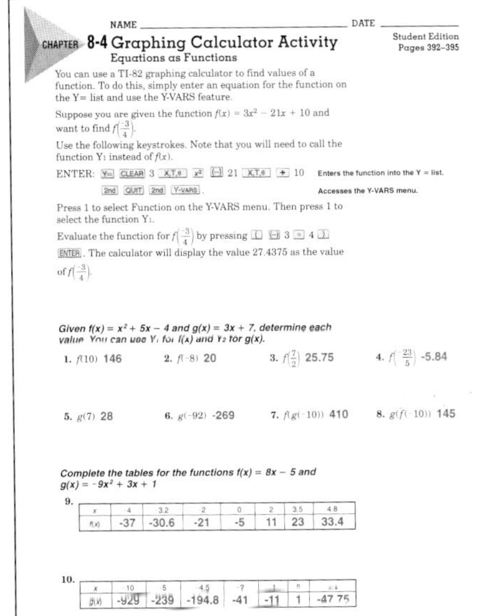

NAME ______ DATE ______

CHAPTER **8-4 Graphing Calculator Activity**
Equations as Functions

Student Edition Pages 392–395

You can use a TI-82 graphing calculator to find values of a function. To do this, simply enter an equation for the function on the Y= list and use the Y-VARS feature.

Suppose you are given the function $f(x) = 3x^2 - 21x + 10$ and want to find $f\left(\frac{-3}{4}\right)$.

Use the following keystrokes. Note that you will need to call the function Y_1 instead of $f(x)$.

ENTER: Y= CLEAR 3 X,T,θ x² (–) 21 X,T,θ + 10 — Enters the function into the Y = list.
2nd QUIT 2nd Y-VARS. — Accesses the Y-VARS menu.

Press 1 to select Function on the Y-VARS menu. Then press 1 to select the function Y_1.

Evaluate the function for $f\left(\frac{-3}{4}\right)$ by pressing ((–) 3 ÷ 4) ENTER. The calculator will display the value 27.4375 as the value of $f\left(\frac{-3}{4}\right)$.

Given $f(x) = x^2 + 5x - 4$ and $g(x) = 3x + 7$, determine each value. You can use Y_1 for $f(x)$ and Y_2 for $g(x)$.

1. $f(10)$ 146
2. $f(-8)$ 20
3. $f\left(\frac{7}{2}\right)$ 25.75
4. $f\left(-\frac{23}{5}\right)$ -5.84
5. $g(7)$ 28
6. $g(-92)$ -269
7. $f(g(-10))$ 410
8. $g(f(-10))$ 145

Complete the tables for the functions $f(x) = 8x - 5$ and $g(x) = -9x^2 + 3x + 1$

9.

x	-4	-3.2	-2	0	2	3.5	4.8
f(x)	-37	-30.6	-21	-5	11	23	33.4

10.

x	-10	-5	-4.5	[illegible]	[illegible]	[illegible]	[illegible]
g(x)	-929	-239	-194.8	-41	-11	1	-47.75

TECH PREP

Tech Prep Applications Masters, pp. 15, 16

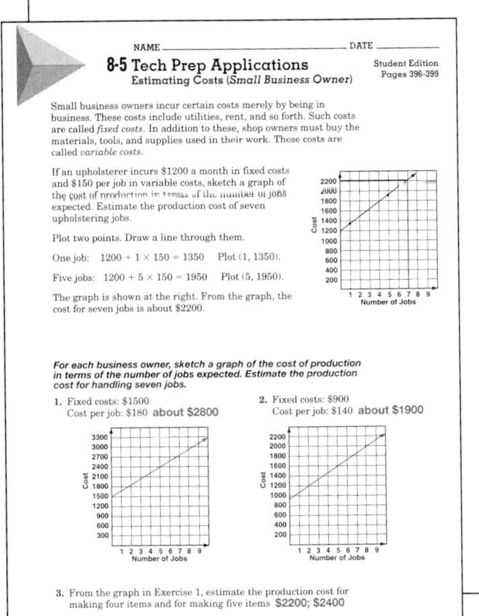

NAME ______ DATE ______

8-5 Tech Prep Applications
Estimating Costs (*Small Business Owner*)

Student Edition Pages 396-399

Small business owners incur certain costs merely by being in business. These costs include utilities, rent, and so forth. Such costs are called *fixed costs*. In addition to these, shop owners must buy the materials, tools, and supplies used in their work. These costs are called *variable costs*.

If an upholsterer incurs $1200 a month in fixed costs and $150 per job in variable costs, sketch a graph of the cost of production in terms of the number of jobs expected. Estimate the production cost of seven upholstering jobs.

Plot two points. Draw a line through them.

One job: $1200 + 1 \times 150 = 1350$ Plot (1, 1350).

Five jobs: $1200 + 5 \times 150 = 1950$ Plot (5, 1950).

The graph is shown at the right. From the graph, the cost for seven jobs is about $2200.

For each business owner, sketch a graph of the cost of production in terms of the number of jobs expected. Estimate the production cost for handling seven jobs.

1. Fixed costs: $1500
   Cost per job: $180 about $2800
2. Fixed costs: $900
   Cost per job: $140 about $1900
3. From the graph in Exercise 1, estimate the production cost for making four items and for making five items $2200; $2400

Activity Masters, pp. 8, 22

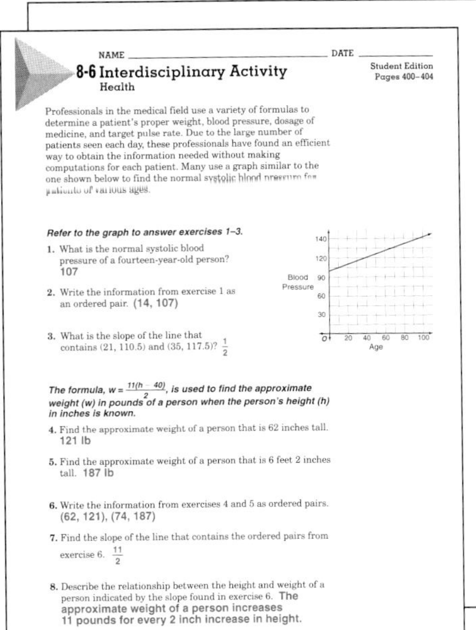

NAME ______ DATE ______

8-6 Interdisciplinary Activity
Health

Student Edition Pages 400–404

Professionals in the medical field use a variety of formulas to determine a patient's proper weight, blood pressure, dosage of medicine, and target pulse rate. Due to the large number of patients seen each day, these professionals have found an efficient way to obtain the information needed without making computations for each patient. Many use a graph similar to the one shown below to find the normal systolic blood pressure for patients of various ages.

Refer to the graph to answer exercises 1–3.

1. What is the normal systolic blood pressure of a fourteen-year-old person? 107
2. Write the information from exercise 1 as an ordered pair. (14, 107)
3. What is the slope of the line that contains (21, 110.5) and (35, 117.5)? $\frac{1}{2}$

The formula, $w = \frac{11(h - 40)}{2}$, is used to find the approximate weight (w) in pounds of a person when the person's height (h) in inches is known.

4. Find the approximate weight of a person that is 62 inches tall. 121 lb
5. Find the approximate weight of a person that is 6 feet 2 inches tall. 187 lb
6. Write the information from exercises 4 and 5 as ordered pairs. (62, 121), (74, 187)
7. Find the slope of the line that contains the ordered pairs from exercise 6. $\frac{11}{2}$
8. Describe the relationship between the height and weight of a person indicated by the slope found in exercise 6. The approximate weight of a person increases 11 pounds for every 2 inch increase in height.

Math Lab and Modeling Math Masters, pp. 66, 83

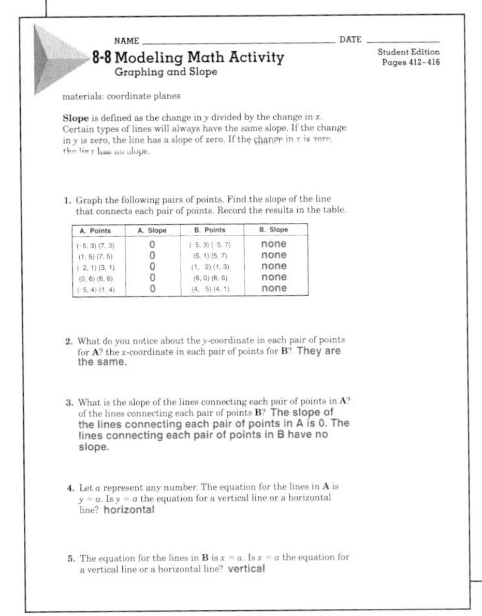

NAME ______ DATE ______

8-8 Modeling Math Activity
Graphing and Slope

Student Edition Pages 412–416

materials: coordinate planes

Slope is defined as the change in y divided by the change in x. Certain types of lines will always have the same slope. If the change in y is zero, the line has a slope of zero. If the change in x is zero, the line has no slope.

1. Graph the following pairs of points. Find the slope of the line that connects each pair of points. Record the results in the table.

A. Points	A. Slope	B. Points	B. Slope
(-5, 3) (7, 3)	0	(-5, 3) (-5, 7)	none
(1, 5) (7, 5)	0	(5, 1) (5, 7)	none
(-2, 1) (3, 1)	0	(1, -2) (1, 3)	none
(0, 6) (6, 6)	0	(6, 0) (6, 6)	none
(-5, 4) (1, 4)	0	(4, -5) (4, 1)	none

2. What do you notice about the y-coordinate in each pair of points for **A**? the x-coordinate in each pair of points for **B**? They are the same.
3. What is the slope of the lines connecting each pair of points in **A**? of the lines connecting each pair of points **B**? The slope of the lines connecting each pair of points in A is 0. The lines connecting each pair of points in B have no slope.
4. Let a represent any number. The equation for the lines in **A** is $y = a$. Is $y = a$ the equation for a vertical line or a horizontal line? horizontal
5. The equation for the lines in **B** is $x = a$. Is $x = a$ the equation for a vertical line or a horizontal line? vertical

MAKING MATHEMATICS RELEVANT

This two-page introduction to the chapter provides students with an opportunity to explore other disciplinary topics and their applications to mathematics.

Background Information

Math and Transportation in the News Honda's first guidance system (1981) used transparent map sheets that the driver had to keep and insert into the system by hand. The company's newest systems are much more sophisticated. The system pioneered by Oldsmobile uses a small display screen attached to the dashboard. It allows the driver to select a specific address, a street intersection, or even a point of interest such as a tourist attraction or hotel. The driver can choose to receive step-by-step instructions along the way or to view an area map that shows the car's location. The map includes a zoom feature for better viewing.

CHAPTER 8 Functions and Graphing

TOP STORIES in Chapter 8

In this chapter, you will:

- determine whether relations are functions,
- graph data in scatter plots,
- graph linear equations using slope and intercepts,
- solve systems of equations, and
- graph inequalities.

MATH AND TRANSPORTATION IN THE NEWS

Finding your way from here to there

Source: Popular Electronics, October, 1994

The days of the map and compass may be coming to an end. Technology is making it easier to find your way in a car. Some electronics companies and car manufacturers are making on-board navigation systems to assist drivers. The car navigation systems rely on computers, databases of map information, and satellites in the Navstar Global Positioning System (GPS). Geometry and precise timing allow the computer to tell you a route to your destination and give visual displays, and computer-generated voice instructions guide you as you go. Although this system is new to consumers in the United States, it has been a popular accessory on cars sold in Japan for a number of years.

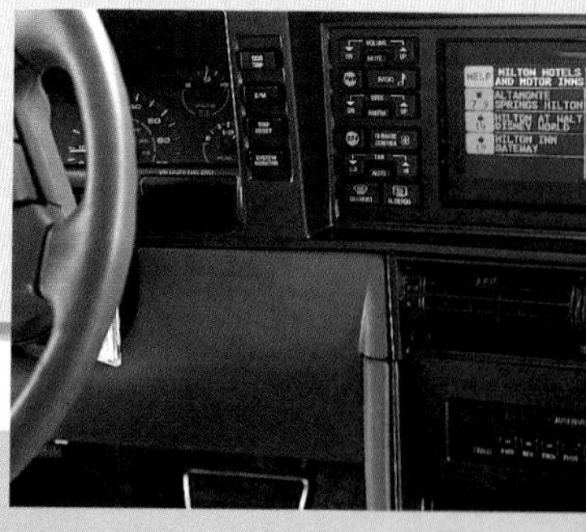

Putting It into Perspective

1200 | 1800 | 1850

1150
Chinese and Mediterranean navigators use magnetic compasses to guide ships.

1769
Steam-powered car is built by French engineer Nicolas-Joseph Cugnot.

Putting It into Perspective

Installation of the Navistar Global Positioning System began with the launch of one satellite into orbit in 1978. The complete system now consists of 21 satellites and three spares. The satellites are positioned in six different orbits about 10,900 miles above Earth. They are positioned such that at least four of them are always available to a navigator located at any point on Earth.

On the Lighter Side

"George and I haven't traveled by car since the map-folding incident of '84."

Statistical Snapshot

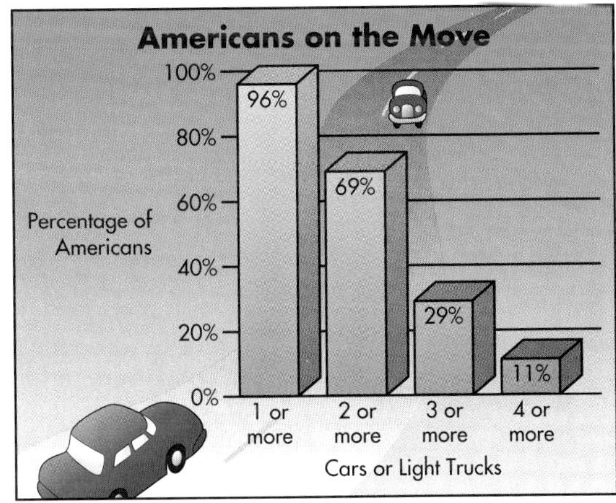

Source: *Ameripoll*, Maritz Marketing Research

interNET CONNECTION For up-to-date information on the Global Positioning System, visit: **www.glencoe.com/sec/math/prealg/mathnet**

1978 The U.S. Government launches the first satellite of the Navstar system.

1994 Oldsmobile is first in North America to offer an onboard navigation system.

1900 1950 2000

o Daimler and nz develop ners of modern e engines.

1971 Leon Sullivan becomes the first African-American to serve on the board of directors of General Motors.

On the Lighter Side

Electronic navigation will ease the burden of traveling for those who find it difficult to use maps or other forms of directions. It will also make navigation easier for drivers traveling alone, and it can be especially useful for travel at night or in bad weather when visibility is limited.

Statistical Snapshot

The diagrams show the two types of visual displays available in the system used by Oldsmobile. The first, displays an arrow and numbers showing the direction of and distance to the next turn along with the remaining distance to the destination. The second, displays a map with an arrow showing the position of the car as it travels. Which format do you prefer? Why?

Investigations and Projects Masters, p. 57

NAME ______________________ DATE ________

CHAPTER 8 Project A

Two-Wheeler Destination

Think of a place you would like to visit that's at least 150 but not more than 500 miles away—a national park, a theme park, the home of a rock star, or even a store you know that has the best ice cream in the world. You will plan a bicycle trip there for you and your friends.

1. First choose your destination. You'll need a road atlas and a map of the local area. Determine your route. It can be the most direct route, the most scenic, or even one that passes by your cousin's house. But, keep in mind that you'll need to stay off highways.

2. How long it will take to ride to your destination? Use one or more resources, interview a long-distance cyclist, or conduct an experiment to find a reasonable average speed. Then complete your route plan by dividing the ride into days and choosing places to spend the night.

3. Next, figure out all costs of your journey. Include meals, lodging, any entrance fees, and anything else you might need such as film for a camera, a flashlight, or a sleeping bag.

4. Now that you've gathered your information, prepare an itinerary for your trip. The itinerary should include
   - a map showing your route including stops along the way;
   - a summary of your time, speed, and distance traveled;
   - an organized list showing all your expenses; and
   - a report explaining how you made your plan, including what was hard to figure out, what was easy to do, and what was the most fun.

5. Present your completed itinerary to the class.

Cooperative Learning

Chapter Projects Two chapter projects are included in the *Investigations and Project Masters*. In Chapter Project A, students extend the topic in the Chapter Opener. In Chapter Project B, students explore graphs in newspapers or magazines. A student page and a parent letter are provided for each Chapter Project.

interNET CONNECTION

Glencoe has made every effort to ensure that the website links for *Pre-Algebra* at www.glencoe.com/sec/math/prealg/mathnet are current and contain appropriate content. However, these website links are not under Glencoe's control.

8-1 LESSON NOTES

NCTM Standards: 1-4, 6, 8, 10, 12

Instructional Resources

- Study Guide Master 8-1
- Practice Master 8-1
- Enrichment Master 8-1
- Group Activity Card 8-1

Transparency 8-1A contains the 5-Minute Check for this lesson; **Transparency 8-1B** contains a teaching aid for this lesson.

Recommended Pacing	
Standard Pacing	Day 1 of 13
Honors Pacing	Day 1 of 12
Block Scheduling*	Day 1 of 7

*For more information on pacing and possible lesson plans, refer to the ***Block Scheduling Booklet***.

1 FOCUS

5-Minute Check *(over Chapter 7)*

Solve each equation or inequality.

1. $3x - 2 = 13$ 5
2. $6 - 5x = 21$ −3
3. $\frac{(2x + 7)}{3} = x + 3$ −2
4. $-12x \geq -24$ $x \leq 2$
5. $7x < 2x - 25$ $x < -5$

Motivating the Lesson

Questioning Ask students how they think the United States compares to other countries in terms of the amount of annual household waste. Given the populations of several countries, have students determine two different ways to order the countries in terms of annual domestic waste.

8-1 Relations and Functions

Setting Goals: *In this lesson, you'll use tables and graphs to represent relations and functions.*

Modeling a Real-World Application: Ecology

Have you ever thought you couldn't make a difference in landfills by recycling? Think again! The table at the right shows how the United States compares to other nations in terms of the amount of annual household waste. These five countries produce the most waste in the world.

Country	Annual Domestic Waste (thousands of tons)	Equivalent per Person (pounds)
France	15,500	634
Great Britain	15,816	620
Japan	40,225	634
United States	200,000	1925
West Germany	20,780	741

Source: *Encarta*, 1994

This data could also be displayed using a set of ordered pairs. Each first coordinate would be the annual domestic waste, and each second coordinate would be the equivalent per person.

{(15,500, 634), (15,816, 620), (40,225, 634), (200,000, 1925), (20,780, 741)}

Learning the Concept

A **relation** is a set of ordered pairs, like the set shown above. The set of first coordinates is called the **domain** of the relation. The set of second coordinates is called the **range** of the relation. So, the domain of the relation above is {15,500, 15,816, 40,225, 200,000, 20,780}, and the range is {634, 620, 1925, 741}. *Notice that 634 is listed only once.*

THINK ABOUT IT

What is the domain and range of this relation?

D: {−3, 0, 4}; R: {1, −2}

A relation can also be modeled by a table or a graph. For example, the relation {(−3, 1), (0, −2), (4, −2)} can be represented in each of the ways shown at the right.

Table

x	y
−3	1
0	−2
4	−2

Graph

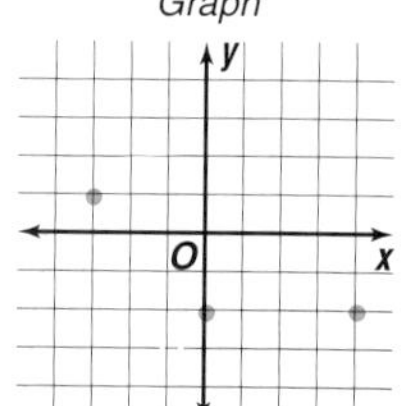

Alternative Learning Styles

Auditory As students work through the lesson, encourage them to orally relate the x value with a move to the right or left and the y value with a move up or down. Thus, the domain determines moves to the left or right while the range tells how far up or down to move.

Example 1 **Express the relation shown in the graph below as a set of ordered pairs. Then determine the domain and range.**

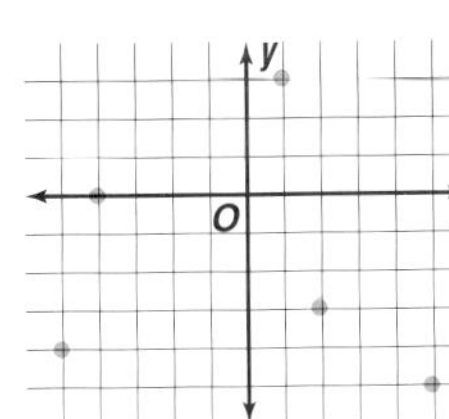

The set of ordered pairs for the relation is $\{(1, 3), (5, -5), (-4, 0), (2, -3), (-5, -4)\}$.

The domain is $\{1, 5, -4, 2, -5\}$.

The range is $\{3, -5, 0, -3, -4\}$.

The relation in Example 1 is a special type of relation called a **function**.

Definition of a Function — A function is a relation in which each element of the domain is paired with exactly one element in the range.

Example 2 **Graph each relation. Then determine whether each relation is a function.**

a.

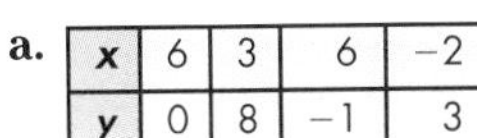

x	6	3	6	−2
y	0	8	−1	3

b.

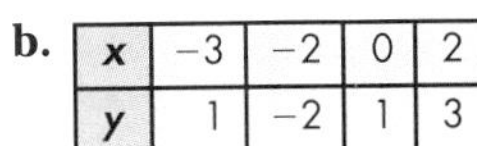

x	−3	−2	0	2
y	1	−2	1	3

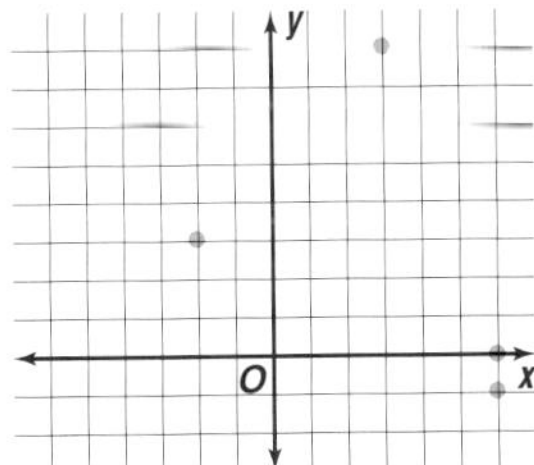

This relation is not a function since the element 6 in the domain is paired with two elements, 0 and -1, in the range.

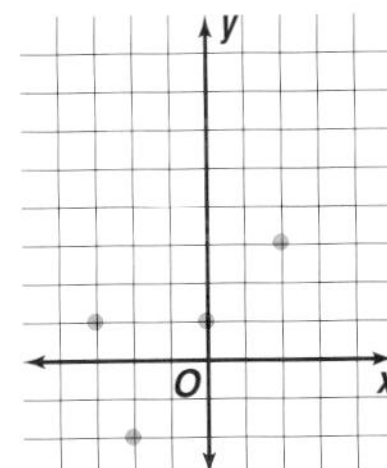

Since each element of the domain is paired with exactly one element in the range, this relation is a function.

You can use the **vertical line test** to determine if a relation is a function. The relation $\{(-2, 5), (0, 4), (4, 2), (6, 1), (8, 0)\}$ is graphed at the right. Place a straightedge at the left of the graph to represent a vertical line. Slowly move the straightedge to the right across the graph.

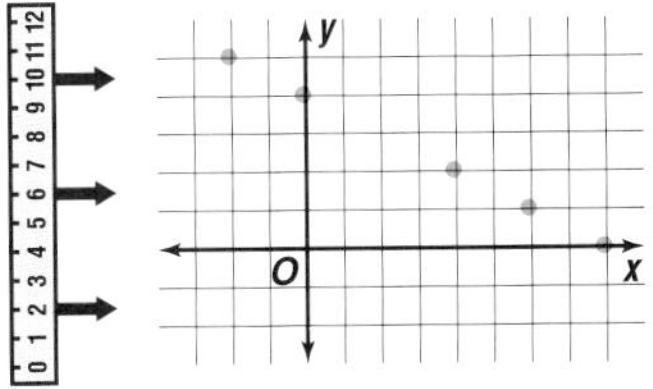

For each value of x, this vertical line passes through no more than one point on the graph. This is true for *every* function.

2 TEACH

In-Class Examples

For Example 1
Express the relation shown in the graph below as a set of ordered pairs. Then determine the domain and range.

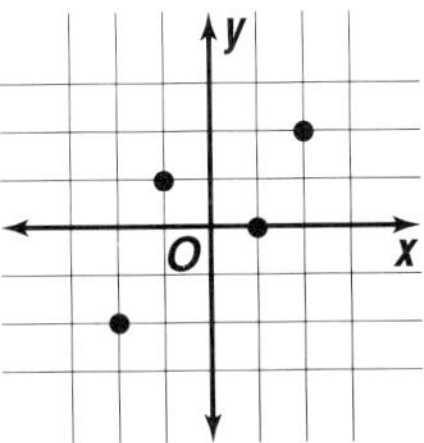

{(2, 2), (-2, -2), (-1, 1), (1, 0)}. The domain is {-2, -1, 1, 2}. The range is {-2, 0, 1, 2}.

For Example 2
Graph each relation. Then determine if each relation is a function.

a.

x	-1	0	4	5
y	-1	0	-4	-5

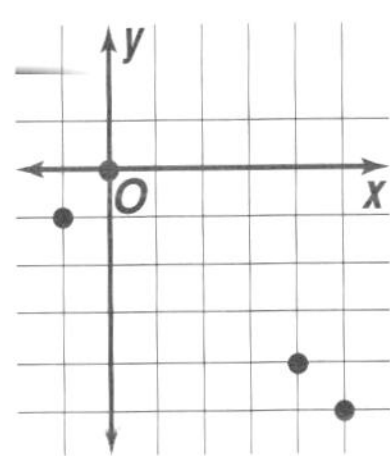

Yes; it is a function because each element of the domain is paired with exactly one element in the range.

b.

x	1	0	1	2
y	4	0	-2	-4

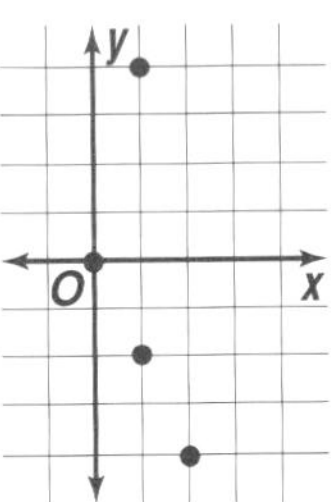

No; it is not a function because the element 1 in the domain maps to both 4 and -2 in the range.

Cooperative Learning

Formations Have students sit in rows and columns, labeled as a grid. Identify seats by (row, column). Have groups of students stand, such as all whose name begins with a vowel or all students wearing red. Students determine whether the points represented by standing students represent a function. Challenge them to find descriptions that will have students stand so a function is formed. These may get complicated, such as all students wearing white tennis shoes and glasses, but not wearing a red shirt. For more information on this strategy, see *Cooperative Learning in the Mathematics Classroom*, p. 30.

In-Class Example

For Example 3

The graph below shows the average time students take to get to school, based on the distance they live from home. Is the relation a function? Explain.

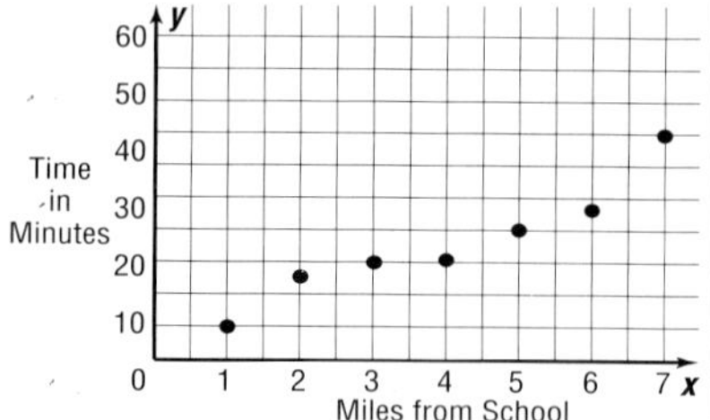

Yes; it is a function because for every *x*-coordinate, there is a unique *y*-coordinate.

3 PRACTICE/APPLY

Checking Your Understanding

Exercises 1–15 are designed to help you assess your students' understanding through reading, writing, speaking, and modeling. You should work through Exercises 1–4 with your students and then monitor their work on Exercises 5–15.

Additional Answers

2. The domain is the set of all first coordinates while the range is the set of all second coordinates.
3. Marita is correct because each *x* is paired with exactly one *y*.
4.

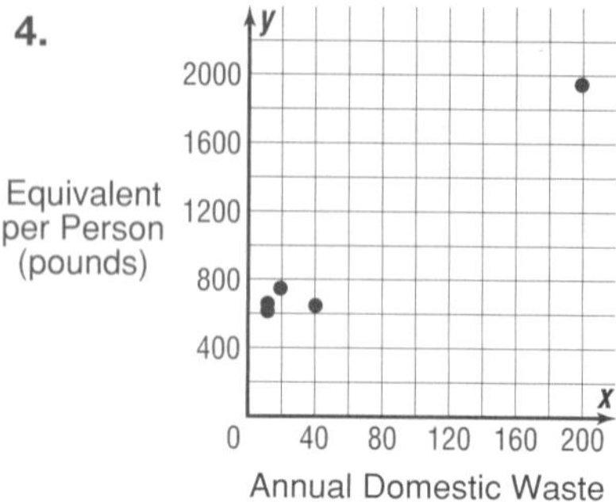

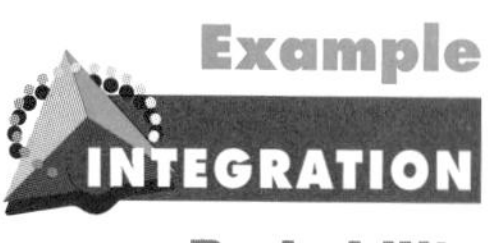

The graph on the left shows the possible outcomes when two six-sided dice are tossed. The graph on the right shows those outcomes when the faces on the dice show a sum of 7. Is either relation a function? Explain.

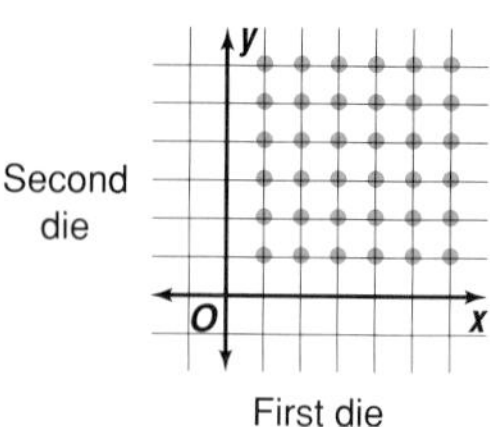

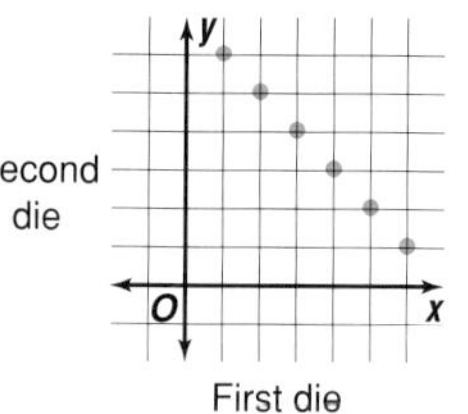

The relation graphed on the left is *not* a function since it does not pass the vertical line test.

The relation graphed on the right is a function since any vertical line passes through no more than one point of the graph of the relation.

Checking Your Understanding

Communicating Mathematics

Read and study the lesson to answer these questions. 2–4. See margin.

1. **Describe** three different ways to show a relation. Give an example of a relation shown in each way. See Solutions Manual.
2. **Explain** the difference between the domain and the range of a relation.
3. **You Decide** Marita says that {(9, 3), (0, 3), (−5, 0), (−1, 11)} is a function. Terrica disagrees. Who is correct and why?
4. **Graph** the relation described by the annual waste figures in the application at the beginning of the lesson.

Guided Practice

State the domain and range of each relation.

5. {(−1.3, 1), (4, −3.9),(−2.4, 3.6)} **{−1.3, 4, −2.4}; {1, −3.9, 3.6}**
6. $\left\{\left(-\frac{1}{2}, -\frac{1}{4}\right), \left(1\frac{1}{2}, -\frac{2}{3}\right), \left(3, -\frac{2}{5}\right), \left(5\frac{1}{4}, 6\frac{2}{7}\right)\right\}$

6. $\left\{-\frac{1}{2}, 1\frac{1}{2}, 3, 5\frac{1}{4}\right\}$; $\left\{-\frac{1}{4}, -\frac{2}{3}, -\frac{2}{5}, 6\frac{2}{7}\right\}$

Express the relation shown in each table or graph as a set of ordered pairs. Then state the domain and range of the relation. 7–8. See margin.

7.

x	y
−1	3
0	6
4	−1
7	2

8.

x	y
2	4
0	2
−2	0
−4	2

Reteaching

Using Manipulatives Have students set up a grid on a geoboard, using small rings or rubber bands to mark points on the grid. Ask them to illustrate a relation that is not a function and a relation that is a function.

Group Activity Card 8-1

Functional Or Not? Group Activity **8-1**

MATERIALS: 2 decks of 16 number cards (see back)

Use one deck of number cards as the domain and the other as the range. Choose a card from each deck and write the numbers as an ordered pair. Repeat this step four more times. Represent the ordered pairs in a table. Graph the ordered pairs. Determine if the relation is a function. Repeat this process four more times.

 Pre-Algebra

9. 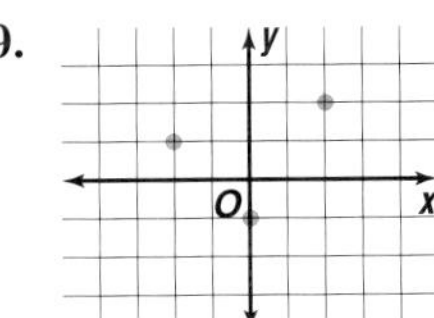

10.

9–10. See Solutions Manual.

Determine whether each relation is a function.

11. $\{(5, -2), (3, 2), (4, -1), (-2, 2)\}$ yes

12.

x	0	2	0
y	4	3	8

no

13. 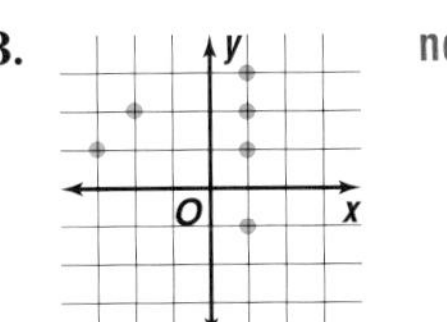 no

14. 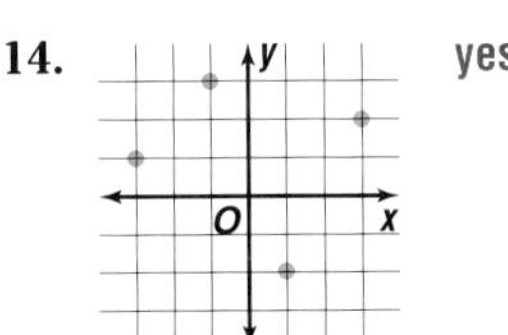 yes

15. **Food** Lisle buys a dozen bagels for a morning meeting. She knows that the staff prefers blueberry bagels over plain bagels. If she buys at least twice as many blueberry bagels as plain bagels, write a relation to show the different possibilities. (*Hint:* Let the relation consist of ordered pairs of the form (blueberry bagels, plain bagels).) $\{(8, 4), (9, 3), (10, 2), (11, 1), (12, 0)\}$

Exercises: Practicing and Applying the Concept

Independent Practice

A

16. $\{5, -2, 2\}$; $\{-4, 3, -1\}$
17. $\{-1, 4, 2, 1\}$; $\{6, 2, 36\}$
18. $\{3.1, 7, -3.9\}$; $\{2, -4.4, -8.8\}$

B

19. $\{1.4, -2, 4, 6\}$; $\{3, 9.6, 4, -2.7\}$
20. $\left\{\frac{2}{5}, 98\frac{3}{5}, -4\right\}$; $\left\{\frac{3}{4}, 37\frac{1}{2}, -\frac{7}{12}\right\}$
21. $\left\{-\frac{1}{2}, 4\frac{2}{3}, -12\frac{3}{8}\right\}$ $\left\{\frac{1}{3}, -17, 66\right\}$

Write the domain and range of each relation.

16. $\{(5, -4), (-2, 3), (5, -1), (2, 3)\}$
17. $\{(-1, 6), (4, 2), (2, 36), (1, 6)\}$
18. $\{(3.1, 2), (7, -4.4), (-3.9, -8.8)\}$
19. $\{(1.4, 3), (-2, 9.6), (4, 4), (6, -2.7)\}$
20. $\left\{\left(\frac{2}{5}, \frac{3}{4}\right), \left(98\frac{3}{5}, 37\frac{1}{2}\right), \left(-4, -\frac{7}{12}\right)\right\}$
21. $\left\{\left(-\frac{1}{2}, \frac{1}{3}\right), \left(4\frac{2}{3}, -17\right), \left(-12\frac{3}{8}, 66\right)\right\}$

Express the relation shown in each table or graph as a set of ordered pairs. Then state the domain and range of the relation.

22.

x	y
0	5
2	3
1	−4
−3	3
−1	−2

23.

x	y
−4	−2
−2	1
0	2
1	−3
3	1

24.

x	y
−1	5
−2	5
−2	4
−2	1
−6	1

25.

x	y
5	4
2	8
−7	9
2	12
5	14

26. 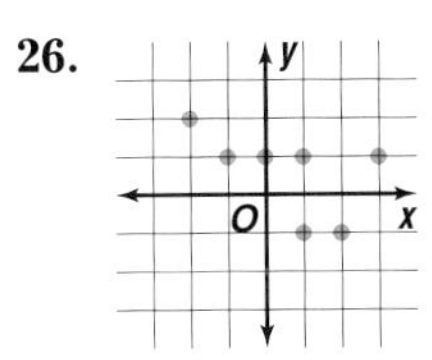

27. 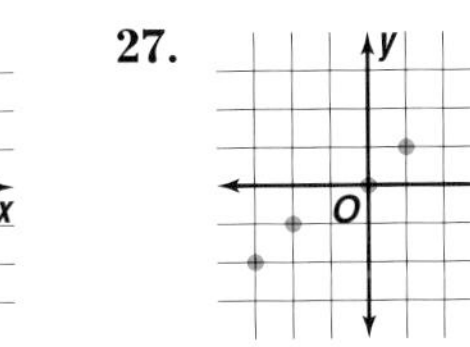

28. 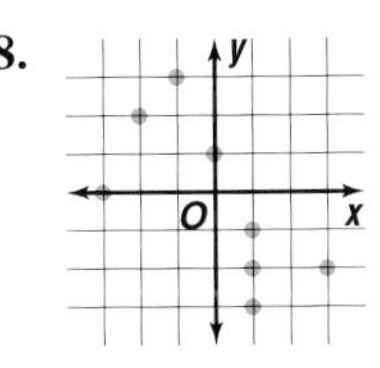

22–28. See margin.

Assignment Guide

Core: 17–43 odd, 44–51
Enriched: 16–38 even, 40–51

For **Extra Practice**, see p. 758.

Additional Answers

7. $\{(-1, 3), (0, 6), (4, -1), (7, 2)\}$; $D = \{-1, 0, 4, 7\}$, $R = \{3, 6, -1, 2\}$
8. $\{(2, 4), (0, 2), (-2, 0), (-4, 2)\}$; $D = \{2, 0, -2, -4\}$, $R = \{4, 2, 0\}$
22. $\{(0, 5), (2, 3), (1, -4), (-3, 3), (-1, -2)\}$; $D = \{0, 2, 1, -3, -1\}$, $R = \{5, 3, -4, -2\}$
23. $\{(-4, -2), (-2, 1), (0, 2), (1, -3), (3, 1)\}$; $D = \{-4, -2, 0, 1, 3\}$, $R = \{-2, 1, 2, -3\}$
24. $\{(-1, 5), (-2, 5), (-2, 4), (-2, 1), (-6, 1)\}$; $D = \{-1, -2, -6\}$, $R = \{5, 4, 1\}$
25. $\{(5, 4,), (2, 8), (-7, 9), (2, 12), (5, 14)\}$; $D = \{5, 2, -7\}$, $R = \{4, 8, 9, 12, 14\}$
26. $\{(-2, 2), (-1, 1), (0, 1), (1, 1), (1, -1), (2, -1), (3, 1)\}$; $D = \{-2, -1, 0, 1, 2, 3\}$; $R = \{2, 1, -1\}$
27. $\{(-3, -2), (-2, -1), (0, 0), (1, 1)\}$; $D = \{-3, -2, 0, 1\}$, $R = \{-2, -1, 0, 1\}$
28. $\{(-3, 0), (-2, 2), (-1, 3), (0, 1), (1, -1), (1, -2), (1, -3), (3, -2)\}$; $D = \{-3, -2, -1, 0, 1, 3\}$, $R = \{0, 2, 3, 1, -1, -2, -3\}$

Study Guide Masters, p. 62

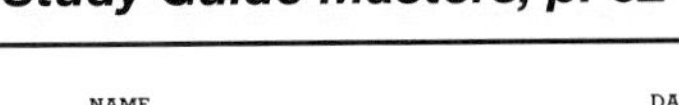

NAME ______________________ DATE ________

8-1 Study Guide
Relations and Functions

Student Edition Pages 372–377

A **relation** is a set of ordered pairs. The set of first coordinates is called the **domain**. The second set of coordinates is called the **range**.

$\{(0, 1), (3, 4), (5, 6)\}$

The domain is $\{0, 3, 5\}$.
The range is $\{1, 4, 6\}$.

x	1	1	4
y	−1	3	7

The domain is $\{1, 4\}$.
The range is $\{-1, 3, 7\}$.

x	y
2	3
6	−1
2	−1

The domain is $\{2, 6\}$.
The range is $\{-1, 3\}$.

A **function** is a relation in which each element of the domain is paired with exactly one element in the range.

x	3	2	−6	−1
y	5	1	2	0

This relation is a function.

Each element of the domain is paired with exactly one element in the range.

$\{(4, 1), (-2, 6), (4, 3), (1, 0)\}$

This relation is not a function.

The element 4 in the domain is paired with two elements, 3 and 1, in the range.

Write the domain and range of each relation.

1. $\{(0, 7), (5, 3), (3, 7), (2, 7)\}$
$D = \{0, 5, 3, -2\}$; $R = \{7, -3\}$

2. $\left\{\left(-\frac{1}{4}, \frac{1}{3}\right), \left(2\frac{2}{3}, -10\right), \left(-6\frac{1}{2}, 42\right)\right\}$ $D = \left\{-\frac{1}{4}, 2\frac{2}{3},\right.$
$R = \left\{\frac{1}{3}, -10, 4\right.$

Express the relation shown in each table or graph as a set of ordered pairs. Then state the domain and range of the relation.

3.

x	y
2	−3
−4	−2
1	5
3	5
2	−1

$\{(2, -3) (-4, -2) (1, 5), (3, 5), (2, -1)\}$
$D = \{2, -4, 1, 3\}$
$R = \{-3, -2, 5, -1\}$

4. 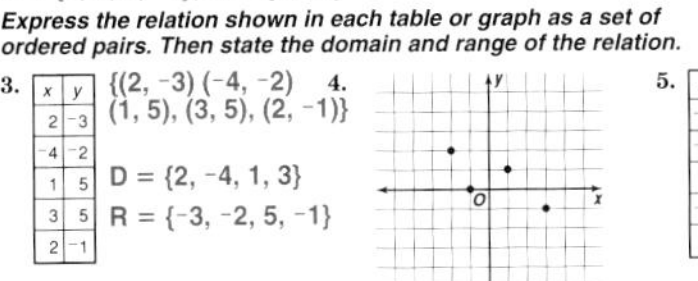

5.

x	y
−1	0
−1	4
3	0
−1	3
2	4

4. $\{(-1, 0), (-2, 2) (1, 1), (3, -1)\}$
$D = \{-1, -2, 1, 3\}$
$R = \{0, 2, 1, -1\}$

5. $\{(-1, 0), (-1, 4) (3, 0), (-1, 3), (2, -$
$D = \{-1, 3, 2\}$
$R = \{0, 4, 3, -4\}$

Determine whether each relation is a function.

6. 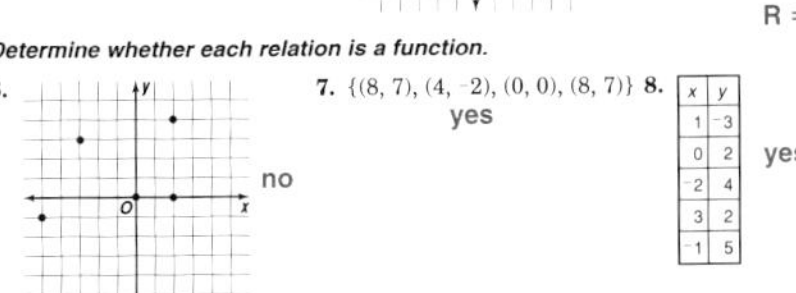 no

7. $\{(8, 7), (4, -2), (0, 0), (8, 7)\}$ yes

8.

x	y
1	−3
0	2
−2	4
3	2
−1	5

yes

Classroom Vignette

"A good extension activity for this lesson is to have students create relations that meet different criteria. For example, have them write relations where the number of elements in the domain and the range are the same, where there are more elements in the domain, and where there are more elements in the range."

Cleo Meek
North Carolina Department of Public Instruction, Retired
Raleigh, NC

Additional Answers

42a. If the cars were considered the domain, then the relation is a function since each number in the domain has exactly one corresponding value in the range.

42b. If the trucks were considered the domain, then the relation would not be a function since 650 trucks is paired with two different numbers of cars.

Practice Masters, p. 62

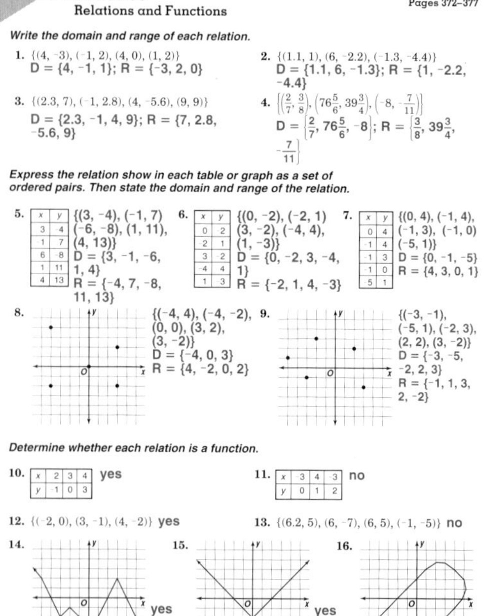

NAME ______________ DATE ______________

8-1 Practice
Relations and Functions

Student Edition Pages 372–377

Write the domain and range of each relation.

1. {(4, −3), (−1, 2), (4, 0), (1, 2)}
D = {4, −1, 1}; R = {−3, 2, 0}

2. {(1.1, 1), (6, −2.2), (−1.3, −4.4)}
D = {1.1, 6, −1.3}; R = {1, −2.2, −4.4}

3. {(2.3, 7), (−1, 2.8), (4, −5.6), (9, 9)}
D = {2.3, −1, 4, 9}; R = {7, 2.8, −5.6, 9}

4. $\left\{\left(\frac{2}{7}, \frac{3}{8}\right), \left(76\frac{5}{6}, 39\frac{3}{4}\right), \left(-8, -\frac{7}{11}\right)\right\}$
D = $\left\{\frac{2}{7}, 76\frac{5}{6}, -8\right\}$; R = $\left\{\frac{3}{8}, 39\frac{3}{4}, -\frac{7}{11}\right\}$

Express the relation show in each table or graph as a set of ordered pairs. Then state the domain and range of the relation.

5.

x	y
3	−4
−1	7
6	−8
1	11
4	13

{(3, −4), (−1, 7) (−6, −8), (1, 11), (4, 13)}
D = {3, −1, −6, 1, 4}
R = {−4, 7, −8, 11, 13}

6.

x	y
0	−2
−2	1
3	−2
−4	4
1	−3

{(0, −2), (−2, 1) (3, −2), (−4, 4), (1, −3)}
D = {0, −2, 3, −4, 1}
R = {−2, 1, 4, −3}

7.

x	y
0	4
−1	4
−1	3
−1	0
−5	1

{(0, 4), (−1, 4), (−1, 3), (−1, 0) (−5, 1)}
D = {0, −1, −5}
R = {4, 3, 0, 1}

8. {(−4, 4), (−4, −2), (0, 0), (3, 2), (3, −2)}
D = {−4, 0, 3}
R = {4, −2, 0, 2}

9. {(−3, −1), (−5, 1), (−2, 3), (2, 2), (3, −2)}
D = {−3, −5, −2, 2, 3}
R = {−1, 1, 3, 2, −2}

Determine whether each relation is a function.

10.

x	2	3	4
y	−1	0	3

yes

11.

x	−3	4	−3
y	0	1	2

no

12. {(−2, 0), (3, −1), (4, −2)} yes

13. {(6.2, 5), (6, −7), (6, 5), (−1, −5)} no

14. yes

15. yes

16. no

Determine whether each relation is a function.

29.

x	1	2	3
y	3	4	5

yes

30.

x	3	5	7
y	−4	−6	−8

yes

31.

x	1	2	1
y	3	5	−7

no

32.

x	−2	0	2
y	3	1	3

yes

33. {(4, 0), (6, 0), (8, 0)} yes

34. {(0, 4), (0, 6), (0, 8)} no

35. {(5, 4), (5, −6), (5, 4), (0, 4)} no

36. {(−2, 3), (4, 7), (24, −6), (5, 4)} yes

37. no

38. no

39. no

Critical Thinking

The *inverse* of any relation is obtained by switching the coordinates in each ordered pair of the relation. Determine whether the inverse of each relation is a function. 41. yes

40. {(4, 0), (5, 1), (6, 2), (6, 3)} yes

41. {(5, 5), (−3, −1), (−2, 1), (−4, 0)}

Applications and Problem Solving

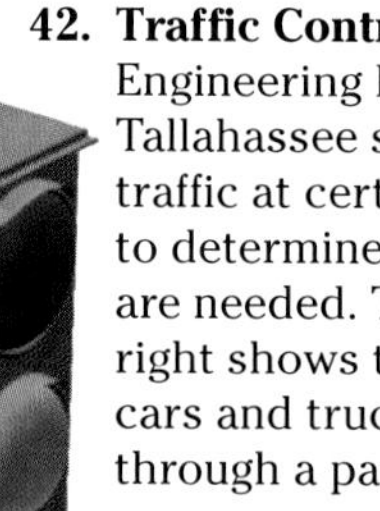

42. Traffic Control The Traffic Engineering Division for Tallahassee studies the flow of traffic at certain intersections to determine if traffic signals are needed. The table at the right shows the actual count of cars and trucks that passed through a particular intersection on one weekday.

Time	Cars	Trucks
3:00P.M.–3:30P.M.	1330	650
3:30P.M.–4:00P.M.	2000	600
4:00P.M.–4:30P.M.	2100	630
4:30P.M.–5:00P.M.	2300	650
5:00P.M.–5:30P.M.	2550	750
5:30P.M.–6:00P.M.	2220	700

42a–b. See margin.

a. Use the information in the second and third columns of the table to illustrate a relation that is a function.

b. Use the information to illustrate a relation that is not a function. Explain.

EARTH WATCH

SAVING THE WHALES

Fishing and whaling have long been important industries in many countries. In recent years, environmentalists have become concerned with the effect that fishing and whaling have had on the populations of these animals. One of the problems associated with protecting whales and other marine animals has been deciding who owns the ocean in which they live. Exclusive Economic Zones, or EEZs, are being established so that countries have rights and responsibilities for parts of the oceans near their shores. Nations can impose quotas or bans on fishing in their EEZ.

Year	Fish catch (metric tons)
1970	66,969,420
1972	63,855,330
1974	68,185,310
1976	71,509,720
1978	73,422,560
1980	75,587,120
1982	80,050,870
1984	87,685,160
1986	96,666,970
1988	103,149,600
1990	101,755,000

Source: World Resources Institute

Extension

Using Cooperative Groups Students explored inverses in Exercises 40–41. Challenge each group of students to identify a relation with a set of ordered pairs so that both the relation and its inverse are functions.

Other possibilities are identifying a relation that is a function with an inverse that is not, a relation that is not a function with an inverse that is, and a non-function relation with a non-function inverse.

43b. more than $30 and less than or equal to $70

✓ Choose

Estimation
Mental Math
Calculator
Paper and Pencil

43. **Shipping Rates** When you order something from a mail-order company, you usually pay a shipping fee. The chart at the right relates shipping costs to the total price of merchandise ordered from *Lands' End*.

Total Price of Merchandise	Shipping Costs
$0.00–$30.00	$4.25
$30.01–$70.00	$5.75
$70.01 and over	$6.95

a. What is the shipping cost for an order of merchandise totaling $75? **$6.95**

b. For what price of merchandise is the shipping cost $5.75?

c. Does the chart represent a function? Explain. **See margin.**

Mixed Review

44. **Consumer Awareness** Which has a greater capacity, a 2-liter bottle of cola or a six-pack of cans each containing 354 mL of the same cola? (Lesson 7-8) **The six-pack has 124 mL more.**

45. Solve $-2.2 < \frac{b}{-10} - 2.4$. (Lesson 7-6) $-2 > b$

46. **Physics** A ball rebounds $\frac{2}{3}$ of its height after every fall. If it is dropped from a height of 48 feet, how high will it bounce at the end of the third bounce? (Lesson 6-8) $14\frac{2}{9}$ **ft**

47. Solve $z - \frac{2}{5} \geq -2$. (Lesson 5-7) $z \geq -1\frac{3}{5}$

48. Simplify $(-7y^2)(3y^3)$. (Lesson 4-8) $-21y^5$

49. **Geometry** Find the perimeter of the rectangle shown below. (Lesson 3-5) **16 miles**

6 miles
2 miles

50. Three subtracted from some number is equal to -7. What is the number? (Lesson 2-5) -4

51. Write a verbal phrase for $2(x + 3)$. (Lesson 1-3) **Sample answer: two times the sum of a number and three.**

See for Yourself

The table at the bottom of page 376 shows the number of tons of fish caught in the world for selected years.

1. Does the table represent a function? **yes**
2. What trend do you see in the size of the catch over these 20 years? **increasing**
3. Research the United States' EEZ. Are there bans or quotas on fishing? How has the state of the marine life changed since the EEZ was established? **See students' work.**

EARTH WATCH

Over 90% of marine life is now living in EEZs. However, fishing still depletes the ocean fish. In the U.S. EEZ, the amount of fish caught by boats from other nations was reduced from 3.8 billion pounds in 1977 to none in 1992. The total catch by boats from the U.S. grew from 1.6 billion pounds to over 6 billion pounds in 1992.

4 ASSESS

Closing Activity

Modeling Have students write two sets of five ordered pairs. One set is to show a relation that is a function. The other set is to show a relation that is not a function.

Additional Answer

43c. Yes, each *x*-value (total price) has exactly one *y*-value (shipping cost).

Enrichment Masters, p. 62

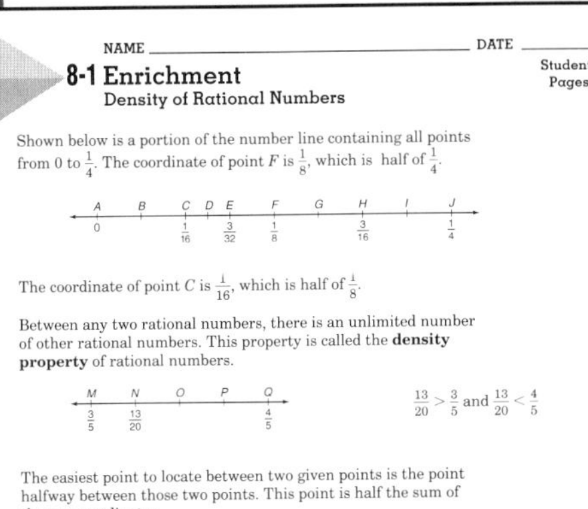

NAME ____________ DATE ____________

8-1 Enrichment
Density of Rational Numbers

Student Edition Pages 372–377

Shown below is a portion of the number line containing all points from 0 to $\frac{1}{4}$. The coordinate of point F is $\frac{1}{8}$, which is half of $\frac{1}{4}$.

The coordinate of point C is $\frac{1}{16}$, which is half of $\frac{1}{8}$.

Between any two rational numbers, there is an unlimited number of other rational numbers. This property is called the **density property** of rational numbers.

$\frac{13}{20} > \frac{3}{5}$ and $\frac{13}{20} < \frac{4}{5}$

The easiest point to locate between two given points is the point halfway between those two points. This point is half the sum of the two coordinates.

$\frac{1}{2}\left(\frac{7}{16} + \frac{8}{16}\right) \rightarrow \frac{1}{2}\left(\frac{15}{16}\right) \rightarrow \frac{15}{32}$

The coordinate of point X is $\frac{15}{32}$.

Using the number lines above, state the coordinates of each point. Assume that each point is halfway between the points to the left and right.

1. B $\frac{1}{32}$
2. D $\frac{5}{64}$
3. G $\frac{5}{32}$
4. I $\frac{7}{32}$
5. O $\frac{7}{10}$
6. P $\frac{3}{4}$
7. W $\frac{29}{64}$
8. Y $\frac{31}{64}$

Name the rational number that is halfway between the two given numbers on a number line.

9. $\frac{1}{2}$ and $\frac{3}{4}$ $\frac{5}{8}$
10. 0 and 5 $2\frac{1}{2}$
11. 4 and $5\frac{1}{4}$ $4\frac{5}{8}$
12. $1\frac{1}{2}$ and $\frac{5}{8}$ $1\frac{1}{16}$
13. $\frac{1}{2}$ and $\frac{2}{3}$ $\frac{7}{12}$
14. $\frac{7}{8}$ and $\frac{8}{9}$ $\frac{127}{144}$

8-2A LESSON NOTES

NCTM Standards: 1-6, 10, 12

Objective

Graph ordered pairs on a coordinate system to determine if there is a relationship.

Recommended Time

Demonstration and discussion: 15 minutes; Exercises: 30 minutes

Instructional Resources

For each student or group of students

- graph paper

Math Lab and Modeling Math Masters

- p. 41 (worksheet)

For teacher demonstration

Overhead Manipulative Resources

1 FOCUS

Motivating the Lesson

Ask students if there are some assumptions people make concerning physical characteristics. For example, tall people have big feet.

2 TEACH

Teaching Tip As each group completes its table, ask students to record their data on a large class graph using a bulletin board, string or yarn, and colored pins or tacks. Then have each student state his or her ordered pair while the other students add this data to their graphs.

3 PRACTICE/APPLY

Assignment Guide
Core: 1–4
Enriched: 1–4

MATHEMATICS LAB

HANDS-ON ACTIVITY

8-2A Scatter Plots

A Preview of Lesson **8-2**

MATERIALS

tape measure

graph paper

Cap size is related to the circumference of your head, and your shoe size is related to the length of your foot. Often it is not easy to establish whether there is a relationship between pairs of numbers by simply looking at them. Graphing ordered pairs on a coordinate system is one way to make it easier to "see" if there is a relationship.

Your Turn Work with a partner.

Name	Shoe Length (in.)	Armspan (in.)

- Measure, to the nearest inch, the length of your partner's shoe and your partner's armspan. Record your data in a table like the one shown at the right.
- Extend the table and combine your data with that of your classmates.
- On a piece of graph paper, draw a large coordinate axes. Label the axes by writing *Shoe Length (in.)* along the horizontal axis and *Armspan (in.)* along the vertical axis.
- Number the vertical lines, called tick marks, on each axis to make an appropriate scale for the data collected.
- Plot a point for each ordered pair (*shoe length, armspan*) using the data your class collected.

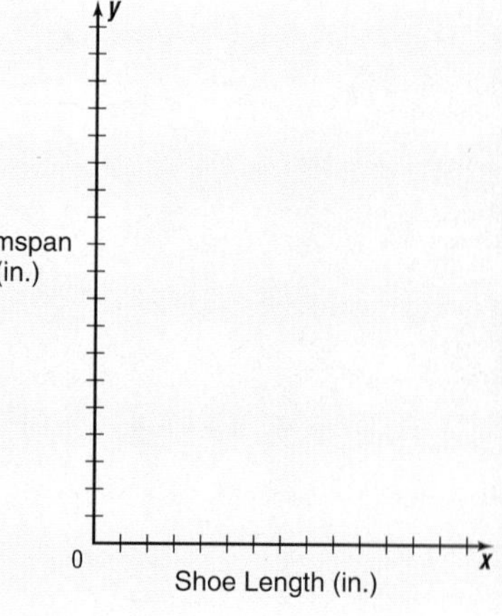

The graph you made is called a **scatter plot** of the data you collected.

TALK ABOUT IT

1. Describe the scatter plot. **1–3. See margin.**
2. Is there a relationship between shoe length and armspan? If so, write a sentence or two that describes the relationship.
3. Why do you suppose this kind of graph is called a scatter plot?

Extension

4. Using similar methods, determine if there is a relationship between another pair of lengths; for example, the circumferences of your head and wrist, or the circumference of your neck and your armspan. **See student's work.**

4 ASSESS

Observing students working in cooperative groups is an excellent method of assessment.

Additional Answers

1. Sample answer: Generally, as shoe length increases, arm span increases.
2. Answers will vary. Sample answer: Yes, arm span is about 6 times shoe length.
3. Sample answer: Because points appear to be scattered around a general area.

8-2 Integration: Statistics
Scatter Plots

Setting Goals: *In this lesson, you'll construct and interpret scatter plots.*

Modeling with Technology

The sales and advertising costs for ten of the largest U.S. industrial corporations are shown in the chart below. Use a TI-82 graphing calculator to create a scatter plot of the data. A **scatter plot** is a graph that shows the general relationship between two sets of data.

Company	Sales (millions of dollars)	Advertising Costs (millions of dollars)
General Motors	132,775	1333.6
Ford	100,786	794.5
IBM	65,096	185.5
General Electric	62,202	250.7
Philip Morris	50,157	2024.1
Chrysler	36,897	756.6
Procter & Gamble	29,890	2165.6
PepsiCo	22,084	928.6
Eastman Kodak	20,577	686.6
Dow Chemical	19,177	186.6

Source: *World Almanac*, 1995

Your Turn Before you create a scatter plot, clear the statistical memories.

Enter: STAT 4 2nd L1 , 2nd L2 ENTER

Next, enter the sales in L1 and the advertising costs in L2.

Enter: STAT ENTER *Accesses the statistical lists.*

132775 ENTER 100786 ENTER . . . 19177 ENTER

▶ 1333.6 ENTER 794.5 ENTER . . . 186.6 ENTER

You can review setting the viewing window in Lesson 2-2B.

After the data is entered, set the range for the graph. Set a viewing window of [0, 150,000] by [0, 2500] with a scale factor of 10,000 on the x-axis and 250 on the y-axis.

Create the scatter plot by pressing 2nd STAT PLOT ENTER and then using the arrow and ENTER keys to highlight "On", the scatter plot, L1 as the Xlist, L2 as the Ylist, and • as the mark. Press GRAPH to see the scatter plot.

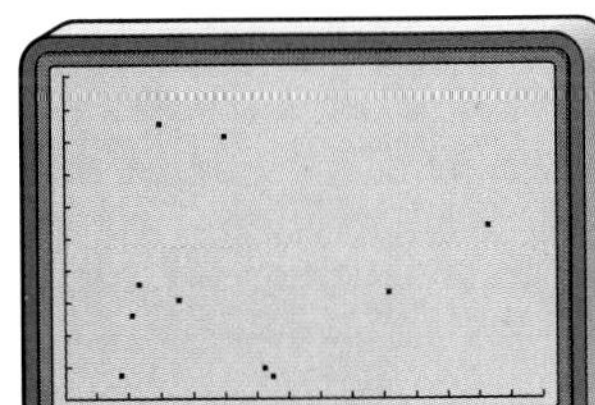

a. Does it appear that there is any relationship between sales and advertising? **It appears that greater sales is related to greater advertising.**

b. Are there any clusters of points that are somewhat separated from the other points? If there are, describe the characteristics of each cluster. **Two points show relatively low sales and high advertising.**

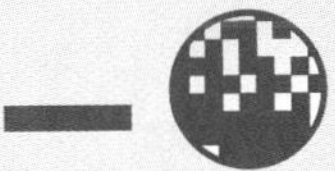

Alternative Teaching Strategies

Student Diversity Have students make the scatter plot at the beginning of the lesson by hand. Ask how students would round the numbers to get a reasonably accurate scatter plot. Help students round the numbers and complete the scatter plot.

8-2 LESSON NOTES

NCTM Standards: 1-5, 8, 10

Instructional Resources

- Study Guide Master 8-2
- Practice Master 8-2
- Enrichment Master 8-2
- Group Activity Card 8-2
- Multicultural Activity Masters, p. 15
- Real-World Applications, 16

Transparency 8-2A contains the 5-Minute Check for this lesson; **Transparency 8-2B** contains a teaching aid for this lesson.

Recommended Pacing	
Standard Pacing	Day 3 of 13
Honors Pacing	Day 2 of 12
Block Scheduling*	Day 2 of 7

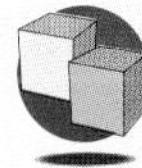

*For more information on pacing and possible lesson plans, refer to the ***Block Scheduling Booklet***.

1 FOCUS

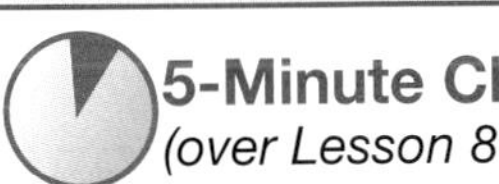

5-Minute Check
(over Lesson 8-1)

Write the domain and range of each relation.

1. {(−2, −1), (0, 1), (2, 3), (4, 5)} **{−2, 0, 2, 4}; {−1, 1, 3, 5}**
2. {(0.5, −4), (0.75, −2), (1.25, −1)} **{0.5, 0.75, 1.25}; {−4, −2, −1}**

Determine whether each relation is a function.

3.

x	1	2	3	4
y	1	2	3	4

yes

4.

x	−1	0	−1	−2
y	−2	0	2	−4

no

5.

x	0	1	2	3
y	3	4	3	4

yes

Motivating the Lesson

Situational Problem From a pediatrician's office, obtain a blank height-weight graph and make a transparency from it. Ask students if they have ever seen such charts. Ask them if there is a relationship between weight and height while growing. If (height, weight) were graphed on a coordinate plane, ask students what they expect a completed graph to look like.

2 TEACH

In-Class Example

For the Example
Make a scatter plot of the following data. Then determine if there is a positive, negative, or no relationship between the time of day and the amount of money spent in a local electronics store. Describe any trends you notice.

Time of Day	Average Amount Spent
9–11 A.M.	\$ 80
11 A.M.–1 P.M.	\$110
1–3 P.M.	\$ 90
3–5 P.M.	\$130
5–7 P.M.	\$150
7–9 P.M.	\$200

See students' graphs. Positive relationship, as time of day gets later, amounts go up, except for period right after lunch.

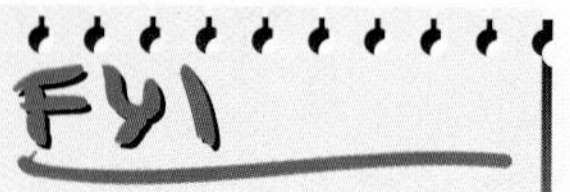

Between 1980 and 1990, the population of Asian-Americans increased by more than 100%.

Learning the Concept

Study the scatter plots below.

If the points appear to suggest a line that slants upward to the right, there is a **positive relationship**.

If the points appear to suggest a line that slants downward to the right, there is a **negative relationship**.

If the points seem to be random, then there is **no relationship**.

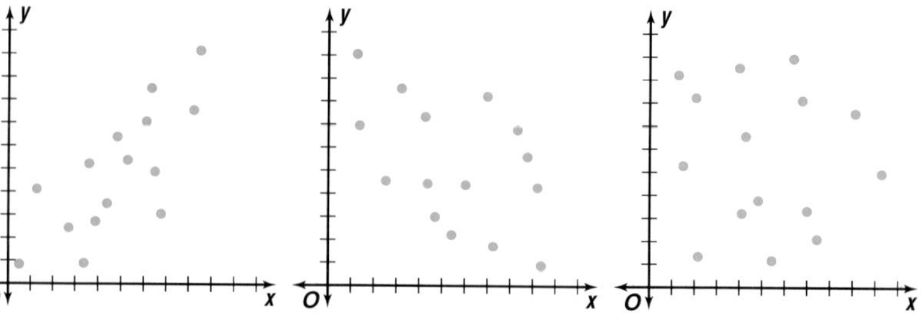

Look at the scatter plot of sales and advertising costs on the graphing calculator. The points in the scatter plot are spread out. There appears to be no relationship between the amount spent on advertising and the amount of sales.

Besides establishing a relationship, scatter plots can be used to study trends or make predictions.

Make a scatter plot of the following data. Then determine if there is a *positive*, *negative*, or *no* relationship. What trends do you see?

FYI

Tagalog is an official language of the Philippines along with Spanish and English.

Languages Other than English and Spanish Spoken at Home by Americans

Language used at home	Total speakers over 5 years old	
	1980	1990
French	1,572,000	1,703,000
German	1,607,000	1,547,000
Italian	1,633,000	1,309,000
Chinese	632,000	1,249,000
Tagalog	452,000	843,000
Polish	826,000	723,000
Korean	276,000	626,000
Vietnamese	203,000	507,000
Portuguese	361,000	430,000

Source: Bureau of the Census, U.S. Dept. of Commerce

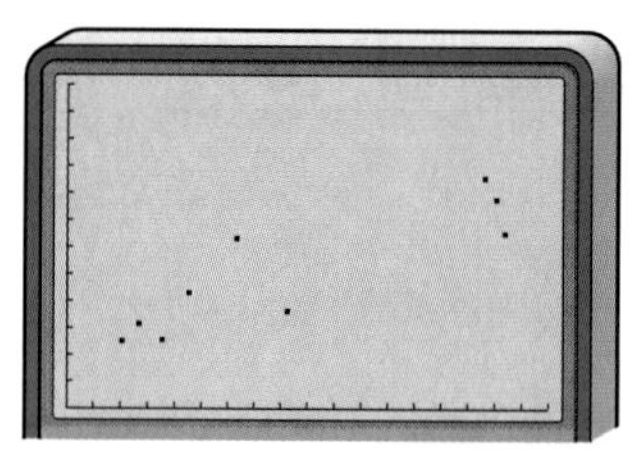

Since the points in the scatter plot are in a pattern that seems to slant upward to the right, the scatter plot shows a *positive* relationship. In general, the greater the number of speakers of a language in 1980, the greater the number of speakers of the same language in 1990.

Sales Associate Students who are interested in a sales career could contact salespersons and ask about the types of data they might analyze to improve sales. For instance, some types of data may be numbers of customers in the store and the time of day, or dollars in sales on non-sale and sale days. For more information on tech prep, see the *Teacher's Handbook*.

Checking Your Understanding

Communicating Mathematics

Read and study the lesson to answer these questions. 1–2. See margin.

1. **Write** a sentence explaining how a scatter plot can be used.
2. **Describe** how to draw a scatter plot for two sets of data.
3. **Draw** a scatter plot with 15 points that shows a negative relationship.
4. **Write** a short paragraph or two describing what you can determine about the data in the scatter plot shown at the right. See margin.

3. See students' work.

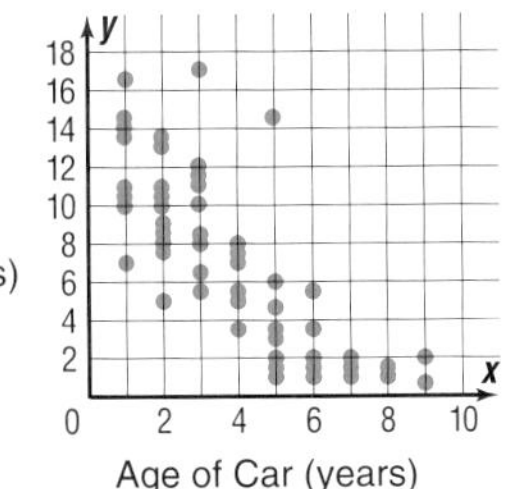

Guided Practice

What type of relationship, *positive*, *negative*, or *none*, is shown by each scatter plot? 5. negative 6. none 7. positive

5.

6.

7.

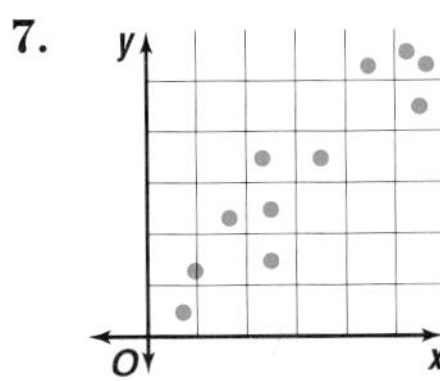

Determine whether a scatter plot of the data for the following might show a *positive*, *negative*, or *no* relationship. Explain your answer.

8. weight, month of birth none
9. playing time, points scored positive
10. **Track and Field** The scatter plot at the right compares the winning time to the year of the race for the women's Olympic 100-meter dash.

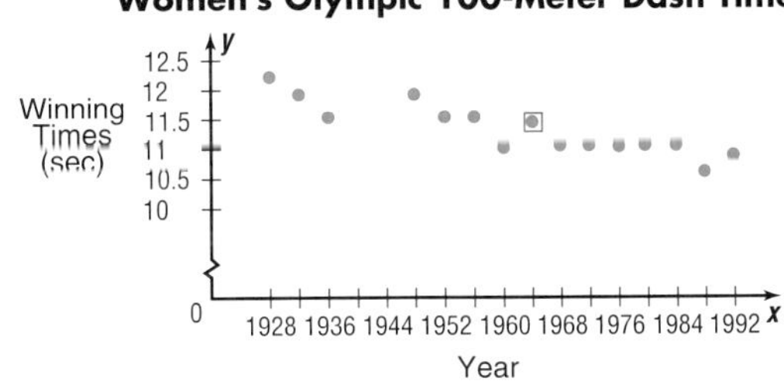

8–9. Sample answers given. See students' explanations.
10b. 12.2 seconds in 1928
10c. about 11.5 seconds
10d. See margin.
10e. See students' work.

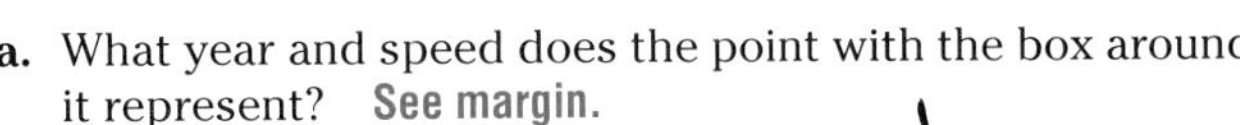

a. What year and speed does the point with the box around it represent? See margin.
b. Find the point(s) that represent(s) the slowest winning time ever run in the Olympic 100-meter dash.
c. If you had trained in the 1950s for the 100-meter race, what time would have been competitive?
d. What conclusion can you make from the scatter plot?
e. Use the scatter plot to predict the winning time in the 1996 Olympics in Atlanta, Georgia. Check your prediction.

3 PRACTICE/APPLY

Checking Your Understanding

Exercises 1–10 are designed to help you assess your students' understanding through reading, writing, speaking, and modeling. You should work through Exercises 1–4 with your students and then monitor their work on Exercises 5–10.

Additional Answers

1. Sample answer: A scatter plot is used to analyze relationships between data.
2. Sample answer: Collect two sets of data and form ordered pairs. Then graph the ordered pairs.
4. Sample answer: The scatter plot shows the cost of cars based on the age of the cars. This scatter plot shows a negative relationship. That is, the age of the car increases as the cost goes down.

10a. It represents the winning time of 11.4 seconds in the 1964 Olympics.

10d. Sample answer: Generally, as the year increases, the associated winning time decreases.

GLENCOE Technology

Interactive Mathematics Tools Software

In this interactive computer lesson, students explore scatter plots. A **Computer Journal** gives students an opportunity to write about what they have learned.

For Windows & Macintosh

Reteaching

Using Manipulatives Have students use a checkerboard and checkers. Label the lines on the board as a grid. Then have students place checkers on the board to show relations that exhibit a positive relationship, a negative relationship, and no relationship.

Error Analysis

Some students may have difficulty differentiating the relationship of a scatter plot. Have students draw a curve enclosing the majority of the plotted points. If the figure is an ellipse, have them draw a straight line lengthwise through the ellipse.

Assignment Guide

Core: 11–23 odd, 24, 25, 27, 29–33
Enriched: 12–22 even, 23–33

For **Extra Practice**, see p. 758.

The red A, B, and C flags, printed only in the Teacher's Wraparound Edition, indicate the level of difficulty of the exercises.

Additional Answers

14a. It represents a study time of 75 minutes with a resulting test score of 87.

14c. The points would be in the upper right portion of the scatter plot.

Exercises: Practicing and Applying the Concept

Independent Practice

What type of relationship, *positive*, *negative*, or *none*, is shown by each scatter plot? 11. none 12. positive 13. negative

11.

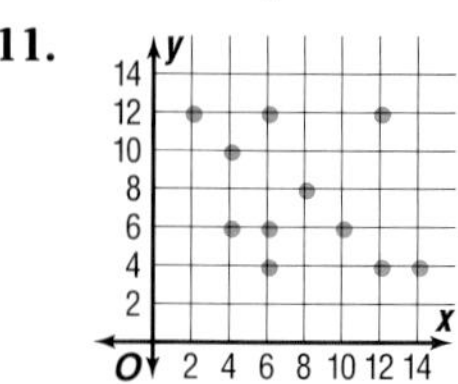

12.

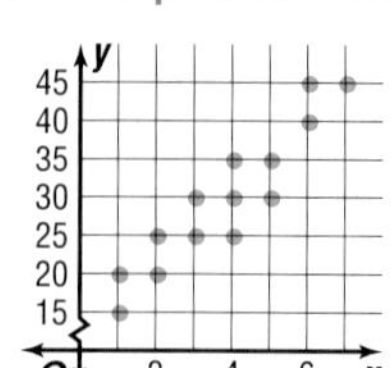

13.

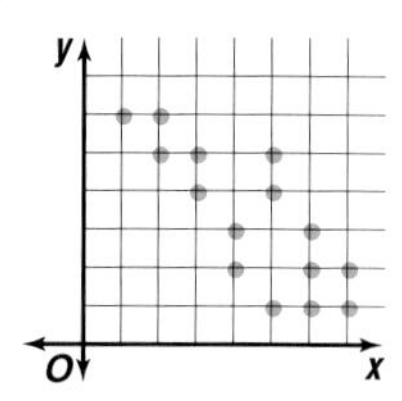

14. A scatter plot of study time and test scores for a class is shown below. 14a. See margin. 14b. yes; positive

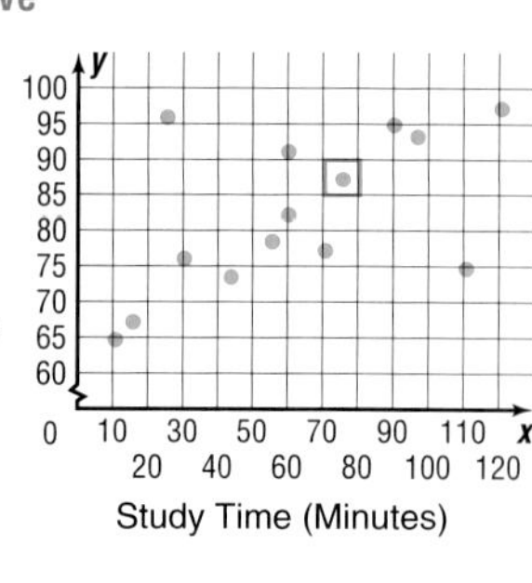

a. What study time and test score does the point with the box around it represent?

b. Do the data show a relationship between study time and test scores? If so, is it positive or negative?

c. Where on the plot are the points for people who studied quite a bit and had a fairly high test score? See margin.

d. How many students are in the class? 14

Determine whether a scatter plot of the data for the following might show a *positive*, *negative*, or *no* relationship. Explain your answer.

15–22. Sample answers given. See students' explanations.

15. height, weight positive

16. armspan of student, test scores

17. hair color, height no

18. outside temperature, heating bill

19. weight of car, miles per gallon negative

20. amount of sales, years of experience positive

21. month of birth, test scores no

22. head circumference, height positive

16. no 18. negative

Graphing Calculator

23. Keyboarding The table below shows the keyboarding speeds in words per minute (wpm) of 12 students. Use a graphing calculator to create a scatter plot of the data. See margin for graph.

Experience (weeks)	4	7	8	1	6	3	5	2	9	6	7	10
Speed (wpm)	38	46	48	20	40	30	38	22	52	44	42	55

a. Do the data show a relationship between speed and experience? If so, is it positive or negative? The data show a positive relationship.

b. Use this scatter plot to estimate the keyboarding speed of a student with 12 weeks experience. Sample answer: about 62 wpm

c. Estimate the experience level of a student whose speed is 33 wpm. Sample answer: between 3 and 4 weeks

d. What conclusion can you make from the scatter plot? The more experience a student has the more words per minute he or she can key.

Additional Answer

23.

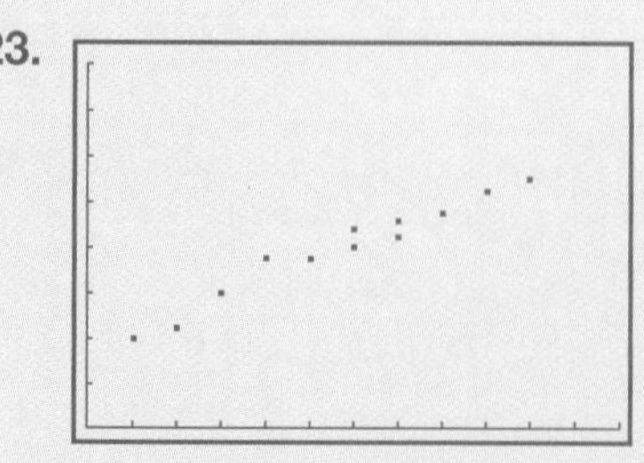

Study Guide Masters, p. 63

NAME ______________________ DATE __________

8-2 Study Guide
Integration: Statistics
Scatter Plots

Student Edition Pages 379–384

A **scatter plot** is a graph of ordered pairs that shows the relationship between two sets of data.

If the points on a graph seem to slant upward to the right, there is a *positive* relationship.

If the points appear to slant downward to the right, there is a *negative* relationship.

If the points seem random, there is *no* relationship.

Example: The scatter plot on the right shows the relationship between the heights of a group of people and their shoe sizes. In this graph, the points go upward to the right. Therefore, there is a positive relationship. In general, the scatter plot seems to show that the taller a person is, the greater the shoe size.

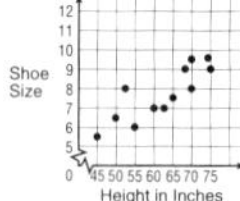

What type of relationship, positive, negative, or none, is shown by each scatter plot?

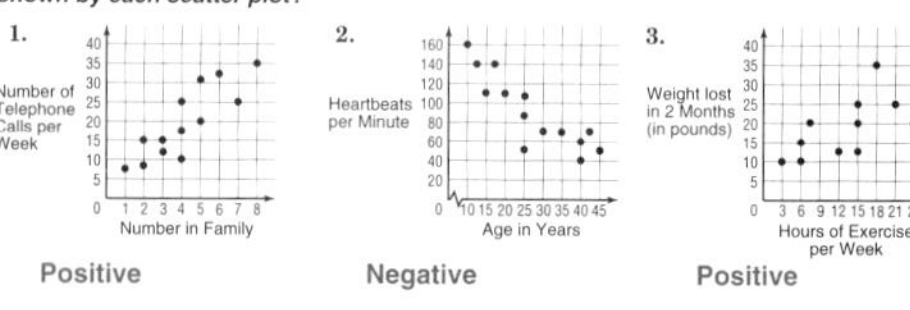

1. Positive
2. Negative
3. Positive

Determine whether a scatter plot of the data for the following might show a positive, a negative, or no relationship. Explain your answer. Sample answers given. See students' explanations.

4. study time, higher grades positive
5. height, intelligence none
6. shoe size, salary none
7. age of car, value of car negative
8. miles per gallon, gas expense negative
9. education, salary positive
10. wrist circumference, appetite none
11. birthdate, ring size none
12. windchill, ice cream sales negative
13. age of tree, number of rings positive
14. amount of snowfall, shovel sales positive
15. hair length, hat size none

Critical Thinking

24. Sales A scatter plot of monthly insulated boot sales and monthly skiing accidents in West Virginia shows a positive relationship.

a. Why might this be true? **24a–b. See margin.**

b. Does a positive relationship necessarily mean that one factor causes the other? Explain.

Applications and Problem Solving

25. Education The table below shows the average tuition for public colleges in the United States for the years 1985 to 1994.

For the latest college costs, visit: **www.glencoe.com/sec/math/prealg/mathnet**

Year	1985	1986	1987	1988	1989
Cost per Year ($)	1386	1536	1651	1726	1846
Year	1990	1991	1992	1993	1994
Cost per Year ($)	2035	2159	2410	2610	2784

a. Make a scatter plot of the data. **See Solutions Manual.**

b. Do the data show a relationship between year and cost per year? If so, what type of relationship is it? **Yes; positive.**

26. Geography Use an almanac to find the area and the average depth of the ten largest oceans and seas.

a. Construct a scatter plot to show the relationship between the average depth and area. **See Solutions Manual.**

b. What can you determine about the relationship between area and depth of oceans from the scatter plot? **See margin.**

27. Sports The number of points scored and rebounds made by members of the 1994 National Basketball Association champion Houston Rockets is given in the table below. **See Solutions Manual.**

Player	Rebounds	Points
Olajuwon	955	2184
Thorpe	870	1149
Maxwell	229	1023
Smith	138	906
Jent	15	31
Horry	440	803
Elie	181	626
Cassell	134	440
Brooks	102	381
Herrera	285	353
Robinson	10	25
Bullard	84	226
Petruska	31	53
Cureton	12	4
Riley	59	88

a. Make a scatter plot of the data to show the relationship between the rebounds and points.

b. Does the scatter plot show any relationship?

c. Could you predict the number of points a player would have if you were given the number of rebounds for that player? Explain.

d. Do you think that the position of a player affects the number of rebounds that player will make? Explain.

e. Do you think the amount of time played affects the number of points scored by a player? Why or why not?

28. Geometry Make a table with ten possible lengths and widths of rectangles with perimeters of 24 centimeters. Then make a scatter plot of the data. Does there appear to be a positive, negative, or no relationship between length and width? **See margin.**

Additional Answers

24a. Sample answer: Both activities are dependent on cold weather

24b. No; buying boots does not cause skiing accidents. The relationship may be that both things are caused by a third factor.

26b. The bodies of water that have the greater surface have the greater depth.

28.

Width	Length
1	11
1.5	10.5
2	10
2.5	9.5
3	9
4	8
4.5	7.5
5	7
5.5	6.5
6	6

See Solutions Manual for graph

Practice Masters, p. 63

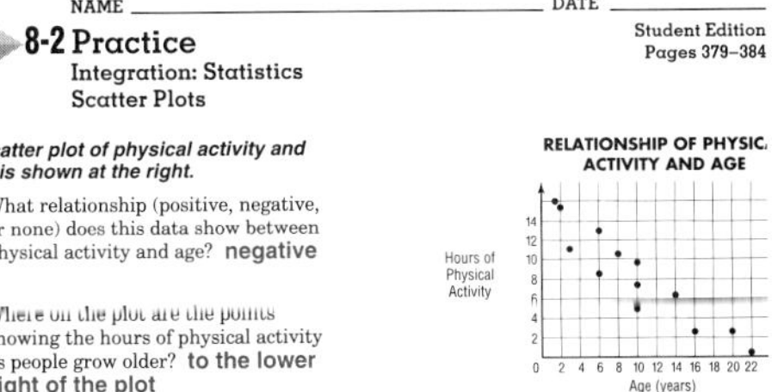

NAME ______________________ DATE __________

8-2 Practice
Integration: Statistics
Scatter Plots

Student Edition
Pages 379–384

A scatter plot of physical activity and age is shown at the right.

1. What relationship (positive, negative, or none) does this data show between physical activity and age? **negative**

2. Where on the plot are the points showing the hours of physical activity as people grow older? **to the lower right of the plot**

3. What happens to the number of hours of physical activity as people grow older? **The number of hours decreases.**

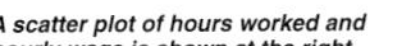

A scatter plot of hours worked and hourly wage is shown at the right.

4. What relationship does this data show between hours worked and hourly wage? **none**

5. How many people are shown on the plot? **14**

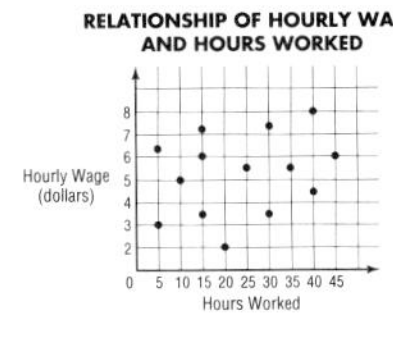

A scatter plot of assisted tackles and solo tackles for each player during a football season is shown at the right.

6. What relationship does this data show between assisted tackles and solo tackles? **positive**

7. What is the greatest number of assists shown on the plot? **64**

8. What is the least number of solo tackles shown on the plot? **3**

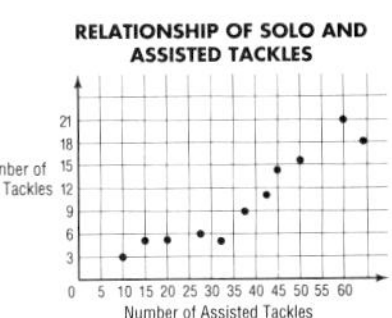

Group Activity Card 8-2

To Tell The Truth Group Activity **8-2**

Is there a relationship between the amount of time students spend watching television and the amount of time they do homework? Investigate this question by collecting data and making a scatter plot.

To begin, ask each classmate how many hours (to the nearest half hour) of homework they do a day, Monday through Thursday. Also, ask each of them how many hours of television they watch on those days. Collect this data as pairs of information, and graph these data on a scatter plot. Use Hours of Studying as your horizontal axis and Hours of Television as your vertical axis.

When you have completed your scatter plot, consider the following questions.

1. Where do most of the data points fall?
2. Does there appear to be a positive or negative relationship between the number of hours spent studying and those spent watching television?

 Pre-Algebra

Extension

Using Cooperative Groups Have students work in groups of four to design an experiment to relate mathematics grades to English grades. Organize the data by using scatter plots. Then, interpret the results.

4 ASSESS

Closing Activity

Writing Have students write a note to a business person of their choice explaining why a scatter plot of particular data would be valuable to that person.

Enrichment Masters, p. 63

NAME ______________________ DATE __________

8-2 Enrichment
Growth Charts

Student Edition Pages 379–384

Scatter plots are often used by doctors to show parents the growth rates of their children. The horizontal scale of the chart at the right shows the ages from 15 to 36 months. The vertical scale shows weight in kilograms. One kilogram is about 2.2 points. The curved lines are used to show how a child's weight compares with others of his or her age.

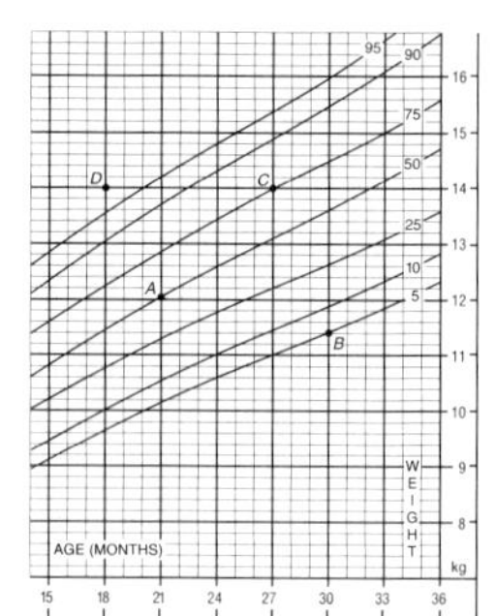

Look at the point labeled *A*. It represents a 21-month-old who weighs 12 kilograms. It is located on the slanted line labeled 50. This means the child's weight is in the "50th percentile." In other words, 50% of all 21-month-olds weigh more than 12 kilograms and 50% weigh less than 12 kilograms.

The location of Point *B* indicates that a 30-month-old who weighs 11.4 kilograms is in the 5th percentile. Only 5% of 30-month-old children will weigh less than 11.4 kilograms.

Solve.

1. Look at the point labeled *C*. How much does the child weigh? How old is he? What percent of children his age will weigh more than he does? **14 kg; 27 months; 25%**
2. Look at the point labeled *D*. What is the child's age and weight? What percent of children her age will weigh more than she does? **18 months; 14 kg; less than 5%**
3. What is the 50th percentile weight for a child 27 months old? **13.2 kg**
4. How much weight would child *B* have to gain to be in the 50th percentile? **2.3 kg**
5. If child *D* did not gain any weight for four months, what percentile would he be in? **90th**
6. How much heavier is a $2\frac{1}{2}$-year-old in the 90th percentile than one in the 10th percentile? **3.7 kg**

Mixed Review

29. Determine the domain and range of the relation {(8, 1), (4, 2), (6, −4), (5, −3), (6, 0)}. (Lesson 8-1) **{8, 4, 6, 5}; {1, 2, −4, −3, 0}**

30. Solve $12x - 24 = -14x + 28$. (Lesson 7-5) **2**

31. Statistics Find the mean, median, and mode for the data set 121, 130, 128, 126, 130, 131. (Lesson 6-6) **127.7; 129; 130**

32. Patterns Write the next three terms of the sequence 5, 6.5, 8, 9.5. (Lesson 5-9) **11, 12.5, 14**

33. Biology Cicadas are sometimes called 17-year locusts because they emerge from the ground every 17 years. The number of a certain type of caterpillar peaks every 5 years. If the peak cycles of the caterpillars and the cicadas coincided in 1990, what will be the next year in which they will coincide? (Lesson 4-7) **2075**

WORKING ON THE

Investigation

Refer to the Investigation on pages 368–369.

Begin working on the Investigation by doing a class experiment. Use a stopwatch to time one student doing the wave. The timer says "go" and starts the stopwatch. The student stands up and then sits down. At this point, the timer stops the stopwatch. Record one person and the time in a table.

Next, time two students doing the wave (each student represents a different section in the stadium). The timer says "go" and starts the stopwatch. The first person stands and then sits. As the first person sits, the second person stands and then sits. The timer stops the stopwatch. Record two people and the time in the table.

Continue the experiment with three people (representing three sections), then eight people, then eleven people, then fourteen people, and so on until you have determined a wave time using every student in the class.

Your table of the class should contain all the data including the number of people (sections) and their corresponding wave times. Use a scatter plot to illustrate the data. The horizontal axis will be the number of people in the wave, and the vertical axis will be the wave time.

Does this data show a relationship between the number of sections and the wave time? Does this scatter plot show a *positive*, *negative*, or *no relationship*?

Write a paper answering the questions raised by the spirit squad on page 369.

Add the results of your work to your Investigation Folder.

WORKING ON THE

Investigation

The Investigation on pages 368 and 369 is designed to be a long-term project that is completed over several days or weeks. Encourage students to keep their materials in their Investigation Folder as they work on the Investigation.

8-3 Graphing Linear Relations

Setting Goals: *In this lesson, you'll find solutions for relations with two variables and graph the solutions.*

Modeling a Real-World Application: Science

Did you know that a cricket is nature's thermometer? You can use the equation $t = c + 40$ to find the temperature t if you know the number of chirps c a cricket makes in 15 seconds.

Chirps in 15 seconds	$c + 40$	Temperature (°F)
21	21 + 40	61
23	23 + 40	63
31	31 + 40	71
50	50 + 40	90
47	47 + 40	87

The chart shows the relationship between various numbers of chirps in 15 seconds and the temperature. The domain is {21, 23, 31, 50, 47}, and the relation is {(21, 61), (23, 63), (31, 71), (50, 90), (47, 87)}. The relation can also be shown in a graph or scatter plot like the one shown at the right.

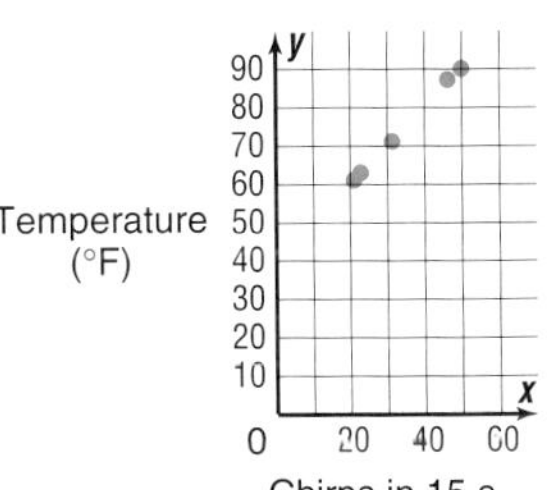

CULTURAL CONNECTIONS
The first attempts at weather forecasting were made by Greek philosophers in the fourth century B.C.

Learning the Concept

Recall that solving an equation means to replace the variable so that a true sentence results. The solutions of an equation with two variables are ordered pairs.

To find a solution of such an equation, choose any value for x, substitute that value into the equation, and find the corresponding value for y.

It is often convenient to organize the solutions in a table.

Example 

a. Find four solutions for the equation $y = -2x + 1$. Write the solutions as ordered pairs.
b. Graph the equation $y = -2x + 1$.

a. Choose four convenient values for x. Substitute each value for x in the expression $-2x + 1$.

Do the computation to find y.

Four solutions are $(-2, 5)$, $(0, 1)$, $(1, -1)$, and $(3, -5)$.

x	$-2x + 1$	y
−2	−2(−2) + 1	5
0	−2(0) + 1	1
1	−2(1) + 1	−1
3	−2(3) + 1	−5

CULTURAL CONNECTIONS
The first meteorological treatise, "Meteorologica," was written by Aristotle in the 4th century B.C. Inventions that improved weather forecasting were the thermometer and weather vane in the 16th century.

8-3 LESSON NOTES

NCTM Standards: 1-5, 8, 9

Instructional Resources

- Study Guide Master 8-3
- Practice Master 8-3
- Enrichment Master 8-3
- Group Activity Card 8-3
- Assessment and Evaluation Masters, p. 211

Transparency 8-3A contains the 5-Minute Check for this lesson; **Transparency 8-3B** contains a teaching aid for this lesson.

Recommended Pacing	
Standard Pacing	Day 4 of 13
Honors Pacing	Day 3 of 12
Block Scheduling*	Day 3 of 7 (along with Lesson 8-4)

*For more information on pacing and possible lesson plans, refer to the ***Block Scheduling Booklet***.

1 FOCUS

5-Minute Check
(over Lesson 8-2)

1. Describe the scatter plot of two sets of data having a positive relationship. **Points slope up to the right.**

Determine whether a scatter plot of the data for the following might show a positive, a negative, or no relationship. Explain.

2. distance traveled, gasoline in tank **Negative; the car uses up the gas as it travels.**
3. color of car, miles per gallon **None; car color has no effect on the mileage.**
4. total cost of a package of meat, pounds in package **Positive; meat is usually priced per pound.**
5. a person's age, that person's weight **Positive; many gain weight as they age.**

Motivating the Lesson

Situational Problem Joanne plans to use 500 feet of fencing to enclose a rectangular garden. The formula for the perimeter of a rectangle is $P = 2\ell + 2w$. Find five possible sizes (in whole numbers only) that Joanne could use.

2 TEACH

In-Class Examples

For Example 1

a. Find four solutions for $y = 4x - 3$. **(0, −3), (1, 1), (−1, −7), (2, 5)**

b. Graph $y = 4x - 3$.

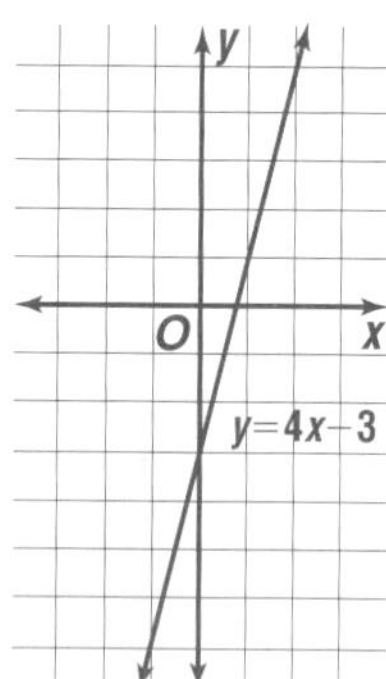

For Example 2

A 20-foot board is sawed into two pieces. If *x* and *y* are the measures of the two pieces, the equation $x + y = 20$ models the conditions.

a. Graph the equation.

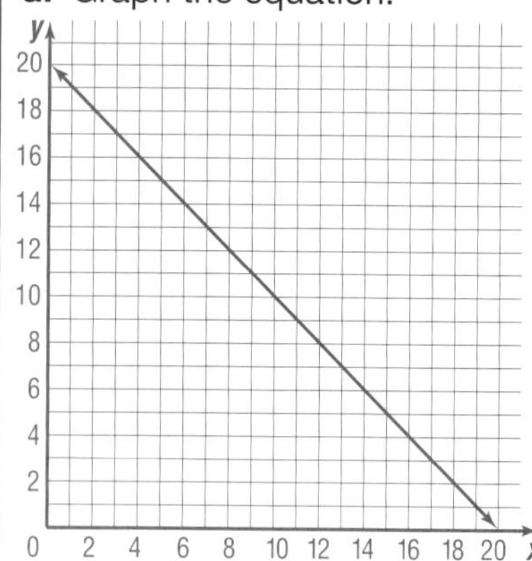

b. Use the graph to name a solution of the equation. **Answers will vary.**

c. Is (−1, 21) a solution of the equation? **Yes; but not the problem.**

THINK ABOUT IT

Is (2, 2) a solution for $y = -2x + 1$? Why or why not?

no; $-2(2) + 1 \neq 2$

b. First, graph the ordered pairs you found in part a.

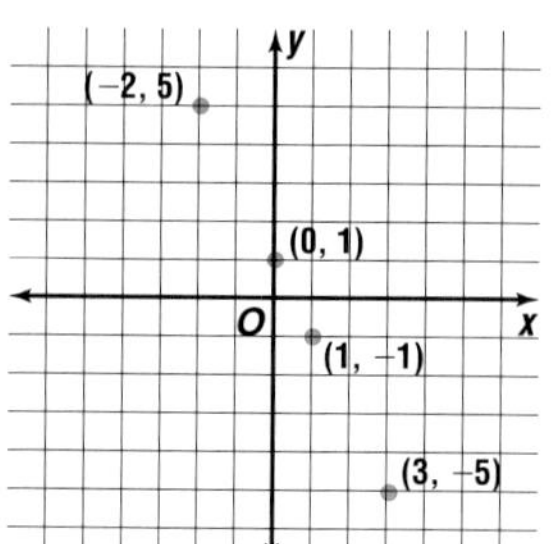

Then draw the line that contains these points.

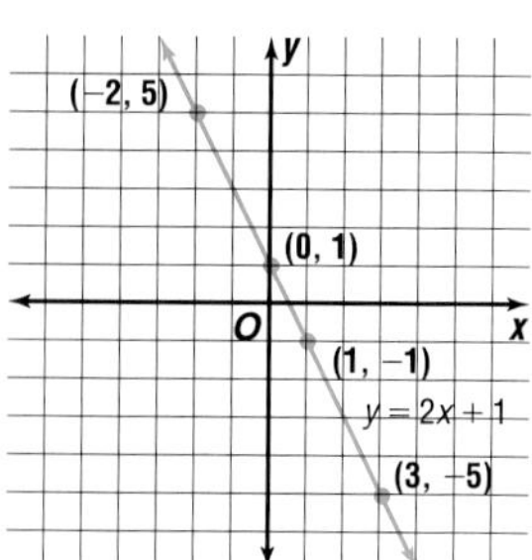

The ordered pair for every point on the line in part b of Example 1 is a solution of $y = -2x + 1$. It appears from the graph that (−1.5, 4) and (2, −3) are also solutions. You can check this by substitution.

Check (−1.5, 4).	Check (2, −3).
$y = -2x + 1$	$y = -2x + 1$
$4 \stackrel{?}{=} -2(-1.5) + 1$	$-3 \stackrel{?}{=} -2(2) + 1$
$4 = 4$ ✓	$-3 = -3$ ✓

An equation like $y = -2x + 1$ is called a **linear equation** because its graph is a straight line. An equation with two variables has an infinite number of solutions.

In any right triangle, the sum of the measures of the two acute angles is 90°. If *x* and *y* represent the measures of the two acute angles, then the equation $x + y = 90$ models this condition.

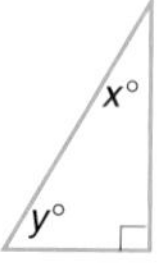

a. Graph the equation.

b. Use the graph to name another solution of the equation.

c. Is (−10, 100) a solution of the equation?

a. First solve the equation for *y* in terms of *x*.

$x + y = 90$

$y = 90 - x$ *Subtract x from each side.*

Now make a table of ordered pairs that satisfy the equation. Since any line can be defined by only two points, only two ordered pairs are needed. Finding a third ordered pair is a good idea to check the accuracy of the first two.

x	90 − x	y	(x, y)
45	90 − 45	45	(45, 45)
10	90 − 10	80	(10, 80)
70	90 − 70	20	(70, 20)

GLENCOE Technology

Interactive Mathematics Tools Software

In this interactive computer lesson, students explore graphing equations. A **Computer Journal** gives students an opportunity to write about what they have learned.

For Windows & Macintosh

Then graph the ordered pairs and connect them with a line.

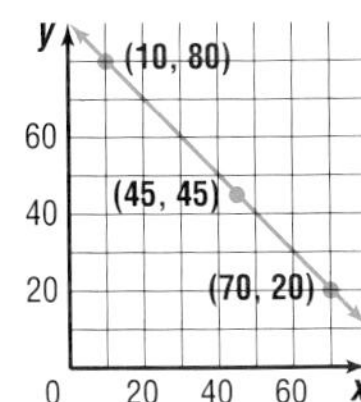

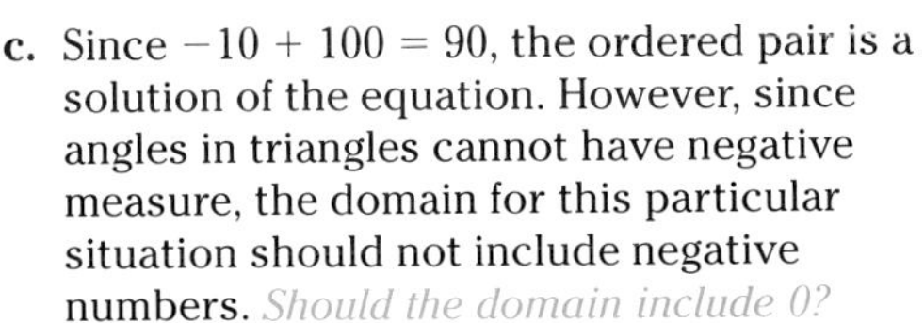

b. Since (50, 40) is on the graph, it is another solution of the equation.

c. Since $-10 + 100 = 90$, the ordered pair is a solution of the equation. However, since angles in triangles cannot have negative measure, the domain for this particular situation should not include negative numbers. *Should the domain include 0?*

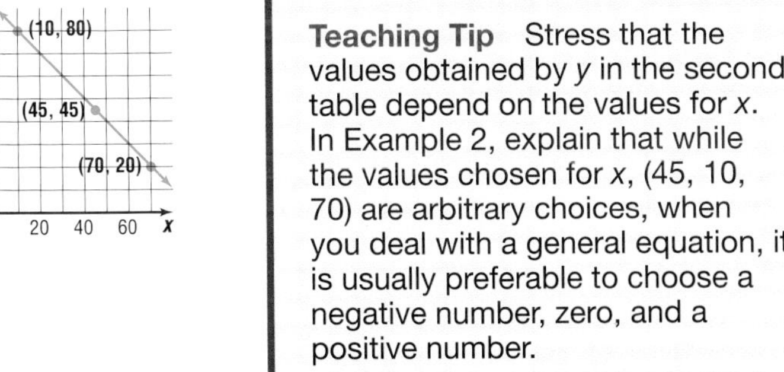

Checking Your Understanding

Communicating Mathematics

Read and study the lesson to answer these questions. 2. See margin.

1. **Determine** which ordered pair, (10, 20), (8, 4), or (300, 600), is a solution of $y = 0.5x$. **(8, 4)**
2. **Explain** why a linear equation has an infinite number of solutions.
3. **Indicate** which equation is graphed at the right. **b**
   a. $x - y = 3$
   b. $x + y = 3$
   c. $y = x + 3$
   d. $y = -x - 3$
4. **Define** a linear equation.
5. **Describe** how you would graph the equation $x + y = 3$. **See margin.**

4. Sample answer: An equation whose graph is a straight line.

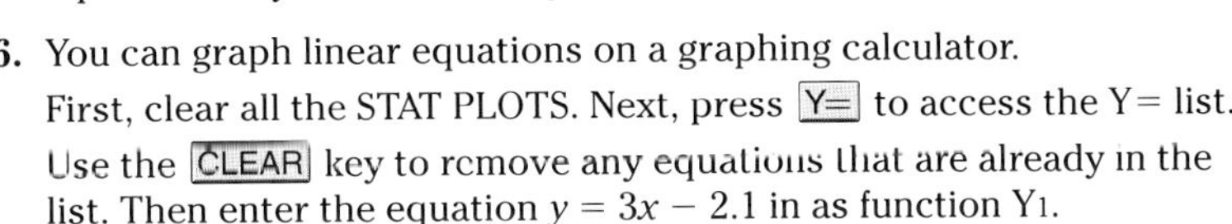

MODELING MATHEMATICS

MATERIALS

graphing calculator

6. You can graph linear equations on a graphing calculator. First, clear all the STAT PLOTS. Next, press [Y=] to access the Y= list. Use the [CLEAR] key to remove any equations that are already in the list. Then enter the equation $y = 3x - 2.1$ in as function Y1.

   Enter: 3 [X,T,θ] [−] 2.1

   Press [ZOOM] 6 to select the standard viewing window and complete the graph.

   Graph each equation in the standard viewing window. **See margin.**
   a. $y = 13 - 5.5x$
   b. $y = 2x + 3.7$

Guided Practice

7. Which ordered pair(s) is a solution of $x - 3y = -7$? **c, d**
   a. (2, 4) b. (2, −1) c. (2, 3) d. (−1, 2)

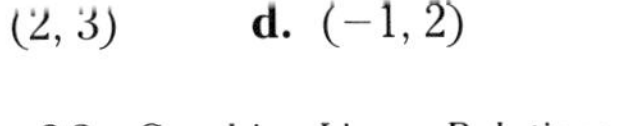

Reteaching

Using Lists Giving students an ordered pair, have students brainstorm a list of equations that the ordered pair will satisfy. For example, if (1, 3) is given, equations might be $y = x + 2$, $y = 3x$, and $y = 2x + 1$.

Teaching Tip Stress that the values obtained by y in the second table depend on the values for x. In Example 2, explain that while the values chosen for x, (45, 10, 70) are arbitrary choices, when you deal with a general equation, it is usually preferable to choose a negative number, zero, and a positive number.

3 PRACTICE/APPLY

Checking Your Understanding

Exercises 1–19 are designed to help you assess your students' understanding through reading, writing, speaking, and modeling. You should work through Exercises 1–6 with your students and then monitor their work on Exercises 7–19.

Additional Answers

2. Sample answer: An equation with two variables has an infinite number of solutions because an infinite number of values can be substituted for x.
5. Sample answer: Find at least four solutions to the equation. Graph the points associated with the ordered pairs. Draw the line containing the four points.

6a.

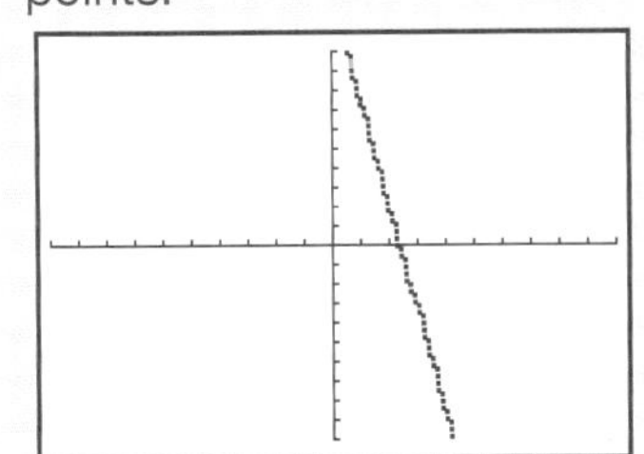

6b.

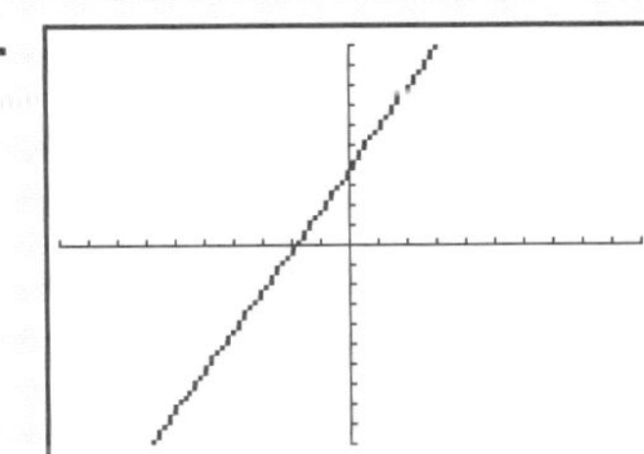

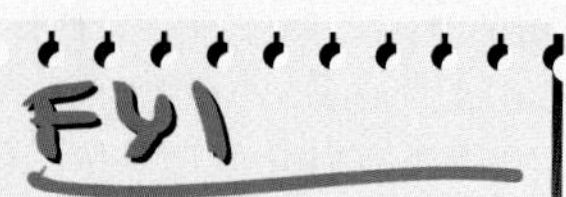

Most of the forestland in the United States belongs to state or federal governments. The Forest Service, a department within the Department of Agriculture, manages the federal forestland.

Additional Answers

10. Answers will vary. Sample answer: (0, −7), (7, 0), (8, 1), (9, 2)

11. Answers will vary. Sample answer: (0, 2.8), (−2, −7.2), (2, 12.8), (−1, −2.2)

12. Answers will vary. Sample answer: (0, 0), (7, −14), (−8, 16), (1, −2)

19b.

x	$y = 0.5x$	y	(x, y)
0	$y = 0.5(0)$	0	(0, 0)
1	$y = 0.5(1)$	0.5	(1, 0.5)
2	$y = 0.5(2)$	1	(2, 1)
3	$y = 0.5(3)$	1.5	(5, 1.5)

Study Guide Masters, p. 64

NAME ______________ DATE ______

8-3 Study Guide
Graphing Linear Relations

Student Edition Pages 385–390

To find a solution of an equation that has two variables, choose any value for x, substitute that value into the equation and find the corresponding value for y.

$y = 7 + x$ Suppose $x = 4$
$y = 7 + 4$ Replace x with 4.
$y = 11$

When $x = 4$, $y = 11$. So the ordered pair (4, 11) is a solution of the equation $y = 7 + x$.

An equation has many ordered pairs that are solutions. For example, four ordered pairs for the equation $y = x - 1$ are (3, 2), (0, −1), (2, 1), and (−1, −2). To graph an equation, graph the ordered pairs and then draw the line that contains the points. An equation is called a **linear equation** if its graph is a straight line.

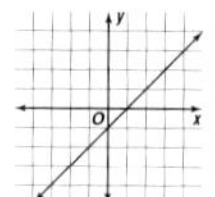

Find four solutions for each equation. Write the solutions as ordered pairs. Answers may vary. Four possible solutions are given.

1. $y = x$ (3,3), (0, 0), (−3, −3), (−1, −1)
2. $y = x + 1$ (0, 1), (3, 4), (−1, 0), (−2, −1)
3. $y = x - 2$ (0, −2), (1,−1), (−2,−4), (2, 0)
4. $2x - y = 3$ (−2, −7), (0, −3), (2, 1), (3, 0)
5. $y = -4x + 1$ (1, −3), (0, 1), (−1, 5), (2, −7)
6. $y = \frac{1}{2}x + 3$ (2, 4), (0, 3), (−2, 2), (4, 5)

Determine whether each relation is linear.

7. $-3x + 2y = 1$ yes
8. $y = x^2 - 1$ no
9. $y = -x - 3$ yes

Graph each equation.

10. $y = 2x - 4$
11. $-3x + y = -5$
12. $y = -\frac{1}{2}x$

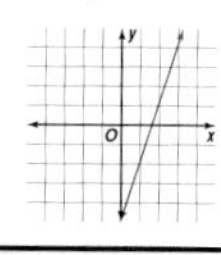
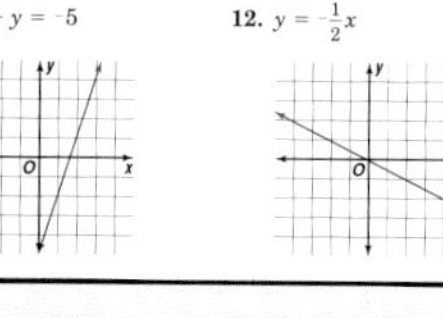

Copy and complete the table for each equation. Then use the results to write four solutions for each equation. Write the solutions as ordered pairs. 8. (−1, 3), (0, 4), (1, 5), (3, 7) 9. (−2, −5), (0, −2), (2, 1), (4, 4)

8. $y = x + 4$

x	$x + 4$	y
−1	−1 + 4	3
0	0 + 4	4
1	1 + 4	5
3	3 + 4	7

9. $y = 1.5x - 2$

x	$1.5x - 2$	y
−2	1.5(−2) − 2	−5
0	1.5(0) − 2	−2
2	1.5(2) − 2	1
4	1.5(4) − 2	4

Find four solutions for each equation. Write the solutions as ordered pairs. 10–12. See margin.

10. $y = x - 7$ **11.** $y = 5x + 2.8$ **12.** $y = -2x$

Determine whether each relation is linear.

13.

x	−2	−1	0	1	2
y	4	5	−1	2	0

no

14.

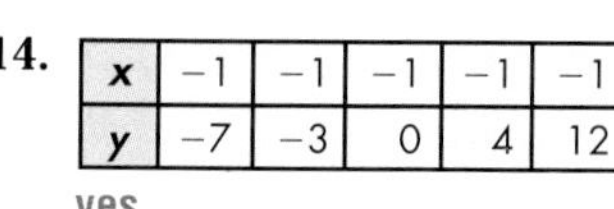

x	−1	−1	−1	−1	−1
y	−7	−3	0	4	12

yes

Graph each equation. See Solutions Manual.

15. $y = x - 7$ **16.** $y = 2.5x + 2$ **17.** $y = -2$ **18.** $y = \frac{1}{3}x$

19. Forestry Since trees are a renewable resource, foresters replace trees that have been cut down. Each year, more than two billion trees are planted in U.S. forests. About half of the seedlings planted survive until they are full grown. Let x represent the number of seedlings planted, and let y represent the number that become full grown trees.

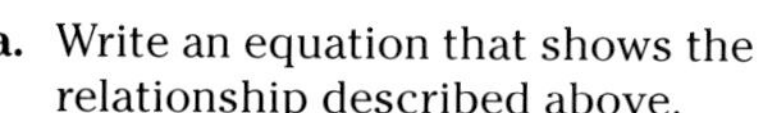

a. Write an equation that shows the relationship described above.

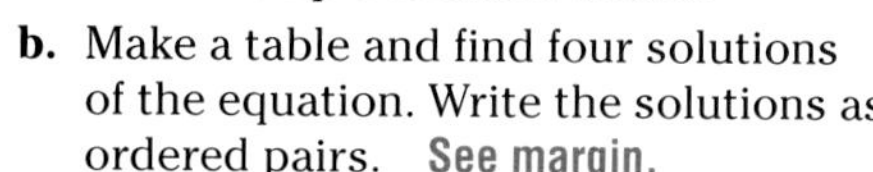

b. Make a table and find four solutions of the equation. Write the solutions as ordered pairs. See margin.

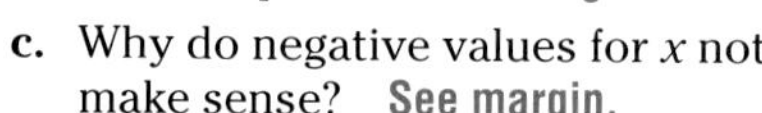

c. Why do negative values for x not make sense? See margin.

d. Graph the equation. See margin.

FYI

The American Forest Council estimates that nearly 70 percent of the forest land that existed in the U.S. when European settlers first landed in the early 1600s is still forested.

19a. $y = 0.5x$

Exercises: Practicing and Applying the Concept

Independent Practice

Which ordered pair(s) is a solution of the equation?

20. $4x + 2y = 8$ a **a.** (2, 0) **b.** (0, 2) **c.** (0.5, −3) **d.** (1, −2)

21. $2a - 5b = 1$ a, d **a.** (−2, −1) **b.** (2, 1) **c.** (7, 3) **d.** (−7, −3)

22. $3x = 8 - y$ b, c **a.** (3, 1) **b.** (2, 2) **c.** (4, −4) **d.** (8, 0)

Cooperative Learning

Round Table Give each group of students a page with a linear equation written at the top. The first student writes an ordered pair that is a solution to the equation. Then the paper is passed to the next student for another solution. After a list of several points is made, have students graph the equation.

For more information on this strategy, see *Cooperative Learning in the Mathematics Classroom*, one of the titles in the Glencoe Mathematics Professional Series, p. 21.

Find four solutions for each equation. Write the solutions as ordered pairs. **23–37. See Solutions Manual.**

23. $y = 2.5x$ 24. $y = 4x$ 25. $y = 3x + 7$
26. $y = x$ 27. $y = -x - 4$ 28. $y = 10x - 1$
29. $x + y = 1$ 30. $2x + y = 10$ 31. $y - x = 1$
32. $y = \frac{1}{3}x$ 33. $y = -\frac{1}{2}x + 3$ 34. $y = \frac{2}{3}x - 1$
35. $x + y = 0$ 36. $y = 5$ 37. $x = -2$

Determine whether each relation is linear.

38. $3x + y = 20$ **yes** 39. $y = x^2$ **no** 40. $y = 5$ **yes**

Graph each equation. **41–49. See Solutions Manual.**

41. $y = x + 1.5$ 42. $y = -\frac{1}{2}x$ 43. $y = \frac{2}{3}x + 1$
44. $y = x$ 45. $2x + y = 5$ 46. $x - y = 4$
47. $x + y = 2$ 48. $y = 5$ 49. $x = -2$

Translate each sentence into an equation. Then find four solutions for each equation. Write the solutions as ordered pairs. Graph each equation. **50–52. See Solutions Manual.**

50. Some number is four more than a second number.
51. Some number is one-fourth a second number.
52. Four times a number minus a second number is 8.

53. A solution of $y = cx + 1$ is (2, 5). Find the value of c. **2**
54. A solution of $2x + ay = 4$ is (−1, 2). Find the value of a. **3**
55. Find a linear equation that has the following solutions. {(0, 0), (1, −3), (2, −6), (−1, 3)} **$y = -3x$ or $3x + y = 0$**

Critical Thinking

56. Find an ordered pair that is a solution for both $x + y = 15$ and $x - y = -1$. **(7, 8)**

Applications and Problem Solving

57. **Geometry** The formula for finding the perimeter of a square with sides s units long is $P = 4s$.

a. Find five ordered pairs of values that satisfy this condition. **See margin.**
b. Draw the graph that contains these points. **See margin.**
c. Why do negative values of s make no sense? **See margin.**

58. **Physics** As a thunderstorm approaches, you see lightning as it occurs, but you hear the accompanying sound of thunder a short time afterward. The distance y, in miles, that sound travels in x seconds, is given by the equation $y = 0.21x$. **58a. See margin.**
a. Create a table to find five ordered pairs of values that relate the time it takes to hear thunder and the distance from the lightning.
b. Graph the equation. **See Solutions Manual.**
c. How far away is lightning when the thunder is heard 2.5 seconds after the light is seen? **about 0.525 miles**

Error Analysis

Watch for students who confuse x and y values. The ordered pair is (x, y), but the equation is $y = ax + b$. One way to prevent this is by emphasizing the use of a table to record values for x and y.

Assignment Guide

Core: 21–55 odd, 56–57, 59–70
Enriched: 20–54 even, 56–70

For **Extra Practice**, see p. 759.

The red A, B, and C flags, printed only in the Teacher's Wraparound Edition, indicate the level of difficulty of the exercises.

Additional Answers

57a. Sample answer: (1, 4), (2, 8), (4, 16), (6, 24), (8, 32)

57b.

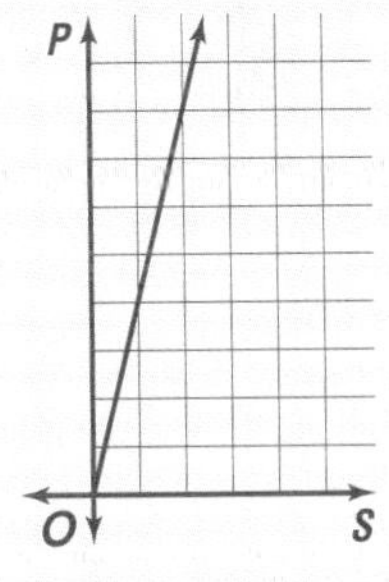

57c. Length cannot have a negative value.

Practice Masters, p. 64

NAME ________________ DATE ________

8-3 Practice
Graphing Linear Relations

Student Edition
Pages 385–390

Which ordered pair(s) is a solution of the equation?

	A.	B.	C.	D.
1. $2a + 3b = 11$ **B, C**	(3, 1)	(1, 3)	(−2, 5)	(4, −1)
2. $2x = 6 - y$ **D**	(4, −4)	(2, −1)	(−4, 3)	(5, −4)
3. $5c - 7d = -4$ **A**	(2, 2)	(−2, −2)	(−2, 2)	(0, 2)

Find four solutions for each equation. Write the solutions as ordered pairs. **Answers may vary. Four possible solutions are given.**

4. $y = 2x$ **(0, 0), (2, 4), (1, 2), (−1, −2)**
5. $y = 5x + 2$ **(2, 12), (1, 7), (4, 22), (−3, −13)**
6. $3x + y = 7$ **(1, 4), (3, −2), (−5, 22), (−2, 13)**
7. $y = -6x + 9$ **(−4, 33), (−1, 15), (1, 3), (5, −21)**
8. $x = -3$ **(−3, 0), (−3, 2), (−3, −1), (−3, 4)**
9. $-2x + y = -4$ **(0, −4), (3, 2) (−1, −6), (−4, −12)**
10. $y = 1$ **(−1, 1), (2, 1), (4, 1), (0, 1)**
11. $y = \frac{1}{4}x$ **(−8, −2), (0, 0), (−4, −1), (8, 2)**
12. $y = \frac{1}{2}x + 5$ **(2, 6), (4, 7), (−2, 4), (−8, 1)**

Determine whether each relation is linear.

13. $y = -4$ **yes** 14. $3 = 2x + y$ **yes** 15. $y = \frac{1}{4}x^2$ **no**

Graph each equation.

16. $y = \frac{1}{4}x$ 17. $y = -2x + 4$ 18. $y = \frac{1}{2}x - 1$

Group Activity Card 8-3

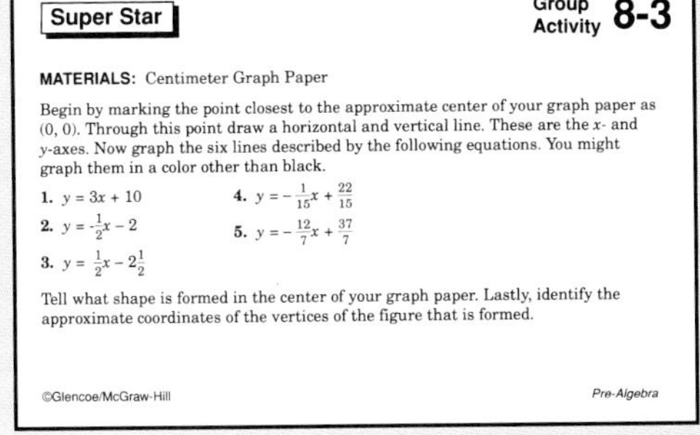

Super Star — Group Activity **8-3**

MATERIALS: Centimeter Graph Paper

Begin by marking the point closest to the approximate center of your graph paper as (0, 0). Through this point draw a horizontal and vertical line. These are the x- and y-axes. Now graph the six lines described by the following equations. You might graph them in a color other than black.

1. $y = 3x + 10$
2. $y = -\frac{1}{2}x - 2$
3. $y = \frac{1}{2}x - 2\frac{1}{2}$
4. $y = -\frac{1}{15}x + \frac{22}{15}$
5. $y = -\frac{12}{7}x + \frac{37}{7}$

Tell what shape is formed in the center of your graph paper. Lastly, identify the approximate coordinates of the vertices of the figure that is formed.

 Pre-Algebra

Additional Answers

19c. Sample answer: There is no such thing as a negative number of seedlings.

19d. **Tree Survival**

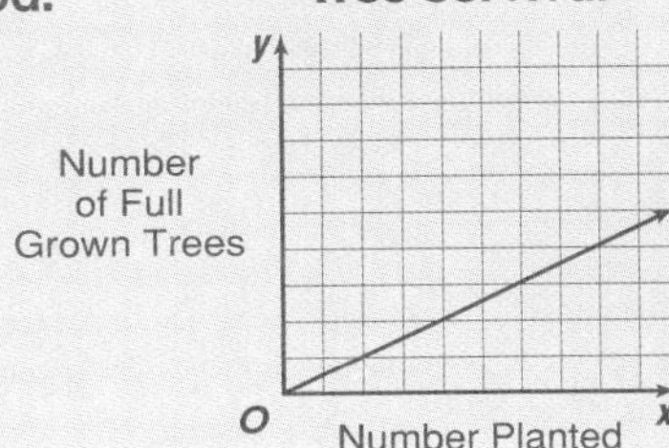

4 ASSESS

Closing Activity

Speaking Have students write a linear equation, find at least four solutions, graph the ordered pairs, and then draw a line through all four points.

Chapter 8, Quiz A (Lessons 8-1 through 8-3) is available in the *Assessment and Evaluation Master*, p. 211.

Additional Answers

58a.

x	y = 0.21x	y	(x, y)
1	y = 0.21(1)	0.21	(1, 0.21)
2	y = 0.21(2)	0.42	(2, 0.42)
3	y = 0.21(3)	0.63	(3, 0.63)
4	y = 0.21(4)	0.84	(4, 0.84)
5	y = 0.21(5)	1.05	(5, 1.05)

67.

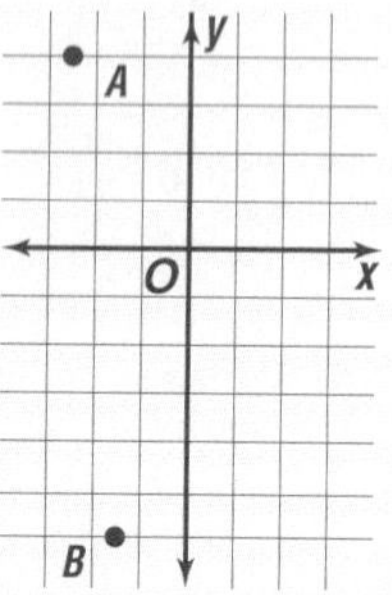

Enrichment Masters, p. 64

NAME ______ DATE ______

8-3 Enrichment
Curves

Student Edition Pages 385–391

Step 1 Make a table of ordered pairs that satisfy the equation.
Step 2 Graph each point.
Step 3 Draw a smooth curve connecting the points.

If you know the graph is going to be a straight line then you only need 2 points (a third one as a check). But if the graph is going to be a curved line, then you need enough points to be able to sketch a nice smooth curve. This may be 5 or more points.

Sketch a smooth curve through the points shown. The right side of the first curve has been done for you.

1.

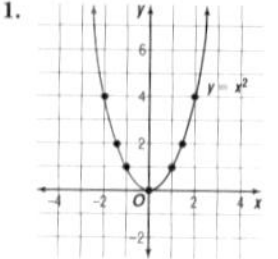

2.

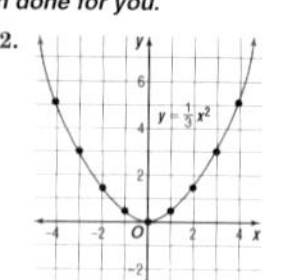

3.

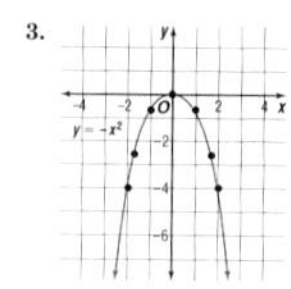

Graph each equation.

4. $y = 2x^2$

x	2	$1\frac{1}{2}$	1	0	1	$1\frac{1}{2}$	2
y	8	$\frac{9}{2}$	2	0	2	$\frac{9}{2}$	8

5. $y = x^2 + 2$

x	3	2	1	0	1	2	3
y	11	6	3	2	3	6	11

6. $y = x^2 - 1$

x	3	2	1	0	1	2	3
y	8	3	0	-1	0	3	8

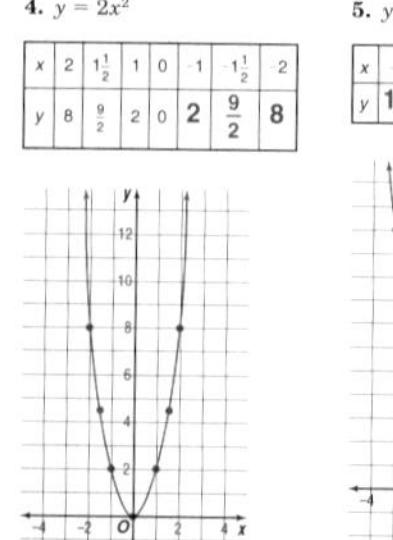

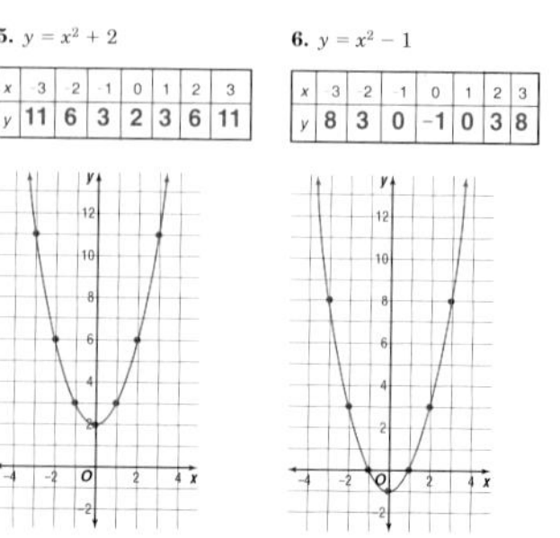

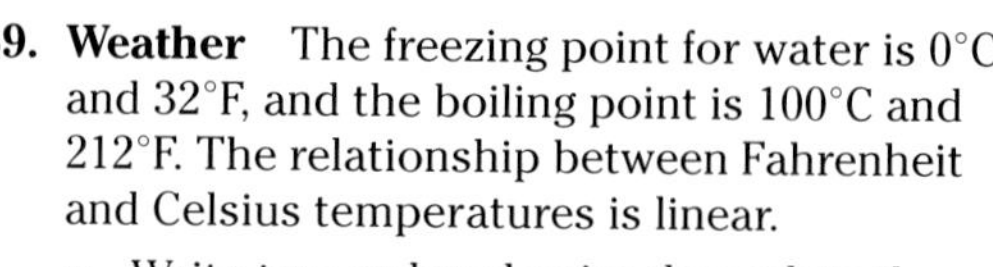

59. Weather The freezing point for water is 0°C and 32°F, and the boiling point is 100°C and 212°F. The relationship between Fahrenheit and Celsius temperatures is linear.

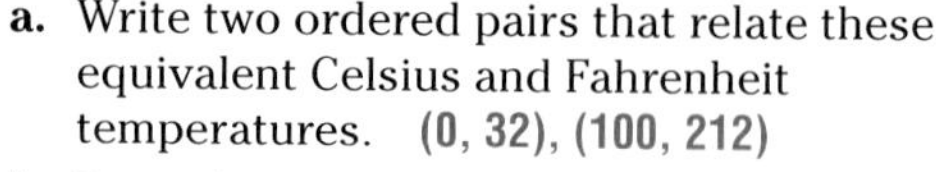

a. Write two ordered pairs that relate these equivalent Celsius and Fahrenheit temperatures. **(0, 32), (100, 212)**

b. Draw the graph that contains these two points. **See Solutions Manual.**

c. How could you use this graph to find other equivalents of other Celsius and Fahrenheit temperatures?

d. Use the graph to estimate two equivalent Celsius and Fahrenheit temperatures. **Sample answer: (20, 68)**

59c. Find the coordinates of other points on the graph.

Mixed Review

60. Health What relationship (*positive*, *negative*, or *none*) do you think a scatter plot of Calorie intake and weight gained might show? (Lesson 8-2) **positive**

61. Solve $2a - 5 > 17$. (Lesson 7-6) **$a > 11$**

62. Express 0.004976 in scientific notation. (Lesson 6-9) **4.976×10^{-3}**

63. Estimate 2.49×1.9 (Lesson 6-2) **4**

64. Carpentry When remodeling a living room, a carpenter needed a $14\frac{3}{4}$-inch piece of molding. How much remained after she cut it from a 36-inch piece of molding? (Lesson 5-5)

64. $21\frac{1}{4}$ inches

65. Find $\frac{7^4}{7^2}$. Express the quotient using an exponent. (Lesson 4-8) **7^2**

66. Geometry The perimeter of any square is 4 times the length of one of its sides. If the perimeter of a square is 56 inches, what is the length of each side of the square? (Lesson 3-5) **14 inches**

67. Graph and label $A(-2.5, 4)$ and $B\left(-1\frac{1}{2}, -6\right)$. (Lesson 2-2) **See margin.**

68. Solve $4b = 36$ mentally. (Lesson 1-6) **9**

69. Simplify $5n + 9n$. (Lesson 1-5) **$14n$**

70. Find the value of $4[12(22 - 19) - 3 \cdot 6]$. (Lesson 1-2) **72**

From the FUNNY PAPERS

In the United States, we measure temperature in degrees Fahrenheit, but most other countries use degrees Celsius.

1. Is it really cooler at "Celsius's Place"? Explain. **See Solutions Manual.**
2. Use the graph you made in Exercise 59 to find the temperature in degrees Fahrenheit if it is 36°C. **96.8°F**

"LET'S GO OVER TO CELSIUS'S PLACE. I HEAR IT'S ONLY 36° OVER THERE."

Extension

Using Research Have students research other equations. Examples from science can be found in a science text. For example, the equation that relates the number of blooms on a cactus (y) to the number of days of sun it gets in a month (x) is $y = 7x - 1$. Have students graph some of the equations they find.

From the FUNNY PAPERS

Have students look in a newspaper to find the temperatures predicted for your area today. Then they can use the graph to convert the temperatures from Celsius to Fahrenheit or vice versa.

HANDS-ON ACTIVITY

8-3B Graphing Parabolas

An Extension of Lesson **8-3**

MATERIALS

graph paper

compass

Not all graphs of equations in two variables are straight lines. In this activity, you will investigate graphs that are nonlinear.

Activity

Work with a partner.

- On a sheet of graph paper, copy the coordinate axes shown at the right.
- Graph and label points $A(0, 1)$, $B(-2, 2)$, and $C(2, 2)$.
- Open the compass to a length of 3 units. With the compass at $F(0, 2)$, draw an arc that intersects the grid line 3 units above the x-axis on each side of the y-axis. Label the points of intersection D and E.
- Open the compass to a length of 4 units. With the compass at point F, draw an arc that intersects the grid line 4 units above the x-axis on each side of the y-axis. Label the points of intersection G and H.
- Repeat this process to plot three more pairs of points.
- Sketch a smooth curve through the points you plotted.

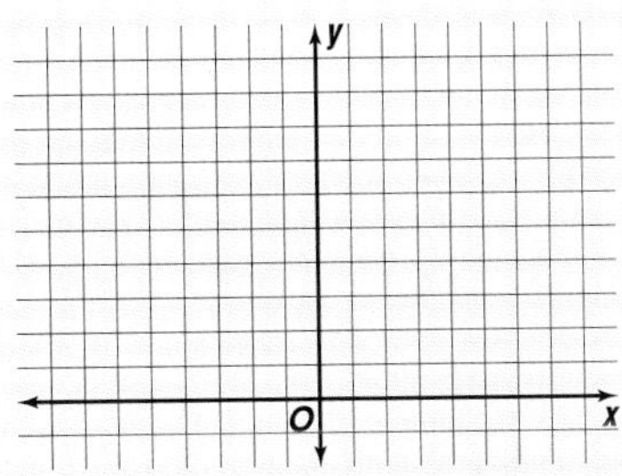

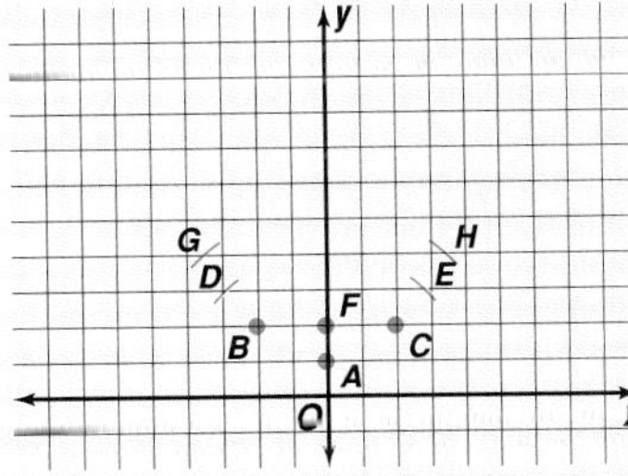

The curve you sketched is called a **parabola**.

TALK ABOUT IT

1. Use the compass to compare the distance above the x-axis and the distance from point F for each point you plotted. **1. Each distance is the same.**
2. Based on your answer to Exercise 1, write a definition for a parabola. (*Hint:* The x-axis is a line.) **See margin.**

Extension

3. **a.** Find solutions for the equation $y = x^2 - 2$ if the replacement set for x is $\{-3, -2, -1, 0, 1, 2, 3\}$. **a–c. See margin.**
   b. Write the solutions as ordered pairs and then graph the equation by drawing a smooth curve through the points.
   c. How does the graph of $y = x^2 - 2$ compare to the graph above?

8-3B LESSON NOTES

NCTM Standards: 1-6, 8

Objective

Investigate graphs that are nonlinear.

Recommended Time

Demonstration and discussion: 15 minutes; Exercises: 30 minutes

Instructional Resources

For each student or group of students

- graph paper
- compass

Math Lab and Modeling Math Masters

- p. 42 (worksheet)

For teacher demonstration

Overhead Manipulative Resources

1 FOCUS

Motivating the Lesson

Discuss with students types of graphs other than line graphs that are possible. Ask a volunteer to describe types of nonlinear graphs that they have seen.

2 TEACH

Teaching Tip Students who may have difficulty distinguishing between the vertical and horizontal lines of graph paper may find it easier to use lined paper instead and mark off one line for the x-axis.

3 PRACTICE/APPLY

Assignment Guide
Core: 1–3
Enriched: 1–3

4 ASSESS

Observing students working in cooperative groups is an excellent method of assessment.

Additional Answers

2. A parabola is the set of all points that are the same distance from a given point and a given line.

3a. {7, 2, −1, −2, −1, 2, 7}

3b. {(−3, 7), (−2, 2), (−1, −1), (0, −2), (1, −1), (2, 2), (3, 7)}; See Solutions Manual for graph.

3c. Sample answer: They both open upward and have their vertex on the y-axis. The graphs pass through different points on the y-axis. The graph of $y = x^2 - 2$ is narrower.

8-4 LESSON NOTES

NCTM Standards: 1-5, 8, 9

Instructional Resources

- Study Guide Master 8-4
- Practice Master 8-4
- Enrichment Master 8-4
- Group Activity Card 8-4
- Graphing Calculator Masters, p. 8

Transparency 8-4A contains the 5-Minute Check for this lesson; **Transparency 8-4B** contains a teaching aid for this lesson.

Recommended Pacing	
Standard Pacing	Day 5 of 13
Honors Pacing	Day 4 of 12
Block Scheduling*	Day 3 of 7 (along with Lesson 8-3)

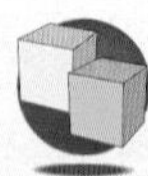

*For more information on pacing and possible lesson plans, refer to the ***Block Scheduling Booklet***.

1 FOCUS

5-Minute Check
(over Lesson 8-3)

Write four solutions for each equation. Write the solutions as ordered pairs.

1. $y = 2x - 4$ **Sample answers: (0, -4), (1, -2), (-1, -6), (2, 0)**
2. $y = \frac{x}{2} - 3$ **Sample answers: (0, -3), (2, -2), (-2, -4), (6, 0)**
3. $x + y = 12$ **Sample answers: (0, 12), (1, 11), (-1, 13), (2, 10)**

Determine whether each relation is linear.

4. $x - y = 3^2$ **yes**
5. $25 - y^2 = x^2$ **no**

Motivating the Lesson

Situational Problem Have students bring in various recipes and explain how they would convert each recipe to serve twice as many or half as many people.

8-4 Equations as Functions

Setting Goals: *In this lesson, you'll determine whether an equation represents a function and find functional values for a given function.*

Modeling with Manipulatives

MATERIALS
index cards
cardboard

Working with a partner, make a function machine using two blank 3-by-5-inch cards. Near the middle of a card, draw three boxes. Cut out the squares on the left and on the right. Label the left "window" INPUT and the right "window" OUTPUT. Write a function rule such as $\times 2 - 3$ in the center box.

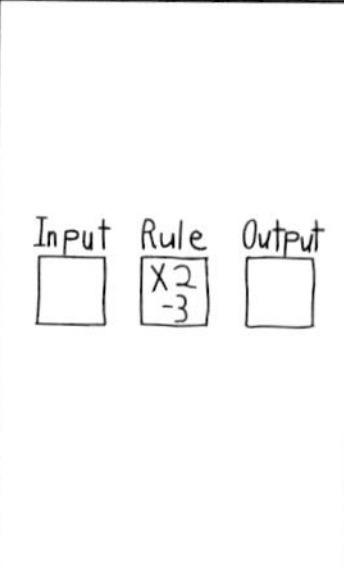

On the other card, list the integers from -5 to 4 in a column about one-half inch from the left edge. Tape this card to a piece of cardboard.

Your Turn Place the function machine over the number column so that -5 is in the left window. Write the output in the right window. Slide the function machine down so that the input is -4. Write the output in the right window. Continue this process for all 10 inputs.

TALK ABOUT IT

a. If the input is represented by x and the output is y, write an algebraic equation that summarizes what the function machine does. $y = 2x - 3$

b. Suppose you were given the output and had to determine the input. How would you go about finding the input? **by working backward**

Learning the Concept

The equation in the function machine above involved using two variables. Since the solutions of an equation in two variables are ordered pairs, such an equation describes a relation. The set of values of x is the domain of the relation. The set of corresponding values of y is the range.

Example 

a. Solve $y = 3x + 4$ if the domain is $\{-2, -1, 0, 1\}$.
b. Graph the equation.
c. Is the equation a function?

a. Make a table of the domain and corresponding range values.

x	y
−2	−2
−1	1
0	4
1	7

b. Use the ordered pairs to graph the equation.

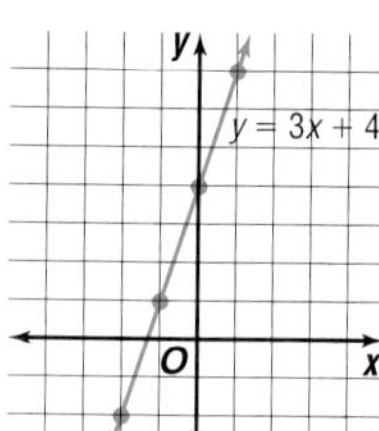

Alternative Learning Styles

Kinesthetic Have students use counters and act as a function machine. For example, for $y = 2x + 3$, a student hands the machine x counters. The machine then doubles that number of counters and adds three more. The result is shown as the value of y for the given number, x.

c. Use the vertical line test to determine if the equation is a function.

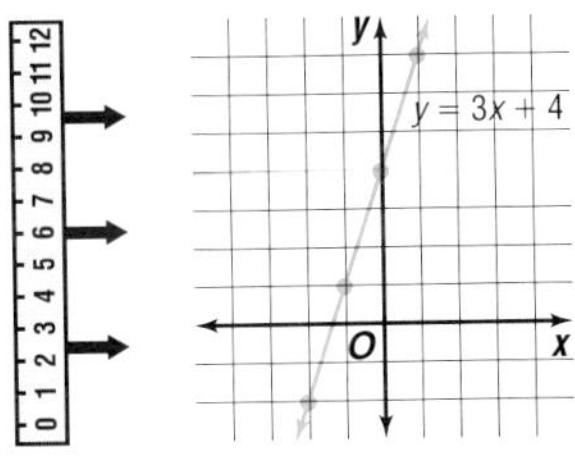

Since the vertical line passes through no more than one point on the graph for each value of x, the equation is a function.

Letters other than f are also used for names of functions.

Equations that represent functions can be written in **functional notation**, $f(x)$. The symbol $f(x)$ is read "f of x" and represents the value in the range that corresponds to the value of x in the domain. For example, $f(8)$ is the element in the range that corresponds to the element 8 in the domain.

You can determine a functional value by substituting the given value for x into the equation. For example, if $f(x) = 7x - 5$ and $x = 2$, then $f(2) = 7(2) - 5$ or 9.

Example 

If $g(x) = 3x + 4$, find each of the following.

a. $g(3)$ **b.** $g(-5)$ **c.** $2g(0)$

$$g(3) = 3(3) + 4 = 9 + 4 = 13$$

$$g(-5) = 3(-5) + 4 = -15 + 4 = -11$$

$$2g(0) = 2[3(0) + 4] = 2[0 + 4] = 2[4] = 8$$

You can use an equation to represent real-world situations.

Example 3

Oceanography

HELP WANTED

Oceanographers are needed in all parts of the world. For more information, contact:

National Ocean Industries Assn.
1050 17th Street NW, Suite 700
Washington, D.C. 20036

Oceanographers use sonar to locate fish and other objects beneath the surface of water. Sonar units emit a pulse of sound and calculate the distance to the object using the time it takes for the pulse to reflect back. You can use the function $vt = d$, where v is the velocity, t is the time, and d is the distance to find the time. The speed or velocity of sound in water is 1454 meters per second. How far below the surface is a school of fish if it takes 0.022 seconds for the sound to return to the sonar?

The sound must go to the fish and return, so the time to the fish is $0.022 \div 2$ or 0.011 seconds.

$$vt = d$$

$$(1454)(0.011) = d \quad \text{Replace } t \text{ with 0.011 and } v \text{ with 1454.}$$

$$15.994 = d \quad \text{Use a calculator.}$$

The fish are about 16 meters from the sonar.

2 TEACH

In-Class Examples

For Example 1

a. Solve $y = 2x - 3$ if the domain is $\{-1, 0, 1, 2\}$. **{(−1, −5), (0, −3), (1, −1), (2, 1)}**

b. Graph the equation.

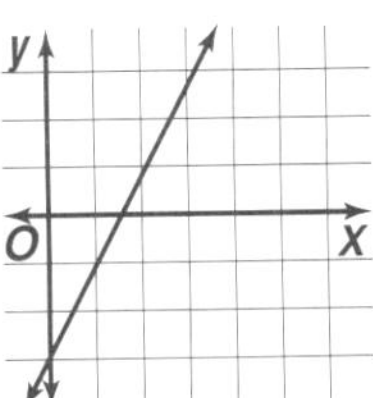

c. Is the equation a function? **yes**

For Example 2

If $h(x) = -4x + 1$, find each of the following.

a. $h(-3)$ **13** **b.** $h(3)$ **−11** **c.** $h(0)$ **1**

For Example 3

A ball is thrown upward with a velocity of 64 feet per second. The height h, in feet, of the ball above the ground after t seconds is given by $h(t) = 64t - 16t^2$. Determine when the ball reaches its maximum height. **in 2 seconds**

Study Guide Masters, p. 65

NAME ______________ DATE ______

8-4 Study Guide
Equations as Functions

Student Edition Pages 392–395

Suppose $y = 3x + 2$ and the domain is $\{^-2, 0, 1\}$. Make a table of the domain and corresponding range values.

x	y
⁻2	⁻4
0	2
1	5

Graph $y = 3x + 2$ using the ordered pairs.

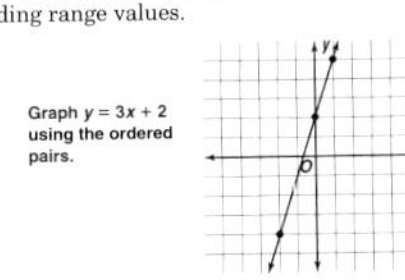

Using the vertical line test, we find that $y = 3x + 2$ is a function.

Equations that represent functions can be written in **functional notation, $f(x)$**. The symbol $f(x)$ represents the value in the range that corresponds to the value of x in the domain. The equation $y = 3x + 2$ can be written as $f(x) = 3x + 2$.

To determine a functional value, substitute the given value for x in the equation. For example, if $f(x) = 3x + 2$ and $x = {}^-3$, then $f(^-3) = 3(^-3) + 2$ or $^-7$.

For each equation,
a. solve for the domain = $\{^-4, 0, 1\}$, and
b. determine whether the equation is a function.

1. $x + y = 12$ 16, 12, 11; yes
2. $x = 3 - y$ 7, 3, 2; yes
3. $y = 8 + x^2$ 24, 8, 9; yes

Given $h(x) = 2x - 9$ and $g(x) = x^2 + 4$, find each value.

4. $h(^-3)$ −3
5. $g(^-5)$ 29
6. $5[h(0)]$ −45
7. $g\left(-\frac{1}{4}\right)$ $4\frac{1}{16}$
8. $g(2.2b)$ $4.84b^2 + 4$
9. $g(3a)$ $9a^2 + 4$
10. $h(^-0.48)$ −9.96
11. $2[g(2)]$ 16
12. $h\left(\frac{1}{2}\right)$ −8

GLENCOE Technology

Interactive Mathematics Tools Software

In this interactive computer lesson, students explore equations and graphs. A **Computer Journal** gives students an opportunity to write about what they have learned.

For Windows & Macintosh

HELP WANTED

Oceanography has traditionally had four major areas of study: research, physical, chemistry, and biology. Newer areas of study include archaeology and conservation.

3 PRACTICE/APPLY

Checking Your Understanding

Exercises 1–11 are designed to help you assess your students' understanding through reading, writing, speaking, and modeling. Work through Exercises 1–3 with your students and then monitor their work on Exercises 4–11.

Assignment Guide

Core: 13–31 odd, 32, 33, 35–40
Enriched: 12–30 even, 32-40

For **Extra Practice**, see p. 759.

The red A, B, and C flags, printed only in the Teacher's Wraparound Edition, indicate the level of difficulty of the exercises.

Additional Answers

1. Sample answer: The graph of the linear equation $x = 1$ is a vertical line but it is not a function because the same element in the domain maps to every element in the range.
2. Sample answer: Substitute 5 for n in the equation and simplify.

Practice Masters, p. 65

NAME ________________ DATE ________

8-4 Practice
Equations as Functions

Student Edition
Pages 392–395

For each equation,
a. solve for the domain = {⁻1, 0, 2, 8}, and
b. determine if the equation is a function.

1. $x + y = 16$ 17, 16, 14, 8; yes
2. $xy = 240$ ⁻240, undefined, 120, 30; yes
3. $y = 8 + 2x$ 6, 8, 12, 24; yes
4. $x = y - 28$ 27, 28, 30, 36; yes
5. $y = {}^-5$ ⁻5, ⁻5, ⁻5, ⁻5; yes
6. $\frac{1}{4}x - 3 = y$ $^-3\frac{1}{4}$, ⁻3, $^-2\frac{1}{2}$, ⁻1; yes
7. $x = 8$ no
8. $x^2 - 4 = y$ ⁻3, ⁻4, 0, 60; yes
9. $y = 6x - 12$ ⁻18, ⁻12, 0, 36; yes

Given $f(x) = 4x + 1$ and $g(x) = x - 3$, find each value.

10. $f(3)$ 13
11. $g(8)$ 5
12. $g(^-2)$ ⁻5
13. $f(^-18)$ ⁻71
14. $f(^-2.5)$ ⁻9
15. $f\left(\frac{1}{4}\right)$ 2
16. $g(2.35)$ ⁻0.65
17. $g\left(\frac{1}{2}\right)$ $^-2\frac{1}{2}$
18. $g(4c)$ $4c - 3$
19. $f(3d)$ $12d + 1$
20. $4[f(a)]$ $16a + 4$
21. $f(0)$ 1
22. $2[g(0)]$ ⁻6
23. $3[f(4)]$ 51
24. $g[g(6)]$ 0

Checking Your Understanding

Communicating Mathematics

Read and study the lesson to answer these questions. 1–2. See margin.

1. **Give an example** of a linear equation that is not a function.
2. **Demonstrate** how to find $f(5)$ if $f(n) = 2n - 6$.
3. **Write** a function for which $f(9) = 13$. Sample answer: $f(x) = 2x - 5$

Guided Practice

For each equation,
a. solve for the domain = {−3, 0, 1, 4}, and
b. determine if the equation is a function.

4. $y = 5x + 8$ −7, 8, 13, 28; yes
5. $g(x) = x^2 + 1$ 10, 1, 2, 17; yes
6. $x = 7$ no

Given $h(x) = 4 - \frac{1}{2}x$, find each value.

7. $h(2)$ 3
8. $h(0)$ 4
9. $h(-4)$ 6
10. $h\left(\frac{1}{2}\right)$ $3\frac{3}{4}$

11. 9 weeks
12. −21, −23, −26, −28; yes
13. 8, 16, 28, 36; yes
14. −180, undefined, 120, 72; yes
15. 2, 2, 2, 2; yes
17. 5, 4, 2.5, 1.5; yes
18. 26, 24, 21, 19; yes
19. 7, 11, 2, −14; yes

11. **Personal Finance** Carlos is saving money to buy a used car for $2200. He already has $1500 in his savings account. He plans to add $80 each week from the money he earns working at the YMCA. The equation $f(x) = 1500 + 80x$ describes Carlos' total savings $f(x)$ after x weeks. After how many weeks will he have enough to purchase the car?

Exercises: Practicing and Applying the Concept

Independent Practice

For each equation,
a. solve for the domain = {−2, 0, 3, 5}, and
b. determine if the equation is a function.

A
12. $x + y = -23$
13. $y = 4x + 16$
14. $xy = 360$
15. $y = 2$
16. $x = -8.8$ no
17. $y = 4 - \frac{1}{2}x$
18. $x = 24 - y$
19. $x^2 + y = 11$

Given $f(x) = 5x + 3$ and $g(x) = x^2 - 2$, find each value. 30. $2b^2 - 4$

B
20. $g(5)$ 23
21. $f(-10)$ −47
22. $f(-4)$ −17
23. $g\left(\frac{1}{2}\right)$ $-1\frac{3}{4}$
24. $g(-3.3)$ 8.89
25. $5[g(3)]$ 35
26. $f\left(\frac{1}{3}\right)$ $4\frac{2}{3}$
27. $f(0.25)$ 4.25

C
28. $f(2a)$ $10a + 3$
29. $g(2b)$ $4b^2 - 2$
30. $2[g(b)]$
31. $f[f(4)]$ 118

Critical Thinking

32. When a relation contains two variables, you can find its *inverse* by interchanging the variables. Use this method to determine whether the inverse of $y = 3x - 1$ is a function. See margin.

Applications and Problem Solving

33a. $\left\{-17\frac{7}{9}, 0, 22\frac{2}{9}, 37\right\}$

33. **Science** A temperature in degrees Fahrenheit has a corresponding Celsius temperature. Let f represent degrees Fahrenheit. Then degrees Celsius can be represented by $C = \frac{5(f - 32)}{9}$.
    a. If the domain is {0, 32, 72, 98.6}, what is the range?
    b. Graph the equation. See Solutions Manual.
    c. Is the relation described by $C = \frac{5(f - 32)}{9}$ a function? Explain. See margin.

Reteaching

Using Tables Provide students with an equation, such as $y = {}^-x - 5$ and a table with given values for x, such as shown below. Have students complete the table and graph the equation.

x	0	2	⁻3	⁻5
y	⁻5	⁻7	⁻2	0

Group Activity Card 8-4

Card Collector

Group Activity **8-4**

MATERIALS: deck of "Equation" cards (see back) • number cube

Shuffle and divide the cards evenly between partners. One partner rolls the number cube. Each partner turns one of his or her cards face up. Using the number indicated on the number cube, each evaluates the function on his or her equation card. The partner having the output with the greatest value collects both equation cards. Roll again, and continue this process until all equation cards have been used. The partner with the greatest number of cards wins the game. Reshuffle and play again. The partner that wins two out of three games wins the match.

Pre-Algebra

34a. See Solutions Manual.

34. Anthropology When a human skeleton is found, an anthropologist uses the length of certain bones to determine the height of the living person. A *femur* is the bone from the knee to the hip. The height h, in centimeters, of a female with a femur of length x is estimated by $h = 61.412 + 2.317x$.

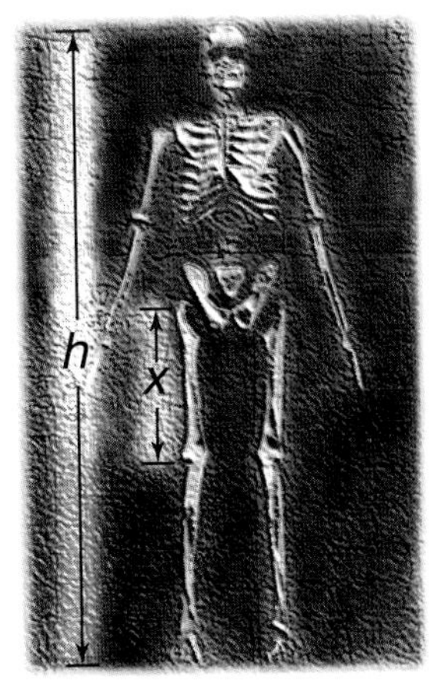

a. Graph the equation $h = 61.412 + 2.317x$.

b. Is the relation described by $h = 61.412 + 2.317x$ a function? Explain. See margin.

c. A woman's femur measuring 49 cm is found in some ruins. What was the height of the person? about 175 cm

Mixed Review

35. Mining The deepest mine in the world is located in Carletonville, South Africa. The temperature of the walls of the mine varies with their depth. The temperature in degrees Fahrenheit, y, is estimated by $y = 18x + 66.5$, where x is the depth in kilometers. (Lesson 8-3)

a. Create a table to find five ordered pairs of values that relate the depth of a mine and the temperature of its walls. See margin.

b. Graph the equation.

c. What would the temperature of the walls be if the mine is 3.6 kilometers deep? 131.3°F

35b. See Solutions Manual.

36. Geometry Find the circumference of a circle with a radius of 14 yards. (Lesson 7-4) about 87.92 yards

37. Find the value of $\left(-\frac{3}{4}\right)^2$. (Lesson 6-3) $\frac{9}{16}$

38. Write $\frac{15rs^2}{50rs}$ in simplest form. (Lesson 4-6) $\frac{3s}{10}$

39. Solve $-3a \geq 18$ and graph the solution. (Lesson 3-7) See margin.

40. Geography The table at the right shows the number of miles of seacoast in selected states. Make a bar graph of the data. (Lesson 1-10) See margin.

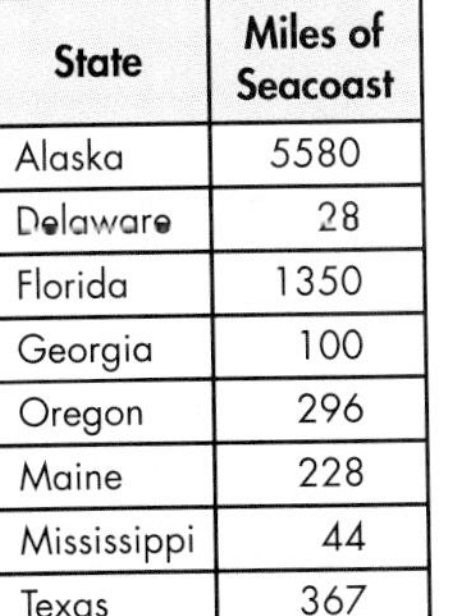

State	Miles of Seacoast
Alaska	5580
Delaware	28
Florida	1350
Georgia	100
Oregon	296
Maine	228
Mississippi	44
Texas	367

4 ASSESS

Closing Activity

Modeling Have students draw a flow chart showing how to graph a linear equation or showing how to determine whether a relation is a function.

Additional Answers

32. Sample answer: The inverse is $x = 3y - 1$. It is a function since no member of the domain is paired with more than one member of the range.

33c. Yes; no member of the domain is paired with more than one member of the range.

34b. Yes; no member of the domain is paired with more than one member of the range.

35a.

x	$y = 18x + 66.5$	y	(x, y)
1	$y = 18(1) + 66.5$	84.5	(1, 84.5)
5	$y = 18(5) + 66.5$	156.5	(5, 156.5)
10	$y = 18(10) + 66.5$	246.5	(10, 246.5)
15	$y = 18(15) + 66.5$	336.5	(15, 336.5)
20	$y = 18(20) + 66.5$	426.5	(20, 426.5)

Enrichment Masters, p. 65

NAME ______________ DATE ______

8-4 Enrichment
Equations with Two Variables

Student Edition Pages 392–395

Complete the table for each equation.

1. $y = 7 + x$

x	y
-4	3
-2	5
-8	-1

2. $y = 2x + 4$

x	y
6	16
4	12
-1	2

3. $y = x - 9$

x	y
3	-6
5	-4
0	-9

4. $y = 3x - 2$

x	y
2	4
-1	-5
3	7

5. $y = \frac{x}{4}$

x	y
8	2
16	4
-24	-6

6. $y = -6x + 1$

x	y
1	-5
2	-11
-2	13

7. $y = 9 - 2x$

x	y
3	3
1	7
4	1

8. $y = \frac{x + 5}{3}$

x	y
4	3
7	4
-8	-1

9. $y = \frac{x}{2} + 5$

x	y
8	9
6	8
10	10

10. $y = x^2$

x	y
2	4
1	1
4	16

11. $y = x^2 - 3$

x	y
3	6
5	22
0	-3

12. $y = 1 - 2x$

x	y
-1	3
-3	7
-5	11

Extension

Using Connections Give students a data chart comparing altitude in feet and temperature in degrees Celsius. The altitudes are: 0, 1500, 6000, and 9000. The corresponding temperatures are 15, 5, -26, and -44. Have students graph the data and determine the altitude at which the temperature is -32°C. 7000 feet

Additional Answers

39. $a \leq -6$;

-6 -4 -2 0

40.

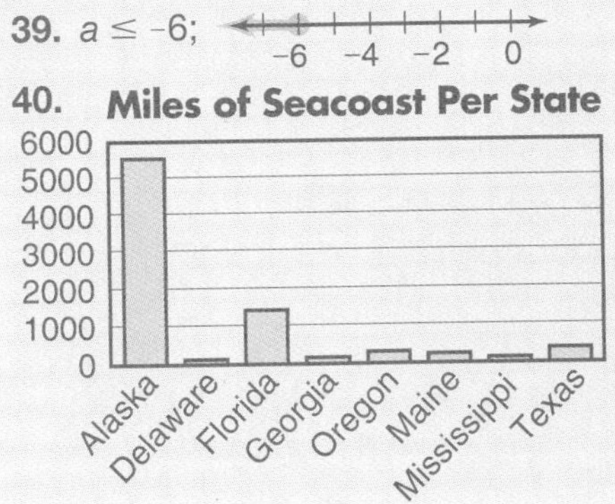

8-5 LESSON NOTES

NCTM Standards: 1-5, 8, 9, 12

Instructional Resources

- Study Guide Master 8-5
- Practice Master 8-5
- Enrichment Master 8-5
- Group Activity Card 8-5
- Assessment and Evaluation Masters, pp. 210, 211
- Tech Prep Applications Masters, p. 15
- Activity Masters, p. 22

Transparency 8-5A contains the 5-Minute Check for this lesson; **Transparency 8-5B** contains a teaching aid for this lesson.

Recommended Pacing	
Standard Pacing	Day 6 of 13
Honors Pacing	Day 5 of 12
Block Scheduling*	Day 4 of 7 (along with Lesson 8-6)

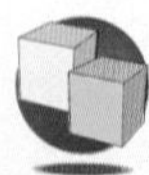

*For more information on pacing and possible lesson plans, refer to the ***Block Scheduling Booklet***.

1 FOCUS

5-Minute Check
(over Lesson 8-4)

Determine whether each relation is a function.

1. $x - y = -3$ **yes**
2. $x - y = -3 - y$ **no**
3. Given $f(x) = 2x - 3$, find the value of $f(-2.5)$. **−8**

Julie and Randy Johnson have $5400 in a savings account. Each week they are going to add $110 to the account. The equation $f(x) = 5400 + 110x$ describes their total savings $f(x)$ after any number of weeks x.

4. How much will they have in savings after 9 weeks? **$6390**
5. After how many weeks will they have enough to buy a used car for $8000? **24 weeks**

8-5 Problem-Solving Strategy Draw a Graph

Setting Goals: *In this lesson, you'll solve problems by using graphs.*

Modeling a Real-World Application: Air Travel

Imagine you are sitting in your summer clothes and sandals eating lunch and it's 74 degrees below zero! It happens every day on high-flying airplanes that are far from the heat of Earth's surface.

Altitude (thousands of feet)	Temp. (°F)
5	41
10	23
15	4
20	−15
25	−33

The temperature at different altitudes above sea level when the temperature at sea level is 60°F is shown in the table at the left.

The table does not include an altitude for −74°. At what altitude would a jet be flying when the thermometer has this reading? *This problem will be solved in Example 1.*

Learning the Concept

A graph represents the same information as its equation. But, because a graph is visual, it allows you to see patterns that may not be obvious from the equation. A graph is a powerful tool in problem solving.

Example 1

APPLICATION

Air Travel

Refer to the application at the beginning of the lesson. Find the altitude where the temperature is −74°F. Assume that the temperature decreases at a constant rate.

Explore You need to estimate an altitude that is not in the table. You know the temperature at five different altitudes and that the temperature decreases at a constant rate.

Plan One way to solve this problem is to graph the given information. Then read the graph to find the altitude that corresponds to a temperature of −74°F.

Solve Let the horizontal axis of the graph represent altitude. Let the vertical axis represent the temperature. Graph each ordered pair (*altitude, temperature*) using the pairs of values taken from the table. Then draw the line that contains these points.

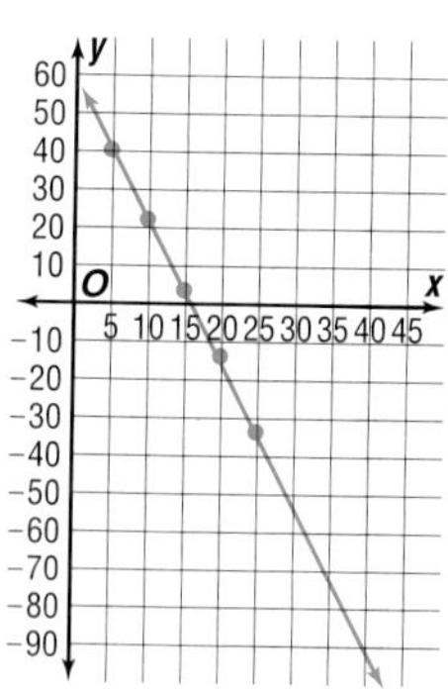

Alternative Teaching Strategies

Reading Mathematics Have students discuss this problem and decide how they could solve it. Encourage students to devise as many strategies as possible. *A swimming pool has been filled with water and the heater turned on. If the temperature rises 2.5°C each hour, by how much will the temperature have risen after 6 hours?* **Sample answers: Use an equation. Use a graph. Make a table.**

An estimate of $-74°$ would be about halfway between -70 and -80 on the vertical axis. Estimate the altitude by reading the corresponding value on the horizontal axis. The ordered pair (36, -74) appears to be on the graph, so $-74°$ corresponds to about 36,000 feet.

Examine Is the solution reasonable? The lowest temperature in the table, $-33°$, corresponds to an altitude of 25,000 feet. The temperature drops about 19 degrees for each 5,000 feet increase in altitude. $-74 - (-33°) = -41°$ and -41 is about $2(-19)$. Therefore, the altitude will be about $2 \times 5{,}000$ or 10,000 feet higher. Since $25{,}000 + 10{,}000 = 35{,}000$, a solution of 36,000 is reasonable.

You can use a TI-82 graphing calculator to plot points and construct a line without erasing the points. Then you can use the trace function to answer the question.

Example 2

Energy

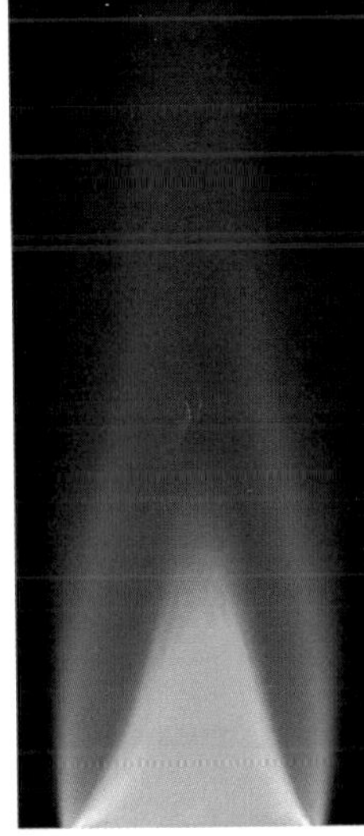

The ordered pairs (50.7, 16.47) and (99.92, 25.42) represent data from two rows of a table comparing cubic meters of natural gas used and the cost. What would the bill for 140 cubic meters of natural gas be?

Use the scatter plot procedure on page 379 to plot the two ordered pairs using □ as the mark.

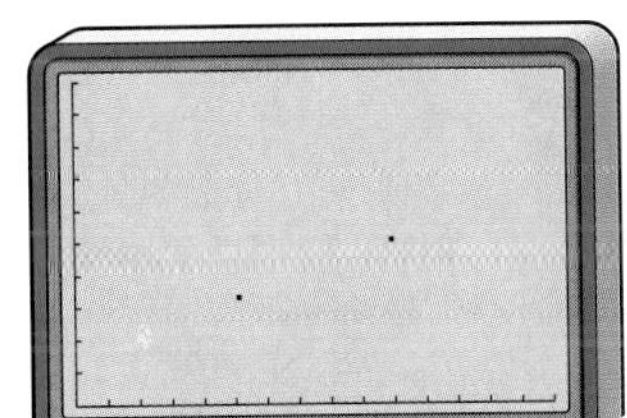

The calculator can draw the graph of the line that passes through the points.

Enter:

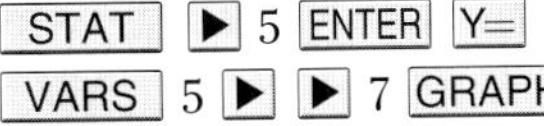

Press TRACE and the up arrow key, to trace the line. Then press the right arrow key until the cursor gets to the point on the line where the x-coordinate is about 140. The corresponding value of y is the cost for 140 cubic meters of gas.

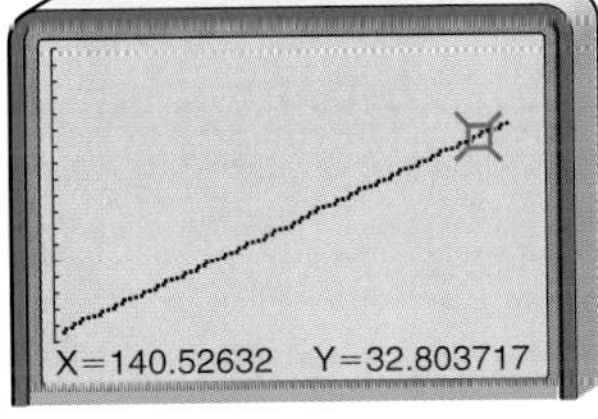

The cost will be about $32.80.

Checking Your Understanding

Communicating Mathematics

Read and study the lesson to answer these questions.

1. **Explain** how a graph can provide the same information as a table and at the same time provide more information. **See margin.**
2. **Describe** how you would make a graph using a table of data. **See margin.**

Motivating the Lesson

Questioning A paper cup is attached to a spring. When a marble is placed in the cup, the spring stretches 0.8 cm. When 5 marbles are placed in the cup, the spring stretches 3.0 cm. How much will the spring stretch if 12 marbles are placed in the cup? 6.85 cm

2 TEACH

In-Class Examples

For Example 1

One way to estimate a dog's age in terms of human years is to add 4 to the animal's age and multiply the result by 4. What is a dog's actual age when its age in terms of human years is 48? **8 years**

For Example 2

Suppose the volume of a certain gas is 380 cm³ at 25°C and 580 cm³ at 75°C. Find the volume of that gas at 40°C. **440°C**

Teaching Tip Discuss the similarities and differences between graphing a line from an equation on a graphing calculator and graphing a line from points.

Study Guide Masters, p. 66

NAME ______________ DATE ______

8-5 Study Guide
Problem-Solving Strategy:
Draw a Graph

Student Edition Pages 396–399

The interest Leon earned on $160 was $8. If he had deposited $200, he would have earned $10. How much money would he need to deposit to earn $12?

Explore Given the interest earned for $160 and $200, you need to determine how much money to deposit to earn $12.

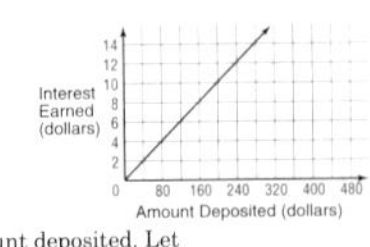

Plan Graph the given information. Then read the graph to find how much money to deposit to earn $12.

Solve Let the horizontal axis represent the amount deposited. Let the vertical axis represent the amount earned. Graph the ordered pairs (160, 8) and (200, 10). Draw a line that contains these points. He will need to deposit $240.

Examine Recall that Leon earned $8 for $160. Since $12 = 8 \times 1.5$, the deposit should be 1.5×160 or $240 to earn $12 in interest.

Use a graph to solve each problem. Assume that the rate is constant in each problem.

1. Scoopers ice cream shop sells one scoop of ice cream for $1.60. They sell three scoops for $4.80. How much money is needed to buy two scoops of ice cream? **$3.20**

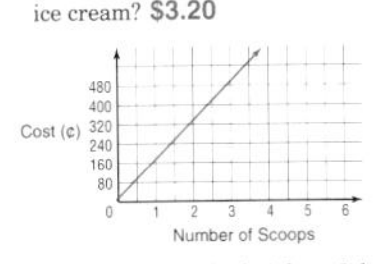

2. Kevin earned $90 for baby-sitting 15 hours. He would have earned $120 for five more hours. How much does he charge per hour? How much will Kevin earn if he works 10 hours? **$6.00; $60.00**

3. Sami measures the heights of the steps going into her house. The 2nd step is 1 foot above ground. The 5th step is $2\frac{1}{2}$ feet above ground. What is the height of the 11th step? **$5\frac{1}{2}$ feet**

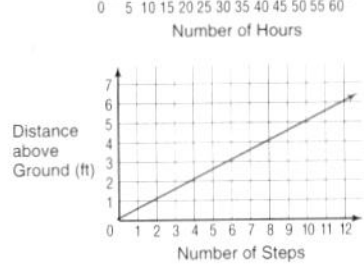

Reteaching

Using Questioning Draw a graph of a linear equation on the chalkboard or overhead. Have students list the ordered pairs that can be identified from the line.

Additional Answers

1. Because a graph is visual, it allows you to see patterns that may not be obvious from the table.
2. Let each axis represent a column of data. Graph points using ordered pairs of values taken from the two columns of data. Then draw a line that contains the points.

3 PRACTICE/APPLY

Checking Your Understanding

Exercises 1–5 are designed to help you assess your students' understanding through reading, writing, speaking, and modeling. You should work through Exercises 1–2 with your students and then monitor their work on Exercises 3–5.

Error Analysis

Watch for students who always assume they are given the value for x in an equation. Have students ask themselves or a partner what information is given. Then, have them point to the axis where that information is represented and read the corresponding horizontal or vertical value.

Assignment Guide

Core: 6–17
Enriched: 6–17
All: Self Test, 1–10

For **Extra Practice**, see p. 759.

The red A, B, and C flags, printed only in the Teacher's Wraparound Edition, indicate the level of difficulty of the exercises.

Practice Masters, p. 66

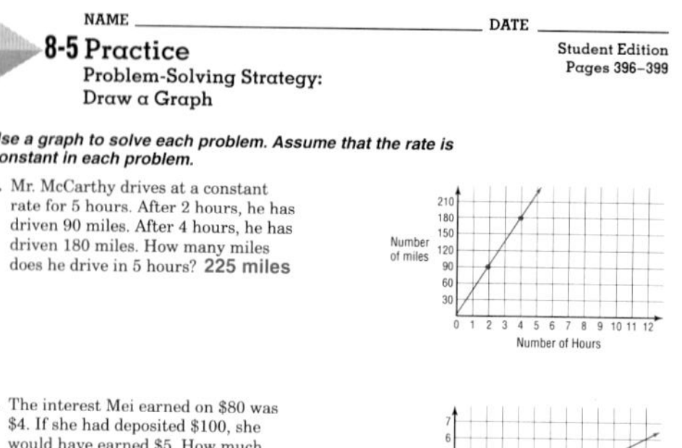

NAME ______ DATE ______

8-5 Practice Student Edition Pages 396–399
Problem-Solving Strategy: Draw a Graph

Use a graph to solve each problem. Assume that the rate is constant in each problem.

1. Mr. McCarthy drives at a constant rate for 5 hours. After 2 hours, he has driven 90 miles. After 4 hours, he has driven 180 miles. How many miles does he drive in 5 hours? **225 miles**

2. The interest Mei earned on \$80 was \$4. If she had deposited \$100, she would have earned \$5. How much would she earn for \$120? **\$6**

3. Larry earned \$150 for working $37\frac{1}{2}$ hours. He would have earned \$160 if he had worked $2\frac{1}{2}$ hours more. What is Larry paid per hour? How much will he earn if he works 30 hours? **\$4; \$120**

4. During a storewide sale, a TV that usually sells for \$450 is on sale for \$360. A stereo that usually sells for \$600 is on sale for \$480. What would the sale price be on a VCR that usually sells for \$500? **\$400**

Guided Practice

4. (5, 2), (11, 3.2)

3. **Estimate** the corresponding value for each item using the graph you created in Example 2. **3b. about \$12.76**
   a. cost: \$21.50 **about 78.2 m³**
   b. gas used: 30 cubic meters

4. **Forestry** An area has been reforested. After five years, one of the trees is 2 meters tall. After six more years, the same tree is 3.2 meters tall. Name two ordered pairs that can be used to graph the information.

5. **Oceanography** The average atmospheric pressure at sea level is 14.7 pounds per square inch. As divers go deeper into the ocean, the pressure increases as shown in the chart. Use a graph to estimate the pressure at 10,000 feet below sea level. Assume that the rate is constant. **about 4500 lb/in²**

Depth (feet)	Pressure (lb/in²)
(Sea level) 0	14.7
500	237
1500	683
4500	2019

Exercises: Practicing and Applying the Concept

Independent Practice

A

6. **Retail** During a storewide sale, a television that usually sells for \$600 is on sale for \$450. A CD player that usually sells for \$200 is on sale for \$150. Name two ordered pairs that can be used to graph the information. **(600, 450), (200, 150)**

Use a graph to solve each problem. Assume that the rate is constant in each problem. **7. \$1.11 a gallon**

B

7. **Business** A gas station sells 87-octane gas for \$1.03 a gallon and 93-octane sells for \$1.28. What is the price of 89-octane gas?

8. **Physical Science** A cup is attached to a spring. When a marble is placed into the cup, the spring stretches 0.6 cm. When 6 marbles are in the cup, the spring stretches 4.0 cm. How much will the spring stretch if 10 marbles are in the cup? **about 6.6 cm**

9. **Energy** Natural gas companies use *degree days* to determine the amount of gas to charge for on a calculated bill. In September, there were 300 degree days, and a homeowner used 85 cubic meters of natural gas. In January, there were 700 degree days, and the homeowner used 265 cubic meters of gas. How much gas would be used if there are 450 degree days? **about 150 cubic meters**

Choose
Estimation
Mental Math
Calculator
Paper and Pencil

Solve using the graph. Explain your method.

10. Tamoko knows that her car can travel about 30 miles on one gallon of gasoline. The graph shows this relationship.
    a. Tamoko is starting on a 250-mile trip. If she has 6 gallons of gasoline in her car, will she have to buy more gasoline sometime during the trip? **yes**
    b. If so, approximately when? **before driving 180 miles**

Driving Distance

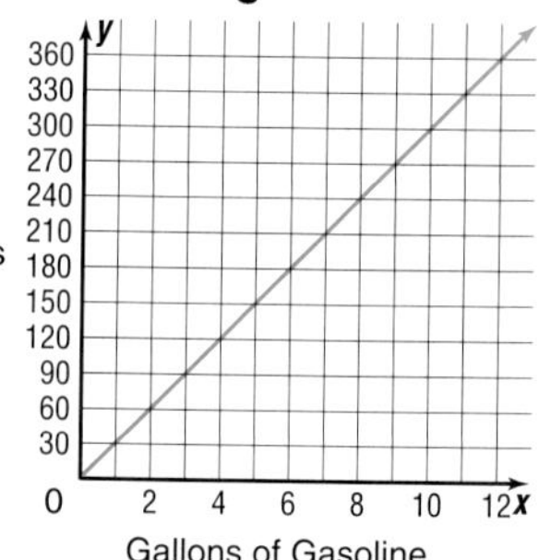

Group Activity Card 8-5

Down Under Group Activity **8-5**

MATERIALS: Graph paper

A scuba diving class was studying the effects of depth on pressure. They learned that at sea level we are under one atmosphere of pressure and at 99 feet we are under four atmospheres pressure. Graph these points and answer the following questions related to diving.

1. If your watch says it is water resistant to 150 feet, how many atmospheres of pressure will it withstand?
2. How far down could you take a flashlight that will withstand $1\frac{1}{2}$ atmospheres of pressure?
3. If you are certified to dive to 60 feet or less, to how many atmospheres pressure are you certified?

 Pre-Algebra

Extension

Business Application A manufacturer knows that it costs \$2.28 per piece plus \$512.50 in set-up costs to manufacture a certain product. How much would it cost to make 500 pieces? **\$1,652.50** Show how you could use a graph to show manufacturing cost of x pieces.

11. **Health** Steve used a chart to see if he was close to the average weight for his height. The chart gave a weight of 130 pounds for a male 60 inches tall. The average weight for a male 66 inches tall was 143 pounds. Steve is 6 feet tall. What should Steve's weight be? **156 pounds**

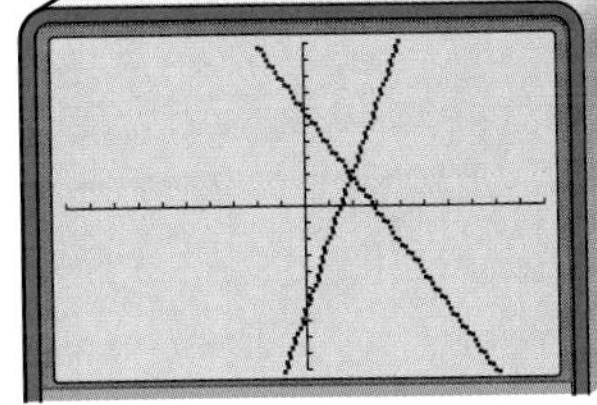

Critical Thinking

12. The graphs of $f(x) = 4x - 6$ and $g(x) = -2x + 6$ are shown at the right. Find an ordered pair that is a solution for both equations. Justify your answer. **(2, 2)**

Mixed Review

13. If $g(x) = 2x - 1$, find $g\left(\frac{5}{2}\right)$. (Lesson 8-4) **4**

14. **Geometry** The area of a trapezoid can be found using $A = h \cdot \frac{1}{2}(b_1 + b_2)$. A trapezoid has an area 64 square inches, a height of 8 inches, and one base 7 inches long. Find the length of the other base. (Lesson 7-2) **9 inches**

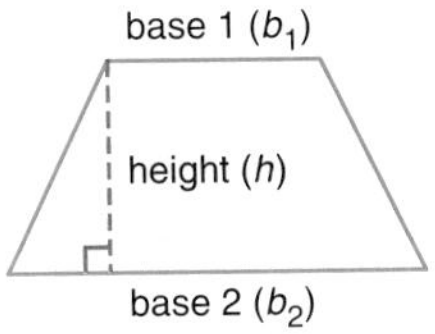

15. Solve $y = 4\frac{3}{4} - 5\frac{1}{6}$. (Lesson 5-5) $-\frac{5}{12}$

16. Evaluate $(a^2 - b)^2$ if $a = 2$ and $b = 4$. (Lesson 4-2) **0**

17. Is the quotient of -42 and 7 positive or negative? (Lesson 2-8) **negative**

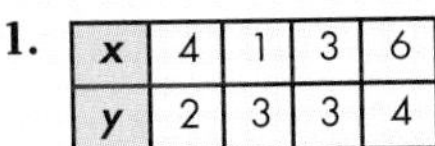

Self Test

Write the domain and range of the relation. Then determine whether each relation is a function. (Lesson 8-1) **1–2. See margin.**

1.

x	4	1	3	6
y	2	3	3	4

2. $\{(4, 2), (-3, 2), (8, 2), (8, 9), (7, 5)\}$

Use the scatter plot to answer each question. (Lesson 8-2)

3. What kind of relationship, if any, is shown by the scatter plot? **positive**

4. What does the point with the triangle around it represent? **185 pound, 6' 2" player**

Weights and Heights of Basketball Players

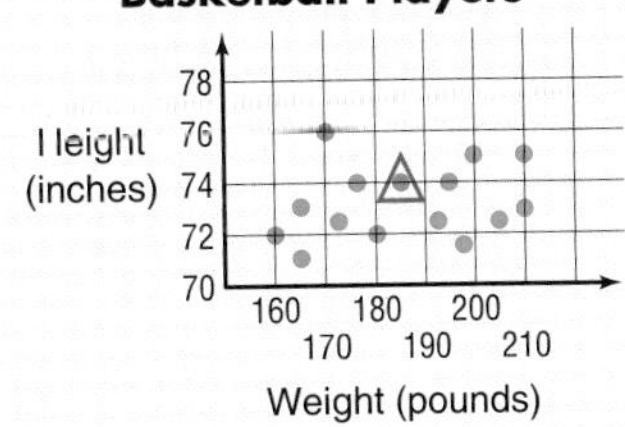

Find four solutions for each equation. Write the solutions as ordered pairs. Then draw the graph. (Lesson 8-3) **5–6. See margin.**

5. $y = 4x - 2$

6. $y = \frac{1}{2}x + 5$

Given $f(x) = 9x - 4$, find each value. (Lesson 8-4)

7. $f(6)$ **50**

8. $f(0)$ **−4**

9. $f(-3)$ **−31**

10. **Physical Science** At 20°C, sound travels 172 meters in 0.5 seconds and 206 meters in 0.6 seconds. Name two ordered pairs that could be used to graph this information. (Lesson 8-5) **(172, 0.5), (206, 0.6)**

4 ASSESS

Closing Activity

Speaking Have students describe a situation in which they would draw a graph to solve a problem. **Answers will vary.**

Chapter 8, Quiz B (Lessons 8-4 and 8-5) is available in the *Assessment and Evaluation Masters*, p. 211.

Mid-Chapter Test (Lessons 8-1 through 8-5) is available in the *Assessment and Evaluation Masters*, p. 210.

Additional Answers Self Test

1. D = {4, 1, 3, 6}; R = {2, 3, 4}; yes
2. D = {4, -3, 8, 7}; R = {2, 9, 5}; no
5. Sample answer: (2, 6), (1, 2), (0, -2), (-1, -6)

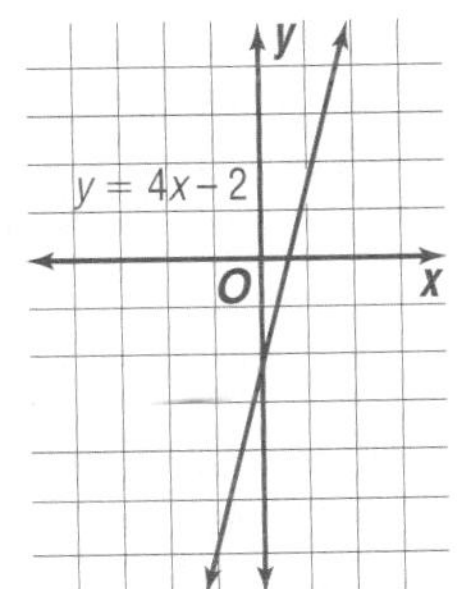

Enrichment Masters, p. 66

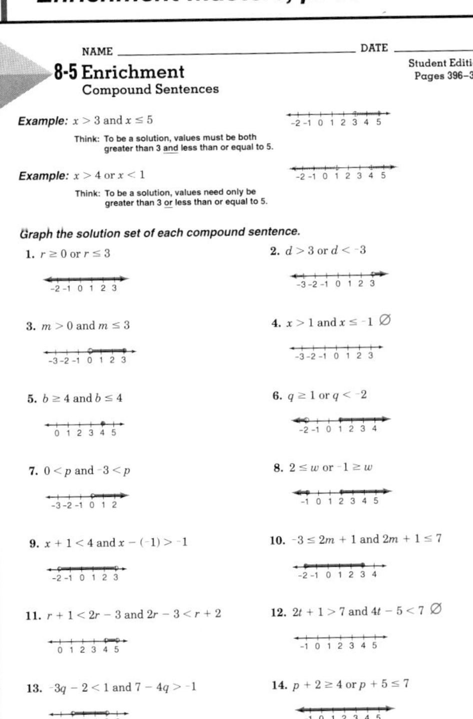

NAME ______ DATE ______

8-5 Enrichment
Compound Sentences

Student Edition Pages 396–399

Example: $x > 3$ and $x \leq 5$

Think: To be a solution, values must be both greater than 3 and less than or equal to 5.

Example: $x > 4$ or $x < 1$

Think: To be a solution, values need only be greater than 3 or less than or equal to 5.

Graph the solution set of each compound sentence.

1. $r \geq 0$ or $r \leq 3$
2. $d > 3$ or $d < -3$
3. $m > 0$ and $m \leq 3$
4. $x > 1$ and $x \leq -1$ ∅
5. $b \geq 4$ and $b \leq 4$
6. $q \geq 1$ or $q < -2$
7. $0 < p$ and $-3 < p$
8. $2 \leq w$ or $-1 \geq w$
9. $x + 1 < 4$ and $x - (-1) > -1$
10. $-3 \leq 2m + 1$ and $2m + 1 \leq 7$
11. $r + 1 < 2r - 3$ and $2r - 3 < r + 2$
12. $2t + 1 > 7$ and $4t - 5 < 7$ ∅
13. $-3q - 2 < 1$ and $7 - 4q > -1$
14. $p + 2 \geq 4$ or $p + 5 \leq 7$

Self Test

The Self Test provides students with a brief review of the concepts and skills in Lessons 8-1 through 8-5. Lesson numbers are given to the right of exercises or instruction lines so students can review concepts not yet mastered.

Additional Answer Self Test

6. Sample answer: (2, 6), (0, 5), (-2, 4), (-4, 3)

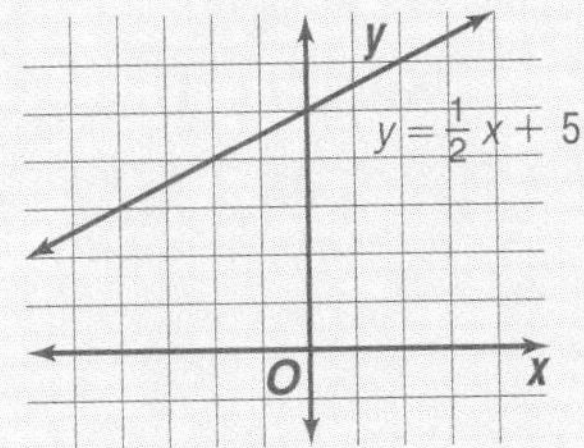

8-6 LESSON NOTES

NCTM Standards: 1-5, 8-10

Instructional Resources

- Study Guide Master 8-6
- Practice Master 8-6
- Enrichment Master 8-6
- Group Activity Card 8-6
- Activity Masters, p. 8

Transparency 8-6A contains the 5-Minute Check for this lesson; **Transparency 8-6B** contains a teaching aid for this lesson.

Recommended Pacing	
Standard Pacing	Day 7 of 13
Honors Pacing	Day 6 of 12
Block Scheduling*	Day 4 of 7 (along with Lesson 8-5)

*For more information on pacing and possible lesson plans, refer to the ***Block Scheduling Booklet***.

1 FOCUS

5-Minute Check
(over Lesson 8-5)

Use a graph to solve the problem below.

A logging truck can carry 20,000 pounds of logs from the forest to the mill. It takes $1\frac{1}{2}$ hours for the truck to make a round-trip from the logging site to the mill. How many pounds of logs can the truck transport in a 6-hour day? **80,000 pounds**

Motivating the Lesson

Questioning Ask student if they have ever been on a roller coaster. Talk about what happens when the roller coaster is steep. What else has a steepness? **a slide on a playground, a street or road, a ski slope**

8-6 Slope

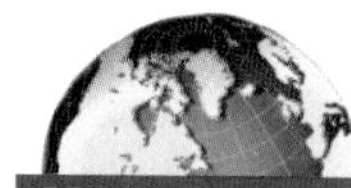

Setting Goals: *In this lesson, you'll find the slope of a line.*

Modeling a Real-World Application: Construction

FYI Rick Hansen is in the Guinness Book of World Records for the longest wheelchair journey. He traveled just under 25,000 miles in 26 months, through 4 continents and 34 countries.

In many buildings, the wheelchair ramps are built next to the walls with stairways in the middle. If you were to compare the handrail on the ramp to the handrail on the stairway, you would notice that the handrail on the ramp is not as steep as the one for the stairs.

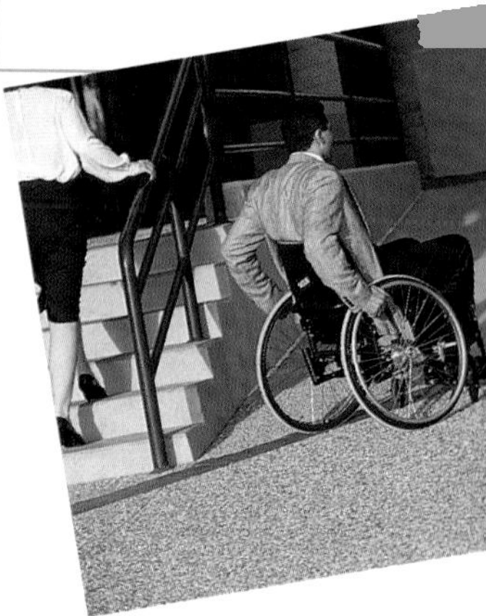

The steepness of the handrails depends on the vertical change and the horizontal change. It can be expressed as a ratio.

$$\text{steepness} = \frac{\text{vertical change}}{\text{horizontal change}}$$

Learning the Concept

In mathematics, the **slope** m of a line describes its steepness. The vertical change is called the **change in *y***, and the horizontal change is called the **change in *x***.

$$\text{slope or } m = \frac{\text{change in } y}{\text{change in } x}$$

Example  **Find the slope of the line graphed below.**

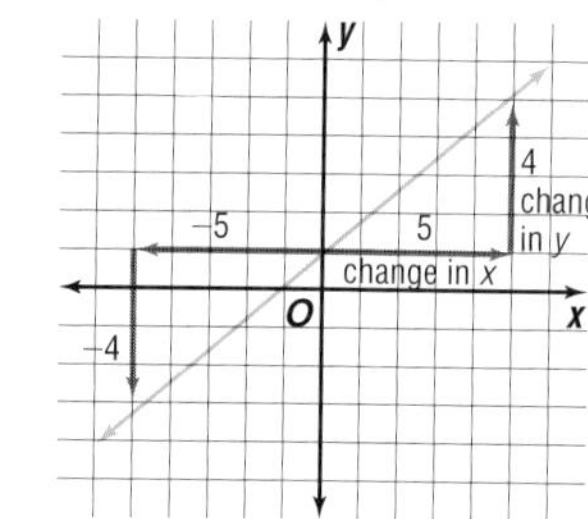

In Quadrant I, the change in y is 4, and the corresponding change in x is 5. Therefore, the slope of the line is $\frac{4}{5}$. Is the slope of the line the same in Quadrant III?

$$\frac{\text{change in } y}{\text{change in } x} \text{ or } \frac{-4}{-5} \text{ or } \frac{4}{5}$$

The slopes are the same.

The slope of a line can be determined by using the coordinates of any two points on the line. The change in y can be found by subtracting the y-coordinates. Likewise, the change in x can be found by subtracting the corresponding x-coordinates.

$$\text{slope} = \frac{\text{change in } y}{\text{change in } x} \text{ or } \frac{\text{difference in } y \text{ coordinates}}{\text{difference in } x \text{ coordinates}}$$

Alternative Learning Styles

Visual Present graphs on the overhead projector. Then help students by tracing graphed lines and counting vertical and horizontal changes. For students with visual limitations, you may wish to furnish extra large grids on which to work.

Example 2 **Find the slope of the line that contains $A(-2, 5)$ and $B(4, -5)$. Then graph the line.**

$$\text{slope} = \frac{\text{difference in } y\text{-coordinates}}{\text{difference in } x\text{-coordinates}}$$

$$\text{slope of line } AB = \frac{5 - (-5)}{-2 - 4} = \frac{10}{-6} \text{ or } -\frac{5}{3}$$

The slope of the line is $-\frac{5}{3}$.

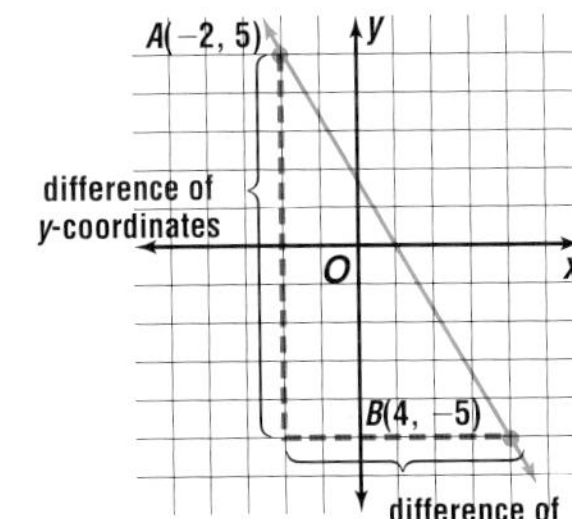

Graph the two points and draw the line.
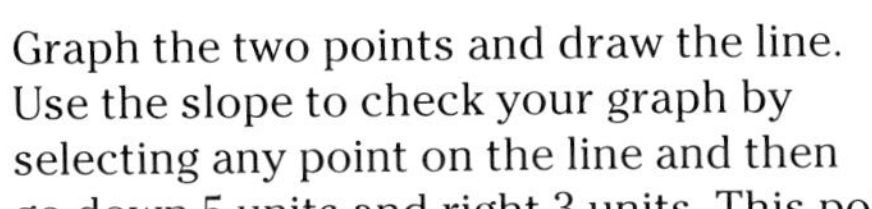
Use the slope to check your graph by selecting any point on the line and then go down 5 units and right 3 units. This point should also be on the line.

Sometimes the vertical change is referred to as the *rise*, and the horizontal change is referred to as the *run*. You can remember slope as *rise over run*.

$$\text{slope} = \frac{rise}{run}$$

Example 3 **APPLICATION** **Highway Safety**

Signs indicating the slope of an upcoming hill are often posted in hilly areas. Find the slope of a road that increases 633.6 feet in 2 miles.

$$\text{slope} = \frac{rise}{run} = \frac{\text{vertical change}}{\text{horizontal change}}$$

Since 1 mile equals 5280 feet, 2 miles equals 10,560 feet.

$$\text{slope} = \frac{633.60}{10{,}560} = \frac{3}{50}$$

The slope of the road is $\frac{3}{50}$. This means that the road rises 3 feet vertically for every 50 feet traveled horizontally.

THINK ABOUT IT

Why is the slope of a horizontal line 0?

Why does a vertical line have no slope?

All y-coordinates are the same; All x-coordinates are the same, so slope denominator is 0.

Connection to Geometry

The figure at the right shows the graphs of two lines that will never intersect. These lines are **parallel**. Is there a special relationship between the slopes of parallel lines?

$$\text{slope of line } CD = \frac{3 - (-3)}{1 - (-1)} = \frac{6}{2} \text{ or } 3$$

$$\text{slope of line } EF = \frac{3 - 0}{3 - 2} = \frac{3}{1} \text{ or } 3$$

The slopes are equal.

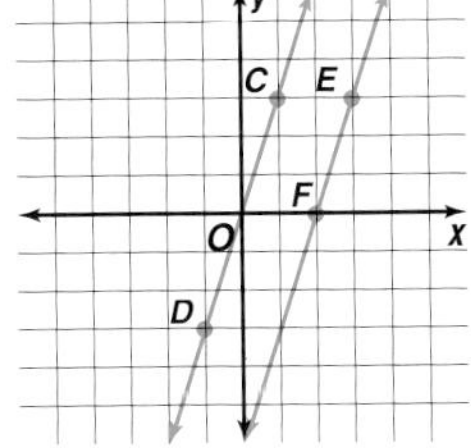

2 TEACH

In-Class Examples

For Example 1

Graph a line that passes through (−2, 3) and (3, −3). Find the slope of the line. **−1**

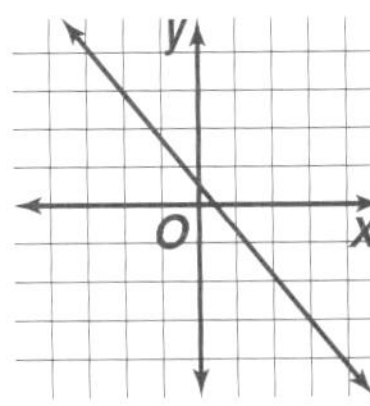

For Example 2

Find the slope of the line that contains $A(2, 6)$ and $B(-3, -8)$. Then graph the line. $\frac{14}{5}$

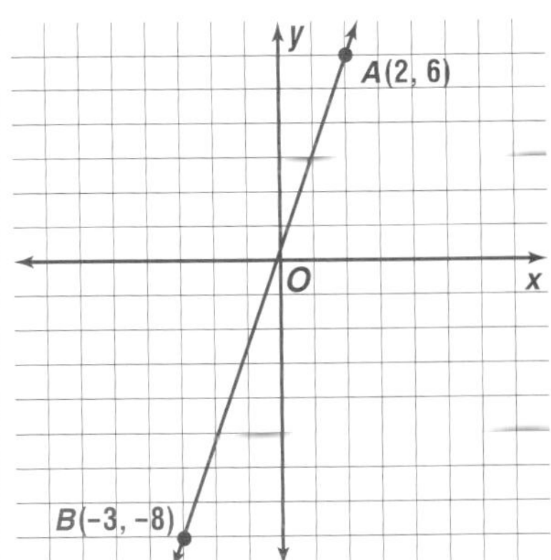

For Example 3

In trapezoid $TRAP$, $\overline{TR}$ is parallel to $\overline{PA}$. If T is at (−2, 3) and R is at (2, 2), what is the slope of side PA? $-\frac{1}{4}$

Teaching Tip Some students may have heard the terms *rise* and *run* when working with slope. Another way to define slope is $\frac{\text{rise}}{\text{run}}$.

Cooperative Learning

Group Discussion Have each group discuss these questions.

- Can you tell if a line has a positive slope by looking at its graph? Explain. **yes, slants up to the right**
- Is it possible for a line to have a slope of zero? Explain. **yes, if y-coordinates are the same**
- What does a line with a zero slope look like? **horizontal**
- What does a line with no slope look like? **vertical**

For more information on this strategy, see *Cooperative Learning in the Mathematics Classroom*, one of the titles in the Glencoe Mathematics Professional Series, p. 31.

3 PRACTICE/APPLY

Checking Your Understanding

Exercises 1–13 are designed to help you assess your students' understanding through reading, writing, speaking, and modeling. Work through Exercises 1–5 with your students and then monitor their work on Exercises 6–13.

Additional Answers

1. Sample answer: If a line has a slope of 3, this means that for every rise of 3 units, it runs to the right 1 unit.
3. Sample answer:

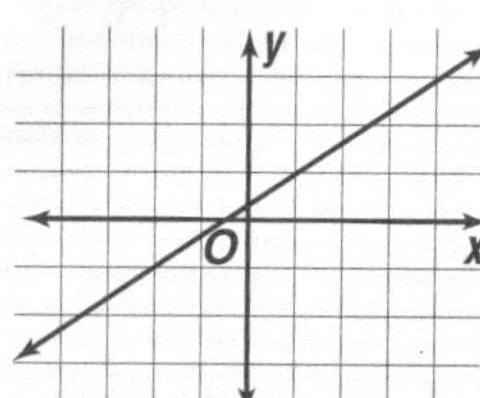

4. Sample answer: Subtract the y-value first. This is the numerator of the slope fraction. Then subtract the x-values. This is the denominator of the slope fraction.
5. Sample answer: A line that has a positive slope slants up to the right. A line that has a negative slope slants down to the right. See Solutions Manual for sample graphs.

Study Guide Masters, p. 67

NAME ______________________ DATE ________

8-6 Study Guide — Student Edition Pages 400–404

Slope

The steepness of a line is called its **slope**. The vertical change is called the **change in y**, and the horizontal change is called the **change in x**.

$$\text{slope} = \frac{\text{change in } y}{\text{change in } x}$$

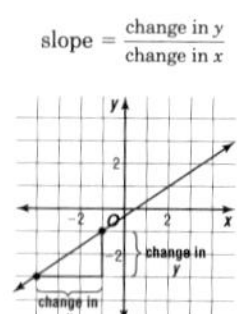

Example: In the graph above, the change in y is 2, and the change in x is 3. Therefore, the slope of the line is $\frac{2}{3}$.

The slope of a line can also be found by using the coordinates of any two points on the line.

$$\text{slope} = \frac{\text{change in } y}{\text{change in } x} \text{ or } \frac{\text{difference in } y\text{-coordinates}}{\text{difference in } x\text{-coordinates}}$$

Example: Find the slope of the line that contains the points $A(^-1, ^-2)$ and $B(^-4, ^-3)$.

$$\text{slope} = \frac{^-2 - (^-3)}{^-1 - (^-4)} = \frac{^-2 + 3}{^-1 + 4} \text{ or } \frac{1}{3}$$

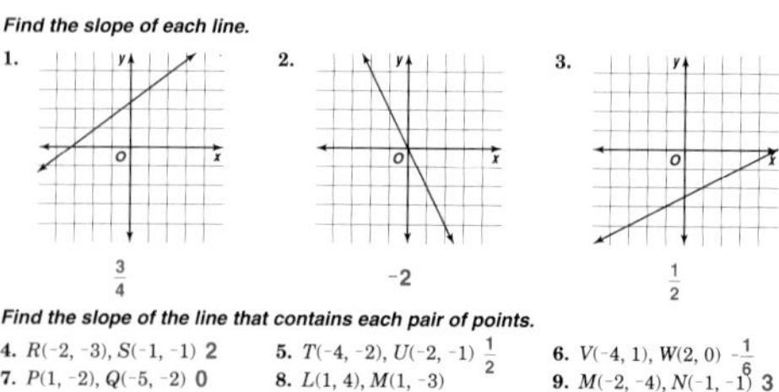

Find the slope of each line.

1. $\frac{3}{4}$ 2. -2 3. $\frac{1}{2}$

Find the slope of the line that contains each pair of points.

4. $R(^-2, ^-3), S(^-1, ^-1)$ **2**
5. $T(^-4, ^-2), U(^-2, ^-1)$ $\frac{1}{2}$
6. $V(^-4, 1), W(2, 0)$ $-\frac{1}{6}$
7. $P(1, ^-2), Q(^-5, ^-2)$ **0**
8. $L(1, 4), M(1, ^-3)$ **no slope**
9. $M(^-2, ^-4), N(^-1, ^-1)$ **3**

Checking Your Understanding

Communicating Mathematics

Read and study the lesson to answer these questions.

1. **Explain** what it means when a line has a slope of 3. **See margin.**

2. Match all the positions of the drawing of a ski slope below with its corresponding slope as shown in the table at the right.

2. a: none, b: negative, c: 0, d: positive

Ski Surface	Slope
A. vertical drop **B.** downhill **C.** flat **D.** uphill	negative none positive 0

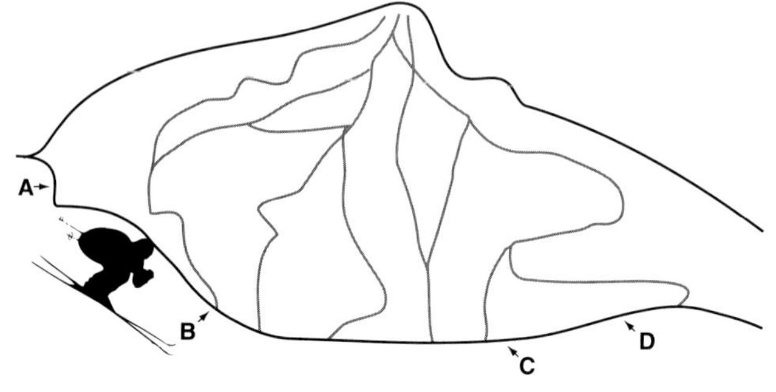

3. **Draw** the graph of a line that has a slope of $\frac{2}{3}$. **See margin.**

4. **Demonstrate** how to find the slope of a line when you know the coordinates of two points on the line. **See margin.**

5. Write a note to someone at home that describes how to tell the difference between the graph of a line with a positive slope and the graph of a line with a negative slope. Include drawings in your note. **See margin.**

Guided Practice

Find the slope of each line.

6.

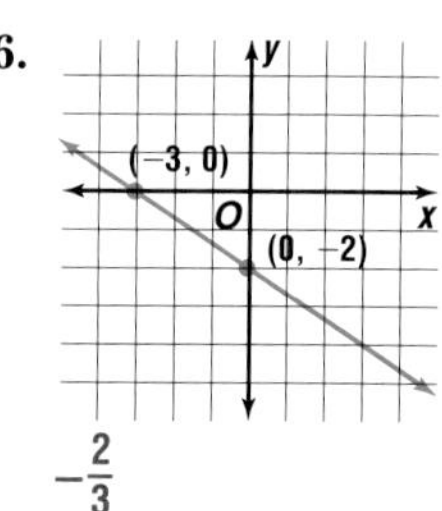

$-\frac{2}{3}$

7.

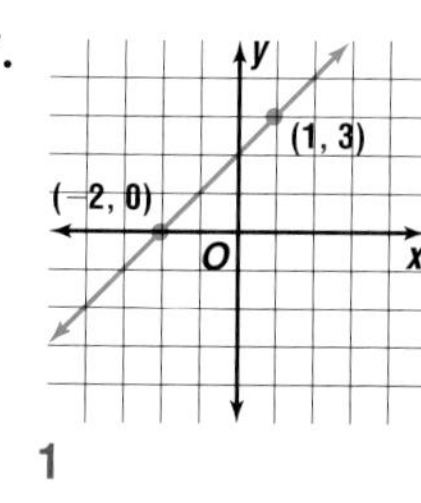

1

8.

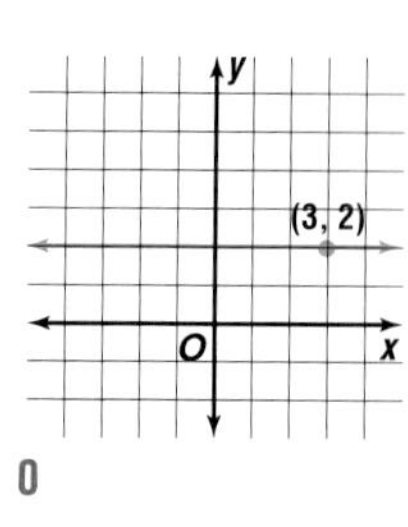

0

Find the slope of the line that contains each pair of points.

9. $R(9, -2), S(3, -5)$ $\frac{1}{2}$
10. $T(14, 3), U(-11, 3)$ **0**
11. $V(-1, -2), X(2, -5)$ **−1**
12. $B(-6, -4), C(-8, -3)$ $-\frac{1}{2}$
13. **Painting** A ladder that reaches a height of 16 feet is placed 4 feet away from the wall. What is the slope of the ladder? **4**

Reteaching

Using Comparison On the board or overhead, draw lines that go through the origin and each of the points: (2, 2), (2, 4), (4, 2), (2, 6), and (6, 2). Have students determine and compare the slopes of the lines.

GLENCOE Technology

Interactive Mathematics Tools Software

In this interactive computer lesson, students explore the slope of a line. A **Computer Journal** gives students an opportunity to write about what they have learned.

For Windows & Macintosh

Exercises: Practicing and Applying the Concept

Independent Practice

Determine the slope of each line named below.

14. a 3
15. b -1
16. c $\frac{3}{2}$
17. d 0
18. e no slope
19. f $-\frac{1}{5}$
20. g $\frac{1}{3}$

21. h -2

22. Look at lines b, f, and h above. These lines have negative slopes. Make a conjecture about how the absolute value of the slope of the line affects how steeply it slants. See margin.

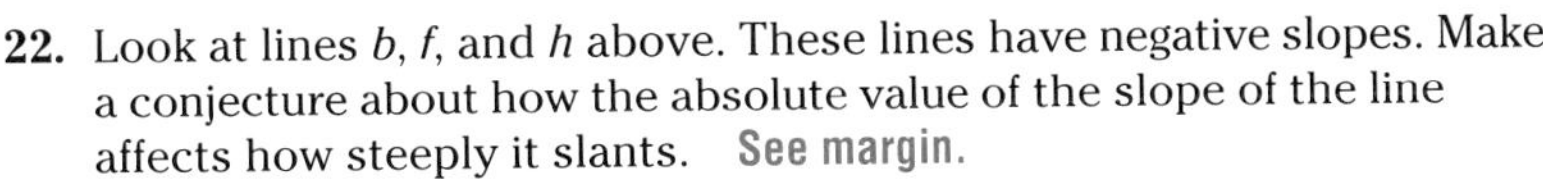

23. **Statistics** What does a scatter plot that shows a positive relationship share with the graph of a line that has a slope of 1? See margin.

Find the slope of the line that contains each pair of points.

24. $C(3, 4), D(4, 6)$ 2
25. $E(-3, 6), F(-5, 9)$ $-\frac{3}{2}$
26. $G(2, 3), H(-1, 3)$ 0
27. $J(1, -3), K(5, 4)$ $\frac{7}{4}$

28. $M(7, -4), N(9, -1)$ $\frac{3}{2}$
29. $P(5, -2), Q(4, -3)$ 1
30. $Y(0, 0), Z(0.5, 0.75)$ 1.5
31. $D(5, -1), E(-3, -4)$ $\frac{3}{8}$
32. $F\left(\frac{3}{4}, 1\right), G\left(\frac{3}{4}, -1\right)$ none
33. $H\left(3\frac{1}{2}, 5\frac{1}{4}\right), J\left(2\frac{1}{2}, 6\right)$ $-\frac{3}{4}$

Critical Thinking

34. **Geometry** The figure at the right shows two lines that are perpendicular. Parallel lines have the same slope. Make a conjecture about the slopes of perpendicular lines. See margin.

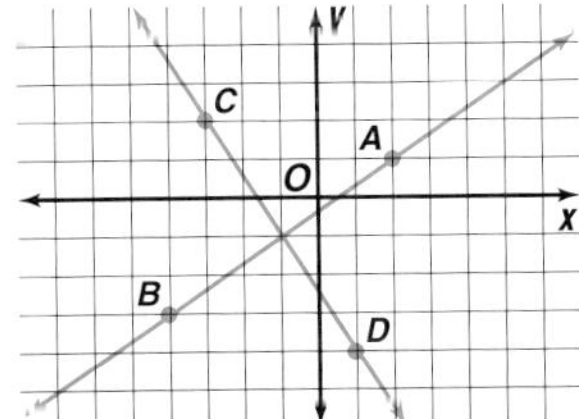

Applications and Problem Solving

35. **Carpentry** In a stairway, the slope of the handrail is the ratio of the riser to the tread. If the tread is 12 inches long and the riser is 8 inches long, what is the slope of the handrail? $\frac{2}{3}$

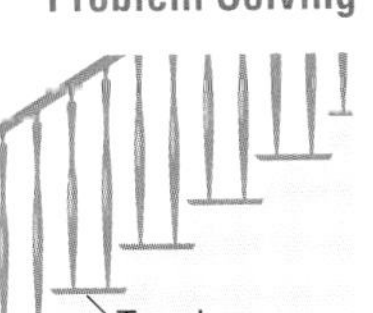

36. **Statistics** The graph at the right shows how fruit juice consumption changed from 1990 through 1994.
    a. Describe the section of the graph that shows when the greatest increase occurred.
    b. Describe what no change in consumption would look like on the graph.
    c. What does a section with a negative slope mean?

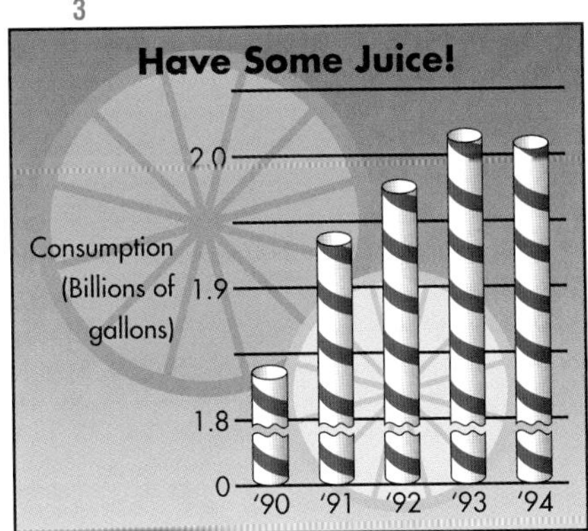

Source: Beverage Marketing Corp.

36a. The section of the graph has the steepest slant up to the right.
36b. horizontal line segment
36c. a decrease in consumption

Error Analysis

When finding slope, students may find the incorrect slope by making any of these errors:

- dividing changes in x-coordinates by changes in y-coordinates
- adding coordinates, rather than subtracting
- not using coordinates in the same order

Monitor student work and correct such errors quickly. Then encourage students to carefully check each step of their work.

Assignment Guide

Core: 15–33 odd, 34–35, 37–44
Enriched: 16–32 even, 34–44

For **Extra Practice**, see p. 760.

The red A, B, and C flags, printed only in the Teacher's Wraparound Edition, indicate the level of difficulty of the exercises.

Additional Answers

22. Sample answer: The larger the absolute value of the slope of a line, the steeper the slant of the line.
23. Sample answer: Both graphs slant up to the right.
34. Sample answer: The slopes are negative reciprocals of each other. That is, the product of their slopes is -1.

Practice Masters, p. 67

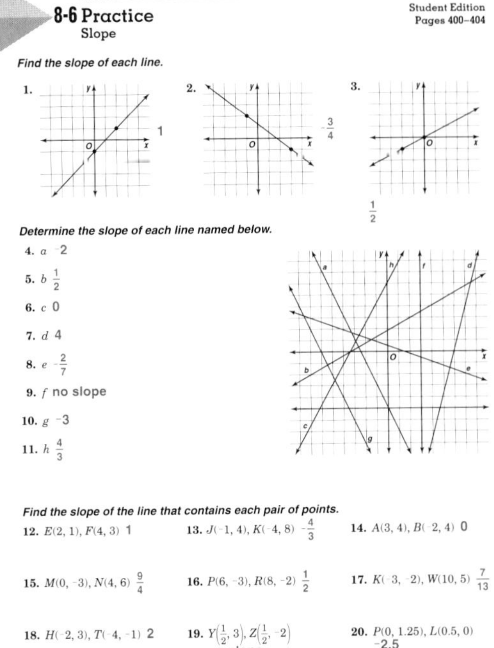

NAME ______ DATE ______

8-6 Practice Slope — Student Edition Pages 400–404

Find the slope of each line.

1. 1
2. $-\frac{3}{4}$
3. $\frac{1}{2}$

Determine the slope of each line named below.

4. a -2
5. b $\frac{1}{2}$
6. c 0
7. d 4
8. e $-\frac{2}{7}$
9. f no slope
10. g -3
11. h $\frac{4}{3}$

Find the slope of the line that contains each pair of points.

12. $E(2, 1), F(4, 3)$ 1
13. $J(-1, 4), K(-4, 8)$ $-\frac{4}{3}$
14. $A(3, 4), B(-2, 4)$ 0
15. $M(0, -3), N(4, 6)$ $\frac{9}{4}$
16. $P(6, -3), R(8, -2)$ $\frac{1}{2}$
17. $K(-3, -2), W(10, 5)$ $\frac{7}{13}$
18. $H(-2, 3), T(-4, -1)$ 2
19. $Y\left(\frac{1}{2}, 3\right), Z\left(\frac{1}{2}, -2\right)$ no slope
20. $P(0, 1.25), L(0.5, 0)$ -2.5

Group Activity Card 8-6

Steep Slopes — Group Activity 8-6

MATERIALS: Four dice • graph paper

The object of this game is to create two ordered pairs from numbers rolled on dice so that the slope of the line containing them meets one of the following requirements.

1. Slope less than 1.1
2. Slope between 1.1 and 2.1
3. Slope between 2.1 and 3.1
4. Slope greater than 3.1

To begin, roll the four dice. All members of the group create their own pairs of ordered pairs from these rolled numbers. They then compute the slope of a line containing the points and determine which of the requirements it meets. On the following rolls they try to create ordered pairs that meet a different requirement.

The winner of the game is the person who first creates ordered pairs that meet each requirement.

 Pre-Algebra

Extension

Using Discussion Have students graph several simple equations, such as $y = -2x + 4$, $y = 3x - 2$, and $y = 5x + 1$. Have students find the slope of each line. Discuss whether there is any relationship between this form of the equation and the slope of the line. slopes -2, 3, 5; yes, the slope is the same as the coefficient of x

4 ASSESS

Closing Activity

Modeling Use two different-colored dice. One is for *x*-values and the other for *y*-values. Roll both and create an ordered pair. Graph the point. Roll the dice again and graph the point. What is the slope determined by the two points? Repeat twice. **Answers may vary.**

Enrichment Masters, p. 67

NAME ________________ DATE ________

8-6 Enrichment
Translations and Reflections

Student Edition
Pages 400–404

The lines on graph paper can help you draw slide images of figures.

1. Graph ΔABC with vertices $A(1, 1)$, $B(-3, 4)$, and $C(-3, -4)$. Draw $\Delta A'B'C'$, the translation image of ΔABC, where the slide is 3 units to the right. Name the coordinates of the image of each vertex. **A' (4, 1), B' (0, 4), C' (0, −4)**

2. Draw ΔJKL with vertices $J(-4, 3)$, $K(0, 2)$, and $L(-2, 0)$. Let $\Delta J'K'L'$ be the image of ΔJKL under a slide of 4 units to the right and then a slide of 3 units up. Graph $\Delta J'K'L'$. Name the coordinates of the vertices of $\Delta J'K'L'$. **J' (0, 6), K' (4, 5), L' (2, 3)**

3. Draw $\overline{A'B'}$, the image formed by reflecting $\overline{AB}$ over the *y*-axis. Then draw $\overline{A''B''}$, the image formed by reflecting $\overline{A'B'}$ over the *x*-axis. What are the coordinates of A'' and B''? What is the relationship between the coordinates of the endpoints of $\overline{AB}$ and those of $\overline{A''B''}$?

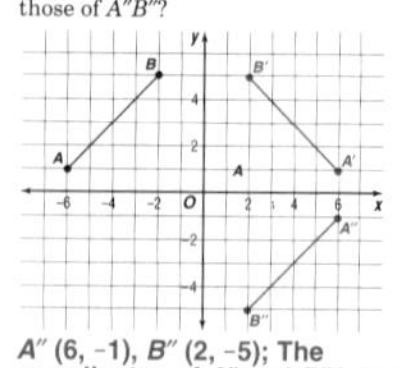

A'' (6, −1), B'' (2, −5); The coordinates of A'' and B'' have the opposite signs of *A* and *B*.

4. Draw $\overline{P'Q'}$, the reflection image of $\overline{PQ}$ over the *y*-axis. Draw $\overline{P''Q''}$, the reflection image of $\overline{P'Q'}$ over the *x*-axis. Find the slopes of $\overline{PQ}$, $\overline{P'Q'}$, an $\overline{P''Q''}$. What is the relationship between the slopes of $\overline{PQ}$ and $\overline{P''Q''}$?

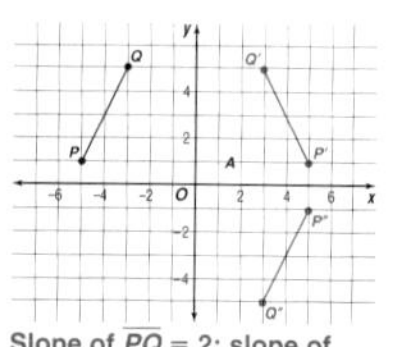

Slope of $\overline{PQ}$ = 2; slope of $\overline{P'Q'}$ = −2; slope of $\overline{P''Q''}$ = 2; They are the same.

37. **Driving** The western entrance to the Eisenhower Tunnel in Colorado is at an elevation of 11,160 feet. The roadway has a downward slope of 0.00895 toward the eastern entrance, and its horizontal distance is 8941 feet. What is the elevation of the eastern end of the tunnel? **11,080 feet**

Mixed Review

38. **Travel** After 2 hours, Andrea checks the odometer in her car. She has traveled 160 kilometers. After 3 more hours, she has traveled a total distance of 400 kilometers. Use a graph to find the number of hours it will take for Andrea to travel a total distance of 560 kilometers. Assume that the rate is constant. (Lesson 8-5) **2 more hours**

39. **Food Preparation** Suppose you are in charge of baking a ham for your family picnic. The directions tell you to bake the ham at 325° for 12 minutes for each pound. The ham weighs 7.45 pounds. To the nearest minute, how long should you bake the ham? (Lesson 6-5)

39. 89 minutes or 1 hour 29 minutes

40. Round 24.692 to the nearest tenth. (Lesson 5-2) **24.7**

41. Is the number 57 *prime* or *composite*? (Lesson 4-4) **composite**

42. Solve $a + (-7) = 8$. (Lesson 3-2) **15**

43. Find the value of $6(5 + 9) \div 7$. (Lesson 1-2) **12**

44. **Problem Solving** After half of the people at a Spanish Club meeting left, one third of those remaining began to plan the club's Cinco de Mayo (5th of May) celebration. The other 18 people were cleaning the room. How many people attended the meeting? (Lesson 1-1) **54**

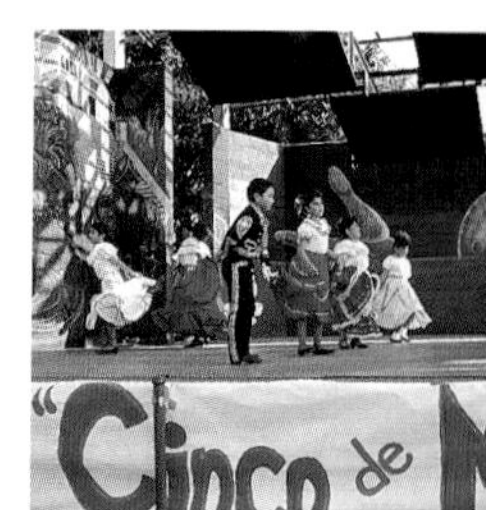

Refer to the Investigation on pages 368–369.

- Transcribe the data from your table in Lesson 8-2 onto a coordinate grid using the same type of axes as your scatter plot. Connect the data points.
- If you compared a graph of the data from another class, how could you tell if 11 sections move faster than 20 sections? Explain.
- Below are three graphs of wave cheer data for a stadium with 26 sections.
- Describe each graph separately. Explain what you think happened. Are all these graphs reasonable?

Add the results of your work to your Investigation Folder.

Graph A

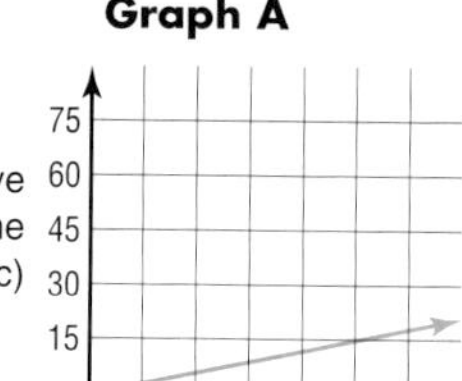

Graph B

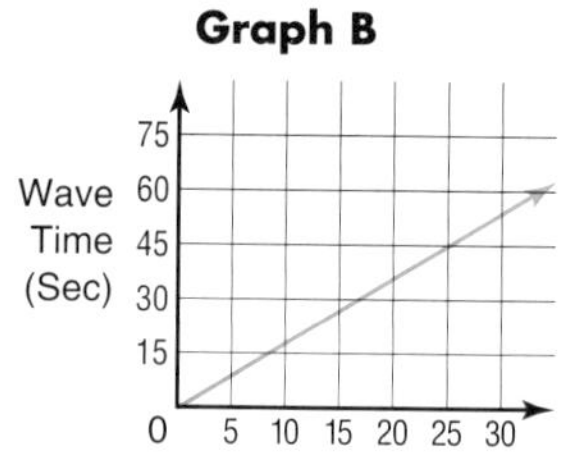

Graph C

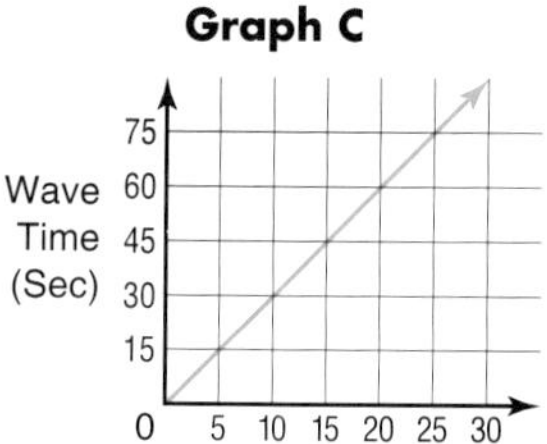

WORKING ON THE

Investigation

The Investigation on pages 368 and 369 is designed to be a long-term project that is completed over several days or weeks. Encourage students to keep their materials in their Investigation Folder as they work on the Investigation.

HANDS-ON ACTIVITY

8-6B Circles and Slope

An Extension of Lesson **8-6**

MATERIALS

- graph paper
- scissors
- tape
- cylindrical objects
- straightedge

Work with a partner.

- Copy the coordinate axes on a sheet of graph paper as shown at the right. Cut off the bottom of the paper so that the horizontal axis becomes the edge of the paper.

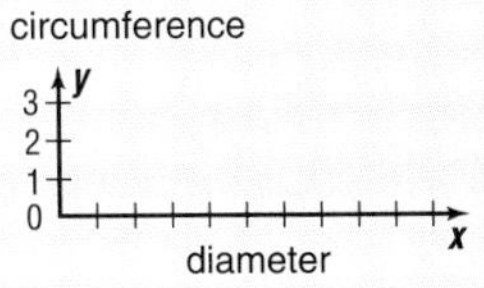

- Affix the graph paper to a wall or chalkboard.
- Hold the cylindrical base of an object against the graph paper so its left edge is at the origin and half of the cylindrical object is on either side of the horizontal axis. Mark the right side of the base where it touches the horizontal axis.

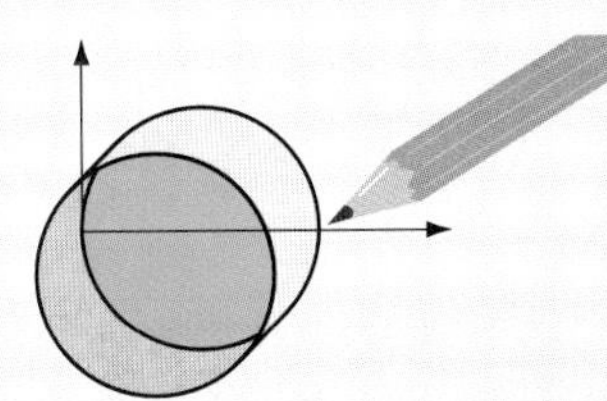

- Roll the cylindrical object up one complete rotation. The mark on the cylindrical object will be at the right side. Mark this point on the graph.

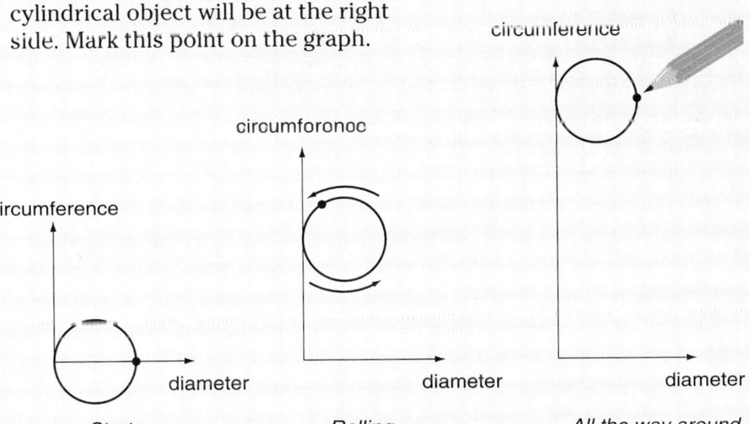

- Repeat this procedure with a different cylindrical object.
- Write the ordered pair for each point.
- Draw a line through the two points marked on the graph paper.

1. Estimate the slope of the line using ordered pairs. **1. See student's work.**
2. Determine the slope of the line. **See margin.**
3. Where would you expect a point to be if you repeated the above procedure with a third cylindrical object? **The point should be on the line.**

Extension

4. Refer to Lesson 7-4 on circles and circumference. What should the slope of the line be? Explain your reasoning. **See margin.**

8-6B LESSON NOTES

NCTM Standards: 1-6, 8

Objective

Use cylindrical objects to discover a connection between algebra and geometry.

Recommended Time

Demonstration and discussion: 15 minutes; Exercises: 30 minutes

Instructional Resources

For each student or group of students

- graph paper
- scissors
- tape
- cylindrical objects
- straightedge

Math Lab and Modeling Math Masters

- p. 43 (worksheet)

For teacher demonstration

Overhead Manipulative Resources

1 FOCUS

Motivating the Lesson

Ask students to predict the slope of a line formed by marking a point on the circle and rotating it one complete revolution to mark that same point.

2 TEACH

Teaching Tip Students must take care to record two points that are directly at "3 o'clock" or two other identical positions. Compare this line to the curve formed if the position of the point is marked several times during its rotation back to the original position.

3 PRACTICE/APPLY

Assignment Guide
Core: 1–4
Enriched: 1–4

Additional Answers

2. The slope should be about 3.
4. Sample answer: The slope should equal π because the change in y is the circumference and the change in x is the diameter and circumference divided by diameter is π. The slope is about 3.14.

4 ASSESS

Observing students working in cooperative groups is an excellent method of assessment.

8-7 LESSON NOTES

NCTM Standards: 1-5, 8, 9

Instructional Resources

- Study Guide Master 8-7
- Practice Master 8-7
- Enrichment Master 8-7
- Group Activity Card 8-7
- Assessment and Evaluation Masters, p. 212
- Math Lab and Modeling Math Masters, p. 66

Transparency 8-7A contains the 5-Minute Check for this lesson; **Transparency 8-7B** contains a teaching aid for this lesson.

Recommended Pacing	
Standard Pacing	Day 8 of 13
Honors Pacing	Day 7 of 12
Block Scheduling*	Day 5 of 7 (along with Lesson 8-8)

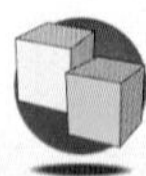

*For more information on pacing and possible lesson plans, refer to the ***Block Scheduling Booklet***.

1 FOCUS

5-Minute Check (over Lesson 8-6)

Find the slope of the line that contains each pair of points.

1. $C(-4, -2), D(5, 3)$ $\frac{5}{9}$
2. $H(1, 4), S(5, -2)$ $-\frac{3}{2}$
3. $G(0, 7), B(2, -7)$ -7
4. $J(0, 0), W(4, -4)$ -1
5. $R(-4, -4), F(-5, -3)$ -1

Motivating the Lesson

Questioning Ask students what the x-coordinate of any point on the y-axis is. What is the y-coordinate of any point on the x-axis? **0, 0**

8-7 Intercepts

Setting Goals: *In this lesson, you'll graph a linear equation using the x- and y-intercepts or using the slope and y-intercept.*

Modeling a Real-World Application: Sports

On your mark, get set, go! Sports analysts have found that the world record for the 10,000-meter run has been decreasing steadily since 1940. The trend can be described by the equation $y = 30.18 - 0.07x$, where x is the number of years since 1940 and y is the record time. The graph of the equation is a line that crosses the x-axis at approximately (431, 0) and crosses the y-axis at (0, 30.18). The points where a graph crosses the axes are called **intercepts**.

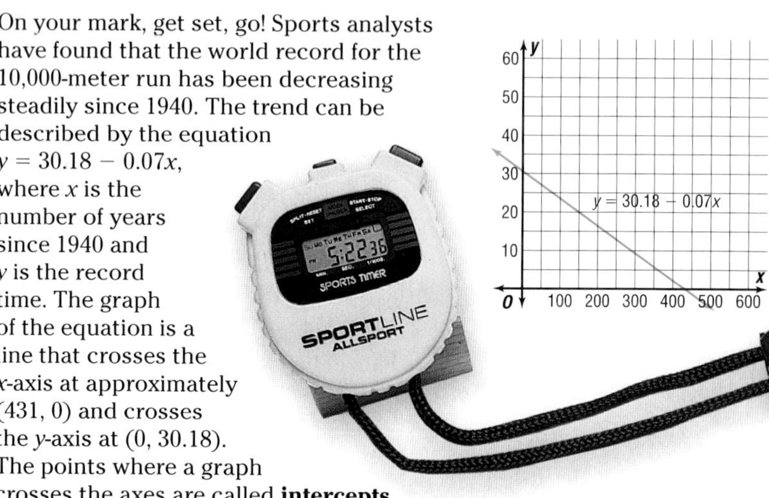

Learning the Concept

THINK ABOUT IT

What kind of line only crosses the x-axis? What kind of line only crosses the y-axis?

vertical; horizontal

The graph of a linear function may cross either the x-axis, the y-axis, or both axes.

The ***x*-intercept** is the x-coordinate of the point where the graph crosses the x-axis.

The ***y*-intercept** is the y-coordinate of the point where the graph crosses the y-axis.

The line graphed at the right crosses both axes.

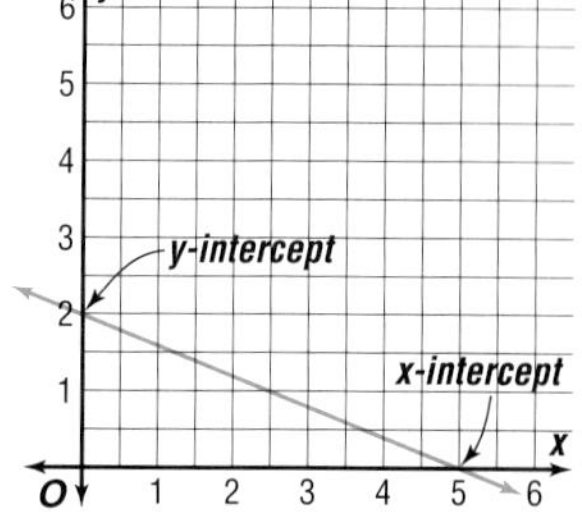

x-intercept The line graphed above crosses the x-axis at (5, 0). Therefore the x-intercept is 5. Note that the corresponding y-coordinate is 0.

y-intercept The line above crosses the y-axis at (0, 2). So the y-intercept is 2. Note that the corresponding x-coordinate is 0.

The y-intercept and the x-intercept can be used to graph a linear equation.

Tech Prep

Manufacturing Cost Analyst When manufacturing certain items there is a base cost that involves fixed costs, such as rent, overhead (lights, telephone), and management fees. The total of these fixed costs are represented by b in the equation $y = mx + b$. The slope, m, represents the cost per item (x) produced and y represents total manufacturing costs. If possible, have students research a business and explain the fixed and unit costs to the class.

For more information on tech prep, see the *Teacher's Handbook*.

Example 1 **Graph $y = \frac{3}{2}x - 6$ using the x- and y-intercepts.**

Step 1 Find the x- and y-intercepts.

To find the x-intercept, let $y = 0$.

$$y = \frac{3}{2}x - 6$$
$$0 = \frac{3}{2}x - 6$$
$$0 + 6 = \frac{3}{2}x - 6 + 6$$
$$6 = \frac{3}{2}x$$
$$4 = x$$

The x-intercept is 4.
The ordered pair is (4, 0).

To find the y-intercept, let $x = 0$.

$$y = \frac{3}{2}x - 6$$
$$y = \frac{3}{2}(0) - 6$$
$$y = -6$$

The y-intercept is -6.
The ordered pair is $(0, -6)$.

Step 2 Graph the intercepts and draw the line that contains them.

Step 3 Check by choosing some other point on the line and determine whether its ordered pair is a solution of $y = \frac{3}{2}x - 6$. Try $(2, -3)$.

$$y = \frac{3}{2}x - 6$$
$$-3 \stackrel{?}{=} \frac{3}{2}(2) - 6$$
$$-3 \stackrel{?}{=} 3 - 6$$
$$-3 = -3 \checkmark$$

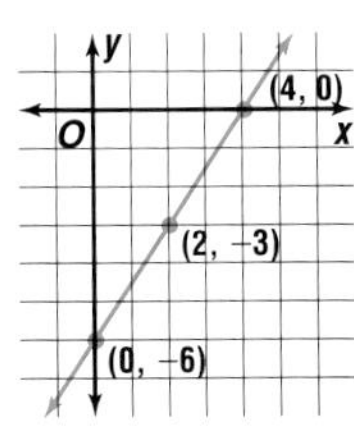

In Example 1, you may have noticed that the y-intercept for the equation was easy to find. Is there a quick way to find the slope? Use the intercepts to find the slope.

$$\text{slope} = \frac{0 - (-6)}{4 - 0} = \frac{6}{4} \text{ or } \frac{3}{2}$$

Notice that $\frac{3}{2}$ is the coefficient of x in the equation $y = \frac{3}{2}x - 6$. When a linear equation is written in the form $y = mx + b$, it is in **slope-intercept form**.

$$y = mx + b$$

↑ slope ↑ y-intercept

You can use the slope-intercept form of an equation to graph a line quickly.

Example 2 APPLICATION **Child Care**

The cost for a child to attend The Learning Station is $19 a day plus a registration fee of $30. Make a graph that can be used to find the cost of any length of attendance.

Explore You know it costs $19 a day and there is a $30 registration fee. Let d represent the number of days. Then $C = 19d + 30$ represents the total cost, C, of attendance.

(continued on the next page)

2 TEACH

In-Class Examples

For Example 1

Graph $y = -\frac{1}{4}x + 3$ using the x- and y-intercepts.

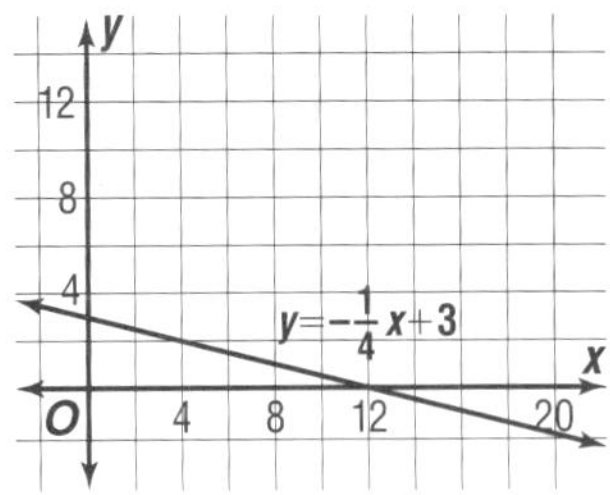

y-intercept is 3; x-intercept is 12.

For Example 2

The cost of a taxi ride is $3 plus $0.20 per tenth of a mile. Make a graph that can be used to find the cost of any length of a ride. **See students' graphs of $y = 0.2x + 3$.**

Teaching Tip You may wish to have students create a flowchart showing how to find the x- and y-intercepts of a given equation.

GLENCOE Technology

Interactive Mathematics Tools Software

In this interactive computer lesson, students explore the intercepts of lines on a coordinate plane. A **Computer Journal** gives students an opportunity to write about what they have learned.

For Windows & Macintosh

3 PRACTICE/APPLY

Checking Your Understanding

Exercises 1–16 are designed to help you assess your students' understanding through reading, writing, speaking, and modeling. Work through Exercises 1–7 with your students and then monitor their work on Exercises 8–16.

Additional Answers

1. Sample answer: The x-intercept is 4. The y-intercept is 2. The slope is $-\frac{1}{2}$.
2. Graph (−3, 0) and (0, 1). Draw the line that contains the two points.
3. To find the x-intercept, let $y = 0$ and solve for x. To find the y-intercept, let $x = 0$ and solve for y. For $y = -\frac{1}{2}x + 8$, the x-intercept is 16 and the y-intercept is 8.
5. In 1940, the time was 431 minutes.
6. Plot the y-intercept on the y-axis. From this point, use the slope to rise and run to find the coordinate on the line. Then connect these two points.

***Study Guide Masters*, p. 68**

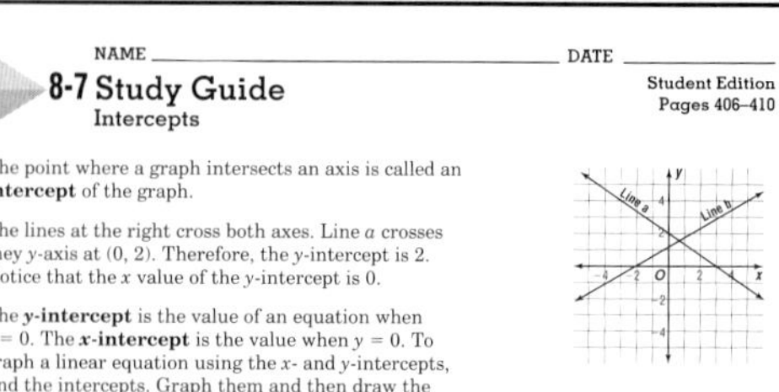
NAME ____ DATE ____
8-7 Study Guide
Intercepts
Student Edition Pages 406–410

The point where a graph intersects an axis is called an **intercept** of the graph.

The lines at the right cross both axes. Line a crosses the y-axis at (0, 2). Therefore, the y-intercept is 2. Notice that the x value of the y-intercept is 0.

The **y-intercept** is the value of an equation when $x = 0$. The **x-intercept** is the value when $y = 0$. To graph a linear equation using the x- and y-intercepts, find the intercepts. Graph them and then draw the line that contains them.

Example: Graph $y = x + 3$ using the x- and y-intercepts.

To find the x-intercept, let $y = 0$.
$y = x + 3$
$0 = x + 3$
$-3 = x$
The x-intercept is −3.
The ordered pair is (−3, 0).

To find the y-intercept, let $x = 0$.
$y = x + 3$
$y = 0 + 3$
$y = 3$
The y-intercept is 3.
The ordered pair is (0, 3).

State the x-intercept and the y-intercept for each line.

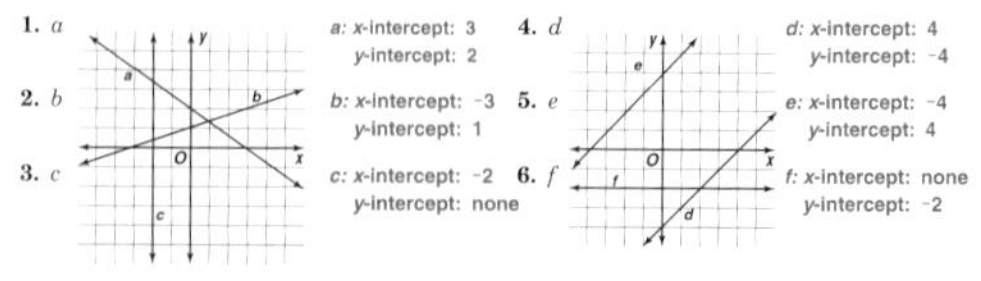
1. a — a: x-intercept: 3, y-intercept: 2
2. b — b: x-intercept: −3, y-intercept: 1
3. c — c: x-intercept: −2, y-intercept: none
4. d — d: x-intercept: 4, y-intercept: −4
5. e — e: x-intercept: −4, y-intercept: 4
6. f — f: x-intercept: none, y-intercept: −2

Use the x-intercept and y-intercept to graph each equation.

7. $y = 1 - 2x$
8. $y = \frac{1}{2}x + 1$
9. $-x - 3y = 3$

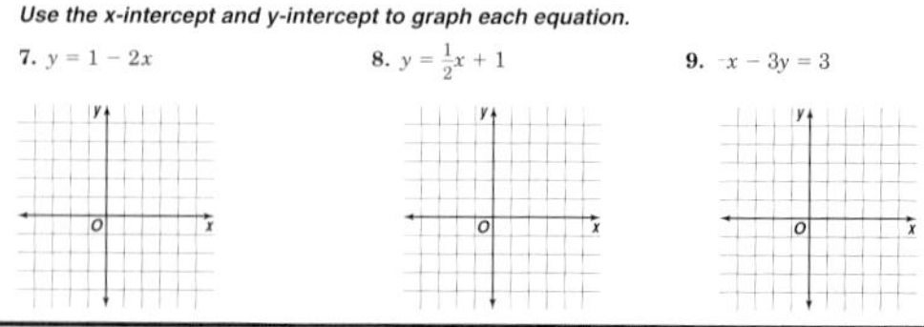

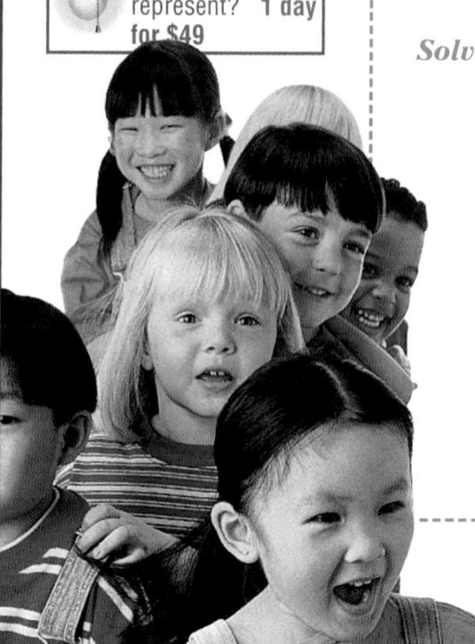

THINK ABOUT IT
What does the ordered pair (1, 49) represent? **1 day for $49**

Plan The y-intercept is 30 so the graph contains the point at (0, 30).

The slope is 19 or $\frac{19}{1}$.

Solve $\frac{19}{1} = \frac{\text{change in } y}{\text{change in } x}$

Starting at (0, 30), go to the right 1 unit and up 19 units. This will be the point at (1, 49). Then draw the line that contains (0, 30) and (1, 49).

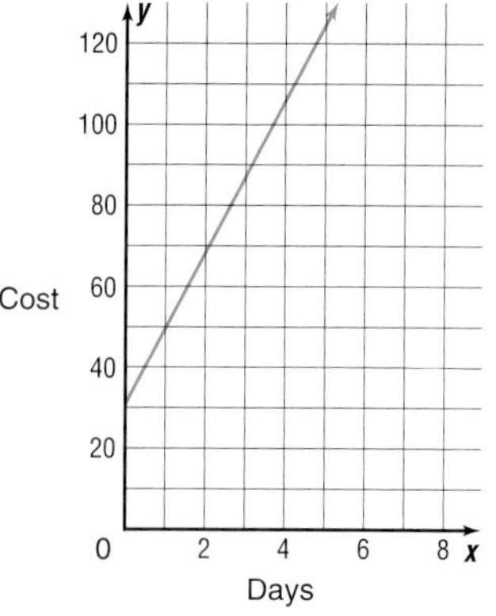

Examine Substitute the coordinates of a third point on the line into the equation $C = 19d + 30$ to determine whether it is a solution of the equation.

Checking Your Understanding

Communicating Mathematics

Read and study the lesson to answer these questions.

1. **Record** all the information you can determine from the graph shown at the right. **See margin.**

2. **Explain** how to graph a line if the x-intercept is −3 and the y-intercept is 1. **See margin.**
3. **Explain** how to find the x- and y-intercept for $y = -\frac{1}{2}x + 8$. **See margin.**
4. **Draw** a graph of a line that has a y-intercept, but no x-intercept. Describe your line. **See students' graphs. It should be a horizontal line.**
5. **Explain** what the intercepts of the graph of the equation involving 10,000-meter run times described at the beginning of the lesson mean.
6. **Illustrate** how to graph a line if you know its y-intercept and its slope.

5–6. See margin.

MATH JOURNAL

7. **Assess Yourself** Design a personal fitness program for yourself. **See students' work.**
   a. Choose an activity from the chart and decide how many hours a week you will do the activity.
   b. Make a graph that shows the number of Calories burned for 0–10 hours.
   c. Use the graph to estimate how many Calories you will burn each week.

Activity	Calories burned per hour
Swimming	288
Walking	300
Jogging	654
Bicycling	174
Aerobics	546
Weight training	756

Reteaching

Using Questioning Ask students what the x-coordinate of any point on the y-axis is. **zero** Ask students what the y-coordinate of any point on the x-axis is. **zero** Ask students how to find the coordinates of the intercepts of any given equation. **Substitute zero for the x-value and solve for y to find the y-intercept; substitute zero for the y-value and solve for x to find the x-intercept.**

Guided Practice

State the x-intercept and the y-intercept for each line.

8. 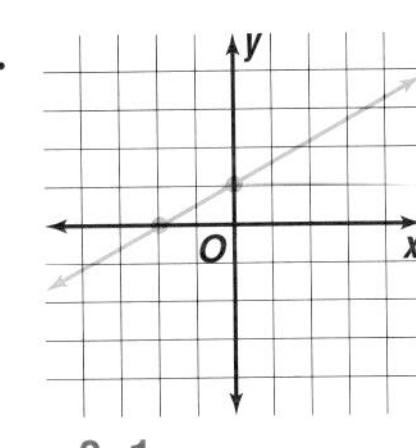

$-2, 1$

9. 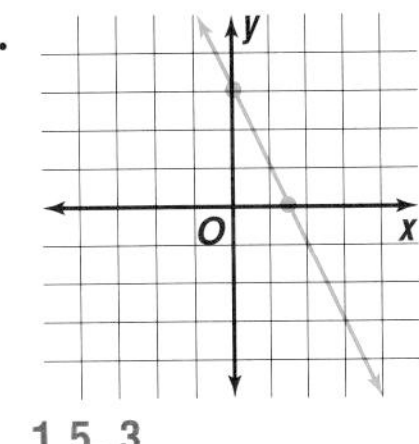

$1.5, 3$

Find the x-intercept and the y-intercept for the graph of each equation. Then graph the line. See Solutions Manual for graphs.

10. $y = x - 8$ $8, -8$

11. $y = x + 6$ $-6, 6$

12. $y = -2x - 1$ $-\frac{1}{2}, -1$

13. $y = 6 - 9x$ $\frac{2}{3}, 6$

Graph each equation using the slope and y-intercept. 14–15. See margin.

14. $y = 2x + 1$

15. $y = -7x - 4$

16. Travel Tom Wishnock is taking a long trip. In the first hour, he drives only 40 miles because of heavy traffic. After that, he averages 60 miles an hour. The equation $y = 60x + 40$, where x represents the number of hours that Tom spent driving 60 miles per hour, and y represents the total distance traveled.

a. Name the x-intercept and the y-intercept of the graph of $y = 60x + 40$. The x-intercept is $-\frac{2}{3}$ and the y-intercept is 40.

b. Graph the equation. See margin.

c. How long will it take Tom to finish his 400-mile trip? 6 hours

Exercises: Practicing and Applying the Concept

Independent Practice

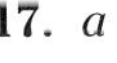

State the x-intercept and the y-intercept for each line.

17. a

18. b

19. c

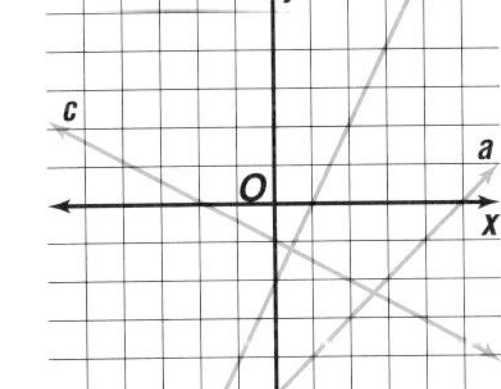

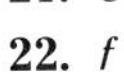

20. d

21. e

22. f

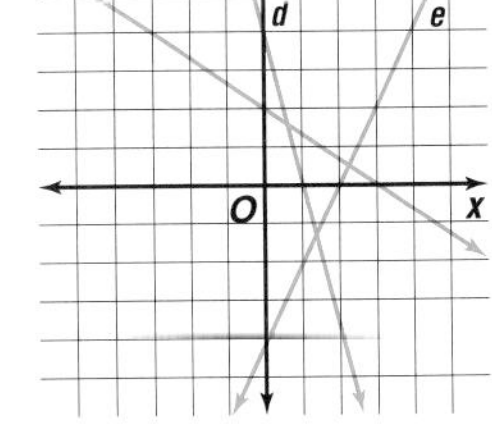

17. $5, -5$
18. $1, -2$
19. $-2, -1$
20. $1, 4$
21. $2, -4$
22. $3, 2$

Use the x-intercept and the y-intercept to graph each equation.

23. $y = x - 2$ $2, -2$

24. $y = x + 1$ $-1, 1$

25. $y = x - 4$ $4, -4$

26. $y = 2x - 1$ $\frac{1}{2}, -1$

27. $y = 3x + 4$ $-\frac{4}{3}, 4$

28. $y = -5x + 10$ $2, 10$

29. $y = \frac{1}{2}x - 5$ $10, -5$

30. $y = 3 - 0.5x$ $6, 3$

31. $y = \frac{1}{3}x + 2$ $-6, 2$

32. $y = \frac{2}{3}x + 4$ $-6, 4$

33. $y = 3x - 4$ $\frac{4}{3}, -4$

34. $y = -5x + 6$ $\frac{6}{5}, 6$

23–34. Intercepts are given. See Solutions Manual for graphs.

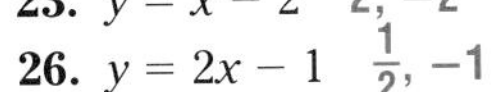

Error Analysis

Students may interchange the values for the x- and y-intercepts. Emphasize that the y-intercept crosses the y-axis and the x-intercept crosses the x-axis.

Assignment Guide

Core: 17–45 odd, 46–53
Enriched: 18–42 even, 43–53

For **Extra Practice**, see p. 760.

The red A, B, and C flags, printed only in the Teacher's Wraparound Edition, indicate the level of difficulty of the exercises.

Additional Answers

14.

15.

Practice Masters, p. 68

NAME ________ DATE ________

8-7 Practice
Intercepts

Student Edition Pages 406–410

State the x-intercept and the y-intercept for each line.

1. a
2. b
3. c

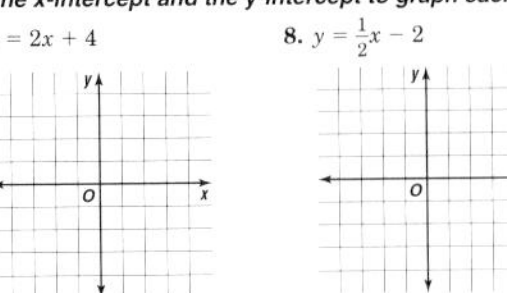

a: x-intercept: 2, y-intercept: 4
b: x-intercept: −2, y-intercept: −1
c: x-intercept: 1, y-intercept: −1

4. d
5. e
6. f

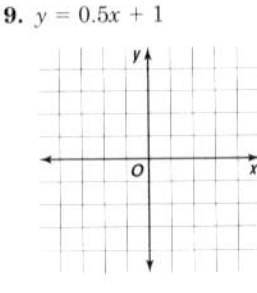

d: x-intercept: none, y-intercept: 2
e: x-intercept: 0, y-intercept: 0
f: x-intercept: −2, y-intercept: 1

Use the x-intercept and the y-intercept to graph each equation.

7. $y = 2x + 4$
8. $y = \frac{1}{2}x - 2$
9. $y = 0.5x + 1$

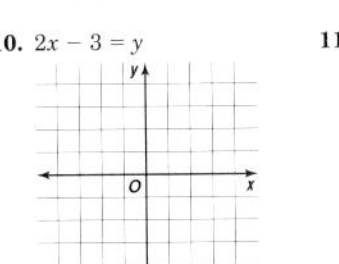

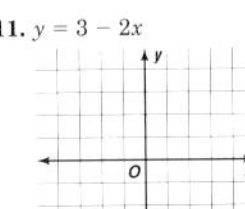

10. $2x - 3 = y$
11. $y = 3 - 2x$
12. $y + 2x = -4$

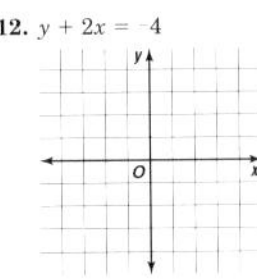

Graph each equation using the slope and y-intercept.

13. $y = \frac{1}{2}x - 3$
14. $x + 2y = 2$
15. $3y - 6 = -x$

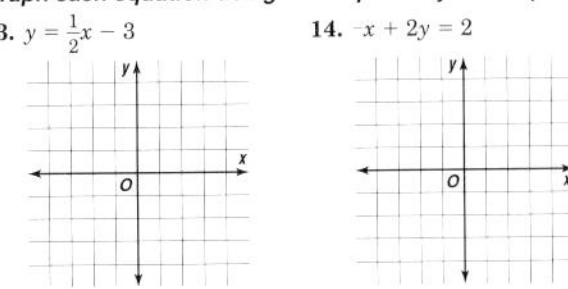

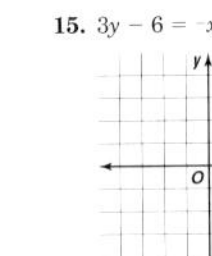

Group Activity Card 8-7

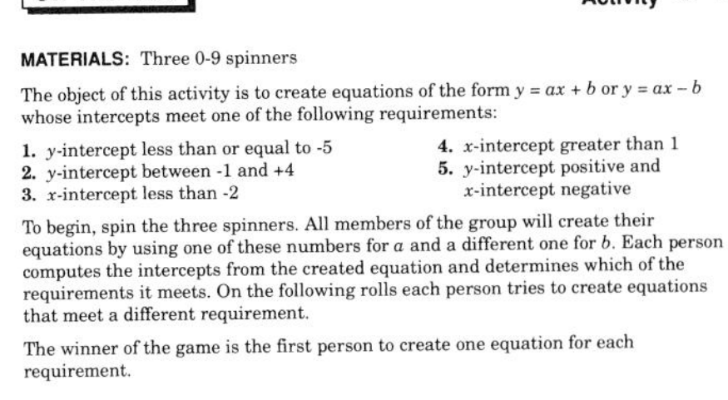

On The Line — Group Activity **8-7**

MATERIALS: Three 0-9 spinners

The object of this activity is to create equations of the form $y = ax + b$ or $y = ax - b$ whose intercepts meet one of the following requirements:

1. y-intercept less than or equal to -5
2. y-intercept between -1 and +4
3. x-intercept less than -2
4. x-intercept greater than 1
5. y-intercept positive and x-intercept negative

To begin, spin the three spinners. All members of the group will create their equations by using one of these numbers for a and a different one for b. Each person computes the intercepts from the created equation and determines which of the requirements it meets. On the following rolls each person tries to create equations that meet a different requirement.

The winner of the game is the first person to create one equation for each requirement.

©Glencoe/McGraw-Hill   Pre-Algebra

Additional Answer

16b.

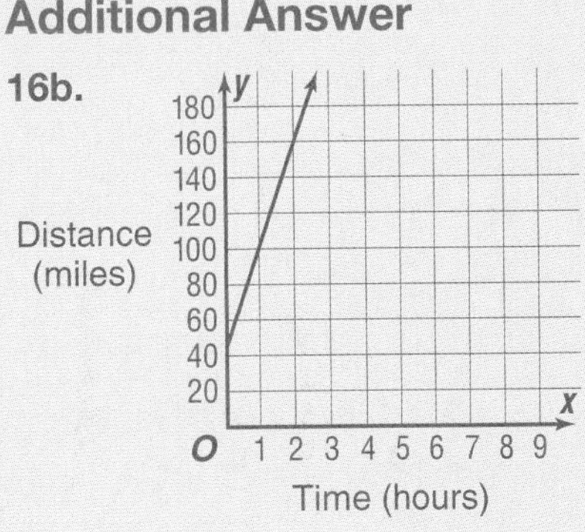

4 ASSESS

Closing Activity

Writing Have students write a sentence or two describing the difference between an x-intercept and a y-intercept.

Chapter 8, Quiz C (Lessons 8-6 and 8-7) is available in the *Assessment and Evaluation Masters*, p. 212.

Additional Answers

43. The x-intercept and y-intercept are both zero. The line, therefore, passes through the origin. Since two points are needed to graph a line, $y = 2x$ cannot be graphed using only the intercepts. See Solutions Manual for graphs.

45a.

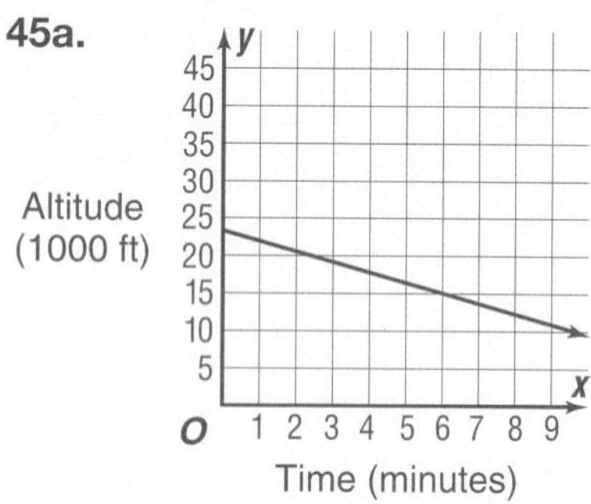

45c. The x-intercept is the time when the plane lands.

Enrichment Masters, p. 68

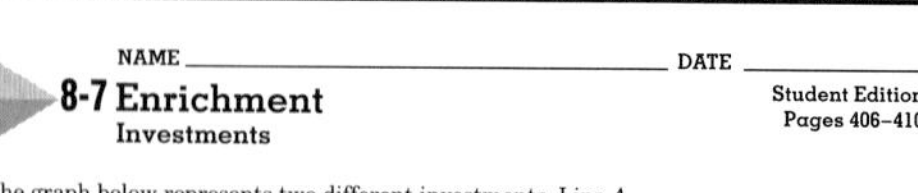

NAME ______ DATE ______

8-7 Enrichment
Investments

Student Edition Pages 406–410

The graph below represents two different investments. Line A represents an initial investment of \$30,000 at a bank paying passbook-savings interest. Line B represents an initial investment of \$5000 in a profitable mutual fund with dividends reinvested and capital gains accepted in shares. By deriving the equation, $y = mx + b$, for A and B, a projection of the future can be made.

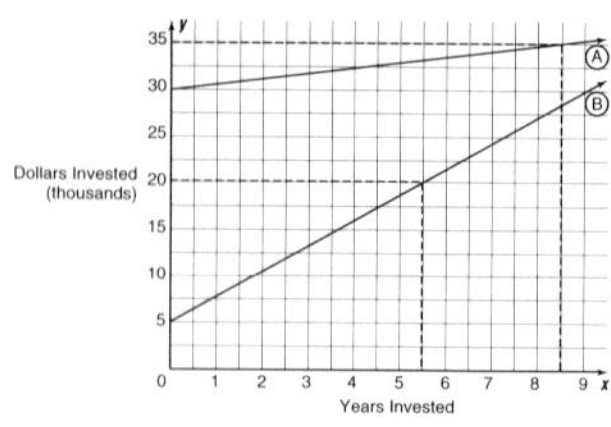

Solve.

1. The y-intercept, b, is the initial investment. Find b for each of the following.
   a. line A 30,000 b. line B 5000
2. The slope of the line, m, is the rate of return. Find m for each of the following.
   a. line A $\frac{35{,}000 - 30{,}000}{8.5 - 0} \approx 588$ b. line B $\frac{20{,}000 - 5000}{5.5 - 0} \approx 2727$
3. What are the equations of each of the following lines?
   a. line A $y = 588x + 30{,}000$ b. line B $y = 2727x + 5000$

Answer each of the following, assuming that the growth of each investment continues in the same pattern.

4. What will be value of the mutual fund after the 11th year?
   $y = 2727(11) + 5000 = \$34{,}997$
5. What will be the value of the bank account after the 11th year?
   $y = 588(11) + 30{,}000 = \$36{,}468$
6. When will the mutual fund and the bank account be of equal value?
   $588x + 30{,}000 = 2727x + 5000 \rightarrow x \approx 11.7$ years
7. In the long term, which investment has the greater payoff? mutual fund

Graph each equation using the slope and y-intercept.

35. $y = \frac{2}{3}x + 3$ **36.** $y = \frac{3}{4}x + 4$ **37.** $y = -\frac{3}{4}x + 4$

C **38.** $-4x + y = 6$ **39.** $-2x + y = 3$ **40.** $3y - 7 = 2x$

35–42. See Solutions Manual.

Given the slope of a line and its y-intercept, graph the line.

41. slope $= 3$, y-intercept $= 1$ **42.** slope $= -\frac{4}{3}$, y-intercept $= -1$

Critical Thinking

43. Explain why you cannot graph the equation $y = 2x$ by using intercepts only. Then draw the graph. See margin.

Applications and Problem Solving

44. Business Suppose you have a lawn-mowing service. You charge a flat fee of \$1.50 per mowing for gas, plus a fee of \$4.50 per hour for labor. The equation $y = 4.5x + 1.5$ represents the total fee, y, for mowing a lawn that takes x hours to mow.
   a. Graph this equation. See Solutions Manual.
   b. What does the y-intercept represent? the flat fee for gas
   c. What is your total fee for a lawn that takes 3 hours to mow? \$15

45. Aviation The equation $a = 24{,}000 - 1500t$, where t is the time in minutes and a is the altitude in feet, represents the steady descent of a jetliner.
   a. Graph this equation. See margin.
   b. Name the x-intercept of the graph. 16
   c. What does the x-intercept represent? See margin.

Mixed Review

46. Find the slope of a line that contains the points $S(-7, -3)$ and $T(-4, -5)$. (Lesson 8-6) $-\frac{2}{3}$

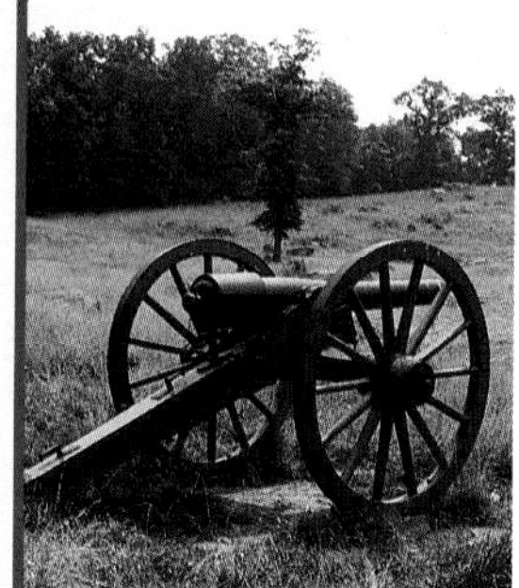

47. Statistics Refer to the graph at the right. Let x represent the number of visitors to Harper's Ferry National Park. The expression $2x + 100{,}000$ is an estimate of the number of visitors at which park? (Lesson 7-5) Gettysburg

Visiting the Past

Most visited in 1989	Visitors
Gettysburg National Military Park, Pa.	1,352,728
Chickamauga and Chattanooga National Military Park, Ga.	851,534
Manassas National Battlefield Park, Va.	767,138
Kennesaw Mountain National Battlefield Park, Ga.	762,422
Harper's Ferry National Historical Park, W. Va.	624,168

48. Solve $y = 15 - 8.7$. (Lesson 5-3) 6.3

49. Find the GCF of 39 and 65. (Lesson 4-5) 13

50. Solve $-41 > r - (-8)$. (Lesson 3-6) $r < -49$

51. Simplify $-3a + 12a + (-14a)$. (Lesson 2-4) $-5a$

52. Sports The Mets are losing to the Astros by a score of 5 to 3. Write an inequality that would tell how many runs would give the Mets the lead. (Lesson 1-9) $3 + x > 5$

53. What property allows you to say that $(9 + 10) + 7 = 9 + (10 + 7)$? (Lesson 1-4) associative, +

Choose
Estimation
Mental Math
Calculator
Paper and Pencil

Extension

Using Cooperative Groups Have students work in groups to solve this problem.

Juan complained to his sister, "I need to find a job closer to home. Yesterday I worked only 2 hours before the machine broke down and they sent me home. So I lost \$3 by going to work."

"How can that be?" asked his sister. Explain what happened. **He had to pay to commute to work. If it costs \$12 to get to work and he earns \$4.50 per hour, he would have lost \$3 by working only 2 hours.**

GRAPHING CALCULATOR ACTIVITY

8-7B Families of Graphs

An Extension of Lesson **8-7**

MATERIALS

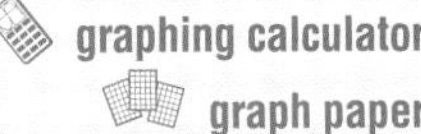

graphing calculator

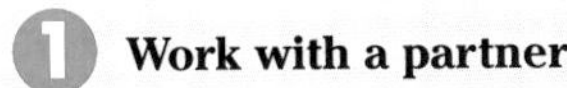

graph paper

A **family of graphs** is a group of graphs that displays one or more similar characteristics. Many linear graphs are related because they have the same slope or the same y-intercept as other functions in the family. You can graph several functions on the same screen and observe if any family traits exist.

Activity 1 **Work with a partner.**

- The TI-82 graphing calculator can graph up to nine functions at one time. Graph $y = 1.5x$, $y = 1.5x + 2$, $y = 1.5x + 4$, and $y = 1.5x - 4$ on the same screen in the standard viewing window.
- Begin by clearing the graphics screen. [Y=] Press and clear any equations in the list using the arrow and the [CLEAR] keys.
- Enter each equation into the Y= list.
- **Enter:** [Y=] 1.5 [X,T,θ] [ENTER] 1.5 [X,T,θ] [+] 2 [ENTER] 1.5 [X,T,θ] [+] 4 [ENTER] 1.5 [X,T,θ] [−] 4

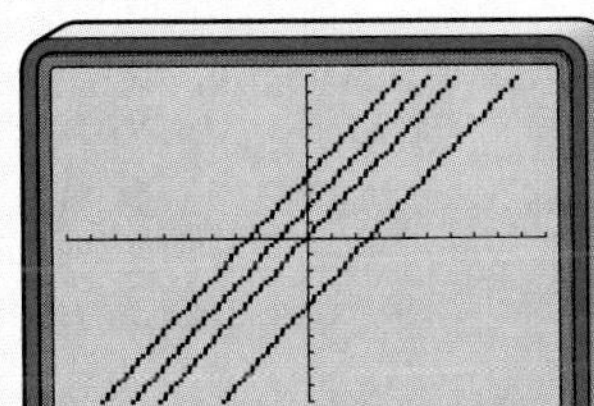

- The standard viewing window can be selected automatically from the zoom menu. Press [ZOOM] 6 and the graphs will appear automatically.

TALK ABOUT IT

1. Describe the family of graphs. **The family of graphs is lines having a slope of 1.5. All are parallel and have different x- and y-intercepts.**

Activity 2 **Work with a partner.**

- Clear the graphics screen. Then graph $y = 4x$ and $y = -4x$ on the same screen in the standard viewing window. Sketch the graphs on a piece of graph paper. **See Solutions Manual.**
- Then graph $y = -\frac{1}{4}x$ on the same screen. Sketch the graph on the graph paper. **See Solutions Manual.**
- Finally graph $y = -6x$ and $y = -1.5x$ on the same screen. Add the graphs to the graph paper. **See Solutions Manual.**

TALK ABOUT IT

2. Explain the difference between the graphs of $y = 4x$ and $y = -4x$.
3. Write the equation of a line whose graph is between the graph of $y = -4x$ and $y = -6x$. **Sample answer: $y = -5x$**

2. $y = 4x$ has a positive slope and $y = -4x$ has a negative slope.

Extension

4. Clear the screen and graph $y = 2x$, $y = 2x + 1$, $y = 2x + 2$, and $y = 2x + 3$. How are the graphs the same? How do they differ? **The slopes are the same. The y-intercepts are different. The lines are parallel.**

Technology

This lesson offers an excellent opportunity for using technology in your pre-algebra classroom. For more information on using technology, see *Graphing Calculators in the Mathematics Classroom*, one of the titles in the Glencoe Mathematics Professional Series.

8-7B LESSON NOTES

NCTM Standards: 1-6, 8

Objective

Use a graphing calculator to graph several functions to determine if any family traits exist.

Recommended Time

15 minutes

Instructional Resources

Graphing Calculator Masters, p. 21

This master provides keystroking instruction for this lesson for the TI-81 and Casio graphing calculators.

1 FOCUS

Motivating the Lesson

Ask students what they think a family of graphs might mean. Ask them how graphs having the same slope might differ.

2 TEACH

Teaching Tip Students having trouble distinguishing the different graphs could enter one equation and record its graph, then enter the equation and record its graph, and continue until all the graphs appear on the screen.

3 PRACTICE/APPLY

Assignment Guide
Core: 1–4
Enriched: 1–4

4 ASSESS

Observing students working with technology is an excellent method of assessment.

NCTM Standards: 1-5, 8, 9

Instructional Resources

- Study Guide Master 8-8
- Practice Master 8-8
- Enrichment Master 8-8
- Group Activity Card 8-8
- Real-World Applications, 17

Transparency 8-8A contains the 5-Minute Check for this lesson; **Transparency 8-8B** contains a teaching aid for this lesson.

Recommended Pacing	
Standard Pacing	Day 9 of 13
Honors Pacing	Day 8 of 12
Block Scheduling*	Day 5 of 7 (along with Lesson 8-7)

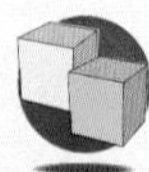

*For more information on pacing and possible lesson plans, refer to the ***Block Scheduling Booklet***.

1 FOCUS

5-Minute Check
(over Lesson 8-7)

Use the *x*- and *y*-intercepts to graph each equation.

1. $y = 6 - x$
Intercepts (0, 6) and (6, 0). See Transparency 8-8A for graph.

2. $y = x - \frac{1}{2}$
Intercepts $(0, -\frac{1}{2})$ and $(\frac{1}{2}, 0)$. See Transparency 8-8A for graph.

Motivating the Lesson

Hands-On Activity Provide each group of students with a straw and grid paper. Tell students to draw a coordinate grid on the paper. Then they are to drop the straw on the paper and draw a line to show the location of the straw. (*Hint:* Be sure the line passes through at least two points on the grid.) Drop the straw again and draw the line. Challenge students to find the solution for that system of equations.

8-8 Systems of Equations

Setting Goals: *In this lesson, you'll solve systems of linear equations by graphing.*

Modeling with Technology

Between 1984 and 1994, the number of cellular phone users doubled more than 7 times.

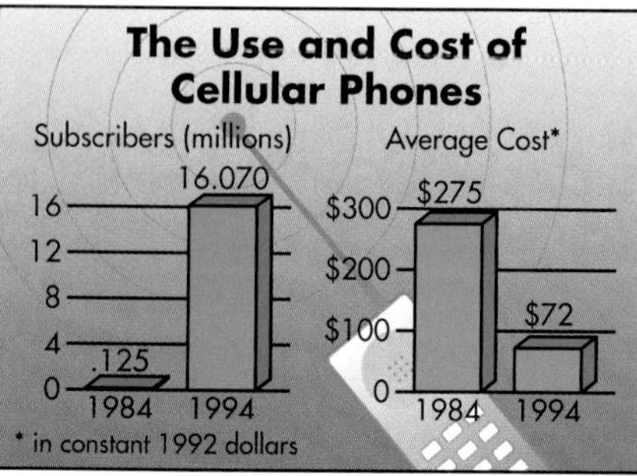

Source: Technology Futures, Inc.

In a questionnaire concerning phone service, potential customers were asked whether they would prefer to pay \$10 a month and 10¢ for each local call or \$5 a month and 20¢ for each local call. Is there a certain number of calls where the two payment plans result in the same total cost?

To solve this problem, let n represent the number of local calls made in a month and let c represent the monthly cost. You can write two equations to represent the situation above.

Plan 1 $c = 10 + 0.1n$ **Plan 2** $c = 5 + 0.2n$

Your Turn You can use a graphing calculator to solve the problem quickly. Enter each equation into the calculator and use the table feature to find the value of n for which the value of n is the same.

Enter: [Y=] 10 [+] .1 [X,T,θ] [ENTER] 5 [+] .2 [X,T,θ] [ENTER] [2nd] [TABLE]

The table on the calculator will show the values of n in the column marked X. Values of c for the plans 1 and 2 will be in columns Y_1 and Y_2 respectively.

TALK ABOUT IT

a. Describe each output. **See margin.**

b. Locate the row where the outputs are the same. Explain what the identical outputs mean and why it happened. **$x = 50$; The cost is the same for both plans when 50 calls are made.**

Learning the Concept

The equations $c = 10 + 0.1n$ and $c = 5 + 0.2n$ together are called a **system of equations**. The **solution** to this system is the ordered pair that is a solution of both equations. Refer to the outputs in the situation above. What is the solution of the system of equations?

Additional Answer
Talk About It

In the Y_1 column the output starts at 10 for x = 0 and each output increases by 0.1. In the Y_2 column, the output starts at 5 for x = 0 and each output increases by 0.2.

Another method for solving a system of equations is to graph the equations on the same coordinate plane. The coordinates of the point where the graphs intersect is the solution of the system of equations.

Example  **Solve the system of equations $y = 2x$ and $y = -x + 3$ by graphing.**

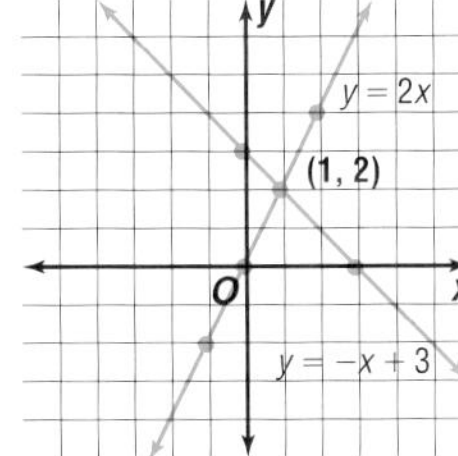

Graph $y = 2x$ and $y = -x + 3$ on the same coordinate plane.

The lines intersect at (1, 2). Therefore, the solution to the system of the equations is (1, 2).

To check, substitute the coordinates into each equation to see whether it is a solution to both equations.

Check:

$y = 2x$	$y = -x + 3$
$2 \stackrel{?}{=} 2(1)$	$2 \stackrel{?}{=} -(1) + 3$
$2 = 2$ ✓	$2 = 2$ ✓

Solving systems of equations is also used to solve problems.

Example 2 **Refer to the application at the beginning of the lesson. Solve the system of equations $c = 10 + 0.1n$ and $c = 5 + 0.2n$ by graphing.**

Communication

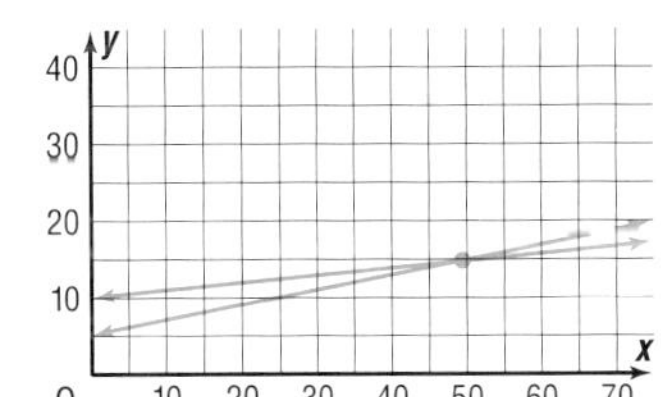

The graphs intersect at the point (50, 15).

Check:

$c = 10 + 0.1n$	$c = 5 + 0.2n$
$15 \stackrel{?}{=} 10 + 0.1(50)$	$15 \stackrel{?}{=} 5 + 0.2(50)$
$15 = 15$ ✓	$15 = 15$ ✓

The solution is (50, 15). This means that two payment plans result in the same total cost, \$15, when 50 local calls are made.

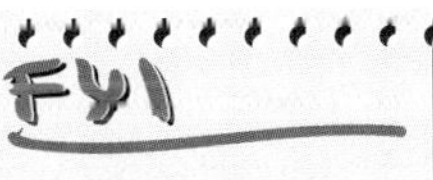

The first test of a cellular phone was in Chicago in 1978.

As you can see in the examples above, when the graphs of two linear equations intersect in exactly one point, the system has exactly one ordered pair as its solution. It is also possible for the two graphs to be parallel lines or to be on the same line. When the graphs are parallel, the system of equations does not have a solution. When the graphs are the same line, the system of equations has infinitely many solutions.

Example  **Solve each system of equations by graphing.**

a. $y = x + 3$ and $y = x + 4$

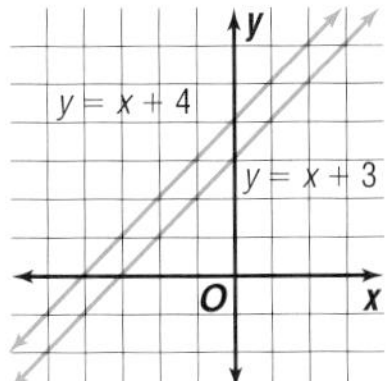

The graphs of the equations are parallel lines. Since they do not intersect, there is no solution to this system of equations.

Notice that the two lines have the same slope but different *y*-intercepts.

2 TEACH

In-Class Examples

For Example 1

Solve the system of equations, $y = -3x$ and $y = 2 - x$, by graphing. **(−1, 3)**

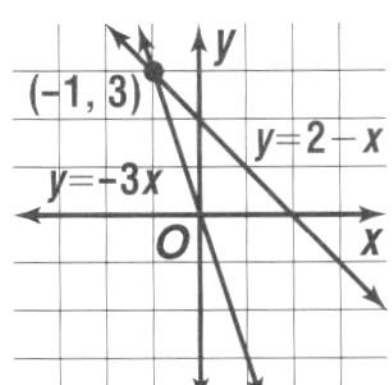

For Example 2

Yari's bookcase has three shelves. The top shelf is $32\frac{1}{4}$ inches above the bottom shelf. The middle shelf is $\frac{1}{4}$ inch thick. The distance between the middle shelf and top shelf is 12 inches more than the distance between the bottom shelf and the middle shelf. Find the distance between the bottom shelf and the middle shelf. Have students use *t* for the distance between the top two shelves and *b* the distance between the bottom two shelves and the equations $t + b + \frac{1}{4} = 32\frac{1}{4}$ and $t = b + 12$ to solve the problem. **The middle shelf is 10 inches above the bottom shelf.**

For Example 3

Solve each system of equations by graphing.

a. $y = 2x - 1$ and $2x - y = 3$ **See students' graphs. The lines are parallel, there are no solutions.**

b. $2y = 5x - 6$ and $y = 2\frac{1}{2}x - 3$ **See students' graphs. The lines coincide, there are infinitely many solutions.**

GLENCOE Technology

Interactive Mathematics Tools Software

In this interactive computer lesson, students use systems of equations in solving problems involving two cyclists. A **Computer Journal** gives students an opportunity to write about what they have learned.

For Windows & Macintosh

It is estimated that there are 574,860,000 telephone lines in use in the world.

3 PRACTICE/APPLY

Checking Your Understanding

Exercises 1–13 are designed to help you assess your students' understanding through reading, writing, speaking, and modeling. You should work through Exercises 1–5 with your students and then monitor their work on Exercises 6–13.

Additional Answers

1. Sample answer: Two or more equations with the same variables form a system of equations. The coordinates of the point where the graphs of the two equations intersect is the solution.
2. Sample answer: Plan 1 costs more if less than 50 calls are made per month. Plan 2 costs more if more than 50 calls are made per month. Preferences will vary.
3. Sample answer: The graphs of the equations are parallel lines. Since they do not intersect, there is no solution.

Study Guide Masters, p. 69

NAME ______________________ DATE ____________

8-8 Study Guide
Systems of Equations

Student Edition Pages 412–416

The equations $y = -\frac{1}{2}x + 2$ and $y = 3x - 5$ together are called a **system of equations**.

The **solution** to this system is the ordered pair that is the solution of both equations. To solve a system of equations, graph each equation on the same coordinate plane. The point where both graphs intersect is the solution of the system of equations.

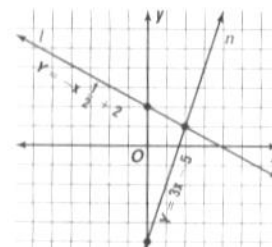

Line l is the graph of $y = -\frac{1}{2}x + 2$.

Line n is the graph of $y = 3x - 5$.

The lines intersect at (2, 1). Therefore, the solution to the system of equations is (2, 1).

Use a graph to solve each system of equations.

1. $y = x - 3$
   $y = -3x + 1$ **(1, −2)**

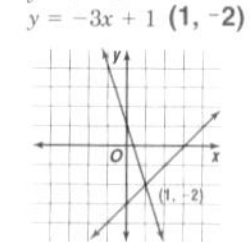

2. $y = 2x$
   $y = x + 1$ **(1, 2)**

3. $y = 4x + 5$ **no**
   $y = 4x - 1$ **solution**

4. $y = x$
   $y = -2$ **(−2, −2)**

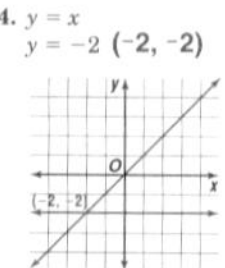

5. $y = -x - 1$
   $y = -3x - 3$ **(−1, 0)**

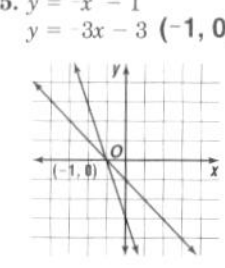

6. $y = 6x - 12$
   $y = 2x - 4$ **(2, 0)**

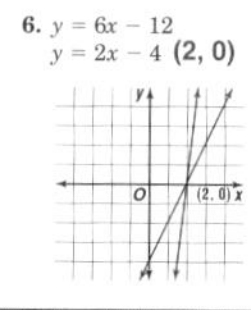

b. $y = -2x + 3$ **and** $3y = -6x + 9$

Each equation has the same graph. Any ordered pair on the graph will satisfy both equations. Therefore, there are infinitely many solutions to this system of equations.

Notice that the graphs have the same slopes and intercepts.

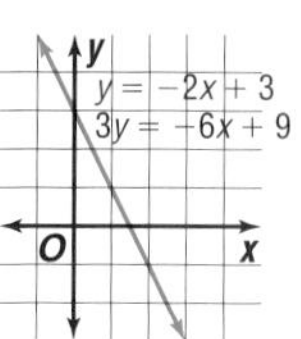

Checking Your Understanding

Communicating Mathematics

Read and study the lesson to answer these questions. **1–3. See margin.**

1. **Explain** what is meant by a system of equations and describe its solution.
2. **Compare and contrast** the two local calling plans at the beginning of the lesson. Which plan would you prefer? Explain.
3. **Show** why there is no solution for the system $x + y = 4$ and $y = 8 - x$.
4. **Create and graph** a system of equations that has one solution, another system that has no solution, and a third that has infinitely many solutions. **See Solutions Manual.**

MATERIALS
- geoboard
- geobands

5. **Geometry** You can use a geoboard to model the first quadrant of the coordinate plane and geobands to model graphs of equations. Let the row of pegs on the bottom represent the x-axis and the column of pegs on the left represent the y-axis. The graph of the equation $y = x + 1$ is shown at the right.

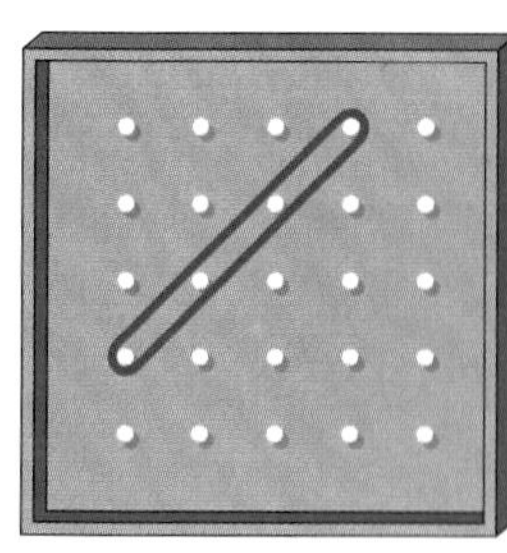

   a. Make the "line" shown at the right on a geoboard. **See students' work.**
   b. Use a geoband to show the graph of $y = -x + 3$ on the same geoboard. **See margin.**
   c. What is the coordinate of the peg that is enclosed by both geobands? **(1, 2)**
   d. What is the solution to the system of equations $y = x + 1$ and $y = -x + 3$? **(1, 2)**

Guided Practice

State the solution of each system of equations.

6. 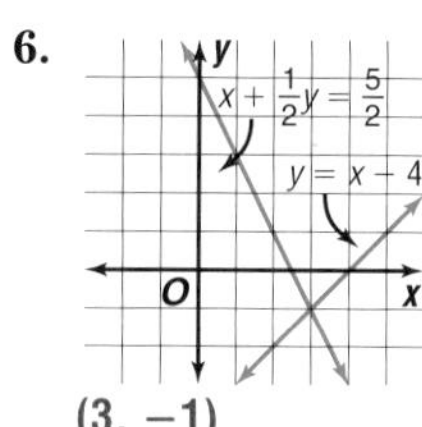

(3, −1)

7. 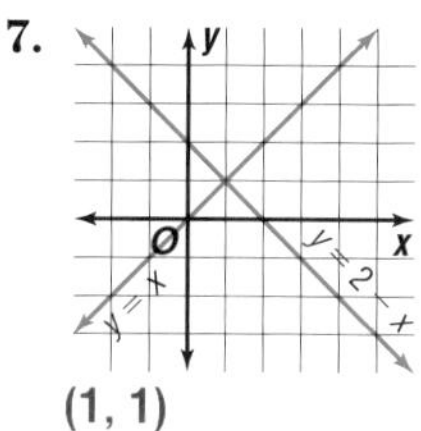

(1, 1)

8. 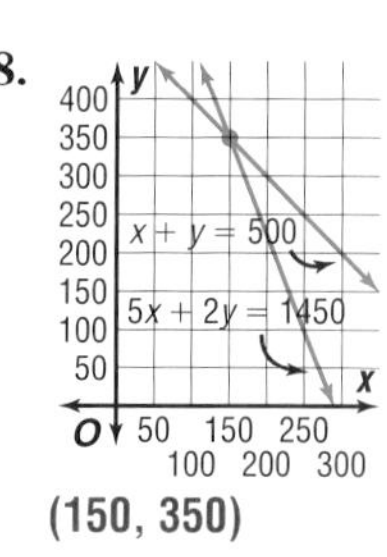

(150, 350)

Reteaching

Using Models Have students graph $y = 4x + 3$ and $y = x - 4$ on a grid. Then have them substitute the coordinates of the point of intersection into both equations to see if it is a solution to both equations. Tell students that any two equations form a system and the intersection of their graphs is the solution of the system.

Additional Answer

5b.

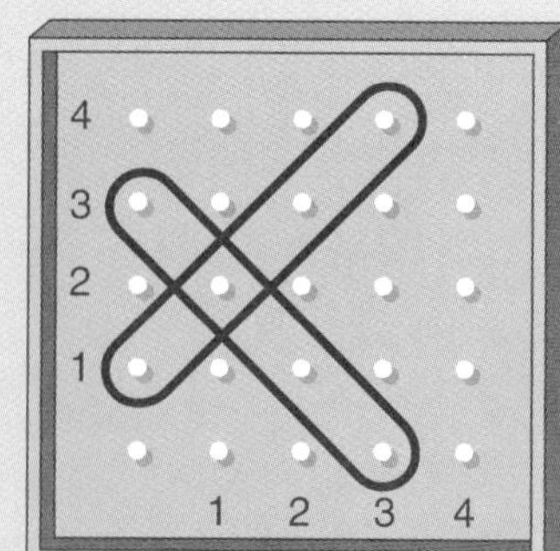

Use a graph to solve each system of equations.

9–12. See Solutions Manual for graphs.

9. $y = x - 8$
$y = -2x - 2$ (2, −6)

10. $y = x + 6$
$y = 6 - 9x$ (0, 6)

11. $y = -\frac{1}{2}x + 2$
$y = 3x - 5$ (2, 1)

12. $y = \frac{3}{8}x - \frac{1}{2}$ no solution
$y = \frac{3}{8}x - \frac{21}{8}$

13. **Art** Kirima bought 12 feet of framing material to make a rectangular frame for her oil painting. She uses all the framing material with no waste and the length is twice the width. $2\ell + 2w = 12$; $\ell = 2w$

a. Write a system of equations that represents this situation.

b. Solve this system of equations by graphing. (4, 2)

c. Explain what the solution means. The length of the frame will be 4 feet and the width will be 2 feet.

Error Analysis

Students may have incorrect solutions because lines are inaccurately drawn. Encourage students to use a sharpened pencil and a straightedge when drawing graphs of systems of equations.

Exercises: Practicing and Applying the Concept

Independent Practice

The graphs of several equations are shown at the right. State the solution of each system of equations.

14. a and c (−3, 3)
15. b and c (0, 0)
16. c and d (5, −5)
17. b and d (2, 1)
18. a and d (1, 3)
19. a and the x-axis no solution
20. b and the y-axis (0, 0)

21. a, b, and d no solution
22. b, c, and the y-axis (0, 0)

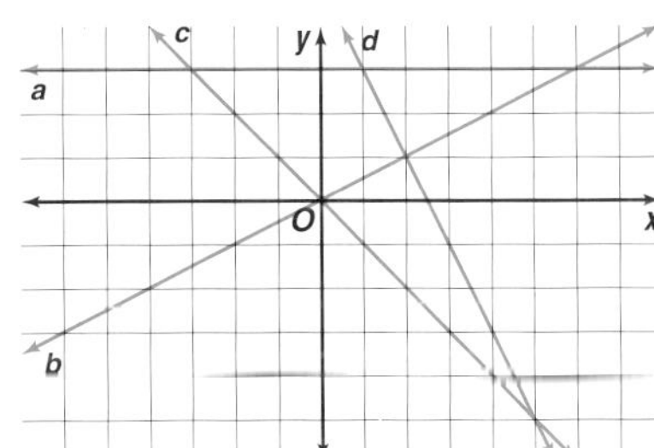

Assignment Guide

Core: 15–39 odd, 40–46
Enriched: 14–34 even, 35–46

For **Extra Practice**, see p. 760.

The red A, B, and C flags, printed only in the Teacher's Wraparound Edition, indicate the level of difficulty of the exercises.

Additional Answer

36a.

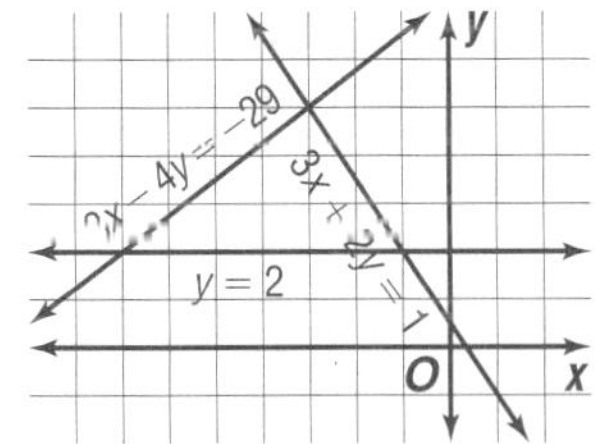

Use a graph to solve each system of equations.

23–34. See Solutions Manual for graphs.

23. $y = x + 2$
$y = -2x + 17$ (5, 7)

24. $y = x$
$y = 3$ (3, 3)

25. $y = -x + 2$
$y = 2x + 5$ (−1, 3)

26. $y = -2x - 5$
$y = -2x - 8$ none

27. $y = x + 6$
$y = 3x$ (3, 9)

28. $y = -x$
$y = 4x$ (0, 0)

29. $y = -5x - 2$
$y = 2x + 12$ (−2, 8)

30. $y = -x$
$y = 0.5x + 3$ (−2, 2)

31. $y = 3x - 8$
$y = 2x - 3$ (5, 7)

32. $\frac{1}{2}x + \frac{1}{3}y = 6$
$y = \frac{1}{2}x + 2$ (8, 6)

33. $\frac{2}{3}x + \frac{1}{4}y = 4$
$x = -\frac{3}{8}y + 6$
infinitely many

34. $3.4x + 6.3y = 4.4$
$2.1x + 3.7y = 3.1$
(5, −2)

Critical Thinking

35. The solution for the system of equations $Ax + y = 6$ and $Bx + y = 7$ is (1, 3). What are the values of A and B? $A = 3$, $B = 4$

Applications and Problem Solving

36. **Geometry** The graphs of the equations $y = 2$, $3x + 2y = 1$, and $3x - 4y = -29$ contain the sides of a triangle. 36a. See margin.

a. Graph the three equations on the same coordinate plane.

b. Find the coordinates of the vertices of the triangle.

36b. (−3, 5), (−1, 2), (−7, 2)

Practice Masters, p. 69

NAME ________ DATE ________

8-8 Practice
Systems of Equations

Student Edition Pages 412–416

The graphs of several equations are shown at the right. State the solution of each system of equations.

1. a and b (−4, 2)
2. c and d (1, 2)
3. c and e (4, −1)
4. b and d (−1, −2)
5. b and e (1, −5)
6. a and f (−4, 2)
7. c and f (−2, 5)
8. a and d (0, 0)
9. a and c (6, −3)
10. b and f (−4, 2)
11. a and the x-axis (0, 0)
12. a, b, and d no solution
13. a, d, and the y-axis (0, 0)

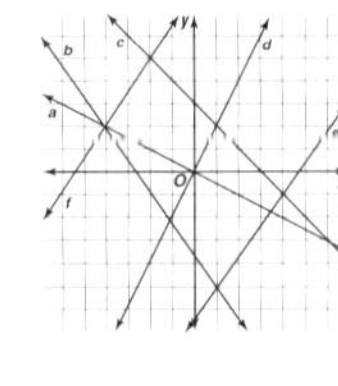

Use a graph to solve each system of equations.

14. $y = -2x + 4$
$y = 2x$ (1, 2)

15. $y = x + 5$
$y = 2x + 6$ (−1, 4)

16. $y = -3x + 3$ no
$y = -3x - 7$ solution

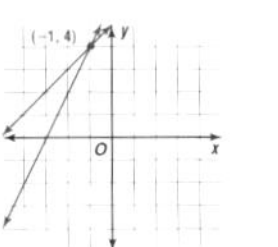

17. $y = -5x$
$y = x$ (0, 0)

18. $x - y = 2$
$x + y = 4$ (1, 3)

19. $2x - y = 3$
$x + y = 3$ (2, 1)

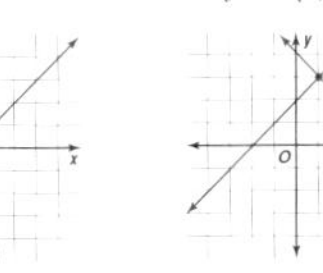

Group Activity Card 8-8

Get The Point Group Activity 8-8

MATERIALS: Graph paper

For this activity, you will consider systems of equations and look for a pattern to determine whether the x-coordinate of their solution will be positive or negative.

Working with a partner, each of you write an equation of the form $y = ax + b$ or $y = ax - b$. Make the equations simple by choosing both a and b to be between -10 and +10.

Next, graph these two equations and locate their point of intersection. After working with a number of these two equations, see if you can explain to another member of the class how he or she may determine where two lines will intersect by looking at their equations.

 Pre-Algebra

4 ASSESS

Closing Activity

Writing Have students write or draw a flowchart showing the steps involved in finding the solution to a system of equation by graphing. **Graph both equations on the same coordinate system. If the lines intersect, locate the point of intersection as the solution.**

Additional Answers

37b. The break-even point would be at point (300, 1500) because that is the solution when the costs equals the income, when 300 items are sold, resulting in an income of \$1500.

39c. After 6 seconds, the dog will catch up to the prowler 90 meters from where the dog started.

Enrichment Masters, p. 69

NAME ______________ DATE ________

8-8 Enrichment
Using Coordinates

Student Edition
Pages 412–416

1. Graph the points (1, 1), (4, 4), and (2, 4) on the coordinate system below. Connect the dots. Name the figure formed. **triangle**
2. Multiply each y-coordinate by −1. Graph the points. How is this triangle related to the one in Exercise 1? **It is a reflection over the x-axis.**
3. Multiply each coordinate in Exercise 1 by −1. Graph the points.
4. What would you have to do to get the coordinates of a triangle in Quadrant II congruent to the ones in Exercises 1–3? **Multiply each x-coordinate in Exercise 1 by −1.**
5. Graph the points (3, 1), (2, 3), (4, 6), and (5, 4) on the coordinate plane at the right. Connect the dots. The figure formed is called a parallelogram.
6. Add 2 to both coordinates of each point and graph the new coordinates. Name the figure formed. **parallelogram**
7. Add −4 to the x-coordinate of each point in Exercise 5 and graph the new coordinates. Is this figure also a parallelogram? **yes**
8. Graph the points (5, −2), (6, −3), (5, −5), and (3, −3) on the coordinate plane at the right. Name the figure formed. **trapezoid**
9. Multiply both coordinates of each point in Exercise 8 by 2 and graph the new coordinates. This is an enlargement.
10. Multiply both coordinates of each point in Exercise 8 by $\frac{1}{2}$ and graph the new coordinates. This is a reduction.

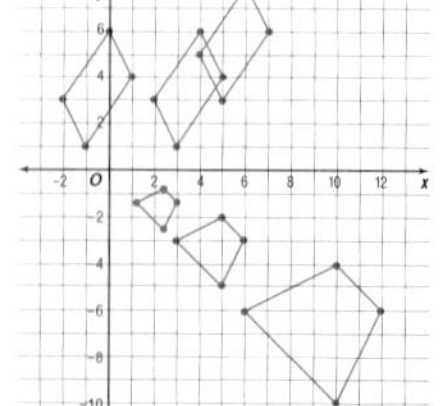

37. **Business** The Howell Company has fixed costs of \$900 per week. Each item produced by the company costs \$2 to manufacture and can be sold for \$5. If x is the number of items produced each week, the cost of producing them can be represented by the equation $y = 900 + 2x$. The weekly income from selling the items can be represented by $y = 5x$. **37b. See margin.**
    a. Solve this system of equations by graphing. **(300, 1500)**
    b. The *break-even point* is the point at which the income is the same as the cost of producing the goods or services. Name the break-even point of the system of equations and explain what it means.
38. **Human Growth** Jordan Gossell was born six weeks premature. He weighed 3 pounds at birth and gained an average of 2 pounds per month. Paige Plesich was born on her due date. She weighed 7 pounds at birth and gained an average of 1 pound per month.
    a. Let a represent each baby's age in months, and let w represent each baby's weight in pounds. Write a system of equations that represents this situation. **Jordan: $w = 3 + 2a$, Paige: $w = 7 + a$**
    b. Solve this system of equations by graphing. **(4, 11)**
    c. Explain what the solution means.

38c. Each baby weighs 11 pounds after 4 months.
39a. $y = 15x$, $y = 5x + 60$

39. **Law Enforcement** A police K-9 unit spots a prowler 60 meters away. The dog runs toward the prowler at a speed of 15 meters per second, and the prowler runs away at a speed of 5 meters per second.
    a. Write a system of equations that represents this situation. (*Hint*: Let x represent time and y represent distance.)
    b. Solve this system of equations by graphing. **(6, 90)**
    c. Explain what the solution means. **See margin.**

Mixed Review

40. Graph $y = -3x + 2$ by using the x-intercept and the y-intercept. (Lesson 8-7) **See margin.**
41. If 4.05 times an integer is increased by 3.116, the result is between 13 and 25. Use an inequality to find the number. (Lesson 7-6) **3, 4, or 5**
42. **Publishing** A page of type is to be divided into three columns. If the page is $6\frac{3}{4}$ inches wide, how many inches wide is each column? (Lesson 6-4) **$2\frac{1}{4}$ inches**
43. **Personal Finance** Leya has \$575.29 in her checking account. She deposits her paycheck for \$125.90. The same day, she receives bills for \$397.28 and \$225.40. If she pays both bills, how much is in her checking account at the end of the day? (Lesson 5-3) **\$78.51**
44. Solve $-15 + t \leq 12$. Check your solution. (Lesson 3-6) **$t \leq 27$**
45. Evaluate $|-17| - |3|$. (Lesson 2-1) **14**
46. Simplify $4(c + 3)$. (Lesson 1-5) **$4c + 12$**

Extension

Consumer Awareness Manufacturer A sells bumper stickers for 22¢ each. Manufacturer B charges a \$50 flat rate plus 11¢ per sticker. Have students graph both plans. Discuss when one manufacturer would be cheaper than the other, and when they would be equal. **Costs are equal at 455 stickers. Less than 455, Manufacturer A is cheaper; more than 455, Manufacturer B is cheaper.**

Additional Answer

40.

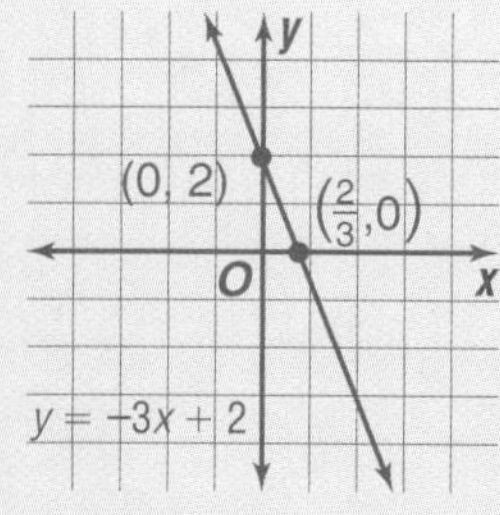

GRAPHING CALCULATOR ACTIVITY

8-9A Graphing Inequalities

A Preview of Lesson **8-9**

MATERIALS

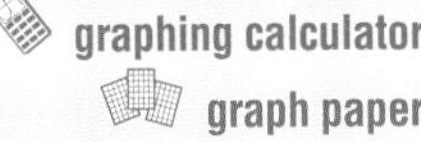

graphing calculator

graph paper

The graph of a linear function separates the coordinate plane into two regions, one above the line and one below it. For the graph of $y = 2x + 5$, shown at the right, points in the region *above* the line are represented by the inequality $y > 2x + 5$. Points in the region *below* the line are represented by the inequality $y < 2x + 5$. You can use a graphing calculator to investigate the graphs of inequalities.

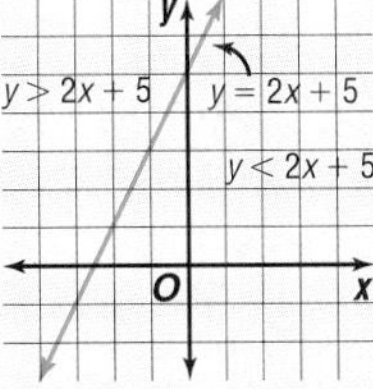

Your Turn Work with a partner.

▶ Graph $y < 2x + 5$ in the standard viewing window.

Enter: [2nd] [DRAW] 7 [(-)] 10 [,] 2 [X,T,θ] [+] 5 [)] [ZOOM] 6 [CLEAR] [ENTER].

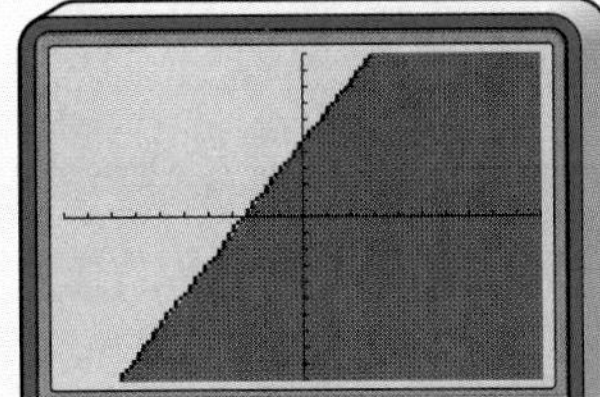

The TI-82 shades between two functions. The first function defines the lower boundary; in this case, Ymin, or −10. The second function defines the upper boundary; in this case, the function y = 2x + 5.

Notice that the graph is a shaded region. This indicates that all ordered pairs in that region satisfy the inequality $y < 2x + 5$.

▶ Clear the drawing that is currently displayed.

Enter: [2nd] [DRAW] [ENTER].

▶ Now graph $y > 2x + 5$ in the standard viewing window.

Enter: [2nd] [DRAW] 7 2 [X,T,θ] [+] 5 [,] 10 [)] [ENTER].

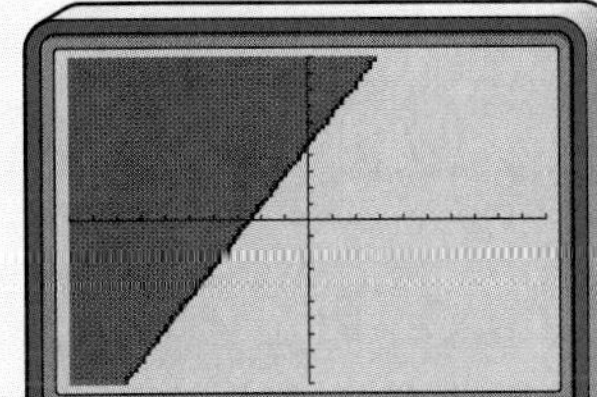

The lower boundary is y = 2x + 5, and the upper boundary is Ymax or 10.

TALK ABOUT IT

1. What is similar about the two graphs?
2. How does the graph of $y < 2x + 5$ differ from the graph of $y > 2x + 5$?
3. How do the graphs differ from the graph of $y = 2x + 5$?
4. Write the keystroke sequence that would graph the following inequalities. Then use a graphing calculator to graph each inequality.
   a. $y < x + 4$ b. $y > 2x - 1$
   c. $y > -x + 3$ d. $y < -2x + 2$

1. Both graphs are shaded regions and the boundary is the same line.
2–4. See Solutions Manual.

8-9A LESSON NOTES

NCTM Standards: 1-6, 8

Objective
Use a graphing calculator to investigate the graphs of inequalities.

Recommended Time
Activity: 15 minutes

Instructional Resources
Graphing Calculator Masters, p. 22

This master provides keystroking instruction for this lesson for the TI-81 and Casio graphing calculators.

1 FOCUS

Motivating the Lesson
Ask students to visualize the graph of an equation and then describe how that graph would change if the equal sign of the equation is changed to ≤, ≥, <, and >.

2 TEACH

Teaching Tip Explain to students who may feel overwhelmed that the only difference between graphing lines and inequalities is the shading and determining whether the line is included in the solution.

3 PRACTICE/APPLY

Assignment Guide
Core: 1–4
Enriched: 1–4

4 ASSESS

Observing students working with technology is an excellent method of assessment.

Technology

This lesson offers an excellent opportunity for using technology in your pre-algebra classroom. For more information on using technology, see *Graphing Calculators in the Mathematics Classroom*, one of the titles in the Glencoe Mathematics Professional Series.

8-9 LESSON NOTES

NCTM Standards: 1-5, 8, 9

Instructional Resources

- Study Guide Master 8-9
- Practice Master 8-9
- Enrichment Master 8-9
- Group Activity Card 8-9
- Assessment and Evaluation Masters, p. 212
- Math Lab and Modeling Math Masters, p. 83
- Multicultural Activity Masters, p. 16
- Tech Prep Applications Masters, p. 16

Transparency 8-9A contains the 5-Minute Check for this lesson; **Transparency 8-9B** contains a teaching aid for this lesson.

Recommended Pacing	
Standard Pacing	Day 11 of 13
Honors Pacing	Day 10 of 12
Block Scheduling*	Day 6 of 7

*For more information on pacing and possible lesson plans, refer to the ***Block Scheduling Booklet***.

1 FOCUS

5-Minute Check
(over Lesson 8-8)

Use a graph to solve each system of equations.

1. $y = x + 1$ and $y = -x + 3$ **(1, 2) See Transparency 8-9A for graph.**
2. $2x - y = 6$ and $x = y + 2$ **(4, 2) See Transparency 8-9A for graph.**

8-9 Graphing Inequalities

Setting Goals: *In this lesson, you'll graph linear inequalities.*

Modeling with Manipulatives

MATERIALS

graph paper

The graph of $y = x + 2$ is shown at the right. Copy the axes and the graph on a piece of graph paper.

Your Turn

Graph the following ordered pairs on your coordinate system.

(3, 4)	(3, −2)	(−2, −1)
(−1, 0)	(4, 4)	(0, −3)

a. Where do these points lie in the plane in relation to the graph of $y = x + 2$?

b. Which sentence, $y = x + 2$, $y > x + 2$, or $y < x + 2$, is true for all of the ordered pairs you graphed?

c. Which sentence, $y = x + 2$, $y > x + 2$, or $y < x + 2$, is true for the points located above the graph of $y = x + 2$?

d. How many points belong to the graph of $y < x + 2$?

a. below the line
b. $y < x + 2$
c. $y > x + 2$
d. infinitely many

Learning the Concept

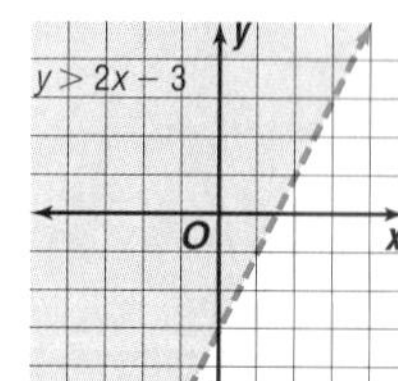

To draw the graph of an inequality such as $y > 2x - 3$, first, draw the graph of $y = 2x - 3$. This graph is a line that separates the coordinate plane into two regions. In the graph at the right, one of the regions is shaded yellow, and the other is shaded blue. The line is the **boundary** of the two regions.

To determine which region is the solution to $y > 2x - 3$, test a point in the blue region. For example, you can test the origin, (0, 0).

$$y > 2x - 3$$
$$0 \overset{?}{>} 2(0) - 3$$
$$0 > -3 \quad \checkmark$$

Since $0 > -3$ is true, (0, 0) *is* a solution of $y > 2x - 3$.

Points on the boundary are *not* solutions of $y > 2x - 3$, so it is shown as a dashed line. Thus, the graph is all points in the blue region.

Now consider the graph of $y \le 2x - 3$. The origin, (0, 0), is not part of the graph of $y \le 2x - 3$ since $0 \le -3$ is not true.

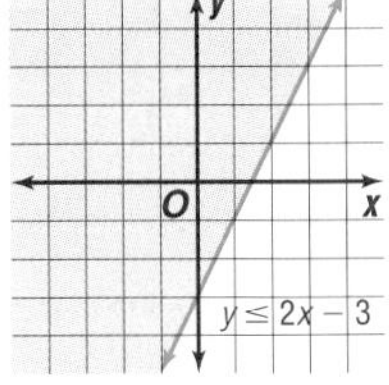

Since the inequality $y \le 2x - 3$ means $y < 2x - 3$ or $y = 2x - 3$, points on this boundary *are* part of the solution of $y \le 2x - 3$. Therefore, the boundary is shown as a solid line on the graph.

Thus, the graph is all the points in the yellow region and the line $y = 2x - 3$.

Example  **Graph $y < -2x + 2$.**

Graph the equation $y = -2x + 2$. Draw a dashed line since the boundary is *not* part of the graph.

The origin, (0, 0), is part of the graph since $0 < -2(0) + 2$ is true. Thus, the graph is all points in the region below the boundary. Shade this region.

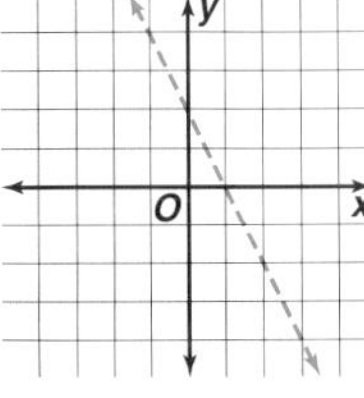

You may have to solve some inequalities for y first to find a solution to the system graphically.

Example  **APPLICATION** **Sales**

Tickets for *West Side Story* are \$66.50 for floor and \$47.50 for balcony. In order to cover the expenses for the show, at least \$66,500 per show must be made from ticket sales.

a. Write an inequality to represent the situation.

b. Use a graph to determine how many of each type of ticket must be sold to cover the expenses.

c. List three possible solutions.

a. Let x represent the number of \$47.50 tickets sold, and let y represent the number of \$66.50 tickets sold.

sales of \$47.50 tickets	*plus*	*sales of \$66.50 tickets*	*is greater than or equal to*	*\$66,500*
$47.5x$	$+$	$66.5y$	$\ge$	66,500

b. To graph the inequality, first solve for y.

$47.5x + 66.5y \ge 66{,}500$

$66.5y \ge 66{,}500 - 47.5x$ *Subtract 47.5x from each side.*

$y \ge 1000 - \frac{5}{7}x$ *Divide each side by 66.5*

Graph $y = 1000 - \frac{5}{7}x$ as a solid line since the boundary is part of the graph. The origin is not part of the graph since $0 \ge 1000 - \frac{5}{7}(0)$ is not true. Thus, the graph is all the points in the region that do not contain the origin.

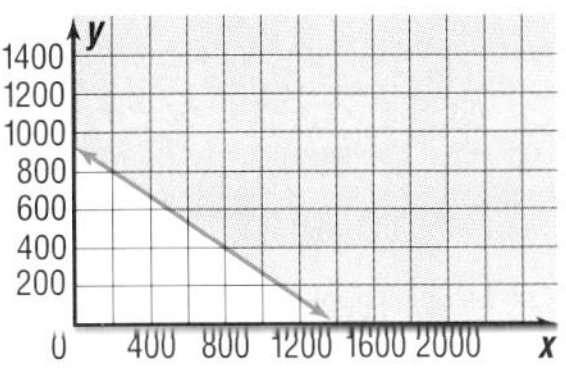

Motivating the Lesson

Questioning Discuss situations where shading is used; for example, a car visor, a baseball cap, sunglasses. Explain that in each case you shade the area you want to use.

2 TEACH

Teaching Tip Have students graph sets of similar inequalities, such as $y > 2x + 2$, $y \ge 2x + 2$, $y < 2x + 2$, and $y \le 2x + 2$ on different coordinate grids. Then have students compare and contrast the graphs.

In-Class Examples

For Example 1

Graph $y > 2x + 5$.

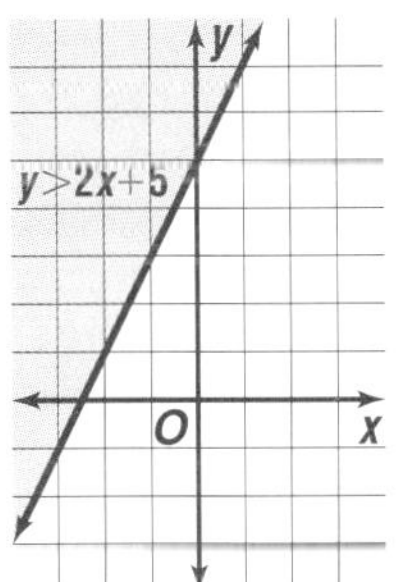

For Example 2

A company has expenses of \$20,000 per month plus monthly salaries of \$2,000 per employee. The company needs to make a profit every month (to make a profit, sales must exceed expenses) Make a graph to show the relation between expenses and the number of employees.

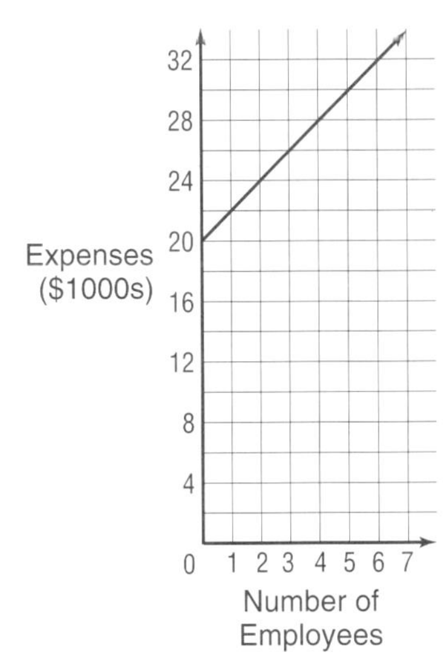

Teaching Tip For Example 2 above, make sure that students understand that whole numbers are necessary since you cannot have a partial employee. Also stress the need to interpret the solution based on the situation. In the example above, the portion of the graph to the left of x is not realistic because if you have no employees, you would have no product or profit.

3 PRACTICE/APPLY

Checking Your Understanding

Exercises 1–18 are designed to help you assess your students' understanding through reading, writing, speaking, and modeling. You should work through Exercises 1–5 with your students and then monitor their work on Exercises 6–18.

Additional Answers

2. Sanura is correct. Dashed lines are for $<$ and $>$. Solid lines are for $\leq$ or $\geq$.
3. $y < 9x + 4$
4. The solution would lie on the graph of the line $47.5x + 66.5y = 66{,}500$.
5. Sample answer: Choose a test point. Substitute its coordinates in the inequality. If the result is true, that side of the boundary is shaded.

Study Guide Masters, p. 70

NAME ______________________ DATE __________

8-9 Study Guide
Graphing Inequalities

Student Edition
Pages 418–422

1. Graph the following ordered pairs on the coordinate system at the right.

   (2, 3) (3, 5) (−2, −1)

   (−3, 2) (3, 4) (−1, 0)

2. Where do these points lie in the plane in relation to the graph of $y = x$? **to the left of, or above, the line**
3. In each ordered pair in Exercise 1, is the x-coordinate less than, equal to, or greater than the y-coordinate? **less than**
4. Which of the following do the ordered pairs in Exercise 1 represent: $y = x$, $y > x$, or $y < x$? **$y > x$**
5. Which of the following represents the points located below the graph of $y = x$: $y = x$, $y > x$, or $y < x$? **$y < x$**
6. To represent all ordered pairs (x, y) where $y > x$, shade the portion of the coordinate plane above the graph of $y = x$. Note that the dashed line means that the graph of $y = x$ is not part of the graph of $y > x$.
7. Which of the following belong to the graph of $y > x$?

     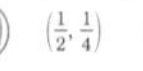 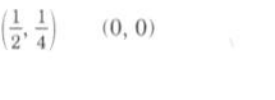

   (10, 20) $\left(\frac{1}{4}, \frac{1}{2}\right)$ $\left(\frac{1}{2}, \frac{1}{4}\right)$ (0, 0)

Graph each inequality.

8. $y < 3x$ 9. $y > 2x + 3$ 10. $x < 1 - y$

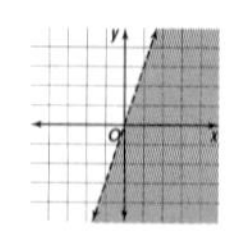
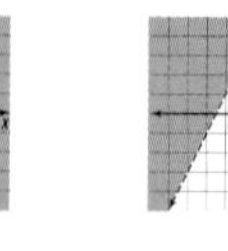
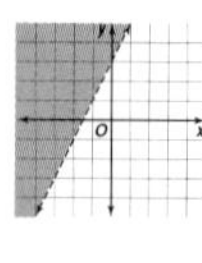
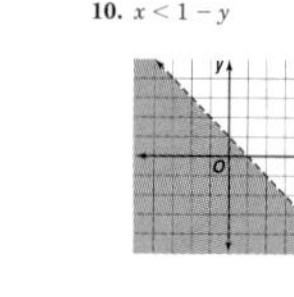

c. Any point in the shaded region represents a solution. However, fractional or negative solutions do not make sense because fractional or negative numbers of tickets cannot be sold. One solution is (500, 700). This corresponds to selling five hundred \$47.50 tickets and seven hundred \$66.50 tickets for a total cost of 47.5(500) + 66.5(700) or \$70,300. (420, 700) and (400, 800) are also solutions.

Checking Your Understanding

Communicating Mathematics

Read and study the lesson to answer these questions. 2–5. See margin.

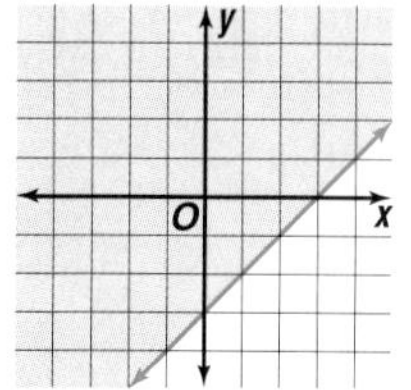

1. **Write** an inequality that describes the graph shown at the right. $y > x - 3$
2. **You Decide** Jeff says that inequalities with $<$ and $\leq$ symbols have dashed lines and those with $>$ and $\geq$ symbols have solid lines. Sanura disagrees. Who is correct and why?
3. **Write** an inequality whose graph is all points below the line $y = 9x + 4$.

MATH JOURNAL

4. **Describe** the location of solutions that result in a total cost of exactly \$66,500 on the graph in Example 2.
5. **Explain** how to determine which side of the boundary line to shade when graphing an inequality.

Guided Practice

State whether the boundary is included in the graph of each inequality.

6. $y \geq x - 13$ yes 7. $y \leq 1$ yes 8. $y > 3x$ no

Determine which ordered pair(s) is a solution to the inequality.

9. $y < 2x + 1$ a. $(-2, 2)$ b. $(4, -1)$ c. $(3, 1)$ b, c
10. $4y > -3x - 2$ a. $(-1, -1)$ b. $(2, -2)$ c. $(-2, 4)$ c

Determine which region is the graph of each inequality.

11. $y > x - 1$ 12. $y \leq \frac{2}{3}x + 2$ 13. $y < 5x$

11. The region above the boundary.
12. The region below the boundary.
13. The region to the right of the boundary.

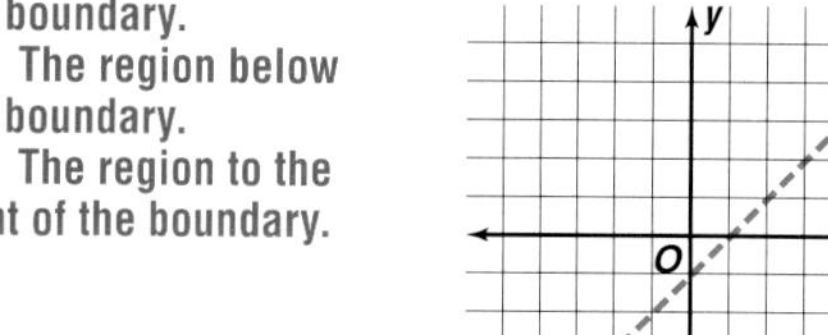

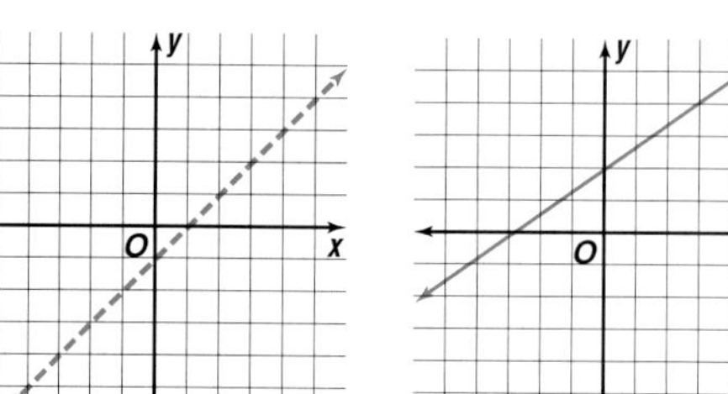

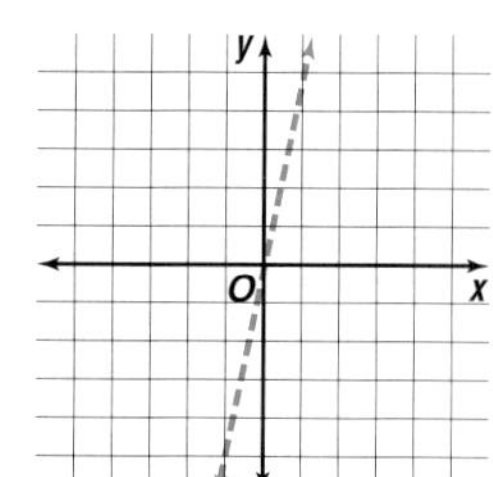

Graph each inequality. 14–17. See Solutions Manual.

14. $y < x + 2$ 15. $y \geq x$ 16. $y < -2x$ 17. $y \geq 3$

Reteaching

Using Charts Display the graph of $y = 2x - 3$ on the chalkboard or overhead. Have students create a chart with three columns. The headings of the columns should be $y < 2x - 3$, $y = 2x - 3$, $y > 2x - 3$. Have students select points on the line and on either side of the line. Then have the students complete the chart by filling in each column with the coordinates of points that make the heading a true sentence.

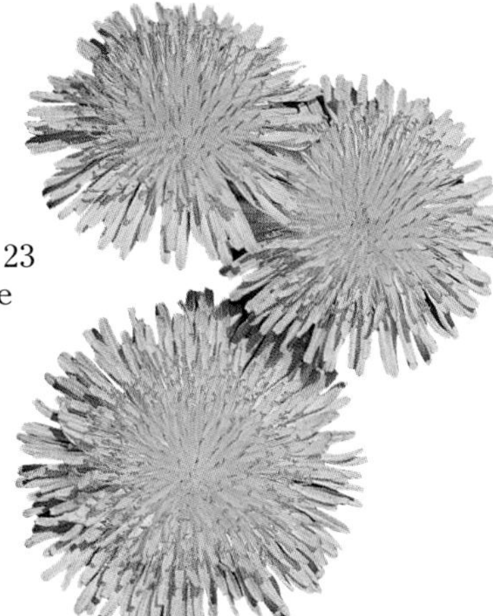

18a. See margin.

18b. Silvina must spend at least 8 hours weeding the garden.

18. Personal Finance Silvina wants to earn at least \$23 this week. Her father has agreed to pay her \$2.25 an hour to weed the garden and \$5.00 to mow the lawn.

a. Graph the inequality $2.25h + 5n \geq 23$ where h is the number of hours she spends weeding the garden and n is the number of times she mows the lawn.

b. If Silvina mows the lawn once, what is the least number of hours she will need to spend weeding the garden in order to make at least \$23?

Exercises: Practicing and Applying the Concept

Independent Practice

Determine which ordered pair(s) is a solution to the inequality.

19. $y < 3 - x$	**a.** $(-2, 2)$	**b.** $(4, -1)$	**c.** $(3, 1)$	a
20. $3y > -1 - 2x$	**a.** $(2, 1)$	**b.** $(-5, 1)$	**c.** $(1, 1)$	a, c
21. $4y - 8 < 0$	**a.** $(0, 2)$	**b.** $(2, 5)$	**c.** $(-2, 0)$	c
22. $3x + 4y \geq 17$	**a.** $(1, 1)$	**b.** $(4, 2)$	**c.** $(-3, 7)$	b, c
23. $2x \geq y - 8$	**a.** $(5, 12)$	**b.** $(4, 8)$	**c.** $(-3, -1)$	a, b, c
24. $5x + 1 \leq 3y$	**a.** $(-2, 3)$	**b.** $(4, 7)$	**c.** $(2, 3)$	a, b

Determine which region is the graph of each inequality.

25. The region to the left of the boundary.
26. The region to the right of the boundary.
27. The region to the left of the boundary.

25. $y > 2x - 6$ **26.** $y \geq 4 - 3x$ **27.** $y > \frac{3}{2}x$

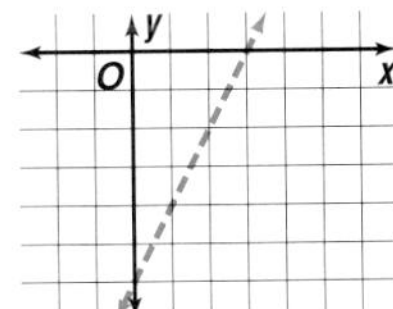

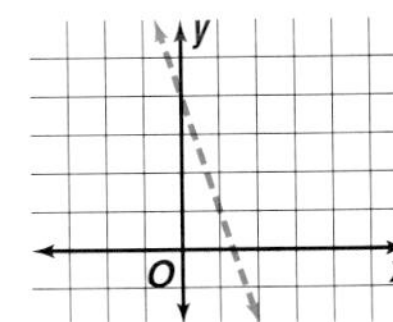

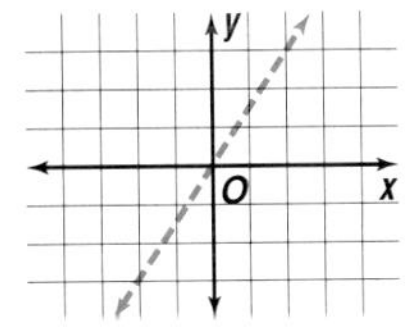

Graph each inequality. 28–43. See Solutions Manual.

28. $y < 2$	**29.** $y \geq -2$	**30.** $x \leq 2$	**31.** $x > -4$
32. $x + y > 2$	**33.** $x + y \geq -3$	**34.** $y \leq 0.5x + 5$	**35.** $x + y < 1$
36. $y > \frac{1}{4}x - 8$	**37.** $y > -x - 6.5$	**38.** $y \leq \frac{1}{3}x$	**39.** $y \geq 4.5x - 0.5$
40. $-y < -x$	**41.** $4y + x \leq 16$	**42.** $2x \geq 3y$	**43.** $-x > -y$

Critical Thinking

44. The solution to a system of inequalities is the set of all ordered pairs that satisfies *both* inequalities.

a. Write a system of inequalities for the graph shown at the right. $y < -x - 1$, $y \geq 1.5x - 3.5$

b. Check your answers by substituting into the inequalities. See students' work.

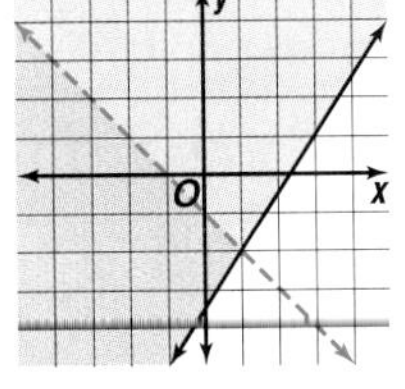

Group Activity Card 8-9

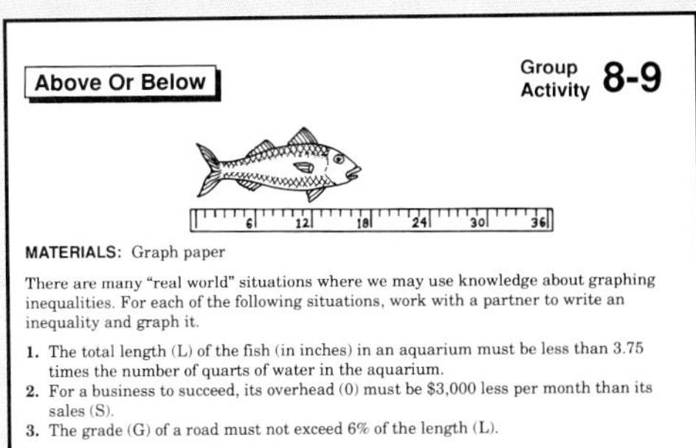

Above Or Below

Group Activity **8-9**

MATERIALS: Graph paper

There are many "real world" situations where we may use knowledge about graphing inequalities. For each of the following situations, work with a partner to write an inequality and graph it.

1. The total length (L) of the fish (in inches) in an aquarium must be less than 3.75 times the number of quarts of water in the aquarium.
2. For a business to succeed, its overhead (0) must be \$3,000 less per month than its sales (S).
3. The grade (G) of a road must not exceed 6% of the length (L).

 Pre-Algebra

Error Analysis

Watch for students who have difficulty reading a graph. Help students with this error, by writing intervals on horizontal axis and using a straightedge to read the graph.

Assignment Guide	
Core:	19–43 odd, 44, 45, 47, 49–56
Enriched:	20–42 even, 44–56

For **Extra Practice**, see p. 761.

The red A, B, and C flags, printed only in the Teacher's Wraparound Edition, indicate the level of difficulty of the exercises.

Additional Answer

18a.

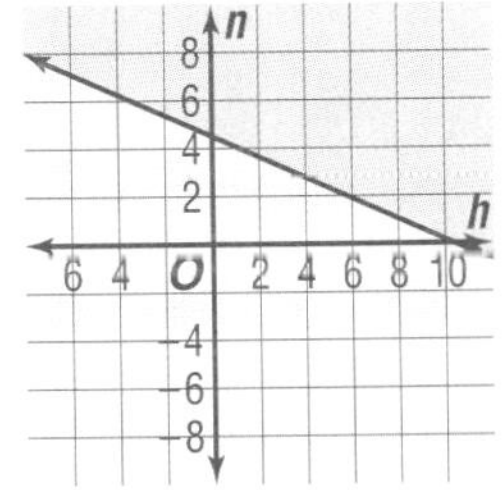

Practice Masters, p. 70

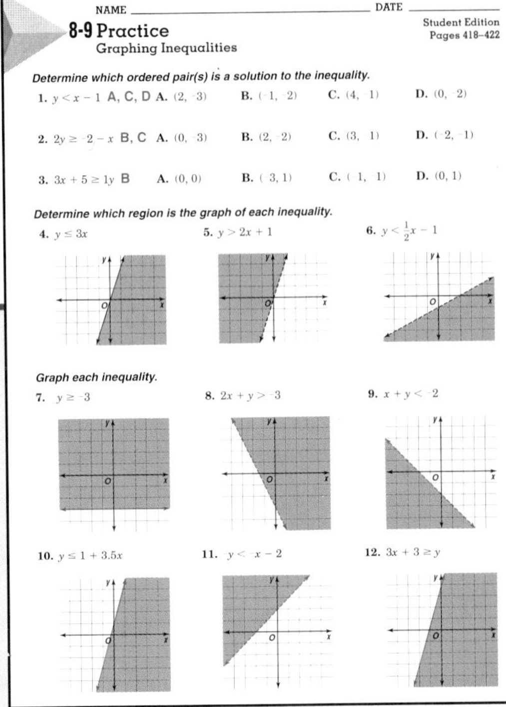

NAME ______ DATE ______

8-9 Practice
Graphing Inequalities

Student Edition Pages 418–422

Determine which ordered pair(s) is a solution to the inequality.

1. $y < x - 1$ A, C, D	**A.** $(2, -3)$	**B.** $(-1, -2)$	**C.** $(4, -1)$	**D.** $(0, -2)$
2. $2y \geq -2 - x$ B, C	**A.** $(0, -3)$	**B.** $(2, -2)$	**C.** $(3, -1)$	**D.** $(-2, -1)$
3. $3x + 5 \geq 1y$ B	**A.** $(0, 0)$	**B.** $(-3, 1)$	**C.** $(-1, -1)$	**D.** $(0, 1)$

Determine which region is the graph of each inequality.

4. $y \leq 3x$ **5.** $y > 2x + 1$ **6.** $y < \frac{1}{2}x - 1$

Graph each inequality.

7. $y \geq -3$ **8.** $2x + y > -3$ **9.** $x + y < -2$

10. $y \leq 1 + 3.5x$ **11.** $-y < -x - 2$ **12.** $3x + 3 \geq y$

4 ASSESS

Closing Activity

Speaking Write several inequalities on the board or overhead. Have students identify and graph the associated equation for each inequality. Then ask students to tell whether to shade above or below the boundary line of each inequality and explain their reasoning.

Chapter 8, Quiz D (Lesson 8-8 and 8-9) is available in *The Assessment and Evaluation Masters*, p. 212.

Additional Answers

45d. No; this would mean that your cost was greater than your sales, which doesn't happen in a successful business.

46c. Sample answer: (12, 4) would mean the total length of the fish is 12 inches in a 4-quart tank. (20, 6) would mean the total length of the fish is 20 inches in a 6-quart tank.

Enrichment Masters, p. 70

NAME ____________ DATE ________

8-9 Enrichment Student Edition Pages 418–422

Graphing Systems of Inequalities

Suppose you are given the following system of inequalities.

$y \geq x + 2$
$y \leq -2x - 1$

The solution of ths system is the set of all ordered pairs that satisfy both inequalities. To find the solution, graph each inequality. The intersection of the graphs represents the solution.

The graphs of the equations $y = x + 2$ and $y = -2x - 1$ are the boundaries of each region. The solution of the system is the region that contains ordered pairs that are solution of *both* inequalities.

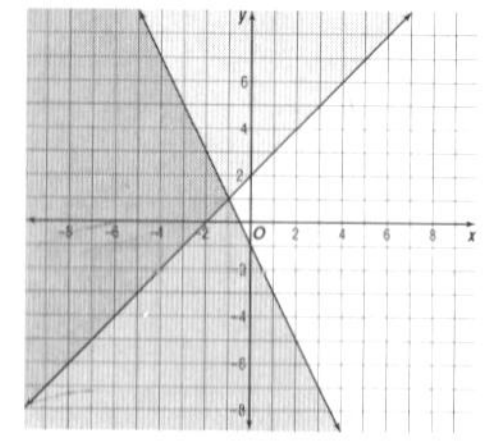

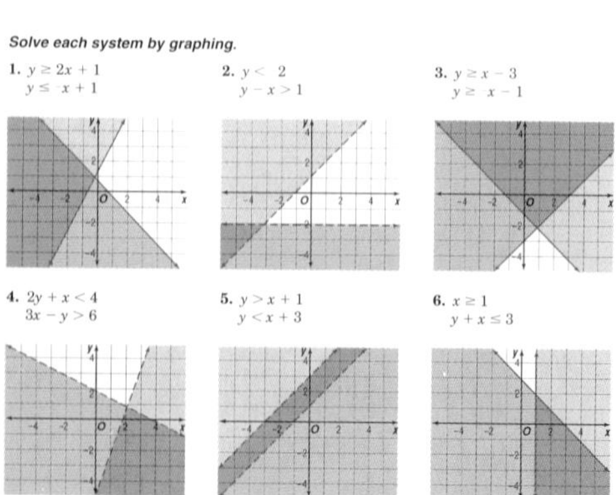

Solve each system by graphing.

1. $y \geq 2x + 1$, $y \leq -x + 1$
2. $y < -2$, $y - x > 1$
3. $y \geq x - 3$, $y \geq -x - 1$
4. $2y + x < 4$, $3x - y > 6$
5. $y > x + 1$, $y < x + 3$
6. $x \geq 1$, $y + x \leq 3$

Applications and Problem Solving

45. Business For a certain business to be successful, its monthly overhead O must be at least \$3000 less than its monthly sales S. **45d. See margin.**

a. Write an inequality to represent this situation. $O < S - 3000$
b. Graph the inequality. **See Solutions Manual.**
c. Do points above or below the boundary line indicate a successful business? Justify your answer. **below the line**
d. Would negative numbers make sense in this problem? Explain.
e. List two solutions. **Sample answer: (6600, 3600), (7000, 4000)**

46. Pets Biologists have found that for fish to thrive, the total length L of all the fish (in inches) in a tank must be no more than 3.75 times the number of quarts Q of water in the tank.

a. Write an inequality to represent this situation. $L < 3.75Q$
b. Graph the inequality. **See Solutions Manual.**
c. List two solutions where at least five 2-inch fish are in a tank. Explain what each solution means. **See margin.**

47. Consumer Awareness You are shopping for cassettes and CDs. Cassettes cost \$7, CDs cost \$14, and you have \$28 to spend. **47a–b. See margin.**

a. Use a graph to determine how many cassettes and CDs you can buy.
b. List two solutions and explain what they mean.

48. Family Activity Investigate your family's eating habits and the U.S. recommended daily allowances (RDAs) of nutrients. Write inequalities for the recommended amounts of at least three nutrients. Does your diet measure up? **See students' work.**

Mixed Review

49. Graph $y - 4x = 3$ and $y = x$. Then find the solution of the system of equations. (Lesson 8-8) **See students' graphs. (−1, −1)**

50. Recreation Angie's bowling handicap is 7 less than half her average. Her handicap is 53. What is Angie's bowling average? (Lesson 7-3) $\frac{a}{2} - 7 = 53$; **120**

51. Order $\frac{3}{5}$, 0.63, and 6.02×10^{-1} from least to greatest. (Lesson 6-1) **51.** $\frac{3}{5}$, 6.02×10^{-1}, 0.63

52. Weather The people of Waynesburg, PA, celebrate July 26 as Rain Day, because it almost always rains on that day. Is this an example of inductive or deductive reasoning? (Lesson 5-8) **52. inductive**

53. Use divisibility rules to determine if 38 is divisible by 2, 3, 5, 6, or 10. (Lesson 4-1) **2**

54. Solve $6b > -31.2$. (Lesson 3-7) **54.** $b > -5.2$

55. Solve $\frac{96}{-8} = m$. (Lesson 2-8) **−12**

56. State whether $3x - 7 > 5$ is *true*, *false*, or *open*. (Lesson 1-9) **open**

Extension

Using Cooperative Groups Have students work in small groups to graph inequalities that contain absolute values. For example, $y \leq |x - 1|$. You may want to give the hint that this inequality is equivalent to the following compound sentence: $y \leq x - 1$ and $y < 1 - x$.

Additional Answers

47a.

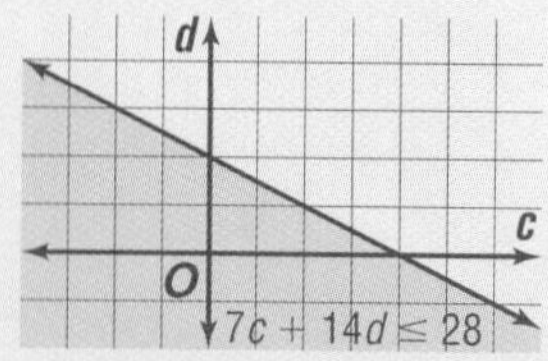

47b. Sample answer: (1, 1) means purchasing 1 cassette and 1 CD for a total cost of \$21. (2, 1) means purchasing 2 cassettes and 1 CD for a total of \$28.

CHAPTER

8 Highlights

Vocabulary

After completing this chapter, you should be able to define each term, property, or phrase and give an example of two of each.

Algebra

boundary (p. 418)
domain (p. 372)
function (p. 373)
functional notation (p. 393)
linear equation (p. 386)
parabola (p. 391)
range (p. 372)
relation (p. 372)
solution of system (p. 412)
system of equations (p. 412)
vertical line test (p. 373)

Statistics

scatter plot (p. 378)
negative relationship (p. 380)
no relationship (p. 380)
positive relationship (p. 380)

Geometry

change in x (p. 400)
change in y (p. 400)
family of graphs (p. 411)
intercepts (p. 406)
parallel lines (p. 401)
slope (p. 400)
slope-intercept form (p. 407)
x-intercept (p. 406)
y-intercept (p. 406)

Problem Solving

draw a graph (p. 396)

Understanding and Using Vocabulary

Choose the term from the list above that best completes each statement or phrase.

1. A __function__ is a relation in which each element of the domain is paired with exactly one element in the range.
2. The set of first coordinates of a relation is called the __domain__.
3. The __y-intercept__ is the y-coordinate of the point where a graph crosses the y-axis.
4. The set of second coordinates of a relation is called the __range__.
5. A __relation__ is a set of ordered pairs.
6. __Parallel__ lines are always the same distance apart.
7. The __x-intercept__ is the x-coordinate of the point where the graph crosses the x-axis.
8. In an equation in slope-intercept form, the b represents the __y-intercept__.

Highlights

Using the Chapter Highlights

The Chapter Highlights begins with an alphabetical listing of the new terms, properties, and phrases that were introduced in this chapter.

Assessment and Evaluation Masters, pp. 199–200

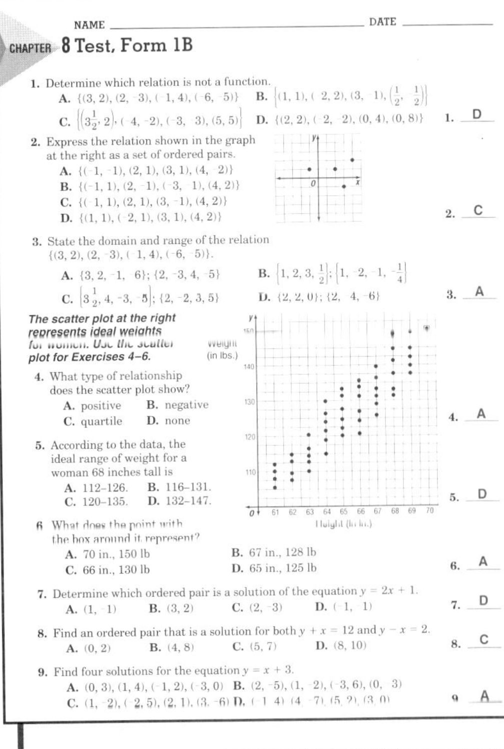

NAME ________ DATE ________

CHAPTER 8 Test, Form 1B

1. Determine which relation is not a function.
   A. $\{(3, 2), (2, -3), (-1, 4), (-6, -5)\}$ B. $\{(1, 1), (-2, 2), (3, -1), (\frac{1}{2}, -\frac{1}{2})\}$
   C. $\{(3\frac{1}{2}, 2), (-4, -2), (-3, -3), (5, 5)\}$ D. $\{(2, 2), (-2, -2), (0, 4), (0, 8)\}$ — 1. D
2. Express the relation shown in the graph at the right as a set of ordered pairs.
   A. $\{(-1, -1), (2, 1), (3, 1), (4, -2)\}$
   B. $\{(-1, 1), (2, -1), (-3, -1), (4, 2)\}$
   C. $\{(-1, 1), (2, 1), (3, -1), (4, 2)\}$
   D. $\{(1, 1), (-2, 1), (3, 1), (4, 2)\}$ — 2. C
3. State the domain and range of the relation $\{(3, 2), (2, -3), (-1, 4), (-6, -5)\}$.
   A. $\{3, 2, -1, -6\}; \{2, -3, 4, -5\}$ B. $\{1, 2, 3, \frac{1}{2}\}; \{1, -2, -1, -\frac{1}{4}\}$
   C. $\{3\frac{1}{2}, 4, -3, -5\}; \{2, -2, 3, 5\}$ D. $\{2, 2, 0\}; \{2, -4, -6\}$ — 3. A

The scatter plot at the right represents ideal weights for women. Use the scatter plot for Exercises 4–6.

4. What type of relationship does the scatter plot show?
   A. positive B. negative
   C. quartile D. none — 4. A
5. According to the data, the ideal range of weight for a woman 68 inches tall is
   A. 112–126. B. 116–131.
   C. 120–135. D. 132–147. — 5. D
6. What does the point with the box around it represent?
   A. 70 in., 150 lb B. 67 in., 128 lb
   C. 66 in., 130 lb D. 65 in., 125 lb — 6. A
7. Determine which ordered pair is a solution of the equation $y = 2x + 1$.
   A. $(1, -1)$ B. $(3, 2)$ C. $(2, -3)$ D. $(-1, -1)$ — 7. D
8. Find an ordered pair that is a solution for both $y + x = 12$ and $y - x = 2$.
   A. $(0, 2)$ B. $(4, 8)$ C. $(5, 7)$ D. $(8, 10)$ — 8. C
9. Find four solutions for the equation $y = x + 3$.
   A. $(0, 3), (1, 4), (-1, 2), (-3, 0)$ B. $(2, -5), (1, -2), (-3, 6), (0, -3)$
   C. $(1, -2), (-2, 5), (2, 1), (3, -6)$ D. $(-1, 4), (4, -7), (5, 2), (3, 0)$ — 9. A

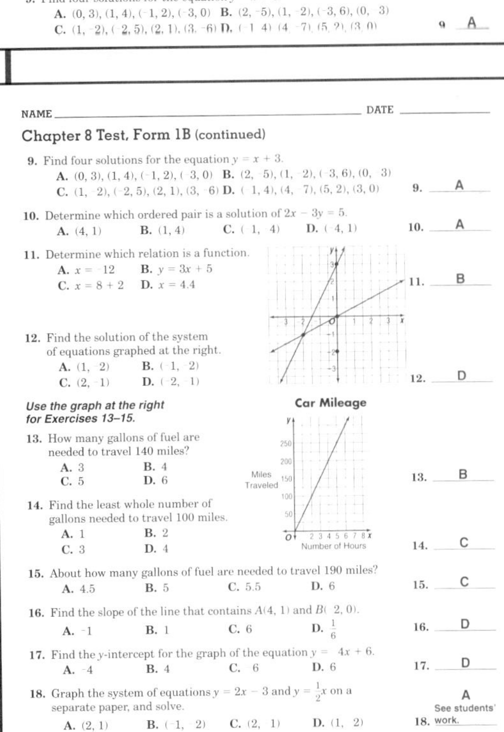

NAME ________ DATE ________

Chapter 8 Test, Form 1B (continued)

9. Find four solutions for the equation $y = x + 3$.
   A. $(0, 3), (1, 4), (-1, 2), (-3, 0)$ B. $(2, -5), (1, -2), (-3, 6), (0, -3)$
   C. $(1, -2), (-2, 5), (2, 1), (3, -6)$ D. $(-1, 4), (4, -7), (5, 2), (3, 0)$ — 9. A
10. Determine which ordered pair is a solution of $2x - 3y = 5$.
    A. $(4, 1)$ B. $(1, 4)$ C. $(-1, -4)$ D. $(-4, 1)$ — 10. A
11. Determine which relation is a function.
    A. $x = -12$ B. $y = 3x + 5$
    C. $x = 8 + 2$ D. $x = 4.4$ — 11. B
12. Find the solution of the system of equations graphed at the right.
    A. $(1, -2)$ B. $(-1, -2)$
    C. $(2, -1)$ D. $(-2, -1)$ — 12. D

Use the graph at the right for Exercises 13–15.

13. How many gallons of fuel are needed to travel 140 miles?
    A. 3 B. 4
    C. 5 D. 6 — 13. B
14. Find the least whole number of gallons needed to travel 100 miles.
    A. 1 B. 2
    C. 3 D. 4 — 14. C
15. About how many gallons of fuel are needed to travel 190 miles?
    A. 4.5 B. 5 C. 5.5 D. 6 — 15. C
16. Find the slope of the line that contains $A(4, 1)$ and $B(-2, 0)$.
    A. -1 B. 1 C. 6 D. $\frac{1}{6}$ — 16. D
17. Find the y-intercept for the graph of the equation $y = -4x + 6$.
    A. -4 B. 4 C. -6 D. 6 — 17. D
18. Graph the system of equations $y = 2x - 3$ and $y = \frac{1}{2}x$ on a separate paper, and solve.
    A. $(2, 1)$ B. $(-1, -2)$ C. $(2, -1)$ D. $(1, -2)$ — 18. A See students' work.

Instructional Resources

Three multiple-choice tests and three free-response tests are provided in the *Assessment and Evaluation Masters*. Forms 1A and 2A are for honors pacing, Forms 1B and 2B are for average pacing, Forms 1C and 2C are for basic pacing, Chapter 8 Test, Form 1B is shown at the right.

Chapter 8 Test, Form 2B is shown on the next page.

CHAPTER 8 Study Guide and Assessment

Using the Study Guide and Assessment

Skills and Concepts Encourage students to refer to the objectives and examples on the left as they complete the review exercises on the right.

Assessment and Evaluation Masters, pp. 205–206

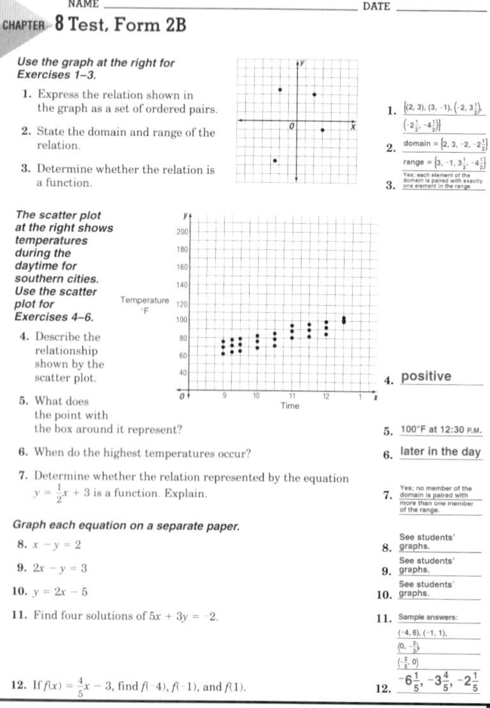

NAME ______________________ DATE ________

CHAPTER 8 Test, Form 2B

Use the graph at the right for Exercises 1–3.

1. Express the relation shown in the graph as a set of ordered pairs. — 1. $\{(2, 3), (3, -1), (-2, 3\frac{1}{2}), (-2\frac{1}{2}, -4\frac{1}{2})\}$
2. State the domain and range of the relation. — 2. domain = $\{2, 3, -2, -2\frac{1}{2}\}$; range = $\{3, -1, 3\frac{1}{2}, -4\frac{1}{2}\}$
3. Determine whether the relation is a function. — 3. Yes; each element of the domain is paired with exactly one element in the range

The scatter plot at the right shows temperatures during the daytime for southern cities. Use the scatter plot for Exercises 4–6.

4. Describe the relationship shown by the scatter plot. — 4. positive
5. What does the point with the box around it represent? — 5. 100°F at 12:30 P.M.
6. When do the highest temperatures occur? — 6. later in the day
7. Determine whether the relation represented by the equation $y = \frac{1}{2}x + 3$ is a function. Explain. — 7. Yes; no member of the domain is paired with more than one member of the range.

Graph each equation on a separate paper.

8. $x - y = 2$ — 8. See students' graphs.
9. $2x - y = 3$ — 9. See students' graphs.
10. $y = 2x - 5$ — 10. See students' graphs.
11. Find four solutions of $5x + 3y = -2$. — 11. Sample answers: $(-4, 6)$, $(-1, 1)$, $(0, -\frac{2}{3})$, $(-\frac{2}{5}, 0)$
12. If $f(x) = \frac{4}{5}x - 3$, find $f(-4)$, $f(-1)$, and $f(1)$. — 12. $-6\frac{1}{5}, -3\frac{4}{5}, -2\frac{1}{5}$

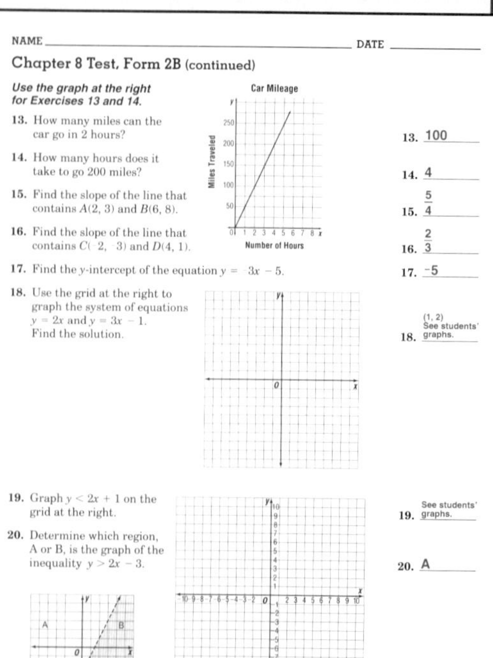

NAME ______________________ DATE ________

Chapter 8 Test, Form 2B (continued)

Use the graph at the right for Exercises 13 and 14.

13. How many miles can the car go in 2 hours? — 13. 100
14. How many hours does it take to go 200 miles? — 14. 4
15. Find the slope of the line that contains $A(2, 3)$ and $B(6, 8)$. — 15. $\frac{5}{4}$
16. Find the slope of the line that contains $C(-2, -3)$ and $D(4, 1)$. — 16. $\frac{2}{3}$
17. Find the y-intercept of the equation $y = -3x - 5$. — 17. -5
18. Use the grid at the right to graph the system of equations $y = 2x$ and $y = 3x - 1$. Find the solution. — 18. (1, 2) See students' graphs.
19. Graph $y < 2x + 1$ on the grid at the right. — 19. See students' graphs.
20. Determine which region, A or B, is the graph of the inequality $y > 2x - 3$. — 20. A

CHAPTER 8 Study Guide and Assessment

Skills and Concepts

Objectives and Examples

Upon completing this chapter, you should be able to:

▶ **use tables and graphs to represent relations and functions.** (Lesson 8-1)

Table

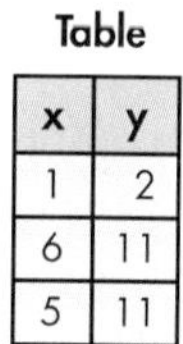

x	y
1	2
6	11
5	11

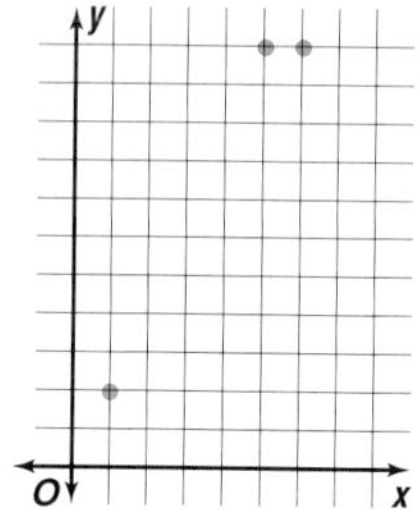

The domain is {1, 6, 5}.

The range is {2, 11}.

Since each element of the domain is paired with exactly one element of the range, this is a function.

▶ **construct and interpret scatter plots** (Lesson 8-2)

Relation of years of experience and amount of money earned

The points in the scatter plot are in a pattern that slants upward to the right. So it shows a positive relationship.

Review Exercises

Use these exercises to review and prepare for the chapter test.

Write the domain and range of each relation. Then determine whether each relation is a function. 9–13. See margin.

9. {(7, 17), (9, 19), (0, 18), (6, 40)}
10. {(4, 5), (4, 10), (4, 22), (4, 36)}
11. {(6, 2), (8, 2), (15, 2)}
12.

x	1	18	18	35	27
y	2	4	6	4	2

13.

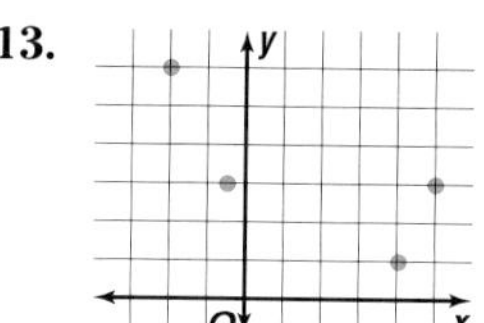

14. What type of relationship, *positive*, *negative*, or *none*, is shown by the scatter plot below? positive

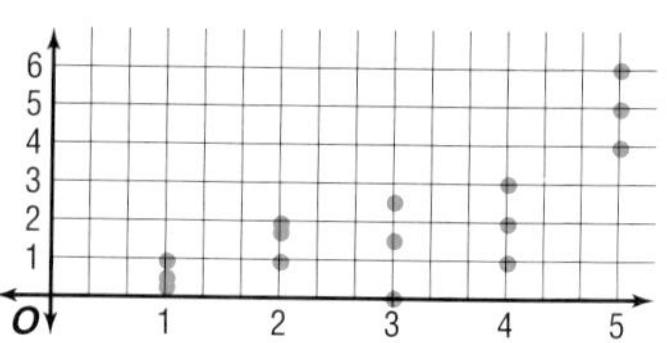

Determine whether a scatter plot of the data might show a *positive*, *negative*, or *no* relationship. Explain your answer.

15. size of TV set, amount of TV watched no
16. number of small children, amount of glass displayed negative
17. age of a person, volume of the radio negative
18. amount of income, value of home positive
19. years of marriage, number of kids no

15–19. Sample answers given. See students' explanations.

GLENCOE Technology

Test and Review Software

You may use this software, a combination of an item generator and item bank, to create your own tests or worksheets. Types of items include free response, multiple choice, short answer, and open ended.

For IBM & Macintosh

Additional Answers

9. {7, 9, 0, 6}; {17, 19, 18, 40}; yes
10. {4}; {5, 10, 22, 36}; no
11. {6, 8, 15}; {2}; yes
12. {1, 18, 35, 27}; {2, 4, 6}; no
13. {4, 5, −0.5, −2}; {3, 6, 1}; yes

33. $\frac{2}{5}$
34. $-2\frac{1}{2}$
35. −1
36. undefined
37. 0
38. $1\frac{1}{2}$

Objectives and Examples

▶ **find solutions for relations with two variables and graph the solution** (Lesson 8-3)

Find four solutions for $y = 2.2x - 1$. Then graph the equation.

x	2.2x − 1	y
0	2.2 (0) − 1	−1
1	2.2 (1) − 1	1.2
2	2.2 (2) − 1	3.4
3	2.2 (3) − 1	5.6

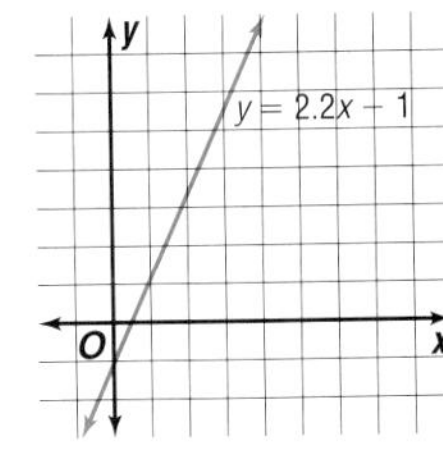

Four solutions are (0, −1), (1, 1.2), (2, 3.4), (3, 5.6).

Review Exercises

Find four solutions for each equation. Write the solution as ordered pairs. Then graph each equation.

20. $y = x - 1$ (1, 0), (2, 1), (3, 2), (4, 3)
21. $y = 3x + 2$ (0, 2), (1, 5), (−1, −1), (2, 8)
22. $x + y = 5$ (0, 5), (1, 4), (−1, 6), (2, 3)
23. $2x + y = 7$ (0, 7), (1, 5), (−1, 9), (2, 3)
24. $x - 2y = -1$ (0, 0.5), (1, 1), (−1, 0), (3, 2)
25. $x = 7$ (7, 0), (7, 2), (7, 4), (7, 6)
26. $y = 1$ (0, 1), (1, 1), (−1, 1), (3, 1)
27. $y = \frac{1}{2}x$ (0, 0.5), (2, 1), (−2, −1), (4, 2)
20–27. See Solutions Manual for graphs.

▶ **find functional values for the given function** (Lesson 8-4)

$$g(x) = 2x^2 - 3$$

$g(3) = 2 \cdot 3^2 - 3$
$= 2(9) - 3$
$= 18 - 3$
$= 15$

$g(-1) = 2 \cdot (-1)^2 - 3$
$= 2(1) - 3$
$= 2 - 3$
$= -1$

Given $f(x) = x + 5$ and $g(x) = 3x^2 + 1$, find each value.

28. $f(2)$ 7
29. $f(0)$ 5
30. $f(-6)$ −1
31. $g(1)$ 4
32. $g(5)$ 76

▶ **find the slope of a line** (Lesson 8-6)

Find the slope of the line that contains $A(4, 4)$ and $B(0, 1)$.

$$\text{slope} = \frac{4 - 1}{4 - 0} = \frac{3}{4}$$

Find the slope of the line that contains each pair of points. 33–38. See margin.

33. $A(-3, 5), B(2, 7)$
34. $J(4, 1), K(2, 6)$
35. $W(4, 0), X(7, -3)$
36. $A(6, 18), B(6, 40)$
37. $P(0, 2), Q(3, 2)$
38. $W(3, 8), X(1, 5)$

▶ **graph a linear equation using the *x*- and *y*-intercepts or using the slope and the *y*-intercept** (Lesson 8-7)

$y = 4x - 1$
slope
$m = 4$
y-intercept
$y = 4(0) - 1$
$y = -1$
$(0, -1)$

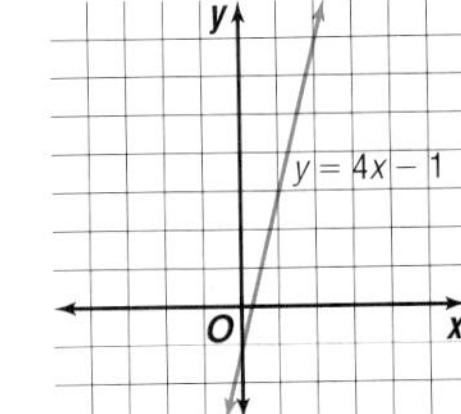

Use the slope and the *y*-intercept to graph each equation. 39–46. See margin.

39. $y = 2x + 1$
40. $y = \frac{2}{3}x - 3$
41. $y = -\frac{1}{2}x + 1$
42. $y = 4$
43. $x = 7$
44. $x + 2y = 6$
45. $y = 2 - \frac{1}{2}x$
46. $5x - y = 9$

Additional Answers

45.
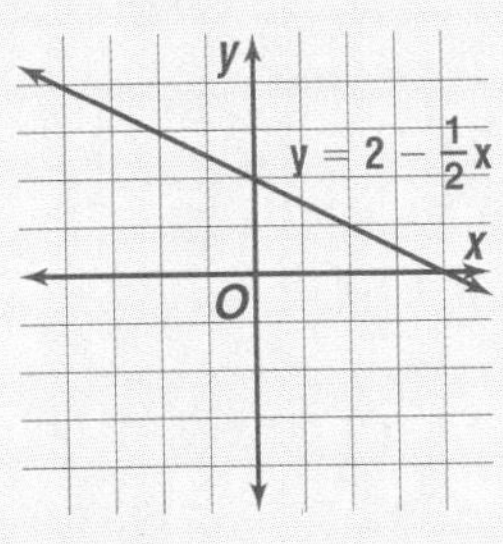

46.
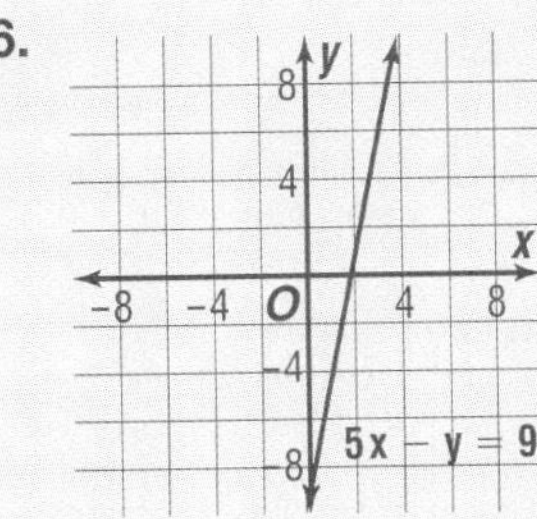

Additional Answers

39.
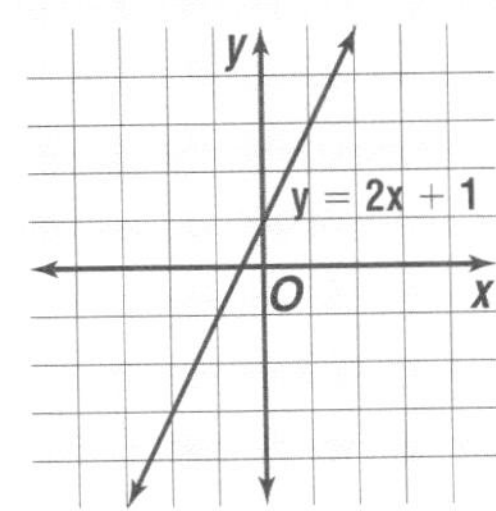

40.
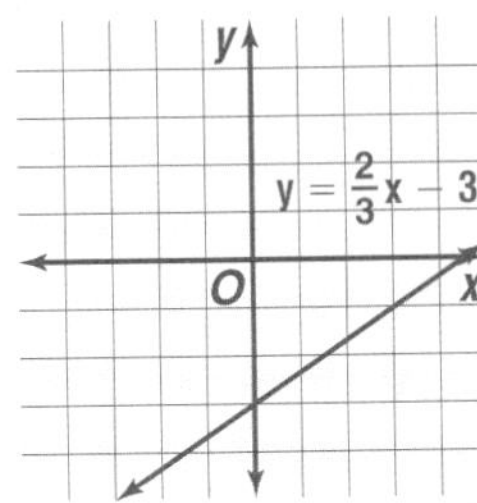

41.
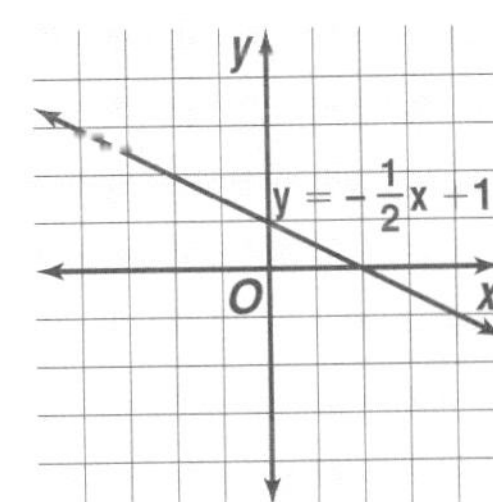

42.
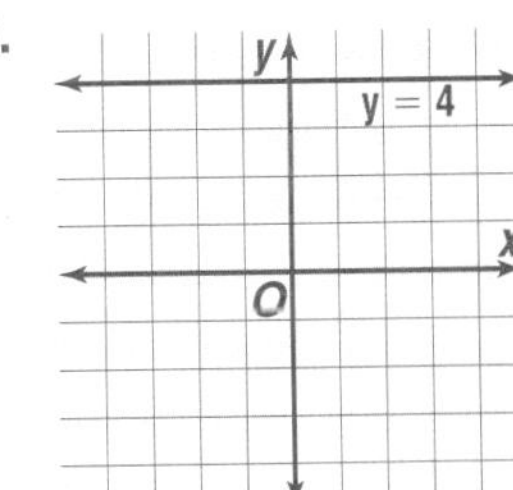

43.
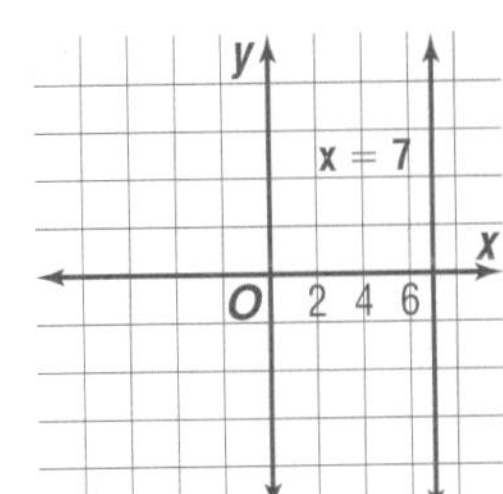

44.
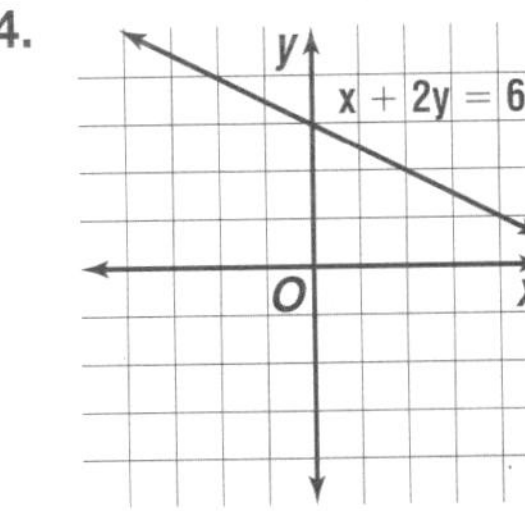

Applications and Problem Solving

Encourage students to work through the exercises in the Applications and Problem Solving section to strengthen their problem-solving skills.

Additional Answers

47–52. See Solutions Manual for graphs.
47. (0, 8)
48. all numbers
49. { }
50. (2, 1)
51. { }
52. (6, -1)
53.

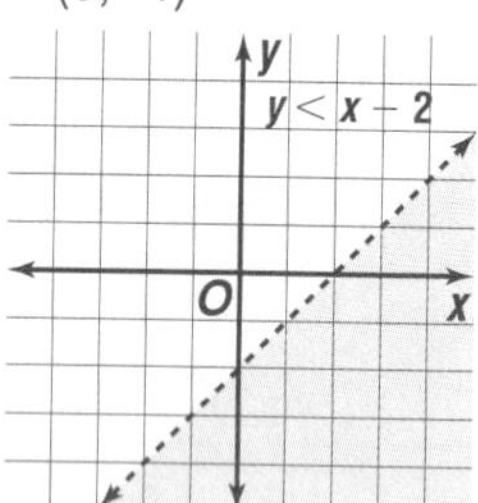

54.

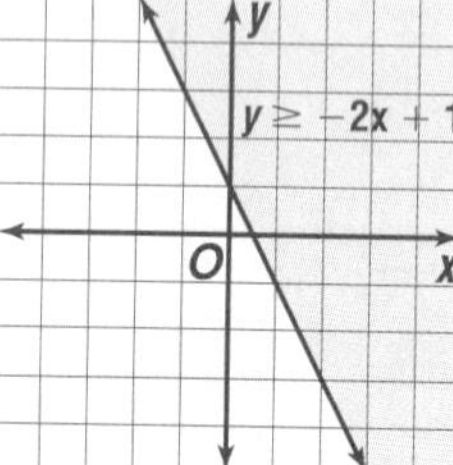

55.

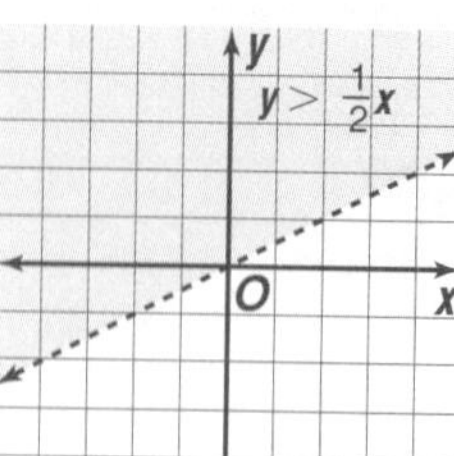

56.

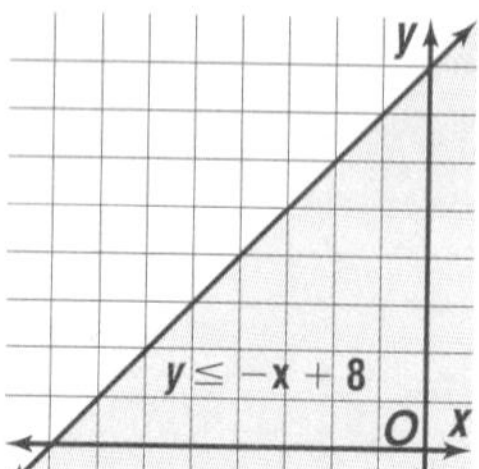

57.

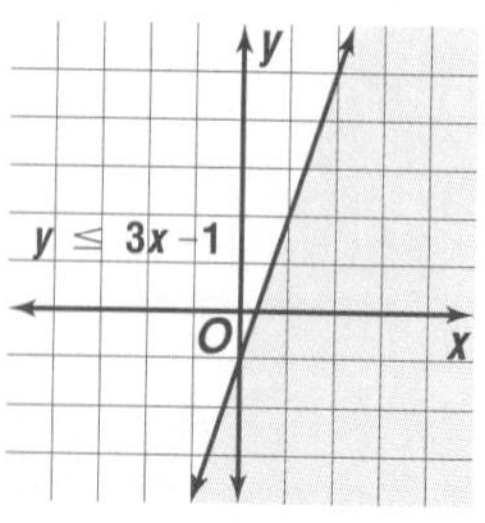

Chapter 8 • Study Guide and Assessment

Objectives and Examples

▶ **solve systems of linear equations by graphing** (Lesson 8-8)

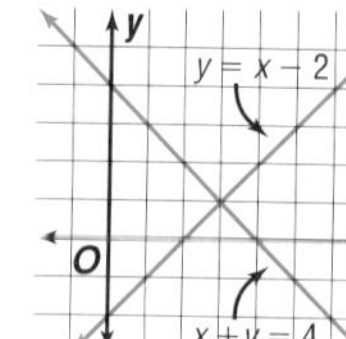

$y = x - 2$
$x + y = 4$

The lines intersect at (3, 1). So the solution to the system of equations is (3, 1).

▶ **graph linear inequalities** (Lesson 8-9)

Graph $y > \frac{1}{3}x + 2$.

The boundary is dashed since the line is not part of the graph. The origin (0, 0) is not part of the graph since $0 < \frac{1}{3}(0) + 2$.

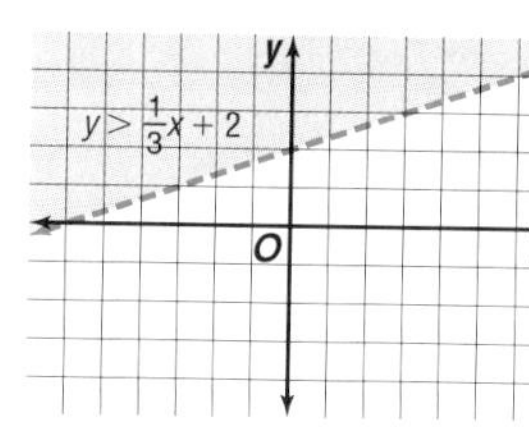

Review Exercises

Use a graph to solve each system of equations. **47–52. See margin.**

47. $x + y = 8$; $x - y = -8$
48. $y = 2x - 4$; $2y = 4x - 8$
49. $y = -x + 1$; $y = -x + 2$
50. $y = \frac{1}{2}x$; $y = \frac{3}{2}x - 2$
51. $y = 3 - x$; $y = 5 - x$
52. $x + 2y = 4$; $-2x - 2y = -10$

Graph each inequality. **53–58. See margin.**

53. $y < x - 2$
54. $y \geq -2x + 1$
55. $y > \frac{1}{2}x$
56. $y \leq -x + 8$
57. $y \leq 3x - 1$
58. $y > \frac{1}{4}x + 2$

Applications and Problem Solving

59. **Nutrition** This table shows nutritional values of selected foods. (Lesson 8-2)

Food	Percentage of Calories from fat	Percentage of Calories from carbohydrate
Apple (medium, delicious)	9	89
Bagel (plain)	6	76
Banana (medium)	2	93
Bran Muffin (large)	40	53
Fig Newtons (four cookies)	18	75
PowerBar	8	76
Snickers bar (regular size)	42	51
Ultra Slim-Fast bar	30	63

a. Draw a scatter plot using Calories from fat as the x values and Calories from carbohydrates as the y values. **See margin.**

b. Does the graph show a positive, negative, or no relationship? **negative**

60. **Living Expenses** An investment company charged $495 a month to rent a 2-bedroom townhouse in 1993. In 1995, the rate was $515 per month. Use a graph to determine how much the townhouse rent will be in 2000. Assume that the rate is constant. (Lesson 8-5) **$565 per month**

61. **Sports** In the 1992–1993 season, Dino Ciccarelli played 82 games and scored 41 goals. The next season, he played 66 games and scored 28 goals. Use a graph to determine about how many goals Dino would score in a 76-game season. Assume that the rate is constant. (Lesson 8-5) **approximately 36 goals scored**

A practice test for Chapter 8 is available on page 781.

Additional Answers

58.

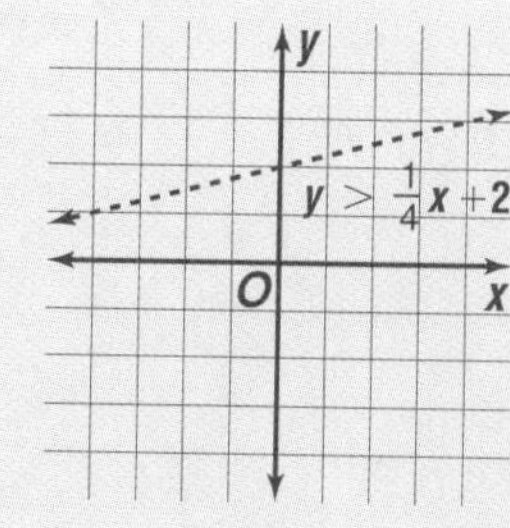

59a.

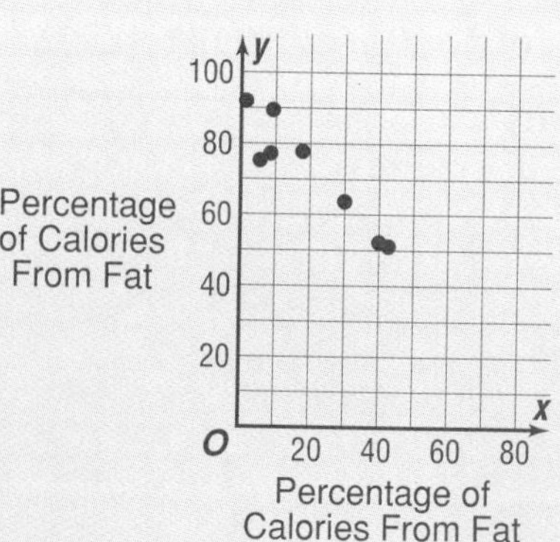

Alternative Assessment

Performance Task

Demonstrate your knowledge by giving a clear, concise solution to each problem. Be sure to include all relevant drawings and justify your answers. You may show your solutions in more than one way or investigate beyond the requirements of the problem.

The table below lists the movies with the top five box-office revenues for 1993.

Rank/Title	Distributor	U.S. Gross Totals (millions)	Foreign Gross Totals (millions)
1. *Jurassic Park*	Universal	$338	$530
2. *The Fugitive*	Warner Brothers	$179	$170
3. *Aladdin*	Buena Vista	$118	$185
4. *The Bodyguard*	Warner Brothers	$45	$248
5. *Indecent Proposal*	Paramount	$107	$151

Source: *Variety*, January 3–9, 1994

1. Make a graph of the information using the rank as the x-coordinate and the world gross totals as the y-coordinate. (*Hint*: The world total is the sum of U.S. and foreign revenues.) **See Solutions Manual.**
2. Give the domain of the relation. **{1, 2, 3, 4, 5}**
3. Give the range of the relation. **{868, 349, 303, 293, 258}**
4. Does the chart represent a function? **yes**
5. Do you think that the film's distributor affected the amount of money it made? **Sample answer: no.**
6. The marketing manager of a film company has determined that a new film must make $700 million in sales between U.S. and foreign sales in order to break even.
   a. Graph an inequality to show this situation. **See Solutions Manual.**
   b. Would the films listed in the table have broken even if this was their goal?

6b. *Jurassic Park* would have made money, but the rest would not.

Thinking Critically

- Give two examples of relations that could be graphed on a scatter plot.
- Choose one of the relations you wrote. Develop sample data and graph the relation. Then determine whether it shows a positive, negative, or no relation. **See students' work.**

Portfolio

Select one of the assignments from this chapter that you found especially challenging. Revise your work if necessary and place it in your portfolio. Explain why you found it to be a challenge. **See students' work.**

Self Evaluation

Never accepting a challenge, you will never feel the excitement of victory. Remember how happy you were when your team won a difficult game? Experience the feeling firsthand in math class.

Assess yourself. Did you find a lesson in the chapter more difficult than the others? How many times did you have to do it in order to solve the problems? How did you feel when you succeeded? What advice would you give to someone faced with a similar challenge? **See students' work.**

Alternative Assessment

The Alternative Assessment section provides students with the opportunity to assess their own work by thinking critically, working with others, keeping a portfolio, and honestly evaluating their own progress. For more information on alternative forms of assessment, see *Alternative Assessment in the Mathematics Classroom*.

Performance Assessment

Performance Assessment tasks for this chapter are included in the *Assessment and Evaluation Masters*. A scoring guide is also provided.

Alternative Assessment

Assessment and Evaluation Masters, pp. 209, 221

NAME ______________________ DATE __________

CHAPTER **8 Performance Assessment**

Instructions: Demonstrate your knowledge by giving a clear, concise solution to each problem. Be sure to include all relevant drawings and justify your answers. You may show your solution in more than one way or investigate beyond the requirements of the problem.

1. Business persons frequently use graphs to communicate and to solve problems.
   a. A corporation has a sales goal of $275,000 this year. This represents an increase of $50,000 over sales last year. Write an equation that describes this situation and solve. Explain the meaning of the answer.
   b. A second goal is to limit the increase in expenses for the year to less than $10,000. Expenses last year were $183,000. Write an inequality that describes this situation and solve. Explain the meaning of the answer.
   c. The rental fee on a VCR for a meeting is given by $y = 8 + 2x$, where x is the number of hours rented and y is the total rental fee in dollars.
      1. Find three solutions to the equation and draw the graph.
      2. Choose a time of rental and use the graph to determine the rental fee. Explain each step.
      3. Find the y-intercept and tell what it means.
      4. Do negative values of x make sense? Why or why not?
   d. Another company also rents VCRs. The rental equation is $y = 8 + 3x$.
      1. Find three solutions to the equation and draw the graph.
      2. Compare and contrast this graph with the graph in part c. Use the words *slope* and *intercept* in your answer.
      3. From which company would you rent a VCR? Explain your reasoning.
2. The home office of a company has limited the number of salespersons and secretaries in a branch office to nine or fewer.
   a. If x is the number of salespersons and y is the number of secretaries, write and graph an inequality.
   b. Does (3, 4) satisfy the inequality? What does (3, 4) represent?
   c. Name an ordered pair that does not satisfy the inequality. What does your ordered pair represent?

CHAPTER **8 Scoring Guide**

Level	Specific Criteria
3 Superior	• Shows thorough understanding of the concepts *equation, inequality, y-intercept, slope*, and *ordered pair*. • Uses appropriate strategies to present data in graphic form. • Computations to solve equations and inequalities are correct. • Written explanations are exemplary. • Graphs are accurate and appropriate. • Goes beyond requirements of some or all problems.
2 Satisfactory, with minor flaws	• Shows understanding of the concepts *equation, inequality, y-intercept, slope*, and *ordered pair*. • Uses appropriate strategies to present data in graphic form. • Computations to solve equations and inequalities are mostly correct. • Written explanations are effective. • Graphs are mostly accurate and appropriate. • Satisfies all requirements of problems.
1 Nearly Satisfactory, with serious flaws	• Shows understanding of most of the concepts *equation, inequality, y-intercept, slope*, and *ordered pair*. • May not use appropriate strategies to present data in graphic form. • Computations to solve equations and inequalities are mostly correct. • Written explanations are satisfactory. • Graphs are mostly accurate and appropriate. • Satisfies most requirements of problems.
0 Unsatisfactory	• Shows little or no understanding of the concepts *equation, inequality, y-intercept, slope*, and *ordered pair*. • May not use appropriate strategies to present data in graphic form. • Computations to solve equations and inequalities are incorrect. • Written explanations are not satisfactory. • Graphs are not accurate or appropriate. • Does not satisfy requirements of problems.

Chapter 8 • Ongoing Assessment

Using the Ongoing Assessment

These two pages review the skills and concepts presented in Chapters 1–8. This review is formatted to reflect new trends in standardized testing.

Assessment and Evaluation Masters, pp. 214–215

NAME ______________________ DATE __________

CHAPTERS 1–8 Cumulative Review

1. Simplify $11a + 3a + 5$. (Lesson 1-5) — 1. $14a + 5$
2. Solve $y = \frac{35}{^-5}$. (Lesson 2-8) — 2. $^-7$
3. Solve $d = rt$ if $r = 54$ and $t = 6$. (Lesson 3-4) — 3. 324
4. Evaluate the expression $(x^2 - y)^2$ if $x = 2$ and $y = 3$. (Lesson 4-2) — 4. 1
5. Determine the prime factorization of 54. (Lesson 4-5) — 5. $2 \cdot 3 \cdot 3 \cdot 3$
6. Simplify $\frac{16a^2b}{24ab}$. (Lesson 4-6) — 6. $\frac{2a}{3}$
7. Write $5x^2y^{-6}$ using positive exponents. (Lesson 4-9) — 7. $\frac{5x^2}{y^6}$
8. Solve $a = 3\frac{5}{6} - 5\frac{2}{3}$. (Lesson 5-5) — 8. $^-1\frac{5}{6}$
9. Solve $\frac{x - 1}{4} > 3$. (Lesson 5-7) — 9. $x > 3\frac{1}{4}$
10. Write the next three terms of 6, 7.5, 9, 10.5, $\cdots$. (Lesson 5-9) — 10. 12, 13.5, 15
11. Solve $y = \frac{6}{7} \cdot \frac{7}{12}$. (Lesson 6-3) — 11. $\frac{1}{2}$
12. Solve $a = 3\frac{1}{3} \div 4\frac{3}{4}$. (Lesson 6-4) — 12. $\frac{40}{57}$
13. Solve $w = 8.64 \div 3.6$. (Lesson 6-5) — 13. 2.4
14. Solve $15 - 6x = {^-45}$. (Lesson 7-2) — 14. 10
15. Solve $^-8(y + 3) = {^-13y} - 4$. (Lesson 7-6) — 15. 4
16. Convert 2.5 liters to milliliters. (Lesson 7-8) — 16. 2500 mL
17. State the domain and range of the relation $\{(^-1.5, 2), (3, ^-4.1), (^-3.7, 2.6)\}$. (Lesson 8-1) — 17. domain = (-1.5, 3, -3.7) range = (2, -4.1, 2.6)

NAME ______________________ DATE __________

Chapters 1–8, Cumulative Review (continued)

18. Graph $x + y = {^-1}$ on the grid (Lesson 8-3) — 18. See students' work.

19. Determine whether the relation described by $f(n) = \frac{9}{5}n + 32$ is a function. (Lesson 8-4) — 19. yes
20. Graph $y < x + 1$ on the grid below. (Lesson 8-9) — 20. See students' work.

CHAPTERS 1–8 Ongoing Assessment

Section One: Multiple Choice

There are eleven multiple-choice questions in this section. After working each problem, write the letter of the correct answer on your paper.

1. If $4x - 8 = 28$, what is the value of x? **B**
   - A. 5
   - B. 9
   - C. 20
   - D. 36
2. An investor bought 80 shares of stock for \$2160 and later sold them for \$1920. Which integer represents the profit or loss per share of stock? **A**
   - A. -3
   - B. -240
   - C. 240
   - D. 3
3. What is the y-intercept for the graph of $y = x + 2$? **D**
   - A. -2
   - B. 1
   - C. 0
   - D. 2
4. A group of divers are at a level of -20 meters. If they descend 15 meters more, at what level will they be? **D**
   - A. 5 m
   - B. -5 m
   - C. 35 m
   - D. -35 m
5. How is the product $5 \times 5 \times 5$ expressed in exponential notation? **C**
   - A. 5^5
   - B. 3^5
   - C. 5^3
   - D. 5^5
6. A mail-order card company charges 50¢ for each greeting card plus a handling charge of \$1.50 per order. Which sentence could be used to find n, the number of cards ordered, if the total charge was \$9? **A**
   - A. $0.5n + 1.5 = 9$
   - B. $0.50n + 1.50n = 9$
   - C. $9 = 50n + 1.50$
   - D. $9 = (0.5 + 1.50)n$
7. Which number should come next in this pattern? **D**

   0.5, 2, 3.5, 5, . . .
   - A. 8
   - B. 7.5
   - C. 7
   - D. 6.5
8. Which is the solution of the system of equations $x + y = 4$ and $y = -3x$? **C**
   - A. (6, 2)
   - B. (1, 3)
   - C. $(-2, 6)$
   - D. (3, 1)
9. One red blood cell is about 7.5×10^{-4} centimeters long. What is another way to express this measure? **B**
   - A. 0.000075 cm
   - B. 0.00075 cm
   - C. 0.0075 cm
   - D. 0.075 cm
10. If $3x + 4(x + 1) = 5x - 8$, what is the value of x? **A**
    - A. -6
    - B. -4.5
    - C. -2
    - D. 6
11. Express 65,000 in scientific notation. **B**
    - A. 6.5×10^3
    - B. 6.5×10^4
    - C. 65×10^3
    - D. 0.65×10^5

Section Two: Free Response

This section contains twelve questions for which you will provide short answers. Write your answer on your paper.

12. John has \$2.50. He spent one third of the money he had this afternoon at lunch. His friend Lakesha borrowed \$2, and he spent 50¢ on a pop. How much money did John have when he left this morning? **\$7.50**
13. Lee wants to set up a baseball tournament. Each team should have 12 different players. How many people does he need to have 8 teams? **96**

Standardized test practice questions are provided in the *Assessment and Evaluation Masters*, p. 213.

14. Yolanda has \$50. She bought jeans for \$25. How much money can she spend for each of 4 gifts and have \$1 left for parking? **$x \leq 6$**

15. The price of Kodak's stock this morning was $87\frac{7}{8}$. At the end of today, the price had increased by $\frac{7}{8}$. What was the closing price of Kodak's stock? **$88\frac{3}{4}$**

16. When practicing during the summer on their own, members of the Lincoln High track team must run at least 18 miles a week. If Jaime wants to run the distance in no more than 6 days each week, write and solve an inequality representing how to find the number of miles he needs to run each day. **$6x > 18$; $x > 3$ miles**

17. Mei used a chart to see if she was close to the average weight for her height. The chart gave a weight of 118 pounds for a female 60 inches tall. The average weight for a female 64 inches tall was 134 pounds. Mei is 5 feet 8 inches tall. What should Mei's weight be? **150 pounds**

18. How many boards, each $3\frac{1}{2}$ feet long, can be cut from a board 21 feet long? **6 boards**

19. Graph the inequality $y \geq 2x + 2$. **See margin.**

20. How can the product $4 \times 4 \times 4 \times 4 \times 4 \times 4$ be written in exponential notation? **4^6**

21. A contractor is going to paint a room with four walls, each 8 feet tall and 12 feet wide. Each wall will be given two coats of paint. If each can of paint will cover up to 400 square feet, how many cans will the job take? **2 cans**

22. The City Marathon had 3542 participants this year. 1054 runners did not finish the race. How many runners finished the City Marathon this year? **2488**

23. A computer catalog offers discount components that customers can add to their computers. Kim chose a printer for \$300, a modem for \$150, and a CD-ROM drive for \$225. If the tax, shipping, and handling was \$71.50, how much was the total bill? **\$746.50**

Test-Taking Tip

Most standardized tests have a time limit, so you must use your time wisely. Some questions will be much easier than others. If you cannot answer a question within a few minutes, go on to the next one. If there is still time left when you get to the end of the test, go back to the questions that you skipped.

Take time out after several problems to freshen your mind.

Section Three: Open-Ended

This section contains two open-ended problems. Demonstrate your knowledge by giving a clear, concise solution to each problem. Your score on these problems will depend upon how well you do the following.

- Explain your reasoning.
- Show your understanding of the mathematics in an organized manner.
- Use charts, graphs, and diagrams in your explanation.
- Show the solution in more than one way or relate it to other situations.
- Investigate beyond the requirements of the problem. **24. See students' work.**

24. Explain the difference between a relation and a function using an example of each.

25. Make a scatter plot of the areas and populations of the states and territories of Australia as shown in the table below. **See Solutions Manual.**

State/Territory	Area (thousands of square miles)	Population (thousands)
New South Wales	310	5732
Victoria	88	4244
Queensland	667	2977
Western Australia	975	1586
South Australia	380	1401
Tasmania	26	452
Australia Capital	1	280
Northern Territory	520	175

Source: *World Almanac*, 1995

Ongoing Assessment

Additional Answer

19.

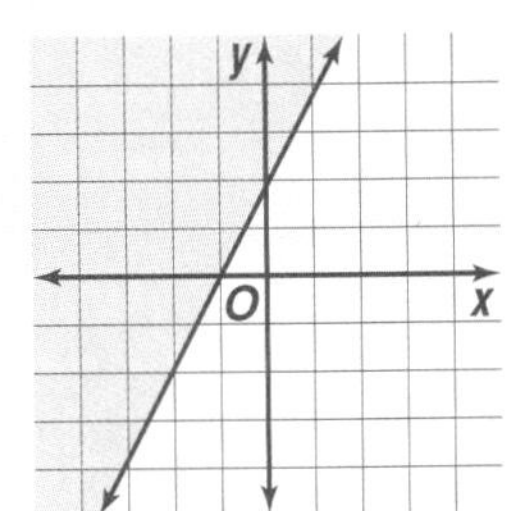

CHAPTER 9

Ratio, Proportion, and Percent

PREVIEWING THE CHAPTER

In this chapter, students study the relationships of fractions, decimals, and percents. In the first lesson, students investigate ratios and rates, including unit rates. The concept of ratios is integrated with the strategy of using a table as students study simple probability. Equal ratios, in the form of proportions are studied next. Then the percent proportion is developed. Next proportions, percent, and statistics are integrated as students use statistics to predict outcomes. The relationship of fractions, decimals, and percents is explored and used as students learn to estimate with percents. The last two lessons in the chapter provide for real-life applications of percents, including the percent of change.

Lesson (pages)	Lesson Objectives	NCTM Standards	State/Local Objectives
9-1 (432–436)	Write ratios as fractions in simplest form and determine unit rates.	1-5, 7, 12	
9-2 (437–438)	Solve problems by making a table, chart, or list.	1-5, 10, 12	
9-3A (439)	Students determine whether a game is fair or unfair.	1-5, 10	
9-3 (440–443)	Find the probability of simple events.	1-4, 10	
9-4 (444–447)	Use proportions to solve problems.	1-5	
9-4B (448)	Use proportions to estimate a population.	1-5, 10	
9-5 (449–453)	Use the percent proportion to write fractions as percents and solve problems involving percents.	1-5, 9	
9-6 (454–457)	Use a sample to predict the actions of a larger group.	1-5, 10	
9-7 (458–461)	Express decimals and fractions as percents and vice versa.	1-5	
9-8 (462–466)	Use percents to estimate.	1-5, 7, 10	
9-9 (467–471)	Solve percent problems using percent equations.	1-5, 7	
9-10 (472–475)	Find percent of increase or decrease.	1-5, 7	

A complete, 1-page lesson plan is provided for each lesson in the ***Lesson Planning Guide***. Answer keys for each lesson are available in the ***Answer Key Masters***.

ORGANIZING THE CHAPTER

You may want to refer to the **Course Planning Calendar** on page T12 for detailed information on pacing.
PACING: Standard—14 days; **Honors**—13 days; **Block**—7 days

LESSON PLANNING CHART

Lesson (Pages)	Materials/ Manipulatives	Extra Practice (Student Edition)	Blackline Masters: Study Guide	Blackline Masters: Practice	Blackline Masters: Enrichment	Blackline Masters: Assessment and Evaluation	Blackline Masters: Math Lab and Modeling Math	Blackline Masters: Multicultural Activity	Blackline Masters: Tech Prep Applications	Blackline Masters: Graphing Calculator	Blackline Masters: Activity	Real-World Applications	Interactive Mathematics Tools Software	Teaching Transparencies	Group Activity Cards
9-1 (432–436)	string	p. 761	p. 71	p. 71	p. 71									9-1A 9-1B	9-1
9-2 (437–438)		p. 761	p. 72	p. 72	p. 72									9-2A 9-2B	9-2
9-3A (439)	two dice*						p. 44								
9-3 (440–443)	graphing calculator deck of playing cards	p. 762	p. 73	p. 73	p. 73	p. 239	p. 84	p. 17		p. 9	p. 39	18		9-3A 9-3B	9-3
9-4 (444–447)	dominoes	p. 762	p. 74	p. 74	p. 74				p. 17		p. 23			9-4A 9-4B	9-4
9-4B (448)	small bowl dry beans						p. 45								
9-5 (449–453)	blank index cards	p. 762	p. 75	p. 75	p. 75	pp. 238, 239		p. 18			p. 9	19	9-5	9-5A 9-5B	9-5
9-6 (454–457)	marbles*	p. 762	p. 76	p. 76	p. 76									9-6A 9-6B	9-6
9-7 (458–461)	grid paper dry beans marbles	p. 763	p. 77	p. 77	p. 77		p. 67–68				p. 40		9-7	9-7A 9-7B	9-7
9-8 (462–466)	water glass	p. 763	p. 78	p. 78	p. 78	p. 240								9-8A 9-8B	9-8
9-9 (467–471)	catalog	p. 763	p. 79	p. 79	p. 79						p. 41	20	9-9	9-9A 9-9B	9-9
9-10 (472–475)	1-inch grid paper	p. 764	p. 80	p. 80	p. 80	p. 240			p. 18			21		9-10A 9-10B	9-10
Study Guide/ Assessment (478–480)						pp. 225–237, 241–243									

*Included in Glencoe's *Student Manipulative Kit* and *Overhead Manipulative Resources*.

ORGANIZING THE CHAPTER

All of the blackline masters in the Teacher's Classroom Resources are available on the ***Electronic Teacher's Classroom Resources*** CD-ROM.

Other Chapter Resources

Student Edition
Chapter Opener, pp. 430–431
Earth Watch, p. 436
From the Funny Papers, p. 466
Working on the Investigation, pp. 453, 471
Closing the Investigation, p. 476

Teacher's Classroom Resources
Investigations and Projects Masters, pp. 61–64
Pre-Algebra Overhead Manipulative Resources, pp. 23–26

Technology
Test and Review Software (IBM & Macintosh)
CD-ROM Activities

Professional Publications
Block Scheduling Booklet
Glencoe Mathematics Professional Series

Outside Resources

Books/Periodicals
The MESA Series: In the Pharmacy, Nancy Cooks
Math/Space Mission, Regional Math Network
Sci-Math: Applications in Proportional Problem Solving (Module 1/Pre-Algebra), Madeline P. Goodstein
Constructing Ideas About Fractions, Decimals and Percent: Grades 6–8. Sandra Ward

Videos/CD-ROMs
Math Vantage Videos: Patterns Unit/Sequences and Ratios, Sunburst
Percents Video, ETA
Real Life Math Percent Series, ETA
Mastering Ratios and World Problem Strategies, Phoenix/BFA Films ad Video

Assessment Resources

Student Edition
Math Journal, pp. 446, 464, 474
Mixed Review, pp. 436, 438, 443, 447, 453, 457, 461, 466, 471, 475
Self Test, p. 457
Chapter Highlights, p. 477
Chapter Study Guide and Assessment, pp. 478–480
Alternative Assessment, p. 481
Portfolio, p. 61
MindJogger Videoquiz, 9

Teacher's Wraparound Edition
5-Minute Check, pp. 432, 437, 440, 444, 449, 454, 458, 462, 467, 472
Checking Your Understanding, pp. 434, 438, 442, 446, 451, 456, 460, 464, 469, 474
Closing Activity, pp. 436, 438, 443, 447, 453, 457, 461, 466, 471, 475
Cooperative Learning, pp. 447, 468

Assessment and Evaluation Masters
Multiple-Choice Tests, Forms 1A (Honors), 1B (Average), 1C (Basic), pp. 225–230
Free-Response Tests, Forms 2A (Honors), 2B (Average), 2C (Basic), pp. 231–236
Performance Assessment, p. 237
Mid-Chapter Test, p. 238
Quizzes A–D, pp. 239,240
Standardized Test Practice, p. 241
Cumulative Review, pp. 242-243

ENHANCING THE CHAPTER

Examples of some of the materials for enhancing Chapter 9 are shown below.

DIVERSITY

Multicultural Activity Masters, pp. 17, 18

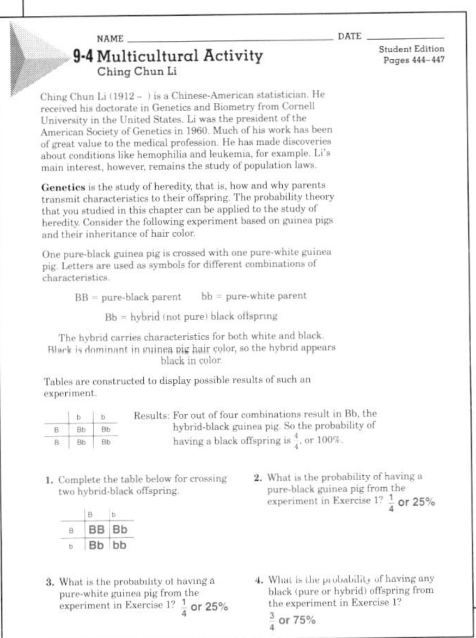

NAME ______________ DATE ______

9-4 Multicultural Activity
Ching Chun Li

Student Edition Pages 444–447

Ching Chun Li (1912 –) is a Chinese-American statistician. He received his doctorate in Genetics and Biometry from Cornell University in the United States. Li was the president of the American Society of Genetics in 1960. Much of his work has been of great value to the medical profession. He has made discoveries about conditions like hemophilia and leukemia, for example. Li's main interest, however, remains the study of population laws.

Genetics is the study of heredity, that is, how and why parents transmit characteristics to their offspring. The probability theory that you studied in this chapter can be applied to the study of heredity. Consider the following experiment based on guinea pigs and their inheritance of hair color.

One pure-black guinea pig is crossed with one pure-white guinea pig. Letters are used as symbols for different combinations of characteristics.

BB = pure-black parent bb = pure-white parent

Bb = hybrid (not pure) black offspring

The hybrid carries characteristics for both white and black. Black is dominant in guinea pig hair color, so the hybrid appears black in color.

Tables are constructed to display possible results of such an experiment.

	b	b
B	Bb	Bb
B	Bb	Bb

Results: For out of four combinations result in Bb, the hybrid-black guinea pig. So the probability of having a black offspring is $\frac{4}{4}$, or 100%.

1. Complete the table below for crossing two hybrid-black offspring.

	B	b
B	BB	Bb
b	Bb	bb

2. What is the probability of having a pure-black guinea pig from the experiment in Exercise 1? $\frac{1}{4}$ or 25%

3. What is the probability of having a pure-white guinea pig from the experiment in Exercise 1? $\frac{1}{4}$ or 25%

4. What is the probability of having any black (pure or hybrid) offspring from the experiment in Exercise 1? $\frac{3}{4}$ or 75%

Real-World Applications, 18, 19, 20, 21

Graphing Calculator Masters, p. 9

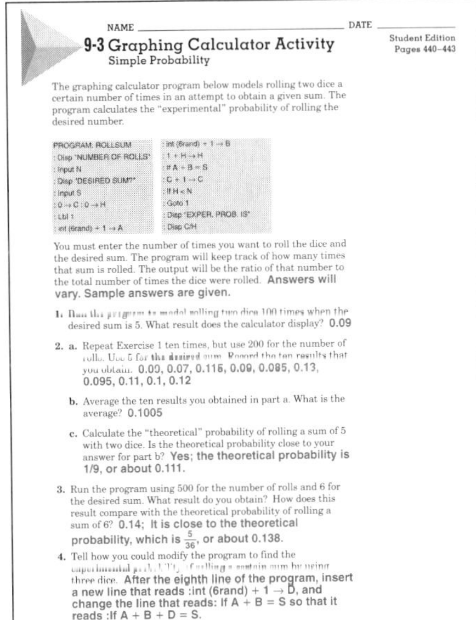

NAME ______________ DATE ______

9-3 Graphing Calculator Activity
Simple Probability

Student Edition Pages 440–443

The graphing calculator program below models rolling two dice a certain number of times in an attempt to obtain a given sum. The program calculates the "experimental" probability of rolling the desired number.

```
PROGRAM: ROLLSUM
:Disp "NUMBER OF ROLLS"
:Input N
:Disp "DESIRED SUM?"
:Input S
:0→C:0→H
:Lbl 1
:int (6rand) + 1 → A
:int (6rand) + 1 → B
:1 + H → H
:If A + B = S
:C + 1 → C
:If H < N
:Goto 1
:Disp "EXPER. PROB. IS"
:Disp C/H
```

You must enter the number of times you want to roll the dice and the desired sum. The program will keep track of how many times that sum is rolled. The output will be the ratio of that number to the total number of times the dice were rolled. **Answers will vary. Sample answers are given.**

1. Run the program to model rolling two dice 100 times when the desired sum is 5. What result does the calculator display? **0.09**

2. a. Repeat Exercise 1 ten times, but use 200 for the number of rolls. Use 5 for the desired sum. Record the ten results that you obtain. **0.09, 0.07, 0.115, 0.09, 0.085, 0.13, 0.095, 0.11, 0.1, 0.12**

   b. Average the ten results you obtained in part a. What is the average? **0.1005**

   c. Calculate the "theoretical" probability of rolling a sum of 5 with two dice. Is the theoretical probability close to your answer for part b? **Yes; the theoretical probability is 1/9, or about 0.111.**

3. Run the program using 500 for the number of rolls and 6 for the desired sum. What result do you obtain? How does this result compare with the theoretical probability of rolling a sum of 6? **0.14; It is close to the theoretical probability, which is $\frac{5}{36}$, or about 0.138.**

4. Tell how you could modify the program to find the experimental probability of rolling a certain sum by using three dice. **After the eighth line of the program, insert a new line that reads :int (6rand) + 1 → D, and change the line that reads: If A + B = S so that it reads :If A + B + D = S.**

Tech Prep Applications Masters, pp. 17, 18

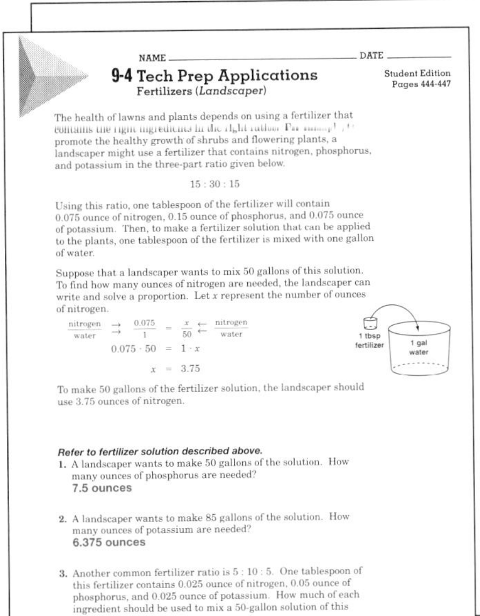

NAME ______________ DATE ______

9-4 Tech Prep Applications
Fertilizers (*Landscaper*)

Student Edition Pages 444-447

The health of lawns and plants depends on using a fertilizer that contains the right ingredients in the right ratios. For example, to promote the healthy growth of shrubs and flowering plants, a landscaper might use a fertilizer that contains nitrogen, phosphorus, and potassium in the three-part ratio given below.

15 : 30 : 15

Using this ratio, one tablespoon of the fertilizer will contain 0.075 ounce of nitrogen, 0.15 ounce of phosphorus, and 0.075 ounce of potassium. Then, to make a fertilizer solution that can be applied to the plants, one tablespoon of the fertilizer is mixed with one gallon of water.

Suppose that a landscaper wants to mix 50 gallons of this solution. To find how many ounces of nitrogen are needed, the landscaper can write and solve a proportion. Let x represent the number of ounces of nitrogen.

$$\frac{\text{nitrogen}}{\text{water}} \rightarrow \frac{0.075}{1} = \frac{x}{50} \leftarrow \frac{\text{nitrogen}}{\text{water}}$$

$$0.075 \cdot 50 = 1 \cdot x$$

$$x = 3.75$$

1 tbsp fertilizer

1 gal water

To make 50 gallons of the fertilizer solution, the landscaper should use 3.75 ounces of nitrogen.

Refer to fertilizer solution described above.

1. A landscaper wants to make 50 gallons of the solution. How many ounces of phosphorus are needed? **7.5 ounces**

2. A landscaper wants to make 85 gallons of the solution. How many ounces of potassium are needed? **6.375 ounces**

3. Another common fertilizer ratio is 5 : 10 : 5. One tablespoon of this fertilizer contains 0.025 ounce of nitrogen, 0.05 ounce of phosphorus, and 0.025 ounce of potassium. How much of each ingredient should be used to mix a 50-gallon solution of this fertilizer? **1.25 oz nitrogen, 2.5 oz phosphorus, 1.25 oz potassium**

Activity Masters, pp. 9, 23, 39, 40, 41

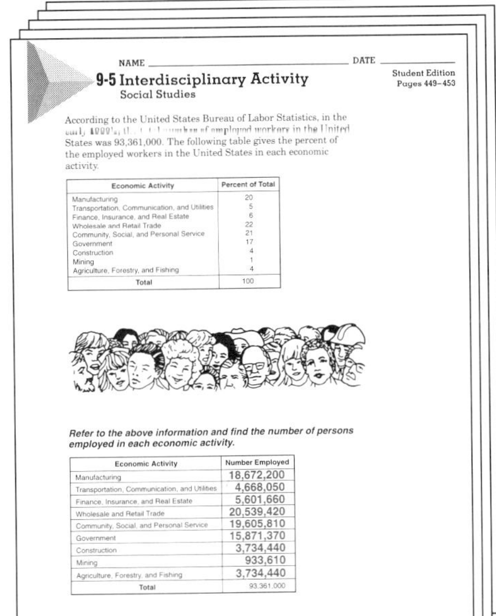

NAME ______________ DATE ______

9-5 Interdisciplinary Activity
Social Studies

Student Edition Pages 449–453

According to the United States Bureau of Labor Statistics, in the early 1990's, the total number of employed workers in the United States was 93,361,000. The following table gives the percent of the employed workers in the United States in each economic activity.

Economic Activity	Percent of Total
Manufacturing	20
Transportation, Communication, and Utilities	5
Finance, Insurance, and Real Estate	6
Wholesale and Retail Trade	22
Community, Social, and Personal Service	21
Government	17
Construction	4
Mining	1
Agriculture, Forestry, and Fishing	4
Total	100

Refer to the above information and find the number of persons employed in each economic activity.

Economic Activity	Number Employed
Manufacturing	18,672,200
Transportation, Communication, and Utilities	4,668,050
Finance, Insurance, and Real Estate	5,601,660
Wholesale and Retail Trade	20,539,420
Community, Social, and Personal Service	19,605,810
Government	15,871,370
Construction	3,734,440
Mining	933,610
Agriculture, Forestry, and Fishing	3,734,440
Total	93,361,000

COOPERATIVE LEARNING

Math Lab and Modeling Math Masters, pp. 67–68, 84

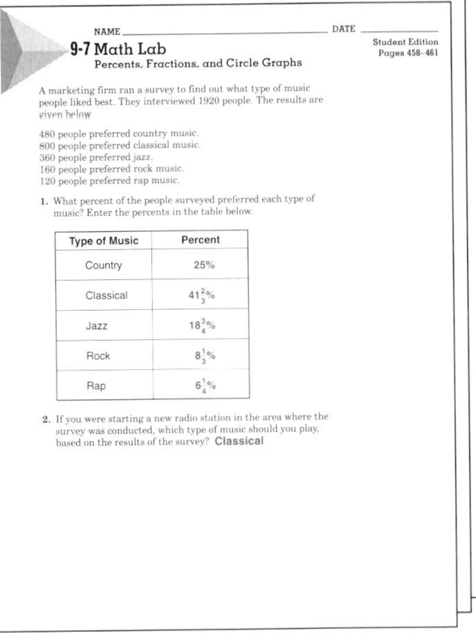

NAME ______________ DATE ______

9-7 Math Lab
Percents, Fractions, and Circle Graphs

Student Edition Pages 458–461

A marketing firm ran a survey to find out what type of music people liked best. They interviewed 1920 people. The results are given below.

480 people preferred country music.
800 people preferred classical music.
360 people preferred jazz.
160 people preferred rock music.
120 people preferred rap music.

1. What percent of the people surveyed preferred each type of music? Enter the percents in the table below.

Type of Music	Percent
Country	25%
Classical	$41\frac{2}{3}$%
Jazz	$18\frac{3}{4}$%
Rock	$8\frac{1}{3}$%
Rap	$6\frac{1}{4}$%

2. If you were starting a new radio station in the area where the survey was conducted, which type of music should you play, based on the results of the survey? **Classical**

MAKING MATHEMATICS RELEVANT

This two-page introduction to the chapter provides students with an opportunity to explore other disciplinary topics and their applications to mathematics.

Background Information

Math and Genetics in the News
Experts believe that dinosaurs dominated Earth in the Triassic Period of the Mesozoic Era, 245 million years ago. In that era, many fish were similar to modern fish. The first turtles, crocodiles, and mammals appeared. It is believed that the first modern humans appeared in the Pleistocene Epoch, a relatively short 2 million years ago.

CHAPTER 9

Ratio, Proportion, and Percent

TOP STORIES in Chapter 9

In this chapter, you will:

- write expressions for ratios, decimals, fractions, and percents,
- find simple probability,
- use ratios and proportions to solve problems,
- use statistics to predict, and
- solve problems by making a table.

MATH AND GENETICS IN THE NEWS

Bone fragments contain what might be dinosaur DNA

Source: USA Today, November 18, 1994

Maybe the movie *Jurassic Park* is not as far from reality as you might think . . . Scientists have found what they believe are genes of a dinosaur.

The genetic material is the first to be taken directly from ancient bones. Researchers believed that complex DNA molecules were too fragile to survive in dinosaur bone for millions of years. But, the team of scientists who studied the bone believe that the dinosaur died in a bog and was covered by peat so that the delicate bones were preserved.

The genetic code found in the bone was 30% similar to birds, 30% similar to reptiles, and 30% similar to mammals. The discovery may help settle whether modern birds are related to ancient dinosaurs.

Some scientists are skeptical about the discovery. If others can repeat the experiment, the conclusions may become accepted.

Putting It into Perspective

1780 | 1820 | 1860

1833
First appearance of a time scale separating prehistoric time into eras of life development.

1859
First oil well is drilled by "Colonel" Edwin Drake in Titusville, Pennsylvania.

1879
The U.S. Geological Survey established.

Putting It into Perspective

The U.S. Geological Survey conducts research in geology, geochemistry, geophysics, hydrology, cartography, and related sciences. It also studies natural hazards such as earthquakes, volcanoes, and flooding. The USGS publishes thousands of technical reports and maps each year.

EMPLOYMENT OPPORTUNITIES

Find Your Place in the Future of Prehistoric Life

Costa Rica • Galapagos • Australia • Alaska • Baja Mongolia • New Zealand

DO YOU LIKE traveling in the most exotic and untamed parts of the world? Career opportunities abound for those interested in the study of the remains of plants and animals preserved in Earth's rocks. Choose your specialty: as a **vertebrate paleontologist**, you will study extinct fishes, amphibians, reptiles, birds, and mammals; **invertebrate paleontologists** deal with fossil insects and shells; and the study of fossil plants is the concentration of the **paleobotanist**. College degree and related experience required.

For more information, contact:
The Paleontological Society
New Mexico Bureau of Mines and Mineral Resources
Socorro, NM 87801

Statistical Snapshot

GEOLOGIC TIME SCALE

ERAS		PERIODS	MAJOR EVENTS
Cenozoic	Present		
	Cenozoic Era	Quaternary	Ice Ages/active faulting in west
	Cenozoic Era	Tertiary	Modern topography; Oldest fossil humans
Mesozoic	65 million years ago		
	Mesozoic Era	Cretaceous	Major mountain-building in Rockies; Giant dinosaurs
	Mesozoic Era	Jurassic	Oldest fossil birds
	Mesozoic Era	Triassic	Beginning of opening of the Atlantic Ocean
Paleozoic	225 million years ago		
	Paleozoic Era	Permian	Last major mountain-building in Appalachians and Quachitas
	Paleozoic Era	Pennsylvanian	Vast coal swamps in North America
	Paleozoic Era	Mississippian	
	Paleozoic Era	Devonian	Oldest forest
	Paleozoic Era	Silurian	Oldest fish
	Paleozoic Era	Ordovician	Major mountain-building in New England and Eastern North America
	Paleozoic Era	Cambrian	Seas covered most of North America
Precambrian	570 million years ago		
	Origin of Earth ±4.5 billion years ago		

interNET CONNECTION For up-to-date information on dinosaurs, visit: **www.glencoe.com/sec/math/prealg/mathnet**

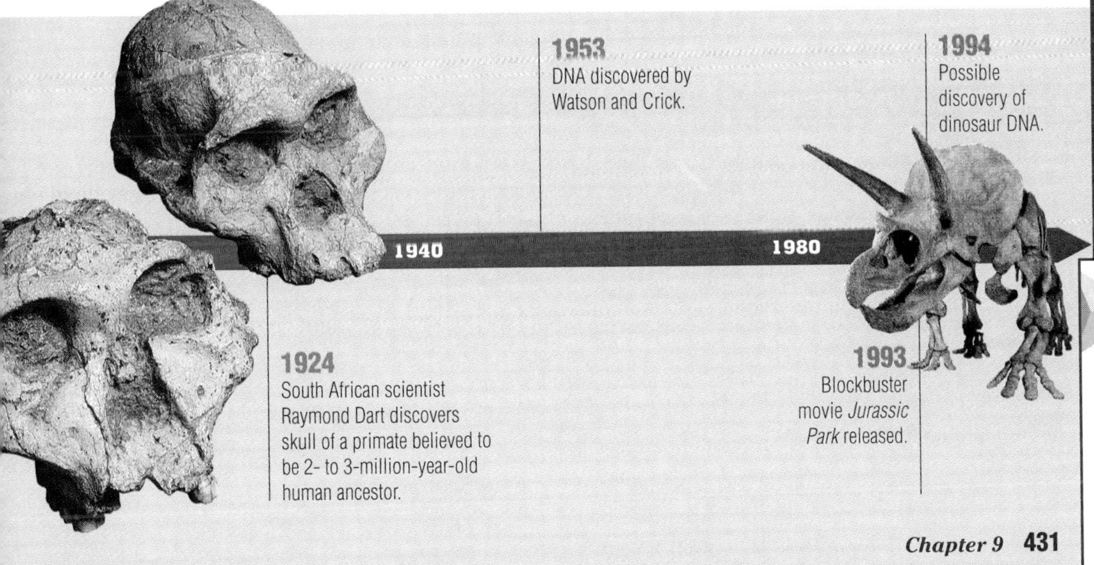

1940
1980
1953
DNA discovered by Watson and Crick.
1994
Possible discovery of dinosaur DNA.
1924
South African scientist Raymond Dart discovers skull of a primate believed to be 2- to 3-million-year-old human ancestor.
1993
Blockbuster movie *Jurassic Park* released.

EMPLOYMENT OPPORTUNITIES

Students might contact universities to see if there are any student expeditions that they can join as volunteers to get an idea of the type of work involved in being a paleontologist.

Statistical Snapshot

Geologists have created the geologic time scale based on a variety of evidence. The relative ages of geologic formations are often determined using radioactive dating techniques. While it is impossible to fix the specific time of an event or the age of a formation, relative ages can be found using different scientific techniques.

Cooperative Learning

Chapter Projects Two chapter projects are included in the *Investigations and Projects Masters*. In Chapter Project A, students extend the topic in the Chapter Opener. In Chapter Project B, students explore the golden ratio. A student page and a parent letter are provided for each Chapter Project.

interNET CONNECTION

Glencoe has made every effort to ensure that the website links for *Pre-Algebra* at www.glencoe.com/sec/math/prealg/mathnet are current and contain appropriate content. However, these website links are not under Glencoe's control.

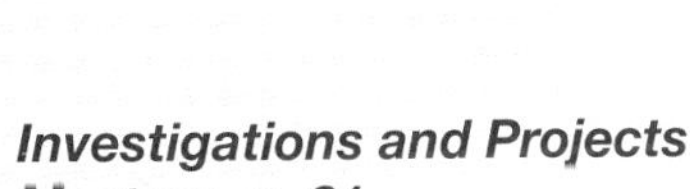

Investigations and Projects Masters, p. 61

NAME ______________________ DATE ____________

CHAPTER 9 Project A
Endangered Species

There are differing theories for the reasons behind the disappearance of the dinosaurs. The impact of an enormous meteorite, the eruptions of several volcanoes, and gradual changes in the Earth's environment are all theories. No matter what caused the end of the dinosaurs, scientists are trying to learn from them and how changes in Earth may effect the species on Earth now.

1. There are many species classified as endangered on both the United States and the International lists of endangered species. Make a list of endangered species. You may wish to consult an almanac or other resource for current information on lesser known endangered species.

2. Go to the library and investigate one of the animals on your list. Look for information on natural habitat, the history of the animal's population, the threats contributing to the animal's endangerment, the outlook for the animal's survival, and if possible the efforts being made to save the animal.

3. If possible, contact a local zoo for more information on your chosen animal. Ask questions like:
   - Are there any of these animals in captivity?
   - If so, are efforts being made to reintroduce some of these animals to the wild?
   - What type of efforts do you know of to help save these animals?
   - What can an average citizen do to help in the efforts?

4. Organize your research and create a book about your chosen animal. Whenever possible, include charts, graphs, or other statistics to illustrate the report.

9-1 LESSON NOTES

NCTM Standards: 1-5, 7, 12

Instructional Resources

- Study Guide Master 9-1
- Practice Master 9-1
- Enrichment Master 9-1
- Group Activity Card 9-1

Transparency 9-1A contains the 5-Minute Check for this lesson; **Transparency 9-1B** contains a teaching aid for this lesson.

Recommended Pacing	
Standard Pacing	Day 1 of 14
Honors Pacing	Day 1 of 13
Block Scheduling*	Day 1 of 7 (along with Lesson 9-2)

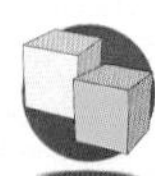

*For more information on pacing and possible lesson plans, refer to the ***Block Scheduling Booklet***.

1 FOCUS

5-Minute Check
(over Chapter 8)

1. Determine the domain and range of the relation {(−1, 2), (−3, 4), (2, 2), (−3, 2)}. Is the relation a function? **D = {−3, −1, 2}; R = {2, 4}; no**
2. Find the slope of the line that contains $A(3, 5)$ and $B(-2, -6)$. $\frac{11}{5}$

Graph each equation or inequality.

3. $y = 3x - 1$

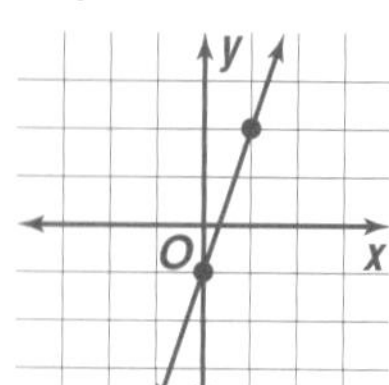

4. $x + y > 6$

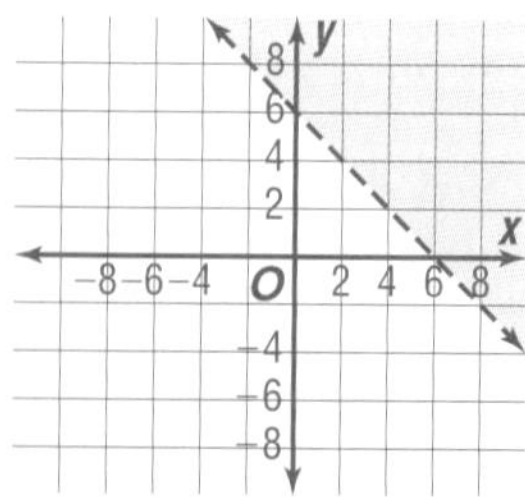

9-1 Ratios and Rates

Setting Goals: *In this lesson, you'll write ratios as fractions in simplest form and determine unit rates.*

MATHEMATICS MINI LAB — Modeling with Manipulatives

MATERIALS
- tape measure
-  grid paper
- ruler

People come in many different shapes and sizes. But with a closer look, you can find a pattern in some body measurements.

Your Turn Work with a partner. Use a tape measure to find the distance around your wrist and around your head to the nearest $\frac{1}{2}$ inch. Copy the table below and record the data for you and your partner. Graph the data as ordered pairs on a graph similar to the one below.

LOOK BACK You can review scatter plots in Lesson 8-2 and slope in Lesson 8-6.

	Head Measurement	Wrist Measurement
You		
Your Partner		

Wrist Measurement (inches) — y; O; x — Head Measurement (inches)

Draw the line between the two points. Then find the slope of the line.

TALK ABOUT IT

a. Divide each wrist circumference by the corresponding head circumference and record your answers. This is the wrist to head *ratio*. How do the ratios compare to the slope of the line? **They are the same.**

b. How do your results compare to the results of other groups? **See students' work.**

Learning the Concept

A **ratio** is a comparison of two numbers by division. If the circumference of someone's wrist is 5 inches and the circumference of his or her head is 20 inches, the ratio comparing wrist and head measurements could be written as follows.

5 to 20   5:20 $\frac{5}{20}$

Ratios are often expressed as fractions in simplest form or as a decimal.

$$\frac{5}{20} = \frac{1}{4}$$ (÷ 5) *The GCF of 5 and 20 is 5.*

or 5 [÷] 20 [=] .25

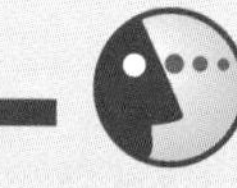

Alternative Learning Styles

Visual On a chalkboard rail, line up rectangles of the following sizes: A, 37.5 cm by 21 cm; B, 29 cm by 24 cm; C, 50 cm by 26 cm; D, 48.5 cm by 30 cm; E, 45 cm by 32 cm; F, 35 cm by 35 cm. Have students choose the rectangle that has the most pleasing appearance to them. Record survey results. Have selected students measure each rectangle. Then have each student find the ratio of length of the height for the rectangle they chose as a decimal. **A 1.857; B 1.208; C 1.923; D 1.617; E 1.406; F 1.0** Rectangle D is a golden rectangle. This shape is frequently chosen by people as the most pleasing shape.

Example 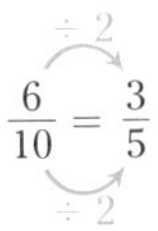 **Express each ratio as a fraction in simplest form and as a decimal.**

a. 6 free throws made out of 10 attempts

$$\frac{6}{10} = \frac{3}{5}$$ (÷ 2)

6 ÷ 10 = *0.6*

$\frac{3}{5}$ or 0.6 of the free-throws were made.

b. \$375 in commission to \$9375 in sales

$$\frac{375}{9375} = \frac{\overset{1}{3} \cdot \overset{1}{5} \cdot \overset{1}{5} \cdot \overset{1}{5}}{\underset{1}{3} \cdot \underset{1}{5} \cdot \underset{1}{5} \cdot \underset{1}{5} \cdot 5 \cdot 5} \text{ or } \frac{1}{25}$$

375 ÷ 9375 = *0.04*

$\frac{1}{25}$ or 0.04 of the sales is commission.

A **rate** is a special ratio. It is a comparison of two measurements with different units of measure, like miles and gallons or cents and grams. The ratio $\frac{89 \text{ cents}}{15 \text{ ounces}}$ compares the price and size of a can of soup. To find the price per ounce, simplify the rate so that the denominator is 1 ounce. A rate with a denominator of 1 is called a **unit rate**.

Example 2 APPLICATION Consumerism

Many grocery stores list the unit price of items on the shelf tag so that customers may compare similar items. The cost of a 12-ounce box of Cheerios is \$3.29. A store brand cereal that is similar to Cheerios costs \$4.89 for an 18-ounce box. Which cereal has a lower per-ounce cost?

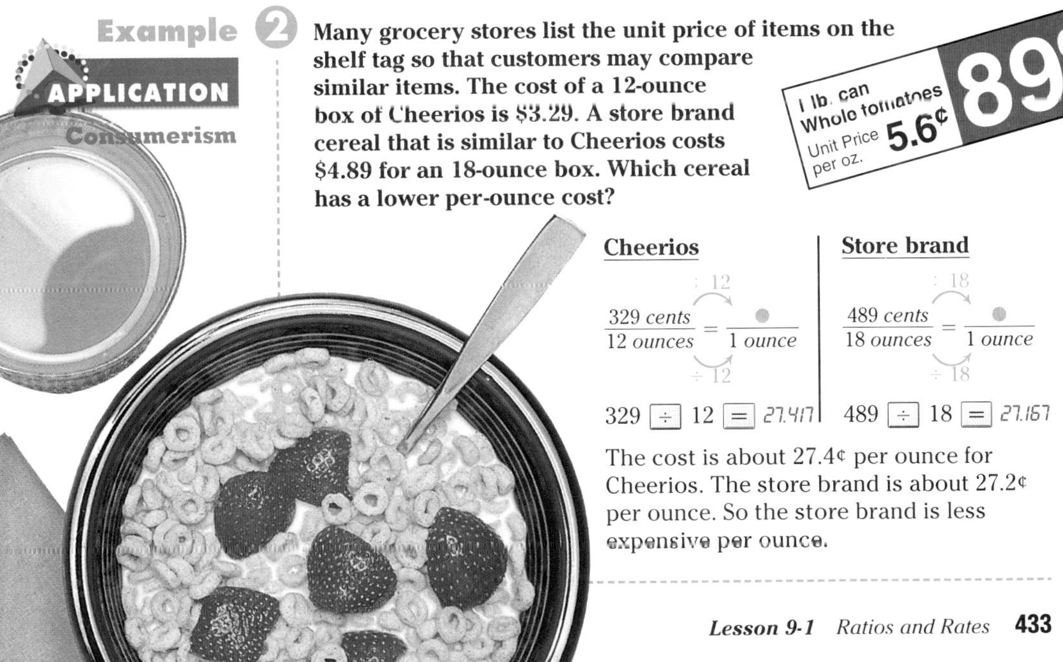

Cheerios

$$\frac{329 \text{ cents}}{12 \text{ ounces}} = \frac{\bullet}{1 \text{ ounce}}$$ (÷ 12)

329 ÷ 12 = *27.417*

Store brand

$$\frac{489 \text{ cents}}{18 \text{ ounces}} = \frac{\bullet}{1 \text{ ounce}}$$ (÷ 18)

489 ÷ 18 = *27.167*

The cost is about 27.4¢ per ounce for Cheerios. The store brand is about 27.2¢ per ounce. So the store brand is less expensive per ounce.

Motivating the Lesson

Hands-On Activity Provide each student with a piece of string about six feet long. Have each student use the string to make a circle on the floor that is the same size as their waist. However, students may not first measure their waist. After all have made their circle, have them check their estimates.

2 TEACH

In-Class Examples

For Example 1

Express each ratio as a fraction in simplest form and as a decimal.

a. 3 broken eggs out of 12 in the carton **$\frac{1}{4}$; 0.25**

b. \$25 tip on a \$125 meal **$\frac{1}{5}$; 0.2**

For Example 2

Sixteen ounces of Planters™ peanuts cost \$3.19 while 12 ounces of Eagle™ peanuts cost \$2.49. Which is the better buy? **Planters at 19.9¢ per ounce is a better buy than Eagle at 20.8¢ per ounce.**

Classroom Vignette

"The golden ratio can be found in art and nature. I like to have students explore the golden ratio more thoroughly by looking through books and magazines to find examples of the golden ratio."

Bonnie Langan

Bonnie Langan
Northern Middle School
Accident, MD

In-Class Example

For Example 3

Segment *AB* is 22 cm long. Place point *C* between *A* and *B* so segment *AC* is 13.6 cm long. Then find each ratio to the nearest thousandth.

a. $\frac{m\overline{AC}}{m\overline{BC}}$ **1.619** b. $\frac{m\overline{AB}}{m\overline{AC}}$ **1.618**

Teaching Tip Using the illustration from Example 3 above, have students explain why the point *C* is called the *golden cut* for segment *AB*. **It divides the segment into two sections such that $\frac{m\overline{AB}}{m\overline{AC}} = \frac{m\overline{AC}}{m\overline{CB}}$ and each ratio is the golden ratio.**

3 PRACTICE/APPLY

Checking Your Understanding

Exercises 1–14 are designed to help you assess your students' understanding through reading, writing, speaking, and modeling. You should work through Exercises 1–4 with your students and then monitor their work on Exercises 5–14.

Study Guide Masters, p. 71

NAME ________ DATE ________

9-1 Study Guide
Ratios and Rates

Student Edition Pages 432–436

A **ratio** is a comparison of two numbers by division. If the two terms of a ratio have no common factors, the ratio is in simplest form.

$\frac{4}{16} = \frac{1}{4}$ (÷4) The GCF of 4 and 16 is 4.

One type of ratio is a **rate**. A rate compares two measurements with different units. Speeds, such as 50 miles per hour or 32 feet per second, are familiar examples of rates. To change a rate to a unit rate, divide both the numerator and denominator by the denominator.

Rate $\frac{5280 \text{ ft}}{4 \text{ min}}$ (÷4) Unit Rate $\frac{1320 \text{ ft}}{1 \text{ min}}$ A unit rate is a rate with a denominator of 1.

Write the ratio that compares each of the following.

1. number of *p*'s to number of *i*'s in *Mississippi* $\frac{2}{4}$ or $\frac{1}{2}$
2. number of *o*'s to total number of letters in *proportion* $\frac{3}{10}$
3. number of months that have an *r* in their name to the number of months in a year $\frac{8}{12}$ or $\frac{2}{3}$

Express each ratio or rate as a fraction in simplest form.

4. 9 to 12 $\frac{3}{4}$
5. 12 to 9 $\frac{4}{3}$
6. 5:20 $\frac{1}{4}$
7. $2.50 for 5 notepads $\frac{50¢}{1}$
8. 60¢ per dozen $\frac{5¢}{1}$
9. $3.00 to rent 2 videotapes $\frac{\$1.50}{1}$

Express each ratio as a unit rate.

10. 120 miles in 2 hours **60 miles per hour**
11. 800 pounds for 40 square inches **20 pounds per square inch**
12. $300 for 5 jackets **$60 per jacket**
13. 45 meters in 3 minutes **15 meters per minute**
14. 10 kilometers in 2 hours **5 kilometers per hour**
15. 30 yards in 15 seconds **2 yards per second**

Connection to Geometry

The group of Greek mathematicians called the Pythagoreans studied the geometric figures and objects in nature that displayed the **golden ratio**. The great painter, Leonardo da Vinci, and many other artists and architects have used the golden ratio to create beautiful designs for over 4000 years. You can find the value of the golden ratio by comparing measurements of the Pyramid of Khufu.

One of the earliest examples of the golden ratio in architecture is the Pyramid of Khufu in Giza, Egypt, built about 2600 B.C. Find the approximate value of the golden ratio by finding the ratio of the slant height of the pyramid to half of the length of one side of the base.

$$\frac{\textit{slant height}}{\textit{half of base length}} = \frac{611.54}{\frac{1}{2}(756.08)} = \frac{611.54}{378.04}$$

611.54 [÷] 378.04 [=] 1.617659507

The golden ratio is approximately 1.618.

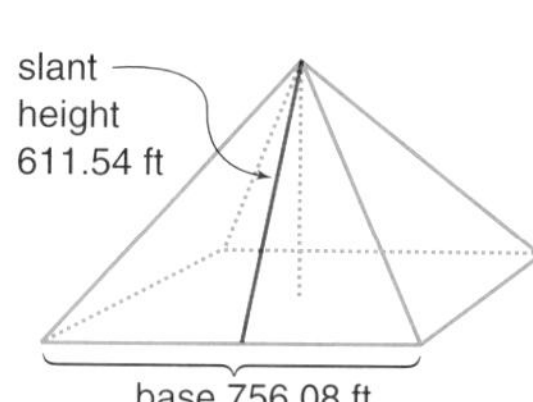

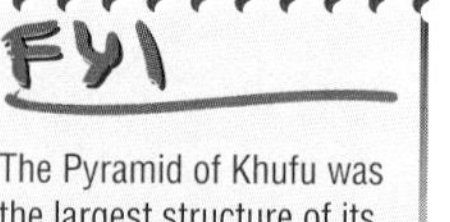

The Pyramid of Khufu was the largest structure of its day. It is the tomb of Pharaoh, Khufu, who died in 2567 B.C.

Checking Your Understanding

Communicating Mathematics

Read and study the lesson to answer these questions. **1. See margin.**

1. **Compare and contrast** ratios and rates.
2. **Write** a ratio that compares the number of students in your school to the number of teachers. Then express this ratio as a unit rate. **2. See students' work.**
3. **Describe** and give an example of a unit rate. **See students' work.**
4. Use a tape measure to measure the distance around your ankle, knee, and wrist to the nearest one-half inch. **a–b. See students' work.**
   a. Write a knee/ankle ratio, wrist/knee ratio, and ankle/wrist ratio.
   b. Compare the ratios in part a to those of a classmate.

MATERIALS

tape measure

Guided Practice

Express each ratio or rate as a fraction in simplest form.

5. 100 to 500 $\frac{1}{5}$
6. 325:25 $\frac{13}{1}$
7. 150 miles in 6 gallons $\frac{25}{1}$
8. 98 cents for 12 ounces $\frac{49}{6}$
9. Juan Gonzalez' 43 home runs in 1992 to 46 home runs in 1993 $\frac{43}{46}$

Express each ratio as a unit rate. **10. 2.4 lb/week** **11. $0.38/1 can**

10. 12 pounds lost in 5 weeks
11. $2.29 for a 6 pack of cola
12. 99¢ for a dozen **8.25¢/1**
13. 6 inches of rain in 24 hours **13. 0.25 in./1 hour**

14. **Consumer Awareness** *Garbage* magazine reported that the average American commuter burns 190 gallons of gasoline a year driving about 4000 miles to and from work. How many miles per gallon does the average commuter get in his or her car? **about 21.1 miles/1 gallon**

FYI

King Khufu's pyramid was 481 feet tall. Kings Khafre and Menkure were King Khufu's son and grandson. Their pyramids were 471 feet tall and 218 feet tall, respectively.

Additional Answer

1. Ratios compare two like units. Rates compare two unlike units.

Exercises: Practicing and Applying the Concept

Independent Practice

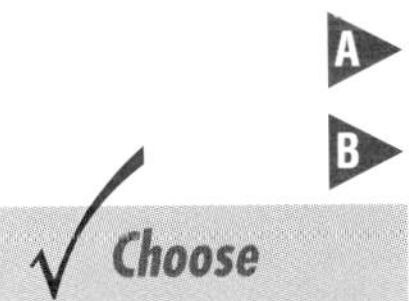

Express each ratio or rate as a fraction in simplest form.

A

15. 13 out of 169 $\frac{1}{13}$
16. 125:50 $\frac{5}{2}$
17. 64 to 16 $\frac{4}{1}$
18. 351:117 $\frac{3}{1}$
19. 156 to 84 $\frac{13}{7}$
20. 35 out of 149 $\frac{35}{149}$

B

21. 2 girls to 15 boys $\frac{2}{15}$
22. 9 Fords to 7 Chevrolets $\frac{9}{7}$
23. 5 out of 7 oranges $\frac{5}{7}$
24. 11 pennies out of 99 pennies $\frac{1}{9}$
25. 24 marbles to 56 marbles $\frac{3}{7}$
26. 15 successes in 35 attempts $\frac{3}{7}$

Estimation
Mental Math
Calculator
Paper and Pencil C

Express each ratio as a unit rate. **27–32. See margin.**

27. 399.3 miles on 16.5 gallons
28. $9.60 for 12 pounds
29. 100 meters in 10.5 seconds
30. 27 inches of snow in 9 hours
31. $210 for 30 tickets
32. 550 mL for $3.30

Critical Thinking

33. Arrange the digits 1 to 9 into two numbers whose ratio is 1:5. **Sample answer: 2697:13,485**

Applications and Problem Solving

34. **Architecture** A golden rectangle is a rectangle in which the ratio of the length to the width is the golden ratio.
    a. Find a golden rectangle in the photo of the Parthenon at the right.
    b. Many everyday objects are golden rectangles. For example, credit cards are approximately golden rectangles. Measure a rectangular object. Is it close to a golden rectangle?

34a–34b. Answers will vary.

35. **Consumer Awareness** Best buys in grocery stores are generally found by comparing unit rates such as cents/oz.
    a. Which is the better buy, a 16-oz bag of candy at $2.49 or a 32-oz bag at $3.69? How do you know? **the 32-oz bag at $3.69**
    b. Are there instances in which the better buy might not necessarily be the one you choose to purchase? **See Solutions Manual.**
36. **Family Activity** Visit a grocery store. Choose a product such as laundry detergent or pretzels. Find the unit price of a number of different brands. Which one would you choose and why?

36. See students' work.

37. **Patterns** Many sequences of numbers follow a pattern in the ratios of a term to the previous term. **37a. All ratios are 2.**
    a. The sequence 3, 6, 12, 24, 48, . . . is geometric. Find the ratio between each pair of consecutive terms. What do you observe?
    b. Recall that the sequence 1, 1, 2, 3, 5, 8, 13, 21, . . . is called the Fibonacci sequence. The ratios between consecutive Fibonacci numbers approach a famous number. Find several ratios and make a conjecture about what number the ratios approach.

37b. See students' work. The ratios approach the golden ratio.

38. **Travel** The distance between two cities on a map is 5 centimeters. The mileage chart with the map shows that the cities are 10 miles apart. The ratio of an actual distance to the distance on a map is called the scale of miles. What is the scale of miles on the map? **0.5 cm/mile**

Assignment Guide

Core: 15–39 odd, 40–46
Enriched: 16–32 even, 33–46

For **Extra Practice**, see p. 761.

The red A, B, and C flags, printed only in the Teacher's Wraparound Edition, indicate the level of difficulty of the exercises.

Additional Answers

27. 24.2 miles/gallon
28. $0.80/lb
29. 9.52 m/s
30. 3 in./h
31. $7/1 ticket
32. 0.6¢/mL

Reteaching

Using Manipulatives Place 4 rolls of pennies (200¢) in a large bowl. Place several cans or boxes of food products near the pennies. Mark each item with its cost. Have students show how many pennies are needed to pay for a particular item and then show how to find the cost per ounce by separating pennies into equal stacks based on the number of ounces.

Group Activity Card 9-1

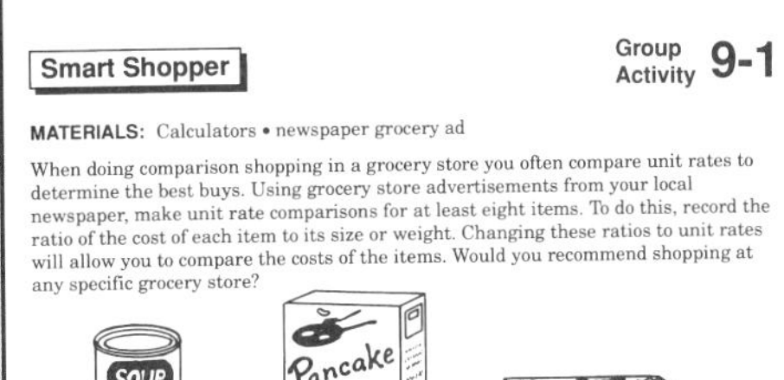

Smart Shopper Group Activity **9-1**

MATERIALS: Calculators • newspaper grocery ad

When doing comparison shopping in a grocery store you often compare unit rates to determine the best buys. Using grocery store advertisements from your local newspaper, make unit rate comparisons for at least eight items. To do this, record the ratio of the cost of each item to its size or weight. Changing these ratios to unit rates will allow you to compare the costs of the items. Would you recommend shopping at any specific grocery store?

 Pre-Algebra

Practice Masters, p. 71

NAME ____________ DATE ______

9-1 Practice
Ratios and Rates

Student Edition
Pages 432–436

Express each ratio or rate as a fraction in simplest form.

1. 50 to 300 $\frac{1}{6}$
2. 800 to 16 $\frac{50}{1}$
3. 425:500 $\frac{17}{20}$
4. 21 out of 91 $\frac{3}{13}$
5. 30:45 $\frac{2}{3}$
6. 18 out of 81 $\frac{2}{9}$
7. 128 to 56 $\frac{16}{7}$
8. 144:36 $\frac{4}{1}$
9. 113 to 339 $\frac{1}{3}$
10. 3 out of 8 automobiles $\frac{3}{8}$
11. 14 dogs to 21 cats $\frac{2}{3}$
12. 16 losses in 40 games $\frac{2}{5}$
13. 9 out of 15 compact discs $\frac{3}{5}$

Express each ratio as a unit rate.

14. $306 for 17 tickets **$18 per ticket**
15. 10 inches of snow in 6 hours **$1\frac{2}{3}$ inches per hour**
16. 300 miles on 12 gallons **25 miles per gallon**
17. $1200 in 3 weeks **$400 per week**
18. 325 words in 5 minutes **65 words per minute**
19. 289 feet in 17 seconds **17 feet per second**
20. $1.32 per dozen **$0.11 each**
21. $7.77 for 3 pounds **$2.59 per pound**
22. 194.8 miles in 4 hours **48.7 miles per hour**
23. 5 kilometers in 8 minutes **0.625 kilometer per minute**

Population of World's Largest Urban Areas (rounded to the nearest million)			
New York, NY	16	Tokyo, Japan	12
Mexico City, Mexico	14	Los Angeles-Long Beach, CA	11
Paris, France	9	Shanghai, China	10
São Paulo, Brazil	8	Buenos Aires, Argentina	10

Use the data above to express the ratio of the populations of the given cities.

24. Paris to Tokyo $\frac{3}{4}$
25. Buenos Aires to São Paulo $\frac{5}{4}$
26. Shanghai to Mexico City $\frac{5}{7}$
27. Shanghai to New York $\frac{5}{8}$
28. New York to São Paulo $\frac{2}{1}$
29. Tokyo to Shanghai $\frac{6}{5}$
30. Los Angeles-Long Beach to Paris $\frac{11}{9}$
31. São Paulo to Mexico City $\frac{4}{7}$

4 ASSESS

Closing Activity

Writing Have students each state a rate in turn while the other students write it as a unit rate.

Additional Answers

40.

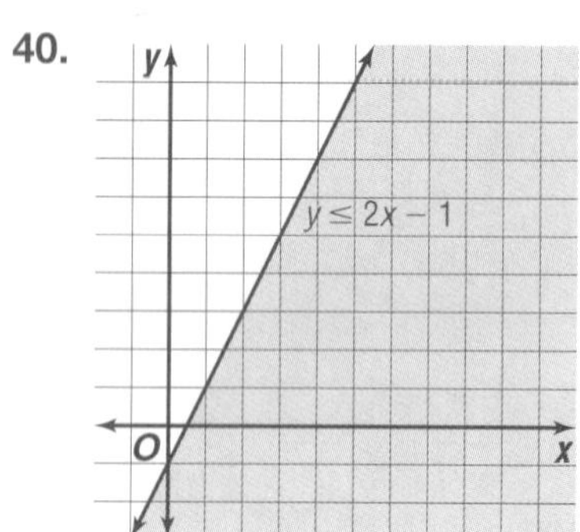

44. $x < 28$

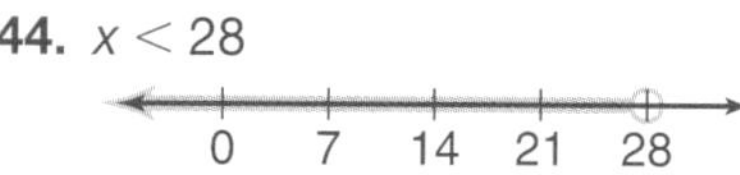

See for Yourself, bullet one: no; This sample has a density of about 0.712 grams per milliliter, so it must not be ethyl alcohol.

See for Yourself, bullet two: yes; The density of saltwater is about 1.04 grams per milliliter. So ethyl alcohol has a lower density and will float.

Enrichment Masters, p. 71

NAME ____________________ DATE __________

9-1 Enrichment
Power Sets

Student Edition
Pages 432–436

The set of possible subsets of a given set is called a **power set** of that set. Note that the set itself and the empty set (∅) are both always subsets.

Set	Number of Elements	Subsets	Number of Subsets
∅	0	∅	1
{25}	1	∅, {25}	2
{25, 50}	2	∅, {25}, {50}, {25, 50}	4
{25, 50, 75}	3	∅, {25}, {50}, {75}, {25, 50} {25, 75}, {50, 75}, {25, 50, 75}	8

Study the pattern for the number of elements and subsets.

1. How many sets are in a power set if the original set has 4 elements? 5 elements? n elements? $2^4 = 16$; $2^5 = 32$; 2^n

Label the eight subsets of {25, 50, 75} as follows.
∅ = ∅ B = {50} D = {25, 50} F = {25, 75}
A = {25} C = {75} E = {50, 75} I = {25, 50, 75}

Complete the table for union (∪) and the table for intersection (∩) for the eight sets in the table above. For example, A ∪ B = {25, 50} = D and A ∩ B = ∅.

2.

∪	∅	A	B	C	D	E	F	I
∅	∅	A	B	C	D	E	F	I
A	A	A	D	F	D	I	F	I
B	B	D	B	E	D	E	I	I
C	C	F	E	C	I	E	F	I
D	D	D	D	I	D	I	I	I
E	E	I	E	E	I	E	I	I
F	F	F	I	F	I	I	F	I
I	I	I	I	I	I	I	I	I

3.

∩	∅	A	B	C	D	E	F	I
∅	∅	∅	∅	∅	∅	∅	∅	∅
A	∅	A	∅	∅	A	∅	A	A
B	∅	∅	B	∅	B	B	∅	B
C	∅	∅	∅	C	∅	C	C	C
D	∅	A	B	∅	D	B	A	D
E	∅	∅	B	C	B	E	C	E
F	∅	A	∅	C	A	C	F	F
I	∅	A	B	C	D	E	F	I

4. Are these eight sets closed under union and intersection; that is, is the result of each union and intersection one of the eight sets? yes

5. What is the identity element for union? for intersection? ∅; I

6. Are the operations of union and intersection commutative? yes

39. **Zoology** On your mark, get set, go! The fastest human can run about 26 miles per hour. That may sound fast, but many animals can leave us in the dust.
    a. A pronghorn antelope is the fastest mammal over long distances. It can run 4 miles in about 6.9 minutes. What is a pronghorn's speed in miles per hour? (*Hint:* Remember that 1 hour is 60 minutes.) **34.8 mph**
    b. The fastest insect, a tropical cockroach, can scurry up to 300 feet in a minute. What is its speed in feet per second? **5 feet /s**

Mixed Review

40. See margin.
42. 6,582,761; 391,000; no mode
43. 2 · 2 · 2 · 3 · 19

40. Graph $y \le 2x - 1$. (Lesson 8-9)
41. Solve $7(a + 2) = 5a - 8$. (Lesson 7-5) **−11**
42. **Geography** The table at the right shows the area in thousands of square miles of the oceans of the world. Find the mean, median, and mode of the areas. (Lesson 6-6)
43. Factor 456 completely. (Lesson 4-4)
44. Solve $x - 9 < 19$. Graph the solution on a number line. (Lesson 3-6) **See margin.**
45. Find $-8 + (-5)$. (Lesson 2-4) **−13**
46. **Pets** In 1988, there were 205 million pets in the United States. By 1995, that number had grown to 235 million. Write an equation for the change in the number of pets from 1988 to 1995. Then solve. (Lesson 1-8) **$205 + c = 235$; 30 million**

Ocean	Area
Pacific Ocean	64,186
Atlantic Ocean	33,420
Indian Ocean	28,351
Arctic Ocean	5106
South China Sea	1149
Caribbean Sea	971
Mediterranean Sea	969
Bering Sea	873
Gulf of Mexico	582
Sea of Okhotsk	537
Sea of Japan	391
Hudson Bay	282
East China Sea	257
Andaman Sea	218
Black Sea	196
Red Sea	175
North Sea	165
Baltic Sea	148
Yellow Sea	114
Persian Gulf	89
Gulf of California	59

Source: *World Almanac,* 1995

EARTH WATCH

KEEPING WATCH OVER OUR OCEANS

You have probably seen news reports about oil spills in the ocean and the effects on the sea life. But oil is just one of the substances that can threaten our oceans.

When dealing with a spill, the first step is for an oceanographer to determine what the material is. One way of doing that is to find its **density.** A material's density is the ratio of its mass to its volume.

$$density = \frac{mass}{volume}$$

Each material has a unique density. So you can compare the density of a sample to the known density of a material to help identify the sample.

See for Yourself

- The density of ethyl alcohol is 0.789 grams per milliliter. An oceanographer has a sample from a chemical spill that has a mass of 136 grams and a volume of 191 milliliters. Is the sample ethyl alcohol? Explain. **See margin.**
- A material that has a lower density will float on one that has a higher density. A 250-milliliter sample of seawater has a mass of 260 grams. Would a spill of ethyl alcohol float on seawater? Justify your answer. **See margin.**
- **Research** ocean spills. Does spilled crude oil float? What chemicals are spilled most often? Describe the methods used to clean up oil and other chemical spills. **See students' work.**

EARTH WATCH

Ethyl alcohol is used for industrial purposes for making photographic film, adhesives, polishes, inks, wood stains, detergents, and personal care products such as lotions and perfumes.

Extension

Using Models Point out to students that 3-by-5 index cards (ratio 0.6) and 5-by-8 index cards (ratio 0.625) are close enough to be used as examples of the golden rectangle. Have students find other examples of rectangular objects that show the golden ratio.

9-2 Problem-Solving Strategy: Make a Table

Setting Goals: *In this lesson, you'll solve problems by making a table.*

Modeling a Real-World Application: Business

On average, 5¢ of every dollar an American spends on food is spent in a vending machine. That's a lot of nickels!

A fruit machine accepts dollars and all the fruit costs 65 cents. If the machine gives only nickels, dimes, and quarters, what combinations of coins could it be programmed to give as change for a dollar?

Learning the Concept

You can make a table to show all of the possible combinations of coins. A table allows you to organize information in an understandable way.

Explore The machine will give back $1.00 − 0.65 or 35 cents in change in a combination that can include nickels, dimes, or quarters.

Plan Organize data in a table using different combinations of nickels, dimes, quarters. Start with the most quarters.

Solve

quarters	dimes	nickels
1	1	0
1	0	2
0	3	1
0	2	3
0	1	5
0	0	7

Examine The total for each combination of the coins is 35 cents. Only nickels, dimes, and quarters are used.

Checking Your Understanding

Communicating Mathematics 1–2. See Solutions Manual.

Read and study the lesson to answer these questions.

1. **Explain** how a table can help you solve problems.
2. What kinds of problems are best solved by making a table or chart?

Guided Practice

Solve by making a table.

3. How many ways can you make change for a 50-cent piece using only nickels, dimes, and quarters? **10**
4. **Number Theory** How many different ways are there to add prime numbers to make a sum of 12? Show all the ways. **See Solutions Manual.**

Reteaching

Using Connections Ms. Pavlova knows that there is 7.5¢ sales tax on each dollar she spends. To help her estimate total costs, make a table that shows the sales tax on whole dollar amounts from $1 to $9 and on ten-dollar amounts from $10 to $100.

Study Guide Masters, p. 72
The Study Guide Master provides a concise presentation of the lesson along with practice problems.

9-2 LESSON NOTES

NCTM Standards: 1-5, 10, 12

Instructional Resources

- Study Guide Master 9-2
- Practice Master 9-2
- Enrichment Master 9-2
- Group Activity Card 9-2

Transparency 9-2A contains the 5-Minute Check for this lesson; **Transparency 9-2B** contains a teaching aid for this lesson.

Recommended Pacing	
Standard Pacing	Day 2 of 14
Honors Pacing	Day 2 of 13
Block Scheduling*	Day 1 of 7 (along with Lesson 9-1)

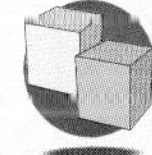

*For more information on pacing and possible lesson plans, refer to the ***Block Scheduling Booklet***.

1 FOCUS

5-Minute Check *(over Lesson 9-1)*

Express each ratio as a fraction in simplest form.

1. 14 out of 35 $\frac{2}{5}$
2. 500:65 $\frac{100}{13}$
3. 16 out of 20 $\frac{4}{5}$
4. 18 dollars to 36 dollars $\frac{1}{2}$
5. 26 successes in 39 attempts $\frac{2}{3}$

2 TEACH

In-Class Example

Jarrett has $0.99 made up of eight U. S. coins. He cannot make change for a dollar, 50¢, 25¢, 10¢, or 5¢. What five coins does Jarrett have? **4 pennies, 2 dimes, 1 quarter, 1 half-dollar**

3 PRACTICE/APPLY

Checking Your Understanding

Exercises 1–5 are designed to help you assess your students' understanding through reading, writing, speaking, and modeling. You should work through Exercises 1–2 with your students and then monitor their work on Exercises 3–5.

Assignment Guide
Core: 7–13 odd, 14–19
Enriched: 6–14 even, 15–19

For **Extra Practice**, see p. 761.

4 ASSESS

Closing Activity

Writing Have students work in small groups and have them write a problem that you can solve by making a table. Then exchange problems among groups and solve.

Enrichment Masters, p. 72
The Enrichment Master extends the concepts in the lessons through new and varied topics.

Practice Masters, p. 72

NAME ______________________ DATE __________

9-2 Practice — Student Edition Pages 437–438
Problem-Solving Strategy: Make a Table

Solve. Use any strategy.

1. Kent Jones has $1.95 consisting of 7 U.S. coins. However, he cannot make change for a nickel or a half-dollar. What 7 coins does Kent have? **2 nickels, 1 dime, 1 quarter, 3 half-dollars**

2. A penny, nickel, dime, quarter, and half-dollar are in a purse. Without looking, Maria picks two coins. How many different amounts of money could she choose? **10 amounts**

3. Rita had 50 stamps in her collection. She traded 6 stamps for 4 from Juana. She traded 4 more for 5 from Mary. She traded another 5 for 3 from Derice. Finally, she traded 12 more stamps for 9 from Mike. How many stamps does Rita have now? **44 stamps**

4. Paul wants to buy a stereo system. The store allows a 15% discount if the purchase is paid for within 30 days. A 20% discount is given if the purchase is paid for within 10 days. If Paul pays $400 at the time of purchase, what was the original price of the stereo system? **$500**

5. Randall Burns bought 50 shares of stock for $2490. When the price per share went up $4, he sold 25 shares. Then the price per share went down $2, so he bought 100 more shares. When the price of the stock went back up $5, he sold 50 shares. How many shares of stock does he have now? How much is each share worth? **75 shares; $56.80**

6. Beth and Janeen both start their jobs at the same time. Beth's starting salary is $16,000 per year with a guaranteed $4000 pay raise per year for a 5-year period. Janeen's starting salary is $18,000 per year with a guaranteed $3000 pay raise per year for a 5-year period. Which person would be making more money during the fifth year? How much money would this person make during the five years? **Beth; $120,000**

5. An octahedron has faces numbered 1 to 8. If two are thrown and the faces down are added, how many ways can you roll a sum less than 8? **21**

Exercises: Practicing and Applying the Concept

Independent Practice

Solve. Use any strategy.

6. Emily had 40 baseball cards. She traded 7 cards for 5 from Dani. She traded 3 more for 4 from Ted and 2 for 1 from Anita. Finally, she traded 11 cards for 8 from Shawna. How many cards does Emily have now? **35**

7. A penny, a nickel, a dime, and a quarter are in a purse. How many amounts of money are possible if you grab two coins at random? **6**

8. Lisa is challenged to find her way through a maze. Each time she walks through a correct opening, she receives $1. In each correct aisle, she is given a reward equal to the total she already has. The correct path passes through 8 openings and 7 aisles. If Lisa takes the correct path, how much money will she have at the end? **$255**

9. Ms. Torres sold 100 shares of stock for $5975. When the price per share went down $5, she bought 200 more shares. When the price per share went back up $3, she sold 100 shares. How much did Ms. Torres gain or lose in these transactions? **She gained $800.**

10. Copy the figure at the right. Place the numbers 1 through 12 at the dots so that the sum of each of the six rows is 26.

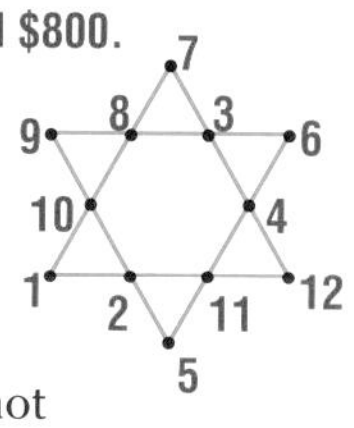

(Figure answer: 7; 9, 8, 3, 6; 10, 4; 1, 2, 11, 12; 5)

11. **Patterns** Find the next number in the pattern 0, $-1, -3, -6, -10, -15, \ldots$ **−21**

12. Tokala has 5 coins with a total value of $1.05. He cannot make change for a dollar, 5 cents, 10 cents, 25 cents, or 50 cents. What five coins does he have?

12. 50-cent piece, quarter, and 3 dimes

13. **Geometry** If the sides of the rectangle are whole numbers of inches, how many different combinations of measures are possible? **5**

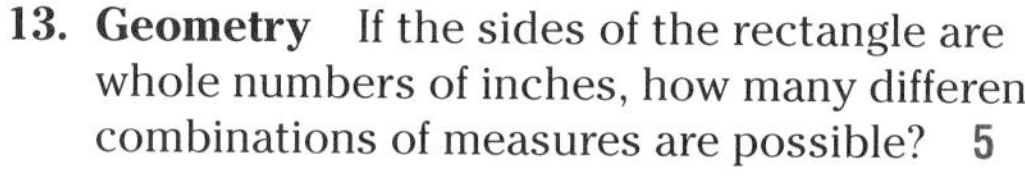
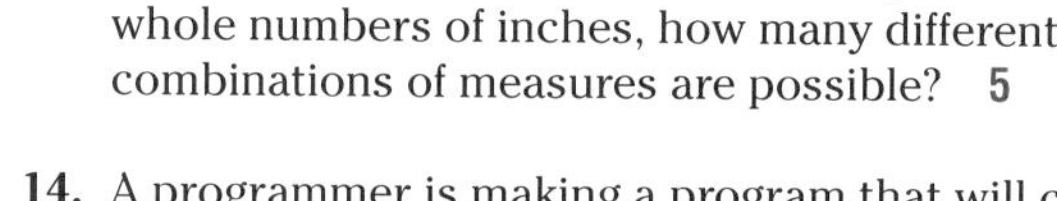

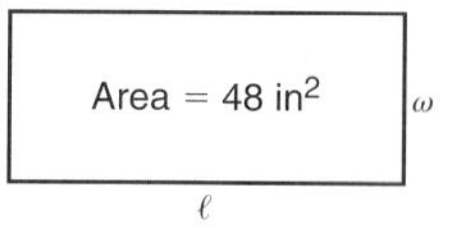

Critical Thinking

14. A programmer is making a program that will create calendars. The program will contain calendar pages with just the days of the week and the dates, no months. The user will insert the name of the month before printing the calendar. If a page will exist for every arrangement of dates in a month, how many pages will need to be designed? **28**

Mixed Review

15. **Aviation** The first non-stop, around-the-world flight was made by Capt. James Gallagher in March 1949. The 23,452-mile flight was made in 94 hours. What was the average rate in miles per hour? (Lesson 9-1) **249.5 mph**

16. Solve $3v + 7 < 19$. (Lesson 7-6) $v < 4$

17. Estimate $90.23 + 89.53$. (Lesson 5-2)

18. Determine whether $6c - d$ is a monomial. Explain why or why not. (Lesson 4-1)

19. Write an inequality using the numbers in the following sentence. *The Montreal Canadiens have won the Stanley Cup 24 times, and the Detroit Red Wings have won it 7 times.* (Lesson 2-3) $24 > 7$

17. Sample answer: 180
18. No; it involves subtraction.

Group Activity Card 9-2

Table Those Ideas — Group Activity **9-2**

Suppose you live on a farm in Iowa. Your father asks you to go out and count the number of cows and chickens in the barnyard, since he is worried that some of the animals got out through a hole in the fence. You count the animals but decide to play a joke on your father. Rather than tell him how many cows and chickens there are, you tell him that you counted 60 animals in all, and there were a total of 210 legs. Can your father figure out how many cows and chickens there are in the barnyard from this information?

With a partner, see if you can solve this problem for the father. It will be helpful to make a table.

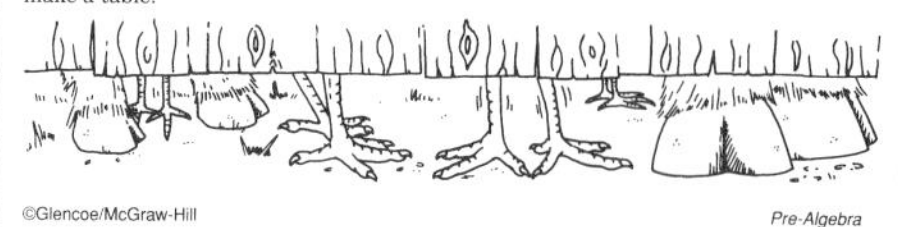

 Pre-Algebra

Extension

Using Technology Have students interview adults to find out how a computer spreadsheet program can be used to solve a problem. Have students explain how using a spreadsheet program and making a table to solve a problem are alike.

HANDS-ON ACTIVITY

9-3A Fair and Unfair Games

A Preview of Lesson **9-3**

MATERIALS

dice or number cubes

Have you ever heard someone say "Hey, no fair!" while you were playing a game? Most games are designed to be fair; that is, each player has an equal chance of winning. If the game is weighted in some way so that one player has a better chance of winning than the other players, the game is unfair.

Suppose you toss a coin 50 times. Each time the coin shows a head, you win and each time the coin shows tails, you lose.

Wins	Losses
卌 卌 卌 卌 III	卌 卌 卌 卌 卌 II

Since there are approximately the same number of wins as losses, the game of tossing a coin seems to be fair.

Your Turn

Work with a partner and play games A and B. Copy the table below and record your results in it.

	Win	Lose
Game A		
Game B		

Roll two dice 50 times.

Game A Find the sum of the numbers on the dice. Record an even sum as a win and record an odd sum as a loss.

Game B Find the product of the numbers on the dice. Record an even product as a win and record an odd product as a loss.

1. Are the games fair or unfair? Explain. **See Solutions Manual.**
2. If one of the games is unfair, modify the rules in order to make it fair. **See students' work.**

Extension

3. Enriqueta is blindfolded. She picks a marble out of a bag containing 6 blue marbles and 6 red marbles. If the marble is red, she wins. If the marble is blue, she loses. Do you think this is a fair game? Why or why not? **Yes; same amount of blue and red marbles.**
4. Two dice are rolled. If both dice show the same number, you win. If the dice show different numbers, you lose. Do you think this is a fair game? Why or why not? **No; there is a greater chance of getting two different numbers.**

9-3A LESSON NOTES

NCTM Standards: 1-4, 10

Objective

Determine whether a game is fair or unfair.

Recommended Time

Demonstration and discussion: 15 minutes; Exercises: 30 minutes

Instructional Resources

For each student or group of students
Student Manipulative Kit
- two dice

Math Lab and Modeling Math Masters
- p. 44 (worksheet)

For teacher demonstration
Overhead Manipulative Resources

1 FOCUS

Motivating the Lesson

Play a game of nim with several pairs of students.

2 TEACH

Teaching Tip Have them continue to play the game until they see a pattern. Discuss what happens when both players understand the winning strategy. Discuss whether this is a fair game in the sense that each player has an equal chance of winning.

3 PRACTICE/APPLY

Assignment Guide
Core: 1–4
Enriched: 1–4

4 ASSESS

Observing students working in cooperative groups is an excellent method of assessment.

9-3 LESSON NOTES

NCTM Standards: 1-5, 10

Instructional Resources

- Study Guide Master 9-3
- Practice Master 9-3
- Enrichment Master 9-3
- Group Activity Card 9-3
- Assessment and Evaluation Masters, p. 239
- Activity Masters, p. 39
- Graphing Calculator Masters, p. 9
- Math Lab and Modeling Math Masters, p. 84
- Multicultural Activity Masters, p. 17
- Real-World Applications, 18

Transparency 9-3A contains the 5-Minute Check for this lesson; **Transparency 9-3B** contains a teaching aid for this lesson.

Recommended Pacing	
Standard Pacing	Day 4 of 14
Honors Pacing	Day 3 of 13
Block Scheduling*	Day 2 of 7

*For more information on pacing and possible lesson plans, refer to the ***Block Scheduling Booklet***.

1 FOCUS

5-Minute Check
(over Lesson 9-2)

Solve. Use any strategy.

1. A rectangle has an area of 60 square inches. If the sides of the rectangle are whole numbers of inches, how many different combinations of measures are possible? **6**
2. Place numbers in each ❑ so the product of each row and column is 60.

❑ ❑ ❑
❑ ❑ ❑
❑ ❑ ❑

Sample answer:
3 4 5
4 5 3
5 3 4

9-3 Integration: Probability
Simple Probability

Setting Goals: *In this lesson, you'll find the probability of simple events.*

Modeling with Technology

In the fall of 1994, WSNY radio station in Columbus, Ohio, ran a contest called "Lucky Bucks". If you had a dollar bill that met the criteria announced by the DJ, you could call in to win \$100, \$1000, or even more. Suppose the DJ said, "If you have a dollar with a serial number ending in 7 or 8, be the first caller and you win!"

Your Turn

Use the graphing calculator program at the right to generate random numbers between 0 and 9 to represent the last digit of your serial number. Generate 50 different digits and record the number of times you qualify to win. Enter 0 as the least integer and 9 as the greatest.

```
PROGRAM: RANDNUM
: ClrHome
: Input "LEAST INTEGER", S
: Input "GREATEST INTEGER", L
: Input "NUMBER OF VALUES", N
: 0 → B
: Lbl 1
: B + 1 → B
: int ((L-S + 1) rand + S) → R
: Disp R
: Pause
: If A ≠ B
: Goto 1
```

TALK ABOUT IT

a. How many times did you qualify?

b. Compare results with your classmates. How many times did someone in the class qualify ?

c. List the digits that qualify to win and the digits that are possible.

d. Compare $\frac{\textit{number of times qualified to win}}{\textit{number of times played}}$ and $\frac{\textit{number of digits that qualify}}{\textit{number of possible digits}}$.

e. If you had 10 dollar bills, do you think at least one would qualify? Explain. **Sample answer: Yes, $\frac{2}{10}$ or 2 should qualify.**

f. Predict the outcome for 1 dollar bill and for 100 dollar bills.

a–b., d. See students' work.

c. 7, 8; 0, 1, 2, 3, 4, 5, 6, 7, 8, 9

f. One bill would probably not qualify; Out of 100 bills, about 20 should qualify.

Learning the Concept

Probability is the chance some event will happen. It is the ratio of the number of ways an event can occur to the number of possible outcomes.

Definition of Probability

$$\text{Probability} = \frac{\textit{number of ways a certain outcome can occur}}{\textit{number of possible outcomes}}$$

Example  **What is the probability that you will roll an even number when you roll a number cube?**

There are 3 possible even outcomes—2, 4, and 6.
There are 6 possible outcomes—1, 2, 3, 4, 5, and 6.

$$P(\text{even}) = \frac{\textit{number of even outcomes}}{\textit{number of outcomes}} = \frac{3}{6} \text{ or } \frac{1}{2}$$

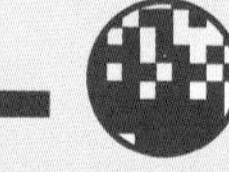

Alternative Teaching Strategies

Student Diversity Have students play a game of Monopoly® with this twist. Before their moves, they must calculate the probability that they will get a "good" move. Good moves include landing on a square that does not cost you money. Landing on unowned property and landing on Chance or Community Chest might be good or bad, so they are neutral moves. If students calculate the probability correctly, they take their turn as usual. If they do not calculate correctly, they lose their turn. Note that other players have to check the calculations and all must agree that it is right or wrong before proceeding.

The set of all possible outcomes is called the **sample space**. The sample space in Example 1 was {1, 2, 3, 4, 5, 6}. When you flip a coin, the sample space is heads or tails, or {heads, tails}.

Example 2 APPLICATION Games

In the game of Backgammon, you get to roll again if you roll doubles. What is the probability of rolling doubles?

Make a table showing all of the combinations that you could roll on a pair of dice as ordered pairs.

	1	2	3	4	5	6
1	(1, 1)	(1, 2)	(1, 3)	(1, 4)	(1, 5)	(1, 6)
2	(2, 1)	(2, 2)	(2, 3)	(2, 4)	(2, 5)	(2, 6)
3	(3, 1)	(3, 2)	(3, 3)	(3, 4)	(3, 5)	(3, 6)
4	(4, 1)	(4, 2)	(4, 3)	(4, 4)	(4, 5)	(4, 6)
5	(5, 1)	(5, 2)	(5, 3)	(5, 4)	(5, 5)	(5, 6)
6	(6, 1)	(6, 2)	(6, 3)	(6, 4)	(6, 5)	(6, 6)

In the sample space, there are 6 outcomes that show doubles and 36 possible outcomes. So the probability of rolling doubles is $\frac{6}{36}$ or $\frac{1}{6}$.

Example 3

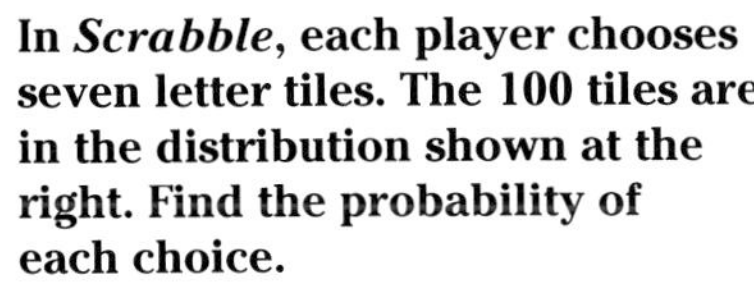

CULTURAL CONNECTIONS
Versions of *Scrabble* in languages other than English have different ratios of letters. For example, in the Dutch version there are 18 Es, 10 Ns, and 2 Js.

In *Scrabble*, each player chooses seven letter tiles. The 100 tiles are in the distribution shown at the right. Find the probability of each choice.

a. an E if no tiles have been chosen
b. an M if 28 tiles have been chosen and 1 of them was an M
c. an X if 42 tiles have been chosen and 1 of them was an X

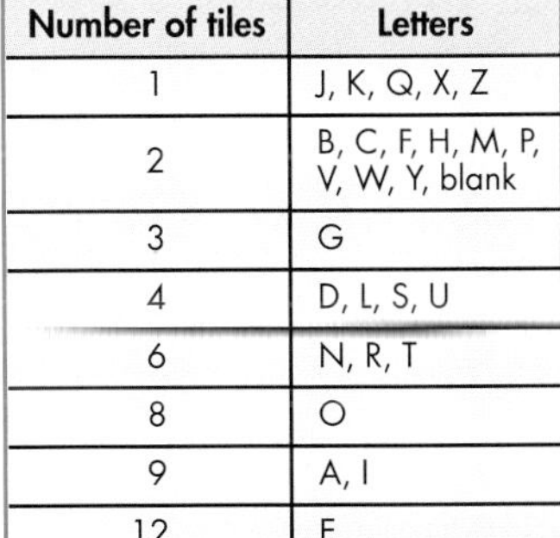

Number of tiles	Letters
1	J, K, Q, X, Z
2	B, C, F, H, M, P, V, W, Y, blank
3	G
4	D, L, S, U
6	N, R, T
8	O
9	A, I
12	E

a. There are 12 E tiles and 100 tiles in all.
$P(\text{choosing an E}) = \frac{12}{100}$ or $\frac{3}{25}$

b. There are 2 M tiles. So, if one has been chosen, 1 remains. There are 100 − 28 or 72 tiles from which to choose.
$P(\text{choosing an M}) = \frac{1}{72}$

c. There is only one X tile, so no Xs remain. The probability of choosing an X is 0.

The probability of an event is always between 0 and 1, inclusive. If something cannot happen, its probability is 0. If something is certain, its probability is 1.

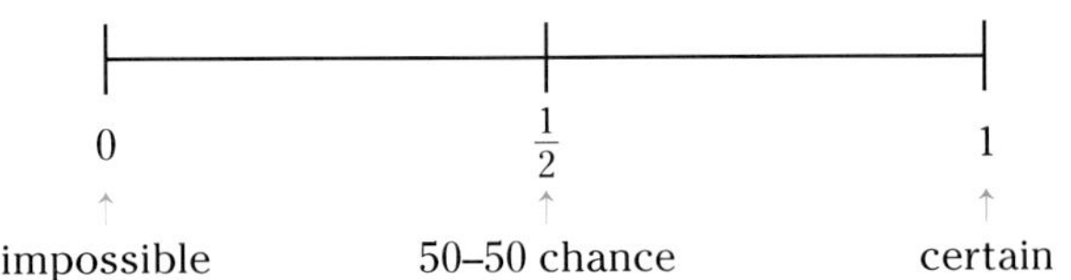

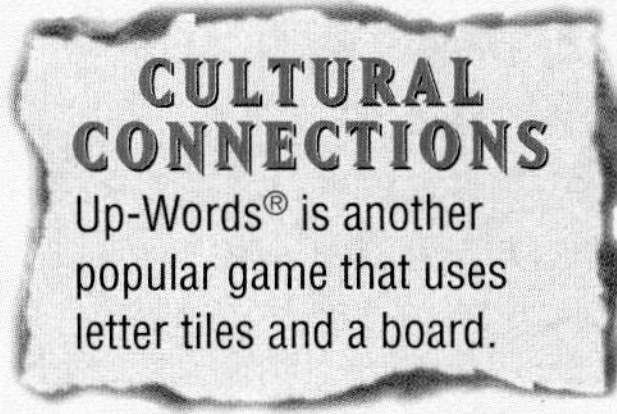

CULTURAL CONNECTIONS
Up-Words® is another popular game that uses letter tiles and a board.

Motivating the Lesson

Questioning Give students questions where the answers "probably" or "probably not" are appropriate; for example, are you going to be in school tomorrow, or will the basketball team win their next game. Discuss how sure they are of the event occurring.

2 TEACH

In-Class Examples

For Example 1
What is the probability that you will roll a 1 or 6 when you roll a number cube? $\frac{1}{3}$

For Example 2
Suppose you must roll a 9 to have a chance to win a game of Trivial Pursuit™. What is the probability of rolling a 9? $\frac{1}{9}$

For Example 3
Find the probability of each choice of Scrabble® tile.

a. an O if no tiles have been chosen $\frac{2}{25}$

b. any vowel if half of all tiles and half of all vowels have been drawn $\frac{21}{50}$

c. a blank if 38 tiles have been chosen, including two blanks 0

Study Guide Masters, p. 73

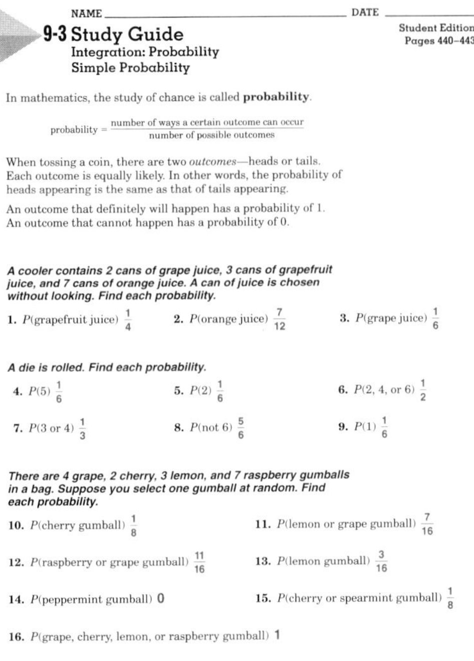

NAME ______ DATE ______

9-3 Study Guide
Integration: Probability
Simple Probability

Student Edition Pages 440–443

In mathematics, the study of chance is called **probability**.

$$\text{probability} = \frac{\text{number of ways a certain outcome can occur}}{\text{number of possible outcomes}}$$

When tossing a coin, there are two *outcomes*—heads or tails. Each outcome is equally likely. In other words, the probability of heads appearing is the same as that of tails appearing.

An outcome that definitely will happen has a probability of 1. An outcome that cannot happen has a probability of 0.

A cooler contains 2 cans of grape juice, 3 cans of grapefruit juice, and 7 cans of orange juice. A can of juice is chosen without looking. Find each probability.

1. P(grapefruit juice) $\frac{1}{4}$
2. P(orange juice) $\frac{7}{12}$
3. P(grape juice) $\frac{1}{6}$

A die is rolled. Find each probability.

4. P(5) $\frac{1}{6}$
5. P(2) $\frac{1}{6}$
6. P(2, 4, or 6) $\frac{1}{2}$
7. P(3 or 4) $\frac{1}{3}$
8. P(not 6) $\frac{5}{6}$
9. P(1) $\frac{1}{6}$

There are 4 grape, 2 cherry, 3 lemon, and 7 raspberry gumballs in a bag. Suppose you select one gumball at random. Find each probability.

10. P(cherry gumball) $\frac{1}{8}$
11. P(lemon or grape gumball) $\frac{7}{16}$
12. P(raspberry or grape gumball) $\frac{11}{16}$
13. P(lemon gumball) $\frac{3}{16}$
14. P(peppermint gumball) 0
15. P(cherry or spearmint gumball) $\frac{1}{8}$
16. P(grape, cherry, lemon, or raspberry gumball) 1

3 PRACTICE/APPLY

Checking Your Understanding

Exercises 1–15 are designed to help you assess your students' understanding through reading, writing, speaking, and modeling. You should work through Exercises 1–5 with your students and then monitor their work on Exercises 6–15.

Additional Answers

1. Probability is the ratio of the number of ways that a certain outcome can happen to the number of possible outcomes that can happen. A sample space is the set of all possible ways an event can happen.
2. Sample: rolling a 5 on a die
3. An event with a probability of 0 cannot happen, an event with a probability of 1 will definitely happen. When the probability is 0 or 1, you know how the event will occur.
4. Manuel is correct because there are an equal number of ways each person can win.

5a. Sample answer:

0 HEADS	T, T, T		
1 HEAD	T, T, H	T, H, T	H, T, T
2 HEADS	H, H, T	H, T, H	T, H, H
3 HEADS	H, H, H		

Practice Masters, p. 73

NAME ______________________ DATE __________

9-3 Practice
Integration: Probability
Simple Probability

Student Edition
Pages 440–443

There are 4 blue marbles, 5 red marbles, 1 green marble, and 2 black marbles in a bag. Suppose you select one marble at random. Find each probability.

1. P(black) $\frac{1}{6}$
2. P(blue) $\frac{1}{3}$
3. P(green) $\frac{1}{12}$
4. P(red) $\frac{5}{12}$
5. P(not blue) $\frac{2}{3}$
6. P(red or green) $\frac{1}{2}$
7. P(blue or black) $\frac{1}{2}$
8. P(neither red nor black) $\frac{5}{12}$
9. P(pink) 0
10. P(not purple) 1

A spinner like the one at the right is used in a game. Determine the probability of spinning each outcome if the spinner is equally likely to land on each section.

11. P(a two) $\frac{1}{4}$
12. P(an odd number) $\frac{1}{2}$
13. P(a one or a four) $\frac{1}{2}$
14. P(the letter A) 0
15. P(a number greater than 1) $\frac{3}{4}$
16. P(prime number) $\frac{1}{2}$
17. P(a number less than one) 0
18. P(not a three) $\frac{3}{4}$

Suppose you roll two dice. Use a chart of possible outcomes to find each probability.

19. How many outcomes are in the sample space? 36
20. What is $P(6, 3)$? $\frac{1}{36}$
21. What is $P(5, 2)$? $\frac{1}{36}$
22. What is P(even number, odd number)? $\frac{1}{4}$
23. What is P(both numbers are odd)? $\frac{1}{4}$

Checking Your Understanding

Communicating Mathematics

Read and study the lesson to answer these questions. 1–3. See margin.

1. **Define** probability and sample space in your own words.
2. Give an example of an event with a probability of $\frac{1}{6}$.
3. **Compare and contrast** an event with a probability of 0 and an event with a probability of 1.
4. **You Decide** Four coins are tossed. Manuel says it will be fair if he wins if the majority of the coins are heads and Paige wins if the majority of the coins are tails. Paige doesn't think the game is fair. Who is correct and why? See margin.

5a. See margin.

MODELING MATHEMATICS

MATERIALS
three coins

5. a. Use coins to help determine the sample space of the toss of three coins. Make a drawing, chart, or table to show your answer.
   b. What is the probability of two heads? $\frac{3}{8}$
   c. What is the probability of at least one tail? $\frac{7}{8}$

Guided Practice

State the probability of each outcome.

6. Today is the 30th of February. 0
7. Randomly turn to a month in a calendar and it is June. $\frac{1}{12}$
8. A 3 is rolled on a die. $\frac{1}{6}$

A spinner like the one at the right is used in a game. Determine the probability of spinning each outcome if the spinner is equally likely to land on each section.

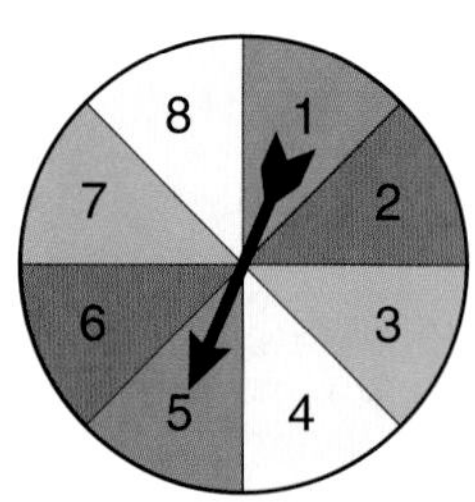

9. a four $\frac{1}{8}$
10. a one $\frac{1}{8}$
11. an even number $\frac{1}{2}$
12. a two or a four $\frac{1}{4}$
13. not a three $\frac{7}{8}$
14. less than a six $\frac{5}{8}$

15. **Government** In 1994, 1529 of the 7424 state legislators in the United States were women. If you chose a state legislator at random for an interview, what is the probability that you would choose a man? $\frac{5895}{7424}$

Exercises: Practicing and Applying the Concept

Independent Practice

There are 5 red marbles, 7 green marbles, 4 black marbles, and 8 blue marbles in a bag. Suppose you select one marble at random. Find each probability.

16. P(red) $\frac{5}{24}$
17. P(orange) 0
18. P(blue) $\frac{1}{3}$
19. P(not green) $\frac{17}{24}$
20. P(green or black) $\frac{11}{24}$
21. P(not yellow) 1

A die is rolled. Find each probability.

22. P(odd) $\frac{1}{2}$
23. P(even) $\frac{1}{2}$
24. P(prime) $\frac{1}{2}$
25. P(greater than 6) 0

Group Activity Card 9-3

Crafty Cooking — Group Activity 9-3

Luigi, the Italian cook, needs grated cheese to make pizza for a birthday party. He has calculated that he needs 18 pounds of grated cheese. However, cheese only comes in 3-pound and 5-pound cans. How many cans of each size should he buy?

Try making a table to solve this problem. You might start by deciding to buy 0, 1, 2, etc. 5-pound cans and then determine the number of 3-pound cans necessary to buy 18 pounds of grated cheese.

Grated Cheese 5 lb
Grated Cheese 3 lb

 Pre-Algebra

Reteaching

Using Models Make or use a spinner divided into four equal sections, numbered 1, 2, 3, and 4. Find the probability of spinning each of the following: a 1, a 2, a 3, a 4, not a 1, an even number, an odd number, a prime number, a composite number. $\frac{1}{4}$, $\frac{1}{4}$, $\frac{1}{4}$, $\frac{1}{4}$, $\frac{3}{4}$, $\frac{1}{2}$, $\frac{1}{2}$, $\frac{1}{2}$, $\frac{1}{4}$

Use the chart of possible outcomes when two dice are rolled in Example 2 to find each probability.

26. P(both odd) $\frac{1}{4}$

27. P(both even) $\frac{1}{4}$

28. P(both prime) $\frac{1}{4}$

29. P(sum greater than 11) $\frac{1}{36}$

30. P(sum of 7) $\frac{1}{6}$

31. P(sum of 1) **0**

Graphing Calculator

32. Refer to the application at the beginning of the lesson. What is the probability that you would have qualified to win the Lucky Bucks contest if the DJ asked for dollars that had serial numbers ending in 24? (*Hint:* Use the graphing calculator program on page 440 with numbers between 0 and 99 to simulate the situation.) $\frac{1}{100}$

Critical Thinking

33. **Complementary events** are two events in which either one or the other must take place, but they cannot both happen at the same time. The sum of their probabilities is 1. Use this definition to solve the following problem. If a batter usually hits 3 out of every 10 pitches, what is the probability that she will *not* hit the next pitch? $\frac{7}{10}$

Applications and Problem Solving

34. **Demographics** The graph at the right shows the population of the United States by age. Suppose that you are taking a telephone poll to determine how popular a political candidate is with different groups of voters.

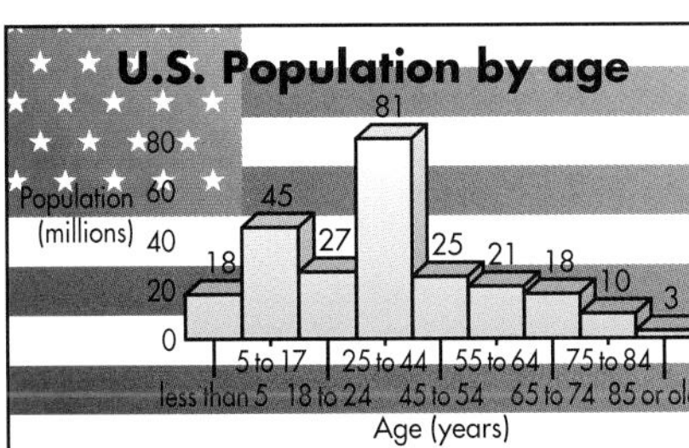

Source: U.S. Census Bureau

a. What is the probability that a randomly chosen person is old enough to vote? (*Hint:* In the United States you must be at least 18 years old to be able to vote.)

34a. $\frac{185}{248}$

b. Find the probability that a randomly-chosen person is 65 years old or older. $\frac{31}{248}$

35. **Space Exploration** What is the probability that humans will explore the moon on foot? What is the probability that humans will explore the Sun on foot? Explain.

35. 1 – humans have been to the Moon; 0 – the Sun is too hot

36. **Family Activity** Devise an experiment using objects in your home such as tossing coins, or choosing objects out of a bag. Write each possible outcome and its probability. Then perform the experiment at least 20 times. How do the probabilities compare with the results? **See students' work.**

Mixed Review

37. **Number Theory** How many whole numbers less than 124 are divisible by 2, 3, and 5? (Lesson 9-2) **4**

38. Find the slope of a line that passes through $A(3, 9)$ and $B(1, -3)$. (Lesson 8-6) **6**

39. Solve $6x + 3 = -9$. (Lesson 7-2) **−2**

40. Solve $n - 9.06 = 2.31$. (Lesson 5-6) **11.37**

41. **Geometry** Find the perimeter and area of a 5.1-by-2.6-meter rectangle. (Lesson 3-5) **15.4 m; 13.26 sq m**

Error Analysis

Students may incorrectly determine probability because they do not consider all possible outcomes. Encourage students to make a summary or list of all possible outcomes before determining the probability.

Assignment Guide

Core: 17–35 odd, 37–41
Enriched: 16–30 even, 32–41

For **Extra Practice**, see p. 762.

The red A, B, and C flags, printed only in the Teacher's Wraparound Edition, indicate the level of difficulty of the exercises.

4 ASSESS

Closing Activity

Modeling Bring in a standard deck of cards. Have students find the probabilities of a number of relationships among the cards. For example, find each probability.

1. P(a four) $\frac{1}{13}$
2. P(a face card) $\frac{3}{13}$

Chapter 9, Quiz A (Lessons 9-1 through 9-3) is available in the *Assessment and Evaluation Masters*, p. 239.

***Enrichment Masters*, p. 73**

NAME ______ DATE ______

9-3 Enrichment
Probability and Tables

Student Edition Pages 440–443

In Rockville High School there are
400 freshmen—60 have A averages and 90 have B averages.
300 sophomores—40 have A averages and 60 have B averages.
200 juniors—10 have A averages and 30 have B averages.
100 seniors—20 have A averages and 60 have B averages.

1. Use the information above to complete the table below. Then use the table to answer Exercises 2–11.

Class \ Grade	A	B	Below B	Total
Freshmen	60	90	250	400
Sophomores	40	60	200	300
Juniors	10	30	160	200
Seniors	20	60	20	100
Total	130	240	630	1000

Suppose a student is selected at random from Rockville High School. Find the probability of selecting each of the following.

2. a freshman $\frac{2}{5}$
3. a senior $\frac{1}{10}$
4. an A student $\frac{13}{100}$
5. a student whose grade is below B $\frac{63}{100}$
6. a sophomore B student $\frac{3}{50}$
7. a junior A student or a senior A student $\frac{3}{100}$
8. a student who is neither a junior A student nor a senior A student $\frac{97}{100}$
9. a B student who is not a junior $\frac{21}{100}$
10. If selecting only from the juniors, what is the probability of picking an A student? $\frac{1}{20}$
11. If selecting only from the students who are neither A nor B students, what is the probability of picking a senior? $\frac{2}{63}$

Extension

Using Manipulatives Have each student in the class write a number between 1 and 5 on a blank index card. Collect all the cards without looking at them. Mix them up and draw a card without looking. Record the results in a table. Make 100 such draws. Have each student record a guess as to how many of each number are in the deck of cards. Look at the cards to determine how many of each there are. Allow time for discussing the results.

9-4 LESSON NOTES

NCTM Standards: 1-5

Instructional Resources

- Study Guide Master 9-4
- Practice Master 9-4
- Enrichment Master 9-4
- Group Activity Card 9-4
- Activity Masters, p. 23
- Tech Prep Applications Masters, p. 17

Transparency 9-4A contains the 5-Minute Check for this lesson; **Transparency 9-4B** contains a teaching aid for this lesson.

Recommended Pacing	
Standard Pacing	Day 5 of 13
Honors Pacing	Day 4 of 14
Block Scheduling*	Day 3 of 7

*For more information on pacing and possible lesson plans, refer to the ***Block Scheduling Booklet***.

1 FOCUS

5-Minute Check *(over Lesson 9-3)*

There are 3 red marbles, 2 white marbles, and 5 blue marbles in a bag. Find each probability.

1. P(red) $\frac{3}{10}$
2. P(blue) $\frac{1}{2}$
3. P(not white) $\frac{4}{5}$
4. P(yellow) 0
5. P(color on the U.S. flag) 1

Motivating the Lesson

Situational Problem Kamil has a computer game that allows more time as the difficulty of the game increases. The number of minutes allowed form a sequence starting with 1, 3, 9. Ask students to find the ratio relating each number to the previous number and find the next three numbers in the sequence. **27, 91, 273**

9-4 Using Proportions

Setting Goals: *In this lesson, you'll use proportions to solve problems.*

Modeling a Real-World Application: Economics

Imagine paying almost \$20 for a 2-liter bottle of Coca-Cola. That's the kind of price that the Russian people have been paying since President Yeltsin opened their markets to free trade in 1992.

The average monthly wage for a Russian is \$110, while the average American takes home about \$2197 each month. So if a Russian pays \$0.44 for a loaf of bread, that is like an American paying \$8.79!

Purchase	Russian cost (U.S. dollars)	Equivalent U.S. cost
eggs (10)	0.69	13.78
blue jeans	48.00	958.56
Coca-Cola (2 liter bottle)	0.95	18.98
gasoline (1 gallon)	1.80	35.96
McDonald's Big Mac	2.08	41.53
Heinz ketchup (17-ounce bottle)	2.31	46.13
American ice cream (2 quarts)	9.24	184.52
oranges (2 pounds)	1.57	31.35

Source: J. Patrick Lewis

The table at the right shows the cost in U.S. dollars for different items purchased in Russia and the equivalent price in terms of the average American income. What is the U.S. equivalent cost for a bottle of aspirin that costs \$6.72 in Russia? *This problem will be solved in Example 3.*

Learning the Concept

You can use a proportion to solve problems that relate to ratios. A **proportion** is a statement of equality of two or more ratios. Consider a general proportion $\frac{a}{b} = \frac{c}{d}$.

$$\frac{a}{b} = \frac{c}{d}$$

$$\frac{a}{b} \cdot bd = \frac{c}{d} \cdot bd \quad \textit{Multiply each side by b and d to eliminate the fractions.}$$

$$ad = cb$$

The products ad and cb are called the **cross products** of a proportion. One way to determine if two ratios form a proportion is to check their cross products.

Property of Proportions

In words: The cross products of a proportion are equal.

In symbols: If $\frac{a}{b} = \frac{c}{d}$, then $ad = bc$. If $ad = bc$, then $\frac{a}{b} = \frac{c}{d}$.

Tech Prep

Travel Planner Various agencies, such as AAA, provide maps and travel plans for clients. They use proportions in dealing with maps, speed rates, and various information required by the client.

For more information on tech prep, see the *Teacher's Handbook*.

Example 1 **Use cross products to determine if each pair of ratios forms a proportion.**

a. $\frac{1}{4}, \frac{4}{16}$

$\frac{1}{4} \stackrel{?}{=} \frac{4}{16}$

$1 \cdot 16 \stackrel{?}{=} 4 \cdot 4$

$16 = 16$

So, $\frac{1}{4} = \frac{4}{16}$.

b. $\frac{1.5}{5.0}, \frac{3}{9}$ *Use a calculator.*

1.5 [×] 9 [=] 13.5

5 [×] 3 [=] 15

$13.5 \neq 15$

So, $\frac{1.5}{5.0} \neq \frac{3}{9}$.

You can use cross products to solve proportions.

Example  **Solve each proportion.**

a. $\frac{c}{35} = \frac{3}{7}$

$c \cdot 7 = 35 \cdot 3$ *Cross products*

$7c = 105$ *Multiply.*

$\frac{7c}{7} = \frac{105}{7}$ *Divide each side by 7.*

$c = 15$

The solution is 15.

b. $\frac{10}{8.4} = \frac{5}{d}$

$10 \cdot d = 8.4 \cdot 5$ *Cross products*

8.4 [×] 5 [÷] 10 [=] 4.2 *Use a calculator.*

Thus, $d = 4.2$.

Proportions can be used to solve many real-life problems.

Example 3 **Refer to the application at the beginning of the lesson. Find the U.S. equivalent cost for a bottle of aspirin.**

Economics

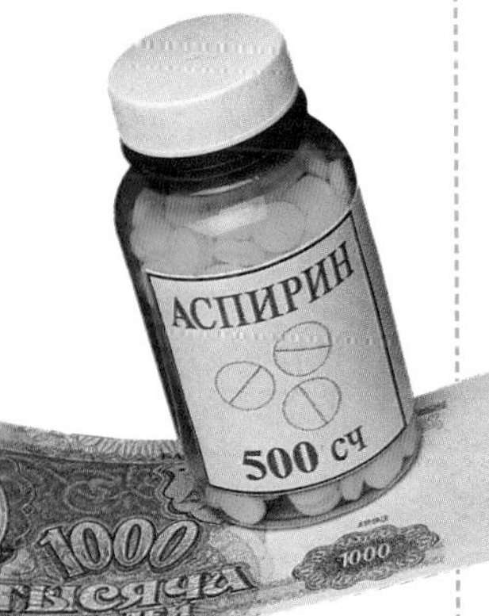

Explore We know the average monthly income for a Russian and an American and the Russian cost for the aspirin. We need to find the U.S. equivalent cost.

Plan Write and solve a proportion with the Russian and American incomes and costs. The table shows that the U.S. equivalents are about 20 times the Russian costs. So an estimate of the U.S. equivalent is 20×7 or \$140.

Solve Let a represent the U.S. cost for aspirin.

$$\frac{\textit{Russian income}}{\textit{Russian cost}} = \frac{\textit{American income}}{\textit{American cost}}$$

$$\frac{110}{6.72} = \frac{2197}{a}$$

$110 \cdot a = 6.72 \cdot 2197$ *Find the cross products.*

6.72 [×] 2197 [÷] 110 [=] 134.2167273

The equivalent American cost is about \$134.22.

Examine The cost is close to the estimate, so it is reasonable.

Connection to Geometry

Proportions are used to relate the size of figures with the same shape.

2 TEACH

In-Class Examples

For Example 1
Use cross products to determine if each pair of ratios forms a proportion.

a. $\frac{3}{7}, \frac{5}{14}$ $\frac{3}{7} \neq \frac{5}{14}$

b. $\frac{6}{3.2}, \frac{9}{4.8}$ $\frac{6}{3.2} = \frac{9}{4.8}$

For Example 2
Solve each proportion.

a. $\frac{b}{64} = \frac{5}{16}$ **20**

b. $\frac{95}{3.6} = \frac{19}{n}$ **0.72**

For Example 3
Find the U.S. equivalent cost for a quart of milk that costs 75¢ in Russia. **\$14.98**

Study Guide Masters, p. 74

NAME ______ DATE ______

9-4 Study Guide
Using Proportions

Student Edition
Pages 444–447

A proportion is a statement of equality of two or more ratios. To determine if two ratios form a proportion, check their cross products. If the cross products are equal, the rations form a proportion.

$\frac{1}{2} \stackrel{?}{=} \frac{2}{4}$ $\frac{2}{5} \stackrel{?}{=} \frac{6}{15}$ $\frac{2}{3} \stackrel{?}{=} \frac{10}{12}$

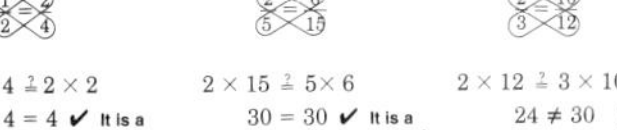

$1 \times 4 \stackrel{?}{=} 2 \times 2$ $4 = 4$ ✔ It is a proportion.

$2 \times 15 \stackrel{?}{=} 5 \times 6$ $30 = 30$ ✔ It is a proportion.

$2 \times 12 \stackrel{?}{=} 3 \times 10$ $24 \neq 30$ It is not a proportion.

Cross products can be used to solve proportions.

Example: Solve the proportion $\frac{3}{4} = \frac{x}{20}$.

$3 \cdot 20 = 4 \cdot x$ Write the cross products.
$60 = 4x$ Multiply.
$15 = x$ Divide each side by 4.

Write = or ≠ in each blank to make a true statement.

1. $\frac{3}{8} \neq \frac{12}{32}$
2. $\frac{15}{20} = \frac{3}{4}$
3. $\frac{4}{7} \neq \frac{16}{49}$
4. $\frac{1}{2} \neq \frac{1}{4}$
5. $\frac{35}{50} = \frac{7}{10}$
6. $\frac{40}{48} = \frac{5}{6}$

Solve each proportion.

7. $\frac{5}{8} = \frac{x}{40}$ $x = 25$
8. $\frac{6}{3} = \frac{10}{t}$ $t = 5$
9. $\frac{n}{5} = \frac{42}{7}$ $n = 30$
10. $\frac{4}{11} = \frac{12}{x}$ $x = 33$
11. $\frac{2}{3} = \frac{0.8}{n}$ $n = 1.2$
12. $\frac{7}{12} = \frac{1.68}{b}$ $b = 2.88$

Write a proportion that could be used to solve each problem. Then solve the proportion.

13. Cole can pick 2 rows of beans in 30 minutes. How long will it take him to pick 5 rows if he works at the same rate? $\frac{2}{30} = \frac{5}{x}$; $x = 75$, 75 minutes
14. A tree casts a shadow 30 meters long. A 2.8-meter pole casts a shadow 2 meters long. How tall is the tree? $\frac{x}{30} = \frac{2.8}{2}$; $x = 42$, 42 meters

Classroom Vignette

"I introduce the concept of scale models to my students as another application of using proportions. My students then use their knowledge of proportion and the cardboard from cereal boxes to make scale models of their bedrooms."

Bonnie Langan

Bonnie Langan
Northern Middle School
Accident, MD

In-Class Example

For Example 4

A set of plans show a building that is $4\frac{1}{4}$ inches long. The key for the plans shows that $\frac{1}{2}$ inch on the plan represents 6 feet. How long will the actual building be? **27 feet**

3 PRACTICE/APPLY

Checking Your Understanding

Exercises 1–14 are designed to help you assess your students' understanding through reading, writing, speaking, and modeling. You should work through Exercises 1–5 with your students and then monitor their work on Exercises 6–14.

Assignment Guide

Core: 15–45 odd, 46–53
Enriched: 16–42 even, 43–53

For **Extra Practice**, see p. 762.

The red A, B, and C flags, printed only in the Teacher's Wraparound Edition, indicate the level of difficulty of the exercises.

Practice Masters, p. 74

NAME ______________________ DATE __________

9-4 Practice
Using Proportions

Student Edition Pages 444–447

Write = or ≠ in each blank to make a true statement.

1. $\frac{4}{6}$ = $\frac{2}{3}$
2. $\frac{16}{4}$ ≠ $\frac{20}{8}$
3. $\frac{21}{28}$ = $\frac{3}{4}$
4. $\frac{4}{5}$ = $\frac{1.2}{1.5}$
5. $\frac{2.1}{4.9}$ ≠ $\frac{6}{1.4}$
6. $\frac{2.6}{4}$ ≠ $\frac{1.6}{0.25}$

Solve each proportion.

7. $\frac{1}{8} = \frac{2}{d}$ $d = 16$
8. $\frac{x}{6} = \frac{15}{18}$ $x = 5$
9. $\frac{0.4}{m} = \frac{2}{4.5}$ $m = 0.9$
10. $\frac{r}{10} = \frac{21}{7}$ $r = 30$
11. $\frac{4}{p} = \frac{8}{11}$ $p = 5.5$
12. $\frac{17}{50} = \frac{x}{25}$ $x = 8.5$
13. $\frac{8}{12} = \frac{b}{48}$ $b = 32$
14. $\frac{0.18}{0.09} = \frac{h}{0.06}$ $h = 0.12$
15. $\frac{0.25}{0.5} = \frac{m}{8}$ $m = 4$
16. $\frac{10}{2.4} = \frac{p}{2.64}$ $p = 11$
17. $\frac{85.8}{d} = \frac{70.2}{9}$ $d = 11$
18. $\frac{0.6}{1.1} = \frac{s}{8.47}$ $s = 4.62$
19. $\frac{9}{15} = \frac{3x}{10}$ $x = 2$
20. $\frac{2}{3} = \frac{x+4}{18}$ $x = 8$
21. $\frac{4.5}{y+5} = \frac{5}{10}$ $y = 4$

Write a proportion that could be used to solve for each variable. Then solve the problem.

22. 1 subscription for $21
28 subscriptions for x dollars
$\frac{1}{21} = \frac{28}{x}$; $x = 588$, $588
23. 20 ounces at $7
17 ounces at x dollars
$\frac{20}{7} = \frac{17}{x}$; $x = 5.95$, $5.95
24. 225 bushels for 3 acres
x bushels for 9.6 acres
$\frac{225}{3} = \frac{x}{9.6}$; $x = 720$, 720 bushels
25. 25 cm by 35 cm enlarged to 150 cm by x cm
$\frac{25}{150} = \frac{35}{x}$; $x = 210$, 210 cm
26. 450 km or 45 liters
1500 km on x liters
$\frac{450}{45} = \frac{1500}{x}$; $x = 150$, 150 liters
27. 3 shirts for $56.85
x shirts for $132.65
$\frac{3}{56.85} = \frac{x}{132.65}$; $x = 7$, 7 shirts

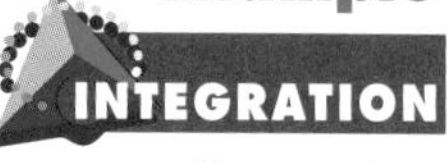

Blueprints for a house show a room that is $2\frac{5}{12}$ inches long. The key to the blueprints says that $\frac{1}{4}$ inch on the blueprints represents 18 inches in the house. How long is the actual room?

Let x represent the actual room length.

$$\begin{array}{lcl} \textit{blueprints} \rightarrow & \dfrac{\frac{1}{4}\text{ inch}}{2\frac{5}{12}\text{ inches}} = \dfrac{18\text{ inches}}{x\text{ inches}} & \leftarrow \textit{actual} \\ \textit{blueprint length} \rightarrow & & \leftarrow \textit{actual length} \end{array}$$

$$\frac{1}{4} \cdot x = 2\frac{5}{12} \cdot 18 \quad \textit{Find the cross products.}$$

$$\frac{1}{4}x = 43\frac{1}{2}$$

$$4 \cdot \frac{1}{4}x = 4 \cdot 43\frac{1}{2} \quad \textit{Multiply each side by 4.}$$

$$x = 174$$

The room is 174 inches or $14\frac{1}{2}$ feet long.

After setting up the proportion, you can estimate to help you decide if your solution is reasonable.

$$\frac{1}{4}x \approx 2 \cdot 20$$
$$\frac{1}{4}x \approx 40$$
$$x \approx 160$$

Checking Your Understanding

Communicating Mathematics

Read and study the lesson to answer these questions.

1–4. See Solutions Manual.

1. **Identify** the relationship between ratios and proportions.
2. **State** in your own words how to solve a proportion.
3. **Describe** three career areas that might use proportions.
4. **Demonstrate** how cross products can be used to tell whether two fractions could form a proportion.

5. **Assess Yourself** Explain how you use proportions in your life. **See students' work.**

Guided Practice

Replace each ● with = or ≠ to make a true statement.

6. $\frac{2}{5}$ ● $\frac{4}{10}$ =
7. $\frac{6.25}{5}$ ● $\frac{2.5}{2}$ =

Solve each proportion. 11. 2.1

8. $\frac{2}{y} = \frac{10}{20}$ 4
9. $\frac{5}{7} = \frac{n}{10.5}$ 7.5
10. $\frac{3}{a} = \frac{18}{24}$ 4
11. $\frac{7}{16} = \frac{x}{4.8}$

Write a proportion that could be used to solve for each variable. Then solve. 12–13. See Solutions Manual.

12. 3 pounds for $1.50
x pounds for $4.50
13. 2.7 liters at m dollars
3 liters at $7.00
14. **Cooking** A recipe calls for $4\frac{1}{2}$ cups of flour for 72 cookies. How many cups of flour would be needed for 48 cookies? **3 cups**

Exercises: Practicing and Applying the Concept

Independent Practice

Replace each ● with = or ≠ to make a true statement.

15. $\frac{16}{5}$ ● $\frac{4}{2}$ ≠
16. $\frac{15}{2}$ ● $\frac{18}{2.4}$ =
17. $\frac{1}{3}$ ● $\frac{8.6}{25.3}$ ≠
18. $\frac{2.1}{3.5}$ ● $\frac{3}{7}$ ≠
19. $\frac{2}{1}$ ● $\frac{15}{7.5}$ =
20. $\frac{3}{8}$ ● $\frac{2.4}{0.64}$ ≠

Group Activity Card 9-4

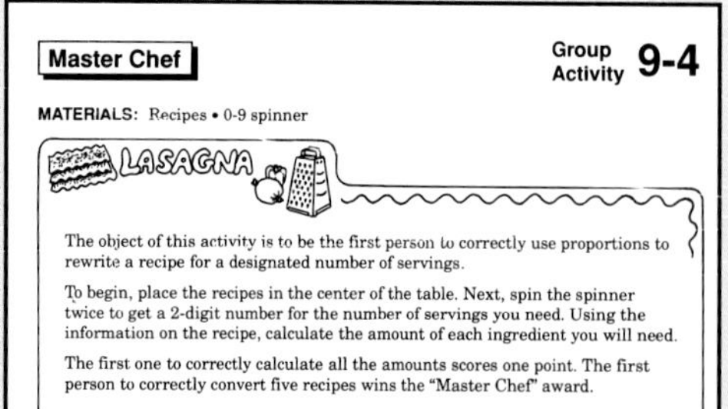

Master Chef

Group Activity **9-4**

MATERIALS: Recipes • 0-9 spinner

LASAGNA

The object of this activity is to be the first person to correctly use proportions to rewrite a recipe for a designated number of servings.

To begin, place the recipes in the center of the table. Next, spin the spinner twice to get a 2-digit number for the number of servings you need. Using the information on the recipe, calculate the amount of each ingredient you will need.

The first one to correctly calculate all the amounts scores one point. The first person to correctly convert five recipes wins the "Master Chef" award.

 Pre-Algebra

Reteaching

Using Cooperative Learning Have students take turns naming a fraction with a denominator of 2, 5, 10, 20, or 25. Improper fractions may be used. The other student uses a proportion to find the equivalent numerator if that fraction is expressed as a fraction with a denominator of 100.

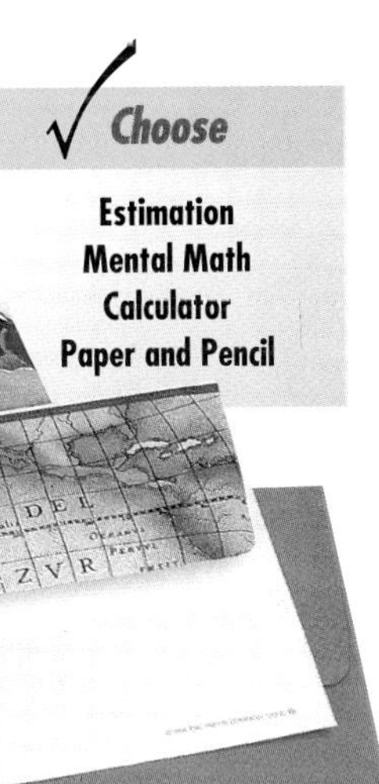

Solve each proportion. 24. 1.0125 32. 0.94

21. $\frac{5}{m} = \frac{25}{35}$ 7
22. $\frac{r}{3.5} = \frac{7.2}{9}$ 2.8
23. $\frac{9.6}{t} = \frac{1.6}{7}$ 42
24. $\frac{6.4}{0.8} = \frac{8.1}{y}$
25. $\frac{5.1}{1.7} = \frac{7.5}{t}$ 2.5
26. $\frac{2.4}{1.6} = \frac{s}{3.4}$ 5.1
27. $\frac{18}{12} = \frac{24}{g}$ 16
28. $\frac{7}{45} = \frac{x}{9}$ 1.4
29. $\frac{7}{16} = \frac{x}{4.8}$ 2.1
30. $\frac{3.5}{6.2} = \frac{7.35}{b}$ 13.02
31. $\frac{3.2}{4.8} = \frac{6.8}{y}$ 10.2
32. $\frac{t}{0.2} = \frac{9.4}{2}$

Write a proportion that could be used to solve for each variable. Then solve the proportion. 33–38. See margin.

33. 5 liters at $6.15
x liters at $8.00
34. 625 bushels for 5 acres
250 bushels for y acres
35. 8 boxes in 2 crates
z boxes in 5 crates
36. 100 candies in 3 bags
300 candies in d bags
37. 25 envelopes in 5 boxes
m envelopes in 25 boxes
38. 12 pens in one package
30 pens in p packages

Solve each proportion. 42. 73

C

39. $\frac{8}{28} = \frac{2x}{7}$ 1
40. $\frac{1.4}{4} = \frac{0.28}{m}$ 0.8
41. $\frac{5}{4} = \frac{x+5}{16}$ 15
42. $\frac{15}{m-3} = \frac{3}{14}$

Critical Thinking

43. Find two integers such that the ratio of the difference to the sum is 1:7 and the ratio of the sum to the product is 7:24. 6 and 8

Applications and Problem Solving

44. **Photography** In simple cameras, like the one shown, light from a subject passes through a lens and makes an image on film. The image and subject are always in proportion.

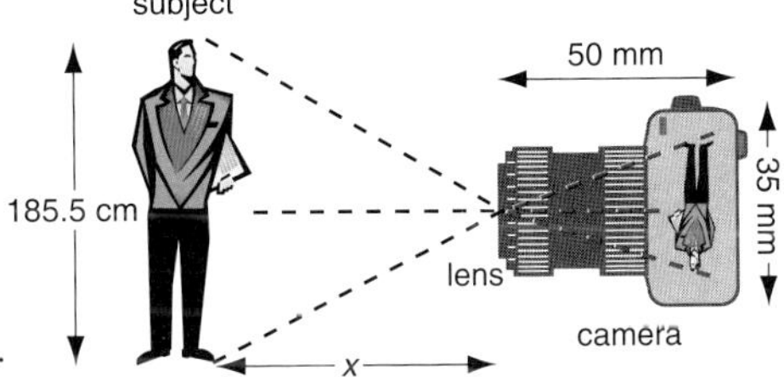

$$\frac{\text{image size}}{\text{subject size}} = \frac{\text{image distance from lens}}{\text{subject distance from lens}}$$

Find the distance represented by x in the diagram. 265 cm

45. **Models** The *Preussen* was the largest sailing ship ever built. Built in Germany in 1902, it could carry 7300 metric tons of cargo. The *Preussen* was 433 feet long and 54 feet wide. How wide should a model that is 26 inches long be? Use estimation to check your answer. 3.24 inches

Mixed Review

46. **Games** Lin needs to get a jack or higher when she cuts a deck of cards in order to win the deal. If an ace is the highest card possible, what is the probability that Lin will get the deal? (Lesson 9-3) $\frac{4}{13}$
47. Find the slope of the line that contains $Q(5, 6)$ and $S(2, -2)$. (Lesson 8-6) $\frac{8}{3}$
48. Solve $-8k - 21 = 75$. (Lesson 7-2) -12
49. Write 9,412,000 in scientific notation. (Lesson 6-9) 9.412×10^6
50. Write $0.\overline{7}$ as a fraction in simplest form. (Lesson 5-1) $\frac{7}{9}$
51. Represent 4^{-3} using positive exponents. (Lesson 4-9) $\frac{1}{4^3}$
52. **Patterns** Find the next number in the pattern 1, 2, 4, 5, 7, 8, 10, 11, . . . (Lesson 2-6) 13
53. Write an expression to represent *$200 more than the budget*. (Lesson 1-3) $b + 200$

4 ASSESS

Closing Activity

Act It Out Have small groups of students use a set of dominoes. Each student draws two dominoes and determines if they form a proportion. If a blank domino is drawn, the student determines the number needed to form a proportion.

Additional Answers

33. $\frac{5}{6.15} = \frac{x}{8.00}$; 6.5
34. $\frac{625}{5} = \frac{250}{y}$; 2
35. $\frac{8}{2} = \frac{z}{5}$; 20
36. $\frac{100}{3} = \frac{300}{d}$; 9
37. $\frac{25}{5} = \frac{m}{25}$; 125
38. $\frac{12}{1} = \frac{30}{p}$; 2.5

Enrichment Masters, p. 74

NAME ______________ DATE ______________

9-4 Enrichment
Direct Variation

Student Edition
Pages 444–447

If the relationship between two quantities is such that when one quantity increases, the other increases, or when one quantity decreases, the other decreases, the quantities are said to **vary directly**.

Example: If 2 loaves of bread cost $1.60, how much will 3 loaves of the same type of bread cost?

The number of loaves and the cost vary directly; that is, as the number of loaves increases, so does the cost.

$\frac{2 \text{ loaves}}{\$1.60} = \frac{3 \text{ loaves}}{x}$ For direct variations, place like quantities [illegible] from each other in a proportion.

$2x = 4.80$
$x = 2.40$

The cost for 3 loaves of bread is $2.40.

Write a proportion to represent each situation. Then solve.

1. Three gallons of gasoline cost $3.36. How much do 5 gallons cost?
$\frac{3}{3.36} = \frac{5}{x}$; $x = 5.6$; $5.60
2. At a rate of 50 mph, a car travels a distance of 600 miles. How far will the car travel at a rate of 40 mph if it is driven the same amount of time?
$\frac{50}{600} = \frac{40}{x}$; $x = 480$; 480 miles
3. If the rent for two weeks is $500, how much rent is paid for 5 weeks?
$\frac{2}{500} = \frac{5}{x}$; $x = 1250$; $1250
4. If 8 newspapers cost $3.30, how much will 6 newspapers cost?
$\frac{8}{3.20} = \frac{6}{x}$; $x = 2.40$; $2.40
5. If 9 fully loaded trucks carry a total of 140,400 pounds, how many pounds can 3 trucks carry?
$\frac{9}{140{,}400} = \frac{3}{x}$; $x = 46{,}800$; 46,800 pounds
6. Twelve floppy disks can hold 16.8 million bytes of data. How many bytes will 20 floppy disks hold?
$\frac{12}{16{,}800{,}000} = \frac{20}{x}$; $x = 28{,}000{,}000$; 28 million bytes
7. If 12 floppy disks hold 16.8 million bytes of data, how many floppy disks are needed to hold 10 million bytes of data?
$\frac{12}{16{,}800{,}000} = \frac{x}{10{,}000{,}000}$; $x \approx 7.14$; 8 disks are needed.
8. There are a total of 2100 Calories in 5 candy bars (all the same kind). How many total Calories are there in 3 dozen of these candy bars?
$\frac{5}{2100} = \frac{36}{x}$; $x = 15{,}120$; 15,120 Calories

Cooperative Learning

Inside-Outside Circle Have students form two circles. The inside student shows two ratios written on an index card to the outside student. The outside student decides whether the two ratios make a proportion. Then students rotate. For more information on this strategy, see *Cooperative Learning in the Mathematics Classroom*, p. 33.

Extension

Using Connections Have each student take his or her pulse for 30 seconds. Have each student use a proportion to tell how many times his or her heart beats in one minute, one hour, and one day.

9-4B LESSON NOTES

NCTM Standards: 1-5, 10

Objective

Use proportions to estimate a population.

Recommended Time

Demonstration and discussion: 15 minutes; Exercises: 30 minutes

Instructional Resources

For each student or group of students

- small bowl
- bag of dry beans

Math Lab and Modeling Math Masters

- p. 45 (worksheet)

For teacher demonstration

Overhead Manipulative Resources

1 FOCUS

Motivating the Lesson

Show students a photograph of a large group of people. Ask students how many people are in the photo.

2 TEACH

Teaching Tip Using the same photo from Motivating the Lesson above, ask students what procedures they would use to determine the number of people in the photo.

Explain that this activity is an example of statistical sampling methods used by naturalists in the field. Discuss why this is done.

3 PRACTICE/APPLY

Assignment Guide
Core: 1–4
Enriched: 1–4

MATHEMATICS LAB

HANDS-ON ACTIVITY

9-4B Capture-Recapture

A Extension of Lesson **9-4**

MATERIALS

- small bowls
- dry beans

Did you know that there is a way to estimate how many deer are in a forest? Often naturalists want to know such a population, but it would be impossible or impractical to make an actual count.

One method of estimating a population is the **capture-recapture** technique. In this activity, you will model this technique using dry beans as "deer" and a bowl as the "forest."

Your Turn

► CAPTURE Fill a small bowl with dry beans. Grab a small handful of beans. Mark each bean with an X on both sides. Count the "tagged" beans and record this number in a chart like the one below. This number is the original number captured. Return the "tagged" beans to the bowl and mix well.

Original Number Captured: ___		
Sample	**Recaptured**	**Tagged**
A		
B		
C		
J		
Total		

► RECAPTURE Grab another small handful of beans. Count the total number of beans you grabbed. This number is the number *recaptured.* Count the number of "tagged" beans. Record these numbers. This is sample A. Return all the beans to the bowl and mix.

► Repeat RECAPTURE nine more times for samples B through J. Find the total tagged and the total recaptured.

► Use the proportion shown below to estimate the number of beans in your bowl.

$$\frac{\textit{original number captured}}{\textit{number in bowl}} = \frac{\textit{total tagged in samples}}{\textit{total recaptured}}$$

1. Why is it a good idea to base your prediction on several samples instead of just one sample? **1–4. See Solutions Manual.**
2. What would happen to your estimate if some of your tags fell or wore off?
3. Count the number of beans in your bowl. How does your estimate compare to the actual number?
4. **Research** Investigate how researchers tag deer and estimate the total number of deer in a forest. Explain how the activity with the bowl and beans simulates their system.

4 ASSESS

Observing students working in cooperative groups is an excellent method of assessment.

9-5 Using the Percent Proportion

Setting Goals: *In this lesson, you'll use the percent proportion to write fractions as percents and to solve problems involving percents.*

Modeling a Real-World Application: Physical Therapy

If you were looking for a bodybuilder, you would probably head for the local gym. But did you think of looking at the medical center? You may not find Arnold Schwarzenegger there, but you will find physical therapists. Physical therapists build bodies by helping people to use exercise to regain their strength after an injury or surgery.

HELP WANTED

If you like to help people, physical therapy may be for you! For more information, contact:
American Physical Therapy Association
1111 N. Fairfax St.
Alexandria, VA 22314

Peggy Harper is a physical therapist for the Findlay City Schools in Findlay, Ohio. She is working with Chris McDaniel, a football player who is recovering from a knee injury. The muscles around Chris's knee are weak from lack of use. Ms. Harper uses a dynamometer to compare the strength of the recovering leg to the strength of the uninjured leg.

Ms. Harper determined that Chris can lift 70 foot-pounds with his weak leg and 125 foot-pounds with his strong leg. Using these numbers, she can find the **percent** of strength of the injured leg.

Learning the Concept

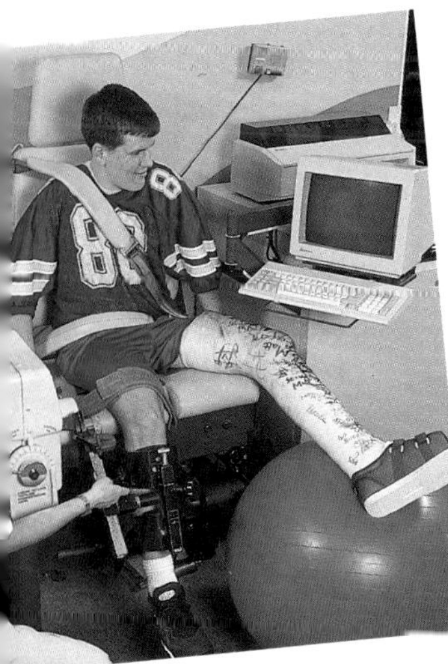

A percent is a ratio that compares a number to 100. Percent also means *hundredths*, or *per hundred*. The symbol for percent is %. To find the percent of strength of the weak leg, set up a proportion.

$$\frac{\text{foot-pounds lifted with injured leg}}{\text{foot-pounds lifted with uninjured leg}} = \frac{\text{strength of injured leg}}{100}$$

$$\frac{70}{125} = \frac{x}{100}$$

$70 \cdot 100 = 125x$ *Write the cross products.*

$\frac{70 \cdot 100}{125} = \frac{125x}{125}$ *Divide each side by 125.*

$56 = x$ *70 [×] 100 [÷] 125 [=] 56*

So, the injured leg has 56% of the strength of the strong leg.

In the proportion that Ms. Harper used, 70 is the **percentage** (P), 125 is the **base** (B), and the ratio $\frac{70}{125}$ is called the **rate** (r).

$$\frac{70}{125} = \frac{56}{100} \rightarrow \frac{\text{Percentage}}{\text{Base}} = \text{Rate}$$

Alternative Learning Styles

Auditory Have students make index study cards of the three types of percent problems as shown in the table on page 450 below Example 1. As you verbally read an exercise similar to Exercises 23–34 on page 452 in the text, have students hold up the correct card to show how to solve the problem.

HELP WANTED

Physical therapists often specialize in an area such as sports medicine, geriatrics, or pediatrics.

9-5 LESSON NOTES

NCTM Standards: 1-5, 9

Instructional Resources

- Study Guide Master 9-5
- Practice Master 9-5
- Enrichment Master 9-5
- Group Activity Card 9-5
- Assessment and Evaluation Masters, pp. 238, 239
- Activity Masters, p. 9
- Multicultural Activity Masters, p. 18
- Real-World Applications, 19

Transparency 9-5A contains the 5-Minute Check for this lesson; **Transparency 9-5B** contains a teaching aid for this lesson.

Recommended Pacing	
Standard Pacing	Day 7 of 14
Honors Pacing	Day 6 of 13
Block Scheduling*	Day 4 of 7 (along with Lesson 9-6)

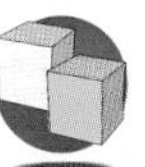

*For more information on pacing and possible lesson plans, refer to the ***Block Scheduling Booklet***.

1 FOCUS

5-Minute Check
(over Lesson 9-4)

Solve each proportion.

1. $\frac{b}{18} = \frac{14}{21}$ 12
2. $\frac{1.5}{3} = \frac{4.9}{n}$ 9.8
3. $\frac{7}{3} = \frac{h}{102}$ 238
4. $\frac{15}{96} = \frac{9}{c}$ 57
5. $\frac{4.5}{m} = \frac{4}{5}$ 5.625

Motivating the Lesson

Situational Problem Pose this situation to students. *You see two jackets on sale. Both originally cost $49.99. One is marked 25% off. The other is marked $\frac{1}{5}$ off.* Ask students which coat is the better buy. (This problem will be solved in In-Class Example 2 on page 450.)

2 TEACH

In-Class Examples

For Example 1
Express each fraction as a percent.

a. $\frac{7}{8}$ **$87\frac{1}{2}\%$**

b. $\frac{92}{80}$ **115%**

For Example 2
You see two jackets on sale. Both originally cost \$49.99. One is marked 25% off. The other is marked $\frac{1}{5}$ off. If you buy the less expensive jacket, how much will it cost? **25% off cost; \$37.49**

If r represents the number per hundred, this equation can be rewritten as $\frac{P}{B} = \frac{r}{100}$. This proportion is called the **percent proportion**. You can use the percent proportion to express fractions as percents and to solve percent problems.

Example  **Express each fraction as a percent.**

a. $\frac{\mathbf{9}}{\mathbf{16}}$

$$\frac{P}{B} = \frac{r}{100} \rightarrow \frac{9}{16} = \frac{r}{100} \quad \textit{Replace P with 9 and B with 16.}$$
$$9 \cdot 100 = 16 \cdot r \quad \textit{Find the cross products.}$$
$$\frac{900}{16} = \frac{16r}{16} \quad \textit{Divide each side by 16.}$$
$$56\frac{1}{4} = r$$

$\frac{9}{16}$ is equivalent to $56\frac{1}{4}\%$.

b. $\frac{\mathbf{15}}{\mathbf{4}}$

$$\frac{P}{B} = \frac{r}{100} \rightarrow \frac{15}{4} = \frac{r}{100} \quad \textit{Replace P with 15 and B with 4.}$$
$$15 \cdot 100 = 4 \cdot r \quad \textit{Find the cross products.}$$
$$\frac{1500}{4} = \frac{4r}{4} \quad \textit{Divide each side by 4.}$$
$$375 = r$$

$\frac{15}{4}$ is equivalent to 375%.

TECHNO TIP

You can find rates on a calculator by dividing the percentage by the base and moving the decimal point two places to the right.

There are three basic types of percent problems. Using the proportion $\frac{1}{2} = \frac{50}{100}$, you can see that the types are related as shown below.

$\frac{\bullet}{2} = \frac{50}{100}$	What number is 50% of 2? *Percentage*	*Find the percentage.*
$\frac{1}{2} = \frac{\bullet}{100}$	1 is what percent of 2? *Rate*	*Find the rate.*
$\frac{1}{\bullet} = \frac{50}{100}$	1 is 50% of what number? *Base*	*Find the base.*

Example  **Sixty-eight is 20% of what number?**

$$\frac{P}{B} = \frac{r}{100} \rightarrow \frac{68}{B} = \frac{20}{100} \quad \textit{Replace P with 68 and r with 20.}$$
$$20 \cdot B = 68 \cdot 100 \quad \textit{Find the cross products.}$$
$$\frac{20 \cdot B}{20} = \frac{68 \cdot 100}{20} \quad \textit{Divide each side by 20.}$$
$$B = 340$$

68 is 20% of 340.

You can also use the percent proportion to find the percentage when the base and the rate are given.

GLENCOE Technology

Interactive Mathematics Tools Software

In this interactive computer lesson, students explore proportions and percents. A **Computer Journal** gives students the opportunity to write about what they have learned.

For Windows & Macintosh

Example 3

APPLICATION

Physical Therapy

Refer to the application at the beginning of the lesson. Physical therapists usually set a goal of bringing an injured leg up to at least 90% of the strength of the strong leg. How many foot-pounds should Chris expect his injured leg to be able to lift when the physical therapy program is complete?

Use the percent proportion to solve for the number of foot-pounds.

$\frac{P}{B} = \frac{r}{100} \rightarrow \frac{P}{125} = \frac{90}{100}$ *Replace B with 125 and r with 90.*

$P \cdot 100 = 125 \cdot 90$ *Find the cross products.*

$\frac{P \cdot 100}{100} = \frac{125 \cdot 90}{100}$ *Divide each side by 100.*

$P = 112.5$ *90 [×] 125 [÷] 100 [=] 112.5*

Chris should expect to be able to lift at least 112.5 foot-pounds with his injured leg when the program is complete.

Checking Your Understanding

Communicating Mathematics

Read and study the lesson to answer these questions. 1–5. See margin.

1. **Explain** in one or two sentences why the value of r in $\frac{P}{B} = \frac{r}{100}$ represents a percent.
2. **Describe** how to write a fraction as a percent.
3. **Name** an advantage of using a percent instead of a fraction.
4. **Write** a proportion that you could use to find your percentage grade on a test that has 25 questions.
5. **Apply** the percent proportion to write a proportion for each.
   a. 52 is 80% of what number?
   b. What number is 80% of 52?
   c. 52 is what percent of 80?

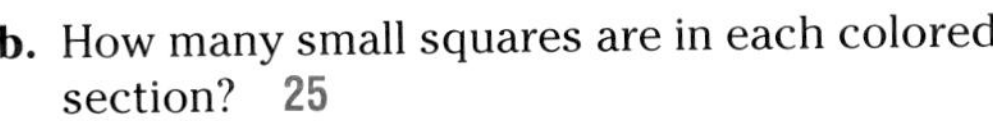

MATERIALS

grid paper

colored pencils

6. Shade a 5 × 5 section in one corner of a 10 by 10 grid. Then use a different color to shade each of the other three 5 × 5 sections.
   a. How many small squares are in the grid? 100
   b. How many small squares are in each colored section? 25
   c. What percent is represented by each colored section? 25%
   d. How many 5 × 5 sections are there? 4
   e. What fraction of the grid is represented by each colored section?

6e. $\frac{1}{4}$

Guided Practice

Express each fraction as a percent.

7. $\frac{44}{100}$ 44%
8. $\frac{1}{20}$ 5%
9. $\frac{8}{250}$ 3.2%
10. $\frac{3}{2}$ 150%

Use the percent proportion to solve each problem.

11. Find 15% of 60. 9
12. Fifty-two is 40% of what number? 130
13. 37 is what percent of 296? 12.5%

Reteaching

Using Connections Choose a topic, and survey the class to find out the fraction represented in each category. A topic might be the number of students owning different types of pets. Use the percent proportion to find the percent equal to the fraction. Round answers to the nearest whole percent.

Additional Answer

5a. $\frac{52}{B} = \frac{80}{100}$

5b. $\frac{p}{52} = \frac{80}{100}$

5c. $\frac{52}{80} = \frac{r}{100}$

In-Class Example

For Example 3

A local grocery store found that about 8% of sales come from items displayed near the cash register. If the sales from those items displayed near the cash register for a week were $18,745, approximately what were the total sales for that week? **$234,312.50**

3 PRACTICE/APPLY

Checking Your Understanding

Exercises 1–14 are designed to help you assess your students' understanding through reading, writing, speaking, and modeling. You should work through Exercises 1–6 with your students and then monitor their work on Exercises 7–14.

Additional Answers

1. $\frac{r}{100}$ is the fractional representation of a percent, r.
2. Set up a proportion and solve.
3. Sample answer: percents are easier to compare than fractions because they always have the same base.
4. $\frac{p}{25} = \frac{r}{100}$; p is the number correct.

Study Guide Masters, p. 75

NAME ______ DATE ______

9-5 Study Guide — Student Edition Pages 449–453

Using the Percent Proportion

The proportion shown at the right is called the **percent proportion**. It can be used to solve problems involving percent.

$\frac{\text{Percentage}}{\text{Base}} = \text{Rate or } \frac{P}{B} = \frac{r}{100}$

The **percentage (P)** is a number that is compared to another number called the **base (B)**. The **rate** is a percent. Always compare r to 100.

Example: Of the 800 tomatoes in a crop, 60% will be used to make ketchup. How many tomatoes will be made into ketchup?

Use the proportion $\frac{P}{B} = \frac{r}{100}$.

$\frac{P}{800} = \frac{60}{100}$

$P \cdot 100 = 800 \cdot 60$

$100P = 48{,}000$

$P = 480$

There will be 480 tomatoes made into ketchup.

Use the percent proportion to solve each problem.

1. Find 70% of 90. 63
2. Find 15% of 400. 60
3. What number is 75% of 600? 450
4. What number is 50% of 96? 48
5. 20% of 140 is what number? 28
6. 45% of 32 is what number? 14.4
7. Find 60% of 60. 36
8. Find 24% of 10.5. 2.52
9. 100% of 8.73 is what number? 8.73
10. What number is 98% of 230? 225.4
11. Joan's income is $190 per week. She saves 20% of her weekly salary. How much does she save each week? $38
12. Ninety percent of the seats of a flight are filled. There are 240 seats. How many seats are filled? 216 seats

Assignment Guide

Core: 15–39 odd, 40–46
Enriched: 16–34 even, 35–46

For **Extra Practice**, see p. 762.

The red A, B, and C flags, printed only in the Teacher's Wraparound Edition, indicate the level of difficulty of the exercises.

14. Statistics In 1992, the National Park Service issued a report saying that visitors spent 18,735 million hours, or about 27% of the total time spent in national parks, on boating. How many millions of hours did visitors spend in national parks in 1992? **about 69,389 million hours**

Exercises: Practicing and Applying the Concept

Independent Practice

Express each fraction as a percent.

A

15. $\frac{1}{4}$ **25%** **16.** $\frac{7}{8}$ **87.5%** **17.** $\frac{3}{5}$ **60%** **18.** $\frac{1}{16}$ **6.25%**

19. $\frac{9}{4}$ **225%** **20.** $\frac{24}{25}$ **96%** **21.** $\frac{45.3}{60}$ **75.5%** **22.** $\frac{24}{3}$ **800%**

Use the percent proportion to solve each problem.

B

23. What is 92% of 116? **106.72**
24. 16 is 40% of what number? **40**
25. 36 is what percent of 80? **45%**
26. 21 is 35% of what number? **60**
27. What is 70% of 50? **35**
28. Find 60% of 48. **28.8**
29. What is 200% of 8? **16**
30. Find 78.5% of 90. **70.65**

C

31. 36 is 45% of what number? **80**
32. 28 is 20% of what number? **140**
33. 126 is 10.5% of what number?
34. Find 0.1% of 450. **0.45**

33. 1200

Critical Thinking

35. Replace the _?_ so that the statement "x is x% of _?_ " is true for all values of x. **100**

36. If $\frac{1}{5}$ is 20% of a number, what percent is $\frac{3}{5}$ of the number? **60%**

Applications and Problem Solving

37. Health In 1994, the U.S. Food and Drug Administration began requiring food manufacturers to label their products with a nutritional label like the one from a bag of pretzels shown at the right.

a. The label states that the bag contains 1.5 grams of fat, which is 3% of the daily value recommended for a 2000-Calorie diet. How many grams of fat are recommended for a 2000-Calorie diet? **50 g**

b. The 760 milligrams of sodium (salt) in the pretzels is 32% of the recommended daily value. What is the recommended daily value of sodium? **about 2375 mg or 2.375 g**

Nutrition Facts
Serving Size 1 package (46.8g)
Servings per container 1

Amount per serving	
Calories 190 Calories from Fat 15	
	% Daily Value*
Total Fat 1.5g	3%
Saturated Fat 0g	0%
Cholesterol 0mg	0%
Sodium 760mg	32%
Total Carbohydrate 37g	12%
Dietary Fiber less than 1g	2%
Sugars 2g	
Protein 5g	
Vitamin A 0% • Vitamin C 0%	
Calcium 0% • Iron 3%	

*Percent Daily Values are based on a 2,000 calorie diet. Your daily values may be higher or lower depending on your calorie needs.
Calories per gram:
Fat 9 • Carbohydrates 4 • Protein 4

Practice Masters, p. 75

NAME ______ DATE ______

9-5 Practice
Using the Percent Proportion
Student Edition Pages 449–453

Express each fraction as a percent.

1. $\frac{7}{25}$ 28%
2. $\frac{97}{100}$ 97%
3. $\frac{13}{50}$ 26%
4. $\frac{9}{4}$ 225%
5. $\frac{7}{8}$ 87.5%
6. $\frac{8}{5}$ 160%
7. $\frac{17}{20}$ 85%
8. $\frac{1}{50}$ 2%

Use the percent proportion to solve each problem.

9. What is 17% of 65? 11.05
10. Find 12.5% of 96. 12
11. What is 6% of 95? 5.7
12. Find 95% of 170. 161.5
13. Find 62.5% of 500. 312.5
14. What is 8% of 17.5? 1.4
15. 42 is what percent of 48? 87.5%
16. 9 is 15% of what number? 60
17. 13 is 5% of what number? 260
18. 24 is what percent of 32? 75%
19. 9% of 2000 is what number? 180
20. 80 is what percent of 300? $26\frac{2}{3}$%
21. 36 is what percent of 24? 150%
22. 76 is what percent of 40? 190%
23. What is 37.5% of 300? 112.5
24. 42 is 63% of what number? $66\frac{2}{3}$
25. 18 is 60% of what number? 30
26. 60 is 75% of what number? 80
27. Find 87.5% of 100. 87.5
28. 39 is 40% of what number? 97.5
29. 96 is what percent of 100? 96%
30. 56 is 1% of what number? 5600
31. Find 6.5% of 250. 16.25
32. 6 is what percent of 5? 120%

Group Activity Card 9-5

Form Fits Group Activity **9-5**

The object of this activity is to determine the type of percent problem that must be solved. These are either percentage, rate, or base problems.

In groups of four or five, take turns making up and telling a problem to the others. The first one to correctly identify the problem as a percentage, rate, or base problem and to correctly solve it scores one point.

Play continues until everyone has three chances to be the problem maker. The person with the highest score wins the game.

EXAMPLES:

1. A nut mix contains 500 nuts. If 85 of them are pecans, what percent are pecans?
2. Twenty percent of a cereal is sugar. If the box weighs 19 ounces, how much sugar does it contain?
3. The state gasoline tax is 8.6%. You paid $1.74 in taxes when you bought gasoline. How much did the gasoline cost before the tax?

 Pre-Algebra

Extension

Using Data Have students study cereal boxes in a local grocery store or their own kitchen. They should observe the total Calories, Calories from fat, grams of fat, and percent of recommended daily amounts. Then have students make up their own problems based upon the information.

39. ME–31,813; NH–8238; WV–18,779; VT–7279; AL–35,071

38. **Personal Finance** Interest on savings accounts is often figured every quarter of a year. One quarter, Louam's savings account earned \$54.84 in interest. This is 2% of her savings.

a. Find Louam's savings. **\$2742**

b. The interest in Louam's account is compounded each quarter, so the interest for a quarter is based on the original balance plus the interest earned the previous quarter. Find the interest Louam will earn next quarter if no deposits or withdrawals are made. **\$55.94**

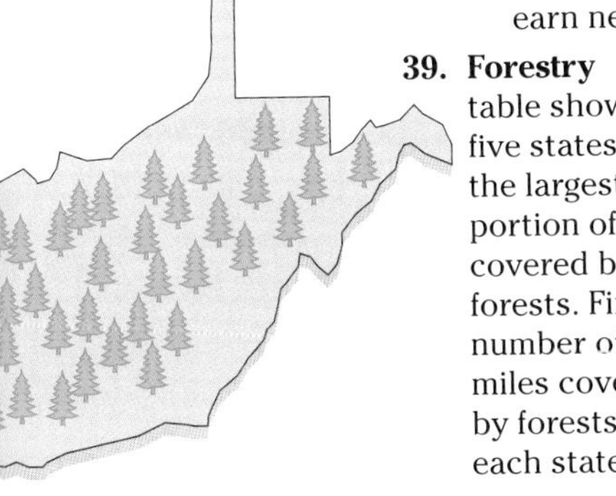

39. **Forestry** The table shows the five states with the largest portion of land covered by forests. Find the number of square miles covered by forests in each state.

State	Percent of land covered by forests	Area of state (square miles)
Maine	89.9%	35,387
New Hampshire	88.1%	9351
West Virginia	77.5%	24,231
Vermont	75.7%	9615
Alabama	66.9%	52,423

Source: The U.S. Forest Service

Mixed Review

40. **Jewelry** Jewelers use karats to describe the amount of gold in a piece of jewelry. A piece of jewelry that is 24-karat gold is all gold. How many grams of gold are in a 15-gram piece of jewelry that is labeled as 18-karat gold? (Lesson 9-4) **11.25 g**

41. $\frac{3}{4}$

41. Express the ratio *6 out of 8 free throws* in simplest form. (Lesson 9-1)

42. Solve the system $y = 8x - 1$ and $3x + y = 21$. (Lesson 8-8) **(2, 15)**

43. **Personal Finance** Yolanda plans to spend no more than \$15 on birthday presents for her brother. She bought a tape for \$7.21 and a keychain for \$3.57. What is the most money Yolanda can spend on other things for her brother? (Lesson 7-1) **\$4.22**

44. Evaluate the expression $4b$ if $b = 3.2$. (Lesson 6-5) **12.8**

45. Find the GCF of 65 and 105. (Lesson 4-5) **5**

46. In which quadrant does the graph of (9, −5) lie? (Lesson 2-2) **IV**

WORKING ON THE Investigation

Refer to the Investigation on pages 368–369.

The ticket prices of the seats are divided into three categories.

Deluxe seats	\$33.50
Spectator seats	\$22.75
Stadium seats	\$15.20

The university has asked you to determine which seats should be designated deluxe, spectator, and stadium. To earn enough gross income, 32% of the seats need to be deluxe, and only 18% should be stadium seats.

They have asked you to prepare a report showing the location of the seats, the price of each seat, the number of seats in each category, and the gross income from a sellout crowd. Prepare this report. Be sure to include any charts or drawings that support your reasoning.

Add the results of your work to your Investigation Folder.

4 ASSESS

Closing Activity

Speaking Have students name a percent and a base. Then have the class estimate what they think the rate will be. Then find the rate. Do this until the class has mastered the concept of estimating the rate.

Chapter 9, Quiz B (Lessons 9-4 and 9-5) is available in the *Assessment and Evaluation Masters*, p. 239.

Mid-Chapter Test (Lessons 9-1 through 9-5) is available in the *Assessment and Evaluation Masters*, p. 238.

WORKING ON THE Investigation

The Investigation on pages 368 and 369 is designed to be a long-term project that is completed over several days or weeks. Encourage students to keep their materials in their Investigation Folder as they work on the Investigation.

Enrichment Masters, p. 75

NAME ______________ DATE ________

9-5 Enrichment
Relative Frequency

Student Edition Pages 449–453

In an English language text, on the average, 1000 letters include the following numbers of particular letters.

E: 131	T: 105	O: 80
R: 68	I: 63	S: 61

The estimate of the likelihood of an event occurring is called **relative frequency**. It is equal to the number of successes divided by the number of trials. In the case of occurrence of letters, it is the number of times the letter occurs (a success) divided by the total number of letters (trials).

From the information above, write the relative frequency of each of the following letters as a decimal.

1. I 0.063 2. R 0.068 3. T 0.105 4. O 0.08

5. E 0.131 6. S 0.061

There are 913 letters in the article at the right. Use the relative frequencies found in Exercises 1–6 to determine the approximate number of times each of the following letters appears.

7. I 58 8. S 56

9. T 96 10. R 62

11. O 73 12. E 120

Find the actual relative frequency of each of the following letters in the article. Round answers to the nearest thousandth.

13. I 0.083 14. S 0.073

15. T 0.108 16. R 0.042

17. O 0.076 18. E 0.117

Music is Everybody's Language was the theme of the special week set aside this month to recognize Music in Our Schools.

Long interested in the attention given to learning about the importance of music in daily life, the sponsoring organization, Music Educators National Conference, consists of a membership of 64,000 music teachers. The influence of this organization has tremendously upgraded the acceptance of music by both students and parents for many years.

All this brings to mind my own experience in Evening High School with the subject of music and its appreciation. As a student I had unfortunately selected two courses with a gap of one period between them. This was a catastrophe in night school because the valuable time was yours to use or waste, and those students attending this kind of schooling were definitely not attracted by waste. Therefore, after a bit of self-condemnation, I decided to take this elective course titled *Music Appreciation*. With my scant knowledge of music and an attentive ear, my curiosity was aroused to the point that I thought it might at least be fun—a sort of vacation.

19. How are the answers for Exercises 13–18 likely to compare with the combined results from ten similar-sized articles? They will be approximately the same.

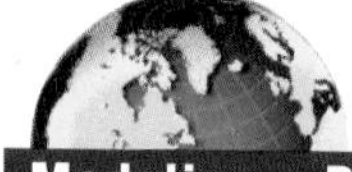

9-6 Integration: Statistics
Using Statistics to Predict

Setting Goals: *In this lesson, you'll use a sample to predict the actions of a larger group.*

Modeling a Real-World Application: Consumerism

How important is the brand name when you choose a product? The International Mass Retail Association surveyed consumers 8 to 17 years old on that subject. The results of the survey are shown in the graph at the right.

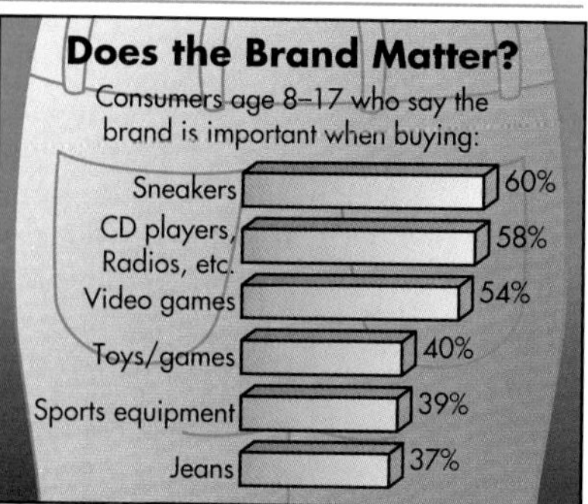

Source: International Mass Retail Association

Learning the Concept

You can use the preferences or behavior of a group to make predictions about a larger or smaller group that is similar. For example, the survey on brand buying was taken using only a portion of the consumers between 8 and 17 years old. But, the results were used to make statements about all consumers in that age group.

Example 1 APPLICATION Consumerism

Refer to the application at the beginning of the lesson. How many of the 423 9th-grade students at Kilbourne High School would you expect to say that they consider brand name when buying jeans?

Use the problem-solving plan to find the expected number.

Explore The graph shows that 37% of the consumers surveyed considered brand name when they purchased jeans. We need to know how many of the freshman class students consider brand name when buying jeans.

Plan Write and solve a proportion using the percent proportion. Estimate that since 37% is a little over $\frac{1}{3}$, the solution should be a little over 423 ÷ 3, or 140.

Solve

$$\frac{\textit{percent considering brand name}}{100} = \frac{\textit{number of Kilbourne freshmen who consider brand name}}{\textit{number of Kilbourne freshmen}}$$

$$\frac{37}{100} = \frac{x}{423}$$

$$37 \cdot 423 = 100 \cdot x \quad 37 \boxed{\times} 423 \boxed{\div} 100 \boxed{=} 156.51$$

$$x = 156.51$$

Expect 157 students to consider brand when buying jeans.

Examine The solution is close to the estimate, so it is reasonable.

9-6 LESSON NOTES

NCTM Standards: 1-5, 10

Instructional Resources

- Study Guide Master 9-6
- Practice Master 9-6
- Enrichment Master 9-6
- Group Activity Card 9-6

Transparency 9-6A contains the 5-Minute Check for this lesson; **Transparency 9-6B** contains a teaching aid for this lesson.

Recommended Pacing	
Standard Pacing	Day 8 of 14
Honors Pacing	Day 7 of 13
Block Scheduling*	Day 4 of 7 (along with Lesson 9-5)

*For more information on pacing and possible lesson plans, refer to the ***Block Scheduling Booklet***.

1 FOCUS

5-Minute Check
(over Lesson 9-5)

Express each fraction as a percent.

1. $\frac{2}{5}$ **40%**
2. $\frac{64}{200}$ **32%**
3. $\frac{633}{300}$ **211%**

Use a percent proportion to solve each problem.

4. What is 52% of 835? **434.2**
5. Twenty-four is 75% of what number? **32**

Motivating the Lesson

Questioning Read the school menu for today. Ask students how they think the person who orders food for the cafeteria knows how much to order.

Example 2

APPLICATION

Entertainment

interNET CONNECTION

For the latest Nielsen ratings, visit: **www.glencoe.com/sec/math/prealg/mathnet**

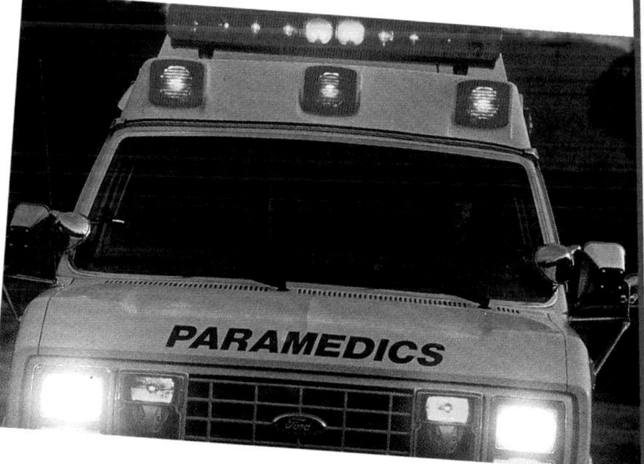

The NBC television program *ER* began the 1995 television season as the number one show in the Nielsen ratings. In the first week of the season, it attracted 41% of the Nielsen viewers. If about 91 million viewers watched television that night, about how many were watching *ER*?

Use a proportion.

$$\frac{x}{91} = \frac{41}{100}$$

$$x \cdot 100 = 91 \cdot 41 \quad \textit{Use a calculator.}$$

$$x = 37.31$$

About 37.31 million viewers watched *ER* that night.

When taking surveys and using the results to predict the actions of a larger population, be sure that your sample is random and is large enough to represent the population. If the sample is too small, it is likely that the items or people that it includes are not typical of the population. For example, suppose you were told that 4 out of 5 athletes chose a particular type of shoe for basketball. The claim would not be meaningful if only 5 athletes were surveyed.

Checking Your Understanding

Communicating Mathematics

Read and study the lesson to answer these questions.

1. **Explain** how to use the results of a survey to predict the characteristics of a population. See margin.
2. **Name** the most important features of a sample. large, random
3. **You Decide** Refer to the application at the beginning of the lesson. Tamika estimates that 152 of the 253 students in her class consider brand name when choosing sneakers. Gloria estimates only 121 of them do. Who is correct and why? Tamika; 152 is about 60% of 253.

Guided Practice

Use the survey on favorite drinks to answer the questions.

4. How large is the sample? 72
5. If this represents a group of 3600, what percent is the sample? 2%
6. What fraction chose fruit drink? What percent is this? $\frac{1}{6}$; $16\frac{2}{3}$%
7. Name a location that would not be a good place to conduct this survey.

7. Sample answer: High school cafeteria because teen's tastes are not representative of all ages.

Favorite Drink	
lemon lime	17
cola	10
root beer	25
fruit drink	12
ginger ale	8

2 TEACH

In-Class Examples

For Example 1

How many of the 516 10th-grade students at Kilbourne High School would you expect to say they consider brand name when buying sneakers? **310**

For Example 2

In a school survey, Alexi found that 32% of the students planned on going to community college after they graduate. If 294 students graduate this year, how many plan to go on to community college? **94 students**

3 PRACTICE/APPLY

Checking Your Understanding

Exercises 1–7 are designed to help you assess your students' understanding through reading, writing, speaking, and modeling. You should work through Exercises 1–3 with your students and then monitor their work on Exercises 4–7.

Study Guide Masters, p. 76

NAME ______________ DATE ______

9-6 Study Guide
Integration: Statistics
Using Statistics to Predict

Student Edition Pages 454–457

Mr. Niles takes a poll of 10 students in his class. Of the 10 students polled, 3 prefer to have the test today, and 7 prefer to have the test tomorrow. The 10 students polled are a **sample** of all the students in the class. The result of the poll can be used to predict the number of students who prefer to have the test tomorrow.

There are 30 students in Mr. Niles's class. How many of these students would you expect to prefer to take the test tomorrow?

Explore What is given? 10 students polled; 3 prefer to have the test today and 7 prefer tomorrow. What is asked? 30 students in the class; how many prefer the test tomorrow?

Plan Assume that the sample is representative of the entire class. Set up a proportion to show two equivalent ratios. Let t represent the total number of students that prefer to have the test tomorrow.

Solve $\frac{7}{10} = \frac{t}{30}$ Solve for t. $7 \times 30 = 10t$; $210 = 10t$; $21 = t$ The solution is 21.

Mr. Niles predicts that 21 students would prefer to have the test tomorrow.

Examine Replacing t in the original equation with 21, we see that $\frac{7}{10} = \frac{21}{30}$. Therefore, the answer is correct.

Solve. Use the poll shown below.

How many days per week should you have physical education?	
one day	13
two days	20
three days	27
four days	20

1. How many people are in the sample? 80
2. What part of the sample chose two days? $\frac{1}{4}$ or 25%
3. What percentage of the sample chose four days? 25%
4. What part of the sample chose three days? $\frac{27}{80}$ or $33\frac{3}{4}$%
5. Suppose there are 240 people in the school. How many do you predict would say three days? 81

Reteaching

Using Manipulatives From a well-mixed bag of marbles of several different colors, draw a representative number of marbles and record the number of each color drawn. Given the number of marbles in the bag, predict the number of marbles of each color. Compare to the actual numbers. Discuss any differences.

Additional Answer

1. Write a proportion with the percentage of the sample and the percentage of the population.

Error Analysis

Students may incorrectly set up the proportion when trying to solve problems. Remind students of the $\frac{P}{B} = \frac{r}{100}$ relationship.

Assignment Guide

Core: 9–17 odd, 19–23
Enriched: 8–14 even, 15–23
All: Self Test 1–11

For **Extra Practice**, see p. 762.

The red A, B, and C flags, printed only in the Teacher's Wraparound Edition, indicate the level of difficulty of the exercises.

Additional Answers

8. Sample answer: This is a valid sample because a large number of customers were surveyed. This is not a valid sample because all the people asked were at the ice cream store, so they all must prefer the type of ice cream served there.

15. Sample answer: No; many of the common two-letter words like *at, in, of,* and *on* do not contain the letter E.

Exercises: Practicing and Applying the Concept

Independent Practice

8. See margin.

B

10. about 200

The owner of a *Dairy Dream* store surveyed 100 people at her store about their favorite ice cream flavor. The results are in the table below.

8. Give some reasons why this is a valid sample and some reasons why it is not.
9. What percent of the people preferred peanut butter fudge? 35%
10. If there were 800 customers on a Saturday, about how many would prefer chocolate?

Favorite Flavors	
vanilla	20
chocolate	25
strawberry	10
peanut butter fudge	35
mint chocolate chip	10

The results of a 1992 survey by Independent Sector on the reasons that teens spend time as volunteers are shown in the table below.

Reason	Percent	Reason	Percent
to help others	47	for a friend	20
enjoy the work	38	religion	19
lots of free time	25	past experience	10
to learn	24	other	7
		don't know	2

11. If you asked 500 teens who volunteer why they do, how many would you expect to answer because they have free time? 125
12. Out of 250 teen volunteers, how many would you expect volunteer because they enjoy the work? 95

13. No; because the sample is too small and it is not random.
14. See students' work.

C

13. If you surveyed 25 volunteers at a homeless shelter on reasons for volunteering, would you expect the same results as this survey? Explain.
14. **Research** Make a frequency table showing the hair color for students in your class. Use the data to predict the number of students in your school with each hair color. Choose a time and place to observe at least 50 students and record their hair color. Compare that data to your predictions. What are your findings?

Critical Thinking

15. Thirteen percent of the letters used in English words are Es. If you were guessing the letters in a two-letter word of a puzzle on *Wheel of Fortune*, would you guess E? Explain. See margin.

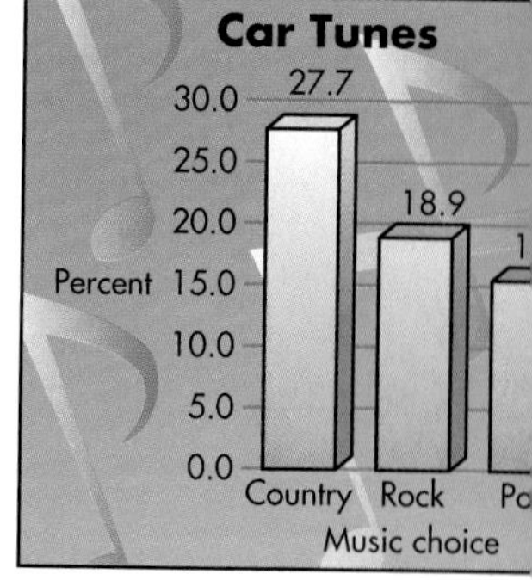

Source: 20/20 Research

Applications and Problem Solving

16a. 84

16. **Music** A *20/20 Research* survey asked what music people listen to in the car.
    a. How many of the 445 cars in a traffic jam do you predict have rock playing?
    b. If you owned a store that specialized in car stereos, what type of music would you have playing? Explain your choice. See students' work.
17. **Fashion** *Cotton Incorporated* surveyed 3600 people about clothing.
    a. 1872 people said they would rather be dressed too casually for an occasion than be dressed too fancy. What percent of the population do you think would feel the same? 52%
    b. What percent of the population would say they were slow to change with the fashions if 2160 of those surveyed said that? 60%

Practice Masters, p. 76

NAME ______________________ DATE ____________

9-6 Practice
Integration: Statistics
Using Statistics to Predict

Student Edition
Pages 454–457

A marketing company surveyed adults in a shopping mall about their favorite radio stations. The results are in the table below.

Favorite Radio Station	
KALM	28
KOOL	32
KLAS	15
KRZY	34
none	11

1. What is the size of the sample? 120
2. What fraction of the sample chose KLAS as a favorite station? $\frac{1}{8}$
3. What is the ratio of KOOL listeners to KALM listeners? 8:7
4. Suppose there are 15,880 people in the listening range of these stations. How many would you expect to listen to KRZY? about 4500 people
5. Suppose KLAS has 780 listeners. How many people would you predict listen to KOOL? about 1664 people

Use the poll at the right to answer each question.

Support of a Senate Bill	
Strongly in favor	3
In favor	18
Opposed	28
Strongly opposed	9
No opinion	5

6. What is the size of the sample? 63
7. What fraction of the sample is in favor or strongly in favor of the bill? $\frac{1}{3}$
8. What fraction of the sample is strongly opposed to the bill? $\frac{1}{7}$
9. Suppose 12,600 people live in the city where the poll was taken. How many would you predict will be opposed or strongly opposed to the bill? about 7400 people
10. How many of the 12,600 people would you predict will have no opinion about the bill? about 1000 people

Group Activity Card 9-6

On Your Feet — Group Activity 9-6

For this activity, collect data about shoes worn by students in your school. First of all, survey your class and record the number of students who wear tennis shoes and the types of tennis shoes students wear. Group these by categories such as Nike, Adidas, Reebok, other tennis shoe brands, and other kinds of shoes. Make a frequency table from this data.

Next, find out the approximate number of students in your school. From the table you have made and your knowledge of the number of students in your school, predict how many would be wearing each type.

Lastly, have every member in your group observe at least 30 other students in your school. Combine your data to see if your predictions are accurate.

 Pre-Algebra

Extension

Using Logical Reasoning Have students work in small groups to research *bias* in sampling. Pose questions such as "If I want to know what the most popular current television program is, can I survey only teenagers? Why or why not?" You may wish to have students take a poll using a biased sample and compare with data obtained from an unbiased poll.

18. **Housekeeping** A *TIME* magazine poll showed that 76% of women and 46% of men surveyed made their bed that morning.
   a. If you used the results of the *TIME* poll, about how many of the 346 females and 362 males at Centennial High school would you predict made their beds this morning? **263 females, 167 males**
   b. The *TIME* poll questioned adults. Do you think that a poll of teens would have the same results? Explain. **See margin.**

Mixed Review

19. 58%
20. yes; 12.5, 14, 15.5

19. **Food** The average American eats 10.28 pounds of chocolate in a year. If the total annual candy consumption of the average American is 17.86 pounds, what percent of the candy is chocolate? (Lesson 9-5)
20. **Patterns** State whether the sequence 2, 3.5, 5, 6.5, 8, 9.5, 11, . . . is arithmetic. Then write the next three terms. (Lesson 5-9)
21. If 15 more than the product of a number and -2 is greater than 12, which of the following could be the number? (Lesson 3-1) **D**
   A. 6 **B.** 4 **C.** 1.5 **D.** 0
22. **Geometry** Use the coordinate grid at the right to name the point for the ordered pair $(-4, 2)$. (Lesson 2-2) **B**

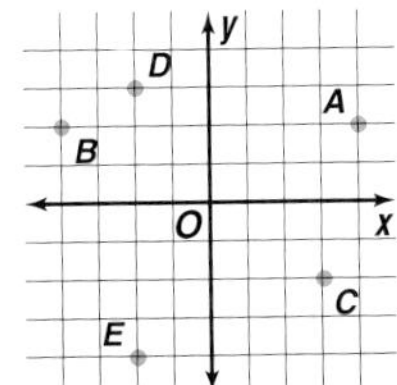

23. State whether $14 \geq 2b + 4$ is true or false if $b = 3$. (Lesson 1-9) **true**

Self Test

Express each ratio as a unit rate. (Lesson 9-1)

1. 300 feet in 5 minutes **60 ft/min**
2. 54 inches of rain in 4 months **13.5 in/mo**
3. How many ways can you make change for a $50 bill using only $5, $10, and $20 bills? (Lesson 9-2) **12**

There are 3 blue pencils, 5 green pencils, 2 black pencils, and 6 red pencils in a drawer. Suppose you grab one pencil at random. Find each probability. (Lesson 9-3)

4. $P(\text{green})$ $\frac{5}{16}$
5. $P(\text{blue or red})$ $\frac{9}{16}$

Solve each proportion. (Lesson 9-4)

6. $\frac{m}{7} = \frac{25}{35}$ **5**
7. $\frac{1.5}{2} = \frac{s}{2.4}$ **1.8**

Use the percent proportion to solve each problem. (Lesson 9-5)

8. Find 75% of 400. **300**
9. 250 is 500% of what number? **50**

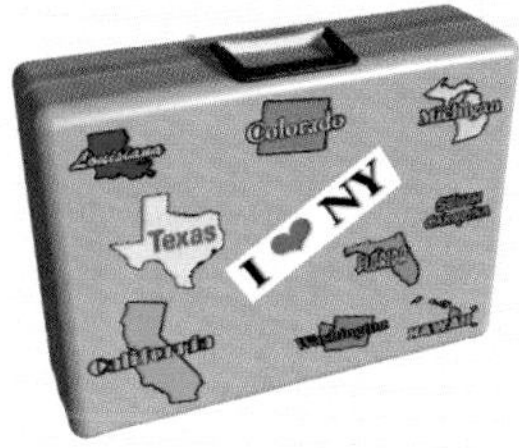

The Travel Industry Association polled 1500 adults on their choice of fall vacation destinations. (Lesson 9-6)

10. How many of those surveyed chose New York as a destination? **165**
11. If a travel agency books vacations for 300 families this fall, how many should they expect will go to Florida? **105**

Destination	Percent
Florida	35
California	30
Hawaii	20
Nevada	14
New York	11
Colorado	9
Arizona	8
Texas	6
South Carolina	6
Washington	6

Source: Travel Industry Assn.

Self Test

The Self Test provides students with a brief review of the concepts and skills in Lessons 9-1 through 9-6. Lesson numbers are given to the right of exercises or instruction lines so students can review concepts not yet mastered.

4 ASSESS

Closing Activity

Speaking Have students name a job and a situation where statistics would be used as a predictor.

Additional Answer

18b. Sample answer: No; many adults are married. So if the woman made the bed, the man would not have to, and vice versa.

Enrichment Masters, p. 76

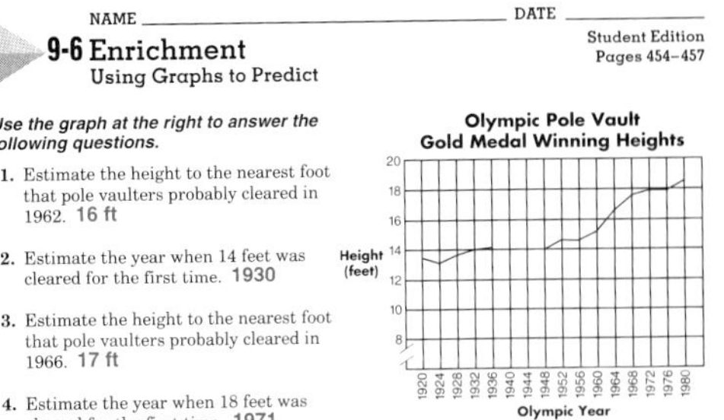

NAME ______ DATE ______

9-6 Enrichment Student Edition Pages 454–457
Using Graphs to Predict

Use the graph at the right to answer the following questions.

Olympic Pole Vault Gold Medal Winning Heights

1. Estimate the height to the nearest foot that pole vaulters probably cleared in 1962. **16 ft**
2. Estimate the year when 14 feet was cleared for the first time. **1930**
3. Estimate the height to the nearest foot that pole vaulters probably cleared in 1966. **17 ft**
4. Estimate the year when 18 feet was cleared for the first time. **1971**
5. If the Olympics had been held in 1940, predict what the winning height would have been (to the nearest foot). **14 ft**
6. Based on the trend from 1960 through 1980, would you predict the winning height in 1984 to have been over or under 19 feet? **over**

Note: Olympics were not held in 1940 and 1944.

Use the graph at the right to answer the following questions.

Personal Savings by Individuals in the U.S.

7. Based on the trend from 1977 through 1981, what level of savings would you predict for 1982? **About $150 billion**
8. How does your prediction for 1982 compare with the actual level of savings for 1982? **About $15 billion too high**
9. Based on the trend from 1981 through 1983, what level of savings would you predict for 1984? **About $100 billion**
10. The actual level of savings in 1984 was $156.1 billion. How does your prediction for 1984 compare with this actual level? **About $55 billion too low**

9-7 LESSON NOTES

NCTM Standards: 1-5

Instructional Resources

- Study Guide Master 9-7
- Practice Master 9-7
- Enrichment Master 9-7
- Group Activity Card 9-7
- Activity Masters, p. 40
- Math Lab and Modeling Math Masters, pp. 67-68

Transparency 9-7A contains the 5-Minute Check for this lesson; **Transparency 9-7B** contains a teaching aid for this lesson.

Recommended Pacing	
Standard Pacing	Day 9 of 14
Honors Pacing	Day 8 of 13
Block Scheduling*	Day 5 of 7 (along with Lesson 9-8)

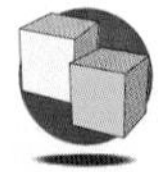

*For more information on pacing and possible lesson plans, refer to the ***Block Scheduling Booklet***.

1 FOCUS

5-Minute Check
(over Lesson 9-6)

A television survey showed these preferences.
situation comedy: 25 people
detective/mystery: 23 people
news/information: 15 people
science fiction: 12 people

1. How many people were involved in the survey? 75
2. What fraction preferred news or information? $\frac{1}{5}$
3. What fraction did not prefer situation comedy? $\frac{2}{3}$
4. What fraction preferred either a detective show or mystery? $\frac{23}{75}$

Motivating the Lesson

Situational Problem *Bill has 10 fish in his aquarium. Two are neon tetras. What part of his fish are neon tetras?* Have students discuss the correct answers and how there can be more than one correct answer.

9-7 Fractions, Decimals, and Percents

Setting Goals: *In this lesson, you'll express decimals and fractions as percents and vice versa.*

MATHEMATICS MINI LAB — Modeling with Manipulatives

MATERIALS
grid paper

You can compare quantities using fractions, percents, or decimals. Each of these can be rewritten in the other two forms. The model at the right represents one unit. Since 36 of the 100 cells are shaded, $\frac{36}{100}$ or 0.36 of the model is shaded. Using the definition of percent, we could also say that 36% of the model is shaded.

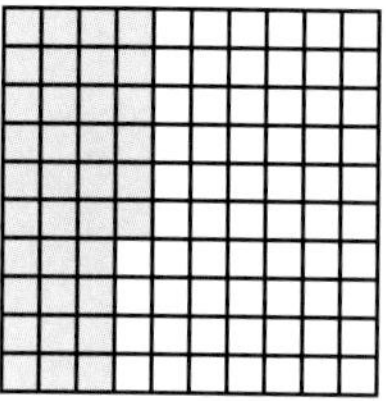

Your Turn Make models to represent $\frac{1}{2}$, 0.42, and 8%.

TALK ABOUT IT

a. How could you write $\frac{1}{2}$ as a decimal and a percent? **0.5, 50%**

b. Write 0.42 as a percent. What do you observe about the decimal point? **42%; it moved 2 places right.**

c. Write a procedure for writing a decimal as a percent and for writing a percent as a fraction. **See margin.**

Learning the Concept

You have written fractions as decimals and percents. Fractions, decimals, and percents are all different names that represent the same number.

$\frac{65}{100}$ ↔ 0.65 ↔ 65%

To express a decimal as a percent, write the decimal as a fraction with 1 as the denominator. Then write that fraction as an equivalent fraction with 100 as the denominator.

Example 1 **Express each decimal as a percent.**

a. 0.36 → $\frac{0.36}{1} = \frac{36}{100}$ or 36% (× 100)

b. 0.07 → $\frac{0.07}{1} = \frac{7}{100}$ or 7% (× 100)

c. 0.004 → $\frac{0.004}{1} = \frac{0.4}{100}$ or 0.4% (× 100)

d. 1.8 → $\frac{1.8}{1} = \frac{180}{100}$ or 180% (× 100)

THINK ABOUT IT

How could you convert a decimal to percent or a percent to a decimal mentally?

See margin.

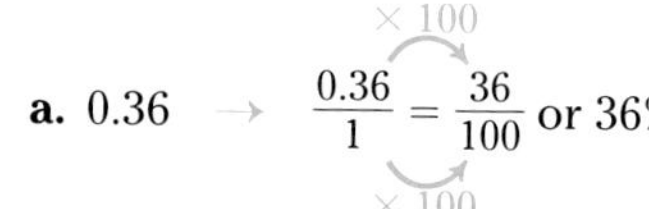

Compare each decimal in Example 1 with its equivalent percent. Notice that, in each case, the decimal point moved two places to the right.

GLENCOE Technology

Interactive Mathematics Tools Software

In this interactive computer lesson, students explore fractions and percents. A **Computer Journal** gives students the opportunity to write about what they have learned.

For Windows & Macintosh

Additional Answers Talk About It

c. Sample answer: move decimal point to the right two places; rewrite fraction as a fraction with a denominator of 100, then the numerator is the percent.

Think About It

Move the decimal point two places to the right or left.

To express a fraction as a percent, first write the fraction as a decimal by dividing numerator by denominator. Then write the decimal as a percent.

Example 2 **Express each fraction as a percent.**

a. $\frac{3}{5} = 0.60 \rightarrow$ 3 [÷] 5 [=] *0.6* $\rightarrow \frac{60}{100} = 60\%$

b. $\frac{7}{4} = 1.75 \rightarrow$ 7 [÷] 4 [=] *1.75* $\rightarrow \frac{175}{100} = 175\%$

c. $\frac{2}{500} = 0.004 \rightarrow$ 2 [÷] 500 [=] *0.004* $\rightarrow \frac{0.4}{100} = 0.4\%$

To express a percent as a fraction, write the percent in the form $\frac{r}{100}$ and simplify.

Example  **Write each percent as a fraction in simplest form.**

LOOK BACK

You can review dividing fractions in Lesson 6-4.

a. $15\% = \frac{15}{100} = \frac{3}{20}$

b. $87\frac{1}{2}\% = \frac{87\frac{1}{2}}{100}$ *The fraction bar indicates division.*

$= 87\frac{1}{2} \div 100$

$= \frac{\overset{7}{\cancel{175}}}{2} \times \frac{1}{\underset{4}{\cancel{100}}}$

$= \frac{7}{8}$

When you want to express a percent as a decimal, write the percent in the form $\frac{r}{100}$ and then write as a decimal.

Example  **Write each percent as a decimal.**

a. $17\% = \frac{17}{100} = 0.17$

b. $250\% = \frac{250}{100} = 2.50$

In Example 4, notice that the decimal point was moved two places to the left to get the equivalent decimal.

APPLICATION

Cosmetics

5 **In a 1995 *Self* magazine survey on favorite perfumes, 26% of those asked used one certain perfume all the time while one-twentieth said that they never use perfume. Which was the larger group?**

Write each number as a decimal to compare.

$26\% = \frac{26}{100} = 0.26$ | $\frac{1}{20} = 0.05$ *1 [÷] 20 [=] 0.05*

Since 0.26 is greater than 0.05, the group that said they use the same perfume each day was larger.

CULTURAL CONNECTIONS

Perfume had its beginning in ancient incense. Greeks and Romans learned about perfumes from the Egyptians. In the 1200s, crusaders took perfumes from Palestine to England and France.

2 TEACH

In-Class Examples

For Example 1
Express each decimal as a percent.

a. 0.83 **83%** b. 0.6 **60%**
c. 0.001 **0.1%** d. 1.2 **120%**

For Example 2
Express each fraction as a percent.

a. $\frac{1}{4}$ **25%** b. $\frac{9}{2}$ **450%**
c. $\frac{9}{200}$ **4.5%**

For Example 3
Write each percent as a fraction in simplest form.

a. 18% $\frac{9}{50}$ b. $66\frac{2}{3}\%$ $\frac{2}{3}$

For Example 4
Write each percent as a decimal.

a. 92% **0.92** b. 180% **1.80**

For Example 5
A fast-food restaurant found in a survey that $\frac{1}{8}$ of customers prefer fish sandwiches and 14% prefer rib sandwiches. Which was the larger group? **14%**

Study Guide Masters, p. 77

NAME ______ DATE ______

9-7 Study Guide
Fractions, Decimals, and Percents

Student Edition Pages 458–461

Fractions, decimals, and percents can all be used to represent the same number.

Example: Express 2.45 as a mixed number and as a percent.

mixed number: $2.45 \rightarrow 2\frac{\overset{9}{\cancel{45}}}{\underset{20}{\cancel{100}}} \rightarrow 2\frac{9}{20}$ | percent: $2.45 \rightarrow 2.45 \rightarrow 245\%$

Example: Express $\frac{1}{4}$ as a decimal and as a percent.

decimal: $\frac{1}{4} \rightarrow 4\overline{)1.00}$ = 0.25; $\frac{1}{4} = 0.25$ | percent: $\frac{1}{4} = \frac{r}{100}$, $100 = 4r$, $25 = r$, $\frac{1}{4} = 25\%$

Express each percent or fraction as a decimal.

1. 49% 0.49
2. 185% 1.85
3. 16.9% 0.169
4. $\frac{2}{5}$ 0.4
5. $\frac{21}{40}$ 0.525
6. $\frac{5}{8}$ 0.625
7. $1\frac{1}{2}$ 1.5
8. 4% 0.04

Express each decimal or fraction as a percent.

9. 5.62 562%
10. 0.327 32.7%
11. 0.007 0.7%
12. $\frac{25}{100}$ 25%
13. $\frac{5}{6}$ $83.\overline{3}\%$
14. $3\frac{2}{5}$ 340%
15. 0.6 60%
16. $2\frac{3}{10}$ 230%

Express each percent or decimal as a fraction or a mixed number.

17. 45% $\frac{9}{20}$
18. 150% $1\frac{1}{2}$
19. 0.3 $\frac{3}{10}$
20. 0.235 $\frac{47}{200}$
21. 4.5 $4\frac{1}{2}$
22. 0.55 $\frac{11}{20}$
23. 0.005 $\frac{1}{200}$
24. 56% $\frac{28}{50}$

CULTURAL CONNECTIONS

Perfumes are generally classified by dominant fragrances. Some common classifications are floral (jasmine, rose, gardenia), spicy (clove, cinnamon), woody (sandalwood and cedarwood), herbal (clover, sweet grass), and mossy (wood moss).

3 PRACTICE/APPLY

Checking Your Understanding

Exercises 1–17 are designed to help you assess your students' understanding through reading, writing, speaking, and modeling. You should work through Exercises 1–4 with your students and then monitor their work on Exercises 5–17.

Error Analysis

Some students may have difficulty with percents that contain decimals. Have them write the percent as 20.34% and then as $\frac{20.34}{100}$, pointing out that percent means *per one hundred*. Then have students move the decimal point to end of the decimal number and move the decimal point after 100 to the right the same number of places, adding zeros as needed.

Assignment Guide

Core: 19–53 odd, 54, 55, 57–67
Enriched: 18–52 even, 54–67

For **Extra Practice**, see p. 763.

The red A, B, and C flags, printed only in the Teacher's Wraparound Edition, indicate the level of difficulty of the exercises.

Practice Masters, p. 77

NAME ______________ DATE ______________

9-7 Practice
Fractions, Decimals, and Percents

Student Edition Pages 458–461

Express each decimal as a percent.

1. 0.5 **50%**	2. 2.72 **272%**	3. 0.65 **65%**
4. 0.08 **8%**	5. 15.7 **1570%**	6. 0.003 **0.3%**
7. 1.076 **107.6%**	8. 0.205 **20.5%**	9. 0.0125 **1.25%**

Express each fraction as a percent.

10. $\frac{3}{8}$ **37.5%**	11. $\frac{5}{100}$ **5%**	12. $\frac{7}{4}$ **175%**
13. $\frac{7}{10}$ **70%**	14. $\frac{13}{16}$ **81.25%**	15. $\frac{7}{8}$ **87.5%**
16. $\frac{1}{6}$ **$16\frac{2}{3}$%**	17. $\frac{11}{12}$ **$91\frac{2}{3}$%**	18. $\frac{5}{2}$ **250%**
19. $\frac{3}{25}$ **12%**	20. $\frac{3}{4}$ **75%**	21. $\frac{6}{5}$ **120%**
22. $\frac{4}{5}$ **80%**	23. $\frac{1}{10}$ **10%**	24. $\frac{5}{16}$ **31.25%**

Express each percent as a fraction.

25. 46% **$\frac{23}{50}$**	26. 9% **$\frac{9}{100}$**	27. 65% **$\frac{13}{20}$**
28. 12.5% **$\frac{1}{8}$**	29. 24.6% **$\frac{123}{500}$**	30. $33\frac{1}{3}$% **$\frac{1}{3}$**
31. 62.5% **$\frac{5}{8}$**	32. $8\frac{1}{8}$% **$\frac{13}{160}$**	33. 2.5% **$\frac{1}{40}$**

Express each percent as a decimal.

34. 6% **0.06**	35. 12% **0.12**	36. 14.6% **0.146**
37. 0.02% **0.0002**	38. 33.3% **0.333**	39. 0.75% **0.0075**

Checking Your Understanding

Communicating Mathematics

Read and study the lesson to answer these questions. **1, 3. See margin.**

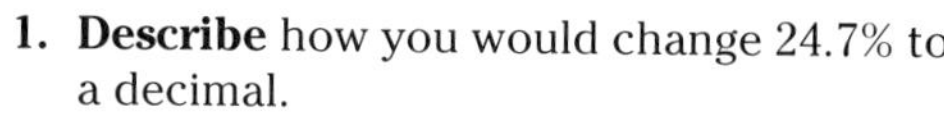
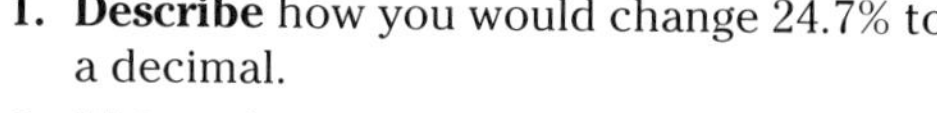

1. **Describe** how you would change 24.7% to a decimal.
2. **Write** a fraction, a decimal, and a percent to represent the model at the right. **$\frac{16}{25}$, 0.64, 64%**

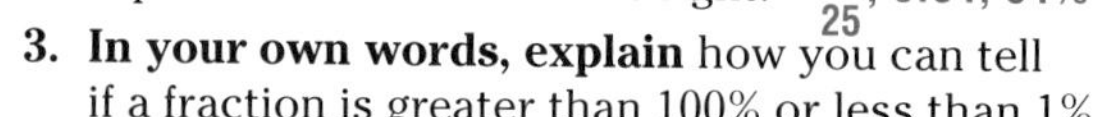

3. **In your own words, explain** how you can tell if a fraction is greater than 100% or less than 1%.

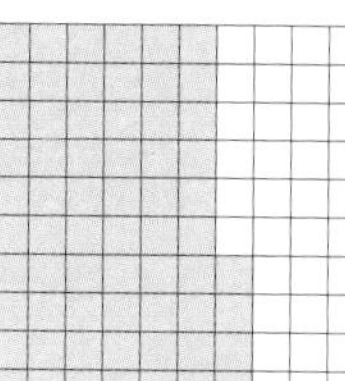

MODELING MATHEMATICS

MATERIALS
dry beans
paper bag

4. Put 25 beans in a bag. Have your partner grab a handful of beans. Express the portion of the beans that were removed as a fraction, a percent, and a decimal. Repeat. Which handful was a greater portion of the beans? **See students' work.**

Guided Practice

Express each decimal as a percent.

5. 0.37 **37%** 6. 0.475 **47.5%** 7. 1.03 **103%**

Express each fraction as a percent.

8. $\frac{27}{100}$ **27%** 9. $\frac{36}{500}$ **7.2%** 10. $\frac{6}{5}$ **120%**

Express each percent as a fraction.

11. 25% **$\frac{1}{4}$** 12. $33\frac{1}{3}$% **$\frac{1}{3}$** 13. 0.3% **$\frac{3}{1000}$**

Express each percent as a decimal.

14. 35% **0.35** 15. $24\frac{1}{2}$% **0.245** 16. 0.4% **0.004**

17. **Sports** In a survey, 48 of 120 people said they preferred hockey as a spectator sport.
 a. What fraction is this? **$\frac{2}{5}$**
 b. What percent is this? **40%**

Exercises: Practicing and Applying the Concept

Independent Practice

Express each decimal as a percent. **25. 0.04%**

18. 0.71 **71%** 19. 0.03 **3%** 20. 0.543 **54.3%** 21. 2.37 **237%**
22. 0.004 **0.4%** 23. 1.32 **132%** 24. 0.035 **3.5%** 25. 0.0004

Express each fraction as a percent.

26. $\frac{3}{16}$ **18.75%** 27. $\frac{7}{12}$ **$58\frac{1}{3}$%** 28. $\frac{5}{4}$ **125%** 29. $\frac{7}{3}$ **$233\frac{1}{3}$%**
30. $\frac{4}{300}$ **$1\frac{1}{3}$%** 31. $\frac{5}{9}$ **$55\frac{5}{9}$%** 32. $1\frac{1}{4}$ **125%** 33. $\frac{3}{40}$ **7.5%**

Choose
Estimation
Mental Math
Calculator
Paper and Pencil

Express each percent as a fraction.

34. 36% **$\frac{9}{25}$** 35. 125% **$\frac{5}{4}$** 36. 58% **$\frac{29}{50}$** 37. 34.5% **$\frac{69}{200}$**
38. 22% **$\frac{11}{50}$** 39. $37\frac{1}{2}$% **$\frac{3}{8}$** 40. $66\frac{2}{3}$% **$\frac{2}{3}$** 41. $16\frac{1}{3}$% **$\frac{49}{300}$**

Group Activity Card 9-7

Conversion Rummy Group Activity **9-7**

SETUP: Make a set of "Conversion Rummy" cards using the numbers given on the back of this card

Shuffle the cards and deal 6 to each player. Place the rest face down with the top card removed and placed face up to begin the discard pile.

Following the usual rules for rummy, play proceeds around the table with each person drawing a card from either pile. Players may keep the drawn card and discard another from their hands, or they may discard the new card. Players may never have more than 6 cards in their hands after a turn.

The object of the game is to get rid of all the cards in your hand by forming sets of three cards. All the cards in a set must name the same number. For example, cards with $\frac{1}{2}$, 0.5, and 50% would form a set. When you have a set, place it on the table in front of you. If someone else has a card that can be used in your set, he or she can place it there.

The first person to play all of his or her cards wins.

 Pre-Algebra

Reteaching

Using Cooperative Learning Have students work in groups of three. The first student names a fraction. The second student names the fraction as a decimal and the third student names it as a percent. The students should rotate roles.

Express each percent as a decimal. 45. 0.398 49. 2.354

42. 35% 0.35 **43.** 75% 0.75 **44.** 93% 0.93 **45.** 39.8%
46. 42.7% 0.427 **47.** 0.4% 0.004 **48.** 127% 1.27 **49.** 235.4%

Choose the greatest number in each set. 51. 22%

50. $\left\{\frac{3}{5}, 0.75, 80\%, 6 \textit{ to } 8\right\}$ 80% **51.** $\left\{22\%, 0.022, \frac{1}{11}, 2 \textit{ out of } 13\right\}$

Write each list of numbers in order from least to greatest.

52. $\frac{1}{2}$, 31%, 0.05 0.05, 31%, $\frac{1}{2}$ **53.** 16%, $\frac{1}{4}$, 0.067 0.067, 16%, $\frac{1}{4}$

Critical Thinking

54. A grocery store carries three different-sized bottles of contact lens solution. Size A is 50% more expensive than size C, but it contains 20% more solution than size B. Size B contains 50% more solution than size C, but it costs 25% more than size A. Which bottle is the most economical choice? A

Applications and Problem Solving

55. Retail Sales A pair of basketball shoes is listed at $33\frac{1}{3}\%$ off. What fraction off is this? $\frac{1}{3}$

56. Health Pediatricians and obstetricians care for newborn babies and their mothers.

a. Pediatricians estimate that 3 babies out of every 1000 are likely to contract a cold their first month. What percent is this? 0.3%

b. Obstetricians have observed that only 4% of babies are born on their due date. What fraction of babies are born on their due dates? $\frac{1}{25}$

57. Business In an interview with David Letterman, an executive of a marshmallow company said "Marshmallows are 80% air. That's how smart we are, we sell air!" What fraction of a marshmallow is air? $\frac{4}{5}$

Mixed Review

58. Cosmetics In the *Self* survey, 34% of those asked prefer floral-scented perfume. Of 50 customers at a department store cosmetics counter, how many would you expect to choose a floral perfume? (Lesson 9-6) 17

59. Find the *x*- and *y*-intercept of the graph of $y = 8x - 12$. (Lesson 8-7) 1.5, −12

60. Solve $\frac{b}{-2} - 12 \leq 11$. (Lesson 7-6) $b \geq -46$

61. Find the product of $\frac{4}{9}$ and $\frac{5}{12}$. (Lesson 6-3) $\frac{5}{27}$

62. Solve $x + 14.7 < 51.2$. (Lesson 5-7) $x < 36.5$

63. Find the product $n^4 \cdot n^{10}$. (Lesson 4-8) n^{14}

64. Science In the metric system, the prefix *giga* means 10^9. Write 10^9 as a product of the same factor. (Lesson 4-2) See margin.

65. Choose the value of *h* if $\frac{h}{0.7} = -2.8$. (Lesson 3-1)

A. −4 **B.** −2.8 **C.** −1.96 **D.** 0.7 C

66. Solve $r = -9889 \div -319$. (Lesson 2-8) 31

67. Solve $4 + x = 12$ using an inverse operation. (Lesson 1-8) 8

4 ASSESS

Closing Activity

Modeling Put 25 marbles in a bag. Have a student take a handful of marbles and count how many were taken. Then have students express the ratio of marbles selected to total marbles as a fraction, a decimal, and a percent. Repeat by having other students take marbles.

Additional Answers

1. Sample answer: Move the decimal point two places to the left.
3. Sample answer: A fraction is greater than 100% if it is greater than 1 and a fraction is less than 1% if it is less than $\frac{1}{100}$.
64. $10 \cdot 10 \cdot 10 \cdot 10 \cdot 10 \cdot 10 \cdot 10 \cdot 10 \cdot 10$

Extension

Using Comparisons Have students compare and contrast fractions, decimals, and percents. Ask them to determine when each would be more appropriate to use in describing different everyday activities.

Enrichment Masters, p. 77

NAME ______ DATE ______

9-7 Enrichment Student Edition Pages 458–461

Finding the Percent One Number Is of Another

Solve. Use the circle graph.

Cost of Eric's Camera Equipment

Camera $244.40; Wide-Angle Lens $140.40; Flash $57.20; Tripod $46.80; Camera Case $31.20

1. How much money did Eric spend on all the camera equipment? $520
2. What percent of the total cost of the equipment did Eric spend on the camera? 47%
3. What percent of the total cost was spent on the camera case? 6%
4. What percent of the total cost was spent on the flash? 11%
5. What percent of the total cost was spent on the wide-angle lens? 27%
6. What percent of the total cost was spent on the tripod? 9%
7. What percent of the total cost was spent on the camera case, tripod, and flash? 26%
8. The salesperson who sold Eric the equipment earned a $6\frac{1}{2}\%$ commission on the sale. How much was the commission? $33.80

9-8 LESSON NOTES

NCTM Standards: 1-5, 7, 10

Instructional Resources

- Study Guide Master 9-8
- Practice Master 9-8
- Enrichment Master 9-8
- Group Activity Card 9-8
- Assessment and Evaluation Masters, p. 240

Transparency 9-8A contains the 5-Minute Check for this lesson; **Transparency 9-8B** contains a teaching aid for this lesson.

Recommended Pacing	
Standard Pacing	Day 10 of 14
Honors Pacing	Day 9 of 13
Block Scheduling*	Day 5 of 7 (along with Lesson 9-7)

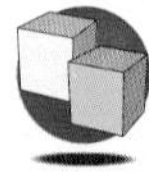

*For more information on pacing and possible lesson plans, refer to the ***Block Scheduling Booklet***.

1 FOCUS

5-Minute Check *(over Lesson 9-7)*

Copy and complete the table.

	Fraction	Decimal	Percent
1.	$\frac{1}{2}$	0.5	50%
2.	$\frac{4}{25}$	0.16	16%
3.	$\frac{1}{250}$	0.004	0.4%
4.	$\frac{3}{8}$	0.375	37.5%
5.	$\frac{49}{200}$	0.245	24.5%

Motivating the Lesson

Hands-On Activity Partially fill a glass with water. Ask students what percent of the glass is full or empty.

9-8 Percent and Estimation

Setting Goals: *In this lesson, you'll use percents to estimate.*

Modeling a Real-World Application: Health

"You are what you eat." So if you want to be healthy, you need to eat healthful food. Nutritionists suggest that people get no more than 30% of their daily Calorie intake from fat. Do any of the popcorn snacks shown in the table at the right contain about 30% fat Calories or less? *This problem will be solved in Example 2.*

Snack	Calories per serving	Calories from fat per serving
Jiffy Pop	140	63
Newman's Own Light	110	27
Pop Secret Original Butter	142	90
Pop Secret by Request	108	18
RedenBudders Light	100	36
Screaming Yellow Zonkers Glazed	130	36
Smartfood White Cheddar Cheese	160	90

Source: *Vitality*

Learning the Concept

Many times when you are working with percents, an exact answer is not needed. In cases like this, you can estimate. We could estimate the percent of the model that is shaded.

Nine of the 20 squares are shaded.

$\frac{9}{20}$ is about $\frac{10}{20}$ or $\frac{1}{2}$

Since $\frac{1}{2} = 50\%$, about 50% of the model is shaded.

There are three methods you can use to estimate a percentage. The table below shows how to estimate 30% of 657 using each method.

Fraction Method	1% Method	Meaning of Percent Method
30% is about 33% or $\frac{1}{3}$.	1% of 660 is 6.6 or about 7.	30% means 30 for every 100 or 3 for every 10. 657 has 6 hundreds and about 6 tens.
$\frac{1}{3}$ of 660 is 220.	30 times 7 is 210.	$(30 \times 6) + (3 \times 6) = 198$
Estimate: 220	Estimate: 210	Estimate: 198

Use a calculator to find the exact amount.

657 [×] 30 [%] [=] *197.1*

The actual percentage is very close to all of the estimates.

Classroom Vignette

"I have students experience real-life discounts by giving students a grocery list, a pricing list, and a number of cents-off coupons. Have them first find the total grocery bill prior to using the coupons and then the final grocery bill using the coupons. Let students explore their savings in terms of percent by finding the discount on their grocery bill when using coupons. Also have them calculate the discount on individual items."

Mrs. Naomi Harwood

Naomi Harwood
Rutherfordton Spindle H.S.
Rutherfordton, NC

Example 1

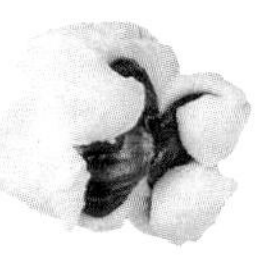

Estimate.

a. 60% of 996
1% of 996 is 9.96 or about 10. So 60% of 996 is about 60 × 10 or 600.

b. 8% of \$58
8% is about 10% or $\frac{1}{10}$.
$\frac{1}{10}$ of \$58 is \$5.80, so
8% of \$58 is about \$6.00

c. 0.5% of 795
0.5% is half of 1%.
795 is almost 800.
1% means 1 out of 100.
So 1% of 800 is 8, and
$\frac{1}{2}$ of 8 is 4.
0.5% of 795 is about 4.

d. 109% of 62
109% is more than 100%,
so 109% of 62 is more than 62.
109% is almost 110%.
110% = 100% + 10%
62(100% + 10%) = 62 + 6.2 = 68.2
109% of 62 is about 68.

Example 2

CONNECTION
Health

Refer to the application at the beginning of the lesson. Do any of the popcorn snacks have less than 30% of their Calories from fat?

Use the meaning of percent method to estimate 30% of the Calories for each snack. 30% means 30 for every 100 Calories and 3 for every 10 Calories.

Snack	Calories	Estimate of 30% of Calories	Fat Calories
Jiffy Pop	140	30 + 3(4) = 42	63
Newman's Own Light	110	30 + 3(1) = 33	27
Pop Secret Original Butter	142	30 + 3(4) = 42	90
Pop Secret by Request	108	30 + 3(1) = 33	18
RedenBudders Light	100	30 + 3(0) = 30	36
Screaming Yellow Zonkers Glazed	130	30 + 3(3) = 39	36
Smartfood White Cheddar Cheese	160	30 + 3(6) = 48	90

FYI

The average American eats 71 quarts of popcorn each year.

Compare the estimate of 30% of the total calories to the number of calories from fat in each snack. Newman's Own Light, Pop Secret by Request, and Screaming Yellow Zonkers Glazed all have about 30% or less of their calories from fat.

Estimating percents is a useful skill for everyday situations.

Example 3

APPLICATION
Consumerism

A tip of 15% is standard for good service in a restaurant. If Simone wants to leave a tip of about 15% on a dinner check of \$23.85, how much should she leave?

Estimate: \$23.85 is about \$24.
10% of \$24 is \$2.40.
5% of \$24 is \$1.20. *5% is half of 10%.*
15% is about 2.40 + 1.20 or \$3.60.

Simone should leave about \$3.60 as a tip.

2 TEACH

In-Class Examples

For Example 1
Estimate. Estimates may vary. Sample answers given.

a. 80% of 524 400

b. 4% of \$29 \$1.20 or \$1.50

c. 0.7% of 213 1.4

d. 248% of 83 200

For Example 2
Leonard estimated that 37% of the Calories in his fast-food meal come from fat. If there are 520 Calories in his meal, about how many Calories come from fat? 185 Calories

For Example 3
Dion treated his entire family to dinner. He wants to leave a tip of about 15% on the bill of \$180.83. How much should he leave? \$27

Teaching Tip You may wish to have students explain orally how to estimate a 15% tip. **Move the decimal point one place to the left. Add half that amount to itself.**

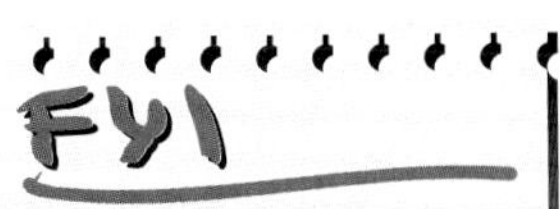

FYI

Rather than throwing away extra popcorn, it can be placed in plastic freezer bags and frozen for later use.

Tech Prep

Financial Advisor When counseling clients about their financial affairs, a financial advisor needs to estimate percents, such as the percent of income being spent on housing expenses, food expenses, and so on.

For more information on tech prep, see the *Teacher's Handbook*.

3 PRACTICE/APPLY

Checking Your Understanding

Exercises 1–17 are designed to help you assess your students' understanding through reading, writing, speaking, and modeling. You should work through Exercises 1–5 with your students and then monitor their work on Exercises 6–17.

Additional Answers

1. Sample answer: 25% of 200 = 50
2. Sample answer: Choose a fraction that is close to the percent like $\frac{1}{4}$ for 28%, and estimate the product.
4. Both; Angeni: $10\% + \frac{1}{2}(10\%) = 15\%$; Darlene: $3\left(5\frac{1}{4}\%\right) \approx 15\%$

Checking Your Understanding

Communicating Mathematics

Read and study the lesson to answer these questions. 1–2. See margin.

1. **Explain** how you would estimate 27% of 198.
2. **Describe** in your own words one method for estimating using percents.
3. **Estimate** the percent of the model at the right that is shaded. 50%
4. **You Decide** Once a month Angeni and Darlene meet for lunch. The food tax where they live is $5\frac{1}{4}\%$. Each of them leaves about a 15% tip. Angeni always finds her tip by taking 10% of the total and then adding that amount to half of that amount. Darlene finds her tip by multiplying the tax by 3. Who is correct and why? See margin.

5. **Describe** a situation that would best use each kind of method for estimating using percents. See students' work.

Guided Practice

Choose the best estimate for the percent shaded.

6. 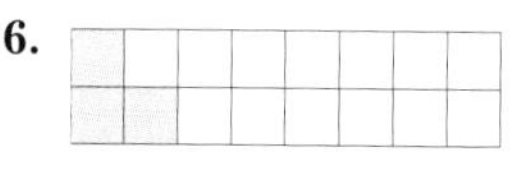

   a. 10% **b.** 25% **c.** 55% b

7. 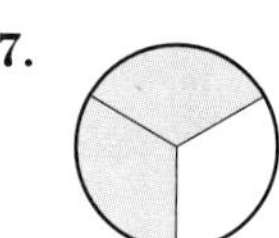

   a. 33% **b.** 50% **c.** 66% c

Write the fraction, mixed number, or whole number you could use to estimate. 8–10. Sample answers given.

8. 24% $\frac{1}{4}$ **9.** 79% $\frac{4}{5}$ **10.** 145% $1\frac{1}{2}$

Estimate. 11–13. Sample answers given.

11. 45% of 430 215 **12.** 112% of 14.5 16 **13.** 0.6% of 325 1.5

Estimate each percent. 14–16. Sample answers given.

14. 8 out of 30 $33\frac{1}{3}\%$ **15.** 9 out of 19 50% **16.** $\frac{5}{7}$ 70%

17. **Retail Sales** Davon bought a pair of shoes on sale for $29. The regular price was $50.
    a. For about what percent of the selling price did he buy them? 60%
    b. About what percent-off was the sale? 40%

Exercises: Practicing and Applying the Concept

Independent Practice

Choose the best estimate.

		a.	**b.**	**c.**	
18.	39% of 300	1.2	12	120	c
19.	47% of 605	3	30	300	c
20.	$\frac{1}{3}\%$ of 240	0.8	8	80	a
21.	129% of 400	50	500	5000	b

Study Guide Masters, p. 78

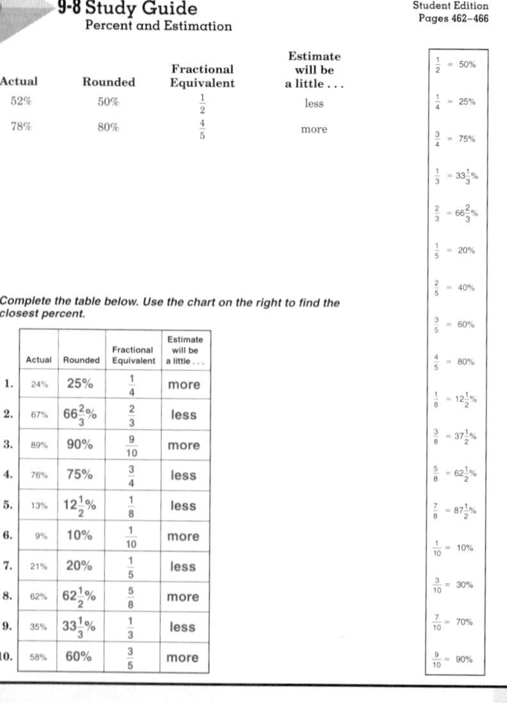

NAME ______________ DATE ______________

9-8 Study Guide
Percent and Estimation

Student Edition Pages 462–466

Actual	Rounded	Fractional Equivalent	Estimate will be a little . . .
52%	50%	$\frac{1}{2}$	less
78%	80%	$\frac{4}{5}$	more

$\frac{1}{2} = 50\%$
$\frac{1}{4} = 25\%$
$\frac{3}{4} = 75\%$
$\frac{1}{3} = 33\frac{1}{3}\%$
$\frac{2}{3} = 66\frac{2}{3}\%$
$\frac{1}{5} = 20\%$
$\frac{2}{5} = 40\%$
$\frac{3}{5} = 60\%$
$\frac{4}{5} = 80\%$
$\frac{1}{8} = 12\frac{1}{2}\%$
$\frac{3}{8} = 37\frac{1}{2}\%$
$\frac{5}{8} = 62\frac{1}{2}\%$
$\frac{7}{8} = 87\frac{1}{2}\%$
$\frac{1}{10} = 10\%$
$\frac{3}{10} = 30\%$
$\frac{7}{10} = 70\%$
$\frac{9}{10} = 90\%$

Complete the table below. Use the chart on the right to find the closest percent.

	Actual	Rounded	Fractional Equivalent	Estimate will be a little . . .
1.	24%	25%	$\frac{1}{4}$	more
2.	67%	$66\frac{2}{3}\%$	$\frac{2}{3}$	less
3.	89%	90%	$\frac{9}{10}$	more
4.	76%	75%	$\frac{3}{4}$	less
5.	13%	$12\frac{1}{2}\%$	$\frac{1}{8}$	less
6.	9%	10%	$\frac{1}{10}$	more
7.	21%	20%	$\frac{1}{5}$	less
8.	62%	$62\frac{1}{2}\%$	$\frac{5}{8}$	more
9.	35%	$33\frac{1}{3}\%$	$\frac{1}{3}$	less
10.	58%	60%	$\frac{3}{5}$	more

Group Activity Card 9-8

Educated Estimates Group Activity **9-8**

MATERIALS: Calculator • two 0-9 spinners • cards with the percents given on the back of this card

Partner's work together to estimate the percent of a number. The percent cards should be shuffled and placed face down in a pile.

The first player looks at the top card on the pile and spins the spinners to obtain a 2-digit number. This player then estimates the percent of the 2-digit number. The other player uses the calculator to find the exact answer.

To earn one point, the first player's estimate must be within five of the exact answer. Players then change roles and play continues.

The first player to earn five points wins the game.

18% 98% 2%

Pre-Algebra

Reteaching

Using Calculators Have students work in small groups and use a calculator to solve problems similar to the following.

A $300 bicycle is on sale for 30% off. The same bicycle can be bought at another store for $250 less a 10% discount. Which is the better buy? **$300 bicycle with 30% off**

Write the fraction, mixed number, or whole number you could use to estimate. 22–29. Sample answers given.

B
22. 67% $\frac{7}{10}$ or $\frac{2}{3}$ 23. 98% 1 24. 18% $\frac{1}{5}$ 25. $2\frac{1}{2}$% $\frac{3}{100}$
26. 148% 1.5 27. $8\frac{5}{9}$% $\frac{1}{10}$ 28. 0.8% $\frac{1}{100}$ 29. 119% $1\frac{1}{5}$

Estimate. 30–38. Sample answers given.

30. 47% of 84 42 31. 28% of 390 120 32. $8\frac{1}{2}$% of 55 5.5
33. 98% of 98 98 34. 126% of 198 250 35. 0.9% of 514 5.15
36. 15% of $34 4.5 37. 116% of 18 20 38. 0.05% of 1180 0.6

Estimate each percent. 39–47. Sample answers given.

39. 12 out of 15 80% 40. 8 out of 35 20% 41. 39 out of 79 50%
42. 57 out of 176 33% 43. 9 out of 95 10% 44. 13 out of 68 21%
C
45. 1 out of 9 10% 46. 3 out of 200 1.5% 47. 7 out of 2445 0.3%

Critical Thinking

48. In a controversial vote, 40 percent of the Democrats and 92.5 percent of the Republicans voted yes. If all the members of the assembly voted and 68 percent of the voters voted yes, what is the ratio of Democrats to Republicans? 7 to 8

Applications and Problem Solving

49. **Nutrition** Estimate the percent of fat Calories of each entree in the chart at the right.

Entree	Calories (3.5 oz)	Calories from fat	
shrimp	90	7	10%
lobster	90	15	20%
chicken wing, with skin	290	171	60%
ham	220	99	45%
ground chuck	222	81	40%

Source: *The Fat Counter*

50. **Research** The land area of Georgia is about 60% of the land area of Great Britain.
a. Find the size (in square miles) of Great Britain.
b. Estimate the size of Georgia.

50a. 94,247 mi^2;
50b. about 54,000 mi^2

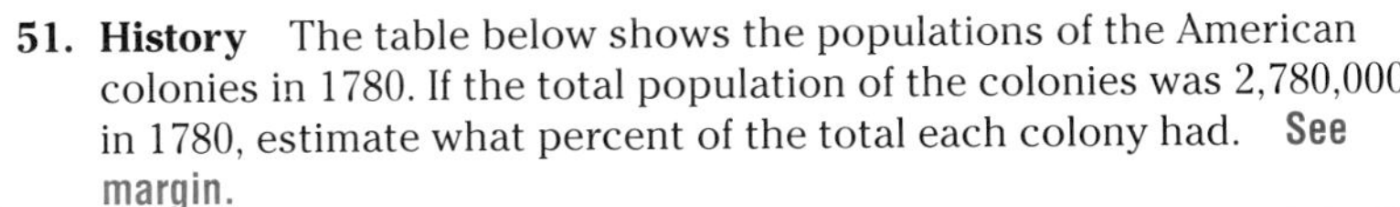

51. **History** The table below shows the populations of the American colonies in 1780. If the total population of the colonies was 2,780,000 in 1780, estimate what percent of the total each colony had. See margin.

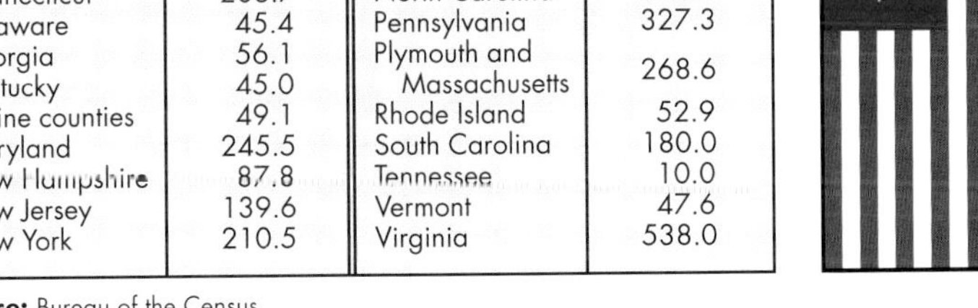

Colony	Population (thousands)	Colony	Population (thousands)
Connecticut	206.7	North Carolina	270.1
Delaware	45.4	Pennsylvania	327.3
Georgia	56.1	Plymouth and Massachusetts	268.6
Kentucky	45.0	Rhode Island	52.9
Maine counties	49.1	South Carolina	180.0
Maryland	245.5	Tennessee	10.0
New Hampshire	87.8	Vermont	47.6
New Jersey	139.6	Virginia	538.0
New York	210.5		

Source: Bureau of the Census

52. **Nutrition** A McDonald's Big Mac has 560 Calories, and 288 of these are fat Calories.
a. About what percent of the Calories are fat Calories? about 50%
b. If a person on a 2000-Calorie-per-day diet eats a Big Mac, how many Calories and fat Calories remain for the rest of the day? 1440; 312

Extension

Using Data Have students use newspaper or store advertisements that contain percentage discounts. If the sale price is given, have students estimate the regular price. If the regular price is given, estimate the sale price. If the regular price and sale price are given, estimate the percent of discount.

Error Analysis

Some students may just randomly choose a fraction to use as an equivalent for a percent. Suggest that students use a calculator to check that the fraction they have chosen to use is indeed close to the given percent.

Assignment Guide

Core: 19–47 odd, 48, 49, 51, 53–60
Enriched: 18–46 even, 48–60

For **Extra Practice**, see p. 763.

The red A, B, and C flags, printed only in the Teacher's Wraparound Edition, indicate the level of difficulty of the exercises.

Additional Answer

51. Sample answers: Connecticut 10%; Delaware 1%; Georgia 2%; Kentucky 1%; Maine 2%; Maryland 10%; New Hampshire 3%; New Jersey 5%; New York 10%; North Carolina 10%; Pennsylvania 15%; Plymouth 10%; Rhode Island 2%; South Carolina 10%; Tennessee 1%; Vermont 2%; Virginia 20%

Practice Masters, p. 78

NAME ______ DATE ______

9-8 Practice
Percent and Estimation

Student Edition Pages 462–466

Choose the best estimate.

1. 19% of 50 A. 1 (B.) 10 C. 100
2. 76% of 240 A. 18 (B.) 180 C. 1800
3. $\frac{3}{4}$% of 90 (A.) 0.9 B. 9 C. 90
4. 193% of 800 A. 16 B. 160 (C.) 1600

Write the fraction, mixed number, or whole number you could use to estimate. Sample answers given.

5. 35% $\frac{1}{3}$ 6. 67% $\frac{2}{3}$ 7. 24% $\frac{1}{4}$ 8. 78% $\frac{3}{4}$
9. 99% 1 10. $9\frac{3}{5}$% $\frac{1}{10}$ 11. 48% $\frac{1}{2}$ 12. $5\frac{1}{5}$% $\frac{1}{20}$
13. 123% $1\frac{1}{4}$ 14. 31.9% $\frac{1}{3}$ 15. 1.2% $\frac{1}{100}$ 16. $\frac{7}{8}$% $\frac{1}{100}$

Estimate. Sample answers given.

17. 9% of 45 4.5 18. 47% of $35.95 18 19. 74% of 40 30
20. 26% of 64 16 21. 66% of $240 $160 22. $9\frac{5}{6}$% of 50 5
23. 98% of 75 75 24. $4\frac{3}{4}$% of $58 $3 25. 126% of 840 1050
26. 1.3% of 97 1 27. $\frac{7}{8}$% of 75 0.75 28. 0.9% of 1500 15

Estimate each percent. Sample answers given.

29. 21 out of 60 $33\frac{1}{3}$% 30. 24 out of 50 50% 31. 21 out of 30 $66\frac{2}{3}$%
32. 7 out of 79 10% 33. 19 out of 80 25% 34. 9 out of 195 5%
35. 12 out of 81 $12\frac{1}{2}$% 36. 53 out of 79 $62\frac{1}{2}$% 37. 73 out of 82 $87\frac{1}{2}$%

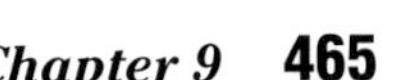

4 ASSESS

Closing Activity

Speaking Have students explain how using the 1% method and the meaning of percent method are both similar to the work backward strategy.

Chapter 8, Quiz C (Lessons 9-6 through 9-8) is available in the *Assessment and Evaluation Masters*, p. 240.

Additional Answers

57. 21;

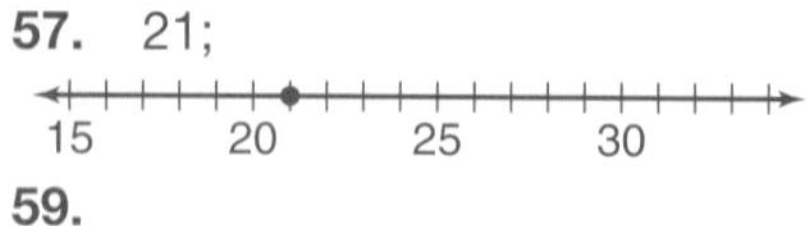

59.

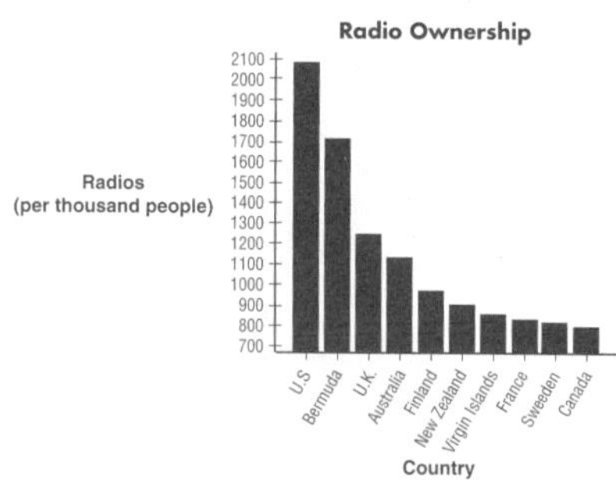

Enrichment Masters, p. 78

NAME ______________________ DATE ________

9-8 Enrichment Student Edition Pages 462–466

Percent Puzzle

Solve each problem.

1. 8% of 75 6
2. 16% of 80 12.8
3. 20% of 85 17
4. 17% of 300 51
5. 40% of 170 68
6. 50% of 380 190
7. 75% of 160 120
8. $33\frac{1}{3}$% of 240 80
9. $62\frac{1}{2}$% of 72 45
10. 30% of 180 54
11. $66\frac{2}{3}$% of 210 140
12. 80% of 160 128
13. 110% of 60 66
14. 95% of 300 285
15. 150% of 75 112.5
16. 400% of 50 200

Answers	
17	R
120	Y
54	T
45	G
6	F
51	A
128	P
200	O
112.5	S
66	I
140	X
190	H
12.8	U
80	N
68	D
285	E

Find the answer to each exercise above and note the letter next to it. Put this letter on the line or lines below that correspond to the exercise number.

HOW MUCH DID THE WORLD'S LARGEST PIZZA WEIGH?

E	I	G	H	T	E	E	N
14	13	9	6	10	14	14	8

T	H	O	U	S	A	N	D
10	6	16	2	15	4	8	5

S	I	X	H	U	N	D	R	E	D
15	13	11	6	2	8	5	3	14	5

S	I	X	T	Y	-	F	O	U	R
15	13	11	10	7		1	16	2	3

P	O	U	N	D	S
12	16	2	8	5	15

Mixed Review

53. Express 64% as a fraction. (Lesson 9-7) $\frac{16}{25}$
54. **World Facts** On German highways, called autobahns, the speed limit is 130 kilometers per hour. What is the speed limit in meters per hour? (Lesson 7-8) **130,000 m per h**
55. **Patterns** State whether 8, 12, 18, 27, 40.5, . . . is a geometric sequence. If so, state the common ratio and list the next three terms. (Lesson 6-8) **geometric; 1.5; 60.75, 91.125, 136.6875**
56. Solve $\frac{9}{16} + \frac{11}{16} = y$. (Lesson 5-4) $1\frac{1}{4}$
57. Solve $5x = 105$ and check your solution. Then graph the solution on a number line. (Lesson 3-3)

57. See margin.

58. Simplify $|-21|$. (Lesson 2-1) **21**
59. **Entertainment** The table at the right shows the number of radios per thousand people in several countries. Make a bar graph of the data. (Lesson 1-10) **See margin.**
60. Translate the phrase *twice the sum of a number and 7* into an algebraic expression. (Lesson 1-3) $2(n + 7)$

Country	Radios per thousand people
United States	2091
Bermuda	1710
United Kingdom	1240
Australia	1144
Finland	984
New Zealand	902
Virgin Islands (U.S.)	884
France	866
Sweden	842
Canada	828

Source: Duncan's American Radio, Inc.

From the FUNNY PAPERS

TALK ABOUT IT

1. Explain why the comic is funny. **See students' work.**
2. If the student missed six out of ten questions on the quiz, how many questions did he answer correctly? **4**
3. What is the student's real percent score on the quiz? **40%**
4. Write a sentence or two to explain to the father how to estimate the percent score on a quiz.

4. See students' work.

From the FUNNY PAPERS

One of baseball's Yogi Berra's quotes is that 99% of the game is half mental. Ask students what is funny about that quote.

9-9 Using Percent Equations

Setting Goals: *In this lesson, you'll solve percent problems using percent equations.*

Modeling with Technology

Maria Morales has saved some money that she wishes to invest. Her banker has provided the terms, interest rates, and minimum investments on the certificates of deposit, or CDs, that they offer. Maria used the information to prepare the spreadsheet below to compare the options.

Option	Minimum Investment	Monthly Interest Rate (percent)	Term (months)	Interest for Term
1	$2500	0.375	6	**$56.25**
2	$2500	0.45	12	**$135.00**
3	$5000	0.54	36	**$972.00**
4	$7500	0.625	60	**$2812.50**

The **principal** is the amount of money in the account. The amount of **interest** earned in a month can be found by using the percent proportion $\frac{\textit{interest}}{\textit{principal}} = \frac{\textit{interest rate}}{100}$. Then multiply the monthly interest by the number of months to find the total interest earned.

Your Turn

a. See margin.

Use a calculator or spreadsheet to find the interest on each investment.

a. Which option earns the most? What factors contribute to this?

b. Let I represent interest, p the principal, r the interest rate as a decimal, and t the time in months. Write a formula for finding the interest on any account. $I = prt$

Learning the Concept

As you discovered in the activity above, you can find the interest earned on an account using the formula $I = prt$ if you write the rate as a decimal. Writing the rate as a decimal in the percent proportion also allows you to write it as an equation and solve percent problems more quickly.

$$\frac{P}{B} = \frac{r}{100}$$

$$\frac{P}{B} \cdot B = \frac{r}{100} \cdot B \quad \text{Multiply each side by } B.$$

$$P = \frac{r}{100} \cdot B$$

Remember $\frac{r}{100}$ is the rate. Let R represent the decimal form of $\frac{r}{100}$.

Therefore, $P = R \cdot B$ or Percentage = Rate · Base.

The form $P = R \cdot B$ is usually easier to use when the rate and base are known.

9-9 LESSON NOTES

NCTM Standards: 1-5, 7

Instructional Resources

- Study Guide Master 9-9
- Practice Master 9-9
- Enrichment Master 9-9
- Group Activity Card 9-9
- Activity Masters, p. 41
- Real-World Applications, 20

Transparency 9-9A contains the 5-Minute Check for this lesson; **Transparency 9-9B** contains a teaching aid for this lesson.

Recommended Pacing	
Standard Pacing	Day 11 of 14
Honors Pacing	Day 10 of 13
Block Scheduling*	Day 6 of 7 (along with Lesson 9-10)

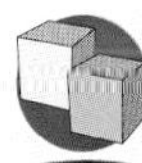

*For more information on pacing and possible lesson plans, refer to the ***Block Scheduling Booklet***.

1 FOCUS

5-Minute Check
(over Lesson 9-8)

Write the fraction or mixed number you should use to estimate.

1. 74% $\frac{3}{4}$

2. 123% $1\frac{1}{4}$

Estimate.

3. 65% of 18 12

4. 18% of 550 110

5. 41% of 25 10

Motivating the Lesson

Questioning Ask the class if anyone has a savings account. Ask students to tell the class what happens to their money if they leave it in the bank for a long time.

Additional Answer Talk About It

a. Option 4 because it has a high interest rate, a long term, and a high initial investment

Interactive Mathematics Tools Software

In this interactive computer lesson, students explore compound interest. A **Computer Journal** gives students the opportunity to write about what they have learned.

For Windows & Macintosh

Alternative Teaching Strategies

Reading Mathematics For students who have difficulty with this lesson, have students identify the parts, P, B, and r, in exercises 6–9 and 14–27 on page 469 before they try to complete the exercises.

2 TEACH

In-Class Examples

For Example 1

a. Find 68% of 125. **85**

b. 78 is what percent of 65? **120%**

For Example 2

In 1995, a town raised \$140,000 for the United Way. Of that money, 42% was used by agencies within the town. How much money was that? **\$58,800**

For Example 3

Joyclyn received a merchandise certificate good for 30% off any coat. The coat she chose was marked \$83.50. The sales tax is 8%. If she uses her certificate to buy the coat, how much will it cost, including tax? **\$63.13**

Teaching Tip You may want students to try each method for each part of Example 4 so they can master both methods.

3 PRACTICE/APPLY

Checking Your Understanding

Exercises 1–13 are designed to help you assess your students' understanding through reading, writing, speaking, and modeling. You should work through Exercises 1–5 with your students and then monitor their work on Exercises 6–13.

Example 

a. Find 36% of 65. *Estimate:* $\frac{1}{3}$ *of 60 is about 20.*

What number is 36% of 65? Write in P = R · B form.

$P = 0.36 \times 65$

0.36 [×] 65 [=] 23.4

23.4 is 36% of 65.

b. 15 is what percent of 125?

$15 = R \cdot 125$ *Replace P with 15 and B with 125.*

$\frac{15}{125} = \frac{125R}{125}$ *Divide each side by 125.*

$0.12 = R$ *Use a calculator.*

15 is 12% of 125.

Example 2

APPLICATION Charity

The Nature Conservancy is among the largest charities involved in protecting the environment. They reported spending \$198,722,500 on conservation in 1993. If this is 72.5% of the money they received, how much money did they receive?

The Nature Conservancy®

198,722,500 is 72.5% of what number?

$198{,}722{,}500 = 0.725 \times B$ *P = R · B*

$\frac{198{,}722{,}500}{0.725} = \frac{0.725B}{0.725}$ *Divide each side by 0.725.*

$274{,}100{,}000 = B$ *Use a calculator.*

The Nature Conservancy raised \$274,100,000 in 1993.

One of the common uses of percents in everyday life is **discounts**.

Example 3

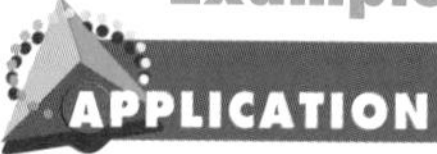

APPLICATION Retail Sales

The regular price of a pair of jeans is \$45.95. If there is a 25% discount and a 6% sales tax (based on the discounted price), how much will the jeans cost?

You can find the sale price in one of two ways.

Method 1

First, find the amount of the discount.

25% of 45.95 = d

0.25 [×] 45.95 [=] 11.475

The discount is \$11.49.

Then, subtract to find the discount price.

45.95 [−] 11.49 [=] 34.46

The discount price is \$34.46.

Method 2

First, subtract the percent discount from 100%.

Since 100% − 25% = 75%, the discount price will be 75% of the original price.

Then, multiply to find the discount price.

0.75 [×] 45.95 [=] 34.4625

The discount price is \$34.46.

Next, find the price including the sales tax.

Cooperative Learning

Trade-A-Problem Have students collect cash register receipts from local businesses. Then either the tax rate, the subtotal, or the amount due is marked out and a word problem is written about that receipt. Attach the receipt to the word problem and trade with another group to solve.

For more information on this strategy, see *Cooperative Learning in the Mathematics Classroom*, one of the titles in the Glencoe Mathematics Professional Series, p. 25.

Method 1

Find the amount of the tax.
6% of 34.46 = t
0.06 [×] 34.46 [=] 2.0676
The tax is \$2.07.

Then, add to find the total price.
34.46 [+] 2.07 [=] 36.53

The total price is \$36.53.

Method 2

First, add the percent of tax to 100%.

Since 100% + 6% = 106%, the total price will be 106% of the discount price.

Then, multiply to find the total price.
1.06 [×] 34.46 [=] 36.5276

The total price is \$36.53.

The jeans will be \$36.53 after the discount and the sales tax.

Checking Your Understanding

Communicating Mathematics

Read and study the lesson to answer each question. 1–3, 5. See margin.

1. **Explain** how r and R are different.
2. **Describe** two ways of finding the price of a \$60 item after a 25% discount.
3. Under what conditions might you use $P = R \cdot B$ instead of $\frac{P}{B} = \frac{r}{100}$?
4. If $R = 0.32$, what is r? **32**
5. **You Decide** Miguel and Jalisa are trying to find the monthly interest on a credit card balance of \$2500 if the monthly percentage rate is 1.5%. Miguel solves using $P = (1.5)(2500)$. Jalisa uses $P = (0.015)(2500)$. Who is correct? Explain.

Guided Practice

Solve each problem by using the percent equation, $P = R \cdot B$.

6. 75 is 50% of what number? **150**
7. 15% of what number is 30? **200**
8. 18 is what percent of 60? **30%**
9. What is 24% of 72? **17.28**

Find the discount or interest to the nearest cent.

10. \$450 TV, 33% off **\$148.50**
11. \$315 suit, 15% off **\$47.25**
12. \$3500 at 12% annually for $2\frac{1}{2}$ years **\$1050.00**
13. **Travel** About 28% of the Japanese tourists who travel abroad each year visit the United States. If 12 million Japanese people went abroad in 1992, about how many visited the United States? **3.36 million**

Exercises: Practicing and Applying the Concept

Independent Practice

Solve each problem by using the percent equation, $P = R \cdot B$.

14. 15 is what percent of 100? **15%**
15. What is 20% of 135? **27**
16. What number is 43% of 15? **6.45**
17. 28 is 40% of what number? **70**

B

18. 15% of what number is 39? **260**
19. 42% of 75 is what number?
20. 45 is what percent of 150? **30%**
21. 42 is what percent of 14? **300%**
22. 110% of what number is 880? **800**
23. 5 is what percent of 300? **$1.\overline{6}$%**
24. 2.04% of 256 is what number?
25. 18.6 is 15.5% of what number?
26. $33\frac{1}{3}$% of 420 is what number? **140**
27. 500% of what number is 100?

19. 31.5
24. 5.2224
25. 120
27. 20

Error Analysis

Students may add discounts or subtract interest when solving problems. Have students write "discount-subtract" and "interest-add" at the top of their paper and use this for reference as they complete their homework.

Assignment Guide

Core: 15–35 odd, 36, 37, 39, 41, 43–51
Enriched: 14–34 even, 36–51

For **Extra Practice**, see p. 763.

The red A, B, and C flags, printed only in the Teacher's Wraparound Edition, indicate the level of difficulty of the exercises.

Additional Answers

1. $R = \frac{r}{100}$
2. Sample answer: Find 75% of \$60 or find 25% of \$60 and subtract from \$60.
3. If the rate and the base are known, then you can just multiply.
5. Jalisa; because 1.5% = 0.015 not 1.5.

Reteaching

Using Flowcharts Help students create flowcharts showing how to decide what information is known and what operations to use to find the unknown.

Group Activity Card 9-9

Percent Puzzle Group Activity **9-9**

MATERIALS: Twenty-four $2\frac{1}{2}$-inch squares cut from index cards

SETUP: The expressions given on the back of this card should be copied onto the squares.

Two players shuffle the cards and place them face up in a 5 by 5 array on the table. Leave a blank space in the lower right corner of the array. Take turns naming correct expressions of the form a% of $b = c$ (for example 20% of 200 = 40) by identifying expressions in the existing array or by moving one square into the blank space to form an expression. Record the expressions you have made. The game continues until no new expressions are made in 10 moves.

The person whose record has the most correct expressions wins. If you incorrectly identify an expression, you subtract one point from your final count.

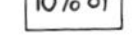

 Pre-Algebra

Study Guide Masters, p. 79

NAME ______ DATE ______

9-9 Study Guide
Using Percent Equations

Student Edition Pages 467–471

To solve percent problems, use the percent equation, $P = R \cdot B$. R represents $\frac{r}{100}$.

Example 1: Find 45% of 36.

What number is 45% of 36?

$P = 45\% \times 36$
$P = 0.45 \times 36$
$P = 16.2$
45% of 36 is 16.2.

Example 2: 8 is what percent of 16?

$8 = R \times 16$
$\frac{8}{16} = \frac{16R}{16}$ Divide each side by 16.
$0.5 = R$ Use a calculator.
8 is 50% of 16.

The amount of **interest *(I)*** earned on an account depends upon the **principal *(p)***, which is the money deposited, the **rate *(r)***, and the **time *(t)*** given in years.

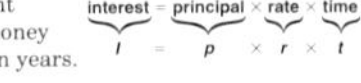

Solve each problem by using the percent equation, P = R · B.

1. Find 40% of 80. **32**
2. Find 15% of 600. **90**
3. What number is 30% of 120? **36**
4. What number is 90% of 50? **45**
5. What percent of 250 is 25? **10%**
6. What percent of 35 is 7? **20%**
7. 100% of 67 is what number? **67**
8. 200% of 67 is what number? **134**

Find the interest to the nearest cent.

9. \$160 at 5.5% for 1.25 years **\$11.00**
10. \$1800 at 6.5% for 2 years **\$234.00**
11. \$350 at 6% for 6 months **\$10.50**
12. \$7050 at 6% for 3 months **\$105.75**
13. \$3500 at 10% for 5 years **\$1750.00**
14. \$75 at 12% for 6 years **\$54.00**

Additional Answers

36. Yes; $x\%$ of $y = \frac{x}{100} \cdot y$ or $\frac{xy}{100}$ and $y\%$ of $x = \frac{y}{100} \cdot x$ or $\frac{xy}{100}$.

41c. You could not add the percents first. For example, if the price was \$100, 10% + 33% = 43% off. Using this method, the price would be 100 − 43 or \$57. Taking 33% off, then 10% would have the price as \$60.30

43. The statements are not equivalent. For example, suppose an item costs \$1.00 on day 1. According to the first statement, it would cost \$2.08, not \$1.00 × 2 or \$2.00, on the fifth day.
    Day 2 1.2 × 1.0 = \$1.20
    Day 3 1.2 × 1.20 = \$1.44
    Day 4 1.2 × 1.44 = 1.728 or \$1.73
    Day 5 = 1.2 × 1.73 = 2.076 or \$2.08

Practice Masters, p. 79

NAME ______________ DATE ________

9-9 Practice
Using Percent Equations

Student Edition Pages 467–471

Solve each problem by using the percent equation, $P = R \cdot B$.

1. 8% of what number is 60.16? 752
2. 64 is what percent of 512? 12.5%
3. What is 15% of \$80. \$12
4. $33\frac{1}{3}\%$ of what number is 30? 90
5. 32 is what percent of 24? $133\frac{1}{3}\%$
6. 125% of what number is 15? 12
7. What number is 19% of \$100? \$19
8. 4 is what percent of 6? $66\frac{2}{3}\%$
9. 14 is 28% of what number? 50
10. Find 200% of 115. 230
11. $87\frac{1}{2}\%$ of 64 is what number? 56
12. 49.2 is 102.5% of what number? 48
13. 2 is what percent of 125? 1.6%
14. 40% of \$9 is what number? \$3.60

Find the discount or interest to the nearest cent.

15. \$500 television, 15% off \$75
16. \$155 bicycle, 20% off \$31
17. \$300 typewriter, 10% off \$30
18. \$35 watch, 15% off \$5.25
19. \$160 violin, $12\frac{1}{2}\%$ off \$20
20. \$125 set of golf clubs, 20% off \$25
21. \$1454 computer, 25% off \$363.50
22. \$15.96 compact disc, $33\frac{1}{3}\%$ off \$5.32
23. \$300 at 6% for 1 year \$18
24. \$750 at 8% for 6 months \$30
25. \$4000 at 10.5% for 6 months \$210
26. \$945 at 11% for 8 months \$69.30
27. \$1200 at 8.5% for 2 months \$17
28. \$30,000 at 12.5% for 25 years \$93,750

Choose
Estimation
Mental Math
Calculator
Paper and Pencil

Find the discount or interest to the nearest cent. 31. \$187.50

28. \$49.95 sweat suit, 28% off \$13.99
29. \$199.99 tools, 35% off \$70
30. \$299.99 VCR, 40% off \$120.00
31. \$1250 refrigerator, 15% off
32. \$500 at 0.3% a month for 15 months \$22.50
33. \$250 at $8\frac{1}{2}\%$ annually for 2 years \$42.50
34. \$1250 at 1.5% monthly for 7 months \$131.25
35. \$945 at 3.5% annually for 15 months \$41.34

Critical Thinking

36. Is $x\%$ of y always equal to $y\%$ of x? Give examples to support your answer. See margin.

Applications and Problem Solving

37. about 93 million

37. Medicine Doctors divide blood into eight different types. Of the approximately 249 million Americans, 37.4% have type O+. How many Americans have O+ blood?

38. Travel Americans went on 43.3 million trips in 1994. The table at the right shows the number of trips made by different modes of transportation. Find the percent of trips made in each mode of transportation.

Mode of Transportation	Number of Trips (millions)
car, not rented	25.7
airplane	12.0
car, rented	3.3
motor home	1.0
bus	0.7
train	0.6

Source: *American Demographics*

39. They would be paying the customers.

39. Advertising A store advertised a 200%-off sale. Explain why you know that their advertisement was incorrect.

40. Business Sales associates are often paid in **commission**. A commission is usually given as a percent of the amount of the associate's sales. If Computer Generation pays its associates 15% commission, how much commission would be earned from the sale of a \$2300 computer? \$345

41. Retail Jackee Johnson receives a 10% discount on all merchandise at the store where she works. She is buying a \$439 television that is on sale at 33% off.

a. Find Jackee's price for the television. \$264.72

b. Does it matter which discount is taken first, the 33% sale price or the 10% employee discount? No; the price is the same either way.

c. Could the discount percents be added together first and then taken off or must they be taken off one at a time? Give an example to support your answer. See margin.

42. Business Many car dealers offer special interest rates as incentives to attract buyers. How much interest would you pay for the first month of a \$5500 car loan with a monthly interest rate of 0.24%? \$13.20

43. Economics When the prices of products in a country rise over time, the country is experiencing inflation. The rate of inflation in the United States is about 5% per year. In 1993, *USA TODAY* reported "The rate of inflation in Yugoslavia is 20% a day. In other words, prices double every five days." Are the two statements made in the article the same? Explain. See margin.

Extension

Using Connections Some businesses use chain discounts, especially when selling through catalogs. In this way, they can keep prices competitive without reprinting the catalog. Suppose a catalog shows a computer with an \$1899 suggested retail price. The catalog's selling price offers a 20% discount on that price at the beginning of the year. By the end of the year, an additional 15% discount is offered on the catalog's selling price. Have students find out whether the two discounts are the same as one discount of 35%. If the amounts differ, by how much and why do they differ? **The single discount is \$664.65 while the chain discount is \$607.68 because the base for the 15% discount is only 80% of the \$1899.**

Mixed Review

44. Sample answer: 140

44. Estimate 68% of 210. (Lesson 9-8)

45. **Statistics** Does the scatter plot at the right show a positive relationship, a negative relationship, or no relationship? (Lesson 8-2) **positive**

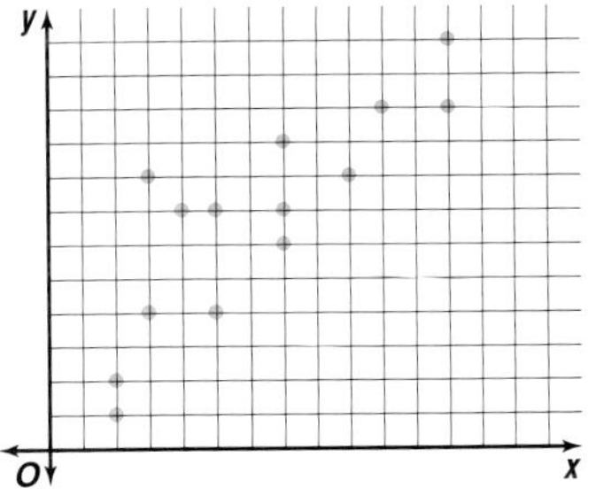

46. Determine whether {(6, 1), (8, 4), (−5, 5), (3, 4)} is a function. (Lesson 8-1) **yes**

50a. $r + 6 = 148$

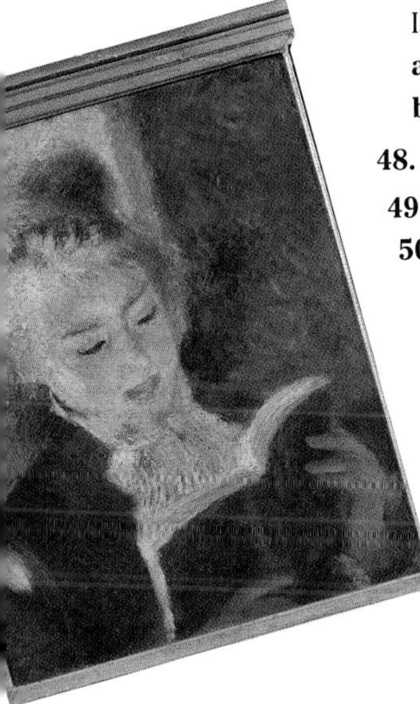

47. **Population** Vatican City is the least populated country in the world. The population of the Falkland Islands is 298 less than three times the population of Vatican City. If the population of the Falkland Islands is 1916, find the population of Vatican City. (Lesson 7-3)
    a. Write an equation that represents this situation. **$3v - 298 = 1916$**
    b. Solve the equation to find the population of Vatican City. **738**

48. Write $\frac{5}{16}$ as a decimal. (Lesson 6-1) **0.3125**

49. Find the GCF of 150 and 345. (Lesson 4-5) **15**

50. **Art** Pablo Picasso is the artist with the most paintings that have sold for over $1 million, with 148 paintings. This is six more than the number of paintings by Auguste Renoir that have sold for more than $1 million. (Lesson 3-2)
    a. Write an equation to represent the number of Auguste Renoir paintings that have sold for more than $1 million.
    b. Find the number of paintings by Auguste Renoir that have sold for more than $1 million. **142**

51. Find 12 + (−5) (Lesson 2-4) **7**

WORKING ON THE Investigation

Refer to the Investigation on pages 368–369.

The Athletic Director of the university called you into her office. She said that the university needs to increase the gross income from the football games. This could be done by either raising the prices of seats or increasing the number of deluxe and/or spectator seats. Conference regulations require that the stadium has at least 18% of its seats priced at under $16.00.

The Athletic Director asked you to prepare three different options to increase the gross income from a game by 15%. Prepare three options and state the advantages and disadvantages of each option.

Add the results of your work to your Investigation Folder.

4 ASSESS

Closing Activity

Modeling Have students choose three items from a catalog, find the total cost, calculate the sales tax, shipping cost (if any) and the amount that should be sent in with the order.

WORKING ON THE Investigation

The Investigation on pages 368 and 369 is designed to be a long-term project that is completed over several days or weeks. Encourage students to keep their materials in their Investigation Folder as they work on the Investigation.

Enrichment Masters, p. 79

NAME ______________________ DATE ____________

9-9 Enrichment
Compound Interest

Student Edition Pages 467–471

Interest may be paid, or compounded, annually (each year), semi-annually (twice per year), quarterly (four times per year), monthly (once per month), or daily.

Example: George had $100 in an account for $1\frac{1}{2}$ years that paid 8% interest compounded semiannually. What was the total amount in his account at the end of $1\frac{1}{2}$ years?

At the end of $\frac{1}{2}$ year: Interest: $\$100 \times 0.08 \times \frac{1}{2} = \$4.00$
New Principal: $\$100 + \$4 = \$104$

At the end of 1 year: Interest: $\$104 \times 0.08 \times \frac{1}{2} = \$4.16$
New Principal: $\$104 + \$4.16 = \$108.16$

At the end of $1\frac{1}{2}$ years: Interest: $\$108.16 \times 0.08 \times \frac{1}{2} = \$4.33$
New Principal: $\$108.16 + \$4.33 = \$112.49$

Find the total amount for each of the following.

	Principal	Rate	Time	Compounded	Total Amount
1.	$200	6%	$1\frac{1}{2}$ years	semiannually	$218.55
2.	$300	5%	2 years	semiannually	$331.15
3.	$100	6%	1 year	quarterly	$106.14
4.	$500	8%	$\frac{3}{4}$ year	quarterly	$530.60
5.	$500	10%	4 months	monthly	$516.88
6.	$800	8%	$\frac{1}{4}$ year	monthly	$816.10
7.	$1000	8%	4 years	annually	$1360.49
8.	$700	6%	$1\frac{1}{2}$ years	semiannually	$764.91

9-10 LESSON NOTES

NCTM Standards: 1-5, 7

Instructional Resources

- Study Guide Master 9-10
- Practice Master 9-10
- Enrichment Master 9-10
- Group Activity Card 9-10
- Assessment and Evaluation Masters, p. 240
- Real-World Applications, 21
- Tech Prep Applications Masters, p. 18

Transparency 9-10A contains the 5-Minute Check for this lesson; **Transparency 9-10B** contains a teaching aid for this lesson.

Recommended Pacing	
Standard Pacing	Day 12 of 14
Honors Pacing	Day 11 of 13
Block Scheduling*	Day 6 of 7 (along with Lesson 9-9)

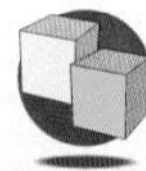

*For more information on pacing and possible lesson plans, refer to the ***Block Scheduling Booklet***.

1 FOCUS

5-Minute Check
(over Lesson 9-9)

Find the discount or interest to the nearest cent.

1. \$28 jeans, 20% off **\$5.60**
2. \$54.95 coat, 15% off **\$8.24**
3. \$3000 at 12.5% for 18 months **\$562.50**
4. \$250 at 9% for 6 months **\$11.25**
5. \$15 CD at 25% off **\$3.75**

Motivating the Lesson

Questioning Ask students to think about the words *increase* and *decrease*. Write each word on the board and make a list of things that can increase or decrease, such as weight, prices, time it takes to do a particular activity, and so on.

9-10 Percent of Change

Setting Goals: *In this lesson, you'll find percent of increase or decrease.*

Modeling a Real-World Application: Electricity Rates

Rate Hike Announced

Source: New York Times, April 22, 1995

The Consolidated Edison Company of New York announced that they will be raising rates on May 1, 1995. The increase will generate \$55.1 million. That represents a 1.1% increase. A typical customer in an apartment that uses 300 kilowatt-hours a month will see an increase in his or her monthly bill from \$50.52 to \$51.26. A Westchester customer using 450 kilowatt-hours a month will pay \$71.05 instead of \$70.05.

Find the percent of increase for each customer and compare your answers to the 1.1% amount the company stated. *This problem will be solved in Example 3.*

Learning the Concept

Newspapers often report changes in prices and rates as a **percent of change**. The percent of change is the ratio of the amount of change to the original amount.

Example  **Find the percent of change from \$120 to \$135.**

There are two methods you can use to find the percent of change.

Method 1

Step 1: Subtract to find the amount of change.
$135 - 120 = 15$

Step 2: Solve the percent equation. Compare the amount of increase to the original amount.

$$P = R \cdot B$$
$$15 = R \cdot 120$$
$$\frac{15}{120} = \frac{120R}{120}$$
$$0.125 = R$$

The percent of change is 12.5%.

Method 2

Step 1: Divide the new amount by the original amount.
135 [÷] 120 [=] *1.125*

Step 2: Subtract 1 from the result and write the decimal as a percent.
$1.125 - 1 = 0.125$ or 12.5%

The percent of change is 12.5%.

Alternative Learning Styles

Kinesthetic Provide students with 1-inch grid paper. Have then cut out a rectangle with a maximum area of 20 square inches. Then have students cut out a larger rectangle. Have them count the squares in each rectangle and find the percent of increase in size.

When an amount increases, like in Example 1, the percent of change is a **percent of increase**. When the amount decreases, the percent of change is negative. You can state a negative percent of change as a **percent of decrease**.

Example  **Find the percent of decrease from a population of 257 thousand to 243 thousand.**

Method 1

Step 1: Subtract.
$243 - 257 = -14$

Step 2: Solve the percent equation. Compare the amount of change to the original amount.

$$P = R \cdot B$$
$$-14 = R \cdot 257$$
$$\frac{-14}{257} = \frac{257R}{257}$$
$$-0.054 = R$$

The percent of change is −5.4%.
The percent of decrease is 5.4%.

Method 2

Step 1: Divide the new amount by the original amount.

243 [÷] 257 [=] .945525

Step 2: Subtract 1 from the result and write the decimal as a percent.

$0.946 - 1 = -0.054$ or -5.4%

The percent of change is −5.4%.
The percent of decrease is 5.4%.

Example 3 **Refer to the application at the beginning of the lesson. Find the percent increase for each customer.**

APPLICATION

Energy

Apartment bill

Use Method 1.

Step 1: Subtract. $51.26 − $50.52 = $0.74 *Estimate:* $\frac{1}{50} = 2\%$

Step 2: Solve the percent equation.

$$P = R \cdot B$$
$$0.74 = R \cdot 50.52$$
$$\frac{0.74}{50.52} = \frac{50.52R}{50.52} \quad \textit{Divide each side by 50.52.}$$
$$0.0146 = R$$

The percent of increase is about 1.5%.

Westchester bill

Use Method 2. *Estimate:* $\frac{71}{70}$ *will be a little over 100%.*

Step 1: Divide the new amount by the original amount.

71.05 [÷] 70.05 [=] 1.014276

Step 2: Subtract 1 from the result and write the decimal as a percent.
$1.014 - 1 = 0.014$ or 1.4%

The percent of increase is 1.4%.

The apartment bill rose 1.5%, and the Westchester bill rose 1.4%, not the 1.1% estimated by the power company.

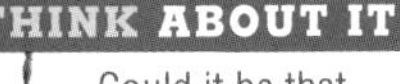

THINK ABOUT IT

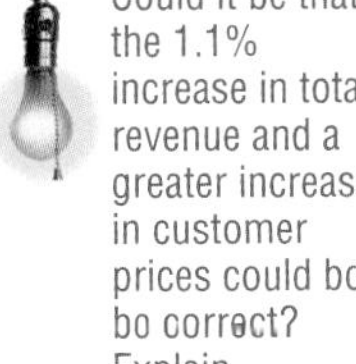

Could it be that the 1.1% increase in total revenue and a greater increase in customer prices could both be correct? Explain.

See margin.

2 TEACH

Teaching Tip You may wish to have students work a few problems using both methods to assure that they understand each method. Then students may choose the method they prefer to use.

In-Class Examples

For Example 1
Find the percent of change from $300 to $396. **32%**

For Example 2
Find the percent of decrease from a population of 120 million to 99 million. **17.5%**

For Example 3
A telephone company announced an increase on long-distance rates. A business had an average long-distance bill of $475 before the increase and $494 after the increase. Find the percent of increase. **4%**

Study Guide Masters, p. 80

NAME ______________ DATE ________

9-10 Study Guide
Percent of Change

Student Edition
Pages 472–475

Example: Al's Sporting Goods Store raised the price of one of its best-selling bicycles from $125 to $140. Find the **percent of increase**.

Subtract to find the amount of change. $140 - 125 = 15$

Solve the percent equation. Compare the amount of increase to the original amount.

$$P = R \cdot B$$
$$15 = R \cdot 125$$
$$\frac{15}{125} = \frac{125R}{125}$$
$$0.12 = R$$

The percent of change is 12%.

Example: Al's Sporting Goods Store reduced the price of one of its less-popular bicycles from $80 to $60. Find the **percent of decrease**.

Divide the new amount by the original amount. $60 \div 80 = 0.75$

Subtract 1 from the result and write the decimal as a percent. $0.75 - 1 = -0.25$ or -25%

The percent of change is −25%.
The percent of decrease is 25%.

State whether each percent of change is a percent of increase or a percent of decrease. Then find the percent of increase or decrease. Round to the nearest whole percent. I = increase; D = decrease

1. a $120 turntable now costs $150 I; 25%
2. a $6 album is now $4.20 D; -30%
2. a $100 digital watch is now $72 D; -28%
4. a $32 sweater now costs $36.80 I; 15%
5. Cheryl weighed 120 pounds. She dieted and lost 12 pounds in three weeks. Find the percent of change in Cheryl's weight. D; -10%
6. The junior high school's enrollment changed from 1200 to 1350 students. Find the percent of change in enrollment. I; 13%
7. Potatoes baked in the oven require 60 minutes to cook. A pressure cooker can do the same job in 20 minutes. Find the percent of change in cooking time. D; -67%
8. A $320 stereo amplifier is on sale for a limited time at $264. Find the percent of change in price. D; -18%

Reteaching

Using Flowcharts Have students create a flowchart that shows how to determine the percent of change.

Additional Answer Think About It

yes; Many utility companies waive fees or reduce rates to customers with limited income. If the rates to these customers were not increased or increased at a lesser rate, both the 1.1% increase and the calculated 1.4 to 1.5% increase could be correct.

3 PRACTICE/APPLY

Checking Your Understanding

Exercises 1–10 are designed to help you assess your students' understanding through reading, writing, speaking, and modeling. You should work through Exercises 1–5 with your students and then monitor their work on Exercises 6–10.

Assignment Guide

Core: 11–29 odd, 30–34
Enriched: 12–22 even, 23–34

For **Extra Practice**, see p. 764.

The red A, B, and C flags, printed only in the Teacher's Wraparound Edition, indicate the level of difficulty of the exercises.

Additional Answer

1. Sample answer: Find the change, then solve for the percent of the change to the original amount; or divide the new amount by the original amount, then subtract 1 and write the result as a percent.

Practice Masters, p. 80

NAME ______________________ DATE ____________

9-10 Practice
Percent of Change

Student Edition
Pages 472–475

State whether each percent of change is a percent of increase or a percent of decrease. then find the percent of increase or decrease. Round to the nearest whole percent. I = increase; D = decrease

1. old: $48.50 new: $38.80 D; −20%
2. old: $15,000 new: $45,000 I; 200%
3. old: $0.80 new: $1.08 I; 35%
4. old: $19.95 new: $23.94 I; 20%
5. old: $0.36 new: $0.60 I; 67%
6. old: $50 new: $35 D; −30%
7. old: 15,200 new: $14,212 D; −7%
8. old: $150 new: $135 D; −10%
9. old: $75 new: $85 I; 13%
10. old: $20.00 new: $15.50 D; −23%
11. old: $2880 new: $3500 I; 22%
12. old: $3.00 new: $3.85 I; 28%
13. old: $58.50 new: $37.50 D; −36%
14. old: $350 new: $311 D; −11%
15. old: $325 new: $375 I; 15%
16. old: $13.50 new: $8.00 D; −41%
17. old: $52.25 new: $78.00 I; 49%
18. old: $16 new: $22 I; 38%
19. old: $135.00 new: $101.25 D; −25%
20. old: $306.25 new: $350.00 I; 14%
21. old: $84.00 new: $205.80 I; 145%
22. old: $533 new: $260 D; −51%
23. old: $1800 new: $1440 D; −20%
24. old: $350 new: $329 D; −6%
25. old: $75.11 new: $72.50 D; −3%
26. old: $16.50 new: $13.55 D; −18%
27. old: $9.75 new: $10.50 I; 8%

Checking Your Understanding

Communicating Mathematics

Read and study the lesson to answer each question.

1. **Explain** how to find a percent of change. See margin.
2. **Estimate** the percent of change if a temperature goes from 20°C to 25°C. (2. +25%)
3. **Name** the amount used as the base when you find a percent of change. the original amount
4. The Cardinals football team scored 14 points against the Vikings. The following week, the Cardinals scored 21 points in their game against the Wolves. The coach said the team had improved their score 150%. Is that correct? Explain. No; the percent of change was 50%.

MATH JOURNAL

5. Write about the least difficult and most difficult concepts for you in this chapter. See students' work.

Guided Practice

State whether each percent of change is a percent of increase or a percent of decrease. Then find the percent of increase or decrease. Round to the nearest whole percent. I = increase; D = decrease

6. old: $10
new: $16 I; 60%
7. old: 25 thousand people
new: 22 thousand people (7. D; 12%)
8. old: 120 pounds
new: 150 pounds I; 25%
9. old: 59°F
new: 55°F D; 7%
10. **Economics** October 19, 1987 is called Black Thursday because the Dow had its biggest decline, 508.32 points, on that day. If the average was 1738.42 points at closing, what was the percent of decrease of the average that day? 22.6%

Tracking the Dow

Exercises: Practicing and Applying the Concept

Independent Practice

State whether each percent of change is a percent of increase or a percent of decrease. Then find the percent of increase or decrease. Round to the nearest whole percent. I = increase; D = decrease

A

11. old: 142 pounds
new: 114 pounds D; 20%
12. old: $0.59
new: $0.30 D; 49%
13. old: 425 people
new: 480 people I; 13%
14. old: $48
new: $44 D; 8%

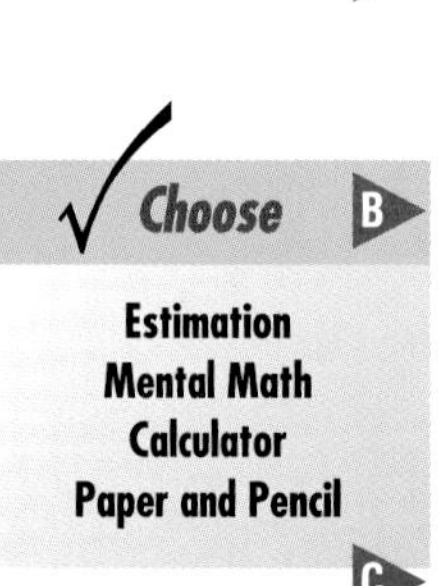

Choose
B
Estimation
Mental Math
Calculator
Paper and Pencil

15. old: 96 minutes
new: 108 minutes I; 13%
16. old: 14.5 liters
new: 12.5 liters D; 14%
17. old: $228
new: $251 I; 10%
18. old: 29.5 ounces
new: 26.5 ounces D; 10%
19. old: $39.99
new: 42.59 I; 7%
20. old: 124 minutes
new: 137 minutes I; 10%

C

21. old: $106
new: $250 I; 136%
22. old: 4329.80 points
new: 4351.50 points I; 1%

Critical Thinking

23. A tennis outfit is marked down 50% and then reduced at the cash register another 30%.
    a. Is this a total reduction of 80%? Why or why not? See margin.
    b. If the outfit is not 80% off, find the actual discount. 65%
24. Explain why a 10% increase followed by a 10% decrease is less than the original amount if the original amount was positive. See margin.

Additional Answers

23a. No; suppose the price of the outfit is $100. The price would be $50 after the first reduction. After the second reduction, the price would be 50 − 0.3(50) = 50 − 15 or $35. A final price of $35 is $\frac{100 - 35}{100}$ or 65% off, not 80% off.

24. Let x represent the original amount. After the increase it is $1.1x$. After the decrease it is $0.9(1.1x)$ or $0.99x$. Since when x is positive, $0.99x < x$, the final amount is less than the original.

Applications and Problem Solving

25. Consumer Awareness Most cars depreciate, or lose value, each year. Kiku paid $7800 for her 1994 Geo Metro. The "blue book" that tells the value of used cars lists the value of a Metro like Kiku's as $6225. What was the percent of decrease of the value of the car? **20%**

26. Merchandising A retail store buys a coat for $85 and marks the price up 65% for sale. Later the price is reduced to $110. **26b. 22%**

26a. 29%

a. If the coat sells for $110, what percent of the selling price is profit?

b. What was the percent of discount from the original selling price?

27. Postage When the U.S. Postal Service changed its rates in 1995, the prices of 1-ounce first class letters and two-pound priority mail packages increased. If the price of a priority package was increased at the same percent of increase as a first class letter, what would priority mail have cost? **$3.20**

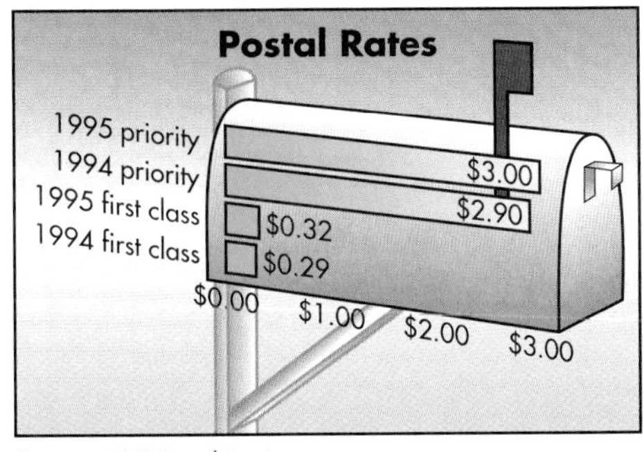

Source: U.S. Postal Service

28. Manufacturing Sometimes manufacturers increase their revenues by selling a product at the same price, but decreasing the quantity sold. For example, a can of fruit juice was changed so that a can which used to hold 6 ounces now contains 5.5 ounces.

a. What was the percent of decrease? **8.3%**

b. If the cans were sold in a 6-pack for $1.59, what was the percent of increase in price per ounce? **9.1%**

29. Pets Looking for an unusual pet? How about a hedgehog! They are becoming so popular that some breeders have reported a 250% increase in sales in recent years. If a breeder sold 50 hedgehogs one year before the increase, how many should he or she expect to sell in a year now? **125**

Mixed Review

30. A pair of earrings usually sell for $15.50. Find the sale price of the earrings if they are on sale for 35% off. (Lesson 9-9) **$10.08**

31. Is the equation whose graph is shown at the right a function? Explain. (Lesson 8-4) **See margin.**

$y = 2x + 1$

32. Energy The United States consumes about 4,921,000,000 barrels of oil each year. The annual production of crude oil in the United States is about half of what is consumed. About how much oil is produced in the United States each year? (Lesson 6-2) **2,500,000,000 barrels**

33. An airline gives each of its flight attendants one red shirt, one white shirt, one blue shirt, a navy blazer, one pair of navy pants, and one pair of navy pin-striped pants. How many different outfits can an attendant wear if the blazer is optional? (*Hint:* Find the outfits possible without the blazer, then add the blazer to each outfit.) (Lesson 4-3)

33. 12 outfits

34. Solve $n = 14 - (-3)$. (Lesson 2-5) **17**

4 ASSESS

Closing Activity

Speaking Use advertisements from newspapers or magazines that show a reduced price for an item (car, dress, shoes, and so on). Have students tell you the steps in finding the percent of change for the price of the item as you write each step on the board.

Chapter 9, Quiz D (Lessons 9-9 and 9-10) is available in the *Assessment and Evaluation Masters*, p. 240.

Additional Answer

31. Yes; it passes the vertical line test.

Group Activity Card 9-10

Why Are They Different? Group Activity **9-10**

MATERIALS: Two 0-9 spinners

Work together in groups of three or four to study percent of increase.

To begin, each member of the group should choose a 2-digit number with which to work. Someone should spin the two spinners to make another 2-digit number. The number formed by the digits on the spinners will be the *amount* of increase for this activity.

Using your own 2-digit number as the original amount, compute the *percent* of increase. Compare your results with other group members. Repeat the process three more times.

With a partner write a short paragraph explaining why all of the percent of increases were not the same.

 Pre-Algebra

Extension

Using Data Have students collect data on one of the following topics and use it to write word problems.

- school enrollments last year and this year
- town/city populations ten years ago and now
- cost of gasoline ten years ago, five years ago, and now

Enrichment Masters, p. 80

NAME ______ DATE ______

9-10 Enrichment
Finance Charges

Student Edition Pages 472–475

PREVIOUS BALANCE	TOTAL PURCHASES	PAYMENTS RETURNS & OTHER CREDITS	FINANCE *CHARGE	THIS IS YOUR NEW BALANCE	PAYMENTS OVERDUE	THIS IS YOUR MINIMUM PAYMENT
245.67	138.56	184.23	2.94	202.94		25.00

*ANNUAL PERCENTAGE RATE OF 15% COMPUTED ON THE AVERAGE DAILY BALANCE OF $235.19.

Eileen Farrell received her monthly statement for her department store charge account. She left part of her balance unpaid last month. Her statement says that her finance charge is $2.94. This is computed at the annual rate of 15% on an average daily balance of $235.19. Is the charge correct?

Example: Compute the monthly finance charge for Eileen's unpaid balance. The annual percentage rate is 15%. To find the monthly rate, divide by 12.

Monthly rate: $15\% \div 12 = 1.25\%$ Compute 1.25% of $235.19.

$B \cdot R = P \rightarrow 235.19 \cdot 0.0125 = P$ 1.25% = 0.0125

$2.939875 = P$

$2.94 = P$ Round to the nearest cent.

The finance charge is $2.94. The statement is correct.

Find the monthly finance charge to the next cent on each average daily balance.

1. Balance: $165.00
Annual percentage rate: 18% **$2.48**

2. Balance: $231.00
Annual percentage rate: 15% **$2.89**

3. Balance: $713.00
Annual rate: 12% **$7.13**

4. Balance: $147.93
Annual rate: 15.5% **$1.91**

5. Balance: $419.60
Annual rate: 14% **$4.90**

6. Balance: $175.14
Annual rate: 16% **$2.34**

7. Balance: $450.00
Annual rate: 17% **$6.38**

8. Balance: $368.50
Annual rate: 18% **$5.53**

9. Balance: $250.50
Annual rate: $18\frac{1}{2}\%$ **$3.86**

10. Balance: $654.90
Annual rate: 12.5% **$6.82**

CLOSING THE

Investigation

This activity provides students an opportunity to bring their work on the Investigation to a close. For each Investigation, students should present their findings to the class. Here are some ways students can display their work.

- Conduct and report on an interview or survey.
- Write a letter, proposal, or report.
- Write an article for the school or local paper.
- Make a display, including graphs and/or charts.
- Plan an activity.

Assessment

To assess students' understanding of the concepts and topics explored in Investigation A and its follow-up activities, you may wish to examine students' Investigation Folders.

The scoring guide provided in the *Investigations and Projects Masters*, p. 19, provides a means for you to score students' work on this Investigation.

Investigations and Projects Masters, p. 19

CHAPTERS 8 and 9 Investigation
Stadium Stampede

Scoring Guide

Level	Specific Criteria
3 Superior	• Shows thorough understanding of *graphing data in scatter plots, graphing linear equations using slope and intercepts, writing expressions for ratios, decimals, fractions, and percents, using ratios and proportions to solve problems, using statistics to predict*, and *solving problems by making a table.* • Uses appropriate strategies to solve problems. • Computations are correct. • Written explanations are exemplary. • Tables and graphs are appropriate and sensible. • Goes beyond the requirement of the Investigation.
2 Satisfactory, with minor flaws	• Shows understanding of *graphing data in scatter plots, graphing linear equations using slope and intercepts, writing expressions for ratios, decimals, fractions, and percents, using ratios and proportions to solve problems, using statistics to predict*, and *solving problems by making a table.* • Uses appropriate strategies to solve problems. • Computations are mostly correct. • Written explanations are effective. • Tables and graphs are appropriate and sensible. • Satisfies all requirements of the Investigation.
1 Nearly Satisfactory, with serious flaws	• Shows understanding of most of the concepts of *graphing data in scatter plots, graphing linear equations using slope and intercepts, writing expressions for ratios, decimals, fractions, and percents, using ratios and proportions to solve problems, using statistics to predict*, and *solving problems by making a table.* • May not use appropriate strategies to solve problems. • Computations are mostly correct. • Written explanations are satisfactory. • Tables and graphs are mostly appropriate and sensible. • Satisfies most requirements of the Investigation.
0 Unsatisfactory	• Shows little or no understanding of *graphing data in scatter plots, graphing linear equations using slope and intercepts, writing expressions for ratios, decimals, fractions, and percents, using ratios and proportions to solve problems, using statistics to predict*, and *solving problems by making a table.* • May not use appropriate strategies to solve problems. • Computations are incorrect. • Written explanations are not satisfactory. • Tables and graphs are not appropriate or sensible. • Does not satisfy requirements of the Investigation

CLOSING THE

Investigation

Refer to the Investigation on pages 368–369.

The spirit squad has an idea for the homecoming game that will increase fan spirit. They want to pass out pompoms in the university colors. There are several items that they have questions about and they have asked you to prepare a presentation for their next meeting.

- How many pompoms should they pass out? Should they go to every fan or only a certain percent of them? How should they determine that percent and which fans should get them?
- How should the pompoms be passed out? Should they be passed out at all of the gates, certain gates, or placed on random seats prior to the game?
- Can a sponsor be obtained to pay for the pompoms? Would the sponsor pay for all of the pompoms, only a certain percent of them, a specified dollar amount, or a certain number of them? If the sponsor cannot cover all the costs, how will the remaining costs be paid?
- Do you charge for the pompoms, and if so, how much do you charge for each one? What would be a good price? At what point will the cost be too much for the fans to want to buy one?
- Of what type of material should the pompoms be made? Should recycling be considered when determining the material? If so, should containers be at the gates for fans to recycle the pompoms on their way out of the stadium after the game?

Reread the list of questions. Are there other issues that need to be addressed? Prepare a written report for the committee.

PORTFOLIO ASSESSMENT

You may want to keep your work on this Investigation in your portfolio.

CHAPTER 9 Highlights

Vocabulary

After completing this chapter, you should be able to define each term, property, or phrase and give an example or two of each.

Algebra
base (p. 449)
commission (p. 470)
cross products (p. 444)
discount (p. 468)
interest (p. 467)
percent (p. 449)
percentage (p. 449)
percent of change (p. 472)
percent of decrease (p. 473)
percent of increase (p. 473)
percent proportion (p. 450)
principal (p. 467)
property of proportions (p. 444)
proportion (p. 444)
rate (p. 433, 449)
ratio (p. 432)
unit rate (p. 433)

Geometry
golden ratio (p. 434)

Probability
capture-recapture (p. 448)
probability (p. 440)
sample space (p. 441)

Problem Solving
make a table (p. 437)

Understanding and Using Vocabulary

Determine whether each statement is *true* or *false*.

1. A rate is a special ratio. true
2. Percentage is a ratio that compares a number to 1. false
3. In the percent proportion $\frac{16}{32} = \frac{50}{100}$, 16 is the base. false
4. A sample space is the area of a closed figure. false
5. When you use the percent equation $P = R \cdot B$, R represents $\frac{r}{100}$ where r is the rate. true
6. A ratio is a comparison of two numbers by multiplication. false
7. The property of proportions allows you to compare two proportions. false

Instructional Resources

Three multiple-choice tests and three free-response tests are provided in the *Assessment and Evaluation Masters*. Forms 1A and 2A are for honors pacing, Forms 1B and 2B are for average pacing, Forms 1C and 2C are for basic pacing, Chapter 9 Test, Form 1B is shown at the right. Chapter 9 Test, Form 2B is shown on the next page.

Highlights

Using the Chapter Highlights

The Chapter Highlights begins with a listing of the new terms, properties, and phrases that were introduced in this chapter.

Assessment and Evaluation Masters, pp. 227–228

NAME ______________ DATE ________

CHAPTER 9 Test, Form 1B

1. Express the ratio of 57 : 76 as a fraction in simplest form.
   A. $\frac{9}{13}$ B. $\frac{19}{26}$ C. $\frac{3}{4}$ D. $\frac{57}{76}$ — 1. C
2. Express the ratio 480 miles in 12 hours as a unit rate.
   A. 0.25 mph B. 40 mph C. 42 mph D. 4 mph — 2. B
3. Angie marks off a garden with a perimeter of 30 meters. If the sides of the garden are whole numbers of meters, how many different combinations of measures are possible?
   A. 7 B. 4 C. 3 D. 2 — 3. A
4. How many different ways are there to add prime numbers to make a sum of 11?
   A. 8 B. 7 C. 5 D. 4 — 4. C
5. Marco has \$50 in \$5 and \$10 bills. How many bills could he have?
   A. 12 B. 9 C. 7 D. 4 — 5. B
6. A small forward usually makes 4 out of 10 hoops he attempts. What is the probability that he will miss his next shot?
   A. $\frac{3}{5}$ B. $\frac{2}{5}$ C. $\frac{1}{3}$ D. $\frac{1}{2}$ — 6. A
7. Two dice are tossed. Find the probability that both are odd.
   A. $\frac{1}{2}$ B. $\frac{1}{3}$ C. $\frac{1}{4}$ D. $\frac{1}{6}$ — 7. C
8. Choose the pair of ratios that form a proportion.
   A. $\frac{3}{2}, \frac{9}{4}$ B. $\frac{3}{2}, \frac{4}{3}$ C. $\frac{2}{3}, \frac{4}{6}$ D. $\frac{2}{3}, \frac{4}{9}$ — 8. C
9. Solve the proportion $\frac{x}{6} = \frac{9}{3}$.
   A. 18 B. 27 C. 12 D. 21 — 9. A
10. Solve the proportion $\frac{5}{2a} = \frac{2}{4}$.
    A. 10 B. 5 C. 20 D. 15 — 10. B
11. A soup recipe calls for 4 pounds of rice for 20 servings. How many pounds of rice are needed for 30 servings?
    A. 4.5 B. 8 C. 6 D. 5 — 11. C
12. Express $\frac{2}{5}$ as a percent.
    A. 2.5% B. 250% C. 0.4% D. 40% — 12. D
13. What is 80% of 20?
    A. 0.16 B. 16 C. 4 D. 0.04 — 13. B
14. Three is what percent of 15?
    A. 45% B. 20% C. 5% D. 50% — 14. B

NAME ______________ DATE ________

Chapter 9 Test, Form 1B (continued)

15. Express 0.305 as a percent.
    A. 0.305% B. 0.00305% C. 30.5% D. 3.05% — 15. C

Use the survey on favorite colors at the right for Exercises 16–17.

Favorite Colors			
Red	26	Green	34
Blue	30	Yellow	10

16. What fraction chose green?
    A. $\frac{13}{50}$ B. $\frac{3}{10}$ C. $\frac{1}{10}$ D. $\frac{17}{50}$ — 16. D
17. For an expected 300 customers, how many pairs of green shorts should a sport shop stock?
    A. 100 B. 102 C. 52 D. 20 — 17. B
18. Brian is ordering ice cream for an auto-racing event. In a random sample, he finds that 18 of 72 people asked prefer coffee-flavored ice cream. For 2000 servings, about how many coffee-flavored ice cream servings should Brian order?
    A. 400 B. 500 C. 600 D. 2000 — 18. B
19. Express 3.4% as a decimal.
    A. 340 B. 0.034 C. 0.34 D. 3.4 — 19. B
20. Estimate 0.4% of 958.
    A. 4 B. 5 C. 3 D. 2 — 20. A
21. Estimate the percent of 13 out of 27.
    A. 40% B. 50% C. 60% D. 70% — 21. B
22. 15 is 30% of what number?
    A. 50 B. 4.5 C. 2 D. 200 — 22. A
23. Akita buys a swimsuit marked \$50. She receives a 20% discount. Find the sale price of the swimsuit.
    A. \$45 B. \$40 C. \$30 D. \$10 — 23 B
24. When an item decreases in value over time, it depreciates. The value of a car changed from \$14,000 to \$11,200. Find the percent of decrease in the value of the car.
    A. 10% B. 14% C. 28% D. 20% — 24. D
25. A dealer buys a coin for \$850 and sells it for \$1190. Find the percent of increase.
    A. 34% B. 340% C. 40% D. 20% — 25. C

Using the Study Guide and Assessment

Skills and Concepts Encourage students to refer to the objectives and examples on the left as they complete the review exercises on the right.

Assessment and Evaluation Masters, pp. 233–234

NAME ______________________ DATE __________

CHAPTER 9 Test, Form 2B

Express each ratio as a fraction in simplest form.

1. 63 to 93 — 1. $\frac{21}{31}$
2. 14 out of 56 — 2. $\frac{1}{4}$

Express each ratio as a unit rate.

3. 202.5 miles in 9 hours — 3. 22.5 mph
4. $495 for 30 tickets — 4. $16.50 per ticket
5. A rectangle has a perimeter of 22 inches. If the sides of the rectangle are whole numbers of inches, how many different combinations of measures are possible? — 5. 5
6. What is the probability of an event that is certain to happen? — 6. 1

A bag contains 5 red, 6 black, 4 blue, and 3 yellow marbles. Suppose you select one marble at random. Find each probability.

7. P (red) — 7. $\frac{5}{18}$
8. P (not black) — 8. $\frac{2}{3}$
9. Solve the proportion $\frac{k}{25} = \frac{3}{5}$. — 9. 15
10. Colleen's car averages 32 miles per gallon of gasoline. How many gallons are needed for a trip of 528 miles? — 10. 16.5 gallons
11. Three ounces of a perfume cost $105. How much should 7 ounces cost? — 11. $245
12. What number is 28% of 425? — 12. 119
13. Six is what percent of 15? — 13. 40%
14. Twelve is 25% of what number? — 14. 48

Use the survey on favorite sports at the right to answer questions 15–16.

Favorite Sport	
Soccer	32
Basketball	26
Football	26
Hockey	12

15. What fraction of the spectators asked prefer soccer? — 15. $\frac{1}{3}$
16. For a town with 1200 sports-club members, how many hockey banners should the town's sports store stock? — 16. 150

NAME ______________________ DATE __________

Chapter 9 Test, Form 2B (continued)

17. Marla is preparing for a weekend trip to Canada. She has $50 dollars to spend on gifts. One U.S. dollar is currently equal to $1.35 Canadian. How much Canadian money will she have for gifts? — 17. $67.50
18. Express 0.4 as a percent. — 18. 40%
19. Express 37.5% as a fraction in simplest form. — 19. $\frac{3}{8}$
20. Express 3.25% as a decimal. — 20. 0.0325
21. Estimate a 15% tip on a restaurant bill of $39.95. — 21. $6
22. Estimate the 6% sales tax on a $635 refrigerator. — 22. $40
23. Dave borrows $800 at an interest rate of 9.5% per year for 3 years. How much interest will he have to pay? — 23. $228
24. Last year, the tuition at a private college was $12,000. This year it will be $12,720. Find the percent of increase. — 24. 6%
25. Find the percent of change from $700 to $784. State whether the change is an increase or a decrease. — 25. 12%, increase

CHAPTER 9 Study Guide and Assessment

Skills and Concepts

Objectives and Examples

Upon completing this chapter, you should be able to:

▶ **write ratios as fractions in simplest form and determine unit rates.** (Lesson 9-1)

Express 15 to 21 as a fraction in simplest form.

$$\frac{15}{21} = \frac{5}{7} \quad (\div 3)$$

Express $1.29 for a dozen as a unit rate.

$$\frac{129 \text{ cents}}{12 \text{ items}} = \frac{\bullet}{1 \text{ item}} \quad (\div 12)$$

129 [÷] 12 [=] *10.75*

The unit rate is 10.75 cents.

Review Exercises

Use these exercises to review and prepare for the chapter test.

Express each ratio as a fraction in simplest form. 8–11. See margin.

8. 10 red to 18 blue
9. 287 yards to 315 yards
10. 15 out of 90
11. 600 to 1000

Express each ratio as a unit rate.

12. 339.2 miles on 10.6 gallons
13. $1.78 for 2 pounds
14. $425 for 17 tickets
15. 142.5 miles in 2.5 hours

12–15. See margin.

▶ **find the probability of simple events.** (Lesson 9-3)

Find the probability of rolling a number that is a factor of 6 on a die.

There are four ways to roll a factor of 6 –1, 2, 3, and 6, and six possible outcomes –1, 2, 3, 4, 5, and 6.

$P(\text{factor of } 6) = \frac{4}{6}$ or $\frac{2}{3}$

There are 6 cans of cola, 4 cans of fruit punch, 9 cans of iced tea, and 2 cans of lemonade in a cooler. Suppose you choose a can randomly. Find each probability. 16–21. See margin.

16. P(cola)
17. P(fruit punch or iced tea)
18. P(not carbonated)
19. P(cola or lemonade)
20. P(iced tea)
21. P(neither cola nor fruit punch)

▶ **use proportions to solve problems.** (Lesson 9-4)

If 2 pounds of pecans cost $13.90, how much will 5 pounds cost?

$\frac{13.90}{2} = \frac{c}{5}$ *Write a proportion.*

$13.90 \cdot 5 = 2 \cdot c$ *Write cross products.*

$\frac{69.50}{2} = \frac{2c}{2}$ *Divide each side by 2.*

$c = 34.75$

Five pounds will cost $34.75.

Solve each proportion.

22. $\frac{8}{6} = \frac{z}{14}$ $18\frac{2}{3}$
23. $\frac{5.1}{1.7} = \frac{7.5}{a}$ 2.5

Write a proportion that could be used to solve for each variable. Then solve. 24–25. See margin.

24. 20 grams for $5.60
    x grams for $9.80
25. 88.4 miles on 3.4 gallons
    161.2 miles on g gallons

GLENCOE Technology

Test and Review Software

You may use this software, a combination of an item generator and an item bank, to create your own tests or worksheets. Types of items include free response, multiple choice, short answer, and open ended.

For IBM & Macintosh

Additional Answers

8. $\frac{5}{9}$
9. $\frac{41}{45}$
10. $\frac{1}{6}$
11. $\frac{3}{5}$
12. $\frac{32 \text{ miles}}{1 \text{ gallon}}$
13. $0.89 per pound
14. $\frac{\$25}{1 \text{ ticket}}$
15. $\frac{57 \text{ miles}}{1 \text{ hour}}$

Objectives and Examples

▸ **use the percent proportion to solve problems involving percents.** (Lesson 9-5)

Sixteen is what percent of 20?

$\frac{P}{B} = \frac{r}{100} \rightarrow \frac{16}{20} = \frac{r}{100}$

$16 \cdot 100 = 20 \cdot r$

$\frac{1600}{20} = \frac{20r}{20}$

$80 = r$

16 is 80% of 20.

Review Exercises

Use the percent proportion to solve each problem.

26. What is 40% of 5? 2
27. Nineteen is what percent of 25? 76%
28. What is 120% of 50? 60
29. What is 0.1% of 4000? 4

▸ **use a sample to predict the actions of a large group.** (Lesson 9-6)

Of the students surveyed, 43% chose pizza as their favorite lunch. How many of the 1294 students in the school would choose pizza?

$\frac{x}{1294} = \frac{43}{100}$

$x \cdot 100 = 1294 \cdot 43$

$x = 556.42$ *Use a calculator.*

About 556 students would choose pizza.

The results of a survey on dessert preferences are shown in the table at the right.

Dessert	Number
cheesecake	7
chocolate cake	48
apple pie	37
ice cream	18

30. How many of the 459 cafeteria customers would choose apple pie? 154
31. If you asked 500 people which of these desserts they prefer, how many would choose cheesecake? 32

▸ **express decimals and fractions as percents and vice versa.** (Lesson 9-7)

Express 0.32 as a percent.

$\frac{0.32}{1} = \frac{32}{100}$ or 32% (× 100)

Express each decimal as a percent.

32. 0.92 92%
33. 0.0056 0.56%

Express each fraction as a percent.

34. $\frac{63}{100}$ 63%
35. $\frac{113}{200}$ 56.5%

Express each percent as a fraction.

36. 90% $\frac{45}{50}$
37. 65% $\frac{13}{20}$

Express each percent as a decimal.

38. 45% 0.45
39. 235% 2.35

▸ **use percents to estimate.** (Lesson 9-8)

Estimate 12% of 54.

12% is about 10%.
10% of 54 is 5.4.
12% of 54 is about 5.4.

Choose the best estimate.

40. 2% of 180 a. 36 b. 3.6 c. 0.36 b
41. 198% of 5 a. 10 b. 1.0 c. 0.10 a
42. 8% of 420 a. 42 b. 5 c. 400 a
43. 73% of 80 a. 60 b. 40 c. 80 a
44. 352% of 20 a. 8 b. 40 c. 70 c
45. 0.6% of 620 a. 2 b. 6 c. 12 b

Chapter 9 • Study Guide and Assessment

Additional Answers

16. $\frac{2}{7}$
17. $\frac{13}{21}$
18. $\frac{5}{7}$
19. $\frac{8}{21}$
20. $\frac{3}{7}$
21. $\frac{11}{21}$
24. $\frac{5.60}{20} = \frac{9.80}{x}$; 35
25. $\frac{88.4}{3.4} = \frac{161.2}{g}$; 6.2

Chapter 9 • Study Guide and Assessment

Applications and Problem Solving

Encourage students to work through the exercises in the Applications and Problem Solving section to strengthen their problem-solving skills.

Additional Answers

52. I; 25%
53. D; 1%
54. D; 5%
55. D; 22%
56. I; 40%
57. D; 12%

Objectives and Examples

▶ **solve percent problems using the percent equation.** (Lesson 9-9)

1.8 is 4% of what number?

$$P = R \cdot B$$
$$1.8 = 0.04 \cdot B \quad \textit{Replace P with 1.8 and R with 0.04.}$$
$$\frac{1.8}{0.04} = \frac{0.04B}{0.04}$$
$$45 = B$$

1.8 is 4% of 45.

Review Exercises

Solve each problem by using the percent equation, $P = R \cdot B$.

46. 54 is 150% of what number? **36**
47. 16% of what number is 3.2? **20**
48. 63 is what percent of 105? **60%**
49. 2600 is 65% of what number? **4000**
50. 16 is what percent of 2? **800**
51. What percent of 16 is 5? **31.25%**

▶ **find percent of increase or decrease.** (Lesson 9-10)

Find the percent of increase from 159 pounds to 168 pounds.

Subtract: 168 − 159 = 9
Use the percent equation.

$$9 = R \cdot 159 \quad P = R \cdot B$$
$$\frac{9}{159} = \frac{R \cdot 159}{159}$$
$$0.0566 = R$$

The percent of increase is about 5.7%.

State whether each percent of change is a percent of increase or a percent of decrease. Then find the percent of increase or decrease. Round to the nearest whole percent.

52. old: \$0.60 new: \$0.75
53. old: 16 ounces new: 15.8 ounces
54. old: 18.4 grams new: 17.5 grams
55. old: \$15.00 new: \$11.75
56. old: \$84 new: \$118
57. old: 210 pounds new: 185 pounds

52–57. See margin.

Applications and Problem Solving

58. **Records** Nippondenso's Micro-Car is listed as the world's smallest car by The Guinness Book of World Records. It is a model of Toyota's first passenger car, the 1936 Model AA sedan. A part of the actual car that is 1 meter long is 0.001 m long on the model. Write the scale of the model as a fraction in simplest form. (Lesson 9-1)

59. **Games** A domino has two squares on its face. Each of the two spaces is marked with 1, 2, 3, 4, 5, or 6 dots, or it is blank. A complete set of dominoes includes one domino for each possible combination of dots and blanks. Doubles, like 5 dots in each space, are included. How many dominoes are in a complete set? (Lesson 9-2) **28**

60. **Safety** Fifty-eight percent of the adults surveyed said they wore their seat belts all the time. If there are 185 million adults in the United States, how many American adults wear their seat belts all the time? (Lesson 9-6) **about 107 million**

58. $\frac{1}{1000}$

A practice test for Chapter 9 is available on page 782.

Alternative Assessment

Cooperative Learning Project

Work as a group. **See students' work.**

1. Develop a multiple-choice questionnaire on any topic. Include at least five questions with three or more choices. Survey 20% of your class for their opinions.
2. Record the data you gather in a table, graph, or scatter plot.
3. What fraction of the people surveyed chose each possible answer? Write a sentence about the probability of randomly choosing a person who chose a certain answer.
4. List the sample space of each question on the questionnaire.
5. Predict the number of people in your class who would choose a certain answer on your questionnaire. Explain how you made your prediction.
6. Write a statement about the percent of your survey group that chose a certain answer.
7. Present the results of your survey in an interesting way. You may wish to make a poster or brochure, including tables or graphs.

Thinking Critically

- When you write a decimal as a percent, the first step is to write it as a fraction with a denominator of 1. Then you multiply by 100. Can you write one step that could replace these two?

  Move the decimal point two places to the right.

- A weather forecaster said "There is a 50% chance of rain on Saturday and a 50% chance of rain on Sunday. So there is a 100% chance of rain this weekend." Do you think this is correct? Explain.

  Sample answer: no; the chance would have to be 100% both days for it to be 100% for the whole weekend.

Portfolio

Select one of the assignments from this chapter that you found especially challenging and place it in your portfolio. **See students' work.**

Self-Evaluation

"It is the province of knowledge to speak and it is the privilege of wisdom to listen." This quote by Oliver Wendell Holmes is saying that it is wise to be a good listener.

Assess yourself. Do you listen well? Studies show that people remember about 20% of what they hear. How much of what you hear do you remember? How can you remember more of what you hear? **See students' work.**

Alternative Assessment

The Alternative Assessment section provides students with the opportunity to assess their own work by thinking critically, working with others, keeping a portfolio, and honestly evaluating their own progress. For more information on alternative forms of assessment, see *Alternative Assessment in the Mathematics Classroom*.

Performance Assessment

Performance Assessment tasks for this chapter are included in the *Assessment and Evaluation Masters*. A scoring guide is also provided.

Alternative Assessment

Assessment and Evaluation Masters, pp. 237, 249

NAME ____________________ DATE __________

CHAPTER 9 Performance Assessment

Instructions: Demonstrate your knowledge by giving a clear, concise solution to each problem. Be sure to include all relevant drawings and justify your answers. You may show your solution in more than one way or investigate beyond the requirements of the problem.

1. Farmers use ratios and proportions to buy supplies, plant crops, care for animals, and determine rates of yield.
   - **a.** The directions on a bag of fertilizer indicate that one bag is enough to fertilize 0.5 acres. Tell how a farmer could find the number of bags of fertilizer he would need for 42 acres. Solve.
   - **b.** A farmer gets about 125 eggs each day from her chickens. She thinks she should be getting about 220 eggs. As a sales representative for chicken feed, compose an advertisement that would be attractive to this farmer. Use percent in your ad.
2. Many photocopiers produce copies that are larger or smaller than the original. The operator determines the rate at which the original is reduced or enlarged. Suppose the buttons on a photocopier read 60%, 100%, and 120%. The 60% button means that photocopies will be 60% as large as the original.
   - **a.** In your own words, tell what the ratio 120% means.
   - **b.** Solve for x and tell what the answer means.

     $\frac{120}{100} = \frac{x}{10}$ inches
   - **c.** Decide on a length and width for an original drawing for which you would like a reduced copy. If you use the copier described above, what would the length and width of a photocopy be?
   - **d.** Describe what a copy would look like if you pressed the 100% button.
   - **e.** Would a 5% button be useful in reproducing printed materials such as the material on this page? Why or why not?
3. A telemarketing group sells subscriptions to 1 of each 35 people it calls.
   - **a.** If the group calls 700 people in one month, how many subscriptions does it sell?
   - **b.** How many calls should the group make to reach a monthly sales goal of 100 subscriptions?

CHAPTER 9 Scoring Guide

Level	Specific Criteria
3 Superior	• Shows thorough understanding of the concepts *percent*, *ratio*, and *proportion*. • Uses appropriate strategies to solve problems. • Computations to solve proportions are correct. • Written explanations are exemplary. • Advertisement is accurate. • Goes beyond requirements of some or all problems.
2 Satisfactory, with minor flaws	• Shows understanding of the concepts *percent*, *ratio*, and *proportion*. • Uses appropriate strategies to solve problems. • Computations to solve proportions are mostly correct. • Written explanations are effective. • Advertisement is mostly accurate. • Satisfies all requirements of problems.
1 Nearly Satisfactory, with serious flaws	• Shows understanding of most of the concepts *percent*, *ratio*, and *proportion*. • May not use appropriate strategies to solve problems. • Computations to solve proportions are mostly correct. • Written explanations are satisfactory. • Advertisement is mostly accurate. • Satisfies most requirements of problems.
0 Unsatisfactory	• Shows little or no understanding of the concepts *percent*, *ratio*, and *proportion*. • May not use appropriate strategies to solve problems. • Computations to solve proportions are incorrect. • Written explanations are not satisfactory. • Advertisement is not accurate. • Does not satisfy requirements of problems.

Investigation

TEACHER NOTES

NCTM Standards: 1-4

This Investigation is designed to be completed over several days or weeks. It may be considered optional. You may want to assign the Investigation and the follow-up activities to be completed at the same time.

Objective

- Design an original game.

Mathematical Overview

This Investigation will use the following skills and concepts from Chapters 10 and 11.

- complete a survey
- display data
- use simulations to test game rules and strategies
- use probabilities to develop a "fair" game

Recommended Time		
Part	**Pages**	**Time**
Investigation	482–483	1 class period
Working on the Investigation	508, 528, 588	20 minutes each
Closing the Investigation	600	1 class period

Instructional Resources

Investigations and Projects Masters, pp. 21–24

A recording sheet, teacher notes, and scoring guide are provided for each Investigation in the *Investigation and Projects Masters*.

1 MOTIVATION

In this Investigation students use research, simulations, and probability to explore games of various types. Allow students to discuss their favorite game to play and favorite game to watch on television. Ask what they like about the games they chose.

LONG-TERM PROJECT

Investigation

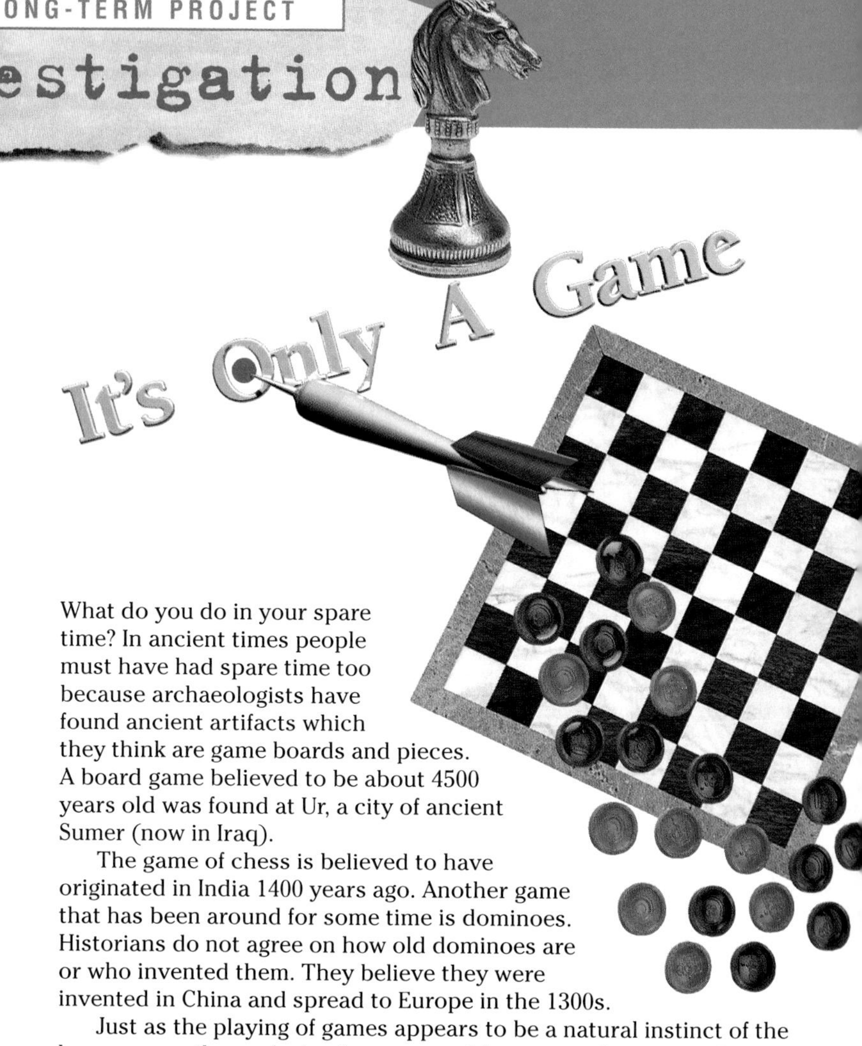

What do you do in your spare time? In ancient times people must have had spare time too because archaeologists have found ancient artifacts which they think are game boards and pieces. A board game believed to be about 4500 years old was found at Ur, a city of ancient Sumer (now in Iraq).

The game of chess is believed to have originated in India 1400 years ago. Another game that has been around for some time is dominoes. Historians do not agree on how old dominoes are or who invented them. They believe they were invented in China and spread to Europe in the 1300s.

Just as the playing of games appears to be a natural instinct of the human race, the analysis of games would appear to be a natural instinct of many mathematicians. Games are most commonly classified according to the equipment they use.

The major types of games are:

- *board games*
- *card games*
- *tile games*
- *target games*
- *dice games*
- *table games*
- *paper and pencil games*
- *electronic games*

Starting the Investigation

Work with your group to design a game. Ask friends, teachers, relatives, and neighbors what their favorite game is and why it is their favorite. Describe, in detail, the qualities of each game that led them to their choice. Determine what qualities are common to the favorite games. Then determine which of these qualities you will use in the design of your game.

Think About . . .

- *What will be the age group of the players who will play your game?*
- *For how many players is the game designed?*
- *What is the object of the game?*
- *Will the game be designed for educational purposes? (Will it teach a player something?)*
- *Where will the game be played? (Is it a party game like charades? a street game, such as hopscotch? a lawn game, such as croquet? or a TV game show?)*
- *What type of game do you want to design? Will it be skill-based, pure chance, or a combination of both?*
- *How will players determine who goes first?*
- *How will players take a turn?*
- *What are the basic rules?*
- *Are there any special rules? What are they based on?*
- *When is the game over? What determines the winner?*

Investigation
It's Only a Game

Working on the Investigation
Lesson 10-4, p. 508

Working on the Investigation
Lesson 10-8, p. 528

Working on the Investigation
Lesson 11-7, p. 588

Closing the Investigation
End of Chapter 11, p. 600

You will work on this Investigation throughout Chapters 10 and 11.

Be sure to keep your materials in your Investigation Folder.

interNET CONNECTION

For current information on game design, visit:
www.glencoe.com/sec/math/prealg/mathnet

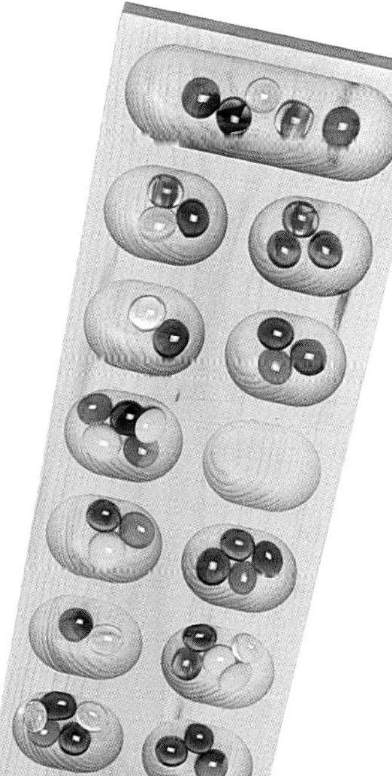

2 SETUP

Have students work in small groups. Each group should complete the surveys and research indicated in the opening discussion on pages 482–483 and in the Working on the Investigation features in Chapters 10 and 11. You may wish to have students consult the following resources for further information.

McNeil, Alex. *Total Television*. New York: Penguin Books

World Wide Web: http://www.cis.ohio-state-edu/hypertext/faq/usenet/tv/game-shows/usa/faw.html

3 MANAGEMENT

Remind students to obtain survey results from a variety of types of people; in other words, various ages, various occupations, and various educational backgrounds.

Sample Answers

Answers will vary as they are based on the research of each group.

Cooperative Learning

This Investigation offers an excellent opportunity for using cooperative learning groups. For more information on cooperative learning strategies and group management, see *Cooperative Learning in the Mathematics Classroom*, one of the titles in the Glencoe Mathematics Professional Series.

Investigations and Projects Masters, p. 24

NAME ____________ DATE ______

CHAPTERS **10 and 11** Investigation
It's Only A Game

Student Edition
Pages 482–483

List three or four of the games you enjoy the most. What type of game is each (board, card, electronic, and so on)? For each game, describe what you like about it.

You will survey friends, neighbors, and family members about their favorite games. Working with your group, plan your survey. What questions will you ask to help you gather information? Write the questions below.

After you complete the survey and gather your information, work with your group. Describe which qualities seem to make games popular. List the qualities which will be part of the game your group will design.

Please keep this page and any other research in your Investigation Folder.

CHAPTER 10

More Statistics and Probability

PREVIEWING THE CHAPTER

In the beginning of the chapter, students study the graphic display of statistics. First, students display and interpret data in stem-and-leaf plots. Then students use measures of variation to compare data. Then box-and-whisker plots, pictographs, and line graphs are used to display data, including the use of a graphing calculator. In the second half of the chapter, students use tree diagrams and the Fundamental Counting Principal to count outcomes. They explore and use permutations and combinations. Probability is extended to finding the odds of a simple event and solving problems using simulations. Students then find experimental probability, the probability of independent and dependent events, and the probability of mutually exclusive or inclusive events.

Lesson (pages)	Lesson Objectives	NCTM Standards	State/Local Objectives
10-1 (486–489)	Display and interpret data in stem-and-leaf plots.	1-4, 10	
10-2 (490–494)	Use measures of variation to compare data.	1-4, 10	
10-3 (495–501)	Use box-and-whisker plots, pictographs, and line graphs to display data.	1-4, 10	
10-3B (502–503)	Use a graphing calculator to construct statistical graphs.	1-4, 10	
10-4 (504–508)	Recognize when statistics are misleading.	1-4, 10	
10-5 (509–513)	Use tree diagrams or the Fundamental Counting Principal to count outcomes.	1-4, 11	
10-6A (514)	Explore and use permutations and combinations.	1-4, 11	
10-6 (515–519)	Use permutations and combinations.	1-4, 11	
10-7 (520–523)	Find the odds of a simple event.	1-4, 11	
10-8 (524–528)	Investigate problems using simulations.	1-4, 11	
10-8B (529)	Conduct experiments to find experimental probability.	1-4, 11	
10-9 (530–534)	Find the probability of independent and dependent events.	1-4, 11	
10-10 (535–538)	Find the probability of mutually exclusive or inclusive events.	1-4, 11	

A complete, 1-page lesson plan is provided for each lesson in the ***Lesson Planning Guide***. Answer keys for each lesson are available in the ***Answer Key Masters***.

ORGANIZING THE CHAPTER

You may want to refer to the **Course Planning Calendar** on page T12 for detailed information on pacing.
PACING: Standard—14 days; **Honors**—13 days; **Block**—7 days

LESSON PLANNING CHART

Lesson (Pages)	Materials/ Manipulatives	Extra Practice (Student Edition)	Blackline Masters: Study Guide	Blackline Masters: Practice	Blackline Masters: Enrichment	Blackline Masters: Assessment and Evaluation	Blackline Masters: Math Lab and Modeling Math	Blackline Masters: Multicultural Activity	Blackline Masters: Tech Prep Applications	Blackline Masters: Graphing Calculator	Blackline Masters: Activity	Real-World Applications	Interactive Mathematics Tools Software	Teaching Transparencies	Group Activity Cards
10-1 (486–489)	tape index cards	p. 764	p. 81	p. 81	p. 81					p. 10		22		10-1A 10-1B	10-1
10-2 (490–494)	graphing calculator	p. 764	p. 82	p. 82	p. 82									10-2A 10-2B	10-2
10-3 (495–501)	stop watch* calculator	p. 764	p. 83	p. 83	p. 83	p. 267		p. 19			p. 10			10-3A 10-3B	10-3
10-3B (502–503)	graphing calculator									p. 23					
10-4 (504–508)		p. 765	p. 84	p. 84	p. 84		p. 69					23		10-4A 10-4B	10-4
10-5 (509–513)		p. 765	p. 85	p. 85	p. 85	pp. 266, 267			p. 19					10-5A 10-5B	10-5
10-6A (514)	blank index cards						p. 46								
10-6 (515–519)		p. 765	p. 86	p. 86	p. 86						p. 42			10-6A 10-6B	10-6
10-7 (520–523)		p. 766	p. 87	p. 87	p. 87									10-7A 10-7B	10-7
10-8 (524–528)	8 pennies cup	p. 766	p. 88	p. 88	p. 88	p. 268		p. 20	p. 20		p. 24	24	10-8	10-8A 10-8B	10-8
10-8B (529)	paper bag marbles*						p. 47								
10-9 (530–534)	graphing calculator	p. 766	p. 89	p. 89	p. 89		p. 85						10-9	10-9A 10-9B	10-9
10-10 (535–538)	colored pencils	p. 767	p. 90	p. 90	p. 90	p. 268								10-10A 10-10B	10-10
Study Guide/ Assessment (540–543)						pp. 253–265, 269–271									

*Included in Glencoe's *Student Manipulative Kit* and *Overhead Manipulative Resources*.

ORGANIZING THE CHAPTER

All of the blackline masters in the Teacher's Classroom Resources are available on the ***Electronic Teacher's Classroom Resources*** CD-ROM.

Other Chapter Resources

Student Edition
Chapter Opener, pp. 484–485
Investigation, pp. 482–483
Working on the Investigation, pp. 508, 528
FYI, pp. 489, 492, 507
Earth Watch, pp. 518–519
From the Funny Pages, p. 523
Cultural Connections, p. 524

Teacher's Classroom Resources
Investigations and Projects Masters, pp. 65–68
Pre-Algebra Overhead Manipulative Resources, pp. 26–29

Technology
Test and Review Software (IBM & Macintosh)
CD-ROM Activities

Professional Publications

Block Scheduling Booklet
Glencoe Mathematics Professional Series

Outside Resources

Books/Periodicals
Exploring Data, James M. Landwehr and Ann E. Watkins
Exploring Probability, C. Newman, T. Obremski, and R. Sheaffer
The Art and Techniques of Simulation, M. Gnandesikan, R. Scheaffer, and J. Swift
What Are My Chances (Book B), Albert P. Schulte and Stuart A. Choate

Videos/CD-ROMs
Mathematical Eye: Probability, Journal Films, Multimedia and Videodisc Compendium

Software
Statistics Workshop, Sunburst
Probability Constructor, Sunburst
MacStat™, Dale Seymour Publications

Assessment Resources

Student Edition
Math Journal, pp. 506, 517, 521
Mixed Review, pp. 489, 494, 501, 508, 513, 519, 523, 528, 534, 538
Self Test, p. 513
Chapter Highlights, p. 539
Chapter Study Guide and Assessment, pp. 540–542
Alternative Assessment, p. 543
Portfolio, p. 543
Ongoing Assessment, pp. 544–545
MindJogger Videoquiz, 10

Teacher's Wraparound Edition
5-Minute Check, pp. 486, 490, 495, 504, 509, 515, 520, 524, 530, 535
Checking Your Understanding, pp. 488, 492, 498, 505, 511, 516, 521, 526, 532, 537
Closing Activity, pp. 489, 494, 501, 508, 513, 519, 523, 528, 534, 538
Cooperative Learning, pp. 499, 515

Assessment and Evaluation Masters
Multiple-Choice Tests, Forms 1A (Honors), 1B (Average), 1C (Basic), pp. 253–258
Free-Response Tests, Forms 2A (Honors), 2B (Average), 2C (Basic), pp. 259–264
Performance Assessment, p. 265
Mid-Chapter Test, p. 266
Quizzes A–D, pp. 267–268
Standardized Test Practice, p. 269
Cumulative Review, pp. 270–271

ENHANCING THE CHAPTER

Examples of some of the materials for enhancing Chapter 10 are shown below.

DIVERSITY

Multicultural Activity Masters, pp. 19, 20

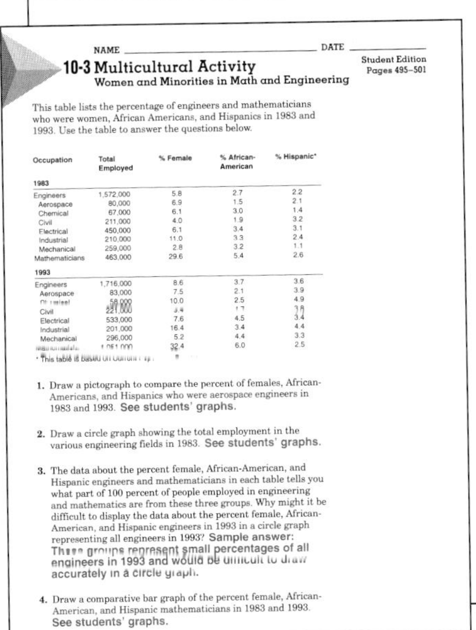

NAME ____________ DATE ________

10-3 Multicultural Activity
Women and Minorities in Math and Engineering

Student Edition Pages 495–501

This table lists the percentage of engineers and mathematicians who were women, African Americans, and Hispanics in 1983 and 1993. Use the table to answer the questions below.

Occupation	Total Employed	% Female	% African-American	% Hispanic*
1983				
Engineers	1,572,000	5.8	2.7	2.2
Aerospace	80,000	6.9	1.5	2.1
Chemical	67,000	6.1	3.0	1.4
Civil	211,000	4.0	1.9	3.2
Electrical	450,000	6.1	3.4	3.1
Industrial	210,000	11.0	3.3	2.4
Mechanical	259,000	2.8	3.2	1.1
Mathematicians	463,000	29.6	5.4	2.6
1993				
Engineers	1,716,000	8.6	3.7	3.6
Aerospace	83,000	7.5	2.1	3.9
[illegible]	58,000	10.0	2.5	4.9
Civil	221,000	[illegible]	[illegible]	3.8
Electrical	533,000	7.6	4.5	3.4
Industrial	201,000	16.4	3.4	4.4
Mechanical	296,000	5.2	4.4	3.3
[illegible]	[illegible]	32.4	6.0	2.5

*This table is based on [illegible]

1. Draw a pictograph to compare the percent of females, African-Americans, and Hispanics who were aerospace engineers in 1983 and 1993. See students' graphs.
2. Draw a circle graph showing the total employment in the various engineering fields in 1983. See students' graphs.
3. The data about the percent female, African-American, and Hispanic engineers and mathematicians in each table tells you what part of 100 percent of people employed in engineering and mathematics are from these three groups. Why might it be difficult to display the data about the percent female, African-American, and Hispanic engineers in 1993 in a circle graph representing all engineers in 1993? Sample answer: These groups represent small percentages of all engineers in 1993 and would be difficult to draw accurately in a circle graph.
4. Draw a comparative bar graph of the percent female, African-American, and Hispanic mathematicians in 1983 and 1993. See students' graphs.

Real-World Applications, 22, 23, 24

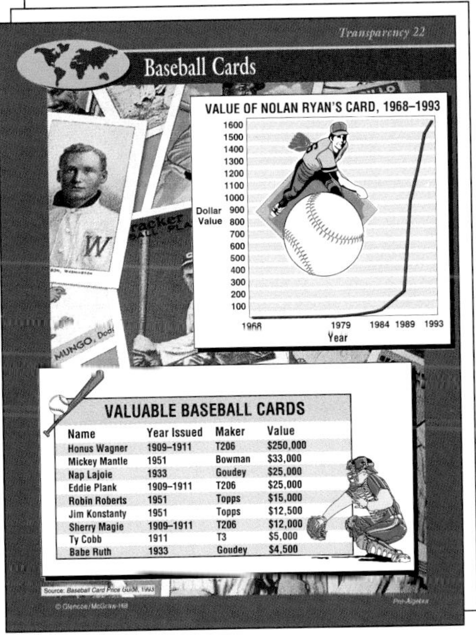

VALUABLE BASEBALL CARDS

Name	Year Issued	Maker	Value
Honus Wagner	1909–1911	T206	$250,000
Mickey Mantle	1951	Bowman	$33,000
Nap Lajoie	1933	Goudey	$25,000
Eddie Plank	1909–1911	T206	$25,000
Robin Roberts	1951	Topps	$15,000
Jim Konstanty	1951	Topps	$12,500
Sherry Magie	1909–1911	T206	$12,000
Ty Cobb	1911	T3	$5,000
Babe Ruth	1933	Goudey	$4,500

Graphing Calculator Masters, p. 10

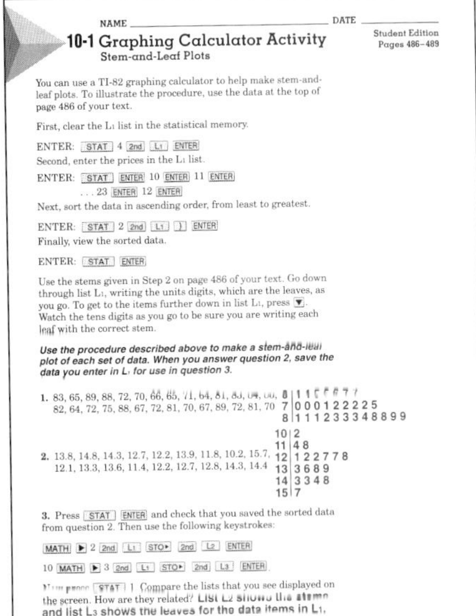

NAME ____________ DATE ________

10-1 Graphing Calculator Activity
Stem-and-Leaf Plots

Student Edition Pages 486–489

You can use a TI-82 graphing calculator to help make stem-and-leaf plots. To illustrate the procedure, use the data at the top of page 486 of your text.

First, clear the L_1 list in the statistical memory.

ENTER: STAT 4 2nd L_1 ENTER

Second, enter the prices in the L_1 list.

ENTER: STAT ENTER 10 ENTER 11 ENTER . . . 23 ENTER 12 ENTER

Next, sort the data in ascending order, from least to greatest.

ENTER: STAT 2 2nd L_1) ENTER

Finally, view the sorted data.

ENTER: STAT ENTER

Use the stems given in Step 2 on page 486 of your text. Go down through list L_1, writing the units digits, which are the leaves, as you go. To get to the items further down in list L_1, press ▼. Watch the tens digits as you go to be sure you are writing each leaf with the correct stem.

Use the procedure described above to make a stem-and-leaf plot of each set of data. When you answer question 2, save the data you enter in L_1 for use in question 3.

1. 83, 65, 89, 88, 72, 70, 66, 65, 71, 64, 81, 83, [illegible], 82, 64, 72, 75, 88, 67, 72, 81, 70, 67, 89, 72, 81, 70
   [illegible]
   7 | 0 0 0 1 2 2 2 2 5
   8 | 1 1 1 2 3 3 3 4 8 8 9 9
2. 13.8, 14.8, 14.3, 12.7, 12.2, 13.9, 11.8, 10.2, 15.7, 12.1, 13.3, 13.6, 11.4, 12.2, 12.7, 12.8, 14.3, 14.4
   10 | 2
   11 | 4 8
   12 | 1 2 2 7 7 8
   13 | 3 6 8 9
   14 | 3 3 4 8
   15 | 7
3. Press STAT ENTER and check that you saved the sorted data from question 2. Then use the following keystrokes:
   MATH ▶ 2 2nd L_1 STO▸ 2nd L_2 ENTER
   10 MATH ▶ 3 2nd L_1 STO▸ 2nd L_3 ENTER
   [illegible] STAT 1. Compare the lists that you see displayed on the screen. How are they related? List L_2 shows the stems and list L_3 shows the leaves for the data items in L_1.

Tech Prep Applications Masters, pp. 19, 20

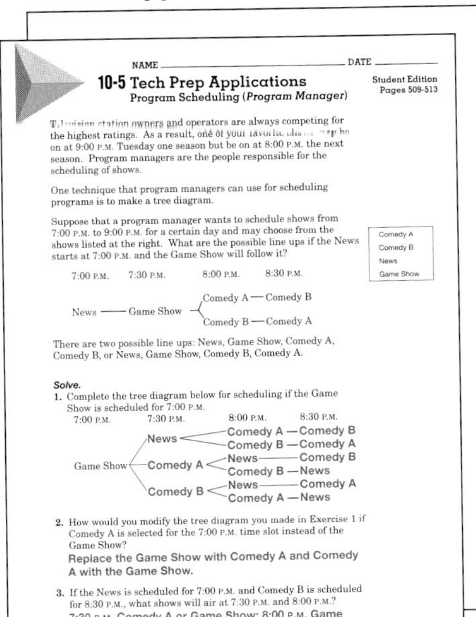

NAME ____________ DATE ________

10-5 Tech Prep Applications
Program Scheduling (*Program Manager*)

Student Edition Pages 509-513

Television station owners and operators are always competing for the highest ratings. As a result, one of your favorite [illegible] on at 9:00 P.M. Tuesday one season but be on at 8:00 P.M. the next season. Program managers are the people responsible for the scheduling of shows.

One technique that program managers can use for scheduling programs is to make a tree diagram.

Suppose that a program manager wants to schedule shows from 7:00 P.M. to 9:00 P.M. for a certain day and may choose from the shows listed at the right. What are the possible line ups if the News starts at 7:00 P.M. and the Game Show will follow it?

Comedy A
Comedy B
News
Game Show

7:00 P.M. 7:30 P.M. 8:00 P.M. 8:30 P.M.

News — Game Show — Comedy A — Comedy B
News — Game Show — Comedy B — Comedy A

There are two possible line ups: News, Game Show, Comedy A, Comedy B, or News, Game Show, Comedy B, Comedy A.

Solve.

1. Complete the tree diagram below for scheduling if the Game Show is scheduled for 7:00 P.M.
   7:00 P.M. 7:30 P.M. 8:00 P.M. 8:30 P.M.
   Game Show — News — Comedy A — Comedy B
   Game Show — News — Comedy B — Comedy A
   Game Show — Comedy A — News — Comedy B
   Game Show — Comedy A — Comedy B — News
   Game Show — Comedy B — News — Comedy A
   Game Show — Comedy B — Comedy A — News
2. How would you modify the tree diagram you made in Exercise 1 if Comedy A is selected for the 7:00 P.M. time slot instead of the Game Show?
   Replace the Game Show with Comedy A and Comedy A with the Game Show.
3. If the News is scheduled for 7:00 P.M. and Comedy B is scheduled for 8:30 P.M., what shows will air at 7:30 P.M. and 8:00 P.M.?
   7:30 P.M. Comedy A or Game Show; 8:00 P.M. Game Show or Comedy A

Activity Masters, pp. 10, 24, 42

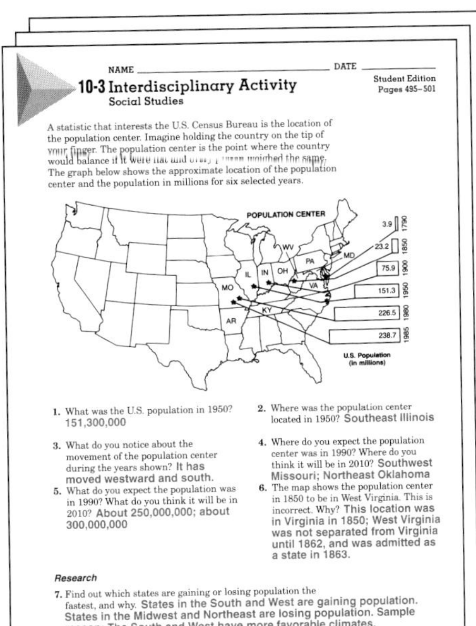

NAME ____________ DATE ________

10-3 Interdisciplinary Activity
Social Studies

Student Edition Pages 495–501

A statistic that interests the U.S. Census Bureau is the location of the population center. Imagine holding the country on the tip of your finger. The population center is the point where the country would balance if it were flat and every person weighed the same. The graph below shows the approximate location of the population center and the population in millions for six selected years.

1. What was the U.S. population in 1950? 151,300,000
2. Where was the population center located in 1950? Southeast Illinois
3. What do you notice about the movement of the population center during the years shown? It has moved westward and south.
4. Where do you expect the population center was in 1990? Where do you think it will be in 2010? Southwest Missouri; Northeast Oklahoma
5. What do you expect the population was in 1990? What do you think it will be in 2010? About 250,000,000; about 300,000,000
6. The map shows the population center in 1850 to be in West Virginia. This is incorrect. Why? This location was in Virginia in 1850; West Virginia was not separated from Virginia until 1862, and was admitted as a state in 1863.

Research

7. Find out which states are gaining or losing population the fastest, and why. States in the South and West are gaining population. States in the Midwest and Northeast are losing population. Sample reason: The South and West have more favorable climates.

COOPERATIVE LEARNING

Math Lab and Modeling Math Masters, pp. 69, 85

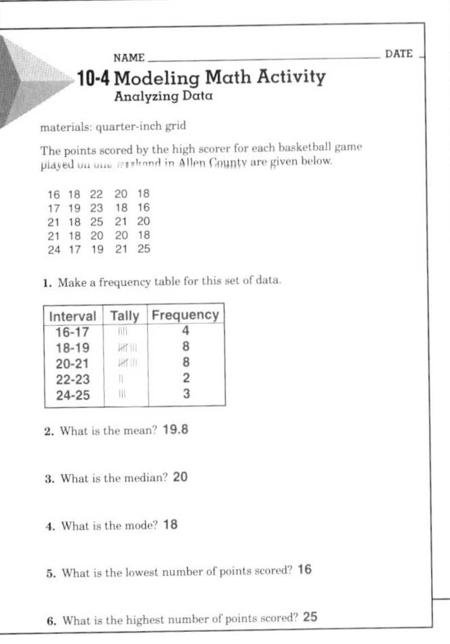

NAME ____________ DATE ________

10-4 Modeling Math Activity
Analyzing Data

materials: quarter-inch grid

The points scored by the high scorer for each basketball game played on one weekend in Allen County are given below.

16 18 22 20 18
17 19 23 18 16
21 18 25 21 20
21 18 20 20 18
24 17 19 21 25

1. Make a frequency table for this set of data.

Interval	Tally	Frequency
16-17	||||	4
18-19	𝍸 |||	8
20-21	𝍸 |||	8
22-23	||	2
24-25	|||	3

2. What is the mean? 19.8
3. What is the median? 20
4. What is the mode? 18
5. What is the lowest number of points scored? 16
6. What is the highest number of points scored? 25

MAKING MATHEMATICS RELEVANT

This two-page introduction to the chapter provides students with an opportunity to explore other disciplinary topics and their applications to mathematics.

Background Information

Math and Marketing in the News
Many advertisers like to know which TV shows are popular with younger viewers. They are willing to pay the networks more for time to show their commercials during these programs. As a result, a show like *Seaquest DSV* could be considered a success by the network even when its overall viewership was declining. This is because the program ranked high in "viewers per viewing household" and the network could charge more for commercial time aimed at reaching these younger households.

CHAPTER 10 More Statistics and Probability

TOP STORIES in Chapter 10

In this chapter, you will:

- display data in stem-and-leaf plots and box-and-whisker plots,
- find measures of variation,
- recognize misleading data,
- count outcomes,
- find permutations and combinations, and
- find probabilities and odds.

MATH AND MARKETING IN THE NEWS

Television ads take aim at young viewers

Source: American Demographics, May, 1995

Many people reach for the TV remote when the commercials come on. But advertisers hope you'll stay tuned, and then go out and buy! The products that are advertised heavily on network television are used primarily by young consumers. Companies compete fiercely for *your* dollars. They hope that if they win you as a customer now, you will be a lifetime customer.

Marketing decisions like when and how to advertise are usually based on surveys and statistics. Attitudes, buying habits, and television viewing patterns are all studied by advertisers as they make decisions. They have found that selling to today's young people is very different from selling to their parents. Advertisers hope that the Nielsen television rating system will soon be updated to give more information about the television viewing habits of consumers of different ages.

Putting It into Perspective

1400 1600 1800

950 Town criers give product information and direct people to shops in European countries.

1440 German inventor Johannes Gutenberg invents movable printing type that makes printing posters, handbills, and newspapers possible.

1704 The first newspaper advertisement in the American colonies appears in *The Boston News-Letter.*

1841 The first advertising agency in the United States opens in Philadelphia.

Putting It into Perspective

Star Broadcasting radio and Channel One television produce programs for classroom use. Many of these programs contain commercials, and the advertisers pay the school system in the form of money or other free goods. Some groups are very critical of this practice, claiming that students are a captive audience, that the school setting adds extra credibility to the messages, that students should not be subjected to commercial pressures in the classroom, and that some of the information being provided is biased or incomplete.

EMPLOYMENT OPPORTUNITIES

The Exciting World of Advertising

ARE YOU CREATIVE? Do you like working with people and ideas? We want you to write advertisements for national magazines, newspapers, and television. If you prefer analyzing data and developing strategies, we want you! Exciting opportunities await you in the field of marketing research and management. Positions for writers and researchers exist in advertising agencies and in marketing departments of larger corporations. College education in journalism, liberal arts, business, marketing, or commercial art needed.

For more information, contact:
American Advertising Federation
1101 Vermont Ave. NW, Suite 500
Washington, D.C. 20005

Statistical Snapshot

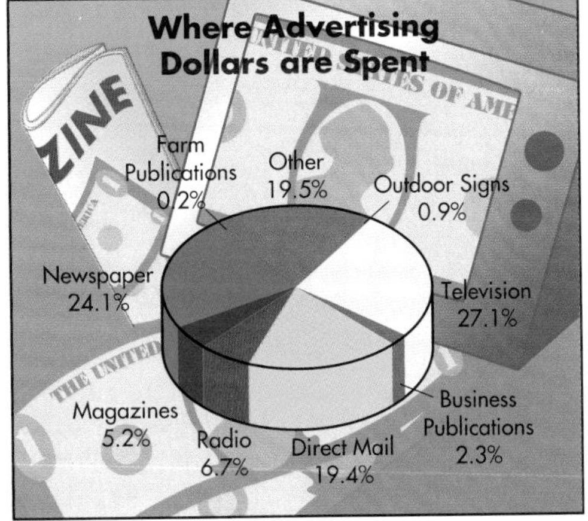

Source: McCann-Erickson, Inc.

interNET CONNECTION For up-to-date information on advertising, visit:
www.glencoe.com/sec/math/prealg/mathnet

I LIKE IKE

1950

1952
Dwight D. Eisenhower uses advertising executives to help plan his successful presidential campaign.

1969
Gail Fisher is the first African-American to have a speaking role in a nationally-televised commercial.

1971
The U.S. Federal Government bans cigarette advertising on radio and television.

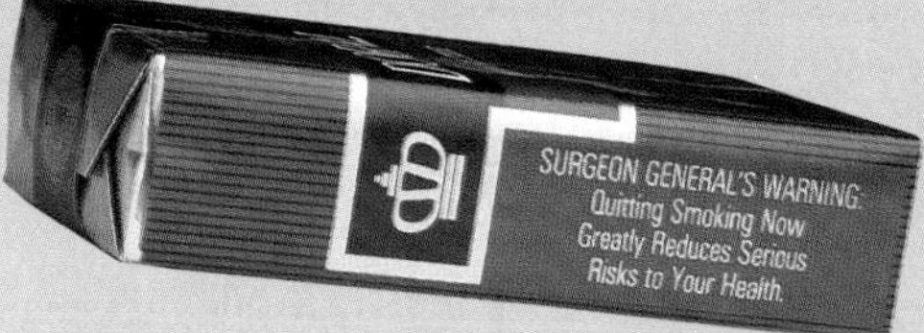

1975

1995
Advertisers consider updating Nielsen system of television ratings.

2000

EMPLOYMENT OPPORTUNITIES

Market researchers can do many tasks. They estimate the demand for new products and services, determine who the potential customers are, provide input into pricing and price changes, and they test the effectiveness of advertising.

Statistical Snapshot

The graph shows the percentages of the money spent on advertising that is spent in different forms of advertising. What types of advertising do students encounter most often? Are they the forms where the most money is spent?

Investigations and Projects Masters, p. 65

NAME ______________________ DATE __________

CHAPTER **10 Project A**

You wouldn't expect to see a car commercial during a morning cartoon show. Nor are you likely to see liquid detergents advertised during a hockey telecast. Advertisers pay a fortune to showcase their products on TV, and they choose time slots very carefully, always keeping the viewing audience in mind. Work collaboratively to find out more about how advertisers think.

1. Have each member of your group be responsible for recording information about the commercials aired during one particular category of programming. Choices might include dramas, situation comedies, news programs, sporting events, game shows, and daytime dramas. Each of you should be responsible for three hours of one kind of program. If possible, use a VCR to tape the shows to make analyzing the commercials easier.
2. As you watch, keep track of the advertisements. Gather information about the number of commercials that air in an hour, the length of each commercial, the kinds of products they advertise, and which products are advertised most often. Record your data in an organized list.
3. Study the commercials you've recorded to analyze them further. Think about what they have in common and the ways in which they differ. For example, which rely on humor? Which provide data? Is the data misleading in any way? How? Which have memorable music, jingles, or slogans? Which feature testimonials from ordinary people or celebrities?
4. Work with your group to record, discuss, and compare findings, and to display the results. For instance, you might use a stem-and-leaf plot to show the number of ads per hour per category of program. Summarize the results of your investigation.
5. Share what you've learned. Present your "Guide to Commercials" to the class.

Cooperative Learning

Chapter Projects Two chapter projects are included in the *Investigations and Projects Masters*. In Chapter Project A, students extend the topic in the Chapter Opener. In Chapter Project B, students explore students' spending habits. A student page and a parent letter are provided for each Chapter Project.

interNET CONNECTION

Glencoe has made every effort to ensure that the website links for *Pre-Algebra* at www.glencoe.com/sec/math/prealg/mathnet are current and contain appropriate content. However, these website links are not under Glencoe's control.

10-1 LESSON NOTES

NCTM Standards: 1-4, 10

Instructional Resources

- Study Guide Master 10-1
- Practice Master 10-1
- Enrichment Master 10-1
- Group Activity Card 10-1
- Graphing Calculator Masters, p. 10
- Real-World Applications, 22

Transparency 10-1A contains the 5-Minute Check for this lesson; **Transparency 10-1B** contains a teaching aid for this lesson.

Recommended Pacing	
Standard Pacing	Day 1 of 14
Honors Pacing	Day 1 of 13
Block Scheduling*	Day 1 of 7 (along with Lesson 10-2)

*For more information on pacing and possible lesson plans, refer to the ***Block Scheduling Booklet***.

1 FOCUS

5-Minute Check *(over Chapter 9)*

Solve.

1. $\frac{5}{16} = \frac{x}{144}$ **45**
2. Find 43% of $62. **$26.66**
3. Fifteen is what percent of 60? **25%**
4. Sixty is 75% of what number? **80**
5. One red marble, two blue marbles, and two green marbles are in a bag. You draw one without looking. What is the probability that you draw a blue marble? $\frac{2}{5}$

Motivating the Lesson

Hands-On Activity Bring a plant to class. Ask students what relationship there is between the stem and leaves. Develop the concept that the leaves may be different but are off the same stem.

10-1 Stem-and-Leaf Plots

Setting Goals: *In this lesson, you'll display and interpret data in stem-and-leaf plots.*

Modeling a Real-World Application: Consumerism

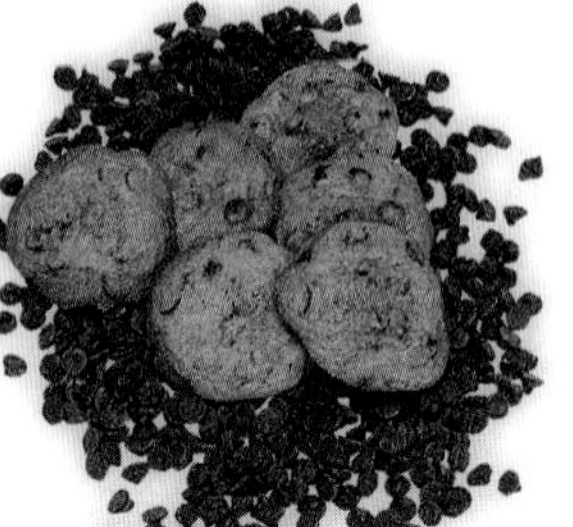

In a taste test reported in *Consumer Reports* and *Zillions* magazines, 35 different chocolate-chip cookies were rated according to flavor, freshness, and texture. For the 27 cookies that were rated good, very good, or excellent, the cost per 30-gram serving is shown below.

10¢	11¢	58¢	10¢	25¢	49¢	24¢	46¢	16¢
13¢	21¢	34¢	16¢	29¢	28¢	13¢	14¢	14¢
20¢	18¢	18¢	21¢	13¢	22¢	11¢	23¢	12¢

If you were asked to organize this set of data and present it in an easy-to-read format, you could construct a **stem-and-leaf plot**.

Learning the Concept

In a stem-and-leaf plot, the greatest place value common to all the data values is usually used for the **stems**. The next greatest place value forms the **leaves**. Follow these steps to construct a stem-and-leaf plot from the data above.

Step 1 Find the least and greatest price.

The least price is 10¢, and the greatest price is 58¢.

Step 2 Find the stems.

The least price, 10¢, has a 1 in the tens place. The greatest price, 58¢, has a 5 in the tens place. Draw a vertical line and write the digits in the tens places from 1 to 5 to the left of the line.

Stem	Leaf
1	
2	
3	
4	
5	

Step 3 Put the leaves on the plot.

Record each of the prices on the plot by pairing the units digit, or leaf, with the corresponding stem. For example, 10 is plotted by placing the units digit, 0, to the right of the stem 1.

Stem	Leaf
1	0 1 0 6 3 6 3 4 4 8 8 3 1 2
2	5 4 1 9 8 0 1 2 3
3	4
4	9 6
5	8

Step 4 Rearrange the leaves so they are ordered from least to greatest.

Step 5 Include an explanation or key of the data.

Stem	Leaf
1	0 0 1 1 2 3 3 3 4 4 6 6 8 8
2	0 1 1 2 3 4 5 8 9
3	4
4	6 9
5	8

1|3 means 13¢.

It is easy to see that nearly all of the prices are between 10¢ and 29¢ because those rows have the most leaves.

Alternative Teaching Strategies

Student Diversity Ask students to name an integer between 10 and 50. List their responses and ask the following questions.

- How could these numbers be organized based on the digits they have in common? **group by tens digits**
- How could you arrange the numbers to make organizing them easier? **from least to greatest**

Sports

In the last ten games of the 1995 NBA playoffs, the points scored by Hakeem Olajuwon of the Houston Rockets were 27, 41, 43, 20, 42, 39, 31, 34, 31, and 35.

a. Make a stem-and-leaf plot of this data.

The stems are 2, 3, and 4.

```
2 | 0 7
3 | 1 1 4 5 9
4 | 1 2 3      2|0 means 20 points.
```

b. In which interval do most of the points scored occur?

From the leaves, you can see that Olajuwon scored in the 30–39 point interval more times than in the other intervals.

A **back-to-back stem-and-leaf plot** is used to compare two sets of data. In this type of stem-and-leaf plot, the leaves for one set of data are on one side of the stem and the leaves for the other set of data are on the other side of the stem. Two keys to the data are needed.

Nutrition

A nutritional analysis was done for several fast food restaurants. The amount of fat, in grams, of their burgers and chicken, is shown below.

Burgers	10, 15, 20, 26, 19, 33, 36, 30
Chicken	9, 13, 13, 19, 18, 20, 15, 15

a. Make a back-to-back stem-and-leaf plot of the data.

The stems are 0, 1, 2, and 3.

```
Burgers |   | Chicken
        | 0 | 9
  9 5 0 | 1 | 3 3 5 5 8 9
    6 0 | 2 | 0
  6 3 0 | 3 |
6|3 = 36          2|0 = 20
```

b. If a lower amount of fat is better, which type of fast food is better?

Chicken is better because it has fewer amounts in the 20–39 range than the burgers.

Checking Your Understanding

Communicating Mathematics

1. See Solutions Manual.

Read and study the lesson to answer these questions.

1. **Explain** why you might use a stem-and-leaf plot.
2. **List** the numbers in the stem for the stem-and-leaf plot at the right. **3, 4, 5, 6**
3. **Explain** how a stem-and-leaf plot is similar to a bar graph. How is it different? **See Solutions Manual.**
4. **Describe** the information you can; obtain from the stem-and-leaf plot in Example 1. **Sample answer: Olajuwon's highest and lowest scores.**

```
3 | 0
4 | 1 5 7
5 | 3 4 4 7
6 | 0 1 8
3|0 = 30
```


Reteaching

Using Manipulatives Provide each student with a blank index card. Tell students to write the first digit of their height in inches on one side of the card and the second digit on the other side. Draw a vertical line on the board and tape cards with numbers 7, 6, 5, 4 in a vertical line on the left of the line to form the stem of a stem-and-leaf plot. Each student in turn, shows the first digit of their height, locates that stem, then turns the card over and tapes it in place on the right side of the line for that corresponding stem. When the stem-and-leaf plot is completed, have students discuss the results.

2 TEACH

In-Class Examples

For Example 1

During Reggie Jackson's first eight full seasons, he had the following numbers of hits: 138, 151, 101, 157, 132, 158, 146, 150.

a. Make a stem-and-leaf plot.

```
10 | 1
11 |
12 |
13 | 2 8
14 | 4
15 | 0 1 7 8      15|0 = 150
```

b. In which interval do most of the seasons occur? **150–159**

For Example 2

The numbers of children at the YMCA in a two-week period are shown below. Boys: 14, 20, 16, 18, 17, 21, 26, 26, 31, 29; Girls: 28, 31, 29, 27, 31, 35, 26, 35, 28, 32

a. Make a back-to-back stem-and-leaf plot of the data.

```
          | 1 | 4 6 7 8
9 8 8 7 6 | 2 | 0 1 6 6 9
5 5 2 1 1 | 3 | 1
5|3 = 35    3|1 = 31
```

b. If you had to plan activities, how could this plot help you? **Answers will vary.**

Study Guide Masters, p. 81

NAME ______________________ DATE ____________

10-1 Study Guide
Stem-and-Leaf Plots

Student Edition
Pages 486–489

The diagram at the right called a **stem-and-leaf plot**. It is one way to organize a list of numbers. This plot shows some of the numbers between 0 and 40.

```
Stems | Leaves
    0 | 3 5
    1 | 2 2 7 8
    2 | 0 1 6 6 9
    3 | 4 5 5 5
3|4 = 34
```

The key indicates that 3|4 represents 34. So, the "stems" are the tens-place digits, and the "leaves" are the ones-place digits. The numbers shown by the plot are listed below.

3, 5, 12, 12, 17, 18, 20, 21, 26, 26, 29, 34, 35, 35, 35

Write the numbers shown by each stem-and-leaf plot.

1.
```
0 | 1 4 6 6        1, 4, 6, 6,
1 | 0 3 4 9 9      10, 13, 14, 19, 19,
2 | 2 3 3 3 4 8    22, 23, 23, 23, 24,
3 | 0 3 5 5        28, 30, 33, 35, 35
0|4 = 4
```

2.
```
5 | 0 0 2 4 8      50, 50, 52, 54, 58,
6 | 1 3 6          61, 63, 66,
7 | 2 5 5 8        72, 75, 75, 78,
8 | 3 7 8 8 9 9    83, 87, 88, 88, 89, 89
6|1 = 61
```

3. Use the stem-and-leaf plot in Exercise 1.
   a. How many numbers are shown on the plot? **19**
   b. Which number(s) appears most frequently? **23**
   c. Which numbers appear least frequently? **1, 4, 10, 13, 14, 22, 24, 28, 30, 33**

4. Use the stem-and-leaf plot in Exercise 2.
   a. Which number(s) appear most frequently? **50, 75, 88, 89**
   b. Are there more numbers greater than 70, or less than 70? **greater than 70**
   c. How many numbers are shown? **18**

5. The following numbers are the results of a survey. A group of ninth-grade students were asked to report the number of hours they spent watching television in one week. Complete the stem-and-leaf plot at the right, using the results of the survey.

   0, 12, 25, 19, 23, 7, 7, 5, 26, 16,
   28, 0, 1, 0, 12, 25, 10, 2, 25, 18,
   23, 1, 14, 0, 26, 19, 14, 21, 25

```
0 | 0 0 0 0 1 1 2 5 7 7
1 | 0 2 2 4 4 6 8 9 9
2 | 1 3 3 5 5 5 5 6 6 8
2|1 = 21
```

3 PRACTICE/APPLY

Checking Your Understanding

Exercises 1–8 are designed to help you assess your students' understanding through reading, writing, speaking, and modeling. You should work through Exercises 1–5 with your students and then monitor their work on Exercises 6–8.

Error Analysis

Students may make careless errors when completing stem-and-leaf plots. Remind them to check that the number of leaves is the same as the original number of data items.

Assignment Guide

Core: 9–19 odd, 20–24
Enriched: 10–14 even, 15–24

For **Extra Practice**, see p. 764.

The red A, B, and C flags, printed only in the Teacher's Wraparound Edition, indicate the level of difficulty of the exercises.

Work with a partner to complete the following activity. **5a–b. See students' work.**

5. Choose a topic that interests you. Research or gather data about the topic. Topics might include: heights of U.S. presidents, ages of U.S. presidents at inauguration, points scored by basketball team members, time it takes to get to school or do homework.
   a. Construct a stem-and-leaf plot of the data you collected.
   b. Write two or three statements that describe the data.

6–7. See Solutions Manual for plots.

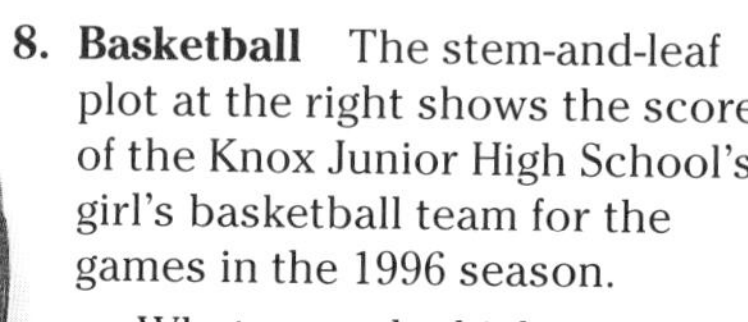

Guided Practice

State the numbers you would use for each stem in a stem-and-leaf plot of each set of data. Then make the plot.

6. 65, 82, 73, 91, 95, 86, 78, 69, 80, 88 **6, 7, 8, 9**
7. 5.7, 5.4, 6.8, 6.3, 7.1, 8.5, 7.5, 6.9, 6.7, 7.7 **5, 6, 7, 8**
8. **Basketball** The stem-and-leaf plot at the right shows the scores of the Knox Junior High School's girl's basketball team for the games in the 1996 season.

Knox		Opponents
3	2	
6 3 1	3	6 7
9 8 6 5 4 4 2	4	0 3 4 5 5 6
9 8 8 6 6 2	5	1 1 3 5 6 6 7
5 0	6	1 3 5 6
0	7	2
3|2 = 23		7|2 = 72

   a. What were the highest and lowest scores for the Knox team? **70, 23**
   b. What were the highest and lowest scores of their opponents? **72, 36**
   c. What is the median of each set of scores? **48.5, 52**

Exercises: Practicing and Applying the Concept

Independent Practice

Make a stem-and-leaf plot of each set of data. **9–13. See Solutions Manual.**

9. 64, 60, 72, 61, 73, 80, 68, 70, 65, 67, 70, 80
10. A 23, 14, 25, 36, 45, 34, 32, 21, 39, 28, 48, 19, 39, 45, 48, 48, 51
11. B 42, 23, 9, 21, 7, 11, 14, 6, 40, 5, 9, 45, 12
12. 83, 94, 54, 92, 85, 54, 96, 89, 75, 117, 68, 99, 116
13. 11.2, 10.4, 12.6, 12.3, 11.1, 14.7, 9.2, 10.8, 12.9, 11.2, 11.7, 13.3, 13.8, 12.8
14. **Architecture** The *World Almanac* lists 15 tall buildings in New Orleans, Louisiana. The number of floors in each of these buildings is listed in the chart at the right.
    a. Make a stem-and-leaf plot of the data. **See margin.**
    b. What is the greatest number of floors? **53**
    c. What is the least number of floors? **23**
    d. How many buildings had 34 or less floors? **8**
    e. How many buildings were in the 30–39 floor range? **6**

51	47	33
53	42	28
45	33	28
39	32	25
36	31	23

Practice Masters, p. 81

NAME ______________________ DATE ____________

10-1 Practice
Stem-and-Leaf Plots

Student Edition
Pages 486–489

Make a stem-and-leaf plot of each set of data.

1. 54, 50, 62, 51, 63, 70, 58, 60, 60, 70
   5 | 0 1 4 8
   6 | 0 0 2 3
   7 | 0 0   5 | 1 = 51
2. 13, 22, 27, 16, 36, 7, 27, 33, 36, 36
   0 | 7
   1 | 3 6
   2 | 2 7 7
   3 | 3 6 6 6   1 | 3 = 13

Thirty-five students took a quiz. The scores were: 7, 10, 7, 10, 6, 7, 2, 9, 6, 0, 20, 10, 16, 18, 14, 10, 18, 10, 6, 18, 16, 20, 20, 24, 18, 27, 21, 12, 15, 24, 15, 12, 21, 30, and 21.

3. Construct a stem-and-leaf plot for the data.
   0 | 0 2 6 6 6 7 7 9
   1 | 0 0 0 0 0 2 2 4 5 5 6 6 8 8 8 8
   2 | 0 0 0 1 1 1 4 4 7
   3 | 0   1 | 5 = 15
4. What was the highest score? 30
5. What was the lowest score? 0
6. What was the mode score? 10
7. Make two or three statements about the data. Answers may vary. Sample answer: More than half of the class scored 10 or better; the highest score was 30 and the lowest score was 0.

The ages of the first thirty people into a concert on Friday were: 19, 21, 24, 18, 20, 20, 19, 17, 20, 23, 18, 20, 21, 20, 24, 25, 22, 21, 25, 18, 19, 20, 21, 19, 22, 23, 17, 22, 25, and 23.

8. Construct a stem-and-leaf plot for the data.
   1 | 7 7 8 8 8 9 9 9 9
   2 | 0 0 0 0 0 0 1 1 1 1 2 2 2 3 3 3 4 4 5 5 5   2 | 1 = 21
9. How old was the oldest person? 25
10. How young was the youngest person? 17
11. What was the median age? 20.5
12. What might account for the limited range of years? Answers may vary. Sample answer: It may have been a concert that attracted persons whose ages were 17 through 25, possibly a rock concert.

Group Activity Card 10-1

Random Numbers?

Group Activity **10-1**

MATERIALS: A table of random numbers

Do tables of random numbers really produce numbers that are random? To help answer this question, you can use a stem-and-leaf plot.

You and your partner should choose any group of 100 digits on the table. From these digits, record the 50 two-digit numbers listed by reading the 100 digits in order.

Now, make a stem-and-leaf plot of these two-digit numbers. From this plot, answer the following questions.

1. What is the range of numbers?
2. Is there one group of numbers that is more frequent than the others?
3. Is there one group of numbers that is less frequent than the others?
4. What would you expect a stem-and-leaf plot of random numbers to look like?
5. Do you think the table you used lists two-digit numbers randomly?

 Pre-Algebra

15. Retail Sales The table below lists the prices, in dollars, for pairs of running shoes.

Men's	83, 70, 70, 75, 70, 82, 70, 125, 82, 45, 70, 120, 110, 133, 100
Women's	124, 70, 72, 70, 65, 90, 60, 80, 55, 70, 67, 45, 55, 75, 85

a. Make a back-to-back stem-and-leaf plot of the prices for men's and women's shoes. **See margin.**

b. What information about the relationship between the prices for men's and women's shoes does it present? **See margin.**

16. See margin.
17. See Solutions Manual.

Critical Thinking

16. Describe any restrictions when using a stem-and-leaf plot.

17. Is it possible to estimate the mean from a stem-and-leaf plot? Explain.

Applications and Problem Solving

18. Meteorology The normal January temperatures of 16 southern cities are shown in the table at the right. **18a–b. See margin.**

a. Make a stem-and-leaf plot of the temperatures.

b. If you were touring the south in January, what range of temperatures would you expect?

Normal January Temperatures

City	Temp	City	Temp
Asheville, NC	37	Memphis, TN	67
Atlanta, GA	42	Mobile, AL	51
Birmingham, AL	42	New Orleans, LA	37
Charleston, SC	49	Norfolk, VA	52
Jackson, MS	46	Raleigh, NC	40
Jacksonville, FL	53	Richmond, VA	40
Miami, FL	38	Savannah, GA	49
Knoxville, TN	40	Tampa, FL	60

Source: *The World Almanac, 1995*

19. Engineering The table below lists the temperatures at launch time for the first 23 space shuttle launches and how many of those times the O-rings on the solid rocket motors experienced thermal distress or no distress. **19a–c. See Solutions Manual.**

Temp (°F)	66	70	69	68	67	72	73	57	63	78	53	75	81	76	79	58
Distress	0	2	0	0	0	0	0	1	1	0	1	1	0	0	0	1
No Distress	1	2	1	1	3	1	1	0	0	1	0	1	1	2	1	0

The last launch of the space shuttle *Challenger* was on January 28, 1986. The temperature was 31°F at the time of the launch.

a. Make a back-to-back stem-and-leaf plot of the temperatures for distress and no distress.

b. What information about the relationship between the temperatures of distress and no distress does your back-to-back stem-and-leaf plot make?

c. If the temperature on the day of the launch was 31°F, would you expect the O-rings to experience thermal distress? Explain.

Mixed Review

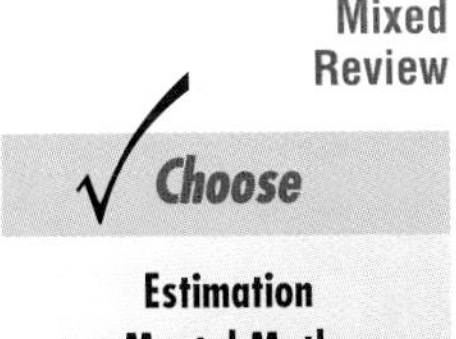

Choose

Estimation
Mental Math
Calculator
Paper and Pencil

20. Economics The Consumer Price Index (CPI) represents the relative costs of goods and services. If the CPI is 233.2 in April and 236.4 in May, what is the percent of increase from April to May? (Lesson 9-10) 1%

21. Express 45% as a fraction in simplest form. (Lesson 9-7) $\frac{9}{20}$

22. Solve $x - 3.4 = 9.2$. Check your solution. (Lesson 5-6) 12.6

23. If 5 more than the product of a number and -2 is greater than 10, which of the following could be that number? (Lesson 3-1) d

a. 3 b. 2 c. -1 d. -4

24. Find the value of $|-9| + 14$. (Lesson 2-1) 23

4 ASSESS

Closing Activity

Speaking List the ages of the members of each student's household. Have students describe what a stem-and-leaf plot of this data would look like.

Additional Answers

15a.

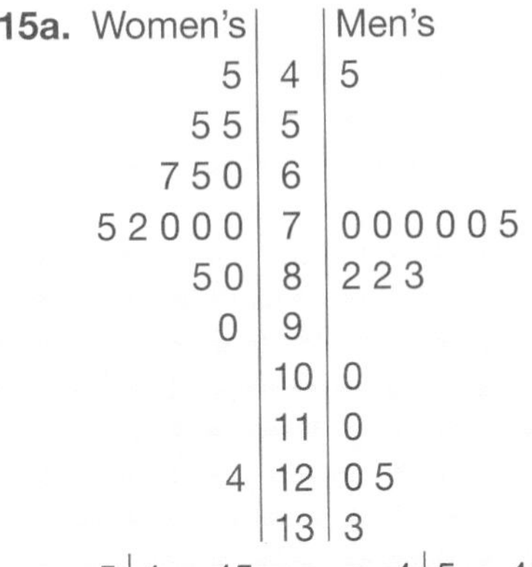

Women's		Men's
5	4	5
5 5	5	
7 5 0	6	
5 2 0 0 0	7	0 0 0 0 0 5
5 0	8	2 2 3
0	9	
	10	0
	11	0
4	12	0 5
	13	3

5|4 = 45   4|5 = 45

15b. Sample answer: Men's running shoes are more expensive than women's. The typical running shoe costs $70.

16. You must have data that are in a relatively small range.

18a.
3 | 6 6 7 9 9
4 | 0 1 2 4 8 9
5 | 0 1 2
6 | 0 7
3|6 = 36°

18b. Sample answer: Somewhere in the 40s.

Enrichment Masters, p. 81

NAME ______ DATE ______

10-1 Enrichment
Statistical Graphs

Student Edition Pages 486–489

Bar graphs and pictographs are used to compare quantities. Line graphs are used to show changes. Circle graphs compare parts to parts, or parts to the whole.

Solve. Use the pictograph.

Principal Languages of the World (to nearest fifty million)

(one symbol = 100 million)

English	
Hindi	
Arabic	
Portuguese	
Chinese	
Russian	
Spanish	
French	
Bengali	

1. How many people speak Portuguese? **150,000,000**
2. What is the ratio of people who speak Spanish to those who speak Russian? **1:1**
3. What three languages are each spoken by about 150 million people? **Arabic, Portuguese, Bengali**
4. How many fewer people speak Arabic than Hindi? **100,000,000**

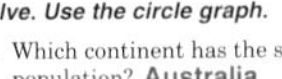

Solve. Use the circle graph.

5. Which continent has the smallest population? **Australia**
6. How does the population of South America compare to that of Africa? **South America has about half as many people.**
7. What is the population of Australia if the world's population is about 4.6 billion? **about 13,800,000**

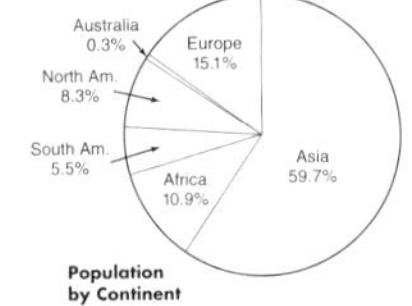

Solve. Use the line graph.

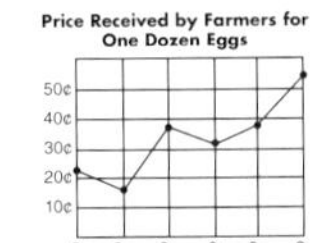

8. During which ten-year period was the increase in the price of eggs greatest? **1940–1950**
9. What was the price of a dozen eggs in 1940? **16¢**
10. By what percent did the price of eggs increase from 1940 to 1950? **about 138%**
11. What was the increase in cents from 1930 to 1980? **33¢**

Extension

Using Research Have students research the average daily temperatures for various U.S. cities. You may wish to assign cities to get a sample from all parts of the country, including Hawaii and Alaska. Have students make a stem-and-leaf plot for each city. Post the plots on a bulletin board and discuss the similarities and differences.

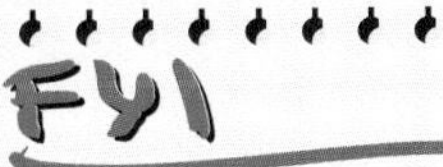

The space shuttle program was authorized on January 5, 1972. The first launch of an occupied shuttle was on April 12, 1981. *Challenger* was first flown on April 4–9, 1983. The January 28, 1986 launch was its 10th flight.

10-2 LESSON NOTES

NCTM Standards: 1-4, 10

Instructional Resources

- Study Guide Master 10-2
- Practice Master 10-2
- Enrichment Master 10-2
- Group Activity Card 10-2

Transparency 10-2A contains the 5-Minute Check for this lesson; **Transparency 10-2B** contains a teaching aid for this lesson.

Recommended Pacing	
Standard Pacing	Day 2 of 14
Honors Pacing	Day 2 of 13
Block Scheduling*	Day 2 of 7 (along with Lesson 10-1)

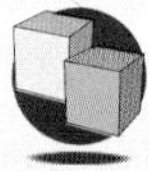

*For more information on pacing and possible lesson plans, refer to the ***Block Scheduling Booklet***.

1 FOCUS

5-Minute Check *(over Lesson 10-1)*

To the nearest dollar, the first twenty-five sales Saturday morning at the jewelry counter of a local department store were \$26, \$32, \$15, \$23, \$54, \$38, \$9, \$45, \$22, \$40, \$14, \$21, \$16, \$28, \$24, \$33, \$7, \$27, \$19, \$20, \$42, \$8, \$59, \$12, and \$29.

1. Make a stem-and-leaf plot for the data. **See Transparency 10-2A.**
2. What was the greatest amount spent? **\$59**
3. What was the least amount spent? **\$7**
4. What price range is most represented? **20s**
5. What is the median price? **\$24**

10-2 Measures of Variation

Setting Goals: *In this lesson, you'll use measures of variation to compare data.*

Modeling with Technology

In Chapter 6, you learned that the mean, median, and mode indicate the center of a set of data. Sometimes measures of center don't fully describe a set of data. In the back-to-back stem-and-leaf plot of bicycle helmet prices at the right, both models have the same mean price, the same median price, and the same mode price. But, it is easy to see that all the prices are not the same.

Adult		Youth
	2	5
7 2 1 1	3	4 7 8
3 3 3	4	2 3 3 8
0	5	

1|3 = \$31 *2|5 = \$25*

You can describe the two lists of helmet prices more accurately by using measures of variation to describe the spread of each set of data. You can use a TI-82 graphing calculator to find common measures of variation.

Your Turn

- ▶ First, clear the statistical memory.

  Enter: [STAT] 4 [2nd] [L1] [,] [2nd] [L2] [ENTER]
- ▶ Enter the adult prices in the L1 list and the youth prices in the L2 list.

  Enter: [STAT] [ENTER] 31 [ENTER] 31 [ENTER] . . . 50 [ENTER] [▶]
  25 [ENTER] 34 [ENTER] . . . 48 [ENTER]
- ▶ Find the median of the lower and upper halves of the adult prices.

  Enter: [STAT] [▶] [ENTER] [2nd] [L1] [ENTER]

  Press the down arrow to scroll through the list. Record the values for Q1 and Q3. Then find the difference between the greatest value, max X, and the least value, min X.
- ▶ Find the median of the lower and upper halves of the youth prices.

  Enter: [STAT] [▶] [ENTER] [2nd] [L2] [ENTER]

  Record the values for Q1 and Q3. Then find the difference between the greatest and least values.

TALK ABOUT IT

a. How do the differences between the greatest and least values of each list compare? **The difference of adult prices is slightly greater.**

b. How do the medians of the lower half (Q1) and upper half (Q3) of each set of data compare? **See margin.**

Learning the Concept

When you found the difference between the greatest and least values of the helmet prices, you calculated a measure of variation called the **range**.

Additional Answers Talk About It

b. The median of the lower half of the youth prices is higher than the lower half of the adult prices. The median of the upper halves are the same.

Definition of Range The range of a set of numbers is the difference between the least and the greatest number in the set.

Example  **Determine the range of each set of data.**

a. 12, 17, 16, 23, 14, 18, 11, 21

First, list the data in order.

11 12 14 16 17 18 21 23

The range is 23 − 11 or 12.

b. (line plot: 12 13 14 15 16 17 18 19)

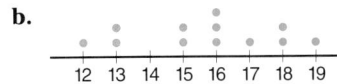

Since the line plot is organized from the least data to the greatest, the range is the difference in the first and last data values of the plot.

The range is 19 − 12 or 7.

The TI-82 graphing calculator uses Q_1 to represent the lower quartile and Q_3 to represent the upper quartile.

In a large set of data, it is helpful to separate the data into four equal parts called **quartiles**. Recall that the median of a set of data separates the data in half. The median of the lower half of a set of data is called the **lower quartile** and is indicated by LQ. Likewise, the median of the upper half of the data is called the **upper quartile** and is indicated by UQ. Quartiles are used in another measure of variation called the **interquartile range**.

Definition of Interquartile Range

In words: The interquartile range is the range of the middle half of a set of numbers.

In symbols: Interquartile range = UQ − LQ

Example 2

Olympics

For the latest on Olympic medals, visit: **www.glencoe.com/sec/math/prealg/mathnet**

In the thirteen summer Olympic Games from 1936–1992, athletes from the United States earned the following numbers of gold medals.

Year	1936	1948	1952	1956	1960	1964	1968
Gold Medals	24	38	40	32	34	36	45

Year	1972	1976	1980	1984	1988	1992
Gold Medals	33	34	0	83	36	37

a. **Find the range and interquartile range for the number of gold medals.**

First, list the data from least to greatest. Then find the median.

0 24 32 33 34 34 36 36 37 38 40 45 83

↑ Median (36)

(continued on the next page)

Motivating the Lesson

Situational Problem Tell students that you are going to open a shoe store that sells only athletic shoes. Ask students what sizes you should order the most of and least of, and why.

2 TEACH

In-Class Examples

For Example 1
Determine the range of each set of data.

a. 94, 75, 84, 86, 96, 77, 88 **21**

b. 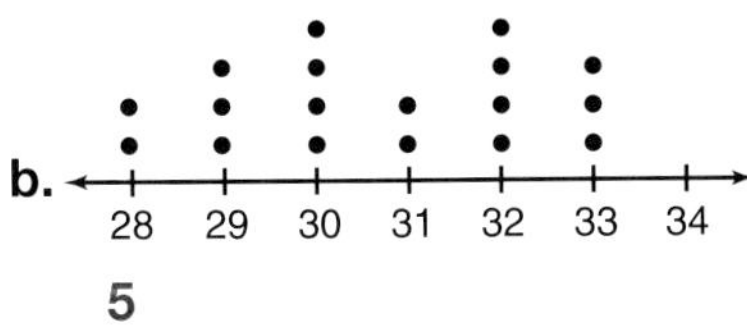

5

For Example 2
Below are the number of assists made by Todd Gill of the Toronto Maple Leafs between the 1985-86 season and the 1994-95 season. 2, 27, 17, 14, 14, 22, 15, 32, 24, 25

a. Find the range and interquartile range for the number of assists. **range 23; interquartile range 11**

b. Which measure of variation better summarizes the typical number of assists made by Gill, the range or the interquartile range? **the interquartile range**

Alternative Learning Styles

Visual Have each student write the last two digits of the year in which one of their parents were born. Have students place the cards on the chalkrail, in order, to make a visual display. Then ask:

- How would you find the lower 25% of the dates? **Sample answer: The lower 25% would be the lowest $\frac{1}{4}$ of the dates.**
- How would you find the range? **Sample answer: Subtract the low from the high.**
- How would you find the middle 50% of the dates? **Take away the upper and lower 25%.**

3 PRACTICE/APPLY

Checking Your Understanding

Exercises 1–9 are designed to help you assess your students' understanding through reading, writing, speaking, and modeling. You should work through Exercises 1–3 with your students and then monitor their work on Exercises 4–9.

Error Analysis

Some students may take one fourth and three fourths of the highest value in the data to find the quartiles. Emphasize the process of counting data and marking the points for the quartiles.

Additional Answers

1. Sample answer: Measures of variation describe the spread of data. Measures of central tendency describe the set as a whole.
2. Sample answer: 10, 12, 16, 18, 18, 20, 22, 24, 26, 26, 28, 40, 60.
3. The data is clustered about the median.

Study Guide Masters, p. 82

NAME ______________________ DATE ____________

10-2 Study Guide
Measures of Variation

Student Edition Pages 490–494

The list below shows the test scores for a sample of 11 high school students.

65 65 (67) 72 75 [75] 75 80 (87) 92 93

The difference between the least and the greatest number in the set is called the **range**. The range above is 93–65 or 28.

When a list of data is arranged in order, the **median** is the middle number. The median above is enclosed in a box.

The circled numbers in the list above are used to analyze the data. The **lower quartile** is the median of the lower half of the data. The **upper quartile** is the median of the upper half.

The difference between the lower quartile and the upper quartile is called the **interquartile range**. The interquartile range above is 87–67 or 20.

Recall that when there are two middle numbers, the median is their mean. This is also true for the upper and lower quartiles.

Find the range, median, upper and lower quartiles, and the interquartile range for each set of data.

1. 48, 50, 53, 50, 44, 52, 45 **9; 50; 52, 45; 7**
2. 32, 0, 6, 20, 0, 12, 15, 25, 18, 15, 24 **32; 15; 24, 6; 18**

3.

Ages of Eastside Health Club Members	
Aimee	19
Lucille	35
Leonard	20
Tiago	52
Nakeisha	43
Hector	27
Robin	44
Haleem	20
Marrissa	32
Tad	27

33; 29.5; 43, 20; 23

4.

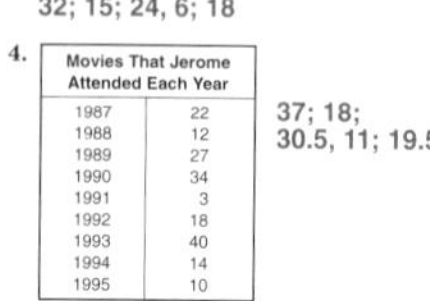

Movies That Jerome Attended Each Year	
1987	22
1988	12
1989	27
1990	34
1991	3
1992	18
1993	40
1994	14
1995	10

37; 18; 30.5, 11; 19.5

FYI: The Summer Olympics were boycotted by the United States and 63 other countries in 1980. The USSR and 13 other Eastern Bloc nations boycotted the 1984 Summer Olympics.

The upper quartile is the median of the upper half and the lower quartile is the median of the lower half.

0 24 32 33 34 34 36 36 37 38 40 45 83

$LQ = \frac{32 + 33}{2}$ or 32.5 Median $UQ = \frac{38 + 40}{2}$ or 39

The interquartile range is 39 − 32.5 or 6.5.

b. Which measure of variation better summarizes the performance of United States athletes, the range or the interquartile range?

Since the range represents values for years in which the Summer Olympics were boycotted, the interquartile range is more representative of the variability of this set of data.

Checking Your Understanding

Communicating Mathematics

Read and study the lesson to answer these questions. **1–3. See margin.**

1. **Compare and contrast** the measures of variation and the measures of central tendency.
2. **Write** a list of at least twelve numbers that has a range of 50 and an interquartile range of 10.
3. What can you say about a group of test scores if it has a small interquartile range?

Guided Practice

Find the range of each set of data.

4. 63, 82, 71, 65, 92, 86, 80, 95, 78, 89 **32**
5. **13**

11	12	13	14	15	16	17	18	19	20	21	22	23	24
×	×		×	×	××	×××		××	××		××	××	×

6. 30.8, 29.9, 30.2, 33.2, 30.1, 30.5, 30.7, 29.8 **2.4**

Use the data in the stem-and-leaf plot shown at the right.

Stem	Leaf
4	2 3 9
5	4 5
6	7
7	3 8
8	2
9	1 3 4

4|2 = 42

7. What is the range? **52**
8. a. Find the median. **70**
   b. What are the upper and lower quartiles? **86.5, 51.5**
   c. Find the interquartile range. **35**
9. **Sales** Discworld's sales of Gloria Estefan's new CD over a two-week period are shown in the chart at the right.
   a. Organize the data into a line plot or a stem-and-leaf plot.
   b. What is the median? **61.5**
   c. Find the upper and lower quartiles.
   d. What is the interquartile range? **28**

M	46	80
T	63	94
W	38	42
T	61	39
F	70	63
S	40	84
S	58	62

9a. See Solutions Manual.

9c. 70, 42

Reteaching

Using Models List 26 test scores on the board. Make a stem-and-leaf plot of the data. Have students determine the median of the data. Then have them find the median of the upper half of the plot and circle it. Have students repeat this procedure for the lower half and find the difference of the two circled numbers. Explain that they found the quartiles and the interquartile range.

FYI In 1992, even though the USSR no longer existed, a United Team represented the 11 former USSR republics in the Olympic Games.

Exercises: Practicing and Applying the Concept

Independent Practice

Find the range, median, upper and lower quartiles, and the interquartile range for each set of data.

A

10. 56, 45, 37, 43, 10, 34, 29
11. 30, 90, 40, 70, 50, 100, 80, 60
12. 19°, 21°, 18°, 17°, 18°, 22°, 36°
13. 8, 11, 23, 7, 2, 4, 16, 2, 4
14. 135, 170, 125, 174, 136, 145, 180, 156, 188 **63; 156; 177, 135.5; 41.5**
15. 211, 225, 205, 207, 208, 213, 180, 200, 210, 229, 199 **49; 208; 213, 200; 13**

10. 46; 37; 45, 29; 16
11. 70; 65; 85, 45; 40
12. 19; 19; 22, 18; 4
13. 21; 7; 13.5, 3; 10.5

B

16.

Stem	Leaf
5	9
6	0 0 1 2 2 3 5
7	1 2

5|9 = 59¢

17.

Stem	Leaf
7	3 5
8	3 8 9
9	0 0 1 2 4 4 7 7 8
10	3 6
11	4

7|3 = 7.3 cm

18.

Stem	Leaf
25	0 3 7 9
26	1 3 4 5 5 6
27	1 5 6 6 9
28	1 2 3 5 8
29	2 5 6 9

29|2 = $292

16. 13; 62; 65, 60; 5
17. 4.1; 9.2; 9.75, 8.85; 0.9
18. 49; 275.5; 283.5, 263.5; 20

19.

World's Busiest Airports in 1994

City	Passengers
Atlanta	54,100,000
Chicago-O'Hare	66,400,000
Dallas-Ft. Worth	52,600,000
Denver	33,100,000
Frankfurt	35,100,000
London Heathrow	51,700,000
Los Angeles	51,100,000
Miami	30,200,000
N.Y.-Kennedy	28,800,000
San Francisco	34,600,000

Source: Airports Council International

20.

Top 10 states with largest Asian-Indian population in 1990

State	Population
California	159,973
Florida	31,457
Illinois	64,200
Maryland	28,330
Michigan	23,845
New Jersey	79,440
New York	140,985
Ohio	20,848
Pennsylvania	28,396
Texas	55,795

Source: U.S. Census Bureau

19. 37,600,000; 43,100,000; 52,600,000; 33,100,000; 19,500,000
20. 139,125; 43,626; 79,440, 28,330; 51,110

21. **Home Runs Hit by League Leaders, 1960–1995**

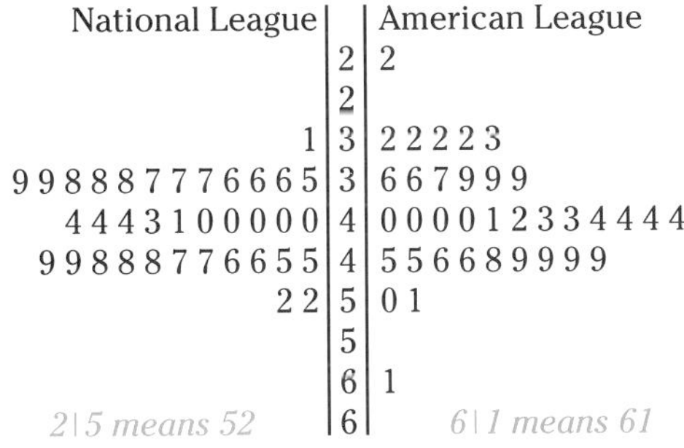

National League		American League
	2	2
	2	
1	3	2 2 2 2 3
9 9 8 8 8 7 7 7 6 6 6 5	3	6 6 7 9 9 9
4 4 4 3 1 0 0 0 0 0	4	0 0 0 0 1 2 3 3 4 4 4 4
9 9 8 8 8 7 7 6 6 5 5	4	5 5 6 6 8 9 9 9 9
2 2	5	0 1
	5	
	6	1
2|5 means 52	6	*6|1 means 61*

21. American League: 39; 42.5; 46, 38; 8
National League: 21; 40.5; 46.5, 38; 8.5

Graphing Calculator

Work with a partner to describe the variation in the populations of capital cities of the U.S.

22. Each person should find the population of half of the state capitals. Enter the data in a TI-82 graphing calculator. Use the sort command to place the populations in order. Find the median, range, and interquartile range of the data. Use this information to describe the population of the capital cities. **See students' work.**

Assignment Guide

Core: 11–25 odd, 26–30
Enriched: 10–22 even, 23–30

For **Extra Practice**, see p. 764.

The red A, B, and C flags, printed only in the Teacher's Wraparound Edition, indicate the level of difficulty of the exercises.

Teaching Tip Point out that in Exercise 21 there are two stems for each number to simplify the plot. Entries with leaves from 0 to 4 are on the first stem and leaves from 5 to 9 are on the second stem.

Additional Answer

23c. Sample answer: The first set of data has a smaller interquartile range, thus the data in the first are more tightly clustered around the median and the data in the second set are spread out over the range.

Practice Masters, p. 82

NAME ______________________ DATE __________

10-2 Practice
Measures of Variation

Student Edition
Pages 490–494

Find the range, median, upper and lower quartiles, and the interquartile range for each set of data.

1. 52, 41, 33, 39, 6, 30, 25
46; 33; 41, 25; 16
2. 25, 85, 35, 45, 95, 75, 55
70; 55; 85, 35; 50
3. 118, 112, 130, 106, 116, 146, 143, 129, 134 **40; 129; 138.5, 114; 24.5**
4. 150, 132, 116, 118, 109, 108, 114, 124
42; 117; 128, 111.5; 16.5

5.

Stem	Leaf
4	8
5	0 1 1 2 2 4 6
6	0 1

4|8 = 4.8 in.
13 in.; 52 in.; 56 in., 51 in.; 5 in.

6.

Stem	Leaf
6	2 4
7	2 7 8
8	0 0 0 1 3 3 6 7 7
9	2 5
10	6

6|2 = $6.20
$4.40; $8.10; $8.70, $7.75; $.95

7.

Stem	Leaf
14	0 2 4 8
15	0 2 3 4 4 5
16	0 6 7 7 8
17	0 4 8 9
18	3 6 7 8

14|0 = 14.0 cm
4.8 cm; 16.6 cm; 17.8 cm, 15.2 cm; 2.6 cm

8.

Words Input Per Minute	
Kalica	64
Celina	53
Sly	51
Marty	90
June	76
Addison	68
Zach	92
Lea	81
Andy	62

41; 68; 85.5, 57.5; 28

9.

Bowling Scores	
Leslie	95
Tyshon	134
Julie	212
Maylin	89
Nate	198
Sloan	107
Nancy	267
Antonio	107
Wes	156

178; 134; 205, 101; 104

The stem-and-leaf plots at the right show the test scores for Kyle and Matt during the first 9-weeks period.

10. How do their medians compare? **Kyle's is greater.**
11. How do their ranges compare? **Matt's is greater.**
12. How do their interquartile ranges compare? **Matt's is less.**
13. Which student is more consistent? Explain your answer. **Matt; his interquartile range is less than Kyle's.**

TEST SCORES FOR FIRST 9 WEEKS

Kyle		Matt
8	4	0
6	5	9
4 0	6	5
5 0	7	0 2 4 4 5 8 9
5 3 2 0	8	0
5 0	9	6

Group Activity Card 10-2

Hot Or Not

Group Activity **10-2**

MATERIALS: Fourteen 3x5 cards: six Statistics Cards and eight Temperature Cards

SETUP: Make the cards using the information given on the back of this card.

The following are high temperatures (in degrees Fahrenheit) for 12 different United States cities on a day in January: 19, 49, 7, 64, 38, 49, 37, 57, 81, 61, 15, 26. Use these temperatures to do the activity.

To begin, shuffle the Statistics Cards and place them face down. Do the same thing with the Temperature Cards. Turn over one Temperature Card and one Statistics Card.

The first person to correctly compute the statistic given on the Statistics Card using the 12 temperatures above plus the additional temperatures on the Temperature Card wins a point. The overall winner is the first person to score five points.

 Pre-Algebra

Critical Thinking

23. a. Produce a set of at least ten test scores that has a median of 60 and an interquartile range of 20. **See students' work.**

b. Produce a set of at least ten test scores that has a median of 60 and an interquartile range of 50. **See students' work.**

c. What conclusions can you draw about the two sets of test scores in parts a and b from comparing the measures of variation? **See margin.**

Applications and Problem Solving

24. Meteorology The table at the right lists the average winter and summer precipitation for selected cities around the world.

a. Construct a back-to-back stem-and-leaf plot for the precipitation amounts.

b. Find the mean and median for each set of data.

c. How do the means compare?

d. How do the medians compare?

e. How do the interquartile ranges compare?

f. Which month has more consistent precipitation? Explain. **See margin.**

January and July Precipitation (in.)

City	Jan.	July
Atlanta, GA	4.9	4.7
Berlin, Germany	1.9	3.1
Dallas, TX	1.7	2.0
Dublin, Ireland	2.7	2.8
Indianapolis, IN	2.7	4.3
Kinshasa, Zaire	5.3	0.1
Miami, FL	2.1	6.0
New Orleans, LA	5.0	6.7
Oklahoma City, OK	5.0	6.7
Oslo, Norway	1.7	2.9
Paris, France	1.5	2.1
Rome, Italy	3.3	0.4
San Diego, CA	2.1	0.0
San Francisco, CA	4.5	0.0
Santiago, Chile	0.1	3.0
Singapore	9.9	6.7
Sydney, Australia	3.5	4.6
Washington, DC	2.8	3.9

Source: Universal Almanac 1995

25. World Cultures Many North American Indians hold conferences called powwows where they celebrate their culture and heritage through various ceremonies and dances.

- The ages of participants in a Menominee Indian powwow were: 20, 18, 12, 13, 14, 72, 65, 23, 25, 43, 67, 35, 68, 13, 56.
- The ages of observers of the powwow were: 43, 55, 70, 63, 15, 41, 9, 42, 75, 25, 16, 18, 51, 80, 75, 39, 23, 55, 50, 54, 60, 43.

a. Find the range of each group. **participants 60; observers 71**

b. Find the interquartile range for each group. **51; 35**

c. Write a paragraph describing the participants. **See margin.**

d. Write a paragraph comparing the participants with the observers. **See margin.**

Mixed Review

26. Statistics The ages of the first twenty people into the museum on Saturday were 17, 9, 12, 25, 8, 39, 27, 14, 29, 40, 36, 8, 15, 41, 28, 29, 30, 31, 29, and 11. Make a stem-and-leaf plot for the data. (Lesson 10-1)

27. Replace the ● in $\frac{2}{3}$ ● $\frac{5}{8}$ with < or > to make a true sentence. (Lesson 6-1) **>**

28. Computers Shantia plans to spend no more than $50 on new software. If she buys a program for $19.95, how much more can she spend? Express your solution as an inequality. (Lesson 5-7)

29. Find the quotient of c^2d^5 and cd. (Lesson 4-8) cd^4

30. *True* or *false*: The graph of a point with one negative coordinate and one positive coordinate is located in Quadrant III. (Lesson 2-2) **False**

√ **Choose**

Estimation
Mental Math
Calculator
Paper and Pencil

24a. See Solutions Manual.
24b. Jan.: 2.8, 2.7; July: 5.3, 4.3
24c. The January mean is about 0.04 higher.
24d. The July median is about 0.3 higher.
24e. The January IQ is 0.3 higher.

26. See margin.

28. $x \leq \$30.05$

4 ASSESS

Closing Activity

Writing Have students write a problem with a given range and interquartile range. For example, give the students a range of 67 and an interquartile range of 33, and have them create data that satisfy the measures.

Additional Answers

24a.

January		July
1	0	0 0 1 4
9 7 7 5	1	
8 7 7 1 1	2	0 1 8 9
5 3	3	0 1 9
9 5	4	3 6 7
3 0 0	5	
	6	0 7 7 7
	7	
	8	
9	9	

1|0 = 0.1 0|1 = 0.1

24f. Sample answer: July; The data are more tightly clustered around the median.

25c. Sample answer: The participants' ages are between 12 and 72. Their ages are spread out over the range of the data. The median age of the participants is 25 years.

Enrichment Masters, p. 82

NAME ______________________ DATE __________

10-2 Enrichment
Variance

Student Edition Pages 490–494

Another way to measure the variation of a set of data is by computing the **variance**. The higher the variance is for a group of numbers, the more "spread out" the data will be.

The table below shows the price of the stock for two companies during one week.

	Monday	Tuesday	Wednesday	Thursday	Friday
Acme Computer Systems	$10	$7	$3	$8	$12
Baker Pencil Company	$7	$8	$7	$9	$9

1. What is the mean average price for the week for each company? **$8 for both companies**

Computing the variance will show which company's stock has the greater variation. To compute the variance, follow these steps:

Step 1 Subtract the mean from each number in the set.
Step 2 Multiply each difference in step 1 by itself.
Step 3 Add these differences.
Step 4 Divide the total by the number of members of the set.

Example: Find the variance for Acme Computer Systems.

$$(10-8)\times(10-8)+(7-8)\times(7-8)+(3-8)\times(3-8)+(8-8)\times(8-8)+(12-8)\times(12-8)$$
$$= 4 + 1 + 25 + 0 + 16$$
$$= 46$$

The variance is $46 \div 5$, or 9.2.

Solve.

2. Do you think the variance for Baker Pencil Company will be higher than the variance for Acme Computer Systems? Why? Compute the variance for Baker Pencil Company to see if you are correct. **No; the prices are closer together; 0.8**

3. Consolidated Airlines also had an average price last week of $48 per share, but its variance was 10.8. Indicate five stock prices that could produce this variance. (*Hint*: Change only the Monday and Tuesday prices for Acme.) **Answers may vary. Sample answer: $11, $6, $3, $8, $12**

4. Sleepy Mattress Company's stock had an average price last week of $8 per share and a variance of 0. What was the price of shares each day last week? **$8 each day**

5. Are there any values that the variance cannot equal? If so, what are these values? **yes; values less than zero**

Extension

Using Sports Connections Have students choose a sport or activity and brainstorm a list of ranges that could be determined from statistics about the sport. For example, in basketball, ranges involving the highest and lowest winning scores would be one possibility. Then have students choose one range and write a sample problem using appropriate data.

Additional Answers

25d. Sample answer: The observers are both younger and older than the participants. The median of the observers is 46.5 while the median of the participants is 25.

26.

Stem	Leaf
0	8 9 9
1	1 2 4 5 7
2	5 7 8 9 9 9
3	0 1 6 9
4	0 1

4|0 = 40

10-3 Displaying Data

Setting Goals: *In this lesson, you'll use box-and-whisker plots, pictographs, and line graphs to display data.*

Modeling a Real-World Application: Safety

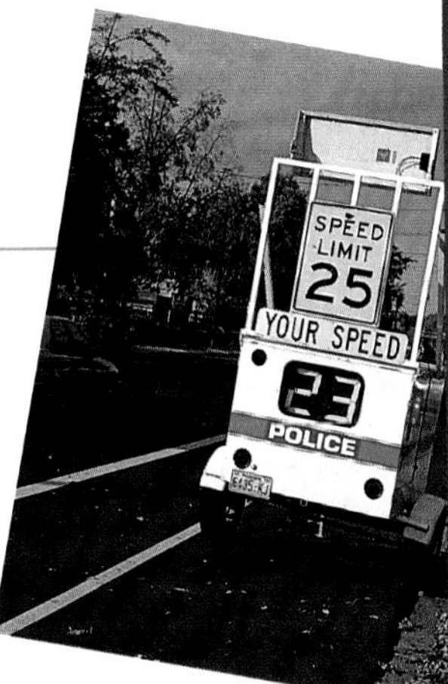

Students at Horace Mann Middle School were concerned about the speed of traffic on the street in front of their school. The student council asked the city's Traffic Engineering Division to place two radar speed sensors with large digital displays near the street. One display allowed vehicles traveling west to see their speed, and the other was positioned for vehicles traveling east to see their speed.

Sample data on the speeds (in mph) of vehicles traveling west were 25, 35, 27, 22, 34, 40, 20, 19, 23, 25, and 30. Speeds of vehicles traveling east were 26, 22, 31, 36, 22, 27, 15, 50, 32, 29, and 30.

The student council voted to ask the city council to erect a stop sign at either end of the block. When they made their presentation, they wanted to make a display of the data so that the city council members would understand it quickly and easily. They chose a **box-and-whisker plot** because it summarizes data using the median, the upper and lower quartiles, and the highest and lowest, or extreme, values.

Learning the Concept

The speeds of the vehicles traveling west are displayed in the stem-and-leaf plot at the right. The median is marked by a box, and the quartiles are circled. The upper extreme is 40, and the lower extreme is 19.

```
1 | 9
2 | 0(2)3 5[5]7
3 | 0(4)5
4 | 0        1|9 = 19
```

Here's how to construct a box-and-whisker plot to display the data.

Step 1 Draw a number line for the range of the speeds. Above the number line, mark points for the extreme, median, and quartile values.

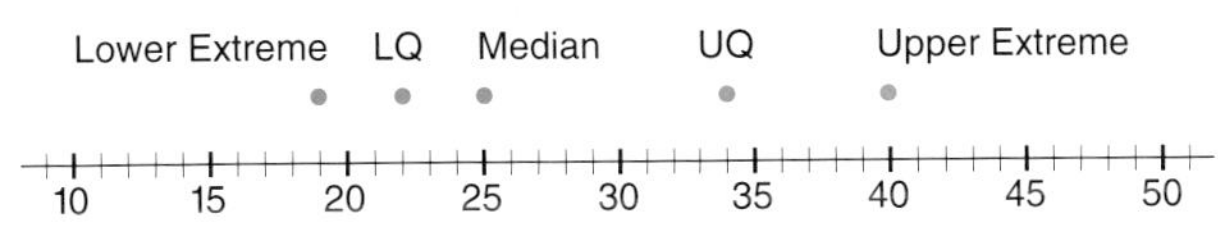

Tech Prep

Banquet Consultant A person who helps plan banquets must be able to make some predictions based on past experience. For instance, knowing that if a banquet is planned for 200 people, the range of people attending the banquet might be as low as 150 and as high as 220, with an average attendance of 190. Using box-and-whisker plots would be a visual way of arranging such data. For more information on tech prep, see the *Teacher's Handbook*.

10-3 LESSON NOTES

NCTM Standards: 1-4, 10

Instructional Resources

- Study Guide Master 10-3
- Practice Master 10-3
- Enrichment Master 10-3
- Group Activity Card 10-3
- Assessment and Evaluation Masters p. 267
- Activity Masters p. 10
- Multicultural Activity Masters p. 19

Transparency 10-3A contains the 5-Minute Check for this lesson; **Transparency 10-3B** contains a teaching aid for this lesson.

Recommended Pacing	
Standard Pacing	Day 3 of 14
Honors Pacing	Day 3 of 13
Block Scheduling*	Day 2 of 7 (along with Lesson 10-4)

*For more information on pacing and possible lesson plans, refer to the ***Block Scheduling Booklet***.

5-Minute Check *(over Lesson 10-3)*

Given the set of data 7, 11, 23, 14, 20, 17, 19, find the following.

1. the range **16**
2. the median **17**
3. the upper and lower quartiles **20, 11**
4. the interquartile range **9**

Motivating the Lesson

Questioning Tell students that you have statistics for the number of people seeking shelter at one homeless shelter in a city each month for one year. Ask how they might display the data.

2 TEACH

In-Class Examples

For Example 1
Use the box-and-whisker plot below to answer the questions.

Number of People Staying Overnight at City Homeless Shelters

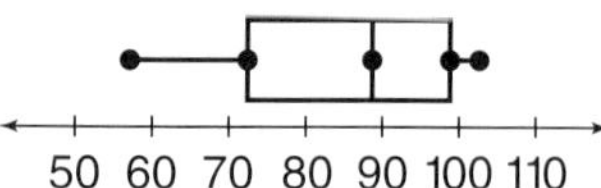

a. What was the fewest people housed? **58**

b. In half of the nights, more than how many people were housed? **89**

c. What percent of the nights were 98 or more people housed? **25%**

For Example 2
Residents living beside a major highway have petitioned the state to reinstate the 55 mph speed limit on the highway near them. They cited the information on sample speeds from state highway patrol records. Speed of cars when speed limit was 55 mph: 50, 54, 55, 55, 58, 58, 59, 60, 66, 69, 85
Speed of cars when speed limit was 65 mph: 58, 60, 65, 65, 70, 70, 71, 74, 79, 85, 88

a. Draw a box-and-whisker plot for the speed data of cars when the speed limit was 55 mph.

Average Speed of Cars on Highway 222

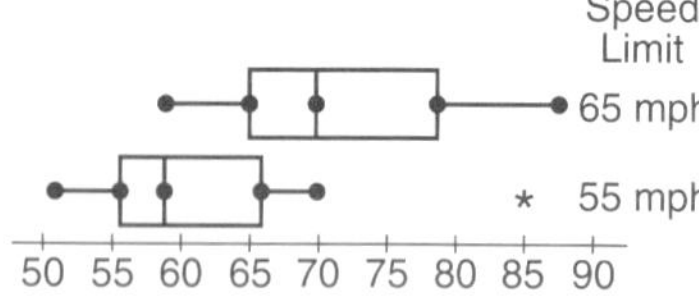

b. Make a double box-and-whisker plot of speeds at 55 mph and 65 mph to convince the state to lower the speed limit.

Step 2 Next, draw a *box* that contains the quartile values. Draw a vertical line through the median value. Finally, extend the *whiskers* from each quartile to the extreme data points.

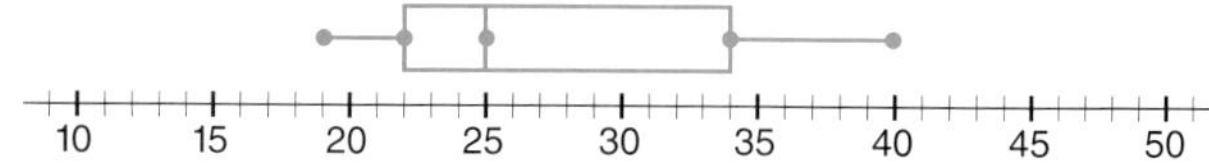

This display, the box-and-whisker plot, separates the data into fourths, each having an equal amount of data. For example, the whisker from 19 to 22 contains 25% of the data.

Example  **Use the box-and-whisker plot below to answer the questions.**

Ages of Volunteers at the Homeless Shelter

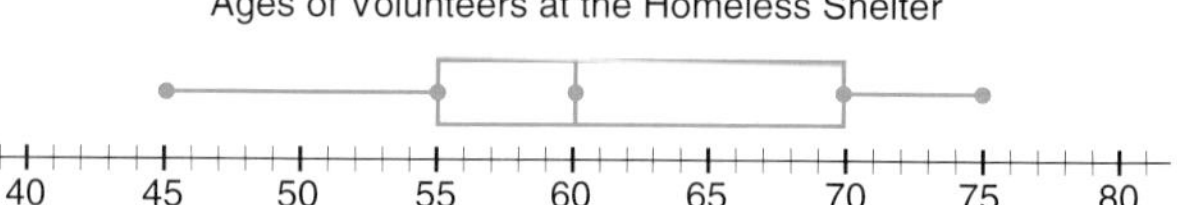

a. What is the age of the youngest volunteer? 45

b. Half of the volunteers are under what age? 60

c. What percent of the volunteers are over 70? 25%

Sometimes the data will have such great variation that one or both of the extreme values will be far beyond the other data. Data that are more than 1.5 times the interquartile range from the quartiles are called **outliers**.

Example  **APPLICATION** **Safety**

Refer to the application at the beginning of the lesson.

a. Draw a box-and-whisker plot for the speed data of the cars traveling east.

15 22 22 26 27 29 30 31 32 36 50

Step 1 The median is 29, the lower quartile is 22, and the upper quartile is 32. Draw a box to show the median and quartiles.

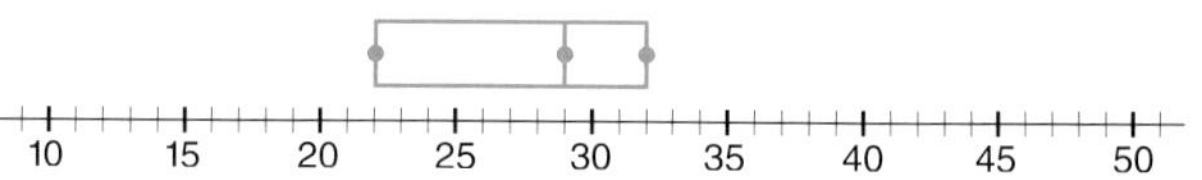

Step 2 The interquartile range is 32 − 22 or 10. So, data more than 1.5 times 10 from the quartiles are outliers.

$$1.5(10) = 15$$

Find the limits of the outliers.

Subtract 15 from the lower quartile $\quad 22 - 15 = 7$

Add 15 to the upper quartile. $\quad 32 + 15 = 47$

The limits for the outliers are 7 and 47. There is one outlier in the data, 50. Plot the outlier with an asterisk. Draw the whiskers to the extreme data that are not outliers.

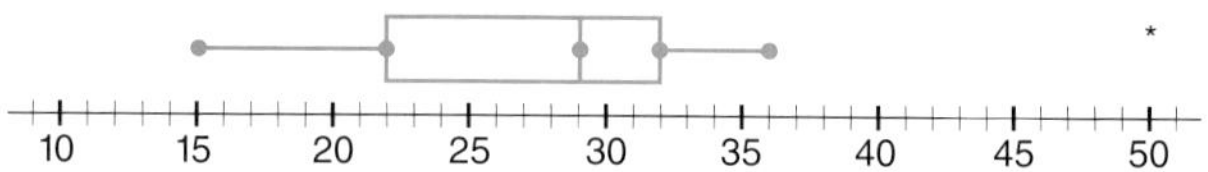

b. The speed limit in front of the school is 20 mph. Make a double box-and-whisker plot of the east and west speeds to convince the city council that a stop sign is needed.

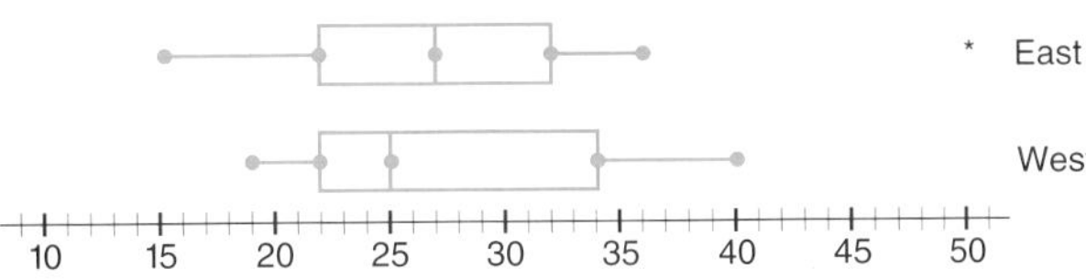

Almost all of the speeds of vehicles traveling west exceed the speed limit, and more than three-fourths of the vehicles traveling east exceed the speed limit. Stop signs at both ends of the block should slow traffic because cars would have to stop twice.

Teaching Tip You may wish to prepare a bulletin board or display of line graphs, pictographs, circle graphs, and comparative graphs found in *USA Today* newspapers. Have students discuss why each particular type of graph was chosen to display the data.

Other statistical graphs that are often used to present data include *line graphs*, *pictographs*, *circle graphs*, and *comparative graphs*.

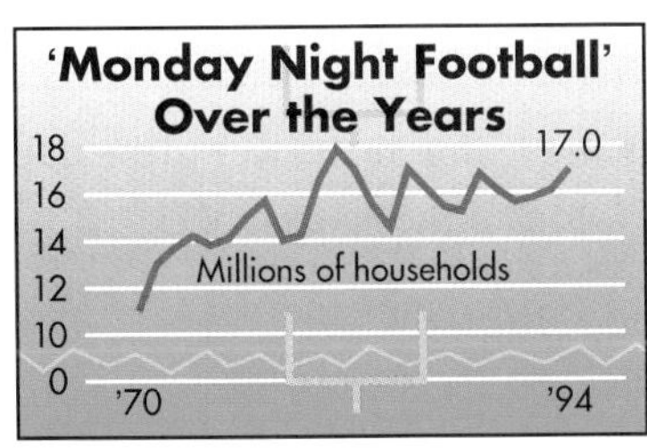

Source: ABC

Line graphs usually show how values change over a period of time. The line graph at the left shows the number of viewers of *Monday Night Football* from 1970 to 1994.

Pictographs use pictures or illustrations to show how specific quantities compare. Each symbol represents a convenient number of items to display the data. The pictograph at the right shows how much different age groups in the United States spend on sporting apparel.

Source: Sporting Goods Manufacturing Association

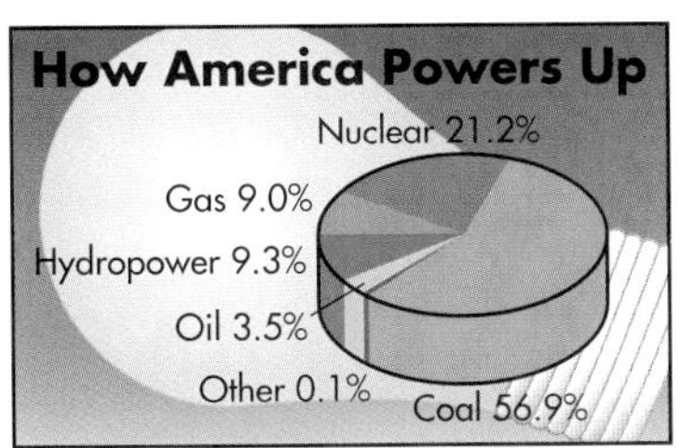

Source: Energy Information Administration

Circle graphs show how parts are related to the whole. The circle graph at the left shows how electricity is generated in the United States.

You can review bar graphs in Lesson 1-10.

Comparative graphs like a double-bar graph are used to show trends. They are usually used to compare results of similar groups. The graph at the right shows how the temperatures in the winter of 1995 compared to the 100-year average.

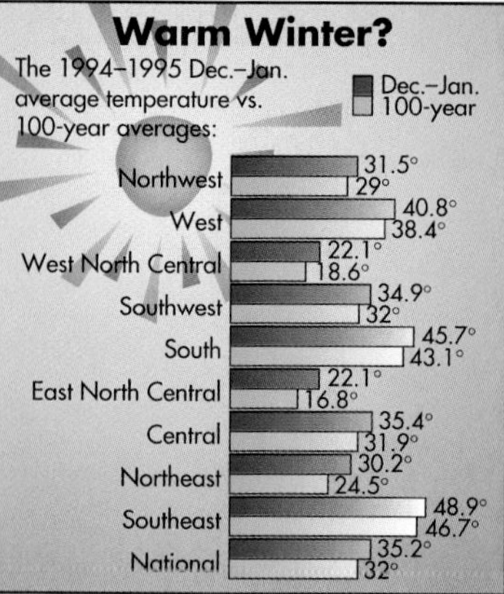

Source: National Climactic Data Center

Example 3

Meteorology

***The Herald Sun* wants to display the high temperatures of the past week. Should they use a box-and-whisker plot, line graph, pictograph, circle graph, or a double-bar graph?**

Since the data would show how values change over a period of time, a line graph would give the reader a clear picture of what temperatures were and the changes in temperature.

Checking Your Understanding

Communicating Mathematics

Read and study the lesson to answer these questions. 1–5. See margin.

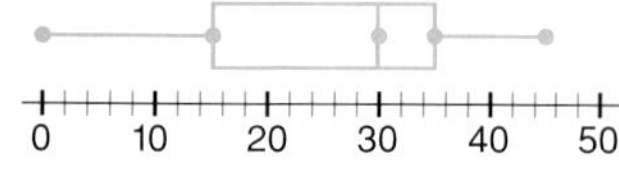

1. **Explain** how to find the median of the data shown in the box-and-whisker plot at the right.
2. **Describe** the data represented by the box.
3. **Write** the five pieces of information you can learn from the plot.
4. **Describe** how to determine if there are any outliers in a set of data.
5. **Explain** how the information you can learn from a set of data shown in a box-and-whisker plot is different from what you can learn from the same set of data shown in a stem-and-leaf plot.

MODELING MATHEMATICS

MATERIALS

stop watch

calculator

Work in groups of 4 or 5 to complete the following activity.

6. Make a table like the one shown below. See students' work.

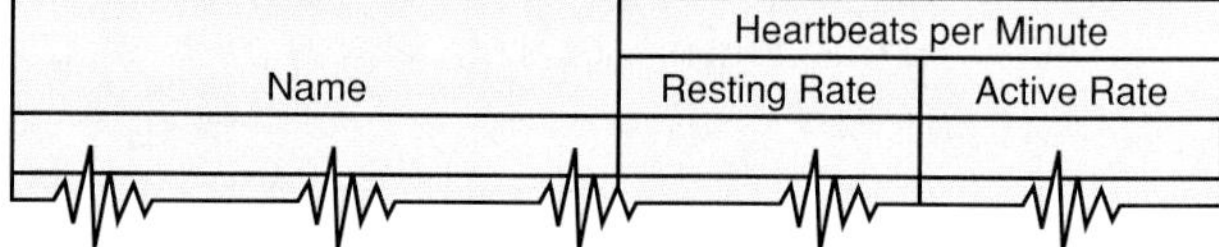

Name	Heartbeats per Minute	
	Resting Rate	Active Rate

a. Find the resting heart rate of each member of your group by taking your pulse for 15 seconds. Then multiply by 4 to find the beats per minute and record the product in your table.

b. In an area designated by your teacher, walk at a brisk rate for 5 minutes without stopping. Then take the pulses again and record them in the table.

c. Extend your table and combine your data of heart rates with that of the other groups.

d. Display the class data in a box-and-whisker plot.

In-Class Example

For Example 3

The *City Telegraph* wants to display how the Granville city government spends its budget. Should they use a box-and-whisker plot, line graph, pictograph, circle graph, or double-bar graph? **Since the data shows how parts of the whole are spent, a circle graph would be a good choice.**

3 PRACTICE/APPLY

Checking Your Understanding

Exercises 1–8 are designed to help you assess your students' understanding through reading, writing, speaking, and modeling. You should work through Exercises 1–6 with your students and then monitor their work on Exercises 7–8.

Additional Answers

1. Sample answer: The median is the point inside the box with the vertical line through it.
2. The data between the upper and lower quartiles.
3. Lower extreme, lower quartile, median, upper quartile, upper extreme.
4. Sample answer: An outlier is an item of data that is more than 1.5 times the interquartile range from the quartiles.
5. Sample answer: It summarizes the data. A stem-and-leaf plot lists all data.

Reteaching

Using Discussion Have students find the least number, lower quartile, median, upper quartile, and greatest number from a set of data that is not in order. Then have them find the same information from a different set of data that is in order. Discuss the differences.

Guided Practice

7d. 20
7f. none

7. Use the box-and-whisker plot below to answer each question.

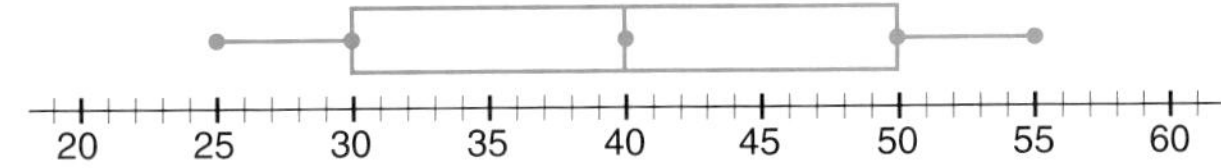

a. What is the median? **40**
b. What is the upper quartile? **50**
c. What is the lower quartile? **30**
d. What is the interquartile range?
e. What are the extremes? **25, 55**
f. What are the outliers, if any?

8. Education From 1987 to 1996, Richton High School had the following number of graduates.

145, 165, 134, 173, 201, 204, 194, 185, 205, 198

a. Find the range, median, upper and lower quartiles, and the interquartile range for the data. **71; 189.5; 201, 165; 36**
b. What are the limits for the outliers? **111, 255**
c. Use the data to make a box-and-whisker plot. **See margin.**

Exercises: Practicing and Applying the Concept

Independent Practice

9. The box-and-whisker plot below shows the heights in inches of the eleven girls on the Towne High girls' volleyball team. **9e. 6**

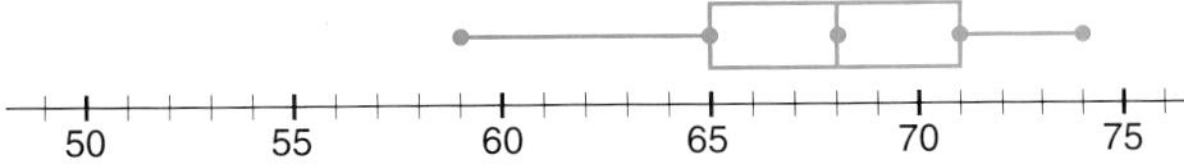

A

a. What is the range? **15**
b. What is the median? **68**
c. What is the upper quartile? **71**
d. What is the lower quartile? **65**
e. What is the interquartile range?
f. Are there any outliers? **no**

10. Use the stem-and-leaf plot at the right to answer each question.

a. Make a box-and-whisker plot of the data. **See margin.**
b. What is the median? **67**
c. What is the upper quartile? **77**
d. What is the lower quartile? **55**
e. What is the interquartile range? **22**
f. What are the extremes? **40, 81**
g. What are the limits for outliers? **22, 110**
h. What are the outliers, if any? **none**

Stem	Leaf
4	0 2
5	1 5 9
6	3 5 7 8
7	2 3 7 8
8	0 1

4|0 = 40

B

11. Use the box-and-whisker plot below to answer each question.

Test Scores

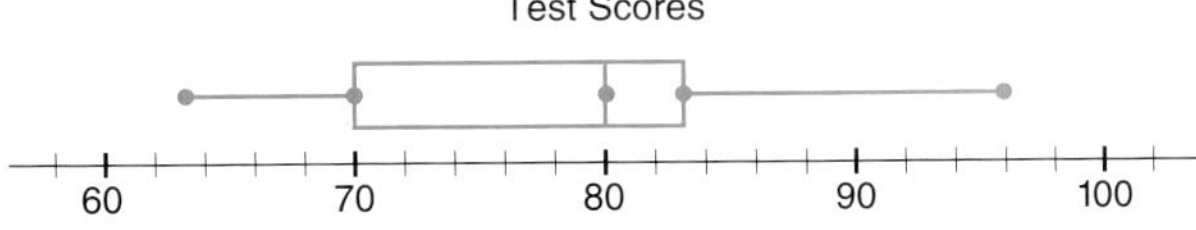

11d. 25%
11e. Scores were closer together between 80–83.
11f. The scores were spread further apart

a. What percent of the test scores are below the median? **50%**
b. What percent of the test scores are represented by the box? **50%**
c. What percent of the test scores are below the lower quartile? **25%**
d. What percent of the test scores are represented by each whisker?
e. Why isn't the median in the middle of the box?
f. Why is one whisker longer than the other?

Error Analysis

Students may fail to identify any outliers or may always identify the extremes of a set of data as outliers. Have students review the mathematical requirements for a piece of data to be an outlier. Then determine the outliers, if any, for Exercise 10h. **There are no outliers.**

Assignment Guide

Core: 9–23 odd, 24–28
Enriched: 10–18 even, 19–28

For **Extra Practice**, see p. 764.

The red A, B, and C flags, printed only in the Teacher's Wraparound Edition, indicate the level of difficulty of the exercises.

Additional Answers

8c.

130 140 150 160 170 180 190 200 210

10a.

40 50 60 70 80 90

Study Guide Masters, p. 83

NAME ______ DATE ______

10-3 Study Guide
Displaying Data

Student Edition Pages 495–501

George surveyed a group of students who walk to school. This list shows the number of minutes it takes each person to walk from his or her house to the school.

5 5 10 (15) 15 20 25 [30] 30 30 30 30 (35) 40 45 50

The median is in the box. The lower and upper quartiles are circled.

A **box-and-whisker plot** is shown at the right. A box is drawn to show the median (30), lower quartile (15), and upper quartile (35). Other key numbers from the data list are the lower extreme (5) and upper extreme (50).

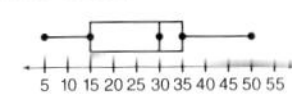

Use the box-and-whisker plots to answer each question below them.

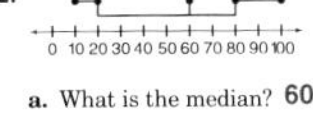

1. a. What is the median? **2**
b. What are the lower quartile and the upper quartile? **1, 5**
c. What are the lower and upper extremes? **0, 8**

2. a. What is the median? **60**
b. What are the lower quartile and the upper quartile? **20, 80**
c. What are the lower and upper extremes? **10, 100**

3. Make a box-and-whisker plot at the right for this list of test scores.
60, 65, 65, 70, 75, 75, 75, 80, 85, 85, 85, 85, 85, 90

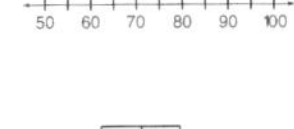

4. Make a box-and-whisker plot at the right for this list of data. It shows the number of pets per family.
0, 0, 0, 1, 1, 1, 2, 2, 2, 2, 2, 3, 3, 3, 5

0 1 2 3 4 5

Cooperative Learning

Jigsaw Have students work in groups of four. Assign each student a letter, A, B, C, or D. Prepare the four corners of the room with signs and examples of the appropriate graphs.
A: Line Graphs; B: Pictographs; C: Circle Graphs; D: Comparative Graphs
Have students go to the corner assigned to their letter, discuss the assigned type of graph, decide for what type of information it can best be used, and discuss how to make the graph. Students then return to their group of four and share the information.

For more information on this strategy, see *Cooperative Learning in the Mathematics Classroom*, pp. 26–27.

Additional Answers

12–17. Sample answers given. See students' explanations.
12. bar graph or pictograph
13. box-and-whisker plot
14. pictograph or bar graph
15. box-and-whisker plot
16. comparative graph
17. line graph or comparative graph
19. The box did not change. The weight 200 is an outlier.
20. Yes; if an extreme value is the same as the upper or lower quartile.
21a.

Physical Science Test Scores.

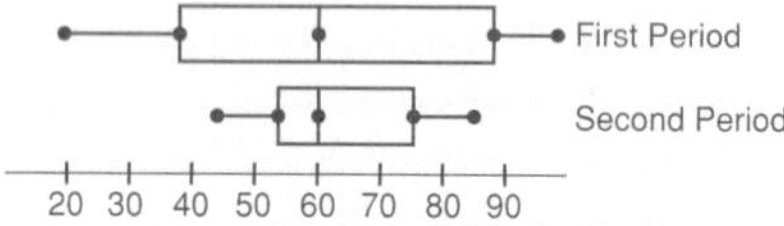

21c. The range and interquartile range are greater for the first period class.

22a.

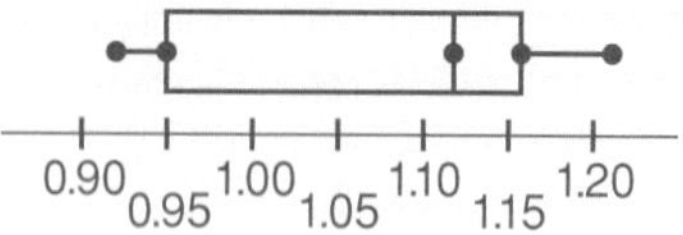

Practice Masters, p. 83

NAME ______ DATE ______

10-3 Practice
Displaying Data

Student Edition Pages 495–501

Use the box-and-whisker plot at the right to answer each question.

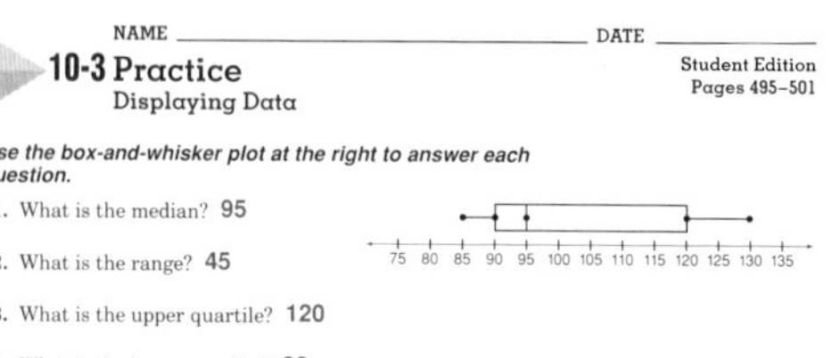

1. What is the median? 95
2. What is the range? 45
3. What is the upper quartile? 120
4. What is the lower quartile? 90
5. What is the interquartile range? 30
6. What are the extremes? 85 and 130
7. What are the limits of the outliers? 45 and 165
8. Are there any outliers? No

Use the stem-and-leaf plot at the right to answer each question.

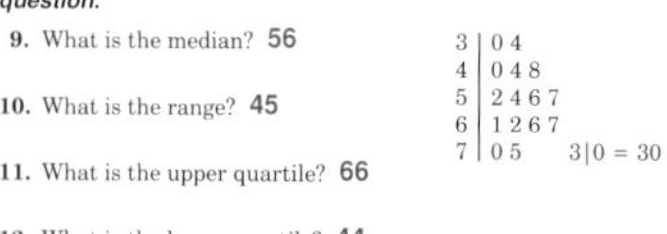

Stem	Leaf
3	0 4
4	0 4 8
5	2 4 6 7
6	1 2 6 7
7	0 5

3|0 = 30

9. What is the median? 56
10. What is the range? 45
11. What is the upper quartile? 66
12. What is the lower quartile? 44
13. What is the interquartile range? 22
14. What are the extremes? 30 and 75
15. What are the limits for the outliers? 11 and 99
16. What are the outliers, if any? none
17. Make a box-and-whisker plot of the data.

20 25 30 35 40 45 50 55 60 65 70 75 80

For each of the following sets of data, determine whether a bar graph, box-and-whisker plot, comparative graph, line graph, or pictograph is the best way to display the data. Explain your reasoning. 12–17. **See margin.**

12. The number of people who have different kinds of pets.
13. Points scored by basketball players in 15 games.
14. Number of movies rented from a video store by different age groups.
15. Shoe size of students in your class.
16. Gender of teachers in your school.
17. Students SAT scores by grade level.

18. **The chart at the right shows the number of sunny days for each city.**
    a. Make a bar graph to represent the number of sunny days. **See margin.**
    b. Make a pictograph to represent the number of sunny days. **See margin.**
    c. Describe any advantages the bar graph has over the pictograph.
    d. Describe any advantages the pictograph has over the bar graph.

18c. Bar graph is easier and more exact.
18d. Pictograph is easy to read at a glance, but not as detailed.

Sunny Days in 1993

City	Number of Days
Albany, NY	60
Burlington, VT	60
Dallas, TX	120
Des Moines, IA	80
Indianapolis, IN	68
Jacksonville, FL	90
Kansas City, MO	100
Lexington, KY	69
Mobile, AL	104
New Orleans, LA	99
Salt Lake City, UT	130

Source: The World Almanac, 1995

Critical Thinking

19. In making a box-and-whisker plot of the weights of the 23 juniors on the varsity football team, Randy found the median to be 170 pounds, the upper quartile to be 178 pounds, the lower quartile to be 168 pounds, and there were no outliers. Then he discovered four more weights: 200, 175, 169, and 167. When he included these in the data, what happened to the box-and-whisker plot? (*Hint:* You may want to sketch two plots.) **See margin.**
20. Is it possible to have a box-and-whisker plot with only one whisker? Explain your answer. **See margin.**

Applications and Problem Solving

21. **Teaching** Mr. Juarez has two classes of Physical Science. He made the stem-and-leaf plot of the scores for the last test shown at the right.
    a. Make a double box-and-whisker plot for each class. (Use the same number line for both; make one plot above the other.) **See margin.**
    b. What is similar about the data in the two plots? **The medians are the same.**
    c. What is different about the data in the two plots? **See margin.**
    d. Which class had scores that were more evenly spread over the range? Explain. **Answers will vary.**

First Period		Second Period
8 0	2	
8 5	3	
5	4	3 5
0	5	2 4 5 6
0 0	6	0 0 0 8
5 2	7	2 5 8
9 8	8	1 5
8 6 1	9	

1|9 = 91 8|1 = 81

Group Activity Card 10-3

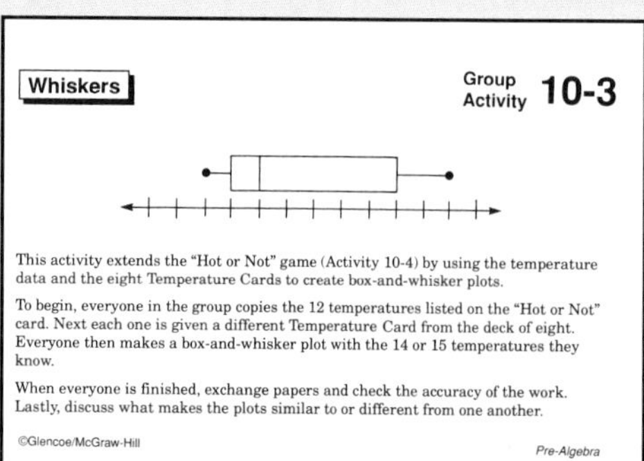

Whiskers Group Activity **10-3**

This activity extends the "Hot or Not" game (Activity 10-4) by using the temperature data and the eight Temperature Cards to create box-and-whisker plots.

To begin, everyone in the group copies the 12 temperatures listed on the "Hot or Not" card. Next each one is given a different Temperature Card from the deck of eight. Everyone then makes a box-and-whisker plot with the 14 or 15 temperatures they know.

When everyone is finished, exchange papers and check the accuracy of the work. Lastly, discuss what makes the plots similar to or different from one another.

 Pre-Algebra

Additional Answers

22d.

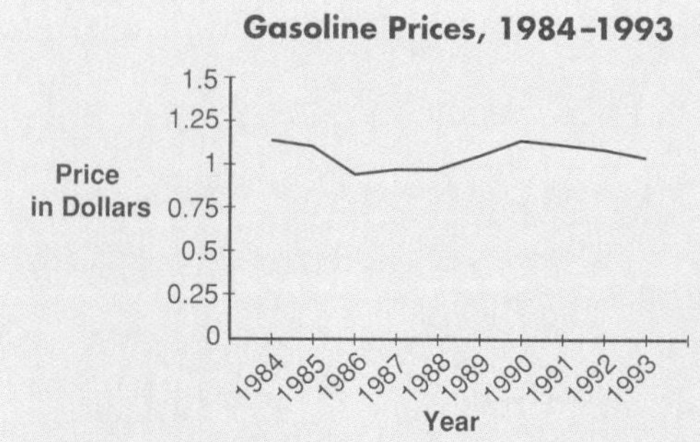

22. **Consumerism** The average retail prices for one gallon of unleaded gasoline from 1984–1993 are shown in the chart below.

Year	1984	1985	1986	1987	1988
Price	$1.21	$1.20	$0.92	$0.95	$0.95

Year	1989	1990	1991	1992	1993
Price	$1.02	$1.16	$1.14	$1.13	$1.11

a. Make a box-and-whisker plot of the prices. **See margin.**
b. What is the median? **$1.12**
c. Are there any outliers? If so, why do you think they occurred?
d. Draw a line graph of the data. **See margin.**
e. Does the line graph present a better picture of the data than the box-and-whisker plot? Explain your answer.

22c. none
22e. Yes; Sample answer: You can see the changes over a period of time.

23. **Employment** The chart below shows the percent of men and women who held two jobs from 1970 to 1995. **a–d. See margin.**

Percent Who Hold Down Two Jobs						
Year	1970	1975	1980	1985	1990	1995
Men	7.0	5.8	5.8	5.9	6.4	5.9
Women	2.2	2.9	3.8	4.7	5.9	5.9

a. Make a comparative bar graph to show the percentage of men and women who hold two jobs.
b. What changes have occurred in men's employment?
c. What changes have occurred in women's employment?
d. In general, what conclusion can you make from the data?

Mixed Review

24. **Statistics** Find the range, median, upper and lower quartiles, and the interquartile range for the set of data in the stem-and-leaf plot shown at the right. (Lesson 10-2) **41; 67; 77; 55; 22**

Words Typed Per Minute

Stem	Leaf
4	0 2
5	1 5 9
6	3 5 7 8
7	2 3 7 8
8	0 1

4|0 = 40

25. Write the domain and range of the relation $\{(5, -3), (-1, 4), (4.5, 0), (-1, -3)\}$ (Lesson 8-1) **$\{5, -1, 4.5\}; \{-3, 4, 0\}$**

26. **Measurement** How many grams are in 0.046 kilograms? (Lesson 7-8) **46**

27. Sample answer: $7

27. Estimate the product of 0.25 and $27.98. (Lesson 6-2)

28. **Nutrition** One cup of skim milk has 80 Calories, and one cup of whole milk has 150 Calories. If you drink one cup of milk per day, about how many Calories would be saved each year by switching from whole milk to skim? (Lesson 1-1)
a. Which method of computation do you think is most appropriate? Justify your choice. **estimation, an exact answer is not needed**
b. Solve the problem using the four-step plan. Be sure to examine your solution. **about 26,000 Calories**

4 ASSESS

Closing Activity

Speaking On the board or overhead, draw the box-and-whisker plot shown below. Then have students state facts from the plot. For example, the median is 7, there is an outlier at 0, the interquartile range is 4, and so on.

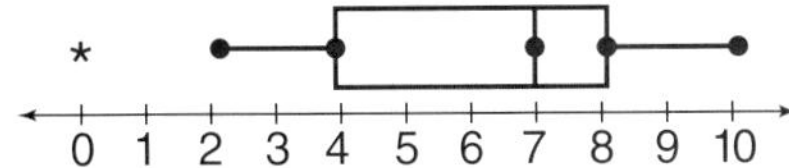

Chapter 10, Quiz A (Lessons 10-1 through 10-3) is available in the *Assessment and Evaluation Masters*, p. 267.

Additional Answer

23a.

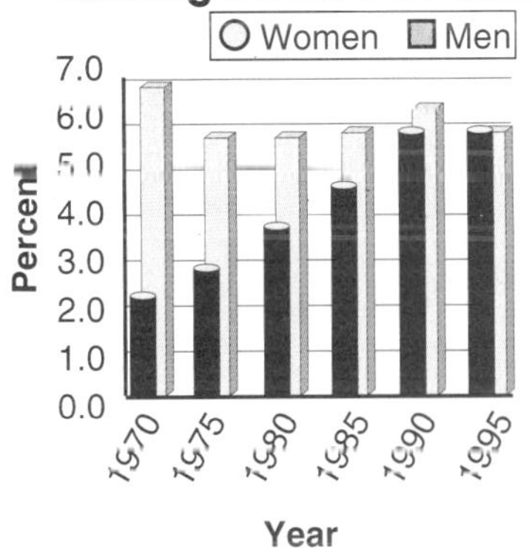

***Enrichment Masters*, p. 83**

NAME ____________ DATE ______

10-3 Enrichment
Normal Curve

Student Edition Pages 495–501

Before actually tossing a coin a number of times, there is a method for determining the probability that a certain number of times heads will result. For example, if you toss a coin 100 times, there is only about a 2.5% chance you will get more than 60 heads.

The **normal curve** is a very useful tool when dealing with problems such as the one above. This curve is also called a *bell curve*. If a teacher says he or she grades "on a curve," he or she is probably referring to the normal curve.

The normal curve at the right shows the percent of the given data that will fall within a certain distance from the mean of the data. These distances are measured in units called *standard deviations*. We will use these units (without actually calculating them) in the exercises below. Also used will be the fact that if a coin is tossed 100 times, the standard deviation for the number of heads is 5 and the mean is 50.

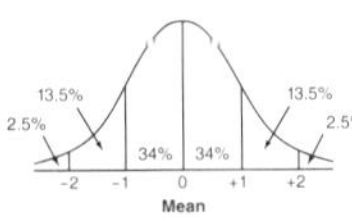

The 34% shown in the area under the curve between 0 and +1 is the percent chance that you will toss between 50 and 55 heads (mean + 1 standard deviation) when tossing the coin 100 times. The 2.5% shown in the area under the curve to the left of −2 is the percent chance that you will toss less than 40 heads (mean − 2 standard deviations) when tossing the coin 100 times.

Use the normal curve shown above to answer each question. The coin is tossed 100 times for each exercise.

1. What is the percent chance you will toss between 45 and 50 heads? **34%**
2. What is the percent chance you will toss more than 55 heads? **16%**
3. What is the percent chance you will toss more than 60 heads? **2.5%**
4. What is the chance you will toss less than 60 heads? **97.5%**

If a coin is tossed 1000 times, the mean for the number of heads is 500 and the standard deviation is approximately 16.

Solve.

5. If a coin is tossed 1000 times, what is the percent chance you will get more than 532 heads? **2.5%**
6. If a coin is tossed 1000 times, there is a 16% chance you will get less than how many heads? **484 heads**

Extension

Work Backward Have students make up data for each condition.
a. The box is very short.
b. The box is very long.
c. The whiskers are very short.
d. The whiskers are very long.
e. The box is very short, and the left whisker is longer than the right whisker.

Have students compare the data.

Additional Answers

23b. Men's employment decreased from 1970 to 1975, but since 1975, men's employment has been relatively consistent.

23c. Women's employment increased steadily from 1970 to 1990 and has remained steady from 1990-95.

23d. Sample answer: Women are as likely as men to have two jobs.

10-3B LESSON NOTES

NCTM Standards: 1-4, 10

Objective
Use a graphing calculator to construct statistical graphs.

Recommended Time
25 minutes

Instructional Resources
Graphing Calculator Masters, p. 23

This master provides keystroking instruction for this lesson for the TI-81 and Casio graphing calculators.

1 FOCUS

Motivating the Lesson
Ask students to discuss the disadvantages of trying to make statistical graphs by hand. Guide the discussion to include the tedious work, the inaccuracy involved when decimals are involved, and so on. Point out that in this lesson, students will use the graphing calculator to make statistical graphs.

2 TEACH

Teaching Tip Have students do research and create graphs from their own data.

3 PRACTICE/APPLY

Assignment Guide
Core: 1–4
Enriched: 1–4

GRAPHING CALCULATOR ACTIVITY

MATHEMATICS LAB

10-3B Making Statistical Graphs

An Extension of Lesson **10-3**

You can use a graphing calculator to construct statistical graphs. A line graph, box-and-whisker plots, and a special kind of bar graph, called a **histogram**, can be created using a TI-82 graphing calculator.

Activity  **Use a TI-82 graphing calculator to construct a line graph of the data on bicycle sales.**

You can review setting the range in Lesson 2-2B and clearing the statistical memory in Lesson 6-6B.

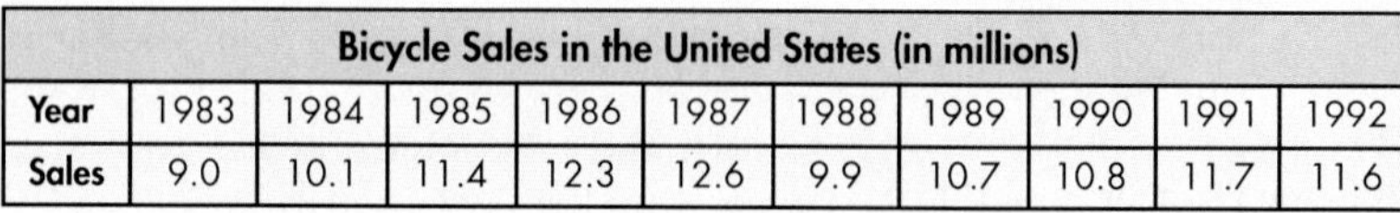

Bicycle Sales in the United States (in millions)										
Year	1983	1984	1985	1986	1987	1988	1989	1990	1991	1992
Sales	9.0	10.1	11.4	12.3	12.6	9.9	10.7	10.8	11.7	11.6

Before you enter any data to graph, you must clear the statistical memory, clear the Y= list, and set the range. Use the range values [1983, 1993] by [0, 15] with a scale factor of 1 on both axes.

Now enter the data into the memory and draw the graph.

Enter: [STAT] [ENTER] *Accesses the statistical lists.*

1983 [ENTER] . . . 1992 [ENTER] *Enters the years in L_1.*

[▶] 9 [ENTER] . . . 11.6 *Enters the sales in L_2.*

Now choose the type of statistical plot you would like to create.

Enter: [2nd] [STAT PLOT] [ENTER] *Accesses the menu for the first statistical graph.*

Use the arrow and [ENTER] keys to highlight your selections for the graph. Select "On", the line graph icon, "L_1" as the Xlist, "L_2" as the Ylist, and "•" for the marks.

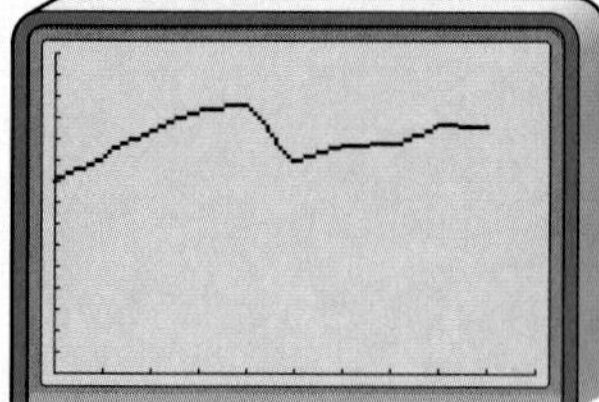

To create the graph, press [GRAPH].

Your Turn

1. Use a TI-82 graphing calculator to create a line graph of the data below. (Enter 1, 2, 3, . . . , 13 in list L_1 to represent the months.) **See Solutions Manual.**

Unemployment Rate, 1993–1994													
Month	Nov.	Dec.	Jan.	Feb.	Mar.	Apr.	May	June	July	Aug.	Sep.	Oct.	Nov.
Percent	6.5	6.4	6.7	6.5	6.5	6.4	6.0	6.0	6.1	6.1	5.9	5.8	5.6

2. Based on your graph, do you think the unemployment rate for December, 1994 increased or decreased? Explain. **See Solutions Manual.**

A *histogram* is a special bar graph that displays the frequency of data that has been organized into equal intervals. The intervals cover all possible values of data. Therefore, there are no spaces between the bars of the histogram.

Activity 2 **Use a TI-82 graphing calculator to create a histogram of the number of graduates for each of the states.**

Public High School Graduates in 1991 (in thousands)

39.0	5.5	31.3	25.7	234.2	31.3	27.3	5.2	87.4	60.1
9.0	12.0	103.3	58.6	28.6	24.4	35.8	33.5	13.2	39.0
52.1	88.2	46.5	23.7	46.9	9.0	16.5	9.4	10.1	67.0
15.2	133.6	62.8	7.6	107.5	33.0	24.6	104.8	7.7	33.1
7.1	44.8	174.3	22.2	5.2	58.4	42.5	21.1	49.3	5.7

When you set the range for a histogram, you determine the number of bars (equal intervals) you want to graph. Use the range values [0, 240] by [0, 25] with a scale factor of 30 on the *x*-axis and 5 on the *y*-axis. These settings will create 240 ÷ 30 or 8 bars.

Now enter the data into the memory and draw the graph.

Enter: STAT ENTER *Accesses the statistical lists.*

39.0 ENTER 5.5 ENTER 31.3 ENTER . . . 5.7 ENTER

Enter: 2nd STAT PLOT ENTER *Accesses the menu for the first statistical graph.*

Use the arrow and ENTER keys to highlight "On", the histogram icon, "L1" as the Xlist, and "1" as the frequency. Press GRAPH to view the histogram.

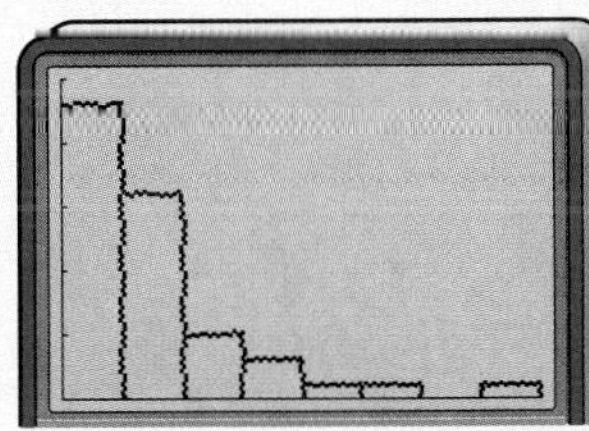

You can make a box-and-whisker plot of the same data.

Enter: 2nd STAT PLOT ENTER

Select the same plot and use the arrow and ENTER keys to highlight the box-and-whisker plot icon. Press GRAPH to view the box-and-whisker plot.

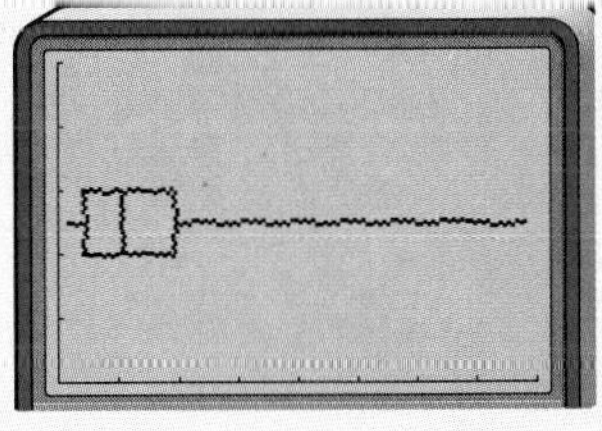

Your Turn

3. See Solutions Manual.

3. Use a TI-82 graphing calculator to create a histogram and a box-and-whisker plot for the heights of the United States presidents.

Heights of U.S. Presidents (in inches)

74	66	74.5	64	72	67	73	66	66	72	68	68	69	70
72	76	70	68.5	68.5	70	74	71	66	66	68	74	71	72
69	71	74	70	70	73	75	71.5	72	70	73	74	74.5	

TALK ABOUT IT

4. If the range of values for a histogram is 35 and you want to create 7 bars, what scale factor should you enter? Explain. **5; 35 ÷ 7 = 5**

4 ASSESS

Observing students working with technology is an excellent method of assessment. You may wish to ask a student at random from each group to explain how to model and solve a problem. Also watch for and acknowledge students that are helping others to understand the concept being taught.

Technology

This lesson offers an excellent opportunity for using technology in your pre-algebra classroom. For more information on using technology, see *Graphing Calculators in the Mathematics Classroom*, one of the titles in the Glencoe Mathematics Professional Series.

10-4 LESSON NOTES

NCTM Standards: 1-4, 10

Instructional Resources

- Study Guide Master 10-4
- Practice Master 10-4
- Enrichment Master 10-4
- Group Activity Card 10-4
- Real-World Applications, 23
- Math Lab and Modeling Math Masters, p. 69

Transparency 10-4A contains the 5-Minute Check for this lesson; **Transparency 10-4B** contains a teaching aid for this lesson.

Recommended Pacing	
Standard Pacing	Day 4 of 14
Honors Pacing	Day 4 of 13
Block Scheduling*	Day 2 of 7 (along with Lesson 10-3)

*For more information on pacing and possible lesson plans, refer to the ***Block Scheduling Booklet***.

1 FOCUS

5-Minute Check *(over Lesson 10-3)*

Use the box-and-whisker plot shown to answer each question.

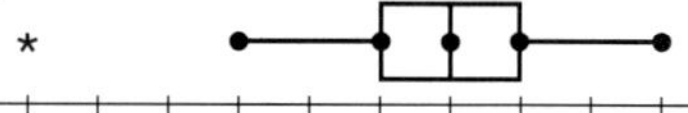

1. What is the median? 80
2. What is the upper quartile? 85
3. What is the lower quartile? 75
4. What is the interquartile range? 10
5. Name any outliers. 50

Motivating the Lesson

Situational Problem Have students discuss times when they were misled by advertisements in newspapers or magazines or on television or radio.

10-4 Misleading Statistics

Setting Goals: *In this lesson, you'll recognize when statistics are misleading.*

Modeling a Real-World Application: Education

Mr. Snyder's sociology class was studying birthrates in the United States. For a project on the history of birthrates from 1960 to 1990, Maria and Jennifer made the two graphs shown below.

- ▶ Do both graphs show the same information?
- ▶ Which graph suggests a baby boom from 1975–1990? Which graph suggests a fairly steady birthrate?

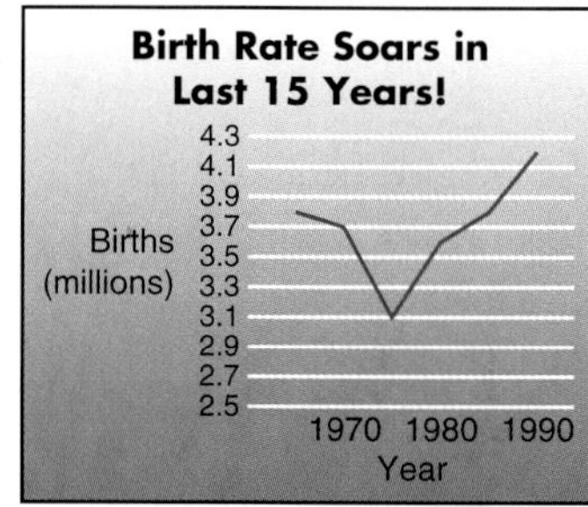

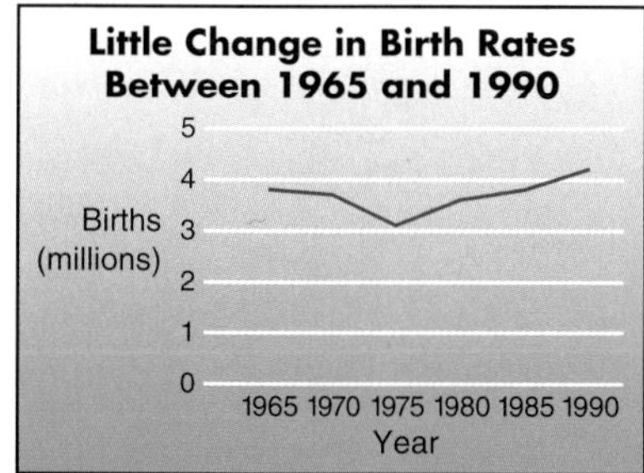

What is the same and what is different about the two graphs?

Learning the Concept

The data used in the graphs above are the same. However, different scales were used to make the data appear to support a particular point of view. The soaring graph at the left is a result of an extended vertical axis and a shortened horizontal axis. A small "break" in the vertical axis is used to show that the axis is not to scale between zero and 2.7. However, it is visually misleading when compared to the complete scale of the graph at the right.

Tech Prep

Political Analyst An analyst must be able to recognize manipulation of statistics, including the use of misleading graphs. For more information on tech prep, see the *Teacher's Handbook*.

Example 1 APPLICATION Entertainment

The graphs at the right display data about movie attendance in the United States.

a. Is the top graph misleading? Explain.

The graph is misleading because there is no title and there are no labels on either scale. Also, the vertical axis does not include zero.

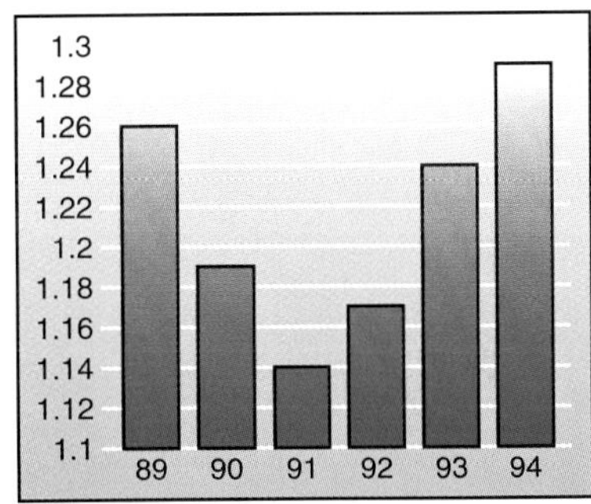

b. Is the bottom graph misleading? Explain.

The graph is not misleading. All the necessary information is present. The scales are labeled, the distance between the units on each axis is uniform, and the vertical axis includes 0.

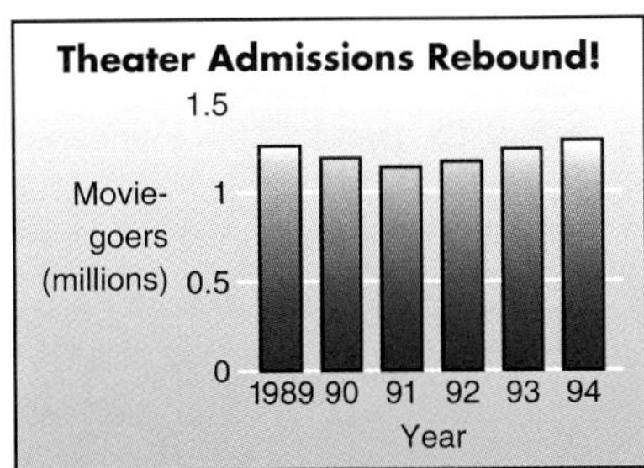

LOOK BACK

You can review measures of central tendency in Lesson 6-6.

Another way statistics can be misleading is by using the inappropriate measure of central tendency.

Example 2 INTEGRATION Statistics

After a semester exam in science, Mr. Myers displayed the results to the class. The scores are shown in the stem-and-leaf plot at the right. Mr. Myers told the principal that the average grade on the exam was 77. Ryan told his parents that the class average was 93. Tai told her grandmother the class average was 80. How can all three people think they are correct?

Scores

9 | 01333
8 | 012345
7 | 023
6 | 0123579

9|0 = 90

Each person used a different measure of central tendency to describe the average grade for the class. Mr. Myers used the mean and Tai used the median. Ryan's use of the mode is misleading because 93 is also the highest score on the semester exam.

Checking Your Understanding

Communicating Mathematics

1–2. See Solutions Manual.

Read and study the lesson to answer these questions.

1. **Describe** two ways graphs can be misleading.
2. **List** some things you might question when reading the results of a survey on employee salaries.

Reteaching

Using Models Show students several misleading graphs such as the one shown at the right. Have students discuss why the graphs are misleading. **In the graph at the right, the bar for Big Company is not just taller, it is wider than the bar for Tiny Company. It makes it seem that the difference is greater.**

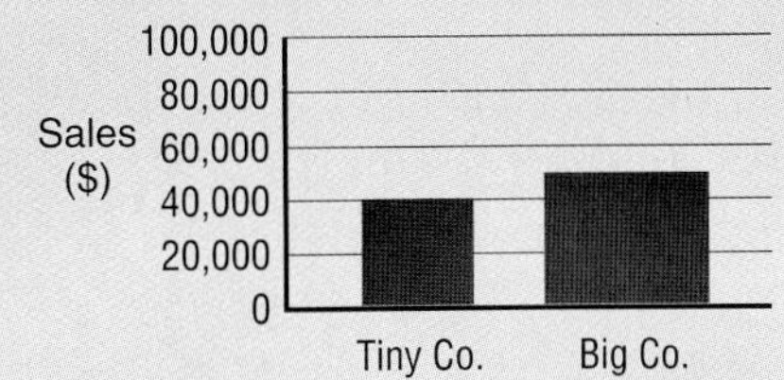

2 TEACH

In-Class Examples

For Example 1

Tell whether each graph is misleading. Explain your answer.

a.

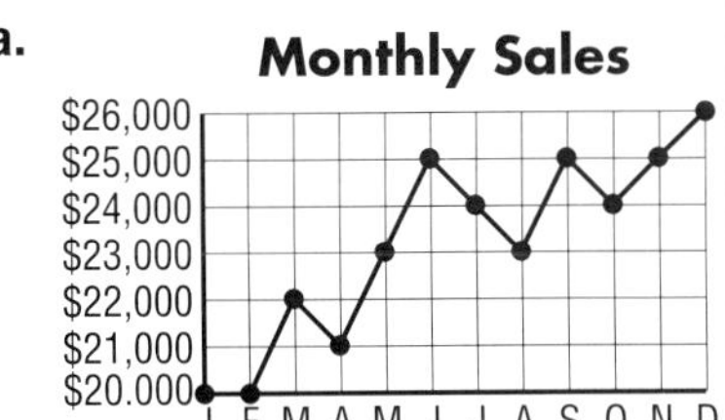

Yes; the scale starts at $20,000.

b.

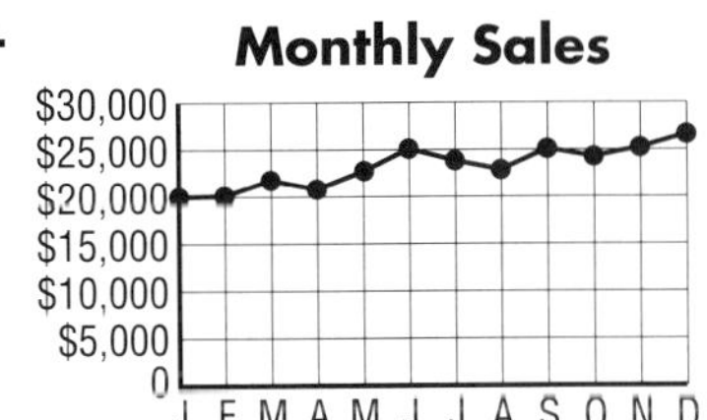

No; the scales are labeled, the distance between the units on each axis is uniform, and the vertical axis includes 0.

For Example 2

Use the data in the table below to answer each question.

Reading Scores	Number of Students
95% or higher	5
80–94%	37
70–79%	75
59–69%	58
below 59%	65

a. Which "average" would the school board want to use? **mode**

b. Which "average" would school critics like to use? **median or mean**

Teaching Tip Tell students that you did a survey and 3 out of 4 students said that math was their favorite subject. Have students discuss what could be misleading about your survey.

3 PRACTICE/APPLY

Checking Your Understanding

Exercises 1–6 are designed to help you assess your students' understanding through reading, writing, speaking, and modeling. You should work through Exercises 1–4 with your students and then monitor their work on Exercises 5–6.

Assignment Guide
Core: 7–13 odd, 14–15, 17–21 **Enriched:** 8–12 even, 14–21

For **Extra Practice**, see p. 765.

The red A, B, and C flags, printed only in the Teacher's Wraparound Edition, indicate the level of difficulty of the exercises.

Additional Answers

5b. The second car is shorter in length and width.

5c. Yes; the actual number of cars sold in 1992 is 1.1 times greater than the number sold in 1991. However, the graph appears to be longer and wider, but both dimensions of the car should not have changed.

Study Guide Masters, p. 84

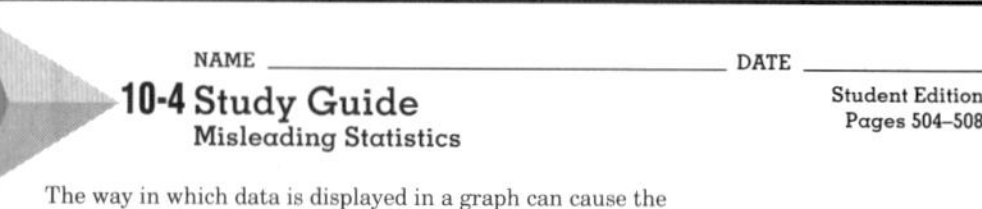
NAME ______ DATE ______

10-4 Study Guide
Misleading Statistics

Student Edition Pages 504–508

The way in which data is displayed in a graph can cause the graph to be visually misleading. The two graphs below show the number of cookies sold each month by a gourmet cookie shop.

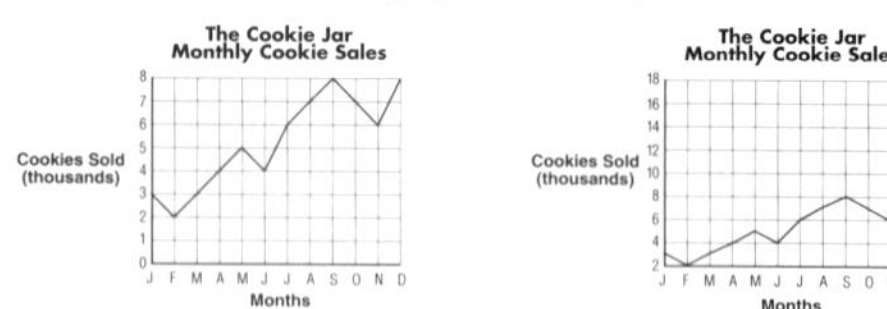

The graphs above contain the same data. However, the graph on the right suggests lower sales than the graph on the left. This is due to the shortened vertical axis. Also, notice that the vertical axis does not include zero.

The graph at the right displays data about money earned by high school students during summer vacation.

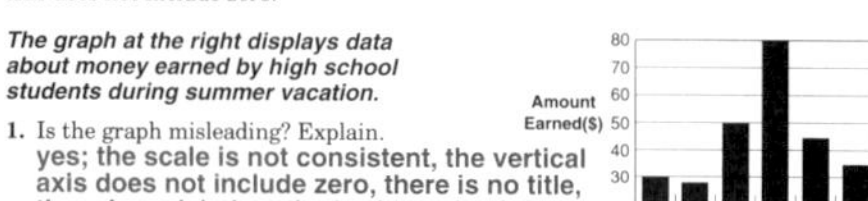

1. Is the graph misleading? Explain. yes; the scale is not consistent, the vertical axis does not include zero, there is no title, there is no label on the horizontal axis

Kimiko made two different graphs showing employee productivity levels for each year in business.

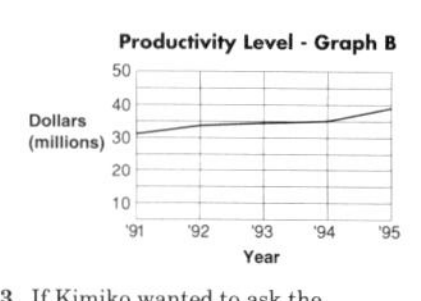

2. Which graph is misleading? Why? Graph A; the extended vertical axis, the shortened horizontal axis, and the vertical axis not to scale between zero and thirty-one all are misleading.

3. If Kimiko wanted to ask the management to give the employees raises in 1996, which graph would be used? Explain your answer. Graph A; it appears to show a higher productivity level.

3. **You Decide** Khandi says that the graph at the right is misleading. Carlos disagrees. Who is correct and why? **Khandi; the scale is not consistent.**

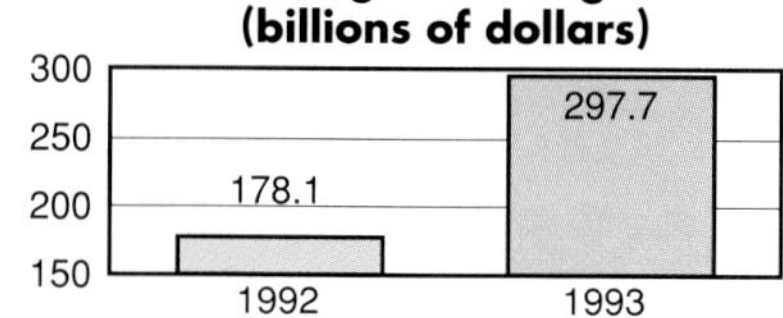

4. **Assess Yourself** Do you understand how to determine whether graphs are misleading? Find three graphs in a newspaper or magazine. Determine if the graphs are misleading and explain why or why not. **See students' work.**

Guided Practice

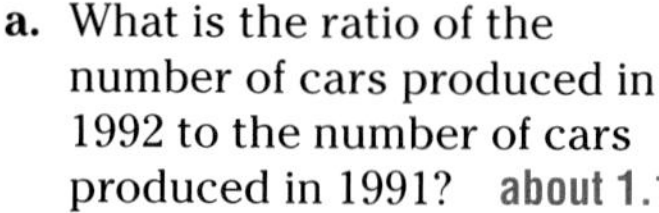

5. **Refer to the graph at the right.**
   a. What is the ratio of the number of cars produced in 1992 to the number of cars produced in 1991? **about 1.1**
   b. How does this ratio compare with the ratio of the sizes of the cars on the graph?
   c. Is this a misleading graph? Explain. **See margin.**

5b. See margin.

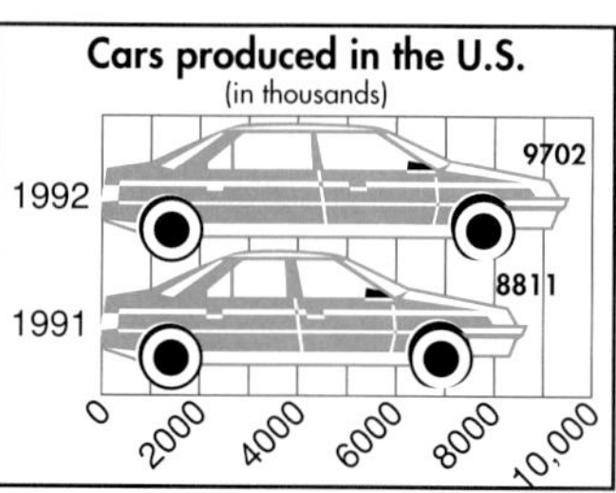

6. **Statistics** Roberta made two graphs of her social studies grades for each grading period.

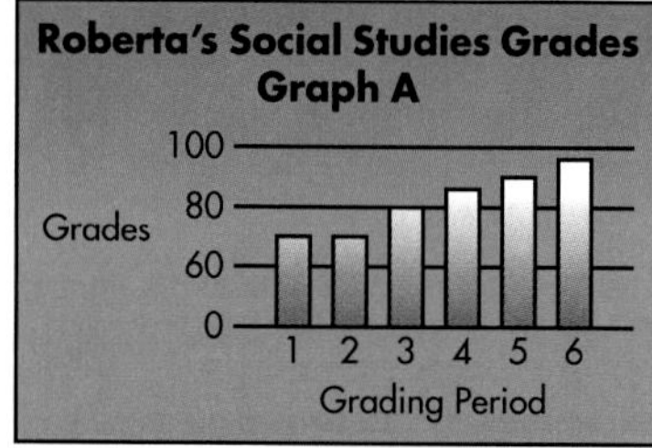

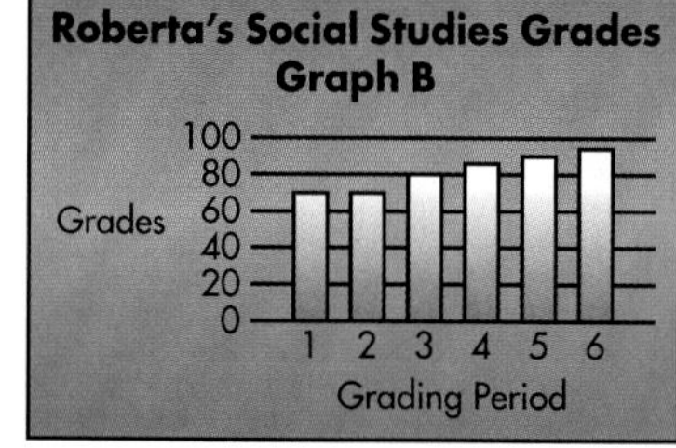

   a. Explain why these graphs made from the same data look different.
   b. If Roberta wanted to convince her parents that her social studies grades have improved dramatically, which graph would she probably show her parents? Explain.

6a. Sample answer: different scales
6b. Graph A because it looks like there is a dramatic rise in social studies grades.

Exercises: Practicing and Applying the Concept

Independent Practice

7. The houses in the Cambridge section of Anderson were assessed for school taxes. The values are shown in the frequency table at the right.
   a. Which measure of central tendency would you use to find the average cost of a house? Why?
   b. Which average might a real estate agent use with a business executive? Explain.

Value	Number
$150,000	2
$130,000	5
$115,000	4
$95,000	10
$85,000	20
$75,000	2

7a. Sample answer: Median; there are large and small values.
7b. Mean; the high values will raise the center of the data.

Group Activity Card 10-4

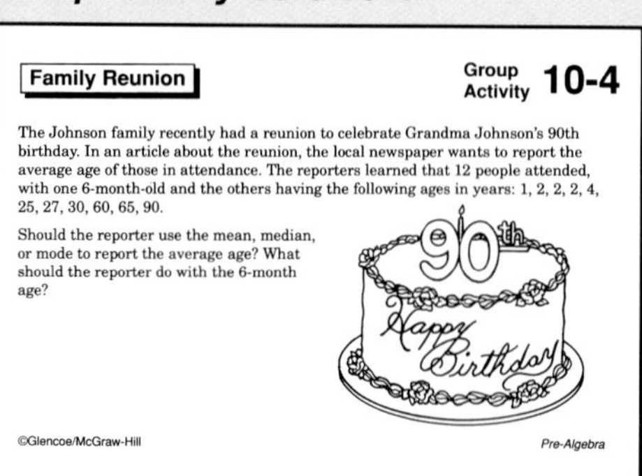
Family Reunion

Group Activity **10-4**

The Johnson family recently had a reunion to celebrate Grandma Johnson's 90th birthday. In an article about the reunion, the local newspaper wants to report the average age of those in attendance. The reporters learned that 12 people attended, with one 6-month-old and the others having the following ages in years: 1, 2, 2, 2, 4, 25, 27, 30, 60, 65, 90.

Should the reporter use the mean, median, or mode to report the average age? What should the reporter do with the 6-month age?

 Pre-Algebra

Use the chart to describe each category. Write the average you used.

	Tashelle	Kim	Dawn	Sita	Bill
Age	40	24	33	28	41
Favorite Entertainment	sports	concerts	theater	movies	sports
Height (cm)	185	163	165	157	193
Salary	$34,700	$18,000	$30,500	$25,500	$47,000

B

8. average age **33; mean**
9. average height **172.6; mean**
10. average salary **30,500; median**
11. average favorite entertainment
12. How can you describe the "average" person in this group?

11. sports; mode
12. See margin.

13. Find the mean, median, mode, and range of the salaries at each company in the table at the right. Which company would you rather work for? Explain your reasoning. **See margin.**

	Frequency	
Salary	Company A	Company B
$15,000	18	
$20,000	4	24
$25,000	4	6
$50,000	2	1
$100,000	1	

Critical Thinking

14. Find a graph in a magazine or newspaper. Redraw the graph so that the data will appear to show different results. **See students' work.**

Applications and Problem Solving

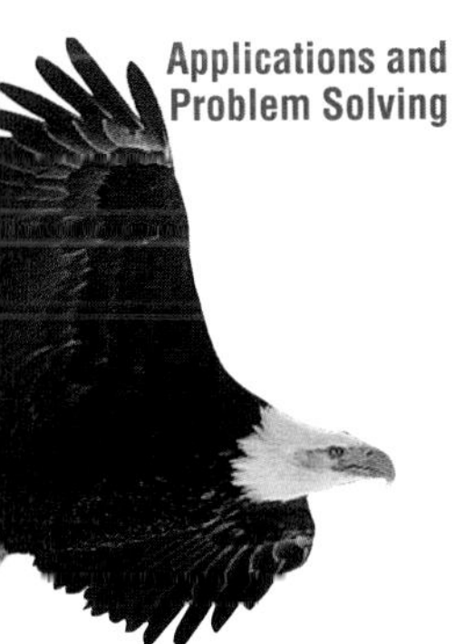

15. **Conservation** In 1963, there were only 417 pairs of bald eagles living in the contiguous 48 states of the United States. DDT, a pesticide used to kill insects, had poisoned many of the birds. In 1973, the U.S. Congress protected the eagle by banning the use of DDT in areas where eagles live. In 1978, the eagle was named an endangered species. As a result, the eagle population has grown.
    a. Do the data in the table represent the growth of the eagle population since the policy changes were made in the United States? Explain. **See margin.**
    b. Draw a graph that you might use to accurately represent the data. **See margin.**

Eagle Pairs in Contiguous 48 States of the U.S.

Year	Number of Pairs
1988	2475
1989	2680
1990	3020
1991	3391
1992	3747
1993	4016

Source: U.S. Fish and Wildlife Service

16. **Employment** The salaries of all the employees of Grenwich Bottling Company are shown in the frequency table at the right. They are negotiating a new contract and want a raise because they learned that the average salary for all hourly employees was only $24,000. The president informed them that the average salary for employees was $31,000.
    a. Which "average" do you think the union would want to use in its negotiations? Explain.
    b. If you were called in to mediate the negotiations, what recommendation would you make and why?

Salary	Number of Employees
$150,000	1
$85,000	2
$60,000	2
$35,000	8
$24,000	25
$15,000	5
$12,000	3

FYI

The hazards of DDT were brought to light in 1962 when marine biologist Rachel Carson wrote *Silent Spring*.

16a–b. See margin.

Additional Answers

12. Sample answer: 33 years old, favortie entertainment: sports; 172 cm tall, $30,500 salary.

13.

	Company A	Company B
mean	$22,413.79	$21,935.48
median	$15,000.00	$20,000.00
mode	$15,000.00	$20,000.00
range	$85,000.00	$30,000.00

You would probably prefer to work for Company B because the salary is higher for more workers.

15a. Answers will vary. Sample answer: no; it only shows the population increases for 6 years.

15b. Sample answer:

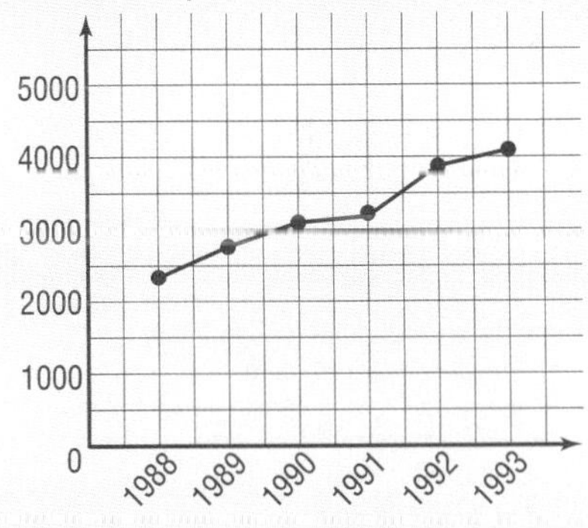

16a. The mode or median because it makes the current salaries look low.

Practice Masters, p. 84

NAME ________________ DATE ________

10-4 Practice
Misleading Statistics

Student Edition
Pages 504–508

Jarred Carson made two graphs of monthly sales for his bakery.

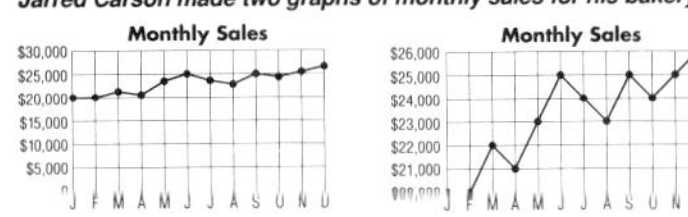

1. Which graph is misleading? Why? **The second; the vertical axis does not include zero.**
2. If Jarred wanted to sell his bakery, which graph would he probably show the buyer? Explain. **The second because it looks like there is a dramatic rise in sales.**

The salaries at the Homecraft Company are shown in the frequency table at the right.

3. Find the mean, median, and mode of the salaries. **$26,000; $24,000; $18,000**
4. Which measure of central tendency would you use to find the average salary? Why? **Median; there are low and high salaries.**
5. Which average might an employer use to attract new employees? Explain. **Mean; the mean average is higher than the median and the mode.**
6. How can you describe the salary of the "average" employee? **$18,000; mode**

Salary	Number of Employees
$12,000	2
$18,000	6
$24,000	5
$28,000	2
$55,000	1
$79,000	1

The table at the right shows shoe sales for the month of March. Use the table to answer the following questions. Write the type of average you used.

7. Which was the most popular size? **size 6: mode**
8. What was the average size sold? **size $6\frac{1}{2}$; median**
9. Which measure of central tendency would be most useful in deciding which sizes to order when the new styles come out? **mode**

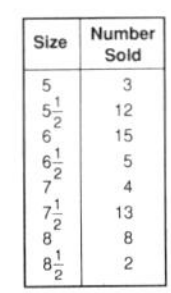

Size	Number Sold
5	3
$5\frac{1}{2}$	12
6	15
$6\frac{1}{2}$	5
7	4
$7\frac{1}{2}$	13
8	8
$8\frac{1}{2}$	2

FYI

DDT is poisonous to humans and animals when swallowed or absorbed through the skin. Residue from DDT has been shown to remain in the ecosystem and food chain long after its original use.

Additional Answer

16b. Answers will vary. Sample answer: Both sides should agree to use the same measure of central tendency.

4 ASSESS

Closing Activity

Modeling Have students make a list of ways that graphs can be misleading. Then have each student choose one way and create a misleading graph.

Mixed Review

17. The box-and-whisker plot below shows the amount students spent on entertainment for one month. (Lesson 10-3)

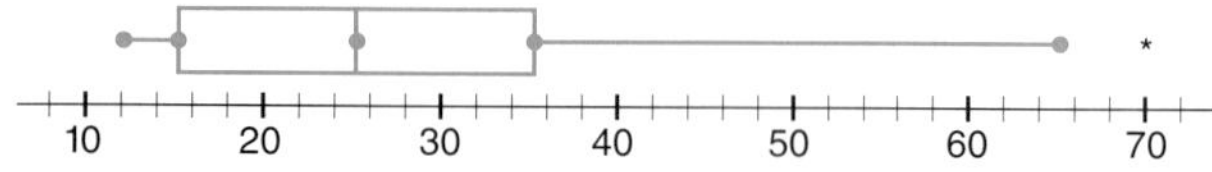

a. What is the range? **\$58**

b. What is the median? **\$25**

c. What is the upper quartile? **\$35**

d. What is the lower quartile?

e. What is the interquartile range?

f. What are the extremes?

g. What are the limits for outliers?

h. What are the outliers, if any?

17d. \$15
17e. \$20
17f. \$12, \$70
17g. \$0 and \$65
17h. \$70

18. Which ordered pair(s) is a solution of $3x + 2y = 12$? (Lesson 8-3) **a, c**

a. (4, 0) **b.** (0, 4) **c.** (5, −1.5) **d.** (−5, 1.5)

19. Geography Multipurpose buildings are common in small towns in Alaska, since the population is small. Between 1980 and 1990, the population of Alaska increased from about 400,000 to 550,000. (Lesson 6-8)

a. Find the ratio of the 1980 population to the 1990 population. $\frac{8}{11}$

b. If the population grows in a geometric pattern over each of the next four decades, what would the population of Alaska be in 2030? **1,965,955**

20. Measurement A millimeter is $\frac{1}{1000}$ of a meter or 10^{-3} meters. A kilometer is 1000 meters. How many millimeters are in a kilometer? (Lesson 4-8) **1,000,000**

21. Patterns Find the next two integers in the pattern 0, −2, 2, −4, 4, . . . (Lesson 2-6) **−6, 6**

WORKING ON THE Investigation

Refer to the Investigation on pages 482–483.

Test your game by having the members of your group play the game several times. Play the game enough times to develop some different strategies. Have each member of your group use a different strategy. Then have players who are in the intended age group play the game several times.

Use the results of your test to:

- determine the best strategy for your game
- revise any rules
- display the scores or winners using a statistical graph
- determine what makes someone playing the game a "good" player
- make an advertisement for your game

Add the results of your work to your Investigation Folder.

Enrichment Masters, p. 84

NAME __________ DATE ________

10-4 Enrichment
Using and Misusing Graphs

Student Edition Pages 504–508

Use the graphs at the right. Complete the following to discover which graph is misleading.

1. Do Graphs A and B give the same information on sales? **Yes**
2. Find the ratio of Hilly's sales to Valley's sales. **3:1**
3. In Graph A, the Hilly van is about 2.5 cm high by 6 cm long. What is its area? **15 cm²**
4. The Valley van is about 0.75 cm high by 2 cm long. What is its area? **1.5 cm²**
5. In Graph B, both vans are about 1.5 cm high. The Hilly van is about 6 cm long. What is its area? **9 cm²**
6. The Valley van is about 2 cm long. What is its area? **3 cm²**
7. Compute the following ratios.

   Graph A: $\frac{\text{Area of Hilly}}{\text{Area of Valley}}$ $\frac{10}{1}$

   Graph B: $\frac{\text{Area of Hilly}}{\text{Area of Valley}}$ $\frac{3}{1}$
8. Compare the results of Exercises 2 and 7. Which graph is misleading? Explain your answer. **A; The actual number of vans sold by Hilly's is 3 times greater, but Graph A appears to show the sales as 10 times greater.**

Use Graphs C and D to answer each question.

9. Which graph is easier to read? **C**
10. Compare the vertical scales. How do they differ? **The units on Graph C increase by 1, while the units on Graph D increase by 10.**
11. Which graph gives a better impression of the trend in sales? Explain. **Sample answer: Neither; the sales increase is overstated by Graph C and understated by Graph D.**

Graph A: Vans Sold In June (Hilly's Van Sales; Valley's Van Sales; Number of Vans 0 2 4 6 8 10)

Graph B: Vans Sold In June (Hilly's Van Sales; Valley's Van Sales; Number of Vans 0 2 4 6 8 10)

Graph C: Total Sales (Sales 35 36 37 38 39 40; Month J F M A M J J A)

Graph D: Total Sales (Sales 0 10 20 30 40 50; Month J F M A M J J A)

Extension

Using Biased Surveys Have students research and report on the ways that statistics can be used to mislead people other than by using misleading graphs. For instance, a survey can be biased by the way samples are selected.

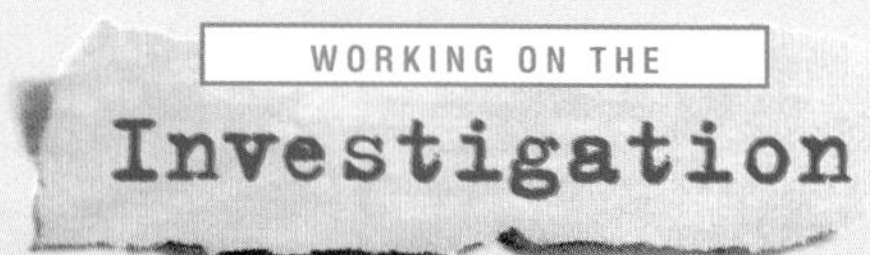

The Investigation on pages 482 and 483 is designed to be a long-term project that is completed over several days or weeks. Encourage students to keep their materials in their Investigation Folder as they work on the Investigation.

10-5 Counting

Setting Goals: *In this lesson, you'll use tree diagrams or the Fundamental Counting Principle to count outcomes.*

Modeling a Real-World Application: Advertising

Bruegger's Fresh Bagel Bakery makes 9 types of bagels and has 10 flavors of cream cheese. They wanted to advertise the number of different bagel with cream cheese combinations that are possible. How many different possible choices are there? *This problem will be solved in Example 1.*

Learning the Concept

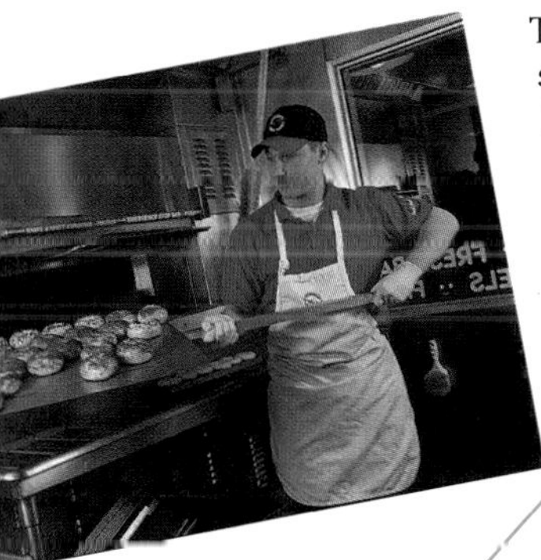

To solve this problem, it may be helpful to solve a simpler problem. Suppose Bruegger's Bagel Bakery offers chicken, tuna, and vegetable sandwiches, on plain or onion bagels. How many possible sandwiches are there?

You can draw a diagram to find the number of possible combinations or **outcomes**.

Deli Item	Bagel	Outcome
chicken	plain	CP
	onion	CO
tuna	plain	TP
	onion	TO
vegetable	plain	VP
	onion	VO

There are 6 possible choices, or outcomes.

The diagram above is called a **tree diagram**. Notice that the product of the number of deli items and the number of bagels, 3×2, also equals 6. Thus, you can find the number of possible outcomes by multiplying. This principle is called the **Fundamental Counting Principle**.

Fundamental Counting Principle

If event M can occur in m ways and is followed by event N that can occur in n ways, then the event M followed by event N can occur in $m \cdot n$ ways.

Alternative Learning Styles

Kinesthetic Bring four different colored socks and four each of three different colors of shoelaces to class. Distribute these to students in the class. Explain that the shoelaces represent different colors of shoes you can wear. Have a student who has a sock put it on the bulletin board. Ask what color of shoes you could wear. One at a time, have students put up the different colored shoelaces. Repeat for the other colors of socks. Ask how many possible outcomes there are. 12

10-5 LESSON NOTES

NCTM Standards: 1-4, 11

Instructional Resources

- Study Guide Master 10-5
- Practice Master 10-5
- Enrichment Master 10-5
- Group Activity Card 10-5
- Assessment and Evaluation Masters, pp. 266-267
- Tech Prep Applications Masters, p. 19

Transparency 10-5A contains the 5-Minute Check for this lesson; **Transparency 10-5B** contains a teaching aid for this lesson.

Recommended Pacing	
Standard Pacing	Day 5 of 14
Honors Pacing	Day 5 of 13
Block Scheduling*	Day 3 of 7 (along with Lesson 10-6)

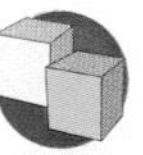

*For more information on pacing and possible lesson plans, refer to the ***Block Scheduling Booklet***.

1 FOCUS

5-Minute Check
(over Lesson 10-4)

Salary	Frequency	
	Lone Star Company	Lincoln Company
$90,000	1	0
$70,000	2	1
$40,000	2	8
$30,000	5	10
$20,000	11	0

Find the following for each company.

1. mean **$32,381; $36,316**
2. median **$20,000; $30,000**
3. mode **$20,000; $30,000**
4. range **$70,000; $40,000**
5. Which company would you rather work for? **Answers may vary.**

Motivating the Lesson

Questioning Show students a copy of a family tree. Ask how many students know about their family trees. Ask how you would determine how many ancestors you have a given number of generations ago.

2 TEACH

In-Class Examples

For Example 1
Palermo's Pizza Palace offers 8 different toppings and 3 types of crust (stuffed, thick, thin). How many different combinations are there for a one-topping pizza? **24**

For Example 2
For an exam in social studies, students had to choose one of twelve essay questions on current political affairs and one essay question on an environmental topic. There were 156 possible combinations of choices. How many environmental topics were there? **13**

For Example 3
A six-sided die is rolled and a fair coin is tossed.

a. How many outcomes are possible? **12**

b. What is the probability of rolling a three and tossing heads. $\frac{1}{12}$

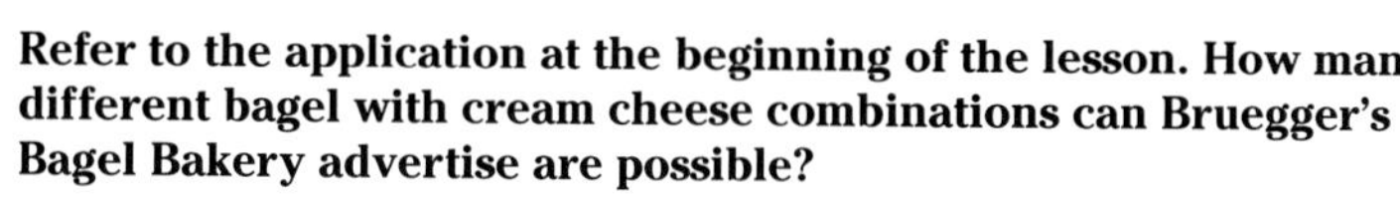

Refer to the application at the beginning of the lesson. How many different bagel with cream cheese combinations can Bruegger's Bagel Bakery advertise are possible?

Explore You know that there are 9 types of bagels and 10 flavors of cream cheese. You want to know how many different possible choices there are.

Plan Use the Fundamental Counting Principle.

Solve $\left(\begin{matrix}\textit{number of types}\\ \textit{of bagels}\end{matrix}\right) \times \left(\begin{matrix}\textit{number of flavors}\\ \textit{of cream cheese}\end{matrix}\right) = \left(\begin{matrix}\textit{number of}\\ \textit{possible outcomes}\end{matrix}\right)$

$9 \times 10 = 90$

There are 90 possible outcomes.

Examine Make a list or draw a tree diagram to verify the solution.

Connection to Algebra

If you already know the total number of possible outcomes, you can write and solve an equation to find the number of ways the events can occur.

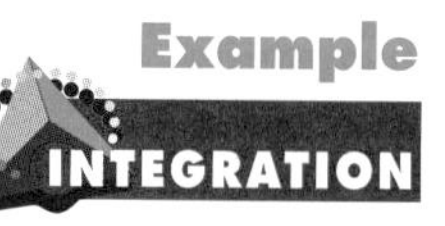

In literature class, each student must choose one short story and one poem to read for homework. The students must choose from a list of 16 short stories and *h* poems. There are 304 different combinations of short stories and poems possible. Find the number of poems on the list.

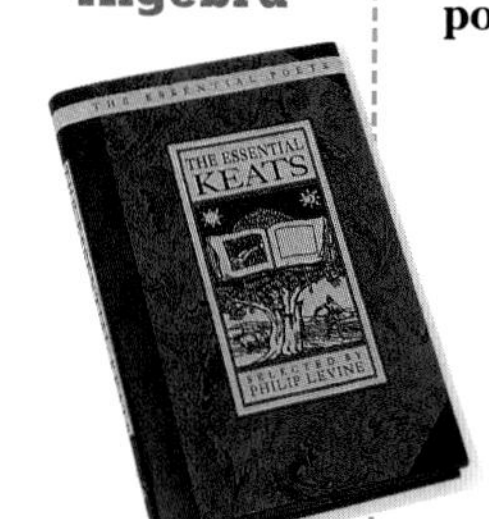

From the Fundamental Counting Principle, you know that 16 times h is 304.

$16h = 304$

$\frac{16h}{16} = \frac{304}{16}$ *Divide each side by 16.*

304 [÷] 16 [=] 19

There are 19 poems on the list.

A die is rolled twice.

a. How many outcomes are possible?

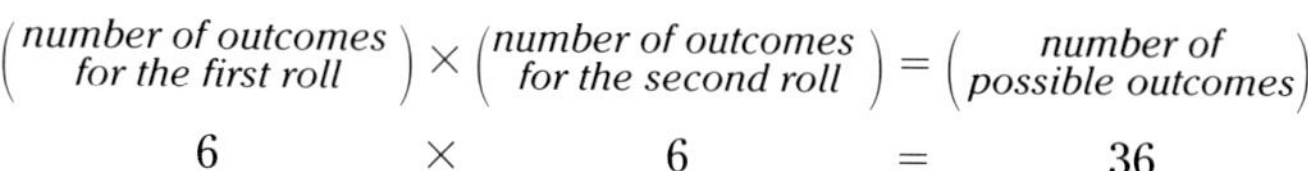

$\left(\begin{matrix}\textit{number of outcomes}\\ \textit{for the first roll}\end{matrix}\right) \times \left(\begin{matrix}\textit{number of outcomes}\\ \textit{for the second roll}\end{matrix}\right) = \left(\begin{matrix}\textit{number of}\\ \textit{possible outcomes}\end{matrix}\right)$

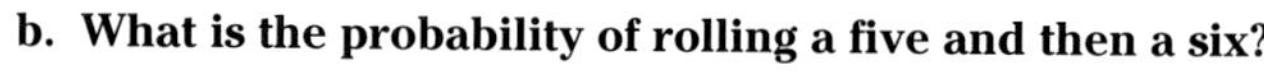

$6 \times 6 = 36$

There are 36 possible outcomes.

b. What is the probability of rolling a five and then a six?

Only one outcome can be a five and then a six.

$P(\text{five and six}) = \frac{\textit{number of favorable outcomes}}{\textit{number of possible outcomes}}$

$= \frac{1}{36}$

The probability of rolling a five and then a six is $\frac{1}{36}$.

LOOK BACK
You can review probability in Lesson 9-3.

Checking Your Understanding

Communicating Mathematics

Read and study the lesson to answer these questions. 1–2. See margin.

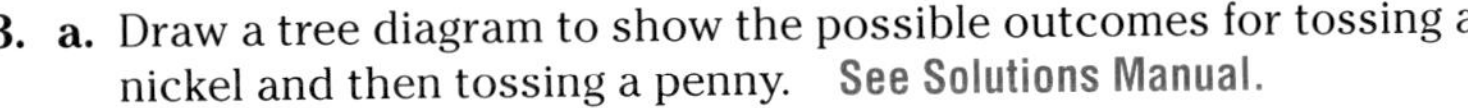

1. **a. Write** a problem that corresponds to the tree diagram at the right.
   b. List the outcomes illustrated in the tree diagram at the right.
2. **Compare and contrast** using the Fundamental Counting Principle and using a tree diagram to find numbers of outcomes.

MODELING MATHEMATICS

MATERIALS

a penny and a nickel

3d. Answers will vary.

3f. Answers will vary.

3. **a.** Draw a tree diagram to show the possible outcomes for tossing a nickel and then tossing a penny. See Solutions Manual.
   b. What is the probability of tossing two heads? 0.25
   c. Make a frequency table listing the outcomes. Toss a nickel and then a penny and tally the outcome in your frequency table. Repeat for 25 trials. See students' work.
   d. What fraction of the time did you toss two heads?
   e. Your answer to part b is the *theoretical probability* of tossing two heads. Your answer to part d is the *experimental probability* of tossing two heads. How does the experimental probability compare to the theoretical probability? Answers will vary.
   f. Toss the coins for 25 more trials. Compare the probabilities again. Did the experimental probability change? If so, how?

Guided Practice

4b. 25 outcomes
4c. 2 outcomes

5a. 12 outcomes

6. 8 outcomes

4. The spinner at the right is spun twice.
   a. Draw a tree diagram that represents the situation. See Solutions Manual.
   b. How many outcomes are possible?
   c. How many outcomes show red and blue?
   d. What is the probability of two oranges? $\frac{1}{25}$
5. A coin is tossed, and a die is rolled.
   a. How many outcomes are possible?
   b. What is the probability of heads and a 4? $\frac{1}{12}$
6. Three coins are tossed. How many outcomes are possible?
7. **Fashion** Reina has three silver necklaces, three pairs of silver earrings, and two silver bracelets. How many combinations of the three types of jewelry are possible? 18 outcomes

Exercises: Practicing and Applying the Concept

Independent Practice

8–9. See Solutions Manual for tree diagrams.

Draw a tree diagram to find the number of outcomes for each situation.

8. Each four-sided die is rolled once.
9. Each spinner is spun once.

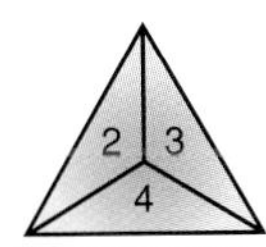

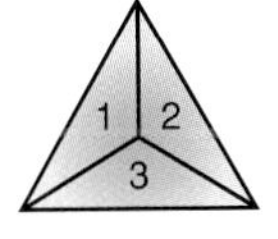

16 outcomes

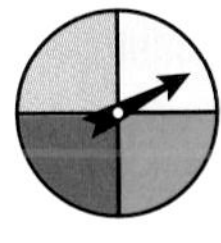

24 outcomes

3 PRACTICE/APPLY

Checking Your Understanding

Exercises 1–7 are designed to help you assess your students' understanding through reading, writing, speaking, and modeling. You should work through Exercises 1–3 with your students and then monitor their work on Exercises 4–7.

Additional Answers

1a. Sample answer: What are the outcomes from two tosses of a three-colored dice? What are the outcomes from spinning a three-choice spinner?

1b. GG, GB, GY, BG, BB, BY, YG, YB, YY

2. Sample answer: The Fundamental Counting Principle is faster and takes less space.

Study Guide Masters, p. 85

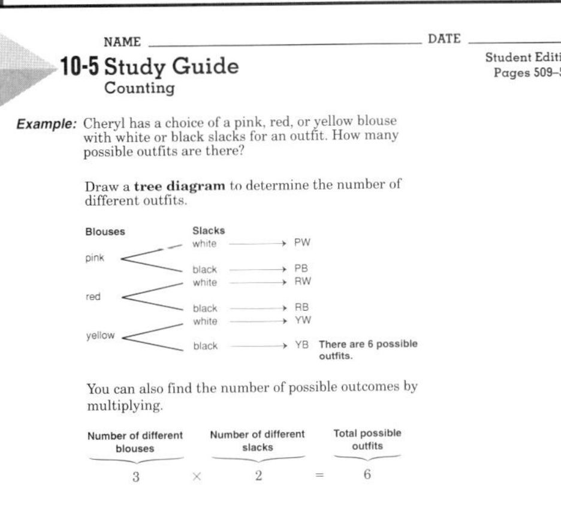

NAME ______________ DATE ________

10-5 Study Guide
Counting

Student Edition Pages 509–513

Example: Cheryl has a choice of a pink, red, or yellow blouse with white or black slacks for an outfit. How many possible outfits are there?

Draw a **tree diagram** to determine the number of different outfits.

You can also find the number of possible outcomes by multiplying.

Number of different blouses		Number of different slacks		Total possible outfits
3	×	2	=	6

Draw a tree diagram to find the number of outcomes for each situation. See student's diagrams.

1. A bag contains one red marble and one white marble. A second bag contains two red marbles and a white one. How many different outcomes are possible if a marble is picked from each bag? The marbles are replaced each time. 6
2. How many different lunches can be made with a choice of hot dog, salad, or chili; pop, milk, or lemonade; and crackers, bread, or a roll? 27

Find the number of possible outcomes for each event.

3. At an ice cream shop, there are 31 flavors and 25 toppings. How many different ways are there to make a one-scoop ice cream sundae with one topping? 775
4. There are 4 quarterbacks and 6 centers on a football team that has 60 players. How many quarterback-center pairings are possible? 24

Reteaching

Act It Out Using four students and four chairs, model the number of different ways the students can be seated. Have a student make a chart or tree diagram on the board and record each student's name as he or she sits in a different chair.

Sample Chart:

Chair 1	Chair 2	Chair 3	Chair 4
Ken	Ari	Marta	Vladmir
Ken	Ari	Vladmir	Marta

Error Analysis

Watch for students who guess at solutions or who omit possibilities when listing solutions. Stress the necessity of using a systematic approach, such as beginning with one item and exploring all possibilities for that item. Then proceeding to the next item.

Assignment Guide

Core: 9–19 odd, 20–21, 23–27
Enriched: 8–18 even, 20–27
All: Self Test, 1–10

For **Extra Practice**, see p. 765.

The red A, B, and C flags, printed only in the Teacher's Wraparound Edition, indicate the level of difficulty of the exercises.

B

10. The continental breakfast at Myriad Hotel has toast, muffin, or bagel with coffee, milk or juice. How many combinations of one bread and one beverage are possible? **9 outcomes**
11. Jarred had five different poses to choose for his senior picture and five different frames. How many different ways could he choose a picture and a frame? **25 ways**

Find the number of possible outcomes for each event.

12. 16 outcomes
13. 36 outcomes

12. Four coins are tossed.
13. A die is rolled twice.
14. Two coins are tossed and a die is rolled. **24 outcomes**
15. School sweatshirts come in two colors, gray or white, and in four sizes, small, medium, large, and extra large.
    a. How many outcomes are possible? **8 outcomes**
    b. What is the probability of selecting a large gray sweatshirt at random? $\frac{1}{8}$

16. 24 autos

16. A sedan comes with two or four doors, a four or six cylinder engine, and six exterior colors. How many different autos are possible?
17. A quiz has five true-false questions. How many outcomes for answering the five questions are possible? (*Hint:* One outcome is FFTFF.) **32 outcomes**

18. An eight-sided die is rolled four times.
    a. How many outcomes are possible? **4096 outcomes**
    b. What is the probability of getting four 7s? $\frac{1}{4096}$
19. A coin is tossed thirty times.
    a. How many outcomes are possible? 2^{30}
    b. What is the probability of getting heads every time? $\frac{1}{2^{30}}$

Critical Thinking

20. A test has all true-false questions. If there are x questions on the test, write an expression for the number of possible answer keys. 2^x

Applications and Problem Solving

21. **Telecommunication** Telephone numbers such as 219-555-1212 are made up of a three-digit area code (219), a three-digit exchange (555), and a four-digit extension (1212). The digits of an exchange cannot be 0 or 1, so it appears that there are $8 \cdot 8 \cdot 8$ or 512 exchanges in an area code. However, there are 26 exchanges that are not available in each area code because of special numbers like 911, 555, and 800. There are actually only 486 exchanges available in each area code.

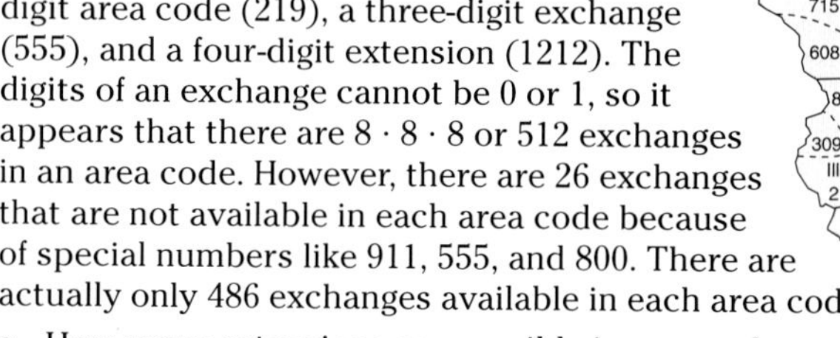

    a. How many extensions are possible in one exchange?
    b. How many telephone numbers are available in each area code?
    c. Many local telephone companies ran out of numbers during the 1990s. They had to extend the three-digit exchange to include 0 and 1 in the second and third digits. (Zero and one still cannot be used as the first digit.) How many new numbers became available in each area code by allowing 0 and 1 in the second and third digits?

21a. 10,000 extensions
21b. 4,860,000
21c. 2,880,000 new numbers

22. 42 toppings

22. **Business** The Yogurt Oasis advertises that there are 1512 ways to enjoy a one-topping sundae. They offer six flavors of frozen yogurt, six different serving sizes, and several toppings. How many toppings do they offer?

Practice Masters, p. 85

NAME ______________________ DATE ____________

10-5 Practice
Counting

Student Edition Pages 509–513

Draw a tree diagram to find the number of outcomes for each situation. **See students' diagrams.**

1. Each spinner is spun once. 8

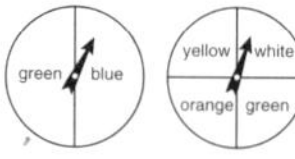

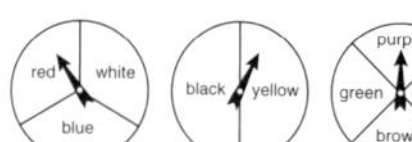

2. Each spinner is spun once. 24

3. The breakfast at Dion's Place has a choice of cereal, eggs, or French toast with a choice of milk or juice. 6
4. Tina has a choice of a sports jersey in blue, white, gray, or black in sizes small, medium, or large. 12

Find the number of possible outcomes for each event.

5. A penny, a nickel, and a dime are tossed. 8
6. Four dice are rolled. 1296
6. Two quarters are tossed. Then a four-sided die is rolled. 16
8. If Alicia has 3 skirts, 2 blouses, and 5 scarves, how many outfits are possible? 30
9. The lunch at Dion's Place has a choice of ham, turkey, or roast beef on rye or white bread with juice, milk, or tea.
   a. How many different lunches are possible?
   b. What is the probability that the lunch special of the day is ham on rye with tea? 18; $\frac{1}{18}$
10. A pizza shop has 6 meat toppings, 5 vegetable toppings, and 3 cheese toppings. How many different pizzas (one meat, one vegetable, and one cheese toppings) can be made? 90

Group Activity Card 10-5

Count The Ways — Group Activity **10-5**

MATERIALS: Seventeen number cards with the numbers 2, 4, 5, 6, 7, 8, 10, 12, 14, 20, 24, 28, 30, 35, 36, 40, 42

Pretend you have one five-color spinner (red, blue, green, yellow, white), two dice, three coins, and a deck of seven cards (with the letters A-G) in a box.

To play, turn over a number card. Your tasks are (1) to decide which combinations of manipulatives were used to create this number of outcomes, and (2) to draw a tree diagram.

Example: If you had to make 10 outcomes you would need to use the spinner and one coin.

H — R, B, G, Y, W   T — R, B, G, Y, W

 Pre-Algebra

Extension

Using Language Connections

The Hawaiian language uses only the vowels a, e, i, o, and u and the consonants h, k, l, m, n, p, and w. How many three-letter Hawaiian words are possible if each contains two consonants separated by a vowel? **245**

Mixed Review

23. See margin.

23. Real Estate A builder has house styles that are priced at \$80,000, \$95,000, \$92,000, and \$170,000. Which "average" price gives the best description of the houses that the builder offers? Why? (Lesson 10-4)

24. \$10.63

24. Finance Use the formula $I = prt$ to find the interest to the nearest cent on \$500 at 4.25% simple interest for 6 months. (Lesson 9-9)

25. Given $f(x) = 2x + 4$, find the value of $f(-5)$. (Lesson 8-4) **−6**

26. $k \leq 7$; See margin for graph.

26. Solve $2(k - 3) \leq 8$ and check your solution. Then graph the solution on a number line. (Lesson 7-6)

27. h = hours, $3.50h < 20$, $h < 5.7$

27. Sarah earned less than \$20 baby-sitting at \$3.50 an hour. How many hours did Sarah baby-sit? Define a variable and translate the sentence into an inequality. Then solve. (Lesson 3-8)

Self Test

The table below shows the normal monthly precipitation, in inches, for selected cities in the United States. Use this data to complete Exercises 1–5. 2. 4.5, 1.8 4–5. See margin.

City	Jan.	Feb.	Mar.	Apr.	May	June	July	Aug.	Sept.	Oct.	Nov.	Dec.
Birmingham, AL	5.1	4.7	6.2	5.0	4.9	3.7	5.3	3.6	3.9	2.8	4.3	5.1
Louisville, KY	2.9	3.3	4.7	4.2	4.6	3.5	4.5	3.5	3.2	2.7	3.7	3.6
Miami, FL	2.0	2.1	2.4	2.9	6.2	9.3	5.7	7.6	7.6	5.6	2.7	1.8
Oklahoma City, OK	1.0	1.3	2.1	2.9	5.5	3.9	3.0	2.4	3.4	2.7	1.5	1.2

1. Make a stem-and-leaf plot of the precipitation for Birmingham. (Lesson 10-1) **See margin.**
2. Find the range and interquartile range of the precipitation for Oklahoma City. (Lesson 10-2)
3. What are the extreme measures of precipitation for Louisville? (Lesson 10-2) **2.7, 4.7**
4. Construct a box-and-whisker plot of the precipitation for Miami. (Lesson 10-3)
5. Determine whether a bar graph, box-and-whisker plot, comparative graph, line graph, or pictograph, is the best way to display the data. Explain your reasoning. (Lesson 10-3)

6–7. See margin. 8. See Solutions Manual 9. 24

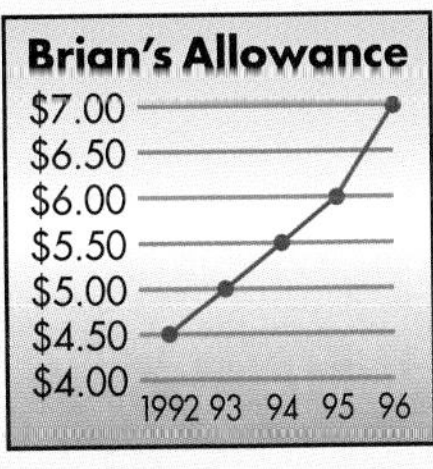

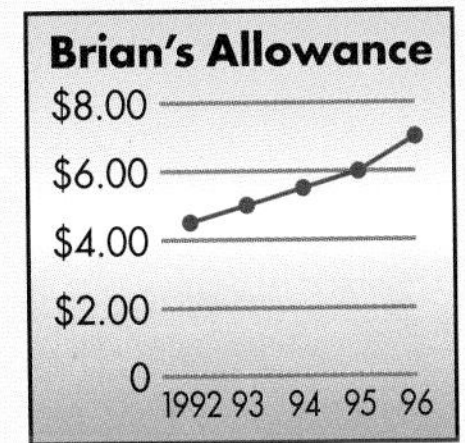

Use the graphs at the right to complete Exercises 6–7. (Lesson 10-4)

6. Explain why these graphs made from the same data look different.
7. Which graph would Brian show to his parents if he wants an increase in his allowance? Explain.
8. Wrapping paper comes in red, pink, green, or blue with white or yellow ribbon. Draw a tree diagram that illustrates how many combinations are possible. (Lesson 10-5)
9. Mark has three pairs of shorts, four shirts, and two pairs of running shoes. How many outcomes are possible? (Lesson 10-5)
10. Three coins are tossed. What is the probability that all 3 coins land heads up? (Lesson 10-5) $\frac{1}{8}$

Self Test

The Self Test provides students with a brief review of the concepts and skills in Lessons 10-1 through 10-5. Lesson numbers are given to the right of exercises or instruction lines so students can review concepts not yet mastered.

Additional Answers Self Test

4.

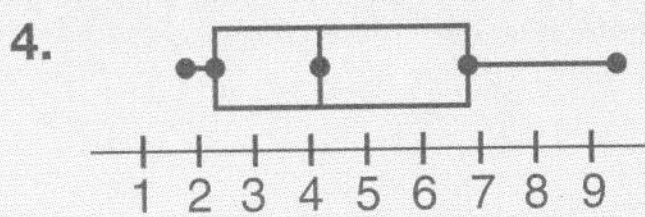

5. Answers will vary. A comparative graph is the best way.
6. Sample answer: They do not have the same scale on the y-axis.
7. Brian would use the bottom graph because it makes the increases appear smaller.

4 ASSESS

Closing Activity

Act It Out Have students write a problem that can be illustrated with cutouts, such as different ice cream cones and different flavors of ice cream. Then have students use the cutouts to model tree diagrams and explain how tree diagrams and the Fundamental Counting Principle are related.

Chapter 10, Quiz B (Lessons 10-4 and 10-5) is available in the *Assessment and Evaluation Masters*, p. 267.

Mid-Chapter Test (Lessons 10-1 through 10-5) is available in the *Assessment and Evaluation Masters*, p. 266.

Additional Answers

23. Sample answer: The median, because it is close to three of four actual prices.

26.

0 2 4 6 8 10

Self Test

1.

```
2 | 8
3 | 6 7 9
4 | 3 7 9
5 | 0 1 1 3
6 | 2        6|2 = 6.2
```

Enrichment Masters, p. 85

NAME ________ DATE ________

10-5 Enrichment Student Edition Pages 509–513

Outcomes

Complete.

1. Complete the spinner so that it will have six different possible outcomes.

1 2 3 4 5 6

The six sections do not have to be the same size.

2. List the numbers that could be placed on the die to provide only four different possible outcomes. **Answers will vary. Sample answer: 1, 1, 1, 2, 3, 4**

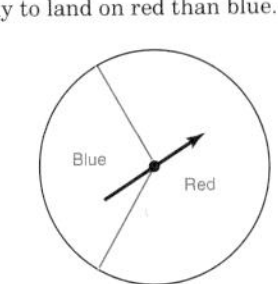

3. Complete the spinner so that it is more likely to land on red than blue.

Blue Red

Red section must be larger than blue section.

4. List the months in which you could choose a date and have 30 possible outcomes. **April, June, September, November**

5. There are white, green, and blue marbles in a bag. What is the minimum number of each so that it is twice as likely that you draw a green one as a white one, and three times as likely that you draw a blue one as a green one? **1 white, 2 green, and 6 blue**

6. A year between 1950 and 2001 is chosen at random. How many possible outcomes are there where the year is a leap year? List them. **13 outcomes; 1952, 1956, 1960, 1964, 1968, 1972, 1976, 1980, 1984, 1988, 1992, 1996, 2000**

10-6A LESSON NOTES

NCTM Standards: 1-4, 11

Objective

Explore and use permutations and combinations.

Recommended Time

Demonstration and discussion: 15 minutes; Exercises: 30 minutes

Instructional Resources

For each student or group of students
- blank index cards
- markers

Math Lab and Modeling Math Masters
- p. 46 (worksheet)

For teacher demonstration
Overhead Manipulative Resources

1 FOCUS

Motivating the Lesson

Have four students stand. Ask how many different groups of three students you could make. 4 (Have students rearrange themselves into groups of three to confirm the answer.)

Place three chairs at the front of the room. Ask how many different ways three of the four students can be seated. 24

Tell students they will explore more about situations like these.

2 TEACH

Teaching Tip Remind students that a "word" is considered any three letter combination, and does not have to be an actual word.

3 PRACTICE/APPLY

Assignment Guide
Core: 1–2
Enriched: 1–2

MATHEMATICS LAB **10-6A**

Permutations and Combinations

A Preview of Lesson **10-6**

MATERIALS

6 index cards
markers

An arrangement of names, objects, or people in a particular order is called a **permutation**. For example, if you use a triangle, square, and circle, you can arrange them in the following permutations.

△□○ △○□ □○△ □△○ ○□△ ○△□

Activity  **Work with a partner.**

Write the letters M, A, T, H on four cards, one letter per card. Shuffle the cards and place them face down. Choose three of the cards and turn them face up. For example, you may have chosen the cards shown at the right. Record the "word" shown; for example, ATM.

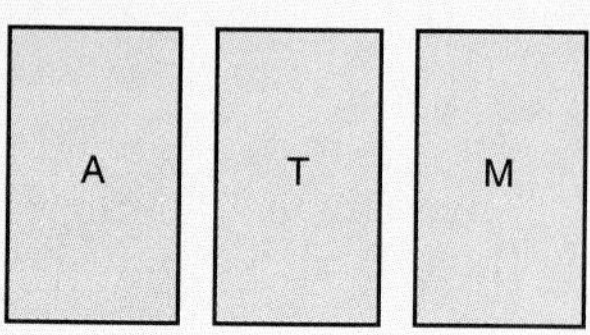

- Rearrange the three cards to make another "word." Record that word.
- Continue rearranging the cards and recording the words until you have listed all the 3-letter words you can with the cards chosen. Record the total number.
- Now use all four cards. Arrange the cards to form all the 4-letter "words" that you can. How many arrangements are there?

Activity  The student council at Northern Potter Junior High School is planning to have a sundae sale. They plan to offer six different toppings.

peanuts	chocolate chips	chocolate syrup
raisins	strawberries	butterscotch syrup

The price depends on the *combinations* of toppings chosen.

Your Turn **Work in groups of two or three.**

- Write the names of the six toppings on six index cards.
- To make a sundae, select any pair of cards. Make a list of all the different combinations of two toppings. Note that the order is not important; that is, *peanuts, raisins*, is the same as *raisins, peanuts*.
- Suppose that the student council decides to offer a deluxe sundae with four toppings on it. Make a list of all the different sundaes that are possible if four toppings of the six are used.

1. In Activity One, suppose that you had five cards with five different letters. How many five-letter words do you think can be formed?
2. What is the difference between a combination (like a combination of toppings) and a permutation? **In a permutation, order is important.**

1. 120 words

4 ASSESS

Observing students working in cooperative groups is an excellent method of assessment.

10-6 Permutations and Combinations

Setting Goals: *In this lesson, you'll use permutations and combinations.*

Modeling a Real-World Application: Swimming

The swim team coach at Clearwater High School is preparing the lineup for the state finals in the 400-meter freestyle relay. She must choose four swimmers from the five that qualified at the district championships. As she looks over the five names, she wonders how many arrangements are possible.

She reasons that any of the 5 swimmers can start the race. Once that swimmer is selected, there are 4 swimmers left who can swim second, 3 who can swim third, and 2 who can swim fourth.

starting swimmers
2nd swimmer
3rd swimmer
4th swimmer
Possible relay teams
5 × 4 × 3 × 2 = 120

Using the Fundamental Counting Principle, as shown above, she finds there are 120 possible relay teams.

Learning the Concept

An arrangement or listing in which order is important is called a **permutation**. For the example above, the symbol $P(5, 4)$ represents the number of permutations of 5 things taken 4 at a time.

$$P(5, 4) = 5 \cdot 4 \cdot 3 \cdot 2$$

If one of the five Clearwater swimmers became ill and was unable to swim, the number of relay teams can be expressed as $P(4, 4)$.

$$P(4, 4) = 4 \cdot 3 \cdot 2 \cdot 1$$

The product $4 \cdot 3 \cdot 2 \cdot 1$ can be written using the mathematical notation 4!. The symbol 4! is read *four **factorial***. The expression $n!$ means the product of all counting numbers beginning with n and counting backward to 1. We define 0! as 1.

Cooperative Learning

Co-op Co-op Have students work in small groups to solve this problem. List all subjects available at your school. Assume that there are no subject requirements and that all subjects are available each period. Have each group determine the permutations and combinations for the number of class periods available.

For more information on this strategy, see *Cooperative Learning in the Mathematics Classroom*, one of the titles in the Glencoe Mathematics Professional Series, p. 32.

10-6 LESSON NOTES

NCTM Standards: 1-4, 11

Instructional Resources

- Study Guide Master 10-6
- Practice Master 10-6
- Enrichment Master 10-6
- Group Activity Card 10-6
- Activity Masters, p. 42

Transparency 10-6A contains the 5-Minute Check for this lesson; **Transparency 10-6B** contains a teaching aid for this lesson.

Recommended Pacing	
Standard Pacing	Day 7 of 14
Honors Pacing	Day 6 of 13
Block Scheduling*	Day 3 of 7 (along with Lesson 10-5)

*For more information on pacing and possible lesson plans, refer to the ***Block Scheduling Booklet***.

1 FOCUS

5-Minute Check *(over Lesson 10-5)*

Find the number of possible outcomes for each event.

1. Three fair coins are tossed. 8
2. A six-sided die is rolled and a fair coin is tossed. 12
3. A six-sided die is rolled and two fair coins are tossed. 24

A fair coin is tossed ten times.

4. How many possible outcomes are there? 2^{10}
5. What is the probability of getting tails every time? $\frac{1}{2^{10}}$

Example  **Find each value.**

a. $\boldsymbol{P(8, 3)}$

$P(8, 3)$ represents the number of permutations or orders of 8 things taken 3 at a time.

$$P(8, 3) = \underbrace{\textit{choices for 1st position}}\ \ \underbrace{\textit{choices for 2nd position}}\ \ \underbrace{\textit{choices for 3rd position}}$$

$$= 8 \cdot 7 \cdot 6$$

$$= 336$$

b. **6!**

$$6! = 6 \cdot 5 \cdot 4 \cdot 3 \cdot 2 \cdot 1$$

$$= 720$$

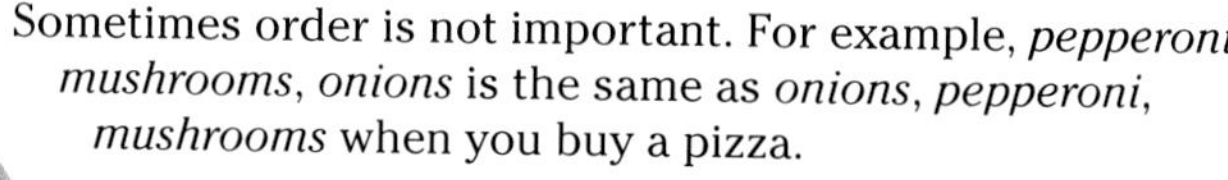

Sometimes order is not important. For example, *pepperoni, mushrooms, onions* is the same as *onions, pepperoni, mushrooms* when you buy a pizza.

Arrangements or listings where order is not important are called **combinations**. To find the number of different three-topping pizzas when there are eight toppings to choose from, divide the number of permutations, $P(8, 3)$, by the number of different ways three things can be ordered. There are 3! or $3 \cdot 2 \cdot 1$ ways to order three toppings. The symbol $C(8, 3)$ means the number of combinations of 8 things taken 3 at a time.

$$C(8, 3) = \frac{P(8, 3)}{3!} = \frac{8 \cdot 7 \cdot 6}{3 \cdot 2 \cdot 1} \text{ or } 56$$

It is possible to make 56 different pizzas with three toppings when there are eight toppings to choose from.

Example 2

Geometry

Polygons can be drawn with any number of sides and vertices. Find the number of line segments that can be drawn between two vertices of a hexagon.

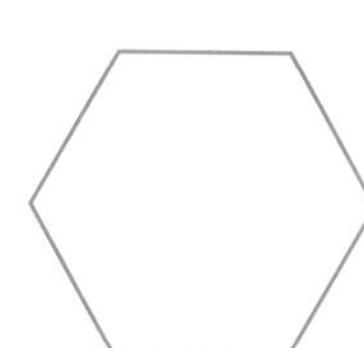

Explore A hexagon has 6 vertices.

Plan Each line segment connects two vertices, so you must find the combination of 6 vertices taken 2 at a time.

Solve

$$C(6, 2) = \frac{P(6, 2)}{2!}$$

$$= \frac{6 \cdot 5}{2 \cdot 1} \text{ or } 15$$

There are 15 line segments that can be drawn.

Examine Draw a hexagon and all the line segments and see if there are 15. Be sure to count the sides of the hexagon.

Motivating the Lesson

Questioning Provide students with a part of the sports section of the newspaper. For whatever sport is in season, have students name the teams in one division. Ask students if there is a way to find out the number of ways first and second place can be determined.

2 TEACH

In-Class Examples

For Example 1
Find each value.

a. $P(7, 4)$ **840**

b. 5! **120**

For Example 2
There are 8 people in a room. Each person is to shake hands with every other person in the room. How many handshakes will there be? $\boldsymbol{C(8, 2) = 28}$

3 PRACTICE/APPLY

Checking Your Understanding

Exercises 1–15 are designed to help you assess your students' understanding through reading, writing, speaking, and modeling. You should work through Exercises 1–4 with your students and then monitor their work on Exercises 5–15.

Error Analysis

Students may confuse combinations and permutations. Emphasize each definition, model various situations, and ask whether order is important.
It may also be helpful to have students make up memory tools for each definition.

Reteaching

Using Manipulatives Have students work in small groups. Give each group seven different books. Have students model permutations and combinations if the books are used five at a time.

Checking Your Understanding

Communicating Mathematics

Read and study the lesson to answer these questions.

1. **Explain** what $P(4, 3)$ means. **See margin.**
2. **Compare** $6 \cdot 5 \cdot 4$ and 6!. **See margin.**
3. **Write** an expression to represent the number of five-person basketball teams that could be formed from all the students in your math class. **Sample answer: $C(25, 5)$.**

4. Write a short summary explaining the difference between a permutation and a combination of 4 things taken 3 at a time. Draw models to illustrate your written explanation. **See students' work.**

Guided Practice

Tell whether each situation represents a *permutation* or *combination*.

5. six people remaining in a game of musical chairs **combination**
6. first, second, third place, and honorable mention awards of six students who are finalists in a science fair **permutation**

How many ways can the letters of each word be arranged? **9. 24**

7. WE **2**
8. STUDY **120**
9. MATH

Find each value.

10. $P(6, 3)$ **120**
11. 5! **120**
12. $C(5, 4)$ **5**
13. $\frac{7!3!}{5!0!}$ **252**

14. 924 combinations

14. How many combinations of 6 different flowers can you choose from one dozen different flowers?
15. **Music** The Riverside High School Orchestra has been invited to play three pieces at the opening of the Summer Music Festival. The orchestra has eight pieces that they could play. How many different programs can they play? **56 programs**

Exercises: Practicing and Applying the Concept

Independent Practice

17. combination
20. permutation
21. combination

Tell whether each situation represents a *permutation* or *combination*.

16. four books in a row **permutation**
17. six CDs from a group of twenty
18. a team of 6 players from 12 **combination**
19. ten people in a line to buy tickets **permutation**
20. position of students at the twelve computers in the computer center
21. ten wooden giraffes placed on display from a large collection

How many ways can the letters of each word be arranged?

22. SHE **6**
23. BOUGHT **720**
24. SANDWICH **40,320**

Additional Answers

1. Sample answer: The number of ways 4 things can be selected 3 at a time.
2. Sample answer: 6! means $6 \cdot 5 \cdot 4 \cdot 3 \cdot 2 \cdot 1$

Assignment Guide

Core: 17–49 odd, 50–58
Enriched: 16–44 even, 45–58

For **Extra Practice**, see p. 765.

The red A, B, and C flags, printed only in the Teacher's Wraparound Edition, indicate the level of difficulty of the exercises.

Teaching Tip Have students divide a sheet of paper into two columns. Head one column "Permutation: order important" and the other "Combination: order not important". As students complete Exercises 16–24 and 41–49, have them write the number of the exercise under the proper column.

Study Guide Masters, p. 86

NAME ______________________ DATE ____________

10-6 Study Guide
Permutations and Combinations

Student Edition Pages 515–519

An arrangement where *order is important* is called a **permutation.**

Example: Mario, Sandy, Fred, and Shana are running for the offices of president, secretary, and treasurer. In how many ways can these offices be filled?

Any of the 4 people can fill the president's position. Once the president has been chosen, there are 3 people to fill the secretary's position. That leaves 2 people to fill the treasurer's position.

The symbol $P(4, 3)$ represents the number of permutations of 4 things taken 3 at a time.

$P(4, 3) = 4 \times 3 \times 2$, or 24. The offices can be filled 24 ways.

An arrangement where *order is not important* is called **combination**.

Example: Charles has four coins in his pocket and pulls out three at a time. How many different amounts can he get? The coins are a penny, a nickel, a dime, and a quarter.

There are $4 \times 3 \times 2$, or 24 different outcomes, but some are the same. To find the number of different combinations, divide the number of permutations $P(4, 3)$, by the number of different ways three items can be arranged.

$C(4, 3) = \frac{P(4, 3)}{3!} = \frac{4 \times 3 \times 2}{3 \times 2 \times 1}$ or 4. 4 different amounts can be chosen.

3! is read "three factorial."

Find each value.

1. 5! **120**
2. $P(4, 2)$ **12**
3. 8! **40,320**
4. $C(15, 6)$ **5005**
5. $P(3, 3)$ **6**
6. 13! **6,227,020,800**
7. $C(5, 5)$ **1**
8. $P(10, 4)$ **5040**
9. In how many ways can five books be arranged on a shelf? (Order is important.) **120**
10. In how many ways can three students council members be elected from five candidates? (Order is not important.) **20**

Classroom Vignette

"To teach combinations, I ask students to generate an advertisement for a new pizza parlor they are opening. Sooner or later, students will suggest advertising a wide variety of pizzas. At this time, I ask students to calculate the possible combinations of pizza toppings given a few parameters."

Denny Wagoner
Cowan High School
Muncie, IN

Dennis L. Wagoner

29. 39,916,800
32. 1680
36. 479,001,600

Find each value.

25. $P(6, 6)$ 720 **26.** 7! 5040 **27.** 9! 362,880 **28.** $C(6, 6)$ 1
29. 11! **30.** $P(5, 1)$ 5 **31.** $C(7, 3)$ 35 **32.** $P(8, 4)$
33. 0! 1 **34.** $C(8, 4)$ 70 **35.** $C(10, 2)$ 45 **36.** $P(12, 12)$
37. $\frac{7!3!}{5!1!}$ 252 **38.** $\frac{10!0!}{6!2!}$ 2520 **39.** $\frac{5!2!}{4!3!}$ $1\frac{2}{3}$ **40.** $\frac{9!5!0!}{11!}$ $1\frac{1}{11}$

Choose

Estimation
Mental Math
Calculator
Paper and Pencil

41. How many different 6-player starting squads can be formed from a volleyball team of 15 players? **5005 teams**

42. How many 3-digit numbers can you write using only the digits 1, 2, and 3 exactly once in each number? **6 numbers**

43. How many ways can 4 members of a family be seated in a theater if the mother is seated on the aisle? **6 ways**

44. How many different 5-card hands is it possible to deal from a standard deck of 52 cards? **2,598,960**

Critical Thinking

45. 10 students
46. 5040 ways

45. There are 720 ways for three students to win first, second, and third place in a debating match. How many students were competing?

46. Ten people are running for chair and vice-chair of the Social Studies Club. After they are selected, three directors will be elected from the remaining candidates. How many different ways can the offices be filled?

Applications and Problem Solving

47. **United Nations** The United States has brought a procedural measure to the United Nations Security Council for a vote. Nine votes are necessary for the measure to pass. There are 5 permanent members (China, France, Russian Federation, United Kingdom, and the United States) and 10 nonpermanent members on the Security Council. How many combinations of votes are possible if the measure passes with 9 votes? **5005 combinations**

48. 66 line segments

48. **Geometry** Twelve points are marked on a circle. How many different line segments can be drawn between any two of the points?

EARTH WATCH

THE NEW GOLD RUSH

You have probably studied the gold rush that occurred in the 1840s and 1850s in the western United States. The news of an easy life of riches made people from all over the world move to the west in search of gold. This westward movement made the United States spread from coast to coast.

But did you know there is a new gold rush occurring in the western United States right now? A new mining process is prompting entrepreneurs to go in search of gold once again.

In the first gold rush, the process of mining was a physical process that was done on a small scale by individual miners. Today, large mining companies use a chemical process called cyanide heap-leaching. This process requires removing millions of tons of rock from the land. Then the rock is pulverized and piled into mounds. For months, the rock is sprinkled with a cyanide solution that draws the gold out of the rock fragments.

Practice Masters, p. 86

NAME ______ DATE ______

10-6 Practice
Permutations and Combinations

Student Edition Pages 515–519

Determine whether each situation represents a permutation or combination.

1. four musical instruments from a group of 12 **combination**
2. seven students in a line to sharpen their pencils **permutation**
3. a choice of three tapes out of 64 **combination**

How many ways can the letters of each word be arranged?

4. RULES 120
5. FOLDERS 5040
6. POWERFUL 40,320

Find each value.

7. 1! 1
8. 7! 5040
9. 9! 362,880
10. 11! 39,916,800
11. $\frac{6!3!}{4!2!}$ 90
12. $\frac{7!3!}{5!1!}$ 252
13. $\frac{8!4!}{5!2!}$ 4032
14. $\frac{9!2!}{6!3!}$ 168
15. $P(4, 3)$ 24
16. $P(6, 4)$ 360
17. $P(7, 2)$ 42
18. $P(8, 5)$ 6720
19. $C(4, 3)$ 4
20. $C(6, 4)$ 15
21. $C(7, 2)$ 21
22. $C(8, 5)$ 56
23. How many ways can a club of 6 members choose a 3-person committee? **20 ways**
24. How many ways can 5 children line up to get on the school bus if Jenny always gets on third? **24 ways**

EARTH WATCH

Have students work in small groups to brainstorm about what some possible advantages and disadvantages of the new gold mining process might be. Then have students complete their research.

Group Activity Card 10-6

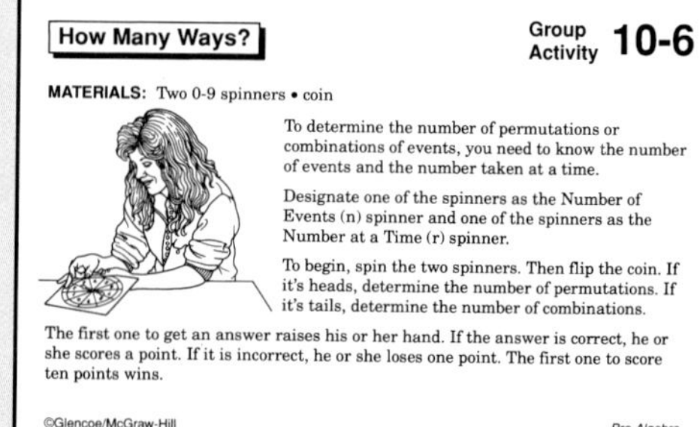

How Many Ways?

Group Activity 10-6

MATERIALS: Two 0-9 spinners • coin

To determine the number of permutations or combinations of events, you need to know the number of events and the number taken at a time.

Designate one of the spinners as the Number of Events (n) spinner and one of the spinners as the Number at a Time (r) spinner.

To begin, spin the two spinners. Then flip the coin. If it's heads, determine the number of permutations. If it's tails, determine the number of combinations.

The first one to get an answer raises his or her hand. If the answer is correct, he or she scores a point. If it is incorrect, he or she loses one point. The first one to score ten points wins.

 Pre-Algebra

49. 140,400,000 license plates

49. License Plates North Carolina issues general license plates with three letters followed by four numerals. (The first numeral cannot be 0.) Numerals can repeat, but letters cannot. How many license plates can North Carolina generate with this format?

Mixed Review

50. Draw a tree diagram to find the number of outcomes for rolling two identical number cubes. (Lesson 10-5) **See Solutions Manual.**

51. about −15° C

51. A temperature of 32° Fahrenheit corresponds to a temperature of 0° Celsius. A temperature of 100° Celsius corresponds to a temperature of 212° Fahrenheit. About what temperature in degrees Celsius corresponds to a temperature of 0° Fahrenheit? (Lesson 8-5)

52. Geometry The perimeter of a rectangle is 38 meters. Find the dimensions if the length is 5 meters less than twice the width. (Lesson 7-5) **8 m; 11 m**

53. −2.5

53. Solve $5.4c = -13.5$. Check your solution. (Lesson 6-7)

54. Estimate $1\frac{3}{5} - \frac{5}{12}$. (Lesson 5-2) **Sample answer: 1**

55. Evaluate $5x^{-2}$ if $x = 2$. (Lesson 4-9) $\frac{5}{4}$

56. Travel Juan and his friends are traveling 120 miles to attend a concert. (Lesson 3-4)

56a. 2.4 hours or 2 hours 24 minutes
56b. 0.6 hours or 36 minutes longer

a. If they average 50 mph, how long will the trip take?

b. If road construction on the trip limits their average speed to 40 mph, how much longer will the trip take?

57. Simplify $3x - 12x$. (Lesson 2-5) $-9x$

58. Name the property shown by the statement $(3 + 6) + 8 = 3 + (6 + 8)$. (Lesson 1-4) **Associative Property of Addition**

See for Yourself See students' work.

- Many of the residents of Colorado and other western states are opposed to the new method of gold mining. They say that the cyanide heap-leaching process frees poisonous heavy metals such as lead, mercury, and arsenic. These metals may then escape into streams and groundwater. Research the effects of these metals on waterways. What are the possible dangers to humans?
- In the United States, the mining of hard-rock minerals such as gold, copper, and silver is regulated by the General Mining Act that was passed in 1892. The act exempts miners from paying royalties to the government and has no provisions for environmental protection or restoration. Investigate the proposals being made to Congress that would update the laws concerning mining.
- Write a letter to your congressional representative giving your opinion of the situation. Include reasons for what you believe and what, if anything, you think should be done.

Extension

Using Research Have students complete research to find the number of members of one of the following groups or organizations: school board, offices of the parent-teacher organization, scout leaders, student governing body, and so on. Have students write problems involving permutations and combinations about the members of the chosen group.

Sample answer: The school board has 7 members. Are there 210 ways or 35 ways of forming a committee of three members?

4 ASSESS

Closing Activity

Writing Give students five permutations verbally and have students write the correct notation and then solve. Then give five combinations and have students write the correct notation and then solve.

Enrichment Masters, p. 86

NAME ______ DATE ______

10-6 Enrichment Student Edition Pages 515–519
Permutations and Combinations

An arrangement of objects *in a given order* is called a **permutation** of the objects. A symbol for the number of permutations is $P(n, x)$, where x represents the number of objects to be arranged in order and n reminds us that these objects are chosen from an original set of n objects.

$$P(n, x) = \frac{n!}{(n-x)!}$$

Example: If gold, silver, and bronze medals are to be awarded to the first three finishers in an 8-person race, in how many ways can the medals be awarded?

$$P(8, 3) = \frac{8!}{5!} = \frac{8 \cdot 7 \cdot 6 \cdot 5 \cdot 4 \cdot 3 \cdot 2 \cdot 1}{5 \cdot 4 \cdot 3 \cdot 2 \cdot 1} = 8 \cdot 7 \cdot 6 = 336$$ ways in which the medals may be awarded

A selection of x objects taken from a set of n objects *without regard for order* of the selection is called a **combination**. A symbol for the number of combinations is $C(n, x)$.

$$C(n, x) = \frac{n!}{x!(n-x)!}$$

Example: In how many ways can you choose 3 people from a group of 12 without regard for order?

$$C(12, 3) = \frac{12!}{3!9!} = \frac{12 \cdot 11 \cdot 10 \cdot 9 \cdot 8 \cdot 7 \cdot 6 \cdot 5 \cdot 4 \cdot 3 \cdot 2 \cdot 1}{(3 \cdot 2 \cdot 1)(9 \cdot 8 \cdot 7 \cdot 6 \cdot 5 \cdot 4 \cdot 3 \cdot 2 \cdot 1)} = \frac{12 \cdot 11 \cdot 10}{3 \cdot 2} = 220$$ possible groups of 3 people

Find each value.

1. $P(7, 2)$ 42
2. $P(7, 5)$ 2520
3. $C(7, 2)$ 21
4. $C(7, 5)$ 21
5. $P(13, 2)$ 156
6. $C(13, 11)$ 78

10-7 LESSON NOTES

NCTM Standards: 1-4, 11

Instructional Resources

- Study Guide Master 10-7
- Practice Master 10-7
- Enrichment Master 10-7
- Group Activity Card 10-7

Transparency 10-7A contains the 5-Minute Check for this lesson; **Transparency 10-7B** contains a teaching aid for this lesson.

Recommended Pacing	
Standard Pacing	Day 8 of 14
Honors Pacing	Day 7 of 13
Block Scheduling*	Day 4 of 7

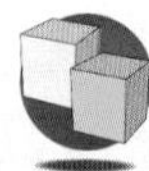

*For more information on pacing and possible lesson plans, refer to the ***Block Scheduling Booklet***.

1 FOCUS

5-Minute Check *(over Lesson 10-6)*

Find each value.

1. $P(6, 3)$ 120
2. $P(7, 4)$ 840
3. $C(5, 2)$ 10
4. $C(9, 3)$ 84
5. $\frac{5!3!}{4!}$ 30

Motivating the Lesson

Situational Problem Pose this situation for students. One of the students in the class will be selected to go to the principal's office. Ask students what the probability is of a girl being chosen. What is the probability that a boy will be chosen?

10-7 Odds

Setting Goals: *In this lesson, you'll find the odds of a simple event.*

MATHEMATICS MINI LAB

Modeling with Manipulatives

MATERIALS
- graph paper
- straight pin

On a piece of graph paper, copy the diagram shown at the right. Shade squares gray or black, as shown.

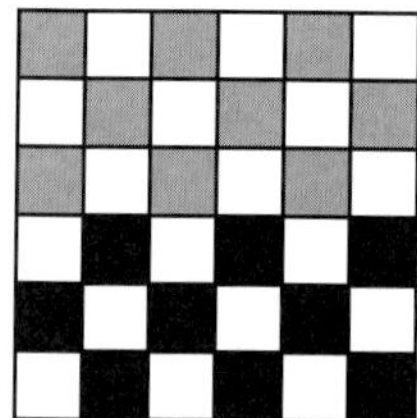

Your Turn

▶ Hold a straight pin about six inches above the center of the diagram. Drop the pin 50 times and record the color of the square where the point of the pin lands. If you miss the diagram, drop the pin again.

TALK ABOUT IT

a. Would you say that the pin is more likely to land on a white square or a shaded square? Explain? *a. White. There are more white squares.*

b. How many ways can the point of the pin land on a black square? **9 ways**

c. How many ways can the point of the pin not land on a gray square? *c. 27 ways*

Learning the Concept

The probability that the point of the pin will land on a gray square is $\frac{9}{36}$ or $\frac{1}{4}$. The probability that the point of the pin will *not* land in a gray square is $\frac{27}{36}$ or $\frac{3}{4}$.

Another way to describe the chance of an event's occurring is with **odds**. The odds in favor of an event is the ratio that compares the number of ways the event can occur to the ways that the event *cannot* occur.

Definition of Odds

The odds in favor of an outcome is the ratio of the number of ways the outcome can occur to the number of ways the outcome cannot occur.

Odds in favor = number of successes: number of failures

The odds against an outcome is the ratio of the number of ways the outcome cannot occur to the number of ways the outcome can occur.

Odds against = number of failures: number of successes

Example 1 **A bag contains 5 green marbles, 2 yellow marbles, and 3 blue marbles. What are the odds of drawing a green marble from the bag?**

There are 10 − 5 or 5 marbles that are not green.

Odds of drawing a green marble

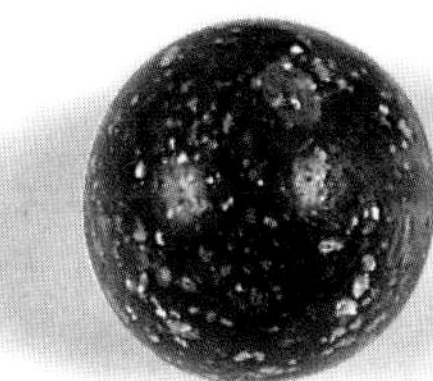

= *number of ways of drawing green marble* : *number of ways of drawing other marbles*

= 5 : 5

= 1:1 *This is read "1 to 1."*

The odds of drawing a green marble are 1:1.

Sometimes when finding odds, you must first find the total number of possible outcomes. This can involve finding permutations or combinations.

Example 2

Finance

The state of Florida has a *Fantasy 5* drawing three times a week in which 5 numbers out of 39 are drawn at random. The proceeds from the lottery help to finance education in the state. What are the odds of winning the *Fantasy 5* jackpot?

Order is not important. You can find the odds using combinations.

There is only one winning combination, so the number of ways to succeed is 1. There are $C(39, 5)$ ways to draw 5 numbers from a group of 39. So the total number of possible outcomes is $C(39, 5)$ or 575,757. This means that the number of failures (not drawing all 5 numbers) must be 575,757 − 1 or 575,756.

odds of drawing 5 correct numbers
= number of successes : number of failures
= 1 : 575,756 *This is read "1 to 575,756."*

The odds of winning the *Fantasy 5* jackpot are 1: 575,756.

Checking Your Understanding

Communicating Mathematics

Read and study the lesson to answer these questions.

1. **Explain** how to find the odds of an event occurring. **See margin.**
2. **You Decide** Ade says that the probability of getting two in one roll of a die is 1 out of 6. Sharla says that the odds of getting two in one roll of a die are 1:5. Who is correct and why? **They are both correct.**
3. Describe a situation in real life that uses odds. **See margin.**

MATH JOURNAL

Guided Practice

Find the odds of each outcome if a die is rolled.

4. a number greater than 5 **1:5**
5. a multiple of 2 **1:1**
6. a number less than 4 **1:1**
7. not a 3 **5:1**

2 TEACH

In-Class Examples

For Example 1
A spinner is half white and half red. What are the odds that the spinner will land on red? **1:1**

For Example 2
To win the Illinois state lottery, you must correctly choose 6 of 54 numbers. Order is not important. What are the odds of winning the Illinois state lottery? **1:25,827,164**

Teaching Tip Be sure students do not confuse probability and odds. If the odds of an event happening are 1:1, the probability is 50%. For odds of 1:3, the probability is 25%.

3 PRACTICE/APPLY

Checking Your Understanding

Exercises 1–8 are designed to help you assess your students' understanding through reading, writing, speaking, and modeling. You should work through Exercises 1–3 with your students and then monitor their work on Exercises 4–8.

Study Guide Masters, p. 87

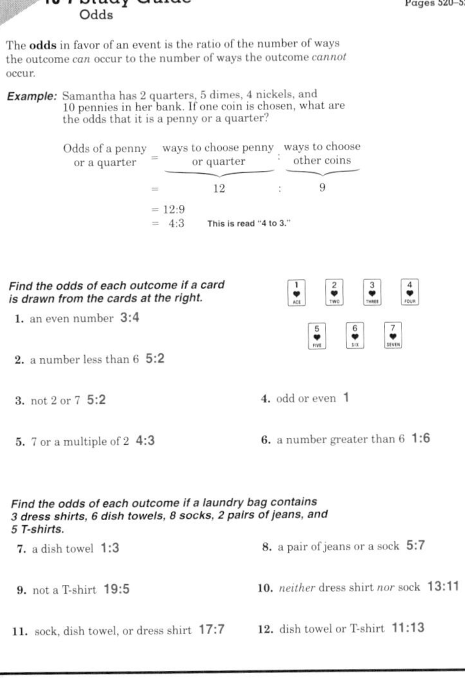

NAME ______ DATE ______

10-7 Study Guide
Odds

Student Edition
Pages 520–523

The **odds** in favor of an event is the ratio of the number of ways the outcome *can* occur to the number of ways the outcome *cannot* occur.

Example: Samantha has 2 quarters, 5 dimes, 4 nickels, and 10 pennies in her bank. If one coin is chosen, what are the odds that it is a penny or a quarter?

Odds of a penny or a quarter = ways to choose penny or quarter : ways to choose other coins
= 12 : 9
= 12:9
= 4:3 This is read "4 to 3."

Find the odds of each outcome if a card is drawn from the cards at the right.

1. an even number **3:4**
2. a number less than 6 **5:2**
3. not 2 or 7 **5:2**
4. odd or even **1**
5. 7 or a multiple of 2 **4:3**
6. a number greater than 6 **1:6**

Find the odds of each outcome if a laundry bag contains 3 dress shirts, 6 dish towels, 8 socks, 2 pairs of jeans, and 5 T-shirts.

7. a dish towel **1:3**
8. a pair of jeans or a sock **5:7**
9. not a T-shirt **19:5**
10. *neither* dress shirt *nor* sock **13:11**
11. sock, dish towel, or dress shirt **17:7**
12. dish towel or T-shirt **11:13**

Reteaching

Using Manipulatives Provide students with marbles of three different colors. One student selects at random some of each color. The other student writes the odds of selecting a specified color from those marbles. Students exchange roles and repeat the procedure.

Additional Answers

1. Sample answer: Write a ratio number of occurrence of a certain outcome and the number of ways that certain outcomes cannot occur.
3. Answers will vary. Sample answer: A newspaper listing the odds of winning some sporting event.

Assignment Guide

Core: 9–25 odd, 26–27, 29, 31–36
Enriched: 10–24 even, 26–36

For **Extra Practice**, see p. 766.

The red A, B, and C flags, printed only in the Teacher's Wraparound Edition, indicate the level of difficulty of the exercises.

8a. 29:21

8. Statistics The United States Census Bureau reported that the number of unmarried adults nearly doubled from 1970 to 1993. The report stated that 58% of unmarried adults had never been married.

a. What are the odds that an unmarried adult was never married?

b. In a group of 200 unmarried adults, how many would you estimate had never been married? **about 116 adults**

Exercises: Practicing and Applying the Concept

Independent Practice

Find the odds of each outcome if the spinner at the right is spun.

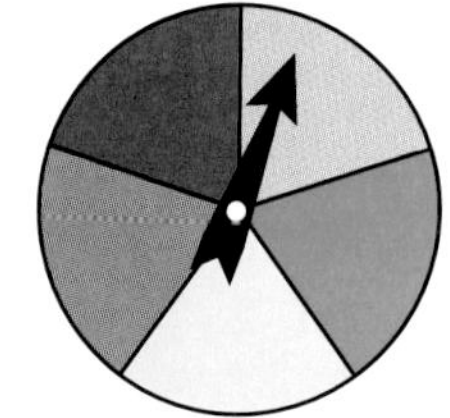

9. blue **1:4**

10. purple **0**

11. not green **4:1**

12. red or orange **2:3**

Find the odds of each outcome if a pair of fair dice are rolled. 18. 1:17

13. an even sum **1:1**

14. an odd sum **1:1**

15. a sum that is a multiple of 3 **1:2**

16. a sum less than 2 **impossible**

17. a composite number **7:5**

18. a prime number greater than 9

19. *not* a sum of 7 **5:1**

20. *not* a sum of 11 or 12 **11:1**

21. a sum of 8 with a 6 on one die **1:17**

22. an even sum or a sum less than 6 **2:1**

23. *neither* an odd sum *nor* a sum greater than 8 **7:11**

Five cards are selected from a standard deck of 52 playing cards.

24. What are the odds of selecting a black king? **1:25**

25. What are the odds of *not* selecting a heart *or* a seven? **9:4**

Critical Thinking

26. A red die and a blue die are rolled. What are the odds that the number showing on the red die is less than the number showing on the blue die? **5:7**

Applications and Problem Solving

27. Entertainment The Durham Bulls baseball team in Durham, North Carolina, invited the community to submit promotion ideas for the opening of their new stadium. One suggestion was that the team call out the last four-digit numbers of a telephone number and award a prize to the first person with those digits in their telephone number to reach the press box. Find the odds that the four digits called out are the last four digits of your telephone number. **1:9999**

28. Television The odds in favor of a person in North America appearing on television sometime in their lifetime is 1:3. If there are 32 students in your class, predict how many will appear on television. **8 students**

Practice Masters, p. 87

NAME ________________ DATE ________

10-7 Practice
Odds

Student Edition
Pages 520–523

Find the odds of each outcome if a bag contains 3 red marbles, 2 black marbles, 4 green marbles, and 1 blue marble.

1. blue marble 1:9

2. brown marble 0

3. red or black marble 1:1

4. not black 4:1

5. black, green, or blue 7:3

6. *neither* blue *nor* red 3:2

Find the odds of each outcome if a die is rolled.

7. a multiple of 5 1:5

8. a prime number 1:1

9. a two digit number 0

10. not a 4 5:1

11. a number less than 3 1:2

12. a number greater than 1 5:1

Find the odds of each outcome if the spinner at the right is spun.

13. a consonant or a number 3:1

14. a prime number or a vowel 1:1

15. not C, 4, or U 5:3

16. a number greater than 1 3:5

17. an even number or a letter 3:1

18. a number less than 1 0

19. *neither* A *nor* 3 3:1

20. a composite number 1:7

21. A cookie jar is filled with the following cookies: 4 chocolate chip, 10 oatmeal raisin, 2 peanut butter and 8 snicker-doodles. What are the odds of getting a snickerdoodle? peanut butter? 1:2; 1:11

22. It usually snows 14 days in December and sleets 4 days. The other days it is cloudy. What are the odds of snow or sleet? cloudy or sleet? 18:13; 17:14

Group Activity Card 10-7

Oddly Enough

Group Activity **10-7**

Copy the table shown at the right. Survey your classmates to find their favorite brand of tennis shoe. Based on the data collected, answer the questions listed below.

Brand of tennis shoe	# of students
Asics	
British Knights	
Fila	
Nike	
Puma	
Reebok	
Other	

1. What are the odds of a student chosen at random picking Nike as their favorite brand?
2. What are the odds that a student will not buy a pair of British Knights?
3. What are the odds of a student buying Fila or Asics tennis shoes?

 Pre-Algebra

Extension

Using Data Have students watch television or listen to the radio to find out the chance of precipitation each day for a week. Then have students use that data to write a summary of the odds for and against precipitation each day. Then have students note what actually happened and compare the actual results to the odds.

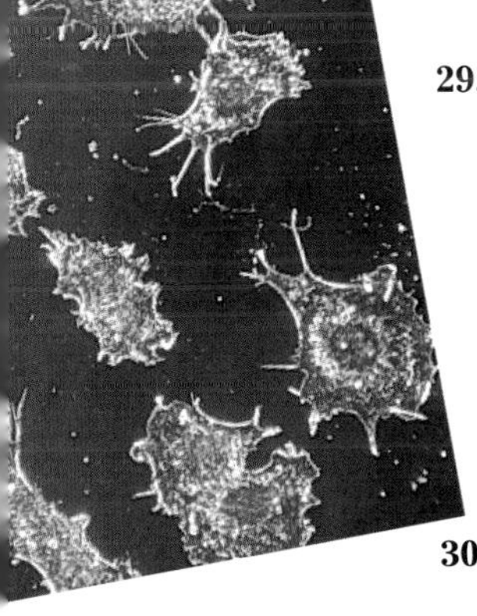

29. **Medicine** A worker at an aluminum plant in Washington state filed a workers' compensation suit against his employer because he developed cancer. He claimed that 8 of 90 workers in his shop had developed the same type of cancer. The medical odds of a person developing this type of cancer is normally 1:1000.
   a. Based on the medical odds, how many people in his shop should have developed the cancer? **Sample answer: less than 1 or none**
   b. Based on the number of workers and the number of cases of cancer in this shop, estimate the number of people who would develop this cancer if 1000 people worked in this shop.

29b. about 89 people

30. **Family Activity** During two hours of watching television or listening to the radio with your family, keep track of the commercials. Record the categories, such as food, cars, cosmetics, and so on, of each commercial break. Find the odds that a randomly-selected commercial is from each category. **See students' work.**

Mixed Review

31. Find the value of 8!. (Lesson 10-6) **40,320**
32. **Sports** The heights (in inches) of members of a volleyball team are 64, 60, 72, 61, 73, 80, 68, 70, 65, 67, 70, and 80. (Lesson 10-1)
   a. Construct a stem-and-leaf plot of the data. **See margin.**
   b. What percent of the heights are between 60 and 70 inches? Why do you think that is true?
33. A penny, a nickel, a dime, and 2 quarters are in a purse. Without looking, Andie picks out three coins. What different amounts of money could she have chosen? (Lesson 9-2)
34. **Geometry** Find the circumference of the circle shown at the right. (Lesson 7-4) **15.71 ft**

5 ft

35. Replace ● with <, >, or = to make $\frac{1}{6}$ ● 0.16 a true statement. Use a number line if necessary. (Lesson 6-1) **>**
36. State whether the inequality $c + 5 < 16$ is *true, false,* or *open.* (Lesson 1-9) **open**

32b. 50%; sample answer: average height for age group < 70 in.
33. $0.16, $0.31, $0.36, $0.40, $0.51, $0.55, $0.60

From the FUNNY PAPERS

1. Express a googol in scientific notation. **1×10^{100}**
2. Do you think the odds are expressed correctly here? Why or why not? **See margin.**
3. Research who defined the term *googol.* **See margin.**

4 ASSESS

Closing Activity

Modeling Provide students with different-colored marbles. Have students show a set of marbles in which the odds of picking a blue marble are 1 to 1. Repeat the procedure for showing the odds of not picking a blue marble being 2 to 3.

Additional Answer

32a.

Stem	Leaf
6	0 1 4 5 7 8
7	0 0 2 3
8	0 0

6 | 0 = 60 inches

From the FUNNY PAPERS

Have students make up another joke that deals with probability or odds. Then have them explain their joke.

From the Funny Papers Additional Answers

2. No; the smaller number (1) should be stated first.
3. Googol is short for googolplex. It was invented in 1938 by Milton Sirotta. He was 9 years old at the time.

Enrichment Masters, p. 87

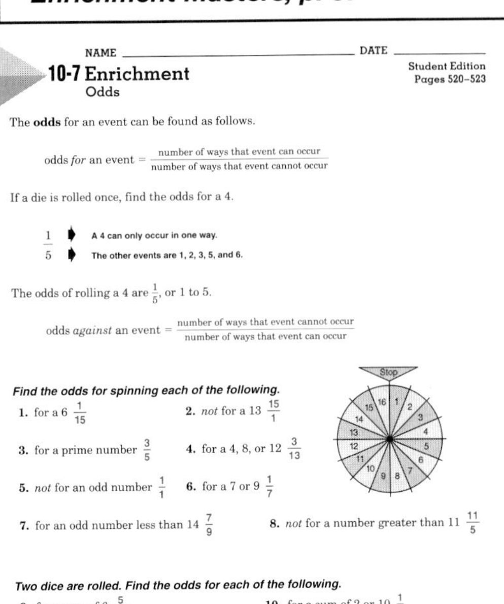

NAME ______________ DATE ______

10-7 Enrichment
Odds

Student Edition
Pages 520–523

The **odds** for an event can be found as follows.

odds *for* an event = $\frac{\text{number of ways that event can occur}}{\text{number of ways that event cannot occur}}$

If a die is rolled once, find the odds for a 4.

1 ➧ A 4 can only occur in one way.
5 ➧ The other events are 1, 2, 3, 5, and 6.

The odds of rolling a 4 are $\frac{1}{5}$, or 1 to 5.

odds *against* an event = $\frac{\text{number of ways that event cannot occur}}{\text{number of ways that event can occur}}$

Find the odds for spinning each of the following.

1. for a 6 $\frac{1}{15}$
2. *not* for a 13 $\frac{15}{1}$
3. for a prime number $\frac{3}{5}$
4. for a 4, 8, or 12 $\frac{3}{13}$
5. *not* for an odd number $\frac{1}{1}$
6. for a 7 or 9 $\frac{1}{7}$
7. for an odd number less than 14 $\frac{7}{9}$
8. *not* for a number greater than 11 $\frac{11}{5}$

Two dice are rolled. Find the odds for each of the following.

9. for a sum of 6 $\frac{5}{31}$
10. for a sum of 2 or 10 $\frac{1}{8}$
11. *not* for a sum of 3 $\frac{17}{1}$
12. for matching numbers (doubles) $\frac{1}{5}$
13. for a sum of 7 and a 2 on one die $\frac{1}{17}$
14. for a sum of 10 and a 4 on one die $\frac{1}{17}$
15. *not* for a sum greater than 9 $\frac{5}{1}$
16. for a sum greater than 9 $\frac{1}{5}$

10-8 LESSON NOTES

NCTM Standards: 1-4, 11

Instructional Resources

- Study Guide Master 10-8
- Practice Master 10-8
- Enrichment Master 10-8
- Group Activity Card 10-8
- Assessment and Evaluation Masters, p. 268
- Activity Masters, p. 24
- Multicultural Activity Masters, p. 20
- Real-World Applications, 24
- Tech Prep Applications Masters, p. 20

Transparency 10-8A contains the 5-Minute Check for this lesson; **Transparency 10-8B** contains a teaching aid for this lesson.

Recommended Pacing	
Standard Pacing	Day 9 of 14
Honors Pacing	Day 8 of 13
Block Scheduling*	Day 5 of 7

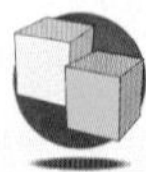

*For more information on pacing and possible lesson plans, refer to the ***Block Scheduling Booklet***.

1 FOCUS

5-Minute Check *(over Lesson 10-7)*

Find the odds of each outcome if a fair pair of dice is rolled.

1. an odd sum **1:1**
2. a sum greater than 12 **impossible**
3. a sum that is a multiple of 6 **1:5**
4. not a sum of 9 **8:1**
5. not a sum of 6 or 8 **13:5**

Motivating the Lesson

Questioning Ask students which would take less time to do—roll a die 100 times or run a computer program that simulates rolling a die 100 times.

10-8 Problem-Solving Strategy: Use a Simulation

Setting Goals: *In this lesson, you'll investigate problems using simulation.*

MATHEMATICS MINI LAB — Modeling with Manipulatives

MATERIALS
- 8 pennies
- cup

The Paiute ('pi- (y)ut) Indians of Pyramid Lake in northwestern Nevada invented the Walnut Shell Game. This game was played with 8 walnut shell halves filled with pitch and powdered charcoal. A player tosses a basket tray containing the walnut shell halves and scores one point if three or five of the walnut shells land with the flat side up. No point is scored for any other combination. The winner is the first player to reach an agreed-on sum. On average, how many times would you have to toss the basket before you get 10 points?

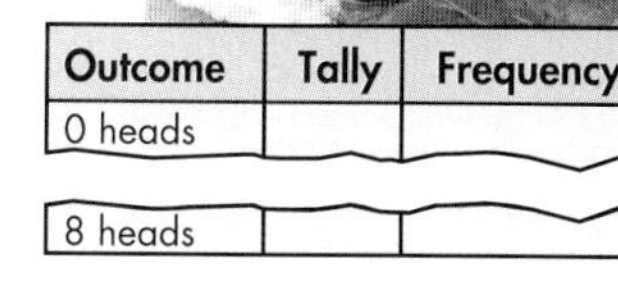

Your Turn Work with a partner to simulate the walnut shell game.

CULTURAL CONNECTIONS Sarah Winnemucca, a member of the Paiute tribe, is famous for her lectures on the U.S. government's treatment of her people.

- ▶ Place eight pennies in a cup. Put your hand over the cup and shake it gently. Then pour the pennies out. Record the number of heads in a frequency table like the one at the right.

Outcome	Tally	Frequency
0 heads		
8 heads		

- ▶ Repeat the process until the sum of the frequencies for three heads and five heads is 10.
- ▶ Compare your results to other pairs of students.

TALK ABOUT IT

a. Do you think this process is a reasonable approximation of the walnut shell game? Explain. **Yes, because you are actually going through the process.**

b. What is an advantage of finding an answer by acting it out? **See margin.**

Learning the Concept

A simulation is an application of the Acting It Out problem-solving strategy.

When you solve a problem by modeling a situation, like you did in the activity above, you are doing a **simulation**. A simulation acts out the event so that you can see outcomes.

You can conduct a simulation of the outcomes of many problems by using manipulatives such as dice, a coin, or a spinner. The manipulative, such as a die, a coin, a spinner, or a combination of them, should have the same number of outcomes as the number of possible outcomes.

Real-World Event	Simulation
having a baby boy or baby girl	flip a coin
history of making 2 free throws for every 3 attempted	spinning a 3-section spinner with 2 sections the same color
win one of twelve prizes	roll a die and flip a coin

Additional Answer Talk About It

b. Sample answer: You may be able to find an approximation for a complicated probability.

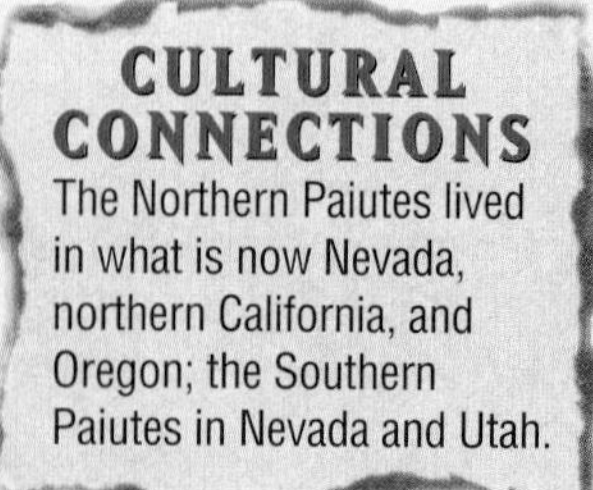

CULTURAL CONNECTIONS The Northern Paiutes lived in what is now Nevada, northern California, and Oregon; the Southern Paiutes in Nevada and Utah.

Example 1

Education

A quiz has 10 true-false questions. The correct answers are T, F, F, T, F, F, T, T, T, F. You need to correctly answer 7 or more questions to pass the quiz. Is tossing a coin to decide the answers a good strategy for taking the quiz?

Explore The quiz has ten questions. Since two choices are available for each answer, tossing a coin is used to simulate guessing the answers.

Plan Toss a coin and record the answer for each question. Write T (true) if tails shows and write F (false) if heads shows. Repeat the simulation three times. Shade the cells with the correct answers.

Solve

Answers	T	F	F	T	F	F	T	T	T	F	Number Correct
Simulation 1	F	T	F	F	T	F	F	F	T	F	4
Simulation 2	F	T	F	F	F	T	T	F	T	F	5
Simulation 3	F	F	T	T	T	F	T	T	T	T	6

Since none of the simulations results in a passing grade, this is probably not a good way to take the quiz.

Examine Try some more simulations to confirm the results.

Sometimes it is necessary to create a manipulative for a simulation.

Example 2

Sports

The Cougars are one point behind in their basketball game. Carmen is fouled as time runs out. If she misses the foul shot, the Cougars lose. If Carmen makes the foul shot, the score is tied and she gets another shot. If she makes the first and second shots, the Cougars win. Carmen has a record of making three out of every four free throws.

a. Conduct a simulation for 25 trips to the foul line.

Construct a spinner with four equal sections like the one shown at the right. Spin the spinner and record the results of several trips to the foul line. A sample simulation for 25 trips to the foul line is shown in the chart below.

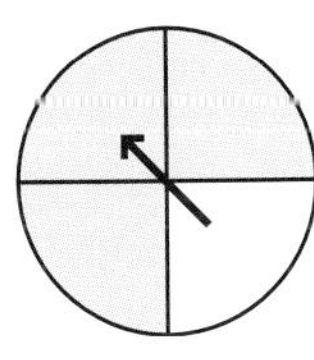

Misses the first shot (0 points)	Makes the first shot, misses the second (1 point)	Makes both shots (2 points)
IIII	IIII	𝍸 𝍸 𝍸 II

b. What is the probability that Carmen makes both shots and the Cougars win the game?

The experiment shows that Carmen will make 2 points $\frac{17}{25}$ or 68% of the time. Based on this simulation, the probability that Carmen scores 2 points and the Cougars win the game is 68%.

2 TEACH

In-Class Examples

For Example 1

A multiple-choice test of 5 questions has 6 choices for each question. The correct choices are A, F, C, D, and E. You need 4 correct answers to pass the test. You roll a die where 1 represents A, 2 represents B, and so on. Is this a good strategy for choosing the answers to the test? **no**

For Example 2

Suppose Carmen had a record of making one out of every two free throws. Toss a coin to simulate the free throws—heads, she makes it; tails, she misses.

a. Conduct a simulation for 25 trips to the foul line.

b. What is the probability that Carmen makes both shots and the Cougars win the game under these new conditions? **Answers will vary. The probability should be about 25%.**

Interactive Mathematics Tools Software

In this interactive computer lesson, students use simulations to explore probability. A **Computer Journal** gives students the opportunity to write about what they have learned.

For Windows & Macintosh

3 PRACTICE/APPLY

Checking Your Understanding

Exercises 1–5 are designed to help you assess your students' understanding through reading, writing, speaking, and modeling. You should work through Exercises 1–3 with your students and then monitor their work on Exercises 4–5.

Assignment Guide

Core: 7–13 odd, 14–19
Enriched: 6–12 even, 13–19

For **Extra Practice**, see p. 766.

The red A, B, and C flags, printed only in the Teacher's Wraparound Edition, indicate the level of difficulty of the exercises.

Teaching Tip You may want to have students devise more than one method of simulating the situations in Exercises 6 and 7.

Study Guide Masters, p. 88

NAME ______________________ DATE ____________

10-8 Study Guide
Problem-Solving Strategy: Use a Simulation

Student Edition Pages 524–528

There are three gumball machines. Each machine contains an equal number of red, blue, and yellow gumballs. If Robin gets one gumball from each machine, what is the probability that two of the gumballs are red?

Explore There are three gumball machines. There are three different colors of gumballs available. We need to find the probability that two out of three gumballs will be red.

Plan Since three colors are available in equal amounts, a spinner like the one at the right can be used to simulate the situation. Spin the spinner and record the results. Repeat the simulation ten times.

red
blue
yellow

Solve

Gumball Machine	Number 1	Number 2	Number 3	
Simulation #1	Y	B	B	
Simulation #2	Y	B	Y	
Simulation #3	Y	B	Y	
Simulation #4	B	Y	Y	
Simulation #5	R	B	R	✓
Simulation #6	R	B	B	
Simulation #7	Y	Y	R	
Simulation #8	B	Y	B	
Simulation #9	Y	B	Y	
Simulation #10	Y	R	B	

One of the simulations results in two red gumballs. You can estimate the probability of getting two red gumballs will be $\frac{1}{10}$.

Examine The actual probability that two of the three gumballs are red is $\frac{1}{3} \times \frac{1}{3}$ or $\frac{1}{9}$. The estimate is reasonable.

Solve. Use a simulation.

1. The aquarium at the pet store contains an equal amount of goldfish that have either orange or white fins. Linda chooses six fish at random. What is the probability that three of the six fish have white fins? See students' work.
2. The M & W Bakery makes homemade white, wheat, rye, pumpernickel, garlic, and Italian bread. The baker chooses a type of bread at random to sell each day. What is the probability that the baker sells the same type of bread four days in one week? See students' work.
3. What is the probability that a family of four children has three girls and one boy? See students' work.

Checking Your Understanding

Communicating Mathematics

Read and study the lesson to answer these questions. 1–2. See margin.

1. **Explain** what a simulation is.
2. **Tell** whether the results of a simulation will be exactly the same as the theoretical probability. Explain your reasoning.
3. **Write** an explanation of how you would modify the spinner in Example 2 to simulate a player who has a record of making one basket out of every three tries. Modify to three sections.

Guided Practice

4. Using a device other than a coin, conduct three more simulations for the problem in Example 1. Do your simulations verify the results obtained while using a coin? See students' work.
5. **Football** Faud Reveiz is a placekicker for the NFL's Minnesota Vikings. In 1994, he made 80% of the field goals he attempted when the distance was between 40 and 49 yards.
   a. How would you simulate his next four field goal attempts between 40 and 49 yards? See margin.
   b. Conduct a simulation for 15 field goal attempts between 40- and 49-yards long. See students' work.
   c. What is the probability that his next two field goal attempts between 40 and 49 yards will be successful? See students' work.

Exercises: Practicing and Applying the Concept

Independent Practice

Solve. Use a simulation to act out the problem.

6. A fast-food restaurant includes prizes with children's meals. During the summer promotional campaign, six different prizes were available. There is an equally likely chance of getting each prize each time.
   a. Use a die to simulate this problem. Let each number represent one of the prizes. Conduct a simulation until you have one of each number. See students' work.
   b. According to your simulation, how many children's meals must be purchased in order to get all six different prizes?

6b. Sample answer: at least 6

7. **Business** Mr. Namura runs a small gourmet restaurant. He has 12 tables. It is a popular restaurant and anyone wanting to eat at 7:00 on Saturday evening must have reservations. Mr. Namura knows that one out of six reservations usually does not show, so he takes reservations for 14 tables for 7:00.
   a. Describe a way to simulate the number of tables that will be filled at 7:00.
   b. Conduct your simulation for part a ten times. See students' work.
   c. From your simulations, how many times is the restaurant overbooked? Are all 12 tables always filled? See students' work.

7a. Roll a die.

Additional Answers

1. Sample answer: A simulation is an imitation of a given problem.
2. Sample answer: The results of a simulation will not be exactly the same as the actual problem because the simulation is just a way of acting out the problem to predict what might happen.

5a. Sample answer: Select a marble out of a bag that has four marbles of the same color and one marble of a different color. Then repeat three more times.

Choose

Estimation
Mental Math
Calculator
Paper and Pencil

Solve. Use any strategy.

8. Unit 2 of a science book starts on page 126 and ends on page 241. How many pages are in the unit? **116 pages**

9. Antonio received a birthday gift of money from his grandmother. He loaned one dollar to his younger brother and spent half of the remaining money. The next day he earned \$6. Later, he lost $\frac{1}{6}$ of his money, but he still had \$15 left. How much money did his grandmother send him? **\$25**

10. There are three traffic lights along the route that Yoshie rides from school to her home. The probability that any one of the lights is green is 0.4, and the probability that it is not green is 0.6. If the three lights operate independently, estimate the probability that these lights will all be green on Yoshie's way home from school. **See margin.**

11. **Hockey** There are 16 hockey teams in a single elimination tournament.
    a. How many games will be played during the tournament? **15 games**
    b. How many ways can the teams win first and second place? **240**

12. Answers will vary. The probability should be about $\frac{5}{8}$. C

12. The diagram below shows a system of power lines that bring electricity from a power plant to three towns. Recently it has been very windy. For the last 10 days, each line has been knocked down about half the time due to the high winds. What is the probability that Altoona will still be able to get power on any given day even if some of the lines are down?

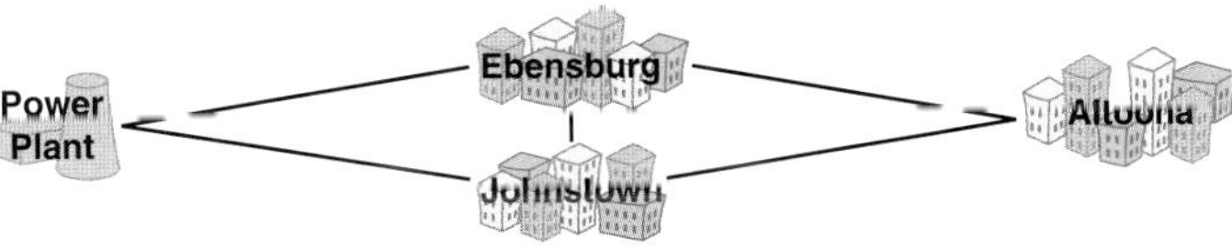

Graphing Calculator

13. When you have a situation involving numbers that cannot be generated from an object like a coin or a spinner, you can use the random number generator function on a computer or a graphing calculator. A random number generator produces numbers that have no pattern.

    The graphing calculator program below will generate 20 random numbers between 1 and the number you enter for T. In order to use the program, you must first enter the program into the calculator's memory. To access the program memory, use the following keystrokes.

    Enter: PRGM ▶ ▶ ENTER

```
PROGRAM: RANDOM
: Input T
: For(N,1,20)
: int((T*rand) + 1) →D
: Disp D
: Pause
: End
```

    Run the program. Press PRGM and choose the program from the list by pressing the number next to its name.

    Press the enter key twenty times. The calculator will display a number from 1 to 6 each time you press the enter key.

    a. Run the program to simulate 20 spins of a spinner that has 7 equal sections. **See student's work.**
    b. How would you alter the program to generate 50 random numbers? **See margin.**

Error Analysis

Watch for students who estimate results or use theoretical probabilities rather than experimental probabilities. Emphasize the need for systematically setting up the experiment and then completing the experiment to find the experimental probability.

Additional Answers

10. Answers will vary. Sample answer: The probability will depend on the results of the simulation. If the simulation is repeated 10 times and one of them results in 3 green light, the probability is 0.1.

13b. Change the second line to "For(N,1,50)"

Practice Masters, p. 88

NAME ______________ DATE ______________

10-8 Practice
Problem-Solving Strategy: Use a Simulation

Student Edition
Pages 524–528

The diagram below shows a system of waterways that bring fresh water from a reservoir to a fish hatchery. Recently there have been many problems with beavers building dams across these waterways. For the last 30 days, each of the waterways has been closed about half of the time due to beaver dams.

What is the probability that the hatchery will still be able to get water even if some of the waterways are closed?

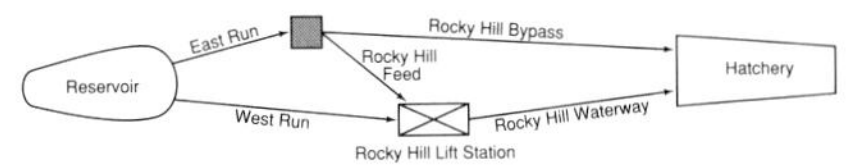

To simulate this situation, you need a procedure to determine if each of the five waterways is open or blocked. Since each waterway is closed about half of the time, you might toss a coin and note heads or tails, or roll a die and note an odd or even number.

Suppose you decide to toss a coin. Let heads mean the water is blocked and tails mean it is open. In the table below, Trial 1 has been completed for you.

1. Complete the table for the remaining trials. **Answers will vary.**

Trial	East Run	West Run	Rocky Hill Feed	Rocky Hill Bypass	Rocky Hill Waterway	Will Water Flow to the Hatchery?
1	T	H	H	T	H	Yes
2						
3						
4						
5						
6						
7						
8						
9						
10						

2. Conduct two more simulations of ten trials each. Based on thirty trials, what is the probability that the hatchery will still be able to get water even if some of the waterways are closed? **Answers will vary; should be about $\frac{5}{8}$.**

Group Activity Card 10-8

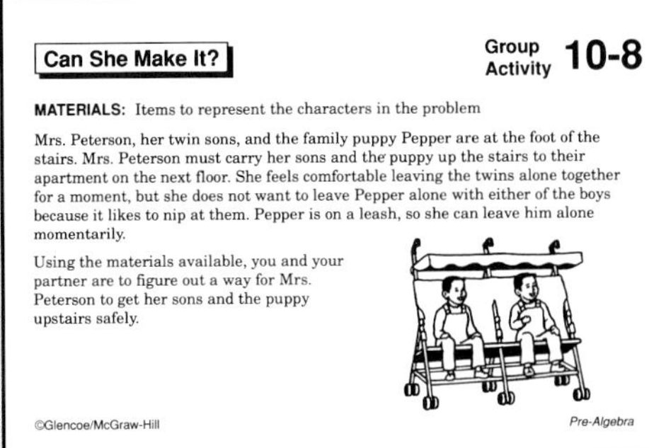

Can She Make It? Group Activity **10-8**

MATERIALS: Items to represent the characters in the problem

Mrs. Peterson, her twin sons, and the family puppy Pepper are at the foot of the stairs. Mrs. Peterson must carry her sons and the puppy up the stairs to their apartment on the next floor. She feels comfortable leaving the twins alone together for a moment, but she does not want to leave Pepper alone with either of the boys because it likes to nip at them. Pepper is on a leash, so she can leave him alone momentarily.

Using the materials available, you and your partner are to figure out a way for Mrs. Peterson to get her sons and the puppy upstairs safely.

 Pre-Algebra

Reteaching

Using Calculators Modify the graphing calculator program in Exercise 13 to simulate 1000 rolls. Have students determine the probability of rolling a one or a two. Run the program and have students count the ones and twos.

4 ASSESS

Closing Activity

Modeling Have students develop a method for simulating any of the following situations.

- selecting the answers to a multiple-choice test with four choices
- finding how many boxes of cereal you will need to purchase before getting one of each of six different prizes available
- finding whether a player will make two free throw shots to win the game if she has previously made 60% of her free throws

Chapter 10, Quiz C (Lessons 10-6 through 10-8) is available in the *Assessment and Evaluation Masters*, p. 268.

Enrichment Masters, p. 88

NAME ________________ DATE ________

10-8 Enrichment — Student Edition Pages 524–528

Experimenting: Coin Tossing

1. Complete the tree diagram for tossing three coins at the same time.

First Coin	Second Coin	Third Coin	Results
heads	heads	heads	heads, heads, heads
		tails	heads, heads, tails
	tails	heads	heads, tails, heads
		tails	heads, tails, tails
tails	heads	heads	tails, heads, heads
		tails	tails, heads, tails
	tails	heads	tails, tails, heads
		tails	tails, tails, tails

You toss three coins at the same time. Express each probability as a fraction.

2. P(3 heads) $\frac{1}{8}$
3. P(2 heads, 1 tail) $\frac{3}{8}$
4. P(2 tails, 1 head) $\frac{3}{8}$
5. P(3 tails) $\frac{1}{8}$

Suppose you toss three coins 40 times each. Predict how many times you would expect to get each of the following results.

6. 3 heads **5**
7. 2 heads, 1 tail **15**
8. 2 tails, 1 head **15**
9. 3 tails **5**

10. Toss three coins 40 times each, and use tally marks to record the results in the table. Count the tally marks to get the number of times each result occurs. **Answers will vary.**

Results	Tally	Number
3 heads		
2 heads, 1 tail		
2 tails, 1 head		
3 tails		

11. Did each result occur exactly the number of times you predicted in Exercises 6–9? **no**

12. Does the probability of a result predict what *will actually happen* or what *will most likely happen*? **what will most likely happen**

Critical Thinking

14. In a survey of 240 students, 100 students said they play soccer, and 120 students said they play basketball. If 40 students play both sports, how many students do not play either basketball or soccer? **60 students**

Mixed Review

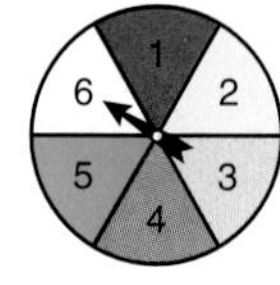

15. **Probability** Find the odds of stopping on 2 if the spinner at the right is spun. (Lesson 10-7) **1:5**

16. **Geometry** Find the slope of the line that contains the points (1, 5) and (3, 3). (Lesson 8-6) **−1**

17. **Sales** Jeff bought a used 10-speed bike for $20 more than one-half its original price. He paid $110 for the bike. (Lesson 7-3)
    a. Write an equation that represents this situation. $110 = \frac{1}{2}x + 20$
    b. What was the original price of the bike? **$180**

18. Write r^4 as the product of the same factor. (Lesson 4-2) $r \cdot r \cdot r \cdot r$

19. **Meteorology** The temperature outside was 20°F. The windchill factor made it feel like −15°F. Find the difference between the real temperature and the apparent temperature. (Lesson 2-5) **35°**

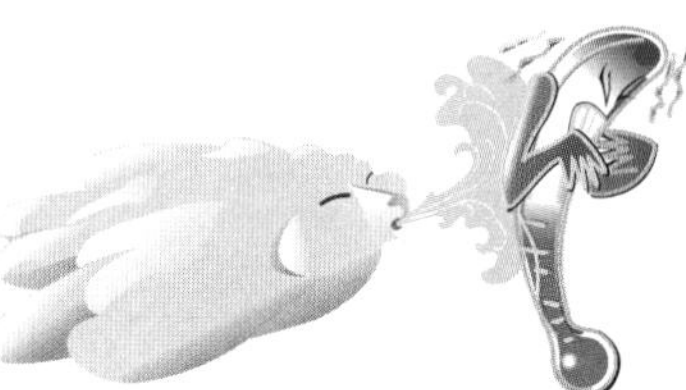

WORKING ON THE

Investigation

Refer to the Investigation on pages 482–483.

Look over the list of favorite games that you placed in your Investigation Folder at the beginning of this Investigation. Some people may have said that their favorite games are TV game shows, such as *The Price is Right.*

For this part of the Investigation, your group has been chosen to test a new TV game show called *Don't Lose It All!* This game is for two players. Each player takes a turn spinning two identical wheels at the same time. Each wheel is divided into six equal regions. Five of the regions say "Win $25" and the sixth region says "Lose Your Cash". The object of the game is to accumulate the most money by the end of the show.

A player's turn ends if one or both of the wheels stop on "Lose Your Cash" or the player decides to let his or her opponent spin. On a spin, one of three things can occur.

1. Both wheels stop on "Win $25" and the player is awarded $50.
2. One wheel stops on "Lose Your Cash," and one stops on "Win $25." In this case, a player loses any money made on this turn but keeps money from any previous turns.
3. Both wheels stop on "Lose Your Cash" and the player loses all the money won so far.

- Create two spinners to simulate the wheels in the game. Play the game enough times to develop some different strategies. Have each member of your group use a different strategy and play the game several more times to determine the best strategy.
- Determine a way to simulate the game your group is designing.

Add the results of your work to your Investigation Folder.

Extension

Using Simulations Have students find methods of creating a simulation other than using dice, coins, spinners, and computers or calculators. For instance, they might choose a page in the telephone book and use the last digit of all the phone numbers on that page. Challenge students to justify that their method truly creates a "fair" random simulation.

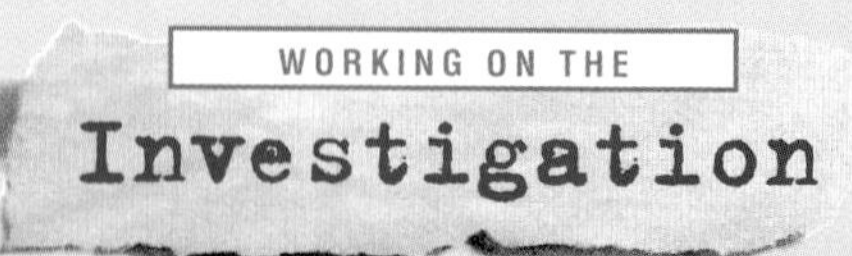

The Investigation on pages 482 and 483 is designed to be a long-term project that is completed over several days or weeks. Encourage students to keep their materials in their Investigation Folder as they work on the Investigation.

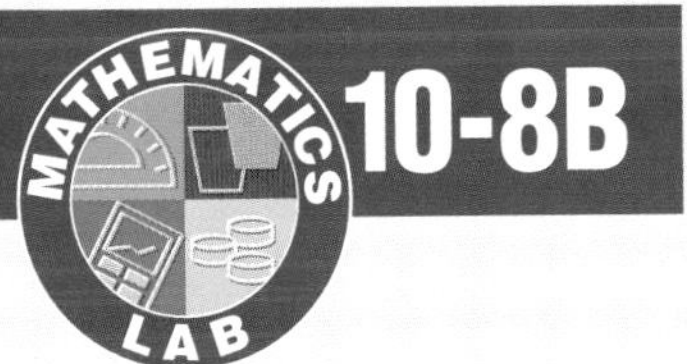

HANDS-ON ACTIVITY

10-8B Making Predictions

An Extension of Lesson **10-8**

MATERIALS

 paper bag

10 marbles of 3 different colors

In this activity, you will conduct and interpret probability experiments. Probability calculated by making observations or experiments is called **experimental probability**. Experimental probability is an estimate based on the **relative frequency**. Relative frequency is the number of times a certain event actually happened.

Activity 1

Work with a partner.

- Get a bag of marbles from your teacher. You will try to determine how many marbles there are of each color.
- Draw one marble from the bag, record its color, and replace it in the bag. Repeat this process 10 times.
- Determine the relative frequency for each color marble.

TALK ABOUT IT

1. Based on the relative frequencies, which color marble do you think is most prevalent in the bag? Predict the colors of the marbles in your bag. **See Solutions Manual.**
2. Based on your experiment, what is the experimental probability of drawing each color marble? **See students' work.**

$$\text{experimental probability} = \frac{\text{relative frequency of color}}{\text{total number of draws}}$$

Activity 2

Repeat both steps above for 20, 30, 40, and 50 draws.

TALK ABOUT IT

3. Is it possible to have a certain color marble in the bag and never draw that color? Is this situation more likely to happen if you make two draws, ten draws, or fifty draws? Explain. **See Solutions Manual.**
4. Compute the experimental probability of drawing each color for each number of draws. Write a paragraph that describes how the experimental probability changed as you increased the number of draws. **See Solutions Manual.**

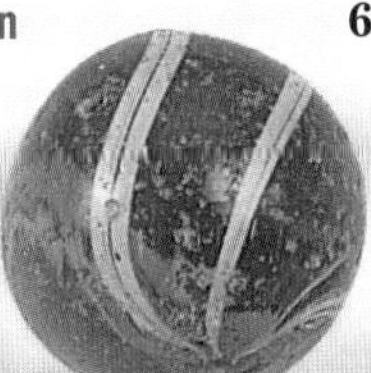

5. Predict the colors of the marbles in your bag. Did your prediction change from the prediction you made in Exercise 1? Explain why or why not. Open the bag and check your prediction against the marbles. **See students' work.**

Extension

6. Make up a fair game for two players to draw marbles from your bag. Assign point values for drawing certain colors of marbles. Make sure each player has an equal chance of winning. **See students' work.**

10-8B LESSON NOTES

NCTM Standards: 1-4, 11

Objective

Conduct experiments to find experimental probability.

Recommended Time

Demonstration and discussion: 15 minutes; Exercises: 30 minutes

Instructional Resources

For each student or group of students

- paper bag
- 10 marbles of 3 different colors

Math Lab and Modeling Math Masters

- p. 47 (worksheet)

For teacher demonstration

Overhead Manipulative Resources

1 FOCUS

Motivating the Lesson

Tell students that you keep your socks in a drawer. Suppose you reach in the drawer and pull out a sock, then replace the sock and draw again. If you pulled black socks out 10 times in a row, could you assume there are only black socks in the drawer?

2 TEACH

Teaching Tip Tell students that part of the fun in this experiment is trying to guess how many of each colored marble they have in the bag, so they should avoid the temptation of looking in the bag.

3 PRACTICE/APPLY

Assignment Guide
Core: 1–6
Enriched: 1–6

4 ASSESS

Observing students working in cooperative groups is an excellent method of assessment.

10-9 LESSON NOTES

NCTM Standards: 1-4, 11

Instructional Resources

- Study Guide Master 10-9
- Practice Master 10-9
- Enrichment Master 10-9
- Group Activity Card 10-9
- Math Lab and Modeling Math Masters, p. 85

Transparency 10-9A contains the 5-Minute Check for this lesson; **Transparency 10-9B** contains a teaching aid for this lesson.

Recommended Pacing	
Standard Pacing	Day 11 of 14
Honors Pacing	Day 10 of 13
Block Scheduling*	Day 6 of 7 (along with Lesson 10-10)

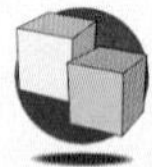

*For more information on pacing and possible lesson plans, refer to the ***Block Scheduling Booklet***.

1 FOCUS

5-Minute Check *(over Lesson 10-8)*

How would you simulate each situation?

1. deciding whether to turn left or right **Sample answer: toss a coin**
2. choosing one of six prizes at random **Sample answer: roll a die**
3. Michelle makes $\frac{3}{4}$ of the free throws she shoots. She is fouled and has two free throws to shoot. What is the probability that she will make both free throws? **Sample answer: use a spinner**

10-9 Probability of Independent and Dependent Events

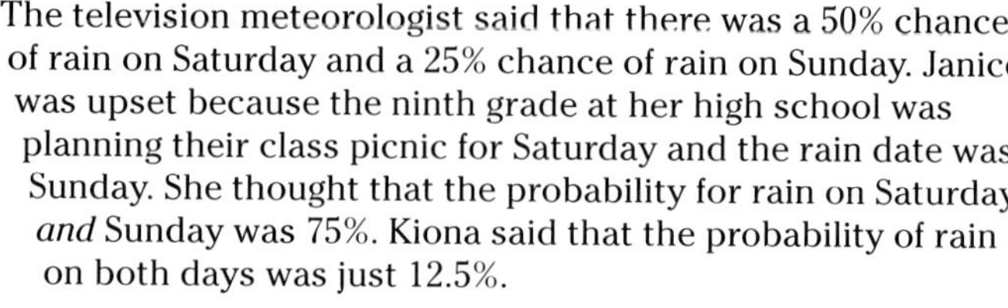

Setting Goals: *In this lesson, you'll find the probability of independent and dependent events.*

Modeling with Technology

The television meteorologist said that there was a 50% chance of rain on Saturday and a 25% chance of rain on Sunday. Janice was upset because the ninth grade at her high school was planning their class picnic for Saturday and the rain date was Sunday. She thought that the probability for rain on Saturday *and* Sunday was 75%. Kiona said that the probability of rain on both days was just 12.5%.

You can use the TI-82 graphing calculator program at the right to simulate this situation. Recall that you must first enter the program into the calculator's memory. Access the program memory using the following keystrokes.

Enter: PRGM ▶ ▶

```
:PROGRAM:RAIN
:0→S: 0→R
:Disp "PROBABILITY",
 "OF RAIN ON"
:Input "SATURDAY", A
:Input "SUNDAY", B
:For (N,1,50)
:int ((100/A*rand)
 +1)→T
:If T=1
:S+1→S
:End
:For (M,1,50)
:int ((100/B*rand)
 +1)→G
:If G=1
:R+1→R
:End
:(S/50)*(R/50)→F
:F*100→F
:Disp "PERCENT OF
 TIMES", "IT RAINS
 BOTH", "DAYS IS", F
:Stop
```

The program generates random numbers for 50 weekends. The first use of the random number function simulates the probability of rain on Saturday. The second use of the random number function simulates the probability of rain on Sunday.

Your Turn Run the program several times and record the percent of times that it rains on both days for each 50-weekend simulation.

TALK ABOUT IT a–b. See margin.

a. Are your results closer to what Janice thought or to what Kiona said?

b. Run the program several times for a weekend when the probability of rain is 100% for Saturday and 20% for Sunday. What do you observe?

Additional Answers Talk About It

a. Sample answer: Kiona

b. The probability is 20%.

Learning the Concept

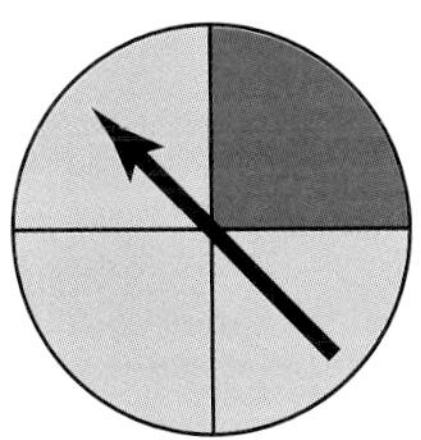

You could also simulate the situation above using manipulatives. This could be done by tossing a coin and spinning the spinner at the right 50 times each. Tossing a head has the same probability as the chance of rain on Saturday and spinning blue on the spinner has the same probability as the chance of rain on Sunday. Does the outcome of tossing the coin affect the outcome of spinning the spinner? no

The weather situation in the activity above is an example of **independent events**. Events are independent when the outcome of one event does *not* influence the outcome of a second event.

Using the Fundamental Counting Principle, you know that there are $2 \cdot 4$ or 8 possible outcomes of tossing a coin and spinning the spinner. Of the 8 outcomes, there is one pair (heads, blue) that matches the meteorologist's prediction (50% chance of rain on Saturday, 25% chance of rain on Sunday). Determine the probability.

$$P(\text{heads, blue}) = \frac{\textit{number of ways heads and blue can occur}}{\textit{number of possible outcomes}}$$
$$= \frac{1}{8}$$

Note that $P(\text{heads})$ is $\frac{1}{2}$, $P(\text{blue})$ is $\frac{1}{4}$, and $\frac{1}{2} \cdot \frac{1}{4} = \frac{1}{8}$, which is $P(\text{heads, blue})$.

This suggests that the probability of two independent events is the product of the probability of each event.

Probability of Two Independent Events

In words:	**The probability of two independent events can be found by multiplying the probability of the first event by the probability of the second event.**
In symbols:	$P(A \text{ and } B) = P(A) \cdot P(B)$

Example 1 APPLICATION Games

In the game of Monopoly, you go to jail if you roll three doubles in a row. What is the probability of rolling three doubles in a row?

The events are independent since each roll of the dice does not affect the outcome of the next roll.

There are six ways to roll doubles, (1, 1), (2, 2), and so on, and there are 36 ways to roll two dice. Thus, the probability of rolling doubles on a toss of the dice is $\frac{6}{36}$ or $\frac{1}{6}$.

LOOK BACK You can review the sample space of rolling two dice on page 441.

$$P\left(\begin{matrix}\textit{doubles on}\\ \textit{three rolls}\end{matrix}\right) = P\left(\begin{matrix}\textit{doubles on}\\ \textit{roll one}\end{matrix}\right) \cdot P\left(\begin{matrix}\textit{doubles on}\\ \textit{roll two}\end{matrix}\right) \cdot P\left(\begin{matrix}\textit{doubles on}\\ \textit{roll three}\end{matrix}\right)$$
$$= \frac{1}{6} \cdot \frac{1}{6} \cdot \frac{1}{6}$$
$$= \frac{1}{216}$$

The probability of rolling all doubles in three rolls of the dice is $\frac{1}{216}$.

Motivating the Lesson

Questioning Write the words *independent* and *dependent* on the board. Ask students to write each pair of events under one of the words.

- having black hair and having blue eyes **independent**
- getting an A in math class and doing your math homework **dependent**
- having movie tickets and getting into the movie **dependent**
- getting an A in science class and doing your English homework **independent**
- drawing a king of clubs from a standard deck of playing cards, replacing the card, and drawing the king of clubs again **independent**

2 TEACH

In-Class Example

For Example 1

You toss a penny, nickel, dime, and quarter. What is the probability of all four landing heads? $\frac{1}{16}$

Alternative Teaching Strategies

Reading Mathematics Be sure that students understand that when the probability of two events is written as *P*(*A* and *B*) it is not possible to tell whether the events are independent or dependent events.

GLENCOE Technology

Interactive Mathematics Tools Software

In this interactive computer lesson, students explore probability. A **Computer Journal** gives students the opportunity to write about what they have learned.

For Windows & Macintosh

Sometimes the outcome of one event affects the outcome of a second event. When this happens, the events are called **dependent events**. For example, suppose you choose a card from a deck, keep it, and then your friend chooses a card. Since you did not replace your card, your friend has fewer cards from which to choose. These events are dependent.

Probability of Two Dependent Events

In words: If two events, A and B, are dependent, then the probability of both events occurring is the product of the probability of A and the probability of B after A occurs.

In symbols: $P(A \text{ and } B) = P(A) \cdot P(B \text{ following } A)$

Example 

Bag lunches are prepared for 50 students on a field trip. There are three different desserts in the bags. Twenty bags contain a brownie, 15 contain a chocolate chip cookie, and 15 contain an oatmeal raisin cookie. The bags are arranged in random order. Connie and Tale were the first and second in line. What is the probability that Connie and Tale both select a bag containing a chocolate chip cookie?

For Connie, $P(\text{chocolate chip}) = \dfrac{\textit{number containing chocolate chip}}{\textit{total number of lunches}}$

$= \dfrac{15}{50}$ or $\dfrac{3}{10}$

Assume that Connie chose a bag containing a chocolate chip cookie.

For Tale, $P(\text{chocolate chip}) = \dfrac{\textit{number containing chocolate chip left}}{\textit{number of lunches left}}$

$= \dfrac{14}{49}$ or $\dfrac{2}{7}$

$P(\text{chocolate chip cookies for both}) = \dfrac{3}{10} \cdot \dfrac{2}{7}$

$= \dfrac{6}{70}$ or $\dfrac{3}{35}$

The probability that both students get a chocolate chip cookie is $\dfrac{3}{35}$.

Checking Your Understanding

Communicating Mathematics

Read and study the lesson to answer these questions. 1, 3. See margin.

1. **Explain** the difference between independent and dependent events.
2. **Write** an example of two independent and two dependent events. List each sample space. See students' work.
3. **Describe** how to find the probability of the second of two dependent events.

In-Class Example

For Example 2

You draw a card from a standard deck of cards. You do not replace the card. Then you draw another card. What is the probability that you will draw two clubs? $\frac{1}{17}$

3 PRACTICE/APPLY

Checking Your Understanding

Exercises 1–11 are designed to help you assess your students' understanding through reading, writing, speaking, and modeling. You should work through Exercises 1–3 with your students and then monitor their work on Exercises 4–11.

Additional Answers

1. Events are independent when neither affects the other and dependent when the first event does affect the following event.
3. Sample answer: Place the number of occurrences of the event in the numerator and reduce the possible occurrences by one.

Study Guide Masters, p. 89

NAME ______________ DATE ________

10-9 Study Guide
Probability of Independent and Dependent Events

Student Edition Pages 530–534

If the outcome of one event does *not* influence the outcome of a second event, the events are **independent**.

Example: A jar contains 12 red bells and 12 silver bells. Pick one, replace it, and pick another. The probability of picking a silver bell twice is

$\frac{1}{2} \times \frac{1}{2}$ or $\frac{1}{4}$.

If the outcome of the second event depends on the outcome of the first event, the events are **dependent**.

Example: A jar contains 12 red bells and 12 silver bells. Pick one, keep it and pick another. The probability of picking two red bells is

$\frac{1}{2} \times \frac{11}{23}$ or $\frac{11}{46}$.

Refer to the ten buttons on the left to find the probability of each outcome. Each button is replaced.

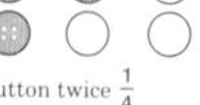

1. a white button twice $\frac{9}{100}$
2. a gray button twice $\frac{1}{4}$
3. a gray button, then a white button $\frac{3}{20}$
4. a white button, then a black button $\frac{3}{50}$
5. a black button twice $\frac{1}{25}$
6. a black button, then a gray button $\frac{1}{10}$

Refer to the ten buttons shown above to find the probability of each outcome. Each button is not replaced.

7. a white button twice $\frac{1}{15}$
8. a gray button twice $\frac{2}{9}$
9. a gray button, then a white button $\frac{1}{6}$
10. a white button, then a black button $\frac{1}{15}$
11. a black button twice $\frac{1}{45}$
12. a black button, then a gray button $\frac{1}{9}$

Reteaching

Using Models Use a bag containing 5 blue and 3 green marbles. To emphasize the difference in dependent and independent events, model problems involving replacement and nonreplacement of items drawn from the bag.

Guided Practice

4–5. See Solutions Manual for explanations.

Determine whether the events are independent or dependent. Explain.

4. rolling a die and then spinning a spinner independent
5. drawing a marble from a bag and then drawing a second marble without replacing the first one dependent

Assignment Guide

Core: 13–27 odd, 28–29, 31–37
Enriched: 12–26 even, 28–37

For **Extra Practice**, see p. 766.

The red A, B, and C flags, printed only in the Teacher's Wraparound Edition, indicate the level of difficulty of the exercises.

The chart below lists the number and type of two soft drinks found in a tub filled with ice. Two cans are selected at random. Find the probability of each outcome.

6. $\frac{7}{95}$
7. $\frac{21}{380}$
8. $\frac{21}{190}$
9. $\frac{9}{190}$

Soda	Regular	Diet
Coca-Cola	3	4
Mountain Dew	7	6

6. a Diet Coke and a regular Mountain Dew
7. a regular Coke and a regular Mountain Dew
8. a diet Mountain Dew and a regular Mountain Dew
9. a regular Coke and a diet Mountain Dew
10. Do the probabilities in Exercises 6–9 represent dependent or independent events? Explain. Dependent; one outcome affects the other.
11. **Economics** In 1993, 44% of full-time, hourly wage earners in the United States were women. If 4% of these women earn the minimum wage, what is the probability that a full-time, hourly wage earner chosen at random for a survey is a woman who is paid the minimum wage? $\frac{11}{625}$ or 1.76%

Exercises: Practicing and Applying the Concept

Independent Practice

12. dependent

14. independent

Determine whether the events are independent or dependent. Explain.

12. choosing a card from a deck of playing cards and then choosing a second card without replacing the first card
13. rolling a die twice independent
14. selecting a name from the Raleigh telephone book and a name from the Miami telephone book
15. tossing a coin and spinning a spinner independent
16. as team captain, selecting someone to be on your basketball team and then selecting a second player dependent

A die is rolled and the spinner is spun. Find each probability.

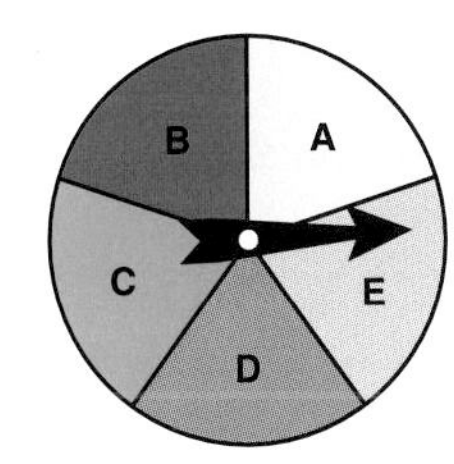

17. P(1 and D) $\frac{1}{30}$
18. P(an odd number and C) $\frac{1}{10}$

19. P(a composite number and a vowel) $\frac{2}{15}$
20. P(an even number and a consonant) $\frac{3}{10}$
21. P(a 6 and F) 0

Practice Masters, p. 89

NAME ____________ DATE ______

10-9 Practice
Probability of Independent and Dependent Events
Student Edition Pages 530–534

Each spinner is spun once. Find each probability.

1. P(A and 1) $\frac{1}{12}$
2. P(C and 2) $\frac{1}{12}$
3. P(B and 3) $\frac{1}{24}$
4. P(A and 4) $\frac{1}{6}$
5. P(C and 3) $\frac{1}{24}$
6. P(B and 2) $\frac{1}{12}$

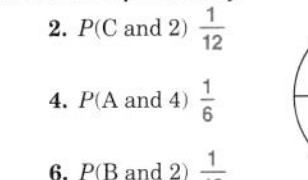
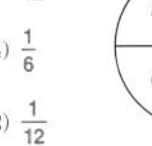
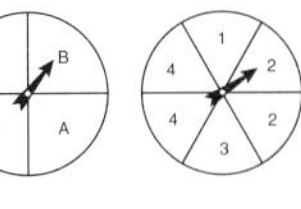

7. P(a consonant and an odd number) $\frac{1}{6}$
8. P(a consonant and a prime number) $\frac{1}{4}$
9. P(a vowel and a 5) 0
10. P(a vowel and a number less than 3) $\frac{1}{4}$

In a bag, there are 4 red marbles, 5 white marbles, and 6 blue marbles. Once a marble is selected, it is not replaced. Find the probability of each outcome.

11. a red marble and then a white marble $\frac{2}{21}$
12. a blue marble and then a red marble $\frac{4}{35}$
13. 2 red marbles in a row $\frac{2}{35}$
14. 2 blue marbles in a row $\frac{1}{7}$
15. a red marble three times in a row $\frac{4}{455}$
16. a white marble three times in a row $\frac{2}{91}$
17. a blue marble, a white marble, and then a red marble $\frac{4}{91}$
18. a blue marble three times in a row $\frac{4}{91}$

Group Activity Card 10-9

In The Bag — Group Activity **10-9**

MATERIALS: Paper bag containing the following pieces from a set of pattern blocks: 3 triangles, 5 hexagons, 5 squares, 6 trapezoids, 7 rhombuses, a set of three cards with the numbers 2, 3, 4

Everyone first should calculate the probability of drawing one specific piece from the bag. Then he or she draws a piece from the bag, shows it to everyone, and returns it to the bag. Do this the number of times indicated on your card.

The other players calculate the probability of the pieces obtained by the first player.

The first player with the correct answer scores the point. Play continues until everyone has had one turn. The person with the most points wins.

 Pre-Algebra

4 ASSESS

Closing Activity

Speaking Use a standard deck of playing cards and have students describe how to determine each outcome. Then find the probability of that outcome.

- Independent events, drawing a king and drawing a jack **Draw a card. Replace it. Draw again. $\frac{1}{169}$**
- Dependent events, drawing a king and drawing a jack **Draw a card. Do not replace it. Draw again. $\frac{4}{663}$**
- Independent events, drawing a heart and drawing another heart **Draw a card. Replace it. Draw again. $\frac{1}{16}$**
- Dependent events, drawing a heart and drawing another heart **Draw a card. Do not replace it. Draw again. $\frac{1}{17}$**

Additional Answers

28a. Sample answer: 3 red, 2 white, and 4 blue.

28b. Sample answer: The numerator must be 24. Any combination of 3, 2, and 4 will have a probability of $\frac{1}{21}$.

Enrichment Masters, p. 89

NAME ______________________ DATE ____________

10-9 Enrichment

Probability of Dependent Events

Student Edition Pages 530–534

Look at the letters in the word MATHEMATICAL. If these letters were placed in a hat, what would be the probability of drawing a vowel and then, without replacing the vowel, drawing a consonant? These are **dependent events** since the letter selected on the first draw affects the probability for the second draw.

$$P(\text{vowel, then consonant}) = \frac{5}{12} \cdot \frac{7}{11} = \frac{35}{132}$$

Find the probability of drawing each of the following from the letters in MATHEMATICAL if the letters are not replaced.

1. two M's $\frac{1}{66}$
2. two A's $\frac{1}{22}$
3. three A's $\frac{1}{220}$
4. three vowels $\frac{1}{22}$
5. five consonants $\frac{7}{264}$
6. the letters MATH in that order $\frac{1}{990}$

Now, think of using variables instead of numbers. This is very useful, since this is the way formulas are developed. Once a formula is found, it can be used for any numbers. Begin by examining the following example.

Example: Three of ten socks in a box are blue. If socks are drawn without looking and not replaced, what is the probability of picking three blue socks in three drawings?

$$\frac{3}{10} \cdot \frac{2}{9} \cdot \frac{1}{8} = \frac{6}{720}, \text{ or } \frac{1}{120}$$

7. If box containing n socks has k blue ones, what is the probability of picking three blue socks in three drawings? $\frac{k}{n} \cdot \frac{k-1}{n-1} \cdot \frac{k-2}{n-2}$
8. If a box containing n socks has k blue ones, what is the probability of picking x blue socks in x drawings? $\frac{k}{n} \cdot \frac{k-1}{n-1} \cdot \frac{k-2}{n-2} \cdots \frac{k-(x-1)}{n-(x-1)}$
9. Use your formula from Exercise 7 to find the probability of picking three blue socks in three drawings from a box containing six socks, four of them blue. $\frac{4}{6} \cdot \frac{3}{5} \cdot \frac{2}{4} = \frac{1}{5}$
10. Use your formula from Exercise 8 to find the probability of picking four blue socks in four drawings from a box containing six socks, five of them blue. $\frac{5}{6} \cdot \frac{4}{5} \cdot \frac{3}{4} \cdot \frac{2}{3} = \frac{1}{3}$

In a bag there are 5 red marbles, 2 green marbles, 4 yellow marbles, and 3 blue marbles. Once a marble is drawn, it is not replaced. Find the probability of each outcome.

23. $\frac{10}{91}$

25. $\frac{3}{91}$

22. two green marbles in a row $\frac{1}{91}$

23. a red and then a yellow marble

24. two red marbles in a row $\frac{10}{91}$

25. a green and then a blue marble

26. a red marble, a green marble, and then a blue marble $\frac{5}{364}$

27. three yellow marbles in a row $\frac{1}{91}$

Critical Thinking

28. There are 9 marbles in a bag. Some of the marbles are red, some are white, and some are blue. The probability of selecting a red marble, a white marble, and then a blue marble is $\frac{1}{21}$.

   a. How many of each color are in the bag? **See margin.**

   b. Explain why there is more than one correct answer to part a. **See margin.**

Applications and Problem Solving

Choose

Estimation
Mental Math
Calculator
Paper and Pencil

29. TV Game Shows On *The Price is Right*, contestants can win a car if they draw the digits in the price of the car from a bag that contains the five digits in the price of the car and three strikes. Once a number or strike is drawn, it is not replaced.

   a. What is the probability that the first draw is a strike? $\frac{3}{8}$

   b. A contestant has drawn three digits in the price of the car without drawing a strike. What is the probability that the next two draws will be strikes? $\frac{3}{10}$

30. Ambulance Service Mercy Hospital has two ambulances that are each available for emergencies 80% of the time.

   a. In an emergency situation, what is the probability that both ambulances will be available? $\frac{16}{25}$

   b. What is the probability that neither will be available? $\frac{1}{25}$

   c. Explain why the probabilities for parts a and b do not have a sum of 1. **See margin.**

31. Aerospace Design The electronic telemetry system on a spacecraft has three motion sensors, a signal conditioner, and a transmitter. The probability of failure for each motion sensor and the signal conditioner is 0.0001, and the probability of failure for the transmitter is 0.001. The systems run independently. What is the probability of *success* for the telemetry system? $\frac{4993}{5000}$ **or 99.86%**

Mixed Review

32. Sports Colin kicks extra points for the football team. He makes 75% of his attempts for extra points. How could you simulate the results of the next six attempts? (Lesson 10-8) **See margin.**

33. Sample answer: $10; about $40

33. Consumer Awareness Keandre is shopping for new tennis shoes. He finds a pair that are priced at $49.99 on a sale rack marked 20% off. Estimate the amount of savings and the sale price. (Lesson 9-8)

34. $\frac{1}{3}$

34. Evaluate $c - a$ if $a = \frac{2}{9}$ and $c = \frac{5}{9}$. Write the result in simplest form. (Lesson 5-4)

35. Solve $14 = b - (-12)$ and check your solution. Then graph the solution on a number line. (Lesson 3-2) **2; See margin for graph.**

36. State whether the product of $-3(5)(-2)$ is positive or negative. Then find the product. (Lesson 2-7) **positive; 30**

37. Write an expression for *twice the sum of a number and 3.* (Lesson 1-3) **$2(n + 3)$**

Extension

Using Connections Post a U.S. map. Have cooperative groups write probability problems relating independent information about the states. Trade and solve. For example, what is the probability that a state name begins with the letter "A" and the state borders the Gulf of Mexico?

Additional Answers

30c. Sample answer: Some of the time one ambulance is available when the other is not.

32. Sample answer: Use a spinner with four equal sections and let one section represent a miss.

35. (number line from 0 to 6 with point at 2)

0 1 2 3 4 5 6

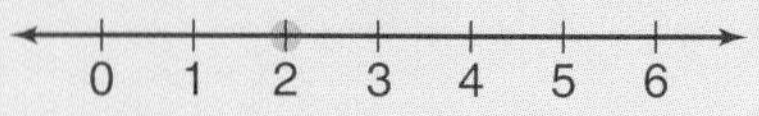

10-10 Probability of Compound Events

Setting Goals: *In this lesson, you'll find the probability of mutually exclusive events or inclusive events.*

Modeling with Manipulatives

MATERIALS

colored pencils

Suppose you are playing a game of *Monopoly* with three friends. It's your turn and you need to roll a 6 or 10 to avoid landing on properties owned by the other players and having to pay them rent. What is the probability that you can avoid paying rent?

Your Turn ▶ The chart at the right is a partial list of the possible sums when two dice are rolled. Copy and complete the chart. Circle the combinations with a sum of 6 with a colored pencil. Then circle those with a sum of 10 using a second color.

Second Die (columns); First Die (rows)

+	1	2	3	4	5	6
1	2	3	4	5	(6)	7
2	3	4	5	(6)	7	8
3	4	5	(6)	7	8	9
4	5	(6)	7	8	9	(10)
5	(6)	7	8	9	(10)	11
6	7	8	9	(10)	11	12

a. What is the probability the sum of the dice will be 6? $\frac{5}{36}$

b. What is the probability the sum of the dice will be 10? $\frac{1}{12}$

c. What is the probability the sum of the dice will be 6 or 10? $\frac{2}{9}$

d. What do you notice about the probabilities in parts a, b, and c? **See margin.**

Learning the Concept

When two events cannot happen at the same time, like rolling a 6 or a 10, they are said to be **mutually exclusive**. The probability that one of two mutually exclusive events will occur is the sum of their individual probabilities.

Probability of Mutually Exclusive Events

In words: The probability of one or the other of two mutually exclusive events can be found by adding the probability of the first event to the probability of the second event.

In symbols: $P(A \text{ or } B) = P(A) + P(B)$

Alternative Learning Styles

Auditory Have students give verbal descriptions of events, such as drawing a four and a heart and have other students respond as to whether the event is mutually exclusive or is inclusive.

Additional Answer Talk About It

d. The sum of the probabilities of pairs a and b equals the probability in part c.

10-10 LESSON NOTES

NCTM Standards: 1-4, 11

Instructional Resources

- Study Guide Master 10-10
- Practice Master 10-10
- Enrichment Master 10-10
- Group Activity Card 10-10
- Assessment and Evaluation Masters, p. 268

Transparency 10-10A contains the 5-Minute Check for this lesson;
Transparency 10-10B contains a teaching aid for this lesson.

Recommended Pacing	
Standard Pacing	Day 12 of 14
Honors Pacing	Day 11 of 13
Block Scheduling*	Day 6 of 7 (along with Lesson 10-9)

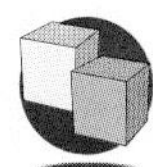

*For more information on pacing and possible lesson plans, refer to the ***Block Scheduling Booklet***.

1 FOCUS

5-Minute Check
(over Lesson 10-9)

There are 5 red marbles, 2 white marbles, and 3 blue marbles in a bag. Find the probability of each outcome.

1. two red marbles in a row if the first is replaced before drawing the second $\frac{1}{4}$
2. a white marble and a blue marble in that order if the first is replaced before drawing the second $\frac{3}{50}$
3. two red marbles in a row if the first is not replaced before drawing the second $\frac{2}{9}$
4. a white marble and a blue marble in that order if the first is not replaced before drawing the second $\frac{1}{15}$
5. a red marble and a green marble if the first is not replaced before drawing the second **0**

Motivating the Lesson

Hands-On Activity Use a standard deck of playing cards. On the board write two headings:

Cannot happen at the same time	Can happen at the same time

Have volunteers show one card, if possible, that would satisfy each pair of conditions and write those conditions under the proper heading.

- a queen and a red card **can**
- a king and a queen **cannot**
- a heart and a red card **can**
- a club and a red card **cannot**

2 TEACH

In-Class Example

For Example 1
You draw a card from a standard deck of playing cards. What is the probability that the card will be a nine or a heart?
$\frac{1}{13} + \frac{1}{4} - \frac{1}{52} = \frac{4}{13}$

Teaching Tip You can use Venn diagrams to show whether two events are mutually exclusive.

Study Guide Masters, p. 90

NAME ______________________ DATE ____________

10-10 Study Guide
Probability of Compound Events

Student Edition
Pages 535–538

When two events *cannot* happen at the same time, they are **mutually exclusive**. For two events that are mutually exclusive, $P(A \text{ or } B) = P(A) + P(B)$.

Example: A die is tossed. Find $P(5 \text{ or } 6)$.

The events are exclusive. The die cannot land on 5 and 6 at the same time.

$$P(5 \text{ or } 6) = P(5) + P(6) = \frac{1}{6} + \frac{1}{6} = \frac{2}{6} \text{ or } \frac{1}{3}$$

When two events *can* occur at the same time, they are **inclusive**. For two events that are inclusive, $P(A \text{ or } B) = P(A) + P(B) - P(A \text{ and } B)$

Example: A die is tossed. Find $P(\text{even or greater than } 4)$.

The events are inclusive. It is possible to land on a number that is both even and greater than 4.

$$P(\text{even or} > 4) = P(\text{even}) + P(> 4) - P(\text{even and} > 4) = \frac{3}{6} + \frac{2}{6} - \frac{1}{6} = \frac{4}{6} \text{ or } \frac{2}{3}$$

Determine whether each event is mutally exclusive or inclusive. The find the probability.

1. The drawing at the right shows the 18 best seats at a concert hall. The tickets for these seats are given away during a raffle.

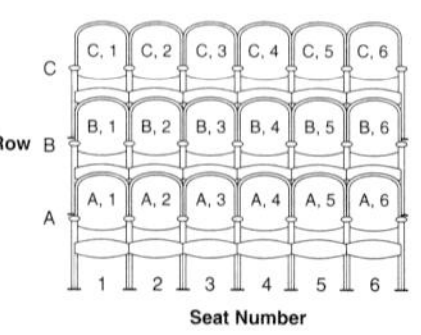

   a. a seat in row A or row C **exclusive; $\frac{2}{3}$**
   b. a seat in row A or a seat numbered 3 **inclusive; $\frac{4}{9}$**
   c. a seat numbered 2 or 5 **exclusive; $\frac{1}{3}$**
   d. a seat in row A, B, or C **exclusive; 1**
   e. a seat numbered 1, 2, or 6 **exclusive; $\frac{1}{2}$**
   f. an aisle seat or a seat in row C **exclusive; $\frac{5}{9}$**
   g. a seat in row C or a prime-numbered seat **inclusive; $\frac{2}{3}$**
   h. an odd-numbered seat or a seat in row C **inclusive; $\frac{2}{3}$**
   i. a seat *not* in row B or a seat numbered 2 **inclusive; $\frac{13}{18}$**

Example 1 **The spinner at the right is spun. What is the probability that the spinner will stop on yellow or on 2?**

THINK ABOUT IT

What is the probability of two events both occurring when they are mutually exclusive? **0**

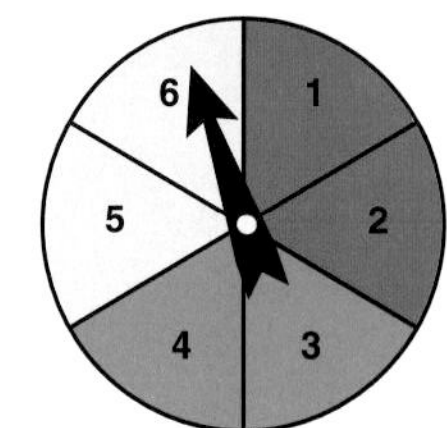

The events are mutually exclusive because the spinner cannot stop on both yellow and 2 at the same time.

$$P(\text{yellow or } 2) = P(\text{yellow}) + P(2) = \frac{1}{3} + \frac{1}{6} = \frac{3}{6} \text{ or } \frac{1}{2}$$

The probability that the spinner stops on yellow or 2 is $\frac{1}{2}$.

When two events are **inclusive**, they can happen at the same time. For example, from a standard deck of 52 playing cards, it is possible to draw a card that is *both* a queen and a red card. What is the probability of drawing a queen or a red card?

$P(\text{queen})$	$P(\text{red card})$	$P(\text{red queen})$
$\frac{4}{52}$	$\frac{26}{52}$	$\frac{2}{52}$
1 queen in each suit	*13 hearts and 13 diamonds*	*queen of hearts and queen of diamonds*

The probability of drawing a red queen is counted twice, once for a queen and once for a red card. To find the correct probability, you must subtract $P(\text{red queen})$ from the sum of $P(\text{queen})$ and $P(\text{red card})$.

$$P(\text{queen or red card}) = \frac{4}{52} + \frac{26}{52} - \frac{2}{52} \text{ or } \frac{7}{13}$$

The probability of drawing a queen or a red card is $\frac{7}{13}$.

Probability of Inclusive Events

In words: **The probability of one or the other of two inclusive events can be found by adding the probability of the first event to the probability of the second event and subtracting the probability of both events happening.**

In symbols: $P(A \text{ or } B) = P(A) + P(B) - P(A \text{ and } B)$

Example 2

Business

An auto dealer finds that of the cars coming in for service, 70% need a tune up, 50% need a new air filter, and 35% need both. What is the probability that a car brought in for service needs either a tune-up or a new air filter?

Since it is possible to get a tune-up *and* a new air filter, these events are inclusive.

$P(\text{tune-up}) = \frac{70}{100}$ $\quad P(\text{air filter}) = \frac{50}{100}$

$P(\text{tune-up and air filter}) = \frac{35}{100}$

$$P(\text{tune-up or air filter}) = \frac{70}{100} + \frac{50}{100} - \frac{35}{100} \text{ or } \frac{85}{100}$$

The probability that a car needs a tune-up or air filter is $\frac{85}{100}$ or 85%.

Reteaching

Using Manipulatives Have each student create a spinner meeting these requirements.

- All sections on the spinner are of the same size. (The number of sections is left to the student.)
- The numerical outcomes are to include at least two and at most all of the digits 1, 2, 3, and 4.
- The colors are to include at least one and at most all of these colors: red, yellow, green, blue.

Have each student show his or her spinner and describe at least one event that is mutually exclusive and one that is inclusive.

Checking Your Understanding

Communicating Mathematics

Read and study the lesson to answer these questions. 1–3. See margin.

1. **Write** an example of two mutually exclusive events in your own life.
2. **Describe** the difference between mutually exclusive and inclusive events.
3. **Explain** why you must subtract the probability of both events happening when finding the probability of two events that are not mutually exclusive.

Guided Practice

Determine whether each event is *mutually exclusive* or *inclusive*.

4. A student is selected at random from a group of male and female students in the seventh, eighth, or ninth grade.
   a. *P*(male or ninth grader)
   b. *P*(male or female) exclusive
   c. *P*(female or eighth grader)
   d. *P*(seventh or eighth grader)

4a. inclusive
4c. inclusive
4d. exclusive

5. The spinner at the right is spun. Find each probability.
   a. *P*(C or D)
   b. *P*(E or vowel)
   c. *P*(B or consonant)

5a. $\frac{1}{3}$
5b. $\frac{1}{3}$
5c. $\frac{2}{3}$
6b. 55%

6. **Advertising** In April 1995, a poll asked 1225 adults across the United States whether they liked a certain television commercial for a long-distance service. Those polled were also asked how effective they thought the commercial was in helping to sell the service. The results of the second question are shown in the graph at the right.

Source: *USA-Today*

   a. If a person is selected at random, what is the probability that he or she would say the commercial was not effective or somewhat effective? 88%
   b. If a person is selected at random, what is the probability that he or she would say the commercial was somewhat or very effective?

Practicing and Applying the Concept

Independent Practice

Determine whether each event is *mutually exclusive* or *inclusive*. Then find the probability.

7. An eight-sided die is tossed.
   a. *P*(even or less than 5)
   b. *P*(6 or odd) exclusive; $\frac{5}{8}$
   c. *P*(4 or prime) exclusive; $\frac{5}{8}$
   d. *P*(greater than 4 or an 8)

7a. inclusive; $\frac{3}{4}$
7d. inclusive; $\frac{1}{2}$

8. A card is drawn from the cards at the right.
   a. *P*(6 or even) inclusive; $\frac{2}{7}$
   b. *P*(9 or even) exclusive; $\frac{3}{7}$
   c. *P*(odd or even) exclusive; 1
   d. *P*(5 or less than 3) exclusive; $\frac{2}{7}$
   e. *P*(4 or greater than 3) inclusive; $\frac{5}{7}$

1 3 4 5
6 7 9

In-Class Example

For Example 2

Of a sample of students, 50% want to play in the band, 35% want to play in the orchestra, and 15% are interested in playing in both. If a student was selected at random, what is the probability that he or she would be interested in playing in the band or in the orchestra? **70%**

3 PRACTICE/APPLY

Checking Your Understanding

Exercises 1–6 are designed to help you assess your students' understanding through reading, writing, speaking, and modeling. You should work through Exercises 1–3 with your students and then monitor their work on Exercises 4–6.

Assignment Guide

Core: 7–21 odd, 22, 23, 25–29
Enriched: 8–20 even, 22–29

For **Extra Practice**, see p. 767.

Practice Masters, p. 90

NAME ______ DATE ______

10-10 Practice
Probability of Compound Events

Student Edition
Pages 535–538

Determine whether each event is mutually exclusive or inclusive.

1. Alicia selects at random from a box of thin and thick crust pizza. Each slice has a topping of mushrooms, pepperoni, or sausage.
   A. *P*(sausage or mushrooms) exclusive
   B. *P*(thin crust or pepperoni) inclusive
   C. *P*(sausage or thick crust) inclusive
   D. *P*(thick or thin crust) exclusive

Determine whether each event is mutually exclusive or inclusive. Then find the probability.

2. A die is rolled.
   A. *P*(odd or greater than 2) inclusive; $\frac{5}{6}$
   B. *P*(odd or prime) inclusive; $\frac{2}{3}$
   C. *P*(even or odd) exclusive; 1
   D. *P*(1 or 6) exclusive; $\frac{1}{3}$
   E. *P*(less than 3 or even) inclusive; $\frac{2}{3}$
   F. *P*(5 or less than 2) exclusive; $\frac{1}{3}$

3. A card is drawn from the bag at the right.
   A. *P*(3 or less than 2) exclusive; $\frac{1}{4}$
   B. *P*(even or prime) inclusive; $\frac{7}{8}$
   C. *P*(6 or 8) exclusive; $\frac{1}{4}$
   D. *P*(1 or odd) inclusive; $\frac{1}{2}$
   E. *P*(odd or greater than 5) inclusive; $\frac{3}{4}$
   F. *P*(8 or less than 8) exclusive; 1

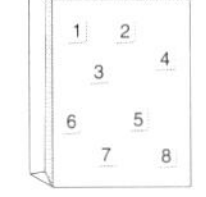

Group Activity Card 10-10

Sweet Probabilities Group Activity **10-10**

MATERIALS: Two bags of candy. One contains four flavors (any combination) to make 35 pieces. The other bag contains three flavors (any combination) to make 21 pieces • one note card with the words "same bag" • one note card with the words "different bags"

One group member draws a card telling him or her to use the same or different bags and draws two pieces of candy from the bag or bags. If one bag is indicated, the member may either replace the candy or draw two pieces without replacement.

The rest of the group then calculates the probability of the draw. When all have finished, the answers are compared and discussed. When all members understand how the correct probability is computed, the cards and bag are passed to the next member.

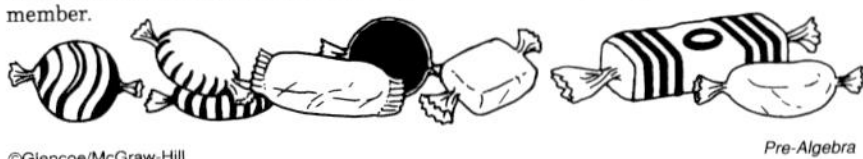

 Pre-Algebra

Additional Answers

1. Answers will vary. Sample answer: walking the dog and practicing the piano.
2. Sample answer: Mutually exclusive events cannot occur at the same time while inclusive events can occur at the same time.
3. Sample answer: The occurrences have already been counted in the probability of the separate events.

4 ASSESS

Closing Activity

Writing Have students write descriptions events that are mutually exclusive and are inclusive. Students are to explain why each event is classified as such.

Chapter 10, Quiz D (Lessons 10-9 and 10-10) is available in the *Assessment and Evaluation Masters*, p. 268.

Additional Answers

22.

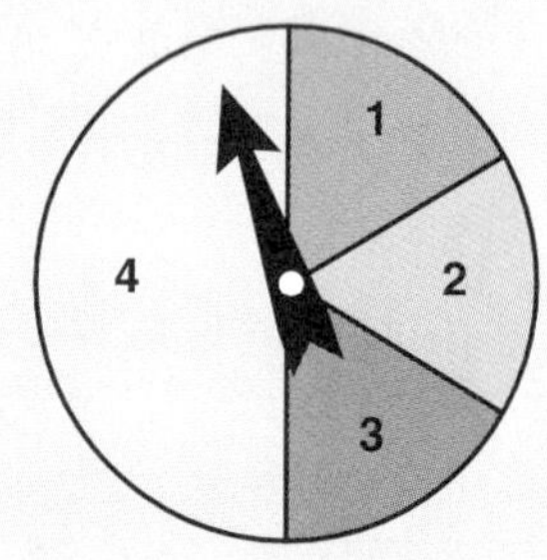

27.

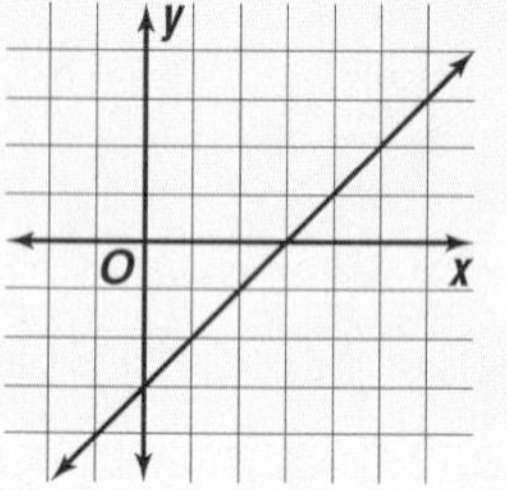

Enrichment Masters, p. 90

NAME ______________________ DATE __________

10-10 Enrichment
Adding Probabilities

Student Edition Pages 535–538

You will need a map of the United States and a list of the 50 states to complete this worksheet.

Assume that the names of the 50 states are written on slips of paper and put into a hat, and each state has the same chance of being picked. Find the probability of selecting each of the following.

1. a state whose name begins with the letter A or the letter W $\frac{4}{50} + \frac{4}{50} = \frac{4}{25}$
2. a state whose name begins with the letter C or the letter N $\frac{3}{50} + \frac{8}{50} = \frac{11}{50}$
2. a state whose name begins with the letter S or a state whose name is two words $\frac{2}{50} + \frac{10}{50} - \frac{2}{50} = \frac{1}{5}$
4. a state whose name begins with the letter I or a state that is entirely east of the Mississippi River $\frac{4}{50} + \frac{26}{50} - \frac{2}{50} = \frac{14}{25}$
5. a state whose name is one syllable or a state whose name is more than two syllables $\frac{1}{50} + \frac{43}{50} = \frac{22}{25}$
6. a state with double letters (side by side) in its name or a state whose name begins with a vowel $\frac{9}{50} + \frac{12}{50} - \frac{1}{50} = \frac{2}{5}$
7. a state that borders on an ocean or a state whose name begins with the letter N $\frac{19}{50} + \frac{8}{50} - \frac{4}{50} = \frac{23}{50}$
8. a state whose name begins and ends with the same letter or a state that is larger in size than Texas $\frac{4}{50} + \frac{1}{50} - \frac{1}{50} = \frac{2}{25}$
9. a state that borders on the Pacific Ocean or a state that was one of the 13 original colonies $\frac{5}{50} + \frac{13}{50} = \frac{9}{25}$
10. a state that borders Wyoming or a state that borders Idaho $\frac{6}{50} + \frac{6}{50} - \frac{2}{50} = \frac{1}{5}$

Suppose $P(A) = \frac{1}{5}$, $P(B) = \frac{2}{3}$, and A and B are mutually exclusive.

9. What is P(A and B)? **0**

10. What is P(A or B)? $\frac{13}{15}$

Suppose $P(A) = \frac{1}{5}$, $P(B) = \frac{2}{3}$, and A and B are independent.

11. What is P(A and B)? $\frac{2}{15}$

12. What is P(A or B)? $\frac{11}{15}$

Two cards are drawn from a standard deck of cards. Find each probability.

14. $\frac{188}{663}$

B

13. P(both aces or both jacks) $\frac{2}{221}$

14. P(both red or both face cards)

15. A bag contains six blue marbles and three red marbles. A marble is drawn, it is replaced and another marble is drawn. What is the probability of drawing a red marble and a blue marble in either order? $\frac{4}{9}$

An eight-sided die is tossed 40 times. Determine how many times you would expect each outcome.

C

16. a 5 or a 7 **about 10 times**

17. a 6 or a 3 **about 10 times**

18. an even number or a 5 **about 25 times**

19. a prime number or an even number **about 35 times**

20. a multiple of 3 or a multiple of 2 **about 25 times**

21. a multiple of 2 or a multiple of 4 **about 20 times**

Critical Thinking

22. Design and draw a spinner containing the numbers 1, 2, 3, and 4 so that the probability of spinning a 3 or a 4 is $\frac{2}{3}$. **See margin.**

Applications and Problem Solving

23. Economics Thirty-one percent of minimum-wage workers are between 16 and 19 years old. Twenty-two percent of minimum-wage workers are between 20 and 24 years old. If a person who makes minimum wage is selected at random, what is the probability that he or she will be between 16 and 24 years old? **53%**

24. $\frac{5}{46}$

24. Government Florida has 23 members in the United States House of Representatives. Eight members are Democrats, and 15 are Republicans. Of the 17 men in the house, 4 are Democrats. The Speaker of the House wants to choose a representative from Florida at random to serve on the agriculture committee. What is the probability that the representative will be a woman or a Democrat?

Mixed Review

25. Probability Suppose you choose a card from a deck of playing cards, returning the card, and then choose another card from the deck. Are these independent or dependent events? Explain. (Lesson 10-9) **independent**

26. No, the unit price is 9.5¢ per ounce.

26. Consumer Awareness Unit prices are often displayed for products at grocery stores. Suppose a 13-ounce box of breakfast cereal costs \$3.69 and a unit price of 0.95¢ per ounce is displayed. Is this the correct unit price? Explain your answer. (Lesson 9-1)

27. Graph $y = x - 3$. (Lesson 8-3) **See margin.**

28. $n \leq 28$; See margin for graph.

28. Write an inequality for "the number of students in each class is less than or equal to 28." Then draw a graph of the inequality. (Lesson 3-6)

29. Find $15 \cdot 20$ mentally using the distributive property. (Lesson 1-5) **300**

Extension

Using Connections Supply students with a list of state representatives, including name, gender, and political affiliation. If a representative is chosen at random, find *P*(man or a Republican), and so on.

Additional Answer

28.

0 7 14 21 28 35

CHAPTER 10 Highlights

Vocabulary

After completing this chapter, you should be able to define each term, property, or phrase and give an example or two of each.

Discrete Mathematics
combination (p. 516)
factorial (p. 515)
Fundamental Counting Principle (p. 509)
outcomes (p. 509)
permutation (pp. 514, 515)
tree diagram (p. 509)

Probability
dependent events (p. 532)
experimental probability (p. 529)
inclusive (p. 536)
independent events (p. 531)
mutually exclusive (p. 535)
odds (p. 520)
relative frequency (p. 529)
simulation (p. 524)

Statistics
back-to-back stem-and-leaf plot (p. 487)
box-and-whisker plot (p. 495)
circle graph (p. 497)
comparative graph (p. 498)
histogram (p. 502)
interquartile range (p. 491)
leaves (p. 486)
line graph (p. 497)
lower quartile (p. 491)
outliers (p. 496)
pictograph (p. 497)
quartile (p. 491)
range (p. 490)
stem-and-leaf plot (p. 486)
stems (p. 486)
upper quartile (p. 491)

Problem Solving
use a simulation (p. 524)

Understanding and Using Vocabulary

Choose the letter of the term that best matches each statement or phrase.

1. two events that can happen simultaneously C
2. the difference between the least number and the greatest number in a set of numbers F
3. an arrangement or listing in which order is important I
4. the median of the upper half of a set of data D
5. two events that cannot happen at the same time A
6. the range of the middle half of a set of data E
7. an arrangement or listing where order is not important G

A. mutually exclusive
B. relative frequency
C. inclusive
D. upper quartile
E. interquartile range
F. range
G. combination
H. outliers
I. permutation

Instructional Resources

Three multiple-choice tests and three free-response tests are provided in the *Assessment and Evaluation Masters*. Forms 1A and 2A are for honors pacing, Forms 1B and 2B are for average pacing, and Forms 1C and 2C are for basic pacing. Chapter 10 Test, Form 1B is shown at the right. Chapter 10 Test, Form 2B is shown on the next page.

Highlights

Using the Chapter Highlights

The Chapter Highlights begins with a listing of the new terms, properties, and phrases that were introduced in this chapter.

Assessment and Evaluation Masters, pp. 255–256

NAME ______ DATE ______

CHAPTER 10 Test, Form 1B

1. Which number would not appear as a stem when drawing a stem-and-leaf plot of these quiz scores: 80, 75, 72, 63, 75, 77, 78, 69?
   A. 5 B. 6 C. 7 D. 8 — 1. A

Use the box-and-whisker plot below for Exercises 2–4.

20 30 40 50 60 70

2. Choose the range.
   A. 55 B. 50 C. 45 D. 70 — 2. C
3. Find the median.
   A. 25 B. 30 C. 45 D. 55 — 3. C
4. Find the lower quartile.
   A. 25 B. 30 C. 40 D. 45 — 4. B
5. On January 7 and 8, 1996, a blizzard struck the eastern United States. The table at the right shows the snowfall in some eastern cities. Which measure of variation would be most accurate in describing the snowfall from the storm over the area?

Cities	Inches of Snow
Washington, D.C.	25
Philadelphia, PA	31
New York, NY	24
Boston, MA	18
Manchester, NH	8

   A. mode B. mean C. median D. range — 5. D
6. Find the number of possible outcomes when a dime and a nickel are tossed.
   A. 2 B. 3 C. 4 D. 8 — 6. C
7. Melvin has packed four pairs of pants and four shirts for a trip. How many different outfits are possible?
   A. 256 B. 16 C. 64 D. 8 — 7. B
8. Find the value of 5!.
   A. 5 B. 25 C. 120 D. 125 — 8. C
9. What is the value of $P(5, 4)$?
   A. 9 B. 20 C. 60 D. 120 — 9. D
10. In English class, each student must select 4 short stories from a list of 5 short stories to read. How many different combinations of short stories could a student read?
   A. 5 B. 20 C. 60 D. 120 — 10. A

NAME ______ DATE ______

Chapter 10 Test, Form 1B (continued)

11. A bag contains 3 oranges, 4 pears, and 5 apples. What are the odds of drawing an orange from the bag?
   A. 1:3 B. 1:5 C. 1:4 D. 1:2 — 11. A
12. What are the odds that the spinner at the right will land on 3 when it is spun?
   A. 1:8 B. 1:7 C. 1:3 D. 3:8 — 12. B

Roland usually makes three free throws of every five attempted. To simulate the probability of Roland's making two free throws in a row, Cathy puts 25 marbles in a bag, red for a basket and blue for a miss. After a marble is drawn and recorded, it is replaced.

13. How many red marbles should Cathy use?
   A. 9 B. 15 C. 12 D. 20 — 13. B
14. In 120 drawings, how many red marbles can Cathy expect to record?
   A. 45 B. 30 C. 72 D. 85 — 14. C
15. A box contains 26 cards. Each card has a different letter of the alphabet on it. Mark draws a card at random. What is the probability of drawing either A or B?
   A. $\frac{1}{13}$ B. $\frac{1}{2}$ C. $\frac{1}{26}$ D. $\frac{1}{24}$ — 15. A
16. Two dice are rolled. Find the probability for rolling an odd number on one die and an even number on the other.
   A. $\frac{1}{6}$ B. $\frac{1}{2}$ C. $\frac{1}{4}$ D. $\frac{1}{36}$ — 16. C
17. Roman has ten cards numbered 1 to 10. What is the probability of picking two even-numbered cards one after the other, if the first card picked is not replaced?
   A. $\frac{2}{5}$ B. $\frac{2}{9}$ C. $\frac{1}{5}$ D. $\frac{1}{2}$ — 17. B
18. A bag contains 4 red, 20 blue, and 6 green marbles. Seth and Amy each pick one at random and keep it. What is the probability that they each select a red marble?
   A. $\frac{2}{300}$ B. $\frac{1}{15}$ C. $\frac{2}{145}$ D. $\frac{1}{870}$ — 18. C
19. Of the people asked to help in an election, 40% will work, 60% will donate money, and 20% will do both. Find the probability that a person chosen at random will either work or donate money, but not both.
   A. 40% B. 80% C. 120% D. 20% — 19. B
20. A die is rolled. What is the probability of rolling a 1, 3, or a 4?
   A. $\frac{1}{3}$ B. $\frac{1}{2}$ C. $\frac{1}{6}$ D. $\frac{2}{3}$ — 20. B

Using the Study Guide and Assessment

Skills and Concepts Encourage students to refer to the objectives and examples on the left as they complete the review exercises on the right.

Assessment and Evaluation Masters, pp. 261–262

NAME ______________________ DATE ________

CHAPTER 10 **Test, Form 2B**

Use the following test scores for Exercises 1–5.
71, 83, 72, 41, 75, 71, 82, 73, 72, 83

1. List the numbers which would appear in the stem of a stem-and-leaf plot of the scores. — 1. 4, 7, 8
2. List the ordered leaves which would appear on stem "7" in a stem-and-leaf plot of the scores. — 2. 1, 1, 2, 2, 3, 5
3. What is the upper quartile of the scores? — 3. 82
4. Find the lower quartile of the scores. — 4. 71
5. What is the interquartile range of the scores? — 5. 11

Use the box-and-whisker plot below for Exercises 6–8.

20 30 40 50 60 70 80 90 100

6. Find the range. — 6. 70
7. What is the upper quartile? — 7. 80
8. Find the median. — 8. 60
9. Which graph, a line graph, a box-and-whisker plot, or a pictograph, would best display data that compares dollars spent on five types of sporting equipment? Justify your answer. — 9. pictograph; shows how specific quantities compare
10. A multiple-choice test has 25 questions with four choices given for each. How many answer keys are possible? — 10. 100

Today's cafeteria menu includes four soups (chicken noodle, chicken rice, tomato, and split pea) and three choices of bread (rye, white, or wheat).

11. How many lunch choices include tomato soup? — 11. 3
12. If dessert is an apple or an orange, how many lunch choices are there that include one soup, one bread, and one dessert? — 12. 24
13. Find the value of 8!. — 13. 40,320
14. What is the value of $P(8, 4)$? — 14. 1680
15. Find the value of $C(8, 4)$. — 15. 70

NAME ______________________ DATE ________

Chapter 10 Test, Form 2B (continued)

16. In a 7-person race, gold, silver, and bronze medals are awarded to the first three finishers. In how many ways can the medals be awarded? — 16. 210
17. Three athletes from the same school compete in a cross-country event. If there is a field of 9 runners, what are the odds that the three athletes from the same school finish in the top three places? — 17. 1:83
18. The State of Massachusetts has a "Mass Cash" lotto drawing twice a week in which 5 numbers out of 36 are drawn at random. What are the odds of winning the "Mass Cash" jackpot? — 18. 1:376,991

Lauretta makes four aces out of five serves in a juniors' tennis tournament. To simulate her chances of making an ace, she puts 40 marbles in a box. A red marble represents an ace, and a blue means a a fault or a returnable serve. After a marble is drawn, it is replaced in the box.

19. How many red marbles does Lauretta need? — 19. 32
20. In 150 drawings, how many times can she expect to draw a blue marble? — 20. 30
21. Five coins are dropped on the floor. Find the probability that they will all land heads up. — 21. $\frac{1}{32}$

Darla has written the numbers 1 to 8 on eight cards of the same size. She picks two cards at random, without replacing the first one. Find the probability of each of the following.

22. drawing two odd-numbered cards — 22. $\frac{3}{14}$
23. drawing cards with numbers that are multiples of 3 — 23. $\frac{1}{28}$
24. A die is rolled. What is the probability of rolling an even number or a number less than 5? — 24. $\frac{5}{6}$
25. Suppose $P(A) = \frac{1}{2}$, $P(B) = \frac{1}{3}$, and A and B are mutually exclusive. What is $P(A \text{ or } B)$? — 25. $\frac{5}{6}$

 262 *Pre-Algebra*

CHAPTER 10 Study Guide and Assessment

Skills and Concepts

Objectives and Examples

Upon completing this chapter, you should be able to:

▶ **display and interpret data in stem-and-leaf plots.** (Lesson 10-1)

Display the data 46, 32, 59, 42, 51, 30, 49 in a stem-and-leaf plot.

3 | 2 0
4 | 6 2 9
5 | 9 1   *5|9 = 59*

Review Exercises

Use these exercises to review and prepare for the chapter test.

Make a stem-and-leaf plot of each set of data. 8–11. See Solutions Manual.

8. 3, 4, 6, 8, 17, 19, 7, 21, 40, 43
9. 10.8, 11.0, 9.8, 10.7, 11.5, 10.7, 9.5, 11.9
10. 18, 16, 20, 25, 30, 23, 30, 29, 21, 15, 18
11. 120, 111, 133, 148, 117, 112, 121, 146

▶ **use measures of variation to compare data.** (Lesson 10-2)

4 | 1 2 0 3
5 | 9 1 8
6 | 6 5   *6|6 = 6.6*

Range: 6.6 − 4.0 or 2.6
Median: 4.0 4.1 4.2 4.3 5.1 5.8 5.9 6.5 6.6
↑ median
Lower Quartile: 4.15
Upper Quartile: 6.2
Interquartile Range: 6.2 − 4.15 or 2.05

Find the range, median, upper and lower quartiles, and the interquartile range.

12. 23 45 16 51 47   12–16. See margin.
13. 6 10 15 21 28 36
14.
9 | 9 7 5
10 | 1 3 6
11 | 2 9 8 3   *11|3 = 113*
15. 205 236 217 214 200 298 240
16.
11 | 0 6 0
12 | 8 5 5
13 | 7 1 3   *12|8 = 12.8*

▶ **use box-and-whisker plots to display data.** (Lesson 10-3)

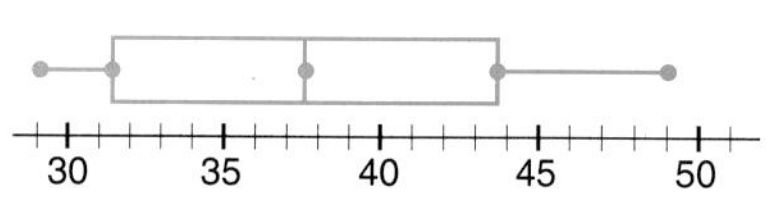

Range: 49 − 29 or 20
Median: 37.5
Upper Quartile: 43.75
Lower Quartile: 31.25
Interquartile Range: 43.75 − 31.25 or 12.50
Extremes: 29 and 49

17. Use the stem-and-leaf plot below to answer each question.

2 | 4 5 5 7 9
3 | 0 3 4 6 8
4 | 1 1 3 5
5 | 4 4   *2|4 = 24*

a. Make a box-and-whisker plot of the data. See Solutions Manual.
b. What is the median? 35
c. What are the outliers, if any? none

GLENCOE Technology

Test and Review Software

You may use this software, a combination of an item generator and an item bank, to create your own tests or worksheets. Types of items include free response, multiple choice, short answer, and open ended.

For IBM & Macintosh

Additional Answers

12. 35; 45; 19.5; 49; 29.5
13. 30; 18; 10; 28; 18
14. 14; 104.5; 95; 119; 24
15. 98; 217; 205; 240; 35
16. 2.7; 12.5; 11.3; 13.2; 1.9

Objectives and Examples

▶ **recognize when statistics are misleading.** (Lesson 10-4)

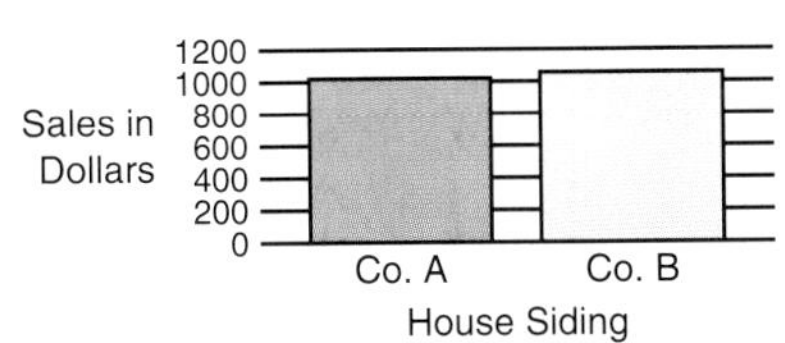

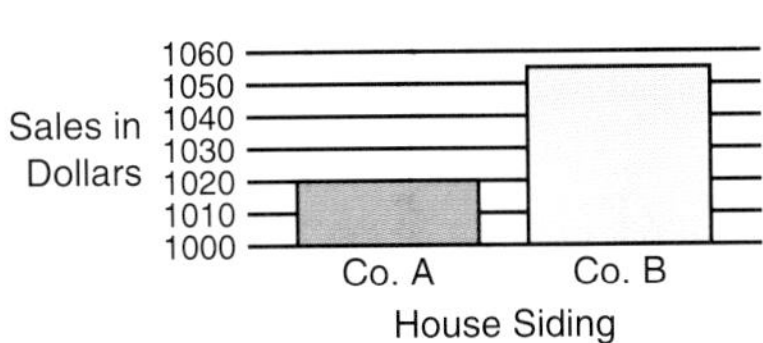

The second graph is misleading because the vertical axis does not include zero.

Review Exercises

Use the following chart to describe each category. Write the type of average you used. 20. 6.5; median

	Tina	Steve	Ryan	Katie
Favorite Sport	football	softball	soccer	soccer
Number of classes	9	10	5	3
Number of family members	3	5	12	6

18. average favorite sport soccer; mode
19. average number of classes. 6.75; mean
20. average number of family members
21. How can you describe the "average" person in this group? See margin.

▶ **use tree diagrams or the Fundamental Counting Principle to count outcomes.** (Lesson 10-5)

How many ways can you choose a shirt and tie?

shirt choices · tie choices

4 · 3 = 12

There are 12 ways to choose a shirt and tie.

Find the number of possible outcomes for each event.

22. An eight-sided die is rolled twice. 64
23. A multiple-choice quiz has 5 questions each with 4 choices. 1024
24. A 6-sided die and a coin are tossed. 12
25. A card is drawn and a coin is tossed. 104

▶ **use permutations and combinations.** (Lesson 10-6)

Seven people are being arranged for a portrait. Does this represent a permutation or a combination?

This is a permutation because the order that the photographer chooses the people determines how the picture looks.

Tell whether each situation represents a *permutation* or a *combination*.

26. five people in a row P
27. 32 cassette tapes from a group of 47 C

Find each value.

28. $P(7,4)$ 840
29. 5! 120
30. $C(8, 2)$ 28
31. $\frac{6!3!}{4!2!}$ 90

▶ **find the odds of a simple event.** (Lesson 10-7)

Green	Red	White	Blue

The odds of picking the green box is 1:3.

Find the odds of each outcome if a card is selected from a standard deck of cards.

32. a queen 1:12
33. not a face card 10:3
34. an ace or six 2:11
35. a red card 1:1

Additional Answer

21. favorite sport: soccer; 6.75 classes in school; 6.5 members in family

Classroom Vignette

"As a review for this chapter, I give each student a bag of M&M's®. They must complete a worksheet that includes a tally chart. They then answer permutation and combination questions about their data. They cannot copy another student's answer, since each bag is different."

Dawn M. Rueter
Amos Alonzo Stagg High School
Palos Hills, IL

Applications and Problem Solving

Encourage students to work through the exercises in the Applications and Problem Solving section to strengthen their problem-solving skills.

Additional Answers

46.

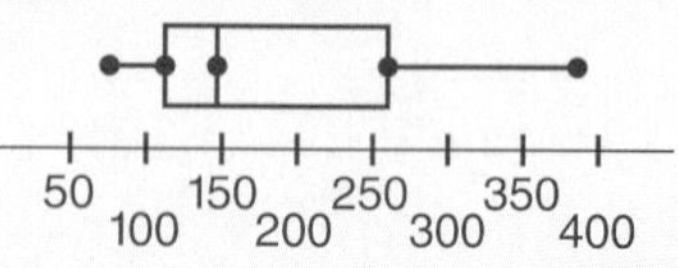

48. Sample Answer: Use a spinner with 4 equal sections. Three of the four spaces means a win. Tally 25 groups of 3 spins as either 3 wins or not.

Objectives and Examples

▸ **find the probability of independent and dependent events.** (Lesson 10-9)

Suppose you are playing a carnival game where each person chooses a plastic duck from a pond. Ten ducks are marked with numbers for prizes. Once a prize is given, the number is removed, and there is one less prize to be won. Is two people winning prizes an example of a dependent or independent event?

It is dependent because the first person winning affects the probability that the second person will win.

Review Exercises

Determine whether the events are independent or dependent. Explain.

36. choosing a recipe from a box and then choosing another without replacement
37. on the first turn you buy a Monopoly property and then on the second turn you buy a property **dependent** **36. dependent**

A bag contains 5 red, 3 yellow, and 3 purple marbles. Once a marble is drawn, it is not replaced. Find each probability.

38. P(two green marbles) **0**
39. P(a red and a yellow) $\frac{3}{22}$
40. P(two yellow) $\frac{3}{55}$
41. P(a purple and a red) $\frac{3}{22}$

▸ **find the probability of mutually exclusive events or inclusive events.** (Lesson 10-10)

Suppose you roll two dice. Are the events *rolling a six* and *rolling a sum of less than seven* mutually exclusive or inclusive?

Since you cannot roll a six and a sum of less than seven in one roll, the events are mutually exclusive.

Determine whether each event is mutually exclusive or inclusive. Then find the probability. **42. inclusive;** $\frac{11}{36}$ **43. inclusive;** $\frac{27}{36}$

42. P(rolling a six or sum greater than ten)
43. P(rolling an even number or sum is odd)

A die is tossed. Find each probability.

44. P(3 or 4) $\frac{1}{3}$
45. P(4 or even) $\frac{1}{2}$

Applications and Problem Solving

46. **Transportation** The numbers of seats in commonly-used passenger airplanes are listed below. Make a box-and-whisker plot of the data. (Lesson 10-3) **See margin.**

398	390	288	281	266	254
248	221	186	185	149	148
144	141	131	124	113	112
100	97	72			

47. **Games** Bruno's Restaurant is giving out game cards as a promotional contest. There are eight different cards available. How many different ways could you get cards on two trips to Bruno's? (Lesson 10-5) **64**

48. **Sports** A soccer team plays 16 games during the season. The team manager predicts that the team has a 75% chance of winning each game. Describe a simulation to determine the probability of the team winning three games in a row. (Lesson 10-8) **See margin.**

A practice test for Chapter 10 is available on page 783.

Alternative Assessment

Chapter 10 • Study Guide and Assessment

Cooperative Learning Activity

Work with your group to write an article for your school newspaper entitled "Are you average?" **See students' work.**

- ▶ Choose the type of information you think students in your school would find interesting. For example, you might investigate students' heights, the numbers of languages they speak, what kind of pets they own, or how long they have lived in the area.
- ▶ Next, conduct a survey in the school cafeteria to gather data.
- ▶ Present the results of your research in an interesting way. Include some graphs and tables. Be sure to explain what the results mean in the article.
- ▶ State the probability that a randomly-chosen student at your school possesses one or more characteristics. For example, how likely is it that a person speaks two languages and owns a hedgehog?

Thinking Critically

In 1994, *First for Women* magazine reported that the average American is a white woman who is 32.7 years old. She is married with children, owns a home, and has two television sets and a VCR. **See students' work.**

- ▶ Does this describe anyone you know?
- ▶ Explain what the magazine meant by "the average American."

Portfolio

Select an item from your work in this chapter that shows how your understanding of the concepts improved and place it in your portfolio. **See students' work.**

Self-Evaluation

"We always have time enough, if we but use it aright."
—Johann Wolfgang Von Goethe

Assess yourself. Do you use your time wisely? Or do you find yourself rushing to complete projects and not giving your best effort? Make a plan for the time that you have to complete the tasks you do each week such as home chores, school work, and community activities. Are there ways you could spend your time more wisely? **See students' work.**

Alternative Assessment

The Alternative Assessment section provides students with the opportunity to assess their own work by thinking critically, working with others, keeping a portfolio, and honestly evaluating their own progress. For more information on alternative forms of assessment, see *Alternative Assessment in the Mathematics Classroom*.

Performance Assessment

Performance Assessment tasks for this chapter are included in the *Assessment and Evaluation Masters*. A scoring guide is also provided.

Assessment and Evaluation Masters, pp. 265, 277

NAME ____________________ DATE __________

CHAPTER 10 Performance Assessment

Instructions: Demonstrate your knowledge by giving a clear, concise solution to each problem. Be sure to include all relevant drawings and justify your answers. You may show your solution in more than one way or investigate beyond the requirements of the problem.

1. a. Name and describe two measures of variation.

b. Produce a set of at least 8 pieces of data that has a mean of 12 and a range of 14.

c. Produce a set of 10 pieces of data that has a median of 25 and an interquartile range of 17.

d. Display the data in part c, using a stem-and-leaf plot or a box-and whisker plot. Tell what your display shows about the data.

2. Construct a misleading graph. Label the axes and give the graph a title. Explain why it is misleading.

3. Suppose you are a salesperson and wish to demonstrate to a customer all of the possible options you have available. Tell how you would do it. Use an example.

4. Probability is important in baseball. Suppose Juan Garcia and Tony Brewer are the first two batters in the ninth inning. Their respective batting averages are .200 or 20% and .250 or 25%.

a. Design a simulation to find the probability of both getting a hit.

b. Tell how to find the probability of independent events. Find the probability of both getting a hit.

c. What would the probability be of Garcia or Brewer getting a hit if they were mutually exclusive events? What would it be if they were not mutually exclusive events? Are the two events mutually exclusive? Why or why not?

CHAPTER 10 Scoring Guide

Level	Specific Criteria
3 Superior	• Shows thorough understanding of the concepts *variation, mean, median, mode, range, interquartile range, stem-and-leaf plot, box-and-whisker plot, combinations, probability, simulation,* and *independent, dependent, and mutually exclusive events.* • Uses appropriate strategies to produce sets of data with given properties. • Computations to count outcomes and determine probabilities are correct. • Written explanations are exemplary. • Graphs and plots are accurate and appropriate. • Goes beyond requirements of some or all problems.
2 Satisfactory, with minor flaws	• Shows understanding of the concepts *variation, mean, median, mode, range, interquartile range, stem-and-leaf plot, box-and-whisker plot, combinations, probability, simulation,* and *independent, dependent, and mutually exclusive events.* • Uses appropriate strategies to produce sets of data with given properties. • Computations to count outcomes and determine probabilities are mostly correct. • Written explanations are effective. • Graphs and plots are mostly accurate and appropriate. • Satisfies all requirements of problems.
1 Nearly Satisfactory, with serious flaws	• Shows understanding of most of the concepts *variation, mean, median, mode, range, interquartile range, stem-and-leaf plot, box-and-whisker plot, combinations, probability, simulation,* and *independent, dependent, and mutually exclusive events.* • May not use appropriate strategies to produce sets of data with given properties. • Computations to count outcomes and determine probabilities are mostly correct. • Written explanations are satisfactory. • Graphs and plots are mostly accurate and appropriate. • Satisfies most requirements of problems.
0 Unsatisfactory	• Shows little or no understanding of the concepts *variation, mean, median, mode, range, interquartile range, stem-and-leaf plot, box-and-whisker plot, combinations, probability, simulation,* and *independent, dependent, and mutually exclusive events.* • May not use appropriate strategies to produce sets of data with given properties. • Computations to count outcomes and determine probabilities are incorrect. • Written explanations are not satisfactory. • Graphs and plots are not accurate and appropriate. • Does not satisfy requirements of problems.

Using the Ongoing Assessment

These two pages review the skills and concepts presented in Chapters 1–10. This review is formatted to reflect new trends in standardized testing.

Assessment and Evaluation Masters, pp. 270–271

NAME ______________________ DATE ____________

CHAPTERS 1 – 10 Cumulative Review

1. Simplify $3 \cdot 5 + 2 \div 2$. (Lesson 1-2) — 1. 16
2. Solve $y = {}^-6 - 8$. (Lesson 2-5) — 2. $^-14$
3. Find the perimeter of a square 12 meters on a side. (Lesson 3-5) — 3. 48 m
4. Evaluate x^4 if $x = {}^-3$. (Lesson 4-2) — 4. 81
5. Find the LCM of 15 and 25. (Lesson 4-7) — 5. 75
6. Mel biked 9.9 miles to the park, 2.1 miles to a store, and 8.1 miles back home. Estimate how far he biked. (Lesson 5-2) — 6. 20 mi
7. Solve $x = {}^-6.2 + 8.379$. (Lesson 5-3) — 7. 2.179
8. Express $\frac{2}{11}$ as a decimal. (Lesson 6-1) — 8. $0.\overline{18}$
9. Solve $({}^-4.07)({}^-6.3) = n$. (Lesson 6-5) — 9. 25.641
10. Express 0.24 in scientific notation. (Lesson 6-9) — 10. 2.4×10^{-1}
11. Solve $3y + 4 = 40$. (Lesson 7-2) — 11. 12
12. Solve $6n - 6 = 5n + 1$. (Lesson 7-5) — 12. 7
13. Complete 3.6 m = __?__ cm. (Lesson 7-8) — 13. 360
14. State the domain and range of the relation $\{({}^-4.2, 3), (3, {}^-5.2), ({}^-6.5, 4.4)\}$. (Lesson 8-1) — 14. domain = $({}^-4.2, 3, {}^-6.5)$ range = $(3, {}^-5.2, 4.4)$
15. Find three solutions for the equation $y = {}^-3x + 3$. Write the solutions as ordered pairs. (Lesson 8-3) — 15. Sample answers: (1, 0), (2, $^-3$), (3, $^-6$)
16. Given $f(x) = 3x + 5$, determine the value for $f(7)$. (Lesson 8-4) — 16. 26
17. Find the slope of the line that contains $M(9, 10)$ and $N(6, 8)$. (Lesson 8-6) — 17. $\frac{2}{3}$
18. Find the x-intercept for the graph of $y = x + 12$. (Lesson 8-7) — 18. $^-12$
19. Write 25:475 as a fraction in simplest form. (Lesson 9-1) — 19. $\frac{1}{19}$
20. Jamal lives across the street from the 9th hole of a golf course and collects golf balls that he finds in the woods nearby. In June he collected 17 golf balls. He gave 6 to his uncle and 6 to his cousin. In July he collected 24 golf balls, and gave a dozen to his neighbor. He collected only 6 golf balls in August, and 24 in September. He gave 6 more to his neighbor and 10 to a friend. How many golf balls did Jamal still have? (Lesson 9-2) — 20. 31

NAME ______________________ DATE ____________

Cumulative Review (continued)

21. Solve $\frac{18}{12} = \frac{9}{y}$. (Lesson 9-4) — 21. 6
22. Express 3.6 as a percent. (Lesson 9-7) — 22. 360%
23. Express 55% as a fraction. (Lesson 9-7) — 23. $\frac{11}{20}$
24. What percent of 12 is 9? (Lesson 9-7) — 24. 75%
25. A used laptop computer sells for \$696. That is a 40% reduction from the original price. What was the original price of the computer? (Lesson 9-10) — 25. \$1160

The stem-and-leaf plot below shows the ages of members of a jury selection pool. (Lesson 10-1)

Stem	Leaf
2	45779
3	33788889
4	255888999
5	00344566799
6	00111355

2 | 4 = 24

26. How old is the youngest member of the pool? — 26. 24
27. What age is the most common? — 27. 38

Use the box-and-whisker plot below. (Lesson 10-3)

30 40 50 60 70 80 90

28. What is the median? — 28. 60
29. What is the upper quartile? — 29. 75
30. Find the value of 7!. (Lesson 10-6) — 30. 5040
31. Find the value of $P(9, 4)$. (Lesson 10-6) — 31. 3024
32. Find the value of $C(7, 3)$. (Lesson 10-6) — 32. 35
33. One bag contains 8 red and 3 white marbles. Another bag has 2 green and 6 black marbles. One marble is drawn from each bag. Find P(white and green). (Lesson 10-9) — 33. $\frac{3}{44}$

CHAPTERS 1–10 Ongoing Assessment

Section One: Multiple Choice

There are ten multiple-choice questions in this section. After working each problem, write the letter of the correct answer on your paper.

1. Theo wants to lose 2 pounds per week for 8 weeks. What will the change in his weight be? D

   A. -6 pounds   B. -8 pounds
   C. $+8$ pounds   D. -16 pounds

2. The high temperatures (°F) for January 15th during the past 10 years are 15°, 18°, 25°, 38°, 6°, 25°, 10°, 30°, 18°, and 25°. What are the lower and upper quartiles? B

   A. 18°, 25°   B. 15°, 25°
   C. 21.5°, 30°   D. 10°, 30°

3. Which fraction is less than $\frac{5}{12}$? D

   A. $\frac{1}{2}$ B. $\frac{3}{4}$
   C. $\frac{5}{8}$ D. $\frac{2}{6}$

4. If $m + 2.1 = 8$, what is the value of m? A

   A. 5.9   B. 8
   C. 6.9   D. 10.1

5. Mr. Garza has 4 black ties, 3 gray ties, 2 blue ties, and 1 brown tie in his closet. If he selects one tie without looking, what is the probability that it will be blue? B

   A. $\frac{1}{2}$ B. $\frac{1}{5}$
   C. $\frac{1}{10}$ D. $\frac{2}{8}$

6. Angie's Place sells made-to-order sandwiches. There are 3 kinds of bread, 3 kinds of cheese, and 4 kinds of meat. How many different combinations of bread, cheese, and meat (one of each) can be ordered? D

   A. 10   B. 18
   C. 12   D. 36

7. If $5y - 4 = 3y + 12$, what is the value of y? D

   A. 1   B. 4
   C. 2   D. 8

8. The number of students in Sam's aerobics class increased from 15 to 24. What was the percent of increase? C

   A. 37.5%   B. 62.5%
   C. 60%   D. 135%

9. A coin is tossed and a die is rolled. What is the probability of tossing tails and rolling a 5? A

   A. $\frac{1}{12}$ B. $\frac{1}{4}$
   C. $\frac{1}{10}$ D. $\frac{1}{3}$

10. Which ordered pairs are solutions for $y = 2x - 3$? A

    A. $(-3, -9), (2, 1), (4, 5)$
    B. $(-3, 0), (-1, -5), (3, 3)$
    C. $(-3, 3), (-2, 1), (-1, -1)$
    D. $(-2, -7), (-1, 1), (0, -3)$

Standardized test practice questions are provided in the *Assessment and Evaluation Masters*, p. 269.

Section Two: Free Response

This section contains eight questions for which you will provide short answers. Write your answer on your paper.

11. Barbara needs at least $7\frac{1}{8}$ pounds of fertilizer for her lawn. She has $4\frac{3}{4}$ pounds. How much more does she need? $2\frac{3}{8}$ pounds.

12. Juanita drove 350 miles and used 14 gallons of gasoline. Express her gas mileage as a unit rate. 25 miles per gallon.

13. The circumference of the circle outlining the boundary of where a discus thrower competes is 17.27 feet. What is the diameter of the circle? 5.5 feet

14. Find the slope of a line that contains (3, −4) and (−2, 2). −1.2

15. As a team, the Eagles made 24 shots in 40 times at the free throw line. What percentage did they make? 60%

16. Matt's parents have agreed to help him buy a car. They will purchase the car and he will make no-interest payments to them over the next three years. Matt can make a down payment of $1500. He cannot afford monthly payments over $200. What is the most he can pay for the car? $8700

17. When rolling a 6-sided die, what are the odds of rolling a number divisible by 3? 1:2

18. In how many different ways can four students be arranged in a row? 24 ways

Test-Taking Tip

As part of your preparation for a standardized test, review basic definitions and formulas such as the ones below.

A number is *prime* if it has no factors except itself and 1. For example, 7 is a prime number. The factors for 8 are 1, 2, 4, and 8. Therefore, 8 is not prime since it has factors other than itself and 1.

The area of a circle with radius r units is πr^2.

Section Three: Open-Ended

This section contains two open-ended problems. Demonstrate your knowledge by giving a clear, concise solution to each problem. Your score on these problems will depend on how well you do the following.

- Explain your reasoning.
- Show your understanding of the mathematics in an organized manner.
- Use charts, graphs, and diagrams in your explanation.
- Show the solution in more than one way or relate it to other situations.
- Investigate beyond the requirements of the problem.

19. Describe how you would calculate the sale price of any item. Use a specific example to demonstrate your method. See students' work.

20. Collect information from your classmates on how long it takes them to get to school. Construct an appropriate statistical graph to display the data. Explain your choice of graph. See students' work.

CHAPTER 11

Applying Algebra to Geometry

PREVIEWING THE CHAPTER

Students study basic geometry as they identify geometric figures and classify angles. Knowledge of angle measures is then used to make circle graphs to illustrate data. Angle relationships are then studied, including angles formed by two parallel lines and a transversal. In the next lesson, the missing angle measure of a triangle is found, and triangles are classified. Students then study congruent and similar triangles, and solve problems involving indirect measurement. Polygons are then explored. The study of transformations is extended to reflections, translations, and rotations.

Lesson (pages)	Lesson Objective	NCTM Standards	State/Local Objectives
11-1 (548–553)	Identify points, lines, planes, rays, segments, angles, and parallel, perpendicular, and skew lines. Classify angles as acute, right, or obtuse.	1-4, 12	
11-1B (554–555)	Construct congruent line segments and angles with a compass and straightedge.	1-4, 12	
11-2 (556–560)	Make a circle graph to illustrate data.	1-4, 10, 12	
11-3 (561–566)	Identify the relationship of vertical, adjacent, complementary, supplementary angles, and angles formed by two parallel lines and a transversal.	1-4, 12	
11-3B (567)	Use a graphing calculator to determine whether two lines are parallel.	1-4, 12	
11-4 (568–572)	Find the missing angle measure of a triangle. Classify triangles by angles and sides.	1-4, 12	
11-5 (573–577)	Identify congruent triangles and corresponding parts of congruent triangles.	1-4, 12	
11-6 (578–583)	Identify corresponding parts and find missing measures of similar triangles. Solve problems involving indirect measurement using similar triangles.	1-4, 12	
11-7 (584–588)	Find the missing angle measure of a quadrilateral. Classify quadrilaterals. Solve problems involving similar quadrilaterals.	1-4, 12	
11-8 (589–593)	Classify polygons and determine the sum of the measures of the interior and exterior angles of a polygon.	1-4, 12	
11-8B (594)	Investigate tessellations using regular polygons.	1-4, 12	
11-9 (595–599)	Identify and draw reflections, translations, and rotations. Identify and draw symmetric figures.	1-4, 12	

A complete, 1-page lesson plan is provided for each lesson in the ***Lesson Planning Guide***. Answer keys for each lesson are available in the ***Answer Key Masters***.

ORGANIZING THE CHAPTER

You may want to refer to the **Course Planning Calendar** on page T12 for detailed information on pacing.
PACING: Standard—13 days; **Honors**—13 days; **Block**—7 days

LESSON PLANNING CHART

Lesson (Pages)	Materials/ Manipulatives	Extra Practice (Student Edition)	Blackline Masters: Study Guide	Practice	Enrichment	Assessment and Evaluation	Math Lab and Modeling Math	Multicultural Activity	Tech Prep Applications	Graphing Calculator	Activity	Real-World Applications	Interactive Mathematics Tools Software	Teaching Transparencies	Group Activity Cards
11-1 (548–553)	protractor*, gum drops, toothpicks	p. 767	p. 91	p. 91	p. 91							25		11-1A 11-1B	11-1
11-1B (554–555)	compass*, ruler*, protractor*						p. 48								
11-2 (556–560)	protractor*, compass*	p. 767	p. 92	p. 92	p. 92				p. 21		p. 11	26	11-2	11-2A 11-2B	11-2
11-3 (561–566)	protractor*	p. 767	p. 93	p. 93	p. 93	p. 295					p. 25			11-3A 11-3B	11-3
11-3B (567)	graph paper									p. 24					
11-4 (568–572)		p. 768	p. 94	p. 94	p. 94								11-4	11-4A 11-4B	11-4
11-5 (573–577)	scissors, graph paper, ruler*	p. 768	p. 95	p. 95	p. 95	pp. 295, 294	p. 86						11-5A	11-5 11-5B	
11-6 (578–583)	geoboards*, graph paper, ruler*, protractor*	p. 768	p. 96	p. 96	p. 96		p. 70		p. 22					11-6A 11-6B	11-6
11-7 (584–588)	straightedge	p. 768	p. 97	p. 97	p. 97	p. 296								11-7A 11-7B	11-7
11-8 (589–593)	computer, LOGO software	p. 768	p. 98	p. 98	p. 98			p. 21			p. 43		11-8	11-8A 11-8B	11-8
11-8B (594)	tracing paper, cardboard, scissors						p. 49								
11-9 (595–599)	cardboard, scissors, ruler*, colored pencils	p. 769	p. 99	p. 99	p. 99	p. 296		p. 22		p. 11	p. 44			11-9A 11-9B	11-9
Study Guide/ Assessment (602–604)						pp. 281–293, 297–299									

*Included in Glencoe's *Student Manipulative Kit* and *Overhead Manipulative Resources*.

ORGANIZING THE CHAPTER

All of the blackline masters in the Teacher's Classroom Resources are available on the ***Electronic Teacher's Classroom Resources*** CD-ROM.

OTHER CHAPTER RESOURCES

Student Edition
Chapter Opener, pp. 546, 547
Working on the Investigation, p. 588
Closing the Investigation p. 600
The Shape of Things to Come: Virtual Reality, p. 560
From the Funny Papers, p. 599.

Teacher's Classroom Resources
Investigations and Projects Masters, pp. 69–72
Pre-Algebra Overhead Manipulative Resources, pp. 30–33

Technology
Test and Review Software (IBM & Macintosh)
CD-ROM Activities

Professional Publications

Block Scheduling Booklet
Glencoe Mathematics Professional Series

OUTSIDE RESOURCES

Books/Periodicals
Geometric Design, Dale Seymour
Informal Geometry Explorations, Margaret J. Kenney, Stanley J. Bezuska, and J. D. Martin
Cooperative Informal Geometry, Wade H. Sherard III
Tessellations Using Logo, Margaret J. Kenney and Stanley J. Bezuska

Videos/CD-ROMs
Basic Geometry Video, ETA
Math Vantage Videos: Patterns Unit/ Tessellations, Sunburst
Math Vantage Videos: Spatial Sense Unit/Trusting Triangles, Sunburst

Software
The Geometric Golfer, Dale Seymour Publications
TesselMania!™, MECC

 ASSESSMENT RESOURCES

Student Edition
Math Journal, pp. 558, 570, 592
Mixed Review, pp. 553, 560, 566, 572, 577, 583, 588, 593, 599
Self Test, p. 572
Chapter Highlights, p. 601
Chapter Study Guide and Assessment, pp. 602–604
Alternative Assessment, p. 605
Portfolio, p. 61
MindJogger Videoquiz, 11

Teacher's Wraparound Edition
5-Minute Check, pp. 548, 556, 561, 568, 573, 578, 584, 589, 595
Checking Your Understanding, pp. 551, 557, 564, 570, 576, 581, 586, 592, 597
Closing Activity, pp. 553, 560, 566, 572, 577, 583, 588, 593, 599
Cooperative Learning, pp. 579, 585

Assessment and Evaluation Masters
Multiple-Choice Tests, Forms 1A (Honors), 1B (Average), 1C (Basic), pp. 281–286
Free-Response Tests, Forms 2A (Honors), 2B (Average), 2C (Basic), pp. 287–292
Performance Assessment, p. 293
Mid-Chapter Test, p. 294
Quizzes A-D, pp. 295, 296
Standardized Test Practice, p. 297
Cumulative Review, pp. 298–299

ENHANCING THE CHAPTER

Examples of some of the materials for enhancing Chapter 11 are shown below.

DIVERSITY

Multicultural Activity Masters, pp. 21, 22

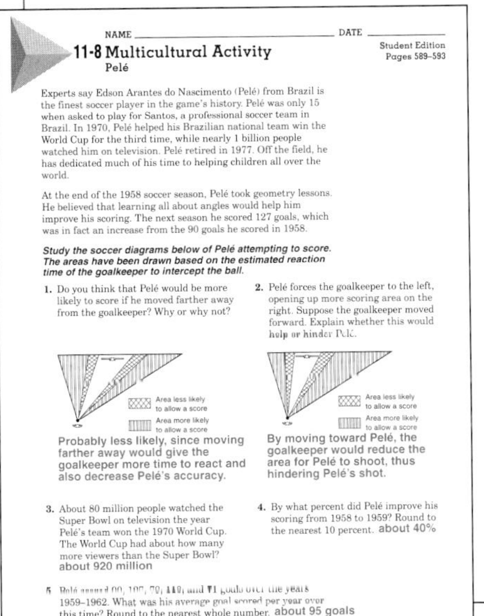

NAME ______________________ DATE __________

11-8 Multicultural Activity
Pelé

Student Edition
Pages 589–593

Experts say Edson Arantes do Nascimento (Pelé) from Brazil is the finest soccer player in the game's history. Pelé was only 15 when asked to play for Santos, a professional soccer team in Brazil. In 1970, Pelé helped his Brazilian national team win the World Cup for the third time, while nearly 1 billion people watched him on television. Pelé retired in 1977. Off the field, he has dedicated much of his time to helping children all over the world.

At the end of the 1958 soccer season, Pelé took geometry lessons. He believed that learning all about angles would help him improve his scoring. The next season he scored 127 goals, which was in fact an increase from the 90 goals he scored in 1958.

Study the soccer diagrams below of Pelé attempting to score. The areas have been drawn based on the estimated reaction time of the goalkeeper to intercept the ball.

1. Do you think that Pelé would be more likely to score if he moved farther away from the goalkeeper? Why or why not?

Probably less likely, since moving farther away would give the goalkeeper more time to react and also decrease Pelé's accuracy.

2. Pelé forces the goalkeeper to the left, opening up more scoring area on the right. Suppose the goalkeeper moved forward. Explain whether this would help or hinder Pelé.

By moving toward Pelé, the goalkeeper would reduce the area for Pelé to shoot, thus hindering Pelé's shot.

3. About 80 million people watched the Super Bowl on television the year Pelé's team won the 1970 World Cup. The World Cup had about how many more viewers than the Super Bowl? about 920 million

4. By what percent did Pelé improve his scoring from 1958 to 1959? Round to the nearest 10 percent. about 40%

5. Pelé scored [illegible] goals over the years 1959–1962. What was his average goal scored per year over this time? Round to the nearest whole number. about 95 goals

APPLICATIONS

Real-World Applications, 25, 26

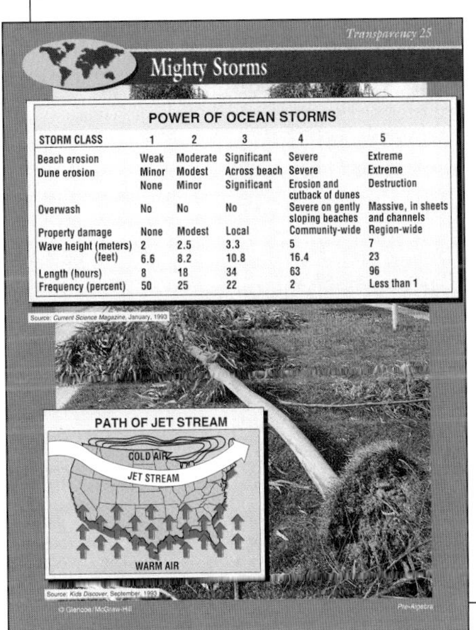

Transparency 25

Mighty Storms

POWER OF OCEAN STORMS

STORM CLASS	1	2	3	4	5
Beach erosion	Weak	Moderate	Significant	Severe	Extreme
Dune erosion	Minor	Modest	Across beach	Severe	Extreme
	None	Minor	Significant	Erosion and cutback of dunes	Destruction
Overwash	No	No	No	Severe on gently sloping beaches	Massive, in sheets and channels
Property damage	None	Modest	Local	Community-wide	Region-wide
Wave height (meters)	2	2.5	3.3	5	7
(feet)	6.6	8.2	10.8	16.4	23
Length (hours)	8	18	34	63	96
Frequency (percent)	50	25	22	2	Less than 1

Source: *Current Science Magazine*, January, 1993

Source: *Kids Discover*, September, 1993

Pre-Algebra

TECHNOLOGY

Graphing Calculator Masters, p. 11

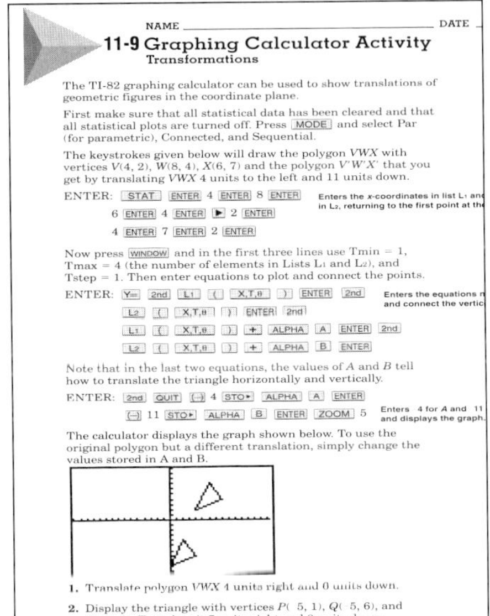

NAME ______________________ DATE

11-9 Graphing Calculator Activity
Transformations

The TI-82 graphing calculator can be used to show translations of geometric figures in the coordinate plane.

First make sure that all statistical data has been cleared and that all statistical plots are turned off. Press MODE and select Par (for parametric), Connected, and Sequential.

The keystrokes given below will draw the polygon *VWX* with vertices $V(4, 2)$, $W(8, 4)$, $X(6, 7)$ and the polygon $V'W'X'$ that you get by translating *VWX* 4 units to the left and 11 units down.

ENTER: STAT ENTER 4 ENTER 8 ENTER 6 ENTER 4 ENTER ▶ 2 ENTER 4 ENTER 7 ENTER 2 ENTER — Enters the *x*-coordinates in list L_1 an[d] in L_2, returning to the first point at th[e]

Now press WINDOW and in the first three lines use Tmin = 1, Tmax = 4 (the number of elements in Lists L_1 and L_2), and Tstep = 1. Then enter equations to plot and connect the points.

ENTER: Y= 2nd L1 (X,T,θ) ENTER 2nd L2 (X,T,θ) ENTER 2nd L1 (X,T,θ) + ALPHA A ENTER 2nd L2 (X,T,θ) + ALPHA B ENTER — Enters the equations [...] and connect the vertic[es]

Note that in the last two equations, the values of *A* and *B* tell how to translate the triangle horizontally and vertically.

ENTER: 2nd QUIT (−) 4 STO▸ ALPHA A ENTER (−) 11 STO▸ ALPHA B ENTER ZOOM 5 — Enters −4 for *A* and −11 and displays the graph.

The calculator displays the graph shown below. To use the original polygon but a different translation, simply change the values stored in A and B.

1. Translate polygon *VWX* 4 units right and 0 units down.
2. Display the triangle with vertices $P(-5, 1)$, $Q(-5, 6)$, and $R(-1, 6)$. Translate it 7 units right and 8 units down.

TECH PREP

Tech Prep Applications Masters, pp. 21, 22

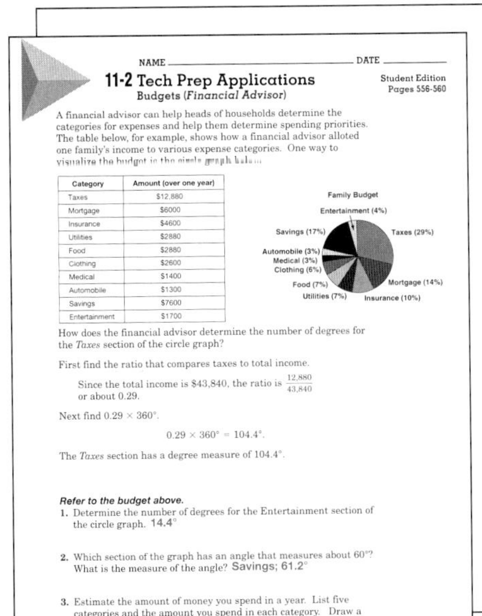

NAME ______________________ DATE __________

11-2 Tech Prep Applications
Budgets (*Financial Advisor*)

Student Edition
Pages 556-560

A financial advisor can help heads of households determine the categories for expenses and help them determine spending priorities. The table below, for example, shows how a financial advisor alloted one family's income to various expense categories. One way to visualize the budget is the circle graph below.

Category	Amount (over one year)
Taxes	$12,880
Mortgage	$6000
Insurance	$4600
Utilities	$2880
Food	$2880
Clothing	$2600
Medical	$1400
Automobile	$1300
Savings	$7600
Entertainment	$1700

How does the financial advisor determine the number of degrees for the *Taxes* section of the circle graph?

First find the ratio that compares taxes to total income.

Since the total income is $43,840, the ratio is $\frac{12,880}{43,840}$ or about 0.29.

Next find $0.29 \times 360°$.

$$0.29 \times 360° = 104.4°.$$

The *Taxes* section has a degree measure of 104.4°.

Refer to the budget above.

1. Determine the number of degrees for the Entertainment section of the circle graph. 14.4°
2. Which section of the graph has an angle that measures about 60°? What is the measure of the angle? Savings; 61.2°
3. Estimate the amount of money you spend in a year. List five categories and the amount you spend in each category. Draw a circle graph for your data. Check students' work.

CONNECTIONS

Activity Masters, pp. 11, 25, 43, 44

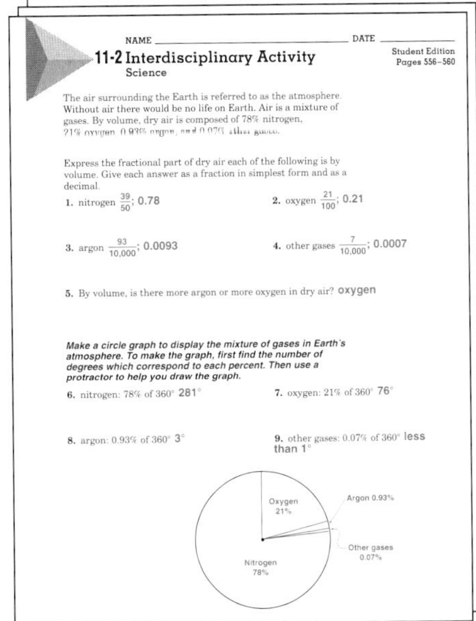

NAME ______________________ DATE __________

11-2 Interdisciplinary Activity
Science

Student Edition
Pages 556–560

The air surrounding the Earth is referred to as the atmosphere. Without air there would be no life on Earth. Air is a mixture of gases. By volume, dry air is composed of 78% nitrogen, 21% oxygen, 0.93% argon, and 0.07% other gases.

Express the fractional part of dry air each of the following is by volume. Give each answer as a fraction in simplest form and as a decimal.

1. nitrogen $\frac{39}{50}$; 0.78
2. oxygen $\frac{21}{100}$; 0.21
3. argon $\frac{93}{10,000}$; 0.0093
4. other gases $\frac{7}{10,000}$; 0.0007
5. By volume, is there more argon or more oxygen in dry air? oxygen

Make a circle graph to display the mixture of gases in Earth's atmosphere. To make the graph, first find the number of degrees which correspond to each percent. Then use a protractor to help you draw the graph.

6. nitrogen: 78% of 360° 281°
7. oxygen: 21% of 360° 76°
8. argon: 0.93% of 360° 3°
9. other gases: 0.07% of 360° less than 1°

COOPERATIVE LEARNING

Math Lab and Modeling Math Masters, pp. 70, 86

NAME ______________________ DATE __________

11-6 Modeling Math Activity
Similar Triangles, Indirect Measurement

Student Edition
Pages 578–583

A **stadia** like the one shown at the right can be used to find the height of an object. Hold the stadia to your eye. Back away from the object until the top and the bottom of the object are in line with the strings. Then record the two readings where the strings appear to cross the stick. Measure the distance from the stadia to the stick and the distance from the stadia to the object. Notice that two similar triangles are formed.

STADIA
side view
stick
0.9 m
2 m
14 m

1. Set up a proportion and find the height of the tree shown in the diagram. $\frac{0.9}{2} = \frac{x}{14}$; 6.3 m
2. Use the stadia and a measuring stick to find the height of the following objects. Remember to measure the distances between yourself and the stick (Distance 1) and between yourself and the object (Distance 2). Record your data in the table below. Answers will vary.

Object	Distance 1	Distance 2	Height
Building			
Tree			
Flagpole			

MAKING MATHEMATICS RELEVANT

This two-page introduction to the chapter provides students with an opportunity to explore other disciplinary topics and their applications to mathematics.

Background Information

Math and Sailing in the News
While three American yachts competed fiercely against each other for the right to defend the America's Cup, the New Zealand team built two boats and spent months racing them against each other so they could learn as much as possible. This enabled them to make a lot of changes and adjustments to one boat and then compare the resulting performance to that of the second boat. This allowed them to "fine-tune" their boats for maximum speed and maneuverability. The American teams relied very heavily upon mathematical and computerized simulations to test their designs, and they had less actual testing time in the water.

CHAPTER 11 Applying Algebra to Geome

TOP STORIES in Chapter 11

In this chapter, you will:

- use some basic terms of geometry,
- construct a circle graph,
- identify the relationships of intersecting, parallel, and skew lines,
- identify properties of congruent and similar figures,
- classify and draw polygons, and
- identify and draw transformations.

MATH AND SAILING IN THE NEWS

It's Magic at the America's Cup

Source: Sail, July, 1995

"Over the past two-and-a-half years, we've done everything we could to prepare for this moment." said *Black Magic 1* team leader Peter Blake. Determination and technology helped Blake and his New Zealand team capture the America's Cup, the most coveted prize in yachting. The team won five races in a row in the final round to earn the victory. The New Zealanders used the lessons they learned in their previous losses to build two boats especially for the competition. Computers and mathematics helped them study the way that the boat and the sails interact with the air and the water. The boat's co-designer says there are literally "hundreds of little refinements." Some of the major refinements are the canoe-style hull, a carbon-fiber mast, innovative sail rigging, and a thin carbon-steel rudder.

Putting It into Perspective
1850
1880
1910
1851
U.S. schooner *America* defeats 17 British yachts to win the Hundred Guinea Cup.
1857
The Hundred Guinea Cup is renamed the America's Cup and becomes an international competition.
1900
Sailing is made an Olympic sport.

Putting It into Perspective

The women's crew eventually added one man to their number during the series of races to decide who would defend for the U.S. Although it did not win, the women's boat, *Mighty Mary*, raced well and came within one victory of being the American defender.

EMPLOYMENT OPPORTUNITIES

Set Sail for Your Future

THERE IS MORE to sailing than just wind and water. And even the most skilled sailor depends upon the boat. The beauty of sailing begins with the designer. If you enjoy creating something from an idea, and are good at visualizing and analyzing, your skills are needed as a **designer**, **engineer**, or a **naval architect**. These jobs require a thorough knowledge of mathematics, especially geometry, and computer skills. A college degree is usually required, but a two-year technical degree with experience will be considered.

For more information, contact:
Society of Naval Architects and Marine Engineers
601 Pavonia Avenue, Suite 400
Jersey City, NJ 07306

Statistical Snapshot

1995 America's Cup Race Course

Leg	1	2	3	4	5	6	Total
Length (nautical miles)	3.275	3.275	3.000	3.000	3.000	3.000	18.550

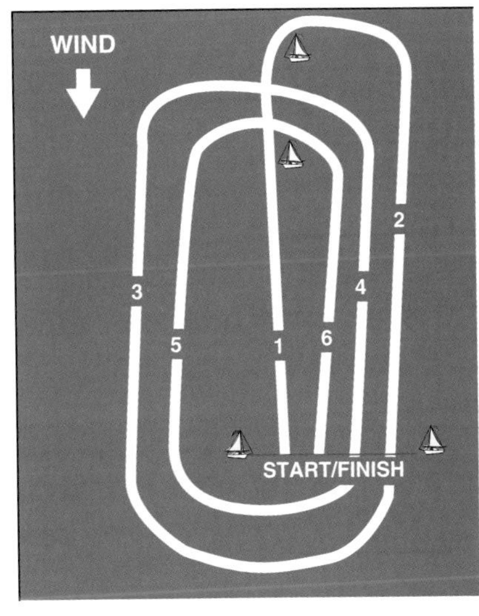

interNET CONNECTION For up-to-date information on boat making, visit: **www.glencoe.com/sec/math/prealg/mathnet**

1980 — **1990** — **2000**

1992
Martin Stephan and Art Prince become the first African Americans to sail in the America's Cup race.

1983
The America's Cup is won by a non-American yacht, *Australia II*, for the first time in 25 competitions.

1995
Black Magic I wins the America's Cup.

1995
The first all-female crew competes to represent the United States as America's Cup defender.

Statistical Snapshot

The course for the 1995 America's Cup race was a straightforward one in which the boats sailed into the wind and away from the wind on each successive leg. The course was estimated to take about $2\frac{1}{2}$ hours to complete. Having just two wind positions to deal with allowed the boats to carry fewer sails so that the cost of competing was lowered.

EMPLOYMENT OPPORTUNITIES

Designers, engineers, and naval architects increasingly need to have skills in computer assisted design (CAD) techniques. Computers can greatly simplify and speed up the design processes. They can simulate behaviors and responses and facilitate the evaluation of alternatives.

Cooperative Learning

Chapter Projects Two chapter projects are included in the *Investigations and Projects Masters*. In Chapter Project A, students extend the topic in the Chapter Opener. In Chapter Project B, students explore automotive design. A student page and a parent letter are provided for each Chapter Project.

interNET CONNECTION

Glencoe has made every effort to ensure that the website links for *Pre-Algebra* at www.glencoe.com/sec/math/prealg/mathnet are current and contain appropriate content. However, these website links are not under Glencoe's control.

Investigations and Projects Masters, p. 69

NAME ______________________ DATE __________

CHAPTER 11 Project A
Ship Shapes

The sleek graceful yachts that you see in the America's Cup races are one of many different kinds of sailing vessels. The first sailing ship, perhaps 8000 years-old or more, probably had a hand-held cloth as a sail. Although there have been numerous improvements since, all sailing ships, from the simple sailboat to the four-masted bark, share many common elements. And the construction and sailing of each relies heavily on geometrical concepts. Find out more about the components of a sailing vessel. Here's what to do:

1. Go to the library. Do research on the different kinds of sailing vessels such as barks, brigs, ketches, yawls, schooners, and so on.
2. Look more closely to examine the features these ships have in common. Study drawings and photos to identify the names, locations, and uses of the key parts, such as the different kinds of sails, masts, and riggings. Record what you discover. Then talk about your findings with classmates to help you better understand what you've learned.
3. Choose one kind of sailing vessel. Make a clear and accurate drawing of it. Label its parts carefully in as much detail as possible. Then write a brief description of the ship, including information about it that someone might find interesting, such as its history, colorful uses, crew size, or top speed.
4. Now look carefully at the picture of your vessel to analyze it mathematically. Prepare a description to accompany the drawing and a report that answers questions such as the following: Identify the line segments—which are parallel or perpendicular? Which are skew? Which intersect? Identify the features that are polygons—what kind are they? Which features are congruent shapes? Which are similar shapes? When you answer these questions, be sure to use the correct ship-shape terminology. For example, you might note that your ship has two lateen sails that are similar acute triangles. Or you might discover that its mainmast is parallel to the mizzenmast and the foremast, but not perpendicular to the bowsprit.
5. Share what you've learned. Prepare a complete "Guide to (your ship)" to display for classmates to read.

11-1 LESSON NOTES

NCTM Standards: 1-4, 12

Instructional Resources

- Study Guide Master 11-1
- Practice Master 11-1
- Enrichment Master 11-1
- Group Activity Card 11-1
- Real-World Applications, 25

Transparency 11-1A contains the 5-Minute Check for this lesson; **Transparency 11-1B** contains a teaching aid for this lesson.

Recommended Pacing	
Standard Pacing	Day 1 of 13
Honors Pacing	Day 1 of 13
Block Scheduling*	Day 1 of 7

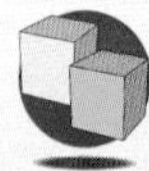

*For more information on pacing and possible lesson plans, refer to the ***Block Scheduling Booklet***.

1 FOCUS

5-Minute Check
(over Chapter 10)

The low temperatures one week in Two Harbors, MN, were 2°, −8°, −3°, 0°, 4°, 6°, and 16°.

1. Name the median. **2**
2. Name the upper and lower quartiles **6, −3**
3. Name the outliers, if any. **There are no outliers.**

Find the probability of each event.

4. rolling a six on a die two times in a row $\frac{1}{36}$
5. drawing a card from a standard deck and getting a five or a heart $\frac{4}{13}$

Motivating the Lesson

Questioning Show students a fishing reel with fishing filament. Ask students why the filament is called a line. Ask what they think the difference is between the fishing line and a geometric line.

11-1 The Language of Geometry

Setting Goals: *In this lesson, you'll identify points, lines, planes, rays, segments, angles, and parallel, perpendicular, and skew lines. You'll also classify angles as acute, right, or obtuse.*

Modeling a Real-World Application: Construction

Students in Todd Scholl's construction trades classes in Fairplay, Colorado, don't just study. They put their learning into practice by building real houses. In fact, they built the town's Silverheels Middle School, a four-classroom structure.

Before construction of a building can begin, the corners of the building site must be level. A laser interferometer can determine the heights of the corners of the site with great accuracy. A laser is visible because light is being reflected from particles in the air.

We can use ideas from the construction of buildings to model geometric ideas.

Learning the Concept

A point has no dimensions.

Each particle visible in the laser beam has a specific location in space and suggests the idea of a geometric **point**. A point is a specific location in space with no size or shape. A point can be represented by a dot and named with a capital letter.

A • point A

B • point B

A line has one dimension, length.

A laser beam travels in a straight path. In geometry, a straight path is called a **line**. A line is a collection of points that extends indefinitely in two directions. Arrowheads are used to show that a line has no endpoints. A line can be named with a single lowercase letter or by two points on the line.

line *CD* or $\overleftrightarrow{CD}$ or line *DC* or $\overleftrightarrow{DC}$ or line ℓ

THINK ABOUT IT

Can you measure the length of a line? Can you measure a line segment?

No; it has no endpoints. Yes.

Each wooden beam used in the construction of a building site is a model of a **line segment**. A line segment is part of a line containing two endpoints and all points between the endpoints. A line segment is named by its endpoints.

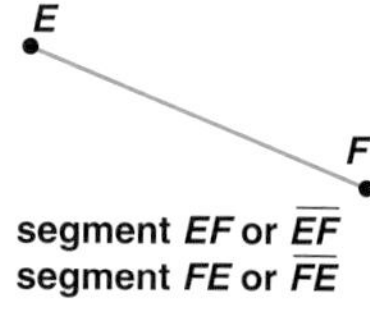

segment *EF* or $\overline{EF}$
segment *FE* or $\overline{FE}$

The path of a laser beam as it begins at the generator and extends in one direction is a model of a **ray**. A ray is a portion of a line that extends from one point infinitely in *one* direction. A ray is named using the endpoint first and then any other point on the ray.

ray *DF* or $\overrightarrow{DF}$

Alternative Learning Styles

Kinesthetic Have the students stand. Call out either *acute angle, obtuse angle,* or *right angle*. Tell students to use their arms to model that kind of angle. Repeat several times. Then have a student model an angle and have the class name the kind of angle.

Example 1 **Name two points, a line, two rays, and a line segment in the figure at the right.**

Two of the points are E and F.

We can use a single lowercase letter or any two of the labeled points to name the line. So we can name it m, $\overleftrightarrow{EF}$, $\overleftrightarrow{EG}$, or $\overleftrightarrow{FG}$.

Two of the rays are $\overrightarrow{FG}$ and $\overrightarrow{FE}$.

One of the line segments shown is $\overline{EG}$.

When two rays have the same endpoint, they form an **angle**. The common endpoint is called the **vertex**, and the rays are called **sides** of the angle. An angle is named using three points with the vertex as the middle letter. If no other angle shares the same vertex, the angle is named with just the vertex letter or with a numeral.

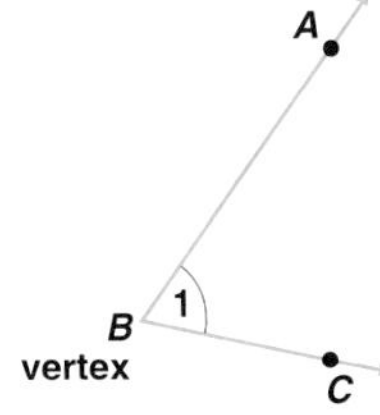

- angle ABC or $\angle ABC$, angle CBA or $\angle CBA$,
- angle B or $\angle B$,
- angle 1 or $\angle 1$

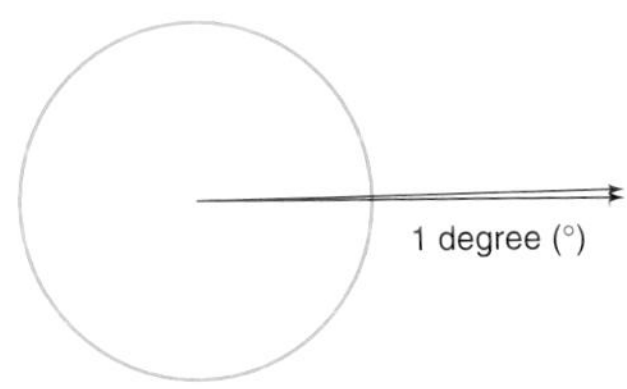

The most common unit of measure for angles is the **degree**. A circle can be separated into 360 arcs of the same length. An angle has a measurement of one degree if its vertex is at the center of the circle and its sides contain the endpoints of one of the 360 equal arcs.

You can use a **protractor** to measure angles.

Example 2 **Use a protractor to measure $\angle PQR$.**

Place the protractor so the center mark is at Q (the vertex of $\angle PQR$) and the straightedge aligns with $\overrightarrow{QR}$ (one side of the angle). Use the scale that begins with 0 at $\overrightarrow{QR}$. Read where $\overrightarrow{QP}$ (the other side of the angle) crosses this scale.

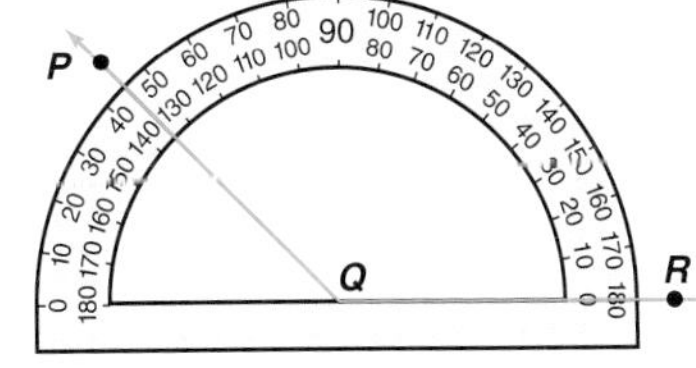

The measure of angle PQR is 135 degrees. Using symbols, $m\angle PQR = 135°$.

You can also use a protractor to draw an angle of a given measure.

2 TEACH

In-Class Examples

For Example 1

Name two points, a line, two rays, and a line segment in the figure below.

Sample answers: Point R, Point J; line b; $\overrightarrow{WR}$ and $\overrightarrow{JW}$; $\overline{RJ}$

For Example 2

Use a protractor to measure $\angle GHJ$. **40°**

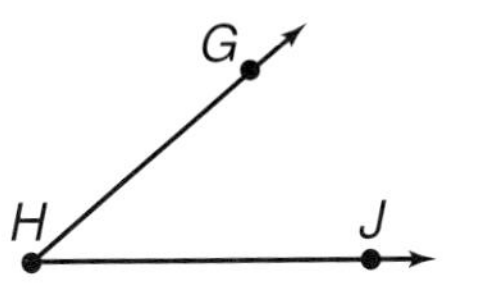

Teaching Tip Using an overhead projector, list the new terms as covered in this lesson. On a separate transparency, give an example of each. After all terms have been covered, show examples in a different order and match each to the correct term.

Classroom Vignette

"I have students measure the angle of their peripheral vision by using a large quarter of a circle, index cards with a dot for the student to follow with their vision, and a string with a heavy object to determine the point on the quarter circle. This helps students see that an angle opens from one side which, in turn, helps them understand which scale to use on the protractor."

Tricia K. Burgess

Tricia Burgess
St. Paul's Episcopal School
Mobile, AL

In-Class Examples

For Example 3
Draw an angle *B* that measures 115°.

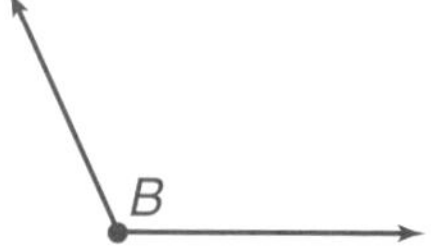

For Example 4
Classify each angle as *acute*, *right*, or *obtuse*. **∠M: obtuse, ∠N: acute, ∠P: right**

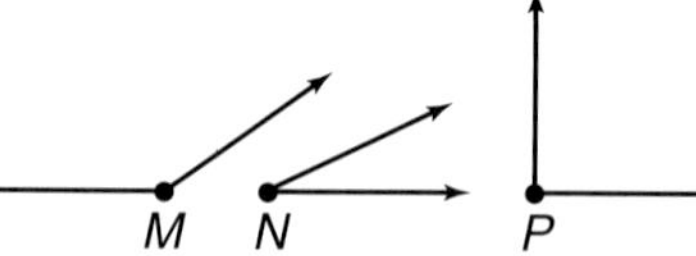

For Example 5
Identify each pair of line segments as *intersecting*, *parallel*, or *skew*.

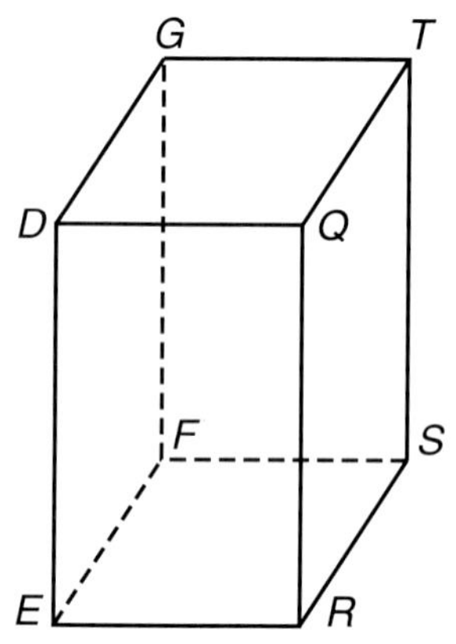

a. $\overline{EF}$ and $\overline{QT}$ **parallel**
b. $\overline{DG}$ and $\overline{TS}$ **skew**
c. $\overline{QT}$ and $\overline{RQ}$ **intersecting**

Example 3 **Draw an angle *Y* that measures 50°.**

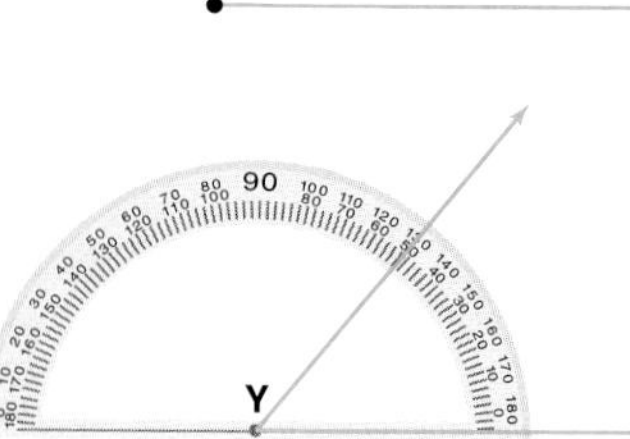

Use a straightedge to draw a ray.

Place the protractor with the center mark at *Y* and the 0-mark aligned with the ray.

Use the scale that begins with 0 at the ray, find the mark for 50, and draw a dot.

Use a straightedge to draw the other side of the angle. Angle *Y* measures 50°.

You can classify angles by their degree measure. **Acute** angles have measures greater than 0° and less than 90°. **Right** angles have measures of 90°. **Obtuse** angles have measures greater than 90° but less than 180°. **Straight** angles measure 180°.

Example 4 APPLICATION Baseball

According to Ted Williams, one of baseball's greatest hitters, the best way to strike a baseball is at a right angle. If you swing too early or too late, the ball will probably go foul. Classify each angle as *acute*, *right*, or *obtuse*.

a.

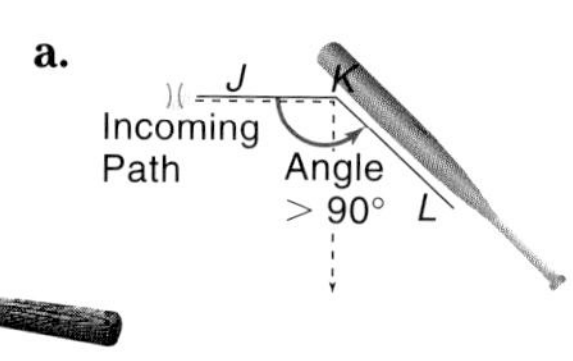

b.

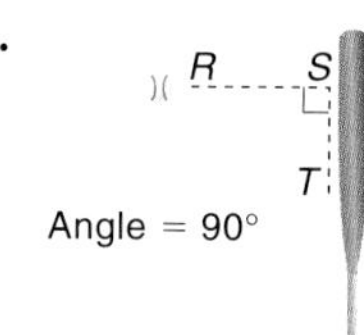

The small square indicates that ∠RST is a right angle.

c.

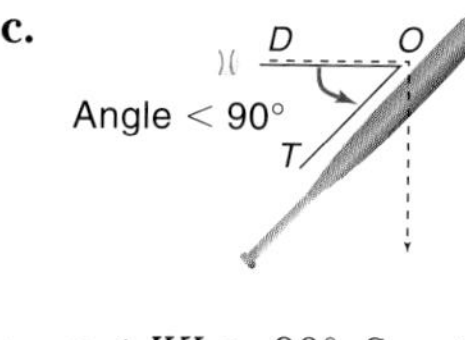

a. $m\angle JKL > 90°$. So, $\angle JKL$ is obtuse. *The swing is too early.*
b. $m\angle RST = 90°$. So, $\angle RST$ is right.
c. $m\angle DOT < 90°$. So, $\angle DOT$ is acute. *The swing is too late.*

A plane has two dimensions, length and width, but no thickness.

The level building site described in the application on page 548 is a model of part of a **plane**. A plane is a flat surface with no edges, or boundaries. A plane can be named by a single uppercase script letter or by using any three points of the plane. *The three points must not lie on the same line.*

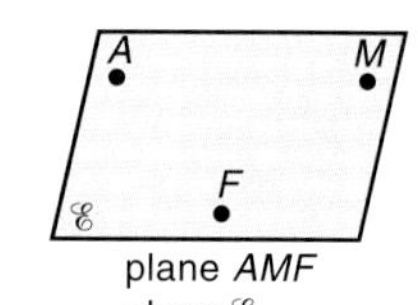

plane *AMF*
plane ℰ

Lines that lie in the *same plane* either intersect or are parallel.

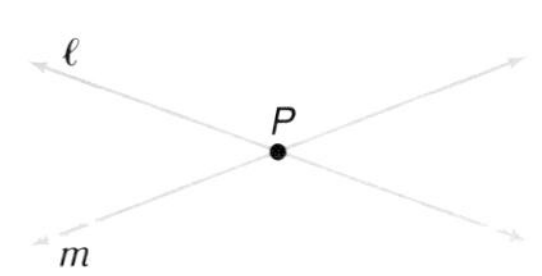

Lines ℓ and m **intersect** at point P.

There is no point of intersection of lines t and n. Lines t and n are **parallel**. Using symbols, $t \parallel n$.

Two lines that intersect to form a right angle are **perpendicular**. Rays and line segments can also be perpendicular. The symbol $\perp$ means *is perpendicular to*.

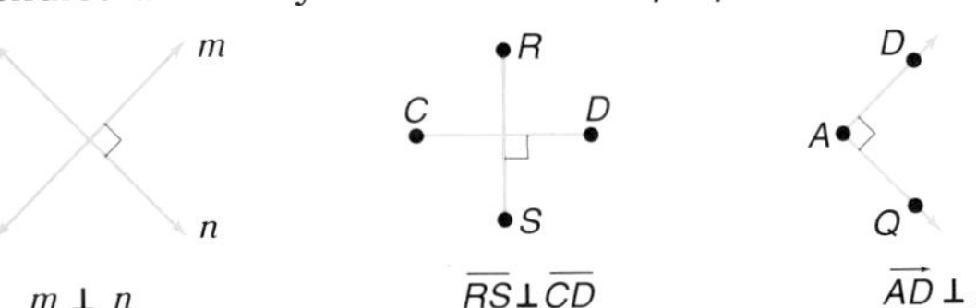

$m \perp n$ $\overline{RS} \perp \overline{CD}$ $\overrightarrow{AD} \perp \overrightarrow{AQ}$

If two lines do not intersect and are not in the same plane, they are called **skew lines**. The roof line along the front of a rectangular building and the line of the base of the side of the building are an example of skew lines.

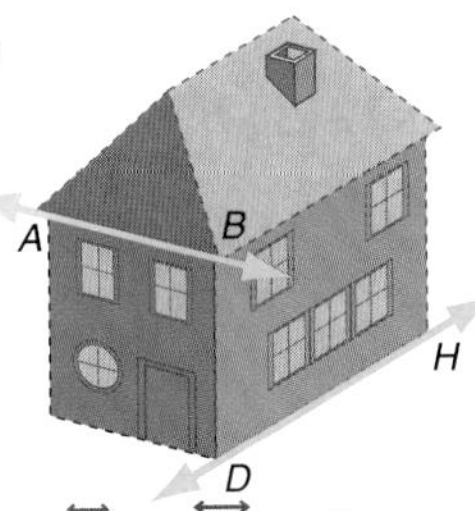

$\overleftrightarrow{AB}$ and $\overleftrightarrow{DH}$ are skew.

Example 5 **Identify each pair of line segments as *intersecting, parallel,* or *skew*.**

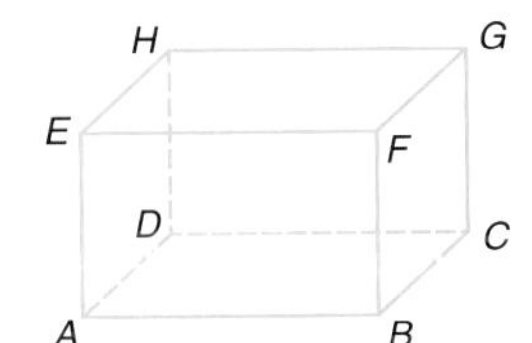

a. $\overline{EH}$ and $\overline{FG}$
These segments are in the same plane (EFG), and they do not intersect. So, $\overline{EH} \parallel \overline{FG}$

b. $\overline{AE}$ and $\overline{FG}$
These segments do not intersect, and they are not in the same plane. So, $\overline{AE}$ and $\overline{FG}$ are skew.

c. $\overline{BC}$ and $\overline{CD}$
These segments are in the same plane (ABC), and they intersect at point C.

Checking Your Understanding

Communicating Mathematics

Read and study the lesson to answer these questions.

1. **Tell** how many endpoints each of the following has.
   a. a line none **b.** a line segment two **c.** a ray one
2. **Compare and contrast** parallel lines and skew lines. See margin.
3. **Demonstrate** how a protractor is used to measure angles and how it is used to draw angles. See students' work.

3 PRACTICE/APPLY

Checking Your Understanding

Exercises 1–17 are designed to help you assess your students' understanding through reading, writing, speaking, and modeling. You should work through Exercises 1–6 with your students and then monitor their work on Exercises 7–17.

Additional Answers

2. Parallel lines never intersect and are in the same plane. Skew lines never intersect but are in different planes.
4.

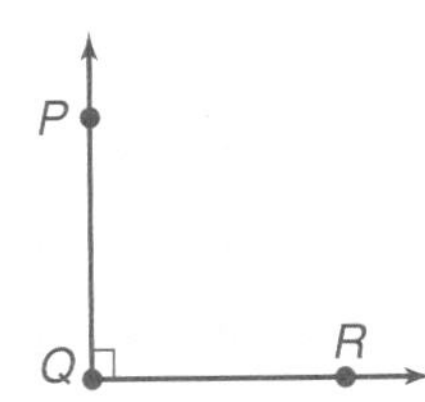

5. Acute angles measure less than 90°. Obtuse angles measure more than 90°, but less than 180°.

Reteaching

Work Backward Have students play Geometric Jeopardy. Write the lesson terms and draw sketches of figures on slips of paper and place in a bag. As the slips of paper are drawn, they are answers, and students must give an appropriate question.

Study Guide Masters, p. 91

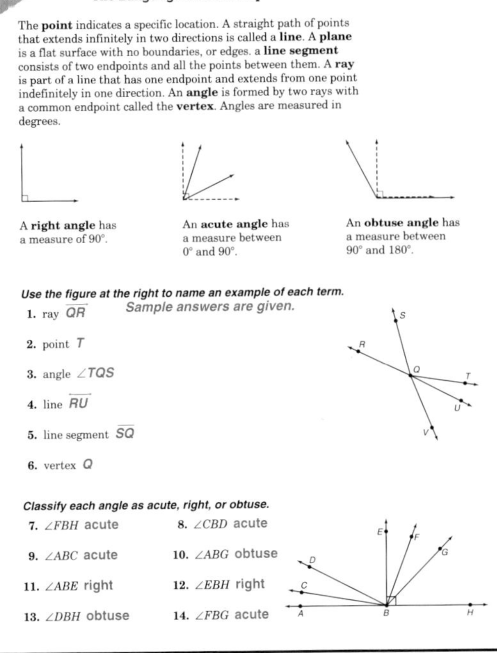

NAME ____________ DATE ________

11-1 Study Guide Student Edition Pages 548–553
The Language of Geometry

The **point** indicates a specific location. A straight path of points that extends infinitely in two directions is called a **line**. A **plane** is a flat surface with no boundaries, or edges. a **line segment** consists of two endpoints and all the points between them. A **ray** is part of a line that has one endpoint and extends from one point indefinitely in one direction. An **angle** is formed by two rays with a common endpoint called the **vertex**. Angles are measured in degrees.

A **right angle** has a measure of 90°. An **acute angle** has a measure between 0° and 90°. An **obtuse angle** has a measure between 90° and 180°.

Use the figure at the right to name an example of each term. *Sample answers are given.*

1. ray QR
2. point T
3. angle $\angle TQS$
4. line RU
5. line segment SQ
6. vertex Q

Classify each angle as acute, right, or obtuse.

7. $\angle FBH$ acute
8. $\angle CBD$ acute
9. $\angle ABC$ acute
10. $\angle ABG$ obtuse
11. $\angle ABE$ right
12. $\angle EBH$ right
13. $\angle DBH$ obtuse
14. $\angle FBG$ acute

Error Analysis

Some students may read the wrong scale on the protractor. Encourage students to estimate whether the measure of an angle is less than or greater than 90° before reading the measure.

Assignment Guide

Core: 19–45 odd, 46, 47, 49–58
Enriched: 18–44 even, 46–58

For **Extra Practice**, see p. 767.

The red A, B, and C flags, printed only in the Teacher's Wraparound Edition, indicate the level of difficulty of the exercises.

Additional Answers

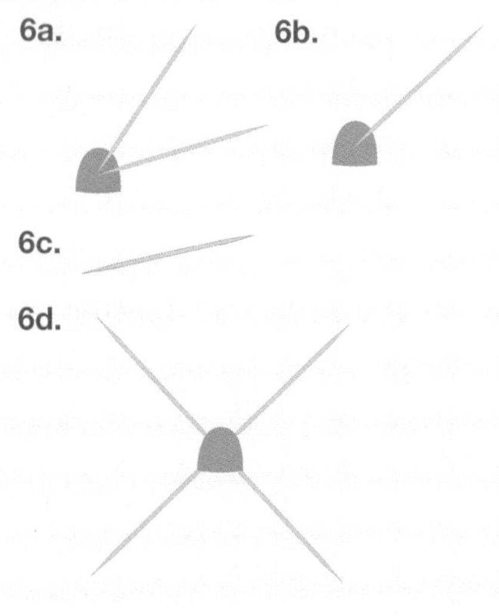

Practice Masters, p. 91

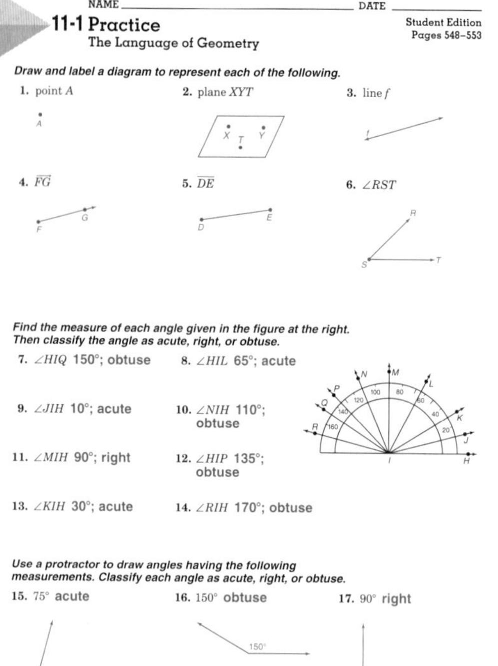

NAME ______________ DATE ______

11-1 Practice
The Language of Geometry

Student Edition Pages 548–553

Draw and label a diagram to represent each of the following.

1. point A
2. plane XYT
3. line f
4. $\overline{FG}$
5. $\overline{DE}$
6. $\angle RST$

Find the measure of each angle given in the figure at the right. Then classify the angle as acute, right, or obtuse.

7. $\angle HIQ$ 150°; obtuse
8. $\angle HIL$ 65°; acute
9. $\angle JIH$ 10°; acute
10. $\angle NIH$ 110°; obtuse
11. $\angle MIH$ 90°; right
12. $\angle HIP$ 135°; obtuse
13. $\angle KIH$ 30°; acute
14. $\angle RIH$ 170°; obtuse

Use a protractor to draw angles having the following measurements. Classify each angle as acute, right, or obtuse.

15. 75° acute
16. 150° obtuse
17. 90° right

4. **Draw** and label a diagram to show a right angle whose sides are $\overrightarrow{QP}$ and $\overrightarrow{QR}$. See margin.
5. **Explain** the difference between an acute angle and an obtuse angle. 5. See margin.
6. You can use gumdrops and toothpicks to model geometric concepts. A gumdrop represents a point, and a toothpick represents a portion of a line. The figure at the right shows a model of a line segment. Use gumdrops and toothpicks to model each of the following.

MATERIALS
toothpicks
gumdrops

   a. an angle **b.** a ray **c.** a line **d.** intersecting lines
   a–d. See margin.

Guided Practice

Tell whether each model suggests a point, a line, a plane, a ray, or a line segment. Explain your answer.

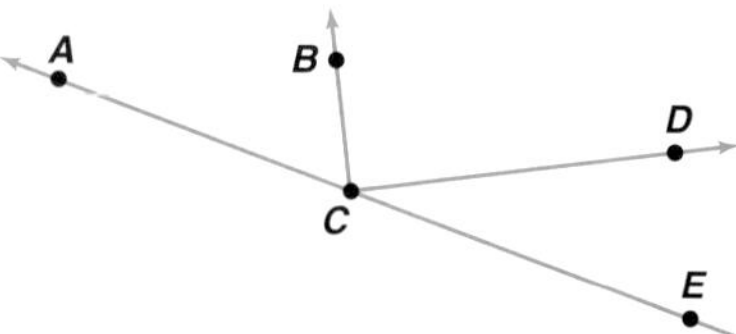

7–8. Sample answers given. See students' explanations.

7. a wall plane
8. the tip of a needle point

9–12. Sample answers are given.

Use the figure at the right to name an example of each term.

9. point A
10. line segment $\overline{AC}$
11. ray $\overrightarrow{CD}$
12. line $\overleftrightarrow{AE}$

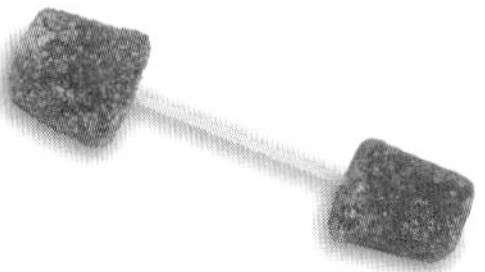

In the figure above, use a protractor to find the measure of each angle. Then classify the angle as *acute*, *right*, or *obtuse*.

13. $\angle ACB$ 63°, acute
14. $\angle BCD$ 90°, right

Use a protractor to draw angles having the following measurements.

15–16. See margin.

15. 78°
16. 145°

17. **Photography** Cameras are often mounted on tripods to give stability. Why do tripods give stability? The three "pods" determine a plane.

Exercises: Practicing and Applying the Concept

Independent Practice

Tell whether each model suggests a point, a line, a plane, a ray, or a line segment. Explain your answer.

18–21. Sample answers given. See students' explanations. A

18. a pencil segment
19. a CD-ROM plane
20. a telephone wire line
21. an arrow's path ray

Draw and label a diagram to represent each of the following.

22–33. See Solutions Manual.

22. point T
23. $\angle Q$
24. line ℓ
25. $\overline{XY}$
26. plane RST
27. $\overrightarrow{EF}$
28. $\angle BOY$
29. $\overleftrightarrow{AB}$
30. lines ℓ and m intersect
31. $\overleftrightarrow{AB} \parallel \overleftrightarrow{CD}$
32. $\overleftrightarrow{XY} \perp \overleftrightarrow{MN}$
33. lines a and b are skew

Group Activity Card 11-1

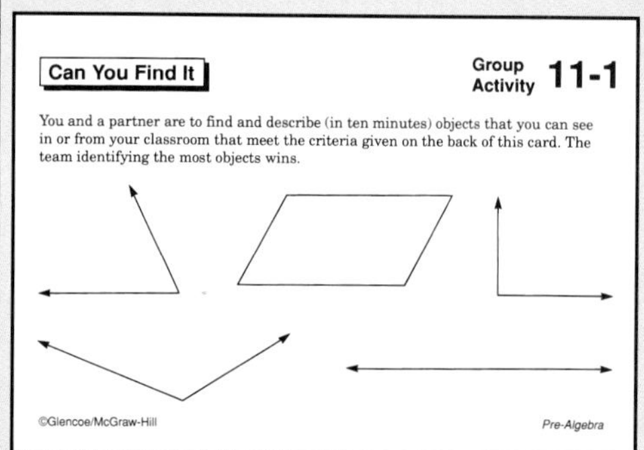

Can You Find It

Group Activity **11-1**

You and a partner are to find and describe (in ten minutes) objects that you can see in or from your classroom that meet the criteria given on the back of this card. The team identifying the most objects wins.

Pre-Algebra

Additional Answer

15.
78°

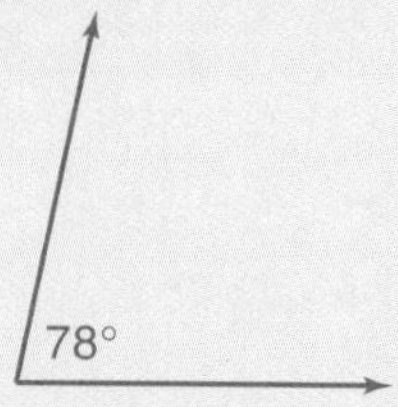

In the figure at the right, use a protractor to find the measure of each angle. Then classify the angle as *acute*, *right*, or *obtuse*.

34. $m\angle MOQ$ 90°; right
35. $m\angle NOT$ 155°; obtuse
36. $m\angle TOP$ 130°; obtuse
37. $m\angle QON$ 65°; acute
38. $m\angle POS$ 90°; right
39. $m\angle ROS$ 15°; acute

Use a protractor to draw angles having the following measurements. Classify each angle as *acute*, *right*, or *obtuse*.

40–45. See margin. See Solutions Manual for drawings.

40. 112° **41.** 47° **42.** 95° **43.** 16° **44.** 162° **45.** 90°

Critical Thinking

46. ***True* or *false*. Explain your answers.**
    a. A line is part of a line segment. False; a segment is part of a line.
    b. The intersection of any two rays is always a point. False; all or part of the rays may coincide, forming a single ray or a line segment.

Applications and Problem Solving

47. **Space Travel** The space shuttle approaches the runway at an angle of 18°, six times as steep as that of a commercial jet. At what angle does a commercial jet make its approach? Draw a diagram showing both approach angles. 3°; See students' work.

48. See margin.

48. **Aviation** Airplane flight paths can be described using angles and compass directions. The path of a particular airplane is described as 36° west of north. Draw a diagram that represents this path.

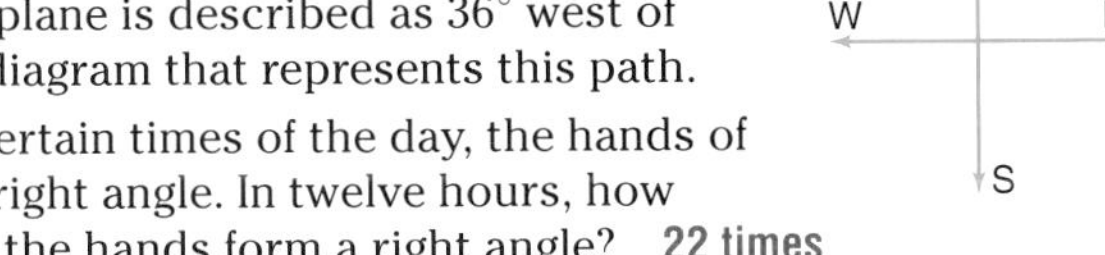

49. **Patterns** At certain times of the day, the hands of a clock form a right angle. In twelve hours, how many times do the hands form a right angle? 22 times

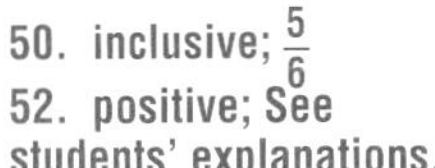

50. inclusive; $\frac{5}{6}$
52. positive; See students' explanations.

Mixed Review

50. **Probability** A die is tossed. Is *P*(even or prime) *mutually exclusive* or *inclusive*? Find the probability. (Lesson 10-10)
51. Replace the ● with = or ≠ to make $\frac{1.5}{2}$ ● $\frac{1.8}{2.4}$ true. (Lesson 9-4) =
52. **Health** What relationship (*positive*, *negative*, or *none*) do you think a scatter plot of number of students with colds and season of year (spring, summer, fall, winter) might show? Explain. (Lesson 8-2)
53. **Number Theory** The sum of an even integer and the next greater even integer is more than 98. Find the least pair of such integers. (Lesson 7-7) 50, 52
54. Solve $n - 6.2 \leq 5.8$. Graph the solution on a number line. (Lesson 5-7) $n \leq 12$; see margin for graph.
55. A volleyball team has 6 members. Suppose each member shakes hands with every other member. How many handshakes take place? (Lesson 4-3) 15
56. **Geometry** Find the perimeter and area of a rectangular floor that is 12 feet wide by 15 feet long. (Lesson 3-5) 54 ft; 180 ft²
57. Evaluate $\frac{d}{8}$ if $d = -96$. (Lesson 2-8) −12
58. Solve $3d = 24$ mentally. (Lesson 1-6) 8

Extension

Using Models Have students look through magazines and newspapers to find examples of each of the lesson terms.

Additional Answer

16. 145°

4 ASSESS

Closing Activity

Modeling Have students draw and label each of the following geometric figures. Then have student measure each angle and state its measurement.

- line *AB*
- ray *CD*
- line segment *EF*
- acute angle *GHI*
- right angle *JKL*
- obtuse angle *MNP*

Additional Answers

40. obtuse
41. acute
42. obtuse
43. acute
44. obtuse
45. right
48.

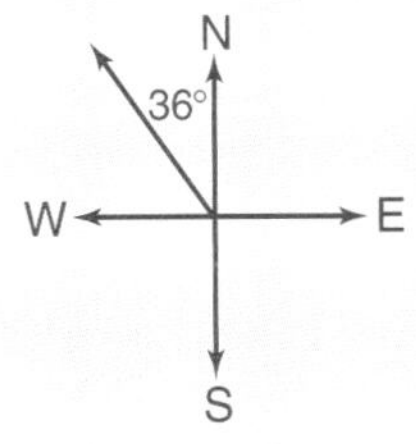

54.

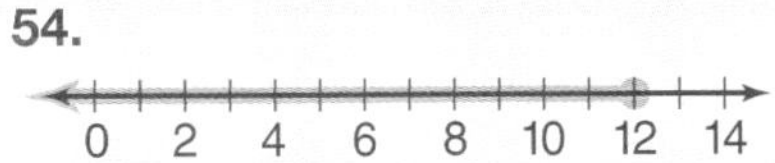

Enrichment Masters, p. 91

NAME ______ DATE ______

11-1 Enrichment
Basic Geometric Terms: Word Search

Student Edition Pages 548–553

R	N	B	A	S	C	V	E	S	R	E	Q	P	O
G	E	O	M	E	T	R	Y	U	D	Z	A	H	B
A	T	E	J	L	S	A	I	P	T	R	U	G	T
T	R	I	A	N	G	L	E	P	A	M	E	H	U
B	A	D	V	M	X	Y	F	L	C	A	G	B	S
E	P	C	P	W	I	B	L	E	F	I	L	N	E
L	E	T	E	S	R	E	S	M	R	D	C	O	C
A	Z	U	R	U	L	Q	V	E	H	G	I	G	W
C	O	M	P	L	E	M	E	N	T	A	R	Y	H
I	I	L	E	A	F	O	W	T	J	N	C	L	G
T	D	X	N	K	R	F	E	A	M	G	L	O	E
R	N	T	D	M	A	X	N	R	O	L	E	P	L
E	E	N	I	L	Y	Z	L	Y	N	E	R	Q	G
V	A	U	C	S	B	P	O	L	L	K	P	I	N
K	S	Q	U	A	R	E	J	H	E	M	L	J	A
T	A	P	L	A	N	E	N	T	O	L	E	Y	T
N	S	V	A	T	C	W	U	X	Q	R	E	S	C
I	K	O	R	P	U	C	T	P	V	M	R	T	E
O	E	Z	Y	Y	A	X	S	U	B	M	O	H	R
P	W	Z	A	D	J	A	C	E	N	T	A	F	S
S	E	G	M	E	N	T	X	Q	W	D	G	Z	I

1. geometry
2. line
3. segment
4. plane
5. point
6. ray
7. angle
8. acute
9. right
10. obtuse
11. adjacent
12. complementary
13. supplementary
14. vertical
15. parallel
16. skew
17. perpendicular
18. triangle
19. polygon
20. rectangle
21. rhombus
22. square
23. trapezoid
24. circle

11-1B LESSON NOTES

NCTM Standards: 1-4, 12

Objective
Construct congruent line segments and angles with a compass and straightedge.

Recommended Time
Demonstration and discussion: 30 minutes; Exercises: 30 minutes

Instructional Resources
For each student or group of students
- compass
- straightedge

Math Lab and Modeling Math Masters
- p. 48 (worksheet)

For teacher demonstration
Overhead Manipulative Resources

1 FOCUS

Motivating the Lesson
Show a diagram of line segment *AB*. Have students solve the following. *A is where Akita lives and B is where Ben lives. Can we show exactly where they would meet if they start at the same time and walk along this segment at the same rate?*

2 TEACH

Teaching Tip A plastic safety compass as provided in the *Overhead Manipulative Resources* is great for overhead demonstrations.

HANDS-ON ACTIVITY

11-1B Constructions

An Extension of Lesson **11-1**

MATERIALS

 compass

ruler

protractor

When two line segments have the same length, we say they are **congruent**. Similarly, two angles with the same measure are congruent. You can use a compass and straightedge to **construct** congruent segments and angles. A **straightedge** is any object you can use to draw a straight line, such as an index card or a ruler. You use a **compass** to draw a circle or part of a circle.

Activity 1 Construct a line segment congruent to a given line segment.

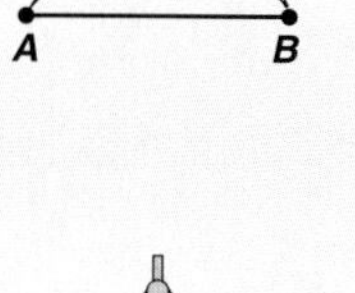

First, draw a segment and label it $\overline{AB}$. This will be the given segment.

Step 1: Use a straightedge to draw $\overrightarrow{PS}$ so it is longer than $\overline{AB}$.

Step 2: Place the steel tip of the compass at *A* and the writing tip at *B*.

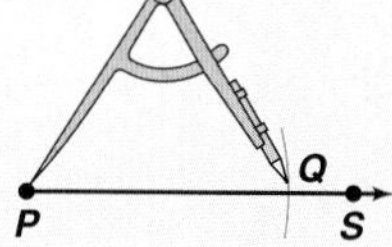

Step 3: Keep the same setting on the compass and place the steel tip at *P*. Draw an arc that intersects $\overrightarrow{PS}$ at *Q*. $\overline{PQ}$ is congruent to $\overline{AB}$. In symbols, $\overline{PQ} \cong \overline{AB}$.

Activity 2 Construct an angle that is congruent to a given angle.

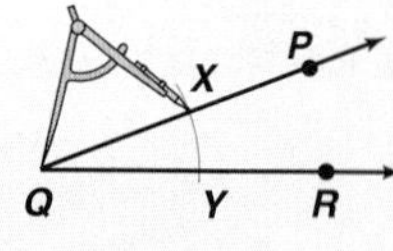

First, draw an angle and label it $\angle PQR$. This will be the given angle.

Step 1: Use a straightedge to draw $\overrightarrow{ST}$.

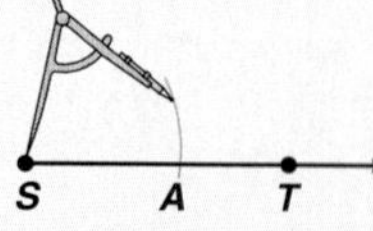

Step 2: Place the steel tip of the compass at *Q*. Draw an arc that intersects both sides of $\angle PQR$ to locate points *X* and *Y*.

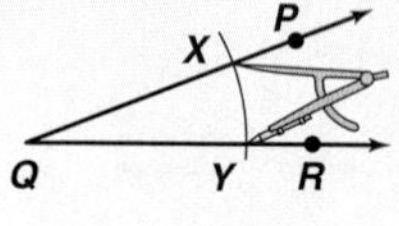

Step 3: Keeping the same compass setting, place the steel tip at *S* and draw an arc that intersects $\overrightarrow{ST}$. Label the intersection point *A*.

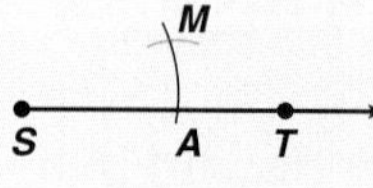

Step 4: Place the compass so one tip is at *X* and the other is at *Y*. Keep that setting and place the steel tip at *A*. Draw an arc that intersects the arc you drew in step 3. Label the intersection point *M*.

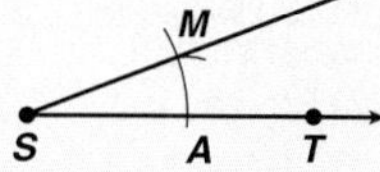

Step 5: With a straightedge, draw $\overrightarrow{SM}$. $\angle MST$ is congruent to $\angle PQR$. In symbols, $\angle MST \cong \angle PQR$.

Cooperative Learning

This lesson offers an excellent opportunity for using cooperative learning groups. For more information on cooperative learning strategies and group management, see *Cooperative Learning in the Mathematics Classroom*, one of the titles in the Glencoe Mathematics Professional Series.

1. See students' work.
2. See students' work.

1. How does the compass work in copying a line segment?
2. Explain how the compass works in copying an angle.

Activity 3 **Construct a line segment that bisects a given line segment.**

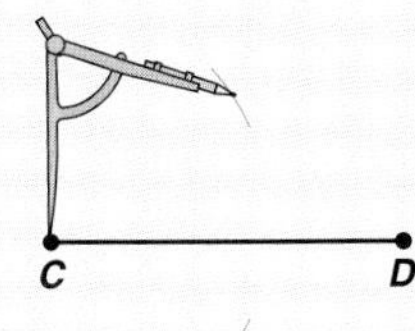

Draw a segment and label it $\overline{CD}$. This is the given line segment.

Step 1: Place the steel tip of the compass at *C*. Extend the compass until the writing tip is more than halfway to *D* and draw an arc "above" and "below" $\overline{CD}$ as shown.

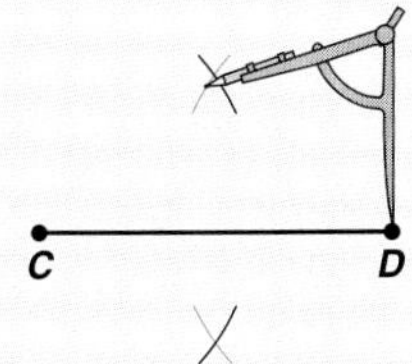

Step 2: With the same setting on the compass, place the steel tip at *D* and draw two arcs as shown to locate two new points, *Q* and *R*.

Step 3: With a straightedge, draw the line segment determined by these two new points. This segment will intersect $\overline{CD}$ at *M*.

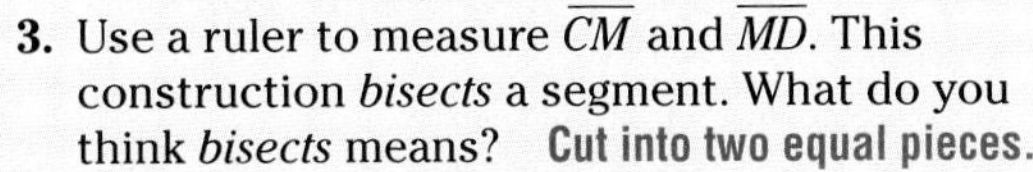

3. Use a ruler to measure $\overline{CM}$ and $\overline{MD}$. This construction *bisects* a segment. What do you think *bisects* means? **Cut into two equal pieces.**
4. In this construction, $\overline{QR}$ is perpendicular to $\overline{CD}$. What type of angles are formed by perpendicular segments? **right angles**

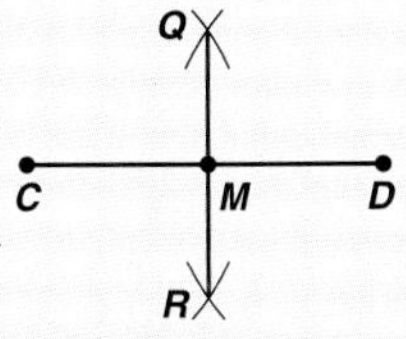

Activity 4 **Construct a ray that bisects a given angle.**

Draw an angle and label it $\angle ABC$. This is the given angle.

Step 1: Place the steel tip of the compass at *B* and draw an arc that intersects both sides of the angle. Label the points of intersection *X* and *Y*.

Step 2: Place the steel tip of the compass at *X* and open the compass to a length more than halfway to *Y*. Draw an arc in the interior of $\angle ABC$.

Step 3: Keeping the same compass setting, place the steel tip at *Y* and draw an arc that intersects the arc you drew in Step 2. Label the point of intersection *T*.

Step 4: With a straightedge, draw $\overrightarrow{BT}$. $\overrightarrow{BT}$ is the bisector of $\angle ABC$.

5. They are the same.

5. Use a protractor to measure $\angle ABT$ and $\angle TBC$. What is true about their measures?
6. Why do we say BT is the bisector of $\angle ABC$? **It separates the angle into two equal angles.**

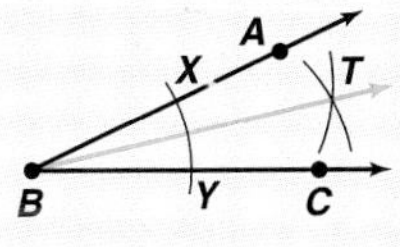

3 PRACTICE/APPLY

Assignment Guide

Core: 1–6
Enriched: 1–6

4 ASSESS

Observing students working in cooperative groups is an excellent method of assessment. You may wish to ask a student at random from each group to explain how to model and solve a problem. Also watch for and acknowledge students that are helping others to understand the concept being taught.

NCTM Standards: 1-4, 10, 12

Instructional Resources
- Study Guide Master 11-2
- Practice Master 11-2
- Enrichment Master 11-2
- Group Activity Card 11-2
- Activity Masters, p. 11
- Real-World Applications, 26
- Tech Prep Applications Masters, p. 21

Transparency 11-2A contains the 5-Minute Check for this lesson; **Transparency 11-2B** contains a teaching aid for this lesson.

Recommended Pacing	
Standard Pacing	Day 3 of 13
Honors Pacing	Day 3 of 13
Block Scheduling*	Day 2 of 7 (along with Lesson 11-3)

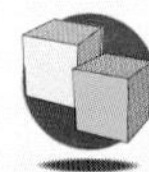

*For more information on pacing and possible lesson plans, refer to the ***Block Scheduling Booklet***.

1 FOCUS

5-Minute Check
(over Lesson 11-1)

Tell whether each model suggests a *point*, a *line*, a *plane*, a *ray*, or a *line segment*. Explain your answers.

1. path of the light from a flashlight **ray**
2. tip of a straight pin **point**
3. edge of a ruler **line segment**

Classify each angle as acute, obtuse, or right.

4. $m\angle ABC = 45°$ **acute**
5. $m\angle PQR = 145°$ **obtuse**

Motivating the Lesson
Questioning Show students a circle graph from a recent newspaper or magazine. Ask what the graph is showing.

11-2 Integration: Statistics
Making Circle Graphs

Setting Goals: *In this lesson, you'll make a circle graph to illustrate data.*

Modeling a Real-World Application: Weather

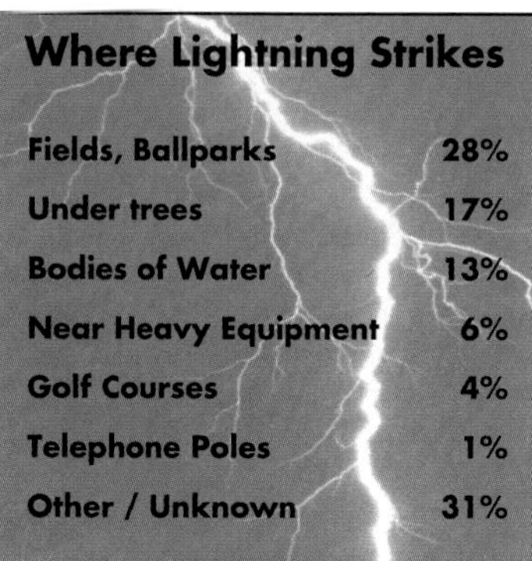

Where Lightning Strikes	
Fields, Ballparks	28%
Under trees	17%
Bodies of Water	13%
Near Heavy Equipment	6%
Golf Courses	4%
Telephone Poles	1%
Other / Unknown	31%

Source: National Weather Service

Have you ever been caught in a thunderstorm and wondered where the safest place to be is? The chart at the left indicates where lightning strikes most often.

A circle graph can be used to display this data. What angles would you use for each section?

Learning the Concept

LOOK BACK You can review percents in Lesson 9-5.

A **circle graph** is used to compare parts of a whole in a way that helps you visualize the information. As you know, there are 360° in a circle. So, you can multiply to find the number of degrees in each section of the circle graph.

THINK ABOUT IT Check the sum of the degree measures. What should it be? **360°**

Fields/Ballparks	28 [%] [×] 360 [=] 100.8
Under Trees	17 [%] [×] 360 [=] 61.2
Bodies of Water	13 [%] [×] 360 [=] 46.8
Near Heavy Equipment	6 [%] [×] 360 [=] 21.6
Golf Courses	4 [%] [×] 360 [=] 14.4
Telephone Poles	1 [%] [×] 360 [=] 3.6
Other/Unknown	31 [%] [×] 360 [=] 111.6

Example 1 shows how to make a circle graph for data that has not been expressed as a percent.

Classroom Vignette

"My students work in cooperative groups to collect data and then use a protractor to construct a circle graph of their data. Then we use a spreadsheet and statistical graphing computer software to enter the same data and make a circle graph. The students compare the two graphs and draw conclusions from the results."

Loren Holthaus
Big Lake Middle School
Big Lake, MN

Example CONNECTION History

You can use spreadsheets and word processing programs to construct circle graphs from data you enter.

To be president of the United States, an individual must be at least 35 years old. Make a circle graph to display the data on presidents' ages at their first inauguration shown at the right.

President's Age on Inauguration Day	
Age Interval	Number of Presidents
35–43	2
44–52	14
53–61	19
62–70	6

Explore You know the number of presidents in each age interval.

Plan Find the total number of presidents. Find the ratio of the number in each interval to the total. Then find the number of degrees for each interval. Finally, draw the graph.

Solve **Step 1:** Find the total number of presidents.
$2 + 14 + 19 + 6 = 41$

Step 2: Find the ratio that compares the number in each interval to the total.

35–43: 2 [÷] 41 [=] 0.048780
44–52: 14 [÷] 41 [=] 0.341463
53–61: 19 [÷] 41 [=] 0.463415
62–70: 6 [÷] 41 [=] 0.146341

Step 3: Multiply to find the number of degrees for each section of the graph. Round to the nearest degree.

35–43: 0.048780 [×] 360 [=] 17.5608 → 18°
44–52: 0.341463 [×] 360 [=] 122.92668 → 123°
53–61: 0.463415 [×] 360 [=] 166.8294 → 167°
62–70: 0.146341 [×] 360 [=] 52.68276 → 53°

Step 4: Draw a circle graph. Use a compass to draw a circle.

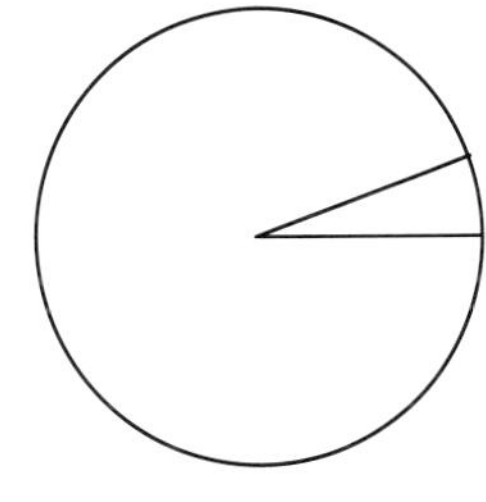

Start with the least number of degrees, in this case 18°. Use a protractor to draw an angle of 18°.

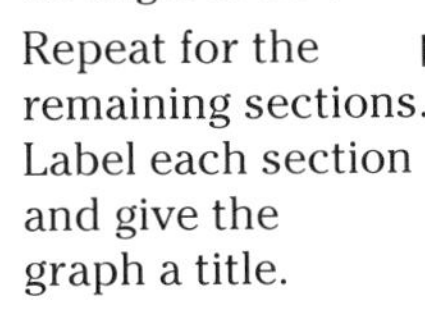

Repeat for the remaining sections. Label each section and give the graph a title.

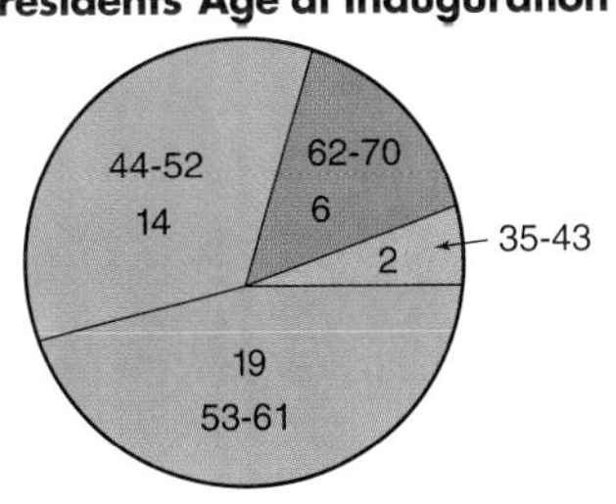

Examine Add the degree measures of each section. The total is 361°. With rounding, this is reasonable.

2 TEACH

In-Class Example

For Example 1
Make a circle graph to display the following data.

Students in Each Class	
Seniors	185
Juniors	105
Sophomores	115
Freshmen	95

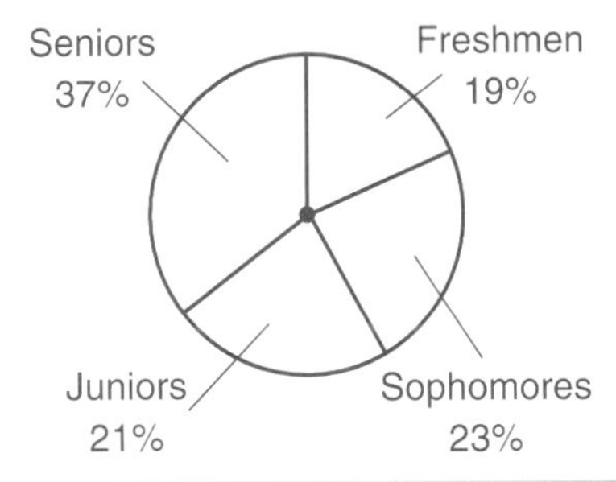

Teaching Tip You may wish to show students that Steps 2 and 3 can be combined into one step by using proportions. For example for 62–70, solve the proportion $\frac{6}{41} = \frac{x}{360}$; $x = 52.68292$ or 53°

3 PRACTICE/APPLY

Checking Your Understanding

Exercises 1–8 are designed to help you assess your students' understanding through reading, writing, speaking, and modeling. You should work through Exercises 1–4 with your students and then monitor their work on Exercises 5–8.

Error Analysis

Some students may not make sure that their data totals 100% before starting the circle graph. Point out that, in that case, they may not have a completed circle graph.

Assignment Guide
Core: 9, 11–13, 15, 17–21
Enriched: 10–21

For **Extra Practice**, see p. 767.

The red A, B, and C flags, printed only in the Teacher's Wraparound Edition, indicate the level of difficulty of the exercises.

Additional Answers

1. A circle graph shows how each piece of data compares to the whole while a bar graph compares data pieces to each other. A line graph shows how data change over time. A circle graph is appropriate when you want to compare parts.
2. Multiply the percent by 360 to find the number of degrees for each section. Draw the angles. Label each section and give the graph a title.
3. Find the total of the data. Write ratios comparing each part to the whole. Multiply each ratio by 360 to find the number of degrees for each angle.

Study Guide Masters, p. 92

NAME ______________________ DATE __________

11-2 Study Guide
Integration: Statistics
Making Circle Graphs

Student Edition
Pages 556–560

A **circle graph** is used to illustrate data. In order to make a circle graph, the circle must be separated into sectors. Each circle is made up of 360°. Therefore, to separate the circle into sectors, multiply the percent needed by 360.

Example: 25% of a family's budget is spent for housing.

$$0.25 \times 360° = 90°$$

Use a protractor and measure 90°.
Then label the sector "Housing" as shown below.

1. The chart shows the percentages allotted for each item of the Johnson's family budget. Make a circle graph to display the data.

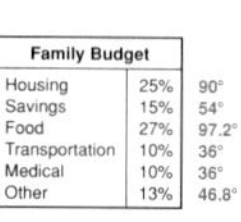

Family Budget		
Housing	25%	90°
Savings	15%	54°
Food	27%	97.2°
Transportation	10%	36°
Medical	10%	36°
Other	13%	46.8°

Johnson Family Budget

2. In a recent poll, a group of people were asked to name their favorite type of television show. Make a circle graph to display the data.

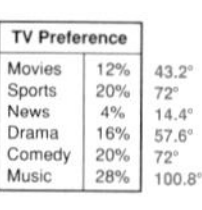

TV Preference		
Movies	12%	43.2°
Sports	20%	72°
News	4%	14.4°
Drama	16%	57.6°
Comedy	20%	72°
Music	28%	100.8°

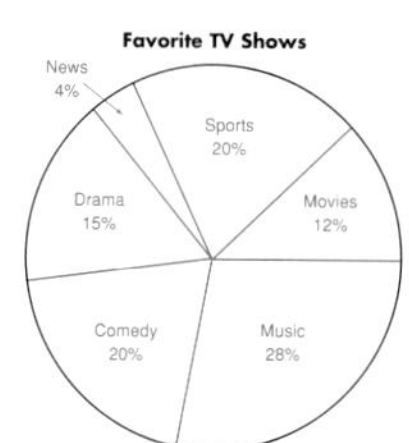

Checking Your Understanding

Communicating Mathematics

Read and study the lesson to answer these questions. 1–3. See margin.

1. **Explain** how a circle graph differs from a bar graph and from a line graph. When is it appropriate to display data in a circle graph?
2. **Tell** the steps to make a circle graph if your data is in percent form.
3. **Explain** how to find the circle graph angles for data that is not in percent form.

4. Find a circle graph in a newspaper or magazine and make a copy to put in your journal. Write a short paragraph that explains the data displayed by the graph. See students' work.

Guided Practice

Use the circle graph at the right to answer each question.

CRUST
TRANSITION ZONE
LOWER MANTLE
LIQUID OUTER CORE
SOLID INNER CORE

Elements of Earth's Crust

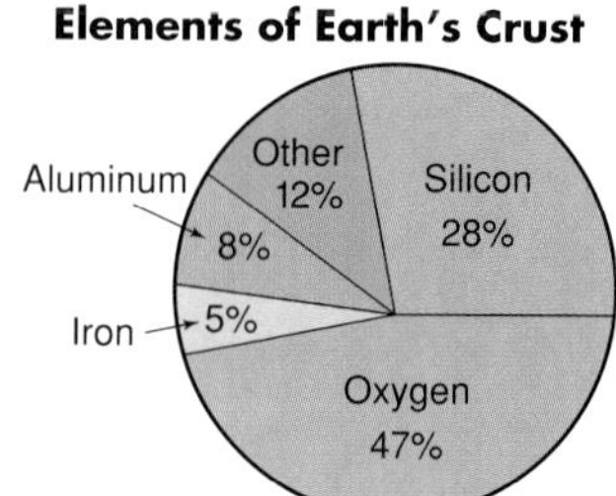

5. Which element makes up approximately $\frac{1}{4}$ of the earth's crust? silicon
6. How much of Earth's crust is composed of oxygen and silicon? 75%
7. The table below shows the Vera family's monthly budget.

Monthly Budget			
Category	**Amount ($)**	**Percent**	**Angle (°)**
Housing	640	31	111.6
Food	400	19	68.4
Transportation	450	22	79.2
Insurance	140	7	25.2
Savings	90	4	14.4
Misc.	360	17	61.2
Totals	2080	100	360.0

a. What total would you expect in the percent column? 100%
b. What total would you expect in the degree column? 360°
c. Copy and complete the table. Round degrees to the nearest tenth. (7c. See the table.)
d. Make a circle graph to display the data. Use spreadsheet or word processing software if it is available. See Solutions Manual.

8. **Probability** What is the probability of tossing a head when you flip a fair coin? What would a circle graph for all the possible results of this experiment be like? $\frac{1}{2}$; half for heads and half for tails.

Exercises: Practicing and Applying the Concept

Independent Practice

9. Count the months of the year that are in each of these three categories: 28 or 29 days long, 30 days long, and 31 days long. Make a circle graph to display your findings. See Solutions Manual.

GLENCOE Technology

Interactive Mathematics Tools Software

In this interactive computer lesson, students make circle graphs. A **Computer Journal** gives students the opportunity to write about what they have learned.

For Windows & Macintosh

Reteaching

Using Data Have students decide upon a topic and take a survey of the class. Then assign each small group of students to find the number of degrees that will be in a specified interval. Finally, have students pool their information and create a class circle graph.

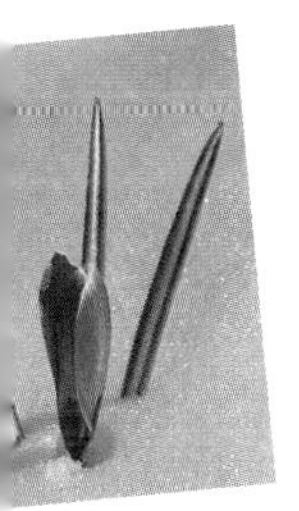

0. See Solutions
Manual.

10. In a recent poll, students were asked to name their favorite season of the year. The results of the survey are shown at the right. Make a circle graph to display the data. Use spreadsheet or word processing software if it is available.

Season Preference	
Summer	27%
Autumn	22%
Winter	18%
Spring	33%

11. Probability Three red marbles and 5 green marbles are placed in a bag. In an experiment, you are to select one marble. If the marble you select is red, you will win a silver dollar. If the marble you select is green, you will lose.

a. What is the probability that you will win? $\frac{3}{8}$

b. Make a circle graph that shows the probability of winning and losing. **See margin.**

Critical Thinking

2a. No, The section
or NBC should only
ave 68°—this shows
more than that.
2b. To make it appear
hat "Friends" had a
arger audience share.

12. Suppose someone drew the graph at the right to display information on the audience share during the NBC show *Friends* on Thursday, October 5, 1995.

CBS 12%
NBC 19%
ALL OTHERS 36%
14% FOX
ABC 19%

a. Does the graph accurately represent the data it presents? Why or why not?

b. Why might someone have drawn the graph this way?

Applications and Problem Solving

Make a circle graph to display each set of data.

13. Geography **See Solutions Manual.** **14. Biology** **See Solutions Manual.**

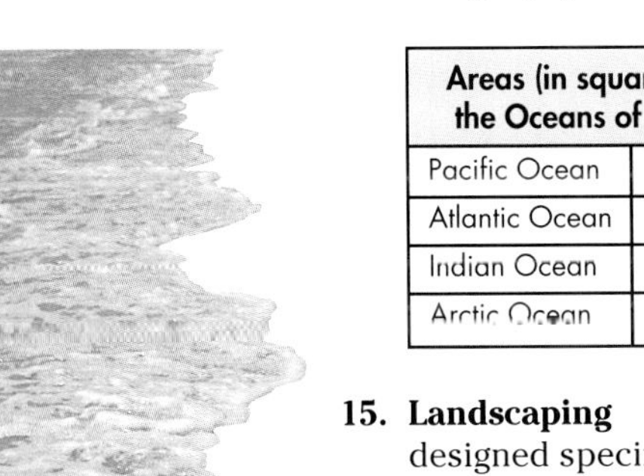

Areas (in square miles) of the Oceans of the World	
Pacific Ocean	64,186,300
Atlantic Ocean	33,420,000
Indian Ocean	28,350,500
Arctic Ocean	5,105,700

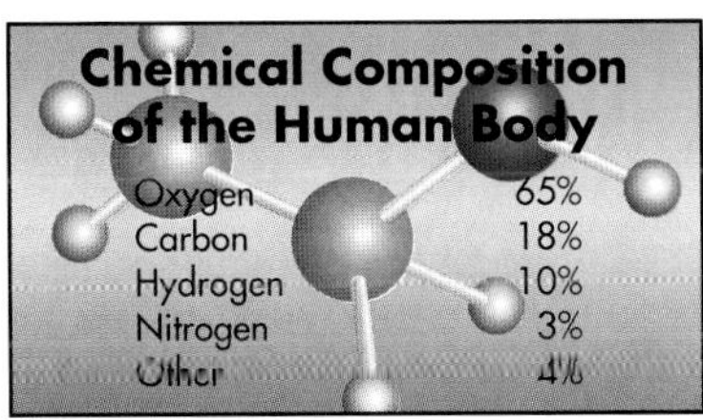

Chemical Composition of the Human Body	
Oxygen	65%
Carbon	18%
Hydrogen	10%
Nitrogen	3%
Other	4%

15. Landscaping A seed company prepares a grass seed mixture that is designed specifically for the Black Hills of South Dakota where the growing season is very short. The following mixture is used.

Black Hills Reclamation Mix	
30% VNS Kentucky Bluegrass	25% VNS Timothy
10% Regar Bromegrass	10% Alsike Clover
10% Lincoln Smooth Brome	10% VNS Tall Fescue
5% Medium Red Clover	

a. Make a circle graph to display the data. **See margin.**

b. Find the amount of each seed that would be needed to make 250 pounds of the mixture. **See margin.**

16. Family Activity Choose a set of data to collect, such as colors of cars in a parking lot, or the number of commercials in TV commercial breaks and record the data in a frequency table. **See students' work.**

a. Make a circle graph and a bar graph to display the data.

b. Which graph better represents the data and why?

Additional Answers

11b.

15a.

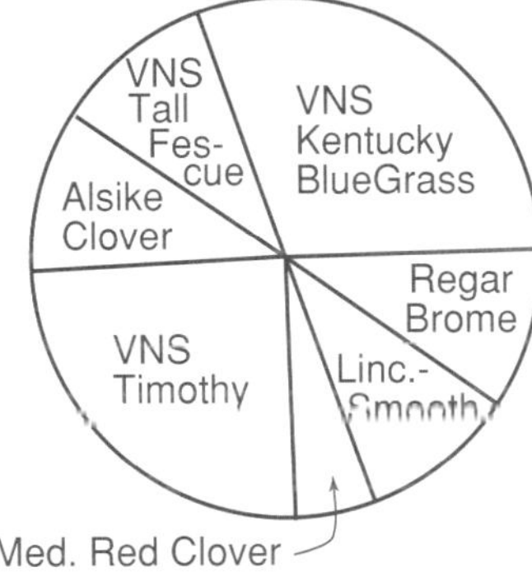

15b. 75 lb VNS Kentucky Bluegrass; 25 lb each of Regar Bromegrass, Lincoln Smooth Brome, Alsike Clover, and VNS Tall Fescue; 12.5 lb Medium Red Clover; 62.5 VNS Timothy

Practice Masters, p. 92

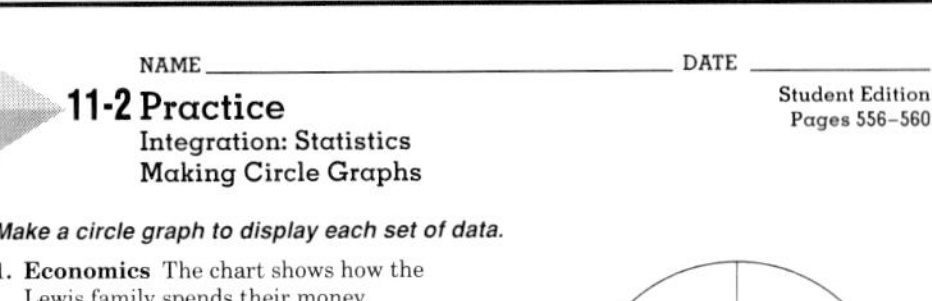

NAME ______________ DATE ______

11-2 Practice
Integration: Statistics
Making Circle Graphs

Student Edition
Pages 556–560

Make a circle graph to display each set of data.

1. **Economics** The chart shows how the Lewis family spends their money. Make a circle graph to display the data.

How the Lewis Family Spends Their Money		
Housing	$10,500	126°
Food	$7500	90°
Clothing	$3000	36°
Investments	$3000	36°
Miscellaneous	$6000	72°

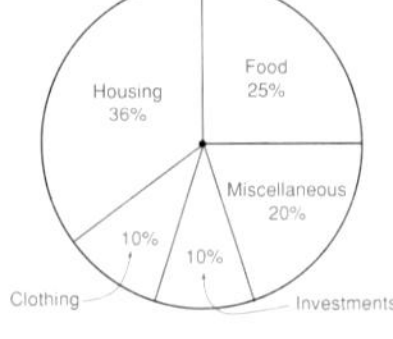

2. **Statistics** The chart shows how many hours are spent in daily activities. Make a circle graph to display the data.

Daily Activities		
Sleeping	8 hours	120°
Eating	3 hours	45°
School	6 hours	90°
Homework	3 hours	45°
Miscellaneous	4 hours	60°

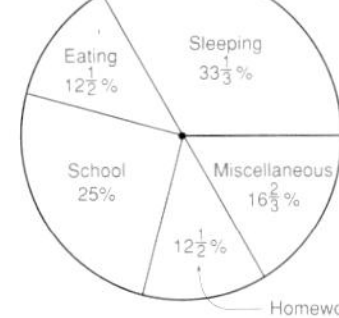

3. **Business** The chart shows four types of consumers who bought computers from a company in one year. Make a circle graph to display the data.

Computer Sales		
Educational	$2,000,000	36°
Personal	$5,200,000	93.6°
Business	$10,400,000	187.2°
Scientific	$2,400,000	43.2°

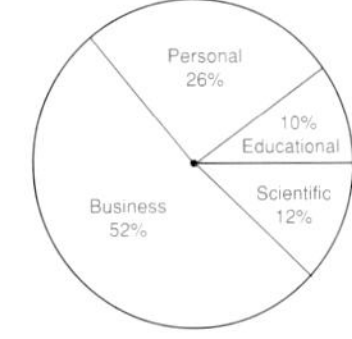

Group Activity Card 11-2

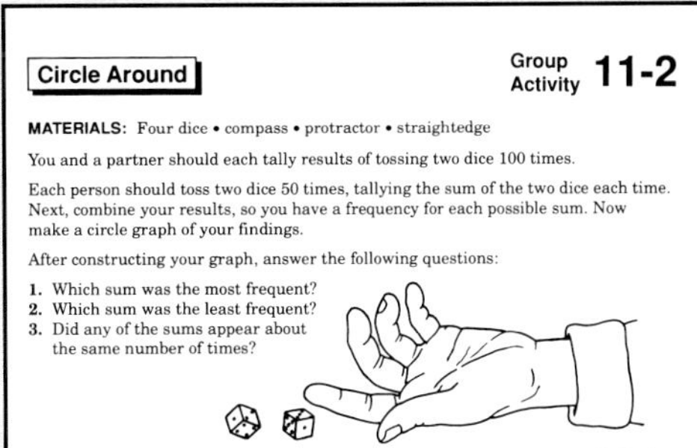

Circle Around Group Activity **11-2**

MATERIALS: Four dice • compass • protractor • straightedge

You and a partner should each tally results of tossing two dice 100 times.

Each person should toss two dice 50 times, tallying the sum of the two dice each time. Next, combine your results, so you have a frequency for each possible sum. Now make a circle graph of your findings.

After constructing your graph, answer the following questions:

1. Which sum was the most frequent?
2. Which sum was the least frequent?
3. Did any of the sums appear about the same number of times?

 Pre-Algebra

Extension

Using Models Have students find circle graphs in other textbooks, magazines, and newspapers. Have them prepare a bulletin board display that includes word problems based upon the circle graphs.

4 ASSESS

Closing Activity

Speaking Display the following data on the board or overhead.

City Zoo Animal Distribution	
Mammals	50%
Fish	25%
Insects	6%
Reptiles	9%
Amphibians	10%

Have students tell how to complete each step in the process of making a circle graph to show the information.

Additional Answers

19.

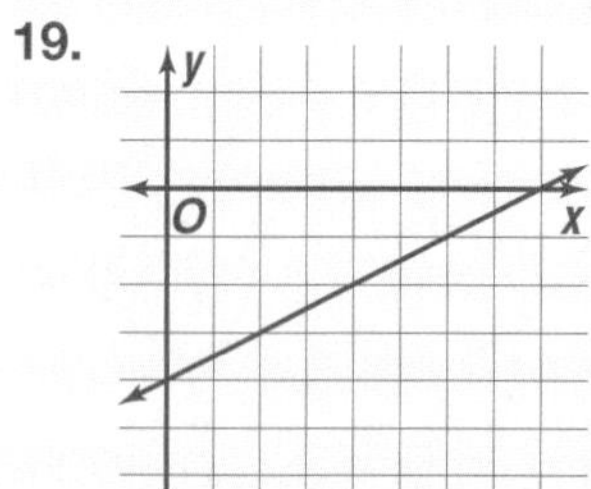

21. Sample answer: Joe earned 3 times as much as Sue. If Joe earned $25, how much did Sue earn?

Enrichment Masters, p. 92

NAME ______ DATE ______

11-2 Enrichment
Secants

Student Edition Pages 556–560

A **secant** is a line that intersects a circle at exactly two points. $\overline{AB}$ and $\overline{BC}$ are secant segments.

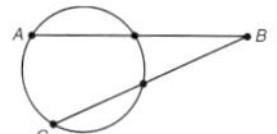

If two secant segments are drawn from a point outside a circle, then the product of the measures of one secant segment and its external part is equal to the product of the measure of the other secant segment and its external part.

Example:

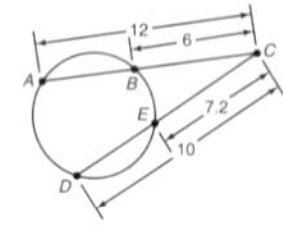

$AC \cdot BC = DC \cdot EC$
$(12)(6) = (10)(7.2)$
$72 = 72$

Substitute the measures from the diagram into the equation. Then multiply.

Find the length of the secant segment or its external part in each of the following.

1. $AC = 15$, $BC = 5$, $EC = 20$; Find DC. 3.75
2. $AC = 25$, $BC = 15$, $DC = 10$; Find EC. 37.5
3. $AC = 75$, $BC = 30$, $EC = 60$; Find DC. 37.5

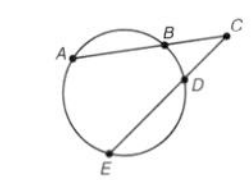
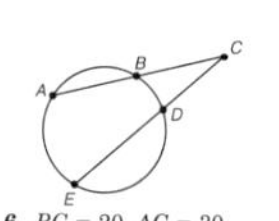

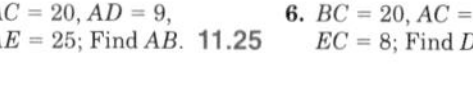

4. $AB = 7$, $AC = 15$, $AE = 8$; Find AD. 13.125
5. $AC = 20$, $AD = 9$, $AE = 25$; Find AB. 11.25
6. $BC = 20$, $AC = 30$, $EC = 8$; Find DC. 75

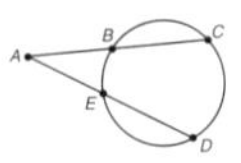
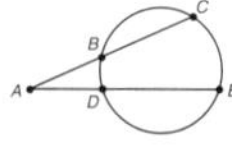
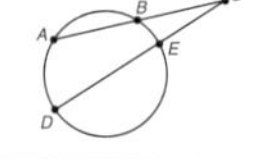

Mixed Review

17. Draw and label a diagram to represent parallel lines a and b. (Lesson 11-1) **See Solutions Manual.**
18. **Business** The phone company assigns the first three digits of a phone number based on the geographic area of the home or business. The remaining four digits are selected from a data base. If any digit from 0 to 9 can be used in any position, how many different combinations of the last four digits are possible? (Lesson 10-6) **10,000**
19. Graph $y = \frac{1}{2}x - 4$ by using the slope and y-intercept. (Lesson 8-7) **19. See margin.**
20. The distance from Kuri's house to school is about 1800 feet. About what part of a mile is this? (Lesson 6-5) **about $\frac{1}{3}$**
21. Write a problem that could be solved using the equation $3x = 25$. (Lesson 1-8) **See margin.**

COOPERATIVE LEARNING PROJECT

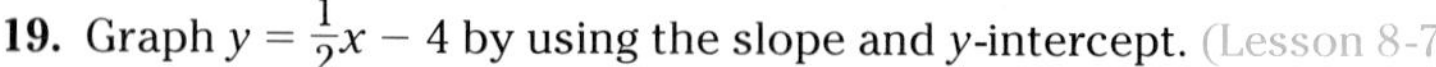

Virtual Reality

Imagine yourself caught up in a daydream so real that you can see and hear the sights and sounds of some faraway reality. Perhaps you are walking through a dense forest. Soon you will be able to simply slip on a headset and enter such a world—a world of *virtual reality*, without ever leaving your room.

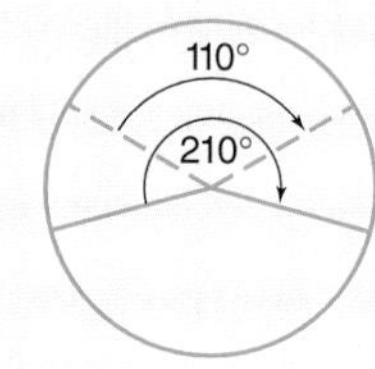

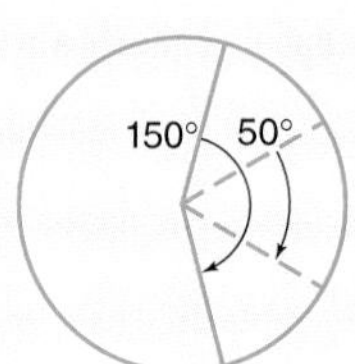

Virtual reality is a three-dimensional computer world. Viewing the images of your virtual reality headset is similar to watching a living, moving graph. Moving your head up and down, the first dimension, allows you to scan whatever images are along the vertical axis. Similarly, moving your head left and right, the second dimension, allows you to scan whatever images are along the horizontal axis. Finally, special lenses cover the screen of your headset and allow you to scan distant objects, the third dimension.

By moving your head from side to side, but not turning your body, your horizontal field of view is about 210°. When you move your head up and down, your vertical field of view is about 150°. Today's virtual reality headsets can only provide a horizontal field of view of about 110° and a vertical field of view of about 50°. But, they should catch up in a few years.

See for Yourself

Research virtual reality systems and computer graphics. **See Solutions Manual.**

- How does the main computer of a virtual reality system know the direction in which you are looking at any given time?
- Identify and name the three dimensions of a virtual reality system.
- How is a virtual reality screen similar to a living, moving graph?
- With the help of two partners, how can you find the number of degrees of your own horizontal and vertical fields of view?
- Why is the reality of virtual reality headsets referred to as "virtual"?

COOPERATIVE LEARNING PROJECT

THE SHAPE OF THINGS TO COME

Virtual Reality *Cyberspace* is another name for virtual reality. Virtual reality is based on the computer's ability to produce images that exhibit the colors, textures, and changing spatial orientations that real objects exhibit. Potential uses range from long-distance manipulation of robot devices to retraining of stroke victims in the use of their limbs.

11-3 Angle Relationships and Parallel Lines

Setting Goals: *In this lesson, you'll identify the relationships of vertical, adjacent, complementary, supplementary angles, and angles formed by two parallel lines and a transversal.*

Modeling with Manipulatives

MATERIALS

protractor

Notice that the writing lines on notebook paper are parallel lines. You can use these lines to investigate angles and parallel lines.

Your Turn

- Use the lines on notebook paper to draw two parallel lines.
- Then draw another line that intersects the parallel lines.
- Use a protractor to measure all the angles formed when this line crosses the parallel lines. Make a sketch that includes the measure of each angle.
- Repeat this activity using a different line.

TALK ABOUT IT

a. Not including straight angles, how many angles are formed when a line intersects two parallel lines?

b. What do you notice about the measures of the angles? **See margin.**

c. Describe the pairs of angles that appear to have the same measure.

d. What do you notice about measures of angles that are side by side?

a. 8
c. See margin.
d. Their sum is 180°.

Learning the Concept

In the activity above, you discovered some of the special relationships that *pairs* of angles can have.

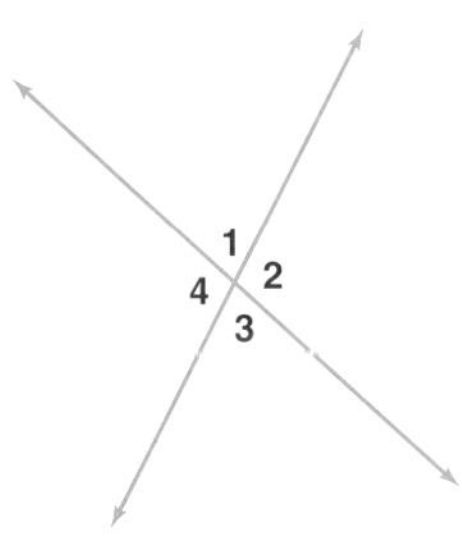

When two lines intersect, they form two pairs of "opposite" angles called **vertical angles**. In the figure at the left, angles 1 and 3 are vertical angles as are angles 2 and 4. If you use a protractor to measure the angles, you will discover that the angles in each pair have the same measure. Angles with the same measure are **congruent**. Vertical angles are always congruent.

When two angles in a plane have the same vertex, share a common side, and do not overlap, they are called **adjacent angles**. In the figure at the right, ∠1 and ∠2 are adjacent angles.

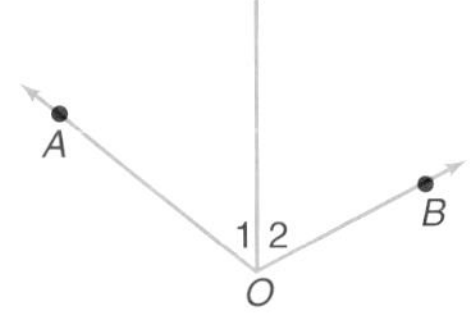

m∠AOB means the degree measure of angle AOB.

We can find the measure of ∠*AOB* by adding the measures of ∠1 and ∠2. That is, $m\angle AOB = m\angle 1 + m\angle 2$.

Additional Answers Talk About It

b. There are just two different measures.

c. Sample answers: the angles across from each other; the angles in the same position on each of the parallel lines; the angles on opposite sides inside the parallel lines; the angles on opposite sides above and below the parallel lines.

11-3 LESSON NOTES

NCTM Standards: 1-4, 10, 12

Instructional Resources

- Study Guide Master 11-3
- Practice Master 11-3
- Enrichment Master 11-3
- Group Activity Card 11-3
- Assessment and Evaluation Masters, p. 295
- Activity Masters, p. 25

Transparency 11-3A contains the 5-Minute Check for this lesson; **Transparency 11-3B** contains a teaching aid for this lesson.

Recommended Pacing	
Standard Pacing	Day 4 of 13
Honors Pacing	Day 4 of 13
Block Scheduling*	Day 2 of 7 (along with Lesson 11-2)

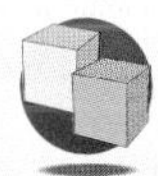

*For more information on pacing and possible lesson plans, refer to the ***Block Scheduling Booklet***.

1 FOCUS

5-Minute Check
(over Lesson 11-2)

Misha kept track of how she spent her evenings during the week. Use the data in the table below to make a circle graph.

How Misha Spent Her Evenings (Monday–Friday)	
Studying	2 hours
Errands/Chores	1 hour
With Family	$\frac{3}{4}$ hour
With Friends	$1\frac{3}{4}$ hours
Other	$\frac{1}{2}$ hour

How Misha Spends Her Evenings (Monday—Friday)

Motivating the Lesson

Situational Problem Sketch a football field on the board. Then pose this situation for students. A football player is here (mark an X on the field) and runs in this direction, crossing the yard-line marker here and running diagonally in a straight line. (From the X, draw an oblique line through the next yard line and stop.) Where should a defensive player be to stop the runner at the next yard-line marker? Point out that the path of the runner is a line segment that crosses the parallel yard lines.

2 TEACH

In-Class Examples

For Example 1
Angle *A* has a measure of 24°.

a. If $\angle A$ and $\angle B$ are complementary, what is the measure of $\angle B$? **66°**

b. If $\angle A$ and $\angle C$ are supplementary, what is the measure of $\angle C$? **156°**

For Example 2
Angles *MNP* and *RST* are supplementary. If $m\angle MNP = 3x - 4$ and $m\angle RST = x - 8$, find the measure of each angle. **$m\angle MNP = 140°$, $m\angle RST = 40°$**

Teaching Tip You may wish to have students make vocabulary cards for *complementary angles* and *supplementary angles* that include a definition and a diagram. Point out that "supplementary" and "straight" both start with "s" and that supplementary angles together form a straight angle. Students can then refer to them if they have trouble remembering which is which.

If the sum of the measures of two angles is 90°, the angles are **complementary**.

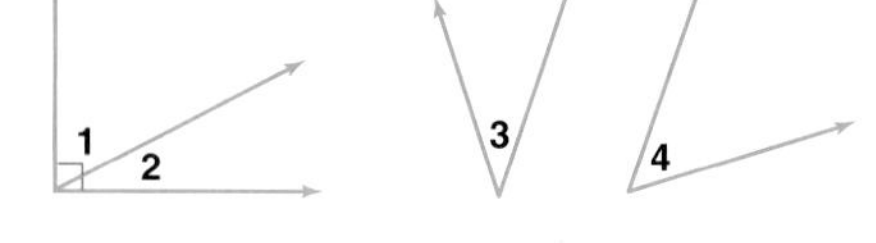

$m\angle 1 + m\angle 2 = 90°$ $m\angle 3 + m\angle 4 = 90°$

If the sum of the measures of two angles is 180°, they are **supplementary**.

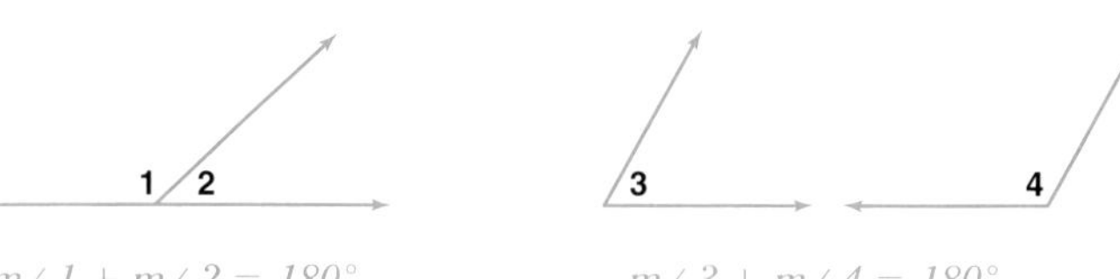

$m\angle 1 + m\angle 2 = 180°$ $m\angle 3 + m\angle 4 = 180°$

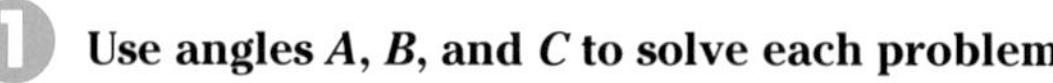

Example 1 **Use angles *A*, *B*, and *C* to solve each problem.**

a. Angles *A* and *B* are complementary. If $m\angle A = 70°$, find $m\angle B$.

$m\angle A + m\angle B = 90$ *Definition of complementary angles*
$70 + m\angle B = 90$ *Substitute.*
$m\angle B = 20$ *Subtract 70 from each side.*

The measure of $\angle B$ is 20°.

b. Angles *A* and *C* are supplementary. Find $m\angle C$.

$m\angle A + m\angle C = 180$ *Definition of supplementary angles*
$70 + m\angle C = 180$ *Replace $m\angle A$ with 70.*
$m\angle C = 110$ *Subtract 70 from each side.*

The measure of $\angle C$ is 110°.

Connection to Algebra

Geometric relationships can be expressed using the tools of algebra.

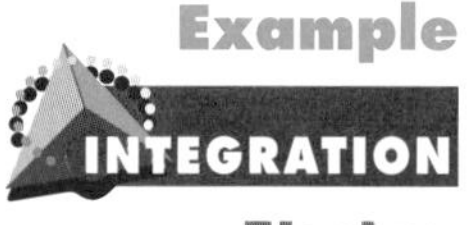

Example 2 **INTEGRATION Algebra**

Angles *PQR* and *RQT* are complementary. If $m\angle PQR = x + 5$ and $m\angle RQT = x - 9$, find the measure of each angle.

$m\angle PQR + m\angle RQT = 90$ *Definition of complementary angles*
$(x + 5) + (x - 9) = 90$ *Substitute.*
$2x - 4 = 90$ *Combine like terms.*
$2x = 94$ *Add 4 to each side.*
$x = 47$ *Divide each side by 2.*

Now replace *x* with 47 in the expression for each angle.

$m\angle PQR = x + 5 = 47 + 5$ or 52 $m\angle RQT = x - 9 = 47 - 9$ or 38

The measure of $\angle PQR$ is 52°, and the measure of $\angle RQT$ is 38°.

Check: $52° + 38° = 90°$ The angles are complementary.

THINK ABOUT IT

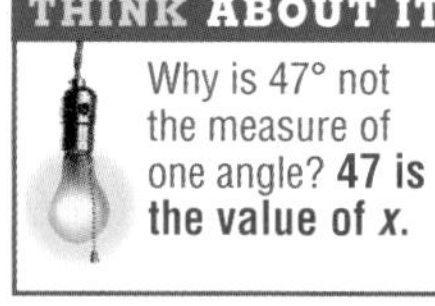

Why is 47° not the measure of one angle? **47 is the value of *x*.**

Alternative Teaching Strategies

Student Diversity Provide each student with four index cards. Have students write these measures, one to a card: 43°; 57°; 137°; 123°. Have students hold up the correct card as you illustrate each situation.

- Draw two intersecting lines and label one angle as 137°. Point to each of the other angles, one at a time.
- Draw two complementary angles. Label one angle 57°. Point to the other angle.
- Draw two supplementary angles. Label one angle 43°. Point to the other angle.
- Draw a set of parallel lines cut by a transversal. Label one angle as 123°. Point to each of the other angles.

THINK ABOUT IT

Why do you think these angles are referred to as interior and exterior angles?

Interior angles are "inside" the parallel lines and exterior angles are "outside."

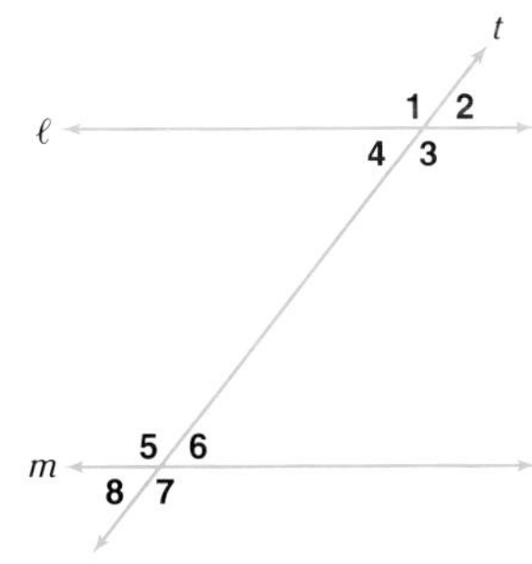

As you discovered in the modeling activity at the beginning of the lesson, when two parallel lines are intersected by a third line, called a **transversal**, eight angles are formed. Four are *interior* angles and four are *exterior* angles.

Interior angles: $\angle 3, \angle 4, \angle 5, \angle 6$
Exterior angles: $\angle 1, \angle 2, \angle 7, \angle 8$

When we study the relationship between the different angles, we can pair the angles as follows.

- **alternate interior angles:** $\angle 4$ and $\angle 6$, $\angle 3$ and $\angle 5$
  Alternate interior angles are nonadjacent interior angles found on opposite sides of the transversal.
- **alternate exterior angles:** $\angle 1$ and $\angle 7$, $\angle 2$ and $\angle 8$
  Alternate exterior angles are nonadjacent exterior angles found on opposite sides of the transversal.
- **corresponding angles:** $\angle 1$ and $\angle 5$, $\angle 2$ and $\angle 6$, $\angle 3$ and $\angle 7$, $\angle 4$ and $\angle 8$
  Corresponding angles are angles that have the same position on two different parallel lines cut by a transversal.

In the activity at the beginning of the lesson, you may have discovered the following relationships.

Parallel Lines Cut by a Transversal

Corresponding angles are congruent.
Alternate interior angles are congruent.
Alternate exterior angles are congruent.

Example 3

A transversal crosses two parallel lines so that $m\angle 4 = 75°$.

a. Which other angles also have a measure of 75°?

b. Find the measures of the other angles.

a. Since $\angle 2$ and $\angle 4$ are vertical angles, they are congruent. So $m\angle 2 = 75°$.
Since $\angle 4$ and $\angle 6$ are alternate interior angles, they are congruent. So, $m\angle 6 = 75°$.
$\angle 4$ and $\angle 8$ are corresponding angles so they are congruent. Thus, $m\angle 8 = 75°$.
Angles 2, 4, 6, and 8 have a measure of 75°.

b. Both angles 1 and 3 are supplementary to $\angle 4$ so the sums of each of their measures and the measure of angle 4 is 180°.
$180 - 75 = 105$. So, $m\angle 1 = 105°$ and $m\angle 3 = 105°$.
Since $\angle 7$ and $\angle 1$ are alternate exterior angles, they are congruent. So, $m\angle 7 = 105°$.
$\angle 5$ and $\angle 1$ are corresponding angles so they are congruent. Thus, $m\angle 5 = 105°$.

Angles 1, 3, 5, and 7 have a measure of 105°.

Teaching Tip You may wish to have students make vocabulary cards for *alternate interior angles*, *alternate exterior angles* and *corresponding angles* that include a definition and a diagram. Students can then refer to them as necessary.

In-Class Examples

For Example 3

A transversal crosses two parallel lines so that $m\angle 1 = 112°$.

a. Which other angles also have a measure of 112°? **3, 6, 8**

b. Find the measures of the other angles. **Angles 2, 4, 5, and 7 have a measure of 78°.**

For Example 4

A rectangular wooden gate has a wire brace attached to two corners.

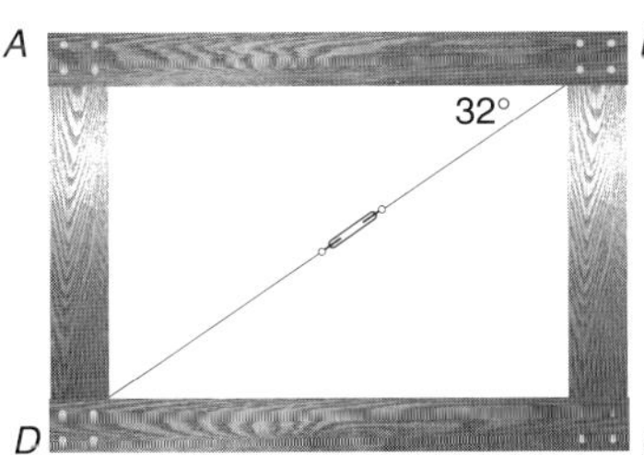

If the wire makes a 32° angle with the top rail of the gate, what angle does it make with the bottom rail of the gate? **32°** What angle does it make with the side rail of the gate? **The two angles formed by the wire at a corner are complementary. So, the angle made with the side rail is 58°.**

3 PRACTICE/APPLY

Checking Your Understanding

Exercises 1–16 are designed to help you assess your students' understanding through reading, writing, speaking, and modeling. You should work through Exercises 1–5 with your students and then monitor their work on Exercises 6–16.

Additional Answers

1. Vertical angles are formed by two intersecting lines. They share only a common vertex. Adjacent angles share a vertex and a common side.
2. Sample answers:

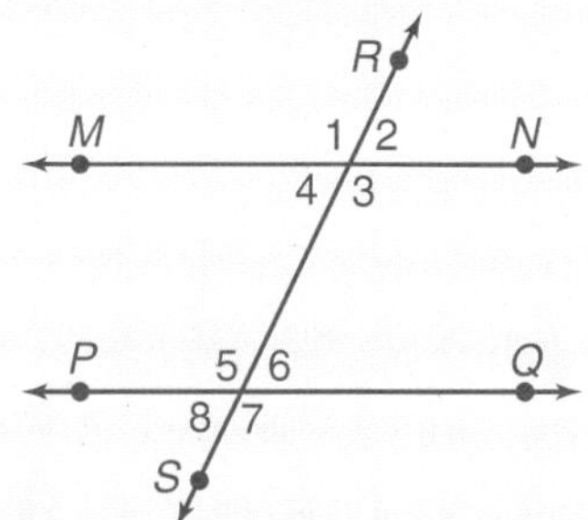

a. ∠1 and ∠5
b. ∠4 and ∠6
c. ∠1 and ∠7
d. ∠1 and ∠2
e. ∠1 and ∠4

Study Guide Masters, p. 93

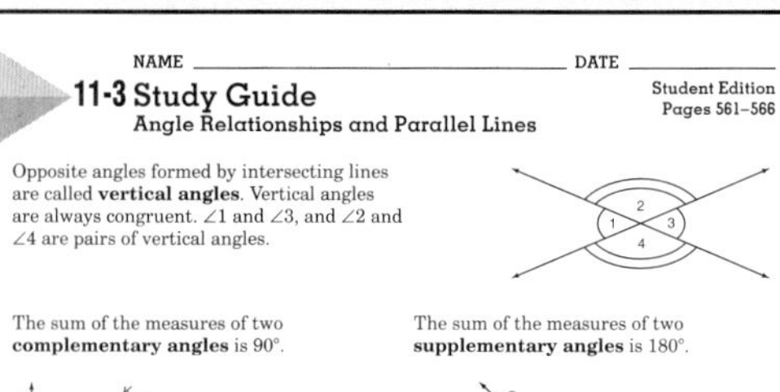

NAME ______________________ DATE ________

11-3 Study Guide
Angle Relationships and Parallel Lines

Student Edition
Pages 561–566

Opposite angles formed by intersecting lines are called **vertical angles**. Vertical angles are always congruent. ∠1 and ∠3, and ∠2 and ∠4 are pairs of vertical angles.

The sum of the measures of two **complementary angles** is 90°.

The sum of the measures of two **supplementary angles** is 180°.

Line n is called a **transversal**. When two parallel lines, ℓ and m, are intersected by a transversal, certain pairs of angles are congruent.

Corresponding angles are congruent. Pairs of corresponding angles are 2 and 6, 4 and 8, 1 and 5, and 3 and 7.
Alternate interior angles are congruent. Pairs of alternate interior angles are 3 and 6, and 4 and 5.
Alternate exterior angles are congruent. Pairs of alternate exterior angles are 1 and 8, and 2 and 7.

Each pair of angles is either complementary or supplementary. Find the value of x in each figure.

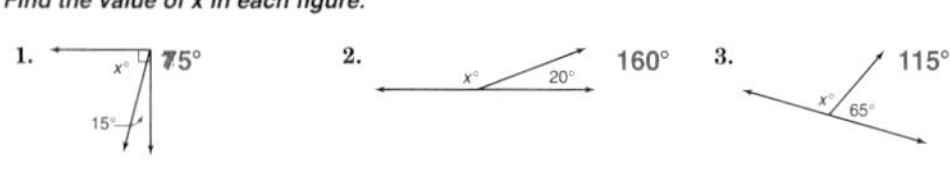

In the figure at the right, r ∥ s. If the measure of ∠4 is 38°, find the measure of each angle.

4. ∠1 38°
5. ∠7 142°
6. ∠2 142°
7. ∠6 142°
8. ∠8 38°
9. ∠3 142°
10. ∠5 38°

Example 4 APPLICATION Carpentry

A carpenter uses a protractor and a plumb line (a string with a weight attached) to measure the angle between a slanted ceiling and the wall. If $m\angle WXY = 65°$, find $m\angle XYZ$.

∠WXY is supplementary to ∠WXV.

$m\angle WXY + m\angle WXV = 180$
$65 + m\angle WXV = 180$
$m\angle WXV = 115$

The plumb line and the wall are parallel lines. So, ∠XYZ is congruent to ∠WXV. *Corresponding angles are congruent.*

$m\angle XYZ = m\angle WXV$
$m\angle XYZ = 115$

The measure of ∠XYZ is 115°.

Checking Your Understanding

Communicating Mathematics

Read and study the lesson to answer these questions. 1–2. See margin.

1. **Compare and contrast** vertical angles and adjacent angles.
2. **Draw** $\overline{MN} \parallel \overline{PQ}$ and transversal $\overline{RS}$. Number the angles and name a pair of each.
   a. corresponding angles
   b. alternate interior angles
   c. alternate exterior angles
   d. adjacent angles
   e. adjacent, supplementary angles

Draw and label a diagram to show each pair of angles and describe their characteristics. 3–5. See margin.

3. ∠WXY and ∠YXZ are adjacent angles.
4. ∠GHI and ∠IHJ are adjacent, complementary angles.
5. ∠KLM and ∠MLN are adjacent, supplementary angles.

Guided Practice

6. Identify the figure in which angles 1 and 2 are supplementary. d
   a. b. c. d.

Find the value of x in each figure.

7. 40
8. 155

9. Angle C and ∠D are complementary. Find $m\angle C$ if $m\angle D$ is 75°. 15°
10. Angle P and ∠Q are supplementary. Find $m\angle P$ if $m\angle Q$ is 90°. 90°
11. Angles X and Y are supplementary. $m\angle X = 2x$ and $m\angle Y = 4x$. Find the measure of each angle. $m\angle X = 60°$, $m\angle Y = 120°$
12. Angles F and G are complementary. $m\angle F = x + 8$ and $m\angle G = x - 10$. Find the measure of each angle. $m\angle F = 54°$; $m\angle G = 36°$

Reteaching

Using Diagrams In a marching band, a certain routine has a drummer crossing two yard lines at an angle of 75° to the line on her left. Have students draw a diagram and give the measures of all eight angles formed.

13. $\angle AFG$, $\angle GFB$, $\angle CGF$, $\angle FGD$
14. $\angle AFE$, $\angle EFB$, $\angle CGH$, $\angle HGD$
15a–b. See margin.

Refer to the diagram at the right to complete Exercises 13–15.

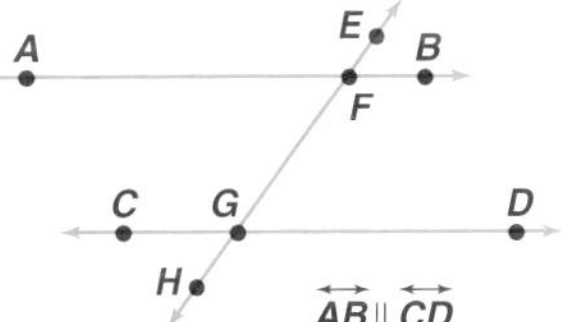

13. Name the interior angles.
14. Name the exterior angles.
15. a. Use a protractor to find the measure of one angle.
    b. Use the measure from **part a** to find all the other angle measures.
16. **Public Transit** The angle at the corner where two streets intersect is 125°. If a bus cannot make a turn at an angle of less than 70°, can bus service be provided on a route that includes turning that corner in both directions? Explain. **No; the corresponding angle has a measure of only 55°.**

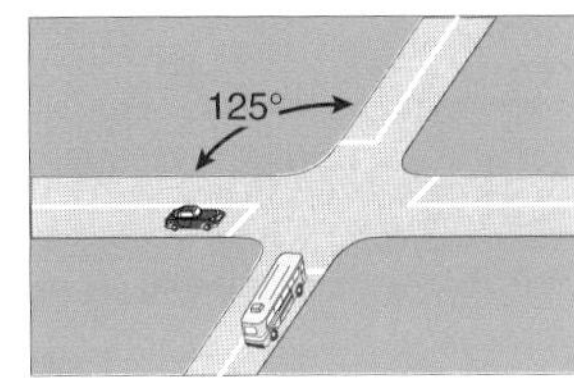

Exercises: Practicing and Applying the Concept

Independent Practice

Find the value of x in each figure.

17. 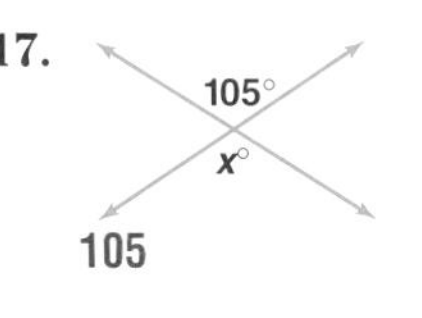

105

18. 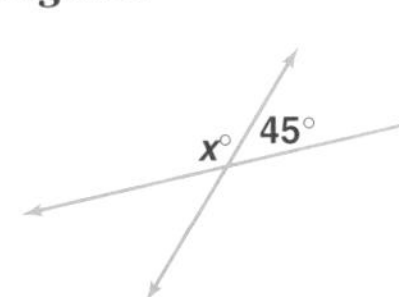

135

19. 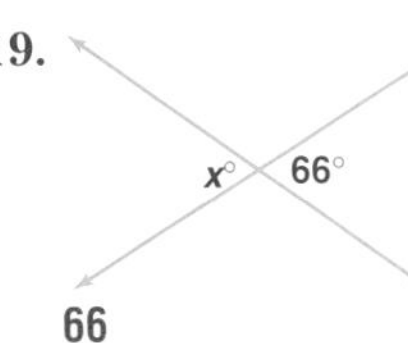

66

20. 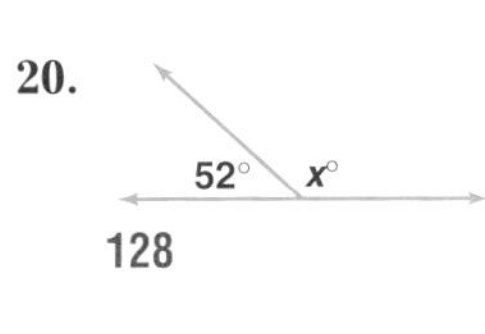

128

21. 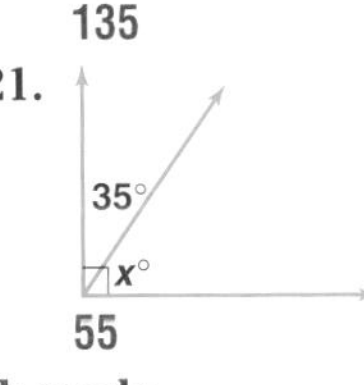

55

22. 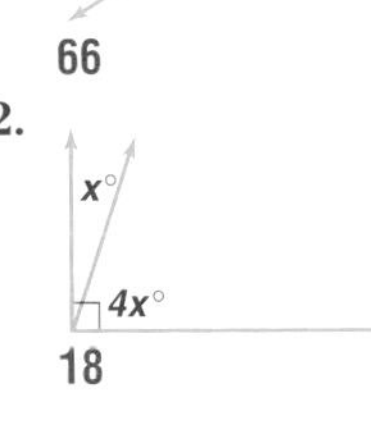

18

Find the measure of each angle.

23. 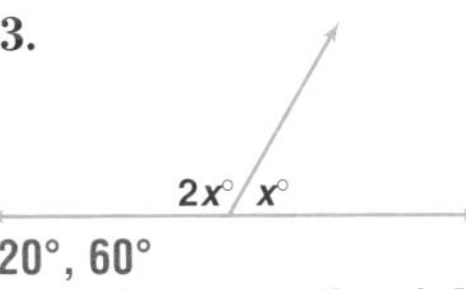

120°, 60°

24. 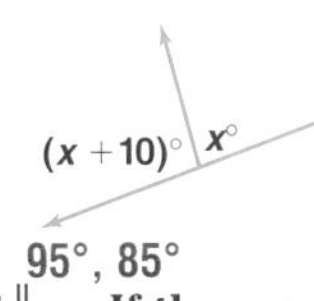

95°, 85°

25. 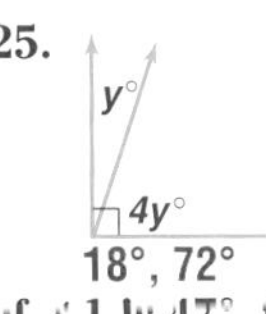

18°, 72°

In the figure at the right, $\ell \parallel m$. If the measure of $\angle 1$ is 47°, find the measure of each angle.

26. $\angle 2$ **133°**
27. $\angle 3$ **47°**
28. $\angle 5$ **47°**
29. $\angle 4$ **133°**
30. $\angle 7$ **47°**
31. $\angle 8$ **133°**

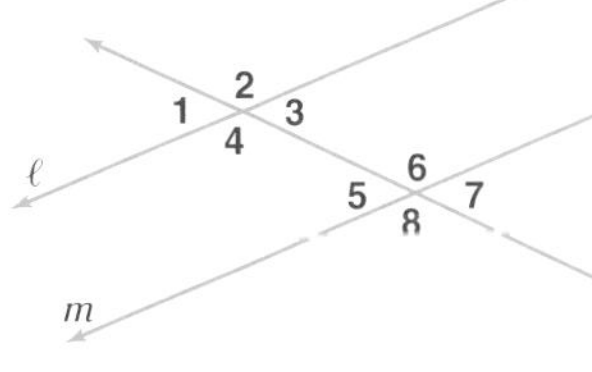

Each pair of angles is complementary. Find each angle measure.

32. $m\angle A = 2x + 15$, $m\angle B = x - 3$
33. $m\angle R = x - 9$, $m\angle S = x + 17$

32. $m\angle A = 67°$; $m\angle B = 23°$
33. $m\angle R = 32°$; $m\angle S = 58°$

Each pair of angles is supplementary. Find the measure of each angle.

34. $m\angle T = 2x + 17$ and $m\angle S = 5x - 40$ **$m\angle T = 75°$; $m\angle S = 105°$**
35. $m\angle F = 3x + 40$ and $m\angle G = 2x + 10$ **$m\angle F = 118°$; $m\angle G = 62°$**

Assignment Guide

Core: 17–43 odd, 44–51
Enriched: 18–38 even, 40–51

For **Extra Practice**, see p. 767.

The red A, B, and C flags, printed only in the Teacher's Wraparound Edition, indicate the level of difficulty of the exercises.

Additional Answers

3. 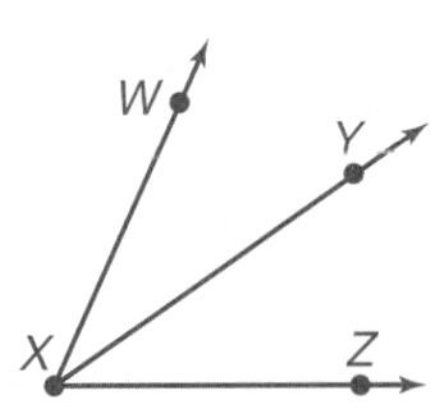

4. 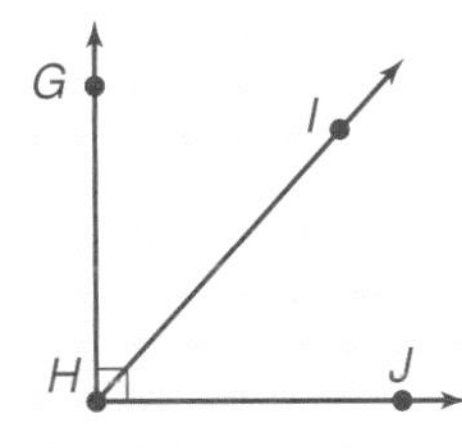

5. 

Practice Masters, p. 93

NAME ______________ DATE ______________

11-3 Practice
Angle Relationships and Parallel Lines

Student Edition Pages 561–566

Find the value of x in each figure.

1. **54°**
2. **122°**
3. **115°**

Each pair of angles is either complementary or supplementary. Find the measure of each angle.

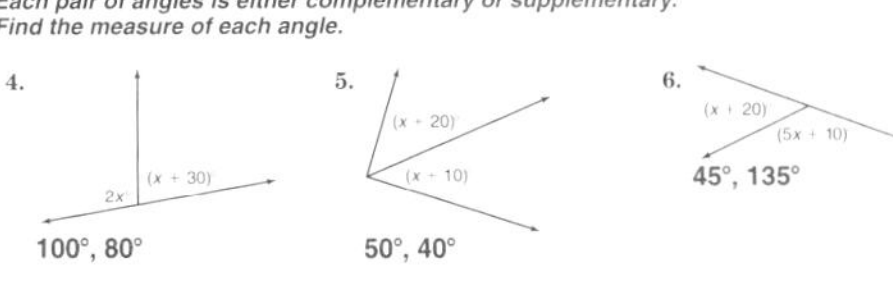

4. **100°, 80°**
5. **50°, 40°**
6. **45°, 135°**

In the figure at the right, m ∥ n. If the measure of ∠3 is 95°, find the measure of each angle.

7. $\angle 1$ **95°**
8. $\angle 4$ **85°**
9. $\angle 5$ **95°**
10. $\angle 6$ **85°**
11. $\angle 7$ **95°**
12. $\angle 8$ **85°**
13. $\angle 2$ **85°**

In the figure at the right ℓ ∥ k. Find the measure of each angle.

14. $\angle 5$ **60°**
15. $\angle 4$ **80°**
16. $\angle 9$ **120°**
17. $\angle 8$ **60°**
18. $\angle 6$ **120°**
19. $\angle 1$ **60°**
20. $\angle 7$ **100°**
21. $\angle 3$ **80°**
22. $\angle 2$ **40°**
23. $\angle 10$ **100°**

Group Activity Card 11-3

Concentrate On Geometry Group Activity **11-3**

MATERIALS: Set of 20 "Geometry Concentration" Cards with information given on the back of this card. Ten cards have geometric terms, and ten have geometric figures.

Geometry Concentration is played like the traditional game of Concentration. To begin, cards should be shuffled and laid face down in five rows of four cards each. Player one is allowed to turn over two cards. If they form a pair by naming and illustrating a geometry term, the player gets to keep that pair and turn over another two cards. If the next two cards do not form a pair, it is then player two's turn.

The game is over when all cards have been paired or when no more pairs can be made. The winner is the player with the most pairs.

Pre-Algebra

Additional Answer

15a–b. $m\angle AFE = 130°$; $m\angle EFB = 50°$; $m\angle BFG = 130°$; $m\angle AFG = 50°$; $m\angle CGF = 130°$; $m\angle FGD = 50°$; $m\angle DGH = 130°$; $m\angle CGH = 50°$

4 ASSESS

Closing Activity

Writing Have students each draw and label two parallel lines and a transversal. Then have them write ten pairs of angles that are congruent and give the reason why they are congruent.

Chapter 11, Quiz A (Lessons 11-1 through 11-3) is available in the *Assessment and Evaluation Masters*, p. 295.

Additional Answers

40.

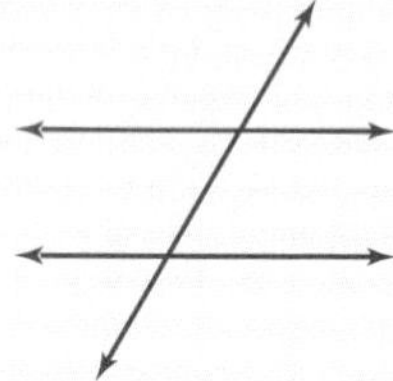

They are supplementary.

41.

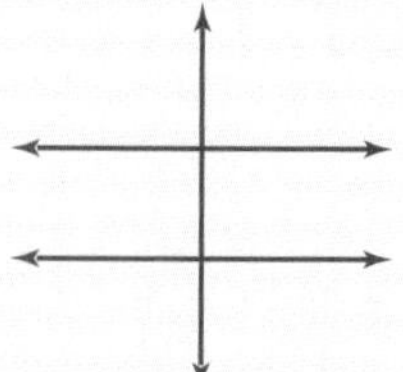

Yes; all the angles formed are right angles.

Enrichment Masters, p. 93

NAME ______ DATE ______

11-3 Enrichment
Angle Relationships

Student Edition Pages 561–566

Angles are measured in degrees (°). Each degree of an angle is divided into 60 minutes ('), and each minute of an angle is divided into 60 seconds (").

60' = 1°
60" = 1'
$67\frac{1}{2}° = 67°30'$
70.4° = 70°24'
90° = 89°60'

Two angles are complementary if the sum of their measures is 90°. Find the complement of each of the following angles.

1. 35°15' **54°45'**
2. 27°16' **62°44'**
3. 15°54' **74°06'**
4. 29°18'22" **60°41'38"**
5. 34°29'45" **55°30'15"**
6. 87°2'3" **2°57'57"**

Two angles are supplementary if the sum of their measures is 180°. Find the supplement of each of the following angles.

7. 120°18' **59°42'**
8. 84°12' **95°48'**
9. 110°2' **69°58'**
10. 45°16'24" **134°43'36"**
11. 39°21'54" **140°38'6"**
12. 129°18'36" **50°41'24"**
13. 98°52'59" **81°7'1"**
14. 9°2'32" **170°57'28"**
15. 1°2'3" **178°57'57"**

In the figure at the right, $\ell \parallel m$. Find the value of x for each of the following. *The figure is not drawn to scale.*

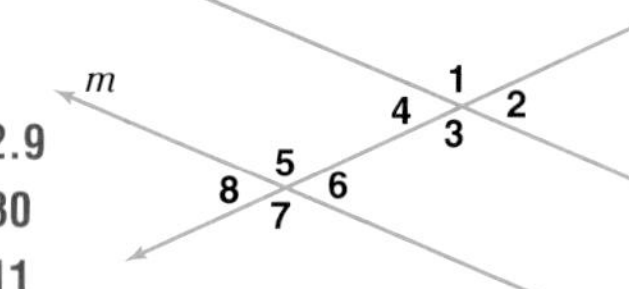

36. $m\angle 1 = 6x - 20$ and $m\angle 2 = x + 40$ **22.9**
37. $m\angle 4 = 3x + 17$ and $m\angle 5 = 2x + 13$ **30**
38. $m\angle 6 = 9x - 10$ and $m\angle 7 = 5x + 36$ **11**

39. The measure of the supplement of an angle is 15° less than four times the measure of the complement. Find the measure of the angle. **55°**

Critical Thinking

Draw a diagram for each situation. Use a protractor to measure the angles formed and answer each question. **40–41. See margin.**

40. If two parallel lines are cut by a transversal, how are the interior angles on the same side of the transversal related? Explain.
41. If a transversal is perpendicular to one of two parallel lines, is it perpendicular to the other parallel line also? Explain.

Applications and Problem Solving

42. **Physics** The drawing at the right shows a beam of light reflected from the surface of a flat mirror. The angle of incidence (the angle between the incident ray and the normal line) and the angle of reflection (the angle between the reflected ray and the normal line) are congruent. The normal line is perpendicular to the surface of the mirror. Suppose a beam of light strikes the mirror at a 52° angle to the mirror. What is the measure of the angle between the incident ray and the reflected ray? **76°**

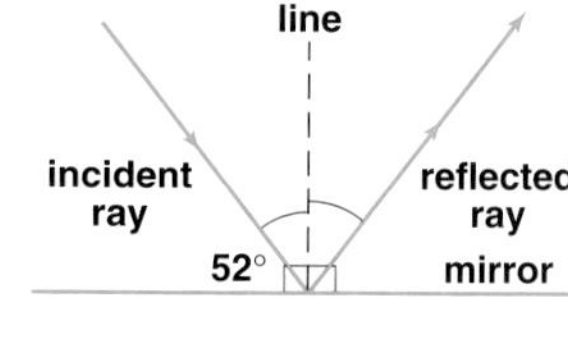

43. **Carpentry** A standard stair rail is designed to make an angle of 45° with the floor. The first vertical post and the lower rail also form an angle of 45°. If the upper rail is parallel to the lower rail, what angle should it make with the first vertical post? **135°**

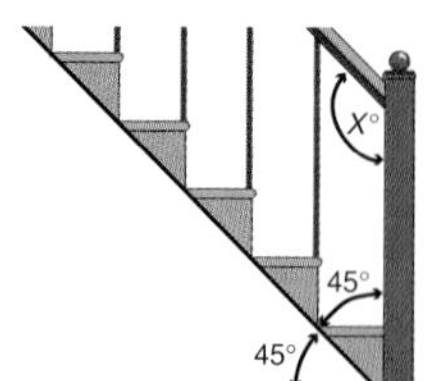

Mixed Review

44. **Retail Sales** 17% of Foodtown's sales come from the produce department. Find the measure of the angle to represent produce sales for a circle graph showing Foodtown's sales. (Lesson 11-2) **61.2°**
45. **Probability** Two dice are thrown. Find $P(\text{sum} > 8)$. (Lesson 9-3) $\frac{5}{18}$
46. **Biology** A certain bacteria doubles its population every 12 hours. After 3 full days, there are 1600 bacteria. How many bacteria were there at the beginning of the first day? (Lesson 7-1) **25**

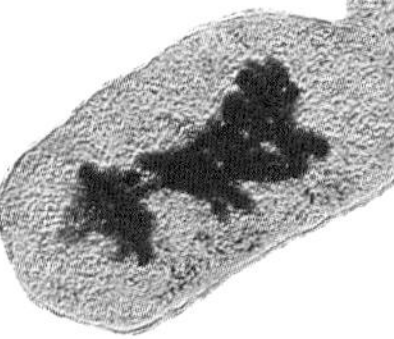

47. Write 56,780 in scientific notation. (Lesson 6-9) 5.678×10^4
48. Write $\frac{12}{54}$ in simplest form. (Lesson 4-6) $\frac{2}{9}$
49. Simplify $9b + (-16)b$. (Lesson 2-4) $-7b$
50. Evaluate $|-5| - |2|$. (Lesson 2-1) **3**
51. Solve $8 + n = 13$ mentally. (Lesson 1-6) **5**

Extension

Using Proofs Have students write an informal proof which shows that if a line is perpendicular to one of two parallel lines, then it is perpendicular to the other parallel line.

11-3B

GRAPHING CALCULATOR ACTIVITY

Slopes of Parallel Lines

An Extension of Lesson **11-3**

MATERIALS

graph paper

You can determine whether two lines are parallel by finding their slopes. In Lesson 8-6, you learned that parallel lines have the same slope. So, if two lines have the same slope, then they are parallel. The graphing calculator program at the right finds the slope of a line when you enter the coordinates of two points on the line.

```
PROGRAM:SLOPE
: Disp "ENTER"
: Disp "COORDINATES"
: Disp "POINT 1"
: Input A
: Input B
: Disp "POINT 2"
: Input C
: Input D
: If A = C
: Then
: Disp "SLOPE UNDEFINED"
: Stop
: End
: (D-B)/(C-A) → M
: Disp "SLOPE = ", M
```

Your Turn Work with a partner.

- Enter the program in a TI-82 graphing calculator.
- Find the coordinates of two points on line m below, Point 1 and Point 2.

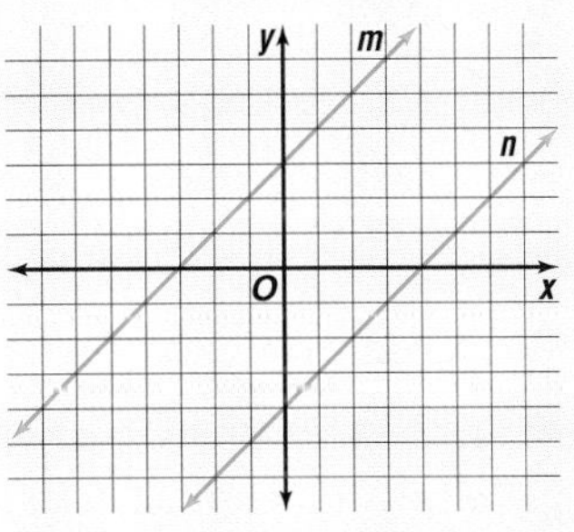
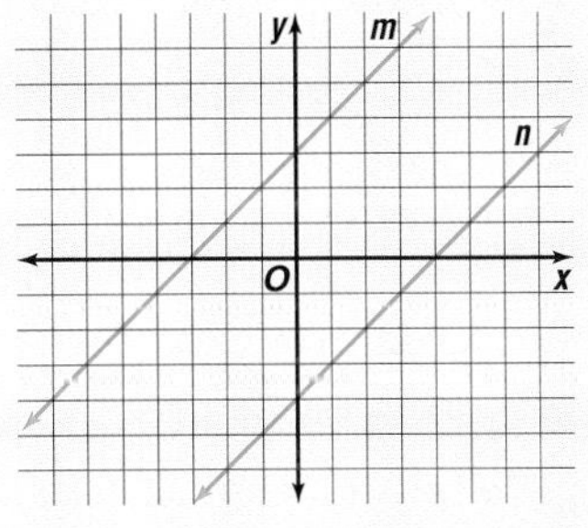

- Enter the coordinates in the program. Record the slope of line m.
- Now find the coordinates of two points on line n above.
- Enter the coordinates in the program. Record the slope of line n.

1. It divides the difference of the y-coordinates by the difference of the x-coordinates.

TALK ABOUT IT

1. How does the program calculate the slope of the line?
2. How do the slopes of lines m and n compare? **They are the same.**
3. Based on their slopes, are lines m and n parallel? **yes**

Your Turn

- Now draw two different lines that you think are parallel on graph paper.
- Find the coordinates of two points on each line.
- Enter the coordinates in the program to find the slope of each line.
- Repeat the steps above for two lines that you think are *not* parallel.

TALK ABOUT IT

4. What were the results for the lines that you drew? **See students' work.**
5. Suppose you wanted to draw two lines that were parallel. How could you draw the lines to be sure that they were parallel? **Draw two lines with the same slope.**

Technology

This lesson offers an excellent opportunity for using technology in your pre-algebra classroom. For more information on using technology, see *Graphing Calculators in the Mathematics Classroom*, one of the titles in the Glencoe Mathematics Professional Series.

11-3B LESSON NOTES

NCTM Standards: 1-4, 12

Objective

Use a graphing calculator to determine whether two lines are parallel.

Recommended Time

15 minutes

Instructional Resources

Graphing Calculator Masters, p. 24

This master provides keystroking instruction for this lesson for the TI-81 and Casio graphing calculators.

1 FOCUS

Motivating the Lesson

Ask students why parallel courses are laid out on a ski slope. **So that the courses are as close to the same as possible.**

2 TEACH

Teaching Tip Review the meaning of slope as taught in Lesson 8-6 on pages 400–404 before proceeding with the exploration.

3 PRACTICE/APPLY

Assignment Guide
Core: 1–5
Enriched: 1–5

4 ASSESS

Observing students working with technology is an excellent method of assessment.

11-4 LESSON NOTES

NCTM Standards: 1-4, 12

Instructional Resources

- Study Guide Master 11-4
- Practice Master 11-4
- Enrichment Master 11-4
- Group Activity Card 11-4

Transparency 11-4A contains the 5-Minute Check for this lesson; **Transparency 11-4B** contains a teaching aid for this lesson.

Recommended Pacing	
Standard Pacing	Day 5 of 13
Honors Pacing	Day 5 of 13
Block Scheduling*	Day 3 of 7

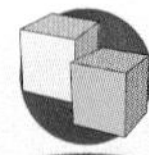

*For more information on pacing and possible lesson plans, refer to the ***Block Scheduling Booklet***.

1 FOCUS

5-Minute Check
(over Lesson 11-3)

Find the measure of $\angle A$ and $\angle B$ if $m\angle A = 2x + 20°$ and $m\angle B = 3x°$ under each set of conditions.

1. $\angle A$ and $\angle B$ are vertical angles. **60°; 60°**
2. $\angle A$ and $\angle B$ are complementary angles. **48°; 42°**
3. $\angle A$ and $\angle B$ are supplementary angles. **84°; 96°**

Line m is parallel to line n and line t crosses lines m and n so that one interior angle has a measure of 102°.

4. What is the measure of the adjacent interior angle? **78°**
5. What is the measure of a corresponding angle? **102°**

11-4 Triangles

Setting Goals: *In this lesson, you'll find the missing angle measure of a triangle. You'll also classify triangles by angles and sides.*

Modeling a Real-World Application: Engineering

Civil engineers who design and build railroad bridges often choose a *truss* bridge to span canyons and rivers. A truss bridge can span more than 1000 feet and requires less building material than other types of bridges.

Learning the Concept

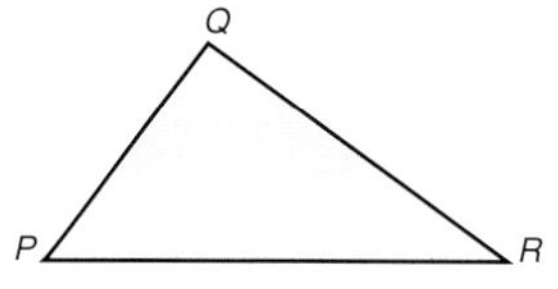

HELP WANTED

Civil engineers use geometry to make beautiful and functional designs. For more information on this field, contact:

American Society of Civil Engineers
345 East 47th Street
New York, NY 10017

Trusses get their strength from braces that form triangles. Because triangles are rigid, they are strong. A **triangle** is formed by three line segments that intersect only at their endpoints. A triangle can be named by its vertices. The triangle shown below is triangle *PQR*, or $\triangle PQR$.

The vertices are P, Q, and R.
The sides are $\overline{PQ}$, $\overline{QR}$, and $\overline{PR}$.
The angles are $\angle P$, $\angle Q$, and $\angle R$.

There is a special relationship among the angles of a triangle.

1. Use a straightedge to draw a triangle on a piece of paper. Cut out the triangle and label the vertices *A*, *B*, and *C*.
2. Fold the triangle so that the point *C* lies on $\overline{AB}$ and the fold is parallel to $\overline{AB}$. Label $\angle C$ as $\angle 2$.
3. Fold again so point *A* meets the vertex of $\angle 2$. Label $\angle A$ as $\angle 1$.
4. Finally, fold so point *B* also meets the vertex of $\angle 2$. Label $\angle B$ as $\angle 3$.

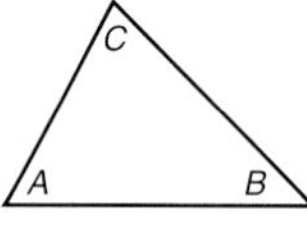

Step 1

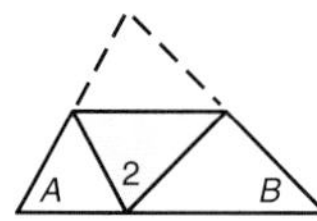

Step 2

Step 3

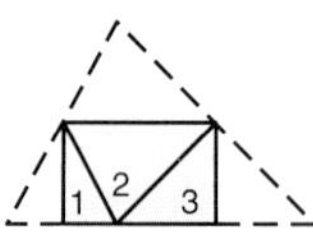

Step 4

What do you notice about the sum of the measures of angles 1, 2, and 3? This activity suggests the following relationship.

Angles of a Triangle — The sum of the measures of the angles of a triangle is 180°.

HELP WANTED

In the mid-18th century, the term "civil" engineering was adopted to differentiate those engineers from military engineers. Transportation civil engineers build highways and bridges and structure engineers concentrate on the design of large buildings.

Example 1 **In $\triangle CAT$, the measure of $\angle C$ is 47°, and the measure of $\angle A$ is 59°. Find the measure of $\angle T$.**

Estimate: *50 + 60 = 110 and 180 − 110 = 70, so $m\angle T$ is about 70°.*

$m\angle C + m\angle A + m\angle T = 180$ *The sum of the measures of the angles of a triangle is 180°.*

$47 + 59 + m\angle T = 180$ *Replace $m\angle C$ with 47 and $m\angle A$ with 59.*

$106 + m\angle T = 180$ *Add 47 + 59.*

$m\angle T = 74$ *Subtract 106 from each side.*

The measure of $\angle T$ is 74°.

Connection to Algebra

The relationships of the angles in a triangle can be represented using algebra.

Example 2 **The measures of the angles of a certain triangle are in the ratio 2:3:4. Find the measure of each angle.**

INTEGRATION **Algebra**

Let $2x$ represent the measure of one angle, $3x$ the measure of a second angle, and $4x$ the measure of the third angle.

$2x + 3x + 4x = 180$ *The sum of the measures is 180°.*

$9x = 180$ *Combine like terms.*

$x = 20$ *Divide each side by 9.*

$2x = 40$, $3x = 60$, and $4x = 80$.

The measures of the angles are 40°, 60°, and 80°. *Check by adding.*

A triangle can be classified by its angles. Every triangle has two acute angles. The third angle can be used to classify the triangle.

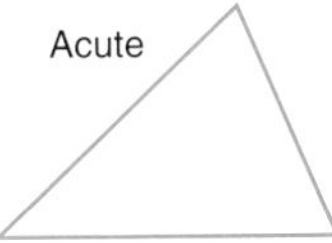

An **acute triangle** has three acute angles.

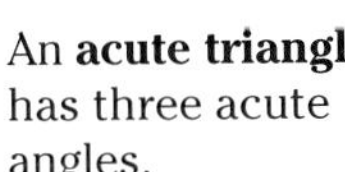

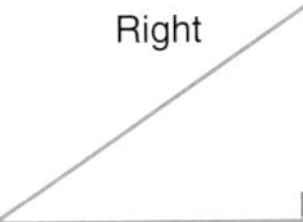

A **right triangle** has one right angle.

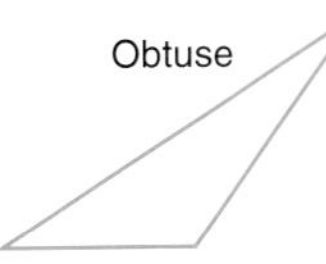

An **obtuse triangle** has one obtuse angle.

Triangles can also be classified by the number of congruent sides.

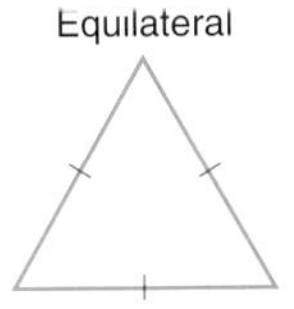

An **equilateral triangle** has three congruent sides.

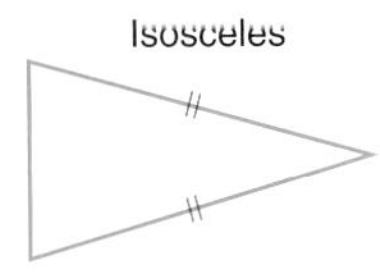

An **isosceles triangle** has at least two congruent sides.

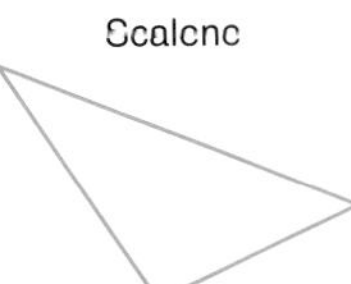

A **scalene triangle** has no congruent sides.

Motivating the Lesson

Hands-On Activity Provide students with a cutout of a right triangle. Have students tear off the acute angles of the triangle and place them so they are adjacent angles. Ask what the sum of the angles is. 90°

Teaching Tip You may wish to have students try the experiment again with a different triangle to show that this works for any triangle.

2 TEACH

In-Class Examples

For Example 1
In $\triangle SUN$, the measure of $\angle S$ is 29°, and the measure of $\angle N$ is 99°. Find the measure of $\angle U$. **52°**

For Example 2
The measures of the angles of a certain triangle are in the ratio 3:4:5. Find the measure of each angle. **40°, 60°, and 80°**

GLENCOE Technology

Interactive Mathematics Tools Software

In this interactive computer lesson, students explore triangles. A **Computer Journal** gives students the opportunity to write about what they have learned.

For Windows & Macintosh

3 PRACTICE/APPLY

Checking Your Understanding

Exercises 1–11 are designed to help you assess your students' understanding through reading, writing, speaking, and modeling. You should work through Exercises 1–5 with your students and then monitor their work on Exercises 6–11.

Additional Answers

1. Subtract the sum of the known angles from 180°.
2. Sample answer: wire hanger
3. Sample answer: pattern block triangle
4. Sample answer: triangle formed by stairs, wall, and floor

Study Guide Masters, p. 94

NAME ______________________ DATE ____________

11-4 Study Guide
Triangles

Student Edition Pages 568–572

A **right triangle** has one right angle. An **obtuse triangle** has one obtuse angle. All other triangles are **acute**. In an acute triangle, each of the three angles is acute.

For any triangle, the sum of the measures of the angles is 180°.

Example: In $\triangle AOD$, the measure of $\angle D$ is 35°, and the measure of $\angle O$ is 107°. Find the measure of $\angle A$.

$m\angle A + m\angle O + m\angle D = 180$	The sum of the measures of the angles is 180.
$m\angle A + 107 + 35 = 180$	Replace $m\angle O$ with 107 and $m\angle D$ with 35.
$m\angle A + 142 = 180$	Add 107 and 35.
$m\angle A = 38$	Subtract 142 from each side.

Find the value of x. Then classify each triangle as acute, right, or obtuse.

1. 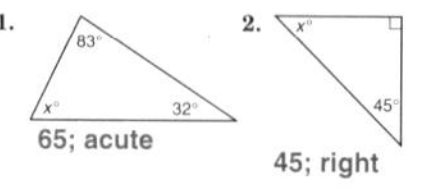

65; acute

2. 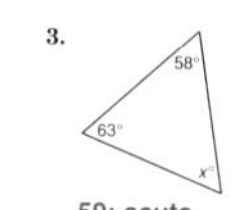

45; right

3.
58°
63°
x°
59; acute

4. 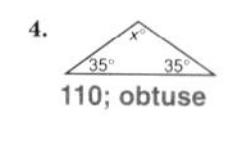

110; obtuse

Use the figure at the right to solve each of the following.

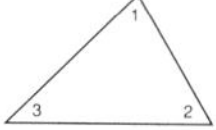

5. Find $m\angle 2$ if $m\angle 3 = 49°$ and $m\angle 1 = 63°$. 68°
6. Find $m\angle 2$ if $m\angle 3 = 45°$ and $m\angle 1 = 50°$. 85°
7. Find $m\angle 3$ if $m\angle 2 = 66°$ and $m\angle 1 = 64°$. 50°
8. Find $m\angle 3$ if $m\angle 2 = 38°$ and $m\angle 1 = 70°$. 72°
9. Find $m\angle 3$ if $m\angle 2 = 42°$ and $m\angle 1 = 58°$. 80°

Checking Your Understanding

Communicating Mathematics

Read and study the lesson to answer these questions.

1. If you know the measures of two of the angles in a triangle, how can you find the measure of the third angle? See margin.

An A-frame house is shaped like an isosceles triangle. Give a real-world example of the following triangles. 2–4. See margin.

2. obtuse
3. equilateral
4. right

5. Make up memory tools you can use to remember the names of triangles that are classified by sides or angles. Record the tool in your journal. See students' work.

Guided Practice

Find the value of *x*. Then classify each triangle as *acute*, *right*, or *obtuse*.

6. 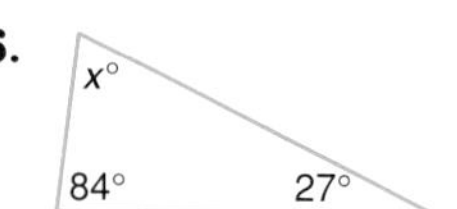

69; acute

7. 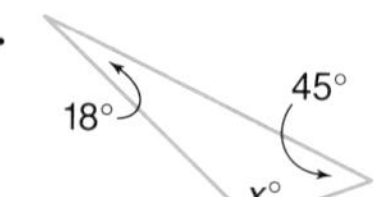

117; obtuse

8. **Algebra** The measures of the angles of a certain triangle are in the ratio 1:3:5. Find the measures of each angle. 20°, 60°, 100°

Use a ruler to determine the number of congruent sides in each triangle. Then classify the triangle as *scalene*, *isosceles*, or *equilateral*.

9. 
equilateral

10. 
isosceles

11. **Construction** Find the measure of the missing angle, $x°$, in the roof truss shown at the right. 119°

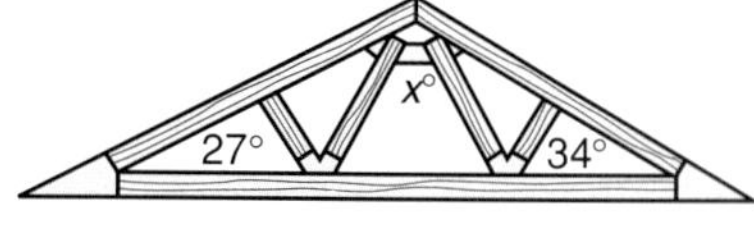

Exercises: Practicing and Applying the Concept

Independent Practice

Estimation
Mental Math
Calculator
Paper and Pencil

Find the value of *x*. Then classify each triangle as *acute*, *right*, or *obtuse*.

12.
36°
x°
32°
112; obtuse

13.
x°
40°
50; right

14. 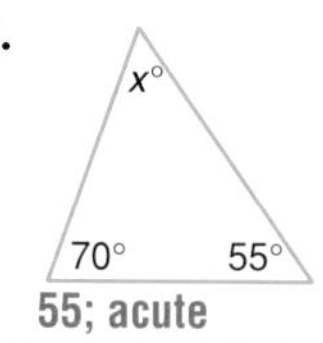

55; acute

15. The measures of the angles of a triangle are in the ratio 2:3:5. Find the measure of each angle. 36°; 54°; 90°
16. The measures of the angles of a triangle are in the ratio 1:4:7. Find the measure of each angle. 15°; 60°; 105°

Reteaching

Using Manipulatives Have students cut out a triangle and color each of its three corners, putting a heavy dot near each vertex. Then have students cut or tear the triangle into three pieces so that each vertex is on a separate piece. Tell students to arrange the pieces so that the vertices are all on the same location and so that the sides of the angles create adjacent angles. Ask what figure is formed by the outer sides of the outer two angles. **a straight line**

Use a ruler to determine the number of congruent sides in each triangle. Then classify the triangle as *scalene*, *isosceles*, or *equilateral*.

17.  isosceles

18.  scalene

19. 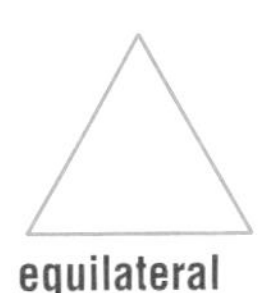 equilateral

Use a protractor to determine the measures of the angles in each triangle. Then classify the triangle as *acute*, *right*, or *obtuse*.

20.  right

21. 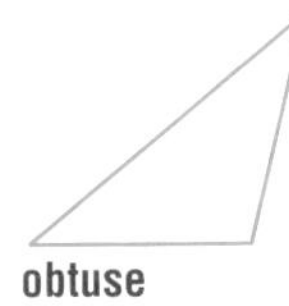 obtuse

22. 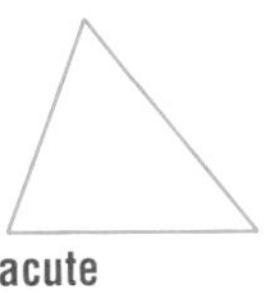 acute

Use the figure at the right to solve each of the following.

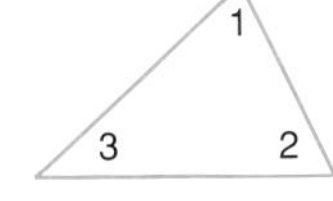

23. Find $m\angle 1$ if $m\angle 2 = 40°$ and $m\angle 3 = 55°$. **85°**

24. Find $m\angle 1$ if $m\angle 2 = 60°$ and $m\angle 3 = 60°$. **60°**

25. Find $m\angle 1$ if $m\angle 2 = 81°$ and $m\angle 3 = 74°$. **25°**

26. Find $m\angle 2$ if $m\angle 1 = 45°$ and $m\angle 3 = 75°$. **60°**

27. Find $m\angle 2$ if $m\angle 1 = 47°$ and $m\angle 3 = 48°$. **85°**

Find the measures of the angles in each triangle.

28. 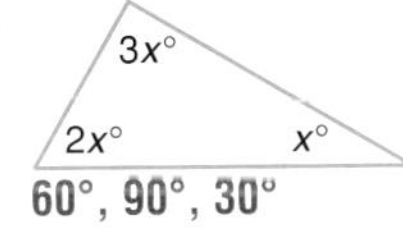

60°, 90°, 30°

29. 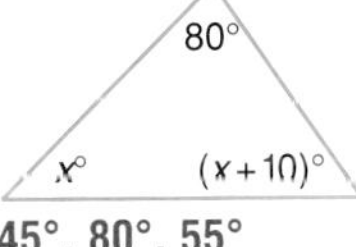

45°, 80°, 55°

30. 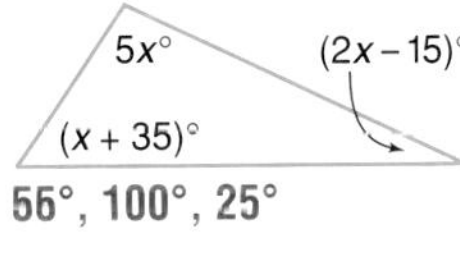

56°, 100°, 25°

Critical Thinking

31. Explain how you could use a piece of string to convince someone that a given triangle was: **a–c. See students' work.**

a. scalene **b.** isosceles **c.** equilateral

32. Are the acute angles of a right triangle complementary? Explain. **See margin.**

Applications and Problem Solving

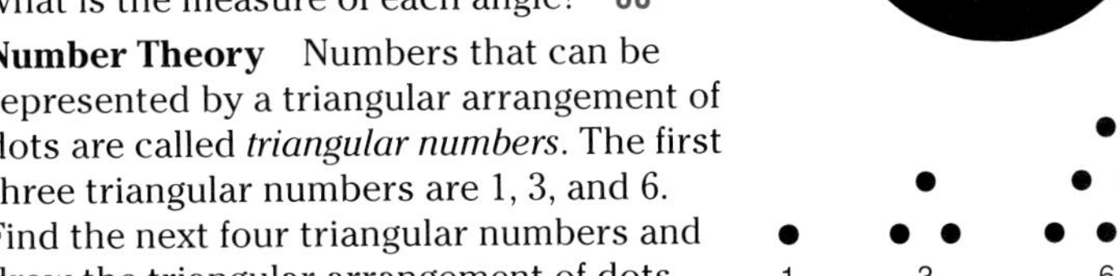

33. Civil Defense The Civil Defense program is a civilian organization created to help people when a disaster occurs. The Civil Defense symbol is shown at the right. If the triangle pictured in the symbol is an equilateral triangle, and all of the angles are congruent, what is the measure of each angle? **60°**

34. Number Theory Numbers that can be represented by a triangular arrangement of dots are called *triangular numbers*. The first three triangular numbers are 1, 3, and 6. Find the next four triangular numbers and draw the triangular arrangement of dots for each one. **See margin.**

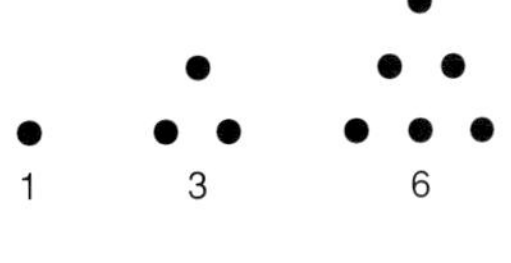

Assignment Guide

Core: 13–35 odd, 36–41
Enriched: 12–30 even, 31–41
All: Self Test, 1–12

For **Extra Practice**, see p. 768.

The red A, B, and C flags, printed only in the Teacher's Wraparound Edition, indicate the level of difficulty of the exercises.

Additional Answers

32. Yes, their sum must be 90° to have the three angles add to 180°.

34.

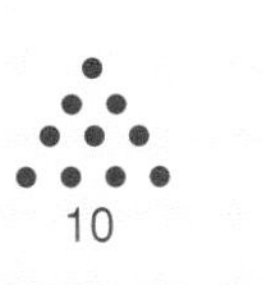
10

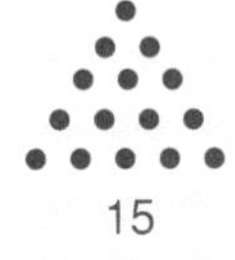
15

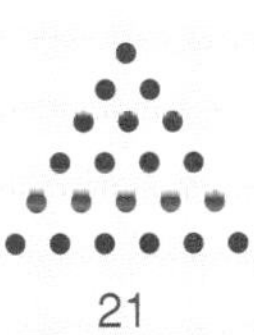
21

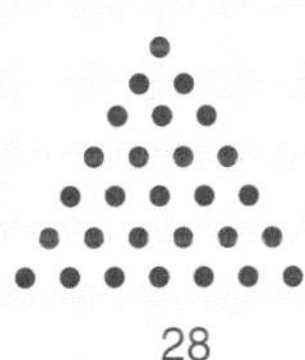
28

Practice Masters, p. 94

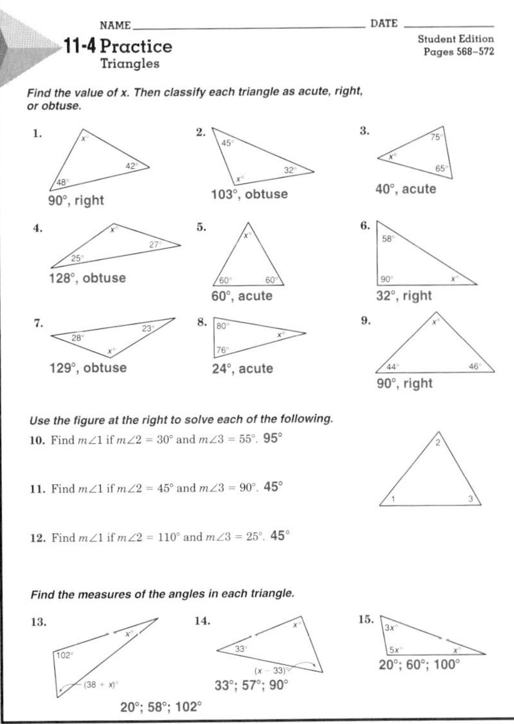

NAME ______________ DATE ______

11-4 Practice
Triangles

Student Edition
Pages 568–572

Find the value of x. Then classify each triangle as acute, right, or obtuse.

1. 90°, right
2. 103°, obtuse
3. 40°, acute
4. 128°, obtuse
5. 60°, acute
6. 32°, right
7. 129°, obtuse
8. 24°, acute
9. 90°, right

Use the figure at the right to solve each of the following.

10. Find $m\angle 1$ if $m\angle 2 = 30°$ and $m\angle 3 = 55°$. **95°**

11. Find $m\angle 1$ if $m\angle 2 = 45°$ and $m\angle 3 = 90°$. **45°**

12. Find $m\angle 1$ if $m\angle 2 = 110°$ and $m\angle 3 = 25°$. **45°**

Find the measures of the angles in each triangle.

13. 20°; 58°; 102°
14. 33°; 57°; 90°
15. 20°; 60°; 100°

Group Activity Card 11-4

Try These Triangles Group Activity **11-4**

MATERIALS: Six dice • nine note cards with the phrases given on the back of this card

Shuffle the cards and place them face down. To begin, one person rolls the six dice, announces the sum of the numbers, and turns over a card. Everyone in the group must then draw a triangle that contains one angle equal to the announced number and satisfies the condition given on the card. Each person should identify the measure of each angle in his or her triangle. Players then check each other's answers for correctness.

The activity continues until everyone has had a chance to roll the dice and turn over a card.

 Pre-Algebra

4 ASSESS

Closing Activity

Modeling Have each student draw a triangle on a piece of paper and measure two of the angles. Write the measurement next to the angle and then exchange papers among the students. Have each student find the missing angle measure without measuring. Then have them check their answers by using a protractor.

Additional Answers

38.

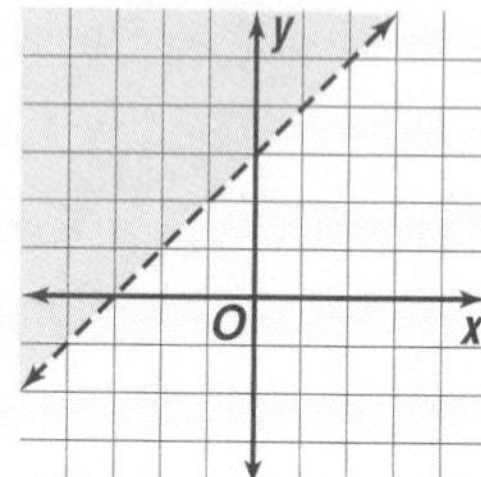

Self Test

4.

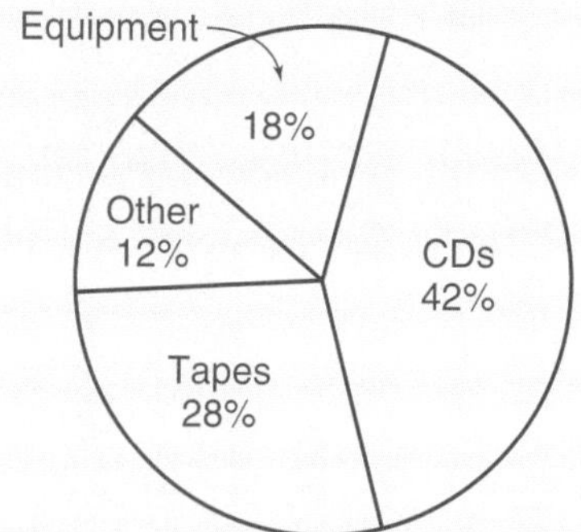

Enrichment Masters, p. 94

NAME ______ DATE ______

11-4 Enrichment Student Edition Pages 568–572

Constructing Designs

Many designs can be made using geometric constructions. Two examples are stained glass rose windows found in churches and Pennsylvania Dutch hex designs found on barns.

Use your compass to draw a circle.

Then, without changing the compass setting, move the point to a point on the circle. Draw a second circle.

Place the point on one of the points where the two circles intersect. Draw a circle. Repeat five more times.

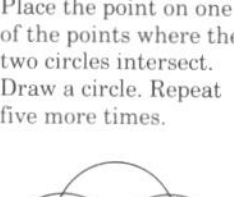

Use your compass and a straightedge to make a design like the one below. (HINT: *See the design above.*) Check students' drawings.

---

35. Public Utilities Above-ground electric wires and transformers, are mounted on utility poles. A support cable, called a guy wire, is sometimes attached to give the pole stability. If the guy wire makes an angle of 65° with the ground, what is the measure of the angle formed by the guy wire and the pole? Assume that the ground is level. **25°**

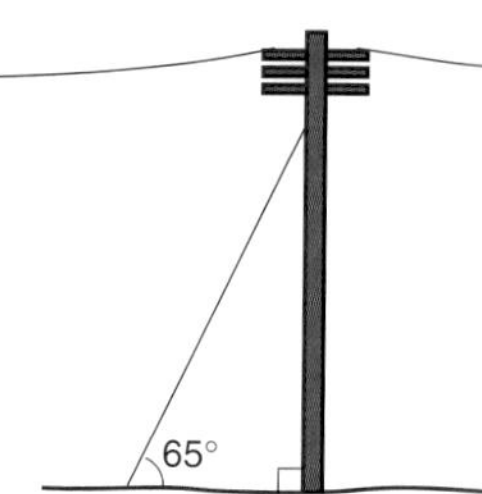

Mixed Review

36. Angles X and Y are supplementary. If $m\angle X = 35°$, find $m\angle Y$. (Lesson 11-3) **145°**

37. Probability Find the odds of rolling a sum less than 6 if a pair of dice are rolled. (Lesson 10-7) $\frac{5}{13}$

38. Graph $y > x + 3$. (Lesson 8-9) **See margin.**

39. $59\frac{1}{4}$ inches

39. Sewing A curtain pattern says to measure the length of the window and add 4 inches for curtain length. Then it says to add 4 inches for a heading, $2\frac{1}{2}$ inches for a rod casting, and 6 inches for a hem. Mary's window is $42\frac{3}{4}$ inches long. How long should the fabric be? (Lesson 5-5)

40. Replace each ● with <, >, or =. (Lesson 2-3)

a. -8 ● -14 **>** **b.** $-|20|$ ● -20 **=** **c.** 0 ● $|-2|$ **<**

41. Simplify $4(x + 3)$. (Lesson 1-5) $4x + 12$

Self Test

Use a protractor to draw an angle with the given measurements. Classify each angle as *acute, right,* or *obtuse.* (Lesson 11-1) **1–3. See Solutions Manual for drawings.**

1. 60° **acute** **2.** 115° **obtuse** **3.** 20° **acute**

4. Retail At Recordtown, 42% of sales are from CDs, 28% are from tapes, 18% are from equipment, and 12% are from other items. Make a circle graph of this data. (Lesson 11-2) **See margin.**

In the figure at the right, ℓ is parallel to m. If the measure of $\angle 3$ is 34°, find the measure of each angle. (Lesson 11-3)

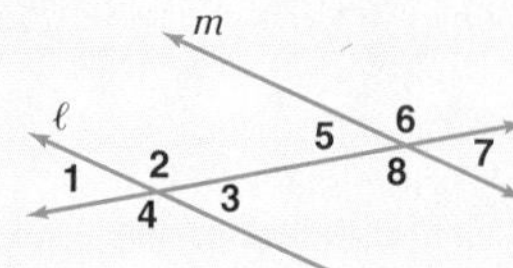

5. $\angle 1$ **34°** **6.** $\angle 4$ **146°** **7.** $\angle 5$ **34°**

Find the measure of each angle using the following information. (Lesson 11-3)

8. $\angle A$ and $\angle B$ are complementary. $m\angle A = x + 25$; $m\angle B = 2x - 10$ $m\angle A = 50°$; $m\angle B = 40°$

9. $\angle F$ and $\angle G$ are supplementary. $m\angle F = 3x - 50$; $m\angle G = 2x - 20$ $m\angle F = 100°$; $m\angle G = 80°$

Find the value of x. Then classify each triangle as *acute, right,* or *obtuse.* (Lesson 11-4)

10.

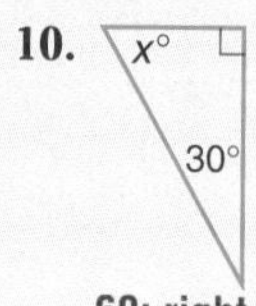

60; right

11.

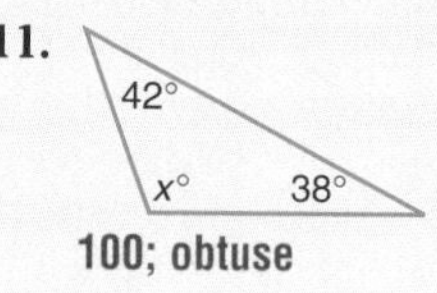

100; obtuse

12.

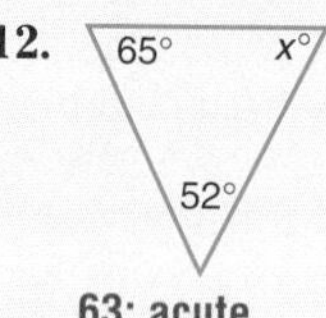

63; acute

Extension

Using Writing Have students write verbal problems using the angle measurements of each type of triangle. Trade and solve. Sample problem: The side of a triangular tent forms a 50° angle with the ground. The poles are perpendicular to the ground. What angle does the side of the tent form with the pole? **40°**

Self Test

The Self Test provides students with a brief review of the concepts and skills in Lessons 11-1 through 11-4. Lesson numbers are given to the right of exercises or instruction lines so students can review concepts not yet mastered.

11-5 Congruent Triangles

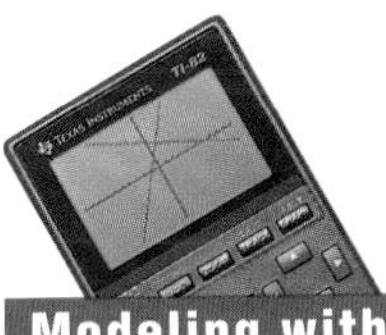

Setting Goals: *In this lesson, you'll identify congruent triangles and corresponding parts of congruent triangles.*

Modeling with Technology

You can use a graphing calculator to plot points and draw line segments connecting them. By plotting three points and the connecting segments, you can graph triangles.

On the TI-82, you will use the line feature from the draw menu. When you enter two points, the calculator graphs the line segment between them.

Your Turn Graph $\triangle XYZ$ with vertices $X(10, 5)$, $Y(30, 5)$, $Z(5, 25)$ and $\triangle X'Y'Z'$ with vertices $X'(-15, 2)$, $Y'(5, 2)$, and $Z'(-20, 22)$ using the steps below.

- Press ZOOM 6 ZOOM 8 ENTER. *Sets the screen to integer values.*
- Press 2nd DRAW 2.
- Use the arrow keys to move the cursor to (10, 5). The coordinates at the bottom of the screen will help you find the point. Press ENTER.
- Then move the cursor to (30, 5). Press ENTER twice.
- Move the cursor to (5, 25) and press ENTER twice. Then move the cursor back to (10, 5) to complete the triangle. Press ENTER.
- Repeat the third, fourth, and fifth steps above using the coordinates of $\triangle X'Y'Z'$ to draw the second triangle.

TALK ABOUT IT

a. How do the size and shape of $\triangle XYZ$ and $\triangle X'Y'Z'$ compare? same

b. If you placed $\triangle X'Y'Z'$ on top of $\triangle XYZ$, which angles would match up?

c. Which sides would match up? $\overline{XY}$ to $\overline{X'Y'}$, $\overline{YZ}$ to $\overline{Y'Z'}$, $\overline{XZ}$ to $\overline{X'Z'}$

b. $\angle X$ to $\angle X'$, $\angle Y$ to $\angle Y'$, $\angle Z$ to $\angle Z'$

Learning the Concept

Figures that have the same size and shape are **congruent**. The triangles shown below are congruent. Slash marks are used to indicate which *sides* are congruent and arcs are used to indicate which *angles* are congruent. Remember, the symbol ≅ means *is congruent to.*

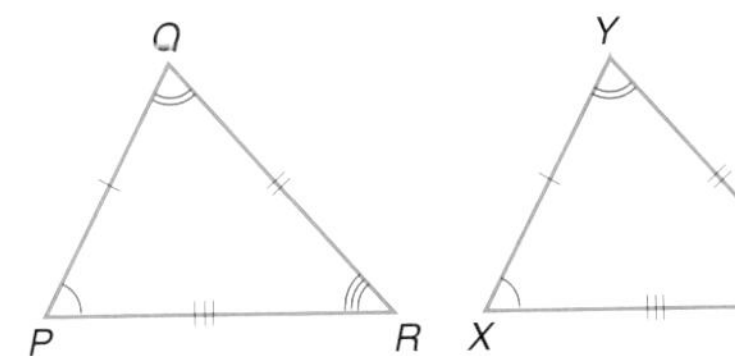

Congruent Angles	Congruent Sides
$\angle P \cong \angle X$	$\overline{PQ} \cong \overline{XY}$
$\angle Q \cong \angle Y$	$\overline{QR} \cong \overline{YZ}$
$\angle R \cong \angle Z$	$\overline{PR} \cong \overline{XZ}$

Parts of congruent triangles that "match" are called **corresponding parts**. For example, in the triangle above, $\angle P$ corresponds to $\angle X$, and $\overline{PQ}$ corresponds to $\overline{XY}$.

Alternative Learning Styles

Visual Have students draw a scalene triangle on a blank sheet of paper. Have them put the labels of vertices A, B, and C *inside* the triangle. Now have them place another sheet of blank paper behind the sheet they drew on and cut out the triangle, thus yielding two triangles of the same shape and size. Have them label the second triangle's vertices R, S, and T. Tell students that corresponding parts have the same measure. Remind students that parts with the same measure are *congruent* and the symbol is ≅. Have them write a list of six pairs of congruent corresponding parts.

11-5 LESSON NOTES

NCTM Standards: 1-4, 12

Instructional Resources

- Study Guide Master 11-5
- Practice Master 11-5
- Enrichment Master 11-5
- Group Activity Card 11-5
- Assessment and Evaluation Masters. pp. 294, 295
- Math Lab and Modeling Math Masters, p. 86

Transparency 11-5A contains the 5-Minute Check for this lesson; **Transparency 11-5B** contains a teaching aid for this lesson.

Recommended Pacing	
Standard Pacing	Day 6 of 13
Honors Pacing	Day 6 of 13
Block Scheduling*	Day 4 of 7 (along with Lesson 11-6)

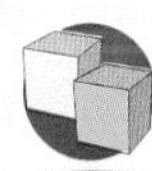

*For more information on pacing and possible lesson plans, refer to the ***Block Scheduling Booklet***.

1 FOCUS

5-Minute Check
(over Lesson 11-4)

In $\triangle ABC$, $m\angle A = 40°$ and $m\angle B = 35°$.

1. Find $m\angle C$. 105°
2. Classify $\triangle ABC$ as *acute*, *right*, or *obtuse*. obtuse

In $\triangle RST$, $m\angle R = (x - 20)°$, $m\angle S = (x + 40)°$, and $m\angle T = (x + 10)°$

3. Find $m\angle R$. 30°
4. Find $m\angle T$. 60°
5. Classify $\triangle RST$ as *acute*, *right*, or *obtuse*. right

Motivating the Lesson

Questioning Show pictures of identical twins. Ask why they are called identical. same characteristics

2 TEACH

In-Class Examples

For Example 1
If $\triangle XYZ \cong \triangle DEF$, name the congruent angles and sides.

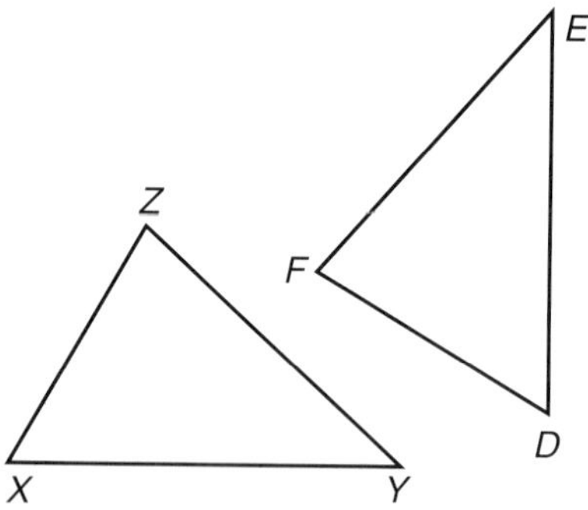

$\angle X \cong \angle D$; $\angle Y \cong \angle E$; $\angle Z \cong \angle F$; $\overline{XY} \cong \overline{DE}$; $\overline{YZ} \cong \overline{EF}$; $\overline{ZX} \cong \overline{FD}$

For Example 2
The corresponding parts of two congruent triangles are marked in the figure. Write a congruence statement for the two triangles.

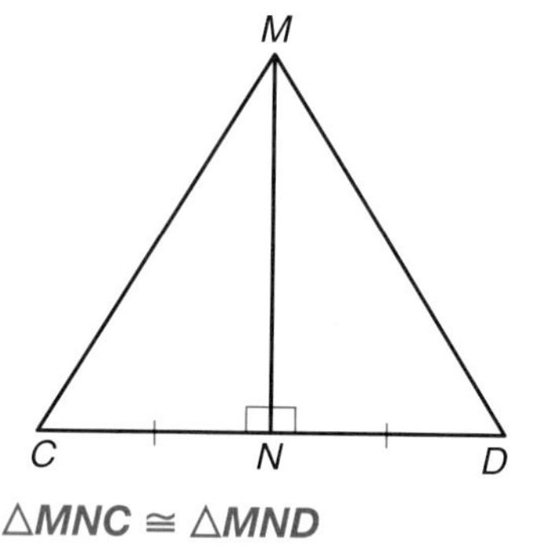

$\triangle MNC \cong \triangle MND$

Teaching Tip Emphasize the fact that vertices are named in order to show congruency.

Corresponding Parts of Congruent Triangles

If two triangles are congruent, their corresponding sides are congruent and their corresponding angles are congruent.

When you write $\triangle PQR \cong \triangle XYZ$, the corresponding vertices are written in order. So, $\triangle PQR \cong \triangle XYZ$ means that vertex P corresponds to vertex X, vertex Q corresponds to vertex Y, and vertex R corresponds to vertex Z.

Example 1 **If $\triangle COD \cong \triangle ATM$, name the congruent angles and sides.**

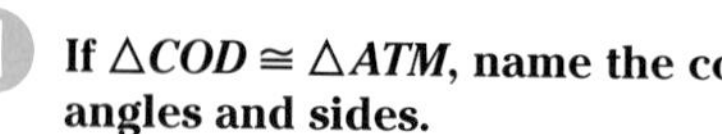

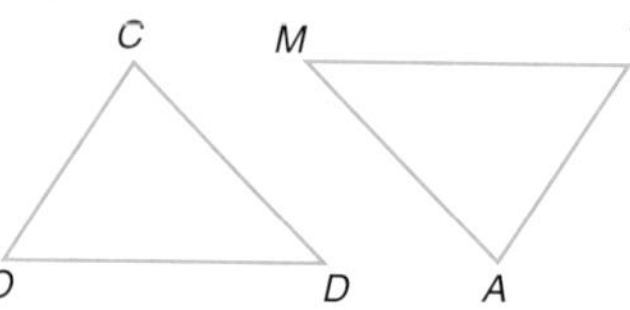

First, name the three pairs of congruent angles by looking at the order of the vertices.

$\angle C \cong \angle A$
$\angle O \cong \angle T$
$\angle D \cong \angle M$

Since C corresponds to A and O corresponds to T, $\overline{CO}$ corresponds to $\overline{AT}$. Similarly $\overline{OD}$ corresponds to $\overline{TM}$, and $\overline{CD}$ corresponds to $\overline{AM}$. The congruent sides are as follows.
$\overline{CO} \cong \overline{AT}$, $\overline{OD} \cong \overline{TM}$, $\overline{CD} \cong \overline{AM}$

You can draw conclusions about congruent triangles based on the congruence statements.

Example 2 **The corresponding parts of two congruent triangles are marked in the figure. Write a congruence statement for the two triangles.**

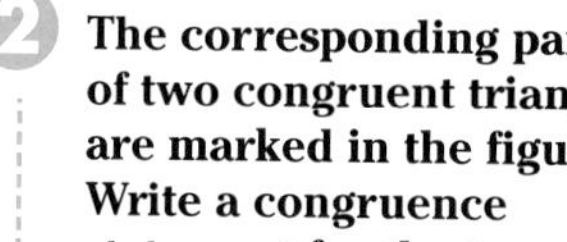

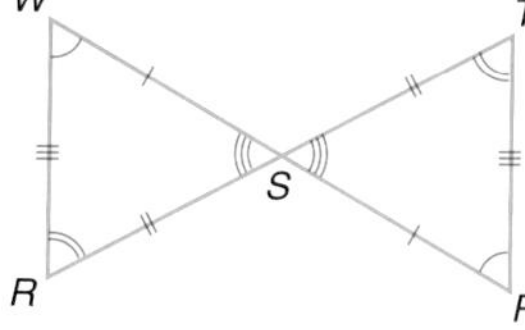

List the congruent angles and sides.

$\angle W \cong \angle P$	$\overline{PS} \cong \overline{WS}$
$\angle R \cong \angle T$	$\overline{TS} \cong \overline{RS}$
$\angle WSR \cong \angle PST$	$\overline{TP} \cong \overline{RW}$

The congruence statement can be written by matching the vertices of the congruent angles. $\triangle TPS \cong \triangle RWS$.

THINK ABOUT IT

How would you know that $\angle WSR \cong \angle PST$ even if they were not marked?

They are vertical angles.

You can use corresponding parts to find measures of angles and sides in a figure that is congruent to a figure with known measures, as shown in Example 3.

Classroom Vignette

"I have students construct congruent triangles on graph paper. The students then cut out the triangles and match the congruent triangles. This activity gives practice in constructions and introduces congruent figures."

Paul Todd
Onondaga Hill Middle School
Syracuse, NY

Example 3 APPLICATION Construction

Roof trusses allow a roof to withstand the stress of heavy loads. A roof truss for a particular building is made so that the angle on the left measures 35°. The triangular sides of the truss are to be congruent as marked below.

a. What should the measure of angle *TSR* be?

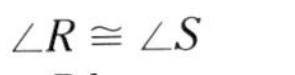

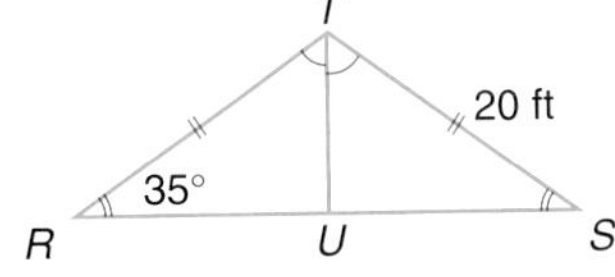

$\angle R \cong \angle S$
$\angle R$ has a measure of 35°, so $\angle S$ also has a measure of 35°.

b. What should the measure of side *TR* be?

$\overline{TR} \cong \overline{TS}$

$\overline{TS}$ has a given measure of 20 feet. So $\overline{TR}$ also has a measure of 20 feet.

Checking Your Understanding

Communicating Mathematics

Read and study the lesson to answer these questions.

1. **Tell** what it means when one triangle is congruent to another triangle. All corresponding parts are congruent.
2. **Name** the side that corresponds to $\overline{ST}$ and name the angle that corresponds to $\angle T$ if $\triangle RST \cong \triangle WXY$. $\overline{XY}$, $\angle Y$
3. **Tell** which figure appears to be congruent to . c

a. 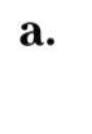 b.  c.  d. none of these

MATERIALS

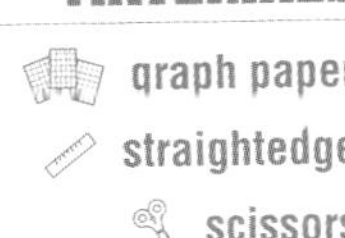
graph paper
straightedge
scissors

4. On a piece of graph paper, draw two triangles like the ones at the right. Label the vertices as shown and cut out the triangles. Place one triangle over the other and turn until the corresponding parts match. Next, slide the top triangle to the right until the two triangles are side-by-side. Now the corresponding parts are in the same relative positions and you can easily identify them.

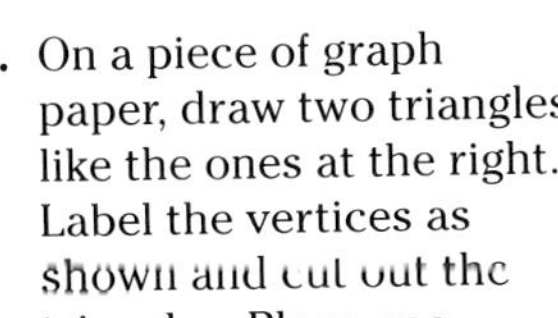
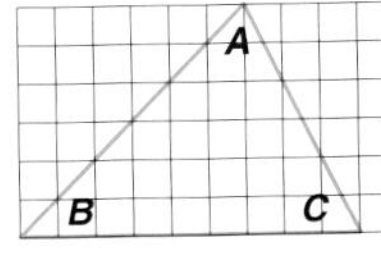

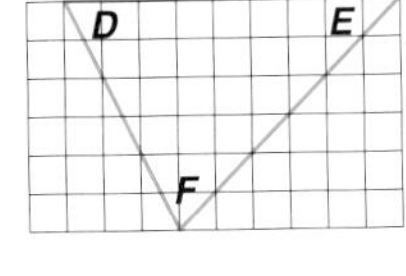

a. Write a congruence statement for the two triangles. $\triangle ABC \cong \triangle FED$
b. Name congruence statements for the congruent angles and sides.

4b. $\angle A \cong \angle F$, $\angle B \cong \angle E$, $\angle C \cong \angle D$; $\overline{AB} \cong \overline{FE}$; $\overline{BC} \cong \overline{ED}$; $\overline{AC} \cong \overline{FD}$

Guided Practice

5. Complete the congruence statement for the triangles shown at the right. Then name the corresponding parts.

a. $\triangle ABC \cong$ ___ $\triangle ZYX$
b. $\overline{AB} \cong$ ___ $\overline{ZY}$ c. $\angle Y \cong$ ___ $\angle B$
d. $\angle C \cong$ ___ $\angle X$ e. $\overline{BC} \cong$ ___ $\overline{YX}$
f. $\overline{XZ} \cong$ ___ $\overline{CA}$ g. $\angle Z \cong$ ___ $\angle A$

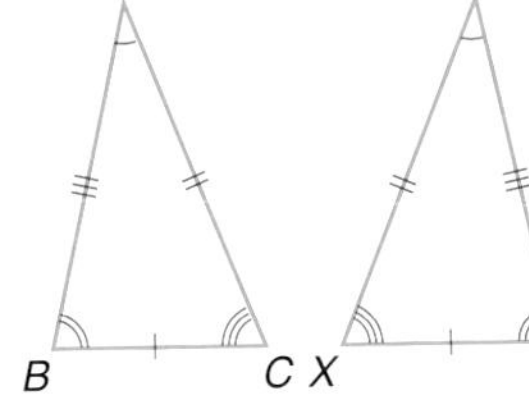

In-Class Example

For Example 3

Lorenzo noticed that the guy wires connected to a telephone pole created two congruent triangles. He found the measurements shown on this sketch.

How far is point *J* from the base of the pole? What is the measure of the angle that guy wire $\overline{JL}$ makes with the ground? **10 feet; 70°**

Study Guide Masters, p. 95

NAME ______ DATE ______

11-5 Study Guide
Congruent Triangles

Student Edition Pages 573–577

The triangles on the right are **congruent**. The parts of congruent triangles that "match" are called **corresponding parts**.

$\angle A$ corresponds to $\angle G$. $\overline{AB}$ corresponds to $\overline{GH}$.
$\angle B$ corresponds to $\angle H$. $\overline{BC}$ corresponds to $\overline{HF}$.
$\angle C$ corresponds to $\angle F$. $\overline{AC}$ corresponds to $\overline{GF}$.

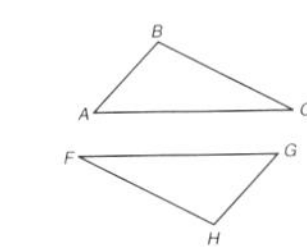

Complete the congruence statement for each pair of congruent triangles. Then name the corresponding parts.

1. $\triangle TRS \cong \triangle$ URX
$\angle T \cong \angle U$, $\angle S \cong \angle X$, $\angle SRT \cong \angle XRU$; $\overline{TS} \cong \overline{XU}$, $\overline{SR} \cong \overline{XR}$, $\overline{TR} \cong \overline{UR}$

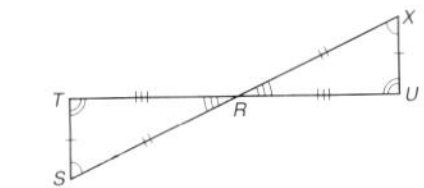

2. $\triangle ABD \cong \triangle$ CBD
$\angle A \cong \angle C$, $\angle ABD \cong \angle CBD$, $\angle ADB \cong \angle CDB$; $\overline{AB} \cong \overline{CB}$, $\overline{AD} \cong \overline{CD}$, $\overline{BD} \cong \overline{BD}$

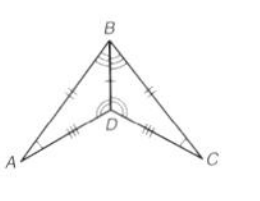

3. $\triangle FHG \cong \triangle$ MGH
$\angle F \cong \angle M$, $\angle FGH \cong \angle MHG$, $\angle FHG \cong \angle MGH$; $\overline{FG} \cong \overline{MH}$, $\overline{FH} \cong \overline{MG}$, $\overline{GH} \cong \overline{GH}$

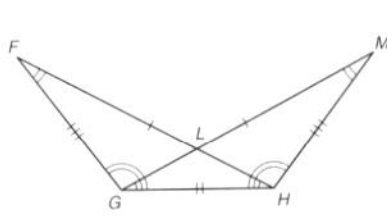

Reteaching

Using Models Have students cut out several different pairs of congruent triangles and mix them up. Then have students match each triangle to its congruent counterpart.

3 PRACTICE/APPLY

Checking Your Understanding

Exercises 1–8 are designed to help you assess your students' understanding through reading, writing, speaking, and modeling. You should work through Exercises 1–4 with your students and then monitor their work on Exercises 5–8.

Assignment Guide

Core: 9–27 odd, 29–33
Enriched: 10–22 even, 24–33

For **Extra Practice**, see p. 768.

The red A, B, and C flags, printed only in the Teacher's Wraparound Edition, indicate the level of difficulty of the exercises.

6. Write a congruence statement for the triangles shown below. **△ABC ≅ △FED**

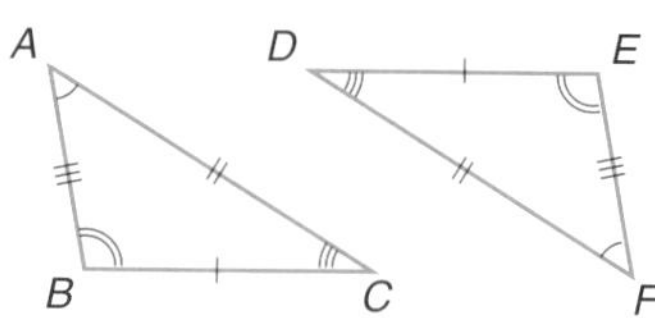

7. If $\triangle DEF \cong \triangle HEG$, what is the measure of $\overline{EH}$? **c**
   a. 6 b. 8
   c. 10 d. not enough information

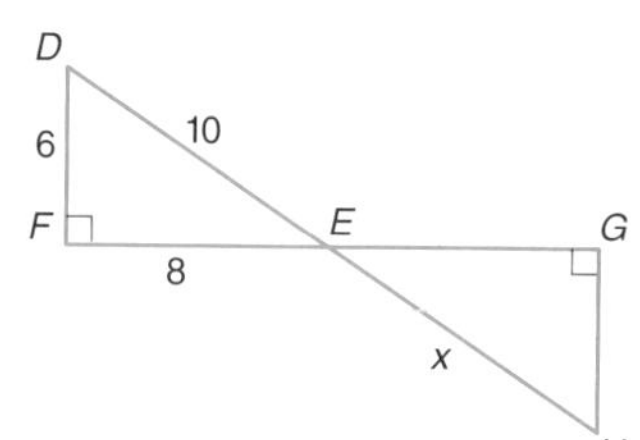

8. **Landscaping** The fence around a triangular garden is 12 meters long. A second garden is congruent to the first. How much fence is needed for the second garden? **12 m**

Exercises: Practicing and Applying the Concept

Independent Practice

Complete the congruence statement for each pair of congruent triangles. Then name the corresponding parts.

9. 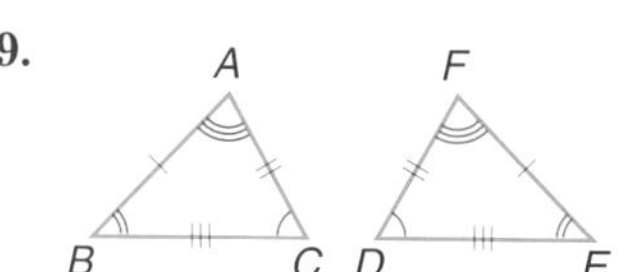

10. 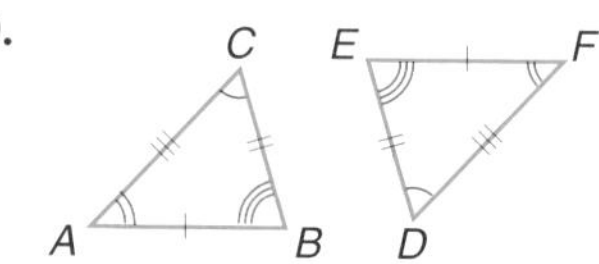

11. 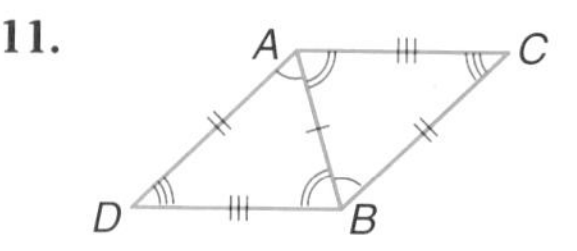

12. 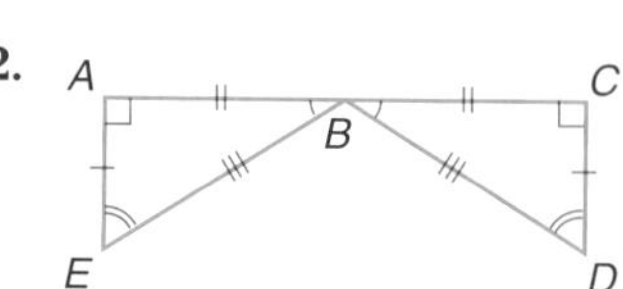

Write a congruence statement for each pair of congruent triangles.

13. 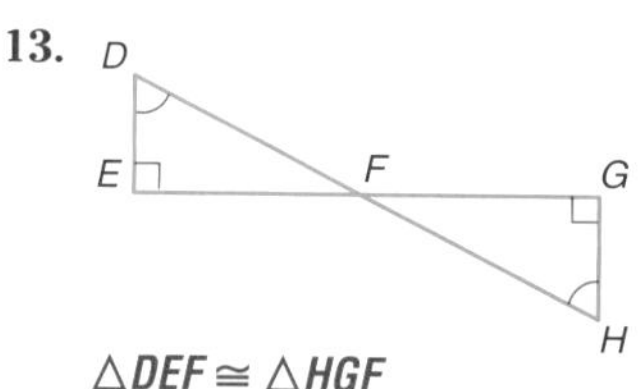

△DEF ≅ △HGF

14. 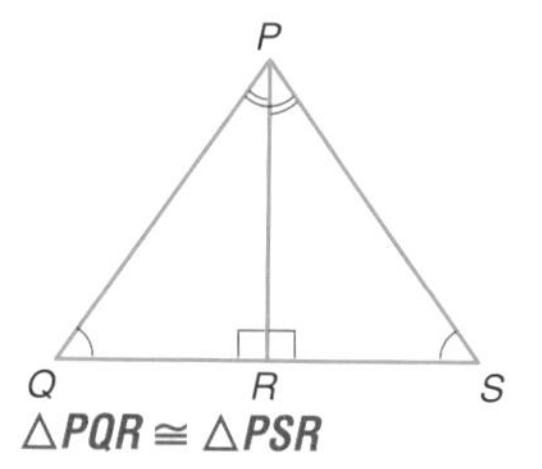

△PQR ≅ △PSR

If $\triangle RST \cong \triangle MAP$, name the part congruent to each angle or segment given. (*Hint*: Make a drawing.)

15. $\overline{RS}$ **$\overline{MA}$** 16. $\angle P$ **∠T** 17. $\overline{MP}$ **$\overline{RT}$** 18. $\overline{ST}$ **$\overline{AP}$** 19. $\angle R$ **∠M** 20. $\angle A$ **∠S**

Answers:

9. △ABC ≅ △FED; ∠A ≅ ∠F, ∠B ≅ ∠E, ∠C ≅ ∠D, $\overline{AB} \cong \overline{FE}$, $\overline{BC} \cong \overline{ED}$, $\overline{AC} \cong \overline{FD}$
10. △CBA ≅ △DEF; ∠C ≅ ∠D, ∠B ≅ ∠E, ∠A ≅ ∠F, $\overline{CB} \cong \overline{DE}$, $\overline{BA} \cong \overline{EF}$, $\overline{CA} \cong \overline{DF}$
11. △BAD ≅ △ABC; ∠BAD ≅ ∠ABC, ∠D ≅ ∠C, ∠ABD ≅ ∠BAC, $\overline{BA} \cong \overline{AB}$, $\overline{AD} \cong \overline{BC}$, $\overline{BD} \cong \overline{AC}$
12. △BCD ≅ △BAE; ∠C ≅ ∠A, ∠D ≅ ∠E, ∠CBD ≅ ∠ABE, $\overline{BC} \cong \overline{BA}$, $\overline{CD} \cong \overline{AE}$, $\overline{BD} \cong \overline{BE}$

Practice Masters, p. 95

NAME ______________ DATE ______________

11-5 Practice
Congruent Triangles

Student Edition Pages 573–577

Complete the congruence statement for each pair of congruent triangles. Then name the corresponding parts.

1. 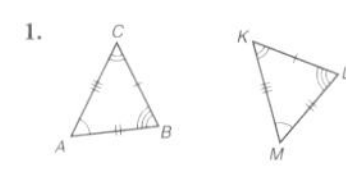
△ABC ≅ △ **MLK**
∠A ≅ ∠M, ∠B ≅ ∠L, ∠C ≅ ∠K; $\overline{AB} \cong \overline{ML}$, $\overline{BC} \cong \overline{LK}$, $\overline{CA} \cong \overline{KM}$

2. 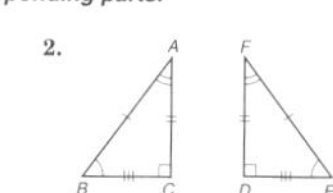
△ABC ≅ △ **FED**
∠A ≅ ∠F, ∠B ≅ ∠E, ∠C ≅ ∠D; $\overline{AB} \cong \overline{FE}$, $\overline{BC} \cong \overline{ED}$, $\overline{CA} \cong \overline{DF}$

3. 
△BAD ≅ △ **DCB**
∠ABD ≅ ∠CDB, ∠DAB ≅ ∠BCD, ∠ADB ≅ ∠CBD; $\overline{BA} \cong \overline{DC}$, $\overline{AD} \cong \overline{CB}$, $\overline{BD} \cong \overline{DB}$

4. 
△XYZ ≅ △ **CAB**
∠X ≅ ∠C, ∠Y ≅ ∠A, ∠Z ≅ ∠B; $\overline{XY} \cong \overline{CA}$, $\overline{YZ} \cong \overline{AB}$, $\overline{XZ} \cong \overline{CB}$

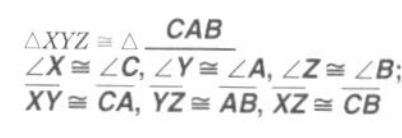

5. 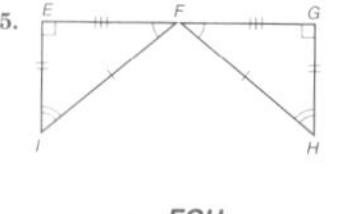
△FEI ≅ △ **FGH**
∠EFI ≅ ∠GFH, ∠E ≅ ∠G, ∠I ≅ ∠H; $\overline{FE} \cong \overline{FG}$, $\overline{EI} \cong \overline{GH}$, $\overline{FI} \cong \overline{FH}$

6. 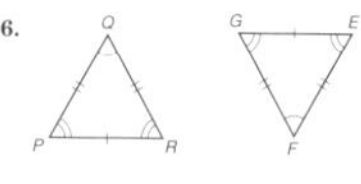
△PQR ≅ △ **GFE**
∠Q ≅ ∠F, ∠P ≅ ∠G, ∠R ≅ ∠E; $\overline{PQ} \cong \overline{GF}$, $\overline{QR} \cong \overline{FE}$, $\overline{PR} \cong \overline{GE}$

If △FGE = △XYZ, name the part congruent to each angle or segment given. (HINT: *Make a drawing.*)

7. ∠X **∠F** 8. $\overline{FE}$ **$\overline{XZ}$** 9. ∠E **∠Z** 10. $\overline{EG}$ **$\overline{ZY}$**

Group Activity Card 11-5

Triangular Trivia Group Activity **11-5**

MATERIALS: Enlarged copies of the geometric figure below • rulers • protractors

How many congruent triangles can you find in the figure below? All right angles have been marked. Work with a partner. Use a protractor and ruler to measure parts of the figure. Discuss why you think certain triangles are congruent. Compare your results with others to see who found the most congruent triangles.

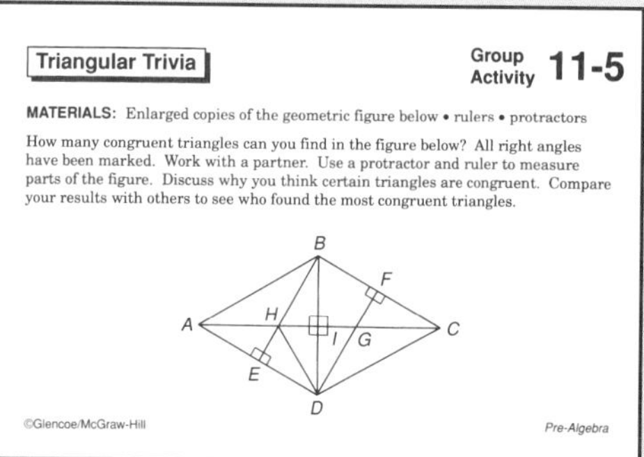

 Pre-Algebra

Find the value of x for each pair of congruent triangles.

21. 3

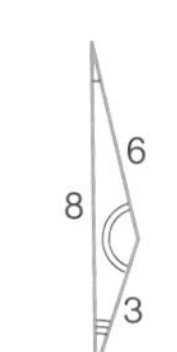

22. 6

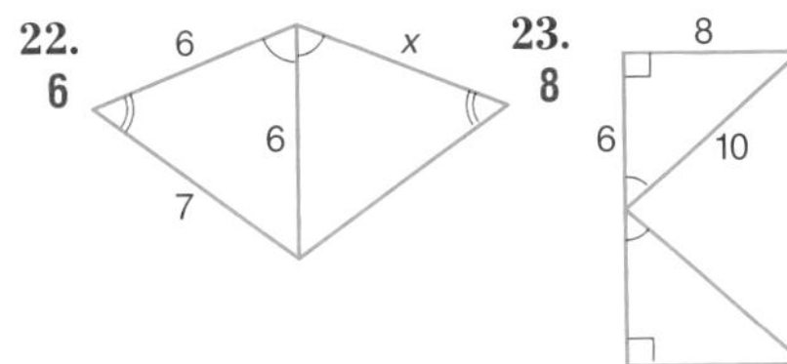

23. 8

Critical Thinking

24. See margin.

24. If two triangles have three pairs of congruent, corresponding angles, are the triangles congruent? Explain your answer by making a drawing.

25. In the figure at the right, two pairs of overlapping triangles appear to be congruent. Write a congruence statement for each pair.
$\triangle ABH \cong \triangle IJG$; $\triangle ACG \cong \triangle IEF$

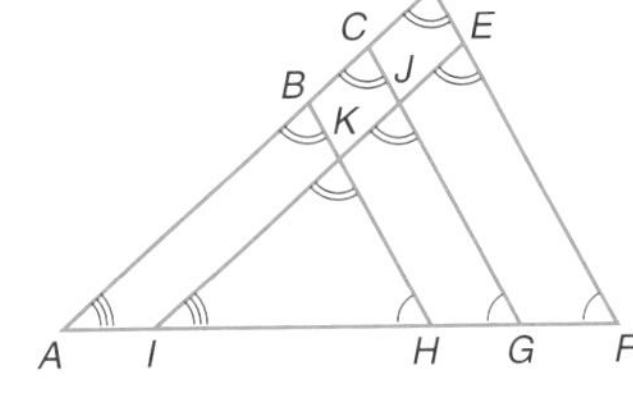

Applications and Problem Solving

26. Construction One pattern for a roof truss is shown at the right. Use the labels in the figure and name the triangle that seems to be congruent to each triangle listed below.

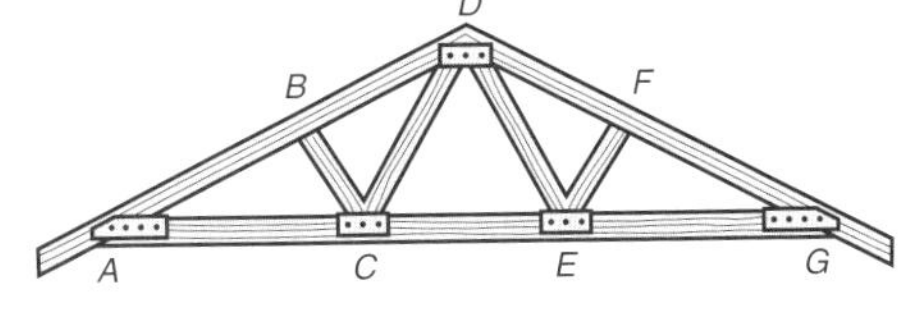

a. $\triangle ABC$ **$\triangle GFE$** **b.** $\triangle BDC$ **$\triangle FDE$** **c.** $\triangle ADC$ **$\triangle GDE$**

27. Geometry Two triangles are congruent and the perimeter of one triangle is 5 feet. What is the perimeter of the second triangle? **5 ft**

28. Recreation Make a paper airplane of your own design from an $8\frac{1}{2}$-by-11 inch piece of paper. When you have finished, open the paper and observe the pattern made by the folds. Look for pairs of congruent triangles. Use a protractor to measure angles if you wish. Label the vertices and write congruence statements for the pairs of congruent triangles. **See students' work.**

Mixed Review

29. 95°, obtuse

29. In $\triangle PQR$, $m\angle P = 45°$, $m\angle Q = 40°$. Find the measure of $\angle R$ and classify the triangle as *acute*, *right*, or *obtuse*. (Lesson 11-4)

30. Consumer Awareness A shirt that normally sells for \$45 is on sale for 15% off. What is the amount of savings? (Lesson 9-5) **\$6.75**

31. Solve $\frac{a}{5} - 6 = 19$. (Lesson 7-2) **125**

32. Name the multiplicative inverse of $1\frac{3}{5}$. (Lesson 6-4) $\frac{5}{8}$

33. Eliminate Possibilities Liz, Renee, and Pablo each either brought a sack lunch, bought a plate lunch, or bought from the snack bar for lunch. Use the clues to find each person's lunch. (Lesson 3-1)
- Liz did not buy a lunch.
- Renee did not have a plate lunch.

Renee – snack bar; Liz – pack; Pablo – plate lunch

4 ASSESS

Closing Activity

Writing Draw and label two congruent triangles on the board or overhead. Have students write the segments and angles that are congruent. Then have them write a congruence statement for the triangles.

Chapter 11, Quiz B (Lessons 11-4 and 11-5) is available in the *Assessment and Evaluation Masters*, p. 295.

Mid-Chapter Test (Lessons 11-1 through 11-5) is available in the *Assessment and Evaluation Masters*, p. 294.

Additional Answer

24. Not necessarily: similar triangles, like $\triangle ABC$ and $\triangle DEF$, have congruent corresponding angles but are not congruent triangles.

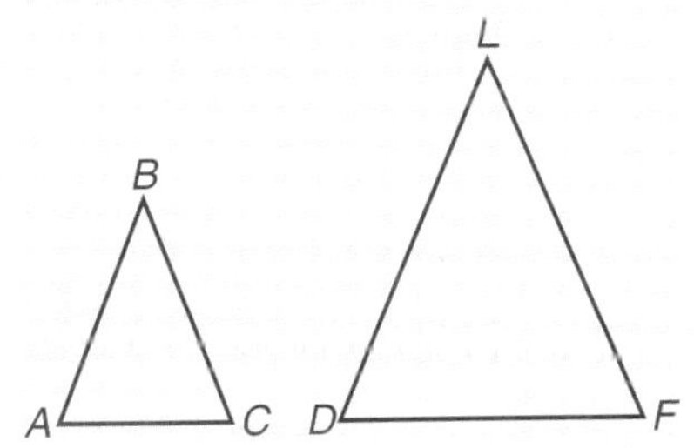

***Enrichment Masters*, p. 95**

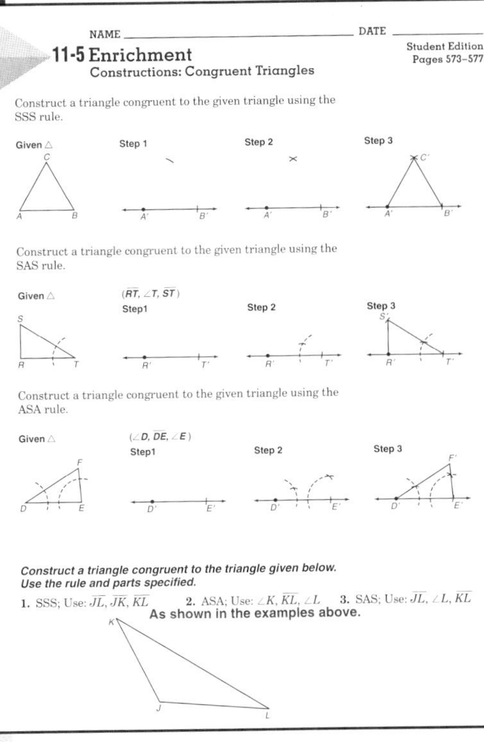
NAME ______ DATE ______

11-5 Enrichment Student Edition Pages 573–577

Constructions: Congruent Triangles

Construct a triangle congruent to the given triangle using the SSS rule.

Given △ Step 1 Step 2 Step 3

Construct a triangle congruent to the given triangle using the SAS rule.

Given △ ($\overline{RT}$, ∠T, $\overline{ST}$) Step1 Step 2 Step 3

Construct a triangle congruent to the given triangle using the ASA rule.

Given △ (∠D, $\overline{DE}$, ∠E) Step1 Step 2 Step 3

Construct a triangle congruent to the triangle given below. Use the rule and parts specified.

1. SSS; Use: $\overline{JL}$, $\overline{JK}$, $\overline{KL}$ 2. ASA; Use: ∠K, $\overline{KL}$, ∠L 3. SAS; Use: $\overline{JL}$, ∠L, $\overline{KL}$

As shown in the examples above.

Extension

Using Manipulatives On a piece of paper, draw two different rectangles and two different parallelograms. Duplicate and distribute to students. Have students draw the diagonals of one of the rectangles. Then have students either measure or cut out the triangles and check whether there are any congruent triangles. **Yes, two pairs**

Have students repeat the procedure for the parallelograms. **Yes, two pairs** Have students use the second rectangle and parallelogram to check their work.

11-6 LESSON NOTES

NCTM Standards: 1-4, 12

Instructional Resources

- Study Guide Master 11-6
- Practice Master 11-6
- Enrichment Master 11-6
- Group Activity Card 11-6
- Math Lab and Modeling Math Masters, p. 70
- Tech Prep Applications Masters, p. 22

Transparency 11-6A contains the 5-Minute Check for this lesson; **Transparency 11-6B** contains a teaching aid for this lesson.

Recommended Pacing	
Standard Pacing	Day 7 of 13
Honors Pacing	Day 7 of 13
Block Scheduling	Day 4 of 7 (along with Lesson 11-5)

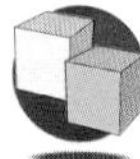

*For more information on pacing and possible lesson plans, refer to the ***Block Scheduling Booklet***.

1 FOCUS

5-Minute Check
(over Lesson 11-5)

Use the figure below to complete the following. The triangles are congruent.

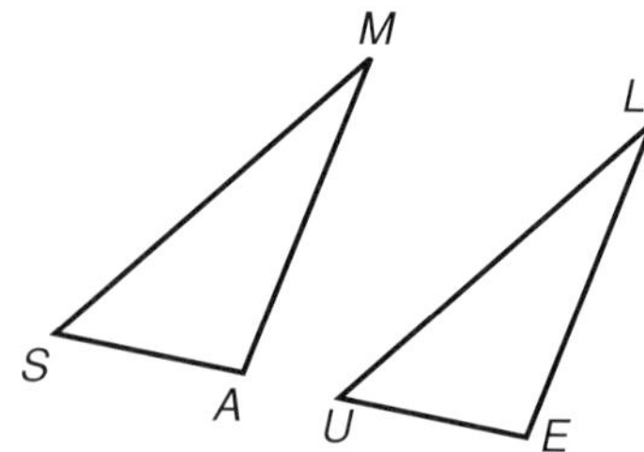

1. Name three pairs of congruent angles. **$\angle S \cong \angle U$; $\angle A \cong \angle E$; $\angle M \cong \angle L$**
2. Write a congruence statement for the two triangles. **$\triangle SAM \cong \triangle UEL$**
3. If side $\overline{AM}$ is 10 cm long, which side of the other triangle is 10 cm long? **side $\overline{EL}$**

11-6 Similar Triangles and Indirect Measurement

Setting Goals: *In this lesson, you'll identify corresponding parts and find missing measures of similar triangles. You'll also solve problems involving indirect measurement by using similar triangles.*

MATHEMATICS MINI LAB

Modeling with Manipulatives

MATERIALS

graph paper
ruler
 protractor

Have you ever used a copy machine to make a reduction or enlargement of something? The original and the copy are the same shape, but different sizes. In mathematics, we call this a **dilation.** You can use graph paper to investigate the relationship of two such figures.

Your Turn

Work with a partner.

- Draw and label a triangle on a coordinate plane.
- Multiply the coordinates of each vertex by 2. Draw the new triangle.

TALK ABOUT IT

a. Measure the angles in the two triangles. How do they compare?

b. Use a ruler to measure the sides of the two triangles. Write a ratio comparing the measures of corresponding sides. What do you notice?

c. Repeat the activity by multiplying the coordinates of the original triangle by $\frac{1}{2}$. Are the results the same?

a. Corresponding angles have the same measure.
b. The ratios are the same.
c. yes

Learning the Concept

The triangles you drew in the activity were **similar**. Figures that have the same shape but not necessarily the same size are similar figures.

As with congruent triangles, write the corresponding vertices of similar triangles in the same order.

In the figure below, $\triangle ABC$ is similar to $\triangle PQR$. This is written as $\triangle ABC \sim \triangle PQR$ in symbols. The symbol $\sim$ means *is similar to.*

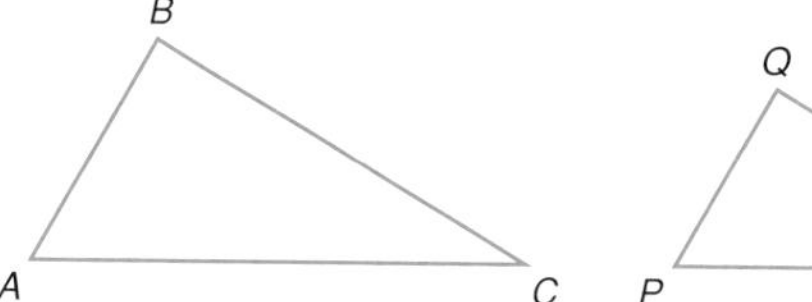

Measure each angle. Compare the corresponding angles. This example and the activity at the beginning of the lesson suggest the following.

Tech Prep

Mural Painter In many towns and cities, you may see a large mural painted on the side of a building or other structure. A mural artist often makes a small drawing of the mural and then uses similar figures and proportions to enlarge the drawing and create the mural.

For more information on tech prep, see the *Teacher's Handbook.*

Corresponding Angles of Similar Triangles

If two triangles are similar, then the corresponding angles are congruent.

It is also true that if the corresponding angles of two triangles are congruent, then the triangles are similar. *You will use this property in Example 2.*

The corresponding sides of congruent triangles also have a special relationship. In the figure below, $\triangle LMN \sim \triangle RST$.

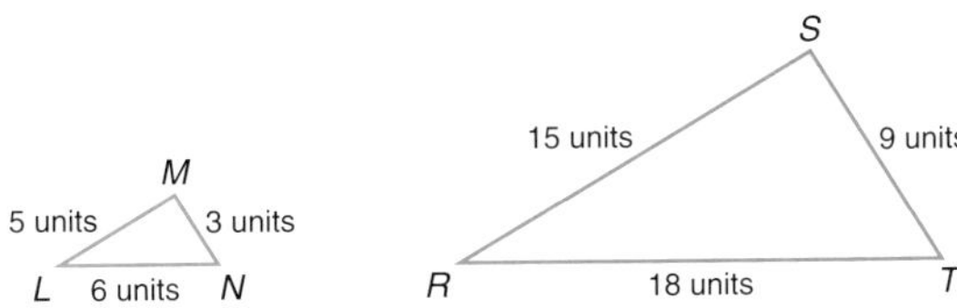

We can compare the measures of corresponding sides by using ratios.

$m\overline{LN}$ means the measure of segment LN.

$\frac{m\overline{LN}}{m\overline{RT}} = \frac{6 \text{ units}}{18 \text{ units}}$ or $\frac{1}{3}$ $\quad \frac{m\overline{MN}}{m\overline{ST}} = \frac{3 \text{ units}}{9 \text{ units}}$ or $\frac{1}{3}$ $\quad \frac{m\overline{LM}}{m\overline{RS}} = \frac{5 \text{ units}}{15 \text{ units}}$ or $\frac{1}{3}$

Notice that the ratios are equivalent. So, the corresponding sides are proportional to each other. This example and the activity at the beginning of the lesson seem to suggest the following.

Corresponding Sides of Similar Triangles

If two triangles are similar, then their corresponding sides are proportional.

It is also true that if the corresponding sides of two triangles are proportional, then the triangles are similar.

You can use proportions to find the measures of the sides of similar triangles when some measures are known.

Example  **If $\triangle GHI \sim \triangle JKL$, find the value of x.**

You can review proportions in Lesson 9-4.

Write a proportion using the known measures.

$\frac{m\overline{GH}}{m\overline{JK}} = \frac{m\overline{HI}}{m\overline{KL}}$ *Corresponding sides are proportional.*

$\frac{12}{4} = \frac{9}{x}$

$12x = 4 \cdot 9$ *Find the cross products.*

$12x = 36$

$x = 3$ *Divide each side by 12.*

The measure of $\overline{KL}$ is 3 cm.

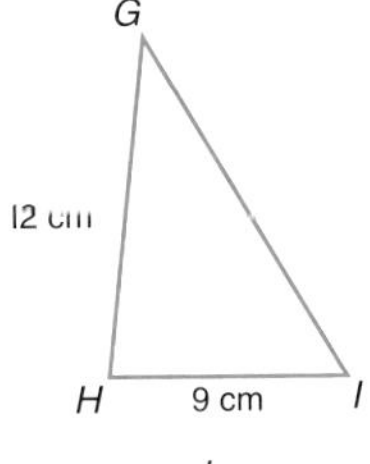

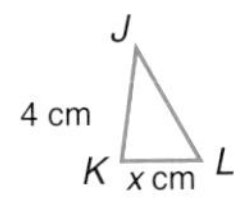

Motivating the Lesson

Situational Problem Read these two paragraphs to students.

One cold, dark night, I found myself alone in the backyard of my home. The trees bent as the cold wind blew from the north. An eerie feeling crept over me, so I went inside.

The very next night I found myself in similar circumstances, when ...

Ask what similar circumstances means.

Teaching Tip If geoboards are not available, have students use dot paper and colored pencils.

2 TEACH

In-Class Example

For Example 1

If $\triangle JUD \sim \triangle ITH$, find the value of x. **8**

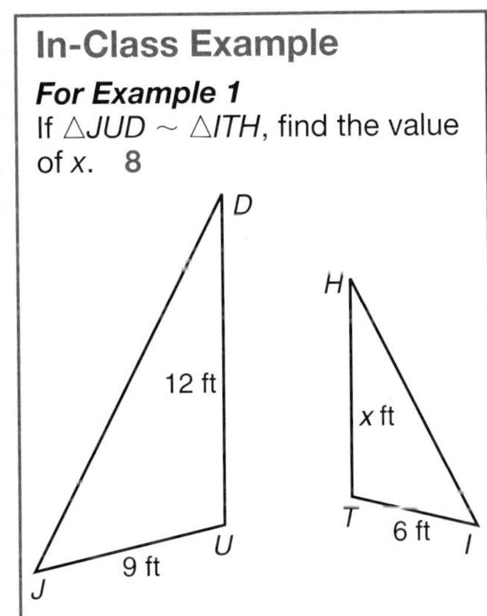

Cooperative Learning

Co-op Co-op Have students work in teams to complete this project. Each team is to draw a simple picture that uses mainly triangles on a piece of typing paper. Then the group is to create a mural or bulletin board that incorporates an enlargement of that picture. You may allow for creativity by allowing captions to be added to the picture.

For more information on this strategy, see *Cooperative Learning in the Mathematics Classroom*, one of the titles in the Glencoe Mathematics Professional Series, p. 32.

In-Class Examples

For Example 2
Use the information in the figure below to find the value of x.

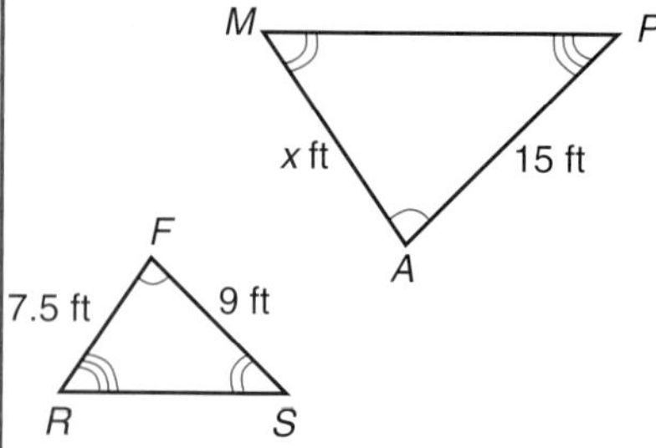

The measure of $\overline{AM}$ is 18 feet.

For Example 3
Quincy and Anna want to know how far it is across a pond. They made the sketch below.

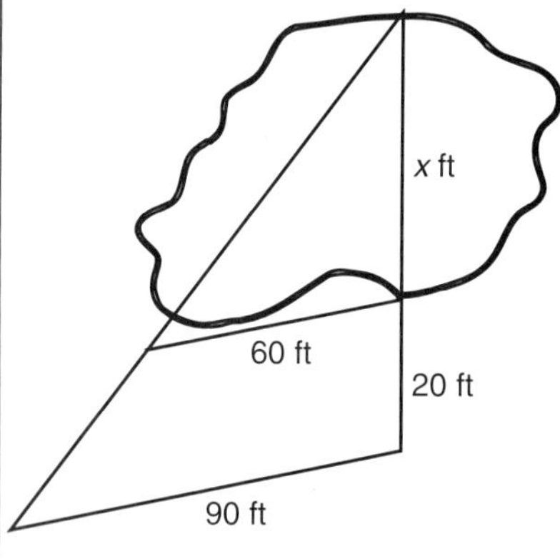

How far is it across the pond?
It is 40 feet across the pond.

Teaching Tip In solving the In-Class Example for Example 3 above, you may need to help students set up the correct proportion, which is $\frac{60}{90} = \frac{x}{x + 20}$.

Additional Answer Think About It

Sample Answer: On a sunny day, measure the shadow cast by the top of the school and the shadow cast by a meterstick. Write and solve a proportion to find the height of the school.

Example  **Use the information in the figure below to find the value of x.**

E x ft D, 12 ft, C, 7.5 ft, A 10 ft B

Since vertical angles are congruent, $\angle ACB \cong \angle DCE$.
Angles A and D are marked congruent.
Angles B and E are congruent because they are both right angles.
So, $\triangle ABC \sim \triangle DEC$, because their corresponding angles are congruent.

$\frac{m\overline{AB}}{m\overline{DE}} = \frac{m\overline{BC}}{m\overline{EC}}$ *Corresponding sides of similar triangles are proportional.*

$\frac{10}{x} = \frac{7.5}{12}$

$10 \cdot 12 = 7.5x$ *Find the cross products.*

$120 = 7.5x$

$\frac{120}{7.5} = \frac{7.5x}{7.5}$ *Divide each side by 7.5.*

120 [÷] 7.5 [=] *16*

$16 = x$

The measure of $\overline{ED}$ is 16 feet.

You can use what you know about similar triangles to find the measure of objects that are too large to measure directly, as shown in Example 3. This kind of measurement is called **indirect measurement**.

Example 3

History

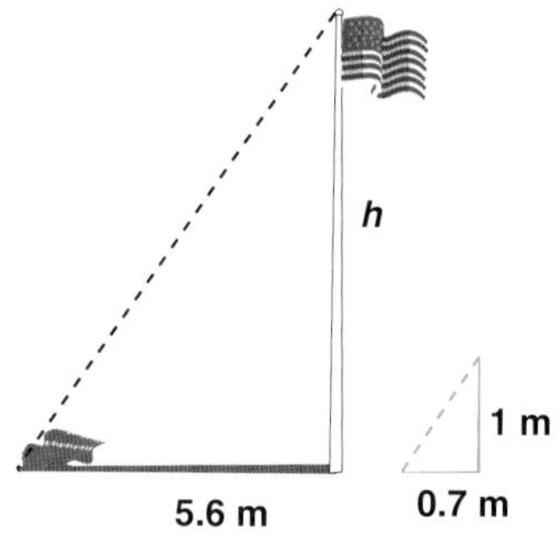

In ancient Egypt, mathematicians used a technique called *shadow reckoning* to determine the heights of tall objects, such as the pyramids. The height of a staff and the length of its shadow are proportional to the height of an object and the length of its shadow. The objects and their shadows form two sides of similar triangles. Use shadow reckoning to find the height of the flagpole in the figure below.

Explore The meterstick in the drawing is like the staff.

Plan In the figure, the shadow of the meterstick is 0.7 meters long, and the shadow of the flagpole is 5.6 meters long. Write a proportion comparing corresponding sides of the similar triangles.

h, 5.6 m, 1 m, 0.7 m

THINK ABOUT IT

How could you use shadow reckoning to measure the height of your school?

See margin.

Solve

$$\frac{\text{length of meterstick shadow}}{\text{length of flagpole shadow}} = \frac{\text{length of meterstick}}{\text{length of flagpole}}$$

$$\frac{0.7}{5.6} = \frac{1}{h}$$

$$0.7h = 5.6 \quad \textit{Find the cross products.}$$

$$\frac{0.7h}{0.7} = \frac{5.6}{0.7} \quad \textit{Divide each side by 0.7.}$$

$$h = 8$$

The flagpole is 8 meters tall.

Examine You can estimate to check your answer. The meterstick is about $1\frac{1}{2}$ times its shadow, so the flagpole is about $1\frac{1}{2}$ times its shadow.

Checking Your Understanding

Communicating Mathematics

Read and study the lesson to answer these questions. **1. See margin.**

1. **Explain** what it means for one figure to be similar to another.
2. **You Decide** Shawn thinks the triangles below are similar. Marta thinks they are not. Who is correct and why? **Marta; the corresponding angles are not congruent.**

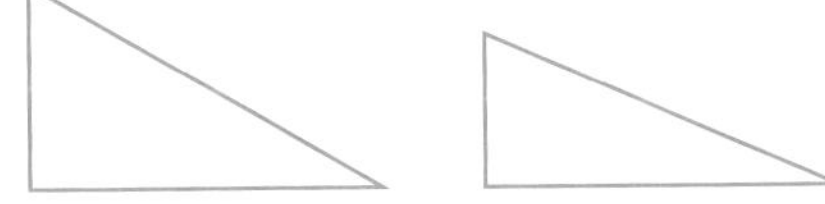

3. **Draw** two similar triangles. Label the vertices and name the corresponding sides and angles. **See students' work.**
4. **Explain** how you can find the height of an object that is too large to measure directly. **See margin.**
5. **Make Up a Problem** Write a problem involving the use of similar triangles. **See students' work.**

MATERIALS

TI-82 graphing calculator

6. Use a TI-82 graphing calculator to graph $\triangle ABC$ with vertices $A(0, 3)$, $B(15, 3)$, and $C(15, 15)$ and $\triangle AXY$ with vertices $A(0, 3)$, $X(30, 3)$, and $Y(30, 27)$. Follow the steps in the procedure on page 573, using the coordinates given in this problem. Do the triangles appear to be similar? **See students' work; yes.**

Guided Practice

7a. $\frac{m\overline{AB}}{m\overline{AC}} = \frac{m\overline{BE}}{m\overline{CD}} = \frac{m\overline{AE}}{m\overline{AD}}$

7. Refer to the similar triangles at the right.

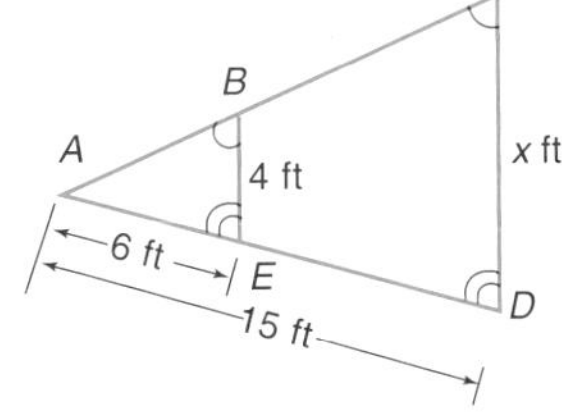

a. List three proportions that can be written for $\triangle ABE$ and $\triangle ACD$.

b. What angle of $\triangle CAD$ corresponds to $\angle BAE$? **$\angle CAD$**

c. What side corresponds to $\overline{BE}$? **$\overline{CD}$**

d. Find the value of x. **10**

3 PRACTICE/APPLY

Checking Your Understanding

Exercises 1–11 are designed to help you assess your students' understanding through reading, writing, speaking, and modeling. You should work through Exercises 1–6 with your students and then monitor their work on Exercises 7–11.

Additional Answers

1. Corresponding angles are congruent and corresponding sides proportional.
4. Measure the shadow cast by the object you can measure, like a meterstick. Write and solve a proportion to find the height of the object.

Study Guide Masters, p. 96

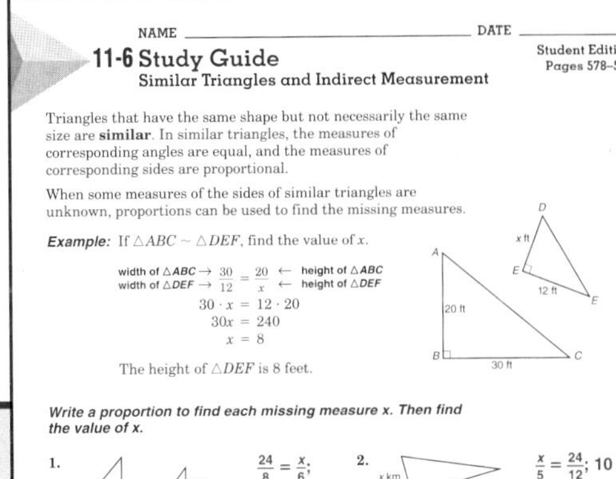

NAME ______ DATE ______

11-6 Study Guide
Similar Triangles and Indirect Measurement

Student Edition Pages 578–583

Triangles that have the same shape but not necessarily the same size are **similar**. In similar triangles, the measures of corresponding angles are equal, and the measures of corresponding sides are proportional.

When some measures of the sides of similar triangles are unknown, proportions can be used to find the missing measures.

Example: If $\triangle ABC \sim \triangle DEF$, find the value of x.

width of $\triangle ABC \rightarrow \frac{30}{12} = \frac{20}{x} \leftarrow$ height of $\triangle ABC$
width of $\triangle DEF \rightarrow$ $\leftarrow$ height of $\triangle DEF$

$$30 \cdot x = 12 \cdot 20$$
$$30x = 240$$
$$x = 8$$

The height of $\triangle DEF$ is 8 feet.

Write a proportion to find each missing measure x. Then find the value of x.

1. $\frac{24}{8} = \frac{x}{6}$; 18 in.
2. $\frac{x}{5} = \frac{24}{12}$; 10 km

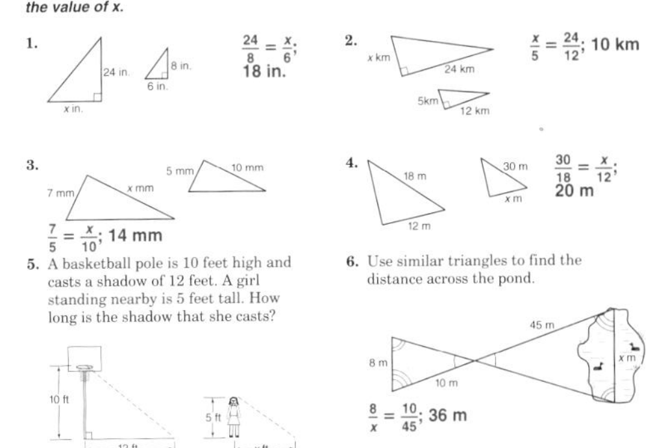

3. $\frac{7}{5} = \frac{x}{10}$; 14 mm
4. $\frac{30}{18} = \frac{x}{12}$; 20 m
5. A basketball pole is 10 feet high and casts a shadow of 12 feet. A girl standing nearby is 5 feet tall. How long is the shadow that she casts? $\frac{10}{5} = \frac{12}{x}$; 6 ft
6. Use similar triangles to find the distance across the pond. $\frac{8}{x} = \frac{10}{45}$; 36 m

Reteaching

Using Models Provide students with pairs of similar triangles. For each pair, have students find the ratios of the corresponding sides. Ask what they notice about the ratios. **They are the same.**

Error Analysis

Students may set up the proportions incorrectly. Have students use colored pencils as an aid in the following manner. Trace one triangle with a blue pencil and one with a red pencil. Then use green to mark the shortest of the sides of each triangle with one green tick mark (I), the longest of the sides of each with three green tick marks (I I I), the remaining sides with two green tick marks (I I). The proportion is then set up using two of three fractions below.

$$\frac{\text{red triangle 1 tick mark}}{\text{blue triangle 1 tick mark}} = \frac{\text{red triangle 2 tick marks}}{\text{blue triangle 2 tick marks}} = \frac{\text{red triangle 3 tick marks}}{\text{blue triangle 3 tick marks}}$$

Assignment Guide

Core: 13–23 odd, 24–28
Enriched: 12–18 even, 19–28

For **Extra Practice**, see p. 768.

The red A, B, and C flags, printed only in the Teacher's Wraparound Edition, indicate the level of difficulty of the exercises.

8. $\frac{4}{8} = \frac{x}{10}$; 5

9. $\frac{1}{x} = \frac{2}{6}$; 3

Write a proportion to find each missing measure x. Then find the value of x.

8. (J, K, L, M, N; 10 ft, 8 ft, 6 ft, x ft, 4 ft)

9. (P, Q, R, S, T, U; x cm, 6 cm, 1cm, 2 cm)

10. Given: triangle with sides 2, 3, 4. Which of the following is similar to the given triangle? **a**

a. 4, 6, 8 **b.** 3, 4, 5 **c.** 4, 5, 6 **d.** none of these

11. Civil Engineering The city of Marion plans to build a bridge across Brandon Lake. Use the information in the diagram at the right to find the distance across Brandon Lake. **121 m**

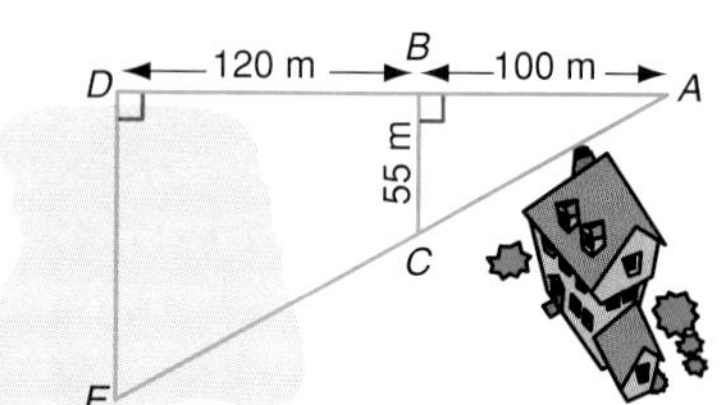

Exercises: Practicing and Applying the Concept

Independent Practice

Write a proportion to find each missing measure x. Then find the value of x.

12. $\frac{4}{10} = \frac{6}{x}$; 15

13. $\frac{3}{2} = \frac{x}{3}$; 4.5

14. $\frac{3}{2} = \frac{5.5}{x}$; $3\frac{2}{3}$

15. $\frac{4}{5} = \frac{8}{x}$; 10

16. $\frac{24}{56} = \frac{21}{x}$; 49

17. $\frac{3}{5} = \frac{4.5}{x}$; 7.5

12.

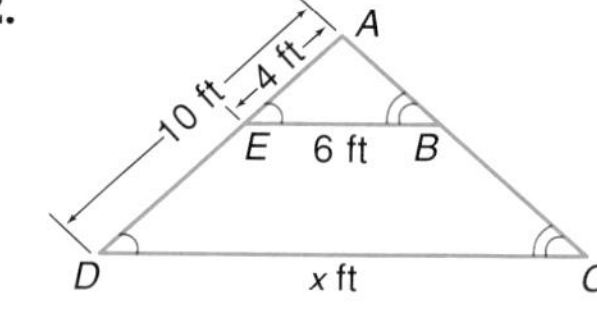

13. (G, H, I, J, K, L; 3 m, x m, 3 m, 2 m)

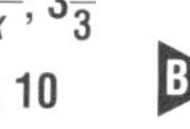

14.

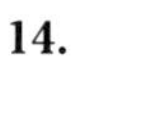
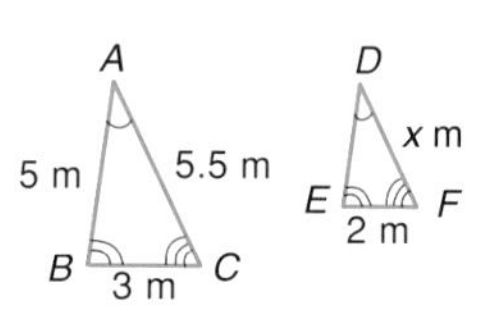

15.

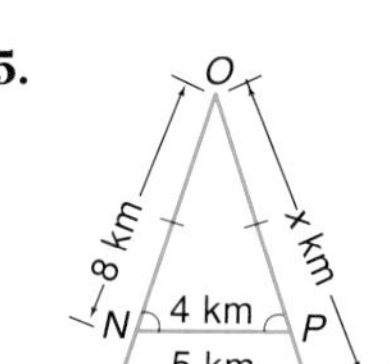

16.

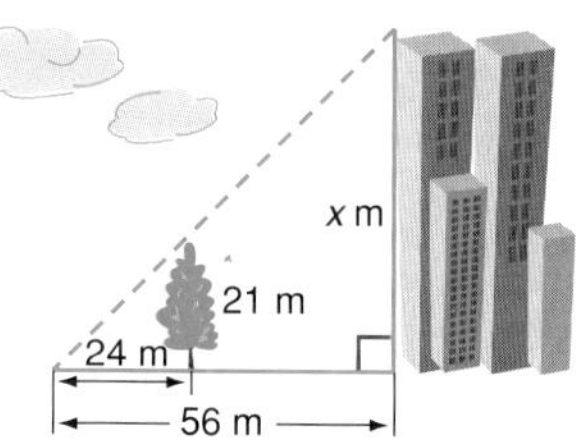

17.

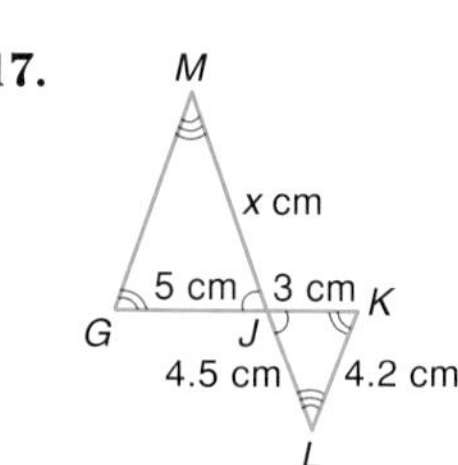

Practice Masters, p. 96

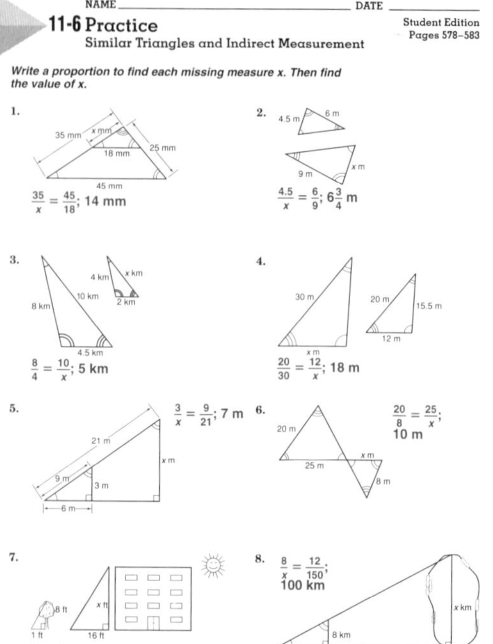

NAME ________ DATE ________

11-6 Practice
Similar Triangles and Indirect Measurement

Student Edition Pages 578–583

Write a proportion to find each missing measure x. Then find the value of x.

1. $\frac{35}{x} = \frac{45}{18}$; 14 mm

2. $\frac{4.5}{x} = \frac{6}{9}$; $6\frac{3}{4}$ m

3. $\frac{8}{4} = \frac{10}{x}$; 5 km

4. $\frac{20}{30} = \frac{12}{x}$; 18 m

5. $\frac{3}{x} = \frac{9}{21}$; 7 m

6. $\frac{20}{8} = \frac{25}{x}$; 10 m

7. $\frac{8}{x} = \frac{1}{16}$; 128 ft

8. $\frac{8}{x} = \frac{12}{150}$; 100 km

Group Activity Card 11-6

Look Alikes Group Activity **11-6**

MATERIALS: Three 0-9 spinners

Work in groups of two to four to make similar triangles.

First, the leader spins the three spinners. If the sum of any two of the numbers is not greater than the third number, the spinner with the least number must be spun again. Use the three numbers for the lengths of the sides of a triangle.

The group's task is to determine the length of the sides of three new triangles similar to the first one. The leader spins one of the spinners again. In making each new triangle, this number replaces one of the sides of the original triangle. An example is given on the back of the card.

 Pre-Algebra

18a. Yes. The corresponding angles are congruent.

18. On a coordinate grid, graph $A(6, 6)$, $B(2, 2)$, $C(8, 2)$, $D(0, 0)$, and $E(9, 0)$. Draw $\triangle ABC$ and $\triangle ADE$.

a. Are the triangles similar? How do you know?

b. Do $\overline{BC}$ and $\overline{DE}$ appear to be parallel? **yes**

c. Will two triangles with a common angle and the sides opposite that angle parallel always be similar? Explain. **Yes, because the corresponding angles are congruent.**

Critical Thinking

19. a. Are all congruent triangles also similar? Explain. **See margin.**

b. Are all similar triangles also congruent? Explain. **See margin.**

20. Photography A photo negative is 52.5 mm wide by 35 mm high. The print is made 15.24 centimeters wide. How many times greater is the area of the print than the area of the negative? **about 8.4 times greater**

Applications and Problem Solving

21. Construction Find the length of the brace in the roof rafter shown at the right. **2.7 ft**

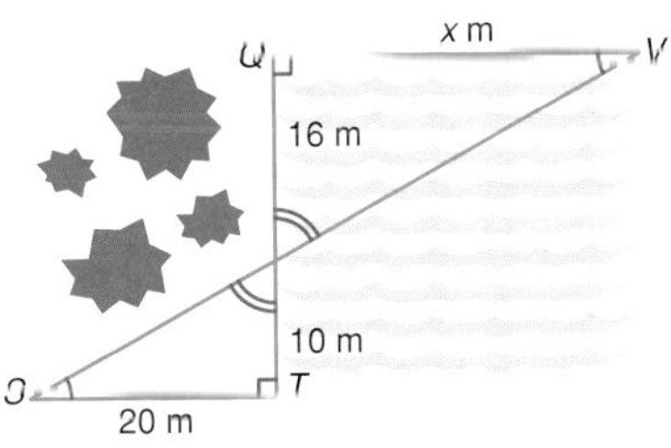

22. History Refer to Example 3. The largest known pyramid is Khufu's pyramid. At a certain time of day, a yardstick casts a shadow 1.5 ft long, and the pyramid casts a shadow 241 ft long. Draw a sketch that shows the two similar triangles and use shadow reckoning to find the height of the pyramid. Estimate to check your answer. **482 ft**

23. Surveying A surveyor needs to find the distance across a river and draws the sketch shown at the right. Explain how the surveyor determined the sketch and the known measurements. Then find the distance across the river. **See margin.**

Mixed Review

24. Geometry Two triangles are congruent, and the area of one triangle is 15 square feet. What is the area of the second triangle? (Lesson 11-5) **15 square feet**

25. A quiz has 5 true-false questions. How many outcomes for giving answers to the five questions are possible? (Lesson 10-5) $\mathbf{2^5}$

26. Solve the system of equations $y = 3x$ and $y = x + 4$ by graphing. (Lesson 8-8) **(2, 6)**

27. Statistics Find the mean, median, and mode for the set of data below. When necessary, round to the nearest tenth. (Lesson 6-6)
26, 32, 54, 34, 36, 48, 39, 40, 54, 89, 78, 45 **47.9; 42.5; 54**

28. Find the GCF of $14n^2$, $22p^2$, and $36n^2p$. (Lesson 4-5) **2**

4 ASSESS

Closing Activity

Modeling Provide five different pairs of cut-out similar triangles. Mix them up. Have a student select one triangle and find the triangle that is similar to the one selected.

Additional Answers

19a. Yes; the ratio of the corresponding sides is 1:1.

19b. No; only those where the ratio of the corresponding sides is 1:1.

23. Sample answer: Stand on the shore (at point *Q*) and sight straight across the river to a point on the other shore, perhaps a tree (a point *V*). Walk along the shore 26 m (to point *T*); then walk perpendicular to the shore 20 m (to point *S*). Sight from *S* across the river to point *V*. Have a second person stand along the shore so that he or she is in the line of sight from *S* to *V*. Measure the distance from that person to *Q* and *T* (16 m and 10 m). The distance across the river is 32 m.

Extension

Logical Thinking Extend Exercise 19 by asking the following questions.

- Are all right triangles similar? **no**
- Are all equilateral triangles similar? **yes**
- Are all isosceles triangles similar? **no**
- Are all isosceles right triangles similar? **yes**

Enrichment Masters, p. 96

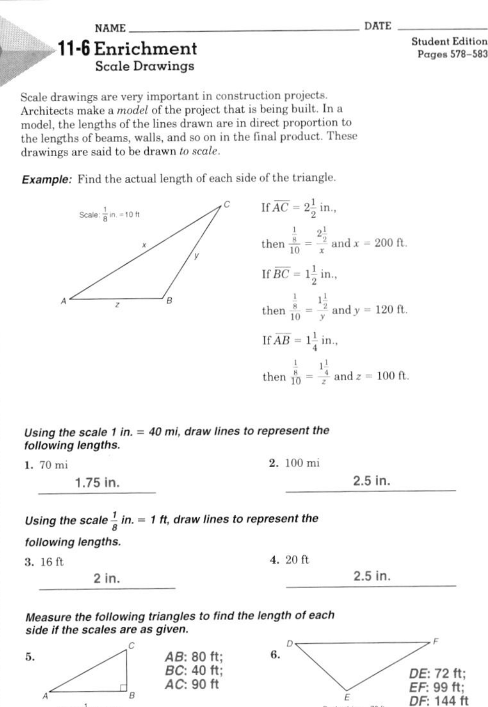

NAME ____________ DATE ________

11-6 Enrichment — Student Edition Pages 578–583

Scale Drawings

Scale drawings are very important in construction projects. Architects make a *model* of the project that is being built. In a model, the lengths of the lines drawn are in direct proportion to the lengths of beams, walls, and so on in the final product. These drawings are said to be drawn *to scale*.

Example: Find the actual length of each side of the triangle.

Scale: $\frac{1}{8}$ in. = 10 ft

If $\overline{AC} = 2\frac{1}{2}$ in.,

then $\frac{\frac{1}{8}}{10} = \frac{2\frac{1}{2}}{x}$ and $x = 200$ ft.

If $\overline{BC} = 1\frac{1}{2}$ in.,

then $\frac{\frac{1}{8}}{10} = \frac{1\frac{1}{2}}{y}$ and $y = 120$ ft.

If $\overline{AB} = 1\frac{1}{4}$ in.,

then $\frac{\frac{1}{8}}{10} = \frac{1\frac{1}{4}}{z}$ and $z = 100$ ft.

Using the scale 1 in. = 40 mi, draw lines to represent the following lengths.

1. 70 mi   1.75 in.

2. 100 mi   2.5 in.

Using the scale $\frac{1}{8}$ in. = 1 ft, draw lines to represent the following lengths.

3. 16 ft   2 in.

4. 20 ft   2.5 in.

Measure the following triangles to find the length of each side if the scales are as given.

5. *AB*: 80 ft; *BC*: 40 ft; *AC*: 90 ft (Scale: $\frac{1}{8}$ in. = 10 ft)

6. *DE*: 72 ft; *EF*: 99 ft; *DF*: 144 ft (Scale: 1 in. = 72 ft)

11-7 LESSON NOTES

NCTM Standards: 1-4, 12

Instructional Resources

- Study Guide Master 11-7
- Practice Master 11-7
- Enrichment Master 11-7
- Group Activity Card 11-7
- Assessment and Evaluation Masters, p. 296

Transparency 11-7A contains the 5-Minute Check for this lesson; **Transparency 11-7B** contains a teaching aid for this lesson.

Recommended Pacing	
Standard Pacing	Day 8 of 13
Honors Pacing	Day 8 of 13
Block Scheduling*	Day 5 of 7 (along with Lesson 11-8)

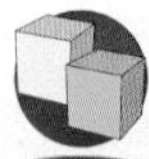

*For more information on pacing and possible lesson plans, refer to the ***Block Scheduling Booklet***.

1 FOCUS

5-Minute Check
(over Lesson 11-6)

1. Name the corresponding sides for the pair of similar triangles below.

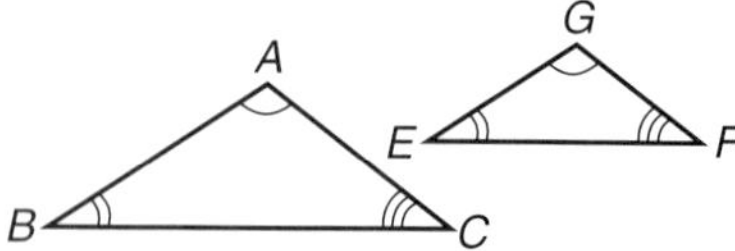

$\overline{BC} \cong \overline{EF}$, $\overline{CA} \cong \overline{FG}$, $\overline{AB} \cong \overline{GE}$

2. Name the corresponding angles. **$\angle B \cong \angle E$; $\angle A \cong \angle G$; $\angle C \cong \angle F$**

3. Find the value of x in the figure below. **$7\frac{1}{2}$**

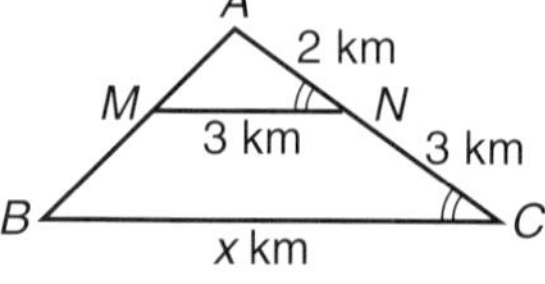

11-7 Quadrilaterals

Setting Goals: *In this lesson, you'll find the missing angle measure of a quadrilateral, classify quadrilaterals, and explore similar quadrilaterals.*

Modeling a Real-World Application: Quilting

FYI

The world's largest quilt was made by 7000 citizens of North Dakota for the state's centennial celebration in 1989. It was 85 feet by 134 feet.

Quilting in America dates back to colonial days. Early quilts were often made using only square or rectangular patches. Later designs used other geometric shapes. The quilt shown at the right was made using a standard one-patch pattern and silk cloth. It is called "Tumbling Blocks" and is on display in the Howell House in Philadelphia, Pennsylvania. Notice how the four-sided patches and the design give the illusion of three dimensions.

Learning the Concept

When you name a quadrilateral, you can begin at any vertex. But it is important to name consecutive vertices in order.

Squares and rectangles are simple quadrilaterals. A **quadrilateral** is a closed figure formed by four line segments that intersect only at their endpoints. As with triangles, a quadrilateral can be named by its vertices. The quadrilateral shown below can be named quadrilateral $ABCD$.

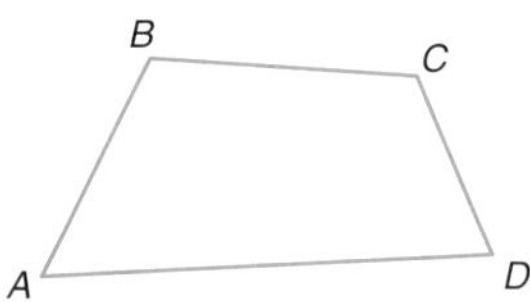

The vertices are A, B, C, and D.
The sides are $\overline{AB}$, $\overline{BC}$, $\overline{CD}$, and $\overline{AD}$.
The angles are $\angle A$, $\angle B$, $\angle C$, and $\angle D$.

If you use a protractor to measure each interior angle of quadrilateral $ABCD$, you will find that $m\angle A = 60°$, $m\angle B = 115°$, $m\angle C = 120°$, and $m\angle D = 65°$. The sum of the measures of the angles of $ABCD$ is 360°.

This example suggests a relationship of the angles of a quadrilateral.

Angles of a Quadrilateral	The sum of the measures of the angles of a quadrilateral is 360°.

Example 

INTEGRATION
Algebra

Find the value of x. Then find the missing angle measures.

The sum of the measures of the angles is 360°.

$72 + 96 + x + 2x = 360$
$168 + 3x = 360$ *Combine like terms.*
$3x = 192$ *Subtract 168 from each side.*
$x = 64$ *Divide each side by 3.*

So $m\angle Q = 64°$, and $m\angle R = 2(64)$ or $128°$.

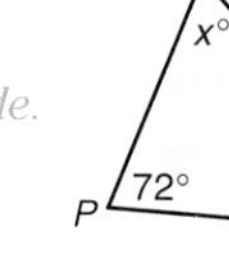

Quilting is a sewing technique used for centuries around the world in many countries, including ancient China and Egypt. Patchwork quilting was a popular form of folk art in preindustrial America.

There are many kinds of quadrilaterals. One way to identify them is to look for pairs of parallel sides.

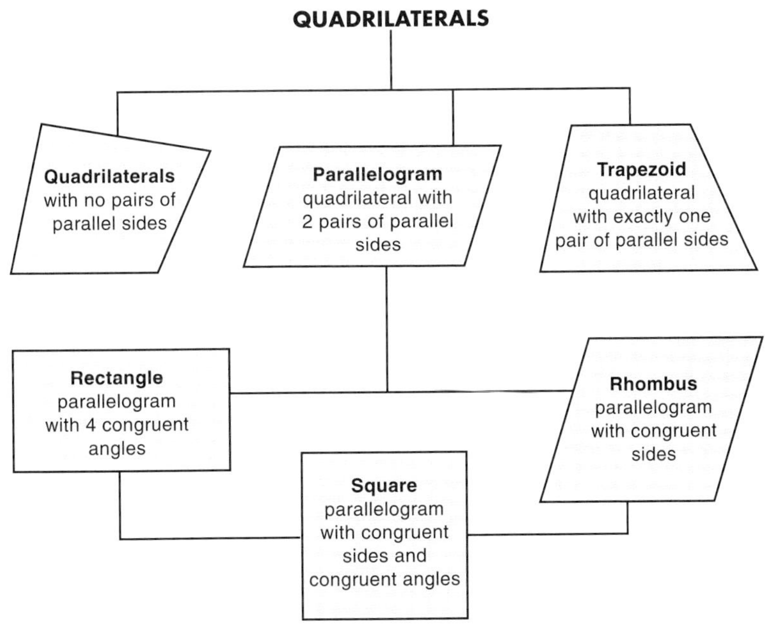

In Lesson 11-6, you learned various properties of similar triangles. These same properties also hold for similar quadrilaterals or any pair of similar figures.

Corresponding Angles and Sides of Similar Figures

If two figures are similar, then the angles of one figure are congruent to the corresponding angles of the other figure. If two figures are similar, then their corresponding sides are porportional.

A **scale drawing** is similar to the actual object, but it is generally either smaller or larger. The scale is the ratio of the lengths on the drawing to the actual lengths of the object.

Example  **On a scale drawing of a room, 1 inch represents 2 feet. In the scale drawing, the room is 11 inches long. How long is the actual room?**

The scale is the ratio 1 in. to 2 ft or $\frac{1 \text{in.}}{2 \text{ft}}$. *This is often written 1 in. = 2 ft.*

Write a proportion.

$$\frac{1 \text{in.}}{2 \text{ft}} = \frac{11 \text{ in.}}{x}$$

$1x = 2(11)$ *Find the cross products.*

$x = 22$

The actual room is 22 feet long.

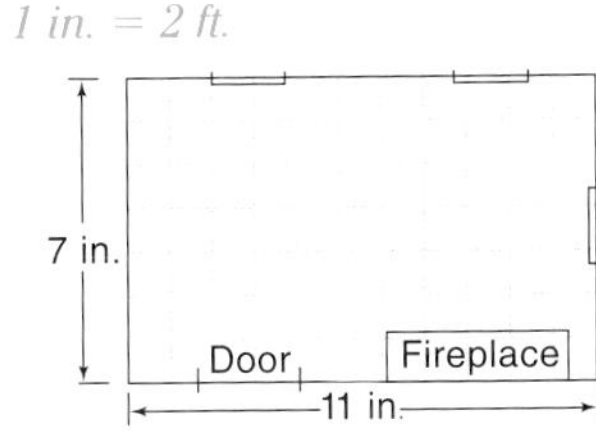

Motivating the Lesson

Hands-On Activity Draw different quadrilaterals on index cards. Have a student pick a card without showing it to anyone else and describe it to the class. Have one student try to draw the figure at the board as the other describes it. Repeat until the cards are gone.

2 TEACH

In-Class Examples

For Example 1
The measures of three angles of a quadrilateral are 87°, 103°, and 52°. What is the measure of the fourth angle? **118°**

For Example 2
On a map, 1 inch represents 80 miles. If two towns are $2\frac{1}{4}$ inches apart on that map, what is the actual distance between the two towns? **180 miles**

Teaching Tip Emphasize that the property stated on page 585 applies to all polygons, not just triangles and quadrilaterals.

Cooperative Learning

Formations Have students work in pairs. As you call out the name of a special quadrilateral, students use their arms to form that figure.
For more information on this strategy, see *Cooperative Learning in the Mathematics Classroom*, one of the titles in the Glencoe Mathematics Professional Series, pp. 31–32.

3 PRACTICE/APPLY

Checking Your Understanding

Exercises 1–9 are designed to help you assess your students' understanding through reading, writing, speaking, and modeling. You should work through Exercises 1–4 with your students and then monitor their work on Exercises 5–9.

Study Guide Masters, p. 97

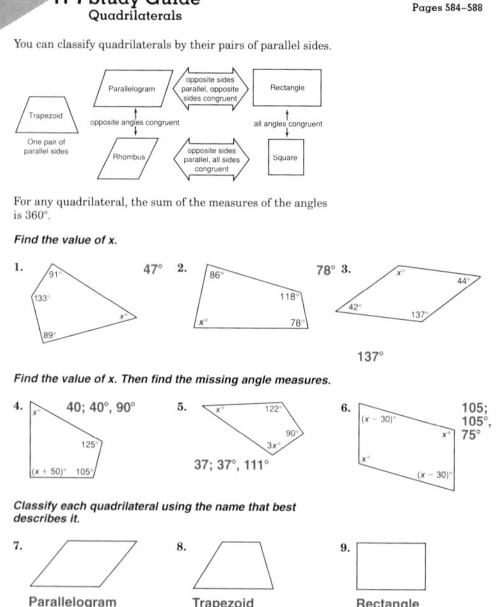

NAME ______________________ DATE ____________

11-7 Study Guide
Quadrilaterals

Student Edition Pages 584–588

You can classify quadrilaterals by their pairs of parallel sides.

For any quadrilateral, the sum of the measures of the angles is 360°.

Find the value of x.

1. 47° 2. 78° 3. 137°

Find the value of x. Then find the missing angle measures.

4. 40; 40°, 90° 5. 37; 37°, 111° 6. 105; 105°, 75°

Classify each quadrilateral using the name that best describes it.

7. Parallelogram 8. Trapezoid 9. Rectangle

Checking Your Understanding

Communicating Mathematics

Read and study the lesson to answer these questions.

1. **Draw** an example of each kind of quadrilateral named in this lesson. Give a real-world example of each. **See students' work.**
2. **You Decide** Susan says that all squares are similar. Lynn says they are not. Who is correct? Explain.
3. Which of the following rectangles is similar to the rectangle at the right? Why?

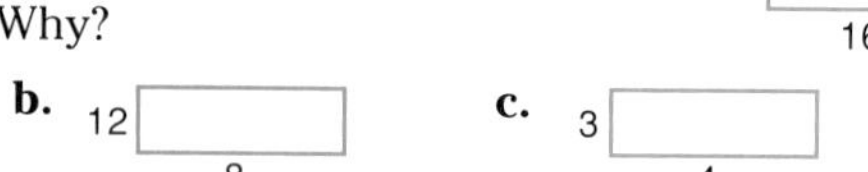

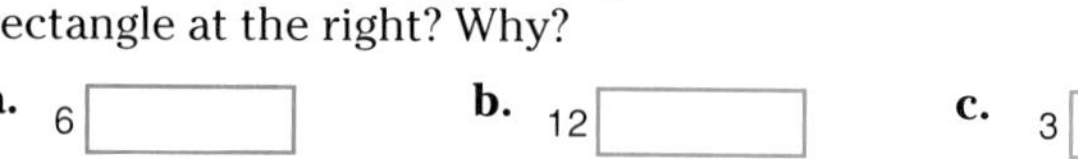

2. Susan. Angles are congruent and sides are proportional.
3. c; the measures of the sides are proportional.

MATERIALS

straightedge

4. Work with a partner.
   a. Use a straightedge to draw a quadrilateral.
   b. Draw a line segment that connects any two nonconsecutive vertices. This is a *diagonal.*
   c. How many triangles are formed by the diagonal? **2**
   d. What is the sum of the measures of the angles of a triangle? of two triangles? **180°; 360°**
   e. How does this activity show that the measures of the angles of a quadrilateral have a sum of 360°? **See Solutions Manual.**

Guided Practice

5. Find the value of x. **115**

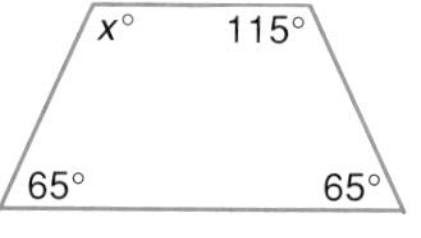

6. Find the value of x. Then find the missing angle measures.

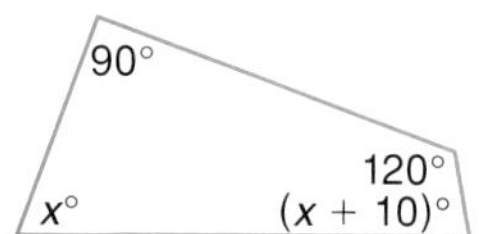

6. 70; 70° and 80°

List every name that can be used to describe each quadrilateral. Indicate the name that *best* describes the quadrilateral.

7.

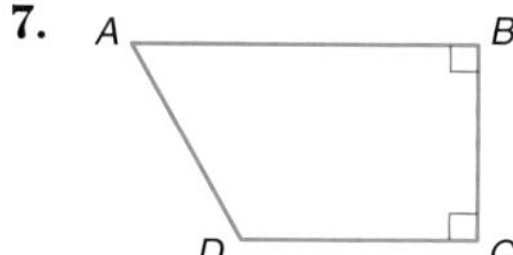

8.

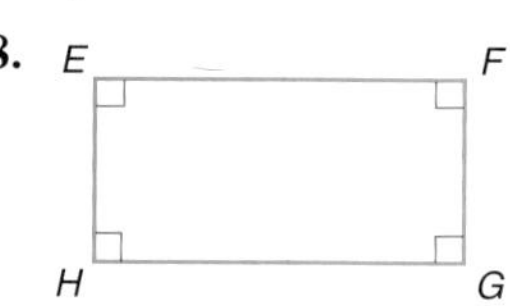

7. quadrilateral, trapezoid; trapezoid
8. quadrilateral, parallelogram, rectangle; rectangle

9. **Interior Design** The scale of the drawing at the right is 1 cm = 0.5 m. What is the length of the actual room? **6 m**

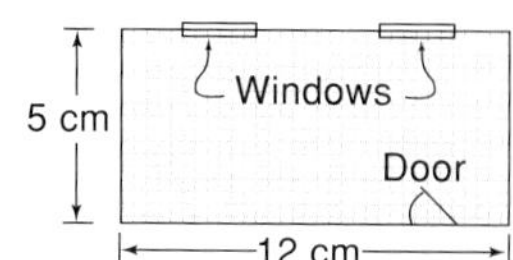

Exercises: Practicing and Applying the Concept

Independent Practice

A

Find the value of x. 10. 150 11. 30 12. 100

10. 150° 30° 30° x°

11.

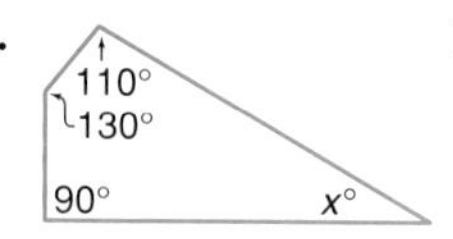

12.

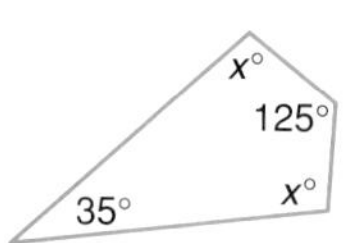

Reteaching

Using Diagrams Have students draw a polygon that has four sides and label it *quadrilateral*. As each special quadrilateral (*parallelogram, rectangle, rhombus, square, trapezoid*) is discussed, have students draw and label a diagram.

Have students choose one of the quadrilaterals they drew and draw one diagonal. Ask what the sum of the measures of the angles of one triangle created by the diagonal is. **180°** Ask what the sum of the measures of the angles of the quadrilateral is. **360°**

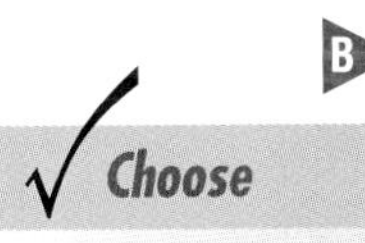

Estimation
Mental Math
Calculator
Paper and Pencil

Find the value of x. Then find the missing angle measures.

13. 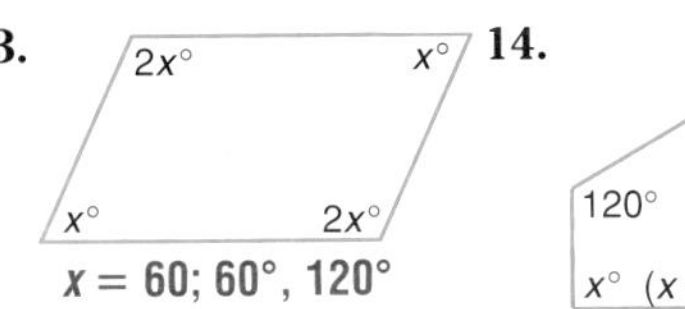

$x = 60; 60°, 120°$

14. 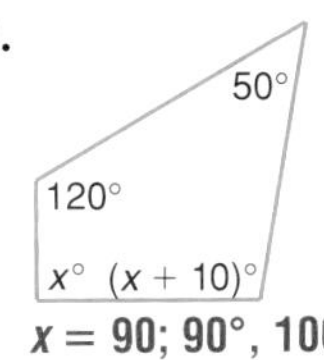

$x = 90; 90°, 100°$

15. 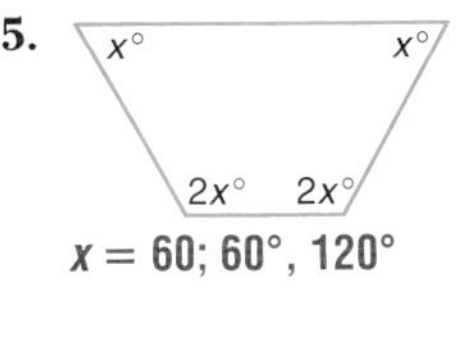

$x = 60; 60°, 120°$

Classify each quadrilateral using the name that *best* describes it.

16.–18. 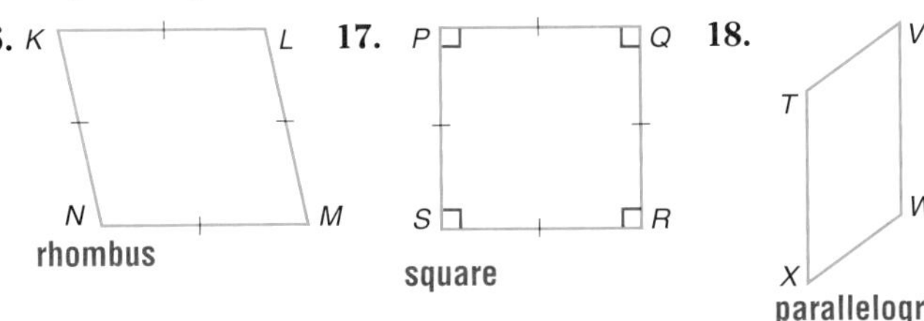

16. rhombus
17. square
18. parallelogram

Make a drawing of each quadrilateral. Then classify each quadrilateral using the name that *best* describes it. See students' drawings.

19. In quadrilateral *WXYZ*, $m\overline{WX} = 3$ in., $m\overline{XY} = 5$ in., $m\overline{YZ} = 3$ in., and $m\overline{WZ} = 5$ in. Angles *W*, *X*, *Y*, and *Z* are right angles. rectangle
20. In quadrilateral *CDEF*, $\overline{CD}$ and $\overline{EF}$ are parallel, and $\overline{CF}$ and $\overline{DE}$ are parallel. Angle *C* is not congruent to angle *D*. parallelogram
21. In quadrilateral *JKLM*, $m\angle J = 90°$, $m\angle K = 50°$, $m\angle L = 90°$, and $m\angle M = 130°$. quadrilateral

Determine whether each statement is *always*, *sometimes*, or *never* true. 22. always 23. sometimes 25. never 26. always

22. A rectangle is a parallelogram.
23. A rectangle is a square.
24. A rhombus is a square. sometimes
25. A trapezoid is a parallelogram.
26. A square is a parallelogram.
27. A square is a rhombus. always

28. See Solutions Manual for enlargements; area is 4 times greater.

28. Use grid paper to draw each figure. Then enlarge each figure by making each segment twice the length of the original. How does the area of the enlargement compare to the area of the original figure?

a. 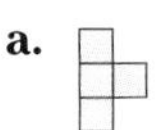
b. 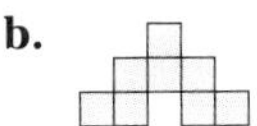
c. 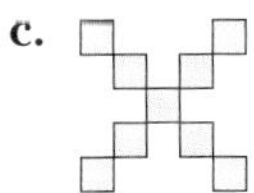

Critical Thinking

29a. Yes; a rhombus is equilateral but may not be equiangular.
29b. Yes; a rectangle is equiangular but may not be equilateral.
31. See students' work.

29. In an **equilateral** figure, all sides have the same measure. In an **equiangular** figure, all angles have the same measure.
   a. Is it possible for a quadrilateral to be equi*lateral* without being equi*angular*? Explain.
   b. Is it possible for a quadrilateral to be equiangular without being equilateral? Explain.
30. One angle of a parallelogram measures 40°. What are the measures of the other angles? 40°, 140°, 140°

Applications and Problem Solving

31. **Sports** Name four sports with playing fields that are quadrilaterals. Find the dimensions of each field and classify the shape.

Error Analysis

Some students may make careless errors when calculating the measures of the angles in a quadrilateral. Encourage students to estimate a reasonable answer both before and after they have completed their calculations. Their answer should be close to the estimates.

Assignment Guide

Core: 11–33 odd, 35–40
Enriched: 10–28 even, 29–40

For **Extra Practice**, see p. 768.

The red A, B, and C flags, printed only in the Teacher's Wraparound Edition, indicate the level of difficulty of the exercises.

Practice Masters, p. 97

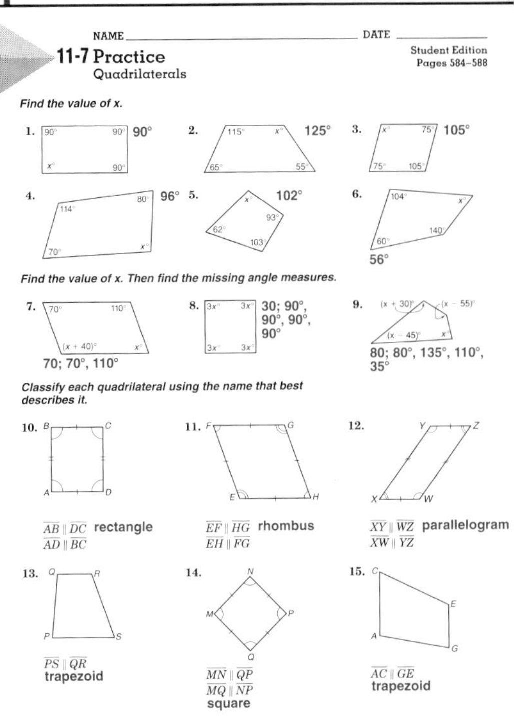

NAME ____ DATE ____

11-7 Practice
Quadrilaterals

Student Edition Pages 584–588

Find the value of x.

1. 90°
2. 125°
3. 105°
4. 96°
5. 102°
6. 56°

Find the value of x. Then find the missing angle measures.

7. 70; 70°, 110°
8. 30; 90°, 90°, 90°, 90°
9. 80; 80°, 135°, 110°, 35°

Classify each quadrilateral using the name that best describes it.

10. $\overline{AB} \parallel \overline{DC}$, $\overline{AD} \parallel \overline{BC}$ rectangle
11. $\overline{EF} \parallel \overline{HG}$, $\overline{EH} \parallel \overline{FG}$ rhombus
12. $\overline{XY} \parallel \overline{WZ}$, $\overline{XW} \parallel \overline{YZ}$ parallelogram
13. $\overline{PS} \parallel \overline{QR}$ trapezoid
14. $\overline{MN} \parallel \overline{QP}$, $\overline{MQ} \parallel \overline{NP}$ square
15. $\overline{AC} \parallel \overline{GE}$ trapezoid

Group Activity Card 11-7

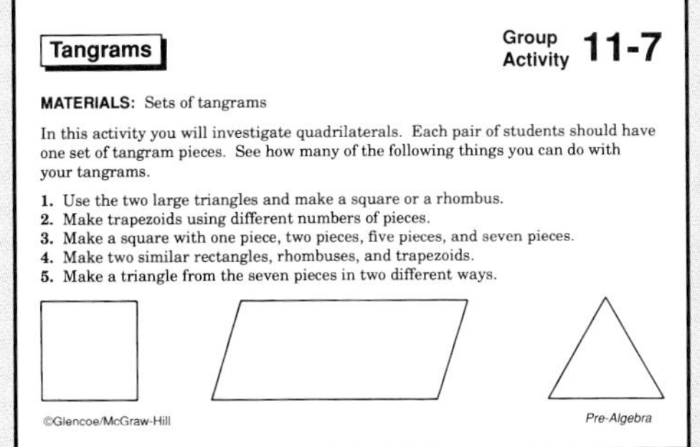

Tangrams Group Activity **11-7**

MATERIALS: Sets of tangrams

In this activity you will investigate quadrilaterals. Each pair of students should have one set of tangram pieces. See how many of the following things you can do with your tangrams.

1. Use the two large triangles and make a square or a rhombus.
2. Make trapezoids using different numbers of pieces.
3. Make a square with one piece, two pieces, five pieces, and seven pieces.
4. Make two similar rectangles, rhombuses, and trapezoids.
5. Make a triangle from the seven pieces in two different ways.

 Pre-Algebra

Extension

Using Analysis Have students list each type of quadrilateral. For each, have them name the minimum number of angles that need to be measured so that the remaining angles can be determined. **any parallelogram, 1; trapezoid, 3; no parallel sides, 3**

4 ASSESS

Closing Activity

Writing Put the six types of quadrilaterals listed on page 585 on index cards. Go around the room and have each of the students draw a card. Then have them draw that quadrilateral on the board. Finally, for the second and each succeeding quadrilateral, the student is to tell how it is different than the one that was drawn just before his or her turn.

Chapter 11, Quiz C (Lessons 11-6 and 11-7) is available in the *Assessment and Evaluation Masters*, p. 296.

Enrichment Masters, p. 97

NAME ______________ DATE ______

11-7 Enrichment — Polygons and Diagonals — Student Edition Pages 584–588

A **diagonal** of a polygon is any segment that connects two nonconsecutive vertices of the polygon. In each of the following polygons, all possible diagonals are drawn.

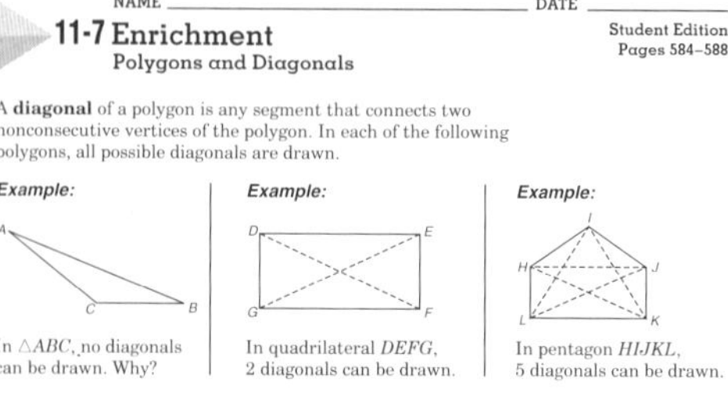

Example: In $\triangle ABC$, no diagonals can be drawn. Why?

Example: In quadrilateral $DEFG$, 2 diagonals can be drawn.

Example: In pentagon $HIJKL$, 5 diagonals can be drawn.

Complete the chart below and try to find a pattern that will help you answer the questions that follow.

	Polygons	Number of Sides	Number of Diagonals From One Vertex	Total Number of Diagonals
	triangle	3	0	0
	quadrilateral	4	1	2
	pentagon	5	2	5
1.	hexagon	6	3	9
2.	heptagon	7	4	14
3.	octagon	8	5	20
4.	nonagon	9	6	27
5.	decagon	10	7	35

Find the total number of diagonals that can be drawn in a polygon with the given number of sides.

6. 6 9
7. 7 14
8. 8 20
9. 9 27
10. 10 35
11. 11 44
12. 12 54
13. 15 90
14. 20 170
15. 50 1175
16. 75 2700
17. n $\frac{n^2 - 3n}{2}$

interNET CONNECTION

For other house blueprints, visit: **www.glencoe.com/sec/math/prealg/mathnet**

32. **Construction** In a house blueprint, the actual height of 8 feet is represented by 2 inches. If the actual length of the house is 60 feet, what is the length of the house in the blueprint? **15 in.**
33. **Make a Drawing** A basketball court is 84 feet by 50 feet. Make a scale drawing on $\frac{1}{4}$-inch grid paper. Use the scale $\frac{1}{4}$ inch = 6 feet.
34. **Publishing** Kiona's school newspaper uses 3-inch columns. If a $3\frac{1}{2}$-by-5 inch vertical photograph is reduced to fit in one column, how long will the reduced photograph be? **about 4.3 inches**

33. **See students' work.**

Mixed Review

35. Find the value of x in the $\triangle DEC$ at the right. (Lesson 11-6) **3 cm**
36. Find the measure of $\angle A$ in $\triangle ABC$ at the right. (Lesson 11-4) **60°**
37. Express 450 miles on 20 gallons of gas as a unit rate. (Lesson 9-1)
38. **Geography** The distance around Earth at the equator is about 25,000 miles. Find the approximate diameter of Earth at the equator. (Lesson 7-4)
39. Viñita noticed the school cafeteria served pizza on the last five Mondays. Viñita decides that the cafeteria always serves pizza on Monday. Is this an example of *inductive* or *deductive* reasoning? (Lesson 5-8)
40. Use mental math, paper and pencil, or a calculator to find at least one number that has four digits and is not divisible by 2, 3, or 5. (Lesson 4-1) **Sample answer: 1001**

37. **22.5 miles per gallon**
38. **about 7957.7 mi**
39. **inductive**

Refer to the Investigation on pages 482–483.

Look over the list of favorite games that you placed in your Investigation Folder at the beginning of Chapter 10. Some people may have said that their favorite games are target games, such as darts or horseshoes.

For this part of the Investigation, your group has been hired to design a new target game using Velcro® balls. The new board should be square and divided into several rectangular regions. Some regions should be marked with an X, and other regions marked with a Y.

- Work with the members of your group to design the board. If a Velcro ball is thrown at random at your board, the game should meet the following requirements.
  1. Player A scores points if the ball lands in a region marked X, and player B scores points if the ball lands in a region marked Y.
  2. The game should be fair.
- Write a report explaining how your board meets the requirements. Be sure to use concepts from Chapter 10 in your explanation.

Add the results of your work to your Investigation Folder.

WORKING ON THE

Investigation

The Investigation on pages 482 and 483 is designed to be a long-term project that is completed over several days or weeks. Encourage students to keep their materials in their Investigation Folder as they work on the Investigation.

11-8 Polygons

Setting Goals: *In this lesson, you'll classify polygons and determine the sum of the measures of the interior and exterior angles of a polygon.*

Modeling with Technology

MATERIALS

computer
LOGO software

Your Turn

If you were to walk a path in the shape of a rectangle, you would make a 90° turn at each corner. In LOGO, the turtle "walks" a path following directions you give. You can use LOGO to make paths of various shapes.

Work with a partner. Type the following commands to make the shapes shown to the right of each set of commands. Sketch each figure and mark the angles. Clear the screen before typing the second set of commands.

FD 100 RT 60
FD 100 RT 90
FD 173 RT 120
FD 173
What angle must you turn to return the turtle to its starting position? **90°**

FD 100 RT 60
FD 80 RT 60
FD 80 RT 60
FD 100 RT 90
FD 139
What angle must you turn to return the turtle to its starting position? **90°**

TALK ABOUT IT

a. In each figure, compare the angle of turn (the exterior angle) to the adjacent angle inside the figure (the interior angle). What pattern do you see? **They always add to 180.**

b. How could you use this pattern to draw figures using LOGO?

c. In each figure, find the sum of the turning (exterior) angles including the final turn to return the turtle to its starting position. What do you notice? **The sum for both figures is 360°.**

b. To draw a particular angle, make a turn equal to the supplement of the angle.

Learning the Concept

THINK ABOUT IT

Can you name a simple, closed figure that is NOT a polygon?

Sample answer: circle

The figures you drew above were simple closed figures. A simple, closed figure can be traced in a continuous path without tracing any point other than the starting point more than once. A **polygon** is a simple, closed figure in a plane that is formed by three or more line segments, called **sides**. The segments meet only at their endpoints. These points of intersection are called **vertices** (plural of **vertex**).

These plane figures are polygons.

These plane figures are not polygons.

Alternative Learning Styles

Kinesthetic If a computer with LOGO software is not available, have students model the program. For instance, for
FD 100 RT 60
FD 100 RT 90,
students would walk forward 10 steps and turn right 60° and then walk forward 10 steps and turn right 90°.

GLENCOE Technology

Interactive Mathematics Tools Software

In this interactive computer lesson, students explore polygons. A **Computer Journal** gives students the opportunity to write about what they have learned.

For Windows & Macintosh

11-8 LESSON NOTES

NCTM Standards: 1-4, 12

Instructional Resources

- Study Guide Master 11-8
- Practice Master 11-8
- Enrichment Master 11-8
- Group Activity Card 11-8
- Activity Masters, p. 43
- Multicultural Activity Masters, p. 21

Transparency 11-8A contains the 5-Minute Check for this lesson; **Transparency 11-8B** contains a teaching aid for this lesson.

Recommended Pacing	
Standard Pacing	Day 9 of 13
Honors Pacing	Day 9 of 13
Block Scheduling*	Day 5 of 7 (along with Lesson 11-7)

*For more information on pacing and possible lesson plans, refer to the ***Block Scheduling Booklet***.

1 FOCUS

5-Minute Check
(over Lesson 11-7)

Find the value of *x*.

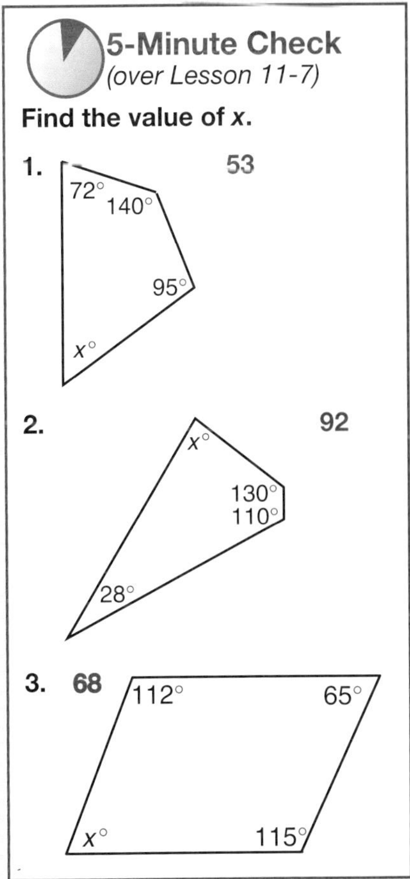

Motivating the Lesson

Questioning Show students pictures of items made in triangular and square shapes. Ask students the total of angle measurements for each of the two shapes. Show a picture of the Pentagon in Washington, D.C. Ask for the total measurement of its angles. **540°**

2 TEACH

In-Class Example

For Example 1

a. Find the sum of the measures of the angles of a heptagon by drawing diagonals. **900°**

b. A dodecagon is a polygon with 12 sides. Find the sum of the measures of the angles of a dodecagon by using the formula. **1800°**

Teaching Tip You may wish to have student make study cards on blank index cards that contain the name and a sketch of each polygon. As the lesson is studied, students can add the sum of the interior angle measures of each polygon.

We can classify polygons by the number of sides. Some of the more common polygons are shown below.

A polygon with n sides is called an ***n-gon***.

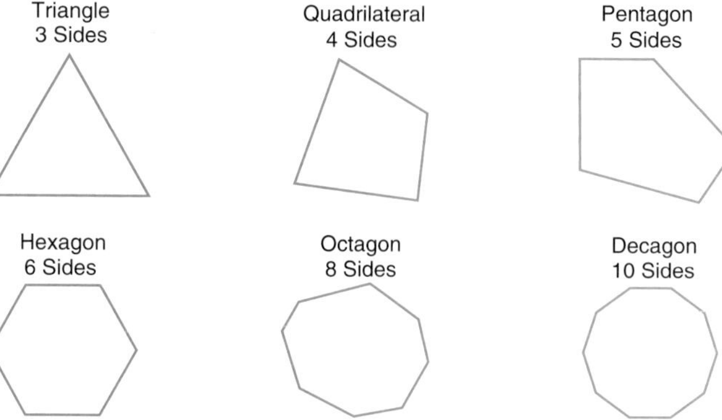

A **diagonal** is a line segment that joins two nonconsecutive vertices. You can draw diagonals in any polygon with more than three sides. In the polygons shown below, all possible diagonals from one vertex are shown. The table shows the number of diagonals drawn, and the number of triangles formed in each polygon.

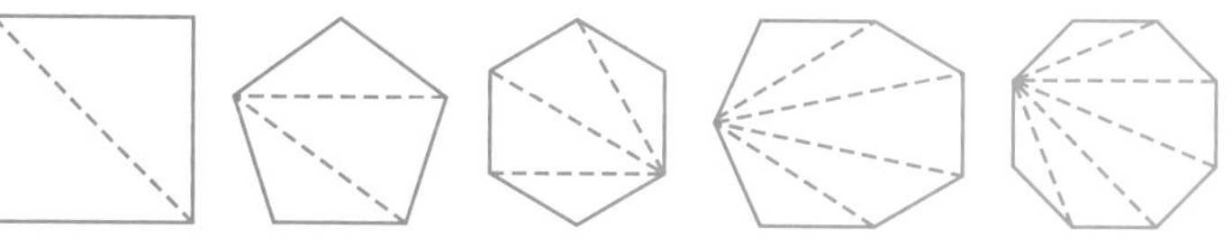

Number of:	Quadrilateral	Pentagon	Hexagon	Heptagon	Octagon
Sides	4	5	6	7	8
Diagonals	1	2	3	4	5
Triangles	2	3	4	5	6

Connection to Algebra

Compare the number of sides to the number of triangles. What pattern do you see? The number of triangles is always 2 less than the number of sides. You can use diagonals and the property of the sum of the angles of a triangle to find the sum of the measures of the angles of any polygon without using a protractor.

Sum of the Interior Angle Measures in a Polygon

If a polygon has n sides, then $n - 2$ triangles are formed, and the sum of the degree measures of the interior angles of the polygon is $(n - 2)180$.

Alternative Teaching Strategies

Reading Mathematics Tell students that Greek prefixes are often used in mathematics. Some of the more common prefixes are listed below.

tri-	3	octa-	8
quadri-	4	nona-	9
penta-	5	deca-	10
hexa-	6	dodeca-	12
hepta-	7		

Have students find and define non-mathematical words that contain these prefixes.

Example 

a. Find the sum of the measures of the interior angles of a pentagon by drawing diagonals.

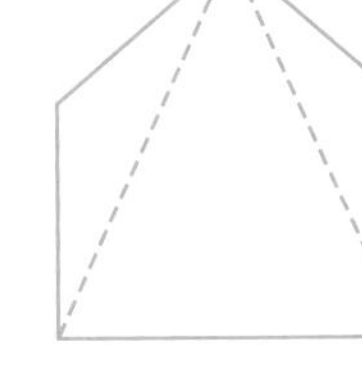

Draw all of the diagonals from one vertex.

Three triangles are formed.

The sum of the measures of the interior angles of a triangle is 180°, so the sum of the measures of the interior angles of a pentagon is 3(180) or 540°. *Check by using the formula.*

b. Find the sum of the measures of the interior angles of a hexagon by using the formula.

A hexagon has 6 sides. Therefore, $n = 6$.

$(n - 2)180 = (6 - 2)180$ *Replace n with 6.*

$= 4(180)$ or 720 *You can check this solution by drawing diagonals.*

The sum of the measures of the interior angles of a hexagon is 720°.

A **regular** polygon is a polygon that is **equilateral** (all sides are congruent) and **equiangular** (all angles are congruent). Since the angles of a regular polygon are congruent, their measures are equal.

Example 2

Architecture

The Pentagon in Washington, D.C., is named for its unusual shape, a regular pentagon. In a scale model of the Pentagon, what is the measure of each angle?

The scale model and the actual building are similar figures. In similar figures, the angles are congruent. From Example 1a, you know that the sum of the measures of the interior angles of a pentagon is 540°. Since the Pentagon is regular, you can find the degree measure of one angle by dividing 540 by 5.

540 [÷] 5 [=] *108*

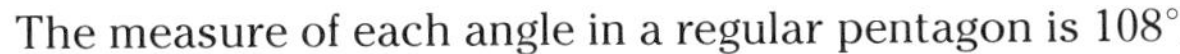

The measure of each angle in a regular pentagon is 108°.

The angles we considered in Examples 1 and 2 are "inside" the polygon. These are the **interior angles**. When a side of a polygon is extended, a special angle is formed, called an **exterior angle**.

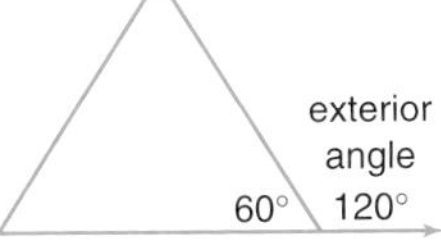

In the activity at the beginning of the lesson, you discovered that the interior and exterior angles of a polygon are supplementary. You also discovered that the sum of the exterior angles of a polygon is 360°. You can use these ideas for another way to find the measure of one angle of a regular polygon.

In-Class Examples

For Example 2

A stop sign is an octagon. For your model railroad village, you use miniature stop signs. What is the measure of each angle of the stop sign? **120°**

For Example 3

A decagon is a polygon that has 10 sides. Find the measure of each interior angle in a regular decagon. **144°**

3 PRACTICE/APPLY

Checking Your Understanding

Exercises 1–10 are designed to help you assess your students' understanding through reading, writing, speaking, and modeling. You should work through Exercises 1–5 with your students and then monitor their work on Exercises 6–10.

Error Analysis

Some students may assume that regular and irregular polygons have different sums of the measures of their angles. Others may assume that all polygons are regular. Provide models for students to examine and measure to clarify their thinking.

Study Guide Masters, p. 98

NAME ______________ DATE ________

11-8 Study Guide
Polygons

Student Edition
Pages 589–593

Polygons are simple, closed figures formed by three or more line segments, called **sides**. If a polygon has n sides, then the sum of the measures of the interior angles is $(n - 2)180$.

Example: Find the sum of the measures of the angles of a heptagon.

A heptagon has 7 sides. Therefore, $n = 7$.
$(n - 2)180 = (7 - 2)180$ Replace n with 7
$= 5(180)$ or 900

The sum of the measures of the angles of a heptagon is 900°.

Regular polygons are figures in which all sides are congruent and all angles are congruent. Since the heptagon above is regular, the measure of one **interior** angle is 900 ÷ 7, or about 129°.

Interior and **exterior** angles of a polygon are supplementary. In a regular heptagon, the measure of the interior angle is about 129°. Therefore, the measure of the exterior angle is 180 − 129, or about 51°.

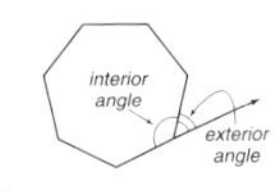

Find the sum of the measures of the interior angles of each polygon.

1. hexagon 720°
2. pentagon 540°
3. quadrilateral 360°
4. octagon 1080°
5. 16-gon 2520°
6. 27-gon 4500°

Find the measure of each exterior angle and each interior angle of each regular polygon.

7. regular octagon 135°, 45°
8. regular decagon 144°, 36°
9. regular 12-gon 150°, 30°
10. regular heptagon 129°, 51°
11. regular 15-gon 156°, 24°
12. regular 24-gon 165°, 15°

Find the perimeter of each regular polygon.

13. regular heptagon with sides 12 ft long 84 ft
14. regular quadrilateral with sides 3.7 m long 14.8 m
15. regular octagon with sides $\frac{1}{2}$ yd long 4 yd
16. regular pentagon with sides $2\frac{4}{5}$ in. long 14 in.

Assignment Guide

Core: 11–29 odd, 30–34
Enriched: 12–26 even, 27–34

For **Extra Practice**, see p. 768.

The red A, B, and C flags, printed only in the Teacher's Wraparound Edition, indicate the level of difficulty of the exercises.

Additional Answers

1. All sides and angles are congruent.
2. The number of triangles is equal to the number of sides minus two.
3. Equiangular means all angles are congruent, while equilateral means all sides are congruent. A rectangle is equiangular but may not be equilateral.
4. 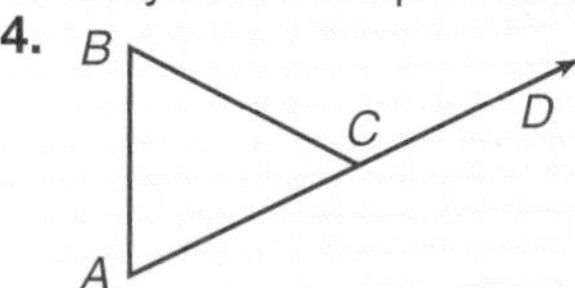

Example  **Find the measure of each interior angle in a regular hexagon.**

Sketch a regular hexagon.
Draw an exterior angle at each vertex.
There are 6 exterior angles. The sum of the measures of the exterior angles is 360°.

360 [÷] 6 [=] 60 *Measure of each exterior angle.*
180 − 60 = 120 *Interior and exterior angles are supplementary.*

The measure of each interior angle in a regular hexagon is 120°.

Checking Your Understanding

Communicating Mathematics

Read and study the lesson to answer these questions. **1–4. See margin.**

1. **Explain** why a square is a regular polygon.
2. **Explain** the relationship between the number of sides in a polygon and the number of triangles formed by the diagonals.
3. **Tell** the difference between equiangular and equilateral. Give an example of a figure that is equiangular or equilateral but not both.
4. **Draw** $\triangle ABC$ with exterior angle BCD.
5. **Assess Yourself** How can knowing the properties of geometric shapes help you in your daily life? **See students' work.**

MATH JOURNAL

Guided Practice

Classify each polygon below and determine whether it appears to be *regular* or *not regular*. **8. 1080°**

6. 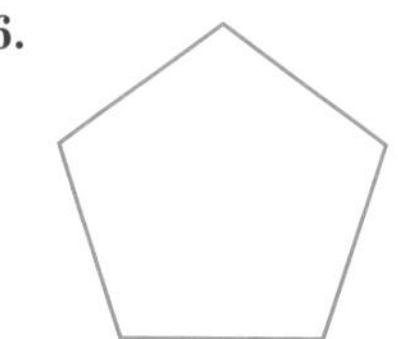

pentagon, regular

7. 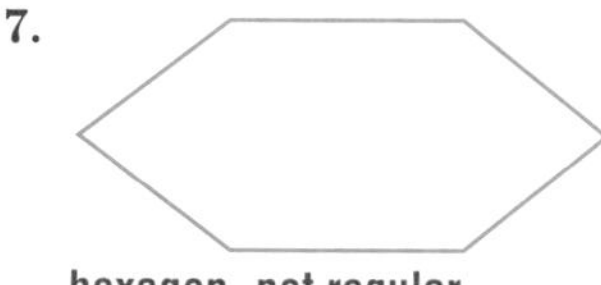

hexagon, not regular

8. Find the sum of the measures of the interior angles of an octagon.
9. Find the measure of each exterior angle and each interior angle in a regular nonagon (9 sides). **40°; 140°**
10. **Scale Models** Refer to Example 2. The scale of the model of the Pentagon is to be 1 in. = 50 ft. Each side of the Pentagon is about 1 mile (5280 ft) long. What length should each side of the scale model be?

10. 105.6 in.

Exercises: Practicing and Applying the Concept

Independent Practice

Find the sum of the measures of the interior angles of each polygon.

A

11. heptagon **900°**
12. decagon **1440°**
13. dodecagon **1800°**
14. nonagon **1260°**
15. 15-gon **2340°**
16. 25-gon **4140°**

Find the measure of each exterior angle and each interior angle of each regular polygon.

17. regular hexagon **60°, 120°**
18. regular decagon **36°, 144°**
19. regular dodecagon **30°, 150°**
20. regular 20-gon **18°, 162°**

Practice Masters, p. 98

NAME ______ DATE ______
11-8 Practice
Polygons
Student Edition Pages 589–593

Find the sum of the measures of the interior angles of each polygon.

1. quadrilateral 360°
2. pentagon 540°
3. octagon 1080°
4. 12-gon 1800°
5. 18-gon 2880°
6. 20-gon 3240°
7. 30-gon 5040°
8. 45-gon 7740°
9. 75-gon 13,140°

Find the measure of each exterior angle and each interior angle of each regular polygon.

10. regular octagon 45°, 135°
11. regular pentagon 72°, 108°
12. regular heptagon 51.43°, 128.57°
13. regular nonagon 40°, 140°
14. regular 18-gon 20°, 160°
15. regular 25-gon 14.4°, 165.6°

Find the perimeter of each regular polygon.

16. regular hexagon with sides 28.5 millimeters long 171 mm
17. regular decagon with sides 2.5 inches long 25 in.
18. regular heptagon with sides 10.75 feet long 75.25 ft
19. regular 12-gon with side 3.25 yards long 39 yd
20. regular 25-gon with sides 6 inches long 150 in.
21. regular 100-gon with sides 9 centimeters long 900 cm

Reteaching

Using Models Bring in pictures that show geometric shapes from a variety of sources, such as newspapers, magazines, and books on crafts, architecture, gardening, science, interior design, and classical paintings. Discuss the shapes found in various pictures. As students study the examples, have them find the angles in one of the polygons in their pictures.

Group Activity Card 11-8

A Puzzling Bird Group Activity **11-8**

See who can be the first in your group to complete the following mathematical statements given on the back of this card. Put the answers on the lines. The letters in the boxes will tell you the answer to the riddle: What did Grandma say when her cat ate her parrot?

 Pre-Algebra

Estimation
Mental Math
Calculator
Paper and Pencil

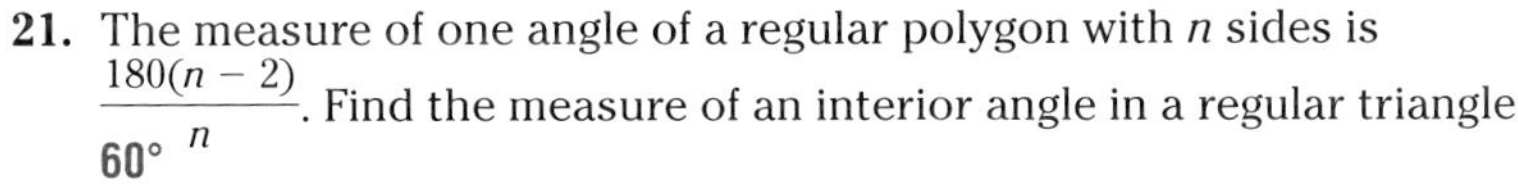

21. The measure of one angle of a regular polygon with n sides is $\frac{180(n-2)}{n}$. Find the measure of an interior angle in a regular triangle. **60°**

Find the perimeter of each regular polygon.

22. equilateral triangle with sides 27 ft long **81 ft**
23. regular quadrilateral with sides 18 cm long **72 cm**
24. regular pentagon with sides 3 m long **15 m**
25. regular hexagon with sides 4.5 in. long **27 in.**
26. regular octagon with sides $5\frac{3}{4}$ yd long **46 yd**

Critical Thinking

27. Trace the dot pattern shown at the right. Without lifting your pencil from the paper, draw four line segments that connect all the points.

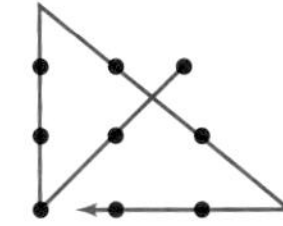

Applications and Problem Solving

28. **Driver's License Exam** Part of a driver's license exam includes identifying road signs by color and by shape. The shapes are polygons and some are regular polygons. Identify the shape of each road sign pictured below and tell what it means. **See margin.**

CULTURAL CONNECTIONS
The English rule of driving on the left of the road has its roots in history. The practice started because horsemen preferred to pass oncoming strangers on their sword-arm side.

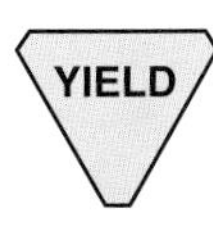

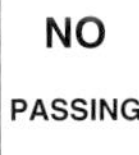

29. **Manufacturing** Some cafeteria trays are designed so that four people can place their trays around a square table without bumping corners. Each tray is shaped like the one shown at the right. The top and bottom of the tray are parallel.

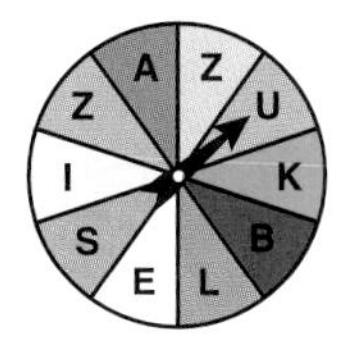

a. What shape does the tray have? **trapezoid**
b. Find the measure of each angle of the tray so that the trays will fit side-to-side around the table. **45°, 45°, 135°, 135°**
c. What polygon will be formed at the center of the table if four trays are placed around the table with their sides touching? **square**

Mixed Review

30. **Geometry** A lot for a new house is shaped like a trapezoid, with two right angle corners. If the third corner is an 80° angle, what is the measure of the fourth angle? (Lesson 11-7) **100°**

31. **Probability** A die is rolled and the spinner is spun. What is P(an even number and a vowel)? (Lesson 10-9)

31. $\frac{1}{5}$

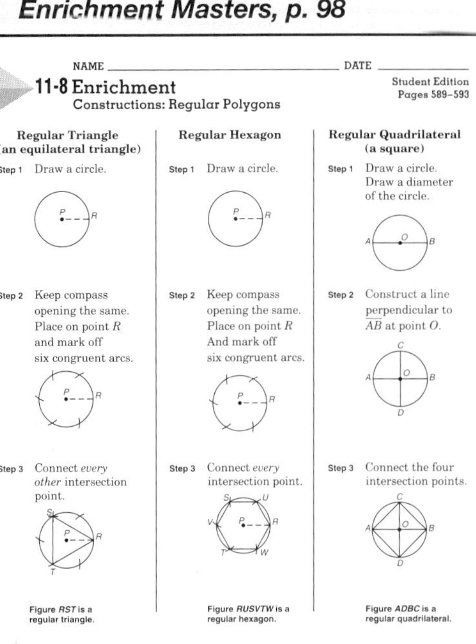

32. Simplify $3.5x + 2.8x + 1.5x$. (Lesson 5-3) **7.8x**
33. Write 1604 in expanded form. (Lesson 4-2)

33. $(1 \times 10^3) + (6 \times 10^2) + (4 \times 10^0)$

34. Solve the equation $-6 - 8 = c$. (Lesson 2-5) **−14**

4 ASSESS

Closing Activity

Speaking Go around the room and have the students think of some object and state the shape of the object. For example, a pentagon is inside a star, a notebook is a quadrilateral. Have each student say one object and its shape.

CULTURAL CONNECTIONS
Men's and women's shirts button on opposite sides because when a man drew his sword from his left side with his right hand, he did not want the sword getting caught in the shirt opening.

Additional Answer

28. square, narrow bridge ahead;
square, curve in the road ahead;
triangle, yield to oncoming traffic;
square, stop sign or light ahead;
octagon, stop;
rectangle, speed limit 50 miles per hour;
rectangle, one way street;
rectangle, no passing allowed

Enrichment Masters, p. 98

NAME ______ DATE ______

11-8 Enrichment Student Edition Pages 589–593
Constructions: Regular Polygons

Regular Triangle (an equilateral triangle)	Regular Hexagon	Regular Quadrilateral (a square)
Step 1 Draw a circle.	Step 1 Draw a circle.	Step 1 Draw a circle. Draw a diameter of the circle.
Step 2 Keep compass opening the same. Place on point R and mark off six congruent arcs.	Step 2 Keep compass opening the same. Place on point R And mark off six congruent arcs.	Step 2 Construct a line perpendicular to $\overline{AB}$ at point O.
Step 3 Connect *every other* intersection point.	Step 3 Connect *every* intersection point.	Step 3 Connect the four intersection points.
Figure RST is a regular triangle.	Figure $RUSVTW$ is a regular hexagon.	Figure $ADBC$ is a regular quadrilateral.

Complete each construction.

1. a regular quadrilateral **See example above.**
2. a regular hexagon **See example above.**

Extension

Using Constructions Have students use a compass and a straightedge to construct a regular hexagon.

- Use the compass to draw a circle.
- Measure its radius with the compass. Keep the length set.
- Place both ends of the compass on the circle and make an arc.
- Place the point on the end of the arc. Make another arc.
- Repeat around.
- Connect the points where the arcs intersect the circle with chords.
- Check that the figure formed is a regular hexagon.

11-8B LESSON NOTES

NCTM Standards: 1-4, 12

Objective
Investigate tessellations using regular polygons

Recommended Time
Demonstration and discussion: 15 minutes; Exercises: 30 minutes

Instructional Resources
For each student or group of students
- tracing paper
- cardboard
- scissors

Math Lab and Modeling Math Masters
- p. 49 (worksheet)

For teacher demonstration
Overhead Manipulative Resources

1 FOCUS

Motivating the Lesson
Show students a honeycomb or a picture of a honeycomb. Ask students to describe the pattern.

2 TEACH

Teaching Tip Have students make a table with three columns headed *Name of Polygon*, *Sum of Angles*, and *Does the Polygon Tessellate?* Tell students to include the hexagon in their table.

3 PRACTICE/APPLY

Assignment Guide
Core: 1–2
Enriched: 1–2

4 ASSESS

Observing students working in cooperative groups is an excellent method of assessment.

HANDS-ON ACTIVITY

MATHEMATICS LAB

11-8B Tessellations

An Extension of Lesson **11-8**

MATERIALS
- tracing paper
- cardboard
- scissors

The repetitive pattern of regular polygons shown at the right is an example of a **tessellation**. In a tessellation, the polygons fit together with no holes or gaps.

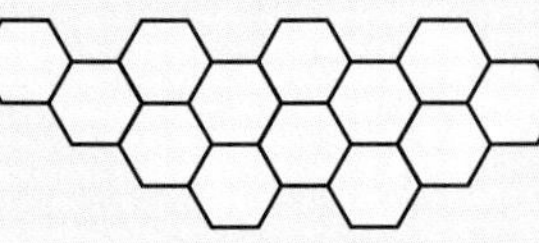

In this Math Lab, you will investigate tessellations with regular polygons.

Your Turn

Work with a partner.

- Trace each regular polygon shown. Cut each shape from the tracing paper and trace it onto a piece of cardboard. Cut each shape from the cardboard.
- Use each cardboard piece to try to draw a tessellation for each regular polygon.
- For each polygon, calculate the sum of the interior angle measures. Then determine the measure of one angle in each polygon. Record your results in a table. **See Solutions Manual.**

Equilateral Triangle

Square

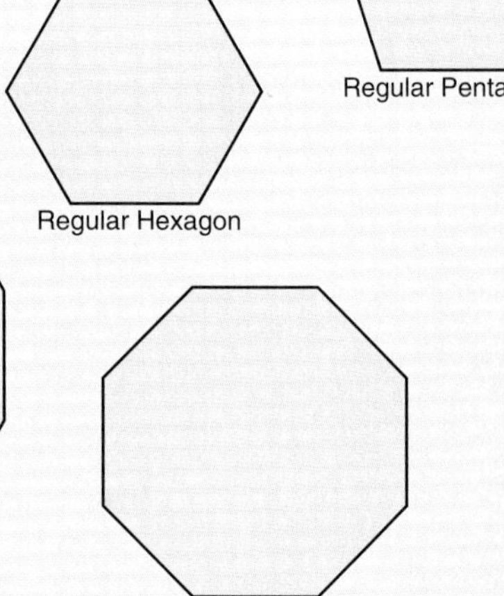

Regular Pentagon

Regular Hexagon

Regular Dodecagon

Regular Octagon

TALK ABOUT IT

1. How can you tell from the measure of one angle whether or not a single polygon will tessellate? **If 360 is a multiple of the angle's measure it will tessellate.**

Your Turn

Try to make tessellations from the combinations below.

a. square and octagon
b. square, triangle, and hexagon
c. square, triangle, and dodecagon
d. another combination you choose

TALK ABOUT IT

2. How can you tell whether a combination of polygons will tessellate?

2. If the sum of the angle measures at each meeting point of the vertices is 360° it will tessellate.

11-9 Transformations

Setting Goals: *In this lesson, you'll identify and draw reflections, translations, and rotations. You'll also identify and draw symmetric figures.*

Modeling with Manipulatives

MATERIALS

cardboard, scissors

If you look around, you will probably see geometric shapes in repeating patterns. Floors and ceilings are often tiled in patterns. In the activity below, you will investigate one way to create a pattern.

Your Turn

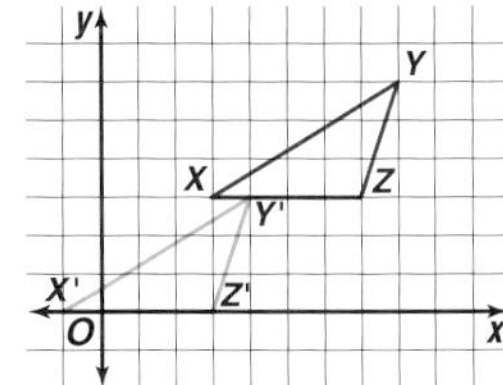

- ▶ Cut a triangle out of cardboard. Place the triangle on a coordinate grid as shown, so one side lies parallel to the *x*-axis. Trace the triangle on the paper. Label the vertices as shown. Write the coordinates of the vertices.
- ▶ Slide the cutout triangle 4 units left and 3 units down as shown. Trace the triangle. Label the vertices X', Y', and Z', so they correspond to the first triangle. Write the coordinates of the vertices of $\triangle X'Y'Z'$.

TALK ABOUT IT How are the coordinates of $\triangle XYZ$ and $\triangle X'Y'Z'$ related? **See margin.**

Learning the Concept

Transformations are movements of geometric figures. A **translation** is one type of transformation. In a translation, you "slide" a figure horizontally, vertically, or both, as you did in the activity.

Two other types of transformations are rotations and reflections.

In a **rotation**, you turn the figure around a point. Computer drawing packages allow the user to create drawings. After a drawing is created, a command called "rotate right" can be used to rotate the figure as shown at the right.

In a **reflection**, you "flip" a figure over a line. You can use a geomirror to draw the reflection of a figure. Each diagram below shows a figure and its reflection over a line. Place a geomirror on the dashed line in each figure and verify each reflection.

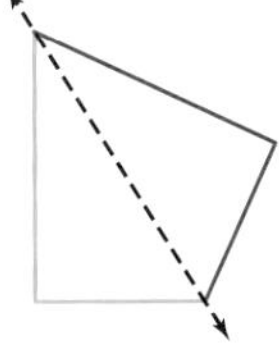

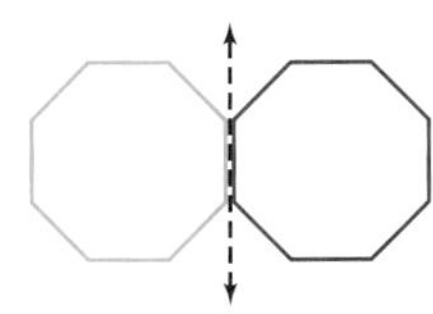

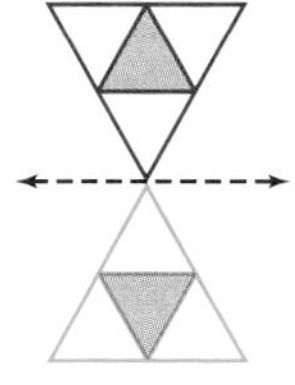

How could you describe the figure and its reflection? **They are mirror images.**

Alternative Learning Styles

Auditory Play a series of single notes on a piano or guitar. Then play the notes one octave higher. Ask what transformation this represents. **Translation up one octave.** Play another series of notes followed by the same notes in reverse order. **Reflection**

Additional Answer Talk About It

Each *x*-coordinate is 4 units less; each *y*-coordinate is 3 units less.

11-9 LESSON NOTES

NCTM Standards: 1-4, 12

Instructional Resources

- Study Guide Master 11-9
- Practice Master 11-9
- Enrichment Master 11-9
- Group Activity Card 11-9
- Assessment and Evaluation Masters, p. 296
- Activity Masters, p. 44
- Graphing Calculator Masters, p. 11
- Multicultural Activity Masters, p. 22

Transparency 11-9A contains the 5-Minute Check for this lesson; **Transparency 11-9B** contains a teaching aid for this lesson.

Recommended Pacing	
Standard Pacing	Day 11 of 13
Honors Pacing	Day 11 of 13
Block Scheduling*	Day 6 of 7

*For more information on pacing and possible lesson plans, refer to the ***Block Scheduling Booklet***.

1 FOCUS

5-Minute Check
(over Lesson 11-8)

Classify each polygon and determine whether it appears to be *regular* or *not regular*.

1. **octagon, regular**

2.  **pentagon, not regular**

3. Find the sum of the measures of the interior angles of a heptagon **900°**

Motivating the Lesson

Hands-On Activity If possible, wear a piece of clothing, such as a shirt or tie that illustrates a pattern that has been formed from a translation. Wallpaper or wrapping paper patterns might also be used. Have students examine the pattern and determine where it repeats. Then ask students if they can explain how the repeated pattern was formed.

2 TEACH

In-Class Examples

For Example 1

Tell whether each transformation is a translation, a rotation, or a reflection.

a.

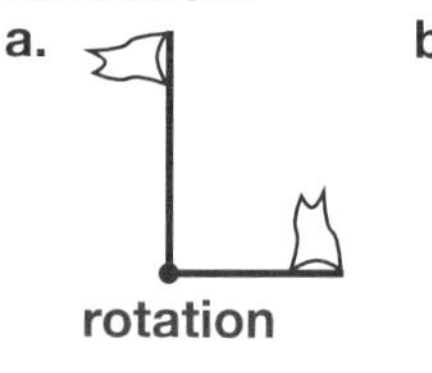

rotation

b.

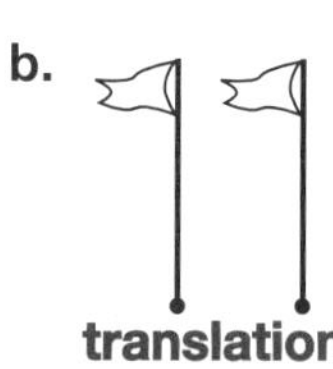

translation

c.

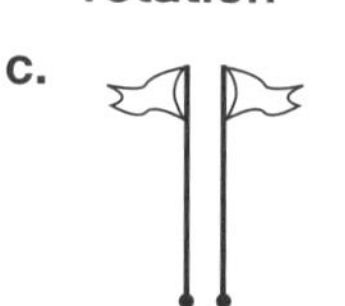

reflection

For Example 2

Draw all lines of symmetry for each figure.

a.

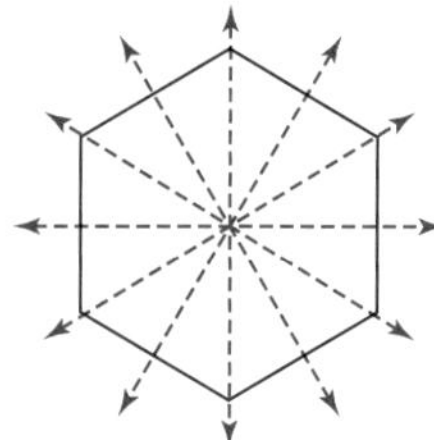

b.

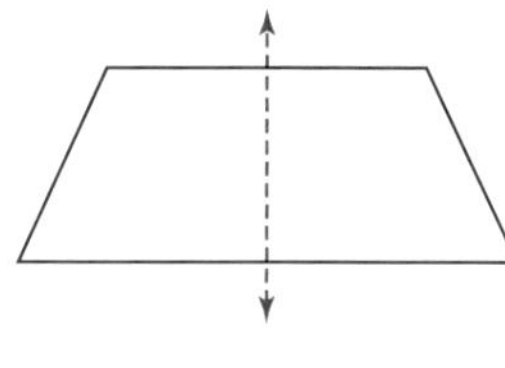

c.

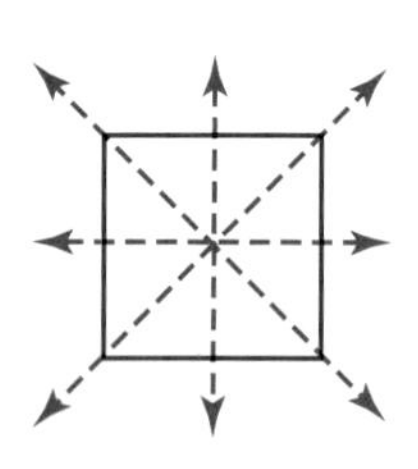

Example 1 **Tell whether each transformation is a translation, a rotation, or a reflection.**

a.

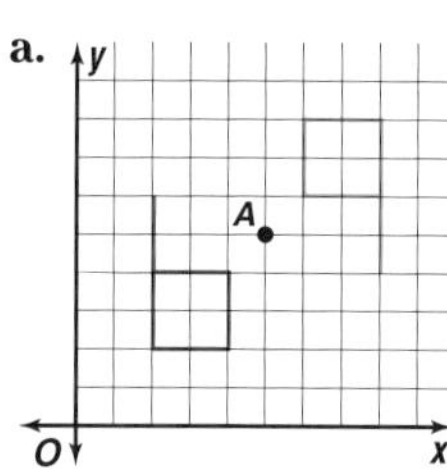

b.

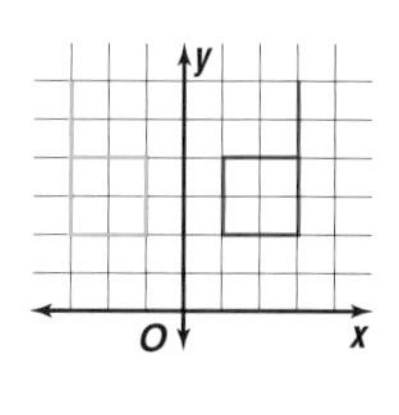

c.

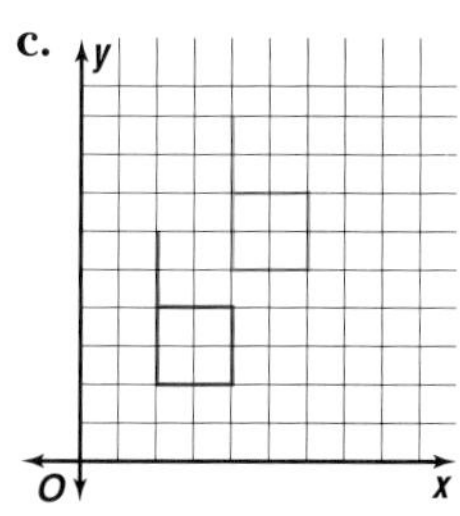

a. The figure is *rotated* about point A.

b. The figure is *reflected* about the y-axis.

c. The figure is *translated* 2 units horizontally and 3 units vertically.

THINK ABOUT IT

How many lines of symmetry does a square have? How many lines of symmetry does a circle have?

four; infinitely many

When you draw a figure and its reflection, you create a figure that is **symmetric**. The line where you placed the geomirror is called a *line of symmetry*. Some common items with line symmetry are shown below.

Some of the items shown above have one line of symmetry. Others have several lines of symmetry. Each line of symmetry separates the figure into *two* congruent parts.

Example 2 **Draw all lines of symmetry for each figure.**

a.

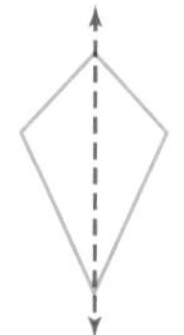

A kite has one line of symmetry.

b.

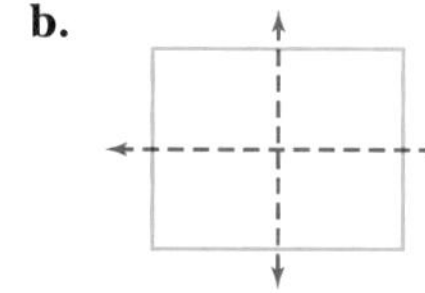

A rectangle has two lines of symmetry.

c.

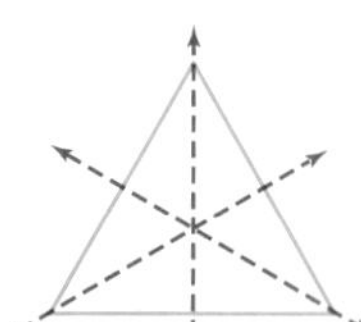

An equilateral triangle has three lines of symmetry.

Artists have used symmetry to create beautiful designs for centuries.

Example 3

World Culture

In 1895, excavators found Hopi pottery among the ruins of the pueblo known as Sikyatki. The pattern below was found on the exterior of a Sikyatki food bowl. Describe the transformations used to make the pottery design.

The original design is translated to the right and rotated 180° or reflected over a horizontal line.

Tech Prep

Textile Designer Many fabric designs use one or more transformations to create the design. Designers must also consider the type of consumer who will be using the fabric. For business shirts, a simple design is more appropriate than a bold flowery design.

For more information on tech prep, see the *Teacher's Handbook*.

Reteaching

Using Manipulatives Allow students to use the overhead projector and congruent geometrical shapes to illustrate each type of transformation.

Checking Your Understanding

Communicating Mathematics

Read and study the lesson to answer these questions. 1–2. See margin.

1. **Compare and contrast** a translation, a reflection, and a rotation of the same figure.
2. **Show** how to complete the figure so that $\overleftrightarrow{XY}$ is a line of symmetry.

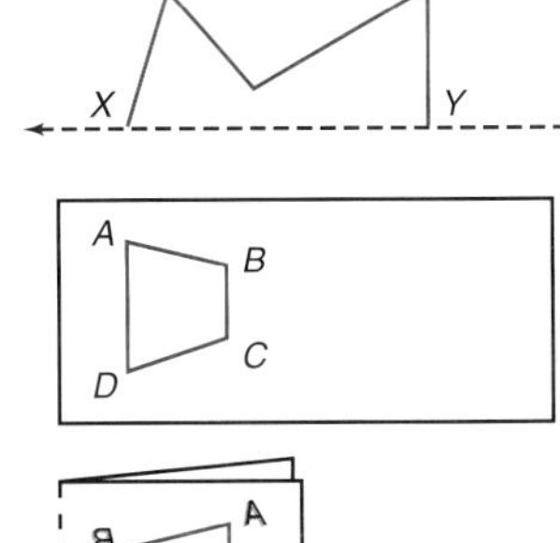

3. a. On one half of a piece of wax paper, use a straightedge and a colored pencil to draw a quadrilateral. Label the vertices *A*, *B*, *C*, and *D*.

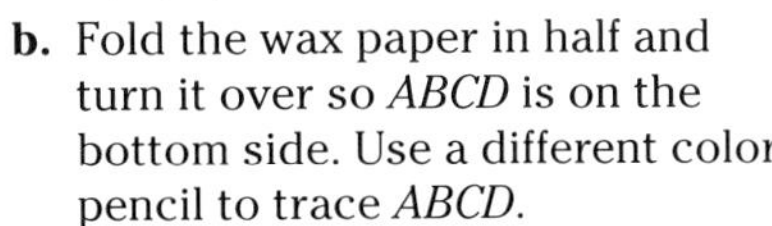

   b. Fold the wax paper in half and turn it over so *ABCD* is on the bottom side. Use a different color pencil to trace *ABCD*.

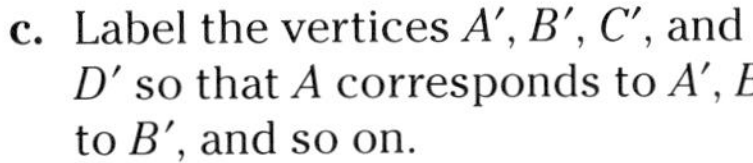

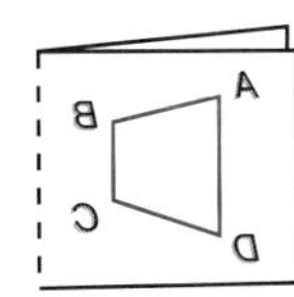

   c. Label the vertices *A′*, *B′*, *C′*, and *D′* so that *A* corresponds to *A′*, *B* to *B′*, and so on.

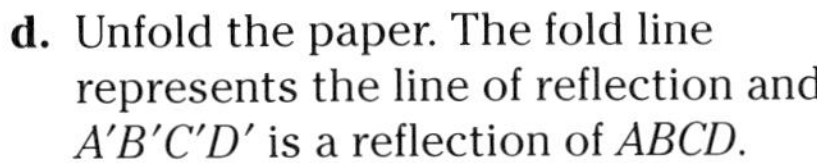

   d. Unfold the paper. The fold line represents the line of reflection and *A′B′C′D′* is a reflection of *ABCD*.

   e. Use a ruler to measure $\overline{AA'}$, $\overline{BB'}$, $\overline{CC'}$, and $\overline{DD'}$. At what point does the line of reflection intersect each segment? midpoint

MODELING MATHEMATICS

MATERIALS

wax paper
colored pencils
ruler

Guided Practice

Tell whether each geometric transformation is a translation, a reflection, or a rotation. Explain your answer.

4–5. Sample answers given. See students' explanations.

4. 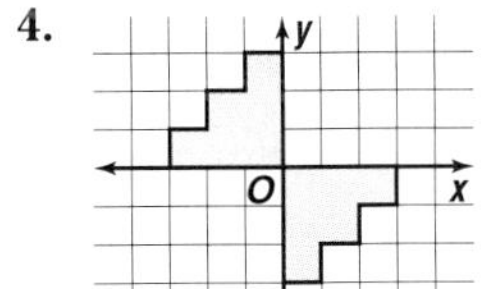 rotation

5. 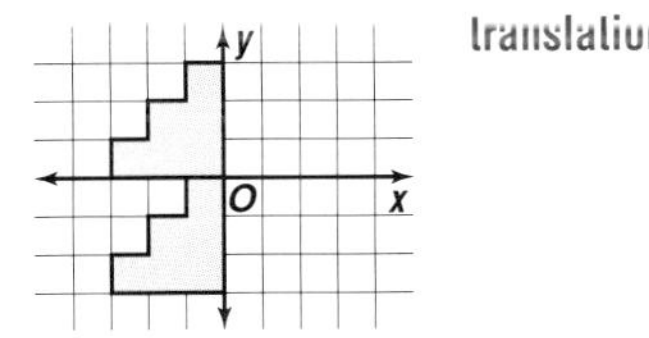 translation

Trace each figure. Draw all lines of symmetry.

6. 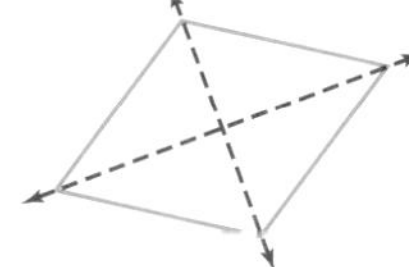

7. 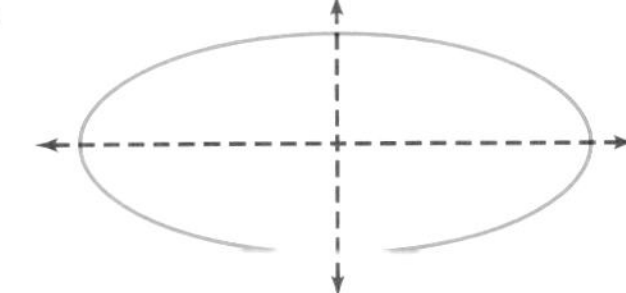

8. **Art** A sample of a Navaho weaving is shown below. Describe the transformations and lines of symmetry that exist in the weaving. Answers may vary. See students' work.

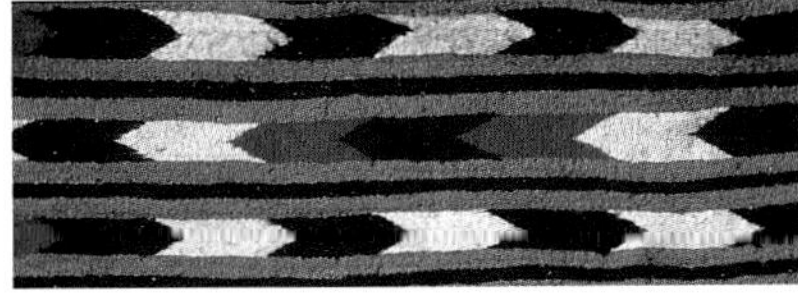

Additional Answers

1. In a translation, the figure changes position but its orientation remains the same; in a reflection, each part of the figure is reversed; in a rotation, the orientation of the figure changes.

2. 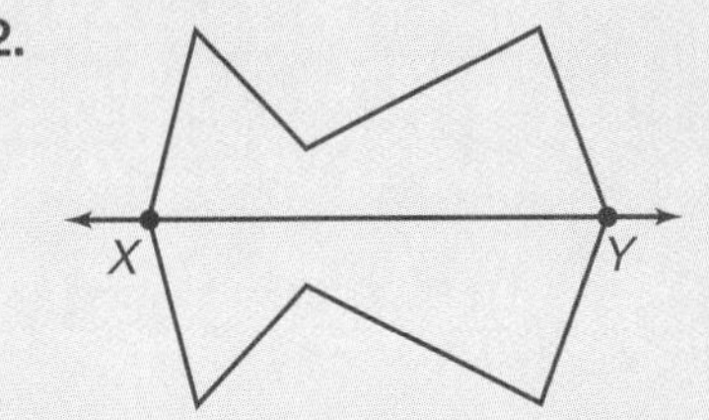

In-Class Examples

For Example 3

Describe the transformations used to make the design below.

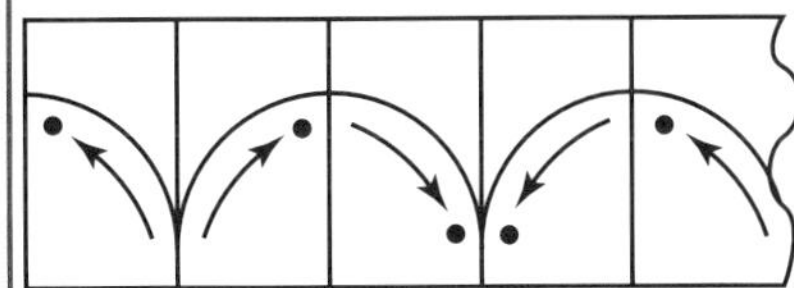

Sample answer: a reflection over a vertical line followed by a 90° rotation or a 90° rotation followed by a reflection over a vertical line

3 PRACTICE/APPLY

Checking Your Understanding

Exercises 1–8 are designed to help you assess your students' understanding through reading, writing, speaking, and modeling. You should work through Exercises 1–3 with your students and then monitor their work on Exercises 4–8.

Study Guide Masters, p. 99

NAME ______________ DATE ______

11-9 Study Guide
Transformations

Student Edition Pages 595–599

Transformations are movements of geometric figures.

When a geometric figure is moved horizontally, vertically, or both, it is called a **translation**. The figure at the right is moved 6 units to the left.

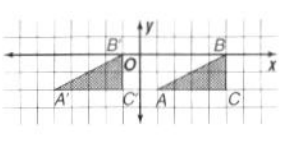

In a **rotation**, a figure is turned about a point.

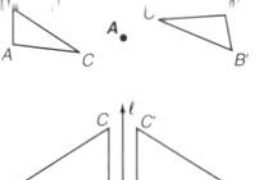

When a figure is "flipped" over a line, it is called a **reflection**. At the right, △*ABC* is *reflected* about line ℓ. Since the figure can be folded over line ℓ so that the two halves correspond, the figure is **symmetric**. Line ℓ is called a *line of symmetry*. A line of symmetry separates a figure into two congruent parts.

The figure at the right has three lines of symmetry.

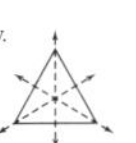

Tell whether each transformation is a translation, a rotation, or a reflection. Explain your answer. Sample answers given.

1. 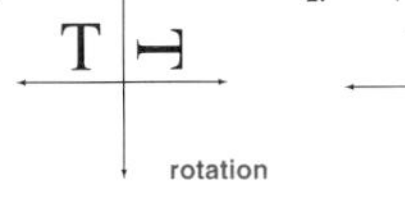 rotation

2.  translation

3. 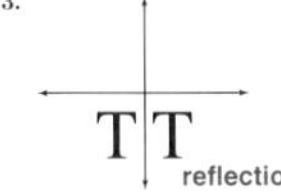 reflection

Draw all lines of symmetry.

4. 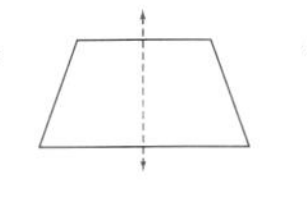

5. 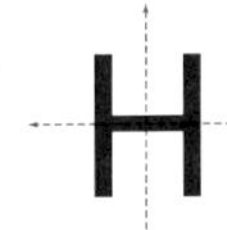

6. 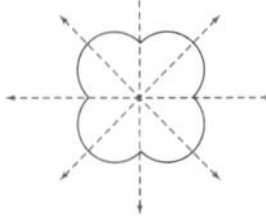

Assignment Guide

Core: 9–19 odd, 20–21, 23–28
Enriched: 10–18 even, 20–28

For **Extra Practice**, see p. 769.

The red A, B, and C flags, printed only in the Teacher's Wraparound Edition, indicate the level of difficulty of the exercises.

Additional Answers

17.

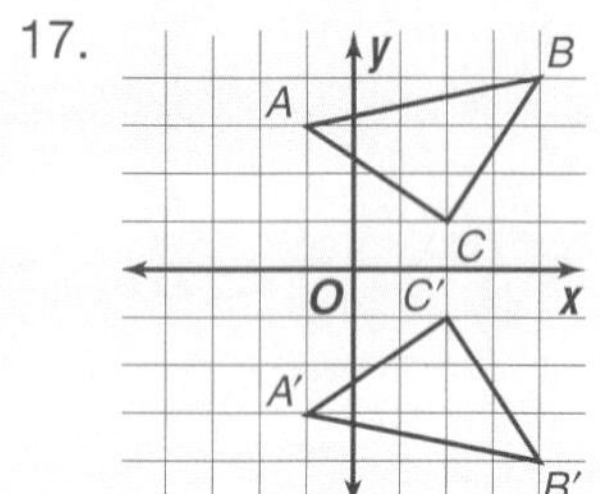

18.

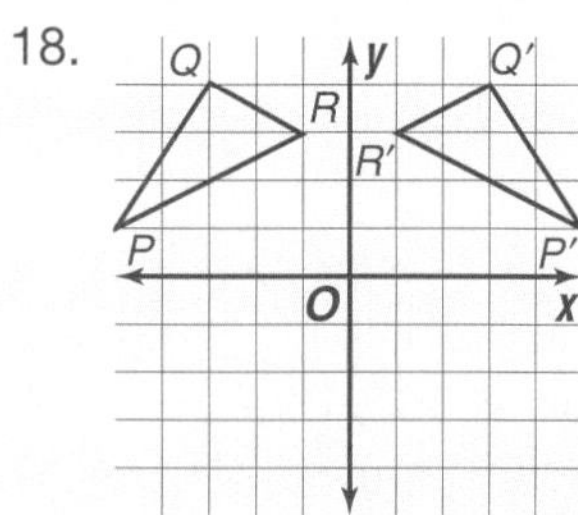

Practice Masters, p. 99

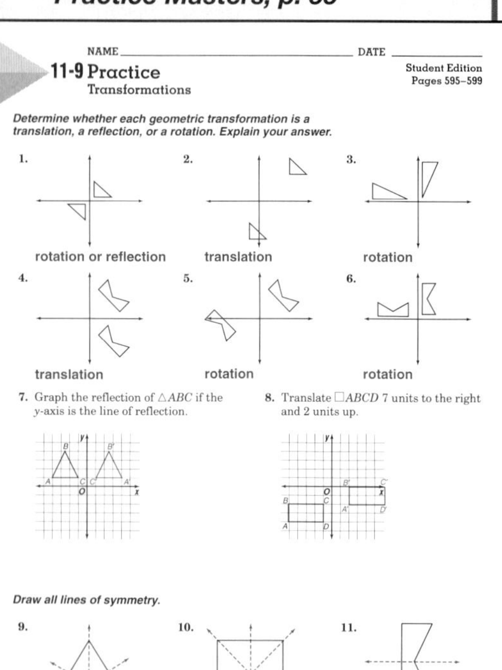

NAME ________________ DATE ________

11-9 Practice
Transformations

Student Edition Pages 595–599

Determine whether each geometric transformation is a translation, a reflection, or a rotation. Explain your answer.

1. rotation or reflection
2. translation
3. rotation
4. translation
5. rotation
6. rotation
7. Graph the reflection of $\triangle ABC$ if the y-axis is the line of reflection.
8. Translate $\square ABCD$ 7 units to the right and 2 units up.

Draw all lines of symmetry.

9.
10.
11.

Exercises: Practicing and Applying the Concept

Independent Practice

Tell whether each geometric transformation is a *translation*, a *reflection*, or a *rotation*. Explain your answer.

9–12. Sample answers given. See students' explanations.

9. 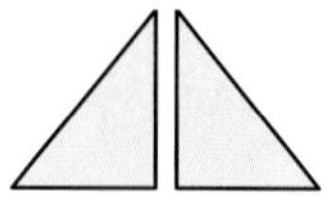 reflection

10.

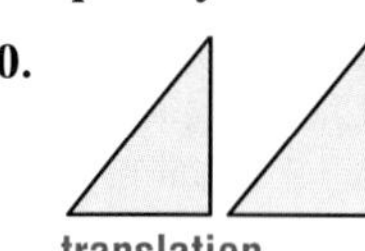

translation

11.

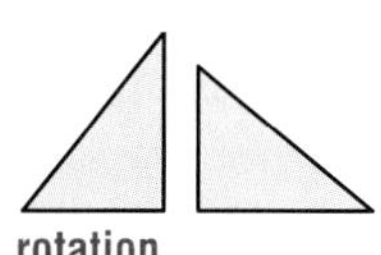

rotation

12. A translation moves $\triangle PQR$ 3 units to the right and 3 units down to form $\triangle P'Q'R'$. Write the coordinates of each vertex of $\triangle P'Q'R'$. $P'(3, 2)$, $Q'(3, -1)$, $R'(6, -1)$

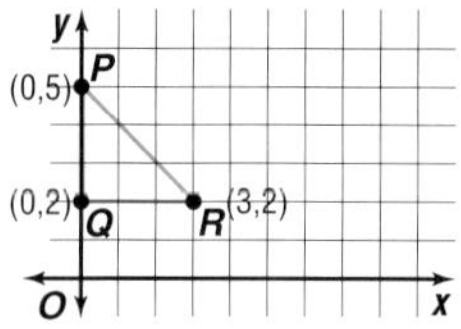

(0,5) P
(0,2) Q
R (3,2)

13. b

13. If the first figure below is rotated about point R, which could be the resulting figure?

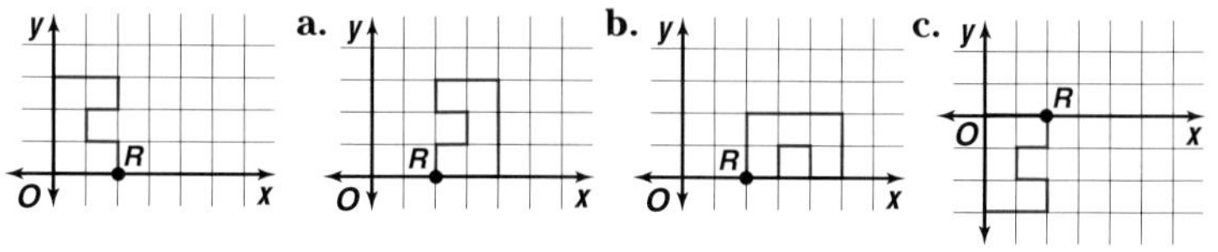

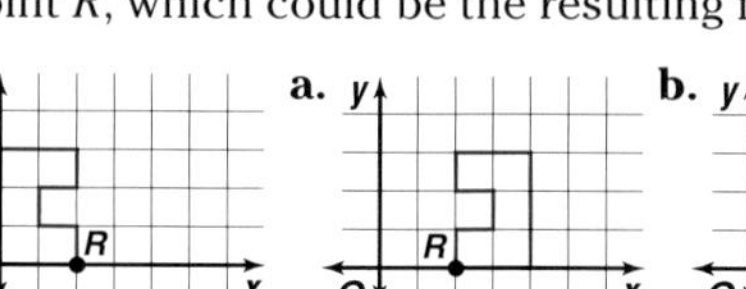

Trace each figure. Draw all lines of symmetry.

14.

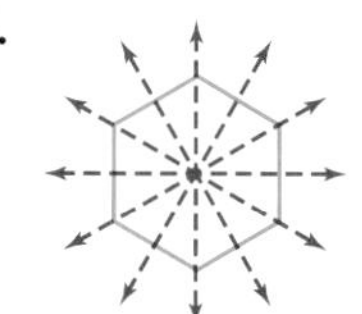

15.

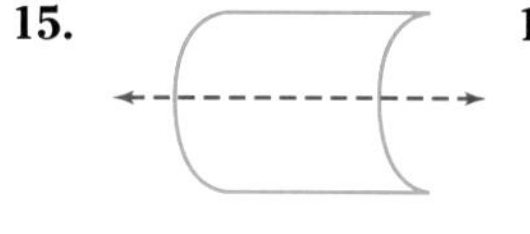

16.

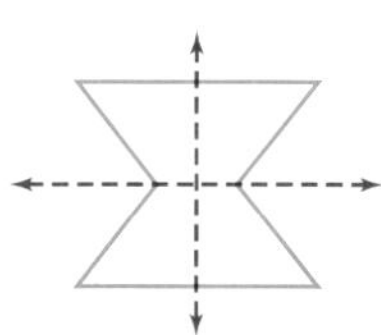

17. Graph $A(-1, 3)$, $B(4, 4)$, and $C(2, 1)$ on a coordinate plane. Draw $\triangle ABC$ and graph its reflection if the x-axis is the line of reflection.

18. Graph $P(-5, 1)$, $Q(-3, 4)$, and $R(-1, 3)$ on a coordinate plane. Draw $\triangle PQR$ and graph its reflection if the y-axis is the line of reflection.

17–19. See margin.

19. Graph $R(3, -1)$, $S(6, 2)$, and $T(1, 4)$ on a coordinate plane. Draw $\triangle RST$ and translate the triangle 5 units to the left and 4 units down.

Critical Thinking

20. Write a formula to find the number of lines of symmetry for any regular polygon with n sides. (*Hint:* Sketch several regular polygons and draw all the lines of symmetry. Record the results in a table and look for a pattern.) There are n lines of symmetry in a regular n-gon.

Application and Problem Solving

21. See margin.

21. Communication International code flags can be used by sailors to send messages at sea. Describe the kinds of symmetry that are displayed by the flags and tell which flags have each kind of symmetry.

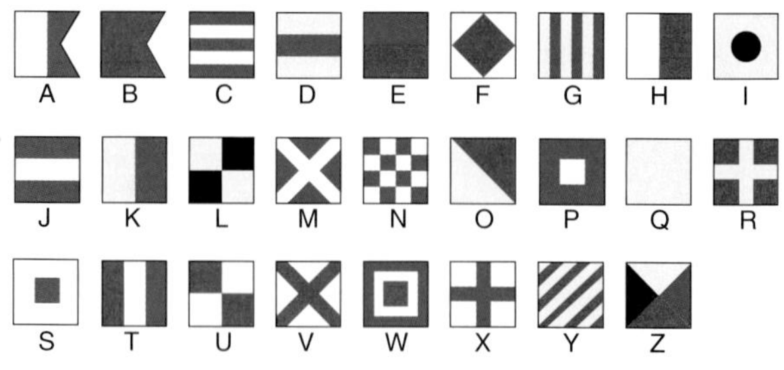

Group Activity Card 11-9

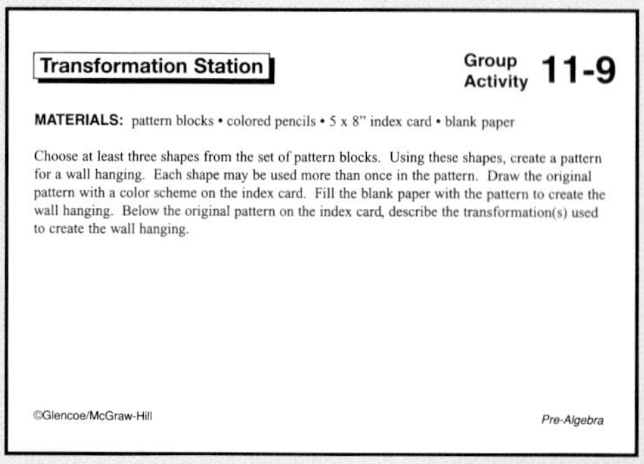

Transformation Station — Group Activity 11-9

MATERIALS: pattern blocks • colored pencils • 5 x 8" index card • blank paper

Choose at least three shapes from the set of pattern blocks. Using these shapes, create a pattern for a wall hanging. Each shape may be used more than once in the pattern. Draw the original pattern with a color scheme on the index card. Fill the blank paper with the pattern to create the wall hanging. Below the original pattern on the index card, describe the transformation(s) used to create the wall hanging.

 Pre-Algebra

Extension

Using Research Have students research ancient uses of transformations as a decorative art. Some suggestions include the To-Ta-Da-Ho belt made by Onondaga tribe of New York, the Washington Covenant and Hiawatha belts used by the Iroquois tribes, and Adinkra cloth made by the Asante people of Ghana, Africa.

22. **Games** Miniature golf is played with a putter and a golf ball on a carpeted "green" surrounded by side rails. On some holes there is a straight path to the hole. Others are "dog legs", and it is not possible to aim directly at the hole. You can, however, get a "hole in one" by mentally reflecting the hole across the line of the wall and aiming for the reflection.

a. Trace the figure at the top right. Draw a path using a reflection that will allow the ball to reach the hole in one stroke.

b. A more difficult "dog leg" is shown at the bottom right. Trace the figure and find the path of the ball for a "hole in one."

FYI

Crossword puzzles first appeared in 1913 in the *New York World* newspaper.

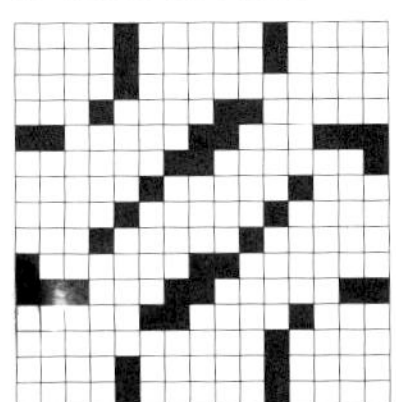

23. **Games** The crossword puzzle at the right is a 15 by 15 crossword puzzle with half-turn symmetry. That is, when the puzzle is turned upside down it looks exactly the same as when it is right-side up. Draw a 9 by 9 grid. Then fill in at least 18 squares to make a puzzle with half-turn symmetry. **See students' work.**

Mixed Review

24. 2340°
25. 200% increase
27. $t < -21$; See margin for graph.
28. Sample answer: (4, 0)

24. Find the sum of the measures of the angles of a 15-gon. (Lesson 11-8)

25. **Health** The average birth weight for full-term female babies is 7 pounds. The average weight of a one-year old girl is 21 pounds. What is the percent of change in weight in the first year? (Lesson 9-10)

26. Solve $\frac{2}{5}x \leq -0.16$. (Lesson 6-7) **−0.4**

27. Solve $-84 > 4t$. Then graph the solution on a number line. (Lesson 3-7)

28. Write the coordinates for a point on the x-axis. (Lesson 1-7)

From the FUNNY PAPERS

© 1995 Bill Amend/Dist. by Universal Press Syndicate
OW!
BONK!
AND PEOPLE THINK TWO-DIMENSIONAL BILLIARDS TAKES SKILL.
I BELIEVE THIS IS YOUR SUPER BALL

TALK ABOUT IT

1. Explain why the comic is funny. **See students' work.**
2. What is shown in the second frame of the comic?
3. What calculations did Jason have to make to get the ball to hit Paige? **He had to calculate the angles of all the bounces.**

2. The ball is bouncing off the walls at various angles.

4 ASSESS

Closing Activity

Modeling Have students create two distinct designs based upon transformations. At least two different transformations should be used for each design.

Chapter 11, Quiz D (Lessons 11-8 and 11-9) is available in the *Assessment and Evaluation Masters*, p. 296.

Additional Answers

19.

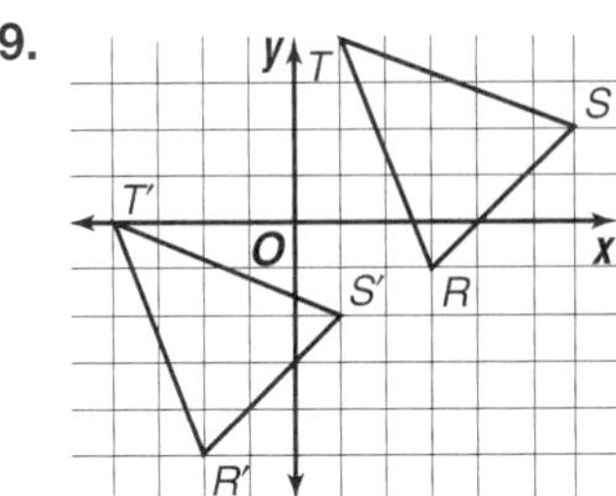

21. top to bottom: all except ELNOUYZ:
side to side: CDEFIJMPQRSVWX
diagonal: FII MNPQRSUVWXY

27.

−35 −28 −21 −14 −7 0 7

Enrichment Masters, p. 99

NAME ______________________ DATE __________

11-9 Enrichment
Nine-Point Circle

Student Edition Pages 595–599

The **nine-point circle**, or **Feuerbach circle**, is a circle on which the following nine points lie. This holds true for any triangle.

Points 1, 2, and 3: the midpoints of the sides of the triangle
Points 4, 5, and 6: the points where the altitudes intersect the sides of the triangle
Points 7, 8, and 9: the midpoints of the segments that join the vertices of the triangle to the point of intersection (orthocenter) of the altitudes

Construct the nine-point circle for $\triangle ABC$ as follows.

Step 1 Construct the perpendicular bisectors of the sides. Locate the midpoints of $\overline{AB}$, $\overline{BC}$, and $\overline{AC}$. Label these points 1, 2, and 3.
Step 2 Label the intersection of the perpendicular bisectors D (circumcenter).
Step 3 Construct a line from each vertex perpendicular to the opposite side. Label the points where these altitudes meet the sides 4, 5, and 6. Label the point of intersection of the altitudes O.
Step 4 Construct the midpoints of $\overline{OA}$, $\overline{OB}$, and $\overline{OC}$. Label these points 7, 8, and 9.
Step 5 Draw $\overline{OD}$. Construct the midpoint of $\overline{OD}$. Label it F.
Step 6 Draw the nine-point circle with center F and radius $F1$. Points 1, 2, 3, 4, 5, 6, 7, 8, and 9 will lie on the circle.

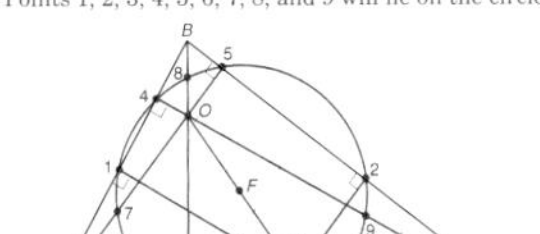

Identify the name of the point of intersection for the following in relation to triangles.

1. perpendicular bisectors **circumcenter** 2. altitudes **orthocenter**

3. The line determined by point O (orthocenter) and point D (circumcenter) is a special line. Do research and find the name of this line. **Euler's line**

FYI

The first crossword puzzle was created by Arthur Wynne, a newspaper editor. The puzzle was diamond-shaped and appeared on December 21, 1913. The *World* was the only newspaper to publish crossword puzzles for more than 10 years.

From the FUNNY PAPERS

Have students discuss the cartoon. Ask students to think of another situation involving the bouncing of a ball off walls and various other objects. The Michael Jordan and Larry Bird "Nothing But Net" McDonald's television commercials.

CLOSING THE

Investigation

Refer to the Investigation on pages 482–483. Add the results of your work to your Investigation folder.

Plan and give an oral presentation to describe what you know about games and strategy. Your presentation should include the following.

- A report of the survey you conducted at the beginning of the Investigation.
- A report of the results of your work at the end of Lesson 10-4. Include any revisions that were the direct result of the test and an explanation of why those revisions were made.
- A report on how your group determined the best strategy for *Don't Lose It All!* at the end of Lesson 10-8.
- A model of the board your group created at the end of Lesson 11-7 and a brief explanation of the accompanying game.
- A final version of the game your group designed. Include an explanation of how the survey at the beginning of the Investigation affected the type of game your group designed.

Extension

Contact a computer game manufacturer and inquire about how a computer game is created and how long it takes to create it.

PORTFOLIO ASSESSMENT

You may want to keep your work on this Investigation in your portfolio.

CLOSING THE

Investigation

This activity provides students an opportunity to bring their work on the Investigation to a close. For each Investigation, students should present their findings to the class. Here are some ways students can display their work.

- Conduct and report upon an interview or survey.
- Write a letter, proposal, or report.
- Write an article for the school or local paper.
- Make a display, including graphs and/or charts.
- Plan an activity.

Assessment

To assess students' understanding of the concepts and topics explored in Investigation A and its follow-up activities, you may wish to examine students' Investigation Folders.

The scoring guide provided in the *Investigations and Projects Masters,* p. 23, provides a means for you to score students' work on this Investigation.

Investigations and Projects Masters, p. 23

CHAPTERS 10 and 11 Investigation
It's Only A Game

Scoring Guide

Level	Specific Criteria
3 Superior	• Shows thorough understanding of the concepts of *displaying data, counting outcomes, finding permutations and combinations, finding probabilities and odds*, and *drawing polygons*. • Determines the best strategies to play the games. • Written explanations are exemplary. • Statistical chart and advertisement are appropriate and sensible. • Goes beyond the requirements of the Investigation.
2 Satisfactory, with minor flaws	• Shows understanding of the concepts of *displaying data, counting outcomes, finding permutations and combinations, finding probabilities and odds*, and *drawing polygons*. • Determines appropriate strategies to play the games. • Written explanations are effective. • Statistical chart and advertisement are appropriate and sensible. • Satisfies all requirements of the Investigation.
1 Nearly Satisfactory, with serious flaws	• Shows understanding of most of the concepts of *displaying data, counting outcomes, finding permutations and combinations, finding probabilities and odds*, and *drawing polygons*. • May not use appropriate strategies to play the games. • Written explanations are satisfactory. • Statistical chart and advertisement are mostly appropriate and sensible. • Satisfies most requirements of the Investigation.
0 Unsatisfactory	• Shows little or no understanding of the concepts of *displaying data, counting outcomes, finding permutations and combinations, finding probabilities and odds*, and *drawing polygons*. • May not use appropriate strategies to play the games. • Written explanations are not satisfactory. • Statistical chart and advertisement are not appropriate or sensible. • Does not satisfy requirements of the Investigation.

CHAPTER 11 Highlights

Vocabulary

After completing this chapter, you should be able to define each term, property, or phrase and give an example or two of each.

Geometry

acute angle (p. 550)
acute triangle (p. 569)
adjacent angles (p. 561)
alternate interior angles (p. 563)
alternate exterior angles (p. 563)
angle (p. 549)
complementary (p. 562)
congruent (pp. 554, 561, 573)
construct (p. 554)
corresponding angles (p. 563)
corresponding parts (p. 573)
degree (p. 549)
diagonal (p. 590)
equiangular (p. 591)
equilateral (p. 591)
exterior angle (p. 591)
indirect measurement (p. 580)
interior angle (p. 591)
intersect (p. 551)
isosceles triangle (p. 569)
line (p. 548)
line segment (p. 548)
obtuse angle (p. 550)
obtuse triangle (p. 569)
parallel (p. 551)
parallelogram (p. 585)
perpendicular (p. 551)
plane (p. 550)
point (p. 548)
polygon (p. 589)
quadrilateral (p. 584)
ray (p. 548)
rectangle (p. 585)
reflection (p. 595)
regular polygon (p. 591)
rhombus (p. 585)
right angle (p. 550)
right triangle (p. 569)
rotation (p. 595)
scale drawing (p. 585)
scalene triangle (p. 569)
sides (pp. 549, 589)
similar (p. 578)
skew lines (p. 551)
straight angle (p. 550)
square (p. 585)
supplementary (p. 562)
symmetric (p. 596)
tessellation (p. 594)
transformation (p. 595)
translation (p. 595)
transversal (p. 563)
trapezoid (p. 585)
triangle (p. 568)
vertex (pp. 549, 589)
vertical angles (p. 561)

Statistics

circle graph (p. 556)

Understanding and Using Vocabulary

Choose the letter of the term that best completes each statement.

1. A(n) _?_ angle measures between 0° and 90°. b
2. Two angles are _?_ if the sum of their measures is 90°. h
3. When a _?_ intersects two parallel lines, the corresponding angles are congruent. g
4. When a line is perpendicular to another line, the angles formed are _?_ angles. f
5. A(n) _?_ angle measures between 90° and 180°. a
6. A parallelogram with four congruent sides is a _?_. e
7. Two angles are _?_ if the sum of their measures is 180°. i

a. obtuse
b. acute
c. vertical
d. trapezoid
e. rhombus
f. right
g. transversal
h. complementary
i. supplementary

Instructional Resources

Three multiple-choice tests and three free-response tests are provided in the *Assessment and Evaluation Masters*. Forms 1A and 2A are for honors pacing, Forms 1B and 2B are for average pacing, Forms 1C and 2C are for basic pacing. Chapter 11 Test, Form 1B is shown at the right. Chapter 11 Test, Form 2B is shown on the next page.

Highlights

Using the Chapter Highlights

The Chapter Highlights begins with a listing of the new terms, properties, and phrases that were introduced in this chapter.

Assessment and Evaluation Masters, pp. 283–284

NAME ____________ DATE ________

CHAPTER 11 Test, Form 1B

1. What is the name for the figure shown at the right? 
   A. segment B. ray C. angle D. line — 1. B
2. The measure of angle A is 92°. Classify angle A.
   A. acute B. straight C. obtuse D. right — 2. C
3. Find the number of degrees for the section of a circle graph that would represent 60%.
   A. 120° B. 240° C. 540° D. 216° — 3. D
4. The pair of angles shown in the figure at the right are supplementary. Find the value of x. 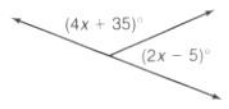
   $(4x + 35)°$ $(2x - 5)°$
   A. 25 B. 35 C. 75 D. 45 — 4. A

In the figure at the right, lines ℓ and m are parallel. Use the figure for Exercises 5 and 6.

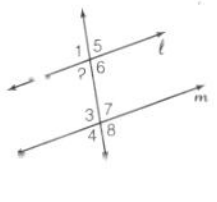

5. If $m\angle 1 = 100°$, find $m\angle 3$.
   A. 160° B. 120° C. 100° D. 80° — 5. C
6. If $m\angle 4 = 80°$, find $m\angle 5$.
   A. 100° B. 80° C. 120° D. 160° — 6. B
7. In $\triangle ABC$, $m\angle A = 18°$ and $m\angle B = 25°$. Find $m\angle C$.
   A. 36° B. 137° C. 143° D. 130° — 7. B
8. If $\triangle ABC \cong \triangle XZY$, which segment is congruent to $\overline{AC}$?
   A. $\overline{XY}$ B. $\overline{XZ}$ C. $\overline{YZ}$ D. $\overline{AB}$ — 8. A
9. If $\triangle LMN \cong VWX$, which angle is congruent to $\angle N$?
   A. $\angle W$ B. $\angle V$ C. $\angle M$ D. $\angle X$ — 9. D

In the figure at the right, $\triangle ABC \sim \triangle FGH$. Use the figure for Exercises 10 and 11.

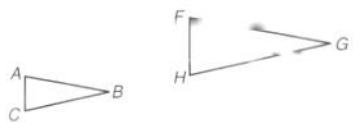

10. If $m\overline{AB} = 10$ cm, $m\overline{FG} = 15$ cm, and $m\overline{BC} = 16$ cm, find $m\overline{GH}$.
    A. 18 cm B. 20 cm C. 24 cm D. 12 cm — 10. C
11. If $m\angle A = 80°$, $m\angle B = 20°$, and $m\angle F = 80°$, find $m\angle G$.
    A. 80° B. 20° C. 100° D. 60° — 11. B

NAME ____________ DATE ________

Chapter 11, Test Form 1B (continued)

12. A photograph measuring 4 inches by 6 inches is enlarged to 12 inches by 18 inches. In the original photo, a tree's image is 1.5 inches high. Find the height of the tree's image in the enlargement.
    A. 4 in. B. 4.5 in. C. 9.5 in. D. 6 in. — 12. B
13. Whitney sees an aerial photograph of her house and the pond behind her house. The images of her house and the pond are 6 centimeters and 9 centimeters wide, respectively. She knows that her house is 35 feet wide. Find the width of the pond.
    A. 44 ft B. 52.5 ft C. 62.5 ft D. 70 ft — 13. B
14. Which quadrilateral does not have two pairs of parallel sides?
    A. trapezoid B. rectangle C. rhombus D. square — 14. A
15. Quadrilateral $ABCD$ is shown at the right. Find $m\angle D$. 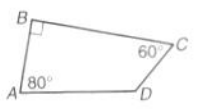
    A. 300° B. 240° C. 130° D. 60° — 15. C
16. Choose the measure of each exterior angle and each interior angle of a regular hexagon.
    A. 60°, 30° B. 60°, 120° C. 6°, 82° D. 6°, 172° — 16. B
17. Find the sum of the measures of the interior angles of a polygon having seven sides.
    A. 1260° B. 490° C. 700° D. 900° — 17. D
18. Find the perimeter of a regular pentagon with sides 2 meters long.
    A. 10 m B. 8 m C. 25 m D. 50 m — 18. A
19. Describe the transformation at the right. 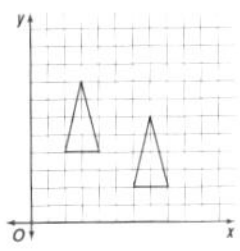
    A. translation
    B. reflection
    C. rotation
    D. cannot tell — 19. A
20. Which letters in addition to A have vertical lines of symmetry?
    A. H, M, V, X B. C, D, E, G
    C. B, C, K, L D. J, N, P, Q — 20. A

Using the Chapter Study Guide and Assessment

Skills and Concepts Encourage students to refer to the objectives and examples on the left as they complete the review exercises on the right.

Assessment and Evaluation Masters, pp. 289–290

NAME ________________ DATE ________

CHAPTER 11 Test, Form 2B

1. What type of geometric figure is shown at the right? — 1. line
2. The measure of angle C is 60°. Classify angle C. — 2. acute
3. Find the number of degrees for the section of a circle graph that would represent 20%. — 3. 72°
4. Angles A and B are supplementary and $m\angle A = 12°$. Find $m\angle B$. — 4. 168°
5. Angles 2 and 4 are vertical and $m\angle 2 = 65°$. Find $m\angle 4$. — 5. 65°

In the figure at the right, lines ℓ and m are parallel. Use the figure for Exercises 6 and 7.

6. If $m\angle 1 = 80°$, find $m\angle 3$. — 6. 100°
7. If $m\angle 4 = 80°$, find $m\angle 5$. — 7. 80°
8. In $\triangle ABC$, $m\angle A = 48°$ and $m\angle B = 26°$. Find $m\angle C$. — 8. 106°
9. If $\triangle ABC \cong \triangle XZY$, what segment is congruent to $\overline{ZY}$? — 9. $\overline{BC}$

In the figure at the right, $\triangle ABC \sim \triangle FGH$. Use the figure for Exercises 10 and 11.

10. If $m\overline{AB} = 20$ cm, $m\overline{FG} = 15$ cm, and $m\overline{BC} = 28$ cm, find $m\overline{GH}$. — 10. 21 cm
11. If $m\angle A = 50°$, $m\angle B = 45°$, find $m\angle H$. — 11. 85°
12. On a small scale drawing, the length of the front of the public library is 12 centimeters, and the length of the front of the high school is 13 centimeters. On a larger scale drawing, the length of the library is 18 centimeters. Find the length of the high school on the larger drawing. — 12. 19.5 cm
13. What is the sum of the measures of the interior angles of a pentagon? — 13. 540°
14. Find the measure of $\angle D$ in the figure at the right. — 14. 70°

NAME ________________ DATE ________

Chapter 11 Test, Form 2B (continued)

15. Find the measure or each exterior angle and each interior angle of a 15-gon. — 15. 24°, 156°
16. Find the sum of the measures of the interior angles of a polygon having ten sides. — 16. 1440°
17. Owen draws two similar 4-sided designs. The smaller one has two short sides each measuring 6 inches and two long sides each measuring 9 inches. The larger design has two short sides each measuring 15 inches. Find the length of the long sides of the larger design. — 17. 22.5 in.
18. Find the perimeter of a regular quadrilateral with sides 2.84 meters long. — 18. 11.36 m
19. Tell whether the transformation at the right is a translation, a rotation, or a reflection. — 19. rotation

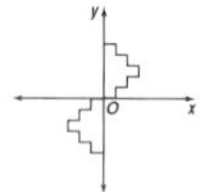

20. Draw all lines of symmetry for the geometric figure shown at the right. — 20. See students' work.

CHAPTER 11 Study Guide and Assessment

Skills and Concepts

Objectives and Examples

Upon completing this chapter, you should be able to:

▶ **identify points, lines, planes, rays, segments, angles, and parallel, perpendicular and skew lines.** (Lesson 11-1)

Draw and label a diagram to represent $\angle ABC$.

▶ **make a circle graph to illustrate data.** (Lesson 11-2)

This circle graph compares the amount of land and water to the whole surface of Earth.

30 [%] [×] 360 [=] 108

70 [%] [×] 360 [=] 252

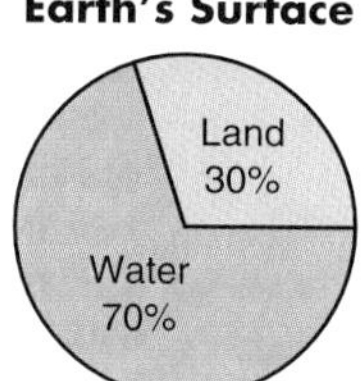

▶ **identify the relationships of vertical, adjacent, complementary, and supplementary angles, and angles formed by two parallel lines and a transversal.** (Lesson 11-3)

Find the measure of each labeled angle.

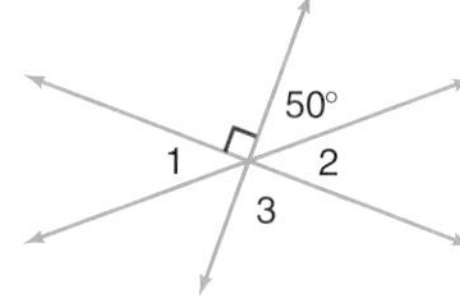

$m\angle 1 = 180 - 90 - 50$ or $40°$

$m\angle 2 = m\angle 1$ or $40°$

$m\angle 3 = 90°$

Review Exercises

Use these exercises to review and prepare for the chapter test.

Draw and label a diagram to represent each of the following. 8–15. See Solutions Manual.

8. point R
9. ray TS
10. plane XYZ
11. acute angle JKL
12. line n
13. $\overline{EF}$
14. $\overleftrightarrow{HG} \parallel \overleftrightarrow{KL}$
15. lines w and y are skew

16. The table below shows the types of jams and jellies that were entered in state fairs in 1994. Make a circle graph of the data. **See margin.**

Type	Entries
Strawberry	266
Raspberry	266
Blackberry	176
Grape	165
Plum	131
Other	592

17. Make a circle graph of the number of hours each week that you study, talk on the telephone, watch television or listen to music, and sleep. **See students' work.**

In the figure below, $a \parallel b$. Find the measure of each angle.

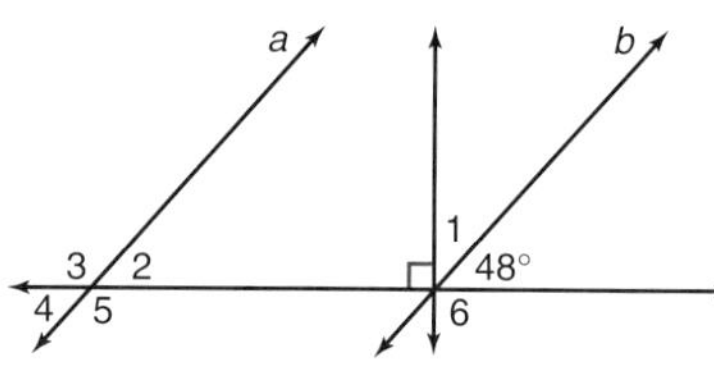

18. $\angle 1$ 42°
19. $\angle 2$ 48°
20. $\angle 3$ 132°
21. $\angle 4$ 48°
22. $\angle 5$ 132°
23. $\angle 6$ 90°

GLENCOE Technology

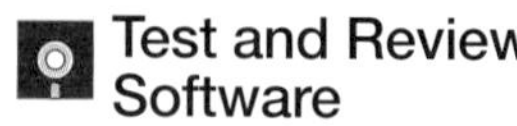

Test and Review Software

You may use this software, a combination of an item generator and an item bank, to create your own tests or worksheets. Types of items include free response, multiple choice, short answer, and open ended.

For IBM & Macintosh

Objectives and Examples

▶ **find the missing angle measure of a triangle.** (Lesson 11-4)

Find $m\angle 1$ if $m\angle 2 = 50°$ and $m\angle 3 = 80°$. Classify the triangle as *acute*, *right*, or *obtuse*.

$m\angle 1 = 180 - 50 - 80$ or $50°$
The triangle is acute.

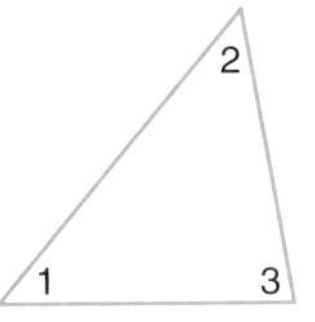

▶ **identify congruent triangles and their corresponding parts.** (Lesson 11-5)

If $\angle ABC \cong \angle FED$, name the parts congruent to angle A and segment AC.

$\angle A$ is congruent to $\angle F$. $\overline{AC}$ is congruent to $\overline{FD}$.

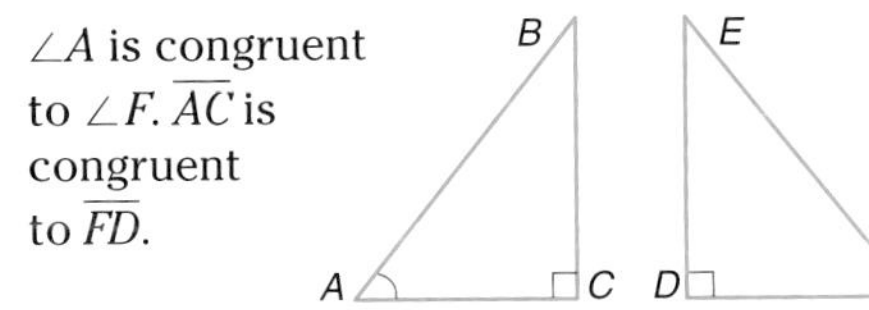

▶ **identify corresponding parts and find missing measures of similar triangles.** (Lesson 11-6)

Find the value of x in the similar triangles.

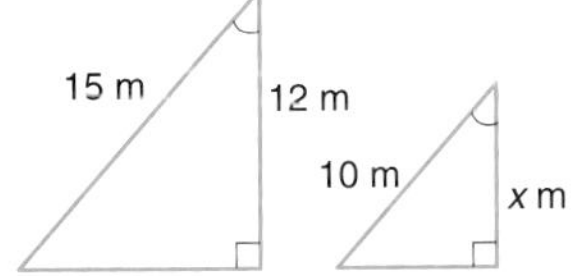

$$\frac{15}{10} = \frac{12}{x}$$
$$120 = 15x$$
$$8 = x$$

▶ **find the missing angle measure of a quadrilateral and classify quadrilaterals.** (Lesson 11-7)

Find the missing angle measures.

x° 135° 75° 2x°

$75 + 135 + x + 2x = 360$
$210 + 3x = 360$
$3x = 150$
$x = 50$

The measures are 50° and $50 \cdot 2$ or 100°.

Review Exercises

Use the figure below to solve each problem.

24. Find $m\angle 1$ if $m\angle 2 = 35°$ and $m\angle 3 = 90°$. Classify the triangle as *acute*, *right*, or *obtuse*. **55°; right**

25. Find $m\angle 2$ if $m\angle 1 = 42°$ and $m\angle 3 = 46°$. Classify the triangle as *acute*, *right*, or *obtuse*. **92°; obtuse**

If $\triangle XYZ \cong \triangle ABC$, name the part congruent to each angle or segment.

26. $\angle X$ **$\angle A$**
27. $\angle B$ **$\angle Y$**
28. $\overline{YZ}$ **$\overline{BC}$**
29. $\overline{AC}$ **$\overline{XZ}$**

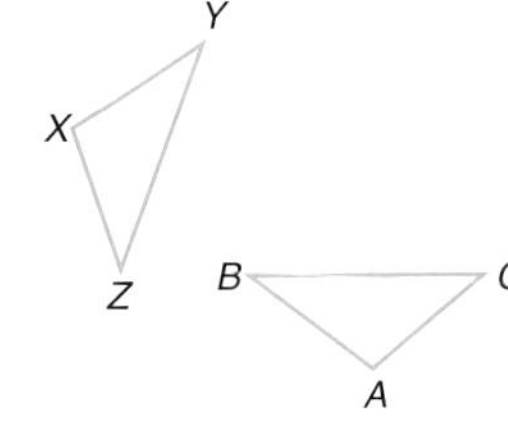

Find the value of x in each pair of similar triangles.

30. x ft, 6 ft, 7 ft, 21 ft — **18**

31. 5 in., 4 in., 5 in., x in., 2 in. — **2**

Find the missing angle measures in each figure.

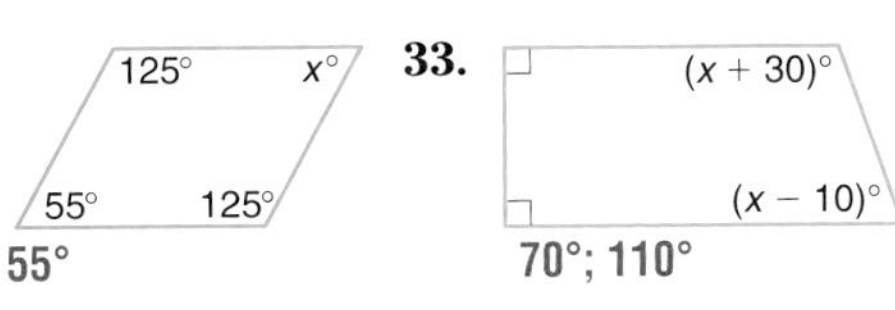

32. **55°**

33. **70°; 110°**

Additional Answers

8. • R

9.
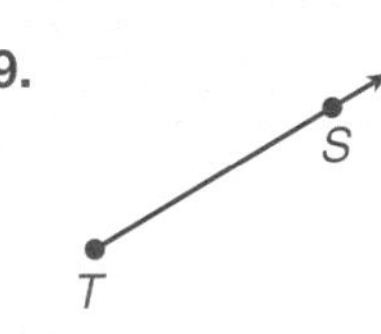
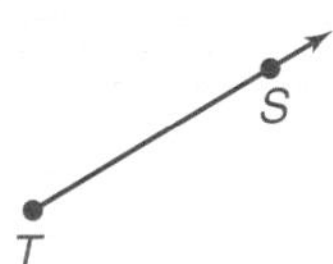

10.
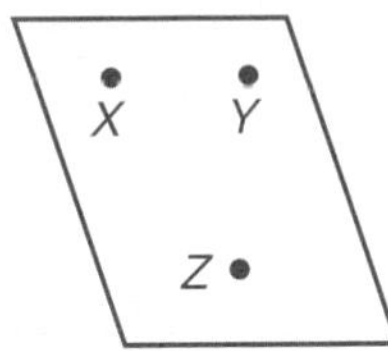

11.
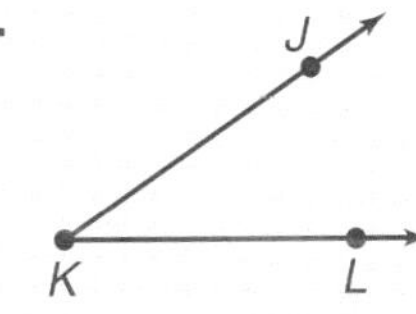

12.

13.

14.
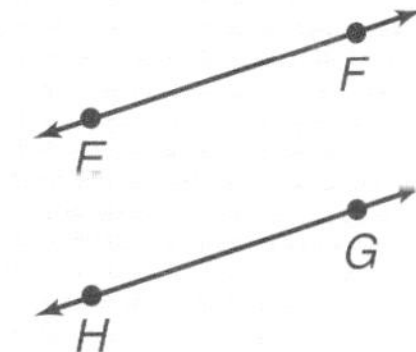

15.
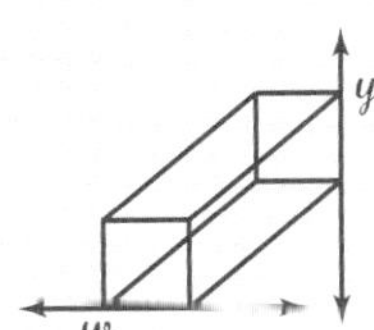

16.
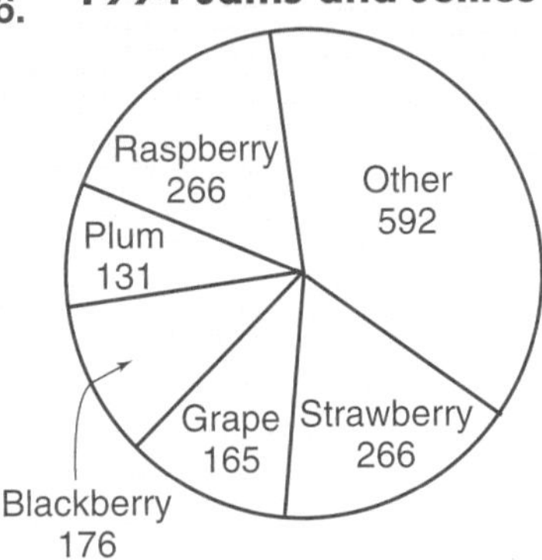

Applications and Problem Solving

Encourage students to work through the exercises in the Applications and Problem Solving section to strengthen their problem-solving skills.

Additional Answer

41. **Water Use in U.S.**

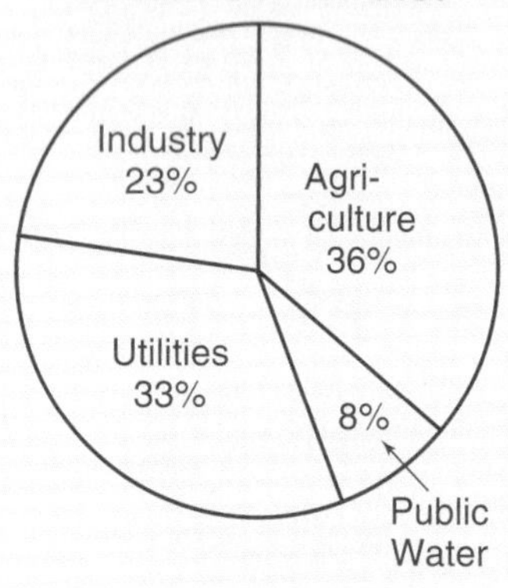

Objectives and Examples

▶ **classify polygons and determine the sum of the measures of the interior and exterior angles of a polygon.** (Lesson 11-8)

Find the sum of the measures of the angles of an octagon.

$(n - 2)180 = (8 - 2)180$
$= 6(180)$
$= 1080$

Review Exercises

34. Find the sum of the degree measures of the angles of a hexagon. **720°**

35. Find the measure of each interior angle in a regular decagon. **144°**

36. Find the measure of an exterior angle in a regular pentagon. **72°**

Objectives and Examples

▶ **identify and draw reflections, translations, and rotations and identify and draw symmetric figures.** (Lesson 11-9)

Draw all lines of symmetry for the figure below.

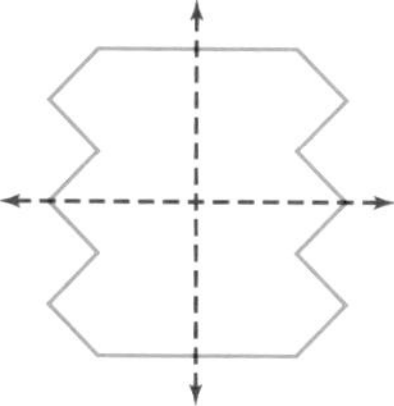

Review Exercises

Trace each figure. Draw all lines of symmetry.

37.

38.

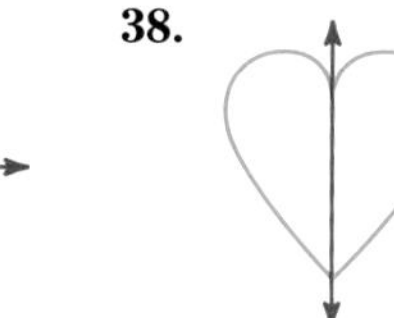

Tell whether each transformation is a *translation*, a *reflection*, or a *rotation*. Explain your answer. Sample answers given. See students' explanations.

39. reflection

40.

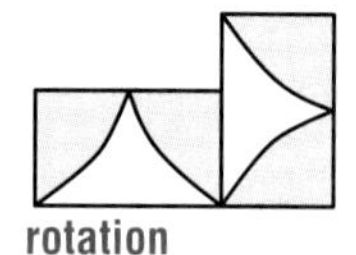

rotation

Applications and Problem Solving

41. Statistics Make a circle graph to display the data in the chart. (Lesson 11-2) **See margin.**

Water Use in the U.S.	
Agriculture	36%
Public Water	8%
Utilities	33%
Industry	23%

42. Home Maintenance When a 6-foot ladder is placed against the side of a house at an angle that is reasonable for climbing, it touches the side of the house at a point 5 feet above the ground. If a 15-foot ladder is placed at the same angle to the house, can it be used to reach a point 12 feet above the ground? How high will it reach? (Lesson 11-6) **yes; 12.5 ft**

A practice test for Chapter 11 is available on page 784.

Alternative Assessment

Cooperative Learning Project

Scale Drawings Measurements of angles and geometric shapes are used to make scale drawings. For this project, you will need to choose a real-life space and complete the following activities.

1. Choose an indoor or outdoor area of which to make a scale drawing. For example, you might choose a room or the exterior of a house.
2. Make a sketch of the area, marking details such as walls, doors, windows, electrical outlets, and any other details in the space.
3. Make a plan for measuring the details of the space. Consider how you will measure the angles in the space as well as the distances.
4. Make your measurements and record them on the sketch.
5. Select an appropriate scale for your drawing.
6. Prepare a final scale drawing from the information on your sketch.

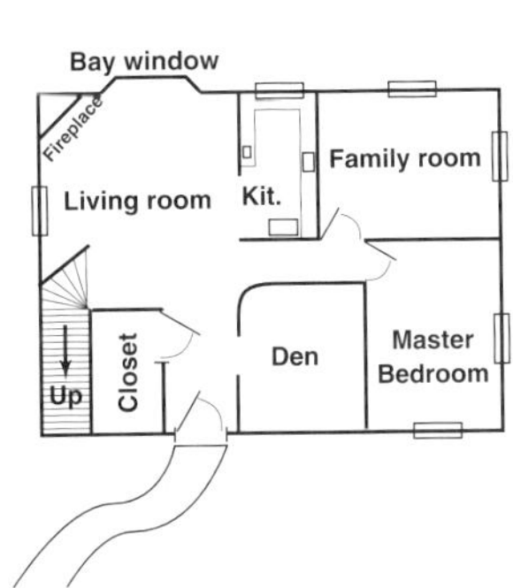

Thinking Critically

- A quadrilateral has two pairs of parallel sides. What kind of figure could it be? **square, rhombus, rectangle, or parallelogram**
- A triangle has angles measuring x, $2x$, and $3x$ degrees. What kind of triangle is it? **right**
- The sum of the degree measures of the angles in a regular polygon is $(n - 2)180$. Explain how you can remember this formula. **See students' work.**

Portfolio

Select your favorite activity from this chapter and write a note explaining what you learned from the activity.

Self Evaluation

Honeycombs are made in the shape of hexagons because hexagons are very strong structures. They distribute the weight and do not collapse.

Assess yourself. How do you distribute the weight of your tasks over the time you have been allowed? Do you wait for the last minute and then rush to finish, or do you spread out your work so you can give it your best attention? Write a paragraph about how you can improve the way you use your time. **See students' work.**

Alternative Assessment

The Alternative Assessment section provides students with the opportunity to assess their own work by thinking critically, working with others, keeping a portfolio, and honestly evaluating their own progress. For more information on alternative forms of assessment, see *Alternative Assessment in the Mathematics Classroom*.

Performance Assessment

Performance Assessment tasks for this chapter are included in the *Assessment and Evaluation Masters*. A scoring guide is also provided.

Alternative Assessment

Assessment and Evaluation Masters, pp. 293, 305

NAME ____________ DATE ________

CHAPTER 11 Performance Assessment

Instructions: Demonstrate your knowledge by giving a clear, concise solution to each problem. Be sure to include all relevant drawings and justify your answers. You may show your solution in more than one way or investigate beyond the requirements of the problem.

1. Find two examples of parallel segments in your classroom. Explain what it means to say that segments are parallel. Find two examples of perpendicular segments in your classroom. Explain what it means to say that segments are perpendicular.
2. Write a plan for measuring the height of your school building by the indirect method of measurement. Include a drawing with your plan.
3. Triangles can be classified by the type of angles they have. All triangles have at least two acute angles but are named according to the measure of the third angle. Sketch and label all possible kinds of triangles according to the measure of their angles.
4. Draw a diagram that classifies different kinds of quadrilaterals according to their characteristics.
5. a. Tell when it is appropriate to graph a set of data in a circle graph.
   b. Why are small groups of data often grouped together under the heading "other" rather than being graphed separately?
   c. Suppose you were given $100 to spend and allowed to purchase 5 items. List each of your purchases by item and price. Group you purchases into similar categories. Draw and label a circle graph that would represent your spending spree.
   d. Give an example of how a business person might use a circle graph.
6. a. Draw a triangle ABC and a line ℓ intersecting the triangle. Draw the reflection of $\triangle ABC$ with respect to ℓ and label the reflection $\triangle A'B'C'$.
   b. Draw a figure with at least one line of symmetry. Show one or more of the line(s) of symmetry.

CHAPTER 11 Scoring Guide

Level	Specific Criteria
3 Superior	• Shows thorough understanding of the concepts *parallel, perpendicular, similar triangles, classifying triangles, classifying quadrilaterals, circle graph, reflection*, and *symmetry*. • Uses appropriate strategies. • Computations to construct circle graphs are correct. • Written explanations are exemplary. • Example of use of circle graph is appropriate and makes sense. • Diagrams and circle graph are accurate. • Goes beyond requirements of some or all problems.
2 Satisfactory, with minor flaws	• Shows understanding of the concepts *parallel, perpendicular, similar triangles, classifying triangles, classifying quadrilaterals, circle graph, reflection*, and *symmetry*. • Uses appropriate strategies. • Computations to construct circle graphs are mostly correct. • Written explanations are effective. • Example of use of circle graph is appropriate and makes sense. • Diagrams and circle graph are mostly accurate. • Satisfies all requirements of problems.
1 Nearly Satisfactory, with serious flaws	• Shows understanding of most of the concepts *parallel, perpendicular, similar triangles, classifying triangles, classifying quadrilaterals, circle graph, reflection*, and *symmetry*. • May not use appropriate strategies. • Computations to construct circle graphs are mostly correct. • Written explanations are satisfactory. • Example of use of circle graph is appropriate and makes sense. • Diagrams and circle graph are mostly accurate. • Satisfies most requirements of problems.
0 Unsatisfactory	• Shows little or no understanding of the concepts *parallel, perpendicular, similar triangles, classifying triangles, classifying quadrilaterals, circle graph, reflection*, and *symmetry*. • May not use appropriate strategies. • Computations to construct circle graphs are incorrect. • Written explanations are not satisfactory. • Example of use of circle graph are not appropriate or sensible. • Diagrams and circle graph are not accurate. • Does not satisfy requirements of problems.

Investigation

TEACHER NOTES

NCTM Standards: 1-4

This Investigation is designed to be completed over several days or weeks. It may be considered optional. You may want to assign the Investigation and the follow-up activities to be completed at the same time.

Objectives

- Investigate types of housing available in the local community.
- Design and make a model of a residential house.

Mathematical Overview

This Investigation will use the following skills and concepts from Chapters 12 and 13.

- find areas, surface areas, and volumes
- make a model or drawing
- use ratios

Recommended Time		
Part	**Pages**	**Time**
Investigation	606–607	1 class period
Working on the Investigation	637, 653, 671	20 minutes each
Closing the Investigation	698	1 class period

Instructional Resources

Investigations and Projects Masters, pp. 25–28

A recording sheet, teacher notes, and scoring guide are provided for each Investigation in the *Investigation and Projects Masters*.

1 MOTIVATION

In this Investigation, students use research to find out about local residential housing. Start with a class brainstorming session about what is involved in designing and building a house.

LONG-TERM PROJECT

Investigation

MATERIALS

- ruler
- protractor
- graph paper
- cardboard or foam board
- glue
- paint (optional)

Look around the room where you are sitting. What factors do you think had to be considered before any walls were built? If you did not know any mathematics, would you be able to construct the building that you are sitting in now?

Few things require the use of mathematics more than the construction of buildings. Even the most complex buildings are made up of arrangements of geometric shapes.

Suppose you are a member of a group of three architects that has been hired to design a home. You will oversee every detail of the design from selecting the location to planting the last shrub in the yard.

To create an attractive and efficient building, you must bring harmony to the following elements.

- ***Appearance***–*The outside appearance of a structure is determined not only by its shape but also by the choice of building materials. Many architects have skillfully blended the rough textures of wood and stone with the smoothness of glass and metal.*
- ***Durability***–*A structure should be well built so that it can withstand all kinds of weather and provide protection from temperature extremes without expensive maintenance.*
- ***Function***–*Every structure is designed for a certain purpose. To be functional, a building should fulfill the needs of its users.*

Starting the Investigation

When someone decides to build a house, the first thing they must do is choose a *lot*, or piece of land. Work with your group to select a location for the dwelling you will design and build. Then check the zoning laws and building codes for that location.

As a group, identify at least four different dwellings in your community. Before you begin, you should familiarize yourself with some styles of houses by referring to an encyclopedia, or consulting with a Realtor. Describe the physical structure of the house and state its purpose. For example, Description: brick 1-story; Purpose: home for a retired couple. Include a sketch of each building.

An *elevation* is a drawing of a building's exterior. An elevation shows the location and size of features such as doors and windows. Typically, four elevation drawings are done, one for each of the four sides of the building. They also show what types of materials will be used on the outside of the building. Choose a style of dwelling that your group would like to build. Then sketch its front elevation.

For current information on architecture, visit:
www.glencoe.com/sec/math/prealg/mathnet

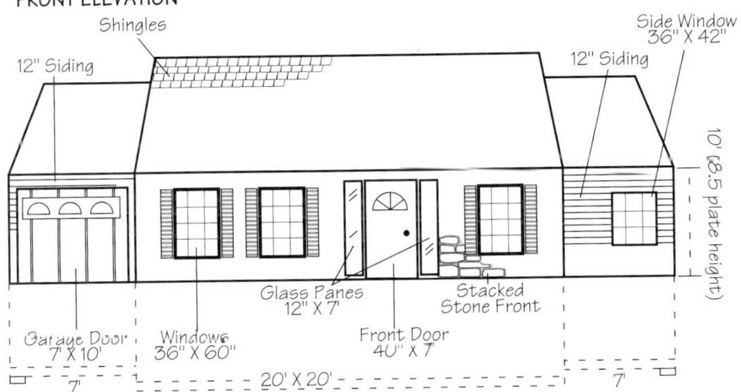

Investigation
Oh Give Me a Home

Working on the Investigation
Lesson 12-5, p. 637

Working on the Investigation
Lesson 12-8, p. 653

Working on the Investigation
Lesson 13-2, p. 671

Closing the Investigation
End of Chapter 13, p. 698

You will work on this Investigation throughout Chapters 12 and 13. Be sure to keep your materials in your Investigation Folder.

2 SETUP

Have students work in small groups. During this Investigation, students will make scale drawings and models. To do this, they will need rulers, protractors, graph paper, cardboard or foam board, a cutting utensil, and glue. Decorative items are optional.

For more information on this topic, consult the following references.

Harms, Henry R., Dennis K. Kroon, and Marlene Weigel. *Experience Technology: Manufacturing and Construction*. New York: Glencoe, 1993.

Salvadori, Mario G. *Why Buildings Stand Up: The Strength of Architecture*. New York: W. W. Norton & Company, 1992

Salvadori, Mario G., and Matthys P. Levy. *Why Buildings Fall Down*. New York: W. W. Norton & Company, 1992

3 MANAGEMENT

Discuss the method that will be used to grade each project *before* students start. For instance, students should know whether an elaborately decorated house will score higher than a plain house.

Investigations and Projects Masters, p. 28

NAME ______________________ DATE ____________

CHAPTERS **12 and 13** Investigation
Oh Give Me A Home!

Student Edition
Pages 606–607

Work with your group of architects.

Name four different house styles that your group has identified. Describe each style.

What style did you select?

Where will you build your house?

In the space below, sketch the front elevation of the house your group chose to build. Include details about materials you will use, and label the dimensions.

Please keep this page and any other research in your Investigation Folder.

Cooperative Learning

This Investigation offers an excellent opportunity for using cooperative learning groups. For more information on cooperative learning strategies and group management, see *Cooperative Learning in the Mathematics Classroom*, one of the titles in the Glencoe Mathematics Professional Series.

Sample Answers

Answers will vary as they are based on the research of each group.

CHAPTER 12

Measuring Area and Volume

PREVIEWING THE CHAPTER

Students begin by using geoboards to find the area of shapes that are not rectangular. Students then find areas of parallelograms, triangles, and trapezoids. Fractals are investigated, and then students find the area of circles. Probability is integrated with area. Students will also use a graphing calculator to experiment with area and probability. Students will solve problems using a model or a drawing. The next portion of the chapter involves finding the surface areas of prisms, cylinders, pyramids, and cones. The chapter concludes with a study of the volume of prisms, cylinders, pyramids, and cones. The final Math Lab allows students to find how changing the dimensions of a solid affects the surface area and volume of the solid.

Lesson (pages)	Lesson Objectives	NCTM Standards	State/Local Objectives
12-1A (610–611)	Use a geoboard to find areas of shapes that are not rectangular.	1-4, 12, 13	
12-1 (612–617)	Find areas of parallelograms, triangles, and trapezoids.	1-4, 12, 13	
12-1B (618)	Explore and construct fractals.	1-4, 12, 13	
12-2 (619–622)	Find areas of circles.	1-4, 12, 13	
12-3 (623–627)	Find probabilities using area models.	1-4, 11-13	
12-3B (628)	Use a graphing calculator to experiment with area and probability.	1-4, 11-13	
12-4 (629–631)	Solve problems using a model or a drawing.	1-4, 12, 13	
12-5 (632–637)	Find surface areas of triangular and rectangular prisms and circular cylinders.	1-4, 12, 13	
12-6 (638–642)	Find surface areas of pyramids and cones.	1-4, 12, 13	
12-7A (643)	Investigate volume by making containers of different shapes and comparing how much each container holds.	1-4, 12, 13	
12-7 (644–648)	Find volumes of prisms and circular cylinders.	1-4, 12, 13	
12-8 (649–653)	Find volumes of pyramids and cones.	1-4, 12, 13	
12-8B (654)	Investigate how changing the dimensions of a solid affects its surface area and volume.	1-4, 12, 13	

A complete, 1-page lesson plan is provided for each lesson in the ***Lesson Planning Guide***. Answer keys for each lesson are available in the ***Answer Key Masters***.

ORGANIZING THE CHAPTER

You may want to refer to the **Course Planning Calendar** on page T12 for detailed information on pacing.
PACING: Standard—13 days; **Honors**—10 days; **Block**—7 days

LESSON PLANNING CHART

Lesson (Pages)	Materials/ Manipulatives	Extra Practice (Student Edition)	Blackline Masters									Real-World Applications	Interactive Mathematics Tools Software	Teaching Transparencies	Group Activity Cards
			Study Guide	Practice	Enrichment	Assessment and Evaluation	Math Lab and Modeling Math	Multicultural Activity	Tech Prep Applications	Graphing Calculator	Activity				
12-1A (610–611)	geoboard* grid paper						p. 50								
12-1 (612–617)	grid paper, scissors, tape	p. 769	p. 100	p. 100	p. 100		p. 71			p. 12	p. 45		12-1.1 12-1.2	12-1A 12-1B	12-1
12-1B (618)	ruler*, compass*, protractor*						p. 51								
12-2 (619–622)	grid paper, compass*, protractor*	p. 769	p. 101	p. 101	p. 101	p. 323	p. 87	p. 23			p. 26	27	12-2	12-2A 12-2B	12-2
12-3 (623–627)	grid paper dry beans	p. 769	p. 102	p. 102	p. 102									12-3A 12-3B	12-3
12-3B (628)	grid paper straightedge									p. 25					
12-4 (629–631)		p. 770	p. 103	p. 103	p. 103	pp. 322, 323	p. 72					28		12-4A 12-4B	12-4
12-5 (632–637)	cereal boxes, scissors, ruler*	p. 770	p. 104	p. 104	p. 104				p. 23					12-5A 12-5B	12-5
12-6 (638–642)		p. 770	p. 105	p. 105	p. 105	p. 324								12-6A 12-6B	12-6
12-7A (643)	5x8 index cards, tape, rice						p. 52								
12-7 (644–648)		p. 770	p. 106	p. 106	p. 106				p. 24				12-7	12-7A 12-7B	12-7
12-8 (649–653)	scissors, tape, rice, compass*	p. 771	p. 107	p. 107	p. 107	p. 324		p. 24			p.12			12-8A 12-8B	12-8
12-8B (654)	sugar cubes						p. 53								
Study Guide/ Assessment (656–658)						pp. 309–321, 325–327									

*Included in Glencoe's *Student Manipulative Kit* and *Overhead Manipulative Resources*.

ORGANIZING THE CHAPTER

All of the blackline masters in the Teacher's Classroom Resources are available on the ***Electronic Teacher's Classroom Resources*** CD-ROM.

Other Chapter Resources

Student Edition
Chapter Opener, pp. 608–609
Working on the Investigation, pp. 637, 653
The Shape of Things to Come: Digital Cameras, p. 627
From the Funny Papers, p. 622

Teacher's Classroom Resources
Investigations and Projects Masters, pp. 73–76
Pre-Algebra Overhead Manipulative Resources, pp. 34–38

Technology
Test and Review Software (IBM & Macintosh)
CD-ROM Activities

Professional Publications

Block Scheduling Booklet
Glencoe Mathematics Professional Series

Outside Resources

Books/Periodicals
The MESA Series: Packaging and the Environment, Nancy Cook
Unit Origami: Multidimensional Transformations, Tomoko Fusé
Middle Grades Mathematics Project Series: Mouse and Elephant/Measuring Growth, Fitzgerald, et al
EQUALS Investigation Project: Flea Sized Surgeons/Surface Area, Volume and Scale, Creative Publications

Videos/CD-ROMs
Math Vantage Videos: Spatial Sense Unit/Containers: Surface Area and Volume, Sunburst
Math Vantage Videos: Spatial Sense Unit/Going Around in Circles, Sunburst

Software

Cabri Geometry, Texas Instruments
The Geometric Supposer, Sunburst

Assessment Resources

Student Edition
Math Journal, pp. 630, 640, 646
Mixed Review, pp. 617, 622, 627, 631, 637, 642, 648, 653
Self Test, p. 642
Chapter Highlights, p. 655
Chapter Study Guide and Assessment, pp. 656–658
Alternative Assessment, p. 659
Portfolio, p. 659
Ongoing Assessment, pp. 660–661
MindJogger Videoquiz, 12

Teacher's Wraparound Edition
5-Minute Check, pp. 612, 619, 623, 629, 632, 638, 644, 649
Checking Your Understanding, pp. 615, 621, 625, 630, 635, 640, 646, 651
Closing Activity, pp. 617, 622, 627, 631, 637, 642, 648, 653
Cooperative Learning, pp. 614, 650

Assessment and Evaluation Masters
Multiple-Choice Tests, Forms 1A (Honors), 1B (Average), 1C (Basic), pp. 309–314
Free-Response Tests, Forms 2A (Honors), 2B (Average), 2C (Basic), pp. 315–320
Performance Assessment, p. 321
Mid-Chapter Test, p. 322
Quizzes A–D, pp. 323–324
Standardized Test Practice, p. 325
Cumulative Review, pp. 326–327

ENHANCING THE CHAPTER

Examples of some of the materials for enhancing Chapter 12 are shown below.

Multicultural Activity Masters, pp. 23, 24

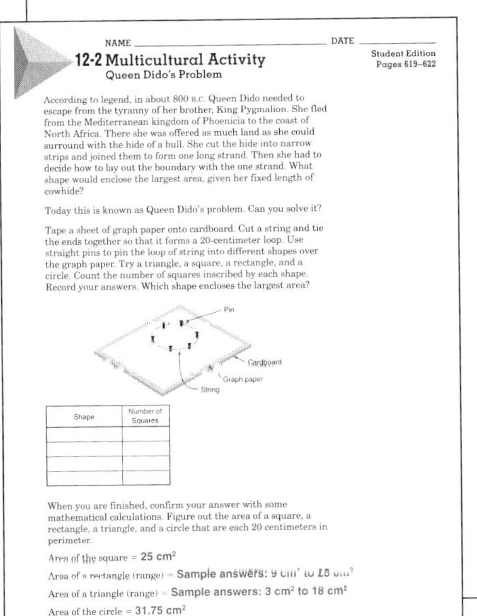

NAME ______________________ DATE __________

12-2 Multicultural Activity
Queen Dido's Problem

Student Edition Pages 619–622

According to legend, in about 800 B.C. Queen Dido needed to escape from the tyranny of her brother, King Pygmalion. She fled from the Mediterranean kingdom of Phoenicia to the coast of North Africa. There she was offered as much land as she could surround with the hide of a bull. She cut the hide into narrow strips and joined them to form one long strand. Then she had to decide how to lay out the boundary with the one strand. What shape would enclose the largest area, given her fixed length of cowhide?

Today this is known as Queen Dido's problem. Can you solve it?

Tape a sheet of graph paper onto cardboard. Cut a string and tie the ends together so that it forms a 20-centimeter loop. Use straight pins to pin the loop of string into different shapes over the graph paper. Try a triangle, a square, a rectangle, and a circle. Count the number of squares inscribed by each shape. Record your answers. Which shape encloses the largest area?

Shape	Number of Squares

When you are finished, confirm your answer with some mathematical calculations. Figure out the area of a square, a rectangle, a triangle, and a circle that are each 20 centimeters in perimeter.

Area of the square = **25 cm²**

Area of a rectangle (range) = **Sample answers: 9 cm² to 25 cm²**

Area of a triangle (range) = **Sample answers: 3 cm² to 18 cm²**

Area of the circle = **31.75 cm²**

Real-World Applications, 27, 28

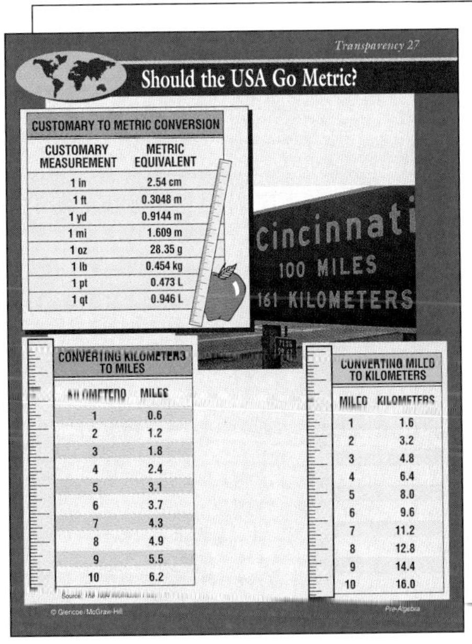

Transparency 27

Should the USA Go Metric?

CUSTOMARY TO METRIC CONVERSION	
CUSTOMARY MEASUREMENT	METRIC EQUIVALENT
1 in	2.54 cm
1 ft	0.3048 m
1 yd	0.9144 m
1 mi	1.609 m
1 oz	28.35 g
1 lb	0.454 kg
1 pt	0.473 L
1 qt	0.946 L

CONVERTING KILOMETERS TO MILES	
KILOMETERS	MILES
1	0.6
2	1.2
3	1.8
4	2.4
5	3.1
6	3.7
7	4.3
8	4.9
9	5.5
10	6.2

Source: [illegible]

CONVERTING MILES TO KILOMETERS	
MILES	KILOMETERS
1	1.6
2	3.2
3	4.8
4	6.4
5	8.0
6	9.6
7	11.2
8	12.8
9	14.4
10	16.0

© Glencoe/McGraw-Hill Pre-Algebra

Graphing Calculator Masters, p. 12

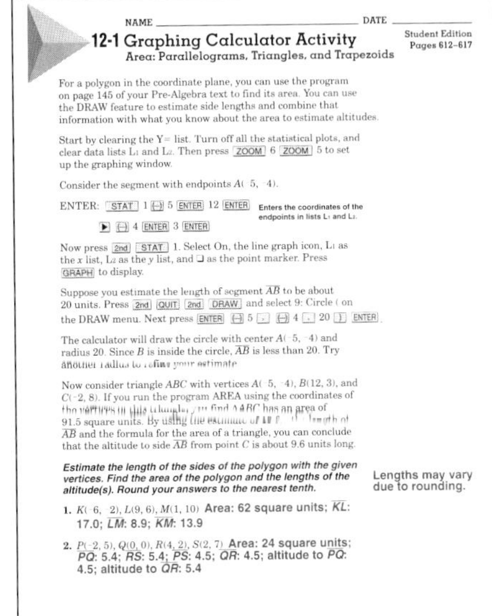

NAME ______________________ DATE __________

12-1 Graphing Calculator Activity
Area: Parallelograms, Triangles, and Trapezoids

Student Edition Pages 612–617

For a polygon in the coordinate plane, you can use the program on page 145 of your Pre-Algebra text to find its area. You can use the DRAW feature to estimate side lengths and combine that information with what you know about the area to estimate altitudes.

Start by clearing the Y= list. Turn off all the statistical plots, and clear data lists L1 and L2. Then press ZOOM 6 ZOOM 5 to set up the graphing window.

Consider the segment with endpoints A(−5, −4).

ENTER: STAT 1 (−) 5 ENTER 12 ENTER ▶ (−) 4 ENTER 3 ENTER — Enters the coordinates of the endpoints in lists L1 and L2.

Now press 2nd STAT 1. Select On, the line graph icon, L1 as the x list, L2 as the y list, and ❑ as the point marker. Press GRAPH to display.

Suppose you estimate the length of segment $\overline{AB}$ to be about 20 units. Press 2nd QUIT 2nd DRAW and select 9: Circle (on the DRAW menu. Next press ENTER (−) 5 , (−) 4 , 20) ENTER.

The calculator will draw the circle with center A(−5, −4) and radius 20. Since B is inside the circle, $\overline{AB}$ is less than 20. Try another radius to refine your estimate.

Now consider triangle ABC with vertices A(−5, −4), B(12, 3), and C(−2, 8). If you run the program AREA using the coordinates of the vertices of this triangle, you find △ABC has an area of 91.5 square units. By using the estimate of [illegible] length of $\overline{AB}$ and the formula for the area of a triangle, you can conclude that the altitude to side $\overline{AB}$ from point C is about 9.6 units long.

Estimate the length of the sides of the polygon with the given vertices. Find the area of the polygon and the lengths of the altitude(s). Round your answers to the nearest tenth.

Lengths may vary due to rounding.

1. K(−6, −2), L(9, 6), M(1, 10) **Area: 62 square units; $\overline{KL}$: 17.0; $\overline{LM}$: 8.9; $\overline{KM}$: 13.9**

2. P(−2, 5), Q(0, 0), R(4, 2), S(2, 7) **Area: 24 square units; $\overline{PQ}$: 5.4; $\overline{RS}$: 5.4; $\overline{PS}$: 4.5; $\overline{QR}$: 4.5; altitude to $\overline{PQ}$: 4.5; altitude to $\overline{QR}$: 5.4**

Tech Prep Applications Masters, pp. 23, 24

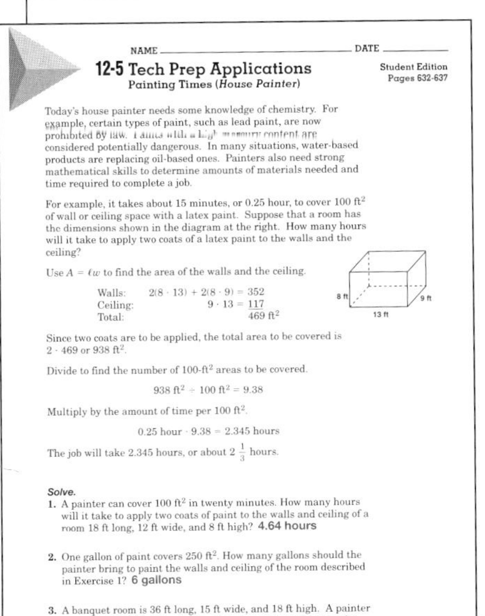

NAME ______________________ DATE __________

12-5 Tech Prep Applications
Painting Times (*House Painter*)

Student Edition Pages 632-637

Today's house painter needs some knowledge of chemistry. For example, certain types of paint, such as lead paint, are now prohibited by law. Paints with a high mercury content are considered potentially dangerous. In many situations, water-based products are replacing oil-based ones. Painters also need strong mathematical skills to determine amounts of materials needed and time required to complete a job.

For example, it takes about 15 minutes, or 0.25 hour, to cover 100 ft² of wall or ceiling space with a latex paint. Suppose that a room has the dimensions shown in the diagram at the right. How many hours will it take to apply two coats of a latex paint to the walls and the ceiling?

Use $A = \ell w$ to find the area of the walls and the ceiling.

Walls: $2(8 \cdot 13) + 2(8 \cdot 9) = 352$
Ceiling: $9 \cdot 13 = 117$
Total: 469 ft²

Since two coats are to be applied, the total area to be covered is $2 \cdot 469$ or 938 ft².

Divide to find the number of 100-ft² areas to be covered.

$$938 \text{ ft}^2 \div 100 \text{ ft}^2 = 9.38$$

Multiply by the amount of time per 100 ft².

$$0.25 \text{ hour} \cdot 9.38 = 2.345 \text{ hours}$$

The job will take 2.345 hours, or about $2\frac{1}{3}$ hours.

Solve.

1. A painter can cover 100 ft² in twenty minutes. How many hours will it take to apply two coats of paint to the walls and ceiling of a room 18 ft long, 12 ft wide, and 8 ft high? **4.64 hours**

2. One gallon of paint covers 250 ft². How many gallons should the painter bring to paint the walls and ceiling of the room described in Exercise 1? **6 gallons**

3. A banquet room is 36 ft long, 15 ft wide, and 18 ft high. A painter can cover 100 ft² in 25 minutes. How long will it take to apply two coats of paint to the walls and ceiling? **19.8 hours**

Activity Masters, pp. 12, 26, 45

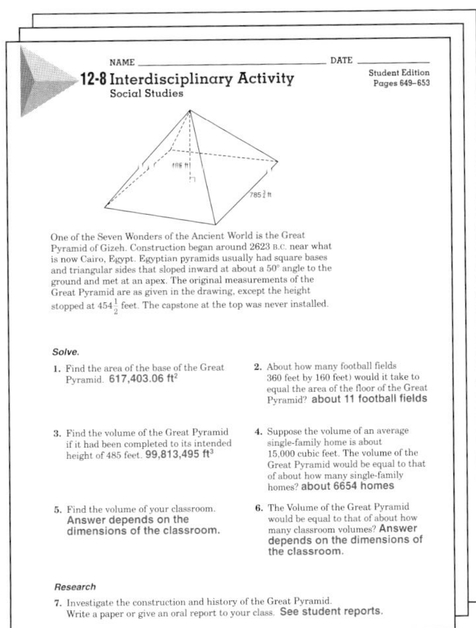

NAME ______________________ DATE __________

12-8 Interdisciplinary Activity
Social Studies

Student Edition Pages 649–653

One of the Seven Wonders of the Ancient World is the Great Pyramid of Gizeh. Construction began around 2623 B.C. near what is now Cairo, Egypt. Egyptian pyramids usually had square bases and triangular sides that sloped inward at about a 50° angle to the ground and met at an apex. The original measurements of the Great Pyramid are as given in the drawing, except the height stopped at $454\frac{1}{2}$ feet. The capstone at the top was never installed.

Solve.

1. Find the area of the base of the Great Pyramid. **617,403.06 ft²**

2. About how many football fields 360 feet by 160 feet) would it take to equal the area of the floor of the Great Pyramid? **about 11 football fields**

3. Find the volume of the Great Pyramid if it had been completed to its intended height of 485 feet. **99,813,495 ft³**

4. Suppose the volume of an average single-family home is about 15,000 cubic feet. The volume of the Great Pyramid would be equal to that of about how many single-family homes? **about 6654 homes**

5. Find the volume of your classroom. **Answer depends on the dimensions of the classroom.**

6. The Volume of the Great Pyramid would be equal to that of about how many classroom volumes? **Answer depends on the dimensions of the classroom.**

Research

7. Investigate the construction and history of the Great Pyramid. Write a paper or give an oral report to your class. **See student reports.**

COOPERATIVE LEARNING

Math Lab and Modeling Math Masters, pp. 71, 72, 87

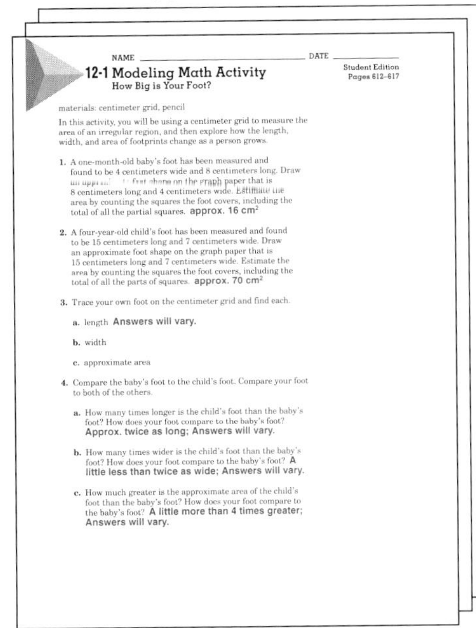

NAME ______________________ DATE __________

12-1 Modeling Math Activity
How Big is Your Foot?

Student Edition Pages 612–617

materials: centimeter grid, pencil

In this activity, you will be using a centimeter grid to measure the area of an irregular region, and then explore how the length, width, and area of footprints change as a person grows.

1. A one-month-old baby's foot has been measured and found to be 4 centimeters wide and 8 centimeters long. Draw an approximate foot shape on the graph paper that is 8 centimeters long and 4 centimeters wide. Estimate the area by counting the squares the foot covers, including the total of all the partial squares. **approx. 16 cm²**

2. A four-year-old child's foot has been measured and found to be 15 centimeters long and 7 centimeters wide. Draw an approximate foot shape on the graph paper that is 15 centimeters long and 7 centimeters wide. Estimate the area by counting the squares the foot covers, including the total of all the parts of squares. **approx. 70 cm²**

3. Trace your own foot on the centimeter grid and find each.
   a. length **Answers will vary.**
   b. width
   c. approximate area

4. Compare the baby's foot to the child's foot. Compare your foot to both of the others.
   a. How many times longer is the child's foot than the baby's foot? How does your foot compare to the baby's foot? **Approx. twice as long; Answers will vary.**
   b. How many times wider is the child's foot than the baby's foot? How does your foot compare to the baby's foot? **A little less than twice as wide; Answers will vary.**
   c. How much greater is the approximate area of the child's foot than the baby's foot? How does your foot compare to the baby's foot? **A little more than 4 times greater; Answers will vary.**

MAKING MATHEMATICS RELEVANT

This two-page introduction to the chapter provides students with an opportunity to explore other disciplinary topics and their applications to mathematics.

Background Information

Math and Archaeology in the News The large tomb was discovered when archaeologist Kent Weeks learned that the Egyptian government planned to expand a road and parking lot near the tomb of Ramses II. He persuaded the government to delay the project until he could investigate to make sure that there were no more caves or tombs under the proposed construction area. Ramses II is believed to be the pharaoh referred to in the Biblical story of the exodus of the Israelites from Egypt.

CHAPTER 12 Measuring Area and Volum

TOP STORIES in Chapter 12

In this chapter, you will:

- find the area of polygons and circles,
- investigate the relationship between area and probability,
- find the surface area and volume of prisms, cylinders, pyramids, and cones, and
- solve problems by making a model or drawing.

MATH AND ARCHAEOLOGY IN THE NEWS

An Amazing Find

Source: U.S. News & World Report, May, 29, 1995

For years, archaeologists believed that the major discoveries had already been made in Egypt's Valley of Kings. But Kent Weeks, an archaeologist from the American University in Cairo, believed there was something still hidden.

When the Egyptian government proposed adding some roadways to accommodate the tourist traffic, Weeks asked them to put off construction so he could check his hunch. After seven summers of digging, Weeks was proved right. He and his research team have discovered the largest and most elaborate tomb in Egypt. It may hold the remains of as many as fifty of the sons of Ramses II, the pharoah who ruled Egypt from 1279 to 1213 B.C. A large hall that is 50 feet square and 12 feet high opens to a T-shaped corridor with many chambers.

Many of the riches of the tomb have been stolen through the centuries, but the knowledge that it holds will be valuable to Egyptologists. Many rooms are yet to be explored, and there may be another level beneath this find.

Putting It into Perspective

1799 The Rosetta Stone is discovered during Napoleon's occupation of Egypt.

1800

1822 Jean François Champollion publishes research that enables archaeologists to translate ancient Egyptian writings.

1850

1879 Prehistoric wall paintings are found in a cave at Altamira, Spain.

Putting It into Perspective

The Rosetta Stone is an ancient Egyptian stone that was a key in deciphering old Egyptian writings (hieroglyphics). It contains the same inscription in three different languages—ancient hieroglyphic script, spoken Egyptian script and Greek.

EMPLOYMENT OPPORTUNITIES

Open a Window to the Past

DO YOU HAVE dreams of being a real-life Indiana Jones? Careers in archaeology are not always as dangerous as his adventures. Archaeologists study the remains of human cultures to learn what life was like and to better understand history. **Field archaeologists** travel to all parts of the world to search for artifacts. Many other archaeologists conduct research or teach in universities. A master's degree is required, a doctorate is preferred.

For more information, contact:
Society for American Archaeology
900 2nd Street NE, No. 12
Washington, D.C. 20002

Statistical Snapshot

Cross Section of the Great Pyramid of Khufu

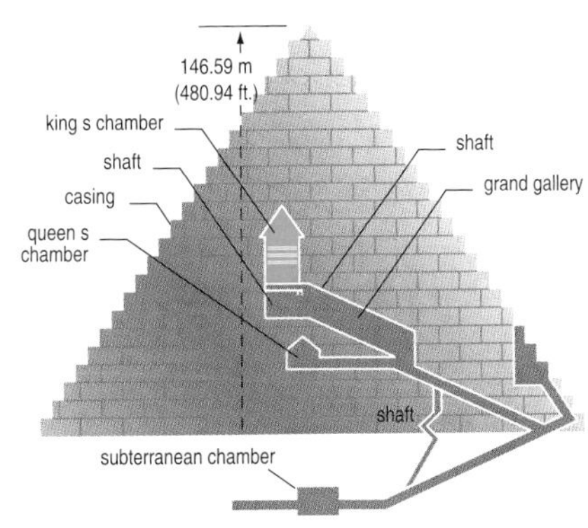

interNET CONNECTION For up-to-date information on archaeology, visit:
www.glencoe.com/sec/math/prealg/mathnet

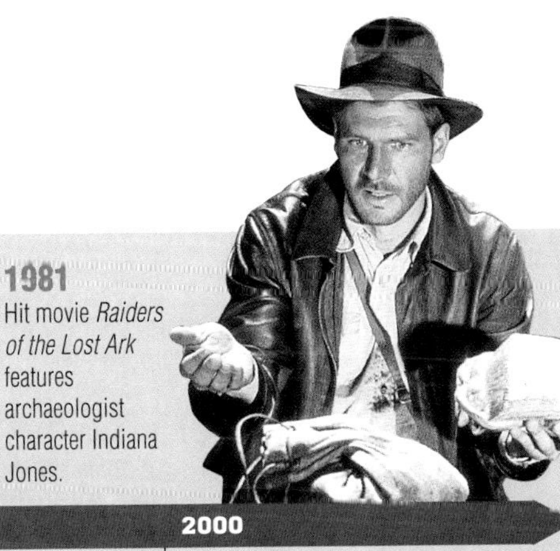

1900 — **1950** — **2000**

1922 Discovery of King Tutankhamen's tomb.

1952 British archaeologist Dame Kathleen Kenyon proves Jericho, Jordan, to be one of the oldest known human communities.

1981 Hit movie *Raiders of the Lost Ark* features archaeologist character Indiana Jones.

1995 Kent Weeks announces discovery of major tomb in Egypt's Valley of Kings.

EMPLOYMENT OPPORTUNITIES

Archaeologists who work for governments study cultural remains on public property and may become involved if public construction projects threaten the remains.

Statistical Snapshot

Ancient archaeological records have allowed scholars to accurately identify the long succession of rulers during much of the Egyptian empire's history. Statues, mummies, and other artifacts have enabled us to view actual physical likenesses of many of the Egyptian rulers. The most well-known leaders were King Tutankhamen (King Tut), Ramses II, and Cleopatra.

Cooperative Learning

Chapter Projects Two chapter projects are included in the *Investigations and Projects Masters*. In Chapter Project A, students extend the topic in the Chapter Opener. In Chapter Project B, students explore the history of mathematics, building, and archaeology. A student page and a parent letter are provided for each Chapter Project.

interNET CONNECTION

Glencoe has made every effort to ensure that the website links for *Pre-Algebra* at www.glencoe.com/sec/math/prealg/mathnet are current and contain appropriate content. However, these website links are not under Glencoe's control.

Investigations and Projects Masters, p. 73

NAME ____________ DATE ______

CHAPTER 12 Project A
The Perfect Home for the Living

The ancient Egyptian rulers spent years and fortunes building elaborate tombs for themselves with many rooms filled with their valuables. Suppose you had the same kinds of resources as the Pharaohs. Imagine designing ideal rooms for yourself that you could fill with things that you treasure. But, imagine that you could use this place while you were still alive! Design the home of your dreams.

1. Decide whether your home will be a house or an apartment, and think about whether it will have traditional rooms or loft-like areas or even odd-shaped rooms. List the kinds of rooms you want and how many you'll need. Think about whether your palace will fit on one floor or whether it will have several floors. Think about whether you like high ceilings, low ceilings, or even vaulted ceilings. Think about some of the features to include—a swimming pool, a movie theater, a bowling alley, a recording studio, and so on. Browse through magazines about luxury homes and estates to get ideas.
2. Once you've decided on the type of home and its features, sketch it on paper. Make changes and get it right. Remember, money is no object!
3. When you are happy with your sketch, draw an accurate floor plan, to scale, using graph paper. Choose a scale that makes sense. You may need to work with a partner on this. Keep in mind that you'll need as many floor plans as you have floors. Label all rooms and dimensions. You may find it useful to check out floor plans in newspapers and real estate flyers to see examples of how this is done.
4. Make a scale model or draw an accurate picture of your dream home. Remember to consider the heights of your ceilings.
5. Using the information in your floor plan and in your picture or model, write an appealing description of the home that includes all its features. This description must also include the area of the living space in square feet as well as its volume in cubic feet. Estimate as needed.
6. Put together a brochure that includes the floor plan, the model or picture, and the description of your ideal home. Display it for classmates to admire.

Chapter 12 **609**

12-1A LESSON NOTES

NCTM Standards: 1-4, 12, 13

Objective

Use a geoboard to find areas of shapes that are not rectangular.

Recommended Time

Demonstration and discussion: 30 minutes; Exercises: 30 minutes

Instructional Resources

For each student or group of students
- grid paper

Student Manipulative Kit
- geoboard
- geobands

Math Lab and Modeling Math Masters
- pp. 7–8 (grid paper)
- p. 50 (worksheet)

For teacher demonstration
Overhead Manipulative Resources

1 FOCUS

Draw an irregularly shaped figure on the chalkboard. Ask students how they would go about finding the area of the figure.

2 TEACH

When looking at the 3–4–5 right triangle, remind students not to assume that a partial square has an area of $\frac{1}{2}$ square unit. In fact, in this case, there are no such partial squares.

MATHEMATICS LAB **12-1A**

HANDS-ON ACTIVITY

Areas and Geoboards

A Preview of Lesson **12-1**

MATERIALS
- geoboard
- geobands
- grid paper

In Lesson 3-5, you learned a formula to find the area of a rectangle. But sometimes you need to find the area of shapes that are not rectangular. In this lab, we'll use a geoboard to find the areas of other polygons.

Activity 1 **Make these figures on a geoboard. If one square represents one square unit, find the area of each figure.**

c. $3\frac{1}{2}$ sq units

a.

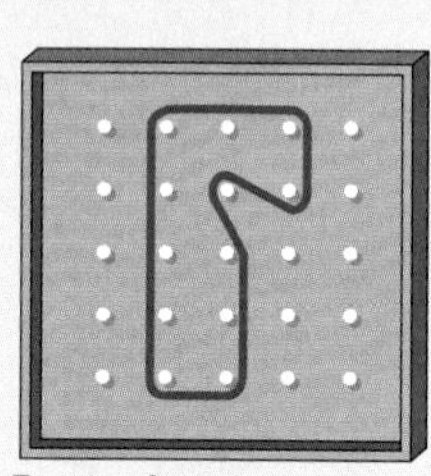

5 sq units

b.

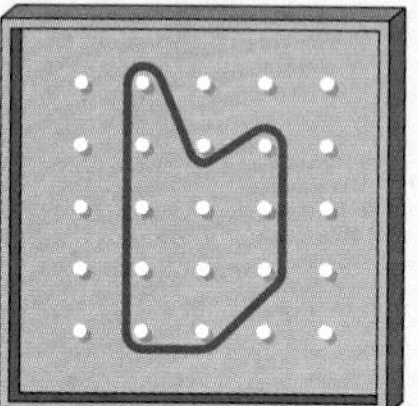

6 sq units

c.

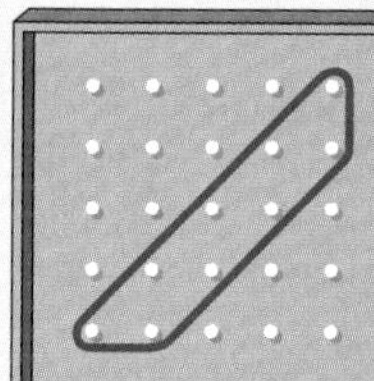

d.

11 sq units

e.

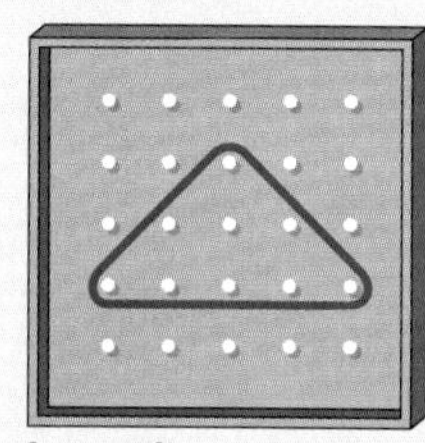

4 sq units

f.

5 sq units

g. Answers will vary.

g. Make a figure on the geoboard and ask your partner to find its area.

1. Sample answer: count whole-squares and half-squares.

1. How did you find the areas of the polygons in a–f?
2. Was your partner's polygon made up of whole- and half-squares, or did it include other parts of squares? **See students' work.**

Activity 2 Look at the triangle below. Suppose you make another triangle just like it to form a rectangle. You can find the area of the rectangle. Then divide by 2 to find the area of each triangle.

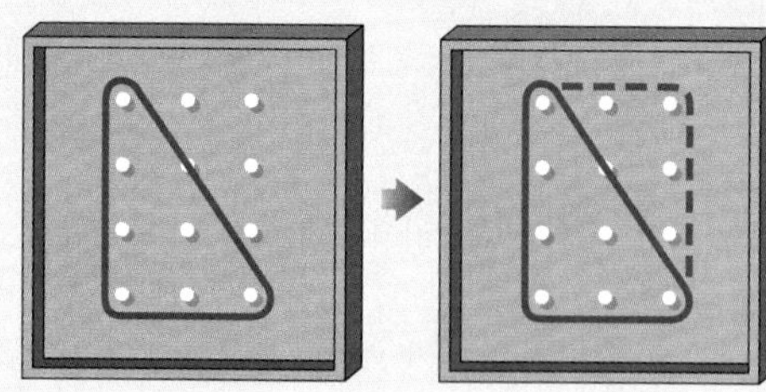

The area of the rectangle is 6 square units. So, the area of each triangle is 3 square units.

Cooperative Learning

This lesson offers an excellent opportunity for using cooperative learning groups. For more information on cooperative learning strategies and group management, see *Cooperative Learning in the Mathematics Classroom*, one of the titles in the Glencoe Mathematics Professional Series.

Your Turn **Find the area of each triangle.**

h. 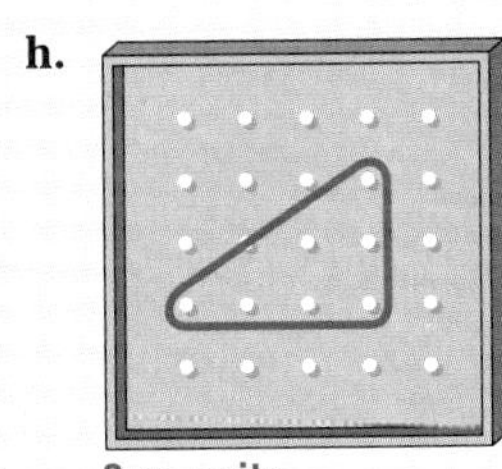

3 sq units

i. 6 sq units

j. 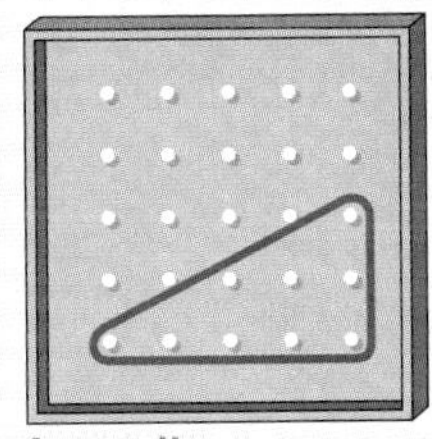

4 sq units

k. Make a right triangle on your geoboard and ask your partner to find the area. **See students' work.**

TALK ABOUT IT

3. Explain how you found the areas of the triangles. **Sample answer: find area of rectangle and divide by 2.**

Extension Here is another strategy for finding areas. Make the triangle at the right on a geoboard. Now build a rectangle around it as shown in step 2. You can subtract to find the area of the original triangle.

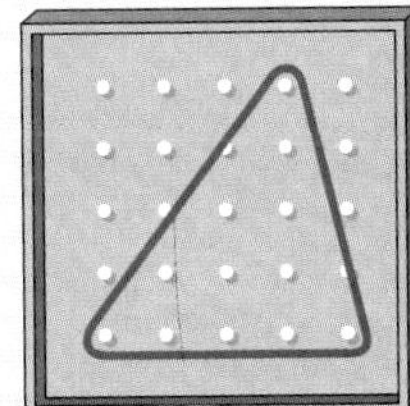

Step 1

The large rectangle has an area of 16 square units. The area of triangle a is half of 4 or 2 square units. The area of triangle b is half of 12 or 6 square units. The area of the original triangle is $16 - 2 - 6$ or 8 square units.

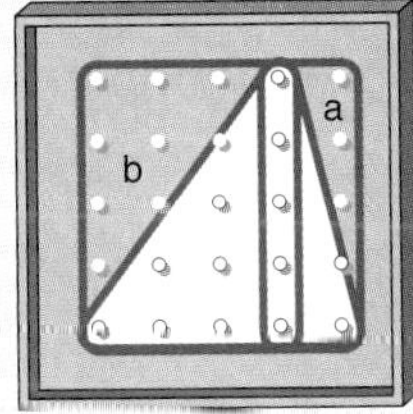

Step 2

Your Turn **Find the area of each polygon.**

l. 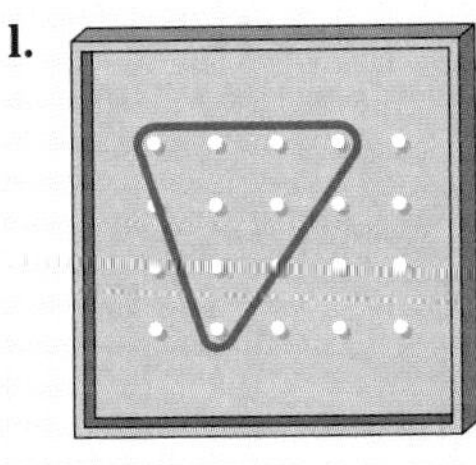

l. $4\frac{1}{2}$ sq units

m. 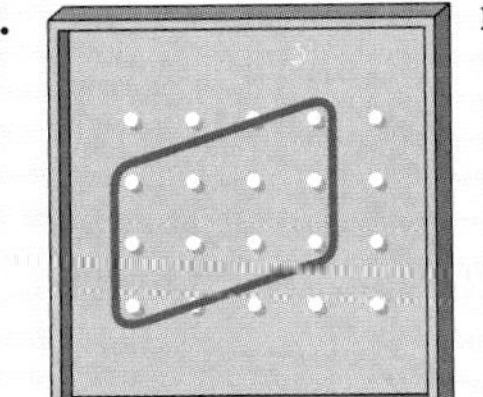

6 sq units

n. 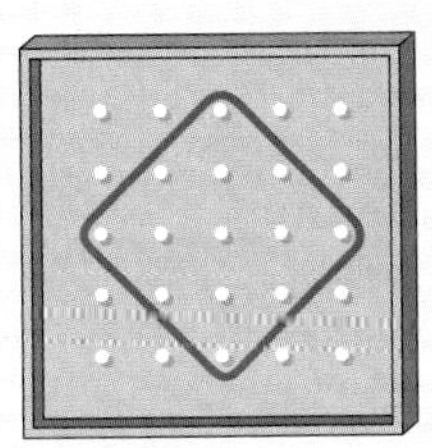

8 sq units

o. 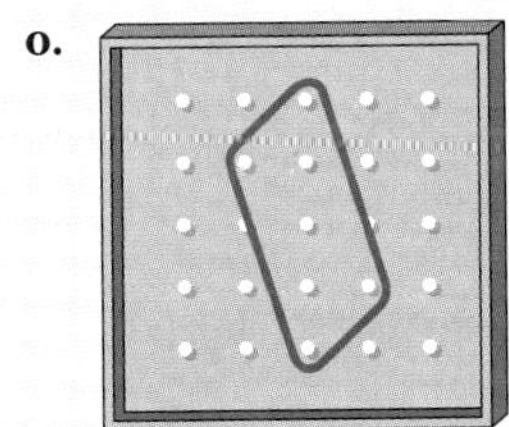

4 sq units

p. 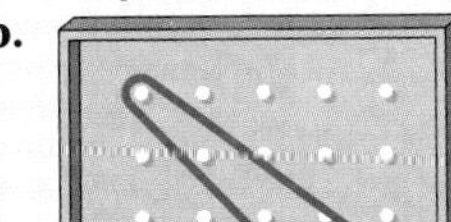

2 sq units

q. 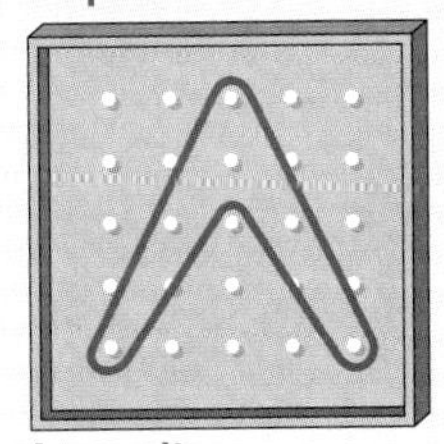

4 sq units

r. Make a polygon on your geoboard and ask your partner to find its area. **See students' work.**

4. Find the area of a rectangle around the polygon and subtract the areas not included in the polygon.

TALK ABOUT IT

4. How did you find the areas of the polygons?

3 PRACTICE/APPLY

Assignment Guide

Core: 1–4
Enriched: 1–4

4 ASSESS

Observing students working in cooperative groups is an excellent method of assessment. You may wish to ask a student at random from each group to explain how to model and solve a problem. Also watch for and acknowledge students who are helping others to understand the concept being taught.

12-1 LESSON NOTES

NCTM Standards: 1-4, 12, 13

Instructional Resources

- Study Guide Master 12-1
- Practice Master 12-1
- Enrichment Master 12-1
- Group Activity Card 12-1
- Activity Masters, p. 45
- Graphing Calculator Masters, p. 12
- Math Lab and Modeling Math Masters, p. 71

Transparency 12-1A contains the 5-Minute Check for this lesson; **Transparency 12-1B** contains a teaching aid for this lesson.

Recommended Pacing	
Standard Pacing	Day 2 of 13
Honors Pacing	Day 1 of 11
Block Scheduling*	Day 1 of 7

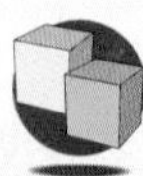

*For more information on pacing and possible lesson plans, refer to the ***Block Scheduling Booklet***.

1 FOCUS

5-Minute Check
(over Chapter 11)

The measure of ∠A is 39°

1. Find the measure of the complement of ∠A. **51°**
2. Find the measure of the supplement of ∠A. **141°**
3. Line *m* is perpendicular to line *n*. What is the measure of the angle formed by the two lines? **90°**

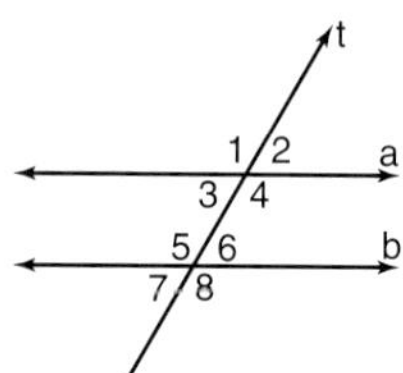

In the figure above, line *a* is parallel to line *b*, and line *t* is a transversal. The measure of ∠4 is 123°.

4. Find the measure of ∠6. **57°**
5. Find the measure of ∠5. **123°**

12-1 Area: Parallelograms, Triangles, and Trapezoids

Setting Goals: *In this lesson, you'll find areas of parallelograms, triangles, and trapezoids.*

Modeling with Manipulatives

MATERIALS

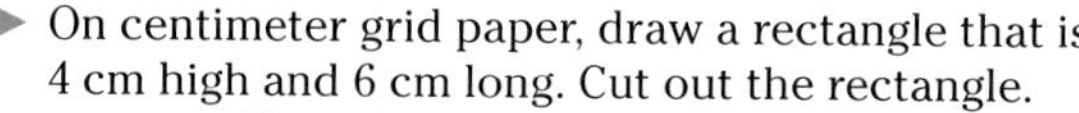

centimeter grid paper
scissors
tape

Remember Transformers™? Several years ago, Transformers were one of the hottest toys around. A Transformer starts out as one toy and can be changed into a different toy by moving its parts. In the activity below, you will transform one geometric shape into another.

Your Turn

Work with a partner.

- On centimeter grid paper, draw a rectangle that is 4 cm high and 6 cm long. Cut out the rectangle. Trace it on a piece of plain paper and label the measures on the sides as shown. Label it rectangle A.
- Use scissors to cut a triangle from one side of the original rectangle as shown.
- Tape the triangle to the other side of the rectangle to form a parallelogram. Trace the parallelogram on the plain paper next to the rectangle. Label it parallelogram A.
- Repeat this activity with two different rectangles (B and C), making two different parallelograms (B and C).

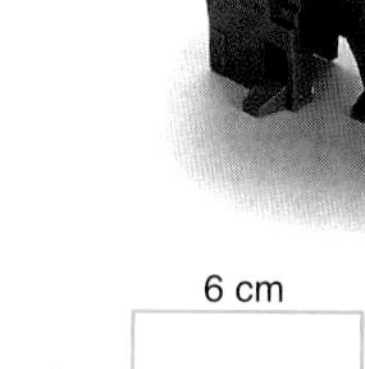

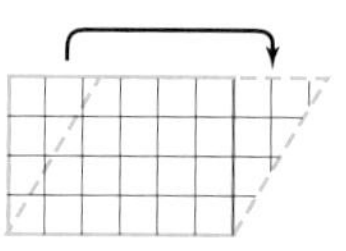

TALK ABOUT IT

a. Compare the area of each rectangle and the parallelogram made from it. **They are the same.**

b. How do the measures of the sides of each rectangle relate to the parallelogram formed? **See margin.**

c. How do you find the area of a rectangle? **multiply length by width**

d. What is the area of a rectangle that is 10 cm long and 2 cm high? **20 cm²**

e. Describe a parallelogram with the same area. **It has a base of 10 cm and a height of 2 cm.**

Learning the Concept

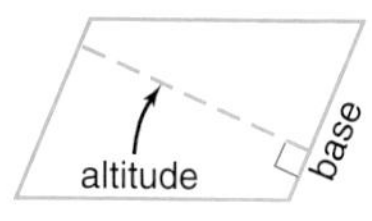

The modeling activity above suggests that you can find the area of a parallelogram by multiplying the measures of the base and the height. The base can be any side of the parallelogram. The height is the length of an **altitude**, a line segment perpendicular to the base with endpoints on the base and the side opposite the base.

Additional Answers Talk About It

b. The length of the rectangle is the same as the base of the parallelogram; the height of the rectangle is the same as the distance from the top to the bottom of the parallelogram.

GLENCOE Technology

Interactive Mathematics Tools Software

In these interactive computer lessons, students explore areas of parallelograms, triangles, and trapezoids. A **Computer Journal** gives students the opportunity to write about what they have learned.

For Windows & Macintosh

Area of a Parallelogram

In words:	If a parallelogram has a base of b units and a height of h units, then the area A is $b \cdot h$ square units.
In symbols:	$A = bh$

Example 1 **Find the area of each parallelogram.**

a.

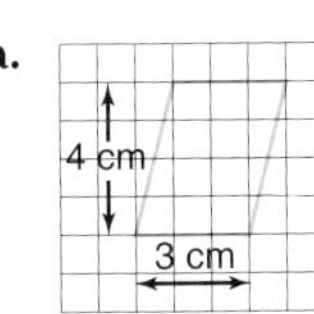

The base is 3 cm. The height is 4 cm.

$A = bh$ *Formula for area of a parallelogram*

$A = 3 \cdot 4$ *Replace b with 3 and h with 4.*

$A = 12$

The area is 12 cm^2.

b.

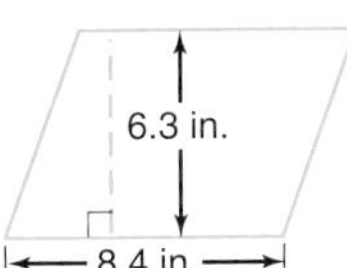

$A = bh$

$A = 8.4 \cdot 6.3$

$A = 52.92$

The area is about 53 in^2.

If you draw a diagonal in a parallelogram, you form two congruent triangles. So the area of each triangle is half the area of the parallelogram. The area of parallelogram $ABCD$ at the right is $8 \cdot 4$ or 32 square units.

So, the area of $\triangle ABD$ is $\frac{1}{2} \cdot 32$ or 16 square units.

Using the formula for the area of a parallelogram, we can find the formula for the area of a triangle.

THINK ABOUT IT

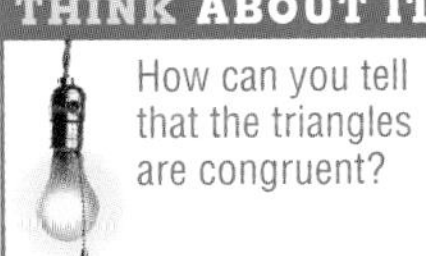

How can you tell that the triangles are congruent?

The three sides are congruent.

Area of a Triangle

In words:	If a triangle has a base of b units and a height of h units, then the area A is $\frac{1}{2} b \cdot h$ square units.
In symbols:	$A = \frac{1}{2}bh$

As with a parallelogram, you can use any side of a triangle as a base. The height is the length of the corresponding altitude, a line segment perpendicular to the base from the opposite vertex.

Notice that an altitude can be outside the triangle.

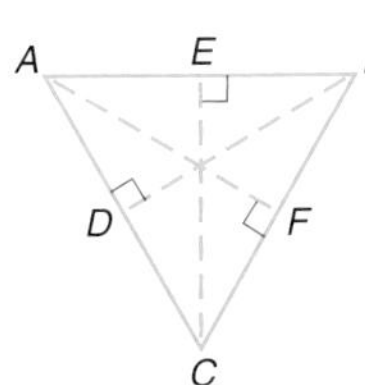

base	altitude
$\overline{AC}$	$\overline{BD}$
$\overline{AB}$	$\overline{CE}$
$\overline{BC}$	$\overline{AF}$

base	altitude
$\overline{NQ}$	$\overline{PR}$
$\overline{NR}$	$\overline{QS}$
$\overline{QR}$	$\overline{NT}$

Side $\overline{RQ}$ is extended to T

Side $\overline{NR}$ is extended to S

Motivating the Lesson

Hands-On Activity Remove the top and bottom of a rectangular box. Tack the bottom to a bulletin board. Hold the box up to the bottom on the bulletin board and change the box's shape to a parallelogram. Ask students whether the area of the bottom of the box when it is shaped like a parallelogram is the same as the area of the bottom tacked to the bulletin board.

2 TEACH

In-Class Example

For Example 1

Find the area of each parallelogram.

a.

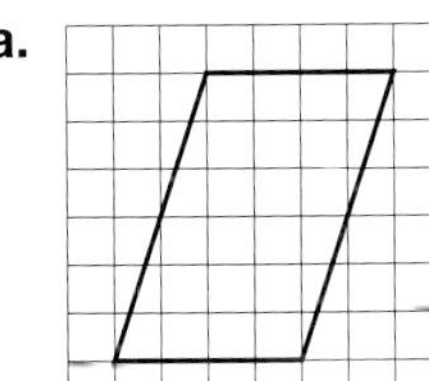

24 square units

b.

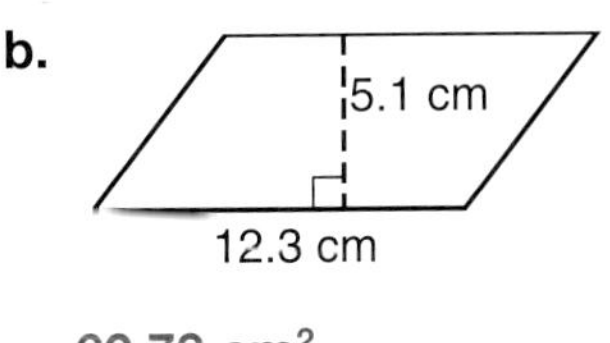

62.73 cm^2

Teaching Tip Point out to students that the length of a segment is measured in units, such as inches, centimeters, meters, and so on. The area of a figure is measured in **square** units, such as square inches (in^2), square centimeters (cm^2), and square miles (mi^2).

Classroom Vignette

"The area of a parallelogram may be introduced by having each student cut a rectangle out of graph paper and then find its area. They then cut the rectangle along a diagonal and rearrange the pieces to form a parallelogram. They now have a different shape and yet the same surface area. Defining the terms *base* and *height* leads them to discover the formula $A = bh$."

Niva Costa
Ambridge Area High School
Ambridge, PA

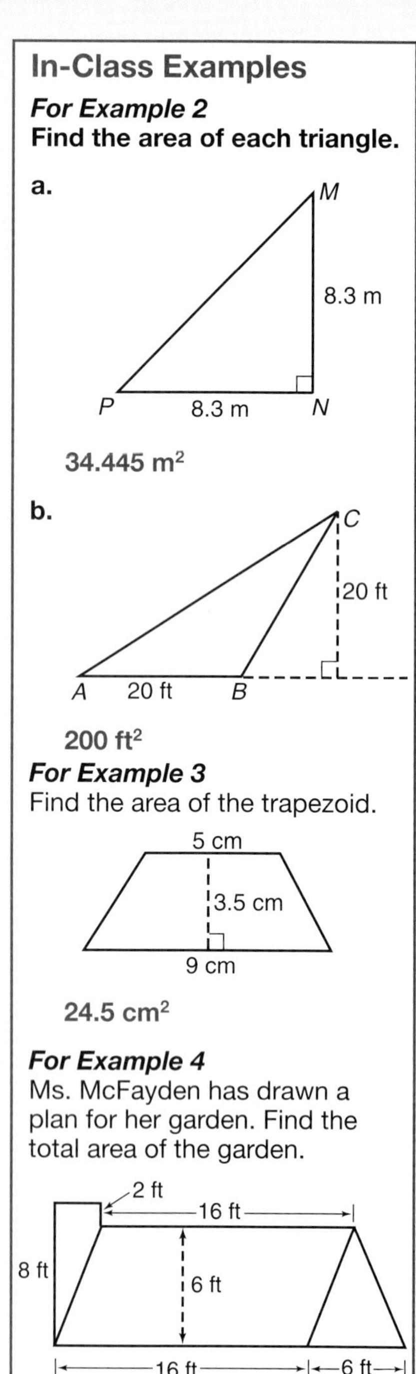

In-Class Examples

For Example 2
Find the area of each triangle.

a.

34.445 m²

b.

200 ft²

For Example 3
Find the area of the trapezoid.

24.5 cm²

For Example 4
Ms. McFayden has drawn a plan for her garden. Find the total area of the garden.

129 square feet

Example  **Find the area of each triangle.**

a. (3 cm, 4 cm)

The base is 4 cm long. The height is 3 cm.

$A = \frac{1}{2}bh$ *Formula for area of a triangle*

$A = \frac{1}{2}(4)(3)$ *Replace b with 4 and h with 3.*

$A = \frac{1}{2}(12)$

$A = 6$

The area of the triangle is 6 cm^2.

b. (6 cm, 5.4 cm)

$A = \frac{1}{2}bh$

$A = \frac{1}{2}(6)(5.4)$ *Replace b with 6 and h with 5.4.*

$A = 16.2$ *$0.5 \times 6 \times 5.4 = 16.2$*

The area of the triangle is 16.2 cm^2.

You can use triangles to find the area of a trapezoid. Remember, a trapezoid is a quadrilateral with exactly two parallel sides called **bases**. The height of a trapezoid is the distance between the bases. The height is the length of an **altitude**, a line segment perpendicular to both bases, with endpoints on the base lines.

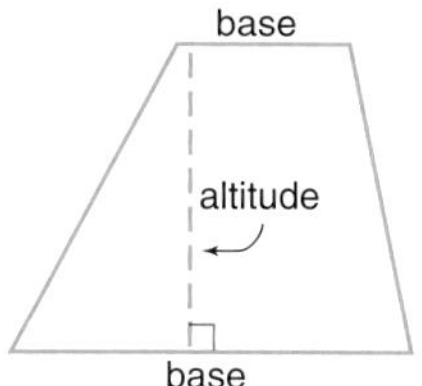

Connection to Algebra

On page 613, we drew a diagonal to separate a parallelogram into two congruent triangles. When you draw a diagonal in a trapezoid, the two triangles formed will *not* be congruent. But, the area of the trapezoid is still the sum of the areas of the triangles.

because the bases are parallel

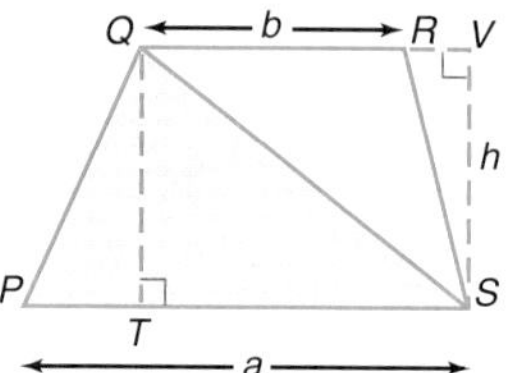

The triangles are $\triangle PQS$ and $\triangle QRS$.
The altitudes of these triangles, $\overline{QT}$ and $\overline{SV}$, are congruent. Both are h units long.
The base of $\triangle PQS$, $\overline{PS}$, is a units long.
The base of $\triangle QRS$, $\overline{QR}$, is b units long.

Area of trapezoid $PQRS$ = area of $\triangle PQS$ + area of $\triangle QRS$

$= \frac{1}{2}ah + \frac{1}{2}bh$

$= \frac{1}{2}h(a + b)$ *Distributive property*

Area of a Trapezoid

In words: If a trapezoid has bases of a units and b units and a height of h units, then the area A of the trapezoid is $\frac{1}{2} \cdot h \cdot (a + b)$ square units.

In symbols: $A = \frac{1}{2}h(a + b)$

Cooperative Learning

Round Table Have students work in groups of four. Each group folds a sheet of plain paper into fourths. The teacher then names an area measurement, such as 24 cm². One student in each group draws a figure with that area in one section of the paper. The paper is passed to other students, who, in turn, draw a different figure with the given area.

For more information on this strategy, see *Cooperative Learning in the Mathematics Classroom*, one of the titles in the Glencoe Mathematics Professional Series, p. 21.

Example 3 **Find the area of the trapezoid.**

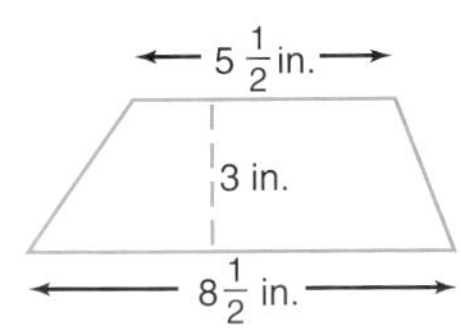

$A = \frac{1}{2}h(a + b)$

$A = \frac{1}{2} \cdot 3\left(8\frac{1}{2} + 5\frac{1}{2}\right)$ *Replace h with 3, a with $8\frac{1}{2}$, and b with $5\frac{1}{2}$.*

$A = \frac{1}{2} \cdot 3 \cdot 14$

$A = 21$

The area of the trapezoid is 21 in^2.

Example 4

Landscaping

Marisa works during the summer taking care of several of her neighbors' lawns. In addition to mowing, she also fertilizes the lawns. The label on the bag of the fertilizer she will be using indicates that one bag fertilizes 2000 square feet. How much fertilizer should she buy for the yard shown below?

The yard is in the shape of a trapezoid. The house and driveway are rectangles within the trapezoid that do not need fertilizer. Subtract the area of the rectangles from the area of the trapezoid to find the area that will be fertilized.

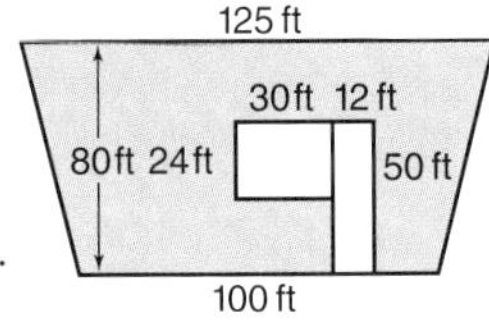

Area of the lot

$A = \frac{1}{2}h(a + b)$ *Formula for area of a trapezoid*

$A = \frac{1}{2} \cdot 80 \cdot (100 + 125)$ *Replace h with 80, a with 100, and b with 125.*

$A = \frac{1}{2} \cdot 80 \cdot 225$ *Use a calculator.*

$A = 9000$

Area of the house

$A = \ell w$ *Formula for area of a rectangle*

$A = 24 \cdot 30$

$A = 720$

Area of the driveway

$A = \ell w$

$A = 12 \cdot 50$

$A = 600$

$9000 - 720 - 600 = 7680$

The area to be fertilized is 7680 square feet. Since each bag of fertilizer covers 2000 square feet, Marisa will need $7680 \div 2000$ or about 4 bags of fertilizer for the yard.

HELP WANTED

Landscape architects design beautiful gardens for homes and businesses. For more information, contact:

American Association of Landscape Architects
4401 Connecticut Ave. NW, Fifth Floor
Washington, D.C. 20008

Checking Your Understanding

Communicating Mathematics

Read and study the lesson to answer these questions. **1–3. See margin.**

1. What is the difference between an altitude in a triangle and an altitude in a parallelogram or trapezoid?
2. **Draw** and label a diagram to show the three altitudes of a right triangle.
3. **Tell**, in your own words, how to find the area of a trapezoid.
4. **Draw** on centimeter grid paper a parallelogram, a triangle, and a trapezoid that each have an area of 16 square units. **See students' work.**

3 PRACTICE/APPLY

Checking Your Understanding

Exercises 1–11 are designed to help you assess your students' understanding through reading, writing, speaking, and modeling. You should work through Exercises 1–4 with your students and then monitor their work on Exercises 5–11.

Additional Answers

1. In a triangle, an altitude is drawn from a vertex and is perpendicular to the opposite side. In a parallelogram or trapezoid, an altitude is any segment perpendicular to the bases.
2. In triangle *ABC* below, the altitudes are $\overline{AB}$, $\overline{BC}$, and $\overline{BD}$.

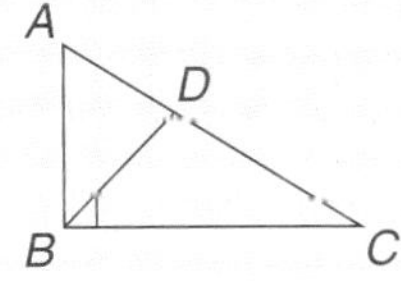

3. Sample answer: Multiply the sum of the bases by half the height.

Study Guide Masters, p. 100

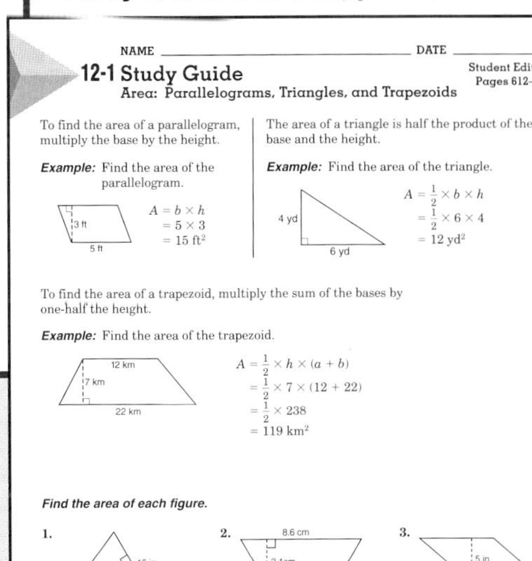

NAME ____ DATE ____

12-1 Study Guide Student Edition Pages 612–617

Area: Parallelograms, Triangles, and Trapezoids

To find the area of a parallelogram, multiply the base by the height.

Example: Find the area of the parallelogram.

$A = b \times h$
$= 5 \times 3$
$= 15$ ft^2

The area of a triangle is half the product of the base and the height.

Example: Find the area of the triangle.

$A = \frac{1}{2} \times b \times h$
$= \frac{1}{2} \times 6 \times 4$
$= 12$ yd^2

To find the area of a trapezoid, multiply the sum of the bases by one-half the height.

Example: Find the area of the trapezoid.

$A = \frac{1}{2} \times h \times (a + b)$
$= \frac{1}{2} \times 7 \times (12 + 22)$
$= \frac{1}{2} \times 238$
$= 119$ km^2

Find the area of each figure.

1. 112.5 in^2
2. 20.615 cm^2
3. 40 in^2
4. 737.5 km^2
5. 34 cm^2
6. 84 cm^2

Reteaching

Using Manipulatives Work in cooperative groups of three. Give each group some corrugated cardboard with a piece of quarter-inch grid paper attached. Using tacks and string, have a student make a parallelogram or triangle on the grid paper. Have another student name the base and altitude and the third student determine the area.

HELP WANTED

Have students discuss various gardens they have visited, such as those found at Cypress Gardens, Disney World, or Niagara Falls.

Error Analysis

In finding the area of a parallelogram or trapezoid, students may use the length of the non-base side as the height or altitude of the figure. Emphasize that the altitude is perpendicular to the base.

Assignment Guide	
Core:	13–25 odd, 28, 29, 31, 33–37
Enriched:	12–26 even, 27–37

For **Extra Practice**, see p. 769.

The red A, B, and C flags, printed only in the Teacher's Wraparound Edition, indicate the level of difficulty of the exercises.

Practice Masters, p. 100

NAME ______________________ DATE ____________

12-1 Practice
Area: Parallelograms, Triangles, and Trapezoids

Student Edition Pages 612–617

Find the area of each figure.

1. 8 ft; 11 ft; 13 ft — 104 ft²
2. 2 m; 4 m; 2.3 m — 2.3 m²
3. 27 ft; 15 ft; 12 ft — 292.5 ft²
4. 3.2 cm; 3.7 cm; 4.0 cm — 5.92 cm²
5. 14.8 cm; 10 cm; 12 cm — 60 cm²
6. 24 m; 24 m; 20 m; 24 m — 240 m²
7. 3.5 m; 4.3 m; 4 m; 4.1 m; 5 m — 17 m²
8. $\frac{5}{16}$ in.; $\frac{5}{8}$ in.; 1 in. — $\frac{5}{16}$ in²
9. 11 km; 7.2 km; 7.9 km — 68.04 km²

Find the area of each figure described below.

10. parallelogram: base, 10 cm; height, 8 cm **80 cm²**
11. parallelogram: base, $2\frac{3}{4}$ ft height, $1\frac{1}{2}$ ft **$4\frac{1}{8}$ ft²**
12. triangle: base, 14 cm height, 10 cm **70 cm²**
13. triangle: base, 8.6 m; height, 6.5 m **27.95 m²**
14. trapezoid: height, 4.6 m; bases, 8.2 m and 8 m **37.26 m²**
15. trapezoid: height, 5.4 km; bases, 13.7 km and 4.6 km **49.41 km²**

Guided Practice

State the measures of the base and height and how to find the area of each figure. Then find each area.

5. 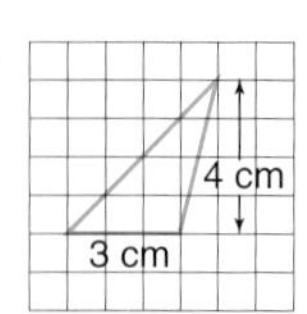

3 cm; 4 cm; 6 cm²

6.
4 cm
6 cm

6 cm; 4 cm; 24 cm²

7. 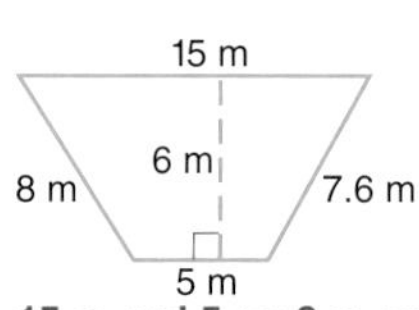

15 m and 5 m; 6 m and 60 m²

Find the area of each figure described below.

8. triangle: base, 10 cm; height, 4.5 cm — 22.5 cm²
9. parallelogram: base, 12 cm; height, 18 cm — 216 cm²
10. trapezoid: bases, $6\frac{1}{2}$ ft and 11 ft; height, 5 ft — $43\frac{3}{4}$ ft²
11. **Geography** The state of Indiana is shaped almost like a trapezoid. The distances along the east and west boundaries are about 200 miles and about 280 miles, and the east-west distance is about 140 miles.
    a. Estimate the area of the state. — 11a. about 33,600 mi²
    b. Look up the actual area in an encyclopedia and compare it to your estimate. The actual area is 35,936 mi².

Exercises: Practicing and Applying the Concept

Independent Practice

Find the area of each figure.

A

12. 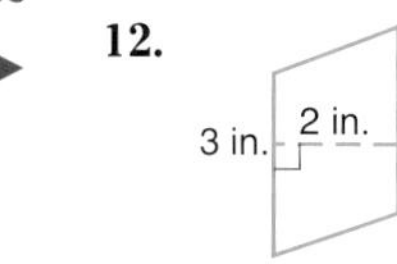

6 in²

13. 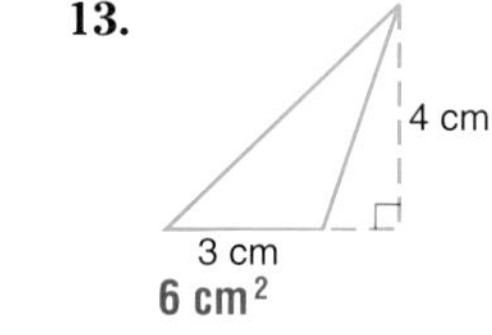

6 cm²

14. 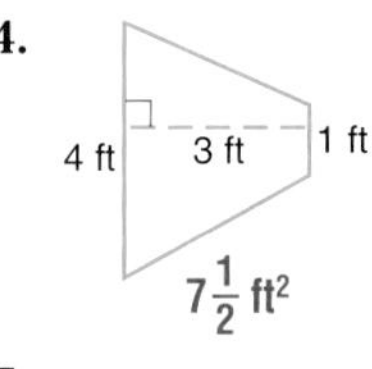

$7\frac{1}{2}$ ft²

B

Estimation
Mental Math
Calculator
Paper and Pencil

15. 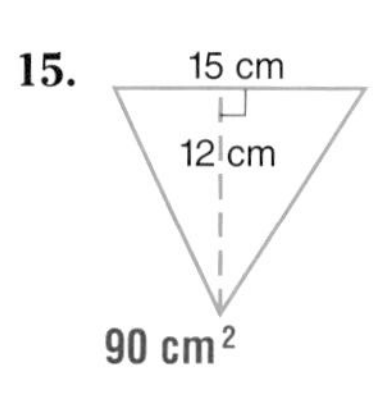

90 cm²

16. 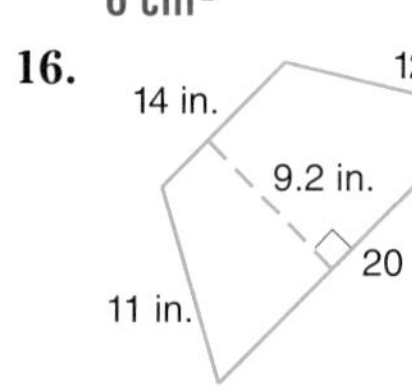

156.4 in²

17. 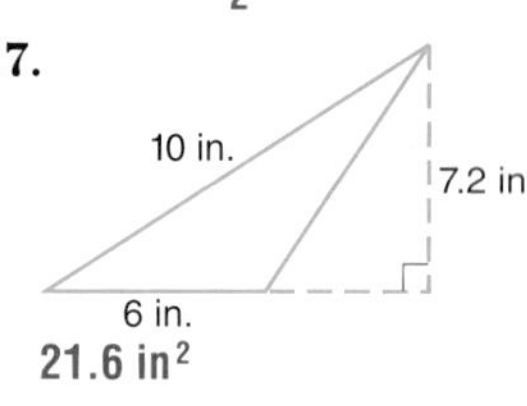

21.6 in²

18. 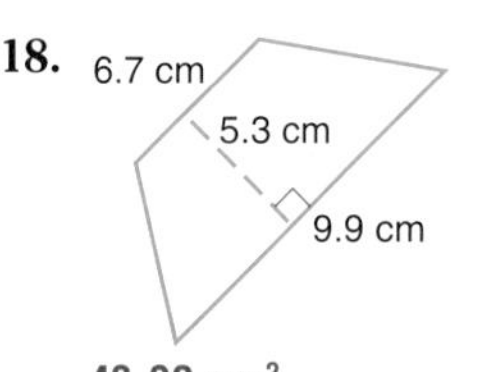

43.99 cm²

19. 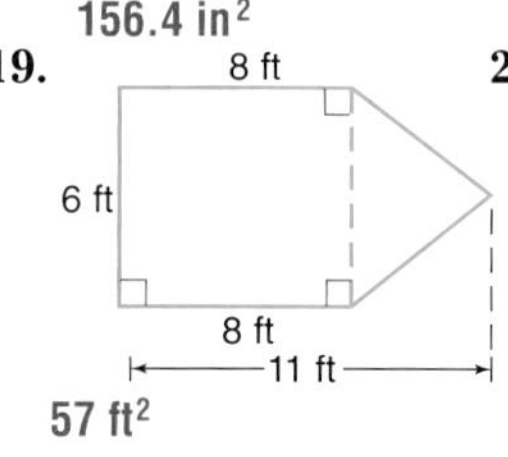

57 ft²

20. 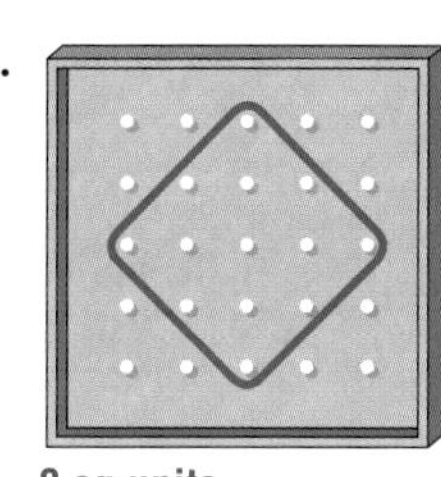

8 sq units

Find the area of each figure described below.

C

21. trapezoid: height, 2 in.; bases, 3 in. and 6 in. **9 in²**
22. triangle: base, 10 in.; height, 7 in. **35 in²**

Tech Prep

Research Assistant Research assistants work with physicists and other scientists as they study various phenomenon. A research assistant may be involved in a number of activities including assisting with experiments, gathering data from various sources, and organizing collected data. For more information on tech prep, see the *Teacher's Handbook*.

Group Activity Card 12-1

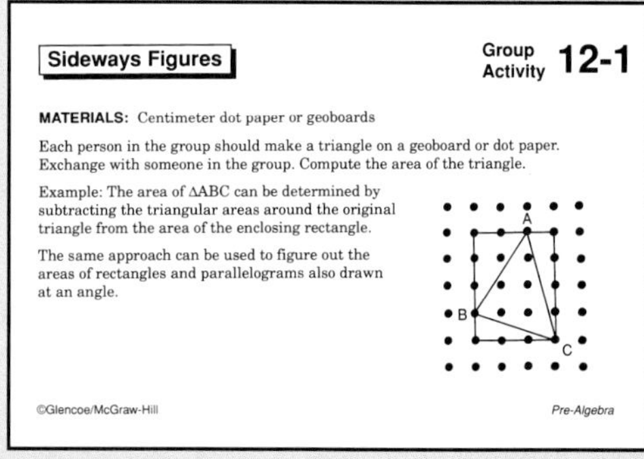

Sideways Figures — Group Activity **12-1**

MATERIALS: Centimeter dot paper or geoboards

Each person in the group should make a triangle on a geoboard or dot paper. Exchange with someone in the group. Compute the area of the triangle.

Example: The area of ΔABC can be determined by subtracting the triangular areas around the original triangle from the area of the enclosing rectangle.

The same approach can be used to figure out the areas of rectangles and parallelograms also drawn at an angle.

©Glencoe/McGraw-Hill — Pre-Algebra

23. parallelogram: base $3\frac{1}{4}$ ft; height, 2 ft **$6\frac{1}{2}$ ft²**

24. triangle: base 0.8 km; height, 2 km **0.8 km²**

25. trapezoid: height, 3.5 m; bases 10 m and 11 m **36.75 m²**

26. parallelogram: base, 25 yd; height, 11 yd **275 yd²**

27. In Exercise 20, you may have found the area of the square by finding the areas of two triangles and adding them. Show that this is the same as $A = \frac{1}{2}d^2$ where d is the length of a diagonal. **See margin.**

Critical Thinking

28. Explain or demonstrate how the formula for the area of a trapezoid can be used to find the areas of parallelograms and triangles. **28. See margin.**

Applications and Problem Solving

29. **Consumerism** Carpeting is sold in several standard widths, but since room sizes are not standard, there frequently is wasted carpeting, for which the home owner must pay.

a. If you are carpeting a room that is 14 feet by 22 feet, what is the area of the room? **308 ft²**

b. If the carpet you have chosen is sold in 12 foot and 15 foot widths, which size will result in less waste? **15 ft width**

30. **Geography** The state of Tennessee is shaped almost like a parallelogram. The distance along the north boundary is 447 miles, and the north-south distance is 116 miles. Estimate the area of the state. **about 51,852 mi²**

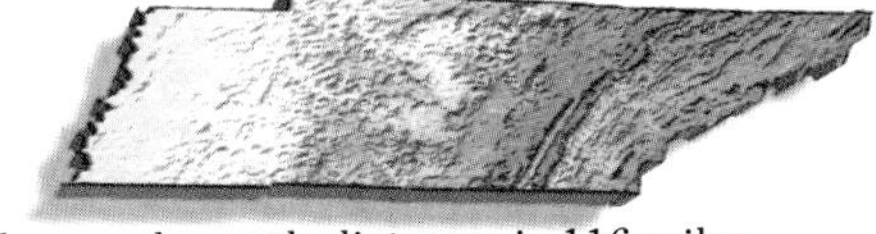

31. **Pet Care** In many communities, pets must be kept in a fenced area, or on a leash. If you had 48 feet of fencing to fence a rectangular area, describe the shape you would choose in order to give your pet the greatest amount of area. Justify your answer. **See margin.**

32. **Advertising** Use the formula in Exercise 27 to determine whether the claim about the viewing area in the advertisement below is accurate. Remember that television screens are measured diagonally and that the diagonals of a square are congruent. **See margin.**

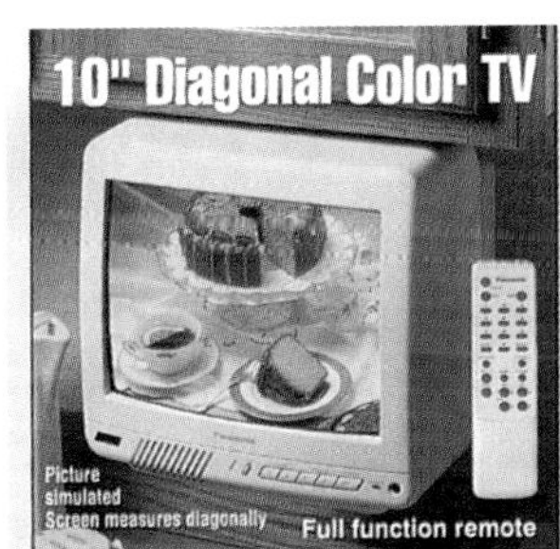

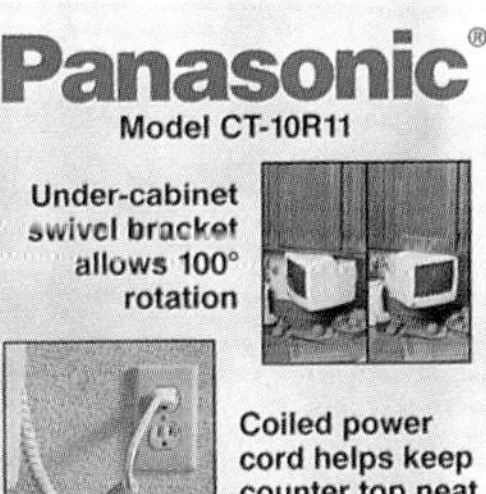

Panasonic
Model CT-10R11
Under-cabinet swivel bracket allows 100° rotation
Coiled power cord helps keep counter top neat
Compact cabinet with 24% more viewing area than a 9" set

Mixed Review

33. **Geometry** How many lines of symmetry does a regular pentagon have? (Lesson 11-9) **5**

34. Estimate 26% of 120. (Lesson 9-8) **30**

35. Solve $n = -9.45 \div -4.5$. (Lesson 6-5) **2.1**

36. Write $(-4)(-4)(-4)$ using exponents. (Lesson 4-2) **$(-4)^3$**

37. Rewrite $3 + (x + 4)$ using the associative and commutative properties. Then simplify. (Lesson 1-4) **$x + 7$**

4 ASSESS

Closing Activity

Speaking Draw different-sized triangles, parallelograms, and trapezoids on the board. Have students name the base and height and then give the area of the figure.

Additional Answers

27. $A_1 = \frac{1}{2}h_1d$; $A_2 = \frac{1}{2}h_2d$
$A_1 + A_2 = \frac{1}{2}d(h_1 + h_2) =$
$\frac{1}{2}d(d) = \frac{1}{2}d^2$

28. $A = \frac{1}{2}h(a + b)$; In a triangle, one "base" = 0, so $A = \frac{1}{2}h(0 + b)$, then $A = \frac{1}{2}hb$ or $A = \frac{1}{2}bh$. In a parallelogram, the bases are congruent, so a = b. Thus, $A = \frac{1}{2}h(b + b)$ or $A = \frac{1}{2}h(2b)$, then $A = bh$.

31. A shape as close as possible to a square gives the greatest area; see students' explanations.

Enrichment Masters, p. 100

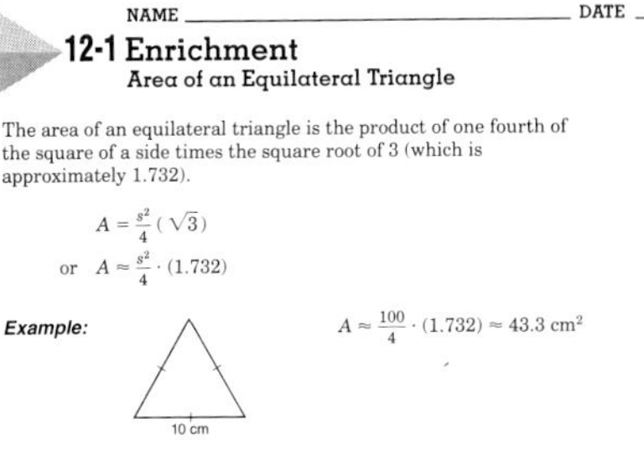

NAME ____________ DATE ____________

12-1 Enrichment Student Edition Pages 612–617

Area of an Equilateral Triangle

The area of an equilateral triangle is the product of one fourth of the square of a side times the square root of 3 (which is approximately 1.732).

$A = \frac{s^2}{4}(\sqrt{3})$

or $A \approx \frac{s^2}{4} \cdot (1.732)$

Example: $A \approx \frac{100}{4} \cdot (1.732) \approx 43.3$ cm²

10 cm

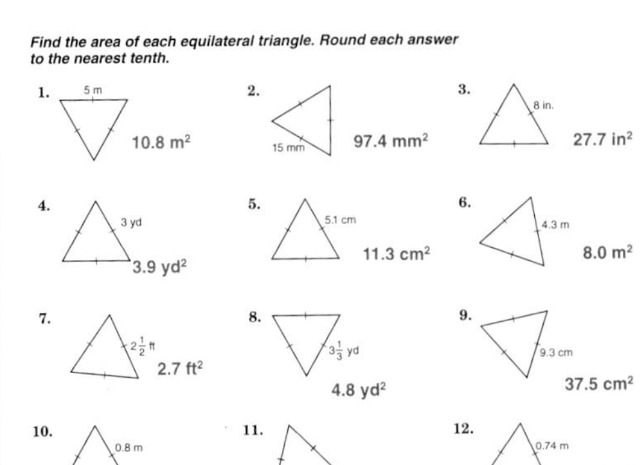

Find the area of each equilateral triangle. Round each answer to the nearest tenth.

1. 5 m — 10.8 m²
2. 15 mm — 97.4 mm²
3. 8 in. — 27.7 in²
4. 3 yd — 3.9 yd²
5. 5.1 cm — 11.3 cm²
6. 4.3 m — 8.0 m²
7. $2\frac{1}{2}$ ft — 2.7 ft²
8. $3\frac{1}{3}$ yd — 4.8 yd²
9. 9.3 cm — 37.5 cm²
10. 0.8 m — 0.3 m²
11. $5\frac{2}{3}$ ft — 13.9 ft²
12. 0.74 m — 0.2 m²

Extension

Using Diagrams A carpenter makes a picture frame from two sets of identical isosceles trapezoids. She cuts the trapezoids from a 2-inch wood strip. Make a diagram to show the picture frame needed for an 8-inch by 10-inch picture. Find the area of the wood in the frame. **88 in²**

Additional Answer

32. $A_1 = \frac{1}{2}(9^2) = 40.5$ in²
$A_2 = \frac{1}{2}(10^2) = 50$ in²
percent of increase
$= \frac{50 - 40.5}{40.5} \approx 23\%$; The advertisement is close to correct.

12-1B LESSON NOTES

NCTM Standards: 1-4, 12, 13

Objective
Explore and construct fractals

Recommended Time
Demonstration and discussion: 15 minutes; Exercises: 30 minutes

Instructional Resources
For each student or group of students
Student Manipulative Kit
- ruler
- compass
- protractor

Math Lab and Modeling Math Masters
- p. 18 (protractors)
- p. 51 (worksheet)

For teacher demonstration
Overhead Manipulative Resources

1 FOCUS

Motivating the Lesson
Draw a large equilateral triangle on the board. Draw three lines connecting the midpoints of the sides to create a figure like the one below.

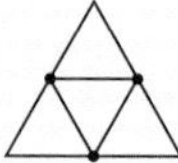

Have different students draw the three lines in each of the three corner triangles you created. Ask students what pattern they see. Do not erase.

2 TEACH

Teaching Tip After completing Lesson 12-1B, return to the large equilateral triangle from the Motivating the Lesson Activity. Repeat the procedure for each of the three-corner triangles drawn by students. Then shade each corner triangle. This design is the start of a fractal called the *Sierpinski triangle*.

HANDS-ON ACTIVITY

12-1B Fractals

An Extension of Lesson **12-1**

MATERIALS
ruler
compass
protractor

The design shown at the right is called the Koch snowflake. It is made up of three rotated copies of the Koch curve, introduced by Helge von Koch, a mathematician who worked on fractal geometry. Fractals are self-similar. That is, if you look at any part of a fractal, you will see a miniature replica of the larger design. A tree of broccoli is an example of a fractal in nature. Each branch of the tree is a miniature replica of the entire tree.

Your Turn Work with a partner to construct a Koch curve.

a. Use a ruler to draw a line segment about 6 inches long. This is stage 0 of the curve.

b. Use the ruler to separate the line into three congruent segments.

c. Set the compass to the length of one segment. Then draw arcs from each end of the middle segment as shown at the right. Use the point where the arcs intersect to draw an equilateral triangle. Then erase the base of the triangle. This is stage 1.

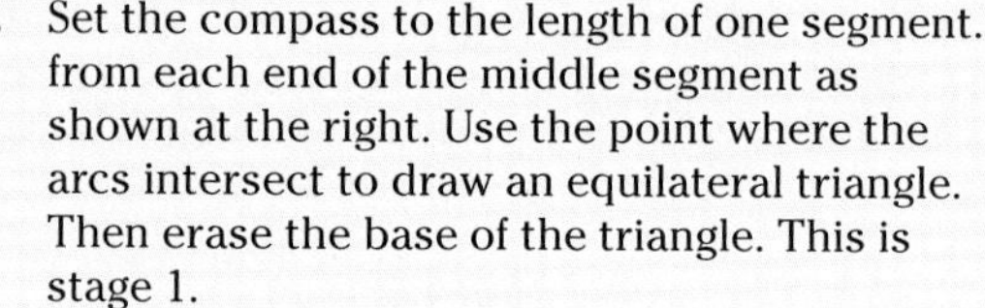

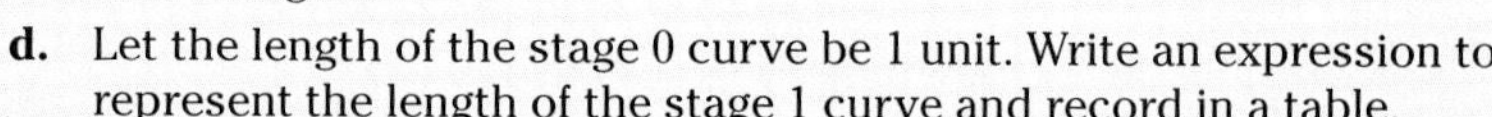

d. Let the length of the stage 0 curve be 1 unit. Write an expression to represent the length of the stage 1 curve and record in a table.

e. Repeat step c on the sides of stage 1 to produce a stage 2 figure. Again determine the length and record in the table.

f. Repeat step c once more to produce a stage 3 figure. Determine the length and record in the table.

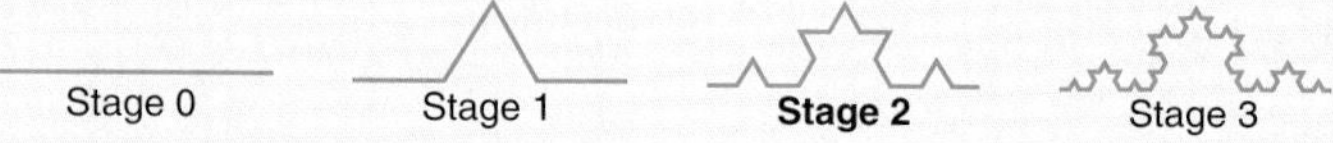

TALK ABOUT IT

1. Each section is a replica of the whole curve.

1. Why do you think the Koch curve is considered to be self-similar?
2. What happens to the length in each stage? **increases by factor of $\frac{4}{3}$**
3. Suppose the curve is continued indefinitely. Make a conjecture about the length. **infinite**
4. Research fractal geometry. Then write your own definition of a fractal. **See students' work.**

Extension

5. The Koch curve is produced by repeating step c above an infinite number of times. The Koch snowflake is made from three rotated copies of the curve. Stage 0 is an equilateral triangle. Use your constructions to produce a stage 1, stage 2, and stage 3 snowflake. Suppose the area of the stage 0 triangle is 1 square unit. Determine the area of each stage. Then make a conjecture about the area of the Koch snowflake.

5. See students' work. The area of the snowflake increases with every stage, but it does approach a limit.

3 PRACTICE/APPLY

Assignment Guide

Core: 1–5
Enriched: 1–5

4 ASSESS

Observing students working in cooperative groups is an excellent method of assessment.

12-2 Area: Circles

Setting Goals: *In this lesson, you'll find areas of circles.*

Modeling with Manipulatives

MATERIALS

- centimeter grid paper
- compass
- ruler
- protractor

Pizza, anyone? Most pizzas are round and come in different sizes, measured by diameter. Have you ever wondered whether one size would give you more pizza for your money? To find out, you need to find the area of a circle. This activity will help you understand how to find the area of a circle.

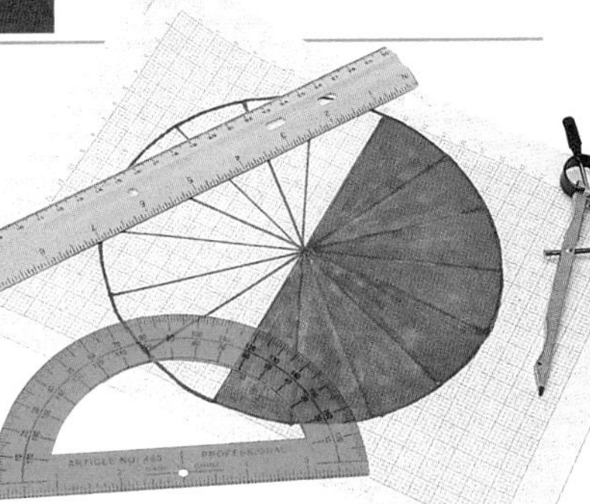

Your Turn Work with a partner.

- ▶ On centimeter grid paper, use a compass to construct a circle with a diameter of 20 centimeters. Do this by putting the point of the compass at the corner of a grid square and adjusting the compass so it is 10 centimeters wide.
- ▶ Count the grid squares and the parts of grid squares that lie inside the circle. Use your judgment on how to count each partial grid square. Record this estimate of the area of the circle.
- ▶ With a protractor, straightedge, and pencil, separate the circle into 16 congruent parts as shown in the photograph above. Shade one half of the circle as shown.
- ▶ Cut out the parts and arrange them in a side-by-side pattern as shown. When you have finished, you should have a shape that looks like a parallelogram.

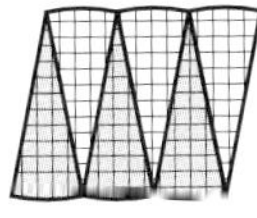

TALK ABOUT IT

a. What part of the circle makes up the base of the parallelogram? **half the circumference**

b. What part of the circle is the height of the parallelogram? **radius**

c. Use a ruler to find the approximate measure of the base and height.

c. 30 cm, 10 cm

d. What is the approximate area of the parallelogram? **$\approx$300 cm^2**

e. Compare to the estimate you found by counting grid squares. **They should be about the same.**

Learning the Concept

In the activity, you separated the circle into parts and made a figure that looks like a parallelogram. The circle has the same area as that figure.

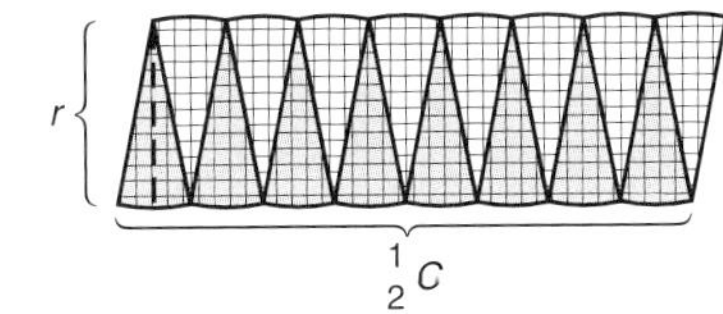

Alternative Teaching Strategies

Student Diversity Allow students to create various artistic circular designs, and then determine the area of various parts of the design. For example, in the design at the right, $\frac{1}{2}$ of the circle is shaded.

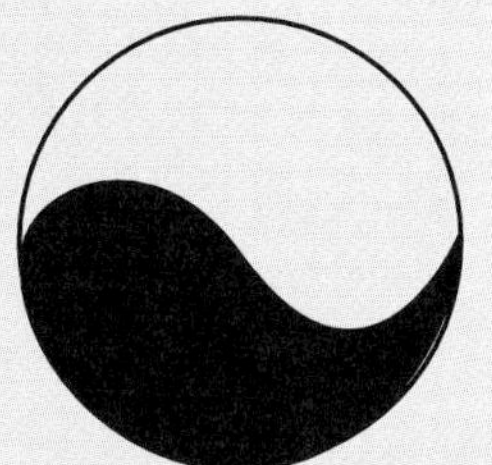

12-2 LESSON NOTES

NCTM Standards: 1-4, 12, 13

Instructional Resources

- Study Guide Master 12-2
- Practice Master 12-2
- Enrichment Master 12-2
- Group Activity Card 12-2
- Assessment and Evaluation Masters, p. 323
- Activity Masters, p. 26
- Math Lab and Modeling Math Masters, p. 87
- Multicultural Activity Masters, p. 23
- Real-World Applications, 27

Transparency 12-2A contains the 5-Minute Check for this lesson; **Transparency 12-2B** contains a teaching aid for this lesson.

Recommended Pacing	
Standard Pacing	Day 4 of 13
Honors Pacing	Day 3 of 11
Block Scheduling*	Day 2 of 7

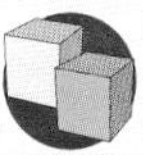

*For more information on pacing and possible lesson plans, refer to the ***Block Scheduling Booklet***.

1 FOCUS

5-Minute Check
(over Lesson 12-1)

Find the area of each figure described below.

1. triangle; base, 8 yd; height, 6 yd **24 yd^2**
2. triangle; base 2.6 km; height 4 km **5.2 km^2**
3. parallelogram; base $6\frac{2}{3}$ ft; height $4\frac{1}{2}$ ft **30 ft^2**
4. rectangle; length $5\frac{1}{2}$ in.; width $2\frac{2}{5}$ in. **$13\frac{1}{5}$ in^2**
5. trapezoid; bases 3 ft and 5 ft; height, 2 ft **8 ft^2**

Motivating the Lesson

Questioning Ask students which has the larger area, a square that is 4 inches on each side or a circle with a 4-inch diameter. **square**

2 TEACH

In-Class Examples

For Example 1

Find the area of each circle. Round to the nearest tenth.

a.

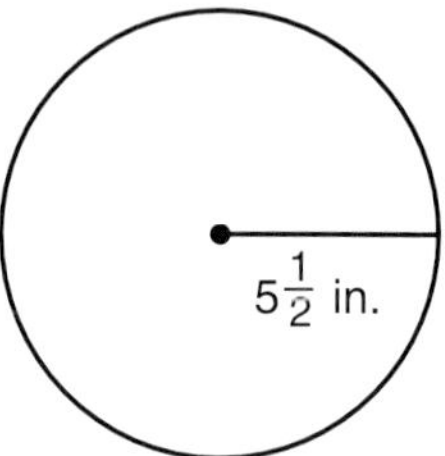

95.0 in²

b.

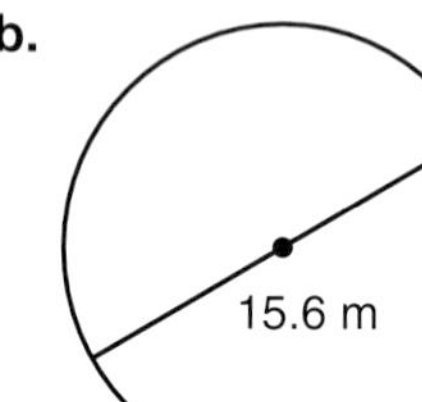

191.1 m²

For Example 2

A 3-foot walkway is to be built around a circular flower garden that has a 20-foot diameter. What will the area of the walkway be? **216.8 ft²**

Study Guide Masters, p. 101

NAME ______ DATE ______

12-2 Study Guide
Area: Circles

Student Edition Pages 619–622

The area of a circle with radius r can be found by using the formula below.

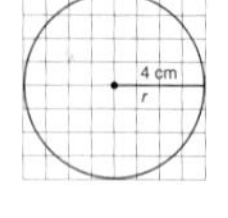

$A = \pi r^2$

Find the area of the circle at the left.

$A = \pi(4)^2$

$= \pi \cdot 16$

≈ 50.27 cm²

Find the area of each circle. Round to the nearest tenth.

1. 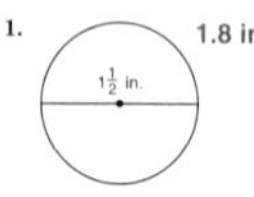

1.8 in²

2. 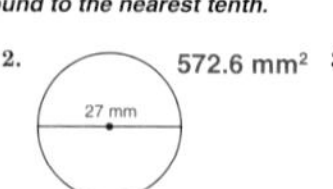

572.6 mm²

3. 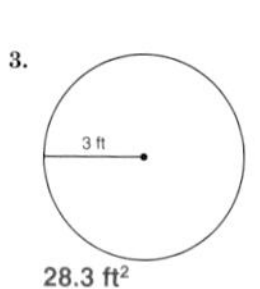

28.3 ft²

4. 15 cm **176.7 cm²**

5. 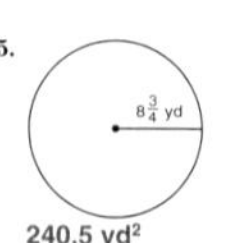

240.5 yd²

6. 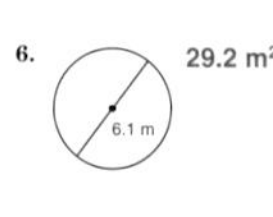

29.2 m²

7. diameter, 22 yd **380.1 yd²**
8. radius, 8.4 in. **221.7 in²**
9. radius $2\frac{1}{4}$ in. **15.9 in²**
10. diameter, 6.5 ft. **33.2 ft²**
11. diameter, 4 cm **12.6 cm²**
12. radius, 4.6 mm **66.5 mm²**

Connection to Algebra

You can use the formula for the area of a parallelogram to find the formula for the area of a circle.

You can review circumference in Lesson 7-4.

$A = b \times h$

$A = \left(\frac{1}{2} \times C\right) \times r$ — *The base of the parallelogram is one-half the circumference.*

$A = \left(\frac{1}{2} \times 2\pi r\right) \times r$ — *Remember, $C = 2\pi r$.*

$A = \pi \times r \times r$

$A = \pi r^2$ — *The formula is $A = \pi r^2$.*

Area of a Circle

In words: If a circle has a radius of r units, then the area A is $\pi \cdot r^2$ square units.

In symbols: $A = \pi r^2$

Example  **Find the area of each circle. Round to the nearest tenth.**

THINK ABOUT IT

You can estimate the area of a circle by squaring the radius and multiplying by 3. So, the area in Example 1 is about $4 \times 4 \times 3$ or 48 cm².

a. 

$A = \pi r^2$ — *Formula for area of a circle*

$A = \pi \cdot 4^2$ — *The radius is 4 in. Estimate: $3 \cdot 16 = 48$*

π [×] 4 [x²] [=] 50.26548246

$A \approx 50.3$ — *Compare with the estimate.*

The area is about 50.3 in².

b. 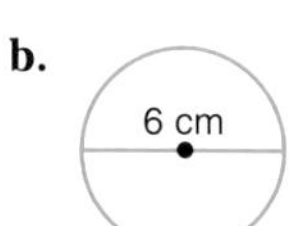

$A = \pi r^2$

$A = \pi \cdot 3^2$ — *The radius is one-half the diameter, so $r = \frac{1}{2}(6)$ or 3.*

$A \approx 28.3$ — ***Estimate:*** *$3 \times 9 = 27$*

The area is about 28.3 cm².

Example  **APPLICATION** **Consumerism**

A 10-inch, two-topping pizza sells for $5.99 and a 14-inch, two-topping pizza sells for $11.19. Which is a better buy, one 14-inch or two 10-inch pizzas?

Use $A = \pi r^2$ to find the area of each size of pizza.

10-inch	14-inch
$A = \pi r^2$	$A = \pi r^2$
$A = \pi \cdot 5^2$	$A = \pi \cdot 7^2$
$A \approx 79$	$A \approx 154$

The area of the 14-inch pizza is almost twice as much as the 10-inch. So two 10-inch pizzas are about the same amount of pizza for 2($5.99) or $11.98, as one 14-inch pizza for $11.19. It appears that the 14-inch pizza is the better buy.

GLENCOE Technology

Interactive Mathematics Tools Software

In this interactive computer lesson, students explore circles. A **Computer Journal** gives students the opportunity to write about what they have learned.

For Windows & Macintosh

Checking Your Understanding

Communicating Mathematics

Read and study the lesson to answer these questions.

1. **Explain** how to find the area of a circle if you know the measure of the radius. Multiply π by the square of the radius.
2. **Estimate** the area of the circle at the right by counting grid squares. Then find the area using the formula. Compare the difference of the two areas to the actual area to find the percent of error. about 50 square units; 50.27; 0.54%

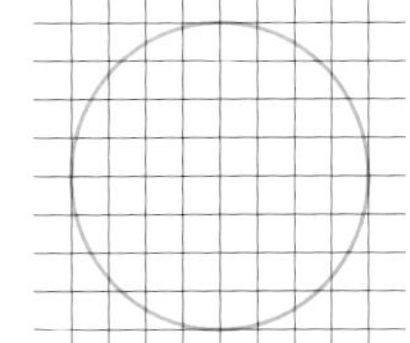

3. **You Decide** Joyce says the area of a circle with a diameter of 10 cm is about 75 cm^2. Sanjay says the area is about 300 cm^2. Use estimation to determine who is correct. Explain.

3. Joyce; you use the radius of 5 cm to calculate the area.

Answers are calculated using π key.

Guided Practice

Find the area of each figure. Round to the nearest tenth.

4. 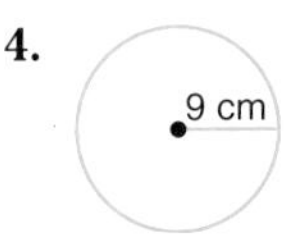

254.5 cm^2

5. 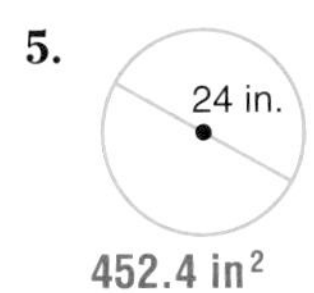

452.4 in^2

6. 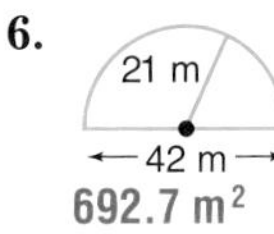

692.7 m^2

7. **City Planning** The circular region inside the streets at DuPont Circle in Washington, DC, is 250 feet across. How much area does the grass and sidewalk cover? 49,087.4 ft^2

Exercises: Practicing and Applying the Concept

Independent Practice

Find the area of each circle. Round to the nearest tenth.

8. 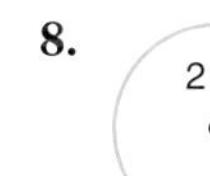

12.6 in^2

9. 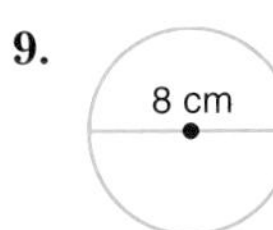

50.3 cm^2

10. 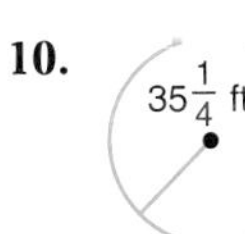

3903.6 ft^2

11. 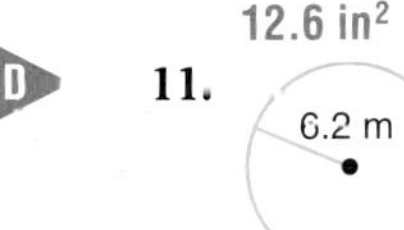

120.8 m^2

12. 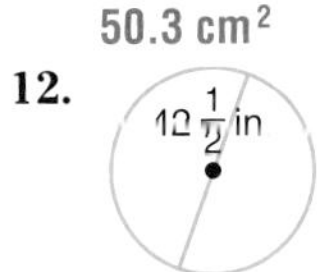

1418.6 in^2

13. 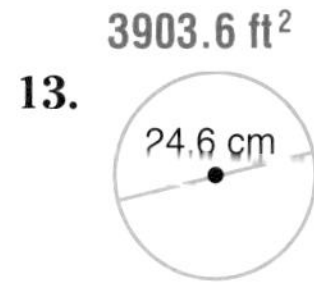

475.3 cm^2

14. radius, 23 cm 1661.9 cm^2
15. diameter, 14.6 in. 167.4 in^2
16. radius, 24.6 m 1901.2 m^2

Find the area of each figure. Round to the nearest tenth.

17. 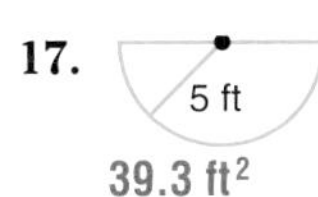

39.3 ft^2

18. 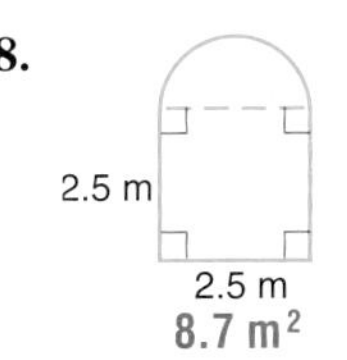

8.7 m^2

19. 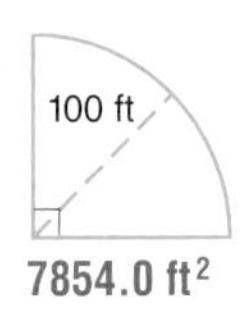

7854.0 ft^2

Critical Thinking

20. The area and the circumference of a circle are the same number.
    a. What is the radius of the circle? 2 units
    b. What is unique about this radius? When $r = 2$, r^2 and $2(r)$ have the same value.

3 PRACTICE/APPLY

Checking Your Understanding

Exercises 1–7 are designed to help you assess your students' understanding through reading, writing, speaking, and modeling. You should work through Exercises 1–3 with your students and then monitor their work on Exercises 4–7.

Assignment Guide
Core: 9–19 odd, 20, 21, 23, 25–30
Enriched: 8–18 even, 20–30

For **Extra Practice**, see p. 769.

The red A, B, and C flags, printed only in the Teacher's Wraparound Edition, indicate the level of difficulty of the exercises.

Practice Masters, p. 101

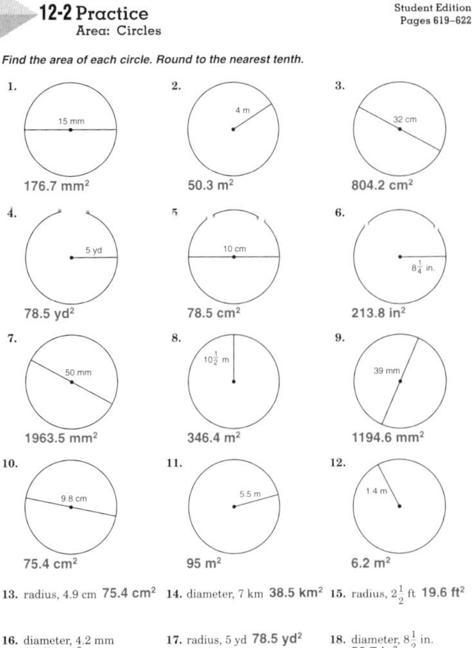

NAME ______ DATE ______

12-2 Practice
Area: Circles

Student Edition Pages 619–622

Find the area of each circle. Round to the nearest tenth.

1. 15 mm — 176.7 mm^2
2. 4 m — 50.3 m^2
3. 32 cm — 804.2 cm^2
4. 5 yd — 78.5 yd^2
5. 10 cm — 78.5 cm^2
6. $8\frac{1}{4}$ in. — 213.8 in^2
7. 50 mm — 1963.5 mm^2
8. $10\frac{1}{2}$ m — 346.4 m^2
9. 39 mm — 1194.6 mm^2
10. 9.8 cm — 75.4 cm^2
11. 5.5 m — 95 m^2
12. 1.4 m — 6.2 m^2
13. radius, 4.9 cm 75.4 cm^2
14. diameter, 7 km 38.5 km^2
15. radius, $2\frac{1}{2}$ ft 19.6 ft^2
16. diameter, 4.2 mm 13.9 mm^2
17. radius, 5 yd 78.5 yd^2
18. diameter, $8\frac{1}{2}$ in. 56.7 in^2

Reteaching

Using Modeling Have students draw circles on $\frac{1}{2}$-inch grid paper. Tell students to be sure that the radius of each circle is a whole number of inches. Remind students that each square on the grid paper has an area of 1/4 in^2. Tell students to count the number of squares and divide by 4 to estimate the area of each circle. Then have students use the formula for the area of a circle to check their estimates.

Group Activity Card 12-2

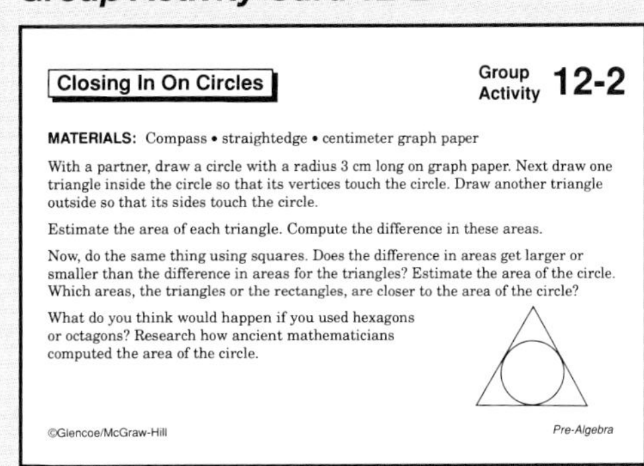

Closing In On Circles Group Activity **12-2**

MATERIALS: Compass • straightedge • centimeter graph paper

With a partner, draw a circle with a radius 3 cm long on graph paper. Next draw one triangle inside the circle so that its vertices touch the circle. Draw another triangle outside so that its sides touch the circle.

Estimate the area of each triangle. Compute the difference in these areas.

Now, do the same thing using squares. Does the difference in areas get larger or smaller than the difference in areas for the triangles? Estimate the area of the circle. Which areas, the triangles or the rectangles, are closer to the area of the circle?

What do you think would happen if you used hexagons or octagons? Research how ancient mathematicians computed the area of the circle.

 Pre-Algebra

4 ASSESS

Closing Activity

Writing Have students write a description of how to find the area of a circle and how to estimate the area of a circle.

Chapter 12, Quiz A (Lessons 12-1 and 12-2) is available in the *Assessment and Evaluation Masters,* p. 323.

Additional Answer

30.

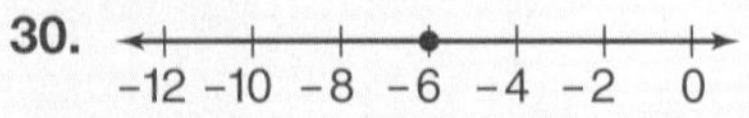

Have students discuss the cartoon. Then have students work in small groups to create their own cartoon relating to some topic covered in this chapter.

Enrichment Masters, p. 101

NAME ______________________ DATE ____________

12-2 Enrichment Student Edition Pages 619–622

Area of a Regular Polygon

The area of a regular polygon is equal to one-half the product of the **apothem** and the **perimeter**. The apothem is the distance from the center of the polygon to a side. The perimeter is the sum of the lengths of all of the sides.

Example:

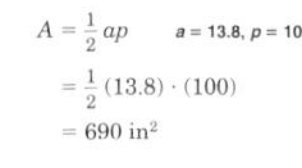

$A = \frac{1}{2}ap$ $a = 13.8$, $p = 100$

$= \frac{1}{2}(13.8) \cdot (100)$

$= 690 \text{ in}^2$

Find the area of each regular polygon.

1. 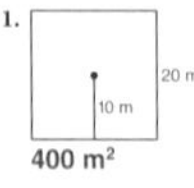

400 m²

2. 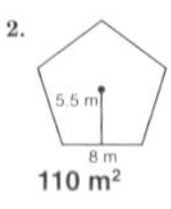

110 m²

3. 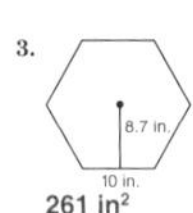

261 in²

4. 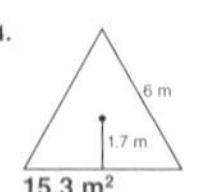

15.3 m²

5. 

275.4 m²

6. 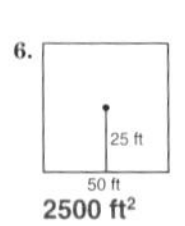

2500 ft²

7. 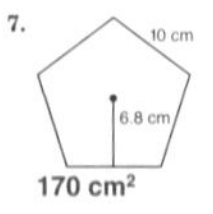

170 cm²

8. 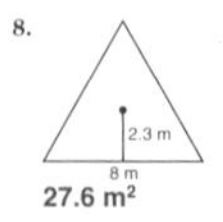

27.6 m²

9. 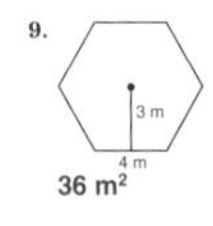

36 m²

Applications and Problem Solving

21. **History** Stonehenge, an ancient monument in England, may have been used as a calendar. The stones are arranged in a circle 30 meters in diameter. Find the area of the circle. **about 707 m²**

22. **Landscaping** Hiroshi needs to match some pipe used in the sprinkling system where he works. He doesn't want to cut the pipe to find the diameter, but he can measure the circumference. He finds that it is about 6.25 inches. What is the diameter of the pipe? **22. about 2 in.**

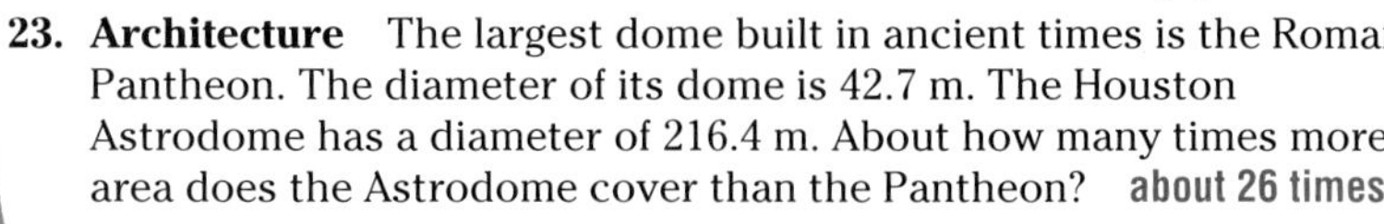

23. **Architecture** The largest dome built in ancient times is the Roman Pantheon. The diameter of its dome is 42.7 m. The Houston Astrodome has a diameter of 216.4 m. About how many times more area does the Astrodome cover than the Pantheon? **about 26 times**

24. **Sports** The Athletic Booster Club for a high school is raising funds for artificial grass for the football playing area in the infield of the track. The grass will cost \$50 per square yard, including installation. To the nearest dollar, how much money must they raise to pay for the grass? **\$348,175**

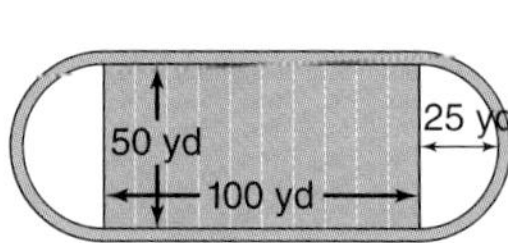

25. **Public Safety** A town has installed a tornado warning system. The sound emitted can be heard for a 2-mile radius. Find the area that will benefit from the system. **about 13 mi²**

Mixed Review

26. **Real Estate** A farmer's land is in the shape of a trapezoid. The north boundary is 220 yards long, the south boundary is 176 yards long, and the east-west distance is 1320 yards. How much land does the farmer have? (Lesson 12-1) **261,360 yd²**

27. Solve $\frac{15}{c} = \frac{8}{9.6}$. (Lesson 9-4) **18**

28. Solve $3x - 5 > 3x + 4$. (Lesson 7-6) **∅**

29. Express $0.\overline{4}$ as a fraction in simplest form. (Lesson 5-1) $\frac{4}{9}$

30. Solve $9c = -54$. Graph the solution on a number line. (Lesson 3-3) **30. −6; See margin for graph.**

From the FUNNY PAPERS

1. What is the formula the man is trying to write? $A = \pi r^2$
2. What does the formula represent? **the area of a circle**
3. Make up a way of remembering the correct formula. **See students' work.**

Extension

Using Constructions Have students draw two line segments with a common endpoint on a sheet of paper. The angle between the line segments must be less than 180°. Have students construct the perpendicular bisector of each segment with a compass and straightedge. Then have students draw a circle with the center at the intersection of the perpendicular bisectors and a radius such that the circle passes through the common endpoint of the segments. What do students notice about the circle? **It passes through the endpoints of the other two segments.** What can this type of construction be used for? **Finding the center of a circle when you only have part of a circle given.**

12-3 Integration: Probability
Geometric Probability

Setting Goals: *In this lesson, you'll find probabilities using area models.*

Modeling a Real-World Application: Games

In a game at the Johnson Middle School Carnival, you toss a beanbag onto a target like the one shown at the right. You get a prize if your beanbag lands in one of the colored circles. Let's assume that all the beanbags land on the board, and that any beanbag is equally likely to land any place on the board. What is the probability that a beanbag will land in one of the circles? *This problem will be solved in Example 2.*

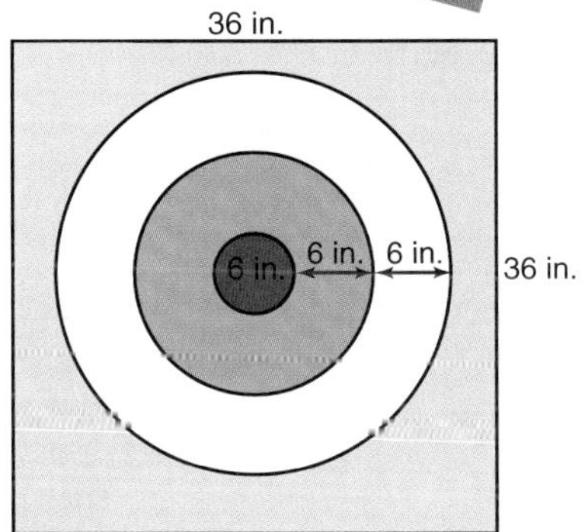

Learning the Concept

In this lesson, for problems involving dartboards, we will assume that the dart will land on the dartboard and that it is equally likely to land anywhere on the dartboard.

Geometric probability uses ideas about area to find the probability of an event. In Lesson 9-3, you calculated probabilities for spinners. A dartboard is similar to a spinner. On the dartboard below, there are ten equally likely outcomes, five of which are shaded. The probability of a dart landing in a shaded section is $\frac{5}{10}$ or $\frac{1}{2}$.

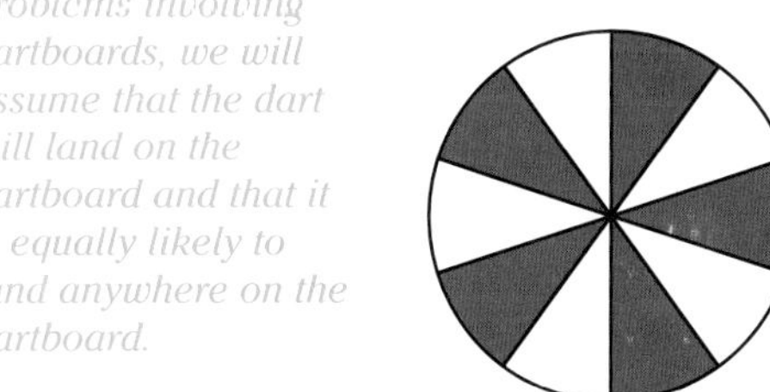

$P(\text{shaded}) = \frac{5}{10}$ or $\frac{1}{2}$

You can also relate probability to the areas of other geometric shapes.

Alternative Learning Styles

Auditory Have students orally explain and review how to find the probability of a particular event happening. Then have students explain their reasoning as they solve various exercises.

12-3 LESSON NOTES

NCTM Standards: 1–4, 11–13

Instructional Resources

- Study Guide Master 12-3
- Practice Master 12-3
- Enrichment Master 12-3
- Group Activity Card 12-3

Transparency 12-3A contains the 5-Minute Check for this lesson; **Transparency 12-3B** contains a teaching aid for this lesson.

Recommended Pacing	
Standard Pacing	Day 5 of 13
Honors Pacing	Day 4 of 11
Block Scheduling*	Day 3 of 7

*For more information on pacing and possible lesson plans, refer to the ***Block Scheduling Booklet***.

1 FOCUS

5-Minute Check
(over Lesson 12-2)

Find the area of each circle described below. Round to the nearest tenth.

1. radius, 9 cm **254.4 cm²**
2. diameter, 24 cm **452.4 cm²**
3. diameter, 6 ft **28.3 ft²**
4. radius, 2.4 m **18.1 m²**
5. diameter, 3 yd **7.1 yd²**

Motivating the Lesson

Questioning Ask students whether they have ever tossed an object at a target to try to win a prize at a carnival or amusement park. Ask whether they think the chances of winning are: very good, average, poor, or impossible.

2 TEACH

In-Class Examples

For Example 1
To win a prize, you are to toss a dart at a board containing 100 small squares. If the dart hits a red square, you win. Twelve of the squares are red. The rest are white.

a. Find the probability of the dart landing in a red square. $\frac{3}{25}$

b. If you threw 75 darts, how many would you expect to land in red squares? **9**

For Example 2
You toss a dart at a large circular area with a diameter of 3 feet that contains 40 small circles, each with a radius of 1 inch. What is the probability that your dart will be inside one of the small circles?
about 12%

Teaching Tip Tell students that for these problems, it is assumed that the object being thrown hits the target each time and that the squares or sections are the same size unless specifically stated otherwise.

Example  **The figure at the right represents a dartboard.**

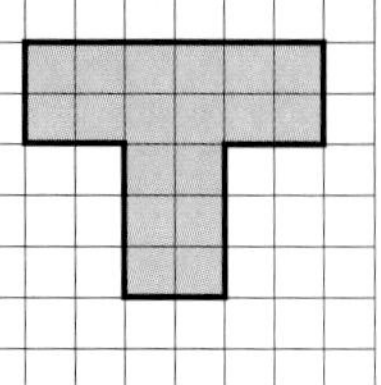

a. Find the probability of landing in the shaded region.

b. Suppose you threw 50 darts. How many would you expect to land in the shaded region?

a. You can count squares to find the area of the shaded region and the area of the whole target.

$$P(\text{shaded}) = \frac{\text{shaded area}}{\text{area of target}}$$
$$= \frac{18}{64} \text{ or } \frac{9}{32}$$

b. Let n represent the darts landing in the shaded region.

$\frac{n}{50} = \frac{9}{32}$ *Write a proportion.*
$50 \cdot 9 = 32n$ *Find the cross products.*
$450 = 32n$
$14.1 \approx n$ *About 14 darts should land in the shaded region.*

  **Refer to the application at the beginning of the lesson. What is the probability of winning a prize on any given toss?**

The area within the circles represents the outcomes of winning a prize and the total area of the board (including the circles) represents all the possible outcomes.

First find the area of the entire board and the area within the circles. Then write a fraction to represent the probability of winning.

The board is a 36-inch by 36-inch square.

$A = s^2$ *Formula for area of a square*
$A = 36^2$ *Replace s with 36.*
$A = 1296$

The board has an area of 1296 in^2.

The area for winning a prize is a circle with a radius of 15 in. *Why?*

$A = \pi r^2$ *Formula for area of a circle*
$A = \pi(15)^2$ *Replace r with 15. Estimate: 3 × 225 = 675*
$A \approx 707$

The area for winning a prize is about 707 in^2.

So, $P(\text{winning a prize}) = \frac{\text{winning area}}{\text{total area}}$
$\approx \frac{707}{1296}$
≈ 0.546 *Use a calculator.*

The probability of winning is about 55% or $\frac{11}{20}$.

Checking Your Understanding

Communicating Mathematics

Read and study the lesson to answer these questions. 1. See margin.

1. Refer to the application at the beginning of the lesson. Tell how you would find the probability of winning a soft drink or a pencil.
2. **Draw** a dart board for which the probability of landing on a shaded area is $\frac{1}{3}$. See students' work.
3. Work with a partner.
   - ▶ Outline two squares as shown at the right.
   - ▶ Hold 25 beans about 3 inches above the paper and drop them onto the paper.
   - ▶ Record the number of beans that land within the small square and the number that land within the large square (including the small square.) If a bean lands on a line, count it in the area where the greater part of the bean lands. Do not count beans that land outside both squares.
   - ▶ Repeat three more times.
   - ▶ Find the ratio $\frac{\text{number of beans in small square}}{\text{number of beans in large square}}$.
   - ▶ Find the ratio $\frac{\text{area of small square}}{\text{area of large square}}$.
   - ▶ Repeat the activity or combine results with other pairs.
   - ▶ Compare the two ratios. What conclusion can you make about the relationship of area and probability? See students' work.

MATERIALS
- grid paper
- dry beans

Guided Practice

Each figure represents a dartboard. Find the probability of landing in the shaded region.

4. $\frac{1}{3}$

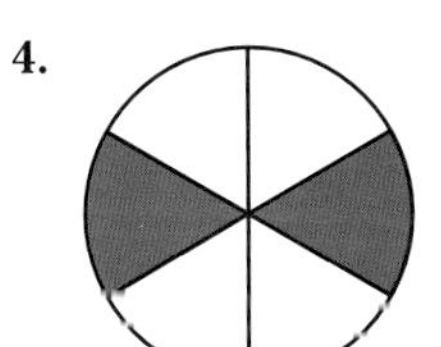

5. $\frac{9}{50}$

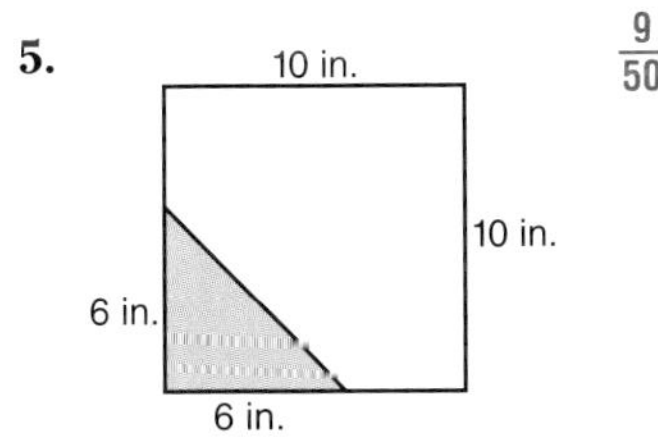

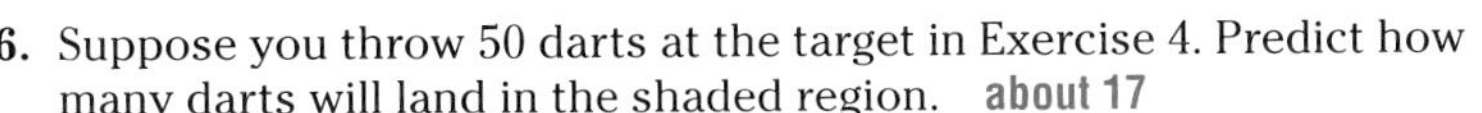

6. Suppose you throw 50 darts at the target in Exercise 4. Predict how many darts will land in the shaded region. about 17
7. Suppose you throw 50 darts at the target in Exercise 5. Predict how many darts will land in the shaded region. about 9
8. **Skydiving** A sky diver parachutes onto a 100-yard by 100-yard square field with a tree in each corner of the field. The diver will not get tangled in tree branches if she stays at least 6 yards away from the tree trunk. Assume the diver lands within the field and that she is equally likely to land anywhere in the field. What is the probability of a clear landing? 98.9%

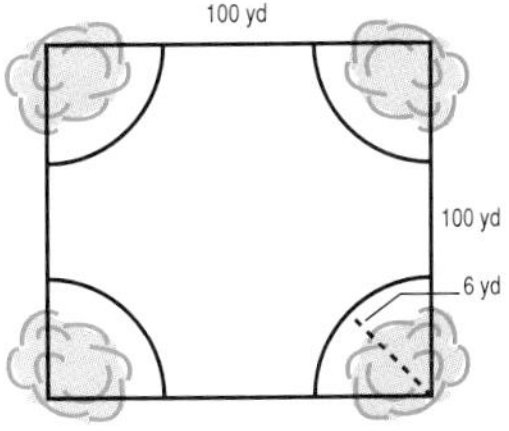

3 PRACTICE/APPLY

Checking Your Understanding

Exercises 1–8 are designed to help you assess your students' understanding through reading, writing, speaking, and modeling. You should work through Exercises 1–3 with your students and then monitor their work on Exercises 4–8.

Teaching Tip Before groups begin dropping their beans in Exercise 3, be sure students understand the rule for what constitutes landing "in" a square. After the experiment, discuss why they needed to make several drops.

Additional Answer

1. Find the area of the circle for winning a soft drink and stuffed animal combined. Then divide this area by the area of the square.
   A (soft drink and stuffed animal) = 254.5 in^2
   A (square) = 1296 in^2
   P (soft drink or stuffed animal) = 19.6%

Study Guide Masters, p. 102

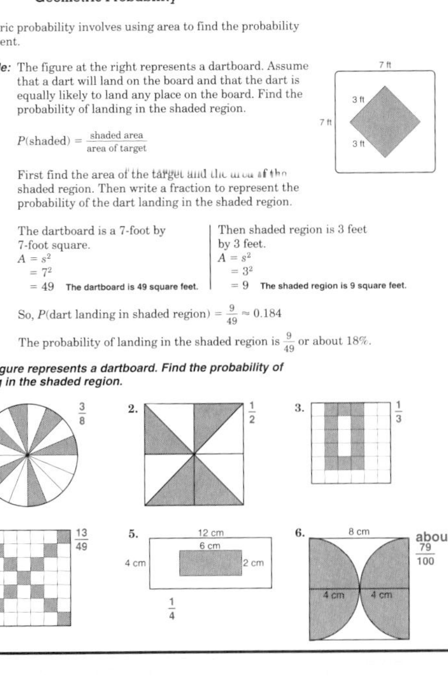

NAME ______________ DATE ______________

12-3 Study Guide
Integration: Probability
Geometric Probability

Student Edition Pages 623–627

Geometric probability involves using area to find the probability of an event.

Example: The figure at the right represents a dartboard. Assume that a dart will land on the board and that the dart is equally likely to land any place on the board. Find the probability of landing in the shaded region.

$P(\text{shaded}) = \frac{\text{shaded area}}{\text{area of target}}$

First find the area of the target and the area of the shaded region. Then write a fraction to represent the probability of the dart landing in the shaded region.

The dartboard is a 7-foot by 7-foot square.
$A = s^2$
$= 7^2$
$= 49$ The dartboard is 49 square feet.

Then shaded region is 3 feet by 3 feet.
$A = s^2$
$= 3^2$
$= 9$ The shaded region is 9 square feet.

So, $P(\text{dart landing in shaded region}) = \frac{9}{49} \approx 0.184$

The probability of landing in the shaded region is $\frac{9}{49}$ or about 18%.

Each figure represents a dartboard. Find the probability of landing in the shaded region.

1. $\frac{3}{8}$
2. $\frac{1}{2}$
3. $\frac{1}{3}$
4. $\frac{13}{49}$
5. $\frac{1}{4}$ (12 cm, 6 cm, 4 cm, 2 cm)
6. about $\frac{79}{100}$ (8 cm, 4 cm, 4 cm)

Reteaching

Using Drawings Tell students to draw a large square on grid paper. Then have them draw a smaller square inside their large square. Have them find the area of each square. Finally, have students find the ratio of the two areas to determine the probability of a dart being thrown at the larger square and landing in the smaller square.

Group Activity Card 12-3

Hand In Foot! Group Activity 12-3

MATERIALS: centimeter grid paper • 2 different-colored pencils

Each partner chooses a color to represent himself or herself throughout the activity. To create a target, Partner A stands with one foot on a sheet of centimeter grid paper while Partner B traces the shoe with Partner A's color. Partner B then lays his or her hand (fingers together) across the middle of the shoe tracing while Partner A traces the hand with Partner B's color. Shade the areas of the shoe and hand that do not overlap. Reverse roles, but not colors, and repeat this procedure on another sheet of grid paper to create a second target. For each target, find the probability that a dart thrown at the target will:

a) land in the shaded region,

b) land in the shoeis shaded region, and

c) land in the region where the hand and shoe overlap.

 Pre-Algebra

Assignment Guide

Core: 9–19 odd, 21–25
Enriched: 10–16 even, 17–25

For **Extra Practice**, see p. 769.

The red A, B, and C flags, printed only in the Teacher's Wraparound Edition, indicate the level of difficulty of the exercises.

Exercises: Practicing and Applying the Concept

Independent Practice

Each figure represents a dartboard. Find the probability of landing in the shaded region.

9. 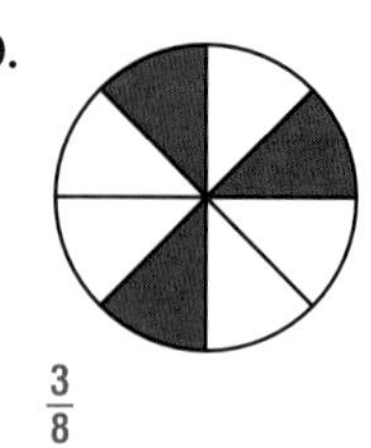 $\frac{3}{8}$

10. 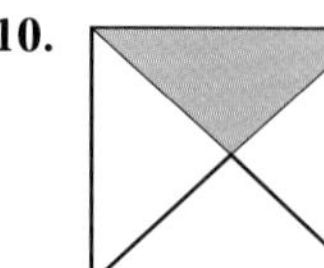 $\frac{1}{4}$

11. 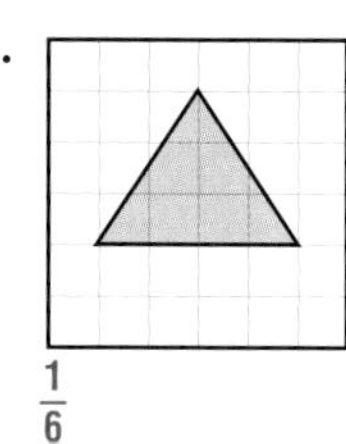 $\frac{1}{6}$

12. 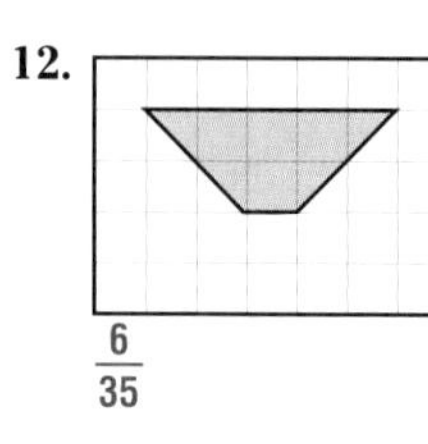 $\frac{6}{35}$

13. 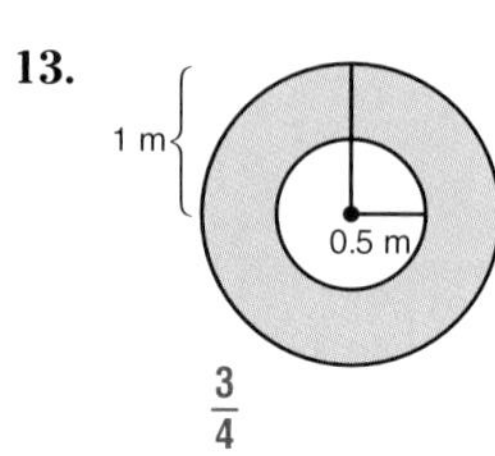

$\frac{3}{4}$

14. 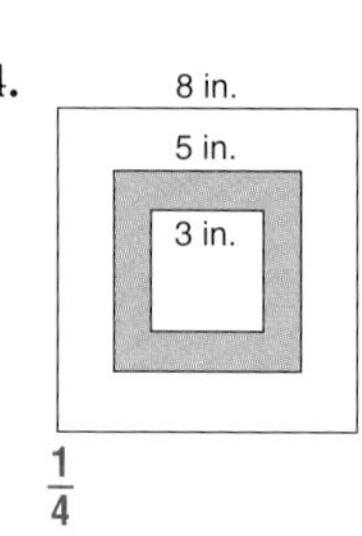

$\frac{1}{4}$

C

15. Suppose you throw 100 darts at the target in Exercise 9. Predict how many darts will land in the shaded region. **about 38**

16. Suppose you throw 75 darts at the target in Exercise 12. Predict how many darts will land in the shaded region. **about 13**

Critical Thinking

17. Games At the Johnson Middle School Carnival, you can win a CD by tossing a quarter onto a grid board so that it doesn't touch a line. The sides of the squares of the grid are 40 millimeters long and the radius of a quarter is 12 millimeters. What is the probability of winning? (*Hint:* Find the area where the center of the coin could land so that the edges don't touch a line.) **0.16**

Applications and Problem Solving

18. Games Refer to the application at the beginning of the lesson. What is the probability of winning a stuffed animal? **about $\frac{9}{400}$**

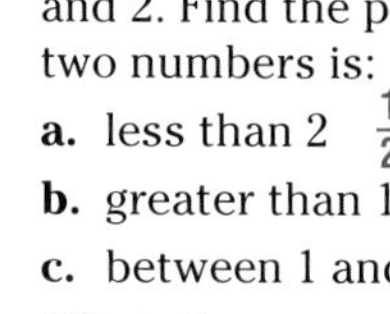

19. Mining A geologist told Sam Antonio that there is oil under 500 acres of his 17,500 acre ranch. If Sam randomly starts drilling, what is the probability that he will strike oil? $\frac{1}{35}$

20. Computers Computers can be used to simulate problem situations using a random number generator. Suppose a computer program randomly selects two real numbers between 0 and 2. Find the probability that the sum of the two numbers is:

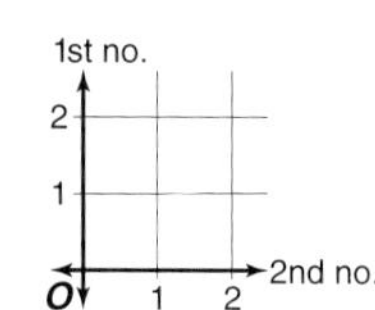

a. less than 2 $\frac{1}{2}$
b. greater than 1 $\frac{7}{8}$
c. between 1 and 2 $\frac{3}{8}$

(*Hint:* Use a coordinate grid.)

Practice Masters, p. 102

NAME ______________ DATE ______________
12-3 Practice
Integration: Probability
Geometric Probability
Student Edition Pages 623–627

Each figure represents a dart board. Find the probability of landing in the shaded region.

1. 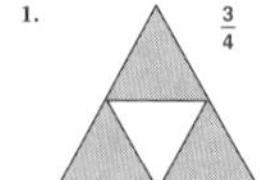 $\frac{3}{4}$
2. 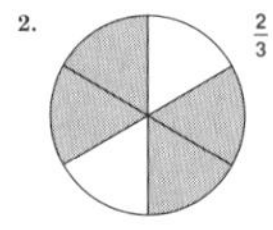 $\frac{2}{3}$
3. 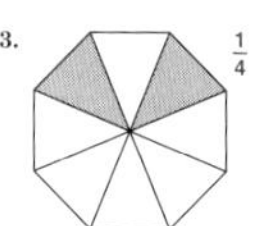 $\frac{1}{4}$
4. 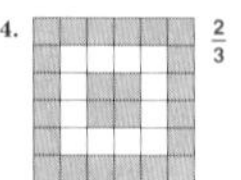 $\frac{2}{3}$
5. 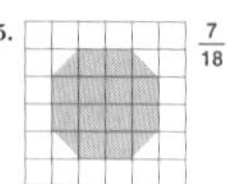 $\frac{7}{18}$
6. 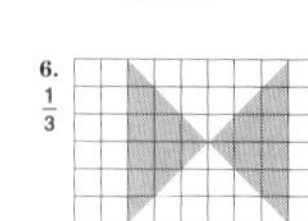 $\frac{1}{3}$
7. 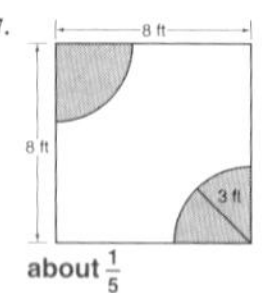

about $\frac{1}{5}$
8. 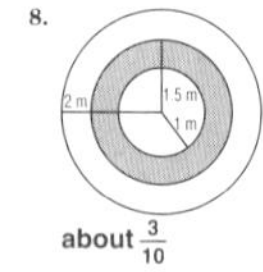

about $\frac{3}{10}$
9. 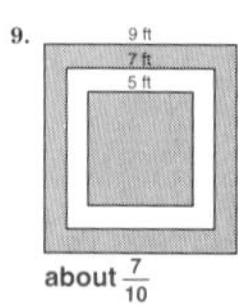

about $\frac{7}{10}$

10. Suppose you throw 25 darts at the target in Exercise 1. How many would you expect to land in the shaded region? **about 18**

11. Suppose you throw 50 darts at the target in Exercise 3. How many would you expect to land in the shaded region? **about 12**

12. Suppose you throw 75 darts at the target in Exercise 5. How many would you expect to land in the shaded region? **about 29**

13. Suppose you throw 100 darts at the target in Exercise 9. How many would you expect to land in the shaded region? **about 70**

Extension

Using Logical Reasoning Place a large piece of paper or cardboard on the floor. This is the target. Have students carefully trace around a quarter somewhere on the target. To win, you must throw a nickel and it must land exactly in the circle, without touching the edges. Have students use logical reasoning to find the probability of winning.

The diameter of the circle is about 2.5 cm. The area of the nickel is 2 cm. So, for the nickel to land completely in the circle, the center of that nickel would have to land in a circular area with a diameter of only 0.5 cm. Thus, the probability can be calculated as $\frac{0.2}{\text{(area of target)}}$, not $\frac{4.9}{\text{(area of target)}}$.

Mixed Review

21. **Geometry** Estimate the area of a circle with a radius of 12 cm. (Lesson 12-2) **about 450 cm²**

22. **Geometry** The measures of the angles of a triangle are in the ratio 1:3:5. Find the measure of each angle. (Lesson 11-4) **20°, 60°; 100°**

23. **Science** An inclined plane is a slanted surface, which may be used for raising objects to higher places. Find the slope of the inclined plane pictured at the right. (Lesson 8-6) **0.1**

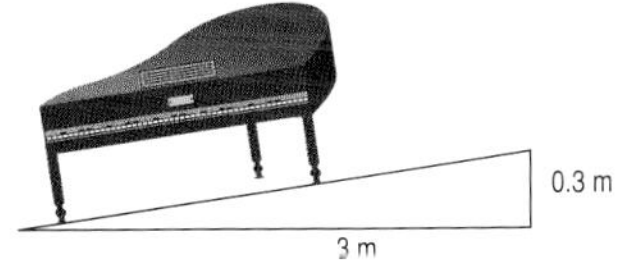

24. Evaluate xy if $x = \frac{1}{5}$ and $y = \frac{3}{5}$. (Lesson 6-3) $\frac{3}{25}$

25. Write an inequality using the numbers in the sentences below. Use the symbols < or >. (Lesson 2-3)
Water boils at 212°F. It freezes at 32°F. **$32 < w < 212$**

COOPERATIVE LEARNING PROJECT

THE SHAPE OF THINGS TO COME

Digital Cameras

Wouldn't it be great to have a camera that allowed you to take a snapshot and instantly view it, before it is developed? Or a camera that, when you viewed a snapshot, allowed you to zoom in on some interesting detail? Such cameras are just now becoming available. They are called digital cameras, cost under $1000, and are about the size of today's film-loaded cameras. Digital cameras can even be plugged into a TV. When you are satisfied with your snapshots, you can transfer them to a videotape, save them on a floppy disk, or print them out on a video printer for placement in an album. By the year 2005, it is expected that the price of digital cameras will come down to about $300.

Digital cameras, like digital computers and digital televisions, produce images by treating the screen as a two-dimensional grid. Each square on the grid is called a pixel. The greater the number of pixels, the higher the "resolution" of the image, and the sharper the image becomes.

To help you understand how image resolution works, think about drawing a circle. But, you can only draw one *straight* line in any given grid square, or pixel. Compare figures 1 and 2. Figure 2 has twice as many squares along each axis.

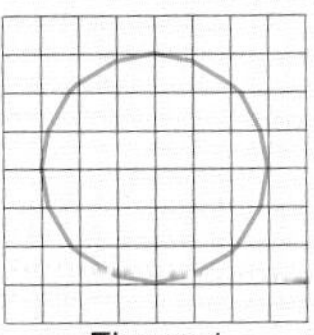
Figure 1

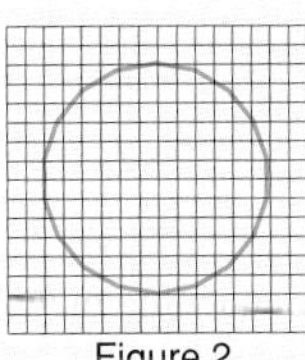
Figure 2

See for Yourself

Research digital photography and the new digital cameras. **See students' work.**

- What are several advantages of digital cameras over today's film-loaded cameras?
- How will you be getting your digital-camera snapshots developed by the year 2005?
- Notice that in figure 2, the drawing comes closer to looking like a circle. As you increase the number of grid squares, you achieve higher resolution and come closer and closer to a sharp, circular image. Reproduce the grid squares of figure 2 on graph paper. Then increase the number of grid squares and, by drawing only one straight line per grid square, see how close you can come to drawing the image of a circle.

4 ASSESS

Closing Activity

Modeling Have students model a target in which the probability of a dart landing in a shaded area is 20%.

Enrichment Masters, p. 102

NAME ______ DATE ______

12-3 Enrichment Student Edition Pages 623–627
Sector of a Circle

The **sector** of a circle is the region bounded by two radii and the arc of the circle. The area of a sector is a fractional part of the area of the circle.

$A = \frac{n}{360} \times \pi \times r^2$ where n is the degree measure of the central angle.

Example: Area of a sector $= \frac{60}{360} \times 3.14 \times 81$
$\approx 42.39 \text{ cm}^2$

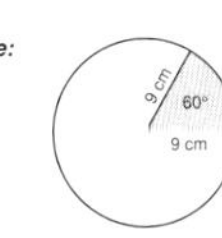

Find the area of each sector. Use 3.14 for π.

1. 12 cm, 45° — **56.52 cm²**
2. 12 cm, 30° — **37.68 cm²**

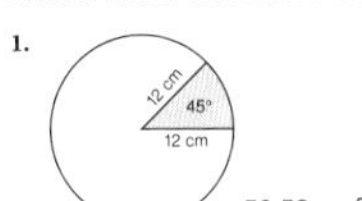

3. 12 cm, 60° — **75.36 cm²**
4. 6 cm, 90° — **28.26 cm²**

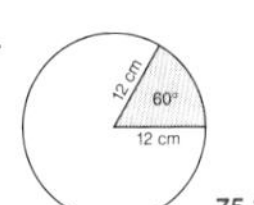

5. 6 cm, 60° — **18.84 cm²**
6. 9 cm, 120° — **84.78 cm²**

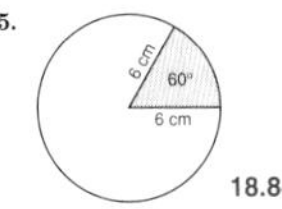

COOPERATIVE LEARNING PROJECT

THE SHAPE OF THINGS TO COME

Digital Cameras Students may be interested in some of the early history of cameras. In 1826, Joseph Nicéphore Niepce used a crude film that captured the first fixed photographic image. In 1888, the Kodak camera made it possible for ordinary people to own and use cameras.

12-3B LESSON NOTES

NCTM Standards: 1-4, 11-13

Objective

Use a graphing calculator to experiment with area and probability.

Recommended Time

15 minutes

Instructional Resources

Graphing Calculator Masters, p. 25

This master provides keystroking instruction for this lesson for the TI-81 and Casio graphing calculators.

1 FOCUS

Motivating the Lesson

Fold a piece of paper into fourths, and shade one of the four sections. Place the paper on the floor and drop a coin so it lands on the paper. Have students discuss the probability of the coin landing in the shaded fourth of the paper.

2 TEACH

Teaching Tip You may wish to review simple probability as taught in Lesson 9-3 before students begin this lesson.

3 PRACTICE/APPLY

Assignment Guide

Core: 1–4
Enriched: 1–4

4 ASSESS

Observing students working with technology is an excellent method of assessment.

MATHEMATICS LAB **12-3B**

Geometric Probability

An Extension of Lesson **12-3**

You can use a T1-82 graphing calculator to experiment with area and probability. The program below generates random ordered pairs.

MATERIALS
- graphing calculator
- grid paper
- straightedge

```
PROGRAM:RANDPRS
: ClrHome
: Fix 1
: Input "LEAST X", A
: Input "GREATEST X", B
: Input "LEAST Y", C
: Input "GREATEST Y", D
: Input "NUMBER OF PAIRS", P
: 0 →I
: Lbl 1
: I +1 → I
: (B–A)rand+A →X
: Disp X
: (D–C)rand+C →Y
: Disp Y
: Disp " "
: Pause
: If I≠P
: Goto 1
: Stop
```

Your Turn Work with a partner.

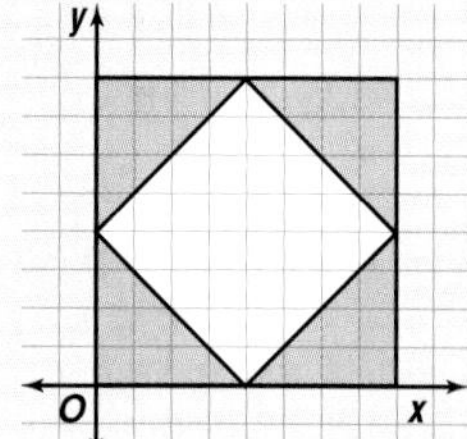

- Use a straightedge to draw a square like the one shown at the right. Shade the figure as shown.
- Enter the program above into a graphing calculator.
- Use the program to generate 50 ordered pairs. Enter 0 as the least value and 8 as the greatest value for both *x* and *y*.
- Graph each pair on the grid paper and keep a tally of how many points are in the shaded and unshaded areas.
- Write a ratio comparing the points in the shaded area to the total number of points.
- Find the areas of the small square and the large squares. Write a ratio comparing the area of the small square to the area of the large square.

TALK ABOUT IT

1. How do the two ratios you wrote compare? **The ratios should be similar.**
2. Repeat the activity above to generate and graph another 50 ordered pairs. Combine the results with the results from the first run and write the ratio. **See students' work.**
3. What effect does more data have on the results? **More data makes the ratios closer.**
4. Predict the results of repeating the activity using the figure at the right. Then check your prediction by running the program and graphing the results. **See students' work.**

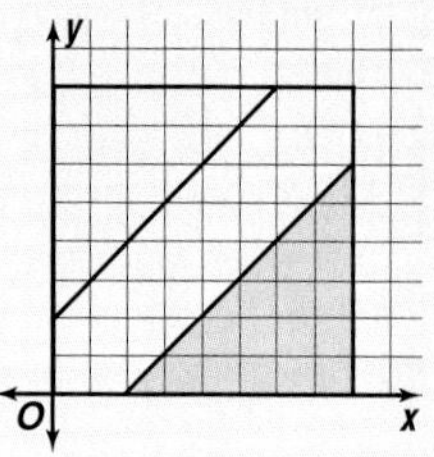

Technology

This lesson offers an excellent opportunity for using technology in your pre-algebra classroom. For more information on using technology, see *Graphing Calculators in the Mathematics Classroom*, one of the titles in the Glencoe Mathematics Professional Series.

12-4 Problem-Solving Strategy: Make a Model or a Drawing

Setting Goals: *In this lesson, you'll solve problems using a model or a drawing.*

Modeling a Real-World Application: Pet Care

June has attached a 20-foot rope to the corner of a 10-foot by 10-foot shed to tie up her horse Harry. How much grazing area does Harry have?

Learning the Concept

Without a drawing, it is difficult to find the area or even visualize the shape of the horse's grazing area. Many problems, especially those that involve geometry, are easier if you make a model or a drawing.

Explore You know the dimensions of the shed and the length of the rope. You need to find the grazing area.

Plan Make a drawing to illustrate the problem. Notice that you have sections of circles. The largest section is $\frac{3}{4}$ of a circle with a radius of 20 feet. The smaller sections are each $\frac{1}{4}$ of a circle with a radius of 10 feet. You can find the areas of each section and add to find the total grazing area.

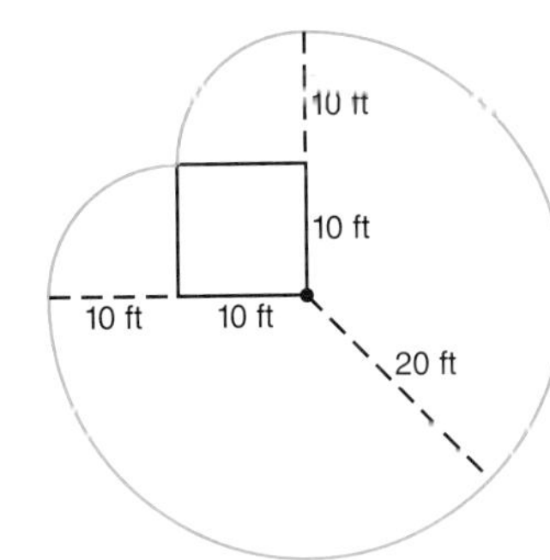

THINK ABOUT IT

How do you know that the large section is three-fourths of a circle and the small sections are one-fourth of a circle?

The building corners are 90° angles and a circle has 360°.

Solve

Area of the large section: $A = \pi r^2 \times \frac{3}{4}$

Estimate: $3 \times 400 \times \frac{3}{4} = 900$

$A = \pi \times 20^2 \times \frac{3}{4}$

$A \approx 942$

Area of each smaller section: $A = \pi r^2 \times \frac{1}{4}$

Estimate: $3 \times 100 \times \frac{1}{4} = 75$

$A = \pi \times 10^2 \times \frac{1}{4}$

$A \approx 79$

Total area = 942 + 2(79) or 1100 *There are 2 smaller sections.*

The horse has a grazing area of about 1100 ft^2.

Examine An answer of 1100 ft^2 seems reasonable, compared to your estimate.

CULTURAL CONNECTIONS

Horses were introduced to North America by the Spanish in 1519. The horses they left behind were probably the ancestors of the American wild horses.

Study Guide Masters, p. 103

The Study Guide Master provides a concise presentation of the lesson along with practice problems.

CULTURAL CONNECTIONS

The earliest horses were of the species *Equus*. The earliest record of the use of Equus as a means of transport comes from the Sumerians between 4000 and 5000 years ago.

12-4 LESSON NOTES

NCTM Standards: 1-4, 12, 13

Instructional Resources

- Study Guide Master 12-4
- Practice Master 12-4
- Enrichment Master 12-4
- Group Activity Card 12-4
- Assessment and Evaluation Masters, pp. 322, 323
- Math Lab and Modeling Math Masters, p. 72
- Real-World Applications, 28

Transparency 12-4A contains the 5-Minute Check for this lesson; **Transparency 12-4B** contains a teaching aid for this lesson.

Recommended Pacing	
Standard Pacing	Day 6 of 13
Honors Pacing	Day 5 of 11
Block Scheduling*	Day 4 of 7 (along with Lesson 12-5)

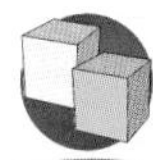

*For more information on pacing and possible lesson plans, refer to the ***Block Scheduling Booklet***.

1 FOCUS

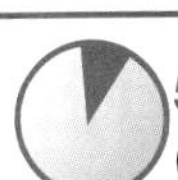

5-Minute Check
(over Lesson 12-3)

Each figure represents a dartboard. Find the probability of a randomly thrown dart landing in the shaded area.

1.  **50%**

2. 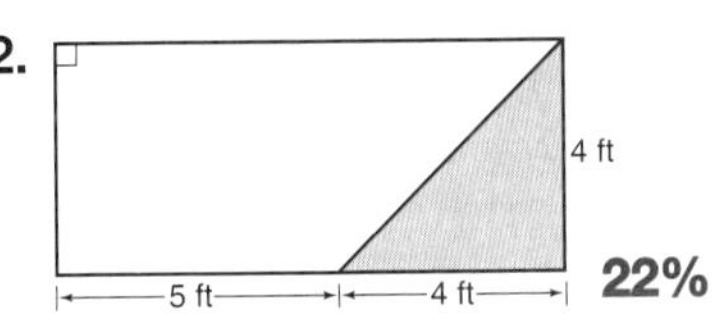

22%

3. Suppose you throw 100 darts at the target in Exercise 1 above. How many would you expect to land in the shaded region? **50**

Motivating the Less

Situational Problem Pose this problem to students. Tell students to listen carefully and take notes if needed. (Read quickly and do not repeat any information.) *Randy jogged 4 blocks north, then 5 blocks south. he stopped to get his breath, turned around and went 3 blocks north. Next, he went 6 blocks east and then 2 blocks south. Where did he end up in reference to where he started?* **6 blocks east from where he started.** Ask whether anyone has an answer. Ask what might be a good way of solving the problem.

2 TEACH

In-Class Example

To start a quilt pattern, Mr. Yon cut out four right triangles, each with sides of 3, 4, and 5 inches. He also cut out a 5-inch square and placed one triangle on each side of the square with the hypotenuse matching the side of the square. What was the shape of his completed pattern? What was the area of the completed pattern? **a square; 49 in^2**

3 PRACTICE/APPLY

Checking Your Understanding

Exercises 1–5 are designed to help you assess your students' understanding through reading, writing, speaking, and modeling. You should work through Exercises 1–3 with your students and then monitor their work on Exercises 4–5.

Assignment Guide
Core: 7–15 odd, 16–21
Enriched: 6–14 even, 16–21

For **Extra Practice**, see p. 770.

Practice Masters, p. 103
The Practice Master provides additional practice problems.

Checking Your Understanding

Communicating Mathematics

Read and study the lesson to answer these questions.

1. Refer to the application at the beginning of the lesson. Draw a picture of the problem if the rope was 25 feet long. Then explain how the problem is different. **See margin.**
2. How can making a model or drawing help in problem solving?
3. Think of a real-life problem where making a model or drawing can help you solve the problem. Write about the problem and its solution. Include a drawing. **See students' work.**

2. It can make the problem easier to visualize.

Guided Practice

Solve by making a model or drawing.

4. Refer to the application at the beginning of the lesson. Suppose the rope is only 15 feet long. Find the grazing area. **≈569 ft^2**
5. **Framing** A painting 15 cm by 25 cm is bordered by a mat that is 3 centimeters wide. The frame around the mat is 2 centimeters wide. What is the area of the picture including the frame and mat? **875 cm^2**

Exercises: Practicing and Applying the Concept

Independent Practice

Solve. Use any strategy.

A 6. A rectangular swimming pool is 15 feet wide and 30 feet long. One-foot-square tiles are to be installed around the pool to form a walkway 3 feet wide all around the pool. What is the area to be covered by the tiles? **306 ft^2**

B 7. There are five ways to place four one-inch squares together so that any two adjoining sides match exactly.

These match: These do not match:

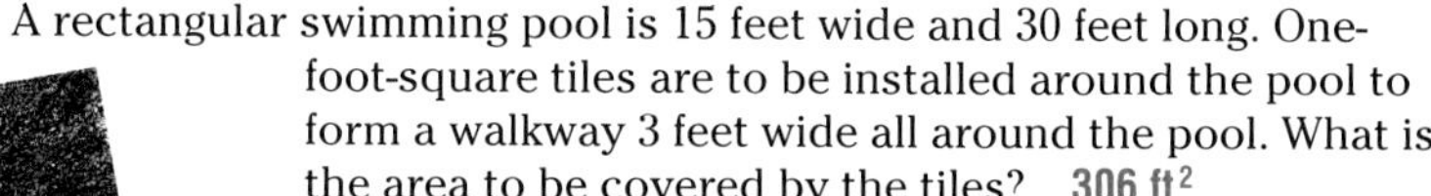

a. Find the five ways. **a–d. See Solutions Manual.**
b. Find the perimeter of each figure.
c. Describe the shape or shapes for which the perimeter is greatest.
d. Describe the shape or shapes for which the perimeter is least.
e. How many ways can you arrange five squares? Do their perimeters differ? **12; yes**

LOOK BACK You can review perimeter in Lesson 3-5.

8. Find the sum of the first 100 *even* positive numbers. **10,100**
9. The perimeter of a rectangle is 32 meters. Its area is 48 square meters. What are the dimensions of the rectangle? **4 m by 12 m**
10. **Sports** A circular track has a diameter of 400 feet. Les starts out jogging around the track at a rate of 600 feet per minute. One minute later, Antonio starts jogging at the same place. He goes in the same direction and jogs at 700 feet per minute. Will Antonio pass Les on the first, second, third, or fourth quarter of the track? **2nd**
11. Sandy has 36 identical cubes. Each edge of a cube is 1 inch long. How can Sandy arrange the cubes to have the smallest surface area?

11. $3 \times 3 \times 4$

Additional Answer

1. The grazing areas overlap.

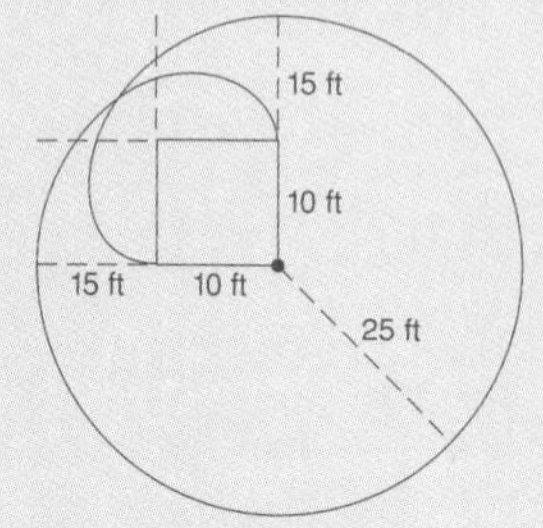

Reteaching

Make a Drawing Have students solve the following problem. *The main part of a garden is a rectangle with a length of 25 feet and a width of 18 feet. On the north side of the rectangle is a triangle with a base of 25 feet and a height of 12 feet. On the south side of the rectangle is a trapezoid with bases of 25 feet and 15 feet and a height of 5 feet. What is the area of the garden?* **700 ft^2**

12. **Geometry** Various views of a solid figure are shown below. The edge of one block represents one unit of length. A dark segment indicates a break in the surface.

    a. Use the views to sketch the solid. **See Solutions Manual.**

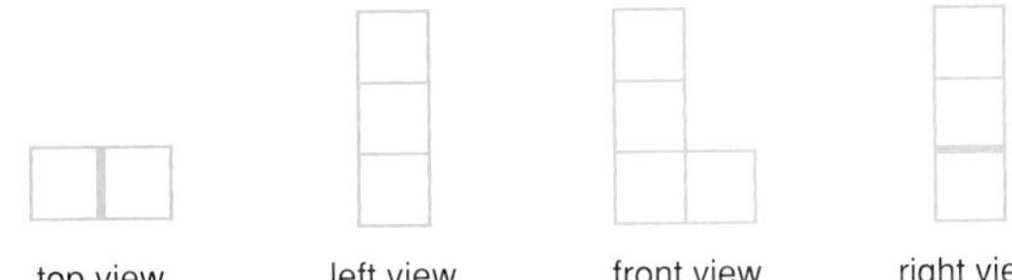

    b. Choose a solid object from your home and draw the top, left, front, and right views. **See students' work.**

13. Ms. Warren sold 100 shares of stock for $2475. When the price per share went down $4, she bought 200 more shares. When the price per share went back up $3, she sold 100 shares. How much did Ms. Warren gain or lose in her transactions?

14. Keiko's little brother fills an 8-quart pail with sand for a sand castle. He also has empty 3-quart and 5-quart pails. How could he use these to divide the sand into two equal portions? **See margin.**

13. $700 gain

15. Three different views of the same cube are shown below. What figure is on the side opposite the X? **A**

Critical Thinking

16. The figures you drew in Exercise 7 are called *nets*. Some nets can be folded to form solid figures. Draw all of the nets containing five one-inch squares that could be folded into open boxes. Mark the square that would be the bottom of the box. **See Solutions Manual.**

Mixed Review

17. **Probability** The figure at the right represents a dartboard. What is the probability of landing in the shaded region? (Lesson 12-3) $\frac{2}{5}$

18. Find the value of $P(5, 4)$. (Lesson 10-6) **120**

19. **Wages** You earn $8 an hour and a $12 bonus. What is the least number of hours you must work to earn at least $100? (Lesson 7-7) **11 hours**

20. Write 2^{-3} using positive exponents. (Lesson 4-9) $\frac{1}{2^3}$

21. **Statistics** The graph at the right shows the number of millions of acres of federally-owned land in several states. (Lesson 1-10)

    a. Which state has the greatest number of acres that are federally owned?

    b. What is the difference between the number of acres owned by the Federal Government in California and Arizona? **17 million acres**

interNET CONNECTION

For the latest on federally-owned land, visit: **www.glencoe.com/sec/math/prealg/mathnet**

21a. Nevada

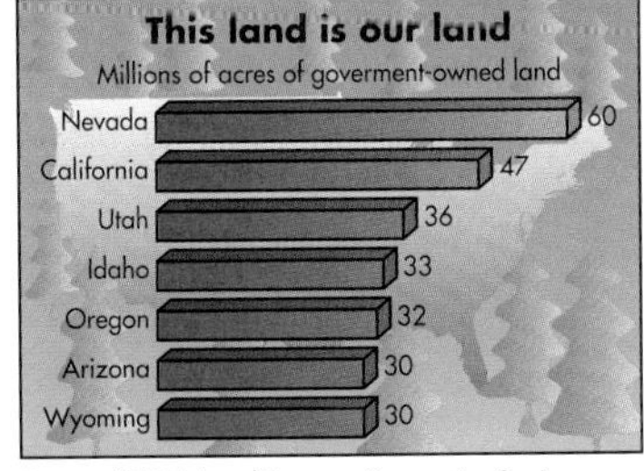

Source: USDA, Natural Resources Conservation Service, National Resources Inventory

4 ASSESS

Closing Activity

Modeling Have students create a drawing to show various ways of arranging the classroom. Then have students calculate the amount of floor space allowed for student seating and for aisles or walking room.

Additional Answer

14. Fill the 5-quart pail from the 8-quart pail. Fill the 3-quart pail from the 5-quart pail, leaving 2 quarts in the 5-quart pail. Empty the 3-quart pail into the 8-quart pail. Pour 2 quarts from the 5-quart pail into the 3-quart pail. Fill the 5-quart pail from the 8-quart pail. Pour from the 5-quart pail into the 3-quart pail until full, leaving 4 quarts in the 5-quart pail. Empty the 3-quart pail into the 8-quart pail, making 4 quarts in the 8-quart pail.

Chapter 12, Quiz B (Lessons 12-3 and 12-4) is available in the *Assessment and Evaluation Masters*, p. 323.

Mid-Chapter Test (Lessons 12-1 through 12-4) is available in the *Assessment and Evaluation Masters*, p. 322.

Enrichment Masters, p. 103

NAME ______________________ DATE ____________

12-4 Enrichment
Problem Solving: Measures

Student Edition Pages 629–631

Solve each problem.

1. A flower bed is in the shape of a triangle. The sides measure 2.5 meters, 3.5 meters, and 4.25 meters. What is the perimeter of the flower bed? **10.25 m**

2. Mindy has a class picture that measures 10 inches by $8\frac{1}{2}$ inches. The wood on the frame she uses is $1\frac{1}{8}$ inches wide. What is the perimeter of the framed picture? **46 in.**

3. The perimeter of a square is 140 meters. Find the length of each side. **35 m**

4. A regular octagon has a perimeter of 10 centimeters. Find the length of each side. **1.25 cm**

5. A security guard walks around a building 115 feet long and 39 feet wide. If her strides average $2\frac{1}{2}$ feet in length, how many strides does she take each trip around? **The perimeter is 308 feet and 308 ÷ 2.5 is about 123 strides. However, since she cannot walk exactly along the edge of the building, the number of strides will be greater than 123.**

6. A regulation baseball diamond, 90 feet square, is laid out on a field 172 feet wide and 301 feet long. How much greater is the distance around the whole field than around the diamond? **586 ft**

7. How many feet of chrome edging do you need to finish a table-top 54 inches long and 30 inches wide? **14 ft**

Group Activity Card 12-4

Can It Be Done? Group Activity **12-4**

MATERIALS: One-inch tiles • graph paper

You will be working with pentominoes in this activity. These are arrangements of five squares. In the arrangements, all squares must share at least one side with another square, and the vertices of the squares must coincide. Arrangement A is allowed, while B is not.

A B

1. Make all possible *different* arrangements of five tiles. Record these on graph paper. "Different" means an arrangement that can't be made from another arrangement by turning or flipping it.
2. Of the arrangements you found in 1, circle the ones that can be folded into a box without a lid.
3. In the arrangements you circled in 2, put an X on the square that would be the bottom of the box.

 Pre-Algebra

Extension

Have students or pairs of students write word problems for the two situations described below.

1. Create a word problem that is best solved by sketching a solid when given the top, side, and front views.
2. Create another word problem that is best solved by drawing a solid or several solids from different perspectives.

Then have students exchange problems and solve.

12-5 LESSON NOTES

NCTM Standards: 1-4, 12, 13

Instructional Resources

- Study Guide Master 12-5
- Practice Master 12-5
- Enrichment Master 12-5
- Group Activity Card 12-5
- Tech Prep Applications Masters, p. 23

Transparency 12-5A contains the 5-Minute Check for this lesson; **Transparency 12-5B** contains a teaching aid for this lesson.

Recommended Pacing	
Standard Pacing	Day 7 of 13
Honors Pacing	Day 6 of 11
Block Scheduling*	Day 4 of 7 (along with Lesson 12-4)

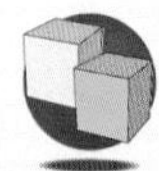

*For more information on pacing and possible lesson plans, refer to the ***Block Scheduling Booklet***.

1 FOCUS

5-Minute Check
(over Lesson 12-4)

Solve each problem by making a model or drawing.

1. Hector liked hexagons. He made an enclosure for his pet that was the shape of two hexagons joined at a common side. How many posts did he use if one is placed at each corner of the enclosure? **10**
2. A 4-inch by 6-inch rectangular miniature painting is to be mounted in a frame made of intricate mosaic tiling. The frame is circular and has a diameter of 8 inches. The tiling goes from the edge of the painting to the outer edge of the circular frame. To the nearest whole square unit, what is the area covered by the mosaic tiling? **26 in²**

12-5 Surface Area: Prisms and Cylinders

Setting Goals: *In this lesson, you'll find surface areas of triangular and rectangular prisms and circular cylinders.*

Modeling a Real-World Application: Postal Service

Maria was buying stamps at a U. S. Post Office when she noticed shipping boxes on display. The price list is shown at the right.

Size (length × width × height)	Price
8 in. × 8 in. × 8 in.	$1.25
15 in. × 10 in. × 12 in.	$2.00
20 in. × 14 in. × 10 in.	$2.50

Maria wondered if the price of each box was related to the amount of material needed to make the box. Is the ratio of price to material about the same for each box? *You will solve this problem in Exercise 27.*

Learning the Concept

THINK ABOUT IT

Rectangles *ABCD* and *EFGH* are bases. Could another pair of regions be the bases of this prism? Does a base always need to be on the bottom?

yes; any two parallel rectangles; no

In geometry, solids like the shipping box shown above are called **prisms**. A prism is a solid figure that has two parallel congruent sides, called **bases**. A prism is named by the shape of its bases. The shipping box and the prism shown at the right are **rectangular prisms**.

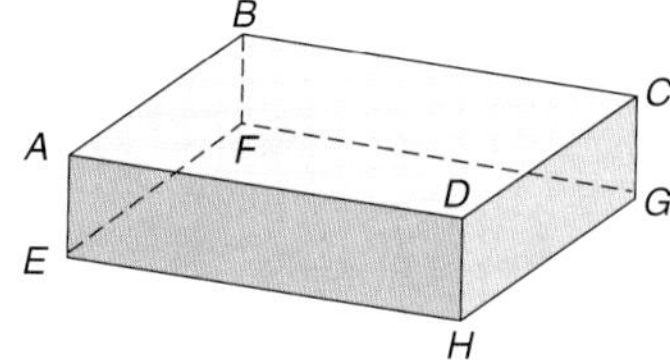

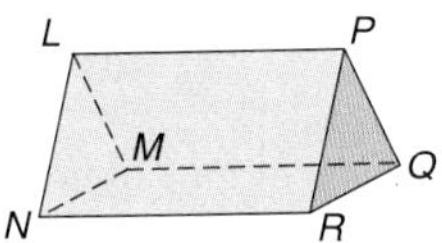

The parallel bases of the prism shown at the left are shaped like triangles, so it is called a **triangular prism**. Can you think of something in everyday life that has the shape of a triangular prism? **Sample answer: a tent**

The amount of material that it would take to cover a geometric solid is the **surface area** of the solid. If you "open up" or "unfold" a prism, the result is a **net**. Nets help us see the regions or **faces** that make up the surface of the prism. The surface area of a prism is the area of the net.

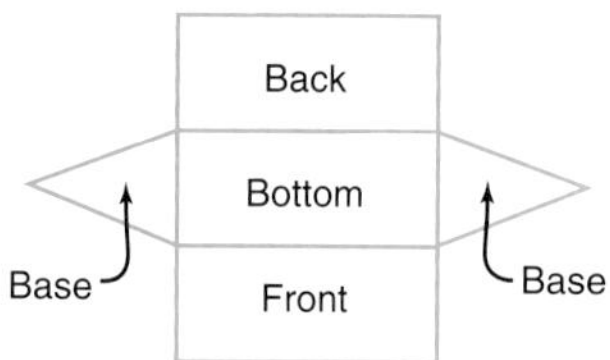

A triangular prism has five faces.

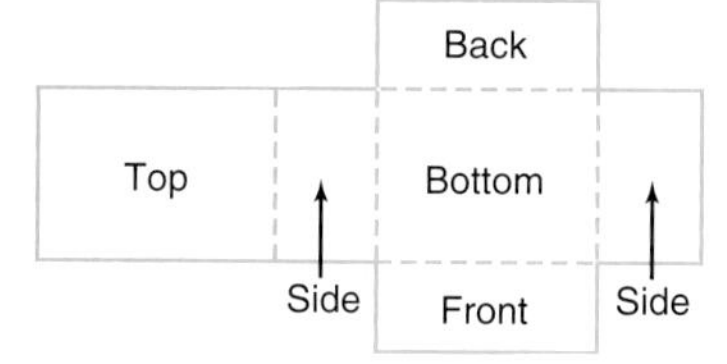

A rectangular prism has six faces.

Alternative Learning Styles

Kinesthetic Provide large cylinders and boxes of various shapes for students to measure and then find the total surface area of each by finding the area of each surface and adding.

Example 1 **Find the surface area of the 15 in. × 10 in. × 12 in. box in the application at the beginning of the lesson.**

Method 1

Sketch the box.
Use the formula $A = \ell w$ to find the areas of the faces.

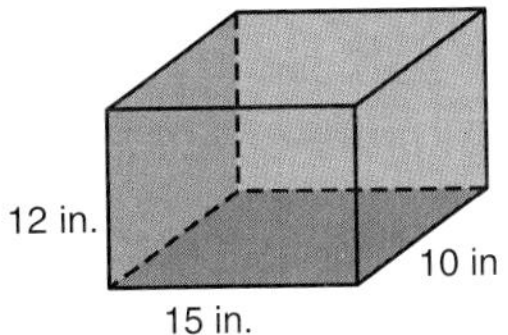

Front and Back	12 [×] 15 [×] 2 = 360
Top and Bottom	10 [×] 15 [×] 2 = 300
Two Sides	12 [×] 10 [×] 2 = 240

Add to find the total surface area. $360 + 300 + 240 = 900$
The surface area of a 15 in. × 10 in. × 12 in. box is 900 in^2.

Method 2

Draw a net for the box and label the dimensions of each face. Find the area of each face. Then add to find the total surface area.

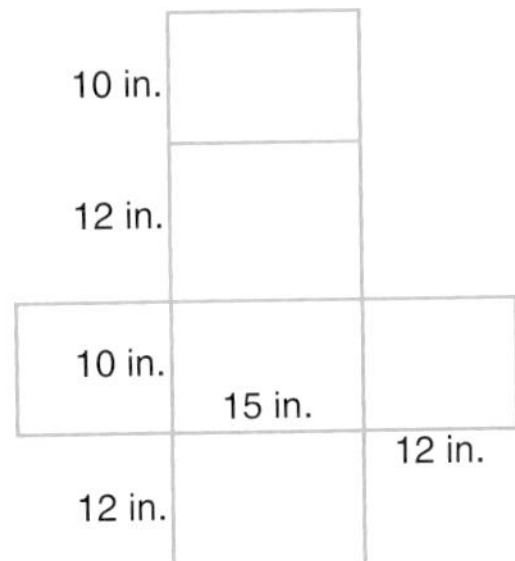

Bottom	10 [×] 15 [=] *150*
Side	12 [×] 10 [=] *120*
Top	10 [×] 15 [=] *150*
Side	12 [×] 10 [=] *120*
Back	12 [×] 15 [=] *180*
Front	12 [×] 15 [=] *180*
Total	150 [+] 120 [+] 150 [+] 120 [+] 180 [+] 180 [=] *900*

The surface area of the box is 900 in^2.

Example 2 **Find the surface area of the triangular prism shown below.**

Use the formula $A = \frac{1}{2}bh$ to find the area of the bases.

$\frac{1}{2} \cdot 6 \cdot 8 = 24$

Use the formula $A = \ell w$ to find the area of the sides.

Front	$9 \cdot 10 = 90$
Bottom	$6 \cdot 9 = 54$
Back	$8 \cdot 9 = 72$

10 m, 8 m, 9 m, 6 m

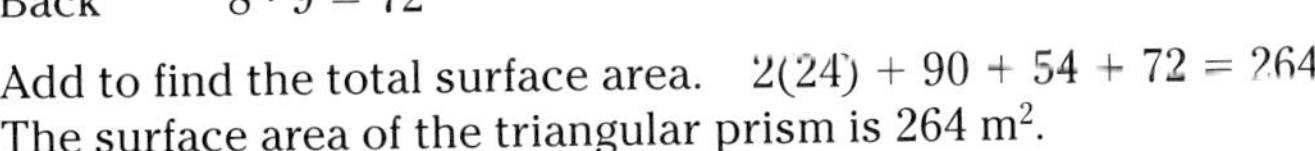

Add to find the total surface area. $2(24) + 90 + 54 + 72 = 264$
The surface area of the triangular prism is 264 m^2.
Check this answer by drawing a net of the prism.

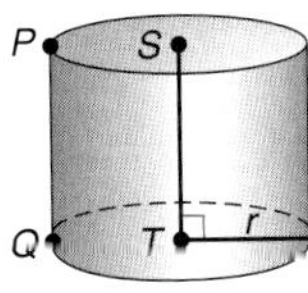

You probably have cans of food in your kitchen at home. These cans are examples of **cylinders**. The bases of a circular cylinder are two parallel, congruent circular regions.
An altitude is any perpendicular line segment joining the two bases. $\overline{ST}$ and $\overline{PQ}$ are altitudes.

Motivating the Lesson

Hands-On Activity Give groups of students models of different prisms. From the models, develop a list of characteristics of a prism. **2 congruent parallel bases connected by rectangles** Have students measure and find the surface area of each.

2 TEACH

In-Class Examples

For Example 1
Find the surface area of a rectangular box that is 12 centimeters wide, 20 centimeters long, and 16 centimeters high. **1504 cm^2**

For Example 2
Find the surface area of the triangular prism shown below. **153 ft^2**

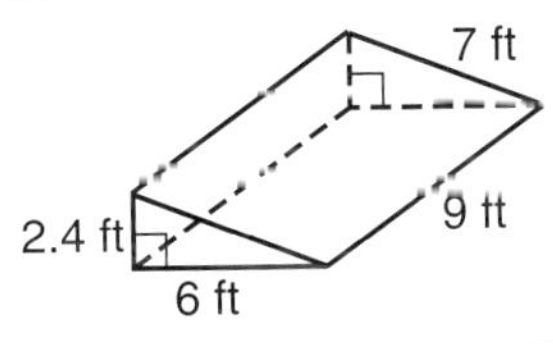

Teaching Tip Show students a drawing of a hexagonal prism. Have students determine how many faces it has. Ask students to develop a formula that tells the number of faces a prism has based on the number of sides of the base. **F = E + 2** Remind students that the area of each face must be added when determining surface area.

In-Class Example

For Example 3

A cylindrical can of peaches is 11.3 centimeters tall and has a diameter of 9.5 centimeters.

a. How much area will the label cover? Round to the nearest tenth. **337.3 cm²**

b. What is the total surface area of the can? Round to the nearest tenth. **801.0 cm²**

Teaching Tip For In-Class Example 3 above, point out to students that the answer is written as 801.0 to indicate that the answer (≈801.97) is rounded to the nearest tenth.

Teaching Tip You may wish to have students use real-life objects to form the net of a cylinder as described in the text.

Additional Answer Think About It

The length of the rectangle coincides with the circumference of a base to form an edge of the cylinder.

If you open the top and bottom of a cylinder and then make a vertical cut in the curved surface, you could lay the cylinder flat in the same way you did with a rectangular prism. You would then have a pattern, or net, for making that cylinder.

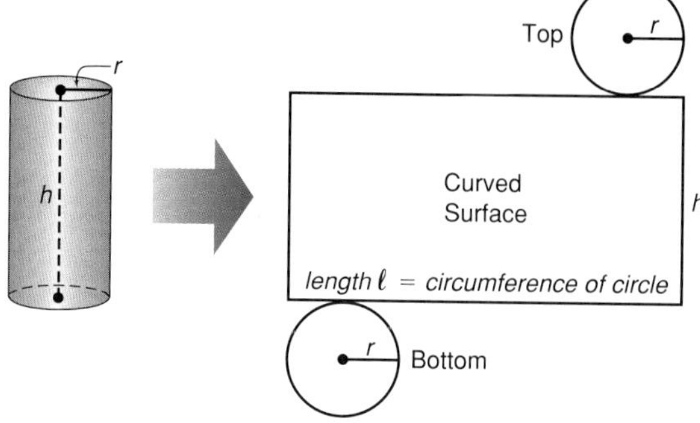

From the diagram above, you can see that:

1. The bases of the cylinder are congruent circular regions.
2. The curved surface of the cylinder opens to form a rectangular region.
3. The width of the rectangle is the height of the cylinder, or h.
4. The length of the rectangle is the circumference of a base, or $2\pi r$.

THINK ABOUT IT

Why is the length of the rectangle the same as the circumference of a base?

See margin.

Connection to Algebra

The area of each base is πr^2. Since the area of a rectangle is ℓw, the area of the curved surface is $2\pi r \cdot h$. So, the surface area of a circular cylinder is $2(\pi r^2) + 2\pi r \cdot h$.

Example 3 **APPLICATION** **Manufacturing**

A Pringles™ can has a diameter of 3 inches and a height of 8.5 inches.

a. How much area must the label cover?

b. What is the total surface area of the can?

Explore You need to find the area of the label and the total surface area of the can. You know the diameter and the height of the can.

Plan The label covers the curved surface of the can. Use $A = \ell w$ or $A = 2\pi r \cdot h$ to find the area of the curved surface.

Estimate: $2 \cdot \pi \cdot 1\frac{1}{2} \cdot 8.5 \rightarrow 2 \cdot 3 \cdot \frac{3}{2} \cdot 9 = 81$

The top and bottom bases of the can are circles. Use $A = \pi r^2$ to find the area of each base.

Estimate: $\pi \cdot 1\frac{1}{2} \cdot 1\frac{1}{2} \rightarrow 3 \cdot 1 \cdot 2 = 6$

Then add to find the total surface area.

Estimate: $81 + 6 + 6 = 93$

Solve	**a.** $A = \ell w$	*Area of a rectangle*
	$A = 2\pi r \cdot h$	*Replace ℓ with the expression for the circumference of a circle.*
	$A = 2 \cdot \pi \cdot 1.5 \cdot 8.5$	*$d = 3$, so $r = 1.5$.*
	$A \approx 80.1$	
	The label covers 80.1 in^2.	
	b. $A = \pi r^2$	*Formula for area of a circle*
	$A = \pi \cdot 1.5^2$	
	$A \approx 7.1$	The area of each base is about 7.1 in^2.
	The total surface area is about 80.1 + 2(7.1) or about 94.3 in^2.	
Examine	Compared to the estimates, the answers make sense.	

Checking Your Understanding

Communicating Mathematics

Read and study the lesson to answer these questions. 1–3. See margin.

1. How is surface area important when you giftwrap a package?
2. **Define** surface area in your own words.
3. **Explain** how to find the surface area of a circular cylinder.
4. **State** an example from your daily life of a rectangular prism, a triangular prism, and a circular cylinder. See students' work.
5. A cube is a rectangular prism with square faces. Which size of box in the application at the beginning of the lesson is a cube? How do you know? $8 \times 8 \times 8$; all sides have the same length.

MODELING MATHEMATICS

MATERIALS
- small cereal boxes
- scissors
- ruler
- calculator

6. Work with a partner.
Measure the edges of a small cereal box. Sketch the box and label its dimensions. Then find a way to cut the box along the seams so you can lay it out flat. Sketch the figure formed.
   a. The figure you formed is a net for the box. What shape is each section of the net? rectangle
   b. Find the area of the cardboard used to make the box. See students' work.

Guided Practice

Name each solid. Then find the surface area. Round to the nearest tenth.

7. triangular prism; 184 ft^2
8. rectangular prism; 192.5 m^2
9. cylinder; 3166.7 m^2

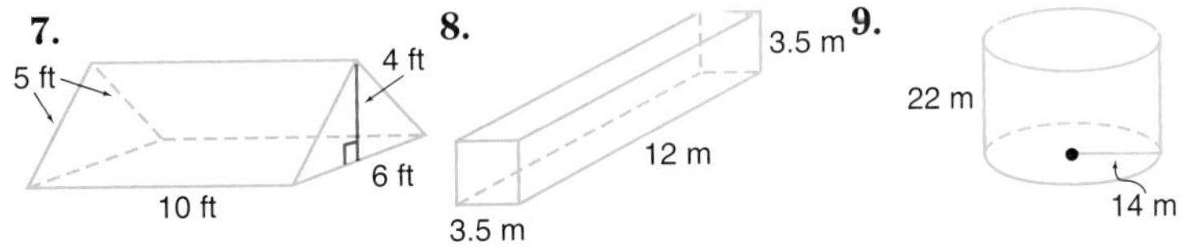

10. 28,274 ft^2; 57 gallons

10. **Building Maintenance** The walls of a cylindrical tank that holds natural gas need to be painted. If the tank is 40 feet high and has a radius of 50 feet, what is the area that needs to be painted? If a gallon of paint covers 500 ft^2, how many gallons of paint are needed?

3 PRACTICE/APPLY

Checking Your Understanding

Exercises 1–10 are designed to help you assess your students' understanding through reading, writing, speaking, and modeling. You should work through Exercises 1–6 with your students and then monitor their work on Exercises 7–10.

Additional Answers

1. You are covering all the surfaces with paper, so the area of the paper must be at least as large as the surface area of the package.
2. Sample answer: The amount of material needed to cover a solid figure.
3. Use πr^2 to find the area of each base; use $2\pi rh$ to find the area of the rectangular region. Then add.

Study Guide Masters, p. 104

NAME ______ DATE ______

12-5 Study Guide
Surface Area: Prisms and Cylinders

Student Edition Pages 632–637

To find the surface area of a solid, find the area of each surface. Then add the areas.

Rectangular Prism

2 cm, 5 cm, 3 cm

Area:
Top and bottom: $(A = \ell \times w)$ $2 \cdot (5 \cdot 3) = 30$ cm^2
Front and back: $(A = \ell \times w)$ $2 \cdot (2 \cdot 5) = 20$ cm^2
Two sides: $(A = \ell \times w)$ $+ 2 \cdot (2 \cdot 3) = 12$ cm^2
Total surface area $= 62$ cm^2

Cylinder

7 in., 14 in., 7 in.
curved surface The length is the circumference of the base. $\ell = 2\pi r$

Area:
Top and bottom: $(A = \pi r^2)$ $2 \cdot \pi \cdot (7)^2 \approx 307.7$ in^2
Curved surface: $(A = 2\pi r \cdot h)$ $+ 2 \cdot \pi \cdot 7 \cdot 14 \approx 615.8$ in^2
Total surface area ≈ 923.5 in^2

Find the surface area of each solid. Round to the nearest tenth.

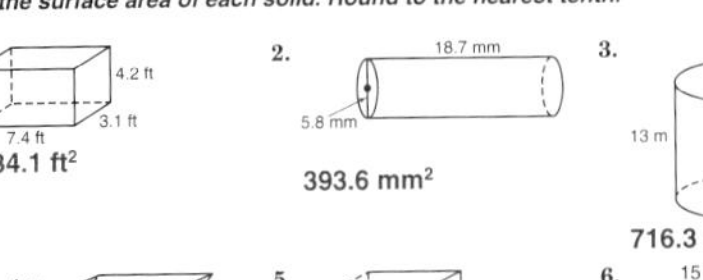

1. 134.1 ft^2
2. 393.6 mm^2
3. 716.3 m^2
4. (4 cm, 3 cm, 5 cm, 5 cm) 72 cm^2

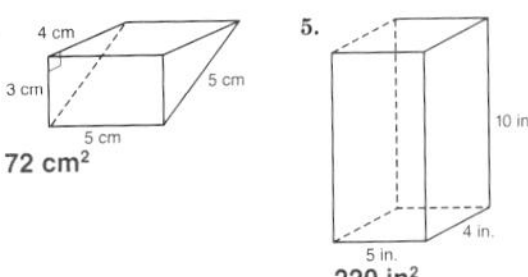

5. 220 in^2

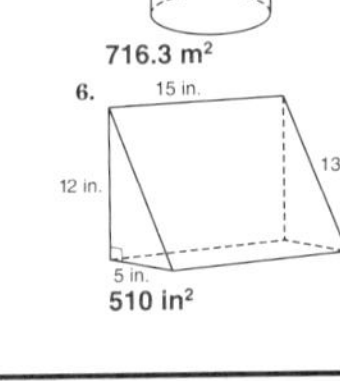

6. 510 in^2

Reteaching

Using Diagrams Provide students with small boxes and cans and 1-inch grid paper. Have students trace around the edges of each face of a box to create a diagram of the entire surface of the box. Then have them count the squares to approximate the surface area. Challenge students to come up with a method that will work for cylinders.

Error Analysis

When finding the surface area of a cylinder, students often become confused as to which formula to use. For these students, you may wish to have them derive and use the formula $A = \pi r(r + 2h)$.

Assignment Guide

Core: 11–29 odd, 30–34
Enriched: 12–22 even, 23–34

For **Extra Practice**, see p. 770.

The red A, B, and C flags, printed only in the Teacher's Wraparound Edition, indicate the level of difficulty of the exercises.

Additional Answer

23b. No, as long as the entire surface of the face of the smaller cube is touching a face of the larger cube, the surface area of the tower will be the same.

Exercises: Practicing and Applying the Concept

Independent Practice

Find the surface area of each solid. Round to the nearest tenth.

11.
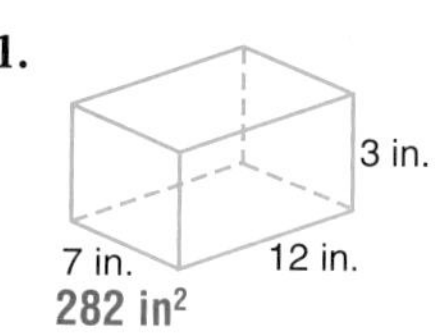

282 in^2

12.
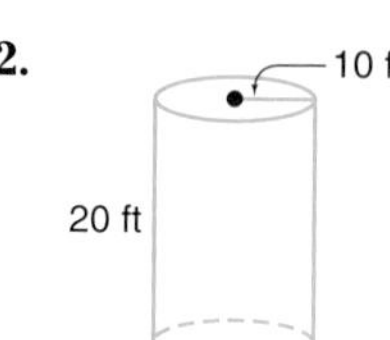

1884.9 ft^2

13.
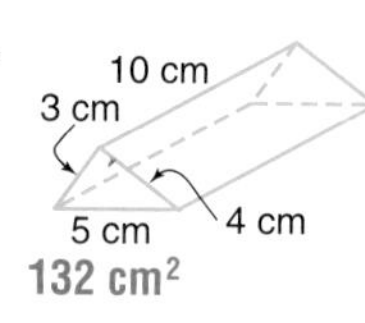

132 cm^2

14.
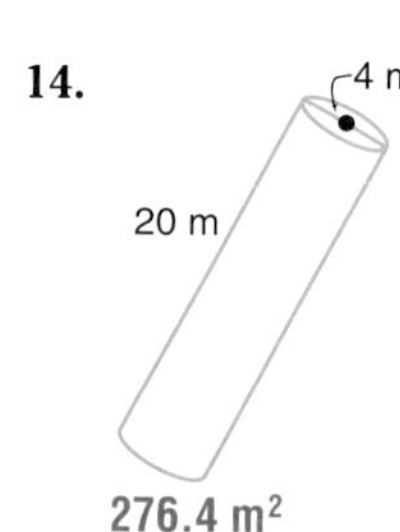

276.4 m^2

15.
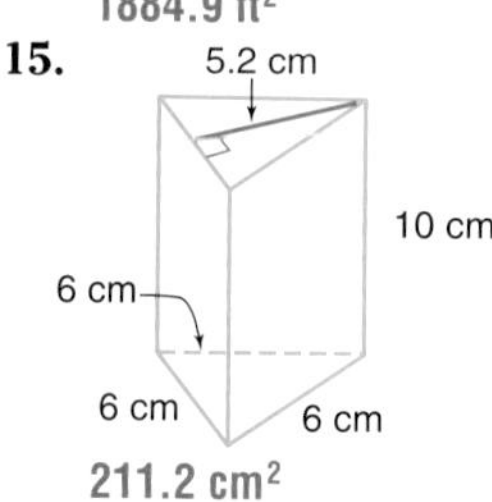

211.2 cm^2

16.
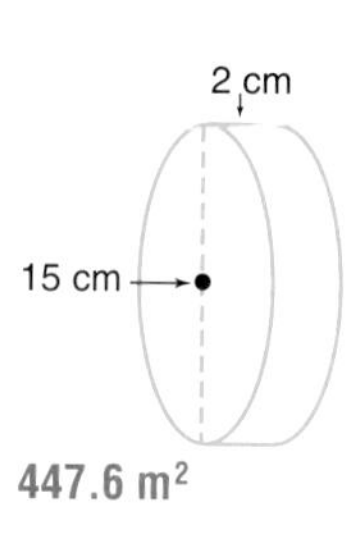

447.6 m^2

17.
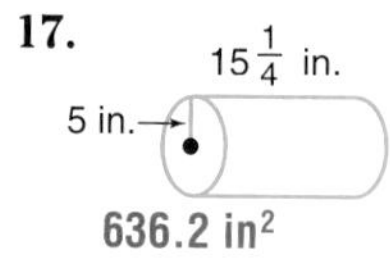

636.2 in^2

18.
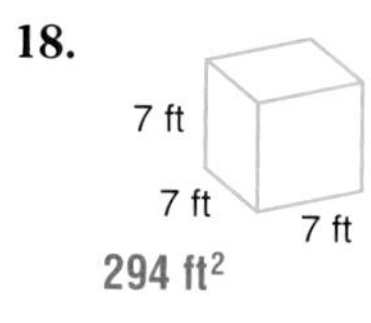

294 ft^2

19.
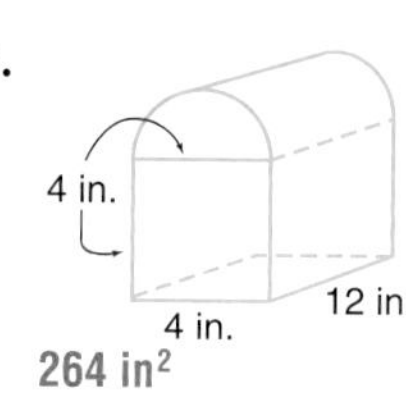

264 in^2

20. A cube has edges measuring a units. Find a formula for the surface area of the cube in terms of a. $S = 6a^2$

21. If you add 1 cm to the length of each side of a cube, which of the following is true about the surface area? d

a. It would increase by 1 cm^2. **b.** It would increase by 6 cm^2.
c. It would remain the same. **d.** need more information

22. Suppose you double the length of the sides of a cube. How is the surface area affected? 4 times greater

Critical Thinking

23. A cube with 12-inch sides is placed on a cube with 15-inch sides. Then a cube with 9-inch sides is placed on the 12-inch cube.

a. What is the surface area of the three-cube tower? 2250 in^2

b. Does the way each cube is placed on the one below it affect the surface area? Why or why not? See margin.

Applications and Problem Solving

24. Business Claire works after school at her family's tent company. One of their best-selling tents is an A-frame tent that is 4 feet high and has a rectangular bottom 4 feet wide by 6 feet long. The sides of the tent are 4.5 feet long. How much canvas is needed to make the tent? 94 ft^2

4.5 ft
4 ft
6 ft
4 ft

25. Manufacturing A carton of canned fruit cocktail holds 24 cans. Each can has a diameter of 7.6 cm and a height of 10.8 cm. How much paper is needed to make the labels for the 24 cans? about 6189 cm^2

Practice Masters, p. 104

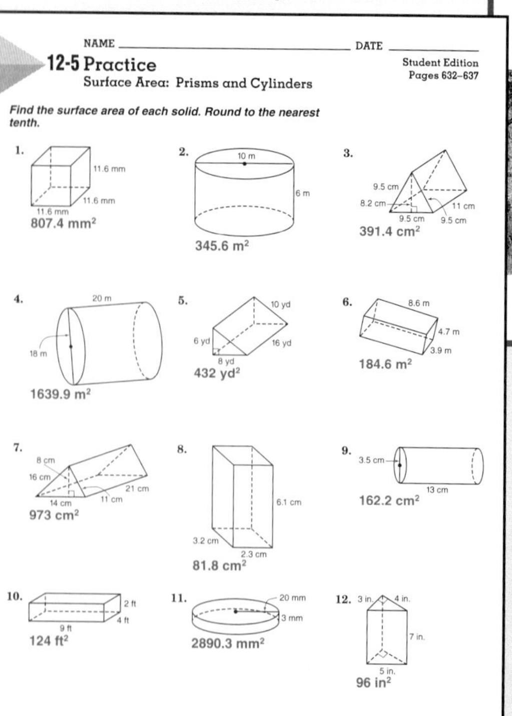

NAME ______ DATE ______

12-5 Practice
Surface Area: Prisms and Cylinders

Student Edition Pages 632–637

Find the surface area of each solid. Round to the nearest tenth.

1. 807.4 mm^2
2. 345.6 m^2
3. 391.4 cm^2
4. 1639.9 m^2
5. 432 yd^2
6. 184.6 m^2
7. 973 cm^2
8. 81.8 cm^2
9. 162.2 cm^2
10. 124 ft^2
11. 2890.3 mm^2
12. 96 in^2

Group Activity Card 12-5

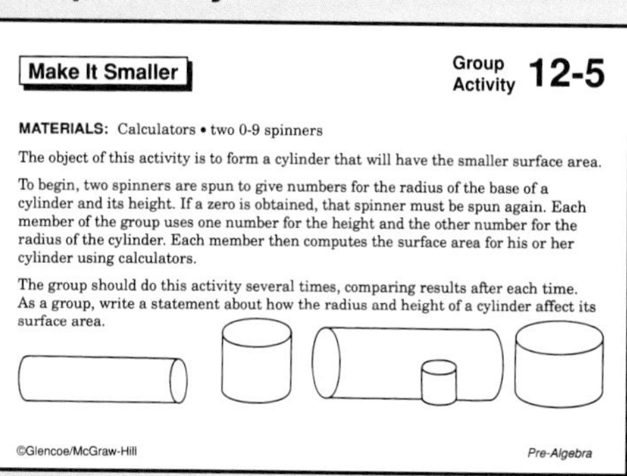

Make It Smaller Group Activity **12-5**

MATERIALS: Calculators • two 0-9 spinners

The object of this activity is to form a cylinder that will have the smaller surface area.

To begin, two spinners are spun to give numbers for the radius of the base of a cylinder and its height. If a zero is obtained, that spinner must be spun again. Each member of the group uses one number for the height and the other number for the radius of the cylinder. Each member then computes the surface area for his or her cylinder using calculators.

The group should do this activity several times, comparing results after each time. As a group, write a statement about how the radius and height of a cylinder affect its surface area.

 Pre-Algebra

26. **Home Improvement** Juan's uncle wants to put drywall on his garage walls and ceiling. The garage is 24 feet by 20 feet, and the walls are $12\frac{1}{2}$ feet high. The drywall he will be using is sold in sheets that are 4 feet wide by 12 feet high by $\frac{1}{2}$ inch thick. How many sheets of drywall will he need to purchase? **33 sheets**

27. **Postal Service** Refer to the application at the beginning of the lesson. How much cardboard does each box take if there are pieces that overlap completely at the top and bottom? Is the ratio of price to material about the same for each box? **512 in^2, 1200 in^2, 1800 in^2; no**

28. **Architecture** A hallway in a Spanish-style house has a semi-circular arch above the walls. The width of the hall and the diameter of the arch are 6 feet. The walls are eight feet from the floor to the beginning of the arch. The hall is 20 feet long. If a gallon of paint covers about 400 square feet, how much paint will be needed to paint the hallway with two coats? **3 gallons**

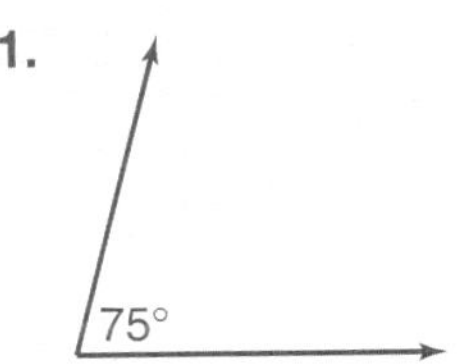

29. **Family Activity** Choose a room in your home. Sketch the room and measure the walls, windows, and doors. **See students' work.**
    a. Find the area of the walls that would be painted or wallpapered.
    b. Visit a paint store and select a paint that would be appropriate for the room. Find the cost for two coats of paint for the room.

Mixed Review

30. **Farming** A farmer attaches a 25-foot rope to the corner of a 10-foot by 15-foot shed to tie up the family's pet goat. How much grazing area does the goat have? (Lesson 12-4) **1727.9 ft^2**

31. **Geometry** Draw an angle that measures 75°. (Lesson 11-1) **31. See margin.**

32. Find the value C(12, 6). (Lesson 10-6) **924**

33. Write 3.605×10^3 in standard form. (Lesson 6-9) **3605**

34. A house sold for more than 3 times its original purchase price of $32,000. Define a variable and translate the sentence into an inequality. Then solve. (Lesson 3-8) **See margin.**

WORKING ON THE

Investigation

OH GIVE ME A HOME

Refer to the Investigation on pages 606–607.

Rooms in a home can be separated into *private*, *service*, and *social zones*. Bedrooms and bathrooms are part of the private zone. The service zone includes the kitchen, laundry, and garage. The social zone is an area where members of the household gather to spend time together.

The three zones don't have to be in different rooms. Many kitchen tables are used to do homework, thus making it a private zone. A kitchen becomes a social room when people gather to eat a meal.

Begin working on the Investigation by doing a group activity to design a model of a 2-bedroom house. The house you design must include all three zones.

Add the results of your work to your Investigation Folder.

4 ASSESS

Closing Activity

Modeling Write measurements from 1 inch to 12 inches on slips of paper and place them in a bag. Draw a number for the height and another number for the radius of a cylinder. Have students sketch the cylinder and determine its surface area. Then draw three slips of paper for the height, width, and length of a rectangular prism. Have students find the surface area of a box with those dimensions.

Additional Answers

31. 75°

34. $v > 3(32{,}000)$; $v > \$96{,}000$

Enrichment Masters, p. 104

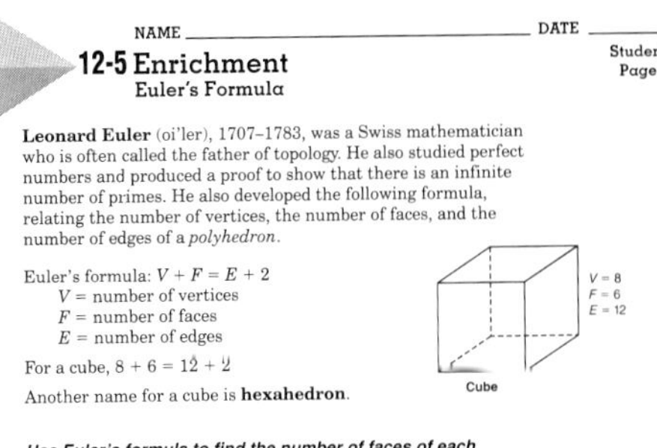

NAME ____________ DATE ________

12-5 Enrichment Student Edition Pages 632–637
Euler's Formula

Leonard Euler (oi'ler), 1707–1783, was a Swiss mathematician who is often called the father of topology. He also studied perfect numbers and produced a proof to show that there is an infinite number of primes. He also developed the following formula, relating the number of vertices, the number of faces, and the number of edges of a *polyhedron*.

Euler's formula: $V + F = E + 2$
V = number of vertices
F = number of faces
E = number of edges

For a cube, $8 + 6 = 12 + 2$

Another name for a cube is **hexahedron**.

Cube: V = 8, F = 6, E = 12

Use Euler's formula to find the number of faces of each polyhedron.

1. tetrahedron; $V = 4$, $E = 6$ **4 faces**
2. octahedron; $V = 6$, $E = 12$ **8 faces**

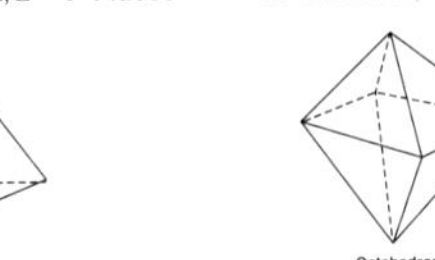

3. icosahedron; $V = 12$, $E = 30$ **20 faces**
4. dodecahedron; $V = 20$, $E = 30$ **12 faces**

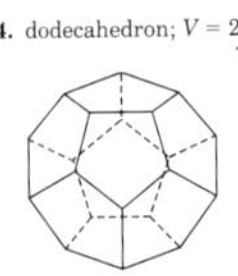

The suffix -hedron comes from the Greek language meaning "face." Find the meaning of each of the following prefixes.

5. hexa- **six** 6. tetra- **four** 7. octa- **eight** 8. icosa- **twenty** 9. dodeca- **twelve**

Extension

Using Consumer Connections

Moesha buys a poster as a gift for a friend. She rolls the poster into a tube 2 feet long and 2 inches in diameter. Find the amount of wrapping paper Moesha needs to buy to wrap the poster, including enclosing the ends. **about 160 square inches**

The Investigation on pages 606 and 607 is designed to be a long-term project that is completed over several days or weeks. Encourage students to keep their materials in their Investigation Folder as they work on the Investigation.

12-6 LESSON NOTES

NCTM Standards: 1-4, 12, 13

Instructional Resources

- Study Guide Master 12-6
- Practice Master 12-6
- Enrichment Master 12-6
- Group Activity Card 12-6
- Assessment and Evaluation Masters, p. 324

Transparency 12-6A contains the 5-Minute Check for this lesson; **Transparency 12-6B** contains a teaching aid for this lesson.

Recommended Pacing	
Standard Pacing	Day 8 of 13
Honors Pacing	Day 7 of 11
Block Scheduling*	Day 5 of 7

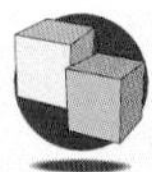

*For more information on pacing and possible lesson plans, refer to the ***Block Scheduling Booklet***.

1 FOCUS

5-Minute Check
(over Lesson 12-5)

Find the surface area of each prism or cylinder. Round to the nearest whole number.

1. 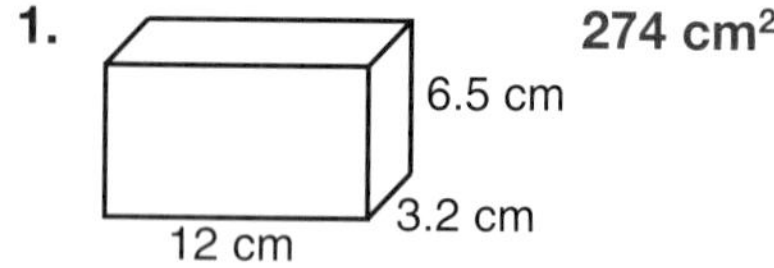

274 cm²

2. 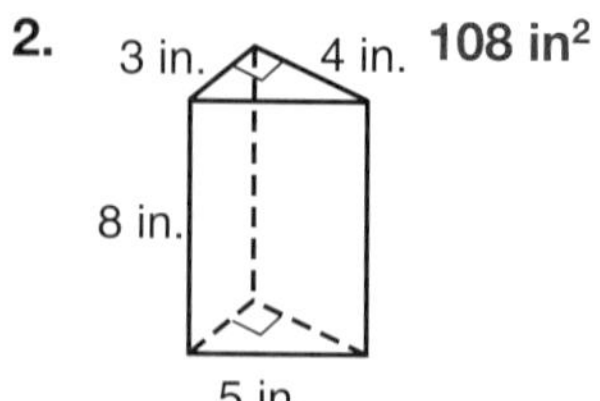

108 in²

3. 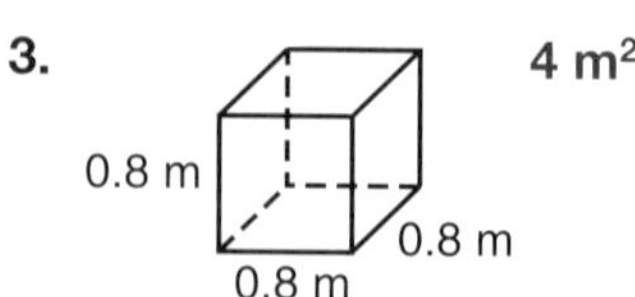

4 m²

4. 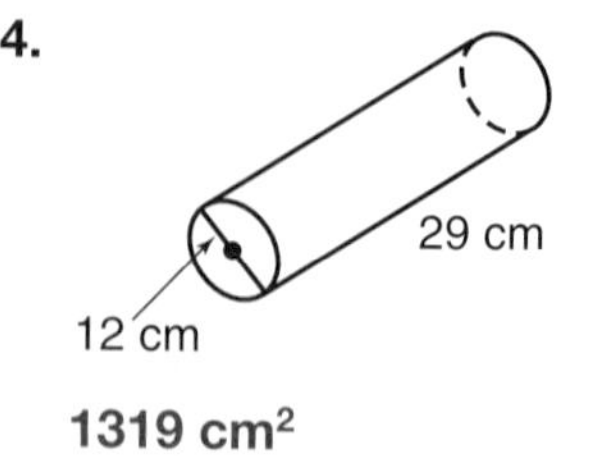

1319 cm²

12-6 Surface Area: Pyramids and Cones

Setting Goals: *In this lesson, you'll find surface areas of pyramids and cones.*

Modeling a Real-World Application: Architecture

FYI

I.M. Pei also designed the Rock-and-Roll Hall of Fame in Cleveland, Ohio, which opened in September, 1995. About the Hall, Pei said, "Architecture is form and space. It's my interpretation of what the music is."

The pyramids of Egypt were built thousands of years ago as burial tombs for the pharaohs and their relatives. Pyramids have also inspired modern architects. The noted Chinese-American architect I. M. Pei completed renovation of the Louvre Museum in Paris in 1990. For the new entrance, he designed a huge square glass pyramid. It has a square base that is 116 feet on each side. How much glass did it take to cover the pyramid? *This problem will be solved in Example 2.*

Learning the Concept

In geometry, a **pyramid** is a solid figure that has a polygon for a base and triangles for the sides. We name pyramids by the shape of their bases. Notice that a pyramid has just one base. All the other faces of a pyramid, the *lateral* faces, intersect at a point called the **vertex**. The pyramid shown at the right is a **square pyramid**.

The pyramid shown at the left is a **triangular pyramid**. Its base is an equilateral triangle, so its lateral faces are congruent triangles.

Let's look at a net of a square pyramid. Notice that the lateral faces are triangles. Because the base is a square, the triangles are all congruent, and their altitudes all have the same length. The length of any one of these altitudes is called the **slant height** of the pyramid.

THINK ABOUT IT

If the base of a pyramid is a pentagon, how many faces would the pyramid have? Would the sides still be shaped like triangles?

6; yes

Square Pyramid

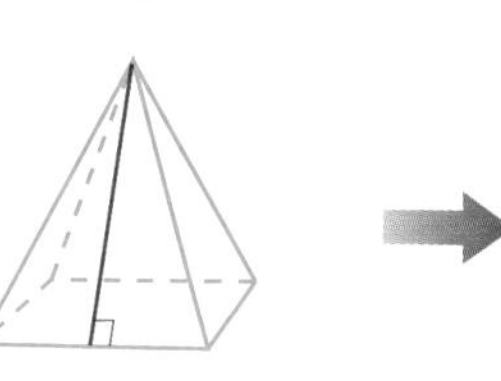

Net of Square Pyramid

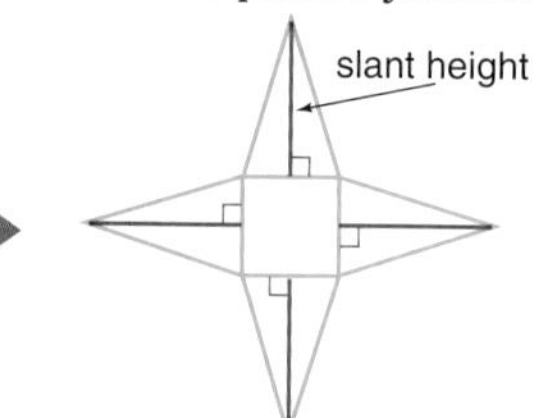

Alternative Learning Styles

Reading Mathematics Have students make index cards for the vocabulary words contained in this chapter. Each card should define one term as well as show a sketch or diagram.

FYI

The Luxor Casino in Las Vegas is another modern-day pyramid, which opened in November, 1993.

The surface area of a pyramid is the sum of the areas of its base and its lateral faces.

Example  **Find the surface area of the pyramid.**

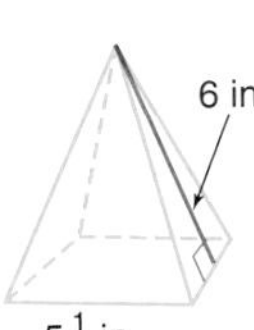

The base is a square. Use the formula $A = s^2$ to find the area. *Estimate: 5 × 5 = 25 and 6 × 6 = 36. The area is about 30 in².*

Use the formula $A = \frac{1}{2}bh$ to find the area of each triangular side.

Estimate: $\frac{1}{2} \cdot 5 \cdot 6 = 15$. The area of the four sides is about 60 in².

Base	Each Triangular Side
$A = s^2$	$A = \frac{1}{2}bh$
$A = \left(5\frac{1}{2}\right)^2$	$A = \frac{1}{2} \cdot 5\frac{1}{2} \cdot 6$
$A = 30\frac{1}{4}$ or 30.25	$A = 16\frac{1}{2}$ or 16.5

Add to find the total surface area.

+
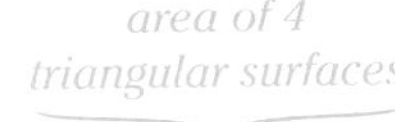

=

30.25 [+]  4 [×] 16.5  [=] 96.25

The surface area is 96.25 in². *Compare with the estimate.*

In the application at the beginning of the lesson, we need to find the **lateral surface** of the pyramid, the surface not including the base.

Example 2 APPLICATION **Architecture**

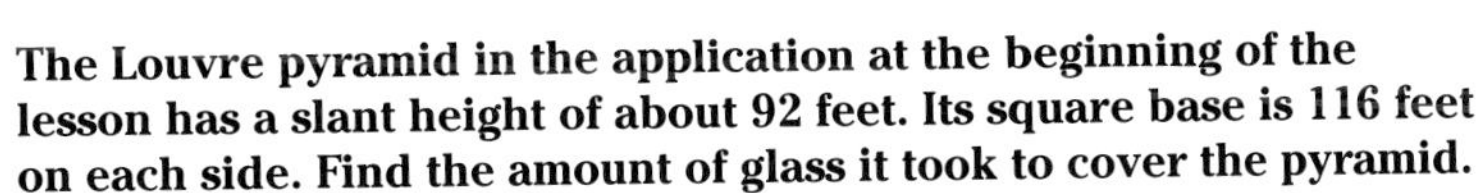

The Louvre pyramid in the application at the beginning of the lesson has a slant height of about 92 feet. Its square base is 116 feet on each side. Find the amount of glass it took to cover the pyramid.

You are covering the lateral surface of the pyramid. Use the formula $A = \frac{1}{2}bh$ to find the area of each triangular face.

$A = \frac{1}{2}bh$

$A = \frac{1}{2} \cdot 116 \cdot 92$ *Replace b with 116 and h with 92.*

$A = 5336$

There are four triangular faces. Multiply to find the lateral surface area.

4 [×] 5336 [=] 21344

It took 21,344 ft² of glass to cover the pyramid.

In this textbook, when we say cone, assume we are talking about a circular cone.

A cone is another three-dimensional shape. You see cones in everyday life such as the cone for some ice cream cones, a holder for cotton candy, or some paper drinking cups. Most of the cones that you see are called **circular cones** because the base is a circle.

Motivating the Lesson

Hands-On Activity Bring in or make a pyramid and a circular cone. Patterns are available in the Math Lab and Modeling Math Masters, pages 22–23. Take them apart and ask students how they would find the surface area.

2 TEACH

In-Class Examples

For Example 1

Find the surface area of this square pyramid. Round to the nearest tenth. **148.8 in²**

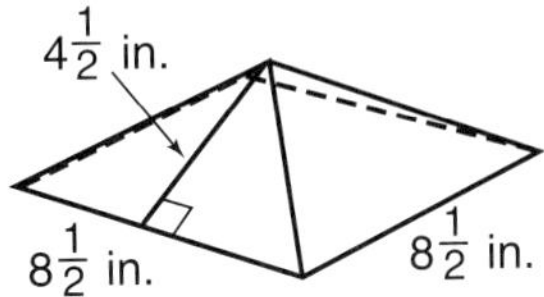

For Example 2

A pyramid with a triangular base is called a tetrahedron. Find the surface area if you make a tetrahedron with four congruent equilateral triangles that have a base of 10 centimeters and an altitude of 11.2 centimeters. **224 cm²**

Teaching Tip You may want to display several prisms and pyramids, including triangular ones of each type. Call out a name and have students hold up an example of that kind of pyramid or prism.

In-Class Example

For Example 3

Find the surface area of this circular cone. Round to the nearest tenth. **251.3 m²**

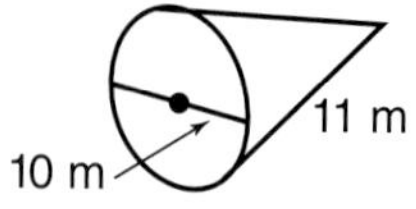

3 PRACTICE/APPLY

Checking Your Understanding

Exercises 1–8 are designed to help you assess your students' understanding through reading, writing, speaking, and modeling. You should work through Exercises 1–4 with your students and then monitor their work on Exercises 5–8.

Study Guide Masters, p. 105

NAME ______________ DATE ________

12-6 Study Guide
Surface Area: Pyramids and Cones

Student Edition
Pages 638–642

To find the surface area of a pyramid, add the areas of each of the faces and the area of the base.

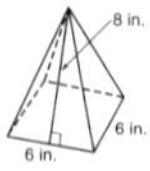

Area:

Square Base
$A = s^2$
$= 6^2$
$= 36 \text{ in}^2$

Four Triangular faces
$A = 4\left(\frac{1}{2} \cdot b \cdot h\right)$
$= 4\left(\frac{1}{2} \cdot 6 \cdot 8\right)$
$= 96 \text{ in}^2$

Surface area is 36 + 96 or 132 in².

To find the surface of a cone, add the area of its circular base and the area of its lateral surface. The area of the lateral surface is πrs.

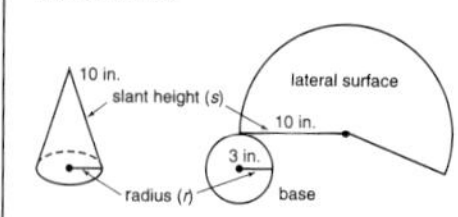

Area:

Circular Base
$A = \pi r^2$
$\approx (3.14)(3)^2$
$\approx 28.26 \text{ in}^2$

Lateral Surface
$A = \pi rs$
$\approx (3.14)(3)(10)$
$\approx 94.2 \text{ in}^2$

Surface area is about 28.26 + 94.2 or 122.46 in².

Find the surface area of each pyramid or cone. Round to the nearest tenth.

1. 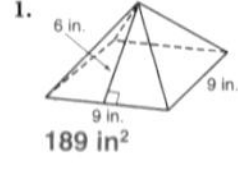

189 in²

2. 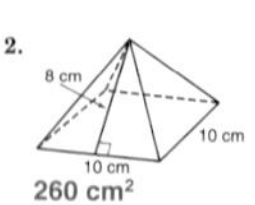

260 cm²

3. 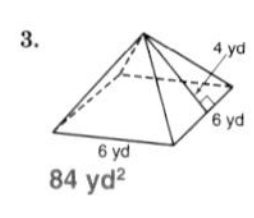

84 yd²

4. 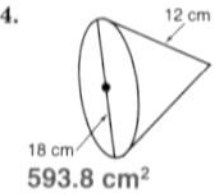

593.8 cm²

5. 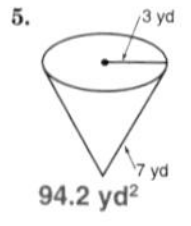

94.2 yd²

6. 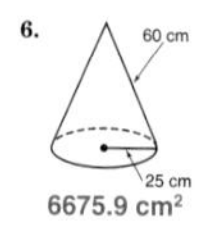

6675.9 cm²

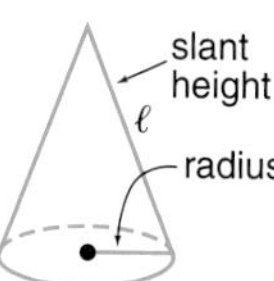

The net of a cone shows the regions that make up the cone. You can see that the base of the cone is a circle, with area πr^2. If you cut the lateral surface into sections, you can form a parallelogram. The base of the parallelogram is half the circumference of the cone, or $\frac{1}{2} \cdot 2\pi r$. Its height is the slant height of the cone, ℓ. So the area of the lateral surface is $A = \frac{1}{2} \cdot 2\pi r \cdot \ell$, or $\pi r\ell$. You can find the total surface area SA of a cone using the formula $SA = \pi r^2 + \pi r\ell$.

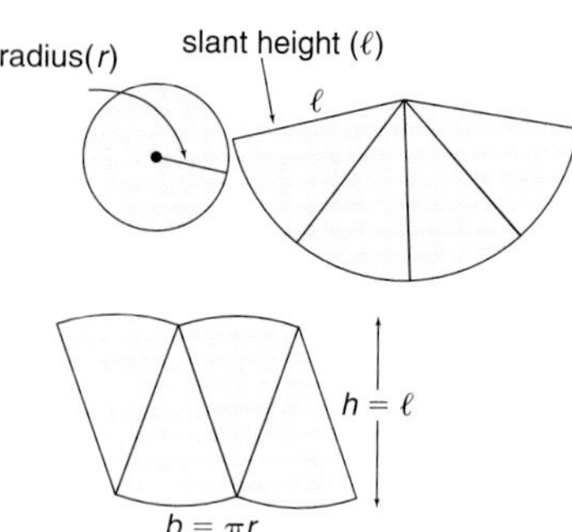

Example  **Find the surface area of a circular cone with a slant height of 10 cm and a base with a radius of 5 cm. Round to the nearest tenth.**

Base	**Lateral Surface**
$A = \pi r^2$	$A = \pi r\ell$
$A = \pi \cdot 5^2$	$A = \pi \cdot 5 \cdot 10$
$A \approx 78.5$	$A \approx 157.1$

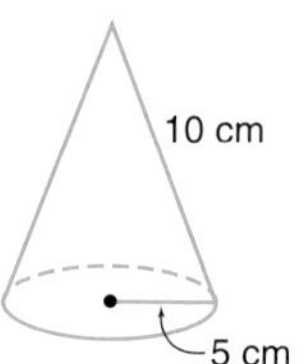

Add to find the total surface area.

78.5 [+] 157.1 [=] *235.6*

The surface area is about 235.6 cm².

Checking Your Understanding

Communicating Mathematics

Read and study the lesson to answer these questions. **1–3. See margin.**

1. **Describe** the slant height of a pyramid and of a cone.
2. **Explain** why the expression $2\pi r$ can be used to find the perimeter of the base of a circular cone.
3. **State** how to find the area of the lateral surfaces of a pyramid.

MATH JOURNAL

4. **Assess Yourself** Write about the meaning of the term *surface area*. Then list some situations in everyday life when you might use surface area. **See students' work.**

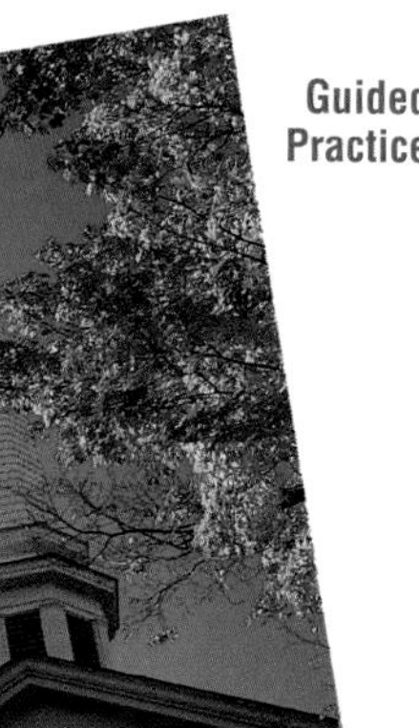

Guided Practice

Name each shape. Then find its surface area. Round to the nearest tenth.

5. (8.2 m; 6 m; 6 m) **pyramid; 134.4 m²**

6. (5 cm; 12 cm) **cone; 267 cm²**

7. (6 in.; 5 in.; 6 in.; 6 in.; 4 in.) **triangular pyramid; 57 in²**

8. **Construction** A church spire is being built in the shape of a regular octagonal pyramid. Each edge of the base is 8 feet long. The slant height of the spire is about 75.3 feet. How much roofing will be needed to cover the spire? **about 2410 ft²**

Reteaching

Using Models Have students construct a pyramid by cutting out four equilateral triangles and a square whose sides are the same length as the sides of the triangles. Have students remove one of the triangular sides and measure the height of the pyramid and an altitude of the triangle. Then have them compare results.

Additional Answers

1. It is the altitude of a triangular face of a pyramid; it is the length from the vertex of a cone to the edge of its base.
2. The base of a circular cone is a circle; the perimeter of a circle is $2\pi r$.
3. Use $A = \frac{1}{2}bh$ to find the area of each side. Add the areas of the sides.

Exercises: Practicing and Applying the Concept

Independent Practice

Find the surface area of each solid. Round to the nearest tenth.

A

9.

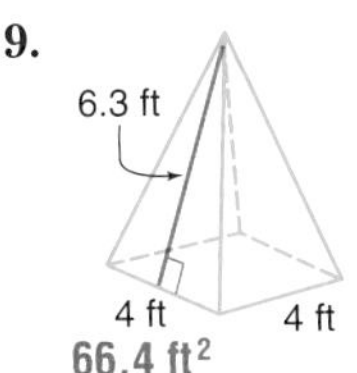

66.4 ft^2

10.

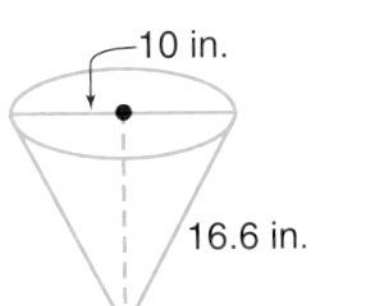

339.3 in^2

11. 11.4 m, 7 m

404.6 m^2

B

12.

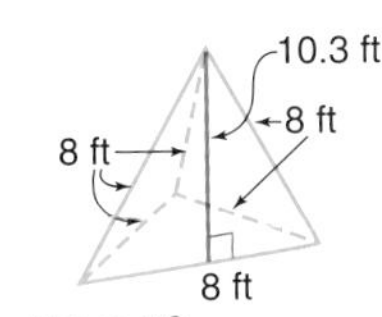

164.8 ft^2

13.

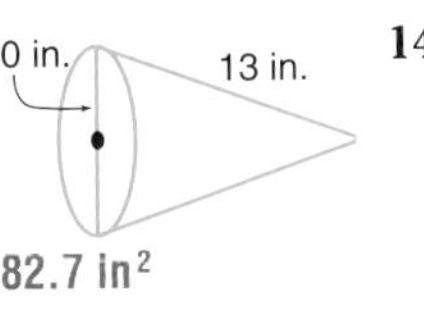

282.7 in^2

14.

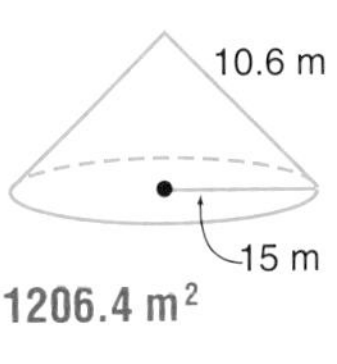

1206.4 m^2

C

15.

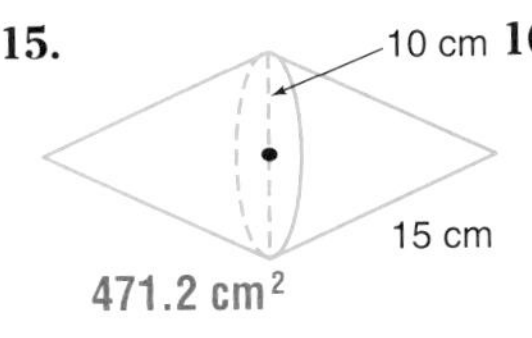

471.2 cm^2

16. 10.6 m, $A = 48$ m^2, 8.3 m

311.9 m^2

17.

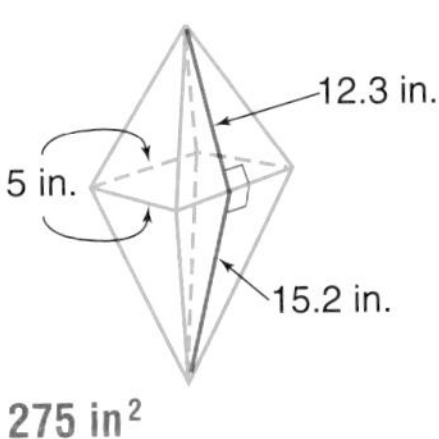

275 in^2

Critical Thinking

18. Manufacturing A bar of lead 13 inches by 2 inches by 1 inch is melted and recast into 100 conical fishing sinkers that have a diameter of 1 inch and a height of 1 inch. The slant height of each sinker is about 1.1 inches. How does the total surface area of all the sinkers compare to the surface area of the original lead bar? Sinkers: about 251 in^2, original bar: 82 in^2; the surface area is about tripled.

Applications and Problem Solving

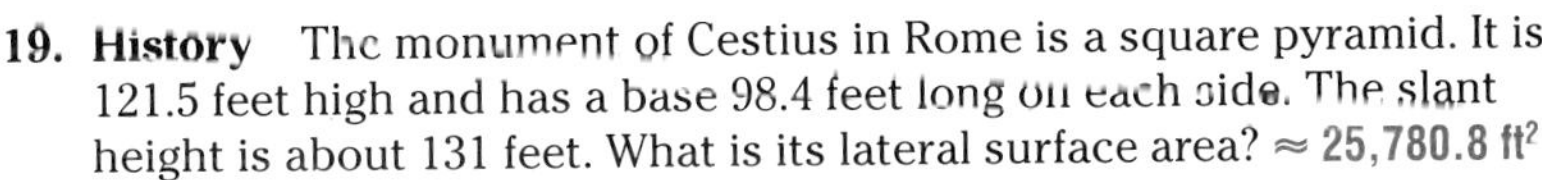

19. History The monument of Cestius in Rome is a square pyramid. It is 121.5 feet high and has a base 98.4 feet long on each side. The slant height is about 131 feet. What is its lateral surface area? $\approx$ 25,780.8 ft^2

20. Architecture I. M. Pei also designed smaller square pyramids that stand near the entrance of the Louvre. These pyramids have bases 26 feet on a side and a slant height of about 20.6 feet. How much glass did it take to cover one of these pyramids? $\approx$ 1071.2 ft^2

21. Tepees The largest tepee in the United States has a diameter of 42 feet and a slant height of about 47.9 feet. The tepee belongs to Dr. Michael Doss of Washington, D.C. Dr. Doss is a member of Montana's Crow Tribe. How much canvas was used to make the cover of the tepee? 3160.1 sq. ft

22. Food An ice cream shop uses sugar cones that are 5 inches long and have a diameter of 2 inches. The manager would like to order special napkins that would fit around the cone twice. Sketch the shape of the napkin and label its dimensions. See margin.

Assignment Guide

Core: 9–17 odd, 18, 19, 21, 23–27
Enriched: 10–16 even, 18–27
All: Self Test, 1–9

For **Extra Practice**, see p. 770.

The red A, B, and C flags, printed only in the Teacher's Wraparound Edition, indicate the level of difficulty of the exercises.

Additional Answer

22.

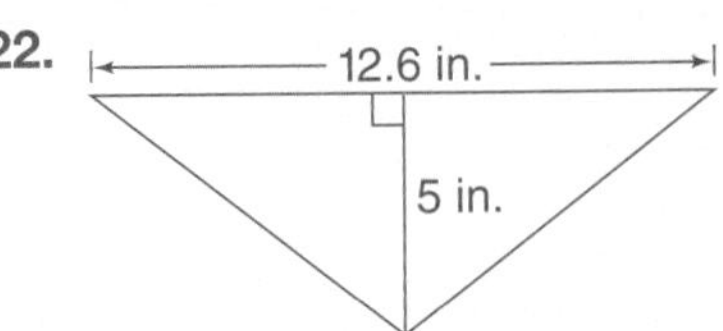

Practice Masters, p. 105

NAME ______ DATE ______

12-6 Practice
Surface Area: Pyramids and Cones

Student Edition Pages 638–642

Find the surface area of each pyramid or cone. Round to the nearest tenth.

1. 18 ft, 10 ft — 879.6 ft^2

2. 8 m, 8 m, 7 m, 8 m, 8 m — 112 m^2

3. 14 cm, 9 cm — 650.3 cm^2

4.

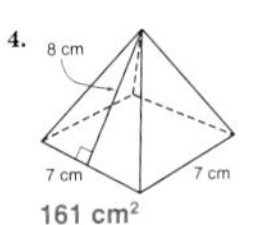

161 cm^2

5.

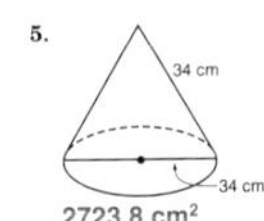

2723.8 cm^2

6.

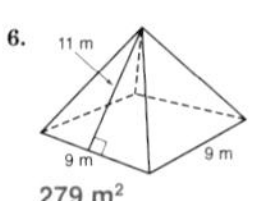

279 m^2

7.

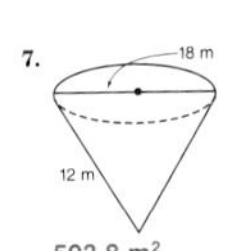

593.8 m^2

8.

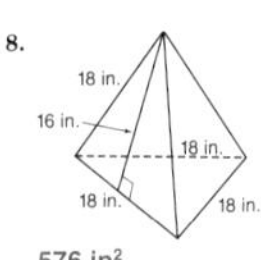

576 in^2

9.

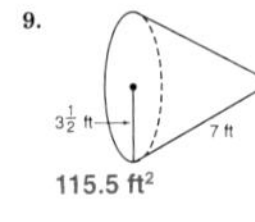

115.5 ft^2

Extension

Using Formulas Find the area of a sector of each circle given the measurements of its radius and central angle. Use $A = \frac{n}{360} \times \pi r^2$ where n is the degree measure of the central angle. Round your answers to the nearest tenth.

1.

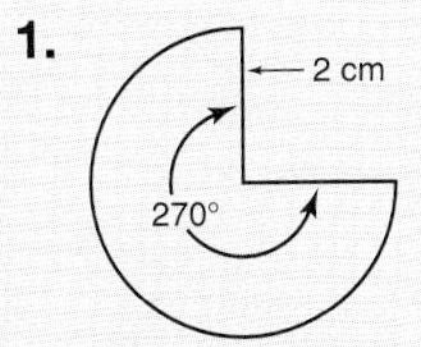

9.4 cm^2

2.

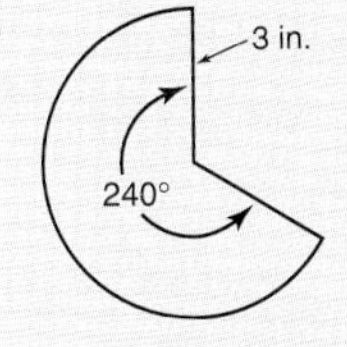

18.8 cm^2

4 ASSESS

Closing Activity

Speaking Have students explain the difference between the slant height and the altitude of a pyramid or cone. Then have students explain how to find the surface area of a square pyramid or a circular cone.

Chapter 12, Quiz C (Lessons 12-5 and 12-6) is available in the *Assessment and Evaluation Masters*, p. 324.

Mixed Review

23. **Geometry** Find the surface area of the rectangular prism at the right. (Lesson 12-5) **900 cm²**

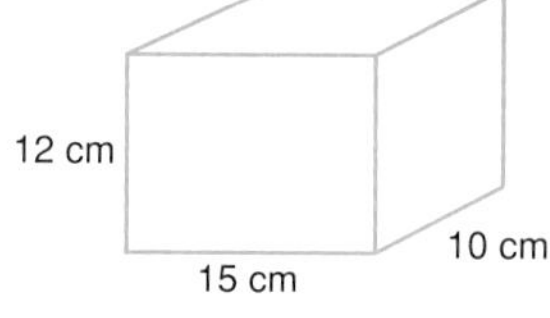

24. **Probability** If two coins are tossed, what is the probability that one will land heads up and the other tails up? (Lesson 9-3) $\frac{1}{2}$

25. Define a variable and write and solve an equation for the following situation. Four times a number plus eight is 20. What is the number? (Lesson 7-3) **$4x + 8 = 20$; 3**

26. Solve $w + \frac{1}{6} = 3\frac{2}{3}$. (Lesson 5-6) $3\frac{1}{2}$

27. Simplify $(5x)(-3y)$. (Lesson 2-7) **$-15xy$**

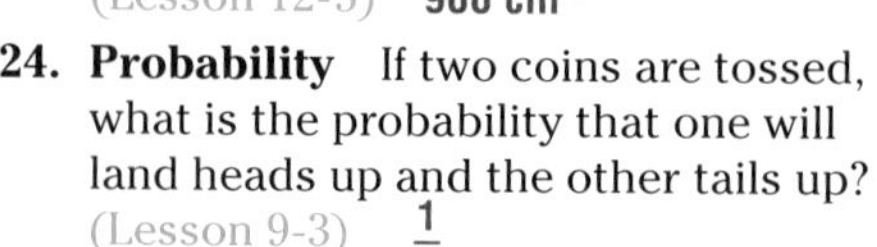

Self Test

Find the area of each figure. (Lessons 12-1 and 12-2)

1. 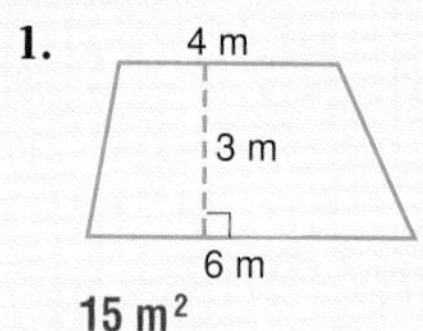

15 m²

2. 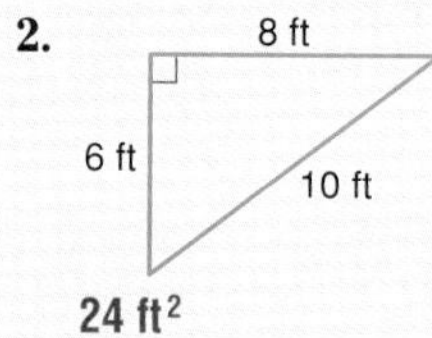

24 ft²

3. 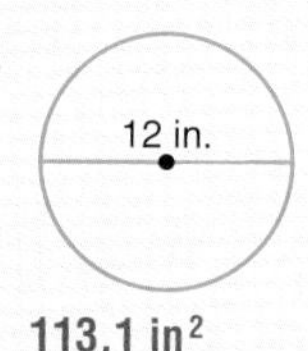

113.1 in²

Name each shape. Then find its surface area. Round to the nearest tenth. (Lessons 12-5 and 12-6)

4. 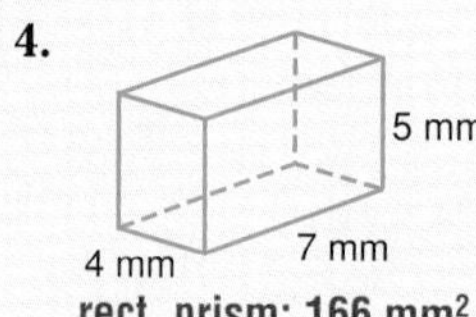

rect. prism; 166 mm²

5. 8 in. 12 in.
cylinder; 402.1 in²

6. 9 m 8 m 8 m
square pyramid; 208 m²

Solve.

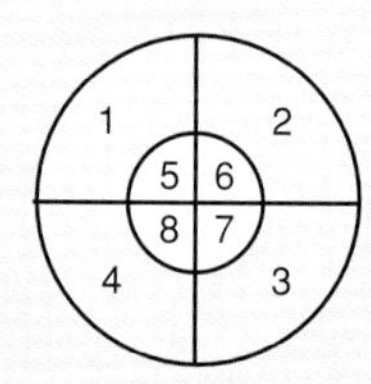

7. **Probability** In a dart game, you win points based on the region in which the dart lands. The diameter of the whole target is 24 inches. The diameter of the inner circle is 8 inches. What is the probability that the dart lands in the 8-point region? (Assume the dart lands on the target.) (Lesson 12-3) **≈ 0.028 or about $\frac{3}{100}$**

8. **Make a Drawing** A circular fountain has a diameter of 8 feet. A flower garden is planted around the fountain. The garden extends 12 feet all around the fountain. If a pound of organic fertilizer covers about 300 square feet, how much fertilizer is needed for the garden? (Lesson 12-4) **3 lbs.**

9. How much paper would be used to make an open paper cone with a diameter of 5 inches and a slant height of 6 inches? (Lesson 12-6) **47.1 in²**

Enrichment Masters, p. 105

NAME ________________ DATE ________

12-6 Enrichment Student Edition Pages 638–642
Line Symmetry

Some geometric figures can be folded in half so that one part will fit exactly on the other. This property is called **line symmetry**. Figures that have line symmetry are often used in designs ranging from building plans to corporate logos.

When a figure is folded, the fold is part of a straight line. Thus, it is referred to as *line of symmetry*. The figure is *symmetrical* with respect to that line.

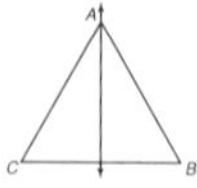

This isosceles triangle is symmetrical with respect to a line that is perpendicular to its base. Notice that if the triangle is folded on this line, point *B* would coincide exactly with point *C*.

For each figure, draw each line of symmetry. If a figure has no line of symmetry, write none.

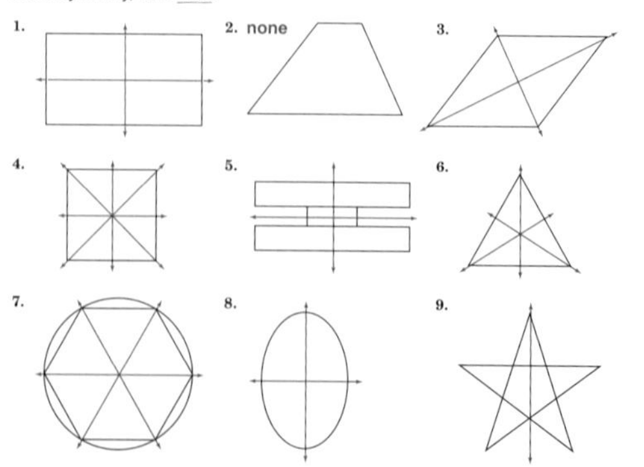

Group Activity Card 12-6

Line Them Up Group Activity **12-6**

MATERIALS: Note cards

In this activity, group members compare the surface areas of pyramids or cones.

A person who is "it" decides if the solids used will be pyramids or cones. Then the other members choose two numbers between 15 and 40, not more than 10 apart. Each member writes the numbers on the front of a note card, identifying one as the slant height and the other as the radius of the circle or the side of the square base.

Each member then computes the surface area of their solid and writes it on the back of their card. When everyone is finished, the cards are put face up on the table. The person who is "it" must place the cards in order so the surface areas go from smallest to largest. He or she gets one point for each card that is placed correctly.

Repeat until each person in the group has been "it." The player with the most points after one round wins the game.

Pre-Algebra

Self Test

The Self Test provides students with a brief review of the concepts and skills in Lessons 12-1 through 12-6. Lesson numbers are given to the right of exercises or instruction lines so students can review concepts not yet mastered.

HANDS-ON ACTIVITY

12-7A Volume

A Preview of Lesson **12-7**

MATERIALS

- 5 × 8 inch index cards
- tape
- rice
- grid paper

In this activity, you will investigate volume by making containers of different shapes and comparing how much each container holds.

Your Turn

Work with a partner.

- Use three 5 × 8 cards to make three containers with a height of 5 inches as shown. Make one with a square base (with 2 in. sides.), one with a triangular base (with sides of 2 in., 3 in., and 3 in.) and one with a circular base.
- Then tape one end of each container to another card as a bottom, but leave the top open.
- Estimate which container would hold the most (have the greatest volume) and which would hold the least (have the least volume). Do you think each container would hold the same amount?
- Use rice to fill the container that you believe holds the least amount. Then put the rice into another container. Does the rice fill this container? Continue the process until you find out which, if any, container has the least volume and which has the greatest.

TALK ABOUT IT

1. Which container held the most? Which held the least?
2. How do the heights of the three containers compare? What is each height? **same height; 5 in.**
3. Compare the perimeters of the bases of each container. What is each base perimeter? **same; 8 in.**
4. Trace the base of each container onto grid paper. Estimate the area of each base. **circular: ≈ 5 in^2; triangular: ≈ 3 in^2; square: 4 in^2**
5. Which container has the greatest base area? **cylinder**
6. Does there appear to be a relationship between the area of the base and the volume? Explain. **Yes; the greater the base area, the greater the volume.**

1. triangular prism; cylinder

12-7A LESSON NOTES

NCTM Standards: 1-4, 12, 13

Objective

Investigate volume by making containers of different shapes and comparing how much each container holds.

Recommended Time

Demonstration and discussion: 15 minutes; Exercises: 30 minutes

Instructional Resources

For each student or group of students

- 5 × 8 inch index cards
- tape
- rice

Math Lab and Modeling Math Masters

- p. 7 (grid paper)
- p. 52 (worksheet)

For teacher demonstration

Overhead Manipulative Resources

1 FOCUS

Motivating the Lesson

On the chalkboard or overhead, draw a circle with 4-inch diameter, a square with sides 3 inches long, and a triangle with sides of 3, 4, and 5 inches. Ask students which one has the greatest area. **circle**

2 TEACH

Teaching Tip Demonstrate how to make the containers by taping only the 5-inch side. To make the cylinder, roll the card up several times to help it stay round. Fold a card in half twice to make a square prism. The triangular prism requires two folds of a card. You may want groups to work with rice inside of a shoe box to contain any that is spilled. If materials are not available for small group work, demonstrate the Your Turn section for the entire class.

3 PRACTICE/APPLY

Assignment Guide

Core: 1–6
Enriched: 1–6

4 ASSESS

Observing students working in cooperative groups is an excellent method of assessment.

12-7 LESSON NOTES

NCTM Standards: 1-4, 12, 13

Instructional Resources

- Study Guide Master 12-7
- Practice Master 12-7
- Enrichment Master 12-7
- Group Activity Card 12-7
- Tech Prep Applications Masters, p. 24

Transparency 12-7A contains the 5-Minute Check for this lesson; **Transparency 12-7B** contains a teaching aid for this lesson.

Recommended Pacing	
Standard Pacing	Day 10 of 13
Honors Pacing	Day 8 of 11
Block Scheduling*	Day 6 of 7 (along with Lesson 12-8)

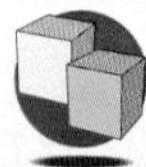

*For more information on pacing and possible lesson plans, refer to the ***Block Scheduling Booklet***.

1 FOCUS

5-Minute Check
(over Lesson 12-6)

Find the surface area of each figure. Round to the nearest tenth.

1. **10.6 m²**

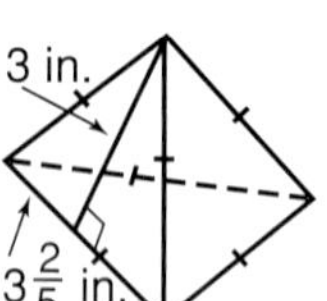

2. **$20\frac{2}{5}$ in²**

3. **82.5 ft²**

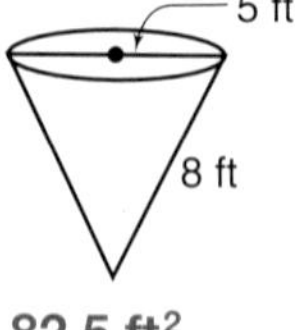

4. **301.6 in²**

12-7 Volume: Prisms and Cylinders

Setting Goals: *In this lesson, you'll find volumes of prisms and circular cylinders.*

Modeling with Technology

The program below computes the volume of a rectangular prism when you enter the dimensions. Use this program to discover how to calculate the volume of a container.

Your Turn Enter the program in a programmable calculator.

```
PROGRAM: VOLUME
: Input "LENGTH =", L
: Input "WIDTH =", W
: Input "HEIGHT =", H
: L*W→A
: L*W*H→V
: Disp "AREA OF BASE ="
: Disp A
: Disp "VOLUME ="
: Disp V
: Stop
```

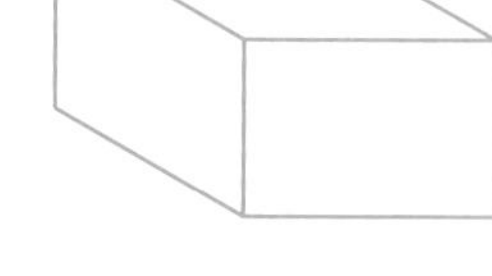

- Sketch a rectangular prism like the one shown at the right. Choose measurements for each side and label the sketch.
- To run the program, press PRGM. Use the down arrow to highlight the program. Press ENTER twice. Enter the data asked for in the program. Record the results on your sketch.
- Repeat for at least three different rectangular prisms. Sketch each prism and record the dimensions and the results of the program. (To run the program again, press ENTER.)

TALK ABOUT IT

a. What dimensions did the program ask for in a rectangular prism? **length, width, height**

b. What is the relationship between the dimensions and the volume? **The volume is the product of the dimensions.**

c. How does the area of the base of the prism relate to its volume? **The area of the base times the height equals the volume.**

Learning the Concept

The amount a container will hold is called its capacity, or **volume**. Volume is usually measured in cubic units. Two common units of measure for volume are the cubic centimeter (cm³) and the cubic inch (in³). The volume of a prism can be found by multiplying the area of the base times the height.

Alternative Learning Styles

Visual Bring in merchandise catalogs and have students find examples of different prisms and cylinders used for packaging various products. Have students discuss why various shapes might have been chosen.

GLENCOE Technology

Interactive Mathematics Tools Software

In this interactive computer lesson, students explore volume and surface area. A **Computer Journal** gives students the opportunity to write about what they have learned.

For Windows & Macintosh

Volume of a Prism

In words: If a prism has a base area of B square units and a height of h units, then the volume V is $B \cdot h$ cubic units.

In symbols: $V = Bh$

Example  **Find the volume of each prism.**

THINK ABOUT IT

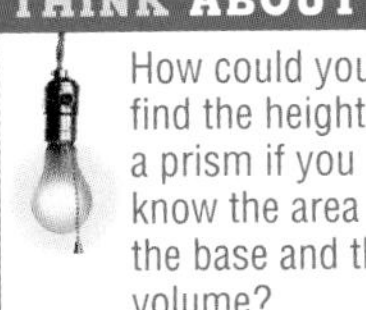

How could you find the height of a prism if you know the area of the base and the volume?

Divide the volume by the area of the base.

a.

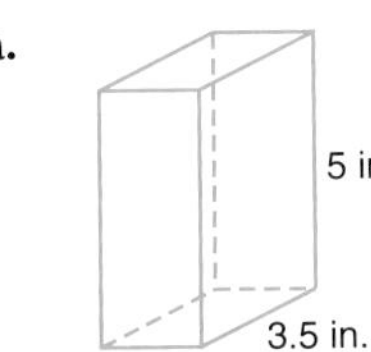

$V = Bh$ *Formula for volume of a prism*

$V = \ell wh$ *Since the base of the prism is a rectangle, $B = \ell w$.*

$V = 2 \cdot 3.5 \cdot 5$ *Replace ℓ with 2, w with 3.5, and h with 5.*

$V = 35$

The volume is 35 in^3.

b. A prism has bases that are right triangles as shown below. The height of the prism is 10 cm.

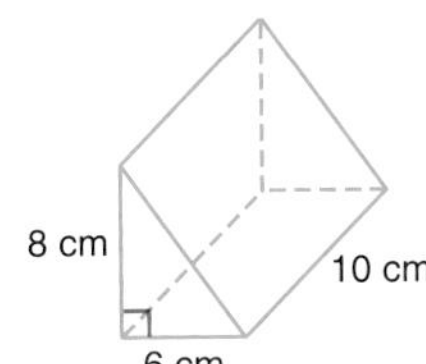

$V = Bh$

$V = \left(\frac{1}{2} \cdot 6 \cdot 8\right) \cdot 10$ *The base of the prism is a triangle. The area of the triangle is $\frac{1}{2} \cdot 6 \cdot 8$.*

$V = 24 \cdot 10$

$V = 240$

The volume is 240 cm^3.

Since the base of a circular cylinder is a circle and its area is equal to πr^2, we can substitute πr^2 for B to find the volume of a cylinder.

Volume of a Circular Cylinder

In words: If a circular cylinder has a base with a radius of r units and a height of h units, then the volume V is $\pi r^2 h$ cubic units.

In symbols: $V = \pi r^2 h$

Example  **A can of tomato juice has a diameter of 14 cm and a height of 20 cm. What is its volume?**

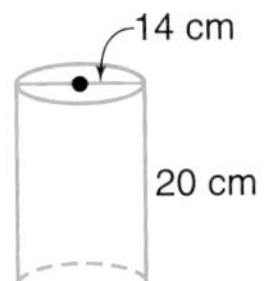

$V = \pi r^2 h$ *Formula for volume*

$V = \pi \cdot 7^2 \cdot 20$ *The diameter of the can is 14 cm; so the radius is 7 cm.*

$V \approx 3079$

The volume is about 3079 cm^3.

Motivating the Lesson

Situational Problem Pose this situation. A man dug a hole that was 3 feet deep, 2 feet wide, and 4 feet long. When he finished, how much dirt was in the hole? **none** How much dirt had he removed? **24 ft^3**

2 TEACH

In-Class Examples

For Example 1

Find the volume of each prism.

a. **288 m^3**

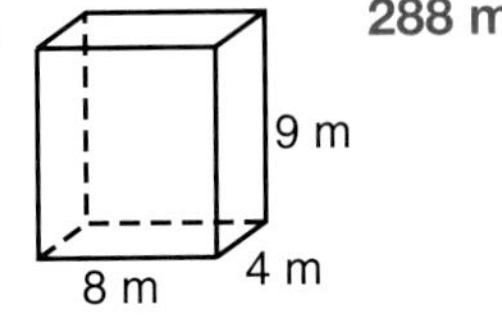

b. **40 cm^3**

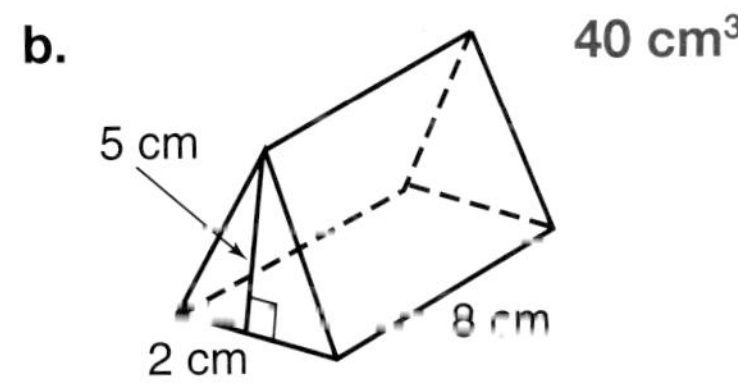

For Example 2

A cylindrical silo is 40 feet tall and has a diameter of 12 feet. What is the volume of the silo? Round to the nearest tenth. **4523.9 ft^3**

Teaching Tip Review with students what kind of units would be used to express the following situations.

- the perimeter of a base of a cylinder **unit**
- the area of the base of cylinder **square units**
- the total surface area of a cylinder **square units**
- the volume of a cylinder **cubic units**

Tech Prep

Air Conditioner Sales Associate
People who sell air conditioners need to know how to find the volume of the space that the air conditioner will be servicing. Although tables containing the needed information may be available, the ability to estimate volumes and compute exact volumes is an advantage to such a salesperson.

For more information on tech prep, see the *Teacher's Handbook*.

In-Class Example

For Example 3

A soft drink can holds 355 milliliters of liquid. Each milliliter has a volume of 1 cm^3. If the diameter of the can is 6.4 cm, what is the height of the can? **11.0 cm**

3 PRACTICE/APPLY

Checking Your Understanding

Exercises 1–12 are designed to help you assess your students' understanding through reading, writing, speaking, and modeling. You should work through Exercises 1–5 with your students and then monitor their work on Exercises 6–12.

Additional Answers

1. Their volumes both depend on the area of the base and the height.
2. V = volume; B = area of the base; h = height
3. Both are measures of three-dimensional objects; surface area covers the outside of a solid while volume fills the inside.

Study Guide Masters, p. 106

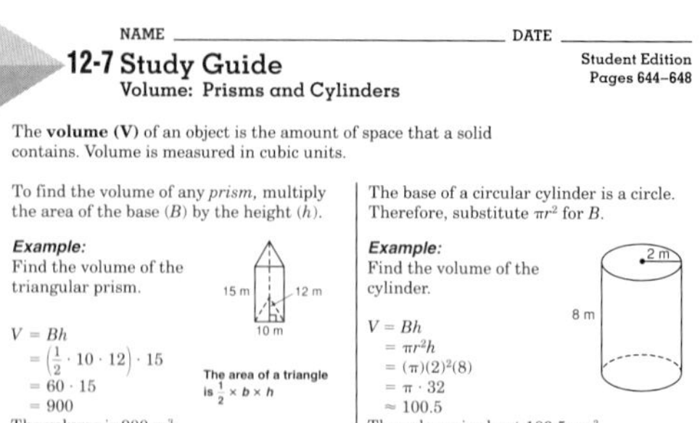

NAME ______________________ DATE __________

12-7 Study Guide
Volume: Prisms and Cylinders

Student Edition Pages 644–648

The **volume (V)** of an object is the amount of space that a solid contains. Volume is measured in cubic units.

To find the volume of any *prism*, multiply the area of the base (B) by the height (h).

Example: Find the volume of the triangular prism.

$V = Bh$
$= \left(\frac{1}{2} \cdot 10 \cdot 12\right) \cdot 15$
$= 60 \cdot 15$
$= 900$

The volume is 900 m^3.

The area of a triangle is $\frac{1}{2} \times b \times h$

The base of a circular cylinder is a circle. Therefore, substitute πr^2 for B.

Example: Find the volume of the cylinder.

$V = Bh$
$= \pi r^2 h$
$= (\pi)(2)^2(8)$
$= \pi \cdot 32$
≈ 100.5

The volume is about 100.5 cm^3.

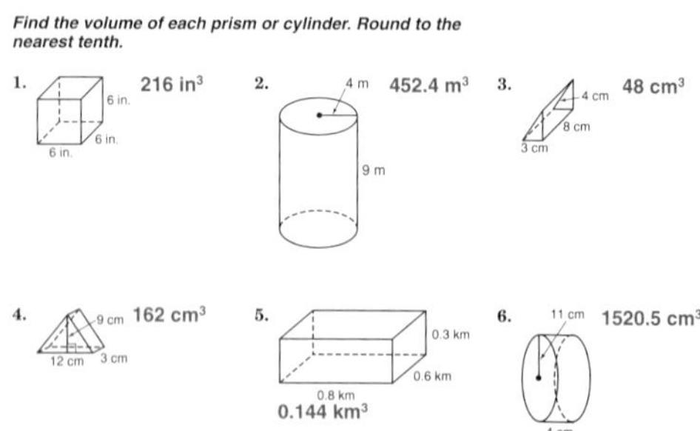

Find the volume of each prism or cylinder. Round to the nearest tenth.

1. 216 in^3
2. 452.4 m^3
3. 48 cm^3
4. 162 cm^3
5. 0.144 km^3
6. 1520.5 cm^3

Connection to Algebra

If you know the volume of a cylinder, you can solve the formula for h or r to find the height or radius of the cylinder.

Example 3

APPLICATION

Engineering

A cylindrical natural gas storage tank is being manufactured to hold at least 1,000,000 cubic feet of natural gas and have a diameter of no more than 80 feet. What height should the tank be?

Explore The tank is a cylinder. You know the volume and the diameter. You must find the height.

Plan The formula for the volume of a cylinder is $V = \pi r^2 h$. Solve the formula for h to find the height.

Solve

$V = \pi r^2 h$
$1{,}000{,}000 = \pi \cdot (40)^2 \cdot h$ *$V = 1{,}000{,}000$, $r = 40$*
$1{,}000{,}000 = \pi \cdot 1600 \cdot h$
1,000,000 [÷] [π] [÷] 1600 [=] *198.9436789*
$198.9 \approx h$

The tank should have a height of at least 199 feet.

Examine If the tank is 199 feet high with a diameter of 80 feet, will it have a volume of at least 1,000,000 cubic feet?

Is $\pi \cdot (40)^2 \cdot 199 \geq 1{,}000{,}000$?

π [×] 40 [x^2] [×] 199 [=] *1000283.101*

The answer checks.

THINK ABOUT IT

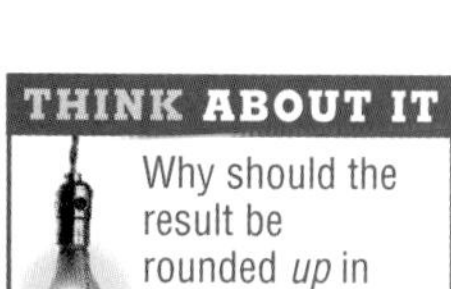

Why should the result be rounded *up* in this case?

So the volume is *at least* 1,000,000 ft^3.

Checking Your Understanding

Communicating Mathematics

Read and study the lesson to answer each question. 1–4. See margin.

1. **Explain** why you can use the same formula to find the volume of a prism or a cylinder.
2. **Tell** what each variable represents in the formula $V = Bh$.
3. **Compare and contrast** surface area and volume.
4. **You Decide** Jake says that doubling the length of each side of a cube doubles the volume. Carlos says the volume is eight times greater. Who is correct? Explain your answer.

MATH JOURNAL

5. Write a problem from an everyday situation in which you will need to find the volume of a cylinder or a rectangular prism. Explain how to solve the problem. See students' work.

Guided Practice

Find the volume of each prism or cylinder. Round to the nearest tenth.

6. 160 cm^3

6.

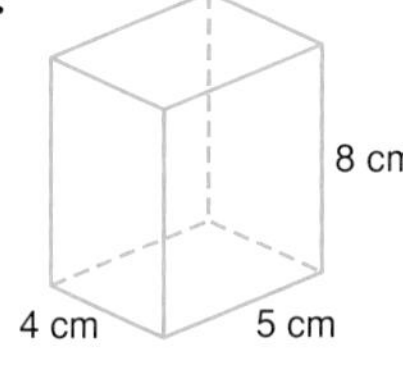

7. 8 in. 15 in. 10 in.
600 in^3

8. 5 cm 15 cm
1178.1 cm^3

Reteaching

Using Manipulatives Have students use centimeter cubes to build as many different rectangular prisms as possible that have a volume of 24 cm^3. Have students make a table of the length, width, and height for each prism. Have students compare the volume to the product of the area of the base times the height.

Additional Answer

4. Carlos is correct; a cube has three dimensions and when you multiply each dimension by 2, you multiply the volume by $2 \times 2 \times 2$ or 8.

Find the volume of each prism or cylinder. Round to the nearest tenth.

9. rectangular prism: length 3 in.; width 5 in.; height 15 in. **225 in²**
10. octagonal prism; base area 25 m²; height of 1.5 m **37.5 m³**
11. circular cylinder: radius 2 ft; height $2\frac{1}{4}$ ft **28.3 ft³**
12. **Pet Care** Tina has an old fish tank that is a circular cylinder. The tank is 2 feet in diameter and 6 feet high. How many cubic feet of water does it hold? **about 19 ft³**

Exercises: Practicing and Applying the Concept

Independent Practice

A

√ Choose

Estimation
Mental Math
Calculator
Paper and Pencil

Find the volume of each solid. Round to the nearest tenth.

13. 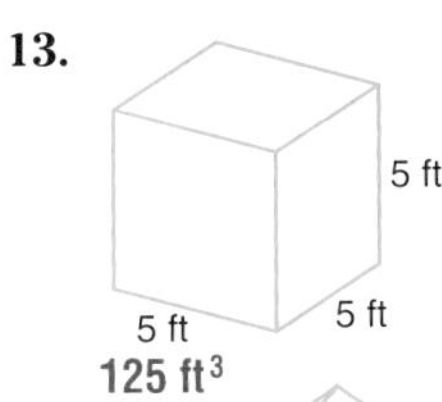

125 ft³

14. 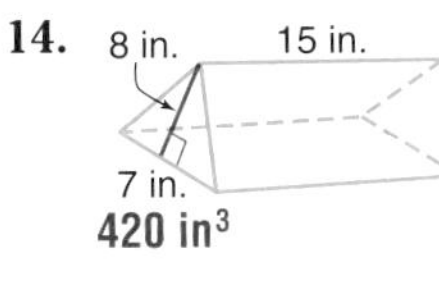

420 in³

15. 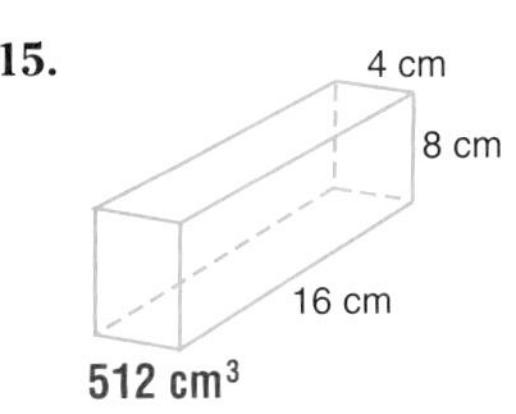

512 cm³

B

16. 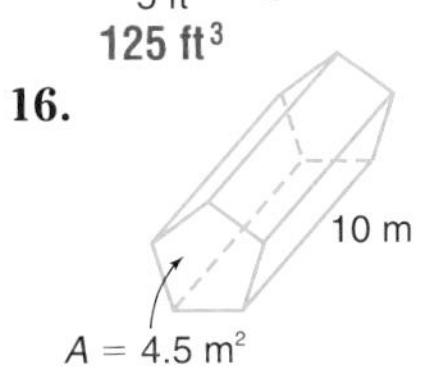

45 m³

17. 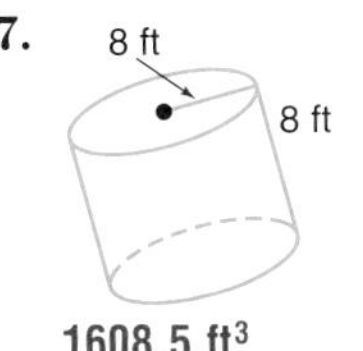

1608.5 ft³

18. 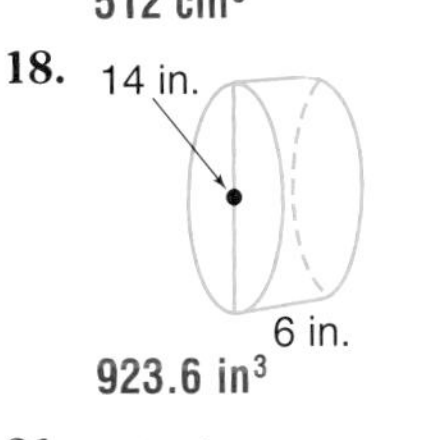

923.6 in³

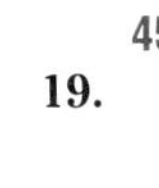

C

24. 2160.6 cm³

19. 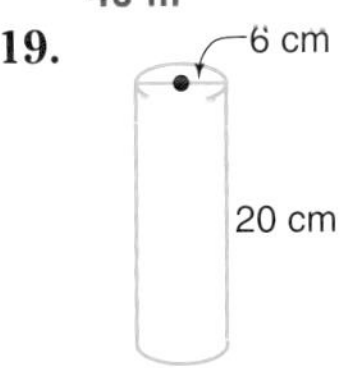

565.5 cm³

20. 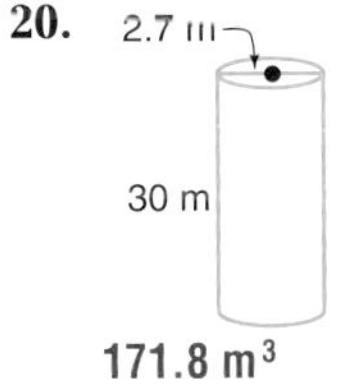

171.8 m³

21. 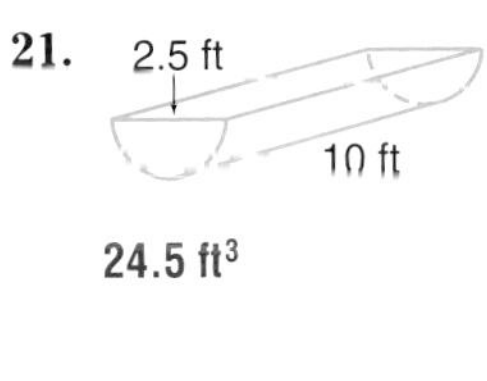

24.5 ft³

22. rectangular prism: length 6 in.; width 6 in.; height 6 in. **216 in³**
23. rectangular prism: length 4.2 cm; width 3.2 cm; height 6.2 cm **83.3 cm³**
24. pentagonal prism: base area 45.2 cm²; height 47.8 cm
25. circular cylinder: radius 10 yd; height 15 yd **4712.4 yd³**
26. triangular prism; base of triangle 8 in.; altitude of triangle 15 in.; height of prism 6.5 in. **390 in³**
27. circular cylinder: diameter 2.6 m; height 3.5 m **18.6 m³**

Graphing Calculator

28a. See margin.
28b. See students' work.

28. Refer to the program in the activity at the beginning of the lesson.
   a. Change the program to find the volume of a circular cylinder.
   b. Sketch three circular cylinders with the same height and different diameters. Run the program to find the volume of each cylinder.
   c. Graph the diameters and volumes on a coordinate plane. What pattern do you observe? **See students' work.**

Assignment Guide

Core: 13–35 odd, 36–42
Enriched: 14–26 even, 28–42

For **Extra Practice**, see p. 770.

The red A, B, and C flags, printed only in the Teacher's Wraparound Edition, indicate the level of difficulty of the exercises.

Additional Answer

28a. Change lines 1–2 to Input "RADIUS=", R
Change line 4 to π*R²→A
Change line 5 to A*H→V

Practice Masters, p. 106

NAME ______ DATE ______

12-7 Practice
Volume: Prisms and Cylinders

Student Edition Pages 644–648

Find the volume of each prism or cylinder. Round to the nearest tenth.

1. 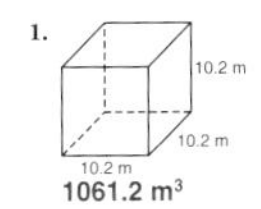

1061.2 m³

2. 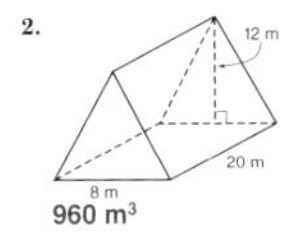

960 m³

3. 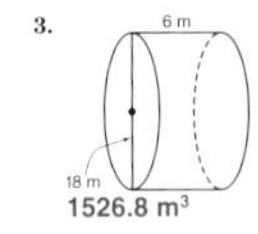

1526.8 m³

4. 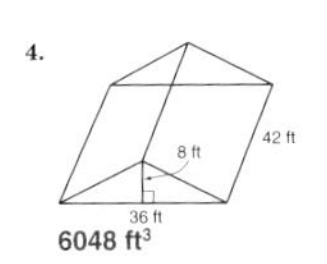

6048 ft³

5. 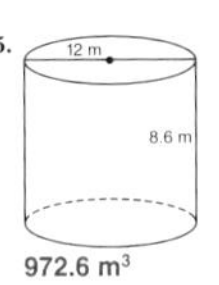

972.6 m³

6. 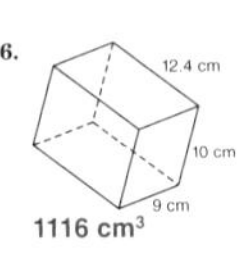

1116 cm³

7. rectangle prism: length, 6 yd; width, 5 yd; height, 3 yd **90 yd³**
8. triangular prism: base of triangle, 6 m; altitude, 4 m; prism height, 3 m **36 m³**
9. circular cylinder: radius, 5 m; height, 10 m **785.4 m³**
10. rectangular prism: length, 16.5 mm; width, 8.4 mm; height, 32. mm **443.5 mm³**
11. circular cylinder: diameter, $5\frac{1}{2}$ yd; height, 13 yd **308.9 yd³**
12. triangular prism: base of triangle, 3 km; altitude, 2 km; prism height, 1 km **3 km³**

Group Activity Card 12-7

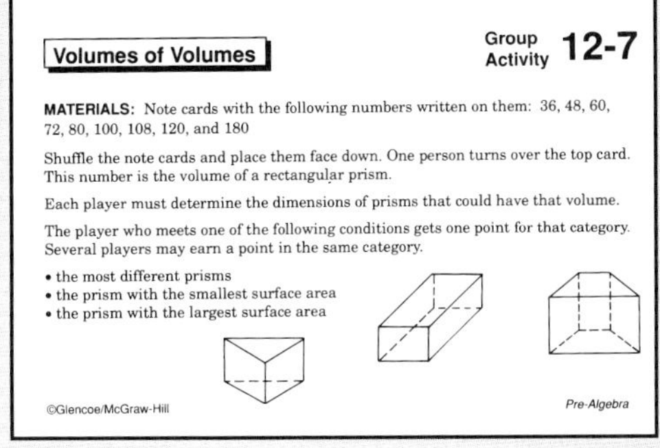

Volumes of Volumes — Group Activity 12-7

MATERIALS: Note cards with the following numbers written on them: 36, 48, 60, 72, 80, 100, 108, 120, and 180

Shuffle the note cards and place them face down. One person turns over the top card. This number is the volume of a rectangular prism.

Each player must determine the dimensions of prisms that could have that volume.

The player who meets one of the following conditions gets one point for that category. Several players may earn a point in the same category.

- the most different prisms
- the prism with the smallest surface area
- the prism with the largest surface area

 Pre-Algebra

4 ASSESS

Closing Activity

Writing Have students draw flowcharts to show how to find the volume of each of the following types of objects.

- rectangular prism with length ℓ, width w, and height h
- triangular prism with the triangular base being a 3–4–5 right triangle and the height of the prism h
- cylinder with a radius r and a height h

Additional Answers

29. The volume is greater with the $8\frac{1}{2}$-inch side as the height.
42. Sample answer: Candy is on sale for 3 pieces for 25 cents. How much is one piece?

Enrichment Masters, p. 106

NAME ______________ DATE ______

12-7 Enrichment — Student Edition Pages 644–648

Spheres

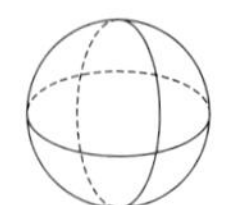

The *great circle* on a sphere is the intersection of the sphere with a plane through the center. Both circles shown to the left are great circles. They are congruent and have circumferences that measure $2\pi r$.

Volume of sphere: $V = \frac{4}{3}\pi r^3$

Surface area of a sphere: $S = 4\pi r^2$

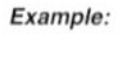

Example:

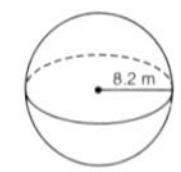

The *area of the great circle* is about 211.1336 m².
The *circumference of the great circle* is about 51.496 m.
The *surface area of the sphere* is about 844.5344 m².
The *volume of the sphere* is about 2308.394 m³.

Solve. Use $\pi \approx 3.14$. Round answers to the nearest hundredth.

1. Find the volume of a sphere with a radius of 4 meters. **267.95 m³**
2. Find the surface of a sphere with a radius of 8 centimeters. **803.84 cm²**
3. A hemispherical dome (half of a sphere) has a height of 32 meters.
   a. What is the volume of the dome? **68,594.35 m³**
   b. What is the surface area? **6430.72 m²**

4. Find the volume of the grain silo shown below. **1138.77 m³**

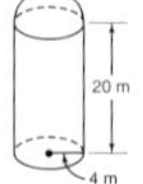

5. A cold capsule is 12 millimeters long and 4 millimeters in diameter. Find the volume of medicine it can contain. **184.21 mm²**

12 mm
4 mm

Critical Thinking

29. Suppose you roll an $8\frac{1}{2}$- by 11-inch piece of paper to form a cylinder. Will the volume be greater if you roll it so the height is $8\frac{1}{2}$ inches or 11 inches, or will the volumes be the same? **See margin.**
30. A cube has a side a units long. What is the formula for the volume of the cube in terms of a? $V = a^3$

Applications and Problem Solving

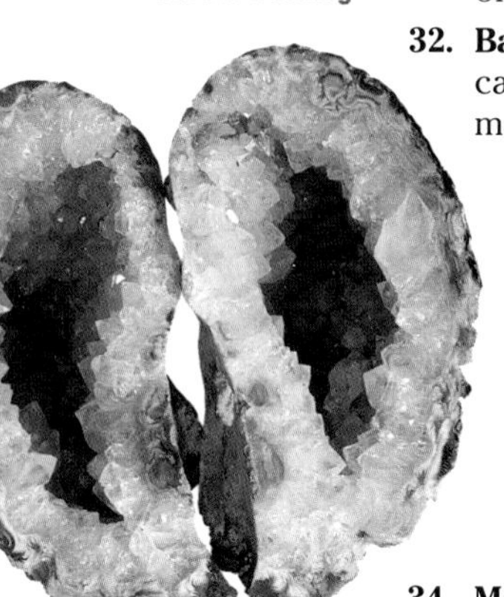

31. **Chemistry** A quartz crystal is a hexagonal prism. It has a base area of 1.41 cm² and a volume of 4.64 cm³. What is its height? **3.29 cm**
32. **Baking** A rectangular cake pan is 30 cm by 21 cm by 5 cm. A round cake pan has a diameter of 21 cm and a height of 4 cm. Which will hold more batter, the rectangular pan or two round pans? **rectangular pan**
33. **Recreation** A cubic foot of water is approximately 7.481 gallons. Which swimming pool below requires about 45,000 gallons of water to fill? **a**

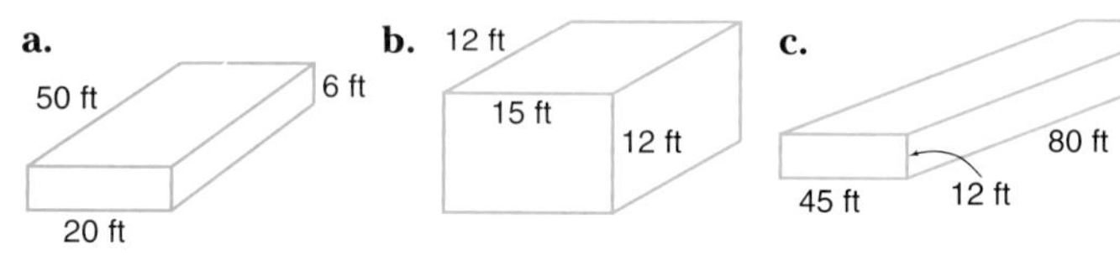

34. **Make a Drawing** A twelve-pack of ice tea cans is sold in a rectangular box. The cans are 6 inch high cylinders with a diameter of 3 inches.
    a. What are the dimensions of the box if the cans are arranged in a 4 by 3 array? **12 by 9 by 6 inches**
    b. How much wasted space is there in the box? **about 140 in³**
35. **Consumer Awareness** Firewood is usually sold by a measure known as a cord. A full cord may be a stack $8 \times 4 \times 4$ feet or $8 \times 8 \times 2$ feet.
    a. What is the volume of a full cord? **128 ft³**
    b. A "short cord" is $8 \times 4 \times$ the length of the logs. What part of a full cord is a short cord of $2\frac{1}{2}$ foot logs? $\frac{5}{8}$

Mixed Review

36. Find the surface area of a square pyramid with a base 14 yards long and a height of 3 yards. (Lesson 12-6) **280 yd²**
37. **Geometry** Find the value of x in $\triangle ABC$ at the right. (Lesson 11-6) **4 mi**

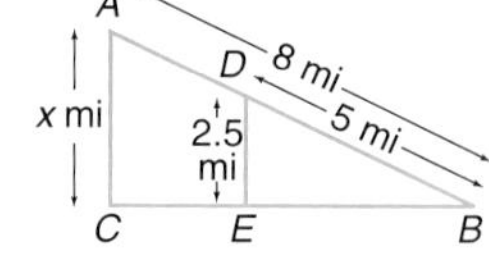

38. **Geometry** Angles X and Y are supplementary. If $m\angle X = 58°$, find $m\angle Y$. (Lesson 11-3) **122°**
39. If $f(x) = 5x - 3$, find $f(2)$. (Lesson 8-4) **7**

40. $8.50

40. **Work Backward** Drew has $2.25. Half of the money he had when he left home this morning was spent on lunch. He loaned Cassie two dollars after lunch. How much money did Drew start with? (Lesson 7-1)
41. **Games** If you toss three 4-sided dice, how many different combinations of tosses are possible? (Lesson 4-3) **64**
42. Write a problem that could be solved using the equation $3x = 25$. (Lesson 1-8) **See margin.**

Extension

Using Data Have students find a prism or cylinder in real life, measure it, and find its volume. Have students use their data to make a class display about prisms, cylinders, and volume.

12-8 Volume: Pyramids and Cones

Setting Goals: *In this lesson, you'll find volumes of pyramids and cones.*

Modeling with Manipulatives

MATERIALS

- centimeter grid paper
- scissors
- tape
- rice
- compass

Look at the nets shown at the right. They will fold into models of a square prism (a cube) with one open side and a square pyramid. The base of both solids is the same size. If you fill each model with rice, you can compare their volumes.

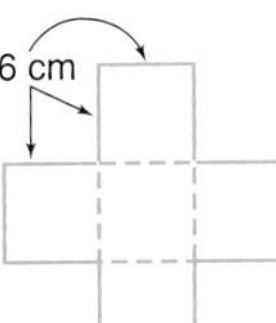

Your Turn Work with a partner.

- ▶ Copy the two nets on centimeter grid paper.
- ▶ Cut them out and fold on the dashed lines.
- ▶ Tape the edges together to form models of the solids.
- ▶ Estimate how much larger the volume of the cube is than the volume of the pyramid.
- ▶ Make an opening in the base of the pyramid so you can put rice into it.
- ▶ Fill the pyramid with rice. Then pour this rice into the cube. Repeat until the cube is filled.

6 cm

a. How many pyramids of rice did it take to fill the cube?

b. What do you know about the areas of the bases of each solid?

c. Compare the heights of the cube and the pyramid.

d. Compare the volume of the cube and the pyramid.

a–d. See margin.

Learning the Concept

In the modeling activity, you found that the volume of the pyramid was $\frac{1}{3}$ the volume of the prism. Since the volume of a prism is Bh, the volume of the pyramid is $\frac{1}{3}Bh$. The height, h, of a pyramid is the length of a segment from the vertex, perpendicular to the base.

Volume of a Pyramid

In words: If a pyramid has a base of B square units and a height of h units, then the volume V is $\frac{1}{3} \cdot B \cdot h$ cubic units.

In symbols: $V = \frac{1}{3}Bh$

Additional Answers Talk About It

a. about 3
b. They are the same.
c. They appear to be the same.
d. The volume of the pyramid is $\frac{1}{3}$ the volume of the cube.

12-8 LESSON NOTES

NCTM Standards: 1-4, 12, 13

Instructional Resources

- Study Guide Master 12-8
- Practice Master 12-8
- Enrichment Master 12-8
- Group Activity Card 12-8
- Assessment and Evaluation Masters, p. 324
- Activity Masters, p. 12
- Multicultural Activity Masters, p. 24

Transparency 12-8A contains the 5-Minute Check for this lesson; **Transparency 12-8B** contains a teaching aid for this lesson.

Recommended Pacing	
Standard Pacing	Day 11 of 13
Honors Pacing	Day 9 of 11
Block Scheduling*	Day 6 of 7 (along with Lesson 12-7)

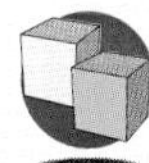

*For more information on pacing and possible lesson plans, refer to the ***Block Scheduling Booklet***.

1 FOCUS

5-Minute Check
(over Lesson 12-7)

Find the volume of each prism.

1. rectangular prism; length 8m, width 5 m, height 10 m **400 m^3**
2. triangular prism; base of triangle 4 ft; altitude 2 ft; prism height 9 ft **36 ft^3**
3. a prism with a base that has an area of 82 m^2 and a height of 17 m **1394 m^3**

Find the volume of each cylinder. Round answers to the nearest tenth.

4. radius, 3.2 cm; height, 10.5 cm **337.8 cm^3**
5. diameter, 4 m; height 8 m **100.5 m^3**

Motivating the Lesson

Questioning Show a picture of the pyramids in Egypt. Those that have been opened have contained a tremendous number of artifacts. How did the Egyptians decide how big to build the pyramids?

2 TEACH

In-Class Examples

For Example 1
Find the volume of this pyramid. **672.8 cm³**

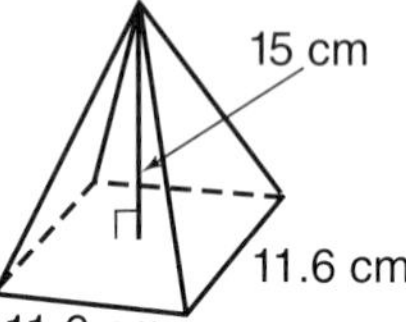

For Example 2
Find the volume of the cone to the nearest whole number. **38 cm³**

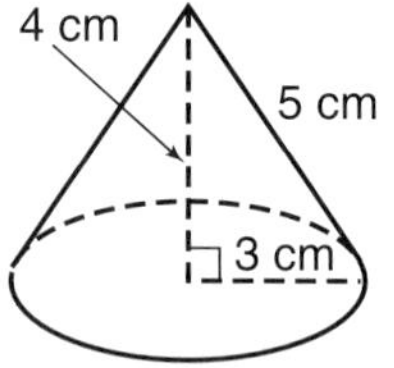

For Example 3
Pyramid Power detergent gave away samples of detergent in packages shaped like a square pyramid. Each edge of the pyramid is 10 centimeters long. The height of the pyramid is about 7 centimeters. Round to the nearest whole unit.

a. Find the volume of the package. **233 cm³**

b. Each pyramid is 80% filled with detergent. This is enough detergent to do two normal wash loads. How much detergent does it take to do one load? **93 cm³**

Example 1 **Find the volume of the pyramid.**

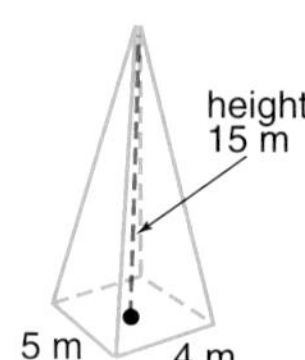

$V = \frac{1}{3}Bh$ *Formula for volume of a pyramid*

$V = \frac{1}{3}\ell wh$ *Replace B with ℓw.*

$V = \frac{1}{3} \cdot 4 \cdot 5 \cdot 15$

$V = 100$

The volume is 100 m³.

The relationship of the volumes of a cone and a cylinder is similar to the relationship of the volumes of a pyramid and a prism. The volume of a cone is $\frac{1}{3}$ the volume of a cylinder. Since the volume of a cylinder is πr^2h, the volume of a cone is $\frac{1}{3}\pi r^2h$. *You'll verify this formula in Exercise 4.*

Volume of a Cone	
In words:	**If a cone has a radius of r units and a height of h units, then the volume V is $\frac{1}{3} \cdot \pi \cdot r^2 \cdot h$.**
In symbols:	$V = \frac{1}{3}\pi r^2h$

Example 2 **Find the volume of the cone to the nearest whole number.**

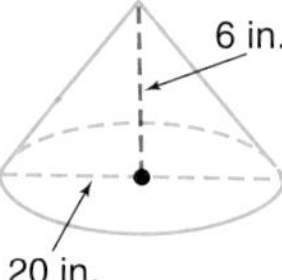

$V = \frac{1}{3}\pi r^2h$ *Formula for volume of a cone*

$V = \frac{1}{3} \cdot \pi \cdot 10^2 \cdot 6$ *Since d = 20, r = 10; h = 6*

$V \approx 628$

The volume is about 628 in³.

Example 3

Highway Maintenance

Many communities use a salt and sand mixture on snowy or icy roads. When the mixture is dumped from a truck, it forms a mound with a conical shape.

a. If the mound has a diameter of 15 feet and a height of 10 feet, how much salt-sand mixture is there? *Estimate: $\frac{1}{3} \cdot 3 \cdot 8^2 \cdot 10 = 640$*

b. If the mixture is spread at a rate of 1 cubic foot per 500 square feet of roadway, how many square feet of roadway can be salted by the mixture in one mound?

a. $V = \frac{1}{3}\pi r^2h$ *Formula for volume of a cone*

$V = \frac{1}{3} \cdot \pi \cdot 7.5^2 \cdot 10$ *Since d = 15, r = 7.5; h = 10*

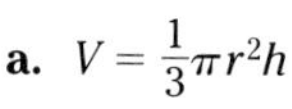

$V \approx 589$

The volume is 589 ft³.

Cooperative Learning

Think-Pair-Share Have students work in pairs to find the surface area and volume of this figure.

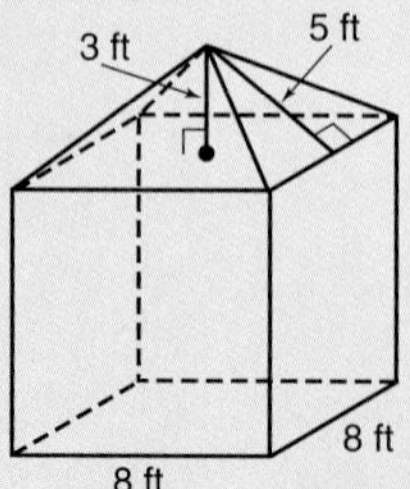

Have one student give the surface area and the other give the volume.
$A = 400$ ft²; $V = 576$ ft³
The activity can be extended by having students create their own figures.
For more on this strategy, see *Cooperative Learning in the Mathematics Classroom*. pp. 24–25.

b. Each cubic foot of mixture will salt 500 square feet of roadway.

589 [×] 500 [=] 294000

294,000 square feet of roadway can be salted by the mixture in the mound.

Checking Your Understanding

Communicating Mathematics

Read and study the lesson to answer these questions. **1–3. See margin.**

1. **Explain** why you can use πr^2 to find the area of the base of a cone.
2. **State** the formula for finding the volume of a pyramid. Then tell what each variable represents.
3. **State** the formula for finding the volume of a cone. Then tell what each variable represents.

MODELING MATHEMATICS

MATERIALS
- compass
- scissors
- tape
- rice

4. Work with a partner.

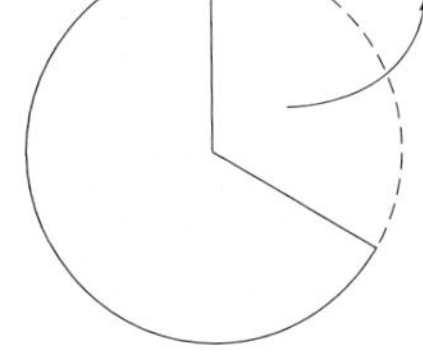

- ▶ Draw a circle with a radius of at least 2 inches. Cut out the circle.
- ▶ Remove about $\frac{1}{3}$ of the circle as shown at the right. Tape the edges of the remaining part of the circle to make an open cone.
- ▶ Trace the base of the cone on another piece of paper and cut out the circle.
- ▶ Cut a strip of paper as wide as the cone is high. Use the strip and the circle to make a model of a cylinder that has the same size base and the same height as the cone.
- ▶ Estimate how many times larger the cylinder is than the cone. Use rice to check your estimate. **See students' work.**

Guided Practice

Explain how to find the volume of each pyramid or cone. Then find the volume. Round to the nearest tenth. **5–7. See students' explanations**

5. 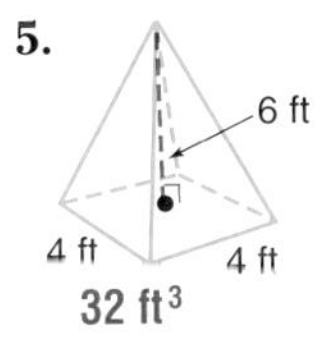

32 ft³

6. 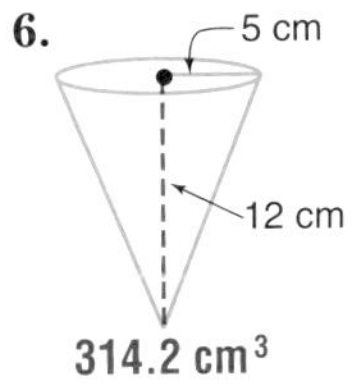

314.2 cm³

7. 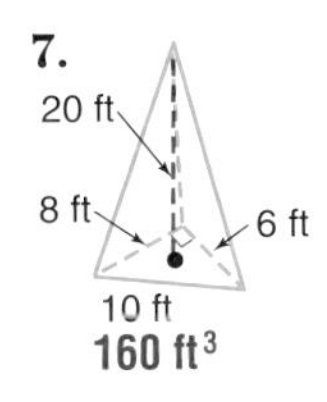

160 ft³

8. 565.5 m³

8. circular cone: diameter 12 m; height 15 m
9. hexagonal pyramid: base area 125 in^2; height $6\frac{1}{2}$ in. **270.8 in³**
10. **Historical Monuments** The top of the Washington Monument is a square pyramid 54 feet high and 34 feet long on each side. What is the volume of the room under the top? **20,808 ft³**

3 PRACTICE/APPLY

Checking Your Understanding

Exercises 1–10 are designed to help you assess your students' understanding through reading, writing, speaking, and modeling. You should work through Exercises 1–4 with your students and then monitor their work on Exercises 5–10.

Additional Answers

1. The base is a circle.
2. $V = \frac{1}{3}Bh$; V represents volume, B represents the area of the base, h represents the height of the pyramid.
3. $V = \frac{1}{3}\pi r^2 h$; V represents volume, r represents the radius of the base, h represents the height of the cone.

Study Guide Masters, p. 107

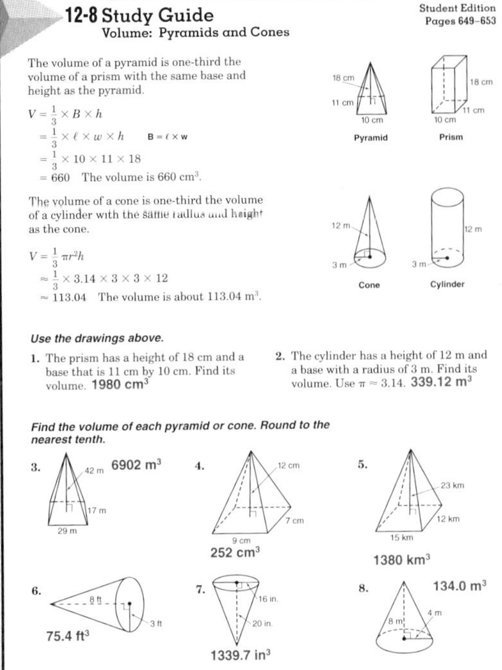

NAME ____________ DATE ________

12-8 Study Guide
Volume: Pyramids and Cones

Student Edition Pages 649–653

The volume of a pyramid is one-third the volume of a prism with the same base and height as the pyramid.

$V = \frac{1}{3} \times B \times h$
$= \frac{1}{3} \times \ell \times w \times h$ $B = \ell \times w$
$= \frac{1}{3} \times 10 \times 11 \times 18$
$= 660$ The volume is 660 cm³.

The volume of a cone is one-third the volume of a cylinder with the same radius and height as the cone.

$V = \frac{1}{3}\pi r^2 h$
$\approx \frac{1}{3} \times 3.14 \times 3 \times 3 \times 12$
≈ 113.04 The volume is about 113.04 m³.

Use the drawings above.

1. The prism has a height of 18 cm and a base that is 11 cm by 10 cm. Find its volume. **1980 cm³**
2. The cylinder has a height of 12 m and a base with a radius of 3 m. Find its volume. Use $\pi \approx 3.14$. **339.12 m³**

Find the volume of each pyramid or cone. Round to the nearest tenth.

3. **6902 m³**
4. **252 cm³**
5. **1380 km³**
6. **75.4 ft³**
7. **1339.7 in³**
8. **134.0 m³**

Reteaching

Using Science Connections The command module of a rocket is basically cone-shaped. Have students use reference materials to find the dimensions of a command module and approximate its volume.

Error Analysis

Some students may use the slant height of a cone rather than the altitude when finding the volume of a pyramid or cone. Model how to find surface area and volume of both a pyramid and a cone.

Assignment Guide
Core: 11–29 odd, 31–36
Enriched: 12–26 even, 27–36

For **Extra Practice**, see p. 771.

The red A, B, and C flags, printed only in the Teacher's Wraparound Edition, indicate the level of difficulty of the exercises.

Additional Answer

27. No; the customer gets only $\frac{1}{3}$ as much popcorn for $\frac{1}{2}$ the price.

Practice Masters, p. 107

NAME ______________ DATE ______

12-8 Practice
Volume: Pyramids and Cones

Student Edition Pages 649–653

Find the volume of each pyramid or cone. Round to the nearest tenth.

1. 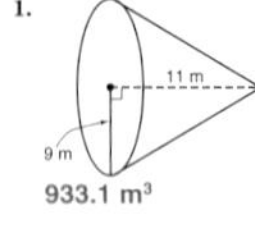

933.1 m³

2. 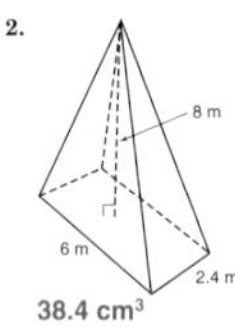

38.4 cm³

3. 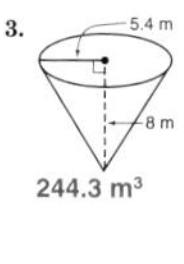

244.3 m³

4. 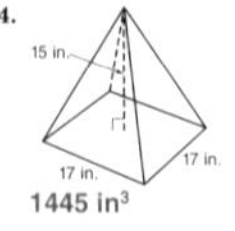

1445 in³

5. 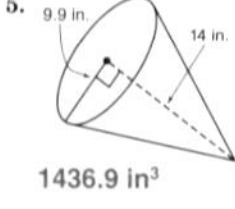

1436.9 in³

6. 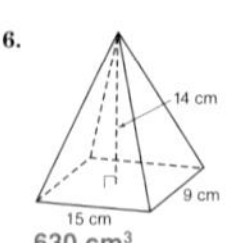

630 cm³

7. rectangular pyramid: length, 8 cm; width, 7 cm; height, 9 cm 168 cm³

8. square pyramid: length, 25 mm; height, 30 mm 6250 mm³

9. circular cone: radius, 12 yd; height, 18 yd 2714.3 yd³

Exercises: Practicing and Applying the Concept

Independent Practice

Find the volume of each solid. Round to the nearest tenth.

11. 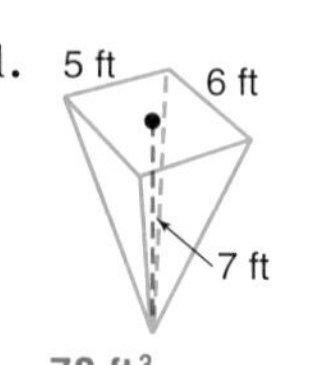

70 ft³

12. 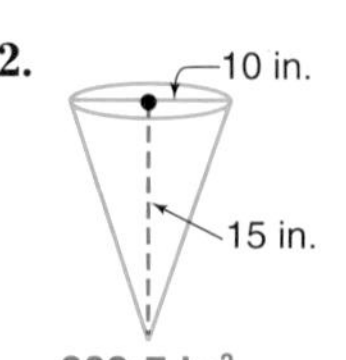

392.7 in³

13. 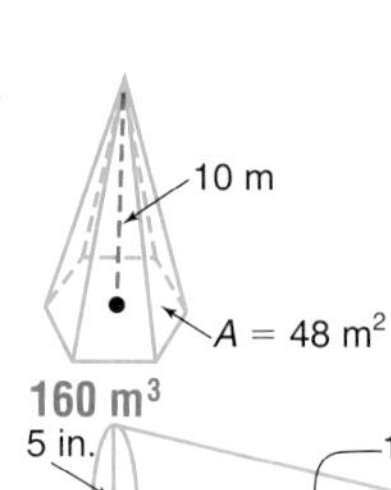

160 m³

14. 9 mm, 14 mm
461.8 mm³

15. 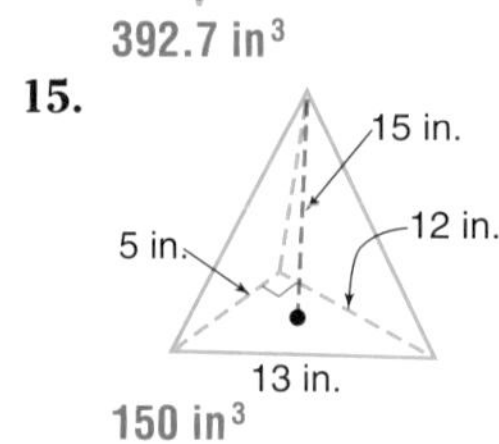

150 in³

16. 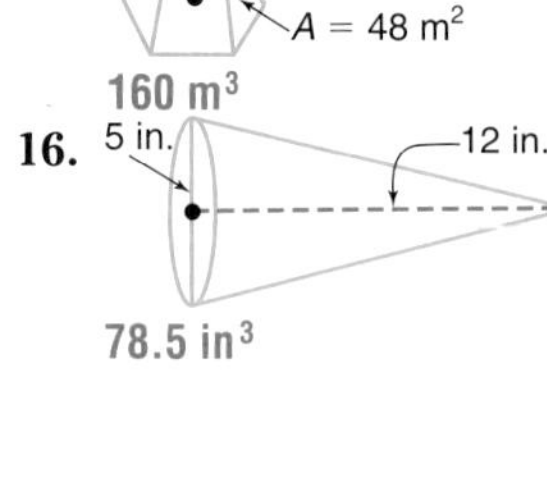

78.5 in³

17. 39 m, 15 m
9189.2 m³

18. 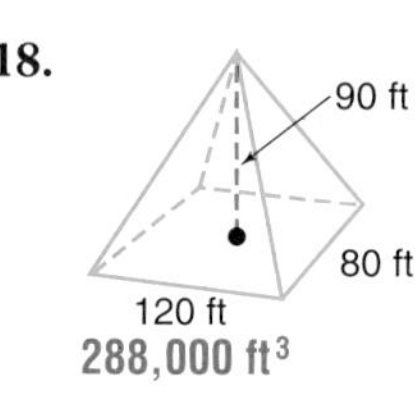

288,000 ft³

19. 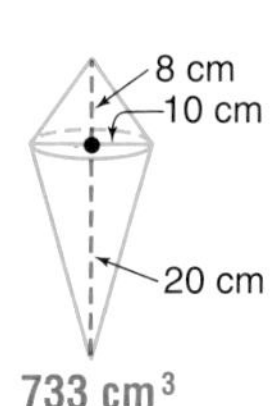

733 cm³

20. rectangular pyramid; length, 9 in.; width, 7 in.; height, 18 in. 378 in³

21. square pyramid; length, 5 cm; height, 6 cm 50 cm³

22. circular cone; radius, 3 ft; height, 14 ft 131.9 ft³

23. circular cone; radius 10 m; height, 18 m 1885.0 m³

24. 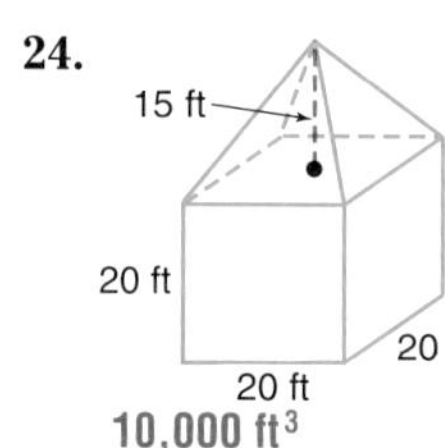

10,000 ft³

25. 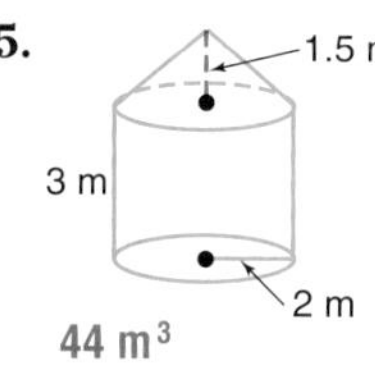

44 m³

26. 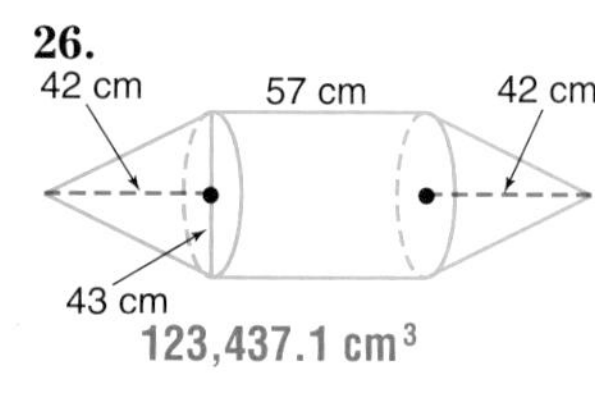

123,437.1 cm³

Critical Thinking

27. See margin.

27. Consumer Awareness Popcorn at the Bijou Theatre used to be sold in rectangular prism containers. Recently the theater changed to a rectangular pyramid container. Both containers have the same height and the same size base. Popcorn in the new box sells for half the former price. Is this a good deal for the customer? Why or why not?

28. a. Suppose you double the height of a cone. How does the volume change? **The volume doubles.**

b. Suppose you double the radius of the base of a cone. How does the volume change? **The volume quadruples.**

Applications and Problem Solving

29. History The Great Pyramid of Khufu in Egypt was originally 481 feet high and had a square base 756 feet on a side. What was its volume? Use an estimate to check your answer. 91,636,272 ft³

Group Activity Card 12-8

Know It All Group Activity **12-8**

MATERIALS: A coin • two 0-9 spinners • four cards (Prism (square base), Pyramid (square base), Cone, and Cylinder)

The leader of each group flips the coin to determine if everyone in the group will compute the surface area (heads) or volume (tails) of a solid.

Next the leader picks a card from the deck to determine which solid they will use. Lastly, the leader spins the two spinners, one to determine the radius of the circle or the side of the square base, and the other to determine the height (for volume) or slant height (for surface area) of the solid. The leader must make sure that the slant height is greater than the radius of the circle or half the side of the square.

All group members then compute the volume or surface area for the solid with the given dimensions.

 Pre-Algebra

Extension

Using Formulas Challenge students to find the formulas for the surface area and the volume of a sphere. Have students make up a problem that requires the use of the formulas. Have students exchange problems and solve. $A = 4\pi r^2$; $V = \frac{4}{3}\pi r^3$

30. **Geology** A stalactite in the Endless Caverns in Virginia is shaped like a cone. It is 4 feet long and has a diameter at the roof of 1.5 feet.

 a. Find the volume of the stalactite to the nearest tenth. **2.4 ft³**

 b. The stalactite is made of calcium carbonate, which weighs 131 pounds a cubic foot. What is the weight of the stalactite? **314.4 lb**

Mixed Review

31. **Construction** Marty's Masonry makes concrete steps for new homes. The steps for one home are to be 5 feet long, with three steps. Each step will be 0.5 foot high and 1 foot wide. How much concrete will be needed? (*Hint:* Make a drawing.) (Lesson 12-7) **15 ft³**

32. **Probability** A die is tossed 30 times. How many times would you expect an outcome of a 1 or a 6? (Lesson 10-10) **about 10 times**

33. Eighteen is what percent of 90? (Lesson 9-5) **20%**

34. Solve $x = 3\frac{3}{5} - 1\frac{3}{10}$. Write the solution in simplest form. (Lesson 5-5)

35. **Geometry** Find the perimeter and area of the rectangle at the right. (Lesson 3-5)

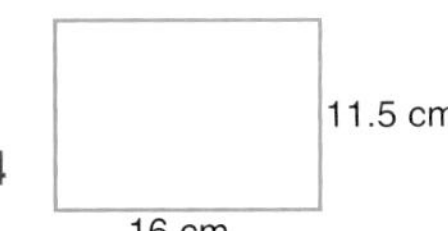

36. Solve the equation $\frac{-144}{6} = k$. (Lesson 2-8) **−24**

34. $2\frac{3}{10}$
35. **55 cm; 184 cm²**

WORKING ON THE
Investigation

Refer to the Investigation on pages 606–607.

Once your group has completed the basic layout of the rooms in your house, you are ready to prepare the *working drawings*. Working drawings are used to show specific information that is necessary to build the structure.

The working drawing that shows the location and size of the rooms is called the *floor plan*. A floor plan shows how the rooms would look if the ceiling were removed and you looked down from above.

Using your basic layout, design your house so that it is no larger than 1200 square feet. Each room should have at least 1 window. Sketch each floor plan on graph paper to a scale of $\frac{1}{4}$ inch = 1 foot.

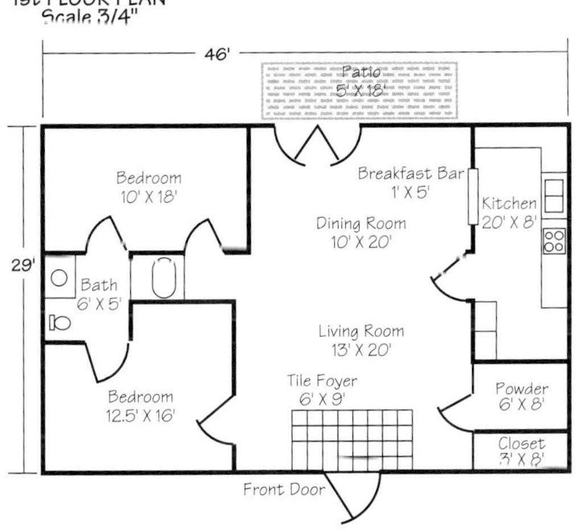

Add the results of your work to your Investigation Folder.

4 ASSESS

Closing Activity

Modeling Which of the following holds more ice cream in the container itself?

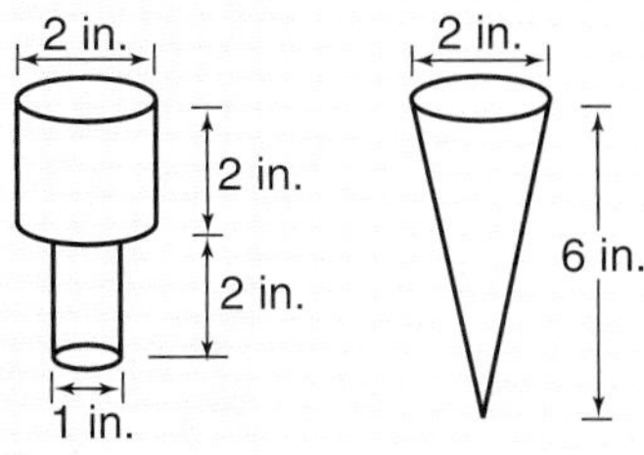

left: 7.85 in³; right: 6.28 in³; The left one holds more.

Chapter 12, Quiz D (Lessons 12-7 and 12-8) is available in the *Assessment and Evaluation Masters*, p. 324.

Enrichment Masters, p. 107

NAME ________________ DATE ________

12-8 Enrichment
Using Area and Volume

Student Edition Pages 649–653

Solve. Use 3.14 for π.

1. Mrs. Bartlett wants to put carpeting in a room that is 15 feet long and 12 feet wide. How many square feet of carpeting does she need? **180 ft²**

2. Tom is making a cover for a box 4.5 feet long and 23 inches wide. What will be the area of the cover, in square feet? **8.625 ft²**

3. How many 2-inch by 3-inch tickets can be cut from a 2-foot by 3-foot sheet of paper? **144 tickets**

4. What is the area of the top of a tree stump that is 42 centimeters in diameter? **1384.74 cm²**

5. Mrs. Stabile has a one-story house that is 32 feet wide and 48 feet long. She plans to build an addition 12 feet by 16 feet. What will the total floor area of the house be then? **1728 ft²**

6. How many cubic meters of dirt must be removed when digging the foundation of a building if the excavation is 32 meters long, 12 meters wide, and 8 meters deep? **3072 m³**

7. Mr. Mendez has a whirling sprinkler for watering his lawn. The sprinkler waters a circle with a radius of 12 meters. Find the area of the lawn that he can water at one time. **452.16 m²**

8. A cylindrical storage tank has a diameter of 6 meters and a height of 5 meters. What is the volume of the storage tank? **141.3 m²**

9. In Mr. Curin's living room there is a circular mirror with a diameter of 18 inches. What is the area of the mirror? **254.34 in²**

10. Find the cost of cementing a driveway 9 meters by 16 meters at $48 per square meter. **$6912**

WORKING ON THE
Investigation

The Investigation on pages 606 and 607 is designed to be a long-term project that is completed over several days or weeks. Encourage students to keep their materials in their Investigation Folder as they work on the Investigation.

NCTM Standards: 1-4, 12, 13

Objective
Investigate how changing the dimensions of a solid affects its surface area and volume.

Recommended Time
Demonstration and discussion: 15 minutes; Exercises: 30 minutes

Instructional Resources
For each student or group of students
- sugar cubes

Math Lab and Modeling Math Masters
- p. 53 (worksheet)

For teacher demonstration
Overhead Manipulative Resources

1 FOCUS

Motivating the Lesson
Show students one box and a second box that is twice as big in length, width, and height as the first box. Ask how they think the volumes of the two boxes compare. Tell students they will compare volumes in this lesson.

2 TEACH

Teaching Tip
You can extend this activity to include other prisms.

3 PRACTICE/APPLY

Assignment Guide
Core: 1–12
Enriched: 1–12

4 ASSESS

Observing students working in cooperative groups is an excellent method of assessment.

HANDS-ON ACTIVITY

12-8B Similar Solid Figures

An Extension of Lesson **12-8**

MATERIALS

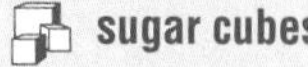
sugar cubes

You can review similar figures in Lesson 11-6.

Have you or someone you know ever built a model car or boat? A scale model is an exact replica of the real object, but is larger or smaller than the original. The dimensions of the model and the original are proportional. The number of times that you increase or decrease the linear dimensions is called the **scale factor**. In this activity, you will find out how changing the dimensions of a solid affects the surface area and volume of the solid.

Activity 1 Work with a partner.

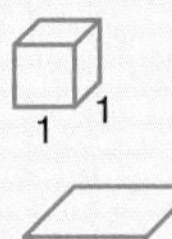

Your Turn
- If each edge of a sugar cube is 1 unit long, then each face is 1 square unit and the volume of the cube is 1 cubic unit.
- Use cubes to make a larger cube that has sides twice as long as the original cube.
- How many small cubes did you use? **8**

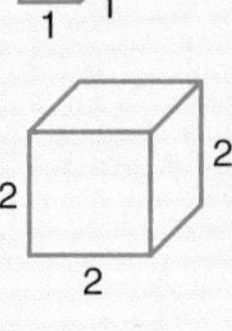

TALK ABOUT IT

1. What is the area of one face of the original cube? **1 unit^2**
2. What is the area of one face of the cube that you built? **4 units^2**
3. What is the volume of the original cube? **1 unit^3**
4. What is the volume of the cube that you built? **8 units^3**

Activity 2 Work with a partner.

Your Turn
- Now use cubes to build a cube that has sides three times as long as the original cube.
- How many small cubes did you use? **27**

TALK ABOUT IT

5. What is the area of one face of the cube? **9 units^2**
6. What is the volume of the cube? **27 units^3**
7. Complete the table below.

Scale Factor	Length of a Side	Area of a Face	Volume
1	1	1	1
2	2	4	8
3	3	9	27

8. Study the table. What happens to the area of a face when the length of a side is doubled? tripled? **× 4; × 9**
9. If the scale factor is x, what is the area of the face? **x^2**
10. What happens to the volume of a cube when the length of a side is doubled? tripled? **× 8; × 27**
11. Let the scale factor be x. Write an expression for the cube's volume. **11. x^3**
12. What would be the area of a face and the volume of a cube with sides 4 times as long as the original cube? **16 units^2; 64 units^3**

CHAPTER 12 Highlights

Vocabulary

After completing this chapter, you should be able to define each term, property, or phrase and give an example or two of each.

Geometry

altitude (pp. 612, 614, 633)
area of a circle (p. 620)
area of a parallelogram (p. 613)
area of a trapezoid (p. 614)
area of a triangle (p. 613)
base (pp. 614, 632)
circular cone (p. 639)
cylinder (p. 633)
face (p. 632)
lateral surface (p. 639)
net (p. 632)
prism (p. 632)
pyramid (p. 638)
rectangular prism (p. 632)
scale factor (p. 654)
slant height (p. 638)
square pyramid (p. 638)
surface area (p. 632)
triangular prism (p. 632)
triangular pyramid (p. 638)
vertex (p. 638)
volume (p. 644)
volume of a circular cylinder (p. 645)
volume of a cone (p. 650)
volume of a prism (p. 645)
volume of a pyramid (p. 649)

Probability

geometric probability (p. 623)

Problem Solving

make a model or a drawing (p. 629)

Understanding and Using Vocabulary

Choose the letter of the term or terms that best completes each statement.

A. altitude	**F.** slant height
B. cubic centimeter	**G.** square centimeter
C. perpendicular	**H.** surface area
D. prism	**I.** volume
E. pyramid	

1. The _?_ is a common unit of measure for area. G
2. The _?_ is a common unit of measure for volume. B
3. The _?_ of a solid is the sum of the areas of its faces. H
4. _?_ is the amount of space that a solid contains. I
5. The _?_ of a parallelogram is a segment perpendicular to the bases with endpoints on the bases. A
6. The _?_ of a pyramid is the length of an altitude of one of its lateral faces. F
7. The height of a pyramid is the length of a segment from the vertex _?_ to the base of the pyramid. C
8. A _?_ is named by the shape of its base. D, E

Instructional Resources

Three multiple-choice tests and three free-response tests are provided in the *Assessment and Evaluation Masters*. Forms 1A and 2A are for honors pacing, Forms 1B and 2B are for average pacing, Forms 1C and 2C are for basic pacing. Chapter 12 Test, Form 1B is shown at the right. Chapter 12 Test, Form 2B is shown on the next page.

Highlights

Using the Chapter Highlights

The Chapter Highlights begins with a listing of the new terms, properties, and phrases that were introduced in this chapter.

Assessment and Evaluation Masters, pp. 311–312

NAME ______________________ DATE ____________

CHAPTER 12 Test, Form 1B

Find the area of each figure shown or described below. Use $\pi \approx 3.14$ with decimal measures and $\pi \approx \frac{22}{7}$ with fractions. Round decimal answers to the nearest tenth and simplify fraction answers.

1. 6 m, 13.4 m, 12 m
   A. $226.1\ m^2$ B. $18\ m^2$ C. $72\ m^2$ D. $36\ m^2$ — 1. D
2. $10\frac{1}{2}$ in., 7 in., 6 in., 8 in., $12\frac{1}{2}$ in.
   A. $66\frac{1}{12}\ in^2$ B. $136\frac{1}{2}\ in^2$ C. $69\ in^2$ D. $81\frac{1}{3}\ in^2$ — 2. C
3. Parallelogram: base, 15 centimeters; height, 21 centimeters
   A. $157.5\ cm^2$ B. $989.1\ cm^2$ C. $315\ cm^2$ D. $105\ cm^2$ — 3. C
4. Circle: diameter, 18 inches
   A. $113.0\ in^2$ B. $254.3\ in^2$ C. $1017.4\ in^2$ D. $56.5\ in^2$ — 4. B
5. 12 cm
   A. $452.2\ cm^2$ B. $144\ cm^2$ C. $113.0\ cm^2$ D. $75.4\ cm^2$ — 5. A
6. The figure at the right represents a dart board. Find the probability that a randomly thrown dart lands in the shaded area.
   A. $\frac{1}{5}$ B. $\frac{2}{5}$ C. $\frac{1}{10}$ D. $\frac{3}{10}$ — 6. A
7. George has 72 identical cubes. Each edge of a cube is 1 inch long. How can George arrange the cubes to have the smallest surface area?
   A. $3 \times 4 \times 6$ B. $3 \times 3 \times 8$ C. $2 \times 4 \times 9$ D. $2 \times 6 \times 6$ — 7. A

Find the surface area of each figure shown or described below. Use $\pi \approx 3.14$. Round decimal answers to the nearest tenth.

8. 10 in., 5 in.
   A. $500\ in^2$ B. $785\ in^2$ C. $207\ in^2$ D. $314\ in^2$ — 8. D

NAME ______________________ DATE ____________

Chapter 12 Test, Form 1B (continued)

9. 2 cm, 3 cm, 5 cm
   A. $43\ cm^2$ B. $86\ cm^2$ C. $62\ cm^2$ D. $30\ cm^2$ — 9. C
10. Rosita is making a lamp shade. It is shaped like a cylinder, with no top or base. The lamp shade will have a diameter of 12 inches and a height of 18 inches. Find the amount of fabric she needs. Use $\pi \approx 3.14$ and round to the nearest tenth.
    A. $226.1\ in^2$ B. $678.2\ in^2$ C. $84.7\ in^2$ D. $254.3\ in^2$ — 10. B
11. Find the surface area of a square pyramid with a slant height of 10 centimeters and a base with an edge of 6 centimeters.
    A. $96\ cm^2$ B. $156\ cm^2$ C. $120\ cm^2$ D. $70\ cm^2$ — 11. B
12. Find the surface area of a circular cone with a slant height of 8 centimeters and a base with a radius of 3 centimeters.
    A. $210.4\ cm^2$ B. $94.2\ cm^2$ C. $103.6\ cm^2$ D. $75.4\ cm^2$ — 12. C
13. Find the volume of the figure below. Use $\pi \approx 3.14$. Round decimal answer to the nearest tenth.
    5 cm, 6 cm, 15 cm
    A. $450\ cm^3$ B. $150\ cm^3$ C. $225\ cm^3$ D. $112.5\ cm^3$ — 13. C
14. The interior of Quyen's refrigerator is shaped like a rectangular prism that is $34\frac{3}{4}$ inches high, 26 inches wide, and $15\frac{1}{2}$ inches deep. Find the volume of the refrigerator. Write the answer in simplest form.
    A. $7052\frac{1}{2}\ in^3$ B. $14{,}004\frac{1}{4}\ in^3$ C. $13{,}260\frac{3}{8}\ in^3$ D. $18{,}806\frac{3}{8}\ in^3$ — 14. B
15. Dean has a small pickle barrel with a radius of 9 inches and a height of 48 inches. Find the volume of the barrel. Use $\pi \approx 3.14$ and round to the nearest tenth.
    A. $6104.2\ in^3$ B. $4069.4\ in^3$ C. $12{,}208.3\ in^3$ D. $2713.0\ in^3$ — 15. C
16. Find the volume of a pyramid with a base area of 23 square inches and a height of 6 inches.
    A. $1058\ in^3$ B. $138\ in^3$ C. $3.2\ in^3$ D. $46\ in^3$ — 16. D
17. Find the volume of a cylinder with a diameter of 28 inches and a height of 10 inches. Use $\pi \approx \frac{22}{7}$ and simplify fraction answer.
    A. $6160\ in^3$ B. $3140\ in^3$ C. $440\ in^3$ D. $250\frac{1}{8}\ in^3$ — 17. A

CHAPTER 12 Study Guide and Assessment

Using the Study Guide and Assessment

Skills and Concepts Encourage students to refer to the objectives and examples on the left as they complete the review exercises on the right.

Assessment and Evaluation Masters, pp. 317–318

NAME ____________ DATE ______

CHAPTER 12 Test, Form 2B

Find the area of each figure. Use $\pi \approx 3.14$ with decimal measures and $\pi \approx \frac{22}{7}$ with fraction measures.

1. 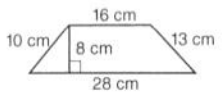

1. 176 cm^2

2. 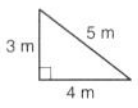

2. 6 m^2

3. 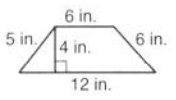

3. 36 in^2

4. 

4. $273\frac{7}{9}$ in^2

5. 

5. 200.96 mm^2

6. Find the area of a circle 600 yards in diameter. Use $\pi \approx 3.14$. 6. 282,600 yd^2

7. Tom cuts a square having 4-inch sides out of a circle having a 9-inch diameter. Find the area of the remaining piece. Use $\pi \approx 3.14$. 7. 47.585 in^2

8. The figure at the right represents a dart board. Suppose you threw 28 darts at random. If they all land on the board, how many would you expect to land in the shaded region? 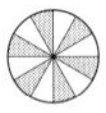 8. 14 darts

Find the surface area of each figure. Use $\pi \approx 3.14$.

9. 

9. 135.648 in^2

10. 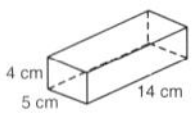

10. 292 cm^2

NAME ____________ DATE ______

Chapter 12 Test, Form 2B (continued)

11. 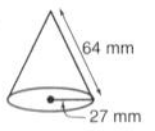

11. 7714.98 mm^2

12. 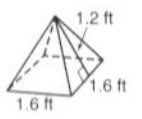

12. 6.4 ft^2

13. Find the volume of a rectangular prism with a length of 6 meters, a height of 4 meters, and a width of 2 meters. 13. 48 m^3

14. Alice has a shell-covered box to hold her rings. It is shaped like a cube measuring $2\frac{1}{2}$ inches on an edge. Find the volume of Alice's shell-covered box. 14. $15\frac{5}{8}$ in^3

Find the volume of each figure. Use $\pi \approx 3.14$. Round decimal answers to the nearest tenth.

15. 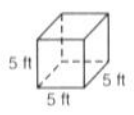

15. 125 ft^3

16. 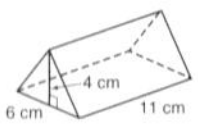

16. 132 cm^3

17. 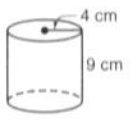

17. 452.2 cm^3

18. 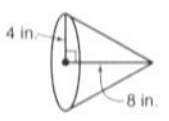

18. 134.0 in^3

19. Find the volume of a triangular pyramid if the base area is 42 square inches and the height is 12 inches. 19. 168 in^3

20. Peter cuts pieces of sod to use in a rectangular play area at a day-care center. The sod is 6 inches thick. Find the volume of sod he needs if the play area is 30 feet wide and 48 feet long. 20. 720 ft^3

CHAPTER 12 Study Guide and Assessment

Skills and Concepts

Objectives and Examples

Upon completing this chapter, you should be able to:

▶ **find areas of parallelograms, triangles, and trapezoids** (Lesson 12-1)

Find the area of the trapezoid.

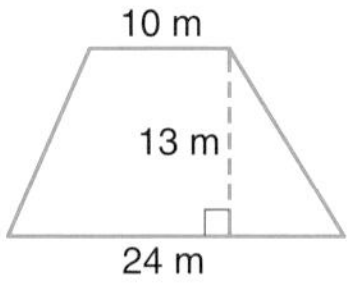

$A = \frac{1}{2}h(a + b)$

$A = \frac{1}{2} \cdot 13(10 + 24)$

$A = 221 \text{ m}^2$

Review Exercises

Use these exercises to review and prepare for the chapter test.

Find the area of each figure.

9. 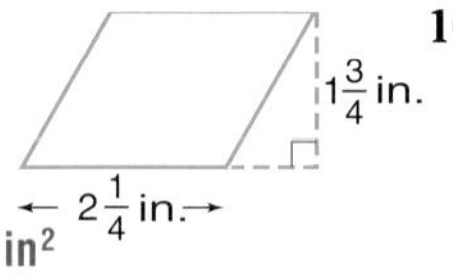

$3\frac{15}{16}$ in^2

10. 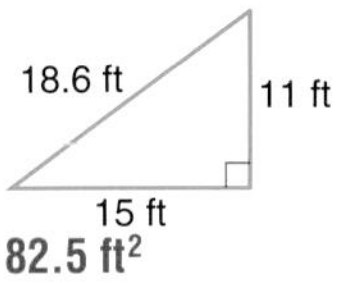

82.5 ft^2

11. 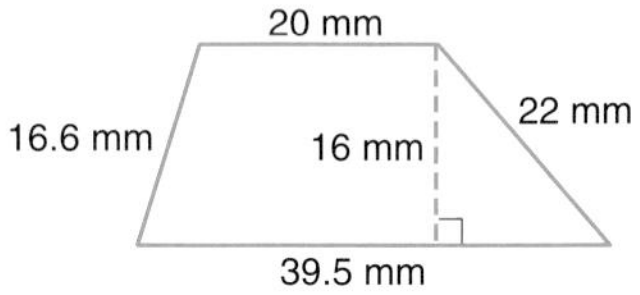

476 mm^2

▶ **find the area of circles** (Lesson 12-2)

Find the area of the circle.

$A = \pi r^2$

$A = \pi \cdot (3)^2$

$A \approx 28.3 \text{ in}^2$

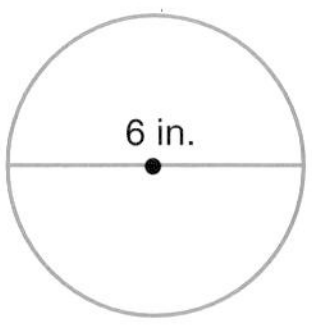

Find the area of each circle described below. Round answers to the nearest tenth.

12. 

254.5 mi^2

13. diameter, 4.2 m 13.9 m^2

▶ **find probabilities using area models** (Lesson 12-3)

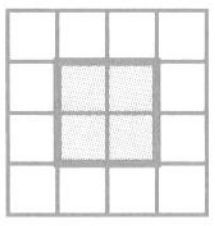

The probability of landing in the shaded region is $\frac{1}{4}$.

14. The figure below represents a dartboard. Find the probability of landing in the shaded region. $\frac{2}{5}$

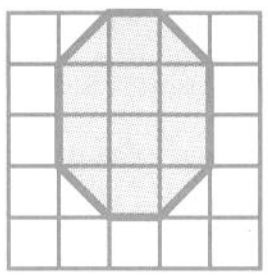

GLENCOE Technology

Test and Review Software

You may use this software, a combination of an item generator and an item bank, to create your own tests or worksheets. Types of items include free response, multiple choice, short answer, and open ended.

For IBM & Macintosh

Chapter 12 •
Study Guide and Assessment

Objectives and Examples	Review Exercises

▶ **solve problems using a model or a drawing** (Lesson 12-4)

Making a model or drawing can help you visualize a problem so you can solve it.

Solve by making a drawing.

15. The radius of a circular flower garden and its border is 2.8 meters. Without the border, the radius is 2.3 meters. Find the area of the border. **8.0 m^2**

▶ **find the surface area of rectangular prisms and circular cylinders** (Lesson 12-5)

Find the surface area of the prism.

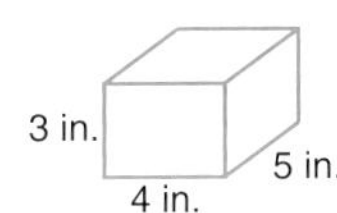

Top and bottom: $2 \times 4 \times 5 = 40$
Front and back: $2 \times 3 \times 4 = 24$
Sides: $2 \times 3 \times 5 = 30$
Surface area $= 40 + 24 + 30$ or 94 in^2

Find the surface area of each figure.

16.

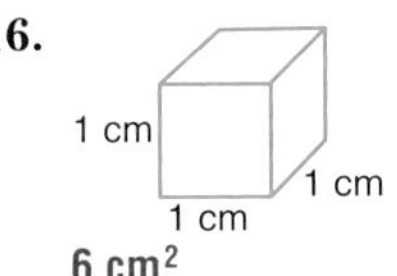

6 cm^2

17.

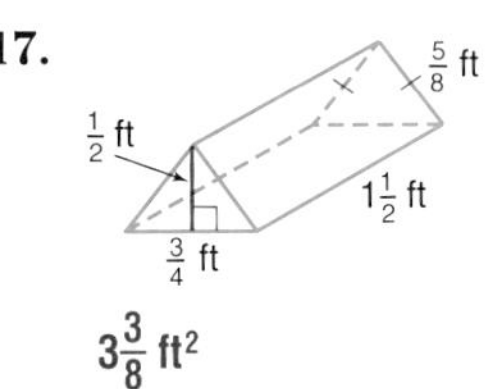

$3\frac{3}{8}$ ft^2

18. circular cylinder: height, 2.3 m; radius, 1 m **21 m^2**
19. circular cylinder: height, 41 ft; diameter, 40 ft **7665 ft^2**

▶ **find the surface areas of pyramids and cones** (Lesson 12-6)

Find the surface area of a cone with radius 5 inches and height 8 inches.

Base:	Curved Surface:
$A = \pi r^2$	$A = \pi r \ell$
$A = \pi \cdot (5)^2$	$A = \pi \cdot 5 \cdot 8$
$A \approx 78.5$	$A \approx 125.7$

Total surface area: $78.5 + 125.7$ or 204.2 in^2

Find the surface area of each figure. Round decimal answers to the nearest tenth.

20.

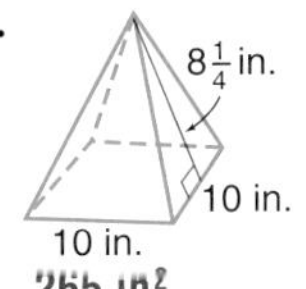

265 in^2

21.

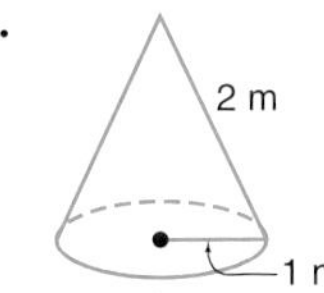

9.4 m^2

▶ **find the volumes of prisms and circular cylinders** (Lesson 12-7)

Find the volume of the cylinder.

$V = \pi r^2 h$
$V = \pi \cdot (2)^2 \cdot 7$
$V = 88.0 \text{ ft}^3$

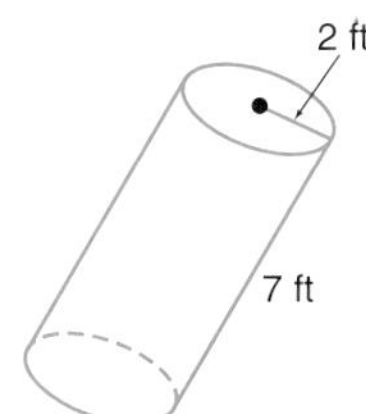

Find the volume of each figure described below. Round decimal answers to the nearest tenth.

22. triangular prism: base of triangle, 5 cm; height, 3.4 cm; prism height, 12 cm **102 cm^3**
23. hexagonal prism: area of base, 166.25 in^2; height, 20 in. **3325 in^3**
24. rectangular prism: length, 4 m; width, 4 m; height, 16 m **256 m^3**
25. cylinder: radius, 7 ft; height, 18 ft **2770.9 ft^3**
26. cylinder: diameter, 10 m; height, 22 m **1727.9 m^3**

Chapter 12 657

Chapter 12 • Study Guide and Assessment

Applications and Problem Solving

Encourage students to work through the exercises in the Applications and Problem Solving section to strengthen their problem-solving skills.

Objectives and Examples	Review Exercises
▶ **find the volumes of pyramids and cones** (Lesson 12-8)	**Find the volume of each figure. Round decimal answers to the nearest tenth.**

Find the volume of the pyramid.

$V = \frac{1}{3}Bh$

$V = \frac{1}{3} \cdot (3 \cdot 3) \cdot 4$

$V = 12\text{m}^3$

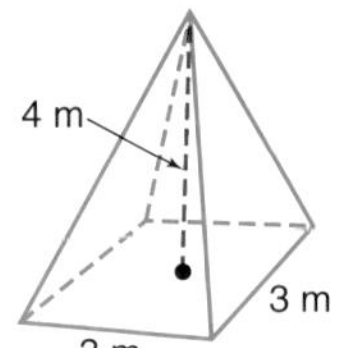

27. 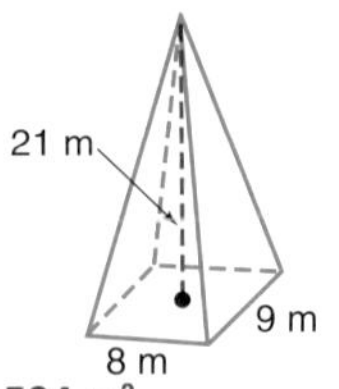

504 m³

28. 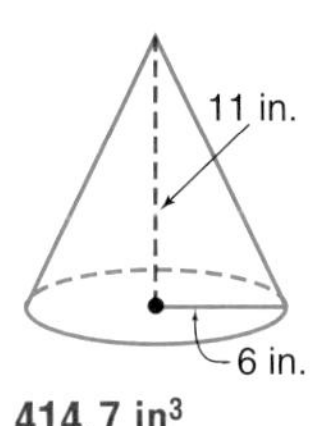

414.7 in³

Applications and Problem Solving

29. Entertainment The Eastside High School Drill Team is having new flags made. Some of the flags will be shaped like triangular pennants. The base of the pennants will be 36 inches and the height will be 48 inches. What is the area of one of these flags? (Lesson 12-1) **864 in²**

30. Agriculture A silo has a cylindrical shape. The radius of the base is about 20 feet, and the silo is 60 feet high. A company has been hired to paint the silo. (Lesson 12-5) **30a. 7540 ft²**

a. What is the area of the curved surface?

b. How many gallons of paint would be needed to apply one coat of paint to the silo if one gallon of paint covers about 450 square feet? **23 gallons**

c. How much would it cost to paint the silo if one gallon costs $10? **$230**

d. How much would it cost to apply two coats of paint to the silo? **$460**

31. Manufacturing A leading manufacturer of sugar cubes packs a rectangular box so that there are six cubes along one edge, eleven cubes along a second edge, and three cubes along the third edge. How many sugar cubes are in the box? (Lesson 12-7) **198**

32. Science Jenny knows that 1 mL of water has a volume of 1 cm³. She needs 225 mL of water for an experiment. She has an unmarked cylindrical container with a diameter of 12 cm. To what height should she fill the container to measure the water for the experiment? (Lesson 12-7) **≈2 cm**

33. Agriculture A water trough is half of a circular cylinder with a radius of 6 feet and a height of 3 feet. If one gallon of water has a volume of about 230 in³, about how many gallons of water will the trough hold? (Lesson 12-7) **about 1275 gallons**

34. Mining An open pit mine in the Elk mountain range is shaped like a cone. The mine is 420 feet across and 250 feet deep. What volume of material was removed? (Lesson 12-8) **≈11,545,353 ft³**

A practice test for Chapter 12 is available on page 785.

Alternative Assessment

Cooperative Learning Project

Work with a group. See students' work.

1. Research how Egyptian pyramids were constructed and what tools were used.
2. Choose one pyramid. Find the dimensions of that pyramid.
3. Plan a way to make a scale model of the pyramid. Include the following information in your plan.
   a. Decide how big to make the model.
   b. Determine the scale factor for the model.
   c. Select materials for making the model.
4. Construct the model.
5. Prepare a report on your findings, your procedures, and any problems you encountered. Include information on the surface area and volume of the original pyramid and of the model. You may wish to make a poster or brochure to accompany your model.

Thinking Critically

- The edge of a cube and the diameter and height of a cylinder all have the same measure. Which has the least surface area? Which has the least volume? **cylinder; cylinder**
- Suppose you are designing a solid figure with two congruent parallel bases and the distance around the base and the height must be the same for whatever solid you design. What shape base would give the greatest surface area? the greatest volume? **circle; circle**
- Design three different nets that could be used to make a rectangular prism 4 cm long, 3 cm wide, and 2 cm high. Which net will allow you to cut the greatest number of patterns from a 20-cm × 30-cm grid? **See students' work.**

Portfolio

Select one of the assignments from this chapter that you found especially challenging and place it in your portfolio. **See students' work.**

Self Evaluation

Good problem solvers often break a problem into several smaller steps that allow them to solve the entire problem.

Assess yourself. Are there any daily problems that you have solved by tackling them one step at a time? Write a paragraph about your experience. How can this strategy be applied to problems about surface area or volume? **See students' work.**

Alternative Assessment

The Alternative Assessment section provides students with the opportunity to assess their own work by thinking critically, working with others, keeping a portfolio, and honestly evaluating their own progress. For more information on alternative forms of assessment, see *Alternative Assessment in the Mathematics Classroom*.

Performance Assessment

Performance Assessment tasks for this chapter are included in the *Assessment and Evaluation Masters*. A scoring guide is also provided.

Alternative Assessment

Assessment and Evaluation Masters, pp. 321, 323

NAME ________________ DATE ________

CHAPTER 12 Performance Assessment

Instructions: Demonstrate your knowledge by giving a clear, concise solution to each problem. Be sure to include all relevant drawings and justify your answers. You may show your solution in more than one way or investigate beyond the requirements of the problem.

1. Mr. Hansen's job is to estimate the labor and materials needed to paint houses and other structures. A part of his job is to determine the area to be painted.
   a. Tell how to find the area, excluding the roof, to be painted on the house at right. Find the area.
   b. Tell how to find the surface area of the silo including the roof. Find the area.
   c. Sometimes he estimates the area to be painted because the actual area is difficult to determine. Would it be better for estimates to be a little too large or a little too small? Why? Give a good estimate for the building at right. Exclude the roof.

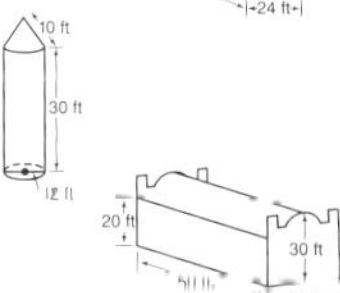

2. Design a target for a target game in which the probability that a random throw wins is $\frac{1}{16}$.
3. The Food and Drug Administration, among other duties, is charged with protecting consumers from misleading packaging. Many schemes have been used through the years to make customers believe they are getting more than they really are.
   a. Find the volume of the cylinder at right. Find the volume if the height of the cylinder is doubled. Find the volume if the radius is doubled. Do you think retailers would tend to use tall bottles or large-diameter bottles to mislead customers? Why or why not?
   b. A way of misleading customers is by creating false bottoms such as those shown at right. Guess which false bottom displaces the most volume. Find the volume of each false bottom to check your answer.
   c. Give other ways distributors may attempt to mislead customers relative to the volume of a container.

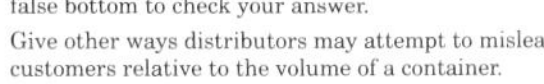

CHAPTER 12 Scoring Guide

Level	Specific Criteria
3 Superior	• Shows thorough understanding of the concepts *area, volume, radius, height, geometric probability,* and *misleading.* • Uses appropriate strategies to find areas and volumes. • Computations to find areas and volumes are correct. • Written explanations are exemplary. • Ways to mislead customers are accurate. • Diagrams are accurate and appropriate. • Goes beyond requirements of some or all problems.
2 Satisfactory, with minor flaws	• Shows understanding of the concepts *area, volume, radius, height, geometric probability,* and *misleading.* • Uses appropriate strategies to find areas and volumes. • Computations to find areas and volumes are mostly correct. • Written explanations are effective. • Ways to mislead customers are mostly accurate. • Diagrams are mostly accurate and appropriate. • Satisfies all requirements of problems.
1 Nearly Satisfactory, with serious flaws	• Shows understanding of most of the concepts *area, volume, radius, height, geometric probability,* and *misleading.* • May not use appropriate strategies to find areas and volumes. • Computations to find areas and volumes are mostly correct. • Written explanations are satisfactory. • Ways to mislead customers are mostly accurate. • Diagrams are mostly accurate and appropriate. • Satisfies most requirements of problems.
0 Unsatisfactory	• Shows little or no understanding of the concepts *area, volume, radius, height, geometric probability,* and *misleading.* • May not use appropriate strategies to find areas and volumes. • Computations to find areas and volumes are incorrect. • Written explanations are not satisfactory. • Ways to mislead customers are not accurate. • Diagrams are not accurate and appropriate. • Does not satisfy requirements of problems.

Chapter 12 • Ongoing Assessment

Using the Ongoing Assessment

These two pages review the skills and concepts presented in Chapters 1–12. This review is formatted to reflect new trends in standardized testing.

Assessment and Evaluation Masters, pp. 326–327

NAME ______________________ DATE __________

CHAPTERS 1–12 Cumulative Review

1. Translate *nine less a number* into an algebraic expression. (Lesson 1-3)	1. $9 - n$
2. Solve $n = -12 + 25$. (Lesson 2-4)	2. 13
3. Solve $y - (-7) = 34$. (Lesson 3-2)	3. 27
4. Find the LCM for $4a$, $6a$, and $18a$. (Lesson 4-7)	4. $36a$
5. Solve $w = 23.5 + (-7.26)$. (Lesson 5-3)	5. 16.24
6. Solve $q = 0.444 \div 0.37$. (Lesson 6-5)	6. 1.2
7. Solve $3a + 4 = 4a - 3$. (Lesson 7-5)	7. 7
8. Find the slope of the line that contains the points $X(2, 2)$ and $Y(5, 11)$. (Lesson 8-6)	8. 3
9. Express 0.5% as a fraction in simplest form. (Lesson 9-7)	9. $\frac{1}{200}$
10. Paula buys a $75 tire and receives a 15% discount. Find the cost of the tire. (Lesson 9-9)	10. $63.75
11. Which average, mode, median, or mean, would best describe hair color of a group of people? (Lesson 10-4)	11. mode
12. Two dice are rolled and a coin is tossed. How many possible outcomes are possible? (Lesson 10-5)	12. 72 outcomes
13. Maria, Li, and Fred are running for class president and Tom, Keli, Miquel, and Don are running for vice president. How many president-vice president pairings are possible? (Lesson 10-5)	13. 12
14. Henri must pick 5 of 7 classes being offered. How many ways can Henri arrange a schedule? (Lesson 10-6)	14. 2520
15. Sasha has 8 antique spoons. She uses 4 in a window display. How many combinations are possible? (Lesson 10-6)	15. 70
16. Write a name for the figure at the right. (Lesson 11-1)	16. $\overline{BA}$

NAME ______________________ DATE __________

Chapters 1–12 Cumulative Review (continued)

17. Classify an angle having a measure of 12 degrees. (Lesson 11-1)	17. acute
18. Aimee must represent 20% on a circle graph. How many degrees are in the measure of the section? (Lesson 11-2)	18. 72°
19. In triangle XYZ, the measure of angle Y is 16° and the measure of angle Z is 81°. Find the measure of angle X. (Lesson 11-4)	19. 83°
Find the area of each figure described below. (Lessons 12-1 and 12-2)	
20. Parallelogram: base, 8 centimeters; height, 6 centimeters	20. 48 cm²
21. Triangle: base, 24 feet; height, 18 feet	21. 216 ft²
22. Trapezoid: base (a), 21 inches; base (b), 12 inches; height, 18 inches	22. 297 in²
23. Circle: radius, 12 feet (Use $\pi \approx 3.14$.)	23. 452.16 ft²
24. Find the volume of a circular cylinder that has a diameter of 12 centimeters and a height of 30 centimeters. (Lesson 12-7)	24. 3391.2 cm³
25. Joyce owns a cone-shaped planter that is 18 inches high and has a diameter of 20 inches. Find the volume of the planter. Use $\pi \approx 3.14$. (Lesson 12-8)	25. 1884 in³

CHAPTERS 1–12 *Ongoing Assessment*

Section One: Multiple Choice

There are eleven multiple-choice questions in this section. After working each problem, write the letter of the correct answer on your paper.

1. **Geometry** What is the area of the trapezoid? **D**

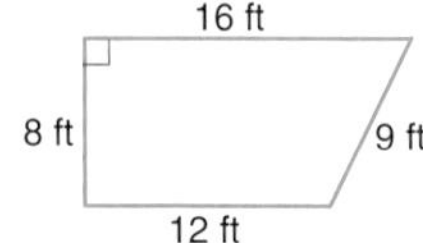

A. 45 ft² **B.** 64 ft² **C.** 96 ft² **D.** 112 ft²

2. **Sports** Sandy's handicap score is found using the formula: $s = g + c$, where s is the handicap score, g is the game score, and c is the handicap. What was Sandy's game score if her handicap score is 186 and her handicap is 15? **B**

A. 161 **B.** 171 **C.** 191 **D.** 201

3. **Geometry** A cylindrical can is 12 cm tall, and the ends each have a diameter of 10 cm. Which expression can be used to find the area the can label covers? **A**

A. $\pi \times 10 \times 12$
B. $\pi \times 5 \times 5 \times 12$
C. $2(\pi \times 5^2) + \pi \times 10 \times 12$
D. $\pi \times 5^2 + 2(\pi \times 10 \times 12)$

4. Which is equivalent to $a^5 \cdot a^2$? **D**

A. $2a^{10}$ **B.** $2a^7$ **C.** a^{10} **D.** a^7

5. **Geometry** What is the volume of the triangular prism? **D**

A. 3600 cm³
B. 4800 cm³
C. 30,000 cm³
D. 36,000 cm³

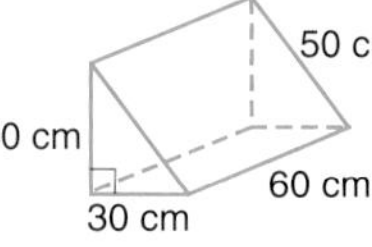

6. Which fractions and decimals are in order from least to greatest? **B**

A. $0.5, \frac{1}{3}, \frac{1}{4}$ **B.** $\frac{3}{5}, 0.7, \frac{9}{12}$
C. $0.75, \frac{5}{8}, \frac{9}{10}$ **D.** $\frac{5}{6}, \frac{5}{8}, 0.5$

7. Lee is practicing for a 2500-meter run. He runs the distance twice a day for 5 days. How many kilometers has he run? **A**

A. 25 km **B.** 25,000 km
C. 250 km **D.** none of these

8. **Geometry** Which shows a step in constructing an angle congruent to a given angle? **B**

A.

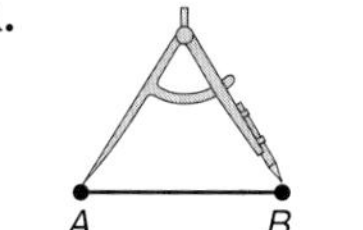

B.

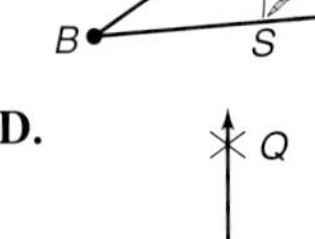

C.

D.

9. In how many different ways can all four students be arranged in a row?

Suki Janet Marcus Bill **C**

A. 4 ways **B.** 12 ways
C. 24 ways **D.** 36 ways

10. **Geometry** $\triangle QRS$ is similar to $\triangle XYZ$. What is the length of $\overline{RQ}$? **B**

A. 16 units
B. 18 units
C. 20 units
D. 22 units

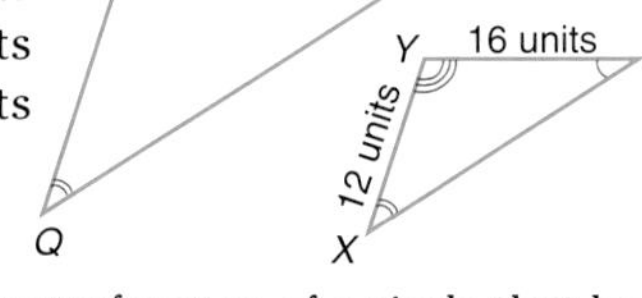

11. Find the circumference of a circle that has a diameter 1.2 meters long. **C**

A. 4.5 meters
B. 7.5 meters
C. 3.8 meters
D. 1.1 meters

Standardized Test Practice Questions are provided in the *Assessment and Evaluation Masters*, p. 325.

Section Two: Free Response

This section contains eight questions for which you will provide short answers. Write your answer on your paper.

12. Geometry Find the value of x. Then classify the triangle as acute, obtuse, or right. 134, obtuse

13. A sweatshirt is on sale for \$16. This is 80% of the regular price. What is the regular price? \$20

14. A rectangular swimming pool is 6 meters wide and 12 meters long. A concrete walkway is poured around the pool. The walkway is 1 meter wide and 0.2 meters deep. What is the volume of the concrete? 8 m^3

15. On a trip, the Changs drive 104 miles in 2 hours. If they continue at the same rate, how long will it take them to complete a 550 mile trip? about 10.5 hours

16. From a 52-card deck, Susanna draws 2 cards. Find the probability they will both be kings. $\frac{1}{221}$

17. Geometry In the figure below, ℓ is parallel to m. If the measure of $\angle 1$ is $108°$, find the measure of the remaining angles. $m\angle 2 = 72°$, $m\angle 3 = 108°$, $m\angle 4 = 72°$, $m\angle 5 = 108°$, $m\angle 6 = 72°$, $m\angle 7 = 108°$, $m\angle 8 = 72°$

18. Geometry Find the value of x. $x = 140$

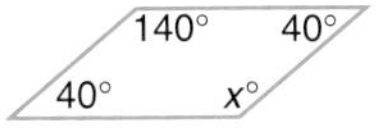

Test-Taking Tip

Although the basic formulas that you need to answer the questions on a standardized test are often given to you in the test booklet, it is a good idea to review the formulas beforehand.

Area of a circle $= \pi r^2$
Circumference of a circle $= 2\pi r$
Area of a square $= s^2$
Perimeter of a rectangle $= 2(\ell + w)$
Area of a rectangle $= \ell w$

19. Sports The Los Angeles Lakers scored 52 points in the first half of a recent game against the Phoenix Suns. Phoenix averages 106 points scored per game. Write and solve an inequality that shows how many points the Lakers need to score in the second half to beat the Suns' average points scored per game. $p + 52 \geq 106$; $p \geq 54$

Section Three: Open-Ended

This section contains two open-ended problems. Demonstrate your knowledge by giving a clear, concise solution to each problem. Your score on these problems will depend upon how well you do the following.

- Explain your reasoning.
- Show your understanding of the mathematics in an organized manner.
- Use charts, graphs, and diagrams in your explanation.
- Show the solution in more than one way or relate it to other situations.
- Investigate beyond the requirements of the problem.

20–21. See students' work.

20. Probability Give two examples where geometric probability can be used.

21. Geometry Explain what the difference is between a line, a line segment, and a ray.

CHAPTER 13

Applying Algebra to Right Triangles

PREVIEWING THE CHAPTER

The chapter begins with finding and using squares and square roots. Students next use Venn diagrams to solve problems. The study of numbers is expanded to include irrational numbers. The Pythagorean theorem is used to find the length of the sides of a right triangle, to solve problems, and to graph irrational numbers. Special triangles are studied as students find missing measures in 30°–60° and 45°–45° right triangles. Trigonometry is studied as students explore the relationships of the sides of right triangles. Finally, students solve problems by using the trigonometric ratios.

Lesson (pages)	Lesson Objectives	NCTM Standards	State/Local Objectives
13-1 (664–668)	Find and use squares and square roots.	1-4, 6, 12, 13	
13-2 (669–671)	Use Venn diagrams to solve problems.	1-4, 12	
13-3 (672–675)	Identify numbers in the real number system and solve equations by finding square roots.	1-4, 6, 12, 13	
13-4 (676–681)	Use the Pythagorean theorem to find the length of the side of a right triangle and to solve problems.	1-4, 12, 13	
13-4B (682)	Use the Pythagorean theorem to graph irrational numbers.	1-4, 12, 13	
13-5 (683–686)	Find missing measures in 30°–60° and 45°–45° right triangles.	1-4, 12, 13	
13-6A (687)	Explore the relationships of the sides of right triangles.	1-4, 12, 13	
13-6 (688–692)	Find missing sides and angles of triangles using the sine, cosine, and tangent ratios.	1-4, 12, 13	
13-6B (693)	Use a graphing calculator to explore the slope of a line.	1-4, 12, 13	
13-7 (694–697)	Solve problems by using the trigonometric ratios.	1-4, 12, 13	

A complete, 1-page lesson plan is provided for each lesson in the ***Lesson Planning Guide***. Answer keys for each lesson are available in the ***Answer Key Masters***.

ORGANIZING THE CHAPTER

You may want to refer to the **Course Planning Calendar** on page T12 for detailed information on pacing.
PACING: Standard—11 days; **Honors**—10 days; **Block**—6 days

LESSON PLANNING CHART

			BLACKLINE MASTERS												
Lesson (Pages)	**Materials/ Manipulatives**	**Extra Practice (Student Edition)**	**Study Guide**	**Practice**	**Enrichment**	**Assessment and Evaluation**	**Math Lab and Modeling Math**	**Multicultural Activity**	**Tech Prep Applications**	**Graphing Calculator**	**Activity**	**Real-World Applications**	**Interactive Mathematics Tools Software**	**Teaching Transparencies**	**Group Activity Cards**
13-1 (664–668)	graphing calculator, tiles*	p. 771	p. 108	p. 108	p. 108				p. 25		p. 46		13-1	13-1A 13-1B	13-1
13-2 (669–671)		p. 771	p. 109	p. 109	p. 109	p. 351	p. 88							13-2A 13-2B	13-2
13-3 (672–675)		p. 771	p. 110	p. 110	p. 110									13-3A 13-3B	13-3
13-4 (676–681)	geoboard	p. 771	p. 111	p. 111	p. 111	pp. 350–351	pp. 73–74	p. 25	p.26		pp. 13, 47	29	13-4	13-4A 13-4B	13-4
13-4B (682)	compass* straightedge						p. 54								
13-5 (683–686)		p. 772	p. 112	p. 112	p. 112									13-5A 13-5B	13-5
13-6A (687)	calculator metric ruler*						p. 55								
13-6 (688–692)		p. 772	p. 113	p. 113	p. 113	p. 352						30	13-6.1 13-6.2	13-6A 13-6B	13-6
13-6B (693)	graphing calculator									p. 26					
13-7 (694–697)	protractor*, index cards, straw, large paper clip, tape, string, calculator	p. 772	p. 114	p. 114	p. 114	p. 352		p. 26		p. 13	p. 27			13-7A 13-7B	13-7
Study Guide/ Assessment (700–702)						pp. 337–349, 353–355									

*Included in Glencoe's *Student Manipulative Kit* and *Overhead Manipulative Resources*.

ORGANIZING THE CHAPTER

All of the blackline masters in the Teacher's Classroom Resources are available on the ***Electronic Teacher's Classroom Resources*** CD-ROM.

OTHER CHAPTER RESOURCES

Student Edition
Chapter Opener, pp. 662–663
Working on the Investigation, p. 671
Closing the Investigation, p. 698
The Shape of Things to Come: Thought-Controlled Computer Technology, p. 692
From the Funny Papers, p. 668

Teacher's Classroom Resources
Investigations and Projects Masters, pp. 77–80
Pre-Algebra Overhead Manipulative Resources, pp. 39–41

Technology
Test and Review Software (IBM & Macintosh)
CD-ROM Activities

Professional Publications
Block Scheduling Booklet
Glencoe Mathematics Professional Series

OUTSIDE RESOURCES

Books/Periodicals
Flying Through Math, Jan de Lange

Videos/CD-ROMs
Algebra Math Video Series: Verbal Problems & Introduction to Trigonometry, ETA

Software
Gliding, Sunburst

ASSESSMENT RESOURCES

Student Edition
Math Journal, pp. 690, 696
Mixed Review, pp. 668, 671, 675, 681, 686, 692, 697
Self Test, p. 681
Chapter Highlights, p. 699
Chapter Study Guide and Assessment, pp. 700–702
Alternative Assessment, p.
Portfolio, p. 61

MindJogger Videoquiz, 13

Teacher's Wraparound Edition
5-Minute Check, pp. 664, 669, 672, 676, 683, 688, 694
Checking Your Understanding, pp. 666, 670, 674, 679, 685, 690, 696
Closing Activity, pp. 668, 671, 675, 681, 686, 692, 697
Cooperative Learning, pp. 673, 695

Assessment and Evaluation Masters
Multiple-Choice Tests, Forms 1A (Honors), 1B (Average), 1C (Basic), pp. 337–342
Free-Response Tests, Forms 2A (Honors), 2B (Average), 2C (Basic), pp. 343–348
Performance Assessment, p. 349
Mid-Chapter Test, p. 350
Quizzes A–D, pp. 351–352
Standardized Test Practice, p. 353
Cumulative Review, pp. 354–355

ENHANCING THE CHAPTER

Examples of some of the materials for enhancing Chapter 13 are shown below.

DIVERSITY

Multicultural Activity Masters, pp. 25, 26

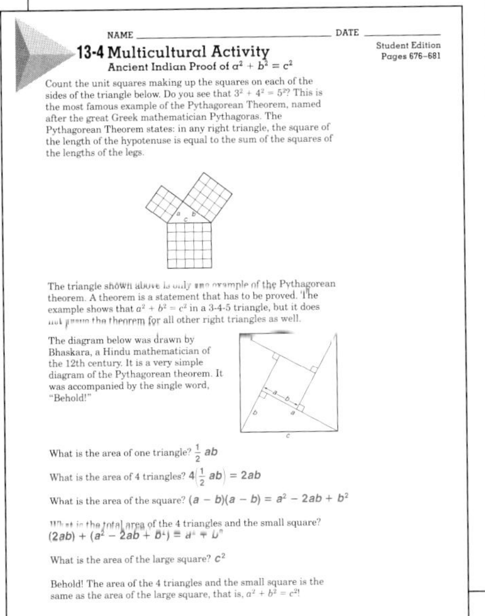

NAME ______________ DATE ______

13-4 Multicultural Activity
Ancient Indian Proof of $a^2 + b^2 = c^2$

Student Edition Pages 676–681

Count the unit squares making up the squares on each of the sides of the triangle below. Do you see that $3^2 + 4^2 = 5^2$? This is the most famous example of the Pythagorean Theorem, named after the great Greek mathematician Pythagoras. The Pythagorean Theorem states: in any right triangle, the square of the length of the hypotenuse is equal to the sum of the squares of the lengths of the legs.

The triangle shown above is only one example of the Pythagorean theorem. A theorem is a statement that has to be proved. The example shows that $a^2 + b^2 = c^2$ in a 3-4-5 triangle, but it does not prove the theorem for all other right triangles as well.

The diagram below was drawn by Bhaskara, a Hindu mathematician of the 12th century. It is a very simple diagram of the Pythagorean theorem. It was accompanied by the single word, "Behold!"

What is the area of one triangle? $\frac{1}{2}ab$

What is the area of 4 triangles? $4\left(\frac{1}{2}ab\right) = 2ab$

What is the area of the square? $(a - b)(a - b) = a^2 - 2ab + b^2$

What is the total area of the 4 triangles and the small square? $(2ab) + (a^2 - 2ab + b^2) = a^2 + b^2$

What is the area of the large square? c^2

Behold! The area of the 4 triangles and the small square is the same as the area of the large square, that is, $a^2 + b^2 = c^2$!

APPLICATIONS

Real-World Applications, 29, 30

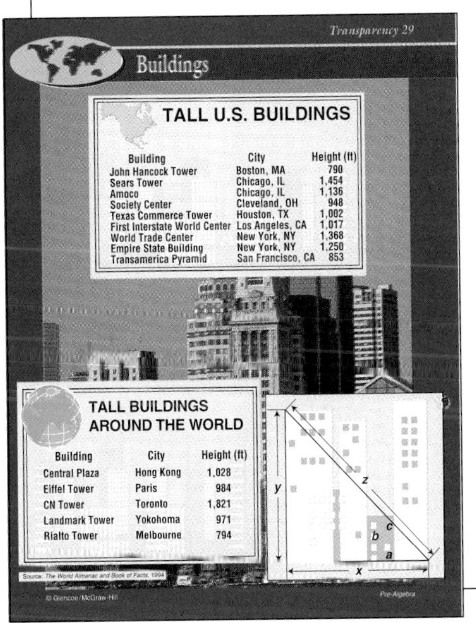

Transparency 29

Buildings

TALL U.S. BUILDINGS

Building	City	Height (ft)
John Hancock Tower	Boston, MA	790
Sears Tower	Chicago, IL	1,454
Amoco	Chicago, IL	1,136
Society Center	Cleveland, OH	948
Texas Commerce Tower	Houston, TX	1,002
First Interstate World Center	Los Angeles, CA	1,017
World Trade Center	New York, NY	1,368
Empire State Building	New York, NY	1,250
Transamerica Pyramid	San Francisco, CA	853

TALL BUILDINGS AROUND THE WORLD

Building	City	Height (ft)
Central Plaza	Hong Kong	1,028
Eiffel Tower	Paris	984
CN Tower	Toronto	1,821
Landmark Tower	Yokohama	971
Rialto Tower	Melbourne	794

Source: *The World Almanac and Book of Facts, 1994*

© Glencoe/McGraw-Hill   Pre-Algebra

TECHNOLOGY

Graphing Calculator Masters, p. 13

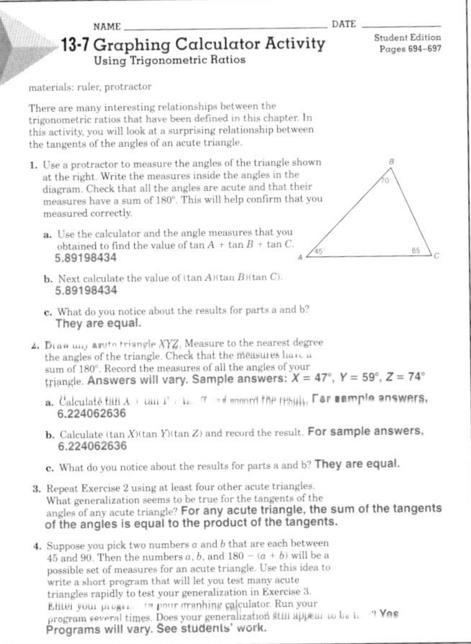

NAME ______________ DATE ______

13-7 Graphing Calculator Activity
Using Trigonometric Ratios

Student Edition Pages 694–697

materials: ruler, protractor

There are many interesting relationships between the trigonometric ratios that have been defined in this chapter. In this activity, you will look at a surprising relationship between the tangents of the angles of an acute triangle.

1. Use a protractor to measure the angles of the triangle shown at the right. Write the measures inside the angles in the diagram. Check that all the angles are acute and that their measures have a sum of 180°. This will help confirm that you measured correctly.
   a. Use the calculator and the angle measures that you obtained to find the value of $\tan A + \tan B + \tan C$. **5.89198434**
   b. Next calculate the value of $(\tan A)(\tan B)(\tan C)$. **5.89198434**
   c. What do you notice about the results for parts a and b? **They are equal.**
2. Draw any acute triangle XYZ. Measure to the nearest degree the angles of the triangle. Check that the measures have a sum of 180°. Record the measures of all the angles of your triangle. **Answers will vary. Sample answers: $X = 47°$, $Y = 59°$, $Z = 74°$**
   a. Calculate $\tan X + \tan Y + \tan Z$ and record the result. **For sample answers, 6.224062636**
   b. Calculate $(\tan X)(\tan Y)(\tan Z)$ and record the result. **For sample answers, 6.224062636**
   c. What do you notice about the results for parts a and b? **They are equal.**
3. Repeat Exercise 2 using at least four other acute triangles. What generalization seems to be true for the tangents of the angles of any acute triangle? **For any acute triangle, the sum of the tangents of the angles is equal to the product of the tangents.**
4. Suppose you pick two numbers a and b that are each between 45 and 90. Then the numbers a, b, and $180 - (a + b)$ will be a possible set of measures for an acute triangle. Use this idea to write a short program that will let you test many acute triangles rapidly to test your generalization in Exercise 3. Enter your program in your graphing calculator. Run your program several times. Does your generalization still appear to be true? **Yes Programs will vary. See students' work.**

TECH PREP

Tech Prep Applications Masters, pp. 25, 26

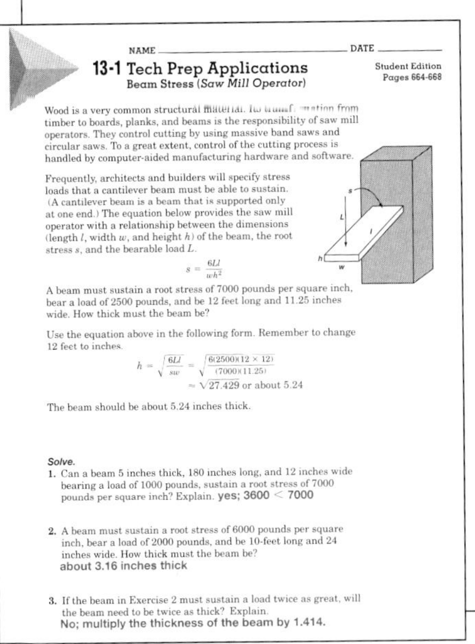

NAME ______________ DATE ______

13-1 Tech Prep Applications
Beam Stress (*Saw Mill Operator*)

Student Edition Pages 664-668

Wood is a very common structural material. Its transformation from timber to boards, planks, and beams is the responsibility of saw mill operators. They control cutting by using massive band saws and circular saws. To a great extent, control of the cutting process is handled by computer-aided manufacturing hardware and software.

Frequently, architects and builders will specify stress loads that a cantilever beam must be able to sustain. (A cantilever beam is a beam that is supported only at one end.) The equation below provides the saw mill operator with a relationship between the dimensions (length l, width w, and height h) of the beam, the root stress s, and the bearable load L.

$$s = \frac{6Ll}{wh^2}$$

A beam must sustain a root stress of 7000 pounds per square inch, bear a load of 2500 pounds, and be 12 feet long and 11.25 inches wide. How thick must the beam be?

Use the equation above in the following form. Remember to change 12 feet to inches.

$$h = \sqrt{\frac{6Ll}{sw}} = \sqrt{\frac{6(2500)(12 \times 12)}{(7000)(11.25)}} \approx \sqrt{27.429} \text{ or about } 5.24$$

The beam should be about 5.24 inches thick.

Solve.

1. Can a beam 5 inches thick, 180 inches long, and 12 inches wide bearing a load of 1000 pounds, sustain a root stress of 7000 pounds per square inch? Explain. **yes; 3600 < 7000**
2. A beam must sustain a root stress of 6000 pounds per square inch, bear a load of 2000 pounds, and be 10-feet long and 24 inches wide. How thick must the beam be? **about 3.16 inches thick**
3. If the beam in Exercise 2 must sustain a load twice as great, will the beam need to be twice as thick? Explain. **No; multiply the thickness of the beam by 1.414.**

CONNECTIONS

Activity Masters, pp. 13, 27, 46, 47

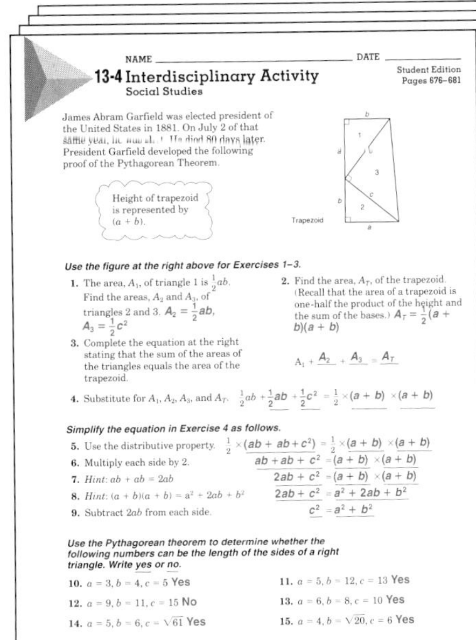

NAME ______________ DATE ______

13-4 Interdisciplinary Activity
Social Studies

Student Edition Pages 676–681

James Abram Garfield was elected president of the United States in 1881. On July 2 of that same year, he was shot. He died 80 days later. President Garfield developed the following proof of the Pythagorean Theorem.

Height of trapezoid is represented by $(a + b)$.

Trapezoid

Use the figure at the right above for Exercises 1–3.

1. The area, A_1, of triangle 1 is $\frac{1}{2}ab$. Find the areas, A_2 and A_3, of triangles 2 and 3. **$A_2 = \frac{1}{2}ab$, $A_3 = \frac{1}{2}c^2$**
2. Find the area, A_T, of the trapezoid. (Recall that the area of a trapezoid is one-half the product of the height and the sum of the bases.) **$A_T = \frac{1}{2}(a + b)(a + b)$**
3. Complete the equation at the right stating that the sum of the areas of the triangles equals the area of the trapezoid. $A_1 + A_2 + A_3 = A_T$
4. Substitute for A_1, A_2, A_3, and A_T. $\frac{1}{2}ab + \frac{1}{2}ab + \frac{1}{2}c^2 = \frac{1}{2} \times (a + b) \times (a + b)$

Simplify the equation in Exercise 4 as follows.

5. Use the distributive property. $\frac{1}{2} \times (ab + ab + c^2) = \frac{1}{2} \times (a + b) \times (a + b)$
6. Multiply each side by 2. $ab + ab + c^2 = (a + b) \times (a + b)$
7. *Hint*: $ab + ab = 2ab$ $\quad 2ab + c^2 = (a + b) \times (a + b)$
8. *Hint*: $(a + b)(a + b) = a^2 + 2ab + b^2$ $\quad 2ab + c^2 = a^2 + 2ab + b^2$
9. Subtract $2ab$ from each side. $c^2 = a^2 + b^2$

Use the Pythagorean theorem to determine whether the following numbers can be the length of the sides of a right triangle. Write yes or no.

10. $a = 3, b = 4, c = 5$ **Yes**
11. $a = 5, b = 12, c = 13$ **Yes**
12. $a = 9, b = 11, c = 15$ **No**
13. $a = 6, b = 8, c = 10$ **Yes**
14. $a = 5, b = 6, c = \sqrt{61}$ **Yes**
15. $a = 4, b = \sqrt{20}, c = 6$ **Yes**

COOPERATIVE LEARNING

Math Lab and Modeling Math Masters, pp. 73, 74, 88

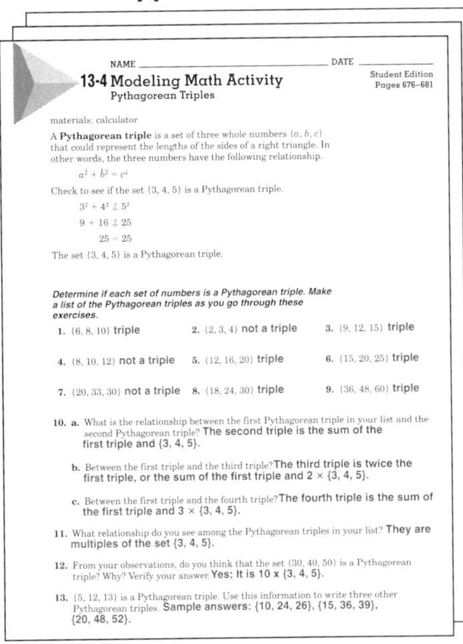

NAME ______________ DATE ______

13-4 Modeling Math Activity
Pythagorean Triples

Student Edition Pages 676–681

materials: calculator

A **Pythagorean triple** is a set of three whole numbers $\{a, b, c\}$ that could represent the lengths of the sides of a right triangle. In other words, the three numbers have the following relationship.

$$a^2 + b^2 = c^2$$

Check to see if the set $\{3, 4, 5\}$ is a Pythagorean triple.

$$3^2 + 4^2 \stackrel{?}{=} 5^2$$
$$9 + 16 \stackrel{?}{=} 25$$
$$25 = 25$$

The set $\{3, 4, 5\}$ is a Pythagorean triple.

Determine if each set of numbers is a Pythagorean triple. Make a list of the Pythagorean triples as you go through these exercises.

1. $\{6, 8, 10\}$ **triple**
2. $\{2, 3, 4\}$ **not a triple**
3. $\{9, 12, 15\}$ **triple**
4. $\{8, 10, 12\}$ **not a triple**
5. $\{12, 16, 20\}$ **triple**
6. $\{15, 20, 25\}$ **triple**
7. $\{20, 33, 30\}$ **not a triple**
8. $\{18, 24, 30\}$ **triple**
9. $\{36, 48, 60\}$ **triple**
10. a. What is the relationship between the first Pythagorean triple in your list and the second Pythagorean triple? **The second triple is the sum of the first triple and {3, 4, 5}.**
    b. Between the first triple and the third triple? **The third triple is twice the first triple, or the sum of the first triple and 2 × {3, 4, 5}.**
    c. Between the first triple and the fourth triple? **The fourth triple is the sum of the first triple and 3 × {3, 4, 5}.**
11. What relationship do you see among the Pythagorean triples in your list? **They are multiples of the set {3, 4, 5}.**
12. From your observations, do you think that the set $\{30, 40, 50\}$ is a Pythagorean triple? Why? Verify your answer. **Yes; It is 10 x {3, 4, 5}.**
13. $\{5, 12, 13\}$ is a Pythagorean triple. Use this information to write three other Pythagorean triples. **Sample answers: {10, 24, 26}, {15, 36, 39}, {20, 48, 52}.**

MAKING MATHEMATICS RELEVANT

This two-page introduction to the chapter provides students with an opportunity to explore other disciplinary topics and their applications to mathematics.

Background Information

Math and Technology in the News Many of the computer systems in use today are old, dating back 20 years or more. At the time many of them were developed, their expected life was about 10 years. Few people expected these systems to still be in operation at the turn of the century. Another complicating factor is that some of the software in these old systems limited the user to a two-digit year indicator. Possible remedies for the situation include the use of special programs that will seek out and fix the defects, the hiring of programmers on a short-term basis to address this specific problem, or the replacement of these old systems with new ones.

CHAPTER 13 Applying Algebra to Right Triangles

TOP STORIES in Chapter 13

In this chapter, you will:

- **find squares and square roots,**
- **solve equations by finding square roots,**
- **use the Pythagorean theorem,**
- **use trigonometric ratios to solve problems, and**
- **use Venn diagrams to solve problems.**

MATH AND TECHNOLOGY IN THE NEWS

My how time flies!

Source: New Scientist, April, 1995

The coming of the year 2000 is making people think about the changes that have taken place in the world in this century and the changes that will take place in the next one. But even though life is very different now than it was one hundred years ago, many computers and appliances with clocks will "think" it's 1900 again. At the stroke of midnight on January 1, 2000, these internal clocks will change from "99" to "00." And unless the programs are changed, that means that the systems will treat the year as 1900. That could mean chaos at banks that determine interest based on time, insurance companies that set rates using their customers' ages, and any other place where the data is affected by time. Experts estimate that the cost of making the necessary software changes before 2000 will cost up to $500 million!

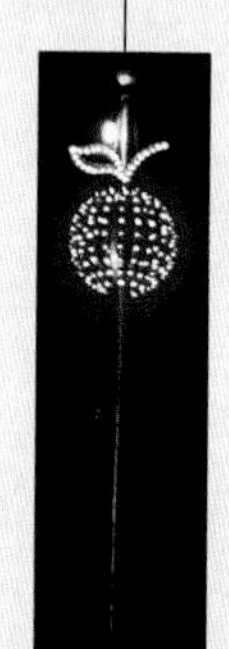

Putting It into Perspective

8000 B.C. A calendar with 12 months of 30 days each is developed in Egypt.

46 Julius Caesar reforms the Roman calendar to use 365 days each year and 1 additional day every four years to adjust the calendar.

1200 A calendar with 18 months of 20 days is developed in Mesomenia by the Olmecs of Mexico.

A.D. 14

Putting It into Perspective

The rotation of Earth is the standard used to set our clocks. Although they are not well-understood, tidal friction and movements of Earth's molten core seem to affect Earth's rotation. Since Earth was formed billions of years ago, it has been slowing down very gradually. At the time of the dinosaurs, Earth's "day" was less than 23 hours long. In order to keep our master clocks accurate, we need to adjust them occasionally for this slowing.

On the Lighter Side

"What time is it when the little hand is on two and the big hand is on the floor?"

interNET CONNECTION For up-to-date information on the calendar, visit:
www.glencoe.com/sec/math/prealg/mathnet

Statistical Snapshot

Comparative Scale of Time

Number of Seconds	Duration
10^{18}	Approximate age of the known universe (12 billion years)
10^{15}	Time since the age of the dinosaurs (135 million years)
10^{12}	Time for light from the center of our galaxy to reach Earth (about 30,000 years)
10^{9}	Human generation span (about 30 years)
10^{7}	Length of a school semester
10^{4}	Length of an average baseball game
10^{2}	Length of a television commercial break
10^{0}	One second
10^{-1}	Time of 1 vibration of lowest-pitched sound audible to humans
10^{-3}	Time for a midge to beat its wings once
10^{-7}	Time for an electron beam to go from source to screen of a television tube
10^{-11}	Time for a visible light wave to pass through a pane of glass
10^{-15}	Time for an electron to revolve once around proton in a Hydrogen molecule

1656
Dutch scientist Christian Huygens designs the first weight-driven clock controlled by a pendulum.

1948
Dr. Willard Frank Libby, professor of chemistry at the University of Chicago, invents the atomic clock.

1600 — 1800 — 2000

A.D. 1582
Pope Gregory XIII uses exact measurements for the solar year to revise the Julian calendar. The Gregorian calendar is the one in use in most of the world today.

1859
Big Ben, the tower clock famous for its accuracy and 13-ton bell, is installed.

1995
The computer industry addresses the year 2000.

Cooperative Learning

Chapter Projects Two chapter projects are included in the *Investigations and Projects Masters*. In Chapter Project A, students extend the topic in the Chapter Opener. In Chapter Project B, students explore hot-air balloons. A student page and a parent letter are provided for each Chapter Project.

interNET CONNECTION

Glencoe has made every effort to ensure that the website links for *Pre-Algebra* at www.glencoe.com/sec/math/prealg/mathnet are current and contain appropriate content. However, these website links are not under Glencoe's control.

On the Lighter Side

It is difficult to write a one-sentence definition of time, even if it's a long sentence. Students can be asked to try it and perhaps asked to talk about any problems they encounter. Does time still exist if clocks are broken and calendars are stuck on one day?

Statistical Snapshot

The table lists time intervals for several time periods and events. The scale is calibrated in seconds and is logarithmic. It shows scales of time from extremely large intervals to extremely small intervals.

Investigations and Projects Masters, p. 77

NAME ____________ DATE ________

CHAPTER **13** Project A
Time for Mathematics

The time line in the Chapter Opener shows that the development of calendars and clocks has taken place over thousands of years. Similarly, mathematics has had a long history. There is evidence that people have been thinking mathematically and doing calculations for at least 5000 years. Surprisingly, it was not until the 18th century that many educated people became acquainted with the mathematics we do today. Work with a partner on the following project, which consists of several related research assignments, to familiarize yourself with some of the key people, ideas, and events in the history of mathematics.

1. Visit the library. Create a list of mathematicians from the past, making sure to include people like Euclid, Pythagoras, Thales, Archimedes, Karl Friedrich Gauss, Leonhard Euler, Fibonacci (Leonardo of Pisa), Rene Descartes, Francois Viete, Blaise Pascal, and Isaac Newton. Don't forget women mathematicians like Hypatia, Sophie Germain, and Emmy Noether. Briefly describe the contribution of each to the field of mathematics. Then choose one person to research in more detail. Write an essay describing his or her life and work.
2. Throughout history, ancient societies developed their own unique numeration systems. Do research so that you can explain the unrelated systems used by the Babylonians and the Mayans. Demonstrate your understanding by writing some small numbers and some greater numbers using these systems.
3. The numerals we use are called Hindu-Arabic numerals. Do research to find out more about the contributions of these cultures to developing and spreading the numerals 0, 1, 2, 3, and so on.
4. In expressing ideas mathematically, we use the symbols $=$, $+$, $\times$, $-$, $\div$, $\sqrt{\ }$, and π. Find out approximately when each of these symbols was first used.
5. Now, at the close of the 20th century, it is a good time to explore some of the mathematics that has been developed in the last 100 years. Research to find out about some of these developments, such as work with computers and calculators and on statistics. Write a brief essay describing what you've learned. *Hint:* You may want to focus on developments that you don't need a grasp of higher mathematics to understand!
6. Make a poster showing what you've learned and display it in the classroom.

13-1 LESSON NOTES

NCTM Standards: 1-4, 6, 12, 13

Instructional Resources

- Study Guide Master 13-1
- Practice Master 13-1
- Enrichment Master 13-1
- Group Activity Card 13-1
- Activity Masters, p. 46
- Tech Prep Applications Masters, p. 25

Transparency 13-1A contains the 5-Minute Check for this lesson; **Transparency 13-1B** contains a teaching aid for this lesson.

Recommended Pacing	
Standard Pacing	Day 1 of 11
Honors Pacing	Day 1 of 10
Block Scheduling*	Day 1 of 6 (along with Lesson 13-2)

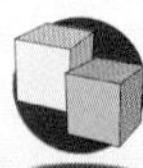

*For more information on pacing and possible lesson plans, refer to the ***Block Scheduling Booklet***.

1 FOCUS

5-Minute Check
(over Chapter 12)

1. Find the area of a trapezoid with bases of 10 cm and 14 cm and a height 8 cm. **96 cm²**
2. Find the probability of a dart landing in the shaded part of the target shown below. $\frac{1}{4}$

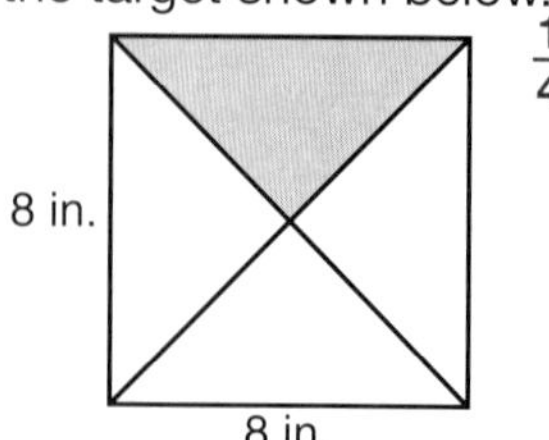

Find the surface area and volume of each figure.

3. a cylinder with a radius of 6 cm and a height of 25 cm **1168.1 cm²; 2826 cm³**
4. a pyramid with a 7-inch square base, a slant height of 12 in., and an altitude of 11.5 in. **217 in²; 187.8 in³**

13-1 Squares and Square Roots

Setting Goals: *In this lesson, you'll find and use squares and square roots.*

Modeling a Real-World Application: Ballooning

Have you ever wondered how far you can see out from an airplane or from the top of a hill? How far you can see depends on the curvature of Earth and your height above it. There is a formula that will help you to estimate how far you can see. The formula states that the view in miles V equals 1.22 times the square root of the altitude in feet A.

$$V = 1.22 \times \sqrt{A}$$

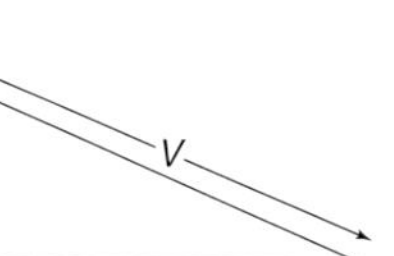

Lucia is riding in a hot air balloon about 158 feet above the ground. She looks off to the horizon and can just see her school in the distance. About how far away is she from the school? *You will solve this problem in Example 4.*

Learning the Concept

A **square root** is one of two equal factors of a number. For example, the square root of 49 is 7 since $7 \cdot 7$ or 7^2 is 49. It is also true that $-7 \cdot (-7) = 49$. So -7 is another square root of 49.

Definition of Square Root
In words: The square root of a number is one of its two equal factors.
In symbols: If $x^2 = y$, then x is a square root of y.

The symbol $\sqrt{}$, called the **radical sign**, is used to indicate a nonnegative square root.

$\sqrt{49}$ indicates the nonnegative square root of 49.

$$\sqrt{49} = 7$$

$-\sqrt{49}$ indicates the negative square root of 49.

$$-\sqrt{49} = -7$$

Example 1 **Find each square root.**

a. $\sqrt{25}$

The symbol $\sqrt{25}$ represents the nonnegative square root of 25. Since $5 \cdot 5 = 25$, $\sqrt{25} = 5$.

Alternative Learning Styles

Kinesthetic Provide each small group of students with about 30 square tiles or cutouts. Ask students whether 20 is a perfect square. Have students count 20 tiles and try to arrange them in a square. Ask what was the largest square they could make. **4 × 4 with 4 tiles left** Ask how many more tiles would be needed to make a 5 × 5 square. **5** Ask whether the square root of 20 is closer to 4 or 5. **4** Ask students to explain why you can say that the square root of 20 is about 4.5.

b. $-\sqrt{64}$

The symbol $-\sqrt{64}$ represents the negative square root of 64. Since $8 \cdot 8 = 64$, $-\sqrt{64} = -8$.

Squaring a number and finding the square root of a number are closely related to the square.

Connection to Geometry

The square shown at the right has sides that are 6 meters long. The area of the square is found by squaring 6. If you know that the area of the square is 36 square meters, then the length of a side is found by finding the square root of 36.

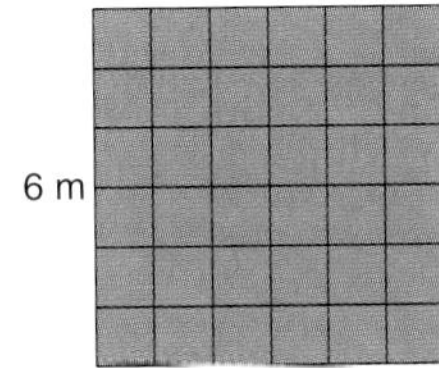

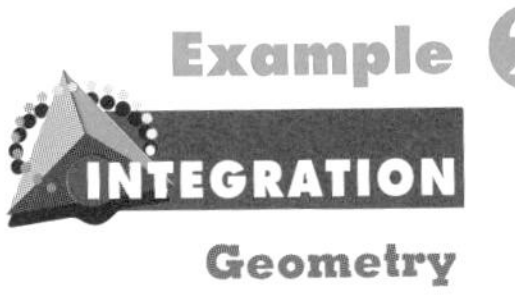

Example 2 INTEGRATION Geometry

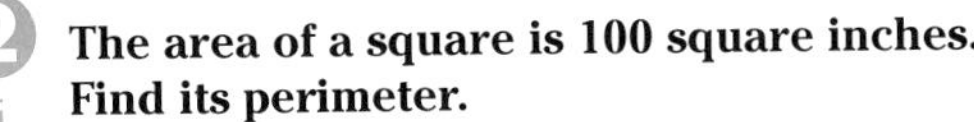

The area of a square is 100 square inches. Find its perimeter.

Area = 100 in^2

First, find the length of each side.

$A = s^2$

$100 = s^2$ *Replace A with 100.*

What number s when multiplied by itself is 100?

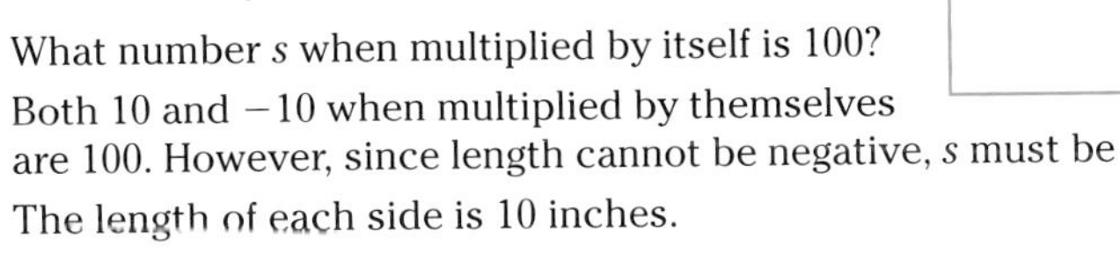

Both 10 and -10 when multiplied by themselves are 100. However, since length cannot be negative, s must be 10. The length of each side is 10 inches.

You can review area and perimeter in Lesson 3-5.

Now find the perimeter.

$P = 4 \cdot s$

$P = 4 \cdot 10$ *Replace s with 10.*

$P = 40$ The perimeter is 40 inches.

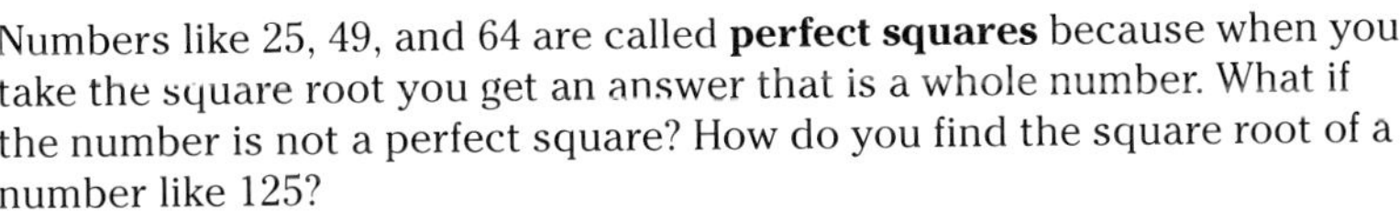

Numbers like 25, 49, and 64 are called **perfect squares** because when you take the square root you get an answer that is a whole number. What if the number is not a perfect square? How do you find the square root of a number like 125?

List some squares of numbers that are close to 125.

$10^2 = 100$ $11^2 = 121$ $12^2 = 144$

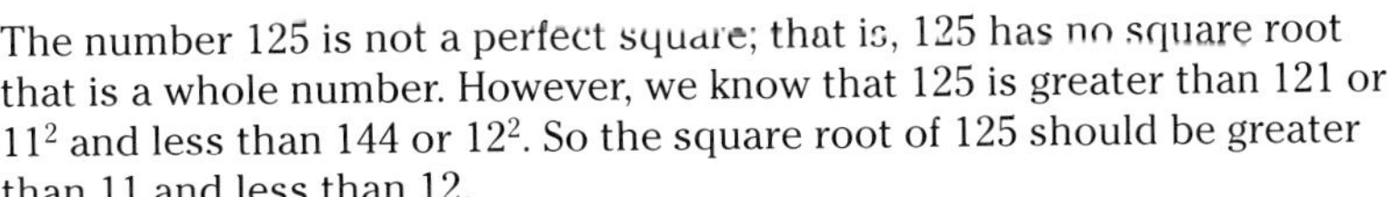

The number 125 is not a perfect square; that is, 125 has no square root that is a whole number. However, we know that 125 is greater than 121 or 11^2 and less than 144 or 12^2. So the square root of 125 should be greater than 11 and less than 12.

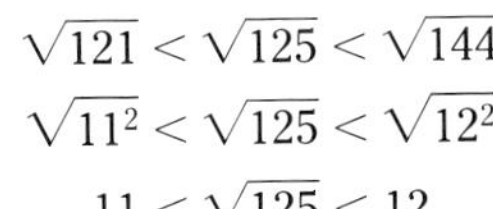

$$\sqrt{121} < \sqrt{125} < \sqrt{144}$$
$$\sqrt{11^2} < \sqrt{125} < \sqrt{12^2}$$
$$11 < \sqrt{125} < 12$$

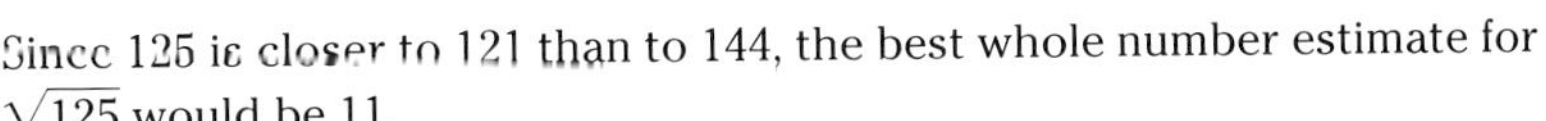

Since 125 is closer to 121 than to 144, the best whole number estimate for $\sqrt{125}$ would be 11.

Motivating the Lesson

Questioning Write the numbers 1, 4, 9, 16, 25, 36, 49, __?__ on the board. Ask what number comes next. **64**

2 TEACH

In-Class Examples

For Example 1

Find each square root.

a. $\sqrt{81}$ **9**

b. $-\sqrt{144}$ **−12**

For Example 2

The area of a square is 169 square inches. Find its perimeter. **52 inches**

Teaching Tip You may want to use a list to keep track on the board of squares and square roots that students discuss in class.

GLENCOE Technology

Interactive Mathematics Tools Software

In this interactive computer lesson, students explore squares and square roots. A **Computer Journal** gives students the opportunity to write about what they have learned.

For Windows & Macintosh

In-Class Examples

For Example 3
Find the best integer estimate for each square root.

a. $\sqrt{79}$ **9**

b. $-\sqrt{14}$ **−4**

For Example 4
The pilot of a jet tells passengers they are flying at an altitude of 28,000 feet. To the nearest mile, how far can the passengers see if it is a clear day?
204 miles

3 PRACTICE/APPLY

Checking Your Understanding

Exercises 1–14 are designed to help you assess your students' understanding through reading, writing, speaking, and modeling. You should work through Exercises 1–4 with your students and then monitor their work on Exercises 5–14.

Study Guide Masters, p. 108

NAME ______________________ DATE ____________

13-1 Study Guide
Finding and Approximating Squares and Square Roots

Student Edition Pages 664–668

Since $6 \times 6 = 36$, a **square root** of 36 is 6.

$$\sqrt{36} = 6$$

Since $-6 \times (-6) = 36$, -6 is also a square root of 36.
A *negative sign* is used to indicate the *negative square root*.

$$-\sqrt{36} = -6$$

6
6
36 is the square of 6.

If the square root of a number is a whole number, the original number is called a **perfect square**.
For example, 81 is a perfect square because $9 \times 9 = 81$.
However, 79 and 80 are not perfect squares.

In cases where the square root of a number is not a whole number, you can estimate the square root by using the two closest perfect squares. Seventy is between 64 and 81.
So, $\sqrt{70}$ is between $\sqrt{64}$ and $\sqrt{81}$ or between 8 and 9. Since 70 is closer to 64 than to 81, $\sqrt{70}$ is closer to 8.

Find each square root.

1. $\sqrt{9}$ **3**
2. $\sqrt{25}$ **5**
3. $-\sqrt{4}$ **−2**
4. $-\sqrt{64}$ **−8**
5. $\sqrt{121}$ **11**
6. $-\sqrt{196}$ **−14**
7. $\sqrt{225}$ **15**
8. $\sqrt{1.44}$ **1.2**
9. $\sqrt{900}$ **30**
10. $-\sqrt{324}$ **−18**
11. $\sqrt{529}$ **23**
12. $\sqrt{100}$ **10**

Find the best integer estimate for each square root. Then check your estimate using a calculator.

13. $-\sqrt{37}$ **−6**
14. $\sqrt{90}$ **9**
15. $-\sqrt{50}$ **−7**
16. $\sqrt{300}$ **17**
17. $-\sqrt{75}$ **−9**
18. $\sqrt{69}$ **8**
19. $\sqrt{1681}$ **41**
20. $\sqrt{27.96}$ **5**
21. $-\sqrt{11.25}$ **−3**

Example 3 **Find the best integer estimate for each square root.**

Some calculators have a *square root key* labeled $\sqrt{\ }$ or $\sqrt{x}$. When you press this key, the number in the display is replaced by its nonnegative square root.

a. $\sqrt{59}$

49 and 64 are the closest perfect squares.

$$\sqrt{49} < \sqrt{59} < \sqrt{64}$$
$$\sqrt{7^2} < \sqrt{59} < \sqrt{8^2}$$
$$7 < \sqrt{59} < 8$$

Since 59 is closer to 64 than 49, the best integer estimate is 8.

b. $-\sqrt{38}$

36 and 49 are the perfect squares closest to 38.

$$-\sqrt{49} < -\sqrt{38} < -\sqrt{36}$$
$$-\sqrt{7^2} < -\sqrt{38} < -\sqrt{6^2}$$
$$-7 < -\sqrt{38} < -6$$

Since 36 is closer to 38 than 49, the best integer estimate is −6.

Many applications in science and other fields use square roots in important formulas.

Example 4 **Refer to the application at the beginning of the lesson. How far away is Lucia from the school?**

Explore You know that Lucia is about 158 feet above the ground. You also know that she can just see her school in the distance.

Plan Find the distance she is from the school. Use the formula $V = 1.22 \times \sqrt{A}$

Solve Substitute values into the formula.
$V = 1.22 \times \sqrt{A}$
$V = 1.22 \times \sqrt{158}$
Use your calculator.

Enter: 1.22 [×] 158 [$\sqrt{x}$] [=] 15.335162

Lucia is about 15.3 miles from the school.

Examine The closest perfect square to 158 is 169 or 13^2. $13 \times 1.22 = 15.86$. The answer is reasonable.

Checking Your Understanding

Communicating Mathematics

Read and study the lesson to answer each question. **1–2, 4. See margin.**

1. **Explain** why a positive number has two different square roots.
2. **Draw** a square that has an area of 64 square centimeters.
3. **Explain** why 49 is a perfect square. **It is the square of 7.**
4. **Explain** why finding the square root and squaring are inverse operations.

Reteaching

Using Models Have students draw a dot grid, using 12 dots horizontally and 12 dots vertically. Connect dots to form as many different squares as possible. For each square, write equations representing the connected dots as a square, then as a square root. For example, for a 3-by-3 square, write $3^2 = 9$, and $\sqrt{9} = 3$.

Additional Answers

1. Sample answer: Each positive number has a positive square root and a negative square root.
2. 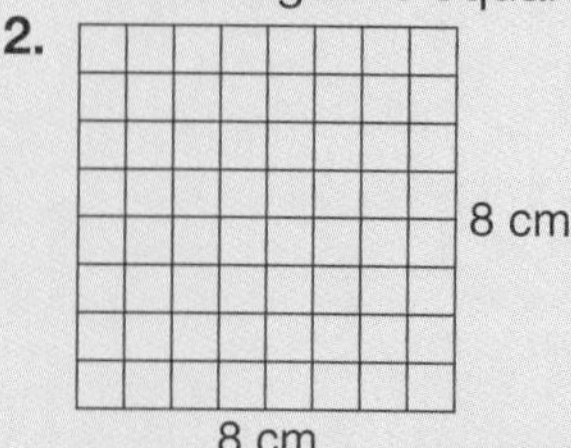

MATERIALS

square tiles

5. Arrange 40 tiles into the largest square possible.
   a. How many tiles did you use? How many are left over? **See margin.**
   b. Add tiles until you have the next larger square. How many tiles did you add? **9 tiles**
   c. Between what two whole numbers is the square root of 40? **6 and 7**

Guided Practice

Find each square root.

6. $-\sqrt{9}$ **−3** 7. $\sqrt{16}$ **4** 8. $-\sqrt{4}$ **−2** 9. $\sqrt{81}$ **9**

Find the best integer estimate for each square root. Then check your estimate using a calculator.

10. $\sqrt{79}$ **9** 11. $-\sqrt{53}$ **−7**
12. $\sqrt{29}$ **5** 13. $\sqrt{120}$ **11**

14. **Aviation** The British Airways Concorde flies from New York to London at an altitude of 60,000 feet. How far can the pilot see when he or she looks out the window? **298.8 miles**

Exercises: Practicing and Applying the Concept

Independent Practice

Find each square root.

15. $\sqrt{4}$ **2** 16. $\sqrt{64}$ **8** 17. $\sqrt{9}$ **3** 18. $-\sqrt{25}$ **−5**
19. $-\sqrt{16}$ **−4** 20. $-\sqrt{81}$ **−9** 21. $\sqrt{100}$ **10** 22. $\sqrt{1}$ **1**
23. $-\sqrt{144}$ **−12** 24. $\sqrt{169}$ **13** 25. $\sqrt{400}$ **20** 26. $\sqrt{2.25}$ **1.5**

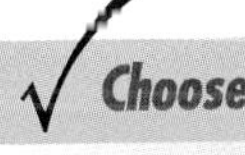

Choose
Estimation
Mental Math
Calculator
Paper and Pencil

Find the best integer estimate for each square root. Then check your estimate using a calculator. **30. −13 38. −12**

27. $\sqrt{89}$ **9** 28. $-\sqrt{44}$ **−7** 29. $\sqrt{200}$ **14** 30. $-\sqrt{170}$
31. $-\sqrt{97}$ **−10** 32. $-\sqrt{6.76}$ **−3** 33. $\sqrt{2600}$ **51** 34. $\sqrt{625}$ **25**
35. $\sqrt{118}$ **11** 36. $\sqrt{13.69}$ **4** 37. $-\sqrt{26.79}$ **−5** 38. $-\sqrt{156.25}$

Simplify. **40. 24**

39. $\sqrt{a^2}$ ***a*** 40. $\sqrt{169} - (-\sqrt{121})$ 41. $\sqrt{\sqrt{81}}$ **3**

Critical Thinking

42. Enter a negative number into your calculator. Press the square root key. What is the result? Why? **See margin.**
43. In $x^2 = y$, could x be negative? Why or why not? **See margin.**

Applications and Problem Solving

44. **Sports** The ability of a hang glider to fly depends on the area of its wing. The relationship between wingspan and wing area is called the aspect ratio. The formula for the aspect ratio, R, is $R = \frac{s^2}{A}$, where s represents the wingspan and A represents the wing area.
   a. Solve $R = \frac{s^2}{A}$ for s to write a formula for wingspan when the aspect ratio and wing area are known. $s = \sqrt{RA}$
   b. An advertisement for a hang glider says the aspect ratio is 2.7 and the wing area is 30 square feet. What is the wingspan of the glider? **9 feet**

Assignment Guide

Core: 15–45 odd, 47–54
Enriched: 16–40 even, 42–54

For **Extra Practice**, see p. 771.

The red A, B, and C flags, printed only in the Teacher's Wraparound Edition, indicate the level of difficulty of the exercises.

Additional Answers

4. Sample answer: Square "undoes" taking a square root.
5a. The square has 36 tiles, with 4 left over.
42. Error message is given; the square root of a negative number is not a real number.
43. Yes; because when a negative number is squared the result is positive.

Practice Masters, p. 108

NAME ______________ DATE ______

13-1 Practice
Finding and Approximating Squares and Square Roots

Student Edition Pages 664–668

Find each square root.

1. $-\sqrt{4}$ −2 2. $\sqrt{81}$ 9 3. $-\sqrt{36}$ −6 4. $\sqrt{1}$ 1
5. $-\sqrt{9}$ −3 6. $\sqrt{144}$ 12 7. $-\sqrt{121}$ −11 8. $\sqrt{36}$ 6
9. $-\sqrt{49}$ −7 10. $-\sqrt{81}$ −9 11. $-\sqrt{169}$ −13 12. $\sqrt{400}$ 20
13. $\sqrt{225}$ 15 14. $-\sqrt{289}$ 17 15. $\sqrt{196}$ 14 16. $\sqrt{324}$ 18
17. $\sqrt{900}$ 30 18. $\sqrt{256}$ 16 19. $\sqrt{625}$ 25 20. $-\sqrt{196}$ −14
21. $\sqrt{361}$ 19 22. $-\sqrt{324}$ −18 23. $-\sqrt{256}$ −16 24. $-\sqrt{900}$ −30
25. $-\sqrt{625}$ −25 26. $\sqrt{961}$ 31 27. $\sqrt{289}$ 17 28. $-\sqrt{361}$ −19

Find the best integer estimate for each square root. Then check your estimate using a calculator.

29. $\sqrt{8}$ 3 30. $-\sqrt{12}$ −3 31. $\sqrt{55}$ 7
32. $-\sqrt{39}$ −6 33. $\sqrt{98}$ 10 34. $\sqrt{500}$ 22
35. $-\sqrt{60}$ −8 36. $\sqrt{19}$ 4 37. $\sqrt{150}$ 12
38. $-\sqrt{70}$ −8 39. $-\sqrt{395}$ −20 40. $\sqrt{200}$ 14
41. $-\sqrt{115}$ −11 42. $\sqrt{1000}$ 32 43. $-\sqrt{40}$ −6
44. $\sqrt{1500}$ 39 45. $\sqrt{196}$ 14 46. $-\sqrt{7.95}$ −3
47. $\sqrt{3.72}$ 2 48. $\sqrt{15.01}$ 4 49. $\sqrt{75.83}$ 9
50. $-\sqrt{60.25}$ −8 51. $-\sqrt{83.91}$ −9 52. $\sqrt{26.17}$ 5

Group Activity Card 13-1

It's Close Group Activity **13-1**

MATERIALS: Calculator • three 0-9 spinners

Form groups of two to four students. The object of the game is to give the closest estimate of the nonnegative square root of a three-digit number.

To begin, the leader spins the three spinners and makes a three-digit number. If desired, the leader may put a decimal point between any of the digits.

Everyone in the group, including the leader, then writes his or her whole number estimate for the square root of the number. When everyone has written an estimate, the leader uses the calculator to get the square root. The player whose estimate is closest to this number is the winner of this round and earns one point.

After everyone has served as the leader, the player with the most points wins.

 Pre-Algebra

Extension

Using Formulas The length of a skid mark can be used to find the approximate speed of the car trying to stop. The formula is $s = 13\sqrt{\ell}$, where s is speed in kilometers per hour and ℓ is the skid mark length in meters. How fast was a car going if a skid mark was 25 meters long? **65 km/h**

4 ASSESS

Closing Activity

Speaking Write the numbers 1–12 on index cards. Go around the room and have the students pick a card and state the square of the given number. Then ask if the given number is a perfect square. If it is, ask for its square root.

Additional Answer

54.

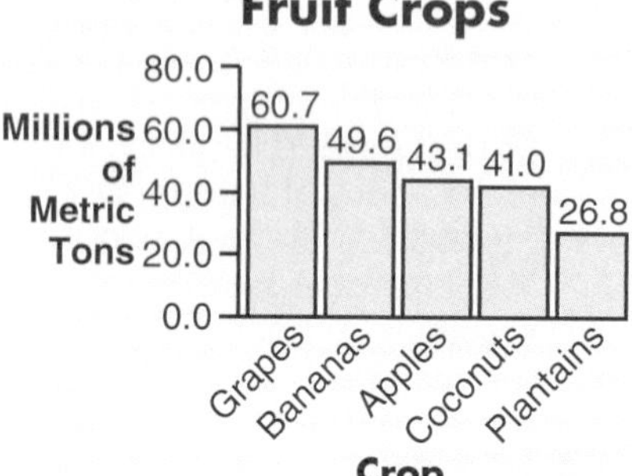

45. Construction Bloomington city code requires that a party house must allow at least 4 square feet for each person on the dance floor. Reston's Hotel Restaurant is going to add a dance floor that is square and is large enough for 100 people. How long should it be on each side? **20 feet**

46. Plumbing Every water heater must have a pipe connecting it to the water meter supplied by the utility company. An engineer who is designing a new water heater has determined that the proper amount of water will be supplied if the opening of the pipe has an area of 0.442 square inches. Pipe is usually described in terms of its interior diameter. Use the formula for the area of a circle, $A = \pi r^2$, to find the proper size of pipe for the water heater. **0.75 inches**

Mixed Review

47. Geometry Find the volume of a rectangular pyramid with length 8 inches, width 8 inches, and height 9 inches. (Lesson 12-8) **192 in^3**

48. How many ways can the letters of the word GYMNAST be arranged? (Lesson 10-6) **5040 ways**

49. Express 28% as a fraction in simplest form. (Lesson 9-7) $\frac{7}{25}$

50. Carpentry How many boards, each $2\frac{1}{2}$ feet long, can be cut from an 18-foot board? (Lesson 6-4) **7**

51. Estimate 9.2×3.85. (Lesson 6-2) **36**

52. Find the prime factorization of 2160. (Lesson 4-4) $2^4 \times 3^3 \times 5$

53. Solve $a = -15 + (-5)$. (Lesson 2-4) **−20**

54. Agriculture Make a bar graph of the data about the fruit production in the table at the right. (Lesson 1-10)

54. See margin.

Annual Fruit Production	
Fruit	**Millions of Metric Tons**
Grapes	60.7
Bananas	49.6
Apples	43.1
Coconuts	41.0
Plantains	26.8

Source: *Top Ten of Everything*

interNET CONNECTION

For the latest U.S. fruit production, visit: **www.glencoe.com/sec/math/prealg/mathnet**

From the FUNNY PAPERS

1. What is wrong with Hobbes's reasoning? **See margin.**
2. Is the figure that Hobbes drew a square? Explain. **No; a square has four equal sides.**
3. Write an explanation for Calvin and Hobbes of how to use square tiles to solve 6 + 3. **See Solutions Manual.**

Enrichment Masters, p. 108

NAME ____________ DATE ______

13-1 Enrichment
Roots

Student Edition Pages 664–668

The symbol $\sqrt{\ }$ indicates a square root. By placing a number in the upper left, the symbol can be changed to indicate higher roots.

$\sqrt[3]{8} = 2$ because $2^3 = 8$
$\sqrt[4]{81} = 3$ because $3^4 = 81$
$\sqrt[5]{100,000} = 10$ because $10^5 = 100,000$

Find each of the following.

1. $\sqrt[3]{125}$ **5**
2. $\sqrt[4]{16}$ **2**
3. $\sqrt[8]{1}$ **1**
4. $\sqrt[3]{27}$ **3**
5. $\sqrt[5]{32}$ **2**
6. $\sqrt[3]{64}$ **4**
7. $\sqrt[3]{1000}$ **10**
8. $\sqrt[3]{216}$ **6**
9. $\sqrt[6]{1,000,000}$ **10**
10. $\sqrt[3]{1,000,000}$ **100**
11. $\sqrt[4]{256}$ **4**
12. $\sqrt[3]{729}$ **9**
13. $\sqrt[6]{64}$ **2**
14. $\sqrt[4]{625}$ **5**
15. $\sqrt[5]{243}$ **3**

From the FUNNY PAPERS

Ask students to tell something funny that has happened when their parents or siblings have tried to help them with their math problems.

Additional Answer From the Funny Papers

1. Sample answer: 6 + 3 can be modeled by adding a length of 6 units and a length of 3 units, but it has nothing to do with the diagonal of a square.

13-2 Problem-Solving Strategy: Use Venn Diagrams

Setting Goals: *In this lesson, you'll use Venn Diagrams to solve problems.*

Modeling a Real-World Application: Surveys

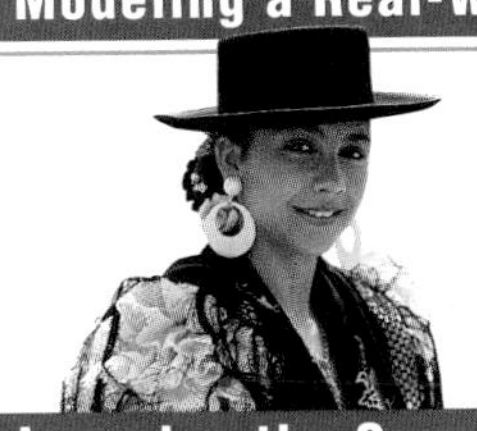

The guidance counselor surveyed 80 eighth-graders to find out their interest in learning Japanese or Spanish. Twenty-two students expressed an interest in both languages. Eighteen expressed interest in Japanese only, and five were not interested in learning any foreign language. How many of the students who were surveyed were interested in Spanish only?

Learning the Concept

You can use a **Venn diagram** to illustrate data. A rectangle is used to represent all the data. A circle inside the rectangle is used to represent one group of data. Data that is common to more than one group is represented by the region where intersecting circles overlap.

Foreign Language Preference

Spanish | Japanese | 22 | 18 | neither 5

Draw two intersecting circles in a rectangle to represent Japanese and Spanish. Write the number of students who are interested in both languages and the number of students who are interested in only Japanese in the appropriate regions. Write the number of students who are not interested in learning either language outside the circles but inside the rectangle. There are 22 + 18 + 5 or 45 students represented in the Venn diagram. There are 80 − 45 or 35 students that are interested in learning only Spanish.

Example

APPLICATION

Music

The Venn diagram at the right shows students' music preferences.

Music Preference

Rock 63 | Oldies 14 | Rap 19 | 7 | 5 | 21 | 9

a. How many students like all three types of music?

The area where all three circles overlap contains a 5. There are 5 students who like all three types.

b. How many students like oldies?

There are 14 students who only like oldies, 9 who like oldies and rap, 7 who like oldies and rock, and 5 who like all three. There are 14 + 9 + 7 + 5 or 35 students who like oldies.

c. How many students like rap and rock?

The area where the circles for rap and rock overlap contains 5 and 21. There are 5 + 21 or 26 students who like rap and rock.

Alternative Teaching Strategies

Study Guide Masters, p. 109 The Study Guide Master provides a concise presentation of the lesson along with practice problems.

Reading Mathematics Be sure students understand what the parts of the overlapping circles represent as well as how to place numbers in the proper sections.

13-2 LESSON NOTES

NCTM Standards: 1-4, 12

Instructional Resources

- Study Guide Master 13-2
- Practice Master 13-2
- Enrichment Master 13-2
- Group Activity Card 13-2
- Assessment and Evaluation Masters, p. 351
- Math Lab and Modeling Math Masters, p. 88

Transparency 13-2A contains the 5-Minute Check for this lesson; **Transparency 13-2B** contains a teaching aid for this lesson.

Recommended Pacing	
Standard Pacing	Day 2 of 11
Honors Pacing	Day 2 of 10
Block Scheduling*	Day 1 of 6 (along with Lesson 13-1)

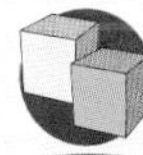

*For more information on pacing and possible lesson plans, refer to the ***Block Scheduling Booklet***.

1 FOCUS

5-Minute Check
(over Lesson 13-1)

Find each square root.

1. $\sqrt{49}$ 7 **2.** $-\sqrt{16}$ −4

3. $\sqrt{121}$ 11

Find the best integer estimate for each square root.

4. $\sqrt{150}$ 12

5. $-\sqrt{54.93}$ −7

Motivating the Lesson

Hands-On Activity Have students tell whether they have each of the following with them in class today: pencils, pens, felt-tipped markers. Record the information on the board to show how many students have only pencils, only pens, only markers, only pens and pencils, only pens and markers, only pencils and marker, or all three. Keep this information for use in In-Class Example 1.

2 TEACH

In-Class Example

For Example 1

Create a Venn diagram for the information obtained in Motivating the Lesson activity.

Sample answer:

Writing Utensils

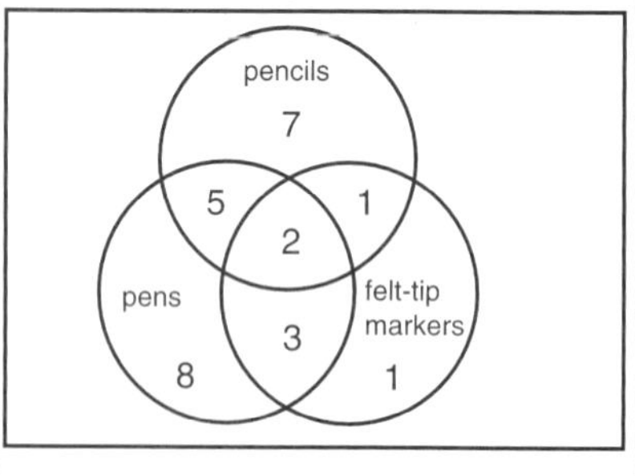

3 PRACTICE/APPLY

Checking Your Understanding

Exercises 1–4 are designed to help you assess your students' understanding through reading, writing, speaking, and modeling. You should work through Exercises 1–2 with your students and then monitor their work on Exercises 3–4.

Practice Masters, p. 109

NAME ______________________ DATE __________

13-2 Practice
Problem-Solving Strategy:
Use Venn Diagrams

Student Edition
Pages 669–671

Solve using Venn Diagrams.

1. There are 90 students participating in winter sports at Whittier School. Forty-five students run track and 67 play basketball. Twenty-two students do both sports. How many students run track only? **23 students**

2. At North High School, there are 413 sophomores. Seventy-five sophomores are taking keyboarding, 115 sophomores are taking computer science, and 33 students are taking both courses. How many sophomores are taking neither keyboarding nor computer science? **256 sophomores**

3. At a breakfast buffet, 93 people chose coffee for their beverage and 47 people chose juice. Twenty-five people chose both coffee and juice. Each person chose at least one of these beverages. How many people visited the buffet? **115 people**

4. One hundred fifty-seven students were surveyed in music class. Ninety-five students prefer rock-and-roll music, and one hundred eight prefer country-western music. Forty-six students prefer both rock-and-roll and country-western music. How many students prefer rock-and-roll music but not country-western music? **49 students**

5. The Olde World Calzone Shoppe sells pepperoni, mushroom and pepperoni-mushroom calzones. On Tuesday, 72 calzones were sold. Thirty-one of the calzones contained mushrooms. If 12 pepperoni-mushroom calzones were sold, how many calzones contained pepperoni? **53 calzones**

6. Nineteen students are in the math club, and 25 students are in the science club. Nine students are in both the math and science clubs. How many students are in one of the two clubs only? **26 students**

Checking Your Understanding

Communicating Mathematics

Read and study the lesson to answer each question. The Venn diagram at the right represents the states that produce more than 100 million bushels of corn, wheat, or soybeans per year.

Grain Production

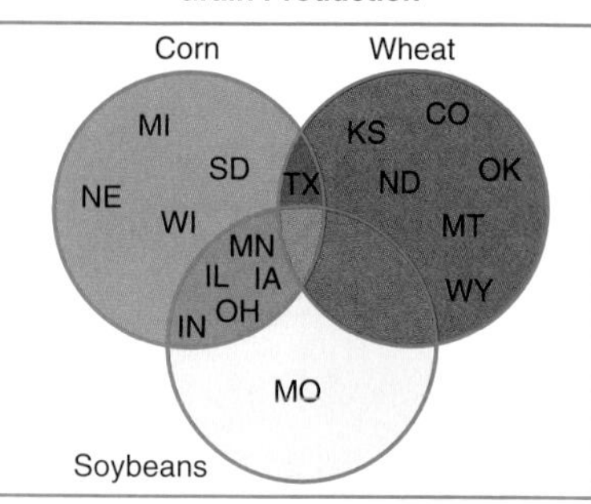

1. **Write** what the section where all three circles overlap represents. **1. States that produce 100 million bushels of all three crops each year.**
2. **Determine** how many states produce more than 100 million bushels of corn or wheat. **16 states**

Guided Practice

Solve using a Venn diagram.

3. At a birthday party, 9 people chose just cake for dessert, and 4 people chose just ice cream. Five people chose both cake and ice cream. If each person chose at least one dessert, how many people were at the birthday party? **18 people**
4. **Recycling** The Venn diagram at the right shows the number of communities serviced by EarthCare Recycling that choose to have curbside recycling of glass, newspaper, and aluminum. How many communities does EarthCare service? **44**

EarthCare Communities

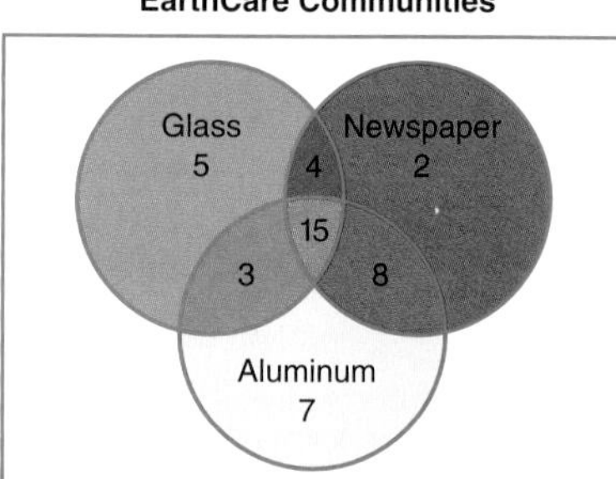

Exercises: Practicing and Applying the Concept

Independent Practice

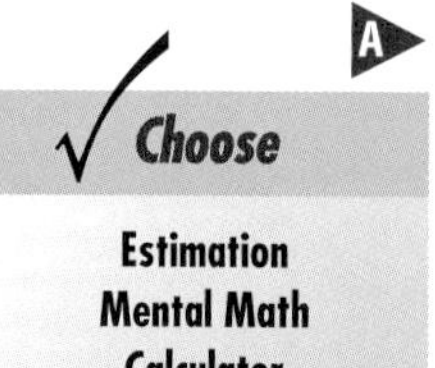

Solve. Use any strategy.

5. **Geometry** There are 2 tiles. One is in the shape of a square, and the other is in the shape of an equilateral triangle. The sides of the square are congruent to the sides of the triangle. The tiles are put together to form a pentagon. If one side of the triangle is 20 cm long, what is the perimeter of the pentagon? **100 cm**
6. Of 150 TV viewers one evening, 64 watched MTV, 36 watched ESPN, and 10 watched both. How many did not watch either? **60 people**
7. Chef Martino made a huge pan of lasagna for a banquet. He makes 6 cuts along the length of the rectangular pan and 10 cuts along the width. How many pieces does he have? **77 pieces**

8. Jaria must read a book for a book report. She read half of her book on Tuesday. On Wednesday she read another 30 pages. On Thursday she read 6 more, and on Friday she read half of what was left. If she still has 20 pages yet to read, how many pages does the book have? **8. 152 pages**
9. A school survey of 250 girls showed that 85 read *Seventeen* and 65 read *Elle*. 110 girls read neither. How many read both? **10 girls**

Reteaching

Using Data Have students take a survey in class and use the data to create a Venn diagram. Possible survey items might include: favorite subject, play in the band, play on an athletic team.

Group Activity Card 13-2

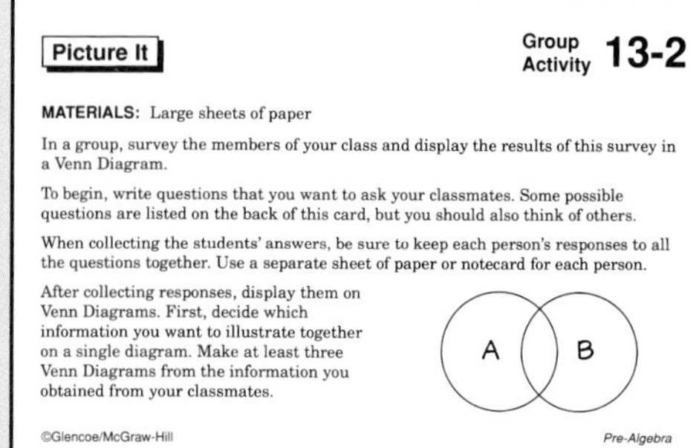

Picture It

Group Activity **13-2**

MATERIALS: Large sheets of paper

In a group, survey the members of your class and display the results of this survey in a Venn Diagram.

To begin, write questions that you want to ask your classmates. Some possible questions are listed on the back of this card, but you should also think of others.

When collecting the students' answers, be sure to keep each person's responses to all the questions together. Use a separate sheet of paper or notecard for each person.

After collecting responses, display them on Venn Diagrams. First, decide which information you want to illustrate together on a single diagram. Make at least three Venn Diagrams from the information you obtained from your classmates.

A B

Pre-Algebra

10. There are 26 students in a math class. The class takes a survey on pets and finds that 14 students have dogs, 10 students have cats, and 5 students have birds. Four students have dogs and cats, 3 students have dogs and birds, and 1 student has a cat and a bird. If no one has all three of these animals, how many students have none of these animals? **5 students**

C 11. In a geography class of 30 students, 16 said they wanted to visit France, 16 wanted to visit Germany, and 11 wanted to visit Italy. Five said they wanted to visit both France and Italy, and of these, 3 wanted to visit Germany as well. Five wanted only Italy, and 8 wanted only Germany. How many students wanted France only? **7 students**

Critical Thinking

12. **Geometry** A geometry teacher drew some quadrilaterals on the chalkboard. There were 5 trapezoids, 12 rectangles, 5 squares, and 8 rhombuses. What is the least number of figures the teacher could have drawn? **20 figures**

13. **Geometry** Draw a Venn diagram to show the relationship between squares and parallelograms. Let the rectangle represent polygons. **See Solutions Manual.**

Mixed Review

14. Find $-\sqrt{121}$. (Lesson 13-1) **−11**

15. **Geometry** Draw an example of a rhombus and describe its characteristics. (Lesson 11-7) **See margin.**

16. $23

16. **Finance** If $800 is deposited in a savings account for six months at 5.75% annual interest, how much interest will it earn? (Lesson 9-9)

17. Solve $2x - 5 = 48$. (Lesson 7-2) **26.5**

18. Solve $n - (-20) \geq 14$ and graph the solution on a number line. (Lesson 3-6) **$n \geq -6$; see margin for graph.**

WORKING ON THE
Investigation

OH GIVE ME A HOME

Refer to the Investigation on pages 606–607.

Look at the floor plan(s) of the house that you placed in your Investigation Folder at the end of Lesson 12-8. Use the floor plan(s) to make elevation drawings from four different directions. Be sure to use the same scale on the elevation drawing that you used on the floor plan. An example of an elevation drawing is shown at the beginning of Chapter 12.

Using your working drawings, construct a cardboard or foam board model of your house. It will probably be easier to work with a scale of $\frac{1}{2}$ inch = 1 foot instead of $\frac{1}{4}$ inch = 1 foot. (*Hint:* Cut doors and windows out before you erect the walls. You can also trace window openings on wax paper. Then draw in the frames and tape the wax paper on the inside wall.)

Make a roof for your model. Don't glue the roof on so that you can remove it to show the arrangement of the rooms.

Place your model on a base that is larger than the house. The excess can be used to show the yard, sidewalks, and driveway. You can also show trees and shrubs.

If possible, take a photograph of your model.

Add the results of your work to your Investigation Folder.

Extension

Using Data

Hair and Eye Color

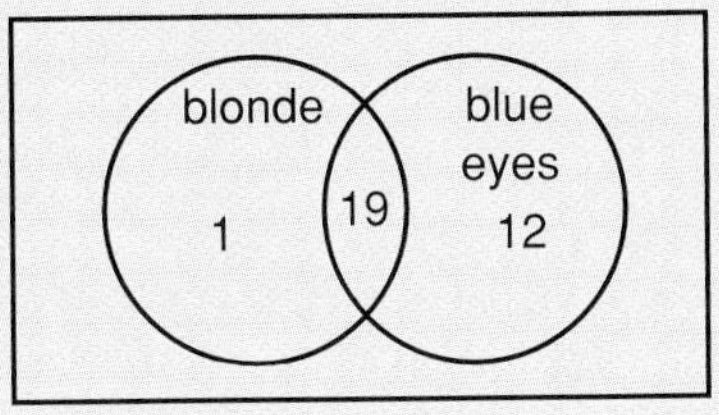
blonde
blue eyes
1
19
12

Based on the information shown in the diagram, ask students whether each of these statements is true or false.

- Most blondes have blue eyes. **true**
- Most blue-eyed people have blonde hair. **false**
- Some blue-eyed people do not have blonde hair. **true**

4 ASSESS

Closing Activity

Writing Separate the class into small groups. Have each group write a problem that can be solved using a Venn diagram.

Chapter 13, Quiz A (Lessons 13-1 and 13-2) is available in the *Assessment and Evaluation Masters*, p. 351.

Additional Answers

15. A rhombus is a quadrilateral with four congruent sides.

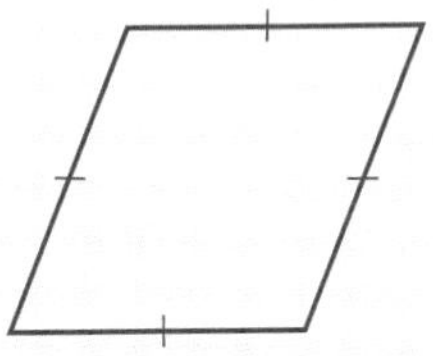

18.

−8
−6
−4
−2
0

WORKING ON THE
Investigation

The Investigation on pages 606 and 607 is designed to be a long-term project that is completed over several days or weeks. Encourage students to keep their materials in their Investigation Folder as they work on the Investigation.

Enrichment Masters, p. 109

NAME ____________ DATE ________

13-2 Enrichment
Hero's Formula

Student Edition
Pages 669–671

Hero of Alexandria (or Heron) was a mathematician and inventor who lived sometime around the first or second century A.D. In his work, there was a strong emphasis toward the applications of mathematics to the real world. He made discoveries in geometry and physics, and he is also believed to have invented a steam engine.

Hero's formula, named after Hero, is a formula for finding the area of a triangle in terms of the lengths of the three sides.

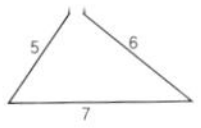
5
6
7

$$\begin{aligned} A &= \sqrt{s(s-a)(s-b)(s-c)} \\ &= \sqrt{9(9-5)(9-6)(9-7)} \\ &= \sqrt{9 \times 4 \times 3 \times 2} \\ &= 6\sqrt{6} \text{ square units} \end{aligned}$$

a, b, c = lengths of sides

s = half the perimeter of the triangle or $\frac{1}{2}(a + b + c)$

Use Hero's formula to find the area of each triangle.

1. 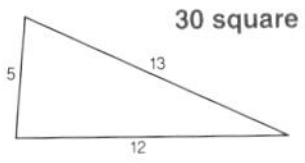
5
13
12
30 square units

2. 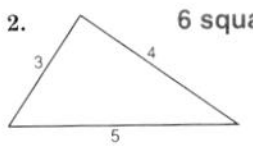
3
4
5
6 square units

3. 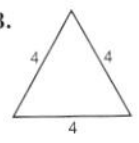
4
4
4
$4\sqrt{3}$ square units

4. 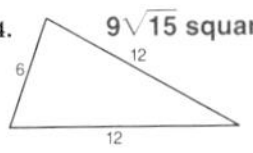
6
12
12
$9\sqrt{15}$ square units

13-3 LESSON NOTES

NCTM Standards: 1-4, 6, 12, 13

Instructional Resources

- Study Guide Master 13-3
- Practice Master 13-3
- Enrichment Master 13-3
- Group Activity Card 13-3

Transparency 13-3A contains the 5-Minute Check for this lesson; **Transparency 13-3B** contains a teaching aid for this lesson.

Recommended Pacing	
Standard Pacing	Day 3 of 11
Honors Pacing	Day 3 of 10
Block Scheduling*	Day 2 of 6 (along with Lesson 13-4)

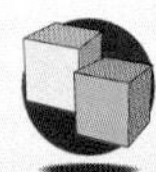

*For more information on pacing and possible lesson plans, refer to the ***Block Scheduling Booklet***.

1 FOCUS

5-Minute Check
(over Lesson 13-2)

Solve. Use any strategy.

1. Of 30 people polled at a resort, 12 were water skiers, 19 were snow skiers, and 7 were both water and snow skiers. How many people were neither water skiers nor snow skiers? **6**
2. At Whitmer High School, 36 freshmen take woodworking, 25 take cooking, and 9 take both courses. There are 108 students in the freshman class. How many freshmen take neither woodworking nor cooking? **56**

Motivating the Lesson

Questioning Draw a chart like the one on page 673, with headings of whole number, integer, perfect square. Have students determine which sets contain 9, $-\sqrt{9}$, 19, -12, -4, and $\sqrt{4}$ **whole 9, 19, $\sqrt{4}$; integer 9, $-\sqrt{9}$, 19, -12, -4, $\sqrt{4}$; perfect square 9**

13-3 The Real Number System

Setting Goals: *In this lesson, you'll identify numbers in the real number system and solve equations by finding square roots.*

Modeling a Real-World Application: Law Enforcement

> **HELP WANTED**
>
> Dedicated individuals are needed as law enforcement officers. If you are interested, contact:
>
> National Fraternal Order of Police
> 2100 Gardiner Lane
> Louisville, KY 40205

When investigating an accident, police officers often need to determine how fast a car was traveling before it skidded to a stop. They use the formula $s = \sqrt{30df}$, where s is the speed in miles per hour, d is the length of the skid marks in feet, and f is the coefficient of friction. What was the approximate speed of a car if the skid marks are 80 feet long and the coefficient of friction is estimated to be 1.2?

$s = \sqrt{30df}$

$s = \sqrt{30 \cdot 80 \cdot 1.2}$ *Replace d with 80 and f with 12.*

$s = \sqrt{2880}$

$s = 53.66563146\ldots$

The car was traveling at about 54 miles per hour.

Learning the Concept

You can review rational numbers in Lesson 5-1.

Recall that numbers that can be expressed as $\frac{a}{b}$, where a and b are integers and b is not 0, are rational numbers. Rational numbers can always be expressed as decimals that terminate or repeat. For example, 7, $\frac{1}{3}$, and -4.9 are rationals. The number $\sqrt{2880}$, which is $53.66563146\ldots$, is not a repeating or terminating decimal. This kind of number is called an **irrational number**.

> **Definition of Irrational Number**
>
> An irrational number is a number that cannot be expressed as $\frac{a}{b}$, where a and b are integers and b does not equal 0.

Other examples of irrational numbers are shown below.

$\sqrt{3} = 1.7320508\ldots$ $\qquad \pi = 3.14159\ldots$

Example 

Determine whether each number is rational or irrational.

a. 0.22222 . . .

The three dots means that the 2s keep repeating. This is a repeating decimal, so it can be expressed as a fraction.

$0.22222\ldots = \frac{2}{9}$

Thus, it is a rational number.

> **HELP WANTED**
>
> You may want to have a local law enforcement officer talk to the class about the ways he or she uses mathematics on the job.

b. 0.35

This is a terminating decimal. It can be expressed as $\frac{35}{100}$ or $\frac{7}{20}$. It is also a rational number.

c. 0.52522522252222 . . .

This decimal does not repeat or terminate and is therefore an irrational number. It does have a pattern to it, but since the 5s are separated by an increasing number of 2s, there is no exact repetition.

The set of rational numbers and the set of irrational numbers together make up the set of **real numbers**. The Venn diagram at the right shows the relationships among whole numbers, integers, rational numbers, irrational numbers, and real numbers.

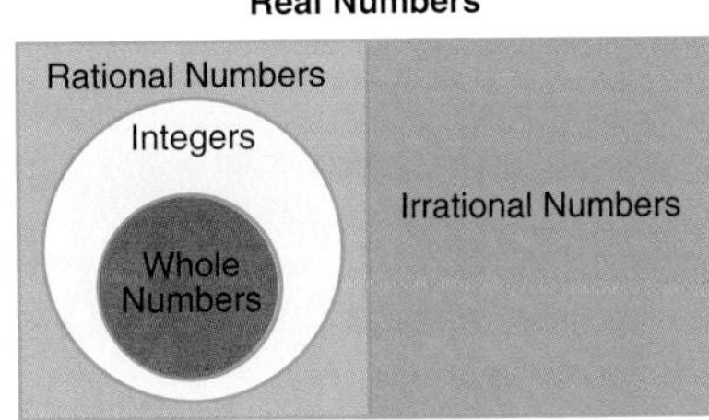

The following chart shows the sets of numbers to which several numbers belong.

Number	Whole Number	Integer	Rational	Irrational	Real
−9		✓	✓		✓
$\sqrt{35}$				✓	✓
$\sqrt{49}$	✓	✓	✓		✓
0.58585858 . . .			✓		✓
0.121231234 . . .				✓	✓
9	✓	✓	✓		✓

The equations you have encountered so far have had rational number solutions. Some equations have irrational number solutions.

You can solve some equations that involve squares by taking the square root of each side.

Example  **Solve each equation. Round decimal answers to the nearest tenth.**

a. $x^2 = 81$

$x^2 = 81$

$x = \sqrt{81}$ or $x = -\sqrt{81}$ *Take the square root of each side.*

$x = 9$ or $x = -9$

Check: $9^2 = 81$ and $(-9)^2 = 81$, so the solution checks.

b. $y^2 = 75$

$y^2 = 75$

$y = \sqrt{75}$ or $y = -\sqrt{75}$ *Take the square root of each side.*

75 [$\sqrt{x}$] 8.660254038

$y \approx 8.7$ or $y \approx -8.7$

2 TEACH

In-Class Examples

For Example 1
Determine whether each number is rational or irrational.

a. 0.55555... **rational**
b. 0.9148 **rational**
c. 0.313113111311113111111...
irrational

For Example 2
Solve each equation. Round decimal answers to the nearest tenth.

a. $x^2 = 144$ **$x = 12$ or $x + -12$**
b. $n^2 = 54$ **$n \approx 7.3$ or $n \approx -7.3$**

Teaching Tip Be sure students understand that the three dots mean that the pattern continues in the same manner. It does not necessarily mean that the last digit continues repeating.

Study Guide Masters, p. 110

NAME ________________ DATE ________

13-3 Study Guide
The Real Number System

Student Edition Pages 672–675

Rational numbers are numbers that can be expressed as a quotient of two integers, where the divisor is not zero.

Irrational numbers are numbers that can be named by nonterminating, non-repeating decimals.

The set of **real numbers** includes both the rational numbers and the irrational numbers.

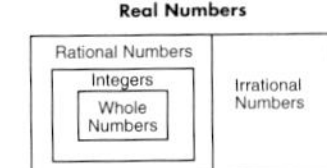

Some equations have solutions that are irrational numbers. To solve an equation that involves squares, take the square root of each side.

Example: Solve $x^2 = 65$

$x^2 = 65$

$x = \sqrt{65}$ or $x = -\sqrt{65}$ Take the square root of each side.

$x \approx 8.1$ or $x \approx -8.1$

Name the sets of numbers to which each number belongs: the whole numbers, the integers, the rational numbers, the irrational numbers, and/or the reals.

1. 6.01001 . . . irrational, real
2. −2 integer, rational, real
3. $-\frac{3}{5}$ rational, real
4. $0.\overline{83}$ rational, real
5. $\sqrt{19}$ irrational, real
6. 0.625 rational, real
7. $-\sqrt{81}$ integer, rational, real
8. 4.2342352 . . . irrational, real
9. $-\sqrt{7}$ irrational, real

Solve each equation. Round decimal answers to the nearest tenth.

10. $c^2 = 49$ $c = 7$ or $c = -7$
11. $x^2 = 4$ $x = 2$ or $x = -2$
12. $w^2 = 14$ $w \approx 3.7$ or $w \approx -3.7$
13. $n^2 = 289$ $n = 17$ or $n = -17$
14. $m^2 = 132$ $m \approx 11.5$ or $m \approx -11.5$
15. $h^2 = 250$ $n \approx 15.8$ or $n \approx -15.8$
15. $k^2 = 3.24$ $k = 1.8$ or $k = -1.8$
16. $r^2 = 1600$ $r = 40$ or $r = -40$
17. $d^2 = 90$ $d \approx 9.5$ or $d \approx -9.5$

Cooperative Learning

Categorizing Give students this group of numbers. 9, −23, $\sqrt{16}$, $\sqrt{-16}$, $-\sqrt{16}$, $\sqrt{7}$, $-\sqrt{7}$, $(-\sqrt{7})^2$, $-(\sqrt{7})^2$, 0.020202..., 0.02002000200002..., 5.55555
Have each group work together to categorize each number in as many ways as possible. Students can record the results as a table or in a Venn diagram.
For more information on this strategy, see *Cooperative Learning in the Mathematic Classroom*, one of the titles in the Glencoe Mathematics Professional Series, pp. 31–32.

In-Class Example

For Example 3

The time it takes for a pendulum to make one complete swing (back and forth) is given by the formula $T = 2\pi\sqrt{\frac{L}{32}}$, where T is the time in seconds and L is the length in feet. How long will it take a 6-foot pendulum to make one complete swing? **2.7 seconds**

3 PRACTICE/APPLY

Checking Your Understanding

Exercises 1–12 are designed to help you assess your students' understanding through reading, writing, speaking, and modeling. You should work through Exercises 1–3 with your students and then monitor their work on Exercises 4–12.

Assignment Guide

Core: 13–45 odd, 46–52
Enriched: 14–40 even, 41–52

For **Extra Practice**, see p. 771.

The red A, B, and C flags, printed only in the Teacher's Wraparound Edition, indicate the level of difficulty of the exercises.

Practice Masters, p. 110

NAME ______________________ DATE ____________

13-3 Practice
The Real Number System

Student Edition
Pages 672–675

Name the sets of numbers to which each number belong: the whole numbers, the integers, the rational numbers, the irrational numbers, and/or the reals.

1. 4 whole, integer, rational, real
2. $\frac{1}{4}$ rational, real
3. $2.\overline{6}$ rational, real
4. $\sqrt{13}$ irrational, real
5. -5.8 rational, real
6. $-\sqrt{36}$ integer, rational, real
7. $-\sqrt{5}$ irrational, real
8. $0.56361345\ldots$ irrational, real
9. $-0.777\ldots$ rational, real
10. $\frac{4}{3}$ rational, real
11. $\sqrt{16}$ whole, integer, rational, real
12. $0.01002003\ldots$ irrational, real
13. -9 integer, rational, real
14. $\frac{-10}{3}$ rational, real
15. $0.583333\ldots$ rational, real
16. $\sqrt{676}$ whole, integer, rational, real

Solve each equation. Round decimal answers to the nearest tenth.

17. $b^2 = 25$ $b = 5$ or $b = -5$
18. $c^2 = 16$ $c = 4$ or $c = -4$
19. $a^2 = 9$ $a = 3$ or $a = -3$
20. $h^2 = 10$ $h \approx 3.2$ or $h \approx -3.2$
21. $p^2 = 20$ $p \approx 4.5$ or $p \approx -4.5$
22. $j^2 = 45$ $j \approx 6.7$ or $j \approx -6.7$
23. $k^2 = 56$ $k \approx 7.5$ or $k \approx -7.5$
24. $n^2 = 70$ $n \approx 8.4$ or $n \approx -8.4$
25. $f^2 = 300$ $f \approx 17.3$ or $f \approx -17.3$
26. $d^2 = 140$ $d \approx 11.8$ or $d \approx -11.8$
27. $h^2 = 190$ $h \approx 13.8$ or $h \approx -13.8$
28. $g^2 = 3.61$ $g \approx 1.9$ or $g \approx -1.9$
29. $a^2 = 2500$ $a = 50$ or $a = -50$
30. $0.0081 = t^2$ $t = 0.09$ or $t = -0.09$
31. $w^2 = 40{,}000$ $w = 200$ or $w = -200$

The solutions to many real-world problems are irrational numbers.

Example 3 APPLICATION Electricity

The voltage, V, in a circuit is given by the formula $V = \sqrt{PR}$. In this formula, V is in volts, P is the power in watts, and R is the resistance in ohms. An electrician has a circuit that produces 1500 watts of power. She wants the voltage in the circuit to be no more than 110 volts. Should she design the circuit with a resistance of a 7.5 ohms or 7.8 ohms?

For this circuit, $P = 1500$ and $V = 110$. To determine which resistance to use, you can use a calculator to evaluate $\sqrt{1500R}$ for $R = 7.5$ and for $R = 7.8$ to find out which one will produce no more than 110 volts.

Evaluate $\sqrt{1500R}$ for $R = 7.5$.

Enter: 1500 [×] 7.5 [=] [√x] 106.0660172

Evaluate $\sqrt{1500R}$ for $R = 7.8$.

Enter: 1500 [×] 7.8 [=] [√x] 108.1665383

She can design the circuit with *either* resistance. If she wants the voltage in the circuit to be as close to 110 volts as possible, then she should design the circuit with a resistance of 7.8 ohms.

Checking Your Understanding

Communicating Mathematics

Read and study the lesson to answer each question.

1. **Compare and contrast** rational and irrational numbers.
2. **You Decide** Jaya says that $\sqrt{16}$ can be both an integer and a rational number. Do you agree? Explain.
3. **Give an example** of an irrational number that is less than -10.

1–3. See Solutions Manual.

Guided Practice

Name the sets of numbers to which each number belongs: the whole numbers, the integers, the rational numbers, the irrational numbers, and/or the real numbers.

4. $0.3333\ldots$ 5. 0 6. $\sqrt{11}$ 7. $-\frac{1}{2}$

Solve each equation. Round decimal answers to the nearest tenth.

8. $m^2 = 49$ 7, −7 9. $r^2 = 361$ 10. $t^2 = 1$ 1, −1 11. $n^2 = 17$
12. **Hang Gliding** Using $R = \frac{s^2}{A}$, find the wingspan s of Kesse's hang glider if the aspect ratio R is 2.5 and the area of the wing A is 120 square feet. about 17.3 feet

4. rational, real
5. whole, integer, rational, real
6. irrational, real
7. rational, real
9. 19, −19
11. 4.1, −4.1

Exercises: Practicing and Applying the Concept

Independent Practice

Name the sets of numbers to which each number belongs: the whole numbers, the integers, the rational numbers, the irrational numbers, and/or the real numbers. 13–18. See Solutions Manual.

13. 8 14. -5 15. $\frac{2}{3}$
16. $\sqrt{5}$ 17. -3.5 18. $-\sqrt{25}$

Group Activity Card 13-3

Simply Simplified Group Activity **13-3**

MATERIALS: Calculators • 12 notecards with the radical expressions given on the back of this card • paper • pencils

The deck of "Radical" cards should be made, shuffled, and placed face down on the table.

One player turns up the top card on the card pile. All players then simplify the same radical expression. After everyone is finished, expressions are checked for accuracy using the calculator, and they are checked to be sure they are in simplified form.

The player who first arrives at the correct, simplified expression scores a point. The overall winner is the player with the most points after all cards have been used.

 Pre-Algebra

Reteaching

Using Lists Have students write the words whole, integer, rational, irrational, and real across a piece of paper. Under each term, have students list five numbers that apply to that term. In cooperative groups, discuss whether each number could also be in other groups.

19–24. See Solutions Manual.

19. 0.121121112... **20.** $\frac{5}{4}$ **21.** $\sqrt{18}$

22. $\frac{-9}{3}$ **23.** $2.\overline{5}$ **24.** 3.14359265...

27. 3.5, −3.5
28. 13, −13
29. 11.0, −11.0
30. 14.1, −14.1
31. 13.4, −13.4
32. 1.2, −1.2
33. 0.02, −0.02
34. 15.5, −15.5
35. 20, −20
36. 300, −300
38. 5, −5

Solve each equation. Round decimal answers to the nearest tenth.

B

25. $r^2 = 36$ **6, −6** **26.** $x^2 = 64$ **8, −8** **27.** $y^2 = 12$ **28.** $169 = m^2$

29. $n^2 = 120$ **30.** $f^2 = 200$ **31.** $180 = j^2$ **32.** $p^2 = 1.44$

33. $0.0004 = s^2$ **34.** $h^2 = 240$ **35.** $400 = q^2$ **36.** $a^2 = 90{,}000$

C

37. $(-b)^2 = 81$ **9, −9** **38.** $c^2 - 3^2 = \sqrt{16^2}$ **39.** $\sqrt{81} = d^2$ **3, −3**

40. Geometry The length of a rectangle is three times its width. What are the dimensions of the rectangle if its area is 192 ft^2? **8 ft by 24 ft**

Critical Thinking

41. Can the product of two irrational numbers be rational? Support your answer with examples. **yes; Sample answer: $\sqrt{2} \times \sqrt{2} = 2$**

Applications and Problem Solving

42. Meteorology You can use the formula $t^2 = \frac{d^3}{216}$ to estimate the amount of time that a thunderstorm will last. In this formula, t is the time in hours, and d is the diameter of the storm in miles. If a thunderstorm is 6 miles wide, how long will the storm last? **1 hour**

In A.D. 60, Heron, an Egyptian mathematician and scientist, invented the *dioptra* for surveying and making astronomical observations. A dioptra could measure horizontal and vertical angles.

43. Geometry You can use Heron's formula to find the area of a triangle if you know the measures of its sides. If the measures of the sides are a, b, and c, the area A equals $\sqrt{s(s-a)(s-b)(s-c)}$, where s is one-half the perimeter. Suppose you have a triangle with sides 80 feet, 60 feet, and 100 feet long. Find the area of the triangle. **2400 ft^2**

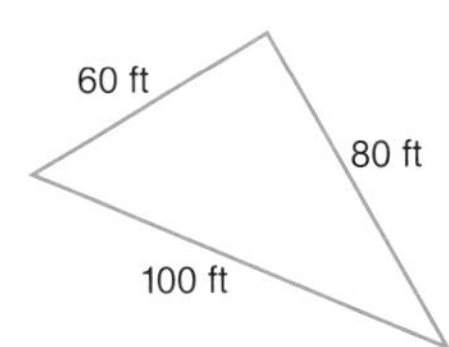

44. Physics If an object is dropped, the time t in seconds it takes to reach a given distance can be estimated by using the formula $d = 0.5gt^2$. In this formula, d is the free-fall distance, and g is the acceleration due to gravity, 32 ft/s^2. If Rachelle drops a ball from the top of a 55-foot building, how long does it take for the ball to hit the ground? **about 1.85 seconds**

45. See students' work.

45. Family Activity Drop a ball from a second or third story window several times and time its fall to the ground. Find the average time. See how that time compares to the answer using the formula given in Exercise 44 by measuring the actual falling distance.

Mixed Review

46. Statistics A survey of 120 people showed that from 7:00 to 8:00 one evening, 46 people watched news programs, 34 watched game shows, and 15 watched both. How many did not watch either? (Lesson 13-2) **55 people**

47. 4908.7 cm^3

47. Geometry Find the volume of a circular cylinder with diameter 25 cm and height 10 cm. Round to the nearest tenth. (Lesson 12-7)

48. Graph $y = 2x + 3$. (Lesson 8-3) **See margin.**

49. Solve $v = \left(\frac{3}{7}\right)\left(-\frac{14}{15}\right)$. (Lesson 6-3) $-\frac{2}{5}$

50. Find the GCF of 36 and $-30ab$. (Lesson 4-5) **6**

51. Evaluate $4cd$ if $c = -2$ and $d = -9$. (Lesson 2-7) **72**

52. Simplify $7(a + b) - 2(3a + 4b)$. (Lesson 1-5) $a - b$

4 ASSESS

Closing Activity

Speaking Have each student name any number. Then have the class state whether the number is rational or irrational.

Heron, or Hero of Alexandra, used his extensive mathematical knowledge in his study of scientific principles. He invented the pneumatic device commonly known as Heron's Fountain, which was a simple form of the steam engine.

Additional Answer

48.

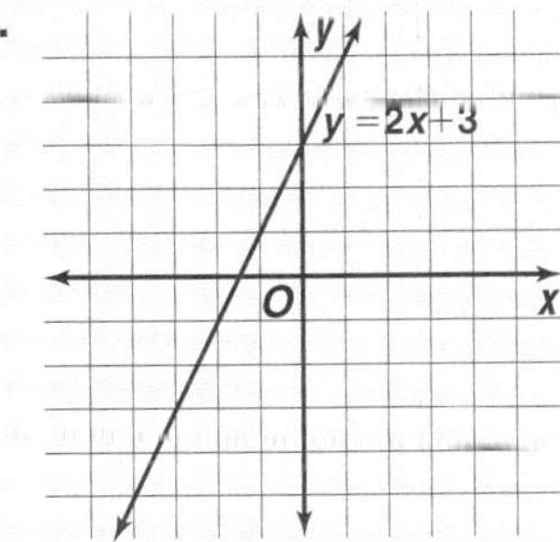

Enrichment Masters, p. 110

NAME ____________ DATE ________

13-3 Enrichment
Diagonals

Student Edition Pages 672–675

To find the length of diagonals in cubes and rectangular solids, a formula can be applied. In the example below, the length of diagonal $\overline{AG}$ or d can be found using the formula

$$d^2 = a^2 + b^2 + c^2 \text{ or } d = \sqrt{a^2 + b^2 + c^2}.$$

Example: The diagonal, d, is equal to the square root of the sum of the squares of the length, a, the width, b, and the height, c.

Example: Find the length of the diagonal of a rectangular prism with length of 8 meters, width of 6 meters, and height of 10 meters.

$d = \sqrt{8^2 + 6^2 + 10^2}$ Substitute the dimensions into the equation.
$= \sqrt{64 + 36 + 100}$ Square each value. Add.
$= \sqrt{200}$ Find the square root of the sum.
$= 14.1$ m Round the answer to the nearest tenth.

Solve. Use $d = \sqrt{a^2 + b^2 + c^2}$. Round answers to the nearest tenth.

1. Find the diagonal of a cube with sides of 6 inches. **10.4 in.**
2. Find the diagonal of a cube with sides of 2.4 meters. **4.2 m**
3. Find the diagonal of a rectangular solid with length of 18 meters, width of 16 meters, and height of 24 meters. **34 m**
4. Find the diagonal of a rectangular solid with length of 15.1 meters, width of 8.4 meters, and height of 6.3 meters. **18.4 m**
5. Find the diagonal of a cube with sides of 34 millimeters. **58.9 mm**
6. Find the diagonal of a rectangular solid with length of 8.9 millimeters, width of 6.7 millimeters, and height of 14 millimeters. **17.9 mm**

Extension

Using Concept Maps Complete a concept map, using the terms *integer, irrational, other, rational, real,* and *whole.*

Additional Answers

13. whole number, integer, rational, real
14. integer, rational, real
15. rational, real
16. irrational, real
17. rational, real
18. integer, rational, real
19. irrational, real
20. rational, real
21. irrational, real
22. integer, rational, real
23. rational, real
24. irrational, real

13-4 LESSON NOTES

NCTM Standards: 1-4, 12, 13

Instructional Resources

- Study Guide Master 13-4
- Practice Master 13-4
- Enrichment Master 13-4
- Group Activity Card 13-4
- Assessment and Evaluation Masters, pp. 350, 351
- Activity Masters, pp. 13, 47
- Math Lab and Modeling Math Masters, pp. 73-74
- Multicultural Activity Masters, p. 25
- Tech Prep Applications Masters, p. 26

Transparency 13-4A contains the 5-Minute Check for this lesson; **Transparency 13-4B** contains a teaching aid for this lesson.

Recommended Pacing	
Standard Pacing	Day 4 of 11
Honors Pacing	Day 4 of 10
Block Scheduling*	Day 2 of 6 (along with Lesson 13-3)

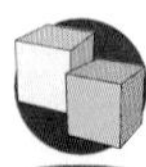

*For more information on pacing and possible lesson plans, refer to the ***Block Scheduling Booklet***.

1 FOCUS

5-Minute Check
(over Lesson 13-3)

Name the set of numbers to which each number belongs: whole numbers, integers, rational numbers, irrational numbers, and/or real numbers.

1. $\sqrt{16}$ **real, rational, integer, whole**
2. $\sqrt{7}$ **real, irrational**

Solve each equation. Round decimal answers to the nearest tenth.

3. $t^2 = 1.4$ **1.2, -1.2**
4. $m^2 = \sqrt{625}$ **5, -5**
5. $(-n)^2 = 121$ **11, -11**

13-4 The Pythagorean Theorem

Setting Goals: *In this lesson, you'll use the Pythagorean theorem to find the length of the side of a right triangle and to solve problems.*

Modeling with Manipulatives

MATHEMATICS MINI LAB

MATERIALS

 dot paper

On dot paper, the horizontal or vertical distance between two adjacent dots is defined as one unit. Therefore, a square with sides 1 unit long has an area of 1 square unit.

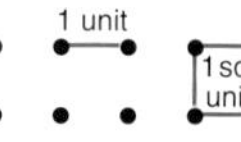

Your Turn **Work with a partner.**

▶ Draw a square like the one shown at the right.

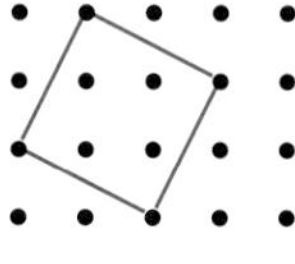

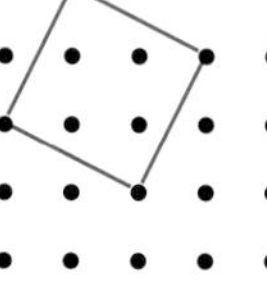

TALK ABOUT IT

a. How can you prove that the area of the square is 5 square units? **See margin.**

Your Turn ▶ Use the side of the square to make a triangle. Then draw squares on the other two sides of the triangle.

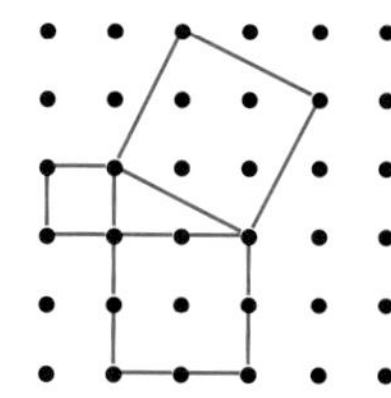

TALK ABOUT IT

b. Classify the triangle you drew. **right triangle**

c. What are the areas of the two smaller squares? **1 square unit, 4 square units**

d. What is the relationship between the areas of the three squares?

d. The area of the larger square equals the sum of the areas of the two smaller squares.

e. If the area of the largest square is 5 square units, what is the length of the side of the triangle opposite the right angle? $\sqrt{5}$

Learning the Concept

In the movie *The Wizard of Oz*, with Judy Garland, when the scarecrow gets his brain he recites the Pythagorean theorem . . . but he says it incorrectly!

The sides of a right triangle that are adjacent to the right angle are called the **legs**. The side opposite the right angle is called the **hypotenuse**.

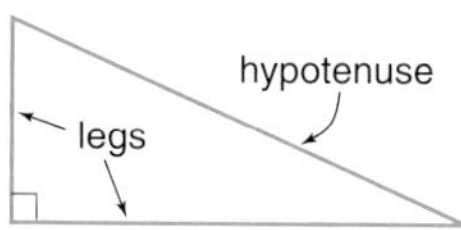

The relationship between the lengths of the legs and the hypotenuse that you noticed in the activity at the beginning of the lesson was proven in the fifth century B.C. by a Greek mathematician, Pythagoras, and his followers. This relationship is called the **Pythagorean theorem**. It is true for *any* right triangle.

In *The Wizard of Oz*, the scarecrow actually says, "the sum of the square roots of any two sides of an isosceles triangle is equal to the square root of the remaining side."

Talk About It

a. Draw a 3-by-3 square around the given square so that the sides of the larger square contain the vertices of the given square. The area of the 3-by-3 square is 9 square units. The four triangles that are formed by the sides of the squares each have an area of $\frac{1}{2}(1 \cdot 2)$ or 1 square unit. Thus, the area of the square is $9 - 4(1)$ or 5 square units.

Pythagorean Theorem

In words: In a right triangle, the square of the length of the hypotenuse is equal to the sum of the squares of the lengths of the legs.

In symbols: $c^2 = a^2 + b^2$

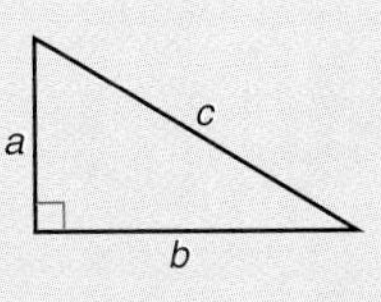

Example 1

Geometry

A rectangle has sides of 6 and 8 inches. How long is a diagonal?

A diagonal and two adjacent sides of the rectangle form a right triangle. The legs of the right triangle are sides of the rectangle, and the hypotenuse of the right triangle is a diagonal of the rectangle.

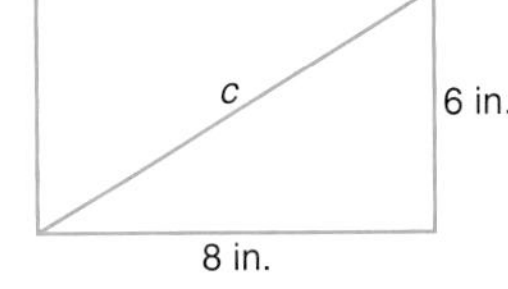

$c^2 = a^2 + b^2$ *Pythagorean theorem*

$c^2 = 6^2 + 8^2$ *Replace a with 6 and b with 8.*

$c^2 = 36 + 64$

$c^2 = 100$

$c = \sqrt{100}$ *Take the square root of each side.*

$c = 10$

The length of the diagonal is 10 inches.

THINK ABOUT IT

Why can you ignore the negative value of *c* in this problem?

The length of the diagonal must be a positive number.

The Pythagorean theorem can also be used to find the length of any side of a right triangle if the lengths of the other two sides are known.

Example 2

Find the length of the third side of the right triangle.

$c^2 = a^2 + b^2$

$27^2 = a^2 + 25^2$

$729 = a^2 + 625$

$729 - 625 = a^2 + 625 - 625$

$104 = a^2$

$\sqrt{104} = a$

104 [$\sqrt{x}$] 10.19803903

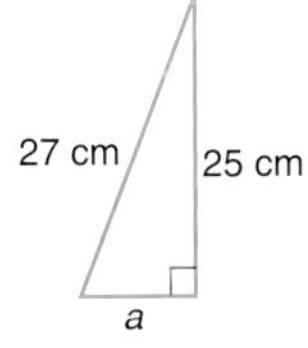

The length of the leg is about 10.2 cm.

You can use the Pythagorean theorem to see whether a triangle is right.

Example 3

The measurements of three sides of a triangle are given. Determine whether each triangle is a right triangle. *Remember, the hypotenuse is the longest side.*

a. 5 ft, 7 ft, 8 ft

$c^2 = a^2 + b^2$

$8^2 \stackrel{?}{=} 5^2 + 7^2$

$64 \stackrel{?}{=} 25 + 49$

$64 \neq 74$

The triangle is *not* right.

b. 10 cm, 24 cm, 26 cm

$c^2 = a^2 + b^2$

$26^2 \stackrel{?}{=} 10^2 + 24^2$

$676 \stackrel{?}{=} 100 + 576$

$676 = 676$

The triangle is right.

Motivating the Lesson

Situational Problem Tell students that last weekend you started at home and drove 3 miles due north then 4 miles due west to get to the mall. Ask how far "as the crow flies" the mall is from your home. Have students brainstorm ways to solve the problem.

2 TEACH

In-Class Examples

For Example 1

A rectangle has sides 10 centimeters and 24 centimeters long. How long is the diagonal? **26 cm**

For Example 2

The hypotenuse of a right triangle is 14 inches long. One of the legs of the triangle is 9 inches long. Find the length of the third side of the triangle. Round to the nearest tenth. **about 10.7 in.**

For Example 3

The measurements of three sides of a triangle are given. Determine whether each triangle is a right triangle.

a. 40 yd, 42 yd, 58 yd **The triangle is a right triangle.**

b. 18 m, 31 m, 35 m **The triangle is not a right triangle.**

GLENCOE Technology

Interactive Mathematics Tools Software

In this interactive computer lesson, students explore the Pythagorean theorem. A **Computer Journal** gives students the opportunity to write about what they have learned.

For Windows & Macintosh

Alternative Teaching Strategies

Student Diversity Create a geoboard on the floor. Have students choose points at which to stand and use string to create various right triangles.

Teaching Tip As you discuss the Examine step of Example 4, remind students that it is also beneficial to make an estimate of the answer before beginning to solve the problem.

> **In-Class Example**
>
> ***For Example 4***
>
> The lid on a grand piano is held open with a 28-inch prop. The prop's base is 55 inches across the piano from the lid's hinges. If the prop forms a right angle with the lid, how far from the hinges must the prop be placed? **about 47 inches**

The Pythagorean theorem can be applied to solve problems that occur in real life.

Example 4 **A Forest Service warehouse facility is used as a supply depot and a fire-fighting center. A security fence is needed to protect a helicopter, some tanker trucks, and other fire-fighting equipment that must be stored outdoors. A diagram of the region to be fenced is shown at the right. How much fencing must be purchased to surround the region?**

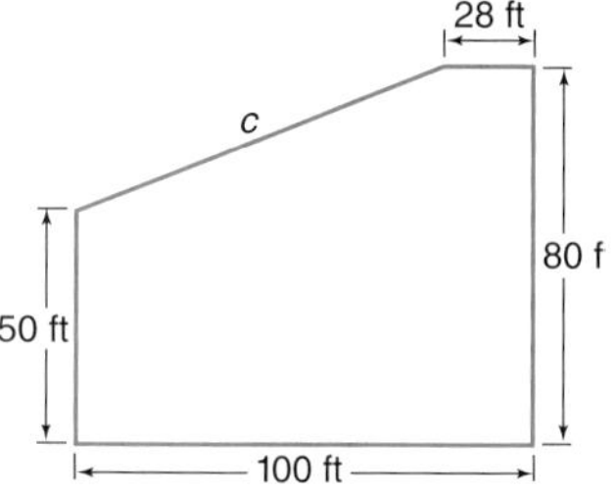

CULTURAL CONNECTIONS

Pythagoras is credited with the proof of the theorem that bears his name. However, the theorem may have been known to Egyptian, Babylonian, and Chinese scholars as much as one thousand years earlier.

Explore You know what the boundary of the region to be fenced looks like. You must find the perimeter of the region.

Plan To find the perimeter, you must find the unknown length, c. You can separate the region into more familiar shapes as shown below.

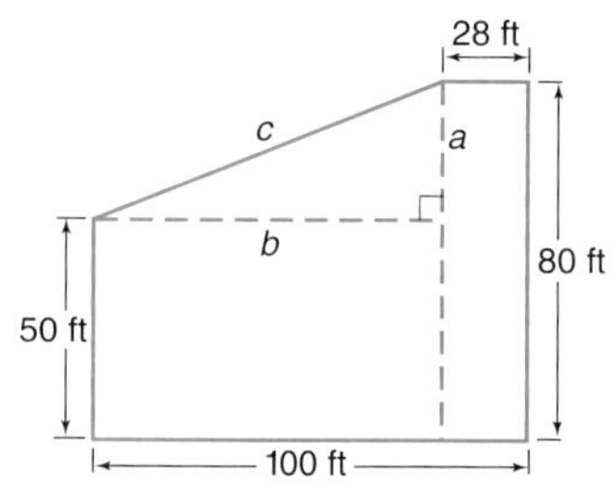

The hypotenuse of the right triangle is the side with length c. Find the measures of the legs of the right triangle and then use the Pythagorean theorem to find the length of the hypotenuse.

Solve First find the measures of the legs.

$a = 80 - 50$ $\quad b = 100 - 28$
$a = 30$ $\quad b = 72$

Next, apply the Pythagorean theorem.

$c^2 = a^2 + b^2$
$c^2 = 30^2 + 72^2$ *Replace a with 30 and b with 72.*
$c^2 = 900 + 5184$
$c^2 = 6084$
$c = \sqrt{6084}$ *Take the square root of each side.*
6084 [$\sqrt{x}$] 78
$c = 78$

Finally, find the perimeter.

$28 + 80 + 100 + 50 + 78 = 336$

So, 336 feet of fencing is needed.

Examine If the region included the corner area, it would be a rectangular shape. Its perimeter would be $2 \cdot 100 + 2 \cdot 80$ or 360 feet. The answer should be less than this. Thus, the answer is reasonable.

Surveyor Surveyors measure distances with various instruments. When distances cannot be measured directly, they often use right triangles and the Pythagorean theorem to find the missing measurements. For more information on tech prep, see the *Teacher's Handbook*.

Checking Your Understanding

Communicating Mathematics

Read and study the lesson to answer each question. **1–4. See margin.**

1. **Explain** how to find the length of a leg of a right triangle if you know the length of the hypotenuse and the length of the other leg.
2. **Draw** and label a triangle with sides that measure 6 meters, 8 meters, and 10 meters. Is it a right triangle? Explain.
3. The roof of a house is 24 feet above the ground. You have a 26-foot ladder. Use the Pythagorean theorem to explain why the base of the ladder must be placed no more than 10 feet from the house in order to reach the roof.
4. Use dot paper to draw squares on each side of a triangle like the one at the right. Write an equation you could use to find the length of the hypotenuse.

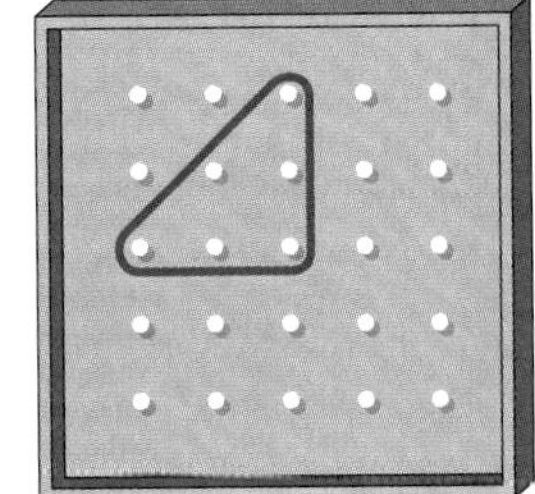

MODELING MATHEMATICS

MATERIALS

dot paper

Guided Practice

Write an equation you could use to solve for x. Then solve. Round decimal answers to the nearest tenth.

5.

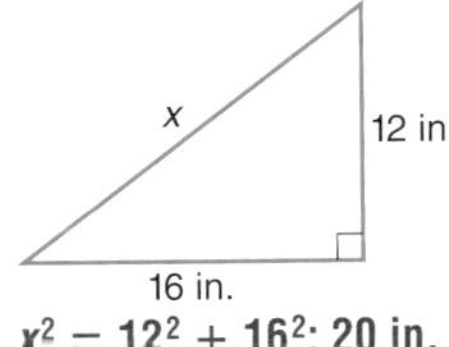

$x^2 = 12^2 + 16^2$; 20 in.

6.

4 ft, 4 ft, x, 4 ft

$4^2 = x^2 + 2^2$; 3.5 ft

Write an equation that can be used to answer each question. Then solve. Round decimal answers to the nearest tenth.

7. How long is the lake?

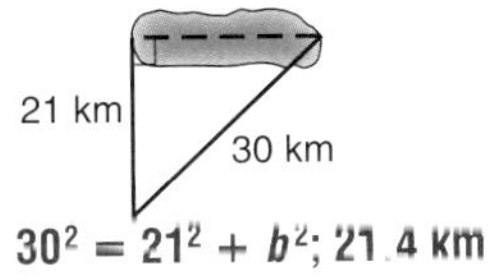

$30^2 = 21^2 + b^2$; 21.4 km

8. How high is the TV screen?

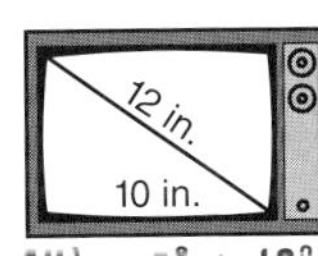

$12^2 = a^2 + 10^2$; 6.6 in.

In a right triangle, if a and b are the measures of the legs and c is the measure of the hypotenuse, find each missing measure. Round decimal answers to the nearest tenth.

9. $a = 3, b = 4$ **5**
10. $a = 3, c = 7$ **6.3**
11. $b = 12, c = 35$ **32.9**
12. $a = 15, b = 16.7$ **22.4**

The measurements of three sides of a triangle are given. Determine whether each triangle is a right triangle.

13. 9 cm, 12 cm, 15 cm **yes**
14. 6 ft, 7 ft, 12 ft **no**
15. **Baseball** A baseball diamond is actually a square. The distance between bases is 90 feet. When a runner on first base tries to steal second, a catcher has to throw from home to second base. How far is it between home and second base? **about 127.3 ft**

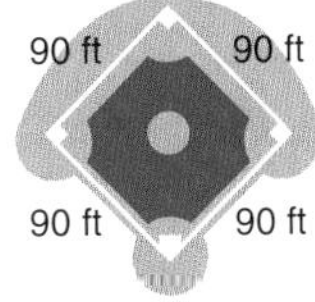

3 PRACTICE/APPLY

Checking Your Understanding

Exercises 1–15 are designed to help you assess your students' understanding through reading, writing, speaking, and modeling. You should work through Exercises 1–4 with your students and then monitor their work on Exercises 5–15.

Additional Answers

1. Sample answer: Square the hypotenuse and subtract the square of the leg, then find the square root of the result.
2. yes; $6^2 + 8^2 = 10^2$

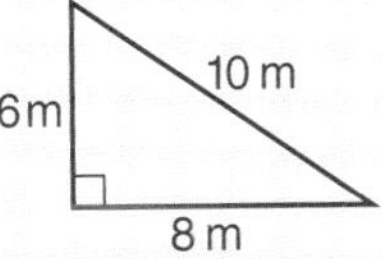

3. Let x represent the greatest distance that the ladder could be from the house.

$$x^2 + 24^2 = 26^2$$
$$x^2 + 576 = 676$$
$$x^2 = 100$$
$$x = 10$$

If the ladder is placed 10 feet from the house, the ladder will just reach the roof.

Study Guide Masters, p. 111

NAME ______________ DATE ______

13-4 Study Guide
The Pythagorean Theorem

Student Edition Pages 676–681

In a right triangle, the square of the hypotenuse, c, is equal to the sum of the squares of the lengths of the other two sides, a and b.

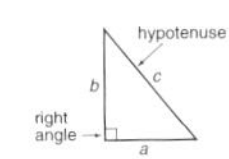

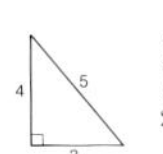

$c^2 = a^2 + b^2$
$5^2 = 3^2 + 4^2$
$5^2 = 9 + 16$
$25 = 25$

Example: How long must a ladder be to reach a window 13 feet above ground? For the sake of stability, the ladder must be placed 5 feet away from the base of the wall.

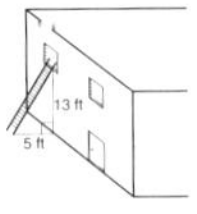

$c^2 = (13)^2 + (5)^2$
$c^2 = 169 + 25$
$c^2 = 194$
$c^2 = \sqrt{194} \approx 13.9$ ft

Solve. Round decimal answers to the nearest tenth.

1. In a softball game, how far must the catcher throw to second base? **84.9 ft**

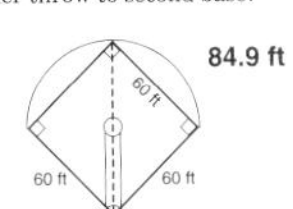

2. How long must the brace be on a closet rod holder if the vertical side is 17 cm and the horizontal side must be attached 30 cm from the wall? **34.5 cm**

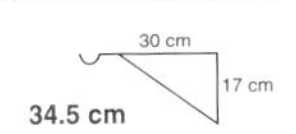

3. If Briny is 32 miles due east of Oxford and Myers is 21 miles due south of Oxford, how far is the shortest distance from Myers to Briny? **38.3 mi**

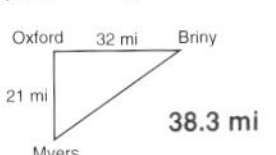

4. In a baseball game, how far must the shortstop (halfway between second base and third base) throw to make an out at first base? **100.6 ft**

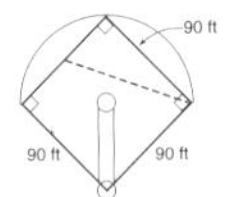

Reteaching

Using Models Using graph paper, have students model right triangles and determine the length of the third side if the lengths of two sides are known. Confirm by counting squares or using a calculator.

Additional Answer

4. See students' diagrams. Sample formula: length of hypotenuse = $\sqrt{(4 + 4)}$

Error Analysis

Some students may always add or always subtract the squares of the given sides, no matter which two sides are given. Remind students to label the right angle and the hypotenuse and then use the Pythagorean theorem correctly.

Assignment Guide

Core: 17–45 odd, 47–54
Enriched: 16–42 even, 43–54
All: Self test, 1–10

For **Extra Practice**, see p. 771.

The red A, B, and C flags, printed only in the Teacher's Wraparound Edition, indicate the level of difficulty of the exercises.

Practice Masters, p. 111

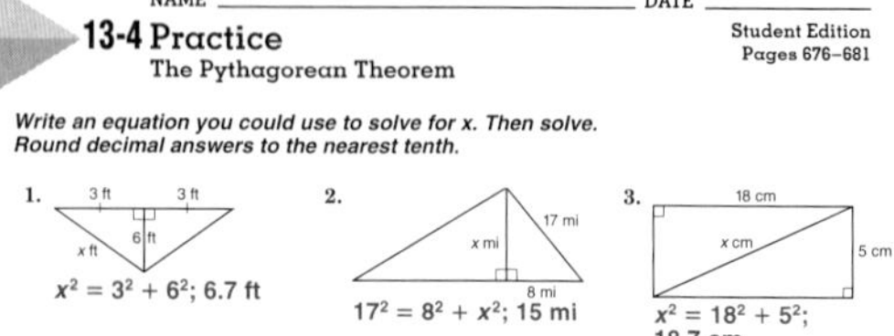

NAME ______________ DATE ______________

13-4 Practice Student Edition Pages 676–681
The Pythagorean Theorem

Write an equation you could use to solve for x. Then solve. Round decimal answers to the nearest tenth.

1. $x^2 = 3^2 + 6^2$; 6.7 ft
2. $17^2 = 8^2 + x^2$; 15 mi
3. $x^2 = 18^2 + 5^2$; 18.7 cm

Solve. Round decimal answers to the nearest tenth.

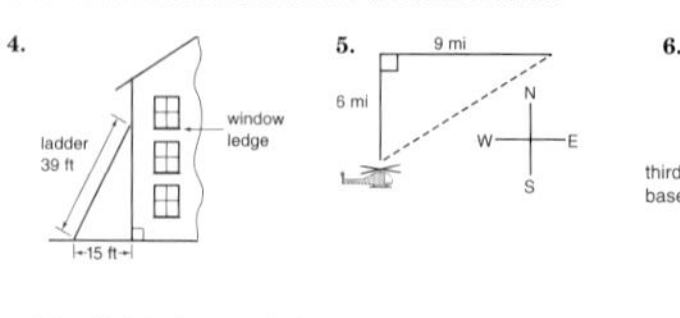

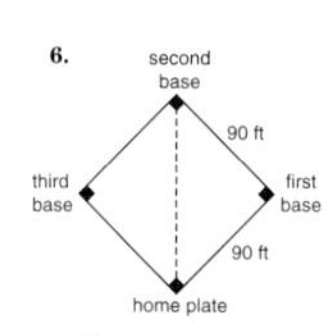

4. How high is the top window ledge above the ground? 36 ft
5. How far is the helicopter from its starting point. 10.8 mi
6. How far does a baseball travel from home plate to second base? 127.3 ft

In a right triangle, if a and b are the measures of the legs and c is the measure of the hypotenuse, find each missing measure. Round decimal answers to the nearest tenth.

7. $b = 16, c = 20$ 12
8. $a = 6, c = 14$ 12.6
9. $a = 9, c = 16$ 13.2
10. $b = 15, c = 20$ 13.2
11. $a = 8, c = 12$ 8.9
12. $b = 5, c = 16$ 15.2

The measurements of three sides of a right triangle are given. Determine whether each triangle is a right triangle.

13. 8 km, 15 km, 17 km yes
14. 15 in., 20 in., 25 in. yes
15. 8 mm, 9 mm, 15 mm no
16. 10 mi, 20 mi, 30 mi no

Exercises: Practicing and Applying the Concept

Independent Practice

Write an equation you could use to solve for x. Then solve. Round decimal answers to the nearest tenth.

16. x ft, 35 ft, 60 ft
$x^2 = 60^2 + 35^2$; 69.5

17.
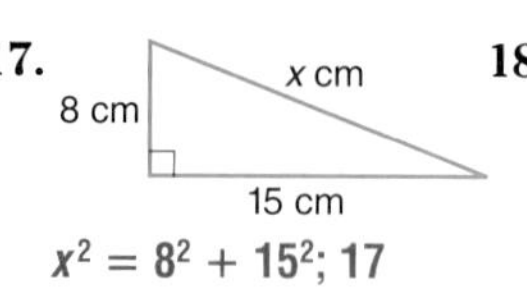

$x^2 = 8^2 + 15^2$; 17

18.
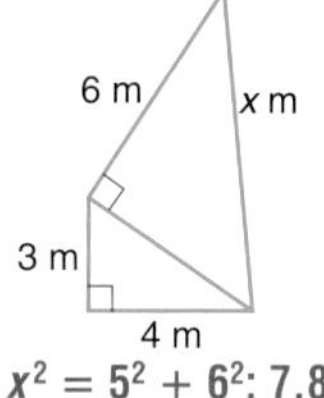

$x^2 = 5^2 + 6^2$; 7.8

Write an equation that can be used to answer each question. Then solve. Round decimal answers to the nearest tenth.

19. How high is the kite?
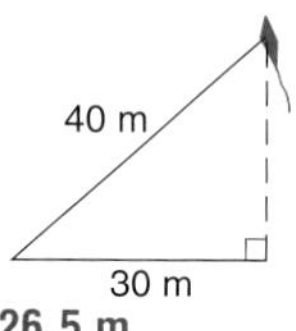

26.5 m

20. How long is a suspension line?
15 ft, 8 ft, suspension lines
11.0 ft

21. How high does the ladder reach?
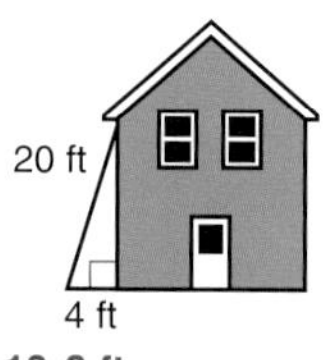

19.6 ft

22. How far apart are the planes?
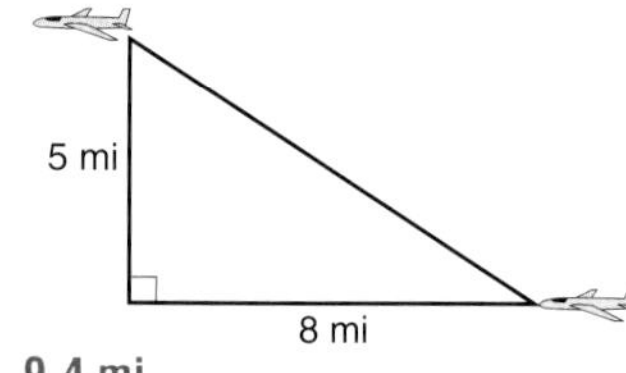

9.4 mi

In a right triangle, if a and b are the measures of the legs and c is the measure of the hypotenuse, find each missing measure. Round decimal answers to the nearest tenth.

23. $a = 12, b = 16$ 20
24. $b = 21, c = 29$ 20
25. $a = 2, b = 5$ 5.4
26. $a = 5, c = 10$ 8.7
27. $a = 7, c = 9$ 5.7
28. $b = 3, c = 7$ 6.3
29. $a = 7, b = 7$ 9.9
30. $b = 36, c = 85$ 77
31. $a = 14, b = 15$ 20.5
32. $a = 180, c = 181$ 19
33. $a = \sqrt{11}, c = 6$ 5
34. $b = 13, c = \sqrt{233}$ 8

The measurements of three sides of a triangle are given. Determine whether each triangle is a right triangle.

35. 8 ft, 9 ft, 10 ft no
36. 10 m, 24 m, 26 m yes
37. 15 in., 12 in., 9 in. yes
38. 6 mi, 7 mi, 8 mi no
39. 18 cm, $\sqrt{24}$ cm, 30 cm no
40. 15 ft, 16 ft, $\sqrt{31}$ ft yes

41. Geometry Find the length of a diagonal of a square if its area is 72 m². 12 m

42. Geometry Find the length of the diagonal of a cube if each side of the cube is 3 feet long. $\sqrt{27}$ or about 5.2 feet

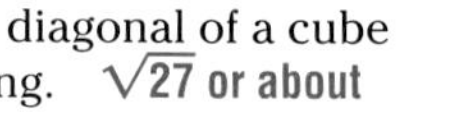

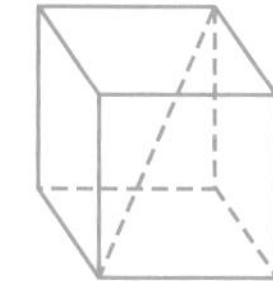

Group Activity Card 13-4

Will You Be Pythagorus? Group Activity **13-4**

MATERIALS: Calculators • "Pythagorean Triangles" cards with the numbers given on the back of this card

This card game for two to four players follows the rules of the "Old Maid" card game. The "Pythagorean Triangles" cards are shuffled and dealt to the players who should make any pairs they have in their hand. Cards are paired when one card contains the legs of a right triangle and the other card contains the hypotenuse for that triangle. When pairs are laid down, the other players must agree that the cards do form a pair.

Play continues around the table to the left, with each player drawing a card from the player on his or her right. If a player can make a pair, he or she does so after drawing a card.

The object of the game is to get rid of all of your cards. The player holding the one card that does not have a pair when the game ends loses.

 Pre-Algebra

Extension

Using Writing Have students work in pairs to write verbal problems using the Pythagorean theorem. Trade and solve.

Critical Thinking

43. The points $C(5, 1)$ and $D(8, 5)$ are graphed at the right. Find the distance between C and D. (*Hint:* Notice the right triangle. First find the measures of $\overline{BC}$ and $\overline{BD}$.) **5**

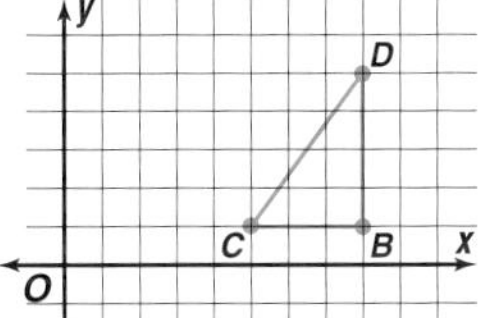

Applications and Problem Solving

Estimation
Mental Math
Calculator
Paper and Pencil

44. **Safety** The National Safety Council recommends placing the base of a ladder one foot from the wall for every three feet of the ladder's length.
    a. How far away from a wall should you place a 15-foot ladder? **5 ft**
    b. How high can the ladder safely reach?

44b. about 14.1 feet

45. **Hiking** Gracia hikes 10 kilometers south, 15 kilometers east, and then 10 km south again. How far is Gracia from the starting point of her hike? **25 km**
46. **Sailing** A rope from the top of a mast on a sailboat is attached to a point 6 feet from the base of the mast. If the rope is 24 feet long, how high is the mast? **about 23.2 feet**

Mixed Review

47. Solve $x^2 = 2.5$ to the nearest tenth. (Lesson 13-3) **1.6**
48. **Geometry** Draw and label a diagram to represent perpendicular lines k and m. (Lesson 11-1) **See margin.**
49. If $g(x) = -x + 4$, find $g(-1)$. (Lesson 8-4) **5**
50. Solve $5 - 2n = 4n + 41$. (Lesson 7-5) **−6**
51. Solve $\left(\frac{2}{3}\right)^2 = x$. (Lesson 6-3) $\frac{4}{9}$
52. **Sports** Willa swam 1 lap on the first day. She swam 2 laps on the second day, 4 laps on the third day, and 8 laps on the fourth day. To continue this pattern, how many laps should she swim on the seventh day? (Lesson 2-6) **64 laps**
53. Solve $6b = 120$ mentally. (Lesson 1-6) **20**
54. Find the value of the expression $3[4(6 - 2) - 5]$. (Lesson 1-2) **33**

Self Test

Find each square root. (Lesson 13-1)

1. $\sqrt{36}$ **6**
2. $\sqrt{121}$ **11**
3. $-\sqrt{100}$ **−10**
4. In a recent survey of 120 students, 60 students said they play tennis and 50 students said they play softball. If 20 students play both sports, how many students do not play either tennis or softball? (Lesson 13-2) **30 students**

Solve each equation. Round decimal answers to the nearest tenth. (Lesson 13-3)

5. $x^2 = 144$ **12, −12**
6. $y^2 = 50$ **7.1, −7.1**
7. $g^2 = 40$ **6.3, −6.3**

In a right triangle, if a and b are the measures of the legs and c is the measure of the hypotenuse, find each missing measure. Round answers to the nearest tenth. (Lesson 13-4)

8. $a = 12, b = 16$ **20**
9. $b = 63, c = 65$ **16**
10. $a = 15, c = 39$ **36**

Self Test

The Self Test provides students with a brief review of the concepts and skills in Lessons 13-1 through 13-4. Lesson numbers are given to the right of exercises or instruction lines so students can review concepts not yet mastered.

4 ASSESS

Closing Activity

Modeling Have students construct right triangles. Then have them construct three squares using each side of the triangle and compare the areas of the squares.

Chapter 13, Quiz B (Lessons 13-3 and 13-4) is available in the *Assessment and Evaluation Masters,* p. 351.

Mid-Chapter Test (Lessons 13-1 through 13-4) is available in the *Assessment and Evaluation Masters,* p. 350.

Additional Answer

48.

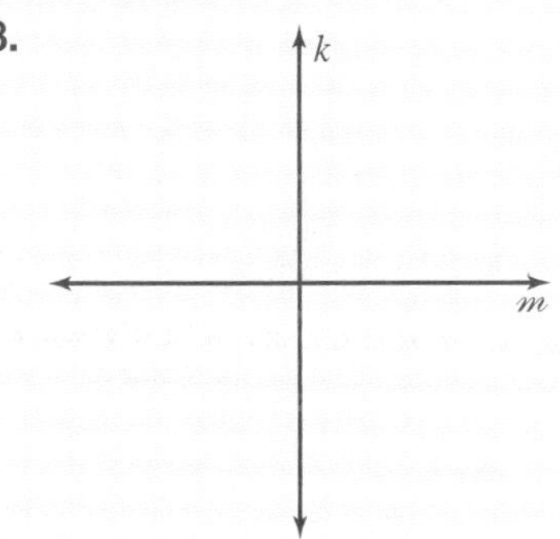

Enrichment Masters, p. 111

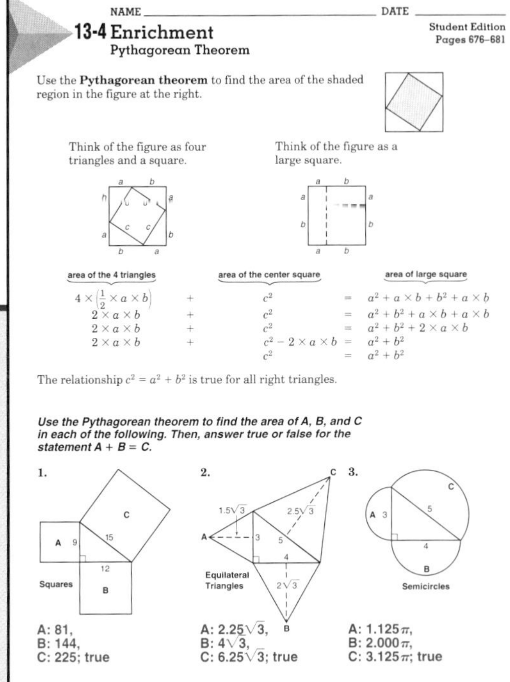

NAME ____________ DATE ____________

13-4 Enrichment
Pythagorean Theorem

Student Edition
Pages 676–681

Use the **Pythagorean theorem** to find the area of the shaded region in the figure at the right.

Think of the figure as four triangles and a square. | Think of the figure as a large square.

area of the 4 triangles		area of the center square		area of large square
$4 \times \left(\frac{1}{2} \times a \times b\right)$	+	c^2	=	$a^2 + a \times b + b^2 + a \times b$
$2 \times a \times b$	+	c^2	=	$a^2 + b^2 + a \times b + a \times b$
$2 \times a \times b$	+	c^2	=	$a^2 + b^2 + 2 \times a \times b$
$2 \times a \times b$	+	$c^2 - 2 \times a \times b$	=	$a^2 + b^2$
		c^2	=	$a^2 + b^2$

The relationship $c^2 = a^2 + b^2$ is true for all right triangles.

Use the Pythagorean theorem to find the area of A, B, and C in each of the following. Then, answer true or false for the statement A + B = C.

1. Squares (A 9, B 12, C 15)
A: 81,
B: 144,
C: 225; true

2. Equilateral Triangles ($1.5\sqrt{3}$, $2.5\sqrt{3}$, $2\sqrt{3}$, 3, 4, 5)
A: $2.25\sqrt{3}$,
B: $4\sqrt{3}$,
C: $6.25\sqrt{3}$; true

3. Semicircles (A 3, B 4, C 5)
A: 1.125π,
B: 2.000π,
C: 3.125π; true

13-4B LESSON NOTES

NCTM Standards: 1-4, 12, 13

Objective

Use the Pythagorean theorem to graph irrational numbers.

Recommended Time

Demonstration and discussion: 15 minutes; Exercises: 30 minutes

Instructional Resources

For each student or group of students
Student Manipulative Kit
- compass
- straightedge

Math Lab and Modeling Math Masters
p. 54 (worksheet)
For teacher demonstration
Overhead Manipulative Resources

1 FOCUS

Motivating the Lesson

Review the construction of a perpendicular, if necessary. Have students construct the right triangle as you model each step on the board or overhead. Ask students to estimate the value of $\sqrt{2}$ based on the graph. Students can check their estimates using a calculator.

2 TEACH

Teaching Tip Have students organize their work in a chart like the one below and use it as they complete the exercises.

Leg 1	Leg 2	Hypotenuse

3 PRACTICE/APPLY

Assignment Guide

Core: 1–5
Enriched: 1–5

HANDS-ON ACTIVITY

MATHEMATICS LAB **13-4B**

Graphing Irrational Numbers

An Extension of Lesson **13-4**

MATERIALS
compass
straightedge

You already know how to graph integers and rational numbers on a number line. How would you graph an irrational number like $\sqrt{10}$?

Activity

- Draw a number line. At 3, construct a perpendicular line segment 1 unit in length. Draw the line segment shown in color. Label it c.

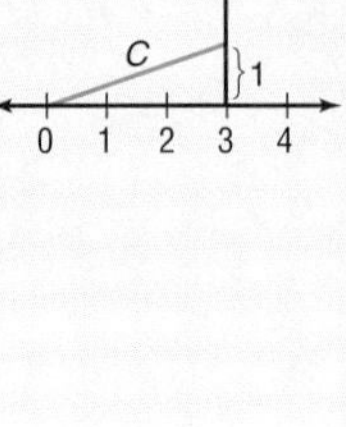

- Using the Pythagorean theorem, you can show that the hypotenuse is $\sqrt{10}$ units long.

$$c^2 = a^2 + b^2$$
$$c^2 = 1^2 + 3^2$$
$$c^2 = 10$$
$$c = \sqrt{10}$$

- Open the compass to the length of the segment you drew in color. With the tip of the compass at 0, draw an arc that intersects the number line at B. The distance from 0 to B is $\sqrt{10}$ units.

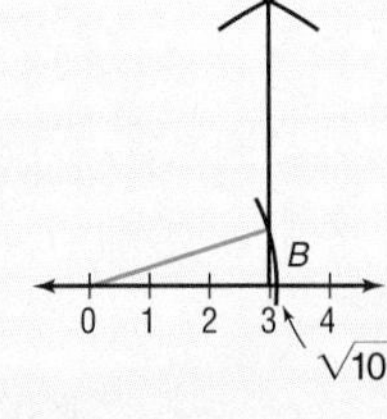

Your Turn Graph $\sqrt{8}$ on a number line. Think of $\sqrt{8}$ as $\sqrt{2^2 + 2^2}$.

TALK ABOUT IT

1–5. See Solutions Manual.

1. Explain how to graph $-\sqrt{10}$.
2. Explain how to graph $\sqrt{2}$.
3. Describe two different ways to graph $\sqrt{5}$.
4. Explain how the graph of $\sqrt{2}$ can be used to locate the point that represents $\sqrt{3}$.

Extension

5. Graph $\sqrt{12}$ on a number line. (*Hint:* Think of $\sqrt{12}$ as $\sqrt{4^2 - 2^2}$.)

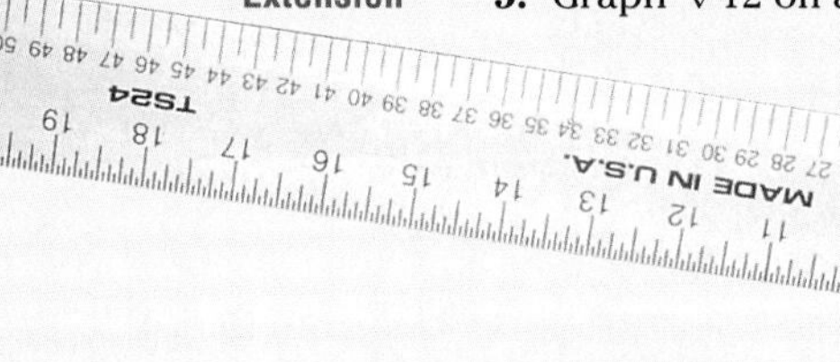

4 ASSESS

Observing students working in cooperative groups is an excellent method of assessment. You may wish to ask a student at random from each group to explain how to model and solve a problem. Also acknowledge students who are helping others to understand the concept being taught.

13-5 Special Right Triangles

Setting Goals: *In this lesson, you'll find missing measures in 30°–60° and 45°–45° right triangles.*

Modeling with Manipulatives

MATERIALS
- compass
- protractor
- scissors
- ruler

Most people know that Alexander Graham Bell invented the telephone. But the drawings that Bell needed to get a patent were done by Lewis Howard Latimer (1848–1928), an African-American drafter and engineer.

Drafters and engineers often use a compass and straightedge to copy measurements accurately from one location to another.

Your Turn **Work with a partner.**

- ▶ Construct and cut out a square.
- ▶ Fold the square along the diagonal to form an isosceles right triangle.
- ▶ Measure each leg and each angle of the triangle.
- ▶ Use the Pythagorean theorem to determine the length of the hypotenuse.
- ▶ Use a calculator to divide the length of the hypotenuse by $\sqrt{2}$.
- ▶ Repeat the above steps for several other squares.

fold line

a. Describe a method for finding the length of the hypotenuse of a right isosceles triangle if you know the length of its legs. **The length of the hypotenuse is $\sqrt{2}$ times the length of a leg.**

Your Turn
- ▶ Construct an equilateral triangle. Cut out the triangle. Fold the triangle to form a right triangle.
- ▶ Measure each side and each angle of the right triangle.
- ▶ Repeat the above steps for several other equilateral triangles.

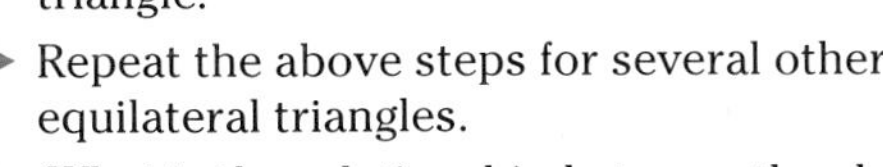

fold line

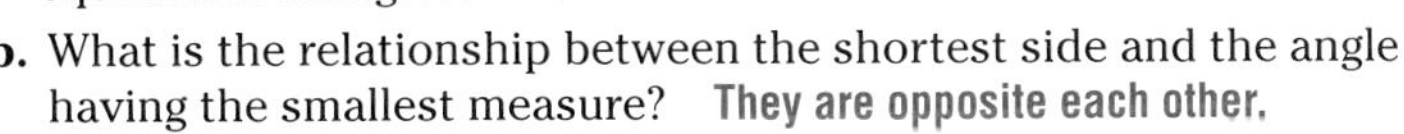

TALK ABOUT IT

b. What is the relationship between the shortest side and the angle having the smallest measure? **They are opposite each other.**

c. What is the relationship between the length of the hypotenuse and the length of the shortest side? **The hypotenuse is twice as long.**

Learning the Concept

The triangle formed in the first activity above is called a **45°–45° right triangle**. The triangle in the second activity is called a **30°–60° right triangle**. The relationships you discovered above are true for any 45°–45° or 30°–60° right triangle.

Alternative Learning Styles

Visual Point out to students that when they see a 45°– 45° right triangle has **two** congruent sides, they know that the length of the hypotenuse is the product of the length of the side and the square root of **two**. For a 30°–60° right triangle, the side adjacent to the **thirty** degree angle has a length of the product of the square root of **three** and the length of the side opposite the 30° angle.

13-5 LESSON NOTES

NCTM Standards: 1-4, 12, 13

Instructional Resources
- Study Guide Master 13-5
- Practice Master 13-5
- Enrichment Master 13-5
- Group Activity Card 13-5
- Real-World Applications, 29

Transparency 13-5A contains the 5-Minute Check for this lesson; **Transparency 13-5B** contains a teaching aid for this lesson.

Recommended Pacing	
Standard Pacing	Day 6 of 11
Honors Pacing	Day 5 of 10
Block Scheduling*	Day 3 of 6

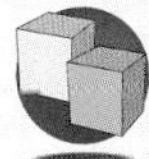

*For more information on pacing and possible lesson plans, refer to the ***Block Scheduling Booklet***.

1 FOCUS

5-Minute Check *(over Lesson 13-4)*

In a right triangle, if *a* and *b* are the measures of the legs and *c* is the measure of the hypotenuse, find each missing measure. Round answers to the nearest tenth.

1. $a = 15$ ft, $b = 17$ ft **8 ft**
2. $a = 300$ m, $c = 500$ m **400 m**
3. $b = \sqrt{7}$, $c = 4$ **3**

The measurements of the three sides of a triangle are given. Determine if each triangle is a right triangle.

4. 9 cm, 5 cm, 12 cm **no**
5. 11 in., 60 in., 61 in. **yes**

Motivating the Lesson

Questioning Draw a square and one of its diagonals on the chalkboard or overhead and ask students to describe the two triangles that are formed. **They are congruent isosceles right triangles.**

2 TEACH

In-Class Examples

For Example 1
Find the length of $\overline{EG}$ in $\triangle EFG$.

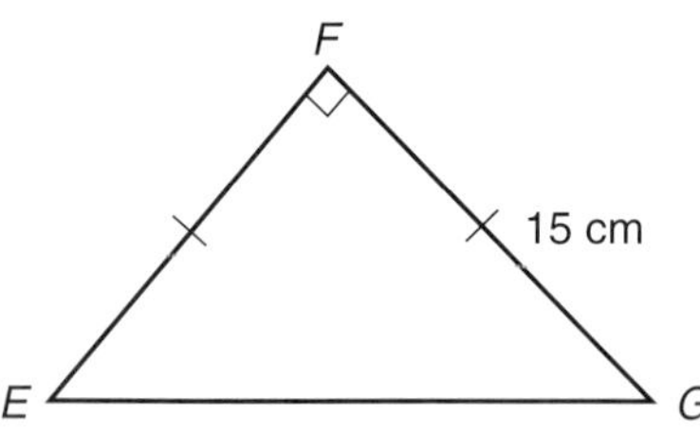

about 21.2 cm

For Example 2
Find the length of $\overline{MN}$ in $\triangle MNP$.

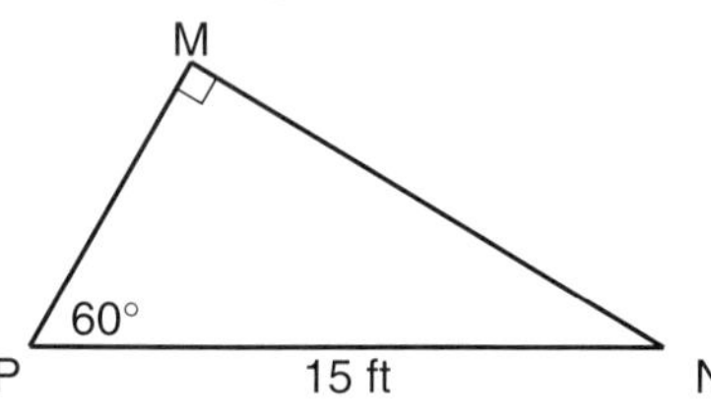

about 13.0 ft

For Example 3
A geologist looking for oil finds a likely rock formation that goes underground at a 30° angle. If the oil well is located 5500 feet from where the rock goes underground, how deep will the driller have to drill to hit the rock formation? **about 3175 feet**

Study Guide Masters, p. 112

NAME ______________________ DATE ____________

13-5 Study Guide
Special Right Triangles

Student Edition Pages 683–686

In any 30°–60° right triangle, the length of the shortest side is one-half the length of the hypotenuse.
The side opposite the 60° angle is $\sqrt{3}$ times the length of the other leg.

Example:

Find the lengths of sides a and b in the triangle below.

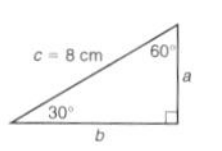

Side a is one-half of side c, 8 cm. Therefore, side $a = \frac{1}{2} \cdot 8$, or 4 cm.
Side b is $\sqrt{3}$ times side a. Thus, side $b = \sqrt{3} \cdot 4$, or about 6.9 cm.

In any 45°–45° right triangle, the length of the hypotenuse is $\sqrt{2}$ times the length of a leg.

Example:

Find the lengths of sides a and b in the triangle below.

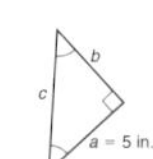

Side b is the same length as side a, 5 in. Therefore, side $c = \sqrt{2} \cdot 5$, or about 7.1 in.

Find the lengths of the missing sides in each triangle. Round decimal answers to the nearest tenth.

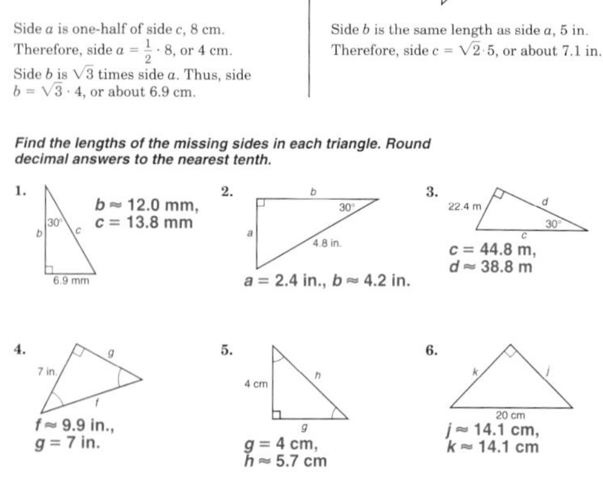

In a 45°–45° right triangle, you can find the length of the hypotenuse by multiplying the length of a leg by $\sqrt{2}$.

Example  **Find the length of $\overline{AC}$ in $\triangle ABC$.**

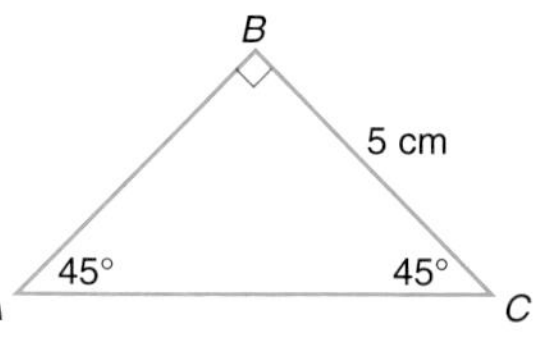

$c = a\sqrt{2}$

$c = 5\sqrt{2}$ *Replace a with 5.*

5 [×] 2 [√x] [=] 7.071067812

The length of $\overline{AC}$ is about 7.1 cm.

In a 30°–60° right triangle, the length of the side opposite the 30° angle is one-half the length of the hypotenuse.

Example  **Find the length of $\overline{PQ}$ in $\triangle PQR$.**

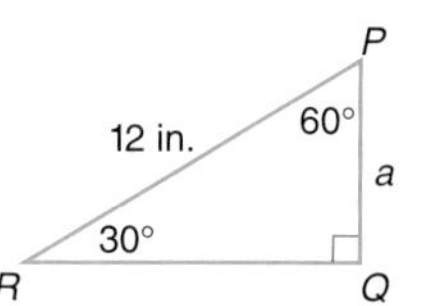

$a = \frac{1}{2}c$

$a = \frac{1}{2}(12)$ *Replace c with 12.*

$a = 6$

The length of $\overline{PQ}$ is 6 inches.

Also, in a 30°–60° right triangle, you can find the length of the side opposite the 60° angle by multiplying the length of the other leg by $\sqrt{3}$.

Example 3

APPLICATION
Construction

Large doors, like those on barns and airplane hangars, are often reinforced with a diagonal brace to prevent warping. A barn door is designed so that the brace forms a 30° angle as shown. If the width of the door is 7 feet, how high is the door?

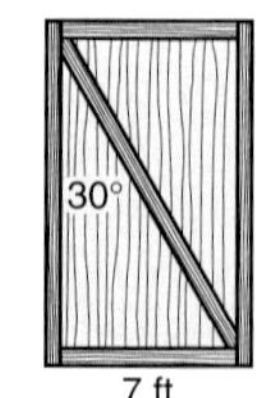

The brace forms two congruent right triangles. Draw a model of one triangle to find the measures.

Let a be the measure of the side opposite the 30° angle and b be the measure of the side opposite the 60° angle.

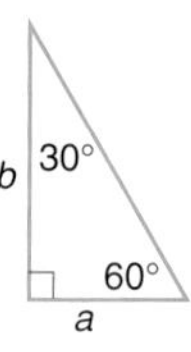

$b = a\sqrt{3}$

$b = 7\sqrt{3}$

$b \approx 12.1$

The door is about 12 feet high.

Checking Your Understanding

Communicating Mathematics

Read and study the lesson to answer each question. **1–2. See margin.**

1. **Explain** how to find the length of the hypotenuse of an isosceles right triangle without using the Pythagorean theorem.
2. **Draw** and label all the sides and angles of a 30°–60° right triangle with the shortest side having a measurement of 5 meters.

Reteaching

Using Diagrams Have students make study cards to use for reference as they complete exercises.

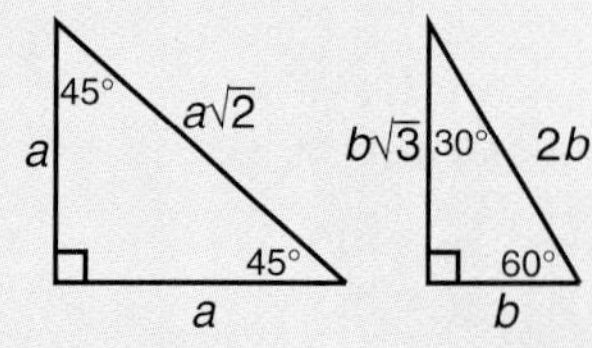

Additional Answers

1. Sample answer: Multiply the measure of a leg by $\sqrt{2}$.
2.

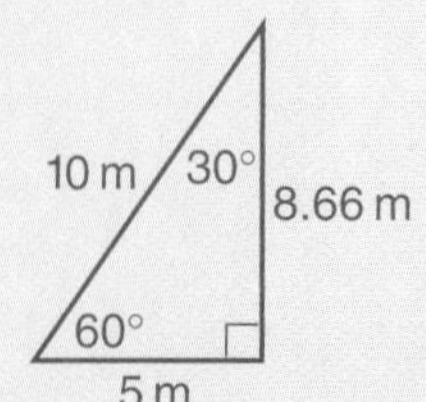

Guided Practice

The length of a leg of a 45°–45° right triangle is given. Find the length of the hypotenuse. Round decimal answers to the nearest tenth.

3. 1 yd **1.4 yd** **4.** 9.5 m **13.4 m**

The length of a hypotenuse of a 30°–60° right triangle is given. Find the length of the side opposite the 30° angle.

5. 18 m **9 m** **6.** $6\frac{1}{2}$ in. $3\frac{1}{4}$ **in.**

Find the lengths of the missing sides in each triangle. Round decimal answers to the nearest tenth.

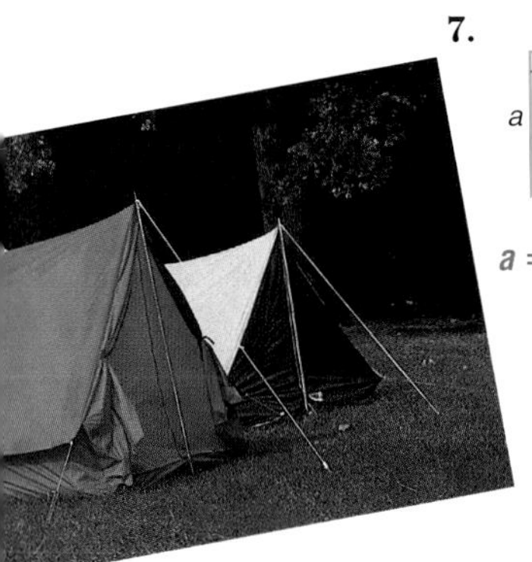

7.

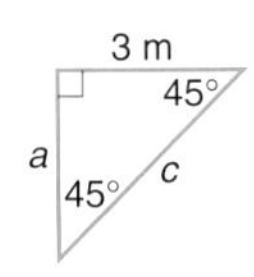

$a = 3$ m, $c \approx 4.2$ m

8.

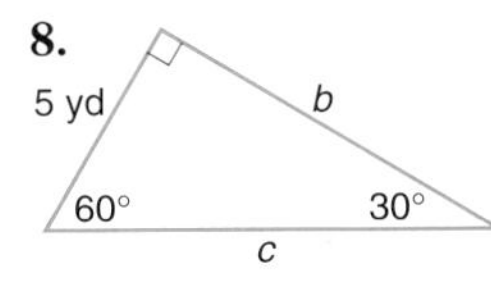

$c = 10$ yd, $b \approx 8.7$ yd

9.

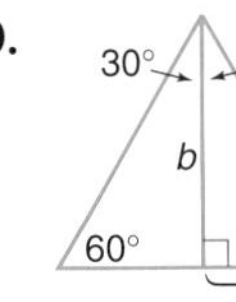

$a = 6$ m, $b = 10.4$ m

10. Camping A rope tied to the top of a tent pole makes a 60° angle with the ground and is anchored 4 feet from the base of the pole.

a. How long is the rope? **8 feet**

b. How tall is the tent pole? **about 7 feet tall**

Exercises: Practicing and Applying the Concept

Independent Practice

11. One leg of a 45°–45° right triangle is 15 centimeters long. What is the length of the other leg? **15 cm**

12. The shorter leg of a 30°–60° right triangle is 3 inches long.

a. What is the length of the hypotenuse? **6 inches**

b. What is the length of the other leg? **about 5.2 inches**

The length of a leg of a 45°–45° right triangle is given. Find the length of the hypotenuse. Round decimal answers to the nearest tenth.

13. 15 cm **21.2 cm** **14.** 7 yd **9.9 yd** **15.** 5.2 m **7.4 m**

16. $2\frac{1}{2}$ ft **3.5 ft** **17.** 6.9 in. **9.8 in.** **18.** 4.1 mm **5.8 mm**

The length of a hypotenuse of a 30°–60° right triangle is given. Find the length of the side opposite the 30° angle.

19. 48 in. **24 in.** **20.** 11 yd **5.5 yd** **21.** $\frac{1}{4}$ mi $\frac{1}{8}$ **mi**

22. 3000 m **1500 m** **23.** 4.63 cm **2.315 cm** **24.** 13 mm **6.5 mm**

Find the lengths of the missing sides in each triangle. Round decimal answers to the nearest tenth.

B

25.

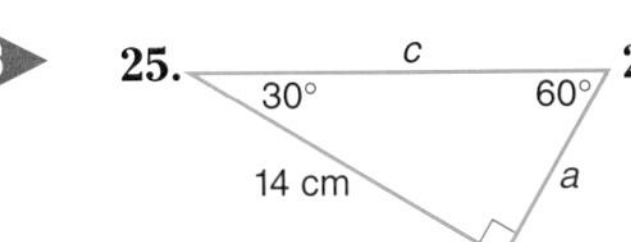

26.

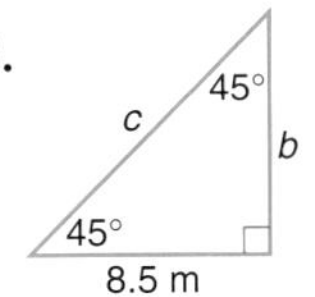

27.

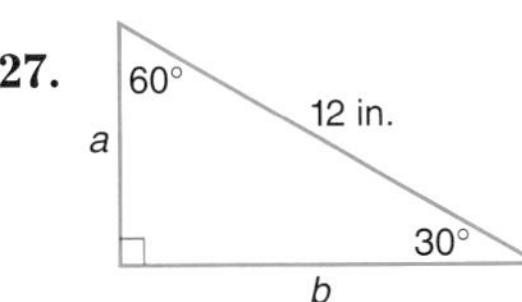

25. $a \approx 8.1$ cm, $c \approx 16.2$ cm
26. $b = 8.5$ m, $c \approx 12.0$ m
27. $a = 6$ in., $b \approx 10.4$ in.

Group Activity Card 13-5

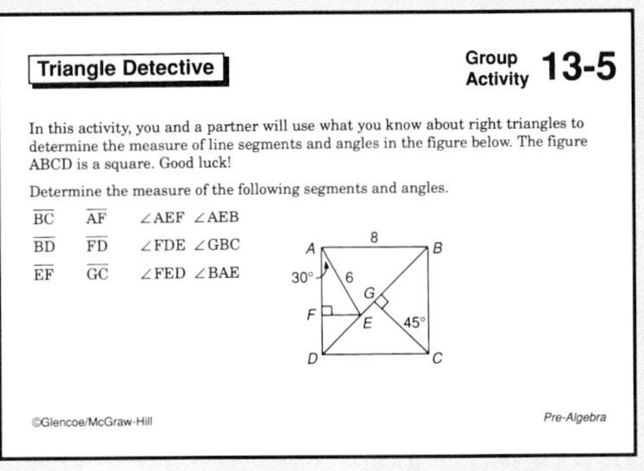

Triangle Detective Group Activity **13-5**

In this activity, you and a partner will use what you know about right triangles to determine the measure of line segments and angles in the figure below. The figure ABCD is a square. Good luck!

Determine the measure of the following segments and angles.

$\overline{BC}$ $\overline{AF}$ ∠AEF ∠AEB

$\overline{BD}$ $\overline{FD}$ ∠FDE ∠GBC

$\overline{EF}$ $\overline{GC}$ ∠FED ∠BAE

 Pre-Algebra

3 PRACTICE/APPLY

Checking Your Understanding

Exercises 1–10 are designed to help you assess your students' understanding through reading, writing, speaking, and modeling. You should work through Exercises 1–2 with your students and then monitor their work on Exercises 3–10.

Assignment Guide

Core: 11–35 odd, 36–37, 39–45
Enriched: 12–34 even, 36–45

For **Extra Practice**, see p. 772.

The red A, B, and C flags, printed only in the Teacher's Wraparound Edition, indicate the level of difficulty of the exercises.

Teaching Tip Remind students that they are working with linear measurements and the answers are in units, not square units.

Practice Masters, p. 112

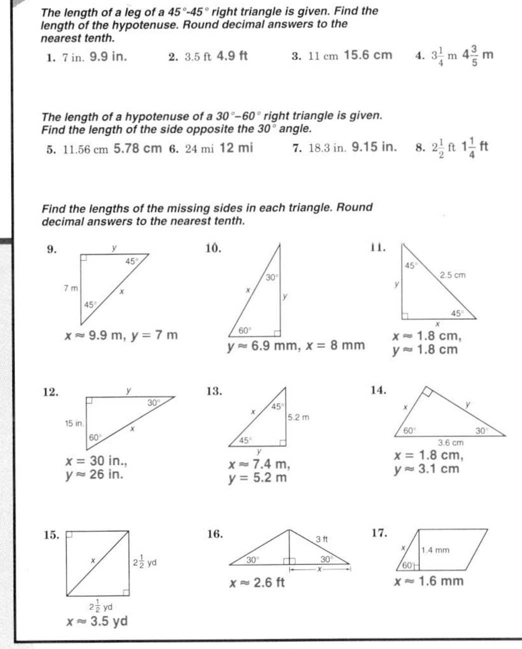

NAME ______________ DATE ________

13-5 Practice
Special Right Triangles

Student Edition Pages 683–686

The length of a leg of a 45°-45° right triangle is given. Find the length of the hypotenuse. Round decimal answers to the nearest tenth.

1. 7 in. **9.9 in.** 2. 3.5 ft **4.9 ft** 3. 11 cm **15.6 cm** 4. $3\frac{1}{4}$ m $4\frac{3}{5}$ m

The length of a hypotenuse of a 30°–60° right triangle is given. Find the length of the side opposite the 30° angle.

5. 11.56 cm **5.78 cm** 6. 24 mi **12 mi** 7. 18.3 in. **9.15 in.** 8. $2\frac{1}{2}$ ft $1\frac{1}{4}$ ft

Find the lengths of the missing sides in each triangle. Round decimal answers to the nearest tenth.

9. $x \approx 9.9$ m, $y = 7$ m
10. $y \approx 6.9$ mm, $x = 8$ mm
11. $x \approx 1.8$ cm, $y \approx 1.8$ cm
12. $x = 30$ in., $y \approx 26$ in.
13. $x \approx 7.4$ m, $y = 5.2$ m
14. $x = 1.8$ cm, $y \approx 3.1$ cm
15. $x \approx 3.5$ yd
16. $x \approx 2.6$ ft
17. $x \approx 1.6$ mm

4 ASSESS

Closing Activity

Writing Go around the room and have each student state something that all 45°–45° right triangle have in common or something that all 30°–60° right triangles have in common. It can be as simple as they all have a right angle. Have students make a list of the answers so no answer is repeated.

Additional Answer

44.

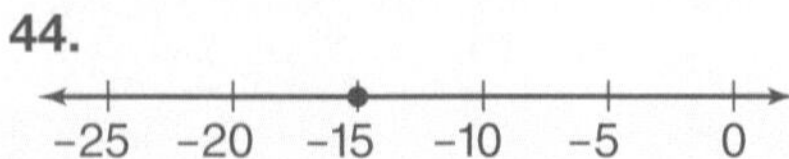

Enrichment Masters, p. 112

NAME ______________ DATE ______________

13-5 Enrichment Student Edition Pages 683–686
Reduced Triangle Principle

The following steps can be used to reduce the difficulty of a triangle problem by converting to easier triangle lengths.

Step 1 Multiply or divide the three lengths of the triangle by the same number.

Step 2 Solve for the missing side of the easier problem.

Step 3 Convert back to the original problem.

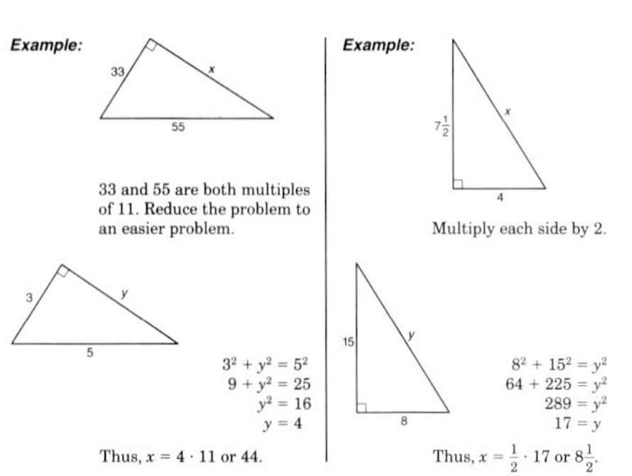
Example:

33 and 55 are both multiples of 11. Reduce the problem to an easier problem.

$3^2 + y^2 = 5^2$
$9 + y^2 = 25$
$y^2 = 16$
$y = 4$

Thus, $x = 4 \cdot 11$ or 44.

Example:

Multiply each side by 2.

$8^2 + 15^2 = y^2$
$64 + 225 = y^2$
$289 = y^2$
$17 = y$

Thus, $x = \frac{1}{2} \cdot 17$ or $8\frac{1}{2}$.

Use the reduced triangle method to find the value of x.

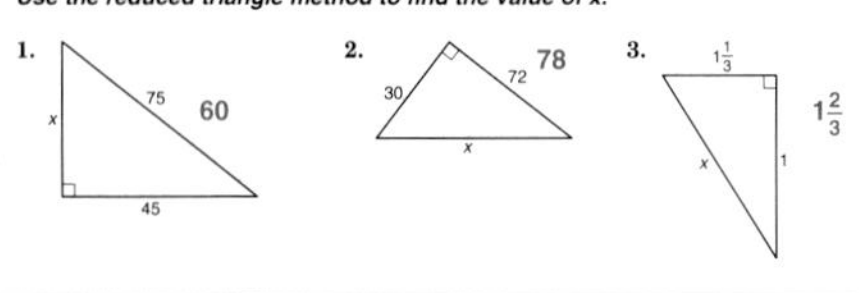

28. $a = 4$ ft, $b \approx 6.9$ ft
29. $b = 5$ yd, $c \approx 7.1$ yd
30. $b = 4$ cm; $c \approx 5.7$ cm

28.

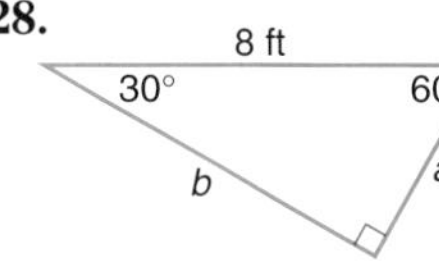

29.

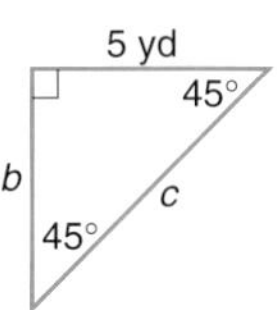

30. b, 45°, 4 cm, c, 45°

31.

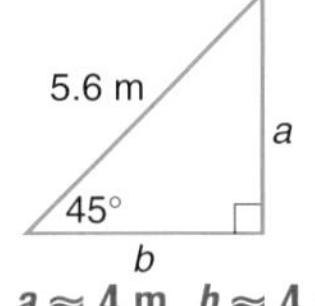

$a \approx 4$ m, $b \approx 4$ m

32. a, 60°, 3 in., b

$a = 1.5$ in., $b \approx 2.6$ in.

33. b, a, 30°, 60°, 11 km

$a = 5.5$ km, $b \approx 9.5$ km

34. The hypotenuse of a 45°–45° right triangle is 12 cm long. Find the length of each of the other sides. **about 8.5 cm each**

35. In triangle *JKL* shown at the right, the measure of $\angle LJK$ is 30°, and the altitude *h* is 5 yards long. Find the length of $\overline{KL}$. **about 5.8 yards**

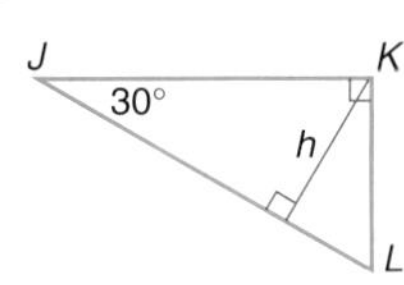

Critical Thinking

36. The area of a square is 900 square feet. Find the length of each of its diagonals. **about 42.4 ft**

Applications and Problem Solving

37. Recreation Two poles at the end of a swing set form a triangle with the ground. Each pole is 10 feet long and forms a 45° angle with the ground. What is the height of the swing set? **about 7.07 feet**

38. Construction An A-frame house is one that looks like the letter A when viewed from the front. An architect is designing an A-frame that has sides that are as long as the width of the front of the house at its base. If the building is to be 26 feet wide, what is the height of the house? **about 22.5 feet**

Mixed Review

39. Construction Harold and Marge are staking an area for their patio. The patio is to be a rectangle 20 feet long by 12 feet wide. To make sure they have the corners square, they measure the diagonals. What should the measure of both diagonals be? (Lesson 13-4) **23.3 feet**

40. Geometry Find the area of a circle with a diameter of $8\frac{1}{2}$ inches. Round to the nearest tenth. (Lesson 12-2) **56.7 in²**

41. Architecture An architect is designing a gazebo in the shape of a regular pentagon. What will be the measure of each interior angle of the gazebo? (Lesson 11-8) **108°**

42. Express $\frac{5}{6}$ as a percent. (Lesson 9-7) **83.3%**

43. Evaluate $3c^2$ if $c = -5$. (Lesson 4-2) **75**

44. Solve $f + 19 = 4$ and graph the solution on a number line. (Lesson 3-2) **−15; See margin for graph.**

45. Translate *a score not less than 50 is a win* into an inequality. (Lesson 1-9) $s \geq 50$

Extension

Using Technology Have students use a calculator to find the area and the perimeter of the figure below. Round your answer to the nearest hundredth.

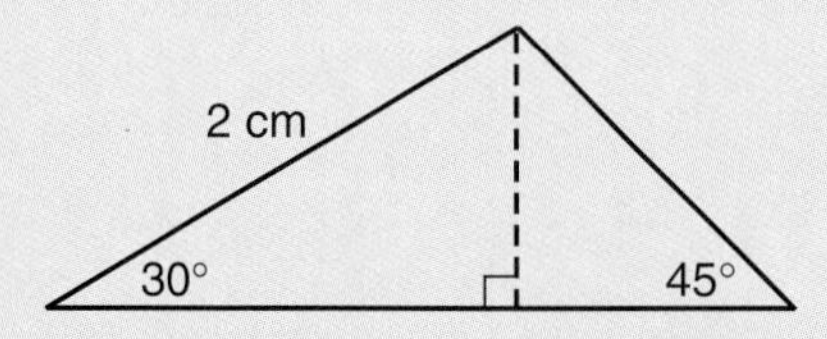

perimeter = 6.15 cm; area = 1.37 cm²

HANDS-ON ACTIVITY

13-6A Ratios in Right Triangles

A Preview of Lesson **13-6**

MATERIALS

protractor

metric ruler

Right triangles and the relationships among their sides have been studied for thousands of years.

When working with right triangles, the side **opposite** an angle is the side that is not part of the angle. In the triangle shown at the right, side r is opposite $\angle R$. The side that is not opposite an angle and not the hypotenuse is called the **adjacent** side. In the triangle at the right, side s is adjacent to $\angle R$.

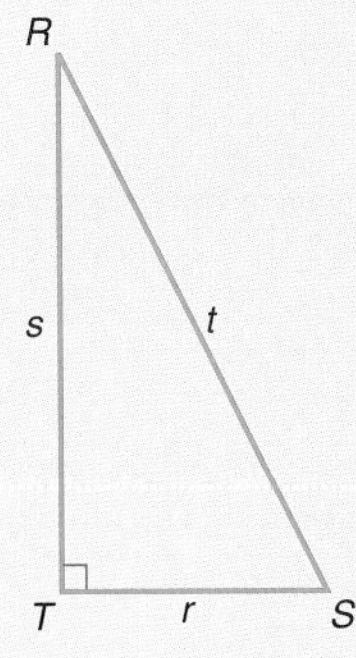

Your Turn

Work in groups of three.

- Each person should copy the table at the right.
- Each person in your group should draw a right triangle XYZ in which $m\angle X = 40°$, $m\angle Y = 50°$, and $m\angle Z = 90°$.
- Measure the leg opposite the 40° angle. Record the measurement to the nearest millimeter.

	40°angle
Length of leg opposite	
Length of leg adjacent	
Length of hypotenuse	
Ratio 1	
Ratio 2	
Ratio 3	

- Measure the leg adjacent to the 40° angle. Record the measurement to the nearest millimeter.
- Measure the hypotenuse and record the measurement to the nearest millimeter.
- Use your measurements and a calculator to find each ratio for the 40° angle. Record each ratio to the nearest hundredth.

ratio 1 = $\frac{\text{opposite leg}}{\text{adjacent leg}}$ ratio 2 = $\frac{\text{opposite leg}}{\text{hypotenuse}}$ ratio 3 = $\frac{\text{adjacent leg}}{\text{hypotenuse}}$

- Add a column to your table and repeat the above procedure for the 50° angle.

TALK ABOUT IT

1. Compare your ratios with the other members of your group. How do they compare? **The ratios are the same.**
2. Make a conjecture about the ratio of the sides of any 40°–50° right triangle. **The ratios will be the same.**
3. Repeat the activity for 25°–75° right triangles. What do you find? **The ratios of the given angles are the same no matter how long the sides are.**

13-6A LESSON NOTES

NCTM Standards: 1-4, 12, 13

Objective

Explore the relationships of the sides of right triangles.

Recommended Time

Demonstration and discussion: 15 minutes; Exercises: 30 minutes

Instructional Resources

For each student or group of students
- calculator

Student Manipulative Kit
- metric ruler

Math Lab and Modeling Math Masters
- p. 55 (worksheet)

For teacher demonstration
Overhead Manipulative Resources

1 FOCUS

Motivating the Lesson

Draw a scalene right triangle on the board. Ask students to tell you everything they know about this triangle. **Sample answers: the other two angles are complementary; the sum of the squares of the legs is equal to the square of the hypotenuse.**

2 TEACH

Teaching Tip Be sure that students understand that the size of the triangle does not affect the ratios.

3 PRACTICE/APPLY

Assignment Guide
Core: 1–3 **Enriched:** 1–3

4 ASSESS

Observing students working in cooperative groups is an excellent method of assessment.

13-6 LESSON NOTES

NCTM Standards: 1-4, 12, 13

Instructional Resources

- Study Guide Master 13-6
- Practice Master 13-6
- Enrichment Master 13-6
- Group Activity Card 13-6
- Assessment and Evaluation Masters, p. 352
- Real-World Applications, 30

Transparency 13-6A contains the 5-Minute Check for this lesson; **Transparency 13-6B** contains a teaching aid for this lesson.

Recommended Pacing	
Standard Pacing	Day 8 of 11
Honors Pacing	Day 7 of 10
Block Scheduling*	Day 4 of 6

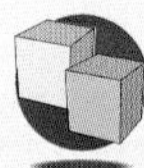

*For more information on pacing and possible lesson plans, refer to the ***Block Scheduling Booklet***.

1 FOCUS

5-Minute Check
(over Lesson 13-5)

Find the lengths of the missing sides in each triangle. Round decimal answers to the nearest tenth.

1. **9 mm**

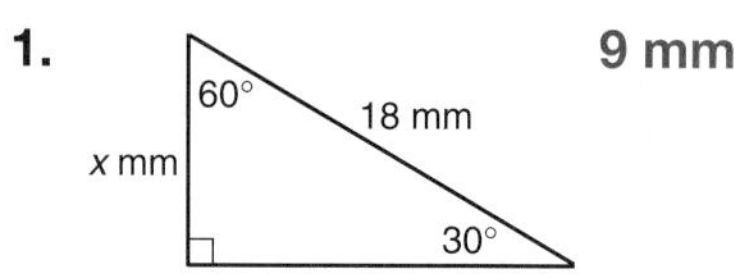

2. **≈ 4.9 ft**

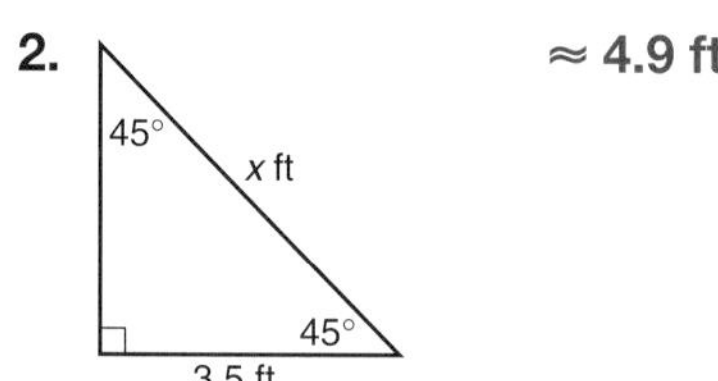

3. **≈ 5.7 m**

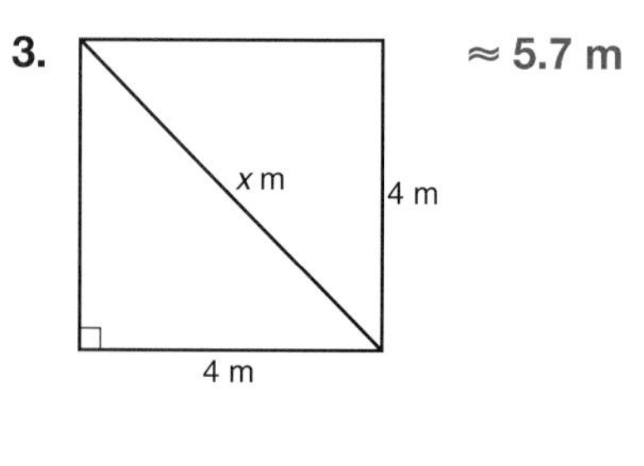

13-6 The Sine, Cosine, and Tangent Ratios

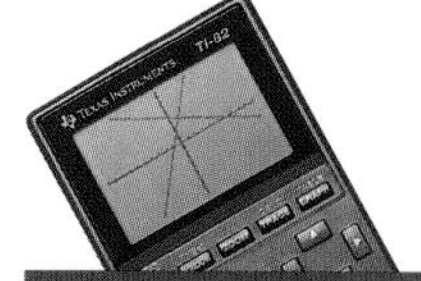

Setting Goals: *In this lesson, you'll find missing sides and angles of triangles using the sine, cosine, and tangent ratios.*

Modeling with Technology

If you know certain measures of a right triangle, ratios can be used to find the measures of the remaining parts. These ratios are used in the study of **trigonometry**.

Your Turn **Work with a partner.**

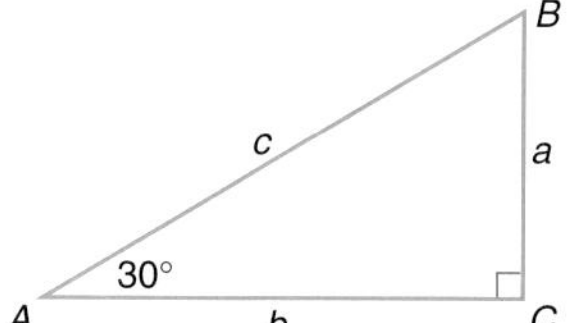

- In triangle ABC, the measure of side a is 1, the measure of side b is 1.732, and the measure of side c is 2.
- Find and record the ratio of a to c. Using a calculator, enter 30 [SIN].
- Find and record the ratio of b to c. Using a calculator, enter 30 [COS].
- Find and record the ratio of a to b. Using a calculator, enter 30 [TAN].

TALK ABOUT IT

a. The first ratio compares the leg opposite $\angle A$ to the hypotenuse. Describe the relationship between the measures of the second ratio and $\angle A$.

a–b. See margin.

b. What is the relationship between the measures of the third ratio and $\angle A$?

c. How did each ratio compare to each calculator display that followed? They were the same.

Learning the Concept

The word *trigonometry* means triangle measurement. The ratios of the measures of the sides of a right triangle are called **trigonometric ratios**.

Three common trigonometric ratios are defined below.

Definition of Trigonometric Ratios

In words: If $\triangle ABC$ is a right triangle and A is an acute angle,

$$\text{sine of } \angle A = \frac{\text{measure of the leg opposite } \angle A}{\text{measure of the hypotenuse}}$$

$$\text{cosine of } \angle A = \frac{\text{measure of the leg adjacent to } \angle A}{\text{measure of the hypotenuse}}$$

$$\text{tangent of } \angle A = \frac{\text{measure of the leg opposite } \angle A}{\text{measure of the leg adjacent to } \angle A}$$

In symbols: $\sin A = \frac{a}{c}$, $\cos A = \frac{b}{c}$, $\tan A = \frac{a}{b}$

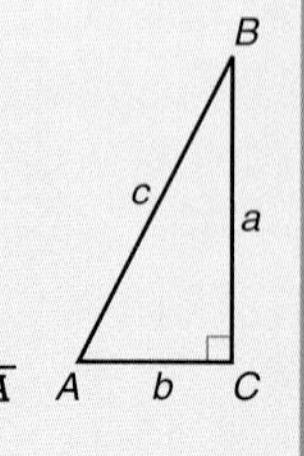

Sine, cosine, and tangent are abbreviated as sin, cos, and tan, respectively.

Additional Answers Talk About It

a. second ratio = $\frac{\text{measure of the leg adjacent to } \angle A}{\text{measure of the hypotenuse}}$

b. third ratio = $\frac{\text{measure of the leg opposite } \angle A}{\text{measure of the leg adjacent to } \angle A}$

Alternative Learning Styles

Auditory Students can use these mnemonic codes as a memory aid for the definitions of trigonometric ratios.
SOH: Sine = Opposite over Hypotenuse
CAH: Cosine = Adjacent over Hypotenuse
TOA: Tangent = Opposite over Adjacent

The value of the trigonometric ratios depends only on the measure of the angle. The size of the triangle does not affect the value of the ratio.

Example  **Find sin S, cos S, and tan S to the nearest thousandth.**

$\sin S = \frac{\text{measure of leg opposite } \angle S}{\text{measure of hypotenuse}}$

$= \frac{12}{13}$ or about 0.923

$\cos S = \frac{\text{measure of leg adjacent to } \angle S}{\text{measure of hypotenuse}}$

$= \frac{5}{13}$ or about 0.385

$\tan S = \frac{\text{measure of leg opposite } \angle S}{\text{measure of leg adjacent to } \angle S}$

$= \frac{12}{5}$ or 2.4

M, L, S; 12, 13, 5

A calculator can be used to find the sine, cosine or tangent ratio for an angle with a given degree measure.

Example  **Find each value to the nearest ten thousandth.**

a. sin 37°

37 [SIN] 0.601815023

Rounded to the nearest ten thousandth, sin 37° ≈ 0.6018.

b. cos 56°

56 [COS] 0.559192903

Rounded to the nearest ten thousandth, cos 56° ≈ 0.5592.

c. tan 72°

72 [TAN] 3.077683537

Rounded to the nearest ten thousandth, tan 72° ≈ 3.0777.

You can also use a calculator to find the degree measure of an angle if the sine, cosine, or tangent ratio is known.

Example 

a. Find the measure of $\angle S$ given that sin S = 0.3569.

.3569 [SIN⁻¹] 20.909936

The measure of $\angle S$ is about 21°.

b. Find the measure of $\angle C$ given that cos C = 0.2831.

.2831 [COS⁻¹] 73.55469014

The measure of $\angle C$ is about 74°.

c. Find the measure of $\angle T$ given that tan T = 4.703.

4.703 [TAN⁻¹] 77.99596131

The measure of $\angle T$ is about 78°.

TECHNO TIP

To use [SIN⁻¹], [COS⁻¹], and [TAN⁻¹], you have to press the [2nd] or [INV] key and then [SIN], [COS], or [TAN].

You can use the trigonometric ratios to find missing measures.

Classroom Vignette

"Have students work with a partner to play a trigonometric game. First, draw $\triangle ABC$ with right angle C, on an index card. Label each side using the letter of the opposite angle, for example, side a is opposite angle A. Then write sin A, sin B, cos A, cos B, tan A, and tan B, each on a separate card. Shuffle the cards and place face down. The first student draws a card and states the ratio for $\triangle ABC$. Continue play until all cards are drawn."

Wm D Leschensky William D. Leschensky
Author

Motivating the Lesson

Situational Problem Sketch a 30°–60° right triangle on the board. Ask a student to write the ratio of the side opposite the 30° angle to the hypotenuse on the board. $\frac{1}{2}$ Now draw a huge 30°–60° right triangle on the board and repeat the activity. **ratio = $\frac{1}{2}$** Finally, drawing a tiny 30°–60° right triangle and repeat. **ratio = $\frac{1}{2}$** Have students explain why that ratios are always the same. **The angles are the same.**

2 TEACH

In-Class Examples

For Example 1

Find sin R, cos R, and tan R. Write each answer as a ratio and then as a decimal rounded to the nearest thousandth.

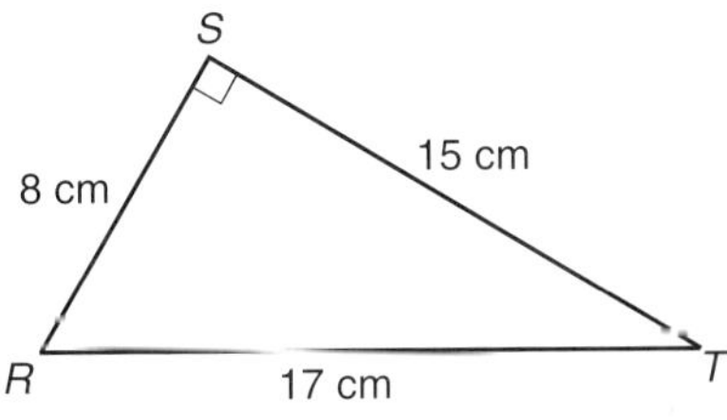

sin $R = \frac{15}{17}$ or about 0.882

cos $R = \frac{8}{17}$ or about 0.471

tan $R = \frac{15}{8}$ or 1.875

For Example 2

Find each value to the nearest ten thousandth.

a. sin 105° **0.9659**
b. cos 83° **0.1219**
c. tan 50° **1.1918**

For Example 3

a. Find the measure of $\angle N$ given that sin N = 0.2588 **≈ 15°**
b. Find the measure of $\angle Q$ given that tan Q = 0.5095 **≈ 27°**
c. Find the measure of $\angle R$ given that cos R = 0.0349 **≈ 88°**

Teaching Tip If students' calculators do not have trigonometric functions, they can use a trigonometric table. Be sure to instruct students on how to use the table.

In-Class Example

For Example 4

For safety reasons, a wheelchair ramp should make an angle of $7\frac{1}{8}°$ or less with the ground. If a ramp makes a 7° angle with the ground and connects to a platform 4 feet high, how long must the ramp be? Round to the nearest tenth. **32.8 ft**

Example 4 **Transportation**

The end of an exit ramp from an interstate highway is 22 feet higher than the highway. If the ramp is 630 feet long, what angle does it make with the highway?

Let $m\angle A = x°$

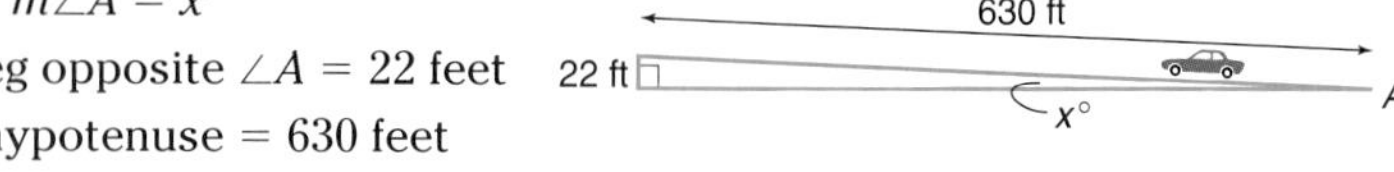

leg opposite $\angle A$ = 22 feet

hypotenuse = 630 feet

Now substitute these values into the definition of sine.

$\sin x° = \frac{22}{630}$ ← *side opposite* ← *hypotenuse*

Use a calculator to find the value of x.

22 [÷] 630 [=] [SIN⁻¹] 2.0012119

The ramp makes an angle of about 2° with the road.

3 PRACTICE/APPLY

Checking Your Understanding

Exercises 1–19 are designed to help you assess your students' understanding through reading, writing, speaking, and modeling. You should work through Exercises 1–6 with your students and then monitor their work on Exercises 7–19.

Additional Answers

4. Sample answer:

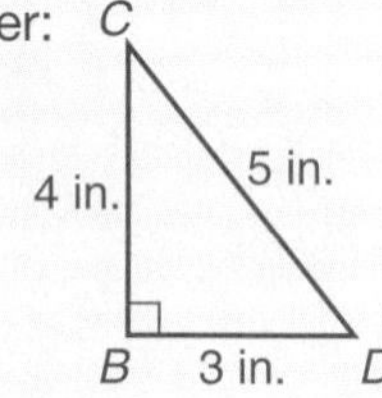

5. Derice. The value of sin 30° is constant.

Study Guide Masters, p. 113

NAME ______________ DATE ________

13-6 Study Guide
The Sine, Cosine, and Tangent Ratios

Student Edition Pages 688–692

In the triangle at the right, the measures of the sides are used to find the following **trigonometric ratios**.

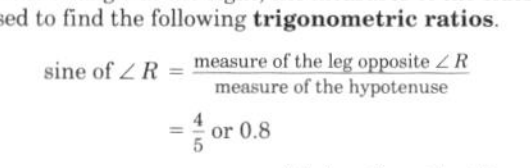
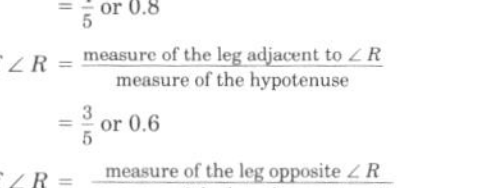

sine of $\angle R = \frac{\text{measure of the leg opposite } \angle R}{\text{measure of the hypotenuse}} = \frac{4}{5}$ or 0.8

cosine of $\angle R = \frac{\text{measure of the leg adjacent to } \angle R}{\text{measure of the hypotenuse}} = \frac{3}{5}$ or 0.6

tangent of $\angle R = \frac{\text{measure of the leg opposite } \angle R}{\text{measure of the leg adjacent to } \angle R} = \frac{4}{3}$ or about 1.333

If a trigonometric ratio is known, then the degree measure of an angle can be determined by using a calculator. For example, in the triangle above, $\sin R = 0.8$. Therefore, the measure of $\angle R$ is 0.8 [sin⁻¹] or about 53.1°.

For each triangle, find sin A, cos A, and tan A to the nearest thousandth.

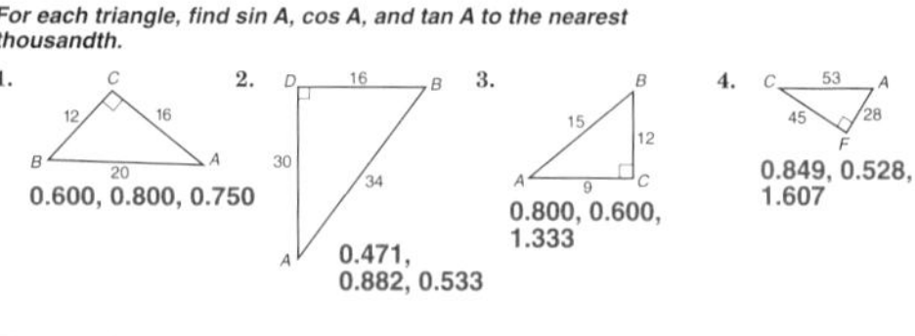

1. 0.600, 0.800, 0.750
2. 0.471, 0.882, 0.533
3. 0.800, 0.600, 1.333
4. 0.849, 0.528, 1.607

For each triangle, find the measure of the marked acute angle to the nearest degree.

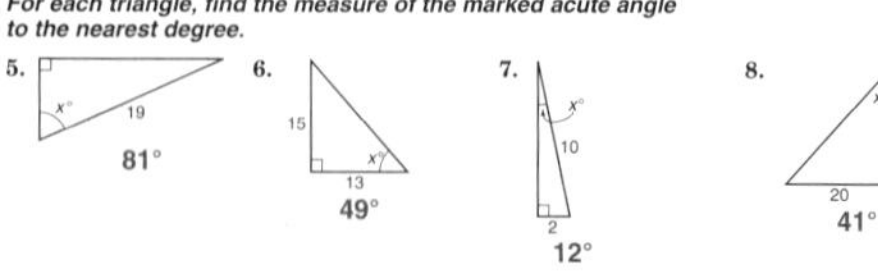

5. 81°
6. 49°
7. 12°
8. 41°

Checking Your Understanding

Communicating Mathematics

Read and study the lesson to answer each question.

1. **Write** a definition of the sine ratio.
2. **Compare and contrast** the sine ratio with the cosine ratio.
3. **Describe** the procedure for using a calculator to find the degree measure of the angle that corresponds to a given tangent ratio.

1–3. See Solutions Manual.

4. **Draw** a right triangle such that $\sin D = \frac{4}{5}$, $\cos D = \frac{3}{5}$, and $\tan D = \frac{4}{3}$.
5. **You Decide** Kelli says that the value of sin A is greater than the value of sin D. Derice says that sin A and sin D are equal. Who is correct and why?

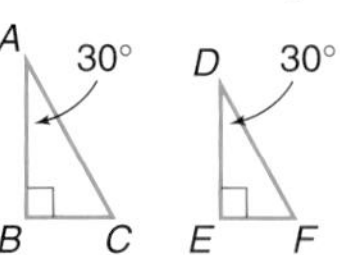

4–5. See margin.

6. *Some Old Horse - Caught A Horse - Taking Oats Away* is a helpful mnemonic device for remembering the trigonometric ratios. S, C, and T represent sin, cos, and tan respectively, while O, H, and A represent opposite, hypotenuse, and adjacent respectively. Develop your own mnemonic device for remembering the ratios. **See students' work.**

Guided Practice

Refer to $\triangle XYZ$. Express each ratio as a fraction in simplest form.

7. $\sin X$ $\frac{24}{25}$
8. $\sin Y$ $\frac{7}{25}$
9. $\cos X$ $\frac{7}{25}$
10. $\cos Y$ $\frac{24}{25}$
11. $\tan X$ $\frac{24}{7}$
12. $\tan Y$ $\frac{7}{24}$

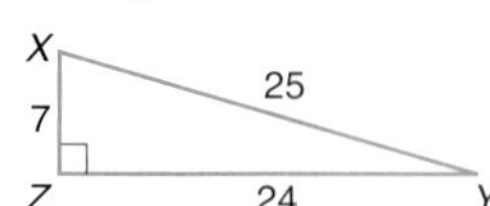

Use a calculator to find each ratio to the nearest ten thousandth.

13. cos 71° **0.3256**
14. tan 2° **0.0349**
15. sin 25° **0.4226**

Use a calculator to find the angle that corresponds to each ratio. Round to the nearest degree.

16. $\sin M = 0.4$ **24°**
17. $\cos N = 0.18$ **80°**
18. $\tan F = 0.64$ **33°**

Reteaching

Using Models Have students use modeling clay, a pencil, and string to model the following problem. A 30-foot tower needs guy wires. What length should they be if their angle with the ground is given? Solve for several different angles.

Group Activity Card 13-6

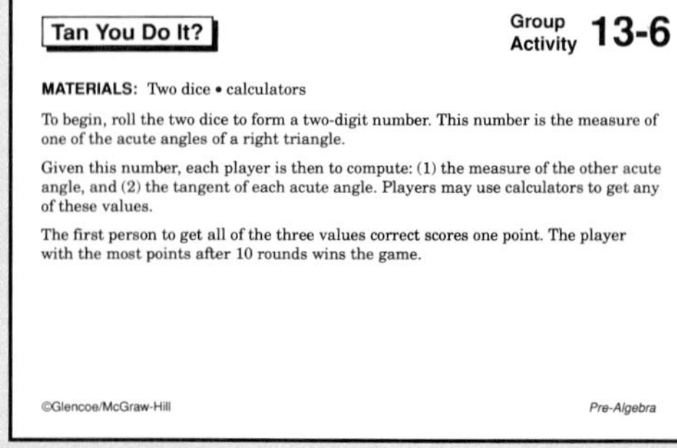

Tan You Do It? Group Activity 13-6

MATERIALS: Two dice • calculators

To begin, roll the two dice to form a two-digit number. This number is the measure of one of the acute angles of a right triangle.

Given this number, each player is then to compute: (1) the measure of the other acute angle, and (2) the tangent of each acute angle. Players may use calculators to get any of these values.

The first person to get all of the three values correct scores one point. The player with the most points after 10 rounds wins the game.

Pre-Algebra

19. **Public Access** The angle that a wheelchair ramp makes with the ground cannot exceed 6°. If a ramp needs to rise 2 feet, how long does the ramp need to be? **about 19 feet**

Exercises: Practicing and Applying the Concept

Independent Practice

For each triangle, find sin *B*, cos *B*, and tan *B* to the nearest thousandth.

20.

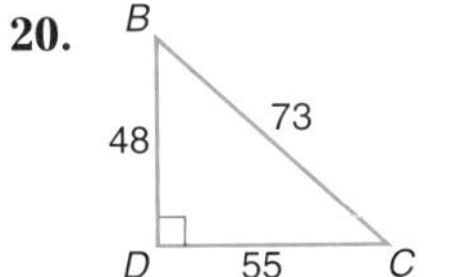

0.753, 0.658, 1.146

21. B, 29, 20, G, 21, A

0.724, 0.690, 1.05

22.

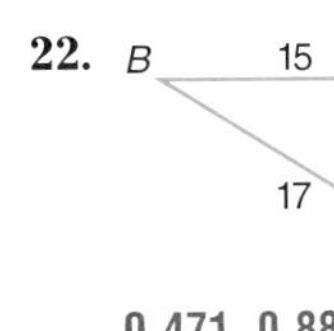

0.471, 0.882, 0.533

23. T, 10, 6, L, 8, B

0.6, 0.8, 0.75

24.

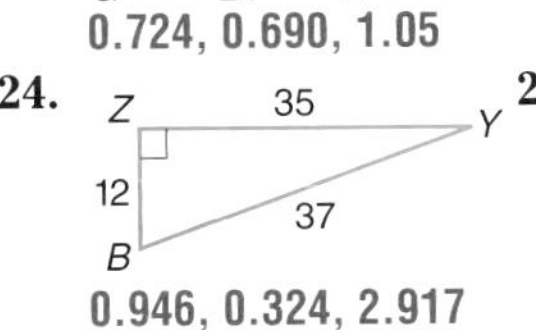

0.946, 0.324, 2.917

25. B, 63, F, 16, 65, D

0.246, 0.969, 0.254

Use a calculator to find each ratio to the nearest ten thousandth.

26. tan 45° **1.0000** 27. sin 30° **0.5000** 28. cos 60° **0.5000**

29. cos 25° **0.9063** 30. tan 31° **0.6009** 31. sin 71° **0.9455**

Use a calculator to find the angle that corresponds to each ratio. Round answers to the nearest degree.

32. tan $J = 0.6$ **31°** 33. sin $R = 0.8$ **53°** 34. cos $F = 0.866$ **30°**

35. sin $E = 0.6897$ **44°** 36. cos $B = 0.4706$ **62°** 37. tan $K = 1.8$ **61°**

For each triangle, find the measure of the marked acute angle to the nearest degree.

38.

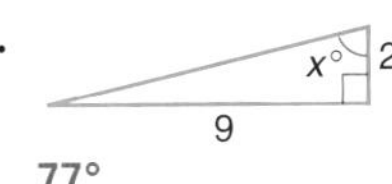

77°

39.

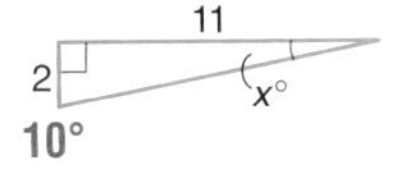

10°

40.

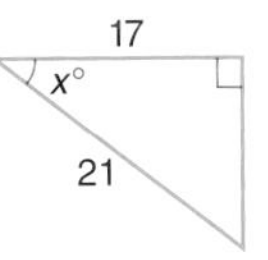

36°

41.

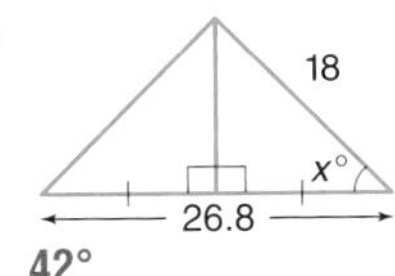

42°

42.

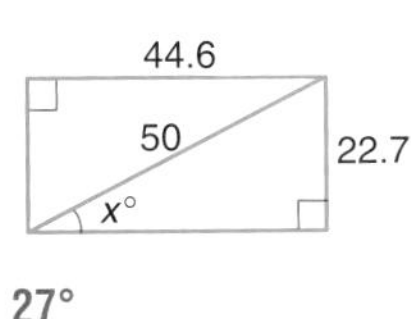

27°

43.

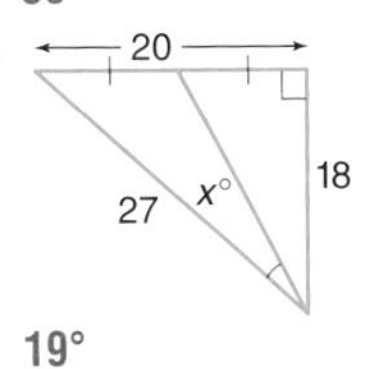

19°

Critical Thinking

44. Write a conjecture about the relationship between the sine and cosine of complementary angles. **See margin.**

Error Analysis

Students may invert or confuse the ratios associated with sine, cosine, and tangent. Have students make study cards for each ratio, including the name, the ratio, and a sketch. These study cards can be used as students complete the exercises.

Assignment Guide

Core: 21–43 odd, 44–45, 47–54
Enriched: 20–42 even, 44–54

For **Extra Practice**, see p. 772.

The red A, B, and C flags, printed only in the Teacher's Wraparound Edition, indicate the level of difficulty of the exercises.

Additional Answer

44. Sample answer: The sine of an angle is equal to the cosine of its complement.

Practice Masters, p. 113

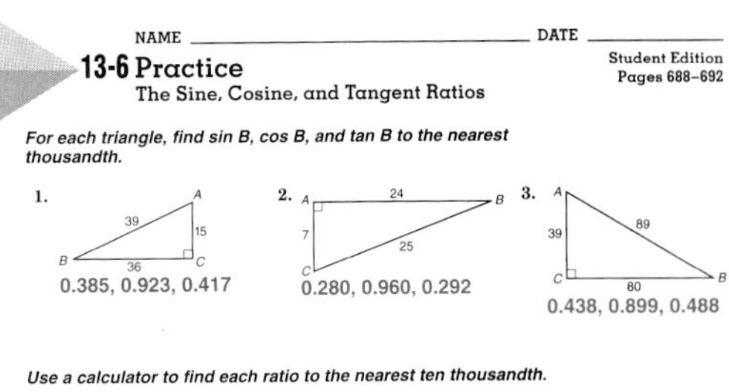

NAME ______ DATE ______

13-6 Practice
The Sine, Cosine, and Tangent Ratios

Student Edition Pages 688–692

For each triangle, find sin B, cos B, and tan B to the nearest thousandth.

1. 0.385, 0.923, 0.417 2. 0.280, 0.960, 0.292 3. 0.438, 0.899, 0.488

Use a calculator to find each ratio to the nearest ten thousandth.

4. sin 35° 0.5736	5. cos 75° 0.2588	6. tan 10° 0.1763
7. cos 34° 0.8290	8. tan 27° 0.5095	9. sin 56° 0.8290
10. tan 48° 1.1106	11. sin 15° 0.2588	12. cos 65° 0.4226
13. sin 89° 0.9998	14. cos 19° 0.9455	15. tan 76° 4.0108

Use a calculator to find the angle that corresponds to each ratio. Round answers to the nearest degree.

16. tan $D = 0.3443$ 19°	17. cos $S = 0.9962$ 5°	18. sin $L = 0.1219$ 7°
19. cos $B = 0.9063$ 25°	20. sin $E = 0.9962$ 85°	21. tan $J = 0.0875$ 5°
22. sin $T = 0.4226$ 25°	23. tan $M = 2.9042$ 71°	24. cos $I = 0.5446$ 57°
25. tan $U = 1.4281$ 55°	26. cos $K = 0.4695$ 62°	27. sin $N = 0.5446$ 33°

For each triangle, find the measure of the marked acute angle to the nearest degree.

28. 12, 17, x° 45°

29.

69°

30.

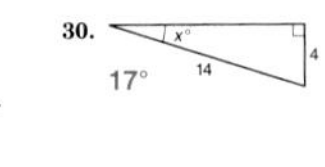

17°

GLENCOE Technology

Interactive Mathematics Tools Software

In these interactive computer lessons, students explore tangent, sine and cosine ratios. A **Computer Journal** gives students the opportunity to write about what they have learned.

For Windows & Macintosh

Extension

Connections Have students assume they are in charge of installing playground equipment at a local elementary school. Have them explain how they would use the sine, cosine, and tangent ratios to determine the height of the slide, the frame of the swings, and the shape of the ball diamond.

4 ASSESS

Closing Activity

Writing Have students draw a right triangle and label it as triangle *QRS* with $\angle R$ being the right angle. Have students write the ratios for the sine, cosine and tangent of $\angle Q$ and then $\angle S$.

Chapter 13, Quiz C (Lessons 13-5 and 13-6) is available in the *Assessment and Evaluation Masters,* p. 352.

Additional Answer

53. 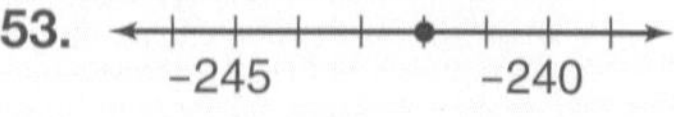

Enrichment Masters, p. 113

NAME ______________________ DATE ____________

13-6 Enrichment
Perspective Drawings

Student Edition
Pages 688–692

The perspective drawing is very realistic. It is based on the observation that objects seem smaller the farther away they are. Parallel lines seem to meet and disappear at a point on the horizon.

The point where all lines seem to meet is called the **vanishing point**. The vanishing point can be in the center of vision, to the right, or to the left of vision.

Perspective drawing with vanishing point to the left

Step 1 Draw the horizon line and locate the vanishing point.

Step 2 Draw vertical line segment representing the height of the object.

Step 3 Complete angular view of the object.

Step 4 Draw light lines toward the vanishing point. Assume depth of the object.

If the vanishing point is in the center of vision, then draw the front view of the object in Step 3.

Locate the vanishing point.

1. 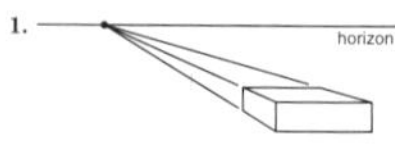

2. 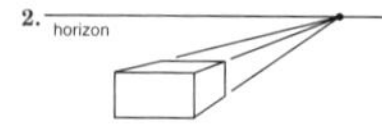

Create your own perspective drawings. Answers will vary.

Applications and Problem Solving

45. **Scale Drawing** The Leaning Tower of Pisa is 55 meters tall and tilts 5 meters off the perpendicular. Maria wants to make a scale drawing for a project in history class. What should be the measure of the angle she draws to represent the tilt of the tower? **about 84.8°**

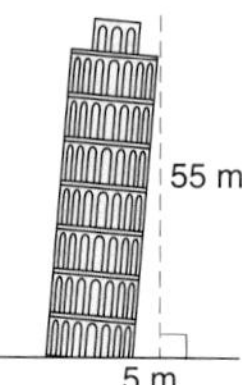

46. **Transportation** The steepest streets in the United States are Filbert Street and 22nd Street in San Francisco. They rise 1 foot for every 3.17 feet of horizontal distance. Find the angle these streets form with the horizontal. **about 17.5°**

Mixed Review

47. The hypotenuse of a 30°–60° right triangle is 20 yards long. What is the length of the side opposite the 30° angle? (Lesson 13-5) **10 yd**

48. **Construction** A rectangular patio is designed to have a concrete border around a tiled area. The tiled area is to be rectangular, and the border is to extend 2 feet around the tile on all sides. The entire patio will be 24 feet by 14 feet. What is the area to be covered by tile? (Lesson 12-4) **200 ft²**

49. There are 3 routes between Jean's house and the school. How many ways can Jean go from her house to the school and back home again? (Lesson 10-5) **9 ways**

50. How many milligrams are in 2.5 grams? (Lesson 7-8) **2500**

51. **Geometry** If two angles of a triangle have the same measure, the triangle is an isosceles triangle. The triangle shown at the right is an isosceles triangle. Is this an example of inductive or deductive reasoning? (Lesson 5-8) **deductive**

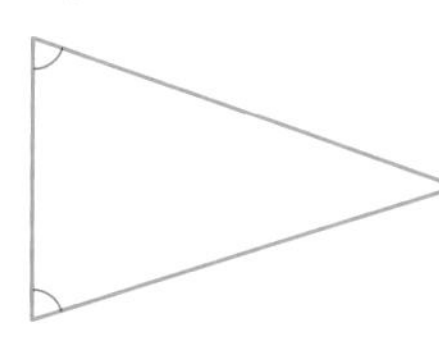

52. Find the product of $(2b^3)$ and $(-6b^4)$. (Lesson 4-8) $\mathbf{-12b^7}$

53. Solve $22 = \frac{n}{-11}$. Then graph the solution on a number line. (Lesson 3-3) **−242; See margin for graph.**

54. Is $4 > 16$ *true*, *false*, or *open*? (Lesson 1-9) **false**

COOPERATIVE LEARNING PROJECT

THE SHAPE OF THINGS TO COME

Thought-Controlled Computer Technology

Have you ever heard the expression "It's the thought that counts?" A new computer accessory will make that expression more than an adage. Experimental work has been done to translate thoughts into digital impulses that can be read by a computer. A sensor is placed on your finger and then attached to a personal computer. Minute physiological responses are measured through the skin and the computer responds onscreen. Some applications of the technology include action games that can be played without a joystick, art and music software that adjust colors and tempos according to thoughts, and memory and learning games.

See for Yourself

Research thought-controlled computer technology. **See students' work.**

- When will the technology be commercially available?
- Is research being done on applications of the technology for things other than games?

COOPERATIVE LEARNING PROJECT

THE SHAPE OF THINGS TO COME

"Mood sensing" devices have been around for a while. Lie detectors or polygraphs analyze a person's physiological responses to questions.

GRAPHING CALCULATOR ACTIVITY

13-6B Slope and Tangent

An Extension of Lesson **13-6**

Be sure to turn off any functions in the Y= list and any statistical plots before starting to draw.

You can use a graphing calculator to draw line segments. The right triangle shown was drawn in the viewing window [0, 14] by [0, 10] with a scale factor of 1 on both axes. The vertices are (0, 3), (6, 3), and (6, 6).

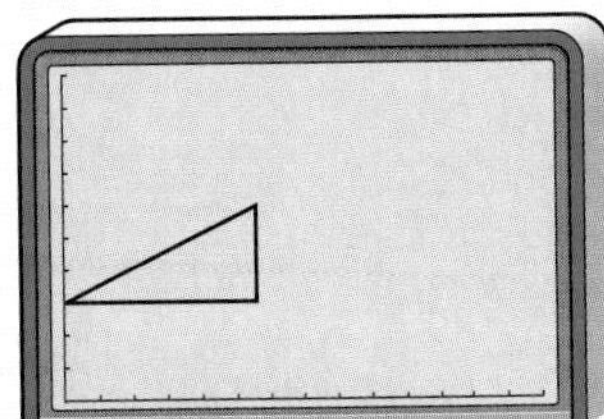

Enter: [2nd] [DRAW] 2 0 [,] 3 [,] 6 [,] 3 [)] [ENTER] [CLEAR]
[2nd] [DRAW] 2 6 [,] 3 [,] 6 [,] 6 [)] [ENTER] [CLEAR]
[2nd] [DRAW] 2 6 [,] 6 [,] 0 [,] 3 [)] [ENTER]

To find the measure of the angle formed by the horizontal line and the hypotenuse, you can use the tangent ratio. The leg opposite the angle is 3 units long, and the leg adjacent to the angle is 6 units long.

$$\tan a = \frac{\text{leg opposite}}{\text{leg adjacent}}$$
$$= \frac{3}{6} \text{ or } 0.5$$

Use a calculator to find a.

Enter: [2nd] [TAN⁻¹] .5 [ENTER] 26.56505118

$a \approx 26.6°$

Your Turn

- Find the slope of the hypotenuse of the triangle shown at the beginning of the lab. $\frac{1}{2}$
- Draw the triangle shown above on a graphing calculator. Then graph the line that contains the hypotenuse, $y = 0.5x + 3$. **See students' work.**

Enter: [Y] .5 [X,T,θ] [+] 3 [GRAPH]

TALK ABOUT IT

1. What is the slope of the line? $\frac{1}{2}$
2. What is the slope of the hypotenuse? $\frac{1}{2}$
3. How does the slope of the hypotenuse compare to the tangent of the angle formed by the hypotenuse and the horizontal leg? **3. They are the same.**
4. Graph $y = 3x + 4$ on your graphing calculator. What is the slope of the line? **3**
5. Sketch the right triangle formed by the graph of $y = 3x + 4$, the x-axis, and the y-axis. Label it $\triangle ABC$, so that $\angle A$ is between the x-axis and the graph of $y = 3x + 4$. What is $\tan A$? **3**
6. What is the measure of the angle that the graph of $y = 3x + 4$ forms with the x-axis? **about 71.6°**

Technology

This lesson offers an excellent opportunity for using technology in your pre-algebra classroom. For more information on using technology, see *Graphing Calculators in the Mathematics Classroom*, one of the titles in the Glencoe Mathematics Professional Series.

13-6B LESSON NOTES

NCTM Standards: 1-4, 12, 13

Objective

Use a graphing calculator to explore the slope of a line.

Recommended Time

15 minutes

Instructional Resources

Graphing Calculator Masters, p. 26

This master provides keystroking instruction for this lesson for the TI-81 and Casio graphing calculators.

1 FOCUS

Motivating the Lesson

Have students draw a right triangle in Quadrant I on graph paper. Tell them to draw any right triangle, but the legs must be parallel to the axes. Have students find the slope of the hypotenuse of their triangle.

2 TEACH

Teaching Tip Remind students that they studied the slope of a line in Lesson 8-6. Review the definition of slope if necessary.

3 PRACTICE/APPLY

Assignment Guide

Core: 1–6
Enriched: 1–6

4 ASSESS

Observing students working with technology is an excellent method of assessment.

13-7 LESSON NOTES

NCTM Standards: 1-4, 12, 13

Instructional Resources

- Study Guide Master 13-7
- Practice Master 13-7
- Enrichment Master 13-7
- Group Activity Card 13-7
- Assessment and Evaluation Masters, p. 352
- Activity Masters, p. 27
- Graphing Calculator Masters, p. 13
- Multicultural Activity Masters, p. 26

Transparency 13-7A contains the 5-Minute Check for this lesson; **Transparency 13-7B** contains a teaching aid for this lesson.

Recommended Pacing	
Standard Pacing	Day 9 of 11
Honors Pacing	Day 8 of 10
Block Scheduling*	Day 5 of 6

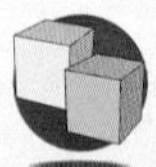

*For more information on pacing and possible lesson plans, refer to the ***Block Scheduling Booklet***.

1 FOCUS

5-Minute Check
(over Lesson 13-6)

Use a calculator to find each ratio to the nearest ten thousandth.

1. tan 76° **4.0108**
2. sin 33° **0.5446**
3. cos 5° **0.9962**

Use a calculator to find the angle that corresponds to each ratio. Round to the nearest degree.

4. sin $M = 0.9925$ **83°**
5. tan $E = 1.8040$ **61°**

Motivating the Lesson

Questioning Show students a picture of the Washington Monument. Ask them how they could determine its height.

13-7 Using Trigonometric Ratios

Setting Goals: *In this lesson, you'll solve problems by using the trigonometric ratios.*

MATHEMATICS MINI LAB

Modeling with Manipulatives

MATERIALS

protractor
index card
straw
paper clip (large)
string
tape

Many problems that can be solved using trigonometric ratios deal with angles of elevation. An **angle of elevation** is formed by a horizontal line and a line of sight above it.

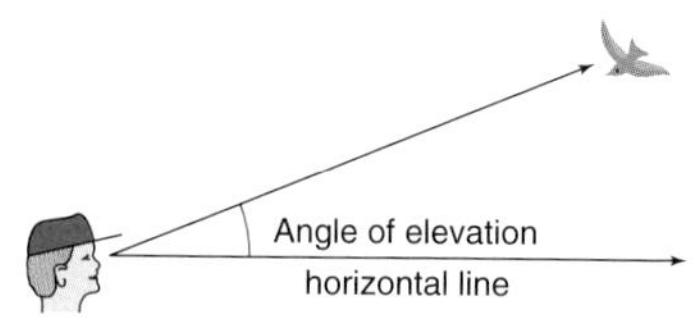

A hypsometer is an instrument that measures angles of elevation. You can make a model of a hypsometer.

Your Turn

Work with a partner.

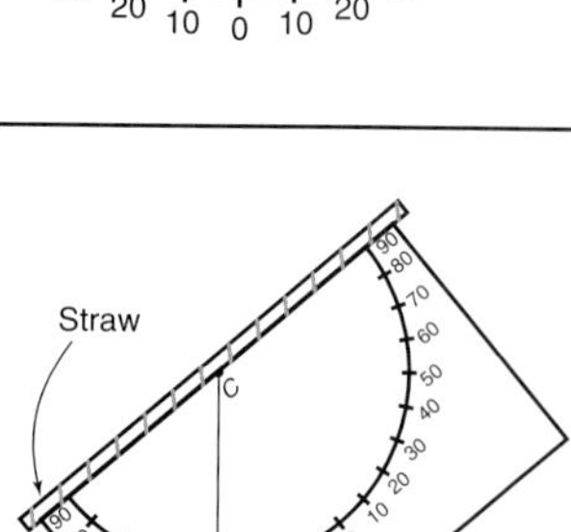

- ▶ Tape a protractor on an index card so that both zero points align with the edge of the card. Mark the center of the protractor on the card and label it *C*. Then mark and label every 10° on the card like the one shown at the right.
- ▶ Tie a piece of string to a large paper clip. Attach the other end of the string to the index card at *C*.
- ▶ Tape a straw to the edge of the index card that contains *C*.
- ▶ To use the hypsometer, look through the straw at the object. Have your partner read the angle from the scale.
- ▶ If possible, use your hypsometer to find the angle of elevation to a tree on your school property. Also measure the distance from the hypsometer to the ground and from the hypsometer to the base of the object.

Straw
String
Paper Clip

a. Draw a diagram of the situation with the hypsometer and the tree. If you do not have your own measurements, use 35° as the angle of elevation, 4 feet as the distance from the hypsometer to the ground, and 46 feet as the distance from the hypsometer to the tree base.

b. How could you find the height of a tree? Solve.

a–b. See margin.

Additional Answers Talk About It

a.

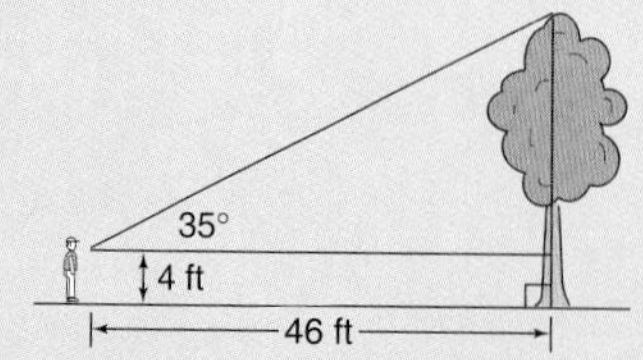

b. See students' work. If the given measures are used, the height of the tree is about 36.2 feet.

Tech Prep

Crop Duster Crop dusters fly small airplanes very close to the ground and must be able to calculate locations accurately. They use a lot of trigonometry when making these calculations.

For more information on tech prep, see the Teacher's Handbook.

Learning the Concept

The trigonometric ratios and the Pythagorean theorem can be used to find the measure of any side or angle of a right triangle if the measure of one side and any other side or acute angle are known.

Example  **Find the measure of $\angle J$ in $\triangle JKL$.**

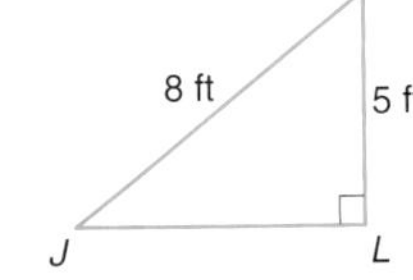

The measure of the leg opposite $\angle J$ and the measure of the hypotenuse are known. Use the sine ratio.

$$\sin J = \frac{\text{measure of leg opposite } J}{\text{measure of hypotenuse}}$$

$$\sin J = \frac{5}{8}$$ *Replace the measure of the leg opposite J with 5 and the measure of the hypotenuse with 8.*

$$\sin J = 0.625$$

$$J \approx 38.7$$ *Use a calculator.*

The measure of $\angle J$ is about 38.7°.

A person on a tower or a cliff must look down to see an object below. An **angle of depression** is formed by a horizontal line and another line of sight below it.

Example 2

APPLICATION

Fire Fighting

A fire is sighted from a fire tower in the Black Hills National Forest. Using a hypsometer, the ranger found the angle of depression to be 2°. If the ranger's eyes are 125 feet above the ground, how far is the fire from the base of the tower?

Let d represent the distance from the fire to the base of the fire tower.

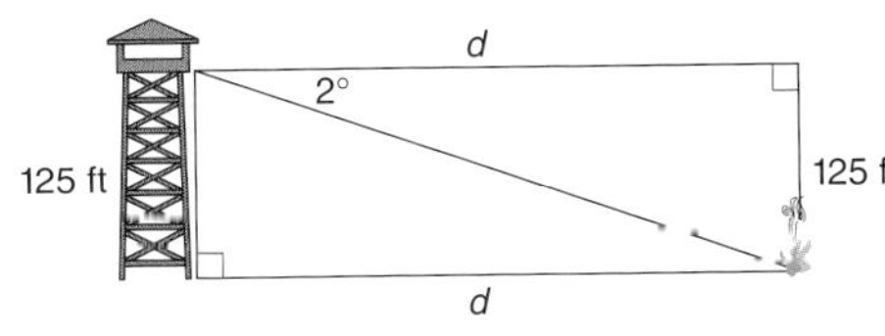

In the diagram, a line from the fire perpendicular to the horizontal line of sight forms a rectangle. Therefore, both lengths can be labeled d. Notice that 125 feet is the length of the leg opposite the 2° angle and d is the length of the leg adjacent to the 2° angle. The tangent ratio should be used to solve this problem.

$$\tan 2° = \frac{\text{measure of leg opposite}}{\text{measure of leg adjacent}}$$

$$\tan 2° = \frac{125}{d}$$ *Replace the measure of the leg opposite with 125 and the measure of the leg adjacent with d.*

$$d \tan 2° = 125$$ *Multiply each side by d.*

$$d = \frac{125}{\tan 2°}$$ *Divide each side by tan 2°.*

$$d \approx 3580$$ *Use a calculator.*

The fire is about 3580 feet from the base of the tower.

2 TEACH

In-Class Examples

For Example 1

a. In $\triangle XYZ$, find the length of $\overline{XY}$. Round to the nearest tenth. **about 1.6 ft**

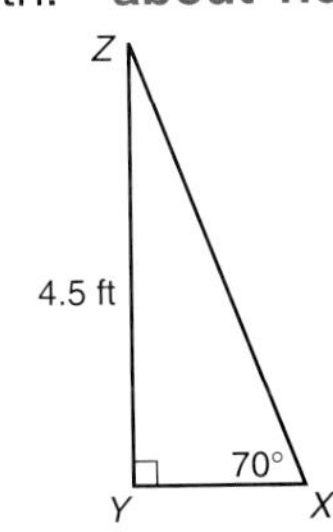

b. Find the measure of $\angle M$ in $\triangle MNP$. Round to the nearest tenth. **about 55.2°**

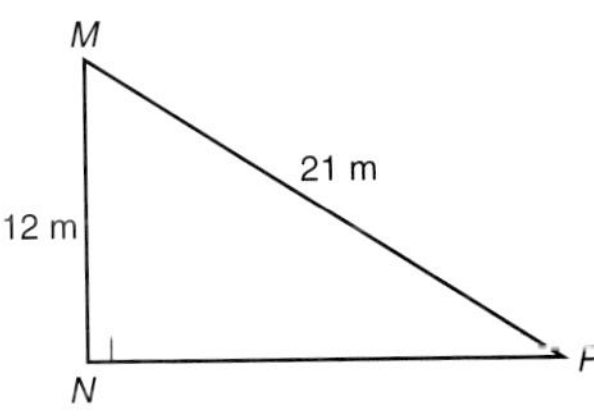

For Example 2

You want to build a walking ramp from your porch to the ground. Safety experts say that the angle should not be more than $16\frac{2}{3}°$. If your porch is 3.5 feet off the ground, how long will your walking ramp have to be? Round to the nearest tenth. **about 12.2 feet**

Study Guide Masters, p. 114

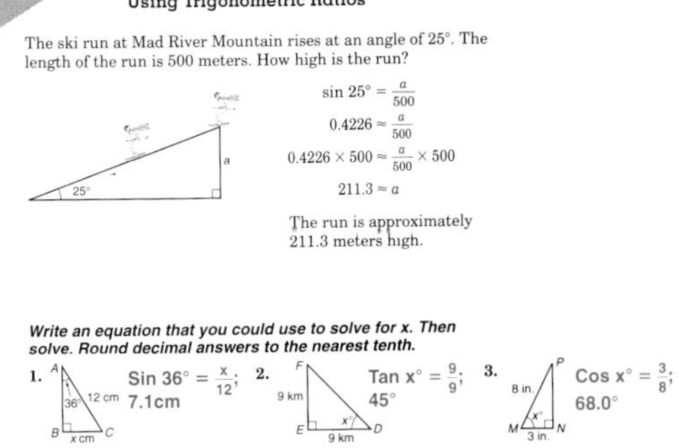

NAME ______________ DATE ______________

13-7 Study Guide
Using Trigonometric Ratios

Student Edition
Pages 694–697

The ski run at Mad River Mountain rises at an angle of 25°. The length of the run is 500 meters. How high is the run?

$$\sin 25° = \frac{a}{500}$$

$$0.4226 \approx \frac{a}{500}$$

$$0.4226 \times 500 \approx \frac{a}{500} \times 500$$

$$211.3 \approx a$$

The run is approximately 211.3 meters high.

Write an equation that you could use to solve for x. Then solve. Round decimal answers to the nearest tenth.

1. $\sin 36° = \frac{x}{12}$; 7.1cm
2. $\tan x° = \frac{9}{9}$; 45°
3. $\cos x° = \frac{3}{8}$; 68.0°

Use trigonometric ratios to solve each problem.

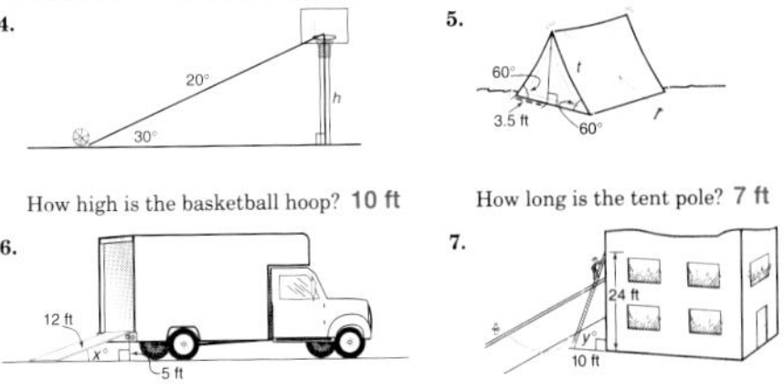

4. How high is the basketball hoop? 10 ft
5. How long is the tent pole? 7 ft
6. What is the angle of the ramp? 25°
7. What is the angle of the ladder? 67°

Cooperative Learning

Brainstorming Have students brainstorm to complete this activity. Use road maps of your state. Find your location and the location of a point of interest in another part of the state. Using the scale, find vertical and horizontal displacements from one location to the other. Calculate the direction and displacement from one location to the other.

For more information on this strategy, see *Cooperative Learning in the Mathematics Classroom*, one of the titles in the Glencoe Mathematics Professional Series, p. 30.

3 PRACTICE/APPLY

Checking Your Understanding

Exercises 1–7 are designed to help you assess your students' understanding through reading, writing, speaking, and modeling. You should work through Exercises 1–3 with your students and then monitor their work on Exercises 4–7.

Assignment Guide

Core: 9–25 odd, 27–31
Enriched: 8–22 even, 23–31

For **Extra Practice**, see p. 772.

The red A, B, and C flags, printed only in the Teacher's Wraparound Edition, indicate the level of difficulty of the exercises.

Additional Answer

2. If you know the measure of the hypotenuse and the measure of the side opposite the angle, use the sine ratio. If you know the measure of the hypotenuse and the measure of the side adjacent to the angle, use the cosine ratio. If you know the measures of the two legs, use the tangent ratio.

Practice Masters, p. 114

NAME ______________________ DATE ____________

13-7 Practice
Using Trigonometric Ratios

Student Edition Pages 694–698

Write an equation that you could use to solve for x. Then solve. Round decimal answers to the nearest tenth.

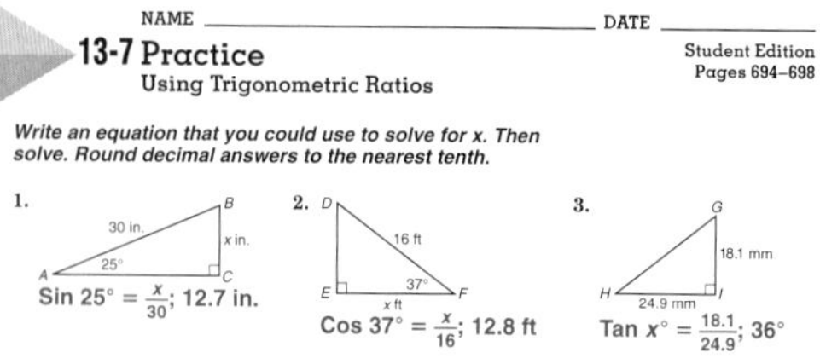

1. $\text{Sin } 25° = \frac{x}{30}$; 12.7 in.
2. $\text{Cos } 37° = \frac{x}{16}$; 12.8 ft
3. $\text{Tan } x° = \frac{18.1}{24.9}$; 36°

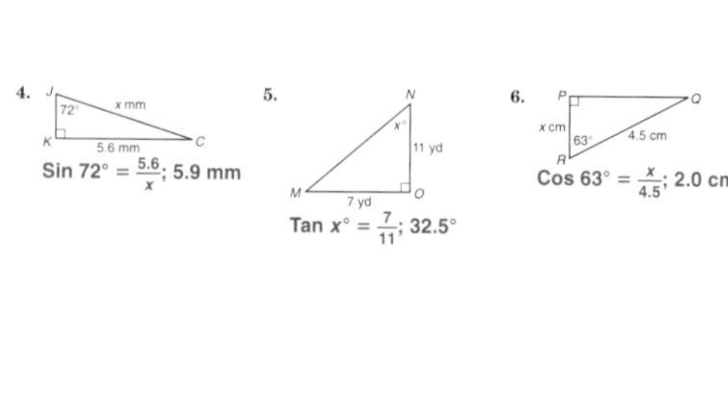

4. $\text{Sin } 72° = \frac{5.6}{x}$; 5.9 mm
5. $\text{Tan } x° = \frac{7}{11}$; 32.5°
6. $\text{Cos } 63° = \frac{x}{4.5}$; 2.0 cm

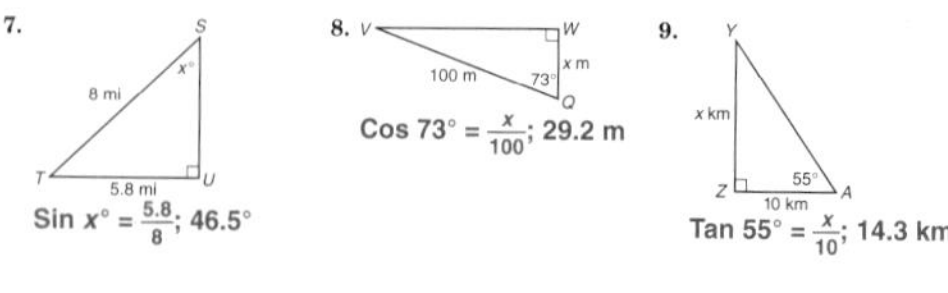

7. $\text{Sin } x° = \frac{5.8}{8}$; 46.5°
8. $\text{Cos } 73° = \frac{x}{100}$; 29.2 m
9. $\text{Tan } 55° = \frac{x}{10}$; 14.3 km

Checking Your Understanding

Communicating Mathematics

Read and study the lesson to answer each question. 2. See margin.

1. **Draw** a diagram and write a few sentences to explain what is meant by an angle of elevation. See students' work.
2. **Explain** how to decide whether to use sine, cosine, or tangent when you are finding the measure of an acute angle in a right triangle.

3. **Assess Yourself** Describe a measure in your home or school that you could find using an angle of elevation. If possible, use the hypsometer you made at the beginning of the lesson to find the measure. Then write a few sentences explaining how you found the height. See students' work.

Guided Practice

Write an equation that you could use to solve for *x*. Then solve. Round decimal answers to the nearest tenth.

4. $\sin 46° = \frac{x}{15}$; 10.8
5. $\tan 39° = \frac{10}{x}$; 12.3
6. $\cos x° = \frac{12}{16}$; 41.4

4.

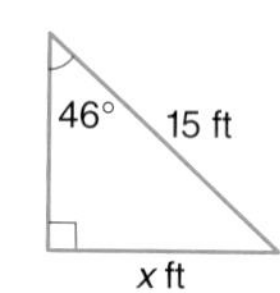

5.

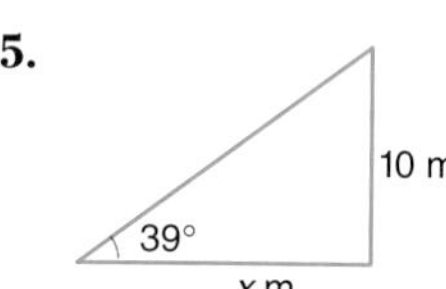

6.

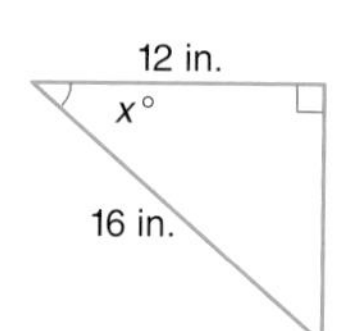

7. **Home Maintenance** A painter props a 20-foot ladder against a house. The angle it forms with the ground is 65°. How far up the side of the house does the ladder reach? about 18.1 feet

Exercises: Practicing and Applying the Concept

Independent Practice

Write an equation that you could use to solve for *x*. Then solve. Round decimal answers to the nearest tenth. 8–16. See margin.

8.

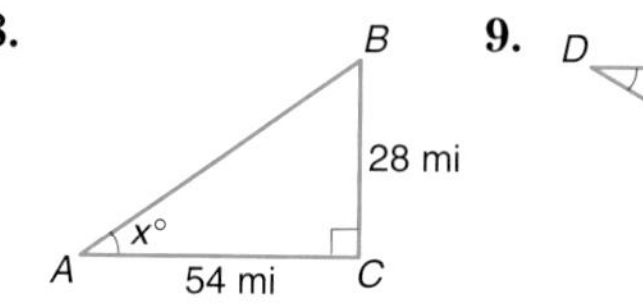

9.

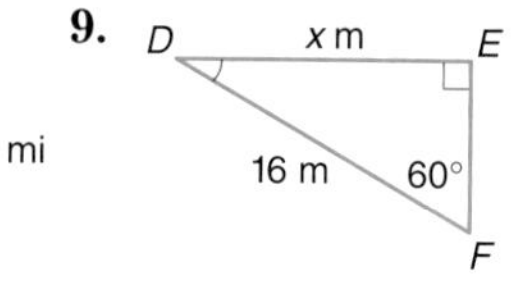

10.

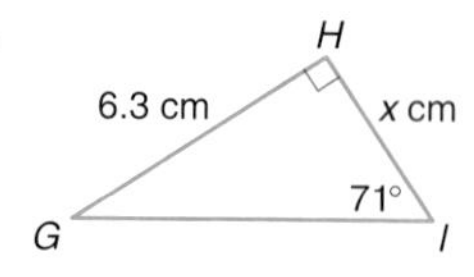

11.

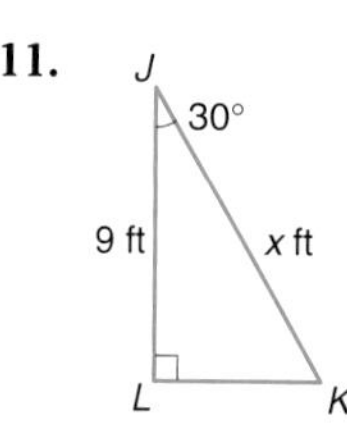

12.

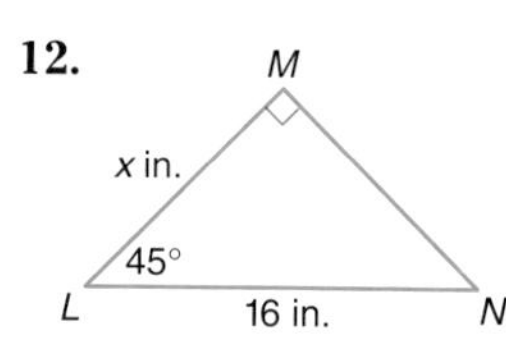

13.

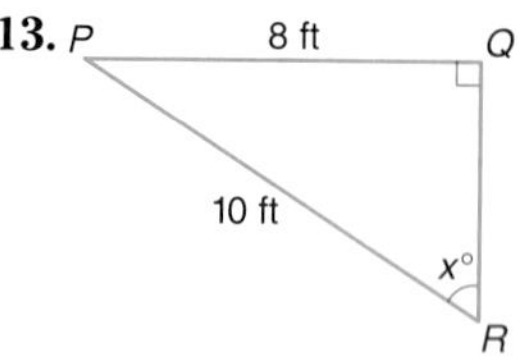

14.

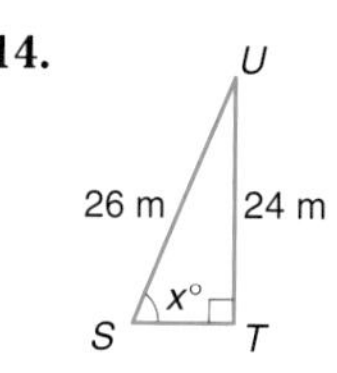

15.

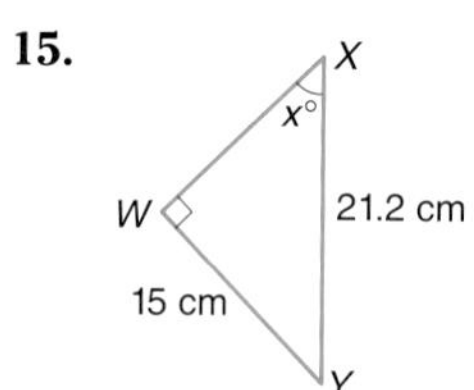

16.

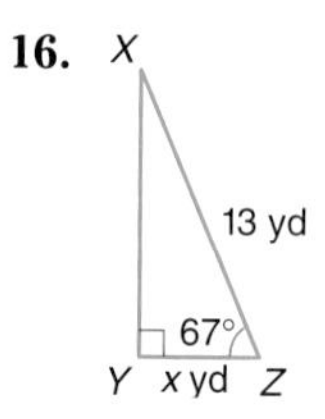

Group Activity Card 13-7

Across The Lot Group Activity **13-7**

MATERIALS: Protractor • meter stick • calculator

You and your partner must use trigonometric ratios to compute the distance across an imaginary parking lot.

To begin, outline a parking lot on the floor of your classroom or on a table. Suppose you cannot walk in the lot because it has just been resurfaced. Draw at least two triangles that you might use to compute the distance diagonally across the parking lot. Then compute the distance using your calculator and trigonometric ratios. Lastly, actually measure the distance across the parking lot to see how accurate your computations are.

 Pre-Algebra

Reteaching

Using Diagrams Have each student draw a right triangle. Label any two sides or an angle and a side. Find the other angles and sides. Repeat using different information on another right triangle.

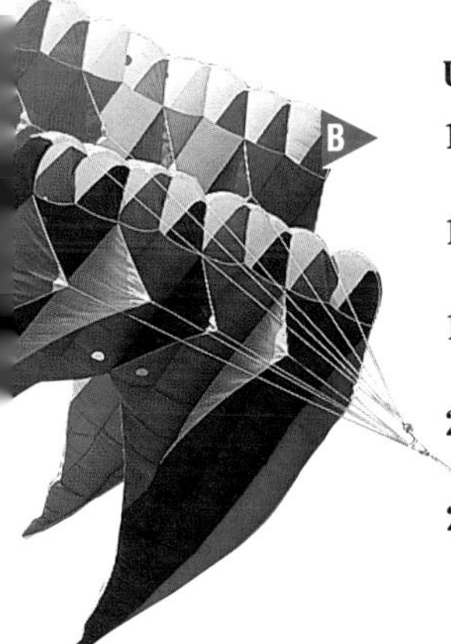

Use trigonometric ratios to solve each problem. 17. 38.3 yards

17. Martina is flying a kite on a 50-yard string. The string is making a 50° angle with the ground. How high above the ground is the kite?
18. A flagpole casts a shadow 25 meters long when the angle of elevation of the sun is 40°. How tall is the flagpole? **about 21 meters**
19. A guy wire is fastened to a TV tower 50 feet above the ground and forms an angle of 65° with the tower. How long is the wire? **about 118 feet**
20. Find the area of a right triangle in which one acute angle measures 35° and the leg opposite that angle is 80 cm long. **4570 cm²**
21. A surveyor is 85 meters from the base of a building. The angle of elevation to the top of the building is 26.5°. If her eye level is 1.6 meters above the ground, find the height of the building. **about 44 m**

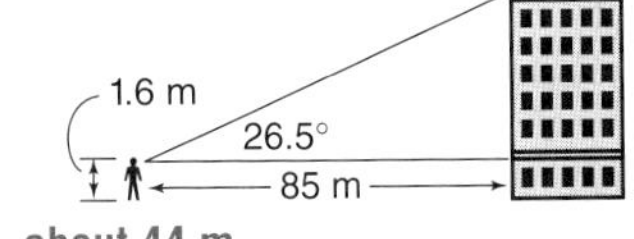

22. Danica is in the observation area of the Sears Tower in Chicago overlooking Lake Michigan. She sights two sailboats going due east from the tower. The angles of depression to the boats are 25° and 18°. If the observation deck is 1353 feet high, how far apart are the boats? **about 1262.6 feet**

Critical Thinking

23. Find the missing measures in the triangle. (*Hint:* Draw an altitude from *B*.) **$a \approx 12.4$ yd, $b \approx 15.9$ yd**

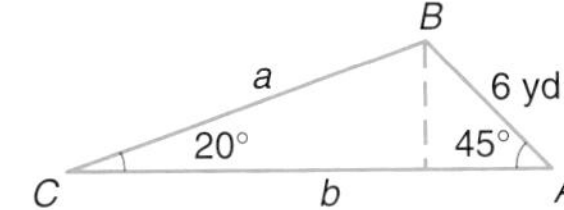

Applications and Problem Solving

24. **Aviation** A plane is 3 miles above the ground. The pilot sights the airport at an angle of depression of 15°. What is the distance between the plane and the airport? **about 11.2 miles**
25. **Mountain Climbing** A mountain climber is planning to scale a vertical rock wall. From a point 10 feet from the base, she estimates the angle of elevation to the top of the wall to be 75°. How high is the vertical wall to the nearest foot? **37 feet**
26. **Meteorology** The cloud ceiling is the lowest altitude at which solid cloud is present. To find the cloud ceiling at night, a searchlight is shone vertically onto the clouds. If a searchlight is 100 meters from a weather office and the angle of elevation to the spot on the clouds is 70°, how high is the cloud ceiling? **about 275 meters**

Mixed Review

27. Use a calculator to find the measure of $\angle B$ given that $\sin B = 0.8829$. (Lesson 13-6) **about 62°**
28. **Collect Data** Find out how many students are in each grade in your school. Make a circle graph to display your findings. (Lesson 11-2)
29. Find the range, median, upper and lower quartiles, and the interquartile range for the set of data {45, 62, 72, 51, 47, 68, 69, 50, 75}. (Lesson 10-2)
30. **Taxes** Use the information in the graph to find the probability that a randomly chosen taxpayer begins his or her tax return in March or April. (Lesson 9-3) **38%**
31. Determine whether {(2, −1), (3, −3), (4, −5)} is a function. (Lesson 8-1)

28. See students' work.
29. 30; 62; 70.5; 48.5; 22
31. yes

When do you start preparing your income tax return?
After Deadline 1%
Year-round 1%
April 13%
January 24%
March 25%
February 36%
APRIL 15

Source: RGA Survey

Extension

Working Backward A pilot has to fly from his present location to a city 200 kilometers east. He must detour around a group of thunderstorms, so he flies a path 20° NNE. When he is directly north of his destination, he flies south to get there. How far out of his way did the pilot have to fly? **85.6 km**

4 ASSESS

Closing Activity

Writing Write as many ratios as you can think of that occur in a right triangle. Then draw and label a picture of a right triangle that corresponds to three of the ratios. Exchange with another student and find the desired measure.

Chapter 13, Quiz D (Lesson 13-7) is available in the *Assessment and Evaluation Masters*, p. 352.

Additional Answers

8. $\tan x° = \frac{28}{54}$; 27.4
9. $\sin 60° = \frac{x}{16}$; 13.9
10. $\tan 71° = \frac{6.3}{x}$; 2.2
11. $\cos 30° = \frac{9}{x}$; 10.4
12. $\cos 45° = \frac{x}{16}$; 11.3
13. $\sin x° = \frac{8}{10}$; 53.1
14. $\sin x° = \frac{24}{26}$; 67.4
15. $\sin x° = \frac{15}{21.2}$; 45.0
16. $\cos 67° = \frac{x}{13}$; 5.1

***Enrichment Masters*, p. 114**

NAME ______ DATE ______

13-7 Enrichment Student Edition Pages 694–697

Trigonometric Applications

The angle of depression is very similar to the angle of elevation. As shown in the figure below, the angle is measured looking down from the horizontal rather than looking up.

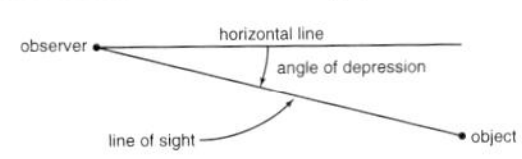

Draw a triangle to model each problem. Then solve. Round answers to the nearest tenth.

1. A lighthouse keeper is in the top of a lighthouse 95 feet above sea level. She notes that the angle of depression to a rock jutting above the water is 6°. How far is the rock from the lighthouse? **903.9 ft**
2. What is the angle of elevation of the sun when a 100-foot water tower casts a shadow 165 feet long? **31.2°**

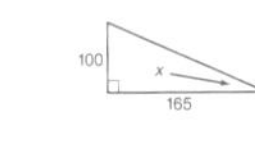

3. A disabled jet can glide at an angle of 11° with the horizontal. If it starts to glide at an altitude of 12,000 feet, can it reach a landing strip 10 miles away? **61,728 ft ≈ 11.7 mi; yes**

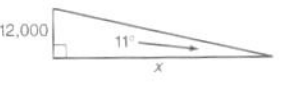

4. From a cliff 150 feet above a lake, Julio saw a boat sailing directly toward him. The angle of depression was 5°. A few minutes later, he measured it to be 11°. Find the distance the boat sailed between the two observations. **942.7 ft**
5. A horizontal road runs due east from Mount Baldy. From two points 235 meters apart on the road, the angles of elevation to the mountaintop are 43° and 30°. How high above the road is the mountaintop? **356.3 m**

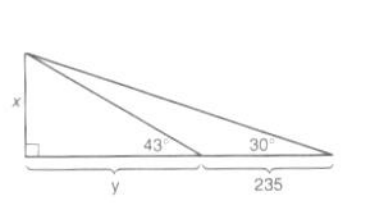

CLOSING THE

Investigation

This activity provides students an opportunity to bring their work on the Investigation to a close. For each Investigation, students should present their findings to the class. Here are some ways students can display their work.

- Conduct and report on an interview or survey.
- Write a letter, proposal, or report.
- Write an article for the school or local paper.
- Make a display, including graphs and/or charts.
- Plan an activity.

Assessment

To assess students' understanding of the concepts and topics explored in Investigation A and its follow-up activities, you may wish to examine students' Investigation Folders.

The scoring guide provided in the *Investigations and Projects Masters*, p. 27, provides a means for you to score students' work on this Investigation.

Investigations and Project Masters, p. 27

CHAPTERS 12 and 13 Investigation
Oh Give Me A Home!

Scoring Guide

Level	Specific Criteria
3 Superior	• Shows thorough understanding of the concepts of *finding the area of polygons and circles*, and *solving problems by making a model or drawing*. • Uses appropriate strategies to solve problems. • Computations are correct. • Written explanations are exemplary. • Working drawings and models are appropriate and sensible. • Goes beyond the requirements of the Investigation.
2 Satisfactory, with minor flaws	• Shows understanding of the concepts of *finding the area of polygons and circles, and solving problems by making a model or drawing*. • Uses appropriate strategies to solve problems. • Computations are mostly correct. • Written explanations are effective. • Working drawings and models are appropriate and sensible. • Satisfies all requirements of the Investigation.
1 Nearly Satisfactory, with serious flaws	• Shows understanding of most of the concepts of *finding the area of polygons and circles, and solving problems by making a model or drawing*. • May not use appropriate strategies to solve problems. • Computations are mostly correct. • Written explanations are satisfactory. • Working drawings and models are mostly appropriate and sensible. • Satisfies most requirements of the Investigation.
0 Unsatisfactory	• Shows little or no understanding of the concepts of *finding the area of polygons and circles*, and *solving problems by making a model or drawing*. • May not use appropriate strategies to solve problems. • Computations are incorrect. • Written explanations are not satisfactory. • Working drawings and models are not appropriate or sensible. • Does not satisfy requirements of the Investigation.

CLOSING THE

Investigation

OH GIVE ME A HOME

Architects often give presentations to potential buyers to show and explain their designs. Plan and give an oral presentation to describe what you know about architectural styles and designing homes. Your presentation should include the following.

- A brief architectural history of your community. You might include old and new photographs, drawings, or maps.
- Examples from your Investigation Folder to help you describe and define building terms you used most often in this Investigation.
- Floor plan(s) and elevation drawings of the house your group designed.
- The model of the house your group designed.
- The reasoning behind your design; that is, an explanation that tells why you designed the house the way you did.
- An explanation of what you would do differently if you had to do this Investigation over again.

Extension

Schedule an interview with an architect. Make a questions to ask him or her.

- Be sure to ask about the importance of mathematics to architecture.
- Include a question about what courses high school students interested in becoming architects should take.

PORTF ASSESS

You may to keep work or Investiga your por

CHAPTER 13 Highlights

Vocabulary

After completing this chapter, you should be able to define each term, property, or phrase and give an example or two of each.

Algebra

irrational number (p. 672)
perfect squares (p. 665)
Pythagorean theorem (p. 676)
radical sign (p. 664)
real numbers (p. 673)
square root (p. 664)

Geometry

adjacent leg (p. 687)
angle of depression (p. 695)
angle of elevation (p. 694)
cosine (p. 688)
45°–45° right triangle (p. 683)
hypotenuse (p. 676)
leg (p. 676)
opposite side (p. 687)
sine (p. 688)
tangent (p. 688)
30°–60° right triangle (p. 683)
trigonometric ratios (p. 688)
trigonometry (p. 688)

Problem Solving

Venn diagram (p. 669)

Understanding and Using Vocabulary

Choose the correct term to complete each sentence.

1. The (square, square root) of 6 equals 36
2. In the expression $\sqrt{25}$, the symbol $\sqrt{\ }$ is called a (radical sign, square).
3. The number $\sqrt{121}$ belongs to the set of (irrational, rational) numbers.
4. The longest side of a right triangle is the (hypotenuse, leg).
5. A right triangle with a leg that is half the length of the hypotenuse is a (30°–60°, 45°–45°) right triangle.
6. The angle formed by a horizontal line and another line of sight above it is an angle of (depression, elevation).

Instructional Resources

Three multiple-choice tests and three free-response tests are provided in the *Assessment and Evaluation Masters*. Forms 1A and 2A are for honors pacing, Forms 1B and 2B are for average pacing, Forms 1C and 2C are for basic pacing, Chapter 13 Test, Form 1B is shown at the right. Chapter 13 Test, Form 2B is shown on the next page.

Highlights

Using the Chapter Highlights

The Chapter Highlights begins with a listing of the new terms, properties, and phrases that were introduced in this chapter.

Assessment and Evaluation Masters, pp. 339–340

NAME ____________________ DATE __________

CHAPTER 13 Test, Form 1B

1. Find the value of $\sqrt{36}$.
   A. -18 B. 1296 C. 18 D. 6 — 1. D
2. Choose the best integer estimate for $\sqrt{51}$.
   A. 8 B. 7 C. 25 D. 102 — 2. B
3. Harriet has a total of 36 red, blue, and red-and-blue socks. If 12 socks are all red and 20 socks have some red, how many socks are all blue?
   A. 16 B. 4 C. 12 D. 24 — 3. B
4. The Venn diagram at the right shows data about 200 music students. How many students are in chorus only?
   A. 22 B. 32
   C. 112 D. 42 — 4. A

Solve each equation. Round decimal answers to the nearest tenth.

5. $y^2 = 50$
   A. 7.1, -7.1 B. -7.0 C. 7.7, -7.7 D. 7.2 — 5. A
6. $z^2 = 63$
   A. -8.1 B. 8.0, -8.0 C. 7.9, -7.9 D. 7.8 — 6. C
7. Use the Pythagorean Theorem to find the length of the hypotenuse of a right triangle if the lengths of the legs are 24 and 32 centimeters.
   A. 28 cm B. 7.48 cm C. 56 cm D. 40 cm — 7. D

In a right triangle, if a and b are the measures of the legs and c is the measure of the hypotenuse, find each missing measure. Round answers to the nearest tenth.

8. $a = 13, b = 15$
   A. 18.3 B. 19.8 C. 28.0 D. 17.2 — 8. B
9. $a = 5, c = 19$
   A. 18.3 B. 19.8 C. 14.0 D. 24.1 — 9. A
10. Rhea bikes 4 miles west, 8 miles north, and 4 miles west. How many miles is she from where she began? Round the answer to the nearest tenth.
    A. 18.3 mi B. 12.4 mi C. 16.0 mi D. 11.3 mi — 10. D

NAME ____________________ DATE __________

Chapter 13 Test, Form 1B (continued)

11. An antenna wire for shortwave radio reception is connected between the tops of two poles. One pole is 40 feet high and the other is 10 feet high. If the poles are 48 feet apart, how long is the wire? Round the answer to the nearest tenth.
    A. 8.8 ft B. 56.6 ft C. 62.5 ft D. 53.9 ft — 11. B
12. Find the length of a diagonal of a square that is 5 centimeters on a side.
    A. $\sqrt{5}$ cm B. $5\sqrt{2}$ cm C. $10\sqrt{2}$ cm D. $25\sqrt{2}$ cm — 12. B
13. Find the length of the shorter leg of a 30°–60° right triangle if the length of the hypotenuse is 12 centimeters.
    A. 6 cm B. $6\sqrt{3}$ cm C. $12\sqrt{3}$ cm D. $6\sqrt{2}$ cm — 13. A
14. Use a calculator to find sin 38° to the nearest ten thousandth.
    A. 0.620 B. 0.7880 C. 0.6157 D. 0.7813 — 14. C
15. Use a calculator to find the angle that corresponds to tan $A = 1.2345$. Round to the nearest degree.
    A. 14° B. 13° C. 76° D. 51° — 15. D
16. Find the measure of the marked acute angle in the triangle at the right. Round to the nearest degree.
    A. 34° B. 44°
    C. 41° D. 36° — 16. D

17. Javier has a kite with a string that is 150 meters long. When he lets out the string fully and the string makes an angle of 60° with the ground, what is the altitude of the kite? Round the answer to the nearest whole number.
    A. 120 m B. 130 m C. 137 m D. 140 m — 17. B
18. Tom rests a 32-foot ladder against a wall. The top of the ladder hits the wall at a point 28 feet high. What is the angle it forms with the ground?
    A. 29° B. 50° C. 90° D. 61° — 18. D
19. Find the measure of A in $\triangle ABC$.
    A. 34.8° B. 20.4°
    C. 55.2° D. 90° — 19. A

20. Kendra can see her house from a height of 80 meters in a hot-air balloon. The angle of depression is 18°. How far is her house from the point directly below the balloon? Round the answer to the nearest whole number.
    A. 218 m B. 236 m C. 246 m D. 264 m — 20. C

CHAPTER 13 Study Guide and Assessment

Using the Chapter Study Guide and Assessment

Skills and Concepts Encourage students to refer to the objectives and examples on the left as they complete the review exercises on the right.

Assessment and Evaluation Masters, pp. 345–346

NAME ____________ DATE ________

CHAPTER 13 Test, Form 2B

1. Find the square root of 169. — 1. 13
2. Find the best integer estimate for $-\sqrt{79}$. — 2. −9
3. Find the best integer estimate for $\sqrt{125}$. — 3. 11
4. Janet has a total of 28 white, blue, and white-and-blue socks. If 8 socks are all blue and 26 socks have some blue, how many socks are all white? — 4. 2
5. Of the French and Spanish classes offered at Knight Middle School, 50 students take Spanish, 28 take French, and 10 take both languages. How many students take French or Spanish classes all together? — 5. 68
6. Solve $y^2 = 222$. Round to the nearest tenth. — 6. 14.9, −14.9
7. Solve $c^2 = 80$. Round to the nearest tenth. — 7. 8.9, −8.9
8. Use the Pythagorean Theorem to find the length of the hypotenuse of a right triangle if the lengths of the legs are 30 meters and 72 meters. — 8. 78 m
9. Find the missing measure of the right triangle at the right. Round the answer to the nearest tenth. — 9. 59.4 ft

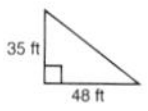

10. An insect crawls 6 inches south, 10 inches east, and 2 inches south. How many inches is it from where it began? Round the answer to the nearest tenth. — 10. 12.8 in.
11. An antenna wire for shortwave radio reception is connected between the tops of two poles. One pole is 50 feet high and the other is 12 feet high. If the poles are 56 feet apart, how many feet long is the wire? Round the answer to the nearest tenth. — 11. 67.7 ft

Round the answers to the nearest tenth.

12. Find the length of a diagonal of a square that is 8 feet on each side. — 12. 11.3 ft
13. The sides of square *DEFG* are 3.4 centimeters long. Find the length of $\overline{DF}$. — 13. 4.8 cm

NAME ____________ DATE ________

Chapter 13 Test, Form 2B (continued)

14. Find the length of the longer leg of a 30°–60° right triangle if the length of the hypotenuse is 14 centimeters. — 14. 12.1 cm

Use a calculator to solve. Round answers to the nearest thousandth.

15. Find sin 48°. — 15. 0.743
16. What is the value of tan 21°? — 16. 0.384

Round answers to the nearest degree.

17. What is the measure of angle *A* if tan *A* = 0.625? — 17. 32°
18. Find the measure of angle *B* if cos *B* = 0.819. — 18. 35°
19. If sin *C* = 0.961, find the measure of angle *C*. — 19. 74°
20. Khandi places a 30-foot ladder against a wall. The top of the ladder makes a 12° angle with the wall. How far from the wall is the foot of the ladder? Round to the nearest tenth. — 20. 6.2 ft

CHAPTER 13 Study Guide and Assessment

Skills and Concepts

Objectives and Examples

Upon completing this chapter, you should be able to:

▶ **find squares and square roots** (Lesson 13-1)

Find $-\sqrt{196}$.

The symbol $-\sqrt{196}$ represents the negative square root of 196. Since $14 \cdot 14 = 196$, $-\sqrt{196} = -14$.

Review Exercises

Use these exercises to review and prepare for the chapter test.

Find each square root.

7. $\sqrt{9}$ **3**
8. $\sqrt{121}$ **11**
9. $-\sqrt{25}$ **−5**
10. $-\sqrt{81}$ **−9**
11. $\sqrt{225}$ **15**
12. $-\sqrt{900}$ **−30**

▶ **find and use squares and square roots** (Lesson 13-1)

Find the best integer estimate for $-\sqrt{29}$.
25 and 36 are the closest perfect squares.

$$-\sqrt{36} < -\sqrt{29} < -\sqrt{25}$$
$$-\sqrt{6^2} < -\sqrt{29} < -\sqrt{5^2}$$
$$-6 < -\sqrt{29} < -5$$

Since −29 is closer to −25 than −36, the best integer estimate is −5.

Find the best integer estimate for each square root. Then check your estimate using a calculator.

13. $\sqrt{83}$ **9**
14. $\sqrt{54}$ **7**
15. $\sqrt{220}$ **15**
16. $-\sqrt{900}$ **−30**
17. $-\sqrt{39}$ **−6**
18. $\sqrt{9.61}$ **3**

▶ **identify numbers in the real number system** (Lesson 13-3)

0.25 can be expressed as $\frac{1}{4}$ so it is a rational number.

The number $\sqrt{6} = 2.449489743\ldots$, which is not a repeating or terminating decimal, is an irrational number.

Name the sets of numbers to which each number belongs: the whole numbers, the integers, the rational numbers, the irrational numbers, and/or the real numbers.

19. 7 **whole, integer, rational, real**
20. $\frac{3}{4}$ **rational, real**
21. −2.6 **rational, real**
22. $\sqrt{11}$ **irrational, real**

▶ **solve equations by finding square roots** (Lesson 13-3)

Solve $b^2 = 49$.

$$b^2 = 49$$
$$b = \sqrt{49} \quad \text{or} \quad b = -\sqrt{49}$$
$$b = 7 \quad \text{or} \quad b = -7$$

Solve each equation. Round decimal answers to the nearest tenth.

23. $x^2 = 196$ **14, −14**
24. $n^2 = 160$ **12.6, −12.6**
25. $t^2 = 15$ **3.9, −3.9**
26. $a^2 = 0.04$ **0.2, −0.2**

GLENCOE Technology

Test and Review Software

You may use this software, a combination of an item generator and an item bank, to create your own tests or worksheets. Types of items include free response, multiple choice, short answer, and open ended.

For IBM & Macintosh

Objectives and Examples

▶ **use the Pythagorean theorem to find the length of the side of a right triangle and to solve problems** (Lesson 13-4)

Find the length of the third side of the right triangle.

$$c^2 = a^2 + b^2$$
$$8^2 = 5^2 + b^2$$
$$64 = 25 + b^2$$
$$39 = b^2$$
$$\sqrt{39} = b$$
$$6.2 \approx b$$

8 in.
b in.
5 in.

The length of the leg is about 6.2 inches.

Review Exercises

In a right triangle, if *a* and *b* are the measures of the legs and *c* is the measure of the hypotenuse, find each missing measure. Round answers to the nearest tenth.

27. $a = 12, b = 16$ **20**

28. $a = 14, b = 40$ **42.4**

29. $a = 30, b = 16$ **34**

30. $a = 8, c = 15$ **12.7**

31. $b = 63, c = 65$ **16**

32. $a = 15, c = 39$ **36**

▶ **find missing measures in 30°–60° and 45°–45° right triangles** (Lesson 13-5)

In a 30°–60° right triangle, the hypotenuse is twice as long as the length of the side opposite the 30° angle.

In a 45°–45° right triangle, the hypotenuse is $\sqrt{2}$ times the length of a leg.

Find the lengths of the missing sides in each triangle. Round decimal answers to the nearest tenth.

33.

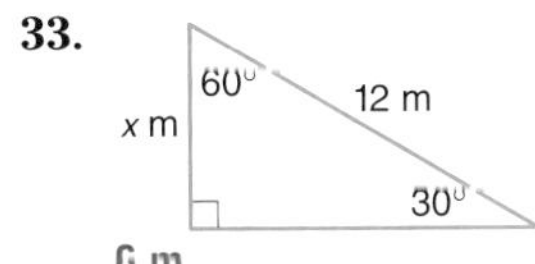

6 m

34.

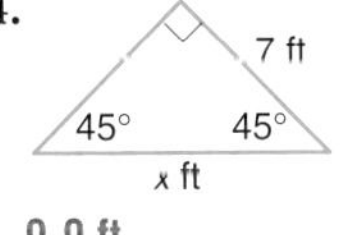

9.9 ft

35.

30°
x cm
60°
1.5 cm

3 cm

▶ **find missing sides and angles of triangles using the sine, cosine, and tangent ratios** (Lesson 13-6)

$$\sin \angle A = \frac{\text{measure of the leg opposite } \angle A}{\text{measure of the hypotenuse}}$$

$$\cos \angle A = \frac{\text{measure of the leg adjacent to } \angle A}{\text{measure of the hypotenuse}}$$

$$\tan \angle A = \frac{\text{measure of the leg opposite } \angle A}{\text{measure of the leg adjacent to } \angle A}$$

For each triangle, find sin *X*, cos *X*, and tan *X* to the nearest thousandth. **36–37. See margin.**

36.

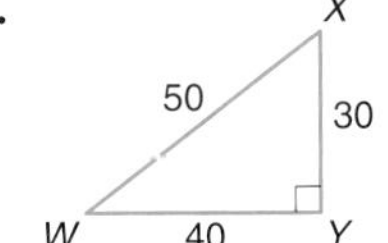

37.

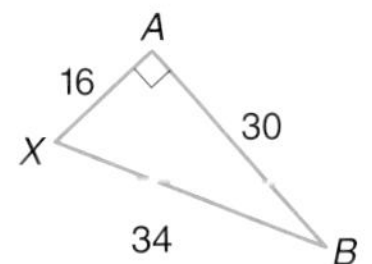

Use a calculator to find the angle that corresponds to each ratio. Round to the nearest degree.

38. $\tan C = 2.145$ **65°**

39. $\sin H = 0.9945$ **84°**

40. $\cos M = 0.2588$ **75°**

Chapter 13 • Study Guide and Assessment

Additional Answers

36. $\sin X = 0.8$, $\cos X = 0.6$, $\tan X = 1.333$

37. $\sin X = 0.882$, $\cos X = 0.471$, $\tan X = 1.875$

Applications and Problem Solving

Encourage students to work through the exercises in the Applications and Problem Solving section to strengthen their problem-solving skills.

Additional Answers

41. $\tan 35^\circ = \frac{x}{15}$

42. $\cos x^\circ = \frac{6}{11}$

43. $\sin 25^\circ = \frac{15}{x}$

Objectives and Examples

▶ **solve problems using the trigonometric ratios** (Lesson 13-7)

In $\triangle ABC$, find the length of $\overline{BC}$.

$$\sin A = \frac{\text{measure of leg opposite } \angle A}{\text{measure of the hypotenuse}}$$

$$\sin 28^\circ = \frac{x}{12}$$

$$(12)(\sin 28^\circ) = x$$

$$5.6 \approx x$$

The measure of $\overline{BC}$ is about 5.6 feet.

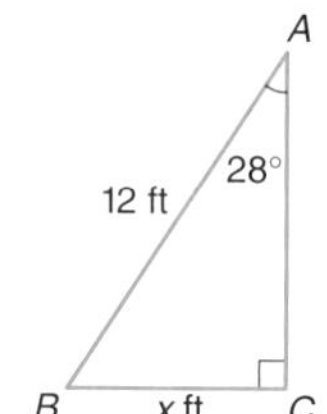

Review Exercises

Write an equation that you could solve for x. Round decimal answers to the nearest tenth. **See margin for equations.**

41.

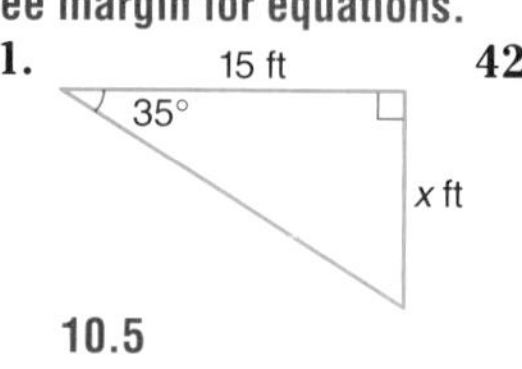

10.5

42.

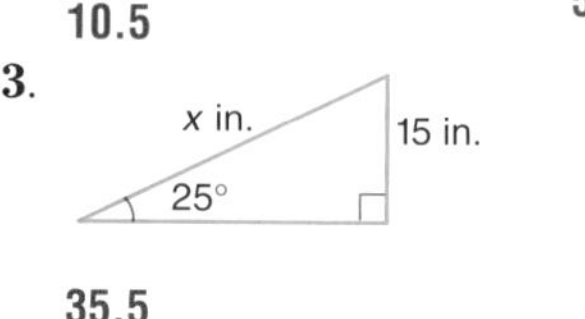

56.9

43.

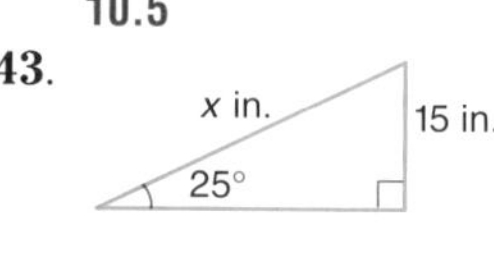

35.5

Applications and Problem Solving

44. Recreation A class survey of 32 students about favorite weekend activities showed that 20 students chose a movie and 15 students chose going to a shopping mall. There were 8 students who chose neither activity. How many students chose both activities? (Lesson 13-2) **11 students**

45. Communication A telephone pole is 28 feet tall. A wire is stretched from the top of the pole to a point on the ground that is 5 feet from the bottom of the pole. How long is the wire? Round to the nearest tenth. (Lesson 13-3) **28.4 ft**

46. Geometry Write a formula for the length of the diagonal of a square if the sides are s units long. (Lesson 13-5) $d = s\sqrt{2}$

47. Navigation From the top of a lighthouse, the angle of depression to a buoy is 25°. If the top of the lighthouse is 150 feet above sea level, find the distance from the buoy to the foot of the lighthouse. (Lesson 13-7) **about 321.68 feet**

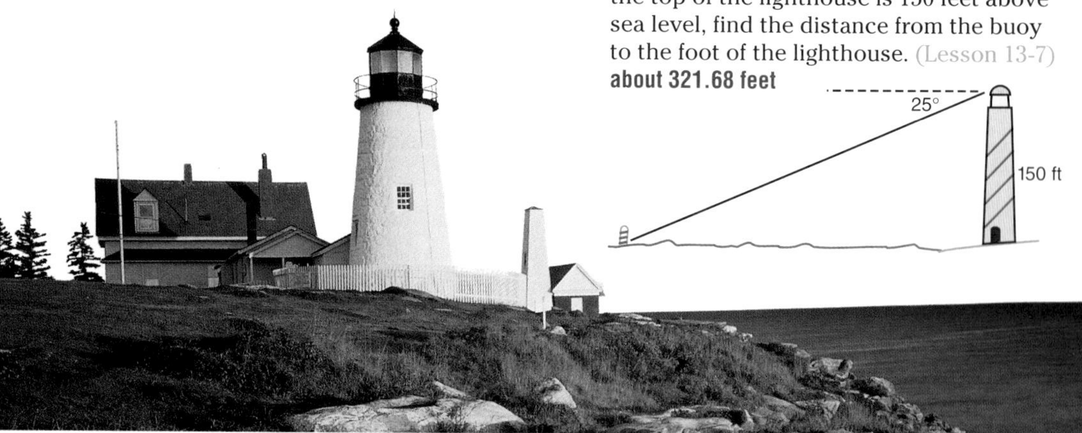

A practice test for Chapter 13 is available on page 786.

Alternative Assessment

Performance Task **Demonstrate your knowledge by giving a clear, concise solution to each problem. Be sure to include all relevant drawings and justify your answers. You may show your solutions in more than one way or investigate beyond the requirements of the problem.** **1. 22° northeast; 16 km from the starting point**

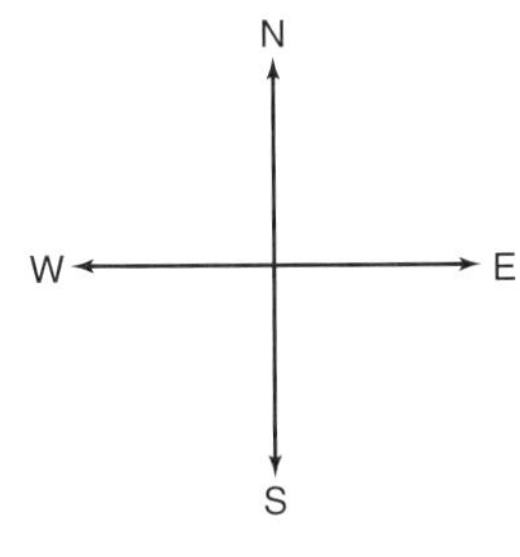

1. You and your group are hiking in the woods. You leave camp and walk 15 kilometers due north. Then you walk 6 kilometers due east. Your job in the group is to chart the group's movement and to find how far you are from the starting point. Vectors are line segments with both magnitude (length) and direction. Draw vectors to represent the group's movement. Then use what you have learned in this chapter to find the direction and distance the hikers are from their starting point.

2. A soccer player kicks the ball 12 feet to the west. Another player kicks the ball 18 feet to the north. What is the direction and distance from the starting point of the ball? Round your answer to the nearest whole number. **56° northwest; 22 ft from the starting point**

Thinking Critically Use a calculator to find the sine, cosine, and tangent of several acute angles.

- When is $\cos A > \sin A$? **when $m\angle A < 45°$**
- When is $\cos A = \sin A$? **when $m\angle A = 45°$**
- Describe the value of $\tan A$ when $m\angle A > 45°$. **$\tan A > 1$**
- Describe the range of values for $\cos A$ and $\sin A$. **$0 \le \sin A \le 1$, $0 \le \cos A \le 1$**

Portfolio Place your favorite word problem from this chapter in your portfolio along with your solution and attach a note explaining why it is your favorite. **See students' work.**

Self Evaluation Euclid's *Elements* is the basis for much of the geometry that is studied today. It was written in about 300 B.C. and was a compilation of most of the geometry known at the time. Euclid taught mathematics at the Library of Alexandria in Alexandria, Egypt. When asked if there was a simpler way to learn geometry than by studying the *Elements*, Euclid replied "There is no 'royal road' to geometry."

Assess yourself. Learning anything well is a building process that takes time. Evaluate the time that you spend studying. Do you invest time each day on study or do you study for tests and forget what you learned? Keep a log of the time that you spend studying for one week, noting when tests and quizzes were given. If you find that you are studying ineffectively, make a new plan for your study time in the next week. **See students' work.**

Alternative Assessment

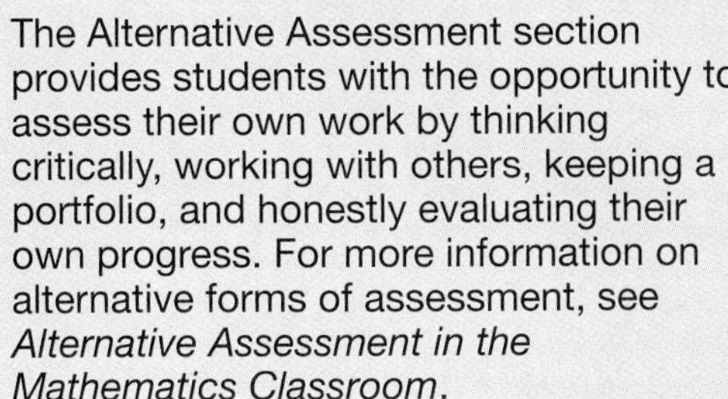

The Alternative Assessment section provides students with the opportunity to assess their own work by thinking critically, working with others, keeping a portfolio, and honestly evaluating their own progress. For more information on alternative forms of assessment, see *Alternative Assessment in the Mathematics Classroom.*

Performance Assessment

Performance Assessment tasks for this chapter are included in the *Assessment and Evaluation Masters*. A scoring guide is also provided.

Alternative Assessment

Assessment and Evaluation Masters, pp. 349, 361

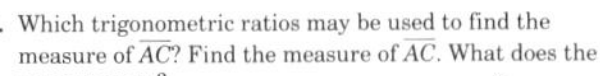

NAME ________________ DATE ________

CHAPTER 13 Performance Assessment

Instructions: Demonstrate your knowledge by giving a clear, concise solution to each problem. Be sure to include all relevant drawings and justify your answers. You may show your solution in more than one way or investigate beyond the requirements of the problem.

1. Carpenters frequently use lines like those shown at right to outline the position of a building they are planning to build. To check that the corners of the house are square, one can mark points 6 feet from the corner on one string and 8 feet from the corner on the other string. If the two points are 10 feet apart, the corner is square.
   a. Explain how the carpenter knows the corner is square.
   b. If the diagonal measure is more than 10 feet, should the string at point A be moved to the right or left? Draw a diagram to demonstrate.
   c. If the diagonal measure is less than 10 feet, should the string at point B be moved to the right or left? Draw a diagram to demonstrate.

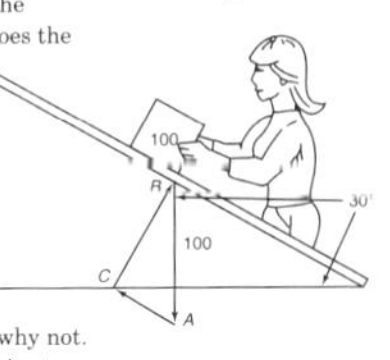

2. An inclined plane is a simple machine often used to load items onto trucks because it is easier to push a heavy box up the inclined plane that to lift it into the truck. The measure of $\overline{BA}$ represents the weight of the box. The measure of $\overline{CB}$ represents the force of the inclined plane on the box. The measure of $\overline{AC}$ represents the force of the person pushing the box.
   a. Which trigonometric ratios may be used to find the measure of $\overline{AC}$? Find the measure of $\overline{AC}$. What does the answer mean?
   b. If the inclined plane was longer, as in the figure at right, would angle A be larger or smaller? Would the force required to push the box be more or less? How do you know?

3. a. Write an example of a terminating decimal. Is this decimal a rational number? Explain why or why not.
   b. Write an example of a repeating decimal. Is this decimal a rational number? Explain why or why not.
   c. Write an example of a decimal that does not terminate or repeat. How do you know it does not repeat? Is this decimal a rational number? Explain why or why not.

CHAPTER 13 Scoring Guide

Level	Specific Criteria
3 Superior	• Shows thorough understanding of the concepts *square corner*, *Pythagorean Theorem*, *inclined plane*, *trigonometric ratios*, *angle measure*, *rational numbers*, and *irrational numbers*. • Uses appropriate strategies to solve problems. • Computations to find the measure of line segments are correct. • Written explanations are exemplary. • Diagrams are accurate and clear. • Goes beyond requirements of some or all problems.
2 Satisfactory, with minor flaws	• Shows understanding of the concepts *square corner*, *Pythagorean Theorem*, *inclined plane*, *trigonometric ratios*, *angle measure*, *rational numbers*, and *irrational numbers*. • Uses appropriate strategies to solve problems. • Computations to find the measure of line segments are mostly correct. • Written explanations are effective. • Diagrams are mostly accurate and clear. • Satisfies all requirements of problems.
1 Nearly Satisfactory, with serious flaws	• Shows understanding of most of the concepts *square corner*, *Pythagorean Theorem*, *inclined plane*, *trigonometric ratios*, *angle measure*, *rational numbers*, and *irrational numbers*. • May not use appropriate strategies to solve problems. • Computations to find the measure of line segments are mostly correct. • Written explanations are satisfactory. • Diagrams are mostly accurate and clear. • Satisfies most requirements of problems.
0 Unsatisfactory	• Shows little or no understanding of the concepts *square corner*, *Pythagorean Theorem*, *inclined plane*, *trigonometric ratios*, *angle measure*, *rational numbers*, and *irrational numbers*. • May not use appropriate strategies to solve problems. • Computations to find the measure of line segments are incorrect. • Written explanations are not satisfactory. • Diagrams are not accurate and clear. • Does not satisfy requirements of problems.

CHAPTER 14

Polynomials

PREVIEWING THE CHAPTER

In this chapter, students use algebra tiles and a hands-on approach as they study polynomials.The chapter begins as students identify and classify polynomials and find the degree of a polynomial. Students then use algebra tiles to model polynomials. The chapter continues as students add and subtract polynomials. In Lesson 14-4, students find powers of monomials. Students then use tiles to multiply a monomial and a polynomial. This is expanded as students use algebra to multiply a polynomial by a monomial and to multiply binomials.

Lesson (pages)	Lesson Objectives	NCTM Standards	State/Local Objectives
14-1 (706–709)	Identify and classify polynomials. Find the degree of a polynomial.	1-4, 9	
14-1B (710)	Use algebra tiles to model polynomials	1-4, 9	
14-2 (711–714)	Add polynomials.	1-4, 9	
14-3 (715–718)	Subtract polynomials.	1-4, 9	
14-4 (719–723)	Find powers of monomials.	1-4, 9	
14-5A (724)	Use algebra tiles to muliply a monomial and a polynomial.	1-4, 9	
14-5 (725–727)	Multiply a polynomial by a monomial.	1-4, 9	
14-6 (728–731)	Multiply binomials.	1-4, 9	
14-6B (732)	Use algebra tiles to factor simple trinomials.	1-4, 9	

A complete, 1-page lesson plan is provided for each lesson in the ***Lesson Planning Guide***. Answer keys for each lesson are available in the ***Answer Key Masters***.

ORGANIZING THE CHAPTER

You may want to refer to the **Course Planning Calendar** on page T12 for detailed information on pacing.
PACING: Honors—9 days

LESSON PLANNING CHART

Lesson (Pages)	Materials/ Manipulatives	Extra Practice (Student Edition)	Blackline Masters										Real-World Applications	Interactive Mathematics Tools Software	Teaching Transparencies	Group Activity Cards
			Study Guide	Practice	Enrichment	Assessment and Evaluation	Math Lab and Modeling Math	Multicultural Activity	Tech Prep Applications	Graphing Calculator	Activity				
14-1 (706–709)		p. 772	p. 115	p. 115	p. 115									14-1A 14-1B	14-1
14-1B (710)	algebra tiles*						p. 56								
14-2 (711–714)	algebra tiles*	p. 773	p. 116	p. 116	p. 116	p. 379			p. 27				14-2	14-2A 14-2B	14-2
14-3 (715–718)	algebra tiles*	p. 773	p. 117	p. 117	p. 117	pp. 378, 379				p. 14			14-3	14-3A 14-3B	14-3
14-4 (719–723)		p. 773	p. 118	p. 118	p. 118			p. 27			p. 48	31		14-4A 14-4B	14-4
14-5A (724)	algebra tiles*						p. 57								
14-5 (725–727)	algebra tiles* product mat*	p. 773	p. 119	p. 119	p. 119	p. 380	p. 75				p. 14			14-5A 14-5B	14-5
14-6 (728–731)	algebra tiles* product mat*	p. 773	p. 120	p. 120	p. 120	p. 380	p. 89	p. 28	p. 28		p. 28	32		14-6A 14-6B	14-6
14-6B (732)	algebra tiles* product mat*						p. 58								
Study Guide/ Assessment (734–736)						pp. 365–377, 381–383									

*Included in Glencoe's *Student Manipulative Kit* and *Overhead Manipulative Resources*.

ORGANIZING THE CHAPTER

All of the blackline masters in the Teacher's Classroom Resources are available on the ***Electronic Teacher's Classroom Resources*** CD-ROM.

Other Chapter Resources

Student Edition
Chapter Opener, pp. 704–705
Earth Watch, p. 723

Teacher's Classroom Resources
Investigations and Projects Masters, pp. 81–84
Pre-Algebra Overhead Manipulative Resources, pp. 42–45

Technology
Test and Review Software (IBM & Macintosh)
CD-ROM Activities

Professional Publications
Block Scheduling Booklet
Glencoe Mathematics Professional Series

Outside Resources

Books/Periodicals
Visualizing Algebra: Algebra Problems & Projects for the Function Analyzer, Dr. Michael Yerushalmy and June Mark
Key to Algebra: Polynomials, Key Curriculum Press

Videos/CD-ROMs
Interactive Mathematics, Ferranti Int.
Math Sleuths, Videodiscovery

Software
Gliding, Sunburst

Assessment Resources

Student Edition
Math Journal, pp. 707, 721
Mixed Review, pp. 709, 714, 718, 723, 727, 731
Self Test, p. 718
Chapter Highlights, p. 733
Chapter Study Guide and Assessment, pp. 734–736
Alternative Assessment, p. 737
Portfolio, p. 737
Ongoing Assessment, pp. 738–739
MindJogger Videoquiz, 14

Teacher's Wraparound Edition
5-Minute Check, pp. 706, 711, 715, 719, 725, 728
Checking Your Understanding, pp. 708, 713, 717, 721, 726, 730
Closing Activity, pp. 709, 714, 718, 723, 727, 731
Cooperative Learning, pp. 711, 720

Assessment and Evaluation Masters
Multiple-Choice Tests, Forms 1A (Honors), 1B (Average), 1C (Basic), pp. 365–370
Free-Response Tests, Forms 2A (Honors), 2B (Average), 2C (Basic), pp. 371–376
Performance Assessment, p. 377
Mid-Chapter Test, p. 378
Quizzes A–D, pp. 379–380
Standardized Test Practice, p. 381
Cumulative Review, pp. 382–383

ENHANCING THE CHAPTER

Examples of some of the materials for enhancing Chapter 14 are shown below.

DIVERSITY

Multicultural Activity Masters, pp. 27, 28

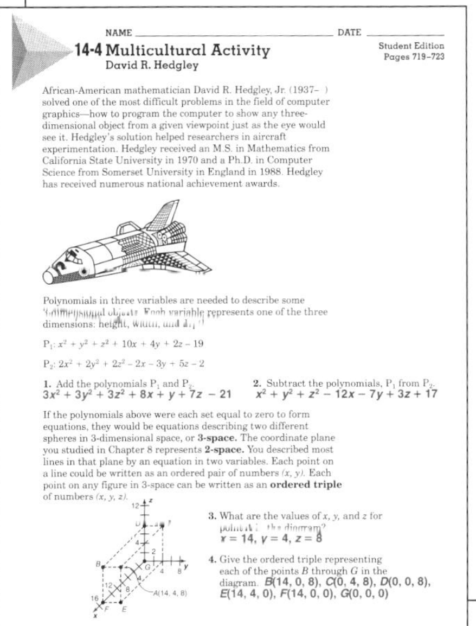

NAME ______________________ DATE __________

14-4 Multicultural Activity
David R. Hedgley

Student Edition Pages 719–723

African-American mathematician David R. Hedgley, Jr. (1937–) solved one of the most difficult problems in the field of computer graphics—how to program the computer to show any three-dimensional object from a given viewpoint just as the eye would see it. Hedgley's solution helped researchers in aircraft experimentation. Hedgley received an M.S. in Mathematics from California State University in 1970 and a Ph.D. in Computer Science from Somerset University in England in 1988. Hedgley has received numerous national achievement awards.

Polynomials in three variables are needed to describe some 3-dimensional objects. Each variable represents one of the three dimensions: height, width, and depth.

P_1: $x^2 + y^2 + z^2 + 10x + 4y + 2z - 19$

P_2: $2x^2 + 2y^2 + 2z^2 - 2x - 3y + 5z - 2$

1. Add the polynomials P_1 and P_2. **$3x^2 + 3y^2 + 3z^2 + 8x + y + 7z - 21$**
2. Subtract the polynomials, P_1 from P_2. **$x^2 + y^2 + z^2 - 12x - 7y + 3z + 17$**

If the polynomials above were each set equal to zero to form equations, they would be equations describing two different spheres in 3-dimensional space, or **3-space.** The coordinate plane you studied in Chapter 8 represents **2-space.** You described most lines in that plane by an equation in two variables. Each point on a line could be written as an ordered pair of numbers (x, y). Each point on any figure in 3-space can be written as an **ordered triple** of numbers (x, y, z).

3. What are the values of x, y, and z for point A in the diagram? **$x = 14, y = 4, z = 8$**
4. Give the ordered triple representing each of the points B through G in the diagram. **$B(14, 0, 8)$, $C(0, 4, 8)$, $D(0, 0, 8)$, $E(14, 4, 0)$, $F(14, 0, 0)$, $G(0, 0, 0)$**

APPLICATIONS

Real-World Applications, 31, 32

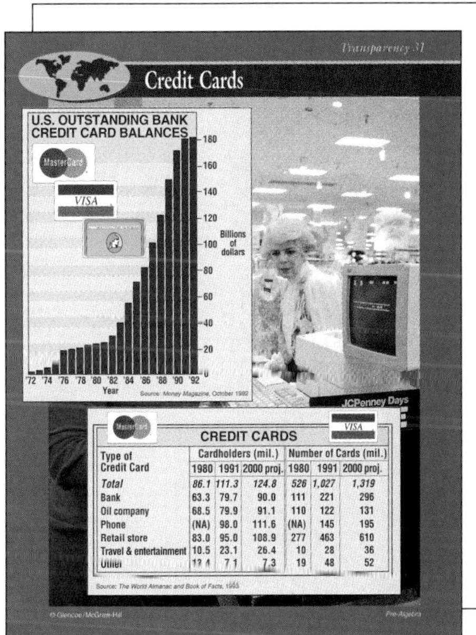

Transparency 31

Credit Cards

CREDIT CARDS

Type of Credit Card	Cardholders (mil.)			Number of Cards (mil.)		
	1980	1991	2000 proj.	1980	1991	2000 proj.
Total	*86.1*	*111.3*	*124.8*	*526*	*1,027*	*1,319*
Bank	63.3	79.7	90.0	111	221	296
Oil company	68.5	79.9	91.1	110	122	131
Phone	(NA)	98.0	111.6	(NA)	145	195
Retail store	83.0	95.0	108.9	277	463	610
Travel & entertainment	10.5	23.1	26.4	10	28	36
Other	12.4	7.1	7.3	19	48	52

Source: The World Almanac and Book of Facts, 1993

Pre-Algebra

TECHNOLOGY

Graphing Calculator Masters, p. 14

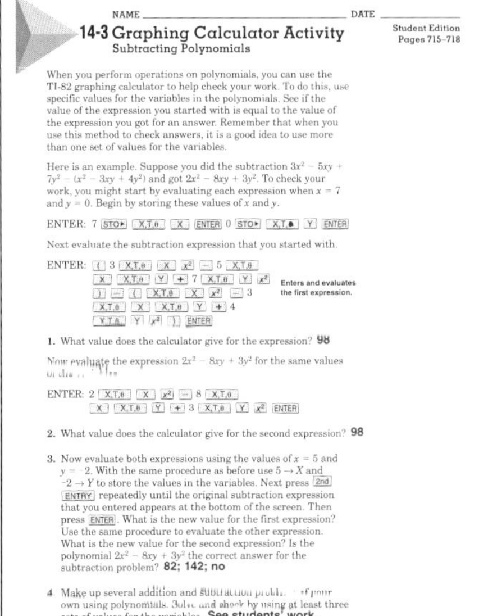

NAME ______________________ DATE __________

14-3 Graphing Calculator Activity
Subtracting Polynomials

Student Edition Pages 715–718

When you perform operations on polynomials, you can use the TI-82 graphing calculator to help check your work. To do this, use specific values for the variables in the polynomials. See if the value of the expression you started with is equal to the value of the expression you got for an answer. Remember that when you use this method to check answers, it is a good idea to use more than one set of values for the variables.

Here is an example. Suppose you did the subtraction $3x^2 - 5xy + 7y^2 - (x^2 - 3xy + 4y^2)$ and got $2x^2 - 8xy + 3y^2$. To check your work, you might start by evaluating each expression when $x = 7$ and $y = 0$. Begin by storing these values of x and y.

ENTER: 7 [STO▸] [X,T,θ] [X] [ENTER] 0 [STO▸] [X,T,θ] [Y] [ENTER]

Next evaluate the subtraction expression that you started with.

ENTER: [(] 3 [X,T,θ] [X] [x^2] [–] 5 [X,T,θ] [X] [X,T,θ] [Y] [+] 7 [X,T,θ] [Y] [x^2] [)] [–] [(] [X,T,θ] [X] [x^2] [–] 3 [X,T,θ] [X] [X,T,θ] [Y] [+] 4 [X,T,θ] [Y] [x^2] [)] [ENTER]

Enters and evaluates the first expression.

1. What value does the calculator give for the expression? **98**

Now evaluate the expression $2x^2 - 8xy + 3y^2$ for the same values of the variables.

ENTER: 2 [X,T,θ] [X] [x^2] [–] 8 [X,T,θ] [X] [X,T,θ] [Y] [+] 3 [X,T,θ] [Y] [x^2] [ENTER]

2. What value does the calculator give for the second expression? **98**
3. Now evaluate both expressions using the values of $x = 5$ and $y = -2$. With the same procedure as before use $5 \rightarrow X$ and $-2 \rightarrow Y$ to store the values in the variables. Next press [2nd] [ENTRY] repeatedly until the original subtraction expression that you entered appears at the bottom of the screen. Then press [ENTER]. What is the new value for the first expression? Use the same procedure to evaluate the other expression. What is the new value for the second expression? Is the polynomial $2x^2 - 8xy + 3y^2$ the correct answer for the subtraction problem? **82; 142; no**
4. Make up several addition and subtraction problems of your own using polynomials. Solve and check by using at least three sets of values for the variables. **See students' work.**

TECH PREP

Tech Prep Applications Masters, pp. 27, 28

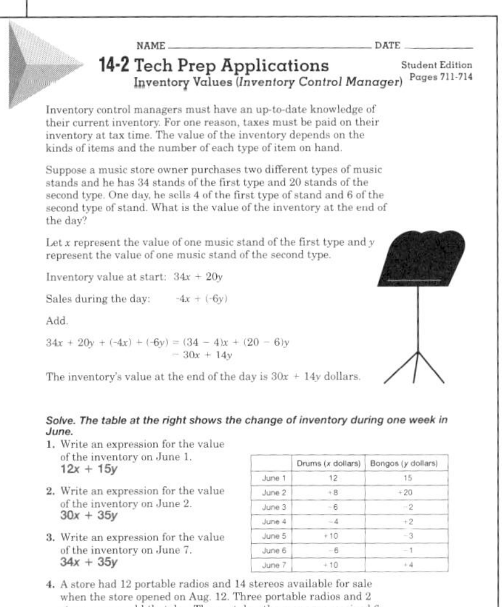

NAME ______________________ DATE __________

14-2 Tech Prep Applications
Inventory Values (*Inventory Control Manager*)

Student Edition Pages 711-714

Inventory control managers must have an up-to-date knowledge of their current inventory. For one reason, taxes must be paid on their inventory at tax time. The value of the inventory depends on the kinds of items and the number of each type of item on hand.

Suppose a music store owner purchases two different types of music stands and he has 34 stands of the first type and 20 stands of the second type. One day, he sells 4 of the first type of stand and 6 of the second type of stand. What is the value of the inventory at the end of the day?

Let x represent the value of one music stand of the first type and y represent the value of one music stand of the second type.

Inventory value at start: $34x + 20y$

Sales during the day: $-4x + (-6y)$

Add.

$$34x + 20y + (-4x) + (-6y) = (34 - 4)x + (20 - 6)y = 30x + 14y$$

The inventory's value at the end of the day is $30x + 14y$ dollars.

Solve. The table at the right shows the change of inventory during one week in June.

	Drums (x dollars)	Bongos (y dollars)
June 1	12	15
June 2	+8	+20
June 3	-6	-2
June 4	-4	+2
June 5	+10	-3
June 6	-6	-1
June 7	+10	+4

1. Write an expression for the value of the inventory on June 1. **$12x + 15y$**
2. Write an expression for the value of the inventory on June 2. **$30x + 35y$**
3. Write an expression for the value of the inventory on June 7. **$34x + 35y$**
4. A store had 12 portable radios and 14 stereos available for sale when the store opened on Aug. 12. Three portable radios and 2 stereos were sold that day. The next day, the manager received 6 portable radios and 4 stereos. During the day, 2 portable radios and 3 stereos were sold. If a represents the value of a portable radio and b represents the value of a stereo, write an expression for the value of the inventory at the end of the second day. **$13a + 13b$**

CONNECTIONS

Activity Masters, pp. 14, 28, 48

NAME ______________________ DATE __________

14-5 Interdisciplinary Activity
Economics

Student Edition Pages 725–727

Last year Mary Beth Hughes started a baby-sitting service. Her price per hour of babysitting was \$5. She found that she was asked to baby-sit 4 hours a week. Wanting to increase her earnings this year she lowered her price to \$4 per hour. She immediately found that her baby-sitting was more in demand. She worked eight hours the first week.

1. The amount Mary Beth earns during a week is her **weekly revenue**. What was her weekly revenue during a week that she charged \$5 per hour? **\$20**
2. What was Mary Beth's weekly revenue when she charged \$4 per hour? **\$32**

Mary Beth's older sister, Ellen, told Mary Beth that the price, p, per hour for baby-sitting can be related by the following equation to the number of hours, x, she is asked to baby-sit each week.

$$p = -\frac{x}{4} + 6$$

This equation is called the **demand equation**.

3. By substituting, show that this equation gives Mary Beth's price when she first started her baby-sitting service and worked only 4 hours per week. Show that it gives the price when she worked 8 hours per week. **$p = -\frac{4}{4} + 6 = 5$; $p = -\frac{8}{4} + 6 = 4$**
4. Describe how to find the weekly revenue if the price per hour and the number of hours worked are known. **Multiply the price per hour times the number of hours worked.**
5. Write a polynomial that represents the revenue when x hours are worked and the price is given by the demand equation. **$-\frac{1}{4}x^2 + 6x$**
6. Write an equation for computing the revenue. Let R represent the revenue and x represent the number of hours worked. **$R = -\frac{1}{4}x^2 + 6x$**
7. Complete the table below. Then plot the points to draw a graph of the revenue equation using the grid at the right.

x	0	4	8	12	16	20
R	**0**	**20**	**32**	**36**	**32**	**20**

R
40
30
20
10
0
Revenue (in dollars)
4 8 12 16 20 x
Hours

8. From the graph in Exercise 7, what value of x gives the greatest value of R? What price would Mary Beth charge for this number of hours? **12; \$3**

COOPERATIVE LEARNING

Math Lab and Modeling Math Masters, pp. 75, 89

NAME ______________________ DATE __________

14-5 Modeling Math Activity
Punnett Squares

Student Edition Pages 725–727

materials: integer mat, algebra tiles, counters

Use the materials to model a Punnett square with two traits. Use the tiles to represent one gene, and the counters to represent the other gene. Let the shaded tile and the shaded counter represent the dominant gene. Let the unshaded tile and the unshaded counter represent the recessive gene.

1. In this square, you will be representing the offspring of peas. The traits are round or wrinkled, and yellow or green. Round and yellow are the dominant genes. Wrinkled and green are the recessive genes. Complete the Punnett square.

	R Y	R y	r Y	r y
R Y	R R Y Y	R R Y y	R r Y Y	R r Y y
R y	R R y Y	R R y y	R r y Y	R r y y
r Y	r R Y Y	r R Y y	r r Y Y	r r Y y
r y	r R y Y	r R y y	r r y Y	r r y y

2. What are the different combinations? **RRYY, RRYy, RRyy, RrYY, RrYy, Rryy, rrYY, rrYy, rryy**
3. How many of the peas are round and yellow? **9**
4. How many of the peas are round and green? **3**
5. How many of the peas are wrinkled and yellow? **3**
6. How many of the peas are wrinkled and green? **1**

MAKING MATHEMATICS RELEVANT

This two-page introduction to the chapter provides students with an opportunity to explore other disciplinary topics and their applications to mathematics.

Background Information

Theme of Chapter The Chinese culture, with its emphasis on family relationships and close human connections, lends itself to the game of bridge in which two partners rely on cooperation and communication. The Chinese language may also play a role in the country's success. It requires students to learn and memorize literally thousands of characters. Compared to this, a game with only 52 cards may not seem so intimidating. The language thus requires accuracy, patience, and perseverance—all qualities of good bridge players.

CHAPTER 14 Polynomials

TOP STORIES in Chapter 14

In this chapter, you will:

- **identify and classify polynomials, and**
- **add, subtract, and multiply polynomials.**

MATH AND GAMES IN THE NEWS

Bridge-building in China

Source: The Atlantic Monthly, February, 1995

"The Chinese are naturals for bridge, a game that brings people together," says Kathie Wei-Sender, a Tennessee-based businesswoman and former world champion bridge player. The Chinese did not begin to compete seriously in international bridge until the 1980s. Nevertheless, both the women's and men's national teams are making strong showings. Young Chinese players are being encouraged to become competitive, and bridge is a part of the curriculum in many schools. Working with a partner and careful bidding are key elements of a winning bridge player. Shanghai-born C. C. Wei invented the "precision system" of bidding that is based on statistics. This system is widely used in China and other countries. Experts on international competition think that it is only a matter of time before the Chinese team wins the championship.

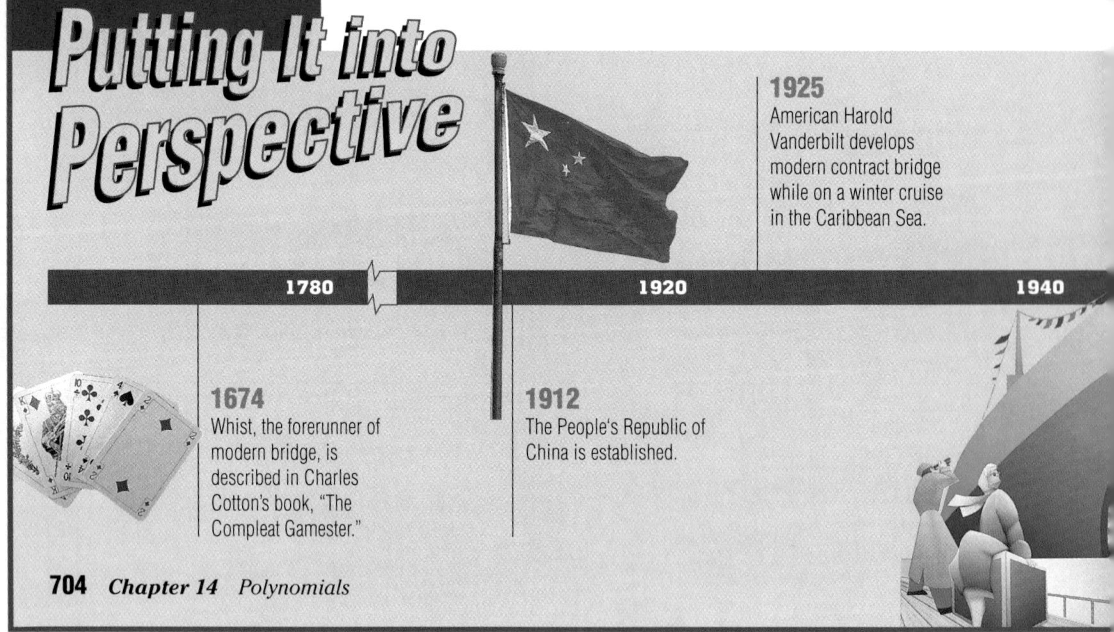

Putting It into Perspective
1925
American Harold Vanderbilt develops modern contract bridge while on a winter cruise in the Caribbean Sea.
1780
1920
1940
1674
Whist, the forerunner of modern bridge, is described in Charles Cotton's book, "The Compleat Gamester."
1912
The People's Republic of China is established.

Putting It into Perspective

As Chinese bridge grew rapidly in China after 1977, differences between the Chinese and American game became evident. The Chinese players are quieter, more polite to each other, and less prone to harsh competitiveness. "Friendship first, competition second" is the rule. Also, because apartments are small, especially in urban areas, the Chinese play bridge mostly away from home, at their workplaces and in clubs.

NORTH
♠ 952
♥ K8642
♦ K54
♣ 63

WEST
♠ AJ864
♥ Q5
♦ 10763
♣ 54

EAST
♠ 103
♥ AJ107
♦ AJ9
♣ Q1097

SOUTH
♠ KQ7
♥ 93
♦ Q82
♣ AKJ82

interNET CONNECTION For up-to-date information on bridge, visit:
www.glencoe.com/sec/math/prealg/mathnet

Statistical Snapshot

Chances of Suit Distributions

Thirteen cards are dealt in a hand of bridge. The distribution of cards that come from different suits is important to the bidding process. A suit distribution of 4-4-3-2 means that 4 cards come from one suit, 4 from a second suit, 3 from a third, and 2 from the fourth suit. The odds of different suit distributions are given in the table below.

Distribution	Approximate Odds Against
4–4–3–2	4 to 1
5–4–2–2	8 to 1
6–4–2–1	20 to 1
7–4–1–1	254 to 1
8–4–1–0	2211 to 1
13–0–0–0	188,753,389,899 to 1

Source: *The World Almanac*

On the Lighter Side

8-30 © 1995 United Feature Syndicate, Inc.

WHO'S THE FELLOW IN THE BLACK HAT?

THAT'S JOE BLACKJACK, THE WORLD FAMOUS RIVER BOAT GAMBLER..

ARE WE GONNA TALK, OR ARE WE GONNA PLAY "OLD MAID"?

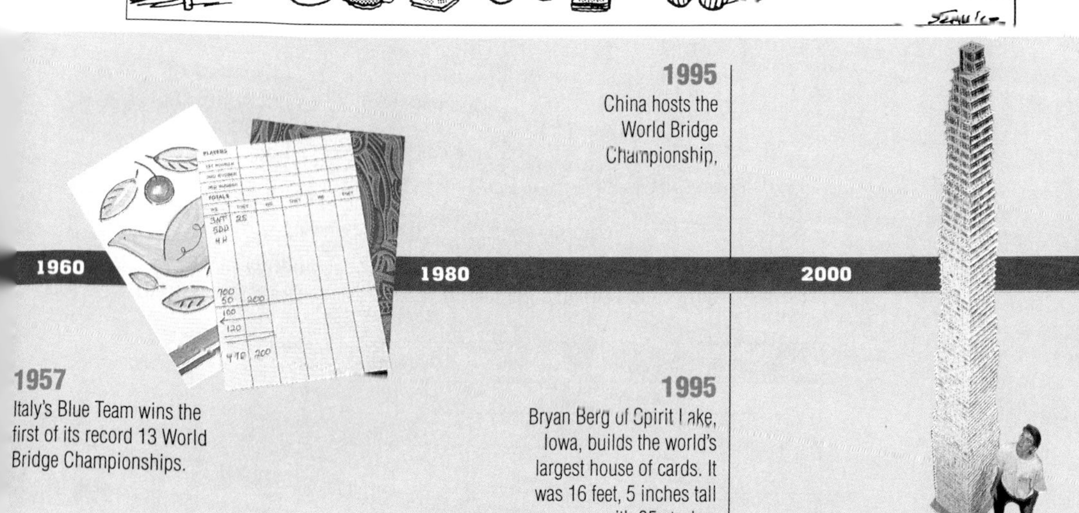

1995
China hosts the World Bridge Championship.
1960
1980
2000
1957
Italy's Blue Team wins the first of its record 13 World Bridge Championships.
1995
Bryan Berg of Spirit Lake, Iowa, builds the world's largest house of cards. It was 16 feet, 5 inches tall with 85 stories.

On the Lighter Side

Temperaments of bridge players vary widely. Some players are fanatically competitive. Others are more easygoing. Some players use tactics to try to intimidate, unsettle, and distract their opponents. To some, this behavior is unsportsmanlike; to others, it's considered to be part of the game.

Statistical Snapshot

Bridge, like most card games, is based heavily on the principles of statistics and chance. The table shows one player's chances of holding various combinations of suits in any given hand of 13 cards.

Cooperative Learning

Chapter Projects Two chapter projects are included in the *Investigations and Projects Masters.* In Chapter Project A, students extend the topic in the Chapter Opener. In Chapter Project B, students explore baseball. A student page and a parent letter are provided for each Chapter Project.

interNET CONNECTION

Glencoe has made every effort to ensure that the website links for *Pre-Algebra* at www.glencoe.com/sec/math/prealg/mathnet are current and contain appropriate content. However, these website links are not under Glencoe's control.

Investigations and Projects Masters, p. 81

NAME ______ DATE ______

CHAPTER 14 Project A
Shuffle the Cards and Cut the Deck

Bridge is a complicated card game with many variations. In addition to "generating kindness" and bringing people together, it calls upon players to use strategy, to understand and use probability concepts, and, in some games, to think as a team. Like other card games, bridge involves a degree of chance. In this project, you will examine several card games mathematically.

1. Create a list of all the card games your classmates know how to play. Your list might include games such as hearts, rummy, pinochle, canasta, crazy eights, war, go fish, and solitaire games.
2. Set up stations where card instructors can teach small groups how to play. Determine who in the class will teach the games you've listed. Having more than one teacher per game may be necessary. You'll also need several decks of cards and perhaps some specialized decks for certain games.
3. Each student should learn how to play three different games. You need not become an expert, but you should know some of the basic concepts behind playing each game. Play several games. Talk about the roles of strategy, chance, and communication. Talk about how statistics and probability affect the game.
4. Get together with a group to analyze the card games. Prepare a brochure that describes the different games and compares and contrasts them. Include a glossary of terms. When formulating your descriptions, address questions such as the following: To what extent does the game involve skill or chance? Is communication needed? How many people play at once? Do you play alone or with a partner? How hard is the game compared to the others? Do you use an understanding of probability? If so, how? What other math skills and concepts do you use? How do you score and win? What special vocabulary, such as *trump*, *shuffle*, and *follow suit*, do you use? What do you like or dislike about the game? Write a summary in which you discuss the essential elements of and draw conclusions about card games.
5. Display your brochures for classmates to examine and read your summaries to the class. As a whole class, talk about what you've learned.

14-1 LESSON NOTES

NCTM Standards: 1-4, 9

Instructional Resources

- Study Guide Master 14-1
- Practice Master 14-1
- Enrichment Master 14-1
- Group Activity Card 14-1

Transparency 14-1A contains the 5-Minute Check for this lesson; **Transparency 14-1B** contains a teaching aid for this lesson.

Recommended Pacing	
Honors Pacing	Day 1 of 9

1 FOCUS

5-Minute Check
(over Chapter 13)

Solve each equation. Round decimal answers to the nearest tenth

1. $r^2 = 121$ **11, -11**
2. $4 = x^2$ **2, -2**
3. $n^2 = 45$ **6.7, -6.7**

In a right triangle, if *a* and *b* are the measures of the legs and *c* is the measure of the hypotenuse, find each missing measure. Round answers to the nearest tenth.

4. $a = 120, b = 160$ **200**
5. $c = 30, b = 14$ **26.5**

Motivating the Lesson

Questioning Use words such as monocle, bicycle, tricycle, and polyester to develop the prefixes mono-, bi-, tri-, and poly-.

Teaching Tip Point out that the degree of a nonzero constant follows the same rules as other monomials because any variable to the zero power has a value of 1. So, $7a^0 = 7 \cdot 1$, or 7.

14-1 Polynomials

Setting Goals: *In this lesson, you'll identify and classify polynomials and find their degree.*

Modeling a Real-World Application: Construction

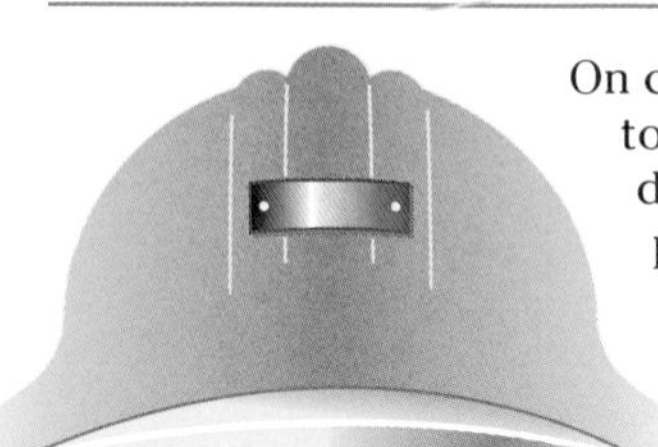

On construction sites, workers and visitors are required to wear hard hats for protection. If an object is dropped from above, it may injure an unprotected person on the ground. The velocity v of a dropped object can be determined by using the formula $v^2 = 2gs$, where g represents the acceleration due to gravity and s represents the distance the object falls.

Learning the Concept

You can review monomials in Lesson 4-1.

As you know, expressions like v^2 and $2gs$ are monomials. Monomials can be numbers, variables, or products of numbers and variables. Expressions that involve variables in the denominator of a fraction or variables with radical signs are not monomials. For example, $3x + 2$, $\frac{5}{m}$, and $\sqrt{ab}$ are not monomials.

An algebraic expression that contains one or more monomials is called a **polynomial**. A polynomial is a sum or difference of monomials. A polynomial with two terms is called a **binomial**, and a polynomial with three terms is called a **trinomial**.

Example 1 **Determine whether the given expression is a polynomial. If it is, classify it as a *monomial*, a *binomial*, or a *trinomial*.**

a. $3x^2 + 7x - 5$

The expression is a polynomial because it is the sum of three monomials. Since there are three terms, it is a trinomial.

b. $3y - \frac{1}{y^2}$

The expression is *not* a polynomial because $\frac{1}{y^2}$ is not a monomial.

The degree of a monomial is the sum of the exponents of its variables. The following chart shows how to find the degree of a monomial.

Monomial	Variables	Exponents	Degree
z	z	1	1
$-3c^2$	c	2	2
$\frac{2}{5}m^3n^4$	m, n	3, 4	3 + 4 or 7
$0.62p^3q^3$	p, q	3, 3	3 + 3 or 6

Remember that $z = z^1$.

Alternative Teaching Strategies

Reading Mathematics After discussion and examples of key terms, separate students into groups of four. Write expressions on cards. For each, have a student in turn tell if it is a monomial or polynomial and name the variable, exponent(s), and degree.

A monomial like 7 that does not have a variable associated with it is called a **constant**. The degree of a nonzero constant is 0. The constant 0 has *no* degree.

A polynomial also has a degree. The degree of a polynomial is the same as that of the term with the greatest degree.

Example  **Find the degree of $x^2 + xy^2 + y^4$.**

First, find the degree of each term.

x^2 has degree 2.
xy^2 has degree 1 + 2 or 3.
y^4 has degree 4. *Greatest degree*

So, the degree of $x^2 + xy^2 + y^4$ is 4.

Example 3

APPLICATION

Medicine

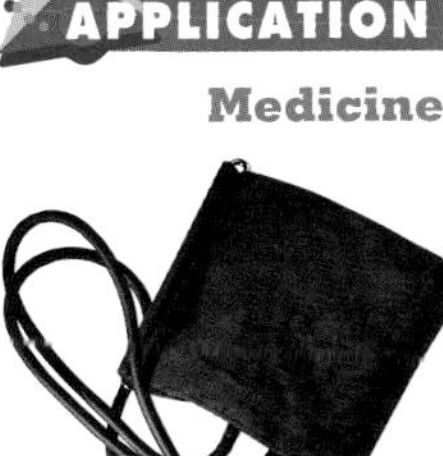

Doctors can study the heart of a potential heart attack patient by injecting dye in a vein near the heart. In a normal heart, the amount of dye in the bloodstream after t seconds is given by the polynomial $-0.006t^4 + 0.140t^3 - 0.53t^2 + 1.79t$. Find the degree of the polynomial.

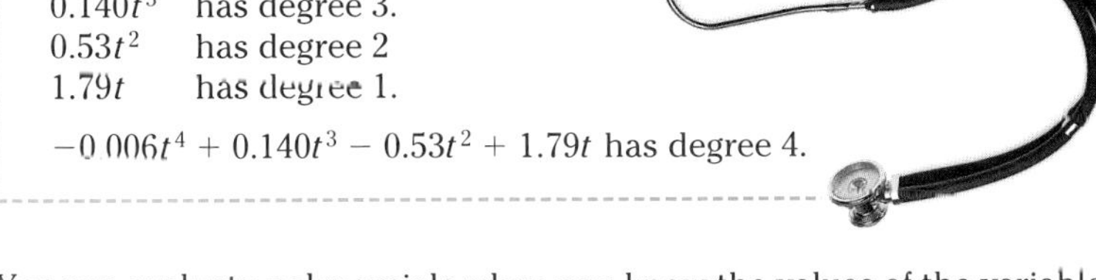

$-0.006t^4$ has degree 4.
$0.140t^3$ has degree 3.
$0.53t^2$ has degree 2
$1.79t$ has degree 1.

$-0.006t^4 + 0.140t^3 - 0.53t^2 + 1.79t$ has degree 4.

You can evaluate polynomials when you know the values of the variables involved.

Example  **Evaluate $6x^2 - 5xy$ if $x = -2$ and $y = -3$.**

$$6x^2 - 5xy = 6 \cdot (-2)^2 - 5(-2)(-3) \quad \textit{Replace x with } -2 \textit{ and y with } -3.$$
$$= 6 \cdot 4 - 5 \cdot 6 \quad \textit{Use the order of operations.}$$
$$= 24 - 30$$
$$= -6$$

Checking Your Understanding

Communicating Mathematics

Read and study the lesson to answer these questions. **1–3. See margin.**

1. **Determine** if $\frac{3w^2}{5y}$ is a monomial. Explain your answer.
2. **Write** three binomial expressions. Explain why they are binomials.
3. **Explain** how to find the degree of a monomial and a polynomial.

4. Write about a way that you can remember the definitions of monomial, binomial, trinomial, and polynomial. **See students' work.**

2 TEACH

In-Class Examples

For Example 1
Determine whether the given expression is a polynomial. If it is, classify it as a *monomial*, a *binomial*, or a *trinomial*.

a. $x^2 - x + \sqrt{n}$ **not a polynomial**

b. $3x - 14$ **binomial**

For Example 2
Find the degree of $5x^2 - 3x^3y - 2x^3y^4 + y^3$. **6**

For Example 3
The total U.S. exports in billions of dollars between 1980 and 1990, can be modeled by the polynomial $T = -0.13t^4 + 2.7t^3 - 14t^2 + 18t - 224$. Find the degree of the polynomial. **4**

For Example 4
Evaluate $5x^2 - 3x + 2y^2$ if $x = 2$ and $y = 5$. **64**

Study Guide Masters, p. 115

NAME ____________ DATE ____________

14-1 Study Guide
Polynomials

Student Edition
Pages 706–709

A **polynomial** is an algebraic expression that contains one or more monomials. A **binomial** is a polynomial with two terms. A **trinomial** is a polynomial with three terms.

A term of a polynomial *cannot* be a quotient with the variable in the denominator or the square root of a variable.

Monimials			
-36	x^3	$0.5y$	$\frac{-z}{10}$

Binomials		
$3x + \sqrt{2}$	$-4a + b^2$	$\frac{c}{3} - \frac{d}{4}$

Trinomials	
$x^2 - 3x + 4$	$4a^2b^3 + 2ab^2 - b$

To find the degree of a monomial, add the exponents on the variables. The monomial $5x^5y^3$ has a degree of 5 + 3 or 8.

The degree of a polynomial the same as that of the term with the highest degree.

Example: Find the degree of $y^3 + 2y^3z + 4x$.

y^3 has degree 3.
$2y^3z$ has degree 3 + 1 or 4. **Greatest degree**
$4x$ has degree 1.

The degree of $y^3 + 2y^3z + 4x$ is 4.

State whether each expression is a polynomial. If it is, classify it as a monomial, bionomial, or trinomial.

1. $r^2 - 2$ **yes, bionomial**
2. $\frac{-10}{c}$ **no**
3. $\frac{w}{2} - 24$ **yes, binomial**
4. $b^2 + 1$ **yes, bionomial**
5. $-14m^2n$ **yes, monomial**
6. $4x^2 - 2xy - 4y^2$ **yes, trinomial**
7. $\frac{1}{x} - \frac{1}{y} - \frac{1}{z}$ **no**
8. $a^2 - b^3 - c^4$ **yes, trinomial**
9. $\sqrt{b - 4ac}$ **no**

Find the degree of each polynomial.

10. $5d^4 + 13a^2b^4c$ **7**
11. $7g^2h - 2h$ **3**
12. $2x^4 - 6xy^3 + 3x^3z^2$ **5**
13. -36 **0**
14. $12a + 3$ **1**
15. $-34r^2s^4t^6$ **12**

Reteaching

Act It Out Have students play a "Tell the Truth" game. Have one student write an expression on the chalkboard. The student then states, "I have a polynomial of degree (integer)". The other students determine whether the statement is true and, if it is not true, why it is false.

Additional Answers

1. No, because there is a variable in the denominator.
2. Sample answers: $4x + 7$; $15y^2 + 22x$; $\frac{1}{2}a - 6$; These are binomials because they are sums or differences of two monomials.
3. The degree of a monomial is the sum of the exponents of its variables. The degree of a polynomial is the same as the degree of the term with the greatest degree.

3 PRACTICE/APPLY

Checking Your Understanding

Exercises 1–17 are designed to help you assess your students' understanding through reading, writing, speaking, and modeling. You should work through Exercises 1–4 with your students and then monitor their work on Exercises 5–17.

Error Analysis

Watch for students who multiply exponents of factors instead of adding them. Suggest writing factors in expanded form, performing the multiplication, and returning to exponential form.

Assignment Guide

Core: 19–53 odd, 54, 55, 57–67
Enriched: 18–52 even, 54–67

For **Extra Practice**, see p. 772.

The red A, B, and C flags, printed only in the Teacher's Wraparound Edition, indicate the level of difficulty of the exercises.

Additional Answers

18. yes, monomial
19. yes, binomial

Practice Masters, p. 115

NAME ______________ DATE ______

14-1 Practice
Polynomials

Student Edition
Pages 706–709

State whether each expression is a polynomial. If it is, classify it as a monomial, binomial, or trinomial.

1. $p^2 - q^2$ yes, binomial
2. $\sqrt{x+1}$ no
3. $9y$ yes, monomial
4. $ax + bx + x^3$ yes, trinomial
5. $\frac{4}{z}$ no
6. -12 yes, monomial
7. $\sqrt{81}$ yes, monomial
8. $6 + d^5$ yes, binomial
9. $e - f$ yes, binomial
10. $\frac{1}{6} + \frac{1}{3}x - x^3$ yes, trinomial
11. $a + \frac{4}{c}$ no
12. $\frac{1}{4}t^3$ yes, monomial

Find the degree of each polynomial.

13. $3x - 1$ 1
14. $2y^3$ 3
15. $3a + 2b$ 1
16. $-2t^2 + s^4$ 4
17. $4c^2 - 9c^2d^3$ 5
18. -2 0
19. $5y^6 - 2xy^3z + x^4yz^2$ 7
20. $2pq^2 - pq^3$ 4
21. $-3ka^4 - 4a^6$ 6

Evaluate each polynomial if a = 1, b = −2, c = −1, and d = −4.

22. $a^2 - 3bd$ −23
23. $b^3 - 4ad$ 8
24. $1 - b^2 + c^3$ −4
25. $2b^4 - 3ad^2$ −16
26. $\sqrt{2bd}$ 4
27. $a^2 - a^4$ 0
28. $4abc - 5ab^2 + 3a^2b$ −18
29. $c^2 - a^2 + 1$ 1
30. $a^2 - c^2 + 1$ 1
31. $6bc^2 - 2ab + 4ab^2$ 8
32. $2b^2 - b + 3$ 13
33. $-d^2 + 3d - 1$ −29

Guided Practice

State whether each expression is a polynomial. If it is, classify it as a *monomial*, *binomial*, or *trinomial*.

5. $\frac{x}{2}$ yes; monomial
6. -5 yes; monomial
7. $\frac{ab}{c} - c$ no
8. $3x^2 + 4x - 2$ yes; trinomial

Find the degree of each polynomial.

9. $-2xy^2$ 3
10. $12x^3y^2$ 5
11. $x^3 + 7x$ 3
12. $x^2 + xy^2 - y^4$ 4

Evaluate each polynomial if $x = 3$, $y = -5$, and $z = -1$.

13. $3x - 5y + z^2$ 35
14. $x^3 + 2y^2 - z$ 78
15. $x^3 - 2xy$ 57
16. $2xy + 6yz^2$ −60
17. **Business** The cost of an order of 12 burgers, 11 fries, and 5 shakes can be represented by the polynomial $12b + 11f + 5s$. Determine the cost of the order if burgers are \$1.79 each, fries are \$0.99 each, and shakes are \$1.29 each. \$38.82

Exercises: Practicing and Applying the Concept

Independent Practice

State whether each expression is a polynomial. If it is, classify it as a *monomial*, *binomial*, or *trinomial*. 18–29. See margin.

18. $-\frac{1}{9}r^2$
19. $c^2 + 3$
20. $x^2 - 4$
21. $1 + 3x + 5x^2$
22. $\frac{6}{a} + b$
23. $5xy$
24. $3x^2y$
25. $ab^2 + 3a - b^2$
26. $x + y$
27. $-\sqrt{49}$
28. $\frac{e}{6}$
29. $x^2 - \frac{1}{2}x + \frac{1}{3}$

B

Find the degree of each polynomial.

30. $11c^2 + 4$ 2
31. $3x + 5$ 1
32. 121 0
33. $4x^3 + xy - y^2$ 3
34. $x^6 + y^6$ 6
35. $d^4 + c^4d^2$ 6
36. $16y^2 + mnp$ 3
37. $x^3 - x^2y^3 + 8$ 5
38. $-7x^3y^4$ 7
39. $-x^2yz^4$ 7
40. $14c^5 - 16c^6d$ 7
41. $2x^5 + 9x + 1$ 5

Evaluate each polynomial if $a = 2$, $b = -3$, $c = 4$, and $d = -5$. 48. 22

42. $a^3 - 2bc$ 32
43. $ab + cd$ −26
44. $2abc + 3a^2b$ −84
45. $b^3 - 2ac + d^2$ −18
46. $b^2 + a^2b$ −3
47. $a^5 + bd^2$ −43
48. $a^2 + b^2 - c^2 + d^2$
49. $abcd - 25 + a$ 97
50. $5ad^2 - 2a + (bc)^2$ 390

53. 82

C

51. $-(-ac)^2 - cd$ −44
52. $\sqrt{c} - b^2d$ 47
53. $d - abc^2 - \sqrt{-27b}$

Critical Thinking

54. Find the degree of the polynomial $a^{x+3} + x^{x-2}b^3 + b^{x+2}$. $x + 3$

Applications and Problem Solving

55. **Landscaping** Lee's Garden Shop wants to place hosta plants along the perimeter of a garden. $2x + 2y + z + xy$
    a. Write a polynomial that represents the perimeter of the garden in feet.
    b. What is the degree of the polynomial? 2

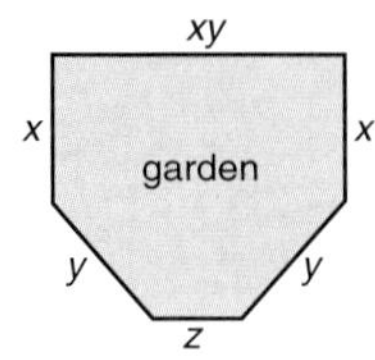

Group Activity Card 14-1

Polynomial Properties Group Activity **14-1**

MATERIALS: Three decks of Polynomial Cards with the expressions given on back of this card

The object of this activity is to correctly compute the degree of a polynomial. The polynomial is obtained by turning over cards from the three decks of the Polynomial Cards until a blank card shows.

To begin, place the three decks on the table with the *number* deck first, the *variable* deck second, and the *operations* deck last.

The leader of the group turns up one card from each deck. If the last card is not blank, the leader turns up another three cards, placing them after the first three cards. When a blank card is turned up, group members compute the degree of the displayed polynomial.

 Pre-Algebra

Additional Answers

20. yes, binomial
21. yes, trinomial
22. no
23. yes, monomial
24. yes, monomial
25. yes, trinomial
26. yes, binomial
27. yes, monomial
28. yes, monomial
29. yes, trinomial

56. **Ecology** In the early 1900s, the deer population of the Kaibab Plateau in Arizona increased rapidly because hunters had reduced the number of predators. However, the food supply was not great enough to support the increased population. So, the population eventually began to decline. The deer population for the years 1905 to 1930 can be approximated by the polynomial $-0.125x^5 + 3.125x^4 + 4000$, where x is the number of years from 1900.
   a. Find the degree of the polynomial. **5**
   b. Evaluate the polynomial to find the deer population in 1920. **104,000**

57. **Family Activity** The amount of money in a savings account can be written as a polynomial. Suppose that your grandparents placed \$100 in your savings account each year on your birthday. On your fifth birthday, there would have been $100x^4 + 100x^3 + 100x^2 + 100x + 100$ dollars, where x is the annual interest rate plus 1. Research the interest rate at your family's bank. How much money would you have on your next birthday at their passbook savings rate? **See students' work.**

Mixed Review

58. **Geometry** Find the length of $\overline{XY}$ in $\triangle WXY$. Round to the nearest tenth. (Lesson 13-7) **11.9 ft**

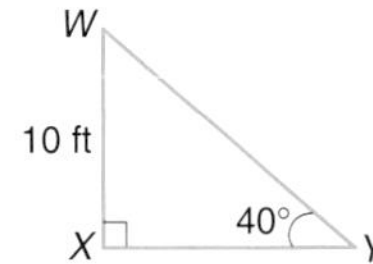

59. **Catering** A caterer is ordering a tent for outdoor parties. The area enclosed by the tent is 8 meters by 8 meters. The top is a square pyramid with a slant height of 4.2 meters. The hanging sides are 2.8 meters tall on each side. If there is no floor, what is the area of the canvas required to make the tent? (Lesson 12-6) **156.8 m²**

60. The stem-and-leaf plot shows the number of canned goods collected by each class for the food drive. (Lesson 10-2)

Stem	Leaf
1	98
2	10 56 78 80 85
3	08 12 60 62 75 80
4	15
5	25

1 | 98 = 198 cans

   a. What is the range? **327**
   b. What is the median? **310**
   c. What is the interquartile range? **97**
   d. Are there any outliers? **yes; 525**

61. **$104.37**
63. **yes; 3; 81, 243, 729**
64. **10.5 m**

61. **Retail Sales** The retail price of a camera is \$98. The sales tax rate is 6.5% of the price. What is the total cost of the camera? (Lesson 9-5)

62. Solve $\frac{2x}{5} - 3 > -5$. (Lesson 7-6) $x > -5$

63. State whether 1, 3, 9, 27, . . . is a geometric sequence. If so, state the common ratio and list the next three terms. (Lesson 6-8)

64. **Geometry** A rectangle has a length of 3.5 meters and a width of 1.75 meters. What is the perimeter of the rectangle? (Lesson 5-3)

65. Write an inequality for the solution set graphed below. (Lesson 3-6) $x > 8$

0 1 2 3 4 5 6 7 8 9 10

66. Find the quotient of 48 and −3. (Lesson 2-8) **−16**

67. Simplify $|-8| + |5|$. (Lesson 2-1) **13**

Extension

Using Science Connections Have students use physical science books to find formulas that involve polynomials. Have students make a list of the polynomials and the degree of each.

4 ASSESS

Closing Activity

Writing Have the students write ten polynomial expressions: three monomials, three binomials, and three trinomials, and one with more than three unlike terms. Then have them write five expressions that are not polynomials.

Enrichment Masters, p. 115

NAME ____________ DATE ________

14-1 Enrichment
The Summation Sign

Student Edition Pages 706–709

The Greek letter Σ (sigma) is called the **summation sign**. It is used to abbreviate the writing of a series.

Consider the series 2 + 4 + 6 + 8 + 10. It can be written as 2(1) + 2(2) + 2(3) + 2(4) + 2(5). Each term is 2 times a number, so $2k$ can be used to represent each term.

The sum of $2k$ from $k = 1$ to $k = 5$ can be written $\sum_{k=1}^{5} 2k$.

This symbol means that you successively replace k by each value 1, 2, 3, 4, and 5, and write the resulting numbers as an indicated sum. The variable k is called the **index**.

Example: Write the series $\sum_{t=1}^{6} (t-1)$ as an indicated sum.

$$\sum_{t=1}^{6} (t-1) = 0 + 1 + 2 + 3 + 4 + 5$$

Write each series as an indicated sum.

1. $\sum_{j=3}^{7} j$ **3 + 4 + 5 + 6 + 7**
2. $\sum_{k=4}^{8} (k+2)$ **6 + 7 + 8 + 9 + 10**
3. $\sum_{t=1}^{5} \frac{1}{2}t$ $\frac{1}{2} + 1 + \frac{3}{2} + 2 + \frac{5}{2}$
4. $\sum_{x=1}^{5} x^2$ **1 + 4 + 9 + 16 + 25**
5. $\sum_{k=2}^{6} (2k+1)$ **5 + 7 + 9 + 11 + 13**
6. $\sum_{k=1}^{5} \frac{k(k+1)}{2}$ **1 + 3 + 6 + 10 + 15**

Use Σ to write each series. **Answers may vary. Sample answers given.**

7. 4 + 7 + 10 + 13 $\sum_{k=1}^{4} (3k+1)$
8. 6 + 10 + 14 + 18 $\sum_{k=1}^{4} (4k+2)$
9. 1 + 8 + 27 + 64 $\sum_{k=1}^{4} k^3$
10. $\frac{1}{4} + \frac{2}{5} + \frac{3}{6} + \frac{4}{7}$ $\sum_{k=1}^{4} \frac{k}{k+3}$

14-1B LESSON NOTES

NCTM Standards: 1-4, 9

Objective
Use algebra tiles to model polynomials.

Recommended Time
Demonstration and discussion: 15 minutes; Exercises: 30 minutes

Instructional Resources
For each student or group of students
Student Manipulative Kit
- algebra tiles

Math Lab and Modeling Math Masters
- p. 1 (algebra tiles)
- p. 56 (worksheet)

For teacher demonstration
Overhead Manipulative Resources

1 FOCUS

Motivating the Lesson
Draw several squares on the board, each with an integer greater than one as the length of a side. Have students name the area of each square. Extend by drawing a square with a side length of 1 and another with a length of x.

2 TEACH

Teaching Tip
Remind students that $2x^2 - 3x + 4$ is the same as $2x^2 + (-3x) + 4$.

3 PRACTICE/APPLY

Assignment Guide
Core: 1–3
Enriched: 1–3

4 ASSESS

Observing students working in cooperative groups is an excellent method of assessment.

HANDS-ON ACTIVITY

14-1B Representing Polynomials with Algebra Tiles

An Extension of Lesson **14-1**

MATERIALS
algebra tiles

You can use the concept of area to model monomials. A model of a monomial can be made using tiles or drawings like the ones shown below.

Let [1 × 1 square] be a 1×1 square. The square has an area of 1.

Let [1 × x rectangle] be a $1 \times x$ rectangle. The rectangle has an area of x.

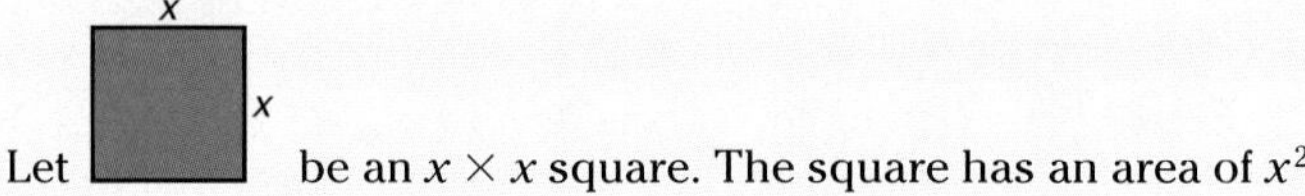

Let [x × x square] be an $x \times x$ square. The square has an area of x^2.

Red tiles are used to represent -1, $-x$, and $-x^2$.

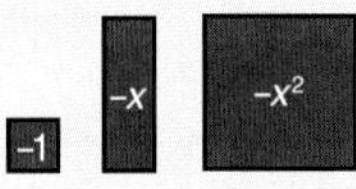

You can use these tiles to make a model of a polynomial.

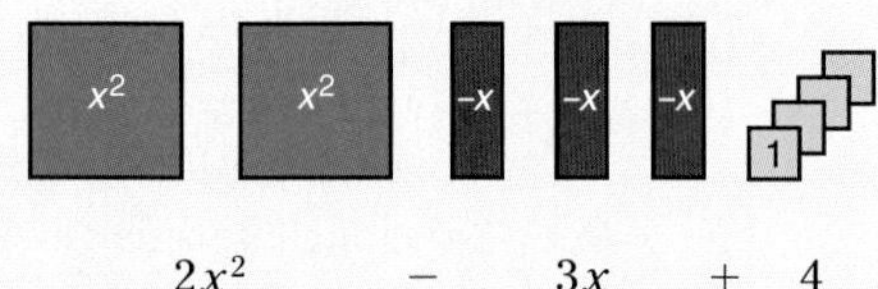

$2x^2 \quad - \quad 3x \quad + \quad 4$

Your Turn ▶ Model $-3x^2$, $5x + 3$, $4x^2 - x$, and $5x^2 + 2x - 3$ using algebra tiles.

TALK ABOUT IT

1. Explain how you can tell if an expression is a monomial, binomial, or trinomial by looking at the area tiles. **See Solutions Manual.**
2. Name the polynomial represented by the model below. $-x^2 + 3x - 5$

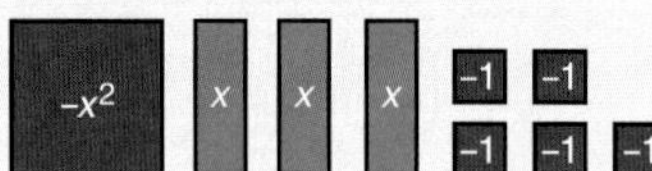

3. Explain how you would find the degree of a polynomial using algebra tiles. **See Solutions Manual.**

See Solutions Manual.

14-2 Adding Polynomials

Setting Goals: *In this lesson, you'll add polynomials.*

Modeling with Manipulatives

MATERIALS

algebra tiles

Algebra tiles can be used to model addition of polynomials.

- The numerical part of a monomial is called the **coefficient**. For example, the coefficient of $5x^2$ is 5.
- When monomials are the same or differ only by their coefficients they are called **like terms**. For example, $2x$ and $7x$, and $-5x^2y$ and $9x^2y$ are like terms; $6a$ and $-5b$, and $7a^3b^2$ and $7a^2b$ are unlike terms. Like terms are represented by tiles that are the same shape and size.
- A zero pair is formed by pairing one tile with its opposite.
- You can remove or add zero pairs without changing the value of the polynomial.

The model below represents $(2x^2 + 2x - 4) + (x^2 + 3x + 6)$.

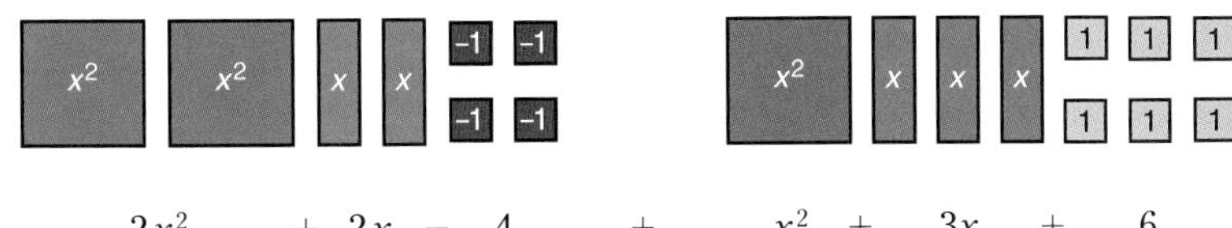

$2x^2 \quad + \; 2x \; - \; 4 \quad + \quad x^2 \; + \; 3x \; + \; 6$

- To add the polynomials, combine like terms and remove all zero pairs.

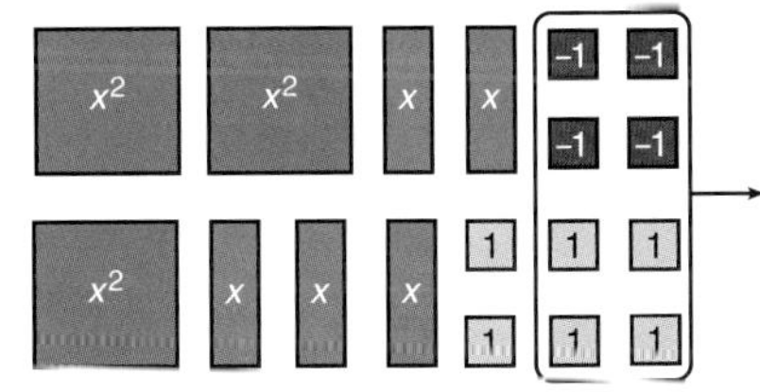

- Write the polynomial for the tiles that remain.
  Therefore, $(2x^2 + 2x - 4) + (x^2 + 3x + 6) = 3x^2 + 5x + 2$.

Your Turn Model $(x^2 + 4x + 2) + (7x^2 - 2x + 3)$ using algebra tiles.
See students' work.

a. What is the sum of $(x^2 + 4x + 2)$ and $(7x^2 - 2x + 3)$? **$8x^2 + 2x + 5$**

b. Compare and contrast finding sums of polynomials using algebra tiles to finding the sum of integers using counters. **See margin.**

Learning the Concept

You can use the principles and properties you learned about integers and rational numbers to add polynomials.

14-2 LESSON NOTES

NCTM Standards: 1-4, 9

Instructional Resources

- Study Guide Master 14-2
- Practice Master 14-2
- Enrichment Master 14-2
- Group Activity Card 14-2
- Assessment and Evaluation Masters, p. 379
- Tech Prep Applications Masters, p. 27

Transparency 14-2A contains the 5-Minute Check for this lesson; **Transparency 14-2B** contains a teaching aid for this lesson.

Recommended Pacing	
Honors Pacing	Day 2 of 9

1 FOCUS

5-Minute Check
(over Lesson 14-1)

Find the degree of each polynomial.

1. $5x^3 - 2y^4 + 6$ **4**
2. $x^3y - x^2y^3 + 7$ **5**

Evaluate each polynomial if $m = 3$, $n = 2$, and $p = 4$.

3. $-12mn + p^3$ **−8**
4. $3n^2 - mp - 2m^2$ **−18**
5. $p^3 - (mn)^2$ **28**

Motivating the Lesson

Situational Problem Show the class a bowl of apples and oranges, and ask them to describe the contents of the bowl using one number. Point out that the word "fruit" must be used when only one number can be used. To accurately describe the apples and oranges, two different phrases must be used. Similarly, in algebra, different variables are used in specific situations.

Additional Answer Talk About It

b. The concept of the zero pairs is the same, but there are tiles that represent different terms in algebra tiles, and counters all represent 1 unit.

Cooperative Learning

Trade-A-Problem Have each student write any rational number from 1 to 5 on an index card. Shuffle the cards. Then each student draws a card and takes it to his or her cooperative learning group. Each group is to use the numbers drawn as coefficients for x and make up problems using the group's monomials. Then groups exchange problems and solve.

For more information on this strategy, see *Cooperative Learning in the Mathematics Classroom*, one of the titles in the Glencoe Mathematics Professional Series, pp. 25–26.

2 TEACH

In-Class Examples

For Example 1
Find each sum.

a. $(9x - 6) + (4x - 5)$ $\mathbf{13x - 11}$

b. $(x^2 - 2x + 3) + (2x^2 - x - 3)$ $\mathbf{3x^2 - 3x}$

c. $(2xy - 3x^2 + 2y^2) + (3y^2 - x^2)$ $\mathbf{-4x^2 + 2xy + 5y^2}$

For Example 2
The length of a rectangular garden is 5 feet more than twice its width.

a. If x is the width, what is the perimeter of the garden? **6x + 10**

b. If x is $4\frac{1}{2}$ feet, what is the perimeter of the rectangle? **37 ft**

Teaching Tip In Example 1c of the In-Class Examples, point out that it is a good idea to rewrite the terms of both polynomials in the same order before trying to add them.

Study Guide Masters, p. 116

NAME ______________________ DATE __________

14-2 Study Guide
Adding Polynomials

Student Edition Pages 711–714

To add two or more polynomials, look for **like terms**. Like terms are those that have exactly the same variables to the same powers.

Method 1: Add Vertically

$$\begin{array}{r} 5x + 1 \\ + \ 4x + 3 \\ \hline 9x + 4 \end{array}$$

Align the terms.
Add.

In Method 1, notice that the terms are aligned and added.

Method 2: Add Horizontally

$$\begin{aligned}(3x + 2y + 1) + (5x + 2) &= (3x + 5x) + (2y) + (1 + 2) \\ &= 8x + 2y + 3\end{aligned}$$

In Method 2, notice that there is nothing to add to the term $2y$. That is, the second polynomial does not include a term having the variable y.

Find each sum.

1. $\begin{array}{r} 2x + 3 \\ + \ 3x + 9 \\ \hline \end{array}$ $\mathbf{5x + 12}$
2. $\begin{array}{r} 8y - 7 \\ + \ x + 13 \\ \hline \end{array}$ $\mathbf{8y + x + 6}$
3. $\begin{array}{r} 12y + 7x \\ + \ 7y - 2x \\ \hline \end{array}$ $\mathbf{19y + 5x}$
4. $\begin{array}{r} 3x^2 - x + 7 \\ + \ 2x^2 + 3x + 3 \\ \hline \end{array}$ $\mathbf{5x^2 + 2x + 10}$
5. $\begin{array}{r} 6x + 12 \\ + \ 8y - 2 \\ \hline \end{array}$ $\mathbf{6x + 8y + 10}$
6. $\begin{array}{r} 8x + 9 \\ + \ x + 4y + 11 \\ \hline \end{array}$ $\mathbf{9x + 4y + 20}$
7. $(4y + 2x) + (6y - x)$ $\mathbf{10y + x}$
8. $(6x + 2y) + (4x - 8y)$ $\mathbf{10x - 6y}$
9. $(11m - 4n) + 3n$ $\mathbf{11m - n}$
10. $(x + y + 8) + (5x + 2y + 3)$ $\mathbf{6x + 3y + 11}$
11. $(3x^2 - 12x + 7) + (3 + 2x - 6x^2)$ $\mathbf{-3x^2 - 10x + 10}$
12. $(5y^2 + 3y + 2) + (2y^2 + 9)$ $\mathbf{7y^2 + 3y + 11}$
13. $(4y^2 - y + 6) + (3y - 4)$ $\mathbf{4y^2 + 2y + 2}$

Example 1 **Find each sum.**

a. $(5x + 7) + (3x + 3)$

Method 1: Add vertically

$$\begin{array}{r} 5x + 7 \\ (+)\ 3x + 3 \\ \hline 8x + 10 \end{array}$$

Notice that the terms are aligned and added.

Method 2: Add horizontally

$$\begin{aligned}(5x + 7) + (3x + 3) &= (5x + 3x) + (7 + 3) && \text{Associative and commutative properties of addition} \\ &= (5 + 3)x + (7 + 3) && \text{Distributive property} \\ &= 8x + 10\end{aligned}$$

The sum is $8x + 10$.

b. $(3x^2 + x + 2) + (x^2 + 2x + 4)$

$$\begin{aligned}(3x^2 + x + 2) + (x^2 + 2x + 4) &= (3x^2 + x^2) + (x + 2x) + (2 + 4) \\ &= 4x^2 + 3x + 6\end{aligned}$$

The sum is $4x^2 + 3x + 6$.

c. $(2x^2 + 5xy + 7y^2) + (4x^2 - 3y^2)$

$$\begin{array}{r} 2x^2 + 5xy + 7y^2 \\ (+)\ 4x^2 \qquad\quad - 3y^2 \\ \hline 6x^2 + 5xy + 4y^2 \end{array}$$

The sum is $6x^2 + 5xy + 4y^2$.

Polynomials are often used to represent measures of geometric figures.

Example 2 INTEGRATION Geometry

The lengths of the sides of golden rectangles are in a special ratio. The rectangle at the right shows the lengths of the sides of a general golden rectangle.

a. Find the perimeter of the rectangle.
b. Find the length, the width, and the perimeter of the rectangle if x is 5.2.

a. $\begin{aligned}x + 1.618x + x + 1.618x &= (1 + 1.618 + 1 + 1.618)x \\ &= 5.236x\end{aligned}$

b. If x is 5.2, then the width is 5.2 units, the length is 1.618(5.2) units, and the perimeter is 5.236(5.2) units.

Estimate the length and perimeter first and then use a calculator.

Length: (1.618)(5.2) is about $2 \cdot 5$ or 10 units.

1.618 [×] 5.2 [=] 8.4136

Perimeter: (5.236)(5.2) is about $5 \cdot 5$ or 25 units.

5.236 [×] 5.2 [=] 27.2272

CULTURAL CONNECTIONS
During the golden age of Greece, artists used the proportions of the golden rectangle to draw human forms. The ratio was rediscovered by European artists during the Renaissance.

CULTURAL CONNECTIONS
The golden section is a line segment that has been divided into two parts (*a* and *b*, $a > b$) in such a way that this proportion is true.

$$\frac{a}{b} = \frac{(a + b)}{a}$$

This ratio is called the *golden ratio.*

Reteaching

Using Diagrams Have students use colored pencils to sketch the algebra tiles of each addend in Exercises 7–12. Have them draw x^2 tiles using blue lines, x tiles using green lines, and units using yellow lines. Negative monomials are shaded red. As students add like terms, remind them to start by removing any zero pairs and then writing the sum of the remaining tiles.

Checking Your Understanding

Communicating Mathematics

Read and study the lesson to answer these questions. 1–3. See margin.

1. **Name** the like terms in $(x^2 + 7x + 1) + (2x^2 - 3x + 5)$.
2. **Demonstrate** how to find the sum of $5x^2 + 6x + 4$ and $2x^2 + 3x + 1$.
3. **You Decide** Caroline says that $4xyz$ and $3zyx$ are like terms. Nikita says that they are not. Who is correct? Explain.

MATERIALS

algebra tiles

Use algebra tiles to find each sum.

4. $(3x^2 - 2x + 1) + (x^2 + 5x - 3)$
5. $(-2x^2 + x - 5) + (x^2 - 3x + 2)$
6. $(x^2 - 3x + 6) + (2x^2 + 5x - 4) + (x^2 + x + 1)$ $4x^2 + 3x + 3$

4. $4x^2 + 3x - 2$ 5. $-x^2 - 2x - 3$

Guided Practice

Find each sum.

7. $(x + 3) + (2x + 5)$ $3x + 8$
8. $(4x + 3) + (x - 1)$ $5x + 2$
9. $(3x - 5) + (x + 9)$ $4x + 4$
10. $(2x - 3) + (x - 1)$ $3x - 4$
11. $\begin{array}{r} 2x^2 + 4x + 5 \\ (+)\ x^2 - x - 3 \\ \hline \end{array}$ $3x^2 + 3x + 2$
12. $\begin{array}{r} 2x^2 - 5x + 4 \\ (+)\ 3x^2 + 8x - 1 \\ \hline \end{array}$ $5x^2 + 3x + 3$

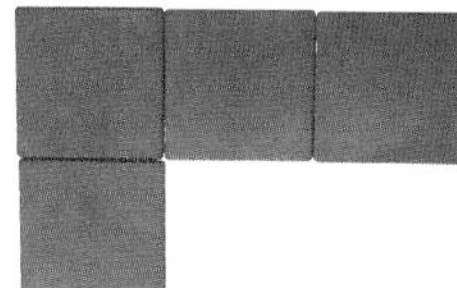

13. **Geometry** Find the perimeter of the rectangle shown at the right. $6x + 24$ units

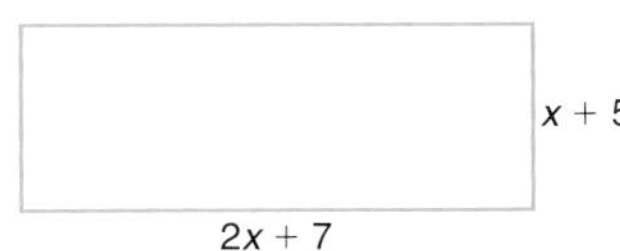

Exercises: Practicing and Applying the Concept

Independent Practice

Find each sum.

14. $(6y - 5r) + (2y + 7r)$ $8y + 2r$
15. $(3x + 9) + (x + 5)$ $4x + 14$
16. $(11x + 2y) + (x - 5y)$ $12x - 3y$
17. $(8m - 2n) + (3m + n)$ $11m - n$
18. $(13x - 7y) + 3y$ $13x - 4y$
19. $(2x^2 + 5x) + (9 - 7x)$
20. $(3r + 6s) + (5r - 9s)$ $8r - 3s$
21. $(5m + 3n) + 12m$ $17m + 3n$
22. $(3x^2 - 9x + 5) + (5x^2 + 5x - 11)$
23. $(5x^2 - 7x + 9) + (3x^2 + 4x - 6)$
24. $(6x^2 + 15x - 9) + (5 - 8x - 8x^2)$
25. $(a^3 - b^3) + (3a^3 + 2a^2b - b^2 + 2b^3)$
26. $\begin{array}{r} -6y^2 + 7b - 5 \\ (+)\ 2y^2 - 9b + 8 \\ \hline \end{array}$
27. $\begin{array}{r} 3a + 5b - 4c \\ 2a - 3b + 7c \\ (+)\ -a + 4b - 2c \\ \hline \end{array}$
28. $\begin{array}{r} 3x^2 - 7x + 9 \\ -2x^2 + x - 4 \\ (+)\ x^2 + 3x - 1 \\ \hline \end{array}$

19. $2x^2 - 2x + 9$

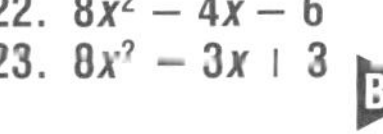

22. $8x^2 - 4x - 6$
23. $8x^2 - 3x + 3$

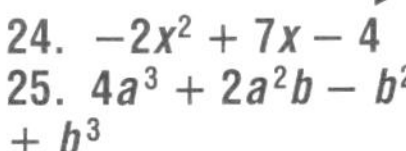

24. $-2x^2 + 7x - 4$
25. $4a^3 + 2a^2b - b^2 + b^3$
26. $-4y^2 - 2b + 3$
27. $4a + 6b + c$
20. $2x^2 - 3x + 4$

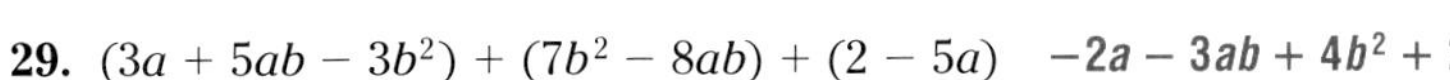

29. $(3a + 5ab - 3b^2) + (7b^2 - 8ab) + (2 - 5a)$ $-2a - 3ab + 4b^2 + 2$
30. $(x^2 + x + 5) + (3x^2 - 4x - 2) + (2x^2 + 2x - 1)$ $6x^2 - x + 2$
31. $(3x^2 + 7) + (4x - 2) + (x^2 - 3x - 6)$ $4x^2 + x - 1$

Find each sum. Then, evaluate if $a = -3$ and $b = 4$.

32. $(3a + 5b) + (2a - 9b)$ $5a - 4b; -31$
33. $(a^2 - 3ab + b^2) + (3a^2 - 2b - 5b^2)$ $4a^2 - 3ab - 2b - 4b^2; 0$
34. $(a^2 + 7b^2) + (5 - 3b^2) + (2a^2 - 7)$ $3a^2 + 4b^2 - 2; 89$

3 PRACTICE/APPLY

Checking Your Understanding

Exercises 1–13 are designed to help you assess your students' understanding through reading, writing, speaking, and modeling. You should work through Exercises 1–6 with your students and then monitor their work on Exercises 7–13.

Error Analysis

When adding polynomials, students may add the exponents when they add the coefficients to combine like terms. Work similar problems involving units of measure instead of variables. Show that only the coefficient changes. Relate this to the like terms with variables.

Assignment Guide

Core: 15–37 odd, 38–45
Enriched: 14–34 even, 35–45

For **Extra Practice**, see p. 773.

The red A, B, and C flags, printed only in the Teacher's Wraparound Edition, indicate the level of difficulty of the exercises.

Group Activity Card 14-2

Add Them Up Group Activity **14-2**

MATERIALS: Three decks of cards with the expressions given on the back of this card.

Shuffle the three decks of cards and place the *number* deck on the table first, the *variable* deck second, and the *operations* last.

The leader turns up one card from each deck. When the card from the last deck is blank, everyone copies the polynomial made. After the leader makes two polynomials in this fashion, everyone adds them.

The first one to get the correct answer wins one point. The cards are shuffled and play starts again. When everyone has been the leader one time, the overall winner is the person with the most points.

 Pre-Algebra

Additional Answers

1. x^2 and $2x^2$; $7x$ and $-3x$; 1 and 5
2. Sample answer: group like terms and add. $(5x^2 + 6x + 4) + (2x^2 + 3x + 1) = (5x^2 + 2x^2) + (6x + 3x) + (4 + 1) = 7x^2 + 9x + 5$
3. Caroline; the variables are the same in both terms, they are just in a different order.

Practice Masters, p. 116

NAME ______________________ DATE ____________

14-2 Practice
Adding Polynomials

Student Edition Pages 711–714

Find each sum.

1. $(5w^2 + 8w) + (2w^2 + 3w)$ $7w^2 + 11w$
2. $(3x^2 - x) + (2x^2 + 3x)$ $5x^2 + 2x$
3. $(x^2 + 3x - 10) + (x^2 + 5x - 14)$ $2x^2 + 8x - 24$
4. $(3x^2 - 4x + 1) + (3x^2 - 5x + 2)$ $6x^2 - 9x + 3$
5. $(x - y) + (x + y)$ $2x$
6. $(9x + 2) + (3x - 6)$ $12x - 4$
7. $(-4a + 2b) + (7a + b)$ $3a + 3b$
8. (4 [illegible]) + (3x + 1) 5
9. $(3x + 6) + (2x - 2)$ $5x + 4$
10. $(3a^2 - 7a - 2) + (5a^2 - 3a - 17)$ $8a^2 - 10a - 19$
11. $(6x^2 + 7x - 2) + (3x^2 - 4x + 10)$ $9x^2 + 3x + 8$
12. $(12x^2 - 8x + 7) + (15x^2 + 4x - 1)$ $27x^2 - 4x + 6$
13. $(3x + 6y + 2) + (-9x - 2y + 8)$ $-6x + 4y + 10$
14. $(4a + 6ab - 2b) + (-5a - 2ab)$ $-a + 4ab - 2b$
15. $(9x^2 + 6x - 5) + (-2x^2 - x + 7)$ $7x^2 + 5x + 2$
16. $(-15x^2 + 5x + 2) + (3x^2 + 7)$ $-12x^2 + 5x + 9$
17. $(-2x + 3y - 4z) + (-5x + 10y)$ $-7x + 13y - 4z$
18. $(4b^2 - 3b + 7) + (-8b^2 - 5b + 16)$ $-4b^2 - 8b + 23$
19. $(6y^2 + 13y - 8) + (4y^2 - 8y + 7)$ $10y^2 + 5y - 1$
20. $(12z^2 - 8z + 17) + (21z - 8)$ $12z^2 + 13z + 9$
21. $\begin{array}{r} 2x + 4 \\ (+)\ x - 7 \\ \hline 3x - 3 \end{array}$
22. $\begin{array}{r} 5x - 7y \\ (+)\ 6x + 8y \\ \hline 11x + y \end{array}$
23. $\begin{array}{r} 2x + 3y \\ (+)\ 6x - 5y \\ \hline 8x - 2y \end{array}$
24. $\begin{array}{r} 7n + 2t \\ (+)\ 4n - 3t \\ \hline 11n - t \end{array}$
25. $\begin{array}{r} 3a + 6c \\ (+)\ 4a - 12c \\ \hline 7a - 6c \end{array}$
26. $\begin{array}{r} 4a^2 - 4b^2 \\ (+)\ 3a^2 + 3b^2 \\ \hline 7a^2 - b^2 \end{array}$
27. $\begin{array}{r} 24x^2 - 14xy - 3y^2 \\ (+)\ 6x^2 - 47xy - 63y^2 \\ \hline 30x^2 - 61xy - 66y^2 \end{array}$
28. $\begin{array}{r} 18a^2 - 9ab - 4b^2 \\ (+)\ 5a^2 + 3ab + 9b^2 \\ \hline 23a^2 - 6ab + 5b^2 \end{array}$

4 ASSESS

Closing Activity

Modeling Have students add polynomials using tiles. Give students two or three polynomial addends and have them find the sum. Be sure to include one problem that contains some unlike terms.

Chapter 14, Quiz A (Lessons 14-1 and 14-2) is available in the *Assessment and Evaluation Masters*, p. 379.

Additional Answer

41a.

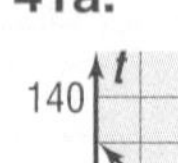

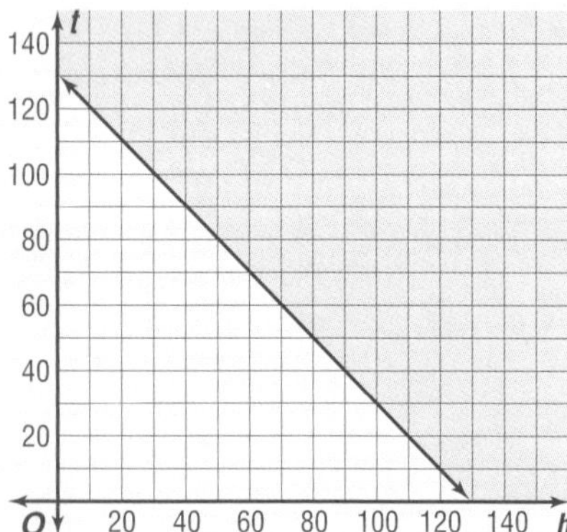

Enrichment Masters, p. 116

NAME ______ DATE ______

14-2 Enrichment Student Edition Pages 711–714

Adding Polynomials

Can you make a sentence using these words?

A FRUIT TIME LIKE AN BUT FLIES BANANA ARROW LIKE FLIES

Add the polynomials. Then find the word in the table at the right that corresponds to the sum. Read the words in order down the column to discover the hidden saying.

	Word
1. $(2x^2 + 3x^2) + (5x^2 + x^2)$ $11x^2$	TIME
2. $(2x^2 + 3x^3) + (5x^2 + x^2)$ $3x^3 + 8x^2$	FLIES
3. $(2x^2 + x) + (xy + x)$ $2x^2 + 2x + xy$	LIKE
4. $(x^3 + 2x^2) + (5x^3 + x)$ $6x^3 + 2x^2 + x$	AN
5. $(x + xy) + (x^2 + xy)$ $x^2 + 2xy + x$	ARROW
6. $(5x^2 + x) + (x + 2x^4)$ $2x^4 + 5x^2 + 2x$	BUT
7. $(xy + y^2 + x^2) + (2xy + x^2)$ $2x^2 + 3xy + y^2$	FRUIT
8. $(3x^2 + 2x^3) + (x^3 + x)$ $3x^3 + 3x^2 + x$	FLIES
9. $(x + x^2) + x^3$ $x^3 + x^2 + x$	LIKE
10. $(x^3 + x^3) + (x^3 + x^3)$ $4x^3$	A
11. $2x^{12} + 2x^{12}$ $4x^{12}$	BANANA

A	$4x^3$
FRUIT	$2x^2 + 3xy + y^2$
TIME	$11x^2$
LIKE	$x^3 + x^2 + x$
AN	$6x^3 + 2x^2 + x$
BUT	$2x^4 + 5x^2 + 2x$
FLIES	$3x^3 + 8x^2$
BANANA	$4x^{12}$
ARROW	$x^2 + 2xy + x$
LIKE	$2x^2 + 2x + xy$
FLIES	$3x^3 + 3x^2 + x$

Critical Thinking

35. If $(4r + 3s) + (6r - 5s) = (10r - 2s)$, find $(10r - 2s) - (6r - 5s)$. $(4r + 3s)$

Applications and Problem Solving

36. **Geometry** The measures of the angles of a triangle are shown in the figure at the right.
    a. Find the sum of the measures of the angles. $4x + 8$
    b. The sum of the measures of the angles in any triangle is 180°. Find the value of x. 43°
    c. Find the measure of each angle. 43°, 53°, 84°

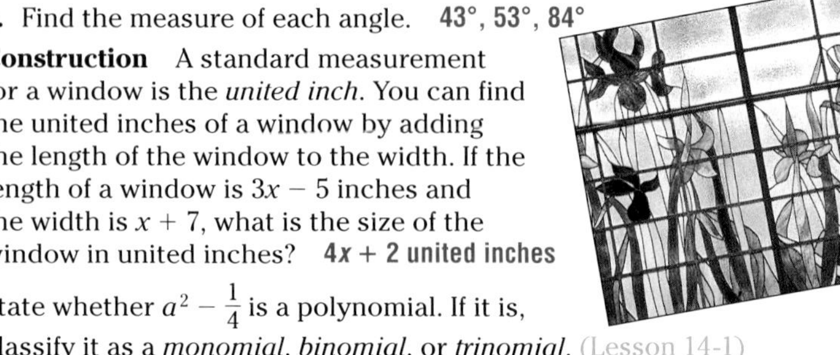

37. **Construction** A standard measurement for a window is the *united inch*. You can find the united inches of a window by adding the length of the window to the width. If the length of a window is $3x - 5$ inches and the width is $x + 7$, what is the size of the window in united inches? $4x + 2$ united inches

Mixed Review

38. State whether $a^2 - \frac{1}{4}$ is a polynomial. If it is, classify it as a *monomial, binomial,* or *trinomial.* (Lesson 14-1)

38. yes; binomial

39. **Geometry** Find the area of the triangle at the right. (Lesson 12-1) 7.5 cm²

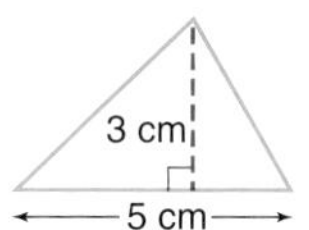

40. **Consumer Awareness** At a 20%-off sale, Antjuan bought a book that regularly sells for \$15.98. How much is the discount? (Lesson 9-9) \$3.20
41. **Weather** When the sum of the temperature (in °F) and humidity exceeds 130, people generally find the weather to be uncomfortable. (Lesson 8-9)
    a. Write an inequality to represent this situation. Then graph the inequality. $t + h > 130$; see margin for graph.
    b. Which area of the graph indicates uncomfortable weather?
    c. Humidity cannot exceed 100%. How does this affect the graph?

41b. the area above the boundary line.
41c. No points with $h > 100$ are included in the graph.

42. Consider the sequence 2, 6, 11, 17, 24, . . . (Lesson 5-9)
    a. Is the sequence arithmetic? no
    b. Write the next three terms of the sequence. 32, 41, 51
43. **Geometry** A rectangle has a perimeter of 50 feet and an area of 150 square feet. Its width is 10 feet. What is its length? (Lesson 3-5) 15 ft
44. **Statistics** The table at the right shows the percent of American homes that have the most popular pets. Make a bar graph of the data. (Lesson 1-10) See margin.

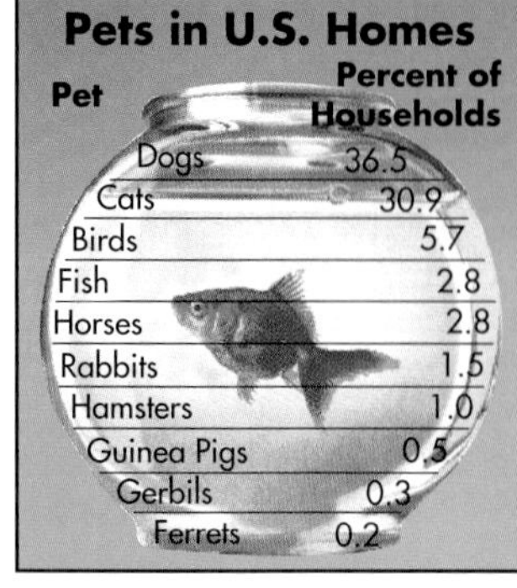

Pets in U.S. Homes

Pet	Percent of Households
Dogs	36.5
Cats	30.9
Birds	5.7
Fish	2.8
Horses	2.8
Rabbits	1.5
Hamsters	1.0
Guinea Pigs	0.5
Gerbils	0.3
Ferrets	0.2

Source: American Veterinary Medical Association

45. **Sports** Two hundred thirty-six women were members of the United States Olympic team for 1992. That number grew to 298 in 1996. How many more women were on the 1996 team than were on the 1992 team? (Lesson 1-6) 62

Extension

Using Health Connections For several meals, have students chart the number of servings he or she eats from each of the four basic food groups. Use *m* for meat; *d* for dairy; *f* for fruit and vegetables; and *b* for bread and cereal. Have each student write a polynomial representing each meal. Find daily and cumulative totals.

Additional Answer

44.

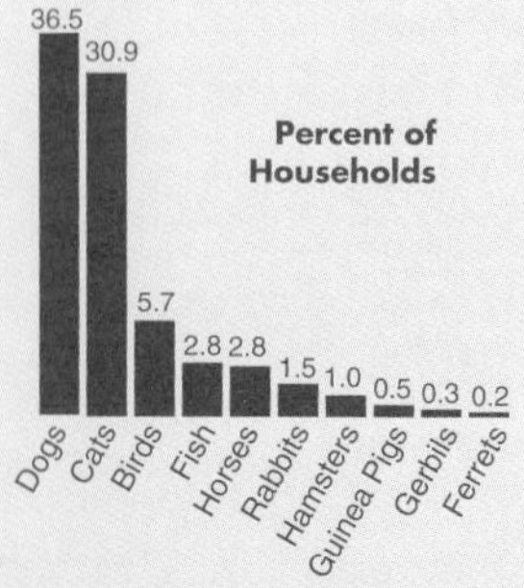

14-3 Subtracting Polynomials

Setting Goals: *In this lesson, you'll subtract polynomials.*

Modeling a Real-World Application: Running

German Silva of Mexico was the 1995 men's winner with a time of 2 hours 11 minutes. He also won the race in 1994.

Tegla Loroupe of Kenya was the first woman to cross the finish line of the 1994 New York Marathon. She completed the 26.2-mile course in 2 hours 27 minutes 37 seconds. In 1995, she won again with a time of 2 hours 28 minutes 6 seconds. How much faster was her 1994 time?

To find the difference of the times you must subtract.

2 hours 28 minutes 6 seconds	→	2 hours 27 minutes 66 seconds
− 2 hours 27 minutes 37 seconds		− 2 hours 27 minutes 37 seconds
		29 seconds

Miss Loroupe's 1994 time was 29 seconds faster.

Learning the Concept

In the application above, like units were subtracted from each other. Similarly, in algebra, when one polynomial is subtracted from another, only like terms can be subtracted.

Algebra tiles were used to help you understand how to add polynomials. They can also be used to help you understand how to subtract polynomials.

Example  **Find $(4x^2 + 7x + 4) - (x^2 + 2x + 1)$ by using algebra tiles.**

Make a model of $4x^2 + 7x + 4$ as shown in the diagram below. Then remove 1 x^2-tile, 2 x-tiles, and 1 1-tile.

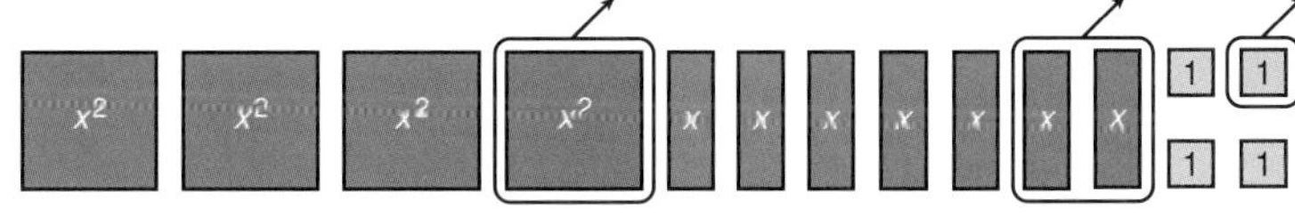

To find the difference, count the number of x^2-tiles, x-tiles, and 1-tiles that remain. There are 3 x^2-tiles, 5 x-tiles, and 3 1-tiles remaining.

$(4x^2 + 7x + 4) - (x^2 + 2x + 1) = 3x^2 + 5x + 3.$

To subtract polynomials, you can use the same methods you used to add them.

Classroom Vignette

"When subtracting polynomials using algebra tiles, remind students that removing positive tiles is equivalent to adding negative tiles. For instance, to solve Example 1, students could add one $-x^2$-tile, two $-x$-tiles, and 1 -1-tile. Then after removing the zero pairs, the result is the same, $4x^2 + 5x + 3$."

Cleo M. Meek
North Carolina Department of Instruction, Retired
Raleigh, NC

14-3 LESSON NOTES

NCTM Standards 1-4, 9

Instructional Resources

- Study Guide Master 14-3
- Practice Master 14-3
- Enrichment Master 14-3
- Group Activity Card 14-3
- Assessment and Evaluation Masters, pp. 378, 379
- Graphing Calculator Masters, p. 14

Transparency 14-3A contains the 5-Minute Check for this lesson; **Transparency 14-3B** contains a teaching aid for this lesson.

Recommended Pacing	
Honors Pacing	Day 3 of 9

1 FOCUS

5-Minute Check
(over Lesson 14-2)

Find each sum.

1. $(6x + 2) + (3x + 5)$ $\mathbf{9x + 7}$
2. $(x^2 + 5x - 4) + (5x^2 + 8x - 6)$ $\mathbf{6x^2 + 13x - 10}$
3. $(7x^2 - x) + (2x^2 + 4x + 16)$ $\mathbf{9x^2 + 3x + 16}$
4. $(6 - x) + (2x^2 + x - 7)$ $\mathbf{2x^2 - 1}$
5. $(5x^2 + 4xy - 7y^2) + (5x^2 - 2xy - 6y^2)$ $\mathbf{10x^2 + 2xy - 13y^2}$

According to legend, in 490 B.C., a Greek soldier named Philippides ran the 36.2 km (22.5 mi) from the site of the battle of Marathon to Athens, where he died after announcing the Greek victory over the Persians. The modern marathon commemorates the feat.

Motivating the Lesson

Questioning Shut the classroom door. Ask students how you can "undo" that action. **Open the door.** If you add 5 to some number, how can you undo that action? **Subtract 5.**

2 TEACH

In-Class Examples

For Example 1
Find $(3x^2 + 6x + 4) - (2x^2 + x + 3)$ by using algebra tiles.
$\mathbf{x^2 + 5x + 1}$

For Example 2
Find $(16x - 7) - (9x + 6)$.
$\mathbf{7x - 13}$

For Example 3
Find each difference.

a. $(2x - 6) - (5x + 4)$
$\mathbf{-3x - 10}$

b. $(2x^2 + xy - 6y^2) - (2x^2 + 4xy + 2y^2)$ $\mathbf{-3xy - 8y^2}$

For Example 4
Abram is painting a mural with a main part surrounded by a border. The main part has an area of $6x^2 + 6$ square feet. The entire mural has an area of $8x^2 - 2$. Find the area of the border. $\mathbf{2x^2 - 8}$

Study Guide Masters, p. 117

NAME ______________ DATE ______

14-3 Study Guide
Subtracting Polynomials

Student Edition
Pages 715–718

A rational number can be subtracted by adding its **additive inverse** or opposite. You can find the additive inverse of a number by multiplying the number by -1. For example, the additive inverse of 3 is $(-1)3$, or -3. You can also subtract binomials and trinomials by adding their additive inverse.

To find the additive inverse or binomials and trinomials, replace each term with its additive inverse.

binomial	additive inverse
$x + 2y$	$-x - 2y$
$2x^2 + 4x$	$-2x^2 - 4x$

trinomial	additive inverse
$2x^2 - 3x + 5$	$-2x^2 + 3x - 5$
$-8x + 5y - 7z$	$8x - 5y + 7z$

Examples

a. $\begin{array}{r} 4x^2 + 5xy \\ -\ 6x^2 + 8xy \\ \hline \end{array}$ additive inverse $-(6x^2 + 8xy)$ $\begin{array}{r} 4x^2 + 5xy \\ (+)\ -6x^2 - 8xy \\ \hline 2x^2 - 3xy \end{array}$

b. $\begin{array}{r} 12a^2 - 3a + 3 \\ -\ 6a^2 - 12a - 4 \\ \hline \end{array}$ additive inverse $-(6a^2 - 12a - 4)$ $\begin{array}{r} 12a^2 - 3a + 3 \\ (+)\ -6a^2 + 12a + 4 \\ \hline 6a^2 + 9a + 7 \end{array}$

Find each difference.

1. $(2x + 6) - (x + 4)$ **$x + 2$**
2. $(6x - 1) - (2x + 5)$ **$4x - 6$**
3. $(14y + 2) - (6y - 1)$ **$8y + 3$**
4. $\begin{array}{r} 9x + y + 8 \\ (-)5x + 3y - 3 \\ \hline \mathbf{4x - 2y + 11} \end{array}$
5. $\begin{array}{r} 6x \qquad + 7 \\ (-)-x - 4y - 12 \\ \hline \mathbf{7x + 4y + 19} \end{array}$
6. $\begin{array}{r} x^2 + 2x + 4 \\ (-)x^2 - 6x + 3 \\ \hline \mathbf{8x + 1} \end{array}$
7. $(5y^2 + 3y + 2) - (2y^2 - 9)$ **$3y^2 + 3y + 11$**
8. $(4y^2 - y + 6) - (-3y - 5)$ **$4y^2 + 2y + 11$**
9. $(9x + 12) - (5y + 3)$ **$9x - 5y + 9$**

Example 2 Find $(5x + 9) - (3x + 6)$.

Method 1: Subtract Vertically

$$\begin{array}{r} 5x + 9 \\ (-)\ 3x + 6 \\ \hline 2x + 3 \end{array}$$

The difference is $2x + 3$.

Method 2: Subtract Horizontally

$$\begin{aligned}(5x + 9) - (3x + 6) &= 5x + 9 - 3x - 6 \\ &= (5x - 3x) + (9 - 6) \\ &= (5 - 3)x + (9 - 6) \\ &= 2x + 3\end{aligned}$$

Recall that you can subtract a rational number by adding its additive inverse. You can also subtract a polynomial by adding its additive inverse.

To find the additive inverse of a polynomial, it is necessary to multiply the entire polynomial by -1. Study the examples in the table below.

Polynomial	Multiply by -1	Additive Inverse
$-y$	$-1(-y)$	y
$4x - 3$	$-1(4x - 3)$	$-4x + 3$
$2xy + 5y$	$-1(2xy + 5y)$	$-2xy - 5y$
$2x^2 - 3x - 5$	$-1(2x^2 - 3x - 5)$	$-2x^2 + 3x + 5$

Example 3 **Find each difference.**

a. $\mathbf{(7x + 5) - (3x + 2)}$

$$\begin{aligned} &= (7x + 5) + (-1)(3x + 2) && \textit{Add the additive inverse of } 3x + 2. \\ &= 7x + 5 + (-3x - 2) && (-1)(3x + 2) = -3x - 2 \\ &= 7x + 5 - 3x - 2 \\ &= 7x - 3x + 5 - 2 && \textit{Commutative property of addition} \\ &= 4x + 3 \end{aligned}$$

b. $\mathbf{(3x^2 - 5xy + 7y^2) - (x^2 - 3xy + 4y^2)}$

Align like terms and add the additive inverse of $x^2 - 3xy + 4y^2$.

$$\begin{array}{r} 3x^2 - 5xy + 7y^2 \\ (-)\ x^2 - 3xy + 4y^2 \\ \hline \end{array} \rightarrow \begin{array}{r} 3x^2 - 5xy + 7y^2 \\ (+)\ -x^2 + 3xy - 4y^2 \\ \hline 2x^2 - 2xy + 3y^2 \end{array}$$

Example 4

Lashonda plans to mat and frame a picture. The area inside the frame is $\mathbf{3x^2 - 5}$ square inches. The area of the picture is $\mathbf{x^2 + 5}$ square inches. How much matting will Lashonda need?

Explore You know the area inside the frame and the area of the picture.

Plan The area of the picture is smaller than the area inside the frame. Subtract the area of the picture from the area inside the frame to find the area needed for the matting.

Solve

$$\begin{aligned}(3x^2 - 5) - (x^2 + 5) &= 3x^2 - 5 + (-1)(x^2 + 5) \\ &= 3x^2 - 5 + (-x^2 - 5) \\ &= 3x^2 - 5 - x^2 - 5 \\ &= 3x^2 - x^2 - 5 - 5 \\ &= 2x^2 - 10\end{aligned}$$

Lashonda will need $2x^2 - 10$ square inches of matting.

Examine Check by aligning like terms and subtracting.

Alternative Learning Styles

Auditory Have students orally describe how to find the additive inverse of a polynomial. Then, as students model subtraction using algebra tiles, have them orally describe what they are doing.

Reteaching

Using Models Students can use algebra tiles to subtract polynomials.
Example: $(2x^2 - 3x - 6) - (x^2 + 5x - 5)$
Model each polynomial. Now model the opposite of the polynomial being subtracted, $-x^2 + (-5x) + 5$.
Finally, add the first polynomial and the opposite of the second polynomial, removing zero pairs where possible.
The result is $\mathbf{x^2 - 8x - 1}$.

Checking Your Understanding

Communicating Mathematics

Read and study the lesson to answer each question. **1–2. See margin.**

1. **Describe** how subtraction and addition of polynomials are related.
2. **Explain** how to find the additive inverse of a given polynomial. What is the additive inverse of $3x^2 - 5x + 7$?
3. **Write** the sum of a polynomial and its additive inverse. **0**
4. **Write** a polynomial subtraction problem with a difference of $3m^2 - 5m + 4$. Check your work by adding.
   Sample answer: $(5m^2 + 2m + 7) - (2m^2 + 7m + 3) = 3m^2 - 5m + 4$

MATERIALS

algebra tiles

Use algebra tiles to find each difference. **5–8. See margin.**

5. $(5x^2 + 6x + 4) - (3x^2 + 2x + 1)$
6. $(2x^2 + 5x + 3) - (x^2 - 2x + 3)$
7. $(3x^2 + 2x - 5) - (2x^2 + x - 4)$
8. $(8x^2 + 2x - 1) - (2x^2 + x + 5)$

Guided Practice

State the additive inverse of each polynomial.

9. $2abc$ **$-2abc$**
10. $3x + 2y$ **$-3x - 2y$**
11. $x^2 + 5x + 1$
12. $3x^2 - 2x + 5$
13. $-8m + 7n$ **$8m - 7n$**
14. $-4h^2 - 5hk - k^2$

11. $-x^2 - 5x - 1$
12. $-3x^2 + 2x - 5$
14. $4h^2 + 5hk + k^2$

Find each difference.

15. $(3x + 4) - (x + 2)$ **$2x + 2$**
16. $(4x + 5) - (2x + 3)$ **$2x + 2$**
17. $(5x - 7) - (3x - 4)$ **$2x - 3$**
18. $(2x - 5) - (3x + 1)$ **$-x - 6$**
19. $\begin{array}{r} 5x^2 + 4x - 1 \\ (-)\ 4x^2 + x + 2 \\ \hline \end{array}$ **$x^2 + 3x - 3$**
20. $\begin{array}{r} 3x^2 + 5x + 4 \\ (-)\ x^2 \qquad - 1 \\ \hline \end{array}$ **$2x^2 + 5x + 5$**

21. **Geometry** Find how much longer the hypotenuse is than the shorter leg of the triangle shown at the right. **$4x - 6$**

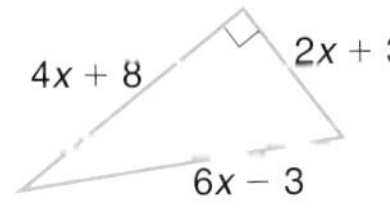

Exercises: Practicing and Applying the Concept

Independent Practice

Find each difference.

A

22. $(9x + 5) - (4x + 3)$ **$5x + 2$**
23. $(2x + 5) - (x + 8)$ **$x - 3$**
24. $(3x - 2) - (5x - 4)$ **$-2x + 2$**
25. $(6x - 5) - (4x + 3)$ **[illegible]**
26. $(2x + 3y) - (x - y)$ **$x + 4y$**
27. $(9x - 4y) - (12x - 9y)$
28. $(5x^2 - 3) - (2x^2 - 7)$ **$3x^2 + 4$**
29. $(x^2 + 6x) - (3x^2 + 7)$

B

30. $\begin{array}{r} 5a^2 + 7a + 9 \\ (-)\ 3a^2 + 4a + 1 \\ \hline \end{array}$
31. $\begin{array}{r} 6m^2 - 5m + 3 \\ (-)\ 5m^2 + 2m - 7 \\ \hline \end{array}$
32. $\begin{array}{r} 5x^2 - 4xy \qquad \\ (-)\qquad - 3xy + 2y^2 \\ \hline \end{array}$
33. $\begin{array}{r} 9m^2 \qquad + 7 \\ (-)\ -6m^2 + 2m - 3 \\ \hline \end{array}$
34. $\begin{array}{r} 15x^2y^2 + 11xy - 9 \\ (-)\ 9x^2y^2 - 13xy + 6 \\ \hline \end{array}$
35. $\begin{array}{r} 14a + 10b - 18c \\ (-)\ 5a + 7b - 11c \\ \hline \end{array}$
36. $(10x^2 + 8x - 6) - (3x^2 + 2x - 9)$ **$7x^2 + 6x + 3$**
37. $(5y^2 + 9y - 12) - (-3y^2 + 5y - 7)$ **$8y^2 + 4y - 5$**

C

38. $(6a^2 + 7ab - 3b^2) - (2a^2 + 3ab - b^2)$ **$4a^2 + 4ab - 2b^2$**
39. $(x^3 - 3x^2y + 4xy^2 + y^3) - (7x^3 - 9xy^2 + x^2y + y^3)$

27. $-3x + 5y$
29. $-2x^2 + 6x - 7$
30. $2a^2 + 3a + 8$
31. $m^2 - 7m + 10$
32. $5x^2 - xy - 2y^2$
33. $15m^2 - 2m + 10$
34. $6x^2y^2 + 24xy - 15$
35. $9a + 3b - 7c$
39. $-6x^3 - 4x^2y + 13xy^2$

3 PRACTICE/APPLY

Checking Your Understanding

Exercises 1–21 are designed to help you assess your students' understanding through reading, writing, speaking, and modeling. You should work through Exercises 1–8 with your students and then monitor their work on Exercises 9–21.

Assignment Guide

Core: 23–39 odd, 40, 41, 43–50
Enriched: 22–38 even, 40–50
All: Self Test, 1–10

For **Extra Practice**, see p. 773.

The red A, B, and C flags, printed only in the Teacher's Wraparound Edition, indicate the level of difficulty of the exercises.

Additional Answers

1. Addition and subtraction are inverse operations.
2. Multiply by −1. The additive inverse of $3x^2 - 5x + 7$ is $-3x^2 + 5x - 7$.
5. $2x^2 + 4x$ [illegible] 3
6. $x^2 + 7x$
7. $x^2 + x - 1$
8. $6x^2 + x - 6$

Practice Masters, p. 117

NAME ______________ DATE ______________

14-3 Practice
Subtracting Polynomials

Student Edition
Pages 715–718

Find each difference.

1. $(7a + 2) - (5a + 1)$ **$2a + 1$**
2. $(3x + 10) - (x + 10)$ **$2x$**
3. $(17x + 13) - (7x - 4)$ **$10x + 17$**
4. $(37y - 17) - (14y + 11)$ **$23y - 28$**
5. $(2x + 3) - (3x - 1)$ **$-x + 4$**
6. $(x^2 + 1) - (x^2 - 1)$ **2**
7. $(x^2 + 7x + 3) - (x^2 + 7x + 3)$ **0**
8. $(5r - 3s) - (7r + 5s)$ **$-2r - 8s$**
9. $(14x^2 - 22) - (14x + 5)$ **$14x^2 - 14x - 27$**
10. $(43xy - 43) - (19xy + 13)$ **$24xy - 56$**
11. $(11x^2 + 5x) - (7x^2 + 3)$ **$4x^2 + 5x - 3$**
12. $(x^2 + 3) - (6 + 4x)$ **$x^2 - 4x - 3$**
13. $(15x - 3y) - (7x - 6y)$ **$8x + 3y$**
14. $(18t^2 + 4t) - (16t^2 - 6t)$ **$2t^2 + 10t$**
15. $(-2a^2 + 4b^2) - (5a^2 - 6b^2)$ **$-7a^2 + 10b^2$**
16. $(-16x - 2y) - (-3x + 7y)$ **$-13x - 9y$**
17. $(6x + 10y) - (13x - 5y + 4)$ **$-7x + 15y - 4$**
18. $(-9x^2 + 7x + 7) - (6x^2 - 2x - 7)$ **$-15x^2 + 9x + 14$**
19. $(4x^2 - 3x - 7) - (-5x^2 - 9x + 12)$ **$9x^2 + 6x - 19$**
20. $(-8y^2 + 7y - 3) - (15y^2 - 9y + 4)$ **$-23y^2 + 16y - 7$**
21. $\begin{array}{r} 6x^2 + 9x + 10 \\ (-)3x^2 + 5x + 4 \\ \hline 3x^2 + 4x + 6 \end{array}$
22. $\begin{array}{r} 8z^2 - 5z + 11 \\ (-)8z^2 + 2z - 8 \\ \hline -7z + 19 \end{array}$
23. $\begin{array}{r} c + 2d - e \\ (-)3c + d + 6e \\ \hline -2c + d - 7e \end{array}$
24. $\begin{array}{r} 4u^2 + 3uv \qquad \\ (-)\qquad - 2uv + v^2 \\ \hline 4u^2 + 5uv - v^2 \end{array}$
25. $\begin{array}{r} 10x^2y^2 + 5xy - 8 \\ (-)-5x^2y^2 - 7xy + 9 \\ \hline 15x^2y^2 + 12xy - 17 \end{array}$
26. $\begin{array}{r} 8k^2 \qquad + 3 \\ (-)5k^2 + 3k - 5 \\ \hline 3k^2 - 3k + 8 \end{array}$
27. $(14n^2 - 8nt + 12t^2) - (-12n^2 + 15nt - 13t^2)$ **$26n^2 - 23nt + 25t^2$**
28. $(54m^3 - 84m^2 - 26m) - (8m^2 - 9m - 2)$ **$54m^3 - 92m^2 - 17m + 2$**

Group Activity Card 14-3

Subtract Them — Group Activity **14-3**

MATERIALS: Three decks of cards made for Activity 14-2

This game is an extension of Activity 14-2. The same decks of cards are used, and polynomials are formed by the leader in the same manner. However, in this game when two polynomials are formed, everyone subtracts the second one from the first one.

The winner is the one who gets the correct answer the fastest. The overall winner is the player with the most points after one round.

15 | A^2 | +

 Pre-Algebra

GLENCOE Technology

Interactive Mathematics Tools Software

In this interactive computer lesson, students explore subtraction of polynomials. A **Computer Journal** gives students the opportunity to write about what they have learned.

For Windows & Macintosh

4 ASSESS

Closing Activity

Modeling Model several polynomial subtraction problems. Have students state the problem that is modeled and how they would solve the problem.

Chapter 14, Quiz B (Lesson 14-3) is available in the *Assessment and Evaluation Masters*, p. 379.

Mid-Chapter Test (Lessons 14-1 through 14-3) is available in the *Assessment and Evaluation Masters*, p. 378.

Additional Answer

46.

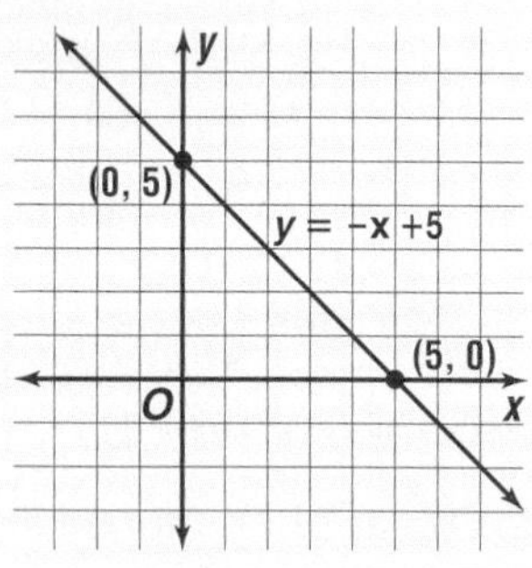

Enrichment Masters, p. 117

NAME ____________ DATE ____________

14-3 Enrichment
Turn Symmetry

Student Edition Pages 715–718

Some geometric figures can be turned about a point and will fit point for point in their original position. This property is called **turn symmetry**. All figures can be turned 360° to fit in their original positions. Turn symmetry refers to turns less than 360°.

For example, a square has turn symmetry. It can be turned 90°, 180°, or 270° about its center of rotation and its appearance will not be changed.

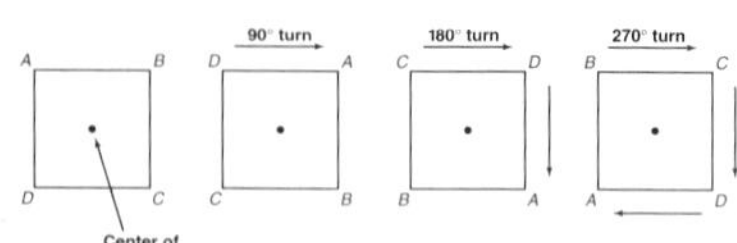

For each figure below, mark the center of rotation with a dot. Then list all possible degrees of turn symmetry so that each figure has the appearance of its original position. If a figure does not have turn symmetry, write none.

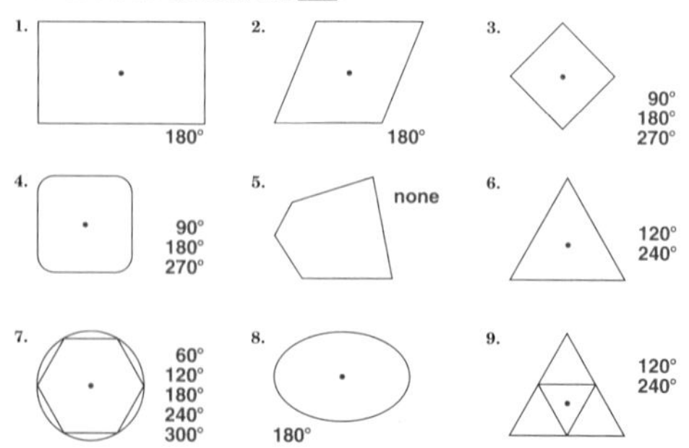

Critical Thinking

40. Suppose that A and B represent polynomials. If $A + B = 3x^2 + 2x - 2$ and $A - B = -x^2 + 4x - 8$, find A and B. **$A = x^2 + 3x - 5$; $B = 2x^2 - x + 3$**

Applications and Problem Solving

interNET CONNECTION
For the latest basketball scores, visit: **www.glencoe.com/sec/math/prealg/mathnet**

41. **Basketball** On December 13, 1983, the Denver Nuggets and the Detroit Pistons broke the record for the highest total score in a basketball game. The total points scored was 2 points more than twice what the Nuggets scored. The Pistons scored 186 points in the game.
   a. What was the Nuggets final score? **184**
   b. How many points did the two teams score? **370**

42. **Geometry** The perimeter of the isosceles trapezoid shown at the right is $16x + 1$ units. Find the length of the missing base of the trapezoid. **$7x + 5$ units**

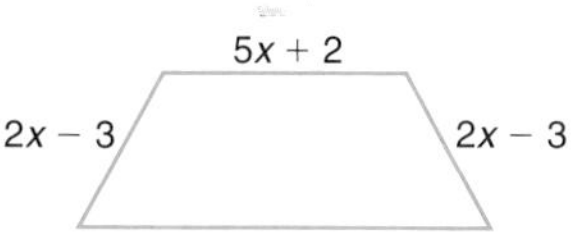

Mixed Review

43. Find the sum $(2a^2 + 3a - 4) + (6a^2 - a + 5)$. (Lesson 14-2)
44. Solve $n^2 = 18$ to the nearest tenth. (Lesson 13-3) **4.2**
45. **Geometry** In $\triangle CDE$, $m\angle C = 35°$ and $m\angle D = 55°$. Find the measure of $\angle E$ and classify the triangle as *acute*, *right*, or *obtuse*. (Lesson 11-4)
46. Graph $y = -x + 5$ by using the x-intercept and the y-intercept. (Lesson 8-7) **See margin.**
47. **Patterns** Two groups share the same meeting room. If the debate team meets every 3rd school day and the dance committee meets every 4th school day, how often will both groups need the room on the same day? (Lesson 4-7) **every 12 days**
48. Solve $x = 32 + 56 + (-18)$. (Lesson 2-4) **70**
49. **Geometry** In which quadrant does the point at $(-5, 3)$ lie? (Lesson 2-2) **II**
50. Solve $8c = 72$ mentally. (Lesson 1-6) **9**

43. $8a^2 + 2a + 1$
45. 90°; right

Self Test

State whether each expression is a polynomial. If it is, classify it as a *monomial*, a *binomial*, or a *trinomial*. Then find the degree of each polynomial. (Lesson 14-1)

1. $ax^2 + 6x$ **yes, binomial; 3**
2. $4b^2 + c^3d^4 + x$ **yes, trinomial; 7**
3. $a^3b^4c^5$ **yes, monomial; 12**

Find each sum. (Lesson 14-2) **4. $11x + 2y$ 5. $-x^2 + 12$ 6. $19m + n$**

4. $(4x + 5y) + (7x - 3y)$
5. $(2x^2 + 5) + (-3x^2 + 7)$
6. $(9m - 3n) + (10m + 4n)$
7. **Geometry** The perimeter of the triangle shown at the right is $7x + 2y$ units. Find the length of the third side of the triangle. (Lessons 14-1 and 14-2) **$2x + 6y$ units**

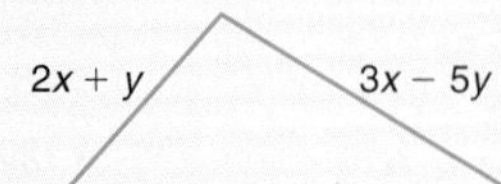

Find each difference. (Lesson 14-3)

8. $(11p + 5r) - (2p + r)$ **$9p + 4r$**
9. $(7a + 6d) - (6a - 7d)$ **$a + 13d$**
10. $(4t + 11r) - (t + 2r)$ **$3t + 9r$**

Extension

Using Critical Thinking Have students develop a model for a term of a polynomial where the power of a variable is three. The model would be a cube.

Self Test

The Self Test provides students with a brief review of the concepts and skills in Lessons 14-1 through 14-3. Lesson numbers are given to the right of exercises or instruction lines so students can review concepts not yet mastered.

14-4 Powers of Monomials

Setting Goals: *In this lesson, you'll find powers of monomials.*

Modeling with Technology

Recall that when you multiply powers that have the same base, you add the exponents. For example, $3^2 \cdot 3^3 = 3^{2+3}$ or 3^5. But how do you find the power of a power like $(3^2)^3$? You can use a calculator to discover a pattern.

Your Turn

TECHNO TIP

The ^ or y^x key allows you to evaluate expressions with exponents.

Copy the table below. Use a calculator to find each value and complete the table.

Power	Value	Power	Value
3^1	3	$(3^2)^1$	9
3^2	9	$(3^2)^2$	81
3^3	27	$(3^2)^3$	729
3^4	81	$(3^2)^4$	6561
3^5	243	$(3^2)^5$	59,049
3^6	729	$(3^2)^6$	531,441

TALK ABOUT IT

a. They are the same.
b. $6 = 2 \cdot 3$.
c. 3^8 or 6561

a. Compare the values of 3^6 and $(3^2)^3$. What do you observe?
b. Compare the exponents of 3^6 and $(3^2)^3$. What do you observe?
c. Guess the value of $(3^4)^2$. Use a calculator to check your answer.
d. Write a rule for finding the value of a power of a power.
Sample answer: Keep the base and multiply the exponents.

Learning the Concept

LOOK BACK

You can review multiplying powers in Lesson 4-8.

Consider the following powers.

$(5^4)^2 = 5^4 \cdot 5^4$ *Definition of exponent*
$= 5^{4+4}$ *Product of powers*
$= 5^8$ *Substitution*

$(x^2)^3 = (x^2) \cdot (x^2) \cdot (x^2)$
$= x^{2+2+2}$
$= x^6$

These and other similar examples suggest that you can find a *power of a power* by multiplying the exponents.

Power of a Power For any number a and positive integers m and n, $(a^m)^n = a^{mn}$.

Alternative Learning Styles

Kinesthetic Provide students with centimeter cubes. Tell students that the length of each side of each cube is x. Ask what the volume of each cube is. x^3 Have students stack centimeter cubes to make a larger cube with sides of $2x$. Ask what the volume of that cube is. $8x^3$ Continue by having students work in groups to make larger cubes and find the volume of each. Relate finding the power of each side measure to finding the power of a power.

14-4 LESSON NOTES

NCTM Standards: 1-4, 9

Instructional Resources

- Study Guide Master 14-4
- Practice Master 14-4
- Enrichment Master 14-4
- Group Activity Card 14-4
- Activity Masters, p. 48
- Multicultural Activity Masters, p. 27
- Real-World Applications, 31

Transparency 14-4A contains the 5-Minute Check for this lesson; **Transparency 14-4B** contains a teaching aid for this lesson.

Recommended Pacing	
Honors Pacing	Day 4 of 9

1 FOCUS

5-Minute Check
(over Lesson 14-3)

Find each difference.

1. $(4x + 7) - (x + 8)$ **$3x - 1$**
2. $(6x^2 + x - 2) - (4x^2 + 7x - 3)$ **$2x^2 - 6x + 1$**
3. $(2x^2 - x) - (5x^2 + 4x - 6)$ **$-3x^2 - 5x + 6$**
4. $(1 - x) - (2x^2 - x)$ **$-2x^2 + 1$**
5. $(2x^2 + 5xy - y^2) - (x^2 - 4xy - 2y^2)$ **$x^2 + 9xy + y^2$**

Motivating the Lesson

Hands-On Activity Show square boxes, one with sides twice as long as those of the other. Ask students how many times the contents of a full smaller box could be poured into the larger box. List student responses. Use rice to test guesses.

2 TEACH

In-Class Examples

For Example 1
Simplify $(r^5)^3$. $\mathbf{r^{15}}$

For Example 2
The formula for the volume of a cone is $V = \frac{1}{3}\pi r^2 h$. A cone has a height of 5x and a circular base with a radius of 2x.

a. Find the measure of the volume of the cone.
$\mathbf{\frac{1}{3}\pi(20x^3)}$

b. If x is 4 centimeters, what is the volume of the pyramid?
about 1340.4 cm³

For Example 3
Simplify $(m^4n)^3$. $\mathbf{m^{12}n^3}$

For Example 4
Simplify $(ab^{-2}c^3)^{-4}$. $\mathbf{a^{-4}b^8c^{-12}}$

3 PRACTICE/APPLY

Checking Your Understanding

Exercises 1–13 are designed to help you assess your students' understanding through reading, writing, speaking, and modeling. You should work through Exercises 1–3 with your students and then monitor their work on Exercises 4–13.

Example  **Simplify $(p^6)^2$.**

$(p^6)^2 = p^{6 \cdot 2}$ *Power of a power*
$= p^{12}$

Check: $(p^6)^2 = p^6 \cdot p^6$
$= p^{6+6}$ or p^{12} ✓

Connection to Geometry

The volume of a cube can be found by multiplying its length, width, and height. The volume of the small cube shown at the right is $x \cdot x \cdot x$ or x^3. If the length of each edge of the cube is doubled, the volume of the new cube would be $(2x)^3$.

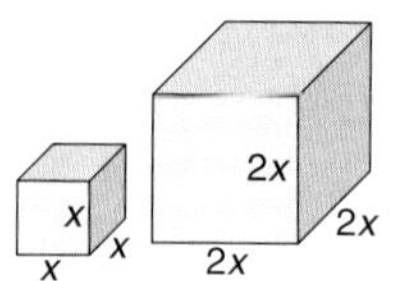

$(2x)^3 = (2x)(2x)(2x)$ *Definition of exponent*
$= (2 \cdot 2 \cdot 2)(x \cdot x \cdot x)$ *Commutative and associative properties*
$= 2^3x^3$ *Exponential notation*
$= 8x^3$ $2^3 = 8$

Notice that when $2x$ was raised to the third power, *both* the 2 and the x were raised to the third power. This example suggests that the power of a product can be found by multiplying the individual powers.

Power of a Product — For all numbers a and b and positive integer m, $(ab)^m = a^mb^m$.

Example 

Geometry

The formula for the volume of a pyramid is $V = \frac{1}{3}\ell wh$.

a. Find the measure of the volume of the pyramid.

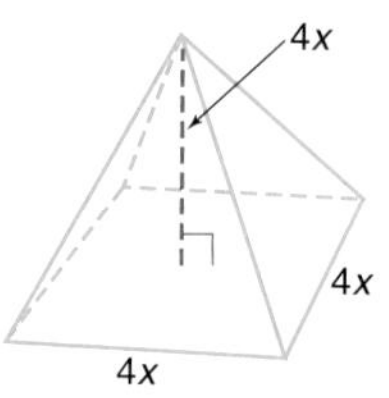

$V = \frac{1}{3}\ell wh$
$= \frac{1}{3}(4x)(4x)(4x)$
$= \frac{1}{3}(4 \cdot 4 \cdot 4)(x \cdot x \cdot x)$
$= \frac{1}{3}(4^3)(x^3)$
$= \frac{1}{3}(4x)^3$ *Power of a product*

The measure of the volume of the pyramid is $\frac{1}{3}(4x)^3$.

b. If x is 5 cm, what is the volume of the pyramid?

$\frac{1}{3}(4x)^3 = \frac{1}{3}(4 \cdot 5)^3$ *Replace x with 5.*
$= \frac{1}{3}(20)^3$ *Estimate: $\frac{1}{3} \cdot 20 \cdot 20 \cdot 20 = 7 \cdot 20 \cdot 20$ or 2800*

20 [y^x] 3 [÷] 3 [=] 2666.6666

The volume is about 2667 cm³. *Compare to the estimate.*

Cooperative Learning

Brainstorming Have students choose replacements for the variables in the examples and then use calculators to evaluate the monomials. Then have students generalize their findings using variables and compare them to the rule.

For more information on this strategy, see *Cooperative Learning in the Mathematics Classroom*, one of the titles in the Glencoe Mathematics Professional Series, p. 30.

Sometimes the rules for the *power of a power* and the *power of a product* are combined into one rule.

Power of a Monomial	**For all numbers a and b and positive integer m, n, and p, $(a^m b^n)^p = a^{mp} b^{mp}$.**

Example  **Simplify $(x^2y^3)^4$.**

$$(x^2y^3)^4 = (x^2)^4(y^3)^4 \quad \text{Power of a monomial}$$
$$= x^{2\cdot 4}y^{3\cdot 4}$$
$$= x^8y^{12}$$

Check:
$$(x^2y^3)^4 = (x^2y^3)(x^2y^3)(x^2y^3)(x^2y^3)$$
$$= (x^2\cdot x^2\cdot x^2\cdot x^2)(y^3\cdot y^3\cdot y^3\cdot y^3)$$
$$= (x^{2+2+2+2})(y^{3+3+3+3})$$
$$= (x^8)(y^{12}) \text{ or } x^8y^{12} \checkmark$$

These rules can also be used with negative exponents.

Example  **Simplify $(x^3y^2z)^{-2}$.**

$$(x^3y^2z)^{-2} = x^{3(-2)}y^{2(-2)}z^{1(-2)} \quad \text{Power of a monomial}$$
$$= x^{-6}y^{-4}z^{-2}$$

Check:
$$(x^3y^2z)^{-2} = \frac{1}{(x^3y^2z)^2}$$
$$= \frac{1}{(x^3y^2z)(x^3y^2z)}$$
$$= \frac{1}{(x^3\cdot x^3)(y^2\cdot y^2)(z\cdot z)}$$
$$= \frac{1}{x^{3+3}y^{2+2}z^{1+1}}$$
$$= \frac{1}{x^6y^4z^2} \text{ or } x^{-6}y^{-4}z^{-2} \checkmark$$

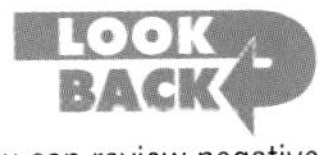

You can review negative exponents in Lesson 4-9.

Checking Your Understanding

Communicating Mathematics

Read and study the lesson to answer each question.

1. **You Decide** Nelia says that $5x^3$ and $(5x)^3$ are equivalent expressions. Julie says they are not. Who is correct? Explain. **See margin.**
2. **Analyze** the square shown at the right. What is its area? What would its area be if the length of each side were doubled or tripled? **x^2; $4x^2$; $9x^2$**

3. **Assess Yourself** How well do you understand the rules given in this lesson? For each of the following, write the rules in your own words and then give an example of it. **See margin.**
   a. Power of a power
   b. Power of a product
   c. Power of a monomial

Additional Answers

1. Julie, $(5x)^3 = 5x \cdot 5x \cdot 5x$ or $125x^3$
3. Sample answers:

   Power of a power: When you raise a power to a power, multiply the exponents and keep the base. Example: $(x^2)^4 = x^8$

   Power of a Product: When you raise a product to a power, distribute the exponent over each term. Example: $(4x)^4 = 256x^4$

   Power of a Monomial. When you raise a product involving powers to a power, multiply each power by the exponent. Example: $(x^2y^3)^4 = x^{2\cdot 4}y^{3\cdot 4}$ or x^8y^{12}

Reteaching

Using Study Aids Have students create study cards by writing an example of each of the following on one side of the card and writing the explanation and example on the back of the card.

Power of a Power
Power of a Product
Power of a Monomial

You may wish to have students create separate cards for situations involving negative exponents.

Study Guide Masters, p. 118

NAME ______________________ DATE ____________

14-4 Study Guide
Powers of Monomials

Student Edition
Pages 719–723

To find a **power of a power**, multiply the exponents.

$(4^2)^3 = 4^{2\cdot 3} = 4^6$ $\qquad$ $(x^3)^4 = x^{3\cdot 4} = x^{12}$

To find a **power of a product**, multiply the individual powers.

$(ab)^4 = a^4b^4$ $\qquad$ $(xy)^{-3} = x^{-3}y^{-3}$

Use the previous rules to find a **power of a monomial**.

$(2a^4b^5)^2 = (2)^2(a^4)^2(b^5)^2 = 2^2a^{4\cdot 2}b^{5\cdot 2} = 4a^8b^{10}$ $\qquad$ $(d^2e^6f^3)^{-3} = (d^2)^{-3}(e^6)^{-3}(f^3)^{-3} = d^{-6}e^{-18}f^{-9}$

Simplify.

1. $(ab)^7$ **a^7b^7**	2. $(2^4)^2$ **2^8**	3. $(6m)^3$ **$216m^3$**
4. $(-4m^2n)^2$ **$16m^4n^2$**	5. $(k^6)^3$ **k^{18}**	6. $[(-2)^4]^2$ **256**
7. $(-3y^5)^2$ **$9y^{10}$**	8. $(-h^4)^5$ **$-h^{20}$**	9. $(d^6e^3)^8$ **$d^{48}e^{24}$**
10. $(4a^2b)^4$ **$256a^8b^4$**	11. $(-pq^2)^9$ **$-p^9q^{18}$**	12. $6x(4x)^3$ **$384x^4$**
13. $(-2x^4y^3)^6$ **$64x^{24}y^{18}$**	14. $-4t(t^4)^5$ **$-4t^{21}$**	15. $[(4)^2]^3$ **4096**
16. $(-w^8)^{-4}$ **w^{-32}**	17. $(-m^9)^{-2}$ **m^{-18}**	18. $7(u^6v)^{-4}$ **$7u^{-24}v^{-4}$**

Evaluate each expression if $x = -1$ and $y = 4$.

19. $3x^2y$ **12**	20. $-2xy^3$ **128**	21. $(x^3y)^3$ **-64**
22. $2x(2y^2)^{-2}$ **$\frac{-1}{512}$**	23. $(x^2y)^4$ **256**	24. $-(xy)^{-2}$ **$-\frac{1}{16}$**

Assignment Guide

Core: 15–41 odd, 42, 43, 45–53
Enriched: 14–40 even, 41–53

For **Extra Practice**, see p. 773.

The red A, B, and C flags, printed only in the Teacher's Wraparound Edition, indicate the level of difficulty of the exercises.

Additional Answers

41a. The numbers in the values columns are the same.
41b. The exponents are distributed across the factors.
41c. When you raise a product to a power, distribute the exponent over each factor.

Guided Practice

Simplify.

4. $(y^5)^3$ **y^{15}**
5. $(2^4)^3$ **2^{12}**
6. $(2m)^4$ **2^4m^4 or $16m^4$**
7. $(3xy)^3$ **7. $3^3x^3y^3$ or $27x^3y^3$**
8. $(a^2b^3)^2$ **a^4b^6**
9. $(-xy)^2$ **x^2y^2**

Evaluate each expression if $a = -1$ and $b = 3$.

10. a^2b **3**
11. $-ab^2$ **9**
12. $(-4a^3b)^2$ **144**

13. **Geometry** Find the volume of the rectangular prism at the right. **$48x^6y^2$**

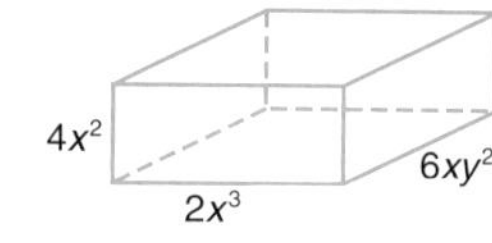

Exercises: Practicing and Applying the Concept

Independent Practice

Simplify.

14. $(7^3)^2$ **7^6**
15. $(yz)^4$ **y^4z^4**
16. $(-3w)^3$ **$-27w^3$**

17. $(-2rs)^4$ **$16r^4s^4$**
18. $[(-4)^2]^2$ **256**
19. $(m^2)^5$ **m^{10}**
20. $(-y^3)^6$ **y^{18}**
21. $(-2x^2)^5$ **$-32x^{10}$**
22. $(x^2y)^3$ **x^6y^3**
23. $(-xy^3)^4$ **x^4y^{12}**
24. $(2a^3b)^5$ **$32a^{15}b^5$**
25. $3x(2x)^2$ **$12x^3$**
26. $-5x(x^3)^2$ **$-5x^7$**
27. $(4x^2y^3)^2$ **$16x^4y^6$**
28. $[(-3)^2]^3$ **729**
29. $(t^3)^{-4}$ **t^{-12}**
30. $(-w^2)^{-5}$ **$-w^{-10}$**
31. $-2(x^3y)^{-2}$ **$-2x^{-6}y^{-2}$**

Evaluate each expression if $a = -2$ and $b = 3$.

32. $2ab^2$ **−36**
33. $-3a^2b$ **−36**
34. $-(ab^2)^2$ **−324**
35. $(-2ab^3)$ **108**
36. $(a^3b)^2$ **576**
37. $-(3a^2b)^2$ **−1296**

38. $-3b(2a)^{-2}$ **$-\frac{9}{16}$**
39. $2b^2(-3ab)^{-3}$ **$\frac{1}{324}$**
40. $2(3a^{-2})^2$ **$1\frac{1}{8}$**

Calculators

41. You can use a calculator to confirm the power of a product rule. Copy and complete the following table using a calculator. **a–c. See margin.**

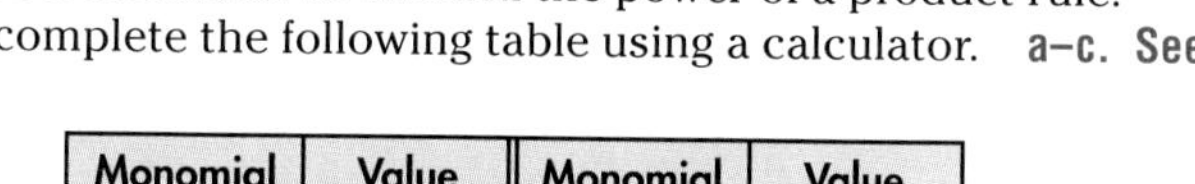

Monomial	Value	Monomial	Value
$(2 \cdot 3)^2$	36	$2^2 \cdot 3^2$	36
$(4 \cdot 5)^3$	8000	$4^3 \cdot 5^3$	8000
$(6 \cdot 7)^4$	3,111,696	$6^4 \cdot 7^4$	3,111,696
$(8 \cdot 3)^5$	7,962,624	$8^5 \cdot 3^5$	7,962,624

a. Compare the values in each row. What do you observe?
b. Compare the monomials in each row. What do you observe?
c. Write a rule for finding the power of a product in your own words.

Critical Thinking

42. Are $(2^6)^4$ and $(4^6)^2$ equal? Explain why or why not. (*Hint:* Write both expressions as a power of 2.) **yes; $(2^6)^4 = 2^{6 \cdot 4}$ or 2^{24}, $(4^6)^2 = ((2^2)^6)^2$ or 2^{24}**

Applications and Problem Solving

43. **Finance** Nuna opens a savings account with $1000. It earns 1% interest every 3 months. So, in one year, the balance is 1000(1.01)(1.01)(1.01)(1.01) or $1000(1.01)^4$. If she keeps the money in the account for 3 years, the deposit is worth $1000(1.01^4)^3$.
   a. Write an expression that represents what the deposit is worth in 6 years. **$1000(1.01^4)^6$**
   b. What is the deposit worth in 6 years? **$1269.73**

Practice Masters, p. 118

NAME ______________ DATE ________

14-4 Practice
Powers of Monomials

Student Edition
Pages 719–723

Simplify.

1. $(4^2)^3$ **4096**
2. $[(-3)^2]^2$ **81**
3. $(x^3)^4$ **x^{12}**
4. $(7d)^3$ **$343d^3$**
5. $(11k)^2$ **$121k^2$**
6. $(-2y)^3$ **$-8y^3$**
7. $(-6m)^4$ **$1296m^4$**
8. $(-t)^{10}$ **t^{10}**
9. $(xy)^4$ **x^4y^4**
10. $(rs)^6$ **r^6s^6**
11. $(ab^2)^4$ **a^4b^8**
12. $(c^3d^5)^2$ **c^6d^{10}**
13. $(-3x^2y^4)^3$ **$-27x^6y^{12}$**
14. $(2d^3f)^4$ **$16d^{12}f^4$**
15. $(3a^2b^3)^2$ **$9a^4b^6$**
16. $(4g^2h)^6$ **$4096g^{12}h^6$**
17. $(-2jk^3)^4$ **$16j^4k^{12}$**
18. $(-6pq)^4$ **$1296p^4q^4$**
19. $(-5x^3z^9)^3$ **$-125x^9z^{27}$**
20. $(-2pj)^5$ **$-32p^5j^5$**
21. $(4x^4y)^3$ **$64x^{12}y^3$**
22. $(x^7y^6)^3$ **$x^{21}y^{18}$**
23. $(cat)^2$ **$c^2a^2t^2$**
24. $(5m^2y)^2$ **$25m^4y^2$**
25. $(-2a^2b^3)^4$ **$16a^8b^{12}$**
26. $(p^{10}x^7)^4$ **$p^{40}x^{28}$**
27. $(-z^3y)^3$ **$-z^9y^3$**
28. $4p(-3p)^2$ **$36p^3$**
29. $2b(2ab)^3$ **$16a^3b^4$**
30. $3y(-2y)^3$ **$-24y^4$**
31. $3m(-2m)^2$ **$12m^3$**
32. $2c^2(-3c)^3$ **$-54c^5$**
33. $-2f(4fg)^2$ **$-32f^3g^2$**

Evaluate each expression if $x = -2$ and $y = -3$.

34. $3xy^2$ **−54**
35. $4x^2y$ **−48**
36. $(xy)^2$ **36**
37. $(-2xy^2)^2$ **1296**
38. $x(y^2)^3$ **−1458**
39. $(2x^2)^3$ **512**

Group Activity Card 14-4

Potent Powers — Group Activity **14-4**

MATERIALS: The deck of *number* and the deck of *variable* cards made for Activity 14-1 • one 0-6 spinner

The object of this game is to be the first one to correctly find the power of a monomial.

To begin, place the two decks on the table with the *number* deck first and the *variable* deck second. The leader turns up one card from the *number* deck and one card from the *variable* deck to form a monomial.

Lastly, the leader spins the spinner. The number obtained is used for the power of the monomial. Everyone must copy the expression.

The first player to simplify the expression correctly scores one point. The winner is the one with the most points after one round.

 Pre-Algebra

Extension

Using Cooperative Groups Have students choose three different one-digit numbers. Use one as the base, one as the exponent, and one as the power of the exponential term. Rearrange the integers until all possible combinations have been written. Evaluate each, compare, summarize, and explain results.

44. **Biology** It takes 2 hours for a culture with 1 bacterium to grow to two bacteria. To grow to 32, or 2^5, bacteria, the culture must double five times. If the culture doubles another five times, there will be $(2^5)^2$ bacteria.

a. Write $(2^5)^2$ as a power of 2. 2^{10}

b. $2^5 = 32$. Use this information to estimate $(2^5)^2$. 1024

45. **Chemistry** Chemicals like salt and magnesium can be extracted from sea water. To determine if such a venture would be profitable, a company would like to know how much salt or magnesium could be extracted from a cubic mile of sea water. How many cubic feet of sea water are in a cubic mile of sea water? (*Hint:* There are 5280 feet in a mile.) 5280^3

Mixed Review

46. **Geometry** The perimeter of the triangle at the right is $5x + 2$ units. Find the length of the base of the triangle. (Lesson 14-3) $2x$

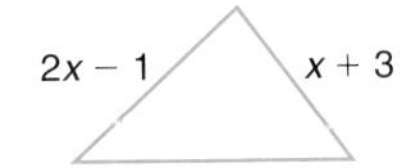

47. 13 in.

47. **Geometry** In a right triangle, if the measures of the legs are 5 inches and 12 inches, find the measure of the hypotenuse. (Lesson 13-4)

48. $\triangle GHI \cong \triangle LKJ$; $\angle G \cong \angle L$; $\angle H \cong \angle K$; $\angle I \cong \angle J$; $\overline{GH} \cong \overline{LK}$; $\overline{HI} \cong \overline{KJ}$; $\overline{GI} \cong \overline{LJ}$

48. **Geometry** Write a congruence statement for the pair of congruent triangles at the right. Then name the corresponding parts. (Lesson 11-5)

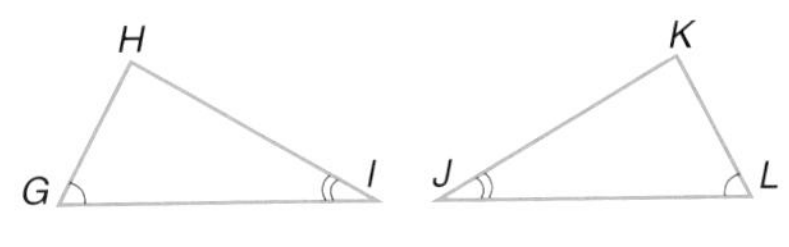

49. **Probability** Betsy plays on a basketball team. Her free throw average is 60%. How could you simulate the probable results of her next five free throw attempts? (Lesson 10-8) See margin.

50. Solve $\frac{2x}{-5} = -24$. (Lesson 7-2) 60

51. See margin for graph.

51. On graph paper, draw coordinate axes. Then graph and label each point. Name the quadrant in which each point is located. (Lesson 2-2)

a. $X(-4, 6)$ II b. $Y(2, 5)$ I c. $Z(-3, 0)$ none

52. Evaluate each expression if $x = -2$, $y = 4$, and $z = -1$. (Lesson 2-1)

a. $|2x| + yz$ 0 b. $\frac{4(x + y)}{2}$ −4 c. $\frac{xyz}{8}$ 1

53. Write an expression that means the same as $\frac{xy}{5}$. (Lesson 1-3) $xy \div 5$

EARTH WATCH

BUILDING WITH RECYCLED MATERIALS

It seems like a house built out of newspaper, sawdust, plastic bags, and old cars would not be the nicest one on the block. But the National Association of Home Builders made a beautiful new home out of recycled materials in Bowie, Maryland. The framing material was steel from old cars and the insulation was made from recycled newspapers.

See for Yourself See students' work.
Research recycled building materials.

- New siding, decking, and asphalt products have been invented that make use of recycled material. What recycled materials do these products use?
- What advantages and disadvantages do products made of recycled materials have as compared to traditional building products?

4 ASSESS

Closing Activity

Writing Have students make up two word problems. One that can be solved by finding the power of a power, and the other by finding the power of a product. Then have students exchange problems and solve.

Additional Answers

49. Sample answer: A spinner divided into 10 sections, six of them being successes.

51.

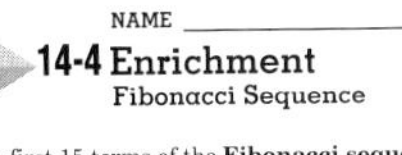

Enrichment Masters, p. 118

NAME ______ DATE ______

Student Edition Pages 719–723

14-4 Enrichment
Fibonacci Sequence

The first 15 terms of the **Fibonacci sequence** follow.

1, 1, 2, 3, 5, 8, 13, 21, 34, 55, 89, 144, 233, 377, 610

In addition to the fact that each term is the sum of the two previous terms, there are other interesting relationships.

Solve.

1. Find the sum of the first four terms. 7
2. Find the sum of the first five terms. 12
3. Find the sum of the first six terms. 20
4. What is the relationship between the sums and terms in the sequence? **The sum is one less than the Fibonacci number after the next Fibonacci number.**
5. Use your answer from Exercise 4 above to predict the sum of the first 10 terms, and of the first 12 terms. Check your answers by finding the sums. 143, 376
6. Divide each of the first 15 terms by 4. Make a list of the remainders. What pattern do you see in the remainder sequence? 1, 1, 2, 3, 1, 0, 1, 1, 2, 3, 1, 0, 1, 1, 2; The six digits 1, 1, 2, 3, 1, 0 repeat.
7. Complete the next three lines of this pattern of the sums of the squares of consecutive terms of the Fibonacci sequence.

$1^2 + 1^2 = 1 \times 2$
$1^2 + 1^2 + 2^2 = 2 \times 3$
$1^2 + 1^2 + 2^2 + 3^2 = 3 \times 5$
$1^2 + 1^2 + 2^2 + 3^2 + 5^2 = 5 \times 8$
$1^2 + 1^2 + 2^2 + 3^2 + 5^2 + 8^2 = 8 \times 13$
$1^2 + 1^2 + 2^2 + 3^2 + 5^2 + 8^2 + 13^2 = 13 \times 21$
$1^2 + 1^2 + 2^2 + 3^2 + 5^2 + 8^2 + 13^2 + 21^2 = 21 \times 34$

EARTH WATCH

You don't have to build a new home to have a "green" home. Many inexpensive improvements can be made to make a house more earth-friendly. For example, replacing standard shower heads with new low-flow heads, adding caulk and weather- stripping, and getting a heating and cooling system tune up are all ways to save energy and improve the environment.

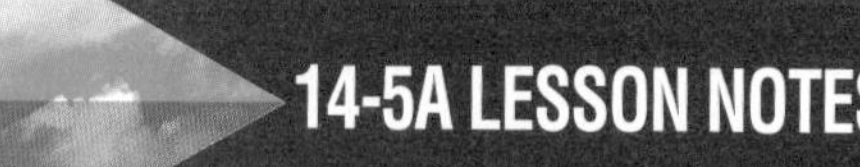

14-5A LESSON NOTES

NCTM Standards: 1-4, 9

Objective

Use algebra tiles to multiply a monomial and a polynomial.

Recommended Time

Demonstration and discussion: 15 minutes; Exercises: 30 minutes

Instructional Resources

For each student or group of students

Student Manipulative Kit

- algebra tiles

Math Lab and Modeling Math Masters

- p. 1 (algebra tiles)
- p. 57 (worksheet)

For teacher demonstration

Overhead Manipulative Resources

1 FOCUS

Motivating the Lesson

Ask students to show you tiles to represent the polynomial $x^2 + 2x + 1$. Ask if they can arrange the tiles in any way to form a rectangle or square.

2 TEACH

Teaching Tip For students who have difficulty working with small pieces, you may wish to provide them with larger, paper models of algebra tiles.

3 PRACTICE/APPLY

Assignment Guide
Core: 1–6 **Enriched:** 1–6

4 ASSESS

Observing students working in cooperative groups is an excellent method of assessment.

HANDS-ON ACTIVITY

14-5A Multiplying Polynomials

A Preview of Lesson **14-5**

MATERIALS

algebra tiles

product mat

Algebra tiles are named based on the area of the rectangle. The area of the rectangle is the product of the width and length.

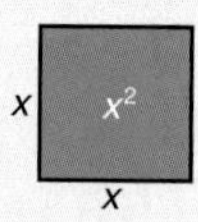

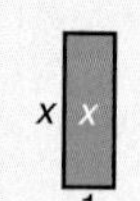

You can use algebra tiles to model more complex rectangles. These rectangles will help you understand how to find the product of simple polynomials. The width and length each represent a polynomial being multiplied. The area of the rectangle represents their product.

Your Turn **Work with a partner to find $x(x + 2)$.**

- Make a rectangle with a width of x and a length of $x + 2$. Use algebra tiles to mark off the dimensions on a product mat.

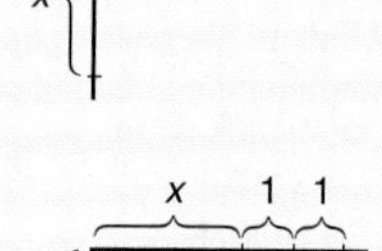

- Using the marks as a guide, fill in the rectangle with algebra tiles.

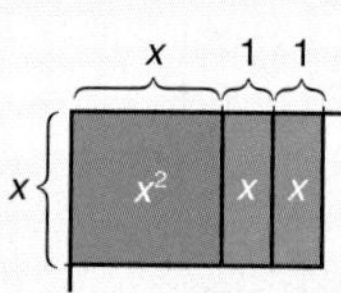

- The area of the rectangle is $x^2 + x + x$. In simplest form, the area is $x^2 + 2x$. Therefore, $x(x + 2) = x^2 + 2x$.

TALK ABOUT IT **Tell whether each statement is *true* or *false*. Justify your answer with algebra tiles.**

1. $x(2x + 3) = 2x^2 + 3x$ **true**
2. $2x(3x + 4) = 6x^2 + 4x$ **false**

Find each product using algebra tiles.

3. $x(x + 5)$ **$x^2 + 5x$**
4. $2x(x + 2)$ **$2x^2 + 4x$**
5. $3x(2x + 1)$ **$6x^2 + 3x$**
6. Suppose you have a square garden plot that measures x feet on a side.
   a. If you double the length of the plot and increase the width by 3 feet, how large will the new plot be? Write two expressions for the area of the new plot. **$2x(x + 3)$; $2x^2 + 6x$**
   b. If the original plot was 10 feet on a side, what is the area of the new plot? **260 ft²**

14-5 Multiplying a Polynomial by a Monomial

Setting Goals: *In this lesson, you'll multiply a polynomial by a monomial.*

Modeling a Real-World Application: Construction

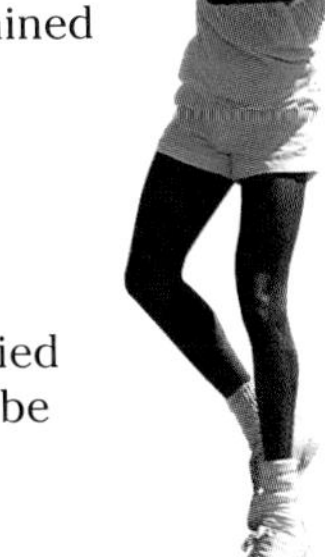

A chain-link fence is going to be installed around the school tennis courts. To determine how much fencing is needed, a representative from the fencing company measures the length and the width of the region that has to be enclosed. The total length of fencing can be determined by finding the perimeter P of the rectangular region.

$P = \ell + w + \ell + w$
$= 2\ell + 2w$ *Combine like terms.*
$= 2(\ell + w)$ *Distributive property*

In the expression $2(\ell + w)$, a polynomial is being multiplied by a monomial. In general, the distributive property can be used to multiply a monomial and a polynomial.

Learning the Concept

As you saw in Lesson 14-5A, you can use algebra tiles to multiply a polynomial and a monomial.

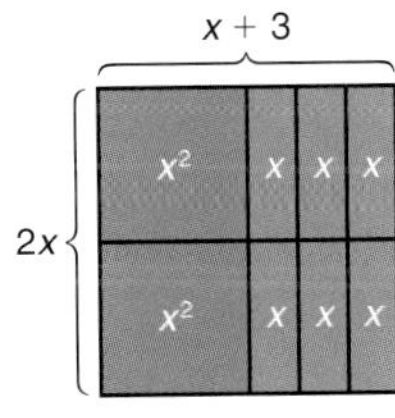

The figure at the right is a rectangle whose length is $x + 3$ and whose width is $2x$.

The area of any rectangle is the product of its length and its width. The area of this rectangle can also be found by adding the areas of the tiles.

Area: Formula	Area: Algebra tiles
$A = \ell w$	$A = x^2 + x^2 + x + x + x + x + x + x$
$A = (x + 3)2x$	$A = 2x^2 + 6x$

Therefore, $(x + 3)2x = 2x^2 + 6x$.

Example 1 **Find each product.**

a. $\mathbf{7(2x + 5)}$

$7(2x + 5) = 7(2x) + 7(5)$ *Distributive property*
$= 14x + 35$

b. $\mathbf{(3x - 7)4x}$

$(3x - 7)4x = (3x)4x - (7)4x$ *Distributive property*
$= 12x^2 - 28x$

c. $\mathbf{2ab(-9a^2 + 5ab - 4b^2)}$

$2ab(-9a^2 + 5ab - 4b^2)$
$= 2ab(-9a^2) + 2ab(5ab) - 2ab(4b^2)$ *Distributive property*
$= -18a^3b + 10a^2b^2 - 8ab^3$ *Multiply monomials.*

14-5 LESSON NOTES

NCTM Standards: 1-4, 9

Instructional Resources

- Study Guide Master 14-5
- Practice Master 14-5
- Enrichment Master 14-5
- Group Activity Card 14-5
- Assessment and Evaluation Masters, p. 380
- Activity Masters, p. 14
- Math Lab and Modeling Math Masters, p. 75

Transparency 14-5A contains the 5-Minute Check for this lesson; **Transparency 14-5B** contains a teaching aid for this lesson.

Recommended Pacing	
Honors Pacing	Day 5 of 9

1 FOCUS

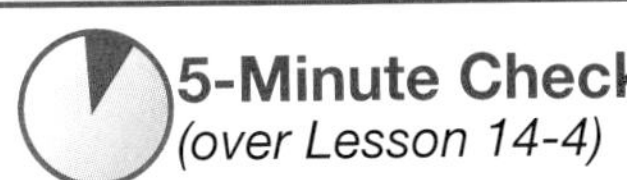

5-Minute Check
(over Lesson 14-4)

Simplify.

1. $(-3^3)^2$ **729**
2. $(6x^4)^3$ $\mathbf{216x^{12}}$
3. $5(x^3y^{-4})^3$ $\mathbf{5x^9y^{-12}}$

Evaluate each expression if $a = -4$ and $b = 2$.

4. $7a^3b$ **−896**
5. $-(a^{-2}b^6)$ **−4**

Motivating the Lesson

Questioning Show students several rectangles. Ask how they would find the perimeter of one of the rectangles. Ask how the area is related to the lengths of the sides.

Study Guide Masters, p. 119 The Study Guide Master provides a concise presentation of the lesson along with practice problems.

Tech Prep

Specialized Architects Some architects specialize in fields such as planning housing accommodations for physically challenged people. They must consider the extra space required for wheelchairs, chairlifts, and other special equipment when planning such areas.

For more information on tech prep, see the *Teacher's Handbook*.

2 TEACH

In-Class Examples

For Example 1
Find each product.

a. $9(5x + 6)$ **$45x + 54$**

b. $(-3m - 5)(5m)$
$-15m^2 - 25m$

c. $(3rs^2)(r - 2s + 5r^2s^3)$
$3r^2s^2 - 6rs^3 + 15r^3s^5$

For Example 2
The maximum length between goal lines of a Rugby Union football field is 40 yards less than twice the width. The maximum perimeter is 370 yards. What are the maximum dimensions of the field? **75 yards by 110 yards**

Additional Answers

1. Multiply $2x$ and -1 by x.
3. The commutative property of multiplication says that the order of the multiplication is not important.

Practice Masters, p. 119

NAME ________________ DATE ________

14-5 Practice
Multiplying a Polynomial by a Monomial

Student Edition
Pages 725–727

Find each product.

1. $3(2x + 3y)$ **$6x + 9y$**
2. $4x(6 - 5m)$ **$24x - 20mx$**
3. $^-2a(3a + 5ab)$ **$^-6a^2 - 10a^2b$**
4. $^-7c(^-4c - 6c^2)$ **$28c^2 + 42c^3$**
5. $10(3x - 4y)$ **$30x - 40y$**
6. $a^4(3a^3 + 4)$ **$3a^7 + 4a^4$**
7. $^-p^2(2p + 3pt - 4p^2)$ **$^-2p^3 - 3p^3t + 4p^4$**
8. $^-3t(5t^3 - 4t)$ **$^-15t^4 + 12t^2$**
9. $12xy(4xy + 6x)$ **$48x^2y^2 + 72x^2y$**
10. $3a^2b(4a + 3b)$ **$12a^3b + 9a^2b^2$**
11. $r(r^2 - 9)$ **$r^3 - 9r$**
12. $^-2y(y + 6)$ **$^-2y^2 - 12y$**
13. $5mp(7m - 2p)$ **$35m^2p - 10mp^2$**
14. $2x(4x + 3)$ **$8x^2 + 6x$**
15. $^-2y(^-5 - 3y)$ **$10y + 6y^2$**
16. $2xy(^-4xy + 3y)$ **$^-8x^2y^2 + 6xy^2$**
17. $2a^2b^3(3a^2b - 4ab^2)$ **$6a^4b^4 - 8a^3b^5$**
18. $y^2(y^2 + y - 2)$ **$y^4 + y^3 - 2y^2$**
19. $^-3xy(5x^4 - 7y^3 + 6x^2y)$
$^-15x^5y + 21xy^4 - 18x^3y^2$
20. $2x(2x^3 + 3xy - 5x)$
$4x^4 + 6x^2y - 10x^2$
21. $3p(7x - 4p)$ **$21px - 12p^2$**
22. $4pt(7pt + 7t)$ **$28p^2t^2 + 28pt^2$**
23. $3(7x + 6y + z)$ **$21x + 18y + 3z$**
24. $5x(7xy + 6x - 8y^2)$
$35x^2y + 30x^2 - 40xy^2$
25. $10a^2(5b^3 + 6a^2b - 8a)$
$50a^2b^3 + 60a^4b - 80a^3$
26. $4a^7(13a^2 - 7)$ **$52a^9 - 28a^7$**

Sometimes problems can be solved by simplifying polynomial expressions.

Example 2 INTEGRATION Algebra

The world's largest swimming pool is the Orthlieb Pool in Casablanca, Morocco. It is 30 meters longer than 6 times its width. If the perimeter of the pool is 1110 meters, what are the dimensions of the pool?

Explore You know the perimeter of the pool. You want to find the dimensions of the pool.

Plan Let w represent the width of the pool. Then $6w + 30$ represents the length. Then write an equation.

Perimeter *equals* *twice* *the sum of the length and width.*

$$P \quad = \quad 2 \quad (\ell + w)$$

Solve

$$P = 2(\ell + w)$$

$1110 = 2((6w + 30) + w)$ *Replace P with 1110 and ℓ with $6w + 30$.*

$1110 = 2(7w + 30)$ *Combine like terms.*

$1110 = 14w + 60$ *Distributive property*

$1050 = 14w$ *Subtract 60 from each side.*

$$\frac{1050}{14} = w$$

$$75 = w$$

The width is 75 meters and the length is $6w + 30$ or 480 meters.

Examine If the width is 75 meters, the length should be 30 meters longer than 6 times 75, or 480 meters. The perimeter is $2(75 + 480)$ or 1110 meters. The solution checks.

Checking Your Understanding

Communicating Mathematics

Read and study the lesson to answer these questions. **1, 3. See margin.**

1. **Explain** how you would find the product of x and $2x - 1$.
2. **State** the product of x and $2x + 3$ using the rectangle at the right. **$2x^2 + 3x$**
3. **Explain** why $x(2x + 3)$ and $(2x + 3)x$ are equivalent expressions.

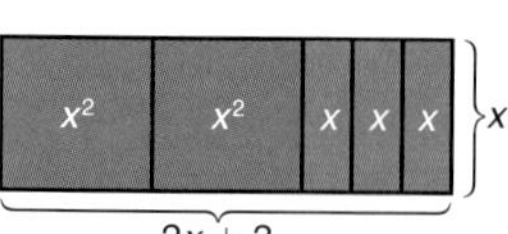

MODELING MATHEMATICS

MATERIALS

algebra tiles
product mat

Use algebra tiles to find each product.

4. $5(x + 2)$ **$5x + 10$**
5. $x(x + 4)$ **$x^2 + 4x$**
6. $2x(x - 1)$ **$2x^2 - 2x$**

Guided Practice

Find each product.

7. $3(x + 4)$ **$3x + 12$**
8. $x(x + 5)$ **$x^2 + 5x$**
9. $3(x - 2)$ **$3x - 6$**
10. $2x(x - 8)$ **$2x^2 - 16x$**
11. $4x(3x + 7)$ **$12x^2 + 28x$**
12. $x(5x - 12)$ **$5x^2 - 12x$**

Reteaching

Using Cooperative Groups Allow students to work together to complete the examples and some exercises. Make sure each student understands how to multiply a polynomial by a monomial before beginning his or her assignment.

Group Activity Card 14-5

Pretty Products Group Activity **14-5**

MATERIALS: Three decks of cards made for Activity 14-1.

First, the leader turns up cards to form first a polynomial (cards are drawn until a blank card is drawn in the *operations* deck) and then a monomial (one card from the *numbers* deck and one card from the *variables* deck).

After these expressions are formed, everyone in the group multiplies the polynomial and the monomial. When all are finished, the answers are checked, and one point is awarded to each player who has the correct answer.

At the end of one round, total points are compared to select the winners.

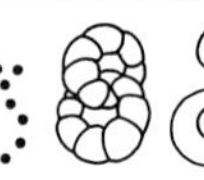

 Pre-Algebra

13. **Sports** The perimeter of a football field is 1040 feet. The length of the field is 120 feet less than 3 times the width. What are the dimensions of the football field? **160 by 360 feet**

Exercises: Practicing and Applying the Concept

Independent Practice

22. $-3x^3 + 21x^2$
23. $-18a + 2a^3$
24. $-14a^2 + 35a - 77$
25. $-15x^2 + 35x - 45$
26. $-18y + 27y^2 - 12y^3$
27. $4c^4 + 28c^3 - 40c$
28. $-15x^5 + 40x^3 + 60x^2$
29. $-12x^5 + 48x^4 - 42x^2$
30. $10x^4 - 8x^3 + 12x^2 - 18x$
31. $-x^5 + x^4 - 3x^3 + 5x^2$

A / B / C

Find each product. 18. $-3x^2 + 15x$ 20. $4m^3 - 4m^2$ 21. $p^2q^2 + 8pq$

14. $7(3x + 5)$ $21x + 35$
15. $-2(x + 8)$ $-2x - 16$
16. $y(y - 9)$ $y^2 - 9y$
17. $3x(2x - 1)$ $6x^2 - 3x$
18. $-3x(x - 5)$
19. $c(a^2 + b)$ $a^2c + bc$
20. $4m(m^2 - m)$
21. $pq(pq + 8)$
22. $-3x(x^2 - 7x)$
23. $-2a(9 - a^2)$
24. $7(-2a^2 + 5a - 11)$
25. $-5(3x^2 - 7x + 9)$
26. $-3y(6 - 9y + 4y^2)$
27. $4c(c^3 + 7c^2 - 10)$
28. $-5x^2(3x^3 - 8x - 12)$
29. $6x^2(-2x^3 + 8x^2 - 7)$
30. $2x(5x^3 - 4x^2 + 6x - 9)$
31. $-x^2(x^3 - x^2 + 3x - 5)$

Solve each equation.

32. $-3(2a - 12) + 48 = 3a + 3$ **9**
33. $2(5w - 12) = 6(-2w + 3) + 2$ **2**

Critical Thinking

34. Use algebra tiles to make a model of $2x + 6$. Then, form a tile rectangle and find the factors of $2x + 6$. Finally, outline a procedure for finding the factors of any binomial. **See margin.**

35. **Manufacturing** The figure at the right shows a pattern for a cardboard box before it has been cut and folded.
    a. Find the area of each rectangular region and add to find a formula for the number of square inches of cardboard needed. $12x^2 + 8x + 1$
    b. Find the surface area if x is 2.5 inches. **96 in²**

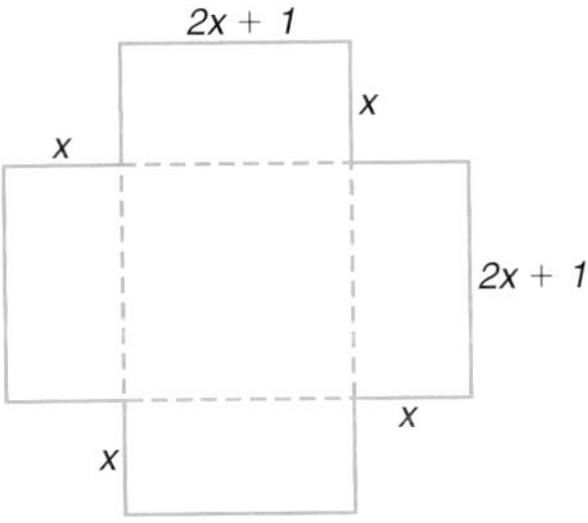

36. **Geometry** Find the measure of the area of the shaded region in simplest terms. $3s^2 - 3s$

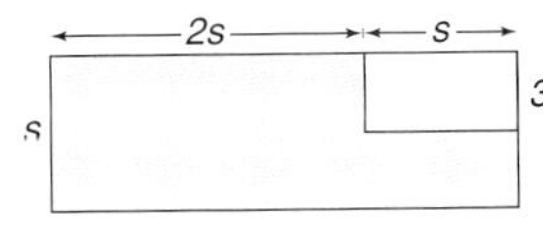

Mixed Review

37. Simplify $-4b(2b)^3$. (Lesson 14-4) $-32b^4$
38. Find the best integer estimate for $-\sqrt{150}$. (Lesson 13-1) **−12**
39. **Statistics** Describe two ways a graph of sales of several brands of cereal could be misleading. (Lesson 10-4) **See margin.**
40. Solve the system of equations $y = x - 3$ and $y = x - 4$ by graphing. (Lesson 8-8) **See Solutions Manual for graph; no solution.**
41. **Draw a Diagram** The Avarillo's backyard is 130 by 90 feet. If their sprinkler can water an area 30 by 30 feet, how many times will it need to be moved to water the entire yard? (Lesson 4-3) **13 times**
42. Is 528 divisible by 2, 3, 5, 6, or 10? (Lesson 4-1) **2, 3, 6**
43. Find the distance traveled if you drive at 55 mph for $3\frac{1}{2}$ hours. (Lesson 3-4) **192.5 miles**

3 PRACTICE/APPLY

Checking Your Understanding

Exercises 1–13 are designed to help you assess your students' understanding through reading, writing, speaking, and modeling. You should work through Exercises 1–6 with your students and then monitor their work on Exercises 7–13.

Assignment Guide

Core: 15–33 odd, 34, 35, 37–43
Enriched: 14–32 even, 34–43

For **Extra Practice**, see p. 773.

4 ASSESS

Closing Activity

Speaking Write several products of a polynomial by a monomial on the chalkboard or overhead. Have students state how to solve the problems.

Chapter 14, Quiz C (Lessons 14-4 and 14-5) is available in the *Assessment and Evaluation Masters*, p. 380.

***Enrichment Masters*, p. 119**

NAME ________________ DATE ________

14-5 Enrichment Student Edition Pages 725–727
The Product Sign

The Greek symbol Π (pi) is called the **product sign**. It can be used to write algebraic or numeric products.

Consider the series 3 × 4 × 5 × 6. To represent it algebraically, the series can be written as

$$\prod_{k=3}^{6} k \quad \text{or} \quad \prod_{k=0}^{3} (k + 3).$$

1. Can you find another way to write the product 3 × 4 × 5 × 6 using Π?
   Answers may vary. Sample answer: $\prod_{k=1}^{4} (k + 2)$

Write each series as an indicated product.

2. $\prod_{k=1}^{3} k$ $1 \times 2 \times 3$
3. $\prod_{i=4}^{7} (i - 3)^2$ $1 \times 4 \times 9 \times 16$
4. $\prod_{j=1}^{4} \frac{j}{j+1}$ $\frac{1}{2} \times \frac{2}{3} \times \frac{3}{4} \times \frac{4}{5}$
5. $\prod_{j=1}^{3} 2^j$ $2 \times 4 \times 8$
6. $\prod_{k=10}^{11} k^2 - 5$ 95×116
7. $\prod_{k=2}^{5} \frac{k+1}{k-1}$ $\frac{3}{1} \times \frac{4}{2} \times \frac{5}{3} \times \frac{6}{4}$ or $3 \times 2 \times \frac{5}{3} \times \frac{3}{2}$

Use Π to write each series. Answers may vary. Sample answers given.

8. $1 \times 2 \times 3$ $\prod_{k=1}^{3} k$
9. $1 \times 4 \times 9 \times 16$ $\prod_{k=1}^{4} k^2$
10. $\frac{1}{2} \times \frac{2}{3} \times \frac{3}{4} \times \frac{4}{5} \times \frac{5}{6}$ $\prod_{k=1}^{5} \frac{k}{k+1}$
11. $15 \times 24 \times 35 \times 48$ $\prod_{k=5}^{8} k(k - 2)$

Extension

Using Diagrams Have students work in groups to draw diagrams of rectangles showing the lengths and widths in terms of x. Then have students make up a problem involving their diagram and the perimeter of the rectangle.

Additional Answers

34. The factors of $2x + 6$ are 2 and $x + 3$. You can find the factors of a binomial by dividing each term by its greatest common factor. Then the greatest common factor and the result are the factors of the binomial.
39. Sample answers: Choose a scale that makes one product look much better than the other; omit some information or labels.

14-6 LESSON NOTES

NCTM Standards: 1-4, 9

Instructional Resources

- Study Guide Master 14-6
- Practice Master 14-6
- Enrichment Master 14-6
- Group Activity Card 14-6
- Assessment and Evaluation Masters, p. 380
- Activity Masters, p. 28
- Math Lab and Modeling Math Masters, p. 89
- Multicultural Activity Masters, p. 28
- Real-World Applications, 32
- Tech Prep Applications Masters, p. 28

Transparency 14-6A contains the 5-Minute Check for this lesson; **Transparency 14-6B** contains a teaching aid for this lesson.

Recommended Pacing	
Honor Pacing	Day 6 of 9

1 FOCUS

5-Minute Check
(over Lesson 14-5)

Find each product.

1. $8(4x - 3)$ $\mathbf{32x - 24}$
2. $-3(-12m - 5)$ $\mathbf{36m + 15}$
3. $fg(-2 - fg^2)$ $\mathbf{-2fg - f^2g^3}$
4. $4y(3y^2 - 2)$ $\mathbf{12y^3 - 8y}$
5. $(-2a)(-3a^3 + 2a^2 - 4)$ $\mathbf{6a^4 - 4a^3 + 8a}$

Motivating the Lesson

Situational Problem Have students verbalize traits that appear frequently within their families. Do some family members not have these traits? Use this discussion to develop definitions of dominant, recessive, and hybrid. Use Punnett squares to show how two parents with brown eyes (dominant) can have a blue-eyed (recessive) child.

14-6 Multiplying Binomials

Setting Goals: *In this lesson, you'll multiply binomials.*

Modeling a Real-World Application: Genetics

Punnett squares are used to show possible ways that genes can combine at fertilization. In a Punnett square, *dominant* genes are shown with capital letters. *Recessive* genes are shown with the lowercase of the same letter.

The Punnett square below represents a cross between two hybrid tall pea plants. A hybrid trait is the result of a combination of a dominant and a recessive gene. The parent plants are tall because the dominant trait masks the recessive trait.

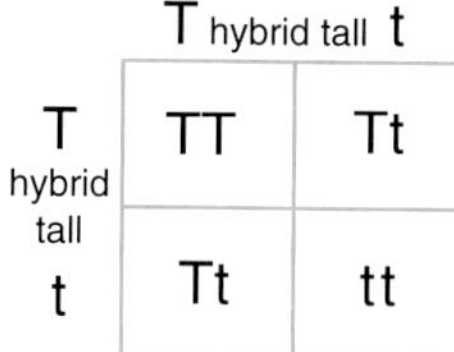

	T (hybrid tall)	t
T (hybrid tall)	TT	Tt
t	Tt	tt

Letters representing the parents' genes are placed on the outer sides of the Punnett square.

Letters inside the boxes of the square show the possible gene combinations for their offspring.

Let T represent the dominant gene for tallness and t represent the recessive gene for shortness.

When the parents' genes are combined, you see all of the possible combinations of the genes of the offspring.

$$(T + t)(T + t) = TT + Tt + Tt + tt$$
$$= TT + 2Tt + tt$$

Learning the Concept

$(T + t)(T + t)$ is an example of the product of two binomials. Algebra tiles can also be used to help you understand how to multiply two binomials.

Consider the binomials $x + 3$ and $x + 2$. To multiply these binomials, mark off a rectangle on a product mat that has dimensions $x + 3$ and $x + 2$. Then fill in the rectangle with algebra tiles.

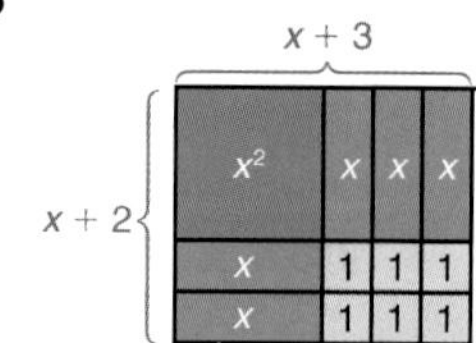

Area: Formula	Area: Algebra Tiles
$A = \ell w$	$A = x^2 + x + x + x + x + x + 1 + 1 + 1 + 1 + 1 + 1$
$= (x + 3)(x + 2)$	$= x^2 + 5x + 6$

The product $(x + 3)(x + 2)$ represents the area of the rectangle. The area is also the sum of the areas of the algebra tiles, $x^2 + 5x + 6$. Therefore, $(x + 3)(x + 2) = x^2 + 5x + 6$.

Tech Prep

Animal Breeder Breeders of purebred and show animals need to know about the genes that animals carry. They use much more complicated combinations than shown in the Punnett square on page 728, but the basic idea is still the same.
For more information on tech prep, see the *Teacher's Handbook*.

You can also use the distributive property to multiply binomials.

Example 1 Find each product.

a. $(2x + 3)(3x + 5)$

$$\begin{aligned}(2x+3)(3x+5) &= (2x+3)3x + (2x+3)5 && \text{Distributive property}\\ &= (2x)3x + (3)3x + (2x)5 + (3)5\\ &= 6x^2 + 9x + 10x + 15 && \text{Multiply polynomials.}\\ &= 6x^2 + 19x + 15 && \text{Combine like terms.}\end{aligned}$$

b. $(2x + 1)(5x - 3)$

$$\begin{aligned}(2x+1)(5x-3) &= (2x+1)5x + (2x+1)(-3) && \text{Distributive property}\\ &= (2x)5x + (1)5x + (2x)(-3) + (1)(-3)\\ &= 10x^2 + 5x + (-6x) + (-3) && \text{Multiply polynomials.}\\ &= 10x^2 - x - 3 && \text{Combine like terms.}\end{aligned}$$

Example 2

APPLICATION

Gardening

A rectangular garden is 5 feet longer than twice its width. It has a sidewalk 3 feet wide on two of its sides. The area of the sidewalk is 213 square feet. Find the dimensions of the garden.

Explore Use algebra tiles to model the situation. Let x represent the width of the garden. Then, $2x + 5$ represents the length of the garden, $x + 3$ the width of the garden and sidewalk, and $2x + 8$ the length of the garden and sidewalk.

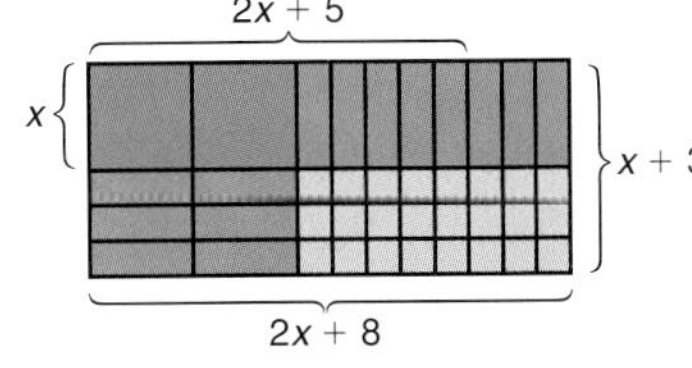

Then, $x(2x + 5)$ = the area of the garden and $(x + 3)(2x + 8)$ = the area of the garden and the sidewalk.

Plan area of garden and sidewalk − area of garden = area of sidewalk

$$(x + 3)(2x + 8) - x(2x + 5) = 213$$

Solve

$$\begin{aligned}2x^2 + 6x + 8x + 24 - 2x^2 - 5x &= 213 && \text{Count the tiles.}\\ (2x^2 - 2x^2) + (6x + 8x - 5x) + 24 &= 213\\ 9x + 24 &= 213 && \text{Simplify.}\\ 9x &= 189\\ x &= 21\end{aligned}$$

The width is 21 feet, and the length is $2x + 5 = 2(21) + 5$ or 47 feet.

Examine

$$\begin{aligned}\text{Total Area} - \text{area of garden} &\stackrel{?}{=} \text{area of sidewalk}\\ (24)(50) - (21)(47) &\stackrel{?}{=} 213\\ 1200 - 987 &\stackrel{?}{=} 213\\ 213 &= 213 \checkmark\end{aligned}$$

2 TEACH

In-Class Examples

For Example 1
Find each product.

a. $(x + 6)(4x + 2)$ $\mathbf{4x^2 + 26x + 12}$

b. $(3x - 5)(2x - 4)$ $\mathbf{6x^2 - 22x + 20}$

For Example 2
Pat makes baby quilts and full-sized quilts. The baby quilts are x-feet by x-feet. The full-sized quilts are 3 feet wider and 4 feet longer than the baby quilts. The difference in the areas of the quilts is 33 square feet. Find the dimensions of each quilt. **baby quilt, 3 ft by 3 ft; full-sized quilt, 6 ft by 7 ft**

***Study Guide Masters*, p. 120**

NAME ______________________ DATE ____________

14-6 Study Guide
Multiplying Binomials

Student Edition Pages 728–731

Use the distributive property to multiply two binomials.

$(a + b)(c + d) = ac + ad + bc + bd$

Example: Find the product $(x + 6)(x + 2)$.

$$\begin{aligned}(x+6)(x+2) &= x(x+2) + 6(x+2) && \text{Distributive property}\\ &= x\cdot x + x\cdot 2 + 6\cdot x + 6\cdot 2 && \text{Distributive property}\\ &= x^2 + 2x + 6x + 12 && \text{Multiply polynomials}\\ &= x^2 + 8x + 12 && \text{Combine like terms}\end{aligned}$$

Find each product.

1. $(x + 1)(x + 2)$ $\mathbf{x^2 + 3x + 2}$
2. $(y + 4)(y + 5)$ $\mathbf{y^2 + 9y + 20}$
3. $(z + 7)(z + 7)$ $\mathbf{z^2 + 14z + 49}$
4. $(y + 2)(y - 9)$ $\mathbf{y^2 - 7y - 18}$
5. $(y + 3)(y - 3)$ $\mathbf{y^2 - 9}$
6. $(x + 1)(x + 1)$ $\mathbf{x^2 + 2x + 1}$
7. $(x + 5)(x - 1)$ $\mathbf{x^2 + 4x - 5}$
8. $(y + 8)(y - 3)$ $\mathbf{y^2 + 5y - 24}$
9. $(3x + 8)(2x + 1)$ $\mathbf{6x^2 + 19x + 8}$
10. $(4x + 3)(x - 6)$ $\mathbf{4x^2 - 21x - 18}$
11. $(x - 7)(3x + 4)$ $\mathbf{3x^2 - 17x - 28}$
12. $(6y + 3)(4y - 6)$ $\mathbf{24y^2 - 24y - 18}$
13. $(3y - 2)(6y + 3)$ $\mathbf{18y^2 - 3y - 6}$
14. $(3x + 8)(3x - 8)$ $\mathbf{9x^2 - 64}$
15. $(2x + 4)(2x + 4)$ $\mathbf{4x^2 + 16x + 16}$

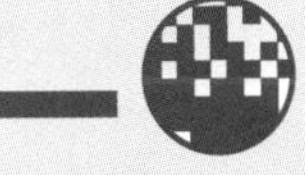

Alternative Teaching Strategies

Student Diversity Some students may have difficulty multiplying binomials without tiles. Provide paper tiles for these students and allow them to use them to complete all exercises.

Reteaching

Using Manipulatives Place algebra tiles in bags—all x^2 in one bag, all x in another, and all 1-unit tiles in a third. Have students draw out tiles without looking. Students then try to make a rectangle with their tiles. If they can make a rectangle, they write the two factors and their product.

3 PRACTICE/APPLY

Checking Your Understanding

Exercises 1–12 are designed to help you assess your students' understanding through reading, writing, speaking, and modeling. You should work through Exercises 1–5 with your students and then monitor their work on Exercises 6–12.

Assignment Guide
Core: 13–33 odd, 34–39
Enriched: 14–30 even, 31–39

For **Extra Practice**, see p. 773.

The red A, B, and C flags, printed only in the Teacher's Wraparound Edition, indicate the level of difficulty of the exercises.

Additional Answer

1.

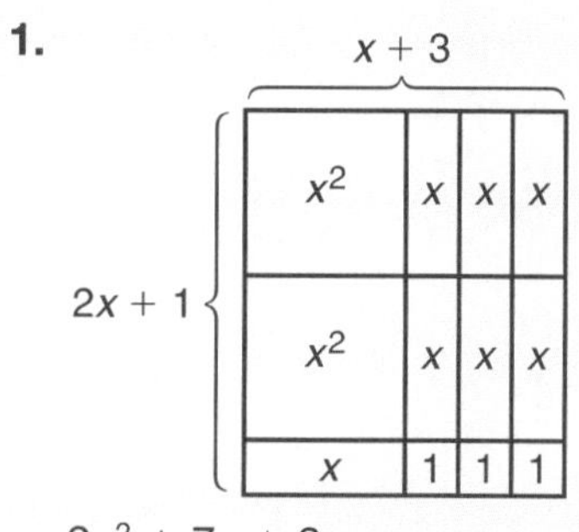

$2x^2 + 7x + 3$

***Practice Masters*, p. 120**

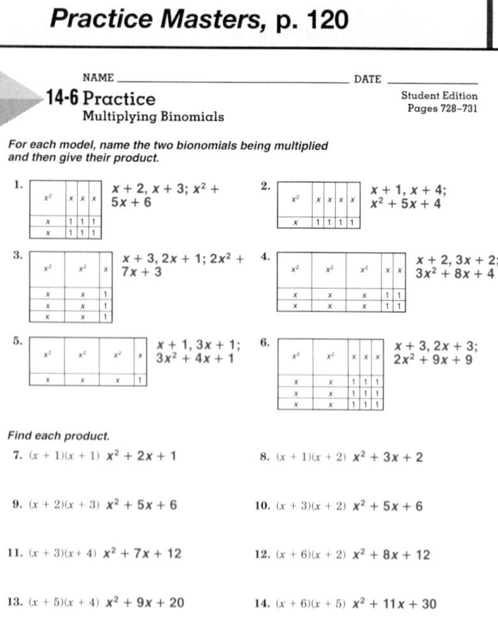

NAME ______ DATE ______

14-6 Practice
Multiplying Binomials

Student Edition Pages 728–731

For each model, name the two bionomials being multiplied and then give their product.

1. $x + 2, x + 3; x^2 + 5x + 6$
2. $x + 1, x + 4; x^2 + 5x + 4$
3. $x + 3, 2x + 1; 2x^2 + 7x + 3$
4. $x + 2, 3x + 2; 3x^2 + 8x + 4$
5. $x + 1, 3x + 1; 3x^2 + 4x + 1$
6. $x + 3, 2x + 3; 2x^2 + 9x + 9$

Find each product.

7. $(x + 1)(x + 1)$ $x^2 + 2x + 1$
8. $(x + 1)(x + 2)$ $x^2 + 3x + 2$
9. $(x + 2)(x + 3)$ $x^2 + 5x + 6$
10. $(x + 3)(x + 2)$ $x^2 + 5x + 6$
11. $(x + 3)(x + 4)$ $x^2 + 7x + 12$
12. $(x + 6)(x + 2)$ $x^2 + 8x + 12$
13. $(x + 5)(x + 4)$ $x^2 + 9x + 20$
14. $(x + 6)(x + 5)$ $x^2 + 11x + 30$
15. $(2x + 1)(x + 2)$ $2x^2 + 5x + 2$
16. $(2x + 1)(2x + 1)$ $4x^2 + 4x + 1$
17. $(x + 4)(2x + 2)$ $2x^2 + 10x + 8$
18. $(3x + 1)(x + 5)$ $3x^2 + 16x + 5$
19. $(x + 6)(3x + 2)$ $3x^2 + 20x + 12$
20. $(2x + 1)(3x + 1)$ $6x^2 + 5x + 1$

Checking Your Understanding

Communicating Mathematics

MATERIALS
algebra tiles
product mat

Read and study the lesson to answer these questions. 1–2. See margin.

1. **Draw** a rectangular region that represents the product of $x + 3$ and $2x + 1$. Then find the product.
2. **Explain** how the procedure used for multiplying two binomials is similar to the procedure for multiplying a binomial and a monomial. Then explain how it is different.

Use algebra tiles to find each product.

3. $(x + 1)(x + 2)$ $x^2 + 3x + 2$
4. $(x + 3)(x + 4)$ $x^2 + 7x + 12$
5. $(2x + 3)(x + 2)$ $2x^2 + 7x + 6$

Guided Practice

For each model, name the two binomials being multiplied and then give their product.

6.

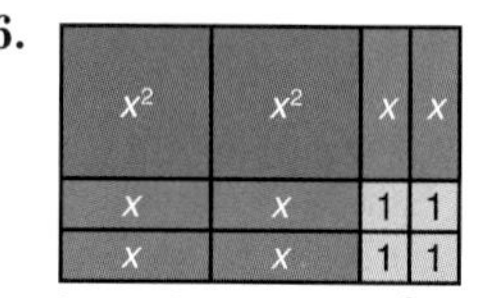

$2x + 2, x + 2; 2x^2 + 6x + 4$

7.

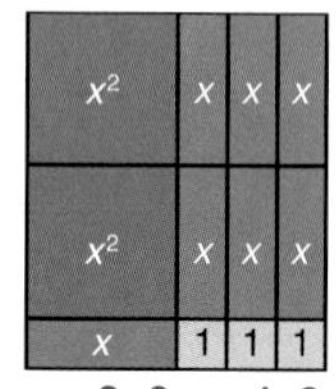

$x + 3, 2x + 1; 2x^2 + 7x + 3$

Find each product.

8. $(x + 3)(2x + 5)$ $2x^2 + 11x + 15$
9. $(2x + 3)(x + 1)$ $2x^2 + 5x + 3$
10. $(2x + 3)(3x + 2)$ $6x^2 + 13x + 6$
11. $(5x - 3)(2x + 1)$ $10x^2 - x - 3$

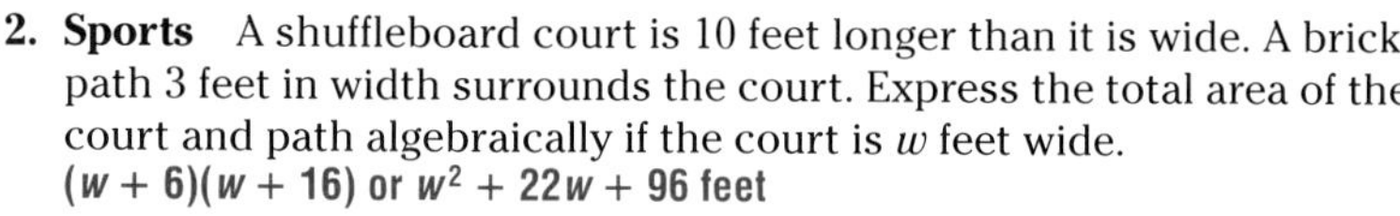

12. **Sports** A shuffleboard court is 10 feet longer than it is wide. A brick path 3 feet in width surrounds the court. Express the total area of the court and path algebraically if the court is w feet wide. $(w + 6)(w + 16)$ or $w^2 + 22w + 96$ feet

Exercises: Practicing and Applying the Concept

Independent Practice

For each model, name the two binomials being multiplied and then give their product.

13.

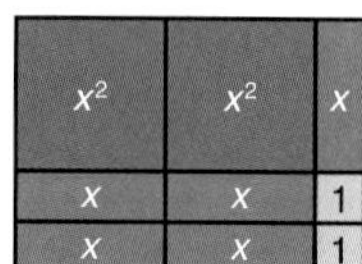

$2x + 1, x + 2; 2x^2 + 5x + 2$

14.

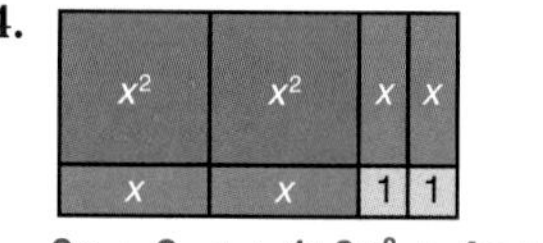

$2x + 2, x + 1; 2x^2 + 4x + 2$

15.

$2x + 3, 2x + 2; 4x^2 + 10x + 6$

16.

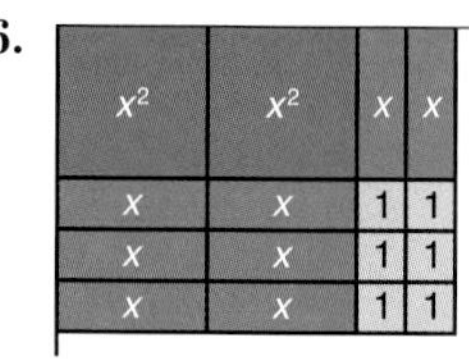

$2x + 2, x + 3; 2x^2 + 8x + 6$

Group Activity Card 14-6

Twice The Fun — Group Activity 14-6

MATERIALS: The three decks of cards used in Activity 14-1

This game is an extension of Activity 14-5.

This time the leader turns up cards to form two binomials. Once again everyone else tries to be the first one to correctly calculate the product.

After the answers are checked, the first player with the correct answer scores one point, and the player with the most points after one round is the overall winner.

 Pre-Algebra

Additional Answer

2. Sample answer: The distributive property is used to multiply both two binomials and a binomial and a monomial. But when you multiply two binomials there are four multiplications, and with a binomial and a monomial there are only two multiplications to perform.

Find each product. **17–30. See margin.**

17. $(x + 4)(x + 3)$	**18.** $(x + 5)(x + 2)$	**19.** $(x - 6)(x + 2)$
20. $(x + 7)(x - 5)$	**21.** $(x - 9)(x + 4)$	**22.** $(x + 2)(x + 2)$
23. $(2x + 3)(x - 4)$	**24.** $(3x - 1)(x + 8)$	**25.** $(x + 3)(x + 3)$
26. $(2x + 5)(3x + 1)$	**27.** $(5x + 2)(2x - 3)$	**28.** $(x - 5)(x + 5)$
29. $\left(3x - \frac{1}{4}\right)\left(6x - \frac{1}{2}\right)$	**30.** $(x - 2)(x^2 + 2x + 4)$	

Critical Thinking

31. Write the multiplication problem modeled by the figure at the right. Name the product. **$(x + 2)(x - 1)$; $x^2 + x - 2$**

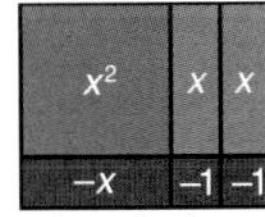

Applications and Problem Solving

32. Volunteer Work As a project, a senior citizens group makes baby quilts and full-size quilts for victims of natural disasters. The baby quilts are $x + 2$ feet long by x feet wide. The full-size quilts are 4 feet longer and 3 feet wider than the baby quilts. **a–d. See margin.**

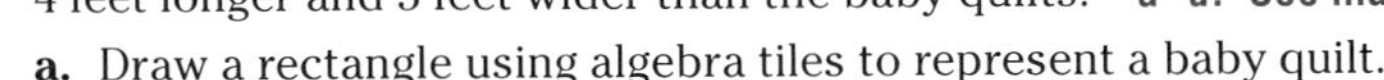

a. Draw a rectangle using algebra tiles to represent a baby quilt.

b. Draw additional tiles to increase the rectangular region to represent a full-size quilt.

c. Write a product for the full-size quilt and another for the baby quilt.

d. What is the difference between the area of the full-size quilt and the area of the baby quilt?

33. Geometry The length of the smaller rectangle at the right is 1 inch less than twice its width. Both the dimensions of the larger rectangle are 2 inches longer than the smaller rectangle. The area of the shaded region is 86 inches. **a. 14 in. and 27 in.**

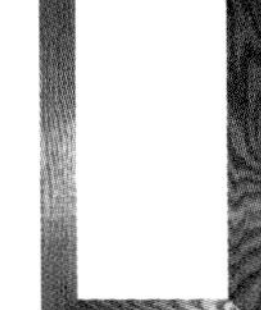

a. What are the dimensions of the smaller rectangle?

b. What is the area of the smaller rectangle? **378 in²**

c. What is the area of the larger rectangle? **464 in²**

Mixed Review

34. Find the product $-2(4c^2 - 3c + 5)$. (Lesson 14-5) **$-8c^2 + 6c - 10$**

35. Measurement On a sunny day, a tree casts a shadow 60 feet long. At the same time, a yardstick casts a shadow 2 feet long. How tall is the tree? (Lesson 11-6) **90 ft**

36. Business One way businesses judge their success is by the ratio of their sales to the total sales of similar products. This is called their market share. The circle graph at the right shows the market shares of ready-to-drink teas. Find the measure of the angle to represent Arizona® brand tea. (Lesson 11-2) **43.2°**

Source: Beverage Marketing Corporation

37. Geometry The radius of a circle is 0.5 m. Find its circumference. (Lesson 7-4) **3.14 m**

38. What is the GCF of 60 and 25? (Lesson 4-5) **5**

39. Solve $6x < -84$. (Lesson 3-7) **$x < -14$**

Extension

Using Patterns Have students complete the exercises below. Then have them summarize and generalize their findings.

$(2 + x)(2 - x)$ **$4 - x^2$**
$(x - 6)(x + 6)$ **$x^2 - 36$**
$(1 + x)(1 - x)$ **$1 - x^2$**
$(1 + 2x)(1 - 2x)$ **$1 - 4x^2$**
$(3x + 4)(3x - 4)$ **$9x^2 - 16$**

The product of the sum and difference of two numbers is equal to the difference of their squares.

4 ASSESS

Closing Activity

Modeling Have students work in pairs. One student models a binomial multiplication problem, and the other states the two binomials and the product. Switch roles and repeat.

Chapter 14, Quiz D (Lesson 14-6) is available in the *Assessment and Evaluation Masters*, p. 380.

Additional Answers

17. $x^2 + 7x + 12$
18. $x^2 + 7x + 10$
19. $x^2 - 4x - 12$
20. $x^2 + 2x - 35$
21. $x^2 - 5x - 36$
22. $x^2 + 4x + 4$
23. $2x^2 - 5x - 12$
24. $3x^2 + 23x - 8$
25. $x^2 + 6x + 9$
26. $6x^2 + 17x + 5$
27. $10x^2 - 11x - 6$
28. $x^2 - 25$
29. $18x^2 - 3x + \frac{1}{8}$
30. $x^3 - 8$
32c. full-sized: $x^2 + 9x + 18$; baby quilt: $x^2 + 2x$
32d. $7x + 18$

Enrichment Masters, p. 120

NAME ______________________ DATE ____________

14-6 Enrichment
Multiplying Binomials

Student Edition Pages 728–731

Two binomials can be multiplied by using a memory device called the **FOIL** method as shown below.

> To multiply two binomials, find the sum of the products of
> **F** the first terms,
> **O** the outer terms,
> **I** the inner terms, and
> **L** the last terms.

Example: Find the product $(2x + 3)(5x + 8)$.

Multiply the **F**irst terms.	$(\mathbf{2x} + 3)(\mathbf{5x} + 8)$ ($2x \cdot 5x$)	$10x^2$
		+
Multiply the **O**uter terms.	$(\mathbf{2x} + 3)(5x + \mathbf{8})$ ($2x \cdot 8$)	$16x$
		+
Multiply the **I**nner terms.	$(2x + \mathbf{3})(\mathbf{5x} + 8)$ ($3 \cdot 5x$)	$15x$
		+
Multiply the **L**ast terms.	$(2x + \mathbf{3})(5x + \mathbf{8})$ ($3 \cdot 8$)	24

$$(2x + 3)(5x + 8) = 10x^2 + 16x + 15x + 24$$
$$= 10x^2 + 31x + 24 \quad \text{Combine like terms.}$$

Use the FOIL method to find each product.

1. $(x + 3)(x + 5)$ **$x^2 + 8x + 15$**	**2.** $(2x + 4)(x + 7)$ **$2x^2 + 18x + 28$**	**3.** $(3x + 1)(3x + 1)$ **$9x^2 + 6x + 1$**
4. $(x + 4)(x + 4)$ **$x^2 + 8x + 16$**	**5.** $(2x + 1)(2x + 3)$ **$4x^2 + 8x + 3$**	**6.** $(x - 3)(x + 2)$ **$x^2 - x - 6$**
7. $(x + 3)(x - 2)$ **$x^2 + x - 6$**	**8.** $(x - 4)(x - 8)$ **$x^2 - 12x + 32$**	**9.** $(2x + 1)(x + 8)$ **$2x^2 + 17x + 8$**

14-6B LESSON NOTES

NCTM Standards: 1-4, 9

Objective

Use algebra tiles to factor simple trinomials.

Recommended Time

Demonstration and discussion: 15 minutes; Exercises: 30 minutes

Instructional Resources

For each student or group of students
Student Manipulative Kit
- algebra tiles

Math Lab and Modeling Math Masters
- p. 1 (algebra tiles)
- p. 58 (worksheet)

For teacher demonstration
Overhead Manipulative Resources

1 FOCUS

Motivating the Lesson

Ask students to name pairs of numbers with a product of 12. Write each different pair on the board. Then have students use 12 1-unit tiles to make rectangles for each pair of factors.

2 TEACH

Teaching Tip Model making rectangles on the overhead projector. If you do not have overhead algebra tiles, just use regular tiles or paper tiles.

3 PRACTICE/APPLY

Assignment Guide
Core: 1–7
Enriched: 1–7

4 ASSESS

Observing students working in cooperative groups is an excellent method of assessment.

HANDS-ON ACTIVITY

14-6B Factoring

An Extension of Lesson **14-6**

MATERIALS
- algebra tiles
- product mat

Consider a rectangle with a length of $x + 1$ and a width of $x + 3$. From the previous lesson, you know that $(x + 1)(x + 3) = x^2 + 4x + 3$. The binomials $(x + 1)$ and $(x + 3)$ are **factors** of the trinomial $x^2 + 4x + 3$.

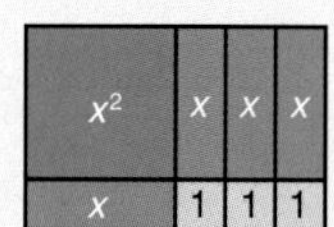

You can use algebra tiles as a model for factoring simple trinomials. If a rectangle with a length and width greater than 1 cannot be formed to represent the trinomial, then the trinomial is not factorable.

Your Turn **Work with a partner to factor $x^2 + 3x + 2$.**

- Model the polynomial.

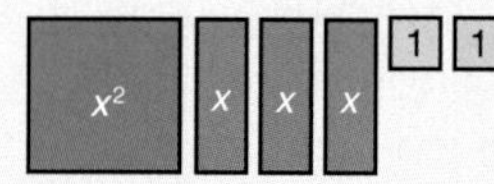

- Place the x^2-tile and the 1-tiles as shown.

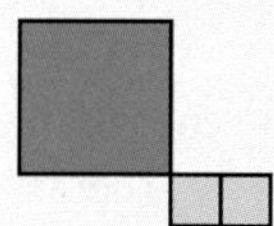

- Complete the rectangle with the x-tiles.

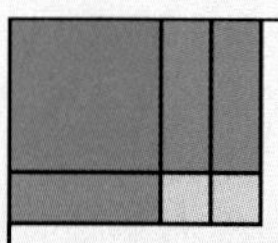

- Since a rectangle can be formed, $x^2 + 3x + 2$ is factorable. The rectangle has a width of $x + 1$ and a length of $x + 2$. So the factors of $x^2 + 3x + 2$ are $x + 1$ and $x + 2$.

TALK ABOUT IT **Tell whether each polynomial is factorable. Justify your answers with algebra tiles.** 1–6. See Solutions Manual for justifications.

1. $x^2 + 6x + 8$ yes
2. $x^2 + 5x + 6$ yes
3. $x^2 + 7x + 3$ no
4. $3x^2 + 8x + 5$ yes
5. $5x^2 - x + 16$ no
6. $8x^2 - 31x - 4$ yes
7. Write a paragraph that explains how you can determine whether a trinomial can be factored. Include an example of one trinomial that can be factored and one that cannot. See margin.

Additional Answer Talk About It

7. Sample answer: A trinomial can be factored if you can make a rectangle with algebra tiles that represent that trinomial. $x^2 + 5x + 6$ can be factored, but $x^2 + 6x + 6$ cannot.

x^2	x	x
x	1	1
x	1	1
x	1	1

CHAPTER 14 Highlights

Vocabulary

After completing this chapter, you should be able to define each term, property, or phrase and give an example or two of each.

Algebra

binomial (p. 706)

coefficient (p. 711)

constant (p. 707)

like terms (p. 711)

polynomial (p. 706)

power of a monomial (p. 721)

power of a power (p. 719)

power of a product (p. 720)

trinomial (p. 706)

Understanding and Using Vocabulary

Choose the correct term to complete each sentence.

1. The degree of a monomial is the (<u>sum</u>, product) of the exponents of its variables.
2. A (binomial, <u>trinomial</u>) is the sum or difference of three monomials.
3. The degree of a polynomial is the same as that of the term that has the (least, <u>greatest</u>) degree.
4. The numerical part of a monomial is called the (<u>coefficient</u>, variable).
5. To subtract two polynomials, you add the (<u>additive inverse</u>, multiplicative inverse) of the second polynomial.
6. The (commutative property, <u>distributive property</u>) is used in multiplying a polynomial by a monomial.

Determine whether each statement is *true* or *false*.

7. Every polynomial is a monomial. false
8. The coefficient of $6ab^2$ is 2. false
9. The additive inverse of $(3x + 6y)$ is $(-3x - 6y)$. true
10. The power of a product property says that $(8m)^3 = 8^3m^3$. true

Instructional Resources

Three multiple-choice tests and three free-response tests are provided in the *Assessment and Evaluation Masters*. Forms 1A and 2A are for honors pacing, Forms 1B and 2B are for average pacing, Forms 1C and 2C are for basic pacing, Chapter 14 Test, Form 1B is shown at the right. Chapter 14 Test, Form 2B is shown on the next page.

Highlights

Using the Chapter Highlights

The Chapter Highlights begins with a listing of the new terms, properties, and phrases that were introduced in this chapter.

Assessment and Evaluation Masters, pp. 367–368

NAME ______________________ DATE __________

CHAPTER 14 Test, Form 1B

1. Choose the expression which is <u>not</u> a monomial.
   A. $\frac{4}{5}$ B. $\frac{4}{x}$ C. $3x^2y$ D. 5 1. B
2. Find the degree of $3ab^6$.
   A. 18 B. 9 C. 6 D. 7 2. D
3. The expression $x^2 + 2x + 4$ is a
   A. monomial. B. binomial. C. trinomial. D. constant. 3. C
4. Find the degree of $ax^2 + xy + 3y^2$.
   A. 1 B. 2 C. 3 D. 4 4. C

Find each sum.

5. $(9f^2 + 4f - 6) + (f^2 - 2f - 2)$
   A. $10f^2 + 2f - 8$ B. $10f^2 + 2f - 4$
   C. $10f^2 + 2f + 8$ D. $9f^2 + 2f + 2$ 5. A
6. $(-2x^2 - 2x - 6) + (-x^2 + 2x - 6)$
   A. $2x + 12$ B. $-3x^2 - 4x - 12$
   C. $3x^2 - 4x - 12$ D. $-3x^2 - 12$ 6. D
7. $(5a^2 + 3a + 2) + (6a^2 - 5a + 6)$
   A. $11a^2 - 2a + 8$ B. $11a^2 + 2a + 8$
   C. $11a^2 + 8a + 8$ D. $11a^2 - 2a - 4$ 7. A
8. $(2x + 3y + 4) + (4x - y - 2) + (3x - 1)$
   A. $9x + 2y + 1$ B. $9x + 4y + 7$
   C. $9x + 3y + 6$ D. $9x - y + 1$ 8. A

Find each difference.

9. $(2x^2 - 7x - 3) - (5x^2 - 9)$
   A. $3x^2 - 7x - 12$ B. $-3x^2 - 7x + 6$
   C. $3x^2 - 7x - 6$ D. $-3x^2 - 7x - 12$ 9. B
10. $(-3z^2 + 4z + 7) - (8z^2 + 4z - 3)$
   A. $-11z^2 + 4$ B. $-11z^2 - 8z + 10$
   C. $5z^2 + 8z + 4$ D. $-11z^2 + 10$ 10. D
11. $(4w^2 + 3w + 6) - (3w^2 + 4w + 7)$
   A. $w^2 - w - 1$ B. $w^2 + 7w + 13$
   C. $w^2 - w + 1$ D. $-w^2 - w + 1$ 11. A
12. $(10y^2 + 4y - 3) - (6y^2 - 8y + 12)$
   A. $4y^2 - 12y - 15$ B. $4y^2 + 12y + 9$
   C. $4y^2 + 12y - 15$ D. $4y^2 - 4y + 9$ 12. C

NAME ______________________ DATE __________

Chapter 14 Test, Form 1B (continued)

Simplify each monomial.

13. $(3^2)^3$
   A. 216 B. 18 C. 27 D. 729 13. D
14. $(-2^3)^2$
   A. -36 B. 36 C. 64 D. -64 14. C
15. $(xy^3)^4$
   A. x^4y^{12} B. xy^7 C. $4xy^{12}$ D. xy^{12} 15. A
16. $(-2x^4)^3$
   A. $8x^{12}$ B. $-8x^{12}$ C. $-8x^7$ D. $-6x^{12}$ 16. B

Find each product.

17. $-2(4x^2 - 8x - 6)$
   A. $-8x^2 - 16x - 12$ B. $-8x^2 - 16x + 12$
   C. $-8x^2 + 16x + 12$ D. $-8x^2 + 16x - 12$ 17. C
18. $3a(4a^2 + 2a - 3)$
   A. $12a^3 + 6a^2 - 9a$ B. $12a^2 + 6a - 9$
   C. $7a^3 + 6a^2 - 9a$ D. $12a^3 + 5a^2 - 9$ 18. A
19. $-4b(20b^4 - 6b^2)$
   A. $-80b^5 - 10b^3$ B. $-80b^5 - 2b^3$
   C. $-80b^5 + 10b^3$ D. $-80b^5 + 24b^3$ 19. D
20. $8x(-4x^2 + 6x - 5)$
   A. $-32x^3 + 48x^2 + 40x$ B. $32x^3 + 48x + 40$
   C. $-32x^3 + 48x^2 - 40x$ D. $32x^2 + 48x - 40$ 20. C
21. $(x + 2)(x + 4)$
   A. $x^2 + 8$ B. $2x + 6$ C. $2x + 8$ D. $x^2 + 6x + 8$ 21. D
22. $(3m + 4)(2m - 7)$
   A. $6m^2 - 29m - 28$ B. $6m^2 - 13m - 28$
   C. $6m^2 + 13m + 28$ D. $6m^2 + 13m - 28$ 22. B
23. $(2x - 3)(2x + 3)$
   A. $4x^2 - 9$ B. $4x^2 - 12x + 9$
   C. $4x^2 - 6$ D. $4x^2 + 9$ 23. A
24. $(2x - 3y)(3x - 2y)$
   A. $6x^2 - 5xy - 6y^2$ B. $6x^2 - 13xy + 6y^2$
   C. $6x^2 + 13xy + 6y^2$ D. $6x^2 - 13xy - 6y^2$ 24. B
25. $(2x - 3y)(2x - 3y)$
   A. $4x^2 + 12xy - 9y^2$ B. $4x^2 + 12xy + 9y^2$
   C. $4x^2 - 12xy - 9y^2$ D. $4x^2 - 12xy + 9y^2$ 25. D

Using the Chapter Study Guide and Assessment

Skills and Concepts Encourage students to refer to the objectives and examples on the left as they complete the review exercises on the right.

Assessment and Evaluation Masters, pp. 373–374

NAME ______ DATE ______

CHAPTER 14 Test, Form 2B

State whether each expression is a polynomial. If it is, classify it as a monomial, binomial, or trinomial.

1. $3x$ — 1. monomial
2. $3 + x + x^2$ — 2. trinomial
3. $\frac{3}{x}$ — 3. no

Find the degree of each polynomial.

4. $5x$ — 4. 1
5. $x^2 + 3x$ — 5. 2
6. 12 — 6. 0

Find each sum or difference.

7. $(3x + 2) + (5x + 3)$ — 7. $8x + 5$
8. $(6a^2 + 4) + (8a^2 - 3a + 6)$ — 8. $14a^2 - 3a + 10$
9. $(5c^2 + 4c + 3) + (3c^2 + 4c + 5)$ — 9. $8c^2 + 8c + 8$
10. $(2x^2 - 3x + 4) + (-6x^2 - 5x - 8)$ — 10. $-4x^2 - 8x - 4$
11. $(9y - 3) - (4y + 7)$ — 11. $5y - 10$
12. $(7y^2 - 3y + 4) - (8y^2 - 6y + 4)$ — 12. $-y^2 + 3y$
13. $(-12x^2 - 3) - (6x^2 - 4x + 8)$ — 13. $-18x^2 + 4x - 11$
14. $(5a^2 - 3a - 6) - (4a^2 + 5a + 7)$ — 14. $a^2 - 8a - 13$

Simplify each monomial.

15. $(3a)^2$ — 15. $9a^2$
16. $(ab^6)^4$ — 16. a^4b^{24}
17. $(x^2y^4z^3)^{-3}$ — 17. $x^{-6}y^{-12}z^{-9}$
18. Evaluate $-3x(x^3)^2$ if $x = 2$. — 18. -384

NAME ______ DATE ______

Chapter 14 Test, Form 2B (continued)

Find each product.

19. $3(2a^2 + 4a - 3)$ — 19. $6a^2 + 12a - 9$
20. $f^3(8f^2 - 6)$ — 20. $8f^5 - 6f^3$
21. $-2a^2(3a^2 - 4a + 1)$ — 21. $-6a^4 + 8a^3 - 2a^2$
22. $(4a + 3)(4a - 3)$ — 22. $16a^2 - 9$
23. $(-2b + 6)(2b + 6)$ — 23. $-4b^2 + 36$
24. $(3x - 4)(4x - 3)$ — 24. $12x^2 - 25x + 12$
25. $(-3c + 7)(-3c + 7)$ — 25. $9c^2 - 42c + 49$

CHAPTER 14 Study Guide and Assessment

Skills and Concepts

Objectives and Examples

Upon completing this chapter, you should be able to:

▶ **identify and classify polynomials** (Lesson 14-1)

The expression $5a^2 - 3a + 4$ is a polynomial. It is a trinomial because it is the sum of three monomials.

The expression $\frac{2}{a}$ is not a polynomial because it has a variable in the denominator.

Review Exercises

Use these exercises to review and prepare for the chapter test.

State whether each expression is a polynomial. If it is, identify it as a *monomial*, *binomial*, or *trinomial*. 11, 14, 15. See margin.

11. $3x^4 - x$
12. $\frac{4}{ax}$ no
13. ax^2 yes; monomial
14. $9b^2 + b - 1$
15. $8x - \frac{5}{3}$
16. $k^3 - \frac{2}{k}$ no

▶ **find the degree of polynomials** (Lesson 14-1)

Find the degree of $x^3y + 3xy - 6y - 1$.

Find the degree of each term.

x^3y has degree 3 + 1 or 4.
$3xy$ has degree 1 + 1 or 2.
$6y$ has degree 1.
1 has degree 0.

So, the degree of $x^3y + 3xy - 6y - 1$ is 4.

Find the degree of each polynomial.

17. $4x$ 1
18. $5a^2b$ 3
19. $3x + y^2$ 2
20. $19m^2n^3 - 14mn^4$ 5
21. $x^2 - 6xy + xy^2$ 3
22. $12rs^2 + 3r^2s + 5r^4s$ 5

▶ **add polynomials** (Lesson 14-2)

Find the sum $(b^2 + 3b + 8) + (2b^2 + 5b - 5)$.

$$\begin{array}{r} b^2 + 3b + 8 \\ (+)\ 2b^2 + 5b - 5 \\ \hline 3b^2 + 8b + 3 \end{array}$$

Find each sum.

23. $(2x^2 - 5x) + (3x^2 + x)$ $5x^2 - 4x$
24. $(a^2 - 6ab) + (3a^2 + ab)$ $4a^2 - 5ab$
25. $(x^2 - 5x + 3) + (4x - 3)$ $x^2 - x$
26. $(-3y^2 + 2) + (4y^2 - 5y - 2)$ $y^2 - 5y$
27. $\begin{array}{r} 4x^2 + 3x + 2 \\ (+)\ x^2 \qquad - 1 \\ \hline \end{array}$ $5x^2 + 3x + 1$
28. $\begin{array}{r} 16x^2y - 2xy + xy^2 \\ (+)\ 4x^2y + 6xy - 8xy^2 \\ \hline \end{array}$ $20x^2y + 4xy - 7xy^2$

GLENCOE Technology

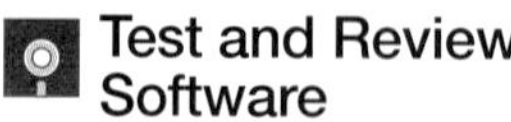

Test and Review Software

You may use this software, a combination of an item generator and an item bank, to create your own tests or worksheets. Types of items include free response, multiple choice, short answer, and open ended.

For IBM & Macintosh

Additional Answers

11. yes, binomial
14. yes, trinomial
15. yes, binomial

Chapter 14 • Study Guide and Assessment

Objectives and Examples	Review Exercises

▶ **subtract polynomials** (Lesson 14-3)

Find the difference $(4y^2 - 2y + 3) - (y^2 + 2y - 4)$.

Align like terms and add the additive inverse of $y^2 + 2y - 4$.

$$\begin{array}{r} 4y^2 - 2y + 3 \\ (-)\ y^2 + 2y - 4 \\ \hline \end{array} \rightarrow \begin{array}{r} 4y^2 - 2y + 3 \\ (+)\ -y^2 - 2y + 4 \\ \hline 3y^2 - 4y + 7 \end{array}$$

Find each difference.

29. $(7a - 11b) - (3a + 4b)$ **$4a - 15b$**
30. $(6y - 8z) - (6y + 4z)$ **$-12z$**
31. $(3a^2 - b^2 + c^2) - (a^2 + 2b^2)$ **$2a^2 - 3b^2 + c^2$**
32. $(14a^2 - 3a) - (6a^2 + 5a + 17)$ **$8a^2 - 8a - 17$**
33. $\begin{array}{r} 18x^2 + 3x - 1 \\ (-)\ 2x^2 + 4x + 6 \\ \hline \end{array}$ **$16x^2 - x - 7$**
34. $\begin{array}{r} 12m^2 - mn + 9n^2 \\ (-)\ 7m^2 + 2mn - 4n^2 \\ \hline \end{array}$ **$5m^2 - 3mn + 13n^2$**

▶ **find powers of monomials** (Lesson 14-4)

$(x^5)^3 = x^{5 \cdot 3}$
$= x^{15}$

$(2d)^4 = 2^4 \cdot d^4$
$= 16d^4$

$(a^3b^2)^3 = a^{3 \cdot 3} \cdot b^{2 \cdot 3}$
$= a^9b^6$

Simplify.

35. $(a^2)^3$ **a^6**
36. $(-2x)^3$ **$-8x^3$**
37. $(p^2q)^3$ **p^6q^3**
38. $-5c(2cd)^3$ **$-40c^4d^3$**
39. $4y(y^2z)^3$ **$4y^7z^3$**
40. $(12a^5)^2b^3$ **$144a^{10}b^3$**
41. $6a(-ab)^7$ **$-6a^8b^7$**
42. $(-2d^2)^6(-3)^3$ **$-1728d^{12}$**

▶ **multiply a polynomial by a monomial** (Lesson 14-5)

$-5y(2y^2 - 4) = -5y(2y^2) - (-5y)(4)$
$= -10y^3 - (-20y)$
$= -10y^3 + 20y$

Find each product. 48. **$-6az^3 - 12a^2z^2 - 3a^3z$**

43. $4d(2d - 5)$ **$8d^2 - 20d$**
44. $x(-5x + 3)$ **$-5x^2 + 3x$**
45. $a^2(2a^3 + a - 5)$ **$2a^5 + a^3 - 5a^2$**
46. $3y(-y^2 - 8y + 4)$ **$-3y^3 - 24y^2 + 12y$**
47. $-2g(g^3 + 6g + 3)$ **$-2g^4 - 12g^2 - 6g$**
48. $-3az(2z^2 + 4az + a^2)$

Applications and Problem Solving

Encourage students to work through the exercises in the Applications and Problem Solving section to strengthen their problem-solving skills.

Objectives and Examples

▶ **multiply binomials** (Lesson 14-6)

$(x - 4)(2x + 3)$
$= (x - 4)2x + (x - 4)(3)$
$= x(2x) + (-4)(2x) + x(3) + (-4)(3)$
$= 2x^2 - 8x + 3x - 12$
$= 2x^2 - 5x - 12$

Review Exercises

49. For the model below, name the two binomials being multiplied and then give their product. **$2x + 1$, $x + 3$; $2x^2 + 7x + 3$**

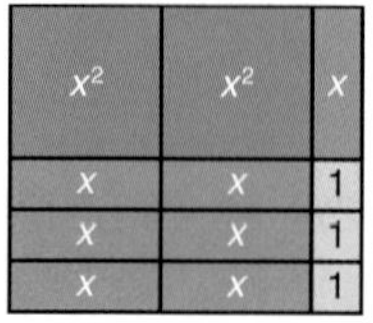

Find each product.

50. $(x + 3)(x + 1)$ **$x^2 + 4x + 3$**

51. $(2x + 1)(x + 1)$ **$2x^2 + 3x + 1$**

52. $(3x + 2)(2x + 2)$ **$6x^2 + 10x + 4$**

Applications and Problem Solving

53. Architecture Norma Merrick Sklarek was the first African-American woman registered as an architect in the United States. Suppose Ms. Sklarek drew the floor plan for the first floor of a house shown below. Using the measurements given, write a polynomial that represents the total area of the first floor. (Lesson 14-1)
$2xy + 2y^2 + 2yz$

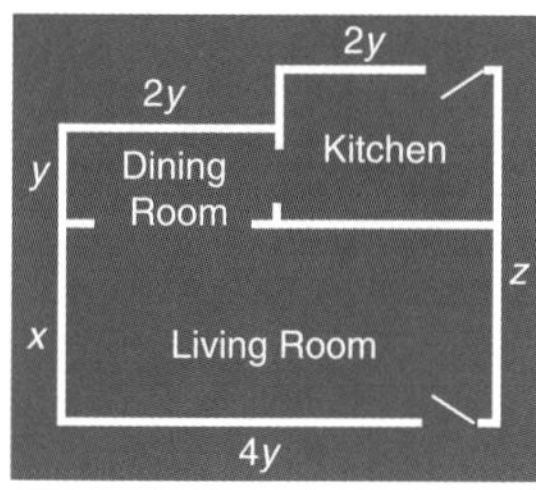

54. Geometry The perimeter of the triangle shown below is $4x - 2y$ units. Find the length of the third side of the triangle. (Lessons 14-1 and 14-2) **$2x - 2y$ units**

55. Geometry Find the volume of the triangular prism below. (Lesson 14-4) **$3x^3$**

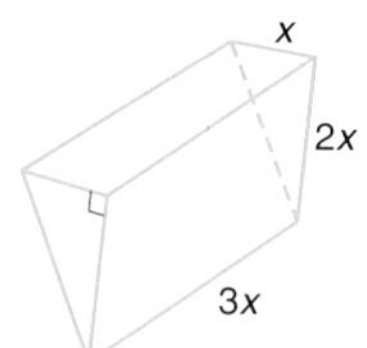

56. Sports The perimeter of a soccer field is 1040 feet. The length of the field is 40 feet more than 2 times the width. What are the dimensions of the field? (Lesson 14-5)
160 ft by 360 ft

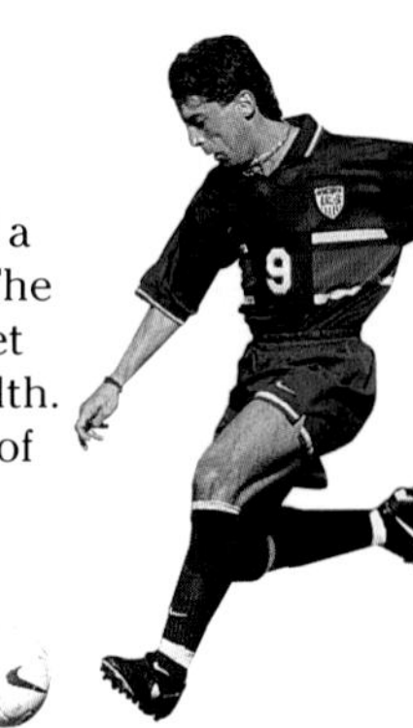

A practice test for Chapter 14 is available on page 787.

Additional Answer Cooperative Learning Project

The sixth row of Pascal's Triangle is the coefficients of the expansion of $(x + y)^5$.
$(x + y)^6 = x^6 + 6x^5y + 15x^4y^2 + 20x^3y^3 + 15x^2y^4 + 6xy^5 + y^6$.

Alternative Assessment

Cooperative Learning Project

The triangular arrangement of numbers below is called Pascal's Triangle. Many patterns exist in Pascal's Triangle. One of the patterns is that the rows in the triangle can be used to write powers of binomials.

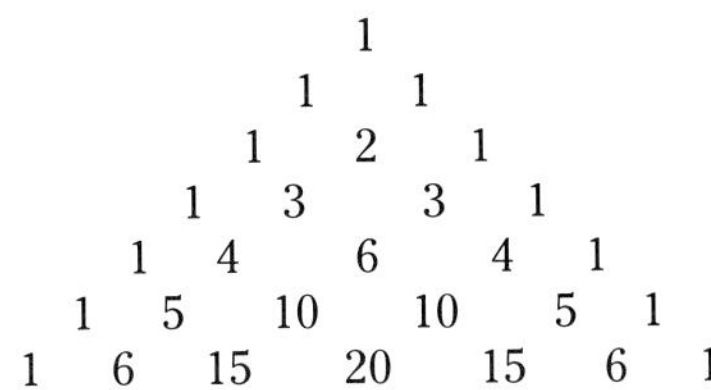

Work with a partner to find the powers of the binomial $(x + y)$ and complete the table below.

Power	Product	Coefficients of the terms
$(x + y)^2$	$x^2 + 2xy + y^2$	1, 2, 1
$(x + y)^3$	$x^3 + 3x^2y + 3xy^2 + y^3$	1, 3, 3, 1
$(x + y)^4$	$x^4 + 4x^3y + 6x^2y^2 + 4xy^3 + y^4$	1, 4, 6, 4, 1
$(x + y)^5$	$x^5 + 5x^4y + 10x^3y^2 + 10x^2y^3 + 5xy^4 + y^5$	1, 5, 10, 10, 5, 1

Compare the last column of the table to the rows of Pascal's Triangle. Describe the relationship you find. Predict the product of $(x + y)^6$. Then multiply to verify your product. **See margin.**

Thinking Critically

- For all numbers a and b such that $b \neq 0$, and any integer m, is $\left(\frac{a}{b}\right)^m = \frac{a^m}{b^m}$ a true sentence? **yes**
- A trapezoid has an area of 425 square inches and a height of 10 inches. The lower base is 5 inches less than twice the upper base. Find the length of each base. **30 in.; 55 in.**

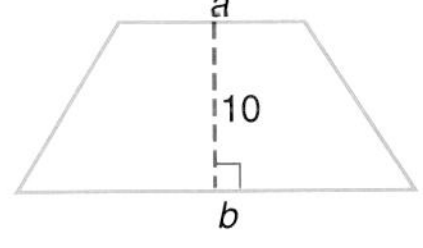

Portfolio

Review the items in your portfolio. Make a list of the items, noting why each item was chosen. Replace any items that are no longer appropriate. **See students' work.**

Self Evaluation

Assess yourself. Look back over your performance in Pre-Algebra this year. Did you meet the goals you set for yourself at the beginning of the course? What could you have done differently to make the year more successful? Write at least one goal that you would like to accomplish in the next mathematics course you take. **See students' work.**

Alternative Assessment

The Alternative Assessment section provides students with the opportunity to assess their own work by thinking critically, working with others, keeping a portfolio, and honestly evaluating their own progress. For more information on alternative forms of assessment, see *Alternative Assessment in the Mathematics Classroom*.

Performance Assessment

Performance Assessment tasks for this chapter are included in the *Assessment and Evaluation Masters*. A scoring guide is also provided.

Alternative Assessment

Assessment and Evaluation Masters, pp. 377, 389

NAME ____________ DATE ________

CHAPTER 14 Performance Assessment

Instructions: Demonstrate your knowledge by giving a clear, concise solution to each problem. Be sure to include all relevant drawings and justify your answers. You may show your solution in more than one way or investigate beyond the requirements of the problem.

1. An algebraic expression that contains more than one monomial is called a polynomial. Complete the table below.

Sample Polynomial	Type of Polynomial	Number of Terms	Degree of Polynomial
$\frac{1}{3}x^2$		1	
$x^3 - x^2$			3
$a^2 + 3ab + b^2$			
$a^4b^3 + 2a^3b^2 - 12a^2b$			

2. Tiles can be used to model polynomials.
   a. Draw tiles to represent $3x^2 + x + 4$.
   b. Draw tiles to find the sum $(2x^2 + 3x + 1) + (x^2 + 2x + 3)$. Explain each step.
   c. Find two polynomials whose difference is $2x^2 + x + 4$. Draw tiles to demonstrate finding their difference.
   d. The tiles below represent the product of two binomials. Use the tiles to form a rectangle and find the two binomials.

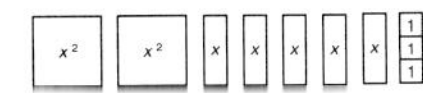

CHAPTER 14 Scoring Guide

Level	Specific Criteria
3 Superior	• Shows thorough understanding of the concepts *polynomial*, and *adding and subtracting polynomials*. • Uses appropriate strategies to solve problems. • Computations to find the difference of polynomials are correct. • Written explanations are exemplary. • Diagrams are accurate and clear. • Goes beyond requirements of some or all problems.
2 Satisfactory, with minor flaws	• Shows understanding of the concepts *polynomial*, and *adding and subtracting polynomials*. • Uses appropriate strategies to solve problems. • Computations to find the difference of polynomials are mostly correct. • Written explanations are effective. • Diagrams are mostly accurate and clear. • Satisfies all requirements of problems.
1 Nearly Satisfactory, with serious flaws	• Shows understanding of most of the concepts *polynomial*, and *adding and subtracting polynomials*. • May not use appropriate strategies to solve problems. • Computations to find the difference of polynomials are mostly correct. • Written explanations are satisfactory. • Diagrams are mostly accurate and clear. • Satisfies most requirements of problems.
0 Unsatisfactory	• Shows little or no understanding of the concepts *polynomial*, and *adding and subtracting polynomials*. • May not use appropriate strategies to solve problems. • Computations to find the difference of polynomials are incorrect. • Written explanations are not satisfactory. • Diagrams are not accurate and clear. • Does not satisfy requirements of problems.

Chapter 14 • Ongoing Assessment

Using the Ongoing Assessment

These two pages review the skills and concepts presented in Chapters 1–14. This review is formatted to reflect new trends in standardized testing.

Assessment and Evaluation Masters, pp. 382–383

NAME ____________ DATE ________

CHAPTERS 1–14 Cumulative Review

1. Evaluate ab if $a = 3$ and $b = 5$. (Lesson 1-3) — 1. 15
2. Simplify $|-5| + |3|$. (Lesson 2-1) — 2. 8
3. Find the perimeter of a rectangle with a length of 8 meters and width of 3 meters. (Lesson 3-5) — 3. 22 m
4. Find LCM of 16 and 24. (Lesson 4-7) — 4. 48
5. Solve $y - 4.7 = 8.3$. (Lesson 5-6) — 5. 13
6. Express $\frac{3}{8}$ as a decimal. (Lesson 6-1) — 6. 0.375
7. Solve $^-5y - 12 = {}^-72$. (Lesson 7-2) — 7. 12
8. Find the slope of the line that contains $A(3, 6)$ and $B(1, 1)$. (Lesson 8-6) — 8. $\frac{5}{2}$
9. A $39.95 radio is on sale for 25% off. Find the discount to the nearest cent. (Lesson 9-9) — 9. $9.99
10. Find the value $C(7, 3)$. (Lesson 10-6) — 10. 35
11. Angles A and B are complementary. The measure of angle A is 37°. Find $m\angle B$. (Lesson 11-3) — 11. 53°
12. In $\triangle ABC$, $m\angle A = 70°$ and $m\angle B = 75°$. Find $m\angle C$. (Lesson 11-4) — 12. 35°
13. Find the area of the trapezoid with bases of 20 centimeters and 12 centimeters, and a height of 10 centimeters. (Lesson 12-1) — 13. 160 cm²
14. Find the area of a circle with a radius of 9 inches. Use $\pi \approx 3.14$. (Lesson 12-2) — 14. 254.34 in²
15. Find the surface area of a circular cylinder with a radius of 7 centimeters and a height of 13 centimeters. (Lesson 12-5) — 15. 879.2 cm²
16. Find the volume of a cone with a radius of 8 millimeters and a height of 15 millimeters. Use $\pi \approx 3.14$. (Lesson 12-8) — 16. 1004.8 mm³

NAME ____________ DATE ________

Chapters 1–14, Cumulative Review (continued)

17. Find the best integer estimate for $\sqrt{200}$. (Lesson 13-1) — 17. 14
18. Is $\sqrt{121}$ rational or irrational? (Lesson 13-3) — 18. rational
19. Find the measure of the hypotenuse for a right triangle with legs 16 meters and 18 meters long. Round to the nearest tenth. (Lesson 13-4) — 19. 24.1 m
20. Find the measure of $\angle A$ if $\tan A = 0.839$. (Lesson 13-6) — 20. 40°
21. Find the sum $(6a - 9c) + (2a + 4c)$. (Lesson 14-2) — 21. $8a - 5c$
22. Find the difference $(9m + 5) - (^-3m + 8)$. (Lesson 14-3) — 22. $12m - 3$
23. Simplify $(^-y^2)^7$. (Lesson 14-4) — 23. $^-y^{14}$
24. Find the product $^-3x(^-5x + 3)$. (Lesson 14-5) — 24. $15x^2 - 9x$
25. Find the product $(x + 5)(x - 5)$. (Lesson 14-6) — 25. $x^2 - 25$

Standardized test practice questions are provided in the *Assessment and Evaluation Masters*, p. 381.

CHAPTERS 1–14 *Ongoing Assessment*

Section One: Multiple Choice

There are ten multiple-choice questions in this section. After working each problem, write the letter of the correct answer on your paper.

1. Which is equivalent to $\frac{3^6}{3^2}$? **B**

   A. 3^3 B. 3^4
   C. 3^8 D. 3^{12}

2. In a right triangle, the tangent ratio is the measure of the leg opposite an angle to the measure of the leg adjacent to the angle. What is the tangent ratio of angle D? **C**

   A. $\frac{3}{5}$
   B. $\frac{4}{5}$
   C. $\frac{4}{3}$
   D. $\frac{5}{3}$

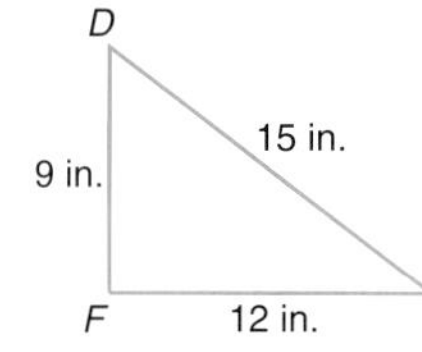

3. Paul Olsen is designing a cereal box. The height of the box is 12 inches, the width is 2 inches, and the length is 6 inches. What is the surface area of the cardboard he needs for one box? **C**

   A. 40 in²
   B. 144 in²
   C. 216 in²
   D. 288 in²

4. Which is equivalent to $(7x^2 + 3y) - (3x^2 + 5y)$? **A**

   A. $4x^2 - 2y$ B. $4x^4 - 2y^2$
   C. $4x^2 + 8y$ D. $4x^4 + 8y^2$

5. A sweater that normally sells for $35 is on sale at 25% off. Which is the best estimate of the sale price? **C**

   A. $9. B. $32.
   C. $26. D. $43.

6. A rectangular window is 40 inches high and 30 inches wide. What is the length of a diagonal? **C**

   A. 60 in. B. 35 in.
   C. 50 in. D. 25 in.

7. Find the surface area of a cone with a diameter of 12 meters and a slant height of 18 meters. Round your answer to the nearest tenth. **B**

   A. 75.4 m² B. 452.2 m²
   C. 94.2 m² D. 678.2 m²

8. Which is equivalent to $(-2x)^3$? **A**

   A. $-8x^3$ B. $-2x^3$
   C. $8x^3$ D. $2x^3$

9. $\triangle EFG$ is similar to $\triangle KML$. What is the measure of $\angle M$? **B**

   A. 40°
   B. 60°
   C. 70°
   D. 80°

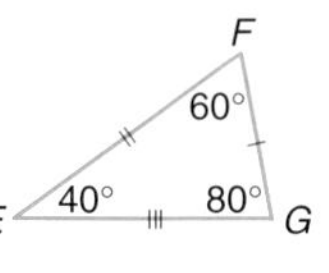

10. $\triangle ABC$ is a right triangle. What is the length of the hypotenuse? **C**

    A. 21 in.
    B. 26 in.
    C. 30 in.
    D. 52 in.

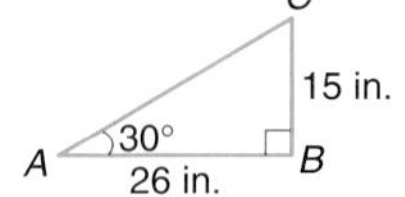

Section Two: Free Response

This section contains twelve questions for which you will provide short answers. Write your answer on your paper.

11. Find the perimeter of a regular hexagon with sides 12.6 meters long. **75.6 m**

12. 15¢, 30¢, 35¢, 55¢, 60¢, 75¢

12. A nickel, a dime, a quarter, and a half dollar are in a purse. Without looking, Alex picks up two coins. What are the different amounts he could choose?

13. Joan and Miguel are flying kites in the park. When Joan is 70 feet from Miguel, her kite forms an angle of 56° with the ground. If the kite is directly over Miguel, how high is the kite to the nearest tenth of a foot? **103.8 feet**

14. Evaluate $a^4 - 2a^3 + 4b^2 - c$ if $a = 2$, $b = -6$, and $c = -2$. **146**

15. Find the area of the circle shown below. **113 cm²**

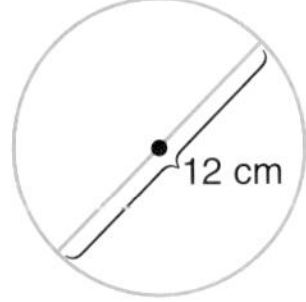

16. There are 25 students in a math class. The class takes a survey and finds that 15 people like pepperoni, 12 people like sausage, and 7 people like both equally. How many people like neither pepperoni nor sausage? **5 people**

17. In a survey of favorite music types, 15 out of 50 people preferred country. In a group of 1000 people, how many would you expect to like country? **300 people**

18. Find the sum of $(4x^3 + 2x^2) + (-2x^3 - 7x^2)$. $\mathbf{2x^3 - 5x^2}$

19. Approximate the value of $\sqrt{45}$. **6.7**

20. The largest country in the world is Russia, with an area of 6,590,876 square miles. The second largest country is Canada. The area of Russia is 110,416 square miles less than twice the area of Canada. What is the area of Canada? **3,350,646 square miles**

21. Write 2.25 as a mixed number. $2\frac{1}{4}$

22. Find the sum of the measures of the interior angles of a trapezoid. **360°**

Test-Taking Tip

On standardized tests, guessing can improve your score if you make educated guesses. First, find out if there is a penalty for incorrect answers. If there is no penalty, a guess can only increase your score, or at worst, leave your score the same. If there is a penalty, try to eliminate enough choices to make the probability of a correct guess greater than the penalty for an incorrect answer. That is, if there is a quarter point penalty for an incorrect response and you have a choice of three responses, the probability of making a correct guess is 1 out of 3, which is better than the penalty.

Section Three: Open-Ended

This section contains three open-ended problems. Demonstrate your knowledge by giving a clear, concise solution to each problem. Your score on these problems will depend on how well you do the following.

- Explain your reasoning.
- Show your understanding of the mathematics in an organized manner.
- Use charts, graphs, and diagrams in your explanation.
- Show the solution in more than one way or relate it to other situations.
- Investigate beyond the requirements of the problem.

23. Show that $4n(-2n^2 + n - 12) = -8n^3 + 4n^2 - 48n$. Use values for n to justify your answer. **See students' work.**

24. Explain the difference between integers and whole numbers. **See students' work.**

25. Write a problem that can be simulated by rolling a die. **See students' work.**

Extra Practice

EXTRA PRACTICE

Lesson 1-1

1. **Postal Service** The U.S. Postal Service offers airmail service to other countries. The rates for International Air Mail letters and packages are shown in the table at the right. Determine the air mail rate for a package that weighs 5.5 ounces.
   a. Write the *Explore* step. What do you know and what do you need to find? **know other rates, need to find rate for 5.5 oz**
   b. Write the *Plan* step. What strategy will you use? What do you estimate the answer to be? **sample answer: look for a pattern**
   c. *Solve* the problem using your plan. What is your answer? **$4.46**
   d. *Examine* your solution. Is it reasonable? Does it answer the question? **See students' work.**

Weight not over (ounces)	Rate
0.5	$0.50
1.0	$0.95
1.5	$1.34
2.0	$1.73
2.5	$2.12
3.0	$2.51
3.5	$2.90
4.0	$3.29

2. **Postal Service** In 1995, the state of Florida celebrated the 150th anniversary of its statehood. The U.S. Postal Service issued a stamp, the first to bear the 32-cent price, to honor the occasion. Ninety million of the commemorative stamps were issued. About how much postage did the stamps represent?
   a. Which method of computation do you think is most appropriate for this problem? Justify your choice. **sample answer: estimation since the problem says "about"**
   b. Solve the problem using the four-step plan. Be sure to examine your solution. **about $30 million**

Lesson 1-2

Find the value of each expression.

1. $8 + 7 + 12 \div 4$ **18**
2. $20 \div 4 - 5 + 12$ **12**
3. $(25 \cdot 3) + (10 \cdot 3)$ **105**
4. $36 \div 6 + 7 - 6$ **7**
5. $30 \cdot (6 - 4)$ **60**
6. $(40 \cdot 2) - (6 \cdot 11)$ **14**
7. $\frac{86 - 11}{11 + 4}$ **5**
8. $\frac{12 + 84}{11 + 13}$ **4**
9. $\frac{5 \cdot 5 + 5}{5 \cdot 5 - 15}$ **3**
10. $(19 - 8)4$ **44**
11. $75 - 5(2 \cdot 6)$ **15**
12. $81 \div 27 \times 6 - 2$ **16**

Lesson 1-3

Evaluate each expression if $a = 2$, $b = 4$, and $c = 3$.

1. $ba - ac$ **2**
2. $4b + a \cdot a$ **20**
3. $11 \cdot c - ab$ **25**
4. $4b - (a + c)$ **11**
5. $7(a + b) - c$ **39**
6. $8a + 8b$ **48**
7. $\frac{8(a + b)}{4c}$ **4**
8. $36 - 12c$ **0**
9. $\frac{9(b + a)}{c - 1}$ **27**
10. $abc - bc$ **12**
11. $28 - bc + a$ **18**
12. $a(b - c)$ **2**

Translate each phrase into an algebraic expression.

13. nine more than a **$a + 9$**
14. eleven less than k **$k - 11$**
15. three times p **$3p$**
16. the product of some number and five **$5n$**
17. twice Shelly's score decreased by 18 **$2s - 18$**
18. the quotient of 16 and n **$16 \div n$**

740 *Extra Practice*

Lesson 1-4

Name the property shown by each statement. 2. associative, +

1. $1 \cdot 4 = 4$ mult. identity
2. $6 + (b + 2) = (6 + b) + 2$
3. $9(6n) = (9 \cdot 6)n$ associative, ×
4. $8t \cdot 0 = 0 \cdot 8t$ commutative, ×
5. $0(13n) = 0$ mult. prop. of 0
6. $7 + t = t + 7$ commutative, +

Find each sum or product mentally.

7. $6 + 8 + 14$ 28
8. $5 \cdot 18 \cdot 2$ 180
9. $0(13 \cdot 6)$ 0
10. $8 + 4 + 12 + 16$ 40
11. $8 \cdot 20 \cdot 10$ 1600
12. $4 \cdot 14 \cdot 5$ 280

13. **History** Many flags have represented the United States since the first colonists arrived. The current flag, called the Stars and Stripes, was originated by the Marine Committee of the Second Continental Congress in a resolution made on June 14, 1777. The flag now has fifty stars in the field to represent the fifty states of the Union. Each row of the flag has either 5 or 6 stars in each row. Rows of 5 stars alternate with rows of 6 stars. How many rows are there with 6 stars? 5 rows with 6 stars

Lesson 1-5

Simplify each expression. 7. $12c + 10$ 8. $13s + 47$

1. $8k + 2k + 7$ $10k + 7$
2. $3 + 2b + b$ $3 + 3b$
3. $t + 2t$ $3t$
4. $9(3 + 2x)$ $27 + 18x$
5. $4(xy + 2) - 2$ $4xy + 6$
6. $(6 + 3e)4$ $24 + 12e$
7. $4 + 9c + 3(c + 2)$
8. $5(7 + 2s) + 3(s + 4)$
9. $9(f + 2) + 14f$ $23f + 18$

10. **Business** The Columbus Dispatch newspaper can be ordered for delivery on weekdays or Sundays. A weekday paper is 35 cents and the Sunday Edition is $1.50. The Stadlers ordered delivery of the weekday papers. The month of March had 23 weekdays and April had 20. How much should the carrier charge the Stadlers for those two months? $0.35(23 + 20) = $15.05

Lesson 1-6

Identify the solution to each equation from the given list.

1. $16 - f = 7$; 5, 7, 9 9
2. $9 = \frac{72}{m}$; 8, 9, 11 8
3. $4b + 1 = 17$; 3, 4, 5 4
4. $17 + r = 25$; 6, 7, 8 8
5. $9 - 7n = 12$; 3, 5, 7 3
6. $67 = 98 - q$; 21, 26, 31 31

Solve each equation mentally.

7. $131 - u = 120$ 11
8. $88 = 11d$ 8
9. $\frac{84}{h} = 12$ 7
10. $5t = 0$ 0
11. $\frac{x}{2} = 8$ 16
12. $88 + y = 96$ 8
13. $13g = 39$ 3
14. $23 = w + 6$ 17
15. $9z = 45$ 5

EXTRA PRACTICE

Extra Practice **741**

EXTRA PRACTICE

Lesson 1-7

Use the grid at the right to name the point for each ordered pair.

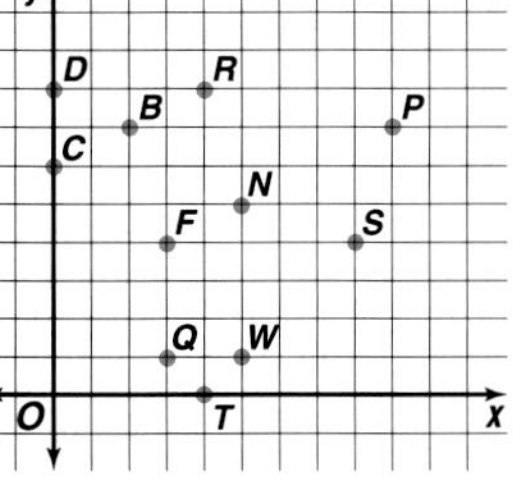

1. (9, 7) P
2. (5, 5) N
3. (3, 1) Q
4. (2, 7) B
5. (8, 4) S
6. (4, 0) T

Use the grid at the right to name the ordered pair for each point.

7. R (4, 8)
8. P (9, 7)
9. W (5, 1)
10. C (0, 6)
11. D (0, 8)
12. F (3, 4)

Lesson 1-8

Solve each equation using the inverse operation. Use a calculator when needed.

1. $37 = b + 22$ 15
2. $6 + u = 14$ 8
3. $p - 19 = 3$ 22
4. $6x = 30$ 5
5. $3v = 99$ 33
6. $48c = 192$ 4
7. $r \div 7 = 4$ 28
8. $12 = \frac{z}{3}$ 36
9. $6.2 = 5.8 + g$ 0.4

Translate each sentence into an equation.

10. The sum of a number and 8 is 14. $x + 8 = 14$
11. Twelve less than a number is 50. $n - 12 = 50$
12. The product of a number and ten is seventy. $10n = 70$
13. A number divided by three is nine. $n \div 3 = 9$

Lesson 1-9

State whether each inequality is *true*, *false*, or *open*.

1. $8 > 0$ true
2. $18 \leq 18$ true
3. $10 < x$ open
4. $24 > 2w$ open
5. $9 < 7$ false
6. $13 \leq 55$ true

State whether each inequality is *true* or *false* for the given value. 10. true

7. $5 \geq 2t - 12; t = 11$ false
8. $7 + n < 25; n = 4$ true
9. $6r - 18 > 0; r = 3$ false
10. $3n + 2 < 26, n = 3$
11. $h - 19 < 13; h = 28$ true
12. $20m \geq 10; m = 0$ false

13. **Literature** William Shakespeare, who lived from 1564 to 1616, was England's best known poet, actor, and playwright. Each year several productions of Shakespeare's plays are performed around the world. Some of his plays have been made into films, including a 1993 version of *Much Ado About Nothing* and a 1990 version of *Hamlet*. There are currently more than 50 versions of *Hamlet* on film. Write an inequality for the number of versions of Hamlet. $n > 50$

742 *Extra Practice*

Lesson 1-10

The table at the right shows the albums that have stayed on the U.S. music charts the longest as of December 31, 1993.

Artist/title	Weeks on chart
Carole King, *Tapestry*	302
Johnny Mathis, *Heavenly*	295
Johnny Mathis, *Johnny's Greatest Hits*	490
Pink Floyd, *Dark Side of the Moon*	741
Tennessee Ernie Ford, *Hymns*	277
Various, Original cast recording of *Camelot*	265
Various, Original cast recording of *My Fair Lady*	480
Various, Original cast recording of *The Sound of Music*	276
Various, Soundtrack of *The King and I*	277
Various, Soundtrack of *Oklahoma!*	305

1. Which album has been on the charts the greatest number of weeks?
2. How many weeks was *Tapestry* by Carole King on the charts? 302
3. How much longer was Johnny Mathis' *Greatest Hits* on the charts than his *Heavenly* album?
4. Make a bar graph of the data on albums. See margin.

1. Pink Floyd, *Dark Side of the Moon* 3. 195 weeks

Lesson 2-1

Simplify.

1. $|-3| + |9|$ 12
2. $|-18| - |5|$ 13
3. $|12 + 7|$ 19
4. $-|6|$ −6
5. $|-8| + |4|$ 12
6. $-|-20|$ −20
7. $|15 - 12|$ 3
8. $|8 + 9|$ 17
9. $-|4| \cdot |-5|$ −20
10. $|-6| \cdot |8|$ 48
11. $-|12| \cdot |9|$ −108
12. $-||-16| + |-22||$ −38

Lesson 2-2

Use the coordinate grid at the right to name the point for each ordered pair.

1. $(-6, 8)$ *D*
2. $(1, -2)$ *J*
3. $(9, 2)$ *C*
4. $(1, 4)$ *L*
5. $(-3, -4)$ *B*
6. $(2, 5)$ *N*

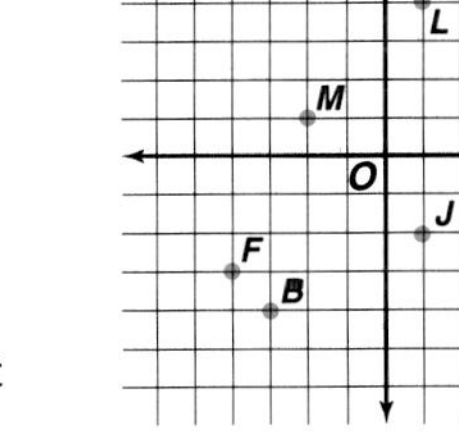

On graph paper, draw coordinate axes. Then graph and label each point. Name the quadrant in which each point is located. 7–12. See margin for graph.

7. $H(-2, -5)$ III
8. $P(1, 5)$ I
9. $R(-3, 1)$ II
10. $M(4, -2)$ IV
11. $K(-4, 5)$ II
12. $G(3, -5)$ IV

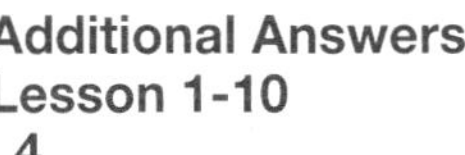

Additional Answers
Lesson 1-10
4.

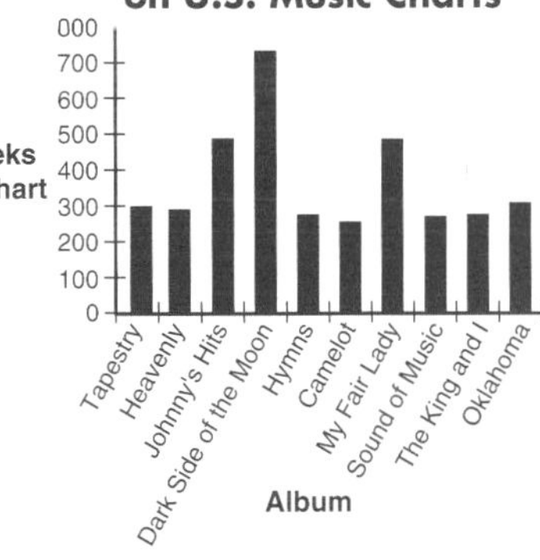

Lesson 2-2
7-12.

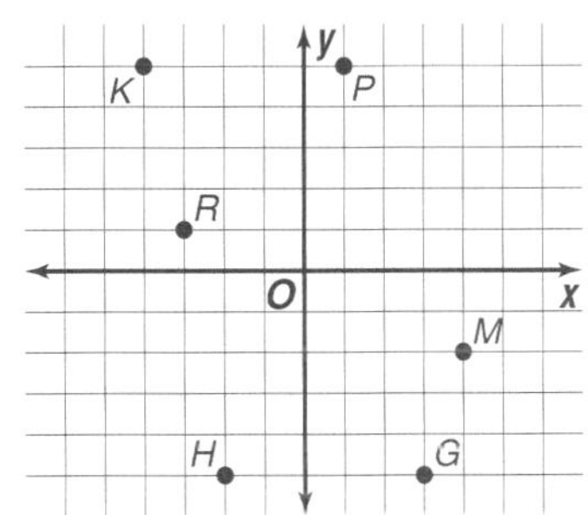

Additional Answers
Lesson 2-3

1. {-5, -1, 2}
2. {-9, -2, 0, 5, 8}
3. {-86, -34, 0, 21, 100}
4. {-43, -40, -1, 8, 16, 27}
5. {-23, -15, 0, 24, 75}
6. {-6, -5, 6, 18}

EXTRA PRACTICE

Lesson 2-3

Order the integers in each set from least to greatest. 1–6. See margin.

1. {−1, 2, −5}
2. {0, −2, 8, 5, −9}
3. {100, −34, −86, 21, 0}
4. {−1, 16, −43, 8, 27, −40}
5. { 0, −23, 75, −15, 24}
6. {−6, 6, −5, 18}

Write an inequality using the numbers in each sentence. Use the symbols $<$ or $>$.

7. World Finland covers 130,119 square miles and is 3 times greater than Ohio which covers 40,953 square miles. **130,119 > 40,953**

8. 1050 > 940 9. 24 < 26

8. Sports Atlanta, Georgia hosted the 1996 Summer Olympics. The point with the highest altitude in the city is 1050 feet above sea level. The lowest point is 940 feet above sea level.

9. Statistics The oldest dog lived to be 24 years old. The oldest cat lived to be 26 years old.

10. Statistics The slowest animal in the world is the garden snail that moves 0.03 mph. The fastest animal is the cheetah that can run 70 mph. **0.03 < 70**

11. Astronomy Mars is 227.9 million kilometers from the Sun. Earth is 149.6 million kilometers from the Sun. **227.9 > 149.6**

Lesson 2-4

Solve each equation. 10. −15 11. 23

1. $n = 5 + (-6)$ **−1**
2. $-17 + 24 = y$ **7**
3. $k = 15 + (-29)$ **−14**
4. $m = -6 + 13$ **7**
5. $50 + (-14) = x$ **36**
6. $w = -21 + (-4)$ **−25**
7. $30 + (-7) = t$ **23**
8. $z = (-3) + (-10)$ **−13**
9. $-15 + 26 = q$ **11**
10. $b = -17 + 4 + (-2)$
11. $d = 50 + (-16) + (-11)$
12. $-17 + 8 + (-14) = f$ **−23**

Lesson 2-5

Solve each equation.

1. $8 - 17 = f$ **−9**
2. $-15 - 3 = r$ **−18**
3. $10 - 21 = a$ **−11**
4. $20 - (-5) = m$ **25**
5. $5 - (-9) = g$ **14**
6. $-12 - (-7) = v$ **−5**
7. $-19 - (-6) = k$ **−13**
8. $-16 - (-23) = b$ **7**
9. $-56 - 32 = z$ **−88**
10. $-49 - (-52) = h$ **3**
11. $-6 - 9 - (-7) = d$ **−8**
12. $6 - (-10) - 7 = n$ **9**

Lesson 2-6

Solve. Look for a pattern. 2. 10 – 15 – 7

1. Find the next two integers in the pattern 2, 3, 5, 9, 17, _?_ , _?_ . **33, 65**
2. If FOG = 6 − 15 − 7 and LOG = 12 − 15 − 7; what does JOG equal by the same logic?
3. The digits 0 to 9 have been used to make the multiplication problem below. Find the missing numbers that make the multiplication correct.

```
   ?02       402
×  ?9      × 39
1???8    15,678
```


744 Extra Practice

Lesson 2-7

Solve each equation.

1. $n = -4(2)$ **−8**
2. $-8(-5) = h$ **40**
3. $r = 13(-4)$ **−52**
4. $-5 \cdot 6 \cdot 10 = c$ **−300**
5. $-6(-2)(-14) = w$ **−168**
6. $u = 18(-3)(6)$ **−324**

Evaluate each expression.

7. $-6t$, if $t = 15$ **−90**
8. $7p$, if $p = -9$ **−63**
9. $-4k$, if $k = -16$ **64**
10. aw, if $a = 0$ and $w = -72$ **0**
11. dk, if $d = -12$ and $k = 11$ **−132**
12. st, if $s = -8$ and $t = -10$ **80**
13. $3hp$, if $h = 9$ and $p = -3$ **−81**
14. $-5bc$, if $b = -6$ and $c = 2$ **60**
15. $-4wx$, if $w = -1$ and $x = -8$ **−32**

Lesson 2-8

Divide.

1. $-36 \div 9$ **−4**
2. $112 \div (-8)$ **−14**
3. $-72 \div 2$ **−36**
4. $-26 \div (-13)$ **2**
5. $-144 \div 6$ **−24**
6. $-180 \div (-10)$ **18**
7. $304 \div (-8)$ **−38**
8. $-216 \div (-9)$ **24**
9. $80 \div (-5)$ **−16**
10. $-105 \div 15$ **−7**
11. $120 \div (-30)$ **−4**
12. $-200 \div (-8)$ **25**

Lesson 3-1

Solve by eliminating possibilities.

1. Lindsay, Lee, Ann, and Marcos formed a study group. Each one has a favorite subject that is different from the others. The subjects are art, math, music, and physics. Use the following information to match each person with his or her favorite subject.
   A. Lindsay likes subjects where she can use her calculator. **Lindsay: PHYSICS**
   B. Lee does not like music or physics. **Lee: MATH**
   C. Ann and Marcos prefer classes in the cultural arts. **Ann: MUSIC**
   D. Marcos plans to be a professional cartoonist. **Marcos: ART**
2. Five friends decided to run in a 5-kilometer run to benefit an education association. Nancy beat Pete. Jeff was not last. Gina was beaten by Tom and Jeff. Jeff crossed the finish line just after Tom did. Tom lost to Pete. Choose the correct order in which the friends finished the race. **c**
   a. Nancy, Pete, Gina, Tom, Jeff
   b. Tom, Jeff, Nancy, Pete, Gina
   c. Nancy, Pete, Tom, Jeff, Gina
   d. Nancy, Gina, Jeff, Tom, Pete

Lesson 3-2 **1–15. See margin for graphs.**

Solve each equation and check your solution. Then graph the solution on a number line.

1. $y + 49 = 26$ **−23**
2. $d - (-31) = -24$ **−55**
3. $q - 8 = 16$ **24**
4. $x - 16 = 32$ **48**
5. $40 = a + 12$ **28**
6. $b + 12 = -1$ **−13**
7. $21 = u - (-6)$ **15**
8. $-52 = p + 5$ **−57**
9. $-14 = 5 - g$ **19**
10. $121 = k + (-12)$ **133**
11. $-234 = m - 94$ **−140**
12. $110 = x + 25$ **85**
13. $f - 7 = 84$ **91**
14. $y - 864 = -652$ **212**
15. $475 + z = -18$ **−493**

EXTRA PRACTICE

Additional Answers
Lesson 3-2

1.

2.
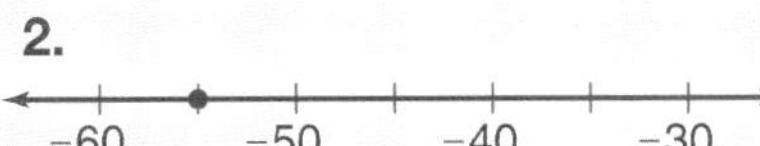

3.
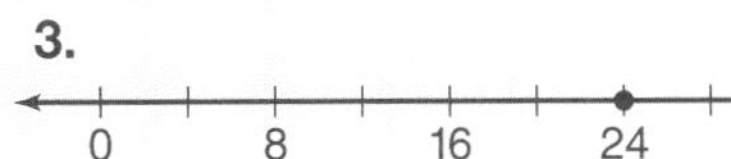

4.
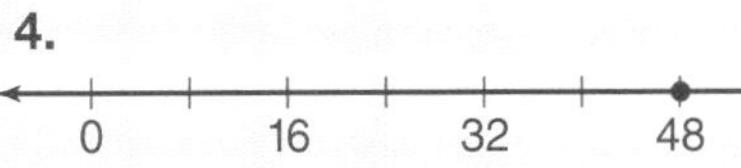

5.
0 14 28 42

6.
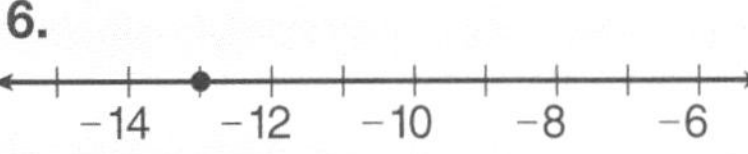

7.
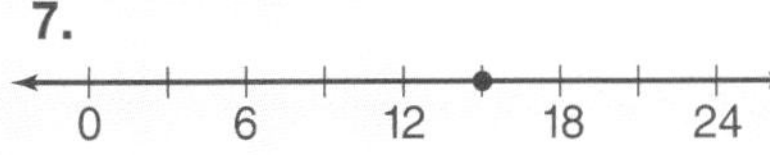

8.
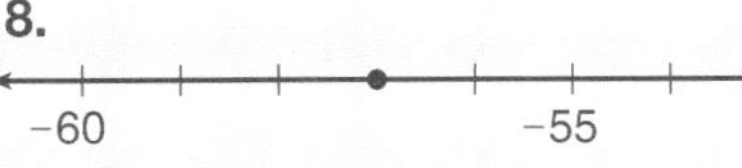

9.
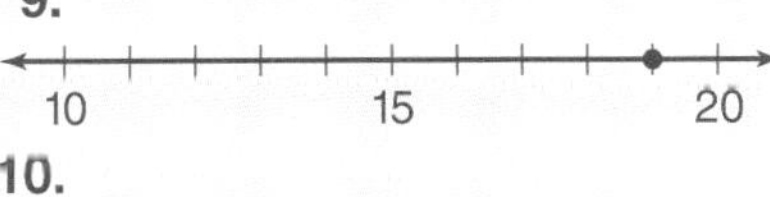

10.

11.
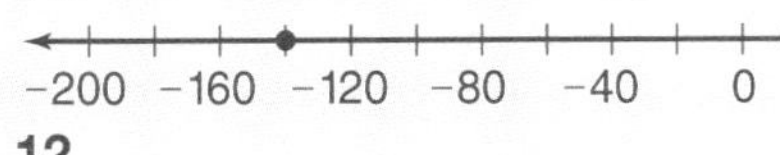

12.
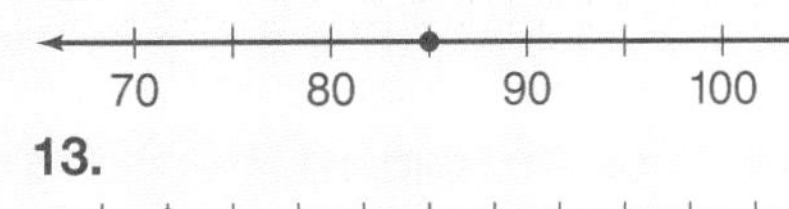

13.
90 95 100

14.
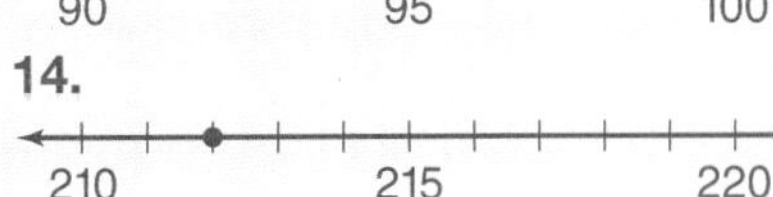

15.
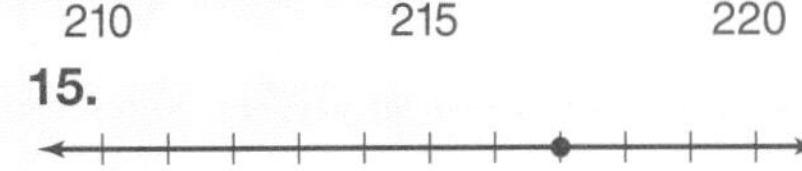

Extra Practice **745**

Additional Answers

Lesson 3-3

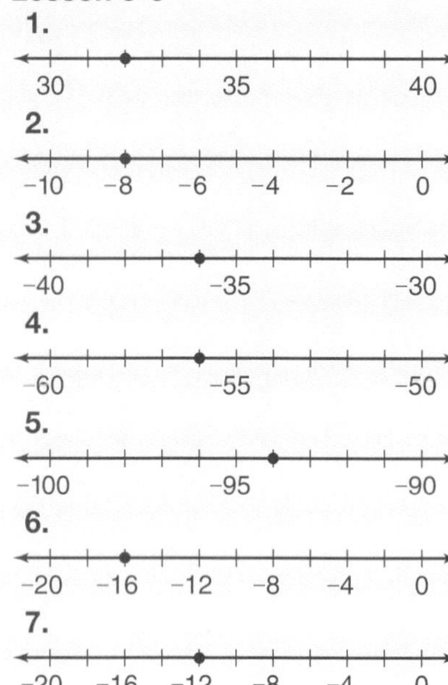

8\.

−1460 −1455 −1450

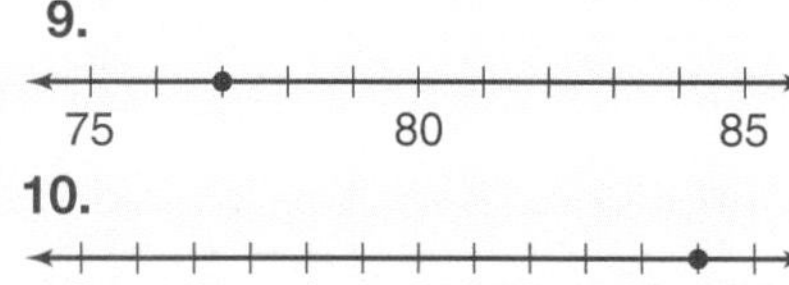

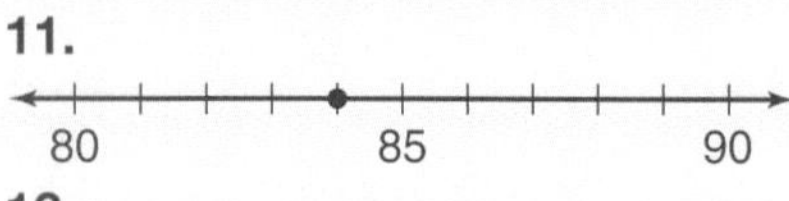

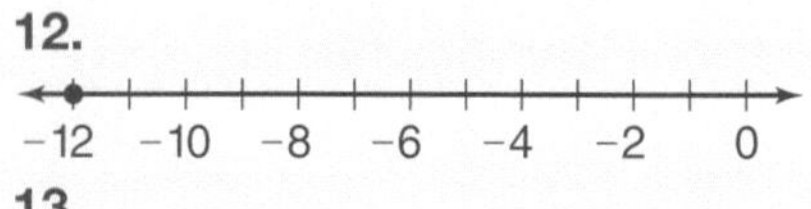

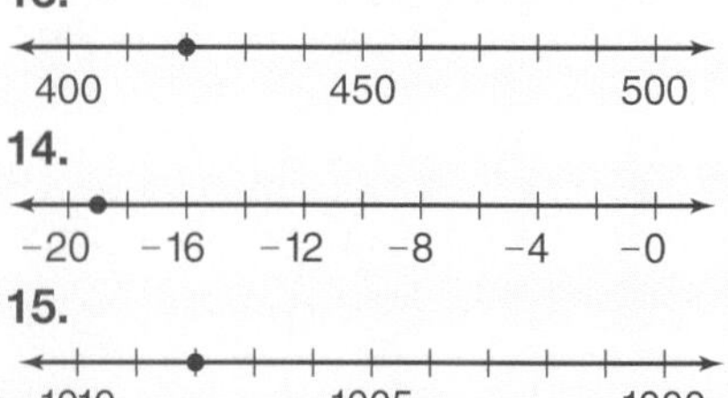

15\.

−1010 −1005 −1000

Lesson 3-3 **1–15. See margin for graphs.**

Solve each equation and check your solution. Then graph the solution on a number line.

1. $-y = -32$ **32**	**2.** $7r = -56$ **−8**	**3.** $\frac{t}{-3} = 12$ **−36**
4. $4 = \frac{s}{-14}$ **−56**	**5.** $\frac{b}{47} = -2$ **−94**	**6.** $64 = -4n$ **−16**
7. $-144 = 12q$ **−12**	**8.** $\frac{r}{11} = -132$ **−1452**	**9.** $-5g = -385$ **77**
10. $-16x = -176$ **11**	**11.** $-21 = \frac{y}{-4}$ **84**	**12.** $-372 = 31k$ **−12**
13. $84 = \frac{k}{5}$ **420**	**14.** $-b = 19$ **−19**	**15.** $\frac{y}{112} = -9$ **−1008**

Lesson 3-4

Solve by replacing the variables with the given values.

1. $A = \pi \cdot r^2$, if $\pi = 3.14$ and $r = 5$ **78.5**
2. $\frac{5}{9}(F - 32) = C$, if $F = 86$ **30**
3. $d = r \cdot t$, if $d = 366$ and $t = 3$ **122**
4. $S = (n - 2) \cdot 180$, if $n = 8$ **1080**
5. $A = \frac{1}{2}bh$, if $A = 36$ and $h = 12$ **6**
6. $P = 4s$, if $P = 108$ **27**
7. $V = \frac{1}{3}(B \cdot H)$, if $B = 27$ and $H = 5$ **45**
8. $h = 69 + 2.2F$, if $F = 42$ **161.4**

Lesson 3-5

Find the perimeter and area of each rectangle.

1. a rectangle 23 centimeters long and 9 centimeters wide **64 cm, 207 cm^2**
2. a 16-foot by 14-foot rectangle **60 ft, 224 ft^2**
3. a rectangle with a length of 31 meters and a width of 3 meters **68 m, 93 m^2**
4. a square with sides 7.5 meters long **30 m, 56.25 m^2**

Find the missing dimension of each rectangle.

	Length	Width	Area	Perimeter
5.	9 ft	**14 ft**	126 ft^2	46 ft
6.	**6 in.**	18 in.	108 in^2	48 in.
7.	13 yd	**21 yd**	273 yd^2	68 yd
8.	**14 cm**	12 cm	168 cm^2	52 cm
9.	**54 m**	3 m	162 m^2	114 m
10.	18 mi	**40 mi**	720 mi^2	116 mi

Lesson 3-6

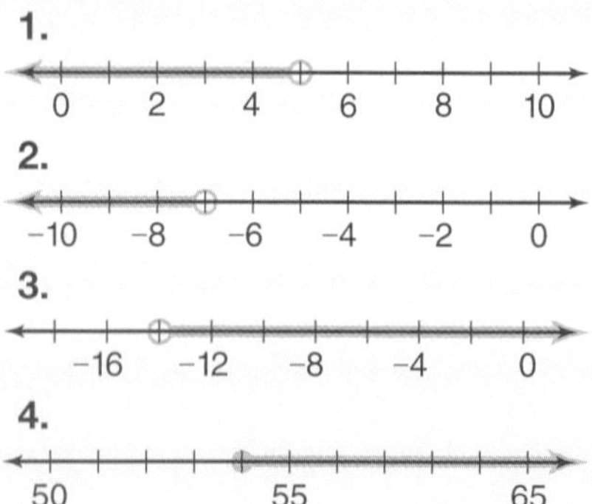

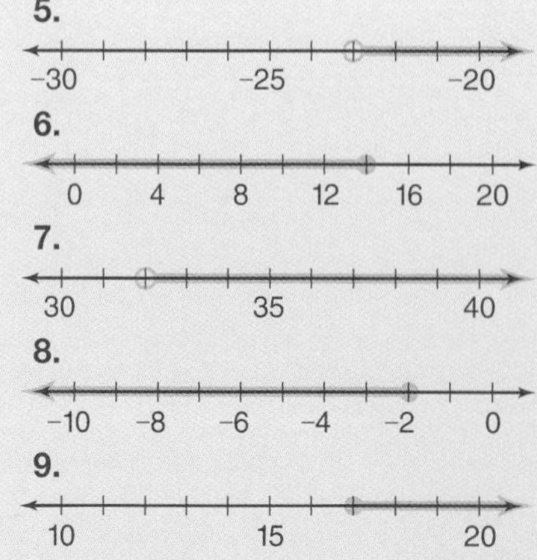

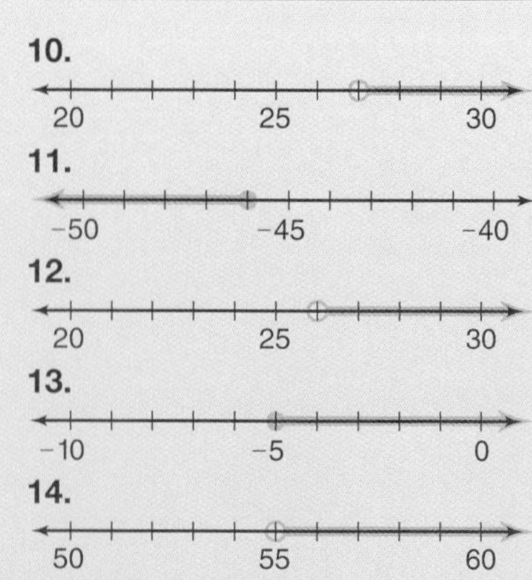

Lesson 3-6 1–18. See margin for graphs. 11. $-46 \geq d$ 13. $n \geq -5$ 16. $p < -26$

Solve each inequality and check your solution. Then graph the solution on a number line.

1. $m + 9 < 14$ $m < 5$
2. $k + (-5) < -12$ $k < -7$
3. $-15 < v - 1$ $-14 < v$
4. $-7 + f \geq 47$ $f \geq 54$
5. $r > -15 - 8$ $r > -23$
6. $18 \geq s - (-4)$ $14 \geq s$
7. $38 < r - (-6)$ $32 < r$
8. $z - 9 \leq -11$ $z \leq -2$
9. $-16 + c \geq 1$ $c \geq 17$
10. $24 < a + -3$ $27 < a$
11. $-52 \geq d + (-6)$
12. $24 < -2 + b$ $26 < b$
13. $n - (-17) \geq 12$
14. $31 < x - 24$ $55 < x$
15. $-20 \leq n + (-3)$ $-17 \leq n$
16. $p + (-11) < -37$
17. $-40 \leq -72 + w$ $32 \leq w$
18. $72 > a + 88$ $a < -16$

Lesson 3-7 1–18. See margin for graphs.

Solve each inequality and check your solution. Then graph the solution on a number line.

1. $6p < 78$ $p < 13$
2. $\frac{m}{-3} > 24$ $m < -72$
3. $-18 < 3b$ $b > -6$
4. $-5k \geq 125$ $k \leq -25$
5. $-75 > \frac{a}{5}$ $-375 > a$
6. $\frac{w}{6} < -5$ $w < -30$
7. $\frac{y}{-13} > -20$ $y < 260$
8. $14t < 266$ $t < 19$
9. $\frac{g}{-25} \geq 8$ $g \leq -200$
10. $-216 \leq 9h$ $-24 \leq h$
11. $\frac{g}{-9} < -8$ $g > 72$
12. $-18d > 108$ $d < -6$
13. $2268 < -63a$ $-36 > a$
14. $52 \leq \frac{p}{4}$ $208 \leq p$
15. $42n \leq -210$ $n \leq -5$
16. $-60 > -4d$ $15 < d$
17. $\frac{k}{-3} \geq -21$ $k \leq 63$
18. $\frac{v}{-8} < 0$ $v > 0$

Lesson 3-8

Define a variable and translate each sentence into an equation or inequality. Then solve.

1. Three times a number is equal to thirty-six. $3x = 36$; 12
2. The sum of a number and 5 is less than 12. $y + 5 < 12$; $y < 7$
3. **Sports** In the 1993–1994 NBA season, David Robinson of the San Antonio Spurs was the leading scorer. Shaquille O'Neal of the Orlando Magic had 6 points less than David Robinson. If O'Neal had 2377 points for the season, how many points did David Robinson get? $r - 6 = 2377$; 2383
4. **College** The enrollment at the Columbus campus of The Ohio State University is more than six times as large as the enrollment at Stanford University. If there are 38,958 students enrolled at Ohio State, how many students are currently enrolled at Stanford? $6s < 38,958$; $s < 6493$

Additional Answers

Lesson 3-7

16.

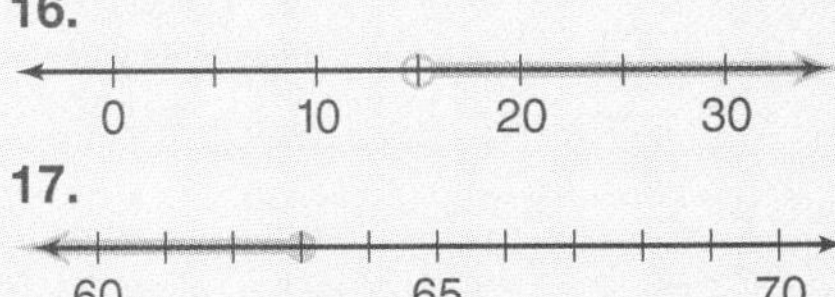

17.

18.

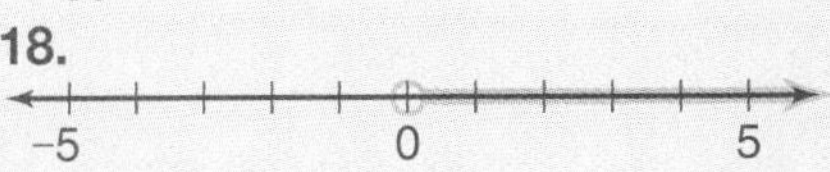

Additional Answers

Lesson 3-6

15. −20 −15 −10

16. −30 −25 −20

17. 30 35 40

18. −20 −16 −12 −8 −4 0

Lesson 3-7

1.

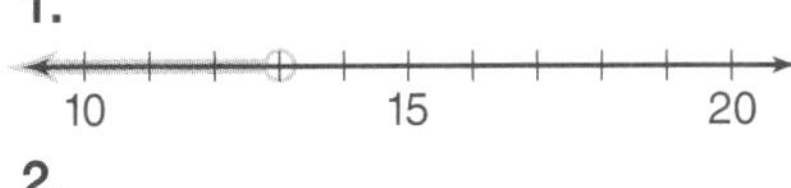

2. −80 −75 −70

3.

4.

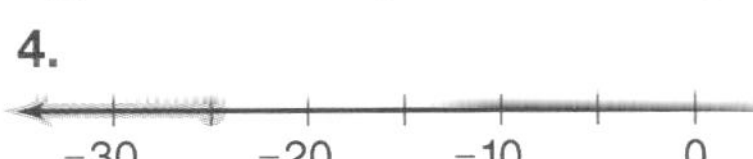

5.

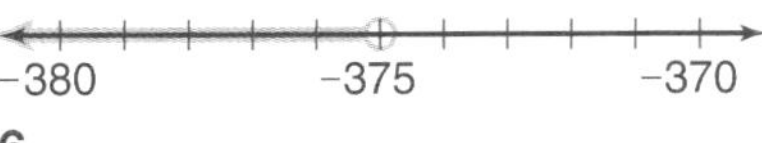

6.

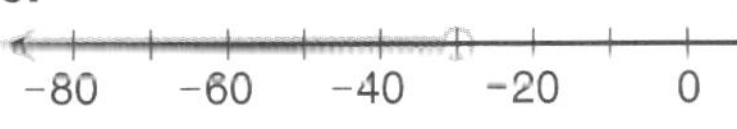

7. 200 250 300

8.

9.

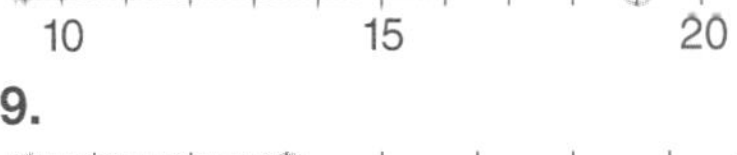

10.

11. 70 75 80

12.

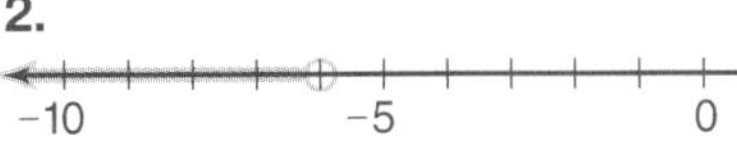

13.

14.

15.

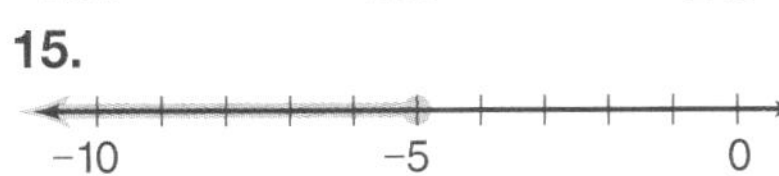

EXTRA PRACTICE

Lesson 4-1

Using divisibility rules, state whether each number is divisible by 2, 3, 5, 6, or 10.

1. 98 **2**
2. 243 **3**
3. 800 **2, 5, 10**
4. 252 **2, 3, 6**
5. 105 **3, 5**
6. 210 **2, 3, 5, 6, 10**
7. 1001 **none**
8. 2016 **2, 3, 6**
9. 2475 **3, 5**

Determine whether each expression is a monomial. Explain why or why not.

10. $-6h$ **yes**
11. $9 - v$ **no**
12. $4g^3$ **yes**

10–12. See Solutions Manual for explanations.

Lesson 4-2

Write each multiplication expression using exponents.

1. $(-6)(-6)(-6)(-6)(-6)$ $\mathbf{(-6)^5}$
2. $(y \cdot y \cdot y) \cdot (y \cdot y \cdot y \cdot y)$ $\mathbf{y^7}$
3. 9 $\mathbf{9^1}$
4. $3q \cdot 3q \cdot 3q \cdot 3q \cdot 3q \cdot 3q$ $\mathbf{(3q)^6}$
5. $\underbrace{n \cdot n \cdot n \cdot \ldots \cdot n}_{\text{17 factors}}$ $\mathbf{n^{17}}$
6. $8 \cdot 8 \cdot 8 \cdot 8$ $\mathbf{8^4}$

Write each power as a multiplication expression of the same factor.

7. 13^3 $\mathbf{13 \cdot 13 \cdot 13}$
8. $(-4)^5$ $\mathbf{(-4)(-4)(-4)(-4)(-4)}$
9. k^9 $\mathbf{k \cdot k \cdot k \cdot k \cdot k \cdot k \cdot k \cdot k \cdot k}$
10. $(2 - w)^2$ $\mathbf{(2 - w)(2 - w)}$
11. $(3m)^{15}$ $\underbrace{\mathbf{3m \cdot 3m \cdot \ldots \cdot 3m}}_{\text{15 factors}}$
12. $(-x)^4$ $\mathbf{(-x)(-x)(-x)(-x)}$

Lesson 4-3

Solve. Use a diagram.

1. Four people are eligible to be officers in the National Honor Society. The positions available are President, Vice President, Treasurer, and Secretary. How many different ways can the offices be filled? **24 ways**

2. Copy the magic square at the right. Use the numbers 0, 4, 8, and 12 to make all the rows, columns, and diagonals add up to 24.

0	12	12	0
8	4	4	8
4	8	8	4
12	0	0	12

3. Beverly and Pam are painting a fence. For every 3 sections Pam paints, Beverly paints 5. If Beverly paints 20 sections in 2 hours how long will it take Pam to paint 18 sections? **3 hours**

4. Determine how many games have to be played to determine a champion from 32 teams in a single elimination softball tournament. **31 games**

748 ***Extra Practice***

Lesson 4-4

Determine whether each number is prime or composite.

1. 59 prime
2. 369 composite
3. 116 composite

Factor each number or monomial completely.

4. 40 $2 \cdot 2 \cdot 2 \cdot 5$
5. $630a$ $2 \cdot 3 \cdot 3 \cdot 5 \cdot 7 \cdot a$
6. 187 $11 \cdot 17$
7. 310 $2 \cdot 5 \cdot 31$
8. 510 $2 \cdot 3 \cdot 5 \cdot 17$
9. 1589 $7 \cdot 227$
10. $-18ab^2$ $-1 \cdot 2 \cdot 3 \cdot 3 \cdot a \cdot b \cdot b$
11. $-117x^3$ $-1 \cdot 3 \cdot 3 \cdot 13 \cdot x \cdot x \cdot x$
12. $435j^2k^5$ $3 \cdot 5 \cdot 29 \cdot j \cdot j \cdot k \cdot k \cdot k \cdot k \cdot k$

EXTRA PRACTICE

Lesson 4-5

Find the GCF of each set of numbers or monomials.

1. 112, 216 8
2. 120, 245 5
3. $84k$, $108k^2$ $12k$
4. $135ab$, $-171b$ $9b$
5. $185fg$, $74f^2g$ $37fg$
6. $44m$, $60n$ 4
7. $90gh$, $225k$ 45
8. 8, $-28h$ 4
9. $-16w$, $-28w^3$ $4w$
10. $24a$, $30ab$, $66a^2$ $6a$
11. $-13z$, $39yz$, $52y$ 13
12. $60a^3b$, $150a^2b^2$, $36a^2b$ $6a^2b$

Lesson 4-6

Write each fraction in simplest form. If the fraction is already in simplest form, write *simplified*.

1. $\frac{3}{54}$ $\frac{1}{18}$
2. $\frac{3}{16}$ simplified
3. $\frac{6}{58}$ $\frac{3}{29}$
4. $\frac{15}{55}$ $\frac{3}{11}$
5. $\frac{10}{90}$ $\frac{1}{9}$
6. $\frac{20}{49}$ simplified
7. $\frac{8}{20}$ $\frac{2}{5}$
8. $\frac{99}{9}$ 11
9. $\frac{18}{54}$ $\frac{1}{3}$
10. $\frac{21}{64}$ simplified
11. $\frac{40}{76}$ $\frac{10}{19}$
12. $\frac{49}{56}$ $\frac{7}{8}$

Lesson 4-7

Find the least common multiple (LCM) of each set of numbers or algebraic expressions.

1. 30, 18 90
2. $3m$, 12 $12m$
3. $6a$, $17a^5$ $102a^5$
4. 2, 5, 7 70

Find the least common denominator (LCD) for each pair of fractions.

5. $\frac{2}{5}, \frac{6}{25}$ 25
6. $\frac{3}{12}, \frac{4}{5}$ 60
7. $\frac{4}{6}, \frac{7}{9}$ 18
8. $\frac{1}{4}, \frac{5}{6}, \frac{2}{9}$ 36

Replace each ● with < or > to make a true statement.

9. $\frac{5}{11}$ ● $\frac{6}{13}$ $<$
10. $\frac{15}{34}$ ● $\frac{4}{8}$ $<$
11. $\frac{8}{18}$ ● $\frac{5}{16}$ $>$
12. $\frac{3}{14}$ ● $\frac{5}{20}$ $<$

Extra Practice **749**

Lesson 4-8

Find each product or quotient. Express your answer in exponential form. 8. $(-2)^8$

1. $r^4 \cdot r^2$ r^6 2. $\frac{2^9}{2^3}$ 2^6 3. $\frac{b^{18}}{b^5}$ b^{13} 4. $12^3 \cdot 12^8$ 12^{11}

5. $x \cdot x^9$ x^{10} 6. $(2s^6)(4s^2)$ $8s^8$ 7. $w^3 \cdot w^4 \cdot w^2$ w^9 8. $(-2)^2(-2)^5(-2)$

9. $\frac{4^7}{4^6}$ 4 10. $3(f^{17})(f^2)$ $3f^{19}$ 11. $(5k)^2 \cdot k^7$ $25k^9$ 12. $\frac{6m^8}{3m^2}$ $2m^6$

Find each missing exponent.

13. $(2^{\bullet})(2^2) = 2^6$ 4 14. $y(y^3)(y^4) = y^{\bullet}$ 8 15. $\frac{5^7}{5^3} = 5^{\bullet}$ 4

16. $12^9 \cdot 12^{\bullet} = 12^{14}$ 5 17. $\frac{b^{10}}{b^{\bullet}} = b$ 9 18. $\frac{2^7}{8^2} = 2^{\bullet}$ 1

Lesson 4-9

Write each expression using positive exponents.

1. y^{-9} $\frac{1}{y^9}$ 2. $3m^{-4}$ $\frac{3}{m^4}$ 3. $5^{-3} \cdot (-a)$ $\frac{-a}{5^3}$ 4. $\left(\frac{2}{x}\right)^{-7}$ $\left(\frac{x}{2}\right)^7$

Write each fraction as an expression using negative exponents.

5. $\frac{1}{p^4}$ p^{-4} 6. $\frac{6a}{b^9}$ $6ab^{-9}$ 7. $\frac{2}{27}$ $2 \cdot 3^{-3}$ 8. $\frac{17}{a}$ $17a^{-1}$

Evaluate each expression.

9. 5^n if $n = -2$ $\frac{1}{25}$ 10. $(2a^{-2}b)^2$ if $a = 3$ and $b = 6$ $1\frac{7}{9}$

11. $15n^{-3}$ if $n = 5$ $\frac{3}{25}$ 12. $9x^{-6}$ if $x = -3$ $\frac{1}{81}$

Find each product or quotient. Express using positive exponents.

13. $(x^8)(x^{-2})$ x^6 14. $(g^{-12})(g^9)$ $\frac{1}{g^3}$ 15. $\frac{r^{14}}{r^{-2}}$ r^{16} 16. $\frac{y^7}{y^{10}}$ $\frac{1}{y^3}$

Lesson 5-1

For exercises 1–3, W = whole numbers, I = integers, and R = rationals.

Name the set(s) of numbers to which each number belongs.

1. $\frac{7}{10}$ R 2. 41 W, I, R 3. −19 I, R

Replace each ● with <, >, or = to make each sentence true. Use a number line if necessary.

4. $\frac{3}{4} \bullet \frac{2}{5}$ > 5. $\frac{-13}{25} \bullet \frac{-3}{5}$ > 6. $\frac{9}{10} \bullet \frac{7}{8}$ >

7. $5\frac{2}{9} \bullet \frac{47}{9}$ = 8. $\frac{5}{26} \bullet \frac{4}{13}$ < 9. $\frac{-11}{4} \bullet -2\frac{1}{2}$ <

Express each decimal as a fraction or mixed number in simplest form.

10. 0.38 $\frac{19}{50}$ 11. 2.346 $2\frac{173}{500}$ 12. $-0.\overline{4}$ $-\frac{4}{9}$

750 *Extra Practice*

Lesson 5-2

Round to the nearest whole number.

1. 6.5 7 **2.** 5.193 5 **3.** 42.09 42

Round each fraction to 0, $\frac{1}{2}$, or 1.

4. $\frac{9}{10}$ 1 **5.** $\frac{1}{14}$ 0 **6.** $\frac{89}{101}$ 1

Estimate each sum or difference. Sample answers are given.

7. $9.9 - 3.2$ 7 **8.** $16.2 + 9.31$ 25 **9.** $1.3 + 2.8 - 3.4$ 1

10. $8\frac{3}{5} - 2\frac{1}{6}$ $6\frac{1}{2}$ **11.** $19\frac{4}{10} + 13\frac{8}{11}$ $33\frac{1}{2}$ **12.** $863\frac{1}{12} - 241\frac{189}{221}$ 621

Lesson 5-3

Solve each equation.

1. $b = 5.8 + 9.3$ 15.1 **2.** $s = 12.4 - 4.52$ 7.88

3. $-4.9 + 8.4 = k$ 3.5 **4.** $-5.2 - 7.8 = q$ −13

5. $p = 14.8 - 29.46$ −14.66 **6.** $-4.25 + 11.2 = t$ 6.95

7. $21.4 - 9.2 = z$ 12.2 **8.** $45.26 - (-6.1) = y$ 51.36

9. $x = -9.27 - 8.1$ −17.37 **10.** $-28.94 + 3.48 = u$ −25.46

Lesson 5-4

Solve each equation. Write the solution in simplest form.

1. $\frac{2}{7} + \frac{3}{7} = g$ $\frac{5}{7}$ **2.** $\frac{8}{15} - \frac{4}{15} = y$ $\frac{4}{15}$ **3.** $\frac{3}{7} + \frac{4}{7} = w$ 1

4. $\frac{5}{6} - \frac{1}{6} = t$ $\frac{2}{3}$ **5.** $\frac{7}{12} - \frac{5}{12} = r$ $\frac{1}{6}$ **6.** $\frac{5}{12} + \frac{11}{12} = w$ $1\frac{1}{3}$

Simplify each expression.

7. $12\frac{7}{8}s - 7\frac{3}{8}s + 2\frac{5}{8}s$ $8\frac{1}{8}s$ **8.** $-6\frac{4}{9}t - \left(-4\frac{5}{9}t\right) + 3\frac{2}{9}t$ $1\frac{1}{3}t$

9. $6\frac{1}{4}g + \left(-6\frac{3}{4}g\right)$ $-\frac{1}{2}g$ **10.** $7\frac{2}{5}n - \left(-4\frac{2}{5}n\right)$ $11\frac{4}{5}n$

Lesson 5-5

Solve each equation. Write each solution in simplest form.

1. $\frac{1}{5} + \frac{2}{7} = d$ $\frac{17}{35}$ **2.** $a = \frac{4}{5} + \frac{7}{9}$ $1\frac{26}{45}$ **3.** $\frac{1}{9} - \frac{7}{12} = n$ $-\frac{17}{36}$

4. $z = \frac{8}{11} - \frac{4}{5}$ $-\frac{4}{55}$ **5.** $\frac{7}{12} - \left(\frac{-4}{11}\right) = y$ $\frac{125}{132}$ **6.** $\ell = -\frac{9}{14} + \frac{15}{16}$ $\frac{33}{112}$

7. $3\frac{2}{5} + 2\frac{4}{7} = k$ $5\frac{34}{35}$ **8.** $r = -4\frac{1}{8} + 2\frac{5}{9}$ $-1\frac{41}{72}$ **9.** $-3\frac{3}{7} - 5\frac{1}{14} = g$ $-8\frac{1}{2}$

EXTRA PRACTICE

Extra Practice **751**

Additional Answers
Lesson 5-7

1.

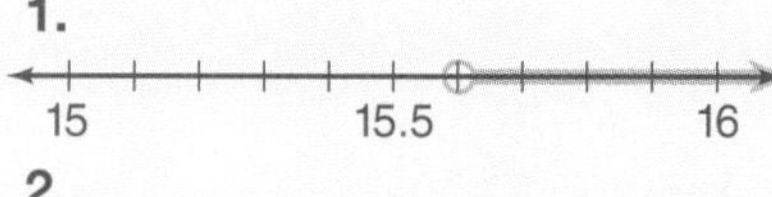

2.

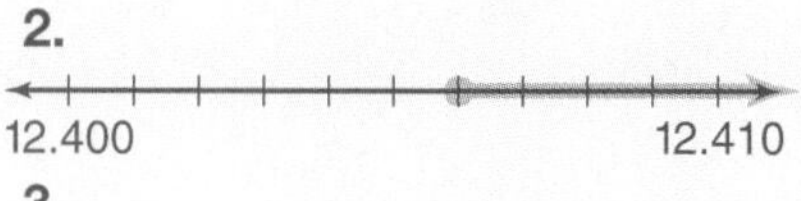

3.

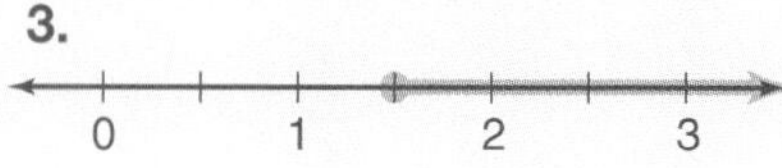

4.

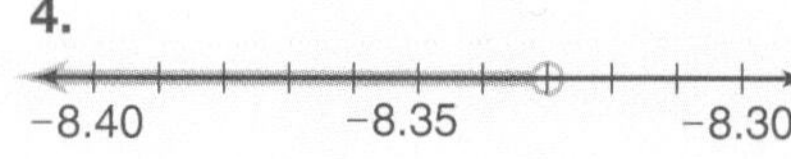

5.

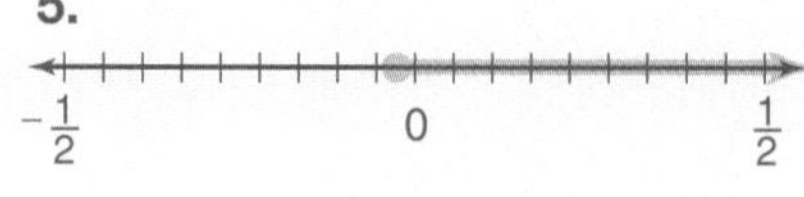

6.

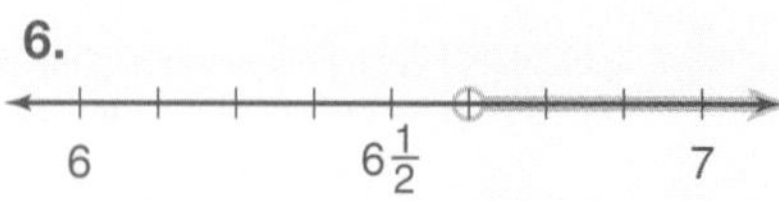

7.

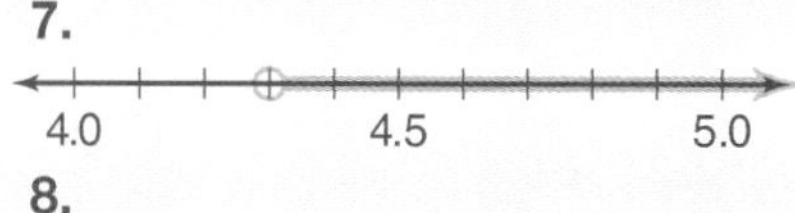

8.

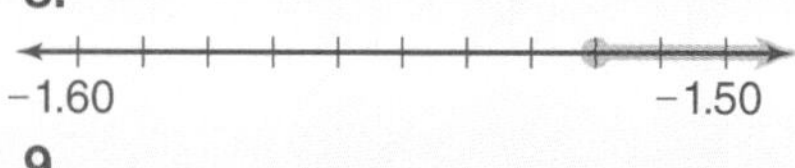

9.

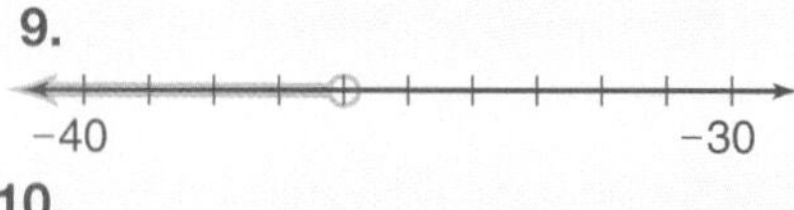

10.

10 11

Lesson 5-6

Solve each equation. Check your solution.

1. $a - 4.86 = 7.2$ **12.06**
2. $n + 6.98 = 10.3$ **3.32**
3. $87.64 = f - (-8.5)$ **79.14**
4. $x - \frac{2}{5} = -\frac{8}{15}$ **$-\frac{2}{15}$**
5. $3\frac{3}{4} + m = 6\frac{5}{8}$ **$2\frac{7}{8}$**
6. $4\frac{1}{6} = r + 6\frac{1}{4}$ **$-2\frac{1}{12}$**
7. $7\frac{1}{3} = c - \frac{4}{5}$ **$8\frac{2}{15}$**
8. $-4.62 = h + (-9.4)$ **4.78**
9. $w - 1\frac{1}{5} = \frac{2}{9}$ **$1\frac{19}{45}$**

Lesson 5-7 1–10. See margin for graphs.

Solve each inequality and check your solution. Graph the solution on a number line.

1. $h + 5.7 > 21.3$ **$h > 15.6$**
2. $78.26 \le v - (-65.854)$ **$12.406 \le v$**
3. $\frac{2}{3} \le a - \frac{5}{6}$ **$1\frac{1}{2} \le a$**
4. $-13.2 > w - 4.87$ **$w < -8.33$**
5. $a + \frac{5}{12} \ge \frac{7}{18}$ **$a \ge -\frac{1}{36}$**
6. $7\frac{1}{2} < n - \left(-\frac{7}{8}\right)$ **$6\frac{5}{8} < n$**
7. $t - 8.5 > -4.2$ **$t > 4.3$**
8. $-7.42 \le d - 5.9$ **$-1.52 \le d$**
9. $m - (-18.4) < -17.6$ **$m < -36$**
10. $s - \frac{2}{3} \ge 9\frac{4}{5}$ **$s \ge 10\frac{7}{15}$**

Lesson 5-8 1–4. Sample answers given. See students' explanations.

State whether each is an example of *inductive* or *deductive* reasoning. Explain your answer.

1. Every day that Felicia takes off from work it rains. She is going to take her birthday off, so she thinks it will rain on her birthday. **inductive**
2. The local radio station calls monthly with a survey. There is less than a week left in the month. They will be calling this week. **inductive**
3. If you have a C or better in all of your classes, you are eligible to play sports. Hank has an A or a B in all of his classes. Hank is eligible to play sports. **deductive**
4. The credit card company charges finance charges if the customer doesn't pay the balance in full. Carla could only afford to pay half of her bill this month. Carla will be charged a finance charge. **deductive**

Lesson 5-9

State whether each sequence is arithmetic. Then write the next three terms of each sequence.

1. 3.5, 4.3, 5.1, . . . **yes; 5.9, 6.7, 7.5**
2. 5, 10, 20, . . . **no; 40, 80, 160**
3. $\frac{1}{2}, \frac{5}{6}, 1\frac{1}{6}, \ldots$ **yes; $1\frac{1}{2}, 1\frac{5}{6}, 2\frac{1}{6}$**
4. $\frac{1}{4}, \frac{1}{2}, 1, 2, \ldots$ **no; 4, 8, 16**
5. 23, 18, 13, . . . **yes; 8, 3, −2**
6. 45, 43, 39, 33, . . . **no; 25, 15, 3**

752 ***Extra Practice***

Lesson 6-1

Write each fraction as a decimal. Use a bar to show a repeating decimal.

1. $\frac{6}{10}$ **0.6** **2.** $-4\frac{7}{12}$ **$-4.58\overline{3}$** **3.** $\frac{8}{11}$ **$0.\overline{72}$** **4.** $3\frac{4}{18}$ **$3.\overline{2}$**

5. $-\frac{3}{16}$ **0.1875** **6.** $8\frac{36}{44}$ **$8.\overline{81}$** **7.** $\frac{9}{37}$ **$0.\overline{243}$** **8.** $\frac{6}{15}$ **0.4**

Replace each ● with <, >, or = to make a true sentence.

9. $\frac{7}{8}$ ● $\frac{5}{6}$ > **10.** 0.04 ● $\frac{5}{9}$ < **11.** $\frac{1}{3}$ ● $\frac{2}{7}$ > **12.** $\frac{3}{5}$ ● $\frac{12}{20}$ =

13. $\frac{1}{2}$ ● 0.75 < **14.** 0.3 ● $\frac{1}{3}$ < **15.** $\frac{2}{3}$ ● 0.64 > **16.** $\frac{2}{20}$ ● 0.10 =

Lesson 6-2

Estimate each product or quotient. Sample answers given.

1. 16.38×1.5 **32** **2.** $35.54 \div 4.1$ **9** **3.** $6\frac{4}{9} \cdot 7.09$ **42** **4.** $18.24 \cdot 3.25$ **54**

5. $\frac{6}{13} \times 150$ **75** **6.** $\left(\frac{1}{4}\right)(15)$ **4** **7.** $78 \div 1\frac{11}{12}$ **39** **8.** $1\frac{7}{8} \cdot 40$ **80**

9. $\frac{1}{3} \cdot 37$ **12** **10.** $75 \div 1\frac{7}{16}$ **50** **11.** $88 \div \frac{3}{8}$ **44** **12.** $71.99 \div 5.7$ **12**

13. Estimate $\frac{4}{9}$ times 20. **10**

14. Estimate the quotient of 65.46 and 5.6. **9**

15. Estimate 32 times $5.49. **$180**

16. Estimate the quotient of 45 and $\frac{6}{13}$. **90**

Lesson 6-3

Solve each equation. Write each solution in simplest form.

1. $d = \frac{2}{5} \cdot \frac{3}{16}$ **$\frac{3}{40}$** **2.** $u = 3\frac{1}{4} \cdot \frac{2}{11}$ **$\frac{13}{22}$** **3.** $\left(\frac{3}{5}\right)\left(-\frac{5}{12}\right) = g$ **$-\frac{1}{4}$** **4.** $t = \left(\frac{4}{5}\right)^3$ **$\frac{64}{125}$**

5. $s = 2\frac{2}{6} \cdot 6\frac{2}{7}$ **$14\frac{2}{3}$** **6.** $2\left(-\frac{7}{12}\right) = r$ **$-1\frac{1}{6}$** **7.** $1\frac{3}{7} \cdot \left(-9\frac{4}{5}\right) = c$ **-14** **8.** $h = \left(-\frac{6}{7}\right)^2$ **$\frac{36}{49}$**

Evaluate each expression if $r = -\frac{1}{5}$, $s = \frac{2}{3}$, $x = 1\frac{1}{4}$, and $y = -2\frac{1}{8}$.

9. rx **$-\frac{1}{4}$** **10.** $5r^2$ **$\frac{1}{5}$** **11.** $s(x + y)$ **$-\frac{7}{12}$**

12. $8y + 12x$ **-2** **13.** $x^2(s + 2)$ **$4\frac{1}{6}$** **14.** $-x(x - 2s)$ **$\frac{5}{48}$**

EXTRA PRACTICE

Extra Practice 753

Lesson 6-4

Name the multiplicative inverse for each rational number.

1. $-\frac{5}{9}$ $-\frac{9}{5}$ or $-1\frac{4}{5}$ **2.** $5\frac{3}{8}$ $\frac{8}{43}$ **3.** 0.7 $\frac{10}{7}$ or $1\frac{3}{7}$

4. 2.35 $\frac{20}{47}$ **5.** -18 $-\frac{1}{18}$ **6.** $\frac{a}{b}$ $\frac{b}{a}$

Estimate the solution to each equation. Then solve. Write the solution in simplest form.

7. $w = \frac{3}{4} \div \frac{15}{16}$ $\frac{4}{5}$ **8.** $16 \div 1\frac{7}{8} = m$ $8\frac{8}{15}$ **9.** $q = 2\frac{1}{6} \div 1\frac{1}{5}$ $1\frac{29}{36}$

10. $y = -11 \div 3\frac{1}{7}$ $-3\frac{1}{2}$ **11.** $a = \frac{8}{45} \div \frac{10}{27}$ $\frac{12}{25}$ **12.** $220 \div \left(-5\frac{1}{2}\right) = p$ -40

Lesson 6-5

Solve each equation. 4. 15.508 7. 0.02298

1. $7.3017 \div 0.57 = a$ 12.81 **2.** $13.42 \div 67.1 = d$ 0.2 **3.** $x = 80 \div (-3.2)$ -25

4. $m = -2.016 \div (-0.13)$ **5.** $3.8 \cdot 2.9 = k$ 11.02 **6.** $85 \cdot 0.07 = w$ 5.95

7. $r = 15.32(0.0015)$ **8.** $(16.2)(0.013) = b$ 0.2106 **9.** $c = 5.4 \cdot 9.7$ 52.38

Evaluate each expression.

10. $4y^3$, if $y = 0.6$ 0.864 **11.** xy, if $x = 0.348$ and $y = -6.4$ -2.2272

12. $\frac{3a}{w}$, if $a = 0.4$ and $w = 2$ 0.6

Lesson 6-6

Find the mean, median, and mode for each set of data. When necessary, round to the nearest tenth.

1. 82, 79, 93, 91, 95 88; 91; none

2. 88, 85, 76, 94, 85, 97 87.5; 86.5; 85

3. 0.57, 12.81, 12.6, 0.96, 6.1, 14.3, 4.1, 12.81, 0.96 7.2; 6.1; 0.96 and 12.81

Use the table to answer Exercises 4–6.

4. What is the mean percent of students who plan to attend college? 68.1

5. What is the median percent of students who plan to attend tech school? 22.45

6. What is the mode percent of students who plan to start a career after high school? 11.5

Percent of Independence High School Seniors with After-High School Plans

Year	College	Tech School	Career
1990	67.7	23.9	11.5
1991	68.2	20.3	11.5
1992	71.1	22.7	6.2
1993	70.4	24.1	5.5
1994	63.9	20.5	10.2
1995	67.4	22.2	10.4

754 *Extra Practice*

Lesson 6-7

Solve each equation or inequality. Check your solution.

1. $3.5a = 7$ **2**

2. $0.8 = -0.8b$ **−1**

3. $8 < \frac{2}{3}c$ **$12 < c$**

4. $\frac{m}{13} \geq 0.5$ **$m \geq 6.5$**

5. $-9 = \frac{3}{4}g$ **−12**

6. $0.4y > -2$ **$y > -5$**

7. $-\frac{1}{2}d \leq -5\frac{1}{2}$ **$d \geq 11$**

8. $-3.5 = 0.07z$ **−50**

9. $-\frac{1}{6}s = 15$ **−90**

Solve each equation or inequality and graph the solution on a number line.

10. $\frac{2}{7}t < 4$ **$t < 14$**

11. $8.37 = 2.7d$ **3.1**

12. $\frac{1}{5}m \geq 4\frac{3}{5}$ **$m \geq 23$**

10–12. See margin for graphs.

Lesson 6-8

State whether each sequence is a geometric sequence. If so, state the common ratio and list the next three terms. **1–6. See margin.**

1. 2, 4, 8, 16, . . .

2. 125, 75, 45, . . .

3. 100, 75, 50, . . .

4. $\frac{1}{5}$, 1, 5, 25, . . .

5. 2401, 49, 7, . . .

6. $-\frac{4}{5}$, 2, −5, $12\frac{1}{2}$, . . .

7. Write the first five terms in a geometric sequence with a common ratio of 3. The first term is −4. **−4, −12, −36, −108, −324**

8. Write the first five terms of a geometric sequence if $a = -12$ and $r = \frac{1}{2}$. **−12, −6, −3, $-\frac{3}{2}$, $-\frac{3}{4}$**

9. In a certain geometric sequence, $a = 1.8$ and $r = -3$. Write the first five terms of the sequence. **1.8, −5.4, 16.2, −48.6, 145.8**

10. Use the expression $ar^{(n-1)}$ to find the eighth term in the geometric sequence −8, 12, −18, ... **$136\frac{11}{16}$**

Lesson 6-9

Write each number in scientific notation.

1. 6,184,000 **6.184×10^6**

2. 27,210,000 **2.721×10^7**

3. 0.00004637 **4.637×10^{-5}**

4. 0.00546 **5.46×10^{-3}**

5. 500,300,100 **5.003001×10^8**

6. 0.00000321 **3.21×10^{-6}**

Write each number in standard form.

7. 9.562×10^{-3} **0.009562**

8. 8.2453×10^{-7} **0.00000082453**

9. 8.2×10^4 **82000**

10. 9.102040×10^2 **910.204**

11. 2.41023×10^6 **2410230**

12. 4.21×10^{-5} **0.0000421**

Additional Answers

Lesson 6-7

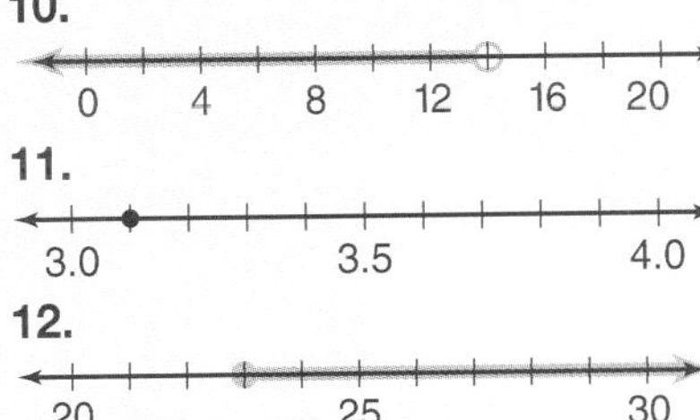

Lesson 6-8

1. yes; 2; 32, 64, 128
2. yes; $\frac{3}{5}$; 27, $16\frac{1}{5}$, $9\frac{18}{25}$
3. no
4. yes; 5; 125, 625, 3125
5. no
6. yes; $-2\frac{1}{2}$; $-31\frac{1}{4}$, $78\frac{1}{8}$, $-195\frac{5}{16}$

Extra Practice **755**

EXTRA PRACTICE

Lesson 7-1

Solve by working backward.

1. Doug brought some milk bottle caps to Antonio's house to trade. Doug traded Antonio a third of his caps in exchange for 2 designer caps. Then Doug gave Antonio's little sister Maria 6 caps as a gift. If Doug left Antonio's house with 42 milk caps, how many caps did he bring to Antonio's house? **69 caps**
2. Kwan had some cookies. She gave a fourth of her cookies to Justin. Justin then gave half of his cookies to Simon. Simon gave a third of his cookies to Sandy. If Sandy has 3 cookies, how many cookies did Kwan have in the beginning? **72 cookies** **3. 25 bacteria**
3. A certain bacteria doubles its population every 12 hours. After 3 full days, there are 1600 bacteria in a culture. How many bacteria were there at the beginning of the first day?
4. Sam and Akita are playing a game. On five turns, Sam's piece advanced 5 spaces, moved back 4 spaces, advanced 2 spaces, moved back 8 spaces, and advanced two spaces. How far did Sam's piece move from its original place on the board in these turns? **The piece moved back 3 places.**

Lesson 7-2

Solve each equation. Check your solution.

1. $3t - 13 = 2$ **5**
2. $-8j - 7 = 57$ **−8**
3. $9d - 5 = 4$ **1**
4. $6 - 3w = -27$ **11**
5. $\frac{k}{6} + 8 = 12$ **24**
6. $-4 = \frac{q}{8} - 19$ **120**
7. $15 - \frac{n}{7} = 13$ **14**
8. $7.25 = 3r - 6.25$ **4.5**
9. $21.63 - h = -32.7$ **54.33**
10. $-19 = 11b - (-3)$ **−2**
11. $6 = 20 + \frac{x}{3}$ **−42**
12. $8.12 + 3a = -3.25$ **−3.79**

Lesson 7-3

Define a variable and write an equation for each situation. Then solve.

1. The sum of 29 and 3 times a number is 44. What is the number? **$29 + 3q = 44$; 5**
2. The opposite of twice a number less 3 is 17. What is the number? **$-2m - 3 = 17$; −10**
3. Find five consecutive numbers whose sum is 95. **17, 18, 19, 20, 21**
4. The lengths of the sides of a quadrilateral are consecutive even integers. The perimeter of the quadrilateral is 108 yards. What are the lengths of the sides of the quadrilateral? **24, 26, 28, 30**

756 ***Extra Practice***

Lesson 7-4

Copy and complete the table below by finding the radius, diameter, or circumference of the circle using the information provided.

	Radius	Diameter	Circumference
1.	5 cm	10 cm	≈ 31.4 cm
3.	7 in.	14 in.	≈ 43.96 in.
5.	9 ft	18 ft	≈ 56.52 ft
7.	8.5 m	17 m	≈ 53.38 m
9.	6 yds	12 yds	≈ 37.68 yds
11.	20 mm	40 mm	≈ 125.6 mm

	Radius	Diameter	Circumference
2.	12.5 mi	25 mi	≈ 78.5 mi
4.	10.5 mi	21 mi	≈ 65.94 mi
6.	8 in.	16 in.	≈ 50.24 in.
8.	16 ft	32 ft	≈ 100.48 ft
10.	11 cm	22 cm	≈ 69.08 cm
12.	13 in.	26 in.	≈ 81.64 in.

Lesson 7-5

Solve each equation. Check your solution.

1. $-7h - 5 = 4 - 4h$ **−3**
2. $5t - 8 = 3t + 12$ **10**
3. $w + 6 = 2(w - 6)$ **18**
4. $m + 2m + 1 = 7$ **2**
5. $3.21 - 7y = 10y - 1.89$ **0.3**
6. $3(b + 1) = 4b - 1$ **4**
7. $\frac{5}{9}g + 8 = \frac{1}{6}g + 1$ **−18**
8. $\frac{s-3}{7} = \frac{s+5}{9}$ **31**

Lesson 7-6 1–12. See margin for graphs. 8. $g >$ 1.75 11. $n \geq 1.625$

Solve each inequality and check your solution. Graph the solution on a number line.

1. $2m + 1 < 9$ $m < 4$
2. $-3k - 4 \leq -22$ $k \geq 6$
3. $-2 > 10 - 2x$ $6 < x$
4. $-6a + 2 \geq 14$ $a \leq -2$
5. $3y + 2 < -7$ $y < -3$
6. $\frac{d}{4} + 3 \geq -11$ $d \geq -56$
7. $\frac{x}{3} - 5 < 6$ $x < 33$
8. $-5g + 6 < 3g + 20$
9. $-3(m - 2) > 12$ $m < -2$
10. $\frac{r}{5} - 6 \leq 3$ $r \leq 45$
11. $\frac{3(n+1)}{7} \geq \frac{n+4}{5}$
12. $\frac{n+10}{-3} \leq 6$ $n \geq -28$

Lesson 7-7 1. $8x - 2 < 15$; $x < \frac{17}{8}$

Define a variable and write an inequality for each situation. Then solve.

1. If 8 times a number is decreased by 2, the result is less than 15. What is the number?
2. George plans to spend at most \$40 for shirts and ties. He bought 2 shirts for \$13.95 each. How much can he spend for ties? **2(13.95) + $t \leq$ 40; \$12.10**
3. Mia is buying a boat. She can make a down payment of \$2,200. She wants to pay off the boat in 3 years. She cannot afford monthly payments over \$350. What is the most she can pay for the boat? **350(36) − 2200 ≥ x; 10,400**

EXTRA PRACTICE

Additional Answers
Lesson 7-6

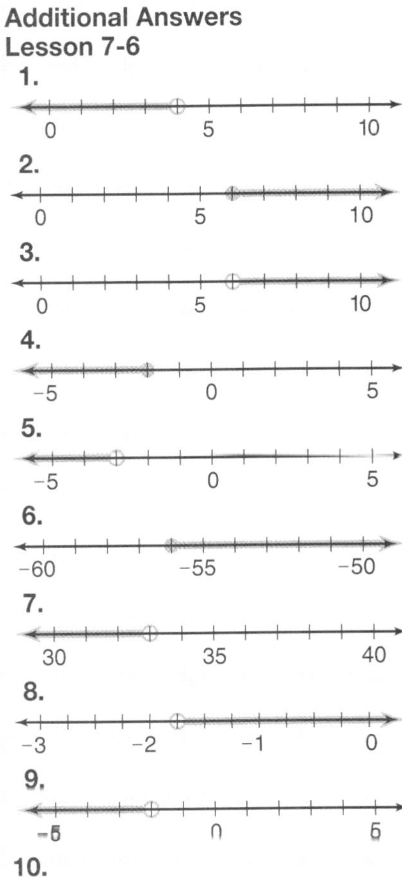

10.

40 45 50

11.

1.620 1.625 1.630

12.

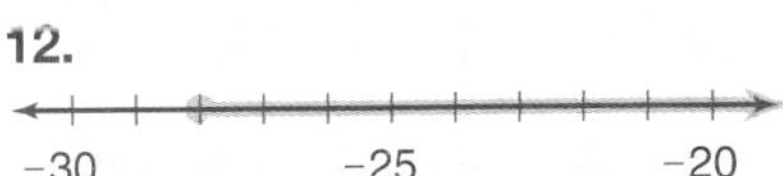

Extra Practice **757**

Additional Answers
Lesson 8-1

1. D = {1, 3, 21, 35};
   R = {1, 7, 6, 64}; yes
2. D = {27, 32, 36, 45}; R = {24}; yes
3. D = {2, 3, 4};
   R = {9, 18, 27, 36}; no
4. D = $\left\{\frac{1}{10}, \frac{1}{8}, \frac{1}{6}, \frac{1}{4}, \frac{1}{2}\right\}$;
   R = {3, 5, 7, 9, 11}; yes
5. D = {1}; R = {0, 9, 18}; no
6. D = {5, 6, 7, 8}; R = {5, 6, 7}; yes
7. D = {8, 15, 22, 29};
   R = {8, 22, 51}; yes
8. D = {−3, 4}; R = {−3, 2.5}; no

EXTRA PRACTICE

Lesson 7-8

Complete each sentence.

1. 8.2 mm = ___?___ cm 0.82
2. 6.7 km = ___?___ m 6700
3. 8.4 kg = ___?___ mg 8,400,000
4. 18 cm = ___?___ m 0.18
5. 250 ml = ___?___ L 0.25
6. 4 g = ___?___ mg 4000

Write which metric unit you would probably use to measure each item.

7. length of a car meters
8. weight of a boy kilograms
9. capacity of a can of motor oil liter
10. weight of a container of deodorant grams

Lesson 8-1 1–8. See margin.

Write the domain and range of each relation. Then determine whether each relation is a function.

1. {(3, 6),(35, 64),(1, 1),(21, 7)}
2. {(32, 24),(27, 24),(36, 24),(45, 24)}
3. {(2, 9),(3, 18),(4, 27),(2, 36)}
4. $\left\{\left(\frac{1}{2}, 3\right),\left(\frac{1}{4}, 5\right),\left(\frac{1}{6}, 7\right),\left(\frac{1}{8}, 9\right),\left(\frac{1}{10}, 11\right)\right\}$
5. {(1, 0),(1, 9),(1, 18)}
6. {(5, 5),(6, 6),(7, 7),(8, 7)}
7.

x	8	15	22	29
y	8	8	51	22

8. 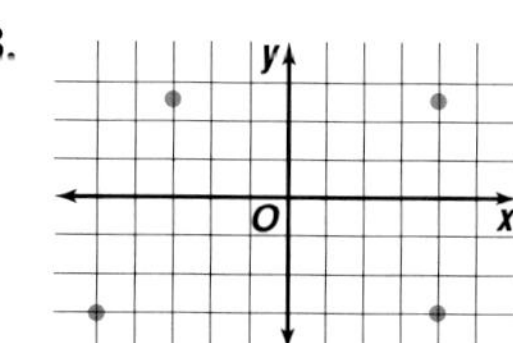

Lesson 8-2

Determine whether a scatter plot of the data for the following might show a *positive*, a *negative*, or *no* relationship. Explain your answer. 1–7. See students' explanations.

1. Size of the bag, number of potato chips positive
2. Number of team members, amount of playing time negative
3. Number of flowers in a garden, number of bees positive
4. Hair color, number of dates none
5. Candy consumed, number of cavities positive
6. Number of meals, weight positive
7. Number of classes, hours studied positive

Lesson 8-3 9. (1, −4), (0, −2), (−1, 0), (−2, 2) 11. (0, −2), (2, −3), (−2, −1), (−4, 0)

Find four solutions for each equation. Write your solutions as ordered pairs.

1. $x = 4$ (4, 2), (4, 3), (4, 5), (4, 6)
2. $y = 0$ (1, 0), (5, 0), (6, 0), (0, 0)
3. $x + y = 2$ (2, 0), (1, 1), (0, 2), (−1, 3)
4. $y = 2x - 6$ (0, −6), (1, −4), (2, −2), (3, 0)
5. $x - y = 5$ (8, 3), (7, 2), (6, 1), (5, 0)
6. $3x - y = 8$ (3, 1), (4, 4), (0, −8), (2, −2)
7. $y = \frac{1}{2}x - 3$ (0, −3), (2, −2), (4, −1), (6, 0)
8. $y = \frac{1}{3}x + 1$ (0, 1), (3, 2), (−3, 0), (6, 3)
9. $2x + y = -2$
10. $2x + 3y = 12$ (6, 0), (0, 4), (3, 2), (−3, 6)
11. $x + 2y = -4$
12. $2x - 4y = 8$ (4, 0), (0, −2), (2, −1), (−2, −3)

EXTRA PRACTICE

Lesson 8-4

Determine whether each equation is a function.

1. $y = 2x - 3$ yes
2. $x = 17$ no
3. $y = |x| - 3$ yes
4. $x + |y| = 2$ no

Given $f(x) = 3x + 2$ and $g(x) = 2x^2 - x$, determine each value.

5. $f(3)$ 11
6. $g(-5)$ 55
7. $g\left(\frac{1}{2}\right)$ 0
8. $f\left(\frac{2}{3}\right)$ 4
9. $g(-1)$ 3
10. $g(2.5)$ 10
11. $f(-2)$ −4
12. $2[f(3)]$ 22

Lesson 8-5

Use a graph to solve each problem. Assume that the rate is constant in each problem.

1. **Climate** The average high temperature in Vancouver, Canada in January is 41°. In July, the average high temperature is 74°. What would the average high temperature be in April? approximately 58°
2. **Economy** A 23 minute phone call from Columbus, Ohio to Dayton, Ohio costs $1.65. A similar phone call lasting 18 minutes costs $1.30. How much would a 20 minute phone call cost? approximately $1.44
3. **Measurement** A distance of 5 miles corresponds to the distance of 8 kilometers. A distance of 30 miles corresponds to the distance of 48.2 kilometers. If you traveled 50 miles, how many kilometers would you have traveled. 80.4 kilometers

Extra Practice **759**

Additional Answers
Lesson 8-7

7.

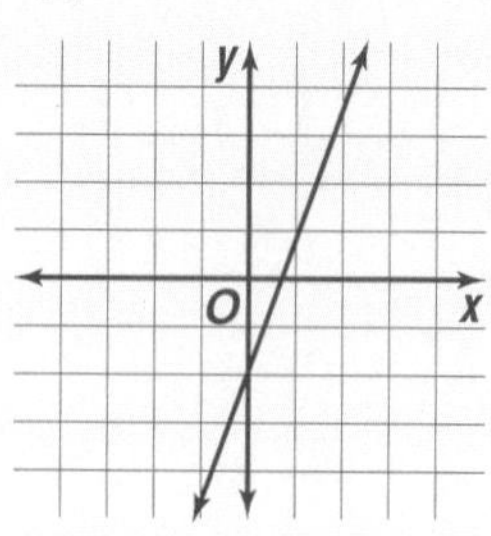

8.

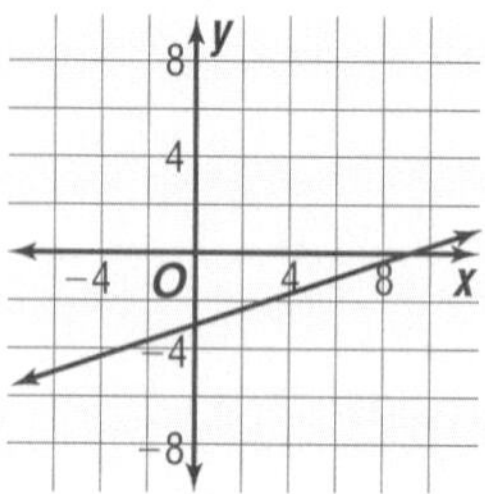

9.

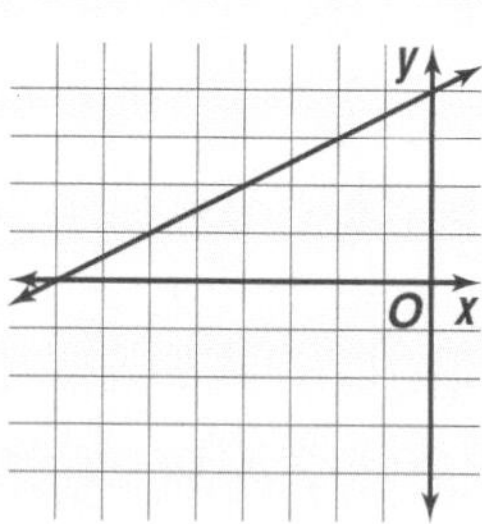

10.

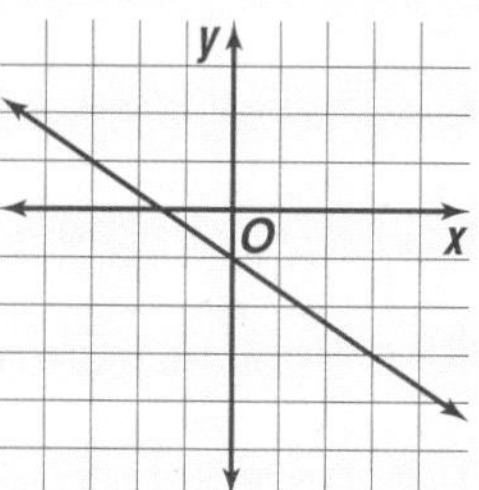

11.

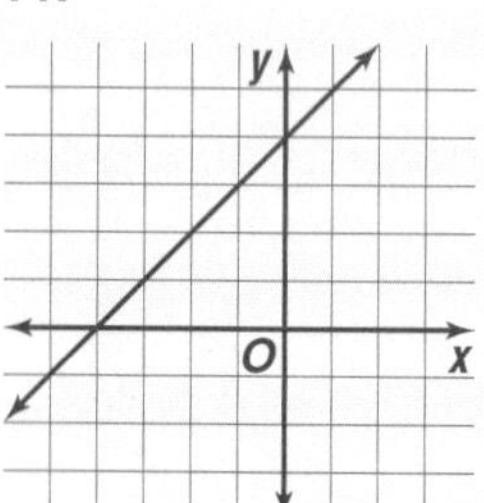

12.

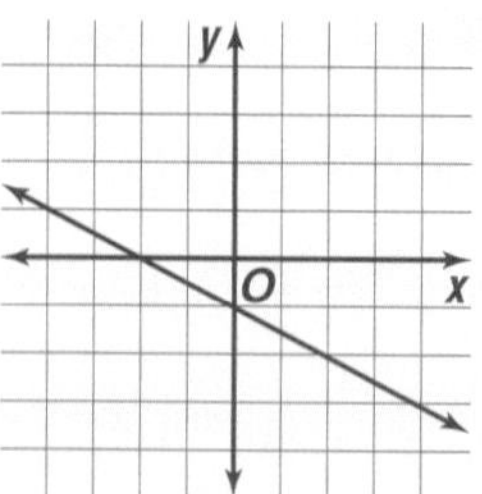

EXTRA PRACTICE

Lesson 8-6

Find the slope of the line that contains each pair of points.

1. $P(3, 8), Q(4, -3)$ −11
2. $D(4, 5), E(-3, -9)$ 2
3. $L(-1, 2), M(0, 5)$ 3
4. $J(6, 2), K(6, -4)$ no slope
5. $B(8, -3), C(-4, 1)$ $-\frac{1}{3}$
6. $D(1, 5), E(3, 10)$ $\frac{5}{2}$
7. $H(7, 2), I(-2, -2)$ $\frac{4}{9}$
8. $K(2, -4), L(5, -19)$ −5
9. $G(5, 6), H(7, 6)$ 0
10. $A(-6, -3), B(-9, 4)$ $-\frac{7}{3}$
11. $P(-1, -6), Q(-5, -10)$ 1
12. $B(5, 9), C(-4, -5)$ $\frac{14}{9}$

Lesson 8-7

Use the *x*-intercept and the *y*-intercept to graph each equation.

1. $2x + y = 6$ 3, 6
2. $-4x + y = 8$ −2, 8
3. $3x - 3y = -12$ −4, 4
4. $x + 2y = -4$ −4, −2
5. $y = -\frac{1}{2}x - 6$ −12, −6
6. $y = \frac{5}{2}x - 1$ $\frac{2}{5}$, −1

1–6. Intercepts are given. See Solutions Manual for graphs.

Graph each equation using the slope and *y*-intercept. 7–12. See margin.

7. $y = 3x - 2$
8. $x - 3y = 9$
9. $y = \frac{1}{2}x + 4$
10. $y = -\frac{2}{3}x - 1$
11. $x - y = -4$
12. $2x + 4y = -4$

Lesson 8-8

Use a graph to solve each system of equations. 1–8. See Solutions Manual for graphs.

1. $x + 2y = 6$
   $y = -0.5x + 3$ (0, 3)
2. $y = -2$
   $4x + 3y = 2$ (2, −2)
3. $2x = 3y$
   $y = 4 + \frac{2}{3}x$ none
4. $y = x - 2$
   $y = -\frac{1}{3}x + 2$ (3, 1)
5. $x + y = 4$
   $y = 2$ (2, 2)
6. $2x + y = 8$
   $2x - y = 0$ (2, 4)
7. $x = 3$
   $y = 4$ (3, 4)
8. $y = x + 2$
   $2y + x = 1$ (−1, 1)

Lesson 8-9

Graph each inequality. 1–8. See margin.

1. $y > 2x - 2$
2. $y \geq x$
3. $y < 1$
4. $x + y \leq -1$
5. $y + 3x \leq 0$
6. $x < -3$
7. $2x + 3y \geq 12$
8. $-2x + y > -1$

Lesson 9-1

Express each ratio or rate as a fraction in simplest form. 4. $\frac{3}{7}$ 8. $\frac{1}{4}$

1. 15 out of 240 $\frac{1}{16}$
2. 140:12 $\frac{35}{3}$
3. $\frac{98}{14}$ $\frac{7}{1}$
4. 6 pens to 14 pens
5. 30 cats to 6 cats $\frac{5}{1}$
6. 18 out of 45 $\frac{2}{5}$
7. 321 to 96 $\frac{107}{32}$
8. 3 cups to 3 quarts

Express each ratio as a unit rate.

9. 343.8 miles on 9 gallons 38.2 mpg
10. \$7.95 for 5 pounds. \$1.59 per pound
11. \$52 for 8 tickets \$6.50 per ticket
12. \$43.92 for 4 CDs \$10.98 per CD

Lesson 9-2

Solve by making a table. 1. 4 dimes, 1 quarter, 1 half-dollar

1. Susan Lee has \$1.15 made up of six United States coins. However, she cannot make change for a dollar, a half-dollar, a dime, or a nickel. What six coins does Susan have?
2. How many ways can you add eight prime numbers to get a sum of 20? You may use a number more than once. 1
3. Anne Zody buys a car for \$100, sells it for \$110, buys it back for \$120, and sells it again for \$130. How much does Ms. Zody make or lose? \$20 gain
4. Dwight bought the books he needed for the fall semester for \$230. He dropped a class and sold the book back for \$18. He added a new class and bought a book for \$35. At the end of the semester, he sold his books back for \$62. How much did Dwight make or lose? \$185 lost
5. A new computer company begins with 48 employees. After the first year, 8 employees leave for more stable jobs, 2 are transferred to other offices, and 17 new employees are hired. Over the next five years the company's profits increase, resulting in the hiring of 21 more people with the loss of only 3 employees. How many employees are employed by the company after 6 years? 73

EXTRA PRACTICE

Additional Answers
Lesson 8-9

1.

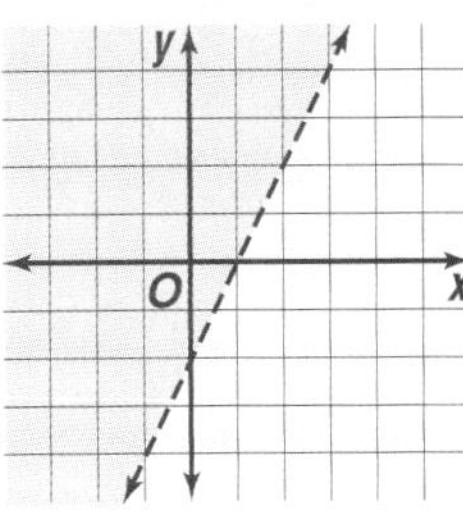

2.

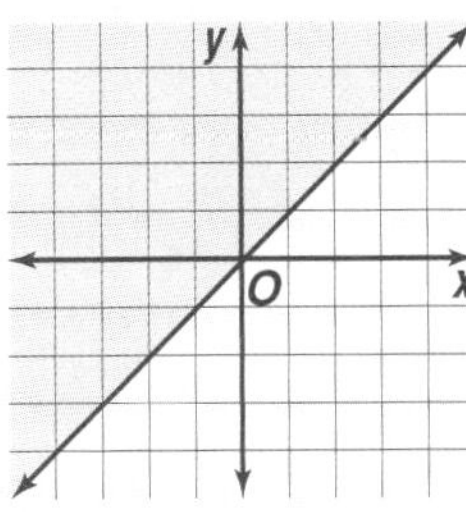

3.

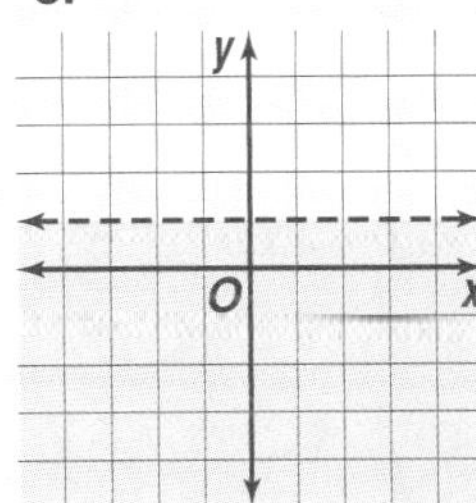

4.

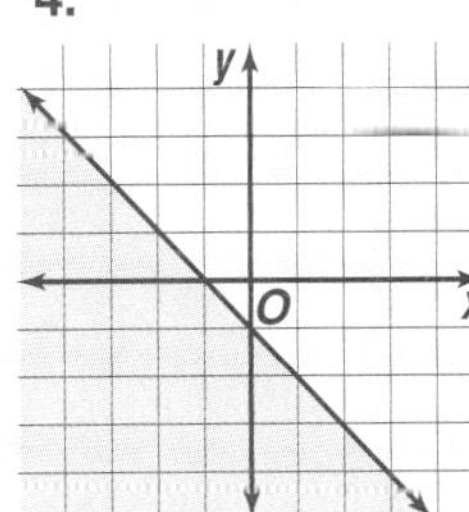

5.

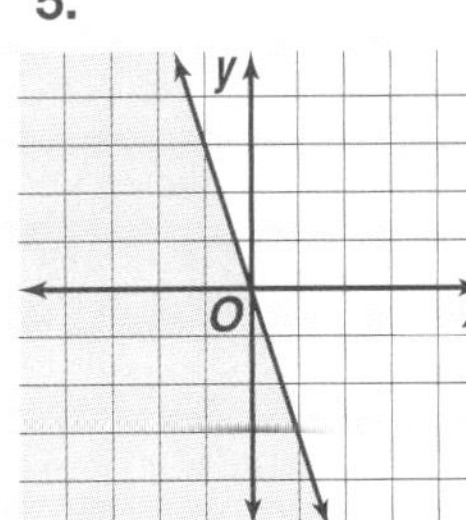

6.

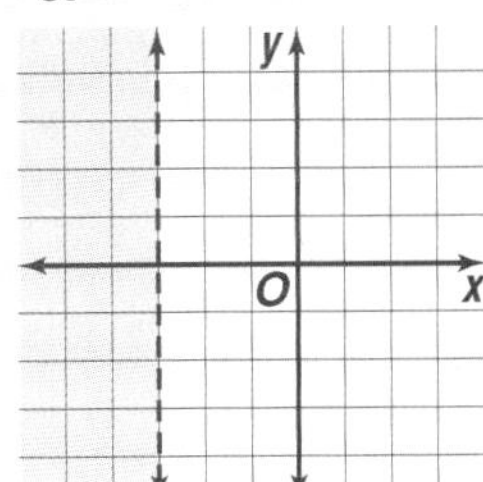

Additional Answers
Lesson 8-9

7.

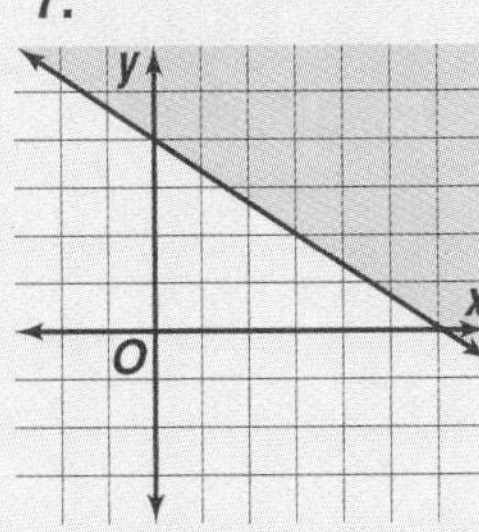

8.

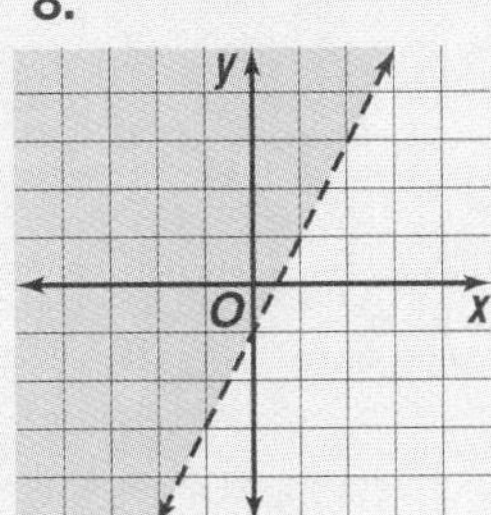

Additional Answers
Lesson 9-4

5. $\frac{6}{1.00} = \frac{10}{d}$
6. $\frac{8}{9.36} = \frac{f}{17.55}$
7. $\frac{3}{53.67} = \frac{7}{m}$
8. $\frac{10}{7.50} = \frac{p}{18}$

Lesson 9-6

1. The people at the candy counter were chosen randomly and the sample was large.
2. The people were chosen at a candy counter, so they will probably prefer the types of candy sold at that candy counter.

EXTRA PRACTICE

Lesson 9-3

There are 4 blue marbles, 5 red marbles, and 3 green marbles in a bag. Suppose you select one marble at random. Find each probability.

1. P(green) $\frac{1}{4}$
2. P(blue) $\frac{1}{3}$
3. P(red) $\frac{5}{12}$
4. P(not green) $\frac{3}{4}$
5. P(white) 0
6. P(blue or red) $\frac{3}{4}$
7. P(neither red nor green) $\frac{1}{3}$
8. P(not orange) 1

Lesson 9-4

Solve each proportion.

1. $\frac{7}{k} = \frac{49}{63}$ 9
2. $\frac{s}{4.8} = \frac{30.6}{28.8}$ 5.1
3. $\frac{6}{11} = \frac{19.2}{g}$ 35.2
4. $\frac{8}{13} = \frac{b}{65}$ 40

Write a proportion that could be used to solve for each variable. Then solve the proportion.

5. 6 plums at \$1
   10 plums at d \$1.67
6. 8 gallons at \$9.36
   f gallons at \$17.55 15
7. 3 packages at \$53.67
   7 packages at m \$125.23
8. 10 cards at \$7.50
   p cards at \$18 24

5–8. See margin for proportions.

Lesson 9-5

Use the percent proportion to solve each problem.

1. What is 81% of 134? 108.54
2. 52.08 is 21% of what number? 248
3. 11.18 is what percent of 86? 13%
4. What is 120% of 312? 374.4
5. 140 is what percent of 400? 35%
6. 430.2 is 60% of what number? 717

Lesson 9-6

William Wonker wants to find out favorite candy bars. He surveyed 100 people at a candy store. 1-2. See margin.

1. Give some reasons why this is a valid sample.
2. Give some reasons why this is not a valid sample.
3. What percent of the people preferred Giggles Bar? 5%
4. If 600 customers were surveyed, how many would prefer a Venus Bar? 126
5. How many would prefer a Galaxy Bar? 222
6. What is the mode? Galaxy Bar

Favorite Candy Bars	
Galaxy Bar	37
Venus Bar	21
Three Amigos Bar	30
Giggles Bar	5
K & K	7

Lesson 9-7

Express each decimal as a percent.

1. 0.06 6% **2.** 0.374 37.4% **3.** 0.0095 0.95% **4.** 56.71 5671%

Express each fraction as a percent.

5. $\frac{3}{4}$ 75% **6.** $3\frac{1}{4}$ 325% **7.** $\frac{45}{50}$ 90% **8.** $\frac{3}{1000}$ 0.3%

Express each percent as a fraction.

9. 17% $\frac{17}{100}$ **10.** 0.8% $\frac{8}{1000}$ **11.** 5268% $52\frac{17}{25}$ **12.** $\frac{15}{4}$% $\frac{375}{10000}$

Lesson 9-8

Choose the best estimate.

1. 28% of 500 **a.** 1.5 **b.** 15 **c.** 150 c
2. 96% of 900 **a.** 90 **b.** 900 **c.** 9000 b
3. 148% of 350 **a.** 5.25 **b.** 52.5 **c.** 525 c
4. $\frac{1}{3}$% of 360 **a.** 1.08 **b.** 10.8 **c.** 108 a

Estimate.

5. 72% of 250 175 **6.** 47% of 198 99
7. 0.8% of 380 3.8 **8.** 98% of 32 32
9. $12\frac{1}{2}$% of 130 13 **10.** 122% of 84 100.8

Estimate each percent. 11–19. Sample answers given.

11. 12 out of 20 60% **12.** 14 out of 40 30% **13.** 3 out of 75 5%
14. 75 out of 179 45% **15.** 19 out of 96 20% **16.** 1.6 out of 88 1%
17. 5 out of 9 50% **18.** 6 out of 210 3% **19.** 12 out of 2250 0.5%

Lesson 9-9

Solve each problem by using the percent equation, $P = R \cdot B$.

1. 9.28 is what percent of 58? 16% **2.** What number is 43% of 110? 47.3
3. 88% of what number is 396? 450 **4.** What number is 61% of 524? 319.64
5. 126 is what percent of 90? 140% **6.** 52% of what number is 109.2? 210

Find the discount or interest to the nearest cent. 10. $18.75

7. $64.98 dress at 32% off. $20.79 **8.** $1000 at 4% annually for 2 years. $80
9. $589 sofa at 20% off. 117.80 **10.** $500 at 2.5% a month for 18 months.
11. $800 at $7\frac{1}{2}$% annually for 15 months. $75 **12.** $2148 bedroom set at 40% off. $859.20

EXTRA PRACTICE

Extra Practice **763**

Lesson 9-10

State whether each percent of change is a percent of increase or a percent of decrease. Then find the percent of increase or decrease. Round to the nearest whole percent. I = increase; D = decrease

1. old: \$56
   new: \$42 D; 25%
2. old: \$26
   new: \$29.64 I; 14%
3. old: \$22
   new: \$37.18 I; 69%
4. old: \$137.50
   new: \$85.25 D; 38%
5. old: \$455
   new: \$955.50 I; 110%
6. old: \$3
   new: \$15 I; 400%
7. old: 750.75
   new: \$765.51 I; 2%
8. old: \$953
   new: \$476.5 D; 50%
9. old: \$101.25
   new: \$379.69 I; 275%
10. old: \$836
    new: \$842.27 I; 3/4%

EXTRA PRACTICE

Lesson 10-1

Make a stem-and-leaf plot of each set of data. 1–5. See margin.

1. 37, 44, 32, 53, 61, 59, 49, 69
2. 3, 26, 35, 8, 21, 24, 30, 39, 35, 5, 38
3. 15.7, 7.4, 0.6, 0.5, 15.3, 7.9, 7.3
4. 172, 198, 181, 182, 193, 171, 179, 186, 181
5. 101.6, 101.8, 100.5, 102.1, 101.0, 100.1, 100.5, 100.5

Lesson 10-2

Find the range, median, upper and lower quartiles, and the interquartile range for each set of data. 1–4. See margin.

1. 44, 37, 23, 35, 61, 95, 49, 96
2. 30, 62, 35, 80, 12, 24, 30, 39, 53, 38
3. 7.15, 4.7, 6, 5.3, 30.1, 9.19, 3.2
4. 271, 891, 181, 193, 711, 791, 861, 818

Lesson 10-3

Use the stem-and-leaf plot at the right to answer each question.

1. Make a box-and-whisker plot of the data. 1–9. See margin.
2. What is the range? 85
3. What is the median? 73
4. What is the upper quartile? 84.5
5. What is the lower quartile? 62.5
6. What is the interquartile range? 22
7. What are the extremes? 12 and 97
8. Are there any outliers? If so, what are they? yes. 12
9. What are the limits for outliers? 29.5 and 117.5

Stem	Leaf
1	2
2	
3	
4	3
5	0
6	2 3 7 9
7	0 3
8	0 2 2 4 5 7
9	0 7

$4|3 = 43$

Additional Answers
Lesson 10-1

1.

Stem	Leaf
3	2 7
4	4 9
5	3 9
6	1 9

$3|2 = 32$

2.

Stem	Leaf
0	3 5 8
1	
2	1 4 6
3	0 5 5 8 9

$0|3 = 3$

3.

Stem	Leaf
0	5 6
1	
2	
3	
4	
5	
6	
7	3 4 9
8	
9	
10	
11	
12	
13	
14	
15	3 7

$0|6 = 0.6$

4.

Stem	Leaf
17	1 2 9
18	1 1 2 6
19	3 8

$17|1 = 171$

5.

Stem	Leaf
100	1 5 5 5
101	0 6 8
102	1

$100|1 = 100.1$

Lesson 10-2

1. 73; 46.5; 78, 36; 42
2. 68; 36.5; 53, 30; 23
3. 26.9; 6; 9.19, 4.7; 4.49
4. 710; 751; 839.5, 232; 607.5

Additional Answer
Lesson 10-3

1.

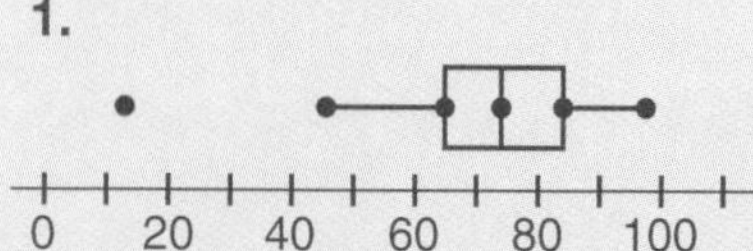

Lesson 10-4

Use the graphs to answer the following questions.

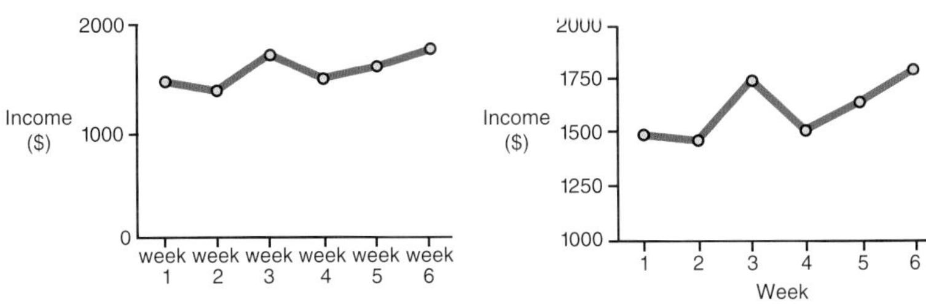

1. Explain why these graphs made from the same data look different. **See margin.**
2. Which graph shows that the family income is consistent? Why? **See margin.**
3. Which graph shows why a son can't have a raise in his allowance? Explain. **See margin.**

Lesson 10-5

Find the number of possible outcomes for each event.

1. Engagement rings come in silver, gold, and white gold. The diamond can weigh $\frac{1}{2}$ karat, $\frac{1}{3}$ karat, or $\frac{1}{4}$ karat. The diamond can have 4 possible shapes. **36**
2. A dress can be long, tea-length, knee-length, or mini. It comes in 2 colors and the dress can be worn on or off the shoulders. **16**
3. The first digit of a 7 digit phone number is a 2. The last digit is a 3. **100,000**
4. A chair can be a rocker, recliner, swivel, straight back, or a combination of any of the first three features. **7**

Lesson 10-6

1. Seven people are running for student council. There will be four people on the board and one alternate. How many ways can the students be elected? **21**
2. How many ways can the letters of the word ISLAND be arranged? **720**
3. How many ways can five candles be arranged in three candlesticks? **10**

Find each value.

4. $7!$ **5040**
5. $P(3, 2)$ **6**
6. $C(9, 4)$ **126**
7. $P(10, 5)$ **30,240**
8. $10!$ **3,628,800**
9. $\frac{6!2!}{5!}$ **12**

Additional Answers
Lesson 10-4

1. The ranges on the vertical axis are different.
2. The graph on the left makes the income look more consistent because the scale has a shorter distance between large numbers.
3. The graph on the right makes it appear that the family expenses have risen sharply.

Additional Answers
Lesson 10-8

1. Sample answer: Roll one die ten times. If the result on the first die is 1, 2, 3, or 4, she has made a shot from the field. Then roll the die five times. If the result is 1, 2, 3, 4, or 5, she has made a free throw. The sum of 2(the number of field shots) + (the number of free throws) is the probable number of points Laneeda will make in the game.
2. Sample answer: Roll a die fourteen times. Each time the die shows 1, Ted wins a soft drink.

EXTRA PRACTICE

Lesson 10-7

Find the odds of each outcome if the spinner below is spun.

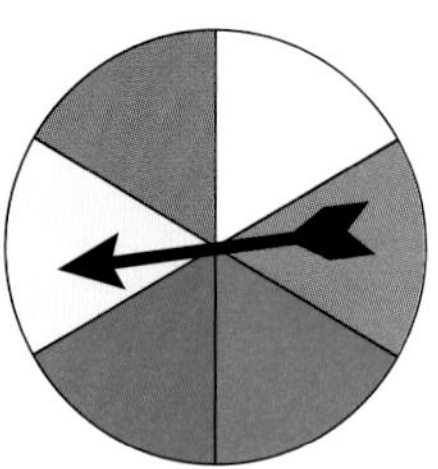

1. blue 1:5

2. a color with less than 5 letters 2:4

3. a color that begins with a consonant 5:1

4. red, yellow, or blue 3:3

Find the odds of each outcome if a 10-sided die is rolled.

5. number less than 7 6:4

6. odd number 5:5

7. composite number 5:5

8. number divisible by 3 3:7

Lesson 10-8

Solve. Use a simulation. 1–2. See margin.

1. Laneeda is a forward on the freshman basketball team. She usually makes $\frac{2}{3}$ of her shots from the field and $\frac{5}{6}$ of her free throws. If she averages 10 shots a game and 5 free throws, describe a simulation that could give her probable number of points in the next game.

2. A certain restaurant gives out game cards. One out of six cards wins a small soft drink. Describe a simulation to predict how many soft drinks Ted will win with 14 game cards.

Lesson 10-9 6. $\frac{1}{506}$ 9. $\frac{15}{322}$

Determine whether the events are independent or dependent. 1. dependent

1. go to the movies on Monday and then going back to the same theater on Tuesday

2. rolling a die and spinning a spinner independent

3. watching a 1 p.m. football game and watching a 4 p.m. football game dependent

4. choosing a person to be quarterback from a roster of 45 people and then choosing a person to be wide receiver dependent

5. selecting a name for your daughter and then selecting a name for your son independent

A deck of Euchre cards consists of 4 nines, 4 tens, 4 jacks, 4 queens, 4 kings, and 4 aces. Once a card is selected, it is not replaced. Find the probability of each outcome.

6. 3 nines in a row

7. a black jack and then a red queen $\frac{1}{138}$

8. a nine of clubs, a black king, and a red ace $\frac{1}{3036}$

9. 4 face cards in a row

10. 2 cards lower than a jack $\frac{7}{69}$

Lesson 10-10

Determine whether each event is *mutually exclusive* or *inclusive*. Then find the probability. A number from 6 to 19 is drawn. 1–8. See margin.

1. P(12 or even)
2. P(13 or less than 7)
3. P(even or odd)
4. P(14 or greater than 20)
5. P(even or less than 10)
6. P(odd or greater than 10)
7. P(divisible by 3 or even)
8. P(prime or even)

Lesson 11-1

In the figure at the right, use a protractor to find the measure of each angle. Then classify the angle as *acute*, *right*, or *obtuse*.

1. $m\angle PQW$ 120°, obtuse
2. $m\angle VQW$ 26°, acute
3. $m\angle TQW$ 90°, right
4. $m\angle SQW$ 140°, obtuse
5. $m\angle SQR$ 40°, acute
6. $m\angle VQR$ 154°, obtuse

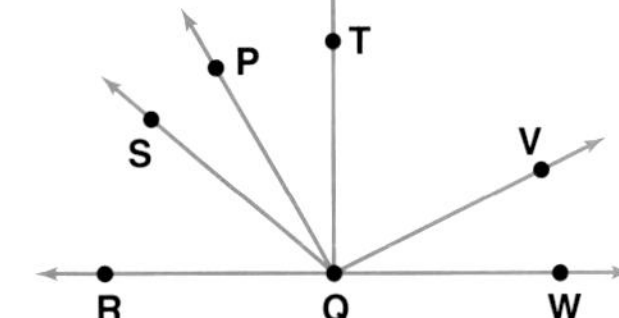

Lesson 11-2

Make a circle graph to display each set of data. 1–2. See margin.

1.

Items Donated for Food Drive	
Canned Goods	169
Pasta	86
Cereal	70
Peanut Butter	42
Condiments	18

2.

Davis Family Monthly Expenditures	
Housing	$800
Taxes	$600
Food	$350
Clothing	$200
Insurance	$150
Savings	$100

Lesson 11-3

In the figure at the right, ℓ is parallel to m. If the measure of $\angle 2$ is 38°, find the measure of each angle.

1. $\angle 1$ 142°
2. $\angle 3$ 38°
3. $\angle 4$ 142°
4. $\angle 5$ 38°
5. $\angle 6$ 142°
6. $\angle 8$ 142°

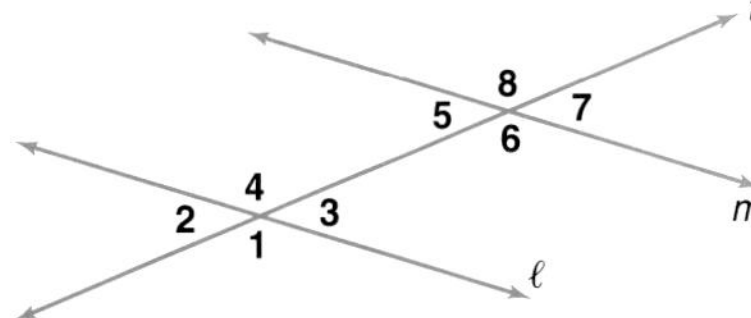

EXTRA PRACTICE

Additional Answers
Lesson 10-10

1. inclusive; $\frac{1}{2}$
2. exclusive; $\frac{1}{7}$
3. exclusive; 1
4. exclusive; $\frac{1}{14}$
5. inclusive; $\frac{9}{14}$
6. inclusive; $\frac{11}{14}$
7. inclusive; $\frac{9}{14}$
8. exclusive; $\frac{6}{7}$

Lesson 11-2

1.

Items Donated for Food Drive

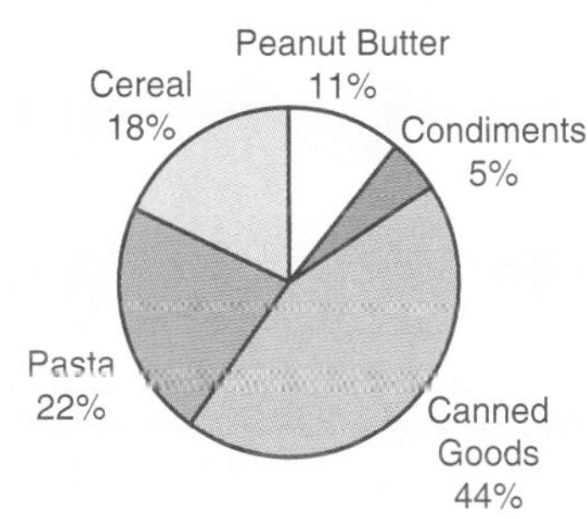

2.

Davis Family Monthly Expenditures

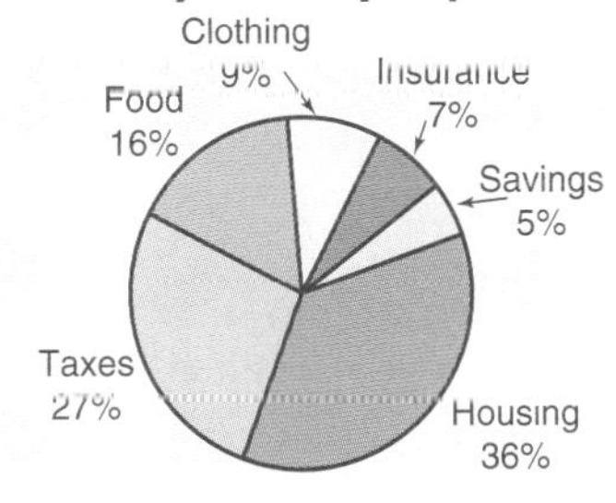

EXTRA PRACTICE

Lesson 11-4

1. Find the measure of each angle of a triangle if they are in the ratio 1:2:3. **30°; 60°; 90°**
2. Find the measure of each angle of a triangle if they are in the ratio 1:1:2. **45°; 45°; 90°**
3. Find the measure of each angle of a triangle if they are in the ratio 1:9:26. **5°; 45°; 130°**

Lesson 11-5

If $\triangle JKL \cong \triangle DGW$, name the part congruent to each angle or segment given.

1. $\angle K$ **$\angle G$** 2. $\overline{WG}$ **$\overline{LK}$** 3. $\angle D$ **$\angle J$** 4. $\overline{KL}$ **$\overline{GW}$** 5. $\overline{DG}$ **$\overline{JK}$**

Lesson 11-6

Write a proportion to find each missing measure x. Then find the value of x.

1.
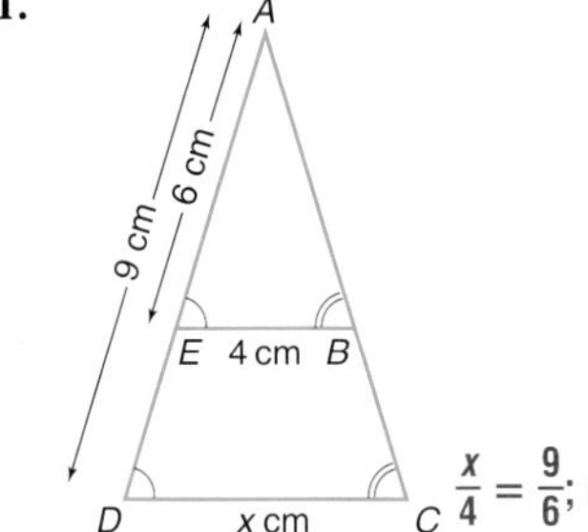

$\frac{x}{4} = \frac{9}{6}$; 6

2.
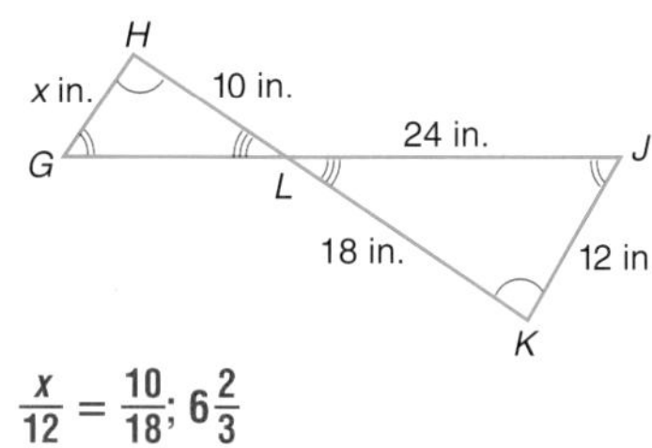

$\frac{x}{12} = \frac{10}{18}$; $6\frac{2}{3}$

Lesson 11-7

Find the value of x. Then find the missing measures.

1.
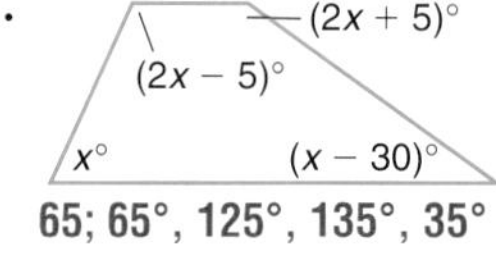

65; 65°, 125°, 135°, 35°

2.
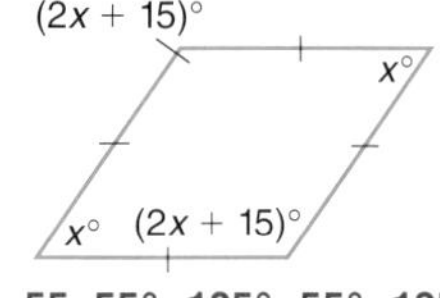

55; 55°, 125°, 55°, 125°

3.
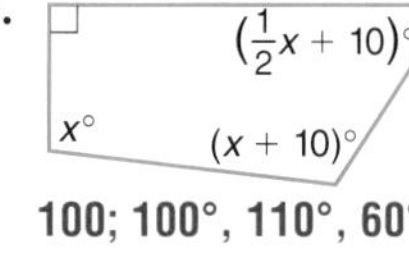

100; 100°, 110°, 60°

Lesson 11-8

Find the measure of each exterior angle and each interior angle of each regular polygon.

1. equilateral triangle **120°, 60°**
2. regular pentagon **72°, 108°**
3. regular octagon **45°, 135°**
4. regular nonagon **40°, 140°**

768 ***Extra Practice***

Lesson 11-9

Trace each figure. Draw all lines of symmetry.

1. 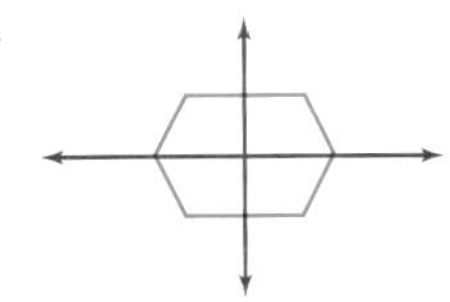

2. 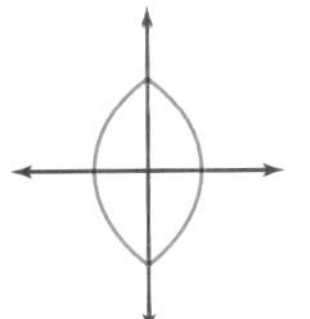

3. 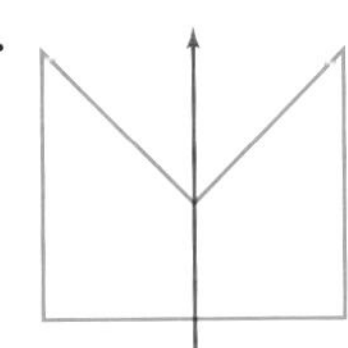

4. 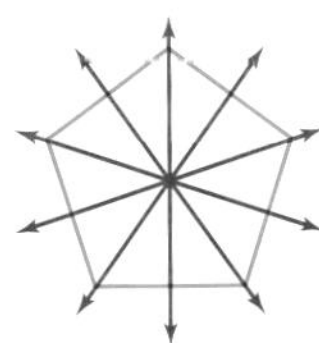

Lesson 12-1

Find the area of each figure.

1. 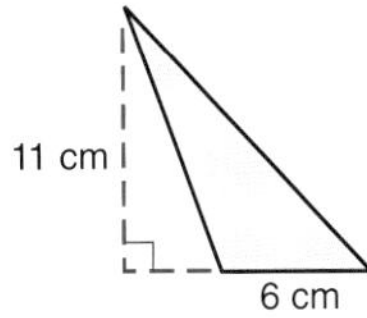

33 cm²

2. 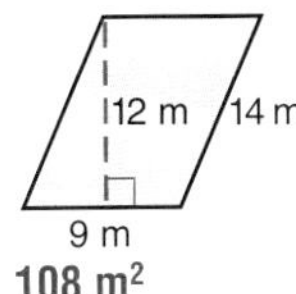

108 m²

3. 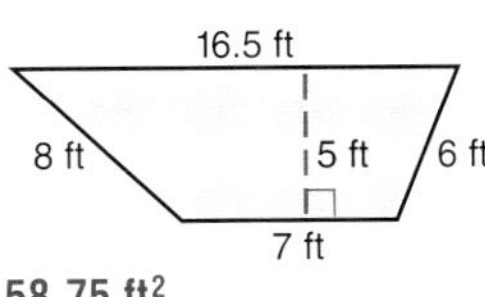

58.75 ft²

Lesson 12-2

Find the area of each circle. Round to the nearest tenth. 1. 78.5 in.²

1. diameter, 10 in.
2. radius, 8 mm 201.1 mm²
3. diameter, 15.4 m 186.3 m²

Lesson 12-3

Each figure represents a dart board. Find the probability of landing in the shaded region.

1. 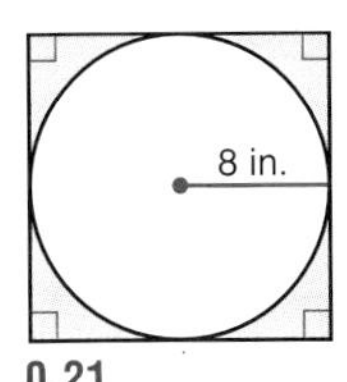

0.21

2. 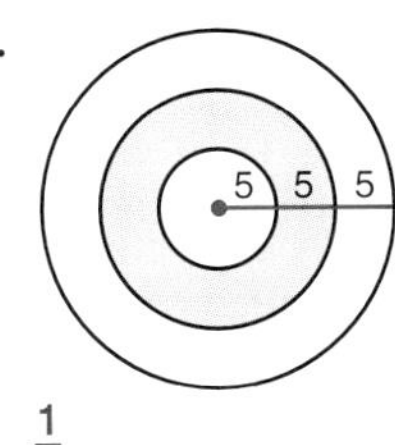

$\frac{1}{3}$

3. 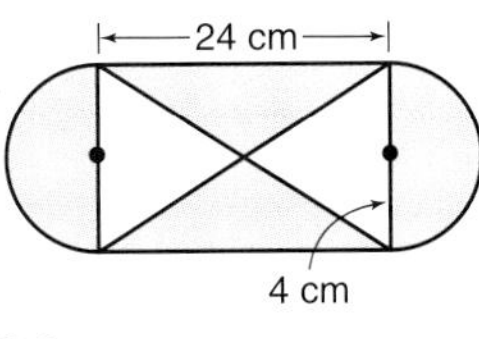

0.6

EXTRA PRACTICE

Extra Practice **769**

Lesson 12-4

Solve by making a model or drawing.

1. A theater section is arranged so that each row has the same number of seats. Andrew is seated in the fifth row from the front and the third row from the back. His seat is sixth from the left and second from the right. How many seats are in this section? 49 seats
2. Sixteen players are competing in a bowling tournament. Each player in the competition bowls against another opponent and is eliminated after one loss. How many games does the winner of the tournament play? 4 games
3. An artist is creating a pyramid piece of art using differently-colored marble cubes. There is one cube on the top, two cubes in the next layer, three cubes in the next layer, and so on until 10 layers are completed. How many cubes did the artist use? 55 cubes

Lesson 12-5

Find the surface area of each solid. Round decimal answers to the nearest tenth.

1. 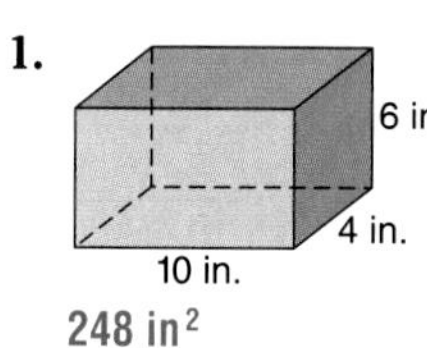

248 in^2

2. 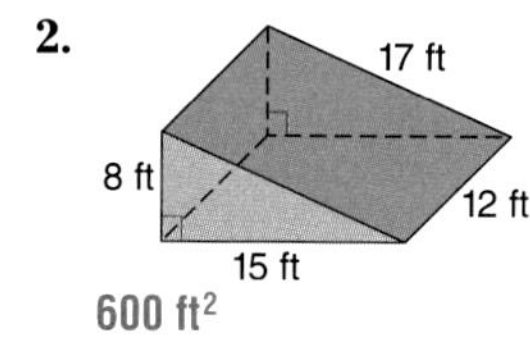

600 ft^2

3. 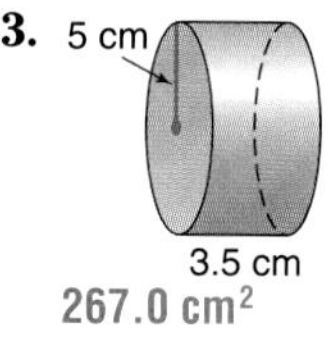

267.0 cm^2

Lesson 12-6

Find the surface area of each pyramid or cone. Round decimal answers to the nearest tenth.

1. 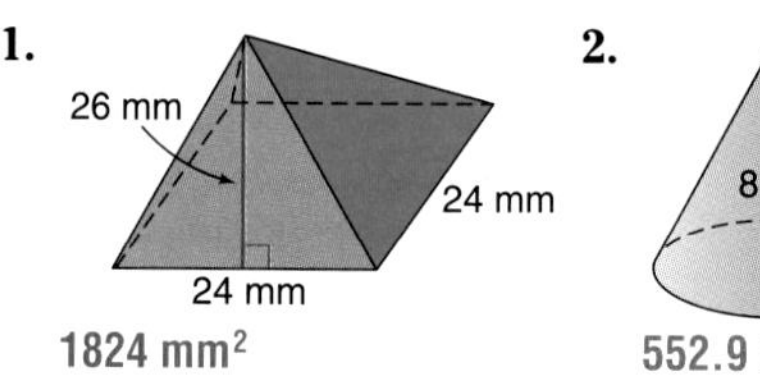

1824 mm^2

2. 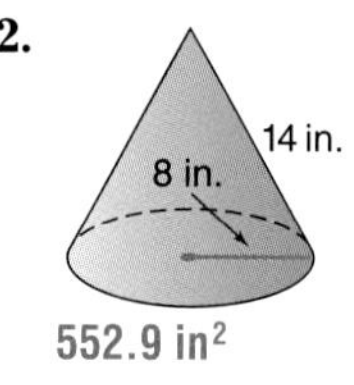

552.9 in^2

3. 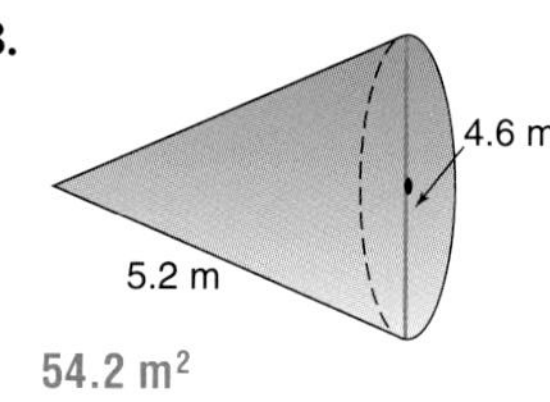

54.2 m^2

Lesson 12-7

Find the volume of each prism or cylinder. Round decimal answers to the nearest tenth.

1. 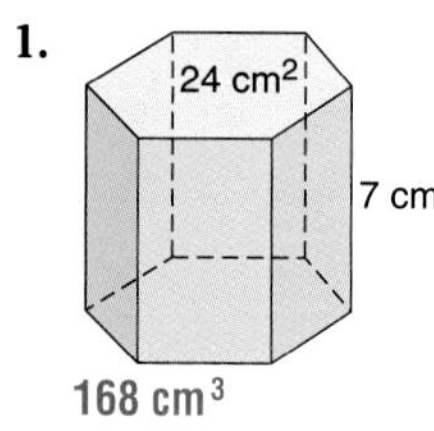

168 cm^3

2. 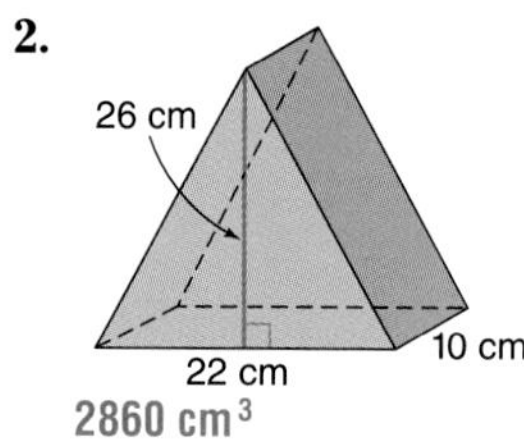

2860 cm^3

3. 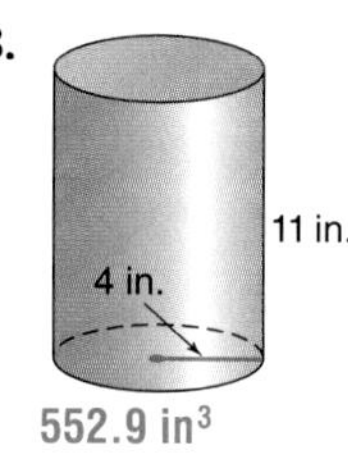

552.9 in^3

770 *Extra Practice*

Lesson 12-8

Find the volume of each pyramid or cone. Round decimal answers to the nearest tenth.

1. 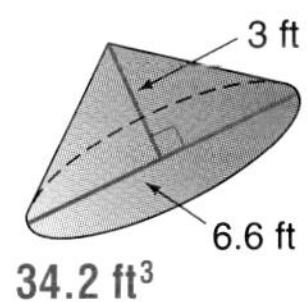

34.2 ft³

2. 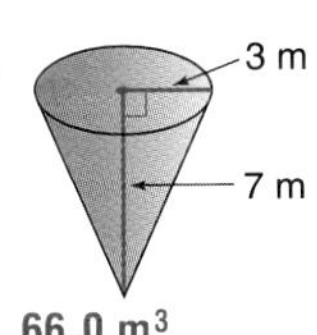

66.0 m³

3. 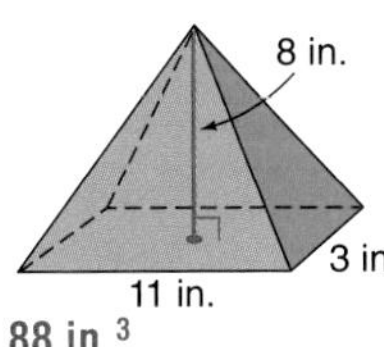

88 in.³

Lesson 13-1

Find the best integer estimate for each square root. Then check your estimate with a calculator.

1. $\sqrt{21}$ 5
2. $-\sqrt{85}$ −9
3. $\sqrt{7.3}$ 3
4. $\sqrt{1.99}$ 1
5. $-\sqrt{62}$ −8
6. $\sqrt{74.1}$ 9
7. $\sqrt{810}$ 28
8. $-\sqrt{88.8}$ −9

Lesson 13-2

Solve using a Venn diagram.

1. A supermarket survey showed that 83 customers chose wheat cereal, 83 chose rice, and 20 chose corn. Six customers bought a box of corn cereal and a box of wheat cereal and 10 others bought a box of rice cereal and a box of corn cereal. Four customers bought all three. How many customers bought a box of rice cereal and a box of wheat cereal if 61 bought just a box of wheat cereal. 12 customers
2. Of the thirty members in a cooking club, 20 like to mix salads, 17 prefer baking desserts, and 8 like to do both.
   a. How many like to do salads, but not bake desserts? 12 people
   b. How many do not like either baking desserts or mixings salads? 1 person

Lesson 13-3

Solve each equation. Round decimal answers to the nearest tenth. 8. 158.1, −158.1

1. $x^2 = 14$ 3.7, −3.7
2. $y^2 = 25$ 5, −5
3. $34 = p^2$ 5.8, −5.8
4. $55 = h^2$ 7.4, −7.4
5. $225 = k^2$ 15, −15
6. $324 = m^2$ 18, −18
7. $d^2 = 441$ 21, −21
8. $r^2 = 25{,}000$

Lesson 13-4

In a right triangle, If a and b are the measures of the legs and c is the measurement of the hypotenuse, find each missing measure.

1. a, 7 m; b, 24 m 25 m
2. a, 18 in.; c, 30 in. 24 in.
3. b, 10 ft; c, 20 ft 17.3 ft
4. a, 3 cm; c, 9 cm 8.5 cm
5. b, 8 m; c, 32 m 31.0 m
6. a, 32 yd; c, 65 yd 56.6 yd

EXTRA PRACTICE

Extra Practice **771**

Lesson 13-5

Find the lengths of the missing sides in each right triangle. Round decimal answers to the nearest tenth.

1.

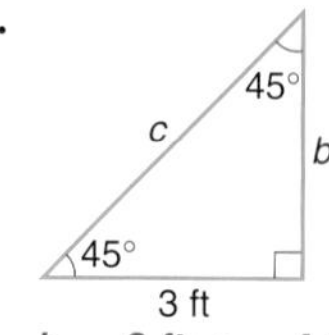

$b = 3$ ft; $c \approx 4.2$ ft

2.

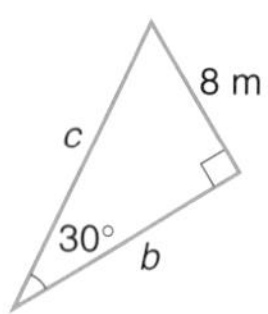

$b \approx 13.9$ m; $c = 16$ m

3. **4.**

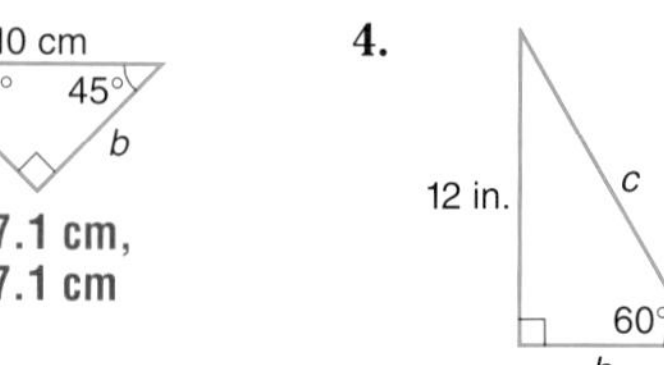

3. $a \approx 7.1$ cm, $b \approx 7.1$ cm

4. $b \approx 6.9$ in. $c \approx 13.9$ in.

Lesson 13-6

For each triangle, find the measure of the marked acute angle to the nearest degree.

1.

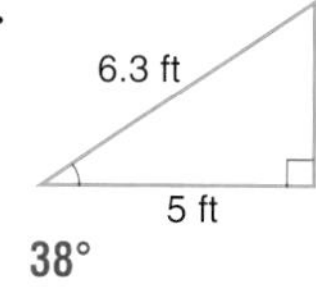

38°

2.

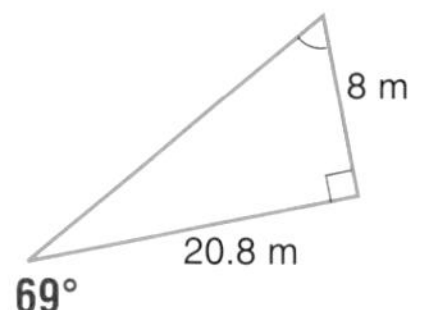

69°

3. 10 cm, 7.7 cm

50°

4.

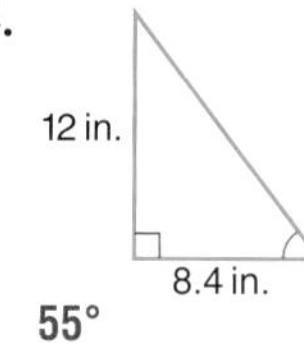

55°

Lesson 13-7

Use trigonometric ratios to solve each problem.

1. In a sightseeing boat near the base of Horseshoe Falls at Niagara Falls, a passenger estimates that angle of elevation to the top of the falls to be 30°. If Horseshoe Falls is 173 feet high, what is the distance from the boat to the base of the falls? **299.6 ft**
2. While picnicking in San Jacinto Battlefield Park, a student 800 feet from the base of the monument estimates the angle of elevation to the top of the monument to be 25°. From this information, estimate the height of the monument. **373.0 ft**
3. From a boat in the ocean, a cliff is sighted through the fog. The angle of elevation is 42° and the height of the cliff is 135 meters. How far is the boat from the cliff? **149.9 m**

Lesson 14-1

Find the degree of each polynomial.

1. $4x$ **1**
2. $a^2 - 6$ **2**
3. $11r + 5s$ **1**
4. $3y^2 4y - 2$ **3**
5. $9cd^3 - 5$ **4**
6. $-5p^3 + 8q^2$ **3**
7. $w^2 + 2x - 3y^3 - 7z$ **3**
8. $\frac{x^3}{6} - x$ **3**
9. $-17n^2p - 11np^3$ **4**

Lesson 14-2

Find each sum. 6. $5a^2 + 2a - 5$

1. $(3a + 4) + (a + 2)$ $4a + 6$
2. $(8m - 3) + (4m + 1)$ $12m - 2$
3. $(5x - 3y) + (2x - y)$ $7x - 4y$
4. $(8p^2 - 2p + 3) + (-3p^2 - 2)$ $5p^2 - 2p + 1$
5. $(-11r^2 + 3s) + (5r^2 - s)$ $-6r^2 + 2s$
6. $(3a^2 + 5a + 1) + (2a^2 - 3a - 6)$

Lesson 14-3

Find each difference.

1. $(3n + 2) - (n + 1)$ $2n + 1$
2. $(-3c + 2d) - (7c - 6d)$ $-10c + 8d$
3. $(4x^2 + 1) - (3x^2 - 4)$ $x^2 + 5$
4. $(5a - 4b) - (-a + b)$ $6a - 5b$
5. $\begin{array}{r} 6x^2 - 4x + 11 \\ (-)\ 5x^2 + 5x - 4 \\ \hline x^2 - 9x + 15 \end{array}$
6. $\begin{array}{r} 8n^2 + 3mn \\ (-)\ 4n^2 + 2mn - 9 \\ \hline 4n^2 + mn + 9 \end{array}$

Lesson 14-4

Simplify.

1. $(a^2)^3$ a^6
2. $(-2x)^3$ $-8x^3$
3. $(p^2q)^3$ p^6q^3
4. $-5c(2cd)^3$ $-40c^4d^3$
5. $4y(y^2z)^3$ $4y^7z^3$
6. $6a(-ab)^7$ $-6a^8b^7$

Lesson 14-5

Find each product.

1. $4n(5n - 3)$ $20n^2 - 12n$
2. $-3x(4 - x)$ $-12x + 3x^2$
3. $6m(-m^2 + 3)$ $-6m^3 + 18m$
4. $-5x(2x^2 - 3x + 1)$ $-10x^3 + 15x^2 - 5x$
5. $7r(r^2 - 3r + 7)$ $7r^3 - 21r^2 + 49r$
6. $-3az(2z^2 + 4az + a^2)$ $-6az^3 - 12a^2z^2 - 3a^3z$

Lesson 14-6

Find each product. 1. $6x^2 + 8x + 2$ 2. $3x^2 + 13x + 4$ 4. $x^2 + 4x + 3$ 5. $2x^2 + 3x + 1$

1. $(2x + 2)(3x + 1)$
2. $(x + 4)(3x + 1)$
3. $(7x + 4)(3x - 11)$ $21x^2 - 65x - 44$
4. $(x + 3)(x + 1)$
5. $(2x + 1)(x + 1)$
6. $(3x + 2)(2x + 2)$ $6x^2 + 10x + 4$

EXTRA PRACTICE

Extra Practice 773

Additional Answer

40.

Registered Dogs

Dogs (thousands)

140
120
100
80
60
40
20
0

Beagle, Chow Chow, Collie, Dalmation, Labrador Retriever, Maltese, Newfoundland, Shar-Pei, Springer Spaniel, Weimaraner

Breeds

Chapter 1 Practice Test

PRACTICE TEST

Solve using the four-step plan.

1. The distance between Jackie's house and Roberta's house is 90 feet. If it takes Jackie 3 seconds to walk 10 feet, how long will it take her to walk to Roberta's house? **27 seconds**

Find the value of each expression.

2. $7 \cdot 4 + 6 \cdot 5$ **58**
3. $8 + 3(16 - 12)$ **20**
4. $6[4 \cdot (72 - 63) \div 3]$ **72**
5. $6[5 \times (41 - 36) - (8 + 14)]$ **18**

Evaluate each expression if $a = 8$, $b = 4$, and $c = 3$.

6. $2c + 6bc - 7$ **71**
7. $15 \div c + 7b$ **33**
8. $6b \div a + 5c$ **18**

Name the property shown by each statement.

9. $4 \cdot (6 \cdot 0) = 4 \cdot (0 \cdot 6)$ **commutative, ×**
10. $5 + (9 + 3) = (5 + 9) + 3$ **associative, +**
11. $0 + h = h$ **identity, +**
12. $8(a + 4) = 8a + 8(4)$ **distributive**
13. $8(6 \cdot 9) = (8 \cdot 6)9$ **associative, ×**
14. $r \cdot 1 = r$ **identity, ×**

Simplify each expression.

15. $(b + 12) + 15$ **$b + 27$**
16. $6(v \cdot 2)$ **$12v$**
17. $18yz + 13yz$ **$31yz$**
18. $5(2w + 3) + 7(w + 13)$ **$17w + 106$**

Solve each equation mentally.

19. $5x = 25$ **5**
20. $8 + g = 8$ **0**
21. $\frac{s}{14} = 3$ **42**
22. $22 - k = 17$ **5**

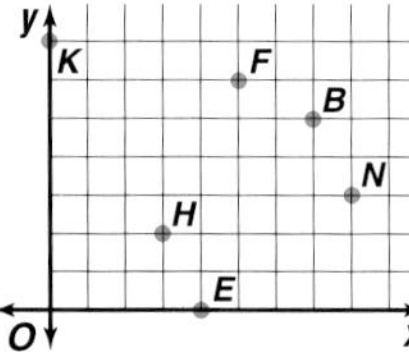

Name the ordered pair for each point graphed on the coordinate plane at the right.

23. B **(7, 5)**
24. E **(4, 0)**
25. F **(5, 6)**
26. H **(3, 2)**
27. N **(8, 3)**
28. K **(0, 7)**

Solve each equation using the inverse operation. Use a calculator when needed.

29. $x + 48 = 55$ **7**
30. $v - 57 = 72$ **129**
31. $\frac{b}{2} = 18$ **36**
32. $672 = 21t$ **32**

State whether each inequality is *true* or *false* for the given value.

33. $16 - t > 7, t = 14$ **false**
34. $4p - 6 \geq 9, p = 4$ **true**
35. $2v + v > 21, v = 7$ **false**
36. $8m - 2m \leq 7m, m = 0$ **true**

The frequency table at the right contains data about the number of dogs of different breeds registered in the American Kennel Club in 1992.

Breed	Dogs (thousands)	Breed	Dogs (thousands)
Beagle	61	Maltese	18
Chow Chow	43	Newfoundland	3
Collie	17	Shar-Pei	90
Dalmatian	39	Springer Spaniel	22
Labrador Retriever	121	Weimaraner	5

Source: American Kennel Club

37. Which breed has the most dogs registered?
38. Write an equation to represent the difference between the number of Maltese and the number of springer spaniels registered. Then find the difference. **$d = 22 - 18$; 4 thousand**
39. How many dogs of these ten breeds are registered? **419,000**
40. Make a bar graph of the data. **See margin.**

37. Labrador retriever

Chapter 2 Practice Test

Write an integer for each situation.

1. a gain of 15 pounds +15
2. a rent increase of \$10 +10
3. a bank overdraft of \$25 −25
4. a loss of 6 yards −6

Simplify.

5. $|-8|$ 8
6. $|-3| + |2|$ 5
7. $|15| - |-3|$ 12
8. $|-20| + |-19|$ 39

Name the ordered pair for each point graphed on the coordinate plane at the right.

9. A (1, −6)
10. C (−7, 5)
11. E (6, 0)
12. R (5, 6)
13. S (−3, −2)
14. M (2, −3)
15. T (−2, 3)
16. V (0, −7)

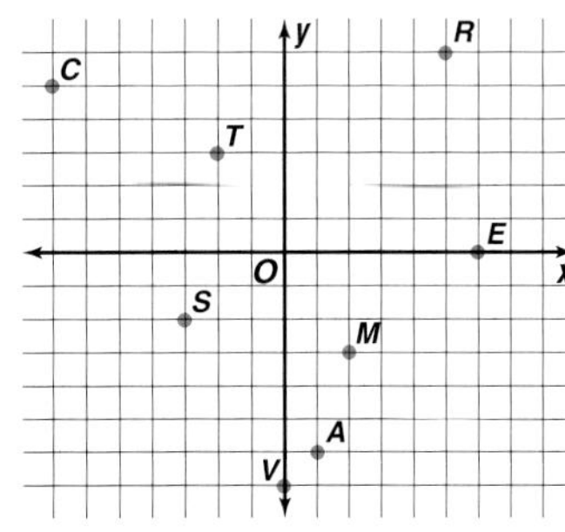

On graph paper, draw coordinate axes. Then graph and label each point. Name the quadrant in which each point is located. 17–22. See margin for graphs.

17. $A(8, -5)$ IV
18. $J(-4, -18)$ III
19. $W(-1, 6)$ II
20. $F(6, 17)$ I
21. $M(-8, 5)$ II
22. $S(-8, -5)$ III

Replace each ● with <, >, or =.

23. −6 ● −12 >
24. 18 ● 18 =
25. −5 ● −4 <
26. $|-7|$ ● 7 =

Simplify each expression.

27. $3x + 5x$ $8x$
28. $6p - (-4p)$ $10p$
29. $-9m - (-7m)$ $-2m$
30. $-8k + (-12k)$ $-20k$
31. $-15a - 6a$ $-21a$
32. $10d + (-3d)$ $7d$

Solve. Look for a pattern. 123,456 × 9 + 7 = 1,111,111

33. Use the pattern at the right to find 123,456 × 9 + 7
34. The following is referred to as an alphametric problem. Each letter stands for one digit 0–9.

1 × 9 + 2 = 11
12 × 9 + 3 = 111
123 × 9 + 4 = 1111

```
  TRIED
+ DRIVE
  RIVET
```

T=1, R=7, I=4, E=6, D=5, V=9

Solve each equation.

35. $q = -13(10)$ −130
36. $-40 \cdot (-6) = h$ 240
37. $-5(-8)(3) = r$ 120
38. $b = \frac{-15}{3}$ −5
39. $\frac{-12}{-6} = a$ 2
40. $115 \div (-5) = k$ −23

PRACTICE TEST

Additional Answers

17–22.

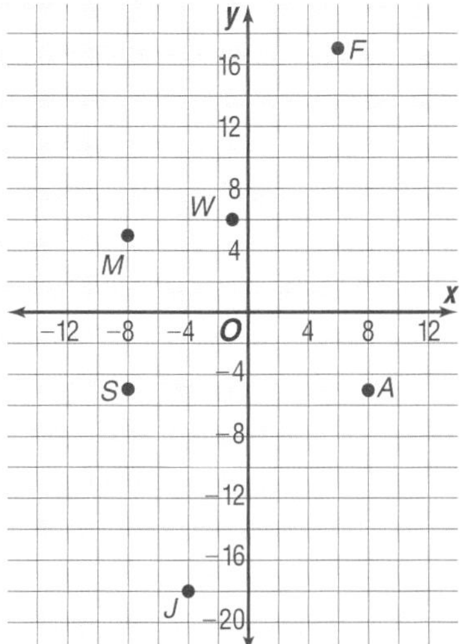

Additional Answers

3.

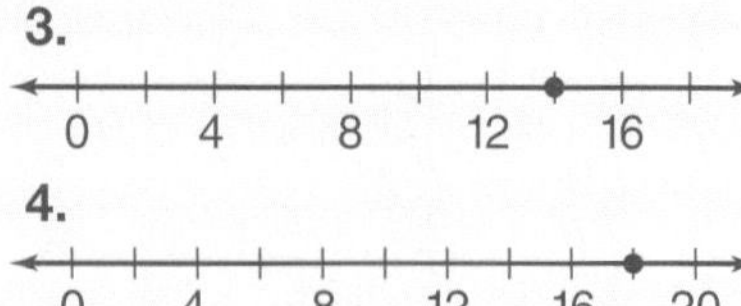

4.

0 4 8 12 16 20

5.

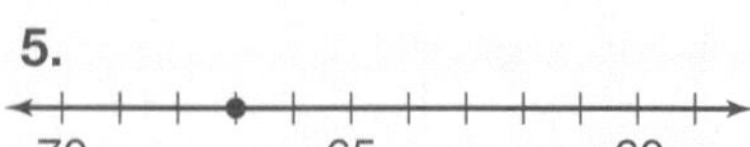

6.

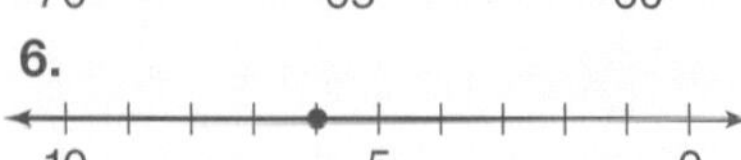

7.

-10 -5 0

8.

0 4 8 12 16 20

9.

-130 -125 -120

10.

0 60 120 180 240 300

18.

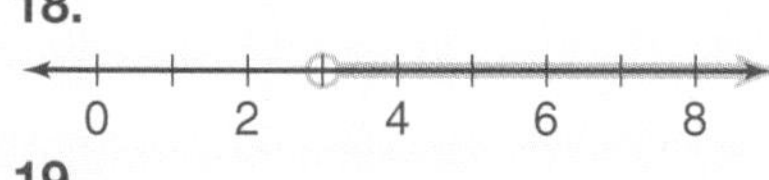

19.

-115 -110 -105

20.

-6 -4 -2 0

21.

-20 -16 -12 -8 -4 0

22.

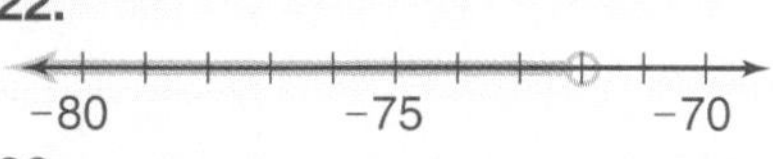

23.

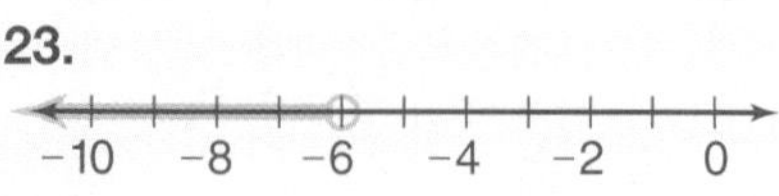

Chapter 3 Practice Test

PRACTICE TEST

Solve by eliminating possibilities.

1. Dave is younger than Angie, but older than Bob and Dwight. Angie is older than Dwight. Dwight is younger than Andy. Donna is older than Angie, but younger than Andy. Andy is older than Dave. Who is the oldest? **Andy**
2. Three people are standing in a straight line as follows: The teenager is to the right of the baby and the adult. The blonde is to the left of the baby. The brunette is to the right of the redhead. State the order of the people and their hair color. **adult, baby, teenager; blonde, redhead, brunette**

Solve each equation and check your solution. Then graph the solution on a number line.

3–10. See margin for graphs.

3. $19 = f + 5$ **14**
4. $-15 + z = 3$ **18**
5. $x - (-27) = -40$ **−67**
6. $j - 7 = -13$ **−6**
7. $-8y = 72$ **−9**
8. $-9k = -162$ **18**
9. $-42 = \frac{p}{3}$ **−126**
10. $\frac{n}{-30} = -6$ **180**

Solve by replacing the variables with the given values.

11. $P = 2(s + b)$, if $s = 4$ and $b = 16$ **40**
12. $S = 2\ell w + 2wh + 2\ell h$, if $\ell = 3$, $w = 5$, and $h = 8$ **158**
13. $E = mc^2$, if $E = 325$ and $c = 5$ **13**
14. $C = \frac{5}{9}(F - 32)$, if $F = 212$ **100**

Find the perimeter and area of each rectangle.

15.

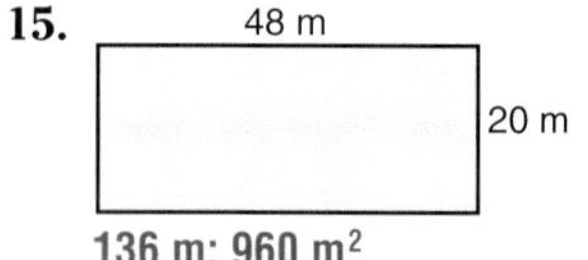

136 m; 960 m²

16.

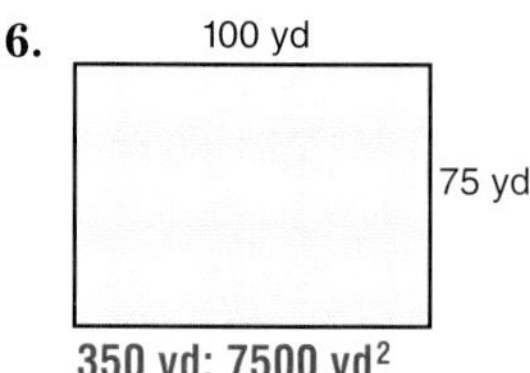

350 yd; 7500 yd²

17.

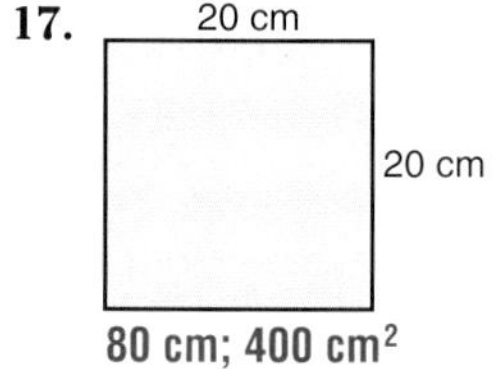

80 cm; 400 cm²

Solve each inequality and check your solution. Then graph the solution on a number line.

18–23. See margin for graphs.

18. $g - 8 > -5$ **$g > 3$**
19. $-61 \le w + 50$ **$-111 \le w$**
20. $17 \ge a - (-19)$ **$-2 \ge a$**
21. $3t \ge -42$ **$t \ge -14$**
22. $\frac{p}{-12} > 6$ **$p < -72$**
23. $-5c > 30$ **$c < -6$**

24. a + 2574 < 6000; There are no more than 3426 vertebrates.

Define a variable and write an equation or inequality. Then solve.

24. **Animals** According to the International Union for Conservation of Nature and Natural Resources, there are just under 6000 types of endangered species. If 2574 are invertebrates, what is the maximum number of vertebrates?
25. **Statistics** In 1990, there were 308,000 Cherokee Indians and 44,000 Creek Indians. How many times more Cherokee Indians are there than Creek Indians? **44,000k = 308,000; There are 7 times more Cherokee Indians than Creek Indians.**

Chapter 4 Practice Test

Using divisibility rules, state whether each number is divisible by 2, 3, 5, 6, or 10.

1. 315 **3, 5**
2. 400 **2, 5, 10**
3. 216 **2, 3, 6**

Write each multiplication expression using exponents. 4. $(-7)^5$

4. $(-7)(-7)(-7)(-7)(-7)$
5. 40 40^1
6. $x \cdot x \cdot x \cdot x \cdot x \cdot x \cdot x \cdot x$ x^8

Solve. Use a diagram.

7. Six teams are to play each other in a round-robin volleyball tournament, that is, each team will play all five of the other teams. How many games will be played? **15 games**

Factor each number or monomial completely.

8. 280 $2 \cdot 2 \cdot 2 \cdot 5 \cdot 7$
9. $882x^3$ $2 \cdot 3 \cdot 3 \cdot 7 \cdot 7 \cdot x \cdot x \cdot x$
10. $-696f^2g$ $-1 \cdot 2 \cdot 2 \cdot 2 \cdot 3 \cdot 29 \cdot f \cdot f \cdot g$

Find the GCF of each set of numbers or monomials.

11. 63, 231 **21**
12. $9xy$, $27x^3y^4$ $9xy$
13. $12z^4$, $72z^3$, $120z^2$ $12z^2$

Write each fraction in simplest form. If the fraction is already in simplest form, write *simplified*.

14. $\frac{95}{155}$ $\frac{19}{31}$
15. $\frac{120}{210}$ $\frac{4}{7}$
16. $\frac{65}{104}$ $\frac{5}{8}$

Replace each ● with <, >, or = to make a true statement.

17. $\frac{3}{7}$ ● $\frac{6}{10}$ $<$
18. $\frac{5}{11}$ ● $\frac{3}{5}$ $<$
19. $\frac{7}{15}$ ● $\frac{1}{3}$ $>$

Find each product or quotient. Express your answer in exponential form.

20. $(2x^4)(-15x^5)$ $-30x^9$
21. $\frac{q^6}{q}$ q^5
22. $18 \cdot 18^7$ 18^8

Write each expression using positive exponents.

23. $3^{-2}xy^{-3}$ $\frac{x}{3^2y^3}$
24. $(st)^{-1}$ $\frac{1}{st}$
25. $\frac{3^{-4}}{2^{-2}}$ $\frac{2^2}{3^4}$

Chapter 5 Practice Test

PRACTICE TEST

Express each decimal as a fraction or mixed number in simplest form.

1. -9.8 $-9\frac{4}{5}$ **2.** 0.125 $\frac{1}{8}$ **3.** $1.\overline{6}$ $1\frac{2}{3}$

Name the set(s) of numbers to which each number belongs.

4. $\frac{2}{3}$ R **5.** -7 I, R **6.** 51 W, I, R

Estimate each sum or difference. Sample answers given.

7. $\$10.03 + \$5.84$ \$16 **8.** $44.03 - 32.9$ 10 **9.** $20.3 + 59.7 + 62.8$ 140

Simplify each expression.

10. $4.75m - 6.2m + 9$ $-1.45m + 9$ **11.** $6.16q - 19 - (-7.32q)$ $13.48q - 19$

Solve each equation. Write the solution in simplest form.

12. $a = \frac{7}{9} + \frac{5}{9}$ $1\frac{1}{3}$ **13.** $\frac{27}{18} - \frac{9}{18} = h$ 1 **14.** $2\frac{3}{8} - 1\frac{5}{8} = p$ $\frac{3}{4}$

15. $\frac{1}{6} + \frac{5}{24} = p$ $\frac{3}{8}$ **16.** $5\frac{1}{2} - 2\frac{2}{3} = r$ $2\frac{5}{6}$ **17.** $\frac{-6}{7} - \left(-2\frac{10}{21}\right) = y$ $1\frac{13}{21}$

Solve each equation or inequality. Check your solution.

18. $u - 7.28 = 14.9$ 22.18 **19.** $a + \frac{2}{3} = 2$ $1\frac{1}{3}$

20. $-9.3 > y - (-4.8)$ $-14.1 > y$ **21.** $j + \frac{7}{12} \leq 7\frac{1}{2}$ $j \leq 6\frac{11}{12}$

22–23. Sample answers given. See students' explanations.

State whether each is an example of *inductive* or *deductive* reasoning. Explain your answer.

22. For the past 3 years of school when Joe wears his black shoes to play football, his team wins. On the nights Joe wears his white shoes, his team loses. Sherry noticed that Joe was wearing his white shoes, so she figured they were going to lose. inductive

23. At 32°F water freezes. It is 15° outside. Kelly knows that it is going to snow instead of rain. deductive

State whether each sequence is arithmetic. Then write the next three terms of each sequence.

24. $\frac{1}{2}, \frac{3}{4}, 1 \ldots$ yes; $1\frac{1}{4}, 1\frac{1}{2}, 1\frac{3}{4}$ **25.** 5040, 720, 120, 24, . . . no; 6, 2, 1

Chapter 6 Practice Test

PRACTICE TEST

Write each fraction as a decimal. Use a bar to show a repeating decimal.

1. $\frac{3}{8}$ 0.375
2. $\frac{-6}{11}$ $-0.\overline{54}$
3. $4\frac{8}{9}$ $4.\overline{8}$

Estimate each product or quotient. Sample answers given.

4. $14.34 \cdot 5.8$ 84
5. $81.2 \div 15.57$ 5
6. $\frac{3}{7} \cdot 98$ 49

Solve each equation. Write each solution in simplest form.

7. $k = -5\left(\frac{-4}{3}\right)$ $6\frac{2}{3}$
8. $1\frac{4}{5} \cdot -1\frac{5}{6} = n$ $-3\frac{3}{10}$
9. $t = \left(\frac{2}{5}\right)^3$ $\frac{8}{125}$

Evaluate each expression.

10. $a \div b$, if $a = \frac{-8}{9}$ and $b = \frac{-4}{3}$ $\frac{2}{3}$
11. $r \div s$, if $r = -1\frac{1}{2}$ and $s = \frac{21}{30}$ $-2\frac{1}{7}$

Solve each equation.

12. $(2.01)(0.04) = d$ 0.0804
13. $2.13 \div (-0.3) = a$ -7.1
14. $a = (81)(0.02)(1.5)$ 2.43
15. $x = 27.9 \div 0.31$ 90
16. $y = 2.3(-0.004)$ -0.0092
17. $-51.408 \div (-5.4) = g$ 9.52

Find the mean, median, and mode for each set of data. When necessary, round to the nearest tenth.

18. 36, 37, 41, 43, 43 40; 41; 43
19. 0.2, 0.4, 0.1, 0.6, 1.2, 1.1 0.6 ;0.5; none
20. 2, 8, 16, 21, 3, 8, 9, 6, 7 8.9; 8; 8
21. 44, 48, 55, 56, 55, 68, 70 56.6; 55; 55

Solve each equation or inequality. Check your solution.

22. $\frac{7}{8}p \geq -4\frac{1}{8}$ $p \geq -\frac{33}{7}$
23. $0.25w = 6\frac{4}{5}$ 27.2
24. $-0.5h < 12.5$ $h > -25$

State whether each sequence is a geometric sequence. If so, state the common ratio and list the next three terms.

25. 384, 96, 24, . . . yes; 0.25; 6, 1.5, 0.375
26. −7, 0, 7, 14, . . . no
27. 20, 16, 12.8, . . . yes; 0.8; 10.24, 8.192, 6.5536

Write each number in scientific notation.

28. 0.0021 2.1×10^{-3}
29. 87,500,000 8.75×10^{7}
30. 0.00000743 7.43×10^{-6}

Write each number in standard form.

31. 3.9×10^3 3900
32. 5.32×10^{-4} 0.000532
33. 7.02×10^0 7.02

Chapter 7 Practice Test

PRACTICE TEST

Solve.

1. Diana makes a salary of \$250 a week. In addition to her weekly salary, she receives a bonus of \$50 for every 25 boxes of books she sells. If she earned a total of \$450 in salary and bonuses this week, how many boxes did she sell? **100 boxes**
2. Frank walks 10 blocks to work every day. It takes him between 15 and 20 minutes to get to work. At this rate, what is the least amount of time it would take for Frank to walk 50 blocks? **75 minutes**

Solve each equation. Check your solution.

3. $25 = 2d - 9$ **17**
4. $4w + (-18) = -34$ **−4**
5. $\frac{p}{6} - (-21) = 8$ **−78**
6. $-7 = \frac{d}{-5} + 1$ **40**

Define a variable and write an equation for each situation. Then solve.

7. The quotient of a number and 8, decreased by 17, is −15.
8. Find the two consecutive odd integers whose sum is 76. **37 and 39**

7. $\left(\frac{x}{8}\right) - 17 = -15$; $x = 16$

Find the circumference of each circle.

9. 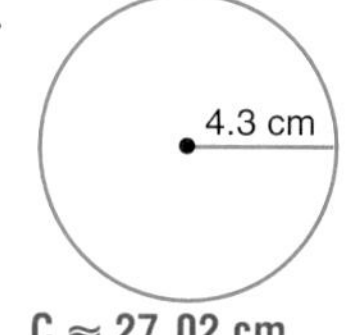

C ≈ 27.02 cm

10. 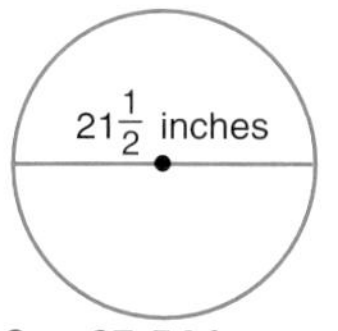

C ≈ 67.54 in.

Solve each equation or inequality. Check your solution. **12. all numbers 16. $14.5 \geq y$**

11. $7y + 6 = 3y - 14$ **−5**
12. $3(x + 2) - 6 = 3x$
13. $\frac{7}{9}d - 7 = \frac{5}{9}d - 3$ **18**
14. $-3j - 4 < -22$ **$j > 6$**
15. $-3.2 + 14s > 15s$ **$-3.2 > s$**
16. $3(y - 2) \geq 5(y - 7)$

Define a variable and write an inequality for each situation. Then solve.

17. Linda plans to spend at most \$85 on jeans and shirts. She bought 2 shirts for \$15.30 each. How much can she spend on jeans? **$2(15.30) + j \leq 85$; \$54.40**

Complete each sentence.

18. 6 m = _?_ cm **600**
19. 8.3 L = _?_ ml **8300**
20. 0.7 kg = _?_ g **700**

Chapter 8 Practice Test

Write the domain and range of each relation. Then determine whether each relation is a function. 1–6. See margin.

1. $\{(8, 9), (7, 6.5), (10, 6)\}$
2. $\{(3, 4), (5, 6), (5, 7), (8, 3)\}$
3. $\{(4, 6), (18, 40), (1.6, 4), (6, 6)\}$
4. $\{(10, 9), (4, 8.5), (17, 3), (4, 7)\}$
5. 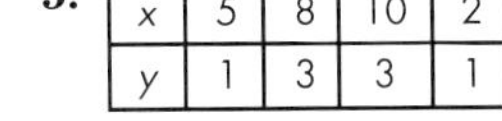

x	5	8	10	2
y	1	3	3	1

6. $\{(-4, 30), (-4, -5), (6, 10)\}$

7–10. Sample answers given. See students' explanations.

Determine whether a scatter plot of the data for the following might show a *positive*, *negative*, or *no* relationship. Explain your answer.

7. income, amount of clothes owned **positive**
8. hair color, weight **none**
9. month of birth, years of college **none**
10. outside temperature, frostbite **negative**

Find four solutions for each equation. Write your solution as ordered pairs. 11–13. See margin.

11. $x + y = 8$
12. $x = -2$
13. $y = \frac{1}{3}x - 2$

For each equation,
a. solve for the domain $\{-4, -2, 0, 2, 4\}$, and
b. determine if the equation is a function.

14. $y = 4x + 5$ −11, −3, 5, 13, 21; yes
15. $x = 9$ no
16. $x^2 - y = 6$ 10, −2, −6, −2, 10; yes

Solve by using a graph. Assume the rate is constant.

17. Of the top 25 collegiate football teams on Nov. 26, 1995, Duke University was rated the number 1 academic football college having a 95% graduation rate. Penn State University was rated 8th having a 77% rate. The Ohio State University was ranked 15th. What was OSU's graduation rate? 59% graduation rate

Find the slope of the line that contains each pair of points.

18. $K(3, 5), L(7, 4)$ $-\frac{1}{4}$
19. $G(9, -3), H(-1, 6)$ $-\frac{9}{10}$
20. $L(-3, -4), M(-5, 2)$ −3
21. $J(-6, 7), K(25, 4)$ $-\frac{3}{31}$

Use the *x*-intercept and *y*-intercept to graph each equation. 22–24. See margin.

22. $y = x - 6$
23. $y = -3 + x$
24. $y = -2x - 9$

Graph each equation using the slope and *y*-intercept. 25–27. See margin.

25. $y = \frac{1}{4}x - 3$
26. $2x - y = -1$
27. $x + 3y = 6$

Use a graph to solve each system of equations. 28–30. See Solutions Manual for graphs.

28. $y = 0.4x$, $y = 0.4x - 3$ no solution
29. $x + y = 2$, $2x - y = -2$ (0, 2)
30. $y = 3x - 6$, $-6x + 2y = -12$ infinitely many

Graph each inequality. 31–33. See Solutions Manual.

31. $y \geq 6$
32. $x + y < 4$
33. $y > \frac{1}{3}x + 3$

PRACTICE TEST

Additional Answers

1. D = {7, 8, 10}; R = {6, 6.5, 9}; yes
2. D = {3, 5, 8}; R = {3, 4, 6, 7}; no
3. D = {1.6, 4, 6, 18}; R = {4, 6, 40}; yes
4. D = {4, 10, 17}; R = {3, 7, 8.5, 9}; no
5. D = {2, 5, 8, 10}; R = {1, 3}; yes
6. D = {-4, 6}; R = {-5, 10, 30}; no

11–13. Sample answers given.

11. {(0, 8), (2, 6), (4, 4), (8, 0)}
12. {(-2, 6), (-2, 2), (-2, 0), (-2, -2)}
13. $\{(3, -1), (0, -2), (1, -1\frac{2}{3}), (-3, -3)\}$

22.

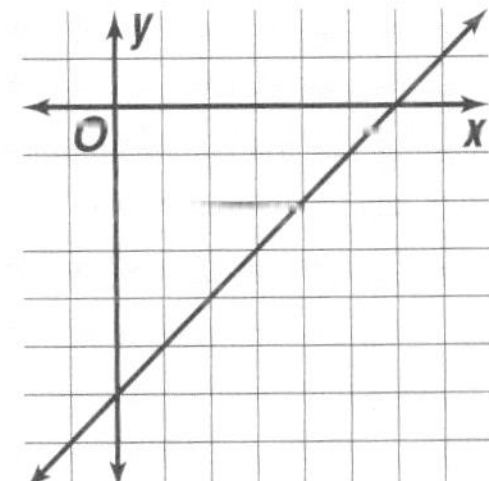

23.

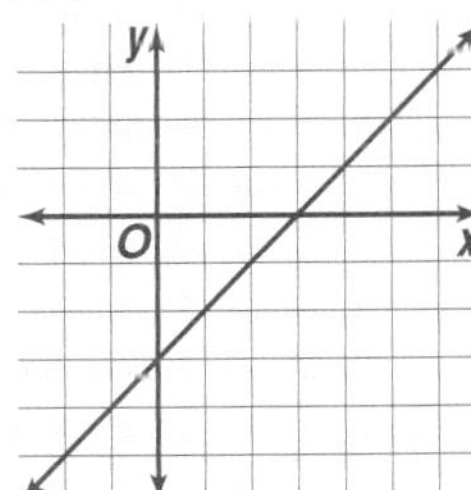

24.

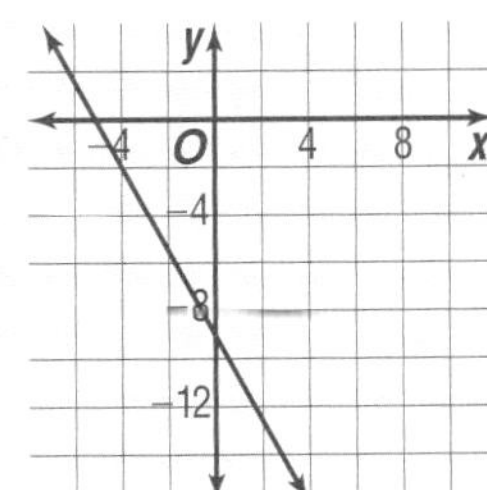

Additional Answers

25.

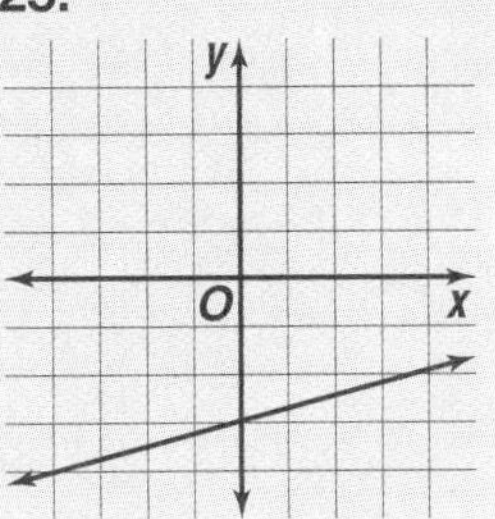

26.

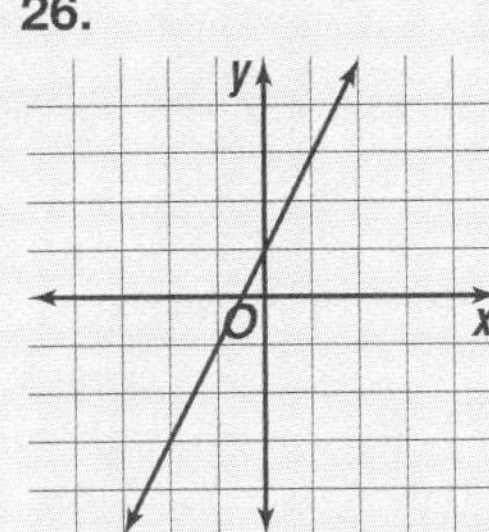

27.

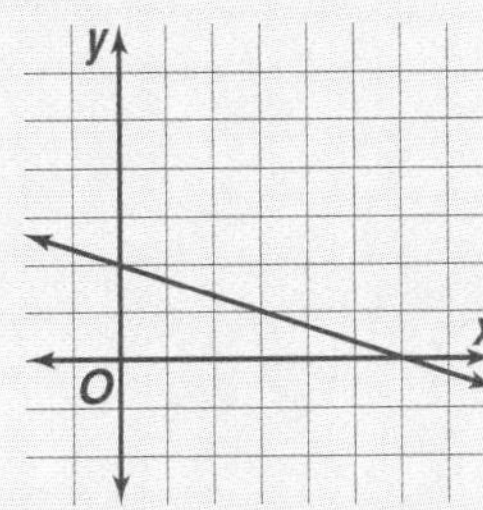

Chapter 9 Practice Test

PRACTICE TEST

Express each ratio or rate as a fraction in simplest form.

1. 16 pencils to 72 pencils $\frac{2}{9}$
2. 2 pints out of 1 gallon $\frac{2}{1}$
3. 88 out of 564 students $\frac{22}{141}$

Solve by making a table.

4. How many ways can you make change for a $50-bill using only $5-, $10-, and $20-bills? 12

A standard deck of playing cards is placed on the table. Suppose you select one card at random. Find each probability.

5. P(a red card) $\frac{1}{2}$
6. P(a six) $\frac{1}{13}$
7. P(a face card) $\frac{3}{13}$

Write a proportion that could be used to solve for each variable. Then solve the proportion.

8. 13 cans for $2.99
   x cans for $4.83 $\frac{13}{2.99} = \frac{x}{4.83}$; 21
9. 3 teaspoons for 24 cookies
   7 teaspoons for y cookies $\frac{3}{24} = \frac{7}{y}$; 56

Use the percent proportion to solve each problem.

10. What is 40% of 60? 24
11. Find 37.5% of 80. 30
12. Twenty-one is 35% of what number? 60
13. Seventy-five is what percent of 250? 30%
14. Fifty-two is what percent of 80? 65%
15. Thirty-six is 45% of what number? 80

Solve.

16. Eight of the 72 students surveyed at Perry Middle School said they would like to take a music appreciation class. The school's policy says there must be at least 30 interested students in the class. If there are 567 students in the school, predict how many students would be interested. Is this enough for the class? 63 students; yes

Express each decimal or fraction as a percent.

17. $\frac{85}{120}$ 71%
18. 0.42 42%
19. $\frac{9}{5}$ 180%
20. 0.086 8.6%

Express each percent as a decimal and as a fraction.

21. 67.5% 0.675; $\frac{27}{40}$
22. $33\frac{1}{3}\%$ $0.3\overline{3}$; $\frac{1}{3}$
23. 10.17% 0.1017; $\frac{1017}{10,000}$

Choose the best estimate.

24. 26% of 49 **a.** 13 **b.** 1.3 **c.** 130 a
25. 47% of 550 **a.** 26 **b.** 2.6 **c.** 260 c

Find the discount or interest to the nearest cent.

26. $1550 at 4% for 2 years $124
27. $138 dress, 15% off $20.70
28. $143 suit, 10% off $14.30
29. $1250 at 10.5% for 3 months $32.81

30–33. I = Increase; D = Decrease

State whether each percent of change is a percent of increase or a percent of decrease. Then find the percent of increase or decrease. Round to the nearest whole percent.

30. old: $32
    new: $19 D; 41%
31. old: $245
    new: $315 I; 29%
32. old: $178
    new: $201 I; 13%
33. old: $1010
    new: $1075 I; 6%

Chapter 10 Practice Test

The ages of Miss Rebar's seventh grade class were 13.3, 12.5, 12.8, 12.5, 12.8, 13.1, 13.2, 12.2, 12.9, 13.3, and 12.6.

1. Make a stem-and-leaf plot of the data. See margin.
2. How young was the youngest student? 12.2 years old

Use the data in the stem-and-leaf plot shown at the right.

2	5 8 8
3	2 3 3 6 9
4	2 3 5 7
5	1 1 3

$5|1 = 51$

3. What is the range? 28
4. Find the median. 39
5. What is the upper quartile? 47
6. What is the lower quartile? 32
7. Find the interquartile range. 15
8. What are the outliers, if any? 8. none
9. Make a box-and-whisker plot of the data. See margin.

The graphs below show how many sports cards Doug buys each month.

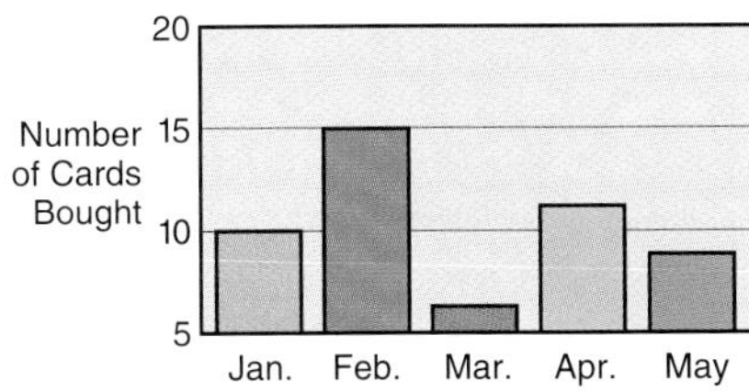

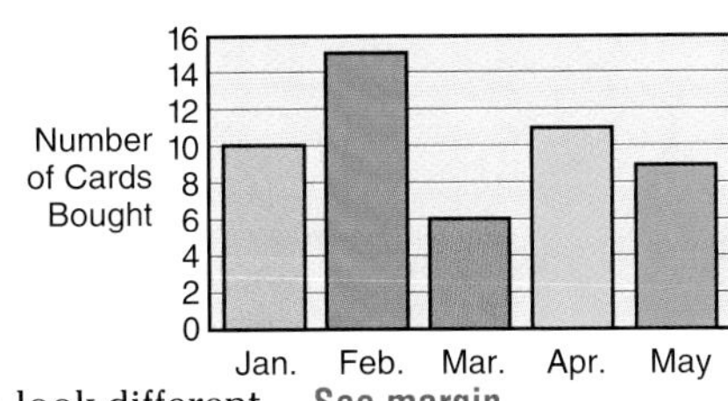

10. Explain why these graphs made from the same data look different. See margin.
11. If Doug wanted to convince his wife that he isn't buying very many cards which graph would he use? Explain. See margin.

Find the number of possible outcomes for each event. 12. 12 outcomes

12. A die is rolled and a coin is tossed.
13. Four dice are rolled. 1296 outcomes

Tell whether each situation represents a *permutation* or a *combination*. 15. combination

14. eight notes in a song permutation
15. eight new CDs from a group of thirty
16. five outfits from 15 to be used for a window display combination

Find the odds of each outcome if a fair eight-sided die is rolled.

17. A number less than five. 1 : 1
18. A composite number. 3 : 5

Heather makes three out of five 10-foot putts. To simulate her chances of making a putt, she puts 40 marbles in a box. A red marble represents a putt made, and a blue marble represents a putt missed. After a marble is drawn, it is replaced in the box.

19. How many blue marbles does Heather need? 16
20. In 120 drawings, how many red marbles can she expect? 72

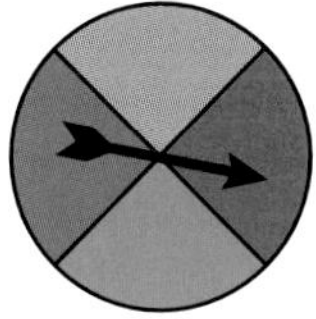

An eight-sided die is rolled and the spinner is spun. Find each probability.

21. P(green and 6) $\frac{1}{32}$
22. P(orange and an odd number) $\frac{1}{8}$

Determine whether each event is *mutually exclusive* or *inclusive*. Then find the probability. A letter of the alphabet is chosen. 23. inclusive; $\frac{6}{13}$

23. P(a vowel or a letter following R)
24. P(a consonant or the letter E) exclusive; $\frac{11}{13}$
25. P(a vowel or a consonant) exclusive; 1

PRACTICE TEST

Additional Answers

1.
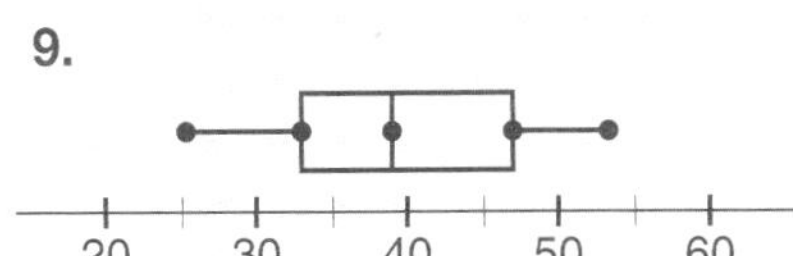

12	2 5 5 6 8 8 9
13	1 2 3 3

$12|2 = 12.2$

9.

10. The vertical axis is condensed in the graph on the right. Also, the vertical axis in the graph on the left starts at 5, so the graph is deceiving.
11. Doug would probably use the graph on the left because the bars are shorter. This makes it appear that he has bought fewer cards.

Additional Answers

1.

C

2.

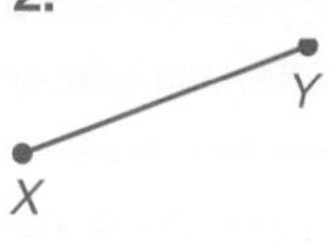

3.

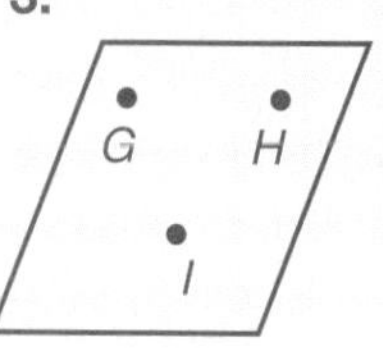

4.

25.

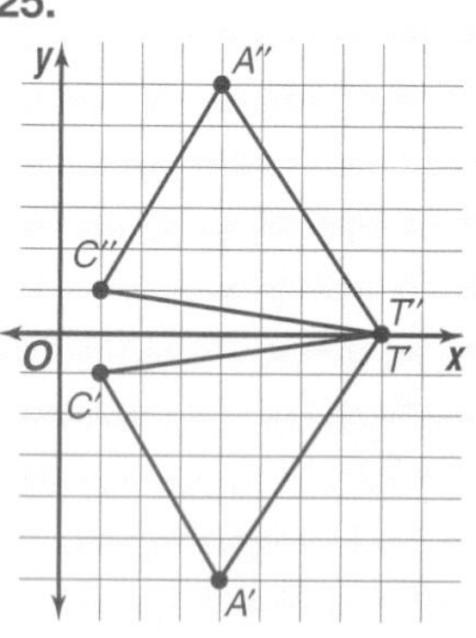

Chapter 11 Practice Test

PRACTICE TEST

Draw and label a diagram to represent each of the following. **1–4. See margin.**

1. point C
2. $\overline{XY}$
3. plane GHI
4. $\overrightarrow{RS}$
5. Make a circle graph to display the data in the chart at the right.

Water Usage in the United States	
Agriculture	36%
Public water	8%
Utilities	33%
Industry	23%

In the figure at the right, $\overline{KG} \parallel \overline{AE}$. If the measure of $\angle HFG$ is 45°, find the measure of each angle.

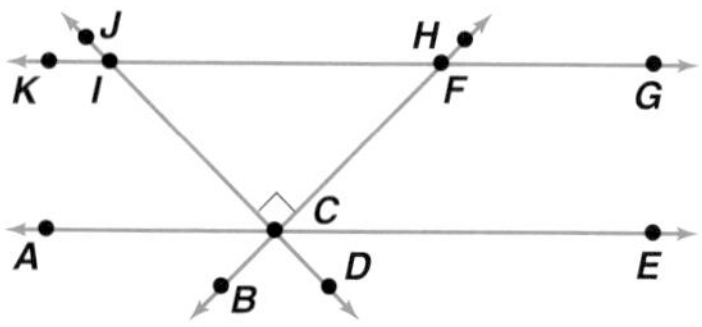

6. $\angle FCE$ **45°**
7. $\angle BCD$ **90°**
8. $\angle JIF$ **135°**
9. $\angle KIC$ **135°**

Find the value of x. Then classify the triangle as *acute*, *right*, or *obtuse*.

10. 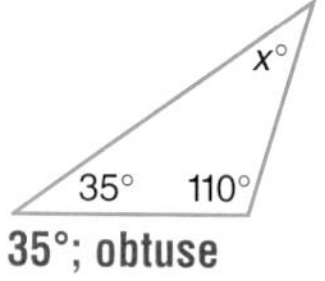

35°; obtuse

11. 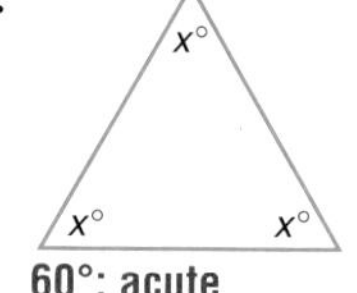

60°; acute

12. 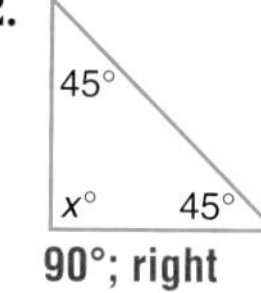

90°; right

If $\triangle XYZ \cong \triangle ABC$, name the part congruent to each angle or segment.

13. $\angle X$ **$\angle A$**
14. $\angle B$ **$\angle Y$**
15. $\overline{YZ}$ **$\overline{BC}$**
16. $\overline{AC}$ **$\overline{XZ}$**

Write a proportion to find the missing measure x. Then find the value of x.

17. **$\frac{12}{15} = \frac{x}{10}$; 8 m**

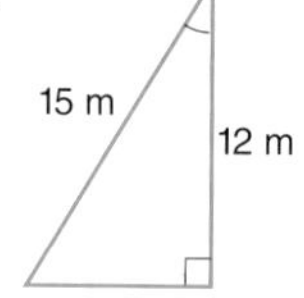

18. **$\frac{7}{6} = \frac{21}{x}$; 18 ft**

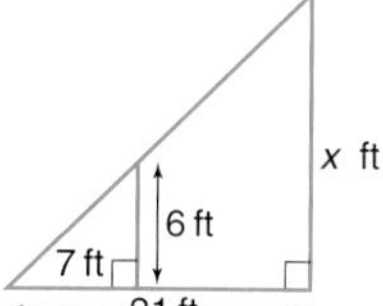

Find the value of x.

19. 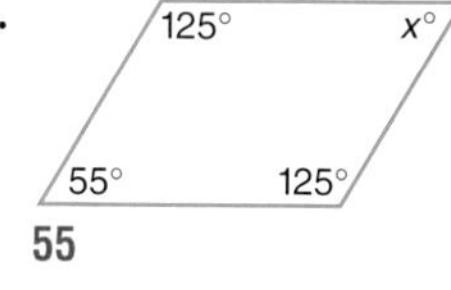

55

20. 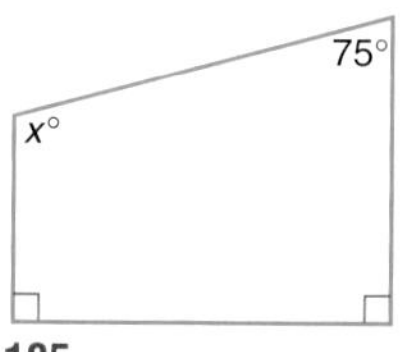

105

21. 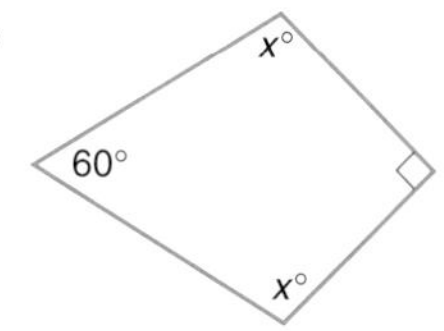

105

Find the sum of the measures of the interior angles of each polygon.

22. hexagon **720°**
23. 20-gon **3240°**
24. A translation moves $\triangle CAT$ 6 units to the right and 3 units up to form $\triangle C'A'T'$. Write the coordinates of each vertex of $\triangle C'A'T'$.
25. Draw $\triangle C'A'T'$ on a coordinate plane and graph its reflection if the x-axis is the line of reflection. **See margin.** **24. $C'(1, 1)$, $A'(4, 6)$, $T'(8, 0)$**

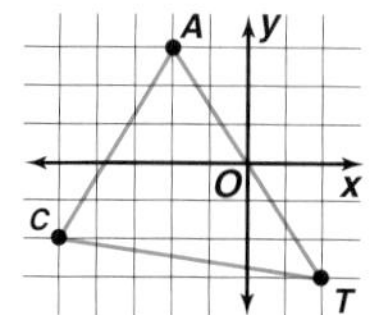

Chapter 12 Practice Test

Find the area of each figure.

1. 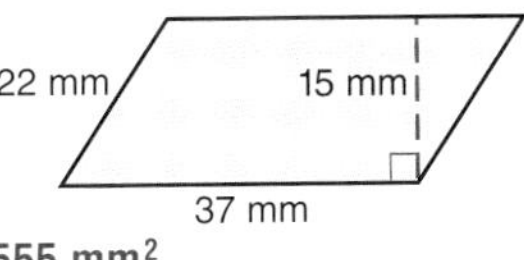

555 mm²

2. 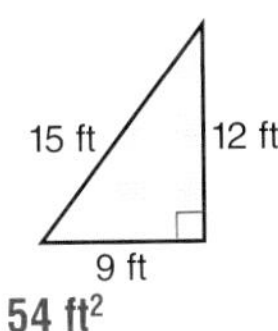

54 ft²

3. 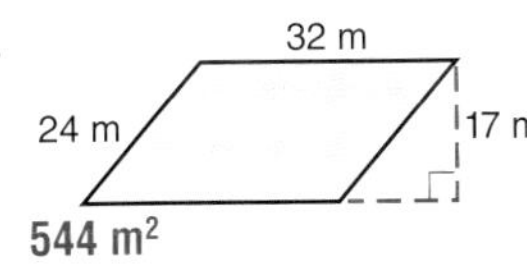

544 m²

Find the area of each circle. Round to the nearest tenth.

4. radius, 16 cm **804.2 cm²**
5. diameter, 42 in. **1385.4 in²**
6. radius, $3\frac{1}{2}$ in. **38.5 in²**

7. The figure at the right represents a dartboard. Find the probability of a dart landing in the shaded region. $\frac{2}{5}$

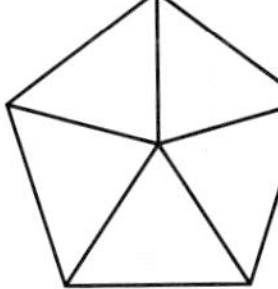

Solve by making a model or drawing.

8. Half of a garden is planted in corn. Half of the remaining garden is planted in strawberries. A third of the part not planted in corn or strawberries is planted in tomatoes. If 6 m² is planted in tomatoes, what is the area planted in corn? **36 m²**

Find the surface area of each solid.

9. 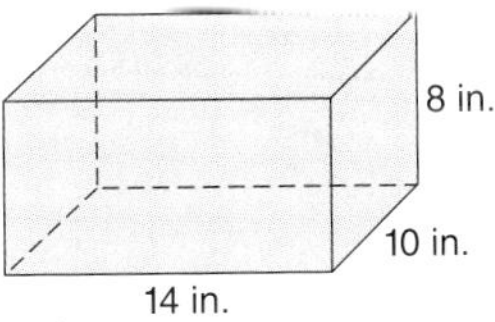

664 in²

10. 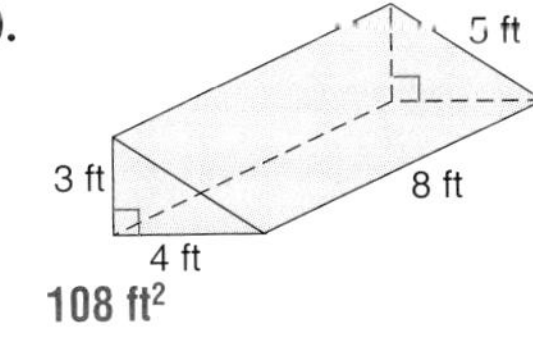

108 ft²

11. 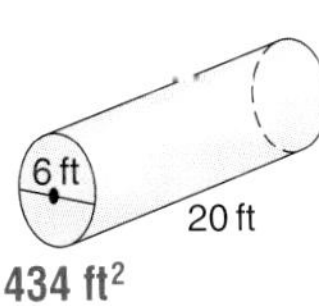

434 ft²

12. 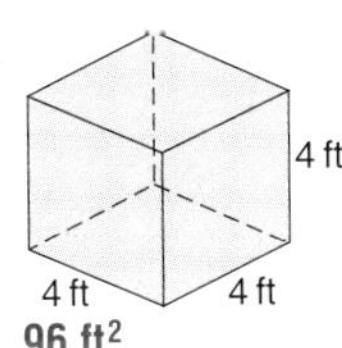

96 ft²

13. 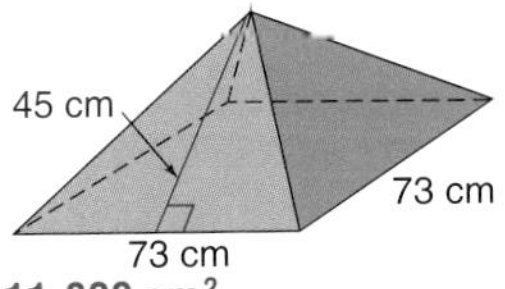

11,899 cm²

14. 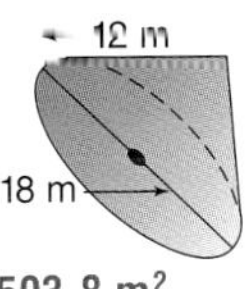

593.8 m²

Find the volume of each solid. Round to the nearest tenth.

15. rectangular prism: length, 4 m; width, 7 m; height, 9 m **252 m³**
16. triangular prism: base of triangle, $2\frac{1}{2}$ ft; altitude, 1 ft; prism height, 2 ft **2.5 ft³**
17. cylinder: diameter, 12 m; height, 9 m **1017.9 m³**
18. rectangular pyramid: length, 18 cm; width, 15 cm; height, 20 cm **1800 cm³**
19. square pyramid: length, 8 in.; height, 9 in. **192 in³**
20. circular cone: radius, 9 cm: height, 16 cm **1357.2 cm³**

Chapter 13 Practice Test

PRACTICE TEST

Find the best integer estimate for each square root. Then check your estimate using a calculator.

1. $\sqrt{90}$ 9 **2.** $\sqrt{20.25}$ 4 **3.** $-\sqrt{62}$ −8 **4.** $-\sqrt{71}$ −8

Solve using a Venn diagram.

5. Exquisite Interiors has 40 sample floor tiles in the shapes of various polygons. Twenty tiles are regular polygons, 14 tiles are quadrilaterals, and 5 tiles are squares. How many tiles are not regular polygons? 11 tiles

Solve each equation. Round decimal answers to the nearest tenth.

6. $s^2 = 121$ 11, −11 **7.** $g^2 = 40$ 6.3, −6.3 **8.** $f^2 = 11$ 3.3, −3.3 **9.** $y^2 = 60$ 7.7, −7.7

In a right triangle, if *a* and *b* are the measures of the legs and *c* is the measure of the hypotenuse, find each missing measure. Round decimal answers to the nearest tenth.

10. *b*, 77 m; *c*, 85 m 36 m **11.** *a*, 15 ft; *c*, 17 ft 8 ft **12.** *b*, 91 yd; *c*, 109 yd 60 yd

Find the lengths of the missing sides in each triangle. Round decimal answers to the nearest tenth.

13.
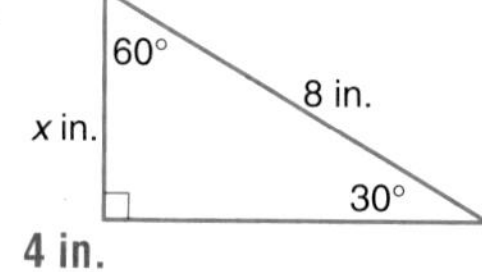

4 in.

14.
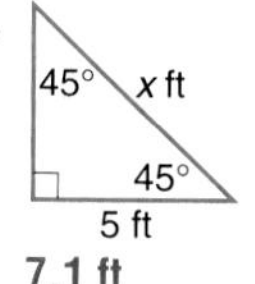

7.1 ft

15.
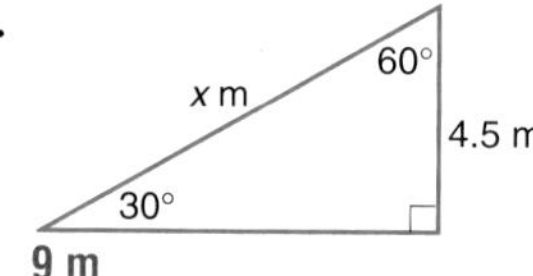

9 m

For each triangle, find the measure of the marked acute angle to the nearest degree.

16.
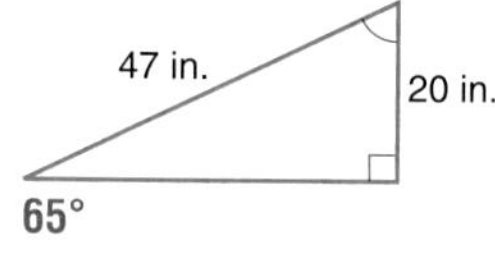

65°

17.
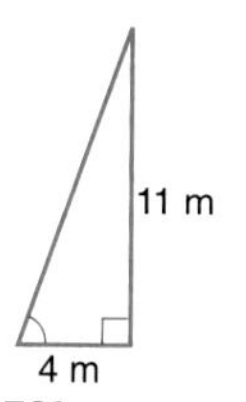

70°

18.
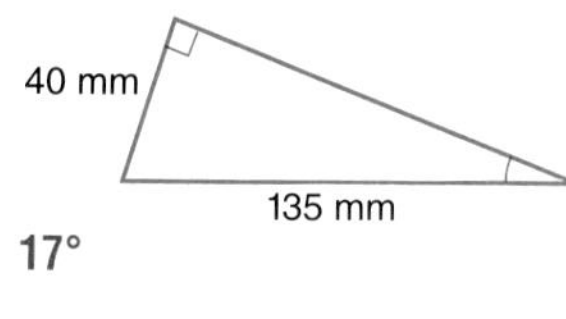

17°

Use trigonometric ratios to solve each problem.

19. What is the angle of elevation of the sun when a 100-foot water tower casts a shadow 165 feet long? 31.2°

20. Building Design A ramp is designed to help people in wheelchairs move more easily from one level to another. If a ramp 16 feet long forms an angle of 6° with the level ground, what is the vertical rise? 1.67 feet

Chapter 14 Practice Test

PRACTICE TEST

State whether each expression is a polynomial. If it is, identify it as a *monomial*, *binomial*, or *trinomial*.

1. $4a^2$ yes; monomial
2. $-2y + 3$ yes; binomial
3. $\frac{a}{5}$ yes; monomial
4. $r^3 - s^2$ yes; binomial
5. $-\frac{3}{(x + y)}$ no
6. $7mt^2 + 3m - 2t$ yes; trinomial

Find the degree of each polynomial.

7. $5a^2b$ 3
8. $12rs^2 + 3r^2s + 5r^4s$ 5
9. $x^2 + 6xy + xy^2$ 3

Find each sum or difference.

10. $(2x^2 + 5x) + (3x^2 + x)$ $5x^2 + 6x$
11. $(7a - 11b) - (3a + 4b)$ $4a - 15b$
12. $(x^2 - 5x + 3) + (4x - 3)$ $x^2 - x$
13. $(3a^2 - b^2 + c^2) - (a^2 + 2b^2)$ $2a^2 - 3b^2 + c^2$

14.
$$\begin{array}{r} -3y^2 + 2 \\ (+)\ 4y^2 - 5y - 2 \\ \hline y^2 - 5y \end{array}$$

15.
$$\begin{array}{r} 14a^2 - 3a \\ (-)\ 6a^2 + 5a + 17 \\ \hline 8a^2 - 8a - 17 \end{array}$$

Simplify.

16. $(5c)^3$ $125c^3$
17. $(4x^2y^3)^2$ $16x^4y^6$
18. $-3x(4xy)^2$ $-48x^3y^2$

Find each product. 20. $2a^5 + a^3 - 5a^2$ 21. $-2g^4 - 12g^2 - 6g$

19. $x(-5x + 3)$ $-5x^2 + 3x$
20. $a^2(2a^3 + a - 5)$
21. $-2g(g^3 + 6g + 3)$
22. $(2x + 1)(3x + 2)$ $6x^2 + 7x + 2$
23. $(3x + 3)(x - 1)$ $3x^2 - 3$
24. $(-x + 3)(3x - 2)$ $-3x^2 + 11x - 6$

Solve.

25. Geometry For the rectangular prism at the right, compute the area of the top, bottom, front, back, left, and right sides. Then add to find the total surface area. $10n^2$, $10n^2$, $5n^2 + 15n$, $5n^2 + 15n$, $2n^2 + 6n$, $2n^2 + 6n$; $34n^2 + 42n$

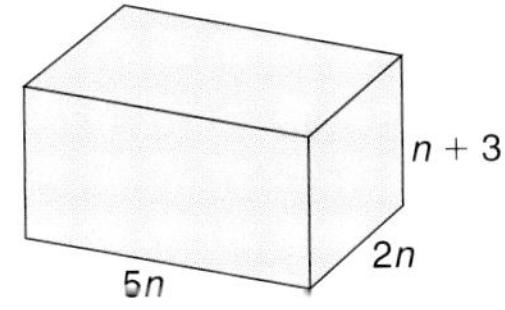

GLOSSARY

absolute value (67) The number of units a number is from zero on the number line.

acute angle (550) An angle with a measure greater than 0° and less than 90°.

acute triangle (569) A triangle that has three acute angles.

addition property of equality (125) If you add the same number to each side of an equation, the two sides remain equal. For any numbers a, b, and c, if $a = b$, then $a + c = b + c$.

addition property of inequality (147) Adding the same number to each side of an inequality does not change the truth of the inequality. For all numbers a, b, and c
1. If $a > b$, then $a + c > b + c$.
2. If $a < b$, then $a + c < b + c$.

additive inverses (89) An integer and its opposite. The sum of an integer and its additive inverse is zero.

adjacent angles (561) Two angles that have a common side and the same vertex, but they do *not* overlap.

adjacent leg (687) The leg of a right triangle that is not opposite an angle and not the hypotenuse.

algebraic expression (16) A combination of variables, numbers, and at least one operation.

algebraic fraction (197) A fraction with variables in the numerator or denominator.

alternate exterior angles (563) Nonadjacent exterior angles found on opposite sides of the transversal.
$\angle 1$ and $\angle 7$, and $\angle 2$ and $\angle 8$ are alternate exterior angles.

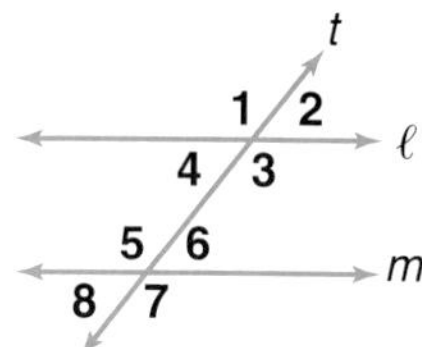

alternate interior angles (563) Nonadjacent interior angles found on opposite sides of the transversal.
$\angle 4$ and $\angle 6$, and $\angle 3$ and $\angle 5$ are alternate interior angles.

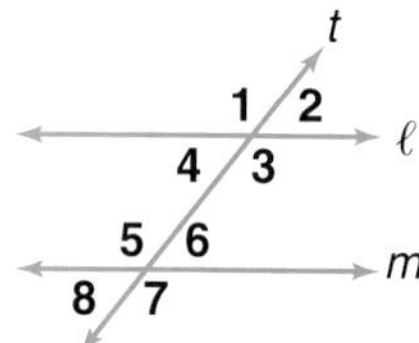

altitude (612, 614, 633) A line segment that is perpendicular to the base of a figure with endpoints on the base and the side opposite the base.

angle (549) Two rays with a common endpoint form an angle. The rays and vertex are used to name an angle.

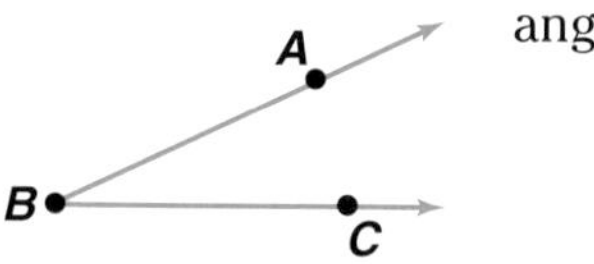

angle ABC or $\angle ABC$

angle of depression (695) An angle of depression is formed by a line of sight along the horizontal and another line of sight below it.

angle of elevation (694) An angle of elevation is formed by a line of sight along the horizontal and another line of sight above it.

area (138) The number of square units needed to cover a surface enclosed by a geometric figure.

arithmetic sequence (258) A sequence in which the difference between any two consecutive terms is the same.

associative property of addition (22, 234) For any rational numbers a, b, and c,
$$(a + b) + c = a + (b + c).$$

associative property of multiplication (22, 296) For any numbers a, b, and c,
$$(ab)c = a(bc).$$

axes (73) Two intersecting number lines that form a coordinate system.

back-to-back stem-and-leaf plot (487) Used to compare two sets of data. The leaves for one set of data are on one side of the stem and the leaves for the other set of data are on the other side.

bar graph (52) A graphic form using bars to make comparisons of statistics.

bar notation (225) In repeating decimals the line or bar placed over the digits that repeat. For example, $2.\overline{63}$ indicates the digits 63 repeat.

base (175) In 5^3, the base is 5. The base is used as a factor as many times as given by the exponent (3). That is,
$$5^3 = 5 \times 5 \times 5.$$

base (449) In a percent proportion, the number to which the percentage is compared.
$$\frac{\text{percentage}}{\text{base}} = \text{rate}$$

base (614) The base of a rectangle, a parallelogram, or a triangle is any side of the figure. The bases of a trapezoid are the parallel sides.

base (632) The bases of a prism are the two parallel congruent sides.

binomial (706) A polynomial with exactly two terms.

boundary (418) A line that separates a graph into half-planes.

box-and-whisker plot (495) A diagram that summarizes data using the median, the upper and lower quartiles, and the extreme values. A box is drawn around the quartile value and whiskers extend from each quartile to the extreme data points.

center (341) The given point from which all points on the circle are the same distance.

circle (341) The set of all points in a plane that are the same distance from a given point called the center.

circle graph (497, 556) A type of statistical graph used to compare parts of a whole.

circular cone (639) A shape in space that has a circular base and one vertex.

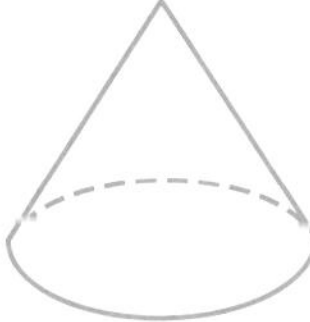

circumference (341) The distance around a circle.

coefficient (711) The numerical part of a monomial.

combination (516) An arrangement or listing in which order is not important.

commission (470) A fee paid to a sales associate based on a percent of the total sales.

common difference (258) The difference between any two consecutive terms in an arithmetic sequence.

common multiples (200) Multiples that are shared by two or more numbers. For example, some common multiples of 2 and 3 are 6, 12, and 18.

common ratio (312) The ratio between any two successive terms in a geometric sequence.

commutative property of addition (22, 234) For any rational numbers a and b,

$$a + b = b + a.$$

commutative property of multiplication (22, 296) For any rational numbers a and b, $ab = ba$.

comparative graph (498) A graph used to compare results of similar groups to show trends.

compass (554) An instrument used for drawing circles or parts of circles.

compatible numbers (280) Numbers that have been rounded so, when the numbers are divided by each other, the remainder is zero.

complementary (562) Two angles are complementary if the sum of their measures is 90°.

composite number (184) Any whole number greater than 1 that has more than two factors.

congruent (554, 561, 573) Line segments that have the same length, or angles that have the same measure, or figures that have the same size and shape.

constant (707) A monomial that does not contain a variable.

coordinate (67, 126) A number associated with a point on the number line.

coordinate system (36, 73) A coordinate plane separated into four equal regions by two intersecting number lines called axes.

corresponding angles (563) Angles that have the same position on two different parallel lines cut by a transversal. $\angle 1$ and $\angle 5$, $\angle 2$ and $\angle 6$, $\angle 3$ and $\angle 7$, and $\angle 4$ and $\angle 8$ are corresponding angles.

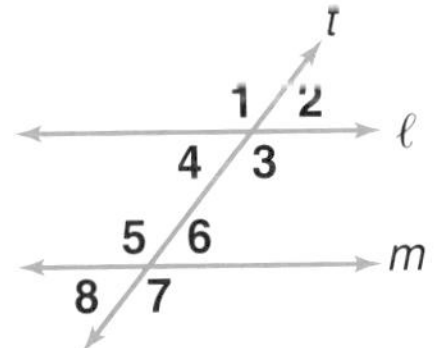

corresponding parts (573) Parts on congruent or similar figures that match.

cosine (688) If $\triangle ABC$ is a right triangle and A is an acute angle,

$$\text{cosine } \angle A = \frac{\text{measure of the side adjacent to } \angle A}{\text{measure of the hypotenuse}}$$

$$\cos A = \frac{b}{c}$$

cylinder (633) A 3-dimensional shape that has two parallel, congruent bases (usually circular) and a curved side connecting the bases.

data (50) Numerical information gathered for statistical purposes.

defining a variable (42) Choosing a variable and a quantity for the variable to represent in an equation.

degree (549) The most common unit of measure for angles.

dependent events (532) Two or more events in which the outcome of one event does affect the outcome of the other event(s).

diagonal (590) A line segment that joins two nonconsecutive vertices of a polygon.

diameter (341) The distance across a circle through its center.

discount (468) The amount deducted from the original price.

discrete mathematics (258) A branch of mathematics that deals with discontinuous or finite sets of numbers. Some fields that deal with discrete mathematics are logic and statistics.

distributive property (26) For any numbers a, b, and c, $a(b + c) = ab + ac$ and $(b + c)a = ba + ca$.

divisible (170) A number is divisible by another if, upon division, the remainder is zero.

division property of equality (130) For any numbers a, b, and c, with $c \neq 0$, if $a = b$, then $\frac{a}{c} = \frac{b}{c}$.

division properties of inequality (152) For all integers a, b, and c, where $c > 0$, if $a > b$, then $\frac{a}{c} > \frac{b}{c}$. For all integers a, b, and c, where $c < 0$, if $a > b$, then $\frac{a}{c} < \frac{b}{c}$.

domain (372) The set of all first coordinates from each ordered pair.

empty set (347) A set with no elements shown by the symbol { } or ∅.

equation (32) A mathematical sentence that contains the equals sign, =.

equiangular (591) All angles are congruent.

equilateral (591) All sides are congruent.

equivalent (275) The same value represented in different forms. For example, $\frac{3}{5}$ and 0.6 are equivalent.

equivalent equations (124) Two or more equations with the same solution. For example, $x + 3 = 5$ and $x = 2$ are equivalent equations.

evaluating an expression (11) Replacing the variables with numbers and finding the numerical value of the expression.

expanded form (176) A number expressed using place value to write the value of each digit in the number.

experimental probability (529) An estimated probability based on the relative frequency of positive outcomes occurring during an experiment.

exponent (175) In 5^3, the exponent is 3. The exponent tells how many times the base, 5, is used as a factor.

$$5^3 = 5 \times 5 \times 5$$

exterior angle (591) The angle formed when a side of a polygon is extended.

face (632) A face of a prism is any surface that forms a side or a base of the prism.

factorial (515) The expression $n!$ is the product of all counting numbers beginning with n and counting backward to 1.

factors (170) Numbers that divide into a whole number with a remainder of zero.

factor trees (185) A diagram showing the prime factorization of a composite number. The factors branch out from the previous factors until all the factors are prime numbers.

Fibonacci sequence (263) A list of numbers in which the first two numbers are both 1 and each number that follows is the sum of the previous two numbers, 1, 1, 2, 3, 5, 8, . . .

formula (134) An equation that states a rule for the relationship between certain quantities.

45°–45° right triangle (683) A right triangle that has two angles that each measure 45°.

frequency table (51) A chart that indicates the number of values in each interval.

function (373) A function is a relation in which each element of the domain is paired with exactly one element in the range.

functional notation (393) Equations that represent functions can be written in *functional notation*, $f(x)$. The symbol $f(x)$ is read "f" of "x" and represents the value in the range that corresponds to the value of x in the domain.

Fundamental counting principle (509) If event M can occur in m ways and is followed by event N that can occur in n ways, then the event M followed by event N can occur in $m \cdot n$ ways.

geometric probability (623) Probability that uses area to find the probability of an event.

geometric sequence (312) A sequence in which the ratio between any two successive terms is the same.

gram (359) The basic unit of mass in the metric system.

graph (73, 126) A dot marking a point that represents a number on a number line or an ordered pair on a coordinate plane.

greatest common factor (GCF) (190) The GCF of two or more integers is the greatest factor that is common to each of the integers.

histogram (502) A bar graph whose columns are next to each other to show how the data are distributed.

hypotenuse (676) The side opposite the right angle in a right triangle.

identity property of addition (23, 234) For any number a, $a + 0 = a$ and $0 + a = a$.

identity property of multiplication (23, 296) For any number a, $a \times 1 = a$.

inclusive (536) When two events can happen at the same time.

independent events (531) Two or more events in which the outcome of one event does *not* affect the outcome of the other event(s).

indirect measurement (580) Finding a measurement by using similar triangles and writing a proportion.

inequality (46) A mathematical sentence that contains $<$, $>$, $\neq$, $\leq$, or $\geq$.

integers (66, 224) The whole numbers and their opposites.

$$\ldots, -3, -2, -1, 0, 1, 2, 3, \ldots$$

intercepts (406) The points where a graph crosses the axes.

interior angle (591) An angle inside a polygon.

interquartile range (491) The range of the middle half of a set of numbers.

Interquartile range = UQ − LQ.

intersect (552) Lines that cross or touch each other at a point.

inverse operations (41) Operations that undo each other, such as addition and subtraction.

inverse property of addition (234) For every rational number a, $a + (-a) = 0$.

inverse property of multiplication (289) For every nonzero number $\frac{a}{b}$, where $a, b \neq 0$, there is exactly one number $\frac{b}{a}$ such that $\frac{a}{b} \times \frac{b}{a} = 1$.

irrational number (672) A number that cannot be expressed as $\frac{a}{b}$, where a and b are integers and b does not equal 0. For example, $\sqrt{3} = 1.7320508\ldots$ is an irrational number.

isosceles triangle (569) A triangle that has at least two congruent sides.

L

lateral surface (639) The lateral surface of a prism, cylinder, pyramid, or cone is all the surface of the figure except the base or bases.

least common denominator (LCD) (201) The least common multiple of the denominators of two or more fractions.

least common multiple (LCM) (200) The least of the nonzero common multiples of two or more numbers. The LCM of 2 and 3 is 6.

leaves (486) The next greatest place value forms the leaves in a stem-and-leaf plot.

leg (676) Either of the two sides that form the right angle of a right triangle.

like terms (27, 711) Expressions that contain the same variables, such as $3ab$ and $7ab$.

line (548) A never-ending straight path. A representation of a line CD ($\overleftrightarrow{CD}$) is shown below.

linear equation (386) An equation for which the graph is a straight line.

line graph (497) A type of statistical graph used to show how values change over a period of time.

line segment (548) Two endpoints and the straight path between them. A line segment is named by its endpoints. A representation of line segment ST ($\overline{ST}$) is shown below.

liter (359) The basic unit of capacity in the metric system.

lower quartile (491) The median of the lower half of a set of numbers indicated by LQ.

mean (300) The sum of the numbers in a set of data divided by the number of pieces of data.

measures of central tendency (300) Numbers or pieces of data that can represent the whole set of data.

median (300) In a set of data, the median is the number in the middle when the data are organized from least to greatest.

meter (359) The basic unit of length in the metric system.

metric system (360) A system of measurement using the basic units: meter for length, gram for mass, and liter for capacity.

mixed number (239) The sum of a whole number and a fraction. For example, $3\frac{1}{2}$.

mode (300) The number or item that appears most often in a set of data.

monomial (172) An expression that is either a real number, a variable, or a product of real numbers and variables.

multiple (200) The product of the number and any whole number.

multiplication properties of inequality (152)

1. For all integers a, b, and c, where $c > 0$, if $a > b$, then $ac > bc$.
2. For all integers a, b, and c, where $c < 0$, if $a > b$, then $ac < bc$.

multiplication property of equality (130) For any numbers a, b, and c, if $a = b$, then $ac = bc$.

multiplicative inverses (289) Two numbers whose product is 1. The multiplicative inverse of $\frac{3}{4}$ is $\frac{4}{3}$.

multiplicative property of zero (23) For any number a, $a \times 0 = 0$.

mutually exclusive (535) Two or more events such that no two events can happen at the same time.

negative (66) A value less than zero.

negative relationship (380) If points in a scatter plot appear to suggest a line that slants downward to the right, there is a *negative relationship*.

net (632) The shape that is formed by "unfolding" a prism. The net shows all the faces that make up the surface area of a prism.

no relationship (380) Points that seem to be random in a scatter plot are said to have no relationship.

null set (347) A set with no elements shown by the symbol { } or ∅.

obtuse angle (550) Any angle that measures greater than 90° but less than 180°.

obtuse triangle (569) A triangle with one obtuse angle.

odds (520) A way to describe the chance of an event occurring.

odds in favor = number of successes:number of failures

odds against = number of failures:number of successes

open sentence (32) An equation that contains variables.

opposite side (687) The side that is not part of the angle in a right triangle.

ordered pair (37) A pair of numbers used to locate a point in the coordinate system.

order of operations (12, 176)

1. Simplify the expressions inside grouping symbols.
2. Evaluate all powers.
3. Then do all multiplications and divisions from left to right.
4. Then do all additions and subtractions from left to right.

origin (36) The point of intersection of the x-axis and y-axis in a coordinate system.

outcomes (509) Possible results of a probability event. For example, 4 is an outcome when a die is rolled.

outliers (496) Data that are more than 1.5 times the interquartile range from the quartiles.

parallel (551) Lines in the same plane that do not intersect. The symbol ∥ means parallel.

parallelogram (585) A quadrilateral with two pairs of parallel sides.

percent (449) A ratio with a denominator of 100. For example, 77% and $\frac{77}{100}$ name the same number.

percentage (449) In a percent proportion, a number that is compared to another number called the base.

$$\frac{\text{percentage}}{\text{base}} = \text{rate}$$

percent of change (472) The ratio of the amount of change to the original amount.

percent of decrease (473) The ratio of an amount of decrease to the previous amount, expressed as a percent.

percent of increase (472) The ratio of an amount of increase to the previous amount, expressed as a percent.

percent proportion (450)

$$\frac{\text{percentage}}{\text{base}} = \text{rate or } \frac{P}{B} = \frac{r}{100}$$

perfect squares (665) Rational numbers whose square roots are whole numbers. 25 is a perfect square because $\sqrt{25} = 5$.

perimeter (138) The distance around a geometric figure.

permutation (515) An arrangement or listing in which order is important.

perpendicular (551) Two lines that intersect to form a right angle.

pi, π (341) The ratio of the circumference of a circle to the diameter of a circle. Approximations for π are 3.14 and $\frac{22}{7}$.

pictograph (497) A type of statistical graph that uses pictures or illustrations to show how specific quantities compare. Each symbol represents a specific number.

plane (550) A flat surface that has no edges or boundaries. A plane can be named by a single uppercase script letter or by using any three noncollinear points of the plane.

point (548) A specific location in space.

polygon (589) A simple closed figure in a plane formed by three or more line segments.

polynomial (706) A monomial or the sum or difference of two or more monomials.

positive relationship (380) In a scatter plot, if the points appear to suggest a line that slants upward to the right, there is a *positive relationship*.

power (175) A number that can be written using an exponent.

power of a monomial (721) For all numbers a and b, and positive integer m, n, and p, $(a^m b^n)^p = a^{mp} b^{np}$.

power of a power (719) For any number a and positive integers m and n, $(a^m)^n = a^{mn}$.

power of a product (720) For all numbers a and b, and positive integer m, $(ab)^m = a^m b^m$.

prime factorization (185) Expressing a composite number as the product of its prime factors.

prime number (184) A whole number greater than 1 that has exactly two factors , 1 and itself.

principal (467) The amount of money in an account.

prism (632) A figure in space that has two parallel and congruent bases in the shape of polygons.

probability (440) The chance that some event will happen. It is the ratio of the number of ways a certain event can occur to the number of possible outcomes.

$$\text{probability} = \frac{\text{number of favorable outcomes}}{\text{number of possible outcomes}}$$

probability of inclusive events (536) The probability of one or the other of two inclusive events can be found by adding the probability of the first event to the probability of the second event and subtracting the probability of both events happening.
$P(A \text{ or } B) = P(A) + P(B) - P(A \text{ and } B)$

probability of mutually exclusive events (535) The probability of one or the other of two mutually exclusive events can be found by adding the probability of the first event to the probability of the second event.
$P(A \text{ or } B) = P(A) + P(B)$

probability of two dependent events (532) If two events A and B are dependent, then the probability of both events occurring is the product of the probability of A and the probability of B after A occurs.
$P(A \text{ and } B) = P(A) \times P(B \text{ following } A)$

probability of two independent events (531) The probability of two independent events can be found by multiplying the probability of the first event by the probability of the second event.
$P(A \text{ and } B) = P(A) \times P(B)$

product of powers (205) For any number a and positive integers m and n, $a^m \times a^n = a^{m+n}$.

property of proportions (444)
If $\frac{a}{b} = \frac{c}{d}$, then $ad = bc$.
If $ad = bc$, then $\frac{a}{b} = \frac{c}{d}$.

proportion (444) A statement of equality of two or more ratios.

protractor (549) An instrument used to measure angles.

pyramid (638) A solid figure that has a polygon for a base and triangles for sides.

Pythagorean theorem (676) In a right triangle, the square of the length of the hypotenuse is equal to the sum of the squares of the lengths of the legs. $c^2 = a^2 + b^2$

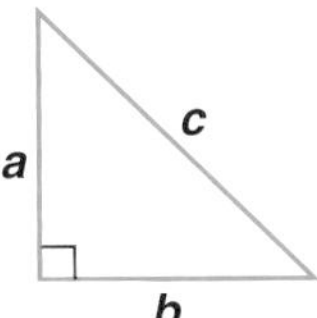

quadrants (73) The four regions into which two perpendicular number lines separate the plane.

quadrilateral (584, 585) A polygon having four sides.

quartile (491) One of four equal parts of data from a large set of numbers.

quotient of powers (206) For any nonzero number a and whole numbers m and n, $\frac{a^m}{a^n} = a^{m-n}$.

radical sign (664) The symbol used to indicate a nonnegative square root. $\sqrt{\ }$

radius (341) The distance from the center of a circle to any point on the circle.

range (372) The range of a relation is the set of all second coordinates from each ordered pair.

range (490) The difference between the least and greatest numbers in the set.

rate (433) A ratio of two measurements having different units.

rate (449) In a percent proportion the ratio of a number to 100.

ratio (196, 432) A comparison of two numbers by division. The ratio of 2 to 3 can be stated as 2 out of 3, 2 to 3, 2:3, or $\frac{2}{3}$.

rational numbers (224) Numbers of the form $\frac{a}{b}$ where a and b are integers and $b \neq 0$.

ray (548) A part of a line that extends indefinitely in one direction.

real numbers (673) The set of rational numbers together with the set of irrational numbers.

reciprocal (289) Another name for a multiplicative inverse.

rectangle (585) A quadrilateral with four congruent angles.

rectangular prism (632) A prism with rectangular bases.

reflection (595) A type of transformation where a figure is flipped over a line.

regular polygon (591) A polygon having all sides congruent and all angles congruent.

relation (372) A set of ordered pairs.

repeating decimal (225) A decimal whose digits repeat in groups of one or more. Examples are 0.181818 . . . and 0.8333. . . .

rhombus (585) A parallelogram with four congruent sides.

right angle (550) An angle that measures 90°.

right triangle (569) A triangle with one right angle.

rotation (595) When a figure is turned around a point.

GLOSSARY

sample (50, 51) A group that is used to represent a whole population.

sample space (441) The set of all possible outcomes.

scale drawing (585) A drawing that is similar but either larger or smaller than the actual object.

scalene triangle (569) A triangle with no congruent sides.

scatter plot (378) In a scatter plot, two sets of data are plotted as ordered pairs in the coordinate plane.

scientific notation (317) A way of expressing a number as the product of a number that is at least 1 but less than 10 and a power of 10. For example, $687{,}000 = 6.87 \times 10^5$.

sequence (258) A list of numbers in a certain order, such as, 0, 1, 2, 3, or 2, 4, 6, 8.

sides (549) The two rays that form an angle are called the sides.

sides (589) The line segments that form a polygon.

similar (578) Figures that have the same shape but not necessarily the same size.

simplest form (28) An expression in simplest form has no like terms and no parentheses.

simplest form (196) A fraction is in simplest form when the GCF of the numerator and the denominator is 1.

simulation (524) The process of acting out a problem.

sine (688) If $\triangle ABC$ is a right triangle and A is an acute angle,

$$\text{sine } \angle A = \frac{\text{measure of the side opposite } \angle A}{\text{measure of the hypotenuse}}$$

$$\sin A = \frac{a}{c}$$

skew lines (551) Two lines that do not intersect and are not in the same plane.

slant height (638) The length of the altitude of a lateral face of a regular pyramid.

slope (400) The slope of a line is the ratio of the change in y to the corresponding change in x.

$$\text{slope} = \frac{\text{change in } y}{\text{change in } x}$$

slope-intercept form (407) The slope-intercept form of a linear equation of a line is $y = mx + b$. The slope of the line is m, and the y-intercept is b.

solution (32) A value for the variable that makes an equation true. The solution for $12 = x + 7$ is 5.

solution of system (412) The ordered pair that is the solution of both equations.

solving an equation (32) The process of finding a solution to an equation.

square (585) A parallelogram with all sides congruent and all angles congruent.

square pyramid (638) A pyramid with a square base.

square root (664) One of the two equal factors of a number. A square root of 144 is 12 since $12^2 = 144$.

standard form (176) The standard form for seven hundred thirty-nine is 739.

statistics (51) Information that has been collected, analyzed, and presented in an organized fashion.

stem-and-leaf plot (486) A system used to condense a set of data where the greatest place value of the data forms the stem and the next greatest place value forms the leaves.

stems (486) The greatest place value common to all the data values is used for the stem of a stem-and-leaf plot.

straightedge (554) Any object that can be used to draw a straight line.

substitution property of equality (17) For all numbers a and b, if $a = b$, then a may be replaced with b.

subtraction property of equality (124) If you subtract the same number from each side of an equation, the two sides remain equal. For any numbers a, b, and c, if $a = b$, then $a - c = b - c$.

subtraction property of inequality (147) Subtracting the same number from each side of an inequality does not change the truth of the inequality.
For all numbers a, b, and c:
1. If $a > b$, then $a - c > b - c$.
2. If $a < b$, then $a - c < b - c$.

supplementary (562) Two angles are supplementary if the sum of their measures is $180°$.

surface area (632) The sum of the areas of all the surfaces (faces) of a 3-dimensional figure.

symmetric (596) A figure created by a shape and its reflection.

system of equations (412) A set of equations with the same variables. For example, $c = 10 + 0.1n$ and $c = 5 + 0.2n$ are called a *system of equations*.

tangent (688) If $\triangle ABC$ is a right triangle and A is an acute angle,

$$\text{tangent } \angle A = \frac{\text{measure of the side opposite } \angle A}{\text{measure of the side adjacent to } \angle A}$$

$$\tan A = \frac{a}{b}$$

term (27) A number, a variable, or a product of numbers and variables.

term (258) Each number within a sequence is called a term.

terminating decimal (225, 275) A decimal whose digits end. Every terminating decimal can be written as a fraction with a denominator of 10, 100, 1000, and so on.

tessellation (594) A repetitive pattern of polygons that fit together with no holes or gaps.

30°–60° right triangle (683) A right triangle that has one angle measuring 30° and another angle measuring 60°.

transformation (595) A movement of a geometric figure.

translation (595) One type of transformation where a figure is slid horizontally, vertically, or both.

transversal (563) A line that intersects two parallel lines to form eight angles.

trapezoid (585) A quadrilateral with exactly one pair of parallel sides.

tree diagram (509) A diagram used to show the total number of possible outcomes in a probability experiment.

triangle (568) A polygon having three sides.

triangular prism (632) A prism with triangular bases.

triangular pyramid (638) A pyramid with an equilateral triangle as a base.

trigonometric ratios (688) Ratios that involve the measures of the sides of right triangles. The tangent, sine, and cosine ratios are three trigonometric ratios.

trigonometry (688) The study of triangle measurement.

trinomial (706) A polynomial with three terms.

truncate (275) An answer being cut off by a calculator at a certain place-value position, ignoring the digits that follow.

unit rate (433) A rate with a denominator of 1.

upper quartile (491) The median of the upper half of a set of numbers.

variable (16) A placeholder in a mathematical expression or sentence.

Venn diagram (669) A diagram consisting of circles inside a rectangle which is used to show the relationships of sets.

vertex (549) A vertex of an angle is the common endpoint of the rays forming the angle.

vertex (75, 589) A vertex of a polygon is a point where two sides of the polygon intersect.

vertex (638) The vertex of a pyramid is the point where all the faces except the base intersect.

vertical angles (561) Congruent angles formed by the intersection of two lines. In the figure, the vertical angles are ∠1 and ∠3, and ∠2 and ∠4.

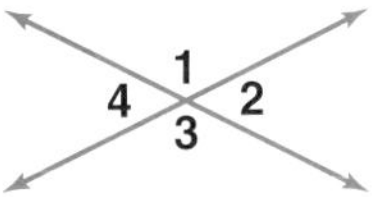

vertical line test (373) If any vertical line drawn on the graph of a relation passes through no more than one point of its graph, then the relation is a function.

volume (644) The number of cubic units needed to fill a container.

whole numbers (224) The set of numbers {0, 1, 2, 3, . . .}. It also includes any number that can be written as a whole number, such as $\frac{5}{5}$ or $\frac{9}{1}$.

***x*-axis** (36) The horizontal number line which helps to form the coordinate system.

***x*-coordinate** (37) The first number of an ordered pair.

***x*-intercept** (406) The *x*-coordinate of the point where the graph crosses the *x*-axis.

***y*-axis** (36) The vertical number line which helps to form the coordinate system.

***y*-coordinate** (37) The second number of an ordered pair.

***y*-intercept** (406) The *y*-coordinate of the point where the graph crosses the *y*-axis.

zero pair (82) The result of pairing one positive counter with one negative counter.

SPANISH GLOSSARY

absolute value/valor absoluto (67) El número de unidades que un número dista de cero en la recta numérica.

acute angle/ángulo agudo (550) Un ángulo cuya medida es mayor que 0° y menor que 90°.

acute triangle/triángulo agudo (569) Un triángulo que tiene tres ángulos agudos.

addition property of equality/propiedad de adición de la igualdad (125) Si sumas el mismo número a ambos lados de una ecuación, los dos lados permanecen igual. Para cualquiera de los números a, b y c, si $a = b$, entonces $a + c = b + c$.

addition property of inequality/propiedad de adición de la desigualdad (147) El sumar el mismo número a cada lado de una desigualdad, no altera la validez de la desigualdad. Para cualquiera de los números a, b y c:
1. Si $a > b$, entonces $a + c > b + c$.
2. Si $a < b$, entonces $a + c < b + c$.

additive inverses/inversos aditivos (89) Un número entero y su opuesto. La suma de un número entero y su inverso aditivo es cero.

adjacent angles/ángulos adyacentes (561) Dos ángulos que tienen un lado común y el mismo vértice, pero que *no* se traslapan.

adjacent leg/cateto adyacente (687) El cateto de un triángulo rectángulo que no es el opuesto a un ángulo y que no es la hipotenusa.

algebraic expression/expresión algebraica (16) Una combinación de variables, números y por lo menos una operación.

algebraic fraction/fracción algebraica (197) Una fracción con variables en el numerador o en el denominador.

alternate exterior angles/ángulos externos alternos (563) Ángulos exteriores no adyacentes que se encuentran en lados opuestos de la transversal. $\angle 1$ y $\angle 7$, y $\angle 2$ y $\angle 8$ son ángulos externos alternos.

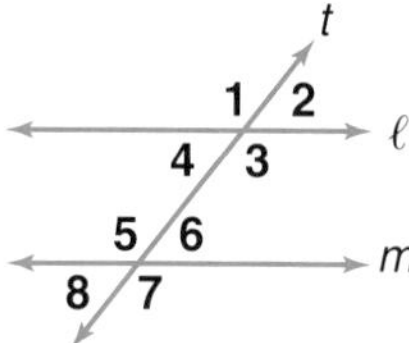

alternate interior angles/ángulos internos alternos (563) Ángulos internos no adyacentes que se encuentran en lados opuestos de la transversal. $\angle 4$ y $\angle 6$, y $\angle 3$ y $\angle 5$ son ángulos internos alternos.

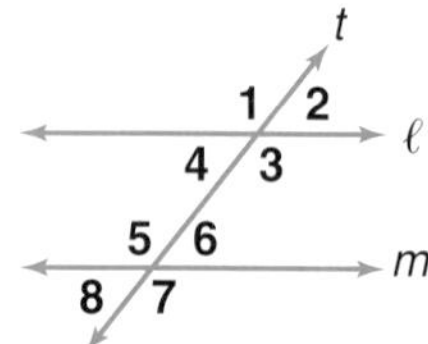

altitude/altitud (612, 614, 633) Un segmento de recta perpendicular a la base de una figura con extremos en la base y el lado opuesto a la base.

angle/ángulo (549) Dos rayos con un punto en común forman un ángulo. Los rayos y los vértices se usan para nombrar el ángulo.

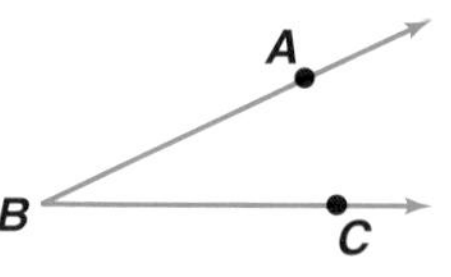

ángulo ABC o $\angle ABC$.

angle of depression/ángulo de depresión (695) El ángulo de depresión se forma por una línea visual a lo largo de la horizontal y otra línea visual por debajo de la misma.

angle of elevation/ángulo de elevación (694) Un ángulo de elevación se forma por una línea visual a lo largo de la horizontal y otra línea visual por encima de la misma.

area/área (138) El número de unidades cuadradas necesarias para cubrir una superficie no encerrada por una figura geométrica.

arithmetic sequence/secuencia aritmética (258) Una secuencia en que la diferencia entre cualquier par de términos consecutivos es la misma.

associative property of addition/propiedad asociativa de la adición (22, 234) Para cualquiera de los números racionales a, b y c,

$$(a + b) + c = a + (b + c).$$

associative property of multiplication/propiedad asociativa de la multiplicación (22, 296) Para cualquiera de los números a, b y c,

$$(ab)c = a(bc).$$

axes/ejes (73) Dos rectas numéricas que se intersecan forman un sistema de coordenadas.

back-to-back stem-and-leaf plot/diagrama de tallo y hojas consecutivo (487) Se usa para comparar dos conjuntos de datos. Las hojas de uno de los conjuntos de datos están a un lado del tallo y las hojas para el otro conjunto de datos están en el otro lado.

bar graph/gráfica de barra (52) Una forma gráfica que usa barras para comparar estadísticas.

bar notation/notación de barra (225) En decimales que se repiten, la línea o barra que se coloca sobre los dígitos que se repiten. Por ejemplo, $2.\overline{63}$ indica que los dígitos 63, se repiten.

base/base (175) En 5^3, la base es 5. La base se usa como un factor tantas veces como lo indique el exponente (3). Es decir,

$$5^3 = 5 \times 5 \times 5.$$

base/base (449) En una proporción de porcentaje, el número con el cual se compara el por ciento.

$$\frac{\text{porcentaje}}{\text{base}} = \text{proporción}$$

base/base (614) La base de un rectángulo, un paralelogramo o un triángulo es cualquier lado de la figura. Las bases de un trapecio son los lados paralelos.

base/base (632) Las bases de un prisma son los dos lados congruentes paralelos.

binomial/binomio (706) Un polinomio con exactamente dos términos.

boundary/frontera (418) Una línea que separa una gráfica en mitades de planos.

box-and-whisker plot/diagrama de caja y patillas (495) Una representación gráfica que resume datos usando la mediana, los cuartillos superior e inferior y los valores extremos. Se dibuja una caja alrededor del valor cuartílico y las patillas se extienden desde cada cuartillo hasta los puntos extremos de los datos.

center/centro (341) El punto dado desde el cual todos los puntos en un círculo están equidistantes.

circle/círculo (341) El conjunto de todos los puntos en un plano que están equidistantes de un punto dado llamado centro.

circle graph/gráfica de círculo (497, 556) Un tipo de gráfica estadística que se usa para comparar partes de un todo.

circular cone/cono circular (639) Una figura en el espacio que tiene una base circular y un vértice.

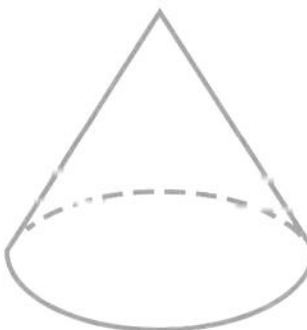

circumference/circunferencia (341) La distancia alrededor de un círculo.

coefficient/coeficiente (711) La parte numérica de un monomio.

combination/combinación (516) Un arreglo o lista en el cual el orden no es importante.

commission/comisión (470) Un estipendio que se le paga a un vendedor o a una vendedora basado en un porcentaje del total de las ventas.

common difference/diferencia común (258) La diferencia entre cualquiera par de términos consecutivos en una secuencia aritmética.

common multiples/múltiplos comunes (200) Múltiplos compartidos por dos o más números. Por ejemplo, algunos múltiplos comunes de 2 y 3 son 6, 12 y 18.

common ratio/proporción común (312) La proporción entre cualquier par de términos consecutivos en una secuencia geométrica.

commutative property of addition/propiedad conmutativa de la adición (22, 234) Para cualquiera de los números racionales a y b,

$$a + b = b + a.$$

commutative property of multiplication/propiedad conmutativa de la multiplicación (22, 296) Para cualquiera de los números racionales a y b,

$$ab = ba.$$

comparative graph/gráfica comparativa (498) Una gráfica que se usa para comparar resultados de grupos similares y mostrar tendencias.

compass/compás (554) Un instrumento que se usa para dibujar círculos o partes de círculos.

compatible numbers/números compatibles (280) Números que han sido redondeados de forma que cuando se dividen uno entre otro, el residuo es cero.

complementary/complementario (562) Dos ángulos son complementarios si la suma de sus medidas es 90°.

composite number/número compuesto (184) Cualquier número entero mayor de uno que posee más de dos factores.

congruent/congruente (554, 561, 573) Segmentos de recta que tienen la misma longitud, o ángulos que tienen la misma medida, o figuras que tienen la misma forma y tamaño.

constant/constante (707) Un monomio que no contiene una variable.

coordinate/coordenada (67, 126) Un número asociado con un punto en la recta numérica.

coordinate system/sistema de coordenadas (36, 73) Un plano de coordenadas separado en cuatro regiones iguales por dos rectas numéricas que se intersecan, llamadas ejes.

corresponding angles/ángulos correspondientes (563) Ángulos que tienen la misma posición en dos rectas paralelas diferentes cortadas por una transversal. $\angle 1$ y $\angle 5$, $\angle 2$ y $\angle 6$, y $\angle 3$ y $\angle 7$, y $\angle 4$ y $\angle 8$ son ángulos correspondientes.

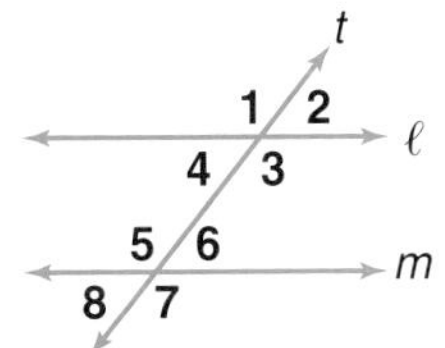

corresponding parts/partes correspondientes (573) Partes en figuras congruentes o similares que encajan.

cosine/coseno (688) Si $\triangle ABC$ es un triángulo rectángulo y A es un ángulo agudo,

$$\text{coseno } \angle A = \frac{\text{medida del lado adyacente a } \angle A}{\text{medida de la hipotenusa}}$$

$$\cos A = \frac{b}{c}$$

cylinder/cilindro (633) Una figura tridimensional que posee dos bases congruentes paralelas (generalmente circulares) y un lado curvo que conecta las bases.

D

data/datos (50) Información numérica que se reúne con propósitos estadísticos.

defining a variable/definir una variable (42) El escoger una variable y una cantidad para la variable para representarla en una ecuación.

degree/grado (549) La unidad de medida más común para ángulos.

dependent events/eventos dependiente (532) Dos o más eventos en los cuales el resultado de un evento afecta el resultado de otro(s) evento(s).

diagonal/diagonal (590) Un segmento de recta que une dos vértices no consecutivos de un polígono.

SPANISH GLOSSARY

diameter/diámetro (341) La distancia a través del centro de un círculo.

discount/descuento (468) La cantidad que se rebaja del precio original.

discrete mathematics/matemáticas discretas (258) Una rama de las matemáticas que tiene que ver con los conjuntos de números finitos o interrumpidos. Algunas disciplinas que tratan con las matemáticas discretas son la lógica y la estadística.

distributive property/propiedad distributiva (26) Para todos los números reales a, b y c, $a(b + c) = ab + ac$ y $(b + c)a = ba + ca$.

divisible/divisible (170) Un número es divisible entre otro si el residuo de la división es cero.

division property of equality/propiedad de división de la igualdad (130) Para cualquiera de los números a, b y c, con $c \neq 0$, si $a = b$, entonces $\frac{a}{c} = \frac{b}{c}$.

division properties of inequality/propiedades de división de la desigualdad (152) Para todos los números enteros a, b y c, en que $c > 0$, si $a > b$, entonces $\frac{a}{b} > \frac{b}{c}$.
Para todos los números enteros a, b y c, en que $c < 0$, si $a > b$, entonces $\frac{a}{c} < \frac{b}{c}$.

domain/dominio (372) El conjunto de todas las coordenadas de cada par ordenado.

empty set/conjunto vacío (347) Un conjunto sin ningún elemento, el cual se representa con el símbolo { } o ∅.

equation/ecuación (32) Un enunciado matemático abierto que contiene el signo de igualdad, =.

equiangular/equiangular (591) Todos los ángulos son congruentes.

equilateral/equilátero (591) Todos los lados son congruentes.

equivalent/equivalente (275) El mismo valor representado de diferentes formas. Por ejemplo, $\frac{3}{5}$ y 0.6 son equivalentes.

equivalent equations/ecuaciones equivalentes (124) Dos o más ecuaciones con la misma solución. Por ejemplo, $x + 3 = 5$ y $x = 2$ son ecuaciones equivalentes.

evaluating the expression/evaluando una expresión (11) Reemplazar las variables con números y hallar el valor numérico de la expresión.

expanded form/forma desarrollada (176) Un número que se expresa usando el valor de posición para escribir el valor de cada dígito en el número.

experimental probability/probabilidad experimental (529) Una posibilidad estimada con base en la frecuencia relativa de los resultados positivos que ocurren durante un experimento.

exponent/exponente (175) En 5^3, el exponente es 3. El exponente indica cuántas veces la base, 5, se usa como factor.

$$5^3 = 5 \times 5 \times 5$$

exterior angle/ángulo externo (591) El ángulo que se forma cuando se extiende un lado de un polígono.

face/cara (632) La cara de un prisma es cualquier superficie que forma un lado o la base del prisma.

factorial/factorial (515) La expresión $n!$ es el producto de todos los números contables comenzando con n y contando al revés hasta 1.

factors/factores (170) Números que se dividen entre un número entero y cuyo residuo es cero.

factor trees/árboles de factores (185) Un diagrama que muestra la factorización prima de un número compuesto. Los factores se ramifican a partir de los factores previos hasta que todos los factores son números primos.

Fibonacci sequence/secuencia de Fibonacci (263) Un listado de números en el que los dos primeros números son ambos 1 y cada número que sigue es la suma de los dos números previos, 1, 1, 2, 3, 5, 8, . . .

formula/fórmula (134) Una ecuación que expresa una regla para la relación entre ciertas cantidades.

45°–45° right triangle/triángulo rectángulo de 45°–45° (683) Un triángulo rectángulo que posee dos ángulos que miden 45° cada uno.

frequency table/tabla de frecuencia (51) Una tabla que indica el número de valores en cada intervalo.

function/función (373) Una función es una relación en que cada elemento del dominio se aparea exactamente con un elemento de la amplitud.

functional notation/notación funcional (393) Las ecuaciones que representan funciones se pueden escribir en *forma de notación*, $f(x)$. El símbolo $f(x)$ se lee "f" de "x" y representa el valor en la amplitud que corresponde al valor de x en el dominio.

fundamental counting principle/principio fundamental de contar (509) Si el evento M puede ocurrir de m maneras y es seguido por el evento N que puede ocurrir de n maneras, entonces el evento M seguido del evento N puede ocurrir en $m \cdot n$ maneras.

geometric probability/probabilidad geométrica (623) Probabilidad que usa el área para hallar la probabilidad de un evento.

geometric sequence/secuencia geométrica (312) Una secuencia en que la proporción entre cualquier par de números consecutivos es la misma.

gram/gramo (359) La unidad básica de masa del sistema métrico.

graph/gráfica (73, 126) Un puntito que representa un número en una recta numérica o un par ordenado en un plano de coordenadas.

greatest common factor (GCF)/máximo factor común (MFC) (190) El MFC de dos o más números enteros es el máximo factor que es común a cada uno de los números.

histogram/histograma (502) Una gráfica de barra cuyas columnas están una al lado de la otra para mostrar cómo están distribuidos los datos.

hypotenuse/hipotenusa (676) El lado opuesto al ángulo recto en un triángulo rectángulo.

identity property of addition/propiedad de identidad de la adición (23, 234) Para cualquier número a, $a + 0 = a$ y $0 + a = a$.

identity property of multiplication/propiedad de identidad de la multiplicación (23, 296) Para cualquier número a, $a \times 1 = a$.

inclusive/inclusivo (536) Cuando dos eventos pueden suceder al mismo tiempo.

independent events/eventos independientes (531) Dos o más eventos en los cuales el resultado de uno de ellos *no* afecta el resultado de ninguno de los otros.

indirect measurement/medida indirecta (580) Hallar una medida mediante el uso de triángulos similares y escribiendo una proporción.

inequality/desigualdad (46) Un enunciado matemático que contiene $<$, $>$, $\neq$, $\leq$ o $\geq$.

integers/números enteros (66, 224) Los números enteros y sus opuestos.

$$\ldots, -3, -2, -1, 0, 1, 2, 3, \ldots$$

intercepts/intersecciones (406) Los puntos en donde una gráfica cruza los ejes.

interior angles/ángulos internos (591) Un ángulo dentro de un polígono.

interquartile range/amplitud intercuartílica (491) La amplitud de la mitad central de un conjunto de datos.

$$\text{Amplitud intercuartílica} = UQ - LQ$$

intersect/intersecar (552) Líneas que se cruzan o se tocan unas a otras en un punto.

inverse operations/operaciones inversas (41) Operaciones que se anulan entre sí, tales como la adición y la sustracción.

inverse property of addition/propiedad del inverso de la adición (234) Para cualquier número racional a, $a + (-a) = 0$.

inverse property of multiplication/propiedad del inverso de la multiplicación (289) Para todo número no cero $\frac{a}{b}$, donde $a, b \neq 0$, hay exactamente un número $\frac{b}{a}$, tal que $\frac{a}{b} \times \frac{b}{a} = 1$.

irrational number/número irracional (672) Un número que no se puede expresar como $\frac{a}{b}$, donde a y b son números enteros y b no es igual a 0. Por ejemplo, $\sqrt{3} = 1.7320508 \ldots$ es un número irracional.

isosceles triangle/triángulo isósceles (569) Un triángulo que tiene por lo menos dos lados congruentes.

lateral surface/superficie lateral (639) La superficie lateral de un prisma, cilindro, pirámide o cono es toda la superficie de la figura excepto la base o las bases.

least common denominator (LCD)/mínimo común denominador (MCD) (201) El mínimo común múltiplo de los denominadores de dos o más fracciones.

least common multiple (LCM)/mínimo común múltiplo (MCM) (200) El menor número diferente de cero de los múltiplos comunes de dos o más números. El MCM de 2 y 3 es 6.

leaves/hojas (486) El valor de posición que le sigue al valor mayor forma las hojas en una gráfica de tallo y hojas.

leg/cateto (676) Cualquiera de los dos lados que forman el ángulo recto de un triángulo rectángulo.

like terms/términos semejantes (27, 711) Expresiones que contienen las mismas variables, tales como $3ab$ y $7ab$.

line/recta (548) Una trayectoria recta sin fin. Una representación de la recta CD ($\overleftrightarrow{CD}$) se muestra a continuación.

linear equation/ecuación lineal (386) Una ecuación cuya gráfica es una recta.

line graph/gráfica lineal (497) Un tipo de gráfica estadística que se usa para mostrar cómo los valores cambian a través de un período de tiempo.

line segment/segmento de recta (548) Dos extremos y la trayectoria recta entre ellos. Un segmento de recta se denomina por sus extremos.

liter/litro (359) La unidad básica de capacidad en el sistema métrico.

lower quartile/cuartillo inferior (491) La mediana de la mitad inferior de un conjunto de números indicados por LQ.

mean/media (300) La suma de los números en un conjunto de datos dividida entre el número de datos.

measures of central tendency/medidas de tendencia central (300) Datos o grupos de datos que pueden representar el conjunto total de datos.

median/mediana (300) En un conjunto de datos, la mediana es el número en el centro cuando los datos se organizan en orden de menor a mayor.

meter/metro (359) La unidad básica de longitud en el sistema métrico.

metric system/sistema métrico (360) Un sistema de medida que usa las unidades básicas: metro para longitud, gramo para masa y litro para capacidad.

mixed numbers/números mixtos (239) La suma de un número entero y una fracción. Por ejemplo, $3\frac{1}{2}$.

mode/modal (300) El número o artículo que aparece con más frecuencia en un conjunto de datos.

monomial/monomio (172) Una expresión algebraica que es un número real, una variable o el producto de números reales y variables.

multiple/múltiplo (200) El producto del número y cualquier número entero.

multiplication properties of inequality/propiedades multiplicativas de la desigualdad (152) **1.** Para todos los números enteros a, b y c, en que $c > 0$, si $a > b$, entonces $ac > bc$. **2.** Para todos los números enteros a, b y c, en que $c < 0$, si $a < b$, entonces $ac > bc$.

multiplication property of equality/propiedad multiplicativa de la igualdad (130) Para cualquiera de los números reales a, b y c, si $a = b$, entonces $ac = bc$.

multiplicative inverses/inversos multiplicativos (289) Dos números cuyo producto es 1. El inverso multiplicativo de $\frac{3}{4}$ es $\frac{4}{3}$.

multiplicative property of zero/propiedad multiplicativa de cero (23) Para cualquier número a, $a \times 0 = 0$.

mutually exclusive/exclusivo mutuamente (535) Dos o más eventos tales que ningún par de eventos pueden suceder al mismo tiempo.

negative/negativo (66) Un valor menor que cero.

negative relationship/relación negativa (380) Si los puntos en un diagrama de dispersión parecen sugerir una recta que se inclina hacia abajo y hacia la derecha, entonces existe una *relación negativa*.

net/red (632) La figura que se forma al "desdoblar" un prisma. La red muestra todas las caras que componen el área de un prisma.

no relationship/no relacionado (380) Se dice que los puntos que parecen ser aleatorios en un diagrama de dispersión son no relacionados.

null set/conjunto vacío (347) Un conjunto sin elementos y el cual se representa con el símbolo { } o $\varnothing$.

obtuse angle/ángulo obtuso (550) Cualquier ángulo que mide más de 90° pero menos de 180°.

obtuse triangle/triángulo obtuso (569) Un triángulo con un ángulo obtuso.

odds/posibilidades (520) Una manera de describir la posibilidad de que un evento ocurra.
posibilidades en favor = número de éxitos: número de fracasos
posibilidades en contra = número de fracasos: número de éxitos

open sentence/enunciado abierto (32) Una ecuación que contiene variables.

opposite side/lado opuesto (687) El lado que no forma parte del ángulo en un triángulo rectángulo.

ordered pair/par ordenado (37) Un par de números utilizados para ubicar un punto en el sistema de coordenadas.

order of operations/orden de operaciones (12, 176)
1. Simplifica las expresiones dentro de los símbolos de agrupación.
2. Evalúa todas las potencias.
3. Luego, realiza todas las multiplicaciones y las divisiones de izquierda a derecha.
4. Finalmente, realiza todas las sumas y las restas de izquierda a derecha.

origin/origen (36) El punto de intersección del eje x y del eje y en un sistema de coordenadas.

outcomes/resultados (509) Posibles resultados de un evento de probabilidad. Por ejemplo, 4 es un resultado cuando se lanza un dado.

outliers/valores atípicos (496) Datos que están a más de 1.5 veces que la amplitud intercuartílica de los cuartillos.

parallel/paralelo (551) Rectas en el mismo plano que no se intersecan. El símbolo $\parallel$ quiere decir paralelo.

parallelogram/paralelogramo (585) Un cuadrilátero con dos pares de lados paralelos.

percent/por ciento (449) Una proporción con un denominador de 100. Por ejemplo, 77% y $\frac{77}{100}$ son el mismo número.

percentage/porcentaje (449) En una proporción de por ciento, un número que se compara con otro llamado base.

$$\frac{\text{porcentaje}}{\text{base}} = \text{tasa}$$

percent of change/porcentaje de cambio (472) La proporción de la cantidad de cambio comparada con la cantidad inicial.

percent of decrease/porcentaje de disminución (473) La proporción de una cantidad de disminución comparada con la cantidad previa, expresada en forma de porcentaje.

percent of increase/porcentaje de aumento (472) La proporción de una cantidad de aumento comparada con la cantidad previa, expresada en forma de porcentaje.

percent proportion/proporción de porcentaje (450)

$$\frac{\text{porcentaje}}{\text{base}} = \text{tasa o } \frac{P}{B} = \frac{r}{100}$$

perfect squares/cuadrados perfectos (665) Números racionales cuyas raíces cuadradas son números enteros. 25 es un cuadrado perfecto porque $\sqrt{25} = 5$.

perimeter/perímetro (138) La distancia alrededor de una figura geométrica.

permutation/permutación (515) Un arreglo o listado en que el orden es importante.

perpendicular/perpendicular (551) Dos rectas que se intersecan para formar un ángulo recto.

pi, π/pi, π (341) La proporción de la circunferencia de un círculo con el diámetro del círculo. Aproximaciones para π son 3.14 y $\frac{22}{7}$.

pictograph/pictografía (497) Un tipo de gráfica estadística que usa láminas o ilustraciones para mostrar comparación entre cantidades específicas. Cada símbolo representa un número específico de artículos.

plane/plano (550) Una superficie plana sin aristas o fronteras. Un plano se puede nombrar con una sola letra mayúscula o mediante el uso de tres puntos no colineales del plano.

point/punto (548) Una ubicación específica en el espacio.

polygon/polígono (589) Una figura cerrada simple en un plano formado por tres o más segmentos de recta.

polynomial/polinomio (706) Un monomio o la suma o diferencia de dos o más monomios.

positive relationship/relación positiva (380) En una gráfica de dispersión, si los puntos parecen sugerir una recta que se inclina hacia arriba y a la derecha, entonces existe una *relación positiva*.

power/potencia (175) Un número que se puede escribir usando un exponente.

power of a monomial/potencia de un monomio (721) Para todos los números a y b, y los números enteros positivos m, n y p, $(a^m b^n)^p = a^{mp} b^{np}$.

power of a power/potencia de una potencia (719) Para cualquier número a y cualquiera de los números enteros m y n, $(a^m)^n = a^{mn}$.

power of a product/potencia de un producto (720) Para todos los números a y cualquier número entero positivo m, $(ab)^m = a^m b^m$.

prime factorization/factorización prima (185) El expresar un número compuesto como un producto de sus factores primos.

prime number/número primo (184) Un número entero mayor que 1 cuyos únicos factores son 1 y el número mismo.

principle/capital (467) La cantidad de dinero en una cuenta.

prism/prisma (632) Figura en el espacio que tiene dos bases paralelas y congruentes en forma de polígonos.

probability/probabilidad (440) La posibilidad de que algún evento pueda suceder. Es la proporción del número de veces que cierto evento puede ocurrir comparada con el número de resultados posibles.

$$\text{Probabilidad} = \frac{\text{número de resultados favorables}}{\text{número de resultados posibles}}.$$

probability of inclusive events/probabilidad de eventos inclusivos (536) La probabilidad de que uno u otro de dos eventos inclusivos ocurra se puede hallar sumando la probabilidad del primero a la probabilidad del segundo evento y restando la probabilidad de que ambos eventos ocurran.

$$P(A \text{ o } B) = P(A) + P(B) - P(A \text{ y } B)$$

probability of mutually exclusive events/probabilidad de eventos exclusivos mutuamente (535) La probabilidad de que el primero o el segundo de dos eventos mutuamente exclusivos ocurra se puede hallar sumando la probabilidad del primero a la probabilidad del segundo evento.

$$P(A \text{ o } B) = P(A) + P(B)$$

probability of two dependent events/probabilidad de dos eventos dependientes (532) Si dos eventos A y B son dependientes, entonces la probabilidad de que ambos eventos ocurran es el producto de la probabilidad de A y de la probabilidad de B después de que A ocurra.

$$P(A \text{ y } B) = P(A) \times P(B \text{ seguida de } A)$$

probability of two independent events/probabilidad de dos eventos independientes (531) La probabilidad de dos eventos independientes se puede hallar multiplicando la probabilidad del primero por la probabilidad del segundo evento.

$$P(A \text{ y } B) = P(A) \times P(B)$$

product of powers/producto de potencias (205) Para cualquiera de los números a y los números enteros positivos m y n, $a^m \times a^n = a^{m+n}$.

property of proportions/propiedad de las proporciones (444)

Si $\frac{a}{b} = \frac{c}{d}$, entonces $ad = bc$.

Si $ad = bc$, entonces $\frac{a}{b} = \frac{c}{d}$.

proportion/proporción (444) Un enunciado de la igualdad de dos o más razones.

protractor/transportador (549) Un instrumento que se usa para medir ángulos.

pyramid/pirámide (638) Una figura sólida que tiene un polígono por base y triángulos por lados.

Pythagorean theorem/teorema de Pitágoras (676) En un triángulo rectángulo, el cuadrado del largo de la hipotenusa es igual a la suma de los cuadrados de los largos de los catetos.

$$c^2 = a^2 + b^2$$

a
c
b

Q

quadrants/cuadrantes (73) Las cuatro regiones en que dos rectas numéricas perpendiculares separan el plano.

quadrilateral/cuadrilátero (584, 585) Un polígono con cuatro lados.

quartile/cuartillo (491) Una de las cuatro partes iguales de datos de un conjunto grande de números.

quotient of powers/cociente de potencias (206) Para todo número a diferente de cero y los números enteros m y n, $\frac{a^m}{a^n} = a^{m-n}$.

R

radical sign/signo radical (664) El símbolo que se usa para indicar una raíz cuadrada no negativa. $\sqrt{\ }$

radius/radio (341) La distancia desde el centro de un círculo hasta cualquier punto en el círculo.

range/amplitud (372) La amplitud de una relación es el conjunto de todas las segundas coordenadas de cada par ordenado.

range/amplitud (490) La diferencia entre los valores mayor y menor en un conjunto de datos.

rate/razón (433) La proporción de dos medidas que tienen diferentes unidades.

rate/tasa (449) En una proporción de por ciento, la razón de un número a 100.

ratio/razón (196, 432) Una comparación de dos números mediante división. La razón de 2 a 3 se puede expresar como 2 de 3, 2 a 3 ó $\frac{2}{3}$.

rational numbers/números racionales (224) Números que se pueden expresar como $\frac{a}{b}$, en que a y b son números enteros y $b \neq 0$.

ray/rayo (548) Una parte de una recta que se extiende infinitamente en una dirección.

real numbers/números reales (673) El conjunto de números racionales unido con el conjunto de números irracionales.

reciprocal/recíproco (289) Otro nombre para el inverso multiplicativo.

rectangle/rectángulo (585) Un cuadrilátero con cuatro ángulos congruentes.

rectangular prism/prisma rectangular (632) Un prisma con bases rectangulares.

reflection/reflexión (595) Un tipo de transformación en que una figura se voltea alrededor de una línea.

regular polygon/polígono regular (591) Un polígono que tiene todos los lados y todos los ángulos congruentes.

relation/relación (372) Un conjunto de pares ordenados.

repeating decimal/decimal periódico (225) Un decimal cuyos dígitos se repiten en grupos de uno o más. Por ejemplo: 0.181818 . . . y 0.8333. . . .

rhombus/rombo (585) Un paralelogramo con cuatro lados congruentes.

right angle/ángulo recto (550) Un ángulo que mide 90°.

right triangle/triángulo rectángulo (569) Un triángulo con un ángulo recto.

rotation/rotación (595) Cuando se hace girar una figura alrededor de un punto.

sample/muestra (50, 51) Un grupo que se usa para representar una población entera.

sample space/espacio muestral (441) El conjunto de todos los resultados posibles.

scale drawing/dibujo a escala (585) Un dibujo que es similar, pero que es o más grande o más pequeño que el objeto en sí.

scalene triangle/triángulo escaleno (569) Un triángulo sin lados congruentes.

scatter plot/diagrama de dispersión (378) En un diagrama de dispersión, se grafican dos conjuntos de datos como pares ordenados en el plano de coordenadas.

scientific notation/notación científica (317) Una manera de expresar un número como el producto de un número que es por lo menos 1, pero menos de 10, y una potencia de 10. Por ejemplo, $687{,}000 = 6.87 \times 10^5$.

sequence/secuencia (258) Una lista de números en cierto orden, como por ejemplo: 0, 1 2 3 ó 2, 4, 6, 8.

sides/lados (549) Los dos rayos que forman un ángulo se llaman lados.

sides/lados (589) Los segmentos de recta que forman un polígono.

similar/similar (578) Figuras que tienen la misma forma, pero no necesariamente el mismo tamaño.

simplest form/forma reducida (28) Una expresión en forma reducida no tiene términos semejantes ni paréntesis.

simplest form/forma reducida (196) Una fracción está en forma reducida cuando el máximo factor común (MFC) del numerador y del denominador es 1.

simulation/simulación (524) El proceso de representar un problema.

sine/seno (688) Si $\triangle ABC$ es un triángulo rectángulo y A es un ángulo agudo,

$$\text{seno } \angle A = \frac{\text{medida del lado opuesto a } \angle A}{\text{medida de la hipotenusa}}$$

$$\text{sen } A = \frac{a}{c}.$$

skew lines/rectas alabeadas (551) Dos rectas que no se intersecan y no están en el mismo plano.

slant height/altura inclinada (638) El largo de la altura de una cara lateral de una pirámide regular.

slope/pendiente (400) La pendiente de la recta es la razón del cambio en y al cambio correspondiente en x.

$$\text{Pendiente} = \frac{\text{cambio en } y}{\text{cambio en } x}$$

slope-intercept form/forma de pendiente-intersección (407) La forma de pendiente-intersección de una ecuación lineal de una recta es $y = mx + b$. La pendiente de la recta es m y la intersección con el eje y es b.

solution/solución (32) El valor de una variable que hace que una ecuación sea válida. La solución de $12 = x + 7$ es 5.

solution of system/solución de un sistema (412) El par ordenado que resuelve ambas ecuaciones en el sistema.

solving an equation/resolviendo una ecuación (32) El proceso de hallar una solución a una ecuación.

square/cuadrado (585) Un paralelogramo con todos los lados congruentes y todos los ángulos congruentes.

square pyramid/pirámide cuadrada (638) Una pirámide con una base cuadrada.

square root/raíz cuadrada (664) Uno de los dos factores iguales de un número. La raíz cuadrada de 144 es 12 puesto que $12^2 = 144$.

standard form/forma estándar (176) La forma estándar de setecientos treinta y nueve es 739.

statistics/estadística (51) Información que ha sido reunida, analizada y presentada de una manera organizada.

stem-and-leaf plot/gráfica de tallo y hojas (486) Un sistema que se usa para condensar un conjunto de datos en el cual el mayor valor de posición de los datos forma el tallo y el siguiente valor de posición forma las hojas.

stems/tallos (486) El mayor valor de posición común a todos los valores de los datos y que forma el tallo en una gráfica de tallo y hojas.

straightedge/regla (554) Cualquier objeto que se puede usar para dibujar una recta.

substitution property of equality/propiedad de sustitución de la igualdad (17) Para todos los números a y b, si $a = b$, entonces, a se puede reemplazar por b.

subtraction property of equality/propiedad de sustracción de la igualdad (124) Si restas el mismo número de cada lado de una ecuación, los dos lados permanecen iguales. Para cualquiera de los números a, b y c, si $a = b$, entonces $a - c = b - c$.

subtraction property of inequality/propiedad de sustracción de la desigualdad (147) Si restas el mismo número de cada lado de una desigualdad, la validez de la desigualdad no cambia.
Para todos los números a, b y c:
1. si $a > b$, entonces $a - c > b - c$;
2. si $a < b$, entonces $a - c < b - c$.

supplementary/suplementario (562) Dos ángulos son suplementarios si la suma de sus medidas es 180°.

surface area/área de superficie (632) La suma de las áreas de todas las superficies (caras) de una figura tridimensional.

symmetric/simétrico (596) Una figura creada por la figura y su reflexión.

system of equations/sistema de ecuaciones (412) Un conjunto de ecuaciones con las mismas variables. Por ejemplo: $c = 10 + 0.1n$ y $c = 5 + 0.2n$ se llaman un *sistema de ecuaciones*.

tangent/tangente (688) Si $\triangle ABC$ es un triángulo rectángulo y A es un ángulo agudo,

$$\text{tangente } \angle A = \frac{\text{medida del lado opuesto a } \angle A}{\text{medida del lado adyacente al } \angle A}$$

$$\tan A = \frac{a}{b}.$$

term/término (27) Un número, una variable o un producto de números y variables.

term/término (258) Cada número en una secuencia se llama un término.

terminating decimal/decimal terminal (225, 275) Un decimal cuyos dígitos terminan. Cada decimal terminal se puede escribir en forma de fracción con un denominador de 10, 100, 1000 y así sucesivamente.

tessellation/teselación (594) Un patrón repetitivo de polígonos que encajan juntos sin dejar espacios vacíos.

30°–60° right triangle/triángulo rectángulo de 30°–60° (683) Un triángulo rectángulo que tiene un ángulo que mide 30° y otro que mide 60°.

transformation/transformaciones (595) Movimientos de figuras geométricas.

translation/traslación (595) Un tipo de transformación en la cual una figura se desliza horizontal, verticalmente o en ambas direcciones.

transversal/transversal (563) Una recta que interseca dos líneas paralelas para formar ocho ángulos.

trapezoid/trapezoide (585) Un cuadrilátero con exactamente un par de lados paralelos.

tree diagram/diagrama de árbol (509) Un diagrama que se usa para mostrar el número total de posibles resultados en un experimento de probabilidad.

triangle/triángulo (568) Un polígono con tres lados.

triangular prism/prisma triangular (632) Un prisma con bases triangulares.

triangular pyramid/pirámide triangular (638) Una pirámide cuya base es un triángulo equilátero.

trigonometric ratios/razones trigonométricas (688) Razones que involucran las medidas de los lados de triángulos rectángulos. La tangente, el seno y el coseno son tres razones trigonométricas.

trigonometry/trigonometría (688) El estudio de la medición de triángulos.

trinomial/trinomio (706) Un polinomio con tres términos.

truncate/truncado (275) Una respuesta que la calculadora interrumpe en una posición determinada, ignorando los dígitos que siguen.

unit rate/razón unitaria (433) Una razón cuyo denominador es 1.

upper quartile/cuartillo superior (491) La mediana de la mitad superior de un conjunto de datos.

variable/variable (16) Símbolo que ocupa un lugar en expresiones matemáticas.

Venn diagram/diagrama de Venn (669) Un diagrama que consiste en círculos dentro de un rectángulo y el cual se usa para mostrar las relaciones de conjuntos.

vertex/vértice (549) El vértice de un ángulo es el extremo común de los rayos que forman el ángulo.

vertex/vértice (75, 589) El vértice de un polígono es un punto en donde se intersecan dos lados del polígono.

vertex/vértice (638) El vértice de una pirámide es el punto en donde se intersecan todas las caras, excepto la base.

vertical angles/ángulos verticales (561) Ángulos congruentes formados por la intersección de dos rectas. En la figura, los ángulos verticales son $\angle 1$ y $\angle 3$, y $\angle 2$ y $\angle 4$.

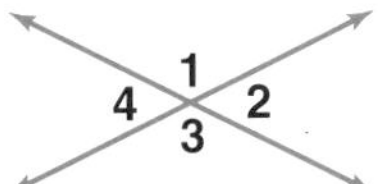

vertical line test/prueba de recta vertical (373) Si cualquier recta vertical dibujada en la gráfica de una relación pasa por un solo punto de esa gráfica, entonces la relación es una función.

volume/volumen (644) El número de unidades cúbicas que se necesitan para llenar un recipiente.

whole numbers/números enteros (224) El conjunto de números {0,1,2,3, . . .}. También incluye cualquier número que se pueda escribir como un número entero, por ejemplo, $\frac{5}{5}$ ó $\frac{9}{1}$.

x-axis/eje x (36) La recta numérica horizontal que ayuda a formar el plano de coordenadas.

x-coordinate/coordenada x (37) El primer número en un par ordenado.

x-intercept/intersección con el eje x (406) La coordenada x del punto sobre el cual una gráfica cruza el eje x.

y-axis/eje y (36) La recta numérica vertical que ayuda a formar el plano de coordenadas.

y-coordinate/coordenada y (37) El segundo número en un par ordenado.

y-intercept/intersección con el eje y (406) La coordenada y del punto sobre el cual una gráfica cruza el eje y.

zero pair/par cero (82) El resultado de aparear un contador positivo con un contador negativo.

SELECTED ANSWERS

Chapter 1 Tools for Algebra and Geometry

Pages 9–10 Lesson 1-1

5a. We know the bonus for 2, 4, 6, and 8 packages over plan. We need to know the bonus for 16 packages over plan. **5b.** Extend the pattern. The bonus should be about $250.

5c.

10	$200
12	$225
14	$250
16	$275

5d. The answer is reasonable and answers the question. **7.** yes **9.** $2.07

11a. 

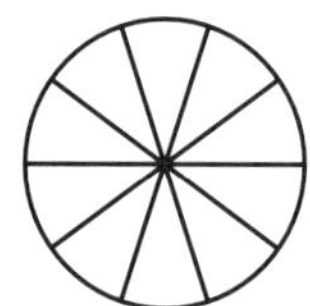 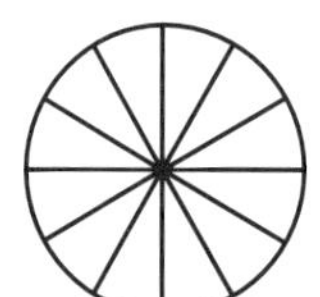

11b. 2, 4, 6, 8, 10, 12 **11c.** 1, 2, 3, 4, 5, 6 **11d.** The number of cuts is half the number of pieces.

Pages 14–15 Lesson 1-2

7. ÷, 61 **9.** ×, 5 **11.** −, 3 **13.** 8 **15.** 9 **17.** 12 **19.** 4 **21.** 4 **23.** 1 **25.** 145 **27.** 60 **29.** 128 **31.** $71 - (17 + 4) = 50$ **33.** $18 \div (3 + 6) + 12 = 14$ **35a.** 3($330) + 2($247) + 4($229) **35b.** $2400 **37.** $252 **39.** 2024, 3024, 7224

Pages 19–20 Lesson 1-3

5. 19 **7.** 12 **9.** 10 **11.** $3 + v$ **13.** $6n$ **15a.** $6x$ **15b.** 204 pounds **17.** 2 **19.** 14 **21.** 4 **23.** 12 **25.** 3 **27.** 10 **29.** $92 \div c$ **31.** $p - 5$ **33.** $u - 10$ **35.** $88 \div b$ **37.** $3k$ **39.** $2s + 2$ **41.** 16 less some number **43.** the quotient of some number and 5 **45.** twice some number plus 1 **47a.** $c \div 4 + 37$ **47b.** 68° **49.** 27 **51.** 29

Page 21 Lesson 1-3B

1. 14 **3.** 7 **5.** 55 **7.** Use REPLAY then change the expression.

Pages 24–25 Lesson 1-4

5. associative addition **7.** commutative addition **9.** 29 **11.** 48 **13.** $(7 \cdot 6)z$; $42z$ **15.** $8.35 **17.** commutative multiplication **19.** multiplicative zero **21.** commutative addition **23.** commutative addition **25.** commutative addition **27.** associative addition **29.** 360 **31.** 0 **33.** 130 **35.** 660 **37.** 9 + 5 **39.** $18w + 9$ **41.** $b(6 \cdot 5)$; $30b$ **43.** $(13 + 11) + m$; $24 + m$ **45.** $p(7 \cdot 4)$; $28p$ **47.** $15 + (4w + w)$; $15 + 5w$ **49b.** Division is not commutative. **53.** 6 **55a.** $w \div 16$ **55b.** 9 pints

Pages 28–30 Lesson 1-5

5. 5(7) + 5(8) **7.** $4(x) + 4(3)$ **9.** $13a + 7$ **11.** $7c + 7$ **13.** $20ab$ **15.** 3(11) + 3(12) **17.** $t(6 + 11)$ **19.** $2(r + 6s)$ **21.** $9(v) + 8(v)$ **23.** $2(4x) + 2(9y)$ **25.** $38a + 45$ **27.** $25c + 16$ **29.** $33y + 35$ **31.** $32b + 60$ **33.** $72 + 25f$ **35.** $11x + 15y$ **37.** 12,000 cubic meters **39.** 8 + (19 + 17) **41.** $n \div 9$ **43.** $362.65 **45.** yes

Page 30 Self Test

1. 1 5 10 10 5 1
1 6 15 20 15 6 1
3. 59 **5.** 2 **7.** commutative addition **9.** $62.50

Page 31 Lesson 1-5B

1. F **3.** T **5.** F

Pages 34–35 Lesson 1-6

7. 13 **9.** 5 **11.** 50 **13a.** $220 + d = 360$ **13b.** 140 people **15.** 39 **17.** 32 **19.** 1 **21.** 6 **23.** 120 **25.** 8 **27.** 9 **29.** 11 **31.** 7 **33.** 7 **35a.** $3973.05 + c = 4003.33$ **35b.** 30.28 **37.** $7x + 1$ **39.** $1943 + n$ **41.** 5

Pages 38–40 Lesson 1-7

5. (3, 2) and (3, 3) **7.** State Capitol **9.** no **11.** M **13.** L **15.** T **17.** J **19.** (6, 4) **21.** (5, 0) **23.** (9, 2) **25.** (10, 8) **27a.** $(a, 0)$ **27b.** $(0, b)$ **27c.** origin

29a.

29b. about 78 years **31a.** 50 cars are waiting at the light at 7:00 A.M. and 10 cars are waiting at 9:00 A.M. **31b.** The workday started at most businesses in the area. **31c.** Sample answer: Add point (5:15, 50). About the same number of cars will take people out from work that took them there in the morning. **33.** $10m + 1$ **35.** 12 **37.** 1

Pages 43–45 Lesson 1-8

5. Sample answer: $12 - 9 = 3$ **7.** Sample answer: $4 + 12 = 16$ **9.** 14 **11.** 11.88 **13.** $n + 27 = 31$; $n = 4$ **15.** Sample answer: Evan missed 7 questions on the quiz. If his score was 28, how many points were possible? **17.** $14 = 11 + 3$ **19.** $14 = 28 \div 2$ **21.** $h = 48 \div 6$ **23.** 22 **25.** 88 **27.** 120 **29.** 0.83 **31.** 3.77 **33.** 0.6 **35.** $3x = 18$; 6 **37.** $n - 7 = 22$; 29 **39.** $n \div 4 = 14$; 56 **41.** Sample answer: Jack scored 16 points in this week's game. That is two more than he scored last week. How

many points did Jack score last week? **43.** Sample answer: The product of any number and 0 is 0 according to the multiplicative property of zero. Thus, we could rewrite $5 = a \cdot 0$ as $5 = 0$. $5 \neq 0$, so division by zero does not make sense.
45. 46 times **47.** 19 **49.** associative property of addition **51.** $5 \cdot t$ **53a.** Sample answer: calculator **53b.** about 317 miles

Pages 48–49 Lesson 1-9

5. false **7.** true **9.** true **11.** $d \geq 27$ **13.** true **15.** open **17.** true **19.** true **21.** open **23.** true **25.** true **27.** true **29.** $t \geq \$100$ **31.** $\ell \geq 3$ **33.** $p < 15$ **35.** $f > 80{,}000$ **37a.** $t \leq 10$ years **37b.** $t \leq 1$ year **39.** 11 **41.** $(8 \cdot 7)c$; $56c$ **43.** 9

Page 50 Lesson 1-10A

1. 9–11 hours

Pages 54–55 Lesson 1-10

7. Miami Metrozoo and Toledo **9.** 13.6 million **11.** 15 **13.** yes, the total of the states is 51 **15.** Most states will issue a driver's license to a 16-year-old, all to an 18-year-old.

17.

Height	Buildings
0–100	0
101–200	0
201–300	0
301–400	4
401–500	8
501–600	4
601–700	1
701–800	3
801–900	0
901–1000	1

19. **Seattle, WA Buildings**

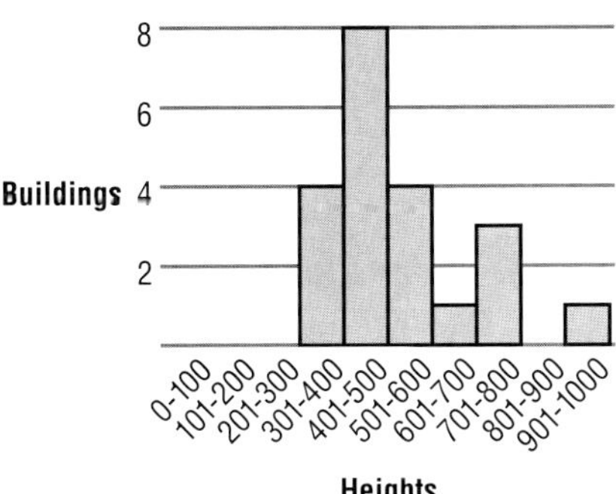

21a. 41 **21b.** 72–74 in. **21c.** You cannot tell from the frequency table how many presidents were exactly six feet tall. The table shows how many presidents fell in different height ranges, not exact heights. **21d.** Leadership is associated with tallness. **23.** $s \geq \$210{,}070$

25.

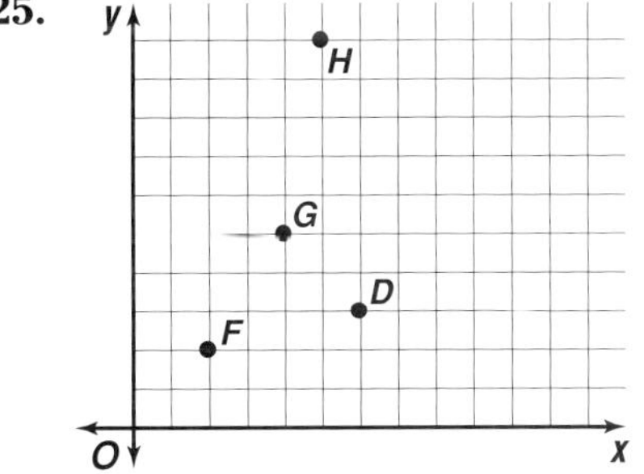

27a. 6 **27b.** 12

Page 57 Chapter 1 Highlights

1. algebraic expression **3.** in simplest form **5.** addition **7.** like terms

Pages 58-61 Study Guide and Assessment

9. 9 **11.** 16 **13.** 6 **15.** 60 **17.** 4 **19.** 10 **21.** 1 **23.** 24 **25.** $10n$ **27.** $b + 10$ **29.** $r - 12$ **31.** commutative, + **33.** associative, × **35.** multiplicative identity **37.** $8n$ **39.** $18x + 8$ **41.** $10x + 26y$ **43.** 6 **45.** 9 **47.** 3 **49.** 5 **51.** N **53.** (5, 6) **55.** (0, 4) **57.** 12 **59.** 28.35 **61.** 108 **63.** false **65.** false **67.** $p < 60$

69.

Representatives	States
1–10	38
11–20	7
21–30	3
31–40	1
41–50	0
51–60	1

71a. Sample answer: Estimation because an exact answer is not needed. **71b.** about 28,000 miles

Chapter 2 Exploring Integers

Pages 69–70 Lesson 2-1

7. −5 **9.** +13 **11.** 5 **13.** 10 **15.** 12 **17a.** −15, −24

17b. −25 −20 −15

19. −5 −4 −3 −2 −1 0 1 2 3 4 5

21. −9 −8 −7 −6 −5 −4 −3 −2 −1 0 1

23. +400 **25.** −3 **27.** +5 **29.** 11 **31.** 0 **33.** 8 **35.** 9 **37.** 12 **39.** −36 **41.** 2 **43.** 6 **45.** 2 **47.** 8, 0, −8 **49.** Sample answer: −1.5

51. −5 −4 −3 −2 −1 0 1 2 3 4 5

53. 17–18

55.

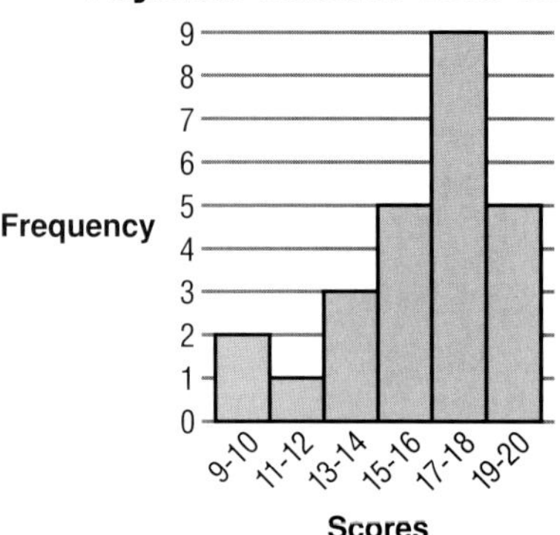

57. 180 square feet **59.** $3x - 8$

Page 71 Lesson 2-1B

1. -1 **3.** Sample answer: There seems to be a cluster between -2 and 1. **5.** Sample answer: Northern states are losing House members while southern states are gaining House members.

Pages 74–76 Lesson 2-2

5. B **7.** D **9.** IV **11.** none **13a.** I **13b.** III **13c.** II **15.** $(4, -2)$ **17.** $(1, -4)$ **19.** $(4, 3)$ **21.** $(2, 2)$ **23.** $(-1, 5)$ **25.** II **27.** III **29.** IV **31.** I **33.** III **35.** none

37.

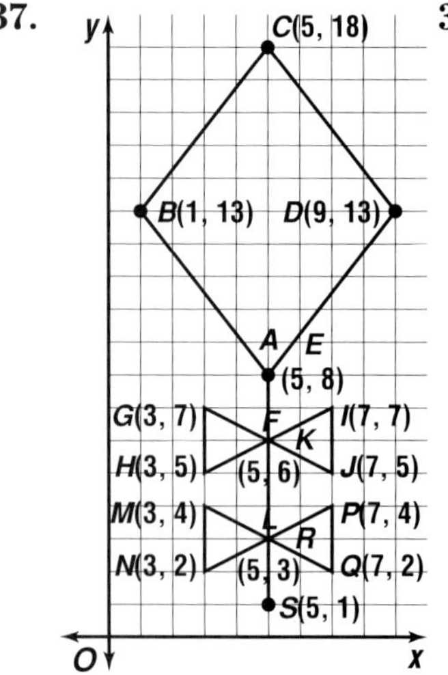

39.

41a. $P(2, 2)$, $Q(2, 5)$, $R(7, 5)$, $S(7, 2)$ **41b.** Rectangle $PQRS$ shifted 4 units down **41c.** A rectangle shaped like $PQRS$ in quadrant III **43.** 12 **45.** $s < 250{,}000$ **47a.** 22 **47b.** 4 **49.** eleven 2-point; eight 1-point

Pages 77–79 Lesson 2-2B

1.

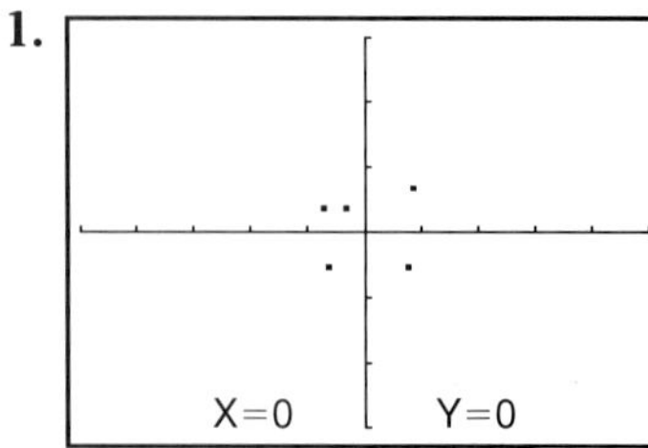

3.

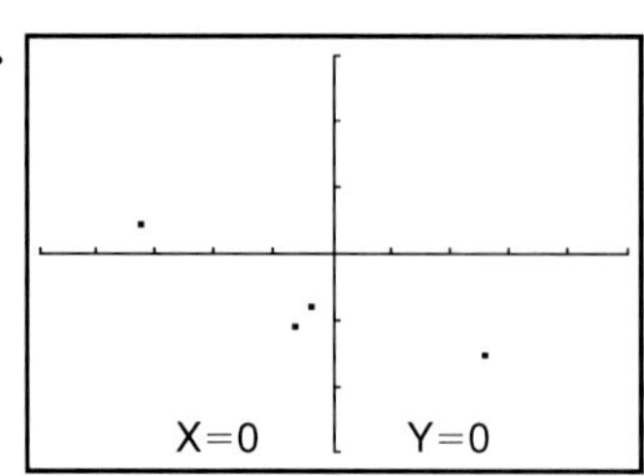

5a. $[-10, 10]$ by $[-10, 10]$ **5b.** $[-47, 47]$ by $[-31, 31]$

Pages 79–81 Lesson 2-3

5. $2 > -5$ **7.** $<$ **9.** $=$ **11.** $\{-88, -9, -4, 0, 3, 43, 234\}$ **13.** $-2 < 3, 3 > -2$ **15.** $>$ **17.** $=$ **19.** $>$ **21.** $<$ **23.** $<$ **25.** $<$ **27.** $3 > 2$ **29.** $60 < 75$ **31.** $565 > 344$ **33.** Sample answer: $66 < 265$ **35.** $-8 < -3; -3 > -8$ **37.** $0 > -8; -8 < 0$ **39.** $\{-11, -3, 8\}$ **41.** $\{-65, -6, 1, 29, 56\}$ **43.** $\{-65, -53, 48, 87, 199\}$ **45b.** Lead **45c.** Helium

47–48.

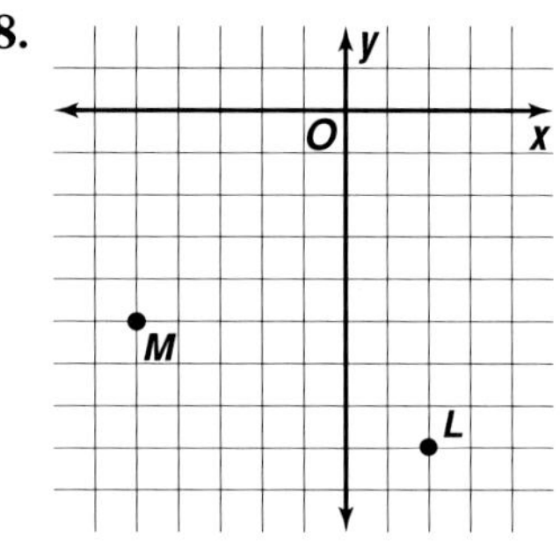

49. -6; 120 **51.** Sample answer: $(6 \times 20) + (6 \times 6)$ **53.** 4 adults

Page 82 Lesson 2-4A

1. 5 **3.** -1 **5.** -3 **7.** -8 **9.** Sample answer: First, place the appropriate number and kind of counters on the mat. Then match all positive counters with all the possible negative counters. These pairs can then be removed because zero does not affect the value of the set on the mat. The remaining counters give you the sum.

Pages 85–87 Lesson 2-4

7. $8 + (-5) = 3$ **9.** $-4 + 9 = 5$ **11.** positive, 9 **13.** negative, -4 **15.** -111 **17.** $-6a$ **19.** 13 **21.** 7 **23.** -4 **25.** 12 **27.** -66 **29.** 26 **31.** -9 **33.** -7 **35.** $-7 + 12$; 5 **39.** 21 **41.** -27 **43.** 63 **45.** $3y$ **47.** $-15m$ **49.** $-20d$ **51.** 7666 **53.** false

55a.

Breed	Change
Akita	−191
Beagle	−390
Chow Chow	+8846
Dachshund	+1473
Labrador Retriever	−4020
Pug	+286

55b. +6004 **57.** false **59.** 3 m/s **61.** $r - 9 = 15$; \$24 **63.** $n + (8 + 9)$; $n + 17$ **65.** false

Page 88 Lesson 2-5A

1. 5 **3.** 1 **5.** −15 **7.** 4 **9.** Sample answer: The answers to each set of exercises are the same. The exercises differ in that the second term of the subtraction exercises is the additive inverse of the second term of the addition exercises.

Pages 91–93 Lesson 2-5

7. 8 **9.** $-9x$ **11.** $-7 + (-11) = x$; −18 **13.** $8 + 3 = b$; 11 **15.** $-14 + 19 = y$; 5 **17.** $-27x$ **19.** −12 **21.** −8 **23.** −14 **25.** −3 **27.** 32 **29.** 9 **31.** 37 **33.** −29 **35.** −63 **37.** −46 **39.** −20 **41.** −27 **43.** −5 **45.** 26 **47.** $-10x$ **49.** $15p$ **51.** $24cd$ **53.** $-8a$ **55.** Sample answer: Every integer has an additive inverse because every integer has an opposite. The sum of an integer and its opposite is zero. Zero is its own additive inverse because $0 + 0 = 0$. **57.** −$2133 **59.** $36m$ **61.** $4 < 6$ **63.** triangle **65.** $3y + 23$ **67.** Sandy Koufax **69.** $k + 3$

Page 93 Self Test

1. 8 **3.** −3, 0, 2, 5

4–6.

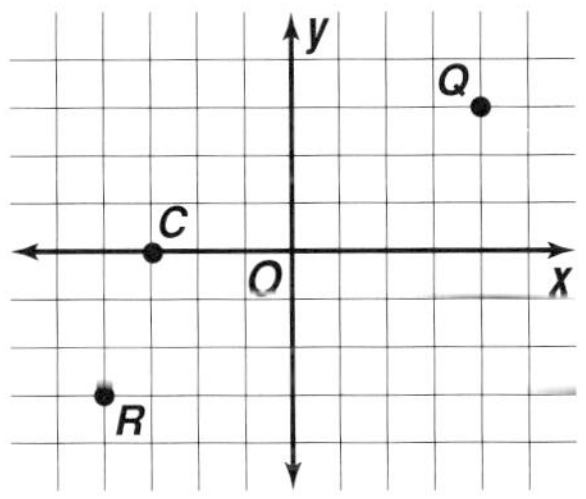

7. > **9.** = **11.** 10 **13.** −15

Pages 95–97 Lesson 2-6

5. 120, 720 **7.** 204 **9.** 123,454,321 **11.** 4 touchdowns **13.** 1:33 P.M.

15a.

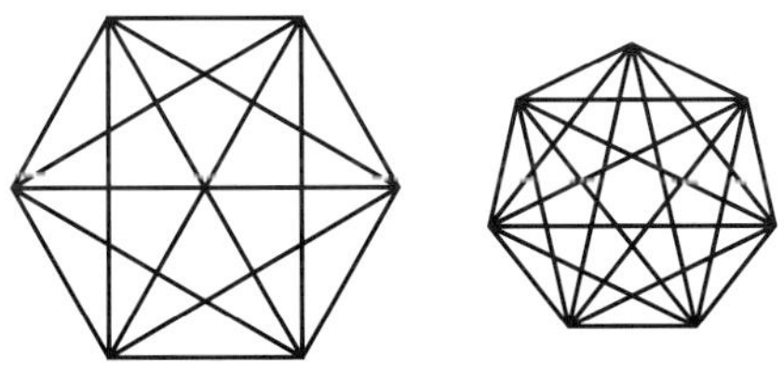

15b. 2, 5, 9, 14; 27 **17.** 36 segments **19.** 55 **21.** 1001 **23.** 6 **25.** $11y + 21$

Page 98 Lesson 2-7A

1. 6 **3.** 6 **5.** −6 **7.** −16 **9.** Sample answer: -2×4 means to remove 2 sets of 4 positive 1-tiles after putting in 8 zero pairs.

Pages 102–103 Lesson 2-7

5. −, −56 **7.** +, 518 **9.** +, 54 **11.** −27 **13.** −320 **15.** −132 **17.** $-40x$ **19.** −56 feet **21.** 28 **23.** −26 **25.** 75 **27.** −585 **29.** 168 **31.** −308 **33.** 56 **35.** −48 **37.** −36 **39.** $-12b$ **41.** $48b$ **43.** $-50rs$ **45.** $-3ab$ **47.** If there is an even number of negative numbers being multiplied, the product will be positive. If there is an odd number of negative numbers being multiplied, the product will be negative. **49.** −36,000 feet or about −6.8 miles **51.** 49°F **53.** Sample answer: 65 > 34 **55.** 39,177 **57.** commutative, +

Pages 107–108 Lesson 2-8

5. −; −7 **7.** −; −6 **9.** −; −15 **11.** 40 **13.** 7 **15.** 13 **17.** −3 **19.** $-60 \div 4 = y$; −15 **21.** −4 **23.** 6 **25.** 13 **27.** −111 **29.** −9 **31.** −12 **33.** −9 **35.** 4 **37.** −19 **39.** −5 **41.** 7 **43.** −9 **45.** −17 **47.** 12 **49.** −7 **51.** −3 **53.** Sample answer: $x = -144$; $y = 12$; $z = -12$ **55.** about $1096 **57.** 10:31, 10:35, 11:21 **59.** 41 **61.** 1014 **63.** 6 **65.** 7 **67.** 27

Page 109 Chapter 2 Highlights

1. F **3.** C **5.** E **7.** A

Pages 110–112 Study Guide and Assessment

9.

11.

13. 24 **15.** −40 **17.** 29 **19.** 12 **21.** III **23.** II **25.** IV **27.** = **29.** > **31.** < **33.** 17 < 35 or 35 > 17 **35.** −16, −4, 2, 3 **37.** −300, −33, 9, 124, 210 **39.** −3 **41.** −6 **43.** −15 **45.** −10 **47.** −30 **49.** 21 **51.** $-7b$ **53.** $6r$ **55.** −8 **57.** 65 **59.** $36cd$ **61.** $-300yz$ **63.** 7 **65.** 15 **67.** −16 **69.** 19 **71.** 18°F **73.** 7623

Chapter 3 Solving One-Step Equations and Inequalities

Pages 120–122 Lesson 3-1

5. 6087 **7.** 24 **9.** 5 people **11.** 501 **13.** You would buy 4 pairs of $70 shoes at a cost of $280 before the $200 pair wore out. It is less expensive to buy the $200 pair of shoes that lasts longer. But, if the shoes are uncomfortable or go out of style, it may make more sense to buy the less expensive shoes. **15.** blue **17.** 21 **19.** $24k + 3$

Page 123 Lesson 3-2A

1. 4 **3.** 3 **5.** 7 **7.** 12 **9.** 11

Pages 127–128 Lesson 3-2

7. −25

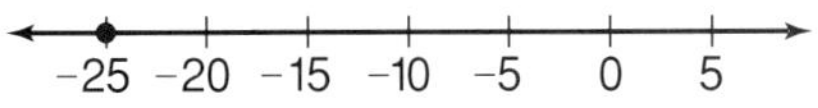

9. 37

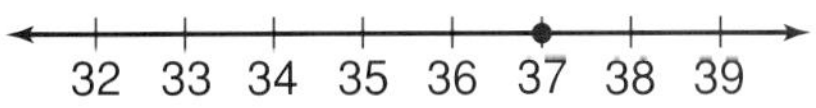

11. −1

13. $322 = 264 + e$, $58 million

15. 38

19. −32

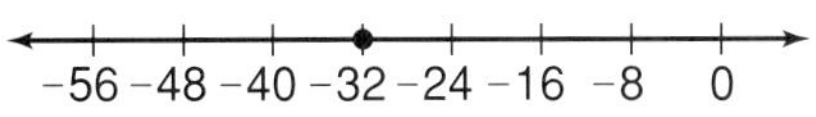

23. −4

27. 50

31. −1384

33. a **35.** 5 **37.** −8 **39.** 12 **41.** 25 years old **43.** −162 **45.** −3, −1, 3, 15 **47a.** v = Calories burned in an hour of volleyball **47b.** $v + 120 = 384$ **47c.** $v = 264$

Pages 131–133 Lesson 3-3

5. 9

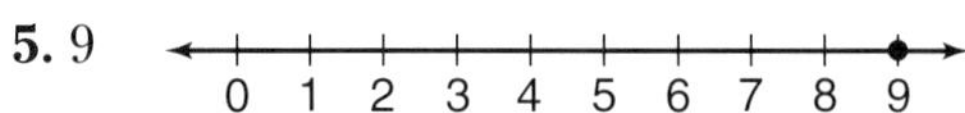

7. −12

9. 17

11. 24 dollars

13. −19

17. 168

21. 8

25. 936

29. 128

33. −13,432

39a. $6x = 300{,}000{,}000$ **39b.** $15x = 300{,}000{,}000$ **39c.** Most species lay between 20 and 50 million eggs. **41.** 18 cm **43.** d **45.** 17 **47.** 10 **49a.** Estimation; an exact answer is not needed. **49b.** Yes, $7 is enough.

Pages 135–137 Lesson 3-4

5. 45 **7.** 165 miles **9.** 360 **11.** 13 **13.** 3.4 **15.** 14 mph **17.** 10.9 hrs **19.** $s = \ell - d$ **21a.** 30 sq in. **21b.** 6 cm **23a.** 0.8 hours or 48 minutes **23b.** 8.9 mph **25a.** $a = \frac{f - s}{t}$ **25b.** 2 m/s^2 **25c.** −2 m/s^2 **25d.** negative **27.** −27 **29.** $x + 11 = 25$; 14

Page 137 Self Test

1. Stiers–Ratcliffe, Gibson–Smith, Means–Powhatan, Hunt–Willow, Bedard–Pocahontas

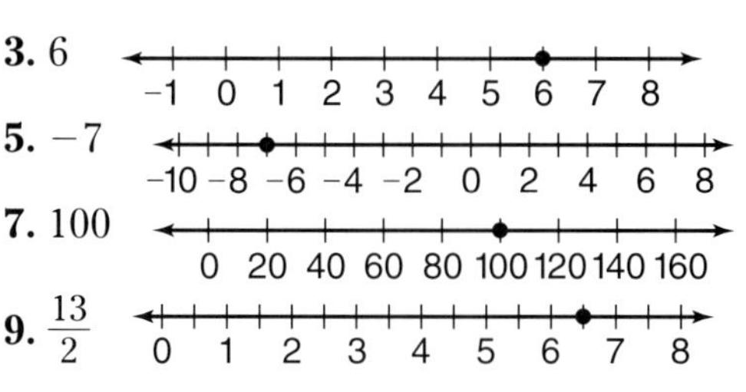

3. 6 **5.** −7 **7.** 100 **9.** $\frac{13}{2}$

Page 138 Lesson 3-5A

1. 18 units

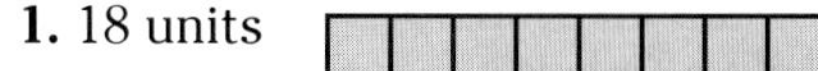

3. 32 units

5. The two figures have the same perimeter because the squares still share the same number of sides.

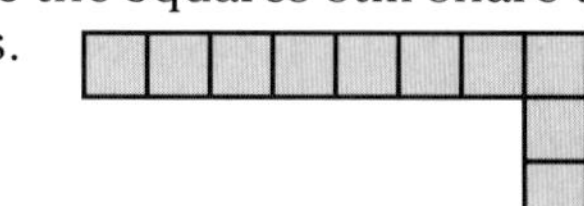

Pages 142–144 Lesson 3-5

5. 14 cm, 12 sq cm **7.** 42 ft; 90 sq ft **9.** 12 m **11.** 70 ft; 264 sq ft **13.** 34 km; 30 sq km **15.** 15 m; 9.86 sq m **17.** 3.6 m; 0.81 sq m **19.** 12.4 cm; 9.61 sq cm **21.** 19.2 mm, 23.04 sq mm **23.** 5 m **25.** 11 yd **27.** 39 ft **29.** 225 sq in.

31a.

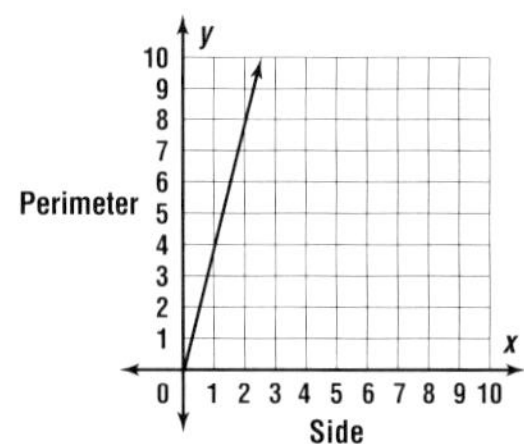

The perimeter is four times the length of a side.

31b. The area is the square of the length of a side.

31c. perimeter: $8x$ units; area: $4x^2$ square units **33a.** 150 ft **33b.** 900 sq ft

37. 20

39. 375 > 210 **41.** $20d + 28$

Page 145 Lesson 3-5B

1. 9 sq units **3.** 32 sq units **5.** Both give the same result.

Pages 148–150 Lesson 3-6

7. $y \leq -1$ **9.** $d < 75$

11. $m < -13$

13. $y < -4$

15. $a \leq -31$

17. $x < 6$ **19.** $y > 1$

21. $t \leq 14$

-2 0 2 4 6 8 10 12 14 16

25. $m < 19$

0 2 4 6 8 10 12 14 16 18 20

29. $x > 27$

0 3 6 9 12 15 18 21 24 27 30

33. $m > -9$

-10 -8 -6 -4 -2 0 2

37. $k < -26$

-28 -24 -20 -16 -12 -8 -4 0

41. $c \geq -25$

-40 -30 -20 -10 0

45. $x + 285 \geq 375$; $x \geq \$90$ **47.** $x \leq 137$ minutes or 2 hours 17 minutes **49.** 100 **51.** $-2a$
53a. 49 million

53b.

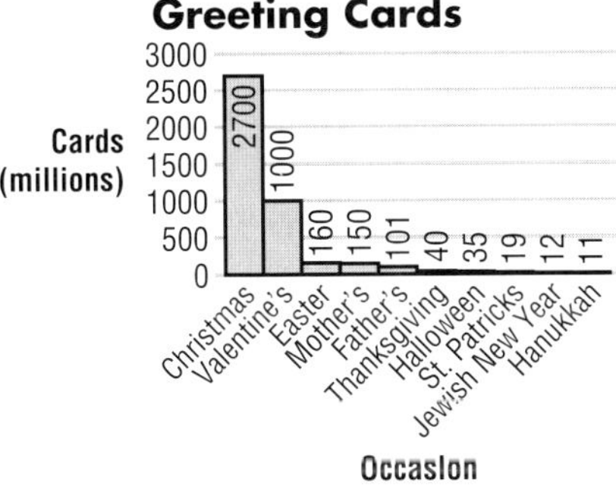

55. Identity, $\times$

Pages 154–155 Lesson 3-7

5. 5, no **7.** -3, yes

9. $x \leq 2$

-3 -2 -1 0 1 2 3

11. $p < 105$

0 20 40 60 80 100

15. $y > 9$

0 10

19. $p \geq -14$

-16 -12 -8 -4 0 4

23. $m \geq 14$

-7 0 7 14 21

27. $r > -21$

-24 -18 -12 -6 0

31. $x < -12$

-12 -10 -8 -6 -4 -2 0

33. 5 **35.** 800 **37.** $t > 6$ years **39.** $m < 28$ **41.** d
43. $-56y$

45a. $[(3 \times 4) + (2 \times 3) + (1 \times 2) + (1 \times 1)] \div 7$
45b. 3.0

Pages 158–159 Lesson 3-8

5. n = number; $4n < 96$; $n < 24$ **7.** Sample answer: The total cost of a car repair was \$187. If the labor cost \$169, how much did the parts cost?
9. n = number; $\frac{n}{-3} > 5$; $n < -15$ **11.** a = account balance; $a + 50 > 400$; $a > \$350$ **13.** p = original purchase price; $4p \leq 85{,}000$; $p \leq 21{,}250$
15. Sample answer: The low temperature was 16° F on Tuesday. This is at least 9° more than the Wednesday low. What was the low temperature on Wednesday?
17. $|x| < 4$; $-3, -2, -1, 0, 1, 2, 3$
19. a = average; $35a < 600$; $a < 17.1$ points per game **21.** c = changes in price; $8c = 32$; $c = 4$ times the 1971 price/oz. **23.** h = height; $5(8)h = 440$; $h = 11$ centimeters **25.** x = sample scores; $80 - 6 \leq x \leq 80 + 6$; $74 \leq x \leq 86$ strokes
27. $h > 13$;

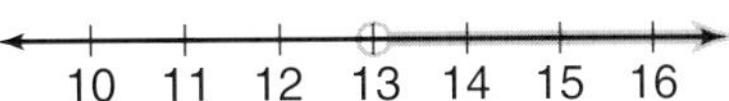

29. $-7y$ **31.** 24

Page 161 Chapter 3 Highlights

1. true **3.** false **5.** true **7.** false

Pages 162–164 Study Guide and Assessment

9. 51

50 55

13. 19

0 5 10 15 20

17. 126

0 25 50 75 100 125

21. 330

300 350

25. 24 **27.** 56.52 **29.** 36 in., 72 sq in. **31.** 40 ft; 75 sq ft

33. $x < -12$

-16 -12 -8 -4 0

37. $f > -4$

-5 0

41. $t < 132$

130 135

45. $f \geq 15$

0 5 10 15 20 25

47. $3x < 36$; less than \$12 **49.** more than 224 boxes **51.** 103

SELECTED ANSWERS

Chapter 4 Exploring Factors and Fractions

Pages 173–174 Lesson 4-1

7. 2 **9.** 2, 3, 5, 6, 10 **11.** yes **13.** yes **15.** 3 **17.** 2, 5, 10 **19.** 5 **21.** 3 **23.** none **25.** 2 **27.** yes **29.** yes **31.** no **33.** no **35.** 0, 4, or 8 **37.** 1035 **39.** 5555 **41b.** 1996, 2008, 2020, 2032, 2044 **43a.** 6 **43b.** 28 **45.** -267 **47.** -8 **49.** 33

Pages 177–179 Lesson 4-2

7. m^3 **9.** $2 \cdot 2 \cdot 2$ **11.** $10 \cdot 10 \cdot 10 \cdot 10 \cdot 10 \cdot 10$ **13.** 54 **15a.** $6(5^2)$ or 150 in^2 **15b.** 5^3 or 125 in^3 **15c.** The surface area is multiplied by 4 and the volume is multiplied by 8. **17.** n^4 **19.** 14^1 **21.** 3^{15} **23.** b^4 **25.** $12 \cdot 12$ **27.** $(-7)(-7)(-7)$ **29.** $\underbrace{6 \cdot 6 \cdot \ldots \cdot 6 \cdot 6}_{\text{20 factors}}$ **31.** $\underbrace{1 \cdot 1 \cdot \ldots \cdot 1 \cdot 1}_{\text{55 factors}}$ **33.** $(-1)(-1)(-1)(-1)$ **35.** 31 **37.** 162 **39.** -78 **41.** $(1 \times 10^2) + (4 \times 10^1) + (9 \times 10^0)$ **43.** 23,405 **45a.** -1 **45b.** $\frac{1}{2}$ **45d.** $\frac{1}{3}$ **47.** The volume of a cube with sides s units long is s^3. **49a.** $2^0, 2^1, 2^2$ **49b.** 2^3 **49c.** 2^{29} cents **49d.** 536,870,912 **49e.** $2^0 + 2^1 + 2^2 + \ldots + 2^{29}$ **49f.** No, the reward will cost \$10,737,418.23 over the thirty-day period.

51. $x > -12$

−14 −12 −10 −8 −6 −4 −2 0 2 4

53. 18°F **55.** $-8p$ **57.** $b + 4$

Page 180 Lesson 4-2B

1. 62 **3.** 2 **5.** 1024 **7.** The values are not the same because the order of operations is different. The value of the first expression is -24 and the value of the second expression is 0.

Pages 182–183 Lesson 4-3

5.

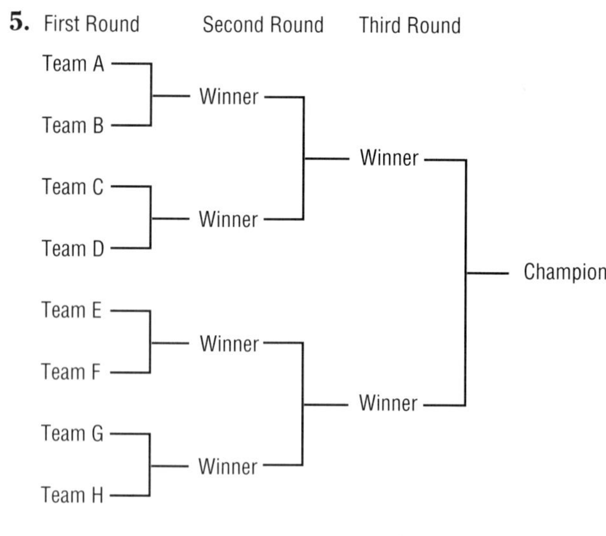

7. $(4 + 3) \times 6 + 3 = 45$ **9.** 1:23 P.M.

11.

4	15	1	14
9	6	12	7
16	3	13	2
5	10	8	11

13. about \$243 **15.** 30 feet **17.** $x \leq -9$ **19.** 4

Pages 187–188 Lesson 4-4

5. composite **7.** $2 \cdot 19$ **9.** $2 \cdot 2 \cdot 2 \cdot 7$ **11.** $-1 \cdot 2 \cdot 3 \cdot 7 \cdot a \cdot b \cdot c$ **13.** Damon is correct that 3067 is prime, but it has two factors, 1 and 3067. **15.** prime **17.** prime **19.** $3 \cdot 17$ **21.** $1 \cdot 41$ **23.** $2 \cdot 5 \cdot 11$ **25.** $3 \cdot 3 \cdot 3 \cdot 3$ **27.** $1 \cdot 13$ **29.** $2 \cdot 3 \cdot 7 \cdot x \cdot y \cdot y$ **31.** $3 \cdot 7 \cdot x \cdot y \cdot y \cdot y$ **33.** $2 \cdot 2 \cdot 7 \cdot f \cdot f \cdot g$ **35.** $2 \cdot 3 \cdot 5 \cdot 7 \cdot m \cdot n \cdot n \cdot n$ **37.** $3 \cdot 5 \cdot 5 \cdot m \cdot m \cdot k$ **39.** 211 **41.** 31 and 53 **43a.** 23 **43b.** yes **45.** 16 **47.** 52 inches; 153 square inches **49.** > **51.** $x + 15$

Page 189 Lesson 4-4B

1. Sample answer: 67, 121, 55, 81, 63, 66

Pages 192–194 Lesson 4-5

5. 2 **7.** 1 **9.** b **11.** 12 inches **13.** 8 **15.** 12 **17.** 36 **19.** $8x$ **21.** $14b$ **23.** 9 **25.** 6 **27.** 6 **29.** $6a$ **31.** yes **33.** no **35.** no **37.** 15 **39.** 18 shelves **41.** $2 \cdot 2 \cdot 2 \cdot 5 \cdot 7 \cdot 11$ **43.** 10 **45.** 135 **47.** 1728

Page 194 Self Test

1. 3 **3.** 3 **5.** 2^3 **7.** 121 **9.** $-1 \cdot 2 \cdot 13$ **11.** 14 **13.** 9

Page 195 Lesson 4-6A

1. Shade 2 sections. **3.** They are equal in size. **5.** Yes; sample answer $\frac{1}{4}$ **7.** yes

Pages 198–199 Lesson 4-6

7. $\frac{1}{7}$ **9.** $\frac{11}{15}$ **11.** $\frac{5}{7}$ **13.** $\frac{9}{22}$ **15.** $\frac{1}{11t}$ **17.** simplified **19.** $\frac{1}{7}$ **21.** $\frac{17}{19}$ **23.** $\frac{1}{3}$ **25.** $\frac{5}{8}$ **27.** simplified **29.** simplified **31.** $\frac{62}{111}$ **33.** $\frac{4x}{5y}$ **35.** $\frac{41}{7}$ **37.** $\frac{z^3}{x}$ **39.** $\frac{10p}{13q}$ **41.** simplified **43.** simplified **45.** $\frac{22}{60 \cdot 24}$ or $\frac{11}{720}$ **47a.** 2:6 **47b.** $\frac{1}{3}$ **47c.** Sample answer: 1 oz. alcohol, 3 oz. water **49.** 108 **51.** $68d$ **53.** 12

Pages 202–204 Lesson 4-7

5. 70 **7.** $60a$ **9.** 630 **11.** 10 **13.** 100 **15.** < **17a.** East Central **17b.** West Central **17c.** New England **19.** 18 **21.** 288 **23.** 84 **25.** $84y$ **27.** 40 **29.** 180 **31.** $105t^2$ **33.** 16 **35.** 35 **37.** 220 **39.** $25b$ **41.** > **43.** < **45.** < **47.** < **49.** $\frac{3}{8} < \frac{5}{12}$ **51.** dog, $\frac{7815}{12} > \frac{4723}{15}$ **53a.** video game **53b.** VCR **53c.** on-line services **55.** $x > 576$ **57.** -98 **59.** Utah

61. **National Park Acreage**

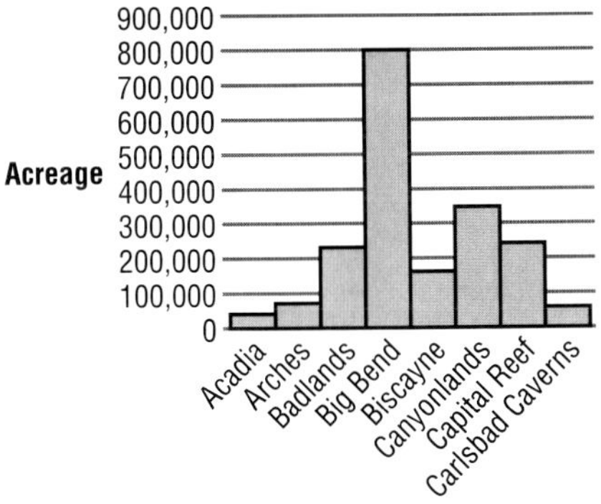

Pages 207–209 Lesson 4-8

5. 10^7 **7.** 3^{10} **9.** 8^6 **11.** y^2 **13.** about 10 times more intense **15.** w^6 **17.** b^7 **19.** 10^7 **21.** a^4 **23.** $20x^7$ **25.** f^{12} **27.** $12a^5$ **29.** a^3b^4 **31.** t **33.** k^2m **35.** w^4y^2 **37.** $3n^8$ **39.** 9 **41.** 3 **43.** 1 **45a.** 100 times **45b.** base **47.** $<$ **49.** 2

51. −7

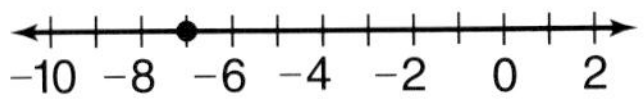

53. II **55.** $2x - 4$

Pages 212–214 Lesson 4-9

5. $\frac{1}{15^5}$ **7.** 10^{-7} **9.** 3^{-2} **11.** $\frac{3}{16}$ **13.** $\frac{1}{4}$ **15.** $\frac{1}{(-2)^5}$ **17.** $\frac{1}{s^2t}$ **19.** $\frac{2}{(xy)^2}$ **21.** 5^{-3} **23.** 2^{-4} or 16^{-1} **25.** $2 \cdot 3^{-2}$ **27.** fg^{-3} **29.** $\frac{1}{27}$ **31.** $\frac{2}{3}$ **33.** a^3 **35.** c^7 **37.** a^{2b-a} **39a.** $\frac{6.67}{10^{11}}$Nm²/kg² **39b.** 0.0000000000667 Nm^2/kg^2 **41.** $-50x^4$ **43.** 7^5 **45.** $2230 + x = 4750$; $2520 **47.** $n - 8 = 3$; 11

Page 215 Chapter 4 Highlights

1. greatest common factor **3.** factors **5.** multiple **7.** least common denominator **9.** ratio

Pages 216–218 Study Guide and Assessment

11. 5 **13.** 3 **15.** 2, 3, 6 **17.** 2, 3, 6 **19.** $(c + 2)^3$ **21.** 1^3 **23.** $k \cdot k \cdot k \cdot k \cdot k \cdot k \cdot k \cdot k \cdot k$ **25.** $32 \cdot 32 \cdot 32 \cdot 32 \cdot 32 \cdot 32 \cdot 32 \cdot 32 \cdot 32 \cdot 32 \cdot 32$ **27.** $-1 \cdot 2 \cdot 3 \cdot 19$ **29.** $-1 \cdot 3 \cdot 5 \cdot 5 \cdot 11 \cdot x \cdot x \cdot y$ **31.** $-1 \cdot 3 \cdot 3 \cdot 7 \cdot 29 \cdot j \cdot k \cdot k \cdot k$ **33.** $2 \cdot 5 \cdot 5 \cdot 11 \cdot m \cdot n \cdot q \cdot q$ **35.** 13 **37.** $35a^2b^2$ **39.** 5 **41.** 1 **43.** $\frac{19}{20}$ **45.** $\frac{8}{11}$ **47.** $\frac{1}{2}$ **49.** simplified **51.** 60 **53.** $30jk$ **55.** 180 **57.** $36abcd$ **59.** $>$ **61.** $>$ **63.** d^5 **65.** $(-4)^4$ **67.** $2x^{22}$ **69.** $110a^{10}b^3$ **71.** $\frac{4d}{c^2}$ **73.** $\frac{4x}{(yz)^2}$ **75.** $3f^9$ **77.** 21 games **79.** home improvement

Chapter 5 Rationals: Patterns in Addition and Subtraction

Pages 227–228 Lesson 5-1

7. $\frac{1}{20}$ **9.** $\frac{2}{9}$ **11.** rationals **13.** none **15.** $=$ **17a.** $\frac{1}{16}$ **17b.** Rational; it is a fraction. **19.** $-\frac{7}{10}$ **21.** $\frac{57}{100}$ **23.** $\frac{3}{100}$ **25.** $-\frac{1}{3}$ **27.** $\frac{25}{99}$ **29.** $2\frac{34}{99}$ **31.** rationals **33.** whole numbers, integers, rationals **35.** none **37.** rationals **39.** rationals **41.** none **43.** $>$ **45.** $=$ **47.** $>$ **49.** $<$ **51.** $\frac{2}{7}$ **53.** Yes; $\frac{6}{2.4} = \frac{5}{2}$ or $\frac{60}{24}$. **55.** $\frac{1}{40}$ **57a.** $\frac{1}{30}$ **57b.** Yes; it can be written as a fraction. **59.** $60x^2$ **61.** $b < -18$ **63.** −19 **65.** 8 servings

Pages 231–233 Lesson 5-2

5. 6 **7.** 803 **9.** 11 **11.** 1 **13.** $\frac{1}{2}$ **15.** a **17.** 40 **19.** 9 **21.** 29 **23.** 60 **25.** 21 **27.** 1 **29.** $7\frac{1}{2}$ **31.** 1 **33.** 5 **35.** 110 **37.** $28\frac{1}{2}$ **39.** 50 **41.** $5 **43.** $3 **45.** No, the change should be about $7. **47.** Sample answer: $11\frac{1}{10} + 1\frac{7}{15}$ **49.** $59 **51.** more than half **53.** $\frac{13}{50}$ **55.** −16 **57.** false **59a.** $2(6.25 + 2.25)$ or $2(6.25) + 2(2.25)$ **59b.** $17

Pages 236–238 Lesson 5-3

5. 3.68 **7.** −4.91 **9.** 77.33 **11.** −6.4 **13.** 43.9 **15.** $-4.1m$ **17.** 14.1 **19.** 1.5 **21.** 63.431 **23.** 2.3 **25.** −26.4 **27.** −51.73 **29.** $12.3m$ **31.** $6.9y$ **33.** $11.5x - 5$ **35.** 29.4 **37.** 12.56 **39.** 12.67 **41.** 24.45 left; no; third side must be less than 20.55 feet **43.** 378.5 **45a.** 24.3 bags per 10,000 passengers **45b.** Sample answer: Bad weather causes changes in flight schedules. **47.** yes; $2 + 4 + 1 + 2 = $9 **49.** $2 < a$ **51.** −11 **53.** 12

Pages 241–243 Lesson 5-4

5. $1\frac{2}{7}$ **7.** $\frac{1}{3}$ **9.** 1 **11a.** $\frac{9}{25}$ **11b.** $\frac{1}{5}$ of each dollar **13.** $\frac{1}{5}$ **15.** $\frac{4}{9}$ **17.** $1\frac{1}{3}$ **19.** $1\frac{1}{3}$ **21.** $2\frac{1}{3}$ **23.** 1 **25.** $\frac{1}{2}$ **27.** $\frac{2}{3}$ **29.** $-\frac{1}{6}$ **31.** $-2r$ **33.** $\frac{2}{5}a$ **35.** $\frac{1}{3}b$ **37.** Sample answer: $\frac{2x}{5} - \frac{1}{5}$ **39.** $5\frac{3}{8}$ inches **41.** $3\frac{1}{2}$ yards **43a.** $\frac{25}{170}$ or $\frac{5}{34}$ **43b.** $\frac{1}{28}$ **45.** $c \geq -13$ **47.** $-22a$

Pages 246–247 Lesson 5-5

7. $\frac{1}{8}$ **9.** $-\frac{1}{8}$ **11.** $\frac{11}{14}$ **13.** $\frac{5}{9}$ **15.** $-\frac{1}{12}$ **17.** $\frac{15}{26}$ **19.** $1\frac{5}{6}$ **21.** $3\frac{1}{6}$ **23.** $4\frac{1}{4}$ **25.** $16\frac{10}{21}$ **27.** $10\frac{1}{15}$ **29.** $-\frac{17}{90}$ **31.** $3\frac{5}{24}$ **33.** $-1\frac{3}{8}$ **35.** $1\frac{5}{24}$ **37.** Sample answer: $\frac{1}{18} + \frac{1}{6}$ **39.** $\frac{11}{102}$ **41.** $\frac{t^3}{s^2}$ **43.** −5 **45.** 13 **47.** $18x$

Page 247 Self Test

1. $-\frac{1}{6}$ **3.** $1\frac{4}{9}$ **5.** 1 **7.** 14.52 **9.** 30.6 **11.** $5\frac{1}{7}$ **13.** $8\frac{3}{4}$ inches

Pages 249–250 Lesson 5-6

5. $-\frac{1}{15}$ **7.** −5.28 **9.** 3.25 **11.** −8.7 **13.** 21.9 miles

15. 6.4 **17.** $15\frac{3}{4}$ **19.** 13.4 **21.** $-\frac{31}{36}$ **23.** $1\frac{39}{40}$ **25.** -2.8 **27.** 8.01 **29.** $12\frac{17}{18}$ **31.** $\frac{19}{312}$ **33.** 688.2 million barrels **35.** 0.39 million or 390,000 metric tons **37.** $4\frac{11}{12}$ **39.** $d \geq -84$ **41.** 18 **43.** 15

Pages 253–254 Lesson 5-7

5. $x \leq 2\frac{23}{30}$

9. $y < 1.92$

13. $y < -\frac{1}{6}$

17. $d < 2\frac{5}{6}$

19. $r \leq 1\frac{7}{8}$

23. $h < 15.96$

27. $t > \frac{5}{108}$

29. x is between 0 and 1 or x is less than -1. **31.** $x > 9.7$ **33a.** Sample answer: The average salary of a high school graduate is slightly less than half of the average salary of a college graduate. **33b.** Sample answer: No, if it was that low the inequality would be written differently. The inequality means that it is less than but close to one-half. **35.** $3\frac{3}{4}$ **37.** 2 **39.** $-30abc$ **41a.** Los Angeles **41b.** 20 cents per mile

Pages 256–257 Lesson 5-8

5. inductive **7.** inductive **9.** 18, 21 **11.** $1111^2 = 1{,}234{,}321$ **13a.** ground you; deductive **13b.** No; there could be another reason. **15a.** 1010 **15b.** 110 **15c.** Inductive reasoning does not always work, so it would not be wise to use it to prove something. **17.** pink **19.** $z \leq 0.91$ **21.** 1 **23.** 10

Pages 260–262 Lesson 5-9

5. yes; 17, 20, 23 **7.** no; 2, -4, -11 **9.** no; 23, 30, 38 **11a.** 9.95, 12.90, 15.85, 18.80, 21.75, 24.70 **11b.** \$36.50 **13.** yes; 2.5, 2.0, 1.5 **15.** no; 25, 36, 49 **17.** no; 32, 64, 128 **19.** yes; 55, 46, 37 **21.** yes; 3, -1, -5 **23.** yes; 9.55, 10.57, 11.59 **25.** no; 46, 59, 74 **27.** no; $\frac{7}{8}, \frac{8}{9}, \frac{9}{10}$ **29.** 7; 11 **31.** 100, 125, 150, 175, 200, 225 **33a.** 50 s **33b.** $d = \frac{1}{5}t$ **35.** 320 feet/second **37.** deductive **39.** 6

41.

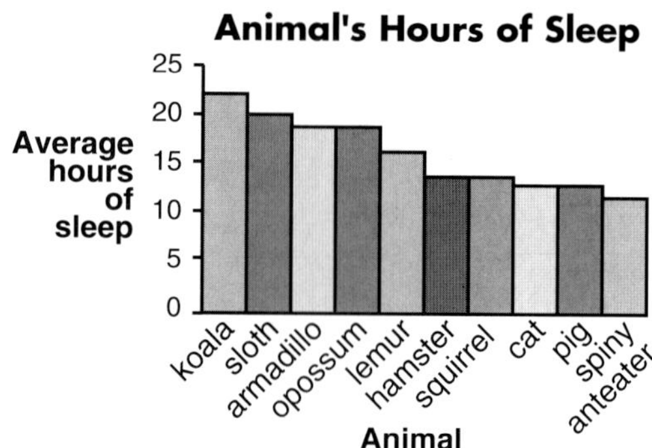

Page 263 Lesson 5-9B

1. 13 **3.** 2, 5, 13, 34, ... The sequence is the 3rd, 5th, 7th, ... terms of the Fibonacci sequence.

Page 265 Chapter 5 Highlights

1. B **3.** I **5.** C **7.** A

Pages 266–268 Study Guide and Assessment

9. integers, rationals **11.** none **13.** $-9\frac{9}{20}$ **15.** $-\frac{13}{100}$ **17.** $4\frac{8}{33}$ **19.** $8\frac{89}{100}$ **21.** 8 **23.** 10 **25.** 1 **27.** $2\frac{1}{2}$ **29.** 11.23 **31.** -7.68 **33.** -17.95 **35.** $13\frac{7}{9}$ **37.** $\frac{1}{4}$ **39.** $-\frac{3}{5}$ **41.** $-\frac{1}{15}$ **43.** $6\frac{11}{12}$ **45.** $5\frac{1}{36}$ **47.** -8.8 **49.** $1\frac{13}{16}$ **51.** $1\frac{5}{6}$

53. $f \leq 4\frac{1}{8}$

55. $q > 5.3$

57. $m > -6.151$

59. $a \leq \frac{9}{10}$

61. yes; 13, 5, -3 **63.** no; $\frac{4}{5}, \frac{2}{3}, \frac{0}{1}$ **65a.** $\frac{35}{1200}$ or $\frac{7}{240}$ **65b.** Rational, it is a fraction. **67.** $6\frac{5}{8}$ inches **69a.** 25 units **69b.** The lengths are the squares of 1, 2, 3, ... It is not an arithmetic sequence because there is no common difference.

Chapter 6 Rationals: Patterns in Multiplication and Division

Pages 277–279 Lesson 6-1

7. $2.\overline{6}$ **9.** 0.125 **11.** $-\frac{3}{4}$ **13.** $-\frac{3}{4}$ **15.** 0.7 **17.** 0.5 **19.** 0.7 **21.** 0.3125 **23.** $-3.\overline{36}$ **25.** -4.3125 **27.** $>$ **29.** $<$ **31.** $<$ **33.** $<$ **35.** $>$ **37.** $<$ **39.** .7777777 **41.** 0.7777778 **43.** $0.\overline{3}$, 0.25, 0.7, 0.8, $0.41\overline{6}$, $0.\overline{571428}$, 0.12 **43a.** terminating: $\frac{1}{4}, \frac{7}{10}, \frac{4}{5}, \frac{3}{25}$;

repeating $\frac{1}{3}, \frac{5}{12}, \frac{4}{7}$ **43b.** $3 = 1 \times 3$; $4 = 2^2$; $10 = 2 \times 5$; $5 = 1 \times 5$; $12 = 2^2 \times 3$; $7 = 1 \times 7$; $25 = 5^2$ **43c.** 2 and 5 **45.** 0.33 **47.** $82.38; $46.88; $36.75 **49.** 64, 58, 52 **51.** b^{11} **53.** 20 feet **55.** $r > 7$ or $r \geq 8$ **57.** 27

Pages 282–283 Lesson 6-2

5. c **7.** d **9.** $18 \div 3 = 6$ **11.** $8 \div 1 = 8$ **13.** $20 \times \frac{1}{4} = 5$ **15.** 160 **17.** 4 **19.** 5 **21.** 2 **23.** 6 **25.** 20 **27.** 150 **29.** 5 **31.** 10 **33.** If the numbers were rounded up the actual answer is less than the estimate. If the numbers were rounded down the product is greater than the estimate. **35.** about 15,000,000 **37.** 5 mph **39.** $\frac{17}{14}$ or $1\frac{3}{14}$ **41.** 27 **43.** 2

Pages 286–288 Lesson 6-3

5. d **7.** $\frac{1}{2}$ **9.** $\frac{4}{9}$ **11.** -24 **13.** $\frac{3}{8}$ **15.** $\frac{9}{25}$ **17a.** 3.8¢ **17b.** 28 years, 1.5 months **19.** $\frac{1}{7}$ **21.** $-1\frac{1}{2}$ **23.** $-4\frac{1}{2}$ **25.** -3 **27.** -24 **29.** $\frac{9}{25}$ **31.** $6\frac{12}{25}$ **33.** $\frac{4}{15}$ **35.** $1\frac{1}{18}$ **37.** $\frac{4}{5}$ **39.** $\frac{11}{36}$ **41.** $\frac{7}{2} \times \frac{6}{3}$ or $\frac{6}{2} \times \frac{7}{3}$; use the greatest factors as numerators and the least as denominators. **43.** $3\frac{3}{4}$ hours **45a.** 6 picas **45b.** $4\frac{1}{2}$ picas **45c.** about 91 characters **47.** 15 **49.** $2^3 \cdot 3^2$ **51.** $-20xy$ **53.** 15

Pages 291–293 Lesson 6-4

7. no **9.** yes **11.** -1 **13.** $-3\frac{3}{4}$ **15.** $\frac{20}{57}$ **17.** $\frac{11}{18}$ **19.** $-\frac{3}{7}$ **21.** $\frac{5}{8}$ **23.** $\frac{2}{3}$ **25.** $\frac{y}{x}$ **27.** $-1\frac{2}{25}$ **29.** -10 **31.** $-1\frac{1}{2}$ **33.** 27 **35.** $34\frac{2}{7}$ **37.** $-1\frac{1}{2}$ **39.** $6\frac{2}{3}$ **41.** $-2\frac{5}{8}$ **43.** $-2\frac{2}{7}$ **45.** $12\frac{2}{3}$ **47.** 21 uniforms plus extra fabric remaining **49.** 8 days **51.** $\frac{2}{5}$ **53.** 6 **55.** II **57.** $13h + 23$

Page 294 Lesson 6-5A

1. 24 squares; 24 hundredths or 0.24 **3.** $0.4 \times 0.2 = 0.08$, each square represents one one-hundredth **5.** 5

Pages 297–299 Lesson 6-5

5. 1.44 **7.** 0.7 **9.** 0.62 **11.** -90 **13.** 25.12 **15.** Inverse, multiplication **17.** 70 pounds **19.** 1.4147 **21.** -5.5875 **23.** -6.592 **25.** 0.32 **27.** -17 **29.** 1460 **31.** 5.6 **33.** -15 **35.** 1.21 **37.** -1.16 **39.** Identity, multiplication **41.** Associative, multiplication **43.** Commutative, multiplication **45.** -2.5; -2.6 **47.** -10; -10.2 **49.** 0.1; 0.2 **51.** Muesli, muesli 18.1¢ cents per oz; frosted corn flakes 18.4¢ per oz. **53.** $3.2 billion **55.** $2\frac{7}{12}$ **57.** integers, rationals **59.** $314 **61.** Commutative, addition

Page 299 Self Test

1. 0.875 **3.** $-0.\overline{72}$ **5.** 16 **7.** 3 **9.** -24

Pages 304–306 Lesson 6-6

5. 29.4, 33, 12 **7.** 100.4, 105, 78 **9.** 49.9, 45, 45 **11.** 6.0, 6.5, 3.6 and 7.2 **13.** 104.4, 105, 113 **15.** 0.92, 0.85, none **17.** 85.4 **19.** 86 **21a.** 86 **21b.** no effect **23a.** 90 **23b.** no effect **25.** 91 **27.** Sample answer: 2, 4, 6, 8, 8, 10, 12, 14 **31.** -12.9824 **33.** $y \leq -24$ **35.** Multiplication property of zero

Page 307 Lesson 6-6B

1. 5.55, 5.75 **3.** -16.36, -15 **5.** 73.6, 49.7 **7.** The mean is affected more by very large or very small numbers. So the median is a better representation if the data has extreme values. If the data are more closely clustered, the mean is a good representation.

Pages 310–311 Lesson 6-7

5. -15

-20 -10 0 10

7. $-9 < f$

-10 -8 -6 -4 -2 0 2

9. $\frac{18}{25}$

0 $\frac{1}{5}$ $\frac{2}{5}$ $\frac{3}{5}$ $\frac{4}{5}$ 1

11. 6 cm **13.** -0.5 **15.** $16\frac{1}{2}$ **17.** $x < -14.4$ **19.** $-6\frac{1}{4}$ **21.** -3.55 **23.** $z \leq -1.3$ **25.** $\frac{-7}{10}$ **27.** 19 **29.** $x < \frac{1}{5}$

31. $y < \frac{15}{32}$

0 $\frac{1}{4}$ $\frac{1}{2}$ $\frac{3}{4}$ 1

35. $16.66 **37.** 88.8 **39.** $1\frac{3}{14}$ **41.** -36

Pages 315–316 Lesson 6-8

7. no **9.** yes, $\frac{1}{3}$; 1, $\frac{1}{3}, \frac{1}{9}$ **11.** yes, 2; 8, 16, 32 **13.** 2048 **15.** yes, $\frac{1}{2}$; $\frac{3}{2}, \frac{3}{4}, \frac{3}{8}$ **17.** yes, $\frac{1}{12}$; $\frac{1}{144}, \frac{1}{1728}, \frac{1}{20736}$ **19.** yes, $\frac{1}{2}$; $\frac{1}{32}, \frac{1}{64}, \frac{1}{128}$ **21.** yes, -2; 112, -224, 448 **23.** no **25.** no **27.** yes, $\frac{1}{4}$; $\frac{3}{32}, \frac{3}{128}, \frac{3}{512}$ **29.** 2, -4, 8, -16, 32 **31.** 80, 100, 125 **33.** 1 **35a.** 1.3 **35b.** 37,129,300 **37.** 40 **39.** 27 **41.** $\frac{1}{3^4}$ or $\frac{1}{81}$ **43.** 5 **45.** 34

Pages 319–320 Lesson 6-9

5. yes **7.** 8.490×10^3 **9.** 8.479×10^2 **11.** 61,000 **13.** 31,557,600 seconds in a year; (31,557,600) (3.00×10^5) or about 9.47×10^{12} km **15.** 5.9×10^{-3} **17.** 1.5×10^{-4} **19.** 4.98×10^{-8} **21.** 7.01×10^8 **23.** 0.0000056 **25.** 0.009001 **27.** 0.00005985 **29.** 1454 **31.** 0.00005 **33.** 700,080,003 **35.** 3.14 **37a.** Bezymianny **37b.** Ngauruhoe **37c.** Bezymianny; Santa Maria; Agung; Mount St. Helens tied with Hekla, 1947;

SELECTED ANSWERS

Hekla, 1970; Ngauruhoe **39.** $x \geq 2\frac{1}{3}$ **41.** −13 **43.** $b + 3$

Page 321 Highlights

1. true **3.** false; Each term in a geometric sequence increases or decreases by a common ratio. **5.** false **7.** true **9.** false; It is possible to have a set of numerical data with no mode.

Pages 322–324 Study Guide and Assessment

11. 0.625 **13.** 0.425 **15.** $0.2\overline{6}$ **17.** 21 **19.** 9 **21.** $-\frac{8}{45}$ **23.** $-1\frac{1}{5}$ **25.** $\frac{64}{125}$ **27.** $\frac{3}{4}$ **29.** $-\frac{1}{9}$ **31.** 28.28 **33.** 0.0629 **35.** $0.291\overline{6}$ **37.** −47.7 **39.** 20.6; 20; 18, 21, and 25 **41.** 14.17; 14; 14 **43.** 9 **45.** −12 **47.** $-4 \geq n$ **49.** 5.3 **51.** no **53.** no **55.** no **57.** 6.74×10^{-6} **59.** 5.81×10^{3} **61.** 57,200 **63.** 202,000,000 **65.** 132¢ or $1.32 **67.** 15.2 times

Chapter 7 Solving Equations and Inequalities

Pages 331–332 Lesson 7-1

5. 3 ounces **7.** $16,200 **9.** $5 **11.** 729 people **13.** 29.8.92 and 2.9.92 **15.** 0.0000006789 **17.** whole numbers, integers, rationals **19.** 31

Page 333 Lesson 7-2A

1. −3 **3.** 4 **5.** 2 **7.** Sample answer: You undo operations to find the value of x.

Pages 336–337 Lesson 7-2

5. 13 **7.** −4 **9.** −3 **11.** 6 ounces **13.** 4 **15.** 17 **17.** −1.2 **19.** 84 **21.** 33 **23.** 51 **25.** −153 **27.** 22.5 **29.** −93 **31.** 16 **33.** Sample answer: $6x + 3 = 2$ **35.** 4, 5 **37.** 90 **39.** > **41.** y^3 **43.** false

Pages 339–340 Lesson 7-3

3. $17 - 2x = 5$; 6 **5a.** $2682 = 4d - 354$ **5b.** 759 thousand **7.** C **9.** E **11.** $\frac{n+6}{7} = 5$; 29 **13.** $\frac{c}{-4} - 8 = -42$; 136 **15.** $5x - 10 = 145$; 31 **17.** 7 yd, 9 yd, 11 yd **19a.** $2w + 50 = 130$ **19b.** 40 feet **21.** 50 **23.** 72 **25.** associative, addition

Pages 343–344 Lesson 7-4

7. 11.31 cm **9.** 10.99 m **11.** 12.57 cm **13.** 34.56 ft **15.** 43.96 mm **17.** 59.032 m **19.** 8.164 yd **21.** $7\frac{1}{3}$ ft **23.** C **25.** B **27.** 0.65 m **29.** 2 units **31.** about 238.76 m **33a.** The perimeter of the square is longer. **33b.** Sample answer: The perimeter of a square is 4 times the length of one side. The circumference of a circle is approximately 3.14 times the diameter. 4 > 3.14 **35.** 3.7 million **37.** $2 \cdot 2 \cdot 5 \cdot 11 \cdot p \cdot q \cdot q$ **39.** >

Page 345 Lesson 7-5A

1. −8 **3.** 4 **5.** 1 **7.** Sample answer: No; the result is the same.

Pages 348–350 Lesson 7-5

5. Sample answer: Add 1 to each side. Subtract $3x$ from each side. The solution is 3. **7.** Sample answer: Simplify the left side to $3a + 66$. Subtract $3a$ from each side. The solution is 4. **9.** Sample answer: Add n to each side. Subtract 4 from each side. The solution is 3. **11.** 21 **13.** 2 **15.** −0.5 **17.** 5 **19.** all numbers **21.** ∅ **23.** 2 **25.** all numbers **27.** 2.48 **29.** 100 m by 130 m **31.** 25 ft by 30 ft **33.** $3y - 14 = y$; 7 **35a.** $199{,}000 - 700x$ **35b.** $165{,}000 + 3700x$ **35c.** $199{,}000 - 700x = 165{,}000 + 3700x$ **35d.** about 8 years **37a.** $20 **39.** 32 **41.** $\frac{2}{7}$ **43.** 52

Page 350 Self Test

1. $569.19 **3.** −12 **5.** $4x - 12 = 18$; 7.5 **7.** 1.02 yd **9.** 65 yards by 120 yards

Pages 353–354 Lesson 7-6

5. $x \leq 7$

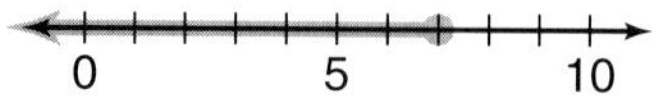

0 5 10

7. $k > 2$

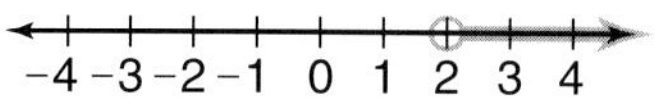

−4 −3 −2 −1 0 1 2 3 4

9. $m < -2$

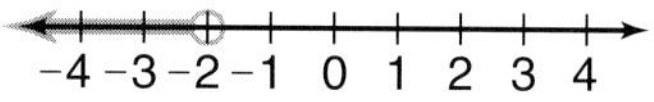

−4 −3 −2 −1 0 1 2 3 4

11. Sample answer: If Heather can find CDs that cost no more than $12.50 each, she can buy the shoes and two CDs.

13. $u \geq -4$

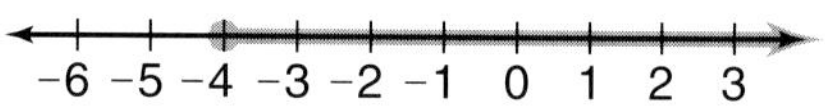

−6 −5 −4 −3 −2 −1 0 1 2 3

17. $j \geq -3$

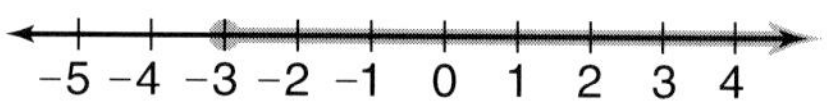

−5 −4 −3 −2 −1 0 1 2 3 4

21. $1.17 > t$

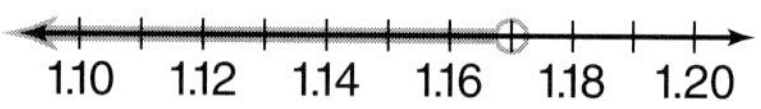

1.10 1.12 1.14 1.16 1.18 1.20

25. $x \leq 4.5$

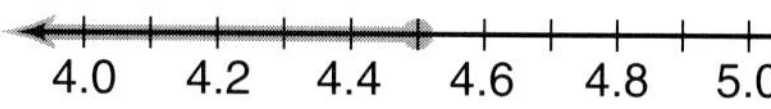

4.0 4.2 4.4 4.6 4.8 5.0

29. $c < -4$

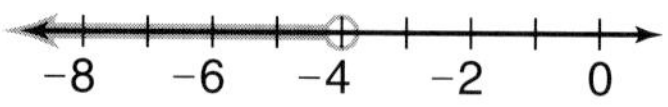

−8 −6 −4 −2 0

31. $-5 < x < 1$ **33.** $s \geq 97$ **35.** 258 **37.** 8 **39.** $n + 3$

Pages 356–357 Lesson 7-7

5. $2n - 9 > 11$; $n > 10$ **7.** $55c + 35 \leq 200$; $c \leq 3$ **9.** $4x + 4 \geq 16$; $x \geq 3$ **11.** Sample answer: A stove and a freezer weigh at least 260 kg. The stove weighs 115 kg. What is the weight of the freezer? **13.** $3(h - 1) \geq 18$; $h \geq 7$ **15.** 17, 18 **17.** $t \geq \frac{11}{3}$ **19.** 3, 4, or 5 **21.** $50 > c > 6$ **23.** $x > -5$ **25a.** $\frac{1.7}{10^{24}}$ **25b.** 0.0000000000000000000000017 **27.** $-126b$

Pages 360–361 Lesson 7-8

5. mL **7.** kg **9.** 0.040 **11.** 12.8 km **13.** cm or m **15.** L **17.** 0.4 **19.** 9.4 **21.** 6 **23.** 8000 **25.** 1000

27. 316 **29.** 946 mL **31.** 603,200 cm **33.** 13 cm **35.** 65 cm, 55 cm **37.** 0.035 m **41.** $b > 7$
43. 17.25 **45.** $-20\frac{4}{5}$

Page 363 Highlights
1. false **3.** true **5.** true **7.** true

Pages 364–366 Study Guide and Assessment
9. 12 **11.** 60 **13.** -20 **15.** 0.2 **17.** 13 **19.** \$250 **21.** 22.0 in. **23.** 30.8 m **25.** -2 **27.** -1 **29.** 14

31. $n > 6$

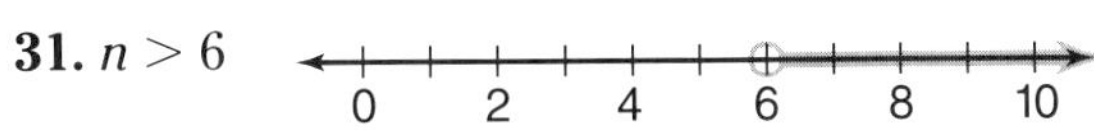

33. $r \leq 98$

35. $c < 7.5$

37. $8n + 2 \geq 18$; $n \geq 2$ **39.** $x + (x + 1) > 47$; $x > 23$ **41.** 0.006 **43.** 43,000 **45.** 300,000 **47.** 8800 **49.** 140.8 beats per minute **51.** less than 16.67 minutes

Chapter 8 Functions and Graphing

Pages 374–377 Lesson 8-1
5. $\{-1.3, 4, -2.4\}$; $\{1, -3.9, 3.6\}$ **7.** $\{(-1, 3), (0, 6), (4, -1), (7, 2)\}$; $D = \{-1, 0, 4, 7\}$, $R = \{3, 6, -1, 2\}$ **9.** $\{(-2, 1), (0, -1), (2, 2)\}$; $D = \{-2, 0, 2\}$, $R = \{-1, 1, 2\}$ **11.** yes **13.** no **15.** $\{(8, 4), (9, 3), (10, 2), (11, 1), (12, 0)\}$ **17.** $\{-1, 4, 2, 1\}$; $\{6, 2, 36\}$ **19.** $\{1.4, -2, 4, 6\}$; $\{3, 9.6, 4, -2.7\}$
21. $\left\{-\frac{1}{2}, 4\frac{2}{3}, -12\frac{3}{8}\right\}$; $\left\{\frac{1}{3}, -17, 66\right\}$ **23.** $\{(-4, -2), (2, 1), (0, 2), (1, -3), (3, 1)\}$; $D = \{-4, -2, 0, 1, 3\}$, $R = \{-2, 1, 2, -3\}$ **25.** $\{(5, 4), (2, 8), (-7, 9), (2, 12), (5, 14)\}$; $D = \{5, 2, -7\}$, $R = \{4, 8, 9, 12, 14\}$ **27.** $\{(-3, -2), (-2, -1), (0, 0), (1, 1)\}$; $D = \{-3, -2, 0, 1\}$, $R = \{-2, -1, 0, 1\}$ **29.** yes **31.** no **33.** yes **35.** no **37.** no **39.** no **41.** yes **43a.** \$6.95 **43b.** more than \$30 and less than or equal to \$70 **43c.** Yes, each x-value (total price) has exactly one y-value (shipping cost).
45. $-2 > b$ **47.** $z \geq -1\frac{3}{5}$ **49.** 16 miles
51. Sample answer: two times the sum of a number and three

Page 378 Lesson 8-2A
1. Sample answer: Generally, as shoe length increases, armspan increases also. **3.** Sample answer: Because points appear to be scattered around a general area.

Pages 381–384 Lesson 8-2
5. negative **7.** positive **9.** positive **11.** none **13.** negative **15.** positive **17.** no **19.** negative **21.** no **23a.** The data show a positive relationship. **23b.** Sample answer: about 62 wpm **23c.** Sample answer: between 3 and 4 weeks **23d.** The more experience a student has, the more words per minute he or she can key.

25a.

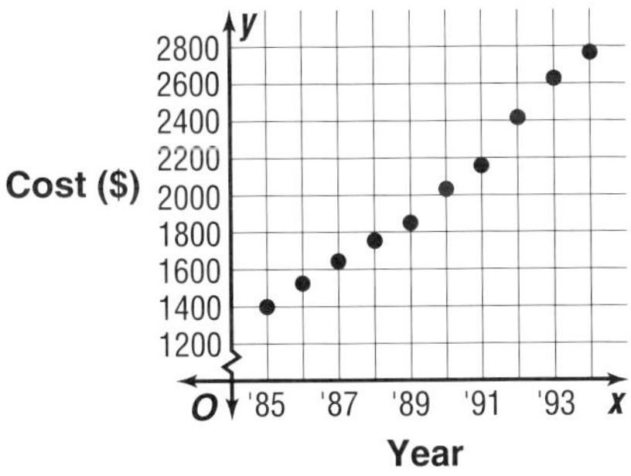

25b. Yes; positive.
27a.

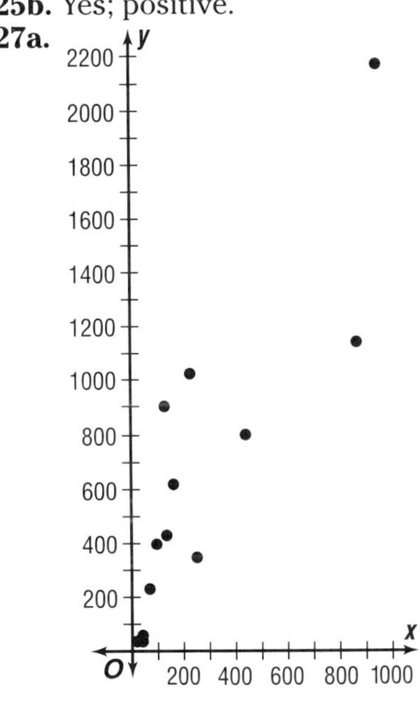

27b. yes; positive **27c.** Sample answer: yes; you could estimate by using the points in the scatter plot. **27d.** Sample answer: Yes; some players have more opportunities to get rebounds because of the position they play for the team. **27e.** Sample answer: Yes, since more playing time provides a player with more opportunity to score points. **29.** $\{8, 4, 6, 5\}$; $\{1, 2, -4, -3, 0\}$ **31.** 127.7; 129; 130 **33.** 2075

Pages 387–390 Lesson 8-3
7. c, d **9.** $(-2, -5)$, $(0, -2)$, $(2, 1)$, $(4, 4)$ **11.** Sample answer: $(0, 2.8)$, $(-2, -7.2)$, $(2, 12.8)$, $(-1, -2.2)$. **13.** no

15.

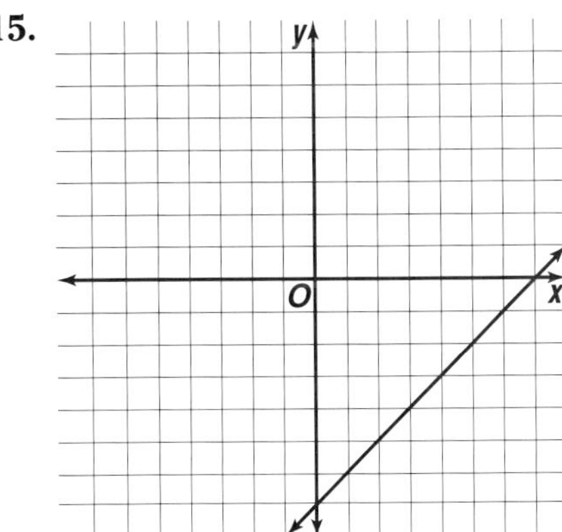

17.

19a. $y = 0.5x$

19b.

x	$y = 0.5x$	y	(x, y)
0	$y = 0.5(0)$	0	(0, 0)
1	$y = 0.5(1)$	0.5	(1, 0.5)
2	$y = 0.5(2)$	1	(2, 1)
3	$y = 0.5(3)$	1.5	(3, 1.5)

19c. Sample answer: There is no such thing as a negative number of seedlings.

19d.

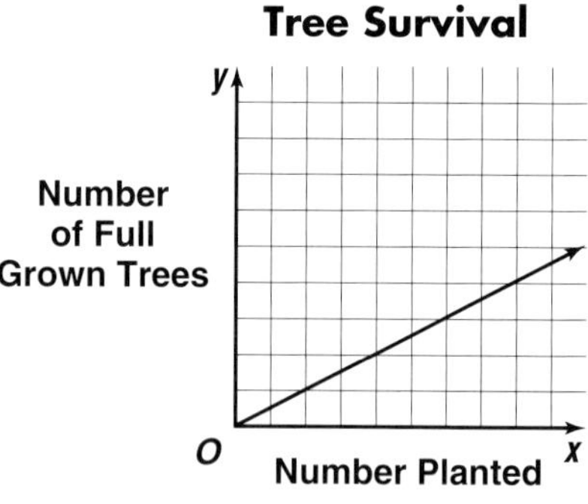

21. a, d **23.–37.** Sample answers given.
23. {(0 ,0), (1, 2.5), (2, 5), (3, 7.5)}
25. {(0, 7), (1, 10), (2, 13), (3, 16)}
27. {(0, −4), (1, −5), (2, −6), (3, −7)}
29. {(0, 1), (1, 0), (2, −1), (3, −2)}
31. {(1, 2), (3, 4), (5, 6), (6, 7)}
33. {(0, 3), (3, 0.5), (4, 1), (6, 0)}
35. {(1, −1), (2, −2), (3, −3), (4, −4)}
37. {(−2, 0), (−2, 1), (−2, 2), (−2, 3)} **39.** no

41.

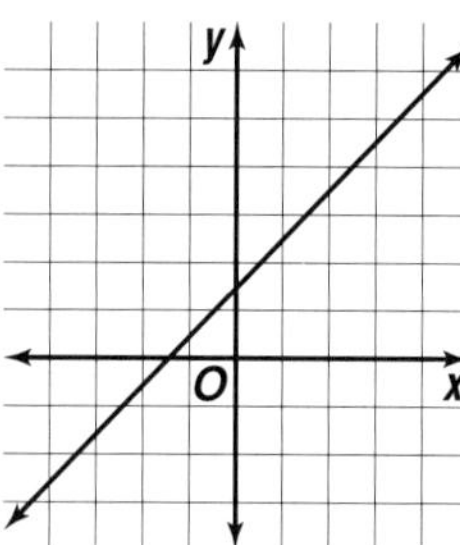

45.

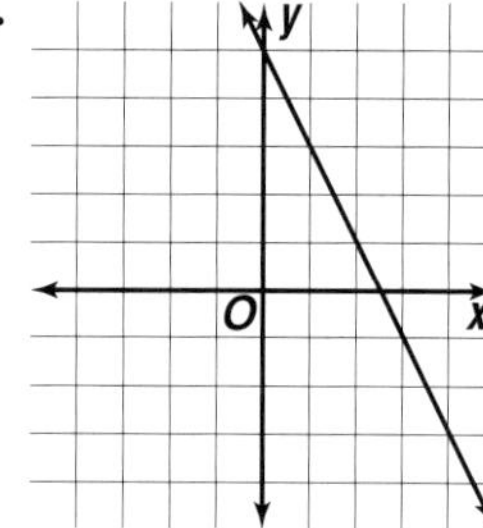

49.

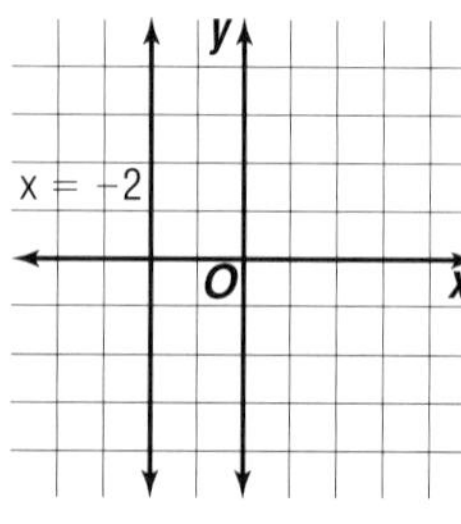

51. $x = \frac{1}{4}y$; {(0, 0), (1, 4), (2, 8), (3, 12)};

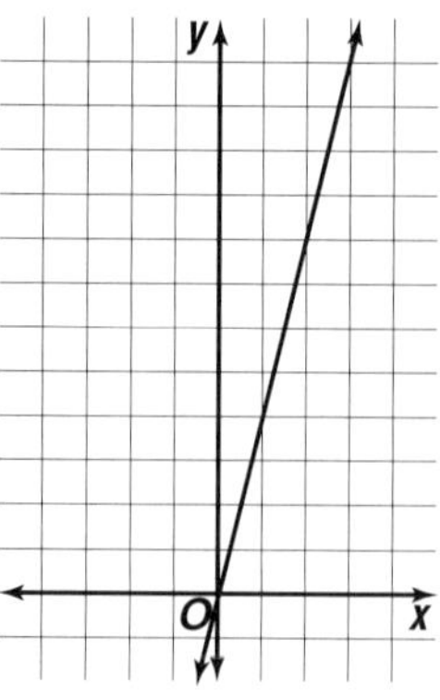

53. 2 **55.** $y = -3x$ or $3x + y = 0$ **57a.** Sample answer: (1, 4), (2, 8), (4, 16), (6, 24), (8, 32)

57b.

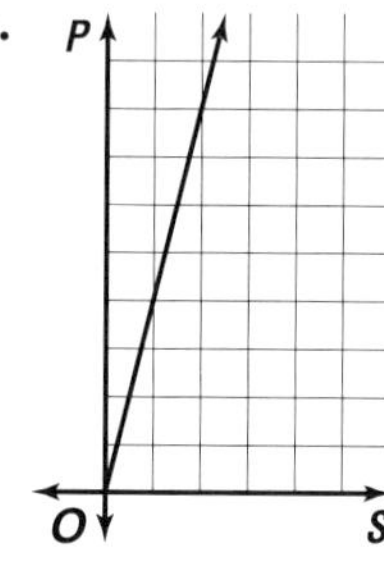

57c. Sample answer: Length cannot have a negative value. **59a.** (0, 32), (100, 212)

59b.

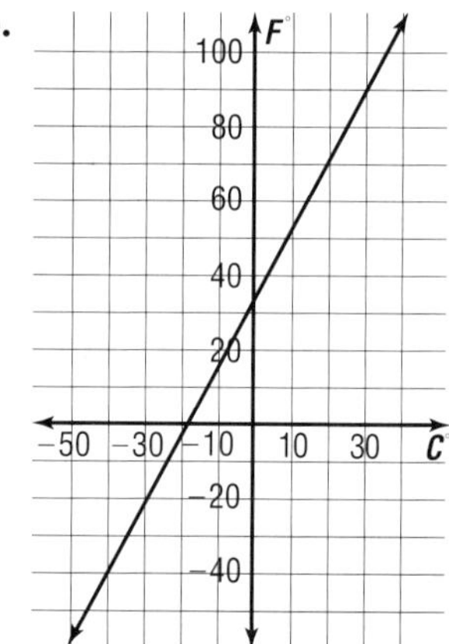

59c. Find the coordinates of other points on the graph. **59d.** Sample answer: (20, 68) **61.** $a > 11$
63. 4 **65.** 7^2

67.

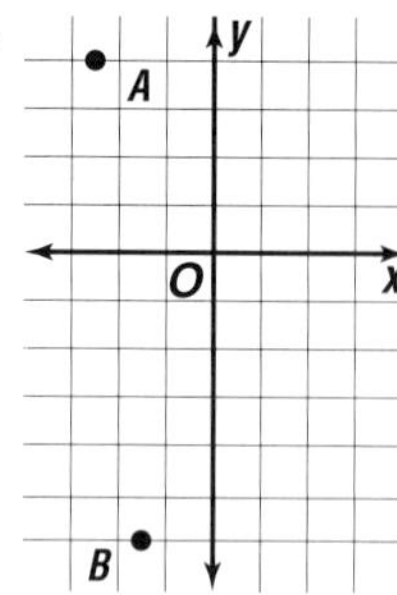

69. $14n$

Page 391 Lesson 8-3B

1. Each distance is the same.
3a. {7, 2, −1, −2, −1, 2, 7}
3b. {(−3, 7), (−2, 2), (−1, −1), (0, −2), (1, −1), (2, 2), (3, 7)};

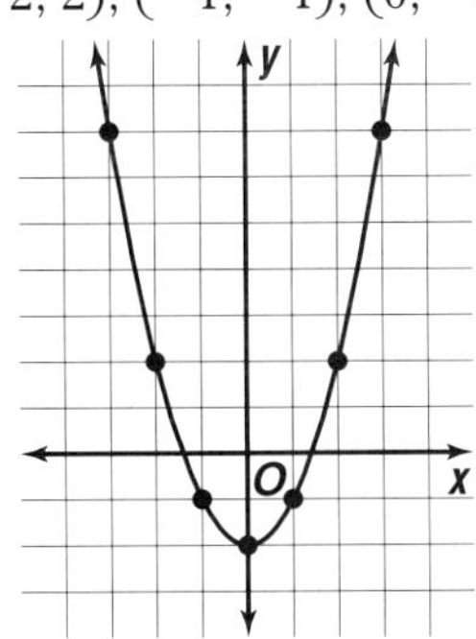

3c. Sample answer: They both open upward and have their vertex on the *y*-axis. The graphs pass through different points on the *y*-axis. The graph of $y = x^2 - 2$ is narrower.

Pages 394–395 Lesson 8-4

5. 10, 1, 2, 17; yes **7.** 3 **9.** 6 **11.** 9 weeks **13.** 8, 16, 28, 36; yes **15.** 2, 2, 2, 2; yes **17.** 5, 4, 2.5, 1.5; yes **19.** 7, 11, 2, −14; yes **21.** −47 **23.** $-1\frac{3}{4}$
25. 35 **27.** 4.25 **29.** $4b^2 - 2$ **31.** 118
33a. $\left\{-17\frac{7}{9}, 0, 22\frac{2}{9}, 37\right\}$
33b.

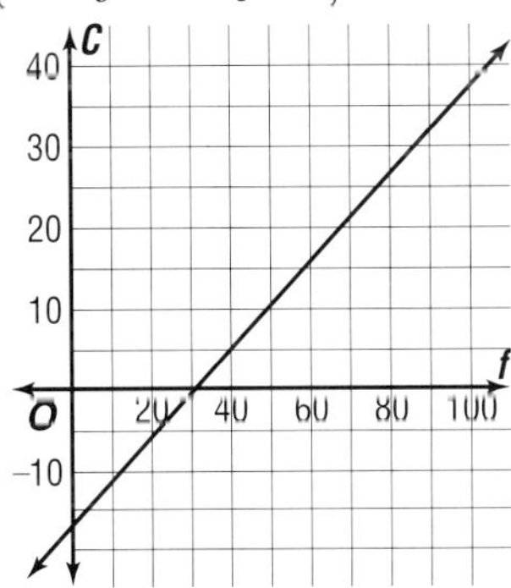

33c. Yes; No member of the domain is paired with more than one member of the range.
35a. Sample answer:

x	y = 18x + 66.5	y	(x, y)
1	y = 18(1) + 66.5	84.5	(1, 84.5)
5	y = 18(5) + 66.5	156.5	(5, 156.5)
10	y = 18(10) + 66.5	246.5	(10, 246.5)
15	y = 18(15) + 66.5	336.5	(15, 336.5)
20	y = 18(20) + 66.5	426.5	(20, 426.5)

35b.

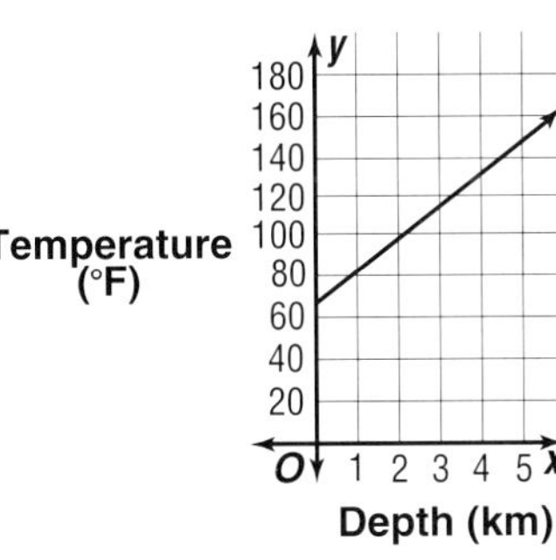

35c. 131.3°F **37.** $\frac{9}{16}$
39. $a \leq -6$;

Pages 398–399 Lesson 8-5

3a. about 78.2 m^3 **3b.** about \$12.76 **5.** about 4500 lb/in^2 **7.** \$1.11 a gallon **9.** about 150 cubic meters **11.** 156 pounds **13.** 4 **15.** $-\frac{5}{12}$
17. negative

Page 399 Self Test

1. D = {4, 1, 3, 6}, R = {2, 3, 4}; yes **3.** positive
5. Sample answer: (2, 6), (1, 2), (0, −2), (−1, −6)

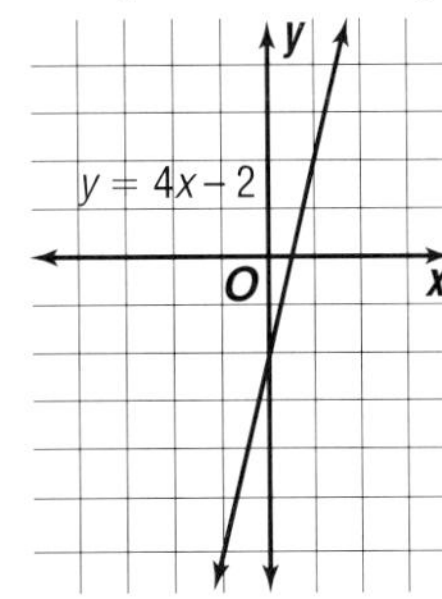

7. 50 **9.** −31

Pages 402–404 Lesson 8-6

7. 1 **9.** $\frac{1}{2}$ **11.** −1 **13.** 4 **15.** −1 **17.** 0 **19.** $-\frac{1}{5}$
21. −2 **23.** Sample answer: Both graphs slant up to the right. **25.** $-\frac{3}{2}$ **27.** $\frac{7}{4}$ **29.** 1 **31.** $\frac{3}{8}$
33. $-\frac{3}{4}$ **35.** $\frac{2}{3}$ **37.** 11,080 feet **39.** 89 minutes or 1 hour 29 minutes **41.** composite **43.** 12

Page 405 Lesson 8-6B

3. The point should be on the line.

Pages 409–410 Lesson 8-7

9. 1.5, 3 **11.** −6, 6

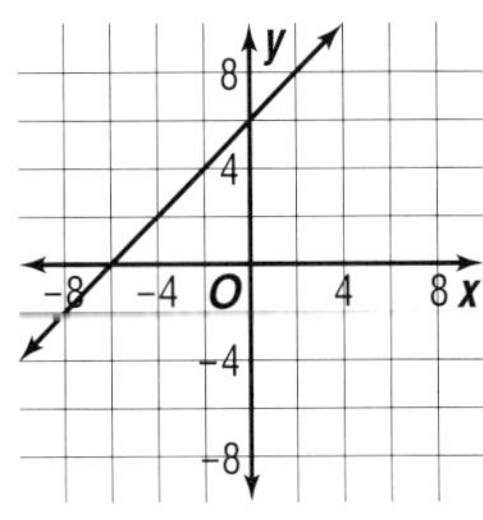

13. $\frac{2}{3}$, 6

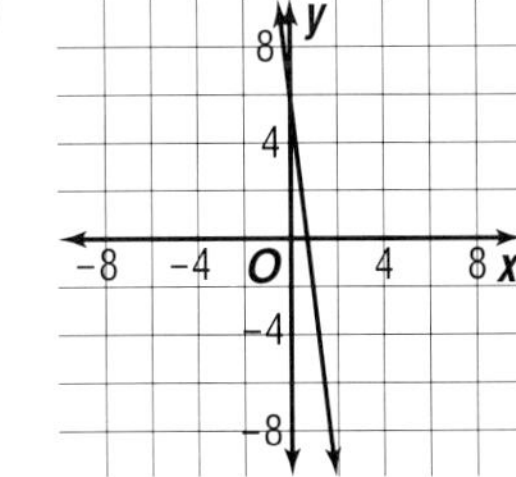

SELECTED ANSWERS

15.

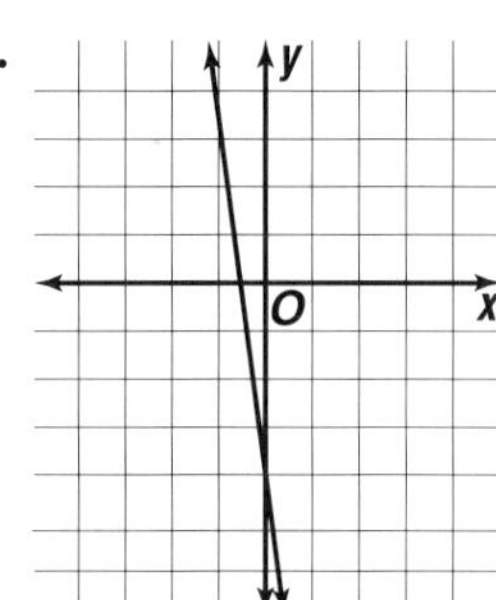

17. 5, −5 **19.** −2, −1 **21.** 2, −4

23.

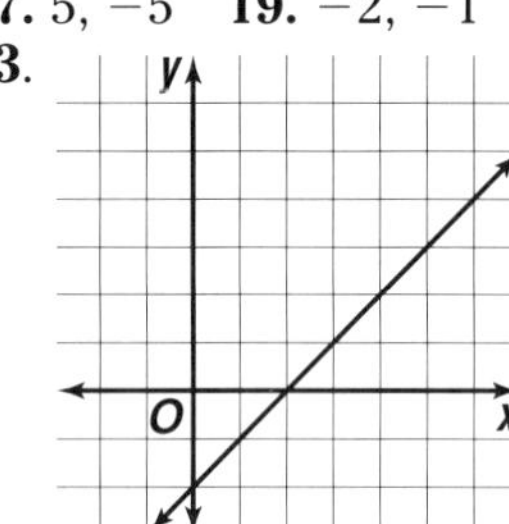

27.

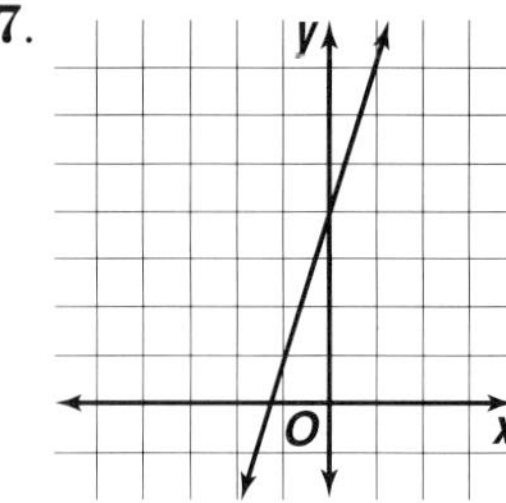

31.

35.

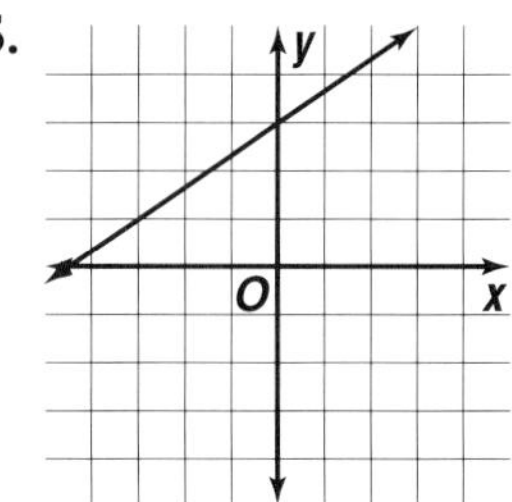

39.

43. The x-intercept and y-intercept are both zero. The line, therefore, passes through the origin. Since two points are needed to graph a line, $y = 2x$ cannot be graphed using only the intercepts.

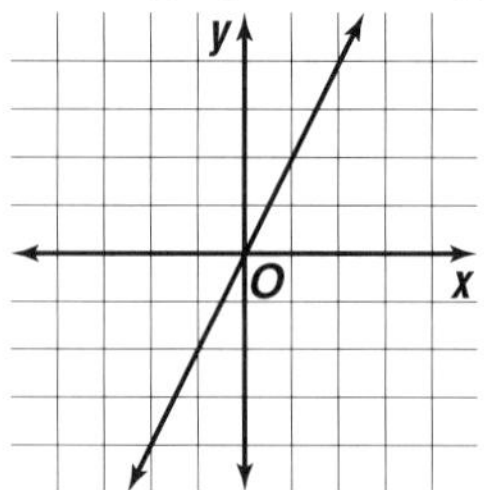

45a.

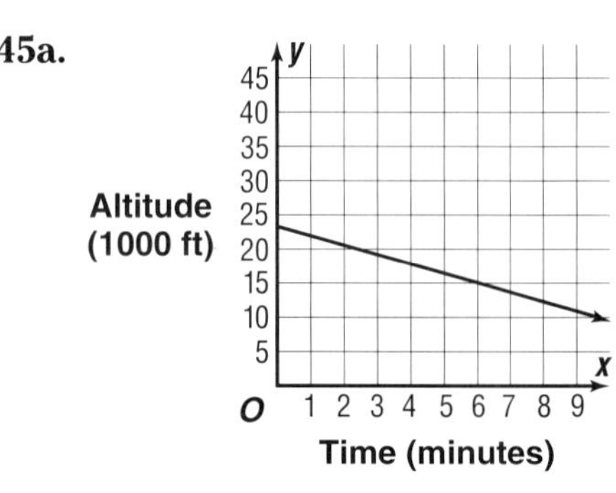

45b. 16 **45c.** The x-intercept is the time when the plane lands. **47.** Gettysburg **49.** 13 **51.** $-5a$ **53.** associative, addition

Page 411 Lesson 8-7B

1. The family of graphs is lines having a slope of 1.5. All are parallel and have different x- and y-intercepts. **3.** Sample answer: $y = -5x$

Pages 414–416 Lesson 8-8

7. (1, 1) **9.** (2, −6) **11.** (2, 1) **13a.** $2\ell + 2w = 12$; $\ell = 2w$ **13b.** (4, 2) **13c.** The length of the frame will be 4 feet and the width will be 2 feet. **15.** (0, 0) **17.** (2, 1) **19.** no solution **21.** no solution **23.** (5, 7) **25.** (−1, 3) **27.** (3, 9) **29.** (−2, 8) **31.** (5, 7) **33.** infinitely many **35.** $A = 3, B = 4$ **37a.** (300, 1500) **37b.** The break-even point would be at point (300, 1500) because that is the solution when the cost equals the income. When 300 items are sold, the income is \$1500. **39a.** $y = 15x$, $y = 5x + 60$ **39b.** (6, 90) **39c.** After 6 seconds, the dog will catch up to the prowler at 90 meters away from where the dog started. **41.** 3, 4, or 5 **43.** \$78.51 **45.** 14

Page 417 Lesson 8-9A

1. Both graphs are shaded regions and the boundary is the same line. **3.** The graph of $y = 2x + 5$ would not be shaded at all.

Pages 420–422 Lesson 8-9

7. yes **9.** b, c **11.** The region above the boundary. **13.** The region to the right of the boundary.

15.

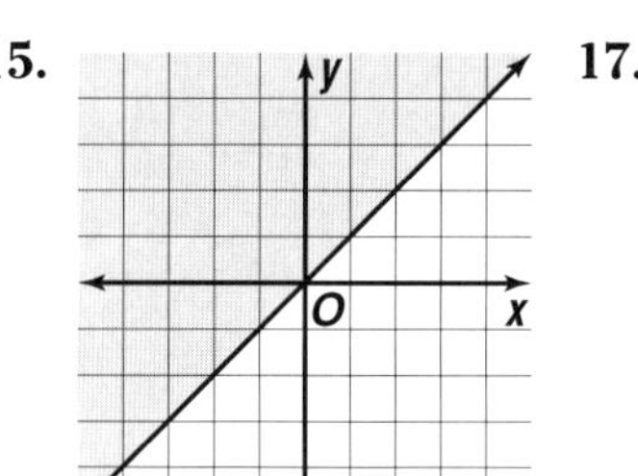

17.

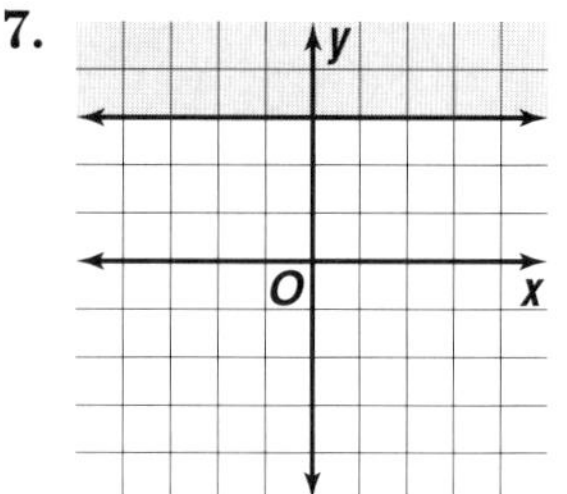

19. a **21.** c **23.** a, b, c **25.** The region to the left of the boundary **27.** The region to the left of the boundary.

29.

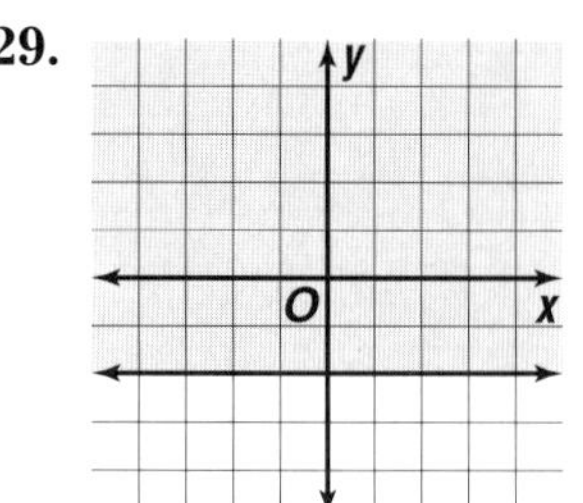

33.

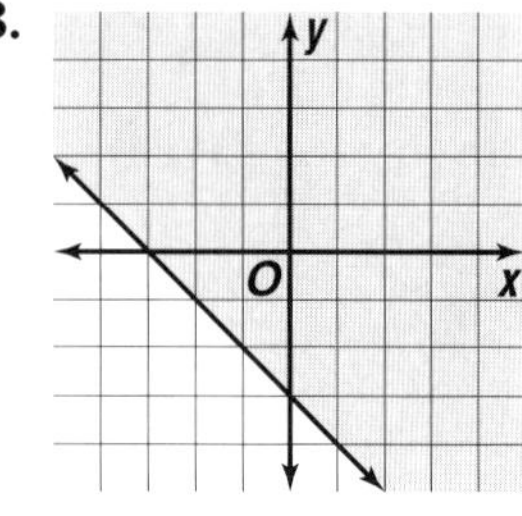

37.

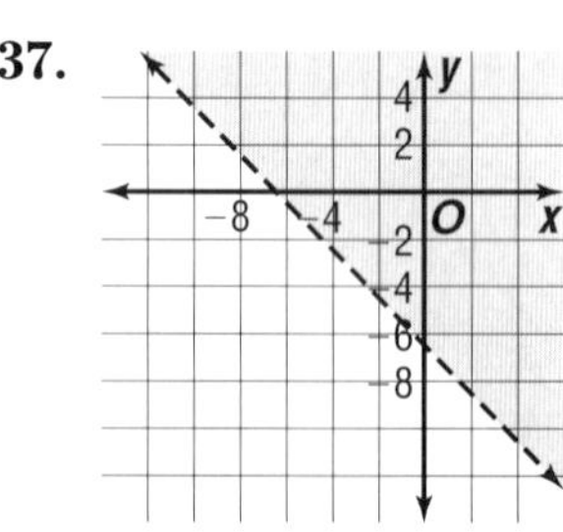

41.

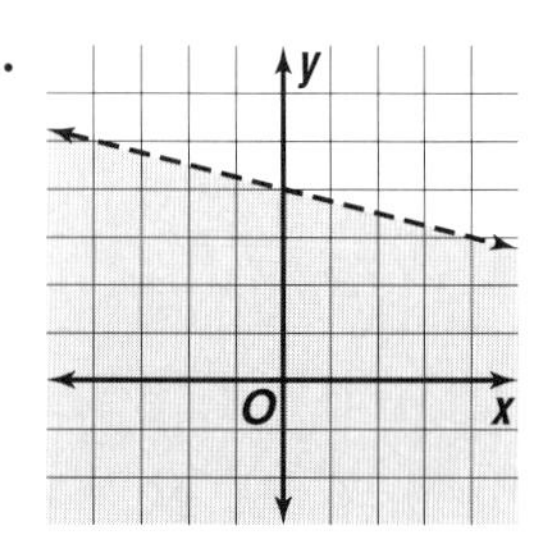

45a. $O < S - 3000$

45b.

5000
4000
3000
2000
1000
O
S
3000 6000

45c. below the line **45d.** No; this would mean that your cost was greater than your sales, which doesn't happen in a successful business.
45e. Sample answer: (6600, 3600), (7000, 4000)

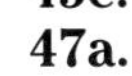

47a.

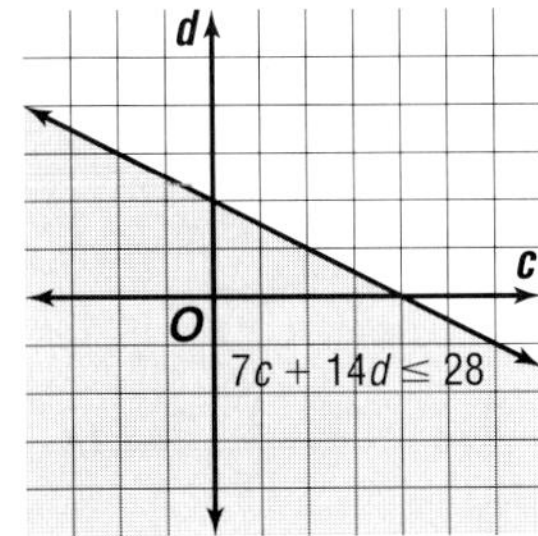

47b. Sample answer: (1, 1) means purchasing 1 cassette and 1 CD for a total cost of \$21. (2, 1) means purchasing 2 cassettes and 1 CD for a total of \$28. **49.** $(-1, -1)$ **51.** $\frac{3}{5}$, 6.02×10^{-1}, 0.63 **53.** 2 **55.** -12

Page 423 Highlights

1. function **3.** *y*-intercept **5.** relation **7.** *x*-intercept

Pages 424–426 Study Guide and Assessment

9. {0, 6, 7, 9}; {17, 18, 19, 40}; yes **11.** {6, 8, 15}; {2}; yes **13.** $\left\{-2, -\frac{1}{2}, 4, 5\right\}$; {6, 3, 1}; yes

15. no **17.** negative **19.** no

21.

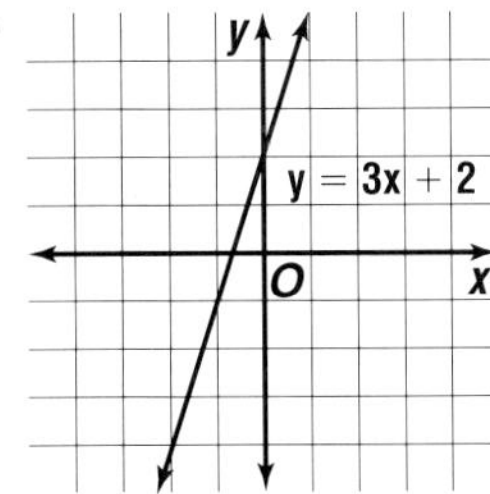

(0, 2), (1, 5), (−1, −1), (2, 8)

25.

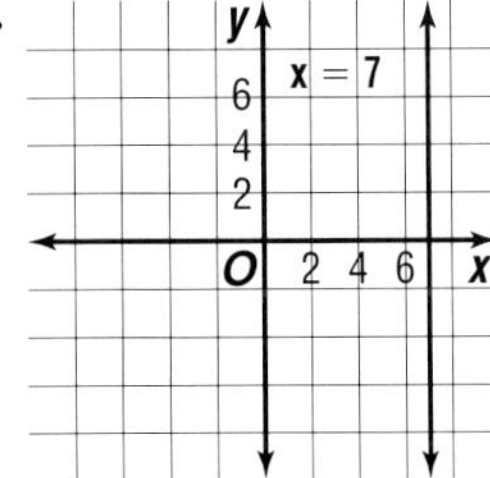

(7, 0), (7, 2), (7, 4), (7, 6)

29. 5 **31.** 4 **33.** $\frac{2}{5}$ **35.** -1 **37.** 0

39.

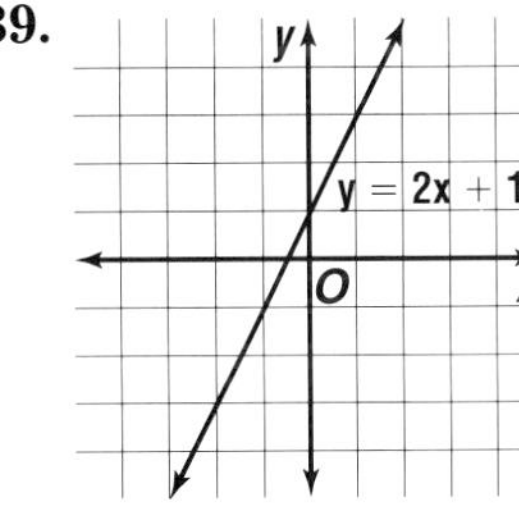

43.

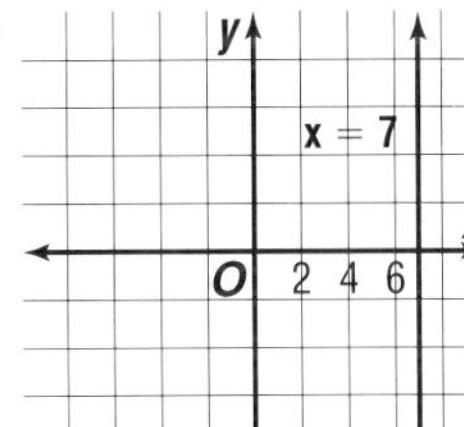

45.

y
$y = 2 - \frac{1}{2}x$
x
O

47. (0, 8) **49.** empty set **51.** empty set

53.

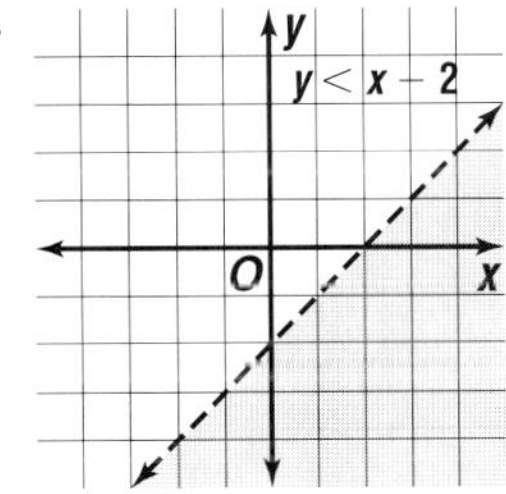

55.

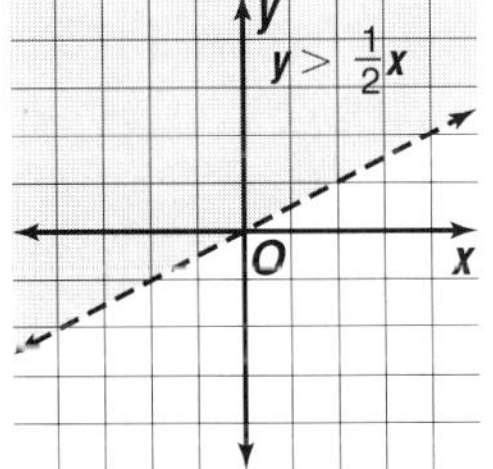

57.

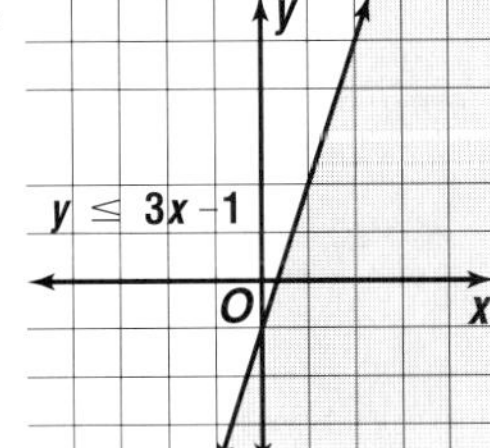

59a.

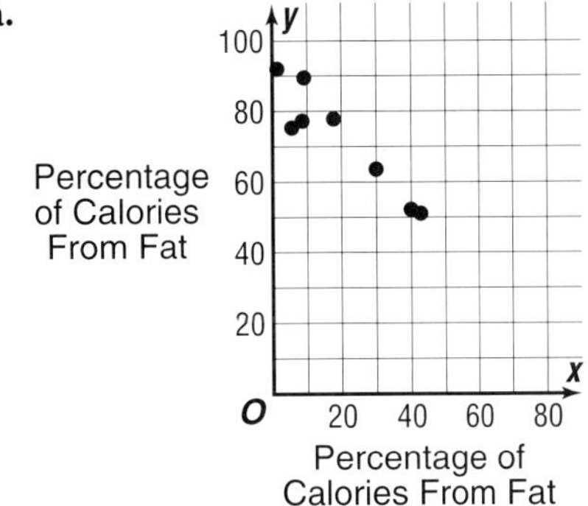

59b. negative **61.** approximately 36 goals scored

Chapter 9 Ratio, Proportion, and Percent

Pages 434–436 Lesson 9-1

5. $\frac{1}{5}$ **7.** $\frac{25}{1}$ **9.** $\frac{43}{46}$ **11.** \$0.38/can **13.** 0.25 in./hour

15. $\frac{1}{13}$ **17.** $\frac{4}{1}$ **19.** $\frac{13}{7}$ **21.** $\frac{2}{15}$ **23.** $\frac{5}{7}$ **25.** $\frac{3}{7}$ **27.** 24.2 miles/gallon **29.** 9.52 m/s **31.** $7/ ticket **33.** Sample answer: 2697:13,485 **35a.** the 32 oz bag at $3.69 **35b.** Sample answer: When the quality of the less expensive product is not as good as that of the more expensive product, it may be better to buy the more expensive product. **37a.** All ratios are 2. **37b.** The ratios approach the golden ratio. **39a.** 34.8 mph **39b.** 5 ft/s **41.** −11 **43.** $2 \cdot 2 \cdot 2 \cdot 3 \cdot 19$ **45.** −13

Page 437–438 Lesson 9-2

3. 10 **5.** 21 **7.** 6 **9.** She gained $800. **11.** −21 **13.** 5 **15.** 249.5 mph **17.** Sample answer: 180 **19.** $24 > 7$

Page 439 Lesson 9-3A

1. Game A is fair. It is fair because there are the same amount of odd and even sums. Game B is unfair because there are many more even products than odd products. **3.** Yes; there are the same number of blue and red marbles.

Pages 442–443 Lesson 9-3

7. $\frac{1}{12}$ **9.** $\frac{1}{8}$ **11.** $\frac{1}{2}$ **13.** $\frac{7}{8}$ **15.** $\frac{5895}{7424}$ **17.** 0 **19.** $\frac{17}{24}$ **21.** 1 **23.** $\frac{1}{2}$ **25.** 0 **27.** $\frac{1}{4}$ **29.** $\frac{1}{36}$ **31.** 0 **33.** $\frac{7}{10}$ **35.** 1–humans have been to the Moon; 0–the Sun is too hot. **37.** 4 **39.** −2 **41.** 15.4 m; 13.26 sq meters

Pages 446–447 Lesson 9-4

7. = **9.** 7.5 **11.** 2.1 **13.** $\frac{2.7}{m} = \frac{3}{7}$; 6.3 **15.** ≠ **17.** ≠ **19.** = **21.** 7 **23.** 42 **25.** 2.5 **27.** 16 **29.** 2.1 **31.** 10.2 **33.** $\frac{5}{6.15} = \frac{x}{8.00}$; 6.5 **35.** $\frac{8}{2} = \frac{z}{5}$; 20 **37.** $\frac{25}{5} = \frac{m}{25}$; 125 **39.** 1 **41.** 15 **43.** 6 and 8 **45.** 3.24 inches **47.** $\frac{8}{3}$ **49.** 9.412×10^6 **51.** $\frac{1}{4^3}$ **53.** $b + 200$

Page 448 Lesson 9-4B

1. Sample answer: A larger sample will lead to a more reliable prediction and result.

Pages 451–453 Lesson 9-5

7. 44% **9.** 3.2% **11.** 9 **13.** 12.5% **15.** 25% **17.** 60% **19.** 225% **21.** 75.5% **23.** 106.72 **25.** 45% **27.** 35 **29.** 16 **31.** 80 **33.** 1200 **35.** 100 **37a.** 50 g **37b.** about 2375 mg or 2.375 g **39.** 31,813; 8238; 18,779; 7279; 35,071 **41.** $\frac{3}{4}$ **43.** $4.22 **45.** 5

Pages 455–457 Lesson 9-6

5. 2% **7.** Sample answer: High school cafeteria; because teen's tastes are not representative of all ages. **9.** 35% **11.** 125 **13.** No; because the sample is too small and it is not random. **15.** Sample answer: No; many of the common two-letter words like *at*, *in*, *of*, and *on* do not contain the letter E. **17a.** 52% **17b.** 60% **19.** 58% **21.** D **23.** true

Page 457 Self Test

1. 60 ft/min **3.** 12 **5.** $\frac{9}{16}$ **7.** 1.8 **9.** 50 **11.** 105

Pages 460–461 Lesson 9-7

5. 37% **7.** 103% **9.** 7.2% **11.** $\frac{1}{4}$ **13.** $\frac{3}{1000}$ **15.** 0.245 **17a.** $\frac{2}{5}$ **17b.** 40% **19.** 3% **21.** 237% **23.** 132% **25.** 0.04% **27.** $58\frac{1}{3}\%$ **29.** $233\frac{1}{3}\%$ **31.** $55\frac{5}{9}\%$ **33.** 7.5% **35.** $\frac{5}{4}$ **37.** $\frac{69}{200}$ **39.** $\frac{3}{8}$ **41.** $\frac{49}{300}$ **43.** 0.75 **45.** 0.398 **47.** 0.004 **49.** 2.354 **51.** 22% **53.** 0.067, 16%, $\frac{1}{4}$ **55.** $\frac{1}{3}$ **57.** $\frac{4}{5}$ **59.** 1.5, −12 **61.** $\frac{5}{27}$ **63.** n^{14} **65.** C **67.** 8

Pages 464–466 Lesson 9-8

7. c **9.** $\frac{4}{5}$ **11.** 215 **13.** 1.5 **15.** 50% **17a.** 60% **17b.** 40% **19.** c **21.** b **23.** 1 **25.** $\frac{3}{100}$ **27.** $\frac{1}{10}$ **29.** $1\frac{1}{5}$ **31.** 120 **33.** 98 **35.** 5.15 **37.** 20 **39.** 80% **41.** 50% **43.** 10% **45.** 10% **47.** 0.3% **49.** 10%, 20%, 60%, 45%, 40% **51.** Sample answers: Connecticut 10%; Delaware 1%; Georgia 2%; Kentucky 1%; Maine 2%; Maryland 10%; New Hampshire 3%; New Jersey 5%; New York 10%; North Carolina 10%; Pennsylvania 15%; Plymouth 10%; Rhode Island 2%; South Carolina 10%; Tennessee 1%; Vermont 2%; Virginia 20% **53.** $\frac{16}{25}$ **55.** geometric; 1.5; 60.75, 91.125, 136.6875 **57.** 21;

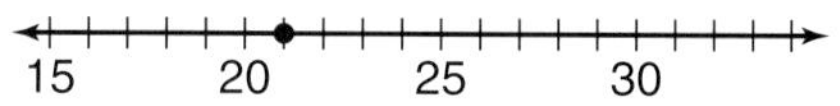

59. Radio Ownership

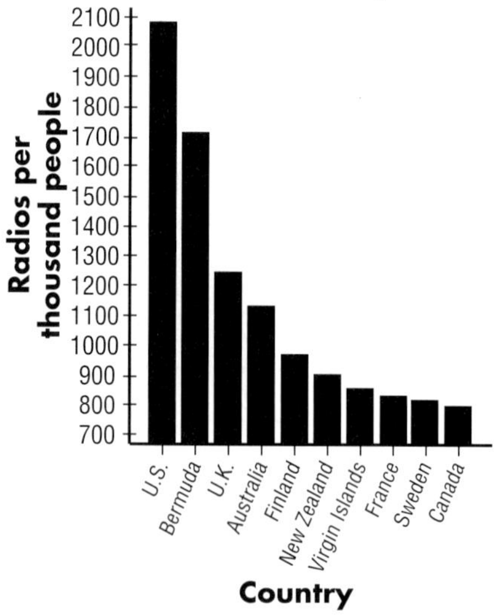

Pages 469–471 Lesson 9-9

7. 200 **9.** 17.28 **11.** $47.25 **13.** 3.36 million **15.** 27 **17.** 70 **19.** 31.5 **21.** 300% **23.** $1.\overline{6}\%$ **25.** 120 **27.** 20 **29.** $70 **31.** $187.50 **33.** $42.50 **35.** $41.34 **37.** about 93 million **39.** They would be paying the customers. **41a.** $264.72 **41b.** No; the price is the same either way. **41c.** You could

not add the percents first. For example, if the price was \$100, 10% + 33% = 43% off. Using this method the price would be 100 − 43 or \$57. Taking 33% off and then 10% would have the price as \$60.30.
43. The statements are not equivalent. For example, suppose an item costs \$1.00 on day 1. According to the first statement, it would cost \$2.08, not \$1.00 × 2 or \$2.00, on the fifth day.

Day 2 1.2 × 1.0 = \$1.20
Day 3 1.2 × 1.20 = \$1.44
Day 4 1.2 × 1.44 = 1.728 or \$1.73
Day 5 1.2 × 1.73 = 2.076 or \$2.08

45. positive **47a.** $3v - 298 = 1916$ **47b.** 738 **49.** 15 **51.** 7

Pages 474–475 Lesson 9-10

7. decrease; 12% **9.** decrease; 7% **11.** decrease; 20% **13.** increase; 13% **15.** increase; 13% **17.** increase; 10% **19.** increase; 7% **21.** increase; 136% **23a.** no; Suppose the price of the outfit is \$100. The price would be \$50 after the first reduction. After the second reduction, the price would be 50 − 0.3(50) = 50 − 15 or \$35. A final price of \$35 is $\frac{100-35}{100}$ or 65% off, not 80% off.
23b. 65% **25.** 20% **27.** \$3.20 **29.** 125 **31.** Yes; it passes the vertical line test. **33.** 12 outfits

Page 477 Highlights

1. true **3.** false **5.** true **7.** false

Pages 478–480 Study Guide and Assessment

9. $\frac{41}{45}$ **11.** $\frac{3}{5}$ **13.** \$0.89 per pound
15. 57 miles/hour **17.** $\frac{13}{21}$ **19.** $\frac{8}{21}$ **21.** $\frac{11}{21}$ **23.** 2.5
25. $\frac{88.4}{3.4} = \frac{161.2}{g}$; 6.2 gal. **27.** 76% **29.** 4 **31.** 32
33. 0.56% **35.** 56.5% **37.** $\frac{13}{20}$ **39.** 2.35 **41.** a
43. a **45.** b **47.** 20 **49.** 4000 **51.** 31.25%
53. decrease; 1% **55.** decrease; 22%
57. decrease; 12% **59.** 28

Chapter 10 More Statistics and Probability

Pages 488–489 Lesson 10-1

7. 5, 6, 7, 8

Stem	Leaf
5	4 7
6	3 7 8 9
7	1 5 7
8	5

5|4 = 5.4

9.

Stem	Leaf
6	0 1 4 5 7 8
7	0 0 2 3
8	0 0

6|0 = 60

11.

Stem	Leaf
0	5 6 7 9 9
1	1 2 4
2	1 3
3	
4	0 2 5

4|0 = 40

13.

Stem	Leaf
9	2
10	4 8
11	1 2 2 7
12	3 6 8 9
13	3 8
14	7

9|2 = 9.2

15.

Women's		Men's
5	4	5
5 5	5	
7 5 0	6	
5 2 0 0 0	7	0 0 0 0 0 5
5 0	8	2 2 3
0	9	
	10	0
	11	0
4	12	0 5
	13	3

5|4 = 45 4|5 = 45

15b. Answers will vary. Sample answer: Men's running shoes are more expensive than women's running shoes. The typical running shoe costs \$70. **17.** Sample answer: Yes; Since the data can be read from a stem-and-leaf plot, you can estimate their sum and then divide by the number of leaves to estimate the mean.

19a.

Distress		No Distress
8 7 3	5	
3	6	6 7 7 7 8 9
5 0 0	7	0 0 2 3 5 6 6 8 9
	8	1

3|5 = 53 7|0 = 70

21. $\frac{9}{20}$ **23.** d

Pages 492–494 Lesson 10-2

5. 13 **7.** 52

9a.

Stem	Leaf
3	8 9
4	0 2 6
5	8
6	1 2 3 3
7	0
8	0 4
9	4

3|8 = 38

9b. 61.5 **9c.** 70, 42 **9d.** 28 **11.** 70; 65; 85, 45; 40 **13.** 21; 7; 13.5, 3; 10.5 **15.** 49; 208; 213, 200; 13 **17.** 4.1; 9.2; 9.75, 8.85; 0.9 **19.** 37,600,000; 43,100,000; 52,600,000; 33,100,000; 19,500,000 **21.** American League: 39; 42.5; 46, 38; 8; National League: 21; 40.5; 46.5, 38; 8.5 **23c.** The first set of data has a smaller interquartile range, thus the data in the first set are more tightly clustered around the median and the data in the second set are spread out over the range. **25a.** participants 60; observers 71 **25b.** 51; 35 **25c.** Sample answer: The participants' ages are between 12 and 72. Their ages are spread out over the range of the data. The median age of the participants is 25 years. **25d.** Sample answer: The observers are both younger and older than the participants, but the range of ages is very close to that of the participants. The median age of the observers is 46.5 while the median ages of the participants is 25. **27.** > **29.** cd^4

Pages 499–501 Lesson 10-3

7a. 40 **7b.** 50 **7c.** 30 **7d.** 20 **7e.** 25, 55 **7f.** none **9a.** 15 **9b.** 68 **9c.** 71 **9d.** 65 **9e.** 6 **9f.** no **11a.** 50% **11b.** 50% **11c.** 25% **11d.** 25% **11e.** Scores were closer together between 80–83. **11f.** The scores were spread further apart. **13.** box-and-whisker plot **15.** box-and-whisker plot **17.** line graph or comparative graph **19.** The box did not change. The weight 200 is an outlier.

21a.

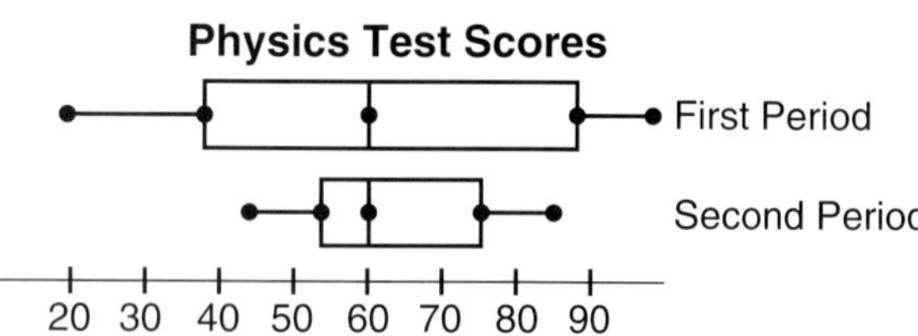

21b. The medians are the same. **21c.** The range and interquartile range are greater for the first period class.

23a.

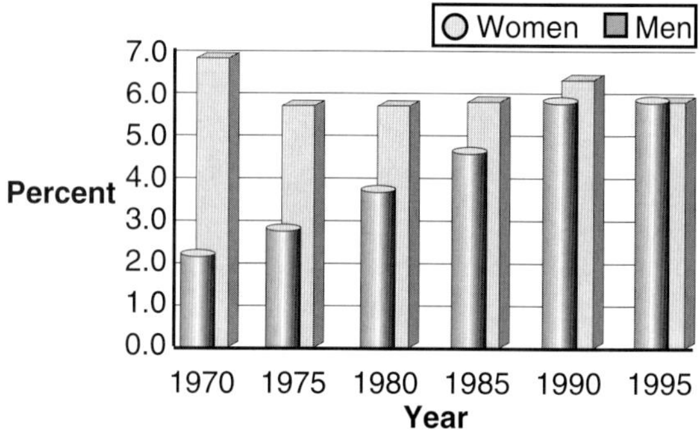

23b. Men's employment decreased from 1970 to 1975, but since 1975 men's employment has been relatively consistent. **23c.** Women's employment increased steadily from 1970 to 1990 and has remained steady from 1990–95. **23d.** Women are as likely as men to have two jobs. **25.** {5, −1, 4.5}; {−3, 4, 0} **27.** Sample answer: $7

Pages 502–503 Lesson 10-3B

1.

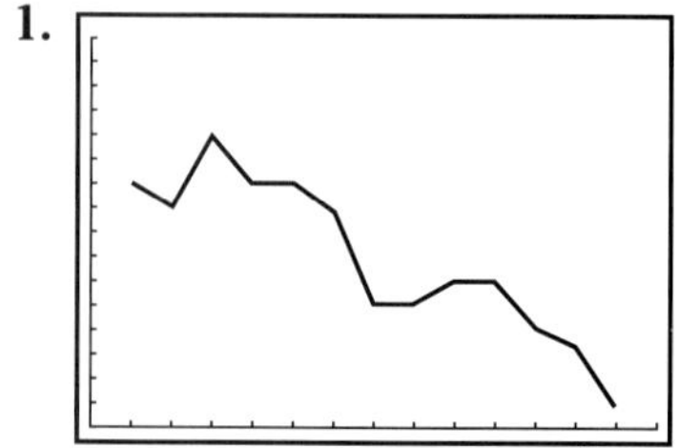

3.

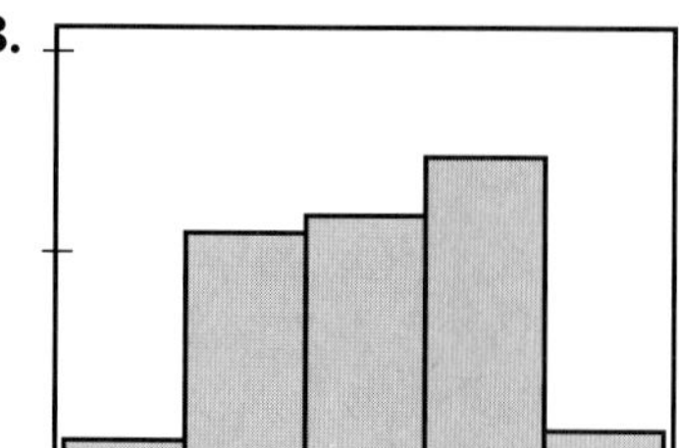

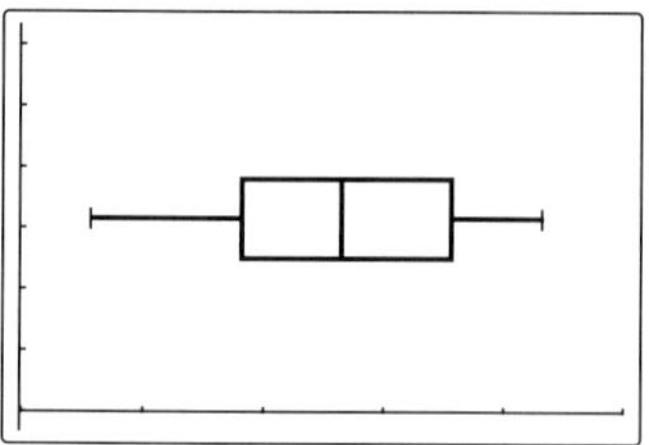

Pages 506–508 Lesson 10-4

5a. about 1.1 **5b.** The second car is shorter in length and width. **5c.** Yes; the actual number of cars sold in 1992 is 1.1 times greater than the number sold in 1991. However, the graph appears to be longer and wider, but both dimensions of the car should not have changed. **7a.** Median; there are large and small values. **7b.** Mean; the high values will raise the center of the data. **9.** 172.6; mean **11.** sports; mode

13.

	Company A	Company B
mean	$22,413.79	$21,935.48
median	$15,000.00	$20,000.00
mode	$15,000.00	$20,000.00
range	$85,000.00	$30,000.00

You would probably prefer to work for Company B. Both the median and mode salaries are greater, so the salary is greater for more employees.
15a. Sample answer: No; it only shows the population increases for 6 years.

15b.

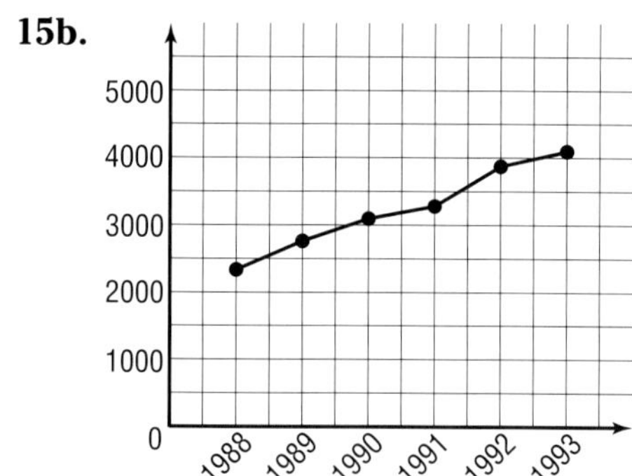

17a. $58 **17b.** $25 **17c.** $35 **17d.** $15 **17e.** $20 **17f.** $12, $70 **17g.** $0 and $65 **17h.** $70 **19a.** $\frac{8}{11}$ **19b.** 1,965,955 **21.** −6, 6

Pages 511–513 Lesson 10-5

5a. 12 outcomes **5b.** $\frac{1}{12}$ **7.** 18 outcomes **9.** 24 outcomes **11.** 25 ways **13.** 36 outcomes

15a. 8 outcomes **15b.** $\frac{1}{8}$ **17.** 32 outcomes

19a. 2^{30} **19b.** $\frac{1}{2^{30}}$ **21a.** 10,000 extensions **21b.** 4,860,000 **21c.** 2,880,000 new numbers **23.** The median; it is close to three of the four actual prices. **25.** −6 **27.** h = hours, $3.50h < 20$, $h < 5.7$

Page 513 Self Test

1.

2	8
3	6 7 9
4	3 7 9
5	0 1 1 3
6	2 *6|2 = 6.2*

3. 2.7, 4.7 **5.** A comparative graph is probably the best way. **7.** Brian would use the bottom graph because it makes the increases he has received appear smaller. **9.** 24 outcomes

Page 514 Lesson 10-6A

1. 120 words

Pages 517–519 Lesson 10-6

5. combination **7.** 2 **9.** 24 **11.** 120 **13.** 252 **15.** 56 programs **17.** combination **19.** permutation **21.** combination **23.** 720 **25.** 720 **27.** 362,880 **29.** 39,916,800 **31.** 35 **33.** 1 **35.** 45 **37.** 252 **39.** $1\frac{2}{3}$ **41.** 5005 teams **43.** 6 ways **45.** 10 students **47.** 5005 combinations **49.** 140,400,000 license plates **51.** about −15° C **53.** −2.5 **55.** $\frac{5}{4}$ **57.** $-9x$

Pages 521–523 Lesson 10-7

5. 1:1 **7.** 5:1 **9.** 1:4 **11.** 4:1 **13.** 1:1 **15.** 1:2 **17.** 7:5 **19.** 5:1 **21.** 1:17 **23.** 7:11 **25.** 9:4 **27.** 1:9999 **29a.** less than 1 or none **29b.** about 89 people **31.** 40,320 **33.** \$0.16, \$0.31, \$0.36, \$0.40, \$0.51, \$0.55, \$0.60 **35.** >

Pages 526–528 Lesson 10-8

5a. Sample answer: Select a marble out of a bag with 4 marbles the same color and one marble a different color. Then repeat three more times. **7a.** Roll a die. **9.** \$25 **11a.** 15 games **11b.** 240 **13b.** Change the second line to "For (N, 1, 50)". **15.** 1:5 **17a.** $110 = \frac{1}{2}x + 20$ **17b.** \$180 **19.** 35°F

Page 529 Lesson 10-8B

1. The most frequently drawn color is probably the one with the greatest number of marbles in the bag. **3.** yes; It is more likely to happen with fewer draws.

Pages 533–534 Lesson 10-9

5. dependent **7.** $\frac{21}{380}$ **9.** $\frac{9}{190}$ **11.** $\frac{11}{625}$ or 1.76% **13.** independent **15.** independent **17.** $\frac{1}{30}$ **19.** $\frac{2}{15}$ **21.** 0 **23.** $\frac{10}{91}$ **25.** $\frac{3}{91}$ **27.** $\frac{1}{91}$ **29a.** $\frac{3}{8}$ **29b.** $\frac{3}{10}$ **31.** $\frac{4993}{5000}$ or 99.86% **33.** \$10; about \$40

35. 2;

0 1 2 3 4 5 6

37. $2(n + 3)$

Pages 537–538 Lesson 10-10

5a. $\frac{1}{3}$ **5b.** $\frac{1}{3}$ **5c.** $\frac{2}{3}$ **7a.** inclusive; $\frac{3}{4}$ **7b.** exclusive; $\frac{5}{8}$ **7c.** exclusive; $\frac{5}{8}$ **7d.** inclusive; $\frac{1}{2}$ **9.** 0 **11.** $\frac{2}{15}$ **13.** $\frac{2}{221}$ **15.** $\frac{4}{9}$ **17.** about 10 times **19.** about 35 times **21.** about 20 times **23.** 53% **25.** independent

27.

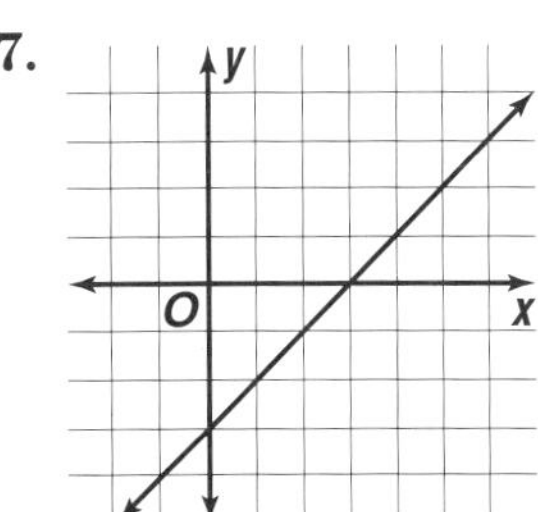

29. 300

Page 539 Highlights

1. C **3.** I **5.** A **7.** G

Pages 540–542 Study Guide and Assessment

9.

9	5 8
10	7 7 8
11	0 5 9 *9|5 = 9.5*

11.

11	1 2 7
12	0 1
13	3
14	6 8 *11|1 = 111*

13. 30; 18; 10, 28; 18 **15.** 98; 217; 205, 240; 35

17a.

20 25 30 35 40 45 50 55

17b. 35 **17c.** none **19.** 6.75; mean **21.** favorite sport: soccer; 6.75 classes; 6.5 family members **23.** 1024 **25.** 104 **27.** C **29.** 120 **31.** 90 **33.** 10:3 **35.** 1:1 **37.** dependent **39.** $\frac{3}{22}$ **41.** $\frac{3}{22}$ **43.** inclusive; $\frac{27}{36}$ **45.** $\frac{1}{2}$ **47.** 64

Chapter 11 Applying Algebra to Geometry

Pages 552–553 Lesson 11-1

7. plane **9.** A **11.** $\overrightarrow{CD}$ **13.** 63°, acute

15.

78°

17. The three "pods" determine a plane. **19.** plane **21.** ray

23. 

25. 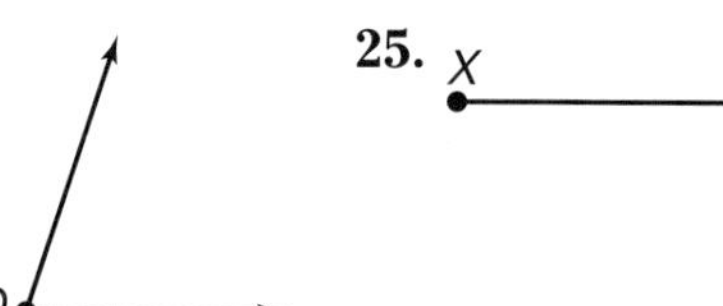

27. 

29. 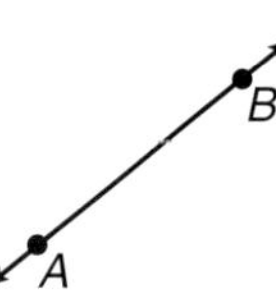

31. A B C D

33. 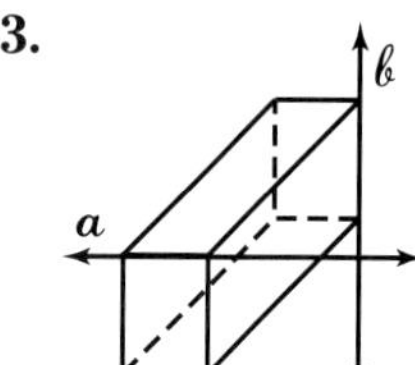

35. 155°; obtuse **37.** 65°; acute **39.** 15°; acute

41. acute; 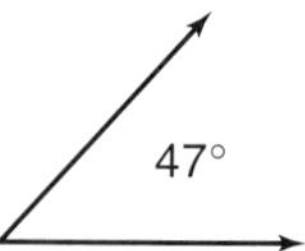

43. acute; 

45. right; 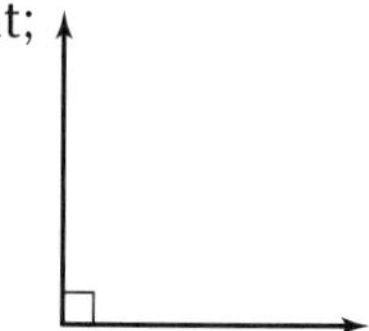

47. 3° **49.** 22 times **51.** = **53.** 50, 52 **55.** 15 **57.** −12

Page 555 Lesson 11-1B

3. Cut in two equal pieces. **5.** They are the same.

Pages 558–560 Lesson 11-2

5. silicon **7a.** 100% **7b.** 360° **7c.** Housing: 31%, 111.6°; Food:19%, 68.4°; Transportation: 22%, 79.2°; Insurance: 7%, 25.2°; Savings: 4%, 14.4°; Misc.: 17%, 61.2°; Totals, \$2080, 100%, 360.0°

7d. 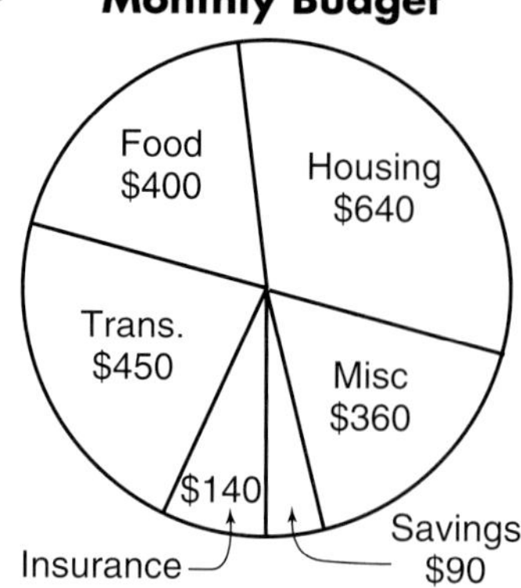

9. 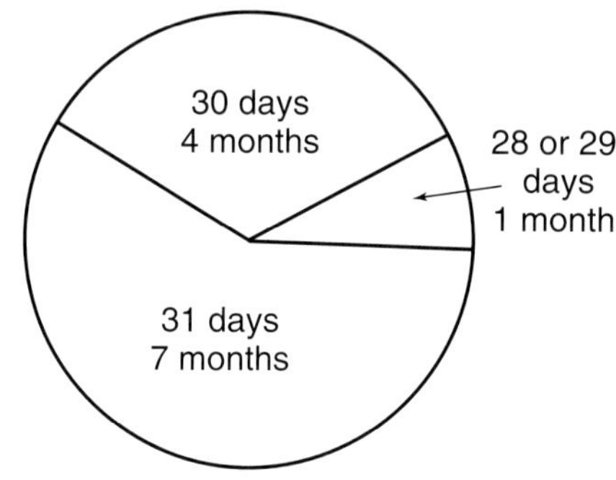

11a. $\frac{3}{8}$

11b. 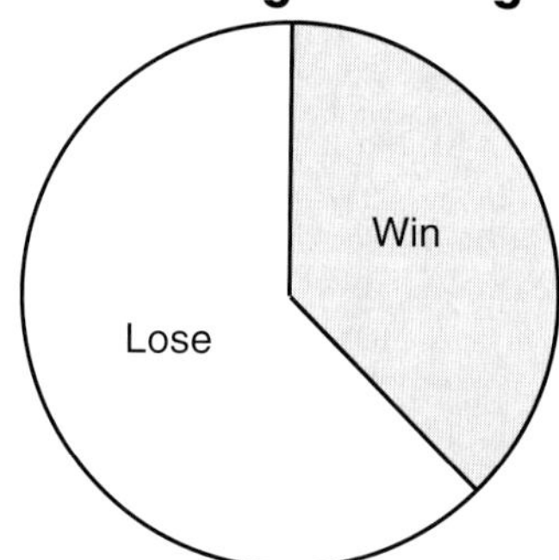

13. 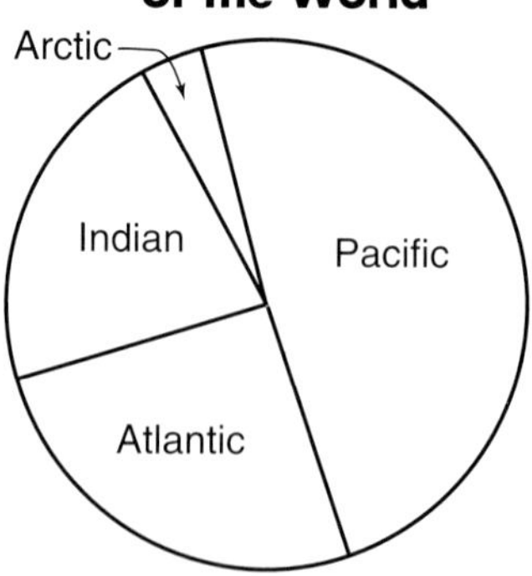

15a.

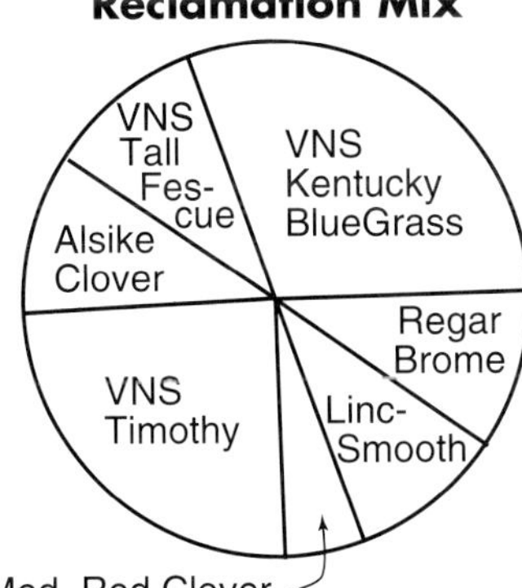

15b. 75 lb. VNS Kentucky Bluegrass; 25 lb each of Regar Bromegrass, Lincoln Smooth Brome, Alsike Clover, and VNS Tall Fescue; 12.5 lb Medium Red Clover; 62.5 lb VNS Timothy

17. **19.**

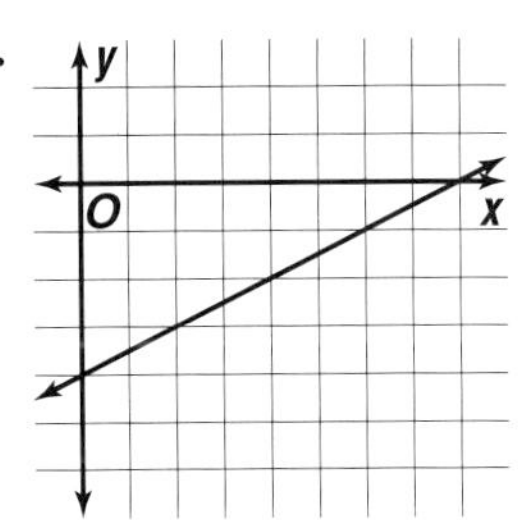

21. Sample answer: Joe earned 3 times as much as Sue. If Joe earned \$25, how much did Sue earn?

Pages 564–566 Lesson 11-3
7. 40 **9.** 15° **11.** $m\angle X = 60°$, $m\angle Y = 120°$
13. $\angle AFG$, $\angle GFD$, $\angle CGF$, $\angle FGD$
15. $m\angle AFE = 130°$; $m\angle EFB = 50°$; $m\angle BFG = 130°$; $m\angle AFG = 50°$; $m\angle CGF = 130°$; $m\angle FGD = 50°$; $m\angle DGH = 130°$; $m\angle CGH = 50°$ **17.** 105 **19.** 66 **21.** 55 **23.** 120°, 60° **25.** 18°, 72° **27.** 47° **29.** 133° **31.** 133° **33.** $m\angle R = 32°$; $m\angle S = 58°$ **35.** $m\angle F = 118°$; $m\angle G = 62°$ **37.** 30 **39.** 55°
41. Yes; all the angles formed are right angles.

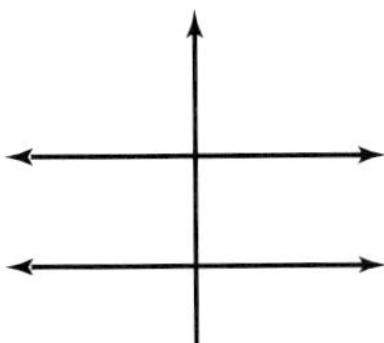

43. 135° **45.** $\frac{5}{18}$ **47.** 5.678×10^4 **49.** $-7b$ **51.** 5

Page 567 Lesson 11-3B
1. It divides the difference of the y-coordinates by the difference of the x-coordinates. **3.** yes
5. Draw two lines with the same slope.

Pages 570–572 Lesson 11-4
7. 117; obtuse **9.** equilateral **11.** 119° **13.** 50; right **15.** 36°; 54°; 90° **17.** isosceles **19.** equilateral **21.** obtuse **23.** 85° **25.** 25° **27.** 85° **29.** 45°, 80°, 55° **33.** 60° **35.** 25° **37.** $\frac{5}{13}$
39. $59\frac{1}{4}$ inches **41.** $4x + 12$

Page 572 Self Test
1. acute **3.** acute **5.** 34° **7.** 34° **9.** $m\angle F = 100°$; $m\angle G = 80°$ **11.** 100; obtuse

Pages 575–577 Lesson 11-5
5a. $\triangle ZYX$ **5b.** $\overline{ZY}$ **5c.** $\angle B$ **5d.** $\angle X$ **5e.** $\overline{YX}$ **5f.** $\overline{CA}$ **5g.** $\angle A$ **7.** c **9.** $\triangle ABC \cong \triangle FED$; $\angle A \cong \angle F$, $\angle B \cong \angle E$, $\angle C \cong \angle D$, $\overline{AB} \cong \overline{FE}$, $\overline{BC} \cong \overline{ED}$, $\overline{AC} \cong \overline{FD}$ **11.** $\triangle BAD \cong \triangle ABC$; $\angle BAD \cong \angle ABC$, $\angle D \cong \angle C$, $\angle ABD \cong \angle BAC$, $\overline{BA} \cong \overline{AB}$, $\overline{AD} \cong \overline{BC}$, $\overline{BD} \cong \overline{AC}$ **13.** $\triangle DEF \cong \triangle HGF$ **15.** $\overline{MA}$ **17.** $\overline{RT}$ **19.** $\angle M$ **21.** 3 **23.** 8 **25.** $\triangle ABH \cong \triangle IJG$; $\triangle ACG \cong \triangle IEF$ **27.** 5 ft **29.** 95°; obtuse **31.** 125
33. Liz–pack; Renee–snack bar; Pablo–plate lunch

Pages 581–583 Lesson 11-6
7a. $\frac{m\overline{AB}}{m\overline{AC}} = \frac{m\overline{BE}}{m\overline{CD}} = \frac{m\overline{AE}}{m\overline{AD}}$ **7b.** $\angle CAD$ **7c.** $\overline{CD}$
7d. 10 **9.** $\frac{1}{x} = \frac{2}{6}$; 3 **11.** 121 m **13.** $\frac{3}{2} = \frac{x}{3}$; 4.5
15. $\frac{4}{5} = \frac{8}{x}$; 10 **17.** $\frac{3}{5} = \frac{4.5}{x}$; 7.5 **19a.** Yes; the ratio of the corresponding sides is 1:1. **19b.** No; only those where the ratio of the corresponding sides is 1:1. **21.** 2.7 ft **23.** 32 m **25.** 2^5 **27.** 47.9; 42.5; 54

Pages 586–588 Lesson 11-7
5. 115 **7.** quadrilateral, trapezoid; trapezoid **9.** 6 m **11.** 30° **13.** $x = 60$; 60°, 120° **15.** $x = 60$; 60°, 120° **17.** square **19.** rectangle
21. quadrilateral **23.** sometimes **25.** never
27. always **29.** Yes; a rhombus is equilateral but may not be equiangular. **29b.** Yes; a rectangle is equiangular but may not be equilateral. **35.** 3 cm
37. 22.5 miles per gallon **39.** inductive

Pages 592–593 Lesson 11-8
7. hexagon, not regular **9.** 40°; 140° **11.** 900° **13.** 1800° **15.** 2340° **17.** 60°, 120° **19.** 30°, 150° **21.** 60° **23.** 72 cm **25.** 27 in.

27.

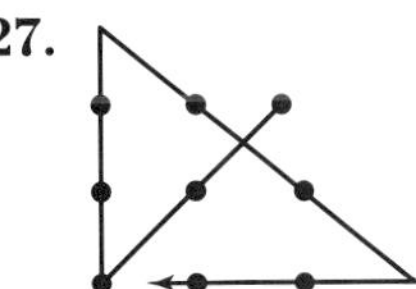

29a. trapezoid **29b.** 45°, 45°, 135°, 135°
29c. square **31.** $\frac{1}{5}$
33. $(1 \times 10^3) + (6 \times 10^2) + (4 \times 10^0)$

Page 594 Lesson 11-8B
1. It will tessellate if 360 is a multiple of the angle's measure.

SELECTED ANSWERS

Pages 597–599 Lesson 11-9

5. translation

7.

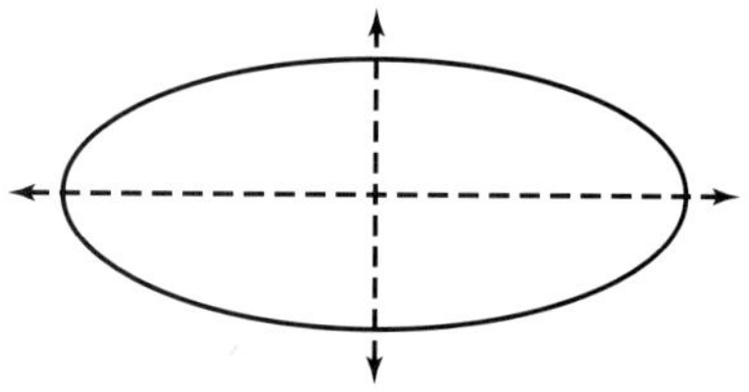

9. reflection **11.** rotation **13.** b

15.

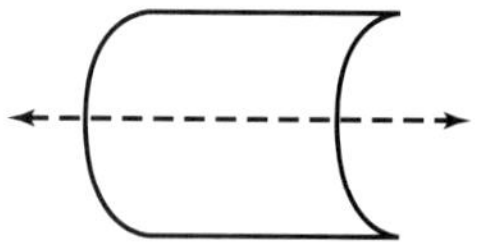

17.

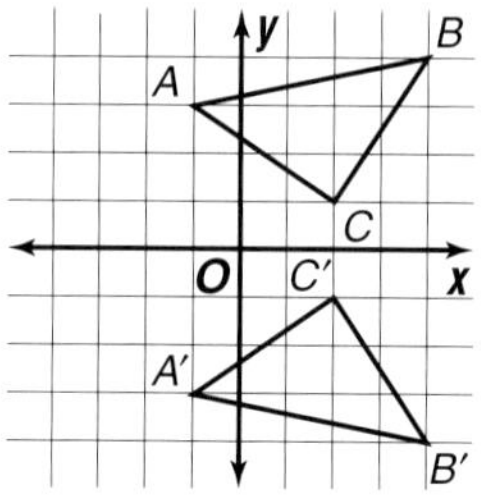

19.

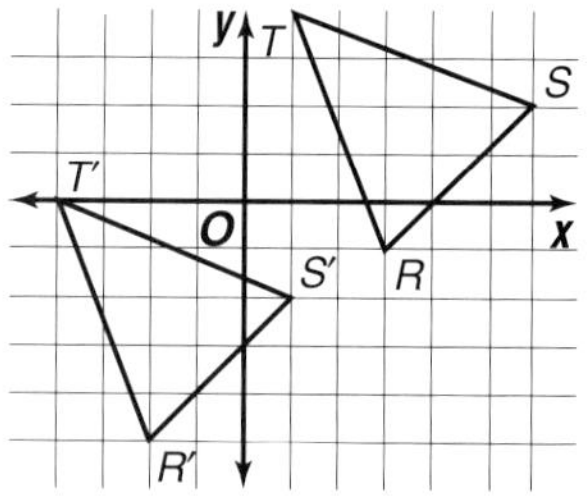

21. top to bottom: *all except* E L N O U Y Z
side to side: C D E F I J M P Q R S V W X
diagonal: F I L M N P Q R S U V W X Y

25. 200% increase

27. $t < -21$

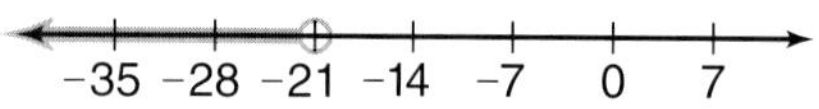

Page 601 Highlights

1. b **3.** g **5.** a **7.** i

Pages 602–604 Study Guide and Assessment

9.

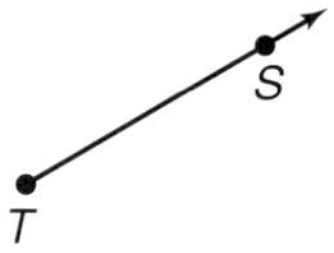

11.

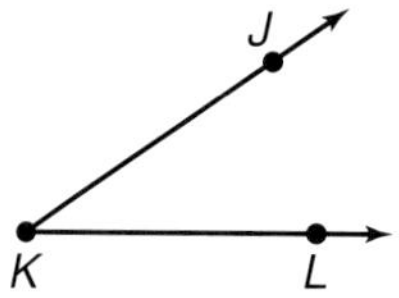

13.

15.

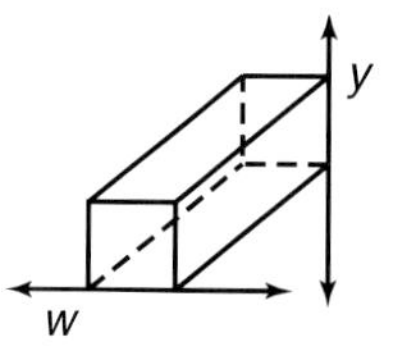

19. 48° **21.** 48° **23.** 90° **25.** 92°; obtuse **27.** $\angle Y$ **29.** $\overline{XZ}$ **31.** 2 in. **33.** 70°; 110° **35.** 144°

37.

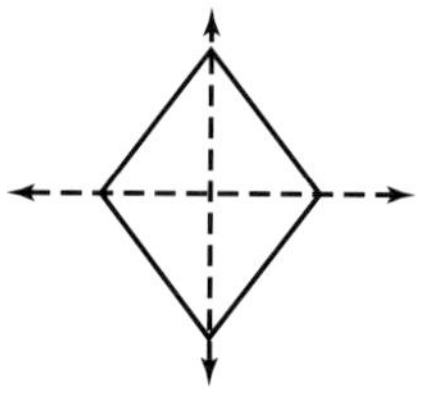

39. reflection

41.

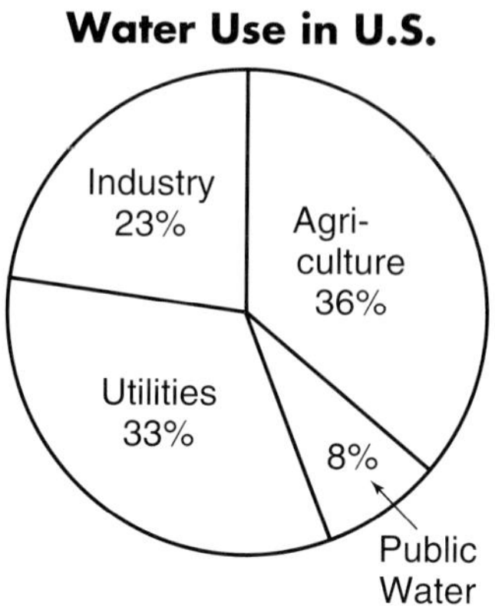

Chapter 12 Measuring Area and Volume

Pages 610–611 Lesson 12-1A

1. Sample answer: Count squares and half-squares.
3. Sample answer: Find the area of the rectangle and divide by 2.

Pages 615–617 Lesson 12-1

5. 3; 4; 6 sq cm **7.** 15 m and 5 m; 6 m; 60 m^2 **9.** 216 cm^2 **11a.** about 33,600 mi^2 **11b.** The actual area is 35,936 mi^2. **13.** 6 cm^2 **15.** 90 cm^2 **17.** 21.6 in^2 **19.** 57 ft^2 **21.** 9 in^2 **23.** $6\frac{1}{2}$ ft^2 **25.** 36.75 m^2 **27.** $A_1 = \frac{1}{2}h_1d$; $A_2 = \frac{1}{2}h_2d$; $A_1 + A_2 = \frac{1}{2}d(h_1 + h_2) = \frac{1}{2}d(d) = \frac{1}{2}d^2$ **29a.** 308 ft^2 **29b.** 15-foot width **31.** A shape as close as possible to a square gives the greatest area. **33.** 5 **35.** 2.1 **37.** $x + 7$

Page 618 Lesson 12-1B

1. Each section is a replica of the whole curve.
3. infinite

Pages 621–622 Lesson 12-2

5. 452.4 in^2 **7.** 49,087.4 ft^2 **9.** 50.3 cm^2 **11.** 120.8 m^2 **13.** 475.3 cm^2 **15.** 167.4 in^2 **17.** 39.3 ft^2 **19.** 7854.0 ft^2 **21.** about 707 m^2 **23.** about 26 times **25.** about 13 mi^2 **27.** 18 **29.** $\frac{4}{9}$

Pages 625–627 Lesson 12-3

5. $\frac{9}{50}$ **7.** about 9 **9.** $\frac{3}{8}$ **11.** $\frac{1}{6}$ **13.** $\frac{3}{4}$

15. about 38 **17.** 0.16 **19.** $\frac{1}{35}$ **21.** about 450 cm^2 **23.** 0.1 **25.** $32 < w < 212$

Page 628 Lesson 12-3B
1. The ratios should be similar. **3.** More data makes the ratios closer.

Pages 630–631 Lesson 12-4
5. 875 cm^2
7b. 10 units; 10 units; 8 units; 10 units; 10 units **7c.** Shapes where two squares share one side with another square and the rest of the squares share two sides with other squares are the greatest in perimeter. **7d.** Square shapes have the least perimeter. **7e.** 12; yes **9.** 4 m by 12 m **11.** $3 \times 3 \times 4$ **13.** $700 gain **15.** A **17.** $\frac{2}{5}$ **19.** 11 hours **21a.** Nevada **21b.** 17 million acres

Pages 635–637 Lesson 12-5
7. triangular prism; 184 ft^2 **9.** cylinder; 3166.7 m^2 **11.** 282 in^2 **13.** 132 cm^2 **15.** 211.2 cm^2 **17.** 636.2 in^2 **19.** 264 in^2 **21.** d **23a.** 2250 in^2 **23b.** No, as long as the entire surface of the face of the smaller cube is touching a face of the larger cube, the surface area of the tower will be the same. **25.** about 6189 cm^2 **27.** 512 in^2, 1200 in^2, 1800 in^2; no

31.

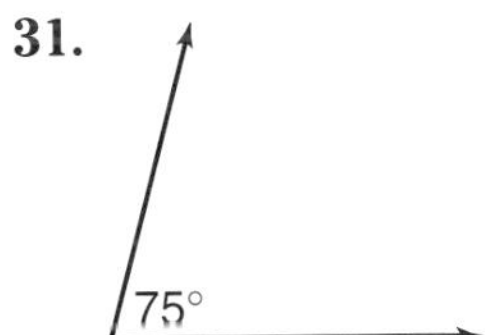

33. 3,605

Pages 641–642 Lesson 12-6
5. pyramid; 134.4 m^2 **7.** triangular pyramid; 57 in^2 **9.** 66.4 ft^2 **11.** 404.6 m^2 **13.** 282.7 in^2 **15.** 471.2 cm^2 **17.** 275 in^2 **19.** $\approx$ 25,780.8 ft^2 **21.** 3160.1 sq ft **23.** 900 cm^2 **25.** $4x + 8 = 20$; 3 **27.** $-15xy$

Page 642 Self Test
1. 15m^2 **3.** 113.1 in^2 **5.** cylinder, 402.1 in^2 **7.** $\approx$ 0.028 or about $\frac{3}{100}$ **9.** 47.1 in^2

Page 643 Lesson 12-7A
1. triangular prism; cylinder **3.** same; 8 in. **5.** cylinder

Pages 646–648 Lesson 12-7
7. 600 in^3 **9.** 225 in^2 **11.** 28.3 ft^3 **13.** 125 ft^3 **15.** 512 cm^3 **17.** 1608.5 ft^3 **19.** 565.5 cm^3 **21.** 24.5 ft^3 **23.** 83.3 cm^3 **25.** 4712.4 yd^3 **27.** 18.6 m^3 **29.** The volume is greater with $8\frac{1}{2}$ as the height. **31.** 3.29 cm **33.** a **35a.** 128 ft^3 **35b.** $\frac{5}{8}$ **37.** 4 mi **39.** 7 **41.** 64

Pages 651–653 Lesson 12-8
5. 32 ft^3 **7.** 160 ft^3 **9.** 270.8 in^3 **11.** 70 ft^3 **13.** 160 m^3 **15.** 150 in^3 **17.** 9189.2 m^3 **19.** 733 cm^3 **21.** 50 cm^3 **23.** 1885.0 m^3 **25.** 44 m^3 **27.** No; the customer gets only $\frac{1}{3}$ as much popcorn for $\frac{1}{2}$ the price. **29.** 91,636,272 ft^3 **31.** 15 ft^3 **33.** 20% **35.** 55 cm; 184 cm^2

Page 654 Lesson 12-8B
1. 1 unit^2 **3.** 1 unit^3 **5.** 9 unit^2
7.

Scale Factor	Length of a Side	Area of a Face	Volume
1	1	1	1
2	2	4	8
3	3	9	27

9. x^2 **11.** x^3

Page 655 Highlights
1. g **3.** h **5.** a **7.** c

Pages 656–658 Study Guide and Assessment
9. $3\frac{15}{16}$ in^2 **11.** 476 mm^2 **13.** 13.9 m^2 **15.** 8.0 m^2 **17.** $3\frac{3}{8}$ ft^2 **19.** 7665 ft^2 **21.** 9.4 m^2 **23.** 3325 in^3 **25.** 2770.9 ft^3 **27.** 504 m^3 **29.** 864 in^2 **31.** 198 **33.** about 1275 gallons

Chapter 13 Applying Algebra to Right Triangles

Pages 667–668 Lesson 13-1
7. 4 **9.** 9 **11.** -7 **13.** 11 **15.** 2 **17.** 3 **19.** -4 **21.** 10 **23.** -12 **25.** 20 **27.** 9 **29.** 14 **31.** -10 **33.** 51 **35.** 11 **37.** -5 **39.** a **41.** 3 **43.** Yes; because when a negative number is squared the result is positive. **45.** 20 feet **47.** 192 in^3 **49.** $\frac{7}{25}$ **51.** 36 **53.** -20

Pages 670–671 Lesson 13-2
3. 18 people **5.** 100 cm **7.** 77 pieces **9.** 10 girls **11.** 7 students
13.

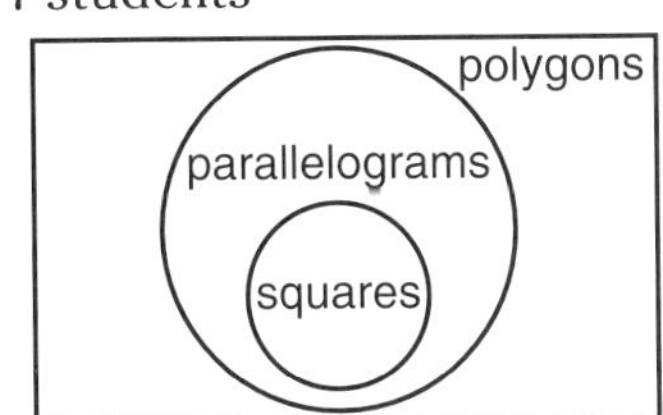

15. A rhombus is a quadrilateral with four congruent sides. **17.** 26.5

Pages 674–675 Lesson 13-3
5. whole, integer, rational, real **7.** rational, real **9.** 19, -19 **11.** 4.1, -4.1 **13.** whole number, integer, rational, real **15.** rational, real

SELECTED ANSWERS

17. rational, real **19.** irrational, real **21.** irrational, real **23.** rational, real **25.** 6, -6 **27.** 3.5, -3.5 **29.** 11.0, -11.0 **31.** 13.4, -13.4 **33.** 0.02, -0.02 **35.** 20, -20 **37.** 9, -9 **39.** 3, -3 **41.** Yes; sample answer: $\sqrt{2} \times \sqrt{2} = 2$ **43.** 2400 ft^2 **47.** 4908.7 cm^3 **49.** $-\frac{2}{5}$ **51.** 72

Pages 679–681 Lesson 13-4

5. $x^2 = 12^2 + 16^2$; 20 in. **7.** $30^2 = 21^2 + b^2$; 21.4 km **9.** 5 **11.** 32.9 **13.** yes **15.** about 127.3 ft **17.** $x^2 = 8^2 + 15^2$; 17 cm **19.** 26.5 m **21.** 19.6 ft **23.** 20 **25.** 5.4 **27.** 5.7 **29.** 9.9 **31.** 20.5 **33.** 5 **35.** no **37.** yes **39.** no **41.** 12 m **43.** 5 **45.** 25 km **47.** 1.6 **49.** 5 **51.** $\frac{4}{9}$ **53.** 20

Page 681 Self Test

1. 6 **3.** -10 **5.** 12, -12 **7.** 6.3, -6.3 **9.** 16

Page 682 Lesson 13-4B

1. Use the procedure in the activity, drawing a right triangle with legs 1 and 3 units long. Then use a compass to find the distance $\sqrt{10}$ on the x-axis. **3.** Use the procedure in the activity, drawing a right triangle with legs 1 unit and 2 units long; or use the procedure in the activity to find the lengths $\sqrt{2}$ and $\sqrt{3}$ and use those lengths as the legs of a right triangle. **5.** Draw a triangle with one leg 2 units long and hypotenuse 4 units long. Transfer $\sqrt{12}$ to the number line.

Pages 685–686 Lesson 13-5

3. 1.4 yd **5.** 9 m **7.** $a = 3$ m, $c \approx 4.2$ m **9.** $a = 6$ m, $b = 10.4$ m **11.** 15 cm **13.** 21.2 cm **15.** 7.4 m **17.** 9.8 in. **19.** 24 in. **21.** $\frac{1}{8}$ mi **23.** 2.315 cm **25.** $a \approx 8.1$ cm, $c \approx 16.2$ cm **27.** $a = 6$ in., $b \approx 10.4$ in. **29.** $b = 5$ yd, $c \approx 7.1$ yd **31.** $a \approx 4.0$ m, $b \approx 4.0$ m **33.** $a = 5.5$ km, $b \approx 9.5$ km **35.** about 5.8 yards **37.** about 7.07 feet **39.** 23.3 feet **41.** 108° **43.** 75 **45.** $s \geq 50$

Page 687 Lesson 13-6A

1. The ratios are the same. **3.** The ratios of the given angles are the same no matter how long the sides are.

Pages 690–692 Lesson 13-6

7. $\frac{24}{25}$ **9.** $\frac{7}{25}$ **11.** $\frac{24}{7}$ **13.** 0.3256 **15.** 0.4226 **17.** 80° **19.** about 19 feet **21.** 0.724, 0.690, 1.05 **23.** 0.6, 0.8, 0.75 **25.** 0.246, 0.969, 0.254 **27.** 0.5000 **29.** 0.9063 **31.** 0.9455 **33.** 53° **35.** 44° **37.** 61° **39.** 10° **41.** 42° **43.** 19° **45.** about 84.8° **47.** 10 yd **49.** 9 ways **51.** deductive

53. -242; 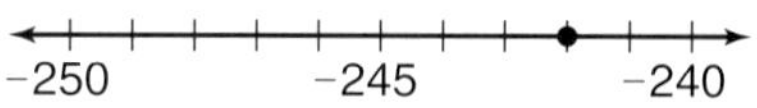

Page 693 Lesson 13-6B

1. $\frac{1}{2}$ **3.** They are the same. **5.** 3

Pages 696–697 Lesson 13-7

5. $\tan 39° = \frac{10}{x}$; 12.3 **7.** about 18.1 feet **9.** $\sin 60° = \frac{x}{16}$; 13.9 **11.** $\cos 30° = \frac{9}{x}$; 10.4 **13.** $\sin x° = \frac{8}{10}$; 53.1 **15.** $\sin x° = \frac{15}{21.2}$; 45.0 **17.** 38.3 yards **19.** about 118 feet **21.** about 44 m **23.** $a \approx 12.4$ yd, $b \approx 15.9$ yd **25.** 37 feet **27.** about 62° **29.** 30; 62; 70.5; 48.5; 22 **31.** yes

Page 699 Highlights

1. square **3.** rational **5.** 30°–60°

Pages 700–702 Study Guide and Assessment

7. 3 **9.** -5 **11.** 15 **13.** 9 **15.** 15 **17.** -6 **19.** whole, integer, rational, real **21.** rational, real **23.** 14, -14 **25.** 3.9, -3.9 **27.** 20 **29.** 34 **31.** 16 **33.** 6 m **35.** 3 cm **37.** $\sin X = 0.882$, $\cos X = 0.471$, $\tan X = 1.875$ **39.** 84° **41.** $\tan 35° = \frac{x}{15}$; 10.5 ft **43.** $\sin 25° = \frac{15}{x}$; 35.5 in. **45.** 28.4 ft **47.** about 321.68 ft

Chapter 14 Polynomials

Pages 708–709 Lesson 14-1

5. yes; monomial **7.** no **9.** 3 **11.** 3 **13.** 35 **15.** 57 **17.** \$38.82 **19.** yes, binomial **21.** yes, trinomial **23.** yes, monomial **25.** yes, trinomial **27.** yes, monomial **29.** yes, trinomial **31.** 1 **33.** 3 **35.** 6 **37.** 5 **39.** 7 **41.** 5 **43.** -26 **45.** -18 **47.** -43 **49.** 97 **51.** -44 **53.** 82 **55a.** $2x + 2y + z + xy$ **55b.** 2 **59.** 156.8 m^2 **61.** \$104.37 **63.** yes; 3; 81, 243, 729 **65.** $x > 8$ **67.** 13

Page 710 Lesson 14-1B

1. A monomial uses just one size of tiles, a binomial uses two different sizes, and a trinomial uses three different sizes. **3.** The degree of the polynomial is determined by the size of the largest algebra tile.

Pages 713–714 Lesson 14-2

7. $3x + 8$ **9.** $4x + 4$ **11.** $3x^2 + 3x + 2$ **13.** $6x + 24$ units **15.** $4x + 14$ **17.** $11m - n$ **19.** $2x^2 - 2x + 9$ **21.** $17m + 3n$ **23.** $8x^2 - 3x + 3$ **25.** $4a^3 + 2a^2b - b^2 + b^3$ **27.** $4a + 6b + c$ **29.** $-2a - 3ab + 4b^2 + 2$ **31.** $4x^2 + x - 1$ **33.** $4a^2 - 3ab - 2b - 4b^2$; 0 **35.** $(4r + 3s)$ **37.** $4x + 2$ united inches **39.** 7.5 cm^2 **41a.** $t + h > 130$

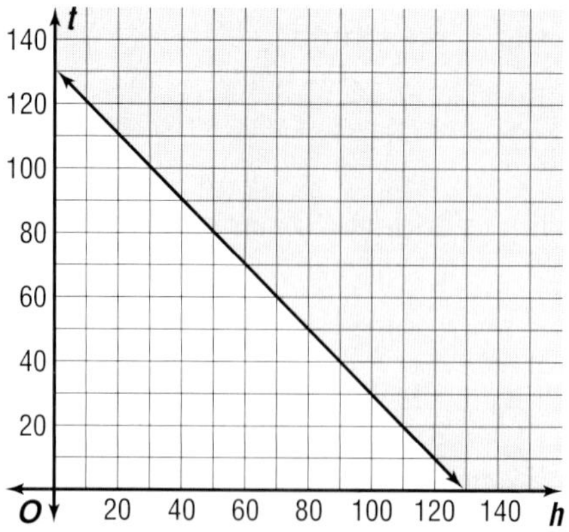

41b. the area above the boundary line

41c. No points with $h > 100$ are included in the graph.
43. 15 ft

Pages 717–718 Lesson 14-3

9. $-2abc$ **11.** $-x^2 - 5x - 1$ **13.** $8m - 7n$ **15.** $2x + 2$ **17.** $2x - 3$ **19.** $x^2 + 3x - 3$ **21.** $4x - 6$ **23.** $x - 3$ **25.** $2x - 8$ **27.** $-3x + 5y$ **29.** $-2x^2 + 6x - 7$ **31.** $m^2 - 7m + 10$ **33.** $15m^2 - 2m + 10$ **35.** $9a + 3b - 7c$ **37.** $8y^2 + 4y - 5$ **39.** $-6x^3 - 4x^2y + 13xy^2$ **41a.** 184 **41b.** 370 **43.** $8a^2 + 2a + 1$ **45.** 90°; right **47.** every 12 days **49.** II

Page 718 Self Test

1. yes, binomial; 3 **3.** yes, monomial; 12 **5.** $-x^2 + 12$ **7.** $2x + 6y$ units **9.** $a + 13d$

Pages 722–723 Lesson 14-4

5. 2^{12} **7.** $3^3x^3y^3$ or $27x^3y^3$ **9.** x^2y^2 **11.** 9 **13.** $48x^6y^2$ **15.** y^4z^4 **17.** $16r^4s^4$ **19.** m^{10} **21.** $-32x^{10}$ **23.** x^4y^{12} **25.** $12x^3$ **27.** $16x^4y^6$ **29.** t^{-12} **31.** $-2x^{-6}y^{-2}$ **33.** -36 **35.** 108 **37.** -1296 **39.** $\frac{1}{324}$ **41a.** The numbers in the values columns are the same. **41b.** The exponents are distributed across the factors. **41c.** When you raise a product to a power, distribute the exponent over each factor. **43a.** $1000(1.01^4)^6$ **43b.** $1269.73 **45.** 5280^3 **47.** 13 in. **49.** A spinner divided into 10 sections, six of them being successes.

51.

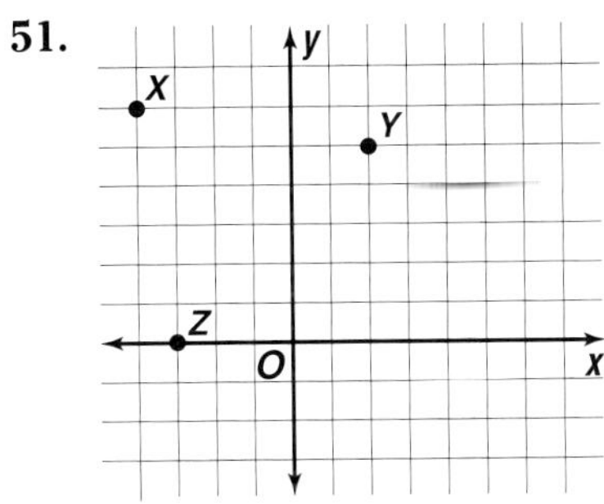

51a. II **51b.** I **51c.** none **53.** $xy \div 5$

Page 724 Lesson 14-5A

1. true **3.** $x^2 + 5x$ **5.** $6x^2 + 3x$

Pages 726–727 Lesson 14-5

7. $3x + 12$ **9.** $3x - 6$ **11.** $12x^2 + 28x$ **13.** 160 ft by 360 ft **15.** $-2x - 16$ **17.** $6x^2 - 3x$ **19.** $a^2c + bc$ **21.** $p^2q^2 + 8pq$ **23.** $-18a + 2a^3$ **25.** $-15x^2 + 35x - 45$ **27.** $4c^4 + 28c^3 - 40c$ **29.** $-12x^5 + 48x^4 - 42x^2$ **31.** $-x^5 + x^4 - 3x^3 + 5x^2$ **33.** 2 **35a.** $12x^2 + 8x + 1$ **35b.** 96 in² **37.** $-32b^4$ **39.** Sample answer: Choose a scale that makes one product look much better than the other; omit some information or labels. **41.** 13 times **43.** 192.5 miles

Pages 730–731 Lesson 14-6

7. $x + 3, 2x + 1; 2x^2 + 7x + 3$ **9.** $2x^2 + 5x + 3$ **11.** $10x^2 - x - 3$ **13.** $2x + 1, x + 2; 2x^2 + 5x + 2$ **15.** $2x + 3, 2x + 2; 4x^2 + 10x + 6$ **17.** $x^2 + 7x + 12$ **19.** $x^2 - 4x - 12$ **21.** $x^2 - 5x - 36$ **23.** $2x^2 - 5x - 12$ **25.** $x^2 + 6x + 9$ **27.** $10x^2 - 11x - 6$ **29.** $18x^2 - 3x + \frac{1}{8}$ **31.** $(x + 2)(x - 1); x^2 + x - 2$ **33a.** 14 in. by 27 in. **33b.** 378 in² **33c.** 464 in² **35.** 90 ft **37.** 3.14 m **39.** $x < -14$

Page 732 Lesson 14-6B

1. yes **3.** no **5.** no

Page 733 Highlights

1. sum **3.** greatest **5.** additive inverse

Pages 734–736 Study Guide and Assessment

11. yes; binomial **13.** yes; monomial **15.** yes; binomial **17.** 1 **19.** 2 **21.** 3 **23.** $5x^2 - 4x$ **25.** $x^2 - x$ **27.** $5x^2 + 3x + 1$ **29.** $4a - 15b$ **31.** $2a^2 - 3b^2 + c^2$ **33.** $16x^2 - x - 7$ **35.** a^6 **37.** p^6q^3 **39.** $4y^7z^3$ **41.** $-6a^8b^7$ **43.** $8d^2 - 20d$ **45.** $2a^5 + a^3 - 5a^2$ **47.** $-2g^4 - 12g^2 - 6g$ **49.** $2x + 1, x + 3; 2x^2 + 7x + 3$ **51.** $2x^2 + 3x + 1$ **53.** $2xy + 2y^2 + 2yz$ **55.** $3x^3$

PHOTO CREDITS

2, 10, 35, 56 (t) Vince Streano/Tony Stone Images; **3** (t) Charles O'Rear/Westlight, (b) Chris Sorensen; **4** (bl) Lois Gervais/Westlight, (bc) Nikolai Zurek/FPG International, (br) Bettmann; **5** (t) Bob Daemmrich/Stock Boston, (bl) Beverly Johnson by Francesco Scavullo August 1974 Vogue. Courtesy Vogue. ©1974 by the Conde Nast Publications, Inc.; (br) Carl Corey/Westlight; **6** (t) Duomo, (b) Mark Burnett; **7** Stacy Pick/Stock Boston; **8** Custom Mcdical Stock Photo; **9** Mark Steinmetz; **11** Mark Burnett; **13** Joe Rimkus Jr./The Miami Herald; **14** Lester Lefkowitz/Tony Stone Images; **15** W. Cody/Westlight; **17** Don Smetzer/Tony Stone Images; **19** (t) Jack Zehrt/FPG International, (b) Dr. E.R. Degginger/Color-Pic; **20** (t) Wide World Photos, (b) FOXTROT ©Bill Amend. Reprinted with permission of UNIVERSAL PRESS SYNDICATE. All rights reserved.; **23** (l) David Allen/LGI Photo Agency, (r) David Allen/LGI Photo Agency; **27** Michael S. Yamashita/Westlight; **28** Katrina De Leon/Still Life Stock; **30** (t) Mark Burnett, (b) E. Sampers/Gamma Liaison; **32** (t) David Young Wolff/Tony Stone Images, (b) Mark Burnett; **33** Doug Martin; **34** Frank Spinelli/Liaison International; **36** Matt Meadows; **38** Mark Burnett; **39** National Optical Astronomy Observatories; **41** Bettmann; **43** Aaron Haupt; **44** Ben Van Hook/Duomo; **45** Photofest; **47** Mark Burnett; **48** Owen Franken/Stock Boston; **50** Matt Meadows; **51** Murray & Associates/Tony Stone Images; **52** Mark Reinstein/FPG International; **54** Lynn Stone; **55** Alexander Gardner, 1863/Bettmann Archive; **56** Chris Sorensen; **61-63** Mark Burnett; **64** (t) Mark Burnett, (bl) Eberhard E. Otto/FPG International, (br) Chromosohm/Joe Sohm/Photo Researchers; **65** FRANK & ERNEST, reprinted by permission of Newspaper Enterprise Association, Inc.; **65** (tl) Steve Gottlieb/FPG International, (b) Bettmann; **66** Buddy Mays/FPG International; **69** Mark Conlin; **70** D.R. Goff, courtesy CAPA; **73** (t) Michael Keller/FPG International, (b) J. Scower/FPG International; **76** Robin Smith/Tony Stone Images; **78** (t) David Scharf/Peter Arnold, Inc., (b) Mark Burnett; **79** Mark Steinmetz; **80** Jose Carrillo; **81** Steve Seiller/Ken Graham Agency; **84** AP/World Wide Photos; **86** Stephen Rose/Gamma International; **87** (t) G. Randall/FPG International, (b) Stan Osolinski/FPG International; **90** Stan Osolinski/FPG International; **92** AP/Wide World Photos; **93** R. Y. Kaufman/Yogi, Inc.; **94** Biblioteca Universitaria, Bologna, Italy/Art Resource; **96** Comstock Inc./Gary Benson; **97** Mark Conlin; **99** (t) Mark Conlin, (l) Painting by Ken Marshall from the book *The Discovery of the Titanic* by Dr. Robert Ballard, (b) Stephen Rose/The Gamma Liaison Network; **101** Josef Beck/FPG International; **102** Ken Graham/Ken Graham Agency; **103** Joseph Pobereski/Tony Stone Images; **106** Remi Benali/Gamma International; **107** John Terence Turner/FPG International; **108** (t) courtesy NASA, (b) Comstock Inc./Jack Clark; **112** Carl R. Sams II/Peter Arnold, Inc.; **113** Wally Dallenbach/Peter Arnold, Inc.; **116** (t) Hoffines/Young, (br) courtesy Perkins School for the Blind, (bl) Bettmann; **117** (t) Ron Rovtar/FPG International, (b) Bettmann; **118** Matt Meadows; **120** Nawrocki Stock Photo; **121** (l) John Elk III/Stock Boston, (r) J. McGuire/Washington Stock Photo; **122** (t) Henry Horenstein/Viesti, (b) Mark Steinmetz; **125** Photofest; **127** Bonnie Sue/Photo Researchers; **128** Photofest; **129, 131** Mark Burnett; **132** Nawrocki Stock Photo; **133** (t) Sharon M. Kurgis, (b) Deborah Denker/Liaison International; **134** Frank Whitney/The Image Bank; **135** Fred Bavendam/Peter Arnold, Inc.; **136** Nawrocki Stock Photo; **137** (t) Joe Viesti/Viesti Associates, (b) Glencoe files; **138** Mark Burnett; **141** Michael Newman/PhotoEdit; **142** Marcel Ehrhard/Liaison International; **144** (t) Gerard Lacz/Peter Arnold, Inc., (b) ZIGGY ©1978 ZIGGY & FRIENDS, INC. Dist. by UNIVERSAL PRESS SYNDICATE. Reprinted with permission. All rights reserved.; **145** Nawrocki Stock Photo; **148** Mark Burnett; **149** Uniphoto; **153** Frank Spinelli/Liaison International; **154** Charlie Trainor/The Miami Herald; **155** (t) Michael Simpson/FPG International, (b) Cliff Beaver; **159** John Lei/Stock Boston; **160** Mark Burnett; **164** Bettmann; **165** Robert Burke/Liaison International; **167** Mark Steinmetz; **168** (br) Frank Driggs Collection, (bl) Painting by Abraham Anderson, c. 1889, National Portrait Gallery, Smithsonian/Bettmann Archives; **169** (t) Comstock Inc./Comstock Inc., (c) Ken Frick, (br) Photofest, (bl) Charles O'Rear/Westlight; **172** Bridgeman/Art Resource, NY; **173** (t) Myrleen Ferguson/PhotoEdit, (b) Nawrocki Stock Photo; **174** G. Randall/FPG International; **175** Aaron Haupt; **177** Comstock Inc./Contrib. Library; **178** Comstock Inc./Comstock Inc.; **179** (t) Alinari/SEAT/Art Resource, NY, (b) THE FAR SIDE ©1985 FARWORKS, INC. Dist. by UNIVERSAL PRESS SYNDICATE. Reprinted with permission. All rights reserved.; **181** (t) Pam Francis/Liaison International, (b) Lawrence Migdale; **183** Lynn Stone; **185** Rosemary Weller/Tony Stone Images; **186** courtesy Dr. Arjen Lenstra, Bell Communications Research; **187** Photofest; **188** (t) Everett Collection, (b) Dick Luria/FPG International; **190** S. Feld/H. Armstrong Roberts, Inc.; **191** Elizabeth Simpson/FPG International; **192** Aaron Haupt; **194** Comstock Inc./Stuart Cohen; **196** courtesy GCI/Atlanta; **199** (t) Matthew Neal McVay/Tony Stone Images, (b) James Blank/FPG International; **200** Stuart Westmorland/Tony Stone Images; **201** NASA/Peter Arnold, Inc.; **203** Matt Meadows; **204** (t) Mike Theiler/Reuters/Bettmann Archive, (b) Larry Ulrich/Tony Stone Images; **207** Paul Richards/UPI/Bettmann Newsphotos; **208, 211** Manfred Kage/Peter Arnold, Inc.; **212** Howard Hughes Medical Institute/Peter Arnold, Inc.; **214** (t) Arthur Tilley/Tony Stone Images, (b) Stock Montage; **219** Comstock Inc./Robert Houser; **222** *American Heritage* magazine; **223** (t)*The Saturday Evening Post* © 1994, (c) Ken Huang/The Image Bank, (br) Mark Steinmetz, (bl) Studiohio; **224** (l) Photofest, (r) Everett Collection; **225** Bob Eighmie/The Miami Herald; **227** Steven Alexander/Uniphoto Picture Agency; **228** (l) courtesy Ingrid Proctor-Fridia, (r) Leonard Lessin/Peter Arnold, Inc.; **229, 230** Mark Burnett; **232** courtesy N.O.A.A.; **233** Robert Landau/Westlight; **236** Eugene Gebhardt/FPG International; **237** Bettmann; **241** G. Schwartz/FPG International; **242** (t) Mark Burnett, (b) David W. Hamilton/The Image Bank; **243** Mark Burnett; **245** Frank Saragnese/FPG International; **248** NASA/Peter Arnold, Inc.; **249, 250** Bettmann; **252** Mark Burnett; **254** (t) William R. Sallaz/Duomo, (b) Mark Steinmetz; **255** PEANUTS reprinted by permission of United Feature Syndicate, Inc.; **256** (t) Stephen Dalton/Photo Researchers, (b) Mark Steinmetz;

257 ©1966 *The Saturday Review*, reproduced by permission of Ed Fisher; **258** A. T. Willett/The Image Bank; **260** Mark Steinmetz; **261** Studiohio; **262** (t) Science and Society Picture Library, (b) Shinichi Kanno/FPG International; **264** Mark Steinmetz; **268** First Image; **269** David M. Grossman; **271** Mark Burnett; **272** (t) Uniphoto, (bl, bc) Bettmann, (br) courtesy International Business Machines Corporation; **273** (t) Drawing by W. Miller, 1994 *The New Yorker* Magazine, Inc., (c) Arthur Tilley/FPG International, (bl) Photofest, (br) Kazu Studio Ltd./FPG International; **274** Mark Burnett; **276** Tim Courlas; **277** courtesy Wendy's International, Inc.; **278** (l) David Zosel, (r) NASA; **279** Raphael. *The Triumph of Galatea.* Palazzo della Farnesina, Rome, Italy./Art Resource; **281** Michal Heron/The Stock Market; **283** (l) Mark Sennet/Onyx, (r) Laurel Canty, Dolphin Research Center, Grassy Key, Florida; **285** Richard Bowditch; **287** Mark Steinmetz; **288** Rudy VonBriel; **290** Bob Daemmrich; **292** Aaron Haupt; **293** Dave Bartruff; **295** Aaron Haupt; **297** Michael Nelson/FPG International; **298** Elaine Shay; **299** Aaron Haupt; **300** Will Vinton/Photofest; **301** U.S. Representative/Westlight; **303** (l) James Schnepf/The Gamma Liaison Network, (r) Thomas Ives/The Stock Market; **304** Mark Steinmetz; **305** Chromosohm/Sohm/The Stock Market; **306** (l) Jim Pickerell/Westlight, (r) Dan Lecca/FPG International; **308** (t) Terence Gili/FPG International, (b) Mark Steinmetz; **310** Photofest; **312** Matt Meadows; **314** Long Photography; **315** Mark Burnett; **317** Manfred Gottschalk/Westlight; **319** (l) Paul W. Nesbit, (r) Stan Osolinski/FPG International; **325** Richard Pasley/Viesti Associates; **328** (t, br) FPG International, (bl) Science & Society Picture Library; **329** (t) THE FAR SIDE ©1990 FARWORKS, INC. Dist. by UNIVERSAL PRESS SYNDICATE. Reprinted with permission. All rights reserved., (b) Photofest; **331** Matthew Klein/Photo Researchers; **332** (t) James Blank/FPG International, (b) Bob Daemmrich; **334** E. Nagele/FPG International; **336** Mark Steinmetz; **337** Aaron Haupt; **338** Focus on Sports; **339** Photofest; **340** from *Not Strictly By the Numbers* ©Carolina Biological Supply Company; **342-3** Jack Zehrt/FPG International; **343** Mark Steinmetz; **344** Ben Simmons/The Stock Market; **349** (t) Elaine Shay, (b) Alan Pitcairn from Grant Heilman; **350** (t) Aaron Haupt, (b) Comstock Inc./Stuart Cohen; **352** Larry Lefever from Grant Heilman; **353** Mark Steinmetz; **354** Warren Morgan/Westlight; **355** Mark Steinmetz; **357** courtesy John Deere Precision Farming; **358** Joachim Messerschmidt/FPG International; **359** Mark Burnett; **360** (t) Bob Daemmrich, (bl) Dr.E.R. Degginger/Color-Pic; **361** (l) Mark Steinmetz, (r) Aaron Haupt; **362** James Marshall/The Stock Market; **366** (l) The Smithsonian Institution, (r) From the Al Rose Collection, Hogan Jazz Archive, Tulane University; **369** (t) Bob Daemmrich, (b) Flip Chalfant/The Image Bank; **370** (t) Ben Van Hook/Black Star Publishing, (b) Archive Photos; **371** (t) *The Saturday Evening Post* ©1995, (c) Cindy Lewis, (bl) Bettmann, (br) Michael R. Brown/Gamma-Liaison; **372** Comstock Inc./Comstock Inc.; **374** MAK-I; **375** Fred Hirschmann/Ken Graham Agency; **377** (t) Henry Groskinsky/Peter Arnold, Inc., (b) Telegraph Colour Library/FPG International; **378** Aaron Haupt; **381** Long Photography; **382** Uniphoto; **383** Brian Drake/Rim Light Photography; **384** J.R. Schnelzer; **388** Comstock Inc./M & C Werner; **390** ©1989 by Sidney Harris - *Einstein Simplified,* Rutgers University Press; **394** Jose Carrillo; **395** (l) J&L Weber/Peter Arnold, Inc., (r) Mark Stephenson/Westlight, (b) Steve Lissau; **397** Ken Skalski/Liaison International; **398** Elaine Shay; **400** B. Busco/The Image Bank; **402** Ken Graham/Ken Graham Agency; **404** Bob Daemmrich; **405** Mark Steinmetz; **406** Aaron Haupt; **408** Penny Gentieu/Tony Stone Images; **410** Warren Motts; **412** Peter Timmerans/Tony Stone Images; **416** James Shaffer/PhotoEdit; **419** Photofest; **421** William J. Weber; **422** (t) Mark Conlin, (b) Bob Daemmrich; **430** (l) Francois Gohier/Photo Researchers, (r) Larry Hamill; **431** (t) Anthro-Photo, (l) John Reader/Science Photo Library/Photo Researchers, (r) Francois Gohier/Photo Researchers; **432** Mark Burnett; 433 Aaron Haupt; **434** Uniphoto; **435** Eberhard Otto/FPG International; **436** John Mitchell/Photo Researchers; **437** Bettmann; **438** (t) Mark Steinmetz, (c) *People Weekly* 1995 Steve Liss, (b) Focus on Sports; **439** Fred Lyon/Photo Researchers; **440** courtesy WSNY; **441** (l) Matt Meadows, (r) Mark Steinmetz; **443** courtesy NASA; **444** (l) P. Le Segretain/Sygma, (r) De Keerle/Sygma; 445 Mark Steinmetz; **446** Doug Martin; **447** Mark Steinmetz; **448** D. Robert Franz/Masterfile; **449** Mark Burnett; **452** (t) Ralph Clevenger/Westlight, (b) Mark Steinmetz; **454** Mark Burnett; **455** W. Hodges/Westlight; **456** Doug Martin; **459** Rosemary Weller/Tony Stone Images; **461** (l)Uniphoto, (r) Paul Morse/LGI Photo Agency; **462, 463** Aaron Haupt; **466** (t) Cindy Lewis Photography, (b) THE BORN LOSER reprinted by permission of Newspaper Enterprises Association, Inc.; **467** Bob Daemmrich; **468** courtesy The Nature Conservancy; **469** Lisa Quinones/Black Star; **470** (t) Larry Mulvehill/Photo Researchers, (b) Aldo Pavan/Gamma Liaison Network; **471** Auguste Renoir. *The Reader,* 1874-1876. Musee d'Orsay, Paris, France. Erich Lessing/Art Resource, NY; **475** Geof du Feu/Masterfile; **476** Brian Bahr/Allsport; **480** courtesy Vehicle Research & Testing Center; **481** Mark Steinmetz; **482, 508, 528** Mark Burnett; **483** KS Studio; **483** Rod Joslin; **483** (t) Mark Steinmetz, (b) Mark Burnett; **484** Bettmann; **485** (t) Gaslight Advertising Archives, (l) Glencoe files, (c) Bettmann, (r) Mark Steinmetz; **486** (l) Mark Burnett, (r) Glencoe files; **488** (l) Mark Burnett, (r) Ed Bock/The Stock Market; **490** Mark Steinmetz; **492** Everett Collection; **492** Mark Burnett; **493** David R. Frazier Photolibrary; **494** James Carlson/David R. Frazier Photolibrary; **495** Trent Steffler/David R. Frazier Photolibrary; **498, 500** Mark Steinmetz; **502** Matt Meadows; **503** Nawrocki Stock Photo; **504** Barbara Campbell/Gamma Liaison International; **505** Matt Meadows; **507** Daniel J. Cox/Gamma Liaison International; **508, 509** Mark Burnett; **510** (l) Mark Steinmetz, (r) David M. Grossman; **513** Comstock Inc./Comstock Inc.; **514** Conway Trefethen/FPG International; **515** Tim Davis/Photo Researchers; **516** Larry Kunkel/FPG International; **517** Nawrocki Stock Photo; **518** Mark Burnett; **519** courtesy Dr. Carol Malloy; **520, 521** Mark Steinmetz; **521** (t) Mark Burnett from the Paul Baumann collection; **523** (t) Cecil Fox/Science Source/Photo Researchers, (b) PEANUTS reprinted by permission of United Feature Syndicate, Inc.; **524** Corbis Media; **525** Mary Ann Evans; **526** James Levin/FPG International; **527** Ken Biggs/The Stock Market; **529** Mark Burnett from the Paul Baumann collection; **530** Thayer Syme/FPG International; **532** Mark Steinmetz; **533** Index Stock Photography; **536** Mark Burnett; **538** Mark Steinmetz; **543** Philippe Poulet/Gamma Liaison International; **546** (l) Bettmann, (c) Stock Newport/Daniel Forster, (r) Thomas Zimmermann/FPG International; **547** (t) G. Whiteley/Photo Researchers, (bl) Stock Newport/Bob Grieser, (br) Stock Newport/Daniel Forster; **548** David Zalubowski/AP Colorado; **550** Bettmann; **551** Mathias

PHOTO CREDITS

Oppersdorff/Photo Researchers; **552** Mark Burnett; **556** Tracy Borland; **557** Bettmann; **559** (t) Comstock Inc./Comstock Inc., (b) Harald Sund/The Image Bank; **560** (t) Mark Burnett, (b) Russell D. Curtis/Photo Researchers; **564** Mark Steinmetz; **565** David R. Frazier Photolibrary; **566** CNRI/Science Photo Library/Photo Researchers; **567, 568** David R. Frazier Photolibrary; **570** David Gorman; **576** Carl Purcell/Photo Researchers; **577, 578** Mark Steinmetz; **580** V.A.P. Neal/Photo Researchers; **583** Mark C. Burnett/Photo Researchers; **584** Janet Adams; **591** Defense Dept.; **594** Pictures Unlimited; **595** David R. Frazier Photolibrary; **596** (l) Phil Degginger, (lc) Matt Meadows, (c) Aaron Haupt, (rc) Mark Steinmetz, (r) Mark Burnett; **597** (l) Dr.E.R. Degginger/Color-Pic, (r) Harald Sund/Image Bank; **599** (t) Audrey Gibson/Mark E. Gibson Photography, (b) FOXTROT ©Bill Amend. Reprinted with permission of UNIVERSAL PRESS SYNDICATE. All rights reserved.; **600** Mark Burnett; **605** Scott Camazine/Photo Researchers; **606** Jose Luis Banus-March/FPG International; **607** David A. Barnes/FPG International; **608** (l) Bridgeman/Art Resource, NY, (r) Scala/Art Resource, NY; **609** (t) Uniphoto, (l) Bettmann, (c) Superstock, (r) Photofest; **612** (l) Glencoe files, (r) Mark Burnett; **615** (l) Mark Steinmetz, (r) Mark Burnett; **617** Joseph Nettis/Photo Researchers; **618, 619** Mark Steinmetz; **621** Mark Segal/Tony Stone Images; **622** (t) Howard Bluestein/ Photo Researchers, (b) FRANK & ERNEST, reprinted by permission of Newspaper Enterprise Association, Inc.; **623** Mark Steinmetz; **624** Mark Burnett; **625** Heinz Fischer/The Image Bank; **626** David R. Stoecklein/ TexStock; **629** Werner Muller/Peter Arnold, Inc.; **630** Ben Mitchell/The Image Bank; **631** Navaswan/FPG International; **632** Mark Steinmetz; **634** Mark Burnett; **636** Gary Cralle/The Image Bank; **638** Marvin E. Newman/The Image Bank; **640** Jules Zalon/The Image Bank; **641** courtesy Dr. Michael R. Doss; **645** Gerard Photography; **646** Ellis Herwig/Stock Boston; **647** E.R.Degginger/Photo Researchers; **648** Herve Berthoule Jacana/Photo Researchers; **650** Bernard Roussel/The Image Bank; **651** Mark Burnett; **653** William E. Ferguson; **654** Mark Steinmetz; **658** David M. Grossman; **659** Ancient Art & Architecture Collection; **662** (t) George Obremski/The Image Bank, (l) Glencoe files, (r) Scala/Art Resource, NY; **663** (t) Reprinted from *The Saturday Evening Post* ©1995, (bl) Peter Beney/ The Image Bank, (br) Bettmann; **664** Superstock; **665** Mark Burnett; **667** Cliff Feulner/The Image Bank; **668** CALVIN AND HOBBES ©1990 Watterson. Dist. by UNIVERSAL PRESS SYNDICATE. Reprinted with permission. All rights reserved.; **669** (t) Travelpix/FPG International, (b) Telegraph Colour Library/FPG International; **670, 672** David R. Frazier Photolibrary; **674** Index; **675** Doug Martin; **678** Leverett Bradley/FPG International; **681, 682** Mark Burnett; **683** Stock Montage; **684** Pictures Unlimited; **685** Mark E. Gibson/Mark E. Gibson Photography; **686** Janine Herron/FPG International; **687** Morton & White Photographic; **690** James Blank/FPG International; **692** John Elk III/Stock Boston; **695** Bill Lea; **697** Aaron Haupt/David R. Frazier Photolibrary; **698** (t) Courtesy Sullivan Gray Bruck Architects, Inc., (b) courtesy Darlene Leshnock; **702** Ron Thomas/FPG International; **704** (l) Mark Burnett, (c) Bettmann, (r) Superstock; **705** (t) PEANUTS reprinted by permission of United Feature Syndicate, Inc., (l) Mark Burnett, (r) courtesy Bryan Berg; **706** Tom Carroll/FPG International; **707** (l) Will & Deni McIntyre/Photo Researchers, (r) Photo Researchers; **708** (t) Mark Burnett, (b) John Kaprielian/Photo Researchers; **709** Audrey Gibson/ Mark E. Gibson Photography; **709** Craig K. Lorenz/Photo Researchers; **712** Ronald Sheridan/Ancient Art & Architecture Collection; **713** Glencoe files; **714** (t) Index, (b) Spencer Grant/Photo Researchers; **715** Steve Allen/Gamma Liaison; **716** Beth Ava/FPG International; **718** John McDonough/Focus on Sports; **720** Superstock; **723** (l) Porterfield-Chickering/Photo Researchers, (r) Platinum Productions; **724** Index; **725** Glencoe files; **726** Franc Shor/©National Geographic Society; **727** (t) Mark Burnett, (b) Mark Steinmetz; **728** Index; **729** Ed Braverman/FPG International; **730** Superstock; **731** Superstock; **736** Tony Quinn Photography; **Cover** Gus Chan; **Cover** Tony Stone Images; **Cover** Tony Stone Images

APPLICATIONS INDEX

A

Academics, 336
Accounting, 92
Advertising, 127, 470, 617
Aerospace, 108
Aerospace Design, 534
Agriculture, 357, 361, 658, 668
Air Travel, 159, 396
Airports, 250
Ambulance Service, 534
Anthology, 395
Architecture, 434, 435, 488, 591, 622, 637, 639, 641, 686
Art, 20, 209, 471
Astronomy, 39, 87, 200, 213, 324
Automotive, 344
Aviation, 92, 366, 410, 438, 553, 697

B

Baking, 648
Banking, 299
Basketball, 718
Biology, 133, 208, 283, 384, 559, 566, 723
Boating, 247
Business, 15, 29, 93, 121, 150, 194, 218, 233, 245, 279, 293, 306, 311, 398, 410, 416, 422, 461, 470, 513, 526, 536, 560, 636, 731

C

Careers, 133
Carpentry, 193, 243, 268, 390, 403, 564, 566, 668
Cartography, 73
Catering, 709
Census, 349
Charity, 49, 468
Chemistry, 80, 208, 336, 648, 723
Civics, 48, 174
Civil Defense, 571
Communication, 150, 598, 702
Computers, 494, 626
Conservation, 507
Construction, 144, 247, 575, 577, 583, 588, 653, 668, 684, 686, 692, 714
Consumer Awareness, 70, 81, 353, 359, 377, 422, 435, 475, 534, 538, 577, 648, 652, 714
Consumerism, 121, 230, 288, 336, 433, 501, 617
Cooking, 144, 288, 350, 446
Cosmetics, 459, 461
Cryptography, 186, 188
Cycling, 237

D

Decorating, 283
Demographics, 45, 188, 199, 279, 443
Desktop Publishing, 290
Driver's License Exam, 593
Driving, 256

E

Ecology, 361, 709
Economics, 108, 243, 299, 324, 445, 470, 489, 533, 538, 709
Education, 118-119, 179, 311, 383, 525
Electricity, 178, 674
Employment, 254, 281, 306, 501, 507
Energy, 397, 398, 473, 475
Engineering, 29, 228, 489, 646
Entertainment, 35, 45, 125, 157, 179, 233, 241, 332, 344, 353, 455, 466, 504, 522
Exercise, 128, 188

F

Family Activity, 144, 188, 254, 306, 361, 422, 435, 443, 523, 559, 637, 675, 709
Family Management, 55
Farming, 637
Fashion, 257, 456
Figure Skating, 254
Finance, 35, 49, 133, 513, 521, 671, 723
Fire Fighting, 695
Food, 228, 233, 276, 283, 293, 311, 337, 466, 457, 642
Food Preparation, 404
Forestry, 196, 324, 453
Fund Raising, 233, 238

G

Games, 441, 447, 480, 531, 542, 599, 626, 648
Game Show, 70
Geography, 76, 87, 92, 174, 316, 383, 395, 436, 508, 559, 588, 617
Geology, 39, 320, 653
Geometry, 10, 30, 35, 39, 45, 55, 93, 159, 177, 178, 183, 188, 243, 288, 293, 340, 342, 344, 357, 354, 357, 361, 377, 395, 399, 403, 414, 415, 438, 443, 457, 518, 519, 523, 528, 577, 583, 593, 599, 617, 627, 631, 637, 642, 653, 668, 670, 671, 675, 681, 686, 692, 702, 709, 714, 718, 723, 727, 731
Grades, 149, 155

H

Health, 49, 60, 155, 288, 293, 366, 390, 399, 452, 461, 463, 553, 599
Highway Maintenance, 650
Highway Safety, 401
Hiking, 681
History, 183, 188, 193, 199, 465, 583, 622, 641, 653
Hobbies, 148, 191, 254, 268
Hockey, 527
Home Improvements, 636
Home Projects, 243
Horse Riding, 122
Horticulture, 658
Housekeeping, 457
Human Growth, 416

J

Jewelry, 453
Jobs, 136

K

Kennels, 143
Keyboarding, 382

L

Landscaping, 144, 150, 559, 615, 622, 708
Law Enforcement, 416
License Plates, 519
Life Science, 350
Lifestyles, 155
Literature, 179, 218
Living Expenses, 426

M

Manufacturing, 133, 228, 475, 593, 634, 637, 658, 727
Measurement, 228, 288, 501, 508, 731
Medicine, 25, 40, 199, 250, 252, 336, 470, 522, 707
Merchandising, 194, 475
Meteorology, 19, 79, 127, 179, 261, 262, 299, 489, 494, 528, 675, 697
Mining, 395, 626
Models, 447
Money, 340
Money Management, 10
Mountain Climbing, 697
Music, 366, 456

Navigation, 702
Number Theory, 571
Nutrition, 15, 247, 300, 426, 465, 487

Oceanography, 93, 306, 393
Oil Production, 250

Paper Recycling, 154
Patterns, 384, 435, 447, 457, 466, 508, 553, 718
Personal Finance, 108, 153, 214, 236, 238, 293, 330, 337, 416, 421, 453
Pet Care, 617
Pets, 87, 89, 155, 422, 436, 475, 714
Photography, 361, 447, 583
Physical Science, 137, 213, 228, 268, 361, 398
Physical Therapy, 136
Physics, 112, 164, 262, 278, 319, 377, 389, 566, 675
Plumbing, 668
Politics, 204, 268
Population, 283, 324, 471, 480
Postage, 159, 260, 475
Postal Service, 10, 637
Probability, 447, 528, 538, 553, 558, 559, 566, 572, 593, 617, 631, 642, 653, 723
Public Safety, 622
Publishing, 416, 588

R

Real Estate, 141, 513, 622
Records, 480
Recreation, 76, 87, 97, 422, 577, 648, 686, 702
Recycling, 232, 237, 297
Research, 344, 465
Retail, 398, 470
Retail Sales, 20, 233, 461, 489, 566, 637

Safety, 480, 496, 681
Sailing, 681
Sales, 350, 419
Scale Drawing, 692
Science, 25, 159, 357, 461, 627, 658
Seismology, 207
Sequences, 293
Sewing, 243, 279, 572
Shipping Rates, 377
Shopping, 133, 150, 243
Social Studies, 55, 173
Space Exploration, 10, 60, 81, 84, 443
Space Flight, 177
Space Travel, 553
Sports, 15, 25, 45, 49, 70, 76, 96, 103, 122, 137, 159, 194, 214, 218, 237, 254, 279, 299, 314, 324, 383, 410, 426, 487, 523, 525, 534, 542, 588, 622, 630, 667
Statistics, 121, 154, 262, 288, 354, 385, 403, 410, 452, 471, 494, 501, 506, 521, 583, 604, 631, 675, 727
Survey, 583

T

Taxes, 150, 697
Teaching, 501
Technology, 204
Teepees, 642
Telecommunication, 512
Television, 522
Theater, 310
Tourism, 344, 352
Tournaments, 159
Traffic, 255
Traffic Control, 40, 376
Transportation, 159, 199, 236, 542, 690, 692
Travel, 128, 144, 237, 404, 435, 470
TV Game Show, 534

United Nations, 518

Veterinary Medicine, 204
Volunteer Work, 731

Wages, 631
Weather, 103, 112, 300, 422, 714
World Cultures, 494

Zoology, 406

INDEX

Red type denotes items only in the Teacher's Wraparound Edition.

A

Absolute value, 67, 76, 78–80, 110
Act it Out, see ***Closing Activity***
Acute angle, 550, 552–553
Acute triangle, 569, 653
Addition
  associative property of, 234
  commutative property of, 234
  of decimals, 234–237
  of fractions, 239–247
  identity property of, 234
  of integers, 83–84
  inverse property of, 234
  phrases, 18
  property of equality, 125
  property of inequality, 147
Additive inverses, 89, 125–126, 716
Adjacent angles, 561–562, 602
Adjacent leg, 687–691
Algebra connections, *see* Connections
Algebraic expression, 16–19, 21, 240–242
  absolute value, 76
  evaluating, 11–15, 68–69, 76, 102, 106, 112, 176–178, 180, 212–213, 286, 287, 291, 297–298
  simplifying, 27–29, 68–69, 85, 86–87, 90–93
Algebraic fraction, 197
Alternate angles, 563–565
Alternative Learning Styles
  Auditory, 42, 96, 139, 175, 259, 285, 334, 372, 449, 535, 595, 623, 688, 715
  Kinesthetic, 7, 36, 73, 119, 181, 196, 239, 290, 351, 392, 472, 509, 548, 589, 632, 664, 719
  Visual, 22, 89, 129, 211, 244, 295, 317, 341, 400, 432, 491, 573, 644, 683, 712
Alternative Teaching Strategies
  Reading Mathematics, 17, 46, 78, 146, 205, 251, 338, 396, 467, 590, 638, 669, 706,
  Student Diversity, 66, 99, 124, 170, 185, 224, 274, 341, 379, 440, 486, 561, 619, 676, 729
Altitude, 612, 614, 634
Angle of depression, 695–697
Angle of elevation, 694–697
Angles, 549–550, 552–554, 602–603
  acute, 550, 552–553
  adjacent, 561, 602
  alternate exterior, 563
  alternate interior, 563
  complementary, 562, 602
  congruent, 561
  corresponding, 563
  of depression, 695–697
  of elevation, 694–697
  exterior, 591
  interior, 591
  measuring, 549, 584, 590–592, 603
  obtuse, 553, 569
  of polygons, 590–592
  of quadrilaterals, 584, 603
  right, 550, 552–553
  supplementary, 562, 602
  vertical, 561, 602
Applications, 4d, 64d, 116d, 168d, 222d, 272d, 328d, 370d, 430d, 484d, 546d, 608d, 662d, 704d
Applications (refer to Applications Index; *see also* Modeling)
Area, 138
  of a circle, 545, 620–622, 629, 656
  of a parallelogram, 612–613, 620, 656
  of a rectangle, 140–143, 150, 634–635
  of a square, 624, 665
  of a trapezoid, 614–616, 656
  of a triangle, 610–611, 613–614, 616, 656
Area model, 286
Arithmetic sequence, 258–261
Assessment
  Alternative Assessment, 61, 113, 165, 219, 269, 325, 367, 427, 481, 543, 605, 659, 703, 737
  Assessment Resources, 4c, 64c, 116c, 168c, 222c, 272c, 328c, 370c, 430c, 484c, 546c, 608c, 662c, 704c
  Performance Assessment, 61, 113, 165, 219, 269, 325, 367, 427, 481, 543, 605, 659, 703, 737
Assessment
  Alternative, 61, 113, 165, 219, 269, 325, 367, 427, 481, 543, 605, 659, 703, 737
  Chapter Tests, 774–787
  Cooperative Learning Project, 113, 481, 543, 605, 659
  math journal, 53, 106, 148, 177, 231, 282, 319, 356, 408, 446, 505, 592, 640, 696, 722
  Ongoing Assessment, 114–115, 326–327, 428–429, 544–545, 660–661, 738–739
  performance task, 61, 165, 219, 269, 325, 367, 427, 703, 737
  portfolio, 61, 113, 165, 219, 269, 325, 367, 427, 481, 543, 605, 659, 703, 737
  self evaluation, 61, 165, 219, 269, 325, 367, 427, 481, 543, 605, 659, 703, 737
  Self Test, 30, 93, 137, 194, 247, 299, 350, 399, 457, 513, 572, 642, 681, 718
  Study Guide and Assessment, 58–60, 110–112, 162–164, 216–218, 266–268, 322–326, 364–366, 424–426, 478–480, 540–542, 602–604, 656–658, 700–702, 734–736
  thinking critically, 61, 113, 165, 219, 269, 325, 367, 427, 481, 543, 605, 659, 703, 737
Associative property
  of addition, 22, 24–25, 234, 236, 712
  of multiplication, 22, 24–25, 30, 101
Auditory, see ***Alternative Learning Styles***
Axes, 73–75

Back-to-back stem-and-leaf plot, 487–489
Bar graph, 52–54, 60, 150
Bar notation, 225
Base, 175, 205, 449–451, 614, 632
Binomials, 706–708
  multiplying, 728–731, 732, 735
Boundary, 418–420
Box-and-whisker plot, 495–496, 498–501, 503, 540
Brackets, 12
Break-even point, 416

Cake method, 185
Capital gain, 166, 264
Capture-recapture, 448
Careers
  employment opportunities
    design, 547
    engineer, 547
    fashion, 5
    field archaeologist, 609
    marketing research and management, 485
    multimedia development, 169
    naval architect, 547
    paleontologist, 431
    for visually impaired, 117
  help wanted
    artist, 297
    automotive service industry, 134
    civil engineer, 568
    landscape architect, 615
    law enforcement officer, 672
    meteorologist, 249
    oceanographer, 393
    physical therapist, 449
    science-related careers, 358
    travel industry, 26
Cell, 209
Center, 341
Central angle, 344

Central tendency, 505
Change in *x*, 400–408
Change in *y*, 400–408
Chapter Project, 5, 65, 117, 169, 223, 273, 329, 371, 431, 485, 547, 609, 663, 705
Circle, 341–344
  area of, 545, 620–622
Circle graph, 497, 556–559, 602
Circular cone, 640–641, 650–652, 657
Circumference, 341–344, 364
Classroom Vignette, 6, 27, 67, 83, 104, 140, 171, 229, 234, 289, 358, 375, 433, 445, 462, 517, 541, 549, 556, 574, 613, 691
Closing Activity
  Act it Out, 25, 70, 204, 238, 361, 447, 513
  Modeling, 35, 45, 76, 93, 133, 144, 155, 179, 188, 194, 247, 254, 288, 299, 306, 311, 337, 344, 354, 377, 395, 404, 443, 461, 471, 507, 523, 528, 553, 572, 583, 599, 627, 631, 637, 653, 681, 714, 718, 731
  Speaking, 15, 40, 55, 81, 97, 103, 128, 137, 150, 159, 183, 199, 214, 233, 250, 262, 316, 320, 332, 350, 390, 399, 422, 453, 457, 466, 475, 489, 501, 534, 560, 593, 617, 642, 668, 675, 727
  Writing, 10, 20, 30, 49, 87, 108, 122, 174, 209, 228, 243, 257, 283, 293, 340, 357, 384, 410, 416, 436, 438, 494, 519, 538, 566, 577, 588, 622, 648, 671, 686, 692, 697, 709, 723
Closing price, 264
Coefficient, 711
Combination, 516, 517, 541
Commission, 470
Common difference, 258, 258–259
Common multiples, 200–204
Common ratio, 312
Commutative property
  of addition, 22–25, 234, 712
  of multiplication, 22, 24–25
Comparative graph, 498, 500
Compass, 554–555
Compatible numbers, 280–282
Complementary angle, 562–565, 602
Composite number, 184–187
Compound inequality, 357
Cones
  surface area, 640–641
  volume of, 650–651
Congruent, 554, 561, 563, 573, 579
  angles, 561, 573–574
  line segments, 554
  regular polygon, 591
  sides, 570–571, 573–575
  triangles, 573–577
Connections, 4d, 64d, 116d, 168d, 222d, 272d, 328d, 370d, 430d, 484d, 546d, 608d, 662d, 704d
Connections
  algebra, 23, 27, 68, 85, 90, 101, 106, 172, 176, 186, 191, 197, 201, 202, 206, 211, 225, 235, 240, 259, 286, 291, 297, 313, 510, 562, 569, 590, 620, 634, 646, 726
  geometry, 134, 401, 405, 434, 445, 677, 712, 716, 720
  interdisciplinary
    biology, 211, 225
    construction, 678
    cultural, 32, 66, 72, 146, 172, 308, 342, 385, 441, 459, 524, 593, 629, 678, 712
    gardening, 729
    geography, 90
    health, 463
    health and exercise, 7
    history, 580
    meteorology, 106
    music, 669
    science, 318
    social studies, 52, 172
Consecutive numbers, 340
Constant, 707
Construct, 554
Consumer price index (CPI), 489
Continuous review, *see* Mixed review
Cooperative Learning
  Brainstorming, 359, 695, 720
  Categorizing, 673
  Co-op Co-op, 515, 579
  Corners, 136
  Formations, 373, 585
  Group Discussion, 276, 401
  Inside-Outside Circle, 191, 447
  Jigsaw, 319, 499
  Math Lab and Modeling Math, 4d, 64d, 116d, 168d, 222d, 272d, 328d, 370d, 430d, 484d, 546d, 608d, 662d, 704d
  Numbered Heads Together, 225
  Pairs Check, 75, 152
  Round Table, 12, 388, 614
  Send-A-Problem, 51, 259, 355
  Think-Pair-Share, 650
  Trade-A-Problem, 100, 176, 468, 711
  Using Cooperative Learning, 3, 5, 50, 65, 71, 117, 167, 223, 273, 329, 371, 431, 554, 610
Cooperative learning projects, *see also* The Shape of Things to Come, 113, 481, 543, 605, 659, 737
Coordinate, 67, 126
Coordinate system, 36–40, 73
Corresponding angles, 563–564, 574, 579
Corresponding parts, 573–576, 579, 603
Cosine, 688–691, 695–697, 701
Cross products, 444–447
Cumulative review, *see* Ongoing Assessments
Cylinder, 634–635, 645–648, 657

Data Collection, 50–51, 56, 128, 160, 174, 209, 238, 264, 337, 362, 384, 404, 439, 453, 471, 486, 508, 524, 588
Data, 50–55
  box-and-whisker plot, 495–501
  cluster, 71
  displaying, 495–501
  gathering, 50–55
  graphing, 379–383, 432
  interpretation, 233
  measures of variation, 490–494
  scatter plots, 379–383
  stem-and-leaf plot, 486–489
Decimals, 296–299, 301–307, 309–311, 317–326
  adding, 234–237
  bar notation, 275, 277
  dividing, 294–298
  estimating differences, 229–232
  estimating sums, 229–232
  to fractions, 225–228
  multiplying, 294–298
  to percents, 458–461
  repeating, 225, 275, 277
  subtracting, 234–237
  terminating, 225, 275, 277
Deductive reasoning, 255–256
Defining a variable, 42
Degree, 549
  of temperature, 128
Demographics, 50, 154, 164
Dependent events, 530–533, 542
Diagonal, 586, 590, 677
  of square, 702
  of polygon, 96
Diameter, 341
Diastolic pressure, 16
Dilation, 578
Dioptra, 675
Discount, 468–470, 475
Discrete mathematics
  arithmetic sequence, 258–261
  combination, 516–517, 541
  common difference, 258–259
  factorial, 515–517
  Fibonacci sequence, 263
  Fundamental counting principle, 509–511, 515, 541
  geometric sequences, 313, 315
  outcomes, 509–512
  permutation, 514–517
  term, 258
  tree diagram, 509, 511, 541
  sequence, 258–262, 313, 315
Distributive property, 26–29, 31, 712, 725–726, 729–731
Diversity, see also ***Alternative Teaching Strategies,*** 4d, 64d, 116d, 168d, 222d, 272d, 328d, 370d, 430d, 484d, 546d, 608d, 662d, 704d
Dividend, 166
Divisibility rules, 170–173, 186, 216
Division
  of decimals, 296–298
  of fractions, 290–292
  of integers, 104–106
  properties of inequality, 152
  property of equality, 130
Domain, 372–376

E

Earth Watch, 80, 214, 232, 376, 436, 518, 723
Eliminating possibilities, 164, 577

INDEX

Employment Opportunities, *see* Careers
Empty set, 348
Enhancing the Chapter, 4d, 64d, 116d, 168d, 222d, 272d, 328d, 370d, 430d, 484d, 546d, 608d, 662d, 704d
Equation, 32–35
  graphing, 126–127, 130–132, 137, 162
  inverse operations, 41–45
  one-step, 124–127, 129–132, 134–136
  percent, 467–470
  with rational numbers, 248–250
  solving, 32–35, 41–45, 84, 85–87, 90–93, 100–103, 105–108, 111, 123–127, 129–132, 134–137, 162, 285, 287, 291, 295–299, 308–311
  two-step, 333–340
  with variables on each side, 345–349, 365
  writing, 338–340, 364
Equiangular, 587, 591
Equilateral, 587, 591
Equilateral triangle, 569–571
Equivalent
  equations, 124
  fractions, 195
  rational numbers, 275
Error Analysis, 10, 15, 44, 69, 75, 80, 86, 96, 102, 107, 127, 132, 136, 143, 173, 193, 198, 203, 208, 213, 231, 246, 261, 277, 287, 305, 336, 343, 349, 357, 382, 389, 398, 403, 409, 415, 421, 443, 456, 460, 465, 469, 488, 492, 499, 512, 517, 527, 552, 558, 582, 587, 592, 616, 636, 652, 680, 691, 708, 713, 717, 727
***Escherichia coli* (*e. coli*),** 208, 320
Estimation
  capture-recapture, 448
  of differences, 229–232
  experimental probability, 529
  hints, 8, 156–157, 177, 233, 281–283, 285, 295–299, 398, 473, 620, 629, 634, 639, 666–667, 700, 712, 720
  of percents, 462–465, 479
  of products, 280–283
  of quotients, 280–283
  of sums, 229–232
Euclid, 703
Evaluating an expression, 16–21, 24
Expanded form, 176
Experimental probability, 511, 529
Experiments, *see* Data collection
Exponent, 175–178, 180, 706, 719–723
Extension
  Analyzing Mazes, 332
  Business Application, 398
  Business Connection, 10
  Comparing and Contrasting, 583
  Connections, 692
  Consumer Awareness, 127, 416
  Make a Drawing, 246
  Make Up a Problem, 340
  Using Analysis, 587
  Using Biased Surveys, 506
  Using Comparisons, 461
  Using Concept Maps, 675
  Using Connections, 76, 81, 137, 257, 336, 349, 395, 447, 470, 534, 538
  Using Constructions, 593, 623
  Using Consumer Connections, 298, 636
  Using Cooperative Groups, 45, 55, 103, 174, 193, 283, 344, 376, 383, 410, 422, 723
  Using Critical Thinking, 718
  Using Data, 452, 465, 475, 522, 648, 671
  Using Diagrams, 97, 143, 617, 631, 727
  Using Discussion, 403
  Using Formulas, 641, 652, 667
  Using Graphs/Graphing, 39, 233
  Using Guess and Check, 150
  Using Health Connections, 714
  Using Language Connections, 512
  Using Lists, 70, 250
  Using Logic/Logical Reasoning, 49, 133, 188, 237, 254, 288, 357, 456, 626
  Using Manipulatives, 87, 108, 122, 178, 213, 228, 353, 361, 443, 577
  Using Mental Math, 155
  Using Models, 93, 183, 204, 279, 293, 436, 553, 559
  Using Number Sense, 25
  Using Partners, 311
  Using Patterns, 208, 243, 731
  Using Problem Solving, 19
  Using Proofs, 566
  Using Puzzles, 34
  Using Research, 159, 390, 489, 519, 598
  Using Science Connections, 262, 320, 709
  Using Simulations, 527
  Using Sports Connections, 494
  Using Surveys, 199
  Using Tables, 316
  Using Technology, 15, 29, 438, 686
  Using Writing, 571, 680
  Working Backward, 306, 501, 697
Exterior angle, 591–592, 604
Extra practice, 740–773

F

Face, 632–633
Factorial, 515
Factor patterns, 189
Factors, 170–173, 175–179, 184–193
  divisibility rule, 170–173
  evaluating expressions, 180
  greatest common factor, 190–193, 196–199
  and monomials, 170–173
  patterns, 189
  powers and exponents, 175–178
  prime factors, 184–187, 196–199
Factor trees, 185–187
Family of graphs, 411
Fibonacci sequence, 95, 263
Floor plan, 653
Formulas, 134–137, 177, 278, 314
  area of circle, 620–622, 629, 635, 661
  area of parallelogram, 612–613, 620
  area of rectangle, 70, 140–143, 163, 661, 725, 728
  area of square, 624, 639, 661
  area of trapezoid, 335, 399, 614–616
  area of triangle, 311, 613–314, 639
  aspect ratio, 667
  car speed according to skid, 672
  circumference, 267, 661
  diagonal of a square, 702
  distance, 35
  energy, 179
  finding Celsius, 346, 394
  finding Fahrenheit, 346
  Heron's, 675
  interest, 467
  men's shoe size, 257
  perimeter of rectangle, 139–140, 142–143, 163, 661, 725
  perimeter of square, 389
  Pythagorean theorem, 676–680
  safe heart rate, 366
  sequence, 314
  slope, 400–401
  slope-intercept form, 407
  surface area of a cone, 640–641
  temperature, 385
  using, 135–137
  velocity of dropped objects, 706
  voltage in a circuit, 674
  volume of circular cylinder, 645–646
  volume of a cone, 650–651
  volume of a pyramid, 649–650, 720
45°–45° right triangle, 683–686, 701
Fractals, 618
Fractions
  adding, 239–247
  to decimals, 274–278
  dividing, 290–292
  equivalent, 195, 276–277
  estimating differences, 230–231
  estimating products, 280
  estimating quotients, 280
  estimating sums, 230–231
  least common denominator, 201–203
  like denominators, 239–243
  mixed, 239–243, 245–247
  multiplying, 284–287
  negative exponents, 210–213
  to percents, 459–461
  simplifying, 217
  unlike denominators, 244–247
Frequency table, 51–55, 60, 70, 79–80, 108, 506
From the Funny Papers, 20, 144, 179, 257, 340, 390, 466, 523, 599, 622
Function
  family of graphs, 411
  graphing, 373–377, 399
  intercepts, 406–410
  linear equations, 392–394, 399
  slope, 400–403
  systems of equations, 412–415

- **Functional notation,** 393–394
- **Fundamental counting principle,** 509–511, 515, 541
- **FYI,** 6, 38, 41, 46, 51, 52, 71, 73, 81, 85, 131, 186, 225, 248, 274, 285, 303, 312, 336, 339, 380, 388, 400, 412, 434, 463, 489, 492, 507, 584, 599, 638

G

- **Geometric probability,** 623–626, 628, 656
- **Geometric sequence,** 312–315
- **Geometry**
  - **acute angle,** 552–553
  - **acute triangle,** 569
  - **adjacent angles,** 561–562, 602
  - **adjacent leg,** 687–691
  - **alternate exterior angles,** 563–565
  - **alternate interior angles,** 563–565
  - **altitude,** 612, 614, 634
  - **angle,** 549–550, 554, 602
  - **angle of depression,** 695–697
  - **angle of elevation,** 694–697
  - **area,** 138
  - **area model,** 286
  - **area of circle,** 620–622, 629
  - **area of parallelogram,** 612–613, 620, 656
  - **area of a rectangle,** 140, 282
  - **area of a trapezoid,** 614–616, 656
  - **area of triangle,** 613–614, 656
  - **axes,** 73–75
  - **base,** 614, 632
  - **center,** 341
  - **change in** x**,** 400–408
  - **change in** y**,** 400–408
  - **circle,** 341
  - **circular cone,** 640–641, 650–652, 657
  - **circumference,** 341–344
  - **compass,** 554–555
  - **complementary,** 562–565, 602
  - **congruent,** 554, 561, 563, 573
  - **construct,** 554
  - **coordinate,** 67, 126
  - **coordinate system,** 36–40, 73
  - **corresponding angles,** 563–564, 574
  - **corresponding parts,** 573–576
  - **cosine,** 688–691, 695–697, 701
  - **cylinder,** 634–635, 645–648, 657
  - **degree,** 549
  - **diagonal,** 586, 590
  - **diameter,** 341
  - **equiangular,** 587, 591
  - **equilateral,** 587, 591
  - **equilateral triangle,** 569–571
  - **exterior angle,** 591–592
  - **face,** 632–633
  - **family of graphs,** 411
  - **45°–45° right triangle,** 683–686, 701
  - **golden ratio,** 434
  - **graph,** 73, 126
  - **hypotenuse,** 676–680, 683–686, 693, 701
  - **indirect measurement,** 580
  - **intercepts,** 406–409
  - **interior angle,** 591–592
  - **intersect,** 550–551
  - **isosceles triangle,** 569–571
  - **lateral surface,** 639–640
  - **leg,** 676–681, 683–686
  - **line,** 548
  - **line segment,** 548–549, 551–552, 554–555
  - **net,** 632
  - **obtuse angle,** 550, 552
  - **obtuse triangle,** 553, 569–571
  - **opposite side,** 687
  - **ordered pair,** 36–40
  - **origin,** 36–40
  - **parallel,** 550–551, 561, 563–565, 602
  - **parallel lines,** 401
  - **parallelogram,** 585–587
  - **perimeter,** 138
  - **perimeter of a rectangle,** 140
  - **perpendicular,** 551
  - **pi (**π**),** 343
  - **plane,** 550, 552, 602
  - **plotting a point,** 73
  - **point,** 548–549, 552, 602
  - **polygon,** 589–593
  - **prism,** 632–633, 645–648, 657
  - **protractor,** 549–552
  - **pyramid,** 638–642, 649–652, 657
  - **quadrants,** 73
  - **quadrilateral,** 584, 603
  - **radius,** 341
  - **ray,** 548–549, 552, 555, 602
  - **rectangle,** 585
  - **rectangular prism,** 632–633, 657
  - **reflection,** 595–598, 604
  - **regular polygon,** 591–592
  - **rhombus,** 585–587
  - **right angle,** 550, 552–553, 569–571, 603
  - **rotation,** 595–598, 604
  - **scale,** 224
  - **scale drawing,** 585
  - **scale factor,** 654
  - **scalene triangle,** 569–571
  - **sides,** 549, 589
  - **similar,** 578
  - **sine,** 688–691, 695–697, 701
  - **skew lines,** 551, 602
  - **slant height,** 638, 640
  - **slope,** 400–403
  - **slope-intercept form,** 407–408
  - **square,** 585–587
  - **square pyramid,** 638–639
  - **straightedge,** 554–555
  - **supplementary,** 562–565, 602
  - **surface area,** 632–636, 657
  - **symmetric,** 596
  - **tangent,** 688–691, 695–697, 701
  - **tessellation,** 594
  - **30°–60° right triangle,** 683–686, 701
  - **transformation,** 595–598
  - **transversal,** 563–566, 602
  - **trapezoid,** 585, 587
  - **triangle,** 568–572
  - **triangular prism,** 632–633
  - **triangular pyramid,** 638
  - **trigonometric ratios,** 688–691, 694–697, 701
  - **trigonometry,** 688
  - **vertex,** 549, 589, 638
    - **of a polygon,** 75
  - **vertical angles,** 561–563, 602
  - **volume,** 643–648
    - **of a circular cylinder,** 645–648, 657
    - **of a cone,** 650–652, 658
    - **of a prism,** 645–648, 657
    - **of a pyramid,** 649–652, 658
  - x**-axis,** 36–40
  - x**-coordinate,** 37–40
  - x**-intercept,** 406–409, 425
  - y**-axis,** 36–40
  - y**-coordinate,** 37–40
  - y**-intercept,** 406–409, 425
- **Geometry connections,** *see* Connections
- **Giga,** 461
- **Golden ratio,** 434, 435
- **Golden rectangles,** 712
- **Gram,** 359–361
- **Graphing calculator**
  - **activities**
    - **area and coordinates,** 145
    - **evaluating expressions,** 21
    - **exploring factors and fractions,** 180
    - **factor patterns,** 189
    - **families of graphs,** 411
    - **finding mean and median,** 307
    - **geometric probability,** 628
    - **graphing inequalities,** 417
    - **slope and tangent,** 693
    - **slopes of parallel lines,** 567
  - **applications, energy cost,** 397
  - **exercises**
    - **graphing linear relations,** 387
    - **measures of variation,** 493
    - **probability,** 443
    - **scatter plot,** 382
    - **solving one-step equations,** 132
    - **use a simulation,** 527
    - **volume,** 647
  - **introduction,** xx, 1
  - **modeling**
    - **congruent triangles,** 573
    - **functions and graphing,** 412
    - **plotting points,** 77
    - **probability,** 440
    - **scatter plots,** 379–380
    - **solving equations,** 39, 124
    - **statistics and probability,** 490, 502–503, 530
  - **programming,** 21, 34, 77, 124, 132, 145, 180, 387, 397, 411, 417, 440, 490, 502–503, 567, 573, 628, 693
- **Graphs,** 73, 126
  - **bar,** 52–54, 150, 395
  - **box-and-whisker plot,** 495–496, 498–501, 503, 540
  - **circle,** 556–557
  - **coordinate systems,** 36–40, 74–76, 93, 110–128, 583
  - **data,** 80, 432
  - **of equations,** 412–415
  - **functions,** 372–376
  - **of inequalities,** 75, 147–149
  - **latitude,** 76
  - **longitude,** 76

INDEX

on number line, 78–80, 110, 112–113, 126–127, 130–132, 162
parabola, 391–394
points, 67–69, 71, 73–76, 78–80
problem solving, 396–399
rectangle, 145
relations, 372–376
Greatest common factor (GCF), 190–193, 196–197, 217, 285

H

Hands-on activities, *see* Mathematics Lab
Hands-On Activity, see ***Motivating the Lesson***
Heat index, 97, 299
Helge von Koch, 618
Help wanted, *see* Careers
Histogram, 502–503
Hypotenuse, 676–680, 686–686, 693, 701
Hypsometer, 694

Identity property
of addition, 23, 234
of multiplication, 23
Inclusive, 536–538, 542
Independent events, 530–533, 542
Indirect measurement, 580–583
Inductive reasoning, 255–256
Inequalities, 46–49, 78–80, 110, 146–159, 203, 217, 227–228
checking solutions, 147–149, 152–154
compound, 357
graphing, 147–149, 152–154, 251–253, 351–353, 417–421, 426
solving multi-step, 351–354, 355–357, 365
solving one-step, 146–149, 151–154, 156–158, 163, 309–311
writing, 355–357, 365
Integers, 66–69, 224
absolute value, 67–69
adding, 82–87, 111
comparing, 78–81
dividing, 104–107
graphing, 73–75, 78–81
multiplying, 98–103, 111
ordering, 78–81
subtracting, 88, 89–92, 104–106, 112
Integration
assessment, *see* Assessment
discrete mathematics
arithmetic sequences, 258–261
geometric sequences, 312–315
geometry, 128, 133
area and perimeter, 139–143
circles and circumference, 341–343
coordinate system, 72–75
ordered pairs, 36–40
Look Back, *see* Look Back
measurement, using the metric system, 358–361
number theory, 15, 30, 95, 174, 188, 218, 287, 437, 443, 553, 571
patterns, *see* Patterns
probability, 374, 440–443, 510
statistics, 53, 93, 121, 164, 262, 288, 384, 403, 452, 471, 494
gathering and recording data, 51–54
measures of central tendency, 301–305
technology, *see* Graphing calculator, Spreadsheets, Technology Tips
Intercepts, 406–409
Interdisciplinary, *see* Connections
Interior angle, 591–592
Internet Connection, 5, 65, 117, 169, 223, 273, 329, 371, 431, 485, 547, 609, 663, 705
Interquartile range, 491–494, 540
Intersecting lines, 550–551
Inverse operations, 41–45
of addition, 234
of multiplication, 289
Investigations (long-term projects)
Bon Voyage, 62–63, 70, 97, 128, 160
It's Only a Game, 482–483, 508, 528, 588, 600
Oh, Give Me a Home, 606–607, 637, 653, 671, 698
Roller Coaster Math, 270–271, 279, 293, 337, 362
Stadium Stampede, 368–369, 384, 404, 453, 471, 476
Taking Stock, 166–167, 174, 209, 238, 264
Up, Up, and Away, 2–3, 10, 35, 56
Irrational numbers, 672–675, 700
Isosceles triangle, 569–571

J

Journal, *see* Math Journal

K

Kinesthetic, see ***Alternative Learning Styles***
Koch curve, 618

L

Lateral surface, 639–640
Least common denominator (LCD), 201–203
Least common multiple (LCM), 200–204, 217
Leaves, 486
Leg, 676–681, 683–686
Lesson Objectives, 4a, 64a, 116a, 168a, 222a, 272a, 328a, 370a, 430a, 484a, 546a, 608a, 662a, 704a
Lesson Planning Guide, 4b, 64b, 116b, 168b, 222b, 272b, 328b, 370b, 430b, 484b, 546b, 608b, 662b, 704b
Like terms, 27, 711, 716
Linear equation, 386–389
Line graph, 497–500
Line plot, 71
Lines, 548, 550–554
intersecting, 550–551
parallel, 561, 563, 567, 602
perpendicular, 551, 602
of symmetry, 550–551, 596–598
transversal, 563
Line segment, 548–549, 551–552, 554–555
bisecting, 555
congruent, 554
perpendicular, 613–614
Liter, 359–361
Logic
matrix, 119–121
reasoning, 70, 74
Long-term project, *see* Investigations
Look Back, 73, 89, 123, 125, 225, 251, 274, 287, 309, 317, 351, 355, 358, 379, 432, 459, 498, 502, 505, 510, 556, 579, 620, 630, 654, 665, 672, 706, 719, 721
Lower quartile, 491–494, 540

Manipulatives
algebra tiles, 664, 710, 724, 729, 730, 732
balance scale, 32, 122, 124, 155, 248, 351
ball, 312
centimeter cubes, 28, 646, 719
checkerboard and checkers, 381
coins, 8, 75, 87, 124, 236, 290, 434, 626, 628
compass, 554, 557, 561, 593, 618, 619, 623, 681, 682
cone models, 639
counters, 18, 23, 45, 47, 85, 91, 98, 155, 304, 351, 392
cups, 18, 45, 351
cylinder models, 632, 635, 648
dice/number cubes, 52, 87, 103, 108, 258, 277, 279, 353, 404
dominoes, 37, 75, 228, 279, 293, 447
drinking straws, 412
geoboard, 374, 579, 610, 621
marbles, 455, 461
masking tape, 36, 643, 676
measuring cups, 241
modeling clay, 690
play money (bills), 93, 236
playing cards, 443
prism models, 632, 635, 639, 648, 654, 719
protractor, 549, 553, 561, 572, 592, 618
pyramid models, 639
rice, 643
rulers/yardsticks/metersticks/ tape measures, 8, 129, 284, 341, 344, 361, 618, 687
spinner, 442
straightedge, 554, 593, 623, 681, 682
string, 70, 284, 341, 378, 432, 676, 690

sugar cubes, 654
thermometers, 346
weights (for balance scale), 32
Manipulatives, *see* Modeling, Mathematics Lab
Market price, 166
Mass, 359–361
Math Journal, 9, 28, 53, 79, 106, 148, 153, 177, 212, 231, 241, 256, 260, 282, 291, 319, 343, 356, 360, 402, 408, 420, 446, 464, 474, 505, 516, 521, 558, 570, 592, 630, 640, 646, 690, 696, 708, 722
Mathematical modeling
charts, 6–7, 11, 51, 71, 78, 118, 139, 175, 189, 200, 205, 210, 234, 251, 258, 280, 301, 308, 330, 341, 372, 385, 396, 432, 437, 444, 462, 502, 535, 719, 728
diagrams, 22, 31, 36, 72, 82, 88, 89, 98, 104, 123, 129, 138, 139, 181, 195, 244, 284, 351, 509, 520, 584, 594, 595, 610, 612, 618, 664, 669, 694
equations, 21, 34, 83, 104, 123, 124, 295, 346, 400, 411, 412, 672, 725
graphs, 21, 71, 77, 145, 151, 180, 229, 248, 274, 372, 385, 391, 405, 406, 411, 417, 418, 454, 495, 502, 504, 556, 567, 573, 693, 694
scatter plots, 379–380
Mathematics Labs, 21, 31, 50, 71, 77, 82, 88, 98, 123, 138, 145, 180, 189, 195, 263, 294, 300, 307, 333, 345, 378, 391, 405, 411, 417, 439, 448, 502, 514, 529, 554–555, 567, 594, 610–611, 618, 628, 643, 654, 682, 687, 693, 710, 724, 732
Mean, 20, 300–301, 303–307, 505
Measurement
gram, 359–361
liter, 359–361
meter, 358–361
metric system, 358–361
Measures of central tendency, 300–307
mean, 300–307, 505
median, 300–307, 491–494, 501, 505, 540
mode, 300–306, 505
Measures of variation, 490–494
Median, 300–307, 491–494, 501, 505, 540
Meter, 358–361
Metric system, 358–361
Mixed number, 239–243
adding, 240–243, 245–247
subtracting, 240–243, 245–247
Mixed review, 20, 25, 29, 35, 40, 45, 49, 55, 70, 75, 80, 87, 93, 97, 103, 108, 122, 128, 133, 137, 144, 150, 155, 159, 174, 179, 194, 199, 204, 209, 214, 228, 233, 238, 243, 247, 250, 254, 257, 262, 279, 283, 288, 299, 306, 316, 320, 332, 337, 340, 344, 350, 354, 357, 361, 377, 384, 395, 399, 404, 410, 416, 422, 436, 438, 447, 453, 457, 461, 466, 471, 475, 494, 501, 507, 519, 523, 528, 534, 553, 560, 566, 572, 577, 588, 593, 617, 622, 627, 631, 637, 648, 653, 668, 671, 681, 686, 692, 697, 709, 714, 718, 723, 727, 731
Mode, 300–306, 505
Modeling
with manipulatives, 22, 31, 36, 89, 98, 104, 123, 126, 138, 139, 146, 170, 184, 284, 289, 333, 341, 345, 351, 392, 405, 418, 432, 458, 520, 524, 535, 561, 578, 595, 612, 619, 649, 676, 694, 711
real-world applications
advertising, 509
air travel, 396
architecture, 638
backpacking, 355
ballooning, 664
baseball, 338
business, 229, 437
cartography, 72–73
child care, 407
comics, 255
construction, 400, 548, 706, 725
consumerism, 334, 454, 486
ecology, 372
economics, 444
education, 118, 504
electricity rates, 472
engineering, 568
entertainment, 32, 51, 181, 339
food, 274
forestry, 196
games, 175, 623
geography, 66
geology, 317
health, 16, 78, 462
health and exercise, 6
history, 190
law enforcement, 672
medicine, 16
meteorology, 248, 301
movies, 224
oceanography, 99
pet care, 629
physical therapy, 449
postal service, 632
quilting, 584
recycling, 151
safety, 495
science, 385
sports, 251, 406
swimming, 515
toys, 129
transportation, 41
travel, 26, 134, 358
weather, 556
world records, 156
with technology, 11, 83, 94, 124, 205, 210, 234, 258, 280, 295, 330, 346, 412, 440, 467, 490, 530, 573, 589, 644, 688, 719
Modeling Mathematics, 14, 21, 22, 31, 36, 47, 50, 54, 62, 68, 71, 77, 82, 88, 89, 91, 98, 102, 104, 120, 123, 135, 138, 139, 145, 146, 170, 173, 180, 182, 184, 189, 192, 195, 198, 202, 227, 239, 244, 263, 277, 284, 288, 289, 294, 297, 300, 304, 307, 315, 331, 333, 335, 341, 345, 348, 351, 353, 378, 387, 391, 392, 404, 405, 411, 414, 417, 418, 432, 434, 439, 442, 448, 451, 458, 460, 488, 498, 502, 511, 514, 520, 524, 529, 535 554–555, 561, 567, 575, 578, 581, 586, 594, 595, 597, 601–611, 618, 619, 625, 628, 635, 643, 649, 651, 654, 667, 676, 679, 682, 687, 693, 694, 710, 711, 713, 717, 724, 726, 730, 732
Modeling, see *Closing Activity*
Monomials, 172–173, 186–187, 205–208, 706, 708, 719–723
dividing, 206–208
factoring, 186–187
multiplying, 205–208
power of, 721–723, 735
power of a power, 719–723
power of a product, 720–723
power of ten, 205, 207
Motivating the Lesson
Hands-On Activity, 23, 42, 72, 104, 124, 129, 171, 190, 196, 234, 248, 258, 280, 284, 312, 346, 358, 412, 432, 462, 486, 535, 568, 584, 595, 613, 632, 639, 669, 710, 719
Labs
Graphing Calculator, 21, 77, 145, 180, 189, 307, 411, 417, 502, 567, 628, 693
Hands-On Activity, 31, 50, 71, 82, 88, 98, 123, 138, 195, 263, 294, 300, 333, 345, 378, 391, 405, 439, 448, 514, 529, 554, 594, 610, 618, 643, 654, 682, 687, 724, 732
Questioning, 7, 11, 51, 78, 90, 100, 134, 147, 151, 181, 184, 211, 224, 244, 274, 301, 308, 317, 334, 338, 355, 372, 397, 400, 406, 418, 440, 454, 467, 472, 495, 504, 509, 515, 524, 530, 548, 556, 573, 589, 619, 623, 649, 664, 672, 683, 694, 706, 715, 725
Situational Problem, 17, 27, 33, 37, 47, 67, 84, 94, 118, 139, 156, 176, 201, 229, 239, 251, 289, 295, 341, 351, 380, 386, 392, 437, 444, 449, 458, 490, 520, 561, 579, 629, 644, 676, 688, 711, 728
Multiple, 200–204
Multiplication
associative property of, 22
of binomials, 728–731, 732, 735
commutative property of, 22
of decimals, 294–298
of fractions, 284–287
of integers, 98–103
of monomials, 205–208
phrases, 18
polynomials by monomials, 725–727
of powers, 719–723
property of equality, 130
properties of inequality, 152
Multiplicative inverse, 289–292
Multiplicative property of zero, 23
Multiplying integers with different signs, 99

INDEX

Multiplying integers with the same sign, 100
Mutually exclusive, 535–538, 542

Negative, 66
Negative exponents, 210–213
Negative relationship, 380–384, 424
Net, 632
Net change, 71
No relationship, 380–384, 424
Null set, 348
Number lines
graphing points, 67, 69, 71, 78–80
graphing solutions to equations, 126–127, 130, 132, 162
graphing solutions to inequalities, 147–149, 152–154, 163
integers, 682
irrational numbers, 682
ordering integers, 78–80
parabolas, 391–394
rational numbers, 682
Numbers, 673–674
composite, 184–187
consecutive, 340
integers, 224–228, 673–674
irrational 673–674
mixed, 239–243, 245–247
perfect squares, 665–667
pi (π), 343
prime, 184–187
rational, 224–228, 672–674, 700
real, 673–674, 700
sequence, 20
whole, 224, 228, 673–674

Objectives, see *Lesson Objectives*
Obtuse angle, 550–553, 569, 603
Obtuse triangle, 553, 569–571
Octagon, 255
Odds, 520–522, 541
Ongoing Assessments, 114–115, 326–327, 428–429, 544–545, 660–661, 738–739
Open-ended problems, 115
Open sentence, 32
Opposite side, 687
Ordered pair, 37, 36–40, 73–77, 103
Order of operations, 11–15, 21, 76
Organizing the Chapter, 4b–4c, 64b–64c, 116b–116c, 168b–168c, 222b–222c, 272b–272c, 328b–328c, 370b–370c, 430b–430c, 484b–484c, 546b–546c, 608b–608c, 662b–662c. 704b–704c
Origin, 36–40
Outcomes, 509–512
Outliers, 71, 496
Outside Resources, 4c, 64c, 116c, 168c, 222c, 272c, 328c, 370c, 430c, 484c, 546c, 608c, 662c, 704c

Parabola, 391–394
Parallel lines, 401, 550–551, 561, 563–565, 602
Parallelogram, 585–587
area of, 612–613, 616
Parentheses, 12–15
Pascal's Triangle, 30
Patterns: problem solving, 20, 25, 35, 49, 95, 96, 103, 122, 178, 183, 189, 193, 205, 210, 256, 312–316, 354, 366, 384, 435, 438, 457, 466, 508, 553
Percent, 449–452
of change, 472–475
to decimals, 459–461
of decrease, 472–475, 480
equations, 467–470
estimating, 462–465, 479
to fractions, 459–461
of increase, 472–475, 480
Percent proportion, 450–451, 499
Perfect number, 174
Perfect squares, 665–667
Performance Assessment, see *Assessment*
Performance task, 61, 165, 219, 269, 325, 367, 427, 703
Perimeter, 138, 678
of a rectangle, 140
of a square, 665
Permutation, 514–517
Perpendicular lines, 551, 602
Pi (π), 255, 341–344
Pictograph, 497, 500
Plane, 550, 552, 602
Plotting a point, 73–75, 77
Point, 548–549, 552, 602
graphing, 77
Polygons, 589–593
angle measures, 590–592
area of, 610–611
exterior angles, 591–592
heptagon, 590
hexagon, 590, 591
interior angles, 591–592
octagon, 590
pentagon, 590–591
quadrilaterals, 590
regular, 591–593
tessellations, 594
Polynomial, 706–708, 710, 734
adding, 711–714, 734
additive inverse, 716–717
binomials, 728–731, 732, 735
multiplying, 724, 725–731, 732, 735
subtracting, 715–718, 734
Portfolio, 61, 113, 160, 165, 219, 269, 325, 427, 476, 481, 543, 600, 605, 659, 698, 703
Positive relationship, 380–384
Powers, 175–178
of a monomial, 706, 720–723, 735
of a power, 719–723
of a product, 720–723
Previewing the Chapter, 4a, 64a, 116a, 168a, 222a, 272a, 328a, 370a, 430a, 484a, 546a, 608a, 662a, 704a
Prime day, 187
Prime factorization, 185–187, 190–191, 196–197, 201–202, 211, 216
Prime number, 545
Principal, 467
Prism, 632–633, 645–648
Probability, 440–443, 478
capture-recapture, 448
dependent events, 530–533, 542
experimental, 511, 529
geometric, 623–626, 628, 656
inclusive, 536–538, 542
independent events, 530–533, 542
mutually exclusive, 535–538, 542
odds, 520–522, 541
relative frequency, 529
sample space, 441, 479
simulation, 524–526
theoretical, 511
Problem-solving strategies
acting it out, *see* Simulations
deductive reasoning, 255–256
draw a diagram, 181–183
draw a graph, 396–398
eliminate possibilities, 118–121
guess and check, 186
inductive reasoning, 255–256
look for a pattern, 94–96, 171
make a drawing, 629–631
make a model, 629–631, 657
make a table, 437, 441
matrix logic, 119–121
method of computation, 8
patterns, 20, 25, 35, 49, 94–97
problem-solving plan, 6–10, 43, 148, 156–157, 171, 181, 290, 314, 330–331, 339, 342, 355–356, 359, 396–397, 407–408, 437, 445, 454, 510, 516, 525, 557, 580–581, 629, 634–635, 646, 678, 726, 729
use a simulation, 524–526
Venn diagram, 669–671, 673
work backward, 314, 330
Product of powers, 205–208
Programming, graphing calculator
area and coordinates, 145
congruent triangles, 573
energy cost, 397
evaluating expressions, 21
exploring factors and fractions, 180
families of graphs, 411
geometric probability, 628
graphing inequalities, 417
graphing linear relations, 387
plotting points, 77
probability, 440
slope and tangent, 693
slopes of parallel lines, 567
solving equations, 34, 124, 132
statistics and probability, 490, 502–503
Projects
cooperative, 113, 481, 543, 605, 659
long-term, *see* Investigations
Properties
of addition, 22, 24–25, 30, 234
additive inverse, 89–91

INDEX

associative, 22, 24–25, 101, 296, 712
commutative, 22–25, 296, 712
distributive, 26–29, 31, 712
of equality, 17, 28, 124, 125
identity, 23, 25, 30, 296
inverse, 296
of multiplication, 22, 24–25, 296
of proportions, 444
substitution, 17, 28
of zero, 23, 25
Proportions, 444–447, 478, 579–583, 585
Protractor, 549–552, 694
Punnett squares, 728
Putting It into Perspective, 4–5, 64–65, 116–117, 168–169, 222–223, 272–273, 328–329, 370–371, 430–431, 484–485, 546–547, 608–609, 662–663, 704–705
Pyramid, 638–642, 649–652, 657
Pythagoras, 676
Pythagorean theorem, 676–680, 682, 694–697, 701

Quadrants, 73–75
Quadrilateral, 584–588, 603
angles, 584, 603
parallelogram, 585–587
rectangle, 585–587
rhombus, 585–587
similar, 585
square, 585–587
trapezoid, 585–587, 614–616
Quartile, 491–494
Questioning, see ***Motivating the Lesson***
Quotient of powers, 206–208

Radical sign, 664–667, 672–675
Radius, 341
Range, 372–376, 490–494, 540
Rate, 433, 449–452
Ratio, 196–199, 432–435, 478, 687, 688–691
cosine, 688–691
sine, 688–691
tangent, 688–691
trigonometric, 688–691, 694–697, 701
Rational numbers, 224–228, 672, 700
adding, 234–237
equivalent, 275
solving equations, 248–250
subtracting, 234–237
Ray, 548–549, 552, 555, 602
Reading Mathematics, see ***Alternative Teaching Strategies***
Real numbers, 673–675, 700
Reciprocal, 289
Rectangle, 585
area of, 140–143, 150, 165, 615, 634–635
graphing, 145
golden, 712
line of symmetry, 596
perimeter, 139–143, 165
volume, 159
Rectangular prism, 632–633
Reflection, 595–598, 604
Regular polygon, 591–593
Relation, 372–376
graphing linear relations, 385–387
inverse, 376
with two variables, 425
Relative frequency, 529
Relatively prime, 193, 218
Repeating decimals, 225–226, 277
Resources, see ***Outside Resources***
Reteaching
Act It Out, 339, 511, 707
Hands-On Activity, 277
Make a Drawing, 630
Making and Using Lists, 674
Paired Partners, 356
Using Calculators, 192, 207, 464
Using Charts, 420
Using Comparisons, 95, 402
Using Computers, 526
Using Connections, 253, 342, 437, 451, 564
Using Cooperative Groups, 8, 43, 314, 726
Using Cooperative Learning, 446, 460
Using Data, 558, 670
Using Diagrams, 586, 635, 685, 696, 713
Using Discussion, 120, 188, 348, 498
Using Drawings, 625
Using Examples, 310
Using Flowcharts, 335, 353, 469, 473
Using Groups of Two, 297
Using Lists, 13, 158, 256, 331, 387
Using Logical Reasoning, 282
Using Manipulatives, 23, 28, 32, 47, 153, 197, 202, 212, 236, 245, 260, 374, 381, 434, 455, 487, 516, 521, 537, 570, 597, 615, 646, 730
Using Mental Math, 230
Using Modeling/Models, 18, 37, 52, 79, 85, 91, 101, 126, 142, 148, 177, 227, 241, 291, 304, 318, 360, 414, 442, 492, 532, 575, 581, 592, 621, 640, 666, 679, 690, 716
Using Questioning, 182, 397, 408
Using Practice Partners, 68
Using Problem Solving, 286
Using Properties, 249
Using Science Connections, 651
Using Study Aids, 721
Using Tables, 135, 173, 394
Using Writing, 74, 106
Working Backward 131, 551
Review
Chapter Tests, 774–787
Extra Practice, 740–773
Highlights, 57, 109, 161, 215, 265, 321, 363, 423, 477, 539, 601, 655, 699, 733
Mixed review, *see* Mixed review
Self Test, 30, 93, 137, 194, 247, 299, 350, 399, 457, 513, 572, 642, 681, 718
Study Guide, 58–60, 110–112, 162–164, 216–218, 266–268, 322–326, 364–366, 424–426, 478–480, 540–542, 602–604, 656–658, 700–702, 734–738
Rhombus, 585–587
Right angle, 550, 552–553, 569–571, 603
Right triangle, 569–571, 676–680, 683–686, 701
angles, 569
cosine ratio, 688–692, 695–697
hypotenuse, 676–681
legs, 676–681
Pythagorean theorem, 676–681
ratios, 687–691, 694–697
sine ratio, 688–692, 695–697
tangent ratio, 688–697
Rotation, 595–598, 604
Rounding
truncate, 275, 281, 296, 298

Sales, 383
Sample, 50, 51
Sample space, 441, 479
Scale, 224
Scale drawing, 585, 605
Scale factor, 654
Scalene triangle, 569–571
Scatter plot, 379–383, 387–384, 424
Scientific notation, 317–319, 324–325
Security, 166
Self Evaluation, 165, 219, 269, 325, 367, 427, 481, 543, 605, 659, 703
Self Test, 30, 93, 137, 194, 247, 299, 350, 399, 457, 513, 572, 642, 681, 718
Sequence, 258–262
geometric, 312–316
Shadow reckoning, 580
Share, 166, 264
Sides, 549, 589
Sieve of Eratosthenes, 193
Similar, 578, 654
Simplest form, 28, 196–199
Simulation, 524–525
Sine, 688–691, 695–697, 701
Situational Problem, see ***Motivating the Lesson***
Skew lines, 551, 602
Slant height, 638, 640
Slope, 400–403, 693
Slope-intercept form, 407–408
Solution, 32–35
graphing, 126–127, 130–132, 147, 152–154
of system, 412–415
Solving an equation, 32
Spatial reasoning
angle relationships and parallel lines, 561–566
area and perimeter of rectangles, 139–143, 163
congruent triangles, 573–577
construction, 554–555
polygons, 589–593

quadrilaterals, 584–587
similar triangles, 578–583
surface area: prisms and cylinders, 632–636
surface area: pyramids and cones, 638–642
tessellations, 594
transformations, 595–598
triangles, 568–572
volume, 643–653, 720
Speaking, see *Closing Activity*
Spiral review, *see* Mixed review
Spreadsheet, 209, 238, 467
Square, 585–587
area, 624, 665
diagonals, 702
Punnett, 728
Square pyramid, 638–639
Square root, 664–667, 672–674
Standard form, 176
Statistical Snapshots, 5, 65, 117, 169, 223, 273, 329, 371, 431, 485, 547, 609, 663, 705
Statistics, 51–55
back-to-back stem-and-leaf plot, 487–489
bar graph, 52–54, 60, 503
box-and-whisker plot, 495–496, 498–501
circle graph, 497, 556–559
comparative graph, 498, 500
data, 50–55
demographics, 50
frequency table, 51–55
histogram, 502–503
interquartile range, 491–494, 540
leaves, 486
line graph, 497, 500
line plot, 71
lower quartile, 491–494, 540
making graph, 502–503
mean, 300–301, 303–307
measures of central tendency, 300–307
median, 300–307
misleading, 504–506, 541
mode, 300–307
negative relationship, 380–384
no relationship, 380–384
odds, 520–522
outliers, 496
pictograph, 497, 500
positive relationship, 382–384, 424
quartile, 491–494
range, 490–494, 540
sample, 50, 51
scatter plot, 378–384, 424
stem-and-leaf plot, 486–489, 540
stems, 486
surveys, 50
upper quartile, 491–494, 540
using to make predictions, 455–456
Stem-and-leaf plot, 486–489, 540
Stems, 486
Stock exchange, 166
Straightedge, 554–555
Substitution property of equality, 17, 28
Subtraction
of integers, 89
like fractions, 240–243
phrases, 18
unlike fractions, 244–247
Subtraction property of equality, 124
Subtraction property of inequality, 147
Supplementary, 288, 562–565, 602
Surface area, 632–636, 657
of cones, 640–641, 657
of cubes, 635–636
of cylinders, 634
of prisms, 632–636, 657
of pyramids, 639–641, 657
Surveys, 50
Symbols
multiplication, 13
division, 13
Symmetric, 596
System of equations, 412–415, 426
Systolic pressure, 16

T

Tangent, 688–691, 695–697, 701
Technology
GLENCOE Technology, Interactive Mathematics Tools, 17, 42, 73, 82, 88, 98, 123, 125, 130, 134, 192, 206, 235, 240, 245, 258, 263, 275, 296, 302, 313, 333, 352, 381, 386, 393, 402, 407, 413, 450, 458, 467, 525, 531, 558, 569, 589, 612, 620, 649, 665, 677, 689, 712, 716
GLENCOE Technology, Test and Review Software, 58, 110, 162, 216, 266, 322, 364, 424, 478, 540, 602, 656, 700, 734
Internet Connection, 5, 65, 117, 169, 223, 273, 329, 371, 431, 485, 547, 609, 663, 705
Using Technology, 21, 77, 145, 180, 189, 307, 411, 417, 438, 502, 567, 628, 693
Technology Masters, 4d, 64d, 116d, 168d, 222d, 272d, 328d, 370d, 430d, 484d, 546d, 608d, 662d, 704d
Technology, *see* Graphing calculator; Modeling; Spreadsheet; Technology Tips
Technology Tips, 12, 83, 177, 235, 276, 318, 450, 491, 666, 689, 719
Tech Prep
Accounting Clerk, 92
Advertising Display Designer, 337
Air Conditioner Salesperson, 645
Animal Breeder, 728
Applications, 4d, 64d, 116d, 168d, 222d, 272d, 328d, 370d, 430d, 484d, 546d, 608d, 662d, 704d
Astronomy Technician, 212
Banquet Consultant, 495
Crop Duster, 694
Electronic Pagination System Operator, 288
Fast Food Supervisor, 231
Financial Advisor, 463
General Contractor, 11
Gymnastics Trainer, 252
Landscaper, 347
Manufacturing Cost Analyst, 406
Mason, 200
Medical Technician, 16
Mural Painter, 578
Political Analyst, 504
Research Assistant, 616
Sales Associate, 380
Specialized Architects, 725
Sports Medicine, 157
Statistician, 302
Surveyor, 678
Textile Designer, 596
Travel Planner, 444
Veterinary Technician, 87
Terminating decimal, 225, 275
Terms, 27, 258
Tessellation, 594
Tests, *see* Assessment
Test-taking tips, 115, 327, 429, 545, 661
Theoretical probability, 511
The Shape of Things to Come
digital cameras, 627
the global positioning system, 40
living batteries, 320
technology in sports, 354
thought-controlled computer technology, 692
virtual reality, 560
Think About It, 7, 12, 17, 18, 23, 24, 27, 33, 37, 42, 43, 52, 67, 104, 171, 172, 197, 201, 202, 226, 252, 281, 341, 352, 372, 386, 401, 406, 458, 536, 548, 556, 562, 563, 574, 580, 589, 596, 613, 614, 629, 632, 634, 638, 645
30°–60° right triangle, 683–686, 701
Timelines, *see* Putting It into Perspective
Transformation, 595–598
Translation, 595, 604
Transversal, 563–566, 602
Trapezoid, 585, 587
area of, 614–616
Tree diagram, 509, 511, 541
Triangle, 568–572
acute, 569–571, 603
angles, 569–577
area of, 311, 610–611, 613–614, 616
congruent, 573–577, 603
equilateral, 569–571, 596
isosceles, 569–571
obtuse, 569–571, 603
right, 569–571, 603
scalene, 569–571
similar, 578–583, 603
special, 683–686
Triangular prism, 632–633
Triangular pyramid, 638
Trigonometric ratios, 688–691, 694–697, 701
Trigonometry, 688
Trinomials, 706–708, 734
Truncate, 275
Twin primes, 188

INDEX

Unit rate, 433–435, 478
Upper quartile, 491–494, 540

Variable, 16–18, 32–35
  defining, 42
Venn diagram, 669, 671, 673
Vertex, 75, 549, 589, 638
Vertical angles, 561–563, 602
Vertical line test, 373–374
Visual, see ***Alternative Learning Styles***
Volume, 643–648
  of a circular cylinder, 645–648, 657
  of a cone, 650–652, 658
  of a cube, 177–178
  of a prism, 645–648, 657
  of a pyramid, 649–652, 658

Whole numbers, 224
Windchill factor, 97
Working drawings, 653
Writing, see ***Closing Activity***

***x*-axis,** 36–40
***x*-coordinate,** 37–40
***x*-intercept,** 406–409

***y*-axis,** 36–40
***y*-coordinate,** 37–40
***y*-intercept,** 406–409

Zero pair, 82